ASTRONOMY AND ASTROPHYSICS ABSTRACTS

A Publication of the Astronomisches Rechen-Institut Heidelberg

Member of the International Council
for Scientific and Technical Information

Astronomy and Astrophysics Abstracts is Prepared
Under the Auspices of the International Astronomical Union

Volume 39
Literature 1985, Part 1

Edited by
S. Böhme U. Esser W. Fricke H. Hefele I. Heinrich W. Hofmann
D. Krahn V. R. Matas L. D. Schmadel G. Zech

Springer Science+Business Media, LLC 1985

Astronomisches Rechen-Institut, Mönchhofstraße 12–14,
D-6900 Heidelberg 1, F.R. Germany
Telex: 461 336 ARIHD D
Director: Prof. Dr. Walter Fricke

Astronomy and Astrophysics Abstracts
Department head: Dr. Lutz D. Schmadel
Editors-in-Chief: Inge Heinrich, Dr. Lutz D. Schmadel

ISBN 978-3-662-12354-6   ISBN 978-3-662-12352-2 (eBook)
DOI 10.1007/978-3-662-12352-2

Library of Congress Catalog Card Number 72-104650.
Media conversion: Daten- und Lichtsatz-Service, Würzburg.

2153/3130-543210

Preface

Astronomy and Astrophysics Abstracts aims to present a comprehensive documentation of the literature concerning all aspects of astronomy, astrophysics, and their border fields. It is devoted to the recording, summarizing, and indexing of the relevant publications throughout the world. *Astronomy and Astrophysics Abstracts* is prepared by a special department of the *Astronomisches Rechen-Institut* under the auspices of the *International Astronomical Union*.

Volume 39 records literature published in 1985 and received before August 15, 1985. Some older documents which we received late and which are not surveyed in earlier volumes are included too. We acknowledge with thanks contributions of our colleagues all over the world. We also express our gratitude to all organizations, observatories, and publishers which provide us with complimentary copies of their publications.

On account of the introduction of an object index the scope of index information will be considerably enlarged beginning with this volume. In connection with the subject index an additional source to satisfy the needs of retrieval is opened up.

Starting with Volume 33, all the recording, correction, and data processing work was done by means of computers. The recording was done by our technical staff members Ms. Helga Ballmann, Ms. Mona El-Choura, Ms. Monika Kohl, Ms. Sylvia Matyssek. Ms. Karin Burkhardt, Ms. Susanne Schlötelburg, and Mr. Stefan Wagner supported our task by careful proofreading. It is a pleasure to thank them all for their encouragement.

Heidelberg, September 1985   **The Editors**

Contents

Stars

Interstellar Matter, Nebulae

Radio Sources, X-ray Sources, Cosmic Rays

Stellar Systems, Galaxy, Extragalactic Objects, Cosmology

Introduction

Astronomical Bibliographies

Astronomy and Astrophysics Abstracts started documentation and abstracting work in 1969 as the direct successor of the Astronomischer Jahresbericht. For information on astronomical literature before this date consultation of one of the following bibliographies is suggested:

(1) J. J. de Lalande, Bibliographie Astronomique, Paris 1803 (this work covers the time from 480 B. C. to the year 1803, VIII + 966 pages).

(2) J. C. Houzeau, A. Lancaster, Bibliographie générale de l'astronomie, Volume I (in two parts), Bruxelles 1887, 1889, Volume II, Bruxelles 1882. The complete title of Volume II is "Bibliographie générale de l'astronomie ou catalogue méthodique des ouvrages, des mémoires et des observations astronomiques, publiés depuis l'origine de l'imprimerie jusqu'en 1880". A new edition of these volumes was prepared by D. W. Dewhirst in 1964.

(3) Bibliography of Astronomy, 1881–1898. The literature of this period was recorded on standard slips by the Observatoire Royal de Belgique. From the material (some 52,000 items) a microfilm version was produced by University Microfilms Limited, Tylers Green, High Wycombe, Buckinghamshire, England, in 1970.

(4) Astronomischer Jahresbericht, 1899 gegründet von Walter Wislicenus, herausgegeben vom Astronomischen Rechen-Institut in Heidelberg (formerly in Berlin), Verlag W. de Gruyter, Berlin. For the period from 1899 to 1968 sixty-eight volumes were published, each of which, in general, covers the literature of one year.

Concept of *Astronomy and Astrophysics Abstracts*

This abstracting service aims to present a comprehensive documentation of the literature in all fields of astronomy and astrophysics and their border fields. It appears in semi-annual volumes. Two of these volumes cover the literature of one calendar year. Every effort will be made to ensure that the average time interval between the receiving date of the original documents and publication of the abstracts will not exceed eight months. This time interval is near to that achieved by monthly abstracting journals, compared to which our system of accumulation of information over six months offers the advantage of greater convenience for the user.

The main characteristics of the concept of *Astronomy and Astrophysics Abstracts* may be summarized as follows:

(1) The subdivision of astronomy and its border fields into subject categories is facilitated by the fact that the astronomical objects appear to be particularly well suited for the formation of categories. It may be assumed that such subdivisions can be maintained for a long period. Experience shows, however, that progress in research might imply minor changes in the classification scheme.

(2) Each paper has been classified into one of 106 numbered subject categories and given a serial number within the category. In this way each item is numbered by six figures: the first three indicate the number of the category, the following three the serial number within the category. Reference to an abstract in Volume 1 is indicated by "01" before the number of the category; for example: 01.074.028, denotes Volume 1, category 074, abstract 028. A paper might be classified into more than one category. In this case, its abstract is placed only in one category, whereas in the other categories only cross references are given. These are listed at the end of each category.

(3) Authors' abstracts are used whenever possible. Popular articles are not abstracted.

(4) If possible, titles of papers and abstracts are given in English. A special reference is made to titles which we have not taken in the original language.

The whole material was recorded by means of modified ITT 3030 microcomputers. All text recording programs and other data processing software were developed by Multicom GmbH, Gröbenzell, F. R. Germany and by our staff members as well. The index computations were carried out on the IBM 3081 D computer of the University of Heidelberg.

Classification Systems

The two most common and widely used classification systems in astronomy and astrophysics are given by Class 9 of the International Classification System for Physics, published by the International Council of Scientific Unions Abstracting Board (Second edition 1978. ICSU-AB, 17 Rue Mirabeau, 75017 Paris, France, ISSN 0305-9618), and the *Astronomy and Astrophysics Abstracts* classification. In order to facilitate literature searches, we introduce a concordance relation between these two very different systems. This solution is only a unilateral one. Starting from the fourth hierarchical level of the ICSU-AB system, the appropriate *Astronomy and Astrophysics Abstracts* chapter numbers are listed. This cannot imply an identical content of the respective chapters in both systems. In many cases there is only a rather partial concordance, and therefore the *Astronomy and Astrophysics Abstracts* numbers are enclosed in parentheses. Considering our objectives, only the astronomical part of Class 9 of the ICSU-AB scheme is covered.

Transliteration Scheme for the Russian Alphabet

The transliteration of the Russian alphabet in use in *Astronomy and Astrophysics Abstracts* is presented here.

А	а	a	П	п	p
Б	б	b	Р	р	r
В	в	v	С	с	s
Г	г	g	Т	т	t
Д	д	d	У	у	u
Е	е	e	Ф	ф	f
Ё	ё	e	Х	х	kh
Ж	ж	zh	Ц	ц	ts
З	з	z	Ч	ч	ch
И	и	i	Ш	ш	sh
Й	й	j	Щ	щ	shch
К	к	k	Ы	ы	y
Л	л	l	Ь	ь	
М	м	m	Э	э	eh
Н	н	n	Ю	ю	yu
О	о	o	Я	я	ya

This transliteration was recommended by the Abstracting Board of the International Council of Scientific Unions in 1969. It corresponds essentially to the transliteration proposed by the Academy of Sciences, Moscow. In this case the letters can be read and printed by usual data processing machines. If the names of Russian authors in the literature are transliterated in a different scheme, we present the names as they are given in the references cited and in addition in brackets according to our transliteration table.

Sources of Information

The majority of sources of information for this volume is given in category **001 Periodicals** and in category **008 Observatories, Institutes.** It may be noted that the titles of the periodicals are given in the original languages, and that Russian titles have been transliterated applying the transliteration scheme given above. Category 008 records publication series of observatories and astronomical institutes. Titles of the periodicals have been given following the recommendations of the "International List of Periodical Title Word Abbreviations" and its additions (see also **Abbreviations**). In most cases they permit recognition of the full title without recourse to the key in category 001. If other secondary sources have been consulted, we cite these papers and give reference to the respective services.

The total number of papers (some do not give names of authors) recorded in this volume amounts to 11,582.

Author, Subject, and Object Indexes

The subject category and the serial number have been used as a reference in all three indexes. These references are more precise than page references and offer considerable advantages in indexing by means of computers.

The author index of this volume contains 12,925 names.

We consider the subject index as an approximation to an optimal index covering all fields of astronomy and astrophysics. Starting with Volume 18, the subject index was enlarged to a certain extent in order to provide a thesaurus of astronomical and astrophysical terms. At present, the *Astronomy and Astrophysics Abstracts Vocabulary* 09.85, containing some 2,303 key words, is in use. This is done not only for the users' convenience, but also with the intention to propose the use of special key words to authors and publishers.

While each volume is scheduled to contain an author index and a subject index, the magnetic tapes containing the index information will be used to produce separate index volumes (authors and subjects) at intervals of five years.

Beginning with Vol. 39 *Astronomy and Astrophysics Abstracts* will additionally contain an object index providing a further key to the documents abstracted. For detailed information concerning selection, standardization, sorting, etc, see introduction to the object index.

The sorting program for the indexes is based on the IBM SORT/MERGE Program. This program sorts blank before hyphen and before letters. Apostrophes are ignored by a special routine.

The users are requested to inform us on spelling errors within the author index in order to assist us in eliminating mistakes in future cumulative indexes.

Astronomy and Astrophysics Abstracts is prepared at the *Astronomisches Rechen-Institut,* Heidelberg under the auspices of the *International Astronomical Union* on a non-profit basis. The editors urge publishers of literature related to astronomy and astrophysics to provide our service in due time with complimentary copies of their material.

Publications should be mailed to:

Astronomy and Astrophysics Abstracts
Astronomisches Rechen-Institut
Moenchhofstrasse 12–14
D 6900 Heidelberg 1
F. R. Germany
Telex: 461 336 ARIHD D

Concordance Relation

between the ICSU-AB International Classification System for Physics
and the *Astronomy and Astrophysics Abstracts* Classification Scheme

ICSU-AB International Classification System for Physics	Astronomy and Astrophysics Abstracts Classification Scheme
0 General	
01.10 Announcements, news, and organizational activities	
01.10.C	006 (010)
01.10.F	011
01.10.H	013 (010)
01.30 Physics literature and publications	
01.30.B	012 (014)
01.30.C	012
01.30.E	003
01.30.K	002 (003)
01.30.M	003
01.30.P	003
01.30.R	014
01.30.T	002
01.40 Education	014
01.50 Educational aids	014
01.60 Biographical, historical, and personal notes	004 (005, 006, 007)
01.65 History of science	004
01.90 Other topics of general interest	015
9 Geophysics, Astronomy, and Astrophysics	
91.10 Geodesy and gravity	
91.10.B	045
91.10.N	044
91.10.Q	081

ICSU-AB International Classification System for Physics	Astronomy and Astrophysics Abstracts Classification Scheme
91.25 **Geomagnetism and paleomagnetism; geoelectricity**	084
91.35 **Earth's interior structure and properties**	081
91.90 **Other topics in solid Earth physics**	081
92.60 **Meteorology**	082
92.65 **Atmospheric optics**	082
94.10 **Physics of the neutral atmosphere**	
94.10.B	082
94.10.D	082
94.10.F	082
94.10.G	082 (063)
94.10.H	082
94.10.L	082
94.10.N	106
94.10.Q	082
94.10.S	084
94.20 **Physics of the ionosphere**	
94.20.B	083
94.20.D	083
94.20.M	083 (084)
94.20.P	083 (084)
94.20.W	083 (062)
94.20.Y	083 (084)
94.30 **Physics of the magnetosphere**	
94.30.C	084
94.30.D	084
94.30.E	084
94.30.F	084 (062)
94.30.G	084 (062)
94.30.H	084
94.30.L	084
94.30.M	084
94.30.S	084
94.30.V	084 (074)
94.30.W	084 (078, 144)

ICSU-AB International Classification System for Physics	Astronomy and Astrophysics Abstracts Classification Scheme
94.40 Cosmic rays	
94.40.C	144
94.40.E	144 (078, 106)
94.40.H	078
94.40.K	144 (085)
94.40.L	144 (078)
94.40.V	105 (144)
94.60 Interplanetary space	
94.60.D	074
94.60.F	074
94.60.G	074 (062)
94.60.K	106
94.60.M	106
94.60.Q	074 (091, 094)
94.60.R	106 (062)
94.80 Aerospace facilities and techniques, space research	
94.80.P	053 (051)
94.80.R	053 (051)
94.80.W	035
95.10 Fundamental astronomy	
95.10.C	042 (043, 052)
95.10.E	042 (052)
95.10.G	041 (079, 095, 096)
95.10.J	041
95.30 Fundamental aspects of astrophysics	
95.30.C	061 (022)
95.30.E	022
95.30.G	022 (061)
95.30.J	063
95.30.L	062
95.30.Q	062
95.30.S	066 (161)
95.45 Observatories	008 (009)

ICSU-AB International Classification System for Physics	Astronomy and Astrophysics Abstracts Classification Scheme
95.55 Astronomical instruments	
95.55.B	032
95.55.C	032 (031)
95.55.E	032
95.55.J	033
95.55.L	035
95.65 Auxiliary and recording instruments	034
95.70 Other instrumentation and techniques **(including clocks, frequency standards, etc.)**	034 (036)
95.75 Techniques of observation and reduction	
95.75.D	036
95.75.F	036
95.75.H	036
95.75.K	036
95.75.M	036
95.75.P	036 (021)
95.80 Catalogues, atlases, etc.	002 (046)
96.10 General, solar nebula, and cosmogony	107 (091)
96.20 Moon	
96.20.B	094
96.20.D	094
96.20.J	094
96.30 Planets and satellites (excluding the moon)	
96.30.D	092
96.30.E	093
96.30.G	097
96.30.H	098
96.30.K	099
96.30.M	100
96.30.T	101

ICSU-AB International Classification System for Physics	Astronomy and Astrophysics Abstracts Classification Scheme
96.50 Other objects in the planetary system	
96.50.D	106
96.50.G	102 (103)
96.50.K	104
96.50.M	105
96.60 Solar physics	
96.60.C	080 (075)
96.60.F	071 (080)
96.60.K	080
96.60.M	071
96.60.N	073 (074)
96.60.P	074
96.60.Q	072
96.60.R	073 (076, 077)
96.60.S	073
96.60.V	078 (074)
97.10 Stellar characteristics and properties	
97.10.B	131
97.10.C	065 (061)
97.10.E	064 (063)
97.10.F	112 (064)
97.10.H	112 (064)
97.10.K	116 (065)
97.10.L	116 (065)
97.10.N	115
97.10.Q	115
97.10.R	115 (113, 114)
97.10.T	114
97.10.V	111
97.10.W	111
97.20 Normal stars (by class): general or individual	
97.20.D	121
97.20.R	126
97.30 Variable and peculiar stars (including novae)	
97.30.E	114 (112)
97.30.F	116

ICSU-AB International Classification System for Physics	Astronomy and Astrophysics Abstracts Classification Scheme
97.30.G	122 (123)
97.30.J	122 (123)
97.30.K	122 (123)
97.30.N	122 (123)
97.30.Q	124 (117, 122, 123)
97.30.S	122 (123)
97.60 Late stages of stellar evolution (including black holes)	
97.60.B	125
97.60.G	126 (067)
97.60.J	067
97.60.L	067
97.60.S	067 (065)
97.80 Binary and multiple stars (including extrasolar planetary systems)	
97.80.D	118 (002)
97.80.F	120
97.80.H	119
97.80.J	142 (117)
97.80.K	118
97.80.M	118
98.10 Stellar dynamics	151
98.20 Stellar clusters and associations	
98.20.C	152
98.20.E	153
98.20.H	154
98.40 Interstellar matter and nebulae	
98.40.B	131
98.40.C	131
98.40.F	131
98.40.H	132
98.40.J	133
98.40.K	131
98.40.M	134
98.40.N	125

ICSU-AB International Classification System for Physics	Astronomy and Astrophysics Abstracts Classification Scheme

98.50   The Galaxy; extragalactic objects and systems

98.50.C	157
98.50.E	157 (161)
98.50.H	157 (158, 161)
98.50.K	160
98.50.L	155
98.50.M	160 (156)
98.50.R	158
98.50.T	160

98.70   Other objects and background radiations of unknown origin or distances

98.70.D	141
98.70.J	159
98.70.L	133
98.70.Q	142 (143)
98.70.S	144
98.70.V	161 (142, 143)

98.80   Cosmology

98.80.B	161
98.80.D	161 (066)
98.80.F	061 (161)

Abbreviations

Abbreviations used in *Astronomy and Astrophysics Abstracts* are primarily based on the 'International List of Periodical Title Word Abbreviations', prepared for the UNISIST/ICSU-AB Working Group on Bibliographic Descriptions (1970).

A.A.B.	Associazione Astrofili Bolognesi
Aarg.	Aargang
AAS	American Astronomical Society
AAVSO	American Association of Variable Star Observers
Abh.	Abhandlung—
Abstr.	Abstract—
Abt.	Abteilung
Acad.	Academi—, Academy
Accad.	Accademi—
Act.	Active, Activit—
Adm.	Administr—
Adv.	Advanc—
Aehron.	Aehronomi—
Aeron.	Aeronom—
Aeronaut.	Aeronauti—
Aerosp.	Aerospace
Afr.	Africa—
AG	Astronomische Gesellschaft
AIAA	American Institute of Aeronautics and Astronautics
AJB	Astronomischer Jahresbericht
Akad.	Akadem—
Ala.	Alabama
Alm.	Almanac—
Am.	America—, Amerika—, Amerique—
Amat.	Amateur—
Amst.	Amsterdam
An.	Anais, Anale—, Anali—, Anals
Anal.	Analis—, Analit—, Analys—, Analyt—
Angew.	Angewandt—
Ann.	Annaes, Annal—
Annu.	Annu—
Anst.	Anstalt
Anu.	Anual—, Anuar—
Anz.	Anzeiger
Appl.	Applied
Arb.	Arbeit
Arch.	Archiv—
Årg.	Årgang
Argent	Argentin—
Ariz.	Arizona
Ark.	Arkiv—
Arkh.	Arkhiv—
Artif.	Artifici—
ASA	Astronomical Society of Australia
Asoc.	Asocia—
ASP	Astronomical Society of the Pacific
ASSA	Astronomical Society of Southern Africa
Assem.	Assembl—
Assoc.	Associ—
Assoz.	Assozi—
Astrofis.	Astrofisic—
Astrofiz.	Astrofizi—
Astrometr.	Astrometr—
Astron.	Astronom—
Astronaut.	Astronauti—, Astronauty—
Astrophys.	Astrophys—
ASV	Astronomical Society of Victoria
ASWA	Astronomical Society of Western Australia
At.	Atom—

Atmos.	Atmosf—, Atmosph—
Aust.	Australi—
BAA	British Astronomical Association
Barc.	Barcelona
Bayer.	Bayerisch—
Beitr.	Beitrag, Beiträge
Belg.	Belge—, Belgi—
Beob.	Beobacht—
Beogr.	Beograd—
Ber.	Bericht—
Bibl.	Bibliot—
Bibliogr.	Bibliograf—, Bibliograph—
BIH	Bureau International de l'Heure
Bimest.	Bimestr—
Bl.	Blatt, Blätter
Bol.	Boletin
Boll.	Bolletino
Br.	British
Bras.	Brasil—
Brun.	Brunens—
Bruss.	Brussel, s
Brux.	Bruxelles
Bul.	Buleten—, Buletin—, Bulten
Bulg.	Bulgar—
Bull.	Bulletin—, Bullettino
Bur.	Bureau—
Byul.	Byuleten—, Byuletin—
Byull.	Byulleten—
C. R.	Comptes Rendus
Cah.	Cahier—
Calif.	California
Camb.	Cambridge
Can.	Canadi—, Canada
Carol.	Carolina—
Cas.	Casopis
Cat.	Catalog—
Celest.	Celestial
Cent.	Center, Central, Centrale, Centrally, Centre
Cercet.	Cercetari
Cesk.	Ceskoslov—
Chem.	Chemi—
Chim.	Chimi—
Chin.	Chinese
Chron.	Chronic—, Chronik, Chronique
Chronom.	Chronometr—
Cie.	Compagnie
Cienc.	Ciencia—
Cient.	Cientific—
Circ.	Circolar—, Circolo, Circolaire—, Circular—, Circulo
Cirk.	Cirkulaer—
Cl.	Clasa, Classe—
Co.	Companies, Company
Coll.	College
Collect.	Collect—
Colloq.	Colloqui—
Colo.	Colorado

Comet.	Cometary
Commentat.	Commentat—
Commun.	Communica—
Comput.	Computation, Computer—, Computing
Comun.	Comunica—
Conf.	Conferen—
Congr.	Congres—
Conn.	Connecticut
Contract.	Contract—
Contrib.	Contribu—
Copenh.	Copenhagen
Cosm.	Cosmic—
Cosmochim.	Cosmochimi—
COSPAR	Committee on Space Research
Crystallogr.	Crystallograph—
CSIRO	Commonwealth Scientific and Industrial Research Organization
Cult.	Cultur—, Cultuur
Curr.	Current
Czech.	Czechoslovak—
D. C.	District of Columbia
DDR	Deutsche Demokratische Republik
Del.	Delaware
Dep.	Departament, Département, Department
Dev.	Development—, Développement—
Dig.	Digest
Dir.	Director—
Diss.	Disserta—
Div.	Divis—
Doc.	Document—
Dok.	Dokument—
Dokl.	Doklad—
Dom.	Dominion
Dtsch.	Deutsch
Ed.	Edit—
Edinb.	Edinburgh
Ehksp.	Ehksperiment—
Eidg.	Eidgenössisch—
Eksp.	Eksperiment—
Electron.	Electroni—
Eng.	Engineer—
Environ.	Environment—
Equip.	Equipement, Equipment
Ergeb.	Ergebnis—
ESA	European Space Agency
ESO	European Southern Observatory
Espec.	Especial—
ESRO	European Space Research Organization
Eur.	Europ—
Eval.	Evaluation—
Exp.	Experiment—
Extraterr.	Extraterrestr—
F. R. Germany	Federal Republic of Germany
Fac.	Facolt—, Faculd—, Facult—
Fak.	Fakult
Fasc.	Fascicul—
Fenn.	Fenni—
Finn.	Finni—
Fis.	Fisic—, Fisik—
Fiz.	Fizic—, Fizik—, Fizyk—
Fla.	Florida
Fluid.	Fluidi—
Fond.	Fondation—, Fondazione
Fortschr.	Fortschritt—
Fotogr.	Fotograf—
Found.	Foundation—
Fr.	Français—
Freq.	Frequen—
Fundam.	Fundamenta—
Fys.	Fysik—, Fysisch, Fysisk—
Fyz.	Fyzik—

G.	Giornale
Ga.	Georgia
Gaz.	Gazeta, Gazette
Gazz.	Gazzetta
Gen.	General
Geochem.	Geochem—
Geochim.	Geochim—
Geod.	Geodaes—, Geodaet—, Geodes—, Geodet—, Geodez—
Geofis.	Geofis—
Geofiz.	Geofiz—
Geofys.	Geofys—
Geogr.	Geograf—, Geograph—
Geokhim.	Geokhim—
Geol.	Geolog—, Geolosk—
Geomagn.	Geomagneti—
Geophys.	Geophys—
Ges.	Gesellschaft
Gesch.	Geschichte
Gl.	Glavno—
Glas.	Glasnik
Gos.	Gosudarst—
Gov.	Government—
Grenzgeb.	Grenzgebiet—
GSFC	Goddard Space Flight Center
H.M.	Her Majesty's, His Majesty's
Hamb.	Hamburg
Handb.	Handbook, Handbuch
Heidelb.	Heidelberg—
Helv.	Helveti—
Her.	Herald—
Hist.	History
Hochsch.	Hochschule
Hoegsk.	Hoegskol—
HR-diagram	Hertzsprung-Russell diagram
Hung.	Hungar—
Hydrogr.	Hydrograf—, Hydrograph—
IAF	International Astronautical Federation
IAU	International Astronomical Union
IBM	International Business Machines Corporation
ICSTI	International Council for Scientific and Technical Information
ICSU	International Council of Scientific Unions
ICSU-AB	International Council of Scientific Unions— Abstracting Board
IEEE	Institute of Electrical and Electronics Engineers
Ill.	Illinois
Inc.	Incorporated
Ind.	Industr—
Inf.	Informat—, Informaz—, Informe—
Ing.	Ingenieur
INIS	International Nuclear Information System
INSPEC	International Information Services for the Physics and Engineering Communities
Inst.	Institut—, Instytut—
Instrum.	Instrument—
Int.	International, Internazional—
Intell.	Intelligenc—
Inter.	Intérieur—, Interior
Interplanet.	Interplanetary
Intez.	Intezet—
Invest.	Investiga—
Ionos.	Ionosfer—, Ionospher—
Ir.	Irish
Iskusstv.	Iskusstvenn—
Isr.	Israel—
Issled.	Issledovan—
Ist.	Istitut
Ital.	Itali—
Izd.	Izdatel—
Izv.	Izvesti—

J.	Joernaal –, Jornal –, Journal –
Jaarb.	Jaarboek –
Jahr.	Jahrbuch, Jahrbücher
Jahresber.	Jahresbericht –
Jahresschr.	Jahresschrift
Jahrg.	Jahrgang
JPL	Jet Propulsion Laboratory
Jpn.	Japan –
K.	Königlich –, Koninkljik –, Kunglig –
Kans.	Kansas
Kartogr.	Kartograf –
Kernforsch.	Kernforschung
Kernphys.	Kernphysik –
Khem.	Khemyi –
Khim.	Khimi –
Kim.	Kimija –, Kimya
Kl.	Klass –
Kolloq.	Kolloquium –
Komet.	Kometnyj
Komm.	Kommission –
Konf.	Konfer –
Kongr.	Kongress
Kosm.	Kosmich –
Kosmog.	Kosmogon –
Kozp.	Kozponti
KPNO	Kitt Peak National Observatory
Ky.	Kentucky
La.	Louisiana
Lab.	Laborato –
LEST	Large European Solar Telescope
Lett.	Letter –, Lettra, Lettre
Libr.	Librair –, Librar –
Madr.	Madrid
Mag.	Magasin, Magazin –
Magn.	Magneti –, Magnitn –
Mar.	Marin –
Mass.	Massachusetts
Mat.	Matemaat –, Matemat –
Mater.	Material –
Math.	Mathemat –
Md.	Maryland
Meas.	Measur –
Mec.	Mecani –
Mech.	Mechani –
Medd.	Meddelande –, Meddelelse
Meded.	Mededeeling, Mededeling –
Mekh.	Mekhani –
Mem.	Memento –, Memoir –, Memori –, Memory –, Memuary
Memo.	Memorand –
Mens.	Mensile, Mensual –, Mensuel –
Messtech.	Messtechni –
Meteorol.	Meteorolog –
Mex.	Mexic –
Mich.	Michigan
Micromec.	Micromecaniq –
Miner.	Mineral, Minerale –, Minerali –
Mineral.	Mineralog –
Minn.	Minnesota
Miss.	Mississippi
MIT	Massachusetts Institute of Technology
Mitt.	Mitteilung –
Mo.	Missouri
Mod.	Modern –
Mol.	Molecul –, Molekul –
Mon.	Monat, Monatlich –, Month –
Monogr.	Monograph –
Mont.	Montana
MPI	Max-Planck-Institut
Mt.	Mount
Muench.	Muenchen
Mus.	Museum
N. C.	North Carolina
N. D.	North Dakota
N. H.	New Hampshire
N. J.	New Jersey
N. M.	New Mexico
N. Y.	New York
N. Z.	New Zealand
Nablyud.	Nablyudeni –
Nac.	Nacion –
Nachr.	Nachricht –
NASA	National Aeronautics and Space Administration
Nat.	Natur –
Natl.	National –
Naturforsch.	Naturforsch –
Naturwiss.	Naturwissenschaft –
Natuurkd.	Natuurkunde
Nauchn.	Nauchny –
Nauk.	Nauka, Naukite, Naukov –, Naukow –
Naut.	Nautic –
Nav.	Naval –
Navig.	Navigat –
Naz.	Nazion –
Nebr.	Nebraska
Ned.	Nederland –
Nev.	Nevada
Newsl.	Newsletter –
Not.	Notationes, Notic –, Notise, Notizi –
Nouv.	Nouveau –, Nouvell –
Nov.	Novoe
Nucl.	Nucléaire –, Nuclear –, Nucl –
Nukl.	Nukle –
Numer.	Numeri –
O-va	Obshchestva
O-vo	Obshchestvo
Obs.	Observ –
Österr.	Österreich –
Off.	Offic –
Okla.	Oklahoma
Opt.	Optic –, Optik –, Optique
Oreg.	Oregon
Oss.	Osserva –
Pa.	Pennsylvania
Pac.	Pacific
Paleontol.	Paleontolog –
Pap.	Paper –, Papier
Part.	Particle
Pekin.	Pekinens –
Perem.	Peremenn –
Period.	Periodi –
Petrol.	Petrolog –
Philos.	Philosoph –
Photogr.	Photograf –, Photograph –
Photogramm.	Photogrammetr –
Photom.	Photometr –
Phys.	Physic –, Physik –, Physique –, Physisch
Pict.	Picture –
Planet.	Planetary
Pol.	Polish, Polon –
Pr.	Prac –
Prelim.	Prelimin –
Prepr.	Preprint
Prib.	Pribor –
Prikl.	Prikladn –
Prilozh.	Prilozhen –
Prir.	Prirodn –
Prirodoved.	Prirodoved –

Probl.	Problem —
Proc.	Proceedings
Prod.	Prodott —, Produc —, Produkt,
Prog.	Progres —
Propag.	Propagation
Prospect.	Prospecting
Prov.	Provinc —, Provints —, Provinz —
Pubbl.	Pubblicazion —
Publ.	Publicac —, Publicas —, Publicat —,
	Publikas —, Publikat —
Q.	Quarterly
Quant.	Quantit —
R.	Royal
R. I.	Rhode Island
Radiat.	Radiati —
Radioact.	Radioactiv —, Radioaktiv —
Radioisot.	Radioisotop —
Rap.	Raport —
Rapp.	Rapport —
RAS	Royal Astronomical Society
Rec.	Record —
Rech.	Recherche —
Ref.	Referat —, Reference —, Referieren
Relat.	Related, Relation —
Relativ.	Relativit —
Rend.	Rendicont —
Rep.	Report —
Repr.	Reprint —
Repub.	Republi —
Res.	Research —
Result.	Resultad —, Resultat —
Rev.	Review —, Revisio, Revista, Revue —
Rezul't.	Rezul'tat —
Ric.	Ricerca, Ricerche
Riv.	Rivist —
Roum.	Roumain —
Rundsch.	Rundschau
S. Afr.	South Africa
S. C.	South Carolina
S. D.	South Dakota
SAF	Société Astronomique de France
SAI	Società Astronomica Italiana
Samml.	Sammlung —
SAO	Smithsonian Astrophysical Observatory
SAS	Société Astronomique de Suisse
Satell.	Satellite
Sb.	Sbornik —
Scand.	Scandinavi —
Sch.	Schul —
Schr.	Schrift —
Schriftenr.	Schriftenreihe
Schweiz.	Schweizer —
Sci.	Scienc —, Scient —, Scienz —
Scr.	Scripta, Scritt —
Secc.	Seccion —
Sect.	Secti —
Sekc.	Sekci —, Sekcj —
Sekt.	Sektion —, Sektor —
Sekts.	Sektsi —
Sel.	Seleccion —, Select —, Selek —, Selezione
Selsk.	Selskab —, Selskap —
Semin.	Séminair —, Seminar —
Sep.	Separat —
Ser.	Seria —, Serie —, Seriya
Serv.	Servic —, Serviz —
Sess.	Sessi —
Signal.	Signalétique —
Simp.	Simpoz —
Sin.	Sinica

Sitzungsber.	Sitzungsbericht —
Skr.	Skrift —
Smithson.	Smithsonian
Soc.	Sociedad —, Societ —
Sol.	Solar
Soln.	Solnechn —
Sonderdr.	Sonderdruck —
Soobshch.	Soobshchen —
South.	Southern
Spacecr.	Spacecraft
Spat.	Spatial —
Spec.	Special —
Spectrosc.	Spectroscop —
Spectrosk.	Spectroskop —
Spets.	Spetsial —
Spez.	Spezial —, Speziell —
SSR	Sovetskaya Sotsialisticheskaya
	Respublika
SSSR	Soyuz Sovetskikh Sotsialisticheskikh
	Respublik
St.	Saint —, Sankt —, Sant —
—St.	—Straße, Street
Stand.	Standard —, Standart —
Sternw.	Sternwarte —
Stiint.	Stiintific —
Stn.	Station, Stazione
Stud.	Studia, Studie —, Studii
Supl.	Suplement —, Supliment —
Suppl.	Supplement —
Surv.	Survey —
Syd.	Sydney
Symp.	Sympos —, Sympoz —
Syst.	System —
Sz.	Szemle
Teach.	Teacher —, Teaching
Tec.	Tecni —
Tech.	Techni —
Technol.	Technolog —
Tecnol.	Tecnolog —
Teh.	Tehnic —, Tehnika, Tehnisk —
Tehnol.	Tehnolog —, Tehnolosk —
Tek.	Tekni —
Tekh.	Tekhni —
Tekhnol.	Tekhnolog —
Teknol.	Teknolog —
Telesc.	Telescop —
Telev.	Television —
Tenn.	Tennessee
Teor.	Teoret —, Teori —
Terr.	Terrestr —
Test.	Testing
Tex.	Texas
TH	Technische Hochschule
Theor.	Theoret —, Theori —
Tidschr.	Tidschrift —
Tidskr.	Tidskrift —
Tidsskr.	Tidsskrift —
Top.	Topic —
Torun.	Torunensis
Tr.	Trudy
Trans.	Transactions, Transazione
Tsentr.	Tsentral —
Tsirk.	Tsirkulyar —
TU	Technical University
Uch.	Uchen —
Uchebn.	Uchebn —
UK	United Kingdom
Umsch.	Umschau
UN	United Nations
Univ.	Universidad —, Universit —, Univerzitet —
Ups.	Upsaliens —

US	United States
USA	United States of America
USSR	Union of Soviet Socialist Republics
Utr.	Utrecht
Va.	Virginia
Var.	Various
Ver.	Verein—, Verenig—
Veränderl.	Veränderlich—
Verh.	Verhandl—
Vermess.	Vermessung—
Vermessungswes.	Vermessungswesen
Veröff.	Veröffentlich—
Vesn.	Vesnik
Vestn.	Vestnik
Vetensk.	Vetenskap—
Vgl.	Vergleich—
Vidensk.	Videnskab—, Videnskap
Vierteljahresschr.	Vierteljahresschrift—
Vierteljahrsschr.	Vierteljahrsschrift
VLB	Very Long Baseline
Volcanol.	Volcanolog—
Vopr.	Vopros—
Vortr.	Vorträge
Vses.	Vsesoyuzn—
Vt.	Vermont

Vyp.	Vypusk—
Vyssh.	Vyssh—
Vyzk.	Vyzkum—
W. Va.	West Virginia
Wash.	Washington
West.	Western
Wet.	Wetenschap—, Wetenskap—
Wis.	Wisconsin
Wiss.	Wissenschaft—
Wyo.	Wyoming
Yad.	Yadern—
Z.	Zeitschrift—
ZA	Zero Age
ZAED	Zentralstelle für Atomkernenergie-Dokumentation
Zap.	Zapisk—, Zapyisk—
Zaved.	Zaveden—
Zent.	Zentral
Zentralbl.	Zentralblatt
Zesz.	Zeszyt
Zh.	Zhurnal—
Zirk.	Zirkular

Periodicals, Proceedings, Books, Activities

001 Periodicals

A.A.O. Newsl.
A.A.O. Newsletter. Anglo–Australian Observatory, PO Box 296, Epping, N.S.W. 2121, Australia.

AAS Photo–Bull.
American Astronomical Society Photo–Bulletin. Working Group on Photographic Materials, 211 Space Sciences Building. University of Florida, Gainesville, Fla. 32611, USA. ISSN 0065–7433.

Abastumanskaya Astrofiz. Obs., Byull.
Abastumanskaya Astrofizicheskaya Observatoriya, Gora Kanobili. Byulleten'. Akademiya Nauk Gruzinskoj SSR. Izdatel'stvo Metsniereba, Tbilisi, USSR. ISSN 0375–6644.

Abh. Akad. Wiss. DDR
Abhandlungen der Akademie der Wissenschaften der DDR. Abteilung Mathematik, Naturwissenschaften, Technik. Akademie–Verlag, Berlin, German Democratic Republic.

Abh. Hamb. Sternw.
Abhandlungen aus der Hamburger Sternwarte. Hamburger Sternwarte, Universität Hamburg, Gojenbergsweg 2, D–2050 Hamburg 80, F.R. Germany. ISSN 0374–1583.

Abstr. Submitted Pap.
Abstracts of Submitted Papers. Institute of Astronomy, The Observatories, Madingley Road, Cambridge CB3 0HA, England.

Acad. R. Belg., Bull. Cl. Sci.
Académie Royale de Belgique, Bulletin de la Classe des Sciences (Koninklijke Academie van België, Mededelingen van de Klasse der Wetenschappen). 5e Série, Palais des Académies, Bruxelles, Belgium.

Acad. R. Belg., Mém. Cl. Sci.
Académie Royale de Belgique, Mémoires de la Classe des Sciences. Collection in 8°, 2e Série, Palais des Académies, Bruxelles, Belgium.

Acad. Sci. Est. SSR, Div. Phys. Math. Tech. Sci., Prepr.
Academy of Sciences of the Estonian SSR, Division of Physical, Mathematical and Technical Sciences, Preprints. Institute of Astrophysics and Atmospheric Physics, Academy of Sciences of the Estonian SSR, 202444 Tartu, Tôravere, Estonian SSR, USSR.

Acta Acust.
Acta Acustica. Science Press, Beijing, People's Republic of China. Subscription address: Guozi Shudian, PO Box 399, Beijing, People's Republic of China.

Acta Astron.
Acta Astronomica. An International Quarterly Journal. Polska Akademia Nauk, Komitet Astronomii, Państwowe Wydawnictwo Naukowe, Warszawa–Kraków, Poland. Subscription address: Ars Polona, 00–068 Warszawa, Krakowskie Przedmieście 7, Poland. ISSN 0001–5237.

Acta Astron. Sin.
Acta Astronomica Sinica. Purple Mountain Observatory, Academia Sinica. Nanjing, People's Republic of China. English translation in Chin. Astron. Astrophys. ISSN 0001–5245.

Acta Astronaut.
Acta Astronautica. Journal of the International Academy of Astronautics. Pergamon Press, Oxford – New York – Toronto – Paris – Frankfurt – Sydney. ISSN 0094–5765.

Acta Astrophys. Sin.
Acta Astrophysica Sinica. Beijing Astronomical Observatory, Academia Sinica, Beijing, People's Republic of China. Subscription address: Guozi Shudian, PO Box 399, Beijing, People's Republic of China. English translation in Chin. Astron. Astrophys.

Acta Cosmologica
Acta Cosmologica. Zeszyty Naukowe Uniwersytetu Jagiellońskiego. Państwowe Wydawnictwo Naukowe, Warszawa–Kraków, Poland. ISSN 0137–2386.

Acta Fac. Rerum Nat. Univ. Comenianae, Astron. Geophys.
Acta Facultatis Rerum Naturalium Universitatis Comenianae, Astronomia et Geophysica. Published for the Komenský University by Slovenské pedagogické nakladatel'stvo, 89112 Bratislava, Czechoslovakia.

Acta Geod. Geophys.
Acta Geodaetica et Geophysica. Institute of Geodesy and Geophysics, Academia Sinica, Beijing, People's Republic of China. Subscription address: Guozi Shudian, PO Box 399, Beijing, People's Republic of China.

Acta Geod. Geophys. Montan.
Acta Geodaetica, Geophysica et Montanistica. Akademiai Kiado, H–1054 Budapest, Alkotmany utca 21, Hungary. ISSN 0374–1842.

Acta Geophys. Pol.
Acta Geophysica Polonica. Subscription address: Ars Polona, 00–068 Warszawa, Krakowskie Przedmieście 7, Poland. ISSN 0001–5725.

Acta Geophys. Sin.
Acta Geophysica Sinica. Department of Geophysical Research, Academia Sinica, Beijing, People's Republic of China. Subscription address: Guozi Shudian, PO Box 399, Beijing, People's Republic of China. ISSN 0001–5733.

Acta Math. Sin.
Acta Mathematica Sinica. Academia Sinica, Beijing, People's Republic of China. Subscription address: Guozi Shudian, PO Box 399, Beijing, People's Republic of China.

Acta Mech. Sin.
Acta Mechanica Sinica. Academia Sinica, Beijing, People's Republic of China. Subscription address: Guozi Shudian, PO Box 399, Beijing, People's Republic of China. ISSN 0254–3060.

Acta Meteorol. Sin.
Acta Meteorologica Sinica. Subscription address: Guozi Shudian, PO Box 399, Beijing, People's Republic of China. ISSN 0577–6619.

Acta Phys. Austriaca
Acta Physica Austriaca. Springer–Verlag, Wien, Austria. ISSN 0001–6713.

Acta Phys. Hung.
Acta Physica Hungarica. Akademiai Kiado, Publishing House of the Hungarian Academy of Sciences, H–1054 Budapest, Alkotmany U. 21, Hungary. ISSN 0231–4428. Formerly entitled Acta Physica Academiae Scientiarum Hungaricae.

Acta Phys. Pol., Ser. A
Acta Physica Polonica, Ser. A. (General Physics, Solid State Physics, Atomic and Molecular Spectroscopy, Applied Physics). Polska Akademia Nauk, Warszawa, Poland. Subscription address: Ars Polona, 00–068 Warszawa, Krakowskie Przedmieście 7, Poland. ISSN 0587–4246.

Acta Phys. Pol., Ser. B
Acta Physica Polonica, Ser. B (Elementary Particle Physics, Nuclear Physics, Theory of Relativity, Field Theory). Polska Akademia Nauk, Warszawa, Poland. Subscription address: Ars Polona, 00–068 Warszawa, Krakowskie Przedmieście 7, Poland. ISSN 0587–4254.

Acta Phys. Sin.
Acta Physica Sinica. Institute of Physics, Academia Sinica, Beijing, People's Republic of China. ISSN 0372–736X.

Acta Phys. Slovaca
Acta Physica Slovaca. VEDA Publishing House of the Slovak Academy of Sciences, 89530 Bratislava, Klemensova 19, Czechoslovakia. ISSN 0323–0465.

Acta Polytech. Scand., Appl. Phys. Ser.
Acta Polytechnica Scandinavia, Applied Physics Series. Finnish Academy of Technical Sciences, Kansakoulukatu 10A, SF–00100 Helsinki, Finland. ISSN 0355–2721.

Acta Sci. Nat. Univ. Pekin.
Acta Scientiarum Naturalium Universitatis Pekinensis. Beijing, People's Republic of China. ISSN 0479–8023.

Acta Sci. Nat. Univ. Sunyatseni
Acta Scientiarum Naturalium Universitatis Sunyatseni (Zhongshandaxue Xuebao). Canton Post Office, Canton, People's Republic of China.

Acta Tech. Acad. Sci. Hung.
Acta Technica Academiae Scientiarum Hungaricae. Akademiai Kiado, Budapest 1363, PO Box 24, Hungary. ISSN 0001–7035.

Acta Tech. ČSAV
Acta Technica Československá akademie věd. Academia, Publishing House of the Czechoslovak Academy of Sciences, Vodičkova 40, 112 29 Praha 1, Czechoslovakia. Subscription address: John Benjamins N.V., Periodical Trade, Warmoesstraat 54, Amsterdam, The Netherlands. ISSN 0001–7043.

Acta Univ. Carol. Math. Phys.
Acta Universitatis Carolinae. Mathematica et Physica. Fakulta matematicko–fyzikální, Karlova universita, Praha, Czechoslovakia. ISSN 0001–7140.

A.D.I.O.N. Bull.
A.D.I.O.N. Bulletin. Association pour le Développement International de l'Observatoire de Nice. Observatoire de Nice, B.P. 252, F–06007 Nice Cedex, France.

Adv. Astronaut. Sci.
Advances in the Astronautical Sciences. American Astronautical Society, Publicatons Office, PO Box 28130, San Diego, Calif. 92128, USA. ISSN 0065–3438.

Adv. Phys.
Advances in Physics. Taylor & Francis Ltd., London, England. ISSN 0001–8732.

Adv. Space Res.
Advances in Space Research. The official journal of the Committee on Space Research (COSPAR). Pergamon Press, Oxford – New York – Toronto – Sydney – Paris – Frankfurt. ISSN 0273–1177.

Aéronaut. Astronaut.
L'Aéronautique et l'Astronautique. Editions Air et Cosmos, 6 Rue Anatole de la Forge, F–75017 Paris, France. ISSN 0001–9275.

AIAA J.
AIAA Journal. American Institute of Aeronautics and Astronautics, 1290 Avenue of the Americas, New York, N.Y. 10019, USA. ISSN 0001–1452.

AIP Conf. Proc.
AIP Conference Proceedings. American Institute of Physics, 335 East 45th Street, New York, N.Y. 10017, USA. ISSN 0094–243X.

Algoritm. Nebesnoj Mekh.
Algoritmy Nebesnoj Mekhaniki (Materialy Matematicheskogo Obespecheniya EhVM). Institut Teoreticheskoj Astronomii Akademii Nauk SSSR. 191187 Leningrad, D–187, nab. Kutuzova, dom–10. USSR.

Alta Freq.
Alta Frequenza. Ufficio Centrale AEI–CEI, Viale Monza 259, I–20126 Milano, Italy. ISSN 0002–6557.

Am. Assoc. Variable Star Obs. Bull.
The American Association of Variable Star Observers Bulletin. The American Association of Variable Star Observers, 187 Concord Avenue, Cambridge, Mass. 02138, USA.

Am. Assoc. Variable Star Obs. Rep.
The American Association of Variable Star Observers Report. The American Association of Variable Star Observers, 187 Concord Avenue, Cambridge, Mass. 02138, USA.

Am. J. Phys.
American Journal of Physics. Published for the American Association of Physics Teachers by the American Institute of Physics, 335 East 45th Street, New York, N.Y. 10017, USA. ISSN 0002–9505.

Am. Mineral.
American Mineralogist. Mineralogical Society of America, 2000 Florida Avenue, N.W., Washington, D.C. 20009, USA. ISSN 0003–004X.

Am. Sci.
American Scientist. Society of Sigma XI, 345 Whitney Avenue, New Haven, Conn. 06510, USA. ISSN 0003–0996.

An. Acad. Bras. Cienc.
Anais da Academia Brasileira de Ciencias. Caixa Postal 229, ZC–00 Rio de Janeiro gb, Brazil. ISSN 0001–3765.

An. Fis., Ser. A
Anales de Fisica, Serie A (Fenomenos e Interacciones). Real Sociedad Espanola de Fisica, Facultades de Ciencias, Ciudad Universitaria, Madrid 3, Spain. ISSN 0211–6243.

An. Fis., Ser. B
Anales de Fisica, Serie B (Aplicaciones, Metodes e Instrumentos). Real Sociedad Espanola de Fisica, Facultades de Ciencias, Ciudad Universitaria, Madrid 3, Spain. ISSN 0211–6251.

Anal. Chem.
Analytical Chemistry. American Chemical Society, 1155 16th Street N.W., Washington, D.C. 20036, USA. ISSN 0003–2700.

Anglo–Aust. Obs., Prepr.
Anglo–Australian Observatory, Preprints (AAO PP). Anglo–Australian Observatory, PO Box 296, Epping, N.S.W. 2121, Australia.

Anglo–Aust. Telesc., Annu. Rep.
Anglo–Australian Telescope, Annual Report. Anglo–Australian Observatory, PO Box 296, Epping, N.S.W. 2121, Australia.

Ann. Acad. Sci. Fenn., Ser. A VI
Annales Academiae Scientiarum Fennicae, Series A VI (Physica). Academia Scientiarum Fennica, Snellmaninkato 9–11, 00 170 Helsinki 17, Finland. ISSN 0066–2003.

Ann. Fond. Louis de Broglie
Annales de la Fondation Louis de Broglie. Fondation Louis de Broglie, 1 rue Montgolfier, F–75003 Paris, France.

Ann. Geophys.
Annales Geophysicae. European Geophysical Society. Gauthier–Villars. Subscription address: C.D.R. Centrale des Revues, 11 rue Gossin, F–92543 Montrouge Cedex, France.

Ann. Inst. Henri Poincaré, Sect. A
Annales de l'Institut Henri Poincaré, Section A (Physique Théoretique). 11 Rue Pierre–Curie, Paris 5, France. ISSN 0020–2339.

Ann. Isr. Phys. Soc.
Annals of the Israel Physical Society. c/o Department of Physics, Bar–Ilan University, Ramat–Gan, Israel. Adam Hilger Ltd., Bristol, England. ISSN 0309–8710.

Ann. Nucl. Energy
Annals of Nuclear Energy. Pergamon Press, Oxford – New York – Toronto – Paris – Frankfurt – Sydney. ISSN 0306–4549.

Ann. N.Y. Acad. Sci.
Annals of the New York Academy of Sciences. The New York Academy of Sciences, 2 East 63rd Street, New York, N.Y., USA. ISSN 0077–8923.

Ann. Obs. Astron. Alger
Annales de l'Observatoire Astronomique d'Alger. Observatoire Astronomique de l'Université d'Alger, Algiers, Algeria.

Ann. Phys. (Leipzig)
Annalen der Physik (Leipzig). Johann Ambrosius Barth, Salomonstr. 18D, Leipzig 701, German Democratic Republic. ISSN 0003–3804.

Ann. Phys. (N.Y.)
Annals of Physics (New York). Academic Press Inc., New York – London. ISSN 0003–4916.

Ann. Phys. (Paris)
Annales de Physique (Paris). Masson et Cie S.A., 120 Boulevard Saint–Germain, F–75280 Paris Cedex 06, France. ISSN 0003–4169.

Ann. Sci.
Annals of Science. Taylor & Francis Ltd., London, England. ISSN 0003–3790.

Ann. Sci. Univ. Besançon, Phys.
Annales Scientifiques de l'Université de Besançon, Physique. Institut des Sciences Naturelles, Place Leclerc, F–25000 Besançon, France. ISSN 0365–6543.

Ann. Shanghai Obs., Acad. Sin.
Annals of Shanghai Observatory, Academia Sinica. Shanghai Scientific and Technical Publishers, Shanghai, Rei Jing Er Street 450, People's Republic of China.

Ann. Soc. Sci. Brux., Sér. I
Annales de la Société Scientifique de Bruxelles, Série I (Sciences Mathématiques, Astronomiques et Physiques). Rue de Bruxelles 61, B–5000 Namur, Belgium. ISSN 0037–959X.

Ann. Tokyo Astron. Obs., Second Ser.
Annals of the Tokyo Astronomical Observatory, Second Series. Tokyo Astronomical Observatory, University of Tokyo, Mitaka, Tokyo 181, Japan. ISSN 0082–4704.

Ann. Univ.–Sternw. Wien
Annalen der Universitäts–Sternwarte Wien. Institut für Astronomie der Universität Wien, Türkenschanzstr. 17, A–1180 Wien, Austria. Published by Ferd. Dümmlers Verlag, Bonn, F.R. Germany. ISSN 0342–4030.

Annu. Rep. Astron. Inst. Greece
Annual Reports of the Astronomical Institutes of Greece. Published by the Greek National Committee for Astronomy, Athens, Greece.

Annu. Rep. (B.I.H.)
Annual Report (B.I.H.). Bureau International de l'Heure, 61, avenue de l'Observatoire, F–75014 Paris, France.

Annu. Rep. Dir., Mt. Wilson Las Campanas Obs.
Annual Report of the Director, The Mt. Wilson and Las Campanas Observatories. Mt. Wilson and Las Campanas Observatories of the Carnegie Institution of Washington, 813 Santa Barbara Street, Pasadena, Calif. 91101–1292, USA.

Annu. Rep. Geophys. Obs.
Annual Report of Geophysical Observations. The International Latitude Observatory of Mizusawa, Mizusawa-Shi, Iwate–Ken, Japan. ISSN 0579–5958.

Annu. Rep. Int. Polar Motion Serv.
Annual Report of the International Polar Motion Service. Central Bureau of the International Polar Motion Service, International Latitude Observatory of Mizusawa, Mizusawa-Shi, Iwate–Ken, Japan. ISSN 0074–7432.

Annu. Rep. Meteorol. Obs. Int. Latitude Obs. Mizusawa
Annual Report of the Meteorological Observations made at the International Latitude Observatory of Mizusawa. The International Latitude Observatory of Mizusawa, Mizusawa-Shi, Iwate–Ken, Japan. ISSN 0303–8378.

Annu. Rev. Astron. Astrophys.
Annual Review of Astronomy and Astrophysics. Annual Reviews Inc., 4139 El Camino Way, Palo Alto, Calif. 94306, USA. ISSN 0066–4146.

Annu. Rev. Earth Planet. Sci.
Annual Review of Earth and Planetary Sciences. Annual Reviews Inc., 4139 El Camino Way, Palo Alto, Calif. 94306, USA. ISSN 0084–6597.

Annu. Rev. Fluid Mech.
Annual Review of Fluid Mechanics. Annual Reviews Inc., 4139 El Camino Way, Palo Alto, Calif. 94306, USA. ISSN 0066–4189.

Annu. Rev. Nucl. Part. Sci.
Annual Review of Nuclear and Particle Science. Annual Reviews Inc., 4139 El Camino Way, Palo Alto, Calif. 94306, USA. ISSN 0163–8998.

Annu. Univ. Sofia
Annuaire de l'Université de Sofia. Faculté de Physique. Bibliothèque de l'Université, Sofia, Bulgaria. ISSN 0584–0279.

Antenna
L'Antenna. Via Monte Generoso 6/a, I–20155 Milano, Italy. ISSN 0003–5386.

Anz. Österr. Akad. Wiss., Math.–Naturwiss. Kl.
Anzeiger der Österreichischen Akademie der Wissenschaften, Mathematisch–Naturwissenschaftliche Klasse. Springer–Verlag, Wien, Austria. ISSN 0065–535X.

Appl. Opt.
Applied Optics. A monthly publication of the Optical Society of America. American Institute of Physics, 335 East 45th Street, New York, N.Y. 10017, USA. ISSN 0003–6935.

Appl. Phys., B
Applied Physics, B (Photophysics and Laser Chemistry). Springer–Verlag, Berlin – Heidelberg – New York – Tokyo. ISSN 0721–7269.

Appl. Phys. Lett.
Applied Physics Letters. American Institute of Physics, 335 East 45th Street, New York, N.Y. 10017, USA. ISSN 0003–6951.

Appl. Spectrosc.
Applied Spectroscopy. Society for Applied Spectroscopy, 428 East Preston Street, Baltimore, Md. 21202, USA. ISSN 0003–7028.

Appl. Spectrosc. Rev.
Applied Spectroscopy Reviews. Marcel Dekker Inc., 270 Madison Avenue, New York, N.Y. 10016, USA. ISSN 0570–4928.

Arch. Elektrotech.
Archiv für Elektrotechnik. Springer–Verlag, Berlin – Heidelberg – New York – Tokyo. ISSN 0003–9039.

Arch. Hist. Exact Sci.
Archive for History of Exact Sciences. Springer–Verlag, Berlin – Heidelberg – New York – Tokyo. ISSN 0003–9519.

Arch. Int. Hist. Sci.
Archives Internationales d'Histoire des Sciences. Istituto dell'Enciclopedia Italiana, fondata da Giovanni Treccani, Roma, Italy. ISSN 0003–9810.

Arch. Sci.
Archives des Sciences. Société de Physique et d'Histoire Naturelle de Genève. Subscription address: Librairie Payot, 6 Rue Grenus, CH–1211 Geneva 11, Switzerland. ISSN 0003–9705.

Archaeoastronomy (U.K.)
Archaeoastronomy. Supplement to Journal for the History of Astronomy. Science History Publications Ltd., Halfpenny Furze, Mill Lane, Chalfont St Giles, Bucks. England, HP8 4NR. ISSN 0142–7253.

Archaeoastronomy (U.S.A.)
Archaeoastronomy. The Bulletin of the Center for Archaeoastronomy, Space Sciences Building, University of Maryland, College Park, Md. 20742, USA. ISSN 0190–9940.

Archaeometry
Archaeometry. Research Laboratory for Archaeology and the History of Art, Oxford University, 6 Keble Road, Oxford OX1 3QJ, England. ISSN 0003–813X.

Archenhold–Sternw. Berlin–Treptow, Sonderdr.
Archenhold–Sternwarte Berlin–Treptow, Sonderdruck. Archenhold–Sternwarte, DDR–1193 Berlin, Alt Treptow 1, German Democratic Republic.

Archenhold–Sternw. Berlin–Treptow, Vortr. Schr.
Archenhold–Sternwarte Berlin–Treptow, Vorträge und Schriften. Archenhold–Sternwarte, DDR–1193 Berlin, Alt Treptow 1, German Democratic Republic. ISSN 0570–6262.

Ark. Fys. Semin. Trondheim
Arkiv for det Fysiske Seminar i Trondheim. c/o Institutt for Teoretisk Fysikk, Universitetet i Trondheim, NTH, N–7034 Trondheim, Norway. ISSN 0365–2459.

Ark. Mat.
Arkiv för Matematik. Institut Mittag–Leffler, Auravägen 17, S–182 62 Djursholm, Sweden. ISSN 0004–2080.

Armagh Obs., Repr.
Armagh Observatory, Reprints. Armagh Observatory, Armagh BT61 9DG, Northern Ireland.

Artif. Satell.
Artificial Satellites. Polish Academy of Sciences, Space Research Committee. Państwowe Wydawnictwo Naukowe, Warszawa–Lódź, Poland. Subscription address: Ars Polona, 00–068 Warszawa, Krakowskie Przedmieście 7, Poland. ISSN 0571–205X.

Astrofiz. Issled. Izv. Spets. Astrofiz. Obs.
Astrofizicheskie Issledovaniya. Izvestiya Spetsial'noj Astrofizicheskoj Observatorii. Akademiya Nauk SSSR. Izdatel'stvo Nauka, Leningradskoe Otdelenie, Leningrad. 199164 Leningrad, V–164, Mendeleevskaya l., 1, USSR. English translation in Bull. Spec. Astrophys. Obs. – North Caucasus. ISSN 0324–1459.

Astrofizika
Astrofizika. Izdatel'stvo Akademii Nauk Armyanskoj SSR, Erevan, USSR. English translation in Astrophysics. ISSN 0571–7132.

Astrofys. Inst., Vrije Univ. Bruss., Overdruk
Astrofysisch Instituut, Vrije Universiteit Brussel, Overdruk. Astrofysisch Instituut, Vrije Universiteit Brussel, Pleinlaan 2, B–1050 Brussel, Belgium.

Astron. Astrophys.
Astronomy and Astrophysics. A European Journal. Springer–Verlag, Berlin – Heidelberg – New York – Tokyo. ISSN 0004–6361.

Astron. Astrophys., Suppl. Ser.
Astronomy and Astrophysics, Supplement Series. A European Journal. Les Editions de Physique, Z.I. de Courtaboeuf, B.P. 112, F–91944 Les Ulis Cedex, France. ISSN 0365–0138.

Astron. Bull., Carter Obs.
Astronomical Bulletin, Carter Observatory. Carter Observatory, PO Box 2909, Wellington 1, New Zealand. ISSN 0373–7268.

Astron. Circ.
Astronomical Circular. Edited by the Chinese Astronomical Society. Compiled by the Editors of Acta Astronomica Sinica, Purple Mountain Observatory, Academia Sinica, Nanjing, People's Republic of China.

Astron. Contrib. Univ. Manchester, Ser. II: Jodrell Bank Repr.
Astronomical Contributions from the University of Manchester, Series II: Jodrell Bank Reprints. Nuffield Radio Astronomy Laboratories, Jodrell Bank, Macclesfield, Cheshire SK11 9DL, England.

Astron. Contrib. Univ. Manchester, Ser. III
Astronomical Contributions from the University of Manchester, Series III. Department of Astronomy, University of Manchester, Oxford Road, Manchester M13 9PL, England.

Astron. Data Cent. Bull.
AstronomicaL Data Center Bulletin. National Space Science Data Center/World Data Center A for Rockets and Satellites. National Aeronautics and Space Administration, Goddard Space Flight Center, Greenbelt, Md. 20771, USA.

Astron. Express
Astronomy Express. Cambridge University Press, London – New York – New Rochelle – Melbourne – Sydney. ISSN 0265–5365.

Astron. Her.
Astronomical Herald. Astronomical Society of Japan, Tokyo Astronomical Observatory, Oosawa Mitaka, Tokyo, Japan. ISSN 0374–2466.

Astron. Inst. "Anton Pannekoek", Univ. Amst., Repr.
Astronomical Institute "Anton Pannekoek", University of Amsterdam, Reprint.

Astron. Inst. Univ. Brno, Contrib.
Astronomical Institute of the University Brno, Contributions. Astronomical Institute, University Brno, Kotlářská 2, Brno, Czechoslovakia.

Astron. Inst. Univ. Brno, Publ.
Astronomical Institute of the University Brno, Publications. Astronomical Institute, University Brno, Kotlářská 2, Brno, Czechoslovakia.

Astron. J.
The Astronomical Journal. Published for the American Astronomical Society by the American Institute of Physics, 335 East 45th Street, New York, N.Y. 10017, USA. ISSN 0004–6256.

Astron. Mitt. Wien
Astronomische Mitteilungen Wien. Institut für Astronomie der Universität Wien, Türkenschanzstraße 17, A–1180 Wien, Austria.

Astron. Nachr.
Astronomische Nachrichten. Akademie–Verlag, DDR–108 Berlin, Leipziger Straße 3–4, German Democratic Republic. ISSN 0004–6337.

Astron. Obs. Trieste, Publ.
Astronomical Observatory Trieste, Publications. Osservatorio Astronomico di Trieste, Via G.B. Tiepolo 11, I–34131 Trieste, Italy.

Astron. Pap.
Astronomical Papers prepared for the use of the American Ephemeris and Nautical Almanac. Published by the Nautical Almanac Office, U.S. Naval Observatory by direction of the Secretary of the Navy and under the authority of Congress. U.S. Government Printing Office, Washington, D.C., USA.

Astron. Q.
The Astronomical Quarterly. Pachart Publishing House, 1130 San Lucas Circle, Tucson, Ariz. 85704, USA. ISSN 0364–9229.

Astron. Raumfahrt
Astronomie und Raumfahrt. Kulturbund der DDR, Zentrale Kommission Astronomie und Raumfahrt. Redaktionssitz: 9630 Crimmitschau, Pionier– und Jugendsternwarte "Johannes Kepler", Strasse der Jugend 8. Available from Zeitungsvertriebsamt, Abt. Export, 1004 Berlin, Strasse der Pariser Kommune 3–4, German Democratic Republic. ISSN 0587–565X.

Astron. Rechen–Inst. Heidelb., Mitt., Ser. A
Astronomisches Rechen–Institut Heidelberg, Mitteilungen, Serie A. Astronomisches Rechen–Institut, Mönchhofstraße 12–14, D–6900 Heidelberg, F.R. Germany.

Astron. Rechen–Inst. Heidelb., Mitt., Ser. B
Astronomisches Rechen–Institut Heidelberg, Mitteilungen, Serie B. Astronomisches Rechen–Institut, Mönchhofstraße 12–14, D–6900 Heidelberg, F.R. Germany.

Astron. Rep.
The Astronomical Reports. Polskie Towarzystwo Miłośników Astronomii, Polish Amateur Astronomical Society, ul. Ludwika Solskiego 30/8, PL–31–027 Kraków, Poland.

Astron. Sch.
Astronomie in der Schule. Verlag Volk und Wissen, DDR–1086 Berlin, Krausenstraße 50, Postfach 1213, German Democratic Republic. ISSN 0004–6310.

Astron. Tidsskr.
Astronomisk Tidsskrift. Astronomisk Selskab, Kobenhavn; Norsk Astronomisk Selskap, Oslo; Svenska Astronomiska Sällskapet, Stockholm. Subscription address: Svenska Astronomiska Sällskapet, Stockholms Observatorium, S–13300 Saltsjöbaden, Sweden. ISSN 0004–6345.

Astron. Tsirk.
Astronomicheskij Tsirkulyar. Izdavaemyj Byuro Astronomicheskikh Soobshchenij Akademii Nauk SSSR. ISSN 0365–7248.

Astron. Vestn.
Astronomicheskij Vestnik. Izdatel'stvo Nauka, 918, Moskva, K–9, USSR. English translation in Sol. Syst. Res. ISSN 0320–930X.

Astron. Zh.
Astronomicheskij Zhurnal. Akademiya Nauk SSSR. Izdatel'stvo Nauka, Moskva, 103717, GSP, Moskva, K–62, Podsosenskij per., 21, USSR. English translation in Soviet Astron. ISSN 0004–6299.

Astronomia
Astronomia. Periodico trimestrale dell'Unione Astrofili Italiani. Subscription address: L. Baldinelli, C.P. 1630, I–40100 Bologna A.D., Italy. ISSN 0392 2308.

Astronomie
L'Astronomie et Bulletin de la Société Astronomique de France. Société Astronomique de France, 3, Rue Beethoven, F–75016 Paris, France. ISSN 0004–6302.

Astronomy
Astronomy. Astro Media Corp., 625 E. St. Paul Avenue, PO Box 92788 Milwaukee, Wis. 53202, USA. ISSN 0091–6358.

Astrophys. J.
The Astrophysical Journal. Published by The University of Chicago Press for the American Astronomical Society. The University of Chicago Press, 5801 S. Ellis Avenue, Chicago, Ill. 60637, USA. ISSN 0004–637X.

Astrophys. J., Lett. Ed.
The Astrophysical Journal, Letters to the Editor. Published by The University of Chicago Press for the American Astronomical Society. The University of Chicago Press, 5801 S. Ellis Avenue, Chicago, Ill. 60637, USA. ISSN 0571–7248.

Astrophys. J., Suppl. Ser.
The Astrophysical Journal, Supplement Series. Published by The University of Chicago Press for the American Astronomical Society. The University of Chicago Press, 5801 S. Ellis Avenue, Chicago, Ill. 60637, USA. ISSN 0067–0049.

Astrophys. Lett.
Astrophysical Letters. Gordon and Breach Science Publishers Inc., New York – London – Paris. ISSN 0004–6388.

Astrophys. Prepr. Ser.
Astrophysics Preprint Series. Astronomy and Astrophysics Group, Faculty of Physics, Pontificia Universidad Católica de Chile, Casilla 6014, Santiago de Chile, Chile.

Astrophys. Relativ., Prepr. Ser.
Astrophysics and Relativity, Preprint Series. Department of Applied Mathematics and Astronomy, University College, PO Box 78, Cardiff CF1 1XL, England.

Astrophys. Space Phys. Rev.
Astrophysics and Space Physics Reviews. Soviet Scientific Reviews, Section E. Harwood Academic Publishers GmbH, PO Box 786, Cooper Station, New York, N.Y. 10276, USA. ISSN 0143–0432.

Astrophys. Space Sci.
Astrophysics and Space Science. An International Journal of Cosmic Physics. D. Reidel Publishing Company, Dordrecht – Boston. ISSN 0004–640X.

Astrophysics
Astrophysics. A cover–to–cover translation of Astrofizika of the Academy of Sciences of the Armenian SSR. Consultants Bureau, 227 West 17th Street, New York, N.Y. 10011, USA. ISSN 0004–6396.

Atmos. Environ.
Atmospheric Environment. Pergamon Press, Oxford – New York – Toronto – Paris – Frankfurt – Sydney. ISSN 0004–6981.

Atomkernenerg. Kerntech.
Atomkernenergie und Kerntechnik. Verlag Karl–Thiemig AG, Postfach 900740, D–8000 München 90, F.R. Germany. ISSN 0171–5747.

Atti Accad. Naz. Lincei, Ser. Ottava, Rend.
Atti della Accademia Nazionale dei Lincei. Serie Ottava, Rendiconti. Classe di Scienze fisiche, matematiche e naturali. Accademia Nazionale dei Lincei, Roma, Italy. ISSN 0001–4435.

Atti Accad. Sci. Torino I
Atti della Accademia delle Scienze di Torino I. Classe di Scienze fisiche, matematiche e naturali. Via Accademia delle Scienze 6, Via Maria Vittoria 3, Torino (208), Italy. ISSN 0001–4419.

Atti Fond. Giorgio Ronchi
Atti della Fondazione Giorgio Ronchi. Largo Enrico Fermi 1, I–50125 Arcetri–Firenze, Italy. ISSN 0015–606X.

Aust. Comput. J.
Australian Computer Journal. Australian Trade Publications, 28 Chippen Street, Chippendale, N.S.W. 2008, Australia. ISSN 0004–8917.

Aust. J. Astron.
Australian Journal of Astronomy. Astral Press, PO Box 107, Wembley, W.A. 6014, Australia. ISSN 0814–5628.

Aust. J. Geod. Photogramm. Surv.
Australian Journal of Geodesy, Photogrammetry, and Surveying. School of Surveying, University of New South Wales, PO Box 1, Kensington, NSW 2033, Australia. ISSN 0313–9220.

Aust. J. Phys.
Australian Journal of Physics. Commonwealth Scientific and Industrial Research Organization (CSIRO), 314 Albert Street, East Melbourne, Victoria 3002, Australia. ISSN 0004–9506.

Aust. Phys.
Australian Physicist. Australian Institute of Physics, Science Centre, 35–43 Clarence Street, Sydney, NSW 2000, Australia. ISSN 0004–9972.

Austrian Pap. Asteroids
Austrian Papers on Asteroids. Institut für Astronomie, Universitätsplatz 5, A–8010 Graz, Austria.

Automatica
Automatica. Pergamon Press, Oxford – New York – Frankfurt – Paris – Sydney – Toronto – Tokyo. ISSN 0005–1098.

BAV Mitt.
BAV Mitteilungen. Berliner Arbeitsgemeinschaft für Veränderliche Sterne e.V., Sternwarte, Munsterdamm 90, D–1000 Berlin 41, F.R. Germany.

BAV Rundbrief
BAV Rundbrief. Mitteilungsblatt der Berliner Arbeitsgemeinschaft für Veränderliche Sterne. BAV Berliner Arbeitsgemeinschaft für Veränderliche Sterne e.V., Sternwarte, Munsterdamm 90, D–1000 Berlin 41, F.R. Germany. ISSN 0405–5497.

BBSAG Bull.
Bedeckungsveränderlichen Beobachter der Schweizerischen Astronomischen Gesellschaft Bulletin. Available from K. Locher, Rebrain 39, 8624 Grüt, Switzerland.

B.I.H. Circ.
Bureau International de l'Heure (B.I.H.) Circulars A, D, E. 61, avenue de l'Observatoire, F–75014 Paris, France.

Biul. Obs. Astron. Uniw. M. Kopernika Toruniu
Biuletyn Obserwatorium Astronomicznego Uniwersytetu M. Kopernika w Toruniu. Institute of Astronomy, Nicolaus Copernicus University, Chopina 12/18, PL–87–100 Toruń, Poland.

Bol. Acad. Cienc. Fis. Mat. Nat.
Boletin de la Academia de Ciencias Fisicas Matematicas y Naturales. Academia de Ciencias Fisicas Matematicas y Naturales, Apartado de Correo 1421, Caracas, Venezuela. ISSN 0366–1652.

Bol. Asoc. Argent. Astron.
Boletin de la Asociación Argentina de Astronomía, La Plata, Argentina. ISSN 0571–3285.

Bol. Astron.
Boletin Astronômico. Observatório do Capricórnio, Prefeitura Municipal de Campinas–SP, Brazil.

Bol. Astron. Obs. Madr.
Boletín Astronómico del Observatorio de Madrid. Instituto Geografico Nacional, General Ibáñez de Ibero, 3. Madrid 3, Spain. ISSN 0373–7101.

Bol. Astron. R Muscae
Boletin Astronomico y R Muscae. Publicación Oficial de la Liga Ibero–Americana de Astronomía, LIADA. Apartado 700, Mérida 5101–A, Venezuela. Casilla de Correos Nro 51 Suc. 48B, 1448 Buenos Aires, Argentina.

Bol. Inst. Tonantzintla
Boletin del Instituto de Tonantzintla. Instituto Nacional de Astrofisica, Optica y Electronica, Apartados Postales Nos. 216 y 51, Puebla, Pue., Mexico. ISSN 0303–7584.

Bol. Obs. Ebro
Boletín del Observatorio del Ebro. Consejo Superior de Investigaciones Científicas, Observatorio del Ebro, Roquetas (Tarragona), Spain. ISSN 0211–5166.

Boll. Geofis. Teor. Appl.
Bollettino di Geofisica Teorica ed Applicada. Osservatorio Geofisico Sperimentale, I–34123 Trieste, Italy. ISSN 0006–6729.

Boundary–Layer Meteorol.
Boundary–Layer Meteorology. D. Reidel Publishing Company, Dordrecht – Boston – London. ISSN 0006–8314.

Boyden Obs., Publ.
Boyden Observatory, Publication. Boyden Observatory, Astronomy Department, University of the Orange Free State, PO Box 339, Bloemfontein 9300, South Africa.

Boyden Obs., Repr.
Boyden Observatory, Reprint. Boyden Observatory, Astronomy Department, University of the Orange Free State, PO Box 339, Bloemfontein 9300, South Africa.

Br. Astron. Assoc. Circ.
British Astronomical Association Circular. British Astronomical Association, Burlington House, Piccadilly, London W1V 0NL, England. ISSN 0264–4185.

Br. J. Philos. Sci.
British Journal for the Philosophy of Science. Cambridge University Press, Cambridge – London – New York – New Rochelle – Melbourne – Sydney. ISSN 0007–0882.

Br. J. Photogr.
British Journal of Photography. Henry Greenwood & Co. Ltd., 28 Great James Street, London WC1N 3HL, England. ISSN 0007–1196.

Bulg. J. Phys.
Bulgarian Journal of Physics. Bulgarian Academy of Sciences, Faculty of Physics, 5 Anton Ivanov Blvd., 1126 Sofia, Bulgaria. ISSN 0323–9217.

Bull. Am. Astron. Soc.
Bulletin of the American Astronomical Society. Published by the American Institute of Physics, 335 East 45th Street, New York, N.Y. 10017, USA. ISSN 0002–7537.

Bull. Assoc. Fr. Obs. Etoiles Variables
Bulletin de l'Association Francaise des Observateurs d'Etoiles Variables. Revue trimestrielle. A.F.O.E.V. Observatoire de Lyon, F–69230 Saint Genis Laval, France. ISSN 0153–9949.

Bull. Astron.
Bulletin Astronomique. Observatoire Royal de Belgique (Astronomisch Bulletin. Koninklijke Sterrenwacht van België). Observatoire Royal de Belgique, 3, avenue Circulaire, Uccle, B–1180 Bruxelles, Belgium.

Bull. Astron. Inst. Czech.
Bulletin of the Astronomical Institutes of Czechoslovakia. Academia, Publishing House of the Czechoslovak Academy of Sciences, Vodičkova 40, 11229 Praha 1, Czechoslovakia. ISSN 0004–6248.

Bull. Astron. Soc. India
Bulletin of the Astronomical Society of India. Astronomical Society of India, Osmania University, Hyderabad 500 007, India. ISSN 0304–9523.

Bull. Aust. Math. Soc.
Bulletin of the Australian Mathematical Society. University of Queensland Press, St. Lucia, Queensland 4067, Australia. ISSN 0004–9727.

Bull. Cl. Sci., Acad. R. Belg.
Bulletin de la Classe des Sciences, Académie Royale de Belgique. Académie Royale des Sciences des Lettres et de Beaux–Arts de Belgique, Brussels, Belgium. ISSN 0001–4141.

Bull. Crimean Astrophys. Obs.
Bulletin of the Crimean Astrophysical Observatory. A cover-to–cover translation of Izv. Krymskoj Astrofiz. Obs. Allerton Press, Inc. 150 5th Avenue, New York, N.Y. 10011, USA.

Bull. Etoiles Tardives Spectre Particulier
Bulletin sur les Etoiles Tardives à Spectre Particulier (Newsletter of Chemically Peculiar Late–type Stars). Observatoire de Strasbourg, 11, rue de l'Université, F–67000 Strasbourg, France.

Bull. Géod.
Bulletin Géodésique. The Journal of the International Association of Geodesy. Bureau Central de l'Association Internationale de Géodésie, 39 rue Gay–Lussac, F–75005 Paris, France. ISSN 0007–4632.

Bull. Geogr. Surv. Inst.
Bulletin of the Geographical Survey Institute. Geographical Survey Institute, Ministry of Construction, Kitasato–1, Yatabe–Machi, Tsukuba–Gun, Ibaraki-ken, Japan. ISSN 0373–7160.

Bull. Geophys.
Bulletin of Geophysics. National Central University, Chung–Li, Taiwan, Republic of China. ISSN 0253–4800.

Bull. Inf. Cent. Données Stellaires
Bulletin d'Information du Centre de Données Stellaires. Observatoire de Strasbourg, 11, rue de l'Université, F–67000 Strasbourg, France. ISSN 0242–6536.

Bull. Inf. Etoiles Be
Bulletin d'Information sur les Etoiles Be (Be Star Newsletter). Observatoire de Strasbourg, 11, rue de l'Université, F–67000 Strasbourg, France.

Bull. Obs. Astron. Belgr.
Bulletin de l'Observatoire Astronomique de Belgrade. Observatoire Astronomique de Belgrade, Beograd, Volgina 7, Yugoslavia. ISSN 0373–3734.

Bull. Res. Inst. Sci. Meas., Tôhoku Univ.
Bulletin of the Research Institute for Scientific Measurements, Tôhoku University, Sendai, Japan. ISSN 0040–8689.

Bull. Soc. R. Sci. Liège
Bulletin de la Société Royale des Sciences de Liège. L'Université, 15 Avenue des Tilleurs, Liège, Belgium. ISSN 0037–9565.

Bull. Spec. Astrophys. Obs. – North Caucasus
Bulletin of the Special Astrophysical Observatory – North Caucasus. A cover–to–cover translation of Astrofiz. Issled. Izv. Spets. Astrofiz. Obs. Allerton Press, Inc., 150 5th Avenue, New York, N.Y. 10011, USA.

Bull., Time Serv. Mizusawa Obs.
Bulletins, Time Service of the Mizusawa Observatory. The International Latitude Observatory of Mizusawa, Mizusawa–Shi, Iwate–Ken, Japan. ISSN 0580–6585.

Bull. Tokyo Gakugei Univ., Ser. IV
Bulletin of Tokyo Gakugei University, Series IV (Mathematics and Natural Sciences) 4–1–1 Nukui–kita–machi, Koganei, Tokyo, Japan. ISSN 0371–6813.

Byull. Inst. Astrofiz.
Byulleten' Instituta Astrofiziki. Akademiya Nauk Tadzhikskoj SSR. Izdatel'stvo Donish, Dushanbe, USSR. ISSN 0568–6865.

Byull. Inst. Teor. Astron.
Byulleten' Instituta Teoreticheskoj Astronomii. Akademiya Nauk SSSR. Leningradskoe Otdelenie, Izdatel'stvo Nauka, Leningrad V–164, Mendeleevskaya l., 1, USSR. ISSN 0002–3302.

C. R. Acad. Sci., Sér. Gén., Vie Sci.
Comptes Rendus de l'Académie des Sciences, Série Générale, La Vie des Sciences. Académie des Sciences, Paris. Subscription address: Gauthier–Villars, C.D.R., Centrale des Revues, 11, rue Gossin, F–92543 Montrouge Cedex, France.

C. R. Acad. Sci., Sér. II
Comptes Rendus de l'Académie des Sciences, Série II: Mécanique, Physique, Chimie, Sciences de l'Univers, Sciences de la Terre. Académie des Sciences, Paris. Subscription address: Gauthier–Villars, C.D.R., Centrale des Revues, 11, rue Gossin, F–92543 Montrouge Cedex, France. ISSN 0249–6305.

Can. J. Earth Sci.
Canadian Journal of Earth Sciences. National Research Council of Canada, Ottawa KIA OR6, Canada. ISSN 0008–4077.

Can. J. Phys.
Canadian Journal of Physics. National Research Council of Canada, Ottawa KIA OR6, Canada. ISSN 0008–4204.

Carter Obs., Repr. Ser.
Carter Observatory, Reprint Series. Carter Observatory, PO Box 2909, Wellington 1, New Zealand.

Cartes Synoptiques
Cartes Synoptiques de la Chromosphère Solaire et Catalogues des Filaments et des Centres d'Activité. Observatoire de Paris, Section d'Astrophysique, F–92190 Meudon, France.

Celest. Mech.
Celestial Mechanics. An International Journal of Space Dynamics. D. Reidel Publishing Company, Dordrecht – Boston. ISSN 0008–8714.

Cent. Astron. Sci. Spat., Obs. Sol.
Centre de l'Astronomie et des Sciences Spatiales, Observations Solaires. Centre de l'Astronomie et des Sciences Spatiales, Academiei Republicii Socialiste România, Bucuresti, Rumania.

Cent. Astrophys., Prepr. Ser.
Center for Astrophysics, Preprint Series. Center for Astrophysics, 60 Garden St., Cambridge, Mass. 02138, USA.

Centaurus
Centaurus. International Magazine of the History of Mathematics, Science, and Technology. Munksgaard Ltd., Copenhagen, Denmark. ISSN 0008–8994.

Česk. Čas. Fyz., Sekce A
Československý časopis pro fyziku, Sekce A. Academia Publishing House of the Czechoslovak Adademy of Sciences, Vodičkova 40, 11229 Praha 1, Czechoslovakia. ISSN 0009–0700.

Chem. Phys.
Chemical Physics. North–Holland Publishing Company, PO Box 211, 1000 AE Amsterdam, The Netherlands. ISSN 0301–0104.

Chem. Phys. Lett.
Chemical Physics Letters. North–Holland Publishing Company, PO Box 211, 1000 AE Amsterdam, The Netherlands. ISSN 0009–2614.

Chin. Astron. Astrophys.
Chinese Astronomy and Astrophysics. A selected translation from the current issues of Acta Astronomica Sinica, Acta Astrophysica Sinica and Chinese Journal of Space Science. Pergamon Press, Oxford – New York – Toronto – Paris – Frankfurt – Sydney. ISSN 0275–1062.

Chin. J. Phys.
Chinese Journal of Physics. Physical Society of the Republic of China, Physics Department, National Taiwan University, Taipei, Taiwan, Republic of China. ISSN 0577–9073.

Chin. J. Space Sci.
Chinese Journal of Space Science. Subscription address: China International Book Trading Corporation, Guozi Shudian, PO Box 2820, Beijing, People's Republic of China.

Chin. Phys.
Chinese Physics. Selected translations from current issues of major Chinese physics and astronomy journals. American Institute of Physics, 335 East 45th Street, New York, N.Y. 10017, USA. ISSN 0273–429X.

Ciel
Le Ciel. Bulletin de la Société Astronomique de Liège. Société Astronomique de Liège, B–4410 Vottem, Belgium.

Ciel Espace
Ciel et Espace. Association Francaise d'Astronomie, 115 rue de Charenton, F–75012 Paris, France.

Ciel Terre
Ciel et Terre. Bulletin de la Société Royale Belge d'Astronomie, de Météorologie et de Physique du Globe. Société Royale Belge d'Astronomie, de Météorologie et de Physique du Globe, 3, avenue Circulaire, B–1180 Bruxelles, Belgium. ISSN 0009–6709.

Circ. Czech. Obs., Time Latitude
Circular of the Czechoslovak Observatories, Time and Latitude. Czechoslovak Academy of Sciences, Astronomical Institute, Prague, Czechoslovakia.

Circ. Inf.
Circulaire d'Information. Union Astronomique Internationale. Commission des Etoiles Doubles. Observatoire de Paris, F–92190 Meudon, France.

Circ. Stn. Astron. Int. Latitudine, Carloforte–Cagliari
Circolari della Stazione Astronomica Internazionale di Latitudine, Carloforte–Cagliari. Series A, B, C. Stazione Astronomica Internazionale di Latitudine, Carloforte–Cagliari, Italy.

Circ. Time Latitude Serv.
Circular Time and Latitude Service. Polish Academy of Sciences, Astronomical Latitude Observatory, Borowiec, Poland.

Classical Quantum Gravity
Classical and Quantum Gravity. Institute of Physics, 47 Belgrave Square, London SW1X 8QX, England.

Clim. Change
Climatic Change. D. Reidel Publishing Company, Dordrecht – Boston – London. ISSN 0165–0009.

CODATA Bull.
CODATA Bulletin. Committee on Data for Science and Technology of the International Council of Scientific Unions. Pergamon Press, Oxford – New York – Toronto – Sydney – Paris – Frankfurt. ISSN 0366–757X.

Coelum
Coelum. Periodico bimestrale per la Divulgazione dell'Astronomia. Osservatorio Astronomico Universitario di Bologna, Cassella Postale 596, I–40100 Bologna, Italy. ISSN 0366–7588.

Comments Astrophys.
Comments on Astrophysics. A Journal of Critical Discussion of the Current Literature. Comments on Modern Physics: Part C. Gordon and Breach Science Publishers, New York – London. ISSN 0146–2970.

Comments At. Mol. Phys.
Comments on Atomic and Molecular Physics. Gordon and Breach Science Publishers, New York – London. ISSN 0010–2687.

Comments Nucl. Part. Phys.
Comments on Nuclear and Particle Physics. Gordon and Breach Science Publishers, New York – London. ISSN 0010–2709.

Comments Plasma Phys. Controlled Fusion
Comments on Plasma Physics and Controlled Fusion. Gordon and Breach Science Publishers, New York – London. ISSN 0374–2806.

Commun. Astron. Dep., Univ. Ankara
Communications from the Astronomical Department, University of Ankara. Fen Fakültesi, Ankara, Turkey.

Commun. Konkoly Obs.
Communications from the Konkoly Observatory of the Hungarian Academy of Sciences, Budapest, Hungary. ISSN 0324–2234.

Commun. Math. Phys.
Communications in Mathematical Physics. Springer–Verlag, Berlin – Heidelberg – New York – Tokyo. ISSN 0010–3616.

Commun. Theor. Phys.
Communications in Theoretical Physics. Huazhong University of Science and Technology Press, Wuhan, People's Republic of China.

Commun. Univ. Obs., St. Andrews
Communications from the University Observatory, St. Andrews. University Observatory, Buchanan Gardens, St. Andrews, Fife KY16 9LZ, Scotland.

Comput. Graphics
Computer Graphics. Association for Computing Machinery, 1133 Avenue of the Americas, New York, N.Y. 10036, USA. ISSN 0097–8930.

Comput. Methods Appl. Mech. Eng.
Computer Methods in Applied Mechanics and Engineering. North–Holland Publishing Company, PO Box 211, 1000 AE Amsterdam, The Netherlands. ISSN 0045–7825.

Comput. Phys. Commun.
Computer Physics Communications. North–Holland Publishing Company, Amsterdam, The Netherlands. ISSN 0010–4655.

Comput. Syst.
Computer Systems. Techpress Publishing Company Ltd., Walton House, 93 High Street, Bromley BR1 1JW, England. ISSN 0264–4193.

Computer
Computer. The Institute of Electrical and Electronic Engineers Computer Society, 5855 Naples Plaza, Suite 301, Long Beach, Calif. 90803, USA. ISSN 0018–9162.

Comun. Obs. Astron. Univ. Coimbra
Comunicações do Observatório Astronomico da Universidade de Coimbra, Portugal.

Contemp. Phys.
Contemporary Physics. Taylor and Francis Ltd., 10–14 Macklin Street, London, WC2B 5NF, England. ISSN 0010–7514.

Contrib. Astron. Obs. Skalnaté Pleso
Contributions of the Astronomical Observatory Skalnaté Pleso. VEDA, vydavateľstvo Slovenskej akadémie vied, Bratislava, Czechoslovakia.

Contrib. Atmos. Phys.
Contributions to Atmospheric Physics. (Beiträge zur Physik der Atmosphäre). Friedrich Vieweg & Sohn Verlagsgesellschaft mbH, Postfach 5829, D–6200 Wiesbaden, F.R. Germany. ISSN 0005–8173.

Contrib. Bosscha Obs.
Contributions from the Bosscha Observatory. Bosscha Observatory, Bandung Institute of Technology, Department of Science, Lembang, Java, Indonesia.

Contrib. Dep. Astron., Univ. Kyoto
Contributions from the Department of Astronomy, University of Kyoto. Department of Astronomy, Faculty of Science, University of Kyoto, Sakyo–ku, Kyoto 608, Japan. ISSN 0388–0230.

Contrib. Dep. Astron., Univ. Tokyo
Contributions from the Department of Astronomy, University of Tokyo. Department of Astronomy, University of Tokyo, Bunko–ku, Tokyo 113, Japan. ISSN 0563–8038.

Contrib. Dep. Geod. Astron., Univ. Thessaloniki
Contributions from the Department of Geodetic Astronomy, University of Thessaloniki. Department of Geodetic Astronomy, University of Thessaloniki, Thessaloniki, Greece.

Contrib. Dom. Astrophys. Obs.
Contributions from the Dominion Astrophysical Observatory. Dominion Astrophysical Observatory, National Research Council of Canada, 5071 West Saanich Road, Victoria, B.C. V8X 4M6, Canada.

Contrib. Dunsink Obs.
Contributions from the Dunsink Observatory. Dunsink Observatory, Castleknock, County Dublin, Republic of Ireland.

Contrib. Geophys. Inst. Slovak Acad. Sci.
Contributions of the Geophysical Institute of the Slovak Academy of Sciences. Slovak Academy of Sciences, Dubravská 9, 84228 Bratislava, Czechoslovakia. ISSN 0586–4607.

Contrib. Inst. Argent. Radioastron.
Contribuciones del Instituto Argentino de Radioastronomia. Instituto Argentino de Radioastronomia, Casilla de Correo No. 5, RA–1894 Villa Elisa, Prov. de Buenos Aires, Argentina.

Contrib. Inst. Astron. Res.
Contributions of the Institute for Astronomical Research. Institute for Astronomical Research, Inc., PO Box 15854, Baton Rouge, La. 70895, USA.

Contrib. Kwasan Hida Obs., Univ. Kyoto
Contributions from the Kwasan and Hida Observatories, University of Kyoto. Kwasan Observatory, University of Kyoto, Yamashina, Kyoto, 607, Japan. Hida Observatory, University of Kyoto, Kamitakara, Gifu-ken, 506–13, Japan. ISSN 0388–2349.

Contrib. La. State Univ. Obs.
Contributions of the Louisiana State University Observatory. Louisiana State University Observatory, Baton Rouge, La. 70803, USA.

Contrib. Lick Obs.
Contributions from the Lick Observatory. Lick Observatory, Board of Studies in Astronomy and Astrophysics, University of California at Santa Cruz, Santa Cruz, Calif. 95064, USA.

Contrib. Nicholas Copernicus Obs. Planetarium Brno
Contributions of the Nicholas Copernicus Observatory and Planetarium in Brno. Nicholas Copernicus Observatory and Planetarium Brno, Czechoslovakia.

Contrib. Nizamiah Japal–Rangapur Obs.
Contributions from Nizamiah and Japal–Rangapur Observatories. Centre of Advanced Study in Astronomy, Osmania University, Hyderabad–500 007, India.

Contrib. Oss. Astron. Torino
Contributi dell'Osservatorio Astronomico di Torino (Pino Torinese). Osservatorio Astronomico di Torino, I–10025 Pino Torinese (Torino), Italy.

Contrib. Van Vleck Obs.
Contributions from the Van Vleck Observatory. Van Vleck Observatory, Middletown, Conn., USA.

Copenh. Univ. Obs., Repr.
Copenhagen University Observatory, Reprints. University Observatory, Øster Voldgade 3, DK–1350 Copenhagen K, Denmark.

Cosmic. Res.
Cosmic Research. A cover-to-cover translation of Kosm. Issled. Consultants Bureau, 227 West 17th Street, New York, N.Y. 10011, USA. ISSN 0010–9525.

Cracow Obs. Repr.
Cracow Observatory Reprints. Cracow Observatory, Jagellonian University, Cracow, Poland.

CSELT Rapp. Tec.
CSELT Rapporti Tecnici. Centro Studi e Laboratori Telecomunicazioni, Via Guglielmo Reiss Romoli 274, Torino, Italy. ISSN 0390–1815.

Curr. Sci.
Current Science. Current Science Association, Raman Research Institute, Bangalore 560006, India. ISSN 0011–3891.

Curr. Top. Chin. Sci., Sect. E: Astron.
Current Topics in Chinese Science, Section E: Astronomy. An annual selection of papers published in Scientia Sinica and Kexue Tongbao. Gordon and Breach Science Publishers, Inc., London – New York – Paris. ISSN 0732–4421.

Czech. J. Phys., Sect. B
Czechoslovak Journal of Physics, Section B. Academia Publishing House of the Czechoslovak Academy of Sciences, Vodičkova 40, 11229 Praha 1, Czechoslovakia. ISSN 0011–4626.

Data Rep. Hydrogr. Obs., Ser. Astron. Geod.
Data Report of Hydrographic Observations, Series of Astronomy and Geodesy. Hydrographic Department of Japan, Tsukiji–5, Chuo–ku, Tokyo 104, Japan.

Debrecen Heliophys. Obs. Hung. Acad. Sci., Repr.
Debrecen Heliophysical Observatory of the Hungarian Academy of Sciences, Reprint. Heliophysical Observatory of the Hungarian Academy of Sciences, 4010 Debrecen, Hungary.

Dějiny Věd Tech.
Dějiny Věd a Techniky. Academia Praha, Praha, Czechoslovakia. ISSN 0300–4414.

Dep. Astron. McDonald Obs. Univ. Tex., Repr.
Department of Astronomy and McDonald Observatory of the University of Texas, Reprints. Astronomy Department, R.L.M. 15.220, University of Texas, Austin, Tex. 78712, USA.

Dep. Astrophys., Univ. Oxford, Publ.
Department of Astrophysics, University of Oxford, Publication. Department of Astrophysics, University of Oxford, South Parks Road, Oxford OX1 3RQ, England.

Diss. Abstr. Int., Sect. B
Dissertation Abstracts International – Section B (The Sciences and Engineering). University Microfilms International, Ann Arbor, Michigan 48106, USA. ISSN 0419–4217.

Dokl. Akad. Nauk BSSR
Doklady Akademii Nauk BSSR. Akademiya Nauk, Minsk, Belorussian SSR. ISSN 0002–354X.

Dokl. Akad. Nauk SSSR. Ser. Mat. Fiz.
Doklady Akademii Nauk SSSR. Seriya Matematika, Fizika. Izdatel'stvo Nauka, Moskva. Editorial address: 117874 GSP–7 Moskva, B–485, Profsoyuznaya Ul., 90 Kom. 533, USSR. ISSN 0002–3264.

Dokl. Bolg. Akad. Nauk
Doklady Bolgarskoj Akademii Nauk. Sofiya, Bulgaria. ISSN 0366–8681.

Dtsch. Geod. Komm. Bayer. Akad. Wiss., Reihe A
Deutsche Geodätische Kommission bei der Bayerischen Akademie der Wissenschaften, Reihe A: Theoretische Geodäsie. Deutsche Geodätische Kommission, Marstallplatz 8, D–8000 München 22, F.R. Germany. ISSN 0065–5309.

Dtsch. Geod. Komm. Bayer. Akad. Wiss., Reihe B
Deutsche Geodätische Kommission bei der Bayerischen Akademie der Wissenschaften, Reihe B: Angewandte Geodäsie. Deutsche Geodätische Kommission, Marstallplatz 8, D–8000 München 22, F.R. Germany. ISSN 0065–5317.

Dtsch. Geod. Komm. Bayer. Akad. Wiss., Reihe C
Deutsche Geodätische Kommission bei der Bayerischen Akademie der Wissenschaften, Reihe C: Dissertationen. Deutsche Geodätische Kommission, Marstallplatz 8, D–8000 München 22, F.R. Germany. ISSN 0065–5325.

Dtsch. Geod. Komm. Bayer. Akad. Wiss., Reihe D
Deutsche Geodätische Kommission bei der Bayerischen Akademie der Wissenschaften, Reihe D: Tafelwerke. Deutsche Geodätische Kommission, Marstallplatz 8, D–8000 München 22, F.R. Germany.

Dtsch. Geod. Komm. Bayer. Akad. Wiss., Reihe E
Deutsche Geodätische Kommission bei der Bayerischen Akademie der Wissenschaften, Reihe E: Geschichte und Entwicklung der Geodäsie. Deutsche Geodätische Kommission, Marstallplatz 8, D–8000 München 22, F.R. Germany. ISSN 0065–5341.

Dtsch. Hydrogr. Inst. Hamb., Zeit–Breitendienst
Deutsches Hydrographisches Institut Hamburg, Zeit– und Breitendienst. Deutsches Hydrographisches Institut, D–2000 Hamburg, F.R. Germany.

Dudley Obs. Rep.
Dudley Observatory Report. The Dudley Observatory, 69 Union Avenue, Schenectady, N.Y. 12308, USA.

Dunsink Obs. Repr.
Dunsink Observatory Reprints. Dunsink Observatory, Castleknock, County Dublin, Republic of Ireland.

Earth, Moon, Planets
Earth, Moon, and Planets. An International Journal of Comparative Planetology. D. Reidel Publishing Company, Dordrecht – Boston. ISSN 0167–9295.

Earth Orientation Bull.
Earth Orientation Bulletin. U.S. Naval Observatory, Time Service Publications, Series 7. U.S. Naval Observatory, Time Service Division (62C), Washington, D.C. 20390, USA.

Earth–Oriented Appl. Space Technol.
Earth–Oriented Applications of Space Technology. Pergamon Press, Oxford – New York – Frankfurt. ISSN 0277–4488.

Earth Planet. Sci. Lett.
Earth and Planetary Science Letters. Elsevier Scientific Publishing Company, PO Box 211, 1000 AE Amsterdam, The Netherlands. ISSN 0012–821X.

Earth–Sci. Rev.
Earth-Science Reviews. Elsevier Scientific Publishing Company, PO Box 330, 1000 AH Amsterdam, The Netherlands. ISSN 0012–8252.

Electron. Lett.
Electronics Letters. Institution of Electrical Engineers, Savoy Place, London, WC2R 0BL, England. ISSN 0013–5194.

Electronics
Electronics. McGraw–Hill Publishing Company, 1221 Avenue of the Americas, New York, N.Y. 10020, USA. ISSN 0013–5070.

Elektronik
Elektronik. Franzis–Verlag GmbH, D–8000 München 37, Postfach 3701 20, Karlstraße 37, F.R. Germany. ISSN 0013–5658.

Elettron. Telecomun.
Elettronica e Telecomunicazioni. Via Arsenale 41, I–10121 Torino, Italy. ISSN 0013–6123.

Endeavour New Ser.
Endeavour New Series. Pergamon Press, Oxford – New York – Toronto – Paris – Frankfurt – Sydney. ISSN 0013–7162.

EOS Trans. Am. Geophys. Union
EOS Transactions of the American Geophysical Union. American Geophysical Union, 2000 Florida Avenue N.W., Washington, D.C. 20009, USA. ISSN 0096–3941.

ESA Brochure
ESA Brochure (ESA BR). European Space Agency, Scientific and Technical Publications Branch, ESTEC, Postbus 299, 2200 AG Noordwijk, The Netherlands. ISSN 0250–1589.

ESA Bull.
ESA Bulletin. European Space Agency, Scientific and Technical Publications Branch, ESTEC, Postbus 299, 2200 AG Noordwijk, The Netherlands. ISSN 0376–4265.

ESA IUE Newsl.
ESA IUE Newsletters. The ESA IUE Observatory. Apartado 540 65, Madrid, Spain. Subscription address: European Space Agency, 8–10 rue Mario–Nikis, F–75738 Paris Cedex 15, France.

ESA J.
ESA Journal. European Space Agency, Scientific and Technical Publications Branch, ESTEC, Postbus 299, 2200 AG Noordwijk, The Netherlands. ISSN 0379–2285.

ESA SCI
European Space Agency SCI. European Space Agency, Scientific and Technical Publications Branch, ESTEC, Postbus 299, 2200 AG Noordwijk, The Netherlands.

ESA Spec. Publ.
ESA Special Publication (ESA SP). European Space Agency, Scientific and Technical Publications Branch, ESTEC, Postbus 299, 2200 AG Noordwijk, The Netherlands. ISSN 0379–6566.

ESO Annu. Rep.
ESO Annual Report. European Southern Observatory, Karl–Schwarzschild–Straße 2, D–8046 Garching bei München, F.R. Germany. ISSN 0531–4496.

ESO Sci. Prepr.
ESO Scientific Preprint. European Southern Observatory, Karl–Schwarzschild–Straße 2, D–8046 Garching bei München, F.R. Germany.

ESO Sci. Rep.
ESO Scientific Report. European Southern Observatory, Karl–Schwarzschild–Straße 2, D–8046 Garching bei München, F.R. Germany.

ESO Tech. Rep.
ESO Technical Report. European Southern Observatory, Karl–Schwarzschild–Straße 2, D–8046 Garching bei München, F.R. Germany.

Eur. J. Phys.
European Journal of Physics. Institute of Physics, 47 Belgrave Square, London SW1X 8QX, England. ISSN 0143–0807.

Europhys. News
Europhysics News. European Physical Society, PO Box 69, CH–1213 Petit–Lancy 2, Switzerland. ISSN 0531–7479.

EXOSAT Express
EXOSAT Express. ESA, EXOSAT Observatory, ESOC, Robert–Bosch–Str. 5, D–6100 Darmstadt, F.R. Germany.

Feingerätetechnik
Feingerätetechnik. VEB Verlag Technik, Oranienburger Straße 13/14, DDR–1020 Berlin, German Democratic Republic. ISSN 0014–9683.

Fis. Tecnol.
Fisica e Tecnologia. Societa Italiana di Fisica, Via Loderingo Degli Andalo 2, I–40124 Bologna, Italy. ISSN 0391–9757.

Fiz. Sz.
Fizikai Szemle. Kiadja a Lapkiado Vallalat, Budapest VII, Lenin korut 9–11, Hungary. ISSN 0015–3257.

Fizika
Fizika. Mladost Export–Import, 41000 Zagreb, Ilica 30, Yugoslavia. ISSN 0015–3206.

Folia Fac. Sci. Nat. Univ. Purkynianae Brun., Phys.
Folia Facultatis Scientiarum Naturalium Universitatis Purkynianae Brunensis, Physica. University J. E. Purkyne, 61137 Brno–Kotlarska 2, Czechoslovakia. ISSN 0323–0287.

Fortschr. Phys.
Fortschritte der Physik. Akademie–Verlag, DDR–108 Berlin, Leipziger Straße 3–4, German Democratic Republic. ISSN 0015–8208.

Found. Phys.
Foundations of Physics. Plenum Publishing Corporation, 227 West 17th Street, New York, N.Y. 10011, USA. ISSN 0015–9018.

Fra Fys. Verden
Fra Fysikkens Verden. Fysisk Institutt, Universitetet i Trondheim, Norges Laererhogskole, N–7000 Trondheim, Norway. ISSN 0015–9247.

Fujitsu Sci. Tech. J.
Fujitsu Scientific and Technical Journal. Fujitsu Ltd., 1015 Kamikodanaka, Nakahara–ku, Kawasaki 211, Kanagawa, Japan. ISSN 0016–2523.

Fundam. Cosmic Phys.
Fundamentals of Cosmic Physics. Gordon and Breach Science Publishers, New York – London – Paris. ISSN 0094–5846.

Funkschau
Funkschau. Franzis–Verlag GmbH, Postfach 370120, Karlstr. 37, D–8000 München 2, F.R. Germany. ISSN 0016–2841.

Fys. Tidsskr.
Fysisk Tidsskrift. Subscription address: Jul. Gjellerups Boghandel, Solvgade 87, DK–1307 Kobenhavn, Denmark. ISSN 0016–3392.

G. A.A.B.
Giornale dell'A.A.B. Associazione Astrofili Bolognesi, Casella Postale 1630, I–40100 Bologna A.D., Italy. ISSN 0392–3932.

G. Astron.
Giornale di Astronomia. Società Astronomica Italiana, Largo E. Fermi, 5, I–50125 Firenze, Italy. ISSN 0390–1106.

Gemini
Gemini. Newsletter of the Royal Greenwich Observatory. Royal Greenwich Observatory, Herstmonceux Castle, Hailsham, East Sussex BN27 1RP, England.

Gen. Relativ. Gravitation
General Relativity and Gravitation. Published under the auspices of the International Committee on General Relativity and Gravitation GRG. Plenum Publishing Corporation, 233 Spring Street, New York, N.Y. 10013, USA. ISSN 0001–7701.

Geochim. Cosmochim. Acta
Geochimica et Cosmochimica Acta. Journal of the Geochemical Society and the Meteoritical Society. Pergamon Press, New York – Oxford – Toronto – Paris – Frankfurt – Sydney. ISSN 0016–7037.

Geod.–Geophys. Arb. Schweiz
Geodätisch–Geophysikalische Arbeiten in der Schweiz. Schweizerische Geodätische Kommission, Institut für Geodäsie und Photogrammetrie, ETH–Hönggerberg, CH–8093 Zürich, Switzerland.

Geod. Geophys. Veröff., Reihe II
Geodätische und Geophysikalische Veröffentlichungen, Reihe II (Solar–terrestrische Beziehungen und Physik der Atmosphäre). Nationalkomitee für Geodäsie und Geophysik bei der Akademie der Wissenschaften der Deutschen Demokratischen Republik, DDR–1500 Potsdam, Telegrafenberg, German Democratic Republic.

Geod. Geophys. Veröff., Reihe III
Geodätische und Geophysikalische Veröffentlichungen, Reihe III (Physik der festen Erde). Nationalkomitee für Geodäsie und Geophysik bei der Akademie der Wissenschaften der Deutschen Demokratischen Republik, DDR–1500 Potsdam, Telegrafenberg, German Democratic Republic. ISSN 0435–6187.

Geod. Kartogr.
Geodezja i Kartografia. Komitet Geodezji Polskiej Akademii Nauk. Publisher: Państwowe Wydawnictwo Naukowe, Warszwa, Poland. ISSN 0016–7134.

Geomagn. Aehron.
Geomagnetizm i Aehronomiya. Akademiya Nauk SSSR. Izdatel'stvo Nauka, Moskva. English translation in Geomagn. Aeron. ISSN 0016–7940.

Geomagn. Aeron.
Geomagnetism and Aeronomy. A cover–to–cover translation of Geomagn. Aehron. American Geophysical Union, 2000 Florida Avenue, N.W., Washington, D. C. 20009, USA. ISSN 0016–7932.

Geophys. Astrophys. Fluid Dyn.
Geophysical and Astrophysical Fluid Dynamics. Gordon and Breach Science Publishers, New York – London – Paris. ISSN 0309–1929.

Geophys. J. R. Astron. Soc.
Geophysical Journal of the Royal Astronomical Society. Published for the Royal Astronomical Society by Blackwell Scientific Publications, Oxford – London – Edinburgh – Boston – Melbourne. ISSN 0016–8009.

Geophys. Norv.
Geophysica Norvegica. Universitetsforlaget, PO Box 2959, Toyen, Oslo 6, Norway. ISSN 0332–5903.

Geophys. Prospect.
Geophysical Prospecting. European Association of Exploration Geophysicists, PO Box 162, 2501 AN The Hague, The Netherlands. ISSN 0016–8025.

Geophys. Res. Lett.
Geophysical Research Letters. American Geophysical Union, 2000 Florida Avenue, N.W., Washington, D.C. 20009, USA. ISSN 0094–8276.

Geophys. Surv.
Geophysical Surveys. D. Reidel Publishing Company, Dordrecht – Boston – London. ISSN 0046–5763.

Geophysics
Geophysics. Society of Exploration Geophysicists, PO Box 3098, Tulsa, Okla. 74101, USA. ISSN 0016–8033.

GEOS Circ.
GEOS Circular. Series: EB (eclipsing binaries), RR (RR Lyrae type variables), SA (small–amplitude variables), SR (red variables). GEOS (Groupe d'Etude et d'Observation Stellaire), 12 rue Bezout, F–75014 Paris, France.

Gerlands Beitr. Geophys.
Gerlands Beiträge zur Geophysik. Akademische Verlagsgesellschaft Geest & Portig KG, DDR–7010 Leipzig, Sternwartenstraße 8, German Democratic Republic. ISSN 0016–8696.

Glas. Mat., Ser. III
Glasnik Matematički, Serija III. Društvo matematičara i fizičara SRH, Marulićev trg 19, YU–41001 Zagreb, P.P. 187, Yugoslavia. ISSN 0017–095X.

Greenwich Time Rep.
Greenwich Time Report. Royal Greenwich Observatory, Time and Latitude Service, Herstmonceux Castle, Hailsham, East Sussex BN27 1RP, England. ISSN 0264–4177.

Hadronic J.
Hadronic Journal. Hadronic Press Inc., Nonantum, Mass. 02195, USA. ISSN 0162–5519.

Heavens
The Heavens. The Oriental Astronomical Association, Ōtsu–shi, Shiga–ken, Japan. In Japanese.

Helv. Phys. Acta
Helvetica Physica Acta. Schweizerische Physikalische Gesellschaft. Birkhäuser Verlag, Elisabethenstraße 19, CH–4000 Basel 10, Switzerland. ISSN 0018–0238.

HHI Sol. Data
HHI Solar Data. Solar Radio Emission. Heinrich–Hertz–Institut, Solare Beobachtungsergebnisse. Akademie der Wissenschaften der DDR, Zentralinstitut für Solar–Terrestrische Physik (Heinrich–Hertz–Institut), DDR–1199 Berlin–Adlershof, German Democratic Republic. ISSN 0323–309X.

HHI STP Rep.
HHI Solar–Terrestrische Physik Reports. Heinrich–Hertz–Institut. Akademie der Wissenschaften der DDR, Zentralinstitut für Solar–Terrestrische Physik (Heinrich–Hertz–Institut), DDR–1199 Berlin–Adlershof, German Democratic Republic.

HHI Suppl. Ser. Sol. Data
HHI Supplement Series of Solar Data. Heinrich–Hertz–Institut. Akademie der Wissenschaften der DDR, Zentralinstitut für Solar–Terrestrische Physik (Heinrich–Hertz–Institut), DDR–1199 Berlin–Adlershof, German Democratic Republic. ISSN 0323–3197.

H.M. Naut. Alm. Off., Libr. Repr.
H.M. Nautical Almanac Office, Library Reprint. H.M. Nautical Almanac Office, Royal Greenwich Observatory, Herstmonceux Castle, Hailsham, East Sussex BN27 1RP, England.

Hvar Obs. Bull.
Hvar Observatory Bulletin. Faculty of Geodesy, University of Zagreb, Kačićeva 26, YU–41000 Zagreb, Yugoslavia. ISSN 0351–2651.

I.A.P.P.P. Commun.
International Amateur–Professional Photoelectric Photometry Communication. Fairborn Observatory, 1247 Folk Road, Fairborn, Ohio 45324, USA.

IAU Circ.
International Astronomical Union, Circular. Central Bureau for Astronomical Telegrams, Smithsonian Astrophysical Observatory, 60 Garden Street, Cambridge, Mass. 02138, USA. ISSN 0081–0304.

IAU Inf. Bull.
International Astronomical Union Information Bulletin. IAU–UAI Secrétariat, 61, Avenue de l'Observatoire, F–75014 Paris, France. Published by D. Reidel Publishing Company, Dordrecht – Boston – London.

Icarus
Icarus. International Journal of Solar System Studies. Academic Press Inc., New York – London. ISSN 0019–1035.

IEE Proc. F
IEE Proceedings F (Communications, Radar and Signal Processing). Institution of Electrical Engineers, PO Box 8, Southgate House, Stevenage, Herts. SG1 1HQ, England. ISSN 0143–7070.

IEEE J. Solid–State Circuits
IEEE Journal of Solid–State Circuits. The Institute of Electrical and Electronics Engineers, 345 East 47th Street, New York, N.Y. 10017, USA. ISSN 0018–9200.

IEEE Spectrum
IEEE Spectrum. The Institute of Electrical and Electronics Engineers, 345 East 47th Street, New York, N.Y. 10017, USA. ISSN 0018–9235.

IEEE Trans. Aerosp. Electron. Syst.
IEEE Transactions on Aerospace and Electronic Systems. The Institute of Electrical and Electronics Engineers, 345 East 47th Street, New York, N.Y. 10017, USA. ISSN 0018–9251.

IEEE Trans. Antennas Propag.
IEEE Transactions on Antennas and Propagation. The Institute of Electrical and Electronics Engineers, 345 East 47th Street, New York, N.Y. 10017, USA. ISSN 0018–926X.

IEEE Trans. Commun.
IEEE Transactions on Communications. The Institute of Electrical and Electronics Engineers, 345 East 47th Street, New York, N.Y. 10017, USA. ISSN 0090–6778.

IEEE Trans. Ind. Electron.
IEEE Transactions on Industrial Electronics. The Institute of Electrical and Electronics Engineers, 345 East 47th Street, New York, N.Y. 10017, USA. ISSN 0278–0046.

IEEE Trans. Instrum. Meas.
IEEE Transactions on Instrumentation and Measurement. The Institute of Electrical and Electronics Engineers, 345 East 47th Street, New York, N.Y. 10017, USA. ISSN 0018–9456.

IEEE Trans. Microwave Theory Tech.
IEEE Transactions on Microwave Theory and Techniques. The Institute of Electrical and Electronics Engineers, 345 East 47th Street, New York, N.Y. 10017, USA. ISSN 0018–9480.

IEEE Trans. Nucl. Sci.
IEEE Transactions on Nuclear Science. The Institute of Electrical and Electronics Engineers, 345 East 47th Street, New York, N.Y. 10017, USA. ISSN 0018–9499.

IEEE Trans. Plasma Sci.
IEEE Transactions on Plasma Science. The Institute of Electrical and Electronics Engineers, 345 East 47th Street, New York, N.Y. 10017, USA. ISSN 0093–3813.

IHW Newsl.
The International Halley Watch Newsletter. Jet Propulsion Laboratory, California Institute of Technology, Pasadena, Calif. 91109, USA.

IMA J. Appl. Math.
IMA Journal of Applied Mathematics. Academic Press Inc., London – New York. ISSN 0272–4960.

Indian East. Eng.
Indian and Eastern Engineer. "Piramal Mansion", 235 Dr. D. Naoroji Road, Bombay 400 001, India. ISSN 0019–4352.

Indian J. Cryog.
Indian Journal of Cryogenics. Indian Cryogenics Council, Jadavpur University, Calcutta 700 032, India. ISSN 0379–0479.

Indian J. Hist. Sci.
Indian Journal of History of Science. Indian National Science Academy, Bahadur Shah Zafar Marg, New Delhi 110 002, India.

Indian J. Phys., Part B
Indian Journal of Physics, Part B. Indian Association for the Cultivation of Science, 2 & 3 Raja Subodh Chandra Mallik Road, Calcutta 700 032, India. ISSN 0374–3330.

Indian J. Pure Appl. Math.
Indian Journal of Pure and Applied Mathematics. Indian National Science Academy, Bahadur Shah Zafar Marg, New Delhi 110 002, India. ISSN 0019–5588.

Indian J. Pure Appl. Phys.
Indian Journal of Pure and Applied Physics. Council of Scientific & Industrial Research. Publications & Information Directorate, Hillside Road, New Delhi 110 012, India. ISSN 0019–5596.

Indian J. Radio Space Phys.
Indian Journal of Radio and Space Physics. Council of Scientific & Industrial Research. Publications & Information Directorate, Hillside Road, New Delhi 110 012, India. ISSN 0367–8393.

Indian J. Theor. Phys.
Indian Journal of Theoretical Physics. Institute of Theoretical Physics, Bognan Kutir, 4–1 Mohan Bagan Lane, Calcutta 700 004, India. ISSN 0019–5693.

Inf. Bull. Variable Stars
Information Bulletin on Variable Stars. Commission 27 of the IAU. Konkoly Observatory, Budapest, Hungary. ISSN 0374–0676.

Infrared Phys.
Infrared Physics. An International Research Journal. Pergamon Press, Oxford – New York – Toronto – Sydney – Paris – Frankfurt. ISSN 0020–0891.

Inst. Astron. Astrophys. Tech. Univ. Berlin, Mitt.
Institut für Astronomie und Astrophysik der Technischen Universität Berlin, Mitteilungen. Institut für Astronomie und Astrophysik der Technischen Universität Berlin, Hardenbergstraße 36, D–1000 Berlin 12, F.R. Germany.

Inst. Astron. Fis. Espacio, Tirada Aparte
Instituto de Astronomía y Fisica del Espacio, Tirada Aparte. Instituto de Astronomía y Fisica del Espacio, Casilla de Correo 67, Sucursal 28, 1428 Buenos Aires, Argentina.

Inst. Astron., Univ. Camb., Annu. Rep.
Institute of Astronomy, University of Cambridge, Annual Report. The Observatories, Madingley Road, Cambridge CB3 0HA, England.

Inst. Astrophys. Paris, Pré–Publ.
Institut d'Astrophysique de Paris, Pré–Publication. Institut d'Astrophysique, 98 bis, Boulevard Arago, F–75014 Paris, France.

Inst. Astrophys., Univ. Liège, Collect. 4°
Institut d'Astrophysique, Université de Liège, Collection in 4°. Institut d'Astrophysique, Université de Liège, Avenue de Cointe, 5, B–4200 Cointe–Ougrée, Belgium.

Inst. Astrophys., Univ. Liège, Collect. 8°
Institut d'Astrophysique, Université de Liège, Collection in 8°. Institut d'Astrophysique, Université de Liège, Avenue de Cointe, 5, B–4200 Cointe–Ougrée, Belgium.

Inst. Obs. Mar., Bol. Astron.
Instituto y Observatorio de Marina, Boletin Astronomico. Instituto y Observatorio de Marina, San Fernando (Cadiz), Spain.

Inst. Obs. Mar., Bol. Ser. B
Instituto y Observatorio de Marina, Boletin Serie B. (Pequeños Planetas). Instituto y Observatorio de Marina, San Fernando (Cadiz), Spain. ISSN 0558–3993.

Inst. Obs. Mar., Bol. Ser. C
Instituto y Observatorio de Marina, Boletin Serie C. (Rotacion de la Tierra). Instituto y Observatorio de Marina, San Fernando (Cadiz), Spain. ISSN 0210–6485.

Inst. Obs. Mar., Bol. Ser. D
Instituto y Observatorio de Marina, Boletin Serie D. (Ocultaciones). Instituto y Observatorio de Marina, San Fernando (Cadiz), Spain.

Inst. Obs. Mar., Mem. Act.
Instituto y Observatorio de Marina, Memoria de las Actividades. Instituto y Observatorio de Marina, San Fernando (Cadiz), Spain.

Inst. Teor. Astrofys., Blindern–Oslo, Småtrykk
Institutt for Teoretisk Astrofysikk, Blindern–Oslo, Småtrykk. Institute of Theoretical Astrophysics, University of Oslo, PO Box 1029, Blindern, N–0315 Oslo 3, Norway.

Inst. Theor. Astrophys., Blindern–Oslo, Rep.
Institute of Theoretical Astrophysics, Blindern–Oslo, Reports. Institute of Theoretical Astrophysics, University of Oslo, PO Box 1029, Blindern, N–0315 Oslo 3, Norway. ISSN 0078–6780.

Inst. Theor. Astrophys., Blindern–Oslo, Repr.
Institute of Theoretical Astrophysics, Blindern–Oslo, Reprints. Institute of Theoretical Astrophysics, University of Oslo, PO Box 1029, Blindern, N–0315 Oslo 3, Norway.

Inst. Theor. Astrophys., Univ. Oslo, Publ. Ser.
Institute of Theoretical Astrophysics, University of Oslo, Publication Series. Institute of Theoretical Astrophysics, University of Oslo, PO Box 1029, Blindern, N–0315 Oslo 3, Norway. ISSN 0800–6652.

Inst. Theor. Phys. Sternw. Univ. Kiel, Repr.
Institut für Theoretische Physik und Sternwarte der Universität Kiel, Reprints. Institut für Theoretische Physik und Sternwarte der Universität Kiel, D–2300 Kiel, F.R. Germany.

Int. Comet Q.
The International Comet Quarterly. Department of Physics and Astronomy, Appalachian State University, Boone, N.C. 28608, USA. ISSN 0736–6922.

Int. J. Gen. Syst.
International Journal of General Systems. Gordon and Breach Science Publishers Inc., New York – London – Paris. ISSN 0308–1079.

Int. J. Heat Mass Transfer
International Journal of Heat and Mass Transfer. Pergamon Press, Oxford – New York – Toronto – Sydney – Paris – Frankfurt. ISSN 0017–9310.

Int. J. Infrared Millimeter Waves
International Journal of Infrared and Millimeter Waves. Plenum Publishing Corporation, 227 West 17th Street, New York, N.Y. 10011, USA. ISSN 0195–9271.

Int. J. Mass Spectrom. Ion Processes
International Journal of Mass Spectrometry and Ion Processes. Elsevier Scientific Publishing Company, PO Box 330, 1000 AH Amsterdam, The Netherlands. ISSN 0020–7381.

Int. J. Non–Linear Mech.
International Journal of Non–Linear Mechanics. Pergamon Press, Oxford – New York – Toronto – Paris – Frankfurt – Sydney. ISSN 0020–7462.

Int. J. Theor. Phys.
International Journal of Theoretical Physics. Plenum Publishing Corporation, 233 Spring Street, New York, N.Y. 10013, USA. ISSN 0020–7748.

Interdisciplinary Sci. Rev.
Interdisciplinary Science Reviews. Heyden & Son Ltd., Spectrum House, Hillview Gardens, London NW4 2JQ, England. ISSN 0308–0188.

Ir. Astron. J.
The Irish Astronomical Journal. A Half–Yearly Publication under the Auspices of the Observatories of Armagh and Dunsink. Armagh Observatory, Armagh BT61 9DG, Northern Ireland. ISSN 0021–1052.

IRIS Bull. A
IRIS Bulletin A (Earth Orientation). Subcommission International Radio Interferometric Surveying (IRIS) Steering Committee. Available from National Geodetic Survey, National Oceanic and Atmospheric Administration, N/CG114, Rockville, Md. 20852, USA.

ISIS
ISIS. An International Review devoted to the History of Science and its Cultural Influences. Department of History and Sociology of Science, University of Pennsylvania, Philadelphia, Pa. 19104, USA. ISSN 0021–1753.

Issled. Solntsa Krasnykh Zvezd
Issledovaniya Solntsa i Krasnykh Zvezd. (Investigations of the Sun and Red Stars). Akademiya Nauk Latvijskoj SSR, Radio-astrofizicheskaya Observatoriya. Izdatel'stvo Zinatne, Riga, USSR. ISSN 0135–1303.

Izv. Akad. Nauk Arm. SSR, Ser. Fiz.
Izvestiya Akademii Nauk Armyanskoj SSR, Seriya Fizika. Akademiya Nauk Armyanskoj SSR, Erevan. 375019 Erevan, ul. Barekamutyan, 24, USSR. ISSN 0002–3035.

Izv. Astron. Ehngel'gardt. Obs.
Izvestiya Astronomicheskoj Ehngel'gardtovskoj Observatorii. Izdatel'stvo Kazanskogo Universiteta, Kazan, ul. Lenina, d. 4/5, USSR.

Izv. Glav. Astron. Obs. Pulkovo
Izvestiya Glavnoj Astronomicheskoj Observatorii v Pulkove. Akademiya Nauk SSSR. Leningradskoe Otdelenie, Izdatel'stvo Nauka, Leningrad. 199164 Leningrad, V–164, Mendeleevskaya l., 1, USSR. ISSN 0367–7966.

Izv. Krymskoj Astrofiz. Obs.
Izvestiya Ordena Trudovogo Krasnogo Znameni Krymskoj Astrofizicheskoj Observatorii. Akademiya Nauk SSSR. Izdatel'stvo 'Nauka', Moskva, USSR. English translation in Bull. Crimean Astrophys. Obs. ISSN 0367–8466.

Izv. Vyssh. Uchebn. Zaved., Radiofiz.
Izvestiya Vysshikh Uchebnikh Zavedenij, Radiofizika. Gor'kovskij Universitet. Gor'kij, ul. Lyadova 25, USSR. English translation in Radiophys. Quantum Electron. ISSN 0021–3462.

J. Am. Assoc. Variable Star Obs.
The Journal of the American Association of Variable Star Observers. The American Association of Variable Star Observers, 187 Concord Avenue, Cambridge, Mass. 02138, USA.

J. Am. Chem. Soc.
Journal of the American Chemical Society. American Chemical Society, 1155 16th Street N.W., Washington, D.C. 20036, USA. ISSN 0002–7863.

J. Appl. Meteorol.
Journal of Applied Meteorology. American Meteorological Society, 45 Beacon Street, Boston, Mass. 02108, USA. ISSN 0021–8952.

J. Appl. Phys.
Journal of Applied Physics. American Institute of Physics, 335 East 45th Street, New York, N.Y. 10017, USA. ISSN 0021–8979.

J. Astron. Soc. Egypt
Journal of the Astronomical Society of Egypt. Astronomical Society of Egypt, Astronomy Department, Faculty of Sciences, Cairo University, Cairo, Egypt.

J. Astron. Soc. West. Aust.
Journal of the Astronomical Society of Western Australia. Astronomical Society of Western Australia, PO Box S1460, Perth, W.A. 6000, Australia.

J. Astronaut. Sci.
Journal of the Astronautical Sciences. American Astronautical Society, 6060 Duke Street, Alexandria, Va. 22304, USA. ISSN 0021–9142.

J. Astrophys. Astron.
Journal of Astrophysics and Astronomy. Indian Academy of Sciences, PO Box 8005, Bangalore 560 080, India. ISSN 0250–6335.

J. Atmos. Sci.
Journal of the Atmospheric Sciences. American Meteorological Society, 45 Beacon Street, Boston, Mass. 02108, USA. ISSN 0022–4928.

J. Atmos. Terr. Phys.
Journal of Atmospheric and Terrestrial Physics. Pergamon Press, Oxford – New York – Frankfurt. ISSN 0021–9169.

J. Aust. Math. Soc., Ser. B
Journal of the Australian Mathematical Society, Series B. (Applied Mathematics). Department of Mathematics, University of Queensland, St. Lucia, QLD 4067, Australia. ISSN 0334–2700.

J. Br. Astron. Assoc.
Journal of the British Astronomical Association. The British Astronomical Association, Burlington House, Piccadilly, London, W1V 0NL, England. ISSN 0007–0297.

J. Br. Interplanet. Soc.
Journal of the British Interplanetary Society. The British Interplanetary Society, 27–29 South Lambeth Road, London, SW8 1SZ, England. ISSN 0007–084X.

J. Chem. Educ.
Journal of Chemical Education. American Chemical Society, 119 West 24th Street, New York, N.Y. 10011, USA. ISSN 0021–9584.

J. Chem. Phys.
Journal of Chemical Physics. American Institute of Physics, 335 East 45th Street, New York, N.Y. 10017, USA. ISSN 0021–9606.

J. Comput. Phys.
Journal of Computational Physics. Academic Press Inc., 111 5th Avenue, New York, N.Y. 10003, USA. ISSN 0021–9991.

J. Differ. Equations
Journal of Differential Equations. Academic Press Inc., New York – London. ISSN 0022–0396.

J. Electron Spectrosc. Relat. Phenom.
Journal of Electron Spectroscopy and Related Phenomena. Elsevier Scientific Publishing Company, PO Box 330, 1000 AH Amsterdam, The Netherlands. ISSN 0368–2048.

J. Electrostat.
Journal of Electrostatics. Elsevier Scientific Publishing Company, PO Box 330, 1000 AH Amsterdam, The Netherlands. ISSN 0304–3886.

J. Environ. Sci.
Journal of Environmental Sciences. Institute of Environmental Sciences, 940 East Northwest Highway, Mt. Prospect, Ill. 60056, USA. ISSN 0022–0906.

J. Fluid Mech.
Journal of Fluid Mechanics. Cambridge University Press, Bentley House, 200 Euston Road, London, NW1 2DB, England. ISSN 0022–1120.

J. Geodyn.
Journal of Geodynamics. Geophysical Press, Brouwersgracht 236, 1013 HE Amsterdam, The Netherlands. ISSN 0264–3707.

J. Geol.
Journal of Geology. The University of Chicago Press, 5801 S. Ellis Avenue, Chicago, Ill. 60637, USA. ISSN 0022–1376.

J. Geomagn. Geoelectr.
Journal of Geomagnetism and Geoelectricity. University of Tokyo Press, c/o Business Centre for Academic Societies, 4–16 Yayoi 2–chome, Bunkyo–ku, Tokyo 113, Japan. ISSN 0022–1392.

J. Geophys.
Journal of Geophysics. Springer–Verlag, Berlin – Heidelberg – New York – Tokyo. ISSN 0340–062X.

J. Geophys. Res.
Journal of Geophysical Research. Section A: Space Physics, Section B: Solid Earth and Planets, Section C: Ocean, Section D: Atmosphere. American Geophysical Union, 2000 Florida Avenue, N.W., Washington, D.C. 20009, USA. ISSN 0148–0227.

J. Guid. Control Dyn.
Journal of Guidance, Control, and Dynamics. American Institute of Aeronautics and Astronautics, 1290 Avenue of the Americas, New York, N.Y. 10104, USA. ISSN 0731–5090.

J. Hist. Astron.
Journal for the History of Astronomy. Published by Science History Publications Ltd., Halfpenny Furze, Mill Lane, Chalfont St Giles, Bucks. HP8 4NR, England. ISSN 0021–8286.

J. Inst. Electron. Commun. Eng. Jpn.
Journal of the Institute of Electronics and Communications Engineers of Japan. Institute of Electronics and Communications Engineers of Japan, Denshi Tsushin Gakkai, Kikai–Shinko–Kaikan, 5–8, Shibakoen 3 Chome Minato–ku, Tokyo 105, Japan. ISSN 0373–6121.

J. Instr. Electron. Telecommun. Eng.
Journal of the Institution of Electronics and Telecommunication Engineers. Institution of Electronics and Telecommunication Engineers, 2 Institutional Area, Lodi Road, New Delhi 110 003, India. ISSN 0377–2063.

J. Korean Astron. Soc.
The Journal of the Korean Astronomical Society. The Korean Astronomical Society, Seoul, Korea.

J. Magn. Magn. Mater.
Journal of Magnetism and Magnetic Materials. North-Holland Publishing Company, PO Box 211, 1000 AE Amsterdam, The Netherlands. ISSN 0304–8853.

J. Math. Phys.
Journal of Mathematical Physics. American Institute of Physics, 335 East 45th Street, New York, N.Y. 10017, USA. ISSN 0022–2488.

J. Math. Phys. Sci.
Journal of Mathematical and Physical Sciences. Indian Institute of Tech., Madras, India. ISSN 0047–2557.

J. Mec. Theor. Appl.
Journal de Mécanique Théorique et Appliquée. Centrale des Revues, 11 rue Gossin, F–92543 Montrouge Cedex, France. ISSN 0750–7240.

J. Mech. Eng. Lab.
Journal of Mechanical Engineering Laboratory. Namiki Sakura–mura, Niihari–gun, Ibaraki, Japan. ISSN 0388–4252.

J. Meteor Res.
The JournaL of Meteor Research. Space News Publishing Co., PO Box 66521, Baton Rouge, La. 70896, USA. ISSN 0277–6057.

J. Microcomput. Appl.
Journal of Microcomputer Applications. Academic Press Inc. Ltd., 24 – 28 Oval Road, London NW1 7DX, England. ISSN 0143–3792.

J. Mol. Spectrosc.
Journal of Molecular Spectroscopy. Academic Press Inc., 111 5th Avenue, New York, N.Y. 10003, USA. ISSN 0022–2852.

J. Nanjing Univ.
Journal of Nanjing University. (Natural Sciences Edition). Nanjing University, Nanjing, People's Republic of China.

J. Navig.
The Journal of Navigation. The Royal Institute of Navigation. Cambridge University Press, Cambridge – London – New York – New Rochelle – Melbourne – Sydney. ISSN 0020–3009.

J. Non–Cryst. Solids
Journal of Non–Crystalline Solids. North–Holland Publishing Company, PO Box 211, 1000 AE Amsterdam, The Netherlands. ISSN 0022–3093.

J. Opt. (Calcutta)
Journal of Optics. Optical Society of India, Department of Applied Physics, University of Calcutta, 92 Acharya Prafulla Chandra Road, Calcutta 700 009, India. ISSN 0304–5811.

J. Opt. (Paris)
Journal of Optics. Masson Editeur, 120 Boulevard Saint-Germain, F–75280 Paris Cedex 06, France. ISSN 0150–536X.

J. Opt. Soc. Am. A
Journal of the Optical Society of America A. Optics and Image Science. Published for the Optical Society of America by the American Institute of Physics, 335 East 45th Street, New York, N.Y. 10017, USA. ISSN 0740–3232.

J. Opt. Soc. Am. B
Journal of the Optical Society of America B. Optical Physics. Published for the Optical Society of America by the American Institute of Physics, 335 East 45th Street, New York, N.Y. 10017, USA. ISSN 0740–3224.

J. Phys.
Journal de Physique. Les Editions de Physique, Z. I. de Courtabœuf, B.P. 112, F–91944 Les Ulis Cedex, France. ISSN 0302–0738.

J. Phys. A
Journal of Physics A. (Mathematical and General Physics). Institute of Physics, 47 Belgrave Square, London SW1X 8QX, England. ISSN 0305–4470.

J. Phys. B
Journal of Physics B. (Atomic and Molecular Physics). Institute of Physics, 47 Belgrave Square, London SW1X 8QX, England. ISSN 0022–3700.

J. Phys. Chem.
Journal of Physical Chemistry. American Chemical Society, 1155 16th Street N.W., Washington, D.C. 20036, USA. ISSN 0022–3654.

J. Phys. Chem. Ref. Data
Journal of Physical and Chemical Reference Data. American Chemical Society, 1155 16th Street, N. W., Washington, D. C. 20036, USA. ISSN 0047–2689.

J. Phys. Colloq.
Journal de Physique Colloque. Les Editions de Physique, Z.I. de Courtabœuf, B.P. 112, F–91944 Les Ulis Cedex, France. ISSN 0449–1947.

J. Phys. D
Journal of Physics D. (Applied Physics). Institute of Physics, 47 Belgrave Square, London SW1X 8QX, England. ISSN 0022–3727.

J. Phys. E
Journal of Physics E. (Scientific Instruments). Institute of Physics, 47 Belgrave Square, London SW1X 8QX, England. ISSN 0022–3735.

J. Phys. Earth
Journal of Physics of the Earth. University of Tokyo Press. Subscription address: Japan Publications Trading Company Ltd., C.P.O. 722, Tokyo, Japan. ISSN 0022–3743.

J. Phys. F
Journal of Physics F. (Metal Physics). Institute of Physics, 47 Belgrave Square, London SW1X 8QX, England. ISSN 0305–4608.

J. Phys. G
Journal of Physics G. (Nuclear Physics). Institute of Physics, 47 Belgrave Square, London SW1X 8QX, England. ISSN 0305–4616.

J. Phys. Lett.
Journal de Physique Lettres. Les Editions de Physique, Z.I. de Courtabœuf, B.P. 112, F–91944 Les Ulis Cedex, France. ISSN 0302–072X.

J. Phys. Soc. Jpn.
Journal of the Physical Society of Japan. Room 211, Kikai Shinko Building, 3–5–8 Shiba Koen, Minato–ku, Tokyo 105, Japan. ISSN 0031–9015.

J. Plasma Phys.
Journal of Plasma Physics. Cambridge University Press, Cambridge – London – New York – New Rochelle – Melbourne – Sydney. ISSN 0022–3778.

J. Proc. R. Soc. N.S.W.
Journal and Proceedings of the Royal Society of New South Wales. Science Centre, 35 Clarence Street, Sydney, N.S.W. 2000, Australia. ISSN 0035–9173.

J. Quant. Spectrosc. Radiat. Transfer
Journal of Quantitative Spectroscopy and Radiative Transfer. Pergamon Press, Oxford – New York – Toronto – Paris – Frankfurt – Sydney. ISSN 0022–4073.

J. R. Astron. Soc. Can.
The Journal of the Royal Astronomical Society of Canada. The Royal Astronomical Society of Canada, 124 Merton Street, Toronto, Ont. M4S 2Z2, Canada. ISSN 0035–872X.

J. Radio Res. Lab.
Journal of the Radio Research Laboratories. Radio Research Laboratories, Ministry of Posts & Telecommunications, Nukui–Kitamachi, Konganei–shi, Tokyo 184, Japan. ISSN 0033–8001.

J. Res. Natl. Bur. Stand.
Journal of Research of the National Bureau of Standards. Subscription address: US Government Printing Office, Washington, D.C. 20402, USA. ISSN 0091 0635.

J. Sci. Ind. Res.
Journal of Scientific and Industrial Research. Council of Scientific & Industrial Research, Publications & Information Directorate, Hillside Road, New Delhi 110 012, India. ISSN 0022–4456.

J. Sci. Res. Banaras Hindu Univ.
Journal of Scientific Research of the Banaras Hindu University. Department of Physics, Faculty of Science, PO Banaras Hindu University, Varnasi 221 005, India. ISSN 0447–9483.

J. Soc. Instrum. Control Eng.
Journal of the Society of Instrument and Control Engineers. Society of Instrument and Control Engineers, Hongo, 1–35–28–303, Bunkyo–ku, Tokyo 113, Japan. ISSN 0453–4662.

J. Spacecr. Rockets
Journal of Spacecraft and Rockets. American Institute of Aeronautics and Astronautics, 1290 Avenue of the Americas, New York, N.Y. 10104, USA. ISSN 0022–4650.

J. Spectrosc. Soc. Jpn.
Journal of the Spectroscopical Society of Japan. Spectroscopical Society of Japan, Room 301, Clean Building, 1–13, Kanda–Awaji–cho, Chiyoda–ku, Tokyo 101, Japan. ISSN 0038–7002.

J. Stat. Phys.
Journal of Statistical Physics. Plenum Publishing Corporation, 227 West 17th Street, New York, N.Y. 10011, USA. ISSN 0022–4715.

J. Toyo Univ., Gen. Educ., Nat. Sci.
Journal of the Toyo University, General Education, Natural Science. Toyo University, 28, Hakusan 5–chôme, Bunkyo–ku, Tokyo, Japan.

Jenaer Rundsch.
Jenaer Rundschau. Jenoptik Jena GmbH. VEB Verlag Technik, Oranienburger Str. 13/14, DDR–1020 Berlin, German Democratic Republic. ISSN 0368–203X.

JETP Lett.
JETP Letters. A cover–to–cover translation of Pis'ma v Zhurnal Ehksperimental'noj i Teoreticheskoj Fiziki. American Institute of Physics, 335 East 45th Street, New York, N.Y. 10017, USA. ISSN 0021–3640.

Johns Hopkins APL Tech. Dig.
Johns Hopkins APL Technical Digest. The Johns Hopkins University Applied Physics Laboratory. Johns Hopkins Road, Laurel, Md. 20707, USA. ISSN 0270–5214.

Jpn. J. Appl. Phys., Part 1
Japanese Journal of Applied Physics, Part 1 (Regular Papers and Short Notes). Publication Office, Daini Toyokaiji Building, 24–8, Shinbashi 4–chome, Minato–ku, Tokyo 105, Japan. ISSN 0021–2922.

Jpn. J. Appl. Phys., Part 2
Japanese Journal of Applied Physics, Part 2 (Letters). Publication Office, Daini Toyokaiji Building, 24–8, Shinbashi 4–chome, Minato–ku, Tokyo 105, Japan. ISSN 0021–4922.

Kandilli Obs., Heliophys. Serv. Publ., Second Ser.
Kandilli Observatory, Heliophysics Service Publications, Second Series. Kandilli Observatory, Istanbul, Turkey.

Kapteyn Astron. Inst., Annu. Rep.
Kapteyn Astronomical Institute, Annual Report. Department of Astronomy Rijksuniversiteit Groningen, Groningen, The Netherlands.

KDD Tech. J.
KDD Technical Journal. International Communications Research Institute, Tokyo, Japan. ISSN 0452–3431.

Kexue Tongbao (Beijing)
Kexue Tongbao. (In Chinese). Academia Sinica, Beijing. Subscription address: Science Press, No. 137, Chaayangmennei Street, Beijing, People's Republic of China. ISSN 0250–7862.

Kexue Tongbao (London)
Kexue Tongbao. (English translation of Kexue Tongbao). Scientific and Technical Books Service Ltd., PO Box 197, London WC2N 4DE, England.

Kinematika Fiz. Nebesn. Tel
Kinematika i Fizika Nebesnykh Tel. Akademiya Nauk Ukrainskoj SSR. Otdelenie Fiziki i Astronomii. Glavnaya Astronomicheskaya Observatoriya Akademii Nauk USSR. Izdatel'stvo Naukova Dumka, 252127 Kiev 127, Goloseevo, USSR. ISSN 0233–7665. Formerly entitled Astrometriya i Astrofizika.

Kodaikanal Obs. Bull., Ser. A
Kodaikanal Observatory Bulletins, Series A. Indian Institute of Astrophysics, Bangalore 560 034, India.

Kodaikanal Obs. Repr.
Kodaikanal Observatory Reprints. Indian Institute of Astrophysics, Bangalore 560 034, India.

Komet. Tsirk.
Kometnyj Tsirkulyar. Gruppa po Issledovaniyu Komet Sektsii 'Solnechnaya Sistema' Astronomicheskogo Soveta AN SSSR. Kievskij Universitet im. T.G. Shevchenko. Glavnaya Astronomicheskaya Observatoriya AN USSR.

Komety Meteory
Komety i Meteory. Akademiya Nauk Tadzhikskoj SSR. Astronomicheskij Sovet Akademii Nauk SSSR. Izdatel'stvo 'Donish', Dushanbe, 29, ul. Ajni, 121, korp. 2, USSR. ISSN 0568–6199.

Kosm. Issled.
Kosmicheskie Issledovaniya. Akademiya Nauk SSSR. Izdatel'stvo 'Nauka', Moskva. Editorial address: 103717 GSP, Moskva, K–62, Podsosenskij per., 21. USSR. English translation in Cosmic Res., Consultants Bureau, New York, N.Y., USA. ISSN 0023–4206.

Kosm. Luchi
Kosmicheskie Luchi. Rezul'taty Issledovanij po Mezhdunarodnym Geofizicheskim Proektam. Mezhduvedomstvennyj Geofizicheskij Komitet pri Prezidiume Akademii Nauk SSSR. Izdatel'stvo Sovetskoe radio, Moskva, Glavpochtamt, a/ya No. 693. USSR.

Kozmos
Kozmos. Popular Astronomical Journal. Slovenské ústredie amatérskej astronomie, Hurbanovo.

Latitude Circ.
Latitude Circular. Astronomic–Geodetical Observatory at Józefosław. Warsaw Technical University, Warsaw, Poland.

LEST Found., Annu. Rep.
Large European Solar Telescope Foundation, Annual Report. Institute of Theoretical Astrophysics, University of Oslo, PO Box 1029, Blindern, N–0315 Oslo 3, Norway. ISSN 0800–7799.

LEST Found., Tech. Rep.
Large European Solar Telescope Foundation, Technical Report. Institute of Theoretical Astrophysics, University of Oslo, PO Box 1029, Blindern, N–0315 Oslo 3, Norway. ISSN 0800–7780.

Lett. Math. Phys.
Letters in Mathematical Physics. D. Reidel Publishing Company, Dordrecht – Boston. ISSN 0377–9017.

Lett. Nuovo Cimento
Lettere al Nuovo Cimento. Società Italiana di Fisica. Editrice Compositori, viale XII Giugno, 1, I–40124 Bologna, Italy. ISSN 0024–1318.

Lick Obs. Bull.
Lick Observatory Bulletin. Lick Observatory, Board of Studies in Astronomy and Astrophysics, University of California at Santa Cruz, Santa Cruz, Calif. 95064, USA.

Lohrmann–Obs., Tech. Univ. Dresden, Zirk.
Lohrmann–Observatorium, Technische Universität Dresden, Zirkular. Lohrmann–Observatorium, Technische Universität Dresden, DDR–8027 Dresden, Mommsenstraße 13, German Democratic Republic.

Lowell Obs. Bull.
Lowell Observatory Bulletin. Lowell Observatory, Flagstaff, Ariz., USA.

Madà
Madà (Science). The Weizmann Science Press of Israel, Jerusalem, Israel. ISSN 0368–833X.

Magy. Geofiz.
Magyar Geofizika. Lapkiado Vallalat, H–1073 Budapest, Lenin korut 9–11, Hungary. ISSN 0025–0120.

Math. Modelling
Mathematical Modelling. Pergamon Press, New York – Oxford – Toronto – Paris – Frankfurt – Sydney. ISSN 0270–0255.

Math. Proc. Camb. Philos. Soc.
Mathematical Proceedings of the Cambridge Philosophical Society. Cambridge University Press, Cambridge – London – New York – New Rochelle – Melbourne – Sydney. ISSN 0305–0041.

Max–Planck–Inst. Radioastron., Bonn, Sonderdr. Ser. A
Max–Planck–Institut für Radioastronomie, Bonn, Sonderdrucke Serie A. Max–Planck–Institut für Radioastronomie, Auf dem Hügel 69, D–5300 Bonn, F.R. Germany.

Meccanica
Meccanica. Pitagora Editrice, via Zamboni 57, I–40126 Bologna, Italy. ISSN 0025–6455.

Meded. K. Acad. Wet., Lett. Schone Kunsten Belg., Kl. Wet.
Mededelingen van de Koninklijke Academie voor Wetenschappen, Letteren en Schone Kunsten van Belgie, Klasse der Wetenschappen. Koninklijke Academie voor Wetenschappen, Letteren en Schone Kunsten van Belgie. Paleis der Academien, Hertogsstraat 1, B–1000 Bruxelles, Belgium.

Mem. Astron. Soc. India
Memoirs of the Astronomical Society of India. Astronomical Society of India, Department of Astronomy, Osmania University, Hyderabad 500 007, India.

Mem. Fac. Eng., Kyoto Univ.
Memoirs of the Faculty of Engineering, Kyoto University. Faculty of Engineering, Kyoto University, Kyoto, Japan. ISSN 0023–6063.

Mem. Fac. Eng., Osaka City Univ.
Memoirs of the Faculty of Engineering, Osaka City University. Faculty of Engineering, Osaka City University, 459 Sugimoto–cho, Sumi Yoshi–kum, Osaka, Japan. ISSN 0078–6659.

Mem. Fac. Sci., Kyoto Univ., Ser. Phys., Astrophys., Geophys., Chem.
Memoirs of the Faculty of Science, Kyoto University, Series of Physics, Astrophysics, Geophysics, and Chemistry. Faculty of Science, University of Kyoto, Kyoto 608, Japan. ISSN 0368–9689.

Mem. Jpn. Astron. Study Assoc.
Memoirs of the Japan Astronomical Study Association. National Science Museum, Ueno Park, Taito–ku, Tokyo, Japan.

Mem. Soc. Astron. Ital.
Memorie della Società Astronomica Italiana. Società Astronomica Italiana, Largo Fermi, 5, I–50125 Firenze, Italy. ISSN 0037–8720.

Mercury
Mercury. The Journal of the Astronomical Society of the Pacific. The Astronomical Society of the Pacific, 1290 24th Avenue, San Francisco, Calif. 94122, USA. ISSN 0047–6773.

Meres Autom.
Meres es Automatika. Lapkiado Vallalat, Budapest VII, Lenin Korut 9 – 11, Hungary. ISSN 0025–9993.

Messenger
The Messenger – El Mensajero. European Southern Observatory, Karl–Schwarzschild–Straße 2, D–8046 Garching bei München, F.R. Germany. ISSN 0722–6691.

Meteoritics
Meteoritics. The Journal of the Meteoritical Society. Center for Meteorite Studies, Arizona State University, Tempe, Ariz. 85287, USA. ISSN 0026–1114

Meteoritika
Meteoritika. Akademiya Nauk SSSR. Komitet po Meteoritam. Izdatel'stvo 'Nauka', Moskva. 117864 GSP-7, Moskva V–485, Profsoyuznaya ul., d. 90, USSR. ISSN 0369–2507.

Meteorol. Mag.
Meteorological Magazine. General Meteorological Office, London Road, Bracknell, Berks. RG12 2SZ, England. ISSN 0026–1149.

Metrologia
Metrologia. Springer–Verlag, Berlin – Heidelberg – New York. ISSN 0026–1394.

Micro Syst.
Micro Systèmes. Société Parisienne d'Edition, 2 à 12 rue de Bellevue, F–75940 Paris Cedex 19, France. ISSN 0183–5084.

Mikrocomput. Z.
Mikrocomputer Zeitschrift. Franzis–Verlag GmbH, Karlstr. 37, D–8000 München 2, F.R. Germany. ISSN 0720–4442.

Mikrowellen Mag.
Mikrowellen Magazin. Sprechsaal–Verlag, PO Box 401, D–8630 Coburg, F.R. Germany. ISSN 0722–8244.

Minor Planet Bull.
The Minor Planet Bulletin. Bulletin of the Minor Planets Section of the Association of Lunar and Planetary Observers. Editorial Office: R. P. Binzel, Department of Astronomy, University of Texas, Austin, Tex. 78712, USA.

Minor Planet Circ.
The Minor Planet Circulars/Minor Planets and Comets. Minor Planet Center, Smithsonian Astrophysical Observatory, 60, Garden Street, Cambridge, Mass. 02138, USA. ISSN 0736–6884.

Mitsubishi Denki Giho
Mitsubishi Denki Giho. Mitsubishi Electric Corporation, Mitsubishi Denki Building, Marunouchi, Tokyo 100, Japan. ISSN 0369–2302.

Mitt. Archenhold–Sternw. Berlin–Treptow
Mitteilungen der Archenhold–Sternwarte Berlin–Treptow. Archenhold–Sternwarte, DDR–1193 Berlin, Alt Treptow 1, German Democratic Republic.

Mitt. Astron. Ges.
Mitteilungen der Astronomischen Gesellschaft, Hamburg. Subscription address: Astronomisches Institut der Universität Bochum, Postfach 102148, D–4630 Bochum, F.R. Germany. ISSN 0172–5483.

Mitt. Astrophys. Obs. Potsdam
Mitteilungen des Astrophysikalischen Wissenschaften der DDR, Zentralinstitut für Astrophysik. Astrophysikalisches Observatorium Potsdam, DDR–1500 Potsdam, Telegrafenberg, German Democratic Republic.

Mitt. Geod. Inst. Rheinischen Friedrich–Wilhelms–Univ. Bonn
Mitteilungen aus den Geodätischen Instituten der Rheinischen Friedrich–Wilhelms–Universität Bonn. Geodätische Institute der Rheinischen Friedrich–Wilhelms–Universität Bonn, Nußallee 17, D–5300 Bonn 1, F.R. Germany. ISSN 0723–4325.

Mitt. Inst. Astron. Phys. Geod. Tech. Hochsch. Münch.
Mitteilungen aus dem Institut für Astronomische und Physikalische Geodäsie der Technischen Hochschule München. Institut für Astronomische und Physikalische Geodäsie der Technischen Hochschule München, D–8000 München, F.R. Germany.

Mitt. Karl–Schwarzschild–Obs. Tautenburg
Mitteilungen des Karl–Schwarzschild–Observatoriums Tautenburg. Akademie der Wissenschaften der DDR, Zentralinstitut für Astrophysik. Karl–Schwarzschild–Observatorium Tautenburg, DDR–6901 Tautenburg, German Democratic Republic.

Mitt. Lohrmann–Obs. Tech. Univ. Dresden
Mitteilungen des Lohrmann–Observatoriums der Technischen Universität Dresden. Lohrmann–Observatorium, Technische Universität Dresden, DDR–8027 Dresden, Mommsenstraße 13, German Democratic Republic.

Mitt. Satell.–Beobachtungsstn. Zimmerwald
Mitteilungen der Satelliten–Beobachtungsstation Zimmerwald. Druckerei der Universität Bern, Switzerland.

Mitt. Sonnenobs. Kanzelhöhe
Mitteilungen des Sonnenobservatoriums Kanzelhöhe. Sonnenobservatorium Kanzelhöhe, A–9520 Sattendorf, Austria.

Mitt. Sternw. Babelsberg, Neue Folge
Mitteilungen der Sternwarte Babelsberg, Neue Folge. Akademie der Wissenschaften der DDR, Zentralinstitut für Astrophysik. Sternwarte Babelsberg, DDR–1502 Potsdam-Babelsberg, Rosa–Luxemburg–Straße 17a, German Democratic Republic.

Mitt. Sternw. Münch.
Mitteilungen der Sternwarte München. Institut für Astronomie und Astrophysik der Universität München, Universitäts–Sternwarte, Scheinerstr. 1, D–8000 München 80, F.R. Germany.

Mitt. Sternw. Sonneberg
Mitteilungen der Sternwarte zu Sonneberg. Akademie der Wissenschaften der DDR, Zentralinstitut für Astrophysik. Sternwarte Sonneberg, DDR–6400 Sonneberg, German Democratic Republic.

Mitt. Univ.–Sternw. Innsb.
Mitteilungen der Universitäts–Sternwarte Innsbruck. Institut für Astronomie der Universität Innsbruck, Universitätsstraße 4, A–6020 Innsbruck, Austria.

Mitt. Univ.–Sternw. Jena
Mitteilungen der Universitäts–Sternwarte zu Jena. Universitätssternwarte Jena, Schillergäßchen 2, DDR–6900 Jena, German Democratic Republic.

Mitt. Universitätssternw. Graz
Mitteilungen der Universitätssternwarte Graz. Institut für Astronomie, Universitätsplatz 5, A–8010 Graz, Austria.

Mitt. Veränderliche Sterne
Mitteilungen über Veränderliche Sterne. Akademie der Wissenschaften der DDR, Zentralinstitut für Astrophysik, Sternwarte Sonneberg, DDR–6400 Sonneberg, German Democratic Republic.

Mitt. Zentralinst. Phys. Erde
Mitteilungen des Zentralinstituts für Physik der Erde. Akademie der Wissenschaften der DDR, Zentralinstitut für Physik der Erde, DDR–1500 Potsdam, Telegrafenberg A17, German Democratic Republic.

Mod. Geol.
Modern Geology. Gordon and Breach Science Publishers, New York – London. ISSN 0026–7775.

Mon. Not. R. Astron. Soc.
Monthly Notices of the Royal Astronomical Society. Published for the Royal Astronomical Society by Blackwell Scientific Publications, Oxford – London – Edinburgh – Boston – Melbourne. ISSN 0035–8711.

Mon. Notes Astron. Soc. S. Afr.
Monthly Notes of the Astronomical Society of Southern Africa. South African Astronomical Observatory, PO Box 9, Observatory, 7935 Cape, South Africa. ISSN 0024–8266.

Mon. Notes Int. Polar Motion Serv.
Monthly Notes of the International Polar Motion Service. Central Bureau of the International Polar Motion Service, International Latitude Observatory of Mizusawa, Mizusawashi, Iwate–ken, Japan. ISSN 0020–8337.

Mons Astrophys. Pap.
Mons Astrophysical Papers. (Communications du Département d'Astrophysique de la Faculté des Sciences de Mons. Departement d'Astrophysique, Université de Mons, B–7000 Mons, Belgium.

MPA Rep.
Max–Planck–Institut für Physik und Astrophysik, Institut für Astrophysik, Reports. Max–Planck–Institut für Physik und Astrophysik, Institut für Astrophysik, Karl–Schwarzschild–Straße 1, D–8046 Garching bei München, F.R. Germany.

MPE Contrib.
Max–Planck–Institut für Physik und Astrophysik, Institut für Extraterrestrische Physik, Contributions. Max–Planck–Institut für Extraterrestrische Physik, Giessenbachstraße, D–8046 Garching bei München, F.R. Germany.

MPE Intern. Rep.
Max–Planck–Institut für Physik und Astrophysik, Institut für Extraterrestrische Physik, Internal Report. Max–Planck–Institut für Extraterrestrische Physik, Giessenbachstraße, D–8046 Garching bei München, F.R. Germany. ISSN 0178–0719.

MPE Prepr.
Max–Planck–Institut für Physik und Astrophysik, Institut für Extraterrestrische Physik, Preprints. Max–Planck–Institut für Extraterrestrische Physik, Giessenbachstraße, D–8046 Garching bei München, F.R. Germany. ISSN 0340–8922.

MPE Rep.
Max–Planck–Institut für Physik und Astrophysik, Institut für Extraterrestrische Physik, Reports. Max–Planck–Institut für Extraterrestrische Physik, Giessenbachstraße, D–8046 Garching bei München, F.R. Germany. ISSN 0173–699X.

MPE Repr.
Max–Planck–Institut für Physik und Astrophysik, Institut für Extraterrestrische Physik, Reprints. Max–Planck–Institut für Extraterrestrische Physik, Giessenbachstraße, D–8046 Garching bei München, F.R. Germany.

MPG Spiegel
MPG Spiegel. Max–Planck–Gesellschaft zur Förderung der Wissenschaften, Residenzstraße 1a, D–8000 München 2, F.R. Germany. ISSN 0341–7727.

MSN Microwave Syst. News
MSN Microwave Systems News. EW Communications Inc., 1170 East Meadow Drive, Palo Alto, Calif. 94303, USA. ISSN 0164–3371.

Mt. Stromlo Siding Spring Obs., Repr.
Mount Stromlo and Siding Spring Observatories, Reprints. Mount Stromlo and Siding Spring Observatories, Research School of Physical Sciences, The Australian National University, Private Bag, Woden PO, ACT 2606, Australia.

Nablyud. Iskusstv. Nebesn. Tel
Nablyudeniya Iskusstvennykh Nebesnykh Tel. Published by Astronomicheskij Sovet Akademii Nauk SSSR, Moskva. Moskva Zh–17, ul. Pyatnitskaya, 48, Astronomicheskij Sovet AN SSSR.

Nachr. Akad. Wiss. Göttingen. II
Nachrichten der Akademie der Wissenschaften in Göttingen. II. Mathematisch–Physikalische Klasse. Vandenhoeck und Ruprecht, Göttingen, F.R. Germany. ISSN 0065–5295.

Nachr. Karten–Vermessungswes., Reihe I
Nachrichten aus dem Karten- und Vermessungswesen, Reihe I: Orginalbeiträge. Verlag des Instituts für Angewandte Geodäsie, Frankfurt a.M., F.R. Germany. ISSN 0469–4236.

Nachr. Olbers–Ges. Bremen
Nachrichten der Olbers–Gesellschaft Bremen. Dr. Walter Stein, Werderstr. 73, Bremen, F.R. Germany.

Nakano wa Kangaeru noda
Nakano wa Kangaeru noda. S. Nakano, PO Box No. 32, Sumoto Post Office, Hyogo–Ken, 656–91, Japan.

Nanjing Univ. Obs., Publ.
Nanjing University Observatory, Publications. Department of Astronomy and Astrophysics, Nanjing University, Nanjing, People's Republic of China.

NASA Conf. Publ.
NASA Conference Publication. National Aeronautics and Space Administration, Scientific and Technical Information Branch, Washington, D.C. 20546, USA. For sale by the National Technical Information Service, Springfield, Va. 22161, USA.

NASA Contract. Rep.
NASA Contractor Report. National Aeronautics and Space Administration, Scientific and Technical Information Branch, Washington, D.C. 20546, USA. For sale by the National Technical Information Service, Springfield, Va. 22161, USA. ISSN 0565–7059.

NASA IUE Newsl.
NASA IUE Newsletter. National Aeronautics and Space Administration, Goddard Space Flight Center, Greenbelt, Maryland 20771, USA. ISSN 0738–2677.

NASA Ref. Publ.
NASA Reference Publication. National Aeronautics and Space Administration, Scientific and Technical Information Branch, Washington, D.C. 20546, USA. For sale by the National Technical Information Service, Springfield, Va. 22161, USA.

NASA Spec. Publ.
NASA Special Publication. National Aeronautics and Space Administration, Scientific and Technical Information Branch, Washington, D.C. 20546, USA. For sale by the National Technical Information Service, Springfield, Va. 22161, USA. ISSN 0091–0805.

NASA Tech. Brief
NASA Technical Brief. NASA Technology Utilization Program, Technology Transfer Division, PO Box 8757, Baltimore/Washington International Airport, Md. 21240, USA. ISSN 0096–7494.

NASA Tech. Memo.
NASA Technical Memorandum. National Aeronautics and Space Administration, Scientific and Technical Information Branch, Washington, D.C. 20546, USA. For sale by the National Technical Information Service, Springfield, Va. 22161, USA. ISSN 0499–9320.

NASA Tech. Note
NASA Technical Note. National Aeronautics and Space Administration, Scientific and Technical Information Branch, Washington, D.C. 20546, USA. For sale by the National Technical Information Service, Springfield, Va. 22161, USA. ISSN 0499–9339.

NASA Tech. Pap.
NASA Technical Paper. National Aeronautics and Space Administration, Scientific and Technical Information Branch, Washington, D.C. 20546, USA. For sale by the National Technical Information Service, Springfield, Va. 22161, USA.

Natl. Acad. Sci. Lett.
National Academy Science Letters. National Academy of Sciences, 5–Lajpat Rai Road, Allahabad 211 002, India. ISSN 0250–541X.

Natl. Astron. Ionos. Cent., Astron. Prepr.
National Astronomy and Ionosphere Center, Astronomy Preprints. National Astronomy and Ionosphere Center, Space Sciences Building, Cornell University, Ithaca, N.Y. 14853, USA.

Natl. Astron. Ionos. Cent., Astron. Publ.
National Astronomy and Ionosphere Center, Astronomy Publications. National Astronomy and Ionosphere Center, Space Sciences Building, Cornell University, Ithaca, N.Y. 14853, USA.

Natl. Astron. Ionos. Cent., Newsl.
National Astronomy and Ionosphere Center, Newsletter. National Astronomy and Ionosphere Center, PO Box 995, Arecibo, P.R. 00613, Puerto Rico.

Natl. Electron. Rev.
National Electronics Reviews. National Electronics Council, Abell House, John Islip Street, London SW1P 4LN, England. ISSN 0305–2257.

Natl. Geogr.
National Geographic. National Geographic Society, 17th and M Sts. N.W., Washington, D.C. 20036, USA. ISSN 0027–9358.

Natl. Radio Astron. Obs., Repr., Ser. A
National Radio Astronomy Observatory, Reprints, Series A. National Radio Astronomy Observatory, PO Box 2, Green Bank, W.Va. 24944, USA.

Natl. Radio Astron. Obs., Repr., Ser. B
National Radio Astronomy Observatory, Reprints, Series B. National Radio Astronomy Observatory, PO Box 2, Green Bank, W.Va. 24944, USA.

Nature
Nature. Macmillan Journals Ltd., London, England. Subscription address: Nature, Circulation Dept., Brunel Road, Basingstoke, Hants RG21 2XS, England. ISSN 0028–0836.

Naturwissenschaften
Die Naturwissenschaften. Springer–Verlag, Berlin – Heidelberg – New York – Tokyo. ISSN 0028–1042.

Nauchn. Inf.
Nauchnye Informatsii. Astronomicheskij Sovet Akademii Nauk SSSR, Moskva, USSR. ISSN 0130–9773.

Naučna Misao
Naučna Misao (Scientific Idea). Society for Promotion and Propagation of Science, 41000 Zagreb, Pavleka Miškine 37, Yugoslavia. ISSN 0467–0468.

Navigation (Paris)
Navigation. Institut Française de Navigation, 3, avenue Octave–Greard, F–75340 Paris Cedex 07, France. ISSN 0028–1530.

Navigation (Wash.)
Navigation. Journal of the Institute of Navigation. Institute of Navigation, Suite 832. 815 15th Street, N.W., Washington, D.C. 20005, USA. ISSN 0028–1522.

Ned. Tijdschr. Natuurkd. A
Nederlands Tijdschrift voor Natuurkunde A. Nederlandse Natuurkundige Vereniging, Princetonplein 5, NL–3508 TA Utrecht, The Netherlands. ISSN 0378–6374.

New Phys., Korean Phys. Soc.
New Physics, Korean Physical Society. Korea Institute for Industrial Economics & Technology, PO Box 205, Chungryang, Seoul, Republic of Korea. ISSN 0374–4914.

New Sci.
New Scientist. New Science Publications, Commonwealth House, 1–19 New Oxford Street, London WC1A 1NG, England. ISSN 0028–6664.

News Lett. Astron. Soc. N.Y.
News Letter of the Astronomical Society of New York. Astronomial Society of New York, 1125 Oxford Place, Schenectady, N.Y. 12308, USA.

Nizamiah Rangapur Obs. Dep. Astron. Osmania Univ., Repr.
Nizamiah and Rangapur Observatories and Department of Astronomy of Osmania University, Reprint. Centre of Advanced Study in Astronomy, Osmania University, Hyderabad 500 007, India.

Nova Acta Leopoldina
Nova Acta Leopoldina. Abhandlungen der Deutschen Akademie der Naturforscher Leopoldina. Deutsche Akademie der Naturforscher Leopoldina, DDR–4010 Halle (Saale), German Democratic Republic. ISSN 0369–5034.

Nova Acta Regiae Soc. Sci. Ups., Ser. V: A
Nova Acta Regiae Societatis Scientiarum Upsaliensis, Ser. V: A (Astronomy and Mathematical Sciences). Royal Society of Sciences of Uppsala. Subscripton address: Almqvist & Wiksell International, Stockholm, Sweden. ISSN 0346–6253.

NSSDC Newsl.
National Space Science Data Center Newsletter. National Space Science Data Center, NASA/Goddard Space Flight Center, Greenbelt, Md. 20771, USA.

NSSDC/WDC–A–R&S, Publ.
NSSDC/WDC–A–R&S, Publication. National Space Science Data Center/World Data Center A for Rockets and Satellites, National Aeronautics and Space Administration, Goddard Space Flight Center, Greenbelt, Md. 20771, USA.

Nucl. Instrum. Methods Phys. Res., Sect. A
Nuclear Instruments and Methods in Physics Research, Section A. Accelerators, Spectrometers, Detectors and Associated Equipment. North–Holland Publishing Company, PO Box 211, 1000 AE Amsterdam, The Netherlands. ISSN 0167–5087.

Nucl. Instrum. Methods Phys. Res., Sect. B
Nuclear Instruments and Methods in Physics Research, Section B. Beam Interactions with Materials and Atoms. North–Holland Publishing Company, PO Box 211, 1000 AE Amsterdam, The Netherlands. ISSN 0168–583X.

Nucl. Phys. A
Nuclear Physics A. North–Holland Publishing Company, PO Box 211, 1000 AE Amsterdam, The Netherlands. ISSN 0029–5582.

Nucl. Phys. B, Part. Phys.
Nuclear Physics B, Particle Physics. North–Holland Publishing Company, PO Box 211, 1000 AE Amsterdam, The Netherlands. ISSN 0550–3213.

Nucl. Sci. Appl., Sect. A
Nuclear Science Applications, Section A. Harwood Academic Publishers GmbH, Poststraße 22, CH–7000 Chur, Switzerland. ISSN 0191–1686.

Nucl. Tracks Radiat. Meas.
Nuclear Tracks and Radiation Measurements. Pergamon Press, Oxford – New York – Toronto – Paris – Frankfurt – Sydney. ISSN 0191–278X.

Numer. Math.
Numerische Mathematik. Springer–Verlag, Berlin – Heidelberg – New York – Tokyo. ISSN 0029–599X.

Nuovo Cimento A
Il Nuovo Cimento A. Società Italiana di Fisica, Editrice Compositori, viale XII Giugno, 1, I–40124 Bologna, Italy. ISSN 0369–4097.

Nuovo Cimento B
Il Nuovo Cimento B. Società Italiana di Fisica, Editrice Compositori, viale XII Giugno, 1, I–40124 Bologna, Italy. ISSN 0369–4100.

Nuovo Cimento C
Il Nuovo Cimento C. Società Italiana di Fisica, Editrice Compositori, viale XII Giugno, 1, I–40124 Bologna, Italy.

N.Z. J. Sci.
New Zealand Journal of Science. New Zealand Department of Scientific and Industrial Research. Subscription address: Science Information Division, DSIR, PO Box 9741, Wellington, New Zealand. ISSN 0028–8365.

Obs. Astron. Antares, Contrib. Cient.
Observatório Astronômico Antares, Contribuição Cientifica. Universidade Estadual de Feira de Santana, Feira de Santana, Brazil.

Obs. Astron. Córdoba, Tirada Aparte
Observatorio Astronomico Córdoba, Tirada Aparte. Observatorio Astronomico, Laprida 854, 5000 Córdoba, Argentina.

Obs. Astron. La Plata, Sep. Astron.
Observatorio Astronómico La Plata, Separata Astronomica. Observatorio Astronómico, Universidad Nacional de La Plata, La Plata, Argentina.

Obs. Astron. Univ. Nac. La Plata, Ser. Astron.
Observatorio Astronómico de la Universidad Nacional de La Plata, Serie Astronómica. Observatorio Astronómico, Universidad Nacional de La Plata, La Plata, Argentina. ISSN 0325–3163.

Obs. Astron. Univ. Nac. La Plata, Ser. Espec.
Observatorio Astronómico de la Universidad Nacional de La Plata, Serie Especial. Observatorio Astronómico, Universidad Nacional de La Plata, La Plata, Argentina. ISSN 0325–3015.

Obs. Astrophys. Lab., Univ. Helsinki, Rep.
Observatory and Astrophysics Laboratory, University of Helsinki, Report. Observatory and Astrophysics Laboratory, University of Helsinki, Tähtitorninmäki, SF–00130 Helsinki 13, Finland. ISSN 0355–9289.

Obs. Haute Provence, Pré–Publ.
Observatoire de Haute Provence, Pré–Publication. Observatoire de Haute Provence, F–04870 St. Michel l'Observatoire, France.

Obs. Lyon Prepr.
Observatoire de Lyon Preprints. Observatoire de Lyon, F–69230 St. Genis Laval, France.

Obs. Lyon Repr.
Observatoire de Lyon Reprint. Observatoire de Lyon, F–69230 St. Genis Laval, France.

Obs. R. Belg., Commun., Sér. A
Observatoire Royal de Belgique, Communications, Série A. (Koninklijke Sterrenwacht van België, Mededelingen, Reeks A). Observatoire Royale de Belgique, 3, avenue Circulaire, Uccle, B–1180 Bruxelles, Belgium.

Obs. R. Belg., Commun., Sér. B
Observatoire Royal de Belgique, Communications, Série B. (Koninklijke Sterrenwacht van België, Mededelingen, Reeks B). Observatoire Royal de Belgique, 3, avenue Circulaire, Uccle, B–1180 Bruxelles, Belgium.

Obs. Satell.
Observations of Satellites. Finnish Meteorological Institute. Subscription address: Government Printing Centre, Marketing Department, PO Box 516, SF–00101 Helsinki 10, Finland. ISSN 0355–2004.

Obs. Variable Stars, Rep.
Observations of Variable Stars, Report. Nederlandse Vereniging voor Weer–en Sterrenkunde. Kapteyn Astronomical Institute, Postbus 800, 9700 AV Groningen, The Netherlands.

Observatory
The Observatory. A Review of Astronomy. Royal Greenwich Observatory, Herstmonceux Castle, Hailsham Sussex, BN27 1RP England, ISSN 0029–7704.

Occas. Rep. R. Obs., Edinb.
Occasional Reports of the Royal Observatory, Edinburgh. Royal Observatory, Blackford Hill, Edinburgh EH9 3HJ, Scotland. ISSN 0309–099X.

Occultation Newsl.
Occultation Newsletter. International Occultation Timing Association (IOTA), PO Box 596, Tinley Park, Ill. 60477, USA.

Österr. Z. Vermessungswes. Photogramm.
Österreichische Zeitschrift für Vermessungswesen und Photogrammetrie. Österreichischer Verein für Vermessungswesen und Photogrammetrie, Friedrich–Schmidt–Platz 3, A–1082 Wien, Austria.

Onsala Space Obs., Prepr.
Onsala Space Observatory, Preprint. Chalmers University of Technology, Onsala Space Observatory, S–43900 Onsala, Sweden.

Opt. Acta
Optica Acta. Taylor and Francis Ltd., 4 John Street, London WC1N 2ET, England. ISSN 0030–3909.

Opt. Commun.
Optics Communications. North–Holland Publishing Company, PO Box 211, 1000 AE Amsterdam, The Netherlands. ISSN 0030–4018.

Opt. Eng.
Optical Engineering. Society of Photo–Optical Instrumentation Engineers (SPIE), PO Box 10, 405 Fieldston Road, Bellingham, Wash. 98225, USA. ISSN 0091–3286.

Opt. Lett.
Optics Letters. Published for the Optical Society of America by the American Institute of Physics, 335 East 45th Street, New York, N.Y. 10017, USA. ISSN 0146–9592.

Opt. Pura Apl.
Optica Pura y Aplicada. Sociedad Española de Optica, Serrano 121, Madrid 6, Spain. ISSN 0030–3917.

Origins Life
Origins of Life. An International Journal Devoted to the Scientific Study of the Origin of Life. D. Reidel Publishing Company, Dordrecht – Boston – London. ISSN 0302–1688.

Orion
Orion. Zeitschrift der Schweizerischen Astronomischen Gesellschaft (SAG). Zentralsekretariat, Hirtenhofstraße 9, CH–6005 Luzern, Switzerland. ISSN 0030–557X.

Orione
Orione. Rivista Trimestrale di Astronomia. Rivista Orione, via Roma 6, I–10025 Pino Torinese, Italy.

Oss. Astrofis. Catania, Pubbl.
Osservatorio Astrofisico di Catania, Pubblicazione. Osservatorio Astrofisico, città universitaria, I–95125 Catania, Italy.

Oss. Astron. Milano–Merate, Contrib.
Osservatorio Astronomico Milano–Merate, Contributo. Osservatorio Astronomico, Via Emilio Bianchi 46, I–22055 Merate, Como, Italy.

Oss. Mem. Oss. Astrofis. Arcetri
Osservazioni e Memorie dell'Osservatorio Astrofisico di Arcetri. Università Degli Studi di Firenze, Firenze, Italy.

Oyo Buturi
Oyo Buturi. Japan Society of Applied Physics, Room No. 209–2, Kikai–Shinko Building, 3 Shiba–Koen Minato–ku, Tokyo, Japan. ISSN 0369–8009.

Pascal Explore
Pascal Explore. E48 (Environnement cosmique terrestre, astronomie et géologie extraterrestre). Centre National de la Recherche Scientifique, Centre de Documentation Scientifique et Technique, 26, rue Boyer, F–75971 Paris Cedex 20, France. ISSN 0761–2109.

Patrika
Patrika. Newsletter of the Indian Academy of Sciences. Indian Academy of Sciences, Bangalore 560 080, India.

Perem. Zvezdy
Peremennye Zvezdy (Variable Stars). Sbornik statej, izdavaemyj Astronomicheskim Sovetom Akademii Nauk SSSR. Editorial address: Sternberg State Astronomical Institute of the Moscow University, Universitetskij prospekt, 13, 117234 Moscow, USSR. ISSN 0373–7683.

Perem. Zvezdy, Prilozh.
Peremennye Zvezdy, Prilozhenie (Variable Stars, Supplement). Sbornik statej, izdavaemyj Astronomicheskim Sovetom Akademii Nauk SSSR. Astronomicheskij Sovet Akademii Nauk SSSR, Moskva, USSR.

Philos. Trans. R. Soc. London, Ser. A
Philosophical Transactions of the Royal Society of London, Series A: Mathematical and Physical Sciences. The Royal Society, 6 Carlton House Terrace, London SW1Y 5AG, England. ISSN 0080–4614.

Photogr. Sci. Eng.
Photographic Science and Engineering. Society of Photographic Scientists and Engineers, Suite 204, 1330 Massachusetts Avenue N.W., Washington, D.C. 20005, USA. ISSN 0031–8760.

Photogramm. Eng. Remote Sensing
Photogrammetric Engineering and Remote Sensing. American Society of Photogrammetry, 105 North Virginia Avenue, Falls Church, Va. 22046, USA. ISSN 0099–1112.

Photonics Spectra
Photonics Spectra. The Magazine of Optical/Electro–Optical/ Laser Technology. Published by the Optical Publishing Co., Inc., PO Box 1146, Pittsfield, Mass. 01201, USA. ISSN 0191–0647.

Photorin
Photorin, Mitteilungen der Lichtenberg–Gesellschaft e.V.

Phys. Abstr.
Physics Abstracts. Science Abstracts Series A. An INSPEC Publication published by the Institution of Electrical Engineers in association with the Institute of Electrical and Electronics Engineers Inc. Subscription address: INSPEC Marketing Department, IEE, Station House, Nightingale Road, Hitchin, Herts. SG5 1RJ, England. ISSN 0036–8091.

Phys. Bl.
Physikalische Blätter. Physik–Verlag GmbH, Postfach 1260/1280, D–6940 Weinheim, F.R. Germany. ISSN 0031–9279.

Phys. Briefs
Physics Briefs. Physikalische Berichte. Edited by Deutsche Physikalische Gesellschaft und Fachinformationszentrum Energie, Physik, Mathematik in cooperation with American Institute of Physics. Physik–Verlag GmbH, Postfach 1260/1280, D–6940 Weinheim, F.R. Germany. ISSN 0170–7434.

Phys. Bull.
Physics Bulletin. Institute of Physics, 47 Belgrave Square, London SW1X 8QX, England. ISSN 0031–9112.

Phys. Chem. Miner.
Physics and Chemistry of Minerals. Springer–Verlag, Berlin – Heidelberg – New York – Tokyo. ISSN 0342–1791.

Phys. Earth Planet. Inter.
Physics of the Earth and Planetary Interiors. Elsevier Scientific Publishing Company, PO Box 211, 1000 AE Amsterdam, The Netherlands. ISSN 0031–9201.

Phys. Educ.
Physics Education. Institute of Physics. 47 Belgrave Square, London SW1X 8QX, England. ISSN 0031–9120.

Phys. Energ. Fortis Phys. Nucl.
Physica Energiae Fortis et Physica Nuclearis. Science Press, Beijing. Subscription address: Guozi Shudian, PO Box 399, Beijing, People's Republic of China. ISSN 0254–3052.

Phys. Fluids
Physics of Fluids. American Institute of Physics, 335 East 45th Street, New York, N.Y. 10017, USA. ISSN 0031–9171.

Phys. Lett. A
Physics Letters A. (General, Atomic, and Solid State Physics). North–Holland Publishing Company, PO Box 211, 1000 AE Amsterdam, The Netherlands. ISSN 0031–9163.

Phys. Lett. B
Physics Letters B. (Nuclear, Elementary Particle, and High–Energy Physics). North–Holland Publishing Company, PO Box 211, 1000 AE Amsterdam, The Netherlands. ISSN 0031–9163.

Phys. Rep.
Physics Reports. North–Holland Publishing Company, PO Box 211, 1000 AE Amsterdam, The Netherlands. ISSN 0370–1573.

Phys. Rev. A
Physical Review A. (General Physics). Published for the American Physical Society by the American Institute of Physics, 335 East 45th Street, New York, N.Y. 10017, USA. ISSN 0556–2791.

Phys. Rev. B
Physical Review B. (Condensed Matter). Published for the American Physical Society by the American Institute of Physics, 335 East 45th Street, New York, N.Y. 10017, USA. ISSN 0163–1829.

Phys. Rev. C
Physical Review C. (Nuclear Physics). Published for the American Phyiscal Society by the American Institute of Physics, 335 East 45th Street, New York, N.Y. 10017, USA. ISSN 0556–2813.

Phys. Rev. D
Physical Review D. (Particles and Fields). Published for the American Physical Society by the American Institute of Physics, 335 East 45th Street, New York, N.Y. 10017, USA. ISSN 0556–2821.

Phys. Rev. Lett.
Physical Review Letters. Published for the American Physical Society by the American Institute of Physics, 335 East 45th Street, New York, N.Y. 10017, USA. ISSN 0031–9007.

Phys. Scr.
Physica Scripta. An International Journal for Experimental and Theoretical Physics. The Royal Swedish Academy of Sciences, Publications Department, Box 50005, S–104 05 Stockholm, Sweden. ISSN 0031–8949.

Phys. Teach.
Physics Teacher. American Institute of Physics, 335 East 45th Street, New York, N.Y. 10017, USA. ISSN 0031–921X.

Phys. Technol.
Physics in Technology. Institute of Physics, 47 Belgrave Square, London SW1X 8QX, England. ISSN 0305–4624.

Phys. Today
Physics Today. American Institute of Physics, 335 East 45th Street, New York, N.Y. 10017, USA. ISSN 0031–9228.

Phys. Unserer Zeit
Physik in unserer Zeit. Verlag Chemie GmbH, Postfach 1260/1280, D–6940 Weinheim, F.R. Germany. ISSN 0031–9252.

Physica A
Physica A. North–Holland Publishing Company, PO Box 211, 1000 AE Amsterdam, The Netherlands. ISSN 0378–4371.

Physica B, C
Physica B and C (Low Temperature and Solid State Physics; Atomic, Molecular and Plasma Physics; Optics). North–Holland Publishing Company, PO Box 211, 1000 AE Amsterdam, The Netherlands. ISSN 0378–4363.

Physica D
Physica D. North–Holland Publishing Company, PO Box 211, 1000 AE Amsterdam, The Netherlands. ISSN 0167–2789.

Pis'ma Astron. Zh.
Pis'ma v Astronomicheskij Zhurnal. Akademiya Nauk SSSR. Izdatel'stvo 'Nauka', Moskva. 103717, GSP, Moskva, K–62, Podsosenskij per., 21. USSR. English translation in Soviet Astron. Lett. ISSN 0320–0108.

Planet. Space Sci.
Planetary and Space Science. Pergamon Press, Oxford – New York – Paris – Frankfurt. ISSN 0032–0633.

Plasma Phys. Controlled Fusion
Plasma Physics and Controlled Fusion. Pergamon Press, Oxford – New York – Toronto – Paris – Frankfurt – Sydney. ISSN 0032–1028.

Pokroky Mat., Fyz. Astron.
Pokroky matematiky, fyziky a astronomie. Academia, Praha, Czechoslovakia. ISSN 0032–2423.

Port. Phys.
Portugaliae Physica. Sociedade Portuguesa de Fisica, Avenida de Republica, 37–40, P–1000 Lisboa, Portugal. ISSN 0048–4903.

Postepy Astron.
Postepy Astronomii. Czasopismo Poświecone Upowszechnianiu Wiedzy Astronomicznej. Polskie Towarzystwo Astronomiczne, Warszawa. Printed in Poland by Pánstwowe Wydawnictwo Naukowe, Warszawa–Lódź, Poland. ISSN 0032–5414.

Postepy Fiz.
Postepy Fiziyki. Polskie Towarzystwo Fizykczne, 00–681 Warszawa, ul. Hoza 69, Poland. ISSN 0032–5430.

Pramāna
Pramāna. Indian Academy of Sciences, Bangalore 560 006, India. ISSN 0304–4289.

Prepr. Astron. Inst. Univ. Basel
Preprint of the Astronomical Institute of the University of Basel. Astronomisches Institut der Universität Basel, Venusstraße 7, CH–4102 Binningen, Switzerland.

Prepr. Steward Obs.
Preprints of the Steward Observatory. Steward Observatory, The University of Arizona, Tucson, Ariz. 85721, USA.

Priroda
Priroda. Izdatel'stvo 'Nauka', Moskva. Editorial address: 117049, Moskva, GSP–1, Maronovskij per., 26, USSR. ISSN 0032–874X.

Probl. Kosm. Fiz.
Problemy Kosmicheskoj Fiziki. Respublikanskij Mezhvedomstvennyj Nauchnyj Sbornik. Izdatel'stvo pri Kievskom Gosudarstvennom Universitete Izdatel'skogo Obedineniya 'Vishcha Shkola'. Kiev, USSR. ISSN 0555–2796.

Proc. Astron. Soc. Aust.
Proceedings of the Astronomical Society of Australia. Astronomical Society of Australia, Sydney. Subscription address: Division of Radiophysics, PO Box 76, Epping, N.S.W. 2121, Australia. ISSN 0066–9997.

Proc. Cosmic–Ray Res. Lab. Nagoya Univ.
Proceedings of the Cosmic–Ray Research Laboratory of Nagoya University. Cosmic Ray Research Laboratory and Department of Physics, Nagoya University, Chikusa–ku, Nagoya, Japan. ISSN 0250 5444.

Proc. IEEE
Proceedings of the IEEE. The Institute of Electrical and Electronics Engineers, 345 East 47th Street, New York, N.Y. 10017, USA. ISSN 0018–9219.

Proc. Indian Acad. Sci., Earth Planet. Sci.
Proceedings of the Indian Academy of Sciences, Earth and Planetary Sciences. Indian Academy of Sciences, Bangalore 560 080, India. ISSN 0370–0089.

Proc. Indian Acad. Sci., Eng. Sci.
Proceedings of the Indian Academy of Sciences, Engineering Sciences. Indian Academy of Sciences, Bangalore 560 080, India. ISSN 0253–4096.

Proc. Indian Natl. Sci. Acad., Part A
Proceedings of the Indian National Science Academy, Part A. (Physical Sciences). The Indian National Science Academy, Bahadur Shah Zafar Marg, New Delhi 110 002, India. ISSN 0370–0046.

Proc. Int. Latitude Obs. Mizusawa
Proceedings of the International Latitude Observatory of Mizusawa. The International Latitude Observatory of Mizusawa, Mizusawa–Shi, Iwate–Ken, Japan. ISSN 0536–3403.

Proc. Jpn. Acad., Ser. A
Proceedings of the Japan Academy, Series A. (Mathematical Sciences). Japan Academy, Ueno Park, Tokyo 110, Japan. ISSN 0386–2194.

Proc. Jpn. Acad., Ser. B
Proceedings of the Japan Academy, Series B. (Physical and Biological Sciences). Japan Academy, Ueno Park, Tokyo 110, Japan. ISSN 0386–2208.

Proc. K. Ned. Akad. Wet., Ser. B
Proceedings of the Koninklijke Nederlandse Akademie van Wetenschappen, Series B. (Physical Sciences). North–Holland Publishing Company, PO Box 211, 1000 AE Amsterdam, The Netherlands. ISSN 0023–3366.

Proc. Natl. Acad. Sci. India, Sect. A
Proceedings of the National Academy of Sciences of India, Section A (Physical Sciences). National Academy of Sciences of India, 5 Lajpatrai Road, Allahabad–2, India. ISSN 0369–8203.

Proc. Natl. Acad. Sci. U.S.A.
Proceedings of the National Academy of Sciences of the United States of America. National Academy of Sciences, 2101 Constitution Avenue, Washington, D.C. 20418, USA. ISSN 0027–8424.

Proc. R. Soc. Edinb., Sect. A
Proceedings of the Royal Society of Edinburgh, Section A (Mathematical and Physical Sciences). Royal Society of Edinburgh, 22 George Street, Edinburgh EH2 2PQ, Scotland. ISSN 0308–2105.

Proc. R. Soc. London, Ser. A
Proceedings of the Royal Society of London, Series A. (Mathematical and Physical Sciences). The Royal Society, 6 Carlton House Terrace, London, SW1Y 5AG, England. ISSN 0080–4630.

Proc. SPIE Int. Soc. Opt. Eng.
Proceedings of the SPIE – The International Society for Optical Engineering. The International Society for Optical Engineering, PO Box 10, Bellingham, Wash. 98227, USA. ISSN 0277–786X.

Prog. Part. Nucl. Phys.
Progress in Particle and Nuclear Physics. Pergamon Press, New York – Toronto – Paris – Frankfurt – Sydney. ISSN 0146–6410.

Prog. Theor. Phys.
Progress of Theoretical Physics. Research Institute for Fundamental Physics and the Physical Society of Japan. Subscription address: Yukawa Hall, Kyoto University, 606 Kyoto, Japan. ISSN 0033–068X.

Prog. Theor. Phys., Suppl.
Progress of Theoretical Physics, Supplement. Research Institute for Fundamental Physics and the Physical Society of Japan. Subscription address: Yukawa Hall, Kyoto University, 606 Kyoto, Japan. ISSN 0375–9687.

PTB Mitt.
PTB Mitteilungen. Forschen + Prüfen. Amts- und Mitteilungsblatt der Physikalisch-Technischen Bundesanstalt Braunschweig – Berlin. Friedr. Vieweg & Sohn Verlagsgesellschaft mbH, Faulbrunnenstraße 13, D–6200 Wiesbaden, F.R. Germany. ISSN 0030–834X.

Pubbl. Stn. Astron. Int. Latitudine, Carloforte–Cagliari, Nuov. Ser.
Pubblicazioni della Stazione Astronomica Internazionale di Latitudine, Carloforte–Cagliari, Nuova Serie. Stazione Astronomica Internazionale di Latitudine, Carloforte–Cagliari, Italy.

Pubbl. Varie Fuori Ser. Oss. Astron. Torino (Pino Torinese)
Pubblicazioni Varie Fuori Serie dell'Osservatorio Astronomico di Torino (Pino Torinese). Osservatorio Astronomico di Torino, I–10025 Pino Torinese (Torino), Italy.

Publ. Astron. Dep. Eötvös Univ.
Publications of the Astronomy Department of the Eötvös University. Department of Astronomy, Eötvös Lorand University, Budapest, Hungary.

Publ. Astron. Inst. Czech. Acad. Sci.
Publications of the Astronomical Institute of the Czechoslovak Academy of Sciences. Astronomical Institute, Czechoslovak Academy of Sciences, Budečská 6, 120 23 Praha 2, Czechoslovakia.

Publ. Astron. Obs. Sarajevo
Publications of the Astronomical Observatory of Sarajevo. Astronomical Observatory of Sarajevo, Sarajevo, M. Tita 44, Yugoslavia. ISSN 0351–4587.

Publ. Astron. Opservatorije Beogr.
Publikacija Astronomske Opservatorije u Beogradu (Publications de l'Observatoire Astronomique de Beograd). Astronomska Opservatorija u Beogradu, YU–11050 Beograd, Volgina 7, Yugoslavia. ISSN 0373–3742.

Publ. Astron. Soc. Jpn.
Publications of the Astronomical Society of Japan. Astronomical Society of Japan, Tokyo Astronomical Observatory. Mitaka–shi, Tokyo 181, Japan. ISSN 0004–6264.

Publ. Astron. Soc. Pac.
Publications of the Astronomical Society of the Pacific. The Astronomical Society of the Pacific, 1290 24th Avenue, San Francisco, Calif. 94122, USA. ISSN 0004–6280.

Publ. Astrophys. Obs. Potsdam
Publikationen des Astrophysikalischen Observatoriums zu Potsdam. Akademie der Wissenschaften der DDR, Zentralinstitut für Astrophysik. Astrophysikalisches Observatorium Potsdam, DDR–1500 Potsdam, Telegrafenberg, German Democratic Republic.

Publ. Beijing Astron. Obs.
Publications of the Beijing Astronomical Observatory. Beijing Astronomical Observatory, Academia Sinica, Beijing, People's Republic of China.

Publ. Bosscha Obs.
Publications of the Bosscha Observatory. Bandung Institute of Technology, Department of Science, Lembang, Indonesia.

Publ. Debrecen Heliophys. Obs.
Publications of Debrecen Heliophysical Observatory. Hungarian Academy of Sciences, H–4010 Debrecen, Hungary. ISSN 0209–7567.

Publ. Dep. Astron., Univ. Beogr.
Publications of the Department of Astronomy, University of Beograd. Department of Astronomy, University of Beograd, Beograd, Yugoslavia. ISSN 0350–3283.

Publ. Dep. Astron., Univ. Cape Town
Publications of the Department of Astronomy of the University of Cape Town. Astronomy Department, University of Cape Town, Rondebosch, 7700 Cape, South Africa.

Publ. Dep. Astron., Univ. Chile
Publicaciones Departamento de Astronomía, Universidad de Chile. Observatorio Astronomico Nacional, Cerro Calan, Santiago de Chile, Chile.

Publ. Dep. Geod. Astron., Univ. Thessaloniki
Publications of the Department of Geodetic Astronomy, University of Thessaloniki. Department of Geodetic Astronomy, University of Thessaloniki, Thessaloniki, Greece.

Publ. Dom. Astrophys. Obs.
Publications of the Dominion Astrophysical Observatory. Dominion Astrophysical Observatory, National Research Center of Canada, 5071 West Saanich Road, Victoria, B.C. V8X 4M6, Canada. ISSN 0078–6950.

Publ. Ege Univ. Obs.
Publications of Ege University Observatory. Ege University Observatory, P.K. 21, Bornova–Izmir, Turkey.

Publ. Espec. Obs. Astron. Quito
Publicacion Especial del Observatorio Astronomico de Quito. Observatorio Astronomico de Quito, Escuela Politécnica Nacional, Quito, Ecuador.

Publ. Finn. Geod. Inst.
Publications of the Finnish Geodetic Institute. Finnish Geodetic Institute, Helsinki, Finland. ISSN 0085–6932.

Publ. Inst. Geophys., Pol. Acad. Sci., Ser. F (Planet Geod.)
Publications of the Institute of Geophysics, Polish Academy of Sciences, Series F (Planetary Geodesy). Państwowe Wydawnictwo Naukowe, Warszawa–Łódź, Poland. ISSN 0318–0222.

Publ. Int. Latitude Obs. Mizusawa
Publications of the International Latitude Observatory of Mizusawa. The International Latitude Observatory of Mizusawa, Mizusawa–shi, Iwate–ken, Japan. ISSN 0386–0779.

Publ. Istanbul Univ. Obs.
Publications of the Istanbul University Observatory. Istanbul University Observatory, Istanbul, Turkey.

Publ. Lick Obs.
Publications of the Lick Observatory. Lick Observatory, Board of Studies in Astronomy and Astrophysics, University of California at Santa Cruz, Santa Cruz, Calif. 95064, USA.

Publ. Obs. Astron. Nac., Univ. Colombia
Publicaciones del Observatorio Astronomico Nacional. Observatorio Astronomico Nacional, Universidad Nacional de Colombia, Apartado Aéreo 2584, Bogotá, Colombia.

Publ. Obs. Astron. "Prof. Manuel de Barros", Fac. Cienc. Porto
Publicações do Observatório Astronómico "Prof. Manuel de Barros", da Faculdade de Ciências do Porto. Observatório Astronómico "Prof. Manuel de Barros", Universidade do Porto, Monte da Virgem – Vila Nova de Gaia, Portugal.

Publ. Obs. Astron. Univ. Cluj
Publicatiile Observatorului Astronomic al Universitatii din Cluj. Astronomical Observatory, University Babes–Bolyai, Cluj–Napoca, Romania.

Publ. Obs. Bordeaux, Nouv. Sér.
Publications de l'Observatoire de Bordeaux, Nouvelle Série. Observatoire de l'Université de Bordeaux I, F–33270 Floirac, France.

Publ. Obs. Ebro
Publicaciones del Observatorio del Ebro (Miscelanea). Consejo Superior de Investigaciones Cientificas, Observatorio del Ebro, Roquetas (Tarragona), Spain.

Publ. Obs. Genève, Sér. A
Publications de l'Observatoire de Genève, Série A. Observatoire de Genève, CH–1290 Sauverny, Switzerland.

Publ. Obs. Genève, Sér. B
Publications de l'Observatoire de Genève, Série B. Observatoire de Genève, CH–1290 Sauverny, Switzerland.

Publ. Purple Mt. Obs.
Publications of the Purple Mountain Observatory. Purple Mountain Observatory, Academia Sinica, Nanking, People's Republic of China.

Publ. R. Obs., Edinb.
Publications of the Royal Observatory, Edinburgh. Science Research Council. Royal Observatory, Blackford Hill, Edinburgh EH9 3HJ, Scotland. ISSN 0305–2001.

Publ. Semin. Astron. Geod. Univ. Complutense Madr.
Publicaciones del Seminario de Astronomia y Geodesia de la Universidad Complutense de Madrid. Seminario de Astronomia y Geodesia, Facultad de Ciencias Matemáticas, Universidad Complutense, Madrid, Spain.

Publ. Shaanxi Astron. Obs.
Publications of the Shaanxi Astronomical Observatory. Shaanxi Astronomical Observatory, Academia Sinica, Lintong, Xian, People's Republic of China.

Publ. Spéc. Cent. Données Stellaires
Publication Spéciale du Centre de Données Stellaires. Observatoire de Strasbourg, 11, rue de l'Université, F–67000 Strasbourg, France.

Publ. U.S. Nav. Obs., Second Ser.
Publications of the United States Naval Observatory, Second Series. U.S. Government Printing Office, Washington, D.C. 20402, USA.

Publ. Variable Star Sect., R. Astron. Soc. N.Z.
Publications of Variable Star Section, Royal Astronomical Society of New Zealand. Astronomical Research Ltd., PO Box 3093, Greerton, Tauranaga, New Zealand.

Publ. Warner Swasey Obs.
Publications of the Warner and Swasey Observatory. Case Western Reserve University, 1975 Taylor Road, East Cleveland, Ohio 44112, USA. ISSN 0160–2500.

Pure Appl. Geophys.
Pure and Applied Geophysics. Birkhäuser Boston Inc., 380 Green Street, Cambridge, Mass. 02139, USA. ISSN 0033–4553.

Q. Appl. Math.
Quarterly of Applied Mathematics. American Mathematical Society, PO Box 1571, Providence, R.I. 02901, USA. ISSN 0033–569X.

Q. Bull. Sol. Act.
Quarterly Bulletin on Solar Activity. International Astronomical Union. Published by the Tokyo Astronomical Observatory, University of Tokyo, Mitaka, Tokyo 181, Japan.

Q. J. R. Astron. Soc.
The Quarterly Journal of the Royal Astronomical Society. Published for the Royal Astronomical Society by Blackwell Scientific Publications, Oxford – London – Edinburgh – Boston – Melbourne. ISSN 0035–8738.

Q. J. R. Meteorol. Soc.
Quarterly Journal of the Royal Meteorological Society. Cromwell House, High Street, Bracknell, Berks., England. ISSN 0035–9009.

R. Greenwich Obs., Annu. Rep.
Royal Greenwich Observatory, Annual Report. Royal Greenwich Observatory, Herstmonceux Castle, Hailsham, East Sussex BN27 1RP, England. ISSN 0308–3322.

R. Greenwich Obs., Bull.
Royal Greenwich Observatory, Bulletins. Royal Greenwich Observatory, Herstmonceux Castle, Hailsham, East Sussex BN27 1RP, England. ISSN 0308–5074.

R. Obs. Ann.
Royal Observatory Annals. Royal Greenwich Observatory, Herstmonceux Castle, Hailsham, East Sussex BN27 1RP, England. ISSN 0080–4371.

R. Obs. Edinb., Res. Facilities
Royal Observatory Edinburgh, Research and Facilities. Royal Observatory, Blackford Hill, Edinburgh EH9 3HJ, Scotland.

Radiant
Radiant. Journal of the Dutch Meteor Society. Federation of European Meteor Astronomers (FEMA), Morssingel 35a, 2312 AZ Leiden, The Netherlands.

Radiat. Eff.
Radiation Effects. Gordon and Breach Science Publishers Inc., New York – London – Paris. ISSN 0033–7579.

Radio Sci.
Radio Science. American Geophysical Union, 2000 Florida Avenue, N.W., Washington, D.C. 20009, USA. ISSN 0048–6604.

Radiochem. Radioanal. Lett.
Radiochemical and Radioanalytical Letters. Elsevier Sequoia S.A., PO Box 851, CH–1001 Lausanne, Switzerland. ISSN 0079–9483.

Rapp. Act. Obs. Paris
Rapport d'Activité de l'Observatoire de Paris. Observatoire de Paris, F–92190 Meudon, France.

Rech. Aerosp.
Recherche Aerospatiale. Office National d'Etudes et de Recherches Aerospatiales, 29 Avenue de la Division Leclerc, F–92320 Chatillon, France. ISSN 0034–1223.

Recherche
Recherche. Société d'Editions Scientifiques, 57, rue de Seine, F–75006 Paris, France. ISSN 0029–5671.

Ref. Zh., 51. Astron.
Referativnyj Zhurnal, 51. Astronomiya. Gosudarstvennyj Komitet SSSR po Nauke i Tekhnike. Vsesoyuznyj Institut Nauchnoj i Tekhnicheskoj Informatsii. Akademiya Nauk SSSR. Moskva. 125219, Moskva, A–219, Baltijskaya ul., 14, USSR. ISSN 0486–2236.

Ref. Zh., 52. Geod. Aehrosemka
Referativnyj Zhurnal, 52. Geodeziya i Aehrosemka. Gosudarstvennyj Komitet SSSR po Nauke i Tekhnike. Vsesoyuznyj Institut Nauchnoj i Tekhnicheskoj Informatsii. Akademiya Nauk SSSR. Moskva. 125219, Moskva, A–219, Baltijskaya ul., 14, USSR. ISSN 0375–9717.

Ref. Zh., 62. Issled. Kosm. Prostranstva
Referativnyj Zhurnal, 62. Issledovanie Kosmicheskogo Prostranstva. Gosudarstvennyj Komitet SSSR po Nauke i Tekhnike. Vsesoyuznyj Institut Nauchnoj i Tekhnicheskoj Informatsii. Akademiya Nauk SSSR. Moskva. 125219, Moskva, A–219, Baltijskaya ul., 14, USSR. ISSN 0034–2408.

Rep. Finn. Geod. Inst.
Reports of the Finnish Geodetic Institute. Finnish Geodetic Institute, Helsinki, Finland. ISSN 0355–1962.

Rep. Inst. Phys. Chem. Res.
Reports of the Institute of Physical and Chemical Research. Institute of Physical and Chemical Research, Rikagaku Kenkyushu, Wako–shi, Saitama 351, Japan. ISSN 0020–3084.

Rep. Obs. Lund
Reports from the Observatory of Lund. Institutionen för Astronomi, Lunds Universitet, Box 1107, 22104 Lund, Sweden. ISSN 0349–4217.

Rep. Prog. Phys.
Reports on Progress in Physics. Published by the Institute of Physics, 47 Belgrave Square, London SW1X 8QX, England. ISSN 0034–4885.

Rep. Ser., Dep. Phys. Sci., Univ. Turku
Report Series, Department of Physical Sciences, University of Turku. Department of Physical Sciences, University of Turku, SF–20500 Turku 50, Finland. ISSN 0356–9896.

Res. Bull. Meisei Univ., Phys. Sci. Eng.
Research Bulletin of Meisei University, Physical Sciences and Engineering. Meisei University, Hino–Shi, Tokyo, Japan. ISSN 0388–130X.

Res. Cent. Astron. Appl. Math., Acad. Athens, Contrib. Ser. I
Research Center for Astronomy and Applied Mathematics, Academy of Athens, Contributions Series I (Astronomy). Research Center for Astronomy and Applied Mathematics, Academy of Athens, Athens, Greece.

Res. Lab. Electron. Onsala Space Obs., Res. Rep.
Research Laboratory of Electronics and Onsala Space Observatory, Research Report. Chalmers University of Technology, Research Laboratory of Electronics, Fack, S–40220 Göteborg 5, Sweden.

Rev. Astron.
Revista Astronomica. Organo de la Asociación Argentina Amigos de la Astronomia. Asociación Argentina Amigos de la Astronomia, C. C. 369–C. Central, 1000–Buenos Aires, Argentina. ISSN 0044–9253.

Rev. Bras. Fis.
Revista Brasileira de Fisica. Sociedade Brasileira de Fisica, Caixa Postal 20553, Sao Paulo SP, Brazil. ISSN 0374–4922.

Rev. Fac. Sci. Univ. Istanbul, Ser. C
Review of Faculty of Science University of Instanbul, Serie C. (Üniversitesi fen Fakültesi Mecmuasi, Seri C). Fen Fakültesi Basimevi, Istanbul, Turkey. ISSN 0444–7298.

Rev. Geophys.
Reviews of Geophysics. American Geophysical Union, 2000 Florida Avenue, N.W., Washington, D.C. 20009, USA. ISSN 8755–1209. Formerly: Rev. Geophys. Space Phys.

Rev. Hist. Sci.
Revue d'Histoire des Sciences. Presses Universitaires de France, Paris, France. ISSN 0151–4105.

Rev. Mex. Astron. Astrofis.
Revista Mexicana de Astronomia y Astrofisica. Instituto de Astronomia, Universidad Nacional Autónoma de México, Apartado Postal 70–264, Mexico 20, D.F., Mexico. ISSN 0185–1101.

Rev. Mex. Fis.
Revista Mexicana de Fisica. Sociedad Mexicana de Fisica, Apartado Postal No. 20–364, Mexico 20, D.F., Mexico. ISSN 0035–001X.

Rev. Mod. Phys.
Reviews of Modern Physics. Published for the American Physical Society by the American Institute of Physics, 335 East 45th Street, New York, N.Y. 10017, USA. ISSN 0034–6861.

Rev. Phys. Appl.
Revue de Physique Appliquée. Editions de Physique, Z.I. de Courtabœuf, B.P. 112, F–91944 Les Ulis Cedex, France. ISSN 0035–1687.

Rev. Quest. Sci.
Revue des Questions Scientifiques. Société Scientifique de Bruxelles, 61, rue de Bruxelles, B–5000 Namur, Belgium. ISSN 0035–2160.

Rev. Radio Res. Lab.
Review of the Radio Research Laboratories. Ministry of Posts & Telecommunications, Nukui–Kitamachi, Konganei-shi, Tokyo 184, Japan. ISSN 0033–801X.

Rev. Roum. Phys.
Revue Roumaine de Physique. Academie Republicii Populare Romine, B.P. 134–135, Bucuresti, Rumania. ISSN 0035–4090.

Rev. Roum. Sci. Tech., Sér. Mec. Appl.
Revue Roumaine des Sciences Techniques, Série de Mécanique Appliquée. Academia R.S.R., Str. Constantin Mille 15, Bucuresti, Rumania. ISSN 0035–4074.

Rev. Sci. Instrum.
Review of Scientific Instruments. American Institute of Physics, 335 East 45th Street, New York, N.Y. 10017, USA. ISSN 0034–6748.

Rezul't. Nablyud. Iskusstv. Sputnikov Zemli
Rezul'taty Nablyudenij Iskusstvennykh Sputnikov Zemli. Published by Astronomicheskij Sovet Akademii Nauk SSSR, Ryazanskij Gosudarstvennyj Pedagogicheskij Institut, Ryazan'. 390000 GSP, Ryazan, ul. Svobody, 46, USSR. ISSN 0131–8586.

Ric. Astron.
Ricerche Astronomiche. Specola Vaticana, I–00120 Città del Vaticano.

Rijksuniv. Gent, Sterrenkundig Obs., Meded.
Rijksuniversiteit Gent, Sterrenkundig Observatorium, Mededeling. Astronomical Observatory, State University of Ghent, Krijgslaan 271, B–9000 Ghent, Belgium.

Říše hvězd
Říše hvězd. Czech popular astronomical journal. Panorama, Praha, Czechoslovakia.

Riv. Nuovo Cimento
La Rivista del Nuovo Cimento della Società Italiana di Fisica. Editrice Compositori, viale XII Giugno 1, I–40124 Bologna, Italy. ISSN 0035–5917.

SAAO Newsl.
SAAO Newsletter. South African Astronomical Observatory, PO Box 9, Observatory 7935, Cape, South Africa.

Sac Peak Update
Sac Peak Update. Sacramento Peak Observatory, Sunspot, N.M. 88349, USA.

S.Afr. Astron. Obs., Annu. Rep.
South African Astronomical Observatory, Annual Report. South African Astronomical Observatory, PO Box 9, Observatory, 7935 Cape, South Africa. ISSN 0250–0671.

S.Afr. Astron. Obs., Circ.
South African Astronomical Observatory, Circulars. South African Astronomical Observatory, PO Box 9, Observatory, 7935 Cape, South Africa.

S.Afr. J. Phys.
South African Journal of Physics. Bureau for Scientific Publications, PO Box 1758, Pretoria 0001, South Africa. ISSN 0379–4377.

Schweiz. Tech. Z.
Schweizerische Technische Zeitschrift. Schweizerischer Technischer Verband, Weinbergstraße 41, CH–8023 Zürich, Switzerland. ISSN 0040–151X.

Sci. Am.
Scientific American. Scientific American, Inc., 415 Madison Avenue, New York, N.Y. 10017, USA. ISSN 0036–8733.

Sci. Atmos. Sin.
Scientia Atmospherica Sinica. Science Press, Peking. Subscription address: Guozi Shudian, PO Box 399, Beijing, People's Republic of China.

Sci. Dimension
Science Dimension. National Research Council of Canada, Ottawa K1A OR6, Canada. ISSN 0036–830X.

Sci. Pap. Inst. Phys. Chem. Res.
Scientific Papers of the Institute of Physical and Chemical Research. Rikagaku Kenkyusho, Wako–shi, Saitama 351, Japan. ISSN 0020–3092.

Sci. Prog.
Science Progress. Blackwell Scientific Publications, Oxford – London – Edinburgh – Boston – Melbourne. ISSN 0036–8504.

Sci. Rep. Tôhoku Univ., Eighth Ser.
The Science Reports of the Tôhoku University, Eighth Series (Physics and Astronomy). Faculty of Science, Tôhoku University, Sendai 980, Japan. ISSN 0388–5607.

Sci. Sin., Ser. A
Scientia Sinica, Series A (Mathematical, Physical, Astronomical and Technical Sciences). Academia Sinica, Beijing. Science Press, No. 137, Chaoyangmennei Street, Beijing, People's Republic of China. Subscription address: Scientific and Technical Books Service Ltd., PO Box 197, London WC2N 4DE, England. ISSN 0253–5831.

Science
Science. American Association for the Advancement of Science, 1515 Massachusetts Avenue, N.W., Washington, D.C. 20005, USA. ISSN 0036–8075.

Scr. Fac. Sci. Nat. Univ. Purkynianae Brun., Phys.
Scripta Facultatis Scientiarum Naturalium Universitatis Purkynianae Brunensis, Physica. University J.E. Purkyne, 61137 Brno–Kotlarská 2, Czechoslovakia. ISSN 0231–6129.

Semin. Astron. Geod., Univ. Madr., Publ.
Seminario de Astronomia y Geodesia, Universidad Complutense, Facultad de Ciencias Matemáticas, Madrid, Publicación. Universidad Complutense, Madrid, Spain.

Sendai Astron. Rap.
Sendai Astronomiaj Raportoj. Astronomical Institute, Tôhoku University, Sendai 980, Japan.

Shaanxi Astron. Obs., Repr.
Shaanxi Astronomical Observatory, Reprints. Shaanxi Astronomical Observatory, Academia Sinica, Lintong, Xian, People's Republic of China.

Shanghai Obs., Acad. Sin., Prepr.
Shanghai Observatory, Academia Sinica, Preprint. Shanghai Observatory, Academia Sinica, 80 Nandan Road, Shanghai, People's Republic of China.

Sitzungsber. Akad. Wiss. DDR, Math. Naturwiss. Tech.
Sitzungsberichte der Akademie der Wissenschaften der DDR, Mathematik – Naturwissenschaften – Technik. Akademie-Verlag, DDR–1086 Berlin, Leipziger Straße 3–4, German Democratic Republic. ISSN 0138–3956.

Sitzungsber. Heidelb. Akad. Wiss.
Sitzungsberichte der Heidelberger Akademie der Wissenschaften. Mathematisch–Naturwissenschaftliche Klasse. Springer-Verlag, Berlin – Heidelberg – New York – Tokyo. ISSN 0371–0165.

Sitzungsber., Österr. Akad. Wiss., Math.–Naturwiss. Kl. Abt. II
Sitzungsberichte, Österreichische Akademie der Wissenschaften, Mathematisch–Naturwissenschaftliche Klasse, Abteilung II (Mathematik, Astronomie, Physik, Meteorologie und Technik). Springer-Verlag, Wien, Austria. ISSN 0029–8816.

Sky Telesc.
Sky and Telescope. Sky Publishing Corporation, 49 Bay State Rd., Cambridge, Mass. 02238–1290, USA. ISSN 0037–6604.

Smithson. Astrophys. Obs., Spec. Rep.
Smithsonian Astrophysical Observatory, Special Report. Available from the Publications Division, Distribution Section, Smithsonian Astrophysical Observatory, Cambridge, Mass. 02138, USA.

Sol. Energy
Solar Energy. Pergamon Press, Oxford – New York – Toronto – Paris – Frankfurt – Sydney. ISSN 0038–092X.

Sol. Maps Act.
Solar Maps and Activity. Manila Observatory, Solar Division, Manila, Philippines.

Sol. Phys.
Solar Physics. A Journal for Solar Research and the Study of Solar Terrestrial Physics. D. Reidel Publishing Company, Dordrecht – Boston. ISSN 0038–0938.

Sol. Syst. Res.
Solar System Research. Translation of Astron. Vestn. Consultants Bureau, 227 West 17th Street, New York, N.Y. 10011, USA. ISSN 0038–0946.

Sol. Terr. Environ. Res. Jpn.
Solar Terrestrial Environmental Research in Japan. Institute of Space and Aeronautical Science, University of Tokyo, 4–6–1, Komaba, Meguro–ku, Tokyo 153, Japan. ISSN 0386–5444.

Solid State Phys.
Solid State Physics. Agne Gijutsu Center, Kitamura Building, 5–1–25, Minamiayama, Minatoku, Tokyo, Japan. ISSN 0454–4544.

Soln. Dannye, Byull.
Solnechnye Dannye, Byulleten' (Solar Data). Akademiya Nauk SSSR. Astronomicheskij Sovet i Glavnaya Astronomicheskaya Observatoriya. Izdatel'stvo 'Nauka', Leningradskoe Otdelenie, Leningrad. 199164 Leningrad, V–164, Mendeleevskaya l. 1, USSR. ISSN 0552–5829.

Sonne
Sonne. Mitteilungsblatt der Amateursonnenbeobachter. Wilhelm–Foerster–Sternwarte, Munsterdamm 90, D–1000 Berlin 41, F.R. Germany. ISSN 0721–0094.

Soobshch. Akad. Nauk Gruz. SSR
Soobshcheniya Akademii Nauk Gruzinskoj SSR. Akademiya Nauk Gruzinskoj SSR. ISSN 0002–3167.

Soobshch. Byurakan. Obs.
Soobshcheniya Byurakanskoj Observatorii. Akademiya Nauk Armyanskoj SSR, Erevan. 375019 Erevan, Barekamutyan, 24, USSR. ISSN 0370–8691.

Soobshch. Gos. Astron. Inst. Shternberg
Soobshcheniya Gosudarstvennogo Astronomicheskogo Instituta im. P.K. Shternberga. Moskovskij Gosudarstvennyj Universitet im. M.V. Lomonosova. Izdatel'stvo Moskovskogo Universiteta, Moskva. Moskva, K–9, ul. Gertsena, 5/7, USSR. ISSN 0038–1489.

Soobshch. Spets. Astrofiz. Obs.
Soobshcheniya Spetsial'noj Astrofizicheskoj Observatorii. Akademiya Nauk SSSR. Izdanie Spetsial'noj Astrofizicheskoj Observatorii, USSR.

South. Stars
Southern Stars. Journal of the Royal Astronomical Society of New Zealand (Inc.). Royal Astronomical Society of New Zealand, PO Box 3181, Wellington, New Zealand. ISSN 0049–1640.

Sov. Astron.
Soviet Astronomy. A translation of Astron. Zh. American Institute of Physics, 335 East 45th Street, New York, N.Y. 10017, USA. ISSN 0038–5301.

Sov. Astron. Lett.
Soviet Astronomy Letters. A translation of Pis'ma Astron. Zh. American Institute of Physics, 335 East 45th Street, New York, N.Y. 10017, USA. ISSN 0360–0327.

Space Educ.
Space Education. A Publication of The British Interplanetary Society. The British Interplanetary Society, 27/29 South Lambeth Road, London SW8 1SZ, England. ISSN 0621–1813.

Space Sci. Rev.
Space Science Reviews. D. Reidel Publishing Company, Dordrecht – Boston. ISSN 0038–6308.

Space Telesc. Sci. Inst., Newsl.
Space Telescope Science Institute, Newsletter. Space Telescope Science Institute, 3700 San Martin Drive, Homewood Campus, Baltimore, Md. 21218, USA.

Space Telesc. Sci. Inst., Prepr. Ser.
Space Telescope Science Institute, Preprint Series. Space Telescope Science Institute, 3700 San Martin Drive, Homewood Campus, Baltimore, Md. 21218, USA.

Spaceflight
Spaceflight. A Publication of The British Interplanetary Society. The British Interplanetary Society, 27/29 South Lambeth Road, London SW8 1SZ, England. ISSN 0038–6340.

Spectrosc. Lett.
Spectroscopy Letters. Marcel Dekker Inc., 270 Madison Avenue, New York, N.Y. 10016, USA. ISSN 0038–7010.

Speculations Sci. Technol.
Speculations in Science and Technology. Elsevier Sequoia, S.A., PO Box 851, CH–1001 Lausanne 1, Switzerland. ISSN 0155–7785.

ST–ECF Newsl.
Space Telescope–European Coordinating Facility Newsletter. Space Telescope–European Facility, European Southern Observatory, Karl–Schwarzschild–Straße 2, D–8046 Garching bei München, F.R. Germany.

Sterne
Die Sterne. Zeitschrift für alle Gebiete der Himmelskunde. Johann Ambrosius Barth, DDR–7010 Leipzig, Salomonstraße 18b, German Democratic Republic. ISSN 0039–1255.

Sterne Weltraum
Sterne und Weltraum. Astronomische Monatsschrift. Verlag Sterne und Weltraum Dr. Vehrenberg GmbH, Portiastraße 10, D–8000 München 90, F.R. Germany, ISSN 0039–1263.

Sternenbote
Der Sternenbote. Österreichische Astronomische Monatsschrift. Astronomisches Büro, Hasenwartgasse 32, A–1238 Wien, Austria. ISSN 0039–1271.

Stockholms Obs., Rep.
Stockholms Observatorium, Report. Stockholms Observatorium, S–13300 Saltsjöbaden, Sweden.

Strolling Astron.
The Strolling Astronomer. The Journal of The Association of Lunar and Planetary Observers. The Strolling Astronomer, Box 3 AZ, University Park, N.M. 88003, USA. ISSN 0039–2502.

Stud. Cercet. Fiz.
Studii si Cercetari de Fizica. Academia Republicii Populare Romine, PO Box 134–5, Calca Victoriei 126, Bucuresti, Rumania, ISSN 0039–3940.

Stud. Geophys. Geod.
Studia Geophysica et Geodaetica. Geophysical Institute of the Czechoslovak Academy of Sciences. Published by Academia, Prague, Czechoslovakia.

Stud. Hist. Philos. Sci.
Studies in History and Philosophy of Science. Pergamon Press, Oxford – New York – Toronto – Paris – Frankfurt – Sydney. ISSN 0039–3681.

Stud. Soc. Sci. Torun., Sect. F
Studia Societatis Scientiarum Torunensis, Sectio F (Astronomia). Państwowe Wydawnictwo Naukowe, Warszawa – Poznań – Toruń, Poland. ISSN 0082–5573.

Surv. High Energy Phys.
Surveys in High Energy Physics. Harwood Academic Publishers GmbH, Poststraße 22, CH–7000 Chur, Switzerland. ISSN 0142–2413.

Swarthmore Astron. Repr.
Swarthmore Astronomical Reprints. Sproul Observatory, Swarthmore College, Swarthmore, Pa. 19081, USA.

Syd. Obs. Pap.
Sydney Observatory Papers. School of Physics, University of Sydney, Sydney, N.S.W. 2006, Australia.

Syst. Int.
Systems International. IPC Electrical Electronic Press Ltd., Quadrant House, The Quadrant, Sutton, Surrey SM2 5AS, England. ISSN 0309–1171.

Tartu Astrofüüs. Obs. Publ.
W. Struve nimelise Tartu Astrofüüsika Observatooriumi, Publikatsioonid. Eesti NSV Teaduste Akadeemia, Tartu.

Tartu Astrofüüs. Obs. Teated
Tartu Astrofüüsika Observatoorium Teated. Eesti NSV Teaduste Akadeemia W. Struve nim. Tartu Astrofüüsika Observatoorium. Valgus, Tallinn.

Tech. Mess. – TM
Technisches Messen – TM. Archiv für Technisches Messen. R. Oldenbourg Verlag GmbH, Rosenheimer Str. 145, D–8000 München 80, F.R. Germany. ISSN 0171–8096.

Tech. Mitt. Krupp Forschungsber.
Technische Mitteilungen Krupp Forschungsberichte. Fachbücherei Krupp, Postfach 917, D–4300 Essen 1, F.R. Germany. ISSN 0494–9382.

Tectonophysics
Tectonophysics. International Journal of Geotectonics and the Geology and Physics of the Interior of the Earth. Elsevier Publishing Company, PO Box 211, 1000 AE Amsterdam, The Netherlands. ISSN 0040–1951.

Telecommun. J.
Telecommunication Journal. (English Edition). International Telecommunications Union. Place des Nations, CH–1221 Genève 20, Switzerland. ISSN 0497–137X.

Tellus, Ser. B
Tellus, Series B. (Chemical and Physical Meteorology). Swedish Geophysical Society, Arrhenius Laboratoriet, S–10691 Stockholm, Sweden. ISSN 0280–6509.

THEOCHEM
THEOCHEM. Elsevier Scientific Publishing Company, PO Box 330, 1000 AH Amsterdam, The Netherlands. ISSN 0166–1280.

Theor. Pap.
Theoretic Papers. The Blindern Theoretic Research Team, P.B. 1029, Blindern, Oslo 3, Norway.

Time Freq. Serv., Bull.
Time and Frequency Services, Bulletin. Shaanxi Astronomical Observatory, Academia Sinica, Lintong, Xian, People's Republic of China.

Time Serv. Annu. Rep.
Time Service Annual Report. Xu–Jia–Hui Section, Shanghai Observatory, Academia Sinica. Subscription address: Shanghai Scientific and Technical Publishers, Shanghai, Rei Jing Er Street 450, People's Republic of China.

Time Serv. Bull.
Time Service Bulletin. Osservatorio Astronomico di Torino, I–10025 Pino Torinese (Torino), Italy.

Tôhoku Geophys. J.
Tôhoku Geophysical Journal. Science Reports of the Tôhoku University, Fifth Series. Faculty of Science, Tôhoku University, Sendai 980, Japan. ISSN 0040–8794.

Tokyo Astron. Bull., Second Ser.
Tokyo Astronomical Bulletin, Second Series. Tokyo Astronomical Observatory, University of Tokyo, Mitaka, Tokyo 181, Japan. ISSN 0082–4690.

Tokyo Astron. Obs. Rep.
Tokyo Astronomical Observatory Report. Tokyo Astronomical Observatory, University of Tokyo, Mitaka, Tokyo 181, Japan. ISSN 0374–4639.

Tokyo Astron. Obs. Repr.
Tokyo Astronomical Observatory Reprints. Tokyo Astronomical Observatory, University of Tokyo, Mitaka, Tokyo 181, Japan. ISSN 0082–4712.

Tokyo Astron. Obs., Kiso Inf. Bull.
Tokyo Astronomical Observatory, Kiso Information Bulletin. Tokyo Astronomical Observatory, University of Tokyo, Mitaka, Tokyo 181, Japan. ISSN 0286–1380.

Tokyo Astron. Obs., Time Latitude Bull.
Tokyo Astronomical Observatory, Time and Latitude Bulletins. Tokyo Astronomical Observatory, University of Tokyo, Mitaka, Tokyo 181, Japan. ISSN 0388–3701.

Top. Astrophys. Astron. Space Sci.
Topics in Astrophysics, Astronomy, and Space Sciences. Centre for Astronomy and Space Sciences, Central Institute of Physics. CIP Press, Cutitul de Argint 5, 75 212 Bucharest, Romania.

Tr. Astrofiz. Inst. Alma–Ata
Trudy Astrofizicheskogo Instituta. Akademiya Nauk Kazakhskoj SSR. Izdatel'stvo Nauka Kazakhskoj SSR, Alma–Ata. 480021, Alma–Ata, ul. Shevchenko, 28, USSR.

Tr. Astron. Obs., Leningrad
Trudy Astronomicheskoj Observatorii. Uchenye Zapiski Gosudarstvennogo Universiteta im. A.A. Zhdanova, Seriya matematicheskikh nauk = Izdatel'stvo Leningradskogo Gosudarstvennogo Universiteta A.A. Zhdanova, Leningrad. 199164, Leningrad V–164, Universitetskij nab., 7/9, USSR. ISSN 0136–8109; ISSN 0136–8141.

Tr. Glav. Astron. Obs. Pulkovo, Ser. 2
Trudy Glavnoj Astronomicheskoj Observatorii v Pulkove, Seriya 2. Akademiya Nauk SSSR. Leningradskoe Otdelenie, Nauka, Leningrad. 199164, Leningrad V–164, Mendeleevskaya l., 1, USSR.

Tr. Gos. Astron. Inst. Shternberg
Trudy Gosudarstvennogo Astronomicheskogo Instituta im. P.K. Shternberga. Izdatel'stvo Moskovskogo Universiteta, Moskva, USSR. ISSN 0371–6791.

Tr. Inst. Teor. Astron., Leningrad
Trudy Instituta Teoreticheskoj Astronomii, Akademiya Nauk SSSR. Izdatel'stvo Nauka, Leningrad. 199164, Leningrad V–164, Mendeleevskaya l., 1, USSR. ISSN 0568–6016.

Tr. Kazan. Gorod. Astron. Obs.
Trudy Kazanskoj Gorodskoj Astronomicheskoj Observatorii. Izdatel'stvo Kazanskogo Universiteta, Kazan. Kazan, ul. Lenina, 2, USSR. ISSN 0371–8247.

Tr. Tashkent. Astron. Obs.
Trudy Ordena Trudovogo Krasnogo Znameni Astronomicheskogo Instituta Akademii Nauk Uzbekskoj SSR. Izdatel'stvo FAN Uzbekskoj SSR, Tashkent. Tashkent, ul. Gololya, 70, USSR.

Trans. Am. Nucl. Soc.
Transactions of the American Nuclear Society. American Nuclear Society, 555 North Kensington Avenue, La Grange Park, Ill. 60525, USA. ISSN 0003-018X.

Trans. Astron. Obs. Yale Univ.
Transactions of the Astronomical Observatory of Yale University. Yale University Observatory, 260 Whitney Avenue, New Haven, Conn. 06511, USA.

Trans. Bose Res. Inst.
Transactions of the Bose Research Institute. Bose Research Institute, 93/1 Acharya Prafulla Chandra Road, Calcutta 9, India. ISSN 0006-7903.

Trans. IAU
Transactions of the International Astronomical Union. Published on behalf of the IAU by D. Reidel Publishing Company, Dordrecht – Boston – London.

Trans. Inst. Electron. Commun. Eng. Jpn., Part C
Transactions of the Institute of Electronic and Communication Engineers of Japan, Part C. Denshi Tsushin Gakkai, Kikai–Shinko–Kaikan, 5–8 Shibakoen 3 Chome, Minato–ku, Tokyo 105, Japan. ISSN 0373-6113.

Transp. Theory Stat. Phys.
Transport Theory and Statistical Physics. Marcel Dekker Inc., 270 Madison Avenue, New York, N.Y. 10016, USA. ISSN 0041-1450.

Tsirk. Astron. Inst. Tashkent
Tsirkulyar Astronomicheskogo Instituta Tashkent. Akademiya Nauk Uzbekskoj SSR. Izdatel'stvo FAN Uzbekskoj SSR, Tashkent. Tashkent, ul. Gololya, 70, USSR. ISSN 0373-7675.

Tsirk., Astron. Obs. L'vov
Tsirkulyar, Astronomicheskaya Observatoriya L'vov. L'vovskij Ordena Lenina Gosudarstvennyj Universitet im. Ivana Franko. Izdatel'stvo L'vovskogo Universiteta, L'vov. 290005, L'vov–5, ul. Lomonosova, 8, USSR. ISSN 0374-0722.

Tsirk. Shemakh. Astrofiz. Obs.
Tsirkulyar Shemakhinskoj Astrofizicheskoj Observatorii. Akademiya Nauk Azerbajdzhanskoj SSR. Izdatel'stvo Ehlm, Baku. 370143 Baku–143, prospekt Narimanova, 31. ISSN 0135-0420.

Turku Univ. Obs., Informeto
Turku University Observatory, Informeto. Turku University Observatory, Tuorla, SF–2150 Piikkiö, Finland.

Turku Univ. Obs., Informo
Turku University Observatory, Informo. Turku University Observatory, Tuorla, SF–21500 Piikkiö, Finland.

UKIRT Rep.
UKIRT Report. United Kingdom Infrared Telescope, 900 Leilani Street, Hilo, Hawaii 96720, USA. ISSN 0260-9983.

Umschau
Die Umschau. Das Wissenschaftsmagazin. Postfach 110262, Stuttgarter Straße 18–24, D–6000 Frankfurt am Main 11, F.R. Germany. ISSN 0722-8562.

Univ. Barc., Dep. Fis. Tierra Cosmos, Publ.
Universidad de Barcelona, Departamento de Fisica de la Tierra y del Cosmos, Publicacion. Departamento de Fisica de la Tierra y del Cosmos, Universidad de Barcelona, Spain.

Univ. Hannover, Astron. Stn., Veröff.
Universität Hannover, Astronomische Station, Veröffentlichungen. Universität Hannover, Astronomische Station, Hannover, F.R. Germany.

Univ. Obs., St. Andrews, Repr.
University Observatory, St. Andrews, Reprint. University Observatory, Buchanan Gardens, St. Andrews, Fife KY16 9LZ, Scotland.

Univ. Tex., Monogr. Astron.
The University of Texas, Monographs in Astronomy. Department of Astronomy, University of Texas, Austin, Tex. 78712, USA.

Univ. Tex., Publ. Astron.
The University of Texas, Publications in Astronomy. Department of Astronomy, University of Texas, Austin, Tex. 78712, USA. ISSN 0276-1106.

Upps. Astron. Obs. Ann.
Uppsala Astronomiska Observatoriums Annaler. Astronomiska Observatoriet, Box 515, S–75120 Uppsala, Sweden.

Upps. Astron. Obs. Rep.
Uppsala Astronomical Observatory, Report. Astronomiska Observatoriet, Box 515, S–75120 Uppsala, Sweden.

Urania (Barc.)
Urania. Revista de Astronomía y Ciencias Afines. Organo de la Sociedad Astronómica de España y America y de la Union Nacional de Astronómia y Ciencias Afines. Sociedad Astronomica de España y America, Avenida Generalisimo, 337, Barcelona, Spain.

Urania (Cracow)
Urania. Miesiecznik Polskiego Towarzystwa Miłośników Astronomii. Polskie Towarzystwo Miłośników Astronomii, Zarzad Gl., 31–027 Kraków, Solskiego 30/8, Poland. ISSN 0042-0794.

U.S. Nav. Obs., Circ.
U.S. Naval Observatory, Circular. U.S. Naval Observatory, Washington, D.C. 20390, USA.

U.S. Nav. Obs., Time Serv. Publ., Ser. 01
U.S. Naval Observatory, Time Service Publications, Series 1 (Worldwide Primary Time and Frequency VLF and HF Transmissions). U.S. Naval Observatory, Time Service Division (62C), Washington, D.C. 20390, USA.

U.S. Nav. Obs., Time Serv. Publ., Ser. 04
U.S. Naval Observatory, Time Service Publications, Series 4 (Daily Time Differences and Relative Phase Values). U.S. Naval Observatory, Time Service Division (62C), Washington, D.C. 20390, USA.

U.S. Nav. Obs., Time Serv. Publ., Ser. 06
U.S. Naval Observatory, Time Service Publications, Series 6 (A.1–UT1 Data). U.S. Naval Observatory, Time Service Division (62C), Washington, D.C. 20390, USA.

U.S. Nav. Obs., Time Serv. Publ., Ser. 10
U.S. Naval Observatory, Time Service Publications, Series 10 (Astronomical Programs). U.S. Naval Observatory, Time Service Division (62C), Washington, D.C. 20390, USA.

U.S. Nav. Obs., Time Serv. Publ., Ser. 11
U.S. Naval Observatory, Time Service Publications, Series 11 (Time Service Report). U.S. Naval Observatory, Time Service Division (62C), Washington, D.C. 20390, USA.

U.S. Nav. Obs., Time Serv. Publ., Ser. 14
U.S. Naval Observatory, Time Service Publications, Series 14
(Time Service Announcement). U.S. Naval Observatory, Time
Service Division (62C), Washington, D.C. 20390, USA.

Utr. Sterrekundige Overdrukken
Utrechtse Sterrekundige Overdrukken. Sonnenborgh Observatory, Zonnenburg 2, NL–3512 Utrecht, The Netherlands.

Uttar Pradesh State Obs., Repr.
Uttar Pradesh State Observatory, Reprint. Uttar Pradesh State
Observatory, Manora Peak, Naini Tal 263 139, India.

Vasiona
Vasiona. Revue d'Astronomie. Bulletin de la Société Astronomique 'R. Bošković'. Vasiona, Narodna opservatorija,
Kalemegdan, Gornji Grad, Beograd, Yugoslavia. ISSN
0506–4295.

Vatican Obs. Publ.
Vatican Observatory Publications. Specola Vaticana, I–00120
Città del Vaticano.

Vatican Obs. Publ., Spec. Ser., Studi Galileiani
Vatican Observatory Publications, Special Series, Studi Gali-
lciani. Specola Vaticana, I–00120 Città del Vaticano.

Veröff. Archenhold–Sternw. Berlin–Treptow
Veröffentlichungen der Archenhold–Sternwarte Berlin–
Treptow. Archenhold–Sternwarte, DDR–1193 Berlin, Alt–
Treptow 1, German Democratic Republic.

Veröff. Astron. Inst. Bonn
Veröffentlichungen der Astronomischen Institute Bonn. Ferd.
Dümmler Verlag, Kaiserstraße 31–37, D–5300 Bonn 1,
F.R. Germany. ISSN 0340–9821.

Veröff. Astron. Rechen–Inst. Heidelb.
Veröffentlichungen des Astronomischen Rechen–Instituts
Heidelberg. Verlag G. Braun, Karl–Friedrich–Straße 14–18,
D–7500 Karlsruhe 1, F.R. Germany.

**Veröff. Bayer. Komm. Int. Erdmessung Bayer. Akad. Wiss.,
Astron.–Geod. Arb.**
Veröffentlichungen der Bayerischen Kommission für die Inter-
nationale Erdmessung der Bayerischen Akademie der Wissen-
schaften, Astronomisch–Geodätische Arbeiten. Bayerische
Akademie der Wissenschaften, Marstallplatz 8, D–8000
München 22, F.R. Germany. ISSN 0340–7691.

**Veröff. Forschungsinst. Dtsch. Mus. Gesch. Naturwiss. Tech.,
Reihe A**
Veröffentlichungen des Forschungsinstituts des Deutschen
Museums für die Geschichte der Naturwissenschaften und der
Technik, Reihe A. Forschungsinstitut des Deutschen Muse-
ums für die Geschichte der Naturwissenschaften und der Tech-
nik, D–8000 München, F.R. Germany. ISSN 0418–9949.

Veröff. Remeis–Sternw. Bamberg
Veröffentlichungen der Remeis–Sternwarte Bamberg. Astro-
nomisches Institut der Universität Erlangen–Nürnberg.
Remeis–Sternwarte, Sternwartstraße 7, D–8600 Bamberg,
F.R. Germany

Veröff. Sternw. Pulsnitz
Veröffentlichungen der Sternwarte Pulsnitz. Sternwarte Pulsnitz, DDR–8514 Pulsnitz, German Democratic Republic.

Veröff. Sternw. Sonneberg
Veröffentlichungen der Sternwarte in Sonneberg. Akademie
der Wissenschaften der DDR, Zentralinstitut für Astrophysik,
Sternwarte Sonneberg, DDR–6400 Sonneberg, German
Democratic Republic. ISSN 0373–6199.

Veröff. Wilhelm–Foerster–Sternw.
Veröffentlichungen der Wilhelm–Foerster–Sternwarte.
Wilhelm–Foerster–Sternwarte, Munsterdamm 90, D–1000
Berlin 41, F.R. Germany.

Veröff. Zentralinst. Phys. Erde
Veröffentlichungen des Zentralinstituts für Physik der Erde.
Akademie der Wissenschaften der DDR, Zentralinstitut für
Physik der Erde, DDR–1500 Potsdam, Telegrafenberg A17,
German Democratic Republic.

Vesmír
Vesmír. Přírodovědecký časopis Čs. akademie věd. Publisher:
Academia, Praha, Czechoslovakia.

Vestn. Khar'kov. Univ.
Vestnik Khar'kovskogo Universiteta. Izdatel'stvo pri
Khar'kovskom Gosudarstvennom Universitete Izdatel'skogo
Obedineniya Vishcha Shkola, Khar'kov. 310003, Khar'kov–3,
ul. Universitetskaya, 16, USSR.

Vestn. Kiev. Univ. Astron.
Vestnik Kievskogo Universiteta. Astronomiya. Izdatel'stvo pri
Kievskom Gosudarstvennom Universitete Izdatel'skogo Obedineniya Vishcha Shkola, Kiev. 252001, Kiev–1, Kreshchatik,
4, USSR. ISSN 0203–7319; ISSN 0321–3927.

Vestn. Leningr. Univ., Mat. Mekh. Astron.
Vestnik Leningradskogo Universiteta Matematika, Mekha-
nika, Astronomiya. Leningradskij Universitet, Universitetskaya Nab. 7/9 Leningrad B–164, USSR. ISSN 0024–0850.

Vestn. Mosk. Univ., Ser. 3. Fiz. Astron.
Vestnik Moskovskogo Universiteta, Seriya 3. Fizika, Astro-
nomiya. Izdatel'stvo Moskovskogo Universiteta, 103009,
Moskva, Ul.Gertsena 5/7, USSR. English translation in Mosc.
Univ. Phys. Bull. ISSN 0579–9392.

Villanova Univ. Obs. Contrib.
Villanova University Observatory Contributions. Department
of Astronomy, Villanova University, Villanova, Pa. 19085,
USA.

Vilniaus Astron. Obs. Biul.
Vilniaus Astronomijos Observatorijos Biuletenis. Vilniaus
Valstybinis V. Kapsuko Universitetas, LTSR Mokslu Aka-
demijos Fizikos Institutas, Vilnius. 232031, Vilnius 31,
Čiurlionio 29, Lithuania, USSR. ISSN 0136–3697.

Vistas Astron.
Vistas in Astronomy. An International Review Journal. Pergamon Press, Oxford – New York – Frankfurt. ISSN 0083–6656.

Warner Swasey Obs., Repr.
Warner and Swasey Observatory, Reprints. Warner and Swasey Observatory, Case Western Reserve University,
1975 Taylor Road, East Cleveland, Ohio 44112, USA.

**Warsaw Univ. Obs. Pol. Acad. Sci., N. Copernicus Astron. Cent.,
Repr.**
Warsaw University Observatory and Polish Academy of Sci-
ences and N. Copernicus Astronomical Center, Reprint. War-
saw University Observatory and N. Copernicus Astronomical
Center, Polish Academy of Sciences, Bartycka 18, PL–00–716
Warsaw, Poland.

Weather
Weather. James Glaisher House, Grenville Place, Bracknell,
Berks, RG12 1BX, England. ISSN 0043–1656.

Werkgroepnieuws
Werkgroepnieuws. WGN, the international circular for meteor
observers. Kontaktblad van meteoorwaarnemers in de Bene-
lux. Tweemaandelijks tijdschrift. P. Roggemans, Delling-
straat 25, B–2800 Mechelen, Belgium.

Wiss. Z. Friedrich–Schiller–Univ. Jena, Math.–Naturwiss. Reihe
Wissenschaftliche Zeitschrift der Friedrich–Schiller–Universität Jena, Mathematisch–Naturwissenschaftliche Reihe. Friedrich–Schiller–Universität Jena. Subscription address: Zeitungsvertriebsamt, Abt. Export, DDR–1004 Berlin, Straße der Pariser Kommune 3/4, German Democratic Republic. ISSN 0043–6836.

Wiss. Z. Humboldt–Univ. Berlin, Math.–Naturwiss. Reihe
Wissenschaftliche Zeitschrift der Humboldt–Universität zu Berlin, Mathematisch–Naturwissenschaftliche Reihe. Humboldt–Universität Berlin. Subscription address: Zeitschriftenvertriebsamt, Abt. Export, DDR–1004 Berlin, Straße der Pariser Kommune 3/4, German Democratic Republic. ISSN 0522–9863.

Wrocław Astron. Obs., Repr.
Wrocław Astronomical Observatory, Reprint. Wrocław University Observatory, Wrocław, Poland.

Wuli
Wuli. Science Press, Beijing. Subscription address: Guozi Shudian, PO Box 399, Beijing, People's Republic of China. ISSN 0379–4148.

Yad. Fiz.
Yadernaya Fizika. Izdatel'stvo Nauka, 103717 GSP Moskva K–62. Podsosenskij Per. 24, SSSR. English translation in Sov. J. Nucl. Phys. ISSN 0044–0027.

Yamamoto Circ.
Yamamoto Circular. Yamamoto Observatory, 289, Kamitanakami–kiryutyo, OTU, Sigaken, 520–21 Japan.

Z. Angew. Math. Mech.
Zeitschrift für angewandte Mathematik und Mechanik. Akademie–Verlag GmbH, DDR–108 Berlin, Leipziger Str. 3–4, German Democratic Republic. ISSN 0044–2267.

Z. Angew. Math. Phys.
Zeitschrift für Angewandte Mathematik und Physik. Verlag Birkhäuser, Postfach 4000, Basel 24, Switzerland. ISSN 0044–2275.

Z. Naturforsch., A
Zeitschrift für Naturforschung, Teil A (Physik, Physikalische Chemie, Kosmophysik). Verlag der Zeitschrift für Naturforschung, PO Box 2645, D–7400 Tübingen, F.R. Germany. ISSN 0340–4811.

Z. Phys., A
Zeitschrift für Physik, A. (Atoms und Nuclei). Springer–Verlag, Berlin – Heidelberg – New York – Tokyo. ISSN 0340–2193.

Z. Phys., B
Zeitschrift für Physik, B (Condensed Matter). Springer–Verlag, Berlin – Heidelberg – New York – Tokyo. ISSN 0722–3277.

Z. Phys., C
Zeitschrift für Physik, C (Particles and Fields). Springer–Verlag, Berlin – Heidelberg – New York – Tokyo. ISSN 0170–9739.

Z. Vermessungswes.
Zeitschrift für Vermessungswesen. Verlag Konrad Wittner KG, Postfach 147, D–7000 Stuttgart 1, F.R. Germany. ISSN 0513–9155.

Zeiss Inf.
Zeiss Information. Carl Zeiss, Oberkochen. F.R. Germany. ISSN 0044–2054.

Zeit–Breitenbestimmungen
Zeit– und Breitenbestimmungen, Empfangszeiten von Zeitsignalen, Präzisionszeitvergleiche. Akademie der Wissenschaften der DDR, Zentralinstitut für Physik der Erde, DDR–1500 Potsdam, Telegrafenberg A 17, German Democratic Republic.

Zemlya Vselennaya
Zemlya i Vselennaya. Astronomiya, Geofizika. Issledovaniya Kosmicheskogo Prostranstva. Nauchno–Populyarnyj Zhurnal Akademii Nauk SSSR. Izdatel'stvo Nauka, Moskva, USSR. ISSN 0044–3948.

Zenit
Zenit. Populair–wetenschappelijk maandblad over sterrenkunde, weerkunde, ruimtevaart, ruimte–onderzoek en aanverwante wetenschappen en technieken. Stichting De Koepel, Nachtegaalstraat 82 bis, Utrecht, The Netherlands. ISSN 0165–0211.

Zentralbl. Math. Grenzgeb. – Math. Abstr.
Zentralblatt für Mathematik und ihre Grenzgebiete – Mathematics Abstracts. Heidelberger Akademie der Wissenschaften und Fachinformationszentrum Energie Physik Mathematik GmbH, Karlsruhe. Springer–Verlag, Berlin – Heidelberg – New York – Tokyo. ISSN 0044–4235.

Zh. Ehksp. Teor. Fiz.
Zhurnal Ehksperimental'noj i Teoreticheskoj Fiziki. Akademiya Nauk SSSR, Leninskij Prosp. 14, Moskva, USSR. English translation in Sov. Phys.–JETP. ISSN 0044–4510.

Zvaigžnota Debess
Zvaigžnota Debess. Latvijas PSR Zinātnu Akadēmijas Radioastrofizikas Observatorijas Populārzinatnisks Rakstu Krāgums. Izdevnieciba Zinātne, Riga. 226004 Riga, Vienibas Gatve 11, USSR. ISSN 0135–129X.

Journals Abstracted Completely

A selected number of journals listed in category 001 (periodicals) are central to the subject scope of *Astronomy and Astrophysics Abstracts*. Depending on their relevance, almost all papers of the journals listed below are abstracted in our service.

AAS Photo-Bull.
Acta Astron.
Acta Astron. Sin.
Acta Astrophys. Sin.
Acta Cosmologica
Algoritm. Nebesnoj Mekh.
Am. Assoc. Variable Star Obs. Bull.
Annu. Rev. Astron. Astrophys.
Archaeoastronomy (U.K.)
Archaeoastronomy (U.S.A.)
Astrofizika
Astron. Astrophys.
Astron. Astrophys., Suppl. Ser.
Astron. Circ.
Astron. Data Cent. Bull.
Astron. J.
Astron. Nachr.
Astron. Q.
Astron. Tidsskr.
Astron. Tsirk.
Astron. Vestn.
Astron. Zh.
Astronomia
Astronomie
Astrophys. J.
Astrophys. J., Lett. Ed.
Astrophys. J., Suppl. Ser.
Astrophys. Lett.
Astrophys. Space Sci.
Aust. J. Astron.

BAV Rundbrief
BBSAG Bull.
Bol. Asoc. Argent. Astron.
Bol. Astron.
Bol. Astron. R Muscae
Br. Astron. Assoc. Circ.
Bull. Am. Astron. Soc.
Bull. Assoc. Fr. Obs. Etoiles Variables
Bull. Astron. Inst. Czech.
Bull. Astron. Soc. India
Bull. Inf. Cent. Données Stellaires

Celest. Mech.
Chin. Astron. Astrophys.
Ciel
Ciel Terre
Circ. Inf.
Coelum
Comments Astrophys.
Curr. Top. Chin. Sci., Sect. E: Astron.

Earth, Moon, Planets

Fundam. Cosmic Phys.

G. Astron.
Gen. Relativ. Gravitation
GEOS Circ.

I.A.P.P.P. Commun.
IAU Circ.
Icarus
IHW Newsl.
Inf. Bull. Variable Stars
Int. Comet Q.
Ir. Astron. J.
Issled. Solntsa Krasnykh Zvezd

J. Am. Assoc. Variable Star Obs.
J. Astrophys. Astron.
J. Br. Astron. Assoc.
J. Hist. Astron.
J. Korean Astron. Soc.
J. Meteor Res.
J. R. Astron. Soc. Can.

Kinematika Fiz. Nebesn. Tel
Komet. Tsirk.
Komety Meteory

Mem. Astron. Soc. India
Mem. Soc. Astron. Ital.
Mercury
Messenger
Meteoritics
Meteoritika
Minor Planet Bull.
Minor Planet Circ.
Mitt. Astron. Ges.
Mon. Not. R. Astron. Soc.
Mon. Notes Astron. Soc. S.Afr.

Nablyud. Iskusstv. Nebesn. Tel
Nauchn. Inf.
News Lett. Astron. Soc. N.Y.

Observatory
Occultation Newsl.
Orion
Orione

Perem. Zvezdy
Perem. Zvezdy, Prilozh.
Pis'ma Astron. Zh.
Postępy Astron.
Probl. Kosm. Fiz.
Proc. Astron. Soc. Aust.
Publ. Astron. Soc. Jpn.
Publ. Astron. Soc. Pac.
Publ. Variable Star Sect., R. Astron. Soc. N.Z.

Q. J. R. Astron. Soc.

Rev. Astron.
Rev. Mex. Astron. Astrofis.

Sky Telesc.
Sol. Phys.
Soln. Dannye Byull.
South. Stars
Space Sci. Rev.
Sterne
Sterne Weltraum
Strolling Astron.

Trans. IAU

Urania (Barcelona)
Urania (Cracow)

Vasiona
Vestn. Kiev. Univ. Astron.
Vistas Astron.

Yamamoto Circ.

Zenit
Zvaigžnota Debess

Publications of Observatories and Astronomical Institutes

Reports, communications, publications, numbered series of reprints and preprints of observatories and astronomical institutes (listed in category 001) which are scanned completely in our service are listed below.

A.A.O. Newsl.
Abastumanskaya Astrofiz. Obs., Byull.
Abh. Hamb. Sternw.
Abstr. Submitted Pap.
Acad. Sci. Est. SSR, Div. Phys. Math. Tech. Sci., Prepr.
A.D.I.O.N. Bull.
Anglo-Aust. Obs., Prepr.
Anglo-Aust. Telesc., Annu. Rep.
Ann. Obs. Astron. Alger
Ann. Shanghai Obs., Acad. Sin.
Ann. Tokyo Astron. Obs., Second Ser.
Ann. Univ.-Sternw. Wien
Annu. Rep. Astron. Inst. Greece
Annu. Rep. (B.I.H.)
Annu. Rep. Dir., Mt. Wilson Las Campanas Obs.
Annu. Rep. Geophys. Obs.
Annu. Rep. Int. Polar Motion Serv.
Annu. Rep. Meteorol. Obs. Int. Latitude Obs. Mizusawa
Archenhold-Sternw. Berlin-Treptow, Sonderdr.
Archenhold-Sternw. Berlin-Treptow, Vortr. Schr.
Armagh Obs., Repr.
Astrofiz. Issled. Izv. Spets. Astrofiz. Obs.
Astrofys. Inst., Vrije Univ. Bruss., Overdruk
Astron. Bull., Carter Obs.
Astron. Contrib. Univ. Manchester, Ser. II: Jodrell Bank Repr.
Astron. Contrib. Univ. Manchester, Ser. III
Astron. Inst., "Anton Pannekoeck", Univ. Amst., Repr.
Astron. Inst. Univ. Brno, Contrib.
Astron. Inst. Univ. Brno, Publ.
Astron. Mitt. Wien
Astron. Obs. Trieste, Publ.
Astron. Pap.
Astron. Rechen-Inst. Heidelb., Mitt., Ser. A
Astron. Rechen-Inst. Heidelb., Mitt., Ser. B
Astron. Zeit-Breitenbestimmungen
Astrophys. Prepr. Ser.
Astrophys. Relativ., Prepr. Ser.
Austrian Pap. Asteroids

BAV Mitt.
B.I.H. Circ.
Biul. Obs. Astron. Uniw. M. Kopernika Toruniu
Bol. Astron. Obs. Madr.
Bol. Inst. Tonantzintla
Bol. Obs. Ebro
Boyden Obs., Publ.
Boyden Obs., Repr.
Bull. Astron.
Bull. Etoiles Tardives Spectre Particulier
Bull. Inf. Cent. Données Stellaires
Bull. Inf. Etoiles Be
Bull. Obs. Astron. Belgr.
Bull., Time Serv. Mizusawa Obs.
Byull. Inst. Astrofiz.
Byull. Inst. Teor. Astron.

Carter Obs., Repr. Ser.
Cartes Synoptiques
Cent. Astron. Sci. Spat., Obs. Sol.
Cent. Astrophys., Prepr. Ser.
Circ. Czech. Obs., Time Latitude
Circ. Stn. Astron. Int. Latitudine, Carloforte-Cagliari
Circ. Time Latitude Serv.
Commun. Astron. Dep., Univ. Ankara
Commun. Konkoly Obs.
Commun. Univ. Obs., St. Andrews
Comun. Obs. Astron. Univ. Coimbra
Contrib. Astron. Obs. Skalnaté Pleso

Contrib. Bosscha Obs.
Contrib. Dep. Astron., Univ. Kyoto
Contrib. Dep. Astron., Univ. Tokyo
Contrib. Dep. Geod. Astron., Univ. Thessaloniki
Contrib. Dom. Astrophys. Obs.
Contrib. Dunsink Obs.
Contrib. Inst. Argent. Radioastron.
Contrib. Inst. Astron. Res.
Contrib. Kwasan Hida Obs., Univ. Kyoto
Contrib. La. State Univ. Obs.
Contrib. Lick Obs.
Contrib. Nicholas Copernicus Obs. Planetarium Brno
Contrib. Nizamiah Japal-Rangapur Obs.
Contrib. Oss. Astron. Torino
Contrib. Van Vleck Obs.
Copenh. Univ. Obs., Repr.
Cracow Obs. Repr.

Data Rep. Hydrogr. Obs., Ser. Astron. Geod.
Debrecen Heliophys. Obs., Hung. Acad. Sci., Repr.
Dep. Astron. McDonald Obs. Univ. Tex., Repr.
Dep. Astrophys., Univ. Oxford, Publ.
Dtsch. Hydrogr. Inst. Hamb., Zeit-Breitendienst
Dudley Obs. Rep.
Dunsink Obs. Repr.

Earth Orientation Bull.
ESO Annu. Rep.
ESO Sci. Prepr.
ESO Sci. Rep.
ESO Tech. Rep.

Gemini
Greenwich Time Rep.

HHI Sol. Data
HHI STP Rep.
HHI Suppl. Ser. Sol. Data
H.M. Naut. Alm. Off., Libr. Repr.
Hvar Obs. Bull.

Inst. Astron. Astrophys. Tech. Univ. Berlin, Mitt.
Inst. Astron. Fis. Espacio, Tirada Aparte
Inst. Astron., Univ. Camb., Annu. Rep.
Inst. Astrophys. Paris, Pré-Publ.
Inst. Astrophys. Univ. Liège, Collect. 4°
Inst. Astrophys. Univ. Liège, Collect. 8°
Inst. Obs. Mar., Bol. Astron.
Inst. Obs. Mar., Bol. Ser. B
Inst. Obs. Mar., Bol. Ser. C
Inst. Obs. Mar., Bol. Ser. D
Inst. Obs. Mar., Mem. Act.
Inst. Teor. Astrofys., Blindern-Oslo, Småtrykk
Inst. Theor. Astrophys., Blindern-Oslo, Rep.
Inst. Theor. Astrophys., Blindern-Oslo, Repr.
Inst. Theor. Astrophys., Univ. Oslo, Publ. Ser.
Inst. Theor. Phys. Sternw. Univ. Kiel, Repr.
Izv. Astron. Ehngel'gardt. Obs.
Izv. Glav. Astron. Obs. Pulkovo
Izv. Krymskoj Astrofiz. Obs.

Kandilli Obs., Heliophys. Serv. Publ., Second Ser.
Kapteyn Astron. Inst., Annu. Rep.
Kodaikanal Obs. Bull., Ser. A
Kodaikanal Obs. Repr.

Latitude Circ.
Lick Obs. Bull.

Lohrmann-Obs., Tech. Univ. Dresden, Zirk.
Lowell Obs. Bull.

Max-Planck-Inst. Radioastron., Bonn, Sonderdr. Ser. A
Messenger
Mitt. Archenhold-Sternw. Berlin-Treptow
Mitt. Astrophys. Obs. Potsdam
Mitt. Geod. Inst. Rheinischen Friedrich-Wilhelms-Univ. Bonn
Mitt. Karl-Schwarzschild Obs. Tautenburg
Mitt. Lohrmann-Obs. Tech. Univ. Dresden
Mitt. Satell.-Beobachtungsstn. Zimmerwald
Mitt. Sonnenobs. Kanzelhöhe
Mitt. Sternw. Babelsberg, Neue Folge
Mitt. Sternw. Münch.
Mitt. Sternw. Sonneberg
Mitt. Universitätssternw. Graz
Mitt. Univ.-Sternw. Innsb.
Mitt. Univ.-Sternw. Jena
Mitt. Veränderliche Sterne
Mitt. Zentralinst. Phys. Erde
Mon. Notes Int. Polar Motion Serv.
Mons Astrophys. Pap.
MPE Contrib.
MPE Intern. Rep.
MPE Prepr.
MPE Rep.
MPE Repr.
Mt. Stromlo Siding Spring Obs., Repr.

Nanjing Univ. Obs., Publ.
NASA IUE Newsl.
Natl. Astron. Ionos. Cent., Astron. Prepr.
Natl. Astron. Ionos. Cent., Astron. Publ.
Natl. Astron. Ionos. Cent., Newsl.
Natl. Radio Astron. Obs., Repr., Ser. A
Natl. Radio Astron. Obs., Repr., Ser. B
Nizamiah Rangapur Obs. Dep. Astron. Osmania Univ., Repr.
NSSDC Newsl.

Obs. Astron. Antares, Contrib. Cient.
Obs. Astron. Córdoba, Tirada Aparte
Obs. Astron. La Plata, Sep. Astron.
Obs. Astron. Univ. Nac. La Plata, Ser. Astron.
Obs. Astron. Univ. Nac. La Plata, Ser. Espec.
Obs. Astrophys. Lab., Univ. Helsinki, Rep.
Obs. Haute Provence, Pré-Publ.
Obs. Lyon Prepr.
Obs. Lyon Repr.
Obs. R. Belg., Commun., Sér. A
Obs. R. Belg., Commun., Sér. B
Obs. Variable Stars, Rep.
Occas. Rep. R. Obs., Edinb.
Onsala Space Obs., Prepr.
Oss. Astrofis. Catania, Publ.
Oss. Astron. Milano-Merate, Contrib.

Prepr. Astron. Inst. Univ. Basel
Prepr. Steward Obs.
Proc. Int. Latitude Obs. Mizusawa
Pubbl. Stn. Astron. Int. Latitudine Carloforte-Cagliari,
   Nuov. Ser.
Pubbl. Varie Fuori Ser. Oss. Astron. Torino (Pino Torinese)
Publ. Astron. Dep. Eötvös Univ.
Publ. Astron. Inst. Czech. Acad. Sci.
Publ. Astron. Obs. Sarajevo
Publ. Astron. Opservatorije Beogr.
Publ. Astrophys. Obs. Potsdam
Publ. Beijng Astron. Obs.
Publ. Bosscha Obs.
Publ. Debrecen Heliophys. Obs.
Publ. Dep. Astron. Univ. Beogr.
Publ. Dep. Astron. Univ. Cape Town
Publ. Dep. Astron., Univ. Chile
Publ. Dep. Geod. Astron., Univ. Thessaloniki
Publ. Dom. Astrophys. Obs.

Publ. Ege Univ. Obs.
Publ. Inst. Geophys., Pol. Acad. Sci., Ser. F (Planet. Geod.)
Publ. Int. Latitude Obs. Mizusawa
Publ. Istanbul Univ. Obs.
Publ. Lick Obs.
Publ. Obs. Astron. Nac., Univ. Colombia
Publ. Obs. Astron. "Prof. Manuel de Barros", Fac. Cienc. Porto
Publ. Obs. Astron. Univ. Cluj
Publ. Obs. Bordeaux, Nouv. Ser.
Publ. Obs. Ebro
Publ. Obs. Genève, Sér. A
Publ. Obs. Genève, Sér. B
Publ. Purple Mt. Obs.
Publ. R. Obs., Edinb.
Publ. Semin. Astron. Geod. Univ. Complutense Madr.
Publ. Shaanxi Astron. Obs.
Publ. Spéc. Cent. Données Stellaires
Publ. U.S. Nav. Obs., Second Ser.
Publ. Warner Swasey Obs.

Q. Bull. Sol. Act.

R. Greenwich Obs., Annu. Rep.
R. Greenwich Obs., Bull.
R. Obs. Ann.
R. Obs. Edinb., Res. Facilities
Rapp. Act. Obs. Paris
Rep. Obs. Lund
Rep. Ser., Dep. Phys. Sci., Univ. Turku
Res. Cent. Astron. Appl. Math., Acad. Athens, Contrib. Ser. I
Res. Lab. Electron. Onsala Space Obs., Res. Rep.
Ric. Astron.
Rijksuniv. Gent, Sterrenkundig Obs., Meded.

SAAO Newsl.
Sac Peak Update
S.Afr. Astron. Obs., Annu. Rep.
S.Afr. Astron. Obs., Circ.
Sci. Rep. Tôhoku Univ., Eighth Ser.
Semin. Astron. Geod., Univ. Madr., Publ.
Sendai Astron. Rap.
Shaanxi Astron. Obs., Repr.
Shanghai Obs., Acad. Sin., Prepr.
Smithson. Astrophys. Obs., Spec. Rep.
Sol. Maps Act.
Soobshch. Byurakan. Obs.
Soobshch. Gos. Astron. Inst. Shternberg
Soobshch. Spets. Astrofiz. Obs.
Space Telesc. Sci. Inst., Newsl.
Space Telesc. Sci. Inst., Prepr. Ser.
ST-ECF Newsl.
Stockholms Obs., Rep.
Stud. Soc. Sci. Torun., Sect. F
Swarthmore Astron. Repr.
Syd. Obs. Pap.

Tartu Astrofüüs. Obs. Publ.
Tartu Astrofüüs. Obs. Teated
Theor. Pap.
Time Freq. Serv., Bull.
Time Serv. Annu. Rep.
Time Serv. Bull.
Tokyo Astron. Bull., Second Ser.
Tokyo Astron. Obs., Kiso Inf. Bull.
Tokyo Astron. Obs. Rep.
Tokyo Astron. Obs. Repr.
Tokyo Astron. Obs., Time Latitude Bull.
Tr. Astrofiz. Inst. Alma-Ata
Tr. Astron. Obs., Leningrad
Tr. Glav. Astron. Obs. Pulkovo, Ser. 2
Tr. Gos. Astron. Inst. Shternberg
Tr. Inst. Teor. Astron., Leningrad
Tr. Kazan. Gorod. Astron. Obs.
Tr. Tashkent. Astron. Obs.
Trans. Astron. Obs. Yale Univ.

Tsirk. Astron. Inst. Tashkent
Tsirk. Astron. Obs. L'vov
Tsirk. Shemakh. Astrofiz. Obs.
Turku Univ. Obs., Informeto
Turku Univ. Obs., Informo

UKIRT Rep.
Univ. Barc., Dep. Fis. Tierra Cosmos, Publ.
Univ. Hannover, Astron. Stn., Veröff.
Univ. Obs., St. Andrews, Repr.
Univ. Tex., Monogr. Astron.
Univ. Tex., Publ. Astron.
Upps. Astron. Obs. Ann.
Upps. Astron. Obs., Rep.
U.S. Nav. Obs., Circ.
U.S. Nav. Obs., Time Serv. Publ., Ser. 1
U.S. Nav. Obs., Time Serv. Publ., Ser. 4
U.S. Nav. Obs., Time Serv. Publ., Ser. 6
U.S. Nav. Obs., Time Serv. Publ., Ser. 10
U.S. Nav. Obs., Time Serv. Publ., Ser. 11
U.S. Nav. Obs., Time Serv. Publ., Ser. 14

Utr. Sterrekundige Overdrukken
Uttar Pradesh State Obs., Repr.

Vatican Obs. Publ.
Vatican Obs. Publ., Spec. Ser., Studi Galileiani
Veröff. Archenhold-Sternw. Berlin-Treptow
Veröff. Astron. Inst. Bonn
Veröff. Astron. Rechen-Inst. Heidelb.
Veröff. Remeis-Sternw. Bamberg
Veröff. Sternw. Pulsnitz
Veröff. Sternw. Sonneberg
Veröff. Wilhelm-Foerster-Sternw.
Veröff. Zentralinst. Phys. Erde
Villanova Univ. Obs. Contrib.
Vilniaus Astron. Obs. Biul.

Warner Swasey Obs., Repr.
Warsaw Univ. Obs. Pol. Acad. Sci.,
    N. Copernicus Astron. Cent., Repr.
Wrocław Astron. Obs., Repr.

002 Bibliographical Publications, Documentation, Catalogues, Atlases

002.001 A catalogue of [Fe/H] determinations, 1984 edition.
G. Cayrel de Strobel, C. Bentolila, B. Hauck, A. Duquennoy.
Astron. Astrophys., Suppl. Ser., Vol. 59, No. 1, p. 145 – 186 (1985).
This is an updated version of the catalogue published by G. Cayrel de Strobel et al. (1981). This version contains 1921 [Fe/H] determinations for 1035 stars, an increase of 48% and 47% respectively. The search in the literature is complete up to December 1983.

002.002 Computer–based catalogue of open–cluster data.
G. Lyngå.
The Milky Way galaxy, p. 143 – 144 (1985). – See Abstr. 012.007 (IAU Symp. No. 106).
The third edition of the computer–based catalogue of open–cluster data has now been produced and is disseminated through the Centre de Données Stellaires, 11 rue de l'Université, F–67000 Strasbourg, France. It is also available through World Data Center A, NASA, Greenbelt, MD 20771, USA.

002.003 Some notes on patterns in citations of papers by American astronomers.
V. Trimble.
Q. J. R. Astron. Soc., Vol. 26, No. 1, p. 40 – 50 (1985).
Average citation rate per person is almost exactly proportional to career length; X–ray and radio astronomers are cited less often than 'old fashioned' optical ones (even after adjusting for shorter average career length); and people who work on stars and galaxies are cited more than twice as often as people who work on the Sun and Solar System.

002.004 Sterrenatlassen vergeleken.
A. Nagel.
Zenit, 12. Jaarg., No. 1, p. 15 – 20 (1985).

002.005 Catalog and bibliography of gamma–ray bursts.
W. A. Baity, G. J. Hueter, R. E. Lingenfelter.
AIP Conf. Proc., No. 115, p. 434 – 484 (1984). – See Abstr. 012.005.
The authors have prepared a list of all reported (until mid–1983) gamma–ray bursts, together with references to the observational and theoretical papers relating to them. All gamma ray bursts which have been reported in the literature, whether confirmed or not, are included, except for events of known solar origin. The catalog lists each burst by date and time, together with the fluence and position, when reported, and relevant references.

002.006 Bibliography of Earth, Moon, and Planets.
Z. Kopal, M. Moutsoulas, F. B. Waranius.
Earth, Moon, Planets, Vol. 32, No. 1, p. 47 – 133 (1985).
Articles entered into the data base and Lunar and Planetary Institute Library, May – October 1983. Concerning: Motion of the Moon and dynamics of the Earth–Moon system: shape and gravity field of the Moon. Physical structure of the Moon; thermal and stress history of the Moon. Morphology of the lunar surface. Origin and stratigraphy of lunar formations: mapping of the Moon. Chemical composition of the Moon: lunar petrology, mineralogy and crystallography. Exploration and utilization of the Moon. Planets. Jupiter. Mars. Mercury. Neptune. Pluto. Saturn. Uranus. Venus.

002.007 Bibliography of Earth, Moon, and Planets.
Z. Kopal, M. Moutsoulas, F. B. Waranius.
Earth, Moon, Planets, Vol. 32, No. 2, p. 193 – 240 (1985).
Articles entered into the data base and Lunar and Planetary Institute Library, May – October 1983. Concerning: asteroids, comets, meteorites, cosmic dust, other particles, etc.

002.008 Improving the nearby star data base.
A. R. Upgren.
News Lett. Astron. Soc. N.Y., Vol. 2, No. 7, p. 15 – 16 (1985).
Abstract. – See Abstr. 010.241.

002.009 Data in astronomy.
C. Jaschek.
Comput. Phys. Commun., Vol. 33, No. 1 – 3, p. 289 – 290 (1984).
Abstr. in Phys. Abstr., Vol. 88, No. 1249, Entry 14489 (1985). – See Abstr. 012.014.

002.010 Photometric catalogue for stars in Selected Areas and other fields in the RGU–system (X). Photometry of fields in and near to the Milky Way: Anticenter 5, Carina (IC 2581), Centaurus III, Aquila II, Cassiopeia (NGC 7654).
W. Becker, R. P. Fenkart, A. Spaenhauer, E. A. Alfaro, G. Kandemir, S. Karaali, C. Fang, J. M. Garcia–Pelayo, H. Steppe, L. Topaktas.
Astronomical Institute of the University of Basle, Venusstr. 7, CH–4102 Binningen, Switzerland. 120 pp. (1984).
The tenth volume of the Basle Photometric Catalogues contains the G–magnitudes, as well as the long– (G–R) and short–wave (U–G) colour indices of the stars in five galactic fields. Additionally, data as obtained from the interpretation of the two–colour diagrams are given, i.e.: if the star in question has been identified as a main–sequence star of either population, or as a late–type giant, or if it is blended on the photographic plates of one or more colours.

002.011 Solar flux atlas from 296 to 1300 nm.
R. L. Kurucz, I. Furenlid, J. Brault, L. Testerman.
National Solar Observatory Atlas No. 1. National Solar Observatory, Sunspot, N.M. 88349, USA. 240 pp. Price $ 13.00 (1984).

002.012 Catalog of CO observations of galaxies.
F. Verter.
Astrophys. J., Suppl. Ser., Vol. 57, No. 2, p. 261 – 285 (1985).
This paper includes all observations of CO isotopes in external galaxies up to spring 1984. Listed for each galaxy are (general) morphological type, apparent magnitude, diameter, recession velocity, H I content, (CO) antenna temperatures of detections and upper limits, detected and extrapolated integrated emission, and distribution of emission in mapped galaxies. For each reference the telescopes used, the features examined, the portion of the galaxy observed, and the uncertainties in the data are given. Calibration and notation conventions are discussed, and conversion factors are given.

002.013 I.A.U. archives of unpublished observations of variable stars: 1981 – 84 data.
M. Breger.
Publ. Astron. Soc. Pac., Vol. 97, No. 587, p. 85 – 88 (1985).

002.014 Desired characteristics of catalogs for cometary astrometry.
C. de Vegt.
Cometary astrometry, p. 87 – 92 (1984). – See Abstr. 012.024.
Reference star catalogs for cometary astrometry have to provide an all–sky coverage with suitable stellar density and limiting magnitude on a fundamental coordinate system. A general catalog meeting all these requirements will not be available for a foreseen timescale. The suitability of various existing catalogs for cometary astrometry is discussed.

002.015   The IHW reference star catalogs for comets Halley and Giacobini–Zinner.
R. S. Harrington.
Cometary astrometry, p. 93 – 95 (1984). – See Abstr. 012.024.

After catalogs in general are discussed, the specific catalogs that should be used for these two comets are described.

002.016   uvbyβ photoelectric photometric catalogue (magnetic tape).
B. Hauck, M. Mermilliod.
Astron. Astrophys., Suppl. Ser., Vol. 60, No. 1, p. 61 – 62 (1985).

This catalogue contains the homogeneous values for 40484 stars observed in the $uvby\beta$ photoelectric photometric system. The first part collects the original data found in the literature (60016 entries). A third file, giving more elaborated star identifications, has been added.

002.017   Conclusion of Viking Lander imaging investigation. Picture catalog of experiment data record.
S. D. Wall, T. C. Ashmore.
NASA Ref. Publ., NASA RP–1137, 10 + 202 pp. (1985).

The images returned by the two Viking Landers during the Viking Survey Mission are presented in this report. Listings of supplemental information which describe the conditions under which the images were acquired are included. Subsets of the images are listed in a variety of sequences to aid in locating images of interest. The format and organization of the digital magnetic tape storage of the images are described. A brief description of the mission and the camera system is also included.

002.018   Halley's comet. A bibliography.
R. S. Freitag (Editor).
Library of Congress, Washington, D.C. For sale by the Superintendent of Documents, U.S. Government Printing Office, Washington, D.C., 20402, USA. 29 + 555 pp. Price US$ 26.00 (1984). ISBN 0–8444–0459–4.
Review in Sky Telesc., Vol. 69, No. 6, p. 519 (1985).

002.019   Sky catalogue 2000.0. Volume 2: Double stars, variable stars and nonstellar objects.
A. Hirshfeld, R. W. Sinnott (Editors).
Cambridge University Press, Cambridge – London – New York – New Rochelle – Melbourne – Sydney and Sky Publishing Corporation, Cambridge, Mass. 57 + 385 pp. Price £ 42.50, US$ 49.50 cloth; £ 17.50, US$ 29.95 paper (1985). ISBN 0–521–25818–9 cloth, ISBN 0–521–27721–3 paper.
Review in Astron. Express, Vol. 1, Nos. 4 – 6, p. 161 (1985).

Contents: Glossary of selected astronomical names. Index to letter names of variable stars. Double and multiple stars. Visual binary stars. Spectroscopic binary stars. Variable stars. Suspected variable stars. Open clusters. Open cluster cross index. Globular clusters. Bright nebulae. Dark nebulae. Planetary nebulae. Galaxies. Quasi–stellar objects (QSO's). Radio sources. X–ray sources.

002.020   100,000 MK types: a mid–course look at the HD reclassification project.
N. Houk.
The MK process and stellar classification, p. 85 – 93 (1984). – See Abstr. 012.033.

002.021   Spectroscopic data bases.
C. Jaschek.
The MK process and stellar classification, p. 94 – 103 (1984). – See Abstr. 012.033.

002.022   Retrieving data for faint stars.
W. Buscombe.
The MK process and stellar classification, p. 104 – 108 (1984). – See Abstr. 012.033.

002.023   General Catalogue of Variable Stars. The Fourth Edition containing information on variable stars discovered and designated till 1982. Volume 1. Constellations Andromeda – Crux.
P. N. Kholopov (Editor).
Nauka Publishing House, Moscow. 376 pp. (1985).

002.024   Atlas of dwarf galaxies in the region of M81 group.
V. E. Karachentseva, I. D. Karachentsev, F. Börngen.
Astron. Astrophys., Suppl. Ser., Vol. 60, No. 2, p. 213 – 227 (1985).

Large scale photographs of dwarf galaxies in the region of the M81 group are presented. A first classification of objects is made according to their common shape, brightness gradient, and degree of resolution into separate stars. A possible new class of irregular amorphous galaxies, which lack the bright stars and H II regions, is described.

002.025   A finding list for the multiplet tables of NSRDS–NBS3, Sections 1 – 10.
C. J. Adelman, S. J. Adelman, D. Fischel, W. H. Warren Jr.
Astron. Astrophys., Suppl. Ser., Vol. 60, No. 2, p. 339 – 341 (1985).

A finding list has been prepared from the multiplet tables contained in NSRDS–NBS3, Sections 1 – 10.

002.026   On the Tartu University astronomical publications issued before 1947.
H. Eelsalu.
Tartu Astrofüüs. Obs. Publ., Tom 50, p. 29 – 37 (1984). In Russian.

Titles of known astronomical disputations published between 1634 and 1709 are tabulated. The list of contents is given for "Astronomische Beyträge" edited by professor J. W. Pfaff in 1806 – 1807. The series of astronomical publications issued by Dorpat (Tartu) University between 1817 and 1946 are featured.

002.027   General Catalogue of Variable Stars.
N. H. Baker.
Inf. Bull. Variable Stars, No. 2709, 1 p. (1985).

002.028   The Guide Star Catalog – structure and publication.
H. Jenkner, B. A. Gillespie, J. J. Harrison,
B. M. Lasker, B. J. McLean, J. A. Pollizzi, D. A. Rosenthal,
J. L. Russell, C. R. Sturch, D. S. Wilson.
Bull. Am. Astron. Soc., Vol. 17, No. 2, p. 549 (1985). Abstract. – See Abstr. 010.065.

002.029   Bibliography of Georges Lemaitre.
The Big Bang and Georges Lemaitre, p. 399 – 407 (1984). – See Abstr. 012.043.

002.030   Systematic errors in Zwicky's magnitudes.
G. Fasano.
Mem. Soc. Astron. Ital., Vol. 55, No. 3, p. 457 – 458 (1984). – See Abstr. 012.049.

The author has compared the magnitudes m_{zw} given by Zwicky in his "Catalogue of Galaxies and of Groups of Galaxies" with the total magnitudes B_T given by Sandage and Tammann in the "Revised Shapley–Ames Catalogue".

002.031   Bibliography of Earth, Moon, and Planets.
Z. Kopal, M. Moutsoulas, F. B. Waranius.
Earth, Moon, Planets, Vol. 32, No. 3, p. 291 – 339 (1985).

Articles entered into the data base at the Lunar and Planetary Institute Library, November 1983 – January 1984. Concerning: Motion of the Moon and dynamics of the Earth–Moon system: shape and gravity field of the Moon. Physical structure of the Moon; thermal and stress history of the Moon. Morphology of the lunar surface: origin and stratigraphy of lunar formations: mapping of the Moon. Chemical composition of the Moon: lunar petrology, mineralogy, and crystallography. Exploration and utilization of the Moon. Planets: Jupiter, Mars, Mercury, Neptune, Pluto, Saturn, Uranus, Venus. Other objects: asteroids, comets, meteorites, cosmic dust, other particles, etc.

002.032 Catalogue of star positions of the USSR Time Services for the epoch and equinox J2000.0.
P. M. Afanas'eva, D. Yu. Belotserkovskij, M. B. Kaufman,
N. V. Shcherbakova, L. I. Yagudin.
Glav. astron. obs. Akad. Nauk SSSR. Leningrad, 35 pp. (1984).
In Russian. Abstr. in Ref. Zh., 51. Astron., 4.51.82 (1985).

002.033 Decametric survey of discrete sources in the northern sky. IX. Source catalogue in the declination range 52° to 60°.
S. Ya. Braude, N. K. Sharykin, K. P. Sokolov,
S. M. Zakcharenko.
Astrophys. Space Sci., Vol. 111, No. 1, p. 1 – 64 (1985).
The region of the celestial sphere between declinations $\delta = 52°3$ and $\delta = 59°$ and right ascensions $\alpha = 1^h$ and $\alpha = 19^h$ has been surveyed with the UTR–2 radiotelescope. The results of the survey are presented. The catalogue contains the positions and flux densities of 313 radio sources detected at 10, 12.6, 14.7, 16.7, 20, and 25 MHz.

002.034 Bibliography and program notes on close binaries, No. 40, (Material received by March 15, 1984).
T. J. Herczceg (Editor), with the collaboration of
K. D. Abhyankar, B. Cester, D. S. Hall, M. Kitamura,
H. Mauder, C. D. Scarfe, A. Schulberg, R. F. Sisteró,
F. van't Veer, M. Vetešnik.
International Astronomical Union, Commission 42, 22 pp. (1984). Available from Department of Physics and Astronomy, The University of Oklahoma, Norman, Oklahoma, USA.

002.035 Bibliography and program notes on close binaries, No. 41. (Material received by September 15, 1984).
T. J. Herczeg (Editor), with the collaboration of
K. D. Abhyankar, B. Cester, D. S. Hall, M. Kitamura,
C. D. Scarfe, A. Schulberg, R. F. Sisteró, F. van't Veer,
M. Vetešnik.
International Astronomical Union, Commission 42, 26 pp. (1985). Available from Department of Physics and Astronomy, The University of Oklahoma, Norman, Oklahoma, USA.

002.036 Documentation for the machine–readable version of the *Revised S201 Catalog of Far–Ultraviolet Objects* (Page, Carruthers and Heckathorn 1982).
W. H. Warren Jr.
NSSDC/WDC–A–R&S, Publ., No. 84–12, 5 + 14 pp. (1984).
A detailed description of the machine–readable revised catalog as it is currently being distributed from the Astronomical Data Center is given. This catalog of star images was compiled from imagery obtained by the Naval Research Laboratory (NRL) Far–Ultraviolet Camera/Spectrograph (Experiment S201) operated from 21 to 23 April 1972 on the lunar surface during the Apollo 16 mission. The documentation includes a detailed data format description, a table of indigenous characteristics of the magnetic tape file, and a sample listing of data records exactly as they are presented in the machine–readable version.

002.037 Documentation for the machine–readable version of *A Finding List for the Multiplet Tables of NSRDS–NBS 3, Sections 1 – 10* (Adelman, Adelman, Fischel and Warren 1984).
W. H. Warren Jr.
NSSDC/WDC–A–R&S, Publ., No. 84–14, 5 + 11 pp. (1984).
A detailed description of the machine–readable finding list, as it is currently being distributed from the Astronomical Data Center, is presented. This version of the list supersedes an earlier one (1977).

002.038 Documentation for the machine–readable version of the *Córdoba Durchmusterung (CD)*.
W. H. Warren Jr.
NSSDC/WDC–A–R&S, Publ., No. 84–15, 5 + 17 pp. (1984).
A detailed description of the machine–readable version of the catalog, as it is currently being distributed from the Astronomical Data Center, is presented. The complete catalog is contained in the magnetic tape file, and corrections published in all corrigenda have been made to the data. The machine version contains 613959 records, but only 613953 stars (six stars were later deleted, but their logical records are retained in the file so that the zone counts are not different from the published catalog).

002.039 Documentation for the machine–readable version of the *Cape Photographic Durchmusterung (CPD)*.
W. H. Warren Jr.
NSSDC/WDC–A–R&S, Publ., No. 84–16, 5 + 11 pp. (1984).
A detailed description of the machine–readable version of the catalog, as it is currently being distributed from the Astronomical Data Center, is presented. The complete catalog is contained in the magnetic tape file, and corrections published in all errata have been made to the data. The machine version contains 454877 records, but only 454875 stars (two stars were later deleted, but their logical records are retained in the file so that the zone counts are not different from the published catalog).

002.040 Computer programs in astronomy.
J. Mosley, A. Fraknoi.
Mercury, Vol. 14, No. 1, p. 27 – 30 (1985).
The authors present a review of astronomical software for microcomputers.

002.041 Pressearchiv zur Geschichte der Astronomie und Raumfahrt.
H. Schienke.
Sterne Weltraum, 24. Jahrg., Nr. 5, p. 276 – 279 (1985).

002.042 A new version of the Abastumani compiled catalogue of galaxies on magnetic tape.
N. G. Kogoshvili.
Astron. Tsirk., No. 1354, p. 5 – 6 (1984). In Russian.

002.043 Star catalogues: their importance, bibliographic control, and a catalogue of the Royal Greenwich Observatory Library Collection.
N. J. Wyatt.
Thesis, City of Birmingham Polytechnic Department of Librarianship, Birmingham, England. 4 + 172 pp. (1984).

002.044 Supplement to the detailed bibliography on the surface photometry of galaxies.
W. D. Pence, E. Davoust.
Astron. Astrophys., Suppl. Ser., Vol. 60, No. 3, p. 517 – 526 (1985).
This is a supplement to the authors' bibliography (Davoust and Pence, 1982) which covers the literature on surface photometry of galaxies from March 1982 to January 1985.

002.045 Bibliography of books and papers published in 1979 on the history of astronomy.
P. G. Kulikovskij (Editor), and compiled by N. B. Lavrova.
Institut istorii estestvoznaniya i tekhniki, Akademiya Nauk SSSR, Moskva. 36 pp. Price 5 Kop. (1984). In Russian and English.

002.046 Messier's catalogue.
M. Jeličić, D. Mikešić.
Vasiona, Année 32, No. 4, p. 83 – 89 (1984). In Croatian.
Extensive tabular review of Messier objects data is presented. The nature and discovery of the objects is particularly emphasized in the text.

002.047 An atlas of selected galaxies, with illustrations of photometric analyses.
B. Takase, K. Kodaira, S. Okamura (Editors).
University of Tokyo Press, Tokyo, Japan and VNU Science Press BV, Utrecht, The Netherlands. 9 + 79 pp. Price DM 241.00 (1984). ISBN 4–13–068103–6 (University of Tokyo Press); ISBN 90–6764–007–7 (VNU Science Press BV).
Part I. General descriptions of the classification and photometric analyses: (1) Classification of galaxies (*B. Takase*). (2) Surface photometry (*K. Kodaira*). (3) Radial luminosity distribution (*S. Okamura*). (4) Luminosity distribution perpendicular to the galactic plane (*M. Hamabe*). (5) Projected profile and

related analysis (*M. Watanabe*). (6) Spectral analysis of spiral patterns (*M. Iye*). (7) Active galaxies and peculiar galaxies (*B. Takase*). (8) Clusters of galaxies (*T. Yamagata*). Part II. The atlas. Part III. Table of data for the atlas.

002.048   A catalog of stellar velocity dispersions. I. Compilation and standard galaxies.
B. C. Whitmore, D. B. McElroy, J. L. Tonry.
Space Telesc. Sci. Inst., Prepr. Ser., No. 40, 36 pp. (1985).
   A catalog of central stellar velocity dispersion measurements is presented. The catalog includes 1096 measurements of 725 galaxies. A set of 51 standard galaxies is defined consisting of galaxies with at least 3 reliable, concordant measurements.

002.049   The bibliography of Calendar and Almanac Bošković (1918 – 1976).
G. Ivanišević.
Publ. Astron. Opservatorije Beogr., No. 33, p. 111 (1985). Abstract. – See Abstr. 012.061.

002.050   Contribution to Yugoslav astronomical terminology. II. The dictionary of extragalactic astronomy terms.
G. Ivanišević.
Publ. Astron. Opservatorije Beogr., No. 33, p. 112 (1985). Abstract. – See Abstr. 012.061.

002.051   The International Portrait Catalogue (IPC) of the Archenhold Observatory.
K. Friedrich.
Veröff. Archenhold–Sternw. Berlin–Treptow, Nr. 11, 42 pp (1984).

002.052   Katalog der ortsfesten Sonnenuhren in der DDR.
A. Zenkert.
Veröff. Archenhold–Sternw. Berlin–Treptow, Nr. 12, 79 + 8 pp. (1984).

002.053   A general catalogue of galactic S stars, second edition.
C. B. Stephenson.
Publ. Warner Swasey Obs., Vol. 3, No. 1, p. 1 – 49 (1984).
   This catalog is intended to list all galactic S stars having known positions of at least roughly the precision of the Henry Draper catalog. The first edition of this catalog, published more than eight years ago, contained little more than half of the present number of stars, but nearly half of those were previously unpublished although lists of Henize's discoveries had been widely circulated. The principal additions since then come from three sources: (1) A low–dispersion infrared survey of the southern Milky Way by Westerlund, the S stars from which are hitherto unpublished; (2) a relatively high–dispersion red survey by MacConnell, also for the southern Milky Way, published in two lists here catalog–coded as MacCon79 and MacCon82; and (3) an unpublished and just–completed intermediate–dispersion red survey by the author, covering the sky north of declination –25° and outside of galactic latitudes ±10°.

002.054   A compilation of UBVRI photometry for galaxies in the ESO/Uppsala catalogue.
A. Lauberts, E. Sadler.
ESO Sci. Rep., No. 3, 52 pp. (1984).
   This catalogue contains all the presently available UBVRI aperture photometry (published up to the end of 1983, and also including some unpublished data) for galaxies in the ESO/Uppsala catalogue. It comprises a total of about 1000 galaxies south of –17° declination.

002.055   A catalogue of quasars and active nuclei (2nd edition).
M.–P. Véron–Cetty, P. Véron.
ESO Sci. Rep., No. 4, 104 pp. (1985).
   Because of the fast increase in the number of published quasars, the authors have prepared an updated version of their catalogue of quasars and active nuclei (Véron–Cetty and Véron, 1984) which now contains 2835 quasars and 739 active galaxies (of which 236 are Seyfert 1). Like the first edition, it includes positions, redshift, photometry and 6 and 11 cm flux densities.

002.056   Computation of the data–base of the star catalogue by EDMS.
T. Ishikawa.
Proc. Int. Latitude Obs. Mizusawa, No. 23, p. 23 – 30 (1984). In Japanese.
   Positions, proper motions and other relevant data of 3467 stars in the FK4 Star Catalogue are compiled in a data–base by using a network–type data–base management system.

002.057   Astronomy and Astrophysics Abstracts. Vol. 38. Literature 1984, Part 2.
S. Böhme, W. Fricke, H. Hefele, I. Heinrich, W. Hofmann, D. Krahn, V. R. Matas, L. D. Schmadel, G. Zech (Editors).
Published for Astronomisches Rechen–Institut by Springer–Verlag, Berlin – Heidelberg – New York – Tokyo. 10 + 919 pp. Price DM 189.00, US$ 68.50 [Subscription Price DM 151.20, US$ 54.80] (1985). ISBN 3–540–15562–7, ISBN 0–387–15562–7 (USA).

002.058   The new catalogue of Tokyo PZT stars (α_{85}/δ_{85} system).
S. Fujii, K. Nakajima.
Ann. Tokyo Astron. Obs., Second Ser., Vol. 20, No. 3, p. 321 – 334 (1985).
   The authors revise the catalogue of Tokyo PZT stars, using 26 years of observational data. The computational method used here is almost the same with that described by Fujii (1981), except the procedure of correcting nutation constants in accordance with the 1980 IAU Theory of Nutation. The fundamental epoch of the new catalogue is correctly changed from B1950.0 to J2000.0, that is the authors first calculate star–place correction for B1950.0 in the system of previous precession constant, and then convert them into J2000.0 using the method of Aoki et al. (1983).

002.059   The Copernicus ultraviolet spectral atlas of Gamma Pegasi.
J. B. Rogerson Jr.
Astrophys. J., Suppl. Ser., Vol. 57, No. 4, p. 751 – 822 (1985).
   An ultraviolet spectral atlas is presented for the B2 IV star γ Pegasi, which has been scanned from 970 to 1501 Å by the Princeton spectrometer aboard the Copernicus satellite. From 970 to 1430 Å the observations have a nominal resolution of 0.05 Å. At the longer wavelengths the resolution is 0.1 Å. The atlas is presented in graphs. Line identifications are also listed.

002.060   Bibliography.
   D. Ballereau, A. Feinstein, H. Hubert, A. M. Hubert–Delplace, J. Jugaku, P. Koubsky, E. Mendoza, M. Ruusalepp, A. Slettebak.
Bull. Inf. Etoiles Be, No. 11, p. 30 – 35 (1985).
   Bibliography of papers on Be stars published 1983/84.

002.061   Le troisième volume d'"Epitome fundamentorum astronomiae".
O. Atanackovic.
Bull. Inf. Cent. Données Stellaires, No. 28, p. 15 – 16 (1985).
   The plans for the third volume of the "Epitome fundamentorum astronomiae" are described. The volume deals with catalogs on radial velocities, parallaxes and proper motions.

002.062   Machine–readable sky surveys available from data centers.
C. Jaschek.
Bull. Inf. Cent. Données Stellaires, No. 28, p. 91 – 94 (1985). – See Abstr. 012.062.

002.063   Guidelines for abstracts and other bibliographic items in the astronomical literature as proposed by the IAU Commission 5 Special Working Group on "Guidelines for Abstracts", second report.
L. D. Schmadel.
Bull. Inf. Cent. Données Stellaires, No. 28, p. 95 – 109 (1985).
   Commission 5 (Documentation and Astronomical Data) of the International Astronomical Union at the XVIIIth General

Assembly Patras 1982 has established a Special Working Group on "Guidelines for Abstracts". It is the aim of this Working Group to elaborate a comprehensive collection of abstracting guidelines. A first draft has been circulated to some distinguished experts of the documentation field. This second report incorporates the results of their common considerations.

002.064 The Bibliographical Star Index, 1979 – 82.
F. Ochsenbein, F. Spite, S. Kirchner, R. Lahmek, M. J. Wagner.
Bull. Inf. Cent. Données Stellaires, No. 28, p. 131 – 132 (1985).
A microfiche of the Bibliographical Star Index (BSI) covering the 4–year period 1979 – 82 is now available. This edition differs slightly from the previous ones; the modifications are described.

002.065 "Physics Briefs": a bibliographic database also for astronomy and astrophysics.
W. Lück, H. Behrens.
Bull. Inf. Cent. Données Stellaires, No. 28, p. 133 – 141 (1985).

002.066 Astronomical data activities in Japan.
S. Nishimura.
Bull. Inf. Cent. Données Stellaires, No. 28, p. 153 – 157 (1985).

002.067 A catalog of small Magellanic Cloud objects in progress.
M. Bischoff, A. Florsch, J. Florsch.
Bull. Inf. Cent. Données Stellaires, No. 28, p. 159 (1985).

002.068 Catalogue of dark nebulae and globules for galactic longitudes 240 to 360 degrees.
J. V. Feitzinger, J. A. Stüwe.
Bull. Inf. Cent. Données Stellaires, No. 28, p. 161 (1985).

002.069 Fourth catalog of orbits of visual binary stars.
C. E. Worley, W. D. Heintz.
Bull. Inf. Cent. Données Stellaires, No. 28, p. 163 (1985).

002.070 The Washington Visual Double Star Catalog, 1984.0.
C. E. Worley, G. G. Douglass.
Bull. Inf. Cent. Données Stellaires, No. 28, p. 165 – 166 (1985).

002.071 Errata in catalogues of spectroscopic binary orbits.
A. Pedoussaut.
Bull. Inf. Cent. Données Stellaires, No. 28, p. 167 – 168 (1985).

002.072 Star catalogs and files available at the Stellar Data Center. Additions.
Bull. Inf. Cent. Données Stellaires, No. 28, p. 173 – 182 (1985).

002.073 Catalogs to be published from Center de Données Stellaires.
Bull. Inf. Cent. Données Stellaires, No. 28, p. 183 (1985).

002.074 Catalogs currently available on microfiche from Centre de Données Stellaires.
Bull. Inf. Cent. Données Stellaires, No. 28, p. 185 – 191 (1985).

002.075 Begriffswörterbuch der Chronologie und ihrer astronomischen Grundlagen.
M. Gossler.
Zweite, verbesserte Auflage, Bibliographische Informationen 12. Universitätsbibliothek Graz, Graz, Austria. 2+174 pp. (1985).

002.076 A catalog of stellar velocity dispersions.
B. C. Whitmore, D. B. McElroy, J. L. Tonry.
Bull. Am. Astron. Soc., Vol. 16, No. 4, p. 890 (1984). Abstract. – See Abstr. 010.062.

002.077 A corrected and uniformly formatted edition of the Abt and Biggs (1972) *Bibliography of Stellar Radial Velocities*.
W. H. Warren Jr.
Bull. Am. Astron. Soc., Vol. 16, No. 4, p. 896 (1984). Abstract. – See Abstr. 010.062.

002.078 Test–case of the Astronomical Mapping Program (AMP): the CPD Star Atlas.
B. N. Rappaport.
Bull. Am. Astron. Soc., Vol. 16, No. 4, p. 907 (1984). Abstract. – See Abstr. 010.062.

002.079 The Palomar Sky Survey II.
J. Schombert, J. R. Mould, W. L. W. Sargent, C. Kowal, A. Maury, R. Brucato.
Bull. Am. Astron. Soc., Vol. 16, No. 4, p. 907 – 908 (1984). Abstract. – See Abstr. 010.062.

002.080 Some notes on patterns in citations of papers by American astronomers.
V. Trimble.
Bull. Am. Astron. Soc., Vol. 16, No. 4, p. 942 – 943 (1984). Abstract. – See Abstr. 010.062.

002.081 A catalog and atlas for the Small Magellanic Cloud.
H. Albers, F. R. Chromey.
Bull. Am. Astron. Soc., Vol. 16, No. 4, p. 948 (1984). Abstract. – See Abstr. 010.062.

002.082 CTI data–management and interrogation.
M. G. M. Cawson, J. T. McGraw.
Bull. Am. Astron. Soc., Vol. 16, No. 4, p. 986 (1984). Abstract. – See Abstr. 010.062.

002.083 A science citation index study of university departments of astronomy.
D. L. Crawford.
Bull. Am. Astron. Soc., Vol. 16, No. 4, p. 999 (1984). Abstract. – See Abstr. 010.062.

002.084 SIMBAD: une base de données pour l'astronomie.
D. Egret, M. Wenger.
Bull. Assoc. Fr. Obs. Etoiles Variables, No. 32, p. 7 – 10 (1985).

002.085 La quatrième édition du G.C.V.S.
E. Schweitzer.
Bull. Assoc. Fr. Obs. Etoiles Variables, No. 32, p. 11 – 12 (1985).

002.086 Catalogue of proper motions of stars relative to galaxies in nine selected areas of southern declination.
K. Umarova.
Astron. inst. Akad. Nauk UzSSR. Tashkent, 115 pp. (1984). In Russian. Abstr. in Ref. Zh., 51. Astron., 5.51.415 (1985).

002.087 Catalogue of proper motions of stars relative to galaxies in nine selected sky areas.
A. G. Rakhimov, K. Umarova.
Astron. inst. Akad. Nauk UzSSR. Tashkent, 122 pp. (1984). In Russian. Abstr. in Ref. Zh., 51. Astron., 5.51.416 (1985).

002.088 Catalogue of proper motions of stars relative to galaxies in nine southern sky areas.
Yu. Baltabaev.
Astron. inst. Akad. Nauk UzSSR. Tashkent, 102 pp. (1984). In Russian. Abstr. in Ref. Zh., 51. Astron., 5.51.417 (1985).

002.089 Catalogue of proper motions of 4423 stars relative to galaxies in 41 areas of the sky.
Yu. Baltabaev.
Astron. inst. Akad. Nauk UzSSR. Tashkent, 174 pp. (1984). In Russian. Abstr. in Ref. Zh., 51. Astron., 6.51.95 (1985).

002.090 Halley's Comet, 1755 – 1984. A bibliography.
B. Morton.
Greenwood Press, Westport, Conn., USA – London, England 16+280 pp. Price £ 34.50, DM 187.50 (1985). ISBN 0–313–24022–1.

002.091   Manuale IHW (International Halley Watch). Guida allo studio scientifico della cometa per astronomi non professionisti.
S. J. Edberg, translated from the English by M. Elena.
Astronomia, Milano and Coelum, Bologna, Italy. 143 pp. (1985).
For the English original see 37.002.101.

002.092   Halley's comet: a bibliography of Canadian newspaper sources, 1835 – 36 and 1910.
J. A. Smith.
J. R. Astron. Soc. Can., Vol. 79, No. 2, p. 54 – 99 (1985).

002.093   Mars: Westliche und östliche Hemisphäre. Kosmos Handkarte 1:23500000.
H. Wolf.
Franckh'sche Verlagshandlung, W. Keller & Co., Stuttgart. F.R. Germany.   1 chart + 38 pp.   Price   DM 16.80   (1985).   ISBN 3–440–05460–8.

002.094   Erdmond: Vorderseite–Rückseite. Kosmos Handkarte 1:12000000.
H. Wolf.
Franckh'sche Verlagshandlung, W. Keller & Co., Stuttgart, F.R. Germany.   1 chart + 73 pp.   Price   DM 16.80   (1985).   ISBN 3–440–05461–6.

002.095   Catalogue of radial velocities of galaxies with emission lines.
R. V. Abdulova, Eh. K. Denisyuk.
Tr. Astrofiz. Inst. Alma–Ata, Tom 40, p. 41 – 65 (1983). In Russian.

002.096   Progress report on the Input Catalogue preparation.
C. Turon.
Processing of scientific data from the ESA astrometry satellite HIPPARCOS, p. 77 – 86 (1985). – See Abstr. 012.087.
The progress accomplished in the preparation of the Input Catalogue from the FAST Thinkshop at Asiago is presented in some details.

002.097   First improvement of the accuracy of the Input Catalogue star coordinates by star mapper data reduction.
F. Donati, E. Canuto, B. Fassino.
Processing of scientific data from the ESA astrometry satellite HIPPARCOS, p. 115 (1985). Abstract. – See Abstr. 012.087.

002.098   Mean stellar radial velocities catalogue.
M. Barbier.
Stellar radial velocities, p. 345 – 350 (1985). – See Abstr. 012.095 (IAU Colloq. No. 88).
A progress report on the preparation at Marseille Observatory of a catalogue of stellar radial velocities published in the period 1970 – 1980 is presented.

002.099   Bibliographic catalogue (*stellar radial velocities*).
M. Barbier.
Stellar radial velocities, p. 351 – 352 (1985). – See Abstr. 012.095 (IAU Colloq. No. 88).

002.100   Bibliography of radial velocity papers (1981 – 1984).
A. G. D. Philip.
Stellar radial velocities, p. 417 – 437 (1985). – See Abstr. 012.095 (IAU Colloq. No. 88).

002.101   The IUE Low–Dispersion Spectra Reference Atlas.
D. Egret, B. J. M. Hassall, A. Heck, C. Jaschek, M. Jaschek, A. Talavera.
Cool stars with excesses of heavy elements, p. 47 – 52 (1985). – See Abstr. 012.101.
The preparation of the second part of the IUE Reference Atlas, which will contain about 650 peculiar stars, is briefly described.

002.102   An analysis of the Basel Star Catalog.
K. U. Ratnatunga, J. N. Bahcall, R. Buser, R. P. Fenkart, A. Spaenhauer.
Bull. Am. Astron. Soc., Vol. 17, No. 2, p. 559 (1985). Abstract. – See Abstr. 010.065.

002.103   Atlas of Infrared Source Cross–Identifications.
M. Schmitz, J. M. Mead, D. Y. Gezari.
Bull. Am. Astron. Soc., Vol. 17, No. 2, p. 572 (1985). Abstract. – See Abstr. 010.065.

002.104   Systematic differences and average errors in the new edition of the Yale Parallax Catalogue.
W. F. van Altena, J. T. Lee.
Bull. Am. Astron. Soc., Vol. 17, No. 2, p. 583 (1985). Abstract. – See Abstr. 010.065.

002.105   A catalog of G5 – M stars in a region at the South Galactic Pole.
R. C. McNeil.
Bull. Am. Astron. Soc., Vol. 17, No. 2, p. 595 (1985). Abstract. – See Abstr. 010.065.

002.106   A galactic center molecular cloud atlas.
J. Bally, A. A. Stark, R. W. Wilson, C. Henkel.
Bull. Am. Astron. Soc., Vol. 17, No. 2, p. 613 – 614 (1985). Abstract. – See Abstr. 010.065.

002.107   List of 3071 high–luminosity stars and Cepheids for meridian observations.
D. K. Karimova, E. D. Pavlovskaya.
Tr. Gos. Astron. Inst. Shternberg, Tom 57, p. 245 – 247 (1985). In Russian.
A list of 3071 high–luminosity stars and Cepheids for meridian observations is compiled. It contains stars for which radial velocities, photoelectric B, V magnitudes (V ⩽ 9.00), two-dimensional spectral classification and astrometric history are available. The list is at Pulkovo.

002.108   MK Spectral Classifications. Sixth General Catalogue.
W. Buscombe.
Astronomy Department, Northwestern University, Evanston, Ill. 60201, USA. 233 pp. (1984). ISBN 0–939160–04–8.
The Sixth General Catalogue of Stellar Spectral Classification is based on literature published up to July 1984. It lists also corrigenda for previous books in this series.

002.109   The UK Schmidt Telescope Objective Prisms. I. Their applicability to some research projects.
A. Savage, S. M. Beard, J. B. Palmer.
Royal Observatory, Edinburgh, 3 + 19 pp. (1985). ISBN 0–902553–34–8.

002.110   The UK Schmidt Telescope Objective Prisms. II. Catalogue of objects and technical data.
A. Savage, J. D. Waldron, D. H. Morgan, S. B. Tritton, R. D. Cannon, J. A. Dawe, M. T. Bruck, S. M. Beard, J. B. Palmer.
Royal Observatory, Edinburgh, 3 + 59 + 4 pp. (1985). ISBN 0–902553–35–6.
The main aim of this report is to provide the user with illustrations of various types of galactic and extragalactic objects as they appear on UKST prism plates so that specific classes of objects can more easily be identified.

002.111   The UK Schmidt Telescope Objective Prisms. III. Illustrations of objective prism spectra.
A. Savage, J. D. Waldron, M. Fretwell, D. H. Morgan, S. B. Tritton, R. D. Cannon, M. T. Bruck, S. M. Beard, J. B. Palmer.
Royal Observatory, Edinburgh, 26 pp. (1985). ISBN 0–902553–36–4.

002.112 Report of IAU Commission 5: Documentation and astronomical data (*Documentation et données astronomiques*).
W. D. Heintz.
Trans. IAU, Vol. XIXA, p. 7 – 11 (1985). – See Abstr. 003.046.

002.113 First supplement to the First Dictionary of the Nomenclature of Celestial Objects.
M.–C. Lortet, F. Spite.
Observatoire de Paris, F–92195 Meudon Principal CEDEX, France, 54 pp. (1985). Accepted by Astron. Astrophys., Suppl. Ser.

002.114 Nomenclature for objects in the galaxy M33.
M.–C. Lortet.
Observatoire de Paris, F–92195 Meudon Principal CEDEX, France, 5 pp. (1985). Accepted by Astron. Astrophys., Suppl. Ser.

002.115 Nomenclature for objects in the Magellanic Clouds.
M.–C. Lortet.
Observatoire de Paris, F–92195 Meudon Principal CEDEX, France, 25 pp. (1985). Submitted to Astron. Astrophys., Suppl. Ser.

002.116 Troisième complément à la bibliographie des publications ayant rapport avec l'astrolabe Danjon ou avec d'autres astrolabes de types nouveaux.
With an introduction by S. Débarbat.
Observatoire de Paris, 61, avenue de l'Observatoire, F–75014 Paris, France. 15 pp. (1985).

002.117 On the RA catalogues F10 and F12 of the Pulkovo Time Service.
V. L. Gorshkov.
Izv. Glav. Astron. Obs. Pulkovo, Astrometr. Astrofiz., No. 202, p. 6 – 16 (1984). In Russian.
The F10 and F12 catalogues have been compiled on the basis of 1971 – 1980 observations with the photoelectric transit instrument. The catalogues contain 476 and 483 stars of the USSR Time Service catalogues, respectively. The catalogues are compared to similar catalogues with similar epochs of observations.

002.118 A preliminary declination catalogue of 96 international programme stars compiled from Kitab PZT observations.
E. A. Litvinenko.
Izv. Glav. Astron. Obs. Pulkovo, Astrometr. Astrofiz., No. 202, p. 17 – 20 (1984). In Russian.
A declination catalogue of 96 stars is given as compiled from Kitab PZT observations. The mean epoch of the catalogue is 1979.6.

002.119 Right ascensions of 194 FKSZ stars reduced for the collimation error using observations of FK4 stars.
A. A. Izvekova.
Izv. Glav. Astron. Obs. Pulkovo, Astrometr. Astrofiz., No. 201, p. 22 – 28 (1985). In Russian.
Transit moments are reduced for the collimation error and other instrumental parameters using a new formula. There is practically no correlation between the coefficients of this formula. The parameters are determined by least squares with high accuracy. The developed method enables to fulfill a closer relation to the reference catalogue system than the classical method.

002.120 On the projection of stellar maps.
A. A. Mikhajlov.
Izv. Glav. Astron. Obs. Pulkovo, Astrometr. Astrofiz., No. 201, p. 135 – 138 (1985). In Russian.
Different solutions of equidistant cylindrical and conical map projections are discussed and it is shown that they have larger scale errors than a zenithal polar projection on a tangent plane. Tables for the construction of stellar maps in conformal projections are given.

002.121 A compilation of *UBVRI* photometry for galaxies in the ESO/Uppsala catalogue.
A. Lauberts, E. M. Sadler.
New aspects of galaxy photometry, p. 335 – 336 (1985). – See Abstr. 012.114.
The authors have compiled a catalogue of all presently available *UBVRI* aperture photometry for galaxies in the ESO/Uppsala catalogue. It comprises a total of about 1000 galaxies south of –17° declination.

002.122 The biographical dictionary of scientists: astronomers.
D. Abbott (Editor).
Blond Educational, London, England. 8 + 204 pp. Price £ 10.95 (1984).
Review in J. Hist. Astron., Vol. 16, Part 1, No. 45, p. 59; 1985 (*O. Gingerich*).

002.123 Atlas of the night sky.
S. Dunlop (Editor).
Newnes Books, London, England. 80 pp. Price £ 4.95 (1984). ISBN 0–600–35113–0.
From Nature, Vol. 314, No. 6013, p. 702 (1985).

002.124 Handbook for Astronomical Societies 1984/85.
B. Jones (Editor).
Federation of Astronomical Societies, c/o B. Jones, 47 St. Blaise Ct., off Manchester Rd., Bradford, W. Yorkshire BD5 0QE, England. 68 pp. Price £ 1.75 (1984).
Review in Sky Telesc., Vol. 69, No. 3, p. 229 (1985).

002.125 Comets: a descriptive catalog.
G. W. Kronk.
Enslow Publishers, Inc., Hillside, N.J., USA. 331 pp. Price US$ 22.50, £ 15.50 (1984). ISBN 0–89490–071–4.
Reviews in Observatory, Vol. 105, No. 1065, p. 59 – 60; 1985 (*N. Eaton*); Orion, 43. Jahrg., Nr. 207, p. 48; 1985 (*A. Tarnutzer*); Strolling Astron., Vol. 31, Nos. 1 – 2, p. 41; 1985 (*J. V. Scotti*).

002.126 Données et définitions fondamentales d'astronomie.
E. Lindemann.
Commission romande de physique. Available from M. Philippe Naudy, Le Chapelet, CH–2208 Les Hauts–Geneveys, Switzerland. 44 pp. Price SFr 6.00 (1984).
Review in Orion, 43. Jahrg., Nr. 206, p. 32 – 33; 1985 (*A. Tarnutzer*).

002.127 Canon of lunar eclipses, –2002 to + 2526.
J. Meeus, H. Mucke.
2. erweiterte Auflage. Astronomisches Büro, Hasenwartgasse 32, A–1238 Wien, Austria. 32 + 256 pp. Price öS 380.00.
Review in Sterne Weltraum, 24. Jahrg., Nr. 3, p. 173; 1985 (*A. Kunert*).

002.128 Canon der Sonnenfinsternisse –2003 bis + 2526.
H. Mucke, J. Meeus.
Astronomisches Büro, Hasenwartgasse 32, A–1238 Wien, Austria. 54 + 910 pp. Price öS 1153.00.
Review in Sterne Weltraum, 24. Jahrg., Nr. 3, p. 174; 1985 (*A. Kunert*).

002.129 A catalogue of the Riche–Covington collection.
Compiled by B. de St. Remy, revised, enlarged and edited by L. H. Morley.
Douglas Library Occasional Paper No. 6. Queen's University, Kingston, Ontario. 63 pp. Price $ 10.00 (1984). Available from Special Collections, Queen's University, Kingston, Ontario, K7L 5C4, Canada.
Review in J. R. Astron. Soc. Can., Vol. 79, No. 3, p. 150 – 151; 1985 (*A. H. Batten*).

002.130   A data base for RS CVn binary systems.
E. R. Nelson, M. Zeilik, D. Hall, R. Genet.
Proceedings of the Southwest Regional Conference for Astronomy and Astrophysics, Vol. 10, p. 27 – 28 (1985). – See Abstr. 012.125.

A catalog of RS CVn binary star systems is being compiled and distributed on a world–wide basis. In this paper the authors report on the nature and status of the catalog.

002.131   Finding charts of radio stars for the HIPPARCOS mission.
R. Hering, H. M. Schwerdtfeger, H. G. Walter.
HIPPARCOS Input Catalogue Consortium Working Paper. Astronomisches Rechen–Institut, Mönchhofstraße 12 – 14, D–6900 Heidelberg, F.R. Germany. 45 pp. (1985).

So far a total of 275 confirmed, suspected and potential radio stars has been proposed to the HIPPARCOS observing list, many of them having insufficient positions for the HIPPARCOS mission. To facilitate reobservations finding charts are provided for a subset of 74 stars with confirmed radio emission. The scale of the charts is 8 arcsec/min; they are produced from Schmidt plates of the POSS and ESO(B) Atlases.

Dokumente aus der Dorpater Zeit Wilhelm Struves in der Charkower Sternwarte.
See Abstr. 004.056.

Kalender– und Astronomie–geschichtliche Publikationen im Tallinner Staatlichen Zentralarchiv.
See Abstr. 004.057.

Rara Astronomica in the Library of the Tartu University: books printed after 1600. Addenda.
See Abstr. 004.058.

Current work on binary and multiple stars.
See Abstr. 013.037.

Programmes d'observation des binaires spectroscopiques et informations du 14ème catalogue complémentaire.
See Abstr. 013.060.

The Astrographic Catalogue and the Carte du Ciel.
See Abstr. 013.061.

"General Catalogue of Variable Stars" and observations of variable stars by amateur astronomers.
See Abstr. 013.089.

Reduction of astrographic catalogues.
See Abstr. 036.052.

Arcsecond positions for milliarcsecond VLBI nuclei of extragalactic radio sources. III. 74 sources.
See Abstr. 041.012.

The effect of astrometric modeling errors of Schmidt plates on the construction of the Guide Star Catalog for Space Telescope.
See Abstr. 041.025.

Chromospheric flare activity in solar cycle 20.
See Abstr. 073.063.

A catalogue of high–speed streams in the solar wind with preference for large–scale and long–term velocity structures.
See Abstr. 074.033.

Refraction Tables of Pulkovo Observatory.
See Abstr. 082.045.

A catalogue of atmospheric gamma ray lines.
See Abstr. 082.078.

The eight–color asteroid survey: results for 589 minor planets.
See Abstr. 098.046.

Statistical catalogue of the parameters of orbits of nearly parabolic comets in Laplacian coordinates.
See Abstr. 102.028.

Photometric observations of long–period comets at large heliocentric distances in the years 1927 to 1955.
See Abstr. 103.007.

Comet observations made at the Skalnaté Pleso Observatory in the years 1972 – 1975.
See Abstr. 103.008.

Catalog of the Collection of Meteorites at the University of California, Los Angeles.
See Abstr. 105.222.

Radial velocities for 28 southern young open clusters.
See Abstr. 111.022.

Radial velocities from S.I.M.B.A.D.
See Abstr. 111.023.

A photometric search for nearby stars in the NLTT catalogue.
See Abstr. 111.049.

A comparison between the MK classification in the Michigan Spectral Catalogue and uvbyβ photometry.
See Abstr. 113.018.

IUE low–dispersion reference atlas.
See Abstr. 114.010.

An IUE high–resolution atlas of O–type spectra.
See Abstr. 114.011.

MK criteria applied to a scanner spectral atlas of the cooler stars.
See Abstr. 114.053.

Micrometer measures of 711 double stars.
See Abstr. 118.044.

Broad band spectral energy distributions of T Tauri stars in the Taurus–Auriga region.
See Abstr. 121.040.

A catalogue of field Type II Cepheids.
See Abstr. 122.117.

Observational studies of Cepheids. III. Catalog of light curve parameters.
See Abstr. 122.164.

Far–infrared survey of the galactic disc. II. The sources.
See Abstr. 133.003.

Infrared sources in the ϱ Ophiuchi dark cloud region.
See Abstr. 133.007.

The 408 MHz all–sky continuum survey.
See Abstr. 141.025.

Source catalogue of γ–rays from SASII data.
See Abstr. 143.074.

Proper motion studies of stars in and around open clusters.
See Abstr. 153.048.

Structure parameters of galactic globular clusters.
See Abstr. 154.069.

The Catalogue of Stellar Groups. II. Cool peculiar stars.
See Abstr. 155.163.

A catalog of dusty elliptical galaxies.
See Abstr. 157.030.

Kiso survey for ultraviolet–excess galaxies. II.
See Abstr. 157.146.

A catalog of radio, optical, and infrared observations of spiral galaxies in clusters.
See Abstr. 160.054.

Morphological and physical characteristics of the Virgo cluster: first results from the Las Campanas photographic survey.
See Abstr. 160.128.

003 Books

003.001 Wind as a geological process on Earth, Mars, Venus and Titan.
R. Greeley, J. D. Iversen.
Cambridge Planetary Science Series, Vol. 4. Cambridge University Press, Cambridge – London – New York – New Rochelle – Melbourne – Sydney. 12 + 333 pp. Price DM 147.50 (1985). ISBN 0–521–24385–8.
Contents: Wind as a geological process. The aeolian environment. Physics of particle motion. Aeolian abrasion and erosion. Aeolian sand deposits and bedforms. Interaction of wind and topography. Windblown dust.

003.002 Problems of cosmic physics, Vypusk 19.
S. K. Vsekhsvyatskij (Editor).
Respublikanskij Mezhduvedomstvennyj Nauchnyj Sbornik. Izdatel'stvo pri Kievskom Gosudarstvennom Universitete Izdatel'skogo Obedineniya "Vishcha Shkola", Kiev. 128 pp. Price 1 Rbl. 60 Kop. (1984). In Russian. ISSN 0555–2796.
The individual contributions are included in their corresponding subject categories – see abstracts 071.002, 072.010, 073.018 – 073.020, 074.019, 074.020, 091.009, 102.005 – 102.007, 103.141, 104.002, 104.003, 113.008, 117.055, 122.018, 157.012.

003.003 The global climate.
J. T. Houghton (Editor).
Cambridge University Press, Cambridge – London – New York – New Rochelle – Melbourne – Sydney. 6 + 233 pp. (1984). ISBN 0–521–25138–9.
Contents: The World Climate Research Programme (*J. T. Houghton, P. Morel*). Global climate research (*C. E. Leith*). Climate variability as estimated from atmospheric observations (*J. K. Angell, G. V. Gruza*). Atmospheric general circulation models: their design and use for climate studies (*A. J. Simmons, L. Bengtsson*). Cloud–radiation interaction and the climate problem (*P. J. Webster, G. L. Stephens*). The sensitivity of numerically simulated climates to land–surface boundary conditions (*Y. Mintz*). On dynamics of deserts and climate (*S. I. Rasool*). The cryosphere (*N. Untersteiner*). The upper ocean and air–sea interaction in global climate (*J. D. Woods*). The ocean circulation in climate (*C. Wunsch*). Strategy of ocean monitoring for climate research (*M. N. Koshlyakov, A. S. Monin*). Biogeochemical processes and climate modelling (*B. Bolin*). The role of carbon dioxide and other minor gaseous components and aerosols in the radiation budget (*K. Ya. Kondratyev, N. I. Moskalenko*).

003.004 The geology of the terrestrial planets.
M. H. Carr (Editor), with a summary and an appendix containing maps of the terrestrial planets.
NASA Spec. Publ., NASA SP–469, 8 + 317 pp. (1984).
The individual contributions are included in their corresponding subject categories – see abstracts 081.012, 092.001, 093.009, 094.009, 097.002, 107.005.

003.005 Theoretical concepts in physics. An alternative view of theoretical reasoning in physics for final–year undergraduates.
M. S. Longair.
Cambridge University Press, Cambridge – London – New York – New Rochelle – Melbourne – Sydney. 13 + 366 pp. Price £ 8.95 paper (1984). ISBN 0–521–25550–3 cloth, ISBN 0–521–27553–9 paper.
Contents: Introduction. The origins of Newton's law of gravitation. Maxwell's equations. Mechanics and dynamics. Thermodynamics and statistical mechanics. The origins of the concept of quanta. Special relativity. General relativity and cosmology.

003.006 Neutron radiative capture.
B. J. Allen, I. Bergqvist, R. E. Chrien, D. Gardner, W. P. Poenitz.
Pergamon Press, Oxford – New York – Toronto – Sydney – Paris – Frankfurt. 14 + 270 pp. (1984). ISBN 0–08–029330–1.
Review in Phys. Abstr., Vol. 88, No. 1251, Entry 19012 (1985). See Abstr. 061.014.

003.007 The universe of galaxies. Readings from Scientific American.
P. W. Hodge (Editor).
W. H. Freeman and Company, New York – Oxford. 10 + 113 pp. Price US$ 20.95 cloth, US$ 10.95 paper (1984). ISBN 0–7167–1675–5 cloth, ISBN 0–7167–1676–3 paper.
Reviews in Nature, Vol. 314, No. 6006, p. 24 – 25; 1985 (*D. W. Hughes*); Sky Telesc., Vol. 69, No. 1, p. 36 (1985); Zenit, 12. Jaarg., No. 3, p. 111 (1985).

003.008 Halley's Comet.
D. Tattersfield.
Basil Blackwell Publisher Limited, Oxford – New York. 9 + 164 pp. Price £ 6.50, US$ 7.50 (1984). ISBN 0–631–13558–8.
Reviews in J. Br. Astron. Assoc., Vol. 95, No. 4, p. 187; 1985 (*A. J. Hollis*); Sky Telesc. Vol. 69, No. 5, p. 422 (1985).

003.009 General relativity.
R. M. Wald.
The University of Chicago Press, Chicago – London. 13 + 491 pp. Price US$ 34.50, £ 31.50 (1984). ISBN 0–226–87032–4 cloth, ISBN 0–226–87033–2 paper.
Review in Nature, Vol. 314, No. 6006, p. 22 – 23; 1985 (*J. D. Barrow*).
Contents: Introduction. Manifolds and tensor fields. Curvature. Einstein's equation. Homogeneous, isotropic cosmology. The Schwarzschild solution. Methods for solving Einstein's equation. Causal structure. Singularities. The initial value formulation. Asymptotic flatness. Black holes. Spinors. Quantum effects in strong gravitational fields. Appendices: Topological spaces. Differential forms, integration, and Frobenius's theorem. Maps of manifolds, Lie derivatives, and Killing fields. Conformal

transformations. Lagrangian and Hamiltonian formulations of Einstein's equation. Units and dimensions.

003.010 A astronomia no Brasil.
A. de Moraes.
Instituto Astronômico e Geofísico, Universidade de São Paulo, São Paulo, Brazil. 8 + 88 pp. (1984). ISBN 85–85047–01–1.

003.011 Cosmologia: teoria cosmogaláctica e galáxcentrismo.
A. Castro Moreira.
A. Castro Moreira, Rua Com. Elias Zarzur, 154 c/4 São Paulo SP. Brasil CEP 04736. 95 pp. (1985).

003.012 Accretion power in astrophysics.
J. Frank, A. R. King, D. J. Raine.
Cambridge University Press, Cambridge – London – New York – New Rochelle – Melbourne – Sydney. 10 + 273 pp. Price £ 30.00, US$ 59.50 (1985). ISBN 0–521–24530–3.
Review in Astron. Express, Vol. 1, Nos. 4 – 6, p. 160 – 161 (1985).
   Contents: Accretion as a source of energy. Gas dynamics. Plasma concepts. Accretion in binary systems. Accretion discs. Accretion on to a compact object. Quasars and active nuclei. Gas flow and line emission in active nuclei. Accretion on to supermassive black holes. Thick discs.

003.013 Radiotelescopes.
W. N. Christiansen, J. A. Högbom.
Second edition. Cambridge Monographs on Physics. Cambridge University Press, Cambridge – London – New York – New Rochelle – Melbourne – Sydney. 9 + 265 pp. Price £ 30.00, US$ 59.50 (1985). ISBN 0–521–26209–7.
Review in Astron. Express, Vol. 1, Nos. 4 – 6, p. 161 (1985).

003.014 Planetary geology in the 1980s.
J. Veverka.
NASA Spec. Publ., NASA SP–467, 14 + 187 pp. (1985).
   Contents: Introduction. Surface features and processes. Chronology of planetary surfaces. Geochemistry in the planetary geology context. Geophysics in the planetary geology context. Geodesy and cartography. The geology of small bodies. Remote sensing and supporting earth–based studies. Summary and recommendations.

003.015 Geschichte der modernen Astronomie.
D. B. Herrmann.
VEB Deutscher Verlag der Wissenschaften, Berlin, GDR. 208 pp. Price DM 47.10 (1984).
Review in Astron. Raumfahrt, 23. Jahrg., Heft 3, p. 71 – 72; 1985 (*W. R. Dick*).

003.016 Investigations of nuclear physics and physics of elementary particles. Collected papers of Samarkand University.
Samarkand. 58 pp. (1984). In Russian.
Review in Ref. Zh., 51. Astron., 3.51.59 (1985).
See abstracts 041.016, 161.142.

003.017 Results from the Solar Maximum Mission.
R. Pallavicini (Editor).
Mem. Soc. Astron. Ital., Vol. 55, No. 4, p. 629 – 829 (1984).
The individual contributions are included in their corresponding subject categories – see abstracts 011.012, 013.047, 051.049, 073.113 – 073.118, 074.076 – 074.078, 076.012 – 076.014, 077.023.

003.018 The Cambridge Atlas of Astronomy.
J. Audouze, G. Israël (Editors).
Cambridge University Press, Cambridge – London – New York – New Rochelle – Melbourne – Sydney and Newnes Books, 84–88 The Centre, Feltham, Middlesex, England TW13 4BH. 432 pp. Price £ 29.95 until 31.12.1985, then £ 40.00, US$ 75.00 (1985). ISBN 0–521–26369–7 (Cambridge University Press), ISBN 0–600–35814–3 (Newnes Books).

   Contents: Introduction. Astronomy today. The Sun. The Solar System. The stars and the Galaxy. The extragalactic domain. The scientific perspective. The history of astronomy. Sky map.

003.019 Astrophysics and Space Physics Reviews. Volume 3 (1984).
R. A. Syunyaev (Editor).
Astrophys. Space Phys. Rev., Vol. 3, 11 + 349 pp. Price US$ 170.00 (1984). ISBN 3–7186–0092–7.
   The papers of this volume are revised and extended English translations of the Russian originals in 34.003.009.
The individual contributions are included in their corresponding subject categories – see abstracts 035.047, 061.035, 061.036, 063.047, 082.041, 117.168, 123.007, 125.050, 161.166.

003.020 Astrophysics and Space Physics Reviews. Volume 4 (1985).
R. A. Syunyaev (Editor).
Astrophys. Space Phys. Rev., Vol. 4, 12 + 315 pp. Price US$ 170.00 (1985). ISBN 3–7186–0125–7.
   The papers of this volume are revised and extended English translations of the Russian originals in 34.003.010.
The individual contributions are included in their corresponding subject categories – see abstracts 107.027, 116.041, 143.041, 144.053, 155.105.

003.021 Cosmology of the early universe.
L. Z. Fang, R. Ruffini (Editors).
Advanced Series in Astrophysics and Cosmology, Vol. 1. World Scientific Publishing Co. Pte. Ltd., Singapore. 7 + 305 pp. Price US$ 56.00 cloth, US$ 32.00 paper (1984). ISBN 9971–950–92–8 cloth, ISBN 9971–950–93–6 paper.
The individual contributions are included in their corresponding subject categories – see abstracts 161.182 – 161.189.

003.022 Investigation of magnetic fields and active formations on the sun.
Collected papers. Sev.–Vost. kompleks. NII. Vladivostok. 91 pp. (1984). In Russian.
Review in Ref. Zh., 51. Astron., 5.51.29 (1985).
See abstracts 036.071, 072.073 – 072.077, 073.158, 073.159, 077.039, 080.111.

003.023 Meteoritic investigations in Siberia: the Tunguska phenomenon – 75 years.
Yu. A. Dolgov (Editor).
Nauka, Novosibirsk. 217 pp. (1984). In Russian.
Review in Ref. Zh., 51. Astron., 6.51.44 (1985).
See abstracts 105.230 – 105.234, 105.236.

003.024 Physics of solar activity.
Eh. I. Mogilevskij (Editor).
Inst. zem. magn., ionos., rasprostr. radiovoln. 203 pp. (1983). In Russian. See also 30.003.107.
Review in Ref. Zh., 51. Astron., 6.51.46 (1985).
See abstracts 034.091, 072.081, 073.166, 073.169, 075.026 – 075.030, 077.040.

003.025 Foundations of radiation hydrodynamics.
D. Mihalas, B. W. Mihalas.
Oxford University Press, Oxford – New York – Tokyo. 17 + 718 pp. Price £ 55.00 (1984). ISBN 0–19–503437–6.
   Contents: Microphysics of gases. Dynamics of ideal fluids. Dynamics of viscous and heat–conducting fluids. Relativistic fluid flow. Waves, shocks, and winds. Radiation and radiative transfer. The equations of radiation hydrodynamics. Radiating flows.

003.026 The Jodrell Bank telescopes.
B. Lovell.
Oxford University Press, Oxford – New York – Tokyo. 16 + 292 pp. Price £ 15.00 (1985). ISBN 0–19–858178–5.
Review in Spaceflight, Vol. 27, No. 6, p. 287 (1985).

003.027   **Chemistry of atmospheres. An introduction to the atmospheres of Earth, the planets, and their satellites.**
R. P. Wayne.
Clarendon Press, Oxford, England. 12 + 361 pp. Price £ 30.00 cloth, £ 14.95 paper (1985). ISBN 0–19–855176–2 cloth, ISBN 0–19–855175–4 paper.

003.028   **Physics and astrophysics. A selection of key problems.**
V. L. Ginzburg, translated by O. Glebov and G. ter Haar.
Pergamon Press, Oxford – New York – Toronto – Sydney – Paris – Frankfurt. 14 + 125 pp. Price US$ 23.50 cloth, US$ 12.50 paper. ISBN 0–08–026498–0 cloth, ISBN 0–08–026499–9 paper (1985). For the Russian original see 27.003.061.

003.029   **Space science – Horizon 2000.**
N. Longdon, H. Olthof (Editors).
ESA Spec. Publ., ESA SP–1070. 5 + 137 pp. Price FF 70.00 (1984).
The individual contributions are included in their corresponding subject categories – see abstracts 013.071 – 013.079, 051.075 – 051.080.

003.030   **Interacting binary stars.**
J. E. Pringle, R. A. Wade (Editors).
Cambridge University Press, Cambridge – London – New York – New Rochelle – Melbourne – Sydney. 10 + 220 pp. Price £ 25.00, US$ 49.50 (1985). ISBN 0–521–26608–4.
Review in Astron. Express, Vol. 1, Nos. 4 – 6, p. 160 (1985).
The individual contributions are included in their corresponding subject categories – see abstracts 065.083, 117.286 – 117.293.

003.031   **Computer simulations of space plasmas.**
H. Matsumoto, T. Sato (Editors).
Advances in Earth and Planetary Sciences. Terra Scientific Publishing Co., Tokyo and D. Reidel Publishing Company, Dordrecht – Boston – Lancaster. 10 + 380 pp. Price Dfl. 190.00, US$ 67.00, £ 48.50 (1985). ISBN 90–277–1952–7.
The individual contributions are included in their corresponding subject categories – see abstracts 062.157 – 062.164, 084.100 – 084.102, 106.066.

003.032   **Annual Review of Earth and Planetary Sciences, Volume 13.**
G. W. Wetherill, A. L. Albee, F. G. Stehli (Editors).
Annu. Rev. Earth Planet. Sci., Vol. 13, 10 + 443 pp. Price US$ 47.00 (1985). ISBN 0–8243–2013–1.
The individual contributions within the subject scope of Astronomy and Astrophysics Abstracts are included in their corresponding categories – see abstracts 081.043, 094.037, 106.069.

003.033   **Handbook of plasma physics. Vol. 2: Basic plasma physics II.**
A. A. Galeev, R. N. Sudan (Editors).
North–Holland Publishing Company, Amsterdam – New York – Oxford. 14 + 850 pp. ISBN 0–444–86645–0.
Review in Phys. Abstr., Vol. 88, No. 1258, Entry 58013 (1985).

003.034   **Infrared and millimeter waves. Vol. 12. Electromagnetic waves in matter, Part II.**
K. J. Button (Editor).
Academic Press Inc., Orlando – San Diego – San Francisco – New York – London – Toronto – Montreal – Sydney – Tokyo – São Paulo. 15 + 323 pp. (1984). ISBN 0–12–147712–6.
Review in Phys. Abstr., Vol. 88, No. 1259, Entry 63596 (1985).
See Abstr. 082.077.

003.035   **Annual Review of Nuclear and Particle Science, Vol. 34.**
J. D. Jackson, H. E. Gove, R. F. Schwitters (Editors).
Annu. Rev. Nucl. Part. Sci., Vol. 34, 7 + 592 pp. (1984). ISBN 0–8243–1534–0.
Review in Phys. Abstr., Vol. 88, No. 1260, Entry 69055 (1985).
See Abstr. 061.112.

003.036   **Jupiter und Saturn. Ergebnisse der Planetenforschung.**
A. Rétyi.
Astrokosmos, Franckh'sche Verlagshandlung, W. Keller & Co., Stuttgart, F.R. Germany. 79 pp. Price DM 19.80 (1985). ISBN 3–440–05491–8.

003.037   **Welcher Stern ist das?**
W. Widmann, K. Schütte.
22nd edition. Kosmos–Naturführer, Franckh'sche Verlagshandlung, W. Keller & Co., Stuttgart, F.R. Germany, 183 pp. Price DM 22.00 (1985). ISBN 3–440–05517–5.

003.038   **Astronomie ganz einfach. Bauen und Beobachten – von der Sonnenuhr zum Spiegelfernrohr.**
P. Seymour.
Kosmos, Franckh'sche Verlagshandlung, W. Keller & Co., Stuttgart, F.R. Germany. 73 pp. Price DM 29.50 (1985). ISBN 7–440–05502–7.

003.039   **Current Topics in Chinese Science, Section E: Astronomy.**
Curr. Top. Chin. Sci., Sect. E: Astron., Vol. 1, 7 + 163 pp. Price US$ 9.00 (1982). ISBN 0–677–31170–2.
A selection of articles from Scientia Sinica, Vol. 24 (1981) and Kexue Tongbao (Beijing), Vol. 26 (1981).
See abstracts 29.065.034, 29.066.031, 30.073.043, 30.052.009, 30.160.034, 30.066.519, 30.065.081, 30.141.145, 30.031.539, 30.142.510, 31.045.004, 31.141.072, 31.080.013, 31.077.005, 31.074.021, 31.101.005, 31.078.005, 32.117.052.

003.040   **Current Topics in Chinese Science, Section E: Astronomy.**
Curr. Top. Chin. Sci., Sect E: Astron., Vol. 2, 7 + 206 pp. Price US$ 24.00 (1984). ISBN 0–677–06260–5.
A selection of articles from Scientia Sinica, Vol. 25 (1982) and Kexue Tongbao (Beijing), Vol. 27 (1982).
Review in Strolling Astron., Vol. 31, Nos. 1 – 2, p. 43; 1985 (*F. D. Miller*).
See abstracts 31.151.046, 31.072.026, 31.151.047, 31.151.088, 31.065.069, 31.045.035, 31.077.062, 32.141.173, 32.084.093, 32.102.090, 32.162.224, 32.162.225, 32.102.091, 32.044.036, 32.044.037, 32.062.096, 39.085.019, 39.073.194, 39.073.195, 33.073.153, 33.119.115, 34.161.010.

003.041   **Calculation of mirror and lens deformations with horizontal and vertical control.**
G. P. Klyueva, N. A. Sukhova, G. G. Telepneva, N. G. Bochkarev.
Izdatel'stvo Moskovskogo Universiteta. 87 pp. Price 25 Kop. (1985). In Russian.

003.042   **Physics of highly charged ions.**
R. K. Janev, L. P. Presnyakov, V. P. Shevelko (*V. P. Shevel'ko*).
Springer Series in Electrophysics, Vol. 13. Springer–Verlag, Berlin – Heidelberg – New York – Tokyo. 10 + 330 pp. Price DM 148.00 (1985). ISBN 3–540–12559–0 (F.R. Germany), ISBN 0–387–12559–0 (USA).
This book is devoted to the basic aspects of the physics of highly charged ions. Its principal aim is to provide a basis for understanding the structure and spectra of these ions, as well as their interactions with other atomic particles (electrons, ions, atoms and molecules). Particular attention is paid to the presentation of theoretical methods for the description of different radiative and collision phenomena involving multiply charged ions. The experimental material is included only to illustrate the validity of theoretical methods or to demonstrate those physical phenomena for which adequate theoretical descriptions are still absent. The general principles of atomic spectroscopy are included to the extent to which they are pertinent to the subject matter.

003.043   Intrinsic geodesy.
A. Marussi, translated by W. I. Reilly.
Springer–Verlag, Berlin – Heidelberg – New York – Tokyo.
17 + 219 pp.   Price DM 160.00 (1985).   ISBN 3–540–15133–8
(F.R. Germany), ISBN 0–387–15133–8 (USA).
Contents: Fundamentals of intrinsic geodesy. Structure of the
gravity field and Laplace's equation. Principles of intrinsic geodesy applied to the normal reference field. Mapping of the actual
gravity field onto the normal reference field. Mapping between
surfaces. Propagation of light in continuous isotropic refracting
media. Posthumous work.

**003.044   Geodetic refraction. Effects of electromagnetic wave
propagation through the atmosphere.**
F. K. Brunner (Editor).
Springer–Verlag, Berlin – Heidelberg – New York – Tokyo.
11 + 213 pp. Price DM 48.00 (1984). ISBN 3–540–13830–7 (F.R.
Germany), ISBN 0–387–13830–7 (USA).
The individual contributions within the subject scope of Astronomy and Astrophysics Abstracts are included in their corresponding subject categories – see abstracts 082.084 – 082.089.

**003.045   Under Newton's shadow. Astronomical practices in the
seventeenth century.**
L. Murdin.
Adam Hilger Ltd., Bristol – Boston. 8 + 152 pp. Price £ 16.75,
US$ 26.00 (1985). ISBN 0–85274–456–0.
Reviews in J. Br. Astron. Assoc., Vol. 95, No. 4, p. 189; 1985
(*E. A. Beet*); Nature, Vol. 315, No. 6018, p. 435 (1985); Sky
Telesc., Vol. 69, No. 6, p. 520 (1985).
This book surveys the characters and personal lives of British
astronomers in the period dominated by the scientific personality
of Isaac Newton. Among these figures are the well known Edmond Halley, Robert Hooke and John Flamsteed; less familiar
are men like Stephen Gray of Canterbury and Thomas Brattle of
New England. This book describes the extent and significance of
their particular contributions to observational astronomy and
discusses how social, educational and economic factors of the
period affected their work.

**003.046   Reports on Astronomy. Transactions of the Inter-
national Astronomical Union, Volume XIXA.**
R. M. West (Editor).
D. Reidel Publishing Company, Dordrecht – Boston – Lancaster.
8 + 726 pp. Price Dfl. 210.00, US$ 69.00, £ 58.25 (1985). ISBN
90–277–2039–8.
The reports of the various IAU commissions are included in their
corresponding subject categories – see abstracts 002.112, 004.127,
013.091, 013.092, 014.052, 015.048, 022.183, 032.046, 041.050,
041.051, 042.114, 044.083, 044.084, 046.022, 051.109, 064.100,
065.096, 072.108, 080.149, 082.096, 082.097, 091.062, 091.063,
098.098, 102.053, 104.055, 106.084, 111.054, 113.070, 114.165,
114.166, 117.376, 118.058, 122.206, 131.356, 141.039, 142.083,
153.060, 155.193, 157.244, 161.358.

003.047   Magnetic fields in space.
N. G. Bochkarev.
Seriya "Problemy Nauki i Tekhnicheskogo Progressa". Glav-
naya Redaktsiya Fiziko–Matematicheskoj Literatury. Nauka,
Moskva. 208 pp. Price 70 Kop. (1985). In Russian.

**003.048   Festschrift dedicated to Hans E. Suess on the occasion of
his 75th birthday 1985.**
K. Marti, H. Wänke (Editors).
Meteoritics, Vol. 20, No. 2, Part 2, p. 287 – 459 (1985).
See abstracts 022.184, 022.185, 061.148, 084.140, 097.024,
105.261 – 105.263, 107.035, 107.036.

**003.049   Main directions of astronomical investigations at the
Moscow University.**
Izdatel'stvo Moskovskogo Universiteta, Moskva. 173 pp. Price
1 Rbl. 80 Kop. (1985). In Russian.
The individual contributions are included in their corresponding
subject categories – see abstracts 009.032 – 009.042, 041.053,
041.054, 042.116, 044.087, 066.195, 081.056, 157.248, 159.143.

**003.050   The cosmic history of the biogenic elements and com-
pounds.**
J. A. Wood, S. Chang (Editors).
NASA Spec. Publ., NASA SP–476, 8 + 80 pp. (1985).
Contents: (1) Introduction. (2) An overview of the cosmic history of the biogenic elements. (3) Nucleosynthesis and ejection to
the interstellar medium of the biogenic elements. (4) Chemical
evolution in the interstellar medium. (5) Protostellar collapse.
(6) Chemical evolution in the solar nebula. (7) The growth of
planetesimals from dust. (8) The accumulation and thermal processing of planetoids. (9) Directions for future research.

003.051   Em busca de outros mundos.
R. R. de Freitas Mourão.
Livraria Francisco Alves Editora S.A., Rua Sete de Setembro,
177, Centro 20050, Rio de Janeiro, Brazil. 231 pp. (1981).

003.052   Astronomia do Macunaíma.
R. R. de Freitas Mourão.
Livraria Francisco Alves Editora S.A., Rua Sete de Setembro,
177, Centro 20050, Rio de Janeiro, Brazil. 85 pp. (1984).

003.053   Explicando o Cosmos. Astronomia ao Seu Alcance.
R. R. de Freitas Mourão.
Editora Tecnoprint S.A., Rio de Janeiro, Brazil. 109 pp. (1984).

**003.054   Mighty is the charm. Lectures on science, literature, and
the arts.**
J. C. Albergotti.
University Press of America, Washington, D.C., USA.
4 + 236 pp. Price US$ 20.95 cloth, US$ 10.95 paper (1982).
Review in J. Hist. Astron., Vol. 16, Part 2, No. 46, p. 152 (1985).

003.055   The sky in X–ray emission.
P. R. Amnuehl'.
Nauka, Glavnaya Redaktsiya Fiziko–Matematicheskoj Litera-
tury, Moskva. 224 pp. Price 80 Kop. (1984). In Russian.
Review in Priroda, No. 2, p. 123 (1985).

**003.056   Springs of scientific creativity: founders of modern sci-
ence.**
R. Aris.
University of Minnesota Press, Minn., USA. 342 pp. Price
US$ 32.50 (1983).
Review in Mercury, Vol. 14, No. 3, p. 90 (1985).

**003.057   The astronomical scrapbook: skywatchers, pioneers, and
seekers in astronomy.**
J. Ashbrook, edited by L. J. Robinson.
Sky Publishing Corporation, 49 Bay State Road, Cambridge,
Mass. 02238–1290, USA and Cambridge University Press, Cam-
bridge – London – New York – New Rochelle – Melbourne –
Sydney. 12 + 468 pp. Price $ 19.95 (1984).
Review in Sci. Am., Vol. 252, No. 6, p. 21 – 22 (1985).

**003.058   The night F region of the ionosphere in the period of
flares on the sun.**
S. V. Avakyan, V. V. Kovalenok, N. F. Solonitsyna.
Nauka, Alma–Ata. 150 pp. (1984). In Russian.
Review in Ref. Zh., 51. Astron., 6.51.45 (1985).

003.059   Kinematics.
J. S. Beggs.
Hemisphere, P.C., Washington and Springer Verlag, Berlin –
Heidelberg – New York – Tokyo. 16 + 223 pp. Price DM 64.00
(1983).
Review in Astron. Nachr., Vol. 306, No. 2, p. 62; 1985
(*H.–J. Schmidt*).

**003.060   The celebrated phaenomena of colours: the early history
of the spectroscope.**
J. A. Bennett.
Whipple Museum of the History of Science, Cambridge, En-
gland. 21 pp. Price £ 1.00 (1984).
Review in J. Hist. Astron., Vol. 16, Part 2, No. 46, p. 153 (1985).

003.061 Astrophysics of cosmic rays.
V. S. Berezinskij, S. V. Bulanov, V. L. Ginzburg,
V. A. Dogel', V. S. Ptuskin.
Nauka, Moskva. 358 pp. (1984). In Russian.
Review in Ref. Zh., 51. Astron., 1.51.52 (1985).

003.062 Three degrees above zero.
J. Bernstein.
Scribner's, New York, USA. 241 pp. Price US$ 17.95 (1984).
ISBN 0–684–18170–3.
Review in Sky Telesc., Vol. 69, No. 3, p. 229 (1985).

003.063 Ferdinand Verbiest, als Oost en West elkaar ontmoeten.
R. A. Blondeau.
Editions Lannoo, Tielt, Belgium. 143 pp. Price BF 495 (1983).
Review in Ciel Terre, Vol. 101, No. 1, p. 23; 1985 (*J. Denoyelle*).

003.064 Astronomy before the telescope, Vol. 1: The Earth–Moon system.
N. T. Bobrovnikoff.
Pachart Publishing House, P.O. Box 35549, Tucson, Ariz. 85740,
USA. 153 pp. Price US$ 38.00 (1984). ISBN 0–88126–201–3.
Review in Sky Telesc., Vol. 69, No. 5, p. 422 (1985).

003.065 Beyond the black hole.
J. Boslough.
Collins, London, England. 127 pp. Price £ 7.95 (1985). ISBN
0–00–217138–4.
From J. Br. Astron. Assoc., Vol. 95, No. 3, p. 139 (1985).

003.066 Celestial navigation.
T. Bottomley.
TAB Books, 241 pp. Price US$ 12.95 paper (1983).
Review in Mercury, Vol. 14, No. 3, p. 94 (1985).

003.067 Astrophysics. Vol. I: Stars. Vol. II: Interstellar matter and galaxies.
R. L. Bowers, T. Deeming.
Jones and Bartlett Publishers, Inc., 20 Park Plaza, Boston, Mass.
02116, USA. Vol. I: 344 pp., Vol. II: 244 pp. Price US$ 37.50,
£ 17.95 each volume (1984). ISBN 0–86720–018–9 (Vol. I), ISBN
0–86720–047–2 (Vol. II).
Reviews in Nature, Vol. 314, No. 6006, p. 24; 1985 (*R. C. Smith*);
Sky Telesc., Vol. 69, No. 1, p. 34 – 35; 1985 (*J. A. Morgan*).

003.068 Aeronomy of the middle atmosphere.
G. Brasseur, S. Soloman.
D. Reidel Publishing Company, Dordrecht – Boston – Lancaster.
16 + 441 pp. Price US$ 44.00 (1984).
Review in Planet. Space Sci., Vol. 33, No. 4, p. 469; 1985
(*D. R. Bates*).

003.069 Die Pfeile der Zeit. Über das Fundamentale in der Natur.
R. Breuer.
Meyster Verlag, München, F.R. Germany. 180 pp. Price
DM 22.00 (1984). ISBN 3–81341–8133–2.
Review in Sterne Weltraum, 24. Jahrg., Nr. 6, p. 354; 1985
(*C. Leinert*).

003.070 Nomenclature of the details of the relief of the Galilean satellites of Jupiter.
G. A. Burba.
Nauka, Moskva. 87 pp. (1984). In Russian.
Review in Ref. Zh., 51. Astron., 2.51.36 (1985).

003.071 Telescopes for the 1980s.
G. Burbidge, A. Hewitt (Editors). Translated from the
English edition by A. A. Tokovinin and edited by Eh. A. Dibaj.
Mir, Moskva. 312 pp. Price 3 Rbl. 30 Kop. (1984). In Russian.
For the original see 32.003.010.
Review in Priroda, No. 4, p. 123 (1985).

003.072 Astrophysics today: readings from Physics Today, No. 1.
A. G. W. Cameron (Editor).
American Institute of Physics, 335 E. 45th St., New York, N.Y.
10017, USA. 337 pp. Price US$ 25.00 (1984). ISBN
0–88318–446–X.
Review in Sky Telesc., Vol. 69, No. 2, p. 132 (1985).

003.073 The story of the earth.
P. Cattermole, P. Moore.
Cambridge University Press, Cambridge – London – New York
– New Rochelle – Melbourne – Sydney. 224 pp. Price £ 12.95
(1985). ISBN 0–521–26292–5.
From J. Br. Astron. Assoc., Vol. 95, No. 3, p. 139 (1985).

003.074 The comet book: a guide for the return of Halley's comet.
R. D. Chapman, J. C. Brandt.
Jones and Bartlett Publishers, Inc., Boston, USA. 168 pp. Price
US$ 15.95 (1984).
Review in Strolling Astron., Vol. 31, Nos. 1 – 2, p. 39 – 41; 1985
(*J. V. Scotti*).

003.075 Stars and physics.
A. D. Chernin.
Nauka, Moskva. 159 pp. (1984). In Russian.
Review in Ref. Zh., 51. Astron., 5.51.30 (1985).

003.076 In the presence of the creator: Isaac Newton and his times.
G. E. Christianson.
The Free P., New York, USA. 623 pp. Price $ 27.50 (1984).
From Phys. Today, Vol. 38, No. 1, p. 104 (1985).

003.077 Starfinder.
H. Couper.
Century Publishing, London, England. 11 + 144 pp. Plus soft-
ware pack for BBC/Model B and Electron Computers. Price
£ 12.95 (1984). ISBN 0–7126–0571–1.
From J. Br. Astron. Assoc., Vol. 95, No. 2, p. 87 (1985).

003.078 Andra världar.
P. Davies.
Akademilitteratur,   Stockholm,   Sweden.   208 pp.   Price
Sv.kr. 124.00 (1984).
Review in Astron. Tidsskr., Årg. 18, Nr. 2, p. 88 – 89 (1985).

003.079 Superforce. The search for a Grand Unified Theory of nature.
P. Davies.
Heinemann, London, England; Simon & Schuster, New York,
USA. 255 pp. Price £ 12.95, US$ 16.95 (1984). ISBN
0–434–17700–8.
Reviews in Mercury, Vol. 14, No. 3, p. 91 – 92 (1985); Sterne
Weltraum, 24. Jahrg., Nr. 6, p. 352; 1985 (*J. Schmid-Burgk*).

003.080 Genesis on planet earth.
W. Day.
Yale University Press, New Haven, Conn., USA. 299 pp. Price
$ 35.00 cloth, $ 12.95 paper (1984). ISBN 0–300–02954–3 cloth;
ISBN 0–300–03202 paper.
Review in Sky Telesc., Vol. 69, No. 5, p. 420 – 422; 1985
(*M. K. Hobish*).

003.081 The brightest stars.
C. de Jager, translated by L. I. Antipova,
L. R. Yungel'son and V. S. Avedisova, edited by Eh. R. Mustel'
and V. S. Strel'nitskij.
Mir, Moskva. 493 pp. Price 5 Rbl. 60 Kop. (1984). For the En-
glish original see 28.003.060.
Review in Priroda, No. 3, p. 125 (1985).

003.082 The hidden universe.
M. Disney.
J. M. Dent, London, England; Macmillan Publishing Co., Inc., New York, USA. 216 pp. Price £ 10.95, US$ 16.95 (1984). ISBN 0–460–04441–9.
Reviews in Nature, Vol. 313, No. 6000, p. 329; 1985 (*G. Efstathiou*); Observatory, Vol. 105, No. 1066, p. 102 – 103; 1985 (*J. Tomkin*).

003.083 The dissolution of the celestial spheres, 1595 – 1650.
W. H. Donahue.
Arno Press, New York, USA. 10 + 330 pp. Price US$ 28.00 (1981).
Review in J. Hist. Astron., Vol. 16, Part 1, No. 45, p. 54 – 55; 1985 (*O. Gingerich*).

003.084 History of the Earth's crust.
D. L. Eicher, A. L. McAlester, M. L. Rottman.
Prentice–Hall, Inc., Englewood Cliffs, N.J. 07632, USA. 197 pp. (1984). ISBN 0–13–389999–3.
From Phys. Abstr., Vol. 88, No. 1252, Entry 24649 (1985).

003.085 Unser Himmel.
J. Ekrutt.
C. Bertelsmann Jugendbuch, München, F.R. Germany. 70 pp. Price DM 19.80 (1983). ISBN 3–570–08129–X.
Review in Sterne Weltraum, 24. Jahrg., Nr. 2, p. 112; 1985 (*M. Sarcander*).

003.086 Rings: discoveries from Galileo to Voyager.
J. Elliot, R. Kerr.
The MIT Press, Cambridge, Mass., USA. 209 pp. Price $ 17.50, £ 19.50 (1985).
Reviews in Nature, Vol. 315, No. 6022, p. 781; 1985 (*C. D. Murray*); Sci. Am., Vol. 252, No. 2, p. 19 – 20; 1985 (*P. Morrison*).

003.087 On Mars: exploration of the red planet 1958 – 1978.
E. C. Ezell, L. Neuman Ezell.
NASA, Washington, D.C., USA. 535 pp. (1984).
From Nature, Vol. 315, No. 6018, p. 437 (1985).

003.088 Space shots: the beauty of nature beyond earth.
T. Ferris.
Pantheon Books, New York, USA. 143 pp. Price US$ 24.95 (1984). ISBN 0–394–53890–0.
Review in Sky Telesc., Vol. 69, No. 4, p. 325 (1985).

003.089 The invisible universe. Probing the frontiers of astrophysics.
G. B. Field, E. J. Chaisson.
Birkhäuser Verlag, Basel – Boston – Stuttgart. 195 pp. Price US$ 19.95 (1985). ISBN 0–8176–3235–2.
Review in Sky Telesc., Vol. 69, No. 5, p. 422 (1985).

003.090 Universe in the classroom: a resource guide for teaching astronomy and instructor's manual for Universe.
A. Fraknoi, W. J. Kaufmann III.
W. H. Freeman & Co., New York – San Francisco – Oxford. 8 + 269 pp. Price US$ 8.95 (1984). ISBN 0–7167–1692–5.
From Nature, Vol. 315, No. 6018, p. 437 (1985); Science, Vol. 228, No. 4704, p. 1192 (1985).

003.091 Probability, statistical optics, and data testing.
B. R. Frieden.
Springer-Verlag, Berlin – Heidelberg – New York –Tokyo. 17 + 404 pp. Price DM 94.00 cloth (1983). ISBN 3–540–11769–5.
Review in Astron. Nachr., Vol. 306, No. 1, p. 42; 1985 (*G. Möstl*).

003.092 The creation of matter: the universe from beginning to end.
H. Fritzsch.
Basic Books, New York, USA. 307 pp. Price US$ 19.95 (1984).
For the German original see 34.003.068.
From Phys. Today, Vol. 38, No. 3, p. 124 (1985).

003.093 Der Halleysche Komet.
R. Froböse.
Verlag Harri Deutsch, Thun – Frankfurt. 131 pp. Price SFr. 15.70. ISBN 3–87144–8370.
Review in Orion, 43. Jahrg., Nr. 208, p. 92; 1985 (*A. von Rotz*).

003.094 Handbook of plasma physics. Vol. 1: Basic plasma physics I.
A. A. Galeev, R. N. Sudan (Editors).
North–Holland Publishing Company, Amsterdam – New York – Oxford. 19 + 751 pp. Price US$ 180.75, Dfl 425.00 (1983).
Review in Astron. Nachr., Vol. 306, No. 3, p. 170; 1985 (*N. Seehafer*).

003.095 Novità celesti e crisi del sapere.
P. Galluzzi (Editor).
Supplemento agli Annali dell'Istituto e Museo di Storia della Scienza, Monografia N. 7, Florence, Italy. 6 + 448 pp. (1983).
Review in J. Hist. Astron., Vol. 16, Part 2, No. 46, p. 152 (1985).

003.096 Wave phenomena in the ionosphere and in cosmic plasma.
B. N. Gershman, L. M. Erukhimov, Yu. Ya. Yashin.
Nauka, Moskva. 392 pp. (1984). In Russian.
Review in Ref. Zh., 62. Issled Kosm. Prostranstva, 1.62.89 (1985).

003.097 Black holes, quasars, and other mysteries of the universe.
S. Gibilisco.
TAB Books, 197 pp. Price US$ 13.50 (1984).
Review in Mercury, Vol. 14, No. 3, p. 89 (1985).

003.098 Kosmisk resa.
B. Gustafsson.
LiberFörlag, Stockholm, Sweden. 128 pp. Price Sv.kr. 55.00 (1984).
Review in Astron Tidsskr., Årg. 18, Nr. 2, p. 89 (1985).

003.099 Friedrich Wilhelm Bessel.
J. Hamel.
Biographien hervorragender Naturwissenschaftler, Techniker und Mediziner, Band 67. BSB B. G. Teubner Verlagsgesellschaft, Leipzig, GDR. 98 pp. Price M 6.80 (1984).
Reviews in Astron. Raumfahrt, 23. Jahrg., Heft 1, p. 24; 1985 (*W. R. Dick*); Sky Telesc., Vol. 69, No. 6, p. 519 (1985); Sterne Weltraum, 28. Jahrg., Nr. 4, p. 232; 1985 (*B. Wöbke*).

003.100 Out of the cradle: exploring the frontiers beyond earth.
W. Hartmann, R. Miller, P. Lee.
Workman Publishing Co., New York, USA. 190 pp. Price $ 11.95 (1984). ISBN 0–89480–770–6.
Review in Sky Telesc., Vol. 69, No. 4, p. 320 – 321; 1985 (*J. Lomberg*).

003.101 Oriens–Occidens II.
W. Hartner, edited by Y. Maeyama.
Georg Olms Verlag, Hildesheim, F.R. Germany. 44 + 423 pp. Price DM 118.00 (1984).
Review in J. Hist. Astron., Vol. 16, Part 2, No. 46, p. 138 – 140; 1985 (*O. Gingerich*).

003.102 Les cinq paliers – comment s'est modifiée notre lecture de l'Univers.
G. Hawkins, translated by J. Guiod.
Editions Mazarine, Paris, France. 336 pp. FB 864 (1985).
Review in Ciel Terre, Vol. 101, No. 3, p. 97; 1985 (*O. Godart*).

003.103 Array signal processing.
S. Haykin, J. H. Justice, N. L. Owsley, J. L. Yen, A. C. Kak (Editors).
Prentice–Hall, Inc., Englewood Cliffs, N.J. 07632, USA. 9+433 pp. (1985). ISBN 0–13–046482–1.
Review in Phys. Abstr., Vol. 88, No. 1256, Entry 46339 (1985).

003.104 Edmond Halley: the man and his comet.
B. H. Heckart.
Childrens Press, Chicago, Ill., USA 111 pp. Price US\$ 8.95 (1984).
Review in Space Educ., Vol. 1, No. 9, p. 432 (1985).

003.105 Islamic cosmology: a study of as–Suyūti's al–Hay'a as–saniya fi l'hay'a as–sunniya.
A. M. Heinen.
Beiruter Texte und Studien, Band XXVII. Franz Steiner Verlag, Beirut, Lebanon. 8+289+81 pp. Price DM 78.00 (1982). In Arabic.
Review in J. Hist. Astron., Vol. 16, Part 2, No. 46, p. 146 – 149; 1985 (*F. J. Ragep*).

003.106 Die neue Astronomie.
N. Henbest, M. Marten.
Birkhäuser Verlag, Basel – Boston – Stuttgart. 240 pp. Price SFr. 60.00 (1984). ISBN 3 7643–1616–0. For the English original see 37.003.094.
Review in Orion, 43. Jahrg., Nr. 206, p. 31; 1985 (*A. Tarnutzer*).

003.107 Satellite sensing of a cloudy atmosphere: observing the third planet.
A. Henderson–Sellers (Editor).
Taylor and Francis Ltd., London – Philadelphia. 336 pp. Price \$ 18.00 paper, \$ 40.00 cloth (1984).
Review in Nature, Vol. 313, No. 5998, p. 165; 1984 (*J. T. Houghton*).

003.108 A hilltop in foggy bottom.
J. Herman.
U.S. Government Printing Office, Washington, D.C. 20402, USA. 84 pp. Price US\$ 3.00 (1984).
Review in Mercury, Vol. 14, No. 3, p. 90 (1985).
A brief, illustrated history of the U.S. Naval Observatory from 1755 through the beginning of the 20th century.

003.109 Přehled astronomie.
O. Hlad, J. Pavlousek.
Státní nakladatelství technické literatury, Praha, Czechoslovakia. 400 pp. Price Kčs 45.00 (1984).
Reviews in Říše hvězd, Vol. 66, No. 5, p. 96 – 97; 1985 (*J. Švestka*), Vesmír, Vol. 64, No. 7, p. 385; 1985 (*J. Bouška*).

003.110 G. J. Rheticus' treatise on holy scripture and the motion of the earth.
R. Hooykaas.
North–Holland Publishing Company, Amsterdam, The Netherlands. 188 pp. (1983).
Review in Astronomie, Vol. 99, p. 211 (1985).

003.111 Le destin ultime de l'univers.
J. N. Islam.
Belfond Sciences. 177 pp. Price FF 89.00 (1984). For the English original see 33.003.050.
Review in Astronomie, Vol. 99, p. 47 (1985).

003.112 The birth of history and philosophy of science: Kepler's "A defence of Tycho against Ursus" with essays on its provenance and significance.
N. Jardine.
Cambridge University Press, Cambridge – London – New York – New Rochelle – Melbourne – Sydney. 9+301 pp. Price \$ 59.50, £ 32.50 (1984). ISBN 0–521–25226–1.
Review in Sky Telesc., Vol. 69, No. 2, p. 132 (1985); Space Sci. Rev., Vol. 39, Nos. 3/4, p. 377; 1984 (*E. G. Forbes*).

003.113 Astronomy laboratory manual.
M. Kafatos.
Kendall/Hunt Publishing Co., PO Box 539, Dubuque, Iowa 52001, USA. 83 pp. Price US\$ 11.95 (1984). ISBN 0–8403–3405–2.
Review in Sky Telesc., Vol. 69, No. 2, p. 132 (1984).

003.114 Galassie e quasars.
W. J. Kaufmann III.
Sansoni, Firenze, Italy. Price L 20,000 (1984). For the English original see 27.003.086.
Review in Coelum, Vol. 53, N. 3, p. 171; 1985 (*F. Gabici*).

003.115 Universe.
W. J. Kaufmann III.
W. H. Freeman & Co., San Francisco – Oxford. 16+594 pp. Price £ 29.95 (1985).
Review in Astron. Express, Vol. 1, Nos. 4 – 6, p. 162 (1985).

003.116 Studies in the Islamic exact sciences.
E. S. Kennedy.
American University of Beirut, Beirut, Lebanon. 16+771 pp. Price \$ 80.00 (1983).
Review in J. Hist. Astron., Vol. 16, Part 1, No. 45, p. 49 – 51; 1985 (*P. Kunitzsch*).

003.117 Mathematical astronomy in Medieval Yemen. A bibliographical survey.
D. A. King.
American Research Center in Egypt Catalogs, Vol. IV. Udena Publications, Malibu, Calif., USA. 14+98 pp. +10 plates. Price US\$ 23.00 cloth; US\$ 16.00 paper (1983).
Review in J. Hist. Astron., Vol. 16, Part 1, No. 45, p. 73 (1985).

003.118 Al–Khwarizmi and new trends in mathematical astronomy in the ninth century.
D. A. King.
Hagop Kevorkian Center for Near Eastern Studies, Occasional Papers of the Near East, 2. New York University, New York 10003, USA. 43 pp. Price US\$ 3.50 (1983).
From J. Hist. Astron., Vol. 16, Part 1, No. 45, p. 73 (1985).

003.119 Relativistic astronomy.
I. A. Klimishin. Translated from Ukrainian and edited by V. S. Imshennik.
Nauka, Moskva. 208 pp. Price 35 Kop. (1983). In Russian.
Review in Astron. Zh., Tom 62, Vyp. 2, p. 413; 1985 (*A. D. Chernin*).

003.120 Shock waves in stellar envelopes.
I. A. Klimishin.
Nauka, Moskva. 215 pp. (1984). In Russian.
Review in Ref. Zh., 51. Astron., 2.51.40 (1985).

003.121 Vesmírní sousedé naší planety.
Z. Kopal.
Academia, Praha, Czechoslovakia. 228 pp. Price Kčs 38.00 (1984).

003.122 Introduction to solar radio astronomy and radio physics.
A. Krüger.
Moskva. 469 pp. (1984). In Russian. For the original see 26.003.078.
Review in Ref. Zh., 51. Astron., 5.51.28 (1985).

003.123 Computational methods in bifurcation theory and dissipative structures.
M. Kubicek, M. Marek.
Springer Series in Computational Physics. Springer Verlag, New York – Berlin – Heidelberg – Tokyo. 11+243 pp. Price DM 108.00 (1983).
Review in Astron. Nachr., Vol. 306, No. 2, p. 62; 1985 (*H.–J. Schmidt*).

003.124   The Universe of motion.
D. B. Larson.
The Structure of the Physical Universe, Vol. 3. 3rd edition. North Pacific Publishers, Portland, Ore., USA. 8+456 pp. Price US$ 19.00 (1984).
From Science, Vol. 228, No. 4704, p 1192 (1985).

003.125   Introduction élémentaire à la physique cosmique et à la physique des relations Soleil–Terre.
J. L. Legrand.
Editions Territoire des terres australes et antarctiques françaises. 306 pp. (1984).
Review in Ciel Terre, Vol. 101, No. 3, p. 98; 1985 (*J.-C. Jodogne*).

003.126   Astro–navigation by calculator.
H. Levison.
David & Charles, Newton Abbot, England. 112 pp. Price £ 7.95 (1984).
Review in Observatory, Vol. 105, No. 1066, p. 102; 1985 (*E. N. Walker*).

003.127   High energy astrophysics.
M. S. Longair.
Mir, Moskva. 396 pp. (1984). In Russian. For the English original see 30.003.093.
Review in Ref. Zh., 51. Astron., 3.51.63 (1985).

003.128   Quantitative aspects of magnetospheric physics.
L. R. Lyons, D. J. Williams.
D. Reidel Publishing Company, Dordrecht – Boston – Lancaster. 15+231 pp. (1984). ISBN 90–277–1663–3.
Review in Phys. Abstr., Vol. 88, No. 1254, Entry 35934 (1985).

003.129   La cometa di Halley.
P. Maffei.
Edizioni Scientifiche e Tecniche Mondadori. Price L 25,000.
Reviews in Coelum, Vol. 53, N. 1, p. 55 – 56; 1985 (*G. Romano, L. Candiano*); Coelum, Vol. 53, N. 2, p. 121; 1985 (*F. Gabici*).

003.130   Historically–astronomical investigations. Vypusk XVII.
L. E. Majstrov (Editor).
Nauka, Moskva. 462 pp. (1984). In Russian.
Review in Priroda, No. 4, p. 120 – 122; 1985 (*S. V. Zhitomirskij*).

003.131   Numerical solutions of the N–body problem.
A. Marciniak.
Mathematics and Its Applications (*East European Series*). D. Reidel Publishing Company, Dordrecht – Boston – Lancaster. 11+242 pp. Price Dfl. 120.00, US$ 39.50, £ 33.25 (1985). ISBN 90–277–2058–4.
Devoted to the study of numerical methods for solving the general N–body problem and related problems, the present volume starts with an overview of the conventional numerical methods for solving the initial value problem. The major part of the book contains original work and features a presentation of special numerical methods conserving the constants of motion in the general N–body problem and methods conserving the Jacobi constant in the problem of motion of N bodies in a rotating frame, as well as an analysis of the applications of both (conventional and special) kinds of methods for solving these problems. For all the methods considered, the author presents algorithms which are easily programmable in any computer language.

003.132   The Galaxy.
L. S. Marochnik, A. A. Suchkov.
Nauka, Moskva. 392 pp. (1984). In Russian.
Review in Ref. Zh., 51. Astron., 2.51.41 (1985).

003.133   Starwatch.
B. Mayer.
Perigee Books, Putnam Publishing Group, New York, USA; Stoddart Publications, Don Mills, Ont., Canada. 144 pp. Price $ 15.95 cloth, $ 7.95 paper (1984). ISBN 0–399–51008–7 cloth, ISBN 0–399–51009–5 paper.

Review in Sky Telesc., Vol. 69, No. 2, p. 130 – 131; 1985 (*G. Lovi*).

003.134   Carl Friedrich Gauss: a bibliography.
Compiled by U. C. Merzbach.
Scholarly Resources, Wilmington, Delaware, USA. 26+551 pp. Price US$ 95.00 (1984).
Review in J. Hist. Astron., Vol. 16, Part 2, No. 46, p. 153 (1985).

003.135   The earth and its rotation.
A. A. Mikhajlov.
Nauka, Glavnaya Redaktsiya Fiziko–Matematicheskoj Literatury, Moskva. 80 pp. Price 15 Kop. (1984). In Russian.
Review in Priroda, No. 2, p. 122 – 123 (1985).

003.136   Le grand tour, voyage à travers le système solaire.
R. Miller, W. K. Hartmann, translated by C. Magnan, B. Magnan.
Editions Robert Laffot, Paris, France. 190 pp. Price F 100.00 (1983). For the English original see 31.003.092.
Review in Astronomie, Vol. 99, p. 157; 1985 (*A. Soubeyran*).

003.137   Astronomy today.
D. Moché, illustrated by H. McNaught.
Kingfisher, London, England. 96 pp. Price £ 5.95 (1984).
Review in Observatory, Vol. 105, No. 1064, p. 16 – 17; 1985 (*J. Mitton*).

003.138   1985 yearbook of astronomy.
P. Moore (Editor).
W. W. Norton and Co., 500 Fifth Avenue, New York, N.Y. 10036, USA. 208 pp. Price US$ 9.95 (1984). ISBN 0–393–30203–2.
From Nature, Vol. 315, No. 6018, p. 437 (1985); Sky Telesc., Vol. 69, No. 3, p. 230 (1985); Strolling Astron., Vol. 31, Nos. 1 – 2, p. 44 (1985).

003.139   En färd i tid och rum.
P. Moore.
Bonnier–Fakta, Stockholm, Sweden, 192 pp. Price Sv.kr. 135.00 (1984).
Review in Astron. Tidsskr., Årg. 18, Nr. 2, p. 89 (1985).

003.140   Het groot Guinness astronomie boek.
P. Moore.
Luitingh, Utrecht, The Netherlands. 248 pp. Price *f*. 34.75 (1984). ISBN 90–245–0805–3.
Review in Zenit, 12. Jaarg., No. 3, p. 111; 1985 (*M. Drummen*).

003.141   The new atlas of the Universe.
P. Moore.
Crown Publishers, Inc., New York, USA; Mitchell Beazley, London, England. 271 pp. Price US$ 40.00, £ 24.95 (1984). ISBN 0–517–55500–X (USA), ISBN 0–85533–537–8 (England).
Review in Sky Telesc., Vol. 69, No. 6, p. 517 – 519; 1985 (*L. A. Marshall*).

003.142   Patrick Moore's armchair astronomy.
P. Moore.
Patrick Stephens, Dennington Estate, Wellingborough, Northants NNB 2QD, England. 185 pp., Price £ 9.95 (1984). ISBN 0–85059–718–8.
Review in J. Br. Astron. Assoc., Vol. 95, No. 2, p. 86; 1985 (*I. Nicolson*).

003.143   Methods of extra–atmospheric astronomy.
E. I. Moskalenko.
Nauka, Moskva. 280 pp. (1984). In Russian.
Review in Ref. Zh., 51. Astron., 2.51.43 (1985).

003.144   Design and origins in astronomy.
G. Mulfinger Jr. (Editor).
Creation Research Society Books, Norcross, Georgia, USA.
152 pp. Price US$ 7.50 (1983).
Review in J. R. Astron. Soc. Can., Vol. 79, No. 1, p. 44 – 45; 1985
(*H. L. Armstrong*).

003.145   The moon's acceleration and its physical origins. Vol. 2:
As deduced from general lunar observations.
R. R. Newton.
Johns Hopkins University Press. Baltimore – London. 320 pp.
Price US$ 35.00 (1984). ISBN 0–8018–2639–X.
Reviews in Johns Hopkins APL Tech. Dig., Vol. 6, No. 2,
p. 176 – 179; 1985 (*S. J. Goldstein Jr.*); Sky Telesc., Vol. 69,
No. 1, p. 36 (1985).

003.146   Foundations of the theory of spectra of atoms and ions.
A. A. Nikitin, Z. B. Rudzikas.
Nauka, Moskva. 320 pp. Price 3 Rbl. 80 Kop. (1983). In Russian.
Review in Astron. Zh., Tom 62, Vyp. 3, p. 614 – 615; 1985
(*A. A. Sapar, T. Kh. Feklistova*).

003.147   Television astronomy.
V. B. Nikonov (Editor).
2nd revised and extended edition. Nauka, Moskva. 272 pp.
(1984). In Russian.
Review in Ref. Zh., 51. Astron., 2.51.44 (1985).

003.148   William Rowan Hamilton. Portrait of a prodigy.
S. O'Donnell.
Boole Press, Dublin, Ireland. 16 + 224 pp. Price US$ 24.95
(1983).
Review in J. R. Astron. Soc. Can., Vol. 79, No 1, p. 30 – 31; 1985
(*A. A. Barrett*).

003.149   Radiation of atoms in cosmic plasma.
I. M. Ojringel', E. B. Klejman.
Nauka, Novosibirsk. 136 pp. (1984). In Russian.
Review in Ref. Zh., 62. Issled. Kosm. Prostranstva, 6.62.58
(1985).

003.150   Contemporary astronomy.
J. M. Pasachoff.
Third edition. Saunders, New York. 477 pp. Price US$ 22.95
(1985).
From Phys. Today, Vol. 48, No. 6, p. 80 (1985).

003.151   Spinors and space–time. Vol. 1: Two–spinor calculus and
relativistic fields.
R. Penrose, W. Rindler.
Cambridge University Press, Cambridge – London – New York
– New Rochelle – Melbourne – Sydney. 458 pp. Price £ 45.00,
$ 89.50 (1984).
Review in Nature, Vol. 313, No. 6003, p. 607 – 608 (1985).

003.152   Ursprung und Zukunft des Weltalls. Pflanzen – Tiere –
Menschen.
J. Pfleiderer (Editor).
Pinguin–Verlag, Innsbruck, Austria. 180 pp. Price öS 385.00
(1983).
Review in Sterne, 61. Band, Heft 2, p. 122; 1985 (*H. Sanke*).

003.153   Les Tables Alphonsines avec les Canons de Jean de
Saxe.
E. Poulle (Editor).
Editions du Centre National de la Recherche Scientifique, Paris,
France. 246 pp. (1984).
Review in J. Hist. Astron., Vol. 16, Part 1, No. 45, p. 51 – 52;
1985 (*J. D. North*).

003.154   Cosmic quest.
M. Poynter, M. J. Klein.
Atheneum Publishers, New York, USA. 124 pp. Price US$ 10.95
(1984). ISBN 0–689–31068–4.
Review in Sky Telesc., Vol. 69, No. 4, p. 325 (1985).

003.155   Cadrans du Soleil; les cadrans peints des Alpes à la
Méditerranée.
P. Ricou, J. M. Homet.
Editions Jeanne Laffitte, Paris, France. 103 pp. Price FF 185,
FB 1.136 (1984).
Review in Ciel Terre, Vol. 101, No. 3, p. 97; 1985 (*J. Donner*).

003.156   Collins guide to stars and planets.
I. Ridpath, W. Tirion.
Collins, London, England. 384 pp. Price £ 8.95 (1984). ISBN
0–00–219071–0.
Review in J. Br. Astron. Assoc., Vol. 95, No. 4, p. 186; 1985
(*R. Knox*).

003.157   Handbuch der Astronomie. Beobachten und verstehen.
I. Ridpath.
Verlag Sauerländer, Aarau – Frankfurt – Salzburg. 224 pp. Price
DM 38.00 (1985).
Review in Sterne Weltraum, 24. Jahrg., Nr. 6, p. 355; 1985
(*G. D. Roth*).

003.158   The center of the Galaxy.
G. Riegler, R. Blandford (Editors).
Translated from the English edition. Mir, Moskva. 271 pp.
(1984). In Russian.
Review in Ref. Zh., 51. Astron., 5.51.31 (1985).

003.159   Le comete.
M. Rigutti.
Rizzoli Editore, Milano, Italy. 166 pp. Price L 32,000 (1984).
Reviews in Coelum, Vol. 53, N. 1, p. 56 – 57; 1985 (*L. Baldinelli*);
Coelum, Vol. 53, N. 2, p. 121; 1985 (*F. Gabici*).

003.160   Indirect imaging: measurement and processing for indi-
rect imaging.
J. A. Roberts (Editor).
Cambridge University Press, Cambridge – London – New York
– New Rochelle – Melbourne – Sydney. 442 pp. Price $ 54.50
(1984). ISBN 0–521–26282–8.
Review in Sky Telesc., Vol. 69, No. 2, p. 132 (1985).

003.161   Scientific writings and astronomical tables in Cracow: a
census of manuscript sources (fourteenth to sixteenth
centuries).
G. Rosinska.
Studia Copernicana, XXII. Ossolineum, Wroclaw, Poland.
561 pp. (1984).
Review in J. Hist. Astron., Vol. 16, Part 2, No. 46, p. 134 – 138;
1985 (*E. Poulle*).

003.162   All the astrolabes.
H. L. Saunders.
Seneico Publishing Co., Oxford, England. 6 + 101 pp. Price
£ 15.95 (1985). ISBN 0–906831–04–0.
From J. Br. Astron. Assoc., Vol. 95, No. 3, p. 139 (1985).

003.163   Wonders of the sky.
F. Schaaf.
Dover Publications, Inc., Mineola, N.Y., USA. 299 pp. Price
US$ 6.95 (1983).
Review in Mercury, Vol. 14, No. 3, p. 88 (1985).

003.164   Tychonic and semi–tychonic world systems.
C. J. Schofield.
Arno Press, New York, USA. 8 + 398 + 11 pp. Price US$ 35.00
(1981).
Review in J. Hist. Astron., Vol. 16, Part 1, No. 45, p. 54 – 55;
1985 (*O. Gingerich*).

003.165   A first course in general relativity.
B. F. Schutz.
Cambridge University Press, Cambridge – London – New York
– New Rochelle – Melbourne – Sydney. 14 + 376 pp. Price
£ 30.00, $ 54.50 cloth, £ 9.95, $ 19.95 paper (1985). ISBN
0–521–25770–0 cloth, ISBN 0–521–27703–5 paper.

Reviews in Astron. Express, Vol. 1, Nos. 4 – 6, p. 159 (1985); Astron. Tidsskr., Årg. 18, Nr. 1, p. 48 (1985); Nature, Vol. 314, No. 6006, p. 22 – 23; 1985 (*J. D. Barrow*); Space Educ., Vol. 1, No. 9, p. 432 (1985).

003.166   Horizons: exploring the universe.
M. A. Seeds.
Second edition. Wadsworth Publishing Co., 10 Davis Dr., Belmont, Calif. 94002, USA. 496 pp. Price US$ 25.95 (1985). ISBN 0–534–04017–9.
Review in Sky Telesc., Vol. 69, No. 6, p. 520 (1985).

003.167   Sternbilder – Blicke in den Nachthimmel.
A. Sfountouris.
Ex Libris Verlag und Grammoclub AG, Zürich, Switzerland. 112 pp. Price SFr. 38.00 (1984).
Review in Sterne Weltraum, 24. Jahrg., Nr. 5, p. 292; 1985 (*M. Sarcander*).

003.168   Precision observations of cosmic rays in Yakutsk.
G. V. Shafer, Yu. G. Shafer.
Nauka, Novosibirsk, 733 pp. (1984). In Russian.
Review in Ref. Zh., 51. Astron., 2.51.37 (1985).

003.169   The cosmic cycle.
T. P. Snow.
Darwin Press, Box 2202, Princeton, N.J. 08540, USA. 109 pp. Price US$ 19.95 (1984). ISBN 0–87850–041–3.
Review in Sky Telesc., Vol. 69, No. 3, p. 229 (1985).

003.170   Maps of the heavens.
G. S. Snyder.
André Deutsch, London – Abbeville – New York. 144 pp. Price £ 19.95, US$ 45.00 (1984).
Review in Nature, Vol. 314, No. 6008, p. 296; 1985 (*C. Stott*).

003.171   Matter of meteorites.
Eh. V. Sobotovich, V. P. Semenenko.
Naukova dumka, Kiev. 191 pp. (1984). In Russian.
Review in Ref. Zh., 51. Astron., 4.51.31 (1985).

003.172   Stars of jade.
D. W. Staal.
Writ Press, P. O. Drawer R, Decatur, Ga. 30031–2244, USA. 204 pp. Price US$ 19.95 (1984). ISBN 0–914635–00–8.
Review in Sky Telesc., Vol. 69, No. 5, p. 422 (1985).

003.173   Möglichkeiten der experimentellen Schwerkraftforschung.
M. Steenbeck, H.–J. Treder.
Akademie der Wissenschaften der DDR, Veröffentlichungen des Forschungsbereichs Kosmische Physik, H. 11. Akademie-Verlag, Berlin, GDR. 46 pp. Price M 10.00 (1984).
Review in Gerlands Beitr. Geophys., Band 94, Heft 3, p. 240; 1985 (*N. Salié*).

003.174   Die Sterne des Abd ar–Rahman as–Sufi.
G. Strohmaier.
Gustav Kiepenheuer Verlag, Leipzig – Weimar. 110 pp. Price M 32.00 (1984). ISBN 3–7833–8447–8.
Reviews in Astron. Raumfahrt, 22. Jahrg., Heft 2, p. 48; 1985 (*W. Häupl*); Sterne Weltraum, 24. Jahrg., Nr. 5, p. 294; 1985 (*R. Gredel*).

003.175   Constrained dynamics. With application to Yang–Mills theory, general relativity, classical spin, dual string model.
K. Sundermeyer.
Lecture Notes in Physics, Vol. 169. Springer-Verlag, Berlin – Heidelberg – New York – Tokyo. Price DM 33.00, US$ 13.20 (1982). ISBN 3–540–11947–7.
Review in Astron. Nachr., Vol. 306, No. 3, p. 128; 1985 (*H. Fuchs*).

003.176   Building the universe.
C. Sutton (Editor).
Basil Blackwell, Oxford, England. 14 + 361 pp. Price £ 19.50 cloth, £ 7.95 paper (1985).
Review in Astron. Express, Vol. 1, Nos. 4 – 6, p. 161 (1985).

003.177   Astronomy & telescopes.
R. Traister, S. Harris.
TAB Books, 200 pp. Price US$ 14.95 (1983).
Review in Mercury, Vol. 14, No. 3, p. 88 (1985).

003.178   Im Augenblick der Schöpfung.
J. S. Trefil.
Birkhäuser Verlag, Basel – Boston – Stuttgart. 256 pp. Price SFr 35.00 (1984). ISBN 3–7643–1606–3.
Review in Orion, 43. Jahrg., Nr. 208, p. 84; 1985 (*K. Städeli*).

003.179   The moment of creation: big bang physics from before the first millisecond to the present universe.
J. S. Trefil.
Macmillan Publishing Co., Inc., New York, USA. 234 pp. Price US$ 6.95 (1984). ISBN 0–02–096770–5.
From Phys. Today, Vol. 38, No. 6, p. 80 (1985); Sky Telesc., Vol. 69, No. 3, p. 230 (1985).

003.180   The science digest book of Halley's comet.
J. Tullius.
Avon, New York, USA. 136 pp. Price US$ 9.95 (1985).
From Phys. Today, Vol. 38, No. 6, p. 80 (1985).

003.181   Le livre du ciel.
J.–P. Verdet.
Editions Gallimard, Paris, France. 94 pp. (1983).
Review in Ciel Terre, Vol. 101, No. 1, p. 22; 1985 (*J.–C. Jodogne*).

003.182   Der zweite Tag der neuen Welt. Die Raumfahrt auf dem Weg ins 3. Jahrtausend.
J. von Puttkamer.
Umschau Verlag, Frankfurt, F.R. Germany. 272 pp. Price DM 39.80 (1985). ISBN 3–524–69054–8.

003.183   Charles Piazzi Smyth, astronomer–artist; his Cape years 1835 – 1845.
B. Warner.
Balkema.   16 + 144 pp.   Price   R 22.50   (1983).   ISBN 0–86961–133–X.
Review in Mon. Notes Astron. Soc. S. Afr., Vol. 44, Nos. 1 – 2, p. 18; 1985 (*I. S. Glass*).

003.184   Gestirnter Himmel über mir. Unverlierbares aus meinem Leben.
D. Wattenberg.
Union–Verlag, Berlin, GDR. 454 pp. Price M 18.80 (1984).
Review in Astron. Raumfahrt, 23. Jahrg., Heft 3, p. 70 – 71; 1985 (*S. Marx*).

003.185   The Nobel Prize.
P. Wilhelm.
Heyden & Son, Inc., 247 S. 41st St., Philadelphia, Pa. 19104, USA. 112 pp. Price US$ 22.00 (1983).
Review in Mercury, Vol. 14, No. 3, p. 91 (1985).

003.186   Theory and experiment in gravitational physics.
C. M. Will.
Cambridge University Press, Cambridge – London – New York – New Rochelle – Melbourne – Sydney. 10 + 342 pp. Price £ 15.00, $ 24.95 (1985). ISBN 0–521–31710–X.
Review in Spaceflight, Vol. 27, No. 6, p. 287 (1985).

003.187 Living the sky: the cosmos of the North American Indian.
R. A. Williamson.
Houghton Mifflin, Boston, USA. 16+366 pp. Price $ 19.95 (1984). ISBN 0–395–35414–5.
Reviews in Archaeoastronomy (U.K.), No. 8, p. S65 – S67; 1985 (*M. Zeilik*); Sky Telesc., Vol. 69, No. 2, p. 132 (1985).

003.188 Astronomie, Theorie und Praxis.
E. Wischnewski.
Available from E. Wischnewski, Heinrich–Heine–Weg 13, 2358 Kaltenkirchen, F.R. Germany. 342 pp. (1984).
Review in Sterne Weltraum, 24. Jahrg., Nr. 2, p. 114; 1985 (*A. M. Quetsch*).

003.189 Astronomy: the evolving universe.
M. Zeilik.
Harper & Row Ltd., London, England. 494 pp. Price $ 31.50 (1985). ISBN 0–06–047374–6.
Review in Sky Telesc., Vol. 69, No. 6, p. 520 (1985).

003.190 Historically–astronomical investigations (Moscow).
Moskva, No. 17, 455 pp. (1984). In Russian.
Review in Ref. Zh., 51. Astron., 2.51.29 (1985).

003.191 Le système solaire et son exploration.
Editions Atlas, Paris, France. Extrait de la série Astronomie. 200 pp. Price FB 1425.
Review in Ciel Terre, Vol. 101, No. 3, p. 97; 1985 (*L. Louys*).

003.192 Mr. Halley's comet.
Editors of Sky and Telescope.
Sky Publishing Corporation, 49 Bay State Road, Cambridge, Mass. 02238–1290, USA. 32 pp. Price US$ 2.00 (1984). ISBN 0–933346–41–7.
Reviews in Observatory, Vol. 105, No. 1066, p. 104 (1985); Sky Telesc., Vol. 69, No. 5, p. 422 (1985).

003.193 Asimov's guide to Halley's comet.
I. Asimov.
Walker, New York, 10+118 pp. Price US$ 12.95 (1985).
From Science, Vol. 228, No. 4704, p. 1192 (1985).

003.194 Evolution of matter and energy on a cosmic and planetary scale.
M. Taube.
Springer–Verlag, New York – Berlin – Heidelberg – Tokyo. 14+289 pp. Price DM 76.00 (1985). ISBN 3–540–13399–2 (F.R. Germany), ISBN 0–387–13399–2 (USA).
Contents: Matter and energy. The interplay of elementary particles and elementary forces. The Universe: how is it observed here and now? Its past and possible future. The origin and nuclear evolution of matter. Chemical evolution and the evolution of life: the cosmic phenomena. The eternal cycle of matter on the earth. The flow of energy on the earth. The biosphere: the coupling of matter and the flow of free energy. Is the future development of mankind on this planet possible? The distant future of mankind – terrestrial or cosmic?

003.195 Das Phänomen des Polarlichtes.
W. Schröder.
Wissenschaftliche Buchgesellschaft, Darmstadt, F.R. Germany. 156 pp. Price DM 35.62 (1984).
Review in J. Atmos. Terr. Phys., Vol. 46, No. 6, p. 621; 1985 (*W. Dieminger*).

003.196 Physics of dense matter.
Y. C. Leung.
Science Press, Beijing, China and World Scientific Publishing Co. Pte. Ltd., Singapore. 4+268 pp. Price DM 143.00 (1984). ISBN 9971–978–10–5.
In this book an attempt is made to provide a theoretical background for the study of the structure of very dense matter. The densities under investigation are those between 10^4 to $10^16g/cm^3$ found in inert stellar objects like the white dwarfs and neutron stars. Even though dense matter physics is also related closely to contemporary investigations in heavy ion collisions and the early universe, the author makes little reference to these issues. Instead emphasis is placed on the stellar objects with the final aim in deriving the appropriate equation of state for matter forming them.

003.197 Atlas des Sonnensystems.
P. Moore, G. Hunt, I. Nicolson, P. Cattermole.
Verlag Herder, Freiburg – Basel – Wien. 462 pp. Price DM 158.00 (1985). ISBN 3–451–19613–1. For the English original see 37.003.128.
Contents: Die Sonne. Der Merkur. Die Venus. Die Erde. Der Mond. Der Mars. Asteroiden. Der Jupiter. Der Saturn. Das äußere Sonnensystem. Geschichte und Beobachtungen.

003.198 Measuring the universe. Cosmic dimensions from Aristarchus to Halley.
A. Van Helden.
University of Chicago Press, Chicago – London. 8+203 pp. Price US$ 30.00 (1985).
From Science, Vol. 228, No. 4705, p. 1308 (1985).

004 History of Astronomy

004.001   Uses for ancient eclipse records.
J. Maddox.
Nature, Vol. 313, No. 6005, p. 733 (1985).
This note comments on the use of ancient eclipse observations to determine the secular evolution of the Earth's rotation.

004.002   G. B. Hodierna's observations of nebulae and his cosmology.
G. F. Serio, L. Indorato, P. Nastasi.
J. Hist. Astron., Vol. 16, Part 1, No. 45, p. 1 – 36 (1985).

004.003   Did Copernicus owe a debt to Aristarchus?
O. Gingerich.
J. Hist. Astron., Vol. 16, Part 1, No. 45, p. 37 – 42 (1985).

004.004   Elijah Burritt and the "Geography of the Heavens".
P. A. Kidwell.
Sky Telesc., Vol. 69, No. 1, p. 26 – 28 (1985).

004.005   The Milky Way from antiquity to modern times.
M. Hoskin.
The Milky Way galaxy, p. 11 – 24 (1985). – See Abstr. 012.007 (IAU Symp. No. 106).
The paper outlines the history of attempts to explain the Milky Way, from antiquity to the early–twentieth century, with special reference to the eighteenth and nineteenth centuries. Also discussed is the relationship of the Galaxy to other star systems, and particularly the question of whether there are other galaxies in the visible universe.

004.006   Kapteyn and statistical astronomy.
E. R. Paul.
The Milky Way galaxy, p. 25 – 42 (1985). – See Abstr. 012.007 (IAU Symp. No. 106).
The author surveys Kapteyn's contributions to the rise of statistical astronomy, principally his studies of systematic stellar motions and his analysis of the sidereal problem, with some attention devoted to his great star cataloguing efforts.

004.007   Studies of the Milky Way 1850 – 1930: some highlights.
R. W. Smith.
The Milky Way galaxy, p. 43 – 58 (1985). – See Abstr. 012.007 (IAU Symp. No. 106).
"The Copernicus of the sidereal system is not to be expected for many generations". So wrote R. A. Proctor in his "Essays in Astronomy" in 1872. Why was this so, and why was there to be such an astonishing transformation of this situation between 1918 and 1930? Certainly these twelve years saw the widespread acceptance of no less than six fundamentally new ways of viewing the Galactic System. The author attempts to describe and analyze what these changes were, what led up to them, as well as to examine the events surrounding them.

004.008   The discovery of the spiral arms of the Milky Way.
O. Gingerich.
The Milky Way galaxy, p. 59 – 70 (1985). – See Abstr. 012.007 (IAU Symp. No. 106).
Attempts in the 1930s and 1940s to determine the spiral structure of the Milky Way by star counting methods, essentially the continuation of the work of the Kapteyn Astronomical Laboratory, failed to reach this goal. A new foundation for the search was laid by Walter Baade in his studies of stellar populations. With the recognition that highly luminous objects, especially H II regions, would outline the spiral structure, W. W. Morgan and his young associates Sharpless and Osterbrock carried out the observational program that first delineated, in 1951, the nearby arms of the Milky Way.

004.009   The mystery of Bayer's *Uranometria*.
L. J. Carter, A. T. Lawton.
Spaceflight, Vol. 27, No. 3, p. 117 – 128 (1985).

004.010   By analogy with the Heavens: Kant's theory of the Earth.
O. Reinhardt, D. R. Oldroyd.
Ann. Sci., Vol. 41, No. 3, p. 203 – 221 (1984). Abstr. in Phys. Abstr., Vol. 88, No. 1250, Entry 14876 (1985).

004.011   The glory of gravity – Halley's comet 1759.
R. Wallis.
Ann. Sci., Vol. 41, No. 3, p. 279 – 286 (1984). Abstr. in Phys. Abstr., Vol. 88, No. 1250, Entry 14879 (1985).

004.012   F. W. Bessel und die Zonenunternehmen der Astronomischen Gesellschaft.
L. Brandt.
Sterne Weltraum, 24. Jahrg., Nr. 1, p. 10 – 15 (1985).

004.013   Die Pekinger Sternwarte in altem Glanz wiederhergestellt.
G. W. E. Beekman.
Sterne Weltraum, 24. Jahrg., Nr. 2, p. 70 – 73 (1985).

004.014   William Huggins und die Anfänge der astronomischen Spektroskopie.
J. B. Hearnshaw.
Sterne Weltraum, 24. Jahrg., Nr. 3, p. 140 – 142 (1985).

004.015   Astronomical scrapbook: the astronomy of Alfonso the Wise.
O. Gingerich.
Sky Telesc., Vol. 69, No. 3, p. 206 – 208 (1985).

004.016   Blaeu's failed celestial globe.
D. J. Warner.
Sky Telesc., Vol. 69, No. 4, p. 294 (1985).

004.017   Archeoastrophysics of supernovae and neutron stars.
K. Brecher.
News Lett. Astron. Soc. N.Y., Vol. 2, No. 4, p. 8 – 9 (1983). Abstract. – See Abstr. 010.242.

004.018   Christiaan Huygens' measurement of the distance to the Sun.
S. J. Goldstein Jr.
Observatory, Vol. 105, No. 1065, p. 32 – 33 (1985).

004.019   Les débuts de la photographie astronomique particulièrement à l'Observatoire de Paris.
T. Weimer.
Astronomie, Vol. 99, 207 – 209 (1985).

004.020   The earliest Pacific observatories – including the world's first (?) high–altitude observatory.
W. R. Beardsley.
Publ. Astron. Soc. Pac., Vol. 97, No. 588, p. 199 (1985). Abstract. – See Abstr. 010.281.

004.021   The Galilean gambit of 1619; the attack on Tycho.
M. Chriss.
Publ. Astron. Soc. Pac., Vol. 97, No. 588, p. 199 – 200 (1985). Abstract. – See Abstr. 010.281.

004.022   Ancient astronomy of East Africa.
L. R. Doyle.
Publ. Astron. Soc. Pac., Vol. 97, No. 588, p. 200 (1985). Abstract. – See Abstr. 010.281.

004.023 Aristotle and parallax: a common error in introductory textbooks.
T. E. Lutz.
Publ. Astron. Soc. Pac., Vol. 97, No. 588, p. 200 – 201 (1985). Abstract. – See Abstr. 010.281.

004.024 A letter from W. H. Pickering to M. M. Bovard, April 30, 1888.
G. Reaves.
Publ. Astron. Soc. Pac., Vol. 97, No. 588, p. 201 (1985). Abstract. – See Abstr. 010.281.

004.025 Naked–eye observations of sunspots in Bohemia in the year 1139.
L. Křivský.
Bull. Astron. Inst. Czech., Vol. 36, No. 1, p. 60 – 61 (1985).
Attention is drawn to naked–eye observations of sunspots in the year 1139, before July 25, which are recorded in the Vyšehrad Canon's chronicle written in Latin.

004.026 The observations of Neptune by Galileo.
G. E. Taylor.
J. Br. Astron. Assoc., Vol. 95, No. 3, p. 116 – 117 (1985).
Suggestions have been made that Galileo's observations of Neptune in 1612 – 1613 indicate that it was as much as 1 arcminute from its predicted position. The author of this paper argues that Galileo's drawing has been misinterpreted and that there is no justification of the adoption of a 1 arcminute error.

004.027 Die astronomische Uhr zu Straßburg.
M. Treiß.
Sterne Weltraum, 24. Jahrg., Nr. 4, p. 222 – 224 (1985).

004.028 Historia de la astronomía.
M. Ruffo.
Rev. Astron., Tomo 56, No. 230, p. 10 – 13 (1984).

004.029 The meaning of celestial mechanics in the Kosmos of Alexander von Humboldt.
F. Schmeidler.
Celest. Mech., Vol. 34, Nos. 1 – 4, p. 7 – 10 (1984). – See Abstr. 012.030.

004.030 Om Tycho Brahe–minnena på Ven.
G. Larsson–Leander.
Astron. Tidsskr., Årg. 18, Nr. 1, p. 1 – 18 (1985).

004.031 Något om Ven–ruinernas tillstånd och behandling i äldre tid.
K. L. Larsson.
Astron. Tidsskr., Årg. 18, Nr. 1, p. 19 – 40 (1985).

004.032 The heliometer principle and some modern applications.
E. H. Geyer.
Astrophys. Space Sci., Vol. 110, No. 1, p. 183 – 192 (1985). – See Abstr. 012.039.
Beside some historical notes about the large Fraunhofer's heliometer used by Bessel and Argelander, some modern applications of the heliometer principle for the geometric and photometric autocalibration of detectors and the determination of absolute radial velocities with slitless field spectrographs are presented.

004.033 Bessel and librations of the Moon.
K. Kozieł.
Astrophys. Space Sci., Vol. 110, No. 1, p. 193 – 196 (1985). – See Abstr. 012.039.
On the occasion of the twohundredth anniversary of F. W. Bessel's birth, his method of heliometric observations for the determination of the Moon's physical libration constants and of the reduction of these observations is presented.

004.034 Evidence from a new site.
C. Ruggles.
Nature, Vol. 314, No. 6007, p. 134 – 135 (1985).
This note comments on the report by MacKie et al. of a newly excavated site of megalithic astronomy at Minard, Argyll (see Abstr. 004.035).

004.035 A prehistoric calendrical site in Argyll?
E. W. MacKie, P. F. Gladwin, A. E. Roy.
Nature, Vol. 314, No. 6007, p. 158 – 161 (1985).
The hypothesis that many prehistoric standing stone sites in Britain were set up, in relation to natural horizon marks, as astronomical observing instruments has been controversial for the past two decades. The authors report here on the findings at Brainport Bay in Argyllshire of artificial features pointing towards the midsummer sunrise that have been dated suitably early and of another feature indicating alignment towards the sunset at the equinox, that was first predicted and then discovered. These results seem to provide strong support for Thom's hypothesis that calendrically useful solar markers existed in Scotland at least as early as the Bronze Age.

004.036 Byzantine calendrical gearing.
F. Maddison.
Nature, Vol. 314, No. 6009, p. 316 – 317 (1985).

004.037 Halley's comet in Babylonia.
C. B. F. Walker.
Nature, Vol. 314, No. 6012, p. 576 – 577 (1985).

004.038 Records of Halley's comet on Babylonian tablets.
F. R. Stephenson, K. K. C. Yau, H. Hunger.
Nature, Vol. 314, No. 6012, p. 587 – 592 (1985).
The late Babylonian texts in the British Museum are shown to contain probable observations of Halley's comet at both its 164 BC and 87 BC apparitions. These texts have important bearing on the orbital motion of the comet in the ancient past.

004.039 Megalithic observatories in Britain: real or imagined?
R. P. Norris.
Proc. Astron. Soc. Aust., Vol. 5, No. 4, p. 428 – 434 (1984).
Over the last two decades there has been an accumulation of exciting evidence that appears to show that, as early as 5000 years ago, people in Britain were making precise observations of the Sun, Moon, and stars, and studying small perturbations in the lunar motion. Recently however a small number of rigorous statistical studies of the sites have cast doubt on the astronomical hypotheses. In this review, the arguments and counter arguments are presented and examined, and we see what can be salvaged from the astronomical hypotheses after the statistical smoke has cleared.

004.040 Status of the Copernican theory before Kepler, Galileo, and Newton.
D. R. Martin.
Am. J. Phys., Vol. 52, No. 11, p. 982 – 986 (1984). Abstr. in Phys. Abstr., Vol. 88, No. 1252, Entry 24660 (1985).

004.041 Stukeley's cosmology and the Newtonian origins of Olbers's paradox.
M. Hoskin.
J. Hist. Astron., Vol. 16, Part 2, No. 46, p. 77 – 112 (1985).

004.042 Solar observations at the Maraghah Observatory before 1275: a new set of parameters.
G. Saliba.
J. Hist. Astron., Vol. 16, Part 2, No. 46, p. 113 – 122 (1985).

004.043 The first predicted return of comet Halley.
P. Broughton.
J. Hist. Astron., Vol. 16, Part 2, No. 46, p. 123 – 133 (1985).

004.044 The first edition of Halley's "Synopsis".
M. Hoskin.
J. Hist. Astron., Vol. 16, Part 2, No. 46, p. 133 (1985).

004.045   **Galen on the astronomers and astrologers.**
G. J. Toomer.
Arch. Hist. Exact Sci., Vol. 32, No. 3 – 4, p. 193 – 206 (1985).

004.046   **Epicycles, eccentrics, and ellipses: the predictive capabilities of Copernican planetary models.**
C. A. Gearhart.
Arch. Hist. Exact Sci., Vol. 32, No. 3 – 4, p. 207 – 222 (1985).

004.047   **Astronomical background to the International Meridian Conference of 1884.**
D. H. Sadler, G. A. Wilkins.
J. Navig., Vol. 38, No. 2, p. 191 – 199 (1985).

004.048   **The International Prime Meridian Conference, Washington, October 1884.**
S. R. Malin.
J. Navig., Vol. 38, No. 2, p. 203 – 206 (1985).

004.049   **Developments of time systems since 1884.**
J. D. H. Pilkington.
J. Navig., Vol. 38, No. 2, p. 207 – 208 (1985).

004.050   **Astronomical navigation since 1884.**
A. E. Fanning.
J. Navig., Vol. 38, No. 2, p. 209 – 215 (1985).

004.051   **Sterrenkunde in Twente: kleine wetenschappers met grote daden.**
A. J. Lindemann.
Zenit, 12. Jaarg., No. 5, p. 171 – 177 (1985).

004.052   **Originele manuscripten van Eise Eisinga ontdekt.**
H. Nieuwenhuis.
Zenit, 12. Jaarg., No. 6, p. 220 – 221 (1985).

004.053   **Newton's lunar mass error.**
N. Kollerstrom.
J. Br. Astron. Assoc., Vol. 95, No. 4, p. 151 – 153 (1985).
   This paper discusses a hitherto undetected error of 100% in the Earth–Moon mass ratio by Newton, and suggests a reason why it has been overlooked.

004.054   **Some historical observations of the nebulae. II. Herschel to photography.**
R. J. Dodd.
South. Stars, Vol. 31, No. 1, p. 23 – 31 (1984).
   This paper begins with the, still visual, observations of William Herschel and traces the progress made in investigating the nature and structure of the nebulae through the nineteenth century including the rapid advances made following the introduction of the spectroscope and the photographic plate.

004.055   **The liquid–mirror telescope – an early example of kiwi ingenuity?**
D. Steel.
South. Stars, Vol. 31, No. 1, p. 32 – 40 (1984).

004.056   **Dokumente aus der Dorpater Zeit Wilhelm Struves in der Charkower Sternwarte.**
W. R. Dick.
Tartu Astrofüüs. Obs. Teated, Nr. 75, p. 5 – 12 (1985).

004.057   **Kalender– und Astronomie–geschichtliche Publikationen im Tallinner Staatlichen Zentralarchiv.**
E. Vendla, U. Urb, H. Eelsalu.
Tartu Astrofüüs. Obs. Teated, Nr. 75, p. 13 – 14 (1985).

004.058   **Rara Astronomica in the Library of the Tartu University: books printed after 1600. Addenda.**
H. Eelsalu.
Tartu Astrofüüs. Obs. Teated, Nr. 75, p. 15 (1985).

004.059   **An attempt to draw up a list of instruments acquired by the Tartu Observatory before 1825: comments.**
H. Eelsalu.
Tartu Astrofüüs. Obs. Teated, Nr. 75, p. 16 – 17 (1985).

004.060   **Briefe von J. D. Sandt an J. W. Pfaff betreffend Troughton. 1804 – 1805.**
M. Raudsepp.
Tartu Astrofüüs. Obs. Teated, Nr. 75, p. 18 – 26 (1985).

004.061   **Old geodetic instruments in the Tartu Astronomical Observatory.**
A. Torim.
Tartu Astrofüüs. Obs. Teated, Nr. 75, p. 27 – 30 (1985).

004.062   **Astronomy in Vojvodina.**
B. D. Jovanović.
Publ. Astron. Opservatorije Beogr., No. 33, p. 96 – 99 (1985). – See Abstr. 012.061.

004.063   **History of astronomy in Bosnia and Herzegovina. Some problems of archaeoastronomy.**
J. Mulaomerović.
Publ. Astron. Opservatorije Beogr., No. 33, p. 103 – 106 (1985). – See Abstr. 012.061.

004.064   **Le Cygne, quel cygne?**
R. Clap, P. Fabre.
Astronomie, Vol. 99, p. 289 – 294 (1985).

004.065   **Antikythera–fyndet – en antik evighetskalender.**
G. Pipping.
Astron. Tidsskr., Årg. 18, Nr. 2, p. 58 – 70 (1985).

004.066   **Which astronomers win prizes?**
N. Sperling.
Bull. Am. Astron. Soc., Vol. 16, No. 4, p. 886 (1984). Abstract. – See Abstr. 010.062.

004.067   **What money could not buy: the Smith fund of the NAS and meteoritic astronomy, 1884 – 1940.**
J. Lankford.
Bull. Am. Astron. Soc., Vol. 16, No. 4, p. 887 (1984). Abstract. – See Abstr. 010.062.

004.068   **Yale contributions to meteoritic astronomy 1742 – 1910.**
D. Hoffleit.
Bull. Am. Astron. Soc., Vol. 16, No. 4, p. 887 (1984). Abstract. – See Abstr. 010.062.

004.069   **Mayan Long Count as a Doomesday prediction.**
W. T. Murawski.
Bull. Am. Astron. Soc., Vol. 16, No. 4, p. 887 (1984). Abstract. – See Abstr. 010.062.

004.070   **The non–astronomical alignment of the descending passage in the pyramid of Kufu.**
R. L. Walker.
Bull. Am. Astron. Soc., Vol. 16, No. 4, p. 887 (1984). Abstract. – See Abstr. 010.062.

004.071   **The early years of astronomy at the University of Arizona.**
G. E. Webb.
Bull. Am. Astron. Soc., Vol. 16, No. 4, p. 887 (1984). Abstract. – See Abstr. 010.062.

004.072   **The Ritchey–Chrétien reflecting telescope: half a century from conception to acceptance.**
J. S. Hall.
Bull. Am. Astron. Soc., Vol. 16, No. 4, p. 888 (1984). Abstract. – See Abstr. 010.062.

004.073 **Ancient apparitions of Halley's comet.**
K. Brecher.
Bull. Am. Astron. Soc., Vol. 16, No. 4, p. 924 (1984). Abstract. –
See Abstr. 010.062.

004.074 **An astronomical howler in Hesiod.**
H. A. T. Reiche.
Bull. Am. Astron. Soc., Vol. 16, No. 4, p. 924 (1984). Abstract. –
See Abstr. 010.062.

004.075 **Greek words for celestial objects prior to 400 BC.**
T. D. Worthen.
Bull. Am. Astron. Soc., Vol. 16, No. 4, p. 924 (1984). Abstract. –
See Abstr. 010.062.

004.076 **Sayers on Lucan.**
B. G. Marsden.
Bull. Am. Astron. Soc., Vol. 16, No. 4, p. 924 (1984). Abstract. –
See Abstr. 010.062.

004.077 **The astronomy of prehistoric Chaco Canyon, New Mexico.**
A. Sofaer, R. M. Sinclair.
Bull. Am. Astron. Soc., Vol. 16, No. 4, p. 924 (1984). Abstract. –
See Abstr. 010.062.

004.078 **Hypotheses for Navajo star ceilings.**
V. D. Chamberlain.
Bull. Am. Astron. Soc., Vol. 16, No. 4, p. 925 (1984). Abstract. –
See Abstr. 010.062.

004.079 **Summer solstice at Casa Rinconada: calendar, heirophony, or nothing?**
M. Zeilik.
Bull. Am. Astron. Soc., Vol. 16, No. 4, p. 925 (1984). Abstract. –
See Abstr. 010.062.

004.080 **A solstitially aligned geoform at Y:3:1 (BLM), Arizona.**
C. T. Hoskinson.
Bull. Am. Astron. Soc., Vol. 16, No. 4, p. 925 (1984). Abstract. –
See Abstr. 010.062.

004.081 **A possible representation of the supernova of 1054 AD in the Quetico Provincial Park.**
G. W. Collins II.
Bull. Am. Astron. Soc., Vol. 16, No. 4, p. 941 (1984). Abstract. –
See Abstr. 010.062.

004.082 **Spica rather than Rigel at the Bighorn and Moose Mountain Medicine Wheels?**
J. H. Robinson.
Bull. Am. Astron. Soc., Vol. 16, No. 4, p. 941 (1984). Abstract. –
See Abstr. 010.062.

004.083 **Korean records of sporadic meteors in the 18th century.**
I.–S. Nha, S.–H. Nha, E.–H. Lee.
Bull. Am. Astron. Soc., Vol. 16, No. 4, p. 941 – 942 (1984). Abstract. – See Abstr. 010.062.

004.084 **Maria Mitchell's observations of double stars.**
B. L. Welther.
Bull. Am. Astron. Soc., Vol. 16, No. 4, p. 942 (1984). Abstract. –
See Abstr. 010.062.

004.085 **The moon – volcanoes or cosmic bombs.**
K. R. Lang.
Bull. Am. Astron. Soc., Vol. 16, No. 4, p. 942 (1984). Abstract. –
See Abstr. 010.062.

004.086 **First infrared observations of a comet.**
A. A. Hoag.
Bull. Am. Astron. Soc., Vol. 16, No. 4, p. 942 (1984). Abstract. –
See Abstr. 010.062.

004.087 **The ethnoastronomy of the historic pueblos. I. Calendrical sun watching.**
M. Zeilik.
Archaeoastronomy (U.K.), No. 8, p. S1 – S24 (1985).

004.088 **A new study of the Aberdeenshire Recumbent Stone Circles. 2. Interpretation.**
C. L. N. Ruggles, H. A. W. Burl.
Archaeoastronomy (U.K.), No. 8, p. S25 – S60 (1985).

004.089 **William Stukeley and the Stonehenge sunrise.**
R. J. C. Atkinson.
Archaeoastronomy (U.K.), No. 8, p. S61 – S62 (1985).

004.090 **Galileo's concept of science: recent manuscript evidence.**
W. A. Wallace.
Vatican Obs. Publ., Spec. Ser., Studi Galileiani, Vol. 1, No. 3, p. 15 – 40 (1985). – See Abstr. 012.064.
Contents: Introduction. The Collegio Romano. Galileo's dependence on Valla. Texts and contents. Galileo and Clavius. Causality and science. Influence in Galileo's scientific writings. Copernican controversies. Reinterpreting Galileo.

004.091 **The rhetoric of proof in Galileo's writings on the Copernican system.**
J. D. Moss.
Vatican Obs. Publ., Spec. Ser., Studi Galileiani, Vol. 1, No. 3, p. 41 – 65 (1985). – See Abstr. 012.064.
Contents: Introduction. The notion of scientific proof in Galileo's letter to Christina. Rhetoric and scientific proof in Galileo's Dialogue.

004.092 **The problem of the annual parallax in Galileo's time.**
J. Casanovas.
Vatican Obs. Publ., Spec. Ser., Studi Galileiani, Vol. 1, No. 3, p. 67 – 74 (1985). – See Abstr. 012.064.
Contents: Introduction. The question of the angular sizes of the stars. Stellar angular sizes after the advent of the telescope. The question of stellar sizes in Galileo's Dialogue. Early proposals for the measurement of the annual parallax.

004.093 **The young Bellarmine's thoughts on world systems.**
G. V. Coyne, U. Baldini.
Vatican Obs. Publ., Spec. Ser., Studi Galileiani, Vol. 1, No. 3, p. 103 – 110 (1985). – See Abstr. 012.064.

004.094 **Galileo's relativity.**
M. Heller.
Vatican Obs. Publ., Spec. Ser., Studi Galileiani, Vol. 1, No. 3, p. 113 – 124 (1985). – See Abstr. 012.064.
Contents: Introduction. Ptolemy's relativity. Copernicus as a relativist. Relativity and inertia according to Galileo. What does really move? From the present perspective.

004.095 **Galileo's views on infinity.**
M. Lubański.
Vatican Obs. Publ., Spec. Ser., Studi Galileiani, Vol. 1, No. 3, p. 125 – 136 (1985). – See Abstr. 012.064.

004.096 **Why Galileo's research program superceded rival programs.**
J. M. Życiński.
Vatican Obs. Publ., Spec. Ser., Studi Galileiani, Vol. 1, No. 3, p. 137 – 148 (1985). – See Abstr. 012.064.

004.097 **Limits to the applicability of Galilean methodology.**
K. Rudnicki.
Vatican Obs. Publ., Spec. Ser., Studi Galileiani, Vol. 1, No. 3, p. 149 – 151 (1985). – See Abstr. 012.064.

004.098 **Galileo and the tyranny of truth.**
P. K. Feyerabend.
Vatican Obs. Publ., Spec. Ser., Studi Galileiani, Vol. 1, No. 3, p. 155 – 166 (1985). – See Abstr. 012.064.

004.099 Myth and imagination in Galileo's discovery.
F. M. Hetzler.
Vatican Obs. Publ., Spec. Ser., Studi Galileiani, Vol. 1, No. 3, p. 167 – 174 (1985). – See Abstr. 012.064.

004.100 Alexander von Humboldt's ideas on a connection between terrestrial and spatial processes and their development in the works of Russian scientists of the 20th century.
A. L. Yanshin.
Vopr. istor. estestvozn. i tekh., No. 4, p. 54 – 57 (1984). In Russian. Abstr. in Ref. Zh., 51. Astron., 6.51.13 (1985).

004.101 Galileo, planetary atmospheres, and prograde revolution.
G. D. Parker.
Science, Vol. 227, No. 4687, p. 597 – 600 (1985).
  Early in March 1610 Galileo was preoccupied with curious brightness variations of the newly discovered satellites of Jupiter. In formulating an incorrect explanation he advanced important generalizations about the existence of planetary atmospheres and counterclockwise circulation within the solar system.

004.102 The First International Halley Watch: worldwide anticipation and observation of comet Halley, 1755 – 1759.
C. B. Waff.
Bull. Am. Astron. Soc., Vol. 17, No. 1, p. 519 – 520 (1985). Abstract. – See Abstr. 010.064.

004.103 The application of historical records to astrophysical problems.
Z. Xi.
High energy astrophysics and cosmology, p. 158 – 169 (1983). – See Abstr. 012.068.
  This paper discusses astrophysical applications of ancient recordings of supernovae, of solar activity, and of solar eclipses.

004.104 The Great Inequality of Jupiter and Saturn: from Kepler to Laplace.
C. Wilson.
Arch. Hist. Exact Sci., Vol. 33, Nos. 1 – 3, p. 15 – 290 (1985).

004.105 The nature of Saturn's rings: James E. Keeler's "Prettiest application of Doppler's principle".
D. E. Osterbrock.
Mercury, Vol. 14, No. 2, p. 46 – 47, 50 – 53, 62 (1985).

004.106 Beyond the planets: early nineteenth–century studies of double stars.
M. Williams.
Br. J. Hist. Sci., Vol. 17, Part. 3, No. 57, p. 295 – 309 (1984). Abstr. in Phys. Abstr., Vol. 88, No. 1253, Entry 30160 (1985).

004.107 On the function and the probable origin of Ptolemy's equant.
J. Evans.
Am. J. Phys., Vol. 52, No. 12, p. 1080 – 1089 (1984). Abstr. in Phys. Abstr., Vol. 88, No. 1254, Entry 35947 (1985).

004.108 Eddington's "Expanding Universe".
S. J. Prokhovnik.
Sir Arthur Eddington Centenary Symposium. Vol. 1: Relativistic astrophysics and cosmology, p. 28 – 44 (1984). – See Abstr. 012.080.
  A review of A. S. Eddington's contributions to the development and dissemination of modern cosmological ideas is presented.

004.109 Eddington's contribution to astrophysics and pulsation theory.
L. D. Chatterji.
Sir Arthur Eddington Centenary Symposium. Vol. 1: Relativistic astrophysics and cosmology, p. 261 – 274 (1984). – See Abstr. 012.080.
  After briefly outlining A. S. Eddington's role in discovering the expansion of the universe, this paper discusses at some length Eddington's theoretical studies of the internal structure and the pulsational behaviour of Cepheid variables.

004.110 Prehistorical evidence for precession.
P. M. James.
Speculations Sci. Technol., Vol. 7, No. 5, p. 303 – 315 (1984). Abstr. in Phys. Abstr., Vol. 88, No. 1256, Entry 51517 (1985).

004.111 Das Werden des Kosmos. Von der Erfahrung der zeitlichen Dimension astronomischer Objekte im 18. Jahrhundert.
F. Krafft.
Ber. Wissenschaftsgesch., Band 8, Heft 2, p. 71 – 85 (1985).
  This paper deals with two of the initial stages through which the dimension of time, in the sense of an irreversible development, found its way into astronomical–cosmological thinking. The one resulted from the first consequential application of Newtonian principles and laws to cosmic entities outside of our solar system found in the "General Natural History or Theory of the Heavens" by Immanuel Kant. The other initial stage is found in the classification of 'nebulae' by William Herschel who introduced the historical time factor as a principle of order in addition to the outward shape, which had become common for all the different elements in natural history during the second half of the 18th century.

004.112 L'astronomia nel Collegio Romano nella prima metà del Seicento.
J. Casanovas.
G. Astron., Vol. 10, N. 3 – 4, p. 149 – 155 (1984).

004.113 Personaggi minori dell'astronomia padovana prima del XIX secolo.
G. Romano.
G. Astron., Vol. 10, N. 3 – 4, p. 157 – 163 (1984).

004.114 Specificità e limiti dell'esperienza di lavoro della specola di Bologna nel XVIII secolo.
A. Braccesi.
G. Astron., Vol. 10, N. 3 – 4, p. 165 – 170 (1984).

004.115 Soluzioni italiane dell'equazione di Keplero nel '700.
E. Badolati.
G. Astron., Vol. 10, N. 3 – 4, p. 171 – 183 (1984).

004.116 La specola Caetani.
G. Monaco.
G. Astron., Vol. 10, N. 3 – 4, p. 185 – 190 (1984).

004.117 Sui primi strumenti di astronomia di posizione della specola di Brera in Milano.
E. Proverbio.
G. Astron., Vol. 10, N. 3 – 4, p. 191 – 200 (1984).

004.118 Origini della specola fiorentina.
M. Miniati.
G. Astron., Vol. 10, N. 3 – 4, p. 209 – 220 (1984).

004.119 La specola pisana (1735 – 1808).
M. di Bono.
G. Astron., Vol. 10, N. 3 – 4, p. 221 – 229 (1984).

004.120 Contatti Brera–Cremona per il rifacimento del quadrante dell'orologio astronomico del Torrazzo ed altre questioni.
A. Leani.
G. Astron., Vol. 10, N. 3 – 4, p. 231 – 239 (1984).

004.121 Connessioni astrologiche con il "Tractatus Sphaerae" di Andalò di Negro.
A. M. Cesari.
G. Astron., Vol. 11, N. 1, p. 91 – 105 (1985).

004.122 Bessels Bibliothek an der Königsberger Sternwarte. Mit einem Bibliogramm der Veröffentlichungen F. W. Bessels.
D. B. Herrmann.
Sterne, 61. Band, Heft 2, p. 96 – 103 (1985).

004.123 The mystery of Bayer's Uranometria – Part 2: The quest for Nathaniel Bliss.
A. T. Lawton, L. J. Carter.
Spaceflight, Vol. 27, No. 6, p. 275 – 282 (1985).

004.124 A possible Roman record of the supernova of A.D. 185.
R. B. Culver, D. MacDonald.
Bull. Am. Astron. Soc., Vol. 17, No. 2, p. 565 (1985). Abstract. – See Abstr. 010.065.

004.125 Hellenistic solstices and Babylonian use of them.
D. Rawlins.
Bull. Am. Astron. Soc., Vol. 17, No. 2, p. 583 (1985). Abstract. – See Abstr. 010.065.

004.126 The Fajada Butte solar marker: a reevaluation.
M. Zeilik.
Science, Vol. 228, No. 4705, p. 1311 – 1313 (1985).
Evalution of the Fajada Butte solar marker in Chaco Canyon, New Mexico, on the basis of its ethnographic context and usefulness for confirmatory and anticipatory solar observations indicates that the site does not function as an accurate solar calendar (accurate in the context of the historic Puebloan culture). The site most likely served as a sun shrine rather than as a calendrical observing station. The interpretation of the site as marking the northern declinations of the lunar 18.6–year cycle is not supported by the ethnographic evidence nor can the site be used to anticipate accurately the year of the standstill.

004.127 Report of IAU Commission 41: History of astronomy (*Histoire de l'astronomie*).
O. Pedersen.
Trans. IAU, Vol. XIXA, p. 581 – 582 (1985). – See Abstr. 003.046.

004.128 Brahe's and Kepler's monument unveiled in Prague.
Říše hvězd, Vol. 65, No. 9, p. 188 – 189 (1984). In Czech.

004.129 Isaac Newton and his Principia.
I. Staríček.
Vesmír, Vol. 64, No. 7, p. 398, 403 – 404 (1985). In Slovak.

004.130 A short history of solar studies. I.
B. M. Ševarlić.
Vasiona, Année 33, No. 1 – 2, p. 9 – 14 (1985). In Serbo-Croatian.
This paper traces the history of solar studies starting from the old Chinese studies.

004.131 Contributions of W. B. Jack and W. F. King to the standardization of early surveying instruments.
J. E. Kennedy.
J. R. Astron. Soc. Can., Vol. 79, No. 3, p. 130 – 133 (1985).

004.132 Rocky sites in Western Rhodopes with a possible astronomic purpose.
V. N. Dermendjiev, S. B. Dermendjieva, L. B. Menkov.
Dokl. Bolg. Akad. Nauk, Tome 37, No. 5, p. 545 – 548 (1984).
The authors report on the first results of the archaeoastronomical investigations of the megalith monument in the Western Rhodope Mountains.

004.133 1884 and Longitude Zero.
D. Howse.
Vistas Astron., Vol. 28, Parts 1/2, p. 11 – 19 (1985). – See Abstr. 012.109.
On October 1 1884, one hundred years ago this year, the International Meridian Conference convened in the Diplomatic Hall of the State Department in Washington, D.C.

004.134 Meridians of reference in pre–Copernican tables.
R. P. Mercier.
Vistas Astron., Vol. 28, Parts 1/2, p. 23 – 27 (1985). – See Abstr. 012.109.

004.135 Prime meridians in medieval Islamic astronomy.
E. S. Kennedy, M. H. Regier.
Vistas Astron., Vol. 28, Parts 1/2, p. 29 – 32 (1985). – See Abstr. 012.109.
Contents: The sources. Zero meridians in the texts. Conversion to Greenwich by a constant. Muslim estimates of the size of the earth. Linear relationships for converting longitudes.

004.136 Prime meridians used by Dutch navigators. A survey of the prime meridians, used by the Dutch for navigation and hydrography, prior to 1884.
W. F. J. Mörzer Bruyns.
Vistas Astron., Vol. 28, Parts 1/2, p. 33 – 39 (1985). – See Abstr. 012.109.

004.137 The determination of early longitudes and base meridians in Hungary.
L. Bartha.
Vistas Astron., Vol. 28, Parts 1/2, p. 41 – 48 (1985). – See Abstr. 012.109.
The author discusses the longitude measurements of base meridians in the Carpathian basin. The increasing accuracy of the measurements reflects progress in instrumental techniques and astronomical and cartographic knowledge during this period. The study is also of interest for the history of cartography.

004.138 Longitudes and meridians on French charts of the Mediterranean in the 17th and 18th centuries.
I. Raynaud–Nguyen.
Vistas Astron., Vol. 28, Parts 1/2, p. 49 – 60 (1985). – See Abstr. 012.109.

004.139 Charles Henry Davis, the foundation of the American Nautical Almanac, and the establishment of an American prime meridian.
C. B. Waff.
Vistas Astron., Vol. 28, Parts 1/2, p. 61 – 66 (1985). – See Abstr. 012.109.

004.140 Time for the astronomer 1484 – 1884.
W. J. H. Andrewes.
Vistas Astron., Vol. 28, Parts 1/2, p. 69 – 86 (1985). – See Abstr. 012.109.
The purpose of this paper is to show how the demands of astronomy influenced the development of the mechanical clock, and, in turn, how the mechanical clock affected the progress of astronomy.

004.141 Guardians of Greenwich time.
B. Hutchinson.
Vistas Astron., Vol. 28, Parts 1/2, p. 87 – 94 (1985). – See Abstr. 012.109.
The author describes some mechanical clocks, built in England, from 1283 onward, especially the Greenwich Tompion's and Graham's clocks.

004.142   The contribution of the mechanical clock to the improvement of navigation.
E. Proverbio.
Vistas Astron., Vol. 28, Parts 1/2, p. 95 – 103 (1985). – See Abstr. 012.109.
Contents: Latitude sailing and the first mechanical clocks at sea. The finding of longitude at sea by astronomical methods. The role of the mechanical clock in longitude determination at sea. The birth of the marine timekeeper.

004.143   Inventing, introducing and objecting to Standard Time.
I. R. Bartky.
Vistas Astron., Vol. 28, Parts 1/2, p. 105 – 111 (1985). – See Abstr. 012.109.
The author focusses on the major events that led to a new time system for the United States, and refers to several analyses that provide a new view of the steps.

004.144   Before Standard Time: distributing time in 19th–century America.
C. E. Stephens.
Vistas Astron., Vol. 28, Parts 1/2, p. 113 – 118 (1985). – See Abstr. 012.109.

004.145   Synodic period relations in Babylonian and Hellenistic astronomy.
K. P. Moesgaard.
Vistas Astron., Vol. 28, Parts 1/2, p. 119 – 121 (1985). – See Abstr. 012.109.

004.146   International coordination and Atomic Time.
H. M. Smith.
Vistas Astron., Vol. 28, Parts 1/2, p. 123 – 128 (1985). – See Abstr. 012.109.
The author reviews the possibilities of time–keeping throughout the World beginning in 1883.

004.147   One hundred years of Standard Time in the United States, 1883 – 1983.
G. Kish.
Vistas Astron., Vol. 28, Parts 1/2, p. 129 (1985). – See Abstr. 012.109.

004.148   The Greenwich meridional instruments (up to and including the Airy Transit Circle).
C. Stott.
Vistas Astron., Vol. 28, Parts 1/2, p. 133 – 145 (1985). – See Abstr. 012.109.

004.149   Meridian astronomy in the private and university observatories of the United Kingdom: rise and fall.
D. W. Dewhirst.
Vistas Astron., Vol. 28, Parts 1/2, p. 147 – 158 (1985). – See Abstr. 012.109.
The later years of the eighteenth century in Europe saw the beginning of a massive programme of observatory building which continued through the next century; a similar epidemic growth started in America fifty years later. The purpose of these observatories was initially to prosecute meridian astrometry: the determination of accurate places of stars, Sun and planets by the observation of their zenith distances and times of meridian transit.

004.150   Greenwich meridian astronomy.
R. H. Tucker.
Vistas Astron., Vol. 28, Parts 1/2, p. 159 – 167 (1985). – See Abstr. 012.109.

004.151   The geodetic link between the Greenwich and Paris Observatories in 1787.
E. G. Forbes.
Vistas Astron., Vol. 28, Parts 1/2, p. 173 – 181 (1985). – See Abstr. 012.109.

004.152   Cassini's meridian.
S. Débarbat.
Vistas Astron., Vol. 28, Parts 1/2, p. 183 – 186 (1985). – See Abstr. 012.109.

004.153   Astronomy versus cartography: late medieval longitudes.
J. Dobrzycki.
Vistas Astron., Vol. 28, Parts 1/2, p. 187 (1985). – See Abstr. 012.109.

004.154   Alexander Dalrymple and John Arnold: chronometers and the representation of longitude on East India Company charts.
A. S. Cook.
Vistas Astron., Vol. 28, Parts 1/2, p. 189 – 195 (1985). – See Abstr. 012.109.

004.155   The longitude and the new science.
J. A. Bennett.
Vistas Astron., Vol. 28, Parts 1/2, p. 219 – 225 (1985). – See Abstr. 012.109.
The Royal Observatory was established in Greenwich Park in 1675, not primarily for the pursuit of astronomy, but to lay down the necessary empirical basis for a solution to a practical problem in navigation. The problem of finding longitude at sea – referred to in the seventeenth century simply as "the longitude" – was the most important technological problem of the day.

004.156   Captain Ortiz–Canelas, a typical Spanish Naval Astronomer in the early 19th century.
A. Orte.
Vistas Astron., Vol. 28, Parts 1/2, p. 227 – 233 (1985). – See Abstr. 012.109.

004.157   Portuguese and Spanish attempts to measure longitude in the 16th century.
W. G. L. Randles.
Vistas Astron., Vol. 28, Parts 1/2, p. 235 – 241 (1985). – See Abstr. 012.109.
Attempts to measure longitude were made in the 16th century for three different purposes: astrology, cartography and navigation. This paper is only concerned with cartography and navigation.

004.158   The problem of longitude at sea in the 18th century in Spain.
A. Lafuente, M. A. Sellés.
Vistas Astron., Vol. 28, Parts 1/2, p. 243 – 250 (1985). – See Abstr. 012.109.
This paper describes the efforts made in 18th century Spain to acquire and adopt technology and knowledge on how to work out a ship's position.

004.159   Ancient geodesy: achievement and corruption.
D. Rawlins.
Vistas Astron., Vol. 28, Parts 1/2, p. 255 – 268 (1985). – See Abstr. 012.109.

004.160   The uncertainties of space and time.
K. R. Lang.
Vistas Astron., Vol. 28, Parts 1/2, p. 277 – 288 (1985). – See Abstr. 012.109.
Contents: Chandler's wobble. Distance to the Sun. The Earth's broken clock. The age of the Universe.

004.161   The Ordnance Survey and the Greenwich Meridian.
A. S. MacDonald.
Vistas Astron., Vol. 28, Parts 1/2, p. 289 – 304 (1985). – See Abstr. 012.109.

004.162   The use of rockets to determine longitude.
M. Hoskin.
Vistas Astron., Vol. 28, Parts 1/2, p. 305 (1985). – See Abstr. 012.109.

004.163 **France, Britain and the resolution of the longitude problem in the 18th century.**
A. J. Turner.
Vistas Astron., Vol. 28, Parts 1/2, p. 315 – 319 (1985). – See Abstr. 012.109.

004.164 **Sir George Airy (1801 – 1892) and the concept of international standards in science, timekeeping and navigation.**
A. Chapman.
Vistas Astron., Vol. 28, Parts 1/2, p. 321 – 328 (1985). – See Abstr. 012.109.

004.165 **The accuracy of ephemerides, 1500 – 1800.**
O. Gingerich.
Vistas Astron., Vol. 28, Parts 1/2, p. 339 – 342 (1985). – See Abstr. 012.109.

This paper brings together into a broad survey the previous results, from pre–Copernican ephemerides to about the end of the eighteenth century.

004.166 **On the history of Kepler's equation.**
E. Badolati.
Vistas Astron., Vol. 28, Parts 1/2, p. 343 – 345 (1985). – See Abstr. 012.109.

004.167 **Abridged multiplication – the architecture of Wilhelm Schickard's calculating machine of 1623.**
F. W. Kistermann.
Vistas Astron., Vol. 28, Parts 1/2, p. 347 – 353 (1985). – See Abstr. 012.109.

004.168 **Planetary observations at the Maraghah Observatory before 1275: a new set of parameters.**
G. Saliba.
Vistas Astron., Vol. 28, Parts 1/2, p. 355 (1985). – See Abstr. 012.109.

This paper introduces an observational notebook written by a thirteenth–century astronomer named Muhyi al Din al Maghribi (died 1283 A.D.), containing a record of the observations that this astronomer had carried out at the Maraghah Observatory.

004.169 **Richard Reeve – the "English Campani" – and the origins of the London telescope–making tradition.**
A. D. C. Simpson.
Vistas Astron., Vol. 28, Parts 1/2, p. 357 – 365 (1985). – See Abstr. 012.109.

004.170 **GMT in Belgium one hundred years ago.**
J. de Graeve.
Vistas Astron., Vol. 28, Parts 1/2, p. 369 – 370 (1985). – See Abstr. 012.109.

004.171 **Greenwich time in Manchester.**
H. Bloch.
Vistas Astron., Vol. 28, Parts 1/2, p. 371 – 372 (1985). – See Abstr. 012.109.

004.172 **Old Scottish meridians.**
D. M. Gavine.
Vistas Astron., Vol. 28, Parts 1/2, p. 373 (1985). – See Abstr. 012.109.

004.173 **The Madras meridian.**
C. K. Ananthasubramaniam.
Vistas Astron., Vol. 28, Parts 1/2, p. 377 – 378 (1985). – See Abstr. 012.109.

004.174 **The observatories movement in India during the 17 – 18th centuries.**
S. M. R. Ansari.
Vistas Astron., Vol. 28, Parts 1/2, p. 379 – 385 (1985). – See Abstr. 012.109.

004.175 **Toronto Magnetic Observatory and international science ca 1850.**
G. A. Good.
Vistas Astron., Vol. 28, Parts 1/2, p. 387 – 390 (1985). – See Abstr. 012.109.

This article discusses collaborative scientific research undertaken by American and Toronto researchers in the 1840s and 1850s.

004.176 **Conservation of the Airy transit circle.**
J. R. Varrall.
Vistas Astron., Vol. 28, Parts 1/2, p. 405 (1985). – See Abstr. 012.109.

004.177 **An astronomical interpretation of the frescoes in the Sala di Galatea in the Villa Farnesina, Rome.**
J. Casanovas.
Vistas Astron., Vol. 28, Parts 1/2, p. 407 (1985). – See Abstr. 012.109.

004.178 **Radio astronomy between Jansky and Reber.**
G. Reber.
Serendipitous discoveries in radio astronomy, p. 71 – 78 (1984). – See Abstr. 012.113.

004.179 **Optical and radio astronomers in the early years.**
J. L. Greenstein.
Serendipitous discoveries in radio astronomy, p. 79 – 88 (1984). – See Abstr. 012.113.

004.180 **Impact of World War II on radio astronomy.**
B. Lovell.
Serendipitous discoveries in radio astronomy, p. 89 – 104 (1984). – See Abstr. 012.113.

004.181 **Early radar research and a beginning in radio astronomy.**
A. E. Covington.
Serendipitous discoveries in radio astronomy, p. 105 – 114 (1984). – See Abstr. 012.113.

004.182 **U.S. radio astronomy following World War II.**
F. T. Haddock.
Serendipitous discoveries in radio astronomy, p. 115 – 120 (1984). – See Abstr. 012.113.

004.183 **Development of aperture synthesis at Cambridge.**
J. W. Findlay.
Serendipitous discoveries in radio astronomy, p. 126 – 128 (1984). – See Abstr. 012.113.

004.184 **The compound interferometer – a precursor of smart arrays.**
N. W. Broten.
Serendipitous discoveries in radio astronomy, p. 129 – 132 (1984). – See Abstr. 012.113.

A compound interferometer for solar radio observations is described.

004.185 **The development of Michelson and intensity long baseline interferometry.**
R. Hanbury Brown.
Serendipitous discoveries in radio astronomy, p. 133 – 145 (1984). – See Abstr. 012.113.

004.186 **Early interferometry at Jodrell Bank.**
A. R. Thompson.
Serendipitous discoveries in radio astronomy, p. 146 – 153 (1984). – See Abstr. 012.113.

004.187 **The almost serendipitous discovery of self–calibration.**
R. D. Ekers.
Serendipitous discoveries in radio astronomy, p. 154 – 159 (1984). – See Abstr. 012.113.

004.188   The discovery of pulsars.
S. J. Bell Burnell.
Serendipitous discoveries in radio astronomy, p. 160 – 170 (1984). – See Abstr. 012.113.

004.189   Discovery of quasars.
M. Schmidt.
Serendipitous discoveries in radio astronomy, p. 171 – 174 (1984). – See Abstr. 012.113.

004.190   Discovery of the 3K radiation.
D. T. Wilkinson, P. J. E. Peebles.
Serendipitous discoveries in radio astronomy, p. 175 – 184 (1984). – See Abstr. 012.113.

004.191   Discovery of the cosmic microwave background.
R. W. Wilson.
Serendipitous discoveries in radio astronomy, p. 185 – 195 (1984). – See Abstr. 012.113.

004.192   Forty years of solar radio astronomy – a history of major advances.
M. R. Kundu.
Serendipitous discoveries in radio astronomy, p. 247 – 251 (1984). – See Abstr. 012.113.

004.193   The discovery of Jupiter bursts.
K. L. Franklin.
Serendipitous discoveries in radio astronomy, p. 252 – 257 (1984). – See Abstr. 012.113.

004.194   Discovery of the Jupiter radiation belts.
F. D. Drake.
Serendipitous discoveries in radio astronomy, p. 258 – 265 (1984). – See Abstr. 012.113.

004.195   Early observations of thermal planetary radio emission.
C. H. Mayer.
Serendipitous discoveries in radio astronomy, p. 266 – 274 (1984). – See Abstr. 012.113.

004.196   Discovery of Mercury's rotation.
G. Pettengill.
Serendipitous discoveries in radio astronomy, p. 275 – 279 (1984). – See Abstr. 012.113.

004.197   The beginnings of molecular radio astronomy.
A. H. Barrett.
Serendipitous discoveries in radio astronomy, p. 280 – 290 (1984). – See Abstr. 012.113.

004.198   The discovery of radio novae.
C. M. Wade.
Serendipitous discoveries in radio astronomy, p. 291 – 293 (1984). – See Abstr. 012.113.

004.199   Serendipity in the Galaxy: the galactic warp and the galactic nucleus.
F. J. Kerr.
Serendipitous discoveries in radio astronomy, p. 294 – 299 (1984). – See Abstr. 012.113.

004.200   Anticipation in Anasazi astronomy.
M. Zeilik.
Proceedings of the Southwest Regional Conference for Astronomy and Astrophysics, Vol. 10, p. 61 (1985). Abstract. – See Abstr. 012.125.

004.201   Die Entwicklung der Astronomie – Stationen der Theoriebildung.
W. Saltzer.
Festschrift für Helmuth Gericke, Mathemata, Reihe "Boethius", Band 12, Franz Steiner Verlag, Wiesbaden – Stuttgart, p. 49 – 68 (1985).

004.202   Kometen und Biosphäre bei Isaac Newton.
W. Petri.
Festschrift für Helmuth Gericke, Mathemata, Reihe "Boethius", Band 12, Franz Steiner Verlag, Wiesbaden – Stuttgart, p. 441 – 457 (1985).

004.203   Über den Einfluß der Biegung auf die Meridiankreisbeobachtung von Bessel, Gauss und Soldner.
F. Schmeidler.
Festschrift für Helmuth Gericke, Mathemata, Reihe "Boethius", Band 12, Franz Steiner Verlag, Wiesbaden – Stuttgart, p. 511 – 521 (1985).

Bibliography of books and papers published in 1979 on the history of astronomy.
See Abstr. 002.045.

Geschichte der modernen Astronomie.
See Abstr. 003.015.

Under Newton's shadow. Astronomical practices in the seventeenth century.
See Abstr. 003.045.

Astronomy before the telescope, Vol. 1: The Earth–Moon system.
See Abstr. 003.064.

The dissolution of the celestial spheres, 1595 – 1650.
See Abstr. 003.083.

Novità celesti e crisi del sapere.
See Abstr. 003.095.

Islamic cosmology: a study of as–Suyūti's al–Hay'a as–saniya fi l'hay'a as–sunniya.
See Abstr. 003.105.

G. J. Rheticus' treatise on holy scripture and the motion of the earth.
See Abstr. 003.110.

The birth of history and philosophy of science: Kepler's "A defence of Tycho against Ursus" with essays on its provenance and significance.
See Abstr. 003.112.

Mathematical astronomy in Medieval Yemen. A bibliographical survey.
See Abstr. 003.117.

Al–Khwarizmi and new trends in mathematical astronomy in the ninth century.
See Abstr. 003.118.

Historically–astronomical investigations. Vypusk XVII.
See Abstr. 003.130.

Les Tables Alphonsines avec les Canons de Jean de Saxe.
See Abstr. 003.153.

Scientific writings and astronomical tables in Cracow: a census of manuscript sources (fourteenth to sixteenth centuries).
See Abstr. 003.161.

Tychonic and semi–tychonic world systems.
See Abstr. 003.164.

Maps of the heavens.
See Abstr. 003.170.

Stars of jade.
See Abstr. 003.172.

Die Sterne des Abd ar–Rahman as–Sufi.
See Abstr. 003.174.

Living the sky: the cosmos of the North American Indian.
See Abstr. 003.187.

Historically–astronomical investigations (Moscow).
See Abstr. 003.190.

Measuring the universe. Cosmic dimensions from Aristarchus to Halley.
See Abstr. 003.198.

Ptolemaic astronomy for an emperor's eyes.
See Abstr. 005.009.

Friedrich Wilhelm Bessel – an appreciation.
See Abstr. 005.010.

Friedrich Wilhelm Bessel (1784 – 1846). In honor of the 200th anniversary of Bessel's birth.
See Abstr. 005.011.

Friedrich Wilhelm Bessel, 1784 July 22 – 1846 March 17.
See Abstr. 005.012.

Nevil Maskelyne, the Nautical Almanac, and G.M.T.
See Abstr. 005.017.

Astronomical work of Stjepan Gradić.
See Abstr. 005.022.

Die geplante Berufung von Johannes Kepler an die Universität Rostock (1629/30).
See Abstr. 005.023.

Galileo's religion.
See Abstr. 005.025.

La dimissione di Boscovich da Brera.
See Abstr. 005.028.

My father (Karl G. Jansky) and his work.
See Abstr. 005.032.

Karl Jansky and the beginnings of radio astronomy.
See Abstr. 005.035.

Karl Guthe Jansky's serendipity, its impact on astronomy and its lessons for the future.
See Abstr. 005.036.

Observatory of the University of Helsinki 150 years old.
See Abstr. 009.028.

The establishment of the U.S. Naval Observatory.
See Abstr. 009.043.

Construction and scientific activities of the new Vienna Observatory 1870 – 1900.
See Abstr. 009.044.

The first years at Parkes.
See Abstr. 009.046.

Radio astronomy at Dover Heights.
See Abstr. 009.047.

International cooperation in monitoring the rotation of the earth.
See Abstr. 044.088.

The geodetic reference systems – the basis of theoretical geodesy.
See Abstr. 045.015.

Greek astronomical calendars. IV. The parapegma of the Egyptians and their "perpetual tables".
See Abstr. 046.003.

The heliometer as a tool for the measure of the Moon.
See Abstr. 094.019.

Far Eastern observations of Halley's comet: 240 BC to AD 1368.
See Abstr. 103.921.

Die Erscheinung des Halleyschen Kometen in den Jahren 1835 und 1836.
See Abstr. 103.932.

005 Biography

005.001   Alfred Wegener (1880 – 1930).
T. Weimer.
Astronomie, Vol. 99, p. 37 – 40 (1985).

005.002   Friedrich Wilhelm Bessel. Un astronome destiné au commerce (1784 – 1846).
F. Van't Veer.
Astronomie, Vol. 99, p. 83 – 86 (1985).

005.003   J. R. Wolf and the Zürich sunspot relative numbers.
A. J. Izenman.
Math. Intell., Vol. 7, No. 1, p. 27 – 33 (1985).

005.004   Wie was Edmond Halley?
G. W. E. Beekman.
Zenit, 12. Jaarg., No. 3, p. 84 – 86, 88 – 91 (1985).

005.005   Giorgio Abetti.
M. G. Fracastoro.
Atti Accad. Naz. Lincei, Ser. Ottava, Rend., Vol. 75, Fasc. 5, p. 259 – 287 (1983).

005.006   Ein bedeutender Astronom und "Entdecker" der Marskanäle.
U. Bach.
Astron. Raumfahrt, 22. Jahrg., Heft 2, p. 31 – 32 (1985).

005.007   Meine Erinnerungen an Cuno Hoffmeister. Teil 1.
R. Kippenhahn.
Sterne Weltraum, 24. Jahrg., Nr. 4, p. 190 – 193 (1985).

005.008   Birkeland and the electromagnetic cosmology.
A. L. Peratt.
Sky Telesc., Vol. 69, No. 5, p. 389 – 391 (1985).

005.009   Ptolemaic astronomy for an emperor's eyes.
O. Gingerich.
Sky Telesc., Vol. 69, No. 5, p. 406 – 408 (1985).

005.010   Friedrich Wilhelm Bessel – an appreciation.
Z. Kopal.
Astrophys. Space Sci., Vol. 110, No. 1, p. 3 – 10 (1985). – See Abstr. 012.039.

005.011   Friedrich Wilhelm Bessel (1784 – 1846). In honor of the 200th anniversary of Bessel's birth.
W. Fricke.
Astrophys. Space Sci., Vol. 110, No. 1, p. 11 – 19 (1985). – See Abstr. 012.039.
This paper is divided into the following parts: (1) Bessel's course of life and family. (2) Education in astronomy in Bremen. (3) Fundamenta Astronomiae. (4) Fundamental observations at the Königsberg Observatory (or according to S. Newcomb (1906): the German School of Astrometry). (5) 61 Cygni. (6) Bibliography of Bessel's original works.

005.012   Friedrich   Wilhelm   Bessel,   1784   July 22 – 1846 March 17.
P. van de Kamp.
Astrophys. Space Sci., Vol. 110, No. 1, p. 103 – 104 (1985). – See Abstr. 012.039.

005.013   M. A. Koval'skij (1821 – 1884).
D. Ya. Martynov.
Zemlya Vselennaya, No. 1, p. 47 – 50 (1985). In Russian.

005.014   Monsignor Georges Lemaitre.
A. Deprit.
The Big Bang and Georges Lemaitre, p. 363 – 392 (1984). – See Abstr. 012.043.

005.015   The scientific work of Georges Lemaitre.
O. Godart.
The Big Bang and Georges Lemaitre, p. 393 – 397 (1984). – See Abstr. 012.043.

005.016   Rudolph Minkowski: observational astrophysicist.
D. E. Osterbrock.
Phys. Today, Vol. 38, No. 4, p. 50 – 57 (1985).

005.017   Nevil Maskelyne, the Nautical Almanac, and G.M.T.
D. Howse.
J. Navig., Vol. 38, No. 2, p. 159 – 177 (1985).

005.018   Harding in Lilienthal.
D. Gerdes.
Nachr. Olbers–Ges. Bremen, Nr. 133, p. 4 – 7 (1985).

005.019   Meine Erinnerungen an Cuno Hoffmeister. Teil 2.
R. Kippenhahn.
Sterne Weltraum, 24. Jahrg., Nr. 5, p. 257 – 259 (1985).

005.020   William Huggins and astronomical spectroscopy.
J. B. Hearnshaw.
South. Stars, Vol. 31, No. 1, p. 17 – 22 (1984).

005.021   N. F. Florya.
P. G. Kulikovskij.
Zemlya Vselennaya, No. 3, p. 48 – 51 (1985). In Russian.

005.022   Astronomical work of Stjepan Gradić.
Ž. Dadić.
Publ. Astron. Opservatorije Beogr., No. 33, p. 100 – 102 (1985). – See Abstr. 012.061.

005.023   Die geplante Berufung von Johannes Kepler an die Universität Rostock (1629/30).
B. Wandt.
NTM–Schriftenreihe für Geschichte der Naturwissenschaften, Technik und Medizin, Band 19, Heft 1, p. 77 – 84 (1982). = Archenhold–Sternw. Berlin–Treptow, Sonderdr., Nr. 27 = Mitt. Archenhold–Sternw. Berlin–Treptow, Nr. 158.

005.024   Giovanni Virginio Schiaparelli: 1835 – 1910.
F. Gabici.
Coelum, Vol. 53, N. 3, p. 129 – 138 (1985).

005.025   Galileo's religion.
O. Pedersen.
Vatican Obs. Publ., Spec. Ser., Studi Galileiani, Vol. 1, No. 3, p. 75 – 102 (1985). – See Abstr. 012.064.

005.026   Seymour Lester Hess (1920 – 1982).
W. A. Baum.
Recent advances in planetary meteorology, p. IX – XI (1985). – See Abstr. 012.071.

005.027   The letters between Titius and Bonnet and the Titius–Bode law of planetary distances.
M. M. Nieto.
Am. J. Phys., Vol. 53, No. 1, p. 22 – 25 (1985). Abstr. in Phys. Abstr., Vol. 88, No. 1257, Entry 51932 (1985).

005.028   La dimissione di Boscovich da Brera.
G. Tagliaferri, P. Tucci.
G. Astron., Vol. 10, N. 3 – 4, p. 201 – 207 (1984).

005.029   Jan Klein (1684 – 1762).
Z. Horský.
Říše hvězd, Vol. 65, No. 7, p. 146 – 147 (1984). In Czech.

005.030   Friedrich Wilhelm Bessel (1784 – 1846).
Vesmír, Vol. 63, No. 12, p. 378 (1984). In Czech.

005.031   Pierre Simon Laplace (1749 – 1827).
J. Beneš.
Vesmír, Vol. 64, No. 6, p. 358 (1985). In Czech.

005.032   My father (Karl G. Jansky) and his work.
D. Jansky.
Serendipitous discoveries in radio astronomy, p. 4 – 21 (1984). – See Abstr. 012.113.

005.033   Personal recollections for the Green Bank symposium honoring Karl G. Jansky.
A. M. Jansky Parsons.
Serendipitous discoveries in radio astronomy, p. 22 – 31 (1984). – See Abstr. 012.113.

005.034   Personal recollections of Karl Jansky.
A. C. Beck.
Serendipitous discoveries in radio astronomy, p. 32 – 38 (1984). – See Abstr. 012.113.

005.035   Karl Jansky and the beginnings of radio astronomy.
W. T. Sullivan III.
Serendipitous discoveries in radio astronomy, p. 39 – 56 (1984). – See Abstr. 012.113.

005.036   Karl Guthe Jansky's serendipity, its impact on astronomy and its lessons for the future.
J. Kraus.
Serendipitous discoveries in radio astronomy, p. 57 – 70 (1984). – See Abstr. 012.113.

The biographical dictionary of scientists: astronomers.
See Abstr. 002.122.

Springs of scientific creativity: founders of modern science.
See Abstr. 003.056.

Ferdinand Verbiest, als Oost en West elkaar ontmoeten.
See Abstr. 003.063.

In the presence of the creator: Isaac Newton and his times.
See Abstr. 003.076.

Friedrich Wilhelm Bessel.
See Abstr. 003.099.

Edmond Halley: the man and his comet.
See Abstr. 003.104.

Carl Friedrich Gauss: a bibliography.
See Abstr. 003.134.

William Rowan Hamilton. Portrait of a prodigy.
See Abstr. 003.148.

Charles Piazzi Smyth, astronomer–artist; his Cape years 1835 – 1845.
See Abstr. 003.183.

Gestirnter Himmel über mir. Unverlierbares aus meinem Leben.
See Abstr. 003.184.

Kapteyn and statistical astronomy.
See Abstr. 004.006.

F. W. Bessel und die Zonenunternehmen der Astronomischen Gesellschaft.
See Abstr. 004.012.

William Stukeley and the Stonehenge sunrise.
See Abstr. 004.089.

The young Bellarmine's thoughts on world systems.
See Abstr. 004.093.

Eddington's "Expanding Universe".
See Abstr. 004.108.

Eddington's contribution to astrophysics and pulsation theory.
See Abstr. 004.109.

Captain Ortiz–Canelas, a typical Spanish Naval Astronomer in the early 19th century.
See Abstr. 004.156.

Sir George Airy (1801 – 1892) and the concept of international standards in science, timekeeping and navigation.
See Abstr. 004.164.

Kometen und Biosphäre bei Isaac Newton.
See Abstr. 004.202.

006 Personal Notes

006.001 Professor S. Bhagavantam is seventy–five.
Patrika, No. 9, p. 4 – 6 (1985).

006.002 Thomas George Cowling received the Catherine Wolfe Bruce Medal 1985 of the Astronomical Society of the Pacific.
J. E. Hesser, A. Fraknoi.
Mercury, Vol. 14, No. 2, p. 40 (1985).

006.003 Thomas George Cowling received the Catherine Wolf Bruce Medal 1985 of the Astronomical Society of the Pacific.
Phys. Today, Vol. 38, No. 6, p. 89 (1985).

006.004 Dale P. Cruikshank received the Muhlmann Prize 1985 of the Astronomical Society of the Pacific.
J. E. Hesser, A. Fraknoi.
Mercury, Vol. 14, No. 2, p. 41 – 44 (1985).

006.005 Dale P. Cruikshank received the Muhlmann Prize 1985 of the Astronomical Society of the Pacific.
Phys. Today, Vol. 38, No. 6, p. 90 (1985).

006.006 Cornélis de Jager received the Prix Janssen 1984.
Astronomie, Vol. 99, p. 102 (1985). In French.

006.007 Angeles I. Diaz and Luis F. Rodriguez received the 1984 Henri Chretien Award of the American Astronomical Society.
Phys. Today, Vol. 38, No. 3, p. 131 (1985).

006.008 Robert Evans and Gregg Thompson received the Amateur Achievement Award of the Astronomical Society of the Pacific.
J. E. Hesser, A. Fraknoi.
Mercury, Vol. 14, No. 2, p. 44 – 45 (1985).

006.009 Robert Evans and Gregg Thompson received the Amateur Achievement Award of the Astronomical Society of the Pacific.
Phys. Today, Vol. 38, No. 6, p. 90 (1985).

006.010 Paul Hertz received the Robert J. Trumpler Award 1985 of the Astronomical Society of the Pacific.
J. E. Hesser, A. Fraknoi.
Mercury, Vol. 14, No. 2, p. 41 (1985).

006.011 Paul Hertz received the Trumpler Award 1985 of the Astronomical Society of the Pacific.
Phys. Today, Vol. 38, No. 6, p. 89 – 90 (1985).

006.012 Jean Kovalevsky received the Prix Joannidès.
C. R. Acad. Sci., Sér. Gén., Vie Sci., Tome 1, No. 6, p. 547 (1984). In French.

006.013 John D. Kraus received the 1985 IEEE Edison Medal.
Phys. Today, Vol. 38, No. 4, p. 98 (1985).

006.014 Presentation of the 1983 Herschel Medal to W. W. Morgan.
P. A. Wayman.
The MK process and stellar classification, p. XIX – XX (1984). –
See Abstr. 012.033.

006.015 L. Perek, 65th birthday.
V. Bumba.
Říše hvězd, Vol. 65, No. 7, p. 148 (1984). In Czech.

006.016 Judith L. Pipher received the Ernest F. Fullam Award of the Dudley Observatory 1984–85.
Phys. Today, Vol. 38, No. 6, p. 92 (1985).

006.017 On the life and work of B. J. Popović on the occasion of his seventieth birthday.
B. D. Jovanović.
Vasiona, Année 33, No. 1 – 2, p. 17 – 20 (1985). In Serbo-Croatian.

006.018 Luis F. Rodriguez and Angeles I. Diaz received the 1984 Henri Chretien Award of the American Astronomical Society.
Phys. Today, Vol. 38, No. 3, p. 131 (1985).

006.019 The seventieth birthday of B. Ševarlić.
D. Teleki.
Vasiona, Année 33, No. 1 – 2, p. 21 – 23 (1985). In Serbo–Croatian.

006.020 Eugene M. Shoemaker received the Barringer Award.
Meteoritics, Vol. 19, No. 4, p. 177 – 182 (1984).

006.021 James Stokley received the Klumpke–Roberts Award 1985 of the Astronomical Society of the Pacific.
J. E. Hesser, A. Fraknoi.
Mercury, Vol. 14, No. 2, p. 44 (1985).

006.022 James Stokley received the Klumpke–Roberts Award 1985 of the Astronomical Society of the Pacific.
Phys. Today, Vol. 38, No. 6, p. 90 (1985).

006.023 Gregg Thompson and Robert Evans received the Amateur Achievement Award of the Astronomical Society of the Pacific.
J. E. Hesser, A. Fraknoi.
Mercury, Vol. 14, No. 2, p. 44 – 45 (1985).

006.024 Gregg Thompson and Robert Evans received the Amateur Achievement Award of the Astronomical Society of the Pacific.
Phys. Today, Vol. 38, No. 6, p. 90 (1985).

006.025 Y. B. Zeldovich, 70th birthday.
J. Štohl.
Kozmos, Vol. 15, No. 5, p. 162 (1984). In Slovak.

007 Obituaries

007.001 M. A. Arakelyan, 1929 – 1983 January 20.
Soobshch. Byurakan. Obs., Vyp. 54, p. 89 – 90 (1983). In Russian and Armenian.

007.002 A. E. H. Bleksley 1908 – 1984.
M. D. Overbeek.
Mon. Notes Astron. Soc. S. Afr., Vol. 44, Nos. 1 – 2, p. 2 – 4 (1985).

007.003 Student memories of Bart Bok, an astronomical godfather.
J. B. Whiteoak.
Proc. Astron. Soc. Aust., Vol. 5, No. 4, p. 608 – 610 (1984).

007.004 In memoriam: Giuseppe Colombo 1920 – 1984.
F. Franklin, M. Grossi, D. Lautman, M. Lecar, I. Shapiro.
Icarus, Vol. 62, No. 1, p. 1 – 3 (1985).

007.005 John P. Cox, 1926 November 4 – 1984 August 19.
C. J. Hansen.
Phys. Today, Vol. 38, No. 1, p. 112 (1985).

007.006 Paul Adrian Maurice Dirac died 1984 October 20.
Patrika, No. 9, p. 7 (1985).

007.007 In memoriam: Cyril Pavlovich Florensky (1915 December 27 – 1982 April 9).
Icarus, Vol. 61, No. 3, p. 351 – 354 (1985).

007.008 E. G. Forbes, 1933 – 1984.
M. Hoskin.
J. Hist. Astron., Vol. 16, Part 1, No. 45, p. 74 (1985).

007.009 La vie et l'œuvre d'Otto Heckmann.
C. Fehrenbach.
C. R. Acad. Sci., Sér. Gén., Vie Sci., Tome 1, No. 6, p. 591 – 593 (1984).

007.010 Henry McKnight Horstman died on 1984 November 9.
G. G. C. Palumbo, G. Pizzichini.
Phys. Today, Vol. 38, No. 3, p. 134 – 135 (1985).

007.011 Martin Christopher Johnson, 1896 October 11 – 1984 November 26.
W. E. Burcham, W. H. McCrea.
Q. J. R. Astron. Soc., Vol. 26, No. 2, p. 219 – 221 (1985).

007.012 D. V. Korol'kov, 1925 – 1984 January 10.
Yu. N. Parijskij.
Astrofiz. Issled. Izv. Spets. Astrofiz. Obs., Tom 19, p. 120 – 122 (1985). In Russian. English translation in Bull. Spec. Astrophys. Obs. – North Caucasus.

007.013 D. V. Korol'kov, 1925 – 1984.
Yu. N. Parijskij.
Radiotekh. Ehlektron., Moskva, Tom 29, No. 9, p. 1851 – 1852 (1984). In Russian. Abstr. in Ref. Zh., 51. Astron., 3.51.28 (1985).

007.014 V. A. Krat, 1911 July 21 – 1983 June 2.
Izv. Glav. Astron. Obs. Pulkovo, Astrometr. Astrofiz., No. 202, p. 3 – 5 (1984). In Russian.

007.015 Anton Kutter, 13. Juni 1903 – 1. Februar 1985.
G. D. Roth.
Mitt. Astron. Ges., Nr. 64, p. 9 – 10 (1985).

007.016 F. Link died 1984 September 23.
Říše hvězd, Vol. 65, No. 11, p. 238 (1984). In Czech.

007.017 Leven en werk van Lukas Plaut (1910 June 5 – 1984 October 4).
A. Blaauw.
Zenit, 12. Jaarg., No. 4, p. 152 – 153 (1985).

007.018 Kalpathi Ramakrishna Ramanathan, 1893 February 28 – 1984 December 31.
Patrika, No. 9, p. 8 – 9 (1985).

007.019 Svein Rosseland, 1894 – 1985.
Ø. Elgarøy.
Astron. Tidsskr., Årg. 18, Nr. 2, p. 71 – 72 (1985).

007.020 In memoriam: A. I. Rybakov, 1908 November 15 – 1984 September 7.
Tr. Gos. Astron. Inst. Shternberg, Tom 57, p. 3 – 4 (1985). In Russian.

007.021 Martin Ryle, pioneer radio astronomer.
F. G. Smith.
Sky Telesc., Vol. 69, No. 2, p. 123 (1985).

007.022 Sir Martin Ryle (1918 – 1984).
J. Bouška.
Vesmír, Vol. 64, No. 7, p. 418 (1985). In Czech.

007.023   **I. S. Shklovskij, 1916 July 1 – 1985 March 3.**
Astron. Zh., Tom 62, Vyp. 3, p. 618 – 619 (1985). In Russian. English translation in Sov. Astron., Vol. 29, No. 3.

007.024   **I. S. Shklovskij died 1985 March 3.**
Kosm. Issled., Tom 23, Vyp. 3, p. 495 (1985). In Russian. English translation in Cosm. Res.

007.025   **I. S. Shklovskij, 1916 July 1 – 1985 March 3.**
Pis'ma Astron. Zh., Tom 11, No. 4, p. 319 – 320 (1985). In Russian. English translation in Sov. Astron. Lett., Vol. 11.

007.026   **J. S. Šklovskij, 1916 – 1985.**
Říše hvězd, Vol. 66, No. 6, p. 113 (1985). In Czech.

007.027   **Hermann Slevogt, 10. Mai 1909 – 13. September 1984.**
H. Schmidt.
Mitt. Astron. Ges., Nr. 64, p. 5 – 6 (1985).

007.028   **Sergej Ivanovich Syrovatskij, 1925 March 2 – 1979 September 26.**
V. Dogiel (*V. A. Dogel'*).
Unstable current systems and plasma instabilities in astrophysics, p. XIII – XV (1985). – See Abstr. 012.002 (IAU Symp. No. 107).

007.029   **Commemorazione di don Giuseppe Tagliaferri.**
S. Piovaneli.
G. Astron., Vol. 10, N. 3 – 4, p. 145 – 146 (1984).

007.030   **In memoriam of Giuseppe Tagliaferri.**
M. G. Fracastoro.
Mem. Soc. Astron. Ital., Vol. 55, No. 3, p. 517 – 520 (1984). – See Abstr. 012.050.

007.031   **K. N. Tavastsherna, 1921 May 1 – 1982 June 24.**
Izv. Glav. Astron. Obs. Pulkovo, Astrometr. Astrofiz., No. 201, p. 3 – 4 (1985). In Russian.

007.032   **Stefan Temesváry 1915 – 13. September 1984.**
R. Kippenhahn.
Mitt. Astron. Ges., Nr. 64, p. 7 – 8 (1985).

007.033   **S. K. Vsechsvjatskij died 1984 October 6.**
Říše hvězd, Vol. 66, No. 5, p. 88 (1985). In Czech.

007.034   **S. K. Vsekhsvyatskij, 1905 June 20 – 1984 October 6.**
Zemlya Vselennaya, No. 2, p. 60 – 63 (1985). In Russian.

007.035   **Professor H. L. Welsh, 1910 – 1984.**
D. A. MacRae.
J. R. Astron. Soc. Can., Vol. 79, No. 3, p. 109 – 112 (1985).

007.036   **Harley Weston Wood, 1911 July 31 – 1984 June 26.**
W. H. Robertson.
Q. J. R. Astron. Soc., Vol. 26, No. 2, p. 225 – 228 (1985).

008 Publications of Observatories, Institutes

Reports, communications, and publications of observatories and astronomical institutes are recorded in this section; included are numbered series of reprints. Whenever possible, the numbers of the abstracts referring to the publications are given. The places of observatories and institutes are listed in alphabetical order. If only the formal name of an observatory or institute is known, its place can be found in the following index list.

Aerospace Corporation — **El Segundo,** Calif.
Algonquin Radio Observatory — **Lake Traverse,** Canada
Allegheny Observatory — **Pittsburgh,** Pa.
Anglo-Australian Observatory — **Epping,** Australia
Antares Observatory — **Feira de Santana,** Brazil
Applied Research Corporation — **Landover,** Md.
Archenhold Observatory — **Berlin,** German Democratic Republic
Argentine Institute of Radio Astronomy — **Villa Elisa,** Argentina
Argentine Radio Astronomy Institute — **Pereyra,** Argentina
Arizona State University — **Tempe,** Ariz.
Astronomisches Rechen-Institut — **Heidelberg,** F.R. Germany

Babelsberg Observatory — **Potsdam,** German Democratic Republic
Bartol Research Foundation — **Newark,** Del.
Battelle Memorial Institute — **Richland,** Wash.
Bell Telephone Laboratories — **Holmdel,** N.J.
Bergedorf Observatory — **Hamburg,** F.R. Germany
Blindern Institute of Theoretical Astrophysics — **Oslo,** Norway
Bosscha Observatory — **Lembang,** Indonesia
Bouzareah Observatory — **Algiers,** Algeria
Boyden Observatory — **Bloemfontein,** South Africa
Bureau International de l'Heure — **Paris,** France

C. E. Kenneth Mees Observatory — **Rochester,** N.Y.
California Institute of Technology — **Pasadena,** Calif.
Cantonal Observatory — **Neuchâtel,** Switzerland
Carter Observatory — **Wellington,** New Zealand
Center for Astrophysics — **Cambridge,** Mass.
Center for High Angular Resolution Astronomy, Georgia State University — **Atlanta,** Ga.
Centre de Données Stellaires — **Strasbourg,** France
Centro Astronomico Hispano-Aleman — **Calar Alto,** Spain
Centro de Investigación de Astronomía (CIDA) — **Mérida,** Venezuela
Cerro Calan National Astronomical Observatory — **Santiago,** Chile
Cerro Tololo Interamerican Observatory — **La Serena,** Chile
Chamberlin Observatory — **Denver,** Colo.
Charles University Astronomical Institute — **Prague,** Czechoslovakia
Climenhaga Observatory — **Victoria,** Canada
Cointe Observatory — **Liège,** Belgium
Computer Sciences Corporation — **Silver Spring,** Md.
Cornell University — **Ithaca,** N.Y.
Crawford Hill Observatory — **Holmdel,** N.J.

David Dunlop Observatory — **Richmond Hill,** Canada
Dearborn Observatory — **Evanston,** Ill.
Department of Astronomy and Space Sciences, Punjabi University — **Patiala,** India
Department of Mathematics, University of Poona — **Poona,** India
Deutsches Hydrographisches Institut — **Hamburg,** F.R. Germany
Dominion Astrophysical Observatory — **Ottawa,** Canada
Dominion Astrophysical Observatory — **Victoria,** Canada
Dominion Radio Astrophysical Observatory — **Penticton,** Canada
Dudley Observatory — **Schenectady,** N.Y.
Dunsink Observatory — **Dublin,** Ireland
Dyer Observatory — **Nashville,** Tenn.

Ebro Observatory — **Roquetas,** Spain
Effelsberg Radio Observatory — **Bonn,** F.R. Germany
Ege University Observatory — **Izmir,** Turkey
Electronics Research Laboratory — **Göteborg,** Sweden
Engelhardt Observatory — **Kazan,** USSR
Erwin W. Fick Observatory — **Ames,** Iowa
European Southern Observatory — **Garching,** F.R. Germany
European Southern Observatory — **La Silla,** Chile

Fabra Observatory	Barcelona, Spain
Felix Aguilar Observatory	San Juan, Argentina
Fernbank Science Center	Atlanta, Ga.
Figl Observatory	Vienna, Austria
Five College Astronomy Department	Amherst, Mass.
Floirac Observatory	Bordeaux, France
Florida State University Radio Observatory	Tallahassee, Fla.
Flower and Cook Observatory	Malvern, Pa.
Fort Skala Station	Cracow, Poland
Franko State University Astronomical Observatory	Lvov, USSR
Georgia State University	Atlanta, Ga.
Goddard Space Flight Center	Greenbelt, Md.
Goethe Link Observatory	Bloomington, Indiana
Hale Observatories	Pasadena, Calif.
Hartebeesthoek Radio Astronomy Observatory	Johannesburg, South Africa
Haute Provence Observatory	Saint Michel, France
Haystack Observatory	Westford, Mass.
Heinrich-Hertz-Institut	Berlin, German Democratic Republic
Heliophysical Observatory	Debrecen, Hungary
Herzberg Institute of Astrophysics	Ottawa, Canada
High Altitude Observatory	Boulder, Colo.
H.M. Nautical Almanac Office	Greenwich, England
Hoher List Observatory	Bonn, F.R. Germany
Hopkins Observatory	Williamstown, Mass.
Horn d'Arturo Observatory	Bologna, Italy
Hvar Observatory	Zagreb, Yugoslavia
IBM Thomas J. Watson Research Center	Yorktown Heights, N.Y.
Indian Institute of Astrophysics	Bangalore, India
Infrared Telescope Facility	Honolulu, Hawaii
Institut Astrofiziki	Dushanbe, USSR
Instituto Nacional de Astrofisica, Optica y Electronica	Puebla, Pue., México
International Latitude Observatory	Carloforte-Cagliari, Italy
International Latitude Observatory	Mizusawa, Japan
IUE Observatory	Villafranca, Spain
James Mims Observatory	Baton Rouge, La.
Jet Propulsion Laboratory	Pasadena, Calif.
Jodrell Bank Radio Observatory	Manchester, England
Judson B. Coit Observatory	Boston, Mass.
Kandilli Observatory	Istanbul, Turkey
Kanzelhöhe Solar Observatory	Graz, Austria
Kapteyn Astronomical Laboratory	Groningen, The Netherlands
Karl Schwarzschild Observatory	Tautenburg, German Democratic Republic
Kiepenheuer Institut	Freiburg, F.R. Germany
Kiso Observatory	Tokyo, Japan
Kitt Peak National Observatory	Tucson, Ariz.
Kodaikanal Observatory	Bangalore, India
Königstuhl Observatory	Heidelberg, F.R. Germany
Konkoly Observatory	Budapest, Hungary
Korean National Astronomical Observatory	Seoul, Korea
Kwasan and Hida Observatories	Kyoto, Japan
Las Campanas Observatory	Pasadena, Calif.
Latitude Station of the Polish Academy of Sciences	Borowiec, Poland
Lawrence Livermore National Laboratory	Livermore, Calif.
Leander McCormick Observatory	Charlottesville, Va.
Lick Observatory	Santa Cruz, Calif.
Lindheimer Astronomical Research Center	Evanston, Ill.
Llano del Hato Observatory	Merida, Venezuela
Lockheed Palo Alto Research Laboratory	Palo Alto, Calif.
Lockheed Solar Observatory	Saugus, Calif.
Lohrmann Observatory	Dresden, German Democratic Republic
Louisiana State University Observatory	Baton Rouge, La.
Lowell Observatory	Flagstaff, Ariz.
Lunar and Planetary Laboratory	Tucson, Ariz.
Main Astronomical Observatory of the USSR Academy of Sciences	Pulkovo, USSR
Max-Planck-Institut für Astronomie	Calar Alto, Spain

Max-Planck-Institut für Astronomie — **Heidelberg, F.R. Germany**
Max-Planck-Institut für Kernphysik — **Heidelberg, F.R. Germany**
Max-Planck-Institut für Physik und Astrophysik — **Garching, F.R. Germany**
Max-Planck-Institut für Radioastronomie — **Bonn, F. R. Germany**
McDonald Observatory — **Fort Davis, Tex.**
McDonnell Center for the Space Sciences — **St. Louis, Mo.**
Meudon Observatory — **Paris, France**
Michigan State University Observatory — **East Lansing, Mich.**
Millstone Hill Radar Observatory — **Westford, Mass.**
Minor Planet Center — **Cambridge, Mass.**
Molonglo Radio Observatory — **Sydney, Australia**
Monterey Institute for Research in Astronomy — **Carmel Valley, Calif.**
Mount Hamilton — **Santa Cruz, Calif.**
Mount John University Observatory — **Lake Tekapo, New Zealand**
Mount Palomar Observatory — **Pasadena, Calif.**
Mount Stromlo Observatory — **Canberra, Australia**
Mount Wilson Observatory — **Pasadena, Calif.**
Mullard Radio Astronomy Observatory — **Cambridge, England**
Mullard Space Science Laboratory — **London, England**
Multiple Mirror Telescope Observatory — **Tucson, Ariz.**

N. Copernicus Astronomical Center — **Warsaw, Poland**
N. Copernicus University Observatory — **Torun, Poland**
Narrabri Observatory — **Sydney, Australia**
NASA Headquarters — **Washington, D.C.**
National Astronomy and Ionosphere Center — **Ithaca, N.Y.**
National Bureau of Standards — **Washington, D.C.**
National Radio Astronomy Observatory — **Green Bank, W.Va.**
National Radio Astronomy Observatory — **Socorro, N.M.**
National Solar Observatory — **Sunspot, N.M.**
Naval Observatory — **San Fernando (Cadiz), Spain**
New Mexico State University Observatory — **Las Cruces, N.M.**
Nicholas Copernicus Observatory — **Brno, Czechoslovakia**
Nizamiah and Japal-Rangapur Observatories — **Hyderabad, India**
Nuffield Radio Astronomy Laboratories — **Manchester, England**

Oak Ridge Observatory — **Cambridge, Mass.**
Observatorio del Ebro — **Roquetas, Spain**
Ole Roemer Observatory — **Aarhus, Denmark**
Onsala Space Observatory — **Göteborg, Sweden**
Owens Valley Radio Observatory — **Big Pine, Calif.**

Pacific Northwest Laboratory — **Richland, Wash.**
Pennsylvania State University — **University Park, Pa.**
Perkins Observatory — **Delaware, Ohio**
Perth Observatory — **Bickley, Australia**
Pic du Midi Observatory — **Toulouse, France**
Pino Torinese Observatory — **Turin, Italy**
Purple Mountain Observatory — **Nanking, China**

Raman Research Institute — **Bangalore, India**
Rattlesnake Mountain Observatory — **Richland, Wash.**
Remeis Observatory — **Bamberg, F.R. Germany**
Rensselaer Observatory — **Troy, N.Y.**
Ritter Astrophysical Research Center — **Toledo, Ohio**
Rosemary Hill Observatory — **Bronson, Fla.**
Rothney Astrophysical Observatory — **Calgary, Canada**
Royal Observatory — **Edinburgh, Scotland**
Royal Observatory — **Greenwich, England**
Royal Observatory of Belgium — **Uccle, Belgium**
Rutgers University — **Piscataway, N.J.**

Sagamore Hill Radio Observatory — **Hanscom, Mass.**
Saltsjöbaden Observatory — **Stockholm, Sweden**
San Vittore Observatory — **Bologna, Italy**
Satelliten-Beobachtungsstation — **Zimmerwald, Switzerland**
Shaanxi Astronomical Observatory — **Lintong, China**
Smithsonian Astrophysical Observatory — **Cambridge, Mass.**
Sommers-Bausch Observatory — **Boulder, Colo.**
Sonnenborgh Observatory — **Utrecht, The Netherlands**
Sonoma State University — **Rohnert Park, Calif.**
South African Astronomical Observatory — **Cape Town, South Africa**
Space Telescope Science Institute — **Baltimore, Md.**
Special Astrophysical Observatory — **Zelenchukskaya, USSR**

Sproul Observatory
Stanford Center for Radar Astronomy
Stellar Data Center
Sternberg State Astronomical Institute
Steward Observatory
Stockert Radio Observatory
Struve Astrophysical Observatory

Tata Institute of Fundamental Research
Tôhoku University Observatory

United Kingdom Infrared Telescope (UKIRT)
University of Sussex
U.S. Naval Observatory
UT Radio Astronomy Observatory
Uttar Pradesh State Observatory

Van Vleck Observatory
Vatican Observatory

Warner and Swasey Observatory
Washburn Observatory
Wesleyan Radio Observatory
Western Ontario University Observatory
Whipple Observatory (MMT)
Wilhelm Foerster Observatory

Yale University Astronomical Observatory
Yerkes Observatory

Zentralinstitut für Physik der Erde

Swarthmore, Pa.
Menlo Park, Calif.
Strasbourg, France
Moscow, USSR
Tucson, Ariz.
Bonn, F.R. Germany
Tartu, USSR

Bombay, India
Sendai, Japan

Hilo, Hawaii
Brighton, England
Washington, D.C.
Marfa, Tex.
Naini Tal, India

Middletown, Conn.
Castel Gandolfo, Vatican City

Cleveland, Ohio
Madison, Wis.
Delaware, Ohio
London, Canada
Tucson, Ariz.
Berlin, F.R. Germany

New Haven, Conn.
Williams Bay, Wis.

Potsdam, German Democratic Republic

008.018   Alma–Ata

Trudy Astrofizicheskogo Instituta.
Tom 40 (39.152.006, 39.151.135, 39.151.136, 39.160.144,
39.157.211, 39.002.095, 39.151.137 – 39.151.139, 39.155.151,
39.043.010, 39.102.049),
Tom 41 (39.072.085, 39.072.086, 39.036.158, 39.072.087,
39.074.128, 39.064.087, 39.034.115, 39.074.129, 39.036.159,
39.036.160, 39.073.182, 39.071.034),
Tom 42 (39.131.302, 39.131.303, 39.064.110, 39.134.064,
39.131.304, 39.132.060, 39.114.133, 39.114.134, 39.082.075,
39.036.177, 39.034.116, 39.031.041),
Tom 43 (39.042.089 – 39.042.091, 39.151.140, 39.161.280,
39.151.141, 39.066.143, 39.159.125, 39.151.142, 39.066.144,
39.066.145, 39.066.146, 39.066.147),
Tom 44 (39.063.082, 39.133.014, 39.134.065, 39.132.061,
39.114.135, 39.034.117, 39.053.011, 39.036.178).

008.021   Ames, Iowa

**Iowa State University, Erwin W. Fick Observatory, Ames, Iowa
50011. Report for the period 1 September 1983 – 31 August
1984.**
Bull. Am. Astron. Soc., Vol. 17, No. 1, p. 216 – 218 (1985).

008.024   Amherst, Mass.

**Five College Astronomy Department: Amherst College,
Amherst, Massachusetts 01002, Hampshire College, Amherst,
Massachusetts 01002, Mount Holyoke College, South Hadley,
Massachusetts 01075, Smith College, Northampton,
Massachusetts 01063, University of Massachusetts, Amherst,
Massachusetts 01003. Report for the academic year 1983 –1984.**
Bull. Am. Astron. Soc., Vol. 17, No. 1, p. 149 – 156 (1985).

008.039   Arecibo, Puerto Rico

**Arecibo Observatory, Arecibo, Puerto Rico 00613. Report for
the period July 1983 – June 1984.**
F. Six.
Bull. Am. Astron. Soc., Vol. 17, No. 1, p. 7 – 13 (1985).

008.051   Atlanta, Ga.

**Georgia State University, Department of Physics and
Astronomy, Atlanta, Georgia 30303. Report for the period
10 October 1982 – 30 September 1984.**
H. A. McAlister.
Bull. Am. Astron. Soc., Vol. 17, No. 1, p. 166 – 169 (1985).

008.054   Austin, Tex.

**University of Texas at Austin, Department of Astronomy,
Austin, Texas 78712. Report for the period 1 September 1983 –
31 August 1984.**
H. J. Smith, F. N. Bash, J. N. Douglas, P. A. Vanden Bout.
Bull. Am. Astron. Soc., Vol. 17, No. 1, p. 425 – 452 (1985).

University of Texas, Publications in Astronomy.
Nos. 22 (39.012.013), 23 (39.151.071).

Department of Astronomy of the University of Texas, Reprints.
Nos. 1112 (22.100.006), 1113 (33.125.201), 1114 (33.114.008),
1115 (33.114.015), 1116 (33.122.006), 1117 (34.121.096),
1118 (33.021.005), 1119 (33.161.014), 1120 (33.161.020),
1121 (33.126.004), 1122 (33.131.042), 1123 (33.101.003),
1124 (33.103.002), 1125 (33.121.007), 1126 (33.132.038), 1127,
1128 (33.131.063), 1129 (33.151.032), 1130 (33.101.015),
1131 (33.153.017), 1132 (33.114.062), 1133 (33.126.024),

1134 (33.103.222), 1135 (33.008.052), 1136 (33.154.055),
1137 (33.034.103), 1138 (33.161.181, 33.161.182),
1139 (33.153.027), 1140 (33.042.033), 1141 (33.100.082),
1142 (33.071.012), 1143 (34.154.005), 1144 (33.157.227),
1145 (34.125.221), 1146 (34.159.003), 1147 (33.131.272),
1148 (33.022.153), 1149 (33.131.275), 1150 (34.114.001).

008.057 **Baltimore, Md.**

**The Johns Hopkins University, Department of Physics and
Astronomy, Baltimore, Maryland 21218. Report.**
Bull. Am. Astron. Soc., Vol. 17, No. 1, p. 219 – 222 (1985).

**Space Telescope Science Institute, Baltimore, Maryland 21218.
Report.**
Bull. Am. Astron. Soc., Vol. 17, No. 1, p. 403 – 419 (1985).

Space Telescope Science Institute, Preprint Series.
Nos. 35 (39.160.073), 36 (39.114.042), 37 (39.160.074),
38 (39.158.168), 39 (39.124.141), 40 (39.002.048),
41 (39.159.065), 42 (39.114.043), 43 (39.131.273),
44 (39.151.072), 45 (39.158.169), 46 (39.132.047).

Space Telescope Science Institute, Newsletter.
Vol. 2, No. 1 (1985).

008.060 **Bamberg**

**Dr.–Remeis–Sternwarte Bamberg, Astronomisches Institut der
Universität Erlangen–Nürnberg. Report for 1984.**
J. Rahe.
Mitt. Astron. Ges., Nr. 64, p. 13 – 18 (1985).

008.069 **Basel**

Astronomisches Institut der Universität Basel. Report for 1984.
G. A. Tammann.
Mitt. Astron. Ges., Nr. 64, p. 19 – 25 (1985).

008.072 **Baton Rouge, La.**

**Louisiana State University Observatory, Baton Rouge,
Louisiana 70803–4001. Report for the period 1 July 1983 –
30 June 1984.**
A. U. Landolt.
Bull. Am. Astron. Soc., Vol. 17, No. 1, p. 259 – 263 (1985).

008.078 **Belgrade**

Publikacije Astronomske opservatorije u Beogradu.
No. 33 (39.012.061).

008.081 **Berkeley, Calif.**

**University of California at Berkeley, Berkeley, California
94720. Report.**
Bull. Am. Astron. Soc., Vol. 17, No. 1, p. 46 – 55 (1985).

**University of California, Ten Meter Telescope, Berkeley,
California 94720. Report for the period July 1983 – June 1984.**
Bull. Am. Astron. Soc., Vol. 17, No. 1, p. 85 (1985).

008.084 **Berlin (East)**

**Jahresberichte der Archenhold–Sternwarte für die Jahre 1982
und 1983.**
D. B. Herrmann.
Mitt. Archenhold–Sternw. Berlin–Treptow, Nr. 140, 14 pp.
(1984).

Archenhold–Sternwarte Berlin–Treptow, Sonderdruck.
Nr. 27 (39.005.023), 28 (39.015.035).

Mitteilungen der Archenhold–Sternwarte Berlin–Treptow.
Nr. 140 (39.008.084), 151 (38.002.028), 156 (38.004.060),
157 (38.004.061), 158 (39.005.023).

Veröffentlichungen der Archenhold–Sternwarte Berlin–Treptow.
Nr. 11 (39.002.051), 12 (39.002.052).

Archenhold–Sternwarte Berlin–Treptow, Vorträge und Schriften.
Nr. 64 (39.009.024).

Blick in die Sternenwelt 1985.
(39.046.013).

**Zentralinstitut für solar–terrestrische Physik
(Heinrich–Hertz–Institut). Report for 1983.**
Akad. Wiss. DDR, Jahrb. 1983, p. 223 – 226 (1985).

Heinrich–Hertz–Institut Supplement Series of Solar Data.
Vol. 5, No. 5 (39.072.058).

Institut für Kosmosforschung. Report for 1983.
Akad. Wiss. DDR, Jahrb. 1983, p. 240 – 241 (1985).

008.087 **Berlin (West)**

**Institut für Astronomie und Astrophysik der Technischen
Universität Berlin. Report for 1984.**
R. Wielen.
Mitt. Astron. Ges., Nr. 64, p. 27 – 35 (1985).

008.093 **Bloemfontein**

Boyden Observatory, Reprint.
No. 57 (38.122.189).

008.096 **Bloomington, Indiana**

**Goethe Link Observatory, Indiana University, Bloomington,
Indiana 47405. Report for the period September 1983 – August
1984.**
R. K. Honeycutt.
Bull. Am. Astron. Soc., Vol. 17, No. 1, p. 170 – 174 (1985).

008.099 **Bochum**

Astronomisches Institut der Ruhr–Universität. Report for 1984.
T. Schmidt–Kaler.
Mitt. Astron. Ges., Nr. 64, p. 37 – 44 (1985).

**Institut für Theoretische Physik, Lehrstuhl IV der
Ruhr–Universität. Report for 1984.**
K. Schindler.
Mitt. Astron. Ges., Nr. 64, p. 45 – 46 (1985).

008.111 **Bonn**

**Astronomisches Institut der Universität Bonn: I. Sternwarte mit
Observatorium Hoher List. Report for 1984.**
H. Schmidt.
Mitt. Astron. Ges., Nr. 64, p. 47 – 58 (1985).

**Astronomisches Institut der Universität Bonn:
II. Radioastronomisches Institut und Teilprojekt A5 und A7 des
Sonderforschungsbereichs 131, Radioastronomie. Report for
1984.**
U. Mebold.
Mitt. Astron. Ges., Nr. 64, p. 59 – 65 (1985).

Astronomisches Institut der Universität Bonn: III. Institut für Astrophysik und Extraterrestrische Forschung. Report for 1984.
W. Priester.
Mitt. Astron. Ges., Nr. 64, p. 66 – 73 (1985).

Max–Planck–Institut für Radioastronomie. Report for 1984.
P. G. Mezger.
Mitt. Astron. Ges., Nr. 64, p. 75 – 83 (1985).

Max–Planck–Institut für Radioastronomie, Bonn, Sonderdrucke Serie A.
Nos. 627 (33.131.251), 628 (39.033.023), 629 (34.159.041), 630 (33.011.032), 631 (37.158.303), 632 (37.160.152), 633, 634 (34.131.022), 635 (34.131.094), 636 (34.132.005), 637 (34.122.013), 638 (34.103.105), 639 (37.142.106), 640 (34.131.109), 641 (34.132.031), 642 (34.131.017), 643 (34.161.030), 644 (39.033.024), 645 (34.159.018), 646 (34.131.029), 647 (34.158.009), 648 (34.159.047), 649 (39.033.025), 650 (34.160.012), 651 (34.155.081), 652 (39.033.026), 653 (39.033.027), 654 (39.033.028), 655 (34.131.120), 656 (34.132.034), 657 (34.157.071), 658 (39.033.029), 659 (34.062.071), 660 (34.157.072), 661 (34.131.123), 662 (34.132.035), 663 (34.131.122), 664 (34.131.145), 665 (37.159.064), 666 (39.033.030), 667 (34.157.074), 668 (34.155.069), 669 (34.131.177), 670 (37.155.039), 671 (37.131.006), 672 (37.112.001), 673 (39.033.031), 674 (37.046.001), 675 (37.158.009), 676 (37.141.001), 677 (37.158.007), 678 (37.158.108), 679 (37.063.015), 680 (37.159.121), 681 (37.159.120), 682 (37.131.030), 683 (37.159.096), 684 (37.157.062), 685 (37.157.107), 686 (37.157.106), 687, 688, 689 (37.158.229), 690 (37.125.033), 691 (37.125.034), 692 (37.158.091), 693 (38.160.125), 694 (38.160.110), 695 (37.141.016), 696 (37.158.144), 697 (37.131.127), 698 (37.131.178), 699, 700 (37.157.199).
This series terminates with No. 700.

Mitteilungen aus den Geodätischen Instituten der Rheinischen Friedrich–Wilhelms–Universität Bonn.
Nr. 66 (39.094.031), 67 (39.083.013).

008.120   Boston, Mass.

Boston University, Astronomy Department, Boston, Massachusetts 02215. Report for the period September 1983 – August 1984.
Bull. Am. Astron. Soc., Vol. 17, No. 1, p. 39 – 42 (1985).

008.123   Boulder, Colo.

High Altitude Observatory, National Center for Atmospheric Research, Boulder, Colorado 80307. Report for the year 1983.
D. K. Watson.
Bull. Am. Astron. Soc., Vol. 17, No. 1, p. 191 – 201 (1985).

008.129   Brno

Astronomical Institute of the University Brno, Contributions.
Nos. 20 (27.123.022), 21 (33.119.111), 22 (33.034.131).

Astronomical Institute of the University Brno, Publications.
Nos. 14 (04.099.012), 22 (11.121.058), 23 (19.121.034), 24 (25.114.023), 26 (37.122.069), 27 (37.122.070).

Contributions of the Nicholas Copernicus Observatory and Planetarium in Brno.
No. 26 (39.119.056 – 39.119.058).

008.138   Bucharest

Bucharest Astronomical Observatory – 75 years of existence.
M. Stavinschi (Editor).
Separate print Centre for Astronomy and Space Sciences, Central Institute of Physics, Bucharest, Romania. 64 pp. (1985).

Centre de l'Astronomie et des Sciences Spatiales, Observations Solaires.
Rotations 1730 – 1743 (39.072.059).

008.147   Byurakan

Soobshcheniya Byurakanskoj Observatorii.
Vyp. 54 (39.141.014, 39.141.015, 39.122.114, 39.122.115, 39.116.033, 39.116.034, 39.114.074, 39.021.010, 39.034.043, 39.036.081, 39.036.082, 39.158.105, 39.007.001).

Byurakan Astrophysical Observatory, Armenia, USSR, Reprints.
Nos. 313 (33.157.131), 314 (33.158.078), 315 (33.114.040), 316 (33.114.041), 317 (33.114.042), 318 (33.121.018), 319 (33.157.132), 320 (33.157.134), 321 (33.158.079), 322 (33.158.080), 323 (33.125.031), 324 (33.158.081), 325 (33.157.135), 326 (33.122.051), 327 (33.158.085), 329 (33.132.044), 330 (33.003.054), 331 (33.159.078), 332 (34.157.091), 333 (34.158.094), 334 (34.160.074), 335 (34.157.094), 336 (34.157.095), 337 (34.158.096), 338 (34.157.097), 339 (34.122.112), 340 (34.063.029), 341 (34.134.021), 342 (34.122.115), 343 (34.158.097), 344 (38.158.041), 345 (38.157.037), 346 (38.158.042), 347 (38.158.044), 348 (38.121.008), 349 (38.114.016), 350 (38.157.039), 351 (38.152.002), 352 (38.131.046), 353 (38.157.049), 354 (38.157.050), 355 (38.002.006), 356 (38.157.051), 357 (38.157.052), 358 (38.157.053), 359 (38.063.023), 360 (38.116.010), 361 (38.122.053), 362 (38.152.003), 364 (38.131.057), 366 (38.122.055), 367 (38.002.007), 368 (38.157.055), 369 (38.157.093), 370 (38.114.023), 371 (38.121.010).

008.156   Cambridge, England

Institute of Astronomy, University of Cambridge, Annual Report 1983 – 1984.
D. Lynden–Bell.
Inst. Astron., Univ. Camb., Annu. Rep., 42 pp. (1984).

008.159   Cambridge, Mass.

Center for Astrophysics: Harvard College Observatory and Smithsonian Astrophysical Observatory, Cambridge, Massachusetts 02138. Report for the period 1 October 1983 – 30 September 1984.
I. Shapiro.
Bull. Am. Astron. Soc., Vol. 17, No. 1, p. 86 – 114 (1985).

The Minor Planet Circulars/Minor Planets and Comets.
Nos. 9315 – 9716 (1985).

IAU Circulars.
Nos. 4025 – 4079 (1985).

008.165   Cape Town

South African Astronomical Observatory. Report for the year ending 1983 December 31.
M. W. Feast.
Q. J. R. Astron. Soc., Vol. 26, No. 2, p. 168 – 190 (1985).

SAAO Newsletter.
No. 5 (1985).

008.171   Cardiff

Astrophysics and Relativity, Preprint Series.
Nos. 113 (39.015.036), 114 (39.015.037), 117 (39.131.187).

008.177   Carmel Valley, Calif.

**The Monterey Institute for Research in Astronomy, Carmel
Valley, California 93924. Report for the period 1 September
1983 – 1 September 1984.**
W. B. Weaver.
Bull. Am. Astron. Soc., Vol. 17, No. 1, p. 308 (1985).

008.180   Castel Gandolfo

Annual Report 1984.
G. V. Coyne.
Vatican Obs. Publ., Vol. 2, No. 8, p. 101 – 111 (1985).

Vatican Observatory Publications.
Vol. 2, No. 8 (39.008.180).

**Vatican Observatory Publications, Special Series, Studi
Galileiani.**
Vol. 1, No. 3 (39.012.064).

008.186   Chapel Hill, N.C.

**University of North Carolina at Chapel Hill, Department of
Physics and Astronomy, Chapel Hill, North Carolina 27514.
Report for the period October 1982 – September 1984.**
B. W. Carney.
Bull. Am. Astron. Soc., Vol. 17, No. 1, p. 366 – 370 (1985).

008.189   Charlottesville, Va.

**National Radio Astronomy Observatory, Charlottesville,
Virginia 22903. Report for the period July 1983 – June 1984.**
R. J. Havlen.
Bull. Am. Astron. Soc., Vol. 17, No. 1, p. 319 – 357 (1985).

008.192   Chicago, Ill.

**University of Chicago, Department of Astronomy and
Astrophysics, Yerkes Observatory, Chicago, Illinois 60637.
Report.**
Bull. Am. Astron. Soc., Vol. 17, No. 1, p. 115 – 122 (1985).

008.198   Cleveland, Ohio

**Warner and Swasey Observatory, Case Western Reserve
University, Cleveland, Ohio 44106. Report for the period 1 July
1983 – 30 June 1984.**
Bull. Am. Astron. Soc., Vol. 17, No. 2, p. 649 – 651 (1985).

Publications of the Warner and Swasey Observatory.
Vol. 3, No. 1 (39.002.053).

Warner and Swasey Observatory, Reprints.
Nos. 303 (29.114.135), 307 (30.151.046), 308 (30.158.197),
319 (27.117.088), 323 (28.118.001), 325 (34.114.006),
327 (27.114.144), 328 (27.114.179), 329 (31.113.013),
330 (28.114.151), 331 (29.120.010), 332 (29.114.150),
333 (29.114.147), 334 (29.114.149), 335 (30.159.008),
336 (30.114.102), 337 (31.117.142), 338 (31.113.087),
339 (31.066.092), 340 (31.123.033), 341 (31.158.296),
342 (32.114.111), 343 (33.002.004), 344 (32.124.003),
345 (33.151.009), 346 (32.116.024), 347 (34.114.064),
348 (34.155.003), 349 (34.134.029), 350 (37.117.055),
351 (34.155.103), 352 (37.156.088), 353 (38.002.001),
354 (37.111.013), 355 (34.114.128), 356 (34.155.131),
357 (34.155.134), 358 (34.111.031), 359 (34.158.076),
360 (37.114.105), 361 (38.111.010), 362 (38.155.064),
364 (37.155.101), 365 (37.013.081), 366 (39.114.027),
367 (37.156.049).

008.207   College Park, Md.

**University of Maryland, College Park, Maryland 20742. Report
for the period 1 October 1983 – 30 September 1984.**
Bull. Am. Astron. Soc., Vol. 17, No. 1, p. 270 – 280 (1985).

008.210   Cologne

I. Physikalisches Institut der Universität. Report for 1984.
G. Winnewisser.
Mitt. Astron. Ges., Nr. 64, p. 195 – 197 (1985).

008.213   Columbia, Mo.

**University of Missouri, Columbia, Missouri 65211 and
St. Louis, Missouri 63121. Report for the period September
1983 – September 1984.**
C. J. Peterson.
Bull. Am. Astron. Soc., Vol. 17, No. 1, p. 306 – 307 (1985).

008.216   Columbus, Ohio

**The Observatories of the Ohio State and Ohio Wesleyan
Universities, Columbus, Ohio 43210. Report.**
E. R. Capriotti.
Bull. Am. Astron. Soc., Vol. 17, No. 1, p. 371 – 375 (1985).

008.228   Crimea

**Izvestiya Ordena Trudovogo Krasnogo Znameni Krymskoj
Astrofizicheskoj Observatorii.**
Tom 69 (39.122.100, 39.117.101, 39.117.102, 39.112.082,
39.112.083, 39.116.022, 39.114.044, 39.143.027, 39.155.080,
39.143.028, 39.116.040, 39.158.073, 39.075.014, 39.073.092,
39.065.029, 39.077.020, 39.080.028, 39.031.005, 39.034.028).

008.234   Delaware, Ohio

**The Observatories of the Ohio State and Ohio Wesleyan
Universities, Delaware, Ohio 43015. Report.**
E. R. Capriotti.
Bull. Am. Astron. Soc., Vol. 17, No. 1, p. 371 – 375 (1985).

008.240   Dresden

**Mitteilungen des Lohrmann–Observatoriums der Technischen
Universität Dresden.**
Nr. 51 (39.012.058).

008.243   Dublin

Dunsink Observatory in 1983. Report for the year 1983.
P. A. Wayman.
Ir. Astron. J., Vol. 17, No. 1, p. 61 – 66 (1985).

008.252   East Lansing, Mich.

Michigan State University, Department of Physics and
Astronomy, East Lansing, Michigan 48824. Report for the
period 1 September 1983 – 31 August 1984.
Bull. Am. Astron. Soc., Vol. 17, No. 1, p. 295 – 297 (1985).

008.255   Edinburgh

Occasional Reports of the Royal Observatory, Edinburgh.
No. 15 (39.031.047).

The UKST objective prisms.
(39.002.109 – 39.002.111).

008.261   Epping

Anglo–Australian Observatory. Report for the year ending 1984
June 30.
D. C. Morton.
Q. J. R. Astron. Soc., Vol. 26, No. 2, p. 191 – 205 (1985).

Anglo–Australian Observatory, Preprints.
Nos. 200 (39.160.075), 201 (39.131.188).

AAO Newsletter.
Nos. 32, 33 (1985).

008.264   Erlangen

Physikalisches Institut, Angewandte Optik. Report for 1984.
G. Weigelt.
Mitt. Astron. Ges., Nr. 64, p. 85 – 86 (1985).

008.267   Evanston, Ill.

Lindheimer Astronomical Research Center and Dearborn
Observatory, Northwestern University, Evanston, Illinois 60201.
Report for the period October 1983 – October 1984.
Bull. Am. Astron. Soc., Vol. 17, No. 1, p. 234 – 236 (1985).

008.269   Fairborn, Ohio

Fairborn Observatory, Fairborn, Ohio 45324. Report for 1983 –
1984.
L. J. Boyd, R. M. Genet.
Bull. Am. Astron. Soc., Vol. 17, No. 1, p. 148 (1985).

008.276   Flagstaff, Ariz.

Lowell Observatory, Flagstaff, Arizona 86002. Report for the
period 1 July 1983 – 30 June 1984.
A. A. Hoag, H. Horstman.
Bull. Am. Astron. Soc., Vol. 17, No. 1, p. 264 – 269 (1985).

008.279   Fort Davis, Tex.

University of Texas at Austin, McDonald Observatory, Fort
Davis, Texas 79737. Report for the period 1 September 1983 –
31 August 1984.
H. J. Smith, F. N. Bash, J. N. Douglas, P. A. Vanden Bout.
Bull. Am. Astron. Soc., Vol. 17, No. 1, p. 425 – 452 (1985).

008.282   Frankfurt

Institut für Theoretische Physik (Astrophysik). Report for 1984.
W. H. Kegel.
Mitt. Astron. Ges., Nr. 64, p. 87 (1985).

008.285   Freiburg

Kiepenheuer–Institut für Sonnenphysik. Report for 1984.
E. H. Schröter.
Mitt. Astron. Ges., Nr. 64, p. 89 – 103 (1985).

008.288   Gainesville, Fla.

University of Florida, Department of Astronomy, Gainesville,
Florida 32611. Report for the period 1982 September – 1984
August.
H. Eichhorn.
Bull. Am. Astron. Soc., Vol. 17, No. 1, p. 157 – 165 (1985).

008.291   Garching

European Southern Observatory. Annual Report 1984.
L. Woltjer.
ESO Annu. Rep. 1984, 83 pp. (1985).

ESO Scientific Report.
Nos. 3 (39.002.054), 4 (39.002.055).

ESO Scientific Preprint.
Nos. 355 (39.158.170), 356 (39.117.187), 357 (39.151.073),
358 (39.125.058), 359 (39.159.066), 360 (39.125.059),
361 (39.157.144), 362 (39.122.137), 363 (39.158.171),
364 (39.158.172), 365 (39.112.087), 366 (39.117.188),
367 (39.125.060), 368 (39.122.108), 369 (39.098.065),
370 (39.112.090), 371 (39.159.067), 372 (39.157.145),
373 (39.042.068).

The Messenger – El Mensajero.
Nos. 39 – 40 (1985).

ST–ECF Newsletter.
No. 1 (1985).

Max–Planck–Institut für Physik und Astrophysik: I. Institut für
Astrophysik. Report for 1984.
R. Kippenhahn.
Mitt. Astron. Ges., Nr. 64, p. 199 – 218 (1985).

MPA Reports.
Nos. 170 (39.062.132), 171 (39.067.098), 172 (39.067.099),
173 (39.158.173), 174 (39.064.061), 175 (39.151.074),
176 (39.064.062), 177 (39.066.092), 178 (39.066.093),
179 (39.065.061), 180 (39.062.133), 181 (39.074.091).

Max–Planck–Institut für Physik und Astrophysik: II. Institut
für extraterrestrische Physik. Report for 1984.
G. Haerendel, J. Trümper.
Mitt. Astron. Ges., Nr. 64, p. 219 – 233 (1985).

Max–Planck–Institut für Physik und Astrophysik. Institut für
extraterrestrische Physik – Tätigkeitsbericht 1984.
MPE Rep., No. 187, 6 + 112 pp. (1985).

MPE Contributions.
Nos. 2163, 2189 (38.078.001), 2190 (37.126.087),
2191 (38.083.046), 2194 – 2197, 2199, 2200, 2201 (38.131.043),
2202, 2204 (38.067.067), 2208 (38.143.019), 2209,
2214 (38.036.154), 2217 (38.125.018), 2218, 2219 (38.117.137),
2221 (39.126.066), 2222, 2223, 2224 (38.097.046), 2225 – 2227,
2228 (38.133.021), 2229 (39.022.152), 2230 (39.084.045),
2231 (39.084.035), 2232 (38.078.013), 2233 (38.074.053),
2234 (38.073.059), 2235 (38.131.291), 2236 (38.107.057),
2238 (39.036.003), 2239 (39.062.019), 2240 (38.124.144),
2242 (39.036.111), 2246 (39.022.188), 2249, 2250, 2252, 2254,
2257, 2259 (39.062.068), 2260 (39.083.029), 2261 (39.084.146),
2262 (39.093.038), 2264 (39.083.006), 2266, 2267 (39.117.189),
2268 (39.084.021), 2269 (39.125.006), 2275, 2286,
2288 (39.062.120), 2289 – 2291, 2293 (39.062.164).

MPE Internal Report.
Nos. 41 (39.034.065), 42.

MPE Reports.
Nos. 186 (39.126.066), 187 (39.008.291), 189 (39.036.111), 190,
191 (39.117.189).

008.297   Ghent

**Rijksuniversiteit Gent, Sterrenkundig Observatorium,
Mededeling.**
Nr. 56 (38.120.017).

008.303   Goeteborg

**Research Laboratory of Electronics and Onsala Space
Observatory, Research Report.**
No. 147 (39.082.048).

Onsala Space Observatory, Preprint.
Nos. 85:1 (39.112.088), 85:2 (39.132.048), 85:12 (39.112.089).

008.306   Göttingen

**Universitätssternwarte Göttingen und Institut für
Sonnenforschung Locarno–Orselina (Tessin). Report for 1984.**
K. J. Fricke.
Mitt. Astron. Ges., Nr. 64, p. 105 – 112 (1985).

008.312   Graz

**Institut für Astronomie der Universität Graz: Institut für
Astronomie (Universitätssternwarte), Observatorium Lustbühel,
Sonnenobservatorium Kanzelhöhe. Report for 1984.**
H. Haupt.
Mitt. Astron. Ges., Nr. 64, p. 113 – 118 (1985).

Mitteilungen der Universitätssternwarte Graz.
Nr. 107 (39.098.063).

008.315   Green Bank, W.Va.

**National Radio Astronomy Observatory, Green Bank, West
Virginia 24944. Report for the period July 1983 – June 1984.**
R. J. Havlen.
Bull. Am. Astron. Soc., Vol. 17, No. 1, p. 319 – 357 (1985).

National Radio Astronomy Observatory, Reprints, Series A.
Nos. 1390 (38.033.016), 1391 (37.008.304), 1392 (39.033.032),
1393 (39.033.033), 1394 (38.141.002).

National Radio Astronomy Observatory, Reprints, Series B.
Nos. 538 (39.033.034), 539 (37.013.053), 540 (39.033.035).

NRAO Workshops.
Nos. 7 (39.012.113), 8 (39.012.115), 9 (39.012.116).

008.318   Greenbelt, Md.

NSSDC Newsletter.
No. 1 (1985).

008.321   Greenwich

Gemini.
No. 13 (1985).

008.327   Hamburg

Hamburger Sternwarte. Report for 1984.
D. Reimers.
Mitt. Astron. Ges., Nr. 64, p. 119 – 125 (1985).

**Deutsches Hydrographisches Institut Hamburg, Zeit– und
Breitendienst.**
Januar – Juni 1984 (39.044.039).

008.333   Hannover

**Universität Hannover, Astronomische Station,
Veröffentlichungen.**
Nr. 16 (39.032.023, 39.115.015, 39.072.060).

008.339   Heidelberg

Astronomisches Rechen–Institut. Report for 1984.
W. Fricke.
Mitt. Astron. Ges., Nr. 64, p. 127 – 137 (1985).

**Astronomisches Rechen–Institut Heidelberg, Mitteilungen,
Serie A.**
Nr. 158 (38.043.007), 159 (38.103.925), 160 (38.103.926),
161 (38.043.011), 162 (38.115.019), 163 (39.098.030),
164 (39.098.032), 165 (39.005.011), 166 (39.046.022),
167 (39.041.051).

**Astronomisches Rechen–Institut Heidelberg, Mitteilungen,
Serie B.**
Nr. 129 (39.098.044), 130 (39.103.932).

Astronomische Grundlagen für den Kalender 1987.
(39.046.024).

Astronomy and Astrophysics Abstracts.
Vol. 38 (39.002.057).

Institut für Theoretische Astrophysik. Report for 1984.
B. Baschek, G. Traving.
Mitt. Astron. Ges., Nr. 64, p. 139 – 142 (1985).

Landessternwarte Heidelberg–Königstuhl. Report for 1984.
I. Appenzeller.
Mitt. Astron. Ges., Nr. 64, p. 149 – 156 (1985).

**Max–Planck–Institut für Astronomie und Deutsch–Spanisches
Astronomisches Zentrum Calar Alto/Almeria. Report for 1984.**
H. Elsässer.
Mitt. Astron. Ges., Nr. 64, p. 157 – 182 (1985).

**Max–Planck–Institut für Astronomie, D–6900
Heidelberg–Königstuhl, F.R. Germany. Centro Astronómico
Hispano–Alemán, Almería, Spain. Report for the year 1983.**
G. Münch.
Bull. Am. Astron. Soc., Vol. 17, No. 1, p. 281 – 288 (1985).

**Max–Planck–Institut für Kernphysik. Abteilung Kosmophysik.
Report for 1984.**
H. Fechtig, H. Völk.
Mitt. Astron. Ges., Nr. 64, p. 143 – 147 (1985).

Max–Planck–Institut für Kernphysik. Jahresbericht 1984.
H. V. Klapdor, E. K. Jessberger.
Max–Planck–Institut für Kernphysik, Heidelberg, F.R.
Germany. 207 pp. (1985).

008.348   **Holmdel, N.J.**

AT&T Bell Laboratories, Crawford Hill Laboratory, Holmdel, New Jersey 07733. Report for the period October 1983 – October 1984.
Bull. Am. Astron. Soc., Vol. 17, No. 1, p. 29 – 34 (1985).

008.351   **Honolulu, Hawaii**

University of Hawaii, Institute of Astronomy, Honolulu, Hawaii 96822. Report for the period 1 July 1983 – 30 June 1984.
Bull. Am. Astron. Soc., Vol. 17, No. 1, p. 175 – 185 (1985).

The Infrared Telescope Facility, Honolulu, Hawaii 96822. Report for the period 1 July 1983 – 30 June 1984.
Bull. Am. Astron. Soc., Vol. 17, No. 1, p. 215 (1985).

008.354   **Houston, Tex.**

William Marsh Rice University, Department of Space Physics and Astronomy, Houston, Texas 77001. Report for the period July 1983 – July 1984.
Bull. Am. Astron. Soc., Vol. 17, No. 1, p. 395 – 397 (1985).

008.357   **Hyderabad**

Contributions from Nizamiah and Japal–Rangapur Observatories.
Nos. 15 (39.117.191), 16 (39.119.059).

Nizamiah and Rangapur Observatories and Department of Astronomy, Osmania University, Reprints.
Nos. 102 (31.151.026), 103 (32.119.073), 104 (32.152.006), 105 (32.119.048), 106 (33.122.143), 107 (33.151.123), 108 (33.091.109), 109 (33.013.098), 110 (34.119.004), 111 (34.091.008), 112 (34.091.009), 113 (34.119.041), 114 (34.119.040), 115 (37.151.002), 116 (39.151.075), 117 (37.153.009), 118 (37.151.018), 119 (34.117.189), 121 (37.153.066), 122 (37.009.037), 123 (37.117.204), 124 (38.074.028), 125 (38.118.018).

008.360   **Innsbruck**

Institut für Astronomie. Report for 1984.
J. Pfleiderer.
Mitt. Astron. Ges., Nr. 64, p. 183 – 186 (1985).

008.375   **Izmir**

Publications of the Ege University Observatory.
No. 25 (38.119.038, 37.119.085, 37.119.087, 38.117.188, 38.119.073, 38.119.076, 38.119.083).

008.378   **Jena**

Mitteilungen der Universitäts–Sternwarte zu Jena.
Nr. 148 (30.161.005) 149 (30.131.124), 150 (31.112.056), 151 (31.131.239), 152 (32.021.028), 153 (32.131.302), 154 (32.031.603), 155 (32.031.608), 156 (32.158.167), 157 (31.021.044), 161 (37.004.037).

008.387   **Kiel**

Institut für Theoretische Physik und Sternwarte der Universität Kiel. Report for 1984.
K. Hunger.
Mitt. Astron. Ges., Nr. 64, p. 187 – 192 (1985).

Institut für Theoretische Physik und Sternwarte der Universität Kiel, Reprints.
Nos. 362 (37.114.002), 363 (37.114.004), 364 (37.126.018), 365 (37.114.032), 366 (37.126.029), 367 (37.065.061), 368, 369 (38.134.020), 370 (38.126.006), 371 (37.114.067), 372 (38.122.025), 373 (38.126.025), 374 (37.126.090), 375 (37.126.092), 376 (38.122.067), 377 (37.126.053), 378 (37.065.115), 379 (38.114.042), 380 (38.134.020), 381 (37.117.013), 382 (37.113.046), 383 (37.126.093), 384 (37.126.085), 385 (37.112.106), 386 (38.064.022), 387 (37.126.091), 388 (38.119.030), 389 (37.114.091), 390 (37.113.041), 391 (38.126.042).

Institut für Reine und Angewandte Kernphysik, Arbeitsgruppe Mathematische Physik. Report for 1984.
K. O. Thielheim.
Mitt. Astron. Ges., Nr. 64, p. 193 – 194 (1985).

008.390   **Kiev**

Kinematika i Fizika Nebesnykh Tel.
Tom 1, Nos. 1 – 3 (1985).

008.399   **Kyoto**

Contributions from the Department of Astronomy, University of Kyoto.
Nos. 158 (34.151.063), 159 (34.062.090), 160 (34.062.091), 161 (34.151.056), 162 (37.154.004), 163 (37.119.055), 164 (34.062.132), 165 (37.158.117), 166 (37.158.010), 167 (37.151.049), 168 (37.062.051), 169 (37.067.062), 170 (37.062.052), 171 (37.112.054), 172, 173 (38.115.009), 174 (38.157.085).

008.400   **La Jolla, Calif.**

University of California at San Diego, Center for Astrophysics and Space Sciences, La Jolla, California 92093. Report.
E. M. Burbidge.
Bull. Am. Astron. Soc., Vol. 17, No. 1, p. 62 – 72 (1985).

008.405   **La Serena**

Cerro Tololo Inter–American Observatory, La Serena, Chile. Report for the period 1 October 1982 – 30 September 1983.
Bull. Am. Astron. Soc., Vol. 16, No. 4, p. 1028 – 1044 (1984).

008.417   **Landover, Md.**

Applied Research Corporation, Landover, Maryland 20785. Report for the period October 1983 – September 1984.
Bull. Am. Astron. Soc., Vol. 17, No. 1, p. 3 – 6 (1985).

008.420   **Las Cruces, N.M.**

New Mexico State University, Department of Astronomy, Las Cruces, New Mexico 88003. Report for the year 1984.
Bull. Am. Astron. Soc., Vol. 17, No. 1, p. 358 – 360 (1985).

008.423   **Lawrence, Kans.**

Clyde W. Tombaugh Observatory, University of Kansas, Lawrence, Kansas 66045. Report for the period August 1983 – August 1984.
Bull. Am. Astron. Soc., Vol. 17, No. 1, p. 453 – 456 (1985).

008.429   Lembang

Contributions from the Bosscha Observatory.
Nos. 76 (32.117.009), 77 (39.151.076), 78 (31.156.010),
79 (37.155.053), 80 (30.064.064), 81 (32.131.037),
82 (37.117.142), 83 (37.120.016).

008.432   Leningrad

Chronicle.
Byull. Inst. Teor. Astron., Tom 15, No. 7 (170),
p. 407 – 408 (1985). In Russian.

Byulleten' Instituta Teoreticheskoj Astronomii.
Tom 15, Nos. 7 (39.103.957, 39.091.066, 39.091.067,
39.052.072, 39.094.045, 39.098.104, 39.008.432),
8 (39.052.073, 39.098.105, 39.094.046, 39.043.009, 39.094.047,
39.052.074, 39.094.048).

Trudy Instituta Teoreticheskoj Astronomii.
Vyp. 19 (39.094.017, 39.041.013).

Ephemerides of minor planets for 1986.
(39.098.096).

008.438   Lintong

Publications of the Shaanxi Astronomical Observatory.
Vol. 7, No. 1 (39.041.021, 39.036.116, 39.044.050 – 39.044.053,
39.034.068, 39.034.069).

Time and Frequency Services, Bulletin.
Nos. 61 – 66 (39.044.040).

008.441   Livermore, Calif.

**Lawrence Livermore National Laboratory, University of
California, Livermore, California 94550. Report.**
Bull. Am. Astron. Soc., Vol. 17, No. 1, p. 237 – 244 (1985).

008.444   London, Canada

**The Observatories of The University of Western Ontario,
London, Ontario N6A 3K7, Canada. Report for the period
1 July 1983 – 30 June 1984.**
W. H. Wehlau.
Bull. Am. Astron. Soc., Vol. 17, No. 1, p. 499 – 501 (1985).

008.447   London, England

**Mullard Space Science Laboratory, University College, London.
Report for the period 1983 November 1 to 1984 October 31.**
J. L. Culhane.
Q. J. R. Astron. Soc., Vol. 26, No. 2, p. 206 – 218 (1985).

008.450   Los Alamos, N.M.

**Los Alamos National Laboratory, Los Alamos, New Mexico
87545. Report for the period 1 July 1983 – 30 June 1984.**
A. N. Cox, J. G. Hills.
Bull. Am. Astron. Soc., Vol. 17, No. 1, p. 245 – 258 (1985).

008.453   Los Angeles, Calif.

**The Aerospace Corporation, Los Angeles, California 90009.
Report for the period 1 October 1983 – 30 September 1984.**
E. E. Epstein.
Bull. Am. Astron. Soc., Vol. 17, No. 1, p. 1 – 2 (1985).

**University of California at Los Angeles, Department of
Astronomy, Los Angeles, California 90024. Report for the
academic year 1983 – 84.**
Bull. Am. Astron. Soc., Vol. 17, No. 1, p. 56 – 61 (1985).

**University of Southern California, Department of Astronomy,
Los Angeles, California 90089. Report.**
Bull. Am. Astron. Soc., Vol. 17, No. 1, p. 401 – 402 (1985).

008.465   Madison, Wis.

**Washburn Observatory, University of Wisconsin at Madison,
Madison, Wisconsin 53706. Report for the period 1 October
1983 – 30 September 1984.**
Bull. Am. Astron. Soc., Vol. 17, No. 1, p. 475 – 480 (1985).

008.468   Madrid

Seminario de Astronomia y Geodesia, Publicación.
Nos. 115 (39.044.043), 116 (39.044.044), 117 – 120.

008.474   Manchester

**Astronomical Contributions from the University of Manchester,
Series II: Jodrell Bank Reprints.**
Nos. 705 (37.117.110), 733 (39.126.040), 734 (38.158.201),
735 (39.158.150), 736 (39.112.021), 737 (39.126.010),
738 (39.159.005), 739 (39.158.084), 740, 741 (38.158.038), 742,
743 (39.126.028), 744 (39.159.013), 745 (39.132.022),
746 (39.158.147), 747 (39.159.045), 748 (39.158.087),
749 (39.158.088), 750 – 753, 754 (39.159.046), 755 (39.158.276),
756, 757 (39.157.084), 758 (39.124.014).

008.477   Manila

Solar Maps and Activity.
(39.072.061).

008.489   Middletown, Conn.

**Van Vleck Observatory, Wesleyan University, Middletown,
Connecticut 06457. Report for the period 1 September 1983 –
31 August 1984.**
A. R. Upgren.
Bull. Am. Astron. Soc., Vol. 17, No. 1, p. 469 – 471 (1985).

008.495   Minneapolis, Minn.

**University of Minnesota, School of Physics and Astronomy,
Minneapolis, Minnesota 55455. Report for the academic year
1983 – 1984.**
Bull. Am. Astron. Soc., Vol. 17, No. 1, p. 298 – 305 (1985).

008.498   Mizusawa

**Proceedings of the International Latitude Observatory of
Mizusawa.**
No. 23 (39.082.049, 39.141.023, 39.002.056, 39.044.042,
39.036.112).

**Publications of the International Latitude Observatory of
Mizusawa.**
Vol. 17, No. 2 (39.082.050, 39.036.113).

Monthly Notes of the International Polar Motion Service.
Nos. 12 (1984), 1 – 4 (1985) (39.044.041).

008.504   Montréal

Université de Montréal, Département de Physique, Montréal,
Quebec H3C 3J7, Canada. Report for the period 1 September
1983 – 31 August 1984.
F. Wesemael, G. Fontaine.
Bull. Am. Astron. Soc., Vol. 17, No. 1, p. 309 – 315 (1985).

008.507   Moscow

Trudy Gosudarstvennogo Astronomicheskogo Instituta im.
P. K. Shternberga.
Tom 57 (39.007.020, 39.081.048, 39.044.075, 39.044.076,
39.052.067 – 39.052.069, 39.042.102, 39.052.070, 39.021.021,
39.042.103 – 39.042.105, 39.094.041, 39.093.053, 39.118.056,
39.002.107).

008.510   Münster

Astronomisches Institut der Universität und Außenstation
Schalkenmehren. Report for 1984.
W. C. Seitter.
Mitt. Astron. Ges., Nr. 64, p. 239 – 245 (1985)

Forschergruppe Erde–Mond–System (Planetologie). Report for
1984.
W. C. Seitter, D. Stöffler.
Mitt. Astron. Ges., Nr. 64, p. 245 – 249 (1985).

008.513   Munich

Institut für Astronomie und Astrophysik der Universität
München, Universitätssternwarte. Report for 1984.
R. P. Kudritzki.
Mitt. Astron. Ges., Nr. 64, p. 235 – 237 (1985).

Mitteilungen aus dem Institut für Astronomische und
Physikalische Geodäsie der Technischen Hochschule München.
Nr. 168 (39.021.015).

008.522   Nanking

Publications of the Purple Mountain Observatory.
Vol. 3, No. 3 (39.073.137, 39.041.007, 39.098.064, 39.103.281,
39.015.038, 39.036.114, 39.034.066, 39.033.036).

008.525   Nashville, Tenn.

Dyer Observatory, Vanderbilt University, Nashville, Tennessee
37235. Report for the period 1 October 1983 – 30 September
1984.
A. M. Heiser.
Bull. Am. Astron. Soc., Vol. 17, No. 1, p. 145 – 147 (1985).

008.528   Neuchâtel

Observatoire cantonal de Neuchâtel. Rapport d'activité pour
l'Exercice 1984.
J. Bonanomi, G. Fischer.
Observatoire cantonal de Neuchâtel, Neuchâtel, Suisse.
23 pp. (1985).

008.531   New Haven, Conn.

Yale University Observatory, New Haven, Connecticut 06511.
Report for the period 1 July 1983 – 30 June 1984.
Bull. Am. Astron. Soc., Vol. 17, No. 1, p. 502 – 510 (1985).

008.534   New York, N.Y.

Columbia University in the City of New York, Department of
Astronomy and Department of Physics, New York, New York
10027. Report for the period September 1983 – August 1984.
Bull. Am. Astron. Soc., Vol. 17, No. 1, p. 127 – 130 (1985).

008.537   Newark, Del.

Bartol Research Foundation and University of Delaware,
Newark, Delaware 19716. Report for the period September
1983 – September 1984.
Bull. Am. Astron. Soc.,Vol. 17, No. 1, p. 35 – 38 (1985).

008.540   Nice

Association pour le Développement International de
l'Observatoire de Nice (A.D.I.O.N.), Bull. Nos. 20 – 21, 1982 –
1984.
Observatoire de Nice, Boîte Postale No. 139, F–06003 Nice,
Cedex, France, 61 pp. (1985).

008.549   Oslo

Theoretic Papers.
Vol. 3, Nos. 2 (39.161.174), 6 (39.107.028).

008.564   Paris

Rapport d'activité de l'Observatoire de Paris 1982.
Observatoire de Paris, 61, avenue de l'Observatoire,
F–75014 Paris, France. 142 pp. (1985).

Institut d'Astrophysique de Paris, Pré–Publication.
Nos. 82 (39.120.017), 83 (39.074.090), 84 (39.079.181),
85 (39.121.035), 86 (39.121.036), 87 (39.103.105),
88 (39.131.189), 89 (39.114.091), 90 (39.061.037),
91 (39.106.056), 92 (39.080.089), 93 (39.117.192),
94 (39.099.060), 95 (39.156.017), 96 (39.102.043),
97 (39.161.175), 98 (39.131.190), 99 (39.079.111),
100 (39.131.191), 101 (39.103.106).

Bureau International de l'Heure (B.I.H.) Circulars.
D218 – D223 (39.044.045).
E13 (39.044.046).

008.606   Potsdam

Zentralinstitut für Astrophysik. Report for 1983.
Akad. Wiss. DDR, Jahrb. 1983, p. 217 – 222 (1985).

Zentralinstitut für Physik der Erde. Report for 1983.
Akad. Wiss. DDR, Jahrb. 1983, p. 227 – 231 (1985).

008.609   Prague

Publications of the Astronomical Institute of the Czechoslovak
Academy of Sciences.
Nos. 58 (39.012.018), 59 (39.104.030, 39.104.031),
61 (39.031.011, 39.031.012).

008.615   Princeton, N.J.

Princeton University Observatory, Princeton, New Jersey 08544.
Report for the period 1 July 1983 – 30 June 1984.
Bull. Am. Astron. Soc., Vol. 17, No. 1, p. 386 – 392 (1985).

008.621   Pulkovo

Izvestiya Glavnoj Astronomicheskoj Observatorii v Pulkove, Astrometriya i Astrofizika.
Nos. 201 (39.007.031, 39.036.216, 39.032.052, 39.002.119,
39.036.217, 39.032.053, 39.032.054, 39.111.055, 39.118.062,
39.044.097, 39.044.098, 39.093.060, 39.097.026, 39.099.089,
39.071.051, 39.075.043, 39.072.121, 39.157.250, 39.117.389,
39.082.103, 39.041.029, 39.031.046, 39.034.175, 39.002.120),
202 (39.007.014, 39.002.117, 39.002.118, 39.032.050,
39.032.051, 39.082.101, 39.082.102, 39.098.106, 39.098.107,
39.071.050, 39.114.172, 39.154.092, 39.153.064, 39.154.093,
39.034.173, 39.034.174).

Refraction Tables of the Pulkovo Observatory.
(39.082.045).

008.636   Richmond Hill

David Dunlap Observatory, University of Toronto, Richmond Hill, Ontario L4C 4Y6, Canada. Report for the period 1 July 1983 – 30 June 1984.
Bull. Am. Astron. Soc., Vol. 17, No. 1, p. 138 – 144 (1985).

008.642   Rochester, N.Y.

C. E. Kenneth Mees Observatory, University of Rochester, Rochester, New York 14627. Report.
Bull. Am. Astron. Soc., Vol. 17, No. 1, p. 289 – 294 (1985).

008.653   Sacramento Peak, N.M.

National Solar Observatory, Sacramento Peak, New Mexico. Report.
J. B. Zirker.
Bull. Am. Astron. Soc., Vol. 16, No. 4, p. 1072 – 1078 (1984).

008.660   San Fernando

Instituto y Observatorio de Marina, Boletin Serie C. (Rotacion de la Tierra).
Nos. 85 (39.044.047), 86 (39.044.048).

008.669   Santa Cruz, Calif.

University of California at Santa Cruz, Board of Studies in Astronomy and Astrophysics, Santa Cruz, California 95064. Report for 1983 – 1984.
Bull. Am. Astron. Soc., Vol. 17, No. 1, p. 73 – 84 (1985).

Lick Observatory, University of California, Santa Cruz, California 95064. Report.
Bull. Am. Astron. Soc., Vol. 17, No. 1, p. 223 – 233 (1985).

008.672   Santiago

Astrophysics Preprint Series.
Nos. 6 (39.160.076), 8 (39.036.115), 9 (39.122.142),
10 (39.113.040), 11 (39.124.010).

008.684   Seattle, Wash.

University of Washington, Astronomy Department, Seattle, Washington 98195. Report for the academic year 1983 – 1984.
B. Margon.
Bull. Am. Astron. Soc., Vol. 17, No. 1, p. 493 – 498 (1985).

008.699   Silver Spring, Md.

Computer Sciences Corporation, Silver Spring, Maryland 20910. Report for the period July 1983 – June 1984.
C. L. Imhoff.
Bull. Am. Astron. Soc., Vol. 17, No. 1, p. 131 – 137 (1985).

008.702   Skalnaté Pleso

Contributions of the Astronomical Observatory Skalnaté Pleso.
Vol. 12 (39.103.007, 39.104.004, 39.119.008, 39.103.008,
39.104.005, 39.074.033, 39.103.009, 39.073.063, 39.117.056,
39.104.006).

008.705   Socorro, N.M.

National Radio Astronomy Observatory, Socorro, New Mexico 87801. Report for the period July 1983 – June 1984.
R. J. Havlen.
Bull. Am. Astron. Soc., Vol. 17, No. 1, p. 319 – 357 (1985).

008.708   Sonneberg

Mitteilungen über Veränderliche Sterne.
Band 10, Heft 4 (39.122.138, 39.121.034, 39.123.012,
39.122.139, 39.119.060, 39.142.035, 39.122.140, 39.123.013,
39.123.014, 39.122.141),
Band 10, Heft 5 (39.121.041, 39.122.145 – 39.122.148,
39.119.061, 39.122.149, 39.122.150).

008.714   St. Louis, Mo.

Washington University, McDonnell Center for the Space Sciences, St. Louis, Missouri 63130. Report for 1983 – 1984.
Bull. Am. Astron. Soc., Vol. 17, No. 1, p. 481 – 492 (1985).

008.715   St. Michel l'Observatoire

Observatoire de Haute Provence, Pré–Publication.
No. 1 (39.158.174).

008.717   Stanford, Calif.

Stanford Center for Radar Astronomy, Stanford University, Stanford, California 94305. Report.
Bull. Am. Astron. Soc., Vol. 17, No. 1, p. 422 – 424 (1985).

008.723   Stony Brook, N.Y.

State University of New York at Stony Brook, Stony Brook, New York 11794. Report for the period 31 August 1983 – 1 September 1984.
D. M. Peterson.
Bull. Am. Astron. Soc., Vol. 17, No. 1, p. 361 – 365 (1985).

008.726   Strasbourg

Centre de Données Stellaires. Annual report of the director for 1983.
C. Jaschek.
Bull. Inf. Cent. Données Stellaires, No. 28, p. 143 – 146 (1985).

Centre de Données Stellaires. Annual report of the director for 1984.
C. Jaschek.
Bull. Inf. Cent. Données Stellaires, No. 28, p. 147 – 152 (1985).

Bulletin d'Information du Centre de Données Stellaires.
No. 28 (1985).

Bulletin d'Information sur les Etoiles Be.
No. 11 (1985).

Bulletin sur les Etoiles Tardives à Spectre Particulier.
No. 2 (1985).

008.732   Swarthmore, Pa.

Department of Astronomy – Sproul Observatory, Swarthmore College, Swarthmore, Pennsylvania 19081. Report for the period 1 July 1983 – 30 June 1984.
J. E. Gaustad.
Bull. Am. Astron. Soc., Vol. 17, No. 1, p. 420 – 421 (1985).

008.744   Tartu

Tartu Astrofüüsika Observatoorium Publikatsioonid.
Tom 50 (39.009.015, 39.002.026, 39.063.028, 39.064.041 –
39.064.043, 39.022.051, 39.064.044, 39.116.027, 39.117.190,
39.122.109, 39.122.110, 39.117.117, 39.131.115, 39.131.116,
39.036.072, 39.036.073, 39.125.030).

Tartu Astrofüüsika Observatoorium Teated.
Nr. 74 (39.117.113 – 39.117.115, 39.114.061, 39.117.116),
75 (39.004.056 – 39.004.061), 76 (39.063.060).

008.750   Tautenburg

Mitteilungen des Karl–Schwarzschild–Observatoriums Tautenburg.
Nr. 105 (29.036.012), 106 (28.158.232), 107 (28.160.068),
108 (30.036.021), 109 (31.098.083), 110 (32.158.165),
111 (33.098.032), 112 (32.031.573).

008.753   Tempe, Ariz.

Arizona State University, Tempe, Arizona 85287. Report for the period September 1983 – September 1984.
D. Burstein.
Bull. Am. Astron. Soc., Vol. 17, No. 1, p. 25 – 28 (1985).

008.759   Tokyo

Annals of the Tokyo Astronomical Observatory, Second Series.
Vol. 20, No. 3 (39.052.030, 39.032.024, 39.157.146, 39.063.062,
39.031.014, 39.002.058).

Tokyo Astronomical Bulletin, Second Series.
No. 272 (39.073.138).

Tokyo Astronomical Observatory Reprints.
Nos. 710 (38.131.231), 711 (38.158.247), 712 (38.117.248),
713 (38.064.063), 714 (38.155.074), 715 (38.119.095),
717 (38.103.161), 718 (38.132.065), 719 (39.157.048),
720 (39.117.029), 721 (39.062.016), 722 (38.032.041),
723 (38.032.059), 724 (39.044.049), 725 (39.073.044),
726 (39.051.007), 727 (39.062.117), 728 (39.065.048),
729 (39.117.160), 730 (39.132.042), 731 (39.073.130),
732 (39.033.020), 733 (39.073.079), 734 (39.114.148),
735 (39.114.156).

Quarterly Bulletin on Solar Activity.
Vol. 22, Part V (39.077.029), Vol. 23, Part III (39.073.139).

008.762   Toledo, Ohio

Ritter Astrophysical Research Center, The University of Toledo, Toledo, Ohio 43606. Report for the period 1 July 1983 – 30 June 1984.
Bull. Am. Astron. Soc., Vol. 17, No. 1, p. 398 – 400 (1985).

008.774   Trieste

Astronomical Observatory Trieste, Publications.
Nos. 910 (37.114.022), 911 (37.114.021), 912 (37.035.003),
913 (37.119.010), 914 (37.160.078), 915 (37.119.004), 916 – 920,
921 (38.117.055), 922 (38.131.078), 923 (38.114.035),
924 (38.114.040), 925 (38.112.018), 926 (38.114.046),
927 (38.117.065), 928 (38.117.068), 929 (38.119.031),
930 (38.036.032), 931 (37.114.012), 932 (37.157.146),
933 (37.119.092), 934 (37.077.062), 935, 936, 937 (39.077.011),
938 (38.117.164), 939 – 952, 953 (38.119.034), 954,
955 (38.114.070), 956 (38.117.136), 957 (38.119.046), 958,
959 (38.114.092), 960 (38.117.169), 961, 962 (34.122.218),
963 (38.160.058), 964 (38.114.094), 965.

008.777   Troy, N.Y.

Rensselaer Polytechnic Institute, Troy, New York 12181. Report.
Bull. Am. Astron. Soc., Vol. 17, No. 1, p. 393 – 394 (1985).

008.780   Tucson, Ariz.

Kitt Peak National Observatory, Tucson, Arizona 85726. Report for the period 1 October 1982 – 30 September 1983.
G. Burbidge.
Bull. Am. Astron. Soc., Vol. 16, No. 4, p. 1045 – 1071 (1984).

National Solar Observatory, Tucson, Arizona. Report.
J. B. Zirker.
Bull. Am. Astron. Soc., Vol. 16, No. 4, p. 1072 – 1078 (1984).

Steward Observatory, University of Arizona, Tucson, Arizona 85721. Report for the period 1 January – 31 December 1983.
P. Strittmatter.
Bull. Am. Astron. Soc., Vol. 16, No. 4, p. 1079 – 1088 (1984).

Preprints of the Steward Observatory.
Nos. 563 (39.157.201), 564 (39.082.053), 565 (39.034.067),
566 (39.155.114), 567 (39.031.013), 568 (39.082.054),
569 (39.158.175), 570 (39.159.068), 571 (39.121.037),
572 (39.158.176), 573 (39.131.192), 574 (39.159.069),
575 (39.159.070), 577 (39.158.177), 578 (39.098.066),
579 (39.117.194), 580 (39.131.193), 581 (39.159.074),
582 (39.022.119), 583 (39.142.036).

University of Arizona, Lunar and Planetary Laboratory, Tucson, Arizona 85721. Report.
Bull. Am. Astron. Soc., Vol. 17, No. 1, p. 14 – 24 (1985).

Multiple Mirror Telescope Observatory, Tucson, Arizona 85721. Report for the period 1 January 1984 – 30 September 1984.
Bull. Am. Astron. Soc., Vol. 17, No. 1, p. 316 – 318 (1985).

National Radio Astronomy Observatory, Tucson, Arizona 85705. Report for the period July 1983 – June 1984.
R. J. Havlen.
Bull. Am. Astron. Soc., Vol. 17, No. 1, p. 319 – 357 (1985).

008.783   Tübingen

Universität Tübingen, Astronomisches Institut. Report for 1984.
M. Grewing.
Mitt. Astron. Ges., Nr. 64, p. 251 – 265 (1985).

008.789   Uccle

Observatoire Royal de Belgique, Communications, Série A.
Nos. 75 (38.071.032), 76 (38.142.051), 77 (32.022.078).

Bulletin Astronomique.
Vol. 9, No. 6 (39.045.024, 39.098.067, 39.098.068, 39.100.032,
39.099.061, 39.103.301, 39.118.037, 39.096.012, 39.036.119,
39.118.038).

008.795   University Park, Pa.

**The Pennsylvania State University, Department of Astronomy,
University Park, Pennsylvania 16802. Report for the period
1 September 1983 – 31 August 1984.**
S. Matsushima.
Bull. Am. Astron. Soc., Vol. 17, No. 1, p. 376 – 385 (1985).

008.798   Uppsala

Uppsala Astronomical Observatory. Annual Report for 1984.
C.-I, Lagerkvist.
Upps. Astron. Obs. Rep., No. 35, 24 pp. (1985).

Uppsala Astronomical Observatory, Report.
Nos. 34 (39.011.023), 35 (39.008.798).

008.801   Urbana, Ill.

**University of Illinois at Urbana–Champaign, Department of
Astronomy, Urbana, Illinois 61801–3000. Report for the period
1 September 1983 – 31 August 1984.**
S. Rosen, J. W. Truran.
Bull. Am. Astron. Soc., Vol. 17, No. 1, p. 205 – 214 (1985).

008.807   Vancouver

**University of British Columbia, Department of Geophysics and
Astronomy, Vancouver, British Columbia V6T 1W5, Canada.
Report.**
Bull. Am. Astron. Soc., Vol. 17, No. 1, p. 43 – 45 (1985).

008.810   Victoria

**Climenhaga Observatory, Physics Department, University of
Victoria, Victoria, British Columbia V8W 2Y2, Canada. Report
for 1983 – 1984.**
Bull. Am. Astron. Soc., Vol. 17, No. 1, p. 123 – 126 (1985).

008.813   Vienna

Institut für Astronomie der Universität Wien. Report for 1984.
M.·Breger, W. M. Tscharnuter.
Mitt. Astron. Ges., Nr. 64, p. 267 – 274 (1985).

008.822   Villanova, Pa.

**Villanova University, Department of Astronomy, Villanova,
Pennsylvania 19085. Report.**
Bull. Am. Astron. Soc., Vol. 17, No. 1, p. 472 – 474 (1985).

008.828   Warsaw

**Warsaw University Observatory and Polish Academy of
Sciences and N. Copernicus Astronomical Center, Reprint.**
Nos. 475 (37.117.027), 476 (37.124.101), 477 (37.117.026).

Latitude Circular.
Nos. 92 – 94 (39.044.054, 39.044.055).

008.831   Washington, D.C.

**U.S. Naval Observatory, Washington, D.C. 20390. Report for
the period 1 July 1983 – 30 June 1984.**
G. Westerhout, C. K. Roberts.
Bull. Am. Astron. Soc., Vol. 17, No. 1, p. 457 – 468 (1985).

**Publications of the United States Naval Observatory, Second
Series.**
Vol. 25, Part I (39.121.040), Part II (39.118.044).

U.S. Naval Observatory, Time Service Publications, Series 4.
Nos. 935 – 960 (39.044.056).

Earth Orientation Bulletin.
Vol. 3, Nos. 1 – 26 (39.044.057).

**U.S. Naval Observatory, Time Service Announcements,
Series 14.**
Nos. 39 – 40 (39.044.058).

008.837   Westford, Mass.

**Haystack Observatory, Northeast Radio Observatory
Corporation (NEROC), Westford, Massachusetts 01886. Report
for the period 1 July 1983 – 30 June 1984.**
Bull. Am. Astron. Soc., Vol. 17, No. 1, p. 186 – 190 (1985).

008.843   Williamstown, Mass.

**Hopkins Observatory, Williams College, Williamstown,
Massachusetts 01267. Report for the 1983 – 84 academic year.**
J. M. Pasachoff.
Bull. Am. Astron. Soc., Vol. 17, No. 1, p. 202 – 203 (1985).

008.852   Würzburg

**Institut für Astronomie und Astrophysik, Lehrstuhl Astronomie.
Report for 1984.**
F.-L. Deubner.
Mitt. Astron. Ges., Nr. 64, p. 275 – 277 (1985).

008.855   Yorktown Heights, N.Y.

**IBM Thomas J. Watson Research Center, Yorktown Heights,
New York 10598. Report.**
B. Elmegreen.
Bull. Am. Astron. Soc., Vol. 17, No. 1, p. 204 (1985).

008.858   Zagreb

Hvar Observatory Bulletin.
Vol. 8, No. 1 (39.072.062, 39.073.141, 39.073.142, 39.098.069).

008.861   Zelenchukskaya

Chronicle.
Astrofiz. Issled. Izv. Spets. Astrofiz. Obs., Tom 19,
p. 118 – 119 (1985). In Russian. English translation in Bull.
Spec. Astrophys. Obs. – North Caucasus.

**Astrofizicheskie Issledovaniya. Izvestiya Spetsial'noj
Astrofizicheskoj Observatorii.**
Tom 19 (39.157.081, 39.159.038, 39.159.039, 39.114.062,
39.116.024, 39.116.025, 39.062.073, 39.067.065, 39.141.005,
39.034.036, 39.033.009, 39.033.010, 39.032.011, 39.033.011 –
39.033.013, 39.036.070, 39.082.062, 39.008.861, 39.007.012).

Soobshcheniya Spetsial'noj Astrofizicheskoj Observatorii.
Vyp. 44 (39.034.168, 39.034.169, 39.033.058).

008.867   Zurich

Institut für Astronomie. Report for 1984.
J. O. Stenflo.
Mitt. Astron. Ges., Nr. 64, p. 279 – 287 (1985).

009 Notes on Observatories, Planetaria, Exhibitions

009.001 **Die Urania–Sternwarte in Zürich.**
E. Peter.
Sternenbote, 28. Jahrg., Nr. 2, p. 22 – 26 (1985).

009.002 **The fantastic voyages of Digistar.**
C. D. Smith.
Sky Telesc., Vol. 69, No. 1, p. 6 – 8 (1985).

009.003 **A visitor's guide to NASA.**
Sky Telesc., Vol. 69, No. 2, p. 102 – 105 (1985).

009.004 **Leiden Observatory: 350 years of astronomy.**
W. Bijleveld, W. B. Burton.
Sky Telesc., Vol. 69, No. 2, p. 119 – 122 (1985).

009.005 **Le planétarium de Strasbourg.**
A. Acker.
Astronomie, Vol. 99, p. 93 – 96 (1985).

009.006 **Neue Sterne über Mannheim.**
W. Wacker.
Sterne Weltraum, 24. Jahrg., Nr. 2, p. 82 – 83 (1985).

009.007 **Why Mount Wilson shouldn't be scrapped.**
L. J. Robinson.
Sky Telesc., Vol. 69, No. 3, p. 197 (1985).

009.008 **W. M. Keck Observatory: when $ 36 million isn't enough.**
L. J. Robinson.
Sky Telesc., Vol. 69, No. 3, p. 223 (1985).

009.009 **The Space Telescope Science Institute.**
W. Tucker.
Sky Telesc., Vol. 69, No. 4, p. 295 – 299 (1985).

009.010 **Spectrographie des structures fines solaires. Un grand pas réalisé au Pic–du–Midi.**
P. Mein.
Astronomie, Vol. 99, p. 167 – 171 (1985).

009.011 **Vatikaanse astronomen verlaten Rome.**
G. W. E. Beekman.
Zenit, 12. Jaarg., No. 2, p. 72 – 73 (1985).

009.012 **Usuda Deep Space Ground Station and its large antenna.**
T. Hayashi, H. Hirosawa, M. Ichikawa.
Astron. Her., Vol. 78, No. 3, p. 82 – 83 (1985).

009.013 **L'observatoire Ras Algethi.**
J. J. Doyle, L. Louys.
Ciel Terre, Vol. 101, No. 2, p. 49 – 50 (1985).

009.014 **Scheduled observations at La Palma.**
P. A. Wayman.
Ir. Astron. J., Vol. 17, No. 1, p. 47 – 50 (1985). – See Abstr. 012.035.

009.015 **Glimpses of the history of Tartu Astronomical Observatory.**
A. Kipper.
Tartu Astrofüüs. Obs. Publ., Tom 50, p. 3 – 28 (1984). In Russian.
   Relied on personal reminiscenses a brief insight into the history of Tartu Astronomical Observatory is presented, covering the scientific directions, activities and achievements during the years 1930 – 1960.

009.016 **Astrophysics at Mount Stromlo: the Woolley era.**
S. C. B. Gascoigne.
Proc. Astron. Soc. Aust., Vol. 5, No. 4, p. 597 – 605 (1984).

009.017 **The observatory at the Lands Department Building, Sydney.**
G. L. White.
Proc. Astron. Soc. Aust., Vol. 5, No. 4, p. 606 – 607 (1984).

009.018 **On the island of La Palma.**
K. Tomita.
Astron. Her., Vol. 78, No. 4, p. 103 – 109 (1985). In Japanese.

009.019 **On the Yonsei University Observatory.**
H. Sato.
Astron. Her., Vol. 78, No. 4, p. 110 – 112 (1985). In Japanese.

009.020 **L'osservatorio astronomico dell'Alta Provenza.**
R. Charvin.
Orione, Vol. 5, N. 2, p. 26 – 31 (1985).

009.021 **Planetari in Italia e nel mondo.**
M. Cavedon.
Orione, Vol. 5, N. 2, p. 32 – 35 (1985).

009.022 **Großplanetarium COSMORAMA für Kanada.**
P. Köhler, G. Nobiling.
Jenaer Rundsch., 30. Jahrg., Heft 1, p. 36 – 37 (1985).

009.023 **Digistar–planetarium in Den Haag.**
G. W. E. Beekman.
Zenit, 12. Jaarg., No. 6, p. 210 (1985).

009.024 **Von der Keilschrifttafel zu Lunochod. Ein Querschnitt durch die museale Sammlung der Archenhold–Sternwarte.**
B. Götz.
Archenhold–Sternw. Berlin–Treptow, Vortr. Schr., Nr. 64, 40 pp. (1984).

009.025 **Der Himmel auf Erden. Der lange Weg vom Riesenglobus zum Projektionsplanetarium.**
G. W. E. Beekman.
Sterne Weltraum, 24. Jahrg., Nr. 6, p. 310 – 314 (1985).

009.026 **Photoelectric astronomy at Mt. Tamborine Observatory.**
A. A. Page.
Aust. J. Astron., Vol. 1, No. 1, p. 41 – 42 (1985).

009.027 **News from observatories: Anglo–Australian Observatory; The Molonglo Radio Observatory; Perth Observatory.**
D. Allen, A. Little, M. Candy.
Aust. J. Astron., Vol. 1, No. 1, p. 43 – 45 (1985).

009.028 **Observatory of the University of Helsinki 150 years old.**
T. Markkanen.
Arkhimedes (Helsinki), 36. vsk, p. 234   240 (1984). In Finnish.
   On September 12th, 1984 the Observatory of the University of Helsinki celebrated its 150th anniversary. The development of astronomy in Finland since 1640 is shortly reviewed.

009.029 **Copernicus Observatory and Planetarium in Brno.**
J. Kohout.
Říše hvězd, Vol. 65, No. 11, p. 226 – 229 (1984). In Czech.

009.030 **Radio Observatory Arecibo.**
J. Grygar, L. Kalašová.
Říše hvězd, Vol. 66, No. 6, p. 108 – 109 (1985). In Czech.

009.031 Planetarium – a window to the universe.
B. Pejović.
Vasiona, Année 33, No. 1 – 2, p. 6 – 9 (1985). In Croatian.
  The paper explains the planetarium as a cultural tool for the popularization of astronomy.

009.032 History of the Moscow Astronomical Observatory and of the Sternberg Astronomical Institute.
E. P. Aksenov.
Main directions of astronomical investigations at the Moscow University, p. 6 – 23 (1985). In Russian. – See Abstr. 003.049.

009.033 Astrophysical investigations at the Sternberg Astronomical Institute.
D. Ya. Martynov.
Main directions of astronomical investigations at the Moscow University, p. 23 – 31 (1985). In Russian. – See Abstr. 003.049.

009.034 Results and perspectives of solar investigations at the Sternberg Astronomical Institute.
Eh. V. Kononovich.
Main directions of astronomical investigations at the Moscow University, p. 39 – 49 (1985). In Russian. – See Abstr. 003.049.

009.035 Radio astronomical investigations at the Sternberg Astronomical Institute.
V. N. Kuril'chik.
Main directions of astronomical investigations at the Moscow University, p. 49 – 58 (1985). In Russian. – See Abstr. 003.049.

009.036 Investigations of the moon at the Sternberg Astronomical Institute.
V. V. Shevchenko.
Main directions of astronomical investigations at the Moscow University, p. 58 – 69 (1985). In Russian. – See Abstr. 003.049.

009.037 On the history of development of stellar–astronomical investigations in Moscow.
P. G. Kulikovskij.
Main directions of astronomical investigations at the Moscow University, p. 69 – 79 (1985). In Russian. – See Abstr. 003.049.

009.038 Investigations of variable stars at the Sternberg Astronomical Institute.
P. N. Kholopov.
Main directions of astronomical investigations at the Moscow University, p. 79 – 90 (1985). In Russian. – See Abstr. 003.049.

009.039 Investigations in the field of celestial mechanics at the Sternberg Astronomical Institute.
A. A. Orlov.
Main directions of astronomical investigations at the Moscow University, p. 102 – 107 (1985). In Russian. – See Abstr. 003.049.

009.040 Astrometry in Moscow in the first 150 years.
A. P. Gulyaev, V. V. Podobed.
Main directions of astronomical investigations at the Moscow University, p. 116 – 121 (1985). In Russian. – See Abstr. 003.049.

009.041 150 years gravimetry at the Sternberg Astronomical Institute.
N. P. Grushinskij, P. A. Stroev.
Main directions of astronomical investigations at the Moscow University, p. 149 – 157 (1985). In Russian. – See Abstr. 003.049.

009.042 Teaching astronomy at the Moscow State University.
V. G. Shamaev.
Main directions of astronomical investigations at the Moscow University, p. 171 – 173 (1985). In Russian. – See Abstr. 003.049.

009.043 The establishment of the U.S. Naval Observatory.
J. K. Herman.
Vistas Astron., Vol. 28, Parts 1/2, p. 391 – 399 (1985). – See Abstr. 012.109.

009.044 Construction and scientific activities of the new Vienna Observatory 1870 – 1900.
M. G. Firneis.
Vistas Astron., Vol. 28, Parts 1/2, p. 401 – 404 (1985). – See Abstr. 012.109.

009.045 Celebrating activities of the 50th anniversary of the founding of the Purple Mountain Observatory and the 1984 symposia on solar physics, celestial mechanics and astronomical instruments.
S.–z. Zhang.
Acta Astron. Sin., Vol. 26, No. 1, p. 98 – 99 (1985). In Chinese.

009.046 The first years at Parkes.
R. M. Price.
Serendipitous discoveries in radio astronomy, p. 300 – 306 (1984). – See Abstr. 012.113.

009.047 Radio astronomy at Dover Heights.
J. G. Bolton.
Serendipitous discoveries in radio astronomy, p. 312 – 321 (1984). – See Abstr. 012.113. Reprinted from the Proc. Astron. Soc. Aust., Vol. 4, No. 4, p. 349 – 358 (1982). – See Abstr. 33.009.009.

009.048 Planetari in Italia e nel mondo.
M. Cavedon.
Orione, Vol. 5, N. 1, p. 31 – 34 (1985).

A hilltop in foggy bottom.
See Abstr. 003.108.

Die Pekinger Sternwarte in altem Glanz wiederhergestellt.
See Abstr. 004.013.

Les débuts de la photographie astronomique particulièrement à l'Observatoire de Paris.
See Abstr. 004.019.

La specola Caetani.
See Abstr. 004.116.

Sui primi strumenti di astronomia di posizione della specola di Brera in Milano.
See Abstr. 004.117.

Origini della specola fiorentina.
See Abstr. 004.118.

La specola pisana (1735 – 1808).
See Abstr. 004.119.

Development of aperture synthesis at Cambridge.
See Abstr. 004.183.

Early interferometry at Jodrell Bank.
See Abstr. 004.186.

The research work at the Central Institute for Physics of the Earth, Potsdam, GDR, in the field of Doppler satellite geodesy.
See Abstr. 013.022.

Searches for short period pulsars at Jodrell Bank.
See Abstr. 126.140.

010 Societies, Associations, Organizations

American Association of Variable Star Observers (AAVSO)

010.021 **The American Association of Variable Star Observers Bulletin.**
No. 48 (1985).

American Astronomical Society (AAS)

010.061 **The 1910 AASA expedition to Hawaii to photograph comet Halley.**
J. Lankford.
Publ. Astron. Soc. Pac., Vol. 97, No. 588, p. 200 (1985). Abstract. – See Abstr. 010.281.

010.062 **The 165th meeting of the American Astronomical Society, held 13 – 17 January 1985 at Tucson, Arizona. Abstracts of presented papers.**
Bull. Am. Astron. Soc., Vol. 16, No. 4, p. 841 – 1025 (1984).

010.063 **Late–paper abstracts from the 16th annual meeting of the Division for Planetary Sciences, held 8 – 12 October 1984 at Kona, Hawaii.**
Bull. Am. Astron. Soc., Vol. 16, No. 4, p. 1026 – 1028 (1984).

010.064 **Late–paper abstracts from the 165th meeting of the American Astronomical Society, held 13 – 17 January 1985 at Tucson, Arizona.**
Bull. Am. Astron. Soc., Vol. 17, No. 1, p. 511 – 522 (1985).

010.065 **The 166th meeting of the American Astronomical Society, held 3 – 7 June 1985 at Charlottesville, Virginia. Abstracts of presented papers.**
Bull. Am. Astron. Soc., Vol. 17, No. 2, p. 525 – 621 (1985).

010.066 **The 16th regular meeting of the Dynamical Astronomy Division, held 28 – 29 March 1985 at Austin, Texas. Abstracts of presented papers.**
Bull. Am. Astron. Soc., Vol. 17, No. 2, p. 622 – 627 (1985).

010.067 **The annual meeting of the Solar Physics Division, held 13 – 15 May 1985 at Tucson, Arizona. Abstracts of presented papers.**
Bull. Am. Astron. Soc., Vol. 17, No. 2, p. 628 – 648 (1985).

010.068 **Bulletin of the American Astronomical Society.**
Vol. 16, No. 4 (1984), Vol. 17, Nos. 1 – 2 (1985).

010.069 **AAS Photo–Bulletin.**
No. 37 (1984).

Association Française des Observateurs d'Etoiles Variables

010.101 **Activité de l'A.F.O.E.V.**
É. Schweitzer.
Astronomie, Vol. 99, p. 41 – 45 (1985).

010.102 **Activité de l'A.F.O.E.V. en 1984.**
E. Schweitzer.
Bull. Assoc. Fr. Obs. Etoiles Variables, No. 32, p. 1 – 6 (1985).
Contents: Activité générale. Variables du type Mira et SRa.

010.103 **La vie de l'Association.**
E. Schweitzer.
Bull. Assoc. Fr. Obs. Etoiles Variables, No. 31, p. 11, No. 32, p. 13 (1985).

010.104 **Bulletin de l'Association Française des Observateurs d'Etoiles Variables.**
Nos. 31 – 32 (1985).

Association of Lunar and Planetary Observers (A.L.P.O.)

010.121 **Meetings of the Association.**
Strolling Astron., Vol. 31, Nos. 1 – 2, p. 1 – 2 (1985).

010.122 **The Minor Planet Bulletin.**
Vol. 12, No. 1 (1985).

010.123 **The Strolling Astronomer. The Journal of the Association of Lunar and Planetary Observers.**
Vol. 31, Nos. 1 – 2 (1985).

Astronomical Society of Australia (ASA)

010.161 **Constitution of the Astronomical Society of Australia.**
Proc. Astron. Soc. Aust., Vol. 5, No. 4, p. 619 – 622 (1984).

010.162 **List of institutions.**
Proc. Astron. Soc. Aust., Vol. 5, No. 4, p. 623 (1984).

010.163 **List of members.**
Proc. Astron. Soc. Aust., Vol. 5, No. 4, p. 623 – 628 (1984).

010.164 **Proceedings of the Astronomical Society of Australia.**
Vol. 5, No. 4 (1984).

Astronomical Society of India

010.201 **Bulletin of the Astronomical Society of India.**
Vol.12, No.4 (1984).

Astronomical Society of Japan

010.221 **Publications of the Astronomical Society of Japan.**
Vol. 37, No. 1 (1985).

010.222 **The Astronomical Herald.**
Vol. 78, Nos. 1 – 6 (1985). In Japanese.

Astronomical Society of New York

010.241 **Abstracts of papers given at the Fall meeting of the Astronomical Society of New York, held November 3, 1984 at Rensselaer Polytechnic Institute in Troy, New York.**
A. G. D. Philip.
News Lett. Astron. Soc. N.Y., Vol. 2, No. 7, p. 1 – 29 (1985).

010.242 **Abstracts of papers given at the Spring meeting of the Astronomical Society of New York, held May 13 – 14, 1983 at Cornell University, Ithaca, New York.**
News Lett. Astron. Soc. N.Y., Vol. 2, No. 4, p. 1 – 34 (1983).

010.243 **Abstracts of papers given at the Fall meeting of the Astronomical Society of New York, held November 5, 1983 at Union College in Schenectady, New York.**
News Lett. Astron. Soc. N.Y., Vol. 2, No. 5, p. 1 – 28 (1984).

010.244 **News Letter of the Astronomical Society of New York.**
Vol. 2, No. 4 (1983), No. 5 (1984), No. 7 (1985).

Astronomical Society of Southern Africa (ASSA)

010.261 **Section reports.**
Mon. Notes Astron. Soc. S. Afr., Vol. 44, Nos. 1 – 2, p. 4 – 9 (1985).

010.262   **Monthly Notes of the Astronomical Society of Southern Africa.**
Vol. 44, Nos. 1 – 2 (1985).

Astronomical Society of the Pacific (ASP)

010.281   **Abstracts of papers presented at the Session on History of Astronomy at the Santa Cruz Summer Meeting of the Astronomical Society of the Pacific, 11 July 1984.**
Publ. Astron. Soc. Pac., Vol. 97, No. 588, p. 199 – 201 (1985).

010.282   **Publications of the Astronomical Society of the Pacific.**
Vol. 97, Nos. 587 – 591 (1985).

010.283   **Mercury. The Journal of the Astronomical Society of the Pacific.**
Vol. 14, Nos. 1 – 3 (1985).

Astronomical Society of Western Australia (ASWA)

010.301   **Journal of the Astronomical Society of Western Australia.**
Vol. 35, January – March, June (1985).

Astronomical Society "Rudjer Bošković"

010.321   **Reminescences on the first days of the Astronomical Society "R. Bošković".**
P. Emanuel.
Vasiona, Année 32, No. 5, p. 91 – 92 (1984). In Serbo–Croatian.

010.322   **Formal meeting commemorating 50 years of the Astronomical Society.**
A. Tomić.
Vasiona, Année 32, No. 5, p. 108 – 110 (1984). In Croatian.

Astronomische Gesellschaft (AG)

010.341   **Mitteilungen des Vorstandes der Astronomischen Gesellschaft.**
H.-U. Keller.
Mitt. Astron. Ges., Nr. 63, p. 203 – 205 (1985).

010.342   **Mitteilungen der Astronomischen Gesellschaft.**
Nr. 63 – 64 (1985).

AURA (The Association of Universities for Research in Astronomy)

010.361   **Chronology of AURA and KPNO 1952 – 1960.**
F. K. Edmondson.
Bull. Am. Astron. Soc., Vol. 16, No. 4, p. 943 (1984). Abstract. – See Abstr. 010.062.

Berliner Arbeitsgemeinschaft für Veränderliche Sterne e.V. (BAV)

010.371   **35 Jahre BAV (1950 – 1985).**
W. Braune.
BAV Rundbrief, 34. Jahrg., Nr. 1, p. 38 – 43 (1985).

British Astronomical Association (BAA)

010.381   **British Astronomical Association Supplementary Circular.**
No. 2 (1985).

010.382   **British Astronomical Association Circular.**
Nos. 647 – 648 (1985).

010.383   **Meetings and activities of the Association.**
J. Br. Astron. Assoc., Vol. 95, Nos. 2 – 4, p. 75 – 81, 126 – 129, 173 – 182 (1985).

010.384   **Section reports.**
J. Br. Astron. Assoc., Vol. 95, Nos. 2 – 4, p. 69 – 73, 110 – 112, 123 – 125, 131 – 132, 143 – 150, 169 –173 (1985).

010.385   **Journal of the British Astronomical Association.**
Vol. 95, Nos. 2 – 4 (1985).

010.386   **The British Astronomical Association Newsletter.**
Nos. 10 – 12 (1985).

British Interplanetary Society (BIS)

010.401   **The BIS in space.**
L. R. Shepherd.
Spaceflight, Vol. 27, No. 1, p. 23 – 27 (1985).

010.402   **JBIS. Journal of the British Interplanetary Society.**
Vol. 38, Nos. 1 – 6 (1985).

010.403   **Spaceflight. A publication of the British Interplanetary Society.**
Vol. 27, Nos. 1 – 6 (1985).

010.404   **Space Education. A publication of the British Interplanetary Society.**
Vol. 1, No. 9 (1985).

European Space Agency (ESA)

010.481   **ESA Bulletin.**
Nos. 41 – 42 (1985).

010.482   **ESA Journal.**
Vol. 9, Nos. 1 – 2 (1985).

010.483   **ESA IUE Newsletter.**
Nos. 22 – 23 (1985).

010.484   **EXOSAT Express.**
Nos. 9 – 11 (1985).

010.485   **ESA Special Publication.**
ESA SP–207 (012.044), SP–213 (012.124), SP–217 (012.021), SP–220 (012.045), SP–224 (012.052), SP–1070 (003.029).

010.486   **ESA Brochure.**
ESA BR–24, (051.114).

International Amateur–Professional Photoelectric Photometry

010.501   **I.A.P.P.P. and its activities.**
D. S. Hall, T. W. Saltsman, R. M. Genet, R. C. Reisenweber.
Bull. Am. Astron. Soc., Vol. 16, No. 4, p. 909 (1984). Abstract. – See Abstr. 010.062.

010.502   **I.A.P.P.P. Communication.**
Nos. 19 – 20 (1985).

International Astronomical Union (IAU)

010.541   **Bibliography and program notes on close binaries.**
Nos. 40 (39.002.034), 41 (39.002.035).

010.542 **IAU Information Bulletin 53.**
R. M. West.
D. Reidel Publishing Company, Dordrecht, Holland. 56 pp. (1985).
 Contents: XIXth General Assembly. Executive Committee. Commissions. International Organisations. IAU Symposia and Colloquia. Meetings co–sponsored by the IAU. Other scientific meetings. IAU Publications. Other publications. Membership. International schemes of support to (young) astronomers. Other matters. Errata.

010.543 **The Minor Planet Circulars/Minor Planets and Comets.**
Nos. 9315 – 9716 (1985).

010.544 **Commission 27 of the IAU, Information Bulletin on Variable Stars.**
Nos. 2650 – 2753 (1985).

010.545 **Circulaire d'Information.**
Nos. 95 – 96 (1985).

010.546 **IAU Circulars.**
Nos. 4025 – 4079 (1985).

010.547 **Reports on Astronomy.**
R. M. West.
Trans. IAU, Vol. XIXA (1985). – See Abstr. 003.046.

010.548 **Quarterly Bulletin on Solar Activity.**
Vol. 22, Part V, Vol. 23, Part V.

010.549 **IAU Symposia.**
Nos. 106 (012.007), 113 (012.070).

010.550 **IAU Colloquia.**
Nos. 82 (012.027), 88 (012.095).

Korean Astronomical Society

010.601 **The Journal of the Korean Astronomical Society.**
Vol. 18, No. 1 (1985).

LEST Foundation

010.621 **LEST Foundation, Annual Report 1984.**
Ø. Hauge.
LEST Found., Annu. Rep., 37 pp. (1985).

010.622 **LEST – Large European Solar Telescope.**
Ø. Hauge.
Astron. Tidsskr., Årg. 18, Nr. 2, p. 49 – 50 (1985).

010.623 **LEST Foundation, Technical Report.**
No. 8 (1984), No. 12 (1985).

Meteoritical Society

010.641 **Abstracts of papers presented at the 47th annual meeting. The Meteoritical Society, July 30 – August 2, 1984, Albuquerque, NM.**
Meteoritics, Vol. 19, No. 4, p. 183 – 348 (1984).

010.642 **Meteoritics. The Journal of the Meteoritical Society.**
Vol. 19, No. 4 (1984), Vol. 20, Nos. 1, 2 (Part 1 – 2) (1985).

National Aeronautics and Space Administration (NASA)

010.681 **History of NASA's airborne astronomy program.**
L. C. Haughney.
Publ. Astron. Soc. Pac., Vol. 97, No. 588, p. 200 (1985). Abstract. – See Abstr. 010.281.

Oriental Astronomical Association

010.701 **The Heavens.**
Vol. 66, Nos. 1 – 6 (1985). In Japanese.

Royal Astronomical Society (RAS)

010.721 **The first results from IRAS. Summaries of papers presented at the RAS specialist discussion, held 1984 May 11 in the Scientific Societies Lecture Theatre, Savile Row, London.**
Observatory, Vol. 105, No. 1064, p. 1 – 7 (1985).

010.722 **The dynamics of stellar and planetary systems. Summary of the RAS specialist discussion, held 1984 October 12 at the Scientific Societies' Lecture Theatre, Savile Row.**
Observatory, Vol. 105, No. 1066, p. 74 – 77 (1985).

010.723 **Meetings of the Society.**
Observatory, Vol. 105, Nos. 1065 – 1066, p. 25 – 29, 61 – 74 (1985).

010.724 **Meetings and activities of the Society.**
Q. J. R. Astron. Soc., Vol. 26, Nos. 1 – 2, p. 115 – 116, 229 – 230 (1985).

010.725 **Monthly Notices of the Royal Astronomical Society.**
Vol. 212, Nos. 1 – 4, Vol. 213, Nos. 1 – 4, Vol. 214, Nos. 1 – 4 (1985).

010.726 **Geophysical Journal of the Royal Astronomical Society.**
Vol. 80, Nos. 1 – 3, Vol. 81, Nos. 1 – 3 (1985).

010.727 **The Quarterly Journal of the Royal Astronomical Society.**
Vol. 26, Nos. 1 – 2 (1985).

Royal Astronomical Society of Canada

010.741 **The Journal of the Royal Astronomical Society of Canada.**
Vol. 79, Nos. 1 – 3 (1985).

010.742 **National Newsletter. Supplement to the Journal of the Royal Astronomical Society of Canada.**
Vol. 79, Nos. 1 – 3 (1985).

Royal Astronomical Society of New Zealand

010.761 **The Royal Astronomical Society of New Zealand (Inc.). The 61st report of council being for the calendar year 1983.**
South. Stars, Vol. 31, No. 1, p. 100 – 112 (1984).

010.762 **Southern Stars. Journal of the Royal Astronomical Society of New Zealand.**
Vol. 31, No. 1 (1984), No. 2 (1985).

Schweizerische Astronomische Gesellschaft (SAG)

010.781 **BBSAG Bulletin.**
Nos. 75 – 76 (1985).

010.782 **Mitteilungen.**
Orion, 43. Jahrg., Nr. 206 – 208, p. 17 – 20, 51 – 57, 89 – 92 (1985).

010.783 **Orion. Zeitschrift der Schweizerischen Astronomischen Gesellschaft. Revue de la Société Astronomique de Suisse.**
Jahrg. 43, Nr. 206 – 208 (1985).

Società Astronomica Italiana (S.A.It.)

010.801   **Memorie della Società Astronomica Italiana.**
Vol. 55, Nos. 3 – 4 (1984).

010.802   **Giornale di Astronomia.**
Vol. 10, N. 3 – 4 (1984), Vol. 11, N. 1 (1985).

Société Astronomique de France

010.821   **Séances, commissions, activités de la Société.**
Astronomie, Vol. 99, p. 96 – 98, 199 – 203, 253 – 254, 295 – 299 (1985).

010.822   **L'Astronomie et Bulletin de la Société Astronomique de France.**
Vol. 99, janvier – juin (1985).

Société Astronomique de Liège

010.841   **Le Ciel.**
Vol. 47, janvier – juin, p. 1 – 180 (1985).

Société Royale Belge d'Astronomie

010.861   **Ciel et Terre. Bulletin de la Société Royale Belge d'Astronomie, de Météorologie et de Physique du Globe.**
Vol. 101, Nos. 1 – 3 (1985).

Vereinigung der Sternfreunde e.V. (VdS)

010.901   **Nachrichten der Vereinigung der Sternfreunde e.V.**
Sterne Weltraum, 24. Jahrg., Nr. 1 – 6, p. 46 – 49, 108 – 111, 166 – 169, 226 – 229, 286 – 289, 346 –349 (1985).

010.902   **Sonne. Mitteilungsblatt der Amateursonnenbeobachter.**
Jahrg. 9, Nr. 33 (1985).

011 Reports on Colloquia, Congresses, Meetings, Symposia, Expeditions

011.001   **Spring MIST meeting 1984.**
P. A. Hadjiry, M. J. Laird.
Q. J. R. Astron. Soc., Vol. 26, No. 1, p. 60 – 67 (1985).

011.002   **Defining the Crab nebula.**
V. Trimble.
Nature, Vol. 313, No. 5998, p. 96 – 97 (1985).
This note reports on the symposium "The Crab nebula and related remnants", held at George Mason University, Virginia, 11 – 12 October 1984.

011.003   **Unsmoothing the Universe.**
V. Trimble.
Nature, Vol. 313, No. 6004, p. 634 – 635 (1985).
This note reports on the Twelfth Texas Symposium on Relativistic Astrophysics, held at Hebrew University, Jerusalem, 17 – 21 December 1984. Emphasis is placed upon recent developments in theories of density fluctuations and galaxy formation in the early Universe.

011.004   **On the Third Asian–Pacific Regional Meeting of the IAU.**
T. Kogure.
Astron. Her., Vol. 78, No. 2, p. 52 – 55 (1985). In Japanese.

011.005   **Gravitational collapse and the problem of singularities in general relativity.**
Patrika, No. 9, p. 2 – 3 (1985).
Academy lecture given by Prof. S. Chandrasekhar on November 21, 1984 at the International Symposium on Theoretical Physics in honour of S. N. Bose.

011.006   **Summary report on the First International Symposium on Space Techniques for Geodynamics, Sopron, Hungary, 9 – 13 July 1984.**
Bull. Géod., Vol. 59, No. 1, p. 103 – 104 (1985).

011.007   **Waves in space plasmas: highlights of a conference held in Hawaii, 7 – 11 February 1983.**
R. L. Dowden, B. J. Fraser.
Space Sci. Rev., Vol. 39, Nos. 3/4, p. 227 – 253 (1984).
The conference was called to bring together investigators of magnetospheric plasma waves having frequencies from VLF whistlers and emissions down through ELF and ULF to Pc5 long period pulsations. The emphasis was on the physics and techniques underlying the entire frequency range. Topics included wave electron interactions and electron precipitation, ray tracing and other methods to track down sources of VLF and ULF waves, VLF–ULF relationships, heavy ion effects in ULF propagation, and long period ULF waves.

011.008   **Cratering theories bombarded.**
P. Weissman.
Nature, Vol. 314, No. 6006, p. 17 – 18 (1985).
This note reports on a conference on "The Galaxy and the Solar System", held at Tucson, Arizona, 10 – 12 January 1985. The main issue of this conference was the hypothetical relation between periodic extinction events in Earth's history and the motion of the solar system in the Galaxy.

011.009   **International conference of directors of planetaria. Stuttgart, West Berlin, Hamburg, F.R. Germany 27 August – 2 September 1984.**
K. A. Portsevskij.
Zemlya Vselennaya, No. 1, p. 57 – 63 (1985). In Russian.

011.010   **Names of landscapes of other worlds.**
P. M. Millman, V. V. Shevchenko.
Zemlya Vselennaya, No. 2, p. 64 – 68 (1985). In Russian.
Report on a conference of the working group in the field of nomenclature of the planetary system held in Tbilisi in April 1984.

011.011   **Materials of the All–Union conference on cosmic rays, held in Yakutsk in July 1984.**
Izv. Akad. Nauk SSSR. Ser. Fiz., Tom 48, No. 11, p. 2066 – 2277 (1984). In Russian. From Ref. Zh., 51. Astron., 3.51.35 (1985).

011.012   **Report on the UK–SMM workshop meetings held in Oxford on 1983 April 11, 12 and 13, September 7, 8, and 9 and 1984 March 26, 27 and 28.**
R. W. P. McWhirter.
Mem. Soc. Astron. Ital., Vol. 55, No. 4, p. 823 – 829 (1984). – See Abstr. 003.017.

011.013 The first IAU Symposium on extraterrestrial life.
M. D. Papagiannis.
J. Br. Interplanet. Soc., Vol. 38, No. 6, p. 276 (1985). Abstract. –
See Abstr. 012.060.

**011.014 The search for extraterrestrial life: recent developments.
A report on IAU Symposium No. 112.**
M. D. Papagiannis.
J. Br. Interplanet. Soc., Vol. 38, No. 6, p. 281 – 285 (1985).
   This paper is a report on the International Symposium "The
search for extraterrestrial life: recent developments" which was
held in Boston, Massachusetts, USA, 18 – 21 June 1984.

011.015 12. Texas–Symposium über relativistische Astrophysik.
H. J. Blome.
Sterne Weltraum, 24. Jahrg., Nr. 5, p. 246 – 247 (1985).

**011.016 Microcomputers in astronomy. II: The Fifth Annual
I.A.P.P.P. Fairborn Symposium and Proceedings.**
R. M. Genet, K. A. Genet, D. S. Hall, E. J. Lurcott.
I.A.P.P.P. Commun., No. 19, p. 1 – 5 (1985).

011.017 The Third European Meeting of the I.A.P.P.P.
E. N. Walker.
I.A.P.P.P. Commun., No. 19, p. 20 – 22 (1985).

**011.018 Some highlights of IAU Colloquium No. 78 on
"Astronomy with Schmidt Telescopes".**
R. J. Dodd.
South. Stars, Vol. 31, No. 1, p. 95 – 99 (1984).
   A report is given of the proceedings of IAU Colloquium
No. 78 on Astronomy with Schmidt Telescopes held at Asiago,
Italy in August/September 1983.

**011.019 All–Union conference of the working group on problems
of "Terrestrial planets and asteroids" and "Giant plan-
ets", held in Gagra, October 16 – 18, 1984.**
L. R. Lisina.
Kinematika Fiz. Nebesn. Tel, Tom 1, No. 3, p. 95 – 96 (1985). In
Russian.

**011.020 IAU symposium in Italy, held in Como, 24 – 29 May
1984.**
I. N. Glushneva.
Zemlya Vselennaya, No. 3, p. 62 – 64 (1985). In Russian.

**011.021 The VIIth national conference of Yugoslav astronomers.
Belgrade, 9 – 11 May 1984.**
M. Jeličić.
Vasiona, Année 32, No. 5, p. 94 – 108 (1984). In Croatian.

**011.022 Compte rendu du Colloque UAI 88 – vitesses radiales
stellaires, Schenectady, 24 – 27 octobre 1985.**
C. Fehrenbach.
Bull. Inf. Cent. Données Stellaires, No. 28, p. 3 – 4 (1985).

011.023 Asteroids, comets, meteors II.
C.-I. Lagerkvist, P. Magnusson, H. Rickman.
Upps. Astron. Obs. Rep., No. 34, 62 pp. (1984).
   This report contains abstracts of papers to be presented at a
symposium, held at Uppsala, 3 – 6 June 1985.

011.024 Berichte aus den Workshops.
Mitt. Astron. Ges., Nr. 63, p. 105 – 109 (1985). – See
Abstr. 012.063.
   Workshop I: Molekülwolken (*T. L. Wilson*), Workshop II:
Planetarische Nebel (*J. Köppen*), Workshop III: Gravitations-
linsen (*P. Schneider*).

011.025 More news from Mars.
I. P. Wright, M. M. Grady.
Nature, Vol. 315, No. 6018, p. 367 – 368 (1985).
   This note comments on new research supporting the hypothesis that some meteorites are of Martian origin. These results were
presented at the 16th Lunar and Planetary Science Conference,
held in Houston, Texas, 11 – 15 March 1985.

**011.026 The G. C. McVittie Eightieth Birthday Celebration
Conference (University of Kent at Canterbury, 1984
June 2).**
J. S. R. Chisholm, W. H. McCrea.
Q. J. R. Astron. Soc., Vol. 26, No. 2, p. 117 – 121 (1985).

**011.027 Conference of the working group on the nomenclature of
the planetary system.**
P. M. Millman.
Astron. Vestn., Tom 18, No. 4, p. 342 – 345 (1984). In Russian.
Abstr. in Ref. Zh., 51. Astron., 5.51.15; 62. Issled. Kosm. Pro-
stranstva, 5.62.13 (1985).

011.028 Making the moon from a big splash.
R. A. Kerr.
Science, Vol. 226, No. 4678, p. 1060 – 1061 (1984).
   Report on a conference on the origin of the moon, held at
Kona, Hawaii, 13 – 16 October 1984.

011.029 Les trous noirs.
J. Demaret.
Ciel Terre, Vol. 101, No. 3, p. 95 – 96 (1985).
   Report on a conference of the Société Royale Belge
d'Astronomie, de Météorologie et de Physique du Globe, held
1985 January 26.

**011.030 All–Union seminar on astrophysics in Leningrad, 4 – 5
October 1984.**
D. I. Nagirner.
Astron. Zh., Tom 62, Vyp. 3, p. 616 – 617 (1985). In Russian.
English translation in Sov. Astron., Vol. 29, No. 3.

011.031 Tsesevich memorial conference of young astronomers.
E. V. Menchenkova, I. L. Andronov.
Sov. Astron., Vol. 28, No. 6, p. 728 – 729 (1984). English transla-
tion of 38.011.035.

**011.032 Astronomy on the 6th Europhysical Conference in Pra-
gue.**
P. Ambrož, J. Grygar.
Říše hvězd, Vol. 66, No. 2, p. 31 – 34 (1985). In Czech.

011.033 Astronomy with the Hubble Space Telescope.
S. van den Bergh.
J. R. Astron. Soc. Can., Vol. 79, No. 3, p. 134 – 142 (1985).
   This paper reports on a Space Telescope Working Group meeting held in Tucson, Arizona, January 17 and 18, 1985 in which
the strategy of Space Telescope use for astronomical research was
discussed. A glossary of Space Telescope acronyms is given in an
appendix.

012 Proceedings of Colloquia, Congresses, Meetings, Symposia

012.001   Future of ultraviolet astronomy based on six years of IUE research. Proceedings of a symposium held at NASA Goddard Space Flight Center, Greenbelt, Maryland, April 3 – 5, 1984.
J. M. Mead, R. D. Chapman, Y. Kondo (Editors).
NASA Conf. Publ., NASA CP–2349. 14 + 544 pp. (1984).
The individual contributions are included in their corresponding subject categories – see abstracts 031.001, 035.001, 036.001, 036.002, 051.001, 065.001, 091.001, 099.001 – 099.003, 101.001, 102.001, 112.001 – 112.019, 114.001 – 114.014, 117.001 – 117.018, 119.001 – 119.005, 120.001, 121.001 – 121.007, 122.001 – 122.005, 124.001, 124.101, 126.001 – 126.009, 131.001 – 131.008, 132.001, 132.002, 134.001, 134.002, 153.001, 154.001, 155.001, 156.001, 156.002, 157.001, 157.002, 158.001 – 158.008, 159.001.

012.002   Unstable current systems and plasma instabilities in astrophysics. Proceedings of the 107th Symposium of the International Astronomical Union, held in College Park, Maryland, U.S.A., August 8 – 11, 1983.
M. R. Kundu, G. D. Holman (Editors), with a summary of the conference by V. M. Vasyliunas.
D. Reidel Publishing Company, Dordrecht – Boston – Lancaster. 22 + 566 pp. Price Dfl. 80.00, US\$ 29.50, £ 20.50 (1985). ISBN 90–277–1886–5 cloth, 90–277–1887–3 paper.
The individual contributions are included in their corresponding subject categories – see abstracts 007.028, 013.001, 062.004 – 062.032, 064.002, 064.003, 073.002 – 073.009, 074.002 – 074.006, 075.001, 076.002, 076.003, 077.001 – 077.004, 084.001 – 084.003, 093.001, 099.004, 117.029, 122.010, 144.001, 144.002, 158.013, 158.014.

012.003   Chapman Conference on collisionless shocks, Parts 1, 2.
J. Geophys. Res., Vol. 90, No. A1, p. 1 – 248, Vol. 90, No. A5, p. 3925 – 3994 (1985).
The individual contributions within the subject scope of Astronomy and Astrophysics Abstracts are included in their corresponding categories – see abstracts 062.033, 062.034, 073.011, 074.007 – 074.010, 074.086, 078.013, 084.084, 099.055, 106.003 – 106.012, 106.052, 106.053.

012.004   Protostars and planets II, conference held in Tucson, Ariz., USA, January 1984.
J. A. Burns (Editor).
Icarus, Vol. 61, No. 1, p. 1 – 59 (1985).
The individual contributions are included in their corresponding subject categories – see abstracts 063.003, 064.005, 065.009, 102.003, 107.003, 107.004, 131.027, 131.028.

012.005   High energy transients in astrophysics. Proceedings of a meeting held at Santa Cruz, Calif., USA, July 11 – 22, 1983.
S. E. Woosley (Editor).
AIP Conf. Proc., No. 115, 15 + 714 pp. Price US\$ 51.25 (1984). ISBN 0–88318–314–5.
The individual contributions are included in their corresponding subject categories – see abstracts 002.005, 032.004, 032.005, 035.003, 035.004, 051.004, 051.005, 062.041, 063.004, 063.005, 065.010, 065.011, 067.007 – 067.029, 073.017, 076.004, 076.005, 082.009, 117.048 – 017.050, 142.004 – 142.020, 143.005 – 143.019.

012.006   Intercomparison of stratospheric/mesospheric data. Proceedings of the Topical Meeting of the COSPAR Interdisciplinary Scientific Commission A (Meeting A1) of the COSPAR Twenty–fifth Plenary Meeting held in Graz, Austria, 25th June – 7th July 1984.
A. Ghazi, R. T. Watson (Editors).
Adv. Space Res., Vol. 4, No. 6, 6 + 147 pp. (1984).

012.007   The Milky Way galaxy. Proceedings of the 106th Symposium of the International Astronomical Union, held in Groningen, The Netherlands, 30 May – 3 June, 1983.
H. van Woerden, R. J. Allen, W. B. Burton (Editors), with a summary and outlook by J. P. Ostriker.
D. Reidel Publishing Company, Dordrecht – Boston – Lancaster. 24 + 660 pp. Price Dfl. 195.00, US\$ 69.00, £ 49.75 cloth; Dfl. 90.00, US\$ 32.50, £ 22.95 paper (1985). ISBN 90–277–1919–5 cloth, ISBN 90–277–1920–9 paper.
The individual contributions are included in their corresponding subject categories – see abstracts 002.002, 004.005 – 004.008, 015.006, 036.009, 112.031, 131.032 – 131.048, 151.011 – 151.034, 155.005 – 155.057, 156.005, 156.006, 157.013 – 157.026, 158.028, 160.008, 161.013.

012.008   Energy transfer during magnetospheric substorms. Papers from CDAW 6 (*Coordinated Data Analysis Workshop 6*).
J. Geophys. Res., Vol. 90, No. A2, p. 1175 – 1374 (1985).
The individual contributions within the subject scope of Astronomy and Astrophysics Abstracts are included in their corresponding categories – see abstracts 084.017 – 084.021.

012.009   Solar maximum analysis. Proceedings of Symposium 2 of the COSPAR Twenty–fifth Plenary Meeting held in Graz, Austria, 25th June – 7th July 1984.
P. A. Simon (Editor).
Adv. Space Res., Vol. 4, No. 7, 8 + 404 pp. (1984).
The individual contributions are included in their corresponding subject categories – see abstracts 013.006, 013.007, 051.007 – 051.009, 073.023 – 073.059, 074.022 – 074.032, 076.006, 076.007, 077.009 – 077.012, 080.011, 106.018 – 106.023.

012.010   Solar–space observations and stellar prospects. Proceedings of the Topical Meeting of the COSPAR Interdisciplinary Scientific Commission E (Meetings E1, E2, and E6) of the COSPAR Twenty–fifth Plenary Meeting held in Graz, Austria, 25th June – 7th July 1984.
J. W. Harvey, H. S. Hudson, R. W. Noyes (Editors).
Adv. Space Res., Vol. 4, No. 8, 6 + 177 pp. (1984).
The individual contributions are included in their corresponding subject categories – see abstracts 035.005 – 035.008, 051.010, 051.011, 065.013, 065.014, 073.060 – 073.062, 075.004, 075.005, 076.008, 076.009, 080.012 – 080.018.

012.011   Classical general relativity. Proceedings of the conference on classical (non–quantum) general relativity, held at City University, London, UK, 21 – 22 December 1983.
W. B. Bonnor, J. N. Islam, M. A. H. MacCallum (Editors).
Cambridge University Press, Cambridge – London – New York – New Rochelle – Melbourne – Sydney. 15 + 269 pp. Price £ 25.00, US\$ 44.50 (1984). ISBN 0–521–26747–1. See also 38.012.069.
Review in Sky Telesc., Vol. 69, No. 6, p. 520 (1985).
The individual contributions are included in their corresponding subject category – see abstracts 066.004 – 066.022.

012.012   Fronts, interfaces and patterns. Proceedings of the Third Annual International Conference of the Centre for Non-linear Studies, held at Los Alamos, N.M., USA, 2 – 6 May 1983.
Physica D, Vol. 12D, No. 1 – 3 (1984).
Review in Phys. Abstr., Vol. 88, No. 1248, Entry 4913 (1985).
See abstracts 021.002, 083.004.

012.013   Report of the optical conference on the 7.6–meter telescope held in Austin, Texas, March 22 – 24, 1982.
H. J. Smith, T. G. Barnes III (Editors), with a summary by R. G. Tull.
Univ. Tex., Publ. Astron., No. 22, 14 + 456 pp. (1984).
Contents: Introductions (*H. J. Smith*). History of the project (*H. J. Smith*). Project constraints (*T. G. Barnes III*). Project con-

straints (*R. G. Tull*). Telescope concept (*R. E. Nather*). Auxiliary instruments (*R. E. Nather*). Paul–Baker prime focus (*R. Angel*). Prime focus and Nasmyth cameras (*A. Meinel*). Nasmyth focal reducers (*M. MacFarlane*). Spectrometry (*R. Angel, R. G. Tull, J. Brault*). Infrared sites (*G. Neugebauer*). IR instrumentation (*F. Gillett*). Prime focus imaging (*E. H. Richardson*). Primary mirror figure control (*R. G. Tull*).

012.014   Data bases and data structures in physics. Proceedings of the Fifth Summer School on Computational Physics, held at Bechyne Castle, Czechoslovakia, 21 – 30 June 1983.
Comput. Phys. Commun., Vol. 33, No. 1 – 3 (1984).
Review in Phys. Abstr., Vol. 88, No. 1249, Entry 9970 (1985).
See abstracts 002.009, 013.014, 013.015.

012.015   Second New Orleans Conference on Quantum Theory and Gravitation, held at New Orleans, La., USA, 25 – 28 May 1983.
A. R. Marlow (Editor).
Int. J. Theor. Phys., Vol. 23, No. 8, 106 pp. (1984).
Review in Phys. Abstr., Vol. 88, No. 1249, Entry 9973 (1985).
See abstracts 080.019, 161.034, 161.035.

012.016   Proceedings of the international conference on glass in planetary and geological phenomena, held at Alfred, N.Y., USA, 14 – 18 August 1983.
J. Non–Cryst. Solids, Vol. 67, No. 1 – 3 (1984).
Review in Phys. Abstr., Vol. 88, No. 1250, Entry 14823 (1985).
See abstracts 081.020, 094.010, 099.018, 105.011 – 105.015.

012.017   Perspectives in nuclear physics at intermediate energies. Proceedings of a workshop, held at Trieste, Italy, 10 – 14 October 1983.
S. Boffi, C. Ciofi degli Atti, M. M. Giannini (Editors).
World Scientific Publishing Co. Pte. Ltd., Singapore, 11 + 546 pp. (1984).
Review in Phys. Abstr., Vol. 88, No. 1250, Entry 14833 (1985).
See Abstr. 067.038.

012.018   Publications of scientific results of the Intercosmos – cooperation Intercosmos/Cospar, Karlovy Vary, Czechoslovakia, September 16 – 21, 1984.
P. Lála (Editor).
Observations of artificial earth satellites, No. 23, 616 pp. (1984) = Publ. Astron. Inst. Czech. Acad. Sci., No. 58.
The individual contributions are included in their corresponding subject categories – see abstracts 013.018 – 013.023, 034.010 – 034.016, 036.017 – 036.029, 044.009, 044.010, 045.002 – 045.008, 052.002 – 052.015, 081.013 – 081.017, 082.012 – 082.016.

012.019   Planets, their origin, interior and atmosphere. Fourteenth Advanced Course of the Swiss Society of Astronomy and Astrophysics, held at Saas–Fee, 26 – 31 March 1984.
P. Bartholdi, P. Bochsler, Y. Chmielewski (Editors).
Published and sold by Geneva Observatory, CH–1290 Sauverny, Switzerland. 10 + 287 pp. Price Sfr 40.00 (1984).
The individual contributions are included in their corresponding subject categories – see abstracts 091.013, 091.014, 107.006.

012.020   Proceedings of the workshop on grand unified theories and cosmology, held at Tsukuba, Japan, 7 – 10 December 1983.
K. Odaka, A. Sugamoto (Editors).
KEK–84–12. Natl. Lab. High Energy Phys., Ibaraki–ken, Japan, 456 pp. (1984).
Review in Phys. Abstr., Vol. 88, No. 1251, Entry 19001 (1985).
See abstracts 061.011, 061.015, 080.021, 080.022, 143.021, 144.019 – 144.021, 155.061, 161.059 – 161.062, 161.071 – 161.073.

012.021   Achievements of the International Magnetospheric Study (IMS). Proceedings of an International Symposium, held at Graz, Austria, 26 – 28 June 1984.
B. Battrick, E. Rolfe (Editors), with an opening address by J. G. Roederer.
ESA Spec. Publ., ESA SP–217, 11 + 760 pp. Price code C5 (1984).

The individual contributions within the subject scope of Astronomy and Astrophysics Abstracts are included in their corresponding categories – see abstracts 082.017, 083.005, 084.031 – 084.060.

012.022   Champs magnétiques stellaires. Comptes rendus de l'Ecole de Goutelas (France), 2 – 7 avril 1984.
A. Baglin, M. Auvergne, C. Caseneuve (Editors).
Société Française des Spécialistes d'Astronomie, Observatoire de Paris, 61, avenue de l'Observatoire, 75014 Paris, France. 489 pp. (1984).
The individual contributions are included in their corresponding subject categories – see abstracts 034.017, 036.037, 062.050 – 062.057, 063.015 – 063.017, 065.022, 073.068, 075.007 – 075.010, 116.009.

012.023   The origin of nonradiative heating/momentum in hot stars. Proceedings of a workshop held at NASA Goddard Space Flight Center, Greenbelt, Maryland, June 5 – 7, 1984.
A. B. Underhill, A. G. Michalitsianos (Editors).
NASA Conf. Publ., NASA CP–2358, 7 + 254 pp. (1985).
The individual contributions are included in their corresponding subject categories – see abstracts 064.013 – 064.024, 065.023, 073.069, 074.035, 112.037 – 112.043, 114.030, 116.010 – 116.014, 121.013, 121.014, 122.026, 122.027, 131.065, 132.015.

012.024   Cometary astrometry. Proceedings of a workshop, held at the European Southern Observatory Headquarters, Garching, F.R. Germany, June 18 – 19, 1984.
D. K. Yeomans, R. M. West, R. S. Harrington, B. G. Marsden (Editors).
Jet Propulsion Laboratory, California Institute of Technology, JPL Publ. 84–82. 5 + 220 pp. (1984).
The individual contributions are included in their corresponding subject categories – see abstract 002.014, 002.015, 013.027 – 013.031, 013.036, 021.004, 034.026, 036.047 – 036.054, 041.006, 042.008, 051.032, 096.008, 102.012 – 102.014, 103.911 – 103.913.

012.025   The Ninth UK Geophysical Assembly, held at the University of East Anglia, Norwich, England, 15 – 17 April 1985.
With a preface by F. J. Vine, P. N. Chroston, R. L. French.
Geophys. J. R. Astron. Soc., Vol. 81, No. 1, p. 307 – 346 (1985).
The individual contributions within the subject scope of Astronomy and Astrophysics Abstracts are included in their corresponding categories – see abstracts 044.013, 044.014, 045.009, 045.010, 084.061, 084.062.

012.026   Atmospheric spectroscopy. The International Workshop held at the SERC Rutherford Appleton Laboratory, Chilton, Didcot, Oxon, England, 19 – 21 July 1983.
G. E. Hunt, J. Ballard (Editors).
J. Quant. Spectrosc. Radiat. Transfer, Vol. 32, No. 5/6, p. 373 – 477 (1984).
The individual contributions are included in their corresponding subject categories – see abstracts 035.015, 036.055, 082.021 – 082.023, 091.018, 099.022, 099.023.

012.027   Cepheids: theory and observations. Proceedings of the IAU Colloquium No. 82, held at Toronto, Canada, 28 May – 1 June 1984.
B. F. Madore (Editor), with a historical preface by J. D. Fernie.
Cambridge University Press, Cambridge – London – New York – New Rochelle – Melbourne – Sydney. 12 + 300 pp. Price £ 20.00, US$ 39.50 (1985). ISBN 0–521–30091–6.
Reviews in Astron. Tidsskr., Árg. 18, Nr. 2, p. 89 (1985); Spaceflight, Vol. 27, No. 6, p. 288 (1985).
The individual contributions are included in their corresponding subject categories – see abstracts 065.027, 122.036 – 122.086, 153.014, 154.010, 156.009, 157.061, 157.062.

012.028 **Effects of variable mass loss on the local stellar environment. Second Trieste–Workshop held in Rupingrande near Trieste, 12 – 20 October 1983.**
R. Stalio, R. N. Thomas (Editors), with a final discussion by C. La Dous, T. Lago, P. Kuin, P. Martens, M. Ramella.
Trieste–Workshop series on nonlinear, nonequilibrium thermodynamics of open, nonthermal systems in astronomy. Osservatorio Astronomico di Trieste, Via G. B. Tiepolo, C. P. Succ. 5, I–34131 Trieste, Italy. 11 + 234 pp. (1984).
The individual contributions are included in their corresponding subject categories – see abstracts 064.030, 064.031, 074.039, 112.049, 112.050, 117.095, 121.018, 124.007, 134.023.

012.029 **Dust in space and comets. Proceedings of the Topical Meeting of the COSPAR Interdisciplinary Scientific Commission B (Meetings B1 and B2) of the COSPAR Twenty–fifth Plenary Meeting held in Graz, Austria, 25th June – 7th July 1984.**
G. E. Morfill, C. T. Russell, M. S. Hanner (Editors).
Adv. Space Res., Vol. 4, No. 9, 7 + 316 pp. (1984). ISBN 0–08–032745–1.
The individual contributions are included in their corresponding subject categories – see abstracts 022.031 – 022.034, 035.016 – 035.020, 062.062, 062.063, 064.033, 091.020 – 091.025, 099.026, 099.027, 100.012 – 100.016, 102.018 – 102.025, 103.015 – 103.019, 103.914 – 103.920, 106.028, 106.029.

012.030 **The stability of planetary systems. Proceedings of the Alexander von Humboldt Colloquium on Celestial Mechanics, held at Ramsau, Styria, March 25 – 31, 1984.**
R. L. Duncombe, R. Dvorak, P. J. Message (Editors).
Celest. Mech, Vol. 34, Nos 1 – 4, p. 1 – 468 (1984). Available also as hardbound edition. Price Dfl. 240.00, $96.00, £ 60.95. ISBN 90–277–1961–6.
The individual contributions are included in their corresponding categories – see abstracts 004.029, 042.013 – 042.032, 043.002 – 043.004, 094.014, 097.005, 098.027 – 098.032, 099.032, 099.033, 100.018, 101.006, 101.007, 107.008.

012.031 **Life sciences and space research XXI(1). Proceedings of the Topical Meeting of the COSPAR Interdisciplinary Scientific Commission F (Meetings F4 and F8) of the COSPAR Twenty–fifth Plenary Meeting held in Graz, Austria, 25th June – 7th July 1984.**
H. P. Klein, G. Horneck (Editors).
Adv. Space Res., Vol. 4, No. 10, 7 + 290 pp. (1984).
The individual contributions within the subject scope of Astronomy and Astrophysics Abstracts are included in their corresponding categories – see abstracts 051.025 – 051.029.

012.032 **Gas in the interstellar medium. Third Rutherford Appleton Laboratory Workshop on Astronomy and Astrophysics, held at the Cosener's House, Abingdon, England, 21 – 23 May 1984.**
P. M. Gondhalekar (Editor).
RAL–84–101. Rutherford Appleton Laboratory, Chilton, Didcot, Oxon OX11 0QX, England. 7 + 171 pp. (1984).
The individual contributions are included in their corresponding subject categories – see abstracts 064.001, 125.025 – 125.028, 131.107 – 131.109, 132.029, 132.030, 134.027, 155.086, 156.012.

012.033 **The MK process and stellar classification. Proceedings of the workshop in honor of W. W. Morgan and P. C. Keenan, held at the University of Toronto, Canada, June 1983.**
R. F. Garrison (Editor), with a general discussion and a summary of the meeting by R. F. Garrison.
David Dunlap Observatory, University of Toronto, Toronto, Canada. 23 + 423 + 11 pp. Price US$ 27.00 paper, US$ 45.00 cloth (1984). ISBN 0–7727–5801–8.
The individual contributions are included in their corresponding subject categories – see abstracts 002.020 – 002.022, 006.014, 034.033 – 034.035, 036.063, 036.064, 064.039, 113.015 – 113.020, 114.046 – 114.059, 115.007, 115.008, 117.109, 118.011, 122.103 – 122.105, 153.017.

012.034 **Proceedings of the meeting held by the Astronomical Science Group of Ireland at University College Galway on 6th April, 1984.**
Ir. Astron. J., Vol. 17, No. 1, p. 11 – 39 (1985).
The individual contributions are included in their corresponding subject categories – see abstracts 035.024, 082.028, 112.064, 117.112, 131.113, 159.037.

012.035 **Proceedings of the meeting held by the Astronomical Science Group of Ireland at University College Dublin on 21st September 1984.**
Ir. Astron. J., Vol. 17, No. 1, p. 40 – 60 (1985).
The individual contributions are included in their corresponding subject categories – see abstracts 009.014, 035.025, 051.030, 102.027, 119.029.

012.036 **Life sciences and space research XXI(2). Proceedings of Workshops VII and XI and of the COSPAR Interdisciplinary Scientific Commission F (Meetings F1, F3, F5, F6, F7 and F9) of the COSPAR Twenty–fifth Plenary Meeting held in Graz, Austria, 25th June – 7th July 1984.**
H. Oser, J. Oró, R. D. MacElroy, H. P. Klein, D. L. DeVincenzi, R. S. Young (Editors).
Adv. Space Res., Vol. 4, No. 12, 7 + 326 pp. (1984). ISBN 0–08–032752–4.
The individual contributions within the subject scope of Astronomy and Astrophysics Abstracts are included in their corresponding categories – see abstracts 022.047 – 022.050, 091.029, 091.030, 100.019, 105.025, 107.009, 131.114.

012.037 **Cosmochemistry and meteoritics. Materials of the 6th All–Union symposium in Kiev.**
Naukova Dumka, Kiev. 220 pp. (1984). In Russian.
Review in Ref. Zh., 51. Astron., 1.51.47 (1985).
The individual contributions are included in their corresponding subject categories – see abstracts 081.027, 091.032, 097.006, 099.037, 105.026, 105.027, 105.029 – 105.035, 105.037 – 105.040, 106.068, 107.012.

012.038 **Scientific ballooning – IV. Proceedings of Symposium 7 of the COSPAR Twenty–fifth Plenary Meeting held in Graz, Austria, 25th June – 7th July 1984.**
W. Riedler, K. Torkar (Editors).
Adv. Space Res., Vol. 5, No. 1, 6 + 131 pp. (1985). ISBN 0–08–032753–2.
The individual contributions within the subject scope of Astronomy and Astrophyscis Abstracts are included in their corresponding subject categories – see abstracts 035.026 – 035.030, 051.033 – 051.041, 079.141.

012.039 **Astrometric binaries. Proceedings of a conference held at the Remeis–Observatory, Bamberg, F.R. Germany, June 13 – 15, 1984.**
Z. Kopal, J. Rahe (Editors), with an introductory address by W. Fricke.
Astrophys. Space Sci., Vol. 110, No. 1, 6 + 210 pp. (1985). Available also as a hardbound edition. Price Dfl. 120.00, $ 39.50, £ 30.50. ISBN 90–277–1979–5.
The individual contributions are included in their corresponding subject categories – see abstracts 004.032, 004.033, 005.010 – 005.012, 013.037, 013.038, 034.038, 036.076, 041.015, 051.042, 094.019, 111.018, 111.019, 117.126 – 117.129, 118.019 – 118.028, 119.034.

012.040 **The XVIth All–Union conference on radio astronomical investigations of the solar system. Zvenigorod, October 1984.**
V. V. Fomichev (Editor).
Moskva. 127 pp. (1984). In Russian.
Review in Ref. Zh., 51. Astron., 2.51.31 (1985).

012.041   **Solar seismology from space. A conference at Snowmass, Colorado, August 17 – 19, 1983.**
R. K. Ulrich, J. Harvey, E. J. Rhodes Jr., J. Toomre (Editors).
JPL Publ. 84–84. NASA. Jet Propulsion Laboratory, California Institute of Technology, Pasadena, California. 8 + 377 pp. (1984).
The individual contributions are included in their corresponding subject categories – see abstracts 034.040 – 034.042, 036.077, 080.034 – 080.057, 082.031.

012.042   **IRAS asteroid workshop number 4: report and recommendations. Workshop held at the Jet Propulsion Laboratory, California Institute of Technology, Pasadena, Calif., 12 – 14 February 1985.**
E. F. Tedesco (Editor).
JPL D–2176. Jet Propulsion Laboratory, California Institute of Technology, Pasadena, Calif. 91109, USA. 8 + 151 pp. (1985).
The IRAS Point Source Catalog with explanation was released in January, 1985 and the asteroid data reduction has been the emphasis since then. A fourth workshop was held at JPL February 12 – 14 to review the asteroid portion of the IRAS project. This document is the report of that workshop. The three areas investigated in detail by the workshop No. 4 were: (1) ADAS design and implementation, (2) asteroid thermal modelling and calibration, and (3) final asteroid data products, their uses and access to them. Three working groups were formed to study and report on these topics. The reports of each of these three working groups is included in this report.

012.043   **The Big Bang and Georges Lemaitre. Proceedings of a symposium in honour of G. Lemaitre fifty years after his initiation of Big–Bang cosmology, Louvain–la–Neuve, Belgium, 10 – 13 October 1983.**
A. Berger (Editor).
D. Reidel Publishing Company, Dordrecht – Boston – Lancaster. 22 + 420 pp. Price Dfl. 150.00, US$ 59.00, £ 38.25 (1984). ISBN 90–277–1848–2.
The individual contributions are included in their corresponding subject categories – see abstracts 002.029, 005.014, 005.015, 021.011, 022.060, 042.059 – 042.061, 061.028, 066.074, 066.075, 080.060, 091.038, 091.039, 098.033, 098.034, 144.037, 151.057, 160.046, 160.047, 161.126 – 161.137.

012.044   **Plasma astrophysics. International School & Workshop on Plasma Astrophysics, held at Varenna, Italy, 28 August – 7 September 1984.**
T. D. Guyenne, J. J. Hunt (Editors).
ESA Spec. Publ., ESA SP–207. 8 + 340 pp. (1984).
The individual contributions are included in their corresponding subject categories – see abstracts 022.061, 062.087 – 062.100, 064.048, 067.079 – 067.081, 073.102 – 073.104, 074.058 – 074.060, 080.061, 084.081, 091.040, 106.039, 106.040, 117.143, 121.028, 121.029, 126.054 – 126.056, 144.038, 144.039, 151.058 – 151.061, 155.097, 158.118 – 158.121, 160.049, 161.138 – 161.140.

012.045   **The hydromagnetics of the sun. Proceedings of the Fourth European Meeting on Solar Physics, Noordwijkerhout, The Netherlands, 1 – 3 October 1984.**
T. D. Guyenne, J. J. Hunt (Editors).
ESA Spec. Publ., ESA SP–220. 12 + 296 pp. (1984).
The individual contributions are included in their corresponding subject categories – see abstracts 034.045, 035.033, 035.034, 062.101 – 062.107, 064.049, 065.040, 071.014 – 071.017, 072.031 – 072.043, 073.105 – 073.112, 074.061 – 074.072, 075.016 – 075.021, 080.064 – 080.078, 116.035.

012.046   **Complex investigations of the sun. 12th Leningrad seminar on cosmophysics. Leningrad, 6 – 8 February 1982.**
V. A. Dergachev, G. E. Kocharov (Editors).
Leningrad. 207 pp. (1982). In Russian.
From Ref. Zh., 51. Astron., 3.51.52 (1985).
See abstracts 073.120 – 073.125, 073.161, 073.163, 073.164, 074.108, 074.110, 077.024, 077.025, 080.112, 085.011, 144.051.

012.047   **Meteor bodies in the interplanetary space and in the earth's atmosphere. All–Union conference held in Suzdal, 17 – 22 September 1984.**
P. B. Babadzhanov (Editor).
Donish, Dushanbe. 64 pp. (1984). In Russian.
Review in Ref. Zh., 51. Astron., 3.51.61 (1985).

012.048   **The IVth symposium on solar–terrestrial physics, held in Sochi in November 1984.**
V. V. Migulin, N. P. Ben'kova.
Moskva. 161 pp. (1984).
Review in Ref. Zh., 51. Astron., 3.51.62 (1985).

012.049   **XXVII annual meeting of the Italian Astronomical Society, Brescia, 21 – 23 October 1983.**
Mem. Soc. Astron. Ital., Vol. 55, No. 3, p. 397 – 514 (1984).
The individual contributions are included in their corresponding subject categories – see abstracts 002.030, 036.086, 074.075, 091.041, 098.040 – 098.042, 102.033, 107.022, 118.029, 144.042, 151.062, 151.063, 153.022, 154.019 – 154.021, 157.114.

012.050   **XXVIII annual meeting of the Italian Astronomical Society, Milano, 19 – 21 October 1984.**
Mem. Soc. Astron. Ital., Vol. 55, No. 3, p. 515 – 628 (1984).
The individual contributions are included in their corresponding subject categories – see abstracts 013.044 – 013.046, 051.047, 051.048, 064.050, 065.041, 065.042, 067.083, 080.080, 106.047, 117.150, 131.149, 158.125, 160.053, 161.146, 161.147.

012.051   **Space debris, asteroids and satellite orbits. Proceedings of Workshops IV and XIII and of the COSPAR Interdisciplinary Scientific Commission P (Meeting P1) of the COSPAR Twenty–fifth Plenary Meeting held in Graz, Austria, 25th June – 7th July 1984.**
D. J. Kessler, E. Grün, L. Sehnal (Editors).
Adv. Space Res., Vol. 5, No. 2, 6 + 229 pp. (1985). ISBN 0–08–033189–0.
The individual contributions within the subject scope of Astronomy and Astrophysics Abstracts are included in their corresponding categories – see abstracts 035.035, 035.036, 036.087, 045.012, 045.013, 051.050, 051.051, 052.037 – 052.048, 098.043, 098.044, 106.048.

012.052   **The Giotto spacecraft impact–induced plasma environment. Proceedings of the Giotto PEWG Meeting, held at Berne, Switzerland, 10 – 12 April 1984.**
E. Rolfe, B. Battrick (Editors), with a preface by R. Reinhard.
ESA Spec. Publ., ESA SP–224. 6 + 104 pp. Price FF 100.00 (1984).
The individual contributions are included in their corresponding subject categories – see abstracts 022.075 – 022.085, 035.037 – 035.040, 051.052.

012.053   **Optical technology for microwave applications. Conference held at Arlington, Va., USA, 1 – 2 May 1984.**
Proc. SPIE Int. Soc. Opt. Eng., Vol. 477 (1984).
Review in Phys. Abstr., Vol. 88, No. 1252, Entry 24588 (1985).

012.054   **Links for the future. Science, systems and services for communications. Proceedings of the International Conference on Communications–ICC 84, held at Amsterdam, The Netherlands, 14 – 17 May 1984.**
P. Dewilde, C. A. May (Editors).
North–Holland Publishing Company, Amsterdam – New York – Oxford (1984).
From Phys. Abstr., Vol. 88, No. 1252, Entry 29819 (1985).
See Abstr. 013.048.

012.055   **1984 International Symposium Digest. Antennas and propagation. Conference held at Boston, Mass., USA, 25 – 29 June 1984.**
IEEE, New York, USA (1984).
From Phys. Abstr., Vol. 88, No. 1252, Entry 29818 (1985).

012.056   **ESCAMPIG 84. Seventh European Sectional Conference on the Atomic and Molecular Physics of Ionized Gases, held at Bari, Italy, 28 – 31 August 1984.**
Eur. Phys. Soc., Petit–Lancy, Switzerland. 23 + 270 pp. (1984).
Review in Phys. Abstr., Vol. 88, No. 1253, Entry 30129 (1985).
See abstracts 022.096, 022.140, 131.298.

012.057   **Comparative planetology. Proceedings of the 27th International Geological Congress held in Moscow, 1984 August 4 – 14, Tom 19.**
M. S. Markov (Editor).
Nauka, Moskva. 142 pp. (1984). In Russian.
Review in Ref. Zh., 51. Astron., 11.51.36 (1984).
See abstracts 015.017, 094.025, 097.008.

012.058   **VIth International Colloquium Geodetic Astrometry. Lohrmann–Observatorium, Dresden 13 – 23 September 1983.**
With an introduction by K.–G. Steinert.
Mitt. Lohrmann–Obs. Tech. Univ. Dresden, Nr. 51 = Wiss. Z. Tech. Univ. Dresden, Band 33, Heft 6, p. 53 – 136 (1984).
The individual contributions are included in their corresponding subject categories – see abstracts 013.050, 032.015 – 032.018, 034.047, 036.092 – 036.095, 041.018, 041.019, 043.005, 044.020 – 044.028, 045.015 – 045.021, 081.032, 081.033.

012.059   **Astronomy from space. Proceedings of the Topical Meeting of the COSPAR Interdisciplinary Scientific Commission E (Meetings E3, E4 and E5) of the COSPAR Twenty–fifth Plenary Meeting held in Graz, Austria, 25th June – 7th July 1984.**
G. G. Fazio, J. A. M. Bleeker, P. A. J. de Korte, J. J. Caldwell (Editors).
Adv. Space Res., Vol. 5, No. 3, 7 + 212 pp. (1985). ISBN 0–08–033192–0.
The individual contributions are included in their corresponding subject categories – see abstracts 035.048, 035.049, 036.104, 051.057 – 051.066, 117.170 – 117.177, 119.048, 119.049, 125.052 – 125.054, 126.062, 131.176, 131.177, 134.036, 142.031 – 142.034, 143.042, 155.106, 155.107, 156.015, 158.156, 160.067.

012.060   **13th International Academy of Astronautics Review Meeting on communication with extraterrestrial intelligence, held in Lausanne, Switzerland, October 1984.**
A. R. Martin (Editor).
J. Br. Interplanet. Soc., Vol. 38, No. 6, p. 276 – 280 (1985).
The individual contributions are included in their corresponding subject categories – see abstracts 011.013, 015.022 – 015.032.

012.061   **Proceedings of the VI. National Conference of Yugoslav Astronomers, Hvar, 1983, May 25 – 27.**
L. Randić, B. Jovanović, G. M. Popović (Editors).
Publ. Astron. Opservatorije Beogr., No. 33, 128 pp. (1985).
The individual contributions are included in their corresponding subject categories – see abstracts 002.049, 002.050, 004.062, 004.063, 005.022, 014.035, 014.036, 015.034, 021.014, 022.118, 032.022, 034.063, 034.064, 036.110, 044.035 – 044.038, 045.023, 063.061, 071.027, 079.161, 082.046, 082.047, 096.011, 104.029, 118.036, 119.055, 141.011, 154.028, 155.113.

012.062   **Proceedings of the Workshop on Sky Surveys, held at Schloss Ringberg, F.R. Germany, November 21 – 23, 1984.**
J. Trümper, R. Wielebinski (Editors).
Bull. Inf. Cent. Données Stellaires, No. 28, p. 39 – 94 (1985).
The individual contributions are included in their corresponding subject categories – see abstracts 002.062, 035.050, 035.051, 051.068, 051.069, 141.024 – 141.031, 143.045, 159.073.

012.063   **Interstellare Materie. Bericht über die wissenschaftliche Tagung in Frankfurt/M., 26. – 29. März 1985.**
With an introduction by W. Seggewiß.
Mitt. Astron. Ges., Nr. 63, 234 pp. Price DM 30.00 (1985). Available from H.–U. Keller, Planetarium Stuttgart, Neckarstr. 47, D–7000 Stuttgart, F.R. Germany.

The individual contributions are included in their corresponding subject categories – see abstracts 011.024, 022.120, 033.037 – 033.039, 041.022, 061.038, 067.102, 106.057, 114.100, 117.210, 119.073, 119.074, 121.042, 121.043, 123.022, 124.011, 124.012, 125.061, 125.062, 126.070, 131.199 – 131.214, 132.049 – 132.052, 134.041 – 134.043, 141.032, 151.078, 151.079, 153.032, 153.033, 155.117 – 155.123, 157.150, 158.183.

012.064   **The Galileo affair: a meeting of faith and science. Proceedings of the conference held at Cracow, Poland, 24 – 27 May 1984.**
G. V. Coyne, M. Heller, J. Życiński (Editors).
Vatican Obs. Publ., Spec. Ser., Studi Galileiani, Vol. 1, No. 3, p. 1 – 180 (1985).
The individual contributions are included in their corresponding subject categories – see abstracts 004.090 – 004.099, 005.025.

012.065   **Third International Conference on Infrared Physics (CIRP 3), held at Zurich, Switzerland, 23 – 27 July 1984.**
F. K. Kneubühl, T. S. Moss (Editors).
Infrared Phys., Vol. 25, No. 1/2, 7 + 529 pp. (1985).
The individual contributions within the subject scope of Astronomy and Astrophysics Abstracts are included in their corresponding categories – see abstracts 022.129, 031.034 – 031.036, 033.044, 034.097 – 034.099, 035.066 – 035.068, 061.047, 082.064, 082.065, 093.038, 158.253.

012.066   **Magnetospheric and ionospheric plasmas. Proceedings of Symposium 9 and of the COSPAR Interdisciplinary Scientific Commission D (Meeting D1) of the COSPAR Twenty–fifth Plenary Meeting held in Graz, Austria, 25 June – 7th July 1984.**
E. R. Schmerling, S. W. H. Cowley, P. H. Reiff (Editors).
Adv. Space Res., Vol. 5, No. 4, 9 + 426 pp. (1985). ISBN 0–08–033193–9.
The individual contributions within the subject scope of Astronomy and Astrophysics Abstracts are included in their corresponding categories – see abstracts 082.068, 083.014, 084.096 – 084.098, 085.015, 093.039 – 093.041, 099.071, 100.038, 100.039.

012.067   **La composition chimique des étoiles dans le voisinage solaire. 7ème journée de Strasbourg tenue le 24 janvier 1985.**
A. Florsch, C. Jaschek, M. Jaschek (Editors).
Comptes rendues sur les journées de Strasbourg, 7ème réunion. Observatoire de Strasbourg, 11, rue de l'Université, F–67000 Strasbourg, France. 98 pp. (1985).
The individual contributions are included in their corresponding subject categories – see abstracts 113.059, 113.060, 114.124 – 114.128, 115.021, 124.243.

012.068   **High energy astrophysics and cosmology. Proceedings of the Academia Sinica – Max–Planck Society workshop on high energy astrophysics, held in Nanjing, China, April 9 – 17, 1982.**
J. Yang, C. Zhu (Editors).
Science Press, Beijing, and Gordon and Breach Science Publishers Inc., New York – London – Paris – Montreux – Tokyo. 5 + 529 pp. Price US$ 65.00 (1983). ISBN 0–677–3134–03.
The individual contributions are included in their corresponding subject categories – see abstracts 004.103, 063.075 – 063.077, 065.082, 066.108, 067.125 – 067.139, 117.282 – 117.284, 126.094 – 126.099, 158.270, 159.118 – 159.120, 161.224 – 161.230.

012.069   **The Virgo cluster of galaxies. Proceedings of the ESO workshop held at Garching, F.R. Germany, 4 – 7 September 1984.**
O.–G. Richter, B. Binggeli (Editors), with an introduction by G. A. Tammann and a summary by D. Lynden–Bell.
ESO Conference and Workshop Proceedings No. 20. European Southern Observatory, Karl–Schwarzschild–Str. 2, D–8046 Garching bei München, F.R. Germany. 8 + 477 pp. Price DM 50.00, US$ 18.00 (1985). ISBN 3–923524–20–X.

The individual contributions are included in their corresponding subject categories – see abstracts 157.205, 158.271, 160.106 – 160.143.

012.070   **Dynamics of star clusters. IAU Symposium No. 113, held in Princeton, New Jersey, U.S.A, 29 May – 1 June, 1984.**
J. Goodman, P. Hut (Editors), with some summary remarks by J. P. Ostriker.
D. Reidel Publishing Company, Dordrecht – Boston – Lancaster. 22 + 622 pp. Price Dfl. 190.00, US$ 69.00, £ 48.50 cloth; Dfl. 85.00, US$ 29.00, £ 21.75 paper (1985). ISBN 90–277–1963–2 cloth; ISBN 90–277–1965–9 paper.
The individual contributions are included in their corresponding subject categories – see abstracts 051.085, 151.100 – 151.133, 153.045 – 153.048, 154.050 – 154.069, 156.025, 157.206.

012.071   **Recent advances in planetary meteorology. Proceedings of the Seymour Hess Memorial Symposium–IUGG General Assembly, held in Hamburg, F.R. Germany, 18 – 19 August 1983.**
G. E. Hunt (Editor).
Cambridge University Press, Cambridge – London – New York – New Rochelle – Melbourne – Sydney. 13 + 161 pp. Price £ 20.00, US$ 39.50 (1985). ISBN 0–521–25866–3.
The individual contributions are included in their corresponding subject categories – see abstracts 005.026, 091.049, 091.050, 093.046, 097.015, 097.016, 099.074, 099.075.

012.072   **Cataclysmic variables and low–mass X–ray binaries. Proceedings of the 7th North American Workshop, held in Cambridge, Mass., USA, 12 – 15 January 1983.**
D. Q. Lamb, J. Patterson (Editors).
Astrophysics and Space Science Library, Vol. 113. D. Reidel Publishing Company, Dordrecht – Boston – Lancaster. 12 + 452 pp. Price Dfl. 165.00, US$ 59.50, £ 41.95 (1985). ISBN 90–277–1947–0.
The individual contributions are included in their corresponding subject categories – see abstracts 064.086, 067.142 – 067.145, 117.300 – 117.336, 124.021, 131.297, 142.063.

012.073   **1984 Annual Conference on Nuclear and Space Radiation Effects, held at Colorado Springs, Colo., USA, 23 – 25 July 1984.**
IEEE Trans. Nucl. Sci., Vol. NS–31, No. 6 (1984).
Review in Phys. Abstr., Vol. 88, No. 1253, Entry 30115 (1985).

012.074   **Proceedings of the national symposium on molecular spectroscopy, held at Calcutta, India, 13 – 15 February 1984.**
Indian J. Phys., Part B, Vol. 58B, No. 4 – 5 (1984).
Review in Phys. Abstr., Vol. 88, No. 1253, Entry 30116 (1985).

012.075   **Proceedings of the second symposium on plasma double layers and related topics, held at Innsbruck, Austria, 5 – 6 July 1984.**
R. Schrittwieser, G. Eder (Editors).
Universität Innsbruck, Innsbruck, Austria. 10 + 411 pp. (1984).
Review in Phys. Abstr., Vol. 88, No. 1253, Entry 30128 (1985).
See abstracts 062.167, 062.169, 062.170, 084.108, 084.115 – 084.119.

012.076   **Proceedings of the international summer school on nucleon–nucleon interaction and nuclear many–body problems, held at Changchun, China, 25 – 31 July 1983.**
S. S. Wu, T. T. S. Kuo (Editors).
World Scientific Publishing Co. Pte. Ltd., Singapore. 15 + 748 pp. (1984).
Review in Phys. Abstr., Vol. 88, No. 1253, Entry 30132 (1985).
See abstracts 061.051, 067.146.

012.077   **Proceedings of the Third International Symposium on Accelerator Mass Spectrometry, held at Zurich, Switzerland, 10 – 13 April 1984.**
Nucl. Instrum. Methods Phys. Res., Sect. B, Vol. 233, No. 2 (1984).
Review in Phys. Abstr., Vol. 88, No. 1254, Entry 35905 (1985).
See abstracts 022.141, 022.142, 022.145, 022.146, 080.117, 105.238, 105.240.

012.078   **Proceedings of the 4th General Conference of the Condensed Matter Division of the EPS, held at The Hague, The Netherlands, 19 – 22 March 1984.**
Physica B, C, Vol. 127B + C, No. 1 – 3 (1984).
Review in Phys. Abstr., Vol. 88, No. 1254, Entry 35913 (1985).

012.079   **High pressure in science and technology. Proceedings of the 9th AIRAPT International High Pressure Conference, held at Albany, N.Y., USA, 24 – 29 July 1983.**
C. Homan, R. K. MacCrone, E. Whalley (Editors).
North–Holland Publishing Company, Amsterdam – New York – Oxford. 3 Vol. 15 + 373 + 13 + 335 + 15 + 397 pp. (1984). ISBN 0–444–00932–9.
Review in Phys. Abstr., Vol. 88, No. 1252, Entry 24619 (1985).

012.080   **Sir Arthur Eddington Centenary Symposium. Vol. 1: Relativistic Astrophysics and Cosmology. Proceedings of a symposium, held at Nagpur, India, 21 – 27 January 1984.**
V. de Sabbata, T. M. Karade (Editors).
World Scientific Publishing Co. Pte. Ltd., Singapore. 9 + 274 pp. Price US$ 34.00 (1984). ISBN 9971–966–99–9.
The individual contributions are included in their corresponding subject categories – see abstracts 004.108, 004.109, 061.057 – 061.059, 066.127 – 066.134, 158.278, 161.255 – 161.257.

012.081   **Proceedings of the XIth International Conference on Neutrino Physics and Astrophysics, held at Nordkirchen, F.R. Germany, 11 – 16 June 1984.**
K. Kleinknecht, E. A. Paschos (Editors).
World Scientific Publishing Co. Pte. Ltd., Singapore. 13 + 811 pp. Price US$ 111.55, £ 105.80 (1984). ISBN 9971–966–70–0.
Review in Astron. Express, Vol. 1, Nos. 4 – 6, p. 160 (1985).
The individual contributions within the subject scope of Astronomy and Astrophysics Abstracts are included in their corresponding categories – see abstracts 034.106 – 034.112, 061.060 – 061.062, 065.086, 080.118, 080.119, 161.258.

012.082   **Conference on the intersections between particle and nuclear physics, held at Steamboat Springs, USA, 23 – 30 May 1984.**
AIP Conf. Proc., No. 123 (1984).
Review in Phys. Abstr., Vol. 88, No. 1255, Entry 40784 (1985).
See abstracts 080.120, 080.121, 144.066.

012.083   **Proceedings of the second workshop on hadronic mechanics, held at Como, Italy, 1 – 3 August 1984.**
Hadronic J., Vol. 7, Nos. 5 – 6 (1984).
Review in Phys. Abstr., Vol. 88, No. 1255, Entry 40799 (1985).
See abstracts 061.063, 061.095, 066.138, 066.141.

012.084   **Proceedings of the conference on neutron–nucleus collisions: a probe of nuclear structure, held at Glouster, Ohio, USA, 5 – 8 September 1984.**
AIP Conf. Proc., No. 124 (1985).
Review in Phys. Abstr., Vol. 88, No. 1256, Entry 46269 (1985).
See abstracts 061.069, 061.071 – 061.074.

012.085   **Physics and physicochemistry of highly condensed matter. Colloque International XXIInd CNRS/EHPRG Meeting, held at Aussois, France, 11 – 14 September 1984.**
J. Phys. Colloq., Vol. 45, No. C-8 (1984).
Review in Phys. Abstr., Vol. 88, No. 1256, Entry 46285 (1985).
See abstracts 091.053, 091.054, 107.010.

012.086   Weather and climate responses to solar variations. Second International Symposium on Solar Terrestrial Influences on Weather and Climate, held at Boulder, Colo, USA, 2 – 6 August 1982.
B. M. McCormac (Editor).
Colorado Associated University Press, Boulder, Colo., USA. 626 pp. (1983). ISBN 0–87081–138–X.
Review in Phys. Abstr., Vol. 88, No. 1256, Entry 46309 (1985).

012.087   Processing of scientific data from the ESA astrometry satellite HIPPARCOS. Second FAST Thinkshop, held in Marseille, France, 21 – 25 January 1985.
J. Kovalevsky (Editor).
Fundamental Astronomy by Space Techniques Consortium. Centre International de Rencontres Mathématiques and Laboratoire d'Astronomie Spatiale, Marseille, France. 5 + 405 pp. (1985).
The individual contributions are included in their corresponding subject categories – see abstracts 002.096, 002.097, 013.084, 031.039, 031.040, 035.078 – 035.080, 036.161 – 036.176, 041.030 – 041.046, 043.007, 051.087 – 051.091, 052.061 – 052.063, 098.090.

012.088   Monopole'83. Proceedings of a NATO Advanced Research Workshop, held at Ann Arbor, Mich., USA, 6 – 9 October 1983.
J. L. Stone (Editor).
Plenum Publishing Corporation, New York, USA. 14 + 699 pp. (1984). ISBN 0–306–41812–6.
Review in Phys. Abstr., Vol. 88, No. 1257, Entry 51854 (1985). See abstracts 061.075 – 061.082, 061.094, 061.096, 061.097, 067.160, 067.161, 067.165, 161.285, 161.303 – 161.305.

012.089   Magnetic monopoles. Proceedings of a NATO Advanced Study Institute, held at Wingspread, Wis., USA, 14 – 17 October 1982.
R. A. Carrigan Jr., W. P. Trower (Editors).
Plenum Publishing Corporation, New York, USA. 10 + 337 pp. (1983). ISBN 0–306–41399–X.
Review in Phys. Abstr., Vol. 88, No. 1257, Entry 51856 (1985). See abstracts 061.083 – 061.091, 061.093.

012.090   International conference on the application, theory and fabrication of periodic structures, diffraction gratings and moiré phenomena II, held at San Diego, Calif., USA, 21 – 23 August 1984.
Proc. SPIE Int. Soc. Opt. Eng., Vol. 503 (1984).
Review in Phys. Abstr., Vol. 88, No. 1258, Entry 57969 (1985). See abstracts 034.118, 034.119, 034.121, 034.122.

012.091   Workshop on instabilities in continuous media, held at Venice, Italy, November – December 1982.
Pure Appl. Geophys., Vol. 121, No. 3 (1983).
Review in Phys. Abstr., Vol. 88, No. 1258, Entry 57972 (1985). See abstracts 091.057, 091.058.

012.092   Antiproton 1984. Proceedings of the VII European Symposium on Antiproton Interactions, held at Durham, England, 9 – 13 July 1984.
M. R. Pennington (Editor).
Adam Hilger Ltd., Bristol, England. 13 + 530 pp. (1985). ISBN 0–85498–164–0.
Review in Phys. Abstr., Vol. 88, No. 1258, Entry 57982 (1985). See abstacts 161.313, 161.314.

012.093   Turbulence and chaotic phenomena in fluids. Proceedings of an international symposium, held at Kyoto, Japan, 5 – 10 September 1983.
T. Tatsumi (Editor).
North–Holland Publishing Company, Amsterdam – New York – Oxford. 13 + 556 pp. (1984). ISBN 0–444–87594.
Review in Phys. Abstr., Vol. 88, No. 1255, Entry 40820 (1985).

012.094   Capture gamma–ray spectroscopy and related topics – 1984. Proceedings of the Fifth International Symposium, held at Knoxville, Tenn., USA, 10 – 14 September 1984.
AIP Conf. Proc., No. 125 (1985).
Review in Phys. Abstr., Vol. 88, No. 1259, Entry 63555 (1985). See abstracts 061.099, 061.101 – 061.107, 061.109 – 061.111.

012.095   Stellar radial velocities. Proceedings of IAU Colloquium No. 88, held at Schenectady, New York, USA, 24 – 27 October 1984.
A. G. D. Philip, D. W. Latham (Editors), with a summary by K. C. Freeman.
L. Davis Press Inc., Schenectady, N.Y., USA. 11 + 454 pp. Price US$ 30.00 (1985). ISBN 0–933485–00–X.
The individual contributions are included in their corresponding subject categories – see abstracts 002.098 – 002.100, 034.125 – 034.127, 036.184 – 036.191, 064.088, 111.032 – 111.046, 120.033 – 120.037, 122.187, 122.188, 152.007, 152.008, 153.051 – 153.053, 154.074 – 154.078, 155.152 – 155.156, 156.026, 157.213.

012.096   18th International Cosmic Ray Conference held at Bangalore, India, August 22 – September 3, 1983.
N. Durgaprasad, S. Ramadurai, P. V. Ramana Murthy, M. V. S. Rao, K. Sivaprasad (Editors).
Tata Institute of Fundamental Research, Homi Bhabha Road, Colaba, Bombay 400 005, India. 13 volumes, Price US$ 108.00 (1983).
Conference papers:
Vol. 1. XG sessions: X rays and gamma rays. 21 + 184 pp.
Vol. 2. OG sessions: origin and galactic phenomena. 37 + 410 pp.
Vol. 3. MG sessions: modulation and geophysical effects. 42 + 570 pp.
Vol. 4. SP sessions: solar particles. 24 + 232 pp.
Vol. 5. HE sessions: high energy physics. 46 + 536 pp.
Vol. 6. EA sessions: extensive air showers. 28 + 269 pp.
Vol. 7. MN sessions: muons and neutrinos. 19 + 138 pp.
Vol. 8. T sessions: techniques. 25 + 210 pp.
Late papers:
Vol. 9. XG, OG, T sessions. 27 + 464 pp.
Vol. 10. MG, SP sessions. 23 + 395 pp.
Vol. 11. HE, EA, MN sessions. 34 + 499 pp.
Invited and rapporteur papers:
Vol. 12. 10 + 515 pp.
General index:
Vol. 13. Author index and list of participants. 7 + 111 pp.
Conference programme and author index. 318 pp.
The individual contributions within the subject scope of Astronomy and Astrophysics Abstracts are included in their corresponding categories – see abstracts 013.085, 022.159 – 022.173, 034.128 – 034.149, 035.089 – 035.122, 036.192 – 036.197, 061.113 – 061.139, 062.179 – 062.181, 066.159, 066.160, 067.177 – 067.181, 072.088, 073.185 – 073.192, 074.130 – 074.132, 075.033, 076.026 – 076.035, 077.045, 077.046, 078.018 – 078.083, 080.123 – 080.129, 082.078 – 082.081, 083.021 – 083.023, 084.123 – 084.129, 094.038, 099.080 – 099.082, 105.243 – 105.246, 106.071 – 106.080, 107.032, 117.338 – 117.341, 118.049, 125.090 – 125.092, 126.105 – 126.113, 131.310, 131.311, 142.064 – 142.071, 143.062 – 143.101, 144.078 – 144.459, 152.009, 155.157 – 155.160, 156.027, 158.281 – 158.285, 159.126 – 159.128, 161.340 – 161.342.

012.097   Results of the ARCAD 3 project and of the recent programmes in magnetospheric and ionospheric physics. Proceedings of a conference held in Toulouse, France, 22 – 25 May 1984.
Cepadues–Editions, 111, rue Nicolas–Vauquelin, 31100 Toulouse, France. 976 pp. Price FF 400.00 (1985). ISBN 2–85428–126–8.
The individual contributions within the subject scope of Astronomy and Astrophysics Abstracts are included in their corresponding categories – see abstracts 083.024, 083.025, 084.130 – 084.135.

012.098   From Spacelab to Space Station. Proceedings of the Fifth AAS/DGLR Symposium, held at Hamburg, F.R. Germany, 3–5 October 1984.
H. Stoewer, P. M. Bainum (Editors).
Adv. Astronaut. Sci., Vol. 56, 10+259 pp. Price US$ 50.00 cloth, US$ 40.00 paper (1985). ISBN 0–87703–209–2 cloth, ISBN 0–87703–210–6 paper.

012.099   Guidance and Control 1985. Proceedings of the Annual Rocky Mountain Guidance and Control Conference, held at Keystone, Colo., USA, 2–6 February 1985.
R. D. Culp, E. J. Bauman, C. A. Cullian (Editors).
Adv. Astronaut. Sci., Vol. 57, 15+601 pp. Price US$ 65.00 cloth, US$ 50.00 paper (1985). ISBN 0–87703–211–4 cloth, ISBN 0–87703–212–2 paper.
The individual contributions within the subject scope of Astronomy and Astrophysics Abstracts are included in their corresponding categories – see abstracts 035.123, 035.124, 052.065, 052.066.

012.100   Permanent presence – making it work. 22nd Goddard Memorial Symposium. Proceedings of an AAS conference, held at Goddard Space Flight Center, Greenbelt, Md., 15–16 March 1984.
I. Bekey (Editor).
Science and Technology Series, Vol. 60. A Supplement to Advances in the Astronautical Siences. Published for the American Astronautical Society by Univelt, Inc., PO Box 28130, San Diego, Calif. 92128, USA. 11+177 pp. Price US$ 40.00 cloth, US$ 30.00 paper (1985). ISBN 0–87703–207–6 cloth, ISBN 0–87703–208–4 paper.

012.101   Cool stars with excesses of heavy elements. Proceedings of the Strasbourg Observatory Colloquium, held at Strasbourg, France, 3–6 July 1984.
M. Jaschek, P. C. Keenan (Editors), with a summary by R. F. Griffin.
Astrophysics and Space Science Library. Vol. 114. D. Reidel Publishing Company, Dordrecht – Boston – Lancaster. 16+398 pp. Price Dfl. 150.00, US$ 54.00, £ 38.25 (1985). ISBN 90–277–1957–8.
The individual contributions are included in their corresponding subject categories – see abstracts 002.101, 064.090 – 064.093, 065.090 – 065.093, 113.062 – 113.064, 114.140 – 114.156, 115.024, 117.343, 117.344, 120.038 – 120.040, 122.189 – 122.191, 155.162 – 155.168, 156.028, 156.029, 157.216, 157.217.

012.102   Asymptotic behavior of mass and spacetime geometry. Proceedings of a conference held at Corvallis, Oregon, USA, 17–21 October 1983.
F. J. Flaherty (Editor).
Lecture Notes in Physics, Vol. 202. Springer–Verlag, Berlin – Heidelberg – New York – Tokyo. 6+213 pp. Price DM 30.00 (1984). ISBN 3–540–13351–8 (F.R. Germany), ISBN 0–387–13351–8 (USA).
The individual contributions are included in their corresponding subject categories – see abstracts 061.143, 066.164 – 066.177, 067.196, 067.197.

012.103   Gravitation, geometry and relativistic physics. Proceedings of the "Journées Relativistes", held at Aussois, France, 2–5 May 1984.
Edited by Laboratoire "Gravitation et Cosmologie Relativistes", Université Pierre et Marie Curie et C.N.R.S., Institut Henri Poincaré, Paris.
Lecture Notes in Physics, Vol. 212. Springer–Verlag, Berlin – Heidelberg – New York – Tokyo. 6+336 pp. Price DM 45.00 (1984). ISBN 3–540–13881–1 (F.R. Germany), ISBN 0–387–13881–1 (USA).
The individual contributions within the subject scope of Astronomy and Astrophysics Abstracts are included in their corresponding categories – see abstracts 034.165 – 034.167, 061.144, 066.178 – 066.188, 067.198, 143.105, 161.354 – 161.357.

012.104   Interacting binaries. Proceedings of a NATO Advanced Study Institute, held at Cambridge, U.K., 31 July – 13 August 1983.
P. P. Eggleton, J. E. Pringle (Editors).
NATO Advanced Science Institutes, Series C, Vol. 150. D. Reidel Publishing Company, Dordrecht – Boston – Lancaster. 6+410 pp. Price Dfl. 160.00, US$ 59.00, £ 41.95 (1985). ISBN 90–277–1966–7.
The individual contributions are included in their corresponding subject categories – see abstracts 036.212, 117.364 – 117.375, 119.103, 120.042.

012.105   Phase transitions in the very early universe. Proceedings of an International Workshop, held at Bielefeld, F.R. Germany, 4–8 June, 1984.
R. Baier, H. Satz (Editors).
Nucl. Phys. B, Part. Phys., Vol. B252, No. 1–2, p. 1–368 (1985).
The individual contributions within the subject scope of Astronomy and Astrophysics Abstracts are included in their corresponding categories – see abstracts 061.146, 066.190, 067.199, 161.359 – 161.375.

012.106   Patterns of change in Earth evolution. Report of a Dahlem Workshop, held at Berlin, 1–6 May, 1983.
H. D. Holland, A. F. Trendall (Editors).
Physical, Chemical, and Earth Sciences Research Report 5. Springer–Verlag, Berlin – Heidelberg – New York – Tokyo. 10+432 pp. Price DM 57.00 (1984). ISBN 3–540–12749–6 (F.R. Germany), ISBN 0–387–12749–6 (USA).
The individual contributions within the subject scope of Astronomy and Astrophysics Abstracts are included in their corresponding categories – see abstracts 081.051 – 081.054, 082.098.

012.107   Supernovae as distance indicators. Proceedings of a workshop held at Cambridge, Mass., USA, 27–28 September 1984.
N. Bartel (Editor).
Lecture Notes in Physics, Vol. 224. Springer–Verlag, Berlin – Heidelberg – New York – Tokyo. 6+226 pp. Price DM 30.00 (1985). ISBN 3–540–15206–7 (F.R. Germany), ISBN 0–387–15206–7 (USA).
The individual contributions are included in their corresponding subject categories – see abstracts 015.043, 122.213, 125.105 – 125.121, 125.262, 157.247.

012.108   Progress in stellar spectral line formation theory. Proceedings of a NATO Advanced Research Workshop, held at Grignano–Miramare, Trieste, Italy, 4–7 September 1984.
J. E. Beckman, L. Crivellari (Editors).
NATO Advanced Science Institutes, Series C, Vol. 152. D. Reidel Publishing Company, Dordrecht – Boston – Lancaster. 16+448 pp. Price Dfl. 165.00, US$ 59.00, £ 45.75 (1985). ISBN 90–277–2007–X.
The individual contributions are included in their corresponding subject categories – see abstracts 022.186, 063.085 – 063.102, 064.101 – 064.109, 071.049, 073.237, 074.162, 131.363.

012.109   Longitude Zero 1884–1984. Proceedings of an International Symposium held at the National Maritime Museum, Greenwich, London, 9–13 July 1984 to mark the centenary of the adoption of the Greenwich Meridian.
S. R. Malin, A. E. Roy, P. Beer (Editors), with an introduction by O. Pedersen.
Vistas Astron., Vol. 28, Parts 1/2, 407 pp. (1985).
The individual contributions within the subject scope of Astronomy and Astrophysics Abstracts are included in their corresponding categories – see abstracts 004.133 – 004.177, 009.043, 009.044, 015.051, 044.088, 051.113.

012.110   Theoretical aspects on structure, activity, and evolution of galaxies: III. Proceedings of a Japanese symposium on galaxies, held at Tokyo, Japan, 21 – 23 January 1985.
S. Aoki, M. Iye, Y. Yoshii (Editors).
Tokyo Astronomical Observatory, University of Tokyo, Mitaka, Tokyo 181, Japan. 3 + 151 pp. (1985).
The individual contributions are included in their corresponding subject categories – see abstracts 062.194, 067.208, 131.364 – 131.366, 151.162 – 151.169, 157.249, 158.320, 161.388 – 161.394.

012.111   Magnetism, planetary rotation, and convection in the solar system: retrospect and prospect. Conference held in the School of Physics at the University of Newcastle upon Tyne, 6 – 8 April 1983, in honour of Prof. S. K. Runcorn. Part 2.
W. O'Reilly (Editor).
Geophys. Surv., Vol. 7, No. 2, p. 125 – 212 (1985). For Part 1 see Abstr. 38.012.082.
The individual contributions are included in their corresponding subject categories – see abstracts 044.090, 044.091, 081.057, 081.058, 084.145.

012.112   Solutions of Einstein's equations: techniques and results. Proceedings of an international seminar held at Retzbach, F.R. Germany, 14 – 18 November 1983.
C. Hoenselaers, W. Dietz (Editors).
Lecture Notes in Physics, Vol. 205. Springer–Verlag, Berlin – Heidelberg – New York – Tokyo. 6 + 439 pp. Price DM 59.00 (1984). ISBN 3–540–13366–6 (F.R. Germany), ISBN 0–387–13366–6 (USA).
The individual contributions are inclued in their corresponding subject categories – see abstracts 066.196 – 066.213, 161.398.

012.113   Serendipitous discoveries in radio astronomy. Proceedings of a workshop held at the National Radio Astronomy Observatory, Green Bank, West Virginia, USA, 4 – 6 May 1983, honoring the 50th anniversary announcing the discovery of cosmic radio waves by Karl G. Jansky on 5 May 1933.
K. Kellermann, B. Sheets (Editors).
NRAO Workshop No. 7. National Radio Astronomy Observatory, PO Box 2, Green Bank, W.Va. 24944–0002, USA. 8 + 321 pp. Price US$ 7.00 (1984).
Review in Sky Telesc., Vol. 69, No. 3, p. 228; 1985 (R. A. Schorn).
The individual contributions are included in their corresponding subject categories – see abstracts 004.178 – 004.199, 005.032 – 005.036, 009.046, 009.047, 013.096, 015.053 – 015.061.

012.114   New aspects of galaxy photometry. Proceedings of the specialized meeting of the Eighth IAU European Regional Astronomy Meeting, held at Toulouse, France, 17 – 21 September 1984.
J.-L. Nieto (Editor).
Lecture Notes in Physics, Vol. 232. Springer–Verlag, Berlin – Heidelberg – New York – Tokyo. 13 + 350 pp. Price DM 52.00 (1985). ISBN 3–540–15657–7 (F.R. Germany), ISBN 0–387–15657–7 (USA).
The individual contributions are included in their corresponding subject categories – see abstracts 002.121, 036.218 – 036.227, 151.174 – 151.180, 156.033, 156.034, 157.253 – 157.291, 158.322, 159.150, 160.158 – 160.162.

012.115   Birth and evolution of neutron stars: issues raised by millisecond pulsars. Proceedings of a workshop held at the National Radio Astronomy Observatory, Green Bank, West Virginia, USA, 6 – 8 June 1984.
S. P. Reynolds, D. R. Stinebring (Editors), with summary remarks by V. Radhakrishnan and a search techniques session.
NRAO Workshop No. 8. National Radio Astronomy Observatory, PO Box 2, Green Bank, W.Va. 24944–0002, USA. 9 + 349 pp. (1984).
Review in Sky Telesc., Vol. 69, No. 6, p. 520 (1985).
The individual contribuiions are included in their corresponding subject categories – see abstracts 033.059, 065.103, 065.104, 067.213 – 067.218, 091.068, 125.122 – 125.124, 126.125 – 126.153.

012.116   Physics of energy transport in extragalactic radio sources. Proceedings of a workshop held at the National Radio Astronomy Observatory, Green Bank, West Virginia, USA, 30 July – 3 August 1984.
A. H. Bridle, J. A. Eilek (Editors).
NRAO Workshop No. 9. National Radio Astronomy Observatory, PO Box 2, Green Bank, W.Va. 24944–0002, USA. 5 + 316 pp. (1984/85).
The individual contributions are included in their corresponding subject categories – see abstracts 062.195 – 062.208, 158.323 – 158.346, 159.151 – 159.155, 160.163.

012.117   Quantum metrology and fundamental physical constants. Proceedings of a NATO Advanced Study Institute, held 16 – 28 November 1981 in Erice, Sicily, Italy.
P. H. Cutler, A. A. Lucas (Editors).
NATO Advanced Study Institute, Ser. B: Physics, Vol. 98. Plenum Press, New York – London. 12 + 658 pp. Price US$ 95.00 (1983).
Review in Astrophys. Space Sci., Vol. 110, No. 2, p. 417 – 418; 1985 (J. Kleczek).

012.118   Supersymmetry and supergravity '82. Proceedings of the Trieste September 1982 School.
S. Ferrara, J. G. Taylor, P. van Nieuwenhuizen (Editors).
World Scientific Publishing Co. Ltd., Singapore. Price US$ 42.00 (1983). ISBN 9971–950–67–7.
Review in Astron. Nachr., Vol. 306, No. 2, p. 90; 1985 (U. Bleyer).

012.119   Supersymmetry and supergravity 1984.
B. De Wit, et al.
World Scientific Publishing Co. Pte. Ltd., Singapore and Taylor and Francis Ltd., London – Philadelphia. 500 pp. Price $ 60.00 (1984).
From IEEE Spectrum, Vol. 22, No. 6, p. 80B (1985).

012.120   Particles and gravity. Proceedings of the Eighth Johns Hopkins Workshop of Current Problems in Particle Theory, held in Baltimore, 20 – 22 June 1984.
G. Domokos, S. Kovesi–Domokos (Editors).
World Scientific Publishing Co. Pte. Ltd., Singapore, distributed by John Wiley & Sons, Chichester – New York – Brisbane – Toronto – Singapore. 10 + 257 pp. Price £ 47.85 (1984).
From Astron. Express, Vol. 1, Nos. 4 – 6, p. 159 (1985).

012.121   Fundamental processes in energetic atomic collisions. Proceedings of a NATO Advanced Study Institute held September 20 – October 1, 1982, in Maratea, Italy.
H. O. Lutz, J. S. Briggs, H. Kleinpoppen (Editors).
NATO Advanced Study Institute, Ser. B: Physics, Vol. 103. Plenum Press, New York – London. 11 + 675 pp. Price US$ 95.00 (1983).
Review in Astrophys. Space Sci., Vol. 110, No. 2, p. 417; 1985 (J. Kleczek).

012.122   Die Sonne. Schrift des 12. Sternfreunde–Seminars 1984 im Planetarium der Stadt Wien.
H. Mucke.
Available from Astronomisches Büro, Hasenwartgasse 32, A–1238 Wien, Austria. 152 pp. Price öS 190.00.
Review in Sterne Weltraum, 24. Jahrg., Nr. 5, p. 292; 1985 (A. Kunert).

012.123   Symposium of the IAGA Working Group I–3, Electromagnetic induction in the earth and moon, held at Hamburg, F.R. Germany, 23 August 1983.
J. Geophys., Vol 55, No. 3 (1984).
Review in Phys. Abstr., Vol. 88, No. 1251, Entry 18983 (1984).

012.124 **Quasat – a VLBI observatory in space. Proceedings of a workshop held at Groß Enzersdorf, Austria, 18 – 22 June 1984.**
W. R. Burke (Editor), with a summary by R. T. Schilizzi and B. F. Burke.
ESA Spec. Publ., ESA SP–213. 12 + 189 pp. (1984).
The individual contributions are included in their corresponding subject categories – see abstracts 013.097, 033.060 – 033.064, 051.115 – 051.120, 052.078, 131.370 – 131.373, 155.197. 158.347 – 158.352, 160.164, 161.400.

012.125 **Proceedings of the Southwest Regional Conference for Astronomy and Astrophysics, held at Los Alamos, New Mexico, USA, 28 May 1984.**
P. F. Gott (Editor).
Southwest Regional Conference for Astronomy and Astrophysics, Vol. 10. Department of Physics, Texas Tech University, Lubbock, Texas, USA. 63 pp. (1985). ISSN 0147–2003.
The individual contributions are included in their corresponding subject categories – see abstracts 002.130, 004.200, 015.064, 032.056, 032.057, 051.121, 065.106, 073.248, 114.173, 117.395, 120.045, 122.220, 142.088, 151.181.

013 Reports on Astronomy in Various Countries and Particular Fields

013.001 **Perspectives on space and astrophysical plasma physics.**
C. F. Kennel, J. Arons, R. Blandford, F. Coroniti, M. Israel, L. Lanzerotti, A. Lightman, K. Papadopoulos, R. Rosner, F. Scarf.
Unstable current systems and plasma instabilities in astrophysics, p. 537 – 552 (1985). – See Abstr. 012.002 (IAU Symp. No. 107).
The authors summarize the discussion of the current status and future prospects of space and astrophysical plasma research prepared by the Panel on Space and Astrophysical plasmas, a part of the study on Physics administered by the National Research Council of the National Academy of Sciences.

013.002 **Spiral arms, comets and terrestrial catastrophism: a discussion.**
A. R. Crawford.
Q. J. R. Astron. Soc., Vol. 26, No. 1, p. 53 – 55 (1985).

013.003 **Physics News in 1984. An American Institute of Physics Report. Astrophysics Section.**
P. F. Schewe (Editor).
Phys. Today, Vol. 38, No. 1, p. S5 – S12 (1985).
The individual contributions are included in their corresponding subject categories – see abstracts 015.005, 106.014, 112.028, 142.003, 143.004, 155.003, 158.023, 161.005.

013.004 **Keck Foundation offers Caltech $70 million for 10–m telescope.**
W. Sweet.
Phys. Today, Vol. 38, No. 2, p. 71 – 75 (1985).

013.005 **All hail the Keck ten–meter telescope project.**
J. L. Greenstein.
Phys. Today, Vol. 38, No. 2, p. 136 (1985).

013.006 **Solar–terrestrial research opportunities – a look to the future.**
D. S. Intriligator.
Adv. Space Res., Vol. 4, No. 7, p. 365 – 373 (1984). – See Abstr. 012.009.

013.007 **Space investigations of the Sun outlined for the end of the 80's – beginning of the 90's.**
S. L. Mandelstam (*S. L. Mandel'shtam*), A. B. Severni (*A. B. Severnyj*).
Adv. Space Res., Vol. 4, No. 7, p. 375 (1984). Abstract. – See Abstr. 012.009.

013.008 **The IRAS project.**
R. E. Jennings.
Observatory, Vol. 105, No. 1064, p. 1 (1985). Abstract. – See Abstr. 010.721.

013.009 **IRAS moving–object search.**
J. K. Davies.
Observatory, Vol. 105, No. 1064, p. 3 – 4 (1985). Abstract. – See Abstr. 010.721.

013.010 **Recent discoveries of comets with the Palomar 46–cm Schmidt camera.**
C. S. Shoemaker, E. M. Shoemaker.
Int. Comet Q., Vol. 7, No. 1, p. 3 – 7 (1985).

013.011 **The Association's plans for observing Halley's Comet.**
M. J. Hendrie.
J. Br. Astron. Assoc., Vol. 95, No. 2, p. 73 – 74 (1985).

013.012 **50 years of radio astronomy.**
P. G. Mezger.
IEEE Trans. Microwave Theory Tech., Vol. MTT–32, No. 9, p. 1224 – 1229 (1984). Abstr. in Phys. Abstr., Vol. 88, No. 1248, Entry 9648 (1985).

013.013 **Radioastronomy – where it has come from – where it is going.**
P. G. Mezger.
Mikrowellen Mag., Vol. 10, No. 4, p. 336 – 338, 340, 342, 344, 346, 348, 350 – 351 (1984). In German. Abstr. in Phys. Abstr., Vol. 88, No. 1248, Entry 9650 (1985).

013.014 **G–EXEC and the World Data Centre.**
B. J. Read.
Comput. Phys. Commun., Vol. 33, No. 1 – 3, p. 235 – 244 (1984). Abstr. in Phys. Abstr., Vol. 88, No. 1249, Entry 14280 (1985). – See Abstr. 012.014.

013.015 **Problems of data bases in geophysics.**
G. K. Hartmann.
Comput. Phys. Commun., Vol. 33, No. 1 – 3, p. 267 – 287 (1984). Abstr. in Phys. Abstr., Vol. 88, No. 1249, Entry 14281 (1985). – See Abstr. 012.014.

013.016 **The present status of space developments and their perspectives.**
S. Saito.
J. Inst. Electron. Commun. Eng. Jpn., Vol. 67, No. 3, p. 248 – 257 (1984). In Japanese. Abstr. in Phys. Abstr., Vol. 88, No. 1249, Entry 14451 (1985).

013.017 **Radio astronomy.**
F. W. Hyde.
Pract. Electron., Vol. 20, No. 8, p. 35 – 39 (1984). Abstr. in Phys. Abstr., Vol. 88, No. 1249, Entry 14491 (1985).

013.018   Interkosmos satellite laser radar in Vietnam Socialist Republic.
Nguyen Ngan, A. Novotný, M. Čech, P. Lála,
I. Yu. Makhalov, S. Schillak.
Publ. Astron. Inst. Czech. Acad. Sci., No. 58, p. 5 – 22 (1984).
Partly in Russian. – See Abstr. 012.018.

After 1982 – 1983 preparation period, the Interkosmos laser radar has been installed in Nha–Trang, Vietnam Socialist Republic, May 1984.

013.019   Upgrading the computer control of the Interkosmos Laser Ranging Station in Helwan.
A. Novotný, I. Procházka.
Publ. Astron. Inst. Czech. Acad. Sci., No. 58, p. 37 – 42 (1984). – See Abstr. 012.018.

To fulfill the requirements of the laser ranging systems, the soft/hardware package of the Helwan station was significantly modified in period 1982 – 84. The max. ranging reprate was increased up to 5 pps. The mount pointing accuracy was increased implementing the mechanical inaccuracies software model. The automatical comparison of the time base to the Loran C signal was put into operation.

013.020   Performance and results of Satellite Ranging Laser Station at Borowiec in 1983.
S. Schillak, E. Butkiewicz, J. Marciniak, J. Offierski,
K. Vorbrich.
Publ. Astron. Inst. Czech. Acad. Sci., No. 58, p. 87 – 92 (1984). – See Abstr. 012.018.

The Interkosmos Satellite Laser Ranging System at Borowiec is described. The system ranging precision has been evaluated. The calculations show the standard deviation associated with the observations to be in the range of ±75 cm. A brief discussion of the influence of the random errors upon calibrations and observations is given.

013.021   Fifteen years of Lunar Laser Ranging and future developments.
J. P. Rozelot, O. Calame.
Publ. Astron. Inst. Czech. Acad. Sci., No. 58, p. 129 – 143 (1984). – See Abstr. 012.018.

The Lunar Laser Ranging experiment is now in operation for about fifteen years. The technical difficulties to get long series of accurate data necessary for scientific objectives are reviewed. These are listed with particular emphasis of what has been done and what can be improved. Some fields of interest for the future are underlined in the case when a centimetric accuracy will be reached quasi–simultaneously from several stations, well distributed around the world. These future prospects concern particularly the terrestrial and celestial reference frames, the extension of the resonances in the Earth–Moon system, the tests of relativistic effects, the questions of lunar frictional energy dissipation and the internal structure of the Earth.

013.022   The research work at the Central Institute for Physics of the Earth, Potsdam, GDR, in the field of Doppler satellite geodesy.
R. Dietrich.
Publ. Astron. Inst. Czech. Acad. Sci., No. 58, p. 593 – 604 (1984). = Mitt. Zentralinst. Phys. Erde, Nr. 1377. – See Abstr. 012.018.

Since 1981 a JMR–4A Doppler receiver is used mostly for stationary observations at the Potsdam observatory. The research activities have three main directions: (1) Methodical investigations of satellite interferometry using Doppler receivers and receiver calibration, (2) computation of regional networks using the orbital program POTSDAM–5, (3) computation of long global arcs for determination of polar motion and station coordinates.

013.023   West–East Doppler Observation Campaign WEDOC–2.
S. Mihály, P. Pesec, K. Rinner.
Publ. Astron. Inst. Czech. Acad. Sci., No. 58, p. 605 (1984). Abstract. – See Abstr. 012.018.

013.024   US experiments and expertise contribute to Soviet Halley probe.
Phys. Today, Vol. 38, No. 3, p. 113 – 114 (1985).

013.025   Millimeter–Astronomie von Köln aus.
G. Winnewisser.
Sterne Weltraum, 24. Jahrg., Nr. 1, p. 16 – 20 (1985).

013.026   The scientific challenge of Space Telescope.
M. Longair.
Sky Telesc., Vol. 69, No. 4, p. 306 – 311 (1985).

013.027   Astrometry at La Silla.
R. M. West.
Cometary astrometry, p. 60 – 65 (1984). – See Abstr. 012.024.

013.028   The astrometry programme at Mt. John Observatory.
A. C. Gilmore, P. M. Kilmartin.
Cometary astrometry, p. 66 – 70 (1984). – See Abstr. 012.024.

013.029   The astrometry network of observers in China.
S. M. Gong.
Cometary astrometry, p. 71 – 73 (1984). – See Abstr. 012.024.

013.030   The astrometry network of observers in Japan.
Y. Kozai.
Cometary astrometry, p. 74 – 75 (1984). – See Abstr. 012.024.

013.031   The astrometry network of observers in U.S.S.R.
S. P. Major, Yu. A. Shokin.
Cometary astrometry, p. 76 – 81 (1984). – See Abstr. 012.024.

013.032   The four great questions of astronomy.
N. Sperling.
Publ. Astron. Soc. Pac., Vol. 97, No. 588, p. 201 (1985). Abstract. – See Abstr. 010.281.

013.033   Ein GIRL wird kaltgemacht. Verzicht auf Spitzentechnologie – Opfer für die Raumstation?
D. Lemke.
Sterne Weltraum, 24. Jahrg., Nr. 4, p. 189 (1985).

013.034   Der Beitrag der Universitäts–Sternwarte Wien zum Projekt MERIT.
P. Jackson.
Sternenbote, 28. Jahrg., Nr. 5, p. 86 – 91 (1985).

013.035   The infrared universe.
J. Davies.
Space Educ., Vol. 1, No. 9, p. 407 – 410 (1985).

013.036   The Astrometry Network of the IHW and the P/Crommelin trial run results.
D. K. Yeomans.
Cometary astrometry, p. 4 – 6 (1984). – See Abstr. 012.024.

013.037   Current work on binary and multiple stars.
M. G. Fracastoro.
Astrophys. Space Sci., Vol. 110, No. 1, p. 105 – 110 (1985). – See Abstr. 012.039.

The author refers on current work carried out by the IAU Commission 26, after having circulated a letter to all members to collect recent information at various places.

013.038   Work on astrometric binaries in China.
L. Yan, L. Wan, S. Gong.
Astrophys. Space Sci., Vol. 110, No. 1, p. 135 – 136 (1985). – See Abstr. 012.039.

The investigations of astrometric binary systems, currently carried out in the People's Republic of China, are described and discussed.

013.039   **Some notes on astronomy in New Zealand.**
R. J. Dodd.
Proc. Astron. Soc. Aust., Vol. 5, No. 4, p. 435 – 437 (1984).

A brief description of astronomy in New Zealand both amateur and professional, past and present is given, with particular emphasis on the work carried out by the two professional establishments; the Carter Observatory, Wellington and the Mt John University Observatory of the University of Canterbury.

013.040   **The International Halley Watch from Australia.**
M. P. Candy.
Proc. Astron. Soc. Aust., Vol. 5, No. 4, p. 438 – 439 (1984).

013.041   **Development of astronomy at Tomsk University.**
G. S. Tyuterev.
Astron. geod., Tomsk, No. 10, p. 3 – 15 (1984). In Russian. Abstr. in Ref. Zh., 51. Astron., 3.51.19 (1985).

013.042   **Development of meteor investigations in Tomsk.**
R. G. Lazarev.
Astron. geod., Tomsk, No. 10, p. 16 – 23 (1984). In Russian. Abstr. in Ref. Zh., 51. Astron., 3.51.20 (1985).

013.043   **Investigations in the field of celestial mechanics at Tomsk University in the years 1968 – 1980.**
T. V. Bordovitsyna.
Astron. geod., Tomsk, No. 10, p. 24 – 30 (1984). In Russian. Abstr. in Ref. Zh., 51. Astron., 3.51.21 (1985).

013.044   **Report on the 1984 activity of the section "Sun and Solar System" of the GNA–CNR (*Gruppo Nazionale di Astronomia*)–(*Research National Council*).**
B. Caccin.
Mem. Soc. Astron. Ital., Vol. 55, No. 3, p. 521 – 525 (1984). – See Abstr. 012.050.

The activities related with solar physics, planetology, fundamental and spherical astronomy are reported.

013.045   **Report on stellar physics researches of the GNA (*Gruppo Nazionale di Astronomia*).**
S. Catalano.
Mem. Soc. Astron. Ital., Vol. 55, No. 3, p. 527 – 536 (1984). – See Abstr. 012.050.

A summary of the stellar physics researches carried out during the years 1983 – 1984 is given.

013.046   **The astrometry network of observers in Italy for International Halley Watch.**
A. Manara.
Mem. Soc. Astron. Ital., Vol. 55, No. 3, p. 557 – 560 (1984). – See Abstr. 012.050.

In this paper astrometric networks of observers and the availability of telescopes in Italy for International Halley Watch are briefly described.

013.047   **Ground–based optical observations during SMM.**
R. Falciani.
Mem. Soc. Astron. Ital., Vol. 55, No. 4, p. 787 – 799 (1984). – See Abstr. 003.017.

The coordination between SMM and optical ground–based observations apparently worked rather poorly. The author discusses the problems; then he reviews new observing methods and the improvement of the existing ones.

013.048   **Communications with outer space – radio astronomy.**
R. Wielebinski.
Links for the future. Science, systems and services for communications, Vol. 1, p. 167 – 170 (1984). Abstr. in Phys. Abstr., Vol. 88, No. 1252, Entry 29819 (1985). – See Abstr. 012.054.

013.049   **Astronomy in Uzbekistan during 60 years.**
V. P. Shcheglov.
Izv. Akad. Nauk UzSSR. Ser. fiz.-mat. nauk, No. 5, p. 53 – 56 (1984). In Russian. Abstr. in Ref. Zh., 51. Astron., 4.51.6 (1985).

013.050   **The programme of observations with the Danjon astrolabe at the Astronomical Latitude Observatory in Borowiec.**
M. Lehmann.
Mitt. Lohrmann–Obs. Tech. Univ. Dresden, Nr. 51, p. 117 – 119 (1984). – See Abstr. 012.058.

The main purpose of observations with the astrolabe OPL 31 in Borowiec is the regular time and latitude service within the framework of the international cooperation of BIH, IPMS and GOSSTANDART (Moscow), as well as bilateral collaboration with the observatories of Potsdam (ZIPE), Ondřejov, Pecný, Bratislava and Irkutsk.

013.051   **The terrestrial coordinate system and international Earth–rotation services.**
G. A. Wilkins.
J. Navig., Vol. 38, No. 2, p. 216 – 217 (1985).

013.052   **Peter van de Kamp en zijn "lieve Barnard's Ster".**
G. Schilling.
Zenit, 12. Jaarg., No. 6, p. 211 – 213 (1985).

013.053   **Halley–specialist Donald K. Yeomans.**
G. Schilling.
Zenit, 12. Jaarg., No. 6, p. 218 – 219 (1985).

013.054   **Developments in meteorite research.**
R. Hutchison.
J. Br. Astron. Assoc., Vol. 95, No. 4, p. 141 – 142 (1985).

013.055   **Automatic photoelectric telescope service.**
L. J. Boyd, R. M. Genet, D. S. Hall.
I.A.P.P.P. Commun., No. 19, p. 41 – 49 (1985).

013.056   **Photoelectric photometry of variable stars from the Pyrenees.**
E. G. Melendo.
I.A.P.P.P. Commun., No. 20, p. 27 – 30 (1985).

013.057   **Moscow astronomers during the years of war.**
G. F. Sitnik.
Zemlya Vselennaya, No. 3, p. 17 – 21 (1985). In Russian.

013.058   **Photometric observing campaign of Be stars. Progress Report No. 8.**
P. Harmanec, J. Horn, P. Koubsky.
Bull. Inf. Etoiles Be, No. 11, p. 18 – 21 (1985).

013.059   **Spectroscopic observing campaign of Be stars.**
P. K. Barker.
Bull. Inf. Etoiles Be, No. 11, p. 22 – 28 (1985).

013.060   **Programmes d'observation des binaires spectroscopiques et informations du 14ème catalogue complémentaire.**
A. Pedoussaut, N. Ginestet, J. M. Carquillat.
Bull. Inf. Cent. Données Stellaires, No. 28, p. 9 – 13 (1985).

The authors present a summary of the spectroscopic binaries observational programs at the Toulouse Observatory. These programs concern the radial velocities of late–type stars, nearby stars, stars with composite spectra, He strong and He weak stars. Some informations about the 14th complementary catalogue of spectroscopic binaries are also given.

013.061   **The Astrographic Catalogue and the Carte du Ciel.**
C. Jaschek.
Bull. Inf. Cent. Données Stellaires, No. 28, p. 169 – 172 (1985).

Answers to a circular letter concerning the Astrographic Catalogue and the Carte du Ciel from Hyderabad (*S. M. Alladin*), Vatican (*G. Coyne*), Mexico (*P. Pismis*), Potsdam (*G. Ruben*), San Fernando (*A. Orte*).

013.062   **Astronomie infrarouge.**
D. Rouan.
Astronomie, Vol. 99, p. 271 – 279 (1985).

013.063   **35 years, together with the solar radio waves.**
T. Takakura.
Astron. Her., Vol. 78, No. 6, p. 163 – 166 (1985). In Japanese.

013.064   **Gamma–ray astronomy and the spirit of Cos–B.**
R. Lüst.
ESA Bull., No. 42, p. 8 – 16 (1985).

013.065   **Reflections on the solar–stellar connection.**
L. Goldberg.
Bull. Am. Astron. Soc., Vol. 16, No. 4, p. 874 (1984). Abstract. – See Abstr. 010.062.

013.066   **Progress report on NASA IUE Regional Data Analysis Facilities.**
C. A. Grady, R. Thompson, N. R. Evans, E. W. Brugel, S. R. Heap.
Bull. Am. Astron. Soc., Vol. 16, No. 4, p. 904 (1984). Abstract. – See Abstr. 010.062.

013.067   **The summer 1984 solar oscillation program of the Mount Wilson 60–foot solar telescope.**
S. Tomczyk, E. J. Rhodes Jr., A. Cacciani, R. K. Ulrich, R. F. Howard.
Bull. Am. Astron. Soc., Vol. 16, No. 4, p. 978 (1984). Abstract. – See Abstr. 010.062.

013.068   **Interference to radio astronomy from satellite transmitters.**
V. Pankonin.
Bull. Am. Astron. Soc., Vol. 16, No. 4, p. 998 – 999 (1984). Abstract. – See Abstr. 010.062.

013.069   **La protection des observatoires astronomiques et géophysiques.**
C. R. Acad. Sci., Sér. Gén., Vie Sci., Tome 2, No. 2, p. 157 – 171 (1985).

013.070   **On perspectives of the development of physics and astrophysics at the end of the XXth century.**
V. L. Ginzburg.
Fiz. XX veka: razvitie i perspectivy. Moskva, p. 281 – 330 (1984). In Russian. Abstr. in Ref. Zh., 51. Astron., 6.51.1 (1985).

013.071   **A long term programme for European space science: actions, proposals, conclusions – Survey Committee report.**
ESA Spec. Publ., ESA SP–1070, p. 2 – 15 (1984). – See Abstr. 003.029.

013.072   **Solar system science – Survey Committee report.**
ESA Spec. Publ., ESA SP–1070, p. 18 – 24 (1984). – See Abstr. 003.029.

013.073   **Report of ESA's topical team on solar and heliospheric physics.**
J. Christensen–Dalsgaard, P. Delache, P. Hoyng, E. R. Priest, R. Schwenn, J. O. Stenflo.
ESA Spec. Publ., ESA SP–1070, p. 26 – 36 (1984). – See Abstr. 003.029.

013.074   **Report of ESA's topical team on space plasma physics.**
M. Dobrowolny, L. Eliasson, R. Gendrin, G. Haerendel, A. Johnstone, S. McKenna, G. Morfill, V. Vasyliunas.
ESA Spec. Publ., ESA SP–1070, p. 38 – 48 (1984). – See Abstr. 003.029.

013.075   **Report of ESA's topical team on planetary science.**
S. J. Bauer, M. Fulchignoni, W.–H. Ip, Y. Langevin, F. W. Taylor, U. von Zahn.
ESA Spec. Publ., ESA SP–1070, p. 50 – 58 (1984). – See Abstr. 003.029.

013.076   **Space astronomy – Survey Committee report.**
ESA Spec. Publ., ESA SP–1070, p. 60 – 67 (1984). – See Abstr. 003.029.

013.077   **Extragalactic astronomy from space in the 1990s.**
A. C. Fabian.
ESA Spec. Publ., ESA SP–1070, p. 68 – 73 (1984). – See Abstr. 003.029.

013.078   **Galactic astronomy from space in the 1990's.**
E. P. J. van den Heuvel.
ESA Spec. Publ., ESA SP–1070, p. 74 – 82 (1984). – See Abstr. 003.029.

013.079   **Space experiments in relativity and gravitation.**
I. W. Roxburgh.
ESA Spec. Publ., ESA SP–1070, p. 84 – 90 (1984). – See Abstr. 003.029.

013.080   **A Soviet plan for exploring the planets.**
M. M. Waldrop.
Science, Vol. 228, No. 4700, p. 698, 703 (1985).

013.081   **Small–telescope photoelectric data acquisition at Pace University.**
S. Engelbrektson, J. B. Dick, K. T. Kwietniak, C. Rosenthal.
Bull. Am. Astron. Soc., Vol. 17, No. 1, p. 511 (1985). Abstract. – See Abstr. 010.064.

013.082   **Compte rendu du stage d'astronomie à Jeunesse et Science, 26 – 30 décembre 1984: Orbite d'une étoile double.**
L. Zimmermann.
Ciel Terre, Vol. 101, No. 3, p. 84 – 86 (1985).

013.083   **Highlights of astronomy in the year 1983.**
J. Grygar.
Říše hvězd, Vol. 65, Nos. 7 – 8, p. 138 – 143, 157 – 164 (1984). In Czech.

013.084   **Exchange of information between FAST and NDAC.**
E. Høg.
Processing of scientific data from the ESA astrometry satellite HIPPARCOS, p. 365 – 366 (1985). – See Abstr. 012.087.
Exchange of information between the two data reduction consortia for HIPPARCOS is planned throughout the development and mission phases. Guidelines are proposed for such exchange with the aim to ensure independence in critical areas and to allow comparison, especially of theory and data. This should ensure the best quality of the final HIPPARCOS catalogue.

013.085   **Cosmic ray data and accelerator data for particle production at very high energies: concurrences and contradictions.**
S. Bhattacharyya.
18th International Cosmic Ray Conference, Vol. 5, p. 202 – 205 (1983). – See Abstr. 012.096.
The sector of multiple production of hadrons provides an area of common interest to both high energy physics and cosmic ray physics. The author tries to demarcate the areas of both agreement and disagreement between the two fields.

013.086   **Il contributo dell'astronomia alle scienze applicate.**
M. Silvestri.
G. Astron., Vol. 11, N. 1, p. 5 – 9 (1985).

013.087   **Amateur astronomer in modern society.**
M. Rigutti.
Astronomia, Suppl. al N. 2, p. 7 – 17 (1985).

013.088   **Amateur astronomy today and tomorrow.**
R. Beck.
Astronomia, Suppl. al N. 2, p. 21 – 23 (1985).

013.089 **"General Catalogue of Variable Stars" and observations of variable stars by amateur astronomers.**
M. S. Frolov.
Astronomia, Suppl. al N. 2, p. 33 – 35 (1985).

013.090 **U.A.I. Saturn section.**
G. Adamoli.
Astronomia, Suppl. al N. 2, p. 41 – 44 (1985).

013.091 **Report of IAU Commission 6: Astronomical telegrams (*Télégrammes astronomiques*).**
M. P. Candy, B. G. Marsden.
Trans. IAU, Vol. XIXA, p. 13 – 14 (1985). – See Abstr. 003.046.

013.092 **Report of IAU Commission 38: Exchange of astronomers (*Echange des astronomes*).**
F. B. Wood.
Trans. IAU, Vol. XIXA, p. 547 (1985). – See Abstr. 003.046.

013.093 **Highlights of astronomy in the year 1984.**
J. Grygar.
Říše hvězd, Vol. 66, Nos. 3 – 6, p. 46 – 47, 70 – 72, 86 – 87, 110 – 112 (1985). In Czech.

013.094 **Czechoslovak astronomy 1945 – 1985.**
M. Kopecký.
Říše hvězd, Vol. 66, No. 5, p. 81 – 83 (1985). In Czech.

013.095 **The NASA Space Station.**
J. M. Beggs.
Johns Hopkins APL Tech. Dig., Vol. 6, No. 2, p. 146 – 148 (1985).
   A brief summary of the basic design considerations for a permanently manned Space Station is presented.

013.096 **Radio astronomy: the progress of a technique–oriented discipline.**
B. F. Burke.
Serendipitous discoveries in radio astronomy, p. 230 – 237 (1984). – See Abstr. 012.113.

013.097 **Opportunities for high angular and temporal resolutions of galactic objects at radio frequencies.**
K. J. Johnston.
ESA Spec. Publ., ESA SP–213, p. 187 – 189 (1984). – See Abstr. 012.124.
   Quasat will extend the spatial and temporal studies of galactic objects such as the sun, nearby flaring stars, collapsed stellar objects with accretion disks, the interstellar medium, and perhaps pulsars. The sensitivity of Quasat should be maximized to allow the maximum number of galactic objects to be studied.

A astronomia no Brasil.
See Abstr. 003.010.

Forty years of solar radio astronomy – a history of major advances.
See Abstr. 004.192.

Why Mount Wilson shouldn't be scrapped.
See Abstr. 009.007.

The satellite ranging system of the Central Institute of Physics of the Earth in Potsdam.
See Abstr. 045.021.

The Solar and Heliospheric Observatory, SOHO – a Phase–A project of the European Space Agency.
See Abstr. 051.008.

On the PSIVA (*Probing Stellar Interiors via Variability and Activity*) approach to stellar seismology and activity from space.
See Abstr. 051.010.

Project HIPPARCOS.
See Abstr. 051.042.

Astrophysical significance of spectral line shape investigations.
See Abstr. 063.061.

Immediate and long–term prospects for helioseismology.
See Abstr. 080.012.

The International Halley Watch.
See Abstr. 103.902.

Searches for short period pulsars at Jodrell Bank.
See Abstr. 126.140.

014 Teaching in Astronomy

014.001 Sterrenkunde op de huiscomputer: helderheden en afstanden.
J. Loonen.
Zenit, 12. Jaarg., No. 1, p. 21 (1985).

014.002 Photographie astronomique.
J. Jonlet.
Ciel, Vol. 47, p. 18 – 25 (1985).

014.003 Orientierungsregel unter der Lupe.
A. Zenkert.
Astron. Raumfahrt, 23. Jahrg., Heft 1, p. 8 – 10 (1985).

014.004 Astronomical computing: How high are lunar mountains?
H. E. Holder.
Sky Telesc., Vol. 69, No. 1, p. 62 – 63 (1985).

014.005 Backyard astronomy – 9. The lure of the variables.
A. MacRobert.
Sky Telesc., Vol. 69, No. 2, p. 124 – 125 (1985).

014.006 Astronomical computing: Short programs, bits and bytes.
G. J. Drobnock, V. J. Slabinski.
Sky Telesc., Vol. 69, No. 2, p. 158 – 159 (1985).
 Photoelectric data reduction. Small angular distances reconsidered. Bits and bytes.

014.007 Spectroscopie solaire a bon marché.
M. Chapelet.
Astronomie, Vol. 99, p. 143 – 146 (1985).

014.008 «Cerchiamo satelliti nel bel Ticino...»
A. Peralta.
Orion, 43. Jahrg., Nr. 206, p. 27 – 30 (1985). In German.

014.009 Fortbildungstagung für Volkshochschuldozenten.
H. Hornung.
Sterne Weltraum, 24. Jahrg., Nr. 1, p. 27 (1985).

014.010 Astrophotographie in der Schule.
L. Laepple.
Sterne Weltraum, 24. Jahrg., Nr. 2, p. 98 – 99 (1985).

014.011 Sternfreund und Heimcomputer.
W. Wepner.
Sterne Weltraum, 24. Jahrg., Nr. 2, p. 104 – 105 (1985).

014.012 Ein Demonstrationsmodell für die Himmelskoordinaten.
J. Sievers.
Sterne Weltraum, 24. Jahrg., Nr. 3, p. 147 (1985).

014.013 Astronomical computing: The waxing and waning Moon.
R. W. Sinnott.
Sky Telesc., Vol. 69, No. 3, p. 254 – 255 (1985).

014.014 Astronomical computing: Stepper–motor control of telescopes – I.
R. M. Genet, L. J. Boyd, M. Trueblood.
Sky Telesc., Vol. 69, No. 4, p. 350 – 351 (1985).

014.015 La lumière raconte...
M. Gros.
Astronomie, Vol. 99, 187 – 197 (1985).

014.016 Astrophysik mit Computern oder: Rechnen ist des Astronomen Lust.
H. U. Fuchs.
Orion, 43. Jahrg., Nr. 207, p. 40 – 45 (1985).

014.017 Sterrenkunde op de huiscomputer: kalender–capriolen.
J. Loonen.
Zenit, 12. Jaarg., No. 2, p. 62 – 63 (1985).

014.018 Sterrenkunde op de huiscomputer: van zonnetijd naar sterrentijd.
J. Loonen.
Zenit, 12. Jaarg., No. 3, p. 104 – 105 (1985).

014.019 Sterrenkunde op de huiscomputer: Paasdatum, Metonische cyclus en andere trucs.
J. Loonen.
Zenit, 12. Jaarg., No. 4, p. 146 – 147 (1985).

014.020 Astronomie pratique.
J. Jonlet.
Ciel, Vol. 47, p. 105 – 110, 134 – 142 (1985).

014.021 Zur Behandlung der Sternentwicklung im Astronomieunterricht der Sekundarstufe II.
O. Zimmermann.
Sterne Weltraum, 24. Jahrg., Nr. 4, p. 205 – 209 (1985).

014.022 Backyard astronomy – 10. Close–up of a star.
A. MacRobert.
Sky Telesc., Vol. 69, No. 5, p. 397 – 398 (1985).

014.023 Astronomical computing: Stepper–motor control of telescopes – II.
R. M. Genet, L. J. Boyd, M. Trueblood.
Sky Telesc., Vol. 69, No. 5, p. 448 – 449 (1985).

014.024 An orbital guide.
R. Christy.
Space Educ., Vol. 1, No. 9, p. 404 – 406 (1985).

014.025 Video for amateur astronomy.
P. Maley.
Space Educ., Vol. 1, No. 9, p. 415 – 417 (1985).

014.026 Halley's comet.
A. Romer.
Phys. Teach., Vol. 22, No. 8, p. 488 – 493 (1984). Abstr. in Phys. Abstr., Vol. 88, No. 1252, Entry 24688 (1985).

014.027 A planetarium in the physics room.
M. M. Payne.
Phys. Teach., Vol. 22, No. 8, p. 518 – 519 (1984). Abstr. in Phys. Abstr., Vol. 88, No. 1252, Entry 24698 (1985).

014.028 Sterrenkunde op de huiscomputer: Coördinatensystemen.
J. Loonen.
Zenit, 12. Jaarg., No. 5, p. 190 (1985).

014.029 Sterrenkunde op de huiscomputer: Massa's van dubbelsterren.
J. Loonen.
Zenit, 12. Jaarg., No. 6, p. 224 (1985).

014.030 Erfahrung mit dem Publikum. Aus dem Alltag der Sternwarte Eschenberg.
M. Griesser.
Orion, 43. Jahrg., Nr. 208, p. 86 – 88 (1985).

014.031 Sinn und Aufgabe der volkstümlichen Astronomie in unserer heutigen Gesellschaft.
J. Wirth.
Orion, 43. Jahrg., Nr. 208, p. 97 – 99 (1985).

014.032   **Bereichsdidaktische Überlegungen zum Unterricht in den Naturwissenschaften.**
W. Winnenburg.
Sterne Weltraum, 24. Jahrg., Nr. 5, p. 266 – 268 (1985).

014.033   **Nachtrag zu "Meridiandurchgang und Kulmination".**
G. Groschopf.
Sterne Weltraum, 24. Jahrg., Nr. 5, p. 268 (1985). See Abstr. 38.014.017.

014.034   **Astronomical computing: Shutter speeds for astrophotography; spectral lines; computer benchmarks.**
M. Neus, R. B. Minton, C. A. Feuchter.
Sky Telesc., Vol. 69, No. 6, p. 544 – 546 (1985).

014.035   **Development of a didactic measuring set "Young astronomer".**
V. Vujnović, M. Šuveljak.
Publ. Astron. Opservatorije Beogr., No. 33, p. 113 – 115 (1985). – See Abstr. 012.061.

014.036   **On astrophysics teaching in secondary schools.**
S. Ninković.
Publ. Astron. Opservatorije Beogr., No. 33, p. 116 – 117 (1985). – See Abstr. 012.061.
The author presents his own three–year–long experience in astrophysics teaching at a Belgrade secondary school.

014.037   **Using science fiction with good science as a teaching tool.**
A. G. Fraknoi.
Bull. Am. Astron. Soc., Vol. 16, No. 4, p. 999 (1984). Abstract. – See Abstr. 010.062.

014.038   **Organizing and conducting astronomy workshops for 2–year and 4–year college science teachers.**
D. B. Hoff, L. A. Kelsey.
Bull. Am. Astron. Soc., Vol. 16, No. 4, p. 999 (1984). Abstract. – See Abstr. 010.062.

014.039   **Zur Pulsation klassischer Cepheiden. Praktische Übungen und Aufgaben.**
F. Gieseking.
Sterne Weltraum, 24. Jahrg., Nr. 6, p. 327 – 329 (1985).

014.040   **Spacecraft navigation and relativity.**
R. W. Flynn.
Am. J. Phys., Vol. 53, No. 2, p. 113 – 119 (1985). Abstr. in Phys. Abstr., Vol. 88, No. 1257, Entry 51906 (1985).

014.041   **When can we treat identical particles as distinguishable? An unfamiliar classical limit.**
R. L. Ingraham.
Am. J. Phys., Vol. 53, No. 2, p. 119 – 121 (1985). Abstr. in Phys. Abstr., Vol. 88, No. 1257, Entry 51907 (1985).

014.042   **The Universe, bit by bit.**
F. Reddy.
Astronomy, Vol. 13, No. 1, p. 6 – 17 (1985). Abstr. in Phys. Abstr., Vol. 88, No. 1257, Entry 57627 (1985).

014.043   **Particle simulation of plasmas and stellar systems.**
T. Tajima, A. Clark, G. G. Craddock, D. L. Gilden, W. K. Leung, Y. M. Li, J. A. Robertson, B. J. Saltzman.
Am. J. Phys., Vol. 53, No. 4, p. 365 – 370 (1985). Abstr. in Phys. Abstr., Vol. 88, No. 1259, Entry 63619 (1985).

014.044   **Comments on B. Davies's 'Derivation of perihelion precession'.**
D. Ebner.
Am. J. Phys., Vol. 53, No. 4, p. 374 (1985). Abstr. in Phys. Abstr., Vol. 88, No. 1259, Entry 63622 (1985).

014.045   **L'eclisse di via di San Michele.**
M. Spadaro.
G. Astron., Vol. 10, N. 3 – 4, p. 299 – 305 (1984).

014.046   **Una unità didattica di astronomia per la I media.**
L. Kustermann.
G. Astron., Vol. 11, N. 1, p. 31 – 39 (1985).

014.047   **Esperienze didattico–sperimentali per l'acquisizione di concetti temporali per alunni della fascia di età fra 6 e 12 anni.**
S. Lai, E. Proverbio.
G. Astron., Vol. 11, N. 1, p. 41 – 49 (1985).

014.048   **Astronomia senza guardare: è possibile?**
M. L. Bruni Ciampini.
G. Astron., Vol. 11, N. 1, p. 51 – 62 (1985).

014.049   **Osservazioni astronomiche a scuola: costruzione sperimentale del piano dell'eclittica.**
L. Camandona, A. Torre.
G. Astron., Vol. 11, N. 1, p. 63 – 77 (1985).

014.050   **The importance of professional astronomers becoming more involved in education at all levels.**
W. J. Bisard.
Bull. Am. Astron. Soc., Vol. 17, No. 2, p. 573 (1985). Abstract. – See Abstr. 010.065.

014.051   **An astronomical laboratory exercise using relativity theory: 3C295.**
A. J. Weitenbeck, D. D. Meisel, R. Mayo.
Bull. Am. Astron. Soc., Vol. 17, No. 2, p. 574 (1985). Abstract. – See Abstr. 010.065.

014.052   **Report of IAU Commission 46: Teaching of astronomy (*Enseignement de l'astronomie*).**
L. N. Houziaux.
Trans. IAU, Vol. XIXA, p. 653 – 654 (1985). – See Abstr. 003.046.

Jupiter und Saturn. Ergebnisse der Planetenforschung.
See Abstr. 003.036.

Welcher Stern ist das?
See Abstr. 003.037.

Astronomie ganz einfach. Bauen und Beobachten – von der Sonnenuhr zum Spiegelfernrohr.
See Abstr. 003.038.

Universe in the classroom: a resource guide for teaching astronomy and instructor's manual for Universe.
See Abstr. 003.090.

Astronomy today.
See Abstr. 003.137.

Teaching astronomy at the Moscow State University.
See Abstr. 009.042.

Compte rendu du stage d'astronomie à Jeunesse et Science, 26 – 30 décembre 1984: Orbite d'une étoile double.
See Abstr. 013.082.

A demonstration of the Sun's motion around the barycenter of the Solar System.
See Abstr. 091.059.

015 Miscellanea (Philosophical Aspects, Extraterrestrial Life, etc.)

015.001   Stellar evolution: motivation for mass interstellar migrations.
B. Zuckerman.
Q. J. R. Astron. Soc., Vol. 26, No. 1, p. 56 – 59 (1985).

The ease and likelihood of interstellar rocket travel is a much-debated issue which is relevant to another controversial topic – the value of 'N' – the number of technological civilizations in our Galaxy. It is argued that even if N is as small as 10 – 100, at least one of these will already have been forced, by the termination of the main–sequence evolutionary phase of their home 'Sun', to carry out a massive interstellar migration. If, as is often argued, $N \sim 10^5$, then as many as 10^4 such migrations may well have taken place. Since each such migration could easily populate the space around more than 10^6 stars, such large values for N imply that our Galaxy is nearly saturated with extraterrestrial creatures.

015.002   Nachlese zum Stern der Weisen: Die Magier/ Geburtsdatum Jesu.
K. Ferrari d'Occhieppo.
Sternenbote, 28. Jahrg., Nr. 1, p. 2 – 6 (1985).

015.003   Observable characteristics of extraterrestrial technological civilisations.
R. A. Freitas Jr.
J. Br. Interplanet. Soc., Vol. 38, No. 3, p. 106 – 112 (1985).

015.004   The zoo we live in.
V. B. Gerard.
J. Br. Interplanet. Soc., Vol. 38, No. 3, p. 137 – 138 (1985).

015.005   Astronomical causes of biological extinctions.
H. L. Shipman.
Phys. Today, Vol. 38, No. 1, p. S10 – S11 (1985). Short review paper. – See Abstr. 013.003.

015.006   Life in the Galaxy?
G. S. Shostak.
The Milky Way galaxy, p. 623 – 632 (1985). – See Abstr. 012.007 (IAU Symp. No. 106).

015.007   Une explication astronomique à la disparition des dinosaures.
A. H. Delsemme.
Astronomie, Vol. 99, p. 3 – 14 (1985).

015.008   Observational search for polarized emission from space vehicles/communication relays near the galactic centre.
J. P. Vallée, M. Simard–Normandin.
Astron. Astrophys., Vol. 143, No. 2, p. 274 – 276 (1985). = NRCC 23979.

The aim of this study is to locate potential communication relays/space vehicles in the interstellar space in the general area where the center of our Galaxy resides, because this could be an ideal location for extraterrestrial purposes in telecommunications within our Galaxy. The authors thus made seven observing runs of roughly 3 d each, from September 1982 until May 1983, in a program to scan slowly a band of sky in search for strongly linearly polarized radio signals from any such extraterrestrial intelligence.

015.009   Das Artensterben aus astronomischer Sicht.
M. Burgdorf, U. Klaas.
Sterne Weltraum, 24. Jahrg., Nr. 2, p. 80 – 81 (1985).

015.010   Astronomie und Öffentlichkeit.
H. Elsässer.
Sterne Weltraum, 24. Jahrg., Nr. 2, p. 83 – 85 (1985).

015.011   The Moon – a second time around?
M. Washburn.
Sky Telesc., Vol. 69, No. 3, p. 209 – 211 (1985).

015.012   Astronomical journalism: responsibility and reflection.
L. Robinson.
News Lett. Astron. Soc. N.Y., Vol. 2, No. 4, p. 20 – 31 (1983). – See Abstr. 010.242.

015.013   "Der Mond ist aufgegangen,..." Plädoyer für die Beschreibung des Vollmondaufganges im "Abendlied" von Matthias Claudius.
K. Rümmler.
Sterne, 61. Band, Heft 1, p. 31 – 38 (1985).

015.014   Búsqueda de vida extraterrestre: desarrollos recientes y nuevas perspectivas.
I. F. Mirabel.
Rev. Astron., Tomo 56, No. 230, p. 2 – 4 (1984).

015.015   Sommaire de 25 années d'observations visant a détecter des êtres intelligents extraterrestres.
J. P. Vallée.
J. R. Astron. Soc. Can., Vol. 79, No. 1, p. 9 – 21 (1985).

The year 1960 will remain in the annals of astronomy as the year in which the first observational Search for ExtraTerrestrial Intelligence (SETI) was conducted. Here the author provides a summary of the techniques and of the fifty or so observational searches made since then – covering a quarter of a century – along with their simplest interpretations.

015.016   Model of "living" and search for extraterrestrial life.
M. D. Nusinov.
Zemlya Vselennaya, No. 1, p. 85 – 90 (1985). In Russian.

015.017   Chemical evolution in space.
J. M. Greenberg.
Comparative planetology, p. 110 – 119 (1984). In Russian. Abstr. in Ref. Zh., 62. Issled. Kosm. Prostranstva, 12.62.298 (1984). – See Abstr. 012.057.

015.018   Frecce del tempo.
A. D'Ercole.
Coelum, Vol. 53, N. 2, p. 69 – 84 (1985).

015.019   Nuove ipotesi astronomiche per spiegare le estinzioni biologiche terrestri.
V. Zappalà.
Orione, Vol. 5, N. 2, p. 48 – 52 (1985).

015.020   A rational goal for mankind: progenitive conception.
B. Lebon.
J. Br. Interplanet. Soc., Vol. 38, No. 6, p. 262 – 264 (1985).

This article presents a planetary model of development which could explain Fermi's paradox on the absence of extraterrestrials on Earth. It also proposes a long term strategy to prepare fruitful and peaceful relations with eventual extraterrestrial civilisations.

015.021   SETI and interstellar migration.
B. R. Finney.
J. Br. Interplanet. Soc., Vol. 38, No. 6, p. 274 – 275 (1985).

015.022   Cosmochemistry and life in the Universe.
C. Ponnamperuma.
J. Br. Interplanet. Soc., Vol. 38, No. 6, p. 276 (1985). Abstract. – See Abstr. 012.060.

015.023   Major chemical evolutionary phases for the emergence of life in planetary systems.
J. Oro.
J. Br. Interplanet. Soc., Vol. 38, No. 6, p. 276 – 277 (1985). Abstract. – See Abstr. 012.060.

015.024   The galactic position of our sun and the optimisation of search strategies for detecting extraterrestrial civilisations.
B. A. Balázs.
J. Br. Interplanet. Soc., Vol. 38, No. 6, p. 277 (1985). Abstract. – See Abstr. 012.060.

015.025   Is mankind unique in the Galaxy?
A. R. Martin, A. Bond.
J. Br. Interplanet. Soc., Vol. 38, No. 6, p. 277 – 278 (1985). Abstract. – See Abstr. 012.060.

015.026   SETI and interstellar migration.
B. R. Finney.
J. Br. Interplanet. Soc., Vol. 38, No. 6, p. 278 (1985). Abstract. – See Abstr. 012.060.

015.027   SETI experiment in Spacelab–1.
U. Merbold, A. Souchier, A. Ducrocq.
J. Br. Interplanet. Soc., Vol. 38, No. 6, p. 278 (1985). Abstract. – See Abstr. 012.060.

015.028   The search for extraterrestrial artifacts (SETA).
F. Valdes, R. A. Freitas Jr.
J. Br. Interplanet. Soc., Vol. 38, No. 6, p. 278 (1985). Abstract. – See Abstr. 012.060.

015.029   A search for the tritium hyperfine line from nearby stars.
F. Valdes, R. A. Freitas Jr.
J. Br. Interplanet. Soc., Vol. 38, No. 6, p. 278 (1985). Abstract. – See Abstr. 012.060.

015.030   Sensitive detection of narrowband pulses.
D. K. Cullers, B. M. Oliver, J. H. Wolfe.
J. Br. Interplanet. Soc., Vol. 38, No. 6, p. 278 – 279 (1985). Abstract. – See Abstr. 012.060.

015.031   Using the Very Large Array (VLA) synthesis radio telescope to perform a parasitic search for extraterrestrial intelligence.
J. Tarter.
J. Br. Interplanet. Soc., Vol. 38, No. 6, p. 279 (1985). Abstract. – See Abstr. 012.060.

015.032   Threats to CETI–SETI, possible solutions and anthropic principle.
M. Subotowicz.
J. Br. Interplanet. Soc., Vol. 38, No. 6, p. 279 – 280 (1985). Abstract. – See Abstr. 012.060.

015.033   Do extraterrestrial civilizations exist?
I. S. Shklovskij.
Zemlya Vselennaya, No. 3, p. 76 – 80 (1985). In Russian.

015.034   Contemporary cosmology and philosophy. (Some aspects of relationships and possibilities of overcoming the present difficulties).
V. Kršljanin.
Publ. Astron. Opservatorije Beogr., No. 33, p. 107 – 110 (1985). – See Abstr. 012.061.

015.035   Amateursternwarte mit zehneckigem Pyramidendach.
H. Pietsch.
Archenhold–Sternw. Berlin–Treptow, Sonderdr., Nr. 28, 8 pp. (1984).

015.036   On the effects of irregularities of internal structure in determining the ultraviolet properties of interstellar grains.
F. Hoyle, N. C. Wickramasinghe, S. Al–Mufti, L. M. Karim.
Astrophys. Relativ., Prepr. Ser., No. 113, 17 pp. (1985).
Whereas data for the extinction of starlight in the visible show the interstellar grains must be partially hollow, data in the ultraviolet show the vesicular interiors must be irregular. The laboratory measurements of absorptions produced by micro–organisms agree very closely with astronomical observations for a large number of early–type stars. Since the interior structures of micro–organisms are indeed highly irregular, the laboratory measurements made with micro–organisms suspended in a fluid, can reasonably be transferred to micro–organisms in vacuo.

015.037   Diatoms on Earth, comets, Europa and in interstellar space.
R. B. Hoover, F. Hoyle, N. C. Wickramasinghe, M. J. Hoover, S. Al–Mufti.
Astrophys. Relativ., Prepr. Ser., No. 114, 54 pp. (1985). Submitted to Astrophys. Space Sci.

015.038   The physiological rhythms and the periods of celestial phenomena.
S.–z. Zhang.
Publ. Purple Mt. Obs., Vol. 3, No. 3, p. 45 – 51 (1984). In Chinese.

015.039   Parasitic, piggyback and opportunistic SETI: it's cheap and it just might work!
J. C. Tarter.
Bull. Am. Astron. Soc., Vol. 16, No. 4, p. 999 (1984). Abstract. – See Abstr. 010.062.

015.040   Mutual help in SETIs.
F. Melia, D. H. Frisch.
Q. J. R. Astron. Soc., Vol. 26, No. 2, p. 147 – 150 (1985).
The probability that we will recognize radio signals sent out purposefully by extraterrestrial sources is parametrized here under the optimistic assumption that one of a few mutually helpful search strategies will be chosen by both senders and receivers.

015.041   Integration processes in science and the problem of search for extraterrestrial civilizations.
L. V. Fesenkova.
Nauchn. dokl. vyshch. shk. Filos. nauk, No. 6, p. 57 – 63 (1984). In Russian. Abstr. in Ref. Zh., 51. Astron., 6.51.2 (1985).

015.042   The astronomy of van Gogh's starry night.
A. Boime.
Bull. Am. Astron. Soc., Vol. 17, No. 1, p. 511 (1985). Abstract. – See Abstr. 010.064.

015.043   Supernovae up close (an after–dinner talk).
B. G. Marsden.
Supernovae as distance indicators, p. 222 – 226 (1985). – See Abstr. 012.107.
A vivid personal account of some jocose highlights of supernova hunting, as registered by the Central Bureau for Astronomical Telegrams, is presented.

015.044   Pioneer, Voyager und die Unendlickeit.
B. Wedel.
Sterne Weltraum, 24. Jahrg., Nr. 6, p. 345 (1985).

015.045   Il principio antropico fra cosmologia e riflessione umanistica.
A. Masani.
G. Astron., Vol. 10, N. 3 – 4, p. 241 – 260 (1984).

015.046   La problematica scientifico–filosofica di Paolo Celesia interessante il principio antropico.
L. Bulferetti.
G. Astron., Vol. 10, N. 3 – 4, p. 261 – 273 (1984).

015.047   Il principio antropico.
V. De Sabbata.
G. Astron., Vol. 10, N. 3 – 4, p. 275 – 298 (1984).

015.048   Report of IAU Commission 51: Search for extraterrestrial life (*Recherche de la vie dans l'univers*).
M. D. Papagiannis.
Trans. IAU, Vol. XIXA, p. 713 – 723 (1985). – See Abstr. 003.046.

015.049   Quantum states of matter in the theory of R. Bošković.
B. Jovanović.
Vasiona, Année 33, No. 1 – 2, p. 14 – 17 (1985). In Croatian.
The ideas of R. Bošković on the quantum states of matter are examined in the light of modern knowledge.

015.050   Supernovae unter der Lupe. Ein Vortrag nach dem Abendessen gehalten am 27. September 1984 in Cambridge, Mass.
B. G. Marsden.
Sterne Weltraum, 24. Jahrg., Nr. 6, p. 324 – 326 (1985). German translation of 015.043.

015.051   The future of the history of astronomy: the Space Telescope history project.
R. W. Smith.
Vistas Astron., Vol. 28, Parts 1/2, p. 375 – 376 (1985). – See Abstr. 012.109.

015.052   A radar for the exploration of extrasolar planets.
F. C. Williams.
Proc. IEEE, Vol. 73, No. 2, p. 355 – 361 (1985).
A radar for detecting and mapping terrestrial–type planets around distant suns is described. Using technology not greatly beyond that presently available, this radar will detect and image terrestrial planets up to a few tens of light years away with a mapping resolution of a few tens of kilometers. The radar is bistatic with the transmit antenna ensemble in solar orbit inside Mercury's orbit and the receive antenna ensemble in an orbit beyond Jupiter. The net motion of the phase center of the two ensembles of antennas generates a circular synthetic array with a diameter of about 10^9km. Such an array will have a synthetic-array angular resolution of 2.5×10^{-13}rad at a carrier frequency of 600 MHz, corresponding to a resolution of 2.5 km per light year of range.

015.053   Impact of news media on scientific research.
W. Sullivan.
Serendipitous discoveries in radio astronomy, p. 121 – 125 (1984). – See Abstr. 012.113.
A talk about radio astronomy seen through the eyes of a journalist.

015.054   Introduction to the panel discussion on the methodology of scientific research.
R. D. Ekers.
Serendipitous discoveries in radio astronomy, p. 196 (1984). – See Abstr. 012.113.
Concerning the history of radio astronomy.

015.055   Observational discovery vs theoretical disvocery.
M. O. Harwit.
Serendipitous discoveries in radio astronomy, p. 197 – 210 (1984). – See Abstr. 012.113.

015.056   Observational innovation and radio astronomy.
R. Hanbury Brown.
Serendipitous discoveries in radio astronomy, p. 211 – 213 (1984). – See Abstr. 012.113.

015.057   Non–standard approaches to astronomical research.
G. R. Burbidge.
Serendipitous discoveries in radio astronomy, p. 214 – 220 (1984). – See Abstr. 012.113.

015.058   The Buffalo Syndrome.
J. Broderick.
Serendipitous discoveries in radio astronomy, p. 221 – 229 (1984). – See Abstr. 012.113.
The author found a sociological phenomenon: he compares the behavior of radio astronomers in the "observational phase space" with that of the buffaloes in the prairie.

015.059   Program–oriented research.
H. van der Laan.
Serendipitous discoveries in radio astronomy, p. 238 – 242 (1984). – See Abstr. 012.113.

015.060   Impact of computers on radio astronomy.
B. Lovell.
Serendipitous discoveries in radio astronomy, p. 243 – 246 (1984). – See Abstr. 012.113.

015.061   SETI – the ultimate serendipitous discovery.
S. von Hoerner.
Serendipitous discoveries in radio astronomy, p. 307 – 311 (1984). – See Abstr. 012.113.

015.062   There is no Fermi Paradox.
R. A. Freitas Jr.
Icarus, Vol. 62, No. 3, p. 518 – 520 (1985).
The "Fermi Paradox", an argument that extraterrestrial intelligence cannot exist because it has not yet been observed, is a logical fallacy. This "paradox" is a formally invalid inference, both because it requires modal operators lying outside the first–order propositional calculus and because it is unsupported by the observational record.

015.063   Nuove ipotesi astronomiche per spiegare le estinzioni biologiche terrestri.
V. Zappalà.
Orione, Vol. 5, N. 1, p. 36 – 42 (1985).

015.064   Comet showers and Nemesis, the death star.
J. G. Hills.
Proceedings of the Southwest Regional Conference for Astronomy and Astrophysics, Vol. 10, p. 39 – 46 (1985). – See Abstr. 012.125.
The author discusses the recently proposed hypothesis that the observed periodic extinctions of terrestrial species is the result of comet showers catalyzed by a hypothetical distant solar companion, Nemesis.

The first IAU Symposium on extraterrestrial life.
See Abstr. 011.013.

The search for extraterrestrial life: recent developments. A report on IAU Symposium No. 112.
See Abstr. 011.014.

Applied Mathematics, Physics

021 Mathematical Papers Related to Astronomy and Astrophysics, Computing

021.001 On period determination methods.
A. Heck, J. Manfroid, G. Mersch.
Astron. Astrophys., Suppl. Ser., Vol. 59, No. 1, p. 63 – 72 (1985).
Two big families of period determination techniques are considered in this paper: those based on the Fourier transform, and the non–parametric algorithms derived from the θ–criterion by Lafler and Kinman (1965). The authors show that these groups of methods are not equivalent to each other under general conditions as asserted by some authors, but that at best all criteria can be expressed as the ratio of two quadratic forms. A comparative study of the performances of some of these methods by numerical simulations is presented.

021.002 Adaptive mesh techniques for fronts in star formation.
K.–H. A. Winkler, M. L. Norman, M. J. Newman.
Physica D, Vol. 12D, No. 1 – 3, p. 408 – 425 (1984). Abstr. in Phys. Abstr., Vol. 88, No. 1248, Entry 9713 (1985). – See Abstr. 012.012.

021.003 Graphic displays of gravitational initial data.
J. M. Bowen.
J. Comput. Phys., Vol. 56, No. 1, p. 42 – 50 (1984). Abstr. in Phys. Abstr., Vol. 88, No. 1249, Entry 14663 (1985).

021.004 Least squares reduction technique.
R. S. Harrington.
Cometary astrometry, p. 151 – 153 (1984). – See Abstr. 012.024.
The algorithm for implementing a standard least squares procedure on a small computer is presented.

021.005 Was ist ein Quadratgrad?
W. Ihle.
Sterne, 61. Band, Heft 1, p. 39 – 40 (1985).

021.006 Nochmals: Wieviel Quadratgrad besitzt die Himmelssphäre? Eine Erwiderung auf die Erwiderungen.
R. Stettler.
Sterne, 61. Band, Heft 1, p. 41 – 42 (1985).

021.007 Himmelsmechanik mit dem PC.
G. Traving.
Sterne Weltraum, 24. Jahrg., Nr. 4, p. 216 – 217 (1985).

021.008 Corrections for biases in slope estimation.
H. L. Marshall.
Astrophys. J., Vol. 289, No. 2, p. 457 – 466 (1985).
The corrections to the estimated slope of the $N(>S)$ relation and of the luminosity function due to Gaussian fluctuation are examined. A method is given for estimating the bias in the slope of the $N(>S)$ relation when several flux–limited samples are combined. The bias in this case is always less than in the case of a single sample and can be extremely small. It is shown that the bias is zero when the measured uncertainties are proportional to the observed flux. The bias in the slope of a luminosity function is determined using similar methods and may be very small.

021.009 On calculation of some data of the astronomical yearbook with EC–computers.
I. S. Guseva, O. V. Kiyaeva, D. D. Polozhnitsev, E. Ya. Stepanova, L. I. Yagudin.
Astron. Geod., Tomsk, No. 12, p. 107 – 109, 145 (1984). In Russian. Abstr. in Ref. Zh., 51. Astron., 2.51.67 (1985).

021.010 Preliminary numerical filtration by median averaging.
R. A. Vardanyan.
Soobshch. Byurakan. Obs., Vyp. 54, p. 60 – 64 (1983). In Russian.
An optimal method of preliminary numerical filtration of two- and three–dimensional numerical matrices is presented. It is shown that preliminary numerical filtration by the median averaging method has some advantages over the other existing numerical methods.

021.011 Finite Euclidean and non–Euclidean geometry with applications to the finite pendulum and the polygonal harmonic motion. A first step to finite cosmology.
R. De Vogelaere.
The Big Bang and Georges Lemaitre, p. 341 – 355 (1984). – See Abstr. 012.043.
Because the universe is both finite and atomic it is desirable to have a geometry which satisfies most of the properties of Euclidean or non–Euclidean geometry for which the number of points on each line is finite. This paper presents essential features of such geometries. The beginning of finite mechanics is present through the application to the finite circular pendulum and to the harmonic polygonal motion, for which time is also discrete.

021.012 Program for constructing high–order Runge–Kutta methods on computers.
Yu. A. Fedyaev.
Astron. geod., Tomsk, No. 10, p. 72 – 83 (1984). In Russian. Abstr. in Ref. Zh., 51. Astron., 3.51.94 (1985).

021.013 TITAN: a computer program for calculating stellar evolution.
C. L. Joseph.
Bull. Am. Astron. Soc., Vol. 17, No. 1, p. 513 – 514 (1985). Abstract. – See Abstr. 010.064.

021.014 Transformation of the rectangular coordinates into heliographic coordinates by the ''Apple II'' computer.
V. Jamičić, J. Ježina.
Publ. Astron. Opservatorije Beogr., No. 33, p. 94 – 95 (1985). – See Abstr. 012.061.

021.015 Observation equations based on expansions into eigenfunctions.
M. Schneider.
Manuscr. Geod., Vol. 9, p. 169 – 208 (1984). = Mitt. Inst. Astron. Phys. Geod. Tech. Hochsch. Münch., Nr. 168.

021.016   Statistical analysis of astronomical data containing upper bounds: general methods and examples drawn from X–ray astronomy.
J. H. M. M. Schmitt.
Astrophys. J., Vol. 293, No. 1, p. 178 – 191 (1985).
Statistical techniques suitable for the analysis of censored data, i.e., data containing both flux measurements and upper limits, are adapted to astronomical usage from the field of survival analysis, and are used in a statistical analysis of X–ray luminosities for single stars and binaries. The conditions of applicability of these methods are discussed and verified, and it is shown that neglect of the information contained in the upper limits can lead to incorrect conclusions. Several methods suitable for linear regression in the presence of censored data are discussed. A new method, based on maximum likelihood estimation in two dimensions, is developed, and is used to determine linear regression curves in the presence of arbitrarily censored data.

021.017   Statistical methods for astronomical data with upper limits. I. Univariate distributions.
E. D. Feigelson, P. I. Nelson.
Astrophys. J., Vol. 293, No. 1, p. 192 – 206 (1985).
The authors discuss statistical techniques applicable when a portion of an object sample is not detected; i.e., when upper limits, or left–censored data, are present. An extensive field of statistics called "survival analysis" of "lifetime data" exists to address these problems. This paper presents the foundations of nonparametric univariate survival analysis and discusses its application to astronomical data. The Kaplan–Meier product–limit estimator, its variance, mean, and standard deviation are presented. It provides a maximum–likelihood–type reconstruction of the true distribution function when upper limits are present. Three nonparametric procedures for testing whether or not two censored samples are drawn from the same distribution – the Gehan, logrank, and Peto–Prentice tests – are described. Some illustrative examples using astronomical data sets are included.

021.018   Planetenpositionen, selbst berechnet.
G. Traving.
Sterne Weltraum, 24. Jahrg., Nr. 6, p. 337 – 339 (1985).

021.019   Applications of non–linear methods in astronomy.
P. C. H. Martens.
Phys. Rep., Vol. 115, No. 6, p. 315 – 378 (1984). Abstr. in Phys. Abstr., Vol. 88, No. 1255, Entry 45930 (1985).

021.020 C^r approximations to the Standard Map.
W. H. Jefferys, A. Sivaramakrishnan.
Bull. Am. Astron. Soc., Vol. 17, No. 2, p. 625 (1985). Abstract. – See Abstr. 010.066.

021.021   A semi–analytical method of solution of systems of linear differential equations with periodic coefficients.
L. G. Luk'yanov.
Tr. Gos. Astron. Inst. Shternberg, Tom 57, p. 160 – 167 (1985). In Russian.
In the well–known structure of the integral matrix there are some unknown periodic functions and roots of the characteristic equation. A semi–analytical method for the determination of these unknown quantities is proposed. This method uses numerical integration on an interval equal to one period of the coefficients of the system. The unknown periodic functions in form of Fourier series are obtained.

021.022   Comments on the summations of spherical harmonics in the geopotential evaluation theories of Deprit and others.
P. J. Melvin.
Celest. Mech., Vol. 35, No. 4, p. 345 – 355 (1985).
Deprit's approach to the summation of the Legendre series in the geopotential evaluation problem is modified to accomodate normalized spherical harmonics and coefficients. Normalization avoids the floating point overflow encountered with high order geopotential models when the computer floating point arithmetic does not provide for large enough exponent. Deprit's algorithm is then appropriate for trajectory generation on–board an autonomous satellite system or for gravity compensation in an inertial navigation system.

021.023   The selection of optimum smoothness and the estimation of accuracy of an observational data series.
M. Zhao.
Acta Astron. Sin., Vol. 26, No. 1, p. 28 – 34 (1985). In Chinese.

021.024   Calculs astronomiques pour amateurs.
J. Meeus.
Astronomie, Vol. 99, p. 87 – 90, 147 – 150, 205 – 206, 249 – 252, 303 – 309 (1985).
Contents: Méridien central de Jupiter. Position des satellites de Jupiter. Magnitudes stellaires. Etoiles doubles. Régression linéaire – corrélation.

Computer programs in astronomy.
See Abstr. 002.040.

Kinematics.
See Abstr. 003.059.

Computational methods in bifurcation theory and dissipative structures.
See Abstr. 003.123.

Numerical solutions of the N–body problem.
See Abstr. 003.131.

The birefringent calibration device: a new approach.
See Abstr. 036.096.

An efficient method for solving Barker's equation.
See Abstr. 042.012.

A Lie integrator program and test for the elliptic restricted three body problem.
See Abstr. 042.045.

Estimate of the efficiency of numerical algorithms for construction of satellite trajectories.
See Abstr. 042.053.

Algorithm and program for numerical averaging of equations of motion of celestial bodies.
See Abstr. 042.066.

On the accuracy of high–speed calculation of orbits of artificial earth satellites by means of numerical integration.
See Abstr. 052.026.

Darstellung des Gravitationspotentials der Erde in kartesischen Basisfunktionen.
See Abstr. 081.034.

Komeet Halley in de computer.
See Abstr. 103.909.

Beobachtungsmöglichkeiten und Positionsberechnungen des Halleyschen Kometen 1985/86.
See Abstr. 103.933.

Application of the Gauss–Halphen–Goryachev method to investigate the orbital evolution of meteor streams.
See Abstr. 104.018.

022 Physical Papers Related to Astronomy and Astrophysics

022.001 Laboratory measurement of the O I 1173/989 Å branching ratio.
M. D. Morrison.
Planet. Space Sci., Vol. 33, No. 1, p. 135 – 139 (1985).

Laboratory measurements of the O I 1173/989 Å branching ratio have been made with a value of 1.5×10^{-4} indicated. This value makes the branching transition at 1173 Å an order of magnitude stronger than the branch at 7990 Å. The 1173 Å branching loss is still too weak a loss process for multiply scattered 989 Å photons to resolve the 989 Å intensity problem in the dayglow.

022.002 Vibrational temperature and excess vibrational energy of molecular nitrogen in the ground state derived from N_2^+ emission bands in aurora.
V. A. Vlaskov, K. Henriksen.
Planet. Space Sci., Vol. 33, No. 2, p. 141 – 145 (1985).

022.003 The dissociative recombination of N_2^+ ($v = 0, 1$) as a source of metastable atoms in planetary atmospheres.
J. L. Queffelec, B. R. Rowe, M. Morlais, J. C. Gomet, F. Vallee.
Planet. Space Sci., Vol. 33, No. 3, p. 263 – 270 (1985).

The yield of metastable nitrogen atoms in dissociative recombination of N_2^+ ($v = 0, 1$) ions has been studied for different experimental conditions. The results are in good agreement with thermospheric models but imply that N_2^+ dissociative recombination is a less important source for nitrogen escape of Mars.

022.004 A Q–rious tale; the origin of the parameter Q in electromagnetism.
B. Jeffreys.
Q. J. R. Astron. Soc., Vol. 26, No. 1, p. 51 – 52 (1985).

One definition of the parameter Q often used to characterize a damped oscillatory process is that Q^{-1} is $(1/2\pi)$ times the fractional energy loss per cycle. The history of the use of Q is traced in fields other than geophysics.

022.005 Einstein A–coefficients for pure rotational transitions in the $H_2^{18}O$–molecule.
S. Chandra.
Astron. Astrophys., Suppl. Ser., Vol. 59, No. 1, p. 59 – 61 (1985).

Einstein A–coefficients for the electric dipole transitions in the $H_2^{18}O$–molecule between the rotational levels of the ground vibrational state up to 1500 cm^{-1} are reported. The coefficients are employed to compute the mean–radiative–life–times of the levels.

022.006 Wert der Lichtgeschwindigkeit endgültig festgelegt.
D. Hantke, J. Tschirnich.
Astron. Raumfahrt, 23. Jahrg., Heft 1, p. 12 – 14 (1985).

022.007 Contribution of large polycyclic aromatic molecules to the infrared emission of the interstellar medium.
J. L. Puget, A. Léger, F. Boulanger.
Astron. Astrophys., Vol. 142, No. 2, p. L19 – L22 (1985).

The IR emission spectrum of a population of graphite grains with a size distribution extending down to polycyclic aromatic molecules of few Å is investigated. The authors use the optical properties of one such molecule, the coronene (Léger and Puget 1984), to evaluate the oscillator strength of lattice vibrations. The calculated spectra account for a variety of astrophysical observations which show extended mid IR emission. The authors predict that the Cirrus Clouds observed by IRAS at 100 µm should also emit at 12 µm.

022.008 Amorphous carbon grains: laboratory measurements in the 2000 Å – 40 µm range.
A. Borghesi, E. Bussoletti, L. Colangeli.
Astron. Astrophys., Vol. 142, No. 2, p. 225 – 231 (1985).

The authors present the results of transmission measurements performed in the range 2000 Å – 40 µm on submicronic amorphous carbon particles. Morphological analysis was performed by using electron microscopy to determine grain properties and size distributions. The extinction of the samples indicates the presence of a pronounced hump at 2400 – 2500 Å with the peak dependent on particle size. Its characteristics are discussed both intrinsically and by comparing them with the interstellar curve. An extinction law $\sim \lambda^{-1}$ is found to hold at wavelengths longward of 4500 Å in agreement with a previous work. Astronomical observations and theoretical calculations are recalled which suggest that amorphous carbon may be produced in space, mainly in regions such as the atmosphere of late–type stars, novae and supernovae.

022.009 A potential extraterrestrial species: the SH^+ molecular ion.
M. Horani, J. Rostas, E. Roueff.
Astron. Astrophys., Vol. 142, No. 2, p. 346 – 350 (1985).

The aim of this paper is to collect the available experimental data which may assist detection of SH^+ in astronomical objects, such as interstellar matter, comet tails or planetary atmospheres, by observation of relevant lines in any part of the electromagnetic spectrum.

022.010 Establishment of internally consistent solar scales of oscillator strenghts and abundances of chemical elements. V. Scandium.
Eh. A. Gurtovenko, R. I. Kostyk, T. V. Orlova.
Kinematika Fiz. Nebesn. Tel, Tom 1, No. 1, p. 75 – 76 (1985). In Russian.

Oscillator strengths for 31 selected Sc I and Sc II lines were determined using the observed equivalent widths of the Fraunhofer lines.

022.011 Moderately accurate oscillator strengths from NBS intensities – a revision.
L. Ward.
Mon. Not. R. Astron. Soc., Vol. 213, No. 1, p. 71 – 74 (1985).

Earlier studies have shown that useful oscillator strengths may be derived from the intensity tabulations in NBS Monograph 145. In the present work recently measured lifetimes are used to obtain an improved formula to be applied to the Nd II spectrum. Oscillator strengths for more than 600 Nd II lines may now be obtained with an uncertainty generally better than 0.17 dex.

022.012 Relative band strengths from the study of intensity distribution in the A–X system of CrO.
S. P. Bagare, N. S. Murthy.
Physica B, C, Vol. 125B + C, No. 2, p. 265 – 267 (1984). Abstr. in Phys. Abstr., Vol. 88, No. 1248, Entry 5964 (1985).

022.013 Observation of the v_1 fundamental band of $DCNH^+$.
T. Amano.
J. Chem. Phys., Vol. 81, No. 7, p. 3350 – 3351 (1984). Abstr. in Phys. Abstr., Vol. 88, No. 1248, Entry 5977 (1985).

022.014 Atomic physics developments for interpretation of spectral line polarization measurements in astrophysics.
S. Sahal–Brechot.
Ann. Phys. (Paris), Vol. 9, No. 4, p. 705 – 712 (1984). Abstr. in Phys. Abstr., Vol. 88, No. 1249, Entry 14462 (1985).

022.015 A theoretical study of the rotation–vibration energy levels and dipole moment functions of CCN^+, CNC^+, and C_3.
W. P. Kraemer, P. R. Bunker, M. Yoshimine.
J. Mol. Spectrosc., Vol. 107, No. 1, p. 191 – 207 (1984). Abstr. in Phys. Abstr., Vol. 88, No. 1250, Entry 15908 (1985).

022.016   The first experimental observation of electronic transitions in C_2^+ and C_2D^+.
A. O'Keefe, R. Derai, M. T. Bowers.
Chem. Phys., Vol. 91, No. 1, p. 161 – 166 (1984). Abstr. in Phys. Abstr., Vol. 88, No. 1250, Entry 15921 (1985).

022.017   Energetics of the protonation of CO: implications for the observation of HOC^+ in dense interstellar clouds.
D. A. Dixon, A. Komornicki, W. P. Kraemer.
J. Chem. Phys., Vol. 81, No. 8, p. 3603 – 3611 (1984). Abstr. in Phys. Abstr., Vol. 88, No. 1250, Entry 17892 (1985).

022.018   Optical spectra of amorphous carbon grains.
G. Maggipinto, A. Minafra, F. Tritto.
Astrophys. Space Sci., Vol. 108, No. 1, p. 101 – 111 (1985).
Extinction coefficient measurements of amorphous carbon grains, produced by hydrocarbon burning, have been carried out at room temperature in the spectral range from 1900 Å to 2.5 μ. The classical KBr pellet technique has been used to obtain quantitative data. Corrections to account for matrix distortion of the spectrum have been determined experimentally by measuring the spectrum of the same grains deposited on a thin quartz plate.

022.019   Use of microwave detected microwave–optical double resonance to assign the 6450 Å band of NH_3.
K. K. Lehmann, S. L. Coy.
J. Chem. Phys., Vol. 81, No. 8, p. 3744 – 3745 (1984). Abstr. in Phys. Abstr., Vol. 88, No. 1251, Entry 20234 (1985).

022.020   Fits for photoionization cross–sections.
K. Butler, C. Mendoza, C. J. Zeippen.
Mon. Not. R. Astron. Soc., Vol. 213, No. 2, p. 345 – 353 (1985).
Fits for the photoionization cross–sections of several ions of astrophysical interest are provided. The fits should facilitate the use of the present radiative data in astrophysical models. The ions considered are N^{2+}, N^{3+}, O^{4+}, Al^+, Si^{2+}, and S^{4+}.

022.021   Erratum: "Collisional excitation of HCN–hyperfine transitions" [Mon. Not. R. Astron. Soc., Vol. 211, No. 2, p. 257 – 266 (1984)].
T. Monteiro.
Mon. Not. R. Astron. Soc., Vol. 213, No. 2, p. 495 (1985). See Abstr. 38.022.085.

022.022   Spectroscopy of CaOH.
P. F. Bernath, C. R. Brazier.
Astrophys. J., Vol. 288, No. 1, p. 373 – 376 (1985).
High–resolution laser spectroscopy of CaOH has produced accurate spectroscopic constants for the $B\,^2\Sigma^+$, $A\,^2\Pi$, and $X\,^2\Sigma^+$ states of CaOH. The pure rotational transition frequencies of the ground state are predicted.

022.023   Oscillator strengths for transitions in N I and the interstellar abundance of nitrogen.
A. Hibbert, P. L. Dufton, F. P. Keenan.
Mon. Not. R. Astron. Soc., Vol. 213, No. 3, p. 721 – 734 (1985).
Oscillator strengths based on configuration interaction wavefunctions are presented for both optically allowed and forbidden transitions in N I. Particular attention is given to the multiplets at 951 Å $(2p^3\,^4S–2p^23d^2D)$, 952 Å $(2p^3\,^4S–2p^23d^4D)$ and 1160 Å $(2p^3\,^4S–2p^23s^2P)$ which have been extensively observed by the *COPERNICUS* satellite. For these transitions, the radiative rates are estimated to have an accuracy of 20 per cent or better. A re–analysis of the *COPERNICUS* observational data indicates there is no depletion of nitrogen towards reddened stars. Possible causes of a small depletion ($\cong 0.2$ dex) towards several nearby unreddened stars are discussed.

022.024   Millimeter and submillimeter wave spectroscopy of the deuterated ethynyl radical.
M. Bogey, C. Demuynck, J. L. Destombes.
Astron. Astrophys., Vol. 144, No. 2, p. L15 – L16 (1985).
The millimeter and submillimeter spectrum of the CCD radical has been observed in a laboratory glow discharge. Rotational, fine and hyperfine structure constants are determined and are used to accurately predict the frequencies of the unobserved components.

022.025   Laboratory photometry of asteroids and atmosphereless bodies. I. The S.A.M. apparatus.
V. D'Ambrosio, R. Burchi, A. Di Paolantonio, C. Giuliani.
Astron. Astrophys., Vol. 144, No. 2, p. 427 – 430 (1985).
The aim of this work is to obtain, by laboratory simulation, information on the geometric and physical characteristics of the asteroids (or other bodies without an atmosphere) such as: their shapes, the orientation of their rotation axes and their surface morphologies. After two years of continuous laboratory tests and experiments the authors' apparatus has reached its almost definitive arrangement. The validity of the results obtained has been verified repeatedly by comparison with numerical data and astronomical observations.

022.026   Laboratory identification of the emission features near 3.5 μm in the pre–main–sequence star HD 97048.
G. P. van der Zwet, L. J. Allamandola, F. Baas, J. M. Greenberg.
Astron. Astrophys., Vol. 145, No. 1, p. 262 – 268 (1985).
This paper presents the results of a laboratory study of formaldehyde (H_2CO) suspended in low temperature molecular matrices with compositions similar to what may be found in the "dirty ice" mantles of grains. It is shown that the emission features near 3.5 μm in the pre–main–sequence star HD 97048 can be matched by a mixture of chemical complexes of H_2CO with surrounding molecules in the grain. Furthermore, a discussion is presented of various possible excitation mechanisms for this emission. The conclusion is, that for the features near 3.5 μm in HD 97048, UV pumped IR fluorescence is the most likely mechanism.

022.027   Redistribution of radiation in the absence of collisions.
G. G. Lombardi, D. E. Kelleher, J. Cooper.
Astrophys. J., Vol. 288, No. 2, p. 820 – 823 (1985).
Redistribution and depolarization of near–resonant radiation was studied for the He $2^1P – 3^1D$ line (668 nm) and some data are also presented for Hα. The measurements of the ratio of Rayleigh to fluorescent intensities confirm the prediction that redistribution of radiation occurs in the *absence* of collisions for transitions having significant lower–level radiative widths. Depolarization rates by collisions with helium were also inferred from the measurements.

022.028   Recombination–cascade X–ray spectra of highly charged helium–like ions.
A. K. Pradhan.
Astrophys. J., Vol. 288, No. 2, p. 824 – 830 (1985).
It is shown that the relative intensity distribution among the X–ray spectral lines of helium–like ions from the $n = 2$ states produced through recombination processes such as radiative and charge transfer recombination may be given by considering in detail the radiative cascades following recombination. Model calculations are presented with predicted line ratios for Ar XVII and Fe XXV in recombination–dominated noncoronal plasmas. Accurate configuration interaction type wave functions are employed to calculate the eigenenergies, transition probabilities, and cascade coefficients.

022.029   On the Z–dependence of lithium–like photoionization cross sections.
J. A. Tully.
Astrophys. J., Vol. 288, No. 2, p. 831 – 832 (1985).
The isoelectronic extrapolation procedure proposed by Pradhan is shown to be invalid since it is based on a paper by Reilman and Manson which contains a printing error. Likewise, the rate coefficients presented by Woods, Shull, and Sarazin for radiative recombination to form Fe^{+22} and Fe^{+23} are incorrect. A simple Z–dependent formula is given which reproduces to within 2% the threshold photoionization cross–sections of Reilman and Manson for the lithium sequence.

022.030   On the singularities of the Newtonian two dimensional N–body problem.
C. Marchioro, M. Pulvirenti.
Atti Accad. Naz. Lincei, Ser. Ottava, Rend., Vol. 75, Fasc. 3 – 4, p. 106 – 110 (1983).

022.031   Sputtering processes: erosion and chemical change.
R. E. Johnson, L. J. Lanzerotti, W. L. Brown.
Adv. Space Res., Vol. 4, No. 9, p. 41 – 51 (1984). – See Abstr. 012.029.

The authors review laboratory data and models on sputter–induced erosion and chemical alterations of ice films and apply the results to icy grains and satellites exposed to magnetospheric ion bombardment. They show that the source of the plasma in the inner magnetosphere of Saturn is likely to be the sputter erosion of the icy objects in this region and consider the sputter erosion and possible stabilization of the E–ring. Ion–induced polymerization is discussed as a source of the darkened rings of Uranus.

022.032   Charges on dust particles.
O. Havnes.
Adv. Space Res., Vol. 4, No. 9, p. 75 – 83 (1984). – See Abstr. 012.029.

The author discusses the potential (charge) on dust particles in various environments. He first considers the classical case of a single isolated dust particle. In conditions which apply to planetary dust rings, the exact value of the dust potential depends critically on several effects (e.g. secondary electron emission, photoelectric efficiency) which are not well known for small dust particles of relevant material and surface conditions. In dust clouds of high dust densities the classical approach fails to give the correct value of the dust potential due to the neglect of collective effects.

022.033   On the importance of fluff on the electric behaviour of cosmic grains.
J.–P. J. Lafon, J. M. Millet.
Adv. Space Res., Vol. 4, No. 9, p. 91 – 94 (1984). – See Abstr. 012.029.

A model is created to describe the effects of "fluff" on the potential and electric field on and close to a charged spherical body embedded in a plasma. The consequences are investigated for dust grains biased at positive or negative potentials, but large enough for electron or ion field emission to be active, especially grains in magnetospheric plasmas. Electron emission reduces the floating potential, whereas ion emission destroys the fluff or even the grain itself. Effects of encounters are discussed. The model also characterizes the levitation of small solid particles from larger bodies.

022.034   The impact of dust grains on fast fly–by spacecraft: momentum multiplication, measurements and theory.
J. A. M. McDonnell, M. Alexander, D. Lyons, W. Tanner, P. Anz, T. Hyde, A.–L. Chen, T. J. Stevenson, S. T. Evans.
Adv. Space Res., Vol. 4, No. 9, p. 297 – 301 (1984). – See Abstr. 012.029.

Energy partitioning during the very high impact speed encountered in a cometary fly–by mission causes a target mass expulsion which leads to a momentum impulse on the target exceeding that of the incident momentum. Theoretical and computational studies are required to provide a basis for predictions of the response at Halley encounter, since experimental data from acceleration of microspheres extends currently only to some 10 km s^{-1}. Such data obtained from the 2 MV Canterbury microparticle accelerator is presented.

022.035   Mean collision frequencies of O^+–He and H^+–He for ionospheric investigations.
A. V. Pavlov.
Kosm. Issled., Tom 23, Vyp. 1, p. 143 – 147 (1985). In Russian. English translation in Cosm. Res.

022.036   Computation of the Franck–Condon factors for a series of astrophysically important molecular systems of N_2.
N. E. Kuz'menko, V. B. Pavlov–Verevkin.
Astrofizika, Tom 22, Vyp. 1, p. 195 – 210 (1985). In Russian. English translation in Astrophysics, Vol. 22, No. 1.

A model quantitative description of the intensities in electronic spectra of diatomic molecules which play an important role in astrophysical investigations of interstellar space and planetary atmospheres has been considered. The results of computations of Franck–Condon factors and other radiative characteristics of a few astrophysically important molecular systems of nitrogen N_2 are given and discussed.

022.037   A mean spherical approximation of the solubility of iron in the internal solar plasma.
I. Ruff, J. Liszi, K. Gombos.
Astrophys. J., Vol. 289, No. 1, p. 409 – 413 (1985).

The model sensitivity of the prediction of the insolubility of iron in internal solar plasma is tested by comparing the mean spherical approximation of a classical plasma of charged hard spheres with the point–charge model of the Debye–Hückel approximation. It is found that the solubility of iron is very insensitive to considerable changes in the model assumptions. Thus the predicted very low solubility of iron must be seriously considered in future models of the Sun, since it may drastically decrease the opacity and, through this effect, provide an explanation for the neutrino dilemma.

022.038   Die Welt im Großen und die heute geltenden Naturgesetze. Gedanken zu einigen fundamentalen Problemen der Physik.
U. Kasper.
Sterne, 61. Band, Heft 1, p. 21 – 27 (1985).

022.039   On the establishment of internally consistent solar scales of oscillator strengths and abundances of chemical elements. VI. Neutral vanadium.
Eh. A. Gurtovenko, R. I. Kostyk, T. V. Orlova.
Kinematika Fiz. Nebesn. Tel, Tom 1, No. 2, p. 62 – 63 (1985). In Russian.

Oscillator strengths of 55 lines of neutral vanadium are determined from equivalent widths of solar Fraunhofer lines using the procedure developed by the authors and reported in the previous papers.

022.040   Dielectronic recombination rates for ions of the magnesium sequence at low energies.
M. P. Dube, R. Rasoanaivo, Y. Hahn.
J. Quant. Spectrosc. Radiat. Transfer, Vol. 33, No. 1, p. 13 – 26 (1985).

The dielectronic recombination (DR) rate coefficient α^{DR} is explicitly calculated for the Si, Ar, Fe and Mo target ions of the Mg isoelectronic sequence with twelve electrons. Both the $3s$, $\Delta n \neq 0$ and $3s$, $\Delta n = 0$ transitions are considered in detail. An explicit LS coupling scheme was applied to all the dominant transitions of these ions. Results of α^{DR} with different free electron temperatures are also discussed.

022.041   The Balmer alpha profile corrected for strong self–absorption.
J. Hernández, F. Torres, S. Mar, M. A. Gigosos.
J. Quant. Spectrosc. Radiat. Transfer, Vol. 33, No. 1, p. 35 – 38 (1985).

The authors have obtained the spectrum of the $H\alpha$ line with strong self–absorption in a pulsed source. The spectrum has been corrected for self–absorption. The corrected profiles allow the authors to obtain plasma electron densities, which are in good agreement with those obtained from the $H\beta$ line.

022.042 Stark–width measurements of argon–ion lines with a Fabry–Pérot interferometer.
P. H. M. Vaessen, J. M. L. van Engelen, J. J. Bleize.
J. Quant. Spectrosc. Radiat. Transfer, Vol. 33, No. 1, p. 51 – 53 (1985).

Stark widths of optically thin line profiles of the 4806 and 4426 Å Ar(II) lines in an atmospheric cascade arc plasma have been measured with a Fabry–Pérot interferometer. These experimental results are significantly lower than all other experimental data and differ about a factor of 2 from theoretical calculations.

022.043 Lifetimes and oscillator strengths of neutral vanadium.
A. Doerr, M. Kock, M. Kwiatkowski, K. Werner, P. Zimmermann.
J. Quant. Spectrosc. Radiat. Transfer, Vol. 33, No. 1, p. 55 – 62 (1985).

Oscillator strengths of 105 V(I) lines in the spectral range 2800 – 6000 Å were obtained by lifetime, emission, and hook measurements.

022.044 Laboratory simulation of meteoritic noble gases. I. Sorption of xenon on carbon: trapping experiments.
J. F. Wacker, M. G. Zadnik, E. Anders.
Geochim. Cosmochim. Acta, Vol. 49, No. 4, p. 1035 – 1048 (1985).

022.045 Laboratory simulation of meteoritic noble gases. II. Sorption of xenon on carbon: etching and heating experiments.
M. G. Zadnik, J. F. Wacker, R. S. Lewis.
Geochim. Cosmochim. Acta, Vol. 49, No. 4, p. 1049 – 1059 (1985).

022.046 Oscillator strengths for C III lines.
R. M. Nasser, Y. P. Varshni.
Astron. Astrophys., Suppl. Ser., Vol. 60, No. 2, p. 325 – 332 (1985).

Oscillator strengths are calculated for all allowed electric dipole transitions (140 in number) between the 40 lowest terms of the C III ion. Configuration–interaction wavefunctions are used. Calculated values are compared with previous results and experimental values, where available.

022.047 Hot atoms in cosmic chemistry.
K. Rössler, H.–J. Jung, B. Nebeling.
Adv. Space Res., Vol. 4, No. 12, p. 83 – 95 (1984). – See Abstr. 012.036.

High energy chemical reactions and atom molecule interactions might be important for cosmic chemistry with respect to the accelerated species in solar wind, cosmic rays, colliding gas and dust clouds and secondary knock–on particles in solids. "Hot" atoms with energies ranging from a few eV to some MeV can be generated via nuclear reactions and consequent recoil processes. Hot atom chemistry may serve for laboratory simulation of the reactions of energetic species with gaseous or solid interstellar matter. Experimental results are given for the systems: C/H_2O (gas), C/H_2O (solid, 77K), N/CH_4 (solid, 77K) and C/NH_3 (solid, 77K). Typical reaction products are: CO, CO_2, CH_4, CH_2O, CH_3OH, HCOOH, NH_3, CH_3NH_2, cyanamide, formamidine, guanidine etc. Products of hot reactions in solids are more complex than in corresponding gaseous systems, which underlines the importance of solid state reactions for the build–up of precursors for biomolecules in space. As one of the major mechanisms for product formation, the simultaneous or fast consecutive reactions of a hot carbon with two target molecules (reaction complex) is discussed.

022.048 Organic chemistry by irradiation in space.
J.–P. Bibring, F. Rocard.
Adv. Space Res., Vol. 4, No. 12, p. 103 – 106 (1984). – See Abstr. 012.036.

The irradiation of grains and/or ices by particles from solar or stellar winds, as well as cosmic rays, induces the synthesis of molecular species. The authors have shown by in–situ infrared spectroscopy of irradiated samples that this chemistry may be responsible for the presence of organic compounds in a large variety of astrophysical sites such as: lunar and asteroidal regoliths, cometary nuclei, rings and satellites of outer planets, circumstellar shells, interstellar clouds. The authors present their experimental results concerning the nature and efficiency of C and N irradiation chemistry, and give plausible astrophysical implications.

022.049 Radiation chemical experiments relevant to studies of cometary nuclei: remarks on working conditions.
I. G. Draganić, Z. D. Draganić.
Adv. Space Res., Vol. 4, No. 12, p. 115 – 119 (1984). – See Abstr. 012.036.

The authors survey some obstacles that a chemist encounters in defining conditions for radiation chemical experiments relevant to cometary nuclei. The choice of working conditions is examined in the light to present knowledge about comets and the facilities available for routine work in radiation chemistry.

022.050 Prebiotic syntheses of purines and pyrimidines.
B. Basile, A. Lazcano, J. Oró.
Adv. Space Res., Vol. 4, No. 12, p. 125 – 131 (1984). – See Abstr. 012.036.

Although several different pathways for the synthesis of purines have been described, they are all variations of the initial mechanism proposed by Oró and Kimball. A number of experiments have shown that purines and pyrimidines can also be obtained from methane, ammonia (nitrogen), and water mixtures, provided an activating source of energy (radiation, electric discharges, etc.) is available. However, in this case the yields are lower by about two orders of magnitude because of the intermediate formation of hydrogen cyanide and cyanoacetylene. The latter two compounds have been found in interstellar space, Titan and other bodies of the solar system. They were probably present in the primordial parent bodies from the solar nebula in concentrations of 10^{-2} to 10^{-3} M as inferred from recent calculations by Miller and coworkers obtained for the Murchison meteorite. These concentrations should have been sufficient to generate relatively large amounts of purine and pyrimidine bases on the primitive Earth.

022.051 Calculation of recombination line intensities for C III ions.
A. A. Nikitin, A. A. Sapar, T. Kh. Feklistova, A. F. Kholtygin.
Tartu Astrofüüs. Obs. Publ., Tom 50, p. 81 – 100 (1984). In Russian.

From the system of equations of statistical equilibrium for singlet and triplet states with main quantum number $n \leqslant 6$ the relative intensities of lines of the recombination spectrum for C III have been found at T_e = 10000, 20000, 30000 and 50000K.

022.052 Energy levels in hydrogen plasmas and the Planck–Larkin partition function – a comment.
W. Ebeling, W. D. Kraeft, D. Kremp, G. Röpke.
Astrophys. J., Vol. 290, No. 1, p. 24 – 27 (1985).

Several comments are presented regarding the Planck–Larkin partition function (PLPF) in connection with a recent article by Rouse. It is shown that in an up–to–date version of the quantum statistics of Coulomb systems with bound states, the discrete energy states of the Bethe–Salpeter equation (BSE) have to be introduced into the PLPF; the latter then becomes temperature– and density–dependent.

022.053 Polycyclic aromatic hydrocarbons and the unidentified infrared emission bands: auto exhaust along the Milky Way!
L. J. Allamandola, A. G. G. M. Tielens, J. R. Barker.
Astrophys. J., Lett. Ed., Vol. 290, No. 1, p. L25 – L28 (1985).

The authors have attributed the unidentified infrared emission features (UIR bands) to a collection of partially hydrogenated, positively charged polycyclic aromatic hydrocarbons (PAHs). This assignment is based on a spectroscopic analysis of the UIR bands. Comparison of the observed interstellar 6.2 and 7.7 µm bands with the laboratory measured Raman spectrum of a collection of carbon–based particulates (auto exhaust) shows a very good agreement, supporting this identification. The infrared emission is due to relaxation from highly vibrationally and electronically excited states. The excitation is probably caused by UV photon absorption.

022.054 Laboratory infrared spectra of predicted condensates in carbon–rich stars.
J. A. Nuth, S. H. Moseley, R. F. Silverberg, J. H. Goebel, W. J. Moore.
Astrophys. J., Lett. Ed., Vol. 290, No. 1, p. L41 – L43 (1985).

Laboratory spectra and mass absorption coefficients of MgS, CaS, FeS, SiS_2, FeS_2, Fe_3C, and a commercial iron carbide are presented over the range 125 μm $\geqslant \lambda \geqslant$ 15 μm. These spectra confirm that MgS is the most likely source of the unidentified 30 μm emission in carbon–rich sources and that FeS, Fe_3C, and "iron carbide" cannot be responsible for this feature although they could contribute to the continuum in this region.

022.055 An apparently first–order transition between two amorphous phases of ice induced by pressure.
O. Mishima, L. D. Calvert, E. Whalley.
Nature, Vol. 314, No. 6006, p. 76 – 78 (1985).

This letter reports that low–density amorphous ice (density $0.94\,\mathrm{g\,cm^{-3}}$) compressed at 77K transforms to high–density amorphous ice ($1.19\,\mathrm{g\,cm^{-3}}$ at zero pressure) at a sharp transition at 6 ± 0.5 kbar. The transition strongly resembles a first order transition in its sharpness and large volume change. It appears to be the first example of an apparently first–order transition between amorphous solids and has implications not only for our understanding of the behaviour of condensed matter, but also for theories of planetary interiors.

022.056 Dependence of Stark widths and shifts on the ionization potential: $np^{k-1}(n+1)s$–np^k resonance transitions.
M. S. Dimitrijević.
Astron. Astrophys., Vol. 145, No. 2, p. 439 – 442 (1985).

Simple relations between the Stark broadening parameters and the ionization potential are derived for $np^{k-1}(n+1)s$–np^k resonance transitions of singly ionized emitters, using the semiclassical approach. The relations obtained are compared with those derived by Purić et al. (1980).

022.057 Rotational spectra of the TiO and ZrO molecules and possibilities of their radio astronomical observations.
M. G. Solov'ev, D. A. Varshalovich.
Astron. Zh., Tom 62, Vyp. 2, p. 268 – 271 (1985). In Russian. English translation in Sov. Astron., Vol. 29, No. 2.

The frequencies and Einstein A–coefficients corresponding to microwave rotational transitions of TiO and ZrO ground electronic states have been calculated. Expected radiation flux densities of the microwave lines are estimated for late–type stars.

022.058 The standard line–formation problem: second–order approximations including continuous absorption.
V. M. Serbin.
Sov. Astron. Lett., Vol. 10, No. 4, p. 231 – 234 (1984). English translation of 38.022.001.

022.059 Line intensity ratios for transitions in O III.
K. M. Aggarwal.
Astron. Astrophys., Vol. 146, No. 1, p. 149 – 158 (1985).

Electron collision strengths have been calculated for the optically allowed transitions among the $1s^22s^22p^2$, $1s^22s2p^3$, and $1s^22p^4$ configurations of O III. The effective collision strengths are tabulated in a temperature range of 5×10^3 to 5×10^5K and are compared with earlier data. These effective collision strengths together with earlier atomic data for energy levels and transition probabilities are employed in calculating level populations of the lowest fifteen energetic levels of O III within the configurations $1s^22s^22p^2$ and $1s^22s2p^3$. The populations are tabulated for these levels in a temperature range of $4\times10^4 - 1\times10^5$K and a density range of $10^5 - 10^{10}\mathrm{cm^{-3}}$. From these populations the line intensity ratios of the astrophysically observable optically allowed transitions are calculated and compared with earlier theoretical and experimental values.

022.060 The critical periodic orbits in the Störmer problem.
R. A. Broucke.
The Big Bang and Georges Lemaitre, p. 257 – 267 (1984). – See Abstr. 012.043.

The author studies some of the periodic orbits, among the families f_0, f_1, f_2, f_3, f_4 and f_5 that were previously found by Goudas and Markellos. The paper concentrates on the periodic orbits which have a special value ($+2$, -2, -1 or 0) of the stability index, because, for these orbits, the period of the periodic solutions of the variational equations is an integer multiple of the original period. Several such solutions are classified according to the Hénon type or the Contopoulos resonant type. Bifurcations or trifurcations of new families are calculated out of these critical orbits.

022.061 Opacity determinations for ICF (*Inertial Confinement Fusion*) materials using an average atom model.
G. Velarde, J. M. Aragonés, C. Cabezudo, J. J. Honrubia, J. M. Martinez–Val, E. Minguez, J. L. Ocaña, J. M. Perlado, J. F. Serrano.
ESA Spec. Publ., ESA SP–207, p. 201 – 204 (1984). – See Abstr. 012.044.

A calculational model to obtain opacity data for the most useful materials in ICF targets is presented. Two modular computer codes coupled have been developed, and their formalisms are explained in detail.

022.062 Statistical mechanics of light elements at high pressure. VII. A perturbative free energy for arbitrary mixtures of H and He.
W. B. Hubbard, H. E. DeWitt.
Astrophys. J., Vol. 290, No. 2, p. 388 – 393 (1985).

The authors present a model free energy which accurately represents results from 45 high–precision Monte Carlo calculations of the thermodynamics of hydrogen–helium mixtures at pressures of astrophysical and planetophysical interest. The free energy is calculated using free–electron perturbation theory (dielectric function theory). Using the new free energy, the authors calculate the phase diagram of mixtures of liquid metallic hydrogen and helium and compare it with earlier results. They use the new free energy expression to compute a theoretical Jovian adiabat and compare the adiabat with results from three–dimensional Thomas–Fermi–Dirac theory.

022.063 Oscillator strengths and collision strengths for neutral sulfur.
Y. K. Ho, R. J. W. Henry.
Astrophys. J., Vol. 290, No. 2, p. 424 – 427 (1985).

Collision strengths for electron–impact excitation of neutral sulfur from the group $3p^4\,^3P$ state to excited states $3p^34s\,^3S^0$, $3p^3(^4S^0)3d\,^3D^0$, and $3p^34s\,^3P^0$ are calculated in a close–coupling approximation for the energy range up to 10^6K. Also, oscillator strengths for various triplet transitions are reported, as well as transitions between $3p^4\,^1D$ and $3p^4\,^1S$ and other singlet excited states.

022.064 Millimeter–wave spectrum of the CCO radical.
C. Yamada, S. Saito, H. Kanamori, E. Hirota.
Astrophys. J., Lett. Ed., Vol. 290, No. 2, p. L65 – L66 (1985).

The pure rotational spectrum of the CCO radical in the $^3\Sigma^-$ ground electronic state has been observed in the laboratory in the region 45 – 185 GHz. The radical was generated in a 3.5 m long free space absorption cell by a DC glow discharge in carbon suboxide (C_3O_2). Seven rotational transitions of $N = 2\leftarrow1$ to $8\leftarrow7$ in each triplet sublevels were precisely measured and the observed frequencies were least–squares analyzed to obtain molecular constants.

022.065 A poorly graphitised carbon contaminant in studies of extraterrestrial materials.
F. J. M. Rietmeijer.
Meteoritics, Vol. 20, No. 1, p. 43 – 48 (1985).

Poorly–graphitised carbon particles are formed during manufacture of sample substrates (holey carbon films) for Analytical Electron Microscopy studies of small particles. The particles form during heat treatment of cellulose acetobutyrate at about 975°C and 1050°C. In AEM studies of fine–grained carbonaceous extraterrestrial materials, these particles are easily recognised.

022.066 High–energy neutron induced prompt gamma–rays: chemical remote sensing of planetary surfaces.
J. Brückner, H. Wänke, P. Englert, R. C. Reedy.
Meteoritics, Vol. 19, No. 4, p. 200 (1984). Abstract. – See Abstr. 010.641.

022.067 Solar wind and cosmic ray irradiation of grains and ices – application to erosion and synthesis of organic compounds in the solar system.
F. Rocard, J. Bénit, J.–P. Bibring, R. Meunier, B. Vassent.
Meteoritics, Vol. 19, No. 4, p. 302 (1984). Abstract. – See Abstr. 010.641.

022.068 Links between astronomical observations of protostellar clouds and laboratory measurements of interplanetary dust: the 6.8 μm carbonate band.
S. A. Sandford, R. M. Walker.
Meteoritics, Vol. 19, No. 4, p. 306 – 307 (1984). Abstract. – See Abstr. 010.641.

022.069 Homogeneous condensation of gaseous mixtures of Si, Fe, O, N, and C in relative cosmic abundance and implications for astronomical condensation.
J. R. Stephens.
Meteoritics, Vol. 19, No. 4, p. 316 (1984). Abstract. – See Abstr. 010.641.

022.070 Hydrogen broadening of vibrational–rotational transitions of ammonia lying near 6450 Å.
C. E. Keffer, C. P. Conner, W. H. Smith.
J. Quant. Spectrosc. Radiat. Transfer, Vol. 33, No. 3, p. 193 – 196 (1985).
The hydrogen broadened half–widths of four ammonia vibrational–rotational transitions near 6450 Å have been measured. The average value of the hydrogen broadening coefficient is 0.101 ± 0.004 cm^{-1}atm with no apparent quantum number dependence.

022.071 The absorption spectrum of beryllium.
J. M. P. Serrão.
J. Quant. Spectrosc. Radiat. Transfer, Vol. 33, No. 3, p. 219 – 226 (1985).
Energy terms, dipole oscillator strengths and photoionization cross–sections from the ground state are calculated. Autoionizing state transition energies and line widths for the $^1P^0$ resonances in the continuum are also obtained. The configuration interaction method for initial and final states is used, and atomic orbitals are generated through angular–momentum–dependent, scaled Thomas–Fermi–Dirac potentials.

022.072 The static approximation in the interpretation of collision broadening.
J. Kielkopf.
J. Quant. Spectrosc. Radiat. Transfer, Vol. 33, No. 3, p. 267 – 274 (1985).
The validity of the static approximation for analyses of spectral line profiles is considered. The Fourier transform of the profile of an isolated line is compared for representative cases with the unified theory, and it is shown that the simple static approximation can be used effectively to interpret low spectral resolution experiments. This result applies to the entire profile in the limit of high density, but it also applies at low density if the instrumental line width is sufficiently large.

022.073 Measurements of nitrogen–, hydrogen– and helium–broadened widths of methane lines at 9030 – 9120 cm^{-1}.
K. Fox, D. E. Jennings.
J. Quant. Spectrosc. Radiat. Transfer, Vol. 33, No. 3, p. 275 – 280 (1985).
Pressure–broadened widths of vibration–rotation lines of both $^{12}CH_4$ and $^{13}CH_4$ have been measured at very high spectral resolution in the R–branch of the $3v_3$ overtone. The broadening gases were N_2, H_2 and He, as well as a mixture of H_2 and He. Results are presented as averages for J–multiplets at ambient temperature.

022.074 Partition functions and dissociation constants for HeH$^+$.
V. P. Gaur, B. M. Tripathi.
J. Quant. Spectrosc. Radiat. Transfer, Vol. 33, No. 3, p. 291 – 292 (1985).
Partition functions and dissociation equilibrium constants for HeH$^+$ have been computed for temperatures between 2,000 and 16,000K in steps of 100K.

022.075 Measurement of charged secondary particle emission under ion and neutral impact.
F. G. Rüdenauer, W. Steiger.
ESA Spec. Publ., ESA SP–224, p. 1 – 10 (1984). – See Abstr. 012.052.
Secondary electron and ion emission yields and energy distributions from Giotto outer surface materials under ion and neutral impact at Giotto–flyby velocities are determined. Bombarding species were N_2^+, Ar^+, Sn^+, Xe^+ (ions) and Ar^0, Xe^0 (neutrals). Sample materials were Al alloy 6061T6 and PCB7 white paint. These experimental data are needed to simulate spacecraft charging during the Halley encounter.

022.076 Contactless determination of the conductivity of the white paint PCB–Z.
H. Arends, R. Schmidt.
ESA Spec. Publ., ESA SP–224, p. 11 – 14 (1984). – See Abstr. 012.052.
The newly developed white conductive paint PCB–Z will be used onboard the spacecraft Giotto, which encounters Halley's comet in 1986. During the encounter the painted part of the dust shield is exposed to the impinging dense flux of molecules and dust. This will give rise to emission of secondary electrons and ions. The knowledge of the conductivity properties of this paint is important to model the general spacecraft/plasma interaction during the fly–by. This paper describes a method to measure the conductivity without mechanical contacts and in vacuo.

022.077 Measurements of integral yields of charged secondary particles using neutral beams simulating a cometary fly–by.
R. Schmidt, H. Arends.
ESA Spec. Publ., ESA SP–224, p. 15 – 19 (1984). – See Abstr. 012.052.
A spacecraft crossing the coma of a comet is exposed to the bombardment of cometary dust and gas particles. In the case of the European cometary probe Giotto, gas molecules and atoms will impinge with a ram energy of 24 eV per amu. Erosion and emission of secondary particles will take place. Secondary electrons and ions will contribute to the electrical charging up of the spacecraft. In order to provide input parameters for model calculations on the general plasma–spacecraft interaction, this work was performed to obtain integral yields for secondary electrons and ions. Different atomic and molecular beams were used to bombard surfaces like gold, aluminum and the newly developed white conductive paint PCB–Z.

022.078 Giotto residual ionization.
F. Arnaudeau, A. de Rouvray, G. Winkelmuller.
ESA Spec. Publ., ESA SP–224, p. 21 – 37 (1984). – See Abstr. 012.052.
The impact of a dust particle at 68.7 km/s on the Giotto aluminum front shield produces a dense plasma with multiply ionized atoms from dust and shield materials. The residual fractional ionization of the plasma resulting from the impact of several foamed silica dust particles of different mass is investigated. The hydrodynamics of the impact is solved by an Arbitrary Lagrangian Eulerian technique (ALE) and an explicit time integration scheme with the ESI finite element code EFHYD–3D.

022.079 Impact ionization from gold, aluminum and PCB–Z.
E. Grün.
ESA Spec. Publ., ESA SP–224, p. 39 – 41 (1984). – See Abstr.
012.052.
Measurements of the impact ionization yield were performed
at Heidelberg dust accelerator for 3 different types of material,
which are representative for the front surface of the Giotto spacecraft: gold, bare aluminum and PCB–Z white paint.

**022.080 Experimental investigations on ion emission with dust
impact on solid surfaces.**
F. R. Krüger, J. Kissel.
ESA Spec. Publ., ESA SP–224, p. 43 – 48 (1984). – See Abstr.
012.052.
Ion types, energy, and angular distributions of ions produced
in dust particle impact are reported and semiempirical yield formulae are given and compared with calculations of other authors.
The data are discussed especially with $v = 69$ km/sec and expected cometary compositions of the dust particles, and the
shield and target materials as used on the Giotto spacecraft.
Some problems for exposed experimental parts, related to neutral
gas impact are also quantitatively discussed.

**022.081 Ion emission from solid surfaces: comparison of dust
impact with other excitations.**
F. R. Krüger.
ESA Spec. Publ., ESA SP–224, p. 49 – 54 (1984). – See Abstr.
012.052.
Cometary dust particles impinging on a spacecraft shield produce ions originating from these particles and, mainly, from the
material of the shield. By means of simulation of dust impact
using the Heidelberg dust accelerator facility the excitation functions have been studied. The results are compared with those of
ionization by short laser pulses applied to solid material, and
some other methods. Some consequences for charging up of
spacecraft in a dust coma are discussed.

**022.082 Electrostatic charging of the Giotto spacecraft due to
neutral gas impact.**
H. Maaßberg.
ESA Spec. Publ., ESA SP–224, p. 55 – 58 (1984). – See Abstr.
012.052.
The stationary charging of Giotto spacecraft is mainly determined by secondary plasma emission generated by high energy
impact of neutral gas onto the front shield. The influence of high
energy tail in the distribution functions of emitted particles on
floating potential and on potential barrier is investigated by
means of a spherically symmetric model.

**022.083 A model of plasma cloud expansion generated by a dust
particle impact.**
H. Maaßberg.
ESA Spec. Publ., ESA SP–224, p. 59 – 63 (1984). – See Abstr.
012.052.
A dust particle impacting on the front sheet of Giotto probe
forms a plasma cloud which expands with more than half the
impact velocity. In case of discrete impacts (not too close to the
nucleus of the comet) the expansion of the dense plasma cloud is
described independently of plasma environment. Electrons which
are released and reach the probe cause a probe potential pulse
lasting up to 10^{-3}s.

**022.084 Giotto plasmasheath and wake charging near Halley's
comet: revised parameters.**
L. W. Parker.
ESA Spec. Publ., ESA SP–224, p. 65 – 69 (1984). – See Abstr.
012.052.
In this continuation of previous work, the author considers
further the Giotto spacecraft charging and electrostatic fields
caused by impact–generated plasma emitted from the Giotto
bumper shield, in the very near vicinity of Halley's comet. In the
previous work, the assumed spacecraft locations were approximately 20,000 km and 10,000 km from the comet. In the present
work, the spacecraft is much closer.

**022.085 Simulations of satellite–plasma interaction, including
electron space charges.**
H. Thiemann.
ESA Spec. Publ., ESA SP–224, p. 71 – 79 (1984). – See Abstr.
012.052.
The impact of dust grains and neutral molecules on the topside
of the Giotto spacecraft results in the generation of secondary
ions and electrons forming a plasma cloud around the body. The
surface potential is controlled by the properties of the secondary
plasma. With a two dimensional particle–in–cell code the author
simulated the time history of this situation by using realistic
electron and ion fluxes from the upper surface of a cylindrical
body. The calculations show the formation of an electron cloud
around the body whereas ions distribute in a wake–like structure.
The surface potential depends on the fluxes and kinetic energy of
the particles. It typically ranges up to several tens of volts.

**022.086 Tabulated optical properties of graphite and silicate
grains.**
B. T. Draine.
Astrophys. J., Suppl. Ser., Vol. 57, No. 3, p. 587 – 594 (1985).
Complex dielectric functions are tabulated for graphite and
"astronomical silicate" for wavelengths λ between 2000 μm and
200 Å, together with Q_{abs}, albedo, and «cos θ» for $a = 0.01$ μm
and $a = 0.1$ μm spheres.

**022.087 Accurate experimental lifetimes of excited levels in
Nd II.**
L. Ward, O. Vogel, A. Arnesen, R. Hallin, A. Wännström.
Phys. Scr., Vol. 31, No. 3, p. 161 – 165 (1985).
Radiative lifetimes of 24 levels in Nd II have been measured
with a laser–ion beam technique using intracavity excitation.
Several levels were found to have a different lifetime than earlier
reported. The measured lifetimes were used to correct old
gf–values deduced from arc spectra. These new gf–values together with 17 equivalent widths, obtained from high resolution
tracings of the solar disc center spectrum, were used in a solar
photosphere abundance analysis which confirmed the authors'
old Nd abundance value.

022.088 On the $S_0(0)$ and $S_1(0)$ spectra of the H_2–H_2 dimer.
G. Danby, D. R. Flower.
J. Phys. B, Vol. 17, No. 24, p. L867 – 870 (1984). Abstr. in Phys.
Abstr., Vol. 88, No. 1253, Entry 31531 (1985).

**022.089 Photodissociation rates of OH, OD, and CN by the
interstellar radiation field.**
J. B. Nee, L. C. Lee.
Astrophys. J., Vol. 291, No. 1, p. 202 – 206 (1985).
The photoabsorption cross sections for OH, OD, and CN in
the vacuum ultraviolet region are measured. The cross sections
for the hydroxyl radicals are of the order of 10^{-17}cm^2, but the
photoabsorption for CN is so low that only an upper limit of
2×10^{-18}cm^2 is obtained. The molecular photodissociative processes are discussed. The photodissociation cross sections are
inferred from the photoabsorption cross sections. On the basis of
the measured data, the photodissociation rates by the interstellar
radiation field are computed and discussed.

**022.090 Experimental phosphorus and sulfur stark widths and
systematic broadening trends for third–row ions.**
M. H. Miller, D. Abadie, A. Lesage.
Astrophys. J., Vol. 291, No. 1, p. 219 – 225 (1985).
Stark broadening parameters for the prominent visible
phosphorus and sulfur lines are measured at
$5 < N_e < 15 \times 10^{16}$cm^{-3} and $10,000 < T < 12,500$K in a well–
calibrated spectroscopic shock tube. Results for 11 of 12 S II lines
and for the two leading visible S I blended multiplets agree within
15% – 30% tolerances with other experiments and with theory.
For the P II and for two of the three P I lines, no prior published
experimental broadening data are available, but results agree
within 15% – 30% tolerances with impact broadening predictions. Systematic broadening trends in the third–row ion sequence Al II, Si II, P II, S II, Cl II, Ar II are examined.

022.091 An update of and suggested increase in calculated radiative association rate coefficients.
E. Herbst.
Astrophys. J., Vol. 291, No. 1, p. 226 – 229 (1985).

A sizable number of radiative association reaction rate coefficients involving polyatomic molecular ions have now been calculated and utilized in ion–molecule models of the chemistry of dense interstellar clouds. Recent evidence suggests that these calculated rate coefficients may be an order of magnitude too low in many cases. The evidence is reviewed and developed and calculated rate coefficients are updated.

022.092 Laboratory measurement of the $S(9)$ pure rotation frequency in H_2.
D. E. Jennings, L. A. Rahn, A. Owyoung.
Astrophys. J., Lett. Ed., Vol. 291, No. 1, p. L15 – L18 (1985).

The frequency of the $\upsilon = 0{\to}0$, $J = 11{\to}9$ transition in molecular hydrogen has been measured in the laboratory using stimulated Raman spectroscopy of H_2 and CO in a methane–air flame. This is the first direct laboratory study of any of the high-J transitions in H_2 which have been observed in the Orion molecular cloud. The laboratory frequency for $S(9)$ is 2130.102(4) cm^{-1}. An improved set of ground–state parameters for H_2 is derived.

022.093 Direct measurement of the fundamental rotational transitions of the OH radical by laser sideband spectroscopy.
J. Farhoomand, G. A. Blake, H. M. Pickett.
Astrophys. J., Lett. Ed., Vol. 291, No. 1, p. L19 – L22 (1985).

The authors report for the first time the direct (zero–field) spectra of the fundamental rotational transitions of the OH radical in its $\Omega = 3/2$ and $1/2$ states at 2509.9 and 1834.7 GHz using a recently developed far–infrared laser sideband spectrometer. These measurements have verified and refined the predictions of previous laser magnetic resonance work, thereby confirming the far–infrared detection of interstellar OH.

022.094 Millimeter wave spectrum of HCS.
M. Bogey, C. Demuynck, J. L. Destombes,
B. Lemoine.
J. Mol. Spectrosc., Vol. 107, No. 2, p. 417 – 418 (1984). Abstr. in Phys. Abstr., Vol. 88, No. 1252, Entry 25963 (1985).

022.095 Electron–ion collisions in high temperature plasmas.
A. Chutjian.
High Temp. Sci., Vol. 17, p. 135 – 153 (1984). Abstr. in Phys. Abstr., Vol. 88, No. 1252, Entry 26921 (1985).

022.096 Determination of the products state in dissociative recombination of molecular ions: N_2^+ and O_2^+ ions.
F. Vallee, J. C. Gomet, M. Morlais, B. R. Rowe,
J. L. Queffelec.
Seventh European Sectional Conference on the Atomic and Molecular Physics of Ionized Gases, p. 79 – 80 (1984). Abstr. in Phys. Abstr., Vol. 88, No. 1252, Entry 26924 (1985). – See Abstr. 012.056.

022.097 Synthesis and sputtering of newly formed molecules by kiloelectronvolt ions.
A. E. de Vries, R. A. Haring, A. Haring, F. S. Klein,
A. C. Kummel, F. W. Saris.
J. Phys. Chem., Vol. 88, No. 20, p. 4510 – 4512 (1984). Abstr. in Phys. Abstr., Vol. 88, No. 1252, Entry 28240 (1985).

022.098 The 1^1S–$n^1P/1^1S$–2^1P emission–line ratios in O VII as temperature diagnostics for solar flares and active regions.
F. P. Keenan, A. E. Kingston, D. L. McKenzie.
Astrophys. J., Vol. 291, No. 2, p. 855 – 857 (1985).

Recent R–matrix calculations of O VII electron excitation rates by Tayal and Kingston are used to determine the theoretical emission–line ratios $R_1 = I(1s^2\,{}^1S{-}1s3p\,{}^1P)/I(1s^2\,{}^1S{-}1s2p\,{}^1P)$ and $R_2 = I(1s^2\,{}^1S{-}1s4p\,{}^1P)/I(1s^2\,{}^1S{-}1s2p\,{}^1P)$. These ratios are found to vary by factors of 3.3 and 4.9, respectively, between $T_e = 6 \times 10^5$ and 2×10^6K. Electron temperatures derived using observed values of R_1 and R_2 from P78–1 satellite spectra of solar flares and active regions are in good agreement.

022.099 Rotational excitation of HCO$^+$ by collisions with H_2.
T. S. Monteiro.
Mon. Not. R. Astron. Soc., Vol. 214, No. 3, p. 419 – 427 (1985).

An ab initio potential surface was obtained for the HCO$^+$–H_2 interaction, making use of C I for part of the surface. This was then used to compute collisional cross–sections and rate coefficients for a range of temperatures (0 – 30K) and between all rotational levels up to and including $j = 4$. Only collisions with para–H_2 were studied here. The effects of different sets of rate coefficients on predicted spectral line intensities were examined.

022.100 Population ratios for the fine structure ground state of Si II applicable to the interstellar medium.
F. P. Keenan, C. T. Johnson, A. E. Kingston, P. L. Dufton.
Mon. Not. R. Astron. Soc., Vol. 214, No. 3, p. 37P – 40P (1985).

Using recent R–matrix calculations of electron excitation rates for the $3s^23p^2P_{1/2}$–$3s^23p^2P_{3/2}$ fine structure transition in Si II, the electron density sensitive population ratio $n({}^2P_{3/2})/n({}^2P_{1/2})$ has been derived for the ranges of temperature (100 – 20000K) and hydrogen density (0 – 1000 cm^{-3}) applicable to H I and H II regions. The results differ appreciably from those of Smeding & Pottasch, and lead to electron density estimates approximately 30 to 40 per cent larger.

022.101 The case for interstellar micro–organisms.
F. Hoyle, N. C. Wickramasinghe, S. Al–Mufti.
Astrophys. Space Sci., Vol. 110, No. 2, p. 401 – 404 (1985).

Recent arguments claiming to disprove the existence of bacteria in interstellar space are shown to be without merit.

022.102 The infrared and ultraviolet absorptions of micro–organisms and their relation to the Hoyle–Wickramasinghe hypothesis.
S. Yabushita, K. Wada.
Astrophys. Space Sci., Vol. 110, No. 2, p. 405 – 411 (1985).

Absorption of yeast and *E. coli* in the infrared and ultraviolet regions and that of diatomaceous soil in the infrared region have been obtained. Electron microscope photographs of aggregates of *E. coli* have also been obtained. These results are discussed in relation to the Hoyle–Wickramasinghe hypothesis regarding the nature of interstellar grains.

022.103 The ultraviolet absorbance of presumably interstellar bacteria and related matters.
F. Hoyle, N. C. Wickramasinghe, S. Al–Mufti.
Astrophys. Space Sci., Vol. 111, No. 1, p. 65 – 78 (1985).

It is shown that the well–known 2200 Å peak in the extinction of starlight is explained by microorganisms. A mixed culture of diatoms and bacteria, which previously the authors found to give excellent fits to astronomical data in the infrared, has a peak absorption slightly shortward of 2200 Å, in very close agreement with the absorptions found in directions towards most early–type stars.

022.104 Calculation of pressure–broadened linewidths for CO in Ar.
S. Green.
J. Quant. Spectrosc. Radiat. Transfer, Vol. 33, No. 4, p. 299 – 305 (1985).

022.105 Ni I oscillator strengths.
A. Doerr, M. Kock.
J. Quant. Spectrosc. Radiat. Transfer, Vol. 33, No. 4, p. 307 – 318 (1985).

Oscillator strengths of 150 Ni I lines in the spectral range 2800 – 6200 Å have been obtained by emission and from hook measurements. Relative sets of f–values were determined by combining emission measurements on a hollow cathode with hook measurements in a high–temperature furnace. No assumption concerning the plasma state is used, and no temperature determination is required. The relative measurements have been placed on an absolute scale by using lifetime data. The uncertainty of the f-values is 13% on average. Comparisons are made with the results of other authors.

022.106 Intensity measurements and self–broadening coefficients in the γ band of O_2 at 628 nm using intracavity laser–absorption spectroscopy (ICLAS).
M. A. Mélières, M. Chenevier, F. Stoeckel.
J. Quant. Spectrosc. Radiat. Transfer, Vol. 33, No. 4, p. 337 – 345 (1985).

022.107 Experimental determination of the temperature dependence of nitrogen–broadened line widths in the $1 \leftarrow 0$ band of HCl.
J. Ballard, W. B. Johnston, P. H. Moffat,
D. T. Llewellyn–Jones.
J. Quant. Spectrosc. Radiat. Transfer, Vol. 33, No. 4, p. 365 – 371 (1985).

022.108 Diode laser measurements of the band strengths of v_3 and v_6 in $^{12}CH_3D$.
R. J. Boyle, G. W. Halsey, D. E. Jennings.
J. Quant. Spectrosc. Radiat. Transfer, Vol. 33, No. 4, p. 411 – 414 (1985).
A diode laser spectrometer has been used to measure line strengths for 143 transitions in the v_6 fundamental band of $^{12}CH_3D$ near 9 µm. These line–strength measurements have been used to derive a band strength for v_6 and v_3. The band strength derived for v_6 is 61.7 ± 1.8 cm^{-2}atm^{-1}, and that for v_3 is 49.3 ± 1.4 cm^{-2}atm^{-1} at 295K.

022.109 Spectral emissivity of the 4.3–µm CO_2 band at high temperature.
A. Coppalle, P. Vervisch.
J. Quant. Spectrosc. Radiat. Transfer, Vol. 33, No. 5, p. 465 – 473 (1985).
The spectral absorptivity of the 4.3–µm CO_2 band was measured at 2900K. A hot CO_2 sample was formed in an oxygen–methane flame. The experimental results are in good agreement with theoretical calculations in the weak– and strong–line limits. However, the statistical model overpredicts the absorptivity at intermediate optical depths.

022.110 Correlation effects in the first–row transition metal atoms and ions: choosing the configurations.
R. Glass.
J. Quant. Spectrosc. Radiat. Transfer, Vol. 33, No. 5, p. 481 – 485 (1985).
A detailed study of core–valence and core–correlation effects using the configuration interaction approach soon becomes unwieldly due to the large number of configurations needed in the expansion. The author examines the possibility of reducing the number of vector–coupled configurational functions in the configuration interaction expansion, formed from different coupling schemes, by coupling the orbital functions in a different order.

022.111 Infrared radiation properties of methane at elevated temperatures.
M. A. Brosmer, C. L. Tien.
J. Quant. Spectrosc. Radiat. Transfer, Vol. 33, No. 5, p. 521 – 532 (1985).
The spectral absorptivity of the v_3 and v_4 fundamental and the $v_1 + v_4$ combination bands of methane have been measured at low resolution for temperatures between 290 and 850K. Spectral mean (narrow–band) parameters for the fundamental bands have been determined from the Elsasser model, while total band absorptance data for all three bands have been correlated using the Edwards exponential wide band model. A total emissivity chart has been developed, based on the wide band absorption models.

022.112 Analyses of experimental observations of electron temperatures in the near wake of a model in a laboratory–simulated solar wind plasma.
D. S. Intriligator, G. R. Steele.
J. Geophys. Res., Vol. 90, No. A5, p. 4027 – 4034 (1985).
Laboratory experiments have been performed that show the effect on the electron temperature of inserting a spherical conducting model, larger than the Debye length, into a free–streaming high–energy (1kv) unmagnetized hydrogen plasma. These experiments are the first electron temperature experiments conducted at energies and compositions directly relevant to solar wind and astrophysical plasma phenomena. The authors discuss their results in the more general context of theoretical studies and of other relevant considerations. The authors also discuss how their findings may be relevant to the maintenance of the nightside ionosphere of Venus and they suggest some specific spacecraft observations that should be carried out using the Pioneer Venus orbiter.

022.113 The role of stimulated processes in cosmic annihilation–line sources. II.
V. V. Zheleznyakov, A. A. Litvinchuk.
Sov. Astron., Vol. 28, No. 5, p. 503 – 510 (1984). English translation of 38.022.029.

022.114 The microwave and far–infrared spectra of the SiH radical.
J. M. Brown, R. F. Curl, K. M. Evenson.
Astrophys. J., Vol. 292, No. 1, p. 188 – 191 (1985).
The frequencies, wavelengths, and line strengths for transitions in the SiH molecule at microwave and far–infrared wavelengths have been calculated from an analysis of its laser magnetic resonance spectrum.

022.115 Total synthesis of interstellar chemical compounds by high energy molecular beam bombardment on pure graphite.
F. M. Devienne, M. Teisseire.
Astron. Astrophys., Vol. 147, No. 1, p. 54 – 60 (1985).
The objective of this paper is to show a possibility of forming interstellar molecules detected in the interstellar space by bombarding a carbon target or graphite grains with high energy neutrals. The authors have bombarded pure graphite in the ultra–vacuum with high energy molecular beams (from 2 to 10 keV) obtained by charge exchange from ion beams of hydrogen, oxygen, or nitrogen. They have observed many organic compounds: binary compounds like hydrogen carbides, ternary compounds containing carbon, nitrogen, oxygen or hydrogen, and finally, quaternary compounds. They also have obtained cyanopolyynes and organic molecules which had previously been observed in the interstellar space. So far, they have identified thirty–two compounds corresponding to molecules observed in the interstellar space and about forty containing only carbon, hydrogen, nitrogen and oxygen.

022.116 CRESU study of the reaction $N^+ + H_2 \rightarrow NH^+ + H$ between 8 and 70K and interstellar chemistry implications.
J. B. Marquette, B. R. Rowe, G. Dupeyrat, E. Roueff.
Astron. Astrophys., Vol. 147, No. 1, p. 115 – 120 (1985).
The reaction $N^+(^3P) + H_2(J = 0,1) \rightarrow NH^+ + H(1)$ has been studied using a newly developed supersonic jet apparatus (CRESU) in the temperature range 8 – 70K. The rate coefficient k_1 decreases dramatically with decreasing temperature following an exponential law. While an activation energy cannot be excluded at the present time, this behavior is probably due to a very slight endothermicity. Implications for the formation of ammonia in interstellar clouds are discussed.

022.117 An updated evaluation of recombination and ionization rates.
M. Arnaud, R. Rothenflug.
Astron. Astrophys., Suppl. Ser., Vol. 60, No. 3, p. 425 – 457 (1985).
The authors present a new evaluation of the ionization and recombination rates by electronic collision for astrophysically

abundant elements. The approach to the ionization rates relies on forty–two measured cross sections and on quantum mechanical calculations using the Coulomb–Born or distorted wave with exchange approximations. The authors determine the equilibrium ionization of elements H, He, C, N, O, Ne, Na, Mg, Al, Si, S, Ar, Ca, Fe, Ni and discuss differences with previous calculations.

022.118   Influence of different line broadening mechanisms on the limb–effect within Na I ($4s^2S-np^2P^0$) series.
I. Vince, M. S. Dimitrijević.
Publ. Astron. Opservatorije Beogr., No. 33, p. 15 – 18 (1985). – See Abstr. 012.061.

Line broadening parameters within Na I ($4s^2S-np^2P^0$) series have been calculated as a function of the position angle of the observed point on the sun. The influence of neutral atom impact– and Stark broadening on the limb–effect are analyzed.

022.119   The spectrum of magnesium hydride.
P. F. Bernath, J. H. Black, J. W. Brault.
Prepr. Steward Obs., No. 582, 24 pp. (1985). To appear in Astrophys. J.

022.120   NLTE–Rechnungen zur Bildung interstellarer Moleküllinien in einem turbulenten Medium.
M. A. Albrecht, W. H. Kegel.
Mitt. Astron. Ges., Nr. 63, p. 151 (1985). – See Abstr. 012.063.

022.121   Laboratory detection of the C_3H radical.
J. M. Vrtilek, C. A. Gottlieb, E. W. Gottlieb, P. Thaddeus.
Bull. Am. Astron. Soc., Vol. 16, No. 4, p. 877 (1984). Abstract. – See Abstr. 010.062.

022.122   Polarization and hyperfine splitting of water maser emission.
S. Deguchi, W. D. Watson.
Bull. Am. Astron. Soc., Vol. 16, No. 4, p. 877 – 878 (1984). Abstract. – See Abstr. 010.062.

022.123   Collision strengths and line strengths for $\Delta n = 0$ transitions with n = 2 in boron–like ions.
D. H. Sampson, G. M. Weaver, G. V. Petrou, R. E. H. Clark, S. J. Goett.
Bull. Am. Astron. Soc., Vol. 16, No. 4, p. 946 (1984). Abstract. – See Abstr. 010.062.

022.124   Simulation of cosmic dust spectra.
J. Hecht, R. Russell, P. Grieve.
Bull. Am. Astron. Soc., Vol. 16, No. 4, p. 1002 (1984). Abstract. – See Abstr. 010.062.

022.125   Infrared spectra of gaseous mononitriles: application to the atmosphere of Titan.
F. Cerceau, F. Raulin, R. Courtin, D. Gautier.
Icarus, Vol. 62, No. 2, p. 207 – 220 (1985).

A list of volatile nitriles, not yet detected in the atmosphere of Titan, but likely to be present in this environment, has been selected: acetonitrile, propionitrile, acrylonitrile, crotononitrile, allyl cyanide, methacrylonitrile, and cyanopropyne. The spectra of these compounds in the gas phase have been systematically studied, in the mid– and far–infrared ranges. Then, in order to estimate the detectability of the selected compounds by infrared spectroscopy in the atmosphere of Titan, the data obtained have been extrapolated to the case of Titan.

022.126   Impact and explosion crater ejecta, fragment size, and velocity.
J. D. O'Keefe, T. J. Ahrens.
Icarus, Vol. 62, No. 2, p. 328 – 338 (1985).

The authors have developed models for the distribution of fragments that are ejected at a given velocity for both impact and explosion cratering. The results from these models have application to the physics of planetary accretion and the origin of meteorites.

022.127   Impact cratering mechanics: relationship between the shock wave and excavation flow.
H. J. Melosh.
Icarus, Vol. 62, No. 2, p. 339 – 343 (1985).

This paper describes the relationship between the shock wave produced by an impact and the excavation flow that opens the crater. The excavation flow velocity is shown to be a nearly constant fraction of the peak particle velocity in the wave. The existence of an excavation flow is due to thermodynamically irreversible processes in the shock.

022.128   The rotational spectra of $HOCO^+$, $HOCS^+$, $HSCO^+$, and $HSCS^+$.
P. R. Taylor, M. Scarlett.
Astrophys. J., Lett. Ed., Vol. 293, No. 1, p. L49 – L51 (1985).

Ab initio molecular electronic structure calculations have been used to predict the equilibrium geometries of protonated CO_2, COS, and CS_2. By correcting for the known inadequacies in calculations of bond lengths and angles, rotational frequencies of considerably higher accuracy than those from the raw ab initio results have been obtained. These frequencies should assist in attempts to identify these species in the laboratory and in tentative assignments of interstellar lines.

022.129   Pure rotational transitions of H_2O molecules in the 8 – 14 μm atmospheric window.
J. Hinderling, M. W. Sigrist, F. K. Kneubühl.
Infrared Phys., Vol. 25, No. 1/2, p. 491 – 496 (1985). – See Abstr. 012.065.

The authors report on the temperature dependence of the water–vapor line absorption near four different CO_2–laser wavelengths at temperatures between 280 and 305K.

022.130   Laboratory measurement on impact ionization by neutrals and floating potential of a spacecraft during encounter with Halley's comet.
R. Schmidt, H. Arends.
Planet. Space Sci., Vol. 33, No. 6, p. 667 – 673 (1985).

A spacecraft penetrating into the dense cloud of ambient gas and dust particles in the coma of Halley's comet, is exposed to a bombardment by these particles having a high kinetic energy due to the large velocity of the spacecraft relative to the cometary coma. The interaction of the spacecraft and cometary neutral particles was simulated by using neutral beams of different species directed towards various target materials such as aluminium, gold and the white conductive paint PCB–Z. Upon impact on the surface of the target, emission of charged as well as neutral secondary particles was initiated. The yields of the charged particles were derived from measurements of the electrical current produced by secondary ions or electrons.

022.131   Role of the gaunt factor in the derivation of dielectronic recombination coefficient.
B. Alam, S. M. R. Ansari.
Sol. Phys., Vol. 96, No. 2, p. 219 – 227 (1985).

A new formula for the coefficient of dielectronic recombination has been derived by substituting excitation cross–section extrapolated below threshold for capture cross–section of Burgess (1964). In this case the excitation cross–section modified by Mewe (1972) through a fitted gaunt factor has been used. The dielectronic recombination coefficients calculated by the new formula are found to be about an order of magnitude smaller than those obtained by Burgess simplified formula. The results for coronal ions of silicon and iron are presented here as our sample calculations.

022.132   Parametric excitation of density waves in circular streams of charged particles (model of the Saturn rings).
P. V. Bliokh, V. V. Yaroshenko.
Izv. Vyssh. Uchebn. Zaved., Radiofiz., Tom 27, No. 11, p. 1471 – 1474 (1984). In Russian. Abstr. in Ref. Zh., 62. Issled. Kosm. Prostranstva, 6.62.326 (1985).

022.133 Approximate and exact photon–Maxwellian electron cross–sections and a Monte Carlo sampling scheme.
B. R. Wienke, J. S. Hendricks, T. E. Booth.
J. Quant. Spectrosc. Radiat. Transfer, Vol. 33, No. 6, p. 555 – 574 (1985).
Fast and accurate approximate methods for computing temperature–corrected photon–Maxwellian electron cross–sections, using distribution averaged electron energies and scattering angles, are presented.

022.134 Relative transition probability measurements in the A–X and B–X systems of CH.
N. L. Garland, D. R. Crosley.
J. Quant. Spectrosc. Radiat. Transfer, Vol. 33, No. 6, p. 591 – 595 (1985).
Relative Einstein emission coefficients have been measured for transitions from the $v' = 0$ and 1 levels in the $A^2\Delta$–$X^2\Pi$ and $B^2\Sigma^-$–$X^2\Pi$ systems of the CH radical. The measurements were made in an atmospheric pressure methane–air flame, and the results are compared with theoretical calculations.

022.135 Integrated infrared intensities in NH_3.
K. Kim.
J. Quant. Spectrosc. Radiat. Transfer, Vol. 33, No. 6, p. 611 – 614 (1985).
The integrated infrared intensities of the fundamental modes of NH_3 have been measured by the Wilson–Wells–Penner–Weber method. The intensities were found to be 43.9 ± 0.6, 567.6 ± 9.4 and 110.5 ± 2.0 atm^{-1}cm^{-2} (standard temperature and pressure) for the $v_1 + v_3$, v_2 and v_4 bands, respectively.

022.136 Variation of the ion dynamics parameter in Stark–broadened helium lines.
H. Richter, A. Piel.
J. Quant. Spectrosc. Radiat. Transfer, Vol. 33, No. 6, p. 615 – 626 (1985).
Stark broadening of quasidegenerate He lines at $\lambda = 447$ nm and 492 nm has been investigated at low electron densities ($N_e = 10^{21}$ and 3×10^{21}m^{-3}). The perturber mass was varied by using H$^+$, He$^+$ and Ar$^+$ ions. Variation of the parameter $(T_0/\mu)^{1/2}$ by a factor of 3 is accompanied by marked ion–dynamical effects in the forbidden component. The results are compared with calculations according to the unified theory and the model microfield method. Characteristic deviations are observed and critically discussed.

022.137 Decomposition of the photoabsorption continuum underlying the Schumann–Runge bands of $^{16}O_2$. I. Role of the $B^3\Sigma_u^-$ state: a new dissociation limit.
B. R. Lewis, L. Berzins, J. H. Carver, S. T. Gibson.
J. Quant. Spectrosc. Radiat. Transfer, Vol. 33, No. 6, p. 627 – 643 (1985).
Extensive measurements are presented of O_2 photoabsorption cross–sections taken at selected minima between rotational lines of the Schumann–Runge band system. Both room temperature and liquid nitrogen temperature results are presented from 1750 to 1800 Å and corrections are applied for the effect of the wings of the rotational lines. Step–like structure in the underlying continuum, due to absorption from rotationally excited levels of the ground state into the $B^3\Sigma_u^-$ state, is verified experimentally for the first time, and the observation is used to deduce a definitive dissociation limit of 57136.0 ± 0.5 cm^{-1} for the B state.

022.138 The splitting of the $2s^2 2p^3 \; ^2P$ term in O II.
M. M. De Robertis, D. E. Osterbrock, C. F. McKee.
Astrophys. J., Vol. 293, No. 2, p. 459 – 462 (1985). = Lick Obs. Bull., No. 1002.
The authors have measured the O II $2p^3 \; ^2P_{1/2}$–$^2P_{3/2}$ and $^2D_{3/2}$–$^2D_{5/2}$ splitting in high–dispersion (~ 3 km s^{-1} resolution), long slit CCD spectra of the bright planetary nebula NGC 7027. Techniques used to deblend the $^2D_{5/2}$–$^2P_{3/2,1/2}$, $^2D_{3/2}$–$^2P_{3/2,1/2}$ pairs of lines are described and applied to the data. Though accurate absolute wavelengths could not be obtained, accurate separations of the ^2P, ^2D levels were determined to be

2.00 ± 0.03 cm^{-1} and 20.11 ± 0.07 cm^{-1}, respectively. Improved–energy level values and improved wavelengths for [O II] are listed.

022.139 The triplet–singlet $\tilde{a}^3A''$–$\tilde{X}^1A'$ emission spectra of hydroxybenzaldehydes and a comparison with CO$^+$ comet tail spectra.
R. K. Baruah, R. P. Dewri, G. D. Baruah.
Indian J. Phys., Part B, Vol. 58B, No. 4 – 5, p. 252 – 261 (1984). Abstr. in Phys. Abstr., Vol. 88, No. 1253, Entry 31544 (1985). – See Abstr. 012.074.

022.140 Reaction of O$^+$, CO$^+$, H$^+$, and CH$_5$$^+$ ions with atomic hydrogen.
W. Federer, H. Villinger, H. Ramler, W. Lindinger.
Seventh European Sectional Conference on the Atomic and Molecular Physics of Ionized Gases, p. 5A (1984). Abstr. in Phys. Abstr., Vol. 88, No. 1253, Entry 34631 (1985). – See Abstr. 012.056.

022.141 ^{205}Pb: accelerator mass spectrometry of a very heavy radioisotope and the solar neutrino problem.
H. Ernst, G. Korschinek, P. Kubik, W. Mayer, H. Morinaga, E. Nolte, U. Ratzinger, W. Henning, W. Kutschera, M. Müller, D. Schull.
Nucl. Instrum. Methods Phys. Res., Sect. B, Vol. 233, No. 2, p. 426 – 429 (1984). Abstr. in Phys. Abstr., Vol. 88, No. 1253, Entry 35551 (1985). – See Abstr. 012.077.

022.142 Production of ^{7}Be, ^{22}Na, ^{24}Na and ^{10}Be from Al in a 4π–irradiated meteorite model.
P. Englert, S. Theis, R. Michel, C. Tuniz, R. K. Moniot, S. Vajda, T. H. Kruse, D. K. Pal, G. F. Herzog.
Nucl. Instrum. Methods Phys. Res., Sect. B, Vol. 233, No. 2, p. 415 – 419 (1984). Abstr. in Phys. Abstr., Vol. 88, No. 1253, Entry 35557 (1985). – See Abstr. 012.077.

022.143 Collision strengths for optically allowed transitions in Ne V and Mg VII.
K. M. Aggarwal.
Astrophys. J., Suppl. Ser., Vol. 58, No. 2, p. 289 – 296 (1985).
Collision strengths for the optically allowed transitions have been calculated among the $1s^2 2s^2 2p^2$, $1s^2 2s 2p^3$, and $1s^2 2p^4$ configurations of Ne V and Mg VII. The R–matrix and Coulomb–Born methods are used to calculate the contribution of partial waves with angular momentum $L \leqslant 9$ and $L > 9$ respectively.

022.144 The structure of linear SiCC: an ab initio SCF CI study including vibrational effects.
F. Pauzat, Y. Ellinger.
Chem. Phys. Lett., Vol. 112, No. 6, p. 519 – 523 (1984). Abstr. in Phys. Abstr., Vol. 88, No. 1254, Entry 36935 (1985).

022.145 Production of long–lived cosmogenic nuclei and their applications.
G. M. Raisbeck, F. Yiou.
Nucl. Instrum. Methods Phys. Res., Sect. B, Vol. 233, No. 2, p. 91 – 99 (1984). Abstr. in Phys. Abstr., Vol. 88, No. 1254, Entry 40377 (1985). – See Abstr. 012.077.

022.146 Accelerator mass spectrometry at the Rehovot Pelletron tandem: measurements of abundances of cosmogenic radioisotopes and future prospects.
D. Fink, O. Meirav, M. Paul, H. Ernst, W. Henning, W. Kutschera, R. Kaim, A. Kaufman, M. Magaritz.
Nucl. Instrum. Methods Phys. Res., Sect. B, Vol. 233, No. 2, p. 123 – 128 (1984). Abstr. in Phys. Abstr., Vol. 88, No. 1254, Entry 40378 (1985). – See Abstr. 012.077.

022.147 Bragg reflection of cosmic neutrinos.
P. F. Smith, J. D. Lewin.
Acta Phys. Pol., Ser. B, Vol. B15, No. 12, p. 1201 – 1214 (1984). Abstr. in Phys. Abstr., Vol. 88, No. 1254, Entry 40456 (1985).

022.148   Electron impact measurement of oscillator strengths for dipole–allowed transitions of atomic oxygen.
J. P. Doering, E. E. Gulcicek, S. O. Vaughan.
J. Geophys. Res., Vol. 90, No. A6, p. 5279 – 5284 (1985).

Optical oscillator strengths for the seven most intense dipole–allowed transitions in the 100–eV incident energy electron scattering spectrum of atomic oxygen have been measured from forward scattering spectra.

022.149   The role of intersystem collisional transfer of excitation in the determination of N_2 vibronic level populations. Application to $B'^3\Sigma_u–B^3\pi_g$ band intensity measurements.
W. Benesch, D. Fraedrich.
J. Chem. Phys., Vol. 81, No. 12, Part 1, p. 5367 – 5374 (1984). Abstr. in Phys. Abstr., Vol. 88, No. 1255, Entry 42326 (1985).

022.150   Analysis of tracks in the stacked film track detector.
M. Hosoe, H. Hasegawa.
Nucl. Instrum. Methods Phys. Res., Sect. A, Vol. 227, No. 3, p. 561 – 564 (1984). Abstr. in Phys. Abstr., Vol. 88, No. 1255, Entry 45926 (1985).

022.151   The millimetre wave spectrum of the $^{13}C^{14}N$ radical in its ground state.
M. Bogey, C. Demuynck, J. L. Destombes.
Can. J. Phys., Vol. 62, No. 12, p. 1248 – 1253 (1984). Abstr. in Phys. Abstr., Vol. 88, No. 1256, Entry 47434 (1985).

022.152   Rotational spectrum of HD at low pressures.
P. Essenwanger, H. P. Gush.
Can. J. Phys., Vol. 62, No. 12, p. 1680 – 1685 (1984). Abstr. in Phys. Abstr., Vol. 88, No. 1256, Entry 47449 (1985).

022.153   High resolution emission spectrum of H_2 between 78 and 118 nm.
J.–Y. Roncin, F. Launay, M. Larzilliere.
Can. J. Phys., Vol. 62, No. 12, p. 1686 – 1705 (1984). Abstr. in Phys. Abstr., Vol. 88, No. 1256, Entry 47482 (1985).

022.154   Difference frequency laser spectroscopy of the v_1 band of $HOCO^+$.
T. Amano, K. Tanaka.
J. Chem. Phys., Vol. 82, No. 2, p. 1045 – 1046 (1985). Abstr. in Phys. Abstr., Vol. 88, No. 1257, Entry 53453 (1985).

022.155   Integrated intensities and the band strengths of the C–X system of LaO.
U. D. Prahllad, N. S. Bapat, N. Sreedhara Murthy.
Physica B, C, Vol. 128B + C, No. 1, p. 123 – 126 (1985). Abstr. in Phys. Abstr., Vol. 88, No. 1257, Entry 53557 (1985).

022.156   Experimental electron scattering spectrum of atomic oxygen.
J. P. Doering, E. E. Gulcicek, S. O. Vaughan.
Chem. Phys. Lett., Vol. 114, No. 3, p. 334 – 337 (1985). Abstr. in Phys. Abstr., Vol. 88, No. 1258, Entry 59382 (1985).

022.157   Proton–proton total cross–section at $E_{c.m.}$ = 15 to 150 TeV from cosmic–ray data.
J. Linsley.
Lett. Nuovo Cimento, Vol. 42, Ser. 2, No. 8, p. 403 – 410 (1985). Abstr. in Phys. Abstr., Vol. 88, No. 1260, Entry 69640 (1985).

022.158 N_2^+ Meinel band quenching.
L. G. Piper, B. D. Green, W. A. M. Blumberg, S. J. Wolnik.
J. Chem. Phys., Vol. 82, No. 7, p. 3139 – 3145 (1985). Abstr. in Phys. Abstr., Vol. 88, No. 1260, Entry 70612 (1985).

022.159   Total and partial inelastic cross sections of proton–nucleus reactions.
C. H. Tsao, R. Silberberg, J. R. Letaw.
18th International Cosmic Ray Conference, Vol. 2, p. 194 – 197 (1983). – See Abstr. 012.096.

An empirical formula is presented for the total inelastic cross sections of protons on nuclei, including the energy dependence.

022.160   Fragmentation of 710 and 1050 MeV/nuc Fe nuclei in CH_2 and C targets – isotopic cross sections for H targets.
W. R. Webber, D. A. Brautigam, J. C. Kish, D. Schrier.
18th International Cosmic Ray Conference, Vol. 2, p. 198 – 201 (1983). – See Abstr. 012.096.

022.161   Fragmentation of ~ 500 MeV/nuc Ne and O nuclei and CH_2, C and H targets – charge and isotopic cross sections.
W. R. Webber, D. A. Brautigam, J. C. Kish, D. A. Schrier.
18th International Cosmic Ray Conference, Vol. 2, p. 202 – 205 (1983). – See Abstr. 012.096.

The authors have studied the fragmentation of ^{20}Ne and ^{16}O nuclei in CH_2 and C targets at several energies between $\sim 400 – 1000$ MeV/nuc. Charge changing and isotopic cross sections for the fragments $Z \geqslant 4$ are obtained. The hydrogen cross sections are obtained by a subtraction procedure.

022.162   A measurement of the high energy muon spectrum by PAIR METER.
I. Nakamura, T. Kitamura, K. Mitsui, Y. Muraki, Y. Ohashi, A. Okada, T. Suda, Y. Kawashima, T. Aoki, T. Takahashi.
18th International Cosmic Ray Conference, Vol. 7, p. 41 (1983). Abstract. – See Abstr. 012.096.

022.163   How sensitive can underground cosmic ray experiments be in measuring neutrino oscillation parameters?
P. V. Ramana Murthy.
18th International Cosmic Ray Conference, Vol. 7, p. 125 – 128 (1983). – See Abstr. 012.096.

022.164   Electronic detection of ultra–heavy nuclei by pyroelectric materials.
J. A. Simpson, A. J. Tuzzolino.
18th International Cosmic Ray Conference, Vol. 8, p. 59 – 62 (1983). – See Abstr. 012.096.

A recent prediction by the authors that pyroelectric materials may be capable of detecting ultra–heavy nuclei has been confirmed.

022.165   An accelerator test of semi–empirical cross–sections.
K. H. Lau, R. A. Mewaldt, M. E. Wiedenbeck.
18th International Cosmic Ray Conference, Vol. 9, p. 255 – 258 (1983). – See Abstr. 012.096.

The authors compare experimentally measured yields of isotopes of elements from $_{12}Mg$ to $_{19}K$ resulting from the fragmentation of ^{40}Ar with calculated yields based on semi–empirical cross–section formulae.

022.166   Interactions of 200 GeV gold nuclei in light elements.
N. R. Brewster, R. K. Fickle, C. J. Waddington, W. R. Binns, M. H. Israel, M. D. Jones, J. Klarmann, T. L. Garrard, B. J. Newport, E. C. Stone.
18th International Cosmic Ray Conference, Vol. 9, p. 259 – 262 (1983). – See Abstr. 012.096.

The authors infer the total and partial cross–sections for $^{197}_{79}Au$ incident on hydrogen. The effects of using these cross–sections in one model of cosmic ray propagation are illustrated.

022.167   The effect of finite temperature on particle energy losses and implications for track formation in SSNTD.
J. Pérez–Peraza, A. Laville, M. Galvez, M. Balcazar–Garcia.
18th International Cosmic Ray Conference, Vol. 9, p. 399 – 402 (1983). – See Abstr. 012.096.

022.168   Determination of coupling coefficients of neutrons incident at arbitrary angles.
Ya. L. Blokh, F. A. Starkov.
18th International Cosmic Ray Conference, Vol. 10, p. 280 – 282 (1983). – See Abstr. 012.096.

022.169   Coupling functions and meteorological coefficients for the detectors of the Elbrus spectrograph.
E. G. Klepach, Kh. M. Khamirzov, V. G. Yanke.
18th International Cosmic Ray Conference, Vol. 10, p. 283 – 286 (1983). – See Abstr. 012.096.

The Elbrus spectrograph of cosmic rays, which consists of three stations is now additionally equipped at Peak Cheget with muon telescopes of plastic scintillators.

022.170   Multiple muons in DUMAND and their interpretation.
P. K. F. Grieder.
18th International Cosmic Ray Conference, Vol. 11, p. 459 – 461 (1983). – See Abstr. 012.096.

The author has studied the major sources of high energy multiple muons and their signatures in the DUMAND detector by means of simulation calculations. His interest focuses on the feasibility to determine the primary composition from the observation of high energy muons. The author presents a first summary of this work.

022.171   Preliminary results on cosmic muons with NUSEX experiment at the Mt. Blanc Laboratory.
G. Battistoni, E. Bellotti, G. Bologna, P. Campana,
C. Castagnoli, A. Castellina, V. Chiarella, D. C. Cundy,
B. D'Ettorre Piazzoli, E. Fiorini, P. Galcotti, R. Iarocci,
C. Liguori, G. Mannocchi, G. P. Murtas, P. Negri,
G. Nicoletti, P. Picchi, A. Pullia, M. Price, S. Ragazzi,
M. Rollier, O. Saavedra, L. Satta, L. Trasatti, S. Vernetto,
L. Zanotti.
18th International Cosmic Ray Conference, Vol. 11, p. 466 – 470 (1983). – See Abstr. 012.096.

With the NUSEX detector, located at a depth of 5000 mwe in the Mt. Blanc Laboratory, the authors observed 6357 single muons and 83 muon bundles during 6094.6 hr of effective running time. The authors have analysed these data in terms of (1) primary spectrum up to 100 TeV, (2) primary cosmic ray composition in the interval 10^{15}–10^{17}eV. Details of this analysis and preliminary results are reported.

022.172   UA1 results from p$\bar{\text{p}}$ collisions at $\sqrt{s}$ = 540 GeV.
W. D. Dau.
18th International Cosmic Ray Conference, Vol. 12, p. 71 – 90 (1983). – See Abstr. 012.096.

The CERN SPS proton–antiproton collider was designed to search for the very massive intermediate vector bosons W^+, W^-, Z^0 and study hadronic interactions. Results from the UA1–experiment on the search for vector bosons and large pt jets are presented.

022.173   The picture of nucleus–nucleus interactions evolving from QCD models and data on hadron–nucleus collisions.
W. V. Jones.
18th International Cosmic Ray Conference, Vol. 12, p. 279 – 298 (1983). – See Abstr. 012.096.

This rapporteur paper is limited to accelerator results relevant to cosmic rays, hadron–nucleus and nucleus–nucleus (B–A) interactions, and recent results on anomalons. Instead of summarizing the reported results, an attempt is made to present the picture of B–A interactions that is evolving from our understanding of the available experimental data and QCD models.

022.174   Laboratory and astronomical identification of C_3H_2.
J. M. Vrtilek, P. Thaddeus, C. A. Gottlieb.
Bull. Am. Astron. Soc., Vol. 17, No. 2, p. 568 (1985). Abstract. – See Abstr. 010.065.

022.175   Laboratory and astronomical detection of the deuterated ethynyl radical CCD.
C. A. Gottlieb, J. M. Vrtilek, P. Thaddeus, W. D. Langer,
R. W. Wilson.
Bull. Am. Astron. Soc., Vol. 17, No. 2, p. 568 (1985). Abstract. – See Abstr. 010.065.

022.176   Rhenium oscillator strengths as a check on technetium oscillator strengths.
R. H. Garstang.
Bull. Am. Astron. Soc., Vol. 17, No. 2, p. 570 (1985). Abstract. – See Abstr. 010.065.

022.177   A new treatment of water vapor opacity.
D. R. Alexander.
Bull. Am. Astron. Soc., Vol. 17, No. 2, p. 600 (1985). Abstract. – See Abstr. 010.065.

022.178   Positrons in a simulated low–density galactic environment: recent experimental results.
B. L. Brown, M. Leventhal.
Bull. Am. Astron. Soc., Vol. 17, No. 2, p. 605 (1985). Abstract. – See Abstr. 010.065.

022.179   Cross sections for excitation of the helium atom by electron impact.
Yu. M. Smirnov.
Sov. Astron., Vol. 28, No. 6, p. 636 – 642 (1984). English translation of 38.022.150.

022.180   Stimulation of H_2O rotational transitions by electron impact.
A. M. Sobolev, S. V. Makarov.
Sov. Astron., Vol. 28, No. 6, p. 720 (1984). English translation of 38.022.151.

022.181   Laboratory simulation of planetesimal collision. 2. Ejecta velocity distribution.
T. Waza, T. Matsui, K. Kani.
J. Geophys. Res., Vol. 90, No. B2, p. 1995 – 2011 (1985).

The velocity distribution of fragments is one of the most important parameters for understanding the quantitative nature of planetary accretion processes and probably controls whether or not planets can be formed. In this paper the authors present high–speed motion picture analysis of low–velocity impact phenomena, with emphasis on the ejecta velocity distributions.

022.182   Clustered impacts: experiments and implications.
P. H. Schultz, D. E. Gault.
J. Geophys. Res., Vol. 90, No. B5, p. 3701 – 3732 (1985).

Impact by clusters of projectiles rather than a single projectile can result from several processes: atmospheric breakup, tidal breakup, and ejecta from a large primary impact. Experiments have been performed in order to establish the characteristics of such events over a wide range of impact velocities (15 m/s to 6 km/s). The authors consider the experimental results, including experimental procedure, cratering efficiency, crater morphology, and ejecta dynamics. They then focus on the relevance and possible implications of the results for planetary–scale processes.

022.183   Report of IAU Commission 14: Atomic and molecular data (*Données atomiques et moléculaires*).
A. H. Gabriel.
Trans. IAU, Vol. XIXA, p. 121 – 165 (1985). – See Abstr. 003.046.

022.184   Suess "wiggles" – a comparison between radiocarbon records.
C. P. Sonett.
Meteoritics, Vol. 20, No. 2, Part 2, p. 383 – 394 (1985). – See Abstr. 003.048.

022.185   Suess "wiggles and deviations" proven by historical and archaeological means.
R. Berger.
Meteoritics, Vol. 20, No. 2, Part 2, p. 395 – 401 (1985). – See Abstr. 003.048.

The veracity of Suess' "wiggles and deviations" was proven in two time segments by radiocarbon dating of known–age buildings or artifacts from the European Middle Ages and Egypt of the first three millennia B.C.

022.186   Line formation in laboratory plasmas.
P. Jaeglé, G. Jamelot, A. Carillon.
Progress in stellar spectral line formation theory, p. 239 – 263 (1985). – See Abstr. 012.108.

Spectral line formation in laboratory plasmas is concerned with plasmas of high density, such that photon escape is not free, though the plasma size may be small. Plasmas produced by laser impact on solid targets, like in inertial confinement experiments are typical examples. This paper reviews new work on the profiles of optically thin lines in these plasmas, discusses frequency redistribution, and presents some results from modeling radiative transfer in high–density plasma. Finally, line formation by simulated emission is briefly considered.

022.187   The lines of carbon, nitrogen and oxygen in the spectra of planetary nebulae. I. Transition probabilities and oscillator strengths.
P. O. Bogdanovich, R. A. Lukoshyavichyus, A. A. Nikitin, Z. B. Rudzikas, A. F. Kholtygin.
Astrofizika, Tom 22, Vyp. 3, p. 551 – 562 (1985). In Russian. English translation in Astrophysics, Vol. 22, No. 3.

The energy levels, wavelengths and transition probabilities for some astrophysical important lines of C, N, and O ions are calculated.

022.188   Frequency selective excitation spectroscopy of the CO intercombination bands.
P. Klopotek, C. R. Vidal.
Can. J. Phys., Vol. 62, No. 12, p. 1426 – 1436 (1984).

Using the method of frequency selective excitation spectroscopy, several bands of different intercombination systems of the CO molecule have been analyzed and measured. Using conventional methods, such as absorption spectroscopy, these bands could not be observed before owing to a severe overlap with the allowed bands of the $A^1\Pi - X^1\Sigma^+$ system.

A finding list for the multiplet tables of *NSRDS–NBS3*, Sections 1 – 10.
See Abstr. 002.025.

Documentation for the machine–readable version of *A Finding List for the Multiplet Tables of NSRDS–NBS 3, Sections 1 – 10* (Adelman, Adelman, Fischel and Warren 1984).
See Abstr. 002.037.

Theoretical concepts in physics. An alternative view of theoretical reasoning in physics for final–year undergraduates.
See Abstr. 003.005.

Foundations of radiation hydrodynamics.
See Abstr. 003.025.

Physics and astrophysics. A selection of key problems.
See Abstr. 003.028.

Physics of highly charged ions.
See Abstr. 003.042.

Foundations of the theory of spectra of atoms and ions.
See Abstr. 003.146.

Physics of dense matter.
See Abstr. 003.196.

Quantum metrology and fundamental physical constants. Proceedings of a NATO Advanced Study Institute, held 16 – 28 November 1981 in Erice, Sicily, Italy.
See Abstr. 012.117.

Fundamental processes in energetic atomic collisions. Proceedings of a NATO Advanced Study Institute held September 20 – October 1, 1982, in Maratea, Italy.
See Abstr. 012.121.

Stability in the double resonance problem.
See Abstr. 042.034.

A note on the integration of the equations of Lie–Deprit's method for unspecified canonical variables.
See Abstr. 042.035.

A variant of the Hori–Lie series method.
See Abstr. 042.037.

Measurement of neutron capture cross sections of s–only isotopes: ^{70}Ge, ^{86}Sr, and ^{87}Sr.
See Abstr. 061.002.

Si II line ratios in the Sun.
See Abstr. 073.093.

The impact of spectroscopic parameters on the composition of the Jovian atmosphere discussed in connection with recent laboratory, Earth and planetary observation programs.
See Abstr. 099.022.

Rate coefficients for the rotational excitation of CO by ortho– and para–H_2.
See Abstr. 131.130.

Polycyclic aromatic hydrocarbons and the diffuse interstellar bands.
See Abstr. 131.133.

Astronomical Instruments and Techniques

031 Astronomical Optics

031.001 **A new generation of spectrometer designs for ultraviolet astronomy.**
M. Hettrick, S. Bowyer.
NASA Conf. Publ., NASA CP–2349, p. 529 (1984). – See Abstr. 012.001.

031.002 **Rendu du contraste et pouvoir séparateur d'instruments utilisés visuellement.**
M. A. M. Van Venrooij.
Astronomie, Vol. 99, p. 125 – 135 (1985).

031.003 **Zur Funktionsweise der Fabry–Linse.**
B.–C. Kämper.
BAV Rundbrief, 34. Jahrg., Nr. 1, p. 28 – 29 (1985).

031.004 **Nützliche Tips für die Okularprojektion.**
P. Hückel.
Sterne Weltraum, 24. Jahrg., Nr. 1, p. 42 – 43 (1985).

031.005 **Optical systems of laser interferometers for testing astronomical optics.**
G. M. Popov, E. G. Popov.
Izv. Krymskoj Astrofiz. Obs., Tom 69, p. 133 – 139 (1984). In Russian. English translation in Bull. Crimean Astrophys. Obs., Vol. 69.
Cemented small laser interferometers with spherical or flat reference surface are described. These interferometers have a reference mirror which coincides with one of the surfaces of a semi–reflecting prism. Two types of interferometers were completed and tested.

031.006 **A comparative study of the optical charateristics of astrograph objectives (Carl Zeiss, Jena). Limiting magnitude.**
G. A. Ivanov, Eh. Rakhmatov, V. A. Yurevich, R. Ya. Inasaridze, I. A. Dautov, A. G. Krylov.
Kinematika Fiz. Nebesn. Tel, Tom 1, No. 2, p. 72 – 77 (1985). In Russian.
Results of a unified study of four–lense objectives of a wide–angle astrograph are given. The magnitude equation is insignificant.

031.007 **Distortion of three astrophotographic objectives.**
K. V. Kuimov.
Astron. Tsirk., No. 1348, p. 6 – 7 (1984). In Russian.

031.008 **A fast relay lens for the next generation of photon–counting systems.**
S. P. Worswick, C. G. Wynne.
Observatory, Vol. 105, No. 1066, p. 95 – 96 (1985).

031.009 **Non–regular circle grating for testing of aspherical surfaces.**
V. Yu. Terebizh.
Astron. Tsirk., No. 1355, p. 1 – 3 (1984). In Russian.

031.010 **Comparison of the star image diameters for wide–angle astrographs.**
I. G. Kolchinskij, L. K. Pakulyak.
Kinematika Fiz. Nebesn. Tel, Tom 1, No. 3, p. 64 – 70 (1985). In Russian.
The dependence of star image diameters on stellar magnitude is given for plates obtained with various wide–angle astrographs installed in the USSR. The diameter ratios for F = 3 m and 2 m are within 1.26 – 1.38, i.e. are smaller than the focal distance ratios. This may indicate that the essential contribution in the increase of image diameters is made not only by the star image tremor but by other factors not connected with the angular fluctuations.

031.011 **Wolter 1 type X–ray mirror system with mean resolution and maximum effective collecting area for the spectral region of 1 – 10 nm.**
R. Hudec, B. Valníček, I. Šolc, V. Landa, L. Svátek, R. Mareček, J. Urban, L. A. Vainshtein (*L. A. Vajnshtejn*), M. E. Plotkin, M. M. Mitropolskii, V. A. Slemzin, N. K. Sukhodrev, V. G. Sumbatov.
Publ. Astron. Inst. Czech. Acad. Sci., No. 61, p. 1 – 41 (1985).
A calculation of parameters, method of manufacturing procedure and results of testing are described for the Wolter 1 X–ray mirror system. It consists of 2 nested pairs of mirrors. Geometrical parameters of the mirrors were chosen to obtain maximum effective area under the limits for overall dimensions.

031.012 **Galvanoplastic grazing incidence mirrors for "vertical" experiments: the tandem tests in soft X–rays.**
R. Hudec, B. Valníček, I. Šolc, M. M. Mitropolskii, I. A. Zhitnik, V. V. Krutov, N. K. Sukhodrev.
Publ. Astron. Inst. Czech. Acad. Sci., No. 61, p. 42 – 76 (1985).
In this paper, the galvanoplastic X–ray mirrors are described and some results of the studies of pictures obtained with these mirrors in the X–ray range are presented and discussed.

031.013 **The Multiple Mirror Telescope as a phased array telescope.**
E. K. Hege, J. M. Beckers, P. A. Strittmatter, D. W. McCarthy.
Prepr. Steward Obs., No. 567, 53 pp. (1985). To be submitted to Appl. Opt.

031.014 **Analytic solutions for three–lens correctors of a paraboloidal primary based on the third–order aberration theory.**
Y. Yamashita, K. Nariai, T. Shigeyama, M. Nakagiri.
Ann. Tokyo Astron. Obs., Second Ser., Vol. 20, No. 3, p. 296 – 320 (1985).
Three–lens correctors for a paraboloidal primary were studied from the third–order aberration theory. Thin–lens solutions were obtained and then, lens thickness was introduced by the variational method. Calculations were made numerically and the optical parameters are presented in a tabular form. Then, uncontrolled higher–order aberrations were investigated by ray traces. It was found from this analytical study that the focal length of the system must be longer than that of the primary and the power distribution of the three lenses must be positive, negative and positive.

031.015 Lacquer coated X–ray optics.
R. C. Catura, E. G. Joki, D. T. Roethig.
Bull. Am. Astron. Soc., Vol. 16, No. 4, p. 885 (1984). Abstract. –
See Abstr. 010.062.

031.016 Design for a low cost 8 m O–IR telescope and enclosure.
J. R. P. Angel, N. J. Woolf, B. L. Ulich.
Bull. Am. Astron. Soc., Vol. 16, No. 4, p. 901 (1984). Abstract. –
See Abstr. 010.062.

031.017 The CTIO 4 m infrared optimized telescope project.
P. S. Osmer.
Bull. Am. Astron. Soc., Vol. 16, No. 4, p. 901 (1984). Abstract. –
See Abstr. 010.062.

031.018 Fabrication of large cellular telescope mirrors by fusing PYREX.
A. P. Odell, P. T. Jones.
Bull. Am. Astron. Soc., Vol. 16, No. 4, p. 901 (1984). Abstract. –
See Abstr. 010.062.

031.019 STRAY, an interactive computer program for the calculation of stray radiation in infared telescopes.
A. S. Dinger.
Bull. Am. Astron. Soc., Vol. 16, No. 4, p. 907 (1984). Abstract. –
See Abstr. 010.062.

031.020 Optical aberrations with an articulating Cassegrain secondary.
A. D. Code.
Bull. Am. Astron. Soc., Vol. 16, No. 4, p. 986 – 987 (1984). Abstract. – See Abstr. 010.062.

031.021 Extrapolated least–squares optimization in optical design.
E. D. Huber.
J. Opt. Soc. Am. A, Vol. 2, No. 4, p. 544 – 554 (1985).

The extrapolated least–squares (ELS) optimization method is a new approach to improving the optimization efficiency of the least–squares techniques used in lens–design computer programs. The ELS method retains information between iterative optimization cycles for the development of second–order extrapolation factors that include up to second–derivative terms. The extrapolation factors are used to update the first–derivative matrix of the residual vector to reflect optimization progress more accurately without requiring the recomputation of the residual vector's first derivatives. For optical–design problems in which the first-derivative matrix is costly and time–consuming to compute, the ELS method may provide great benefit.

031.022 Some ideas about representations of aspheric optical surfaces.
D.–Q. Su, Y.–N. Wang.
Appl. Opt., Vol. 24, No. 3, p. 323 – 326 (1985).

031.023 Aspherical mirror testing using a CGH with small errors.
A. Ono, J. C. Wyant.
Appl. Opt., Vol. 24, No. 4, p. 560 – 563 (1985).

A method for reducing errors in aspherical mirror testing using a computer–generated hologram (CGH) is described. By using a modified filtering method the carrier frequency in the CGH can be reduced by two–thirds, and the resulting error due to distortion is only one–half of that of a conventional CGH. By adopting a Fizeau–type optical setup, only the surface quality of the reference affects the measured results.

031.024 Surface analysis of an actively controlled telescope primary mirror under static loads.
F. B. Ray, Y.–T. Chung.
Appl. Opt., Vol. 24, No. 4, p. 564 – 569 (1985).

The performance of an actively controlled mirror with a diameter/thickness ratio of 63:1 is modeled by a large scale software system. The model allows generalized static disturbances to be applied to the mirror, generates a corrective force field, and tests the result through a finite element simulation.

031.025 Aberrations of an IR chopping secondary: comment 1.
M. Bottema.
Appl. Opt., Vol. 24, No. 7, p. 942 – 943 (1985). See Abstr. 38.031.047.

031.026 Aberrations of an IR chopping secondary: authors' reply to comment 1.
A. B. Meinel, M. P. Meinel.
Appl. Opt., Vol. 24, No. 7, p. 943 (1985).
Concerning Abstr. 031.025. For the original paper see Abstr. 38.031.047.

031.027 Aberrations of an IR chopping secondary: comment 2.
H. van de Stadt.
Appl. Opt., Vol. 24, No. 7, p. 943 (1985). See Abstr. 38.031.047.

031.028 Aberrations of an IR chopping secondary: authors' reply to comment 2 and continuation of authors' reply to comment 1.
A. B. Meinel, M. P. Meinel.
Appl. Opt., Vol. 24, No. 7, p. 944 (1985).
Concerning abstracts 031.025 and 031.027. For the original paper see Abstr. 38.031.047.

031.029 Optical aberration functions: derivatives with respect to surface parameters for symmetrical systems.
T. B. Andersen.
Appl. Opt., Vol. 24, No. 8, p. 1122 – 1129 (1985).

Formulas are presented for the computation of derivatives of the optical aberration functions S, T, V, W, and K with respect to the surface parameters for symmetrical optical systems. The important case of conic–section surfaces of revolution is considered separately. The formulas are suitable for use in computer programs for automatic computation of optical aberration coefficients.

031.030 Catadioptric lens with aberrations balanced by aspherizing one surface.
K. M. Bystricky, P. R. Yoder Jr.
Appl. Opt., Vol. 24, No. 8, p. 1206 – 1208 (1985).

The authors describe an optical system with one surface aspherized to give a specific level of performance and to equalize the intensity profiles of star images over the field. The system was a 25.4–cm EFL, 5.8° field, $f/1.53$ catadioptric Cassegrain. Aberrations were controlled over a 0.5 – 1.1–μm spectral region.

031.031 Grazing incidence echelle spectrometers using varied line–space gratings.
M. C. Hettrick.
Appl. Opt., Vol. 24, No. 9, p. 1251 – 1255 (1985).

The author previously suggested a new two–element echelle spectrometer composed of varied line–space (VLS) gratings at grazing incidence. The absence of other correcting or collimating/camera optics permits high reflection efficiency in comparison to standard approaches. In combination with a two–element grazing incidence telescope, a sample design for use in space astronomy has been calculated to deliver an optics efficiency greater than 5%. In this letter the author provides quantitative estimates of the imaging performance expected from a VLS echelle spectrometer.

031.032 Conical imaging mirrors for high–speed X–ray telescopes.
R. Petre, P. J. Serlemitsos.
Appl. Opt., Vol. 24, No. 12, p. 1833 – 1837 (1985).

The conical X–ray imaging mirror represents the long focal length limit of a Wolter type I grazing incidence mirror, in which the curved surfaces have been replaced by simple cones. When many thin–walled cones are nested, such a mirror affords the relatively high aperture filling factor needed for telescopes well suited to broadband X–ray astronomy. The authors describe the spatial resolution and filling factor of conical optics as a function of various design parameters as characterized using a Monte Carlo ray tracing procedure.

031.033 Transverse ray aberrations for paraboloid–hyperboloid telescopes.
T. T. Saha.
Appl. Opt., Vol. 24, No. 12, p. 1856 – 1863 (1985).
Transverse ray aberration expansions are derived for a paraboloid–hyperboloid telescope. The expansions are valid for glancing incidence Wolter type II and normal incidence Cassegrain telescopes. The analysis gives all third–order aberration terms except distortion and four fifth–order aberration terms as a function of the system parameters and entrance pupil coordinates. The spot diagrams derived from exact ray tracing and the aberration expansions for a Wolter type II design and Cassegrain design agree well. The importance of fifth–order terms is discussed for the Wolter type II telescope.

031.034 Infrared fibers.
J. Lucas.
Infrared Phys., Vol. 25, No. 1/2, p. 277 – 281 (1985). – See Abstr. 012.065.
A review of this new field of materials and optics science is presented.

031.035 Polarization of materials and optical components at near–millimetre wavelengths.
A. Blanco, S. Fonti, M. Mancarella.
Infrared Phys., Vol. 25, No. 1/2, p. 283 – 284 (1985). – See Abstr. 012.065.
Knowledge of the degree of polarization of materials and optical components widely used in the near–millimetre region is important in many applications. In this work the authors present some results, in the wavelength interval 0.5 – 3 mm, concerning materials and optical components in common use. The polarized emission of a well–known FIR source (Philips HPK 125 W) is also measured.

031.036 A test of diffraction structures at near–millimetre wavelengths.
A. Blanco, S. Fonti, M. Mancarella.
Infrared Phys., Vol. 25, No. 1/2, p. 285 – 288 (1985). – See Abstr. 012.065.
The problem of diffraction effects is particularly important at millimetre wavelengths. The geometrical theory of diffraction (GTD) has been a good aid in coping with this kind of problem. The application of GTD, however, has been limited to monochromatic radiation. In this paper the authors show by means of experimental tests that the same method, with some minor computational modifications, also can be applied successfully in non–monochromatic systems.

031.037 Ultraviolet resolution of large mirrors via Hartmann tests and two–dimensional fast Fourier transform analysis.
L. Dame, F. Vakili.
Opt. Eng., Vol. 23, No. 6, p. 759 – 765 (1984). Abstr. in Phys. Abstr., Vol. 88, No. 1254, Entry 37461 (1985).

031.038 Zero–coma condition for decentered and tilted secondary mirror in Cassegrain/Nasmyth configuration.
A. B. Meinel, M. P. Meinel.
Opt. Eng., Vol. 23, No. 6, p. 801 – 805 (1984). Abstr. in Phys. Abstr., Vol. 88, No. 1254, Entry 40484 (1985).

031.039 Optical model for the HIPPARCOS telescope in its real expected configuration.
D. Bertani, M. Cetica, D. Iorio Fili.
Processing of scientific data from the ESA astrometry satellite HIPPARCOS, p. 43 – 48 (1985). – See Abstr. 012.087.
This paper presents a model for the HIPPARCOS optical system in its real expected configuration. Each surface is described by an expression that takes into account possible thermomechanical deformations and manufactoring errors. Misalignments and tilts are also considered. These features allow tolerance analysis and simulation of the HIPPARCOS real output.

031.040 The signal of the HIPPARCOS nominal system.
M. Amoretti, M. Badiali, D. Cardini, A. Emanuele.
Processing of scientific data from the ESA astrometry satellite HIPPARCOS, p. 49 – 58 (1985). – See Abstr. 012.087.
The authors describe the software which computes the light propagation through the HIPPARCOS nominal system and obtains all parameters of the output signal for three stars of different colours. The authors also give some results about amplitudes and phases of the Fourier components of the signal in 25 points of the field of view.

031.041 Fast–speed four–mirror system.
E. G. Popov, B. A. Bulibekov.
Tr. Astrofiz. Inst. Alma–Ata, Tom 42, p. 133 – 140 (1983). In Russian.
A fast-speed four–mirror system with flat field and corrected for aberration in a large field angle and wide spectral region is considered.

031.042 On the problem of application of liquid mirrors in astronomy.
V. P. Vasil'ev.
Astron. Zh., Tom 62, Vyp. 3, p. 598 – 601 (1985). In Russian. English translation in Sov. Astron., Vol. 29, No. 3.
Zonation interferometry and astrophotographical tests show that a parabolic mirror being formed by a rotating reflecting liquid in the system of capacities free–floating in each other meets the standards of optical telescope objectives.

031.043 Polychromatic adaptive optics for infrared telescopes.
J. M. Beckers, P. Eisenhardt, L. Goad, F. Roddier.
Bull. Am. Astron. Soc., Vol. 17, No. 2, p. 571 (1985). Abstract. – See Abstr. 010.065.

031.044 A normal–incidence telescope for solar EUV studies.
R. J. Thomas, G. L. Epstein, R. A. M. Keski–Kuha.
Bull. Am. Astron. Soc., Vol. 17, No. 2, p. 592 (1985). Abstract. – See Abstr. 010.065.

031.045 Optical–shop testing of liquid mirrors.
E. F. Borra, M. Beauchemin, R. Arsenault, R. Lalande.
Publ. Astron. Soc. Pac., Vol. 97, No. 591, p. 454 – 464 (1985).
Borra (1982) has argued that it is scientifically useful and should be technically feasible to build very large optical telescopes (diameters over 15 meters) having as primary mirror a rotating container filled with mercury. The authors report here the results of optical tests on prototype liquid mirrors. Knife–edge and Hartmann tests, as well as direct imagery, show that optical quality surfaces are attained. The largest mirror produced has a diameter of 1.65–m but the authors tested thoroughly only a 1–m diameter mirror.

031.046 Vignetting in three–mirror telescopes.
N. N. Mikhel'son.
Izv. Glav. Astron. Obs. Pulkovo, Astrometr. Astrofiz., No. 201, p. 127 – 131 (1985). In Russian.
Radial heights of various field angles have been determined in the plane of each of the three mirrors and in the focal plane. Four possible cases are considered.

031.047 Optical/IR telescope arrays: a study of their potential for satisfying observational requirements in astronomy for a large aperture system in the 1990s and beyond.
C. M. Humphries, D. J. Walshaw, V. C. Reddish, A. H. Greenaway.
Occas. Rep. R. Obs., Edinb., No. 15, 6 + 127 pp. (1984). ISBN 0–902553–32–1.
Contents: Preface. (1) The scientific case for very large aperture optical/IR telescopes and the advantages of an array telescope. (2) Cost optimisation in the construction of optical telescopes: cost–based design strategy for an array telescope. (3) Incoherent operation of an optical/IR array telescope: SNR

(*signal–to–noise ratio*) considerations. (4) Coherent operation of an optical/IR array telescope: long baseline interferometry. (5) Design solutions for an 18 m array telescope. (6) Maintenance and operation of an array: system reliability and staffing requirements. Appendix A: telescope cost scaling laws and their origin. Appendix B: project cost comparisons.

Calculation of mirror and lens deformations with horizontal and vertical control.
See Abstr. 003.041.

Report of the optical conference on the 7.6–meter telescope held in Austin, Texas, March 22 – 24, 1982.
See Abstr. 012.013.

Quality of star images and determination of dimensions of the well–corrected field on plates of wide–angle astrographs.
See Abstr. 032.009.

Angular measurements by means of optical multiple reflection.
See Abstr. 036.112.

032 Astronomical Instruments

032.001   Plans for new technology telescope come into sharper focus.
Phys. Today, Vol. 38, No. 1, p. 91 – 92 (1985).

032.002   Erratum: 'Measuring electron density in coronal active regions. II: A multichannel coronagraph with a photoelectric spectrograph and a reflex monitor at λ5303 Å' [Sol. Phys., Vol. 94, No. 1, p. 117 – 131 (1984)].
J. C. Noëns, J. Pageault, G. Ratier.
Sol. Phys., Vol. 95, No. 1, p. 199 (1985). See Abstr. 38.032.030.

032.003   De Jones–Bird telescoop.
H. Feijth.
Zenit, 12. Jaarg., No. 1, p. 28 – 29 (1985).

032.004   The Explosive Transient Camera (ETC): an instrument for the detection of gamma–ray burst optical counterparts.
G. R. Ricker, J. P. Doty, J. V. Vallerga, R. K. Vanderspek.
AIP Conf. Proc., No. 115, p. 669 – 686 (1984). – See Abstr. 012.005.
The Explosive Transient Camera (ETC) is a wide field (~ 3 steradians) electronic camera array which can detect coincident optical flashes with durations of $\sim 10^{-1}$ to 10^{+2} seconds. Each array element is a $20° \times 30°$ FOV, cooled CCD detector, developed at MIT. An optical transient as faint as $B = +11$ (1 second duration) can be detected with S/N > 20, and its position determined to an accuracy of ± 10 arcsec.

032.005   The rapidly moving telescope: an instrument for the precise study of optical transients.
B. J. Teegarden, T. T. von Rosenvinge, T. L. Cline, R. Kaipa.
AIP Conf. Proc., No. 115, p. 687 – 693 (1984). – See Abstr. 012.005.
The authors have initiated at the Goddard Space Flight Center the development of a small telescope with a very rapid pointing capability whose purpose is to search for and study fast optical transients that may be associated with gamma–ray bursts and other phenomena. The telescope will have the capability of rapidly acquiring any target in the night sky within 0.7 second and locating the object's position with ± 1 arcsec accuracy. The initial detection of the event will be accomplished by the MIT Explosive Transient Camera.

032.006   The control of an astronomical telescope using a multimicroprocessor system.
B. J. Rooney.
J. Microcomput. Appl., Vol. 7, No. 2, p. 189 – 201 (1984). Abstr. in Phys. Abstr., Vol. 88, No. 1249, Entry 14472 (1985).

032.007   An astrometric lens caper.
A. Hoag.
Sky Telesc., Vol. 69, No. 3, p. 214 – 215 (1985).

032.008   Das Sonnenteleskop der Sternwarte Hubelmatt.
A. Tarnutzer.
Sonne, Jahrg. 9, Nr. 33, p. 21 – 23 (1985).

032.009   Quality of star images and determination of dimensions of the well–corrected field on plates of wide–angle astrographs.
G. A. Ivanov.
Kinematika Fiz. Nebesn. Tel, Tom 1, No. 2, p. 64 – 68 (1985). In Russian.
The radius of the well–corrected field on a plate is determined from seeing estimates and measurement errors of stellar positions using different methods.

032.010   New chromospheric telescope.
V. G. Banin.
Priroda, No. 3, p. 111 (1985). In Russian.

032.011   Investigation of the rotation characteristics and position stability of the first (vertical) axis of a telescope.
V. Ya. Vajnberg.
Astrofiz. Issled. Izv. Spets. Astrofiz. Obs., Tom 19, p. 82 – 87 (1985). In Russian. English translation in Bull. Spec. Astrophys. Obs. – North Caucasus.
The rotation characteristics of the vertical axis of the 6–meter telescope have been investigated for two years. A conclusion has been drawn on the high level of the mounting accuracy and temperature deformations of the telescope design.

032.012   Investigation of an automatic stabilization system of compound mirrors of an adaptive telescope.
V. I. Kryukov, A. K. Rodionov, P. V. Sergeev, V. V. Sychev.
Opt.–mekh. prom–st', No. 6, p. 1 – 4 (1984). In Russian. Abstr. in Ref. Zh., 51. Astron., 1.51.1275 (1985).

032.013   Preliminary results of an investigation of the eccentricity of the Odessa meridian circle.
M. I. Myalkovskij, V. V. Lyakhovich.
Mater. konf. mol. uchenykh Odes. astron. nauchn.–proizv. akad.–univ. kompleksa, Odessa, 15 fevr., 1983. Odessa, p. 76 – 79 (1983). In Russian. Abstr. in Ref. Zh., 51. Astron., 3.51.954 (1985).

032.014   Investigation of the ocular micrometer of the Odessa meridian circle.
M. I. Myalkovskij.
Mater. konf. mol. uchenykh Odes. astron. nauchn.–proizv. akad.–univ. kompleksa, Odessa, 15 fevr., 1983. Odessa, p. 70 – 75 (1983). In Russian. Abstr. in Ref. Zh., 51. Astron., 3.51.955 (1985).

032.015   The Sopron–Vienna zenith camera.
K. Bretterbauer, J. Somogyi, G. Szádeczky–Kardoss.
Mitt. Lohrmann–Obs. Tech. Univ. Dresden, Nr. 51, p. 110 – 111 (1984). In German. – See Abstr. 012.058.
Das Institut für Theoretische Geodäsie und Geophysik der TU Wien und das Geodätische und Geophysikalische Forschungsinstitut der Ungarischen Akademie der Wissenschaften in Sopron haben gemeinsam eine transportable Zenitkammer entwickelt um sie für Zwecke der geodätischen Astronomie und Astrometrie einzusetzen. Das Instrument wird beschrieben.

032.016   The application of the PZT 2 photographic zenith telescope in the Potsdam geodetic–astronomical observatory.
M. Meinig.
Mitt. Lohrmann–Obs. Tech. Univ. Dresden, Nr. 51, p. 120 – 126 (1984). = Mitt. Zentralinst. Phys. Erde, Nr. 1277. In German. – See Abstr. 012.058.
Das photographische Zenitteleskop PZT 2, das seit Juli 1980 im Zentralinstitut für Physik der Erde (ZIPE) im Einsatz ist, ist eine Weiterentwicklung des PZT 1, das in Ondrejov/ČSSR arbeitet. Durch die instrumentellen Verbesserungen gelang es, die Leistungsfähigkeit des Gerätes zu steigern, die Möglichkeiten zur Gestaltung des Beobachtungsprogramms zu erweitern, sowie die Funktionssicherheit und den Bedienungskomfort zu erhöhen. Im folgenden werden eine Beschreibung des PZT 2 gegeben, sein Einsatz im geodätisch–astronomischen Observatoriumsprogramm erläutert und Ergebnisse mitgeteilt.

032.017   Further development of the photographic zenith telescope (PZT) into a photoelectric zenith tube (PEZR).
J. Dittrich, H. Fischer, G. Knischewski, P. Notni.
Mitt. Lohrmann–Obs. Tech. Univ. Dresden, Nr. 51, p. 126 – 127 (1984). = Mitt. Zentralinst. Phys. Erde, Nr. 1290. In German. – See Abstr. 012.058.

032.018   Progress report on vacuum meridian marks.
I. Pakvor.
Mitt. Lohrmann–Obs. Tech. Univ. Dresden, Nr. 51, p. 127 – 129 (1984). – See Abstr. 012.058.
More than ten years ago, the Large Transit Instrument of Belgrade Astronomical Observatory was equipped, as the first and only one up to now, with vacuum meridian marks, which were supposed to ensure a stable image of the meridian mark, at all times and all meteorological conditions. Application of the vacuum meridian marks in absolute R. A. determinations is described.

032.019   Chromatic aberration of the wide–angle astrograph of the Sternberg Institute.
V. A. Eliseev, K. V. Kuimov, O. D. Solov'eva.
Astron. Tsirk., No. 1348, p. 5 – 6 (1984). In Russian.

032.020   A method for the daily azimuth control of the PZT.
Z. M. Malkin.
Astron. Tsirk., No. 1349, p. 7 – 8 (1984). In Russian.

032.021   Investigation of the alidade levels of the Wanschaff vertical circle.
A. V. Bakhonskij.
Kinematika Fiz. Nebesn. Tel, Tom 1, No. 3, p. 85 – 90 (1985). In Russian.
A new comparator with optical caliper was used to study the two–second levels of the Wanschaff vertical circle. The results of investigation are presented. The influence of scale division change on measured zenith distance is estimated. A number of practical recommendations are given to decrease this influence. The temperature coefficients are obtained.

032.022   Correction to the angular value of the screw revolution of the Belgrade zenith–telescope micrometer derived from latitude observations.
M. Djokić.
Publ. Astron. Opservatorije Beogr., No. 33, p. 70 – 74 (1985). – See Abstr. 012.061.

032.023   Transportable Zenitkameras: Reduktion, Konstruktion, Anwendung.
K. Pilowski.
Univ. Hannover, Astron. Stn., Veröff., Nr. 16, p. 3 – 14 (1985).

032.024   Graduation–error determination of the Tokyo photoelectric meridian circle.
M. Miyamoto, S. Suzuki.
Ann. Tokyo Astron. Obs., Second Ser., Vol. 20, No. 3, p. 209 – 236 (1985).
In order to establish the feasibility of an efficient measuring scheme for the graduation–error determination, proposed by Miyamoto and Kühne (1982), the conventional and newly proposed measurements of divisions by the reading microscopes have been carried out with the photoelectric meridian circle at Toyko Astronomical Observatory. Comparing two independent results, it is found that the two sets of the circle correction thus obtained coincide with each other within the accuracy of $\pm 0\rlap{.}''018$ (r.m.s.), whereby the feasibility as well as the reliability of the proposed scheme are proven.

032.025   LEST telescope tower study deformation analysis.
J. Sjölund.
LEST Found., Tech. Rep., No. 8, 9 pp. (1984).

032.026   Den tekniske løsning for LEST.
O. Engvold.
Astron. Tidsskr., Årg. 18, Nr. 2, p. 51 – 57 (1985).

032.027   Open loop telescope automation.
N. L. Markworth, J. B. Rafert.
Bull. Am. Astron. Soc., Vol. 16, No. 4, p. 900 (1984). Abstract. – See Abstr. 010.062.

032.028   Measurement of atmospheric seeing with Ronchi telescopes.
A. Buffington.
Bull. Am. Astron. Soc., Vol. 16, No. 4, p. 900 (1984). Abstract. – See Abstr. 010.062.

032.029   A new life for an old telescope.
G. W. Wolf.
Bull. Am. Astron. Soc., Vol. 16, No. 4, p. 901 (1984). Abstract. – See Abstr. 010.062.

032.030   The CCD/transit instrument – scientific programs.
D. V. Arganbright, J. T. McGraw, J. T. Stocke, J. R. P. Angel, M. J. Keane.
Bull. Am. Astron. Soc., Vol. 16, No. 4, p. 902 (1984). Abstract. – See Abstr. 010.062.

032.031   The CCD/transit instrument – operation and data management.
M. J. Keane, J. T. McGraw, J. T. Stocke, J. R. P. Angel, D. V. Arganbright.
Bull. Am. Astron. Soc., Vol. 16, No. 4, p. 902 (1984). Abstract. – See Abstr. 010.062.

032.032   Braeside Observatory – a computer controlled 16–inch Cassegrain and photometer.
R. E. Fried, D. C. Morse.
Bull. Am. Astron. Soc., Vol. 16, No. 4, p. 908 (1984). Abstract. – See Abstr. 010.062.

032.033   Software for an automated supernova search.
E. C. Pearce, K. B. Meier, S. A. Colgate.
Bull. Am. Astron. Soc., Vol. 16, No. 4, p. 909 (1984). Abstract. – See Abstr. 010.062.

032.034   A telescope for automatic photometry.
F. M. Melsheimer, R. M. Genet.
Bull. Am. Astron. Soc., Vol. 16, No. 4, p. 909 (1984). Abstract. – See Abstr. 010.062.

032.035   Fairborn Observatory automatic photoelectric telescopes.
L. J. Boyd, R. M. Genet, D. J. Sauer, P. S. Hawthorn,
L. W. Slonaker Jr., J. Chatto.
Bull. Am. Astron. Soc., Vol. 16, No. 4, p. 909 (1984). Abstract. –
See Abstr. 010.062.

032.036   Ein Spiegelteleskop mit quarzgesteuerter Nachführung.
E. Bauer.
Astron. Raumfahrt, 23. Jahrg., Heft 3, p. 58 – 62 (1985).

032.037   On the scale of micrometric measurements of a Zeiss zenith telescope.
R. I. Popova, L. S. Otkidach, E. I. Podshchipkova.
Vrashchenie i priliv. deformatsii Zemli, Kiev, No. 16, p. 37 – 42 (1984). In Russian. Abstr. in Ref. Zh., 51. Astron., 5.51.814 (1985).

032.038   Determination of the errors of adjustment of a zenith telescope in zenith observations.
A. V. Gozhij.
Vrashchenie i priliv. deformatsii Zemli, Kiev, No. 16, p. 53 – 56 (1984). In Russian. Abstr. in Ref. Zh., 51. Astron., 5.51.815 (1985).

032.039   Influence of instrumental errors on micrometric meridian measurements in the zenith.
V. K. Budz'ko.
Vrashchenie i priliv. deformatsii Zemli, Kiev, No. 16, p. 56 – 60 (1984). In Russian. Abstr. in Ref. Zh., 51. Astron., 5.51.816 (1985).

032.040   On the use of dynamical dampers for reducing the vibration level of a solar telescope.
B. G. Korenev, A. K. Kitov.
Issled. Geomagn., Aehron. Fiz. Solntsa, Moskva, No. 69, p. 197 – 203 (1984). In Russian. Abstr. in Ref. Zh., 51. Astron., 5.51.824 (1985).

032.041   Air–support concept for a large telescope.
H. W. Babcock.
Appl. Opt., Vol. 24, No. 9, p. 1248 – 1250 (1985).
The air–support concept eliminates the external dome and accommodates conventional focal ratios such as $f/3$ or $f/4$ for the primary mirror. It also permits realization of the operational advantages of the equatorial mounting, with a major reduction in size and cost of heavy, precisely machined parts. The basic design features are explored and an analysis of feasibility is presented.

032.042   Die deutschen Sonnenteleskope des Observatorio del Teide auf Teneriffa.
E. H. Schröter, E. Wiehr.
Sterne Weltraum, 24. Jahrg., Nr. 6, p. 319 – 323 (1985).

032.043   Distributing computing for astronomical data acquisition at McDonald Observatory.
P. W. Kelton.
Proceedings of the real–time systems symposium, held at Austin, Tex., USA, 4 – 6 December 1984. IEEE Comput. Soc. Press, Silver Spring, Md., USA (1984). p. 83 – 88.

032.044   Astrometric performance of the Fan Mountain 1 m astrometric reflector.
B. McNamara, L. W. Fredrick, P. A. Ianna.
Bull. Am. Astron. Soc., Vol. 17, No. 2, p. 573 (1985). Abstract. –
See Abstr. 010.065.

032.045   A helium filled astrometric telescope.
G. D. Gatewood, J. H. Kiewiet de Jonge, J. W. Stein,
L. A. Breakiron.
Bull. Am. Astron. Soc., Vol. 17, No. 2, p. 582 – 583 (1985). Abstract. – See Abstr. 010.065.

032.046   Report of IAU Commission 9: Instruments and techniques (*Instruments et techniques*).
W. C. Livingston.
Trans. IAU, Vol. XIXA, p. 41 – 56 (1985). – See Abstr. 003.046.

032.047   The Leuschner Observatory automated 30–inch telescope.
R. R. Treffers.
Publ. Astron. Soc. Pac., Vol. 97, No. 591, p. 446 – 450 (1985).
The automation of the Leuschner Observatory 30–inch (76–cm) telescope is discussed. All of the telescope control functions, the telescope positioning, the angle encoders, and the dome alignment have been interfaced to the IEEE–488 instrument bus and can be controlled by a personal computer. The same bus is used to control and collect data from a *UBVRI* photometer, an intensified Reticon spectrograph, and a Fabry–Perot spectrometer. This telescope is currently being used for the Berkeley Automated Supernova and Nemesis star search.

032.048   Infrared secondary chopping mirror for a 60 cm reflecting telescope.
Y.–h. Gu, J. Zhu, Z.–l. Xia.
Acta Astron. Sin., Vol. 25, No. 4, p. 431 – 436 (1984). In Chinese.
The design and construction of a servo controlled chopper driver for the infrared secondary mirror of the 60 cm reflecting telescope of Beijing Observatory is described. The system allows square wave modulation for a chopping amplitude of 60 arcsec at 11.7 Hz, the 10 – 90% rise time will be 4 – 5 ms with 5 – 10% overshoot without causing vibration or acoustic noise. The system has been used during the infrared observation with a InSb photometer.

032.049   The astrometric properties of the Schmidt telescope of the Beijing Astronomical Observatory.
Y. Su.
Acta Astron. Sin., Vol. 26, No. 1, p. 22 – 27 (1985). In Chinese.
The astrometric properties of the 60/90/180 cm Schmidt telescope of the Beijing Astronomical Observatory have been investigated. One plate of the Pleiades cluster and another of the α Per moving cluster were measured by the Zeiss ASCORECORD measuring machine. The standard errors of measurement were $\pm 1.4\,\mu$ or $\pm 0.''15$ in x and $+ 1.5\,\mu$ or $\pm 0.''17$ in y. The standard deviation of the computed values using 10 plate constants in the entire field of $4.°8$ diameter from the catalogue positions was $\pm 0.''38$ and in the field of $3.°2$ diameter was $\pm 0.''34$ (averaged for right ascension and declination). No visible magnitude dependence was found.

032.050   Note on the meridian mark stability.
V. A. Varina.
Izv. Glav. Astron. Obs. Pulkovo, Astrometr. Astrofiz., No. 202, p. 21 – 24 (1984). In Russian.
Some results of the study of meridian marks of the Pulkovo Large Transit Instrument installed in the southern hemisphere are given. A conclusion is made on high stability of these short–foci marks (about 60 m) and their use for the azimuth derivation in absolute RA determinations.

032.051   The effect of wind in observations with the Pulkovo photoelectric transit instrument PPI–I during 1961 – 1971.
G. V. Staritsyn.
Izv. Glav. Astron. Obs. Pulkovo, Astrometr. Astrofiz., No. 202, p. 25 – 31 (1984). In Russian.
The systematic error in astronomical observations called the effect of winds is discussed. The results of an analysis of a 10 year observational series with the Pulkovo photoelectric transit instrument PPI–I confirm the existence of the above mentioned effect. The calculations reveal that the instrument, pavilion, ventilation of the pavilion have no influence on the effect of winds.

032.052   The pentag meridian circle.
A. A. Nemiro.
Izv. Glav. Astron. Obs. Pulkovo, Astrometr. Astrofiz., No. 201,
p. 13 – 21 (1985). In Russian.
The main drawbacks of the existing meridian instruments are
given. An optical–mechanical scheme of a pentag meridian circle
is presented. The instrument can be used as meridian circle, as a
vertical circle, as a zenith telescope and a reversible transit instrument. Formulae for the reduction of observations are given.

**032.053   A study of the errors of limb divisions of the Pulkovo
photographic vertical circle.**
B. K. Bagil'dinskij, E. G. Zhilinskij, S. P. Pulyaev,
O. E. Shornikov.
Izv. Glav. Astron. Obs. Pulkovo, Astrometr. Astrofiz., No. 201,
p. 31 – 35 (1985). In Russian.
A study of the errors of the two limbs of the Pulkovo photographic vertical circle was done in steps of 3°. The method of the
study is described, which is a modified method of Bruns' rosettes.

**032.054   On the problem of protection of rays passing from merid-
ian marks to a transit instrument.**
G. M. Petrov, R. T. Fedorova, P. N. Fedorov.
Izv. Glav. Astron. Obs. Pulkovo, Astrometr. Astrofiz., No. 201,
p. 36 – 38 (1985). In Russian.
Two methods of protection of rays passing from meridian
marks to the instrument were tested using the Nikolaev transit
instrument. In the first method the light passed in a tube with
good thermal isolation and ventilation; in the second the light
passed through vacuum tubes. The latter method enabled a considerable improvement in determinations of the azimuth and
collimation errors of the instrument.

**032.055   Il telescopio nazionale e le sue relazioni con l'astronomia
spaziale degli anni '90.**
C. Barbieri.
Orione, Vol. 5, N. 1, p. 20 – 30 (1985).

**032.056   Canada–France–Hawaii Observatory 3.6 meter tele-
scope.**
P. R. Engle.
Proceedings of the Southwest Regional Conference for Astronomy and Astrophysics, Vol. 10, p. 51 – 53 (1985). – See Abstr.
012.125.

**032.057   The ARC (*Astrophysical Research Consortium*) tele-
scope project.**
K. S. Anderson.
Proceedings of the Southwest Regional Conference for Astronomy and Astrophysics, Vol. 10, p. 55 – 59 (1985). – See Abstr.
012.125.
A consortium of universities intends to construct a 3.5 meter
optical–infrared telescope at a site in south–central New Mexico.
The use of innovative mirror technology, a fast primary, and an
alt–azimuth mounting results in a compact and lightweight instrument. This telescope will be uniquely well–suited for addres-
sing certain observational programs by virtue of its capability for
fully remote operation and rapid instrument changes.

CTI data–management and interrogation.
See Abstr. 002.082.

**The UK Schmidt Telescope Objective Prisms. II. Catalogue of
objects and technical data.**
See Abstr. 002.110.

Telescopes for the 1980s.
See Abstr. 003.071.

All the astrolabes.
See Abstr. 003.162.

The liquid–mirror telescope – an early example of kiwi ingenuity?
See Abstr. 004.055.

**The Ritchey–Chrétien reflecting telescope: half a century from
conception to acceptance.**
See Abstr. 004.072.

**Report of the optical conference on the 7.6–meter telescope held in
Austin, Texas, March 22 – 24, 1982.**
See Abstr. 012.013.

Keck Foundation offers Caltech $70 million for 10–m telescope.
See Abstr. 013.004.

All hail the Keck ten–meter telescope project.
See Abstr. 013.005.

**A comparative study of the optical charateristics of astrograph
objectives (Carl Zeiss, Jena). Limiting magnitude.**
See Abstr. 031.006.

The Multiple Mirror Telescope as a phased array telescope.
See Abstr. 031.013.

**Zero–coma condition for decentered and tilted secondary mirror in
Cassegrain/Nasmyth configuration.**
See Abstr. 031.038.

Vignetting in three–mirror telescopes.
See Abstr. 031.046.

**Optical/IR telescope arrays: a study of their potential for satisfying
observational requirements in astronomy for a large aperture system in the 1990s and beyond.**
See Abstr. 031.047.

Fibre optics at the UK Schmidt Telescope.
See Abstr. 034.044.

**Progress in the development of a durable silver–based high–
reflectance coating for astronomical telescopes.**
See Abstr. 034.095.

Science with the Solar Optical Telescope (SOT).
See Abstr. 035.033.

**Notes on telescope drive automation. I. The precalculation of slew
command strings.**
See Abstr. 036.107.

**Photoelectric meridian circles for planetary theory and stellar dy-
namics.**
See Abstr. 041.020.

Latest instruments and methods in the astrogeodetic field.
See Abstr. 045.023.

033 Radio Telescopes and Equipment

033.001 Calibration of the instrumental polarization of radio telescopes.
Z. Turlo, T. Forkert, W. Sieber, W. Wilson.
Astron. Astrophys., Vol. 142, No. 1, p. 181 – 188 (1985).
  A calibration scheme is described which allows the determination of the instrumental polarization characteristics of radio telescopes with good accuracy, reliability and repeatability. The method is based on an estimate of the instrumental Müller matrix made by observing known calibration sources.

033.002 First results of a 36 GHz SIS receiver.
D. Crete, J. C. Pernot, C. Letrou, P. Encrenaz.
Int. J. Infrared Millimeter Waves, Vol. 5, No. 7, p. 965 – 970 (1984). Abstr. in Phys. Abstr., Vol. 88, No. 1248, Entry 9602 (1985).

033.003 A cross–correlation receiver for radio astronomy employing quadrature channel generation by computed Hilbert transform.
W. F. Lo, P. E. Dewdney, T. L. Landecker, D. Routledge, J. F. Vaneldik.
Radio Sci., Vol. 19, No. 5, p. 1413 – 1421 (1984). Abstr. in Phys. Abstr., Vol. 88, No. 1249, Entry 14473 (1985).

033.004 A very low–noise receiver for 80 – 120 GHz.
J. W. Archer, M. T. Faber.
Int. J. Infrared Millimeter Waves, Vol. 5, No. 8, p. 1069 – 1081 (1984). Abstr. in Phys. Abstr., Vol. 88, No. 1250, Entry 18646 (1985).

033.005 Receivers for RT–3 Torun radio telescope.
A. J. Kus, S. Gorgolewski, A. Kepa, B. Krygier, J. Mazurek, E. Pazderski.
Postepy Astron., Vol. 31, No. 4, p. 303 – 314 (1983). In Polish. Abstr. in Phys. Abstr., Vol. 88, No. 1250, Entry 18651 (1985).

033.006 The very long baseline interferometry. III. VLBI data processing.
K. M. Borkowski.
Postepy Astron., Vol. 31, No. 4, p. 256 – 277 (1983). In Polish. Abstr. in Phys. Abstr., Vol. 88, No. 1250, Entry 18675 (1985).

033.007 Low–noise heterodyne receivers for near–millimeter–wave radio astronomy.
J. W. Archer.
Proc. IEEE, Vol. 73, No. 1, p. 109 – 130 (1985).
  Increased interest in the near–millimeter wavelength region, covering the range 3 mm to 300 μm, during the past decade has stimulated the development of sensitive heterodyne receivers for a wide range of applications. This review paper considers current low–noise receiver technology with emphasis on applications in radio astronomy. A brief discussion of the astrophysical importance of radio astronomy at millimeter wavelengths is presented. The concepts of receiver design and the particular problems associated with this region of the spectrum are discussed.

033.008 El radioespectrómetro.
A. E. Osorio.
Rev. Astron., Tomo 56, No. 230, p. 14 – 15 (1984).

033.009 The peculiarities of using radioholography for testing the radio telescope RATAN–600.
G. A. Pinchuk, A. A. Stotskij.
Astrofiz. Issled. Izv. Spets. Astrofiz. Obs., Tom 19, p. 71 – 75 (1985). In Russian. English translation in Bull. Spec. Astrophys. Obs. – North Caucasus.
  The peculiarity of reconstruction of the amplitude and phase distribution on the aperture of a wide–angle focusing system by the radioholography method is considered. It is shown that when the Fourier transform is used, errors in the reconstruction of the amplitude appear. The results of mathematical modelling of the radioholography process for the RATAN–600 radio telescope in autocollimation mode are given.

033.010 ”Slipping” method at the radio telescope RATAN–600.
M. G. Mingaliev, Z. E. Petrov, V. I. Filipenko, L. N. Cherkov.
Astrofiz. Issled. Izv. Spets. Astrofiz. Obs., Tom 19, p. 76 – 81 (1985). In Russian. English translation in Bull. Spec. Astrophys. Obs. – North Caucasus.
  A ”slipping” method of observations at RATAN–600 allowing to increase the integration time is described. Within a tolerable range of aberration a source can be tracked by simple shift of the feed horns along the focal line of the secondary reflector. The results of effectiveness evaluation are given.

033.011 On the choice of a polarimeter for measuring linear polarization of radio emission with the radio telescope RATAN–600.
V. I. Abramov, D. V. Korol’kov.
Astrofiz. Issled. Izv. Spets. Astrofiz. Obs., Tom 19, p. 88 – 92 (1985). In Russian. English translation in Bull. Spec. Astrophys. Obs. – North Caucasus.
  The radio telescope RATAN–600 has large instrumental circular polarization. Since the polarimeter is not ideal the circular polarization can transform into linear one that restricts the accuracy of the linear polarization measurements. Three schemes of polarimeters are considered; the demands to the construction accuracy of the main elements of the polarization tract from the view point of the instrumental linear polarization decrease are defined.

033.012 Investigation of the instrumental linear polarization of the southern sector with a flat reflector of RATAN–600 at 13 cm.
V. I. Abramov, E. N. Vinyajkin.
Astrofiz. Issled. Izv. Spets. Astrofiz. Obs., Tom 19, p. 93 – 100 (1985). In Russian. English translation in Bull. Spec. Astrophys. Obs. – North Caucasus.
  The instrumental linear polarization of the south sector with a flat reflector of RATAN–600 has been measured at 13 cm in intervals of elevation angles of 17 – 87°. Its value amounts to 1.9 – 2.7% and the angle dependence is non–monotonic. The effect is considered in detail of the extension of cosmic sources used for the measurements of the instrumental linear polarization and partial linear polarization of the radio emission of some of these sources on the results of measurements.

033.013 Experimental research of the southern sector with a flat mirror of the RATAN’s patterns.
A. V. Temirova.
Astrofiz. Issled. Izv. Spets. Astrofiz. Obs., Tom 19, p. 101 – 108 (1985). In Russian. English translation in Bull. Spec. Astrophys. Obs. – North Caucasus.
  The results of an experimental research of the RATAN’s patterns at different transverse shifts of the primary feed from the focus antenna are discussed. Measurement of patterns was carried out using radiation of intense point sources. The measured patterns were compared with calculated ones.

033.014 The Australia Telescope – the radio astronomer’s highway to the future.
R. H. Frater.
Proc. Astron. Soc. Aust., Vol. 5, No. 4, p. 440 – 445 (1984).

033.015 The Molonglo Observatory transient event recorder.
M. I. Large, A. E. Vaughan, J. M. Durdin, A. G. Little.
Proc. Astron. Soc. Aust., Vol. 5, No. 4, p. 569 – 574 (1984).
  In its normal synthesis mode of operation, the Molonglo Observatory Synthesis Telescope (MOST) tracks a region of sky for a period of 12 hours with 64 real–time fan beams having high

sensitivity at 843 MHz (Mills 1981). It thus provides an excellent opportunity to monitor the sky at the same time for transient radio events. This technique for recognizing extra–terrestrial sources was of considerable value in the first Molonglo pulsar search when only two beams were used. The transient event recorder has been designed to exploit the characteristics of the MOST over the widest possible range of transient time scales.

033.016   The FST – a 20 arc second synthesis telescope.
       I. G. Jones, A. Watkinson, P. C. Egau,
T. M. Percival, D. J. Skellern, G. R. Graves.
Proc. Astron. Soc. Aust., Vol. 5, No. 4, p. 574 – 578 (1984).
The baseline of East–West array of the Fleurs Synthesis Telescope (FST) has been extended from 786 metres to 1585 metres by the addition of two 13.7 metre dishes. A digital receiver has been built to accommodate the extra delay and correlation requirements, low noise FET preamplifiers have been installed on the large antennas, and a software package has been developed for processing observation data on a VAX 11/780. The FST is now capable of producing accurate wide field maps at 1415 MHz with a resolution of 20 arc seconds and a sensitivity of several milliJansky.

033.017   Acousto–optical spectrometer for radio astronomy.
       N. A. Esepkina, N. F. Ryzhkov,
S. V. Pruss–Zhukovskij, Yu. A. Kotov, A. I. Shishkin,
V. G. Grachev, I. A. Kochergina, L. Eh. Abramyan.
Spets. astrofiz. obs. Akad. Nauk SSSR, Prepr., No. 11L, 35 pp. (1984). In Russian. Abstr. in Ref. Zh., 51. Astron., 3.51.995 (1985).

033.018   The Arecibo–Los Caños spectral line interferometer.
       S. R. Kulkarni, K. C. Turner, C. Heiles,
J. M. Dickey.
Astrophys. J., Suppl. Ser., Vol. 57, No. 3, p. 631 – 642 (1985).
The authors describe the technical aspects and the astronomical performance of a new 21 cm spectral line interferometer that is now operational at the Arecibo Observatory. The interferometer consists of the 305 m Arecibo spherical reflector at one end and the 30.5 m Los Caños paraboloid at the other end. At 20 cm, the interferometer has a gain of 1 K Jy^{-1} and a fringe spacing of 4″; these factors, together with the good velocity resolution of the spectrometer, make the interferometer ideal for galactic H I absorption spectroscopy.

033.019   Observation system for cosmic radio telescope.
       A. Kurokawa, H. Oosawa, T. Kawata.
Fujitsu, Vol. 35, No. 4, p. 455 – 464 (1984). In Japanese. Abstr. in Phys. Abstr., Vol. 88, No. 1252, Entry 29845 (1985).

033.020   The radiometer and polarimeters at 80, 35, and 17 GHz for solar observations at Nobeyama.
H. Nakajima, H. Sekiguchi, M. Sawa, K. Kai, S. Kawashima, T. Kosugi, N. Shibuya, N. Shinohara, Y. Shiomi.
Publ. Astron. Soc. Jpn., Vol. 37, No. 1, p. 163 – 170 (1985).
An 80 GHz whole sun radiometer with a new technique for reducing the atmospheric effect has been completed at Nobeyama. The radiometer has been proved to have the capability of detecting small bursts with the flux density of down to ~10 sfu (solar flux unit) under usual weather conditions; the detection limit has been lowered by one and a half orders of magnitude compared with that achieved with radiometers of a conventional type.

033.021   VLBA – a continent–size radio telescope.
       M. A. Gordon.
Sky Telesc., Vol. 69, No. 6, p. 487 – 490 (1985).

033.022   The Hamilton radio telescope.
       K. R. Gledhill, B. S. Liley.
South. Stars, Vol. 31, No. 1, p. 41 – 52 (1984).
A radio telescope with a three metre diameter parabolic antenna has been built at the Physics Department of the University of Waikato. It operates in the 1400 – 1427 MHz radio astronomy frequency band, and can detect densities of radio flux as low as 100 Jy. The instrument was tested by observing transits of the Sun and the Galactic plane.

033.023   Fundamental mode waveguide at submillimetre wavelengths.
A. R. Gillespie, A. Schulz.
Electron. Lett., Vol. 19, No. 12, p. 440 – 441 (1983). = Max–Planck–Inst. Radioastron., Bonn, Sonderdr. Ser. A, Nr. 628.
Experiments with a TE$_{10}$ waveguide at 460 GHz are described using a simple cross–guide multiplier and a Golay cell. Electroforming production techniques give components with high attenuation, but unconventional methods give acceptable losses.

033.024   Mott–barrier mixer diodes with improved semiconductor profiles for cooled operation at submillimetre wavelengths.
N. J. Keen.
IEE Proc., Vol. 130, Part 1, No. 4, p. 171 – 174 (1983). = Max–Planck–Inst. Radioastron., Bonn, Sonderdr. Ser. A, Nr. 644.
The use of thick, low–doped epitaxial profiles permits lower noise, lower capacitance Mott–barrier mixer diodes to be fabricated without complicating diode manufacturing technology. The paper reports model predictions for metal–GaAs diodes made from such materials.

033.025   Design features of a 10 m telescope for submillimeter astronomy.
J. W. M. Baars, P. G. Mezger, B. L. Ulich, W. F. Hoffmann, R. E. Parks.
Proc. SPIE Int. Soc. Opt. Eng., Vol. 444, p. 65 – 71 (1984). = Max–Planck–Inst. Radioastron., Bonn, Sonderdr. Ser. A, Nr. 649.
The Max–Planck–Institut für Radioastronomie and Steward Observatory are jointly planning the establishment of a submillimeter observatory on Mount Lemmon, Arizona. The telescope will have a diameter of 10 meter and enable diffraction–limited operation at a shortest wavelength of 350 µm. Thus the reflector surface accuracy will be about 15 µm and the pointing accuracy 1″. This paper describes the design of the telescope.

033.026   Desgin and software aspects for the control system of the 30 m MRT.
J. Schraml, W. Brunswig, G. Juen.
Proc. SPIE Int. Soc. Opt. Eng., Vol. 444, p. 122 – 126 (1984). = Max–Planck–Inst. Radioastron., Bonn, Sonderdr. Ser. A, Nr. 652.
The 30 m Millimeter Radiotelescope (MRT) will have a beamwidth of less than 10″ when operated at a wavelength of 1.2 mm. It is an open air telescope located on a mountain ridge in southern Spain. There the instrument is exposed to severe environmental influences, especially wind and temperature changes. The pointing and tracking accuracy required is of the order of a few seconds of arc. Simulations have shown that these specifications cannot be met with conventional servo design. An improvement of performance can be obtained applying the modern concept of state control.

033.027   Low–cost enclosure for the sub–millimeter telescope.
       B. L. Ulich, W. F. Hoffmann, W. B. Davison,
J. W. M. Baars, P. G. Mezger.
Proc. SPIE Int. Soc. Opt. Eng., Vol. 444, p. 72 – 76 (1984). = Max–Planck–Inst. Radioastron., Bonn, Sonderdr. Ser. A, Nr. 653.
The University of Arizona and the Max–Planck–Institut für Radioastronomie are collaborating to construct a sub–millimeter wavelength radio telescope facility at the summit of Mt. Lemmon (2791 m above sea level) near Tucson, Arizona. The authors have designed a corotating building to protect the 10 m diameter Sub–Millimeter Telescope (SMT) against storm damage, to provide large instrumentation rooms at the Nasmyth foci, and to minimize degradation of the reflector profile accuracy and pointing errors caused by wind forces and solar radiation.

033.028 High resolution position and velocity measurement with incremental encoders.
U. Beckmann, R. Bardenheuer.
Proc. SPIE Int. Soc. Opt. Eng., Vol. 444, p. 127 – 131 (1984). = Max–Planck–Inst. Radioastron., Bonn, Sonderdr. Ser. A, Nr. 654.

The use of optical telescopes for infrared and radioastronomy, where observations can also be made in daytime, requires reproducible positioning. Exceptionally good tracking is demanded particularly for radioastronomy, when the observation range extends to the millimeter– and submillimeter wavelengths. For radio telescopes with a small antenna beamwidth (below 1 arcmin) position errors should not exceed a few arcsec. The measurement of magnitudes in the arcsec region can be done with commercially available incremental encoders. For these encoders electronic circuits have been developed to guarantee the required resolution. The design of the electronic circuits also provide possibilities to correct encoder errors and interface to the host computer.

033.029 Technology of large radio telescopes for millimeter and submillimeter wavelengths.
J. W. M. Baars.
Max–Planck–Inst. Radioastron., Bonn, Sonderdr. Ser. A, Nr. 658, 41 pp. (1983). = Chapter 5 of ”Infrared and millimeter waves”, Vol. 9 (1983).

033.030 Hot–electron noise generation in gallium–arsenide Schottky–barrier diodes.
N. J. Keen, H. Zirath.
Electron. Lett., Vol. 19, No. 20, p. 853 – 854 (1983). = Max–Planck–Inst. Radioastron., Bonn, Sonderdr. Ser. A, Nr. 666.

Measurements of excess (hot–electron) noise in a GaAs Schottky–barrier diode at room temperature are reported. A monotonic decrease with increasing frequency is found. It is concluded that only diode hot–electron noise at intermediate frequency contributes to the excess noise of a mixer.

033.031 Investigation of a heterodyne receiver with open structure mixer at 324 GHz and 693 GHz.
H. P. Röser, E. J. Durwen, R. Wattenbach, G. V. Schultz.
Int. J. Infrared Millimeter Waves, Vol. 5, No. 3, p. 301 – 314 (1984). = Max–Planck–Inst. Radioastron., Bonn, Sonderdr. Ser. A, Nr. 673.

033.032 A novel quasi–optical frequency multiplier design for millimeter and submillimeter wavelengths.
J. W. Archer.
IEEE Trans. Microwave Theory Tech., Vol. 32, No. 4, p. 421 – 426 (1984). = Natl. Radio Astron. Obs., Repr., Ser. A, No. 1392.

The paper gives a description of the waveguide and stripline structures which bring the power to the diode at the pump frequency. The practical constraints on this aspect of the design are discussed using the tripler mount for 200 – 280 GHz output, as an example.

033.033 An efficient 200 – 290 GHz frequency tripler incorporating a novel stripline structure.
J. W. Archer.
IEEE Trans. Microwave Theory Tech., Vol. 32, No. 4, p. 416 – 420 (1984). = Natl. Radio Astron. Obs., Repr., Ser. A, No. 1393.

This paper describes a broadly tuneable frequency tripler which can provide more than 2–mW output power at any frequency between 200 and 290 GHz. It is derived from an earlier narrow–band prototype design.

033.034 A very low–noise receiver for 80 – 120 GHz.
J. W. Archer, M. T. Faber.
Int. J. Infrared Millimeter Waves, Vol. 5, No. 8, p. 1069 – 1081 (1984). = Natl. Radio Astron. Obs., Repr., Ser. B, No. 538.

A new dual–polarized cryogenic Schottky barrier mixer receiver for radio astronomy applications is described. Novel features include a very broadband, fixed tuned mixer design and a compact, low–loss, linear polarization diplexing scheme.

033.035 An SIS mixer for 90 – 120 GHz with gain and wide bandwidth.
L. R. D'Addario.
Int. J. Infrared Millimeter Waves, Vol. 5, No. 11, p. 1419 – 1442 (1984). = Natl. Radio Astron. Obs., Repr., Ser. B, No. 540.

An SIS mixer for the 3 mm wavelength band has been developed. It has sufficient RF bandwidth to allow double–sideband operation at an IF of 1.4 GHz.

033.036 The overall calibration for a special equatorial radio telescope.
J.-s. Wang.
Publ. Purple Mt. Obs., Vol. 3, No. 3, p. 91 – 106 (1984). In Chinese.

Using the solar radio emission to measure the cm–wave atmospheric refraction, a special equatorial radio telescope is to be calibrated so that the measurement errors are less than 20 arcsec. This paper states the overall calibration of that radio telescope in detail.

033.037 Das Kölner 3 m–Radioteleskop: I. Die Empfänger.
M. Bester, W. Hilberath, K. Jacobs, B. Vowinkel, G. Winnewisser.
Mitt. Astron. Ges., Nr. 63, p. 111 – 115 (1985). – See Abstr. 012.063.

Auf Grund der hohen Oberflächengenauigkeit des Hauptreflektors von 30 µm (rms) ist das Teleskop bis ca. 600 GHz (0.5 mm) einsetzbar. Die entsprechenden Empfänger sind im Sekundärfokus des Cassegrain–Systems untergebracht. Momentan steht ein Heterodynempfänger mit Schottky–Mischer für den Bereich 70 – 90 GHz zur Verfügung. Ein zweiter Empfänger mit SIS–Mischer für den Bereich 140 – 150 GHz ist zur Zeit im Aufbau. Weitere Empfänger für 230 GHz und für den Submm–Bereich sind in Planung.

033.038 Das Kölner 3 m–Radioteleskop: II. Die Spektrometer.
R. Ewald, G. Winnewisser, W. Zensen.
Mitt. Astron. Ges., Nr. 63, p. 116 – 121 (1985). – See Abstr. 012.063.

Akusto–optische Spektrometer und Filterspektrometer werden beschrieben, sowie deren Anwendung bei verschiedenen Linienprofilen.

033.039 Das Kölner 3 m–Radioteleskop: III. Steuerung und Datenerfassung.
M. Olberg, G. Rau, G. Winnewisser.
Mitt. Astron. Ges., Nr. 63, p. 122 – 125 (1985). – See Abstr. 012.063.

Der Artikel beschreibt die Hardwarekonfiguration des neuen Kölner 3 m–Radioteleskops. Die einzelnen Komponenten dienen zur Steuerung und Datenerfassung.

033.040 Development of low–noise receivers at shorter-millimeter wavelengths.
H. Ogawa, Y. Fukui.
Astron. Her., Vol. 78, No. 6, p. 152 – 155 (1985). In Japanese.

033.041 Time series modeling of atmospherically disturbed VLA phase data.
J. P. Basart, Y. Zheng.
Bull. Am. Astron. Soc., Vol. 16, No. 4, p. 886 (1984). Abstract. – See Abstr. 010.062.

033.042 The UA/MPIfR submillimeter telescope: a progress report on work at the University of Arizona.
B. L. Ulich, W. B. Davison, W. F. Hoffmann, P. A. Strittmatter, R. E. Parks, R. R. Shannon.
Bull. Am. Astron. Soc., Vol. 16, No. 4, p. 986 (1984). Abstract. – See Abstr. 010.062.

033.043 Investigation of the instrumental system of the northern sector of RATAN–600.
P. M. Afanas'ev, G. A. Pinchuk, A. L. Pozhakov, V. A. Fomin.
Spets. astrofiz. Obs. Akad. Nauk SSSR. Prepr., No. 14 L, 13 pp. (1984). In Russian. Abstr. in Ref. Zh., 51. Astron., 5.51.842 (1985).

033.044 Measurements with a Schottky–barrier waveguide mixer at 460 GHz.
N. J. Keen, K.–D. Mischerikow, G. A. Ediss, R. Engelhardt, E. Perchtold, M. Vester.
Infrared Phys., Vol. 25, No. 1/2, p. 353 – 356 (1985). – See Abstr. 012.065.

033.045 Low–noise fixed tuned, broadband mixer for 200 – 270 GHz.
J. W. Archer, M. T. Faber.
IEEE MTT–S International Microwave Symposium Digest 1984. Expanding Microwave Horizons, San Francisco, Calif., USA, 29 May – 1 June 1984, p. 557 – 559 (1984). Abstr. in Phys. Abstr., Vol. 88, No. 1253, Entry 35547 (1985).

033.046 The UK/NL millimeter wave telescope.
R. W. Newport.
Endeavour New Ser., Vol. 8, No. 4, p. 159 – 165 (1984). Abstr. in Phys. Abstr., Vol. 88, No. 1254, Entry 40481 (1985).

033.047 Radio astronomy application of Josephson junctions.
J. Inatani.
Oyo Buturi, Vol. 53, No. 6, p. 526 – 531 (1984). In Japanese. Abstr. in Phys. Abstr., Vol. 88, No. 1254, Entry 40483 (1985).

033.048 Spatial filtering radio astronomical data: one–dimensional case.
H. E. Rowe.
AT T Bell Lab. Tech. J., Vol. 63, No. 9, p. 1997 – 2031 (1984). Abstr. in Phys. Abstr., Vol. 88, No. 1254, Entry 40500 (1985).

033.049 A continuous comparison radiometer at 97 GHz.
C. R. Predmore, N. R. Erickson, G. R. Huguenin, P. F. Goldsmith.
IEEE Trans. Microwave Theory Tech., Vol. MTT–33, No. 1, p. 44 – 51 (1985). Abstr. in Phys. Abstr., Vol. 88, No. 1255, Entry 45922 (1985).

033.050 Interferometer measurements of atmospheric phase noise at 86 GHz.
J. H. Bieging, J. Morgan, W. J. Welch, S. N. Vogel, M. C. H. Wright.
Radio Sci., Vol. 19, No. 6, p. 1505 – 1509 (1984). Abstr. in Phys. Abstr., Vol. 88, No. 1255, Entry 45931 (1985).

033.051 Coherence limits in VLBI observations at 3–millimeter wavelength.
A. E. E. Rogers, A. T. Moffet, D. C. Backer, J. M. Moran.
Radio Sci., Vol. 19, No. 6, p. 1552 – 1560 (1984). Abstr. in Phys. Abstr., Vol. 88, No. 1255, Entry 45932 (1985).

033.052 Heterodyne receiver for short millimeter wave radio-astronomy.
E. Ogawa, Y. Hayashi.
Proc. Infrared Soc. Jpn., No. 9, p. 22 – 28 (1984). In Japanese. Abstr. in Phys. Abstr., Vol. 88, No. 1257, Entry 57616 (1985).

033.053 85–115 GHz receivers for radio astronomy.
D. P. Woody, R. E. Miller, M. J. Wengler.
IEEE Trans. Microwave Theory Tech., Vol. MTT–33, No. 2, p. 90 – 95 (1985). Abstr. in Phys. Abstr., Vol. 88, No. 1258, Entry 63216 (1985).

033.054 Simultation of a radiotelescope with the aid of the block orientated simulation language ISRSIM.
G. Juen, V. Maass, M. Zeitz.
Simulationstechnik. 2. Symposium Simulationstechnik, held at Wien, Austria, 25 – 27 September 1984. Springer–Verlag, Berlin – Heidelberg – New York – Tokyo (1984). p. 636 – 640. In German.

033.055 Quasioptical components for 230 GHz and 460 GHz.
G. A. Ediss, S. J. Wang, N. J. Keen.
IEE Proc. H, Vol. 132, No. 2, p. 99 – 106 (1985). Abstr. in Phys. Abstr., Vol. 88, No. 1259, Entry 68677 (1985).

033.056 Halley's comet exploration and Usuda's large antenna.
T. Nomura, T. Hayashi, H. Matsuo, H. Hirosawa, M. Ichikawa.
J. Inst. Electron. Commun. Eng. Jpn., Vol. 67, No. 10, p. 1039 – 1044 (1984). In Japanese. Abstr. in Phys. Abstr., Vol. 88, No. 1259, Entry 68734 (1985).

033.057 First observations with a new pulsar signal averager.
L. A. Rawley, J. H. Taylor, J. M. Weisberg, M. M. Davis.
Bull. Am. Astron. Soc., Vol. 17, No. 2, p. 555 (1985). Abstract. – See Abstr. 010.065.

033.058 Geodesic adjustment of RATAN–600: increasing the accuracy of the main reflector.
Yu. K. Zverev.
Soobshch. Spets. Astrofiz. Obs., Vyp. 44, p. 53 – 68 (1984). In Russian.
The geodesic measurements at the north sector of the round reflector of the RATAN–600 radio telescope, carried out for the purpose of the determination of the main errors of the reflector elements in the working position, are described.

033.059 Proposed U. C. Berkeley fast pulsar search machine.
S. R. Kulkarni, D. C. Backer, D. Werthimer, C. Heiles.
Birth and evolution of neutron stars: issues raised by millisecond pulsars, p. 245 – 251 (1984). – See Abstr. 012.115.
With the discovery of 1937 + 21 by Backer et al. (1982) there is much renewed interest in an all sky survey for fast pulsars. At U. C. Berkeley the authors have designed and are in the process of building an innovative and powerful, stand–alone, real–time, digital signal–processor to conduct an all sky survey for pulsars with rotation rates as high as 2000 Hz and dispersion measures less than 120 $cm^{-3}pc$ at 800 MHz. The machine is anticipated to be completed in the Fall of 1985.

033.060 Potential affiliate stations to work with the EVN (*European VLBI Network*).
P. N. Wilkinson.
ESA Spec. Publ., ESA SP–213, p. 111 – 113 (1984). – See Abstr. 012.124.
It is vital to demonstrate that EVN–based ground arrays working in concert with Quasat can produce high quality images. The author made a study of how various potential affiliate antennas would extend the u, v coverage of the EVN and tested the imaging performance of EVN–based arrays.

033.061 Correlator and data processing requirements for Quasat.
D. H. Roberts.
ESA Spec. Publ., ESA SP–213, p. 115 – 118 (1984). – See Abstr. 012.124.
The use of a space–borne antenna for VLBI will place new requirements on the correlator and data–processing systems. The large orbit and rapid motion of the satellite lead to delays, delay rates, fringe rates, fringe accelerations, and Doppler bandshifts substantially greater than those encountered with ground–based antennas alone. The scheduled updating of the Mark III correlator, or small modifications to the planned Very Long Baseline Array (VLBA) correlator, will enable the direct processing of Quasat data.

033.062   Limitation of the VLBI angular resolution.
L. I. Matveyenko (*L. I. Matveenko*).
ESA Spec. Publ., ESA SP–213, p. 119 – 121 (1984). – See Abstr. 012.124.

The limit of angular resolution in VLBI is determinated by scattering and by sensitivity. The maximum baseline is equal to 100000 km at cm wavelengths and the angular resolution may reach $\sim 40\ \mu as$.

033.063   Processing system for space VLBI.
S. V. Pogrebenko, M. V. Popov, G. S. Tzarevsky (*G. S. Tsarevskij*).
ESA Spec. Publ., ESA SP–213, p. 123 – 125 (1984). – See Abstr. 012.124.

The main characteristics of the ground/space VLBI preliminary data processing (PDP) system are derived from consideration of the assumed values of delay and fringe rate and their expected errors.

033.064   Quasat and the Canadian Long Baseline Array.
E. R. Seaquist.
ESA Spec. Publ., ESA SP–213, p. 127 (1984). – See Abstr. 012.124.

The proposed Canadian Long Baseline Array (CLBA) should be considered as a potential contributor to the ground arrays for Quasat. The scope and status of the CLBA proposal are briefly outlined.

Radiotelescopes.
See Abstr. 003.013.

The Jodrell Bank telescopes.
See Abstr. 003.026.

The compound interferometer – a precursor of smart arrays.
See Abstr. 004.184.

The development of Michelson and intensity long baseline interferometry.
See Abstr. 004.185.

Early interferometry at Jodrell Bank.
See Abstr. 004.186.

Usuda Deep Space Ground Station and its large antenna.
See Abstr. 009.012.

50 years of radio astronomy.
See Abstr. 013.012.

Radioastronomy – where it has come from – where it is going.
See Abstr. 013.013.

Communications with outer space – radio astronomy.
See Abstr. 013.048.

35 years, together with the solar radio waves.
See Abstr. 013.063.

Interference to radio astronomy from satellite transmitters.
See Abstr. 013.068.

Opportunities for high angular and temporal resolutions of galactic objects at radio frequencies.
See Abstr. 013.097.

A radar for the exploration of extrasolar planets.
See Abstr. 015.052.

Receiver technology for the submillimeter–wave region.
See Abstr. 034.098.

A solution to the short spacing problem in radio interferometry.
See Abstr. 036.035.

Phase–defined ground–orbital aperture synthesis system.
See Abstr. 036.140.

Space astronomy: technical comments – interferometry missions.
See Abstr. 051.078.

The Quasat mission: an overview.
See Abstr. 051.115.

Some prospects of space VLBI.
See Abstr. 051.116.

Space VLBI studies in Japan.
See Abstr. 051.117.

Quasat: technical aspects of the proposed mission.
See Abstr. 051.118.

A study of the imaging potential of Quasat.
See Abstr. 051.119.

The wide–range space radiointerferometer.
See Abstr. 051.120.

A short note on the high–precision navigation of Quasat.
See Abstr. 052.078.

Spatio–temporal features of the development of microwave emission of active regions and flares.
See Abstr. 077.004.

Astrophysik interstellarer Moleküle.
See Abstr. 131.200.

The relevance of Quasat to the star formation problem.
See Abstr. 131.370.

Galactic OH and H_2O masers.
See Abstr. 131.371.

H_2O masers and distance measurements: the impact of Quasat.
See Abstr. 131.373.

Determination of flux densities of radio sources using broad–band radiometers of the radio telescope RATAN–600.
See Abstr. 141.005.

Report of IAU Commission 40: Radio astronomy (*Radio astronomie*).
See Abstr. 141.039.

The environment of active galactic nuclei.
See Abstr. 158.348.

The compact radio emission regions in active galactic nuclei.
See Abstr. 158.349.

Combined MERLIN/VLA observations of the superluminal quasar 3C 179.
See Abstr. 159.046.

Quasat, gravitational lenses, and studies of the interstellar, intracluster and intergalactic medium.
See Abstr. 160.164.

034 Auxiliary Instrumentation, Photographic Materials, Clocks

034.001 Comparative performances of an objective grating concept in UV spectroscopic observations of solar system objects. I. Imaging capabilities.
P. Bruston, M. Choucq–Bruston.
Astron. Astrophys., Vol. 142, No. 1, p. 59 – 70 (1985).

Among different optical designs proposed for a far UV spectroscopic observatory, an objective grating instrument has interesting performances for the study of solar system objects and extended sources since it combines imaging capability with high spectral and spatial resolution. In the present paper, the authors give a schematic description of such an instrument and demonstrate its capabilities through the discussion of the method of data analysis to be used. Application of data simulation programs to a simple model of the Io torus is presented in order to clearly define the basic concept of the instrument which calls for integrated spectral and spatial data. These simulations demonstrate instrument performances and confirm the large advantage of its imaging capabilities when compared to a slit spectrograph.

034.002 On the equivalence of electromagnetic and clock–transport synchronization in noninertial frames and gravitational fields.
H. Rumpf.
Z. Naturforsch., A, Band 40a, Heft 1, p. 92 – 93 (1985).

Synchronization by slow clock transport is shown to be equivalent to that by electromagnetic signals for clocks moving along the trajectories of a timelike Killing vector field, provided the gravitational redshift is corrected for and the synchronization paths are the same.

034.003 La rétrogradation de l'ombre dans les cadrans solaires analemmatiques.
J.–P. Parisot.
J. Hist. Astron., Vol. 16, Part 1, No. 45, p. 43 – 48 (1985).

034.004 Quick–look photo–astrometry with a linear micrometer.
C. Townsend, J. S. Hanssen.
J. Br. Astron. Assoc., Vol. 95, No. 2, p. 62 – 66 (1985).

The purpose of this article is to describe an astrometry technique developed by the authors as part of a self–sponsored sky patrol programme known as "Project Sky Scan". Since the technique simply entails the integration of an inexpensive microcomputer (or programmable calculator) and a precisison linear micrometer with photographic equipment already available to many amateur astronomers, it is expecially easy to implement. If 125 × 175–millimetre (or larger) negatives, positives or prints are utilized, the resultant comet, asteroid or nova co–ordinates are usually accurate enough for reporting to the IAU.

034.005 The design and construction of a charge–coupled device imaging system.
D. H. Gudehus, D. J. Hegyi.
Astron. J., Vol. 90, No. 1, p. 130 – 138 (1985).

The design and construction of a charge–coupled device (CCD) imaging system optimized for the detection of low surface brightness objects is discussed in detail. In particular, the authors describe the design of the Dewar and the CCD electronics, as well as the choice of CCD operating voltages, clock timings, and operating temperature. In addition, they discuss the measurement of readout and spatial noise for an RCA SID53612 thinned CCD (320 by 512 pixels).

034.006 The implications of photomultiplier fatigue for pulse–counting photometry.
W. A. Rosen, F. R. Chromey.
Astron. J., Vol. 90, No. 1, p. 139 – 143 (1985).

Fatigue is an unavoidable effect in all photomultiplier tubes. The authors have conducted laboratory experiments to investigate the importance of fatigue as a source of error in astronomical pulse–counting photometry. It is found that astronomers using common photometric practices can expect fatigue to introduce unpredictable errors which may approach the 1% level. The authors suggest one possible scheme that observers might use to monitor for the presence of fatigue during an observing run.

034.007 Observations of small air showers at Mt. Norikura.
S. Sakakibara, T. Yamada, M. Orito, M. Hayase.
Proc. Cosmic–Ray Res. Lab. Nagoya Univ., Vol. 27, No. 1, p. 1 – 50 (1984). In Japanese. Abstr. in Phys. Abstr., Vol. 88, No. 1248, Entry 9623 (1985).

034.008 A two kiloton liquid argon detector for solar neutrinos and proton decay.
K. L. Giboni.
Nucl. Instrum. Methods Phys. Res., Sect. A, Vol. 225, No. 3, p. 579 – 582 (1984). Abstr. in Phys. Abstr., Vol. 88, No. 1250, Entry 18656 (1985).

034.009 Review of the Monte Bianco, Frejus and Gran–Sasso detectors.
J. Ernwein.
Nucl. Instrum. Methods Phys. Res., Sect. A, Vol. 225, No. 3, p. 583 – 591 (1984). Abstr. in Phys. Abstr., Vol. 88, No. 1250, Entry 18657 (1985).

034.010 Interkosmos laser radar, version mode locked train.
B. Kvasil, K. Hamal, H. Jelínková, A. Novotný, I. Procházka, M. Fahim, B. B. Baghos, A. G. Masevich, S. K. Tatevyan.
Publ. Astron. Inst. Czech. Acad. Sci., No. 58, p. 23 – 27 (1984). – See Abstr. 012.018.

The exploitation of a mode locked train laser pulse for satellite ranging has been used. The Nd YAG oscillator/amplifier/frequency doubler/laser generates a mode locked train at 0.53 µm, most of the energy is contained in three pulses, the individual pulse duration is 70 psec. During the 1983 Merit Campaign, within 4 months, 100 low satellite passes and 31 Lageos passes have been ranged. When the mode locked substructure has been resolved for the received signal (50% of passes), the RMS was 6 – 8 cm.

034.011 Microprocessor oriented laser radar electronics.
M. Čech, P. Hirsl.
Publ. Astron. Inst. Czech. Acad. Sci., No. 58, p. 51 – 55 (1984). – See Abstr. 012.018.

This paper deals with a newly developed electronic system used in the INTERKOSMOS satellite tracking stations. The system consists of microprocessor based single board microcomputer, laser clock board with 100 ns accuracy and external synchronization logic, time gate board (fully programmable – delay 100 ns to 1 s, gate 100 ns to 1 ms, both with steps 100 ns) with laser trigger logic and two communication boards. The system is very flexible by means of software. It may be used in the first generation laser radar or the second generation with reprate 10 Hz.

034.012 Mode–locked train laser transmitter.
H. Jelínková.
Publ. Astron. Inst. Czech. Acad. Sci., No. 58, p. 57 – 67 (1984). – See Abstr. 012.018.

The passively mode locked frequency doubled oscillator amplifier Nd YAG laser radar transmitter with the train of 3 – 5 70 psec long pulses, 30 mJ output energy in green and 0.2 mrad divergence in beam with reprate 1 – 2.5 Hz is described.

034.013 Start discriminator for mode locked train laser radar.
M. Čech.
Publ. Astron. Inst. Czech. Acad. Sci., No. 58, p. 69 – 73 (1984). – See Abstr. 012.018.

A problem of start detector is very topical in laser radar station working with a train of pulses generated by mode locked Nd:YAG laser. This article deals with a newly developed start detector used in Intercosmos laser radar station in Helwan, Egypt. The time resolution of the detector is better than 150 ps.

034.014   Intelligent step motor control unit for continuous tracking.
M. Čech.
Publ. Astron. Inst. Czech. Acad. Sci., No. 58, p. 75 – 80 (1984). – See Abstr. 012.018.

The object of this paper is a microprocessor based control unit of step motors. This single–board unit controls simultaneously two step motors. Required data are supplied by the supervisor computer. A new continuous satellite tracking algorithm is used. The maximum speed of the motors (with 800 steps/turn) is 1.2 turns/sec. This unit may be used as inherent part of the laser radar electronic system. The unit allows small operator corrections in aiming during satellite tracking.

034.015   Mode locked train YAG laser calibration experiment.
K. Hamal, I. Procházka, J. Gaignebet.
Publ. Astron. Inst. Czech. Acad. Sci., No. 58, p. 81 – 86 (1984). – See Abstr. 012.018.

To range the satellite, the authors are using the train of picosecond pulses generated by Nd YAG oscillator/amplifier/second harmonic generator laser system. To establish an optimum discriminator/timing system, the indoor and the short baseline outdoor calibration experiments were used. The experimental results indicate a limit single short uncertainty of 6 cm RMS.

034.016   Improving the SLR measurement accuracy by determination of the pulse center of mass.
Z. Neumann.
Publ. Astron. Inst. Czech. Acad. Sci., No. 58, p. 175 – 180 (1984). – See Abstr. 012.018.

In the case of the 30 ns long laser pulses the jitter of distances of the pulse centers of mass is less than the jitter of distances when using the threshold discriminator. The paper describes an analogous case of determination of the pulse center of mass.

034.017   THEMIS: le nouveau magnétographe solaire français pour l'Observatoire des Iles Canaries.
J. Rayrole.
Champs magnétiques stellaires, p. 443 – 452 (1984). – See Abstr. 012.022.

034.018   Zwei preiswerte Sonnenphotometer.
G. Hirth.
Sterne Weltraum, 24. Jahrg., Nr. 2, p. 96 – 97 (1985).

034.019   Photographie extrem schwacher H II–Regionen. Teil 2.
W. E. Celnik, P. Riepe.
Sterne Weltraum, 24. Jahrg., Nr. 2, p. 100 – 103 (1985).

034.020   Photographische Hochkontrastverstärkung astronomischer Negative.
B. Koch.
Sterne Weltraum, 24. Jahrg., Nr. 3, p. 156 – 157 (1985).

034.021   A fibre–optic four–channel photometer.
E. N. Walker.
Vistas Astron., Vol. 27, Part 4, p. 421 – 432 (1984).

034.022   New results with the COSMOS machine.
H. T. MacGillivray, R. S. Stobie.
Vistas Astron., Vol. 27, Part 4, p. 433 – 475 (1984).

The COSMOS system in its present form is described, and some of the more important discoveries that have been made from COSMOS data over the past few years discussed. These discoveries serve to illustrate the usefulness of high–speed, plate-measuring machines as powerful tools in the extraction and analysis of the vast quantity of information contained on astronomical photographs.

034.023   Astronomical imaging with a new infrared sensitive CCD imaging system.
A. Moneti, W. J. Forrest, J. Pipher, C. Woodward.
News Lett. Astron. Soc. N.Y., Vol. 2, No. 4, p. 14 (1983). Abstract. – See Abstr. 010.242.

034.024   Un miroir plan de référence.
J. Funel.
Astronomie, Vol. 99, 199 – 203 (1985).

034.025   Telescoop richten met computer.
A. Westerveld.
Zenit, 12. Jaarg., No. 4, p. 179 – 180 (1985).

034.026   Astrometry of comets using hypersensitized type 2415 film.
E. Everhart.
Cometary astrometry, p. 41 – 44 (1984). – See Abstr. 012.024.

Kodak Technical Pan Film 2415 should be known to those doing cometary astrometry. It has exceedingly fine resolution (320 lines/mm) and, when properly hypersensitized, it is almost as fast as treated IIIa–J plates and reaches fainter stars. Reciprocity failure with the treated film is practically zero, and the shelf life of treated film sheets is about a month at 2°C stored in a nitrogen atmosphere. Over 120 astrometric measures of negatives on this film have shown a median residual error in comet positions of 1.1″, a value that compares favorably with those of most observatories reporting positions.

034.027   The new near–infrared array camera at the University of Rochester.
W. J. Forrest, A. Moneti, C. E. Woodward, J. L. Pipher, A. Hoffman.
Publ. Astron. Soc. Pac., Vol. 97, No. 588, p. 183 – 198 (1985).

A new near–infrared array camera for use in the 1 μm – 5 μm region of the spectrum has been constructed at the University of Rochester. The camera is based on a 32 × 32 element InSb array on loan from the Santa Barbara Research Center. The authors give a thorough description of the camera system and of the image processing, and present some images of various astronomical objects that are presently under investigation by their research group.

034.028   The blaze of concave diffraction gratings with profiled strokes.
A. V. Bruns.
Izv. Krymskoj Astrofiz. Obs., Tom 69, p. 139 – 145 (1984). In Russian. English translation in Bull. Crimean Astrophys. Obs., Vol. 69.

034.029   Heliostats, siderostats, and coelostats: a review of practical instruments for astronomical applications.
A. A. Mills.
J. Br. Astron. Assoc., Vol. 95, No. 3, p. 89 – 99 (1985).

Members of the 'stats family of astronomical instruments have in common the function of feeding light from a celestial object to a fixed telescope. Particular groups are known as heliostats, siderostats, uranostats and coelostats, considerable variations in design being possible within each group. This paper explains their general construction, differences, and respective advantages, concluding with survey of those two–mirror forms currently applied to solar astronomy.

034.030   New infrared photometer and F/35 chopping secondary at the 3.6 m telescope.
A. Moorwood, A. van Dijsseldonk.
Messenger, No. 39, p. 1 – 3 (1985).

034.031   Der 3M 1000 in der Astrophotographie.
W. E. Celnik, J. Plagge, P. Riepe, I. Schmidt.
Sterne Weltraum, 24. Jahrg., Nr. 4, p. 218 – 220 (1985).

034.032 Interferometer for high–resolution Fourier transform spectroscopy.
L. B. Masleev.
Kinematika Fiz. Nebesn. Tel, Tom 1, No. 2, p. 69 – 71 (1985). In Russian.
Variants of a Fourier spectroscopy interferometry system are suggested.

034.033 Electronic–detector arrays for spectral classification.
D. F. Gray.
The MK process and stellar classification, p. 112 – 124 (1984). – See Abstr. 012.033.

034.034 Spectral classification of OB stars with digital detectors.
N. R. Walborn.
The MK process and stellar classification, p. 125 – 135 (1984). – See Abstr. 012.033.
The advantages of digital detectors for spectral classification studies are discussed from a practical viewpoint, together with evidence that several such devices can now provide the required observational quality. Observations of galactic and Magellanic–Cloud OB stars obtained at 2.5 Å resolution with the new CTIO 1.5–meter SIT–Vidicon system are illustrated, with references to corresponding montages of photographic spectrograms for comparison.

034.035 Electronic detectors and computer–assisted spectral classification.
P. M. Rybski.
The MK process and stellar classification, p. 153 – 174 (1984). – See Abstr. 012.033.
Properties which affect the quality of digital data gathered by intensified–dissector–scanner systems – such as resolution, counting statistics, and flat–field precision – are reviewed. Particular attention is given to problems of acquiring high–quality MK–classification spectra. The change from a "many–standards–per–box" to a "one–standard–per–box" definition of spectral types, which occurred between the MKK Atlas (Morgan et al. 1943) and the MAT Atlas (Morgan et al. 1978), is discussed to clarify why certain numerical algorithms cannot be used to assign MK–78 spectral types with computers. A brief review is given of methodological problems which remain in current efforts to automate the classification process.

034.036 A polarimeter–magnetometer for hydrogen lines.
V. G. Shtol', V. D. Bychkov, N. A. Vikul'ev, O. Yu. Georgiev, Yu. V. Glagolevskij, S. V. Drabek, I. D. Najdenov, I. I. Romanyuk.
Astrofiz. Issled. Izv. Spets. Astrofiz. Obs., Tom 19, p. 66 – 70 (1985). In Russian. English translation in Bull. Spec. Astrophys. Obs. – North Caucasus.
A polarimeter–magnetometer was constructed on the basis of the spectrograph UAGS for measuring stellar magnetic fields from hydrogen lines.

034.037 Sounding system for determination of the intensity of solar cosmic radiation.
Kh. N. Gajnanov, E. I. Zykov, V. A. Matvienko, Yu. F. Kazantsev, V. A. Vorob'ev, M. I. Bukej.
Ural. politekh. inst. Sverdlovsk, 26 pp. (1984). In Russian. Abstr. in Ref. Zh., 62. Issled. Kosm. Prostranstva, 1.62.106 (1985).

034.038 SECAM – a seeing–controlled camera for high–resolution photographs.
W. Schlosser.
Astrophys. Space Sci., Vol. 110, No. 1, p. 75 – 76 (1985). – See Abstr. 012.039.
The basic design of a seeing–controlled camera (SECAM) is described. The camera is currently under construction, and will be used in combination with the 61 cm telescope of the Ruhr–University Bochum at La Silla, Chile.

034.039 Laser interferometer enhanced positional precision of a scanning microdensitometer for astrometric plate scanning.
J. H. Kinsey, R. J. Denman, A. Evzerov.
Rev. Sci. Instrum., Vol. 55, No. 12, p. 1960 – 1963 (1984). Abstr. in Phys. Abstr., Vol. 88, No. 1256, Entry 51509 (1985).

034.040 The magneto–optical filter, working principles and recent progress.
A. Cacciani, E. J. Rhodes Jr.
Solar seismology from space, p. 115 – 123 (1984). – See Abstr. 012.041.
Solar oscillation observations are in progress at the 60–foot Solar Tower of the Mt. Wilson Observatory. This note is devoted to a brief description of the working principle of the magneto–optical filter and a few possible optical arrangements.

034.041 Evaluation of a magneto–optical filter and a Fabry–Perot interferometer for the measurement of solar velocity fields from space.
E. J. Rhodes Jr., A. Cacciani, J. Blamont, S. Tomczyk, R. K. Ulrich, R. F. Howard.
Solar seismology from space, p. 125 – 155 (1984). – See Abstr. 012.041.
A magneto–optical filter and a molecular adherence Fabry–Perot interferometer have been installed in a newly–constructed observing system located at the 60–foot tower telescope at the Mt. Wilson Observatory. Two–dimensional k_h–ω power spectra which show clearly the well–known p–mode ridges are presented and compared with similar power spectra obtained recently with the 13.7–m McMath spectrograph at Kitt Peak. Future plans for continued evaluation of these instruments at Mt. Wilson and for the establishment of a second observing station which will operate in conjunction with Mt. Wilson are discussed. Finally, the operation of the Fabry–Perot as a pressure–scanned narrow band interference filter is described.

034.042 The Fourier Tachometer II – an instrument for measuring global solar velocity fields.
T. M. Brown.
Solar seismology from space, p. 157 – 163 (1984). – See Abstr. 012.041.
The High Altitude Observatory and Sacramento Peak Observatory have jointly constructed a second version of the Fourier Tachometer, which is now undergoing final integration and testing. This is an interferometric instrument for measuring the Doppler shift of solar spectrum lines. The principal features and performance goals of this instrument are: simultaneous velocity observations over a 2–dimensional, 100 × 100 pixel field of view; measurement of absolute Doppler shifts with 1 m/s accuracy; noise level for moderate–l oscillation modes of 1 cm/s for a 1–day observing run; flexibility and ease of use.

034.043 The new electronic equipment of the stellar photometer of the Byurakan Astrophysical Observatory.
Eh. N. Kyurinyan, A. S. Melkonyan, A. V. Mironov, A. V. Oskanyan Jr., V. S. Oskanyan.
Soobshch. Byurakan. Obs., Vyp. 54, p. 65 – 71 (1983). In Russian.
The description and the parameters of the new pulse counting electronic equipment of the photoelectric photometer mounted on the AZT–14 telescope of the Byurakan Observatory are given.

034.044 Fibre optics at the UK Schmidt Telescope.
F. G. Watson, J. A. Dawe.
Proc. Astron. Soc. Aust., Vol. 5, No. 4, p. 579 – 580 (1984).

034.045 The scanning heliometer.
J. Rösch, R. Yerle.
ESA Spec. Publ., ESA SP–220, p. 217 – 218 (1984). – See Abstr. 012.045.
The description of a new heliometer operating by fast photoelectric scans of opposite limbs simultaneously is given. It has been designed primarily for solar diameter measurements, but it turned out to provide also data for helioseismology, since it

measures the fluctuations of the maximum brightness gradient all around the solar limb. An example of the first scans obtained is presented.

034.046   A new multidetector system with magnetic spectrograph for study of cosmic ray extensive air shower components.
D. K. Basak, N. Chakrabarty, B. Ghosh, G. C. Goswami, N. Chaudhuri, N. Mukherjee, M. Gosh, S. K. Sarkar.
Nucl. Instrum. Methods Phys. Res., Sect. A, Vol. 227, No. 1, p. 167 – 172 (1984). Abstr. in Phys. Abstr., Vol. 88, No. 1252, Entry 29850 (1985).

034.047   Automatic determination of the direction angle with the Kern E 1 electronic theodolite by means of zenith distances of the sun.
N. Solarić.
Mitt. Lohrmann–Obs. Tech. Univ. Dresden, Nr. 51, p. 114 – 117 (1984). In German. – See Abstr. 012.058.
Die automatische Zuführung der Daten vom Horizontal– und Vertikalkreis des elektronischen Theodolits Kern E 1 mit dem programmierbaren Rechner HP 41 CV, wird beschrieben.

034.048   L'horloge astronomique d'Orly–Ouest.
P. Muller.
Astronomie, Vol. 99, p. 245 – 248 (1985).

034.049   Stability of an uncooled Reticon detector system at Mauna Loa Solar Observatory: a simple calibration scheme.
D. G. Sime, K. A. Rock, R. R. Fisher.
AAS Photo–Bull., No. 37, p. 10 – 12 (1984).
The authors demonstrate a simple calibration scheme used to quantify the stability of the photoelectric detector system and the end–to–end instrument function of the Mark III K–coronameter at Mauna Loa Solar Observatory. With this scheme the authors derive limits on the performance fluctuations, both day–to–day and over a 3–year period. These confirm the correctness of engineering decisions made at fabrication not to cool the detector device.

034.050   Il quadrante analemmatico.
S. Brunetto.
Orione, Vol. 5, N. 2, p. 38 – 47 (1985).

034.051   Photographische Bildverstärkung an astronomischen Negativen.
T. Langthaler.
Sternenbote, 28. Jahrg., Nr. 6, p. 105 – 107 (1985).

034.052   TV solar images in the He I 10830 Å line.
L. D. Parfinenko.
Soln. Dannye, Byull., 1985, No. 2, p. 87 – 91 (1985). In Russian.
The idea of a modified classical spectroheliograph has been used for getting solar images in He I 10830 Å. A sum of images in the red and blue wings of the helium line can be obtained by the TV method.

034.053   On the influence of hydrogen hypersensitization on DQE of plates.
I. I. Brejdo, O. M. Mikhajlova.
Astron. Tsirk., No. 1344, p. 7 – 8 (1984). In Russian.

034.054   Kleinbildoptiken in der Astrophotographie.
P. Riepe.
Sterne Weltraum, 24. Jahrg., Nr. 5, p. 280 – 283 (1985).

034.055   Starlight–1 solutions.
R. C. Wolpert.
I.A.P.P.P. Commun., No. 19, p. 28 – 31 (1985).

034.056   An improved ac stellar photometer.
J. H. DeWitt Jr.
I.A.P.P.P. Commun., No. 19, p. 32 – 40 (1985).

034.057   A portable digital data acquisition system for asteroid occultation photometry.
T. D. Oswalt, J. B. Rafert.
I.A.P.P.P. Commun., No. 20, p. 18 – 23 (1985).

034.058   On the influence of hydrogen hypersensitization on the spectral sensitivity of photographic plates and films.
I. I. Brejdo.
Astron. Tsirk., No. 1351, p. 6 – 8 (1984). In Russian.

034.059   On the choice of the method of investigation of errors of divided circle diameters.
M. I. Myalkovskij.
Astron. Tsirk., No. 1355, p. 4 – 5 (1984). In Russian.

034.060   On accidental and systematic errors of run determination.
M. I. Myalkovskij.
Astron. Tsirk., No. 1355, p. 5 – 6 (1984). In Russian.

034.061   Comparator with optical caliper.
A. V. Bakhonskij, V. S. Samojlov.
Kinematika Fiz. Nebesn. Tel, Tom 1, No. 3, p. 83 – 84 (1985). In Russian.

034.062   Infrared J, H, K, L–photometer.
V. P. Kuz'kov.
Kinematika Fiz. Nebesn. Tel, Tom 1, No. 3, p. 91 – 94 (1985). In Russian.
A small–size infrared photometer with PbS detector cooled by liquid nitrogen is described.

034.063   The photodiode PS–100–30 as a position sensor in astronomy.
A. Kubičela, M. Rakić.
Publ. Astron. Opservatorije Beogr., No. 33, p. 23 – 27 (1985). – See Abstr. 012.061.

034.064   Photon counter construction with microcomputer data processing.
M. Muminović, M. Stupar, B. Stupar.
Publ. Astron. Opservatorije Beogr., No. 33, p. 44 (1985). Abstract. – See Abstr. 012.061.

034.065   Balloon–borne, He–cooled Fourier transform spectrometer for far infrared astronomy.
M. F. Campbell, S. Drapatz.
MPE Intern. Rep., No. 41, 55 pp. (1985).

034.066   A method for absolute calibration of an ionization chamber in the ultraviolet and for a proportional counter at X–ray wavelengths.
Y.–l. Xu, P. Xie, R.–j. Zhang.
Publ. Purple Mt. Obs., Vol. 3, No. 3, p. 83 – 90 (1984). In Chinese.

034.067   The lost photon problem in image intensifiers.
R. H. Cromwell.
Prepr. Steward Obs., No. 565, 6 pp. (1985). Paper presented at the 165th meeting of the American Astronomical Society in Tucson, January 13 – 16, 1985.

034.068   A simple direct TV satellite receiver for time comparison.
H.–y. Wang, J.–a. Song.
Publ. Shaanxi Astron. Obs., Vol. 7, No. 1, p. 48 – 51 (1984). In Chinese.

034.069   An automatic multiple switch constructed by means of a single board microprocessor.
B.–t. Zhu, Y.–l. Sang.
Publ. Shaanxi Astron. Obs., Vol. 7, No. 1, p. 52 – 55 (1984). In Chinese.

034.070   **Two multi–object spectroscopic options at the ESO 3.6 m telescope.**
S. D'Odorico.
Messenger, No. 40, p. 11 – 12 (1985).

034.071   **Construction d'un spectrocoronographe à protubérances.**
F. Costard.
Astronomie, Vol. 99, p. 295 – 299 (1985).

034.072   **A possible space–research dedicated accelerator facility.**
C. S. Zaidins.
Bull. Am. Astron. Soc., Vol. 16, No. 4, p. 884 – 885 (1984). Abstract. – See Abstr. 010.062.

034.073   **A portable digital data acquisition system for asteroid occultation photometry.**
J. B. Rafert, T. D. Oswalt.
Bull. Am. Astron. Soc., Vol. 16, No. 4, p. 886 (1984). Abstract. – See Abstr. 010.062.

034.074   **Arms control for the MX spectrometer.**
M. P. Lesser, J. M. Hill.
Bull. Am. Astron. Soc., Vol. 16, No. 4, p. 902 – 903 (1984). Abstract. – See Abstr. 010.062.

034.075   **A faint object spectrometer for the infrared.**
R. G. Smith, A. T. Tokunaga, D. Toomey, D. L. DePoy.
Bull. Am. Astron. Soc., Vol. 16, No. 4, p. 903 (1984). Abstract. – See Abstr. 010.062.

034.076   **Improvements for the PDS 1010a photometric system.**
D. A. Klinglesmith III, G. D. Harris.
Bull. Am. Astron. Soc., Vol. 16, No. 4, p. 903 – 904 (1984). Abstract. – See Abstr. 010.062.

034.077   **Calibration of photographic plates from trailed stellar images.**
A. Warnock III, D. A. Klinglesmith III.
Bull. Am. Astron. Soc., Vol. 16, No. 4, p. 904 (1984). Abstract. – See Abstr. 010.062.

034.078   **The lost photon problem in image intensifiers.**
R. H. Cromwell.
Bull. Am. Astron. Soc., Vol. 16, No. 4, p. 904 (1984). Abstract. – See Abstr. 010.062.

034.079   **Observational problems with the EMI Starlight–1 photometer.**
W. Osborn, W. Bisard.
Bull. Am. Astron. Soc., Vol. 16, No. 4, p. 909 (1984). Abstract. – See Abstr. 010.062.

034.080   **An observatory to study 10^{10} to 10^{16} eV gamma rays.**
T. C. Weekes.
Bull. Am. Astron. Soc., Vol. 16, No. 4, p. 963 (1984). Abstract. – See Abstr. 010.062.

034.081   **The application of imaging to very high energy gamma ray astronomy.**
K. Gibbs, T. C. Weekes, P. N. Gorham, V. J. Stenger, R. C. Lamb, D. F. Liebing, M. F. Cawley, D. J. Fegan, N. A. Porter, P. K. MacKeown, K. E. Turver.
Bull. Am. Astron. Soc., Vol. 16, No. 4, p. 963 (1984). Abstract. – See Abstr. 010.062.

034.082   **A new photomultiplier test facility at Kitt Peak National Observatory.**
D. S. Hayes, A. D. Grauer.
Bull. Am. Astron. Soc., Vol. 16, No. 4, p. 985 (1984). Abstract. – See Abstr. 010.062.

034.083   **A 3–channel microdensitometer for interactive analysis of plate spectra.**
C. S. Sutton, S. Biyabani.
Bull. Am. Astron. Soc., Vol. 16, No. 4, p. 985 (1984). Abstract. – See Abstr. 010.062.

034.084   **A spectrometer for accurate radial accelerometry of stars.**
R. S. McMillan, P. H. Smith, J. E. Frecker, W. J. Merline, M. L. Perry.
Bull. Am. Astron. Soc., Vol. 16, No. 4, p. 985 (1984). Abstract. – See Abstr. 010.062.

034.085   **Four modes of CCD operation.**
T. Gehrels, J. E. Frecker, R. S. McMillan, M. L. Perry, J. V. Scotti.
Bull. Am. Astron. Soc., Vol. 16, No. 4, p. 986 (1984). Abstract. – See Abstr. 010.062.

034.086   **A near–infrared camera using a hybrid CCD.**
M. J. Lebofsky, W. F. Kailey.
Bull. Am. Astron. Soc., Vol. 16, No. 4, p. 986 (1984). Abstract. – See Abstr. 010.062.

034.087   **Solar disk sextant at GSFC.**
H. Y. Chiu, S. Sofia, E. J. Maier, K. H. Schatten.
Bull. Am. Astron. Soc., Vol. 16, No. 4, p. 993 (1984). Abstract. – See Abstr. 010.062.

034.088   **Apparatus for recording velocity fields in prominences.**
I. S. Kim, O En Den, A. I. Stepanov.
Inst. zem. magn., ionos. i rasprostr. radiovoln Akad. Nauk SSSR. Prepr., No. 34/508, 8 pp. (1984). In Russian. Abstr. in Ref. Zh., 51. Astron., 5.51.825 (1985).

034.089   **Peculiarities of the operation of an interference–polarization filter in the optical scheme of a telescope.**
Yu. A. Klevtsov.
Issled. Geomagn., Aehron. Fiz. Solntsa, Moskva, No. 69, p. 183 – 189 (1984). In Russian. Abstr. in Ref. Zh., 51. Astron., 5.51.826 (1985).

034.090   **Magnetograph with line–profile scanning.**
G. M. Nikol'skij, I. S. Kim, V. Yu. Klepikov, S. Koutchmy, O En Den, A. I. Stepanov.
Inst. zem. magn., ionos. i rasprostr. radiovoln Akad. Nauk SSSR. Prepr., No. 33/507, 10 pp. (1984). In Russian. Abstr. in Ref. Zh., 51. Astron., 5.51.827 (1985).

034.091   **Limiting accuracy of an integral–interference solar tachometer.**
I. E. Kozhevatov.
Physics of solar activity, p. 182 – 185, 203 (1983). In Russian. Abstr. in Ref. Zh., 51. Astron., 6.51.765 (1985). – See Abstr. 003.024.

034.092   **Geometrical invariance property of gratings.**
D. Maystre, M. Neviere, W. R. Hunter.
Appl. Opt., Vol. 24, No. 2, p. 215 – 216 (1985).
The invariance of the direction of a given order diffracted by a grating used in conical diffraction (i.e., in off–plane mounting) when the grating is rotated about a suitable direction is established. An explanation of this amazing property is given in terms of Fermat's principle and applies to any reflecting surface.

034.093   **Speckle imaging with the PAPA detector.**
C. Papaliolios, P. Nisenson, S. Ebstein.
Appl. Opt., Vol. 24, No. 2, p. 287 – 292 (1985).
A new 2–D photon–counting camera, the PAPA (precision analog photon address) detector has been built, tested, and used successfully for the acquisition of speckle imaging data. The camera has 512×512 pixels and operates at count rates of at least 200,000/sec. The authors present technical details on the camera and include some of the laboratory and astronomical results which demonstrate the detector's capabilities.

034.094 Boron and silicon: filters for the extreme ultraviolet.
S. Labov, S. Bowyer, G. Steele.
Appl. Opt., Vol. 24, No. 4, p. 576 – 578 (1985).
Thin films of boron and silicon have been developed using electron beam deposition. The transmissions of these filters were measured from soft X–ray wavelengths to the far ultraviolet and at optical wavelengths. The boron filter transmission peaks near 66 Å and the silicon filter peaks near 136 Å. The peak transmission of these filters does not change with time, but the width of the silicon filter bandpass is reduced slightly as the filter ages.

034.095 Progress in the development of a durable silver–based high–reflectance coating for astronomical telescopes.
D.–Y. Song, R. W. Sprague, H. A. Macleod, M. R. Jacobson.
Appl. Opt., Vol. 24, No. 8, p. 1164 – 1170 (1985).
Silver–based high–reflectance coatings that can withstand the humid and polluted conditions common in the open air have been developed for astronomical telescope optics. The successful designs incorporate a silver reflective layer with a copper underlayer and a stack of dielectric overlayers.

034.096 Far–infrared dichroic bandpass filters.
D. D. Nolte, A. E. Lange, P. L. Richards.
Appl. Opt., Vol. 24, No. 10, p. 1541 – 1545 (1985).
Filters which serve as dichroic beam dividers are useful for certain types of far–infrared photometer. The authors have used inductive cross metal mesh as reflective coatings on solid dielectric Fabry–Perot etalons to produce dichroic filters with quality factors in the range $3 < Q < 10$. These filters reflect strongly at frequencies below the bandpass and retain their well–defined bandpass at angles of incidence as large as 45°. They have been used to make an efficient far–infrared dichroic photometer with six frequency bands between 10 and 100 cm^{-1}.

034.097 Physics and design of advanced IR bolometers and photoconductors.
E. E. Haller.
Infrared Phys., Vol. 25, No. 1/2, p. 257 – 266 (1985). – See Abstr. 012.065.
Spaceborne IR telescopes and other very low photon flux experiments require photoconductors and bolometers with ever increasing sensitivity working at the background limit. Improvements in the understanding of the physics of these devices, advances in semiconductor materials synthesis and device technology, and the introduction of new device concepts such as uniaxially–stressed photoconductors can, in combination, fulfill these new demands.

034.098 Receiver technology for the submillimeter–wave region.
B. J. Clifton.
Infrared Phys., Vol. 25, No. 1/2, p. 267 – 272 (1985). – See Abstr. 012.065.
The development of heterodyne receivers with good sensitivity in the submillimeter–wave region has resulted in a wide range of applications in radio astronomy, plasma physics, atmospheric measurements and laboratory spectroscopy.

034.099 Development of high–responsivity Ge:Ga photoconductors.
N. M. Haegel, M. R. Hueschen, E. E. Haller.
Infrared Phys., Vol. 25, No. 1/2, p. 273 – 276 (1985). – See Abstr. 012.065.
Czochralski–grown Ga–doped Ge (Ge:Ga) single–crystal samples, with a compensation of 10^{-4} have been modified by the indiffusion of Cu to produce photoconductors which provide NEPs comparable to current optimum Ge:Ga detectors, but exhibit responsivities a factor of 5 – 6 times higher when tested at a background photon flux of 10^8photon/s at $\lambda = 93$ μm. The introduction of Cu, a triple acceptor in Ge which acts as a neutral scattering center, reduces carrier mobility and extends the breakdown field significantly in this ultra–low compensation material.

034.100 Two kinds of "blueward shift" in the NEP spectral curves of InSb detectors.
Y.–X. Zhang.
Infrared Phys., Vol. 25, No. 3, p. 579 – 582 (1985).
It is well–known that as the operating temperature of an InSb detector with a given sensitive area is reduced, the shape of its NEP spectral curve remains the same but shifts towards shorter wavelengths. This is called the "blueward shift". The author presents another kind of blueward shift in the InSb NEP spectral curves. It can be seen when the temperature of the detector remains constant and the equivalent zero–bias resistance of the detector increases. It is suggested that the two kinds of blueward shift can be interpreted by different physics mechanisms.

034.101 Far–infrared astronomical spectrometers.
J. W. V. Storey.
Infrared Phys., Vol. 25, No. 3, p. 583 – 590 (1985).
FIR spectral line astronomy has now matured to the point where an overview of instrumental capabilities is desirable. This review presents a summary of existing instrumentation and attempts to place the different approaches to spectrometer design into perspective. It is shown that the optimum type of spectrometer for a given situation depends on the desired scientific objectives.

034.102 A solar chromospheric variability monitor from Mauna Loa.
R. R. Fisher, D. G. Sime.
Bull. Am. Astron. Soc., Vol. 17, No. 1, p. 514 (1985). Abstract. – See Abstr. 010.064.

034.103 Die CCD–Kamera.
J. Fried.
Sterne Weltraum, 24. Jahrg., Nr. 6, p. 318 (1985).

034.104 Initiation à la photographie astronomique.
Y. Thirionet.
Ciel Terre, Vol. 101, No. 3, p. 77 – 80 (1985).

034.105 New procedures for processing and storage of Kodak spectroscopic plates, type IIIa–J.
W. E. Lee, F. J. Drago, A. T. Ram.
J. Imaging Technol., Vol. 10, No. 1, p. 22 – 28 (1984). Abstr. in Phys. Abstr., Vol. 88, No. 1253, Entry 30497 (1985).

034.106 Solar neutrino detectors based on indium neutrino reaction.
M. Cribier.
Proceedings of the XIth International Conference on Neutrino Physics and Astrophysics, p. 524 – 529 (1984). – See Abstr. 012.081.

034.107 The gallium solar neutrino detector and neutrino oscillations.
W. Hampel.
Proceedings of the XIth International Conference on Neutrino Physics and Astrophysics, p. 530 – 535 (1984). – See Abstr. 012.081.
The current status of the Gallium Solar Neutrino Experiment is briefly described. The paper then addresses the question of neutrino oscillations and discusses their possible impact on solar neutrino experiments, especially on the expected signal from the gallium detector.

034.108 Neutral current detector for neutrino physics, astronomy, and geology.
L. Stodolsky.
Proceedings of the XIth International Conference on Neutrino Physics and Astrophysics, p. 536 – 542 (1984). – See Abstr. 012.081.
Due to the coherent scattering of MeV range neutrinos on nuclei by the neutral current, the cross section is unusually large by neutrino standards, varying as the square of the number of neutrons in the nucleus and the square of the incident energy. A

neutral current detector based on this process is briefly outlined here.

034.109   Status of DUMAND neutrino detector.
V. Peterson.
Proceedings of the XIth International Conference on Neutrino Physics and Astrophysics, p. 543 – 549 (1984). – See Abstr. 012.081.
A Deep Underwater Muon and Neutrino Detector (DUMAND) is being developed by a 10–institution collaboration. The site is at 4.5 km depth near the island of Hawaii. The primary scientific goal is neutrino astrophysics in the TeV energy range. The present status of DUMAND is described.

034.110   Progress report on Lake Baikal neutrino experiment: site studies and stationary string.
L. B. Bezrukov, B. A. Borisovets, E. V. Bugaev,
Zh.-A. M. Jilckibaev, G. V. Domogatsky (*G. V. Domogatskij*),
M. D. Gal'perin, A. M. Klabukov, S. I. Klimushin,
B. K. Lubsandorzhiev, A. I. Panfilov, I. A. Sokal'sky,
L. N. Stepanov, I. I. Trofimenko, N. M. Budnev,
L. V. Bumazhkin, N. P. Butin, V. I. Dobrynin, V. A. Efremov,
A. P. Koshechkin, G. A. Kushnarenko, M. I. Nemchenko,
S. A. Nickiforov, Yu. V. Parfenov, V. A. Poleschuk,
A. A. Schestakov, B. A. Taraschansky, V. L. Zurbanov,
V. A. Fialkov, E. B. Karabanov, P. P. Scherstyankin,
S. B. Ignat'ev, V. B. Kabikov, L. A. Kuzmichöv, G. N. Dudkin,
A. G. Kohomsky, M. A. Novikov.
Proceedings of the XIth International Conference on Neutrino Physics and Astrophysics, p. 550 – 555 (1984). – See Abstr. 012.081.
Results of two–year observation of water luminescence in Lake Baikal are reported. The Stationary Short String (SSS) experiment operating since April 1984 at Lake Baikal is described and some preliminary results are given.

034.111   Neutrino physics and astrophysics with the liquid scintillation detector in the Mont Blanc Laboratory.
G. Badino, G. Bologna, C. Castagnoli, B. D'Ettorre Piazzoli,
W. Fulgione, P. Galeotti, G. Mannocchi, P. Picchi,
O. Saavedra, V. L. Dadykin, V. B. Korchaguin
(*V. B. Korchagin*), P. V. Korchaguin (*P. V. Korchagin*),
A. S. Malguin (*A. S. Mal'gin*), O. G. Ryazhskaya,
A. L. Tziabuk, V. P. Talochkin, G. T. Zatsepin,
V. F. Yakushev.
Proceedings of the XIth International Conference on Neutrino Physics and Astrophysics, p. 556 – 567 (1984). – See Abstr. 012.081.
A variety of topics on elementary particle physics and astrophysics will be studied with a new, massive liquid scintillation detector (LSD) recently put in operation in the Mont Blanc Laboratory. A brief description of the LSD experiment is presented. The main aim of LSD is to search for neutrino bursts from collapsing stars.

034.112   Superhigh energy neutrino radio detectors in ice.
L. G. Dedenko, G. A. Gusev, I. M. Zheleznykh.
Proceedings of the XIth International Conference on Neutrino Physics and Astrophysics, p. 568 – 576 (1984). – See Abstr. 012.081.
The possibility for the detection of neutrinos and muons using radio emission of cascades in natural dielectric medium (ice, atmosphere, etc.) is discussd. It is possible to construct a radio detector of muons and neutrinos (RAMAND) with an effective volume of 10^9–$10^{11} m^3$ in ice where the absorption of radio waves in the decimeter frequency band is relatively small. The detection of neutrinos in the energy range 10^{18}–10^{19}eV may be possible with such a detector.

034.113   Sensitivity analysis of a two–mode resonant gravitational–wave detector with arbitrary tuning.
M. Bassan.
Nuovo Cimento C, Vol. 7C, Ser. 1, No. 3, p. 303 – 316 (1984). Abstr. in Phys. Abstr., Vol. 88, No. 1256, Entry 46513 (1985).

034.114   Initial operation at liquid–helium temperature of the M = 2270 kg Al 5056 gravitational–wave antenna of the Rome Group.
E. Amaldi, E. Coccia, C. Cosmelli, Y. Ogawa, G. Pizzella,
P. Rapagnani, F. Ricci, P. Bonifazi, M. G. Castellano,
G. Vannaroni, F. Bronzini, P. Carelli, V. Foglietti,
G. Cavallari, R. Habel, I. Modena, G. C. Pallottino.
Nuovo Cimento C, Vol. 7C, Ser. 1, No. 3, p. 338 – 354 (1984). Abstr. in Phys. Abstr., Vol. 88, No. 1256, Entry 46514 (1985).

034.115   Two–channel electrophotometer with integral signal accumulation for observations of "Ellermann bombs" on a large coronograph.
A. K. Ajmanov, A. S. Zubtsov, S. S. Shumilin.
Tr. Astrofiz. Inst. Alma–Ata, Tom 41, p. 60 – 69 (1983). In Russian.

034.116   Investigation of the photoelectric block of the spectrometer SEJA–Namioka and its improvement.
V. A. Bandrin, P. N. Bojko, S. S. Shumilin.
Tr. Astrofiz. Inst. Alma–Ata, Tom 42, p. 116 – 132 (1983). In Russian.

034.117   Fabry–Perot scanning spectrometer for observation of extended objects with low surface luminosity.
L. I. Shestakova.
Tr. Astrofiz. Inst. Alma–Ata, Tom 44, p. 62 – 76 (1984). In Russian.

034.118   Unusual diffraction grating mountings for the vacuum ultraviolet.
W. R. Hunter.
Proc. SPIE Int. Soc. Opt. Eng., Vol. 503, p. 78 – 85 (1984). Abstr. in Phys. Abstr., Vol. 88, No. 1258, Entry 58440 (1985). – See Abstr. 012.090.

034.119   Laboratory evaluation of conical diffraction spectrographs.
D. L. Windt, W. C. Cash.
Proc. SPIE Int. Soc. Opt. Eng., Vol. 503, p. 98 – 105 (1984). Abstr. in Phys. Abstr., Vol. 88, No. 1258, Entry 58441 (1985). – See Abstr. 012.090.

034.120   Operation of a microchannel plate counting system in a mass spectrometer.
D. M. Murphy, K. Mauersberger.
Rev. Sci. Instrum., Vol. 56, No. 2, p. 220 – 226 (1985). Abstr. in Phys. Abstr., Vol. 88, No. 1258, Entry 58462 (1985).

034.121   Interference methods in the testing and fabrication of new–design grazing incidence gratings.
M. C. Hettrick, C. Martin.
Proc. SPIE Int. Soc. Opt. Eng., Vol. 503, p. 106 – 113 (1984). Abstr. in Phys. Abstr., Vol. 88, No. 1258, Entry 59689 (1985). – See Abstr. 012.090.

034.122   A flat field spectrograph for solar X–ray spectroscopy.
S. Mrowka.
Proc. SPIE Int. Soc. Opt. Eng., Vol. 503, p. 86 – 91 (1984). Abstr. in Phys. Abstr., Vol. 88, No. 1258, Entry 63215 (1985). – See Abstr. 012.090.

034.123   Double grating prisms and their use for absolute radial velocity measurements of astronomical objects.
B. Nelles.
Optik, Vol. 69, No. 4, p. 153 – 159 (1985). Abstr. in Phys. Abstr., Vol. 88, No. 1258, Entry 63228 (1985).

034.124   Semi–automated, three–dimensional measurement of etched tracks in solid–state nuclear track detectors.
P. B. Price, W. Krischer.
Nucl. Instrum. Methods Phys. Res., Sect. A, Vol. 234, No. 1, p. 158 – 167 (1985). Abstr. in Phys. Abstr., Vol. 88, No. 1259, Entry 64824 (1985).

034.125   The DAO radial velocity spectrometer and recent results.
R. D. McClure, J. M. Fletcher, W. A. Grundman,
E. H. Richardson.
Stellar radial velocities, p. 49 – 62 (1985). – See Abstr. 012.095
(IAU Colloq. No. 88).

A photoelectric radial–velocity spectrometer has been in use at the coudé focus of the 1.2 m DAO telescope for several years. New improvements, resulting in a lowering of background noise, have allowed observations of stars significantly fainter than previously possible. Recent observations with the instrument at the coudé focus of the CFH telescope in Hawaii have shown that velocity measurements to better than 1 km s^{-1} accuracy are possible at 19th B magnitude. Various observing programs being carried out at DAO and CFHT are discussed.

034.126   The LPL radial accelerometer.
R. S. McMillan, P. H. Smith, J. E. Frecker,
W. J. Merline, M. L. Perry.
Stellar radial velocities, p. 63 – 86 (1985). – See Abstr. 012.095
(IAU Colloq. No. 88).

The LPL stellar accelerometer is an interferometrically calibrated spectrometer. Light is fed from the telescope focal plane to the spectrometer entrance aperture by a single optical fiber. Wavelength calibration is imposed on the starlight by a Fabry–Perot interferometer used in transmission. This interferometer is in turn calibrated by observing independent references. The spectrum is dispersed by an echelle grating crossed with a conventional plane diffraction grating. The two–dimensional spectrum is focused on a charge–coupled–device (CCD) used as a detector.

034.127   The McDonald Observatory high precision radial velocity spectrometer.
W. D. Cochran, B. W. Young.
Stellar radial velocities, p. 109 – 120 (1985). – See Abstr. 012.095
(IAU Colloq. No. 88).

The authors are developing a prototype instrument for McDonald Observatory designed to measure stellar radial velocity variations to a precision of a few meters per second. A fixed gap Fabry–Perot etalon, used in reflection, imposes a set of fixed reference absorption lines on the stellar spectrum before it enters the McDonald Observatory 2.7 m coudé spectrograph. The spectrum is recorded on a set of eight Reticon arrays. Doppler shifts of the stellar spectral lines with respect to the fixed Fabry–Perot orders are measured by cross–correlation techniques. Calibration methods have been developed to measure any long–term drifts within the system.

034.128   Application of two dimensional imaging of atmospheric Cerenkov light to very high energy gamma ray astronomy.
M. F. Cawley, J. Clear, D. J. Fegan, K. Gibbs, P. Gorham,
R. C. Lamb, I. MacRae, P. K. MacKeown, N. A. Porter,
V. J. Stenger, K. E. Turver, T. C. Weekes.
18th International Cosmic Ray Conference, Vol. 1, p. 118 – 121
(1983). – See Abstr. 012.096.

A camera description is given as well as operation methods. Then the results are shown and discussed.

034.129   Nagoya cosmic–ray muon spectrometer.
S. Shibata, Y. Kamiya, R. Tatsuoka, K. Munakata,
S. Iida.
18th International Cosmic Ray Conference, Vol. 7, p. 40 (1983).
Abstract. – See Abstr. 012.096.

034.130   A precision measurement of the cosmic ray muon angular distribution underground.
M. F. Crouch, C. B. Bratton, J. S. Hansen, W. R. Kropp,
F. Reines.
18th International Cosmic Ray Conference, Vol. 7, p. 42 – 45
(1983). – See Abstr. 012.096.

A unique horizontal–vertical array of multi–wire proportional counters termed the "underground fly's eye" has been used for a precise measurement of the angular distribution of cosmic ray muons.

034.131   The investigation of the cosmic muon energy spectrum, charge ratio and HAS (*horizontal showers*) with the Aragats complex installation.
T. L. Asatiani, A. V. Abrahamian, S. V. Alchudjian,
V. A. Ivanov, L. I. Kozliner, G. S. Martirossian
(*G. S. Martirosyan*), V. A. Melkumiants, A. K. Pogossian,
S. V. Ter–Antonian (*S. V. Ter–Antonyan*).
18th International Cosmic Ray Conference, Vol. 7, p. 47 – 50
(1983). – See Abstr. 012.096.

034.132   10,000 m^2 COSmic ray MUon Detection system (*COSMUD*) located at rather shallow underground.
T. Aoki, S. Higashi, Y. Kamiya, F. Kajino, Y. Kawashima,
T. Kitamura, K. Kobayakawa, K. Mitsui, Y. Minorikawa,
Y. Muraki, I. Nakamura, A. Okada, Y. Ohashi, S. Ozaki,
H. Shibata, N. Shibata, T. Suda, T. Takahashi.
18th International Cosmic Ray Conference, Vol. 7, p. 58 (1983).
Abstract. – See Abstr. 012.096.

034.133   The investigation of muon groups at large zenith angles with the Aragats spark calorimeter.
T. L. Asatiani, A. V. Abrahamian, S. V. Alchudjian,
V. A. Ivanov, L. I. Kozliner, G. S. Martirossian
(*G. S. Martirosyan*), A. K. Pogossian, S. V. Ter–Antonian
(*S. V. Ter–Antonyan*).
18th International Cosmic Ray Conference, Vol. 7, p. 60 – 63
(1983). – See Abstr. 012.096.

The experimental installation consisting of a spark calorimeter and a magnetic spectrometer is described. The estimation of muon energy by the spark calorimeter is given. The experimental results on space–energy characteristics of muon groups with energies above 5 GeV at the zenith angles 60° – 90° are reported.

034.134   An experiment to detect UHE γ–ray sources.
B. C. Raubenheimer, G. van Urk, E. J. de Villiers.
18th International Cosmic Ray Conference, Vol. 8, p. 24 – 27
(1983). – See Abstr. 012.096.

The design of a new ground based Cerenkov detector to be erected in South Africa is described.

034.135   Scintillator–fiber charged–particle track–imaging detector.
W. R. Binns, M. H. Israel, J. Klarmann.
18th International Cosmic Ray Conference, Vol. 8, p. 89 – 92
(1983). – See Abstr. 012.096.

This paper presents details of the detector technique, properties of the tracks obtained, and range measurements of 15 MeV protons stopping in the fiber bundle.

034.136   A small air shower array with angular resolution system.
Z.–h. Liu, G. Bai, G. Li, Q. Geng, J. Liu.
18th International Cosmic Ray Conference, Vol. 8, p. 141 (1983).
Abstract. – See Abstr. 012.096.

034.137   A new 24 NM–64 neutron supermonitor on Musala.
S. P. Kavlakov, L. M. Georgiev.
18th International Cosmic Ray Conference, Vol. 8, p. 151 – 154
(1983). – See Abstr. 012.096.

The experimental arrangement of the whole apparatus is presented and results of preliminary measurements are discussed.

034.138   Studies of coded aperture gamma–ray optics using an Anger camera.
P. M. Charalambous, A. J. Dean, J. B. Stephen,
N. G. S. Young, A. R. Gourlay.
18th International Cosmic Ray Conference, Vol. 8, p. 156 – 159
(1983). – See Abstr. 012.096.

An experimental arrangement using an Anger camera as a position sensitive focal plane in conjunction with a series of coded aperture masks has been employed to generate laboratory γ–ray images. It is shown that by proper design the major sources of image defects may be reduced to a level compatible with the production of good quality γ–ray sky images.

034.139 A multi–channel pulse height recorder for EAS arrays.
G. K. D. Mazumdar, P. K. Barua, K. M. Pathak.
18th International Cosmic Ray Conference, Vol. 8, p. 204 – 206 (1983). – See Abstr. 012.096.

034.140 A Bragg crystal flux concentrator for annihilation radiation.
N. Lund, R. K. Smither.
18th International Cosmic Ray Conference, Vol. 9, p. 339 – 342 (1983). – See Abstr. 012.096.

034.141 Progress report on the Utah Fly's Eye.
R. Cady, G. L. Cassiday, J. W. Elbert,
P. R. Gerhardy, E. C. Loh, Y. Mizumoto, M. H. Salamon,
P. Sokolsky, D. Steck.
18th International Cosmic Ray Conference, Vol. 9, p. 351 – 354 (1983). – See Abstr. 012.096.
All 67 mirrors of the Fly's Eye have been operational for about a year. At a site 3.4 km from the first eye, a second eye has been started, with 8 mirrors operational. A prototype higher resolution eye (with 256 photomultiplier tubes/mirror unit) is also being tested.

034.142 Low budget extensions to the Haverah Park shower array.
G. Brooke, J. Linsley, J. Lloyd–Evans, P. A. Ogden,
R. J. O. Reid, A. A. Watson.
18th International Cosmic Ray Conference, Vol. 9, p. 420 – 423 (1983). – See Abstr. 012.096.
The authors describe two methods which may be used to extend the existing Haverah Park array (sensitive area 10 km^2 at 4×10^{19}eV) to enhance the collecting rate for cosmic ray primaries above 5×10^{18}eV.

034.143 Design and performance of a thin scintillation counter.
M. S. Darjazi, H. F. Masjed, F. Ashton.
18th International Cosmic Ray Conference, Vol. 9, p. 428 – 431 (1983). – See Abstr. 012.096.
Thin scintillation counters of moderate area (in the present case 0.4 m^2) are useful in a variety of cosmic ray experiments. The design of the scintillation counter is given.

034.144 Theoretical and experimental study of a large temporal scintillation counter of ionizing radiation.
A. N. Avlentyev, A. A. Chaikin, Ya. L. Blokh, L. I. Dorman.
18th International Cosmic Ray Conference, Vol. 9, p. 433 – 436 (1983). – See Abstr. 012.096.

034.145 Relevance of the parameters of multiplicity distribution in neutron monitor to cosmic ray energy spectrum at observation level.
L. I. Dorman, V. K. Korotkov.
18th International Cosmic Ray Conference, Vol. 9, p. 437 – 440 (1983). – See Abstr. 012.096.

034.146 The apparatus for visual investigation of certain spatial characteristics of extensive air showers.
D. M. Kotlyarevski, I. V. Morosov, A. A. Novalov.
18th International Cosmic Ray Conference, Vol. 9, p. 452 – 455 (1983). – See Abstr. 012.096.

034.147 Western Washington University BATISS Neutrino Telescope description.
J. R. Alber, W. Brown, P. Kotzer, R. Lindsay, S. Kondratick,
R. Lord, A. S. Rupaal, J. Turner.
18th International Cosmic Ray Conference, Vol. 11, p. 480 (1983). Abstract. – See Abstr. 012.096.

034.148 Feasibility studies of underwater neutrino astrophysical detectors.
J. R. Albers, W. Brown, P. Kotzer, R. H. Lindsay, J. Pullen,
A. S. Rupaal, E. V. Kolomeets, V. S. Murzin.
18th International Cosmic Ray Conference, Vol. 11, p. 489 (1983). Abstract. – See Abstr. 012.096.

034.149 Some recent developments in particle detectors.
G. Charpak.
18th International Cosmic Ray Conference, Vol. 12, p. 61 – 69 (1983). – See Abstr. 012.096.
Amongst the recent developments in particle detectors, the emphasis is on those based on the detection of ultraviolet or vacuum ultraviolet photons. The combination of wire chambers and photoionizable vapours, or photocathodes obtained by the condensation of such vapours, permits the building of a new class of Ring–Imaging Cherenkov detectors and BaF$_2$ calorimeters.

034.150 A far infrared photometer for ground based astronomical observations.
G.–z. Xie, C. Ceccarelli, L. Pietranera, G. Dall'Oglio, G. Ferri,
S. Radford.
Acta Astron. Sin., Vol. 26, No. 2, p. 180 – 186 (1985). In Chinese.
The authors have developed a photometer which is suitable for astronomical observations through the millimetric atmospheric windows. It consists of a Ge–Ga bolometer operating at 1.0K, with an input AW = 0.3 cm^{2}sr and a NET = 1 mk/Hz$^{1/2}$. Two metalic meshs set the windows at about 10 cm^{-1} and 5 cm^{-1}. It is sensitive enough to be limited by quantum fluctuations of atmospheric radiation.

034.151 Germanium photodiodes with integrating preamplifiers for use at 0.8 μm to 1.6 μm.
C. Philips–Walker, G. H. Rieke, E. F. Montgomery.
Bull. Am. Astron. Soc., Vol. 17, No. 2, p. 571 (1985). Abstract. – See Abstr. 010.065.

034.152 Berkeley infrared camera.
J. F. Arens.
Bull. Am. Astron. Soc., Vol. 17, No. 2, p. 571 (1985). Abstract. – See Abstr. 010.065.

034.153 The AFGL infrared mosaic spectrometer.
P. D. LeVan, T. L. Murdock, P. C. Tandy,
S. J. Little.
Bull. Am. Astron. Soc., Vol. 17, No. 2, p. 571 (1985). Abstract. – See Abstr. 010.065.

034.154 A post–dispersion infrared detection system for Fourier transform spectrometers.
G. Wiedemann, D. Jennings, R. Hanel, G. Lamb, V. Kunde.
Bull. Am. Astron. Soc., Vol. 17, No. 2, p. 571 (1985). Abstract. – See Abstr. 010.065.

034.155 A 803 GHz (374 micron) laser heterodyne receiver.
C. L. Bennett, G. Chin, S. Petuchowski, D. Buhl.
Bull. Am. Astron. Soc., Vol. 17, No. 2, p. 571 – 572 (1985). Abstract. – See Abstr. 010.065.

034.156 A shaped aperture for microdensitometer scanning.
B. M. Lasker, M. Damashek, P. Garnavich.
Bull. Am. Astron. Soc., Vol. 17, No. 2, p. 573 – 574 (1985). Abstract. – See Abstr. 010.065.

034.157 A high precision, high resolution intensified Reticon detector for stellar spectroscopy.
B. W. Bopp, P. V. Noah, R. A. Jones.
Bull. Am. Astron. Soc., Vol. 17, No. 2, p. 574 (1985). Abstract. – See Abstr. 010.065.

034.158 The Penn State fiber coupled/CCD spectrograph system.
L. W. Ramsey, C. Brungardt, D. P. Huenemoerder,
S. Rosenthal.
Bull. Am. Astron. Soc., Vol. 17, No. 2, p. 574 (1985). Abstract. – See Abstr. 010.065.

034.159 The Wisconsin dual etalon CCD imaging spectrometer.
J. Brinkmann, F. Scherb, R. J. Reynolds,
F. L. Roesler.
Bull. Am. Astron. Soc., Vol. 17, No. 2, p. 574 (1985). Abstract. – See Abstr. 010.065.

034.160 Theoretical internal precision of the multi–channel astrometric photometer.
J. H. Kiewiet de Jonge.
Bull. Am. Astron. Soc., Vol. 17, No. 2, p. 582 (1985). Abstract. – See Abstr. 010.065.

034.161 A Fabry–Perot etalon for differential spectral imaging.
D. M. Rust, C. Burton, R. Abell.
Bull. Am. Astron. Soc., Vol. 17, No. 2, p. 642 – 643 (1985). Abstract. – See Abstr. 010.067.

034.162 Hydrogen masers for Rumanian atomic time.
O. C. Gheorghiu, C. M. Mandache, C. M. Szekely, M. P. Dinca, L. C. Giurgiu, V. M. Stavinschi.
Top. Astrophys. Astron. Space Sci., Vol. 1, p. 77 – 86 (1985).
The hydrogen maser is a frequency standard with high accuracy and frequency stability. Its very high long term frequency stability makes the hydrogen maser suitable for atomic time keeping.

034.163 La realizzazione di un misuratore di lastre ad un solo asse per la determinazione di accurate posizioni di asteroidi e comete.
U. Quadri.
Astronomia, N. 2, p. 3 – 5 (1985).

034.164 An experience in hypersensityzing emulsion for technical uses with pure hydrogen: preliminary results of laboratory tests.
C. Frisoni, D. Marani, S. Orlandi, E. Pancaldi, G. Sette.
Astronomia, Suppl. al N. 2, p. 45 – 51 (1985). In English and Italian.

034.165 The interferometric detection of gravitational waves.
A. Brillet.
Gravitation, geometry and relativistic physics, p. 195 – 203 (1984). – See Abstr. 012.103.
The principles of operation and a noise analysis are presented for a gravitational wave interferometer to be constructed at Orsay, France.

034.166 The development of long baseline gravitational radiation detectors at Glasgow University.
J. Hough, S. Hoggan, G. A. Kerr, J. B. Mangan, B. J. Meers, G. P. Newton, N. A. Robertson, H. Ward, R. W. P. Drever.
Gravitation, geometry and relativistic physics, p. 204 – 212 (1984). – See Abstr. 012.103.

034.167 Improved sensitivities in laser interferometers for the detection of gravitational waves.
R. Schilling, L. Schnupp, D. H. Shoemaker, W. Winkler, K. Maischberger, A. Rüdiger.
Gravitation, geometry and relativistic physics, p. 213 – 221 (1984). – See Abstr. 012.103.
A status report of the interferometer detector for gravitational radiation at Garching is presented.

034.168 Investigation of some types of photomultipliers and the possibility of their use for broadband astronomical photometry.
S. I. Neizvestnyj, V. G. Debur, G. A. Georgieva.
Soobshch. Spets. Astrofiz. Obs., Vyp. 44, p. 5 – 37 (1984). In Russian.
The results of the investigations of the photomultipliers FEU–106, –128, –130, –140 are presented. The characteristics of a new photomultiplier with a GaAs–cathode are studied in detail. It has been shown that on the basis of FEU–130 and –140 one can realize a UBV photometric system.

034.169 On the use of AMD–1 in the program "Optical identification of radio sources" carried out with the radio telescope RATAN–600.
L. S. Ugol'kova.
Soobshch. Spets. Astrofiz. Obs., Vyp. 44, p. 39 – 52 (1984). In Russian.

A method of the use of the automatized microdensitometer AMD in the program "Optical identification of radio sources" is described. It allows to automatize partly the process of identification made with devices of Ascorecord type.

034.170 Multiple object spectroscopy: the MX Spectrometer design.
J. M. Hill.
Diss. Abstr. Int., Sect. B, Vol. 45, No. 4, p. 1217 (1984). Thesis, University of Arizona, 218 pp. (1984). Order No. DA8415068.

034.171 Application of an intensified silicon vidicon to astronomical spectrophotometry.
W. G. Weller.
Diss. Abstr. Int., Sect. B, Vol. 45, No. 8, p. 2581 (1985). Thesis, York University (Canada) (1984).

034.172 The No. 1 CCD system at Yunnan Observatory.
B.–x. Ye, X.–m. Meng, C.–j. Wang.
Acta Astron. Sin., Vol. 26, No. 2, p. 172 – 179 (1985). In Chinese.

034.173 Properties of the new infrared film I–1060B.
I. I. Brejdo.
Izv. Glav. Astron. Obs. Pulkovo, Astrometr. Astrofiz., No. 202, p. 118 – 120 (1984). In Russian.
The results of testing the infrared film I–1060B are given. This film is by about an order more sensitive in the region $\lambda = 1060 - 1080$ nm than other films of the same region.

034.174 Application of the method of unsharp masking in astrophotography.
O. M. Mikhajlova.
Izv. Glav. Astron. Obs. Pulkovo, Astrometr. Astrofiz., No. 202, p. 121 – 123 (1984). In Russian.
A method of printing from astronomical negatives with the use of unsharp masking is given which provides for the reproduction of maximum information.

034.175 Converter of solar mean time signals into sidereal time signals.
V. A. Krat.
Izv. Glav. Astron. Obs. Pulkovo, Astrometr. Astrofiz., No. 201, p. 132 – 134 (1985). In Russian.
A simple digital apparatus is described permitting to convert impulse signals of the mean solar time with a frequency up to 10 MHz into impulse signals of sidereal time. Several computational schemes are considered which differ by the algorithm and the value of the systematic variation of the obtained approximation to the signals of the sidereal time.

Katalog der ortsfesten Sonnenuhren in der DDR.
See Abstr. 002.052.

Cadrans du Soleil; les cadrans peints des Alpes à la Méditerranée.
See Abstr. 003.155.

Les débuts de la photographie astronomique particulièrement à l'Observatoire de Paris.
See Abstr. 004.019.

Die astronomische Uhr zu Straßburg.
See Abstr. 004.027.

Contatti Brera–Cremona per il rifacimento del quadrante dell'orologio astronomico del Torrazzo ed altre questioni.
See Abstr. 004.120.

Guardians of Greenwich time.
See Abstr. 004.141.

Spectrographie des structures fines solaires. Un grand pas réalisé au Pic–du–Midi.
See Abstr. 009.010.

Interkosmos satellite laser radar in Vietnam Socialist Republic.
See Abstr. 013.018.

Upgrading the computer control of the Interkosmos Laser Ranging Station in Helwan.
See Abstr. 013.019.

Performance and results of Satellite Ranging Laser Station at Borowiec in 1983.
See Abstr. 013.020.

Fifteen years of Lunar Laser Ranging and future developments.
See Abstr. 013.021.

Astrophotographie in der Schule.
See Abstr. 014.010.

A fast relay lens for the next generation of photon–counting systems.
See Abstr. 031.008.

Grazing incidence echelle spectrometers using varied line–space gratings.
See Abstr. 031.031.

Report of IAU Commission 9: Instruments and techniques (*Instruments et techniques*).
See Abstr. 032.046.

The Leuschner Observatory automated 30–inch telescope.
See Abstr. 032.047.

A new method for the determination of the counting loss of photon counting photometers.
See Abstr. 036.034.

Steuerung eines Amateurfernrohres durch einen Mikrocomputer.
See Abstr. 036.039.

A possible nonlinearity in IDS (*Image Dissector Scanner*) data.
See Abstr. 036.059.

Submillimetre spectroscopy on La Silla.
See Abstr. 036.120.

Electronography: performances, results, expectations.
See Abstr. 036.220.

Food for the photometrists – faint galaxies revealed.
See Abstr. 036.221.

Report of IAU Commission 31: Time (*L'heure*).
See Abstr. 044.084.

Report of IAU Commission 25: Stellar photometry and polarimetry (*Photométrie et polarimétrie stellaires*).
See Abstr. 113.070.

Simultaneous linear and circular polarimetry of EF Eri.
See Abstr. 117.027.

Narrow–band observations of planetary nebulae with a photon–counting imaging detector.
See Abstr. 134.043.

Search for light flashes from a gamma ray burst source.
See Abstr. 143.034.

Very high–energy gamma–ray astronomy.
See Abstr. 143.041.

035 Space Instrumentation

035.001   The Multi–Anode Microchannel Array detector system: current status and future prospects.
J. G. Timothy.
NASA Conf. Publ., NASA CP–2349, p. 530 – 533 (1984). – See Abstr. 012.001.

The Multi–Anode Microchannel Arrays (MAMAs) are a family of photoelectric pulse–counting array detectors that are being developed specifically for use in space. MAMA detectors with formats as large as 256×1024 pixels are currently under evaluation in the laboratory, and a (24×1024)–pixel extreme-ultraviolet MAMA detector was recently flown successfully on a sounding rocket. A (256×1024)–pixel ultraviolet MAMA detector system is now being prepared for flight on the Balloon–Borne Ultraviolet Stellar Spectrograph. The performance characteristics of this detector system are briefly described in this paper and the implications for the design of the detectors for the FUSE (now Columbus) mission are discussed.

035.002   Photometric calibrations of an MCP space qualified photographic–camera.
S. Koutchmy, A. Verlhac.
Astron. Astrophys., Vol. 142, No. 2, p. 355 – 360 (1985).

The authors describe the methods developed to guarantee quantitative calibrations of a high sensitivity micro–channel plate photo–camera designed to obtain large scale night sky pictures of astronomical and geophysical interest in the optical and near infrared spectral range. In–flight calibrations were obtained with a sensitometer and absolute calibrations were performed in laboratory with a quasi–monochromatic black body source. Results

obtained for calibrating the camera are given with emphasis on the comparison of measurements deduced in ground based and orbital condition.

035.003   BATSE/GRO observational capabilities.
G. J. Fishman, C. A. Meegan, T. A. Parnell, R. B. Wilson, W. Paciesas.
AIP Conf. Proc., No. 115, p. 651 – 664 (1984). – See Abstr. 012.005.

The Burst and Transient Source Experiment (BATSE) will be a sensitive, all–sky monitor for the Gamma–Ray Observatory (GRO). The eight scintillation detector modules of BATSE are positioned around the GRO spacecraft to provide a complete view of the entire sky above the horizon. Rough burst locations can be derived from the relative counting rates of the detectors; longer–lived (>100 min) transient sources can be detected and located by Earth occultations. Details of the design and capabilities of the BATSE are presented.

035.004   All sky high resolution cameras for hard and soft X–rays.
P. Gorenstein, C. W. Mauche.
AIP Conf. Proc., No. 115, p. 694 – 708 (1984). – See Abstr. 012.005.

An all sky camera concept for the detection and localization of gamma–ray bursts from a single spacecraft in near–Earth orbit is described. The system consists of three units each containing a position sensitive detector plus a coded aperture and an imaging modulation collimator. Monte Carlo simulations indicate that the instrument will provide positions based upon measurements in the 20 – 100 keV band that are precise to 10 arcsec or better for bursts as strong as the Apollo 16 event. The authors also describe a more sensitive all sky soft (2 – 6 keV) X–ray camera which is capable of monitoring the X–ray afterglows of gamma–ray bursts as well as all X–ray bursts, transients, and steady sources.

035.005   Solar high resolution balloon spectra obtained in the 190 – 300 nm wavelength band.
P. Lemaire, D. Samain.
Adv. Space Res., Vol. 4, No. 8, p. 37 – 41 (1984). – See Abstr. 012.010.
A new solar instrumentation and an equatorial mounting platform is described and results obtained during the first flight are presented and discussed.

035.006   The SOUP (*Solar Optical Universal Polarimeter*) and CIP (*Coordinated Instrument Package*) instruments.
A. Title.
Adv. Space Res., Vol. 4, No. 8, p. 67 – 74 (1984). – See Abstr. 012.010.
The scientific goals, instrument characteristics, and data handling properties of the SOUP and the Solar Optical Telescope CIP are described.

035.007   Solar coronal studies using normal–incidence X–ray optics.
L. Golub.
Adv. Space Res., Vol. 4, No. 8, p. 75 – 82 (1984). – See Abstr. 012.010.
The author describes the progress which has been made in constructing a new type of X–ray telescope, which operates at normal incidence in the soft X–ray region by the use of multilayer coatings. The principles involved in state–of–the–art multilayer technology and some recent high–resolution imaging results are discussed. The scientific program for solar coronal studies and future instrumental developments are also discussed.

035.008   A compact Dopplergraph/magnetograph suitable for space–based measurements of solar oscillations and magnetic fields.
E. J. Rhodes Jr., A. Cacciani, S. Tomczyk, R. K. Ulrich, J. Blamont, R. F. Howard, P. Dumont, E. J. Smith.
Adv. Space Res., Vol. 4, No. 8, p. 103 – 112 (1984). – See Abstr. 012.010.
By combining a unique magneto–optical resonance filter with CID and CCD cameras the authors have been able to obtain full– and partial–disk Dopplergrams and magnetograms. Time series of the velocity images are converted into k–ω power spectra which show clearly the solar nonradial p–mode oscillations. Magnetograms suitable for studying the long–term evolution of solar active regions have also been obtained with this instrument. The results have been encouraging enough to warrant a continued series of tests with this device.

035.009   A new eye on the universe.
J. J. Burger.
Spaceflight, Vol. 27, No. 2, p. 76 – 78 (1985).

035.010   A balloon–borne high–resolution spectrometer for observations of gamma–ray emission from solar flares.
C. J. Crannell, R. Starr, A. R. Stottlemyer, J. I. Trombka.
Nucl. Instrum. Methods Phys. Res., Sect. A, Vol. 225, No. 1, p. 195 – 208 (1984). Abstr. in Phys. Abstr., Vol. 88, No. 1248, Entry 9600 (1985).

035.011   A lightweight shield–detector combination for use in hard X–ray telescopes.
H. Griffiths, R. R. Hillier.
Nucl. Instrum. Methods Phys. Res., Sect. A, Vol. 225, No. 2, p. 418 – 422 (1984). Abstr. in Phys. Abstr., Vol. 88, No. 1248, Entry 9601 (1985).

035.012   Effect of the Mott cross section on charge identification in the HEAO–3 Heavy Cosmic Ray Experiment.
J. H. Derrickson, P. B. Eby, J. W. Watts Jr.
Nucl. Instrum. Methods Phys. Res., Sect. A, Vol. 225, No. 1, p. 185 – 194 (1984). Abstr. in Phys. Abstr., Vol. 88, No. 1248, Entry 9620 (1985).

035.013   Maxwell's last frontier.
M. Washburn.
Sky Telesc., Vol. 69, No. 3, p. 212 (1985).

035.014   Astronomers, Congress, and the Large Space Telescope.
P. A. Hanle.
Sky Telesc., Vol. 69, No. 4, p. 300 – 305 (1985).

035.015   A balloon–borne microwave limb sounder for stratospheric measurements.
J. W. Waters, J. C. Hardy, R. F. Jarnot, H. M. Pickett, P. Zimmermann.
J. Quant. Spectrosc. Radiat. Transfer, Vol. 32, No. 5/6, p. 407 – 433 (1984). – See Abstr. 012.026.
The balloon–borne microwave limb sounder (BMLS) measures atmospheric thermal emission from millimeter wavelength spectral lines to determine vertical profiles of stratospheric species. The instrument flown to date operates at 205 GHz to measure ClO, O_3, and H_2O_2. A 63 GHz radiometer will be added to test the technique for determining tangent point pressure from the MLS experiment on the upper atmosphere research satellite (UARS). Many additional species could also be measured by the BMLS.

035.016   The instrument IKS and its calibration.
J.–P. Bibring, S. Cazes, J. Charra, M. Combes, N. Coron, B. Cougrand, J.–F. Crifo, J. Crovisier, C. Emerich, T. Encrenaz, R. Gispert, B. Gondet, G. Guyot, D. Harduin, J.–M. Lamarre, G. Levanti, C. Maurel, D. Parisot, F. Rocard, P. Salvetat, A. Soufflot.
Adv. Space Res., Vol. 4, No. 9, p. 273 – 276 (1984). – See Abstr. 012.029.
The IKS infrared spectro–photometer will fly on board the VEGA platforms. It is designed to characterize the size, temperature and emissivity of the Comet Halley nucleus, to identify the major gaseous components of the inner coma and to detect the emission of the cometary grains. This paper presents the "calibration" experiments required to reduce the raw data.

035.017   The Infrared Imaging Channel of the IKS Instrument for detection of Comet Halley nucleus.
J.–M. Lamarre, B. Gondet, R. Gispert, C. Emerich, F. Rocard.
Adv. Space Res., Vol. 4, No. 9, p. 277 – 281 (1984). – See Abstr. 012.029.
The Imaging Channel of the IKS Instrument placed on board the Vega fly–by probes will perform measurements of the infrared emission of the central region of Comet Halley at distances in the $10^4 – 10^5$ km range. An encoding wheel analyses one spatial frequency of the infrared image during the whole fly–by. Inversion of this measurement will give low resolution brightness profiles of the nucleus and its immediate surroundings, in two wavelength bandpasses and in two directions of analysis.

035.018   Oriented platform for the Vega–probe to the Halley comet, position detection, working program.
B. Valníček, J. Reček.
Adv. Space Res., Vol. 4, No. 9, p. 283 – 285 (1984). – See Abstr. 012.029.

035.019   A spectral photopolarimeter for Giotto: Halley Optical Probe Experiment.
A. C. Levasseur-Regourd, J. L. Bertaux, R. Dumont, M. C. Festou, R. H. Giese, G. Giovane, P. Lamy, A. Llebaria, J. L. Weinberg.
Adv. Space Res., Vol. 4, No. 9, p. 287 – 290 (1984). – See Abstr. 012.029.
The Halley Optical Probe Experiment (HOPE) is designed to provide in situ photopolarimetric data on both the dust cloud

and the gaseous atmosphere in Halley's coma. The optical probe concept is presented, together with a description of the instrumentation and with the possibilities for cross–checks between HOPE results and those of other space and ground–based experiments.

035.020 First calibration measurements with the dust impact detector DIDSY–IPM.
E. Grün, E. Bussoletti, A. Minafra, H. Kuczera, J. A. M. McDonnell.
Adv. Space Res., Vol. 4, No. 9, p. 291 – 295 (1984). – See Abstr. 012.029.

035.021 The Giotto spacecraft configuration and its achievement.
D. C. Clayton, P. Truss.
J. Br. Interplanet. Soc., Vol. 38, No. 5, p. 222 – 230 (1985).

The considerations leading to the final design of the European Space Agency Giotto Halley's comet probe are discussed, including scientific payload, launch vehicle, thermal control, and spacecraft balance requirements.

035.022 Das Space–Teleskop Observatorium.
M. Woche.
Astron. Raumfahrt, 22. Jahrg., Heft 2, p. 26 – 31 (1985).

035.023 HST (*Hubble Space Telescope*): astronomy's greatest gambit.
J. K. Beatty.
Sky Telesc., Vol. 69, No. 5, p. 409 – 414 (1985).

035.024 EXOSAT – present status and future prospects.
R. M. Redfern.
Ir. Astron. J., Vol. 17, No. 1, p. 31 – 39 (1985). – See Abstr. 012.034.

035.025 The Space Telescope Project.
P. Jakobsen.
Ir. Astron. J., Vol. 17, No. 1, p. 50 – 55 (1985). – See Abstr. 012.035.

035.026 A balloon–borne sun–tracking multichannel photometer for atmospheric aerosol measurements.
Y. B. Acharya, A. Jayaraman, B. H. Subbaraya.
Adv. Space Res., Vol. 5, No. 1, p. 65 – 68 (1985). – See Abstr. 012.038.

A balloon–borne multichannel photometer for measurement of atmospheric scattering in the near ultraviolet and the visible wavelength regions has been developed for study of the size distribution and number density of aerosols at tropospheric and lower stratospheric altitudes. The instrumentation involves tracking the sun in elevation and scanning in azimuth. The payload was recently flown from the Hyderabad Balloon Facility on 18 April 1984. The paper details the instrument design and presents a few illustrations of the instrument performance from this flight.

035.027 Balloon–borne, high–altitude gravimetry.
A. R. Lazarewicz.
Adv. Space Res., Vol. 5, No. 1, p. 79 – 82 (1985). – See Abstr. 012.038.

Gravity measurements from a high–altitude balloon can verify global and upward–continued gravity models. A gravimeter suspended beneath a balloon is in a dynamic, and largely unpredictable, environment sensing accelerations due to gravity and balloon motions. Independent measurements of balloon motions using inertial navigation data combined with ground tracking data will allow for separation of balloon–induced accelerations from gravitational accelerations. The first engineering test flight occurred on 11 October 1983, during the seasonal wind reversal and was very successful. Flight duration was approximately seven hours, with two hours of data collected at each of 30 km and 26 km altitudes. The results include gravity estimates, design criteria for future flights and feasibility analysis for vertical gravity profiles during ascent and descent.

035.028 Microgravity experiment system utilizing a balloon.
M. Namiki, S. Ohta, T. Yamagami, Y. Koma, H. Akiyama, H. Hirosawa, J. Nishimura.
Adv. Space Res., Vol. 5, No. 1, p. 83 – 86 (1985). – See Abstr. 012.038.

A system for microgravity experiments by using a stratospheric balloon has been planned and developed in ISAS since 1978. A rocket–shaped chamber mounting the experiment apparatus is released from the balloon around 30 km altitude. The microgravity duration is from the release to opening of parachute, controlled by an on–board sequential timer. Test flights were performed in 1980 and in 1981. In September 1983 the first scientific experiment, observing behaviors and brain activities of fish in the microgravity circumstance, have been successfully carried out. The chamber is specially equipped with movie cameras and sub-transmitters, and its release altitude is about 32 km. The microgravity observed inside the chamber is less than $2.9 \times 10^{-3}G$ during 10 sec. Engineering aspects of the system used in the 1983 experiment are presented.

035.029 A two–dimensional position–sensitive spectroscopic proportional counter for balloon–borne hard X–ray astronomy.
P. Ubertini, A. Bazzano, L. Boccaccini, N. A. Dipper, L. Iafrate, C. LaPadula, M. Mastropietro, R. Patriarca, V. Polcaro, M. L. Urciuoli.
Adv. Space Res., Vol. 5, No. 1, p. 105 – 108 (1985). – See Abstr. 012.038.

A new design of position–sensitive spectroscopic proportional counter is described, for use in a balloon–borne hard X–ray telescope. Initial position and spectral resolution data from a one–dimensional laboratory prototype are reported. With this device, the final telescope will have an angular resolution of better than 10 minutes of arc.

035.030 Design study for a three–meter balloon–borne telescope for far infrared and submillimeter astronomy.
W. F. Hoffmann, G. G. Fazio, D. A. Harper.
Adv. Space Res., Vol. 5, No. 1, p. 117 – 120 (1985). – See Abstr. 012.038.

A NASA supported design study is being carried out for a three–meter balloon–borne far infrared and submillimeter telescope. The goal of this project is to provide a facility for frequent flights for photometry, spectroscopy, and imaging in the spectral region 30 micrometers to 1 millimeter. It is intended to provide a scientific and technical step on the way to a large submillimeter telescope in space in the future. The study is concentrating on areas where technical advances are required: materials and fabrication techniques for lightweight primary mirrors, telescope and gondola structure, and pointing and stabilization. Innovative approaches to the telescope support and stabilization are being explored for achieving the required 1 arcsecond pointing stability.

035.031 The use of proportional counters for X–ray radiometric analysis of the composition of planetary surfaces.
V. I. Chesnokov.
Geofiz. apparatura, Leningrad, No. 80, p. 65 – 69 (1984). In Russian. Abstr. in Ref. Zh., 51. Astron., 2.51.165 (1985).

035.032 X–ray experiment aboard the artificial earth satellite Prognoz 9.
M. I. Kudryavtsev, S. I. Svertilov.
Vestn. Mosk. Univ., Ser. 3. Fiz. Astron., Tom 25, No. 5, p. 81 – 88 (1984). In Russian. Abstr. in Ref. Zh., 62. Issled. Kosm. Prostranstva, 2.62.110 (1985).

035.033 Science with the Solar Optical Telescope (SOT).
S. D. Jordan.
ESA Spec. Publ., ESA SP–220, 165 – 175 (1984). – See Abstr. 012.045.

The Solar Optical Telescope (SOT) will provide the necessary data for attacking several fundamental problems in the energetics and dynamics of the solar atmosphere, including the origin and evolution of the Sun's magnetic field, the structure of solar subsurface convection, the heating of the outer solar atmosphere, and sources of the solar wind in the lower lying regions of the outer atmosphere. This telescope, of a Gregorian configuration, will have 1.3–meter–diameter primary mirror that will be capable of achieving close to 0.1–arc–second angular resolution on the Sun in both visible and ultraviolet wavelengths. This paper gives further details on the telescope and how it will be operated in space to achieve the scientific objectives.

035.034   Science with SOHO.
M. C. E. Huber, M. Malinovsky–Arduini.
ESA Spec. Publ., ESA SP-220, 177 – 183 (1984). – See Abstr. 012.045.
The scientific goals of the Solar and Heliospheric Observatory (SOHO) comprise the basic and interconnected questions regarding the interior structure of the Sun, the heating of the corona and its expansion as solar wind. The observations needed to pursue these questions will be taken from a halo orbit around Lagrange point L_1 on the Earth–Sun line, which offers uninterrupted observations of the Sun and – lying outside the magnetosphere – of its solar–wind streams. In this paper the authors outline some patent solar and heliospheric problems, define the resulting observing needs and present the SOHO model payload (helioseismology, optical plasma diagnostics of the corona at extreme ultraviolet (XUV) and visible wavelengths, in–situ particle and field measurements) in the light of these requirements.

035.035   Satellite–based instrument concepts for the measurement of orbital debris.
J. A. Sanguinet.
Adv. Space Res., Vol. 5, No. 2, p. 59 – 62 (1985). – See Abstr. 012.051.
Small orbiting particles which pose a potential hazard to spacecraft have cross sections too small to be measured by conventional ground–based radars; thus, the magnitude of the hazard is uncertain. It has therefore been suggested that an orbiting sensor be deployed for the purpose of gathering statistical data on debris population. A number of sensor approaches for fulfilling this mission are discussed, including active millimeter wave radar, passive/active IR and visible sensors, and passive sensors. The performance capability of each is compared in terms of system complexity, range performance, and the expected debris particle detection rate.

035.036   Two–stage acoustic penetration and impact detector for micrometeoroid and space debris application.
H. Kuczera, H. Iglseder, U. Weishaupt, E. Igenbergs.
Adv. Space Res., Vol. 5, No. 2, p. 91 – 94 (1985). – See Abstr. 012.051.
An active impact detector is described which has recently been developed and tested in the Munich Plasma Accelerator Facility. Piezo microphones are used for detection of penetration and impact events in a set–up which consists of a thin front foil and an impact plate. The elastic–wave propagation times between the impact location and the microphones will be measured in both the foil and the target and, by correlation of these time sequences and the signal amplitudes, the impact event time, the impact location, the projectile velocity, the flight path direction and the mass of the impacting particle can be evaluated. This measurement principle can be applied to a wide range of projectile sizes and velocities. Results of impact simulation experiments are presented and discussed with respect to a flight experiment application.

035.037   Anticipated impact plasma problems for the Copernic–RPA experiment in the cometary environment.
C. d'Uston, H. Rème.
ESA Spec. Publ., ESA SP-224, p. 81 – 86 (1984). – See Abstr. 012.052.
The flight through the cometary environment at a very high relative velocity will present a new set of spacecraft/plasma interactions that may influence the charged particle measurements by the Copernic–RPA experiment onboard Giotto in the last two minutes of the mission. The most sensitive areas are the electrostatic environment of the S/C and the secondary particle population. It will be necessary to simulate the trajectories of the particles coming into the detectors in order to know their distortion by the electrostatic environment of the S/C.

035.038   An impact plasma monitor for cometary missions.
R. Grard.
ESA Spec. Publ., ESA SP–224, p. 87 – 88 (1984). – See Abstr. 012.052.
The emission of plasma associated with high velocity impacts of molecules (70 km/s) is a source of interference for the experiments which will study the neutral and ionized gas environment of Comet Halley. The impact plasma monitor is a simple device which will detect this effect on–board Giotto, Vega–1 and Vega–2 by measuring the saturation current of secondary electrons extracted from the surface of a probe exposed to the molecular bombardment. It is expected that these in–situ observations will complement the information obtained both from theory and laboratory investigations and help to characterize the immediate vicinity of the spacecraft during their flyby through the cometary coma.

035.039   Plasma environment effects on the Giotto ion mass spectrometer.
R. Goldstein.
ESA Spec. Publ., ESA SP–224, p. 89 – 90 (1984). – See Abstr. 012.052.
Spacecraft–environment interactions are discussed relative to the performance of the Giotto Ion Mass Spectrometer (IMS). Ion trajectory distortions due to space charge regions forward of the spacecraft do not seriously disturb the measurements of sensor IMS–1, but may affect sensor IMS–2, which looks in the forward direction. Charging of the spacecraft itself can distort ion spectra measured by both sensors. Intercomparison of data from other instruments may allow estimation of the amount of charging and therefore the extent of distortion.

035.040   Thermal design of the Giotto spacecraft.
F. Felici, P. Lo Galbo.
ESA Spec. Publ., ESA SP–224, p. 91 – 98 (1984). – See Abstr. 012.052.
During its nine months long mission to reach Halley's Comet in March 1986, Giotto will be subject to a wide range of sun fluxes illuminating the spacecraft at various angles. These excursions of the thermal environment, together with restrictions on thermal control materials and on the general spacecraft configuration have led to a design that makes use of both passive and active elements (among the latter newly developed thermal shutters) to keep all equipment and the scientific payload within specified temperature ranges during all the phases of the mission.

035.041   NASA applications for acousto–optic spectrometers.
G. Chin.
Proc. SPIE Int. Soc. Opt. Eng., Vol. 477, p. 128 – 131 (1984). Abstr. in Phys. Abstr., Vol. 88, No. 1252, Entry 24967 (1985). – See Abstr. 012.053.

035.042   Measuring planetary hydrogen by remote gamma–ray sensing.
E. L. Haines, A. E. Metzger.
Nucl. Instrum. Methods Phys. Res., Sect. A, Vol. 226, No. 2 – 3, p. 509 – 516 (1984). Abstr. in Phys. Abstr., Vol. 88, No. 1252, Entry 29846 (1985).

035.043   Measuring planetary neutron albedo fluxes by remote gamma–ray sensing.
E. L. Haines, A. E. Metzger.
Nucl. Instrum. Methods Phys. Res., Sect. A, Vol. 226, No. 2 – 3, p. 517 – 523 (1984). Abstr. in Phys. Abstr., Vol. 88, No. 1252, Entry 29847 (1985).

035.044 A balloon–borne instrumentation for cosmic gamma–ray burst detection and measurement.
G. Ventura, H. M. Horstman, A. Brighenti, C. Cavani, M. Camprini, P. Cazzola, G. Giovannini, C. Labanti, J. M. Poulsen.
Nucl. Instrum. Methods Phys. Res., Sect. A, Vol. 226, No. 2–3, p. 524–533 (1984). Abstr. in Phys. Abstr., Vol. 88, No. 1252, Entry 29848 (1985).

035.045 The Hopkins Ultraviolet Telescope.
A. F. Davidsen, G. H. Fountain.
Johns Hopkins APL Tech. Dig., Vol. 6, No. 1, p. 28–37 (1985).
A new capability in ultraviolet astronomy will be realized in 1986 when the Hopkins Ultraviolet Telescope is carried aloft by the space shuttle. Designed to measure far ultraviolet and extreme ultraviolet radiation from a broad range of astronomical objects, this telescope will complement the Space Telescope's observations of the same objects at longer wavelengths.

035.046 An alternate fine guidance sensor for the Space Telescope.
M. D. Griffin, T. E. Strikwerda, D. G. Grant.
Johns Hopkins APL Tech. Dig., Vol. 6, No. 1, p. 51–59 (1985).
A study was conducted to develop a preliminary design of an alternate fine guidance sensor for the National Aeronautics and Space Administration's Space Telescope; an electrostatically focused silicon diode, quadrant array detector was selected. Simulations of the Space Telescope control system showed that the recommended sensor could meet the required performance specifications.

035.047 Prospects for developing satellite–borne gamma–ray telescopes.
A. M. Gal'per, V. G. Kirillov–Ugryumov, V. E. Nesterov, O. F. Prilutskij.
Astrophys. Space Phys. Rev., Vol. 3, p. 301–328 (1984). – See Abstr. 003.019. Revised and extended English translation of 34.035.035.
Contents: (1). Introduction. (2). Present status of observational gamma–ray astronomy. (3). Gamma–ray telescope with wide–gap spark chambers. (4). Aperture encoding method.

035.048 The German Infrared Laboratory (GIRL) – a progress report.
D. Lemke, M. Grewing, P. Preussner, W. Martin, D. Offermann, G. Lange, S. Drapatz, R. Katterloher, H. Denner, G. Klipping, F. Dahl, K. Proetel.
Adv. Space Res., Vol. 5, No. 3, p. 11–17 (1985). – See Abstr. 012.059.
The liquid helium cooled 50 cm–IR–telescope is equipped with a 4–band–camera, a photopolarimeter, a grating spectrometer and a Michelson interferometer. These focal plane instruments allow measurements with high spectral resolution, high sensitivity and diffraction limited spatial resolution in the wavelength region 2.5–200 μm. – The "thermal model" phase of the project was successfully completed in 1983. This phase included the development and construction of a full size cryostat, which was thoroughly tested by an experiment simulator. Prototypes of all focal plane instruments have been developed and several have already been tested under the expected flight conditions.

035.049 Astrophysical and geophysical observations with PIRAMIG/Salyut 7 experiment.
A. C. Levasseur–Regourd, G. Courtes, M. Heise, S. Koutchmy, P. Lamy, T. M. Muliarchik, B. Rocca–Volmerange, S. A. Savchenko, B. Secher, H. M. Tovmassian (*G. M. Tovmasyan*), A. N. Beresovoy (*A. N. Berezovoj*), J. L. Chretien, A. S. Ivanchenkov, V. A. Djanibekov, V. V. Lebedev, L. I. Popov, S. E. Savitskaya, A. A. Serebrov.
Adv. Space Res., Vol. 5, No. 3, p. 27–30 (1985). – See Abstr. 012.059.
The facility offered by the Salyut 7 vehicle has allowed teams of scientists of 3 CNRS French laboratories to develop an instrument optimized for several disciplines in astrophysics and geo-

physics. P.I.R.A.M.I.G. (Photography Infra–Red Atmosphere, Interplanetary Medium, Galaxies) is a wide field (10° and 40°) camera devoted to high sensitivity photographic photometry. The spectral range is limited to the visible and near–infrared (400–850 nm), the main advantage being to observe above the absorbing and emitting layers of the atmosphere.

035.050 The EGRET (*Energetic Gamma Ray Experiment Telescope*) high energy gamma–ray telescope on GRO (*Gamma Ray Observatory*).
G. Kanbach.
Bull. Inf. Cent. Données Stellaires, No. 28, p. 69–71 (1985). – See Abstr. 012.062.

035.051 The observation program of COMPTEL.
V. Schönfelder.
Bull. Inf. Cent. Données Stellaires, No. 28, p. 77–80 (1985). – See Abstr. 012.062.
COMPTEL is one of the four gamma ray telescopes onboard of the NASA gamma–ray observatory GRO which is to be launched in 1988 by the Space Shuttle. The four experiments cover the entire gamma ray energy range.

035.052 The study of astrophysical interactions with the ESA Photon Counting Detector.
S. di Serego Alighieri, M. A. C. Perryman, F. Macchetto.
ESA Bull., No. 42, p. 17–21 (1985).
The authors have used the ESA Photon Counting Detector, a scientific model of the Faint–Object Camera to be flown on the Space Telescope, to study various aspects of the interactions between astrophysical objects. Examples include the relationship between a planetary nebula and a nearby star, the effects of shock waves in the interstellar medium of a young galaxy, the interaction of the jet in a radio galaxy with the surrounding intergalactic medium, and the perturbation in a quasar gas halo caused by a gravitationally interacting galaxy in the same group.

035.053 On the viability of exploiting L–shell fluorescence for X–ray polarimetry.
M. C. Weisskopf, P. G. Sutherland, R. F. Elsner, B. D. Ramsey.
Bull. Am. Astron. Soc., Vol. 16, No. 4, p. 885 (1984). Abstract. – See Abstr. 010.062.

035.054 An analysis of the Hettrick–Bowyer Type I grazing incidence telescope.
J. Green, S. Bowyer.
Bull. Am. Astron. Soc., Vol. 16, No. 4, p. 885 (1984). Abstract. – See Abstr. 010.062.

035.055 Development of a quantum calorimeter for X–ray spectroscopy.
R. L. Kelley, J. C. Mather, S. H. Moseley, R. F. Mushotzky, A. E. Szymkowiak, D. McCammon.
Bull. Am. Astron. Soc., Vol. 16, No. 4, p. 902 (1984). Abstract. – See Abstr. 010.062.

035.056 The Ultraviolet Imaging Telescope for the Astro missions.
T. P. Stecher, R. C. Bohlin, R. W. O'Connell, M. S. Roberts, A. M. Smith, R. A. Parise.
Bull. Am. Astron. Soc., Vol. 16, No. 4, p. 905 (1984). Abstract. – See Abstr. 010.062.

035.057 Satellite–borne, UV, Fourier transform spectrometer for high–resolution, spatially–imaged spectroscopy.
P. L. Smith, W. H. Parkinson.
Bull. Am. Astron. Soc., Vol. 16, No. 4, p. 905–906 (1984). Abstract. – See Abstr. 010.062.

035.058   Multiband imaging photometer for SIRTF.
G. H. Rieke, C. J. Lada, M. J. Lebofsky, F. J. Low,
P. A. Strittmatter, E. T. Young, J. R. Mould, G. Neugebauer,
S. Gaalema, C. A. Beichman, T. N. Gautier III, M. W. Werner,
J. Arens, E. Haller, P. Richards.
Bull. Am. Astron. Soc., Vol. 16, No. 4, p. 906 (1984). Abstract. –
See Abstr. 010.062.

035.059   SIRTF: the Space Infrared Telescope Facility.
M. W. Werner, F. C. Witteborn, J. Murphy,
G. Thorley, L. Manning, W. Brooks.
Bull. Am. Astron. Soc., Vol. 16, No. 4, p. 906 (1984). Abstract. –
See Abstr. 010.062.

**035.060   A wide field and diffraction limited array camera for the
Space Infrared Telescope Facility (SIRTF).**
G. G. Fazio, D. G. Koch, G. J. Melnick,
R. M. Tresch–Fienberg, S. P. Willner, D. Y. Gezari, G. Lamb,
P. Shu, G. Chin, R. Silverberg, J. C. Mather, W. F. Hoffmann,
N. J. Woolf, J. Pipher, W. J. Forrest, C. R. McCreight.
Bull. Am. Astron. Soc., Vol. 16, No. 4, p. 906 (1984). Abstract. –
See Abstr. 010.062.

**035.061   Moderate resolution spectroscopy from the Space Infra-
red Telescope Facility (SIRTF).**
J. R. Houck, S. V. W. Beckwith, T. Herter, E. E. Salpeter,
W. J. Forrest, K. Matthews, B. T. Soifer, D. M. Watson,
T. Roellig, D. Weedman.
Bull. Am. Astron. Soc., Vol. 16, No. 4, p. 907 (1984). Abstract. –
See Abstr. 010.062.

**035.062   The Hopkins Ultraviolet Telescope – an FUV/EUV in-
strument for the Space Shuttle.**
K. S. Long, A. F. Davidsen, S. T. Durrance, H. A. Weaver.
Bull. Am. Astron. Soc., Vol. 16, No. 4, p. 985 (1984). Abstract. –
See Abstr. 010.062.

**035.063   The prime focus spectrograph of the Hopkins Ultraviolet
Telescope.**
R. A. Kimble, C. W. Bowers, K. C. Chambers, H. C. Ferguson,
A. F. Davidsen.
Bull. Am. Astron. Soc., Vol. 16, No. 4, p. 985 – 986 (1984). Ab-
stract. – See Abstr. 010.062.

**035.064   An imaging optical/UV monitor for X–ray astronomy
observatories.**
F. A. Córdova, K. O. Mason, W. C. Priedhorsky, B. Margon,
J. B. Hutchings, P. Murdin.
Astrophys. Space Sci., Vol. 111, No. 2, p. 265 – 290 (1985).
The authors argue that simultaneous X–ray, optical and ultraviolet observations could be achieved more logically, cheaply,
and effectively by mounting a small boresighted optical/UV–
telescope alongside future X–ray telescopes. A 12″ optical/UV
monitor could, for instance, be incorporated into X–ray facilities
such as the American AXAF on the European XMM missions
with minimal impact on the total cost, weight, size, and telemetry
requirements.

035.065   Extreme Ultraviolet Explorer spectrometer.
M. C. Hettrick, S. Bowyer, R. F. Malina, C. Martin,
S. Mrowka.
Appl. Opt., Vol. 24, No. 12, p. 1737 – 1756 (1985).
The design and calculated performance is described for a spectrometer included on the Extreme Ultraviolet Explorer (EUVE)
astronomical satellite. The instrument is novel in design, consisting of three plane reflection gratings mounted in the converg-
ing beam behind a grazing incidence telescope. This configuration is based on new varied line–space (VLS) gratings which have
recently been proposed. The spectrometer has an inherent resolu-
tion of $\lambda/\Delta\lambda \sim 300$, but if combined with a worst–case satellite
performance will yield a spectral resolution of $\lambda/\Delta\lambda = 110 - 240$
and a spatial resolution of $1 - 2$ min of arc. For a 40,000–sec
observation, the average $3\,\sigma$ sensitivity to continuum flux is
$\sim 2 \times 10^{-27} \mathrm{erg/cm^2/sec/Hz}$.

**035.066   Studies of doped IR detectors (Si:In, Si:Ga, Si:As, Si:Sb,
Si:P, Ge:Be and Ge:Ga) for low–background astronomi-
cal applications.**
J. Wolf, D. Lemke.
Infrared Phys., Vol. 25, No. 1/2, p. 327 – 328 (1985). – See Abstr.
012.065.
The authors report on investigations on low–background pho-
toconductors for the photopolarimeter E2 of GIRL.

**035.067   An FIR cooled grating spectrometer for the Kuiper Air-
borne Observatory.**
E. F. Erickson, J. R. Houck, M. O. Harwit, D. M. Rank,
M. R. Haas, D. J. Hollenbach, J. P. Simpson, G. C. Augason.
Infrared Phys., Vol. 25, No. 1/2, p. 513 – 515 (1985). – See Abstr.
012.065.
A liquid–He–cooled spectrometer is being developed as a facility instrument for the Kuiper Airborne Observatory, primarily to
study far–infrared (FIR) lines originating in the interstellar medium. It achieves a maximum resolving power ~ 6000 by means
of a 45 cm long echelle grating and is optically capable of operating in the spectral range from 25 to 300 µm. An array of detectors
is used to measure simultaneously a line and the adjacent continuum from astronomical sources. Currently six detectors allow
mesurements in the $30 - 120$ µm spectral band. This summary
briefly describes the instrument and its performance.

**035.068   A long–wavelength spectrometer for the Infrared Space
Observatory (ISO).**
I. Furniss, W. M. Glencross, P. Cruvellier, M. Joubert,
J.–Y. LeGall, G. Chanin, J.–P. Torre, J. J. Wijnbergen,
G. Serra.
Infrared Phys., Vol. 25, No. 1/2, p. 517 – 520 (1985). – See Abstr.
012.065.
A combination of Fabry–Pérots and a grating monochromator
working in a parallel beam has been designed as a possible long–
wavelength spectrometer for the ISO. It is a very flexible instrument and is capable of working at a high sensitivity and a range
of resolving powers in the FIR, up to 10^4. The principles of the
instrument's design are discussed and a brief outline of its capabilities given.

035.069   Empfängeranordnungen für gekühlte Infrarotteleskope.
J. Wolf.
Diss. Naturwiss.–Math. Gesamtfak. Ruprecht–Karls–Univ.,
Heidelberg, F.R. Germany. $5 + 72$ pp. (1985).

035.070   Space Telescope low–scattered–light camera: a model.
J. B. Breckinridge, T. G. Kuper, R. V. Shack.
Opt. Eng., Vol. 23, No. 6, p. 816 – 820 (1984). Abstr. in Phys.
Abstr., Vol. 88, No. 1254, Entry 40494 (1985).

035.071   Infrared–Fourier–spectrometers in the Venusian orbit.
D. Oertel, W. M. Linkin.
Feingerätetechnik, Vol. 33, No. 8, p. 339 – 340 (1984). In German. Abstr. in Phys. Abstr., Vol. 88, No. 1256, Entry 51496
(1985).

**035.072   Construction of the optical block of the infrared–
Fourier–spectrometer FS 1/4.**
H. Becker–Ross, W. Stadthaus.
Feingerätetechnik, Vol. 33, No. 8, p. 341 – 344 (1984). In Ger-
man. Abstr. in Phys. Abstr., Vol. 88, No. 1256, Entry 51497
(1985).

**035.073   Thermal regime of the optical block of the infrared–
Fourier–spectrometer FS 1/4.**
V. Adam, W. Stadthaus, H. Becker–Ross.
Feingerätetechnik, Vol. 33, No. 8, p. 350 – 353 (1984). In Ger-
man. Abstr. in Phys. Abstr., Vol. 88, No. 1256, Entry 51498
(1985).

035.074   **A large area telescope for balloon–borne hard X–ray astronomy.**
R. E. Baker, G. Barbaglia, A. Bazzano, L. Boccaccini,
A. Bussini, A. Carzaniga, A. Court, A. J. Dean, N. A. Dipper,
G. Ferrandi, N. Haskell, C. La Padula, R. A. Lewis,
D. Maccagni, M. Mastropietro, R. Patriarca, F. Perotti,
V. F. Polcaro, E. Quadrini, D. Ramsden, S. Sembay, R. Spicer,
P. Ubertini, G. Villa, D. Whatley.
Nucl. Instrum. Methods Phys. Res., Sect. A, Vol. 228, No. 1,
p. 183 – 192 (1984). Abstr. in Phys. Abstr., Vol. 88, No. 1256,
Entry 51499 (1985).

035.075   **Balloon–borne detection system for solar infrared brightness temperature.**
Q.–w. Guo, H.–c. Zou.
Chin. J. Infrared Res., Vol. 3, No. 3, p. 176 – 179 (1984). In Chinese. Abstr. in Phys. Abstr., Vol. 88, No. 1256, Entry 51500 (1985).

035.076   **Energetic particle detectors for space research.**
A. Balogh.
Sci. Prog., Vol. 69, No. 275, p. 359 – 374 (1985). Abstr. in Phys. Abstr., Vol. 88, No. 1256, Entry 51502 (1985).

035.077   **Drive unit of an interferometer mirror for satellite–infrared–Fourier–spectrometers.**
H. Driescher, H. Hirsch.
Feingerätetechnik, Vol. 33, No. 8, p. 347 – 349 (1984). In German. Abstr. in Phys. Abstr., Vol. 88, No. 1256, Entry 51505 (1985).

035.078   **HIPPARCOS payload modelling.**
J. Y. Le Gall.
Processing of scientific data from the ESA astrometry satellite HIPPARCOS, p. 29 – 42 (1985). – See Abstr. 012.087.
This paper presents the latest model of the instrument. One will find the modulation factors of the main experiment and the corresponding chromaticities. Results about the star mapper are also presented.

035.079   **Some reflexions about the payload calibration.**
J. Y. Le Gall.
Processing of scientific data from the ESA astrometry satellite HIPPARCOS, p. 65 – 71 (1985). – See Abstr. 012.087.
This paper presents the work done up to now about the payload calibrations. One will find the description of the device allowing the in–flight calibration of the chromaticity which has been approved by ESA.

035.080   **Possible optical grid calibration.**
R. S. Le Poole.
Processing of scientific data from the ESA astrometry satellite HIPPARCOS, p. 73 – 76 (1985). – See Abstr. 012.087.
A means of calibrating the HIPPARCOS grid optically is proposed. The technique uses the demagnified image of a detector array as "reference ruler", and is expected to be capable of such accuracy, that errors in gridgeometry can be eliminated from the total HIPPARCOS error budget.

035.081   **The influence of disturbing effects on the performance of a wide field coded mass X–ray camera.**
M. R. Sims, M. J. L. Turner, R. Willingale.
Nucl. Instrum. Methods Phys. Res., Sect. A, Vol. 228, No. 2 – 3, p. 512 – 531 (1985). Abstr. in Phys. Abstr., Vol. 88, No. 1257, Entry 57615 (1985).

035.082   **Physical–technical conception of the Infrared–Fourier–Spectrometers for 'Venera 15' and 'Venera–16'.**
D. Oertel, H. Jahn, R. Schuster, G. Fellberg, H. Becker–Ross,
W. Barwald, W. Stadthaus.
Exp. Tech. Phys., Vol. 33, No. 1, p. 41 – 59 (1985). In German. Abstr. in Phys. Abstr., Vol. 88, No. 1259, Entry 68705 (1985).

035.083   **Television system of the VEGA experiment.**
A. Balazs, G. Bango, M. Gardos, E. Hamza,
M. Kanyo, G. Kovacs, Z. Nyitrai, R. Redl, P. Rusznyak,
B. Szabo, L. Szabo, S. Szalai, K. Szucs.
Meres Autom., Vol. 33, No. 1 – 2, p. 9 – 13 (1985). In Hungarian. Abstr. in Phys. Abstr., Vol. 88, No. 1260, Entry 74057 (1985).

035.084   **The on–board software of the VEGA TV–system.**
E. Denes, I. Manno, G. Pinter, I. Renyi, L. Varhalmi,
M. Zsenei.
Meres Autom., Vol. 33, No. 1 – 2, p. 13 – 18 (1985). In Hungarian. Abstr. in Phys. Abstr., Vol. 88, No. 1260, Entry 74058 (1985).

035.085   **Data collecting equipment of the VEGA space probes.**
L. Drimusz, T. Hetenyi, Z. Koros, I. Papp,
J. Selmeczi, L. Balogh.
Meres Autom., Vol. 33, No. 1 – 2, p. 19 – 26 (1985). In Hungarian. Abstr. in Phys. Abstr., Vol. 88, No. 1260, Entry 74059 (1985).

035.086   **The equipment TUNDE–M of the VEGA programme.**
M. Farago, T. Gombosi, K. Kecskemety, G. Kozma,
L. Lohonyai, A. Somogyi, L. Szabo, A. Szepesvari, I. Szucs,
A. Varga, J. Windberg, A. Zarandy, A. Banfalvi, R. Redl,
J. Szabo.
Meres Autom., Vol. 33, No. 1 – 2, p. 27 – 31 (1985). In Hungarian. Abstr. in Phys. Abstr., Vol. 88, No. 1260, Entry 74060 (1985).

035.087   **The charged particle analyzer PLAZMAG.**
I. Apathy, A. Banfalvi, G. Endroczy, R. Redl,
I. Szemerey, S. Szendro.
Meres Autom., Vol. 33, No. 1 – 2, p. 32 – 37 (1985). In Hungarian. Abstr. in Phys. Abstr., Vol. 88, No. 1260, Entry 74061 (1985).

035.088   **Imaging photon counting, Starlab and Australian space astronomy.**
A. W. Rodgers.
Aust. Phys., Vol. 22, No. 1, p. 2 – 4 (1985). Abstr. in Phys. Abstr., Vol. 88, No. 1260, Entry 74084 (1985).

035.089   **Coded aperture imaging of X–ray and gamma–ray sources.**
R. Kroeger, D. Müller.
18th International Cosmic Ray Conference, Vol. 8, p. 1 – 4 (1983). – See Abstr. 012.096.
The authors discuss coded aperture telescopes employing arrays of a small number of discrete detector elements for hard X–rays or gamma–rays. Aperture patterns are described that permit a unique reconstruction of the image with high contrast, and that exhibit a specific rotational antisymmetry to suppress systematic distortions. High flux sensitivity can be achieved in this fashion as well as good angular resolution.

035.090   **A large area X–ray telescope for study of discrete hard X–ray sources.**
A. R. Rao, P. C. Agrawal, R. K. Manchanda.
18th International Cosmic Ray Conference, Vol. 8, p. 5 – 8 (1983). – See Abstr. 012.096.
The authors describe briefly the details of a balloon borne large area detector system and associated instrumentation for spectral studies of hard X–ray sources. The average detection efficiency in 20 – 100 keV band is about 50% and typical energy resolution at 44 keV is 10% FWHM. The telescope can be preprogrammed to track any celestial source.

035.091   **The FIGARO experiment for low–energy gamma–ray astronomy: a technical description and an evaluation of performances.**
G. Agnetta, B. Agrinier, A. Bui–Van, J. P. Chabaud,
J. C. Christi, E. Costa, R. Di Raffaele, P. Frabel, G. Gerardi,

P. Mandrou, J. L. Masnou, E. Massaro, J. Narbonne, M. Niel,
G. Rouaix, B. Sacco, M. Salvati, L. Scarsi, G. Vedrenne.
18th International Cosmic Ray Conference, Vol. 8, p. 10 (1983).
Abstract. – See Abstr. 012.096.

035.092   The ZEBRA telescope – preliminary laboratory tests.
R. E. Baker, P. M. Charalambous, A. J. Dean,
M. Drane, A. Gil, J. B. Stephen, L. Barbareschi, G. Boella,
A. Bussini, F. Perotti, G. Villa, C. Butler, E. Caroli,
G. Di Cocco, E. Morelli, A. Spizzichino, M. Badiali,
A. Bazzano, C. D. La Padula, F. Polcaro, P. Ubertini.
18th International Cosmic Ray Conference, Vol. 8, p. 11 – 14
(1983). – See Abstr. 012.096.

An improved design of the low energy γ–ray imaging telescope,
ZEBRA, is described. Results of a number of laboratory tests are
presented and their implications discussed in relation to the capabilities of the astronomical telescope.

035.093   The GSFC advanced Compton telescope (ACT).
R. C. Hartman, C. E. Fichtel, D. A. Kniffen,
G. Stacy, J. I. Trombka.
18th International Cosmic Ray Conference, Vol. 8, p. 16 – 18
(1983). – See Abstr. 012.096.

A new telescope is being developed at GSFC for the study of
point sources of gamma rays in the energy range 1 – 30 MeV.

**035.094   EGRET: the high energy gamma ray telescope for
NASA's Gamma Ray Observatory.**
C. E. Fichtel, D. L. Bertsch, R. C. Hartman, D. A. Kniffen,
D. J. Thompson, R. Hofstadter, E. B. Hughes,
L. E. Campbell–Finman, K. Pinkau, H. Mayer–Hasselwander,
G. Kanbach, H. Rothermel, M. Sommer, A. J. Favale,
E. J. Schneid.
18th International Cosmic Ray Conference, Vol. 8, p. 19 – 22
(1983). – See Abstr. 012.096.

The EGRET high energy γ–ray telescope will have an energy
range of approximately 12 to 30,000 MeV, energy resolution of
about 15% FWHM over most of that range, an effective area of
about 2000 cm^2 at high energies, and single photon angular accuracy of ~2° at 100 MeV, <0.1° above 5 GeV. This instrument
can locate strong sources to an accuracy of about 5 arc min.

**035.095   An imaging spectrometer for cosmic gamma–ray astron-
omy.**
J. L. Matteson, R. M. Pelling, L. E. Peterson.
18th International Cosmic Ray Conference, Vol. 8, p. 23 (1983).
– See Abstr. 012.096.

035.096   Sources of background noise in γ–ray telescopes.
P. Charalambous, A. J. Dean, N. A. Dipper,
R. Lewis, J. B. Stephen.
18th International Cosmic Ray Conference, Vol. 8, p. 28 – 31
(1983). – See Abstr. 012.096.

**035.097   A large area experiment to determine cosmic ray isotopic
abundances.**
B. G. Mauger, V. K. Balasubrahmanyan, J. F. Ormes,
R. E. Streitmatter, W. Heinrich, M. Simon, H. O. Tittel.
18th International Cosmic Ray Conference, Vol. 8, p. 36 – 39
(1983). – See Abstr. 012.096.

A 1.2 m^2–sr instrument to study the isotopic composition of
the elements from oxygen through argon is being constructed.
This instrument uses two scintillators and two Cerenkov detectors (n = 1.33 and 1.4).

**035.098   Cerenkov detectors for cosmic ray telescopes employing
the Cerenkov X total energy technique of mass identification.**
W. R. Webber, J. C. Kish.
18th International Cosmic Ray Conference, Vol. 8, p. 40 – 43
(1983). – See Abstr. 012.096.

The authors consider the characteristics of a large area
(~0.5 m^2–ster) cosmic ray isotope telescope being developed for
use on balloons or spacecraft.

**035.099   Soviet–Roumanian experiments on the registration of
the cosmic ray nuclear component aboard the
"Salyut–6" scientific station.**
K. Blaj, E. V. Gorchakov, N. L. Grigorov, V. V. Kovalyonok,
A. Marin, R. A. Nymmik, L. I. Popov, D. Prunariu,
V. Savinykh, M. Haiduc, D. Hasagan, M. Giobanu.
18th International Cosmic Ray Conference, Vol. 8, p. 44 – 46
(1983). – See Abstr. 012.096.

**035.100   Čerenkov calorimeter to detect antiprotons at low ener-
gies.**
V. K. Balasubrahmanyan, B. G. Mauger, S. A. Stephens,
R. E. Streitmatter.
18th International Cosmic Ray Conference, Vol. 8, p. 47 (1983).
Abstract. – See Abstr. 012.096.

035.101   The Bristol University BUGS 4 detector.
P. H. Fowler, M. R. W. Masheder, J. A. Clifford,
M. Grande, J. A. L. Shearer, G. R. Moss.
18th International Cosmic Ray Conference, Vol. 8, p. 48 – 51
(1983). – See Abstr. 012.096.

The main contribution and aim of BUGS 4 is to measure the
charge spectrum of cosmic rays as a function of measured γ
between 1.4 and 1000 for species between oxygen and the iron
group.

**035.102   On the generation of delta–rays in detectors for high–
energy cosmic–ray nuclei.**
S. P. Swordy, D. Müller, A. Ten Have.
18th International Cosmic Ray Conference, Vol. 8, p. 55 – 58
(1983). – See Abstr. 012.096.

A common problem in detectors for cosmic–ray nuclei is the
generation of energetic knock–on electrons by an incident nucleus. These electrons may produce additional signals which can
obscure the true nature of the signal from the nucleus itself. The
authors discuss Monte Carlo simulations of these effects in the
University of Chicago CRNE instrument for Spacelab–2, and
show that they do not impair the performance of the transition
radiation detector as designed. Some qualitative properties of
delta–ray effects are identified, which are of use to other experi-
menters in this field.

**035.103   Tests and design of a driftchamber to be used in a cosmic
ray isotope experiment.**
M. Simon, M. Henkel, R. Hundt, K. D. Mathis, G. Schieweck,
T. A. Suck.
18th International Cosmic Ray Conference, Vol. 8, p. 93 (1983).
Abstract. – See Abstr. 012.096.

**035.104   Results from a gas–scintillation–driftchamber exposed
to α–particles and heavy ions at LBL (USA).**
K. D. Mathis, M. Simon, M. Henkel.
18th International Cosmic Ray Conference, Vol. 8, p. 94 (1983).
Abstract. – See Abstr. 012.096.

**035.105   Results of a Transmediterranean balloon flight experi-
ment designed to evaluate the high resolution characteristics of CR–39 in cosmic ray studies.**
D. O'Sullivan, A. Thompson, A. Vidal–Quadras, F. Fernandez,
C. Baixeras, M. Casas.
18th International Cosmic Ray Conference, Vol. 8, p. 117 (1983).
Abstract. – See Abstr. 012.096.

**035.106   Cellulose acetate butyrate as a detector for cosmic ray
heavy nuclei.**
V. S. Bhatia, G. Singh, M. Puri.
18th International Cosmic Ray Conference, Vol. 8, p. 118 – 121
(1983). – See Abstr. 012.096.

035.107 **Detector module for Indian cosmic ray experiment aboard space shuttle Spacelab–3.**
S. Biswas, N. Durgaprasad, P. J. Kajarekar, M. N. Vahia,
J. S. Yadav, L. M. Kukreja, D. D. Bhawalkar,
U. K. Chatterjee, C. Basu, J. N. Goswami.
18th International Cosmic Ray Conference, Vol. 8, p. 122 – 125 (1983). – See Abstr. 012.096.

The authors present the technique and also their study of effect of laser cutting on track properties of CR–39 (DOP).

035.108 **A versatile data processing unit (DPU) for satellite–borne composition instrumentation.**
J. B. Blake, J. F. Fennell, R. Koga, N. Katz.
18th International Cosmic Ray Conference, Vol. 8, p. 186 – 189 (1983). – See Abstr. 012.096.

A data processing unit (DPU) for use with a charge–composition experiment to be flown aboard the VIKING auroral research satellite is described. The design of this DPU is such that it can be readily adapted to a variety of cosmic–ray composition experiments.

035.109 **Si detector and an alpha source setup for accurate measurement of the grammage at balloon altitude.**
V. Sinha, S. D. Verma.
18th International Cosmic Ray Conference, Vol. 8, p. 200 – 203 (1983). – See Abstr. 012.096.

035.110 **A high energy resolution experiment in the hard X–ray range.**
W. Paciesas, R. Baker, D. Boclet, T. Cline, P. Durouchoux,
C. Ehrmann, N. Gehrels, J. M. Hameury, R. Haymes,
B. Teegarden, J. Tueller.
18th International Cosmic Ray Conference, Vol. 9, p. 335 – 338 (1983). – See Abstr. 012.096.

The Low Energy Gamma–Ray Spectrometer (LEGS), a joint project among NASA/GSFC, CEN–SACLAY and Rice University, is designed to perform fine energy resolution measurements of astrophysical sources.

035.111 **High sensitivity high resolution double scatter 1 – 30 MeV gamma ray telescope.**
A. D. Zych, B. Dayton, O. T. Tümer, R. S. White.
18th International Cosmic Ray Conference, Vol. 9, p. 343 – 346 (1983). – See Abstr. 012.096.

The use of large Na I(Tl) scintillators for energy, position and timing measurements in gamma ray astronomy can provide the much needed increases in sensitivity and angular and energy resolutions. The new medium energy (1 – 30 MeV) double scatter balloon–borne telescope incorporates long bars of Na I(Tl) and plastic scintillator so as to preserve the wide field–of–view (π ster.), large area (1 m^2) and excellent background rejection inherent in the technique while providing much improved energy and angular resolutions.

035.112 **A rotation modulation collimator for imaging of point X–ray sources up to 100 keV.**
R. Staubert, J. Theinhardt, E. Kendziorra.
18th International Cosmic Ray Conference, Vol. 9, p. 347 – 350 (1983). – See Abstr. 012.096.

035.113 **A large–area instrument to measure the charge and energy spectrum of the anomalous cosmic rays including a possible molecular ion component.**
P. H. Fowler, N. A. McGowan, A. Worley.
18th International Cosmic Ray Conference, Vol. 9, p. 355 – 358 (1983). – See Abstr. 012.096.

An instrument, suitable for a Space Shuttle exposure, has been designed to study several aspects of the anomalous component of the cosmic rays.

035.114 **The high energy (1 – 100 MeV) proton telescope for the CRRES mission.**
D. R. Parsignault.
18th International Cosmic Ray Conference, Vol. 9, p. 359 – 362 (1983). – See Abstr. 012.096.

The author describes the high energy proton telescope to be flown on the CRRES satellite toward the end of 1985 which will measure the trapped proton environment and the ion fluxes, from He^4 to Fe^{56}, in the 1 – 100 MeV and 1.5 – 200 MeV energy ranges, respectively.

035.115 **EPONA, an energetic particle detector system for the Giotto mission to Halley's comet.**
S. McKenna–Lawlor, A. Thompson, D. O'Sullivan, E. Kirsch,
K.-P. Wenzel, D. Melrose.
18th International Cosmic Ray Conference, Vol. 9, p. 363 – 366 (1983). – See Abstr. 012.096.

The particle detector EPONA (Energetic Particle Onset Ad-monitor) is designed to measure energetic particles during the Cruise Phase and during the encounter with Halley. This paper describes the present technical design and scientific objectives of the experiment.

035.116 **The non–Z^2 response of the heavy nuclei cosmic ray detector on HEAO–3.**
T. L. Garrard, B. J. Newport, E. C. Stone, W. R. Binns,
M. H. Israel, M. D. Jones, J. Klarmann, R. K. Fickle,
C. J. Waddington.
18th International Cosmic Ray Conference, Vol. 9, p. 367 – 370 (1983). – See Abstr. 012.096.

035.117 **Use of relativistic rise in ionization chambers for measurement of high energy heavy nuclei.**
S. D. Barthelmy, M. H. Israel, J. Klarmann, J. S. Vogel.
18th International Cosmic Ray Conference, Vol. 9, p. 371 – 374 (1983). – See Abstr. 012.096.

A balloon borne instrument has been constructed to measure the energy spectra of cosmic ray heavy nuclei in the range of ~ 0.3 to ~ 100 GeV/amu. The main emphasis of the instrument is the determination of the change of the ratio of iron (26) to the iron secondaries (21 – 25) in the energy range of 10 to 100 GeV/amu. Preliminary data from a balloon flight in the fall of 1982 from Palestine, Texas is presented.

035.118 **An instrument employing electronic counters and an emulsion chamber for studying heavy cosmic ray interactions (JACEE–3).**
R. W. Austin, T. H. Burnett, S. Dake, M. Fuki, J. C. Gregory,
T. Hayashi, R. Holynski, J. Iwai, W. V. Jones, A. Jurak,
J. J. Lord, C. A. Meegan, O. Miyamura, T. A. Parnell,
T. Ogata, H. Oda, T. Saito, W. J. Selig, S. C. Strausz,
T. Tabuki, Y. Takahashi, T. Tominaga, J. W. Watts,
B. Wilczynska, R. J. Wilkes, W. Wolter, B. Wosiek.
18th International Cosmic Ray Conference, Vol. 9, p. 375 – 378 (1983). – See Abstr. 012.096.

The JACEE–3 instrument was designed to study the properties of nucleus–nucleus interactions above 20 GeV/amu from Z = 6 to Z = 26.

035.119 **Measurement of heavy cosmic rays above 20 GeV/n in passive detectors: preliminary results from JACEE–3.**
T. H. Burnett, S. Dake, W. Fountain, M. Fuki, T. Hayashi,
R. Holynski, J. C. Gregory, J. Iwai, W. V. Jones, A. Jurak,
J. J. Lord, C. A. Meegan, O. Miyamura, T. Ogata, H. Oda,
T. A. Parnell, T. Saito, S. C. Strausz, M. Szarska, T. Tabuki,
Y. Takahashi, T. Tominaga, J. W. Watts, B. Wilczynska,
R. J. Wilkes, W. Wolter, B. Wosiek.
18th International Cosmic Ray Conference, Vol. 9, p. 379 – 382 (1983). – See Abstr. 012.096.

The JACEE–3 experiment combined electronic counters and an emulsion chamber to study nucleus–nucleus interactions between 20 and 100 GeV/n. Here, the authors describe the use of the passive detector stack.

035.120 Analytic calculation of differential apertures of telescopes consisting of one horizontal and one vertical rectangular detector.
M. Crouch.
18th International Cosmic Ray Conference, Vol. 9, p. 387 – 390 (1983). – See Abstr. 012.096.

Analytical expressions are derived for the differential aperture $dA/d\theta$ of the "Horizontal–Vertical" telescope configuration. The analytical nature of the relations makes them especially useful for telescope array design studies.

035.121 The ultra heavy cosmic ray experiment on the NASA Long Duration Exposure Facility (LDEF).
D. O'Sullivan, A. Thompson, J. Daly, C. O'Ceallaigh, K.–P. Wenzel, V. Domingo, A. Smit.
18th International Cosmic Ray Conference, Vol. 9, p. 403 – 406 (1983). – See Abstr. 012.096.

A large array ($\cong 20$ m^{2}sr) of solid state nuclear track detectors is being prepared for a twelve month exposure in earth orbit aboard the NASA LDEF. The experiment is designed to study the charge spectrum of cosmic ray nuclei with $Z > 30$, particularly of those above $Z \sim 70$.

035.122 The variation of track response with registration temperature in solid state nuclear track detectors and its implications for cosmic ray composition studies.
A. Thompson, D. O'Sullivan, J. H. Adams, L. P. Beahm.
18th International Cosmic Ray Conference, Vol. 9, p. 407 – 410 (1983). – See Abstr. 012.096.

035.123 Acquisition and track algorithms for the ASTROS star tracker.
E. Shalom, J. W. Alexander, R. H. Stanton.
Adv. Astronaut. Sci., Vol. 57, p. 375 – 398 (1985). – See Abstr. 012.099.

ASTROS is a sub–arcsecond CCD star tracker that will be a critical part of the set of Astro missions starting in March 1986. This payload will be mounted in the Shuttle bay and will be used to observe Halley's comet and other astronomical targets. ASTROS has demonstrated the ability to acquire and track stars from fields over a 2000 to 1 range in star brightness (magnitude range: –0.8 to 8.2) in less than 30 seconds. These stars are tracked with a noise–equivalent–angle of less than 0.3 arcsec (0.1 arcsec for stars brighter than magnitude 7.0), and small image motions measured to better than 1/100 pixel accuracy.

035.124 Guidung performance of a Quadrant Digicon sensor.
R. O. Ginaven, L. L. Acton, R. D. Smith, J. G. McCoy.
Adv. Astronaut. Sci., Vol. 57, p. 399 – 411 (1985). – See Abstr. 012.099.

The operation and performance of a Digicon–based guiding system recently developed are discussed. The photosensor is a twelve–channel, photon–counting Digicon. The design and implementation of the system are discussed with an emphasis on the objective of developing a general purpose guiding system for a wide range of applications.

035.125 Zeroth order of "observing efficiency" of Space Telescope.
R. C. Nigam.
Bull. Am. Astron. Soc., Vol. 17, No. 2, p. 549 (1985). Abstract. – See Abstr. 010.065.

035.126 The optimization of the Rowland circle grating for high–resolution XUV spectroscopy.
J. F. Meekins, R. G. Cruddace, H. Gursky, G. G. Fritz.
Bull. Am. Astron. Soc., Vol. 17, No. 2, p. 549 (1985). Abstract. – See Abstr. 010.065.

035.127 Diffuse infrared background experiment (DIRBE) prototype photometric performance.
H. J. Wood, T. Magner, T. French.
Bull. Am. Astron. Soc., Vol. 17, No. 2, p. 549 – 550 (1985). Abstract. – See Abstr. 010.065.

035.128 A hybrid PIN X–ray imaging device.
J. G. Jernigan, J. F. Arens, M. C. Peck, S. D. Gaalema.
Bull. Am. Astron. Soc., Vol. 17, No. 2, p. 572 (1985). Abstract. – See Abstr. 010.065.

035.129 The Johns Hopkins University UVX experiment.
P. D. Tennyson, P. D. Feldman, R. C. Henry, J. Murthy.
Bull. Am. Astron. Soc., Vol. 17, No. 2, p. 572 (1985). Abstract. – See Abstr. 010.065.

035.130 Large format microchannel detectors for EUV and UV astronomy.
M. Lampton, O. Siegmund.
Bull. Am. Astron. Soc., Vol. 17, No. 2, p. 572 – 573 (1985). Abstract. – See Abstr. 010.065.

035.131 Point spread functions and resolving power of the high resolution spectrograph for Space Telescope.
D. Ebbets, M. Erickson.
Bull. Am. Astron. Soc., Vol. 17, No. 2, p. 574 (1985). Abstract. – See Abstr. 010.065.

035.132 Laboratory calibration of the high resolution spectrograph for Space Telescope: absolute sensitivity.
K. G. Carpenter, G. Cushman, D. Ebbets, S. Heap, J. Brandt.
Bull. Am. Astron. Soc., Vol. 17, No. 2, p. 574 (1985). Abstract. – See Abstr. 010.065.

035.133 New developments in UV and EUV detectors.
D. Weistrop.
Bull. Am. Astron. Soc., Vol. 17, No. 2, p. 590 (1985). Abstract. – See Abstr. 010.065.

035.134 The status of absolute flux calibration in the 1050 – 3200 Å spectral region.
R. C. Bohlin.
Bull. Am. Astron. Soc., Vol. 17, No. 2, p. 590 (1985). Abstract. – See Abstr. 010.065.

035.135 First results from the UNH directional gamma–ray telescope.
M. L. McConnell, P. P. Dunphy, D. J. Forrest, E. L. Chupp.
Bull. Am. Astron. Soc., Vol. 17, No. 2, p. 604 (1985). Abstract. – See Abstr. 010.065.

035.136 An analysis of the structure calibration and system test data obtained from the GIOTTO Structural Model Spacecraft at ESTEC, Noordwijk, 22 October to 5 November 1983.
S. Evans, A. Ridgeley.
RAL–85–019, Rutherford Appleton Laboratory, Chilton, Didcot, Oxon, OX11 0QX, England, 4 + 43 pp. (1985).

The results of tests on the GIOTTO Structural Model Spacecraft, using a variety of stimulation techniques to imitate the effect of dust particle impacts, are presented.

035.137 Concept and design studies for the exploration of high energy gamma–ray astronomy on the Gamma–Ray Observatory.
L. C. Finman.
Diss. Abstr. Int., Sect. B, Vol. 45, No. 3, p. 900 (1984). Thesis, Stanford University, 277 pp. (1984). Order No. DA8412841.

035.138 X–ray polarimetry: the measurement of the polarization of solar flare X–rays and the design of a Compton polarimeter.
L. J. Tramiel.
Diss. Abstr. Int., Sect. B, Vol. 45, No. 10, p. 3257 – 3258 (1985). Thesis, Columbia University, 147 pp. (1984). Order No. DA8427487.

The heliometer principle and some modern applications.
See Abstr. 004.032.

Astronomy with the Hubble Space Telescope.
See Abstr. 011.033.

Electronic detection of ultra–heavy nuclei by pyroelectric materials.
See Abstr. 022.164.

A new generation of spectrometer designs for ultraviolet astronomy.
See Abstr. 031.001.

Wolter 1 type X–ray mirror system with mean resolution and maximum effective collecting area for the spectral region of 1 – 10 nm.
See Abstr. 031.011.

Galvanoplastic grazing incidence mirrors for "vertical" experiments: the tandem tests in soft X–rays.
See Abstr. 031.012.

Conical imaging mirrors for high–speed X–ray telescopes.
See Abstr. 031.032.

Transverse ray aberrations for paraboloid–hyperboloid telescopes.
See Abstr. 031.033.

Optical model for the HIPPARCOS telescope in its real expected configuration.
See Abstr. 031.039.

The signal of the HIPPARCOS nominal system.
See Abstr. 031.040.

A normal–incidence telescope for solar EUV studies.
See Abstr. 031.044.

Report of IAU Commission 9: Instruments and techniques (*Instruments et techniques*).
See Abstr. 032.046.

Il telescopio nazionale e le sue relazioni con l'astronomia spaziale degli anni '90.
See Abstr. 032.055.

Correlator and data processing requirements for Quasat.
See Abstr. 033.061.

Processing system for space VLBI.
See Abstr. 033.063.

Evaluation of a magneto–optical filter and a Fabry–Perot interferometer for the measurement of solar velocity fields from space.
See Abstr. 034.041.

Balloon–borne, He–cooled Fourier transform spectrometer for far infrared astronomy.
See Abstr. 034.065.

Geometrical invariance property of gratings.
See Abstr. 034.092.

Physics and design of advanced IR bolometers and photoconductors.
See Abstr. 034.097.

Operation of a microchannel plate counting system in a mass spectrometer.
See Abstr. 034.120.

Interference methods in the testing and fabrication of new–design grazing incidence gratings.
See Abstr. 034.121.

The International Ultraviolet Explorer (IUE) point spread function at low resolution.
See Abstr. 036.045.

The X–ray Timing Explorer.
See Abstr. 051.004.

On the PSIVA (*Probing Stellar Interiors via Variability and Activity*) approach to stellar seismology and activity from space.
See Abstr. 051.010.

Wetenschappelijke ruimtevloot zet koers naar komeet Halley.
See Abstr. 051.022.

LDEF–1 and beyond.
See Abstr. 051.030.

First results of a stratospheric experiment using a Montgolfière Infra–Rouge (MIR).
See Abstr. 051.034.

Hard X–ray astronomy from balloons.
See Abstr. 051.038.

Coronal physics with the Solar and Heliospheric Observatory.
See Abstr. 051.048.

The Solar Maximum Mission.
See Abstr. 051.049.

The Exosat mission.
See Abstr. 051.057.

The Tenma mission.
See Abstr. 051.058.

X–ray sky surveys and the ROSAT mission.
See Abstr. 051.068.

Detection of X–ray sources with ROSAT.
See Abstr. 051.069.

Space astronomy: technical comments – ultraviolet and optical instrumentation.
See Abstr. 051.075.

Space astronomy: technical comments – X– and gamma–ray astronomy.
See Abstr. 051.076.

Space astronomy: technical comments – infrared, submillimetre and radio astronomy.
See Abstr. 051.077.

Space astronomy: technical comments – interferometry missions.
See Abstr. 051.078.

Science mission planning for AXAF.
See Abstr. 051.105.

Report of IAU Commission 44: Astronomy from space (*L'astronomie à partir de l'espace*).
See Abstr. 051.109.

Ad astra HIPPARCOS. The European Space Agency's Astrometry Mission.
See Abstr. 051.114.

Quasat: technical aspects of the proposed mission.
See Abstr. 051.118.

The Galileo scan platform pointing control system – a modern control theoretic viewpoint.
See Abstr. 052.066.

Organisation of a unified system of energetic calibration of X–ray experiments.
See Abstr. 076.006.

Results from the transition region camera.
See Abstr. 076.008.

Infrared experiment aboard the automatic interplanetary stations Venera 15 and Venera 16. I. Methods and first results.
See Abstr. 093.026.

036 Methods of Observation and Reduction, Data Processing

036.001   Calibrations of wavelengths in SWP echelle spectra.
M. D. De La Pena, T. R. Ayres.
NASA Conf. Publ., NASA CP–2349, p. 521 – 524 (1984). – See Abstr. 012.001.

In order to obtain high quality radial velocity measurements of emission lines in the far ultraviolet with IUE, it is crucial to minimize sources of random and systematic errors in the echelle wavelength scales. Here the authors analyze wavelength calibration spectra secured from their own observing programs and from the IUE archives.

036.002   Improved continuum definition in high–background IUE images.
R. L. Hackney, K. R. H. Hackney, Y. Kondo.
NASA Conf. Publ., NASA CP–2349, p. 525 – 528 (1984). – See Abstr. 012.001.

A technique is described for automatically filtering low–dispersion IUE images of untrailed point sources during extraction from the guest observer tape, for the purpose of reducing the effects of radiation events and "bright pixels" and obtaining improved definition of the continuum in long duration or high–background exposures. The procedure results in a significant reduction of discrepancies in the fitted power–law indices for repeated exposures of faint BL Lac objects, compared with the results of previously attempted automatic filtering techniques.

036.003   A high accuracy algorithm for maximum entropy image restoration in the case of small data sets.
R. Wilczek, S. Drapatz.
Astron. Astrophys., Vol. 142, No. 1, p. 9 – 12 (1985).

An iterative algorithm for the solution of maximum entropy problems containing a χ^2–constraint is presented. Convergence and high accuracy are certain even for "tough cases". Applications are discussed with respect to the maximum entropy image restoration approach suggested by Gull and Daniell. Computational results for a series of test cases and for image restoration of scanning experiments in the field of infrared astronomy are given.

036.004   Point and interval estimation of the true unbiased degree of linear polarization in the presence of low signal–to–noise ratios.
J. F. L. Simmons, B. G. Stewart.
Astron. Astrophys., Vol. 142, No. 1, p. 100 – 106 (1985).

Four estimators which attempt to correct for biasing in the degree of linear polarization are discussed for the specific case when the errors on an observation of the normalised Stokes parameters q and u are known and assumed equal to σ. These estimators are the maximum likelihood, the median, Serkowski's mean estimator used by optical astronomers and Wardle and Kronberg's "most probable" which is often applied in radio astronomy. The authors also briefly discuss particular areas where applications of these results are important, namely in polarimetric standards, stellar populations, binary stars, interstellar polarization and extinction measurements at high galactic latitudes.

036.005   Surface photometry from the ESO/SRC Atlas.
R. Schindler, J. Isserstedt.
Astron. Astrophys., Suppl. Ser., Vol. 59, No. 3, p. 497 – 504 (1985). In German.

The authors examine whether the on–film copies of the IIIa–J plates of the ESO/SRC Survey are suitable for the determination of relative intensities within galaxies. For the transformation of the densities measured to relative magnitudes they used the photometric wedges on the films. By comparing the brightness distributions of galaxies in overlapping regions of each two films it is found that the accuracy of the magnitude scales and the accidental errors due to grain structure on the films are similar to good original plates.

036.006   On a possible astrometric method of reducing star plates.
Ya. S. Yatskiv, A. N. Kur'yanova.
Kinematika Fiz. Nebesn. Tel, Tom 1, No. 1, p. 18 – 26 (1985). In Russian.

The scheme of confluent analysis known in mathematical statistics is used as a basis when suggesting a method of reducing star plates in which the plate constants and the corrections to the spherical star coordinates are the adjustment parameters. The method is tested using a model of star plates.

036.007   On the block–adjustment reduction of differential meridian observations.
V. V. Tel'nyuk–Adamchuk.
Kinematika Fiz. Nebesn. Tel, Tom 1, No. 1, p. 27 – 32 (1985). In Russian.

The advantage of block–adjustment reduction of meridian observations is discussed. The conditional and normal equations are obtained for the case when the solution is found in the form of corrections to the coordinates. The weights of the instrument constants and coordinate corrections in the case of block–adjustment and in the case of traditional reduction are compared for each night separately. Taking into account the possibility of obtaining the statistically substantial model of the instrument the block–adjustment reduction is more preferable in respect to result accuracy.

036.008   On the FOKAT programme observations with the 40–cm astrograph of the Gissar Observatory.
N. N. Matveev.
Kinematika Fiz. Nebesn. Tel, Tom 1, No. 1, p. 94 – 95 (1985). In Russian.

Results are given of the application of circular diaphragms for improving the stellar image quality on the plates obtained with the 40–cm astrograph. The two–exposure method used for observations of southern stars (zone $-16°$ to $-30°$) is described.

036.009 The production of a 16–mm film of M31.
G. S. Shostak, E. Brinks.
The Milky Way galaxy, p. 443 – 444 (1985). – See Abstr. 012.007
(IAU Symp. No. 106).

036.010 Detection of weak molecular lines.
R. D. Brown, P. D. Godfrey, E. H. N. Rice.
Observatory, Vol. 105, No. 1064, p. 12 – 15 (1985).

036.011 Repérage des coordonnées sur les photos.
R. Behrend.
Orion, 43. Jahrg., Nr. 206, p. 9 – 11 (1985).

**036.012 Die Bestimmung von Farbindices bei der Veränder-
lichenbeobachtung.**
C. Peitscher.
BAV Rundbrief, 34. Jahrg., Nr. 1, p. 24 – 27 (1985).

036.013 The observation of bodies in close proximity to the sun.
J. E. Bortle.
Int. Comet Q., Vol. 7, No. 1, p. 7 – 11 (1985).

036.014 Maximum entropy method in image processing.
S. F. Gull, J. Skilling.
IEE Proc. F, Vol. 131, No. 6, p. 646 – 650 (1984). Abstr. in Phys.
Abstr., Vol. 88, No. 1248, Entry 9636 (1985).

**036.015 Methods of remote surface chemical analysis for aster-
oid missions.**
R. Z. Sagdeev, G. G. Managadze, I. Yu. Shutyaev, K. Szego,
P. P. Timofeev.
Report KFKI–1984–82, Hungarian Acad. Sci., Budapest, Hun-
gary, 18 pp. (1984). Abstr. in Phys. Abstr., Vol. 88, No. 1248,
Entry 9644 (1985).

036.016 Molecular spectroscopy and planetary atmospheres.
C. de Bergh.
Ann. Phys. (Paris), Vol. 9, No. 4, p. 575 – 584 (1984). Abstr. in
Phys. Abstr., Vol. 88, No. 1249, Entry 14479 (1985).

036.017 Two wavelength picosecond ranging of ground target.
K. Hamal, I. Procházka, J. Gaignebet.
Publ. Astron. Inst. Czech. Acad. Sci., No. 58, p. 29 – 35 (1984).
– See Abstr. 012.018.
One of the limiting factors in decreasing the systematic error of
laser ranging is the influence of the atmospheric refraction. Two
colour ranging may contribute useful information for more precise refraction factor modelling and calculation. The authors
describe two wavelength experiments using a streak camera as a
high resolution detector for ground target distance measurement.

036.018 Mode locked train YAG laser ranging data processing.
I. Procházka.
Publ. Astron. Inst. Czech. Acad. Sci., No. 58, p. 43 – 49 (1984).
– See Abstr. 012.018.
The method of processing collected laser ranging data using
the passively mode locked YAG train laser is described. The
algorithm for resolving individual peaks in measured ranges histogram and system internal noise determination is explained together with the crosscorrelation methods for system calibration
constant evaluation. The low/Lageos satellite and calibration
ranging results are included.

**036.019 Simulation studies on the statistics of a multipulse laser
radar working at the single photoelectron level.**
R. Neubert.
Publ. Astron. Inst. Czech. Acad. Sci., No. 58, p. 93 – 102 (1984).
= Mitt. Zentralinst. Phys. Erde, Nr. 1351. – See Abstr. 012.018.
The residual ambiguity of a laser radar system using pulse
trains from a mode locking laser has been investigated by computer simulation. Numerical results are given for pulse groups
with a Gaussian envelope. A slight modification of this signal
shape results in strong reduction of the ambiguity.

036.020 A simple calibration link for a laser radar.
R. Neubert, L. Grunwaldt.
Publ. Astron. Inst. Czech. Acad. Sci., No. 58, p. 103 – 109 (1984).
= Mitt. Zentralinst. Phys. Erde, Nr. 1350. – See Abstr. 012.018.
A very simple calibration link based on dual diffuse scattering,
which can be easily attached in front of any laser radar, was
introduced at the Potsdam station one year ago. No optical
components other than diffuse reflectors are used. The inherent
attenuation is about 10^{13}, so that a 1000:1 additional filter is
sufficient to reach the single photoelectron level. The method has
been used for routine operations during the MERIT campaign
and for investigating some error sources as well.

**036.021 Estimation of phase errors on laser observations of
artificial earth satellites.**
D. T. Matveev.
Publ. Astron. Inst. Czech. Acad. Sci., No. 58, p. 145 – 148 (1984).
– See Abstr. 012.018.
Error estimations arising from phase–wave fluctuations on
laser beam propagation through a turbulent atmosphere are
given. The phase error value increases with the distance of the
satellite and is up to 8 cm for satellites with synchronous orbit in
the daytime.

**036.022 An effective method for on–line corrections of two–axes
laser mount.**
P. Lála, K. Pivo.
Publ. Astron. Inst. Czech. Acad. Sci., No. 58, p. 149 – 153 (1984).
– See Abstr. 012.018.
An algorithm for correction of two–axes laser mount direction
and gates in case of atmospheric drag induced deviations from
the predicted satellite motion is described. Results of numerical
simulations are shown.

**036.023 Precise computation of the refractive index correction
for laser observations.**
E. Pedrero Gonzales.
Publ. Astron. Inst. Czech. Acad. Sci., No. 58, p. 155 – 161 (1984).
In Russian. – See Abstr. 012.018.
On the basis of the model of Hopfield the correction due to
refraction is given with a precision of the order of 1 cm for laser
observations with zenith distances $z \leqslant 80°$.

**036.024 Single shot accuracy improvement by proper filtration
and fraction value in some processing methods.**
W. Kiełek.
Publ. Astron. Inst. Czech. Acad. Sci., No. 58, p. 181 – 190 (1984).
– See Abstr. 012.018.
Formulae were developed for the range error due to discrete
generation and gain of photoelectrons in multiphotoelectron case
for some signal processing methods, as constant threshold and
constant fraction of photomultiplier current or charge and some
others. In many methods, including near optimum estimation,
the error decreases when filter response width increases. Simple
estimation methods can give nearly as good results as optimum
methods, when using proper filtration and fraction values. Simulation and experimental results are in fair agreement with the
theory. The results can be used to improve many existing laser
stations.

**036.025 Atmospheric turbulence influence on the range error and
maximum range.**
J. Siuzdak, W. Kiełek.
Publ. Astron. Inst. Czech. Acad. Sci., No. 58, p. 191 – 196 (1984).
– See Abstr. 012.018.
Atmospheric turbulence can be a source of some systematic
and random range errors and of a decrease of the maximum
range measured. Formulae are collected for energy density decrease and for some of the errors, including a systematic photon
trajectory length increase, not mentioned before according to the
authors' knowledge.

036.026   Calibration of Doppler receivers.
L. Bányai.
Publ. Astron. Inst. Czech. Acad. Sci., No. 58, p. 487 – 497 (1984).
– See Abstr. 012.018.

Geodetic and geodynamic applications of satellite Doppler observations require a proper knowledge of receiver parameters corresponding to available reduction software. In this paper different methods used for calibration of Doppler receivers are summarized. One of them based on GEODOP–III program and broadcast ephemeris is presented and checked by test computations.

036.027   Results of the observations with Doppler receiver DOG–2.
P. Fraczyk, W. Jaks.
Publ. Astron. Inst. Czech. Acad. Sci., No. 58, p. 517 – 528 (1984).
– See Abstr. 012.018.

In this paper a general description of the operation of the Doppler receiver DOG–2, produced in Poland, is presented as well as the actual results of determination of the Borowiec geocentric coordinates obtained from observations made by the use of that receiver. Processing of data was performed with the program GEODOP.

036.028   Doppler receiver performance evaluations.
I. Fejes, F. J. Lohmar.
Publ. Astron. Inst. Czech. Acad. Sci., No. 58, p. 561 – 571 (1984).
– See Abstr. 012.018.

A series of common antenna observations using different Doppler receiver pairs has been evaluated with the Cross Doppler Count method. Results for four pairs of Magnavox MX 1502 and one pair of JMR1A type receiver evaluations are presented.

036.029   Interferometric analysis of Doppler measurements for differential receiver calibration.
R. Dietrich, K. Lehmann.
Publ. Astron. Inst. Czech. Acad. Sci., No. 58, p. 587 – 592 (1984).
= Mitt. Zentralinst. Phys. Erde, Nr. 1376. – See Abstr. 012.018.

The interferometric approach to the calibration problem consists in the use of observation differences of two receivers instead of single observations. Introducing the satellite positions and one receiver position as known the authors determine delay differences, frequency differences and coordinate differences (baseline components in a horizontal system) in their adjustment procedure. Numerical results are presented and optimum strategies for calibration parameter determination (e.g. delay differences, antenna phase center differences) are discussed.

036.030   Interactive reduction of echelle spectrograms.
C. Rossi, R. Lombardi, S. Gaudenzi, G. A. De Biase.
Astron. Astrophys., Vol. 143, No. 1, p. 13 – 18 (1985).

The authors describe a fully interactive approach to the reduction and analysis of the echelle spectrograms, totally independent of the spectrograph characteristics and of the image sampling direction. The method developed provides a modular procedure which uses an interactive image display and takes into account the recorded image defects, the tilt of the spectral lines and the non–uniform widening of the orders.

036.031   A simple maximum entropy deconvolution algorithm.
T. J. Cornwell, K. F. Evans.
Astron. Astrophys., Vol. 143, No. 1, p. 77 – 83 (1985).

A simple maximum entropy image deconvolution algorithm, now implemented in the Astronomical Image Processing System AIPS as task VM, is described. VM uses a simple Newton–Raphson approach to optimise the relative entropy of the image subject to constraints upon the rms error and total power enforced by Lagrange multipliers. Some examples of the application of VM to VLA data are given.

036.032   Narrowband spatial filtering in differential measurements of the line–of–sight velocity of the solar atmosphere.
N. I. Kobanov.
Astron. Astrophys., Vol. 143, No. 1, p. 99 – 101 (1985).

A further improvement upon the differential method is reported, which permits to achieve narrowband filtering of the spatial harmonics in the $K = 0.1 - 5 \,(\mathrm{Mm})^{-1}$ range with an accuracy better than $0.2 \,\mathrm{m\,s}^{-1}$. Preliminary observations have demonstrated that the amplitude of the spatial harmonics falls off rapidly as the wavelength λ decreases from 20″ down to 8″. A maximum for $\lambda = 24″$ is observed, that is most likely associated with supergranulation. At the center of the solar disk, no large–scale meridional flows with oscillation amplitude $\overline{V}_\mathrm{m} \geqslant 0.2 \,\mathrm{km\,s}^{-1}$ were detected.

036.033   The correlation between normalised Stokes parameters in a sinusoidally modulated intensity polarimeter.
B. G. Stewart.
Astron. Astrophys., Vol. 143, No. 1, p. 235 – 240 (1985).

For a polarimeter employing a continuously rotating half–wave plate positioned in front of a polarizer it is shown that the presence of noise introduces a correlation between simultaneous measurements of the normalised Stokes parameters q and u. Using a particular measurement procedure which integrates over specific parts of the modulated intensity, and the application of photon counting techniques, a mathematical expression for the correlation is derived in terms of polarization, signal strength, photon shot noise and scintillation noise. It is also shown that the projected confidence regions for the true normalised Stokes parameters on the q–u plane are different for uncorrelated and correlated measurements, the correlated regions generally being rotated and increased in size relative to the uncorrelated situation.

036.034   A new method for the determination of the counting loss of photon counting photometers.
G. Hildebrandt, D. Lange.
Astron. Nachr., Vol. 306, No. 2, p. 97 – 100 (1985). In German.

A simple method for the determination of statistical pulse losses of a photon counting photometer in dependence on the pulse intensity was developed. The slowly decreasing brightness of the clear sky in the dusk serves as light source. With the help of a neutral filter one gets two different pulse rates for two intensities. From a series of such measurements with different intensities one can derive the correction in dependence on pulse density in form of a power sequence.

036.035   A solution to the short spacing problem in radio interferometry.
R. Braun, R. A. M. Walterbos.
Astron. Astrophys., Vol. 143, No. 2, p. 307 – 312 (1985).

An efficient and reliable method is developed for the derivation of short spacing data missing from a radio interferometer observation, without the use of single dish data. The same approach is shown to apply to other instances of localized missing or damaged spatial frequency data. The method is based upon a direct fit of the missing Fourier coefficients to the (non–linearly) isolated map plane response.

036.036   Optimal enhancement of features in digital spectra.
J. W. Sulentic, H. Arp, J. Lorre.
Astron. J., Vol. 90, No. 3, 522 – 532, 569 – 572 (1985).

The use of median–filter algorithms as a tool for removing background contamination from digital spectra, and for enhancing the detectivity of faint spectra features, is discussed. Two examples of the application of this technique and its efficacy are described. In the first, an optimal procedure is described to search for features in the spectrum of knot C in the jet of M87. In the second, emission–line structure is mapped across NGC 5679ab, a pair of (apparently) interacting spiral galaxies with $\Delta V \cong 1200 \,\mathrm{km\,s}^{-1}$. The median filter algorithm as a sky–subtraction tool is found to be about 25% more efficient than traditional sky template subtraction techniques. A factor of 3 – 5 improvement in detectivity of faint emission–line structure results from median–filter processing.

036.037   Mesure des champs magnétiques à grande échelle.
C. Megessier.
Champs magnétiques stellaires, p. 107 – 120 (1984). – See Abstr. 012.022.

036.038   Simultan–Beobachtungsverfahren in der Astronomie.
C. de Loore.
Umschau, 85. Jahrg., Nr. 4, p. 238 – 242 (1985).

036.039   Steuerung eines Amateurfernrohres durch einen Mikro-computer.
S. Ritt, A. Weber.
Sterne Weltraum, 24. Jahrg., Nr. 1, p. 39 – 41 (1985).

036.040   Ein photometrisches Zwei–Blenden–Verfahren.
K. Güssow.
Sterne Weltraum, 24. Jahrg., Nr. 1, p. 44 – 45 (1985).

036.041   Mondbeobachtung nach dem "Apollo–Schock".
H. Hilbrecht.
Sterne Weltraum, 24. Jahrg., Nr. 2, p. 93 (1985).

036.042   Die Sonnenphotographie und ihre Probleme. Teil 1.
E. Remmert.
Sterne Weltraum, 24. Jahrg., Nr. 3, p. 158 – 159 (1985).

036.043   FODS: a system for faint object detection and classi-fication in astronomy.
M. L. Malagnini, F. Pasian, M. Pucillo, P. Santin.
Astron. Astrophys., Vol. 144, No. 1, p. 49 – 56 (1985).
The structure of FODS, the system for faint object detection and classification developed at the ASTRONET pole of Trieste Observatory, is presented. The main characteristic of such a system is its flexibility, achieved through a suitable catalogue structure. Details on the fundamental operations, i.e. plate digitization, object detection, preliminary screening and analysis of peculiarities, parameter computation and object classification, are illustrated. An example of application to star/galaxy discrimination is given.

036.044   An analysis of scattered light in low dispersion IUE spectra.
G. Basri, J. T. Clarke, B. M. Haisch.
Astron. Astrophys., Vol. 144, No. 1, p. 161 – 170 (1985).
The authors have constructed a detailed computer simulation of the scattering of light from the low–resolution grating in the short wavelength spectrograph of the IUE Observatory to estimate quantitatively the effects of scattering on both continuum and line emission spectra.

036.045   The International Ultraviolet Explorer (IUE) point spread function at low resolution.
A. Cassatella, J. Barbero, P. Benvenuti.
Astron. Astrophys., Vol. 144, No. 2, p. 335 – 342 (1985).
The resolution performances obtainable with IUE low resolution images are reviewed. It is shown that the spatial resolution (perpendicular to the dispersion) is strongly dependent on the telescope focussing conditions while the spectral resolution is not. After determining the telescope optimum focussing conditions, information is provided on the wavelength dependence of the spectral and spatial resolution at optimum focus for the three operational cameras (SWP, LWP, and LWR). The analytical shape of the PSF perpendicular to the dispersion is then analysed.

036.046   General structure of regularization procedures in image reconstruction.
D. M. Titterington.
Astron. Astrophys., Vol. 144, No. 2, p. 381 – 387 (1985).
Regularization procedures are portrayed as compromises between the conflicting aims of fidelity with the observed image and perfect smoothness. The selection of an estimated image involves the choice of a prescription, indicating the manner of smoothing, and of a smoothing parameter, which defines the degree of smoothing. Prescriptions of the minimum–penalized–distance type are considered and are shown to be equivalent to maximum–penalized–smoothness prescriptions. These include, therefore, constrained least–squares and constrained maximum entropy methods. The formal link with Bayesian statistical analysis is pointed out. Two important methods of choosing the degree of smoothing are described.

036.047   Cometary astrometry with Schmidt cameras.
K. S. Russell.
Cometary astrometry, p. 34 – 40 (1984). – See Abstr. 012.024.
The techniques used at the UK Schmidt to observe and derive astrometric positions of comets are described.

036.048   Astrometric plates obtained at the primary focus of large aperture reflectors.
A. Mrkos.
Cometary astrometry, p. 45 – 47 (1984). – See Abstr. 012.024.

036.049   Kitt Peak measurements of P/Halley positions.
M. J. S. Belton.
Cometary astrometry, p. 48 – 57 (1984). – See Abstr. 012.024.
Techniques used for the acquisition and reduction of imaging data for astrometric positions of comet Halley at Kitt Peak National Observatory are described. They yield positions that are uncertain by ± 0.9 arcsec. The reliability and consistency of the positions the author has already derived could be improved by as much as a factor of four in a more ambitious astrometric program.

036.050   Problems and procedures of cometary astrometry.
R. E. McCrosky.
Cometary astrometry, p. 58 – 59 (1984). – See Abstr. 012.024.

036.051   Obtaining accurate comet positions despite some inaccurate catalog stars.
E. Everhart.
Cometary astrometry, p. 96 – 98 (1984). – See Abstr. 012.024.
From an astrographic negative a grid is determined from measurement of all the reference stars and using their catalog positions, and other grids from selections of reference stars. These grids are determined from many stars, and individual stars will have errors with respect to the grid. It is the author's impression that 15 – 20% of the stars in the S.A.O. catalog are 1.5″ or more off their catalog positions. An interactive session with a computer can find and eliminate these errant stars and result in more accurate comet positions.

036.052   Reduction of astrographic catalogues.
J. Stock, F. Della Prugna, J. Cova.
Cometary astrometry, p. 99 – 104 (1984). – See Abstr. 012.024.
An automatic program for the reduction of overlapping plates is described.

036.053   Measuring – the quantifying art.
J. Gibson.
Cometary astrometry, p. 125 – 150 (1984). – See Abstr. 012.024.
This paper deals with the use of one– and two–coordinate measuring engines to measure the positions of images which may be elongated, comatic, or both, as well as round images, and the training of novice measurers to cope with such images. The grading of plates for potential accuracy of positions, preparation of plates for measuring, the accuracy of measurement, effects of telescope focal length and reduction technique upon the accuracy of positions, and the procedures and checks which may prevent erroneous positions also are discussed.

036.054   Reduction of astrometric plates.
J. Stock.
Cometary astrometry, p. 154 – 159 (1984). – See Abstr. 012.024.
A rapid and accurate method for the reduction of comet or asteroid plates is described.

036.055   Passive microwave remote sensing in meteorology and atmospheric physics.
K. F. Künzi.
J. Quant. Spectrosc. Radiat. Transfer, Vol. 32, No. 5/6, p. 435 – 438 (1984). – See Abstr. 012.026.
The basic principles in microwave remote sensing are given and the state of the art of ground–based, air– and spaceborne sensors is briefly reviewed. Selected references are listed, with emphasis on recent publications. Some subjective recommendations for future research, and possible trends in operational and experimental applications are discussed.

036.056   Candidates for astrometric or direct–imaging detection of extra–solar planets, and comments on the photometric method.
D. R. Soderblom.
Icarus, Vol. 61, No. 2, p. 343 – 345 (1985).
Knowledge of a star's rotation period and $v \sin i$ can be used to select stars that are seen pole–on, and thus are well suited to planetary searches by astrometric or direct–imaging means. A table of such stars is presented. This method is not suitable for discriminating equator–on systems and so cannot be used to select candidates for the photometric method of W. J. Borucki and A. L. Summers (1984).

036.057   On the expected performance of a solar oscillation network.
F. Hill, G. Newkirk Jr.
Sol. Phys., Vol. 95, No. 2, p. 201 – 219 (1985).
The authors have estimated the performance of several hypothetical ground–based networks intended to provide continuous observations of solar oscillations for one year. These networks were composed of from 2 to 6 stations distributed both in longitude and between the northern and southern hemispheres. Comparison of an existing 6 station network with the authors' model of the same network suggests that the modelling procedure is realistic provided that the estimates of the climate parameters are accurate.

036.058   The effect of spatial smearing on solar Doppler measurements. I. Mathematical formulation and application to measurements of solar rotation.
F. Albregtsen, B. N. Andersen.
Sol. Phys., Vol. 95, No. 2, p. 239 – 249 (1985).
A mathematical method for calculating the influence of scattered light on solar Doppler measurements is presented. It is shown that the main contribution to the error signal is caused by the long range scattering component of the stray–light. The method is applied on Doppler measurements of solar rotation, and the results are compared with the observations from Stanford and Mt. Wilson. Some of the large scatter in the published results on the solar rotation rate is shown to be caused by a too simple parameterization of the straylight. Within the uncertainties of the current measurements it is shown that the Doppler rotation rate of the solar photosphere is equal to the sunspot tracer rotation rate.

036.059   A possible nonlinearity in IDS (*Image Dissector Scanner*) data.
M. Rosa.
Messenger, No. 39, p. 15 – 17 (1985).

036.060   Some thoughts on limiting visual stellar and cometary magnitudes with various apertures.
D. W. E. Green.
Int. Comet Q., Vol. 7, No. 2, p. 40 – 46 (1985).

036.061   An approximate method for elimination of the apparatus function from the observed profiles of solar lines.
S. I. Gandzha.
Kinematika Fiz. Nebesn. Tel, Tom 1, No. 2, p. 53 – 61 (1985). In Russian.
An approximate method is suggested for eliminating the apparatus response function from the observed profiles of solar lines. The method permits to obtain the true line profile with a higher

accuracy than the Burger and Van Sittert method in the case when the spectral lines have half–widths at least twice as large as the half–width of the apparatus response function.

036.062   The method of plate measurements of a photographic catalogue at the Hissar Observatory.
N. N. Matveev.
Kinematika Fiz. Nebesn. Tel, Tom 1, No. 2, p. 93 – 95 (1985). In Russian.
The method of photographic star position measurements on FOKAT catalogue plates is described. The computational algorithm for calculations of star ephemerides and for identification of measured coordinates is briefly described.

036.063   Progress in automation techniques for MK classification.
M. J. Kurtz.
The MK process and stellar classification, p. 136 – 152 (1984). – See Abstr. 012.033.
The development of "quantitative" techniques for spectral classification is briefly reviewed, up to the onset of their automated implementation. Methods used in achieving automation are discussed; examples, caveats, and suggestions are given. The recent use of pattern–recognition techniques is reviewed. The work of Kurtz (1982) is extended, resulting in a new classification methodology, capable of giving a rigorous definition, in the sense of statistical pattern recognition, to numerical MK classification. Finally some suggestions for future research are made.

036.064   A program for two–dimensional classification.
H. Zekl.
The MK process and stellar classification, p. 175 – 188 (1984). – See Abstr. 012.033.
A computer program is presented for the determination of spectral class and absolute luminosity of normal stars between MK classes B0 and K3. With the use of slit spectra with a dispersion of 112 Å/mm and a mean calibration curve, the accuracy in spectral class is 1.4 subclasses, while the results for the absolute luminosities show considerable scatter.

036.065   A relationship between the Jurkevich periodogram and the Fourier transform spectral estimator.
D. N. Swingler.
Astron. J., Vol. 90, No. 4, p. 675 – 679 (1985).
An interesting mathematical link is derived between the Jurkevich periodogram and a conventional Fourier analysis of the same data. The only approximation involved is that the data points should be distributed approximately equally across the M Jurkevich bins, which is felt to be reasonable in practice. In particular, the effect of the number of bins M is readily apparent, and various effects related to the phase of the Fourier spectrum are explored.

036.066   An algorithm for significantly reducing the time necessary to compute a Discrete Fourier Transform periodogram of unequally spaced data.
D. W. Kurtz.
Mon. Not. R. Astron. Soc., Vol. 213, No. 4, p. 773 – 776 (1985).
A technique for reducing the computation time for a Discrete Fourier Transform by a factor of 4 to 6 is presented. There is no significant loss of accuracy. A FORTRAN code which will directly replace Deeming's FORTRAN code is given.

036.067   Correlation analyses of deep galaxy samples – III.
P. R. F. Stevenson, T. Shanks, R. Fong, H. T. MacGillivray.
Mon. Not. R. Astron. Soc., Vol. 213, No. 4, p. 953 – 969 (1985).
Estimates of the two–point angular correlation function, $w(\theta)$, are presented for galaxy samples obtained from COSMOS machine measurements of 1.2–m UK Schmidt telescope (UKST) and 4–m Anglo–Australian telescope (AAT) plates. All of the estimated $w(\theta)$ are consistent with a –0.8 power–law slope at small scales. At larger angular scales a break from the power–law behaviour is seen in the UKST $w(\theta)$ corresponding to a spatial separation of $3\,h^{-1}$Mpc ($H_0 = 100\,h$ km s^{-1}Mpc^{-1}) in agree-

ment with the earlier results of Shanks et al. The AAT plates allow the correlation analyses to be carried out to 24 mag in the blue (b_J) passband and 22 mag in the red (r_F) passband. It is observed that the correlation function amplitude scaling relation in both passbands is very similar. This allows the authors to add an important additional constraint to the modelled scaling relation as well as to the number–magnitude, $n(m)$, count models.

036.068 Automated analysis of objective–prism spectra. I. Quasar detection.
P. C. Hewett, M. J. Irwin, P. Bunclark, M. T. Bridgeland, E. J. Kibblewhite, X. T. He, M. G. Smith.
Mon. Not. R. Astron. Soc., Vol. 213, No. 4, p. 971 – 989 (1985).

A fully automated system for the location, measurement and analysis of large numbers of low–resolution objective–prism spectra is described. The system allows processing of objective–prism, grens or grism data. Particular emphasis is placed on techniques to obtain the maximum signal–to–noise ratio from the data, both in the initial spectral estimation procedure and for subsequent feature identification. Precise wavelength scales for each spectrum and the definition of magnitude–limited samples simplify subsequent analysis considerably. Comparison of a high–quality visual catalogue of faint quasar candidates with an equivalent automated sample demonstrates the ability of the APM system to identify all the visually selected quasar candidates. In addition, a large population of new, faint ($m_J \sim 20$) candidates is identified.

036.069 Comparison of brightnesses of lunar features by three methods of observations.
W. S. Cameron.
Strolling Astron., Vol. 31, Nos. 1 – 2, p. 33 – 40 (1985).

036.070 Processing of non–linear stellar spectra.
I. I. Nazarenko, V. S. Shergin.
Astrofiz. Issled. Izv. Spets. Astrofiz. Obs., Tom 19, p. 109 – 114 (1985). In Russian. English translation in Bull. Spec. Astrophys. Obs. – North Caucasus.

Software for the photometric complex of SAO is described, which includes the following programs: AMD–1 scanning of the stellar spectra with S–distortion, automatical plot of the characteristic and dispersion curves, summation of intensities of linearized wavelength spectra, correction for the curve of the system reaction, calculation of the spectral line radial velocities.

036.071 Visualization of atmospheric inhomogeneities during solar observations.
M. I. Fisenko.
Investigation of magnetic fields and active formations on the sun, p. 70 – 87 (1984). In Russian. Abstr. in Ref. Zh., 51. Astron., 6.51.113 (1985). – See Abstr. 003.022.

036.072 Instrumentation and methods for determination of radial velocities of galaxies from absorption spectra.
A. Kaasik.
Tartu Astrofüüs. Obs. Publ., Tom 50, p. 296 – 308 (1984). In Russian.

Observations of absorption spectra of low surface brightness galaxies and the algorithm for determination of radial velocities are described.

036.073 Error analysis for the cross–correlation technique of determining radial velocities of galaxies.
A. Kaasik, E. Saar.
Tartu Astrofüüs. Obs. Publ., Tom 50, p. 309 – 326 (1984). In Russian.

An error estimate is derived for the radial velocities of galaxies obtained by the cross–correlation technique. Different strategies to determine the radial velocities are compared. It appears that to reduce the errors it is better to break the spectrum up into non–overlapping intervals and analyze them separately.

036.074 Main dependences of photogrammetry at the reduction of a radar survey.
Yu. S. Tyuflin.
Geod. kartogr., No. 9, p. 28 – 36 (1984). In Russian. Abstr. in Ref. Zh., 52. Geod. Aehrosemka, 1.52.195 (1985).

036.075 Relativistic corrections in data of radio interferometric observations of distant cosmic sources.
N. V. Pavlov.
Sovrem. teor. i ehksp. probl. teor. otnositel'nosti i gravitatsii. Tez. dokl. Vses. konf. 6 Sov. gravitats., Moskva, 3 – 5 iyulya, 1984. Moskva, p. 260 – 261 (1984). In Russian. Abstr. in Ref. Zh., 51. Astron., 1.51.150 (1985).

036.076 Modern methods in the astrometry of double stars.
L. W. Fredrick.
Astrophys. Space Sci., Vol. 110, No. 1, p. 111 – 115 (1985). – See Abstr. 012.039.

The techniques for studying double stars continue to evolve in a predictable way. The most recent major breakthroughs have been the area scanner and speckle interferometry. On the horizon looms the application of large format CCDs which will replace the photographic plate for many astrometric problems.

036.077 Spectral information from gapped data: a comparison of techniques.
J. R. Kuhn.
Solar seismology from space, p. 293 – 303 (1984). – See Abstr. 012.041.

Three methods for estimating the power spectrum of gapped time series are briefly discussed. Numerical experiments, using several synthetic data sets, are used to compare these techniques.

036.078 High dispersion spectroscopy trials using an echelle spectrograph with CCD camera.
B. Bates, C. D. McKeith, P. R. Jorden, I. G. van Breda.
Astron. Astrophys., Vol. 145, No. 2, p. 321 – 323 (1985).

High dispersion ($\sim 1.0 \, \text{Å mm}^{-1}$) spectral observations of several bright stars obtained with a Queen's University Belfast echelle spectrograph and a Royal Greenwich Observatory CCD camera are discussed. The signal–to–noise in the recorded spectra are compared with predicted performance, and the results of these trials are examined with a view to longer term applications of this combination for programmes of high dispersion spectroscopy at resolving power $\sim 10^5$.

036.079 The division corrections of a graduated circle. A comparison between Høg's, Benevides–Soares' and Boczko's methods.
M. Rapaport.
Astron. Astrophys., Vol. 145, No. 2, p. 377 – 379 (1985).

The author considers the two methods currently used to determine the division corrections of declination circles. He first obtains the expression of the covariance matrix of the solution using Høg's method, and remarks that this matrix can be very easily compared with the covariance matrix of the solution using Benevides–Soares and Boczko's method. The author also gives a lower limit for the variance of the unknowns using the rosette method of order m, and compares this theoretical limit with the variance of his choice of rosette apertures.

036.080 Automation of geodetic and astronomical measurements.
V. Ya. Dotsenko.
Vses. zaoch. inzh.–stroit. inst. Moskva, 7 pp. (1984). In Russian. Abstr. in Ref. Zh., 52. Geod. Aehrosemka, 2.52.48 (1985).

036.081 Comparative analysis of the informativeness of different methods of astronomical observations.
K. G. Arutyunyan.
Soobshch. Byurakan. Obs., Vyp. 54, p. 72 – 76 (1983). In Russian.

An attempt is made to compare the efficiencies of different detectors of radiation used in astronomy from an informative point of view. The results for three methods of astronomical observations – direct photography, spectral photography and observations with a photomultiplier tube – are compared. It is shown that the direct photography is the most informative of the considered methods.

036.082   Automatic plotting of identification charts.
G. G. Tovmasyan.
Soobshch. Byurakan. Obs., Vyp. 54, p. 77 – 79 (1983). In Russian.
A computer program for plotting of identification charts is developed. The computer selects the stars from the SAO catalogue located in the given area of sky and prints out the SAO numbers of the selected stars, their equatorial coordinates, magnitudes, spectral types and linear coordinates computed in the given scale.

036.083   Iterative method of the atmospheric extinction reduction in multicolor broadband stellar photometry.
V. G. Moshkalev, Kh. F. Khaliullin.
Astron. Zh., Tom 62, Vyp. 2, p. 393 – 403 (1985). In Russian. English translation in Sov. Astron., Vol. 29, No. 2.
A new method of determination of the spectral extinction coefficient p_λ and atmospheric extinction reduction is developed for using in broadband stellar photometry. The method is based on modeling computations of p_λ as a function of a few parameters, which can be determined from multicolor observations themselves. It permits to exclude systematic errors exceeding $0^{\text{m}}005$ from broadband and integral measurements. Noticeable variations of selectivity factors in atmospheric extinction were discovered from *WBVR* observations. These variations can be taken into account by this method.

036.084   Techniques for mapping galactic SNRs with the MOST (*Molonglo Observatory Synthesis Telescope*).
R. S. Roger, D. K. Milne, K. J. Wellington, R. F. Haynes, M. J. Kesteven.
Proc. Astron. Soc. Aust., Vol. 5, No. 4, p. 560 – 562 (1984).

036.085   Radio astronomical absolute scale for fluxes.
V. P. Ivanov, K. S. Stankevich.
Gork'k nauchn.–issled. radiofiz. inst., Prepr., No. 183, 52 pp. (1984). In Russian. Abstr. in Ref. Zh., 51. Astron., 3.51.985 (1985).

036.086   A proposal for shape analysis in astronomy.
F. Pasian, P. Santin.
Mem. Soc. Astron. Ital., Vol. 55, No. 3, p. 459 – 468 (1984). – See Abstr. 012.049.
A series of techniques and methodologies, already tested and stable in the pattern recognition field, are described for the solution of the shape analysis problem for images of astronomical interest. Examples, referring to shape analysis of galaxies at low signal level and in the case of multiple objects, and to information extraction on the direction of the orders of echelle spectrograms, are given.

036.087   Methods of remote surface chemical analysis for asteroid missions.
R. Z. Sagdeev, G. G. Managadze, I. Yu. Shutyaev, K. Szegö, P. P. Timofeev.
Adv. Space Res., Vol. 5, No. 2, p. 111 – 120 (1985). – See Abstr. 012.051.
In this paper different active remote sensing methods are discussed which can be applied to investigate the composition of minor solar bodies. Methods using ion, laser and electron beams are treated in detail.

036.088   Imagery in radioastronomy.
F. Biraud.
Onde Electr., Vol. 64, No. 6, p. 92 – 94 (1984). In French. Abstr. in Phys. Abstr., Vol. 88, No. 1252, Entry 29854 (1985).

036.089   Representation of the interferometric delay and of the interference frequency in coordinate–independent form in terms of measured values.
Spets. astrofiz. obs. Akad. Nauk SSSR. Prepr., No. 10L, 18 pp. (1984). In Russian. Abstr. in Ref. Zh., 51. Astron., 4.51.147 (1985).

036.090   Radio interferometric observations of cosmic sources in the decameter range.
A. V. Men'.
Visn. Akad. Nauk USSR, No. 11, p. 28 – 41 (1984). In Ukrainian. From Ref. Zh., 51. Astron., 4.51.851 (1985).

036.091   Recording system of superlong interferometry data at the SM 5305 storage.
L. R. Kogan, E. V. Lakutina, O. V. Metlova, L. S. Chesalin.
Inst. kosm. issled. Akad. Nauk SSSR. Prepr., No. 952, 23 pp. (1984). In Russian. Abstr. in Ref. Zh., 51. Astron., 4.51.852 (1985).

036.092   The determination of absolute proper motions on Schmidt plates at Tautenburg.
E. Schilbach.
Mitt. Lohrmann–Obs. Tech. Univ. Dresden, Nr. 51, p. 70 – 72 (1984). In German. – See Abstr. 012.058.
Der Vergleich mit den Untersuchungsergebnissen an Lick– und Pulkowo–Observatorien erlaubt, eine sichere Bestimmung der absoluten Eigenbewegungen mit Tautenburger Schmidt– Platten zu erwarten. Der Vorteil der Bestimmung der absoluten Eigenbewegungen mit dem Tautenburger Schmidt–Teleskop liegt darin, daß es auch für sehr schwache Objekte möglich ist, diese mit guter Genauigkeit zu erhalten.

036.093   Model studies to adjust overlapping plate pairs.
H. Sefkow.
Mitt. Lohrmann–Obs. Tech. Univ. Dresden, Nr. 51, p. 72 – 74 (1984). In German. – See Abstr. 012.058.

036.094   Collocation on Schmidt plates.
C. Witschas.
Mitt. Lohrmann–Obs. Tech. Univ. Dresden, Nr. 51, p. 75 – 76 (1984). In German. – See Abstr. 012.058.

036.095   The effect of the dark–slide on the accuracy of astrometric positions used for the 2 m Schmidt camera.
D. Böhme.
Mitt. Lohrmann–Obs. Tech. Univ. Dresden, Nr. 51, p. 112 – 114 (1984). In German. – See Abstr. 012.058.
Die astrometrische Brauchbarkeit von Platten wurde untersucht bei Verwendung von quadratischen bzw. Ringkassetten am Tautenburger Schmidtspiegel. Es zeigte sich, daß beide Platten gleich gute Resultate liefern. Als Nebenprodukt der Untersuchungen entstand ein Katalog relativer Sternpositionen hoher Genauigkeit, der als astronomischer Standard für die Konstantenbestimmung bei Aufnahmen mit anderen Instrumenten nützlich sein kann.

036.096   The birefringent calibration device: a new approach.
U. Peppel, U. Finkenzeller.
AAS Photo–Bull., No. 37, p. 3 – 9 (1984).
The authors report on their experience with the classical method of obtaining characteristic curves with a birefringent calibration device and describe an alternative, more accurate calibration procedure.

036.097   Manuale di osservazione delle comete.
A. Milani.
Astronomia, Suppl. al N. 1, 20 pp. (1985).

036.098   Detectability of extrasolar planetary transits.
W. J. Borucki, J. D. Scargle, H. S. Hudson.
Astrophys. J., Vol. 291, No. 2, p. 852 – 854 (1985).
Precise stellar photometry can be used to detect other planetary systems. However, the intrinsic variability of stellar luminosity imposes a fundamental limit on the sensitivity of this

method. Based on recent precise solar observations made from the SMM satellite, it appears that the detection of Earth–sized planets will be marginal during periods of high stellar activity. However, with a suitable photometer larger planets should be readily detectable even in the presence of stellar activity equal to that of the sun at the peak of its sunspot cycle.

036.099   Prospects for determining the cosmological helium–3 to helium–4 ratio via absorption–line studies of the local interstellar medium.
M. Hurwitz, S. Bowyer.
Publ. Astron. Soc. Pac., Vol. 97, No. 589, p. 214 – 218 (1985).

The authors consider the prospects for determining the interstellar helium–3 to helium–4 ratio via absorption–line studies in the extreme ultraviolet. For the purpose of this study the authors have assumed a high–resolution extreme ultraviolet spectrometer fed by a 1–meter class telescope. They find that detection of helium–3 may be possible with less than three days of observing time. To measure the helium–3 to helium–4 ratio with sufficient precision to determine the number of light neutrino species will require over 50 days of integration time for each object studied.

036.100   A digital technique for the separation of the eclipses of a white dwarf and an accretion disc.
J. H. Wood, M. J. Irwin, J. E. Pringle.
Mon. Not. R. Astron. Soc., Vol. 214, No. 3, p. 475 – 479 (1985).

A method of extracting the white–dwarf component of an eclipse from the light curve of an eclipsing cataclysmic variable is presented. The method also gives the times of ingress and egress of the white dwarf and the times of mid–eclipse.

036.101   Absolute astronomical accelerometry.
P. Connes.
Astrophys. Space Sci., Vol. 110, No. 2, p. 211 – 255 (1985).

Two distinct but fully compatible novel concepts are proposed here for solar/stellar velocity measurements.The two principal fields of application are celestial seismology (a seismometer is nothing but an accelerometer), and the search for extra–solar planetary systems. In both cases a large number of objects will be accessible with a small telescope. One may also look for solar system accelerations (relative to some system of reference stars) due to any cause, for instance a faint solar companion, or even gravitational waves.

036.102   Restoration of digital spectra: a comparison between two methods.
C. Bonoli.
Astrophys. Space Sci., Vol. 110, No. 2, p. 311 – 319 (1985).

Two different methods for restoring digital data are compared from the point of view of the astronomical application. The results obtained for the same sets of spectroscopic data, both simulated and experimentally observed, are discussed.

036.103   Generalized projection theorem for mapping with radio interferometers.
K. M. Borkowski.
Astrophys. Space Sci., Vol. 111, No. 1, p. 203 – 205 (1985).

A theorem, which provides a relationship between the one-dimensional Fourier transform of a line section across two-dimensional (spatial) spectrum and that across the corresponding (brightness) distribution function, is proved. The theorem is then shown to be relevant in some problems in radio astronomy and possibly in other fields connected with image reconstructions from one–dimensional scans through objects or their spectra.

036.104   Roll deconvolution of Space Telescope data.
M. Walter, G. Weigelt.
Adv. Space Res., Vol. 5, No. 3, p. 169 – 171 (1985). – See Abstr. 012.059.

Roll deconvolution is a speckle method that can improve the resolution of the 2.4 m Space Telescope at short UV wavelengths. In digital simulations the authors have investigated the dependence of the signal–to–noise ratio of the reconstruction on photon noise (10^4 to 10 photons per pixel), the object size, the telescope point spread function and guiding errors.

036.105   On a possibility of observations of variable stars in visual binary systems with the area scanning technique.
S. Yu. Gorda.
Astron. Tsirk., No. 1341, p. 7 – 8 (1984). In Russian.

036.106   A Jurkevich period search program.
D. L. DuPuy, G. A. Hoffman.
I.A.P.P.P. Commun., No. 20, p. 1 – 17 (1985).

036.107   Notes on telescope drive automation. I. The precalculation of slew command strings.
J. B. Rafert.
I.A.P.P.P. Commun., No. 20, p. 24 – 26 (1985).

036.108   Spatial frequency filtering of images.
S. G. Ryan.
South. Stars, Vol. 31, No. 1, p. 53 – 60 (1984).

Spatial frequency filtering is a technique whereby the information structure of an image is changed by obtaining its two-dimensional spatial Fourier transform either digitally or optically, and attenuating selected spatial frequencies. Astronomical applications include the removal of mosaic lines from composite photographs, and the reduction of instrumental noise from stellar spectra.

036.109   A model for four–fold overlapping of a $6° \times 6°$ sky field using plates taken with the Zeiss wide–angle astrograph.
A. I. Yatsenko.
Kinematika Fiz. Nebesn. Tel, Tom 1, No. 3, p. 46 – 52 (1985). In Russian.

On the basis of a mathematical model it is shown that four-fold overlapping of a limited sky field by plates of the Zeiss wide–angle astrograph makes it possible to reduce the r.m.s. errors 1.8 times in comparison with single overlapping.

036.110   Correction to the solar line–of–sight velocities for topocentric motion of the observer.
Z. Ivanović, I. Vince.
Publ. Astron. Opservatorije Beogr., No. 33, p. 19 – 22 (1985). – See Abstr. 012.061.

A method for determining the radial velocity of an observer moving around the sun is described. The Julian date (JD) as a unique independent variable is used. The method is developed for the reduction of data obtained at Belgrade Astronomical Observatory in the course of the solar spectral line shift observations. The accuracy is about ± 0.1 m/s.

036.111   Besondere Themen der Infrarot–Interferometrie.
R. A. Hanel.
MPE Rep., No. 189, 39 pp. (1985).

036.112   Angular measurements by means of optical multiple reflection.
C. Kakuta, S. Takano, S. Tsuruta.
Proc. Int. Latitude Obs. Mizusawa, No. 23, p. 51 – 56 (1984). In Japanese.

An applicability of a method of optical multiple reflection for angular measurements in the astrometric observations is discussed. An incident angle of star light beam on the objective is changed to displacements by using multiple reflection between two plane parallel mirrors, such as Fabry–Perot plates. The multiple reflection method is equivalent to lengthening the focal length of the objective and has a merit of decreasing atmospheric disturbances in the optical path of the objective and mechanical instability of a long focal telescope.

036.113   Guidelines for the reduction of optical astrometry observations during the main campaign in Project MERIT.
K. Yokoyama.
Publ. Int. Latitude Obs. Mizusawa, Vol. 17, No. 2, p. 47 – 61 (1984).

In accordance with the adoption of the FK5 system and the IAU (1976) System of Astronomical Constants from 1984 Jan. 1 onwards, the method for reducing the optical astrometry observations at each station, as well as at the international centers, should be modified. This paper presents a full description of the new method.

036.114 Barycentric method for positioning of PDS scan data.
C.-j. Sheng, Y.-y. Zhang.
Publ. Purple Mt. Obs., Vol. 3, No. 3, p. 52 – 60 (1984). In Chinese.

A barycentric method for positioning of PDS scan data was developed and applied to several practical problems. The resulting precision was analysed and compared with the traditional methods.

036.115 A simple method to calibrate intensities of photographic slit spectrograms.
N. Vogt, L. H. Barrera.
Astrophys. Prepr. Ser., No. 8, 15 pp. (1985). To appear in Astron. Astrophys.

036.116 Improvement of a method for determining the photoelectric delay and a design of photoelectric equipment with zero temperature drift.
H.-x. Cai.
Publ. Shaanxi Astron. Obs., Vol. 7, No. 1, p. 9 – 25 (1984). In Chinese.

036.117 Logiciel de pointage des raies spectrales et mesures des vitesses radiales.
J. Chauville.
Bull. Inf. Cent. Données Stellaires, No. 28, p. 29 – 33 (1985).

An algorithm for accurate measurements of radial velocities from stellar spectra is presented. It may be used for many kinds of spectroscopic data; here the author considers only the case of spectrographic plates recorded by means of a digitalized microdensitometer. Line centers are determined by a least square fit, after a special treatment consisting of an apodized Fourier transform smoothing for which a criterion is given.

036.118 Le programme de vitesses radiales de Strasbourg au prisme–objectif de Fehrenbach.
A. Florsch.
Bull. Inf. Cent. Données Stellaires, No. 28, p. 35 – 37 (1985).

The objective–prism technique for radial velocities remains efficient, especially for detection of high–velocity stars. Such programmes are under way at Strasbourg in the direction of the Small Magellanic Cloud and in galactic fields.

036.119 Algorithme pour analyse des images d'étoiles doubles visuelles fournies par un équipement de télévision.
C. Kint.
Bull. Astron., Vol. 9, No. 6, p. 322 – 324 (1984).

036.120 Submillimetre spectroscopy on La Silla.
E. Krügel, A. Schulz.
Messenger, No. 40, p. 12 – 14 (1985).

036.121 LEST polarimetry with large two–dimensional detector arrays.
J. O. Stenflo, H. Povel.
LEST Found., Tech. Rep., No. 12, 9 pp. (1985).

Using an optical demodulation scheme it is possible to combine fast (50 – 100 kHz) piezoelastic modulation of the Stokes vector with large (e.g. 1000×1000) CCD arrays with long integration times. The optical demodulator is a piezoelastic modulator locked in frequency and phase to the corresponding modulation frequency. In combination with an optical phase switch the influence of the large pixel–to–pixel sensitivity variations is eliminated from the polarization images of Q/I, U/I, and V/I.

036.122 Speckle image reconstruction: weighted shift–and–add analysis.
J. C. Christou, E. K. Hege, J. Freeman, P. A. Strittmatter.
Bull. Am. Astron. Soc., Vol. 16, No. 4, p. 885 (1984). Abstract. – See Abstr. 010.062.

036.123 Identification of speckles by matched filtering.
E. Ribak, E. K. Hege, J. C. Christou.
Bull. Am. Astron. Soc., Vol. 16, No. 4, p. 885 – 886 (1984). Abstract. – See Abstr. 010.062.

036.124 Differential speckle interferometry using the MMT.
J. Hebden, E. K. Hege, J. Beckers.
Bull. Am. Astron. Soc., Vol. 16, No. 4, p. 886 (1984). Abstract. – See Abstr. 010.062.

036.125 Noise statistics of Pioneer 11 VLF gravitational wave search data.
J. W. Armstrong, F. B. Estabrook, J. M. Rotenberry.
Bull. Am. Astron. Soc., Vol. 16, No. 4, p. 886 (1984). Abstract. – See Abstr. 010.062.

036.126 The interpretation of astronomical speckle interferometry results.
E. K. Hege, J. Drummond, A. Y. S. Cheng, J. C. Christou, P. A. Strittmatter.
Bull. Am. Astron. Soc., Vol. 16, No. 4, p. 903 (1984). Abstract. – See Abstr. 010.062.

036.127 Investigation of random and fixed pattern noise in high dispersion IUE spectra.
S. J. Adelman, D. S. Leckrone.
Bull. Am. Astron. Soc., Vol. 16, No. 4, p. 903 (1984). Abstract. – See Abstr. 010.062.

036.128 IUE autographic spectral features – setting the limits.
K. R. H. Hackney, R. L. Hackney, Y. Kondo.
Bull. Am. Astron. Soc., Vol. 16, No. 4, p. 904 (1984). Abstract. – See Abstr. 010.062.

036.129 Taming the Medusa: data reduction.
J. M. Hill, J. Eisenhamer, D. R. Silva.
Bull. Am. Astron. Soc., Vol. 16, No. 4, p. 905 (1984). Abstract. – See Abstr. 010.062.

036.130 Interactive digital image registration software.
M. D. Potter, M. M. Shara.
Bull. Am. Astron. Soc., Vol. 16, No. 4, p. 905 (1984). Abstract. – See Abstr. 010.062.

036.131 A microwave data link between the spacewatch camera on Kitt Peak and the spacewatch computer in Tucson.
M. L. Perry, R. S. McMillan.
Bull. Am. Astron. Soc., Vol. 16, No. 4, p. 908 (1984). Abstract. – See Abstr. 010.062.

036.132 Photometric detection of open cluster binaries.
J. P. Dabrowski.
Bull. Am. Astron. Soc., Vol. 16, No. 4, p. 913 (1984). Abstract. – See Abstr. 010.062.

036.133 Coronal–hole detectability on solar–type stars.
R. C. Altrock.
Bull. Am. Astron. Soc., Vol. 16, No. 4, p. 939 – 940 (1984). Abstract. – See Abstr. 010.062.

036.134 The Feinheit method: a phase–independent scheme for calibrating the period–luminosity relation of classical Cepheids.
B. F. Madore.
Bull. Am. Astron. Soc., Vol. 16, No. 4, p. 970 (1984). Abstract. – See Abstr. 010.062.

036.135 Chondrule volume percents: statistical behavior and visual–estimation guides produced by numerical simulation.
A. Woronow, E. A. King.
Meteoritics, Vol. 20, No. 1, p. 103 – 112 (1985).

Estimates of the volume percents of chondrules in petrographic thin sections are notoriously unreliable, ranging widely among different observers for the same meteorite or even for the same thin section. The set of visual–estimation guides presented herein should help narrow the uncertainties in such estimates and lead to a more consistent picture of chondrule abundances than is now apparent from the literature.

036.136 Prospects of an experiment in astrophysics with a Compton polarimeter.
G. M. Gorodinskij, B. I. Ermolaev, S. V. Zajtsev, E. M. Kruglov, V. V. Khmylko.
Izv. Akad. Nauk SSSR. Ser. fiz., Tom 48, No. 11, p. 2090 – 2092 (1984). In Russian. Abstr. in Ref. Zh., 62. Issled. Kosm. Prostranstva, 3.62.82 (1985).

036.137 Astrometric and photometric estimators for TYCHO photon counts.
M. Yoshizawa, G. K. Andreasen, E. Høg.
Astron. Astrophys., Vol. 147, No. 2, p. 227 – 236 (1985).
The performance of different astrometric and photometric estimators is studied with the purpose of optimizing the evaluation of photon counts from the TYCHO space astrometry experiment on board the HIPPARCOS satellite. Theoretical values of the performance are confirmed by numerical simulations. A numerical filter method is developed to solve the normal equations of the Maximum–Likelihood method. This method requires orders of magnitude less computing time than the exact method.

036.138 Evaluation of Compton contamination events in an actively shielded Ge gamma–ray detector.
A. Owens.
Astrophys. Space Sci., Vol. 112, No. 1, p. 75 – 81 (1985).
An expression has been developed which gives the flux of unvetoed Compton–contamination events for a particular source measurement using an actively shielded gamma–ray telescope. By use of suitable approximations, this expression reduces to simpler functions which have been evaluated for some recent measurements of the Crab nebula. It has been shown that for the exposures typical of balloon–borne observations, and a reasonably efficient anti–coincidence system ($> 50\%$), the non–photopeak component of the detected source energy–loss spectrum can be neglected.

036.139 Gamma–ray astronomy and the study of electrons $> 10^{13}$eV and bursts using an array of γ–ray detectors in space platform.
S. A. Stephens.
Astrophys. Space Sci., Vol. 112, No. 1, p. 145 – 155 (1985).
The author has discussed the possibility of employing an array of high–energy γ–ray detectors in space platform in order to achieve higher sensitivity than by using a single detector. It is shown that such a detector system can be utilized for the study of faint galactic and extragalactic sources, and for deep survey of complex regions. The array can also be operated in the γ–ray burst mode to detect bursts. Further, the array can be used for the study of cosmic–ray electrons above 10^{13}eV, for which there is no other technique to detect them up to now.

036.140 Phase–defined ground–orbital aperture synthesis system.
L. I. Gurvits, S. V. Pogrebenko.
Astrophys. Space Sci., Vol. 112, No. 2, p. 337 – 346 (1985).
The problem of reconstructing a visibility phase from *VLBA* data is discussed. It is shown that a multi–element system including four radiotelescopes, among them one orbital radiotelescope, can contribute much to the "phase problem" solution. A method of completely reconstructing the visibility phase is proposed and numerical simulation results are given.

036.141 Peculiarities of reduction of pulse radio signals distorted when propagating in the interstellar medium.
A. A. Bocharov.
Inst. kosm. issled. Akad. Nauk SSSR. Prepr., No. 962, 42 pp. (1984). In Russian. Abstr. in Ref. Zh., 51. Astron., 6.51.508 (1985).

036.142 Image restoration by the shift–and–add algorithm.
W. G. Bagnuolo Jr.
Opt. Lett., Vol. 10, No. 5, p. 200 – 202 (1985).
A new method for image restoration based on the shift–and–add (SAA) algorithm is presented, the main advantages of which appear to be speed and simplicity. The SAA pattern produced by an object is given by the object correlated by a nonlinear replica of itself whose intensity distribution is strongly weighted toward the brighter pixels. A method of successive substitutions can then be used to decorrelate the SAA pattern and recover the object. The method is applied to the case of the extended chromosphere of Betelgeuse.

036.143 Bispectral analysis at low light levels and astronomical speckle masking.
B. Wirnitzer.
J. Opt. Soc. Am. A, Vol. 2, No. 1, p. 14 – 21 (1985).
The bispectrum is the Fourier transform of the triple correlation, sometimes also referred to as the triple–product integral. The influence of photon noise on the bispectrum of an image-intensity distribution is discussed. As an example, the astronomical speckle–masking method is considered. It is shown theoretically that bispectral analysis in speckle masking should yield true, diffraction–limited images in all those cases in which the speckle-interferometry process has been successful in reconstructing the object autocorrelation.

036.144 Fast iterative solution to exact equations for the two–dimensional phase–retrieval problem.
K. Chalasinska–Macukow, H. H. Arsenault.
J. Opt. Soc. Am. A, Vol. 2, No. 1, p. 46 – 50 (1985).

036.145 Effects of finite spectral bandwidth and focusing error on the transfer function in stellar speckle interferometry.
J. Ohtsubo.
J. Opt. Soc. Am. A, Vol. 2, No. 5, p. 667 – 673 (1985).
The normal model of light propagation through a turbulent atmosphere is used to investigate the telescope–atmosphere transfer function in stellar speckle interferometry. Some observational results are presented and compared with theory. The normal model leads to an analytically simple solution and properly describes light scattering from turbulent atmospheres. The results for the normal model are compared with those for the log-normal model.

036.146 Detection of nebulosity around stars using a Fabry–Perot spectrometer: a new technique.
K. C. Sahu, R. Gupta, M. Srinivasan, J. N. Desai.
Appl. Opt., Vol. 24, No. 1, p. 10 – 12 (1985).
The finding of nebulosity around certain stars, particularly the WR type, has created considerable interest in the last few years, and the search for nebulosity has been carried out by employing various methods, e.g. narrowband photometry, photography, and spectroscopy. The authors offer a simple technique for carrying out such a search along with the principle used and the results of some laboratory experiments. The method uses a photoelectric Fabry–Perot spectrometer with a centrally masked multiple–zone aperture and has certain advantages over other methods.

036.147 Stellar speckle reconstruction by the shift–and–add method.
N. Baba, S. Isobe, Y. Norimoto, M. Noguchi.
Appl. Opt., Vol. 24, No. 10, p. 1403 – 1405 (1985).

036.148 The basic equations for scanning heliometer measurement of solar diameters.
J. Rösch.
Sol. Phys., Vol. 96, No. 2, p. 213 – 217 (1985).
The basic equations for measurement of solar diameters by means of the scanning heliometer are given, and the first processing of the rough data is described, either for oblateness evaluation, for medium term oscillations detection, or for investigation of secular diameter changes.

036.149 Infrared speckle imaging.
R. R. Howell.
Bull. Am. Astron. Soc., Vol. 17, No. 1, p. 515 (1985). Abstract. –
See Abstr. 010.064.

036.150 Spektroskopische Veränderlichenbeobachtung.
E. Pollmann.
Sterne Weltraum, 24. Jahrg., Nr. 6, p. 340 – 344 (1985).

036.151 A new statistic for the analysis of circular data with applications in ultra–high energy gamma–ray astronomy.
R. J. Protheroe.
Astron. Express, Vol. 1, Nos. 4 – 6, p. 137 – 142 (1985).

A new statistic for the analysis of circular data is proposed. The power function of the test based on this statistic has been evaluated and the test has been found to be powerful against alternatives in the form of a uniform background plus a narrow peaked pulse. Critical values of the statistic are tabulated. A single peak plus background is expected in γ–ray observations of pulsars and binary X–ray sources at ultra–high energies and the proposed statistic should prove useful in analysing such observations.

036.152 Changes in personal equations of circumzenithal observers.
G. Karský.
Bull. Astron. Inst. Czech., Vol. 36, No. 3, p. 144 – 149 (1985).

Deviations of Universal Time (UT 1) and latitude from BIH results, determined for 1481 observation series measured with a visual circumzenithal at the Geodetical Observatory Pecný in the years 1970 – 1983, were analysed.

036.153 Automatic analysis of crowded fields.
M. J. Irwin.
Mon. Not. R. Astron. Soc., Vol. 214, No. 4, p. 575 – 604 (1985).

There are many important 2D data reduction problems in astronomy that are not amenable to conventional automatic analysis. Typically these problem areas arise in fields where the number density of images is high and where the local sky background may vary rapidly. Within such regions the image number density becomes so high that the majority of images overlap, even at relatively high isophotes, and simple image parameter estimation algorithms become confused. The author's goal has been to examine the potential for a fully automatic method that is both robust and efficient in terms of computer requirements, capable of dealing with complex multiple overlaps and able to generate the optimum estimates of image parameters.

036.154 Spectrophotometry from uncalibrated Schmidt plates.
W. B. Weaver.
Publ. Astron. Soc. Pac., Vol. 97, No. 590, p. 307 – 309 (1985).

A technique for the recovery of spectrophotometric data from uncalibrated Schmidt objective–prism plates is presented. This method could be applied to any of millions of such spectra which have been obtained over the last 35 years. It is demonstrated on a flare–star spectrum obtained by N. Sanduleak; an interesting absorption feature is noted.

036.155 Kitt Peak multi–tasking FORTH–11.
T. E. McGuire.
J. Forth Appl. Res., Vol. 2, No. 2, p. 57 – 67 (1984). Abstr. in Phys. Abstr., Vol. 88, No. 1254, Entry 40499 (1985).

036.156 Long–baseline Michelson interferometry with large ground–based telescopes operating at optical wavelengths. II. Interferometry at infrared wavelengths.
F. Roddier, P. Lena.
J. Opt. (Paris), Vol. 15, No. 6, p. 363 – 374 (1984). Abstr. in Phys. Abstr., Vol. 88, No. 1254, Entry 40501 (1985).

036.157 Composite pictures of recentred short exposures to improve the angular resolution.
J. Hecquet, G. Coupinot.
J. Opt. (Paris), Vol. 15, No. 6, p. 375 – 383 (1984). In French. Abstr. in Phys. Abstr., Vol. 88, No. 1254, Entry 40502 (1985).

036.158 On reduction of signals of a solar magnetograph.
N. N. Morozov.
Tr. Astrofiz. Inst. Alma–Ata, Tom 41, p. 12 – 18 (1983). In Russian.

036.159 Attenuation of solar radiation in an attachment with two screens.
R. Kh. Gajnullina.
Tr. Astrofiz. Inst. Alma–Ata, Tom 41, p. 83 – 90 (1983). In Russian.

036.160 Method of mathematical reduction of results of observations of spicule occultations by the moon.
B. I. Demchenko, G. K. Ajmanova.
Tr. Astrofiz. Inst. Alma–Ata, Tom 41, p. 91 – 108 (1983). In Russian.

036.161 Report on the task raw data treatment.
F. Donati, E. Canuto.
Processing of scientific data from the ESA astrometry satellite HIPPARCOS, p. 117 (1985). Abstract. – See Abstr. 012.087.

036.162 Double star analysis strategy.
W. Delaney.
Processing of scientific data from the ESA astrometry satellite HIPPARCOS, p. 153 – 158 (1985). – See Abstr. 012.087.

A strategy is presented for recognizing double stars and estimating their parameters. Some features of a least squares recognition – estimation algorithm are outlined, with particular regard for the problem of automatic evaluation of first approximations for the parameters.

036.163 Double star detection.
L. Borriello, W. Delaney.
Processing of scientific data from the ESA astrometry satellite HIPPARCOS, p. 159 – 167 (1985). – See Abstr. 012.087.

In this work the analysis of two indicators suitable for double star detection is made. The indicators considered are the amplitude ratio and the phase difference between the first two harmonics of the star modulation signal. A complete analysis of the signals without noise is presented, while for signals with noise only preliminary results are given.

036.164 Estimation of double star parameters.
A. Farilla.
Processing of scientific data from the ESA astrometry satellite HIPPARCOS, p. 169 – 173 (1985). – See Abstr. 012.087.

A model useful for least squares estimation of double star parameters on the basis of HIPPARCOS measurements is defined. Results obtained with simulated data are presented.

036.165 Imaging approach to multiple star recognition.
L. Borriello.
Processing of scientific data from the ESA astrometry satellite HIPPARCOS, p. 175 – 181 (1985). – See Abstr. 012.087.

The observations of the astrometric satellite HIPPARCOS can be considered as an incomplete sampling of a two dimensional Fourier space; thus, by means of Fourier transform techniques, it is possible to reconstruct a degraded image of the observed objects. This paper is devoted to demonstrate that the techniques developed in VLBI field can be applied to such images in order to estimate the magnitudes, the separation, the positions and in some cases the orbital motion of multiple stars. Examples and preliminary results are given.

036.166 Reception and preparation task.
J. Kovalevsky, J. L. Falin.
Processing of scientific data from the ESA astrometry satellite HIPPARCOS, p. 309 – 313 (1985). – See Abstr. 012.087.

After a general presentation of the reasons for which this task has been introduced in the data reduction chain, its connections with the Interface Document are given. This task is indeed an interface between ESOC tapes and the Input Catalogue on one side, the reduction subsystems on the other side.

036.167 Architecture de la tâche "Réception et Préparation".
M. Villenave.
Processing of scientific data from the ESA astrometry satellite HIPPARCOS, p. 315 – 322 (1985). – See Abstr. 012.087.

Après un rappel des rôles essentiels de la tâche "Réception et Préparation du Flot de données en provenance de l'ESOC", cet article présente une version préliminaire de l'architecture de cette tâche ainsi que la démarche qui a permis à l'auteur d'élaborer celle-ci à partir de l'analyse fonctionnelle.

036.168 Le logiciel de simulation. Les données de la grille principale.
M. Froeschlé, J. L. Falin, J. Kovalevsky, F. Mignard.
Processing of scientific data from the ESA astrometry satellite HIPPARCOS, p. 335 – 350 (1985). – See Abstr. 012.087.

The authors describe the processor developed to simulate the data generated by the passage of a star on the main grid of HIPPARCOS. In a first section they present the software required for the preparation of the simulation, namely the attitude motion of the satellite as well as its relative displacement with respect to the barycenter of the solar system. In particular, allowance is made in the authors' routine for the interruptions of observations because of the presence of the earth or the moon in an extended FOV. Then the main lines of the observation strategy are described.

036.169 Observations with the star–mapper.
M. Froeschlé, J. L. Falin, F. Mignard.
Processing of scientific data from the ESA astrometry satellite HIPPARCOS, p. 351 – 358 (1985). – See Abstr. 012.087.

The authors describe the different steps in the simulation of the star mapper data. After a description of the star mapper they consider the transit of a typical star on the two grids. The authors then present the two softwares used in this simulation along with the format of the output file.

036.170 Simulation for the sphere reconstitution.
M. Lattanzi, B. Bucciarelli.
Processing of scientific data from the ESA astrometry satellite HIPPARCOS, p. 359 – 364 (1985). – See Abstr. 012.087.

The general features of the simulation devoted to produce the input data for the sphere reconstitution are presented. Some aspects related to the simulation methods employed are discussed in detail. Finally the main properties of the simulation are given.

036.171 A first draft of the simulation task, exploitation scheme.
J. Dupic.
Processing of scientific data from the ESA astrometry satellite HIPPARCOS, p. 367 – 374 (1985). – See Abstr. 012.087.

This is a presentation of a proposed simulation exploitation scheme and a request for a final definition of the simulation part within the reduction qualification.

036.172 First–look processing.
J. Schrijver.
Processing of scientific data from the ESA astrometry satellite HIPPARCOS, p. 375 – 378 (1985). – See Abstr. 012.087.

The FAST first-look processing is discussed. A number of possible tests is identified. Some thoughts about the software implementation are presented.

036.173 Review of the data processing task.
J. L. Pieplu.
Processing of scientific data from the ESA astrometry satellite HIPPARCOS, p. 379 – 385 (1985). – See Abstr. 012.087.

The aim of this paper is to present the main topics of the work achieved or undertaken since the last thinkshop in Asiago. At the software system level, these topics deal with the software compatibility, the computer work load assessment, the Data Management and Command System (DMCS) software requirements and the software test policy. At the subsystem level, they concern the simulation architecture and exploitation policy, the design of the reception and preparation software and several implementations of institute software mock ups.

036.174 The general structure of the data reduction and its consequences on the exploitation phase.
C. Huc.
Processing of scientific data from the ESA astrometry satellite HIPPARCOS, p. 387 – 392 (1985). – See Abstr. 012.087.

The new general structure of the HIPPARCOS astrometric data reduction is presented. The representation used enables the author to make an analysis of the frozen and flexible aspects of the data reduction.

036.175 A first approach of the software requirements relating to the D.M.C.S.
C. Huc.
Processing of scientific data from the ESA astrometry satellite HIPPARCOS, p. 393 – 396 (1985). – See Abstr. 012.087.

The software requirements relating to a given system allow to define what this system should do before describing how it should do it. The author tries to give a first idea of what could be the software requirements relating to the Data Management and Command System (D.M.C.S.).

036.176 Present estimate of the computer workload.
A. Vargas.
Processing of scientific data from the ESA astrometry satellite HIPPARCOS, p. 397 – 402 (1985). – See Abstr. 012.087.

The aim of this paper is to present an estimate of the CPU time and magnetic tape needs for all the FAST–HIPPARCOS computer processings which are the simulation, the qualification of the reduction with simulated data and the reduction.

036.177 Use of TV electronics in astronomical observations.
A. V. Didenko, V. A. Kachmin.
Tr. Astrofiz. Inst. Alma–Ata, Tom 42, p. 111 – 115 (1983). In Russian.

036.178 On the identification of stars in observations of artificial celestial bodies. I.
V. S. Matyagin, T. P. Nosova, N. V. Sinyaeva, L. A. Usol'tseva.
Tr. Astrofiz. Inst. Alma–Ata, Tom 44, p. 83 – 93 (1984). In Russian.

036.179 An astronomical application of deconvolution by the maximum entropy method.
D. O'Donoghue.
Proceedings of the Second South African Symposium on Digital Image Processing, Durban, South Africa, 25 – 26 July 1984. A. G. Sartori–Angus (Editor). Univ. Natal, Durban, South Africa (1984). p. 24 – 33. Abstr. in Phys. Abstr., Vol. 88, No. 1257, Entry 57621 (1985).

036.180 Evaluation of an experimental system for spaceborne processing of multispectral image data.
B. D. Meredith, N. D. Murray, R. J. LaBaugh, J. V. Aanstoos.
Opt. Eng., Vol. 24, No. 1, p. 189 – 196 (1985). Abstr. in Phys. Abstr., Vol. 88, No. 1258, Entry 59493 (1985).

036.181   **Analysis of variance: a statistical approach to pattern recognition and motion detection.**
G. X. Ritter, J. Ahmad.
Proc. SPIE Int. Soc. Opt. Eng., Vol. 504, p. 264 – 267 (1984).
Abstr. in Phys. Abstr., Vol. 88, No. 1258, Entry 63227 (1985).

036.182   **Automatic classification of galaxies into morphological types.**
M. Thonnat, M. Berthod.
Seventh International Conference on Pattern Recognition, held at Montreal, Que., Canada, 30 July – 2 August 1984. IEEE Comput. Soc. Press, Silver Spring, Md., USA (1984), p. 844 – 846.
Abstr. in Phys. Abstr., Vol. 88, No. 1259, Entry 68683 (1985).

036.183   **Analysis of a time–varying scene observed with a digital CCD camera.**
N. Kämpfer, W. A. Schöchlin.
IEEE Trans. Geosci. Remote Sensing, Vol. GE–23, No. 2, p. 150 – 157 (1985).
The analysis of digital solar flare images obtained with a charge–coupled device (CCD) camera, their representation in graphical form, and the various corrections (calibrations, image motion, atmospheric conditions) to be applied to the data are presented.

036.184   **Stellar radial velocities of high precision: techniques and results.**
B. Campbell, G. A. H. Walker.
Stellar radial velocities, p. 5 – 19 (1985). – See Abstr. 012.095 (IAU Colloq. No. 88).
Relative radial velocities can now be determined with internal errors as small as ~ 10 m s^{-1}. A variety of techniques for achieving this level of precision are either presently running, or will soon be on–line. These could be used on several astronomical frontiers; the most demanding application is the search for extra–solar planetary systems. The authors briefly review these techniques, and highlight the results obtained to date.

036.185   **Digital stellar speedometry.**
D. W. Latham.
Stellar radial velocities, p. 21 – 34 (1985). – See Abstr. 012.095 (IAU Colloq. No. 88).
Many areas of research have been opened up by the advent of instruments that can mass–produce radial velocities for large numbers of faint stars with precisions better than ± 1 km s^{-1}. This paper summarizes the instruments and techniques in use at Center for Astrophysics, and the performance that they achieve for the digital speedometry of stars.

036.186   **Cross–correlation spectroscopy using CORAVEL.**
M. Mayor.
Stellar radial velocities, p. 35 – 48 (1985). – See Abstr. 012.095 (IAU Colloq. No. 88).
Several studies are reviewed to illustrate the possibilities of cross–correlation spectroscopy to derive rotational velocities, macro–turbulence, metallicities, magnetic fields and information on oscillation modes. As a result of different radial–velocity surveys carried out with CORAVEL, the eccentricity–period relations are discussed, showing the concurrent influence of the tidal circularization, circularization by mass–exchange and the dependence of the initial eccentricity distribution with the orbital period.

036.187   **Radial velocity information in solar–type spectra.**
W. J. Merline.
Stellar radial velocities, p. 87 – 98 (1985). – See Abstr. 012.095 (IAU Colloq. No. 88).
The author has developed a criterion for determining the amount of radial velocity information theoretically available at the Earth's surface from a star as functions of wavelength and spectral resolution. The extent to which this information can be utilized depends upon the experimental technique and the efficiency of the measurement system. In addition, astrophysical effects intrinsic to the star will alter the signature of kinematic accelerations.

036.188   **Radial velocities from CCD detectors.**
D. W. Willmarth, H. A. Abt.
Stellar radial velocities, p. 99 – 108 (1985). – See Abstr. 012.095 (IAU Colloq. No. 88).
The authors have developed a simple system to judge the suitability and accuracy of CCD detectors for determining radial velocities. While earlier studies determine the Doppler shift in Fourier space, the method described here analyses the spectrum in data space, using a straightforward technique that has been demonstrated to give accurate reliable results.

036.189   **Radial–velocity work at Cambridge and Palomar.**
R. F. Griffin.
Stellar radial velocities, p. 121 – 122 (1985). – See Abstr. 012.095 (IAU Colloq. No. 88).

036.190   **The CFA system for digital correlations.**
W. F. Wyatt.
Stellar radial velocities, p. 123 – 130 (1985). – See Abstr. 012.095 (IAU Colloq. No. 88).
The Optical and Infrared Division at the Center for Astrophysics has developed an extensive set of powerful and flexible software to perform spectral data reduction and analysis. The system's flexibility and power allows analysis of spectra from all spectrographs (high dispersion to low), easy creation of templates, many correlations per object, and a very useful display of the correlation results. Several changes in filtering and correlation techniques have resulted in the improved reanalysis of old observations.

036.191   **Absolute astronomical accelerometry.**
P. Connes.
Stellar radial velocities, p. 131 – 137 (1985). – See Abstr. 012.095 (IAU Colloq. No. 88).
A new technique is proposed, leading to two closely related instruments, the solar and stellar accelerometers. The principal fields of application are solar/stellar seismology and the search for extrasolar planets. The technique uses a variable path Fabry–Pérot interferometer whose channelled spectrum is constrained by a servo loop to track the Doppler shifted astronomical spectrum. Then a tunable laser tracks the interferometer. The ultimately measured quantity is the beat frequency between tunable laser and a stabilized one.

036.192   **Optical lightnings as a background in searches for cosmic X– and gamma–ray bursts through atmospheric fluorescence.**
C. L. Bhat, R. K. Kaul, M. L. Sapru, H. Razdan.
18th International Cosmic Ray Conference, Vol. 8, p. 32 – 35 (1983). – See Abstr. 012.096.

036.193   **A method to detect ultra high energy electrons using Earth's magnetic field as a radiator.**
S. A. Stephens, V. K. Balasubrahmanyan.
18th International Cosmic Ray Conference, Vol. 8, p. 81 – 84 (1983). – See Abstr. 012.096.
The authors examine the possibility of detecting electrons above a few TeV by the photons emitted through synchrotron radiation in the Earth's magnetic field. This technique would open up an important window of energy for the study of primary cosmic ray electrons.

036.194   **Absolute calibration of photomultipliers for the measurement of atmospheric Čerenkov light intensities.**
A. G. Gregory, J. R. Patterson, B. R. Dawson.
18th International Cosmic Ray Conference, Vol. 8, p. 145 – 148 (1983). – See Abstr. 012.096.

036.195   **Microprocessor control of a cosmic–ray hadron spectrometer.**
M. Lumme, M. Nieminen, J. Peltonen, J. J. Torsti, E. Vainikka, E. Valtonen.
18th International Cosmic Ray Conference, Vol. 8, p. 182 – 185 (1983). – See Abstr. 012.096.
The microprocessor control system of a hadron spectrometer is described.

036.196 **Precision of the spectrographic method.**
L. I. Dorman, A. V. Zaikin, V. G. Yanke.
18th International Cosmic Ray Conference, Vol. 10, p. 140 – 143 (1983). – See Abstr. 012.096.
The spectrographic method is applied to cosmic ray spectrograph data to find the variation spectrum parameters, the geomagnetic cutoff rigidity variation, and the variations of atmospheric temperature distribution.

036.197 **The simplest versions of the global spectrographic method.**
A. V. Belov, L. I. Dorman, V. G. Yanke.
18th International Cosmic Ray Conference, Vol. 10, p. 144 – 147 (1983). – See Abstr. 012.096.
The global–spectroscopy method is used to specify the variations of cosmic ray density and anisotropy during a deep Forbush–decrease in August 1972.

036.198 **Some methods for searching for weak spectra.**
J. I. Katz.
Bull. Am. Astron. Soc., Vol. 17, No. 2, p. 550 – 551 (1985). Abstract. – See Abstr. 010.065.

036.199 **Accurate differential magnitudes of binary star components as obtained using the SAA algorithm.**
D. J. Hutter, H. A. McAlister, W. I. Hartkopf.
Bull. Am. Astron. Soc., Vol. 17, No. 2, p. 551 (1985). Abstract. – See Abstr. 010.065.

036.200 **Reduction and analysis techniques developed for the GSU ICCD speckle camera.**
W. I. Hartkopf, H. A. McAlister, D. J. Hutter.
Bull. Am. Astron. Soc., Vol. 17, No. 2, p. 551 (1985). Abstract. – See Abstr. 010.065.

036.201 **Statistical methods of astronomical data with upper limits.**
T. Isobe, E. D. Feigelson, P. I. Nelson.
Bull. Am. Astron. Soc., Vol. 17, No. 2, p. 573 (1985). Abstract. – See Abstr. 010.065.

036.202 **PSAIPS – the Penn State astronomical image processing system.**
R. J. Truax, J. A. Nousek, D. W. Weedman, G. Weaver.
Bull. Am. Astron. Soc., Vol. 17, No. 2, p. 575 (1985). Abstract. – See Abstr. 010.065.

036.203 **Performance AIPS on a low budget.**
C. J. Lonsdale, E. D. Feigelson, C. Center.
Bull. Am. Astron. Soc., Vol. 17, No. 2, p. 575 (1985). Abstract. – See Abstr. 010.065.

036.204 **CCD–echelle reduction with the Charles River supermicro.**
D. P. Huenemoerder, L. W. Ramsey.
Bull. Am. Astron. Soc., Vol. 17, No. 2, p. 575 (1985). Abstract. – See Abstr. 010.065.

036.205 **A C implementation of FORTH.**
S. Gajar, R. J. Truax, D. P. Huenemoerder, J. A. Nousek.
Bull. Am. Astron. Soc., Vol. 17, No. 2, p. 575 (1985). Abstract. – See Abstr. 010.065.

036.206 **Galaxies seen through gravitational lenses.**
E. E. Falco, M. H. Schneps.
Bull. Am. Astron. Soc., Vol. 17, No. 2, p. 580 (1985). Abstract. – See Abstr. 010.065.

036.207 **A preliminary look at astrometric accuracy as a function of photon count.**
L. A. Breakiron, G. D. Gatewood, J. W. Stein, C. DiFatta, J. H. Kiewiet de Jonge.
Bull. Am. Astron. Soc., Vol. 17, No. 2, p. 583 (1985). Abstract. – See Abstr. 010.065.

036.208 **The solar radial velocity and implications for spectroscopic planetary detection.**
D. Deming, F. Espenak, D. Jennings, J. W. Brault.
Bull. Am. Astron. Soc., Vol. 17, No. 2, p. 588 (1985). Abstract. – See Abstr. 010.065.

036.209 **Statistical properties of phase dispersion minimization period finding techniques.**
A. F. Linnell Nemec, J. M. Nemec.
Bull. Am. Astron. Soc., Vol. 17, No. 2, p. 597 (1985). Abstract. – See Abstr. 010.065.

036.210 **Tests of the Fourier quotient method.**
H. F. Levison, J. B. Laird.
Bull. Am. Astron. Soc., Vol. 17, No. 2, p. 612 (1985). Abstract. – See Abstr. 010.065.

036.211 **Astrometria fotografica (con scala graduata).**
F. Cerchio.
Astronomia, N. 2, p. 23 – 26 (1985).

036.212 **Maximum entropy reconstruction of accretion disk images from eclipse data.**
K. Horne.
Interacting binaries, p. 327 – 348 (1985). – See Abstr. 012.104.
The eclipsing cataclysmic variables provide a unique opportunity for the observational testing of accretion disk theory. This article treats the eclipse of an accretion disk as an image reconstruction problem. The numerical methods used to synthesize light curves for accretion disk eclipses are reviewed. The paper then develops techniques for reconstructing the intensity distribution on the face of the disk (an image of the disk) from an observed eclipse light curve. Results of simulation tests in which disk images were reconstructed from synthetic light curve data are discussed. Finally, the present status and future prospects of the eclipse mapping method are summarized.

036.213 **Infrared astronomy.**
P. Mayer.
Vesmír, Vol. 63, No. 7, p. 205 – 208, 213 – 214 (1984). In Czech.

036.214 **On using a space telescope to detect faint galaxies.**
E. L. Wright.
Publ. Astron. Soc. Pac., Vol. 97, No. 591, p. 451 – 453 (1985).
Observations using space telescopes should be optimized for conditions that prevail in space. Since the sky at 1 µm is very dark above the OH nightglow, and because distant galaxies are brightest at $\lambda > 1$ µm, the HST should be very good at detecting and measuring faint, distant galaxies in the R and I bands. For later-generation space telescopes the optimal wavelength for detecting high–redshift faint galaxies is the 3 µm "window" in the zodiacal light.

036.215 **A possible method for the determination of the magnitude and inclination of a sunspot magnetic field.**
J.–h. Jin.
Chin. Astron. Astrophys., Vol. 9, No. 1, p. 15 – 19 (1985). English translation of 38.036.170.

036.216 **Peculiarities of absolute determinations of the coordinates of radio sources with the VLBI method.**
V. S. Gubanov, N. D. Umarbaeva.
Izv. Glav. Astron. Obs. Pulkovo, Astrometr. Astrofiz., No. 201, p. 5 – 12 (1985). In Russian.
On the basis of a numerical model of VLBI observations of extragalactic radio sources, the drawbacks of the generally adopted method for absolute determinations of coordinates of the sources are shown. These drawbacks are due to a strong correlation between the unknowns. For the interferometer baselines, parallel to the equator plane when this correlation is particularly strong, two modifications of the observational method are proposed free from the drawbacks.

036.217   On systematic errors of two methods of determination of declinations of stars.
S. A. Tolchel'nikova–Murri.
Izv. Glav. Astron. Obs. Pulkovo, Astrometr. Astrofiz., No. 201, p. 29 – 30 (1985). In Russian.

Equations of observations with zenith telescopes are more accurate than those with the vertical circle, but lead to an unsolvable system. An approximate method used for declination determinations causes most significant systematic errors of declination catalogues of zenith telescopes. Several methods are given for revealing the origin of declination systematic errors of observations made with vertical circles and zenith telescopes.

036.218   New trends in CCD photometry of galaxies.
B. Fort.
New aspects of galaxy photometry, p. 3 – 12 (1985). – See Abstr. 012.114.

This paper summarizes some possibilities of CCDs for photometry of galaxies. After a quick–look at CCD development, and at some basic information for data reduction, it discusses the relevance of CCD/focal reducer techniques.

036.219   A new filter system for CCD surface photometry.
L. Vigroux.
New aspects of galaxy photometry, p. 13 – 20 (1985). – See Abstr. 012.114.

Classical surface photometry of galaxies made with photographic plate has been restricted to wide band filter, such as the *UBV* of the Gunn system. In contrast the very high sensitivity of CCD allows the use of narrow filters centered on emission or absorption features which are characteristic of the physical state of the observed object. This paper gives a list of absorption features which can be used for surface photometry, and presents some results obtained in early type galaxies using intermediate band photometry.

036.220   Electronography: performances, results, expectations.
J.–P. Picat.
New aspects of galaxy photometry, p. 21 – 26 (1985). – See Abstr. 012.114.

Properties related to the spatial continuity of electronographic data are shown. Performances are compared to those of sampling detectors such as CCDs. Applications are proposed where, resolution, large field and low noise are in favour of electronography.

036.221   Food for the photometrists – faint galaxies revealed.
D. F. Malin.
New aspects of galaxy photometry, p. 27 – 32 (1985). – See Abstr. 012.114.

036.222   Computer–linked microdensitometers: performance, results and expectations.
H. T. MacGillivray.
New aspects of galaxy photometry, p. 33 – 38 (1985). – See Abstr. 012.114.

The development of fast, computer–controlled microdensitometers has provided a whole new dimension in the high–speed extraction of data from astronomical photographs, enabling the potential of these photographs to be fully exploited by digital means for the first time. This paper summarises the main features of these machines, describes their advantages and disadvantages and discusses their suitability for galaxy photometry.

036.223   High resolution in galaxy photometry and imaging.
J.–L. Nieto, G. Lelièvre.
New aspects of galaxy photometry, p. 43 – 48 (1985). – See Abstr. 012.114.

The authors present briefly results from high–resolution observations of galaxies either with the CFH telescope or at Pic du Midi Observatory. They mainly correspond to direct acquisition data relying only upon the high resolution of the site. To improve the resolution, several possibilities have been already or are being exploited. The authors discuss a few of them, in the light of the expectations derived from atmospheric turbulence theory.

036.224   Surface photometry with the Space Telescope.
F. D. Macchetto.
New aspects of galaxy photometry, p. 49 (1985). Abstract. – See Abstr. 012.114.

036.225   Comparison of photographic and CCD surface photometry of galaxies.
L. Vigroux, J.–L. Nieto.
New aspects of galaxy photometry, p. 67 – 72 (1985). – See Abstr. 012.114.

All classical works on galaxy surface photometry have been performed with photographic plates. During the seventies, several other panoramic detectors such as electronographic cameras or intensified televisions have been used. They never challenged, in this particular field of surface photometry, the photographic plates. The appearance of CCDs in late seventies has reversed the situation and despite their small size, they are ideally suited for surface photometry. In this paper the authors review the good and bad sides of plates and CCDs for surface photometry, and compare the results obtained with the two methods for selected galaxies.

036.226   Photographic photometry of 16000 galaxies on ESO blue and red survey plates.
A. Lauberts, E. A. Valentijn.
New aspects of galaxy photometry, p. 73 – 78 (1985). – See Abstr. 012.114.

A brief description of the automated photometry of galaxies in the ESO/Uppsala survey of southern galaxies and an example for the automated parameter extraction are presented.

036.227   Some methods for photometry of galaxies in clusters.
A. Bijaoui, G. Mars, E. Slezak, O. Lefèvre.
New aspects of galaxy photometry, p. 297 – 298 (1985). – See Abstr. 012.114.

The authors have developed digital methods to process photometric observations of clusters of galaxies on photographic and electronographic emulsions. These methods allow to obtain global magnitudes and radial profiles of all detectable objects.

Array signal processing.
See Abstr. 003.103.

Methods of extra–atmospheric astronomy.
See Abstr. 003.143.

Television astronomy.
See Abstr. 003.147.

Indirect imaging: measurement and processing for indirect imaging.
See Abstr. 003.160.

Bessel and librations of the Moon.
See Abstr. 004.033.

The almost serendipitous discovery of self–calibration.
See Abstr. 004.187.

Über den Einfluß der Biegung auf die Meridiankreisbeobachtung von Bessel, Gauss und Soldner.
See Abstr. 004.203.

The astrometry programme at Mt. John Observatory.
See Abstr. 013.028.

On period determination methods.
See Abstr. 021.001.

Statistical analysis of astronomical data containing upper bounds: general methods and examples drawn from X–ray astronomy.
See Abstr. 021.016.

Statistical methods for astronomical data with upper limits. I. Univariate distributions.
See Abstr. 021.017.

The selection of optimum smoothness and the estimation of accuracy of an observational data series.
See Abstr. 021.023.

STRAY, an interactive computer program for the calculation of stray radiation in infared telescopes.
See Abstr. 031.019.

Transportable Zenitkameras: Reduktion, Konstruktion, Anwendung.
See Abstr. 032.023.

Graduation–error determination of the Tokyo photoelectric meridian circle.
See Abstr. 032.024.

Distributing computing for astronomical data acquisition at McDonald Observatory.
See Abstr. 032.043.

Report of IAU Commission 9: Instruments and techniques (*Instruments et techniques*).
See Abstr. 032.046.

The astrometric properties of the Schmidt telescope of the Beijing Astronomical Observatory.
See Abstr. 032.049.

High resolution position and velocity measurement with incremental encoders.
See Abstr. 033.028.

Spatial filtering radio astronomical data: one–dimensional case.
See Abstr. 033.048.

Interferometer measurements of atmospheric phase noise at 86 GHz.
See Abstr. 033.050.

Coherence limits in VLBI observations at 3–millimeter wavelength.
See Abstr. 033.051.

Simultation of a radiotelescope with the aid of the block orientated simulation language ISRSIM.
See Abstr. 033.054.

Correlator and data processing requirements for Quasat.
See Abstr. 033.061.

Comparative performances of an objective grating concept in UV spectroscopic observations of solar system objects. I. Imaging capabilities.
See Abstr. 034.001.

Mode locked train YAG laser calibration experiment.
See Abstr. 034.015.

Improving the SLR measurement accuracy by determination of the pulse center of mass.
See Abstr. 034.016.

Electronic detectors and computer–assisted spectral classification.
See Abstr. 034.035.

On the choice of the method of investigation of errors of divided circle diameters.
See Abstr. 034.059.

On accidental and systematic errors of run determination.
See Abstr. 034.060.

Speckle imaging with the PAPA detector.
See Abstr. 034.093.

Double grating prisms and their use for absolute radial velocity measurements of astronomical objects.
See Abstr. 034.123.

Application of the method of unsharp masking in astrophotography.
See Abstr. 034.174.

Report of IAU Commission 24: Photographic astrometry (*Astrométrie photographique*).
See Abstr. 041.051.

About the error analysis of prime vertical–meridian observations.
See Abstr. 041.056.

A differential method for latitude determination by observations in the first vertical.
See Abstr. 044.036.

Peculiarities of a study of the earth's rotation with radio astrometrical methods.
See Abstr. 044.097.

Latest instruments and methods in the astrogeodetic field.
See Abstr. 045.023.

High–resolution spectroscopy of active regions. 1. Observing procedures.
See Abstr. 072.005.

Optimized response functions for two–dimensional observations of solar oscillations.
See Abstr. 080.045.

The effects of a nearly 100% duty cycle on observations of solar oscillations.
See Abstr. 080.048.

Analysis and interpretation of synthetic time strings of oscillation data.
See Abstr. 080.049.

Theory of the two wavelength dispersion method for determination of refractive index for laser observations.
See Abstr. 082.012.

Atmospheric extinction in relation to stellar photometry.
See Abstr. 082.018.

Satellite–earth range measurements. I. Correction of the excess path length due to atmospheric water vapour by ground based microwave radiometry.
See Abstr. 082.048.

Programs for determination of the coordinates of planetary surface features from photos made with space instrumentation.
See Abstr. 091.043.

The heliometer as a tool for the measure of the Moon.
See Abstr. 094.019.

Sternbedeckungen durch den Mond und ihre astrophysikalische Bedeutung.
See Abstr. 096.015.

Les astronomes amateurs et la campagne "PHEMU85".
See Abstr. 099.021.

On a method for determining the sensitivity of a radar station during meteor observations.
See Abstr. 104.021.

Geminiden 1984. A telescopic ZHR (*Zenit Hourly Rate*).
See Abstr. 104.054.

Report of IAU Commission 25: Stellar photometry and polarimetry (*Photométrie et polarimétrie stellaires*).
See Abstr. 113.070.

Stellar ultraviolet flux distributions in the 912 – 1200 Å wavelength range.
See Abstr. 114.014.

Comparison of observations of some stars in the infrared region with catalogue data.
See Abstr. 114.121.

R 136a and the central object in the giant H II region NGC 3603 resolved by holographic speckle interferometry.
See Abstr. 115.016.

Development and investigation of an algorithm of determination of Ap stars magnetic field geometry by means of solving the inverse problem.
See Abstr. 116.003.

On the determination of stellar rotation and differential rotation from chromospheric activity data.
See Abstr. 116.039.

Images of accretion discs – I. The eclipse mapping method.
See Abstr. 117.063.

More on the $\lambda 2800$ Å 'interstellar extinction' feature.
See Abstr. 131.125.

Spectral indices.
See Abstr. 141.029.

Diffuse component of gamma–ray radiation with $E_\gamma \geqslant 5\,\mathrm{MeV}$ measured by high altitude balloons.
See Abstr. 143.083.

Methodical principles for studying cosmic ray variations.
See Abstr. 144.431.

New EAS project to study primary composition at $\geqslant 10^{15}\mathrm{eV}$.
See Abstr. 144.439.

Accuracy and improvements in galaxy photometry: why and how.
See Abstr. 157.254.

Galaxy photometry in the infrared.
See Abstr. 157.256.

On the extendedness of faint ultraviolet excess quasar candidates.
See Abstr. 159.034.

Quasars on plates taken with a Schmidt telescope.
See Abstr. 159.073.

Positional Astronomy, Celestial Mechanics

041 Astrometry

041.001 General precession computed by numerical integration of ordinary differential equations.
T. B. Andersen.
Astron. Nachr., Vol. 306, No. 1, p. 29 – 34 (1985).
It is suggested that the effects of change in the celestial coordinates due to general precession be computed by solving numerically the differential equations for the instantaneous rates of change in right ascension and declination. An approximate solution in closed form is given. Similar differential equations may be solved for the complete rotation matrix, requiring only the knowledge of the variations of the two functions m and n with time. This suggests the existence of a relation between the classical precessional elements z, ζ_0 and Θ, and this relation is derived. Some numerical examples are also presented.

041.002 Combined determination of the instrumental system and corrections for stellar coordinates.
P. F. Lazorenko.
Kinematika Fiz. Nebesn. Tel, Tom 1, No. 1, p. 5 – 10 (1985). In Russian.
Some procedures are considered for compiling absolute and differential catalogues using the least–squares method for determination of all unknown parameters.

041.003 Consideration of the magnitude equation using exposures of different duration. II. Investigation of the magnitude equation for the astrograph (40/550) of the Main Astronomical Observatory of the Ukrainian Academy of Sciences.
G. A. Ivanov, A. B. Onegina, A. I. Yatsenko.
Kinematika Fiz. Nebesn. Tel, Tom 1, No. 1, p. 11 – 17 (1985). In Russian.
Investigation of the magnitude equation using the catalogue of reference stars has shown that photography with two exposures of different duration permitted to determine the effect of this error on the star position. Magnitude errors are $+0''05$ and $-0''03$ per stellar magnitude along the coordinates x and y, respectively.

041.004 Allowance for instrumental errors in meridian micrometric measurements in the zenith.
V. K. Budz'ko.
Kinematika Fiz. Nebesn. Tel, Tom 1, No. 1, p. 33 – 40 (1985). In Russian.
The effect of instrumental errors on observations of zenith stars by the Talcott method is analysed. It is recommended to register the time of star transits through the lateral thread in the two positions of the instrument.

041.005 Pulsar astrometry via VLBI.
N. Bartel, M. I. Ratner, I. I. Shapiro, R. J. Cappallo, A. E. E. Rogers, A. R. Whitney.
Astron. J., Vol. 90, No. 2, p. 318 – 325 (1985).
The authors demonstrate the capability of the Mark III VLBI system for pulsar astrometry. Their observations of the pulsars PSR 0329 + 54 and PSR 1133 + 16 yielded their respective celestial coordinates (J2000.0) for the epoch 1981.21 in the reference frame determined by VLBI observations of compact extragalactic radio sources: $\alpha = 03^h32^m59^s3484 \pm 0^s0005$, $\delta = 54°34'43''663 \pm 0''005$ for PSR 0329 + 54; and $\alpha = 11^h36^m03^s2756 \pm 0^s0021$, $\delta = 15°51'02''65 \pm 0''23$ for PSR 1133 + 16. The uncertainties represent the root–sum–square of the statistical standard deviations and the errors associated with the authors' estimates of the uncertainties of the values assumed for other parameters related to atmospheric delays, Earth orientation, and source structure.

041.006 Remarks on cometary astrometry.
B. G. Marsden.
Cometary astrometry, p. 27 – 33 (1984). – See Abstr. 012.024.
A statistical examination of the astrometric observations made during the past 20 years indicates that there are generally sufficient data on bright and well–publicized comets. On the other hand, astrometric coverage of fainter objects has recently become a severe problem. In spite of an increase in the number of participating observatories, fewer observatories are making significant contributions. Automated methods have generally improved the quality of observations.

041.007 Formulae for calculating the corrections for the light deflection to apparent places.
D.–z. Xian.
Publ. Purple Mt. Obs., Vol. 3, No. 3, p. 18 – 21 (1984). In Chinese.
In the Chinese Astronomical Ephemeris for the year 1984 (and 1985), all the apparent places do not include the effect of the light deflection due to solar gravitation. They should be corrected for this effect if more precise coordinates are required. Formulae which apply to both planets and stars are given in this paper for calculating the increments to be added to right ascension and declination.

041.008 A determination of precise heliographic coordinates at the Kislovodsk Station of the Pulkovo Observatory. I.
Yu. A. Nagovitsyn, E. Yu. Nagovitsyna.
Soln. Dannye, Byull., 1984, No. 11, p. 76 – 81 (1985). In Russian.
The errors in measurements of heliographic coordinates are considered. A method of correcting the coordinates for the position angle of the references thread is proposed.

041.009 A determination of precise heliographic coordinates at the Kislovodsk Mountain Station of the Pulkovo Observatory. II.
Yu. A. Nagovitsyn, E. Yu. Nagovitsyna.
Soln. Dannye, Byull., 1984, No. 12, p. 54 – 59 (1985). In Russian.
The algorithm "Helicor–5" for determination of precise heliographic coordinates is described.

041.010 Experience of studying the systematic errors of star catalogues from the analysis of photographic observations of selected minor planets.
S. P. Major.
Kinematika Fiz. Nebesn. Tel, Tom 1, No. 2, p. 3 – 8 (1985). In Russian.
424 observations of selected minor planets have been reduced to the systems of Yale, SAO and AGK3 catalogues. The mean $O-C$ residuals are tabulated for small rectangular fields. Systematic errors of these catalogues are estimated attributing the $O-C$ residuals to errors in the star positions only. A comparison is made with the Yale corrections determined by Pierce.

041.011   Experience of using orthogonal polynomials for establishing the relation between measured and ideal coordinates.
A. N. Kur'yanova, L. N. Kizyun.
Kinematika Fiz. Nebesn. Tel, Tom 1, No. 2, p. 9 – 14 (1985). In Russian.

A procedure of using orthogonal polynomials to establish the dependence between measured and ideal coordinates is briefly discussed. The method is applied to reduction of astronomical negatives. It is shown that the coefficient for preliminary accounting for the distortion in the measured coordinates may be estimated from the analysis of cubic polynomial terms.

041.012   Arcsecond positions for milliarcsecond VLBI nuclei of extragalactic radio sources. III. 74 sources.
D. D. Morabito, A. E. Wehrle, R. A. Preston, R. P. Linfield, M. A. Slade, J. Faulkner, D. L. Jauncey.
Astron. J., Vol. 90, No. 4, p. 590 – 594 (1985).

VLBI measurements at 2290 and 8420 MHz on baselines of 10^4 km have been used to determine the positions of milliarcsecond nuclei in 74 extragalactic radio sources. Estimated accuracies range from $0''.1$ to $4''.3$ in both right ascension and declination with typical accuracies of $\sim 0''.3$. The observed sources are part of an all–sky VLBI catalog of milliarcsecond radio sources. Arcsecond positions have now been determined for 819 sources. These positions are presently being used to identify optical counterparts in the southern hemisphere.

041.013   The determination of the FK4 orientation from Washington observations of the sun and planets.
M. L. Sveshnikov.
Tr. Inst. Teor. Astron., Leningrad, Vyp. 19, p. 31 – 74 (1985). In Russian.

The paper deals with the theoretical analysis of the accuracy of determination of catalogue orientation from planet observations as well as the determination of the FK4 equinox and the equator by observations of the sun, Mercury and Venus (Washington, 1911 – 1971). New formulae have been suggested for the phase corrections in the planet observations. The ephemeris positions are computed using the analytical theories by S. Newcomb and G. A. Krasinsky (AT-1; for the sun). The value of velocity of the equinox motion ($+ 1''05 \pm 0''54$ per century) has been obtained by the simultaneous treatment of all observations, which is in agreement with the value of the empirical E–term ($+ 1''23 \pm 0''16$ per century) derived on the basis of proper motions of the FK4 stars. It has been shown that one of the possible causes of this motion is a methodical error of determination of the equinox established from short sets of planet observations. As an example the system of the proper motions of the stars is constructed which is independent from the effect of the E–term. The results of the analysis of the modern Washington observations are in accordance with the ones obtained from treatment of T. Hornsby's planet observations (1774 – 1798).

041.014   Declinations of the sun and major planets obtained from observations with the Wanschaff vertical circle at the Goloseevo Observatory in 1977 – 1979.
P. F. Lazorenko, N. F. Minyajlo, E. M. Nenakhova, A. S. Kharin.
Glav. astron. obs. Akad. Nauk USSR. Kiev, 17 pp. (1984). In Russian. Abstr. in Ref. Zh., 51. Astron., 1.51.164 (1985).

041.015   Relativistic astrometry.
M. Soffel, J. Schastok, H. Ruder, M. Schneider.
Astrophys. Space Sci., Vol. 110, No. 1, p. 95 – 101 (1985). – See Abstr. 012.039.

Concepts of relativistic astrometry – such as Weyl's stellar compass or the concept of 'flat–space plus forces' – are discussed. To visualize effects from light deflection pictures showing the stellar sky as seen from the vicinity of a strongly gravitating source are presented.

041.016   Radiointerferometric astrometry and general theory of relativity.
R. K. Kadyev.
Investigations of nuclear physics and physics of elementary particles, p. 20 – 24 (1984). In Russian. Abstr. in Ref. Zh., 51. Astron., 3.51.119 (1985). – See Abstr. 003.016.

041.017   Graphical dependences of differential variations of the horizontal coordinates of heavenly bodies.
N. S. Yusupov.
Kazan. inzh.–stroit. inst. Kazan', 18 pp. (1984). In Russian. Abstr. in Ref. Zh., 51. Astron., 3.51.138 (1985).

041.018   The calculation of apparent positions of stars.
W. Major.
Mitt. Lohrmann–Obs. Tech. Univ. Dresden, Nr. 51, p. 62 – 64 (1984). = Mitt. Zentralinst. Phys. Erde, Nr. 1287. In German. – See Abstr. 012.058.

Contents: Umwandlung der B1950,0 mittleren Sternörter und Eigenbewegungen auf das Äquinox und die Epoche J2000,0; Berechnung scheinbarer Sternörter; Berücksichtigung der Lichtstrahlkrümmung im Gravitationsfeld der Sonne; Die jährliche Aberration unter Beachtung der speziellen Relativitätstheorie.

041.019   Observations of minor planets at the Dresden Lohrmann Observatory.
E. Asenjo.
Mitt. Lohrmann–Obs. Tech. Univ. Dresden, Nr. 51, p. 77 – 79 (1984). In German. – See Abstr. 012.058.

Von 1949 bis 1974 wurden mehr als 21000 photographische Beobachtungen von zehn durch das Institut für Theoretische Astronomie (ITA) der AdW der UdSSR in Leningrad ausgewählten kleinen Planeten in verschiedenen Sternwarten der Welt durchgeführt. Eine Tabelle zeigt eine Zusammenstellung der Ergebnisse zur Bestimmung der Verbesserung von Äquinoktium und Äquator des FK4.

041.020   Photoelectric meridian circles for planetary theory and stellar dynamics.
M. Miyamoto.
Astron. Her., Vol. 78, No. 5, p. 120 – 127 (1985). In Japanese.

041.021   On the error equations for determining the corrections to the zero points of a star catalogue.
Y.–f. Xia, B.–x. Xu, C.–z. Zhang.
Publ. Shaanxi Astron. Obs., Vol. 7, No. 1, p. 1 – 8 (1984). In Chinese.

041.022   Das Radialgeschwindigkeitssystem der Fundamentalsterne.
H.–P. Frantzen.
Mitt. Astron. Ges., Nr. 63, p. 177 (1985). – See Abstr. 012.063.

041.023   Berichtigung: Rotationsmatrizen aus der alten Fundamentalepoche B1950.0 [Mitt. Astron. Ges., Nr. 62, p. 154 (1984)].
N. P. Wieth–Knudsen.
Mitt. Astron. Ges., Nr. 63, p. 206 (1985). See Abstr. 38.041.043.

041.024   Astrometry using the CCD system of the Spacewatch Camera.
J. V. Scotti, J. E. Frecker, T. Gehrels, R. S. McMillan.
Bull. Am. Astron. Soc., Vol. 16, No. 4, p. 902 (1984). Abstract. – See Abstr. 010.062.

041.025   The effect of astrometric modeling errors of Schmidt plates on the construction of the Guide Star Catalog for Space Telescope.
J. L. Russell, B. McLean.
Bull. Am. Astron. Soc., Vol. 16, No. 4, p. 986 (1984). Abstract. – See Abstr. 010.062.

041.026 Analysis of some methods of determination of errors of a reference catalogue from meridian observations.
I. A. Strelkova, S. A. Tolchel'nikova–Murri.
Vestn. Leningr. Univ., Mat. Mekh. Astron., No. 1, p. 95 – 102 (1985). In Russian. Abstr. in Ref. Zh., 51. Astron., 6.51.97 (1985).

041.027 Main reductions of heliometrical observations.
V. S. Borovskikh, Yu. A. Nefed'ev.
Kazan. inzh.–stroit. inst. Kazan', 10 pp. (1985). In Russian. Abstr. in Ref. Zh., 51. Astron., 6.51.103 (1985).

041.028 The equinox correction, based on current observations of the sun, Venus, and Mars.
A. S. Kharin, N. F. Minyajlo, Yu. I. Safronov.
Sov. Astron. Lett., Vol. 10, No. 6, p. 392 – 394 (1984). English translation of 38.041.049.

041.029 On the problem of photographic irradiation.
I. I. Brejdo.
Izv. Glav. Astron. Obs. Pulkovo, Astrometr. Astrofiz., No. 201, p. 122 – 126 (1985). In Russian.

It is proved that photographic irradiation in astrophotographic observations is most frequently determined not by light scattering by the photographic layer, but by peculiarities of the actual distribution of light in the optical image of a star which is due to the optics and atmospheric conditions.

041.030 Weighting of phases and grid coordinates.
J. Kovalevsky, M. Froeschlé, J. L. Falin, F. Mignard.
Processing of scientific data from the ESA astrometry satellite HIPPARCOS, p. 119 – 123 (1985). – See Abstr. 012.087.

Two different weighting factors may be proposed to define the mean phase that is used to determine the grid coordinates of a star. They give to a good approximation the same results for a single star. But for a double star, they may differ by an amount much larger than the expected error and none of them corresponds to a physical point in a double star system. It is concluded that there is no overwhelming argument in favour of any of the definitions and it is proposed that the simplest be used, namely: $\bar{\Psi} = C_1\psi_1 + C_2\psi_2$ where C_1 and C_2 are well chosen numerical values, constant throughout the mission.

041.031 Great–circle reduction by the Northern Data Analysis Consortium (NDAC).
L. Lindegren, C. Petersen.
Processing of scientific data from the ESA astrometry satellite HIPPARCOS, p. 183 – 191 (1985). – See Abstr. 012.087.

In a 12 hour interval, the HIPPARCOS satellite will observe some 1800 stars in a circular strip around the sky with a maximum width of 3°. The great circle reduction will produce a single normal point, called abscissa, for each star in the interval. A least–squares solution takes as input the instantaneous object coordinates on the modulating grid and determines, in addition to the abscissae, the precise attitude motion in the scanning direction and the grid–to–field coordinate transformation. This paper describes the basic observation equations, solution method, and data management system used by NDAC.

041.032 The current state of the reduction on circles.
D. T. van Daalen.
Processing of scientific data from the ESA astrometry satellite HIPPARCOS, p. 193 – 196 (1985). – See Abstr. 012.087.

This paper surveys the progress made in the FAST consortium on the reduction on great circles since the Asiago Thinkshop. Results of studies on the following topics are presented: apparent and geometric positions; attitude representations; grid coordinate and field coordinates; smoothing and adjustment organization; variance computation; reordering of star unknowns. Then the results of a real scale simulation experiment are shown.

041.033 Star abscissae improvement by smoothing of attitude data.
H. van der Marel.
Processing of scientific data from the ESA astrometry satellite HIPPARCOS, p. 197 – 208 (1985). – See Abstr. 012.087.

Smoothing of the along scan attitude during reduction on circles improves the precision of the star abscissae. In this paper formulae for the updating of the star abscissae are given and the results of some simulation experiments are shown. Specifically the author discusses numerical smoothing with spline functions. Simulation experiments showed that the standard deviation of stars brighter than magnitude 7 is improved by a factor 1.45. Also a number of simulation experiments with different observation strategies has been carried out.

041.034 Large scale calibration during reduction on circles.
H. van der Marel.
Processing of scientific data from the ESA astrometry satellite HIPPARCOS, p. 209 – 219 (1985). – See Abstr. 012.087.

An important byproduct of reduction on circles is the large scale instrumental calibration. The mathematical model of the instrumental perturbation and the estimability of instrumental coefficients is discussed. It is shown from a simulated reduction on circles that most components of the large scale deformation and chromaticity are very well estimable, without a perceptible decrease in the precision of the star abscissae. Some attention is given to the solution methods, and their optimization, for the instrumental parameters.

041.035 Grid step inconsistency correction during reduction on circles.
F. A. van der Heuvel, D. T. van Daalen.
Processing of scientific data from the ESA astrometry satellite HIPPARCOS, p. 221 – 228 (1985). – See Abstr. 012.087.

The angles measured by the HIPPARCOS main grid are determined up to an integer number of grid periods, to be estimated on the basis of a priori attitude and star positions. Observations with a wrongly estimated grid step number can give rise to a grid step inconsistency which should be removed during reduction on circles. The nature and the distribution of grid step inconsistencies are discussed. Two methods to correct inconsistencies – pre– and post–adjustment – are discussed, and simulation results are presented.

041.036 Continuous experiments on the great circle reduction.
T. Tommasini Montanari, B. Bucciarelli, M. Lattanzi.
Processing of scientific data from the ESA astrometry satellite HIPPARCOS, p. 229 – 235 (1985). – See Abstr. 012.087.

The authors present a detailed analysis of the application of a conjugate gradient type method to the least squares problem arising from the great circle reduction. It is a large scale problem involving many thousands of unknowns. Particular attention is given to the preconditioning technique and to the quality of the statistical informations of this method.

041.037 Sphere reconstitution by the Northern Data Analysis Consortium (NDAC).
L. Lindegren, S. Söderhjelm.
Processing of scientific data from the ESA astrometry satellite HIPPARCOS, p. 237 – 242 (1985). – See Abstr. 012.087.

The sphere reconstitution or "global solution" combines the intermediate quantities (abscissae) derived from HIPPARCOS data, to obtain the five astrometric parameters for all virtually single stars (perhaps 85 – 90,000 out of the 100,000 programme stars). The solution allows one abscissa zero–point correction per reference great circle and a few tens of "global parameters" representing large–scale distortions of the system of positions e.g. from thermal/instrumental effects and optical chromaticity. However, terms for the gravitational light deflection in a generalized metric for the solar system may also be included.

041.038   Two different methods for the solution of the large least squares problem in the sphere reconstitution.
T. Tommasini Montanari, B. Bucciarelli, C. Tramontin.
Processing of scientific data from the ESA astrometry satellite HIPPARCOS, p. 243 – 254 (1985). – See Abstr. 012.087.

The authors propose and analyse two different algorithms for the solution of the large scale least squares problem in the sphere reconstitution for the HIPPARCOS data processing scheme. In order to measure their computational complexity, particular attention is given to the storage requirement and the computing time of the two methods. Some alternative algorithms are given to compute the variance–covariance matrix in the direct approach. Results have been obtained until a problem involving ten thousand stars.

041.039   A continuous model for the arcwise sphere reconstitution.
B. Betti, F. Sansó.
Processing of scientific data from the ESA astrometry satellite HIPPARCOS, p. 255 – 262 (1985). – See Abstr. 012.087.

The aim of this work is to derive some approximate information on the covariance matrix of the star's unknown parameters. The authors use a continuous model of the least square principle.

041.040   Experiments with the arcwise sphere reconstitution.
B. Betti, L. Mussio, F. Sansó.
Processing of scientific data from the ESA astrometry satellite HIPPARCOS, p. 263 – 279 (1985). – See Abstr. 012.087.

The proposal of the arcwise sphere reconstitution in the HIPPARCOS Project is discussed in comparison with the three steps approach. The feasibility of this proposal is studied by suitable, simplified numerical tests. A new approach to the sphere reconstitution is presented.

041.041   Astrometric parameter determination, methods, algorithms and program implementation.
H. G. Walter, R. Hering, U. Bastian, H. H. Bernstein.
Processing of scientific data from the ESA astrometry satellite HIPPARCOS, p. 281 – 290 (1985). – See Abstr. 012.087.

The objectives of the task astrometric parameter determination are explained with due consideration of the interfaces with preceding sections of the data reduction chain. In the case of routine treatment of star abscissae the observation equations are set up, and the methods and algorithms employed for parameter adjustment and variance–covariance estimation of the unknowns are presented. The numerical performance of the processor is discussed in terms of accuracy and computer time. Specific processors are likely to become indispensable for double stars and stars of irregular behaviour. The steps for refined analysis and result verification are briefly outlined.

041.042   Numerical results on astrometric parameters and accuracy assessment.
H. H. Bernstein.
Processing of scientific data from the ESA astrometry satellite HIPPARCOS, p. 291 – 294 (1985). – See Abstr. 012.087.

On the basis of simulated coordinates referred to reference great circles for a given star, the program for astrometric parameter adjustment was tested. Some typical results on accuracies and correlation coefficients are presented.

041.043   Treatment of the grid step ambiguity problem within task 6000.
U. Bastian.
Processing of scientific data from the ESA astrometry satellite HIPPARCOS, p. 295 – 298 (1985). – See Abstr. 012.087.

A very simple and fast method of solution for the grid step ambiguity problem within the astrometric parameter determination task is described and test results with simulated HIPPARCOS measurements are presented.

041.044   Two remarks about the final reduction.
F. Mignard, M. Froeschlé, J. L. Falin.
Processing of scientific data from the ESA astrometry satellite HIPPARCOS, p. 299 – 307 (1985). – See Abstr. 012.087.

In the first part of the paper the authors stress the need of selecting the epoch of the HIPPARCOS catalogue as close as possible of the mean epoch of observation of each star so as to avoid correlations between the coordinate of position and the proper motion. In the second part, the authors investigate the influence of the coordinate system used in the reduction on the covariance matrix. They show that right ascension and declination are correlated variables whereas longitude and latitude are not.

041.045   Generation of geometric and apparent star positions.
H. G. Walter.
Processing of scientific data from the ESA astrometry satellite HIPPARCOS, p. 323 – 328 (1985). – See Abstr. 012.087.

In the framework of the HIPPARCOS data reduction the meaning of geometric and apparent positions is defined, and those effects are treated which contribute to the transformation from mean to apparent positions with more than 0.2 milliseconds of arc. The formulae accounting for these effects are implemented on a computer and the numerical performance is discussed.

041.046   Apparent places of stars as seen by the HIPPARCOS satellite.
V. Catullo.
Processing of scientific data from the ESA astrometry satellite HIPPARCOS, p. 329 – 334 (1985). – See Abstr. 012.087.

A vectorial formulation is exploited in the transformation from catalogue coordinates of stars to their apparent places (APS), in accordance with IAU resolutions. A software program has been developed which provides a relative accuracy of $0\rlap{.}''001$ in APS prediction, referred to both a reference great circle and an osculating great circle.

041.047   General Catalogue of positions and proper motions of 4949 geodetic stars from $+90°$ to $-90°$.
E. V. Khrutskaya.
Astron. Zh., Tom 62, Vyp. 3, p. 605 – 607 (1985). In Russian. English translation in Sov. Astron., Vol. 29, No. 3.

In 1983 at the Pulkovo Observatory was completed the compilation of the general catalogue of positions and proper motions of 4949 geodetic stars from $+90°$ to $-90°$. The obtained proper motions permitted to determine the centennial precessional corrections $\Delta p = 1\rlap{.}''10 \pm 0\rlap{.}''07$, $\Delta E = 1\rlap{.}''21 \pm 0\rlap{.}''07$, Oort's constants $P = A/4{,}74 = +0\rlap{.}''34 \pm 0\rlap{.}''03$, $Q = B/4{,}74 = -0\rlap{.}''20 \pm 0\rlap{.}''02$, and the coordinates of the solar apex $A_\odot = 271\rlap{.}°2 \pm 3\rlap{.}°8$. $D_\odot = 33\rlap{.}°8 \pm 3\rlap{.}°2$ on the basis of new material on the nearest stars.

041.048   Effects of source structure on position determinations made with VLBI observations.
J. S. Ulvestad, J. B. Thomas, F. R. Bletzacker.
Bull. Am. Astron. Soc., Vol. 17, No. 2, p. 550 (1985). Abstract. – See Abstr. 010.065.

041.049   Very long baseline astrometry.
N. Drake, W. R. Kubinec, R. J. Dukes Jr.
Bull. Am. Astron. Soc., Vol. 17, No. 2, p. 550 (1985). Abstract. – See Abstr. 010.065.

041.050   Report of IAU Commission 8: Positional Astronomy (*Astronomie de position*).
J. A. Hughes.
Trans. IAU, Vol. XIXA, p. 29 – 40 (1985). – See Abstr. 003.046.

041.051   Report of IAU Commission 24: Photographic astrometry (*Astrométrie photographique*).
W. Gliese.
Trans. IAU, Vol. XIXA, p. 253 – 258 (1985). – See Abstr. 003.046.

041.052 Epoch B1950.0 or J2000.0?
J. Bouška.
Říše hvězd, Vol. 65, No. 7, p. 133 – 138 (1984). In Czech.

041.053 Perspectives of present–day astrometry.
V. V. Podobed, A. P. Gulyaev.
Main directions of astronomical investigations at the Moscow University, p. 121 – 129 (1985). In Russian. – See Abstr. 003.049.

041.054 The problem of compilation of the FK5 Fundamental Catalogue.
M. S. Zverev.
Main directions of astronomical investigations at the Moscow University, p. 130 – 148 (1985). In Russian. – See Abstr. 003.049.

041.055 A comparison of the AGK3 with the $NPZT_{74}$ catalog.
C.-l. Lu, G.-c. Shi.
Acta Astron. Sin., Vol. 25, No. 4, p. 426 – 430 (1984). In Chinese.

Systematic differences in relation to right ascension, declination, magnitude and spectral type between the two catalogs are presented. The effects of errors of AGK3 on photographic determination of positions of extragalactic radio sources are discussed. The conclusion is that it is necessary to establish a homogeneous, precise faint reference stars system.

041.056 About the error analysis of prime vertical–meridian observations.
B. Tang.
Acta Astron. Sin., Vol. 26, No. 1, p. 35 – 41 (1985). In Chinese.

This paper analyses and discusses the influence of instrumental errors of prime vertical–meridian observations by which absolute azimuth a and φ, the latitude, can be determined. It is obvious that the accuracy of a single observation for the azimuth a is much higher than that of φ. In order to get a highly accurate and independent declination system as early as possible, the paper presents a method, that is used to determine absolutely the stars as fundamental stars in order to determine the equator point of graduated circle.

Desired characteristics of catalogs for cometary astrometry.
See Abstr. 002.014.

The IHW reference star catalogs for comets Halley and Giacobini–Zinner.
See Abstr. 002.015.

Catalogue of star positions of the USSR Time Services for the epoch and equinox J2000.0.
See Abstr. 002.032.

Documentation for the machine–readable version of the *Córdoba Durchmusterung (CD)*.
See Abstr. 002.038.

Documentation for the machine–readable version of the *Cape Photographic Durchmusterung (CPD)*.
See Abstr. 002.039.

The new catalogue of Tokyo PZT stars (α_{85}/δ_{85} system).
See Abstr. 002.058.

Progress report on the Input Catalogue preparation.
See Abstr. 002.096.

First improvement of the accuracy of the Input Catalogue star coordinates by star mapper data reduction.
See Abstr. 002.097.

Troisième complément à la bibliographie des publications ayant rapport avec l'astrolabe Danjon ou avec d'autres astrolabes de types nouveaux.
See Abstr. 002.116.

Right ascensions of 194 FKSZ stars reduced for the collimation error using observations of FK4 stars.
See Abstr. 002.119.

Finding charts of radio stars for the HIPPARCOS mission.
See Abstr. 002.131.

Meridian astronomy in the private and university observatories of the United Kingdom: rise and fall.
See Abstr. 004.149.

Greenwich meridian astronomy.
See Abstr. 004.150.

Über den Einfluß der Biegung auf die Meridiankreisbeobachtung von Bessel, Gauss und Soldner.
See Abstr. 004.203.

Astrometry in Moscow in the first 150 years.
See Abstr. 009.040.

Cometary astrometry. Proceedings of a workshop, held at the European Southern Observatory Headquarters, Garching, F.R. Germany, June 18 – 19, 1984.
See Abstr. 012.024.

Astrometry at La Silla.
See Abstr. 013.027.

The astrometry programme at Mt. John Observatory.
See Abstr. 013.028.

The astrometry network of observers in China.
See Abstr. 013.029.

The astrometry network of observers in Japan.
See Abstr. 013.030.

The astrometry network of observers in U.S.S.R.
See Abstr. 013.031.

The Astrometry Network of the IHW and the P/Crommelin trial run results.
See Abstr. 013.036.

The Astrographic Catalogue and the Carte du Ciel.
See Abstr. 013.061.

Exchange of information between FAST and NDAC.
See Abstr. 013.084.

Optical model for the HIPPARCOS telescope in its real expected configuration.
See Abstr. 031.039.

The signal of the HIPPARCOS nominal system.
See Abstr. 031.040.

Progress report on vacuum meridian marks.
See Abstr. 032.018.

The astrometric properties of the Schmidt telescope of the Beijing Astronomical Observatory.
See Abstr. 032.049.

Astrometry of comets using hypersensitized type 2415 film.
See Abstr. 034.026.

HIPPARCOS payload modelling.
See Abstr. 035.078.

Possible optical grid calibration.
See Abstr. 035.080.

Acquisition and track algorithms for the ASTROS star tracker.
See Abstr. 035.123.

Guidung performance of a Quadrant Digicon sensor.
See Abstr. 035.124.

On a possible astrometric method of reducing star plates.
See Abstr. 036.006.

On the block–adjustment reduction of differential meridian observations.
See Abstr. 036.007.

Cometary astrometry with Schmidt cameras.
See Abstr. 036.047.

Astrometric plates obtained at the primary focus of large aperture reflectors.
See Abstr. 036.048.

Kitt Peak measurements of P/Halley positions.
See Abstr. 036.049.

Problems and procedures of cometary astrometry.
See Abstr. 036.050.

Obtaining accurate comet positions despite some inaccurate catalog stars.
See Abstr. 036.051.

Reduction of astrographic catalogues.
See Abstr. 036.052.

Measuring – the quantifying art.
See Abstr. 036.053.

Reduction of astrometric plates.
See Abstr. 036.054.

The method of plate measurements of a photographic catalogue at the Hissar Observatory.
See Abstr. 036.062.

Modern methods in the astrometry of double stars.
See Abstr. 036.076.

The division corrections of a graduated circle. A comparison between Høg's, Benevides–Soares' and Boczko's methods.
See Abstr. 036.079.

Model studies to adjust overlapping plate pairs.
See Abstr. 036.093.

A model for four–fold overlapping of a $6° \times 6°$ sky field using plates taken with the Zeiss wide–angle astrograph.
See Abstr. 036.109.

Guidelines for the reduction of optical astrometry observations during the main campaign in Project MERIT.
See Abstr. 036.113.

Astrometric and photometric estimators for TYCHO photon counts.
See Abstr. 036.137.

Double star analysis strategy.
See Abstr. 036.162.

Double star detection.
See Abstr. 036.163.

Estimation of double star parameters.
See Abstr. 036.164.

Imaging approach to multiple star recognition.
See Abstr. 036.165.

Reception and preparation task.
See Abstr. 036.166.

Architecture de la tâche "Réception et Préparation".
See Abstr. 036.167.

Le logiciel de simulation. Les données de la grille principale.
See Abstr. 036.168.

Observations with the star–mapper.
See Abstr. 036.169.

Simulation for the sphere reconstitution.
See Abstr. 036.170.

A first draft of the simulation task, exploitation scheme.
See Abstr. 036.171.

First–look processing.
See Abstr. 036.172.

Review of the data processing task.
See Abstr. 036.173.

The general structure of the data reduction and its consequences on the exploitation phase.
See Abstr. 036.174.

A first approach of the software requirements relating to the D.M.C.S.
See Abstr. 036.175.

Present estimate of the computer workload.
See Abstr. 036.176.

A preliminary look at astrometric accuracy as a function of photon count.
See Abstr. 036.207.

Astrometria fotografica (con scala graduata).
See Abstr. 036.211.

Linking the HIPPARCOS catalog to the VLBI inertial reference system, high angular resolution structures and VLBI positions of 10 radio stars.
See Abstr. 043.007.

Calculation of declination corrections for Dresden Talcott observations from 1963 to 1982 in comparison with the ITB system.
See Abstr. 044.021.

Project HIPPARCOS.
See Abstr. 051.042.

The Hipparcos satellite's mission: the objectives and their implementation.
See Abstr. 051.070.

HIPPARCOS satellite, project status.
See Abstr. 051.087.

The nominal scanning law and the interruptions by the moon.
See Abstr. 051.088.

Proposals of observations with the Space Telescope in the domain of astrometry.
See Abstr. 051.089.

Space based radio astronomy arrays.
See Abstr. 051.103.

The HIPPARCOS satellite.
See Abstr. 051.113.

Ad astra HIPPARCOS. The European Space Agency's Astrometry Mission.
See Abstr. 051.114.

Attitude reconstitution by the Northern Data Analysis Consortium (NDAC).
See Abstr. 052.061.

Atmospheric effects and selection of places for astrometric stations.
See Abstr. 082.025.

On the sensitive measurement of horizontal temperature gradients of air near an astrometric instrument for correcting anomalous refraction.
See Abstr. 082.050.

Atmospheric refraction effects in time and latitude observations using classical techniques.
See Abstr. 082.088.

A demonstration of the Sun's motion around the barycenter of the Solar System.
See Abstr. 091.059.

Photographic position observations of Venus at the Main Astronomical Observatory of the Academy of Sciences of the Ukrainian SSR in 1980.
See Abstr. 093.017.

Photographic position observations of Venus in 1980.
See Abstr. 093.053.

An analysis of lunar occultations in the years 1955 – 1980 using the new lunar ephemeris ELP 2000.
See Abstr. 094.015.

The modulation curve of a minor planet.
See Abstr. 098.090.

Observations astrométriques et comparaison à la théorie des satellites de Saturne effectuées à l'Observatoire de Bordeaux sur le réfracteur de 38 cm durant les oppositions de 1981 et 1982.
See Abstr. 100.001.

On the orbit improvement results obtained by referring old cometary observations to the SAO Star Catalog.
See Abstr. 102.014.

The work on comet Halley's orbit at ESOC.
See Abstr. 103.913.

U.S. Naval Observatory parallaxes of faint stars. List VII.
See Abstr. 111.005.

Rigorous treatment of the heliocentric motion of stars.
See Abstr. 111.009.

Potential of astrographic catalog star first–epoch positions for proper–motion determination.
See Abstr. 111.020.

HIPPARCOS astrometric binaries.
See Abstr. 118.019.

Statistical models for HIPPARCOS binaries.
See Abstr. 118.021.

Astrometry of millisecond pulsar 1937 + 214 and binary pulsar 1913 + 160.
See Abstr. 126.075.

Submilliarcsecond VLBI observations of the close pair GC 1342 + 662 and GC 1342 + 663.
See Abstr. 159.107.

042 Celestial Mechanics, Figures of Celestial Bodies

042.001 Fission sequences of self–gravitating and rotating fluid with internal motion.
Y. Eriguchi, I. Hachisu.
Astron. Astrophys., Vol. 142, No. 2, p. 256 – 262 (1985).
The authors have computed equilibrium sequences of a self–gravitating and rotating fluid with uniform vorticity. These sequences are extensions of the Riemann sequences because no restrictions to the shape are imposed. Equilibrium solutions bifurcate from the Riemann ellipsoidal sequences into dumb–bell like sequences. The waist of the dumb–bell slims more and more and finally it leads to fission. The authors have also computed rotating binary sequences with uniform vorticity. They smoothly connect to the dumb–bell like sequences for the relatively small vorticity case, but for the large vorticity case the separation of the binary reaches a minimum value and the two fluids cannot approach each other closer than this value. This is the Roche limit for the case where the internal motion is included.

042.002 On the determination of the asymptotic solutions in the restricted elliptical three–body problem.
L. G. Luk'yanov.
Astron. Zh., Tom 62, Vyp. 1, p. 153 – 159 (1985). In Russian. English translation in Sov. Astron., Vol. 29, No. 1.
It is shown that for every collinear libration point there exist two planar one–dimensional families of asymptotic solutions. In the vicinity of a libration point these families completely fill some curvilinear sectors in the plane of special coordinates. With the transition to the restricted circular problem the families degenerate to four well–known trajectories.

042.003 Stability of the triangular libration points of the photogravitational three–body problem.
A. L. Kunitsyn, A. T. Tureshbaev.
Pis'ma Astron. Zh., Tom 11, No. 2, p. 145 – 148 (1985). In Russian. English translation in Sov. Astron. Lett., Vol. 11.
The stability of relative equilibrium positions (the triangular libration points) of a gas–dust cloud particle in the gravitational field of a binary system with light repulsion taken into consideration is investigated. The photogravitational restricted circular three–body problem is accepted as a dynamical model. A simple geometrical interpretation is given for the derived necessary stability conditions.

042.004 On the best mean–square approximations for the gravitational potential of a planet.
N. I. Lobkova.
Pis'ma Astron. Zh., Tom 11, No. 2, p. 149 – 155 (1985). In Russian. English translation in Sov. Astron. Lett., Vol. 11.
The continuous problem of approximating the gravitational potential of a planet in the form of polynomials of solid spherical functions is considered. The best mean–square polynomials, referred to different parts of space, are compared with each other.

The harmonic coefficients corresponding to the surface of a planet are shown to be unstable with respect to the degree of the polynomial and to differ from the Stokes constants.

042.005   Regular motions of a satellite and some small effects in the motion of the moon and Phobos.
Yu. V. Barkin.
Kosm. Issled., Tom 23, Vyp. 1, p. 26 – 36 (1985). In Russian. English translation in Cosm. Res.

042.006   Accurate methods in general planetary theory.
J. Laskar.
Astron. Astrophys., Vol. 144, No. 1, p. 133 – 146 (1985).

The author develops a formalism and an algorithm which allow to compute with great relative precision (10^{-6}) the autonomous system which regulates the very long period ($> 50,000$ yr) variations of the 8 planets' orbital elements, up to degree 5 in eccentricity–inclination and up to 2nd order with respect to the masses. In particular, the author studies the convergence of the series giving the coefficients of this system. The computations are optimized thanks to a particular organization of the series which enables to establish beforehand a unique address table for all the products to be made in the computations of the second order. The author also presents here the first results he obtained by the application of the algorithm in the computation of a Lagrange solution up to the second order with respect to the masses.

042.007   The orbital evolution of Pluto–like objects.
D. I. Steel.
Observatory, Vol. 105, No. 1065, p. 40 – 42 (1985).

042.008   The orbits of comets Halley and Giacobini–Zinner.
D. K. Yeomans.
Cometary astrometry, p. 167 – 175 (1984). – See Abstr. 012.024.

The ongoing activities to improve the orbits and ephemerides of comets Halley and Giacobini–Zinner are outlined. The non-gravitational acceleration model of Marsden et al. (1973) is used to represent the motion of these two comets over their observed intervals, and recent orbital updates are presented.

042.009   Universal time, lunar tidal deceleration and relativistic effects from observations of transits, eclipses and occultations in the XVIIIth – XXth centuries.
G. A. Krasinsky (*G. A. Krasinskij*), E. Y. Saramonova, M. L. Sveshnikov, E. S. Sveshnikova.
Astron. Astrophys., Vol. 145, No. 1, p. 90 – 96 (1985).

Lunar and planetary observations for the time span 1717 – 1973 (transits of Mercury and Venus, solar eclipses, occultations of Venus, Mars and Aldebaran) are discussed. Two modern planetary and lunar theories are used. It is shown that the adopted system of differences beween the ephemeris (dynamic) and universal time requires the correction $(-12\overset{s}{.}9 \pm 1\overset{s}{.}3)$/cy before 1960. With this new UT time scale the evaluation of the observed secular motions of the perigee and the node of Mercury is made and the consistency with the relativistic values is confirmed within 1″ limits. The results do not support the hypotheses of the secular decrease of the gravitational constant.

042.010   Conservative fields derived from two monoparametric families of planar orbits.
G. Bozis, G. Tsarouhas.
Astron. Astrophys., Vol. 145, No. 1, p. 215 – 220 (1985).

With the aid of Szebehely's partial differential equation for the inverse problem in dynamics, the authors prove that, in general, two preassigned monoparametric families of planar curves $f(x, y) = c_1$ and $g(x, y) = c_2$ are not compatible, i.e., there does not exist a potential field $U = U(x, y)$ which can generate both these families. Only if the given functions $f(x, y)$ and $g(x, y)$ satisfy a certain condition, the families are compatible. The corresponding potential field, as well as the total energies along each orbit, are, in general, determined uniquely, apart from a multiplicative and an additive constant. In some special cases, up to three additional arbitrary constants may enter into the expression for the potential.

042.011 z–solutions of the symmetric restricted spatial three–body problem.
E. A. Grebenikov, A. K. Egizbaeva.
Pis'ma Astron. Zh., Tom 11, No. 3, p. 227 – 230 (1985). In Russian. English translation in Sov. Astron. Lett., Vol. 11.

z–solutions of the symmetric restricted spatial three–body problem are found. These solutions form a two–parametric family of functions. For all real–time values the motion of the non-gravitating point turns out to be along the Oz axis, which is perpendicular to the plane of motion of the two other gravitating masses.

042.012   An efficient method for solving Barker's equation.
R. Meire.
J. Br. Astron. Assoc., Vol. 95, No. 3, p. 113 – 115 (1985).

In a parabolic orbit, the true anomaly v (as a function of the time) can be obtained by solving a cubic equation for tan v/2, the so–called Barker equation. A modified method of solution is described and it is shown that this new form is very efficient from a computational point of view: it needs less program statements, is faster, and it has greater accuracy than the normally used trigonometric solution.

042.013   Review of concepts of stability.
V. Szebehely.
Celest. Mech., Vol. 34, Nos. 1 – 4, p. 49 – 64 (1984). – See Abstr. 012.030.

Concepts of stability, associated nomenclature and names of originators are reviewed emphasizing some global aspects as well as specific applications to dynamics and to celestial mechanics. Details concerning a few fundamental concepts (Hills's, Liapunov's, Poincaré's stability, etc.) are offered. Short definitions and descriptions are also given for about 50 concepts of stability in the form of a dictionary.

042.014   Three–body problem.
C. Marchal, J. Yoshida, Y.–s. Sun.
Celest. Mech., Vol. 34, Nos. 1 – 4, p. 65 – 93 (1984). – See Abstr. 012.030.

In the first part of this paper (Marchal et al., 1985) the authors have analyzed three–body systems satisfying the condition $r \leqslant kR$ where k is a suitable constant, r the mutual distance of the two masses of the "binary" and R the distance between the center of mass of the binary and the "third mass". In this second part the authors look for initial conditions under which the inequality $r \leqslant kR$ will remain forever satisfied and develop the corresponding tests of escape and their applications. This leads to a major improvement of the knowledge of the nature of three-body motions especially in the vicinity of triple close approaches.

042.015   The Lyapunov characteristic exponents – applications to celestial mechanics.
C. Froeschlé.
Celest. Mech., Vol. 34, Nos. 1 – 4, p. 95 – 115 (1984). – See Abstr. 012.030.

After a presentation of Lyapunov characteristic exponents (LCE) the author recalls their basic properties and numerical methods of computation. He reviews some numerical computations which are concerned with LCEs and mainly related to celestial mechanics problems.

042.016   Kolmogorov entropy as a measure of disorder in some non–integrable Hamiltonian systems.
R. Gonczi, C. Froeschlé, C. Froeschlé.
Celest. Mech., Vol. 34, Nos. 1 – 4, p. 117 – 124 (1984). – See Abstr. 012.030.

This paper is devoted to a quantitative study of the stochastic behaviour of some Hamiltonian systems with closed velocity curves. The authors investigate Hamiltonians already studied by Ali and Somorjai. For each energy they compute the Lyapunov characteristic exponents of fifty orbits chosen at random, in order to calculate the Kolmogorov entropy by Pesin's formula. The results are in agreement with those of Ali and Somorjai.

042.017   Generalizations of the Jacobian integral.
V. Szebehely, A. L. Whipple.
Celest. Mech., Vol. 34, Nos. 1 – 4, p. 125 – 133 (1984). – See Abstr. 012.030.

The restricted problem of three bodies is generalized to the restricted problem of $2+n$ bodies. The system is modified so that there are several gravitationally interacting bodies with small masses. Their motions are influenced by the primaries but they do not influence the motions of the primaries. The separate Jacobian integrals of the minor bodies are lost but a conservative (time-independent) Hamiltonian of the system is obtained. For the case of two minor bodies, the five Lagrangian points of the classical problem are generalized and fourteen equilibrium solutions are established. The four linearly stable equilibrium solutions which are the generalizations of the triangular Lagrangian points are once again stable but only for considerably smaller values of the mass parameter of the primaries than in the classical problem.

042.018   Application of Lie–series to regularized problems in celestial mechanics.
A. Hanslmeier.
Celest. Mech., Vol. 34, Nos. 1 – 4, p. 135 – 143 (1984). – See Abstr. 012.030.

In this paper the author has tried to apply the Lie–formalism to the regularized restricted three body problem. It is shown that this algorithm leads to a very simple structure program which is also fast.

042.019   Integration of the elliptic restricted three–body problem with Lie series.
M. Delva.
Celest. Mech., Vol. 34, Nos. 1 – 4, p. 145 – 154 (1984). – See Abstr. 012.030.

The method of Lie series is used to construct a solution for the elliptic restricted three–body problem. In a synodic pulsating coordinate system, the Lie operator for the motion of the third infinitesimal body is derived as function of coordinates, velocities and true anomaly of the primaries. The terms of the Lie series for the solution are then calculated with recurrence formulae which enable a rapid successive calculation of any desired number of terms. This procedure gives a very useful analytical form for the series and allows a quick calculation of the orbit.

042.020   The stability of our solar system.
P. J. Message.
Celest. Mech., Vol. 34, Nos. 1 – 4, p. 155 – 163 (1984). – See Abstr. 012.030.

Most investigations of the stability of the solar system have been concerned with the question as to whether the very long term effect of the gravitational attractions of the planets on each other will be to alter the nearly coplanar, nearly circular nature of the orbits in which they move. Analytical investigations in the traditions of Laplace, Lagrange, Poisson and Poincaré strongly indicate stability. Numerical integration experiments have thrown considerable light on possible types of motions, especially in fictitious solar systems in which the planetary masses have been increased to enhance the perturbations, and in testing how critical are stability boundary estimates given by Hill surface type methods.

042.021   Termes séculaires dans la théorie des perturbations.
T. Inoue.
Celest. Mech., Vol. 34, Nos. 1 – 4, p. 185 – 191 (1984). – See Abstr. 012.030.

The convergence property is examined for the series which appear in the theory of the planetary motions. With another point of view, the author obtains a result different from that already given by Gyldén.

042.022   Amélioration des théories planétaires analytiques.
P. Bretagnon.
Celest. Mech., Vol. 34, Nos. 1 – 4, p. 193 – 201 (1984). – See Abstr. 012.030.

The VSOP82 and TOP82 theories intend to represent the motion of planets, with a satisfactory accuracy, over an interval of 1000 years from and after J2000.0. The precision of the Newtonian part of the solutions for the system of the Sun and eight point masses is given. The author presents the construction of complements in order to keep this accuracy over one thousand years for the real motion: the relativistic perturbations, the perturbations by the minor planets, the perturbations by the Moon. Besides, he has undertaken the improvement of the solutions through lengthening the interval of validity up to six thousand years from and after J2000.0.

042.023   Progress in general planetary theory.
J. Laskar.
Celest. Mech., Vol. 34, Nos. 1 – 4, p. 219 – 221 (1984). Abstract. – See Abstr. 012.030.

042.024   Tidal effects and the motion of a satellite.
J. Kovalevsky.
Celest. Mech., Vol. 34, Nos. 1 – 4, p. 243 – 244 (1984). Abstract. – See Abstr. 012.030.

042.025   Stability of L_4 and L_5 against radiation pressure.
F. Mignard.
Celest. Mech., Vol. 34, Nos. 1 – 4, p. 275 – 287 (1984). – See Abstr. 012.030.

The restricted three–body problem is generalized with the inclusion of solar radiation pressure. For small particles (typically 1 μm to 1 mm) the familiar equilibrium triangular points L_4 and L_5 no longer exist. However libration orbits are not completely destroyed, although an effect of resonance causes their amplitude to be very large, for a particle initially at rest at either of the triangular point. Finally the results of a study of the linearized equations of motion, supplemented by a numerical integration, rule out the possibility of an accumulation of dust at the Earth–Moon lagrangian triangular points.

042.026   The dynamics of bodies with variable masses.
H. Lichtenegger.
Celest. Mech., Vol. 34, Nos. 1 – 4, p. 357 – 368 (1984). – See Abstr. 012.030.

The Lie–series provide a convenient and simple method to solve systems of differential equations, especially in problems dealing with variable masses. A system of n–bodies moving inside a cloud and collecting mass is considered. The equations of motion are derived whereby the interchange of momentum is treated as a perturbation. It is shown that the solutions, represented by Lie–series, can be expressed by binomial expansions plus a perturbation which can be solved by iteration. In addition, the motion of massless bodies experiencing frictional forces is briefly discussed.

042.027   Numerical experiments on planetary orbits in double stars.
R. Dvorak.
Celest. Mech., Vol. 34, Nos. 1 – 4, p. 369 – 378 (1984). – See Abstr. 012.030.

This is a numerical study of orbits in the elliptic restricted three–body problem concerning the dependence of the critical orbits on the eccentricity of the primaries. They are defined as being the separatrix between stable and unstable single periodic orbits. As the results are adapted to the existence of planetary orbits in double stars the author concentrated first on the P–orbits (defined to surround both primaries). Using the results of some 300 integrated orbits for 10^3 to 3×10^3 periods of the primaries he established lower and upper bounds for the critical orbits for different values of the eccentricity.

042.028   Periodic orbits.
J. D. Hadjidemetriou.
Celest. Mech., Vol. 34, Nos. 1 – 4, p. 379 – 393 (1984). – See Abstr. 012.030.

Recent results on periodic orbits are presented. Planetary systems can be studied by the model of the general 3–body problem and also some satellite systems and asteroid orbits can be studied by the model of the restricted 3–body problem. Triple stellar systems and planetary systems with two Suns are close to periodic systems. Finally, the motion of stars in various types of galaxies can be studied by finding families of periodic orbits in several galactic models.

042.029   The problem of small divisors in planetary motion.
C. A. Williams.
Celest. Mech., Vol. 34, Nos. 1 – 4, p. 395 – 410 (1984). – See Abstr. 012.030.

The author focuses on small divisors arising from near commensurability of two or more frequencies of the motion.

042.030   The Ideal Resonance problem – a comparison of two formal solutions. I.
A. H. Jupp, A. Y. Abdulla.
Celest. Mech., Vol. 34, Nos. 1 – 4, p. 411 – 423 (1984). – See Abstr. 012.030.

Garfinkel's solution of the Ideal Resonance problem derived from a Bohlin–von Zeipel procedure, and Jupp's solution, using Poincaré's action and angle variables and an application of Lie series expansions, are compared. Two specific Hamiltonians are chosen for the comparison and both solutions are compared with the numerical solutions obtained from direct integrations of the equations of motion. It is found that in deep resonance the second–mentioned solution is generally more accurate, while in the classical limit the first solution gives excellent agreement with the numerical integrations.

042.031   A note on resonance in regular variables and averaging.
S. Ferraz–Mello, W. Sessin.
Celest. Mech., Vol. 34, Nos. 1 – 4, p. 453 – 457 (1984). – See Abstr. 012.030.

A point relative to the application of the method of Hori to resonant systems is considered: For systems having one degree of freedom the topology of the phase plane of the auxiliary Hori's system is unaltered in the process of construction of a formal solution. The transformed Hamiltonian may not lead to singular points other than those included in the auxiliary Hori's Hamiltonian.

042.032   On the Brown conjecture.
B. Garfinkel.
Celest. Mech., Vol. 34, Nos. 1 – 4, p. 459 – 463 (1984). – See Abstr. 012.030.

In 1911 E. W. Brown conjectured that the family of long–periodic orbits in the Trojan case of the restricted problem of three bodies terminates in an asymptotic orbit that passes through the Langrangian point L_3 for $t = \pm\infty$. The paper refutes this conjecture analytically, thereby confirming the previously published numerical refutation by Henrard (1983).

042.033   A transformation of the two–body problem.
V. R. Bond.
Celest. Mech., Vol. 35, No. 1, p. 1 – 7 (1985).

A transformation of the differential equations of motion of the two–body problem in the spherical coordinates to oscillator form is derived. It is shown that the independent variable transformation $dt/ds = r^2$ is a transformation which makes the oscillator form possible.

042.034   Stability in the double resonance problem.
M. H. C. F. Lacaz.
Celest. Mech., Vol. 35, No. 1, p. 9 – 17 (1985).

The stability of the equilibrium points and the behavior near the equilibrium points of an Ideal Double Resonance Problem are studied. In the case where the characteristic roots are purely imaginary and such that the stability cannot be decided with linear terms, the nonlinear terms are considered and some theorems of Arnold and of Khazin are used.

042.035   A note on the integration of the equations of Lie–Deprit's method for unspecified canonical variables.
W. Sessin.
Celest. Mech., Vol. 35, No. 1, p. 19 – 21 (1985).

In this note it is shown that the equations generated by Lie–Deprit's method for unspecified canonical variables could be solved in the same way as Hori did in his method. Here the notations follow those used by Deprit in his paper.

042.036   Virial oscillations of celestial bodies. IV. The Lyapunov stability of motion.
V. I. Ferronsky, S. A. Denisik, S. V. Ferronsky.
Celest. Mech., Vol. 35, No. 1, p. 23 – 43 (1985).

The discussed integral approach to the solution of dynamics problems of a gravitating system is the most general approach to the n–body problem, where the role of the perturbation function is played by the energy dissipation function. In this connection the classical n–body problem from a physical point of view should be considered as a problem of evolution of integral characteristics (total, potential, and kinetic energies and moment of inertia) of dynamic system at dissipation of its electromagnetic energy (mass defect).

042.037   A variant of the Hori–Lie series method.
D.–x. Cui, B. Garfinkel.
Celest. Mech., Vol. 35, No. 1, p. 89 – 94 (1985).

If the undisturbed Hamiltonian F_0 is a function of the momenta only, then a variant of the Hori–Lie series perturbation method achieves a solution to any order with a single canonical transformation, without the use of the pseudo–time.

042.038   Stability of binary asteroids.
A. L. Whipple, L. K. White.
Celest. Mech., Vol. 35, No. 1, p. 95 – 104 (1985).

The restricted problem of 2 + 2 bodies is applied to the study of the stability and dynamics of binary asteroids in the solar system. Numerical investigation of the behavior of the orbital elements and the maximal Lyapunov characteristic number of binary asteroids reveal extensive regions where bounded quasiperiodic motion is possible. These regions are compared to the bounded regions which are predicted by the classical restricted problem of three bodies. Regions of bounded chaotic solutions are also found.

042.039   On the collinear libration points in the photo–gravitational three–body problem.
A. L. Kunitsyn, A. T. Tureshbaev.
Celest. Mech., Vol. 35, No. 2, p. 105 – 112 (1985). See Abstr. 34.042.017.

The existence of the three–parametric family of the collinear libration points in the photo–gravitational three–body problem (differing from the classical one by the addition to the gravitational field the light repulsion force–field) is proved. The number and situation of these points are determined with respect to the system parameters. Their stability to a first approximation is investigated. It is shown that contrary to the classical problem the internal collinear libration points may be stable in some domain of parameter–space.

042.040   Ejection and collision orbits of the spatial restricted three–body problem.
J. Llibre, J. Martínez Alfaro.
Celest. Mech., Vol. 35, No. 2, p. 113 – 128 (1985).

The authors begin by describing the global flow of the spatial two body rotating problem, $\mu = 0$. The remainder of the work is devoted to a study of the ejection and collision orbits when $\mu \geqslant 0$. The authors make use of the "blow up" techniques to show that for any fixed value of the Jacobian constant the set of these orbits is diffeomorphic to $S^2 \times R$. Some particular collision–ejection orbits are found.

042.041   Universal Keplerian state transition matrix.
S. W. Shepperd.
Celest. Mech., Vol. 35, No. 2, p. 129 – 144 (1985).

A completely general method for computing the Keplerian state transition matrix in terms of Goodyear's universal variables is presented. This includes a new scheme for solving Kepler's problem which is a necessary first step to computing the transition matrix. The Kepler problem is solved in terms of a new independent variable requiring the evaluation of only one transcendental function. Furthermore, this transcendental function may be conveniently evaluated by means of a Gaussian continued fraction.

042.042 The restricted 3–body problem with radiation pressure.
J. F. L. Simmons, A. J. C. McDonald, J. C. Brown.
Celest. Mech., Vol. 35, No. 2, p. 145 – 187 (1985).

The restricted 3–body problem is generalised to include the effects of an inverse square distance radiation pressure force on the infinitesimal mass due to the large masses, which are both arbitrarily luminous. A complete solution of the problems of existence and linear stability of the equilibrium points is given for all values of radiation pressure of both luminous bodies, and all values of mass ratios. It is shown that the inner Lagrange point, L_1, can be stable, but only when both large masses are luminous. Four equilibrium points, L_6, L_7, L_8, and L_9 can exist out of the orbital plane when the radiation pressure of the smaller mass is very high. Although L_8 and L_9 are always linearly unstable, L_6 and L_7 are stable for a small range of radiation pressures provided that both large masses are luminous.

042.043 Asymptotic solutions of the restricted problem near the equilateral Lagrangian points.
R. Cid, S. Ferrer, J. A. Caballero.
Celest. Mech., Vol. 35, No. 2, p. 189 – 200 (1985).

The restricted problem in the vicinity of the Lagrangian point L_4 is studied by finding a convergent binomial expansion of the disturbing function. Using a Hamiltonian formulation if Delaunay variables and removing the short–period terms, a resonance problem (already considered by Giacaglia (1970) in an attempt of enlarging the Ideal Resonance) is obtained. It is shown that this extension is reducible to Garfinkel's ideal resonance in the libration region.

042.044 On tidal evolution to spin–orbit resonance.
M. Šidlichovský.
Bull. Astron. Inst. Czech., Vol. 36, No. 2, p. 65 – 70 (1985).

The analytical solution to the equations of tidal evolution for an equatorial circular orbit of a mass point around a tidally deformed triaxial body is presented. The slowing down of the rotation of the body due to momentum transfer to the revolving point is taken into account. Effect of resonances is briefly investigated using non–singular variables.

042.045 A Lie integrator program and test for the elliptic restricted three body problem.
M. Delva.
Astron. Astrophys., Suppl. Ser., Vol. 60, No. 2, p. 277 – 284 (1985).

A new algorithm for the calculation of orbits of massless bodies in the elliptic restricted three body problem is presented and tested. It uses a Lie series development and avoids the integration of the orbit of the primary bodies. The computer version of the algorithm is straightforward and comprehensible. It is shown that large stepsizes still give enough accuracy. Compared with another high accuracy program, the need of CPU–time is about only one third up to one half of the time normally necessary.

042.046 Bifurcating periodic orbits in the three–body problem.
S. Kasperczuk, S. Gaska.
Acta Astron., Vol. 34, No. 4, p. 469 – 472 (1984).

The authors discuss the bifurcation of periodic orbits in the restricted three–body problem near resonant tori. They show the creation and annihilation of periodic orbits of the second kind of Poincaré.

042.047 Stochasticity in dynamical systems and the curvature of the associated Riemannian manifold.
H. Varvoglis.
Astrophys. Space Sci., Vol. 109, No. 2, p. 395 – 397 (1985).

It is shown that no conflicts need to arise between results on the stochastic properties of a dynamical system obtained through the method of the surface of section mapping and results obtained through the Hedlund-Hopf-Lobachevsky-Hadamard theorem.

042.048 Numerical investigation of the efficiency of some regularized transformations in calculating orbits with large eccentricities.
V. A. Shefer, A. V. Kardash.
Astron. Geod., Tomsk, No. 12, p. 92 – 94, 144 (1984). In Russian. Abstr. in Ref. Zh., 51. Astron., 2.51.65 (1985).

042.049 On the efficiency of application of iteration methods to improvement of orbital parameters.
A. M. Chernitsov, S. S. Kraev.
Astron. Geod., Tomsk, No. 12, p. 95 – 104, 144 – 145 (1984). In Russian. Abstr. in Ref. Zh., 51. Astron., 2.51.66 (1985).

042.050 Programs for investigation of the motion of the outer satellites of planets.
L. E. Bykova, V. V. Shikhalev, V. A. Yurga.
Astron. Geod., Tomsk, No. 12, p. 110 – 114, 145 – 146 (1984). In Russian. Abstr. in Ref. Zh., 51. Astron., 2.51.68 (1985).

042.051 Uniform approximation in problems of celestial mechanics and numerical integration of equations of motion of comet Halley.
E. Z. Khotimskaya.
Astron. Geod., Tomsk, No. 12, p. 54 – 58, 142 – 143 (1984). In Russian. Abstr. in Ref. Zh., 51. Astron., 2.51.73 (1985).

042.052 Results of an All–Union experiment on investigation of the efficiency of algorithms and programs of numerical integration of equations of motion of celestial bodies.
T. V. Bordovitsyna.
Astron. Geod., Tomsk, No. 12, p. 5 – 17, 141 (1984). In Russian. Abstr. in Ref. Zh., 51. Astron., 2.51.76 (1985).

042.053 Estimate of the efficiency of numerical algorithms for construction of satellite trajectories.
Yu. V. Surnin, S. V. Kuzhelev.
Astron. Geod., Tomsk, No. 12, p. 18 – 26, 141 (1984). In Russian. Abstr. in Ref. Zh., 51. Astron., 2.51.77 (1985).

042.054 Construction of periodic orbits and problems of capture.
C. Edelman.
Astron. Astrophys., Vol. 145, No. 2, p. 455 – 460 (1985). In French.

The method of construction of periodic orbits from an asynchronous or from a non–periodic generating family is extended to systems other than perturbed integrable systems. Some numerical examples are presented within the framework of the elliptical restricted three–body problem in space, in order to study the correspondence between the initial orbit and the periodic orbit obtained. In the capture case most of these orbits present some of the peculiar features of the observed short period comets.

042.055 Secular perturbations of the third order in the motion of a satellite of a nonspherical planet.
V. A. Tamarov.
Astron. Zh., Tom 62, Vyp. 2, p. 380 – 384 (1985). In Russian. English translation in Sov. Astron., Vol. 29, No. 2.

Formulae for calculation of secular perturbations of the third order due to the combined effect on the satellite from harmonics of the geopotential with number $n > 2$ are obtained.

042.056 Coplanar solutions in the photogravitational, restricted, circular three–body problem.
L. G. Luk'yanov.
Sov. Astron., Vol. 28, No. 4, p. 462 – 465 (1984). English translation of 38.042.001.

042.057 On the structure of absolutely rigid bodies admitting of plane motions.
V. V. Vidyakin.
Sov. Astron., Vol. 28, No. 4, p. 465 – 467 (1984). English translation of 38.042.002.

042.058   On 6π–periodic solutions of the plane restricted elliptical three–body problem.
A. Ikromov.
Sov. Astron., Vol. 28, No. 4, p. 467 – 469 (1984). English translation of 38.042.003.

042.059   Dynamics of orbiting dust under radiation pressure.
A. Deprit.
The Big Bang and Georges Lemaitre, p. 151 – 180 (1984). – See Abstr. 012.043.
For a three–dimensional Keplerian system in the presence of a homogeneous field possibly in uniform rotation, action and angle variables are introduced by canonical transformation in the averaged Hamiltonian truncated at the first order. After substitution, the first order averaged system proves to be integrable. It is shown how the orbit space decomposes into a pair of spheres in a three–dimensional space, on which the representative curves are the small circles induced by a finite rotation about a fixed axis.

042.060   Generalizations of the restricted problem of three bodies.
V. Szebehely, A. L. Whipple.
The Big Bang and Georges Lemaitre, p. 195 – 205 (1984). – See Abstr. 012.043.
This paper offers several generalizations of the restricted problem of three bodies from an analytical and dynamical point of view. First, a short review of the classical restricted problem is offered which is followed by the most general reformulation of the problem. In this most general formulation the authors consider a dynamical system consisting of several large bodies and of several smaller masses. The influence of the large masses on the small ones can be arbitrary but in most practical cases only gravitational forces are considered. The authors also allow forces acting between the small bodies influencing their motion. The restriction comes in that no influence of the small bodies on the motion of the large ones is allowed.

042.061   A relativistic approach to the Kepler problem.
J. G. Bryant.
The Big Bang and Georges Lemaitre, p. 231 – 237 (1984). – See Abstr. 012.043.
The classical Kepler problem can be modified so that the velocity is always bounded. This is done by using a new time variable. A new Hamiltonian function can be constructed, and it receives a relativistic interpretation: Newtonian mass and energy are replaced by relativistic (proper) mass and energy. A classical model for the photon can also be formulated.

042.062   Studies on orbital resonance. I.
L. Liu, K. A. Innanen, S.–P. Zhang.
Astron. J., Vol. 90, No. 5, p. 877 – 886 (1985).
A discriminant designed to be suitable for assessing resonant stability for certain three–body configurations is described. Its utility is illustrated by its application to a selection of 71 asteroid orbits.

042.063   Studies on orbital resonance. II.
L. Liu, K. A. Innanen.
Astron. J., Vol. 90, No. 5, p. 887 – 891 (1985).
The effects of several terms neglected in the preceding paper on orbital resonance are investigated, and illustrated by application to minor bodies.

042.064   Motion in the region of the triangular libration points with variable ratio of the masses of components.
B. B. D'yakov, B. I. Reznikov.
Fiz.–tekh. inst. Akad. Nauk SSSR., Prepr., No. 904, 32 pp. (1984). In Russian. Abstr. in Ref. Zh., 51. Astron., 3.51.88 (1985).

042.065   On the application of methods of extension to problems of improvement of orbital parameters.
A. M. Chernitsov, S. S. Kraev.
Astron. geod., Tomsk, No. 10, p. 137 – 142 (1984). In Russian. Abstr. in Ref. Zh., 51. Astron., 3.51.89 (1985).

042.066   Algorithm and program for numerical averaging of equations of motion of celestial bodies.
A. Yu. Vol'fengaut.
Astron. geod., Tomsk, No. 10, p. 114 – 120 (1984). In Russian. Abstr. in Ref. Zh., 51. Astron., 3.51.96 (1985).

042.067   Translational–rotational motion of a rigid body in the restricted problem of three rigid bodies.
E. N. Eremenko.
Sov. Astron., Vol. 28, No. 5, p. 590 – 594 (1984). English translation of 38.042.026.

042.068   Evolution of families of double– and triple–periodic orbits.
B. Barbanis.
ESO Sci. Prepr., No. 373, 22 pp. (1985). To appear in Celest. Mech.

042.069   Initial orbit determination: final word.
B. G. Marsden.
Center for Astrophysics, Prepr. Ser., No. 2132, 14 pp. (1985). Submitted to Astron. J.

042.070   Effect of perturbations on the stability of triangular points in the restricted problem of three bodies with variable mass.
J. Singh, B. Ishwar.
Celest. Mech., Vol. 35, No. 3, p. 201 – 207 (1985).
The effect of small perturbations ε and ε' in the coriolis and the centrifugal forces respectively on the stability of the triangular points in the restricted problem of three bodies with variable mass has been studied. It is found that the range of stability of triangular points increases or decreases depending upon whether the perturbation point $(\varepsilon, \varepsilon')$ lies in one or the other of the two parts in which the $(\varepsilon, \varepsilon')$ plane is divided by the line $J_8\varepsilon - J_9\varepsilon' = 0$ where J_8 and J_9 depend upon β, the constant due to the variation in mass governed by Jeans' law.

042.071   Resonance in regular variables. I. Morphogenetic analysis of the orbits in the case of a first–order resonance.
S. Ferraz–Mello.
Celest. Mech., Vol. 35, No. 3, p. 209 – 220 (1985).
A morphogenetic analysis of the orbits of the ideal first–order resonance problem in the neighbourhood of the origin is presented. It is shown that for problems involving central and near–central resonance it is necessary to consider as parameter the cube root of the perturbation instead of the square root used in classical non–central resonance problems.

042.072   Resonance in regular variables. II. Formal solutions for central and non–central first–order resonance.
S. Ferraz–Mello.
Celest. Mech., Vol. 35, No. 3, p. 221 – 234 (1985).
The method of Hori (1966) is applied to the study of motions in the neighbourhood of the origin in the case of a first–order resonance. Lie–series expansions about the origin and Hori's averaging principles and auxiliary equations are discussed, with emphasis on the introduction of the auxiliary parameter t^*.

042.073   Some manifolds of periodic orbits in the restricted three–body problem.
G. Gómez, M. Noguera.
Celest. Mech., Vol. 35, No. 3, p. 235 – 255 (1985).
The author gives some numerical results about natural families of periodic orbits, which emanate from limiting orbits around the equilateral equilibrium points of the restricted three–body problem, when the mass ratio is greater than Routh's critical one.

042.074   Three–dimensional periodic solutions around equilibrium points in Hill's problem.
C. Zagouras, V. V. Markellos.
Celest. Mech., Vol. 35, No. 3, p. 257 – 267 (1985).
The three–dimensional periodic solutions originating at the equilibrium points of Hill's limiting case of the restricted three–body problem, are studied. Fourth–order parametric expansions

by the Lindstedt–Poincaré method are constructed for them. The two equilibrium points of the problem give rise to two exactly symmetrical families of three–dimensional periodic solutions. The family $_HL_{2v}{}^e$ originating at L_2 is continued numerically and is found to extend to infinity. The family originating at L_1 behaves in exactly the same way. All orbits of the two families are unstable.

042.075   Resonant structure of the outer solar system.
A. Milani, A. M. Nobili.
Celest. Mech., Vol. 35, No. 3, p. 269 – 287 (1985).

Hierarchical stability of the outer solar system is monitored through its 3–body subsystems by using numerically computed ephemerides for 5×10^6yr. It is found that the stability parameters of Sun–Jupiter–Saturn and Sun–Uranus–Neptune oscillate in anti–phase in $\sim 1.1 \times 10^6$yr. The mechanism responsible for this locking is a secular resonance between Uranus' perihelion and Jupiter's aphelion: the difference between the two librates within $\sim \pm 70°$ with the same period of $\sim 1.1 \times 10^6$yr.

042.076   Periodic orbits of the general three–body problem for the Sun–Jupiter–Saturn system.
J. H. Kwok, P. E. Nacozy.
Celest. Mech., Vol. 35, No. 3, p. 289 – 303 (1985).

Two families of symmetric periodic orbits of the planar, general, three–body problem are presented. The masses of the three bodies include ratios equal to the Sun–Jupiter–Saturn system and the periods of the orbits of Jupiter and Saturn are in a 2:5 resonance. The (linear) stability of the orbits is studied in relation to eccentricity and mass variations.

042.077   The determination of the gravitational potential of axisymmetric mass distribution in spherical coordinates. II.
R. Tschaepe.
Astron. Nachr., Vol. 306, No. 3, p. 129 – 136 (1985).

The numerical computation of gravitational potential of rotational–symmetric mass distribution in spherical coordinates requires a lot of computing time. The author discusses direct methods and presents a reduction method which is superior to methods used so far. Results for accuracy which is similar to that of other methods and computing times are given.

042.078   Expansion theory for the elliptic motion of arbitrary eccentricity and semi–major axis. VII. Elliptic expansions in terms of the sectorial variables for the seventh and eighth categories.
M. A. Sharaf.
Astrophys. Space Sci., Vol. 112, No. 1, p. 51 – 68 (1985).

In this paper of the series, elliptic expansions in terms of the sectorial variables $\theta f^{(i)}$ introduced recently in Paper IV (Sharaf, 1982) to regularise highly oscillating perturbations force of some orbital systems will be established analytically and computationally for the seventh and eighth categories. For each of the elliptic expansions belonging to a category, literal analytical expressions for the coefficients of its trigonometric series representation are established. Moreover, some recurrence formulae satisfied by these coefficients are also established to facilitate their computations, numerical results are included to provide test examples for constructing computational algorithms.

042.079   The Roche problem in celestial mechanics.
V. V. Pavlovskaya.
Iz istor. mat. estestvozn., Kiev, p. 70 – 75 (1984). In Russian. Abstr. in Ref. Zh., 51. Astron., 6.51.9 (1985).

042.080   On an investigation of the motion of a satellite relative to the mass center near a libration point.
V. N. Shinkin.
IXth Mezhdunar. konf. po nelinejn. kolebaniyam, Kiev, 30 avg. – 6 sent., 1981. Tom 3. Kiev, p. 303 – 305 (1984). In Russian. Abstr. in Ref. Zh., 51. Astron., 6.51.76 (1985).

042.081   Intermediate motion of a resonance satellite in the gravitational field of a rotating nonspherical planet.
Yu. V. Barkin, V. A. Kitova.
Izv. Akad. Nauk TSSR. Ser. fiz.–tekh., khim., geol. nauk, No. 4, p. 32 – 36 (1984). In Russian. Abstr. in Ref. Zh., 51. Astron., 6.51.83 (1985).

042.082   On the definition of planeoids.
M. Burša.
Bull. Astron. Inst. Czech., Vol. 36, No. 3, p. 142 – 144 (1985).

The problem of determining the vertical datum of planets and satellites has been discussed and its solution outlined. It has been suggested to choose the best–fitting equipotential surface as reference for heights. Different conditions to be imposed have been discussed.

042.083   On periodic motions of a satellite in a circular orbit.
A. P. Markeev.
Kosm. Issled., Tom 23, Vyp. 3, p. 323 – 330 (1985). In Russian. English translation in Cosm. Res.

042.084   Influence of vortical currents on the rotation and orientation of a satellite.
Yu. G. Martynenko.
Kosm. Issled., Tom 23, Vyp. 3, p. 347 – 357 (1985). In Russian. English translation in Cosm. Res.

042.085   Stationary translatory–rotational motions of resonance satellites taking into account the influence of a third body.
S. G. Zhuravlev, A. T. Tagaev.
Kosm. Issled., Tom 23, Vyp. 3, p. 371 – 381 (1985). In Russian. English translation in Cosm. Res.

042.086   Evolution of orbital elements of outer planets in a time interval of 800,000 years.
A. A. Sukhotin.
Astron. Geod., Tomsk, No. 12, p. 80 – 91, 144 (1984). In Russian. Abstr. in Ref. Zh., 51. Astron., 2.51.70 (1985).

042.087   The two–body problem: a diagrammatic discussion.
F. Remy, F. Mignard.
Am. J. Phys., Vol. 52, No. 12, p. 1116 – 1121 (1984). Abstr. in Phys. Abstr., Vol. 88, No. 1254, Entry 35957 (1985).

042.088   Energy conserving, arbitrary order numerical solutions of the N–body problem.
A. Marciniak.
Numer. Math., Vol. 45, No. 2, p. 207 – 218 (1984). Abstr. in Phys. Abstr., Vol. 88, No. 1255, Entry 40889 (1985).

042.089   On nonstationary model problems of celestial mechanics.
M. D. Minglibaev, T. B. Omarov.
Tr. Astrofiz. Inst. Alma–Ata, Tom 43, p. 3 – 11 (1984). In Russian.

042.090   On energy increase in the n–body problem with decreasing masses at the same rate.
M. D. Minglibaev.
Tr. Astrofiz. Inst. Alma–Ata, Tom 43, p. 12 – 14 (1984). In Russian.

042.091   A case of generalization of the restricted collinear problem of three bodies of variable mass.
A. A. Bekov.
Tr. Astrofiz. Inst. Alma–Ata, Tom 43, p. 15 – 22 (1984). In Russian.

042.092   Second–order p–iterative solution of the Lambert/Gauss problem.
F. W. Boltz.
J. Astronaut. Sci., Vol. 32, No. 4, p. 475 – 485 (1984). Abstr. in Phys. Abstr., Vol. 88, No. 1257, Entry 57583 (1985).

042.093   The manifold structure for collision and for hyperbolic–parabolic orbits in the *n*–body problem.
D. G. Saari.
J. Differ. Equations, Vol. 55, No. 3, p. 300 – 329 (1984). Abstr. in Phys. Abstr., Vol. 88, No. 1259, Entry 63671 (1985).

042.094   A new method of initial orbit determination.
L. G. Taff.
Bull. Am. Astron. Soc., Vol. 17, No. 2, p. 622 – 623 (1985). Abstract. – See Abstr. 010.066.

042.095   Initial orbit determination: final word.
B. G. Marsden.
Bull. Am. Astron. Soc., Vol. 17, No. 2, p. 623 (1985). Abstract. – See Abstr. 010.066.

042.096   Motion in the restricted problem to three bodies for small values of the mass parameter.
R. G. Hopkins.
Bull. Am. Astron. Soc., Vol. 17, No. 2, p. 625 (1985). Abstract. – See Abstr. 010.066.

042.097   A sequence of canonical transformations for planetary theory.
C. A. Williams.
Bull. Am. Astron. Soc., Vol. 17, No. 2, p. 626 (1985). Abstract. – See Abstr. 010.066.

042.098   Secular variation of planetary elements.
J. Laskar.
Bull. Am. Astron. Soc., Vol. 17, No. 2, p. 626 (1985). Abstract. – See Abstr. 010.066.

042.099   Birkhoff normalization and resonances in dynamics.
W. Presler, R. A. Broucke.
Bull. Am. Astron. Soc., Vol. 17, No. 2, p. 626 (1985). Abstract. – See Abstr. 010.066.

042.100   On Szebehely's equation for the determination of the potential by satellite observations.
M. Stavinschi.
Top. Astrophys. Astron. Space Sci., Vol. 1, p. 41 – 46 (1985).
Studying the inverse problem of celestial mechanics and implicitly Szebehely's equation, the author found that this equation was independently established by C. Dramba (1963).

042.101   On the determination of the potential in the three–dimensional case.
M. Stavinschi.
Top. Astrophys. Astron. Space Sci., Vol. 1, p. 47 – 49 (1985).
The article describes the extension of Dramba–Szebehely's equation in the three–dimensional case and in an inertial system.

042.102   Von Zeipel's method.
I. A. Gerasimov.
Tr. Gos. Astron. Inst. Shternberg, Tom 57, p. 118 – 129 (1985). In Russian.
The paper presents the translation of that part of von Zeipel's memoir which contains the idea of his method.

042.103   The use of a semi–analytical method for the solution of linear differential equations in the restricted elliptical problem of three bodies.
A. V. Zakharov, L. G. Luk'yanov.
Tr. Gos. Astron. Inst. Shternberg, Tom 57, p. 168 – 176 (1985). In Russian.
The solution of the variational equations of the considered problem is constructed in form of trigonometric series with numerical values of the frequencies and amplitudes which are obtained with the aid of a semi–analytical method using numerical integration of the time interval equal to the period of the coefficients of the system.

042.104   Periodic rotational motions of a rigid body having a triaxial ellipsoid of inertia.
L. Yu. Khorseva.
Tr. Gos. Astron. Inst. Shternberg, Tom 57, p. 177 – 185 (1985). In Russian.
The problem of translatory – rotational motion of a rigid body which has three mutually perpendicular planes of geometric and dynamical symmetries is considered. The mass centre of the satellite lies near the triangular libration point of the restricted problem of three rigid bodies. Each of the main bodies is assumed to be an axially symmetric rigid body having the plane of symmetry normal to the axis of symmetry. Periodic solutions are found by Poincaré's small–parameter method. The small parameter is the dynamical compression of the body. Necessary conditions of stability of the solutions are obtained in first approximation.

042.105   Asymptotical trajectories to the collinear points in the restricted elliptical three–body problem.
V. M. Zuev.
Tr. Gos. Astron. Inst. Shternberg, Tom 57, p. 186 – 193 (1985). In Russian.

042.106   A new simple method for the analytical solution of Kepler's equation.
N. I. Ioakimidis, K. E. Papadakis.
Celest. Mech., Vol. 35, No. 4, p. 305 – 316 (1985).
A new simple method for the closed–form solution of non–linear algebraic and transcendental equations through integral formulae is proposed. This method is applied to the solution of the famous Kepler equation in the two–body problem for elliptic orbits. The resulting formulae are quite elementary and, beyond their analytical interest, they can also provide quite accurate numerical results by using Gauss–type quadrature rules.

042.107   Cas special du problème restreint des plusieurs corps.
G. N. Duboshin.
Celest. Mech., Vol. 35, No. 4, p. 317 – 328 (1985).
Nous considerons ici un cas particulier du problème restreint des n + 1 corps. Les n points matériels actifs des masses unitaires sont situés dans les sommets d'un polygone équilatéral, qui tourne uniformément autour son centre. Ces n corps agissent sur un point matériel passif par une loi quelconque. On trouve les points de libration correspondants et on recherche le problème de la stabilité de ces points au sens de Lyapunov.

042.108   Comments on "About an unsuspected integrable problem" by F. Mignard and M. Henon.
B. Garfinkel.
Celest. Mech., Vol. 35, No. 4, p. 343 – 344 (1985). See Abstr. 38.042.029 for the original paper.

042.109   Bifurcation at complex instability.
D. C. Heggie.
Celest. Mech., Vol. 35, No. 4, p. 357 – 382 (1985).

042.110   The central drop configurations.
E. M. Nezhinskij.
Celest. Mech., Vol. 35, No. 4, p. 383 – 396 (1985).
This paper deals with the investigation of central configurations consisting of a point body, a homogeneous sphere and some drops of homogeneous ideal fluid. The existence of such central configurations, as well as the stability of the drops in linear approximation, has been proved by using the virial method (Chandrasekhar, 1969).

042.111   The equatorial equilibrium–configurations of the Magnetic–Binary problem.
T. Kalvouridis, A. Mavraganis.
Celest. Mech., Vol. 35, No. 4, p. 397 – 408 (1985).
The equilibrium–configurations of the Magnetic–Binary problem are investigated in the case of equatorial motion. The law of this constrained motion is derived and then the procedure for localizing the equilibrium points is developed. The type of equilibrium is also studied by means of the known method of variational equations.

042.112 Evolution of the rotation of a viscoelastic planet on a circular orbit in a central force field.
V. G. Vil'ke, S. A. Kopylov, Yu. G. Markov.
Sov. Astron., Vol. 28, No. 6, p. 702 – 705 (1984). English translation of 38.042.084.

042.113 Structure of coefficients of the series representing the solution of the plane, circular three–body problem.
E. I. Timoshkova, V. B. Titov.
Sov. Astron., Vol. 28, No. 6, p. 709 – 712 (1984). English translation of 38.042.085.

042.114 Report of IAU Commission 7: Celestial mechanics (*Mécanique céleste*).
J. Kovalevsky.
Trans. IAU, Vol. XIXA, p. 15 – 28 (1985). – See Abstr. 003.046.

042.115 The restricted problem of 2 + n bodies.
A. L. Whipple.
Diss. Abstr. Int., Sect. B, Vol. 45, No. 7, p. 2201 (1985). Thesis, University of Texas, 117 pp. (1984). Order No. DA8421817.

042.116 Problems of present–day celestial mechanics.
G. N. Duboshin.
Main directions of astronomical investigations at the Moscow University, p. 107 – 116 (1985). In Russian. – See Abstr. 003.049.

042.117 The Roche limit based on vibrational stability.
G.–x. Song.
Chin. Astron. Astrophys., Vol. 9, No. 2, p. 106 – 113 (1985). English translation of Acta Astron. Sin., Vol. 25, No. 4, p. 365 – 375 (1984).
Starting from the concept of vibrational stability, the Roche limit for a uniform incompressible liquid sphere is derived. It is shown how this method can be generalized to the case where there is a non–uniform matter distribution in the radial direction.

042.118 Behavior of the motion of an isolated body in the three–body problem.
Y.–s. Sun, C. Marchal.
Acta Astron. Sin., Vol. 26, No. 1, p. 1 – 12 (1985). In Chinese.

042.119 Surfaces of zero velocity in the restricted problem of three bodies.
J. Lundberg, V. Szebehely, S. Nerem, B. Beal.
Bull. Am. Astron. Soc., Vol. 17, No. 2, p. 625 (1985). Abstract. – See Abstr. 010.066.

042.120 On the last Brown conjecture.
B. Garfinkel.
Bull. Am. Astron. Soc., Vol. 17, No. 2, p. 625 – 626 (1985). Abstract. – See Abstr. 010.066.

042.121 Evolution of the moment of inertia in the many–body problem.
C. Marchal, Y.–s. Sun.
Sci. Sin., Ser. A, Vol. 28, No. 6, p. 638 – 647 (1985).
In this paper the authors discuss the attainable domain and the forbidden domain for the "mean quadratic distance" in the many–body problem and obtain the bifurcation curves between these two domains. The properties of these bifurcation curves are studied.

Numerical solutions of the N–body problem.
See Abstr. 003.131.

The Great Inequality of Jupiter and Saturn: from Kepler to Laplace.
See Abstr. 004.104.

Soluzioni italiane dell'equazione di Keplero nel '700.
See Abstr. 004.115.

Investigations in the field of celestial mechanics at the Sternberg Astronomical Institute.
See Abstr. 009.039.

The dynamics of stellar and planetary systems. Summary of the RAS specialist discussion, held 1984 October 12 at the Scientific Societies' Lecture Theatre, Savile Row.
See Abstr. 010.722.

Investigations in the field of celestial mechanics at Tomsk University in the years 1968 – 1980.
See Abstr. 013.043.

Particle simulation of plasmas and stellar systems.
See Abstr. 014.043.

Comments on B. Davies's 'Derivation of perihelion precession'.
See Abstr. 014.044.

Himmelsmechanik mit dem PC.
See Abstr. 021.007.

C^r approximations to the Standard Map.
See Abstr. 021.020.

The critical periodic orbits in the Störmer problem.
See Abstr. 022.060.

Photoelectric meridian circles for planetary theory and stellar dynamics.
See Abstr. 041.020.

The Earth's precession and nutations in geological epochs.
See Abstr. 044.066.

On normal places, obtained by matched polynomials.
See Abstr. 052.002.

Some modifications of the Hori–Deprit method in constructing a third–order semianalytical theory for resonance satellites.
See Abstr. 052.004.

Axisymmetric flow around a gravitating body.
See Abstr. 062.194.

Equilibrium models of differentially–rotating polytropic cylinders.
See Abstr. 065.017.

Equilibrium structures of rotating isothermal gas clouds. I.
See Abstr. 065.020.

A general computational method for obtaining equilibria of self–gravitating and rotating gases.
See Abstr. 065.050.

Equilibrium structures of rotating isothermal gas clouds. II. Dependence on the angular momentum distribution.
See Abstr. 065.056.

Equilibrium models of differentially rotating polytropes and the collapse of rotating stellar cores.
See Abstr. 065.057.

Self–gravitating configurations with magnetic fields.
See Abstr. 065.085.

Non–gravitational forces in the evolution of the solar system.
See Abstr. 091.038.

On the stability of the Solar System as hierarchical dynamical system.
See Abstr. 091.039.

Resonances in the solar system.
See Abstr. 091.047.

Newtonian N–body calculations of the advance of Mercury's perihelion.
See Abstr. 092.002.

Numerical models for the study of motion of lunar satellites.
See Abstr. 094.003.

Numerical investigation of collision orbits of lunar satellites.
See Abstr. 094.004.

Planetary perturbations on the libration of the Moon.
See Abstr. 094.014.

Relativistic effects in the earth–moon dynamics.
See Abstr. 094.017.

Untersuchungen zur Verbesserung Brownscher Mondentfernungen aufgrund von Variationen der Integrationskonstanten und anderer Parameter.
See Abstr. 094.031.

The depletion of the outer asteroid belt.
See Abstr. 098.021.

Trojan orbits in secular resonances.
See Abstr. 098.030.

Critical inclination of Trojan asteroids.
See Abstr. 098.031.

An application of Labrouste's method to quasi–periodic asteroidal motion.
See Abstr. 098.032.

Secular perturbations of asteroids with commensurable mean motions.
See Abstr. 098.033.

A mechanism of depletion for the Kirkwood's gaps.
See Abstr. 098.034.

The effect of the dynamical parameters in the motion of the Galilean satellites of Jupiter.
See Abstr. 099.032.

Libration of Laplace's argument in the Galilean satellites theory.
See Abstr. 099.033.

Some dynamical models for the Saturn co–orbital satellites.
See Abstr. 100.043.

Approximation methods in celestial mechanics. Application to Pluto's motion.
See Abstr. 101.006.

The dispersion of the Geminid stream by planetary perturbations.
See Abstr. 104.007.

The three–body problem in stellar dynamics.
See Abstr. 151.057.

Bifurcations and stability in three–dimensional systems.
See Abstr. 151.073.

Jacobi integral for the system of two extended configurations (binary galaxies) and an infinitesimal particle (a star).
See Abstr. 151.075.

Numerical simulations of encounters of hard binaries.
See Abstr. 151.120.

Collisional relaxation: a new approach.
See Abstr. 151.121.

On the number of effective integrals in galactic models.
See Abstr. 151.158.

043 Astronomical Constants, Reference Systems

043.001   Die Schiefe der Ekliptik und ihre zeitliche Variation.
    A. D. Wittmann.
Sterne Weltraum, 24. Jahrg., Nr. 1, p. 24 – 26 (1985).

043.002   Inertial systems – definitions and realizations.
    H. Eichhorn.
Celest. Mech., Vol. 34, Nos. 1 – 4, p. 11 – 18 (1984). – See Abstr. 012.030.
It is pointed out that within the framework of Newtonian mechanics the concept of an inertial frame of reference is global, but within the framework of general relativity only local. It is further pointed out that none of the practically determined approximations to an inertial frame of reference (e.g., the dynamical reference frame of celestial mechanics, the FK5) must be regarded as a definition of an inertial system itself. The zero longitude direction of the equatorial coordinate system is critiqued, and because a rigorous conceptual definition of the ecliptic is impossible, it is suggested to specify the intersection of the invariable plane with the equator as the direction of longitude zero.

043.003   Minor planet observations and the fundamental reference system.
R. L. Duncombe, P. D. Hemenway, A. L. Whipple.
Celest. Mech., Vol. 34, Nos. 1 – 4, p. 19 – 36 (1984). – See Abstr. 012.030.
A 15 year project to establish a dynamical reference system utilizing ground–based and Space Telescope observations of 34 minor planets is being undertaken. The orbits of these minor planets will be knit into a common system through the use of "crossing point" observations. The system of orbits thus established can be used to measure long arcs in the sky (similar to the function of a transit circle) and can be used to detect individual star errors as well as residual periodic effects in the fundamental reference system. The minor planet dynamical reference system will also provide an independent method to establish the zero point and the solid–body rotation of the HIPPARCOS reference system.

043.004   The empirical inertial system determined in FK5 by the dynamics of the planetary system.
W. Fricke.
Celest. Mech., Vol. 34, Nos. 1 – 4, p. 37 (1984). Abstract. – See Abstr. 012.030.

043.005   Relations between terrestrial reference systems as exemplified by the MERIT short campaign.
R. Dietrich.
Mitt. Lohrmann–Obs. Tech. Univ. Dresden, Nr. 51, p. 65 – 67 (1984). = Mitt. Zentralinst. Phys. Erde, Nr. 1289. In German. – See Abstr. 012.058.

043.006   Reference coordinate systems and frames: concepts and realization.
I. I. Mueller.
Bull. Géod., Vol. 59, No. 2, p. 181 – 188 (1985).

In view of the unprecedented progress in the ability of geodetic observational systems to measure crustal movements and the rotation of the earth, as well as in theory and model development, there is a great need for the theoretical definition, practical realization, and international acceptance of suitable coordinate system(s) to facilitate such work. This article deals with certain aspects of the establishment and maintenance of such a coordinate system.

043.007   Linking the HIPPARCOS catalog to the VLBI inertial reference system, high angular resolution structures and VLBI positions of 10 radio stars.
J. F. Lestrade, R. A. Preston, R. L. Mutel, A. E. Niell, R. B. Phillips.
Processing of scientific data from the ESA astrometry satellite HIPPARCOS, p. 87 – 95 (1985). – See Abstr. 012.087.

The HIPPARCOS coordinate system will be inertial if tied to an extragalactic VLBI reference frame. This tie would also have the advantage of unifying the optical and radio celestial coordinate systems. In principle, this tie could be achieved by observing radio stars both with the HIPPARCOS satellite and with the VLBI technique. To assess this method, the authors have conducted a VLBI survey of 18 radio stars and detected 11 of them since December 1982. They are reporting on their spatial radio structures and on initial VLBI position determinations.

043.008   A note on upper limits for hypothetic variation in the Newtonian constant of gravitation.
M. Burša.
Stud. Geophys. Geod., Vol. 28, No. 4, p. 360 – 365 (1984).

It has been demonstrated on the basis of recent astronomical, satellite and LLR data that the variations in the Newtonian constant of gravitation, if any, do not exceed $\sim 5 \times 10^{-15} \mathrm{cy}^{-1}$ of its relative value.

043.009   The obliquity of the ecliptic from occultations of ε Geminorum by Mars (1976) and α Leonis by Venus (1959).
G. A. Krasinskij.
Byull. Inst. Teor. Astron., Tom 15, No. 8 (171), p. 440 – 448 (1985). In Russian.

From 14 photoelectric observations of the ε Geminorum occultation by Mars (1976) a normal place in the form of differences between the equatorial coordinates of the planet and the star was deduced. Combined treatment with observations of the occultation of α Leonis by Venus (1959) made it possible to determine accurate corrections to the parameters of orientation of the FK4 reference frame. For the obliquity ε of the ecliptic at J2000.0 the estimation $\varepsilon = 23°26'21.''30 \pm 0.''015$ was found. Thus, the adopted IAU 1976 value ε has to be diminished by $0.''13$. The mean ecliptic was determined by means of the analytical theory of the sun of the Bureau des Longitudes (France).

043.010   On inertial reference systems.
Z. Kh. Kurmakaev.
Tr. Astrofiz. Inst. Alma–Ata, Tom 40, p. 91 – 97 (1983). In Russian.

Fifteen years of Lunar Laser Ranging and future developments.
See Abstr. 013.021.

General precession computed by numerical integration of ordinary differential equations.
See Abstr. 041.001.

The calculation of apparent positions of stars.
See Abstr. 041.018.

Universal time, lunar tidal deceleration and relativistic effects from observations of transits, eclipses and occultations in the XVIIIth – XXth centuries.
See Abstr. 042.009.

Clarifications concerning the definition and determination of the celestial ephemeris pole.
See Abstr. 044.031.

The geodetic reference systems – the basis of theoretical geodesy.
See Abstr. 045.015.

Survey of relativistic effects in geodesy and fundamental astronomy.
See Abstr. 066.186.

Observations of minor planets with the Very Large Array.
See Abstr. 098.027.

044 Time and Latitude Determination, Earth Rotation, Polar Motion

044.001   On the interpretation of non–polar latitude variations.
A. A. Korsun'.
Pis'ma Astron. Zh., Tom 11, No. 1, p. 78 – 80 (1985). In Russian.
English translation in Sov. Astron. Lett., Vol. 11.

After eliminating the influence of errors of declinations and proper motions of observed stars from non–polar latitude variations these variations are presented as a function of fluctuations of geoid deformations. Estimates of such variations of the z–term are given.

044.002   The secular acceleration of the earth's spin.
R. R. Newton.
Geophys. J. R. Astron. Soc., Vol. 80, No. 2, p. 313 – 328 (1985).

The power spectrum of the earth's spin has important components with periods ranging from a few days to at least a few thousand years, and probably to the age of the earth. The secular acceleration, as the term is used in the paper, refers to the components with periods longer than three centuries. In the year 600, the secular acceleration was -19.9 ± 0.8 parts in 10^9 per century, while the value at the present time is less than half this size. The spin acceleration has important contributions from tidal friction and from an effect that is proportional to the square of the magnetic dipole moment. When these contributions are subtracted from the observed acceleration, we are left with a contribution that amounts to $+41$ parts in 10^9 per century. This amount probably results from an unknown combination of changes in the size of the core, in the amount of glaciation, and in the size of the gravitational constant.

044.003   Discrete polar motion equations.
C. R. Wilson.
Geophys. J. R. Astron. Soc., Vol. 80, No. 2, p. 551 – 554 (1985).

A digital filter equation is derived which is appropriate for predicting polar motion from excitation axis displacements, or for inferring the excitation axis changes from observed polar motion. The result differs from previously published equations in its phase response. Two additional equations are presented which are useful if samples of the excitation and polar motion functions are required to be at the same time values.

044.004   On the arrangement of the Time Service at the Crimea – Pushchino radio interferometer.
N. S. Blinov, V. E. Zharov, L. R. Kogan, L. I. Matveenko, E. N. Fedoseev.
Astron. Zh., Tom 62, Vyp. 1, p. 160 – 165 (1985). In Russian.
English translation in Sov. Astron., Vol. 29, No. 1.

The possibility of application of the very–long–baseline interferometer Crimea – Pushchino to the determination of Universal Time is considered.

044.005   A comparison of different methods of adjustment of time and latitude observations.
Z. M. Malkin.
Astron. Zh., Tom 62, Vyp. 1, p. 166 – 171 (1985). In Russian.
English translation in Sov. Astron., Vol. 29, No. 1.

On the basis of the observations obtained with the PZT at Kitab a comparison is made of four modifications of the chain method and two modifications of the least–squares method. The results of the comparison show that all the methods are equivalent. An approximate criterion of the a priori estimation of the efficiency of the methods of adjustment taking into account the evening linear variation of the parameters is proposed.

044.006   On estimation of the period of the Chandler wobble.
V. A. Olevskij.
Kinematika Fiz. Nebesn. Tel, Tom 1, No. 1, p. 95 – 96 (1985). In Russian.

A more accurate estimation of the period of the Chandler wobble (431.65 days) is given using the hypothesis of regular beating in polar motion.

044.007   Precession–nutation torque in terms of the Stokes constants.
M. Šidlichovský.
Stud. Geophys. Geod., Vol. 28, No. 1, p. 1 – 8 (1984).

Precession–nutation torque from the Moon is calculated in terms of the Stokes constants of the Earth. The relation of individual terms of this torque to tidal forming force function is discussed.

044.008   Earth tides and polar motion.
P. Lanzano.
J. Geodyn., Vol. 1, No. 2, p. 121 – 142 (1984). Abstr. in Phys. Abstr., Vol. 88, No. 1248, Entry 9144 (1985).

044.009   An analysis of classical and modern methods of the Earth's rotation determination.
N. Georgiev, A. Khadzhijskij, V. Kotseva.
Publ. Astron. Inst. Czech. Acad. Sci., No. 58, p. 549 – 559 (1984). In Russian. – See Abstr. 012.018.

The authors compare and analyze classical and space methods for determining the Earth rotation irregularities.

044.010   Comparison of polar motion as determined by classical and space techniques.
J. Vondrák.
Publ. Astron. Inst. Czech. Acad. Sci., No. 58, p. 573 – 586 (1984). – See Abstr. 012.018.

The coordinates of the Earth's axis of rotation as obtained from different observation techniques are intercompared in three frequency windows. It is shown that the best accordance between different techniques is achieved in the medium frequency band (i.e. between 0.5 and 1.0 cpy) and that it is neither possible to give absolute preference nor to abandon any of the techniques used.

044.011   Schwankungen der Erdrotation und elektromagnetische Kern–Mantel–Kopplung.
M. Stix.
Sterne Weltraum, 24. Jahrg., Nr. 2, p. 87 – 88 (1985).

044.012   Influence of time variation in the second zonal harmonic on polar motion.
M. Burša, M. Šidlichovský.
Bull. Astron. Inst. Czech., Vol. 36, No. 1, p. 24 – 27 (1985).

The time variations in the amplitude and frequency of the polar motion caused by variations in the second zonal harmonic in the Earth's gravitational field, have been derived.

044.013   Progress report on project MERIT–COTES and on related research.
G. A. Wilkins.
Geophys. J. R. Astron. Soc., Vol. 81, No. 1, p. 314 (1985). Abstract. – See Abstr. 012.025.

044.014   Earth rotation and polar motion by laser ranging to satellites.
V. Ashkenazi, T. Moore.
Geophys. J. R. Astron. Soc., Vol. 81, No. 1, p. 314 (1985). Abstract. – See Abstr. 012.025.

044.015   Predictability of the Earth's polar motion.
B. F. Chao.
Bull. Géod., Vol. 59, No. 1, p. 81 – 93 (1985).

The present paper is an experimental study of the predictability of polar motion based on a homogeneous BIH data set for the period 1967 – 1983. It is shown that with a floating–period numerical model and a capable estimation method, predictions accurate to within $0\overset{''}{.}012$ to $0\overset{''}{.}024$ can be achieved for polar motion, depending on the prediction length up to one year. The superiority of this "floating–period predictor" to other predictors that are based on critically different numerical models (including the conventional fixed–period predictors) is demonstrated.

044.016   Zur Frage der Reform des Nullmeridians.
M. Gossler.
Sterne, 61. Band, Heft 1, p. 28 – 30 (1985).

044.017   A possible estimate of the beginning of secular deceleration of the angular velocity of the earth.
V. T. Rykov.
Sovrem. teor. i ehksp. probl. teor. otnositel'nosti i gravitatsii. Tez. dokl. Vses. konf. 6 Sov. gravitats. Moskva, 3 – 5 iyulya, 1984. Moskva, p. 263 (1984). In Russian. Abstr. in Ref. Zh., 51. Astron., 1.51.178 (1985).

044.018   On the study of the parameters of the earth's nearly diurnal free nutation based on latitude observations.
L. D. Kovbasyuk.
Astron. Zh., Tom 62, Vyp. 2, p. 385 – 392 (1985). In Russian. English translation in Sov. Astron., Vol. 29, No. 2.
The results of comparison are presented for the parameters of the harmonics with frequencies close to the frequency of the nearly diurnal free polar motion predicted by the theory of rotation of the liquid–core earth. These parameters were obtained by different authors from the astronomical latitude observations in Pulkovo, Paris, Washington, Greenwich, Poltava, the International Latitude Stations, and Gorky. In spite of the remoteness of the epochs compared, all the determinations are in a satisfactory agreement. The harmonic with the period $23^h56^m55^s$ and amplitude $0\overset{''}{.}008$, coinciding with the predicted period for the earth's model I by Molodenskij, is sufficiently steady, but has a direction opposite to the predicted one.

044.019   Spectral analysis of the diurnal nonuniformity of the earth's rotation.
G. P. Pil'nik.
Sov. Astron., Vol. 28, No. 4, p. 470 – 473 (1984). English translation of 38.044.004.

044.020   Synchronization and propagation of time by satellites.
G. Hemmleb, R. Stecher.
Mitt. Lohrmann–Obs. Tech. Univ. Dresden, Nr. 51, p. 59 – 62 (1984). = Mitt. Zentralinst. Phys. Erde, Nr. 1288. In German. – See Abstr. 012.058.
Contents: Präzisionsverfahren zum Vergleich von Zeitskalen; Einwegmethoden der Zeitübertragung und des Zeitkalendervergleichs über Satelliten; Zweiwegmethoden des Zeitskalenvergleichs über Satelliten; Das Experiment LASSO; Geplante Experimente zum Präzisionszeitskalenvergleich über Satelliten.

044.021   Calculation of declination corrections for Dresden Talcott observations from 1963 to 1982 in comparison with the ITB system.
S. Wächter.
Mitt. Lohrmann–Obs. Tech. Univ. Dresden, Nr. 51, p. 79 – 80 (1984). In German. – See Abstr. 012.058.

044.022   A contribution to the determination of polar motion.
F. Chollet, S. Débarbat.
Mitt. Lohrmann–Obs. Tech. Univ. Dresden, Nr. 51, p. 86 – 88 (1984). In French. – See Abstr. 012.058.

044.023   Capabilities of the proposed POPSAT satellite system for monitoring polar motion and earth rotation.
K. Kaniuth, H. Müller, C. Reigber.
Mitt. Lohrmann–Obs. Tech. Univ. Dresden, Nr. 51, p. 88 – 90 (1984). – See Abstr. 012.058.

044.024   On the possibility of determining tidal variations of the Earth's rotation using various observational methods.
J. Melicher, J. Hefty.
Mitt. Lohrmann–Obs. Tech. Univ. Dresden, Nr. 51, p. 91 – 93 (1984). – See Abstr. 012.058.

044.025   On some geophysical effects in latitude variations at Józefosław.
M. Barlik, J. B. Rogowski.
Mitt. Lohrmann–Obs. Tech. Univ. Dresden, Nr. 51, p. 95 – 98 (1984). – See Abstr. 012.058.

044.026   The inverse solution of the differential equation of Earth rotation as an aid for interpreting cyclic geophysical phenomena.
H. Jochmann.
Mitt. Lohrmann–Obs. Tech. Univ. Dresden, Nr. 51, p. 98 – 101 (1984). = Mitt. Zentralinst. Phys. Erde, Nr. 1198. In German. – See Abstr. 012.058.

044.027   Coefficients for two–week nutation terms and for the Oppolzer term with the (S–2L) argument derived from determinations in Potsdam.
J. Höpfner.
Mitt. Lohrmann–Obs. Tech. Univ. Dresden, Nr. 51, p. 102 – 104 (1984). = Mitt. Zentralinst. Phys. Erde, Nr. 1279. In German. – See Abstr. 012.058.

044.028   Annual term of local latitude variations.
B. Napiórkowska, B. Kołaczek.
Mitt. Lohrmann–Obs. Tech. Univ. Dresden, Nr. 51, p. 106 – 109 (1984). – See Abstr. 012.058.
Local annual and semiannual changes of φ and λ are represented by systematic corrections R and S calculated by BIH for every station and instrument every year.

044.029   The IRIS Earth orientation parameters.
IRIS Bull. A, Nos. 11 – 16 (1985).

044.030   Les temps en astronomie.
Y. Ottelet.
Ciel, Vol. 47, p. 49 – 54 (1985).

044.031   Clarifications concerning the definition and determination of the celestial ephemeris pole.
N. Capitaine, J. G. Williams, P. K. Seidelmann.
Astron. Astrophys., Vol. 146, No. 2, p. 381 – 383 (1985).
The adoption of the 1980 IAU theory of nutation and the celestial ephemeris pole implies a change in the reference pole of $0\overset{''}{.}0087$. In this paper the interpretation of the differences and the significance of the change from the instantaneous pole of rotation to the celestial ephemeris pole are discussed.

044.032   Anomalies of some tidal waves of UT1.
N. Capitaine, B. Guinot.
Geophys. J. R. Astron. Soc., Vol. 81, No. 3, p. 563 – 568 (1985).
The M_f and M_m waves of UT1 have been analysed from the BIH data during the period 1967.0 to 1984.0 in order to derive the Love number k.

044.033   The influence of the systematic errors $\Delta\alpha_\alpha$ of the Catalogue of the Time Service on the determination of the scale of the "Universal Time" of the USSR.
N. S. Blinov, G. M. Blank, V. L. Molchanova.
Astron. Tsirk., No. 1342, p. 5 – 7 (1984). In Russian.

044.034   Détemination d'un lieu à l'aide de deux observations au sextant.
R. Behrend.
Orion, 43. Jahrg., Nr. 208, p. 82 – 83 (1985).

044.035   First determination of the latitude of Hvar Observatory.
P. Terzić.
Publ. Astron. Opservatorije Beogr., No. 33, p. 53 – 56 (1985). – See Abstr. 012.061.
The adopted latitude of the Hvar Observatory is: $43°10'39\overset{''}{.}05 \pm 0\overset{''}{.}04$.

044.036   A differential method for latitude determination by observations in the first vertical.
B. Kilar.
Publ. Astron. Opservatorije Beogr., No. 33, p. 57 – 61 (1985). – See Abstr. 012.061.

044.037   "Night error" in the Belgrade latitude observations made in the period 1960 – 1980.
R. Grujić, R. Krga, Z. Stančić.
Publ. Astron. Opservatorije Beogr., No. 33, p. 62 – 65 (1985). – See Abstr. 012.061.
The "night error" (a systematic error produced by meteorological and other effects) in the Belgrade latitude observations, made in the period 1960 through 1980, has been investigated.

044.038   Effects of the level bubble length variation on the latitude values.
G. Teleki, R. Grujić.
Publ. Astron. Opservatorije Beogr., No. 33, p. 66 – 69 (1985). – See Abstr. 012.061.
The level bubble length variations during observations are found to affect the latitude values derived by the Talcott method.

044.039   Zeit– und Breitendienst. Januar – September 1984.
Dtsch. Hydrogr. Inst. Hamb., Zeit–Breitendienst, 7 + 7 + 7 pp. (1985).

044.040   Time and Frequency Services Bulletin, 1984 October – 1985 March.
Time Freq. Serv., Bull., Nos. 61 – 66, 40 pp. (1984/85).

044.041   Monthly Notes of the International Polar Motion Service.
Mon. Notes Int. Polar Motion Serv., Nos. 12, 1 – 4, p. 199 – 214, 1 – 68 (1984/85).

044.042   On the mean pole and earthquakes.
E. Onodera.
Proc. Int. Latitude Obs. Mizusawa, No. 23, p. 46 – 50 (1984). In Japanese.
It is found that the refraction points in the pole path coincide with the active periods of earthquakes.

044.043   Aplicación de la V.L.B.I. al estudio del movimiento del polo.
M. A. Montull, M. J. Sevilla, Y. A. Gonzalez–Camacho.
Semin. Astron. Geod., Univ. Madr., Publ., No. 115, 13 pp. (1981).
In this paper an analysis of some equations under different conditions of the V.L.B.I. problem is carried out. The more appropiate methods of adjustment are also studied.

044.044   Algunas relaciones entre diferentes ejes que se consideran en la rotación de la Tierra.
A. G. Camacho, M. J. Sevilla.
Semin. Astron. Geod., Univ. Madr., Publ., No. 116, 21 pp. (1981).
In this paper some relations between the "Celestial Ephemeris Pole" and several axes are considered for an elementary elastic earth model. The authors give a geometrical and geodynamical interpretation.

044.045   Bureau International de l'Heure (B.I.H.). Circular D.
B.I.H. Circ., D218 – D223 (1985).
Contents: UTC, TAI, UT and coordinates of the pole, rotation of the Earth. Data for 1984 November – 1985 April.

044.046   UTC time step of the 1st of July 1985.
B. Guinot.
B.I.H. Circ., E13 (1985).

044.047   Rotación de la tierra año 1982. Resultados obtenidos en San Fernando con el Astrolabio Impersonal Danjon OPL No. 37.
Inst. Obs. Mar., Bol. Ser. C, No. 85, 19 pp. (1985).
This bulletin contains definitive results of time and latitude observations made at San Fernando with the Danjon astrolabe during 1982. Corrections for internal regularization of stars inside each fundamental group (CLI) have been applied, but not "group corrections". In tables the BIH corrections and astronomical coordinates of the astrolabe are shown.

044.048   Rotación de la tierra año 1983. Resultados obtenidos en San Fernando con el Astrolabio Impersonal Danjon OPL No. 37.
Inst. Obs. Mar., Bol. Ser. C, No. 86, 15 pp. (1985).
This bulletin contains definitive results of time and latitude observations made at San Fernando with the Danjon astrolabe during 1983. Corrections for internal regularization of stars inside each fundamental group (CLI) have been applied, but not "group corrections". In tables the BIH corrections and astronomical coordinates of the astrolabe are shown.

044.049   On the decade fluctuation of the acceleration of the Earth's rotation.
N. Sekiguchi.
J. Geod. Soc. Jpn., Vol. 30, No. 4, p. 249 – 253 (1984). = Tokyo Astron. Obs. Repr., No. 724.
The decade fluctuation of the acceleration of the Earth's rotation is repeated almost periodically with period about 25 years. Its semi–amplitude is about 0.1 s yr^{-2}. The power spectrum of its variation has low values for the frequencies between $(3 \, yr)^{-1} = 0.33 \, yr^{-1}$ and $(12.5 \, yr)^{-1} = 0.08 \, yr^{-1}$. Therefore, one can imagine that the causes of the fluctuation are divided into two parts by this frequency interval.

044.050   An experiment for active TV time synchronization.
H.–q. Zheng, X.–y. Han, S.–y. Lian.
Publ. Shaanxi Astron. Obs., Vol. 7, No. 1, p. 26 – 32 (1984). In Chinese.
Experimental results have indicated that the accuracy of active TV synchronization is better than $0.2 \, \mu s$.

044.051   Measurement of the absolute time delay of a TV receiving system.
H.–q. Zheng, X.–y. Han, S.–y. Lian.
Publ. Shaanxi Astron. Obs., Vol. 7, No. 1, p. 33 – 36 (1984). In Chinese.
The measurement of the absolute time delay of a TV receiving system is very important for TV active time synchronization. A method for the measurement of time delay is proposed. The accuracy is better than 20 ns.

044.052   Measurement of the time delay of a longwave reception system for the synthetic atomic time.
H.–q. Chen, C.–y. He.
Publ. Shaanxi Astron. Obs., Vol. 7, No. 1, p. 37 – 40 (1984). In Chinese.

044.053   Preliminary results of passive time comparison via broadcasting satellite (714 MHz).
J.–a. Song, D.–c. Lo, Z.–f. Shi, S.–h. Lu, B. Huang.
Publ. Shaanxi Astron. Obs., Vol. 7, No. 1, p. 41 – 47 (1984). In Chinese.

044.054   Results of determination of latitude in Józefosław, July – December 1984.
L. Pieczyński.
Latitude Circ., Nos. 92, 94 (1984/85).

044.055   The observational programme of 12 groups of Horrebow–Talcott pairs for latitude determinations at Józefosław for the years 1984 – 1995.
M. Dukwicz–Łatka, K. Borkowski.
Latitude Circ., No. 93, 10 pp. (1985).

044.056 Daily time differences and relative phase values. 1985 January – June.
U.S. Nav. Obs., Time Serv. Publ., Ser. 04, Nos. 935 – 960 (1985).

044.057 Earth Orientation. 1985 January – June.
Earth Orientation Bull., Vol. 3, Nos. 1 – 26 (1985).

044.058 Time Service Announcement.
G. M. R. Winkler.
U.S. Nav. Obs., Time Serv. Publ., Ser. 14, Nos. 39 – 40, 40.1 (1985).

044.059 Estimability of astronomical longitude and latitude only from theodolite observations within three–dimensional networks of terrestrial type.
E. W. Grafarend.
Bull. Géod., Vol. 59, No. 2, p. 124 – 138 (1985).

044.060 Using VLBI and CEI to determine the earth's orientation.
A. K. Babcock.
Bull. Am. Astron. Soc., Vol. 16, No. 4, p. 908 (1984). Abstract. – See Abstr. 010.062.

044.061 The flow at the core mantle boundary as a possible source of excitation of the Chandler wobble.
J.-L. Le Mouël, C. Gire, J. Hinderer.
C. R. Acad. Sci., Sér. II, Tome 301, No. 1, p. 27 – 32 (1985). In French.

The flow of the core fluid at the core–mantle boundary is generally accepted as the primary cause of secular variation of the geomagnetic field. The overpressure field varies with time in conjunction with the geomagnetic secular variation field and with the core flow itself. The authors show that the order of magnitude and geometry of this overpressure field are sufficient to alter the products of inertia of the elastic mantle in such a way as to excite the Chandler wobble.

044.062 The law of distribution of residual errors of determination of time and latitude with the Danjon astrolabe.
I. V. Dzhun', A. A. Slavinskaya.
Vrashchenie i priliv. deformatsii Zemli, Kiev, No. 16, p. 69 – 74 (1984). In Russian. Abstr. in Ref. Zh., 51. Astron., 5.51.98 (1985).

044.063 On possibilities of increasing the accuracy of latitude observations with zenith telescopes.
V. P. Shlyakhovyj, A. P. Stehpa.
Vrashchenie i priliv. deformatsii Zemli, Kiev, No. 16, p. 74 – 83 (1984). In Russian. Abstr. in Ref. Zh., 51. Astron., 5.51.99 (1985).

044.064 UTC time step on the 1st of July 1985.
Yamamoto Circ., No. 2036 (1985). In Japanese.

044.065 Time adjustment on 1985 June 30.
IAU Circ., No. 4046 (1985).

044.066 The Earth's precession and nutations in geological epochs.
M. Burša.
Bull. Astron. Inst. Czech., Vol. 36, No. 3, p. 139 – 141 (1985).

On the basis of the observed secular variations in the Earth's rotation and its principal moments of inertia, a rough estimate of the amplitudes and periods of the Earth's precession–nutation motions during the last 10^6 years has been carried out. It is assumed that the volume as well as the total mass of the Earth have been preserved during the period given. The Earth's polar flattening might be larger in the past as well as the departures from the hydrostatic equilibrium state.

044.067 The components of polar motion.
Z. Li, W.-z. Ma, H.-z. Zhang, Y. Han.
Kexue Tongbao (Beijing), Vol. 29, No. 9, p. 1210 – 1215 (1984). Abstr. in Phys. Abstr., Vol. 88, No. 1253, Entry 35111 (1985).

044.068 The possibility of world wide astro/geodetic datum.
B. P. Lambert.
Aust. J. Geod. Photogramm. Surv., No. 40, p. 69 – 75 (1984). Abstr. in Phys. Abstr., Vol. 88, No. 1256, Entry 50915 (1985).

044.069 Earth's rotation and polar motion from NAVSAT.
E. S. Colquitt, R. J. Anderle, C. A. Malyevac.
J. Astronaut. Sci., Vol. 32, No. 4, p. 393 – 405 (1984). Abstr. in Phys. Abstr., Vol. 88, No. 1257, Entry 57085 (1985).

044.070 Time between physics and astronomy.
M. Stavinschi.
Stud. Cercet. Fiz., Vol. 36, No. 5, p. 448 – 455 (1984). In Rumanian. Abstr. in Phys. Abstr., Vol. 88, No. 1258, Entry 63221 (1985).

044.071 The Earth's rotation rate.
J. Wahr.
Am. Sci., Vol. 73, No. 1, p. 41 – 46 (1985). Abstr. in Phys. Abstr., Vol. 88, No. 1260, Entry 73697 (1985).

044.072 On the deformation of the earth and inelasticity of its mantle.
J. Höpfner.
Gerlands Beitr. Geophys., Band 94, Heft 3, p. 200 – 204 (1985). In German. = Mitt. Zentralinst. Phys. Erde, Nr. 1348.

Using the mean Chandler period obtained from latitude observations at the Potsdam astrolabe, the deformation factor and from this the Love number were computed. The contribution to the Chandler period for inelasticity in the earth's mantle was determined by comparison with a theoretical Chandler period corrected for the influence of the polar tide.

044.073 On the slow changes in the earth's rotation.
S. J. Goldstein Jr.
Bull. Am. Astron. Soc., Vol. 17, No. 2, p. 624 (1985). Abstract. – See Abstr. 010.066.

044.074 On some opinions concerning the Earth's rotation theory.
M. Ciobanu.
Top. Astrophys. Astron. Space Sci., Vol. 1, p. 25 – 30 (1985).

The paper reminds that in the case of a rigid Earth, the luni-solar nutation changes the position of the Earth's rotation axis. This is in contrast to the usual opinion which maintains the independence of these two phenomena.

044.075 Differential equations of the n–body problem for a numerical research of the earth's rotational motion.
A. I. Rybakov.
Tr. Gos. Astron. Inst. Shternberg, Tom 57, p. 20 – 68 (1985). In Russian.

Differential equations of the n–body problem are exhibited, the latter being obtained provided that all harmonic polynomials of the fourth degree inclusive are retained in the expansion of the earth's potential of attraction. The above equations will be used for a research of the earth's rotation around its axis by means of numerical integration.

044.076 A numerical research of the earth's rotation in the framework of the three–body problem.
A. I. Rybakov, E. P. Kalinina.
Tr. Gos. Astron. Inst. Shternberg, Tom 57, p. 69 – 82 (1985). In Russian.

Differential equations of motion in the earth–sun–moon system were integrated numerically on a 100–years interval of time. Tables of the angle of the earth's proper rotation are obtained and then fitted in correspondence with the least–squares method provided that the angular acceleration is constant. The numerical value of this constant is found. Amplitudes of the periodic components of the angular velocity corresponding to the period of nutation have been determined.

044.077   Extreme cycles of motion of the earth's pole.
L. D. Kostina, V. I. Sakharov.
Sov. Astron., Vol. 28, No. 6, p. 698 – 701 (1984). English translation of 38.044.071.

044.078   Global character of the spectrum of nearly diurnal latitude variations.
L. D. Kovbasyuk.
Sov. Astron., Vol. 28, No. 6, p. 712 – 717 (1984). English translation of 38.044.072.

044.079   True polar wander and plate–driving forces.
D. M. Davis, S. C. Solomon.
J. Geophys. Res., Vol. 90, No. B2, p. 1837 – 1841 (1985).

A net torque on the global lithosphere can be exerted both by "ridge push" and "trench pull" forces. A net ridge push torque is present if there is an asymmetry in the age distribution of seafloor about one or more ocean ridges. A net torque due to trench forces is present if the forces per trench length acting on the subducting and overthrust plates are not equal in magnitude. The authors suggest that the most likely explanation for negligible true polar wander in the Cenozoic is that the net ridge and trench torques nearly cancel one another. The net ridge and trench torque vectors are in nearly opposite directions and have been so through the Cenozoic, so the condition of near cancellation amounts to a restriction on the magnitude of the force imbalance along trench boundaries. Such a near cancellation of net torques and the resulting negligible true polar wander are likely to be consequences of present plate geometries and need not be representative of earlier eras.

044.080   Transient polar motions and the nature of the asthenosphere for short time scales.
E. Boschi, R. Sabadini, D. A. Yuen.
J. Geophys. Res., Vol. 90, No. B5, p. 3559 – 3568 (1985).

The role of viscoelastic flows excited by earthquakes in driving short–term polar motions is investigated analytically by means of a four–layer model consisting of an elastic lithosphere, a thin asthenosphere, a mean mantle, and an inviscid core.

044.081   Geodetic radio interferometric surveying: applications and results.
W. E. Carter, D. S. Robertson, J. R. MacKay.
J. Geophys. Res., Vol. 90, No. B6, p. 4577 – 4587 (1985).

Very long baseline interferometry observations collected primarily under projects POLARIS (Polar–Motion Analysis by Radio Interferometric Surveying) and IRIS (International Radio Interferometric Surveying) since 1980 have been used to derive polar motion and UT1 time series.

044.082   A spectral analysis of the earth's angular momentum budget.
T. M. Eubanks, J. A. Steppe, J. O. Dickey, P. S. Callahan.
J. Geophys. Res., Vol. 90, No. B7, p. 5385 – 5404 (1985).

Changes in the circulation of the earth's atmosphere cause fluctuations in the length of day by exchanging angular momentum with the surface of the earth. In this paper the authors examine the earth's polar angular momentum budget using earth rotation data from optical astrometry and lunar laser ranging together with simultaneous estimates of the atmospheric angular momentum derived from meteorological data.

044.083   Report of IAU Commission 19: Rotation of the earth (*Rotation de la terre*).
Ya. S. Yatskiv.
Trans. IAU, Vol. XIXA, p. 193 – 205 (1985). – See Abstr. 003.046.

044.084   Report of IAU Commission 31: Time (*L'heure*).
G. Hemmleb.
Trans. IAU, Vol. XIXA, p. 383 – 396 (1985). – See Abstr. 003.046.

044.085   Project MERIT and orbit dynamics of the satellites.
J. Klokočník.
Říše hvězd, Vol. 66, No. 3, p. 62 – 63 (1985). In Czech.

044.086   Analysis and modeling of variations in length of day.
A. K. Babcock.
Diss. Abstr. Int., Sect. B, Vol. 45, No. 12, p. 3847 (1985). Thesis, University of Virginia, 123 pp. (1984). Order No. DA8503494.

044.087   Work and problems of the Time Service.
N. S. Blinov.
Main directions of astronomical investigations at the Moscow University, p. 98 – 102 (1985). In Russian. – See Abstr. 003.049.

044.088   International cooperation in monitoring the rotation of the earth.
G. A. Wilkins.
Vistas Astron., Vol. 28, Parts 1/2, p. 329 – 335 (1985). – See Abstr. 012.109.

044.089   The Moon's zonal tidal effect in the variation of the Earth's spin rate.
Y.–z. Chu.
Chin. Astron. Astrophys., Vol. 9, No. 1, p. 35 – 38 (1985). English translation of 38.044.073.

044.090   On the excitation of short–term variations in the length of the day and polar motion.
R. Hide.
Geophys. Surv., Vol. 7, No. 2, p. 163 – 167 (1985). – See Abstr. 012.111.

Variations in the distribution of mass within the atmosphere and changes in the pattern of winds produce fluctuations in all three components of the angular momentum of the atmosphere on time–scales upwards of a few days. It has been shown that variations in the axial component of atmospheric angular momentum during the Special Observing Periods in the recent "First GARP Global Experiment" are well correlated with short–term changes in the length of the day. It has also been shown that fluctuations in the equatorial components of atmospheric angular momentum make a major contribution to the observed wobble of the instantaneous pole of the Earth's rotation with respect to the Earth's crust.

044.091   Non–tidal changes in the length of the day: 700 BC to AD 1982.
F. R. Stephenson, L. V. Morrison.
Geophys. Surv., Vol. 7, No. 2, p. 201 – 210 (1985). – See Abstr. 012.111.

Occultations and eclipses from ancient times down to the present are analysed to determine changes in the length of the day. By subtracting the expected tidal contribution from the observed changes, the non–tidal variations are obtained. The non–tidal variations are shown to occur on time–scales of decades and millennia.

044.092   Predictability of the Earth's polar motion.
B. F. Chao.
NASA Tech. Memo., NASA TM–86095, 3 + 17 pp. (1984).

The present paper is an experimental study of the predictability of the polar motion based on a homogeneous BIH data set for the period 1967 – 1983. It is shown that with a floating–period numerical model and a capable estimation method, predictions accurate to within $0\overset{"}{.}012$ to $0\overset{"}{.}024$ can be achieved for the polar motion, depending on the prediction length up to one year. The superiority of this "floating–period predictor" to other predictors that are based on critically different numerical models (including the conventional fixed–period predictor) is demonstrated.

044.093   The secular acceleration of the Earth's spin.
R. R. Newton.
Johns Hopkins APL Tech. Dig., Vol. 6, No. 2, p. 120 – 129 (1985).

The spin rate of the Earth varies constantly. Daily changes are associated with atmospheric winds; long–term changes are related to lunar and solar tidal friction and other slowly changing geophysical parameters. The changes in the Earth's spin rate reported here have occurred over periods measured in centuries and are based on observations in historical astronomical texts. The (negative) secular spin acceleration was −19.8 parts per billion per century around the year 600 AD and is now −8.6 parts per billion. These changes in spin rate are due to contributions from tidal friction and from an effect proportional to the square of the time–varying magnetic dipole of the Earth. When these contributions are subtracted from the observed acceleration, a residual contribution of +41 parts per billion per century remains that is probably due to variations in the diameter of the Earth's core and other geophysical changes.

044.094   Wobbles of the reference axes of the elastic earth with a liquid core.
N.–y. Xiao.
Acta Astron. Sin., Vol. 25, No. 4, p. 409 – 418 (1984). In Chinese.

In this paper, for the earth model with an elastic mantle and a liquid core the author discusses the solution to both free wobble equations and forced wobble equations of rotation axis, axis of angular momentum and figure axis, respectively. The main conclusion is as follows: for the earth with a liquid core the three axes mentioned above are no more coplanar and the deviation of the axis of angular momentum is remarkable.

044.095   On the comparison and choice between the formulae of the earth's tidal correction in astronomical time and latitude observations.
J.–y. Xia.
Acta Astron. Sin., Vol. 25, No. 4, p. 419 – 425 (1984). In Chinese.

The author collects three formulae of the earth's tidal correction in astronomical time and latitude observations and he calculates and compares two of them, i.e. the equilibrium tidal formula and the Wahr's one.

044.096   Effect of oceanic tides on astronomical latitude and longitude observations.
H.–z. Xu, F.–m. Guo, Z.–b. Chen, Y.–x. Sun.
Acta Astron. Sin., Vol. 26, No. 2, p. 162 – 171 (1985). In Chinese.

044.097   Peculiarities of a study of the earth's rotation with radio astrometrical methods.
V. S. Gubanov.
Izv. Glav. Astron. Obs. Pulkovo, Astrometr. Astrofiz., No. 201, p. 52 – 59 (1985). In Russian.

Possibilities of determining the parameters of the rotation of the earth using VLBI observations of extragalactic radio sources and artificial earth satellites are considered. The influence of the polar motion and tidal deformations of the earth on the orientation of the vector of the interferometer baseline is noted. It is shown that both components of the quasi–diurnal nutation of the earth's mantle can be practically obtained from round–the–clock observations with one interferometer, while at least two interferometers are required for the measurement of both components of long–period polar motions.

044.098   An analysis of long–period variations of amplitudes of the Chandler and annual components of the polar motion of the earth.
L. D. Kostina, V. I. Sakharov.
Izv. Glav. Astron. Obs. Pulkovo, Astrometr. Astrofiz., No. 201, p. 60 – 63 (1985). In Russian.

Variations of the amplitudes of the Chandler and annual polar motions have been calculated for the period 1890.00 1980.85. There are polycyclic oscillations with the cycles 46.2, 20.0, 10.6, 9.0, 6.9, 6.0, 4.6 yrs and 42.0, 19.0, 14.0, 10.0, 6.8, 5.1, 4.0 yrs, respectively. Mean values of the above amplitudes are equal to 0".147 and 0".085.

044.099   The Antarctic Circumpolar Current and its influence on the Earth's rotation.
P. Brosche, J. Sündermann.
Deutsche Hydrographische Zeitschrift (Hamburg), Band 38, Heft 1, p. 1 – 6 (1985).

The main question is: how much of the oceanic angular momentum is temporarily stored within the oceans and what is the time scale of the transfer to the solid Earth. As an example, the authors have estimated the phase and the amplitude of the angular momentum which is stored in the Antarctic Circumpolar Current. Its phase resembles the one of the whole observed semiannual discrepancy in the angular momentum budget of the solid Earth plus the atmosphere; the amplitudes are comparable.

On the RA catalogues F10 and F12 of the Pulkovo Time Service.
See Abstr. 002.117.

The earth and its rotation.
See Abstr. 003.135.

Uses for ancient eclipse records.
See Abstr. 004.001.

Developments of time systems since 1884.
See Abstr. 004.049.

Prehistorical evidence for precession.
See Abstr. 004.110.

International coordination and Atomic Time.
See Abstr. 004.146.

The research work at the Central Institute for Physics of the Earth, Potsdam, GDR, in the field of Doppler satellite geodesy.
See Abstr. 013.022.

Der Beitrag der Universitäts–Sternwarte Wien zum Projekt MERIT.
See Abstr. 013.034.

The programme of observations with the Danjon astrolabe at the Astronomical Latitude Observatory in Borowiec.
See Abstr. 013.050.

The terrestrial coordinate system and international Earth–rotation services.
See Abstr. 013.051.

The application of the PZT 2 photographic zenith telescope in the Potsdam geodetic–astronomical observatory.
See Abstr. 032.016.

Correction to the angular value of the screw revolution of the Belgrade zenith–telescope micrometer derived from latitude observations.
See Abstr. 032.022.

A simple direct TV satellite receiver for time comparison.
See Abstr. 034.068.

An automatic multiple switch constructed by means of a single board microprocessor.
See Abstr. 034.069.

Hydrogen masers for Rumanian atomic time.
See Abstr. 034.162.

Guidelines for the reduction of optical astrometry observations during the main campaign in Project MERIT.
See Abstr. 036.113.

Improvement of a method for determining the photoelectric delay and a design of photoelectric equipment with zero temperature drift.
See Abstr. 036.116.

Changes in personal equations of circumzenithal observers.
See Abstr. 036.152.

Universal time, lunar tidal deceleration and relativistic effects from observations of transits, eclipses and occultations in the XVIIIth – XXth centuries.
See Abstr. 042.009.

Reference coordinate systems and frames: concepts and realization.
See Abstr. 043.006.

Further improvements of the orbital program system POTSDAM–5 and their utilization in geodetic – geodynamic investigations.
See Abstr. 045.005.

Geodetic and geophysical results from Lageos.
See Abstr. 045.013.

Latest instruments and methods in the astrogeodetic field.
See Abstr. 045.023.

The "Prognoz" routine, its structure and first results.
See Abstr. 052.008.

Sensitivity analysis and optimal design of range, range–difference, and range–rate measurements for the determination of pole coordinates and variation of Earth's rotation.
See Abstr. 052.011.

Around–the–world relativistic Sagnac experiment.
See Abstr. 066.107.

Growth rhythms, evolution of the Earth's interior, and origin of the Metazoa.
See Abstr. 081.058.

Atmospheric refraction effects in time and latitude observations using classical techniques.
See Abstr. 082.088.

045 Astronomical Geodesy, Satellite Geodesy, Navigation

045.001   Determining highly elliptical earth orbits with VLBI and ΔVLBI.
R. B. Frauenholz, J. Ellis.
J. Astronaut. Sci., Vol. 32, No. 2, p. 159 – 174 (1984). Abstr. in Phys. Abstr., Vol. 88, No. 1250, Entry 18658 (1985).

045.002   Altimetry–gravimetry functional boundary value problem.
L. D. Stoyanov.
Publ. Astron. Inst. Czech. Acad. Sci., No. 58, p. 265 – 279 (1984). In Russian. – See Abstr. 012.018.
The functional conditions referring to the anomalous potential have been worked out by the integral formula of Green. The values of the functionals over the continental areas have been calculated by the measured values of the gravity anomaly and the components of deflection of the vertical or of the geopotential number and for ocean areas by the measured values of the gravity disturbance or of the boundary values of the potential. The functional problem has been reduced to the restoration of the finite–dimensional and unique geopotential which satisfies the functional conditions best (in least–squares sense). The author proposes four spherical solutions of the altimetry–gravimetry boundary problem by a discrete description and two in a closed form by a generalisation of the function of Stokes and Neumann (Hotine kernel).

045.003   Determination of the position of the Station Borowiec No. 7811 by satellite laser observations.
W. Dobaczewská, A. Drozyner, M. Rutkowska, S. Schillak, J. B. Zieliński.
Publ. Astron. Inst. Czech. Acad. Sci., No. 58, p. 281 – 293 (1984). – See Abstr. 012.018.
Laser observations were performed in Borowiec in three years 1977 – 79 of the satellites Geos A and Geos C. These data were processed by means of the program ORBITA and station coordinates were calculated by dynamical methods. Another solution was found with the processing by the program GRIPE of SAO. These two dynamical solutions are compared with the translocation solution Wettzel–Borowiec.

045.004   An analysis of the optimal transformation of coordinate systems.
M. Mojzeš, J. Pecár.
Publ. Astron. Inst. Czech. Acad. Sci., No. 58, p. 295 – 308 (1984). – See Abstr. 012.018.
The paper deals with an analysis of optimal determination of transformation parameters between two coordinate systems when using D–optimality and Σ–optimality. The analysis is applied to the network of geodynamic satellite stations GEOS–REA II.

045.005   Further improvements of the orbital program system POTSDAM–5 and their utilization in geodetic – geodynamic investigations.
G. Gendt.
Publ. Astron. Inst. Czech. Acad. Sci., No. 58, p. 421 – 427 (1984). = Mitt. Zentralinst. Phys. Erde, Nr. 1374. – See Abstr. 012.018.
The program system POTSDAM–5 was developed in the last years on the basis of POTSDAM–4. It is implemented on the computer EC 1040 and will provide a model accuracy of some cm. This high precision can only be achieved by improvements of the reference system, force model and parameter estimation. Results are presented using laser data for the determination of station coordinates, baselines and polar motion.

045.006   An indirect way to determine the geocentric coordinates of the Hvar Doppler station in PE–system starting from two new MPBE–solutions for the project IDOC–82.
K. Colić, F. J. Lohmar, M. Solarić.
Publ. Astron. Inst. Czech. Acad. Sci., No. 58, p. 477 – 486 (1984). – See Abstr. 012.018.
The authors have tried to determine as accurately as possible the geocentric coordinates of the Hvar Doppler station in the coordinate systems of broadcast and precise ephemerides.

045.007   Results of the Photodoppler method from observations of the satellite NOVA.
T. Borza, O. A. Dorodnitsyna.
Publ. Astron. Inst. Czech. Acad. Sci., No. 58, p. 499 – 504 (1984). In Russian. – See Abstr. 012.018.
In 1983 a special campaign was organized in the framework of program Photodoppler, with observations of satellite NOVA. About 400 directions were measured at station Riga. The data were processed and Riga coordinates obtained with an accuracy similar to those of Single Point Positioning. In this method the photographic camera can be considered as a stationary Doppler receiver.

045.008 Preliminary results of Finnish–Hungarian Doppler Observation Campaign.
A. Czobor, J. Ádám, S. Mihály, T. Vass, T. Parm, M. Ollikainen.
Publ. Astron. Inst. Czech. Acad. Sci., No. 58, p. 529 – 548 (1984). – See Abstr. 012.018.

The Finnish–Hungarian Doppler Observation Campaign was carried out in Finland in August 1983 during 13 days. The campaign was organized by the Finnish Geodetic Institute. Three Hungarian JMR–1A receivers and one Finnish JMR–4 occupied 9 stations together with the 1st order triangulation network points of Finland. The data processing was performed in the Satellite Geodetic Observatory, Penc. The observation strategy, the methods of data processing and results obtained by GEODOP and SADOSA programs as well as the S–transformations are presented in this paper.

045.009 Analysis of satellite laser ranging observations of LAGEOS to derive observing station coordinates and interstation baselines.
A. T. Sinclair, G. M. Appleby, J. Xia.
Geophys. J. R. Astron. Soc., Vol. 81, No. 1, p. 314 (1985). Abstract. – See Abstr. 012.025.

045.010 The Global Positioning System and geophysical applications.
V. Ashkenazi, J. Yau.
Geophys. J. R. Astron. Soc., Vol. 81, No. 1, p. 314 – 315 (1985). Abstract. – See Abstr. 012.025.

045.011 Application of differential superlong baseline radio interferometry to astronavigation.
L. R. Kogan, L. I. Matveenko, V. I. Kostenko.
Kosm. Issled., Tom 23, Vyp. 1, p. 167 – 174 (1985). In Russian. English translation in Cosm. Res.

045.012 Geophysical model improvement using TRANET data.
R. Anderle, E. Colquitt.
Adv. Space Res., Vol. 5, No. 2, p. 211 – 217 (1985). – See Abstr. 012.051.

Doppler observations of U.S. Navy Navigation Satellites have been used to strengthen terrestrial networks in many areas of the world and to connect the networks and isolated sites to an earth–centered coordinate system. Parameters of the model used in the computation of the satellite ephemerides used for this purpose result in geodetic positions which are displaced from the center of mass of the earth by about 4 m with respect to the equator, and in longitudes which are rotated from the BIH conventional longitude by about 0.80 seconds of arc. Third order ionospheric effects neglected in the orbit computations in the calculation of site coordinates result in radial displacements of as much as several meters for high levels of solar activity when the right ascension of the node of the satellite is near the right ascension of the sun.

045.013 Geodetic and geophysical results from Lageos.
D. E. Smith, D. C. Christodoulidis, R. Kolenkiewicz, M. H. Torrence, S. M. Klosko, P. J. Dunn.
Adv. Space Res., Vol. 5, No. 2, p. 219 – 228 (1985). – See Abstr. 012.051.

Seven years of laser tracking of the Lageos spacecraft have been used to derive geodetic quantities describing the earth and its rotational motion. The dynamical motions of the solid–earth on its axis have been derived continuously since launch and changes in the length–of–day show very high correlation with variations in the atmospheric zonal winds between 1000 and 50 mbars. A significant improvement in the determination of the product of the earth's mass and the gravitational constant has been made. The high accuracy of the orbit determination of Lageos over the 7 years since launch has permitted the identification of a small deceleration in the nodal precession of the orbit. Measurements of the distances between the tracking stations over several years are showing changes consistent with tectonic plate motion and with general ideas of vertical movements.

045.014 On an estimate of the accuracy of determinations of astronomical longitudes at high latitudes.
V. Z. Khalkhunov.
Geod. kartogr., No. 10, p. 4 – 6 (1984). In Russian. Abstr. in Ref. Zh., 52. Geod. Aehrosemka, 4.52.82 (1985).

045.015 The geodetic reference systems – the basis of theoretical geodesy.
L. Stange.
Mitt. Lohrmann–Obs. Tech. Univ. Dresden, Nr. 51, p. 55 – 58 (1984). In German. – See Abstr. 012.058.

045.016 Calculation of ephemerides for nautical purposes.
J. Böhme.
Mitt. Lohrmann–Obs. Tech. Univ. Dresden, Nr. 51, p. 68 – 70 (1984). In German. – See Abstr. 012.058.

045.017 A contribution to solving semidynamic problems in satellite geodesy.
J. Kabeláč.
Mitt. Lohrmann–Obs. Tech. Univ. Dresden, Nr. 51, p. 80 – 84 (1984). In German. – See Abstr. 012.058.

045.018 Preliminary results of the Doppler observation campaign in Poland during 1983.
J. B. Rogowski.
Mitt. Lohrmann–Obs. Tech. Univ. Dresden, Nr. 51, p. 94 (1984). – See Abstr. 012.058.

045.019 The investigation of geodynamic parameters by means of laser ranging to artificial Earth satellites within the framework of the MERIT project.
G. Gendt, H. Montag.
Mitt. Lohrmann–Obs. Tech. Univ. Dresden, Nr. 51, p. 104 – 105 (1984). In German. – See Abstr. 012.058.

045.020 State of development of satellite laser ranging.
R. Neubert, L. Grunwaldt.
Mitt. Lohrmann–Obs. Tech. Univ. Dresden, Nr. 51, p. 105 – 106 (1984). In German. – See Abstr. 012.058.

045.021 The satellite ranging system of the Central Institute of Physics of the Earth in Potsdam.
H. Fischer, L. Grunwaldt, R. Neubert, C. Selke, R. Stecher.
Mitt. Lohrmann–Obs. Tech. Univ. Dresden, Nr. 51, p. 130 – 131 (1984). In German. – See Abstr. 012.058.

045.022 A decimal system of celestial navigation based on the second.
K. K. White.
J. Navig., Vol. 38, No. 2, p. 293 – 295 (1985).

045.023 Latest instruments and methods in the astrogeodetic field.
P. Terzić.
Publ. Astron. Opservatorije Beogr., No. 33, p. 50 – 52 (1985). – See Abstr. 012.061.

045.024 Résultats des observations Doppler effectuées à Uccle de 1972 à 1983.
P. Pâquet, R. Verbeiren, V. Dehant.
Bull. Astron., Vol. 9, No. 6, p. 266 – 282 (1984).

045.025 On a possibility of determining the astronomical azimuth from photographic observations of the sun.
A. N. Denisov, V. A. Kovalenko.
Geod., kartogr. aehrosemka, L'vov, No. 41, p. 20 – 22 (1985). In Russian. Abstr. in Ref. Zh., 52. Geod. Aehrosemka, 6.52.64 (1985).

045.026 On the exact transmission of the grid bearing and plumb–line deflections from synchronous observations of celestial bodies at two points of the earth surface.
A. E. Filippov.
Geod., kartogr. aehrosemka, L'vov, No. 41, p. 114 – 119 (1985). In Russian. Abstr. in Ref. Zh., 52. Geod. Aehrosemka, 6.52.65 (1985).

045.027 Space geodynamics.
J. Klokočník.
Říše hvězd, Vol. 66, No. 1, p. 6 – 12 (1985). In Czech.

045.028 Astronomy, geodesy and geophysics.
J. Vondrák.
Říše hvězd, Vol. 66, No. 3, p. 59 – 61 (1985). In Czech.

Intrinsic geodesy.
See Abstr. 003.043.

Celestial navigation.
See Abstr. 003.066.

Astro–navigation by calculator.
See Abstr. 003.126.

Astronomical navigation since 1884.
See Abstr. 004.050.

Fifteen years of Lunar Laser Ranging and future developments.
See Abstr. 013.021.

The research work at the Central Institute for Physics of the Earth, Potsdam, GDR, in the field of Doppler satellite geodesy.
See Abstr. 013.022.

West–East Doppler Observation Campaign WEDOC–2.
See Abstr. 013.023.

Observation equations based on expansions into eigenfunctions.
See Abstr. 021.015.

Calibration of Doppler receivers.
See Abstr. 036.026.

Results of the observations with Doppler receiver DOG–2.
See Abstr. 036.027.

Estimability of astronomical longitude and latitude only from theodolite observations within three–dimensional networks of terrestrial type.
See Abstr. 044.059.

On computation of satellite orbits for geodynamical research.
See Abstr. 052.001.

The "Prognoz" routine, its structure and first results.
See Abstr. 052.008.

Problems of application of the analytical theory of the earth artificial satellite motion in geodynamics and satellite geodesy.
See Abstr. 052.009.

Modelling and optimization of Doppler satellite observations.
See Abstr. 052.015.

Altimetry, orbits and tides.
See Abstr. 052.055.

Survey of relativistic effects in geodesy and fundamental astronomy.
See Abstr. 066.186.

Local approximation of the quasi–geoid from astrogeodetic data.
See Abstr. 081.032.

Overview of geodetic refraction studies.
See Abstr. 082.084.

Two wavelength angular refraction measurement.
See Abstr. 082.085.

Zur Berechnung ionosphärischer Refraktionskorrekturen für VLBI–Beobachtungen aus simultanen Dopplermessungen nach Satelliten.
See Abstr. 083.013.

046 Ephemerides, Almanacs, Calendars, Chronology

046.001 1986 Japanese Ephemeris. Pub. No. 684.
Hydrographic Department, Maritime Safety Agency, Tokyo, Japan. 6 + 475 + 43 pp. (1985). ISSN 0373–3696.

046.002 Het jaar van de stier.
O. Namba.
Zenit, 12. Jaarg., No. 2, p. 44 – 46, 48 – 50 (1985).

046.003 Greek astronomical calendars. IV. The parapegma of the Egyptians and their "perpetual tables".
B. L. van der Waerden.
Arch. Hist. Exact Sci., Vol. 32, No. 2, p. 95 – 104 (1985).

046.004 Astronomical Calendar of the Sofia Observatory for the year 1985.
B. Kovachev, D. Rajkova, Z. Krajcheva, V. Ivanova, A. Antov; edited by A. Bonov.
Izdatelstvo na Blgarskata Akademiya na Naukite, Sofiya, Bulgaria. 109 pp. Price 1.24 Lv. (1984). ISSN 0224–1408. In Bulgarian.

046.005 Posizioni dei satelliti di Giove per l'anno 1985.
P. Gregorio.
Orione, Vol. 5, N. 2, p. 56 – 59 (1985).

046.006 Anuário Astronômico 1985.
Instituto Astronômico e Geofisico, Universidade de São Paulo, Caixa Postal 30627, São Paulo, Brasil. 12 + 279 pp. (1984). ISSN 0080–6412.

046.007 Anuario del Observatorio Astronómico de Madrid para 1985.
Instituto Geográfico Nacional, Madrid, Spain. 502 pp. Price Pts. 400.00 (1984). ISBN 84–505–1088–0, ISSN 0373–5125.

046.008 Tables of sunrise, sunset, twilight, moonrise, & moonset 1985.
Prepared by the Astronomical Observation Division, National Geophysical and Astronomical Office.
Philippine Atmospheric, Geophysical and Astronomical Services Administration, Quezon City, Philippines, 10 + 57 pp. (1984). ISSN 0115–3307.

046.009   The Star Almanac for Land Surveyors for the year 1986.
Prepared by H.M. Nautical Almanac Office.
Her Majesty's Stationery Office, London, England. 16 + 80 pp.
Price £ 2.50 (1985). ISBN 0–11–886924–8.

046.010   Nautički Godišnjak 1985.
Hidrografski Institut Jugoslavenske Ratne Mornarice,
Split, Yugoslavia. HI–N–31, Godina 43, 11 + 213 + 69 pp. (1984).

046.011   Jupiter's satellites.
J. Br. Astron. Assoc., Vol. 95, No. 4, p. 167 – 168
(1985).

046.012   Astronomical ephemeris for the year 1985.
N. Čabrić.
Vasiona, Année 32, No. 4, p. 69 – 82 (1984). In Croatian.

**046.013   Blick in die Sternenwelt 1985. Astronomischer Kalender
der Archenhold–Sternwarte.**
E. Rothenberg.
Archenhold–Sternw. Berlin–Treptow, 48 pp. (1984).

046.014   Quelques phénomènes astronomiques remarquables.
J. Meeus.
Astronomie, Vol. 99, p. 300 – 302 (1985).

**046.015   Generalized equations for Julian day numbers and calen-
dar dates.**
D. A. Hatcher.
Q. J. R. Astron. Soc., Vol. 26, No. 2, p. 151 – 155 (1985).
A distinction between given and computational calendars allows a single set of equations to be used for calendars as diverse
as the Gregorian solar calendar and the schematic Islamic lunar
calendars.

046.016   The Astronomical Almanac for the year 1986.
Data for astronomy, space sciences, geodesy, survey-
ing, navigation and other applications.
Issued by the Nautical Almanac Office, United States Naval
Observatory (Washington) and Her Majesty's Nautical Almanac
Office, Royal Greenwich Observatory (London).
US Edition: Superintendent of Documents, US Government
Printing Office, Washington, D.C. 20402, USA. UK Edition:
Her Majesty's Stationery Office, London, England. 9 + 532 pp.
Price £ 18.00 (1985). ISBN 0–11–886923–X.

046.017   Almanac for Geodetic Engineers 1985.
Prepared by the Astronomical Observation Division of
the National Geophysical and Astronomical Office.
Philippine Atmospheric, Geophysical and Astronomical Services
Administration, Quezon City, Philippines. 10 + 26 pp. (1984).
ISSN 0569–0838.

**046.018   Connaissance des Temps. Ephémérides Astronomiques
pour l'An 1986.**
Bureau des Longitudes, 77 Avenue Denfert Rochereau, 75014
Paris, France. 54 + 124 pp. Price FF 150,00 (1985). ISBN
2–11–080454–8, ISSN 0181–3048.

046.019   Philippine Astronomical Handbook 1985.
Prepared by the Astronomical Observation Division of
the National Geophysical and Astronomical Office.
Philippine Atmospheric, Geophysical and Astronomical Services
Administration, Quezon City, Philippines. 12 + 64 pp. (1984).
ISSN 0115–1207.

**046.020   On the Japanese ephemeris for the year 1985 – reforma-
tion of the astronomical ephemeris.**
K. Inoue.
Navigation (Tokyo), No. 81, p. 67 – 71 (1984). In Japanese.
Abstr. in Phys. Abstr., Vol. 88, No. 1255, Entry 45899 (1985).

046.021   International Geophysical Calendar for 1985.
J. Atmos. Terr. Phys., Vol. 46, No. 12, p. 1233 – 1235
(1984).

**046.022   Report of IAU Commission 4: Ephemerides (*Ephémé-
rides*).**
T. Lederle.
Trans. IAU, Vol. XIXA, p. 1 – 6 (1985). – See Abstr. 003.046.

046.023   Efemérides Astronômicas 1985.
CNPq Conselho Nacional de Desenvolvimento Cientí-
fico e Tecnológico, Observatório Nacional, Rio de Janeiro, Brazil. 7 + 529 pp. (1984).

046.024   Astronomische Grundlagen für den Kalender 1987.
Compiled by T. Lederle, edited by Astronomisches
Rechen–Institut, Heidelberg.
Verlag G. Braun, Karl–Friedrich–Straße 14 – 18, D–7500 Karls-
ruhe 1, F.R. Germany. 84 pp. Price DM 59.00 (1985). ISBN
3–7650–0186–4, ISSN 0067–0014.

046.025   Der Sternenhimmel 1985.
W. Burgat.
Annuaire astronomique de la Schweizerischen Astronomischen
Gesellschaft. 45e année. Verlag Sauerländer, Postfach 570,
CH–5001 Aarau, Switzerland. 192 pp. Price SFr 29.80.
Review in Ciel Terre, Vol. 101, No. 3, p. 99; 1985 (*W. Willems*).

046.026   Sterrengids 1985.
Compiled by W. Gielingh, J. Meeus.
Nederlandse Vereniging voor Weer– en Sterrenkunde en Stich-
ting De Koepel, Utrecht, The Netherlands. 157 pp. Price f 29.50;
BF 530.00. ISBN 90–6638–004–7.
Review in Zenit, 12. Jaarg., No. 3, p. 110; 1985 (*A. Mak*).

046.027   Skywatcher's Almanac 1985.
R. L. Mansfield.
Astronomical Data Service, 3922 Leisure Lane, Colorado
Springs, Colo. 80917, USA. 38 pp. Price US$ 13.00 (1985).
Reviews in Mercury, Vol. 14, No. 3, p. 88 (1985); Sky Telesc.,
Vol. 69, No. 2, p. 132 (1985).

046.028   Hemelkalender 1985.
J. Meeus.
Numéro spécial de Heelal (Bulletin de la VVS). 100 pp. Price
BF 200.00.
Review in Ciel Terre, Vol. 101, No. 1, p. 23 – 24; 1985
(*R. Charles*).

046.029   Astronomical Calendar 1985.
G. Ottewell.
Department of Physics, Furman University, Greenville,
S.C. 29613, USA. 66 pp. Price US$ 10.00 (1984). ISBN
0–934546–13–4.
Reviews in Sci. Am., Vol. 252, No. 1, p. 14 – 15; 1985
(*P. Morrison*); Sky Telesc., Vol. 69, No. 3, p. 229 (1985); Strolling
Astron., Vol. 31, Nos. 1 – 2, p. 41; 1985 (*J. R. Smith*).

046.030   Astronomisk årsbok 1985.
A. L. J. Schildt (Editor).
Bokförlaget INOVA, Stockholm, Sweden. 96 pp. Price
Sv.kr. 30.00 (1984).
Review in Astron. Tidsskr., Årg. 18, Nr. 2, p. 83 – 84; 1985
(*S. Söderhjelm*).

046.031   Stjärnhimlen 1985.
P. Schlyter, M. Malmort, T. Jürisoo, P. Aulin.
Bokförlaget INOVA, Stockholm, Sweden. 112 pp. Price
Sv.kr. 45.00 (1984).
Review in Astron. Tidsskr., Årg. 18, Nr. 2, p. 82 – 83; 1985
(*S. Söderhjelm*).

046.032   Almanacka för 500 år – från år 1500 till år 2000.
K.–G. Segland.
Förlaget Kalendator, Enköping, Sweden. 79 pp. Price
Sv.kr. 55.00 (1984).
Review in Astron. Tidsskr., Årg. 18, Nr. 2, p. 89 (1985).

046.033   Calendrier astronomique 1985.
Edité par l'Association Française d'Astronomie, 17, rue Emile–Deutsch–de–la Meurthe, F–75014 Paris, France. Price FF 40.00.
Review in Ciel Terre, Vol. 101, No. 1, p. 24; 1985 (*J. Sauval*).

046.034   Ephémérides des satellites de Jupiter, Saturne et Uranus pour 1986. (Ephemerides of the satellites of Jupiter, Saturn and Uranus for 1986).
Supplément à la Connaissance des Temps – Bureau des Longitudes.
Les Editions de Physique, avenue du Hoggar, Zone Industrielle de Courtabœuf, B.P. 112, F–91944 Les Ulis Cedex, France. 93 pp. Price FF 190.00 (1985). ISBN 2–902731–98–1.

The accuracy of ephemerides, 1500 – 1800.
See Abstr. 004.165.

On calculation of some data of the astronomical yearbook with EC–computers.
See Abstr. 021.009.

Formulae for calculating the corrections for the light deflection to apparent places.
See Abstr. 041.007.

Millisecond pulsars and the location of the solar system barycenter.
See Abstr. 091.068.

Ephemerides of minor planets for 1986.
See Abstr. 098.096.

Space Research

051 Extraterrestrial Research Related to Astronomy and Astrophysics

051.001 Future ultraviolet experiments, including FUSE/ COLUMBUS.
A. Boggess.
NASA Conf. Publ., NASA CP–2349, p. 80 (1984). Abstract. – See Abstr. 012.001.

051.002 Giotto's enkele reis naar komeet Halley.
P. Lindhout.
Zenit, 12. Jaarg., No. 1, p. 4 – 6, 8 – 14 (1985).

051.003 An asteroid for the asking.
J. K. Beatty.
Sky Telesc., Vol. 69, No. 2, p. 127 (1985).

051.004 The X–ray Timing Explorer.
J. E. McClintock, A. M. Levine.
AIP Conf. Proc., No. 115, p. 642 – 650 (1984). – See Abstr. 012.005.

Near the end of this decade, the X–ray Timing Explorer (XTE) will provide X–ray astronomers with a tool of unprecedented sensitivity for the study of compact X–ray sources. The principal instruments, a 1 m^2 proportional counter array and a 2000 cm^2 phoswich array, will view the sky through coaligned one–degree collimators and provide high sensitivity over a 2 – 200 keV energy range. During the course of each satellite orbit, the entire sky will be scrutinized by the sky monitor experiment to a sensitivity of 0.05 Crab.

051.005 The High Energy Transient Explorer (HETE).
E. L. Chupp, T. L. Cline, W. D. Evans, E. Fenimore, P. Gorenstein, J. Grindlay, D. Q. Lamb, W. H. G. Lewin, R. E. Lingenfelter, J. L. Matteson, R. Ramaty, G. R. Ricker, G. Share, B. Teegarden, S. E. Woosley.
AIP Conf. Proc., No. 115, p. 709 – 714 (1984). – See Abstr. 012.005.

051.006 A preliminary spectroscopic assessment of the Spacelab 1/Shuttle optical environment.
M. R. Torr, D. G. Torr.
J. Geophys. Res., Vol. 90, No. A2, p. 1683 – 1690 (1985).

The Spacelab 1/Shuttle mission which orbited for 10 days following launch on November 28, 1983, carried amongst the complement of instruments, an array of imaging spectrometers known as the Imaging Spectrometric Observatory. Because the instrument covers a broad wavelength range extending from the vacuum ultraviolet to the near infrared and has a relatively high spectral resolution (3 – 6 Å), it has provided some of the first spectral information on the Shuttle optical environment. Shuttle flights to date have shown that surfaces directed into the velocity vector develop a bright orange–red glow. As there are extensive plans for optical studies from the Shuttle, as well as other orbiting structures such as the Space Telescope and Space Station, there has been considerable concern as to whether such glows and other sources of optical contamination might prevent or limit observations. In this paper the authors present spectral data from 1150 to 8000 Å measured with the instrument looking tangentially away from the earth and into the velocity vector.

051.007 The HESP/R satellite project.
K. Tanaka, E. Hiei.
Adv. Space Res., Vol. 4, No. 7, p. 377 – 381 (1984). – See Abstr. 012.009.

The aim of the HESP/R (High Energy Solar Physics/ Radiation) satellite project is to obtain data of γ–ray, hard X–ray soft X–ray, EUV, and visible radiation of solar flares at the next solar maximum in order to study the physics of flares.

051.008 The Solar and Heliospheric Observatory, SOHO – a Phase–A project of the European Space Agency.
M. Malinovsky–Arduini, C. Fröhlich.
Adv. Space Res., Vol. 4, No. 7, p. 383 – 392 (1984). – See Abstr. 012.009.

In this paper the present understanding of coronal heating, solar wind generation and solar oscillations is described. The proposed model SOHO instrument payload is outlined and it is shown how it would contribute to our understanding in the three fields mentioned.

051.009 Initial results from the repaired Solar Maximum Mission and future prospects.
B. E. Woodgate.
Adv. Space Res., Vol. 4, No. 7, p. 393 – 402 (1984). – See Abstr. 012.009.

Goals of the recently repaired Solar Maximum Mission Observatory are outlined, including continued emphasis on diagnosing impulsive phase of flares, studies of prominence and coronal plasmas, solar cycle variations of flares, the corona and solar irradiance, and comets. Some preliminary observations taken after the repair are shown, particularly of the X13 flare of April 1984.

051.010 On the PSIVA (*Probing Stellar Interiors via Variability and Activity*) approach to stellar seismology and activity from space.
F. Praderie, A. Mangeney, P. Lemaire.
Adv. Space Res., Vol. 4, No. 8, p. 163 – 168 (1984). – See Abstr. 012.010.

The authors propose a modest payload, using a simple telescope and well known technology, with the aim of placing the probing of stellar interiors on an experimental basis. Two types of observations are sought: The pulsation spectrum and the active phenomena of magnetic origin in the atmosphere, at various altitudes simultaneously, for a representative sample of stars over the Hertzsprung–Russell diagram. The basic requirements are long series of continuous and homogeneous observations with high photometric accuracy in white light and in four spectroscopic indices, three of them being in the UV range (Ca II λ3933 Å, Mg II λ2800 Å, He II λ1640 Å, C IV λ1550 Å). The motives to perform these observations from space are analyzed.

051.011 A survey for photometric variability from space.
H. S. Hudson.
Adv. Space Res., Vol. 4, No. 8, p. 169 – 175 (1984). – See Abstr. 012.010.

A survey for photometric variability in a wide variety of astronomical objects would produce much new information about their interiors and dynamics. This paper discusses reasons for such a survey, showing the example of the recent precise SMM observations of total solar irradiance variations as a guide to what might be expected from main–sequence stars, and proposes a concept for a satellite dedicated to a survey of photometric variability.

051.012 Space Report.
Spaceflight, Vol. 27, No. 1, p. 6 – 9 (1985).
1985 January review of space news and events.

051.013 Space Report.
Spaceflight, Vol. 27, No. 2, p. 65 – 68 (1985).
1985 February review of space news and events.

051.014 Space at JPL.
W. I. McLaughlin.
Spaceflight, Vol. 27, No. 2, p. 71 – 74 (1985).
Contents: Hypervelocity impact. Possible new solar system. NSCAT (*NASA Scatterometer*) starts up. Uranian rings. Shuttle imaging radar.

051.015 Space Report.
Spaceflight, Vol. 27, No. 3, p. 110 – 113 (1985).
1985 March review of space news and events.

051.016 Space Report.
Spaceflight, Vol. 27, No. 4, p. 150 – 153 (1985).
1985 April review of space news and events.

051.017 Space at JPL.
W. I. McLaughlin.
Spaceflight, Vol. 27, No. 4, p. 163 – 167 (1985).
Contents: Technology. Nuclear–electric power. The Hypercube. Artificial intelligence. Viking technology. IRAS catalogue delivered. A weekend in Brighton.

051.018 The Pioneer 10/11 program: from 1969 to 1994.
C. F. Hall.
IEEE Trans. Reliab., Vol. R–32, No. 5, p. 414 – 416 (1983). Abstr. in Phys. Abstr., Vol. 88, No. 1249, Entry 14450 (1985).

051.019 Geopotential Research Mission (GRM).
T. Keating.
J. Astronaut. Sci., Vol. 32, No. 2, p. 145 – 158 (1984). Abstr. in Phys. Abstr., Vol. 88, No. 1250, Entry 18621 (1985).

051.020 Cassini – a concept for a Titan probe.
G. E. N. Scoon.
ESA Bull., No. 41, p. 12 – 20 (1985).
In April 1984, the Joint Science Working Group on Cooperation for Planetary Exploration recommended three candidate projects as possible joint missions to Saturn. This article outlines the novel features of this interesting mission and presents a preliminary design concept for the "Titan probe" which might represent ESA's contribution.

051.021 Soho and Cluster: Europe's possible contribution to the ISTP (*International Solar Terrestrial Physics Programme*).
A. Atzei, J. Ellwood, G. Whitcomb.
ESA Bull., No. 41, p. 21 – 27 (1985).

051.022 Wetenschappelijke ruimtevloot zet koers naar komeet Halley.
P. van Nes.
Zenit, 12. Jaarg., No. 2, p. 52 – 60 (1985).

051.023 Comet coma sample return via Giotto II.
P. Tsou, D. E. Brownlee, A. L. Albee.
J. Br. Interplanet. Soc., Vol. 38, No. 5, p. 232 – 239 (1985).
A comet coma sample return is possible with a low–cost flyby mission. Collecting coma materials and returning them to Earth can be accomplished in a free–return trajectory. This paper focuses on the sample return aspects, including sample return objectives, sample collection techniques, experimental work to verify collection concepts, and some of the characteristics of the cometary targets for sample return.

051.024 The International Cometary Explorer Mission to P/Giacobini–Zinner and P/Halley.
J. C. Brandt.
Int. Comet Q., Vol. 7, No. 2, p. 35 – 39 (1985).

051.025 The radiation situation in space and its modification by geomagnetic field and shielding.
W. Heinrich, J. Beer.
Adv. Space Res., Vol. 4, No. 10, p. 133 – 142 (1984). – See Abstr. 012.031.
The fluxes of the nuclear component of the galactic cosmic radiation are discussed in terms of energy spectra for the different elements. Influences of shielding by the earth's magnetic field on these spectra are described. Then energy spectra behind absorbing matter are calculated considering energy loss and fragmentation. Based on these energy spectra LET–spectra are calculated. The form of the LET–spectra and their dependence on the composition of the shielding material are discussed.

051.026 LET–distributions and doses of HZE radiation components at near–earth orbits.
R. Silberberg, C. H. Tsao, J. H. Adams Jr, J. R. Letaw.
Adv. Space Res., Vol. 4, No. 10, p. 143 – 151 (1984). – See Abstr. 012.031.
Among cosmic rays, the heavy nuclei ranging from carbon to iron provide the principal contribution to the dose equivalent. The LET–distributions and absorbed dose and dose equivalent have been calculated and are presented as a function of shielding and tissue self–shielding.

051.027 Summary of current radiation dosimetry results on manned spacecraft.
E. V. Benton.
Adv. Space Res., Vol. 4, No. 10, p. 153 – 160 (1984). – See Abstr. 012.031.

051.028 Cosmic ray exposure factors for shuttle altitudes derived from calculated cut–off rigidities.
D. F. Smart, M. A. Shea.
Adv. Space Res., Vol. 4, No. 10, p. 161 – 164 (1984). – See Abstr. 012.031.
The allowed cosmic radiation flux accessible to an earth–orbiting spacecraft is a complex function of the satellite position and the geomagnetic cutoff characteristics at each zenith and azimuth angle at each position. The authors have determined cosmic ray exposure factors for the galactic cosmic ray spectrum for typical shuttle altitudes and inclinations up to 50°.

051.029 The monitoring and prediction of solar particle events – an experience report.
G. Heckman, J. Hirman, J. Kunches, C. Balch.
Adv. Space Res., Vol. 4, No. 10, p. 165 – 172 (1984). – See Abstr. 012.031.

051.030 LDEF–1 and beyond.
D. O'Sullivan.
Ir. Astron. J., Vol. 17, No. 1, p. 40 – 47 (1985). – See Abstr. 012.035.

051.031 The X–ray view of the universe.
Space Educ., Vol. 1, No. 9, p. 428 – 431 (1985).

051.032 Astrometric needs for the ISEE–3/ICE mission to comet Giacobini–Zinner.
D. W. Dunham.
Cometary astrometry, p. 207 – 213 (1984). – See Abstr. 012.024.
International Cometary Explorer's flyby of comet Giacobini–Zinner will provide valuable experience for the astrometry needed to target other spacecraft to encounter Halley's comet.

051.033 Vols de longue durée sous Montgolfière Infrarouge MIR.
P. Malaterre.
Adv. Space Res., Vol. 5, No. 1, p. 23 – 26 (1985). – See Abstr. 012.038.

Des vols de longue durée viennent d'avoir lieu utilisant comme ballon des Montgolfières Infrarouges volant dans la stratosphère. La Montgolfière Infrarouge est un ballon à air chaud qui utilise le rayonnement infrarouge de la terre comme source de chaleur. L'article relate les caractéristiques des différents vols effectués et les performances de l'engin.

051.034 First results of a stratospheric experiment using a Montgolfière Infra–Rouge (MIR).
J. P. Pommereau, F. Dalaudier, J. Barat, J. L. Bertaux, F. Goutail, A. Hauchecorne.
Adv. Space Res., Vol. 5, No. 1, p. 27 – 30 (1985). – See Abstr. 012.038.

An original stratospheric experiment in the tropics has been performed on board the new long duration balloon system developped by the Centre National d'Etudes Spatiales in France. The experiment together with the performances reached in flight are presented.

051.035 The University of Wyoming's small scientific balloon program.
D. J. Hofmann, J. M. Rosen, N. T. Kjome, G. L. Olson, D. W. Martell.
Adv. Space Res., Vol. 5, No. 1, p. 31 – 34 (1985). – See Abstr. 012.038.

Over 500 small scientific balloons have been launched by the University of Wyoming's Atmospheric Physics Group from 26 locations over the globe in a study of stratospheric aerosol physics and chemistry which began in 1971. These flights have led to a basic understanding of the evolution of sulfurous gases, injected into the stratosphere by major volcanic eruptions, into sulfuric acid aerosol droplets. The recent use of new, thin film balloon technology, to reduce cost and simplify launch techniques, has been a major advantage to the program.

051.036 Balloon system and balloon–borne experiments in China.
Y.–d. Gu.
Adv. Space Res., Vol. 5, No. 1, p. 35 – 38 (1985). – See Abstr. 012.038.

The Chinese scientific balloon project was started in 1979 by the Academia Sinica which has now established a permanent balloon facility. It consists of 1500 m^2 launching site complete with telecontrol and PCM and FM telemetry, and meteorological and communications equipment. A series of 5×10^2 to 5×10^4 m^3 zero–pressure natural shape balloons produced in China have served for scientific observations with a maximum payload weight of 250 kg and with flight durations up to 18 hrs. For astronomical observations an attitude system is available. Successful scientific observations using this balloon facility included: Observations of the primary cosmic ray nuclei and cross–section measurements of high energy heavy nuclei interactions. Measurement of the vertical distribution of aerosols in the atmosphere. Hard X–ray astronomy observations. Remote sensing in the infra–red band at balloon altitudes. Observations of solar far infra–red radiation.

051.037 Feasibility studies of "Polar Patrol Balloon".
J. Nishimura, M. Kodama, K. Tsuruda, H. Fukunishi.
Adv. Space Res., Vol. 5, No. 1, p. 87 – 90 (1985). – See Abstr. 012.038.

Engineering and meteorological feasibilies of a circum-south–polar ballooning project, called "Polar Patrol Balloon (PPB)", for space and geophysical researches are studied.

051.038 Hard X–ray astronomy from balloons.
A. J. Dean.
Adv. Space Res., Vol. 5, No. 1, p. 93 – 103 (1985). – See Abstr. 012.038.

The region of the electromagnetic spectrum between 15 keV and 300 keV is the first high energy photon band available to the astronomer at balloon altitudes. Thin, actively shielded sodium iodide detectors form the mainstream X–ray detection units. Directionality is achieved by means of honeycomb and modulation collimators. However as suitable position sensitive planes are developed it is possible to anticipate the increasing usage of the coded aperture mask as the key element for fine angular imaging. Balloon–borne hard X–ray telescopes tend to deploy larger sensitive areas than their satellite cousins, and for this reason, with suitably fast timing of the data, may be used to study the classes of objects which exhibit rapid temporal X–ray intensity variations. Spectral studies are also of great astrophysical importance in this range. The usage of balloons and related astrophysical problems are reviewed.

051.039 Solar and cosmic X– and gamma–ray studies with long duration balloon flights.
K. Hurley.
Adv. Space Res., Vol. 5, No. 1, p. 109 – 116 (1985). – See Abstr. 012.038.

The scientific goals of balloon–borne X– and gamma–ray transient studies are reviewed. Several scientific objectives are common to both solar and cosmic investigations, specifically: high time and energy resolution measurements of line and continuum emission, and log N–log S studies. A brief survey of six instruments is presented. It is shown that current proven detector and gondola technologies are sufficient to advance significantly our understanding of transient phenomena via these investigations within the decade and within the current and future constraints of proposed long duration (10 – 30 days) balloon facilities.

051.040 The role of scientific ballooning for exploration of the magnetosphere.
L. P. Block, L. L. Lazutin, W. Riedler.
Adv. Space Res., Vol. 5, No. 1, p. 121 – 128 (1985). – See Abstr. 012.038.

The magnetosphere is explored in situ by satellites, but measurements near the low altitude magnetospheric boundary by rockets, balloons and groundbased instruments play a very significant role. The geomagnetic field provides a frame with anisotropic wave and particle propagation effects, enabling remote sensing of the distant magnetosphere by means of balloon–borne and groundbased instruments. Examples are given of successful studies, with coordinated satellite and balloon observations, of substorm, pulsation and other phenomena propagating both along and across the geomagnetic field.

051.041 Balloon measurements in future exploration of auroral zone phenomena.
S. Ullaland.
Adv. Space Res., Vol. 5, No. 1, p. 129 – 130 (1985). – See Abstr. 012.038.

In order to study auroral zone phenomena small stratospheric balloons (5000 – 50000 m^3) have been used since 1958. During these last 25 years there have been balloon campaigns in the auroral zone consisting of single or multiple flights launched from one station and of up to 40 flights launched from 4 stations. This paper limits itself to review planned auroral zone balloon programs.

051.042 Project HIPPARCOS.
P. L. Bernacca.
Astrophys. Space Sci., Vol. 110, No. 1, p. 21 – 45 (1985). – See Abstr. 012.039.

The HIPPARCOS satellite project of the European Space Agency is described in detail. The paper focusses on the basic rationale, on the performance of the scientific instrumentation and on an explanation of how scientific objectives can be achieved. The purpose is to provide quantitative visibility of basic project features to astronomers and astrophysicists not familiar with astrometric techniques by exploiting updated technical information on project parameters not generally available in the literature.

051.043 Prospects for Giotto and Halley.
M. K. Wallis.
Nature, Vol. 314, No. 6010, p. 400 (1985).

051.044 Sample return from a comet flyby.
D. E. Brownlee.
Meteoritics, Vol. 19, No. 4, p. 199 (1984). Abstract. – See Abstr. 010.641.

051.045 The Giotto mission to Halley's comet.
A. D. Johnston, R. Reinhard.
Meteoritics, Vol. 19, No. 4, p. 247 (1984). Abstract. – See Abstr. 010.641.

051.046 A proposed GIOTTO/comet coma sample return mission.
B. L. Swenson, A. C. Mascy, P. Tsou, A. Friedlander.
Meteoritics, Vol. 19, No. 4, p. 318 (1984). Abstract. – See Abstr. 010.641.

051.047 Space programs of solar physics.
E. Antonucci.
Mem. Soc. Astron. Ital., Vol. 55, No. 3, p. 543 – 548 (1984). – See Abstr. 012.050.
Contents: The Solar and Heliospheric Observatory (SOHO), a European space program. NASA programs in solar physics: Solar Maximum Mission (SMM), results and future programs; the Solar Optical Telescope (SOT).

051.048 Coronal physics with the Solar and Heliospheric Observatory.
R. Pallavicini.
Mem. Soc. Astron. Ital., Vol. 55, No. 3, p. 549 – 556 (1984). – See Abstr. 012.050.
A brief overview of the main results in coronal physics obtained in the last decade is given, and some of the key problems which may be addressed quite effectively by a future space mission such as the proposed Solar and Heliospheric Observatory (SOHO) are outlined.

051.049 The Solar Maximum Mission.
R. Pallavicini.
Mem. Soc. Astron. Ital., Vol. 55, No. 4, p. 633 – 652 (1984). – See Abstr. 003.017.
Contents: Introduction. The spacecraft. The experiment payload: Gamma Ray Experiment, Hard X–ray Burst Spectrometer, Hard X–ray Imaging Spectrometer, Soft X–ray Polychromator, Ultraviolet Spectrometer and Polarimeter, Coronograph/Polarimeter, Active Cavity Radiometer Irradiance Monitor. Data acquisition and analysis. Highlights of results. The Italian participation.

051.050 The J.W.G. (*Joint Working Group*) asteroid mission.
A. Brahic.
Adv. Space Res., Vol. 5, No. 2, p. 97 – 106 (1985). – See Abstr. 012.051.
By the early 1990's, asteroids will be the only major class of objects in the solar system still awaiting the visit of a spaceprobe. After several studies of space missions towards the asteroids over the past few years both in the U.S. and Europe, the N.A.S. and the E.S.F. have formed a Joint Working Group to undertake a study for a joint U.S.-European space mission programme for the exploration of the planets. Following the advice of the Primitive Bodies Study Team, a multiple asteroid orbiter with fly–bys has been recommended with a Saturn–Titan and a Mars mission. The main conclusions of the Study Team final report are presented.

051.051 Fly–by missions to near–earth asteroids.
D. Hartke, D. Möhlmann.
Adv. Space Res., Vol. 5, No. 2, p. 107 – 110 (1985). – See Abstr. 012.051.
Based on the TRIAD–catalogue and the Ephemerides of the Leningrad Institute for Theoretical Astronomy, orbits of 26 asteroids with small inclinations and one (or both) nodal points inside 1.5 AU are investigated to find objects appropriated for a near–asteroid fly–by in course of a Hohmann–transfer to Venus or Mars. Up to 1998 especially 2340 Hathor, 2062 Aten and 1982 XB could be visited on this way. Furthermore, the near–Earth positions of seven asteroids could be a cause for direct missions to these objects while a near–Mercury fly–by of 2340 Hathor in April 1991 has been found to be appropriated for a direct Hathor–Mercury mission.

051.052 Spacecraft environment during the Giotto–Halley encounter: a summary.
D. T. Young.
ESA Spec. Publ., ESA SP–224, p. 99 – 104 (1984). – See Abstr. 012.052.
A summary of the Giotto PEWG Meeting is presented in which the separate disciplines are drawn together to give an overview of the spacecraft environment during the Giotto Halley interaction. Specific recommendations are made as to how the work of the Plasma Environment Working Group might continue to contribute to the Giotto program during encounter and post–encounter data analysis.

051.053 Proposed design and performance analysis of NASA/JPL 70–m dual reflector antennas.
D. A. Bathker, A. G. Cha, W. A. Imbriale, W. F. Williams.
Antennas and propagation, Vol. 2, p. 767 – 770 (1984). Abstr. in Phys. Abstr., Vol. 88, No. 1252, Entry 29818 (1985). – See Abstr. 012.055.

051.054 Japanse vluchten naar komeet Halley.
O. Namba.
Zenit, 12. Jaarg., No. 5, p. 164 – 170 (1985).

051.055 Ruimtestations hebben de toekomst.
D. de Hoop.
Zenit, 12. Jaarg., No. 6, p. 204 – 209 (1985).

051.056 Some questions of planning flights to comets.
R. K. Kazakova, V. A. Kotin, O. V. Papkov, A. K. Platonov, K. G. Sukhanov.
Kosm. Issled., Tom 23, Vyp. 2, p. 296 – 307 (1985). In Russian. English translation in Cosm. Res.

051.057 The Exosat mission.
B. G. Taylor.
Adv. Space Res., Vol. 5, No. 3, p. 35 – 44 (1985). – See Abstr. 012.059.
Exosat, the European X–ray Observatory, was placed in orbit on 26 May 1983. The spacecraft, stabilized in three axes to a few arc seconds, carries four instruments, two one–metre focal length imaging telecopes, a large area proportional counter array and a gas scintillation proportional counter spectrometer. The salient features of the instrumentation, the sensitivities achieved in orbit and the status after the first year of orbital operation are described. Three specific observations, V0332+53, Sco X–1 and M83 are discussed to demonstrate the power of the Exosat instrumentation and the operational flexibility of the spacecraft and ground system.

051.058 The Tenma mission.
Y. Tanaka.
Adv. Space Res., Vol. 5, No. 3, p. 81 – 89 (1985). – See Abstr. 012.059.
Tenma, the second X–ray astronomy satellite of Japan launched in February 1983, is outlined. The main instrument of Tenma is a large–area gas scintillation proportional counter array. Some of the highlights of the results thus far obtained are briefly discussed.

051.059 Status of the Hubble Space Telescope Observatory.
C. R. O'Dell.
Adv. Space Res., Vol. 5, No. 3, p. 151 – 152 (1985). – See Abstr. 012.059.
The Hubble Space Telescope activity by NASA and ESA began in 1971 and is now nearing completion. The scientific

instruments are complete and their performance confirmed. The optical telescope assembly integration is proceeding well, with the crucial fine guidance sensors expected to perform as designed. Telescope delivery to the spacecraft contractor is expected in December, 1984. Completion of the verification and assembly program will then be followed by launch in the summer of 1986.

051.060   Space Telescope science operations.
E. J. Schreier, R. Doxsey.
Adv. Space Res., Vol. 5, No. 3, p. 153 – 155 (1985). – See Abstr. 012.059.
The activities which will take place at the Space Telescope Science Institute in support of science operations are summarized. These activities follow proposal acceptance and range over planning and scheduling, observation implementation, data processing and archiving, and support of astronomers in all these.

051.061   Science at the performance limits of the Hubble Space Telescope.
D. J. Schroeder.
Adv. Space Res., Vol. 5, No. 3, p. 157 – 167 (1985). – See Abstr. 012.059.
As with any large telescope on the ground, HST will often be used at the limits of its performance capabilities. The actual performance will only be determined once HST is in orbit, but detailed modeling based on mirror, structure, and control system characteristics gives a reasonable assessment of the expected performance. The purpose of this paper is to present (1) selected image characteristics expected for HST, (2) performance limits for selected instrument characteristics, and (3) examples of observations for which the performance limits are approached.

051.062   Infrared astronomy on the Hubble Space Telescope.
J. L. Pipher, S. P. Willner, G. G. Fazio.
Adv. Space Res., Vol. 5, No. 3, p. 173 – 179 (1985). – See Abstr. 012.059.
The Hubble Space Telescope offers enormous advantages to infrared astronomy in certain situations. The advantages of being above the atmosphere include an increase in spatial resolution, a much wider range of wavelengths available, and lower background radiation. Compared to proposed cooled telescopes, HST offers higher spatial resolution and increased collecting area. HST is particularly well suited to obervations at wavelengths less than ~ 5 µm, where the diffraction limit is less than the seeing limit from the ground and thermal emission does not seriously compromise the sensitivity of the detectors. HST is also favorable for observations requiring high spectral resolution at all wavelengths not accessible from the ground.

051.063   Space Telescope studies of Jupiter and Saturn simulated by Voyager observations.
G. E. Hunt, V. Moore.
Adv. Space Res., Vol. 5, No. 3, p. 181 – 188 (1985). – See Abstr. 012.059.
Space Telescope (ST) observations of Jupiter and Saturn will offer a unique opportunity for monitoring their changing meteorological characteristics. They will provide higher spatial and temporal resolution for composition and vertical structure studies than have been available to date. The authors have simulated the planetary camera observations of Jupiter and Saturn by Voyager images of the appropriate spatial scale. With this data set the authors have investigated the meteorological properties of these atmospheres which can be studied at these scales. In addition they have considered the advances obtainable with the high resolution spectrometer on ST compared with observations from ground–based and other Earth–orbiting satellites.

051.064   Space Telescope observations of aurorae on the giant planets.
R. Wagener, J. Caldwell.
Adv. Space Res., Vol. 5, No. 3, p. 189 – 193 (1985). – See Abstr. 012.059.
For all four giant planets, Space Telescope can improve upon the quality of current optical observations. For spectroscopy, the low resolution mode of the High Resolution Spectrograph is particularly well suited to auroral observations because of its spectral range, adequate resolution and high sensitivity. For ultraviolet imaging through appropriate filters, the ST spatial resolution, expected to be of order 5 hundredths of an arc second, is also well suited to determine the spatial properties of the aurorae.

051.065   Space Telescope observations of the Earth's upper atmosphere by stellar occultations.
J. Caldwell.
Adv. Space Res., Vol. 5, No. 3, p. 195 – 199 (1985). – See Abstr. 012.059.
Stellar occultations provide a useful means of measuring the trace gas composition of the Earth's mesosphere with a sensitivity of order one part per billion. The operational details will differ from those of other astronomical observations by ST, because of the difficulties in guiding near the Earth's limb. Two specific trace gases of interest to atmospheric studies, Cl and ClO, are discussed in the paper.

051.066   Photometric calibration and first results obtained with the UFT experiment on board the ASTRON satellite.
S. T. Hua, G. Courtes, P. Cruvellier, D. Huguenin, A. B. Severny (*A. B. Severnyj*), A. Boyarchuk, R. Gershberg, P. Petrov, G. Pronik, R. Tovmassian (*G. M. Tovmasyan*).
Adv. Space Res., Vol. 5, No. 3, p. 201 – 205 (1985). – See Abstr. 012.059.
The UV telescope of 80 cm diameter equipped with a three channel scanner spectrometer aboard the highly eccentric orbit space station "ASTRON" is described.

051.067   Through the system of Saturn to the history of the matter of planets.
O. V. Nikolaeva.
Priroda, No. 4, p. 63 – 71 (1985). In Russian.

051.068   X–ray sky surveys and the ROSAT mission.
J. Trümper.
Bull. Inf. Cent. Données Stellaires, No. 28, p. 81 – 86 (1985). – See Abstr. 012.062.

051.069   Detection of X–ray sources with ROSAT.
G. Hasinger.
Bull. Inf. Cent. Données Stellaires, No. 28, p. 87 – 90 (1985). – See Abstr. 012.062.

051.070   The Hipparcos satellite's mission: the objectives and their implementation.
M. Schuyer.
ESA Bull., No. 42, p. 22 – 29 (1985).

051.071   Possible SIRTF observations of infrared–bright galaxies.
M. Jura.
Bull. Am. Astron. Soc., Vol. 16, No. 4, p. 906 (1984). Abstract. – See Abstr. 010.062.

051.072   Infrared observations of distant galaxies using SIRTF.
E. L. Wright, A. Choksi.
Bull. Am. Astron. Soc., Vol. 16, No. 4, p. 907 (1984). Abstract. – See Abstr. 010.062.

051.073   Capabilities of proposed NASA missions to observe below Lyman alpha.
E. J. Weiler, R. E. Stencel.
Bull. Am. Astron. Soc., Vol. 16, No. 4, p. 984 (1984). Abstract. – See Abstr. 010.062.

051.074   Planning a probing voyage to Jupiter.
R. F. Draper, T. V. Johnson.
IEEE Spectrum, Vol. 22, No. 6, p. 70 – 76 (1985).

051.075   Space astronomy: technical comments – ultraviolet and optical instrumentation.
J. M. Deharveng.
ESA Spec. Publ., ESA SP–1070, p. 92 – 95 (1984). – See Abstr. 003.029.

051.076   Space astronomy: technical comments – X– and gamma-ray astronomy.
H. W. Schnopper.
ESA Spec. Publ., ESA SP–1070, p. 96 – 103 (1984). – See Abstr. 003.029.

051.077   Space astronomy: technical comments – infrared, submillimetre and radio astronomy.
G. Winnewisser.
ESA Spec. Publ., ESA SP–1070, p. 104 – 107 (1984). – See Abstr. 003.029.

051.078   Space astronomy: technical comments – interferometry missions.
R. T. Schilizzi.
ESA Spec. Publ., ESA SP–1070, p. 108 – 115 (1984). – See Abstr. 003.029.

051.079   Trend analysis of mission concepts – ESA executive.
ESA Spec. Publ., ESA SP–1070, p. 118 – 121 (1984). – See Abstr. 003.029.

051.080   Industrial benefits derived from the European Space Science programmes – ESA executive.
ESA Spec. Publ., ESA SP–1070, p. 122 – 125 (1984). – See Abstr. 003.029.

051.081   Space research in the era of the space station.
K. J. Frost, F. B. McDonald.
Science, Vol. 226, No. 4681, p. 1381 – 1385 (1984).
With the continuing flights and increasing capabilities of the space shuttle, and with the design and development of a space station, there will be a significant increase in our space research capabilities during the 1990's. Ways in which the American space science program may evolve in response to these developments are described.

051.082   A Giacobini–Zinner Watch (GZW).
R. L. Newburn Jr., J. Rahe.
IHW Newsl., No. 6, p. 2 – 3 (1985).

051.083   The Spartan Halley Mission.
C. A. Barth, S. A. Stern, A. I. F. Stewart.
IHW Newsl., No. 6, p. 2 – 11 (1985).
Shortly before Comet Halley reaches perihelion on 9 February 1986 a University of Colorado ultraviolet spectrometer experiment will make unique measurements of the coma and tail. The experiment will be transported into space by the Shuttle Orbiter Challenger in late January 1986. Two ultraviolet spectrometers and associated instrumentation will be aboard a Spartan carrier that will be deployed to function as an autonomous subsatellite for about 48 hours.

051.084   Space missions to comet Halley: Giotto, Vega, and Planet–A.
T. Encrenaz.
IHW Newsl., No. 6, p. 11 – 18 (1985).

051.085   Space Telescope observations of globular clusters.
J. N. Bahcall.
Dynamics of star clusters, p. 481 – 498 (1985). – See Abstr. 012.070 (IAU Symp. No. 113).
Six proposed Space Telescope programs involving globular clusters are described. The projects appropriate for galactic clusters are: the detection of white dwarfs, the study of the faint end of the Population II luminosity function, the measurement of mass segregation, and the search for a cusp in the density distribution caused by core collapse or by a massive black hole. The two programs that involve extragalactic globular clusters are: the determination of the luminosity function of clusters around different galaxies and the measurement of tidal radii of clusters surrounding elliptical galaxies.

051.086   Galileo spacecraft integration: international cooperation on a planetary mission in the Shuttle era.
R. J. Spehalski.
Earth–Oriented Appl. Space Technol., Vol. 4, No. 3, p. 139 – 150 (1984). Abstr. in Phys. Abstr., Vol. 88, No. 1254, Entry 40462 (1985).

051.087   HIPPARCOS satellite, project status.
M. Schuyer.
Processing of scientific data from the ESA astrometry satellite HIPPARCOS, p. 11 – 27 (1985). – See Abstr. 012.087.
After FAST Thinkshop No. 1, at the time of which most design aspects of the HIPPARCOS mission and satellite design had been frozen and were reported to the FAST community, further refinements and tradeoffs activities took place, extending into the current design and development phase of the Project. The present paper reports selected features of to–day's status of the design, together with findings following production and evaluation of the first hardware items, and an updated assessment of achievable accuracy for the main mission and Tycho.

051.088   The nominal scanning law and the interruptions by the moon.
J. Kovalevsky, J. L. Falin, M. Froeschlé, F. Mignard.
Processing of scientific data from the ESA astrometry satellite HIPPARCOS, p. 59 – 64 (1985). – See Abstr. 012.087.
The present nominal scanning law has the disadvantage to undergo frequent interruptions due to the presence of the moon that may involve more than two consecutive days. Some ecliptic stars will loose up to 7% of the observing time. The authors propose to shift the axis of rotation by a sufficient amount so as to let the moon by–pass the scanning circles, then to return to the nominal law. It is shown that this shift must always be made in order to decrease the inclination of the scanning circle on the ecliptic. The analysis of the interruptions shows also that it is possible to optimize the initial phase of the revolving angle.

051.089   Proposals of observations with the Space Telescope in the domain of astrometry.
A. Fresneau.
Processing of scientific data from the ESA astrometry satellite HIPPARCOS, p. 97 – 102 (1985). – See Abstr. 012.087.
The use of the Hubble Space Telescope for astrometry is advertised at the same level as for photometry, spectroscopy, or polarimetry. The prime instrument to be used for that goal is one of the three fine guidance sensors. The interferometric design of the stellar sensor is adequate for stellar diameter measurements (>0.01 arcsec) close binaries separation determination (<0.1 arcsec) and differential astrometry on targets in a field of view of 60 square arcmin and in the visual magnitude range from 3 to 18. Moving targets brighter than 14 with an apparent motion slower than 150 arcsec per hour can be tracked at the same level of accuracy.

051.090   Precision and method for HIPPARCOS photometry.
E. Høg, L. Lindegren.
Processing of scientific data from the ESA astrometry satellite HIPPARCOS, p. 133 – 141 (1985). – See Abstr. 012.087.
The expected precision of HIPPARCOS photometry is about 0.013 mag for a star of blue magnitude B = 9 from one crossing of the main field. The mission will give ten million observations of similar quality. A strategy for calibration of the modulation factor M_1 and the background is proposed. The background determination will be enhanced by the use of Star Mapper data obtained at the reference stars since these give ten times as frequent and more accurate determinations than obtained by the IDT data.

051.091 About the accuracy in magnitude determination.
F. Mignard, J. L. Falin, M. Froeschlé, P. Granès.
Processing of scientific data from the ESA astrometry satellite
HIPPARCOS, p. 143 – 151 (1985). – See Abstr. 012.087.

The reduction process of the HIPPARCOS data will allow the intensity of the signal to be determined and then interpreted in terms of magnitude. In this paper the authors investigate the random and systematic errors of such a determination. The detection of variable stars of various periods is a promising goal at the level of 0.03 m for a star magnitude 9.

051.092 Interplanetary navigation through the year 2005: the inner solar system.
L. J. Wood, J. F. Jordan.
J. Astronaut. Sci., Vol. 32, No. 4, p. 357 – 376 (1984). Abstr. in Phys. Abstr., Vol. 88, No. 1257, Entry 57579 (1985).

051.093 Results obtained from the IRAS satellite.
H. J. Habing.
Ned. Tijdschr. Natuurkd. A, Vol. A51, No. 1, p. 8 – 11 (1985). In Dutch. Abstr. in Phys. Abstr., Vol. 88, No. 1258, Entry 63232 (1985).

051.094 USSR leads the way to Halley's comet.
N. Henbest.
New Sci., Vol. 104, No. 1434, p. 10 – 11 (1984). Abstr. in Phys. Abstr., Vol. 88, No. 1259, Entry 68645 (1985).

051.095 VEGA – the international space experiment to observe Halley's comet.
I. Apathy, P. Bereczki, G. Endroczy, T. Gombosi,
A. Geschwind, A. Lohonyai, G. Kozma, I. Naday, I. Renyi,
A. Somogyi, L. Szabo, S. Szalai, K. Szego.
Meres Autom., Vol. 33, No. 1 – 2, p. 1 – 8 (1985). In Hungarian. Abstr. in Phys. Abstr., Vol. 88, No. 1260, Entry 74056 (1985).

051.096 Space Report.
Spaceflight, Vol. 27, No. 5, p. 198 – 201 (1985).
1985 May review of space news and events.

051.097 The Mars dual orbiter.
F. Taylor.
Spaceflight, Vol. 27, No. 5, p. 202 – 205 (1985).

The next phase in the exploration of Mars, after the Viking landers, can be accomplished with relatively simple orbiting spacecraft. A complementary pair in radically different orbits is the best way to address the outstanding scientific objectives; a possibility currently under discussion is for the US and Europe to provide one each.

051.098 The ESA science programme.
G. Whitcomb.
Spaceflight, Vol. 27, No. 5, p. 206 – 209 (1985).

051.099 Space at JPL.
W. I. McLaughlin.
Spaceflight, Vol. 27, No. 5, p. 210 – 213 (1985).
Contents: Saturn and SETI. Venus Radar Mapper Mission. Two encounters for Voyager. Worlds in miniature. Micro-spacecraft design.

051.100 The Vega missions.
P. Clark.
Spaceflight, Vol. 27, No. 5, p. 218 – 220 (1985).

051.101 Space Report.
Spaceflight, Vol. 27, No. 6, p. 244 – 248 (1985).
1985 June review of space news and events.

051.102 Space at JPL.
W. I. McLaughlin.
Spaceflight, Vol. 27, No. 6, p. 257 – 259 (1985).
Contents: A peculiar ring about Neptune. Asteroid option for Galileo. Electro–optics and lasers. Keck ten–metre telescope. Acronymese.

051.103 Space based radio astronomy arrays.
J. H. Spencer, K. J. Johnston, L. J. Rickard,
R. S. Simon.
Bull. Am. Astron. Soc., Vol. 17, No. 2, p. 550 (1985). Abstract. – See Abstr. 010.065.

051.104 A suggested future space mission to the low–luminosity star Ross 248 = Gliese 905.
F. R. West.
Bull. Am. Astron. Soc., Vol. 17, No. 2, p. 552 (1985). Abstract. – See Abstr. 010.065.

051.105 Science mission planning for AXAF.
M. S. Davis.
Bull. Am. Astron. Soc., Vol. 17, No. 2, p. 572 (1985). Abstract. – See Abstr. 010.065.

051.106 The Astro Mission: an update.
T. R. Gull.
Bull. Am. Astron. Soc., Vol. 17, No. 2, p. 573 (1985). Abstract. – See Abstr. 010.065.

051.107 Scientific observing plans for the SOUP (*Solar Optical Universal Polarimeter*) instrument on Spacelab 2 in July, 1985.
T. D. Tarbell, M. L. Finch, A. M. Title.
Bull. Am. Astron. Soc., Vol. 17, No. 2, p. 641 (1985). Abstract. – See Abstr. 010.067.

051.108 Biophysical significance of heavy cosmic rays.
D. Hasegan.
Top. Astrophys. Astron. Space Sci., Vol. 1, p. 51 – 75 (1985).

Long duration manned space flights as well as the experiments aboard artificial satellites have shown that one of the main problems, which have to be taken into account, is the protection against cosmic rays.

051.109 Report of IAU Commission 44: Astronomy from space (*L'astronomie à partir de l'espace*).
M. Oda.
Trans. IAU, Vol. XIXA, p. 607 – 643 (1985). – See Abstr. 003.046.

051.110 Astronautics in the year 1983.
M. Grün, P. Koubský.
Říše hvězd, Vol. 65, No. 11, p. 221 – 225 (1984). In Czech.

051.111 Space research of the Sun.
M. Rybanský.
Kozmos, Vol. 16, No. 1, p. 12 – 14 (1985). In Slovak.

051.112 Astronautics in the year 1984.
A. Vítek.
Vesmír, Vol. 64, No. 4, p. 189 – 193 (1985). In Czech.

051.113 The HIPPARCOS satellite.
C. A. Murray.
Vistas Astron., Vol. 28, Parts 1/2, p. 169 (1985). – See Abstr. 012.109.

In 1988 the European Space Agency will launch the astrometric satellite HIPPARCOS. This is designed primarily to measure very precise astrometric data for about 10^5 stars.

051.114 Ad astra HIPPARCOS. The European Space Agency's Astrometry Mission.
M. A. C. Perryman.
ESA Brochure, ESA BR–24, 4 + 45 pp. (1985).
Contents: Some pertinent questions and brief answers. The history of astrometry. The HIPPARCOS Mission. The HIPPARCOS Scientific Consortia. Description of the satellite. Operations. Industrial development.

051.115 The Quasat mission: an overview.
R. T. Schilizzi, B. F. Burke, J. F. Jordan,
A. Hawkyard.
ESA Spec. Publ., ESA SP–213, p. 13 – 18 (1984). – See Abstr.
012.124.
The Quasat mission is reviewed from the historical, scientific
and technical viewpoints and the main conclusions from the as-
sessment studies of NASA and ESA are presented.

051.116 Some prospects of space VLBI.
R. Z. Sagdeev.
ESA Spec. Publ., ESA SP–213, p. 19 – 21 (1984). – See Abstr.
012.124.
Basic concepts of space very long base radio interferometry are
discussed. Two Soviet ground–space interferometer projects are
considered: one with a low–orbit space telescope (very complete
coverage of the uv–plane) and one with a high orbit space telescope (very good angular resolution).

051.117 Space VLBI studies in Japan.
M. Morimoto.
ESA Spec. Publ., ESA SP–213, p. 23 (1984). Abstract. – See
Abstr. 012.124.

051.118 Quasat: technical aspects of the proposed mission.
H. Ames, S. Bolton, B. F. Burke, S. Burks,
C. Christensen, V. Dhawan, B. Freland, D. Grippi,
C. Hamilton, D. Hansen, N. Jacobi, K. Johnson, J. Jones,
F. Jordan, K. I. Kellermann, C. R. Lawrence, G. Levy, R. Lin,
R. Linfield, D. Meier, A. T. Moffet, J. M. Moran, M. Nein,
D. L. Potts, B. Preston, N. Renzetti, M. Reid, T. Readhead,
D. Roberts, A. Rogers, I. Shapiro, J. Springett, P. Theisinger,
C. Uphoff, C. Walker, A. Whitney, Y. Rahmat–Samii,
B. Anderson, A. van Ardenne, A. Boischot, R. S. Booth,
P. Buia, A. Davidson, P. Erichsen, A. Festa, W. Flury,
D. A. Graham, J. Hammer, A. Hawkyard, N. Jensen,
P. Kamoun, K. van 't Klooster, K. A. Langenhoff, H. Laue,
E. Lijphart, H. Olthof, M. Perryman, J. Ponsonby, E. Preuss,
G. Reibaldi, B. O. Rönnäng, C. Savage, R. T. Schilizzi,
P. N. Wilkinson, R. Wilson
ESA Spec. Publ., ESA SP–213, p. 27 – 99 (1984). – See Abstr.
012.124.
This report summarizes the results of parallel assessment stud-
ies of the Quasat mission concept carried out by NASA and ESA
in early 1984. The fundamental requirement on the Quasat mission is to provide an earth orbiting radio telescope, which serves
as an extension of the ground based VLBI antenna arrays, and
thereby provides images of total intensity and polarized emission
of radio source of higher resolution and quality than those obtainable with the ground arrays only.

051.119 A study of the imaging potential of Quasat.
A. C. S. Readhead, R. A. Preston, D. L. Meier,
R. P. Linfield, C. R. Lawrence, V. Dhawan, R. S. Simon,
P. N. Wilkinson, S. Bolton, B. F. Burke, T. J. Cornwell,
K. F. Evans, K. J. Johnston, F. Jordan, A. T. Moffet,
A. E. Niell, T. J. Pearson, J. H. Spencer, D. H. Roberts,
S. C. Unwin.
ESA Spec. Publ., ESA SP–213, p. 101 – 110 (1984). – See Abstr.
012.124.
The authors have made a study of the imaging potential of
Quasat. The capabilities of an orbiting VLBI observatory depend
chiefly on the orbit, the ground array to be used in conjunction
with the space antenna, and the sensitivities of the space and
ground systems. The authors have looked at all of these parameters, and chosen a set of standard parameters which is realistic
and which will yield high quality images, i.e. images in which the
noise is within a factor of three of the thermal noise limit.

051.120 The wide–range space radiointerferometer.
V. V. Andrejanov (*V. V. Andreyanov*), L. I. Gurvits,
N. S. Kardashev, S. V. Progrebenko, V. A. Rudakov,
R. Z. Sagdeev, G. S. Tsarevsky (*G. S. Tsarevskij*).
ESA Spec. Publ., ESA SP–213, p. 161 – 164 (1984). – See Abstr.
012.124.

The concept of a high–orbit wide–band space radiointerferometer is proposed and discussed. It is shown that observations in one octave (or more) in the wavelength range essentially spread the uv–plane coverage. Some problems of image
restoration in the wide–band space aperture synthesis system are
considered.

051.121 The January 1986 Voyager–Uranus encounter.
R. F. Beebe.
Proceedings of the Southwest Regional Conference for Astron-
omy and Astrophysics, Vol. 10, p. 47 – 48 (1985). Abstract. – See
Abstr. 012.125.

051.122 EXOSAT, European X–Ray Astronomy Satellite.
EXOSAT Express, Nos. 9 – 11, 59 + 62 + 90 pp. (1985).

The Guide Star Catalog – structure and publication.
See Abstr. 002.028.

Progress report on the Input Catalogue preparation.
See Abstr. 002.096.

Em busca de outros mundos.
See Abstr. 003.051.

Solar–terrestrial research opportunities – a look to the future.
See Abstr. 013.006.

**Space investigations of the Sun outlined for the end of the
80's – beginning of the 90's.**
See Abstr. 013.007.

The present status of space developments and their perspectives.
See Abstr. 013.016.

The infrared universe.
See Abstr. 013.035.

Astronomie infrarouge.
See Abstr. 013.062.

Gamma–ray astronomy and the spirit of Cos–B.
See Abstr. 013.064.

**A long term programme for European space science: actions, pro-
posals, conclusions – Survey Committee report.**
See Abstr. 013.071.

Solar system science – Survey Committee report.
See Abstr. 013.072.

Report of ESA's topical team on solar and heliospheric physics.
See Abstr. 013.073.

Report of ESA's topical team on space plasma physics.
See Abstr. 013.074.

Report of ESA's topical team on planetary science.
See Abstr. 013.075.

Space astronomy – Survey Committee report.
See Abstr. 013.076.

Extragalactic astronomy from space in the 1990s.
See Abstr. 013.077.

Galactic astronomy from space in the 1990's.
See Abstr. 013.078.

Space experiments in relativity and gravitation.
See Abstr. 013.079.

A Soviet plan for exploring the planets.
See Abstr. 013.080.

The NASA Space Station.
See Abstr. 013.095.

The impact of dust grains on fast fly–by spacecraft: momentum multiplication, measurements and theory.
See Abstr. 022.034.

Measurement of charged secondary particle emission under ion and neutral impact.
See Abstr. 022.075.

Contactless determination of the conductivity of the white paint PCB–Z.
See Abstr. 022.076.

Measurements of integral yields of charged secondary particles using neutral beams simulating a cometary fly–by.
See Abstr. 022.077.

Giotto residual ionization.
See Abstr. 022.078.

Impact ionization from gold, aluminum and PCB–Z.
See Abstr. 022.079.

Experimental investigations on ion emission with dust impact on solid surfaces.
See Abstr. 022.080.

Ion emission from solid surfaces: comparison of dust impact with other excitations.
See Abstr. 022.081.

Electrostatic charging of the Giotto spacecraft due to neutral gas impact.
See Abstr. 022.082.

A model of plasma cloud expansion generated by a dust particle impact.
See Abstr. 022.083.

Giotto plasmasheath and wake charging near Halley's comet: revised parameters.
See Abstr. 022.084.

Simulations of satellite–plasma interaction, including electron space charges.
See Abstr. 022.085.

Laboratory measurement on impact ionization by neutrals and floating potential of a spacecraft during encounter with Halley's comet.
See Abstr. 022.130.

Potential affiliate stations to work with the EVN (*European VLBI Network*).
See Abstr. 033.060.

Science with SOHO.
See Abstr. 035.034.

Satellite–based instrument concepts for the measurement of orbital debris.
See Abstr. 035.035.

An analysis of the structure calibration and system test data obtained from the GIOTTO Structural Model Spacecraft at ESTEC, Noordwijk, 22 October to 5 November 1983.
See Abstr. 035.136.

Methods of remote surface chemical analysis for asteroid missions.
See Abstr. 036.087.

Roll deconvolution of Space Telescope data.
See Abstr. 036.104.

Report on the task raw data treatment.
See Abstr. 036.161.

The effect of astrometric modeling errors of Schmidt plates on the construction of the Guide Star Catalog for Space Telescope.
See Abstr. 041.025.

Very long baseline astrometry.
See Abstr. 041.049.

Linking the HIPPARCOS catalog to the VLBI inertial reference system, high angular resolution structures and VLBI positions of 10 radio stars.
See Abstr. 043.007.

Probability of collisions of artificial bodies in the Earth environment.
See Abstr. 052.037.

Contribution of explosion and future collision fragments to the orbital debris environment.
See Abstr. 052.038.

Collision probabilities at geosynchronous altitudes.
See Abstr. 052.039.

Space research and the new approach to the mechanics of fluid media in cosmos.
See Abstr. 062.087.

UV and X–ray spectroscopy as a tool for the diagnostics of non–thermal processes in the atmospheres of hot stars.
See Abstr. 064.050.

Finalmente un asteroide da vicino.
See Abstr. 098.052.

Saturne, après Voyager, avant le vaisseau spatial Cassini.
See Abstr. 100.040.

Dust environment models for comet P/Halley: support for targeting of the GIOTTO S/C.
See Abstr. 103.914.

Space missions to Halley's comet.
See Abstr. 103.916.

L'esplorazione ravvicinata della cometa di Halley.
See Abstr. 103.928.

La cometa di Halley e la missione spaziale Giotto.
See Abstr. 103.929.

IUE observations of faint standard stars for calibration of the Space Telescope.
See Abstr. 114.013.

HIPPARCOS astrometric binaries.
See Abstr. 118.019.

Statistical models for HIPPARCOS binaries.
See Abstr. 118.021.

The Galaxy scene and Quasat.
See Abstr. 155.197.

052 Astrodynamics, Navigation of Space Vehicles

052.001 On computation of satellite orbits for geodynamical research.
V. K. Taradij, M. L. Tsesis.
Kinematika Fiz. Nebesn. Tel, Tom 1, No. 1, p. 55 – 60 (1985). In Russian.
The most efficient computation of satellite orbits for determining geodynamical parameters may be obtained by a modern algorithm of numerical integration combined with the regularized equations of satellite motion (KS–theory). As for Lageos observations they may be also analysed using direct integration of equations in rectangular coordinates by the Adams algorithm of variable order, for example, based on divided differences.

052.002 On normal places, obtained by matched polynomials.
B. B. Baghos.
Publ. Astron. Inst. Czech. Acad. Sci., No. 58, p. 111 – 127 (1984). – See Abstr. 012.018.
A method is given for constructing the proper spline. The necessary expressions for the variances of the normal places taken at two neighbouring polynomials and their covariance are found and given.

052.003 Model of conjugated point masses of a smoothed field for LAGEOS orbit determination.
A. N. Marchenko.
Publ. Astron. Inst. Czech. Acad. Sci., No. 58, p. 233 – 241 (1984). In Russian. – See Abstr. 012.018.
The solution of the problem of external gravitational field simulation by means of the potential of a system of reciprocal point masses is described. The problem is considered as that of nonlinear programming with Tykhonov's regularization. A model of 85 reciprocal point masses, which represents the LAGEOS orbit on 5–day arcs with the same accuracy as of GEM–L2, is constructed on the basis of the full set of GEM–L2 harmonics. The usage of a conjugated point–mass model instead of GEM–L2 harmonics leads to a reduction of computer time by 50%.

052.004 Some modifications of the Hori–Deprit method in constructing a third–order semianalytical theory for resonance satellites.
I. V. Tupikova.
Publ. Astron. Inst. Czech. Acad. Sci., No. 58, p. 249 – 261 (1984). – See Abstr. 012.018.
To construct a third–order theory of satellite motion, the analytical expression for the mixed perturbations caused by the second zonal harmonic and the tesseral harmonics is obtained. Explicit expressions for the third–order transformed Hamiltonian and the generating function of the Hori–Deprit method are given. A major part of the third–order short–period perturbations may be taken into account by a simple iterative procedure.

052.005 Recurrence relations for the normalized inclination function.
J. Kostelecký.
Publ. Astron. Inst. Czech. Acad. Sci., No. 58, p. 263 (1984). Abstract. – See Abstr. 012.018.

052.006 The use of mixed observations from one station to determine the preliminary orbits.
B. B. Baghos.
Publ. Astron. Inst. Czech. Acad. Sci., No. 58, p. 347 – 363 (1984). – See Abstr. 012.018.

052.007 Analysis of non–gravitational effects in satellite motion.
N. Georgiev, V. Kotseva.
Publ. Astron. Inst. Czech. Acad. Sci., No. 58, p. 365 – 378 (1984). In Russian. – See Abstr. 012.018.
The effects of some non–gravitational perturbations on the intermediate coordinates ξ, η, ω of the artificial Earth's satellites were investigated. The analytical theory used is based on the asymmetrical method of two fixed centres and powers of regularized time τ.

052.008 The "Prognoz" routine, its structure and first results.
N. A. Sorokin, E. A. Samus', S. K. Tatevyan.
Publ. Astron. Inst. Czech. Acad. Sci., No. 58, p. 379 – 397 (1984). In Russian. – See Abstr. 012.018.
The "Prognoz" routine has been worked out at the Astronomical Council permitting high–accuracy prediction of artificial satellite motion in an interval of several days. Numerical integration of equations of the motion of artificial satellites was performed; the geopotential, gravitational influence of the Moon and the Sun, solar pressure, the influence of solid Earth tides and upper atmosphere drag are taken into account. The routine can be used to improve orbital elements, station positions, determination of coordinates of the Earth's pole and instability of the Earth's rotation. Using laser ranging data of Lageos, during preliminary MERIT campaign, an accuracy of less than 1 meter for orbital elements was obtained.

052.009 Problems of application of the analytical theory of the earth artificial satellite motion in geodynamics and satellite geodesy.
N. V. Emel'yanov.
Publ. Astron. Inst. Czech. Acad. Sci., No. 58, p. 399 – 411 (1984). In Russian. – See Abstr. 012.018.
A method is described of the development of an analytical theory by means of integration in the first and second orders of differential equations for elements of E. P. Aksenov intermediate orbit. The earth's oblateness is fully taken into account in the intermediate orbit. The theory takes into account combined perturbations with respect to different perturbing factors. A comparison of the theory with the results of numerical integration is made for the case of perturbations caused by the terrestrial gravitational field. The comparison shows that the accuracy of the new theory turns out to be not worse than 18 cm in satellite coordinates.

052.010 An analysis of the accuracy of the determination of luni–solar perturbations in the analytical theory of artificial Earth satellites' motion.
L. P. Nasonova.
Publ. Astron. Inst. Czech. Acad. Sci., No. 58, p. 413 – 419 (1984). In Russian. – See Abstr. 012.018.
An analytical method of determination of luni–solar perturbations of satellite motion is described. The motion of perturbing body is described by trigonometrical series with the necessary accuracy. The determination of perturbations is made by analytical transformations of the series by means of computer. As a result of these transformations the original accuracy is lost because of the limitation of the number of terms in series. The loss of accuracy is subjected to an analysis by means of numerical tests and initial accuracy of original series is estimated for fixed required accuracy of the calculation of satellite coordinates.

052.011 Sensitivity analysis and optimal design of range, range–difference, and range–rate measurements for the determination of pole coordinates and variation of Earth's rotation.
H. Montag.
Publ. Astron. Inst. Czech. Acad. Sci., No. 58, p. 429 – 440 (1984). = Mitt. Zentralinst. Phys. Erde, Nr. 1375. – See Abstr. 012.018.
By means of analytical relations the influences of orbital errors (along–track, across–track and radial component) on the range, range–difference, and range–rate measurements are studied and compared. In a similar manner the design matrices for the determination of pole coordinates and variation of Earth rotation by means of these measurements are discussed.

052.012 On orbit determination with lumped coefficients.
J. Klokočnik, J. Kostelecký.
Publ. Astron. Inst. Czech. Acad. Sci., No. 58, p. 441 – 451 (1984).
– See Abstr. 012.018.

For some purposes, such as for accurate orbit prediction, it seems to be useful to replace the large fields of harmonic coefficients in the Earth gravity field models by a limited set of the lumped geopotential coefficients, tailored to the individual orbits. The lumped values absorb the orbital information to a high degree for the particular order of the harmonics, and a small set of the lumped coefficients, written for various orders, should be sufficient to describe the orbit as perturbed by the Earth. Some problems and questions concerning the application of this method for orbits similar to those of SEASAT or TOPEX are discussed and answered.

052.013 Some problems connected with the determination of artificial satellite orbits.
A. Drożyner.
Publ. Astron. Inst. Czech. Acad. Sci., No. 58, p. 453 – 459 (1984).
– See Abstr. 012.018.

The purpose of this paper is to give a short conceptual description of the following three problems: actual status of the ORBITA computer system, computation of the motion of the satellite–subsatellite system, and mathematical formulation of the unmodeled accelerations in the satellite motion as coded in ORBITA system.

052.014 Improvement of the shadow function and its influence on the orbital elements of satellites.
J. Kabeláč.
Publ. Astron. Inst. Czech. Acad. Sci., No. 58, p. 461 – 473 (1984).
– See Abstr. 012.018.

The task is to improve the shadow function. For that purpose the Earth is regarded as a sphere and the shadows as two cones. The former is taken for the umbra and the latter for the penumbra. After that the influence of the Earth's atmosphere is taken into account: the astronomical refraction and the atmospheric absorption. The theory is brought to completion by numerical applications.

052.015 Modelling and optimization of Doppler satellite observations.
Ts. Gergov, N. Georgiev.
Publ. Astron. Inst. Czech. Acad. Sci., No. 58, p. 505 – 515 (1984).
In Russian. – See Abstr. 012.018.

A model experimental investigation of the accuracy of ground station coordinates, determined by Doppler satellite observations is made, depending on the number of passages, satellite altitude, orbit inclination, frequency of Doppler transmitter, intervals of Doppler counts, etc. By using satellite passages at altitudes of 1000, 3000 and 5000 km, optimization of the Doppler observations is also achieved.

052.016 Implementation of a semianalytic satellite theory with recovery of short period terms.
B. Kaufman, W. H. Harr.
Acta Astronaut., Vol. 11, No. 6, p. 279 – 286 (1984). Abstr. in Phys. Abstr., Vol. 88, No. 1251, Entry 24262 (1985).

052.017 Precise orbit determination of a geosynchronous satellite by ΔVLBI method.
T. Shiomi, S.-I. Kozono, Y. Arimoto, S. Nagai, M. Isogai.
J. Radio Res. Lab., Vol. 31, No. 133, p. 111 – 132 (1984). Abstr. in Phys. Abstr., Vol. 88, No. 1251, Entry 24264 (1985).

052.018 On the evolution of nearly circular orbits of 12–hour artificial earth satellites.
M. A. Vashkov'yak.
Kosm. Issled., Tom 23, Vyp. 1, p. 3 – 15 (1985). In Russian. English translation in Cosm. Res.

052.019 Comparison of conditionally–periodic solutions with results of a numerical integration in the problem of translatory–rotational motion of a satellite.
A. A. Zlenko.
Kosm. Issled., Tom 23, Vyp. 1, p. 16 – 25 (1985). In Russian. English translation in Cosm. Res.

052.020 Application of the relativistic theory to problems of trajectory measurements of space vehicles.
V. S. Chaplinskij.
Kosm. Issled., Tom 23, Vyp. 1, p. 49 – 62 (1985). In Russian. English translation in Cosm. Res.

052.021 Mathematical model of the motion of a planetary vehicle.
E. I. Grigor'ev, S. N. Ermakov.
Kosm. Issled., Tom 23, Vyp. 1, p. 92 – 99 (1985). In Russian. English translation in Cosm. Res.

052.022 Numerical determination of artificial earth satellite trajectories by the Adams variable order method. II.
V. K. Taradij, M. L. Tsesis.
Kinematika Fiz. Nebesn. Tel, Tom 1, No. 2, p. 15 – 23 (1985). In Russian.

Algorithms and a programme are suggested for the Adams variable order numerical integration for systems of ordinary differential equations of the first, second and mixed orders.

052.023 Orbit manoeuvres with finite thrust.
J. Weiß.
ESA J., Vol. 9, No. 1, p. 49 – 63 (1985).

This paper describes a catalogue concept for the presentation of finite–thrust effects, such as the additional propellant consumption and the drift of the apsidal radius or of the line of apsides compared with impulsive manoeuvres. Only inplane manoeuvres are considered and the impulsive velocity increment must be imparted at one of the apsidal points. Three thrust policies with different degrees of freedom and four cases specifying the final orbit are allowed.

052.024 Influence of the errors of a model of motion on the accuracy of determination of the position of an artificial earth satellite along its orbit.
R. R. Nazirov.
Inst. kosm. issled. Akad. Nauk SSSR. Prepr., No. 875, 44 pp. (1984). In Russian. Abstr. in Ref. Zh., 62. Issled. Kosm. Prostranstva, 1.62.231; 51. Astron. 2.51.81 (1985).

052.025 Semi–analytical method for calculation of the motion of artificial earth satellites of the "Navstar"–type in Eulerian elements.
T. S. Boronenko, L. A. Moskovkina, V. A. Tamarov, Yu. B. Shmidt.
Astron. Geod., Tomsk, No. 12, p. 68 – 72, 143 (1984). In Russian. Abstr. in Ref. Zh., 51. Astron., 2.51.75 (1985).

052.026 On the accuracy of high–speed calculation of orbits of artificial earth satellites by means of numerical integration.
N. A. Sorokin.
Astron. Geod., Tomsk, No. 12, p. 27 – 36, 141 – 142 (1984). In Russian. Abstr. in Ref. Zh., 51. Astron., 2.51.78 (1985).

052.027 A procedure of integration of equations for elements of the intermediate orbit of a satellite.
N. V. Emel'yanov.
Astron. Zh., Tom 62, Vyp. 3, p. 590 – 597 (1985). In Russian. English translation in Sov. Astron., Vol. 29, No. 3.

Under arbitrary perturbing factors of gravitational nature a procedure is considered based on consecutive approximations and used for the integration of equations for the elements of an artificial earth satellite's intermediate orbit.

052.028   Some numerical estimates of perturbations of orbital elements of an artificial satellite with total dynamical symmetry around the rotating non–spherical earth.
D. Z. Koenov.
Dokl. Akad. Nauk TadzhSSR, Tom 27, No. 6, p. 314–317 (1984). In Russian. Abstr. in Ref. Zh., 51. Astron., 2.51.83; 62. Issled. Kosm. Prostranstva, 2.62.153 (1985).

052.029   Calculation of trajectories of artificial earth satellites. Construction of algorithms of numerical integration of ordinary differential equations.
V. K. Taradij, M. L. Tsesis.
Inst. teor. fiz. Akad. Nauk USSR. Prepr., No. 60R, 16 pp. (1984). In Russian. Abstr. in Ref. Zh., 62. Issled. Kosm. Prostranstva, 2.62.150 (1985).

052.030   Toward the optimum selection of the orbit of a Japanese space VLBI satellite.
M. Tsuboi, H. Hirabayashi.
Ann. Tokyo Astron. Obs., Second Ser., Vol. 20, No. 3, p. 191–208 (1985).

The authors present a method to calculate the UV coverage of the space VLBI and show the UV coverages in some selected cases. It is shown that space VLBI with one space antenna gives fine coverage of spatial Fourier components of source brightness distribution. But, there are some problems to be solved in the optimum selection for the orbits, such as "the low declination problem" and "the coincidence problem" for example.

052.031   Calculation of the trajectories of artificial earth satellites. Comparison of programs of numerical integration.
M. L. Tsesis.
Inst. teor. fiz. Akad. Nauk USSR. Prepr., No. 91R, 32 pp. (1984). In Russian. Abstr. in Ref. Zh., 62. Issled. Kosm. Prostranstva, 2.62.155 (1985).

052.032   A packet of subroutines for calculation of trajectories of artificial earth satellites and accompanying reference information.
A. K. Kerimov.
Inst. zem. magn. i rasprostr. radiovoln Akad. Nauk SSSR., Prepr., No. 10/484, 36 pp. (1984). In Russian. Abstr. in Ref. Zh., 51. Astron., 3.51.90 (1985).

052.033   Algorithm for constructing the analytical theory of motion of artificial earth satellites in the Eulerian elements using Lie series.
T. S. Boronenko, V. A. Tamarov, Yu. B. Shmidt.
Astron. geod., Tomsk, No. 10, p. 49–56 (1984). In Russian. Abstr. in Ref. Zh., 51. Astron., 3.51.93 (1985).

052.034   Approximation of results of observations of the topocentric trajectory of an artificial earth satellite by orthogonal Chebyshev polynomials.
Yu. V. Surnin, Yu. V. Dement'ev.
Astron. geod., Tomsk, No. 10, p. 84–93 (1984). In Russian. Abstr. in Ref. Zh., 51. Astron., 3.51.95 (1985).

052.035   On the accuracy of forecasting the orbital motion of artificial earth satellites.
K. V. Kholshevnikov, L. L. Sokolov, E. I. Timoshkova, V. B. Titov.
Vestn. Leningr. Univ., Mat. Mekh. Astron., No. 19, p. 68–71 (1984). In Russian. Abstr. in Ref. Zh., 51. Astron., 3.51.105; 62. Issled. Kosm. Prostranstva, 2.62.154 (1985).

052.036   Oscillations and stability of a gravitationally oriented satellite perturbed by light pressure.
V. V. Beletskij, E. L. Starostin.
Inst. prikl. mat. Akad. Nauk SSSR, Prepr., No. 126, 28 pp. (1984). In Russian. Abstr. in Ref. Zh., 51. Astron., 3.51.108 (1985).

052.037   Probability of collisions of artificial bodies in the Earth environment.
L. Sehnal.
Adv. Space Res., Vol. 5, No. 2, p. 21–24 (1985). – See Abstr. 012.051.

The probability of the collisions of the artificial bodies orbiting the Earth is computed by analogy to the kinetic theory of gases; the input parameters are analysed to the end of 1982. The number of bodies is estimated as ~ 10000 till the $10^{-6} m^2$ size and their distribution is compared to that determined by the same way for the epoch of April 30, 1978. The probability of collisions is determined from the analysis of collision velocities and density distribution. The effect of the upper atmosphere drag and its "cleaning" consequences on the close Earth environment are discussed in connection with the solar activity variations.

052.038   Contribution of explosion and future collision fragments to the orbital debris environment.
S.–Y. Su, D. J. Kessler.
Adv. Space Res., Vol. 5, No. 2, p. 25–34 (1985). – See Abstr. 012.051.

The time evolution of the near–Earth man–made orbital debris environment modeled by numerical simulation is presented in this paper.

052.039   Collision probabilities at geosynchronous altitudes.
M. Hechler.
Adv. Space Res., Vol. 5, No. 2, p. 47–57 (1985). – See Abstr. 012.051.

A considerable collisional hazard to operational geostationary satellites is induced by a continuously increasing population of abandoned objects and related debris. During the past 5 years first measures have been taken to remove geostationary spacecraft from the geostationary altitude at the end of their operational life. Another concern is the crowding of active satellites at some preferred longitude positions. This paper analyses the hazard due to abandoned objects and the probability of a collision between satellites maintained within the same longitudinal slot.

052.040   Orbit determination error analyses for POPSAT and ERS–1.
K. F. Wakker, B. A. C. Ambrosius.
Adv. Space Res., Vol. 5, No. 2, p. 137–146 (1985). – See Abstr. 012.051.

The missions of both the future ESA geodetic POPSAT and remote–sensing ERS-1 satellites require that the orbits are known very accurately. The paper summarizes some results of POPSAT and ERS–1 orbit determination error analyses for 1–5 day arcs of tracking data acquired by global tracking networks, and of error analyses to estimate the accuracy with which the positions of ground stations and of the earth's poles can be determined from POPSAT tracking data. In addition, the accuracy of tailored gravity models for POPSAT is addressed and some simulation results are summarized. Finally, the application of POPSAT–to–ERS tracking data is discussed and it is shown that this is a very–attractive tracking concept for a follow–on spacecraft in the ERS–family.

052.041   Orbit determination for ISRO satellite missions.
C. Sreehari Rao, S. K. Sinha.
Adv. Space Res., Vol. 5, No. 2, p. 147–153 (1985). – See Abstr. 012.051.

052.042   Precise Seasat ephemeris from laser and altimeter data.
B. E. Schutz, B. D. Tapley, C. K. Shum.
Adv. Space Res., Vol. 5, No. 2, p. 155–168 (1985). – See Abstr. 012.051.

052.043   Determination of highly accurate orbits for altimeter satellites over limited geographic areas.
R. Kolenkiewicz, C. F. Martin.
Adv. Space Res., Vol. 5, No. 2, p. 169–174 (1985). – See Abstr. 012.051.

052.044 Analytical generation of orbits at the Royal Aircraft Establishment.
R. H. Gooding.
Adv. Space Res., Vol. 5, No. 2, p. 175 – 183 (1985). – See Abstr. 012.051.

The paper traces the development of the main computer programs for analytical orbit determination and orbit evolution carried out at the RAE.

052.045 Computation of precise satellite orbits by analytical and numerical computer programs.
P. Lála.
Adv. Space Res., Vol. 5, No. 2, p. 185 – 191 (1985). – See Abstr. 012.051.

The computer program, used for the computation of precise satellite orbits by an analytical method, has been modified to increase its accuracy. The modification consists mainly in the implementation of an orbit generator based on the numerical integration of perturbations in spherical coordinates. The theoretical background is given and both methods are compared to the real case of Lageos observations.

052.046 Satellite orbit determination from radio interferometry data.
J. B. Zieliński, E. Butkiewicz.
Adv. Space Res., Vol. 5, No. 2, p. 193 – 197 (1985). – See Abstr. 012.051.

052.047 An analysis of the influences of gravity nature by using an analytical theory for artificial satellite movement.
N. Georgiev, V. Kotseva.
Adv. Space Res., Vol. 5, No. 2, p. 199 – 203 (1985). – See Abstr. 012.051.

On the basis of the generalized problem of two immobile centres, an analytical theory for high–accuracy forecast of artificial satellite movement is elaborated. Polynomials obtained by the authors for determination of gravity perturbations by means of the power series on the regularized time are analysed in the paper. Perturbation values from zonal, tesseral and sectorial harmonics of the geopotential, from the Moon and Sun, from the Earth's tidal deformation, from attraction of the atmosphere and other planets and relativistic effects are performed for various cases.

052.048 Effect of rotational deformations of the Earth on satellite orbits.
M. Burša, J. Kostelecký.
Adv. Space Res., Vol. 5, No. 2, p. 205 – 209 (1985). – See Abstr. 012.051.

Because of variations in direction, as well as in modulus of the rotation vector of the Earth, mass deformations occur. For a perfectly elastic Earth, they can be expressed in harmonics both on the boundary surface and in the outer space. The perturbations in orbital elements of the Earth satellites caused by the deformations in question are derived. Possibilities of resonance effects are discussed.

052.049 On first order perturbations in the problem of motion of a space vehicle to the moon in a parabolic orbit.
V. D. Petelina.
Vopr. mat. i mekh. splosh. sred., Moskva, p. 118 – 128 (1984). In Russian. Abstr. in Ref. Zh., 51. Astron., 4.51.71; 62. Issled. Kosm. Prostranstva, 4.62.122 (1985).

052.050 On a case of determining the elements of an intermediate orbit.
E. L. Lukashevich.
Kosm. Issled., Tom 23, Vyp. 2, p. 308 – 310 (1985). In Russian. English translation in Cosm. Res.

052.051 Analytical estimates of the accuracy of determination and forecast of the parameters of motion of AES from data of altimeter measurements.
M. P. Nevol'ko, E. L. Mosin.
Kosm. Issled., Tom 23, Vyp. 2, p. 310 – 314 (1985). In Russian. English translation in Cosm. Res.

052.052 Expansion of the perturbation function due to the asphericity of the earth.
N. V. Emel'yanov, L. P. Nasonova.
Sov. Astron., Vol. 28, No. 5, p. 594 – 598 (1984). English translation of 38.052.011.

052.053 Use of nonstationary models of the atmosphere for the description of motion of the IKB 1300 artificial earth satellite.
T. A. Timokhova, P. E. Ehl'yasberg.
Inst. kosm. issled. Akad. Nauk SSSR, Prepr., No. 916, 28 pp. (1984). In Russian. Abstr. in Ref. Zh., 62. Issled. Kosm. Prostranstva, 5.62.144 (1985).

052.054 Solution of canonical differential equations of the disturbed motion of an artificial earth satellite with the method of Lie transformations taking the attraction of the sun (or the moon) into account.
D. Z. Koenov.
Izv. Akad. Nauk TadzhSSR. Otd–nie fiz.–mat., khim., geol. nauk, No. 1, p. 22 – 27 (1984). In Russian. Abstr. in Ref. Zh., 62. Issled. Kosm. Prostranstva, 5.62.145 (1985).

052.055 Altimetry, orbits and tides.
O. L. Colombo.
NASA Tech. Memo., NASA TM–86180, 9 | 173 pp. (1984).

This report explains the nature of the orbit error and its effect on the sea surface heights calculated with satellite altimetry. It includes an introduction of the elementary concepts of celestial mechanics required to follow a general discussion of the problem. This leads to a detailed consideration of errors in the orbits of satellites with precisely repeating ground tracks (SEASAT, TOPEX, ERS–1, POSEIDON, amongst past and future altimeter satellites). The theoretical conclusions are illustrated with the numerical results of computer simulations. The report ends with a presentation of some elements of tidal theory, showing how these principles can be combined with those pertinent to the orbit error to make direct maps of the tides using altimetry.

052.056 The determination and analysis of the orbit of Nimbus 1 Rocket, 1964–52B. Part 1: the orbits.
W. J. Boulton.
Planet. Space Sci., Vol. 33, No. 6, p. 721 – 734 (1985).

052.057 The determination and analysis of the orbit of Nimbus 1 Rocket, 1964–52B. Part 2: variations in orbital eccentricity.
W. J. Boulton.
Planet. Space Sci., Vol. 33, No. 6, p. 735 – 756 (1985).

052.058 Perturbations of the coordinates of artificial earth satellites calculated with the Laplace method.
S. A. Gasanov.
Gos. astron. inst. MGU, 84 pp. (1985). In Russian. Abstr. in Ref. Zh., 62. Issled. Kosm. Prostranstva, 6.62.146 (1985).

052.059 Determination of the orientation of AES with the use of measurements of angular velocities.
M. L. Pivovarov.
Kosm. Issled., Tom 23, Vyp. 3, p. 331 – 334 (1985). In Russian. English translation in Cosm. Res.

052.060 Analysis of the accuracy of determination of ephemerides of AES in the presence of a random stream of pulse perturbations.
A. I. Zverev, V. I. Karlov, M. N. Krasil'shchikov.
Kosm. Issled., Tom 23, Vyp. 3, p. 382 – 387 (1985). In Russian. English translation in Cosm. Res.

052.061 Attitude reconstitution by the Northern Data Analysis Consortium (NDAC).
L. Lindegren, F. van Leeuwen.
Processing of scientific data from the ESA astrometry satellite HIPPARCOS, p. 103 – 109 (1985). – See Abstr. 012.087.

The attitude of the HIPPARCOS satellite is defined by three "heliotropic" angles, relating the axes of the telescope to the mean ecliptic of J2000.0 and a nominal Sun direction. In the quiet intervals between jet firings, these angles are represented as continuous spline functions of time. Spline coefficients are obtained by least–squares fitting to star mapper transit times and angular rotation rates from three gyroscopes. Early in the mission, the positions of bright programme stars are progressively improved by use of the transit time residuals of the attitude solution.

052.062   Report on the task attitude reconstitution.
F. Donati, E. Canuto.
Processing of scientific data from the ESA astrometry satellite HIPPARCOS, p. 111 (1985). Abstract. – See Abstr. 012.087.

052.063   Modelling the HIPPARCOS attitude motion in solar eclipse conditions.
F. Donati, E. Canuto, P. Belforte.
Processing of scientific data from the ESA astrometry satellite HIPPARCOS, p. 113 – 114 (1985). Abstract. – See Abstr. 012.087.

052.064   Satellitenstart und Satellitenbahn.
P. Dahms.
Astron. Raumfahrt, 23. Jahrg., Heft 3, p. 50 – 53 (1985).

052.065   Venus Radar Mapper attitude control system.
G. W. Francis.
Adv. Astronaut. Sci., Vol. 57, p. 253 – 267 (1985). – See Abstr. 012.099.
The Venus Radar Mapper mass expulsion control system has been designed to perform a series of distinct functions. The system provides Thrust Vector Control during Trajectory Correction Maneuvers, Venus Orbit Insertion and Orbit Trajectory Maneuvers. The system was also designed to establish the vehicle in the initial cruise attitude after separation from the Centaur stage and prior to handing over to the reaction wheel control system. These varied functions are accomplished with common digital phase plane logic by selection of attitude rate boundaries and switching thruster selection logic for the various functions. The performance of the system has been verified through simulation and shown to meet mission requirements.

052.066   The Galileo scan platform pointing control system – a modern control theoretic viewpoint.
G. E. Sevaston, G. A. Macala, G. K. Man.
Adv. Astronaut. Sci., Vol. 57, p. 297 – 321 (1985). – See Abstr. 012.099.
In this paper the authors review the current Galileo scan platform pointing control system, and discuss ways in which modern control concepts could contribute toward a better result. Moreover, the authors describe in detail a robust modern control system architecture and the associated design procedures.

052.067   The calculation of normalized inclination functions for high degrees and orders.
N. V. Emel'yanov.
Tr. Gos. Astron. Inst. Shternberg, Tom 57, p. 83 – 91 (1985). In Russian.
Methods of calculation of normalized inclination functions are exposed to analysis. It is proposed an effective algorithm for calculation of normalized inclination functions and their derivatives for high degrees and orders.

052.068   Calculation of perturbations of a satellite's motion due to the tidal effect of the earth.
N. V. Emel'yanov.
Tr. Gos. Astron. Inst. Shternberg, Tom 57, p. 92 – 98 (1985). In Russian.
A method to calculate perturbations of elements of an intermediate orbit of a satellite due to the tidal effect of the earth is proposed. The tidal friction inside the terrestrial body is neglected. The Love numbers are assumed to be constants.

052.069   Expansion of the coordinates of intermediate motion of a satellite in Poisson series with the help of a computer.
Huan Nguen Din, N. V. Emel'yanov.
Tr. Gos. Astron. Inst. Shternberg, Tom 57, p. 99 – 117 (1985). In Russian.
For some functions of satellite coordinates being contained in the development of the perturbing function a method of development in trigonometric series is considered. The coordinates of a satellite are described with formulas of the intermediate orbit based on the solution of the generalised two fixed centers problem. For the solution of the problem the UPP–system of programming of analytical operations with Poisson series is used. The series are obtained and printed up to the 9th power of the satellite's eccentricity inclusive.

052.070   The use of Laplace's method in the theory of motion of artificial earth satellites.
S. A. Gasanov.
Tr. Gos. Astron. Inst. Shternberg, Tom 57, p. 130 – 159 (1985). In Russian.
The paper describes the application of the Laplace method to the problem of satellite motion under the influence of the earth's oblateness and lunar perturbations. The analytical expressions for the elements are obtained in the first approximation.

052.071   Radial variations of a satellite orbit due to gravitational errors: implications for satellite altimetry.
C. A. Wagner.
J. Geophys. Res., Vol. 90, No. B4, p. 3027 – 3036 (1985).
The linear perturbations of the radius of a satellite orbit due to the geopotential are derived. From these, estimates are made of the radial orbit error due just to geopotential errors from current models employing only conventional satellite–tracking data.

052.072   The luni–solar perturbations and the motion of high satellites.
A. S. Sochilina.
Byull. Inst. Teor. Astron., Tom 15, No. 7 (170), p. 383 – 395 (1985). In Russian.
Two possible procedures have been investigated for the construction of analytical expressions of the luni–solar perturbations in the elements of high orbit satellites. The first procedure consists in the construction of an intermediate solution by the Krylov–Bogolyubov method of averaging. In the second case the Laplacian plane has been adopted as a reference plane for the satellite orbit. It has been shown that the last procedure is more effective for determining the luni–solar perturbations of the first order for any inclinations of a satellite orbit except the critical one.

052.073   On normal places obtained by matched polynomials.
B. B. Baghos.
Byull. Inst. Teor. Astron., Tom 15, No. 8 (171), p. 413 – 418 (1985). In Russian.
A method is given for constructing the spline which represents the laser range measurements of the AES during one passage over the station. For normal places taken at two neighbouring polynomials of the spline the expressions for the variances and their covariance have been obtained.

052.074   Analytical theory of motion of a satellite of a nonspherical planet.
V. A. Tamarov.
Byull. Inst. Teor. Astron., Tom 15, No. 8 (171), p. 457 – 472 (1985). In Russian.
A theory of the artificial satellite motion in the earth's gravitational field is constructed on the basis of a symmetrical intermediate orbit of the generalized two–fixed centres problem including all perturbations mixed with the second zonal harmonics of the geopotential. The series of the theory are obtained using Kaula's inclination functions and Hansen's coefficients. They are suitable for any orbits with the exception of critical inclination and commensurability between the periods of satellite revolution and planetary rotation.

052.075 Orbital variation of the synchronous satellite and its calculation.
L. Liu.
Chin. Astron. Astrophys., Vol. 9, No. 1, p. 27 – 34 (1985). English translation of 38.052.044.

052.076 The orbit of Lageos and solar eclipses.
D. P. Rubincam, N. R. Weiss.
NASA Tech. Memo., NASA TM–86076, 5 + 10 pp. (1984).

The authors point out the importance of the effect of solar eclipses on the orbit of the Lageos satellite. They examine how the eclipses that occurred between launch on 4 May 1976 and the end of 1983 affected the semimajor axis. It is shown that some eclipses have perturbed the orbit at the one centimeter level. This is significant, since a 1 cm change in the semimajor axis translates into an along–track error of 9 m over a period of 15 days.

052.077 Some ideas on applying a sequential estimation algorithm to satellite dynamics.
W.–y. Zhu, Z.–y. Cheng.
Acta Astron. Sin., Vol. 26, No. 1, p. 13 – 21 (1985). In Chinese.

052.078 A short note on the high–precision navigation of Quasat.
G. Tang.
ESA Spec. Publ., ESA SP–213, p. 185 – 186 (1984). – See Abstr. 012.124.

This note discusses a method whereby the astronomical observations of a space VLBI system are used for determining high–precision positions of the observing system itself.

Upgrading the computer control of the Interkosmos Laser Ranging Station in Helwan.
See Abstr. 013.019.

Performance and results of Satellite Ranging Laser Station at Borowiec in 1983.
See Abstr. 013.020.

Intelligent step motor control unit for continuous tracking.
See Abstr. 034.014.

Estimation of phase errors on laser observations of artificial earth satellites.
See Abstr. 036.021.

An effective method for on–line corrections of two–axes laser mount.
See Abstr. 036.022.

Doppler receiver performance evaluations.
See Abstr. 036.028.

Estimate of the efficiency of numerical algorithms for construction of satellite trajectories.
See Abstr. 042.053.

Project MERIT and orbit dynamics of the satellites.
See Abstr. 044.085.

Determination of the position of the Station Borowiec No. 7811 by satellite laser observations.
See Abstr. 045.003.

Further improvements of the orbital program system POTSDAM–5 and their utilization in geodetic – geodynamic investigations.
See Abstr. 045.005.

Results of the Photodoppler method from observations of the satellite NOVA.
See Abstr. 045.007.

Astrometric needs for the ISEE–3/ICE mission to comet Giacobini–Zinner.
See Abstr. 051.032.

Individual geopotential coefficients of order 15 and 30, from resonant satellite orbits.
See Abstr. 081.003.

The use of geosynchronous satellites for determining parameters of the geopotential.
See Abstr. 081.014.

Further comparisons of Earth gravity models by means of lumped coefficients.
See Abstr. 081.022.

Density scale height determined from the motion of Dash 2 satellite.
See Abstr. 082.013.

Models of the thermosphere for usage in satellite dynamics.
See Abstr. 082.016.

Gravity field of the Jovian system from Pioneer and Voyager tracking data.
See Abstr. 099.017.

The work on comet Halley's orbit at ESOC.
See Abstr. 103.913.

Geomagnetic transmission functions for a 400 km altitude satellite.
See Abstr. 144.277.

053 Artificial Satellites, Space Probes

053.001   **The Soviet Venera programme.**
P. S. Clark.
J. Br. Interplanet. Soc., Vol. 38, No. 2, p. 74 – 93 (1985).

053.002   **Satellite Digest–179.**
R. D. Christy.
Spaceflight, Vol. 27, No. 1, p. 43 (1985).
1985 January listing of satellite and spacecraft launches.

053.003   **Satellite Digest–180.**
R. D. Christy.
Spaceflight, Vol. 27, No. 2, p. 91 (1985).
1985 February listing of satellite and spacecraft launches.

053.004   **Satellite Digest–181.**
R. D. Christy.
Spaceflight, Vol. 27, No. 3, p. 140 (1985).
1985 March listing of satellite and spacecraft launches.

053.005   **Satellite Digest–182.**
R. D. Christy.
Spaceflight, Vol. 27, No. 4, p. 184 (1985).
1985 April listing of satellite and spacecraft launches.

053.006   **Influence of the aerodynamical moment on the gravita-
tional orientation of the Salyut 6–Soyuz orbital complex.**
V. A. Sarychev, V. V. Sazonov.
Kosm. Issled., Tom 23, Vyp. 1, p. 63 – 83 (1985). In Russian.
English translation in Cosm. Res.

053.007   **Visual observations, horizontal and equatorial coordinates of Bolgariya 1300 1981–75–1.**
Rezul't. Nablyud. Iskusstv. Sputnikov Zemli, Vyp. 102 (242),
72 pp. (1983). In Russian.
Concerning Bolgariya 1300 1981–75–1. Horizontal coordinates, January – October 1982. Equatorial coordinates, July –
November 1982.

053.008   **Visual observations, horizontal and equatorial coordinates (1950.0) of the artificial earth satellite
Intercosmos 19 1979–20–1.**
Rezul't. Nablyud. Iskusstv. Sputnikov Zemli, Vyp. 105 (245),
63 pp. (1984). In Russian.

Concerning Intercosmos 19 1979–20–1. January – September
1983. April – September 1983.

053.009   **Visual observations, horizontal and equatorial coordi-
nates (1950.0) of the artificial earth satellite Oreol 3.**
Rezul't. Nablyud. Iskusstv. Sputnikov Zemli, Vyp. 106 (246),
62 pp. (1984). In Russian.
Concerning Oreol 3 1981–94–1. January – August 1983.
April – November 1983.

053.010   **Artificial satellites.**
H. Miles.
J. Br. Astron. Assoc., Vol. 95, No. 4, p. 164 – 166 (1985).

053.011   **On standardization of the brightness of artificial celestial bodies.**
A. V. Didenko.
Tr. Astrofiz. Inst. Alma–Ata, Tom 44, p. 77 – 82 (1984). In Russian.

053.012   **Three–colour observations of a variable geostationary
satellite.**
A. V. Didenko.
Astron. Tsirk., No. 1354, p. 6 – 8 (1984). In Russian.

053.013   **Satellite Digest–183.**
R. D. Christy.
Spaceflight, Vol. 27, No. 5, p. 236 (1985).
1985 May listing of satellite and spacecraft launches.

053.014   **Satellite Digest–184.**
R. D. Christy.
Spaceflight, Vol. 27, No. 6, p. 284 (1985).
1985 June listing of satellite and spacecraft launches.

**Accuracy of Earth gravity field models as a function of order of
harmonic coefficients and a draft of a laser geodynamic twin–
satellite.**
See Abstr. 081.016.

Theoretical Astrophysics

061 General Aspects (Nucleosynthesis, Elementary Particles, Neutrino Astronomy, etc.)

061.001 Suche nach Neutrino–Oszillationen an Kernreaktoren.
V. Zacek, G. Zacek.
Naturwissenschaften, 72. Jahrg., Heft 2, p. 62 – 69 (1985).
One of the most challenging problems in todays physics remains the question, whether the neutrino possesses a finite restmass. Experiments searching for neutrino oscillations provide important information on neutrino masses and on the possibility of mixing of different neutrinos states. In this paper the principles of a neutrino oscillation experiment at a nuclear power reactor are described and the latest results are presented.

061.002 Measurement of neutron capture cross sections of s–only isotopes: ^{70}Ge, ^{86}Sr, and ^{87}Sr.
G. Walter, H. Beer.
Astron. Astrophys., Vol. 142, No. 2, p. 268 – 272 (1985).
In order to improve the s–process data basis in the mass range $A < 100$ the neutron capture cross sections of ^{86}Sr, ^{87}Sr and, for the first time, ^{70}Ge have been measured by time–of–flight in the energy range from 3.5 to 240 keV. The Maxwellian averaged capture cross sections have been calculated from the data for $kT = 20$ keV up to $kT = 50$ keV. At $kT = 30$ keV the authors obtained the values 92 ± 5, 74 ± 5, and 100 ± 7 mb for ^{70}Ge, ^{86}Sr, and ^{87}Sr, respectively.

061.003 Neutrons in science and technology.
D. A. Bromley.
Nucl. Instrum. Methods Phys. Res., Sect. A, Vol. 225, No. 1, p. 240 – 279 (1984). Abstr. in Phys. Abstr., Vol. 88, No. 1248, Entry 5480 (1985).

061.004 The steady–state free–electron population of free space.
H. Aspden.
Lett. Nuovo Cimento, Vol. 41, Ser. 2, No. 7, p. 252 – 256 (1984). Abstr. in Phys. Abstr., Vol. 88, No. 1249, Entry 10206 (1985).

061.005 Exact solution to the Einstein–Yang–Mills equation.
Y. Hosotani.
Phys. Lett. B, Vol. 147B, No. 1 – 3, p. 44 – 46 (1984). Abstr. in Phys. Abstr., Vol. 88, No. 1249, Entry 10522 (1985).

061.006 Mirror neutrinos in standard cosmology.
D. Fargion, M. Roos.
Phys. Lett. B, Vol. 147B, No. 1 – 3, p. 34 – 38 (1984). Abstr. in Phys. Abstr., Vol. 88, No. 1249, Entry 10551 (1985).

061.007 Long–range antigravity.
K. I. Macrae, R. J. Riegert.
Nucl. Phys. B, Part. Phys., Vol. B244, No. 2, p. 513 – 522 (1984). Abstr. in Phys. Abstr., Vol. 88, No. 1249, Entry 14469 (1985).

061.008 Origin of the chemical elements.
R. J. Tayler.
Proc. R. Soc. London, Ser. A, Vol. 396, No. 1810, p. 21 – 54 (1984). Abstr. in Phys. Abstr., Vol. 88, No. 1249, Entry 14807 (1985).

061.009 Effective Lagrangian for massless pair production by an oscillating field.
D. E. Neville.
Phys. Rev. D, Vol. 30, No. 8, p. 1695 – 1706 (1984). Abstr. in Phys. Abstr., Vol. 88, No. 1250, Entry 15030 (1985).

061.010 Global and gauge symmetries in finite temperature sypersymmetric theories.
M. Mangano.
Phys. Lett. B, Vol. 147B, No. 4 – 5, p. 307 – 310 (1984). Abstr. in Phys. Abstr., Vol. 88, No. 1250, Entry 18968 (1985).

061.011 Issues in monopole–catalyzed proton decay.
Y. Kazama.
Grand unified theories and cosmology, p. 61 – 77 (1984). Abstr. in Phys. Abstr., Vol. 88, No. 1251, Entry 19510 (1985). – See Abstr. 012.020.

061.012 On the determination of neutrino mass – critical status report.
C.–r. Ching, T.–h. Ho.
Phys. Rep., Vol. 112, No. 1, p. 1 – 51 (1984). Abstr. in Phys. Abstr., Vol. 88, No. 1251, Entry 19611 (1985).

061.013 Electrodisintegration and electrocapture in primordial nucleosynthesis.
H. S. Picker.
Phys. Rev. C, Vol. 30, No. 5, p. 1751 – 1752 (1984). Abstr. in Phys. Abstr., Vol. 88, No. 1251, Entry 19641 (1985).

061.014 Applications to stellar nucleosynthesis (*neutron capture*).
B. J. Allen.
Neutron radiative capture, p. 176 – 186 (1984). Abstr. in Phys. Abstr., Vol. 88, No. 1251, Entry 19660 (1985). – See Abstr. 003.006.

061.015 Astrophysical constraints on the monopole abundance.
M. Fukugita.
Grand unified theories and cosmology, p. 38 – 60 (1984). Abstr. in Phys. Abstr., Vol. 88, No. 1251, Entry 24580 (1985). – See Abstr. 012.020.

061.016 Radiative formation of positronium in a vacuum.
A. Erdas, G. Mezzorani, P. Quarati, G. Puddu.
Astron. Astrophys., Vol. 144, No. 2, p. 295 – 297 (1985).
The authors report highly precise numerical results of e^+e^- total radiative recombination cross section $\sigma(\varepsilon) = \Sigma \sigma_n(\varepsilon)$. Using a scaling law for $\sigma_n(\varepsilon) = \Sigma \sigma_{nlj}(\varepsilon)$, valid at any n and ε, they give analytical expressions, valid at any T, of the total recombination coefficient α. They calculate the recombination coefficients α_n and the total recombination coefficient α in the temperature range of astrophysical interest. The radiative energy produced in the recombination process is also calculated.

061.017 **The properties and effects on stellar burning of fractionally charged nuclei.**
R. N. Boyd, R. E. Turner, L. Rybarcyk, C. Joseph.
Astrophys. J., Vol. 289, No. 1, p. 155 – 164 (1985).

The consequences of unconfined quarks which may have been left over from the big bang, especially as to how they might participate in nucleosynthesis, are examined. Possible properties of the fractionally charged nuclei (Q–nuclei) thus produced, including β–decay half–lives, binding energies, energy level densities, and thermonuclear reaction rates, are studied. Stellar burning cycles are suggested by these considerations in which the Q–nuclei could contribute significantly to stellar nucleosynthesis. Possible implications of the existence of Q–nuclei for stellar evolution are considered, and the results of a calculation are presented which confirm that no obvious conflicts with the known parameters of the Sun are encountered.

061.018 **Self–duality in Euclidean supergravity.**
G. M. O'Brien, D. H. Tchrakian.
Gen. Relativ. Gravitation, Vol. 17, No. 1, p. 55 – 61 (1985).

The compatibility of Euclidean supergravity with curvature satisfying (1) self–duality and (2) a double duality Ansatz are investigated. The Minkowskian version of (2) is also considered.

061.019 **Energy loss by slow magnetic monopoles in a thermal plasma.**
N. Meyer–Vernet.
Astrophys. J., Vol. 290, No. 1, p. 21 – 23 (1985).

When calculating the deceleration of a magnetic monopole in a thermal plasma, one cannot neglect the nonlocal dispersive properties if the plasma is conductive and the monopole's velocity is subthermal. This yields values of the stopping power in astrophysical plasmas smaller than previously found.

061.020 **Neutrino production from discrete high–energy gamma–ray sources.**
H. Lee, S. A. Bludman.
Astrophys. J., Vol. 290, No. 1, p. 28 – 32 (1985).

Using cross section formulae from Tan and Ng and from Hillas, the authors perform a cascade calculation of neutrino and photon production from a point cosmic–ray source surrounded by matter. The cascade effect on photon production is important in thick matter. It is shown that ultra–high–energy neutrinos from observed ultra–high–energy gamma–ray sources are just detectable in a DUMAND–size detector. Nearby sources of lower energy neutrinos, which are blanketed by thick matter and therefore unidentified electromagnetically, may be detected in underground detectors.

061.021 **Unconventional ^{12}C production in Population III stars.**
R. Mitalas.
Astrophys. J., Vol. 290, No. 1, p. 273 – 275 (1985).

The production of ^{12}C by unconventional nuclear reactions is investigated. The reactions of α–particles with minor constituents of the p–p chain, ^{7}Be and ^{8}B, establish equilibrium abundances of ^{11}B and ^{11}C. For $\varrho = 10\,\mathrm{g\,cm}^{-3}$ and $(X, Z) = (0.739, 0)$ the carbon–producing reactions ^{11}B$(p, \gamma)^{12}$C and ^{11}C$(p, \gamma)^{12}$N$(e^+v)^{12}$C dominate the triple–α reaction for $T_6 < 76$; the nonresonant triple–α reaction dominates the triple–α reaction for $T_6 < 61$. The relative importance of these reactions in massive Population III stars is discussed.

061.022 **Experimental location of Gamow–Teller strength for astrophysical calculations in the region of $A = 54$–58.**
F. Ajzenberg–Selove, R. E. Brown, E. R. Flynn, J. W. Sunier.
Phys. Rev. C, Vol. 30, No. 6, p. 1850 – 1854 (1984). Abstr. in Phys. Abstr., Vol. 88, No. 1253, Entry 30839 (1985).

061.023 **The importance of impatience.**
J. Maddox.
Nature, Vol. 314, No. 6010, p. 399 (1985).

Unifying gravity with the other three fundamental forces has been brought a large step nearer. But there is a long way still to go.

061.024 **Unification of forces and particles in superstring theories.**
M. B. Green.
Nature, Vol. 314, No. 6010, p. 409 – 414 (1985).

Superstring field theories have emerged as potentially consistent quantum field theories that unify gravity with the other fundamental forces in an almost unique manner. They are based on the dynamics of string–like fundamental quanta rather than the point–like quanta of more familiar relativistic "point field theories" such as Yang–Mills gauge theory or general relativity. In these theories the observed fundamental particles, such as the leptons and quarks, may arise as the ground states of a string.

061.025 **The shadow world of superstring theories.**
E. W. Kolb, D. Seckel, M. S. Turner.
Nature, Vol. 314, No. 6010, p. 415 – 419 (1985).

Recent attempts to construct a superstring theory that unifies all the interactions of nature, including gravity, in a finite anomaly–free quantum theory have led to the speculation that there may exist another form of matter ("shadow matter") in the Universe, which only interacts with "ordinary matter" (for example, quarks, leptons) through gravity or gravitational–strength interactions. The existence of shadow matter would have any astrophysical and cosmological implications, some of which are discussed here.

061.026 **The production and survival of ^{205}Pb in stars, and the ^{205}Pb – ^{205}Tl s–process chronometry.**
K. Yokoi, K. Takahashi, M. Arnould.
Astron. Astrophys., Vol. 145, No. 2, p. 339 – 346 (1985).

The s–process production ratio of ^{204}Pb and ^{205}Pb, which is of importance for the ^{205}Pb – ^{205}Tl chronometry, is calculated in two schematic s–process models. Special emphasis is put on the role of the bound state β⁻–decay of ^{205}Tl which has been totally overlooked in previous studies. It is shown that the ^{205}Pb production under certain s–process conditions might be large enough to justify a renewed search for extinct ^{205}Pb in meteorites. The influence of uncertainties in the input quantities (electron capture Q–value and ft values) on the results is also discussed.

061.027 **Helium detonation in pancake stars.**
B. Pichon.
Astron. Astrophys., Vol. 145, No. 2, p. 387 – 390 (1985).

The author examines, within the framework of the adiabatic affine star model, the conditions of helium detonation in the core of a star highly compressed by external tidal forces when it penetrates deeply into the Roche radius of a giant black hole. The author determines also for the more extreme events the principal elements synthesized during the brief hot "pancake" configuration.

061.028 **Quasar energy from frozen fusion via massive neutrinos?**
J. Steyaert.
The Big Bang and Georges Lemaitre, p. 133 – 138 (1984). – See Abstr. 012.043.

This paper speculates on the hypothesis that the nuclear reactions H⁺ + H⁻ → D + v and H + H → D + v with finite–mass neutrinos may be important mechanisms of energy generation in astrophysical sources.

061.029 **Astration of cosmological deuterium.**
D. D. Clayton.
Astrophys. J., Vol. 290, No. 2, p. 428 – 432 (1985).

The author reconsiders the degree of astration of primordial deuterium by the continuous galactic processes of star formation and chemical evolution because of its importance to cosmology. The calculations investigate the dependence on the value of the return fraction and on galactic chemical evolution when infall of matter with constant composition occurs. The results suggest that big bang D/H was at least 3 times larger than the largest values observed in today's solar neighborhood and even larger yet if matter falling onto the disk is already astrated.

061.030 Anisotropic emission of neutrinos originating in beta–decay processes under the influence of a magnetic field of high intensity.
O. F. Dorofeev, V. N. Rodionov, I. M. Ternov.
Pis'ma Astron. Zh., Tom 11, No. 4, p. 302 – 309 (1985). In Russian. English translation in Sov. Astron. Lett., Vol. 11.

Estimates of the maximal momentum obtained by a star as a result of anisotropic emission of neutrinos in URCA processes on free nucleons in a strong magnetic field under conditions of high temperatures and densities are given.

061.031 Stellar neutron capture rates for ^{46}Ca and ^{48}Ca.
F. Käppeler, G. Walter, G. J. Mathews.
Astrophys. J., Vol. 291, No. 1, p. 319 – 327 (1985).

Stellar neutron capture rates for ^{46}Ca and ^{48}Ca have been measured by the activation technique. Both $kT = 25$ keV Maxwellian–like incident neutron spectra and non–Maxwellian higher energy spectra have been utilized to study the possible role of individual capture resonances. The possibility of a neutron capture origin for ^{46}Ca and ^{48}Ca is discussed in the light of the new cross sections, as well as a mechanism for the production of the observed isotopic anomalies in inclusion EK–1–4–1 from the Allende meteorite.

061.032 Cosmological and experimental constraints on the tau neutrino.
S. Sarkar, A. M. Cooper.
Phys. Lett. B, Vol. 148B, No. 4 – 5, p. 347 – 354 (1984). Abstr. in Phys. Abstr., Vol. 88, No. 1252, Entry 25113 (1985).

061.033 A difficulty with evasion of a cosmological limit on massive neutrinos.
M. Gronau, R. Yahalom.
Phys. Rev. D, Vol. 30, No. 11, p. 2422 – 2423 (1984). Abstr. in Phys. Abstr., Vol. 88, No. 1252, Entry 25180 (1985).

061.034 Equilibration of ^{176}Lu^{g,m} during the s–process.
E. B. Norman, T. Bertram, S. E. Kellogg, S. Gil, P. Wong.
Astrophys. J., Vol. 291, No. 2, p. 834 – 837 (1985).

The effects of photoexcitation and positron annihilation–excitation of ^{176}Lu^g to ^{176}Lu^m have been investigated. It is found that as a result of these two processes alone, ^{176}Lu^g and ^{176}Lu^m are in thermal equilibrium at temperatures $\geq 3.5 \times 10^8$K. This implies that ^{176}Lu is not a reliable s–process chronometer.

061.035 Thermal cyclotron radiation in astrophysics.
V. V. Zheleznyakov.
Astrophys. Space Phys. Rev., Vol. 3, p. 157 – 195 (1984). – See Abstr. 003.019. Revised and extended English translation of 34.061.084.

The main characteristics of cyclotron radiation of electrons and of the attendant absorption of electromagnetic waves by an equilibrium plasma in a magnetic field are presented. Examples of a classical plasma, of a quantized plasma, and of a plasma in which vacuum polarization in strong magnetic fields is taken into account are analyzed. The discussion is focused on specific application of the cyclotron radiation mechanism in astrophysics, specifically, the origin of the microwave radio emission of the Sun (the S component of radiation), the possibility of detecting cyclotron lines in this radiation, and the structure of cyclotron lines observed in the spectra of the X–ray sources Her X–1 and 4U 0115+63 and in the spectra of gamma–ray bursts.

061.036 Vacuum polarization by a magnetic field and its astrophysical manifestations.
G. G. Pavlov, Yu. N. Gnedin.
Astrophys. Space Phys. Rev., Vol. 3, p. 197 – 253 (1984). – See Abstr. 003.019. Revised and extended English translation of 34.061.085.

In strong magnetic fields of neutron stars and white dwarfs the electron–positron vacuum behaves as an anisotropic medium that has birefringent properties. Thus vacuum influences the generation and propagation of electromagnetic radiation and changes the spectrum, angular distribution and polarization of the radiation from objects having a strong magnetic field. The study of specific vacuum effects in the radiation from neutron stars and white dwarfs yields additional information on the emitting plasma (magnetic field, density, etc.), which is essential for constructing models of these objects.

061.037 Effects of nuclear uncertainties and chemical evolution on the Standard Big Bang nucleosynthesis.
P. Delbourgo–Salvador, C. Gry, G. Malinie, J. Audouze.
Inst. Astrophys. Paris, Pré–Publ., No. 90, 35 pp. (1985). To appear in Astron. Astrophys.

061.038 Zeitabhängige Lösungen für die Stoßbeschleunigung von energetischen Teilchen.
L. O'C. Drury, R. Beck.
Mitt. Astron. Ges., Nr. 63, p. 172 (1985). – See Abstr. 012.063.

061.039 Neutrinoruhemassen – der aktuelle Stand.
E. Dreisigacker.
Phys. Bl., 41. Jahrg., Heft 6, p. 151 (1985).

061.040 De Sitter superalgebras and supergravity.
K. Pilch, P. van Nieuwenhuizen, M. F. Sohnius.
Commun. Math. Phys., Vol. 98, No. 1, p. 105 – 117 (1985).

A general analysis of all possible super–extensions of anti–de Sitter and de Sitter algebras O(3,2) and O(4,1) is presented. It is shown that actions with de Sitter local supersymmetry exist, but contain vector–ghosts.

061.041 Numerical models of expanding fireballs.
L. J. Caroff, J. A. Eilek, P. D. Noerdlinger, M. Doye.
Bull. Am. Astron. Soc., Vol. 16, No. 4, p. 954 (1984). Abstract. – See Abstr. 010.062.

061.042 Improved astronomical limits on the neutrino mass.
R. Epstein, J. Madsen.
Bull. Am. Astron. Soc., Vol. 16, No. 4, p. 1015 (1984). Abstract. – See Abstr. 010.062.

061.043 Positivity of energy in five–dimensional classical unified field theories.
A. H. Taub.
Lett. Math. Phys., Vol. 9, No. 3, p. 243 – 253 (1985).

This letter contains the outline of the proof that a positive energy theorem holds in the classical unified field theories of Kaluza–Klein, Jordan–Thiry, and Veblen as well as in Yang–Mills theory with a gauge group U(1). The theorem is analogous to that of the Einstein theory of general relativity but differs from it in that it involves the geometry of a five–dimensional space with a Lorentzian metric, which admits a space–like Killing vector.

061.044 The Jordan pair content of the Magic Square and the geometry of the scalars in $N = 2$ supergravity.
P. Truini, G. Olivieri, L. C. Biedenharn.
Lett. Math. Phys., Vol. 9, No. 3, p. 255 – 261 (1985).

The close connection between Jordan and Lie algebras makes these Jordan structures of interest to physicists. The Freudenthal–Tits Magic Square, which exemplifies this connection, has recently entered into constructing supergravity. The authors show how Jordan pairs – which are, from several points of view, a most natural Jordan structure – are imbedded in the Magic Square.

061.045 Neutral fermions as the missing mass matter in galactic halos.
B. Datta, C. Sivaram, S. K. Ghosh.
Astrophys. Space Sci., Vol. 111, No. 2, p. 413 – 417 (1985).

The authors point out that several independent considerations rule out the hypothesis that the missing mass in galactic halos is dominated by massive neutral fermions such as neutrinos, gravitinos or photinos.

061.046   Stellar weak interaction rates for intermediate–mass nuclei. IV. Interpolation procedures for rapidly varying lepton capture rates using effective log (ft) – values.
G. M. Fuller, W. A. Fowler, M. J. Newman.
Astrophys. J., Vol. 293, No. 1, p. 1 – 16 (1985).

Simple expressions for continuum electron and positron capture phase space factors and the associated neutrino energy loss integrals are presented in terms of standard Fermi integrals. Continuous approximations to the relevant Fermi integrals and their first derivatives are made. These allow the computation of effective $\log(ft)$ – values, at each temperature and density point, for the continuum lepton capture rates considered in the earlier papers in this series. Generalization of the Fermi integral expressions for the lepton continuum capture phase space factors are given for astrophysical environments where there exists an equilibrium distribution of electron–type neutrinos.

061.047   Enhancement of forbidden nuclear β–decay by low–frequency electromagnetic fields.
H. R. Reiss.
Infrared Phys., Vol. 25, No. 1/2, p. 525 – 529 (1985). – See Abstr. 012.065.

Nuclear β–decays that are forbidden by angular momentum and/or parity selection rules can be significantly enhanced by external electromagnetic fields. The effect is largely nonperturbative in nature, with a governing intensity parameter that becomes larger at long wavelengths. FIR radiation is a focus of interest for inducing the effect. An examination is done of the field–dependent terms which lead to forbiddenness removal in β–decay. These terms come primarily from field interaction with the initial and final nuclear charges, and from the longitudinal field interaction term for the β–particle. Relevant astrophysical environments are discussed in which the effect can occur.

061.048   The quest for the origin of the elements.
W. A. Fowler.
Science, Vol. 226, No. 4677, p. 922 – 935 (1984).

061.049   Interaction between antiprotons and helium and astrophysics.
M. G. Sapozhnikov.
Priroda, No. 6, p. 70 – 81 (1985). In Russian.

061.050   Upper bound on entropy.
I. Khan, A. Qadir.
Lett. Nuovo Cimento, Vol. 41, Ser. 2, No. 15, p. 493 – 496 (1984). Abstr. in Phys. Abstr., Vol. 88, No. 1253, Entry 35538 (1985).

061.051   Nuclear solutions to astrophysical problems.
S. E. Koonin.
Nucleon–nucleon interaction and nuclear many–body problems, p. 167 – 177 (1984). Abstr. in Phys. Abstr., Vol. 88, No. 1253, Entry 35540 (1985). – See Abstr. 012.076.

061.052   The lifetime of an elementary particle in a field.
L. C. B. Ryff.
Gen. Relativ. Gravitation, Vol. 17, No. 6, p. 515 – 519 (1985).

By assuming that one can associate a proper frame to a particle in a field and using relativistic considerations, a generalization of Apsel's prediction that electromagnetic potentials can alter the lifetime of a charged particle is obtained. It is shown, by a new procedure, that these same assumptions lead to the Euler–Lagrange equations of motion for the particle.

061.053   Incoherent radiation in an n–dimensional space.
S. Giddings.
Am. J. Phys., Vol. 52, No. 12, p. 1125 – 1127 (1984). Abstr. in Phys. Abstr., Vol. 88, No. 1254, Entry 35959 (1985).

061.054   Non–gauge real vector multiplets coupled to the $N = 1$ supergravity.
A. Kakuto.
Prog. Theor. Phys., Vol. 72, No. 3, p. 594 – 605 (1984). Abstr. in Phys. Abstr., Vol. 88, No. 1254, Entry 36080 (1985).

061.055   The magnetic monopole.
P. Galeotti.
G. Fis., Vol. 25, No. 1, p. 3 – 28 (1984). In Italian. Abstr. in Phys. Abstr., Vol. 88, No. 1254, Entry 36416 (1985).

061.056   Cluster model of $A = 7$ nuclei and the astrophysical S factor for $^3\mathrm{He}(\alpha,\gamma)^7\mathrm{Be}$ at zero energy.
B. Buck, R. A. Baldock, J. A. Rubio.
J. Phys. G, Vol. 11, No. 1, p. L11 – L16 (1985). Abstr. in Phys. Abstr., Vol. 88, No. 1254, Entry 36632 (1985).

061.057   A remark on the possibility of an extension of Heisenberg's uncertainty principle to the electric charge.
V. de Sabbata, M. Gasperini.
Sir Arthur Eddington Centenary Symposium. Vol. 1: Relativistic astrophysics and cosmology, p. 224 – 227 (1984). – See Abstr. 012.080.

This note points out that in the framework of the classical Kaluza–Klein theory the introduction of a relation between the electric charge and an extra dimension of space suggests a naive extension of Heisenberg's uncertainty relations.

061.058   Large number coincidences and unification of the parameters underlying elementary particles astrophysics and cosmology.
C. Sivaram.
Sir Arthur Eddington Centenary Symposium. Vol. 1: Relativistic astrophysics and cosmology, p. 228 – 243 (1984). – See Abstr. 012.080.

The ubiquitous occurrence of the Dirac–Eddington large dimensionless numbers when relating the physical parameters such as mass, radius, angular momentum etc. of typical astrophysical objects like stars and galaxies to the fundamental constants of atomic physics is reviewed. Several other interesting coincidences and relationships connecting the parameters of cosmology and elementary particle physics are pointed out. The significance of these relations is explored especially in connection with the time variation of the fundamental constants and the unification of cosmology and quantum physics.

061.059   Extension of Eddington's ideas to quark–lepton level.
R. C. Verma.
Sir Arthur Eddington Centenary Symposium. Vol. 1: Relativistic astrophysics and cosmology, p. 253 – 260 (1984). – See Abstr. 012.080.

The author extends an empirical relation between proton–electron masses and coupling ratios to the quark–lepton level. He finds that the ratios of average current quark mass and charged lepton mass bears a constant ratio for each family. The top quark mass is predicted to be 25 GeV.

061.060   Lepton–number violation in cosmology and astrophysics.
E. W. Kolb.
Proceedings of the XIth International Conference on Neutrino Physics and Astrophysics, p. 243 – 253 (1984). – See Abstr. 012.081.

The cosmological and astrophysical implications of lepton number violation are discussed. The lepton number violation considered is due to the interactions of neutrinos with Nambu–Goldstone bosons.

061.061   High energy neutrino astrophysics ($10^2 – 10^7 \mathrm{GeV}$) with emphasis on Cyg X–3.
V. S. Berezinsky (*V. S. Berezinskij*), G. T. Zatsepin.
Proceedings of the XIth International Conference on Neutrino Physics and Astrophysics, p. 589 – 605 (1984). – See Abstr. 012.081.

061.062   Supergravity and the unification of fundamental interactions.
H. Nicolai.
Proceedings of the XIth International Conference on Neutrino Physics and Astrophysics, p. 769 – 779 (1984). – See Abstr. 012.081.

This paper reviews some of the recent developments in $N = 8$ supergravity and Kaluza–Klein supergravity.

061.063 **A stochastic Lie–isotopic approach to the foundations of classical electrodynamics.**
T. L. Gill.
Hadronic J., Vol. 7, No. 5, p. 1224 – 1258 (1984). Abstr. in Phys. Abstr., Vol. 88, No. 1255, Entry 40915 (1985). – See Abstr. 012.083.

061.064 **Soliton supermultiplets and Kaluza–Klein theory.**
G. W. Gibbons, M. J. Perry.
Nucl. Phys. B, Part. Phys., Vol. B248, No. 3, p. 629 – 646 (1984). Abstr. in Phys. Abstr., Vol. 88, No. 1255, Entry 41005 (1985).

061.065 **Magnetic monopoles and duality in Kaluza–Klein theory.**
A. Iwazaki.
Prog. Theor. Phys., Vol. 72, No. 4, p. 834 – 840 (1984). Abstr. in Phys. Abstr., Vol. 88, No. 1255, Entry 41012 (1985).

061.066 **A model for neutrino decays.**
A. Kumar, R. N. Mohapatra.
Phys. Lett. B, Vol. 150B, No. 1 – 3, p. 191 – 195 (1985). Abstr. in Phys. Abstr., Vol. 88, No. 1255, Entry 41477 (1985).

061.067 **A new mechanism of gauge symmetry breaking induced by gravity and a naturally vanishing cosmological constant.**
F.-x. Dong, T.-s. Tu, P.-y. Xue, X.-j. Zhou.
Commun. Theor. Phys., Vol. 3, No. 5, p. 653 – 656 (1984). Abstr. in Phys. Abstr., Vol. 88, No. 1256, Entry 46788 (1985).

061.068 **Superconducting strings in axion models.**
G. Lazarides, Q. Shafi.
Phys. Lett. B, Vol. 151B, No. 2, p. 123 – 126 (1985). Abstr. in Phys. Abstr., Vol. 88, No. 1256, Entry 46795 (1985).

061.069 **Isovector vibrational modes in heavy nuclei.**
J. Wambach.
AIP Conf. Proc., No. 124, p. 146 – 170 (1985). Abstr. in Phys. Abstr., Vol. 88, No. 1256, Entry 46957 (1985). – See Abstr. 012.084.

061.070 **Magnetic monopoles in stellar interiors.**
L. Bracci, G. Fiorentini.
Lett. Nuovo Cimento, Vol. 42, Ser. 2, No. 3, p. 123 – 128 (1985). Abstr. in Phys. Abstr., Vol. 88, No. 1256, Entry 47334 (1985).

061.071 **r– and s–processes: chronometers, thermometers and neutron dosimeters.**
R. R. Winters.
AIP Conf. Proc., No. 124, p. 484 – 503 (1985). Abstr. in Phys. Abstr., Vol. 88, No. 1256, Entry 51803 (1985). – See Abstr. 012.084.

061.072 **Neutron capture processes of heavy element synthesis.**
J. W. Truran.
AIP Conf. Proc., No. 124, p. 504 – 510 (1985). Abstr. in Phys. Abstr., Vol. 88, No. 1256, Entry 51804 (1985). – See Abstr. 012.084.

061.073 **A parametric study of dynamic s–process neutron–capture nucleosyntheses: nuclear data needs.**
G. J. Mathews, W. M. Howard, K. Takahashi, R. A. Ward.
AIP Conf. Proc., No. 124, p. 511 – 514 (1985). Abstr. in Phys. Abstr., Vol. 88, No. 1256, Entry 51805 (1985). – See Abstr. 012.084.

061.074 **Neutron capture processes in astrophysics.**
B. S. Meyer, D. N. Schramm.
AIP Conf. Proc., No. 124, p. 515 – 525 (1985). Abstr. in Phys. Abstr., Vol. 88, No. 1256, Entry 51806 (1985). – See Abstr. 012.084.

061.075 **Kaluza–Klein theories and the Dirac monopole.**
M. J. Perry.
Monopole '83, p. 29 – 38 (1984). Abstr. in Phys. Abstr., Vol. 88, No. 1257, Entry 52084 (1985). – See Abstr. 012.088.

061.076 **A 5–dimensional monopole.**
R. Sorkin.
Monopole '83, p. 39 – 45 (1984). Abstr. in Phys. Abstr., Vol. 88, No. 1257, Entry 52085 (1985). – See Abstr. 012.088.

061.077 **Magnetic monopoles in grand unified and Kaluza–Klein theories.**
Q. Shafi.
Monopole '83, p. 47 – 49 (1984). Abstr. in Phys. Abstr., Vol. 88, No. 1257, Entry 52086 (1985). – See Abstr. 012.088.

061.078 **Monopoles and grand unification.**
E. J. Weinberg.
Monopole '83, p. 1 – 16 (1984). Abstr. in Phys. Abstr., Vol. 88, No. 1257, Entry 52610 (1985). – See Abstr. 012.088.

061.079 **Superheavy magnetic monopoles and the standard cosmology.**
M. S. Turner.
Monopole '83, p. 61 – 83 (1984). Abstr. in Phys. Abstr., Vol. 88, No. 1257, Entry 52614 (1985). – See Abstr. 012.088.

061.080 **Galactic magnetic fields and magnetic monopoles.**
E. N. Parker.
Monopole '83, p. 125 – 136 (1984). Abstr. in Phys. Abstr., Vol. 88, No. 1257, Entry 52615 (1985). – See Abstr. 012.088.

061.081 **Monopole catalyzed nucleon decay: the astrophysical connection.**
E. W. Kolb.
Monopole '83, p. 239 – 249 (1984). Abstr. in Phys. Abstr., Vol. 88, No. 1257, Entry 52623 (1985). – See Abstr. 012.088.

061.082 **Monopoles in 1983.**
J. Preskill.
Monopole '83, p. 663 – 686 (1984). Abstr. in Phys. Abstr., Vol. 88, No. 1257, Entry 52624 (1985). – See Abstr. 012.088.

061.083 **Monopole compatibility with cosmology.**
G. Lazarides.
Magnetic monopoles, p. 71 – 80 (1983). Abstr. in Phys. Abstr., Vol. 88, No. 1257, Entry 52628 (1985). – See Abstr. 012.089.

061.084 **Reducing the monopole abundance.**
A. H. Guth.
Magnetic monopoles, p. 81 – 96 (1983). Abstr. in Phys. Abstr., Vol. 88, No. 1257, Entry 52629 (1985). – See Abstr. 012.089.

061.085 **Monopoles and astrophysics.**
M. S. Turner.
Magnetic monopoles, p. 127 – 140 (1983). Abstr. in Phys. Abstr., Vol. 88, No. 1257, Entry 52632 (1985). – See Abstr. 012.089.

061.086 **Monopoles and the galactic magnetic field.**
E. M. Purcell.
Magnetic monopoles, p. 141 – 149 (1983). Abstr. in Phys. Abstr., Vol. 88, No. 1257, Entry 52633 (1985). – See Abstr. 012.089.

061.087 **The plasma physics of magnetic monopoles in the Galaxy.**
I. Wasserman.
Magnetic monopoles, p. 151 – 158 (1983). Abstr. in Phys. Abstr., Vol. 88, No. 1257, Entry 52634 (1985). – See Abstr. 012.089.

061.088 **Monopolonium.**
C. T. Hill.
Magnetic monopoles, p. 159 – 173 (1983). Abstr. in Phys. Abstr., Vol. 88, No. 1257, Entry 52635 (1985). – See Abstr. 012.089.

061.089   Binding of monopoles in matter and search in large quantities of old iron ore.
D. B. Cline.
Magnetic monopoles, p. 245 – 258 (1983). Abstr. in Phys. Abstr., Vol. 88, No. 1257, Entry 52636 (1985). – See Abstr. 012.089.

061.090   Monopole energy loss and detector excitation mechanics.
S. P. Ahlen.
Magnetic monopoles, p. 259 – 290 (1983). Abstr. in Phys. Abstr., Vol. 88, No. 1257, Entry 52637 (1985). – See Abstr. 012.089.

061.091   Electronic cosmic ray monopole searches.
E. C. Loh.
Magnetic monopoles, p. 291 – 305 (1983). Abstr. in Phys. Abstr., Vol. 88, No. 1257, Entry 52638 (1985). – See Abstr. 012.089.

061.092   The ^{13}N(p, γ)^{14}O reaction at stellar energies.
K. Langanke, O. S. van Roosmalen, W. A. Fowler.
Nucl. Phys. A, Vol. A435, No. 2, p. 657 – 668 (1985). Abstr. in Phys. Abstr., Vol. 88, No. 1257, Entry 52971 (1985).

061.093   Acoustic detection of monopoles.
B. C. Barish.
Magnetic monopoles, p. 219 – 243 (1983). Abstr. in Phys. Abstr., Vol. 88, No. 1257, Entry 57569 (1985). – See Abstr. 012.089.

061.094   Monopolar contamination of normal sequence stars.
A. K. Drukier, G. G. Raffelt.
Monopole '83, p. 153 – 160 (1984). Abstr. in Phys. Abstr., Vol. 88, No. 1257, Entry 57613 (1985). – See Abstr. 012.088.

061.095   Some possible tests of the inapplicability of Pauli's exclusion principle.
Y. Chang.
Hadronic J., Vol. 7, No. 6, p. 1469 – 1473 (1984). Abstr. in Phys. Abstr., Vol. 88, No. 1258, Entry 58131 (1985). – See Abstr. 012.083.

061.096   Neutrino oscillations and neutrino astronomy in a large flat detector.
J. C. van der Velde.
Monopole '83, p. 431 – 438 (1984). Abstr. in Phys. Abstr., Vol. 88, No. 1258, Entry 59095 (1985). – See Abstr. 012.088.

061.097   First results from the Chicago–Fermilab–Michigan cosmic ray magnetic monopole detector.
J. R. Incandela, M. Campbell, H. Frisch, S. Somalwar, M. Kuchnir, H. R. Gustafson.
Monopole '83, p. 461 – 470 (1984). Abstr. in Phys. Abstr., Vol. 88, No. 1258, Entry 59098 (1985). – See Abstr. 012.088.

061.098   Nontrivial anisotropic supergravity cosmological solution.
D. Lorenz–Petzold.
Lett. Nuovo Cimento, Vol. 42, Ser. 2, No. 6, p. 309 – 312 (1985). Abstr. in Phys. Abstr., Vol. 88, No. 1259, Entry 63811 (1985).

061.099   Capture processes and element synthesis in the Universe.
H. V. Klapdor.
AIP Conf. Proc., No. 125, p. 732 – 747 (1985). Abstr. in Phys. Abstr., Vol. 88, No. 1259, Entry 64398 (1985). See Abstr. 012.094.

061.100   Beta decay far from stability and its role in nuclear physics and astrophysics.
H. V. Klapdor.
Fortschr. Phys., Vol. 33, No. 1, p. 1 – 55 (1985). Abstr. in Phys. Abstr., Vol. 88, No. 1259, Entry 64400 (1985).

061.101   Neutron capture and total cross sections for ^{48}Ca: astrophysical implications.
R. F. Carlton, J. A. Harvey, N. W. Hill, R. L. Macklin.
AIP Conf. Proc., No. 125, p. 774 – 777 (1985). Abstr. in Phys. Abstr., Vol. 88, No. 1259, Entry 64474 (1985). See Abstr. 012.094.

061.102   The Dy163–Ho163 branching: an s–process barometer.
H. Beer, G. Walter, R. L. Macklin.
AIP Conf. Proc., No. 125, p. 778 – 781 (1985). Abstr. in Phys. Abstr., Vol. 88, No. 1259, Entry 64475 (1985). See Abstr. 012.094.

061.103   Target thermalization effect in ^{187}Os neutron capture.
G. Reffo.
AIP Conf. Proc., No. 125, p. 782 – 784 (1985). Abstr. in Phys. Abstr., Vol. 88, No. 1259, Entry 64476 (1985). See Abstr. 012.094.

061.104   Thermonuclear reaction rates from (p,n) reaction.
S. Kailas.
AIP Conf. Proc., No. 125, p. 789 – 792 (1985). Abstr. in Phys. Abstr., Vol. 88, No. 1259, Entry 64477 (1985). See Abstr. 012.094.

061.105   Investigation of capture reactions far off stability by β–delayed neutron emission.
M. Wiescher, B. Leist, W. Ziegert, H. Gabelmann, B. Steinmuller, H. Ohm, K.–L. Kratz, F.–K. Thielemann, W. Hillebrandt.
AIP Conf. Proc., No. 125, p. 908 – 911 (1985). Abstr. in Phys. Abstr., Vol. 88, No. 1259, Entry 64481 (1985). See Abstr. 012.094.

061.106   Status of helium burning of ^{12}C.
H. P. Trautvetter, A. Redder, C. Rolfs.
AIP Conf. Proc., No. 125, p. 748 – 752 (1985). Abstr. in Phys. Abstr., Vol. 88, No. 1259, Entry 64506 (1985). See Abstr. 012.094.

061.107   Indirect investigation of proton capture reactions at stellar energies.
P. Schmalbrock, T. R. Donoghue, H. J. Hausman, M. Wiescher, V. Wijekumar, C. P. Browne, A. A. Rollefson, C. Rolfs.
AIP Conf. Proc., No. 125, p. 785 – 788 (1985). Abstr. in Phys. Abstr., Vol. 88, No. 1259, Entry 64507 (1985). See Abstr. 012.094.

061.108   The E2 contribution to the ^{12}C(α,γ)^{16}O reaction at stellar energies in a coupled channel approach.
C. Funck, K. Langanke, A. Weiguny.
Phys. Lett. B, Vol. 152B, No. 1 – 2, p. 11 – 16 (1985). Abstr. in Phys. Abstr., Vol. 88, No. 1259, Entry 64510 (1985).

061.109   Neutron capture reactions in astrophysics.
F. Kappeler.
AIP Conf. Proc., No. 125, p. 715 – 731 (1985). Abstr. in Phys. Abstr., Vol. 88, No. 1259, Entry 68663 (1985). See Abstr. 012.094.

061.110   Some effects of high temperature and density of neutron–capture nucleosynthesis.
E. B. Norman, S. E. Kellogg.
AIP Conf. Proc., No. 125, p. 753 – 765 (1985). Abstr. in Phys. Abstr., Vol. 88, No. 1259, Entry 68664 (1985). See Abstr. 012.094.

061.111   Dynamic stellar neutron capture nucleosynthesis: the need for more nuclear data for the s–process.
G. J. Mathews, W. M. Howard, K. Takahashi, R. A. Ward.
AIP Conf. Proc., No. 125, p. 766 – 773 (1985). Abstr. in Phys. Abstr., Vol. 88, No. 1259, Entry 68665 (1985). See Abstr. 012.094.

061.112 **Nucleosynthesis.**
J. W. Truran.
Annu. Rev. Nucl. Part. Sci., Vol. 34, p. 53 – 97 (1984). Abstr. in Phys. Abstr., Vol. 88, No. 1260, Entry 74076 (1985). – See Abstr. 003.035.

061.113 **The effect of strange particles on the cosmic ray processes and energy distribution between particles of unlike nature.**
N. A. Kruglov, A. S. Proskuryakov, L. I. Sarycheva, L. N. Smirnova.
18th International Cosmic Ray Conference, Vol. 5, p. 13 – 16 (1983). – See Abstr. 012.096.

061.114 **Evolution of monopoles in big bang universe.**
K. Sato.
18th International Cosmic Ray Conference, Vol. 5, p. 44 – 46 (1983). – See Abstr. 012.096.
Grand Unified Theories predict that monopoles are produced copiously in the very early universe. The author investigates how these monopoles evolve in the expanding universe assuming various values of the abundance.

061.115 **Monopolonium.**
C. T. Hill.
18th International Cosmic Ray Conference, Vol. 5, p. 47 (1983). Abstract. – See Abstr. 012.096.

061.116 **Hadroproduction of heavy quarks extrapolated from accelerators to cosmic rays.**
A. Yu. Khodjamirian, A. G. Oganessian (*A. G. Oganesyan*).
18th International Cosmic Ray Conference, Vol. 5, p. 48 – 51 (1983). – See Abstr. 012.096.

061.117 **Search for slowly moving penetrating particles at Baksan underground telescope.**
E. N. Alexeyev (*E. N. Alekseev*), M. M. Boliev, A. E. Chudakov, S. P. Mikheyev (*S. P. Mikheev*), O. Yu. Shkvorets.
18th International Cosmic Ray Conference, Vol. 5, p. 52 – 55 (1983). – See Abstr. 012.096.
No candidates have been recorded during 230 days of observation. This corresponds to the upper limit for superheavy magnetic monopoles flux $5.6 \times 10^{-6} \mathrm{m}^{-2} \mathrm{st}^{-1} \mathrm{d}^{-1}$ (90% confidence level).

061.118 **A search for slowly moving magnetic monopoles.**
F. Kajino, T. Kitamura, K. Mitsui, Y. Ohashi, A. Okada, Y. K. Yuan, T. Aoki, S. Matsuno.
18th International Cosmic Ray Conference, Vol. 5, p. 56 – 59 (1983). – See Abstr. 012.096.
A search for slowly moving magnetic monopoles has been performed using scintillation counters and proportional chambers which are situated on the ground at sea level. No candidate for the magnetic monopoles is obtained.

061.119 **A hybrid detector cosmic ray experiment to search for GUT monopoles.**
D. J. Fegan, G. C. MacNeill.
18th International Cosmic Ray Conference, Vol. 5, p. 61 – 64 (1983). – See Abstr. 012.096.

061.120 **A limit on the flux of heavily–ionizing penetrating particles in cosmic rays at large zenith angles.**
V. D. Ashitkov, V. V. Borog, T. M. Kirina, A. P. Klimakov, R. P. Kokoulin, A. A. Petrukhin.
18th International Cosmic Ray Conference, Vol. 5, p. 65 – 68 (1983). – See Abstr. 012.096.
A limit on the flux of the heavily–ionizing penetrating particles at sea level over the ionization range ≥ 100 minimum ionizing particles has been derived from the data collected during a long experiment with an ionization calorimeter. The flux of the particles with the ionizing power corresponding to that of the relativistic magnetic monopole does not exceed $2.8 \times 10^{-13} \mathrm{cm}^{-2} \mathrm{s}^{-1} \mathrm{sr}^{-1}$ (95% confidence level).

061.121 **Search for magnetic monopoles using proportional counters.**
S. Higashi, S. Ozaki, T. Takahashi, K. Tsuji.
18th International Cosmic Ray Conference, Vol. 5, p. 69 – 72 (1983). – See Abstr. 012.096.
Slowly moving magnetic monopoles in cosmic rays have been searched at sea level. No reliable candidates due to monopoles were found.

061.122 **Search for cosmic monopoles by plastic nuclear track detector with large area.**
T. Doke, T. Hayashi, R. Hamasaki, T. Akioka, K. Ito, T. Yanagimachi, S. Kobayashi, T. Takenaka, K. Nagata.
18th International Cosmic Ray Conference, Vol. 5, p. 73 (1983). Abstract. – See Abstr. 012.096.

061.123 **Search for baryon decay catalysed by the Rubakov interaction of slow magnetic monopoles.**
N. A. Porter, J. Clear, D. J. Fegan, G. C. MacNeill, K. Gibbs, T. C. Weekes.
18th International Cosmic Ray Conference, Vol. 5, p. 74 – 77 (1983). – See Abstr. 012.096.
24 pairs of events were recorded, but do not constitute evidence for the presence of monopoles.

061.124 **Limits on production cross–section of heavy stable leptons and sleptons in hadron–nucleus collisions.**
P. V. Ramana Murthy.
18th International Cosmic Ray Conference, Vol. 5, p. 89 – 92 (1983). – See Abstr. 012.096.

061.125 **Cosmic ray constraints on neutron–antineutron and hydrogen–antihydrogen oscillation time scales.**
C. Sivaram.
18th International Cosmic Ray Conference, Vol. 5, p. 515 – 516 (1983). – See Abstr. 012.096.
The phenomenon of neutron–antineutron ($n\bar{n}$) oscillations has been postulated as another observable manifestation of grand unified theories.

061.126 **The development of statistical methods in cosmic ray physics.**
A. A. Chilingarian.
18th International Cosmic Ray Conference, Vol. 5, p. 524 – 526 (1983). – See Abstr. 012.096.

061.127 **Neutrino oscillations and the atmospheric neutrino fluxes.**
G. V. Dass, K. V. L. Sarma.
18th International Cosmic Ray Conference, Vol. 7, p. 96 – 99 (1983). – See Abstr. 012.096.

061.128 **Neutrino oscillations, parity violation and solar neutrinos.**
P. Raychaudhuri.
18th International Cosmic Ray Conference, Vol. 7, p. 100 – 103 (1983). – See Abstr. 012.096.
It is shown that the left handed and right handed ν may be represented as the positive and negative orientations of ν and thus the nonconservation of parity is assigned to the asymmetry of the internal structure of the particles. If one takes into account this fact then ν oscillates between left handed and right handed ν's and this type of oscillation may explain the solar ν_e experiment and reactor ν_e experiment.

061.129 **The search for antineutrino fluxes from collapsing stars at the Artyomovsk scientific station in 1981 – 1982.**
F. F. Khalchukov, V. G. Ryassny (*V. G. Ryasnyj*), O. G. Ryazhskaya, G. T. Zatsepin.
18th International Cosmic Ray Conference, Vol. 7, p. 112 – 115 (1983). – See Abstr. 012.096.

061.130 Production of high–energy cosmic neutrinos in pγ–collisions.
V. S. Berezinsky (*V. S. Berezinskij*), A. Z. Gazizov.
18th International Cosmic Ray Conference, Vol. 7, p. 116 – 119 (1983). – See Abstr. 012.096.

For the purposes of high–energy neutrino astronomy the neutrino yields Y_v are calculated for neutrinos produced in collisions of high–energy protons (cosmic rays) with low–energy ambient photons.

061.131 Neutrino energy spectra from relativistic plasma.
S. Karakula, W. Tkaczyk, F. Giovannelli.
18th International Cosmic Ray Conference, Vol. 7, p. 121 – 124 (1983). – See Abstr. 012.096.

The authors determined the neutrino and antineutrino energy production spectra in a relativistic plasma. The considered plasma temperature range was $10^{11} – 10^{13}$K. Spherical accretion of matter onto massive objects (Schwarzschild black holes) was considered.

061.132 The late evolution of WC stars: weak s–process and neutron–rich nuclei in cosmic rays.
N. Prantzos, M. Arnould, M. Cassé.
18th International Cosmic Ray Conference, Vol. 9, p. 155 – 158 (1983). – See Abstr. 012.096.

The authors reexamine and extend the study of nucleosynthesis in the most massive WC stars, since this unique class of objects seems to offer a natural explanation for the ^{12}C, ^{22}Ne, ^{25}Mg and ^{26}Mg excesses inferred at the cosmic–ray sources w.r.t. solar system.

061.133 The possible production of globs associated with gammaization processes in cosmic ray interaction.
J. N. Capdevielle, J. Gawin, B. Grochalska, J. Wdowczyk.
18th International Cosmic Ray Conference, Vol. 11, p. 8 – 11 (1983). – See Abstr. 012.096.

It is shown that the possible production of globs of ultra condensed matter near of 10^6GeV can explain several cosmic ray phenomena.

061.134 Results from a new search for GUT monopoles with plastic scintillators.
G. Tarlé, T. M. Liss, S. P. Ahlen.
18th International Cosmic Ray Conference, Vol. 11, p. 12 – 15 (1983). – See Abstr. 012.096.

061.135 Limits on the flux of GUT monopoles from the KGF experiment.
M. R. Krishnaswamy, M. G. K. Menon, N. K. Mondal, V. S. Narasimham, B. V. Sreekantan, Y. Hayashi, N. Ito, S. Kawakami, S. Miyake.
18th International Cosmic Ray Conference, Vol. 11, p. 39 – 42 (1983). – See Abstr. 012.096.

The authors have exploited following properties of massive monopoles in the search for monopoles: large ionization and slow velocity. In addition a new method of monopole search has opened up with the possibility of catalysis of baryon non-conserving nucleon decays. Such a chain of nucleon decays along the monopole track is looked for in the KGF detector as an additional search for monopoles.

061.136 Atmospheric neutrinos, astrophysical neutrinos and proton decay experiments.
A. Dar.
18th International Cosmic Ray Conference, Vol. 11, p. 169 – 172 (1983). – See Abstr. 012.096.

The author presents simple analytical formulae for the spectra of both atmospheric and interstellar neutrinos.

061.137 Deep underground nuclear processes of small energy exchanges: possible evidence for magnetic monopoles or monopolonium.
S. N. Anderson, R. E. Gibbs, P. Kotzer, R. J. Wilkes, S. C. Strausz, J. J. Lord, J. Iwai, T. A. Koss, R. J. Davission.
18th International Cosmic Ray Conference, Vol. 11, p. 178 (1983). Abstract. – See Abstr. 012.096.

061.138 Upper limit on high energy extraterrestrial neutrinos (Baksan data).
M. M. Boliev, A. E. Chudakov, S. P. Mikheyev (*S. P. Mikheev*), V. N. Zakidyshev.
18th International Cosmic Ray Conference, Vol. 11, p. 481 – 484 (1983). – See Abstr. 012.096.

No indication for existing of point–like sources of high energy neutrino have been found during 4 years of operation of Baksan Underground Scintillation Telescope. The neutrino induced muon flux, as measured at Baksan is in agreement with expected from atmospheric neutrinos only.

061.139 Limits on astrophysical v_e flux at $E_v > 10^{19}$.
B. Cady, G. Cassiday, J. Elbert, P. Gerhardy, E. Loh, Y. Mizumoto, P. Sokolsky, D. Steck.
18th International Cosmic Ray Conference, Vol. 11, p. 485 – 488 (1983). – See Abstr. 012.096.

The authors report on a search for upward EAS using the University of Utah Fly's Eye detector. No events have been found in 3.9×10^6sec of running time.

061.140 Radioactive potassium–40 as a cosmic chronometer.
V. I. Slysh.
Pis'ma Astron. Zh., Tom 11, No. 4, p. 310 – 318 (1985). In Russian. English translation in Sov. Astron. Lett., Vol. 11.

Production of isotopes ^{21}Ne, ^{36}S, ^{37}Cl, ^{40}K and ^{40}Ar is calculated in a standard s–process which provides observed solar system abundance of elements heavier than iron. The age of the radioactive ^{40}K is $6.7(+0.5; –0.6) \times 10^9$ years in single–event nucleosynthesis and $(10.3 \pm 2) \times 10^9$ years in a continuous constant rate nucleosynthesis and coincides with the age of r–process elements uranium and thorium as well as with the age of iron in the Galaxy.

061.141 Observational tests of light element nucleosynthesis: D in SNR's, Li in the galactic center and IR sources.
D. A. Lubowich.
Bull. Am. Astron. Soc., Vol. 17, No. 2, p. 582 (1985). Abstract. – See Abstr. 010.065.

061.142 Pion production in proton–proton and proton–helium reactions.
C. Dermer.
Bull. Am. Astron. Soc., Vol. 17, No. 2, p. 604 – 605 (1985). Abstract. – See Abstr. 010.065.

061.143 Causality of classical supergravity.
Y. Choquet–Bruhat.
Asymptotic behavior of mass and spacetime geometry, p. 61 – 84 (1984). – See Abstr. 012.102.

061.144 Supergravities.
Y. Choquet–Bruhat.
Gravitation, geometry and relativistic physics, p. 88 – 106 (1984). – See Abstr. 012.103.

A brief review of the basic principles of both simple and extended supergravity theories is presented.

061.145 Fourth generation massive neutrinos.
R. J. N. Phillips.
RAL–85–008. Rutherford Appleton Laboratory, Chilton, Didcot, Oxon, OX11 0QX, England, 8 pp. (1985). Presented at the Telemark Conference on Neutrino Mass and Low Energy Weak Interactions, held at Telemark Lodge, Cable, Wis., USA, 25 – 27 October 1984.

There may be a massive, multi–GeV fourth generation neutrino. If so, it could have striking signatures and could appear in present or near–future experiments.

061.146 Prospects and problems of locally supersymmetric Kaluza–Klein theories.
P. Fré.
Nucl. Phys. B, Part. Phys., Vol. B252, No. 1 – 2, p. 331 – 342 (1985). – See Abstr. 012.105.
This paper briefly reviews the role of supersymmetric Kaluza–Klein theories in the general quest for a unified field theory.

061.147 Weakons, the unification of interactions and their significance for astrophysics.
V. Čelebonović.
Vasiona, Année 33, No. 1 – 2, p. 1 – 6 (1985). In Croatian.
An account is presented of the recent discovery of the W, Z and t particles and of the related investigations of the unification of interactions and their possible significance for astrophysics.

061.148 Solar abundances and the role of nucleogenesis in low-to–medium mass stars in the Galaxy.
L. H. Aller.
Meteoritics, Vol. 20, No. 2, Part 2, p. 321 – 330 (1985). – See Abstr. 003.048.
The pattern of solar elemental abundances agrees well with that shown by CI chondrites for nonvolatile elements. For metals of the iron peak, the chief source of uncertainty seems to be the structure of the solar atmosphere. Lines of rare elements are frequently masked by atomic and molecular lines of abundant species. The vast majority of stars (including the sun) will do little to change the bulk composition of the interstellar medium from which new stars are formed. He, C, and N in small quantities are supplied by stars from 1 to 8 solar masses as they evolve and produce nebular envelopes that dissipate into the interstellar medium, but as has long been recognized, oxygen, heavier elements, and all r–process and proton–rich nuclides are made in massive stars.

061.149 Molecular hydrogen and thermal phases in astrophysics.
S. H. Lepp.
Diss. Abstr. Int., Sect. B, Vol. 45, No. 7, p. 2200 (1985). Thesis, University of Colorado, 109 pp. (1984). Order No. DA8422623.

Evolution of matter and energy on a cosmic and planetary scale.
See Abstr. 003.194.

Particles and gravity. Proceedings of the Eighth Johns Hopkins Workshop of Current Problems in Particle Theory, held in Baltimore, 20 – 22 June 1984.
See Abstr. 012.120.

Fundamental processes in energetic atomic collisions. Proceedings of a NATO Advanced Study Institute held September 20 – October 1, 1982, in Maratea, Italy.
See Abstr. 012.121.

When can we treat identical particles as distinguishable? An unfamiliar classical limit.
See Abstr. 014.041.

Bragg reflection of cosmic neutrinos.
See Abstr. 022.147.

Neutral current detector for neutrino physics, astronomy, and geology.
See Abstr. 034.108.

Status of DUMAND neutrino detector.
See Abstr. 034.109.

Progress report on Lake Baikal neutrino experiment: site studies and stationary string.
See Abstr. 034.110.

Neutrino physics and astrophysics with the liquid scintillation detector in the Mont Blanc Laboratory.
See Abstr. 034.111.

Superhigh energy neutrino radio detectors in ice.
See Abstr. 034.112.

Helium plasma: pycnonuclear triple alpha reaction rates in stars.
See Abstr. 062.046.

Electron–positron pair equilibrium in strongly magnetized plasmas.
See Abstr. 062.095.

Electron–positron pairs in a mildly relativistic plasma in active galactic nuclei.
See Abstr. 062.096.

An expanding vortex site for the r–process in rotating stellar collapse.
See Abstr. 065.046.

Presupernova core structure and explosive nucleosynthesis.
See Abstr. 065.068.

Neutrinos from collapsing stars.
See Abstr. 065.086.

Analysis of Zr and Tc abundances from S–stars using the s–process with an exponential distribution of neutron exposures.
See Abstr. 065.092.

The s–process within the R Coronae Borealis star U Aquarii.
See Abstr. 065.093.

The synthesis of heavy elements by charged particle neutron producing processes.
See Abstr. 065.106.

Spontaneously broken symmetries in gravitational fields.
See Abstr. 066.051.

Cylindrically symmetric Einstein–Yang–Mills–Higgs gauge configurations.
See Abstr. 066.052.

Gravity and grand unified theories.
See Abstr. 066.056.

The limit of mass for gravitational collapse and nuclear forces at short distance.
See Abstr. 066.061.

Planck length as the lower bound to all physical length scales.
See Abstr. 066.064.

Time and singularity.
See Abstr. 066.074.

Einstein's gravity with massive gravitons.
See Abstr. 066.106.

Gravitational lens effects of neutrino celestial objects of uniform density.
See Abstr. 066.108.

A constraint on physical quantum states in cosmological space–times.
See Abstr. 066.118.

On quantum gravity with dynamical torsion.
See Abstr. 066.130.

Unified field theories: from Eddington and Einstein up to now.
See Abstr. 066.133.

Classical aspects of Yang–Mills theories.
See Abstr. 066.206.

Explicit and hidden symmetries of dimensionally reduced (super–) gravity theories.
See Abstr. 066.208.

Projective relativity and exact solutions.
See Abstr. 066.212.

Monopoles in pulsar PSR 1929+10.
See Abstr. 067.160.

Neutron star physics and monopole flux limits.
See Abstr. 067.161.

Black holes in compactified supergravity.
See Abstr. 067.165.

Consequences of neutrino production in the Cygnus X–3 system.
See Abstr. 067.193.

Properties of bosonic black holes.
See Abstr. 067.199.

Chlorine and gallium solar neutrino experiments.
See Abstr. 080.091.

Solar neutrinos: prospects for detection and implications.
See Abstr. 080.118.

An extremely metal–poor star with r–process overabundances.
See Abstr. 114.037.

Relative isotopic abundances of zirconium in R Cygni and V Cancri.
See Abstr. 114.041.

Les étoiles pauvres en métaux.
See Abstr. 114.127.

Neutrino binaries.
See Abstr. 117.350.

Constraints on the sites of nitrogen nucleosynthesis from $^{15}NH_3$–observations.
See Abstr. 131.093.

Gamma–ray observations of recently synthesized interstellar ^{26}Al.
See Abstr. 131.224.

A plan of a monopole search experiment using the 100 m² calorimeter of the Akeno air shower array.
See Abstr. 144.019.

Antiprotons from thick cosmic–ray sources.
See Abstr. 144.055.

Carbon deflagrating supernovae and the chemical history of the solar neighbourhood.
See Abstr. 155.002.

Galactic chemical evolution and nucleocosmochronology: analytic quadratic models.
See Abstr. 155.076.

Far–ultraviolet background observations at high galactic latitude. I. The Coma Cluster.
See Abstr. 160.072.

Cosmology of magnetic monopoles.
See Abstr. 161.008.

Does cosmology imply a Dirac neutrino mass?
See Abstr. 161.016.

A model for the early universe and the connection between gravitation and the quantum nature of matter.
See Abstr. 161.020.

Restriction on amount of antimatter in the early Universe from $\bar{p}$–^4He reaction data.
See Abstr. 161.021.

Cosmological problems for spontaneously broken supergravity.
See Abstr. 161.030.

Nonequilibrium processes and primordial nucleosynthesis.
See Abstr. 161.038.

Supersymmetric inflation, baryon asymmetry and the gravitino problem.
See Abstr. 161.044.

Nucleosynthesis in anisotropic cosmologies revisited.
See Abstr. 161.048.

Dark matter and inflation.
See Abstr. 161.101.

Heavy neutrinos and the evolution of primordial gravitational perturbations.
See Abstr. 161.124.

The primordial nucleosynthesis.
See Abstr. 161.129.

Numerical simulation of evolution of a multi–dimensional Higgs field in the new inflationary scenario.
See Abstr. 161.134.

Primordial nucleosynthesis and nuclear reaction rates uncertainties.
See Abstr. 161.135.

Cosmology, galactic astronomy and elementary particle physics.
See Abstr. 161.138.

How can heavy neutrinos dominate the Universe?
See Abstr. 161.150.

Quantum mechanics in the tunneling universe.
See Abstr. 161.151.

Early universe and GUT.
See Abstr. 161.182.

Phase transitions in supersymmetric grand unified models.
See Abstr. 161.184.

Massive neutrino and cosmology.
See Abstr. 161.186.

The rest–mass of the neutrino and the cosmic clustering phenomena.
See Abstr. 161.227.

On neutrino halos of galaxies or clusters of galaxies and their restriction on neutrino mass.
See Abstr. 161.228.

On massive neutrino halos and galactic structure.
See Abstr. 161.229.

Cosmological production of Kaluza–Klein monopoles.
See Abstr. 161.238.

Abundance of light elements and lepton asymmetry.
See Abstr. 161.242.

Can primordial black holes solve the overproduction problem of monopoles?
See Abstr. 161.266.

Kaluza–Klein cosmology.
See Abstr. 161.285.

Primordial inflation with flat supergravity potentials.
See Abstr. 161.286.

A new mechanism for baryogenesis.
See Abstr. 161.288.

Thermodynamic fluctuations and the monopole density of the early Universe.
See Abstr. 161.302.

Particle physics and cosmology.
See Abstr. 161.314.

Cosmology with decaying particles.
See Abstr. 161.318.

Global–symmetry evolution in axion cosmologies.
See Abstr. 161.320.

Cosmological models in eleven–dimensional supergravity.
See Abstr. 161.324.

Primordial helium abundance and grand unification schemes.
See Abstr. 161.340.

Cosmic nucleosynthesis and nonlinear inhomogeneities.
See Abstr. 161.345.

Report of IAU Commission 47: Cosmology (*Cosmologie*).
See Abstr. 161.358.

On the nature of the baryon asymmetry.
See Abstr. 161.361.

Massive relic neutrinos – a status report.
See Abstr. 161.364.

Inflation–driving scalar field.
See Abstr. 161.367.

The invisible axion, the gravitino and cosmology.
See Abstr. 161.372.

Cosmic supergravity.
See Abstr. 161.373.

Kaluza–Klein cosmology and the inflationary universe.
See Abstr. 161.374.

Improved astronomical limits on the neutrino mass.
See Abstr. 161.382.

062 Hydrodynamics, Magnetohydrodynamics, Plasma

062.001   Diffusive shock acceleration in modified shocks.
   T. J. Bogdan, I. Lerche.
Mon. Not. R. Astron. Soc., Vol. 212, No. 2, p. 413 – 423 (1985).
   The authors investigate the effects of shock substructure on the diffusive shock acceleration of energetic particles by solving the steady–state transport equation for a unidirectional, inhomogeneous, flow through a stationary shock front, in the test–particle approximation. Far downstream from the shock front, the accelerated particle distribution function is a power law at high momentum. The spectral index is a smoothly varying function of the ratio l/λ, where l is the length scale of the shock substructure, and λ is the accelerated particle scattering mean free path. Possible application to the non–linear (accelerated particle back reaction) diffusive shock acceleration problem, and implications for particle (cosmic ray) acceleration in radiating shocks are briefly discussed.

062.002   Pair production, Comptonization and dynamics in astrophysical plasmas.
P. W. Guilbert, S. Stepney.
Mon. Not. R. Astron. Soc., Vol. 212, No. 3, p. 523 – 544 (1985).
   Electron–positron pair production is an important cooling mechanism for plasmas at mildly relativistic temperatures. The authors investigate thermal plasmas with temperatures $kT \cong m_e c^2$ and optical depths $1 \lesssim \tau \lesssim 5$, including pair processes, Comptonization and bremsstrahlung. Results are presented for equilibrium and impulsively heated models. It is found that, in the former case, the observed spectrum is featureless, but in the latter case it can show a broad, flat annihilation feature. The authors then discuss non–thermal pair production in plasmas confined by strong magnetic fields, such as those thought to exist at the surface of neutron stars. Finally the authors consider simple dynamical systems and the effect of expansion on the spectrum of a gas in which pair production is important.

062.003   An upper limit to the growth rate of instabilities in constant specific angular momentum tori.
O. M. Blaes.
Mon. Not. R. Astron. Soc., Vol. 212, No. 3, p. 37P – 40P (1985).
   The author derives an exact upper limit for the growth rate of any unstable mode of a polytropic, non–self–gravitating torus rotating with constant specific angular momentum. He also derives a lower limit for the ratio of real to imaginary parts of the complex frequency of the mode. These results depend only on the slenderness of the torus and the azimuthal wavenumber m of the mode.

062.004   Magnetic field reconnection in cosmic plasmas.
   B. U. Ö. Sonnerup.
Unstable current systems and plasma instabilities in astrophysics, p. 5 – 23 (1985). – See Abstr. 012.002 (IAU Symp. No. 107).
   A brief review is presented of the concept of magnetic field reconnection or merging. This process occurs whenever an electric field is present along a separator line in the magnetic field. The basic properties of reconnection are discussed in the context of the classical MHD models by Sweet and Parker and by Petschek. Attention is then focussed on reconnection in collision–free plasmas. The energization of charged particles during their interaction with the current layers associated with the reconnection geometry is discussed and the nature of the processes occurring in the so–called diffusion region which surrounds the separator is considered. Finally, comments are made on the nonsteady aspects of reconnection at the earth's magnetopause.

062.005   Patchy reconnection and magnetic ropes in astrophysical plasmas.
C. T. Russell.
Unstable current systems and plasma instabilities in astrophysics, p. 25 – 42 (1985). – See Abstr. 012.002 (IAU Symp. No. 107).
   Reconnection is clearly observed at the terrestrial magnetopause but seldom in the simple geometry originally proposed.

Most often reconnection is patchy, forming tubes of twisted flux. The passage of one of these twisted tubes has been called a flux transfer event. Similar twisted tubes, or flux ropes, are formed at Venus by velocity shear. These tubes become so highly twisted that they become kink unstable. The presence of the kink instability suggests a way of creating compound flux ropes as have been postulated to be necessary to explain photospheric magnetic structure.

062.006   Laboratory experiments on current sheet disruptions, double layers, turbulence and reconnection.
R. L. Stenzel, W. Gekelman.
Unstable current systems and plasma instabilities in astrophysics, p. 47 – 60 (1985). – See Abstr. 012.002 (IAU Symp. No. 107).

The role of laboratory experiments to the understanding of current systems in space plasmas is reviewed. It is shown that laboratory plasmas are uniquely suited to make detailed investigations of basic physical processes in current–carrying plasmas. Examples are given for double layers, current–driven instabilities, and the plasma dynamics at magnetic neutral points during reconnection. Observations of current sheet disruptions show the coupling between local plasma phenomena (double layers) and global circuit properties (magnetic energy storage).

062.007   Reconnection in sheared magnetic fields in space and astrophysics.
J. F. Drake.
Unstable current systems and plasma instabilities in astrophysics, p. 61 – 81 (1985). – See Abstr. 012.002 (IAU Symp. No. 107).

The current theoretical understanding of the linear and nonlinear evolution of resistive tearing instabilities in sheared magnetic fields is reviewed. The physical mechanisms underlying this instability are emphasized. Some of the problems which are encountered in developing a model of magnetic energy dissipation in coronal loops are discussed and possible solutions are suggested.

062.008   Formation, equilibrium and stability of jets.
C. A. Norman.
Unstable current systems and plasma instabilities in astrophysics, p. 85 – 94 (1985). – See Abstr. 012.002 (IAU Symp. No. 107).

Consideration of the many observed types of jets on scales ranging from parsecs to megaparsecs seen in radio, optical, infrared and X–ray wavebands with a variety of morphologies both in galactic and extragalactic systems leads to some constraints on their fundamental nature. Jet formation is introduced with the concept of the Laval nozzle. Current ideas on jet formation at the black hole and accretion disk are given. Stability of jet propagation is reviewed with emphasis on magnetised and unmagnetised Kelvin–Helmholtz instabilities and the various dominant modes. The particle acceleration physics of shocks, wave–particle interactions and turbulence is summarised. Jet equilibrium associated with the non–linear saturation of instabilities, the formation of cocoons, shock stabilisation and magnetic fields is discussed. Detailed plasma physics studies that could significantly clarify jet physics are indicated.

062.009   On the role of double layers in astrophysical plasmas.
R. A. Smith.
Unstable current systems and plasma instabilities in astrophysics, p. 113 – 123 (1985). – See Abstr. 012.002 (IAU Symp. No. 107).

Limitations of current knowledge of plasma double layers create difficulties in extrapolating double–layer concepts for application to astrophysical models. Some problems of this sort are described, and some central issues in structure and dynamics of double layers are identified, which must be addressed in astrophysical contexts. These include the determination of kinetic boundary conditions, and the relations of time and length scales of local dynamics and structure to those of the global circuit in which the double layer is contained.

062.010   Particle energization in stochastic double layers.
W. Lotko.
Unstable current systems and plasma instabilities in astrophysics, p. 125 – 129 (1985). – See Abstr. 012.002 (IAU Symp. No. 107).

Electrostatic turbulence develops in current carrying plasmas when the relative electron–ion drift exceeds the critical value for laminar current flow. Recent 2D computer experiments (Barnes, 1982) indicate that many weak ion acoustic double layers form in such turbulence. The double layers emerge from the incoherent spectrum of electrostatic ion cyclotron and ion acoustic waves as intense localized electric field structures propagating subsonically relative to the ion bulk flow. An important question concerns the effect of these electric fields on plasma transport properties such as bulk heating and acceleration. For instance, one might expect nonlinear diffusion processes, manifested as distinct non–thermal features in the particle spectra, to accompany the quasilinear diffusion of ions as they traverse turbulent regions in space. This idea motivates the work presented here.

062.011   Links between jet instabilities, radiation and propagation in astrophysics.
G. Benford.
Unstable current systems and plasma instabilities in astrophysics, p. 131 – 138 (1985). – See Abstr. 012.002 (IAU Symp. No. 107).

At the head of a jet the confining medium of plasma frequency v_p is compressed, so that streaming instabilities between relativistic electrons and this plasma produce waves at $v_p' > v_p$. Considerable power can be lodged in these electrostatic waves, and conversion to electromagnetic waves allows them to propagate far beyond the jet. Emission at $v \approx v_p'$ or Compton boosted radiation at $v \lesssim \gamma^2 v_p'$ yields a cone of radiation of angle $\sim 1/\gamma$, which illuminates the region directly in front of the jet. This emission is not absorbed by the surrounding plasma unless a cloud blocks the jet. Absorption in a cloud can lead to tunneling through large clouds, or propelling of smaller clouds out of the jet path. In this fashion jets may clear their way through an inhomogeneous medium, avoiding lateral disturbances and preheating their path.

062.012   Laboratory experiments on reconnection in current sheets.
A. Bratenahl, P. J. Baum.
Unstable current systems and plasma instabilities in astrophysics, p. 147 – 166 (1985). – See Abstr. 012.002 (IAU Symp. No. 107).

Laboratory reconnection experiments dedicated to problems of space and astrophysics are briefly reviewed with the purpose of demonstrating that such experiments can provide important insights of considerable value to the development of reconnection theory. Moreover, many of these insights are of a kind not likely to be perceived either when working directly with space observations or while pursuing a course of pure theoretical reasoning without reference to laboratory results.

062.013   3D simulation of externally driven reconnection.
T. Sato.
Unstable current systems and plasma instabilities in astrophysics, p. 211 – 215 (1985). – See Abstr. 012.002 (IAU Symp. No. 107).

A 3D magnetohydrodynamic simulation is presented. The essential conclusions obtained by the previous 2D model, such as the slow shock formation and the strong plasma jet generation, are reconfirmed. In addition, several new findings pertinent to three dimensionality are obtained. Among them, particularly interesting and important is the generation of field aligned currents at the slow shock associated with a local interruption of the neutral sheet current. It is also interesting to observe the generation of super–magnetosonic flows with the Mach number of 2.

062.014   Nonlinear evolution of the resistive tearing mode.
R. S. Steinolfson, G. Van Hoven.
Unstable current systems and plasma instabilities in astrophysics, p. 273 – 276 (1985). – See Abstr. 012.002 (IAU Symp. No. 107).

Numerical solutions of the MHD equations are used to investigate the nonlinear behavior of the tearing instability. The mode evolves from a linearly growing excitation, followed by a period of greatly reduced nonlinear growth. Constant-Ψ solutions evolve much more slowly than comparable nonconstant-Ψ modes. The nonconstant-Ψ computations indicate a reduction by approximately 20% of the energy in the initial shear layer. For long–wavelength solutions, secondary–flow vortices, opposite in direction to the linear vortices, generate a new magnetic island centered at the initial x–point.

062.015 Resistive instabilities in astrophysical conditions: a critical discussion.
P. Batistoni, G. Einaudi, F. Rubini, C. Chiuderi, G. Torricelli.
Unstable current systems and plasma instabilities in astrophysics, p. 277 – 280 (1985). – See Abstr. 012.002 (IAU Symp. No. 107).

Resistive instabilities have often been indicated as the possible cause of rapid release of energy in astrophysical situations. A correct assessment of the validity of this idea requires a detailed analysis of the theory of resistive instabilities in the regimes of astrophysical interest. In particular, effects as the presence of asymmetries due to current gradients, the influence of geometry and of shear flows must be explicitly evaluated. The authors have started a program of investigation on this subject, the preliminary results of which are reported here.

062.016 "Sweeping pinch" mechanism and the acceleration of jets in astrophysics.
Y. Uchida, K. Shibata.
Unstable current systems and plasma instabilities in astrophysics, p. 287 – 291 (1985). – See Abstr. 012.002 (IAU Symp. No. 107).

A magnetodynamic mechanism of jet formation, in which a packet of the toroidal component of the magnetic field B_φ plays a role, is proposed. Such a packet of toroidal field, produced by the rotational motion in the $\beta = p_g/p_m \gg 1$ region, relaxes itself in the $\beta \ll 1$ region when brought up into such a region, for example, by the process of flux emergence due to magnetic buoyancy. In the $\beta \ll 1$ region, a progressive pinch is caused by this relaxation and the mass is swept out by the pinch near the axis and also by the $j \times B$ force in the twisted field region surrounding the axis.

062.017 Plasma instabilities generated by streaming particles.
M. André.
Unstable current systems and plasma instabilities in astrophysics, p. 309 – 312 (1985). – See Abstr. 012.002 (IAU Symp. No. 107).

It is well known that particles streaming along the ambient magnetic field in space plasmas may generate waves with frequencies of the order of the local ion gyrofrequency (ion waves), (Kindel and Kennel 1971). In this study the author analyzes the dispersion relation of these waves numerically and discusses mechanisms for damping and instability. All numerical results in this report are obtained with the computer code WHAMP (Rönnmark 1982), which solves the dispersion relation of linear waves in a homogeneous plasma for a complex frequency as a function of a real wavevector. As a specific example the author considers the S3–3 satellite observations of banded electrostatic ion waves, associated with streaming particles (Kintner et al. 1979, Cattell 1981).

062.018 Anomalous transport in current sheets.
J. D. Huba.
Unstable current systems and plasma instabilities in astrophysics, p. 315 – 328 (1985). – See Abstr. 012.002 (IAU Symp. No. 107).

A review of several microinstabilities that have been suggested as possible anomalous transport mechanisms in current sheets is presented. The specific application is to a "field reversed plasma" which is relevant to the so–called "diffusion region" of a reconnection process. The linear and nonlinear properties of the modes are discussed, and each mode is assessed as to its importance in reconnection processes based upon these properties. It is concluded that the two most relevant instabilities are the ion acoustic instability and the lower–hybrid–drift instability. However, each instability has limitations as far as reconnection is concerned, and more research is needed in this area.

062.019 Anomalous transport induced by field aligned currents and its relation to electromagnetic coupling.
C. T. Dum.
Unstable current systems and plasma instabilities in astrophysics, p. 329 – 340 (1985). – See Abstr. 012.002 (IAU Symp. No. 107).

The wealth of detailed observations on transport in turbulent plasmas that has become available over the last decade from laboratory experiments, in situ spacecraft observations and computer simulation demonstrates that in contrast to classical transport, the various steps in the analysis of anomalous transport, involving microscopic processes and the global dynamics, are closely coupled. It also points to many new, exciting, and truly anomalous phenomena. These statements apply especially to the highly dynamic processes connected with field aligned currents. The aim of this paper is primarily to describe some of the main issues and to provide some key references.

062.020 Nonlinear effects and the limitation of electron streaming instabilities in astrophysics.
S. R. Spangler, J. P. Sheerin.
Unstable current systems and plasma instabilities in astrophysics, p. 355 – 359 (1985). – See Abstr. 012.002 (IAU Symp. No. 107).

Nonlinear effects, such as soliton collapse, will result in evolution of hydromagnetic waves excited by a field–aligned charged particle beam. If the time scale for such evolution is comparable to, or shorter than, linear time scales such as those for wave growth or pitch angle isotropization, then nonlinear effects may limit the instability. For conditions appropriate to relativistic electron streaming in a radio galaxy, the nonlinear time scale may be comparable to the linear time scales.

062.021 Phase mixing of propagating Alfvén waves.
L. Nocera, B. Leroy, E. R. Priest.
Unstable current systems and plasma instabilities in astrophysics, p. 365 – 369 (1985). – See Abstr. 012.002 (IAU Symp. No. 107).

Among MHD waves, Alfvén waves have been proved to be the best candidates to reach the solar corona and, eventually, to be responsible for the heating of this outer part of the solar atmosphere. The problem concerning the mechanism able to transform the energy stored in the waves into heat is considered.

062.022 Kelvin–Helmholtz instabilities in a magnetised compressible plasma (sheared flow).
T. P. Ray, A. I. Ershkovich.
Unstable current systems and plasma instabilities in astrophysics, p. 375 – 377 (1985). – See Abstr. 012.002 (IAU Symp. No. 107). See Abstr. 34.062.013.

The authors have investigated Kelvin–Helmholtz (K–H) instabilities for a homogeneous compressible plasma containing a uniform magnetic field and a linear velocity shear. A derivation of the relevant K–H dispersion equation and details regarding method of solution are given elsewhere. The authors present here an outline of their results.

062.023 Plasma heating by Alfvén waves – kinetic properties of magnetohydrodynamic disturbances.
A. Hasegawa.
Unstable current systems and plasma instabilities in astrophysics, p. 381 – 389 (1985). – See Abstr. 012.002 (IAU Symp. No. 107).

Mechanisms of Alfvén wave heating in space–astrophysical plasmas are presented with particular emphasis on the parallel electric field generated in the magnetohydrodynamic perturbations due to the finite Larmor radius effects.

062.024 Kinetic theory of Alfvén wave heating.
S. M. Mahajan.
Unstable current systems and plasma instabilities in astrophysics, p. 391 – 392 (1985). Abstract. – See Abstr. 012.002 (IAU Symp. No. 107).

062.025 MHD equilibrium and instabilities in extragalactic jets.
A. Ferrari.
Unstable current systems and plasma instabilities in astrophysics, p. 393 – 412 (1985). – See Abstr. 012.002 (IAU Symp. No. 107).

Morphologies, energetics and nonthermal radiation emission of extragalactic radio sources can be explained in the framework of models based on the physics of supersonic jets interacting with the surrounding intergalactic medium. In this review current physical interpretation of acceleration, collimation, modulation and nonthermal radiation of these jets are discussed.

062.026   Bursting particle acceleration in radio jets.
R. N. Henriksen.
Unstable current systems and plasma instabilities in astrophysics, p. 413 – 423 (1985). – See Abstr. 012.002 (IAU Symp. No. 107).

This work follows on the papers by Henriksen, Bridle and Chan (1982) and by Eilek and Henriksen (1984). These papers introduced a comprehensive model of (hydrodynamic) turbulence driven Alfvén wave, resonant, particle acceleration. In this paper the author studies how the theory applies in the presence of convection and of axial variations in the local properties of the turbulence. The well studied sources NGC 315 and NGC 6251, are referred to for illustration.

062.027   Instabilities in astrophysical jets: disease and cure.
D. Eichler.
Unstable current systems and plasma instabilities in astrophysics, p. 425 – 431 (1985). – See Abstr. 012.002 (IAU Symp. No. 107).

The general criterion for the various instabilities is that internal stress, thermal or magnetic, supports the disruption of the flow. Following an argument that instabilities are a serious problem for jet models that invoke dynamically significant internal stresses, a model is presented for the hydrodynamic collimation of jets having no significant internal stresses.

062.028   Current systems in radio jets.
J. A. Eilek.
Unstable current systems and plasma instabilities in astrophysics, p. 433 – 437 (1985). – See Abstr. 012.002 (IAU Symp. No. 107).

The structure of the magnetic field in radio jets is a topic of recent interest, especially due to the possibility that some high pressure jets are confined by a magnetic pinch. Several such jets have been found. Recent radio interferometer observations of surface brightness and polarization allow the possibility of determining the magnetic field structure. The author presents some basic considerations of the current and field structure required if the observed jets are to be magnetically confined.

062.029   Laboratory plasma processes of astrophysical interest.
C. S. Liu.
Unstable current systems and plasma instabilities in astrophysics, p. 447 – 452 (1985). – See Abstr. 012.002 (IAU Symp. No. 107).

The author reviews some of the recent progress in fusion plasma physics bearing on astrophysical and space plasma processes: (1) turbulent relaxation towards the state of minimum magnetic energy; (2) resonant radial diffusion of energetic ions; (3) wave acceleration of electrons to sustain a current in a plasma.

062.030   Theory of strongly turbulent two–dimensional cross field convection of current carrying space plasmas.
M. J. Keskinen.
Unstable current systems and plasma instabilities in astrophysics, p. 475 (1985). Abstract. – See Abstr. 012.002 (IAU Symp. No. 107).

062.031   Bending waves on current sheets.
G. Bertin, B. Coppi.
Unstable current systems and plasma instabilities in astrophysics, p. 491 – 496 (1985). – See Abstr. 012.002 (IAU Symp. No. 107).

Current sheets are found to be subject to bending waves described by a dispersion relation indicating that these are, essentially, modified surface Alfvén waves. Applications to the observed magnetic polarity sectors in the solar wind and to other astrophysical environments, such as planetary magnetospheres, are suggested.

062.032   Quasi–two–dimensional cosmic jets.
K. Tsinganos, A. Ferrari, R. Rosner.
Unstable current systems and plasma instabilities in astrophysics, p. 497 – 501 (1985). – See Abstr. 012.002 (IAU Symp. No. 107).

In this article, the authors use the experience gained by studying the nearest known astrophysical jet – high–speed solar wind streams – to address some of the problems of astrophysical jet acceleration and collimation associated with objects as diverse as SS433, star–forming molecular clouds and, in particular, jets associated with galaxies and quasars.

062.033   Shock drift acceleration in the presence of waves.
R. B. Decker, L. Vlahos.
J. Geophys. Res., Vol. 90, No. A1, p. 47 – 56 (1985). – See Abstr. 012.003.

Charged particle acceleration via the shock drift mechanism at quasi–perpendicular shocks has generally been analyzed by assuming uniform, time–independent conditions at and near the shock. The authors present results from a model designed to study how the shock drift mechanism is modified when wave activity is included in the shock's upstream and downstream vicinities. The technique involves numerically following test particle trajectories in the wave–shock system for predefined wave fields. In order to compare these results with those obtained in the scatter–free (i.e., nonwave) case, the authors restricted particles to a single shock encounter. As a particular example, they injected ensembles of ions into a system consisting of a quasi-perpendicular shock moving through the interplanetary spectrum of ambient Alfvén waves.

062.034   The second–order theory of electromagnetic hot ion beam instabilities.
S. P. Gary, R. L. Tokar.
J. Geophys. Res., Vol. 90, No. A1, p. 65 – 72 (1985). – See Abstr. 012.003.

The wave–particle interactions of a hot ion beam streaming along a magnetic field $\mathbf{B}$ are studied. A second–order theory of electromagnetic instabilities in a homogeneous, collisionless plasma at propagation parallel to $\mathbf{B}$ is used. The two instabilities most likely to be driven by a hot beam are the right–hand and left–hand resonant ion beam instabilities. If the conditions necessary for the validity of the theory are met, the two modes are found to reinforce one another. Thus this theory predicts that whenever sufficiently hot or "diffuse" ions are found at a collisionless shock and the plasma is sufficiently homogeneous that significant wave growth is possible, ion beam instabilities will act to produce a mixture of both right– and left–hand polarized MHD–like waves.

062.035   Perturbations in the velocity distribution in a collisionless plasma.
J. W. Dungey.
J. Geophys. Res., Vol. 90, No. A1, p. 370 – 376 (1985).

062.036   Lower hybrid waves in finite–β plasmas, destabilized by electron beams.
S. Migliuolo.
J. Geophys. Res., Vol. 90, No. A1, p. 377 – 385 (1985).

The linear and quasilinear theories of lower hybrid waves in a finite–β inhomogeneous plasma with electrons streaming parallel to the magnetic field are examined. These waves are found to be unstable for very low–β and to be stabilized in finite–β plasmas. The quasilinear feedback of unstable waves is shown to result in heating of ions perpendicular to the equilibrium magnetic field. Thus, unstable lower hybrid waves may play a role in producing ion conics in the auroral zone, and in heating ions in the solar plasma.

062.037   The excitation of plasma waves by a current source moving in a magnetized plasma: the MHD approximation.
C. E. Rasmussen, P. M. Banks, K. J. Harker.
J. Geophys. Res., Vol. 90, No. A1, p. 505 – 515 (1985).

062.038   Kosmische gassen.
M. Nepveu.
Zenit, 12. Jaarg., No. 1, p. 26 – 27 (1985).

062.039   Relativistic plasmas and active galactic nuclei.
M. Kusunose, F. Takahara.
Astron. Her., Vol. 78, No. 2, p. 46 – 49 (1985). In Japanese.

062.040   Magnetized viscous jets.
M. Nepveu.
Astron. Astrophys., Vol. 142, No. 2, p. 375 – 377 (1985).

Jets with a turbulent viscosity and a magnetic field are studied. The fields are parallel to the velocities everywhere. It is shown that in some cases the formal calculations on such jets can be reduced to those performed earlier on viscous hydro–jets. Brightness distributions can be computed; they are in fair correlation with the local velocities in the jet. The range of validity for the jet solutions is given.

062.041 Equilibria in stronlgy magnetized pair plasmas.
A. K. Harding.
AIP Conf. Proc., No. 115, p. 615 – 618 (1984). – See Abstr. 012.005.

Positron–electron pair densities for a thermal plasma in the steady–state equilibrium where pair production balances pair annihilation are found as a function of temperature $kT/mc^2 \lesssim 1$ source size R and magnetic field strength B. When the plasma is strongly magnetized, $B > 10^{12}$G, the important processes are synchrotron radiation, one–photon (magnetic) pair production and two–photon pair annihilation.

062.042 Surface emissivity of an optically thick, magnetized plasma and luminosities of accreting neutron stars.
G. G. Pavlov, Yu. A. Shibanov.
Astron. Zh., Tom 62, Vyp. 1, p. 43 – 53 (1985). In Russian. English translation in Sov. Astron., Vol. 29, No. 1.

The authors calculate the total energy flux radiated from an optically thick, isothermal, magnetized plasma with either homogeneous or exponentially growing inward density. Analytical dependences of the flux on temperature, magnetic field and density are obtained in limiting cases. For plasma parameters adopted usually for accreting neutron stars the plasma emissivity appears to be smaller than the black–body one. Observed luminosities of X–ray pulsars and X–ray bursters are discussed on the basis of the results obtained.

062.043 Non–axisymmetric instability in thin discs.
P. Goldreich, R. Narayan.
Mon. Not. R. Astron. Soc., Vol. 213, No. 1, p. 7P – 10P (1985).

Thin discs of arbitrary specific angular momentum are shown to have unstable non–axisymmetric modes provided there is at least one good reflecting edge.

062.044 Boundary layers in space plasmas: a kinetic model of tangential discontinuities.
M. Roth.
XVI International Conference on Phenomena in Ionized Gases, p. 139 – 147 (1983). Abstr. in Phys. Abstr., Vol. 88, No. 1248, Entry 9595 (1985). – See Abstr. 38.012.017.

062.045 Propagation of weak discontinuities in a vibrationally–excited gas flow.
A. Rai, M. Gaur.
Astrophys. Space Sci., Vol. 108, No. 1, p. 175 – 186 (1985).

Propagation of weak discontinuities headed by wavefronts of arbitrary shape in three dimensions are studied in vibrationally relaxing gas flow. The transport equations representing the rate of change of discontinuities in the normal derivatives of the flow variables are obtained, and it is found that the nonlinearity in the governing equations does not contribute anything to the vibrationally relaxing gas. An explicit criterion for the growth and decay of weak discontinuities along bi–characteristic curves in the characteristic manifold of the governing differential equations is given. A special case of interest is also discussed.

062.046 Helium plasma: pycnonuclear triple alpha reaction rates in stars.
A. E. M. Khairozzaman.
Astrophys. Space Sci., Vol. 108, No. 2, p. 221 – 226 (1985).

The plots of dimensionless parameters for helium plasma are given, as they are of great importance for the study of physical conditions prevalent in stellar interiors consuming helium. The relevant temperature and density leading to weak and strong screening can be obtained from the plots to compute the effect of screening to enhance the thermonuclear reaction rates. To make a realistic approach, the pycnonuclear triple alpha reaction rates are found for the 7.66 (O^+) MeV level for pair creation. The equation used here is suitable for white dwarfs satisfying

$0 < 3\Gamma/\tau \ll 1$. The rates are enhanced approximately 10^{42} times, which indicates that the mean life will be reduced by the same factor.

062.047 Nonlinear wave conversion in electron–positron plasmas.
M. E. Gedalin (*M. Eh. Gedalin*), J. G. Lominadze, L. Stenflo, V. N. Tsytovich.
Astrophys. Space Sci., Vol. 108, No. 2, p. 393 – 400 (1985).

Wave conversion mechanisms causing large–frequency shifts are considered for an electron–positron plasma in a strong magnetic field. In particular, the authors discuss the effects of the nonlinear Čerenkov as well as the cyclotron resonances in order to associate pulsar radio–emissions with their present model for nonlinear conversion of high–frequency radiation into the low frequency region.

062.048 Effects of ion collisions on the interaction of transverse waves with half–space plasma.
R. K. Singh, D. R. Phalswal.
Astrophys. Space Sci., Vol. 108, No. 2, p. 409 – 414 (1985).

For half–space ($Z > 0$), homogeneous, collisional and warm plasma, the expressions for fields and penetration depth δ/δ_e are derived and discussed numerically. It is concluded that the propagation of transverse waves is only slightly affected by the ion collisions and the applied magnetic field when the plasma frequency is greater than the wave frequency ($\omega_{pe} > \omega$). For the case of $\omega_{pe} \leqslant \omega$, the damping of the wave is not affected by the changes in the ion collision frequency and the ion temperature. However, in this case, the propagation of the wave is drastically affected by the applied magnetic field and the wave damps quickly as the magnetic field strength or the gyrofrequency increases.

062.049 Local potential analysis of MHD instability.
K. K. Sen, S. J. Wilson.
Astrophys. Space Sci., Vol. 109, No. 1, p. 131 – 143 (1985).

In many astrophysical problems, the study of the stability of an atmosphere in the presence of a magnetic field is of importance. In most cases the MHD instabilities of atmospheres are studied by energy principle of Bernstein et al. (1958). In this paper, a general method for studying the stability of a system subject to MHD equations of conditions has been proposed. This is based on the local potential concept put forward by Glansdorff and Prigogine (1964). The scheme for securing stability criteria has been demonstrated in two particular cases.

062.050 Introduction à la magnétohydrodynamique.
B. Leroy.
Champs magnétiques stellaires, p. 123 – 178 (1984). – See Abstr. 012.022.

062.051 Magnétohydrostatique.
J. Heyvaerts.
Champs magnétiques stellaires, p. 179 – 219 (1984). – See Abstr. 012.022.

062.052 Méthodes aux différences finies pour la résolution numérique des équations aux dérivées partielles de la dynamique des fluides et de la magnétohydrodynamique dans le cadre de problèmes à conditions aux limites et conditions initiales.
J.–M. Malherbe.
Champs magnétiques stellaires, p. 221 – 251 (1984). – See Abstr. 012.022.

062.053 Aspects of dynamo theory.
H. K. Moffatt.
Champs magnétiques stellaires, p. 253 – 268 (1984). – See Abstr. 012.022.

062.054 Où en est la turbulence dévelopée?
U. Frisch.
Champs magnétiques stellaires, p. 269 – 293 (1984). – See Abstr. 012.022.

062.055 Le déclin de la turbulence MHD.
J. Leorat.
Champs magnétiques stellaires, p. 295 – 303 (1984). – See Abstr. 012.022.

062.056 Simulation numérique de l'effet dynamo dans un fluide conducteur turbulent.
M. Meneguzzi.
Champs magnétiques stellaires, p. 305 – 324 (1984). – See Abstr. 012.022.

062.057 Deux problèmes numériques de la MHD tri-dimensionnelle.
D. Galloway.
Champs magnétiques stellaires, p. 325 – 337 (1984). – See Abstr. 012.022.

062.058 Electromagnetic ion beam instabilities: hot beams at interplanetary shocks.
S. P. Gary.
Astrophys. J., Vol. 288, No. 1, p. 342 – 352 (1985).

This paper considers Vlasov instabilities driven by a very hot ion beam streaming parallel to a magnetic field. The linear theory of electromagnetic instabilities driven by such a beam in a homogeneous, collisionless, nonrelativistic plasma is used. If the thermal speed of the beam is greater than its drift speed, the two dominant modes are the right–hand and left–hand resonant ion beam instabilities. Arguments are presented that these two modes are driven by the suprathermal ion component at interplanetary shocks, and that the growth of these modes is sufficient to account for the amplitudes of MHD–like waves observed at such shocks.

062.059 Magnetic vortex tubes, jets and nonthermal sources.
P. F. Browne.
Astron. Astrophys., Vol. 144, No. 2, p. 298 – 314 (1985).

The generation and dissipation of magnetic fields are discussed. Due to induction the time scale for both processes will be less than the Hubble age only if the field has scale length $< 10^{13}$cm. A hierarchy of vortices is postulated in order to reconcile this scale and the large scale of fields observed in radio jets. The specific viscous forces on electrons and ions in a differentially rotating plasma generate a current whose magnetic field is proportional to vorticity, at least whilst the field is weak enough for gyroradius to exceed mean free path for electrons. For an array of vortices of scale 10^{13}cm fields of 4×10^{-5}G are predicted. Energy in large scale differential rotation is transferred through the hierarchy to small scale vorticity and thence into the magnetic field. During collapse of a spherical rotating gas cloud toward disc geometry, a vortex can form along the rotation axis, being maintained by pressure driving equatorial inflow and bipolar outflow of gas. Stellar bipolar nebulae and jet–lobe radio structures are examples.

062.060 Evolution of diamagnetic material in a nonuniform magnetic field.
G. W. Pneuman, P. J. Cargill.
Astrophys. J., Vol. 288, No. 2, p. 653 – 664 (1985).

The basic properties of isothermal diamagnetic bodies injected into a nonuniform magnetic field are studied. Such bodies are subject to two sets of forces: the gradient in the external magnetic field tends to accelerate them to regions of weaker field strength (the melon–seed effect), and lateral forces on their surface tend to compress them, while, at the same time, internal forces tend to expand them. Although these two sets of forces have been considered independently in the past, the present paper outlines their combined effect. The results of this study may be of importance to many solar phenomena such as spicules, macrospicules, and coronal bullets, and also in the acceleration of the solar wind.

062.061 Magnetohydrodynamic stability of an axisymmetric, line–tied, diamagnetic plasmoid embedded in a uniform magnetic field.
T. J. Bogdan.
Astrophys. J., Vol. 288, No. 2, p. 672 – 678 (1985).

The stability of a line–tied, axisymmetric, hemispherical plasmoid embedded in a uniform magnetic field is investigated by using the MHD energy principle. The equilibrium configuration studied resembles the magnetic field topology of an isolated sunspot or of a newly emerged region of magnetic flux in the solar photosphere, and provides insight into the properties of these features.

062.062 Dust in magnetised plasmas: basic theory and some applications.
T. G. Northrop, G. E. Morfill.
Adv. Space Res., Vol. 4, No. 9, p. 63 – 73 (1984). – See Abstr. 012.029.

The authors review the theory of charged test particle motion in magnetic fields. This theory is then extended to charged dust particles, for which gravity and charge fluctuations play an important role. It is shown that systematic drifts perpendicular to the magnetic field and stochastic transport effects may then have to be considered – none of which occur in the case of atomic particles (with the exception of charge exchange reactions). Some applications of charged dust particle transport theory to planetary rings are then briefly discussed.

062.063 Numerical model of time–dependent hydrodynamic flows with mass loading.
N. I. Kömle, H. I. M. Lichtenegger.
Adv. Space Res., Vol. 4, No. 9, p. 249 – 252 (1984). – See Abstr. 012.029.

The authors study the influence of mass loading on the formation of shocks employing a time–dependent hydrodynamic model. Numerical examples illustrate in which way the formation and propagation of shocks in the plasma is affected by variations of the ionization rate (due to changes of solar wind parameters or of neutral gas production). It is found that mass loading may be an efficient mechanism to prevent the formation of discontinuities in hydrodynamic flows.

062.064 Generalized solutions for magnetic configurations of force with given boundary conditions for curves planes.
V. A. Romanov.
Soln. Dannye, Byull., 1984, No. 12, p. 59 – 62 (1985). In Russian.

A generalized solution for Neyman's boundary problem has been obtained using the energetic method. A mathematical substantiation of the obtained solution and the results of the tests of the solution of the reverse problem by the Fourier series method are given.

062.065 On a mechanism of excitation of Alfvén waves in a cosmic plasma.
R. G. Dzhangiryan, F. A. Kostanyan.
Astrofizika, Tom 22, Vyp. 1, p. 189 – 194 (1985). In Russian. English translation in Astrophysics, Vol. 22, No. 1.

The problem of radiation of Alfvén waves in the solar wind from a dense beam of cosmic rays while crossing the contact jump of the wind has been considered. The spectral distribution of radiated energy both qualitatively and quantitatively is shown to coincide with that of the observed one.

062.066 Force–free equilibria of magnetized jets.
A. Königl, A. R. Choudhuri.
Astrophys. J., Vol. 289, No. 1, p. 173 – 187 (1985).

The authors study the force–free equilibria of magnetic–pressure–dominated, supersonic jets which are confined by a slowly varying external pressure. They demonstrate that, if the jets possess an internal dissipation mechanism, then the lowest–energy magnetic field configuration will in general be a superposition of two modes, one axisymmetric and one helical of wavelength $\lambda \approx 5R$ (R is the jet's radius). It is shown that the nonaxisymmetric mode dominates when either the net magnetic flux in the jet is small or the external pressure is low, and that the synchrotron emission then exhibits the nonaxisymmetric features that are observed in jets like NGC 6251. The authors are thus able to account for these phenomena in a natural way as a consequence of the equilibrium structure of the jet, without having to invoke any instabilities.

062.067 Radiative gas dynamics in the transonic regime.
C. J. Cannon.
Astrophys. J., Vol. 289, No. 1, p. 363 – 372 (1985).

The three equations specifying conservations of mass, momentum, and energy in radiating gases, coupled to the equation of radiative transfer, are examined for flow velocities approaching the local speed of sound. It is found that time–dependent terms are essential to any analysis of this transonic regime. In particular, wave–type perturbations start to amplify, rather than damp, when the flow velocity exceeds a critical value v_{crit} which is approximately 80% of the local thermal speed. Further, the work done by pressure forces during a cycle of these wave disturbances generates a nonzero heating of the gas when the temperature fluctuations lead the velocity fluctuations as is the case when the gas radiates.

062.068 Radiation from charges driven by large–amplitude longitudinal plasma waves.
K. H. Strobl, C. Leubner.
Astrophys. J., Vol. 289, No. 2, p. 467 – 474 (1985).

The detailed spectral and angular characteristics of the radiation emitted by electrons streaming at relativistic velocities within an electron ion plasma and being driven by a large–amplitude longitudinal plasma wave are calculated with a view to the possible occurrence of such processes in pulsar wind zones.

062.069 Power–law X–ray and gamma–ray emission from relativistic thermal plasmas.
A. A. Zdziarski.
Astrophys. J., Vol. 289, No. 2, p. 514 – 525 (1985).

The author considers pair equilibrium in thermal plasmas emitting power–law photon spectra by repeated Compton scatterings of a soft photon source. He studies the pair equilibrium equation for the maximum luminosity of a source of a given temperature $\Theta \equiv kT/m_e c^2$. Analytical solutions to the equation are found for the sub–relativistic region, $\Theta \lesssim 0.3$, and for the ultrarelativistic region, $\Theta \gtrsim 3$. In the transrelativistic region the solutions can be expressed by single integrals over the pair production cross sections, which are performed numerically. The results of these calculations are compared with observational data for active galactic nuclei.

062.070 The dynamical stability of differentially rotating discs. II.
J. C. B. Papaloizou, J. E. Pringle.
Mon. Not. R. Astron. Soc., Vol. 213, No. 4, p. 799 – 820 (1985).

The authors extend their investigation of the dynamical stability of differentially rotating fluid tori of uniform entropy to the case in which there is a gradient of specific angular momentum. This case is much more complicated and in this paper the authors limit themselves to an analytic approach. The overall conclusions are that (1) the dynamical instabilities found to exist in constant specific angular momentum tori persist in this case; (2) additional unrelated Kelvin–Helmholtz–like instabilities are introduced by allowing a gradient in specific angular momentum and (3) the general unstable mode is a mixture of these two.

062.071 Validity of electron temperature measurements using continuum plasma emission.
M. Lamoureux, C. Möller, P. Jaeglé.
J. Quant. Spectrosc. Radiat. Transfer, Vol. 33, No. 2, p. 127 – 131 (1985).

062.072 Changes in spectral intensities of thermal plasmas in the presence of fast charged particles.
R. Jayakumar, H. H. Fleischmann.
J. Quant. Spectrosc. Radiat. Transfer, Vol. 33, No. 2, p. 177 – 191 (1985).

The influence of monoenergetic, fast, charged particles present in a Maxwellian plasma on excitation and ionization processes is studied in the context of plasmas and fast particle diagnostics.

062.073 Stratification of radiation behind interstellar shock waves.
K. V. Bychkov.
Astrofiz. Issled. Izv. Spets. Astrofiz. Obs., Tom 19, p. 41 – 53 (1985). In Russian. English translation in Bull. Spec. Astrophys. Obs. – North Caucasus.

The cooling of gas behind a shock wave propagating at a velocity of $30 – 100$ km s^{-1} in a neutral medium is dominanted by hydrogen excitation unless the electron temperature exceeds 1 eV. The Balmer lines are formed in rather hot regions while the forbidden lines of heavy elements are formed in the cold ones which are farther from the shock wave. The intensity of any spectral line depends only on its contribution to the cooling function.

062.074 Sound and thermal waves in a fluid with an arbitrary heat–loss function.
M. H. Ibáñez S.
Astrophys. J., Vol. 290, No. 1, p. 33 – 46 (1985).

The propagation of sound and thermal waves in a nonreacting arbitrary fluid is studied, using the linear approximation. The fluid is assumed to be cooled or heated by nonspecified processes, which can be represented by a heat–loss function $L(\varrho, T)$. Thermal conduction is taken into account, but magnetic fields are neglected. The phase speed and the characteristic scale lengths for damping of sound and thermal waves are found. The condition for sound wave amplification and the corresponding characteristic scale length are obtained.

062.075 Erratum: ”Bremsstrahlung spectra at relativistic temperatures” [Acta Astron., Vol. 31, No. 4, p. 457 – 469 (1981)].
A. Górecki, W. Kluźniak.
Acta Astron., Vol. 34, No. 4, p. 494 (1984). See Abstr. 32.062.030.

062.076 Nonlinear Kelvin–Helmholtz instability in hydromagnetics.
S. K. Malik, M. Singh.
Astrophys. Space Sci., Vol. 109, No. 2, p. 231 – 239 (1985).

By taking into account the temporal as well as the spatial effects, a weakly nonlinear theory of the propagation of wave packets in the Kelvin–Helmholtz instability problem in the presence of uniform magnetic fields, acting along the surface of separation of two moving superposed fluids, is presented. With the use of the method of multiple scaling, the evolution of the amplitude of the two–dimensional wave packets, which is governed by a nonlinear Klein–Gordon equation, is derived. The various stability criteria arising out of this equation are examined. The nonlinear cut–off wavenumber, which separates the region of stability from that of instability, is determined.

062.077 Free molecular flow of rarefied gas over an oscillating plate under a periodic external force.
A. G. El–Sakka, R. A. Abdellatif, S. A. Montasser.
Astrophys. Space Sci., Vol. 109, No. 2, p. 259 – 270 (1985).

The free molecular flow over an infinite oscillating plane wall under external periodic force is considered. The Boltzmann equation is solved by using moments method with two–stream distribution functions. The boundary condition is obtained by assuming that the reflection of the particles from the solid surface takes place with complete energy accommodation. An analytical form for the velocity (X) and shear stress (Y) at any point is obtained. The results show that the amplitude of both the velocity change (X_1) and the shear stress change (Y_1) due to the periodic external force at the boundary ($y = 0$) is an increasing function of time (t).

062.078 Free convection and mass transfer in the hydromagnetic oscillatory flow past an infinite vertical porous plate with internal heat generation due to radiation absorption.
P. Stogianidis, D. P. Georgiou, G. A. Georgantopoulos.
Astrophys. Space Sci., Vol. 109, No. 2, p. 309 – 326 (1985).

The influence of radiation absorption on the flow–field of an unsteady laminar boundary layer due to free convection is considered. The flow is that of an incompressible viscous dissipative and electrically conducting fluid past an infinite vertical porous plate, when the flow is subjected to the action of a transverse magnetic field and the mainstream is oscillating around a mean value and an oscillating suction. The radiation is absorbed by a second material in a small concentration within the fluid and the absorption rate is proportional to the local concentration. The solution of the problem is obtained in the form of power series of Eckert number E, analytical expression for the velocity, temperature and the induced magnetic field are given for both the steady and the unsteady flows. In addition, the influence of the absorption on the skin friction of the heat flux is given.

062.079 Alfvén surface waves along coronal streamers.
A. S. Narayanan, K. Somasundaram.
Astrophys. Space Sci., Vol. 109, No. 2, p. 357 – 364 (1985).

The dispersive characteristics of Alfvén surface waves along a moving plasma surrounded by a stationary plasma are discussed. The stability curves for the symmetric and the asymmetric modes are also discussed.

062.080 On the formation of current layers near the surfaces of a magnetic field.
S. V. Bulanov, A. M. Zaborov, M. A. Ol'shanetskij.
Inst. teor. ehksp. fiz. Moskva, Prepr., No. 76, 58 pp. (1984). In Russian. Abstr. in Ref. Zh., 51. Astron., 2.51.125 (1985).

062.081 Turbulence and angular momentum in extragalactic radio jets. I. Linear stability analysis.
P. Londrillo.
Astron. Astrophys., Vol. 145, No. 2, p. 353 – 368 (1985).

Linear stability analysis of a stationary, supersonic turbulent jet, both in the hydrodynamical and M.H.D. regimes, has been performed by a numerical computation of the relevant dispersion relations. The stability analysis has been restricted to the inviscid limit and parallel flow approximation, the main interest being for a preliminary overview of some dynamical effects which are considered to be important for extragalactic radio jets structure. Three configurations, modelling the radio jet physical regimes in different phases of its evolution, have been analysed. The relevance of the results to the problem of wiggle formation in radio jets is discussed.

062.082 On hydrodynamics of astrophysical jets – I. Basic equations.
L. Nobili, M. Calvani, R. Turolla.
Mon. Not. R. Astron. Soc., Vol. 214, No. 2, p. 161 – 176 (1985).

A model for jet acceleration in the funnels of a geometrically thick, radiation–supported disc is constructed. The dynamics of matter lost from the funnels' surface is studied in a general relativistic background and by means of a fully hydrodynamical treatment, including viscosity. Hydrodynamic equations governing the gas flow are derived either for an optically thick or for an optically thin medium. The general behaviour of the solutions is discussed with particular emphasis to their existence and uniqueness.

062.083 Three–dimensional time–dependent field line reconnection.
V. S. Semenov, E. P. Vasil'ev.
Astron. Zh., Tom 62, Vyp. 2, p. 291 – 300 (1985). In Russian. English translation in Sov. Astron., Vol. 29, No. 2.

A solution of the three–dimensional time–dependent reconnection problem was obtained, which is determined by the following functions of time: (1) shape of a reconnection line, (2) electric field in the diffusion region. A qualitative scheme of the process was developed. It included a generation of the electric field in the diffusion region, propagation of an Alfvén wave along the current sheet and a disruption of an arbitrary MHD discontinuity.

062.084 Multi–mode study of time–dependent thermal convection with hexagonal planforms.
J. M. Lopez, J. O. Murphy.
Proc. Astron. Soc. Aust., Vol. 5, No. 4, p. 483 – 487 (1984).

062.085 Fragmentation of gas clouds.
J. J. Monaghan, J. C. Lattanzio.
Proc. Astron. Soc. Aust., Vol. 5, No. 4, p. 493 – 494 (1984).

The calculations described in this paper use a refined SPH method which provides a resolution twice as fine as that used by Gingold and Monaghan (1981). The main improvement is a grid–based algorithm for determining the gravitational potential, typically with 30^3 grid points.

062.086 Photospheric flux changes and the MHD approximation.
P. R. Wilson.
Proc. Astron. Soc. Aust., Vol. 5, No. 4, p. 500 – 502 (1984).

062.087 Space research and the new approach to the mechanics of fluid media in cosmos.
H. Alfvén, F. Čech.
ESA Spec. Publ., ESA SP–207, p. 3 – 24 (1984). – See Abstr. 012.044.

Space research has increased our knowledge of fluid media in space in two ways: (1) It has made it possible to see our cosmic environment not only in visual light and radio waves but also in infrared, ultraviolet, X–rays and gamma rays. (2) In situ measurements in the magnetospheres – including heliosphere – combined with laboratory experiments have given us new information about the properties of cosmic plasmas. The new results are applied to cosmogony (origin and evolutionary history of the solar system). Magnetospheric physics has matured to such an extent that it is possible to treat certain aspects of cosmogony as an extrapolation of magnetospheric physics. It is shown that certain events 4 – 5 billion years ago can be reconstructed with an accuracy of a few percent.

062.088 On the spectrum of longitudinal waves in the hot degenerate plasma.
J. G. Lominadze, G. I. Melikidze, V. P. Silin, V. N. Ursov.
ESA Spec. Publ., ESA SP–207, p. 101 – 103 (1984). – See Abstr. 012.044.

The longitudinal wave spectra in the hot degenerate plasma, having Fermi distribution – both one and three dimensional, are studied in the whole range of wave vectors. In the frames of classical mechanics, the relativistic case as well as nonrelativistic one are examined. The real and imaginary parts of the solutions of the dispersion equation are found. It is shown that the spectra asymptotics are not connected with strong Landau damping.

062.089 Nonlinear waves in the magnetosphere of a pulsar.
A. D. Pataraya, B. B. Chargeishvili
(*B. B. Chargejshvili*).
ESA Spec. Publ., ESA SP–207, p. 105 – 107 (1984). – See Abstr. 012.044.

The paper deals with the dispersion relations and growth–rates of linear waves propagating across the external magnetic field in an electron–positron plasma (EPP). The basic distribution function is presumed to be a one–dimensional power function. The effects of polarization plane rotation of linear waves due to Coriolis force and mixture of heavy particles in EPP are investigated. Weak nonlinear waves are studied as well.

062.090 X–ray astronomy and plasma astrophysics.
C. R. Canizares.
ESA Spec. Publ., ESA SP–207, p. 159 – 168 (1984). – See Abstr. 012.044.

The author reviews current research in X–ray astronomy with emphasis on studies of thin, thermal plasmas. These are found in stellar coronae, supernova remnants and clusters of galaxies. The author describes plasma diagnostics for density, temperature and elemental abundance, as well as diagnostics for departures from ionization equilibrium. Each of these has been used in recent analyses of data from imaging and spectroscopic instruments on the Einstein satellite.

062.091   Self–similar modified shock structures.
R. Beck, L. O'C. Drury.
ESA Spec. Publ., ESA SP–207, p. 181 – 184 (1984). – See Abstr. 012.044.

Time dependent shock structures for a "three fluid" system including gas, mildly relativistic particles and ultrarelativistic particles are obtained by considering self–similar solutions of the ideal hydrodynamical equations. The effective diffusion coefficient of the ultrarelativistic particles is assumed to be spatially constant but to increase linearly with time. The diffusive particle acceleration process turns out to be very efficient, with overall compressions in the range 10 – 20 although no energy loss from the system occurs.

062.092   Forced reconnection by nonlinear MHD waves and sequential triggering of solar flares.
J. Sakai, T. Tajima, F. Brunel.
ESA Spec. Publ., ESA SP–207, p. 185 – 187 (1984). – See Abstr. 012.044.

Magnetic reconnection processes may play a significant role in the fast release of magnetic energy stored in current–carrying plasmas. The old steady reconnection and non–steady models of tearing modes are insufficient to explain the impulsive phase of solar flares or sequential triggering phase observed by Vorpahl (1976). This observation hints a mechanism of triggering of solar flares by fast magnetosonic waves. In order to explain the triggering phase, the authors propose a theoretical model of forced reconnection caused by fast magnetosonic waves. They examine and visualize the forced reconnection by means of a MHD particle simulation. Some conditions for fast reconnection are discussed with applications to solar flares.

062.093   Explosive transverse electric field and particle acceleration during magnetic collapse.
J. Sakai, T. Tajima, R. Sugihara.
ESA Spec. Publ., ESA SP–207, p. 189 – 192 (1984). – See Abstr. 012.044.

It is shown that in streams across the magnetic field moving with electrons cause an explosive transverse electrostatic field during magnetic collapse such as coalescence instability of current loops. The electrostatic field can explosively accelerate ions and electrons perpendicular both to the magnetic field and ion streams. Ions and electrons are almost simultaneously accelerated to the opposite direction, respectively. The results obtained in the paper well explain the simulation results of collisionless coalescence instability of current loops. The simultaneous acceleration mechanism of ions and electrons is applied to the origin of explosive high energy particles in cosmic plasmas.

062.094   Nonlinear evolution of astrophysical Alfvén waves.
S. R. Spangler.
ESA Spec. Publ., ESA SP–207, p. 197 – 200 (1984). – See Abstr. 012.044.

The author reports numerical studies of nonlinear Alfvén waves, using the Derivative Nonlinear Schrödinger Equation as a model. He has studied the evolution of a variety of initial conditions, such as envelope solitons, amplitude–modulated waves, and band–limited noise. The last two furnish models for naturally occurring Alfvén waves in an astrophysical plasma. The author has studied a collapse instability in which a wave packet becomes more intense and of smaller spatial extent. He conjectures that this instability will lead to enhanced plasma heating. The author has begun studies in which the waves are amplified by an electron beam. The aforementioned instability tends to modestly inhibit the growth of the waves.

062.095   Electron–positron pair equilibrium in strongly magnetized plasmas.
A. K. Harding.
ESA Spec. Publ., ESA SP–207, p. 205 – 208 (1984). – See Abstr. 012.044.

Steady states of thermal electron–positron pair plasmas at mildly relativistic temperatures and in strong magnetic fields are investigated. The pair density in steady–state equilibrium where pair production balances annihilation is found as a function of temperature, magnetic field strength and source size, by means of a numerical calculation which includes pair production attenuation and Compton scattering of the photons. It is found that there is a maximum pair density for each value of temperature and field strength, and also a source size above which optically thin equilibrium states do not exist.

062.096   Electron–positron pairs in a mildly relativistic plasma in active galactic nuclei.
F. Takahara, M. Kusunose.
ESA Spec. Publ., ESA SP–207, p. 209 – 212 (1984). – See Abstr. 012.044.

The authors investigate the electron–positron pair concentration in an optically thin mildly relativistic plasma. Firstly the equilibrium pair concentration is calculated when copious soft photons are supplied by the cyclotron higher harmonics. It is shown that the attainable states of a plasma are strongly restricted by the pair equilibrium condition. Secondly the authors investigate pair production in the hot plasma produced by accretion onto a supermassive black hole, comparing relevant time scales. The authors show that the plasma becomes pair dominated if the accretion rate is moderately high and accretion flow is slow compared to free fall. The unsaturated Comptonization leads to the formation of a pair dominated plasma in most cases.

062.097   Hydrodynamics of astrophysical jets.
R. Turolla, M. Calvani, L. Nobili.
ESA Spec. Publ., ESA SP–207, p. 251 – 253 (1984). – See Abstr. 012.044.

The authors present a model for the acceleration of jets in the framework of geometrically thick accretion disks. The hydrodynamical equations for a non–perfect, heat–conducting fluid are discussed.

062.098   Viscous jets.
M. Nepveu.
ESA Spec. Publ., ESA SP–207, p. 255 – 257 (1984). – See Abstr. 012.044.

The starting flow is calculated around an object that imparts momentum unidirectionally to its surroundings. This object represents a central engine of an active galaxy. Viscosity is used to mimick turbulence. A steady jet results for which the rate of entrainment of gas can be numerically computed. A comparison is made with earlier analytical calculations.

062.099   An analytic model for hydromagnetospheres.
K. Tsinganos, B. C. Low.
ESA Spec. Publ., ESA SP–207, p. 289 – 292 (1984). – See Abstr. 012.044.

The authors outline the main steps needed to isolate a general solution of the complete hydromagnetic equations. In this solution, the axisymmetric magnetosphere that surrounds the central object consists of two regions; the first, around the magnetic equator, is controlled by closed and dipole–like magnetic fieldlines wherein the plasma is in static equilibrium with the magnetic and gravitational fields, while the second around the magnetic axis, is controlled by flows along the open and monopole–like magnetic fieldlines. The total pressure is continuous everywhere and the solution is globally consistent. Several astrophysical applications are sketched.

062.100   The continuous spectrum of an axisymmetric, self–gravitating equilibrium in the presence of a poloidal magnetic field.
D. Hermans, M. Goossens, S. Poedts.
ESA Spec. Publ., ESA SP–207, p. 297 – 300 (1984). – See Abstr. 012.044.

The continuous spectrum of oscillation frequencies is examined for an axisymmetric, self–gravitating equilibrium in the presence of a purely poloidal magnetic field. It is shown that the continuous spectrum is given by an eigenvalue problem of two uncoupled ordinary second–order differential equations along the magnetic field lines. The two decoupled continuous spectra have modes that are polarized either perpendicular or parallel to the magnetic field lines. Curvature and toroidicity influence the two continua, but only the cusp continuum is affected by gravity and compressibility. Variational expressions for the continuum frequencies are derived and discussed.

062.101   Kinematic dynamo theory for an arbitrary mean flow.
P. Hoyng.
ESA Spec. Publ., ESA SP–220, p. 97 – 100 (1984). – See Abstr. 012.045.

The problem of an arbitrary, incompressible mean flow v_0 in kinematic dynamo theory is amenable to a systematic treatment leading to an intuitively simple result. The author's approach is based on the theory of stochastic differential equations.

062.102   Waves in inhomogeneous media.
B. Roberts.
ESA Spec. Publ., ESA SP–220, 137 – 145 (1984). – See Abstr. 012.045.

Recent developments in the theory of magnetohydrodynamic waves in a magnetically structured atmosphere are reviewed. Both continuously and discretely structured media are considered, with applications to coronal loops, open field regions, sunspots, and Hα fibrils. Analogies with other subjects (e.g. oceanography, fibre optics) are drawn. The propagation of an impulsively generated fast magnetoacoustic wave in a density duct leads to periodicities of about a second in the corona and tens of seconds in the chromosphere.

062.103   Linear incompressible magnetoconvection in a planar layer with a non–uniform horizontal magnetic field.
D. Hermans, M. Goossens, R. Polfliet.
ESA Spec. Publ., ESA SP–220, p. 199 – 200 (1984). – See Abstr. 012.045.

The linear stability of an incompressible planar layer in the presence of a non–uniform horizontal magnetic field is studied for three sequences of equilibrium configurations. It is found that the Alfvén continuum is fundamental for the understanding of the linear stability.

062.104   The continuous spectrum of an axisymmetric equilibrium with a mixed poloidal and toroidal magnetic field and with gravity included.
S. Poedts, M. Goossens, D. Hermans.
ESA Spec. Publ., ESA SP–220, p. 201 – 202 (1984). – See Abstr. 012.045.

062.105   The generation of magnetic fields by a flow with chaotic streamlines.
D. J. Galloway, U. Frisch.
ESA Spec. Publ., ESA SP–220, p. 221 – 222 (1984). – See Abstr. 012.045.

Kinematic dynamo action due to a spatially periodic flow with regions of chaotic streamlines is investigated numerically, using a spectral method.

062.106   Dynamo action of magnetostrophic waves.
D. Schmitt.
ESA Spec. Publ., ESA SP–220, p. 223 – 224 (1984). – See Abstr. 012.045.

The stability of magnetostrophic waves is considered in a horizontal thin plane layer of a perfectly conducting fluid, which is stratified according to gravitation and permeated by a variable toroidal magnetic field. The layer rotates rigidly around an axis inclined to the horizontal plane. It is shown that unstable magnetostrophic waves, driven by magnetic buoyancy, are capable of inducing an electromotive force parallel to the toroidal magnetic field.

062.107   Energy balance in resistive instabilities.
G. Torricelli–Ciamponi.
ESA Spec. Publ., ESA SP–220, p. 271 – 272 (1984). – See Abstr. 012.045.

An uni–dimensional current–sheet model, where magnetic reconnection takes place, is considered. The analysis is particularly devoted as to how the perturbation will change the balance between magnetic and thermal energy both in the thin resistive region and in the wide ideal one.

062.108   Nonlinear mathematical problems of the dynamics of a radiating gas.
K. Yu. Kazakov, A. V. Kalinin, S. F. Morozov.
Tr. nauchn.–metod. konf. prep. i sotr. mekh.–mat. fak. Gor'k. univ. Gor'kij, 23 – 24 apr. 1984. Gor'kij, p. 63 – 68 (1984). In Russian. Abstr. in Ref. Zh., 51. Astron., 3.51.156 (1985).

062.109   Electron temperature determination in LTE and non–LTE plasmas.
T. L. Eddy.
J. Quant. Spectrosc. Radiat. Transfer, Vol. 33, No. 3, p. 197 – 211 (1985).

A method is presented for calculating electron temperatures (T_e) in dense plasmas, which does not assume equivalence with the excited level distribution temperatures (T_{ex}). The method involves the upper–level Saha ionization equation at the ionization limit, the limiting weighted population density obtained from measured population densities and the experimentally obtained electron density. Electron temperatures calculated for 0.1–bar hydrogen and 1–atm helium and argon arcs are found to be up to twice as large as excited level distribution temperatures. For subatmospheric argon arcs, the calculated T_e are equivalent to the excitation temperature of the middle levels, but are two to three times smaller than the quoted T_{ex} for the highest levels. Reasons are discussed for the apparent invisibility of true electron temperatures and for differences between them and the excitation temperatures.

062.110   Free–free radiative transitions for the screened Coulomb potential.
R. Lange, D. Schlüter.
J. Quant. Spectrosc. Radiat. Transfer, Vol. 33, No. 3, p. 237 – 242 (1985).

Free–free transitions form a substantial part of the thermal emission of plasmas at high temperatures and long wavelengths. The modifications caused by charged perturbing particles, especially electrons, are estimated for the case of a hydrogen plasma by the use of the Debye–Hückel potential and compared with other approximations.

062.111   Effects of a finite plasma temperature on electron–cyclotron maser emission.
R. M. Winglee.
Astrophys. J., Vol. 291, No. 1, p. 160 – 169 (1985).

Auroral kilometric radiation, Jupiter's decametric radio emission, and microwave spike bursts have all been attributed to the semirelativistic maser instability. The effect of a finite plasma temperature on the emission from this instability is investigated. Temperature effects reduce the frequency of the x mode and thereby enable fundamental x–mode radiation to occur at higher ω_p/Ω_e (where ω_p is the plasma frequency and Ω_e is the electron–cyclotron frequency). The z–mode radiation is subject to electron–cyclotron damping, and this damping can cause heating of the plasma in the vicinity of the source region.

062.112   Nonlinear spiral density waves: an inviscid theory.
F. H. Shu, C. Yuan, J. J. Lissauer.
Astrophys. J., Vol. 291, No. 1, p. 356 – 376 (1985).

The authors develop an inviscid fluid theory for nonlinear self–gravitating spiral density waves that adopts, for simplicity, the most stringent of the asymptotic approximations normally used, but relaxes the assumption of small amplitude. This treatment allows the consideration of both free wave trains and those excited by resonant forcing. In a frame which corotates with the pattern, the pressure–free problem reduces in a Lagrangian description to the solution of a complex nonlinear integral equation. Numerical solutions are obtained for forced density waves of the strengths typically seen in Saturn's rings. Although meaningful detailed comparisons with observations require the inclusion of wave damping by induced viscous stresses, several important implications are derived for the density waves in Saturn's rings.

062.113   The stability of confined radio jets: the role of reflection modes.
D. G. Payne, H. Cohn.
Astrophys. J., Vol. 291, No. 2, p. 655 – 667 (1985).

The linear stability of a confined radio jet is reinvestigated. The roles of both absolute (temporal) and convected (spatial) instability are considered, and it is demonstrated that the two are related through the group velocity. The dispersion relation is analyzed asymptotically for the fundamental and reflection modes. Numerical results are presented for pinching modes. A geometrical interpretation of the modes is presented in terms of the propagation angle and is visualized by contour plots of the pressure perturbation. A confined jet is seen to be an acoustic waveguide that supports a set of self–excited modes. The implications of the calculations for the development of large–scale features in extragalactic radio jets are discussed.

062.114   Reconnections in force–free sheared magnetic fields. I.
A. A. Solov'ev.
Soln. Dannye, Byull., 1985, No. 1, p. 60 – 66 (1985). In Russian.

A qualitative analysis of the magnetic field lines reconnection in a force–free sheared magnetic field is given. The existence of a singular magnetic surface is shown to be closely connected with the shear of the field lines. The statement is formulated that the resistive evolution (reconnections) of the force–free field has to be developed in the direction of shear decreasing.

062.115   Thermal–mode stabilization of the resistive tearing instability of a current sheet.
B. I. Meerson.
Sov. Astron. Lett., Vol. 10, No. 5, p. 299 – 301 (1984). English translation of 38.062.048.

062.116   Reconnections in force–free sheared magnetic fields. II.
A. A. Solov'ev.
Soln. Dannye, Byull., 1985, No. 2, p. 47 – 52 (1985). In Russian.

A number of formulae for the parameters of the reconnection process in a force–free sheared field is derived. The velocity, time and space scales are expressed through the average values which could be obtained from the observations.

062.117   A magnetodynamic mechanism for the formation of astrophysical jets. I. Dynamical effects of the relaxation of nonlinear magnetic twists.
K. Shibata, Y. Uchida.
Publ. Astron. Soc. Jpn., Vol. 37, No. 1, p. 31 – 46 (1985).

A magnetodynamical mechanism for the formation of jets in astrophysical situations is presented. The effect of the relaxation of the accumulated nonlinear magnetic twist is calculated in an axisymmetric quasi–three dimensional simulation, and it is shown that a jet is formed in the direction of the axis parallel to the large–scale magnetic field along which the packet of the magnetic twist relaxes. One of the characteristic features of the jet produced in the present mechanism is its helical motion which is favorable in explaining some of the observations of astrophysical jets.

062.118   Magnetic flux loss in gaseous layers.
T. Nakano.
Publ. Astron. Soc. Jpn., Vol. 37, No. 1, p. 69 – 95 (1985).

The magnetic flux loss due to the drift of plasma and magnetic field is investigated for gaseous layers with the magnetic field parallel to the layer surfaces. When the magnetic force is initially much stronger than gravity and the pressure of the external medium is negligible, a large drift velocity of ions is induced. However, because the cloud expands nearly freely, the magnetic flux loss is limited to the early phase of the expansion. The magnetic flux loss is also investigated for a gaseous layer which is kept in quasi–equilibrium by the gas pressure and the magnetic field parallel to the surfaces by taking into account the friction of charged grains. The density fluctuation in such a layer grows in a time scale of t_f and the layer breaks into fragments with negligible flux loss.

062.119   A note on the stability of helical velocity and magnetic fields.
R. Kant, S. K. Trehan.
Astrophys. Space Sci., Vol. 110, No. 2, p. 297 – 300 (1985).

The stability of nonparallel helical velocity and magnetic fields in a plasma cylinder with free boundaries is examined. It is shown that, in case of weak helicity the stability of the system remains unaffected.

062.120   The propagation of a dense quasi–neutral ion beam across a magnetized plasma.
R. A. Treumann, B. Häusler.
Astrophys. Space Sci., Vol. 110, No. 2, p. 371 – 378 (1985).

Propagation of a quasi–neutral narrow ion beam across a magnetised cold plasma is investigated in slab geometry. This problem is of interest in connection with artificial beam injection experiments and with naturally appearing plasma injections into magnetic fields as astrophysical jets. Several different cases are discussed briefly where the beam is assumed either slow or fast.

062.121   Driven magnetic field reconnection.
B. K. Shivamoggi.
Astrophys. Space Sci., Vol. 110, No. 2, p. 397 – 399 (1985).

The author considers the magnetic field reconnection in a plasma induced by perturbing the boundaries of a slab of incompressible plasma with a magnetic neutral surface inside. He assumes that the boundaries of the plasma slab are perturbed at a rate which is fast compared with the hydromagnetic evolution rate; and investigates the ensuing adjustments in the plasma and the magnetic field threading through it.

062.122   Erratum: ''The critical layers and other singular regions in ideal hydrodynamics and magnetohydrodynamics'' [Astrophys. Space Sci., Vol. 105, No. 2, p. 401 – 412 (1984)].
J. A. Adam.
Astrophys. Space Sci., Vol. 110, No. 2, p. 419 (1985). See Abstr. 38.062.093.

062.123   Self–similar solutions for the blast wave problem in radiative gasdynamics. I.
J. B. Singh, K. S. Singh.
Astrophys. Space Sci., Vol. 111, No. 1, p. 79 – 85 (1985).

Similarity solutions, describing the flow of a perfect gas behind spherical shock waves, are investigated including the radiation heat flux. The shock is assumed to be propagating in a medium at rest. Shock radius varies exponentially with time and density is inversely proportional to fifth power of the shock radius immediately ahead of the shock front.

062.124   Unsteady hydromagnetic free convection flow past an accelerated infinite vertical porous plate.
M. A. Hossain, A. C. Mandal.
Astrophys. Space Sci., Vol. 111, No. 1, p. 87 – 95 (1985).

Unsteady laminar free convection flow of a viscous incompressible and electrically conducting fluid past an accelerated vertical infinite porous plate subjected to a suction velocity proportional to $(time)^{-1/2}$ is studied in presence of a uniform horizontal magnetic field. Results are discussed with the effects of the

Grashof number Gr, and the magnetic field parameter M for Pr (the Prandtl number) = 0.71 and 7.0 representing air and water respectively at 20°C.

062.125   The onset of shock wave in an electrically conducting and radiating gas.
A. Kumar, L. P. Singh, R. Shyam.
Astrophys. Space Sci., Vol. 111, No. 1, p. 131 – 137 (1985).

The present work applies the method of characteristics to study the behaviour of planar and cylindrical wave–heads propagating through a perfectly electrically conducting and thermally radiating inviscid gas under the optically thin limit in the presence of a transverse magnetic field. The true nonlinear progress of the flow variable gradients at the wavefront is predicted and the critical distance at which the characteristics pile up at the wavefront to form a shock wave is obtained. It is investigated as to how the effects of radiative flux, the magnetic field strength and the specific heat ratio influence the process of steepening or flattening of the characteristic wavefront.

062.126   Electromagnetic mass model of a static charged perfect fluid sphere.
D. D. Dionysiou.
Astrophys. Space Sci., Vol. 111, No. 1, p. 207 – 209 (1985).

Using the results, which have been already published in two earlier papers (Dionysiou, 1982; Dionysiou and Kostakis, 1983), one can easily see that the total gravitational mass, the density and pressure of a static charged perfect fluid sphere are of electromagnetic origin.

062.127   The loss–cone driven instability for Langmuir waves in an unmagnetized plasma.
R. G. Hewitt, D. B. Melrose.
Sol. Phys., Vol. 96, No. 1, p. 157 – 179 (1985).

The authors evaluate the absorption coefficient for Langmuir waves due to an axially symmetric distribution of suprathermal electrons; they also develop a geometric interpretation of the resonance condition in terms of a resonance hyperboloid in momentum space, and use this to treat an idealized loss–cone distribution. They discuss the evolution of initial Maxwellian and loss–cone distributions in a trap–plus–precipitation model, showing how a 'gap' distribution develops. Numerical results for the growth rate are presented and discussed.

062.128   Ohmic dissipation and relaxation of force–free magnetic fields.
A. A. Solov'ev.
Astrofizika, Tom 21, Vyp. 3, p. 627 – 639 (1984). In Russian. English translation in Astrophysics, Vol. 21, No. 3.

The passive ohmic dissipation of a force–free field in a low–pressure plasma with inhomogeneous conductivity has been considered. The process of the topological resistive relaxation of the force–free field in the twisted magnetic flux loop to the state of minimum magnetic energy is also investigated. It is found that the magnetic flux rope is uniformly twisted in this state if the external pressure is fixed and the torque on the ends of the loop is constant. A critical analysis of Taylor's model of relaxation has been given.

062.129   Use of the 447 nm He I line to determine electron concentrations.
A. Czernichowski, J. Chapelle.
J. Quant. Spectrosc. Radiat. Transfer, Vol. 33, No. 5, p. 427 – 436 (1985).

The helium 447 nm complex line has been excited in a wall–stabilized arc fed at atmospheric pressure by pure He, He–H_2, He–Ne–H_2 or He–Ar–H_2 mixtures. Photoelectric end–on observations of the central part of the arc channel were made with high spatial (1/600) and spectral (53,000) resolution. A collection of 88 helium 447 nm line profiles, of which 75 were recorded simultaneously with H_β and Ne I or Ar II line profiles, yielded information about the electron concentration (2×10^{20}–2×10^{22}m^{-3}), temperature and relative ion composition of the plasma. The authors propose simple formulae, which may be useful for practical determinations of the electron concentrations in helium–containing plasmas with an accuracy of $\pm 15\%$ and without taking into account either the chemical composition of the plasma or the temperature.

062.130   A perturbation theory for the free oscillations of a self–gravitating sphere with phase boundaries.
S. V. Vorontsov.
Sov. Astron., Vol. 28, No. 5, p. 500 – 503 (1984). English translation of 38.062.070.

062.131   The back action of accelerated particles on shock–front structure.
S. V. Bulanov, I. V. Sokolov.
Sov. Astron., Vol. 28, No. 5, p. 515 – 520 (1984). English translation of 38.062.071.

062.132   Baroclinic instability in the presence of a strong horizontal shear. I. Instability conditions.
E. Knobloch, H. C. Spruit.
MPA Rep., No. 170, 25 pp. (1984). Submitted to Geophys. Astrophys. Fluid Dyn.

062.133   The hydrodynamics of clouds overtaken by supernova remnants. I. Cloud crushing phenomena.
G. Tenorio–Tagle, M. Rozyczka.
MPA Rep., No. 180, 20 pp. (1985). Submitted to Astron. Astrophys.

062.134   Cosmic jets.
L. L. Smarr.
Bull. Am. Astron. Soc., Vol. 16, No. 4, p. 918 (1984). Abstract. – See Abstr. 010.062.

062.135   A variational principle for MHD based upon magnetic helicity.
G. Field.
Bull. Am. Astron. Soc., Vol. 16, No. 4, p. 928 (1984). Abstract. – See Abstr. 010.062.

062.136   Spatial and temporal growth of surface waves on an expanding jet.
P. E. Hardee.
Bull. Am. Astron. Soc., Vol. 16, No. 4, p. 932 (1984). Abstract. – See Abstr. 010.062.

062.137   Wave propagation in a radiating medium.
B. W. Mihalas.
Bull. Am. Astron. Soc., Vol. 16, No. 4, p. 947 (1984). Abstract. – See Abstr. 010.062.

062.138   Theoretical studies of the linear polarization of synchrotron radiation from relativistic jets.
W. K. Cobb, J. F. C. Wardle, D. H. Roberts.
Bull. Am. Astron. Soc., Vol. 16, No. 4, p. 954 (1984). Abstract. – See Abstr. 010.062.

062.139   Polarization transport in turbulent synchrotron sources.
J. A. Eilek.
Bull. Am. Astron. Soc., Vol. 16, No. 4, p. 954 (1984). Abstract. – See Abstr. 010.062.

062.140   Formation and propagation of radio jets in more realistic galactic potentials.
J. Siah, P. J. Wiita.
Bull. Am. Astron. Soc., Vol. 16, No. 4, p. 954 – 955 (1984). Abstract. – See Abstr. 010.062.

062.141   Vertical helices in deep convection.
K. L. Chan, S. Sofia.
Bull. Am. Astron. Soc., Vol. 16, No. 4, p. 992 (1984). Abstract. – See Abstr. 010.062.

062.142 On the numerical computation of nonlinear force–free magnetic fields.
S. T. Wu, H. M. Chang, M. J. Hagyard, E. A. West.
Bull. Am. Astron. Soc., Vol. 16, No. 4, p. 992 – 993 (1984). Abstract. – See Abstr. 010.062.

062.143 Nonlinear development of convective instability within slender flux tubes. II. The effect of radiative heat transport.
P. Venkatakrishnan.
J. Astrophys. Astron., Vol. 6, No. 1, p. 21 – 34 (1985).
Inclusion of radiative heat transport in the energy equation for a slender flux tube leads to oscillations of the tube. The amplitude of the oscillations depends on the radius of the tube when lateral heat exchange alone is considered. Longitudinal heat transport has a greater influence on the evolution of the instability than lateral heat exchange for the particular value of tube radius considered in the calculation. Heat transport is seen to reduce the efficiency of concentration of magnetic fields by convective collapse in the case of polytropic tubes.

062.144 Electron temperatures of astrophysical plasmas from the [O III] line ratio.
S. M. V. Aldrovandi, R. B. Gruenwald.
Astron. Astrophys., Vol. 147, No. 2, p. 331 – 334 (1985).
It is shown that the introduction of K–shell photoionization of O^0, followed by Auger effect, as a population mechanism of O^{2+} metastable levels can modify the behaviour of $R_{\text{OIII}} =$ [O III]$\lambda 4363/\lambda\lambda(5007+4959)$ intensity ratio with electron temperature. The results indicate that one should be cautious in using this line ratio in temperature determinations for the emission line region of active galactic nuclei.

062.145 Spherical shock waves in viscous magnetogasdynamics.
B. G. Verma, R. C. Srivastava, V. K. Singh.
Astrophys. Space Sci., Vol. 111, No. 2, p. 253 – 263 (1985).
The point–source, spherical magnetogasdynamics shock wave moving into a constant density γ–law gas is considered in the limit of infinite shock strength, from the point of view of the Richtmyer–Von Neumann viscosity technique. Numerical solutions of this problem has been obtained in viscous and non-viscous regions. A similarity solution of this problem is shown to exist. The authors have shown that field variables change rapidly when the magnetic field is imposed in both the viscous and non-viscous regions.

062.146 Double layers in strong turbulent plasmas.
X.–q. Li.
Astrophys. Space Sci., Vol. 112, No. 1, p. 13 – 24 (1985).
On the basis of two magneto–fluid models for two time–scales, the double layer formation in strong turbulent plasma is examined; it is shown that double layer is a moving nonlinear entity in this case: soliton and wave–packet shock. It turns out that subsonic moving double layer is a layer–double layer and might be accounted as electric detonator in cosmic bodies; supersonic moving double layer possesses soliton pattern of shock type and it can be broken up into a single double layer within the bounds of possibility.

062.147 Effect of anisotropy and viscous dissipation on turbulence production.
N. Kishore, S. R. Singh.
Astrophys. Space Sci., Vol. 112, No. 1, p. 209 – 212 (1985).
The paper is concerned with turbulent flow of incompressible, spatially homogeneous viscous fluid. A model for turbulence energy equation is obtained, ignoring the pressure redistribution term in dynamical equations for the Reynolds stresses. The mechanism of dissipation on turbulence production is discussed and shown that the turbulence kinetic energy decays up to a constant value as time becomes infinitely large, i.e., for isotropy, dissipation inhibits the production process and if $\overline{u^2} > \overline{v^2}$ initially then dissipation causes reduction in anisotropy.

062.148 Free–convection flow past an impulsively started vertical plate in a porous medium by finite difference method.
A. A. Raptis, A. K. Singh.
Astrophys. Space Sci., Vol. 112, No. 2, p. 259 – 265 (1985).
A numerical solution for the effects of the free–convection currents of a viscous fluid through a porous medium bounded by a vertical moving infinite vertical plate is considered, when the flow is unsteady. The graphs of velocity profiles for different values of permeability parameter of the porous medium and the Grashof number are discussed.

062.149 Convection flow over a uniform–flux vertical surface with variable viscosity.
F. S. Ibrahim, A. H. Khater, F. M. Hady.
Astrophys. Space Sci., Vol. 112, No. 2, p. 303 – 310 (1985).
A regular perturbation analysis is presented for natural convection flow over a uniform–flux vertical surface with temperature dependent viscosity. Numerical calculations are presented for $P_r = 6.7$ which show that the first–order correction to the local temperature difference and to the local skin–fraction are negative whereas it is positive for the local Nusselt number. The effects of variable viscosity on the temperature, velocity profiles, the local temperature difference, the local Nusselt number and the local skin fraction are discussed.

062.150 Effect of a transverse magnetic field and porosity on the Falkner–Skan flows of a non–Newtonian fluid.
F. M. Hady, I. A. Hassanien.
Astrophys. Space Sci., Vol. 112, No. 2, p. 381 – 390 (1985).
This paper provides an analysis of the effects of a uniform transverse magnetic field and porosity on the boundary layer flow of a homogeneous incompressible fluid of the second grad past a wedge placed symmetrically with respect to the flow direction. The variation of the skin friction with respect to porosity and magnetic field parameters are discussed. Also the first order velocities of the fluid are given.

062.151 Hall effects on thermal hydromagnetic instability of a partially ionized medium.
K. C. Sharma, S. Sharma.
Astrophys. Space Sci., Vol. 112, No. 2, p. 391 – 395 (1985).
Thermal instability of a finitely conducting hydromagnetic composite medium is considered including the effects of Hall currents and the collisions with neutrals. The equilibrium magnetic field is assumed to be uniform and vertical.

062.152 A class of exact solutions of the unsteady magnetohydrodynamic free–convection flows.
J. N. Tokis.
Astrophys. Space Sci., Vol. 112, No. 2, p. 413 – 422 (1985).
The unsteady free–convection flow of an electrically-conducting fluid near a moving vertical plate of infinite extent is investigated in the presence of uniform transverse magnetic field fixed to the fluid or to the plate. Exact solution of this problem is obtained with the aid of the Laplace transform technique, when the plate is moving with a velocity which is an arbitrary function of time. The solution is exemplified for three particular cases of physical interest; the non–magnetic case is also dicussed.

062.153 The turbulent viscosity of a gravitating gaseous disk.
A. G. Morozov, Yu. M. Torgashin, A. M. Fridman.
Pis'ma Astron. Zh., Tom 11, No. 3, p. 231 – 238 (1985). In Russian. English translation in Sov. Astron. Lett., Vol. 11.
It is shown that for a gaseous gravitationally stable disk having a region being near the border of its gravitational stability state an anomalous high coefficient of turbulent viscosity would arise.

062.154 Axisymmetric expansion of a rotating adiabatic gas.
B. C. Low.
Astrophys. J., Vol. 293, No. 1, p. 44 – 51 (1985).
A differentially rotating adiabatic gas can exhibit axisymmetric, time–dependent, self–similar expansion. Analytic hydrodynamic solutions describing such an expansion are derived. These solutions are the generalization of well–known spherically symmetric self–similar solutions for which rotation is not included.

Basic properties are discussed and illustrated with explicit solutions for rotating gaseous bodies expanding freely in vacuum. The discussion relates these self–similar solutions to others found recently for time–dependent magnetohydrodynamic flows in the presence of gravity.

062.155   Possible mechanism of the alternating jets (flip–flop jets) by the evaporating disk model.
S.–U. Choe, R. N. Henriksen.
Bull. Am. Astron. Soc., Vol. 17, No. 1, p. 517 (1985). Abstract. – See Abstr. 010.064.

062.156   Knots, wiggles, and flaring in ballistic jets.
D. Roberts.
Bull. Am. Astron. Soc., Vol. 17, No. 1, p. 517 (1985). Abstract. – See Abstr. 010.064.

062.157   Introduction to particle simulation models and their application to electrostatic plasma waves in space.
H. Okuda.
Computer simulation of space plasmas, p. 3 – 41 (1985). – See Abstr. 003.031.

A review of plasma simulation models using finite–size particles is given with an emphasis on the electrostatic models. Physical properties of such simulation models as well as numerical techniques are studied in some detail. Modification to plasma kinetic theory, collisions and fluctuations, Vlasov–Landau equations are derived along with an introduction to a spatial grid, charge–force interpolation and finite–difference schemes. A study of electrotastic ion cyclotron waves driven by a field-aligned current is also studied by numerical simulations.

062.158   Particle simulation of electromagnetic waves and its application to space plasmas.
H. Matsumoto, Y. Omura.
Computer simulation of space plasmas, p. 43 – 102 (1985). – See Abstr. 003.031.

The authors present a basic description of particle simulation codes developed at RASC of Kyoto University for the study of wave phenomena in space plasmas. Out of the available codes at RASC, two most–frequently–used codes are presented. Principles and basic techniques of the codes as well as examples of their applications to space plasma problems are presented.

062.159   Relativistic code applied to radiation generation.
A. T. Lin.
Computer simulation of space plasmas, p. 103 – 116 (1985). – See Abstr. 003.031.

A general algorithm for simulating a relativistic electromagnetic plasma is given. Emphasis is placed on how to handle the electron mass dependent on its energy and to treat boundary conditions for outgoing electromagnetic waves. The results of mode conversion from an extraordinary wave into an electrostatic Bernstein wave and radiation generation which exploits the relativistic effects of electrons such as gyrotron and auroral kilometric radiation are presented.

062.160   Modern development in particle simulation.
J. C. Adam.
Computer simulation of space plasmas, p. 117 – 130 (1985). – See Abstr. 003.031.

The author reviews some of the extension that has been made recently to particle codes to allow them to handle low frequency phenomena economically.

062.161   Principles of magnetohydrodynamic simulation in space plasmas.
T. Sato.
Computer simulation of space plasmas, p. 133 – 153 (1985). – See Abstr. 003.031.

The aim of this article is to place a special emphasis on the philosophical as well as physical principles which are essential in the establishment of the magnetohydrodynamic (MHD) simulation study in the solar terrestrial plasma research. Taking the limits of the abilities of the present–day computers for granted, the author emphasizes the importance of the "local" MHD simulation. The local MHD simulation is defined as a self–contained MHD simulation in each elementary region. Differences in the roles between the local and present–day global MHD simulations are briefly discussed. The importance and difficulties of the boundary condition are also discussed in some detail. Finally, a couple of notes on the finite–difference method and the importance of the diagnostics are pointed out.

062.162   Anomalous transport by Kelvin–Helmholtz instabilities.
A. Miura.
Computer simulation of space plasmas, p. 203 – 224 (1985). – See Abstr. 003.031.

Simulation of magnetohydrodynamic Kelvin–Helmholtz-instabilities has been performed for parallel ($B_0 \parallel V_0$) and transverse ($B_0 \perp V_0$) configurations, modeling high latitude (or downstream flanks) and dayside low latitude magnetospheric boundaries.

062.163   Vlasov simulations of ion acoustic double layers.
G. Chanteur.
Computer simulation of space plasmas, p. 279 – 301 (1985). – See Abstr. 003.031.

The author presents a detailed description of a Vlasov code which has been used to study the formation of weak ion acoustic double layers and he discusses the physical process responsible for the formation of these structures.

062.164   Simulation models for space plasmas and boundary conditions as a key to their design and analysis.
C. T. Dum.
Computer simulation of space plasmas, p. 303 – 375 (1985). – See Abstr. 003.031.

The basic types of simulation models as they derive from a variety of space plasma problems are examined. The design of these models and the evaluation of their potential in advancing physical theories, including also the caveats in their physical interpretation and generalization, are emphasized rather than a review of existing simulation results. Boundary condition play the key role in this analysis. The analysis of physical problems and corresponding simulation models presented in the paper puts some emphasis on double layers, anomalous resistivity and reconnection, not only because these processes are fundamental and thus also actively studied in simulations, but also because they are outstanding examples for a strong dependence on boundary conditions. Related laboratory experiments and corresponding simulations are also presented.

062.165   Radiative theory of MHD wave propagation (Review).
A. V. Gul'el'mi.
Geomagn. Aehron., Tom 25, No. 3, p. 356 – 370 (1985). In Russian. English translation in Geomagn. Aeron.

062.166   Jets with entrained clouds – I. Hydrodynamic simulations and magnetic field structure.
C. S. Coleman, G. V. Bicknell.
Mon. Not. R. Astron. Soc., Vol. 214, No. 3, p. 337 – 355 (1985).

The formation of bow shocks in freely expanding supersonic jets is simulated using the method of Smoothed Particle Hydrodynamics, suitably reformulated for application to flows with axial symmetry. It is found that a stable bow shock is formed when a jet encounters an obstruction of diameter d less than about half the jet diameter D. Results are presented for flow simulations with adiabatic indices of 4/3 and 5/3, representing the two cases in which the pressure due to relativistic particles is dominant and negligible respectively, and for pre–shock Mach numbers of 3 and 5. In each of these simulations the jet has an initial diameter of $D = 5d$. The frozen–in magnetic field configuration is calculated for each flow simulation.

062.167   Global effects of double layers.
M. A. Raadu.
Second symposium on plasma double layers and related topics, p. 3 – 31 (1984). Abstr. in Phys. Abstr., Vol. 88, No. 1253, Entry 32525 (1985). – See Abstr. 012.075.

062.168 Yang–Mills magnetohydrodynamics: nonrelativistic theory.
D. D. Holm, B. A. Kupershmidt.
Phys. Rev. D, Vol. 30, No. 12, p. 2557 – 2560 (1984). Abstr. in Phys. Abstr., Vol. 88, No. 1254, Entry 36371 (1985).

062.169 The relationship between the electric fields associated with plasma expansion and double layers.
N. Singh, R. W. Schunk.
Second symposium on plasma double layers and related topics, p. 272 – 277 (1984). Abstr. in Phys. Abstr., Vol. 88, No. 1255, Entry 43201 (1985). – See Abstr. 012.075.

062.170 Current fluctuations and dynamical features of double layer potential profiles.
N. Singh, H. Thiemann, R. W. Schunk.
Second symposium on plasma double layers and related topics, p. 278 – 283 (1984). Abstr. in Phys. Abstr., Vol. 88, No. 1255, Entry 43202 (1985). – See Abstr. 012.075.

062.171 Shock formation and decay in hydrodynamic flows with mass loading.
N. I. Komle, H. I. M. Lichtenegger.
Comput. Phys. Commun., Vol. 34, No. 1 – 2, p. 47 – 55 (1984). Abstr. in Phys. Abstr., Vol. 88, No. 1255, Entry 45890 (1985).

062.172 Review of the critical ionization velocity effect in space.
P. T. Newell.
Rev. Geophys., Vol. 23, No. 1, p. 93 – 104 (1985).

Laboratory experiments have shown under a variety of conditions that when a neutral gas passes through a magnetized plasma with a relative velocity perpendicular to the magnetic field that is greater than a critical velocity, anomalously high ionization of the neutrals occurs. The conditions under which the same effect is to be expected in space plasmas is still unclear. The experimental evidence for the occurrence of the critical ionization velocity effect in space is summarized, and various areas in which it has been proposed that the effect should be significant are discussed.

062.173 Solitons in weakly nonlinear electron–positron plasmas and pulsar microstructures.
U. A. Mofiz, U. de Angelis, A. Forlani.
Phys. Rev. A, Vol. 31, No. 2, p. 951 – 955 (1985). Abstr. in Phys. Abstr., Vol. 88, No. 1257, Entry 54444 (1985).

062.174 Nonlinear dynamos: a complex generalization of the Lorenz equations.
C. A. Jones, N. O. Weiss, F. Cattaneo.
Physica D, Vol. 14D, No. 2, p. 161 – 176 (1985). Abstr. in Phys. Abstr., Vol. 88, No. 1257, Entry 57711 (1985).

062.175 Dissipation by thermal forces in quantum plasmas.
R. R. Burman, D. E. McClelland.
J. Plasma Phys., Vol. 32, Part 3, p. 369 – 385 (1984). Abstr. in Phys. Abstr., Vol. 88, No. 1258, Entry 60290 (1985).

062.176 A parametric survey of the first critical Mach number for a fast MHD shock.
J. P. Edmiston, C. F. Kennel.
J. Plasma Phys., Vol. 32, Part 3, p. 429 – 441 (1984). Abstr. in Phys. Abstr., Vol. 88, No. 1258, Entry 60301 (1985).

062.177 Spontaneous reconnection of magnetic field lines in a collisionless plasma.
A. A. Galeev.
Handbook of plasma physics. Vol. 2: Basic plasma physics II, p. 305 – 335 (1984). Abstr. in Phys. Abstr., Vol. 88, No. 1258, Entry 60309 (1985). – See Abstr. 003.033.

062.178 Longitudinal friction and intermediate Mach number collisionless transverse magnetosonic shocks.
W. M. Manheimer, D. S. Spicer.
Phys. Fluids, Vol. 28, No. 2, p. 652 – 665 (1985). Abstr. in Phys. Abstr., Vol. 88, No. 1258, Entry 60373 (1985).

062.179 The X–radiation of optically thick plasma: calculation of the comptonization contribution.
S. R. Kelner, Yu. D. Kotov, E. S. Shihovtseva
(*E. S. Shikhovtseva*).
18th International Cosmic Ray Conference, Vol. 1, p. 16 – 19 (1983). – See Abstr. 012.096.

062.180 Coulomb losses in first–order Fermi shock acceleration.
D. C. Ellison.
18th International Cosmic Ray Conference, Vol. 4, p. 1 (1983). Abstract. – See Abstr. 012.096.

062.181 Further properties of the motion of an electron in a neutral sheet.
S.–y. An.
18th International Cosmic Ray Conference, Vol. 4, p. 223 – 226 (1983). – See Abstr. 012.096.

This paper is concerned with the properties of the motion of an electron in the neighbourhood of a neutral sheet, in their dependence upon the sheet thickness. It is shown that some new quantum states of momentum may be allowed for larger sheet thickness, and that the state density of momentum space tends toward a certain distribution with increasing the sheet thickness.

062.182 MHD beam solitons: incompressible flow.
R. L. Fiedler.
Bull. Am. Astron. Soc., Vol. 17, No. 2, p. 578 (1985). Abstract. – See Abstr. 010.065.

062.183 Three dimensional hydrodynamic structure of a supersonic jet.
C. N. Arnold.
Bull. Am. Astron. Soc., Vol. 17, No. 2, p. 578 (1985). Abstract. – See Abstr. 010.065.

062.184 3–dimensional models of the evolution of radio jets.
J. J. Mitteldorf, P. J. Wiita.
Bull. Am. Astron. Soc., Vol. 17, No. 2, p. 609 (1985). Abstract. – See Abstr. 010.065.

062.185 Solutions for magnetic fields and surface electric fields produced by charged particle beams in which electrons (or electrons and positrons) have drift velocities with respect to protons.
W. K. Rose.
Bull. Am. Astron. Soc., Vol. 17, No. 2, p. 609 (1985). Abstract. – See Abstr. 010.065.

062.186 A variational approach to the non–linear force–free magnetic fields.
G. A. Gary.
Bull. Am. Astron. Soc., Vol. 17, No. 2, p. 641 (1985). Abstract. – See Abstr. 010.067.

062.187 Calculation on nonlinear force–free magnetic fields.
P. A. Sturrock, W. Yang.
Bull. Am. Astron. Soc., Vol. 17, No. 2, p. 641 (1985). Abstract. – See Abstr. 010.067.

062.188 Viscous dissipation in the tearing instability.
T. Tachi, R. S. Steinolfson.
Bull. Am. Astron. Soc., Vol. 17, No. 2, p. 645 (1985). Abstract. – See Abstr. 010.067.

062.189 Almost parallel electromagnetic wave propagation in a hot anisotropic plasma.
S. S. Sazhin.
J. Atmos. Terr. Phys., Vol. 47, No. 6, p. 517 – 522 (1985).

When using the approach previously suggested by Sazhin et al. (1981) a formula is derived for almost parallel R and L wave propagation in a hot anisotropic plasma. It is further analysed for magnetospheric conditions.

062.190   Particle acceleration in space and laboratory plasmas.
R. Bingham, D. A. Bryant, D. S. Hall.
RAL–84–057. Rutherford Appleton Laboratory, Chilton, Didcot, Oxon, OX11 0QX, England, 17 pp. (1984). Presented at the Canadian Association of Physicists Annual Meeting, Sherbrook, Quebec, Canada, 18 June 1984.

The general principle of charged particle acceleration in space and laboratory plasmas is illustrated by a discussion of particular types of acceleration mechanisms which can be classified as either deterministic processes or stochastic processes. Acceleration by parallel electric fields, produced in double layers is an example of a deterministic process. Fermi acceleration and acceleration by turbulent wave fields are examples of stochastic processes. The physical acceleration mechanism involved in each type of process is discussed and examples given for space and laboratory plasmas.

062.191   The calculation of the opacity of hot dense plasmas.
S. J. Rose.
RAL–85–022, Rutherford Appleton Laboratory, Chilton, Didcot, Oxon, OX11 0QX, England, 20 pp. (1985). Invited paper at the Second International Conference on the Radiative Properties of Hot Dense Matter.

A review of the calculation of the radiative opacity of hot dense material is presented. An attempt is made to describe some of the approximations and assumptions implicit in current calculations and to indicate areas in which improvements may be made.

062.192   Simulations of high–Mach–number collisionless perpendicular shocks in astrophysical plasmas.
K. B. Quest.
Phys. Rev. Lett., Vol. 54, No. 16, p. 1872 – 1874 (1985).

A problem of critical importance to space physics and astrophysics is the existence and properties of high–Mach–number shocks. The author presents the preliminary results of a simulation of a perpendicular shock with Alfvén Mach number 22. He shows that for sufficiently small electron resistivity the dissipation for this shock is provided by a periodic, rather than time–stationary, reflection of ions. The author discusses the problem of electron heating and the extension to higher Mach numbers.

062.193   Periodic and aperiodic dynamo waves.
N. O. Weiss, F. Cattaneo, C. A. Jones.
Geophys. Astrophys. Fluid Dyn., Vol. 30, No. 4, p. 305 – 341 (1984).

In order to show that aperiodic magnetic cycles, with Maunder minima, can occur naturally in nonlinear hydromagnetic dynamos, the authors have investigated a simple nonlinear model of an oscillatory stellar dynamo. Including the nonlinear interaction between the magnetic field and the velocity shear results in a system of seven coupled nonlinear differential equations.

062.194   Axisymmetric flow around a gravitating body.
H. Takeda, E. Shima, T. Matsuda.
Theoretical aspects on structure, activity, and evolution of galaxies: III, p. 67 – 73 (1985). – See Abstr. 012.110.

Gravitational interaction between a cosmic object and the surrounding medium has been a matter of interest for astronomers. To find the essential points of the gravitational interactions and hydrodynamical characteristics, the authors restrict themselves to axisymmetric flow around a gravitating body and assume the medium to be a perfect ideal gas. They briefly review former works and show the results of their computation.

062.195   Constraints on the properties of bent beams.
C. P. O'Dea.
Physics of energy transport in extragalactic radio sources, p. 64 – 75 (1984). – See Abstr. 012.116.

The physics of bent radio luminous plasma beams is reviewed and constraints on the momentum and kinetic energy flux are examined. Expressions for the bulk velocity, particle density, and efficiency of conversion of bulk kinetic and internal energy into radio luminosity are given. VLA data on the intensity ratios of opposing jets in a sample of narrow angle tail (NAT) sources are used to set an upper limit of $0.2\,c$ to the bulk velocity of the beams/plasmons. Within the context of models for NATs, order of magnitude estimates are made of the bulk velocities, particle densities, efficiencies and mass loss rates of the beams/plasmons in 19 NATs.

062.196   Linear analysis of jet stability.
P. E. Hardee.
Physics of energy transport in extragalactic radio sources, p. 144 – 149 (1984). – See Abstr. 012.116.

The linear stability analysis of jets with cylindrical cross section is reviewed. In a linear stability analysis perturbations to a jet are analyzed in terms of Fourier components which in cylindrical geometry split into Fourier modes with different configuration. In fact it is the pinch Fourier mode whose effects are directly observable in numerical models of axisymmetric jets and whose effects may be responsible for knots in the structure of some extragalactic jets. It also seems likely that the twisting seen in many extragalactic jets which allow non–axisymmetric Fourier modes can be described by the propagation properties of the twist Fourier mode.

062.197   Knot production and jet disruption via nonlinear Kelvin–Helmholtz pinch instabilities.
M. L. Norman, K.–H. A. Winkler, L. Smarr.
Physics of energy transport in extragalactic radio sources, p. 150 – 167 (1984). – See Abstr. 012.116.

The authors summarize the results of time–dependent numerical hydrodynamical simulations which investigate the behavior of pinching modes of the Kelvin–Helmholtz instability in the nonlinear regime. Ordinary mode pinch instabilities, important only at Mach numbers of order unity, are shown to be disruptive. Reflection mode pinch instabilities, important in supersonic jets with parameters of astrophysical interest, are shown to be not disruptive, but rather saturate at finite amplitude through the formation of oblique internal shockwaves. The effect of pinch instabilities on energy transport in extragalactic jets is discussed.

062.198   Why dominant cluster jets are different.
D. M. Sumi, L. L. Smarr.
Physics of energy transport in extragalactic radio sources, p. 168 – 181 (1984). – See Abstr. 012.116.

The authors have discussed possible atmospheres a jet might encounter as it emerges from an elliptical galaxy. Wind type atmospheres tend to keep a jet in its initial shock regime. Cooling inflow atmospheres will tend to drive a jet toward the planar shock regime and degrading it into a subsonic flow. This type of jet instability may make it possible to understand the observed association between short jet – hotspot – plume sources and dominant central galaxies.

062.199   Vortex–rings in extended doubles.
L. Rudnick.
Physics of energy transport in extragalactic radio sources, p. 182 – 184 (1984). – See Abstr. 012.116.

Vortex rings appear to be a commonly excited mode in the fluid transport of relativistic material to radio source lobes. They may serve as both passive probes of the transport physics and as active dynamical contributors to the source evolution.

062.200   Current–carrying jets.
G. Benford.
Physics of energy transport in extragalactic radio sources, p. 185 – 192 (1984). – See Abstr. 012.116.

The best evidence for current–carrying jets comes from a need to invoke a B_θ to confine high–pressure jets. One can learn how such jets form by studying the electrodynamics of existing laboratory beams. The author proposes that magnetically dominated nozzles account for Bridle's observed correlations in weak vs. strong jets. If knots and hot spots emit through coherent plasma mechanisms, this greatly relieves the pressure problem. This radical suggestion might be testable for small, nearby lobes such as those of Sco X–1. Observations of amplitude jitter and long autocorrelation times would be clear evidence of non–synchrotron emission.

062.201 Formation and propagation of magnetized radio jets.
J. Siah, P. J. Wiita.
Physics of energy transport in extragalactic radio sources,
p. 193 – 199 (1984). – See Abstr. 012.116.
Numerical models of the boundaries of radio jets consisting of
a relativistic fluid flowing into a confining gas cloud are discussed. In these models a fixed fraction of the total energy of the
jets is assumed to be in the form of ordered magnetic fields.

062.202 Force–free equilibria of magnetized jets.
A. Königl.
Physics of energy transport in extragalactic radio sources,
p. 200 – 201 (1984). Abstract. – See Abstr. 012.116.

**062.203 Collimation of intermediate scale motion entrained by
nuclear jets.**
R. N. Henriksen.
Physics of energy transport in extragalactic radio sources,
p. 211 – 215 (1984). – See Abstr. 012.116.
The relation between the VLBI nuclear jets and the much
larger scale VLA jets remain a crucial question. The author
adopts the view that the VLBI jets are the "prime movers" for the
larger scale VLA motion. They may not in fact be the "first jets"
to the VLA and single dish "last jets", but they are sufficiently
different in scale to be regarded as a major step along the causal
chain.

062.204 The physics of particle acceleration in radio galaxies.
J. A. Eilek.
Physics of energy transport in extragalactic radio sources,
p. 216 – 228 (1984). – See Abstr. 012.116.
Particle acceleration in the diffuse plasma environment of extragalactic radio sources probably occurs through turbulence
and/or shocks in the plasma. This paper reviews the basic physics
of each process and attempts some discussion of the efficiency,
spectral output and sites of occurrence of each process.

**062.205 The relationship between laboratory and astrophysical
jets.**
G. V. Bicknell.
Physics of energy transport in extragalactic radio sources,
p. 229 – 244 (1984). – See Abstr. 012.116.
The author presents the view, first suggested by Fanti et al.
(1982) that the slow surface brightness decline observed in a
number of jets is due to adiabatic processes associated with entrainment. He presents arguments that the jets in the edge-
darkened (Fanaroff–Riley Class I) sources are of low Mach number along a large fraction of their lengths, that they are largely
non–dissipative (except near the cores), and that their morphology and surface brightness is governed by turbulent processes.

**062.206 Global invariants of a mean fluid flow and local turbulent
substructure.**
R. N. Henriksen.
Physics of energy transport in extragalactic radio sources,
p. 245 – 254 (1984). – See Abstr. 012.116.
Fluid turbulence is presumably fully described by exact solutions of the Navier–Stokes equations which contain a sufficient
array of spatial and temporal scales. Unfortunately such solutions are not known and if they were known they would be so
complex in general as to be useless in practice. There is one
exception to this state of affairs, which is to be the subject of this
paper, and that is when the various scales are coupled by a
renormalizing constraint such as local self–similarity. The real
significance of the discovery of the universality of the Kolmogorov Cascade, is that such coupling does occur, at least asymptotically.

**062.207 Magnetic energy dissipation as the source of synchrotron
emission in jets.**
A. Königl.
Physics of energy transport in extragalactic radio sources,
p. 260 – 264 (1984). – See Abstr. 012.116.
It is proposed that the dissipation of internal magnetic energy,
rather than of bulk kinetic energy, could be the main source of

power for the synchrotron emission in certain radio jets. This
possibility arises naturally in the context of the force–free model
of magnetized jets from the requirement that the outward-
convected field adjust continuously to maintain a minimum-
energy configuration. The rate of energy dissipation calculated
from this model is shown to depend only on the nonaxisymmetric component of the magnetic field. A rough estimate
of this rate in the inner jet of NGC 6251 is found to be consistent
with the observations.

**062.208 Laboratory electron beam simulation of cosmic radio
jets.**
R. G. Spulak Jr., J. O. Burns.
Physics of energy transport in extragalactic radio sources,
p. 265 – 271 (1984). – See Abstr. 012.116.
Astrophysical jets are injections of particles, magnetic fields,
and possibly large–scale currents into a background plasma, the
intergalactic medium. In principle, they are therefore somewhat
similar to the laboratory injection of an electron beam into a
plasma. The authors consider scaling between the astrophysical
and laboratory cases and discuss the similarities and differences
between them. They discuss what aspects of the physics of astrophysical jets might be investigated with electron beam experiments; in general, the laboratory will be most useful to study
purely electrodynamic effects.

Foundations of radiation hydrodynamics.
See Abstr. 003.025.

Handbook of plasma physics. Vol. 2: Basic plasma physics II.
See Abstr. 003.033.

Magnetic fields in space.
See Abstr. 003.047.

Handbook of plasma physics. Vol. 1: Basic plasma physics I.
See Abstr. 003.094.

Wave phenomena in the ionosphere and in cosmic plasma.
See Abstr. 003.096.

Radiation of atoms in cosmic plasma.
See Abstr. 003.149.

**Waves in space plasmas: highlights of a conference held in Hawaii,
7 – 11 February 1983.**
See Abstr. 011.007.

Perspectives on space and astrophysical plasma physics.
See Abstr. 013.001.

Report of ESA's topical team on space plasma physics.
See Abstr. 013.074.

Adaptive mesh techniques for fronts in star formation.
See Abstr. 021.002.

**Energy levels in hydrogen plasmas and the Planck–Larkin partition
function – a comment.**
See Abstr. 022.052.

Line formation in laboratory plasmas.
See Abstr. 022.186.

**Fission sequences of self–gravitating and rotating fluid with inter-
nal motion.**
See Abstr. 042.001.

Energy loss by slow magnetic monopoles in a thermal plasma.
See Abstr. 061.019.

Redistribution functions for astrophysical problems.
See Abstr. 063.001.

Cyclotron line profiles and photon diffusion in a hot plasma.
See Abstr. 063.021.

Spectral line inversion as a diagnostic tool.
See Abstr. 063.025.

The anisotropic radiative transfer problem in optically thick, strongly magnetized plasma: a comparison of results.
See Abstr. 063.044.

Induced emission of radiation near $2\omega_e$ by a synchrotron–maser instability.
See Abstr. 063.083.

A self–consistent model of the magnetosphere with centrifugal wind. I.
See Abstr. 064.007.

A self–consistent model of the magnetosphere with centrifugal wind. II.
See Abstr. 064.008.

Magnetic stellar winds. A 2–D generalization of Weber–Davis model.
See Abstr. 064.061.

Euler potential method in three–dimensional stellar wind problems.
See Abstr. 064.077.

Three–dimensional structures of magnetostatic atmospheres. I. Theory.
See Abstr. 064.080.

QUIP: a time–dependent standard model.
See Abstr. 065.031.

Numerical simulations of stellar convective dynamos. II Field propagation in the convection zone.
See Abstr. 065.045.

A general computational method for obtaining equilibria of self–gravitating and rotating gases.
See Abstr. 065.050.

An Eulerian variational principle and a criterion for the occurrence of nonaxisymmetric neutral modes along rotating axisymmetric sequences.
See Abstr. 065.059.

Density bifurcation in a homogeneous isotropic collapsing star.
See Abstr. 065.084.

Non–static fluid spheres in general relativity.
See Abstr. 066.032.

Magnetohydrodynamical model of pulsar rotation. Exact periodic solutions.
See Abstr. 067.030.

Analytic structure of cosmic radio jets: a preliminary investigation.
See Abstr. 067.054.

On the possibility of numerical modelling of MHD–processes in solar active regions. I.
See Abstr. 072.113.

On the possibility of numerical modelling of MHD–processes in solar active regions. II.
See Abstr. 072.117.

A unified treatment of the filament and flare instabilities.
See Abstr. 073.006.

Energy dissipation mechanisms in the solar corona.
See Abstr. 074.002.

A unified theory of coronal heating.
See Abstr. 074.003.

Velocity shear instabilities in the anisotropic solar wind.
See Abstr. 074.004.

Alfvén waves in a nonuniform layer in a gravity field.
See Abstr. 074.159.

Numerical simulations of stellar convective dynamos. III. At the base of the convection zone.
See Abstr. 075.041.

Quenching of the beam–plasma instability by large–scale density fluctuations in 3 dimensions.
See Abstr. 077.027.

Magnetohydrodynamical model of some instationary phenomena on the sun.
See Abstr. 080.033.

Non–stochastic acceleration of protons in the magnetic neutral sheet.
See Abstr. 084.003.

Particle simulation in a dipole field.
See Abstr. 084.115.

Evolution of ion holes and large potential structures in simulations of bounded current–driven magnetized plasma systems.
See Abstr. 084.116.

A simple model of convection in the terrestrial planets.
See Abstr. 091.065.

Similarity solutions for the structure of supernova blast waves driven by clumped ejecta. I. Undecelerated clumps.
See Abstr. 125.045.

Self–consistent models for the X–ray emission from supernova remnants: an application to Kepler's remnant.
See Abstr. 125.046.

Disc–like magneto–gravitational equilibria.
See Abstr. 131.014.

Classical thermal evaporation of clouds: an electrostatic analogy.
See Abstr. 131.156.

Predicted long–slit, high–resolution emission–line profiles from interstellar bow shocks.
See Abstr. 131.299.

Cosmic ray generation by plasma turbulence – development of a new paradigm in physics.
See Abstr. 144.038.

Application of a new gasdynamics code to gas flow problems in disk galaxies.
See Abstr. 151.008.

Oscillations of rotating gas disks: p–modes, g–modes, and r–modes.
See Abstr. 151.022.

Rossby solitons in a galactic disk.
See Abstr. 157.116.

Is the jet in M87 magnetically confined?
See Abstr. 158.013.

A model of the polarization position–angle swings in BL Lacertae objects.
See Abstr. 158.072.

Hotspots in radio galaxies: a comparison with hydrodynamic simulations.
See Abstr. 158.089.

Jets in galaxies.
See Abstr. 158.120.

Extragalactic variable radio sources.
See Abstr. 158.248.

Turbulence, entrainment and magnetic fields.
See Abstr. 158.341.

A theorist's perspective – I.
See Abstr. 158.345.

A theorist's perspective – II.
See Abstr. 158.346.

063 Radiative Transfer, Scattering

063.001   Redistribution functions for astrophysical problems.
C. Magnan.
Astron. Astrophys., Vol. 142, No. 1, p. 117 – 123 (1985).

Since the general problem of frequency redistribution is almost impossible to solve exactly, the author offers an approximation that appears convenient for astrophysical plasmas. This approximation takes advantage of the fact that the actual line broadening is always much larger than the pure thermal Doppler broadening. In such case, it is shown that the thermal velocity distribution of the atoms cannot depend upon their atomic state and is not determined by the line radiative processes. As a result the combination of velocity effects and level population effects becomes simpler, and the approximation leads to useful practical redistribution functions covering all cases of interest.

063.002   Hydrogen lines Stark broadening tables in the presence of a magnetic field.
G. Mathys.
Astron. Astrophys., Suppl. Ser., Vol. 59, No. 2, p. 229 – 253 (1985).

Tables of Stark broadened profiles of the lines $L\alpha$, $L\beta$ and $H\alpha$ of hydrogen, in the presence of a magnetic field are presented. Unpolarized intensities observed in directions parallel (longitudinal profiles) and perpendicular (transversal profiles) to the magnetic field have been computed on the basis of the unified theory, for electron densities higher than or equal to $10^{15}\mathrm{cm}^{-3}$, temperatures ranging from 5×10^3 to 4×10^4K, and magnetic fields up to 4×10^4G.

063.003   Radiative transfer in disk geometry: emergent intensity from cool gray disks.
G. F. Spagna Jr., Chun Ming Leung.
Icarus, Vol. 61, No. 1, p. 27 – 35 (1985). – See Abstr. 012.004.

As part of a continuing project to develop self–consistent radiative transfer solutions in disk geometry, numerical models for disk shaped dust clouds are presented. Models are computed with the assumption of gray (Rosseland mean) opacity for cool, externally heated disks. For a given temperature distribution, based on a reasonable estimate for molecular clouds, or precollapse protoplanetary disks, the effect on the emergent intensity of disk geometry (i.e., degree of flattening), optical depth, and inclination to the observer's line of sight is examined. The goal of this research is to develop a computational tool for interpretation of observations of disk–shaped objects.

063.004   Theory of optical flashes.
R. A. London.
AIP Conf. Proc., No. 115, p. 581 – 589 (1984). – See Abstr. 012.005.

The theory of optical flashes created by X– and γ–ray burst heating of stars in binaries is reviewed. Calculations of spectra due to steady–state X–ray reprocessing and estimates of the fundamental time scales for the non–steady case are discussed. The results are applied to the extant optical data from X–ray and γ–ray bursters.

063.005   Non–thermal synchrotron radiation in a strong magnetic field.
R. W. Bussard.
AIP Conf. Proc., No. 115, p. 611 – 614 (1984). – See Abstr. 012.005.

The synchrotron radiation spectra of mildly relativistic electrons injected into strong fields ($B \sim 10^{12}$G) are calculated by a Monte Carlo technique. The energy losses, extremely rapid for such field strengths, are taken into account self–consistently. For photon energies below 1 MeV, the results are well described by the product of a power law and an exponential, and the exponent and characteristic energy are tabulated with field strength and injection energy.

063.006   Multimode radiative transfer in finite optical media. I. Fundamentals.
R. W. Preisendorfer, G. L. Stephens.
J. Atmos. Sci., Vol. 41, No. 5, p. 709 – 724 (1984). Abstr. in Phys. Abstr., Vol. 88, No. 1248, Entry 9310 (1985).

063.007   Multimode radiative transfer in finite optical media. II. Solutions.
G. L. Stephens, R. W. Preisendorfer.
J. Atmos. Sci., Vol. 41, No. 5, p. 725 – 735 (1984). Abstr. in Phys. Abstr., Vol. 88, No. 1248, Entry 9311 (1985).

063.008   Polarization of intrinsic radiation of tidally distorted stars with electron–scattering atmospheres. I. The case of conservative scattering.
N. G. Bochkarev, E. A. Karitskaya, N. I. Shakura.
Astrophys. Space Sci., Vol. 108, No. 1, p. 1 – 14 (1985).

The variable linear polarization of the intrinsic radiation emitted by tidally distorted stars was computed with approximation of pure electron or Rayleigh scattering in the atmospheres. A precise form of the equipotential surfaces within the binary system Roche model, which approximates the form of the star surface, was taken into account. Calculations were made using the Sobolev (1949) and Chandrasekhar (1950) solutions of transport equations for an optically thick medium without true absorption. The behaviour of the Stokes polarization parameters Q/I and U/I and the dependence of these on the parameters of the model were subjected to detailed investigation. It is concluded that the amplitude of polarization variations is small, not generally exceeding 0.4%; however, in a number of cases it remains at the observable level.

063.009   Distribution of polarization and intensity of radiation across the stellar disk and numerical values of atmospheric characteristics governing this distribution.
N. G. Bochkarev, E. A. Karitskaya, N. A. Sakhibullin.
Astrophys. Space Sci., Vol. 108, No. 1, p. 15 – 29 (1985).

Computations of polarization and intensity of radiation from a unit stellar surface area are presented, as well as a study of the numerical characteristics of atmospheres – single–scattering albedo Ω_λ and the initial source function $B_\lambda(\tau_\lambda)$, which define the polarization behaviour of atmospheres. The models of stellar atmospheres presented by Kurucz et al. (1974) and Kurucz (1979) have been used for calculations. The plot of Ω_λ is characterized by discontinuities at the boundaries of spectral series for hydrogen and, sometimes, for helium. Maximum Ω_λ are attained in the Lyman region of $\lambda = 912 - 1200$ Å. For stars with $T_{\rm eff} \gtrsim 35000$K, high values of Ω_λ also are attained for $\lambda < 912$ Å. Within the infrared region, Ω_λ is always small because of bremsstrahlung absorption. A rapid growth of the source function B_λ with the optical depth τ_λ, typical for ultraviolet range, together with high values of Ω_λ results in the strong polarization of emission from a unit stellar surface element. For longer wavelengths the plane of polarization abruptly turns 90° in the central parts of the visible stellar disk.

063.010   The transfer of polarized radiation in spectral lines: formalism and solutions in simple cases.
S. J. McKenna.
Astrophys. Space Sci., Vol. 108, No. 1, p. 31 – 66 (1985).

A general treatment of the transfer of polarized radiation in spectral lines assuming a Rayleigh phase function and a general law of frequency redistribution is derived. It is shown how nine families of coupled integral equations for the moments of the radiation field arise which are necessary to fully describe the state of polarization of the emergent radiation from a plane–parallel, semi–infinite atmosphere. The special case of angle independent redistribution functions is derived from the general formalism, and it is shown how the nine families of integral equations reduce to the six linearly independent integral equations derived by Collins (1972). To serve as a test of the formulation, solutions for isothermal atmospheres are given.

063.011   Solution of radiative transfer equation with spherical–symmetry in partially scattering medium.
A. Peraiah, B. A. Varghese.
Astrophys. Space Sci., Vol. 108, No. 1, p. 67 – 80 (1985).

The authors have solved the equation of radiative transfer in spherical symmetry with scattering and absorbing medium. They have set the albedo for single scattering to be equal to 0.5. They have set the Planck function constant throughout the medium in one case and in another case the Planck function has been set to vary as r^{-2}. The geometrical extension of the spherical shell has been taken as large as one stellar radius. Two kinds of variations of the optical depth are employed (1) that remains constant with radius and (2) that varies as r^{-2}. In all these cases the internal source vectors and specific intensities change depending upon the type of physics employed in each case.

063.012   Polarization of intrinsic radiation of tidally distorted stars with electron–scattering atmospheres. II. An account of true absorption.
N. G. Bochkarev, E. A. Karitskaya.
Astrophys. Space Sci., Vol. 109, No. 1, p. 1 – 14 (1985).

Linear polarization of radiation emitted by tidally distorted stars as a function of the binary system phase is computed, taking into account true absorption and the scattering of light on free and bound electrons within hot stellar atmospheres. Computations are made both for the linear distribution of true sources across the atmospheres and for radiative–stable model atmospheres presented by Kurucz et al. (1974) and Kurucz (1979). Polarization variability was investigated as a function of wavelength λ. In a number of cases, polarization variability was found to be at an observable level. For fairly long waves where the limb–darkening coefficient falls below a certain critical value $u_{\rm cr} \sim 0.5$, the plane of polarization is found to be turned by 90° as compared to the case of a pure electron atmosphere. For limb–darkening coefficients far from the value of $u_{\rm cr}$ the form of the polarization phase curves, as well as dependence on the parameters of a binary system, remain approximately the same as those in the case of pure electron scattering.

063.013   A new representation of Güttler's theory of electromagnetic scattering by a composite sphere.
G. A. Shah.
Astron. Nachr., Vol. 306, No. 2, p. 63 – 66 (1985).

The multipole expansion coefficients for electromagnetic scattering by a stratified core–mantle sphere, suitable for fast and accurate evaluation of the radiation scattering parameters, have been recast.

063.014   A viable mechanism to establish relativistic thermal particle distribution functions in cosmic sources.
R. Schlickeiser.
Astron. Astrophys., Vol. 143, No. 2, p. 431 – 434 (1985).

It is shown that the combination of first and second order Fermi acceleration with synchrotron and inverse Compton radiation losses explains the existence of relativistic Maxwellian electron distribution functions in cosmic objects.

063.015   Théorie de la polarisation des raies: interaction matière–rayonnement en présence de champ magnétique et de collisions.
S. Sahal–Bréchot.
Champs magnétiques stellaires, p. 5 – 78 (1984). – See Abstr. 012.022.

063.016   Transfert de rayonnement en lumière polarisée. Application à la mesure des champs magnétiques par effet Zeeman.
M. Semel.
Champs magnétiques stellaires, p. 79 – 96 (1984). – See Abstr. 012.022.

063.017   Méthodes de mesure des champs magnétiques: mérites respectifs de l'effet Zeeman et de l'effet Hanle.
J.–L. Leroy.
Champs magnétiques stellaires, p. 97 – 105 (1984). – See Abstr. 012.022.

063.018   Hyperfine selective collisional excitation of interstellar molecules.
J. Stutzki, G. Winnewisser.
Astron. Astrophys., Vol. 144, No. 1, p. 1 – 12 (1985).

The authors discuss in detail the hyperfine selective collisional excitation of linear rotor and symmetric top molecules and show that the commonly used simpler approximations for handling the hyperfine selective collisional excitation are not sufficient. The scattering process is described in the framework of formal scattering theory, which also allows for a concise derivation of the CS and IOS approximations. The authors give numerical results for the scattering of para NH_3/He and HCN/He. The correct description of the hyperfine selective collisional excitation is important for the interpretation of radioastronomical observations of these molecules, which show that the hyperfine states are not thermally populated in many molecular clouds.

063.019   On the interpretation of hyperfine–structure intensity anomalies in the NH_3 $(J, K) = (1, 1)$ inversion transition.
J. Stutzki, G. Winnewisser.
Astron. Astrophys., Vol. 144, No. 1, p. 13 – 26 (1985).

For NH_3 the authors present a simple radiative transfer model with selective trapping in the far IR $(J, K) = (2, 1) \rightarrow (1, 1)$ transition, which together with hyperfine selective collisional excitation is able to reproduce the observed hyperfine intensity anomalies in the molecules $(J, K) = (1, 1)$ inversion transition. The model leads to the interpretation, that the anomalous hyperfine intensities emitting clouds are composed of many small $(10^{-2}$pc), high density $(10^6 - 10^7 {\rm cm}^{-3})$ clumps of $0.3 - 1\ M_\odot$, which may be the precursors of low mass star formation.

063.020 Line formation in the presence of turbulence.
C. Magnan.
Astron. Astrophys., Vol. 144, No. 1, p. 186 – 190 (1985).

The problem of line formation in the presence of a random velocity field only requires direct computation of the mean value of the radiative intensity. The author shows how the layer addition formalism both provides a physical solution to this problem and can deal with any spatially correlated velocity field. Specific formulae are derived for the LTE case that agree with previous classical results. A solution to the non–LTE case is given in terms of an approximation that extends a solution already proposed in the rigid cell model.

063.021 Cyclotron line profiles and photon diffusion in a hot plasma.
J. Ventura, M. Soffel, H. Herold, H. Ruder.
Astron. Astrophys., Vol. 144, No. 2, p. 479 – 484 (1985).

The structure of cyclotron line profiles in approximate treatments of radiative transfer in strongly magnetized ($B \sim 10^{12}$G) plasma systems is discussed. It is shown how the uncritical use of scattering and absorption cross sections in coherent transfer calculations may lead to unphysical cyclotron line structures. A new concept for the treatment of coherent radiative transfer in the frame of the diffusion approximation is introduced. By means of this concept the principle of micro–reversibility can be ensured and physically reasonable photon fluxes emitted e.g. from the accretion column of X–ray pulsars can be obtained.

063.022 Non–LTE line transfer with partial redistribution. II. An equivalent–two–level–atom approach.
I. Hubený.
Bull. Astron. Inst. Czech., Vol. 36, No. 1, p. 1 – 9 (1985).

A formulation of equations for radiative transfer in a gas of multilevel atoms, taking into account recent developments of theoretical description, is presented. It is shown that for a simple case where one chosen transition is allowed to depart from complete redistribution, the global multilevel problem may be solved by suitably modified complete–redistribution numerical techniques. In particular, the author has formulated a modification of the equivalent–two–level–atom approach that enables a multilevel transfer to be solved by a simple iteration scheme. Various approximate forms of the line source function are also discussed.

063.023 On asymptotic formulae of the theory of radiative transfer in a sphere and a spherical envelope.
A. K. Kolesov.
Astrofizika, Tom 22, Vyp. 1, p. 177 – 188 (1985). In Russian. English translation in Astrophysics, Vol. 22, No. 1.

Integral constraints for the radiation intensities in homogeneous, absorbing and anisotropically scattering media with radial symmetry are obtained. Asymptotic formulae for the radiation intensities in the outer layers of a sphere of large optical radius and an optically thick spherical envelope are found by means of the integral constraints and physical considerations. Cases of a central point source, uniformly distributed sources and conic sources on the outer boundary surface of the media are considered.

063.024 Circular polarization of interstellar absorption lines at radio frequencies.
S. Deguchi, W. D. Watson.
Astrophys. J., Vol. 289, No. 2, p. 621 – 629 (1985).

The formation of linearly and circularly polarized line radiation at radio frequencies is investigated when radiation from a distant continuum source propagates through a diffuse gas such as the interstellar medium. Radiative transfer equations for the Stokes parameters (I, Q, U, and V) are obtained for the commonly encountered circumstance in which the Zeeman splitting of the particle (atoms or molecules) states is much greater than the inverse lifetime of the state, but much less than the breadth of the line. If the particles are partially aligned, magnetorotation converts linear polarized radiation into circularly polarized radiation. The authors perform numerical calculations for a static, homogeneous gas with a uniform magnetic field and for a static, homogeneous gas with a twisted magnetic field.

063.025 Spectral line inversion as a diagnostic tool.
A. Kavetsky, B. J. O'Mara.
J. Quant. Spectrosc. Radiat. Transfer, Vol. 33, No. 2, p. 93 – 100 (1985).

A general technique for inverting spectral line profiles to obtain physical information about a semi–infinite plasma is described. The technique allows for distinct line and continuum source functions, the presence of both stochastic and systematic velocity fields, and permits all spectral lines in a mutiplet to be inverted simultaneously.

063.026 Multiple scattering in a two–dimensional rectangular medium exposed to collimated radiation.
A. L. Crosbie, R. G. Schrenker.
J. Quant. Spectrosc. Radiat. Transfer, Vol. 33, No. 2, p. 101 – 125 (1985).

Two–dimensional multiple scattering in a rectangular medium exposed to uniform collimated radiation is studied using an exact radiative transfer formulation. The two–dimensional integral equation is solved numerically by subtracting out the singularity of the kernel. Graphical and tabulated results of the flux and scattered intensity at the boundary of the rectangular medium, as well as the source function, are presented for a wide variety of geometries exposed to normal incident, collimated radiation. Comparisons are made to existing solutions of two–dimensional radiative transfer in a one–dimensional geometry. Results of the source function using the first term of a Taylor series expansion are also presented. A method is presented for extending these results to the problem of a strongly anisotropic scattering phase function which is made up of a spike in the forward direction superimposed on an isotropic phase function.

063.027 Efficiency of optical pumping in layered media.
P. B. Kunasz.
J. Quant. Spectrosc. Radiat. Transfer, Vol. 33, No. 2, p. 155 – 165 (1985).

In pursuit of a mechanism to produce a population inversion at far ultraviolet or X–ray wavelengths, the author has carried out detailed calculations for optical pumping of a target resonance line by radiation in a second, nearly coincident, resonance line belonging to a different species.

063.028 Solutions of some problems in radiative transfer.
T. Viik.
Tartu Astrofüüs. Obs. Publ., Tom 50, p. 38 – 49 (1984). In Russian.

The solution of some model problems for a semi–infinite homogeneous isotropically scattering plane–parallel medium can be written explicitly in a compact form if one defines two functions h and g using the resolvent function. The differential and integral equations for the h and g functions are derived and the first moments of the g function, which are essentially the moments of the point–direction gain in the sense of van de Hulst, are found.

063.029 Intensity of radiation reflected by a vertically inhomogeneous medium through multiple scattering.
K. Kawabata, R. Hirata.
Astrophys. Space Sci., Vol. 109, No. 2, p. 345 – 356 (1985).

To facilitate the computation of the radiative intensity reflected, upon multiple scattering, by a vertically inhomogeneous medium, an implicit formula for integrating the invariant imbedding equation for Fourier–decomposed reflection function is derived starting with a formal solution. The integration involved in the formal solution is then carried out analytically, yielding a corrector–type formula for finding the reflection function at each step of τ (optical thickness). It is expected that this formula is capable of handling general cases of inhomogeneous media where both single–scattering albedo and phase function are allowed to vary continuously with height. Similar, but 'explicit' expressions are also derived for the single and the second–order scattering solutions, with which the higher–order Fourier terms of reflection function are to be approximated.

063.030 A treatment of opacity suitable for media of low density and temperature. I.
R. Capuzzo–Dolcetta, A. Di Fazio, F. Palla.
Astron. Astrophys., Vol. 145, No. 2, p. 290 – 295 (1985).

The paper is concerned with the determination of mean absorption coefficients for the most important physical processes, suitable for a wide range of physical conditions. In particular, they do not need any particularly restrictive assumption about the thermodynamical state of matter, and about the coupling between matter and radiation. All these reasons, and the choice of the Planck mean, make the authors' mean absorption coefficients especially useful in those particular low density and temperature conditions, that are typical of protogalaxies and protostars. Furthermore, the opacities are given in an explicit functional form which allows to avoid interpolation of large multi–parameter tables.

063.031 A modified Rybicki method and the partial coherent scattering approximation.
I. Hubený.
Astron. Astrophys., Vol. 145, No. 2, p. 461 – 474 (1985).

It is the purpose of the paper to demonstrate that it is possible to combine the advantages of the partial coherent scattering (PCS) approximation with the favorable numerical properties of the Rybicki method. To accomplish this goal, the author carries out an analysis which indicates the way of accounting for Doppler diffusion within the framework of the PCS approximation. The essence of the present method consists in considering the frequency that sets the boundary between the complete redistribution core region and the coherent wing region to be depth–dependent. From the mathematical point of view, the present modification is purely at the computational level and thus retains the ease of formulation and programming of the original Rybicki method.

063.032 Numerical solution of the radiative transfer equation in a magnetized medium.
K. N. Nagendra, A. Peraiah.
Mon. Not. R. Astron. Soc., Vol. 214, No. 2, p. 203 – 218 (1985).

A numerical method of solution based on the discrete space theory of radiative transfer as applied to the transfer problems in an anisotropic medium is discussed. Two simple applications, namely the scattering in the atmosphere of a hot magnetic white dwarf and in a plasma slab immersed in a superstrong magnetic field are discussed. The normal wave transfer equations for the scattering and absorption of radiation are used for this purpose. The solutions are compared with those obtained for the non–magnetic Thomson scattering in the same media. A comparative study is made of the normal wave and Stokes vector equations for a Zeeman active gas.

063.033 Transfer of line radiation: approximate solutions. Atmospheres of finite optical thickness.
V. M. Serbin.
Astron. Zh., Tom 62, Vyp. 2, p. 272 – 282 (1985). In Russian. English translation in Sov. Astron., Vol. 29, No. 2.

Approximate methods for calculating the line source function in plane homogeneous static atmospheres of finite optical thickness are studied. Complete frequency redistribution is assumed.

063.034 On a long–standing problem of the radiative transfer theory.
V. V. Ivanov.
Astron. Zh., Tom 62, Vyp. 2, p. 283 – 290 (1985). In Russian. English translation in Sov. Astron., Vol. 29, No. 2.

The standard problem of the theory of line formation (i.e., isothermal semi–infinite atmosphere, two–level atom, complete frequency redistribution upon scattering, no continuum absorption) is considered.

063.035 Rayleigh scattering in a thin spherical envelope.
J. Freimanis.
Astron. Zh., Tom 62, Vyp. 2, p. 314 – 322 (1985). In Russian. English translation in Sov. Astron., Vol. 29, No. 2.

The model of a geometrically thin homogeneous spherical envelope with an isotropically radiating unpolarized point source in its centre has been discussed. It is assumed that Rayleigh scattering and true absorption take place in the shell. The distribution of Stokes parameters of the outcoming radiation on the surface of the envelope, as well as the intensity and the degree of polarization of the total radiation coming from a partially obscured envelope, has been calculated as a function of the single scattering albedo, optical thickness and degree of obscuration.

063.036 The transfer of line radiation. I. General analysis of approximate solutions.
V. V. Ivanov, V. M. Serbin.
Sov. Astron., Vol. 28, No. 4, p. 405 – 409 (1984). English translation of 38.063.016.

063.037 Polarized light in planetary atmospheres for perpendicular directions.
J. W. Hovenier, J. F. de Haan.
Astron. Astrophys., Vol. 146, No. 1, p. 185 – 191 (1985).

A plane–parallel atmosphere with perpendicularly incident light or scattered light travelling in perpendicular directions is considered. It is shown on the basis of general geometrical arguments that substantial simplifications arise compared with situations in which no perpendicular directions are involved. The azimuth dependence of the radiation field is explicitly derived and shown to require only one or two terms in a Fourier series expansion. When the incident light illuminates the atmosphere perpendicularly, the scattered light travelling in perpendicular directions up and down can be described by matrices which have the same simple form as the scattering matrix of a volume–element has for strictly backward and forward scattering, respectively.

063.038 Radiative transfer equation in spherical symmetry.
A. Peraiah, B. A. Varghese.
Astrophys. J., Vol. 290, No. 2, p. 411 – 423 (1985).

The authors present a numerical solution of the radiative transfer equation in spherically symmetric geometry using integral operators within the framework of the discrete space theory and expressing the specific intensity in terms of the nodal values of the radius–angle mesh. The solution obtained satisfies the following basic tests: (1) the invariance of the specific intensity in a medium in which radiation is neither absorbed nor emitted, (2) the continuity of the solution in both angle and radial distribution, (3) a numerical proof showing the uniqueness of the solution, and (4) the condition of zero net flux in a scattering medium with one boundary having a specular reflector, and global conservation of energy.

063.039 Saturation and beaming in astrophysical masers.
C. Alcock, R. R. Ross.
Astrophys. J., Vol. 290, No. 2, p. 433 – 444 (1985).

The authors have constructed a "four–stream" model system of equations that are closely analogous to the equations of radiative transfer in a maser spectral line and are amenable to numerical solution. This model is used to investigate radiative transfer in asymmetric maser clouds. Competition between different beams of radiation for available pump photons is important in both unsaturated and saturated masers. Consequently, the radiation from a typical cloud will be highly beamed, even if the maser is saturated, because the shape of a typical cloud is not symmetric.

063.040 Excitation of the hyperfine transitions of atomic hydrogen, deuterium, and ionized helium 3 by Lyman–alpha radiation.
S. Deguchi, W. D. Watson.
Astrophys. J., Vol. 290, No. 2, p. 578 – 586 (1985).

The profile of Lyman–alpha radiation in an expanding gas cloud is calculated in detail in order to determine the color temperature of the radiation scattered by an H I atom within the cloud. Scattering of Lyα is likely to dominate in the excitation of the 21 cm transition of H I in astronomical gas clouds with very low densities. The spin temperature will then become equal to this color temperature. The method involves a Sobolov–like treatment and the application of a redistribution function for the scattering that preserves detailed balance when the recoil of the atom is included. Analogous color temperatures are calculated that are relevant for the excitation of the hyperfine transitions of atomic deuterium and ionized ^{3}He.

063.041   Validity of band–model calculations for CO_2 and H_2O applied to radiative properties and conductive–radiative transfer.
A. Soufiani, J. M. Hartmann, J. Taine.
J. Quant. Spectrosc. Radiat. Transfer, Vol. 33, No. 3, p. 243 – 257 (1985).
A previously presented line–by–line (LBL) calculation is used to test the validity of various approximate models. Narrow–band model parameters are generated from the lines used by the LBL approach in the $150 – 8000$ cm^{-1} and $300 – 1500$K ranges. The accuracy of narrow–band models and of approximations for nonuniform paths applied to transmissivities and intensities of columns is studied.

063.042   Integrals involving an exponential integral function and exponentials arising in the solution of radiation transfer.
S. T. Thynell, M. N. Özisik.
J. Quant. Spectrosc. Radiat. Transfer, Vol. 33, No. 3, p. 259 – 266 (1985).
The solution of the equation of radiative transfer for a participating medium generally results in the evaluation of integrals involving a product of an exponential integral function or an exponential with a polynomial. Although expressions are available in the literature for the evaluation of such integrals, which appear to be structurally easy and simple to program, they are found to be not so accurate for certain parameters involved. To overcome such difficulties, alternative, computationally more accurate, analytic expressions are presented, and numerical techniques for their evaluation are discussed.

063.043   Probability of photon scattering by electrons. A case of chaotically moving monoenergetic electrons.
G. A. Arutyunyan, V. A. Dzhrbashyan.
Astrofizika, Tom 22, Vyp. 2, p. 379 – 386 (1985). In Russian. English translation in Astrophysics, Vol. 22, No. 2.
The question of finding the probability of photon scattering by a free electron ensemble is investigated for cases of rather general assumptions concerning the physics and geometry of the problem. A formula describing the given probability calculated for unit time is obtained in the case of chaotically moving monoenergetic electrons.

063.044   The anisotropic radiative transfer problem in optically thick, strongly magnetized plasma: a comparison of results.
G. G. Pavlov, Yu. A. Shibanov, N. A. Silant'ev, W. Nagel.
Astrophys. J., Vol. 291, No. 1, p. 170 – 177 (1985).
Recently developed methods for solving the coherent radiative transfer problem in a strongly magnetized plasma are compared and analyzed for the case of a semi–infinite, homogeneous plasma with the magnetic field perpendicular to the surface. The work of Mészáros and Bonazzola is shown to contain some errors. The accuracy of numerical methods proposed by Silant'ev and Nagel is investigated for various plasma parameters and photon energies. The coupled diffusion approximation developed by Nagel and Kaminker et al. appears to give quite satisfactory results.

063.045   Line fluorescence in astrophysics.
M. Elitzur, H. Netzer.
Astrophys. J., Vol. 291, No. 2, p. 464 – 467 (1985).
The authors develop a formalism to calculate emission–line strengths in the case of partial or complete overlapping of two spectral lines (so–called "line fluorescence"). The procedure is based on the escape probability method and is particularly suitable for use in numerical codes where line and continuum transfer is treated in this way. The method is applied to two cases of astrophysical interest: Fe II fluorescence with Lyα, and Lyβ fluorescence with O I λ1025 in quasar clouds.

063.046   Bowen fluorescence mechanism in X–ray binaries.
S. Deguchi.
Astrophys. J., Vol. 291, No. 2, p. 492 – 504 (1985).
The Bowen fluorescence mechanism is investigated for the ionized gas cloud surrounding X–ray binaries. It is found that the helium–oxygen photon–conversion efficiency is about $0.5 – 0.8$ at a Lyα optical depth of 10^7 in an expanding–cloud model or in a rotating–disk model. The helium–nitrogen photon–conversion efficiency through the oxygen–nitrogen resonance is about 0.1. The characteristic scale length of the cloud and the number density of He II are found to be $\lesssim 10^{12}$cm and $\gtrsim 10^8$cm^{-3} respectively. Two possibilities are considered for the gas cloud responsible for the optical emission lines: one is the gas of the stellar wind which is illuminated on one side by X–rays, and the other is the gas which is accumulated at the Langrangian triangular points.

063.047   Theory of radiation transfer in spectral lines.
D. I. Nagirner.
Astrophys. Space Phys. Rev., Vol. 3, p. 255 – 300 (1984). – See Abstr. 003.019. Revised and extended English translation of 34.063.036.
Theoretical studies of the spectral line formation in stationary isotropic media under the assumptions of complete and partial frequency redistribution upon scattering in a line are reviewed. The main attention is given to the analytical methods and to the analytical results. After formulating the problem and giving a brief historical background, the article describes the exact solutions of the simplest problems, asymptotic and approximate theoretical methods, probabilistic interpretation of multiple scattering and certain numerical methods. The application of the theory to the study of astrophysical objects – nebulae, atmospheres of stars and planets, solar phenomena – is discussed.

063.048   Statistical description of a radiation field on the basis of the invariance principle. II. The mean number of scatterings in a medium containing energy sources.
A. G. Nikogosyan.
Astrofizika, Tom 21, Vyp. 3, p. 595 – 607 (1984). In Russian. English translation in Astrophysics, Vol. 21, No. 3.
Under general assumptions concerning the elementary act of scattering equations for determining the mean number of scatterings of photons in a three–dimensional semi–infinite medium are obtained. For illustration the case of complete frequency redistribution, including the absorption and emission in a continuum, is examined in more detail. In this case a number of new formulas for the mean numbers of scatterings in semi–infinite and infinite media are given.

063.049   Processes of energy exchange between electrons and photons at intense radiation fields encountered in some astrophysical objects. I.
G. T. Ter–Kazaryan.
Astrofizika, Tom 21, Vyp. 3, p. 609 – 625 (1984). In Russian. English translation in Astrophysics, Vol. 21, No. 3.
Three classes of basic problems and its importance in the interpretation of dates of some astrophysical objects (nuclei of Seyfert galaxies, quasars, interstellar masers, pulsar NP 0532, radio pulsars) are enduced on the basis of analysis of works dedicated to the single photon Compton interaction between electrons and photons. The existing physical conditions in the mentioned astrophysical objects show that it is necessary to take into account also processes of multi–photon comptonization, because in their efficiency the latter may essentially exceed single photons (for example, for pulsar NP 0532 and radio pulsars). The particular problem of the relaxation of the nonequilibrium isotropic radiation interacting with nondegenerate nonrelativistic electrons via the multi–photon Compton scattering is considered. The equations describing the time evolution of energy exchange, the heating and cooling of electron gas are also obtained.

063.050 Radiation transfer in an isotropically scattering homogeneous solid sphere.
S. T. Thynell, M. N. Özisik.
J. Quant. Spectrosc. Radiat. Transfer, Vol. 33, No. 4, p. 319 – 330 (1985).
Radiative heat transfer in an absorbing, emitting, isotropically scattering sphere with a diffusively reflecting and emitting boundary is solved by the Galerkin method.

063.051 Computational techniques for radiative transfer by spherical harmonics.
W. H. Wells, J. J. Sidorowich.
J. Quant. Spectrosc. Radiat. Transfer, Vol. 33, No. 4, p. 347 – 363 (1985).

063.052 The Neumann solution of the multiple scattering problem in a plane–parallel medium. I. The infinite medium.
B. Rutily, J. Bergeat.
J. Quant. Spectrosc. Radiat. Transfer, Vol. 33, No. 4, p. 373 – 380 (1985).
This paper deals with the study of the radiation field arising from a plane distribution of isotropic sources in a plane–parallel, infinite medium. The classical Neumann method is used together with the theory of singular integrals of the Cauchy type. The main advantage of this technique is that it is available for a general treatment of radiative transfer problems in semi–infinite and finite spaces.

063.053 Dirac–delta function approximations to the scattering phase function.
A. L. Crosbie, G. W. Davidson.
J. Quant. Spectrosc. Radiat. Transfer, Vol. 33, No. 4, p. 391 – 409 (1985).
Dirac–delta function approximations are used to represent the single scattering phase function of large spherical particles or voids. The Dirac–delta function accounts for strong forward scattering. Particular attention is given to large ice spheres and spherical voids in ice. The Dirac–delta function is shown effective in reducing the number of terms needed to describe the phase function.

063.054 Spatial differencing of the discrete ordinate equations in optically thick media.
P. A. Brown, J. G. Hill, G. C. Pomraning.
J. Quant. Spectrosc. Radiat. Transfer, Vol. 33, No. 5, p. 437 – 452 (1985).
The authors present a modification to the linear characteristic method used to finite difference the discrete ordinate (S–N) equations. This modification is designed to produce the correct diffusion limit in the radiative transfer context in the limit of optically thick mesh cells. Numerical tests suggest that this finite difference method is a practical calculational scheme for radiative transfer problems.

063.055 Radiative transfer of resonance lines with internal sources.
G. R. Gladstone.
J. Quant. Spectrosc. Radiat. Transfer, Vol. 33, No. 5, p. 453 – 458 (1985).
The modeling of resonance lines in the aurorae and dayglow usually requires a large computer code to solve the radiative transfer problem. The author presents a simple theory that gives reasonable estimates of the emergent intensities, including the angular distribution, for very optically thick lines with internal sources. The theory is compared with detailed numerical calculations for resonance lines with Voigt emission line profiles, including the cases of both partial and complete frequency redistribution. The effects of frequency redistribution on the angular distribution of the emergent radiation and on the line profiles are also shown.

063.056 Two–dimensional linearly anisotropic scattering in a finite–thick cylindrical medium exposed to a laser beam.
A. L. Crosbie, R. L. Dougherty.
J. Quant. Spectrosc. Radiat. Transfer, Vol. 33, No. 5, p. 487 – 520 (1985).

063.057 The transfer of line radiation. II. Approximate solutions for semi–infinite atmospheres.
V. V. Ivanov, V. M. Serbin.
Sov. Astron., Vol. 28, No. 5, p. 524 – 531 (1984). English translation of 38.063.036.

063.058 Bremsstrahlung spectra from thick–target electron beams with noncollisional energy losses.
J. C. Brown, A. L. MacKinnon.
Astrophys. J., Lett. Ed., Vol. 292, No. 1, p. L31 – L34 (1985).
The authors consider what can be learned from the bremsstrahlung radiation of fast electrons in a thick target, generalized to include electron energy losses additional to collisions. They show that the observed photon spectrum can, in principle, be inverted to yield an integral functional of the electron spectrum and the effective energy loss rate. In the light of this result, there seems no reason to suppose, in the absence of a priori information to the contrary, that the photon spectrum is symptomatic more of the fast electron distribution than of the energy loss processes.

063.059 Resonance lines in dusty gaseous nebulae.
R. Wehrse, W. Kalkofen.
Astron. Astrophys., Vol. 147, No. 1, p. 71 – 83 (1985).
The authors study the effect of absorption and scattering by dust grains on resonance lines formed in static gaseous nebulae with plane–parallel stratification, assuming that the dust is either homogeneously mixed throughout the whole volume or confined to one half or one quarter (front or rear) of the space. They compare the total emergent flux in the line from the dust–filled nebula with the emergent flux from a dust–free medium for many values of the optical thickness in the line and several combinations of the opacity ratio of dust and gas, dust albedo, and collision parameter. The transfer equation for a spectral line in the presence of scattering and absorbing dust is solved analytically by means of a new method as a set of coupled first–order difference equations.

063.060 A method of the resolvent function approximation in radiative transfer.
T. Viik, R. Rôôm, A. Heinlo.
Tartu Astrofüüs. Obs. Teated, Nr. 76, 131 pp. (1985). In Russian.

063.061 Astrophysical significance of spectral line shape investigations.
M. Dimitrijević.
Publ. Astron. Opservatorije Beogr., No. 33, p. 11 – 14 (1985). – See Abstr. 012.061.
The astrophysical implications of the spectral line shape investigations are briefly reviewed. A short survey of results reached by Yugoslav researchers is also given.

063.062 Accuracy of quadrature formulae for the intensity and the flux integrals.
K. Yoshioka, K. Nariai.
Ann. Tokyo Astron. Obs., Second Ser., Vol. 20, No. 3, p. 282 – 295 (1985).
The authors have examined the accuracy of several quadrature formulae for the intensity and the flux integrals. As the source function for the test, Kourganoff's best solution for the gray model which is precise almost to the sixth decimal place was used. The best quadrature formula gives precise values of J and F for the gray model down to the fourth decimal place.

063.063 Comptonization in a trapped, divergent flow.
P. A. Becker, M. C. Begelman.
Bull. Am. Astron. Soc., Vol. 16, No. 4, p. 899 (1984). Abstract. – See Abstr. 010.062.

063.064 Effects of source geometry on continuum radiation transport.
G. F. Spagna Jr., C. M. Leung.
Bull. Am. Astron. Soc., Vol. 16, No. 4, p. 961 (1984). Abstract. – See Abstr. 010.062.

063.065 Nonthermal, optically thin e^+e^- pair production and the universal X–ray spectrum of active galactic nuclei.
A. A. Zdziarski, A. P. Lightman.
Bull. Am. Astron. Soc., Vol. 16, No. 4, p. 1009 (1984). Abstract. – See Abstr. 010.062.

063.066 Theoretical models of polarization structure in compact radio jets.
T. W. Jones.
Bull. Am. Astron. Soc., Vol. 16, No. 4, p. 1009 (1984). Abstract. – See Abstr. 010.062.

063.067 Radiation from relativistic extragalactic jets interacting with interstellar matter.
W. K. Rose.
Bull. Am. Astron. Soc., Vol. 16, No. 4, p. 1010 (1984). Abstract. – See Abstr. 010.062.

063.068 H– and auxiliary functions for phase functions of the type $p(\cos\theta) = \omega_0 + \omega_1 P_1(\cos\theta) + \omega_2 P_2(\cos\theta)$.
R. K. Bhatia, K. D. Abhyankar.
Bull. Astron. Soc. India, Vol. 12, No. 4, p. 364 – 383 (1984).

063.069 Multiple scattering and the dependence of the phase variation of equivalent widths on line profiles.
R. K. Bhatia, K. D. Abhyankar.
Bull. Astron. Soc. India, Vol. 12, No. 4, p. 384 – 392 (1984).
An explanation for the inverse phase effect for absorption lines formed in a Rayleigh scattering atmosphere is given in terms of multiple scattering. It is pointed out that the inverse phase effect is a function of the line profile chosen, with the Lorentz profile giving the maximum change and the Doppler profile the least.

063.070 Moment method for multiple scattering of solar Lα radiation in the nearby interstellar medium.
S. J. Wilson.
Astrophys. Space Sci., Vol. 112, No. 1, p. 69 – 74 (1985).
The problem of solar Lα (1216 Å) photons scattered coherently by interplanetary medium is solved for a realistic density distribution using a simple three–stream division of the radiation field.

063.071 Mean number of photon scatterings in a spherical shell.
S. Karanjai, M. Karanjai.
Astrophys. Space Sci., Vol. 112, No. 2, p. 407 – 412 (1985).
An expression for the mean number of photon scattering in an absorbing and scattering medium, when the medium is a spherical shell of finite thickness, has been derived. Also the expression of probability of such emergence of photons absorbed is given.

063.072 On a method of solution of the integral equation of radiative transfer.
V. V. Zharkova.
Kiev. univ. Kiev, 9 pp. (1984). In Russian. Abstr. in Ref. Zh., 51. Astron., 5.51.116 (1985).

063.073 The Sobolev approximation for line formation with continuous opacity.
D. G. Hummer, G. B. Rybicki.
Astrophys. J., Vol. 293, No. 1, p. 258 – 267 (1985).
The Sobolev approximation for line–formation problems in atmospheres with high–speed flows is generalized to include the effects of continuum absorption and emission in the region of the line. A comparison with accurate numerical solutions for simple problems in plane–parallel geometry is presented. A three–dimensional version of the theory is given that applies to general geometries.

063.074 A simple one–per–cent approximation of the Voigt function.
V. Dobrichev.
Dokl. Bolg. Akad. Nauk, Vol. 37, No. 8, p. 991 – 993 (1984). In Russian. Abstr. in Ref. Zh., 51. Astron., 6.51.119 (1985).

063.075 The Cerenkov microwave line emission of the hydroxyl radical.
J. You, S. Li.
High energy astrophysics and cosmology, p. 497 – 508 (1983). – See Abstr. 012.068.
In this paper, the possibility of microwave emission lines produced by the Cerenkov effect is discussed using OH as an example. The intensity and other characteristics of OH Cerenkov microwave line emission are calculated.

063.076 Correlations in the synchrotron–self Compton spectra.
J. You, G. Xie, M. Bao, K. Li.
High energy astrophysics and cosmology, p. 509 – 521 (1983). – See Abstr. 012.068.
The production of high–energy photons by the synchrotron-self Compton effect for relativistic electrons is investigated and correlations between the synchrotron radio emission and the Compton UV and X–ray radiation are derived.

063.077 Motion and radiation of electrons in an inhomogeneous magnetic field.
S. An, L. Luo.
High energy astrophysics and cosmology, p. 522 – 527 (1983). – See Abstr. 012.068.

063.078 The polarization properties of magnetic accretion columns. III. A grid of uniform temperature and shock front models.
D. T. Wickramasinghe, S. M. A. Meggitt.
Mon. Not. R. Astron. Soc., Vol. 214, No. 4, p. 605 – 618 (1985). With microfiche MN 214/1.
The authors present calculations of cyclotron and free–free emission for a wide range of values of the electron temperature (T_e), frequency (ω/ω_c), viewing angle with respect to the magnetic field (θ) and size parameter (Λ) both for constant temperature models and models incorporating a temperature structure (shock front models). The conditions under which the temperature structure within the shock front can play an important role in determining the properties of the radiation emerging from AM Herculis–type systems are discussed. It is suggested that the narrow range of field strengths observed in AM Herculis–type systems is most probably a selection effect due to the emission properties of cyclotron radiation.

063.079 On the singular components of the solution to the searchlight problem in radiative transfer.
C. E. Siewert.
J. Quant. Spectrosc. Radiat. Transfer, Vol. 33, No. 6, p. 551 – 554 (1985).
Explicit expressions for the singular components of the solution to the searchlight problem are reported.

063.080 Green's function formulae for the internal intensity in radiative transfer computations by matrix–vector methods.
S. Twomey.
J. Quant. Spectrosc. Radiat. Transfer, Vol. 33, No. 6, p. 575 – 579 (1985).
In radiative transfer computations, Green's functions are particularly useful and have a straightforward physical interpretation, as discussed by Cogley and co–workers. By recognizing that, in thermal emission problems, the angular distribution of emission is prescribed a priori, one can obtain a more rapid computational procedure in which most computations are in discrete formalism one–dimensional or vector formulae.

063.081   The standard problem of line formation in moving atmospheres.
V. M. Serbin.
Astrofizika, Tom 22, Vyp. 2, p. 387 – 409 (1985). In Russian. English translation in Astrophysics, Vol. 22, No. 2.

Under the standard assumptions of the theory of line formation (plane–parallel isothermal atmosphere, two–level atom, complete frequency redistribution in comoving frame, Doppler absorption profile) the line source functions (LSF) are found numerically from the basic integral equation for the LSF. It is assumed that the atmosphere expands with small constant velocity gradient. The emergent profiles are also found. Both semi–infinite atmospheres and atmospheres of finite optical thickness are considered. The numerical results are compared with the available analytical information.

063.082   One–dimensional scattering as a possible model for interpretation of emission of globules and dust nebulae. I.
D. A. Rozhkovskij.
Tr. Astrofiz. Inst. Alma–Ata, Tom 44, p. 3 – 19 (1984). In Russian.

063.083   Induced emission of radiation near $2\omega_e$ by a synchrotron–maser instability.
C. S. Wu, G. C. Zhou, J. D. Gaffey Jr.
Phys. Fluids, Vol. 28, No. 3, p. 846 – 853 (1985). Abstr. in Phys. Abstr., Vol. 88, No. 1259, Entry 66003 (1985).

063.084   The production of soft radiation by photon–electron scattering in a strong magnetic field.
R. W. Bussard, P. Mészáros, S. Alexander.
Bull. Am. Astron. Soc., Vol. 17, No. 2, p. 554 (1985). Abstract. – See Abstr. 010.065.

063.085   General aspects of partial redistribution and its astrophysical importance.
I. Hubeny.
Progress in stellar spectral line formation theory, p. 27 – 58 (1985). – See Abstr. 012.108.

A review is given of new developments in the theory of partial redistribution in radiative transfer problems. Emphasis is on the transfer of unpolarized radiation in plane parallel static media, but effects of velocity fields and of geometrical structure are also briefly discussed. Applications to solar and stellar spectral line formation are outlined.

063.086   Kinetic aspects of redistribution in spectral lines.
J. Oxenius.
Progress in stellar spectral line formation theory, p. 59 – 71 (1985). – See Abstr. 012.108.

This paper considers the question of how atomic velocity distributions affect the line profile coefficients of the radiative transfer equation. To this end, it discusses in turn the two–level atom and the three–level atom.

063.087   Non local effects on the redistribution of resonant scattered photons.
E. Simonneau.
Progress in stellar spectral line formation theory, p. 73 – 86 (1985). – See Abstr. 012.108.

In resonance lines of the most abundant elements non–LTE radiative transfer can induce a convective transport of excited atoms. This transport can strongly modify the excited atom's density and therefore the intensity of the spectral line. The author develops here the first truly self–consistent non–LTE solution to the two–level atom transfer problem.

063.088   Asymptotic properties of complete and partial frequency redistribution.
H. Frisch.
Progress in stellar spectral line formation theory, p. 87 – 100 (1985). – See Abstr. 012.108.

Radiative transfer problems with frequency redistribution may be investigated by asymptotic methods when the mean number of scatterings undergone by photons is very large. These methods provide scaling laws for characteristic parameters of the line radiation field. These methods also provide asymptotic transfer equations which describe the large scale behaviour of the radiation field away from boundaries. Complete redistribution and the four standard types of partial redistribution is discussed. Implications for numerical calculations are briefly considered.

063.089   Some comments upon the line emission profile Ψ_v.
R. Freire Ferrero.
Progress in stellar spectral line formation theory, p. 101 – 108 (1985). – See Abstr. 012.108.

The mathematical expression of the emission profile Ψ_v is given in the case of a two–level atom plus continuum: the result is that Ψ_v is independent of populations and abundances, depending only on T_e, N_e and J_v.

063.090   A modified Rybicki method with partial redistribution.
I. Hubeny.
Progress in stellar spectral line formation theory, p. 109 – 114 (1985). – See Abstr. 012.108.

A new approximate numerical method is presented that retains the basic computational advantages of the Rybicki method while still being capable of handling partial redistribution transfer problems. The crucial point of this method is to consider the frequency which separates the complete redistribution core region from the coherent wing region, to be depth–dependent. The present method yields excellent agreement with exact calculations and gives much better results than any depth–independent version of the partial coherent scattering approximation.

063.091   Redistribution functions: a review of computational methods.
P. Heinzel.
Progress in stellar spectral line formation theory, p. 115 – 124 (1985). – See Abstr. 012.108.

This paper reviews and compares the existing computational methods developed to evaluate various types of redistribution functions applicable in astrophysics. It discusses in detail several numerical aspects (codes) of calculating both the angle–dependent and angle–averaged laboratory–frame functions.

063.092   Progress in stellar spectral line formation theory. Panel discussion on partial redistribution.
R. Freire Ferrero, H. Frisch, J. Linsky, J. Oxenius, E. Simonneau.
Progress in stellar spectral line formation theory, p. 143 – 151 (1985). – See Abstr. 012.108.

063.093   Numerical methods in radiative transfer.
W. Kalkofen.
Progress in stellar spectral line formation theory, p. 153 – 168 (1985). – See Abstr. 012.108.

The author discusses the operator perturbation method for the solution of radiative transfer problems in the integral equation formulation. The example given is that of line transfer in complete redistribution for a two–level atom in statistical equilibrium. The essence of the method is the separation of the calculation into two parts: the calculation of corrections to a solution with the aid of an approximate integral operator; and the calculation of the error with which the solution satisfies the conservation equation.

063.094   Partial versus complete linearization.
W. Kalkofen.
Progress in stellar spectral line formation theory, p. 169 – 174 (1985). – See Abstr. 012.108.

The convergence properties of the partially or completely linearized equations for a grey model atmosphere in radiative equilibrium are compared. The completely linearized equations show the quadratic convergence properties of Newton–Raphson equations. When the opacity depends strongly on temperature, the convergence of the partially linearized equations is very slow initially but improves once the maximal error has moved to the lower boundary.

063.095 Radiative transfer diagnostics: Understanding multi-level transfer calculations.
A. Skumanich, B. W. Lites.
Progress in stellar spectral line formation theory, p. 175 – 187 (1985). – See Abstr. 012.108.

The authors present a method of interpreting the solution to a multi-level, multi-transition non-LTE transfer problem. The method respresents the solutions in terms of equivalent two-level forms with a scattering and a source term. The resulting individual quenching probability, i.e. the difference of the scattering albedo from one, and source term are then decomposed by a perturbation method into their principal dependence on collisional and/or radiative rates. The method is illustrated by considering the excitation and ionization of hydrogen in the VAL 3C model of the quiet sun chromosphere.

063.096 A new method for solving multi-level non-LTE problems.
G. B. Scharmer, M. Carlsson.
Progress in stellar spectral line formation theory, p. 189 – 198 (1985). – See Abstr. 012.108.

A new scheme for solving multi-level non-LTE problems is described. This method uses an approximate operator for the relation between the intensity and the source function. This operator results in a matrix equation for the population numbers which has a simple and characteristic structure. Solutions are obtained such that the results are "exact", irrespective of the choice of the approximate operator.

063.097 Escape probability methods.
G. B. Rybicki.
Progress in stellar spectral line formation theory, p. 199 – 206 (1985). – See Abstr. 012.108.

The physical foundations of escape probability methods, and methods derived from them, are briefly reviewed. First-order escape probability methods, the core saturation method, second-order escape probability methods, and Scharmer's method are discussed.

063.098 Numerically stable discrete ordinate solutions of the radiative transfer equation.
R. Wehrse.
Progress in stellar spectral line formation theory, p. 207 – 213 (1985). – See Abstr. 012.108.

A new numerically stable and highly accurate method for the computation of the radiation field in a spectral line is presented. It involves the following three main steps: First by means of a discretisation of the angle and frequency space the transfer equation is transformed into a system of coupled first order differential equations. Next the system is solved analytically and finally it is cast into a form, which gives the emergent intensities as a function of the inflowing radiation and which contains only negative eigenvalues.

063.099 NLTE spectral line formation in a three-dimensional atmosphere with velocity fields.
Å. Nordlund.
Progress in stellar spectral line formation theory, p. 215 – 224 (1985). – See Abstr. 012.108.

A method to solve the "two-level-atom-with-overlapping-continuum" problem in a three-dimensional atmosphere is presented. The method is based on treating the radiative transfer along a number of rays through the models as separate sub-problems. In each iteration, the error in the source function is evaluated along all the rays through the model, and an estimate of the necessary correction is obtained for each ray. The converged solution is an exact solution to the problem. As an application, the method is used on the case of a neutral iron line in the solar photosphere.

063.100 A code for line blanketing without local thermodynamic equilibrium.
L. S. Anderson.
Progress in stellar spectral line formation theory, p. 225 – 232 (1985). – See Abstr. 012.108.

A numerical code has been written which is designed to calculate radiation transport and atmospheric structure under the constraints of statistical equilibrium in atomic transitions and radiative and hydrostatic equilibrium in the medium. It uses a multi-frequency/multi-grey algorithm which admits the inclusion of many spectral lines in full statistical equilibrium. The program can comfortably accept up to about 300 specific lines arising from about 30 lower states and any number of continua. By way of example, the author presents a model of a stellar atmosphere with effective temperature 35000K and surface gravity 10^4cm s^{-2}.

063.101 Progress in stellar spectral line formation theory. Panel discussion on radiative transfer methods.
W. Kalkofen, J. Linsky, G. Rybicki, G. Scharmer, R. Weherse.
Progress in stellar spectral line formation theory, p. 233 – 237 (1985). – See Abstr. 012.108.

063.102 Hydrogen line formation in dense plasmas in the presence of a magnetic field.
G. Mathys.
Progress in stellar spectral line formation theory, p. 381 – 388 (1985). – See Abstr. 012.108.

The transfer equation of arbitrarily polarized radiation in the hydrogen lines is formulated. The effect of a large scale, uniform magnetic field, the linear Stark effect due to the charged particles of a surrounding dense plasma and the Doppler broadening due to the thermal motions of the radiating atoms are included. The theory is suitable to conditions typical of atmospheres of magnetic chemically peculiar stars or of active solar regions.

063.103 The radiation field in media with radial symmetry.
A. K. Kolesov.
Astrofizika, Tom 22, Vyp. 3, p. 571 – 583 (1985). In Russian. English translation in Astrophysics, Vol. 22, No. 3.

The structure of Green's functions in problems of radiative transfer theory for media with radial symmetry is investigated. Linear integral equations for the light reflection coefficients are found for the cases of a sphere and an infinite space with a spherical black-body hollow. The kernel functions and the free terms of the equations are expressed in terms of the infinite space Green's function.

063.104 On the problem of noncoherent scattering in a one-dimensional medium.
M. S. Gevorkyan, A. Kh. Khachatryan.
Astrofizika, Tom 22, Vyp. 3, p. 599 – 612 (1985). In Russian. English translation in Astrophysics, Vol. 22, No. 3.

The problem of noncoherent isotropic scattering radiation in one-dimensional approximation in a semi-infinite medium and in a layer of finite thickness is effectively solved. Results of numerical calculations in the fourth and eighth approximations for the case of Doppler line broadening are given. Results of certain numerical calculations performed by two different methods are compared.

063.105 The thermodynamical proof of Rosseland's theorem.
A. M. Sobolev, V. S. Strel'nitskij, N. N. Chugaj.
Astrofizika, Tom 22, Vyp. 3, p. 613 – 618 (1985). In Russian. English translation in Astrophysics, Vol. 22, No. 3.

The proof of Rosseland's theorem for any area of spectra has been rendered based on the use of brightness temperature for the description of a nonequilibrium radiation field. Such a proof elucidates a simple thermodynamic sense of the theorem.

Foundations of radiation hydrodynamics.
See Abstr. 003.025.

Radiation of atoms in cosmic plasma.
See Abstr. 003.149.

Redistribution of radiation in the absence of collisions.
See Abstr. 022.027.

The standard line–formation problem: second–order approximations including continuous absorption.
See Abstr. 022.058.

Opacity determinations for ICF (*Inertial Confinement Fusion*) materials using an average atom model.
See Abstr. 022.061.

The role of stimulated processes in cosmic annihilation–line sources. II.
See Abstr. 022.113.

NLTE–Rechnungen zur Bildung interstellarer Moleküllinien in einem turbulenten Medium.
See Abstr. 022.120.

Line formation in laboratory plasmas.
See Abstr. 022.186.

Neutrino production from discrete high–energy gamma–ray sources.
See Abstr. 061.020.

Thermal cyclotron radiation in astrophysics.
See Abstr. 061.035.

Vacuum polarization by a magnetic field and its astrophysical manifestations.
See Abstr. 061.036.

Equilibria in stronlgy magnetized pair plasmas.
See Abstr. 062.041.

Force–free equilibria of magnetized jets.
See Abstr. 062.066.

Effects of a finite plasma temperature on electron–cyclotron maser emission.
See Abstr. 062.111.

Wave propagation in a radiating medium.
See Abstr. 062.137.

Theoretical studies of the linear polarization of synchrotron radiation from relativistic jets.
See Abstr. 062.138.

Polarization transport in turbulent synchrotron sources.
See Abstr. 062.139.

Relativistic code applied to radiation generation.
See Abstr. 062.159.

Radiative theory of MHD wave propagation (Review).
See Abstr. 062.165.

The calculation of the opacity of hot dense plasmas.
See Abstr. 062.191.

Molecular emission from expanding circumstellar envelopes: polarization and profile asymmetries.
See Abstr. 064.004.

Comptonization of low–frequency radiation in accretion disks: angular distribution and polarization of hard radiation.
See Abstr. 064.011.

Multiline transfer and the dynamics of stellar winds.
See Abstr. 064.032.

On models of extended stellar atmospheres.
See Abstr. 064.041.

The Chandrasekhar *H*–function for exponentially varying atmosphere.
See Abstr. 064.057.

Effects of multiquantum transitions on molecular populations in grain–forming circumstellar environments
See Abstr. 064.085.

Radiation transfer in accretion disk coronae.
See Abstr. 064.086.

Observed and computed stellar line profiles: the roles played by partial redistribution, geometrical extent and expansion.
See Abstr. 064.101.

The effect of abundance values on partial redistribution line computations.
See Abstr. 064.102.

The theory of line transfer in expanding atmospheres.
See Abstr. 064.106.

Computed He II spectra for Wolf–Rayet stars.
See Abstr. 064.107.

Partial redistribution in the winds of red giants.
See Abstr. 064.108.

Modeling lines formed in the expanding chromospheres of red giants.
See Abstr. 064.109.

Inverse comptonization vs. thermal synchrotron.
See Abstr. 067.027.

Physics of the synchrotron model of cosmic gamma–ray bursts.
See Abstr. 067.028.

Radiative transfer in the accretion column of X–ray pulsars: effects from the hot spot.
See Abstr. 067.057.

Spectra of gamma–ray bursts.
See Abstr. 067.120.

Pressure broadening and solar limb effect.
See Abstr. 071.049.

Partial redistribution interlocking in the solar chromosphere.
See Abstr. 073.237.

Transfer of Lyman–α radiation in solar coronal loops.
See Abstr. 074.162.

Application of a radiative transfer model to bright icy satellites.
See Abstr. 100.010.

Predicted continuum spectra of Type II supernovae: LTE results.
See Abstr. 125.023.

A preliminary discussion of Bose–Einstein diffusion in supernovae.
See Abstr. 125.121.

The interstellar radiation field and the production of inverse–Compton gamma rays in the Galaxy.
See Abstr. 131.128.

Shielding of CO from dissociating radiation in interstellar clouds.
See Abstr. 131.144.

On a model construction of interstellar scattering processes with dust formation in the Galaxy.
See Abstr. 131.302.

A review of line formation in molecular clouds.
See Abstr. 131.363.

Models of H II regions: heavy element opacity, variation of temperature.
See Abstr. 132.021.

One–photon and two–photon annihilation lines in gamma–bursts. II.
See Abstr. 143.057.

Annihilation radiation from the galactic center: positrons in dust?
See Abstr. 155.083.

A model of the polarization position–angle swings in BL Lacertae objects.
See Abstr. 158.072.

Consequences of hot gas in the broad–line region of active galactic nuclei.
See Abstr. 158.161.

Diffuse shock acceleration and quasar photospheres.
See Abstr. 159.020.

064 Stellar Atmospheres, Stellar Envelopes, Mass Loss, Accretion

064.001   Stellar winds and the interstellar medium.
J. E. Dyson.
Gas in the interstellar medium, p. 107 – 113 (1984). – See Abstr. 012.032.

The author first discusses some of the general principles of the interaction of stellar winds with their surroundings and how they can be used to obtain information on the stellar wind properties, specifically, the mass loss rates and wind velocities. He then discusses another wind interaction process, namely, the production of Herbig–Haro objects, and presents a preliminary account of yet another model for their formation. Finally, the author briefly mentions recent observational work on what at first glance appear to be Herbig–Haro objects and indicates why models of "classical" H–H objects appear inapplicable.

064.002   Accretion disk electrodynamics.
F. V. Coroniti.
Unstable current systems and plasma instabilities in astrophysics, p. 453 – 469 (1985). – See Abstr. 012.002 (IAU Symp. No. 107).

Accretion disk electrodynamic phenomena are separable into two classes: (1) disks and coronae with turbulent magnetic fields; (2) disks and black holes which are connected to a large–scale external magnetic field. Turbulent fields may originate in an α–ω dynamo, provide anomalous viscous transport, and sustain an active corona by magnetic buoyancy .The large–scale field can extract energy and angular momentum from the disk and black hole, and be dynamically configured into a collimated relativistic jet.

064.003   Magnetic field reconnection in differentially rotating accretion disks.
D. Gilden, T. Tajima.
Unstable current systems and plasma instabilities in astrophysics, p. 477 – 482 (1985). – See Abstr. 012.002 (IAU Symp. No. 107).

Differentially rotating accretion disks threaded by a uniform magnetic field have been numerically simulated. Fast reconnection followed by coalescence allows the magnetic field to drive small amplitude radial oscillations in the disk. These oscillations may be observable as the viscous stresses cause the disk to brighten and fade as the disk expands and contracts. Episodes of reconnection may also be observable as hot spots produced locally at the sites of coalescence. Cataclysmic variables, and in particular dwarf novae, provide a natural interpretation for these calculations.

064.004   Molecular emission from expanding circumstellar envelopes: polarization and profile asymmetries.
M. Morris, R. Lucas, A. Omont.
Astron. Astrophys., Vol. 142, No. 1, p. 107 – 116 (1985).

Two aspects of the transfer of radiation from molecular lines arising in expanding circumstellar envelopes are investigated.

First, the authors propose an explanation for the recently observed asymmetries in radio line profiles from IRC +10216. Second, they investigate the relative alignment of the rotational axes of those molecules whose rotational levels are excited predominantly by absorption and re–emission of infrared vibration–rotation line photons.

064.005   Mass loss rates from protostars and O I (63 μm) shock luminosities.
D. Hollenbach.
Icarus, Vol. 61, No. 1, p. 36 – 39 (1985). – See Abstr. 012.004.

The high–velocity ejection of material from protostars results in a wind shock which may be observable in O I (63 μm) emission. It is shown that for a wide range of conditions, the O I (63 μm) luminosity is proportional to the mass loss rate from the protostar. Application is made to shock O I (63 μm) emission observed around IRc 2 in the BN–KL region of Orion.

064.006   Transfer calculations of dusty gray disks: Eddington approximation.
G. Spagna.
News Lett. Astron. Soc. N.Y., Vol. 2, No. 7, p. 21 – 22 (1985). Abstract. – See Abstr. 010.241.

064.007   A self–consistent model of the magnetosphere with centrifugal wind. I.
S. Shibata, O. Kaburaki.
Astrophys. Space Sci., Vol. 108, No. 1, p. 203 – 217 (1985).

Under the purely centrifugal approximation, stellar magnetospheres are classified into three main types of different physical properties in the two–dimensional parameter space. They are characterized essentially by the strength of the magnetic field and the plasma density, at the base of the magnetosphere. Among the three types, the type II magnetosphere has moderate surface densities for a given field strength, and is expected to possess a centrifugal wind blowing across the magnetic field lines without affecting them appreciably. Such a situation may be realized through a modification of the electric field from that under the ideal–MHD condition, owing to the inertia of a plasma. In order to illustrate this mechanism, the type II magnetosphere is taken up for a numerical simulation.

064.008   A self–consistent model of the magnetosphere with centrifugal wind. II.
S. Shibata.
Astrophys. Space Sci., Vol. 108, No. 2, p. 337 – 362 (1985).

In order to construct an axisymmetric model of magnetospheres with centrifugal wind, especially of the type II magnetosphere in Paper I (Shibata and Kaburaki, 1984), the author presents a numerical iterative scheme, in which a tenuous plasma with conspicuous trans–field motion is treated self–consistently with the electromagnetic field. The obtained properties of type II magnetosphere are as follows. (1) Plasma particles in fact flow out across the closed magnetic field lines. (2) The centrifugal force is exerted powerfully on the positive particles to form a disk–like structure, and the strong electric force makes the negative particles drift to the disk. (3) There appears the electric field parallel to the magnetic field, which is shown to be necessary for the steady wind to exist.

064.009   Stochastic electron acceleration in stellar coronae.
T. J. Bogdan, R. Schlickeiser.
Astron. Astrophys., Vol. 143, No. 1, p. 23 – 28 (1985).

The authors consider the acceleration of electrons by the second order Fermi mechanism in a quasi–stationary turbulent plasma of dimension l, mean magnetic field strength B, and mean number density n. They determine criteria for the existence of, and give exact analytic solutions for, the steady–state electron energy spectra. Implications of the solutions for electron acceleration associated with flare–like microwave emission in late–type stars are briefly discussed.

064.010   Two–dimensional models of stellar wind bubbles.
I. Numerical methods and their application to the investigation of outer shell instabilities.
M. Różyczka.
Astron. Astrophys., Vol. 143, No. 1, p. 59 – 71 (1985).

This paper is the first of a series on the evolution and stability of stellar wind bubbles. Interactions between winds and the interstellar medium are dealt with, but processes leading to wind generation and acceleration are not studied. In the present paper a second order two–dimensional hydrodynamical code is developed and shown to have clear advantages over widely used first-order schemes. Its application range is discussed and research perspectives are outlined.

064.011   Comptonization of low–frequency radiation in accretion disks: angular distribution and polarization of hard radiation.
R. A. Sunyaev (*R. A. Syunyaev*), L. G. Titarchuk.
Astron. Astrophys., Vol. 143, No. 2, p. 374 – 388 (1985).

In accretion disks the angular distribution and polarization of hard radiation forming via Comptonization, i.e. multiple scatterings in the disk, depend only on the optical thickness of the disk and are independent of either the photon frequency or the geometric distribution of low–frequency photons. This paper presents calculations of the angular distribution and polarization made for several values of the optical thickness of disks, for the case of the Thomson scattering. Specific astrophysical applications of the results obtained are discussed. A simplified theory of electron temperature dependence on the vertical coordinate in the disk is given. The density distribution over the vertical coordinate in the radiation–dominated disks is derived.

064.012   The scaling of the shear stress in an accretion disk.
H. Burm.
Astron. Astrophys., Vol. 143, No. 2, p. 389 – 392 (1985).

In most accretion disk models, the viscosity is scaled to the pressure with a parameter α. There are three pressure terms of interest: gas, magnetic and radiation pressure. If α is kept constant throughout the disk, scaling with the total pressure results in an unstable disk; scaling with the gas pressure results in a stable disk. It is demonstrated that scaling with the gas pressure is only justified, when the amplification of the magnetic field in the disk is limited by buoyancy. However, in the case of reconnection limited hydromagnetic turbulence, the shear stress scales with the total pressure.

064.013   The scaling of coronal models from one star to another.
R. Hammer.
NASA Conf. Publ., NASA CP–2358, p. 121 – 124 (1985). – See Abstr. 012.023.

The purpose of this paper is to discuss the requirements that must be met in order that stationary numerical corona models can be scaled from one star to another.

064.014   Overheated open coronal regions.
R. Hammer.
NASA Conf. Publ., NASA CP–2358, p. 125 – 129 (1985). – See Abstr. 012.023.

This paper studies the stability of stellar coronal shells when the coronal heating flux exceeds a certain limit, which depends on the damping length over which the energy is dissipated in the corona.

064.015   Co–rotating interaction regions in stellar winds: particle acceleration and non–thermal radio emission in hot stars.
D. J. Mullan.
NASA Conf. Publ., NASA CP–2358, p. 130 – 135 (1985). – See Abstr. 012.023.

A co–rotating interaction region (CIR) forms in a stellar wind when a fast stream from a rotating star overtakes a slow stream. The author points out the usefulness of CIR's in OB star winds to explain two properties of such winds: deposition of non–radiative energy in the wind far from the stellar surfaces and acceleration of non–thermal particles.

064.016   Synchrotron emission from chaotic stellar winds.
R. L. White.
NASA Conf. Publ., NASA CP–2358, p. 136 – 141 (1985). – See Abstr. 012.023.

A new model is presented for the radio emission from hot stars. Electrons are accelerated to relativistic energies by shocks in the wind near the star and emit radio radiation through the synchrotron mechanism. The particle energy spectrum and radio spectrum for this model are derived; the model can account for many of the observed characteristics of some recently discovered stars with peculiar radio emission.

064.017   Winds from rotating, magnetic, hot stars.
D. B. Friend.
NASA Conf. Publ., NASA CP–2358, p. 142 – 147 (1985). – See Abstr. 012.023.

This paper studies the question whether a magnetic field, coupled with stellar rotation, can enhance the wind from a hot star, which is driven primarily by line radiation pressure.

064.018   Non–radiative energy from differential rotation in hot stars.
K. L. Chan, H. G. Mayr, I. Harris.
NASA Conf. Publ., NASA CP–2358, p. 164 – 168 (1985). – See Abstr. 012.023.

A model of differential rotation is proposed for the upper radiative envelopes of hot stars, and it is suggested that the actions of such motions can be the source of energy for heating the coronae.

064.019   Radiatively driven winds and what they imply: a review of radiative driven instabilities and their importance in hot stars.
A. G. Hearn.
NASA Conf. Publ., NASA CP–2358, p. 188 – 198 (1985). – See Abstr. 012.023.

The mass loss and acceleration of the mass flow from hot stars is driven by the radiative forces of the resonance lines. These radiative forces can also cause instabilities to grow in the flow, and this is probably an important source of non–radiative heating and momentum in hot stars. In this paper the physical origin of these instabilities is described, and some of the problems of hot stars where radiative driven instabilities may play a part are briefly discussed.

064.020   Magnetic effects in the heating and modification of flows in the outer stellar atmospheres with application to early type stars.
Y. Uchida.
NASA Conf. Publ., NASA CP–2358, p. 199 – 220 (1985). – See Abstr. 012.023.

Possible magnetic effects in the heating and modification of flows in early type star atmospheres are discussed by referring to the physically related phenomena in late type stars, young stars, and close binary systems. It is pointed out that magnetic fields may play an important role also in early type star atmospheres in converting the energy of the radiatively driven outflow into heat, or in modifying the outflow by nozzling or by initial modification of the temperature and/or momentum distribution.

064.021 Effect of scattering on instabilities in line–driven stellar winds.
S. P. Owocki, G. B. Rybicki.
NASA Conf. Publ., NASA CP–2358, p. 221 – 225 (1985). – See Abstr. 012.023.

064.022 New instabilities in line driven winds.
P. C. H. Martens.
NASA Conf. Publ., NASA CP–2358, p. 226 – 232 (1985). – See Abstr. 012.023.

The author proposes a general three–dimensional treatment of the stability problem of line–driven stellar winds, which leads to the general dispersion equation. From this dispersion equation a new instability in stellar winds is derived: the 'thermal drift instability'. It is related to changes in absorption of radiation caused by temperature perturbations. This mechanism results in growing, inwardly propagating sound waves.

064.023 Radiative amplification of acoustic waves in hot stars.
B. E. Wolf.
NASA Conf. Publ., NASA CP–2358, p. 233 – 235 (1985). – See Abstr. 012.023.

The author proposes a model for the transport of mechanical energy into the outer atmospheres of hot stars by means of acoustic waves, which are forming from random fluctuations and are amplified by the radiation field.

064.024 Multi–line transfer and the dynamics of stellar winds.
D. C. Abbott, L. B. Lucy.
NASA Conf. Publ., NASA CP–2358, p. 241 – 242 (1985). Abstract. – See Abstr. 012.023.

064.025 Rapid non–LTE calculations of Balmer lines and hydrogen ionization: the solar case.
S. Dumont, S. Collin–Souffrin.
Astron. Astrophys., Vol. 144, No. 1, p. 245 – 248 (1985).

The authors propose an extremely rapid approximation method to determine a model of a plane–parallel atmosphere in hydrostatic equilibrium given the temperature distribution, the gravity, and the chemical composition. The populations of the energy levels of hydrogen, the principal component, as well as its degree of ionization, are determined without the assumption of local thermodynamic equilibrium (non–LTE) by an approximation method based on the escape probability. The comparison of the results thus obtained with those of Vernazza et al. (1981) for their Model C is satisfactory for the Balmer lines and for ionization in the low chromosphere and in the photosphere.

064.026 Does nucleation theory apply to the formation of refractory circumstellar grains?
B. Donn, J. A. Nuth.
Astrophys. J., Vol. 288, No. 1, p. 187 – 190 (1985).

The authors point out that kinetic factors inherent in cosmic systems will prevent the attainment of an approximate equilibrium precondensation cluster distribution and therefore such systems will violate one of the most fundamental assumptions of nucleation theory. It is shown that of the 11 refractory metal/metal oxide systems which have been studied experimentally, none are consistent with the predictions of either classical nucleation theory or the Lothe–Pound modification. The authors conclude that nucleation theory cannot be made to work in expanding circumstellar shells.

064.027 Properties and spectra of extended static model photospheres of M giants.
M. Scholz.
Astron. Astrophys., Vol. 145, No. 1, p. 251 – 261 (1985).

Extended model photospheres in hydrostatic equilibrium have been computed for M type giants with effective temperatures $3800K \geqslant T_{eff} \geqslant 2500K$. Surface gravities g_s range from 0.1 to $6\ cm\ s^{-2}$, and photospheric extensions d range from plane–parallel stratifications to more than 50% corresponding to different mass–luminosity combinations. Solar, metal–rich and metal–poor compositions are considered as well as modifications of CNO abundances representing dredge–up of processed material. The properties of the models and their spectra are discussed. Various narrow band color indices are tested as indicators of T_{eff}, g_s and d. Metal abundance variations as small as a factor of 3 seriously affect most color indices whereas CNO abundance modifications are hardly detectable. Wavelength dependent radius determinations are briefly discussed.

064.028 Approximate formulae for radiative acceleration in stars.
G. Alecian.
Astron. Astrophys., Vol. 145, No. 1, p. 275 – 277 (1985).

The author proposes an analytical expression for the radiative acceleration due to absorption of photons through bound–bound transitions, which is valid in optically thick media, when pressure broadening is dominant. He then derives approximate formulae for the numerical evaluation of radiative acceleration in Ap star envelopes.

064.029 Erratum: "Acoustic waves in early–type stars. I. An efficient method for the computation of thermodynamic quantities in time–dependent stellar atmosphere calculation" [Astron. Astrophys., Vol. 127, No. 1, p. 93 – 96 (1983)].
B. E. Wolf.
Astron. Astrophys., Vol. 145, No. 1, p. 278 (1985). See Abstr. 034.064.029.

064.030 Empirical–theoretical structural patterns for stellar atmospheres and their local environment relative to variable mass–loss.
R. N. Thomas.
Effects of variable mass loss on the local stellar environment, p. 3 – 21 (1984). – See Abstr. 012.028.

064.031 The interaction of the stellar wind with the local environment.
S. Kwok.
Effects of variable mass loss on the local stellar environment, p. 175 – 186 (1984). – See Abstr. 012.028.

Contents: Introduction. Interstellar bubbles. Planetary nebulae. Ring nebulae around Wolf–Rayet stars. Supernovae. Novae. T Tauri stars. High velocity flows near young stars. Thermal structure of the wind interaction regions. Ionization fronts. Conclusions.

064.032 Multiline transfer and the dynamics of stellar winds.
D. C. Abbott, L. B. Lucy.
Astrophys. J., Vol. 288, No. 2, p. 679 – 693 (1985).

A Monte Carlo technique for treating multiline transfer in stellar winds is decribed and tested. With a line list containing many thousands of transitions and with fairly realistic treatments of ionization, excitation, and line formation, the resulting code allows the dynamical effects of overlapping lines to be investigated quantitatively, as well as providing the means of directly synthesizing the complete spectrum of a star and its wind. As a result of this improved treatment of multiline transfer, the computed mass loss rate for ζ Puppis and the synthesized spectrum of ζ Puppis agree with observational data.

064.033 Orbit perturbation: evolution of a Keplerian disk.
R. Pellat, P. Barge, P. Hornung, J. M. Millet.
Adv. Space Res., Vol. 4, No. 9, p. 95 – 105 (1984). – See Abstr. 012.029.

The dynamical evolution of a Keplerian disk of matter around a star or around a planet is a crucial point for the understanding of the formation of solar systems or for the formation of planetary rings. In this paper the authors adopt a general point of view: they describe the disk by a distribution function in phase space and follow its evolution in presence of a stochastic force or binary collisions.

064.034   Model atmosphere for the bright component of the binary system υ Sgr.
V. V. Leushin, G. P. Topil'skaya.
Astrofizika, Tom 22, Vyp. 1, p. 121 – 135 (1985). In Russian. English translation in Astrophysics, Vol. 22, No. 1.

A set of model atmospheres for the bright component of the system υ Sgr has been computed with $10000° \leqslant T_e \leqslant 14000°$ and $1.0 \leqslant \log g \leqslant 4.0$. The chemical composition for the models was obtained from analysis of spectra. The helium abundance was taken in the ranges 0.70 – 0.95, hydrogen $0.01 – 10^{-6}$ and heavy elements 0.05 – 0.30 by number of atoms. The theoretical continuum and the character of the models have been compared with the observed spectrum. The atmosphere of υ Sgr is in agreement with the models: $T_e = 13500° \pm 200°$, $\log g = 1.5 \pm 0.25$, and $T_e = 14000° \pm 200°$, $\log g = 2.0 \pm 0.25$.

064.035   Stellar winds driven by super–Eddington luminosities.
T. Quinn, B. Paczyński.
Astrophys. J., Vol. 289, No. 2, p. 634 – 643 (1985).

The authors constructed over 100 models of a steady–state, spherically symmetric outflow of gas from a star of $1.4\ M_\odot$ with a pure helium envelope and a luminosity somewhat exceeding the Eddington limit. Super–Eddington luminosities were made possible by the decrease of the electron–scattering opacity at the high temperature at which accreted helium burns on neutron stars. The authors explored a large region in the mass–loss rate $(\dot{M})$ – energy–loss rate $(\dot{E})$ diagram, and they found self–consistent models for $1.01 < \dot{E}/L_{\mathrm{Edd}} < 1.11$, and for $16.5 < \log \dot{M} < 19.5$ (in grams per second). The models have photospheric radii much larger than neutron star radii, and they are faint in the X–ray region of the spectrum. Applications to sources of X–ray bursts with precursors are discussed.

064.036   The structures and spectra of magnetic, line–blanketed model atmospheres.
K. G. Carpenter.
Astrophys. J., Vol. 289, No. 2, p. 660 – 668 (1985).

Magnetic, line–blanketed model atmospheres for upper–main-sequence stars with normal elemental abundances and a slightly distorted dipolar magnetic field have been constructed. These were computed with a modified version of the Kurucz ATLAS6 model atmosphere code and newly computed opacity distribution functions (ODFs), which take into account the Zeeman splitting of the contributing atomic lines. The enhanced blanketing represented by the "magnetic" ODFs in combination with the structure changes and Zeeman broadening of individual lines causes the emergent spectrum to vary with viewing inclination and to differ from the nonmagnetic case. The calculations are compared with observations of Ap stars.

064.037   Synchrotron emission from chaotic stellar winds.
R. L. White.
Astrophys. J., Vol. 289, No. 2, p. 698 – 708 (1985).

A new model is presented for the radio emission from hot stars. Electrons are accelerated to relativistic energies by shocks in the wind near the star and emit radio radiation through the synchrotron mechanism. The particle energy spectrum and radio spectrum for this model are derived; the model can account for many of the observed characteristics of some recently discovered stars which have peculiar radio emission. These particles may make hot stars significant sources of γ–rays, cosmic rays, or both.

064.038   Extreme–ultraviolet emission from cool star outer atmospheres.
M. Landini, B. C. Monsignori Fossi, F. Paresce, R. A. Stern.
Astrophys. J., Vol. 289, No. 2, p. 709 – 720 (1985).

Extreme–ultraviolet line and continuum emission spectra are computed for cool stars with varying levels of coronal and chromospheric activity. The stars' visible surfaces are covered by a network of identical solar–like magnetic loop structures. The density and thermal structures of the loops are determined by stationary solutions of the appropriate coupled mass, momentum, and energy balance equations including self–consistent convective, conductive, and radiative input and loss terms. The model's boundary conditions are constrained by available IUE and Einstein data. Calculated loop models and associated EUV emission spectra are given for α Cen, the Hyades stars BD +15°640 and BD +14°693, and the very active triple star HD 165590.

064.039   On the relevance of the MK system and process to the theory of stellar atmospheres.
D. Mihalas.
The MK process and stellar classification, p. 4 – 16 (1984). – See Abstr. 012.033.

064.040   Derivation of atmospheric structure from emission–line fluxes.
C. Jordan.
Mon. Not. R. Astron. Soc., Vol. 214, No. 1, p. 1P – 4P (1985).

It is shown how the pressure and temperature gradient in a stellar chromosphere and corona are related to the emission measure distribution. The conditions under which a simple approximation for the temperature versus height suggested by Lago, Penston & Johnstone is appropriate are critically discussed.

064.041   On models of extended stellar atmospheres.
T. Kipper.
Tartu Astrofüüs. Obs. Publ., Tom 50, p. 50 – 53 (1984).

Discussed is the applicability of the method suggested by Peraiah and Grant (1973) for calculation of models of extended atmospheres of late spectral type stars for solution of the equation of radiative transfer in spherical envelopes. The program realizing this method is briefly described.

064.042   The solution of the NLTE problem for magnesium in the atmosphere of an M2 giant.
Ya. V. Pavlenko.
Tartu Astrofüüs. Obs. Publ., Tom 50, p. 54 – 67 (1984). In Russian.

The methods and the results of the solution of the NLTE problem for magnesium in the atmosphere of an M2 giant $(T_{\mathrm{eff}} = 3800\mathrm{K},\ \lg g = 1.5)$ are described. The five–level model of the magnesium atom was used. The NLTE problem was solved by the partial linearization method. The depth dependence of the source functions and the thermalization processes of the transitions of magnesium were investigated.

064.043   Radiative balance in transitions of magnesium in the atmosphere of an M2 giant.
Ya. V. Pavlenko.
Tartu Astrofüüs. Obs. Publ., Tom 50, p. 68 – 80 (1984). In Russian.

The NLTE problem was solved for magnesium in the atmosphere of an M2 III giant $(T_{\mathrm{eff}} = 3800\mathrm{K},\ \lg g = 1.5)$. The five–level model of the Mg I atom was used. It was assumed that the bound–bound radiative transitions are balanced. Three forms of radiative balance were investigated. The radiative transfer equation was solved for the frequencies of bound–bound transitions. The population of the levels and the source functions were compared with the solution of the "complete" NLTE problem, when the radiative transfer equation was solved for both bound–bound and bound–free transitions.

064.044   Mass loss rates of Wolf–Rayet stars and the velocity structure of their winds.
T. Nugis.
Tartu Astrofüüs. Obs. Publ., Tom 50, p. 101 – 151 (1984).

From the study of the line profiles and the IR flux distributions of ten WR stars it is concluded that their matter flows are accelerated up to about two stellar radii, then there follows a decelerating region where the flow velocity decreases 2 – 3 times and further on the velocity remains constant. The $\dot{M}$ values found for 24 WR stars range from 2×10^{-5} up to $2 \times 10^{-4} M_\odot \mathrm{y}^{-1}$.

064.045   Computed He II spectra for Wolf–Rayet stars.
W.–R. Hamann.
Astron. Astrophys., Vol. 145, No. 2, p. 443 – 448 (1985).

Synthetic spectra of spherically symmetric, expanding stellar atmospheres of pure helium are calculated in non–LTE for a given atmospheric structure. The radiation transfer is solved in the "comoving frame", and consistency with the multilevel atom rate equations is achieved iteratively by an "equivalent two level atom approach". A small grid of models is calculated. He II emission lines as typically observed in WR stars are obtained, if the effective temperature of the star lies above 35kK and the mass loss rate exceeds $10^{-5} M_\odot$/yr.

064.046   Two–photon cyclotron emission in accretion columns.
J. G. Kirk, D. B. Melrose, J. G. Peters.
Proc. Astron. Soc. Aust., Vol. 5, No. 4, p. 478 – 480 (1984).
The authors have argued in this paper that two–photon emission (and absorption) may play an important role in forming the spectrum near the cyclotron line in the emission from an accretion column above a magnetized neutron star.

064.047   The stability of stellar atmospheres.
J. R. Auman.
Proc. Astron. Soc. Aust., Vol. 5, No. 4, p. 489 – 493 (1984).
The question of the stability of stellar atmospheres has essentially a non–local character and must be considered globally for the atmosphere as a whole. The present paper describes the start of a program to make such an analysis.

064.048   Equilibrium model of thin magnetic flux tubes.
G. Bodo, A. Ferrari, S. Massaglia, W. Kalkofen, R. Rosner.
ESA Spec. Publ., ESA SP–207, p. 277 – 280 (1984). – See Abstr. 012.044.

064.049   Stellar granulation.
D. F. Gray.
ESA Spec. Publ., ESA SP–220, p. 211 – 212 (1984). – See Abstr. 012.045.
Properties of stellar granulation are inferred from observed line bisectors.

064.050   UV and X–ray spectroscopy as a tool for the diagnostics of non–thermal processes in the atmospheres of hot stars.
R. Stalio.
Mem. Soc. Astron. Ital., Vol. 55, No. 3, p. 569 – 574 (1984). – See Abstr. 012.050.
The author summarizes the following points: (a) The thermal model of stellar atmospheres and their inadequacy in explaining an important part of the observations. (b) The role of UV and X–ray spectroscopy in fixing the non–thermal structure of the atmospheres of hot stars. (c) The kind of information of these non–thermal processes that one expects from the future FUV (100 – 912 Å) and FUV (912 – 1200 Å) missions. The "EUV Spectral Imager" is presented.

064.051   Non–LTE effects in the atmospheres of F type supergiants. I. Over–ionization of Fe I atoms.
A. A. Boyarchuk, L. S. Lyubimkov, N. A. Sakhibullin.
Astrofizika, Tom 22, Vyp. 2, p. 339 – 356 (1985). In Russian. English translation in Astrophysics, Vol. 22, No. 2.
Non–LTE computations of the Fe I–Fe II ionization balance are performed for a sample of F type supergiants and dwarfs. It is shown that the departures from LTE lead to over–ionization of Fe I atoms in upper layers of the atmospheres.

064.052   Molecular structure in the envelopes of metal–rich red giants.
S. F. C. Rossi, W. J. Maciel.
Rev. Bras. Fis., Vol. 13, No. 4, p 735 – 746 (1983). Abstr. in Phys. Abstr., Vol. 88, No. 1255, Entry 46065 (1985).

064.053   The gravity dependence of the Hα width in late–type stars.
D. M. Zarro.
Astrophys. J., Vol. 291, No. 1, p. 297 – 299 (1985).
The author develops a theoretical gravity–scaling law for the Hα absorption width in late–type stars. The derivation is based upon (1) the hydrostatic thickening of stellar chromospheres with decreasing surface gravity, and (2) a dependence of the Hα width

upon opacity and Doppler width in a region subject to a chromospheric temperature rise. The scaling relation is approximately consistent with the mean gravity dependence deduced from the empirical correlation between Hα and Ca II K Wilson–Bappu widths. The calculations suggest that gravity variations in chromospheric–mass column density may, in addition to Doppler velocity enhancements, control the width–luminosity broadening of the Hα profile in late–type stars.

064.054   Magnetohydrodynamic thermal instabilities in cool inhomogeneous atmospheres.
G. Bodo, A. Ferrari, S. Massaglia, R. Rosner, G. S. Vaiana.
Astrophys. J., Vol. 291, No. 2, p. 798 – 805 (1985).
The stability of magnetic loops to current–driven filamentation instabilities is investigted. The unperturbed atmosphere is assumed to be composed of an (upper) isothermal, optically thin, low–density portion and a (lower) higher density portion which is in radiative equilibrium. Conditions appropriate for the surface of a solar–like star are adopted. The authors perform a linear stability analysis; numerical results show that physically plausible current densities, which would be generated by typical loop foot-point motions, are effective in driving magnetohydrodynamic instabilities in such plasma.

064.055   Occurrence of optically thick winds.
M. Kato.
Publ. Astron. Soc. Jpn., Vol. 37, No. 1, p. 19 – 30 (1985).
Surface regions of stars are examined in relation to the occurrence of steady mass loss in which the gas is accelerated inside the photosphere. Static solutions for the inside of the photosphere are found to exist only if the ratio of the luminosity to the Eddington luminosity at the surface, l_s, is smaller than the critical value, l_s^{max}, which is very near to unity. In the envelopes with large l_s ($\leqslant l_s^{max}$), the thermal energy in the surface region is as large as the gravitational energy. Solutions with the critical value l_s^{max} have structures very similar to those of the steady mass–loss solutions. This indicates that, when the luminosity increases, the envelope begins to lose its mass immmediately after static solutions cease to exist. Several envelope models are calculated for neutron stars and white dwarfs and compared with mass–loss solutions.

064.056   Cooling of magnetic flux tubes as a mechanism for suppression of magnetic buoyant escape of the flux tubes from the Sun and stars.
H. Yoshimura.
Publ. Astron. Soc. Jpn., Vol. 37, No. 1, p. 171 – 181 (1985).
A new mechanism to suppress the magnetic buoyancy of flux tubes in the solar and stellar convection zones is proposed as a solution of the enigma that the very convective stratification, that drives plasma fluid motions to generate magnetic fields, is unfavorable to keep the generated fields because of the magnetic buoyancy. The mechanism is cooling of flux tubes by inhibition of convective heat transport by magnetic fields inside the tubes that could also be responsible for the cooling of sunspots, starspots and pores. It is suggested that the cooling mechanism as a stabilizing mechanism of the magnetic buoyancy is plausible.

064.057   The Chandrasekhar H–function for exponentially varying atmosphere.
M. H. Haggag, H. M. Machali.
Astrophys. Space Sci., Vol. 111, No. 1, p. 189 – 195 (1985).
The Chandrasekhar H–function for exponentially–varying atmosphere is calculated. The calculations have been carried out on the assumption that the scattering function behaves also exponentially. Numerical results are given for H and for the emergent flux at the surface.

064.058   Improved methods of predicting stellar atmospheres from known atmospheres in the presence of convection.
A. W. Irwin.
Astron. Astrophys., Vol. 146, No. 2, p. 282 – 292 (1985).
The well–known temperature scaling technique has often been used to predict stellar atmospheric models from a known model. This paper investigates two alternatives to temperature scaling

which predict model atmospheres using pre–determined corrections to the diffusion approximation or the grey approximation. Diffusion prediction gives the same and grey prediction nearly the same results as temperature scaling in the outer radiative zone. The real improvement over the temperature scaling technique occurs in the convection zone, where diffusion prediction gives better results, and grey prediction gives the best results.

064.059   Hydrogen–deficient atmospheres for cool carbon stars.
H. R. Johnson, D. R. Alexander, C. D. Bower, D. A. Lemke, D. G. Luttermoser, J. P. Petrakis, M. D. Reinhart, K. A. Welch, J. H. Goebel.
Astrophys. J., Vol. 292, No. 1, p. 228 – 230 (1985).

Motivated by recent work which hints at a possible deficiency of hydrogen in non–Mira N–type carbon stars, the authors have computed a series of hydrogen–deficient models for carbon stars. For these models T_{eff} = 3000K, and log g = 0.0. Solar abundances are used for all elements except for carbon (which is enhanced to give C/O = 1.05), hydrogen, and helium. As the fractional abundance of hydrogen is decreased, being replaced by helium, the temperature–optical depth relation is affected only slightly, but the temperature–pressure relation is changed.

064.060   Linear and quadratic limb–darkening coefficients for a large grid of LTE model atmospheres.
R. A. Wade, S. M. Rucinski.
Astron. Astrophys., Suppl. Ser., Vol. 60, No. 3, p. 471 – 484 (1985).

The authors tabulate linear and quadratic limb–darkening coefficients for the model atmosphere grid of Kurucz (1979). They note some self–consistency problems with the Kurucz fluxes as published. They discuss the applicability of low–order polynomial approximations to the limb–darkening law, especially for low effective temperatures at short wavelengths.

064.061   Magnetic stellar winds. A 2–D generalization of Weber–Davis model.
T. Sakurai.
MPA Rep., No. 174, 33 pp. (1985). Submitted to Astron. Astrophys.

064.062   Two–dimensional models of stellar wind bubbles. V. The efficiency of momentum transfer between winds and dense clumps in free wind regions.
M. Rozyczka, G. Tenorio–Tagle.
MPA Rep., No. 176, 23 pp. (1985). To appear in Acta Astron.

064.063   Infrared excesses in Be stars: an axisymmetric model.
K. S. Bjorkman, L. B. F. M. Waters, H. J. G. L. M. Lamers.
Bull. Am. Astron. Soc., Vol. 16, No. 4, p. 894 (1984). Abstract. – See Abstr. 010.062.

064.064   A periodic shock wave model for Mira variable atmospheres.
E. Bertschinger, R. A. Chevalier.
Bull. Am. Astron. Soc., Vol. 16, No. 4, p. 898 – 899 (1984). Abstract. – See Abstr. 010.062.

064.065   A non–radial radiation force in hot star winds.
D. B. Friend, K. B. MacGregor.
Bull. Am. Astron. Soc., Vol. 16, No. 4, p. 899 (1984). Abstract. – See Abstr. 010.062.

064.066   Kinetic efficiencies of stellar wind bubbles.
D. Van Buren.
Bull. Am. Astron. Soc., Vol. 16, No. 4, p. 938 (1984). Abstract. – See Abstr. 010.062.

064.067   Disk accretion by magnetic stars.
G. J. Zylstra, F. K. Lamb, J. J. Aly, H. Cohn.
Bull. Am. Astron. Soc., Vol. 16, No. 4, p. 944 (1984). Abstract. – See Abstr. 010.062.

064.068   Continuum spectral flux from Type II supernova model atmospheres.
S. Hershkowitz, E. Linder, R. V. Wagoner.
Bull. Am. Astron. Soc., Vol. 16, No. 4, p. 971 (1984). Abstract. – See Abstr. 010.062.

064.069   Effect of scattering on the instability of radiation–driven stellar winds.
G. B. Rybicki, S. P. Owocki.
Bull. Am. Astron. Soc., Vol. 16, No. 4, p. 993 (1984). Abstract. – See Abstr. 010.062.

064.070   On the influence of density and temperature fluctuations on the formation of spectral lines in stellar atmospheres.
J. Stahlberg.
Astron. Nachr., Vol. 306, No. 3, p. 137 – 144 (1985).

A method taking into account the influence of temperature and density fluctuations generated by the velocity field in stellar atmospheres on the formation of spectral lines is presented. The influenced line profile is derived by exchanging the values in a static atmosphere by a mean value and a fluctuating one. The correlations are calculated with the help of the well–known hydrodynamic equations. It results that in normal stellar atmospheres the visual lines are only very weakly influenced by such fluctuations due to the small values of the gradients of the pressure and density and of the velocity dispersion.

064.071   Optically thick radial accretion onto main sequence stars and degenerate dwarfs.
R. Tylenda.
Astron. Astrophys., Vol. 147, No. 2, p. 197 – 201 (1985).

This paper presents a simple method which allows to determine the radius and temperature of the effective photosphere for an optically thick accretion flow. If one defines $\dot{M}_T$ as an accretion rate at which a transition between the optically thin and optically thick regimes occurs then for main sequence stars $\dot{M}_T \approx 10^{-3} \dot{M}_{Edd}$ has been found. In the case of degenerate dwarfs $\dot{M}_T$ is a function of the dwarf mass and varies from $5 \times 10^{-3} \dot{M}_{Edd}$ for 0.15 $M_\odot$ to 0.5 $\dot{M}_{Edd}$ for 1.35 $M_\odot$. It is suggested that YY Orionis stars accrete at a rate close to $\dot{M}_T$. In binary systems the radial accretion onto a main sequence star or a degenerate dwarf from a strong, variable, anisotropic wind of an M giant companion may lead to phenomena similar to those observed in symbiotic stars.

064.072   Two–dimensional models of stellar wind bubbles. II. Variable mass–loss rates and the possibility of outer-shell fragmentation in relation to the origin of interstellar bullets.
M. Różyczka, G. Tenorio–Tagle.
Astron. Astrophys., Vol. 147, No. 2, p. 202 – 208 (1985).

Models of stellar wind bubbles are examined for stability against shell disruption due to increasing luminosity of the wind. The disruption advances most quickly in a high density model whose shocked wind region is degenerated to a thin and cool layer of gas. In low density models, where shocked wind regions are broad and hot, the development of shell instabilities is markedly slower. The possibility is discussed that fragments obtained in the high density model evolve into interstellar bullets which could account for some of Herbig–Haro objects.

064.073   Two–dimensional models of stellar wind bubbles. III. Self–confining flows in media with strong density gradients.
M. Różyczka, G. Tenorio–Tagle.
Astron. Astrophys., Vol. 147, No. 2, p. 209 – 219 (1985).

Hydrodynamic models of stellar wind bubbles evolving in a medium with strong density gradients are presented. Wind parameters are representative of T Tauri stars. The nonuniformity of the medium is caused by the presence of compact circumstellar discs which are simulated by plane–parallel stratified slabs. The bubbles break out of the discs and evolve into secondary bubbles and/or into hollow cylinders. In the latter case self–confining flows are established due to combined effects of cooling and momentum conservation and a significant degree of collimation of originally isotropic winds can be achieved. During breakout the outer shock front of the bubble may separate from the shell, propagate to large distances from the wind source and drive a bipolar molecular outflow.

064.074 Two–dimensional models of stellar wind bubbles. IV. Properties of bow shocks around dense clumps in free wind regions.
M. Ròżyczka, G. Tenorio–Tagle.
Astron. Astrophys., Vol. 147, No. 2, p. 220 – 226 (1985).

Numerical models of stationary bow shocks produced by cool stellar winds flowing past dense cloudlets are presented. A velocity range representative of Herbig–Haro objects $(25 - 400$ km s$^{-1})$ is explored, and the winds are highly supersonic (Mach number > 50). In agreement with the bow shock theory shapes, dimensions and density distributions of all adiabatic models are identical. However, when allowances for cooling are made, they become very sensitive to the cooling efficiency which peaks in the 200 km s^{-1} model. Maps of the emission integrated over all frequencies are obtained and the efficiency of the emission is calculated. The results are discussed in relation to Herbig–Haro objects.

064.075 The reflection effect in model stellar atmospheres. I. Grey atmospheres with convection.
L. P. R. Vaz, Å. Nordlund.
Astron. Astrophys., Vol. 147, No. 2, p. 281 – 299 (1985).

The effects of the mutual illumination of the components of binary systems are investigated, by introducing an external radiation field in a model for plane–parallel stellar atmospheres in radiative + convective equilibrium. For grey atmospheres in radiative equilibrium, the results are verified against exact solutions. Changes in the limb–darkening due to illumination are also discussed and it is shown that, at least for grey atmospheres, a convenient numerical expression may be given for the reduction of the limb–darkening as a function of frequency, angle of incidence, and relative incident flux.

064.076 Formation of bipolar flow by the stellar wind embedded in a molecular disk.
S. Sakashita, H. Hanami, M. Umemura.
Astrophys. Space Sci., Vol. 111, No. 2, p. 213 – 223 (1985).

The interaction of the isotropic stellar wind with the rotating isothermal cloud surrounding the young star is investigated. The density distribution of the cloud is taken as that for the equilibrium state of the rotating isothermal cloud modified by adding the rarefied interstellar gas in the polar region. The development of the shock envelope and the structure of the shell induced by the stellar wind are obtained. It is shown that the envelope of the shock front elongates and opens to the polar direction with half opening angle of about 20 degrees resulting the bipolar flow which is able to reproduce well the observed properties for the outflow in the bipolar sources.

064.077 Euler potential method in three–dimensional stellar wind problems.
O. Kaburaki.
Astrophys. Space Sci., Vol. 112, No. 1, p. 157 – 174 (1985).

A theoretical scheme is developed to deal with the problems of stellar winds in three–dimensional situations, and relativistic fluid equations are integrated formally under isentropic and quasi–stationary conditions, in a flat space–time.

064.078 On the initial phase of interaction between expanding stellar envelopes and surrounding medium.
D. K. Nadyozhin (*D. K. Nadezhin*).
Astrophys. Space Sci., Vol. 112, No. 2, p. 225 – 249 (1985).

The initial stages of deceleration in the circumstellar medium of a stellar envelope, thrown off by a shock wave, are investigated. The equations of spherical–symmetric adiabatic hydrodynamics are shown to have a similarity solution in the case of the density of the expanding envelope being approximated by a reasonable power law. The results obtained can be employed to describe the interaction between the exploding core of a red giant star and its rarefied envelope. This is of interest for explosive nucleosynthesis. The similarity solution is applied to the envelopes expelled both by type–II supernovae and by rapid novae.

064.079 Self–consistent electromagnetic field in a nearly co–rotating magnetosphere.
O. Kaburaki.
Astrophys. Space Sci., Vol. 112, No. 2, p. 287 – 301 (1985).

An example of the self–consistent solution which belongs to the non–trivial solution, obtained in a previous paper (Kaburaki, 1985), is found in a nearly co–rotating inner magnetosphere. Though the stellar wind is neglected there compared with the co–rotational velocity, drift motion around the magnetic axis, which is a manifestation of inertial effects, is determined self–consistently with the electromagnetic field. In this process, explicit expressions for the energy integral in the rotating frame and for the density distribution are also obtained.

064.080 Three–dimensional structures of magnetostatic atmospheres. I. Theory.
B. C. Low.
Astrophys. J., Vol. 293, No. 1, p. 31 – 43 (1985).

This paper presents two families of magnetostatic equilibrium states for two atmospheres, one with a uniform gravity and the other with a gravity due to a point mass in spherical geometry. Taking the electric current to be everywhere perpendicular to the gravitational force, the fully three–dimensional problem is shown to be amenable to analytic treatment. A subset of the new magnetostatic solutions have magnetic topologies identical to those of the set of all potential magnetic fields. This subset of solutions are analyzed as illustrative examples, with an application to model the three–dimensional magnetic structure of cool plasma loops, often observed in the EUV corona.

064.081 Carbon–enriched stellar envelopes: nuclei of planetary nebulae and R Coronae Borealis stars.
I.–J. Sackmann, A. I. Boothroyd.
Astrophys. J., Vol. 293, No. 1, p. 154 – 164 (1985).

Envelopes rich in carbon were computed, taking envelope carbon opacities into account that were hitherto unavailable. All effects of carbon partial ionizations were fully included. The authors investigated stars with envelope carbon content $X_C = 0.1$ and $X_C = 0.31$ (fraction by weight), and compared them to stars of normal composition. They investigated R Coronae Borealis stars and nuclei of planetary nebulae, considering high–luminosity objects of $\log(L/L_\odot) = 4.1$, of total stellar mass $0.815\ M_\odot$, and ranging in effective temperature from $\log T_e = 3.5$ to 5.3. The effects of carbon on the total opacity and on the structure of the convective zone are discussed in detail.

064.082 Amplification and propagation of sound waves in early–type stars.
B. E. Wolf.
Diss. Naturwiss.–Math. Gesamtfak. Ruprecht–Karls–Univ., Heidelberg, F.R. Germany. 2 + 262 pp. (1985).

064.083 Instabilities in non–alpha accretion disk models.
B. A. Fryxell, R. E. Taam.
Bull. Am. Astron. Soc., Vol. 17, No. 1, p. 517 (1985). Abstract. – See Abstr. 010.064.

064.084 Formation of silicate crystals and/or glasses in the circumstellar environment.
J. Svatoš.
Bull. Astron. Inst. Czech., Vol. 36, No. 3, p. 172 – 174 (1985).

A quantitative approach to distinguish the formation of silicate glasses and crystals under circumstellar conditions is given. It is shown that the formation of amorphous (glassy) silicate grains in oxygen–rich stars is much more probable than the crystalline ones.

064.085   Effects of multiquantum transitions on molecular populations in grain–forming circumstellar environments
J. A. Nuth, M. Wiant, J. E. Allen Jr.
Astrophys. J., Vol. 293, No. 2, p. 463 – 469 (1985).

To examine the effects of multiquantum translation to vibration transitions in expanding circumstellar envelopes, vibrational populations of the lowest 20 levels of CO have been calculated as a function of pressure and radiation density for H atom–CO collisions. Significant departure from local thermodynamic equilibrium is indicated, which implies lower dissociation rates for molecular components and a subsequent enhancement in the rate of grain formation by many orders of magnitude. Stabilization of intermediate species before they can dissociate may facilitate the formation of refractory grain cores in very hot, dilute outflows.

064.086   Radiation transfer in accretion disk coronae.
R. A. London.
Cataclysmic variables and low–mass X–ray binaries, p. 121 – 132 (1985). – See Abstr. 012.072.

The physics of X–ray excited accretion disk coronae is described. A numerical model for the hydrostatic inner region, including radiation transfer effects, is presented. The resulting appearance is compared to recent observations of X–ray dips in several X–ray binaries. The author also discusses the dynamical effects of X–ray induced mass loss, which originates at large radius in the disk.

064.087   Outflow of matter from stars at high radiative pressure.
L. V. Tambovtseva.
Tr. Astrofiz. Inst. Alma–Ata, Tom 41, p. 37 – 59 (1983). In Russian.

064.088   Stellar lineshifts induced by photospheric convection.
D. Dravins.
Stellar radial velocities, p. 311 – 320 (1985). – See Abstr. 012.095 (IAU Colloq. No. 88).

Effects of stellar atmospheres on measured radial velocities are examined. Surface convection ("stellar granulation") causes photospheric line asymmetries and wavelength shifts of $\cong$ 100 – 500 m/s. Cyclic changes in the convection patterns, such as observed during the solar 11–year cycle, may mimic radial velocity variations of perhaps 30 m/s. The study of stellar atmospheres would benefit from accurate ($<$ 100 m/s) differential radial velocity measurements among lines of different parameters (strength, excitation potential, wavelength region) in the same star.

064.089   A method for determination of the electron density in stellar atmospheres.
V. G. Cholakyan.
Pis'ma Astron. Zh., Tom 11, No. 6, p. 458 – 462 (1985). In Russian. English translation in Sov. Astron. Lett., Vol. 11.

Using the ultraviolet resonance lines Mg I 2852 and Mg II 2795 the mean electron density is determined in the atmospheres of 32 (A – G)–type stars. The electron density drops quickly from A stars towards late spectral types.

064.090   Model atmospheres for peculiar red giant stars.
H. R. Johnson.
Cool stars with excesses of heavy elements, p. 271 – 292 (1985). – See Abstr. 012.101.

An overview of model atmospheres available for peculiar red giant stars is given. After listing the assumptions of the classic stellar atmosphere, the author examines the prospects for generalizing each of these. An examination of the parameter space in effective temperature, surface gravity, and chemical composition covered by published atmospheres is presented. Tests of the adequacy of current models as regards both self–consistency and agreement with stellar observations are dicussed. Several applications of models, especially in calibrating the temperature scale, are described. Finally, spherical models and chromospheres for peculiar red giants are discussed.

064.091   Model atmospheres for M (super–) giants with different abundances of the heavy metals and the CNO group.
R. Wehrse.
Cool stars with excesses of heavy elements, p. 293 – 294 (1985). – See Abstr. 012.101.

064.092   HCN in stellar atmospheres – a quantum mechanical calculation.
U. G. Jørgensen.
Cool stars with excesses of heavy elements, p. 301 – 305 (1985). – See Abstr. 012.101.

CI–CASSCF calculations of transitions between the 360 lowest vibrational energy levels in HCN are discussed, together with results of its application to model stellar atmospheres of cool carbon stars.

064.093   Dust formation in stellar winds.
H.–P. Gail, E. Sedlmayr.
Cool stars with excesses of heavy elements, p. 307 – 312 (1985). – See Abstr. 012.101.

Classical nucleation theory is applied to grain formation in C–rich cool stellar winds of heavily obscured objects. The effectivity of the formation process and its implications for the structure of the shells are discussed.

064.094   A disk model of M8E–IR.
D. M. Peterson, M. Simon.
Bull. Am. Astron. Soc., Vol. 17, No. 2, p. 557 (1985). Abstract. – See Abstr. 010.065.

064.095   Radio spectra of collimated, ionized stellar winds.
S. P. Reynolds.
Bull. Am. Astron. Soc., Vol. 17, No. 2, p. 564 (1985). Abstract. – See Abstr. 010.065.

064.096   Chromospheric dust formation and mass loss.
R. E. Stencel.
Bull. Am. Astron. Soc., Vol. 17, No. 2, p. 569 (1985). Abstract. – See Abstr. 010.065.

064.097   The infra–red colors of cool model stellar atmospheres.
R. A. Bell.
Bull. Am. Astron. Soc., Vol. 17, No. 2, p. 569 (1985). Abstract. – See Abstr. 010.065.

064.098   The effect of gravity on the X–ray and UV emission of cool stars.
S. K. Antiochos.
Bull. Am. Astron. Soc., Vol. 17, No. 2, p. 570 (1985). Abstract. – See Abstr. 010.065.

064.099   3D structures in turbulent convection and how to read them.
K. L. Chan, S. Sofia.
Bull. Am. Astron. Soc., Vol. 17, No. 2, p. 592 – 593 (1985). Abstract. – See Abstr. 010.065.

064.100   Report of IAU Commission 36: Theory of stellar atmospheres (*Théories des atmosphères stellaires*).
B. Gustafsson.
Trans. IAU, Vol. XIXA, p. 503 – 519 (1985). – See Abstr. 003.046.

064.101   Observed and computed stellar line profiles: the roles played by partial redistribution, geometrical extent and expansion.
J. L. Linsky.
Progress in stellar spectral line formation theory, p. 1 – 26 (1985). – See Abstr. 012.108.

The author reviews partial redistribution (PRD) radiative transfer with emphasis on the complex interaction of observations and theoretical predictions of spectral line shapes. He summarizes the work that has led to "realistic" plane parallel static chromospheric models for the Sun and other late–type stars. The author then discusses the various roles played by atmospheric extension and expansion (winds) in determining resonance line profile shapes, and summarizes the existing PRD calculations for late–type stars.

064.102   The effect of abundance values on partial redistribution line computations.
R. Freire Ferrero, P. Gouttebroze.
Progress in stellar spectral line formation theory, p. 125 – 136 (1985). – See Abstr. 012.108.

This paper discusses line formation theory for so–called chromospheric indicators, in particular the cores of Ca II, Mg II, C II and Si II resonance lines in the case of A dwarf stars. Some of these lines must be interpreted as computed partial redistribution (PR) line profiles and others as complete redistribution (CR) ones. The role of abundances in this theoretical interpretation is analyzed.

064.103   Observational problems in spectral line formation.
M. Hack.
Progress in stellar spectral line formation theory, p. 265 – 277 (1985). – See Abstr. 012.108.

The main problem for the interpretation of stellar spectral lines is to explain superionization and stellar winds. Results giving information on these two phenomena in solar–type stars, hot dwarfs and supergiants, A–type dwarfs and supergiants are reviewed.

064.104   Current problems of line formation in early–type stars.
D. C. Abbott.
Progress in stellar spectral line formation theory, p. 279 – 304 (1985). – See Abstr. 012.108.

Extensive observations of UV P Cygni lines in the winds of early–type stars reveal the following four major problems with line formation theory: (1) the evidence for photon destruction even in resonance lines; (2) the presence of strong wind shocks and the resultant nonmonotonic flow; (3) the mystery of the UV discrete components, and (4) line blending of P Cygni profiles and the multiple scattering of photons. A first comparison of CCD observations of an O star (ζ Pup) to model atmospheres, both with and without wind–blanketing, shows that the spectral classification can depend as strongly on mass loss as on the usual parameters of T_{eff} and $\log g$.

064.105   Stellar surface inhomogeneities and the interpretation of stellar spectra.
M. S. Giampapa.
Progress in stellar spectral line formation theory, p. 305 – 318 (1985). – See Abstr. 012.108.

The author discusses manifestations of stellar surface thermal inhomogeneities in stellar spectra. He illustrates the effects of multi–components in stellar atmospheres on the interpretation of line diagnostics and single–component models as well as on the treatment of line transfer problems. The examples offered involve metal abundance determinations, chromospheric line diagnostics, the realistic representation of pre–main sequence atmospheres and stellar magnetic fields.

064.106   The theory of line transfer in expanding atmospheres.
P. B. Kunasz.
Progress in stellar spectral line formation theory, p. 319 – 333 (1985). – See Abstr. 012.108.

The observational motivation for the development of line transfer in moving atmospheres is discussed. This is followed by a review of methods. Special attention is given to techniques developed within the past few years, and to their application and extension to problems of current astrophysical interest, with an emphasis on those involving multi–dimensional geometries, and formidable atomic coupling.

064.107   Computed He II spectra for Wolf–Rayet stars.
W.–R. Hamann.
Progress in stellar spectral line formation theory, p. 335 – 342 (1985). – See Abstr. 012.108.

Synthetic spectra of spherically symmetric, expanding stellar atmospheres of pure helium are calculated in non–LTE for a given atmospheric structure. The radiation transfer is solved in the "comoving frame". A small grid of models is calculated. He II emission lines as typically observed in WR stars are obtained, if the effective temperature of the star lies above 35000K and the mass loss rate exceeds $10^{-5} M_\odot$/yr.

064.108   Partial redistribution in the winds of red giants.
K. Hempe.
Progress in stellar spectral line formation theory, p. 343 – 349 (1985). – See Abstr. 012.108.

Most late–type giants and supergiants are losing mass in cool stellar winds. The only evidence for mass loss in single stars is the presence of a characteristic asymmetry deep in the core of photospheric lines. This paper studies the effects of partial redistribution functions and complete redistribution on line profiles. The equation of transfer is solved for a two level atom in the comoving frame.

064.109   Modeling lines formed in the expanding chromospheres of red giants.
S. A. Drake.
Progress in stellar spectral line formation theory, p. 351 – 357 (1985). – See Abstr. 012.108.

This paper discusses the application of radiative transfer techniques to the study of the physical conditions in the extended, expanding chromospheres of cool giants and supergiants. The important diagnostic feature indicating the outflow of chromospheric material in such stars is the presence of blue–shifted absorption components in the h and k resonance lines of Mg II. To model these lines, the author uses a spherically symmetric co–moving frame solution of the equation of radiative transfer that takes proper account of partial redistribution effects. Preliminary results of the study of the Mg II lines in α Boo are given.

064.110   Theory of radiatively driven stellar winds.
Eh. Ya. Vil'koviskij, L. V. Tambovtseva.
Tr. Astrofiz. Inst. Alma–Ata, Tom 42, p. 28 – 38 (1983). In Russian.

A gas dynamical mass loss theory from stars with large radiative pressure due to ion lines of different elements is presented. It is given the analysis of thermal balance, and numerical solutions are obtained.

Accretion power in astrophysics.
See Abstr. 003.012.

Shock waves in stellar envelopes.
See Abstr. 003.120.

Partition functions and dissociation constants for HeH$^+$.
See Abstr. 022.074.

A new treatment of water vapor opacity.
See Abstr. 022.177.

Neutron capture reactions in astrophysics.
See Abstr. 061.109.

Evolution of diamagnetic material in a nonuniform magnetic field.
See Abstr. 062.060.

Radiative gas dynamics in the transonic regime.
See Abstr. 062.067.

Baroclinic instability in the presence of a strong horizontal shear. I. Instability conditions.
See Abstr. 062.132.

Axisymmetric expansion of a rotating adiabatic gas.
See Abstr. 062.154.

Hydrogen lines Stark broadening tables in the presence of a magnetic field.
See Abstr. 063.002.

Radiative transfer in disk geometry: emergent intensity from cool gray disks.
See Abstr. 063.003.

Polarization of intrinsic radiation of tidally distorted stars with electron–scattering atmospheres. I. The case of conservative scattering.
See Abstr. 063.008.

Distribution of polarization and intensity of radiation across the stellar disk and numerical values of atmospheric characteristics governing this distribution.
See Abstr. 063.009.

The transfer of polarized radiation in spectral lines: formalism and solutions in simple cases.
See Abstr. 063.010.

Polarization of intrinsic radiation of tidally distorted stars with electron–scattering atmospheres. II. An account of true absorption.
See Abstr. 063.012.

Line formation in the presence of turbulence.
See Abstr. 063.020.

The standard problem of line formation in moving atmospheres.
See Abstr. 063.081.

General aspects of partial redistribution and its astrophysical importance.
See Abstr. 063.085.

Partial versus complete linearization.
See Abstr. 063.094.

Radiative transfer diagnostics: Understanding multi–level transfer calculations.
See Abstr. 063.095.

NLTE spectral line formation in a three–dimensional atmosphere with velocity fields.
See Abstr. 063.099.

A code for line blanketing without local thermodynamic equilibrium.
See Abstr. 063.100.

Hydrogen line formation in dense plasmas in the presence of a magnetic field.
See Abstr. 063.102.

Theoretical constraints on far UV radiation from old Population II stellar systems.
See Abstr. 065.006.

Non–radiative heating and enhanced mass–loss due to shear turbulence in massive hot stars.
See Abstr. 065.023.

Examination of wave behaviors in differentially rotating systems.
See Abstr. 065.048.

Accretion onto magnetized neutron stars.
See Abstr. 067.078.

Numerical simulation of the growth of thick accretion disks.
See Abstr. 067.141.

A unified treatment of the filament and flare instabilities.
See Abstr. 073.006.

Observational constraints on heating processes.
See Abstr. 074.065.

Effects of CO molecules on the outer solar atmosphere: a time–dependent approach.
See Abstr. 080.001.

New insights into the solar–stellar connection.
See Abstr. 080.107.

Is the ratio of observed X–ray luminosity to bolometric luminosity in early–type stars really a constant?
See Abstr. 112.041.

On the role played by lines in radiatively driven stellar winds depending on the position of the stars in the HR diagram.
See Abstr. 112.042.

Summary of the origin of nonradiative heating/momentum in hot stars.
See Abstr. 112.043.

The Be stars.
See Abstr. 112.049.

The geometric extent of C II (UV 0.01) emitting regions around luminous, late–type stars.
See Abstr. 112.061.

Three–component analysis of ultraviolet emission lines of solar–type stars.
See Abstr. 112.108.

The surface morphology of solar–type Hyades stars.
See Abstr. 112.109.

The filling factor of active regions of non–dMe stars.
See Abstr. 112.113.

Theoretical models for the polarization of astronomical masers.
See Abstr. 112.146.

Brightness distribution over the disk of μ Geminorum, derived from lunar occultations.
See Abstr. 113.035.

A model atmosphere analysis of the F5 IV – V subgiant Procyon.
See Abstr. 114.020.

SIT Vidicon and IDS spectra of central stars of planetary nebulae.
See Abstr. 114.022.

Spectrum analysis of the giant Am star HR 178.
See Abstr. 114.079.

C, N, O, and their isotope abundances in coolest stars of the red giant branch.
See Abstr. 114.156.

Beginnings of asteroseismology.
See Abstr. 116.030.

Solar and stellar activities.
See Abstr. 116.059.

Investigation of a wind model for cataclysmic variable ultraviolet resonance line emission.
See Abstr. 117.065.

On the nature of envelopes surrounding W UMa–type stars.
See Abstr. 117.117.

Disk–instability model for outbursts of dwarf novae. II. Full–disk calculations.
See Abstr. 117.159.

Hydrodynamical modeling of mass transfer from cataclysmic variable secondaries.
See Abstr. 117.196.

X–ray emission from cataclysmic variables with accretion disks. I. Hard X–rays.
See Abstr. 117.197.

X–ray emission from cataclysmic variables with accretion disks. II. EUV/soft X–ray radiation.
See Abstr. 117.198.

Helium in the atmospheres of binary stars.
See Abstr. 119.036.

Energy dissipation mechanisms in flare stars.
See Abstr. 122.010.

Sur l'origine des ondes de choc dans l'atmosphère des étoiles Mira.
See Abstr. 122.019.

Bipolar flows and X–ray emission from young stellar objects.
See Abstr. 131.065.

Shapes of planetary nebulae.
See Abstr. 134.003.

Cosmic–ray acceleration and transport in stellar winds with terminal shocks.
See Abstr. 144.144.

065 Stellar Structure and Evolution

065.001   Mass distribution and evolutionary scheme for central stars of planetary nebulae.
S. R. Heap, H. G. Augensen.
NASA Conf. Publ., NASA CP–2349, p. 258 – 261 (1984). – See Abstr. 012.001.

Nearly all investigators agree on the general scheme for the late stages of evolution of stars having low or intermediate masses. According to this scheme, a star ascends the asymptotic giant branch all the while losing mass via a strong stellar wind. At some point, the star enters a brief "superwind" phase, during which the planetary nebula is formed, and the remnant central star then evolves toward its final fate as a white dwarf. In this paper, the authors present new estimates of the initial mass and final mass of central stars of planetary nebulae.

065.002   Mechanics of the affine star model.
B. Carter, J. P. Luminet.
Mon. Not. R. Astron. Soc., Vol. 212, No. 1, p. 23 – 55 (1985).

In a recent pioneering study of the phenomena occurring when a star is disrupted by passage through the tidal field of a large black hole, it was found convenient to make use of a simplified affine star model as a preliminary approximation pending more accurate hydrodynamical investigations. The present work consists of a general study of the intrinsic mechanical properties of a wide class of such affine star models (including allowance for non–adiabatic effects such as thermonuclear energy release and viscous dissipation) independently of any particular physical context.

065.003   Tidal squeezing of stars by Schwarzschild black holes.
J.–P. Luminet, J.–A. Marck.
Mon. Not. R. Astron. Soc., Vol. 212, No. 1, p. 57 – 75 (1985).

The authors present a relativistic generalization of the problem of tidal disruption of a star deeply plunging within the Roche radius of a massive Schwarzschild black hole, on the basis of the affine star model developed by Carter & Luminet in a Newtonian context. It is shown that new specific relativistic effects occur. In particular, tidal compression acting on the direction orthogonal to the orbital plane of the star may induce formation of many successive strong squeezings of the star when the periastron of its orbit is sufficiently close to the horizon of the black hole. Illustrative numerical results are displayed in the last section.

065.004   Stellar evolution with the Roxburgh criterion for convection.
C. Doom.
Astron. Astrophys., Vol. 142, No. 1, p. 143 – 149 (1985).

The author shows that the Roxburgh criterion of convection gives an approximation of the flux of turbulent kinetic energy in a stellar convection zone. He presents evolutionary computations of $1.2 - 120\ M_\odot$ stars, including overshooting from convective cores according to this criterion. He shows that these models represent the observational constraints much better than the classical models.

065.005   The life and times of an intermediate mass star – in isolation/in a close binary.
I. Iben Jr.
Q. J. R. Astron. Soc., Vol. 26, No. 1, p. 1 – 39 (1985).

A review is given of the properties of intermediate mass $(1 < M/M_\odot < 9)$ single stars as they evolve through the initial stages of gravitational contraction on to the main sequence and then through various stages at rates controlled by hydrogen– and helium–burning reactions. Colourful events which punctuate the life of such stars include: the core helium flash $(M \lesssim 2.3\ M_\odot)$; acoustical pulsations as RR Lyrae or Cepheid stars during core helium burning; thermal pulses (also known as helium shell flashes) on the asymptotic giant branch where carbon and s–process isotopes are produced and thereupon brought to the surface whence they are expelled into the interstellar medium by a strong wind; ejection of a planetary nebula and excitation of this nebula by a final remnant which either (1) ultimately becomes a DA white dwarf exhibiting hydrogen at its surface (75 per cent of the time) or (2) experiences a final helium shell flash (25 per cent of the time), briefly becomes a born–again asymptotic giant branch star, re–excites the surrounding nebular shell, and loses the final vestiges of surface hydrogen to become ultimately a non–DA white dwarf.

065.006   Theoretical constraints on far UV radiation from old Population II stellar systems.
V. Caloi, V. Castellani, R. Nesci, L. Rossi.
Astron. Astrophys., Suppl. Ser., Vol. 59, No. 3, p. 505 – 509 (1985).

The UV flux from Population II systems is evaluated on the basis of globular cluster population distribution, ZAHB structures, and model atmospheres. The results may be used as constraints in stellar population analysis when interpreting far UV observations of globular clusters and galaxies.

065.007   Collapse and formation of stars.
A. P. Boss.
Sci. Am., Vol. 252, No. 1, p. 28 – 33 (1985).

Hidden from observation, this process can nonetheless be modeled on high–speed computers. Pictures that emerge yield insight into the formation of our own solar system.

065.008 A dynamical instability as a driving mechanism for stellar oscillations.
M. Auvergne, A. Baglin.
Astron. Astrophys., Vol. 142, No. 2, p. 388 – 392 (1985).

Nonlinear effects in some cases lead to overstable pseudo periodic, irregular or even chaotic motions in dynamical systems, where catastrophic ones (i.e. dynamical instabilities) are expected from the linear theory. In a rough approximation, the motion of an ionization zone in a star is described by a simple nonlinear differential equation. This equation, analogous to the one obtained by Moore and Spiegel (1966) in the treatment of radiating convective elements, exhibits behaviours sensitive to small perturbations due to the competition between several equilibrium configurations: a dynamically unstable one, and two stable ones. Astrophysical applications to Miras and white dwarfs are tentatively proposed.

065.009 Do we understand rotating isothermal collapses yet?
J. E. Tohline.
Icarus, Vol. 61, No. 1, p. 10 – 21 (1985). – See Abstr. 012.004.

In this paper, the results of many different studies of rotating isothermal clouds are couched in a unified, global model of equilibrium cloud structures. The scalar virial equation is used here to describe the global structural parameters of these various models, and the apparent contradiction between the results of Stahler (1983) and those of Hayashi, Narita, and Miyama (1982) is resolved. Finding the parameter regimes over which stable equilibrium models exist provides a good framework in which dynamical evolutions of isothermal clouds can be discussed.

065.010 Evolution of binary systems into transient sources.
R. E. Taam.
AIP Conf. Proc., No. 115, p. 1 – 10 (1984). – See Abstr. 012.005.

The author examines the evolutionary paths of primordial binary systems to the X–ray binary stage. Although mass and angular momentum losses from the system are likely, the massive X–ray binaries can be qualitatively described in terms of conservative evolution. On the other hand, low mass X–ray binaries could not have evolved to their present state by normal evolutionary processes alone. These systems have suffered severe mass and angular momentum losses in a common envelope phase.

065.011 The evolution of binaries of moderate primordial mass ($\lesssim 10\,M_\odot$) and the formation of transient and explosive objects due to accumulation instabilities in such binaries.
I. Iben Jr., A. V. Tutukov.
AIP Conf. Proc., No. 115, p. 11 – 30 (1984). – See Abstr. 012.005.

Possible scenarios are discussed for the evolution of close binaries characterized primordially by: masses of individual components less than $\sim 10\,M_\odot$, semimajor axes A in the range $10 \lesssim A/R_\odot \lesssim 2000$, and orbital periods less than 25 yrs. Final products include double degenerate dwarfs with $A \gtrsim (3-5)\,R_\odot$, single rapidly rotating dwarfs ($P_{rot} \gtrsim 10$ sec), or neutron stars ($P_{rot} \gtrsim 10^{-3}$ sec), and supernovae of Type I. The authors suggest that binary systems consisting of a low mass non–degenerate component and a neutron star or black hole may be primarily the result of inelastic capture collisions between single neutron stars and single low mass stars or of exchange captures involving the replacement of one component of a low mass binary by a passing neutron star. A simple conservation law for "accumulation" instabilities accompanying the evolution of low mass binaries is presented. Several accumulation instabilities leading to SN I and novae, to dwarf novae and X–ray recurrent bursters, and to X–ray and γ–ray bursters are discussed.

065.012 Accretion with uniform eddy diffusivity. I. Polytropic–tube models.
G. Rüdiger, R. Tschäpe.
Astron. Nachr., Vol. 306, No. 1, p. 21 – 28 (1985).

The authors consider accretion onto a two–dimensional polytropic tube. In contrast to the standard papers they apply spatially uniform eddy viscosity (and eddy conductivity). In principle, the resulting rotation law is non–Keplerian and proves to be rather flat. Whether the angular velocity increases or decreases inwardly depends on the strength of the torque exerted on the outer surface which is due to the inflow of angular momentum. Large (small) torque causes super–(sub–) rotation. The frictionally originated luminosity varies for various models. It lies one or two orders of magnitude below the total available accretion energy. For certain combinations of the Mach numbers the radial inflow profile possesses a sonic point.

065.013 Theory and application of stellar seismology.
W. Dziembowski.
Adv. Space Res., Vol. 4, No. 8, p. 143 – 150 (1984). – See Abstr. 012.010.

The author evaluates prospects of using oscillation data as a probe of stellar internal structure. The results obtained for the sun are quite encouraging.

065.014 Stellar magnetic activity, rotation and convection.
R. W. Noyes.
Adv. Space Res., Vol. 4, No. 8, p. 151 – 161 (1984). – See Abstr. 012.010.

The study of stellar magnetic activity, rotation, and convection is an important current research topic which is relevant to the study of stellar oscillations. The properties of both activity and oscillations depend on the structure and dynamics of stellar interiors, so that information gained in the study of one is applicable to the other. As an important example, stellar oscillation data can in principle yield a depth–averaged value of the star's rotation; by combining this with a measure of the surface rotation rate information about the star's differential rotation with depth may be obtained.

065.015 L'évolution des étoiles massives.
M. Gabriel.
Ciel Terre, Vol. 101, No. 1, p. 3 – 12 (1985).

First the evolution of massive stars without mass loss is described. Secondly the observational H–R diagram of bright stars is presented and the main contradictions with the theoretical predictions are pointed out. Thirdly the main properties of the Wolf–Rayet stars are described. Finally the evolution of massive stars is described and the evidences of mass loss from luminous stars for two laws of mass loss rates and the link with the WR stars is discussed.

065.016 Shock propagation in supernovae: concept of net ram pressure.
J. Cooperstein, H. A. Bethe, G. E. Brown.
Nucl. Phys. A, Vol. A429, No. 3, p. 527 – 555 (1984). Abstr. in Phys. Abstr., Vol. 88, No. 1249, Entry 14645 (1985).

065.017 Equilibrium models of differentially–rotating polytropic cylinders.
P. Veugelen.
Astrophys. Space Sci., Vol. 109, No. 1, p. 45 – 55 (1985).

The equilibrium structure of differentially rotating polytropic cylinders is determined numerically. The author sets $n = 3$ and uses a quadratic function for the law of differential rotation. He constructs different models by varying the angular velocity at the axis and the ratio of the angular velocity at the surface to the angular velocity at the axis. By taking a decreasing function for the rotation law he is able to treat models with an angular velocity at the axis greater than the break–up velocity of uniformly rotating cylinders. He also determines whether a Richardson–like criterion for stability is violated in the models.

065.018 The rapidly oscillating Ap stars.
M. Gabriel, A. Noels, R. Scuflaire, G. Mathys.
Astron. Astrophys., Vol. 143, No. 1, p. 206 – 208 (1985).

Frequency separation $\Delta\nu$ between consecutive modes have been computed for rapidly oscillating stars. It is shown that $\Delta\nu$ is a good indicator of the star's position in the main sequence band. These results are applied to Kurtz observations of HD 101065 and HD 24712. The problem of HD 83368 is also discussed.

065.019   Linear oscillations of differentially rotating self–gravitating cylinders.
P. Veugelen.
Astron. Astrophys., Vol. 143, No. 2, p. 256 – 266 (1985).

The influence of a differential rotation on the oscillations and the dynamical stability of self–gravitating cylinders is investigated. The author discusses the solutions associated with the continuous spectrum of eigenfrequencies of cylinders with an arbitrary equation of state and an arbitrary law for the differential rotation. To investigate the influence of the differential rotation on the normal modes with discrete eigenfrequencies, the author determines non–axisymmetric oscillations of polytropic cylinders with a parabolic law for the rotation. All non–axisymmetric modes with real eigenfrequencies can be identified as p–, g–, and f–modes modified by the differential rotation.

065.020   Equilibrium structures of rotating isothermal gas clouds. I.
I. Hachisu, Y. Eriguchi.
Astron. Astrophys., Vol. 143, No. 2, p. 355 – 364 (1985).

The authors have computed equilibrium sequences of rotating isothermal gas clouds in order to predict the outcomes of the collapse. For simplicity, they assume that the initial state of the gas cloud is a uniformly rotating homogeneous sphere and that the cloud is axisymmetric during the contraction and, therefore, that the angular momentum is conserved locally. The final states for the collapsing gas cloud are: a stable cloud, a pre–solar type system, and a multiple system.

065.021   Classification of normal modes of rotating cylindrical systems.
P. Veugelen.
Astron. Astrophys., Vol. 143, No. 2, p. 458 – 460 (1985).

Modes of rotating cylindrical systems are investigated by considering non–axisymmetric oscillations of uniformly rotating polytropic cylinders. The classification of the modes is clarified by the use of propagation diagrams. When the Brunt–Väisälä frequency tends to zero, the eigenfrequencies of backward propagating g–modes of rotating cylinders tend to zero, and the eigenfrequencies of forward propagating g–modes tend to frequencies of r–modes. The forward propagating g–modes of rotating cylindrical systems can thus be considered as gravitational–rotational modes.

065.022   Champ magnétique stellaire à grande échelle. Le modèle du rotateur oblique.
C. Megessier.
Champs magnétiques stellaires, p. 455 – 466 (1984). – See Abstr. 012.022.

065.023   Non–radiative heating and enhanced mass–loss due to shear turbulence in massive hot stars.
S. R. Sreenivasan, W. J. F. Wilson.
NASA Conf. Publ., NASA CP–2358, p. 177 – 182 (1985). – See Abstr. 012.023.

Rotation in massive stars generates shear turbulence. This has the consequence of heating the outer layers of massive stars and enhancing mass–loss in the early spectral types. Model calculations are presented and discussed in support of these effects.

065.024   Time–dependent models of rotating magnetic stars. II. The displaced dipole.
D. Moss.
Mon. Not. R. Astron. Soc., Vol. 213, No. 3, p. 575 – 589 (1985). With a correction in Vol. 215, No. 1, p. 159 – 160 (1985).

Earlier studies of time–dependent models of rotating magnetic stars are extended to the case where the field is no longer strictly antisymmetric with respect to an equatorial plane, thus modelling the "displaced dipole" configuration. Results are presented for models which are rotationally dominated, with the meridional flow approximately given by the Eddington Sweet circulation through most of the radiative region. The angle χ between the rotation and magnetic axes is taken as either 0 or $\pi/2$. With increasing rotation speed the ratio of the surface field at the stronger pole to that at the weaker declines with time.

065.025   Erratum: "Linear radial pulsations in stars in thermal imbalance" [Astron. Astrophys., Vol. 141, No. 2, p. 297 – 303 (1984)].
P. Smeyers, P. Bruggen.
Astron. Astrophys., Vol. 144, No. 2, p. 516 (1985). See Abstr. 38.065.093.

065.026   Evolution of massive stars with semiconvective diffusion.
N. Langer, M. F. El Eid, K. J. Fricke.
Astron. Astrophys., Vol. 145, No. 1, p. 179 – 191 (1985).

Evolutionary sequences at constant mass are presented for Pop I stars of 15 $M_\odot$ and 30 $M_\odot$ until the onset of core carbon burning. A time–dependent diffusion model based on a vibrational instability analysis is adopted in order to describe semiconvection. The fully–convective layers are determined according to the Ledoux criterion. For comparison, evolutionary sequences are also calculated for the same masses without assuming semiconvection but using the Schwarzschild criterion for convective instability. It is found that the vibrational instability does not establish the Schwarzschild–Härm criterion for semiconvection during the fast phases of evolution such as shell hydrogen burning. The modification of the chemical profile by semiconvection during this phase does strongly affect the evolutionary tracks in the H–R diagram during core helium burning.

065.027   Current problems in horizontal–branch theory: some implications for RR Lyrae variables.
P. Demarque.
Cepheids: theory and observations, p. 268 – 271 (1985). – See Abstr. 012.027. (IAU Colloq. No. 82).

065.028   Type II supernova energetics.
J. M. Lattimer, A. Burrows, A. Yahil.
Astrophys. J., Vol. 288, No. 2, p. 644 – 652 (1985).

This paper exploits the fact that the collapsing core of a massive star, at the endpoint of its thermonuclear life, rapidly becomes hydrostatic after bouncing at nuclear densities. The energy transferred by pdV work from the inner unshocked core as it becomes hydrostatic to the bounce shock in the outer core is, to a good approximation, the initial energy of the supernova. It can be shown semi–analytically that this energy is equal, with minor qualifications, to the binding energy of the hydrostatic remnant. The authors present analytical formulae for the structure and binding energy of a hydrostatic residue modeled as a composite of two nested polytropes. It is demonstrated that the results derived from considering the nested double polytrope are in fact very general and are only weakly dependent on the behavior of the nuclear equation of state.

065.029   On the method of numerical calculation of nonlinear radial pulsations of stars.
A. G. Kosovichev.
Izv. Krymskoj Astrofiz. Obs., Tom 69, p. 108 – 122 (1984). In Russian. English translation in Bull. Crimean Astrophys. Obs., Vol. 69.

Some features of using the finite difference method for numerical investigation of nonlinear radial pulsations of stars were considered. The numerical method can be used for calculation of large–amplitude radial pulsations of stars.

065.030   Erratum: "Do monopoles keep white dwarfs hot?" [Astrophys. J., Vol. 286, No. 1, p. 216 – 220 (1984)].
K. Freese.
Astrophys. J., Vol. 289, No. 2, p. 858 (1985). See Abstr. 38.065.146.

065.031   QUIP: a time–dependent standard model.
A. Munier.
Astrophys. J., Vol. 290, No. 1, p. 47 – 65 (1985).

A time–dependent solution representing the monotonic expansion or contraction of a self–gravitating gas sphere with polytropic index $n = 3$, the QUIP (quasi–invariant polytrope) is obtained from group theory. Radiation is introduced into the model through a dynamic diffusion description. Compatibility between the equations of hydrodynamics and radiation leads to the definition of a (time–dependent) standard QUIP generalizing Eddington's standard (static) model. All quantities such as the ratio of the mean to the central density, the central density, pressure and temperature, etc., are calculated. A mass–radius–hydrodynamical time scale relation is given, together with a luminosity formula.

065.032   Deflagration of white dwarfs as a model for type–I supernovae.
D. Jeffery, P. Sutherland.
Astrophys. Space Sci., Vol. 109, No. 2, p. 277 – 285 (1985).

Carbon–oxygen white dwarfs may be the progenitors of type–I supernovae. Spherically–symmetric models of such dwarfs have been evolved from an artificial core incineration. The convectively unstable incinerated region was allowed to grow at a velocity prescribed by the mixing–length theory of convection. The mixing length can be varied to give different cases. In all the cases considered the dwarfs exploded and were totally disrupted. The calculations were stopped after the dwarf matter had gone into homologous expansion. The model with the best estimated mixing length incinerated 0.8 $M_\odot$. The energy released in burning this amount of carbon–oxygen to ^{56}Ni provides a disrupted dwarf with velocities suitable for type–I supernovae.

065.033   The evolution of stars and origin of chemical elements.
D. Sugimoto.
Perspektivy kvant. fiz. Kiev, p. 274 – 287 (1982). In Russian. Abstr. in Ref. Zh., 51. Astron., 2.51.376 (1985).

065.034   On the central explosion of a gravitating rotating body with a magnetic field.
V. A. Startsev.
Dokl. Akad. Nauk SSSR. Ser. Mat. Fiz., Tom 278, No. 2, p. 316 – 321 (1984). In Russian. Abstr. in Ref. Zh., 51. Astron., 2.51.391 (1985).

065.035   Pion stars.
Yu. L. Vartanyan, G. S. Adzhyan, G. B. Alaverdyan.
Sov. Astron., Vol. 28, No. 4, p. 396 – 401 (1984). English translation of 38.065.007.

065.036   Collapse anisotropy for massive stars.
V. G. Kornilov, V. M. Lipunov.
Sov. Astron., Vol. 28, No. 4, p. 402 – 404 (1984). English translation of 38.065.008.

065.037   Detonation–wave structure and nucleosynthesis in exploding carbon–oxygen cores and white dwarfs.
V. S. Imshennik, A. M. Khokhlov.
Sov. Astron. Lett., Vol. 10, No. 4, p. 262 – 265 (1984). English translation of 38.065.015.

065.038   Low mass asymptotic giant branch evolution.
J. C. Lattanzio.
Proc. Astron. Soc. Aust., Vol. 5, No. 4, p. 498 – 500 (1984).

065.039   The evolution of massive stars losing mass and angular momentum: supergiants.
S. R. Sreenivasan, W. J. F. Wilson.
Astrophys. J., Vol. 290, No. 2, p. 653 – 659 (1985).

Evolutionary sequences have been computed to central helium exhaustion for massive Population I star models with initial masses of 40, 60, 80, and 100 $M_\odot$. Mass loss occurs through radiatively driven winds and through the effects of nonthermal winds generated both by rotational shear turbulence during the main–sequence phase and by the acoustic energy flux from the convective envelope during the red supergiant phase. In these models the mass loss and evolutionary patterns for a star depend on its inital rotational speed. The numerical results are discussed in terms of internal structure and the formation of blue loops in the H–R diagram, blue–to–red supergiant ratios, and Wolf-Rayet stars.

065.040   Energy relations on the small oscillations in the stars.
W. G. Gavryusev (*V. G. Gavryusev*),
E. A. Gavryuseva.
ESA Spec. Publ., ESA SP–220, p. 195 – 196 (1984). – See Abstr. 012.045.

The formula of the work integral given by A. Eddington is used often in the calculation of the small oscillations excitation of stars. It is shown that some derivations of the Eddington formula are incorrect. However the well known formula of the work integral is correct because it is the consequence of the linear equations of oscillations.

065.041   The present status of stellar activity theory.
G. Belvedere.
Mem. Soc. Astron. Ital., Vol. 55, No. 3, p. 575 – 578 (1984). – See Abstr. 012.050.

065.042   Effects of convective overshooting on intermediate mass stars.
C. Chiosi.
Mem. Soc. Astron. Ital., Vol. 55, No. 3, p. 579 – 583 (1984). – See Abstr. 012.050.

The author summarizes the results of a study on the effects of convective overshooting on the evolution of intermediate mass stars, from the main sequence until the AGB phase. Several major observational problems that cannot find an easy solution in the context of classical models are discussed in the light of the new ones.

065.043   Recurrent novae as a consequence of the accretion of solar material onto a 1.38 $M_\odot$ white dwarf.
S. Starrfield, W. M. Sparks, J. W. Truran.
Astrophys. J., Vol. 291, No. 1, p. 136 – 146 (1985).

The authors have computed three evolutionary sequences which treat the accretion of hydrogen–rich material (solar composition) onto 1.38 $M_\odot$ white dwarfs. In the first sequence they utilized an accretion rate of $1.7 \times 10^{-8} M_\odot \mathrm{yr}^{-1}$ onto a white dwarf with an initial luminosity of 0.1 $L_\odot$. It took this sequence ~ 33 yr to reach the peak of the thermonuclear runaway which resulted in an outburst that ejected $3 \times 10^{-8} M_\odot$ of material. The light curve, the time to outburst, and the amount of mass ejected during the evolution are in excellent agreement with the observed outburst of Nova U Sco 1979. The second study involved an accretion rate of $1.7 \times 10^{-9} M_\odot \mathrm{yr}^{-1}$ onto a white dwarf with an initial luminosity of $10^{-2} L_\odot$. It took nearly 1600 yr to reach the burst phase of the evolution. This sequence ejected $3 \times 10^{-7} M_\odot$, only 13% of the accreted envelope.

065.044   Convective heating of the inner core of red giants prior to the peak of the core helium flash.
P. W. Cole, P. Demarque, R. G. Deupree.
Astrophys. J., Vol. 291, No. 1, p. 291 – 296 (1985).

The effects of convective overshooting across the temperature inversion in the cores of red giants are investigated from the onset of the core convection zone to the peak of the core helium flash using a model for overshooting in stellar evolution, based on two–dimensional and three–dimensional hydrodynamic simulations of the core helium flash. A major effect of the overshooting is the substantial heating of the material interior to the temperature inversion, producing a smoother temperature profile. The implications of this interior heating for the subsequent evolution through the peak of the core helium flash are discussed.

065.045   Numerical simulations of stellar convective dynamos. II Field propagation in the convection zone.
G. A. Glatzmaier.
Astrophys. J., Vol. 291, No. 1, p. 300 – 307 (1985).

The author presents numerical simulations of nonlinear, three–dimensional, time–dependent, giant–cell stellar convection and magnetic field generation. The velocity, magnetic field, and thermodynamic variables satisfy the anelastic magnetohydrodynamic equations for a stratified, rotating, spherical shell of ionized gas. The interaction of rotation and convection produces a nonlinear transport of angular momentum that maintains a differential rotation in radius and latitude. Magnetic fields are generated by the shearing and twisting effects of the differential rotation and helical motions and are destroyed by eddy diffusion. The simulations are compared with the solar magnetic cycle and it is suggested that, instead of operating in the turbulent convective region, the solar dynamo may be operating at the base of the convection zone.

065.046   An expanding vortex site for the r–process in rotating stellar collapse.
E. M. D. Symbalisty, D. N. Schramm, J. R. Wilson.
Astrophys. J., Lett. Ed., Vol. 291, No. 1, p. L11 – L14 (1985).

It is shown from two–dimensional stellar collapse calculations with rotation that ejected jets form which meet the conditions for dynamical r–process nucleosynthesis. The mass yield of such neutron–rich ejecta per supernova is $\sim 4 \times 10^{-4} M_\odot$ which is shown to be consistent with the yields per supernova necessary to explain the present r–process abundances.

065.047   CNO abundances resulting from diffusion in accreting nova progenitors.
A. Kovetz, D. Prialnik.
Astrophys. J., Vol. 291, No. 2, p. 812 – 821 (1985).

The evolution of white dwarfs accreting hydrogen–rich matter toward a nova outburst has been followed quasi–statically. The effect of diffusion on the composition profiles in the accreted matter and in the white dwarf core has been investigated. For each of the three initial parameters involved, namely, the accretion rate $\dot{M}$, the white dwarf mass M_{WD}, and its initial luminosity L_{WD}, two values have been chosen within the range relevant to nova outbursts: $\dot{M} = 10^{-9}$ and $10^{-11} M_\odot \text{yr}^{-1}$; $M_{WD} = 1.25$ and $0.9\ M_\odot$; and $L_{WD} = 1.5 \times 10^{-2}$ and $7.5 \times 10^{-3} L_\odot$. The range of resulting CNO abundances in the convective envelope at the onset of the nova outburst was from $Z = 0.08$ to 0.40. The masses accreted prior to the development of a nuclear runaway ranged from 4×10^{-6} to $3 \times 10^{-4} M_\odot$.

065.048   Examination of wave behaviors in differentially rotating systems.
H. Ando.
Publ. Astron. Soc. Jpn., Vol. 37, No. 1, p. 47 – 67 (1985).

Characteristics and excitation mechanisms of nonaxisymmetric oscillations in differentially rotating systems are investigated. A realistic solar model is used as the equilibrium state, and wave solutions are obtained by numerical analysis. Solutions are classified by analyzing the fractional contributions of kinetic energy, gravity potential energy, and so on to the total pulsation energy and its kinetic energy distribution in the radial direction.

065.049   A simple model for binary star evolution.
C. A. Whyte, P. P. Eggleton.
Mon. Not. R. Astron. Soc., Vol. 214, No. 3, p. 357 – 378 (1985).

A simple model for calculating the evolution of binary stars is presented. Detailed stellar evolution calculations of stars undergoing mass and energy transfer at various rates are reported and used to identify the dominant physical processes which determine the type of evolution. These detailed calculations are used to calibrate the simple model and a comparison of calculations using the detailed stellar evolution equations and the simple model is made. Results of the evolution of a few binary systems are reported and compared with previously published calculations using normal stellar evolution programs.

065.050   A general computational method for obtaining equilibria of self–gravitating and rotating gases.
Y. Eriguchi, E. Müller.
Astron. Astrophys., Vol. 146, No. 2, p. 260 – 268 (1985).

A general computational method is presented to compute equilibria of rapidly rotating self–gravitating gases. The superior capabilities of this method as compared to previously published methods stem from the fact that the equation of hydrostatic equilibrium and Poisson's equation are solved simultaneously with a Newton–Raphson iteration scheme. To demonstrate the capabilities of the method the authors have computed equilibrium sequences of differentially rotating polytropes for several rotation laws, and one equilibrium sequence using the ideal Fermi gas equation of state. Although in these examples the pressure is a function of the density only, a simple extension of the method allows to model closely more realistic temperature–dependent equations of state.

065.051   On oscillations of a star with phase transition.
G. S. Bisnovatyj–Kogan, Z. F. Seidov.
Astrofizika, Tom 21, Vyp. 3, p. 563 – 571 (1984). In Russian. English translation in Astrophysics, Vol. 21, No. 3.

The mechanism of damping of star oscillations is considered which is connected with phase transition in matter. The equation of motion and its solutions are obtained for different kinds of radial oscillations of the star in the approximation of an incompressible fluid. Estimates of the maximal amplitude of the oscillations are given when phase damping is absent in the pion condensation phase transition.

065.052   Lower limits for the central pressure of a star.
R. J. Tayler.
Observatory, Vol. 105, No. 1066, p. 93 – 94 (1985).

065.053   On the nonradial pulsations of general relativistic stellar models.
S. Detweiler, L. Lindblom.
Astrophys. J., Vol. 292, No. 1, p. 12 – 15 (1985).

The authors present a nonsingular fourth–order system of equations which describe the nonradial pulsations of general relativistic stellar models. These equations represent an improvement over previous discussions in the literature which presented fifth–order systems or fourth–order systems to describe these pulsations. The nonsingular power–series solutions to these equations near $r = 0$ are also presented.

065.054   Stellar evolution at high mass with convective core overshooting.
R. B. Stothers, C.–w. Chin.
Astrophys. J., Vol. 292, No. 1, p. 222 – 227 (1985).

Convective overshooting beyond the formal boundary of a convectively unstable stellar core is studied here in terms of a free overshoot parameter d/H_P (overshoot distance divided by local pressure scale height) in order to determine in more detail than previously the effect of widespread convective core mixing on the main–sequence evolution of stars in the mass range $15 - 120\ M_\odot$. Comparison of theoretically constructed isochrones for very young star clusters with observed main–sequence turnups in the H–R diagram strongly suggests that a significant amount of convective core overshooting (or some other form of mixing) increases the luminosities and decreases the effective temperatures of evolving stars of high mass.

065.055   Observable quantities of nonradial pulsations in the presence of slow rotation.
W. D. Pesnell.
Astrophys. J., Vol. 292, No. 1, p. 238 – 248 (1985).

Theoretical power spectra of both light and radial velocity variations of a nonradially pulsating star that include the effects of rotation are presented. Three classes of motion are defined, depending on the symmetry of the mode of interest and its orientation with respect to the rotating axis. The use of tabulated model atmospheres to predict color variations is compared to the blackbody approximation for two cases, a hydrogen white dwarf (DA) and a main–sequence B Star.

065.056   Equilibrium structures of rotating isothermal gas clouds. II. Dependence on the angular momentum distribution.
I. Hachisu, Y. Eriguchi.
Astron. Astrophys., Vol. 147, No. 1, p. 13 – 21 (1985).

Using a thin disk approximation model, the authors have computed equilibrium sequences of rotating isothermal gas clouds with angular momentum distributions $j_r = (1+\varepsilon)(J/M)\{1-(1-M_r/M)^{1/\varepsilon}\}$. Here M is the total mass, J the total angular momentum, M_r the mass within the cylindrical radius r, and j_r, the specific angular momentum at r. Three cases ($\varepsilon = 2$, 1, and 1/2) have been computed (a uniformly rotating homogeneous sphere corresponds to $\varepsilon = 3/2$). There are two phases which are stable against the compression mode – a pressure dominant phase and a rotation dominant phase for all angular momentum distributions. When $\varepsilon \geqslant 1$, there exist equilibria with infinite central density, however, when $\varepsilon = 1/2$, a ring structure appears in the rotation dominant phase. When a spherical homogeneous isothermal gas cloud, which is not in equilibrium, is considered, the cloud will begin to collapse.

065.057   Equilibrium models of differentially rotating polytropes and the collapse of rotating stellar cores.
Y. Eriguchi, E. Müller.
Astron. Astrophys., Vol. 147, No. 1, p. 161 – 168 (1985).
The collapse of rotating degenerate cores of massive stars $M \geqslant 8\,M_{sol}$ is studied. Extending the analysis of Tohline (1984), equilibrium models of differentially rotating axisymmetric polytropes are used to estimate the properties of the core at the endpoint of its collapse without performing a detailed collapse calculation. Slowly and rapidly rotating equilibrium models have been calculated with two kinds of angular momentum distributions and with adiabatic indices $\gamma = 4/3$, 1.32, 1.29, and 1.25, respectively.

065.058   The evolution of massive stars losing mass and angular momentum: rotational mixing in early–type stars.
S. R. Sreenivasan, W. J. F. Wilson.
Astrophys. J., Vol. 292, No. 2, p. 506 – 510 (1985).
Rotation is shown to produce an extended core due to mixing in early–type stars, in a manner consistent with the mixing–length theory employed to study other properties in stellar evolution. Whereas this effect is seen to have consequences similar to convective overshoot, it is capable of producing a greater amount of core enlargement, as appears to be required by observations. However, while it extends the main–sequence lifetime, widens the main–sequence band, and affects the blue–to–red supergiant ratios, it is not found to produce a blue loop.

065.059   An Eulerian variational principle and a criterion for the occurrence of nonaxisymmetric neutral modes along rotating axisymmetric sequences.
J. R. Ipser, R. A. Managan.
Astrophys. J., Vol. 292, No. 2, p. 517 – 521 (1985).
A new method is developed, in the Newtonian framework, for locating Dedekind bifurcation points, at which an axisymmetric configuration of rotating fluid possesses a nonaxisymmetric neutral, or zero–frequency, mode in the inertial frame. The method is based on an "Eulerian" variational principle, which involves integrals whose integrands depend on only the Eulerian change, the difference between values at the same location in the perturbed and unperturbed configurations, of the sum of enthalpy and gravitational potential. Hence the complications associated with principles based on the Lagrangian displacement are avoided automatically.

065.060   The action of inexorable meridional circulation on a stellar magnetic field.
M. J. Roberts, W. P. Wood.
Astrophys. J., Vol. 292, No. 2, p. 578 – 588 (1985).
Results for the behavior over time of an initial magnetic field in the presence of prescribed meridional circulation within a rotating upper–main–sequence star are presented. Various model flows are studied, and it is found that the presence of mass motions leads to a more rapid depletion of the magnetic flux which emerges from the stellar surface than for the situation when mass motions are absent. It is also shown that, for the circulation models studied here, the e–folding time of decay is not a reliable measure of the decay rate of the surface magnetic field.

065.061   Equilibrium models of differentially rotating, completely catalyzed, zero–temperature configurations with central densities intermediate to white dwarf and neutron star densities.
E. Müller, Y. Eriguchi.
MPA Rep., No. 179, 33 pp. (1985). Submitted to Astron. Astrophys.

065.062   The formation and evolution of carbon stars.
J. R. Mould, M. Aaronson.
Bull. Am. Astron. Soc., Vol. 16, No. 4, p. 875 (1984). Abstract. – See Abstr. 010.062.

065.063   On the observed properties and long term structure and evolution of white dwarfs in cataclysmic variables.
E. M. Sion.
Bull. Am. Astron. Soc., Vol. 16, No. 4, p. 945 (1984). Abstract. – See Abstr. 010.062.

065.064   Evolution of the nonradial pulsation properties of hot pre–white dwarf stars.
S. D. Kawaler, C. J. Hansen, D. E. Winget.
Bull. Am. Astron. Soc., Vol. 16, No. 4, p. 945 (1984). Abstract. – See Abstr. 010.062.

065.065   Effective log(ft) values for accurate interpolation of rapidly varying lepton capture rates in the stellar interior.
M. J. Newman, G. M. Fuller, W. A. Fowler.
Bull. Am. Astron. Soc., Vol. 16, No. 4, p. 946 (1984). Abstract. – See Abstr. 010.062.

065.066   The evolution of massive stars losing mass and angular momentum. VII. Rotational mixing in early–type stars.
S. R. Sreenivasan, W. J. F. Wilson.
Bull. Am. Astron. Soc., Vol. 16, No. 4, p. 947 (1984). Abstract. – See Abstr. 010.062.

065.067   Conditions for very massive protostellar accretion.
M. G. Wolfire, J. P. Cassinelli.
Bull. Am. Astron. Soc., Vol. 16, No. 4, p. 960 (1984). Abstract. – See Abstr. 010.062.

065.068   Presupernova core structure and explosive nucleosynthesis.
T. A. Weaver, S. E. Woosley.
Bull. Am. Astron. Soc., Vol. 16, No. 4, p. 971 (1984). Abstract. – See Abstr. 010.062.

065.069   Horizontal–branch sequences for globular cluster RR Lyrae stars.
A. V. Sweigart, A. Renzini, A. Tornambé.
Bull. Am. Astron. Soc., Vol. 16, No. 4, p. 996 (1984). Abstract. – See Abstr. 010.062.

065.070   Rotational evolution of hot stars.
K. B. MacGregor, D. B. Friend, R. L. Gilliland.
Bull. Am. Astron. Soc., Vol. 16, No. 4, p. 996 (1984). Abstract. – See Abstr. 010.062.

065.071   The effect of opacity on theoretical models of RR Lyrae stars.
O. Hubickyj.
Bull. Am. Astron. Soc., Vol. 16, No. 4, p. 1012 (1984). Abstract. – See Abstr. 010.062.

065.072   A simple model of the convection–pulsation interaction.
W. D. Pesnell.
Bull. Am. Astron. Soc., Vol. 16, No. 4, p. 1012 (1984). Abstract. – See Abstr. 010.062.

065.073   Anomalous magnetic field diffusion during star formation.
C. Norman, J. Heyvaerts.
Astron. Astrophys., Vol. 147, No. 2, p. 247 – 256 (1985).
The authors discuss the physics of magnetic field dissipation and reconnection during protostellar formation. In their analysis of quasi–equilibrium collapse they have incorporated both ambipolar diffusion and magnetic dissipation processes. Both the final flux and angular momentum problems can be resolved in their model which incorporates some of the detailed physics that was previously lacking in such theories. The energy input during flux destruction is considerable and the authors have speculated that

consequences may include the formation of a hot corona with an observable X–ray flux, and the pumping of OH masers. Radio emission may be observable in this phase.

065.074   Vibrational instability of Wolf–Rayet stars.
A. Maeder.
Astron. Astrophys., Vol. 147, No. 2, p. 300 – 308 (1985).

The radial vibrational stability of massive star models evolving with mass loss at observed rates is tested numerically throughout the various evolutionary stages of initial stellar masses 85 and 120 $M_\odot$: O–stars, Hubble–Sandage variables, Wolf–Rayet stars of types WNL (late), WNE (early), WC, and WO.

065.075   Hydrodynamic models for population–II cepheids.
Yu. A. Fadeyev (*Yu. A. Fadeev*), A. B. Fokin.
Astrophys. Space Sci., Vol. 111, No. 2, p. 355 – 374 (1985).

The nonlinear self–excited pulsations of population–II stars with mass 0.6 $M_\odot$ and luminosities from 128 to 1280 $L_\odot$ are studied. The pulsation periods are found to be in the range of 1.3 to 19 days. An increase of the stellar luminosity is shown to be accompanied by an increasing nonadiabaticity and decreasing efficiency of the radiative damping region. This leads to both an increase of the growth rate while pulsations are exciting and an increase of the oscillation amplitude of the limit cycle. The theoretical period–luminosity relation proposed for population–II cepheids is in good agreement with that obtained from observations.

065.076   Radial pulsation of less–massive yellow supergiants. I. 1 $M_\odot$, 3200 $L_\odot$ models.
T. Aikawa.
Astrophys. Space Sci., Vol. 112, No. 1, p. 125 – 131 (1985).

Linear pulsation models of less–massive yellow supergiants with $M = 1\,M_\odot$ and $L = 3200\,L_\odot$ were investigated with the aim of understanding FG Sge's pulsation. The author has found that the instability of the third overtone extends to at least $T_e = 8500K$, while the blue edge of the fundamental mode is near $T_e = 5500K$. Nonlinear hydrodynamic simulations were performed for models with $T_e = 6000K$ and 5400K with the same mass and luminosity as the linear models. It is suggested that the pulsation of FG Sge must have undergone mode–switching from the third overtone to the fundamental mode during its evolution toward later spectral types.

065.077   Thermal imbalance effects on pulsational stability in stellar models.
N. S. Chauhan.
Astrophys. Space Sci., Vol. 112, No. 1, p. 185 – 192 (1985).

The treatment followed by Cox et al. (1973) for the estimation of the effects of termal imbalance on the pulsational stability of stars is applied to Roche and point source (Chandrika Prashad) model. The calculations suggest that the eigenvalues obtained for the above model are correct to $0(t_{ff}/t_s)$. The expressions for stability coefficients and corresponding eigenfunctions for the stars in thermal imbalance for the above models have also been derived.

065.078   An analysis of nonradial pulsations of the central star of the planetary nebula K1–16.
S. Starrfield, A. N. Cox, R. B. Kidman, W. D. Pesnell.
Astrophys. J., Lett. Ed., Vol. 293, No. 1, p. L23 – L27 (1985).

The authors have analyzed stellar models on the high-luminosity branch of the evolutionary track from the AGB to the white dwarfs. Those stars with effective temperatures in the range of $(1.50 – 1.65) \times 10^5 K$ and masses of 0.6 $M_\odot$ are shown to be unstable to nonradial oscillations at periods close to those observed for the central star of the planetary nebula K1–16. The stellar composition utilized in this study consisted of half carbon and half oxygen by mass. The analysis shows that K1–16 is in the same instability strip as the PG 1159–035 class of variable stars because the pulsation driving mechanism in this star is the cyclical ionization of carbon and oxygen just as it is for the PG variables. As K1–16 evolves, it will pulsate at shorter and shorter periods.

065.079   Energetic relations in excitation of small oscillations in stars.
V. G. Gavryusev, E. A. Gavryuseva.
Inst. yader. issled. Akad. Nauk SSSR. Prepr., No. 0374, 13 pp. (1984). In Russian. Abstr. in Ref. Zh., 51. Astron., 6.51.348 (1985).

065.080   On stars, their evolution and their stability.
S. Chandrasekhar.
Science, Vol. 226, No. 4674, p. 497 – 505 (1984).

065.081   Gravitational collapse revisited.
W. D. Arnett.
Bull. Am. Astron. Soc., Vol. 17, No. 1, p. 518 – 519 (1985). Abstract. – See Abstr. 010.064.

065.082   Supernova theory.
W. Hillebrandt.
High energy astrophysics and cosmology, p. 133 – 157 (1983). – See Abstr. 012.068.

The present status of core–collapse models for type II supernovae is reviewed. It is demonstrated that stars of about 10 $M_\odot$ on the main sequence can eject some matter in a supernova explosion, leaving behind a neutron star of around 1.5 $M_\odot$. For these models the explosion mechanism is simply a reflected hydrodynamical shock wave from the central neutron star. This simple and straightforward mechanism probably does not work for more massive stars ($M \gtrsim 15\,M_\odot$). In fact, recent numerical studies indicate, that those stars are more likely to form black holes if they do not rotate sufficiently fast or possess strong magnetic fields, or both.

065.083   Stellar evolution and binaries.
R. F. Webbink.
Interacting binary stars, p. 39 – 70 (1985). – See Abstr. 003.030.

Contents: Introduction. The evolution of isolated stars. The mass–losing star. The accreting star. Systematic mass and angular momentum loss. Mass transfer remnants. Statistics of zero–age binaries. Close binary evolution.

065.084   Density bifurcation in a homogeneous isotropic collapsing star.
S. Bouquet, M. R. Feix, E. Fijalkow, A. Munier.
Astrophys. J., Vol. 293, No. 2, p. 494 – 503 (1985).

The collapse of an isotropic nonradiating sphere of perfect gas submitted to a polytropic equation of state is studied from the point of view of similarity solutions. The equations representing the evolution of the sphere throughout the collapse are shown to have a critical point. An analytic study of this critical point is made possible in the simple case of the homogeneous solution. In this special case, a mathematical and numerical bifurcation is shown to exist.

065.085   Self–gravitating configurations with magnetic fields.
S. K. Trehan.
Arabian J. Sci. Eng., Vol. 9, No. 2, p. 177 – 186 (1984). Abstr. in Phys. Abstr., Vol. 88, No. 1254, Entry 40477 (1985).

065.086   Neutrinos from collapsing stars.
W. Hillebrandt, E. Müller.
Proceedings of the XIth International Conference on Neutrino Physics and Astrophysics, p. 229 – 242 (1984). – See Abstr. 012.081.

Weak interaction processes leading to the production and emission of neutrinos from collapsing stars are reviewed. Based on recent numerical simulations of type II supernova explosions average electron neutrino energies and luminosities are computed. It is shown that typically a fast burst of neutrinos of energy 10 to 20 MeV and a luminosity of around $10^{54} \mathrm{erg\,s^{-1}}$ with a typical duration of a few ms is expected when the supernova shock reaches the neutrino sphere. Most of the neutrinos, however, are leaving the star on a much longer timescale with significantly lower average energies and luminosities.

065.087 Effects of general relativity on the mass ejection of neutrino–trapping supernovae.
M. Takahara, K. Sato.
Prog. Theor. Phys., Vol. 72, No. 5, p. 978 – 988 (1984). Abstr. in Phys. Abstr., Vol. 88, No. 1257, Entry 57606 (1985).

065.088 The problem of non–radial oscillations of highly centrally condensed stellar models.
C. Mohan, K. Singh.
Indian J. Theor. Phys., Vol. 30, No. 4, p. 303 – 313 (1982). Abstr. in Phys. Abstr., Vol. 88, No. 1260, Entry 74216 (1985).

065.089 On initial masses of stars that become Wolf–Rayet stars.
A. V. Tutukov, L. R. Yungel'son.
Astron. Zh., Tom 62, Vyp. 3, p. 604 – 605 (1985). In Russian. English translation in Sov. Astron., Vol. 29, No. 3.

It is shown that the spatial distribution and number ratio of massive O and B stars, found by Conti et al. (1983) is consistent with the statement that Wolf–Rayet stars are produced by components of close binaries with $M \gtrsim 20\ M_\odot$ and by all other stars with $M \gtrsim 40\ M_\odot$.

065.090 Formation of helium stars with extended envelopes in binaries.
A. V. Tutukov, I. Iben.
Cool stars with excesses of heavy elements, p. 345 – 347 (1985). – See Abstr. 012.101.

The authors present a short summary of results related to the formation of some helium rich stars.

065.091 Stellar evolution, nucleosynthesis and dredge–up in cool giants.
P. R. Wood.
Cool stars with excesses of heavy elements, p. 357 – 371 (1985). – See Abstr. 012.101.

The results of calculations of helium shell flashes in stars on the asymptotic giant branch (AGB) are discussed. These results when compared with observations of cool giants with peculiar abundances in the Magellanic Clouds unambiguously show that there is an evolutionary sequence up the AGB in the sense M→S→C (N and late R) which is caused by dredge–up of carbon and s–process elements at helium shell flashes. The Ba and CH stars, which may all have white dwarf companions, are probably the result of mass transfer from N stars in binary systems. The results of calculations of helium core flashes in stars at the tip of the first giant branch are also discussed.

065.092 Analysis of Zr and Tc abundances from S–stars using the s–process with an exponential distribution of neutron exposures.
H. Beer, G. Walter.
Cool stars with excesses of heavy elements, p. 373 – 377 (1985). – See Abstr. 012.101.

The abundances of Zr, Mo, Tc and Ru in S–stars were studied in the frame of the s–process with an exponential neutron fluence distribution. Estimates for the average time integrated neutron flux and the s–process neutron density and temperature have been derived from the abundances of the Zr–isotopes and from Tc, respectively.

065.093 The s–process within the R Coronae Borealis star U Aquarii.
R. A. Malaney.
Cool stars with excesses of heavy elements, p. 379 – 383 (1985). – See Abstr. 012.101.

This paper attempts to show why a detailed study of U Aquarii, with its peculiar abundance characteristics of an extreme hydrogen deficiency coupled with a strontium and yttrium overabundance of about 100 relative to the sun, could lead to an improved understanding of some aspects of current asymptotic giant branch theory. Calculations using nuclear reaction networks suggest some outstanding problems in U Aquarii with respect to this theory.

065.094 The evolution of very low–mass stars and the nature of VB 8B.
L. A. Nelson, S. A. Rappaport, P. C. Joss.
Bull. Am. Astron. Soc., Vol. 17, No. 2, p. 553 (1985). Abstract. – See Abstr. 010.065.

065.095 Blue stragglers: are they rotationally transformed members of a cluster?
S. R. Sreenivasan, W. J. F. Wilson.
Bull. Am. Astron. Soc., Vol. 17, No. 2, p. 600 (1985). Abstract. – See Abstr. 010.065.

065.096 Report of IAU Commission 35: Stellar constitution (*Constitution des étoiles*).
A. N. Cox.
Trans. IAU, Vol. XIXA, p. 479 – 502 (1985). – See Abstr. 003.046.

065.097 Spherically symmetric radiation in gravitational collapse.
D. J. Bridy.
Diss. Abstr. Int., Sect. B, Vol. 45, No. 2, p. 583 (1984). Thesis, Syracuse University, 137 pp. (1983). Order No. DA8410705.

065.098 Thermal effects in stellar pulsations.
W. D. Pesnell.
Diss. Abstr. Int., Sect. B, Vol. 45, No. 4, p. 1218 (1984). Thesis, University of Florida, 137 pp. (1983). Order No. DA8415150.

065.099 Theoretical and observational studies of stellar activity.
J. H. M. M. Schmitt.
Diss. Abstr. Int., Sect. B, Vol. 45, No. 9, p. 2957 – 2958 (1985). Thesis, Harvard University, 211 pp. (1984). Order No. DA8419457.

065.100 A generalized time–dependent local convection theory.
D.–r. Xiong.
Chin. Astron. Astrophys., Vol. 9, No. 1, p. 1 – 7 (1985). English translation of 38.065.153.

065.101 A numerical model for type II supernovae – the collapse stage.
C. Xu, Z.–w. Li, Y.–z. Ge.
Chin. Astron. Astrophys., Vol. 9, No. 2, p. 154 – 158 (1985). English translation of Acta Astrophys. Sin., Vol. 5, No. 1, p. 50 – 58 (1985).

The authors use a numerical model for the collapse stage of a type II supernova of 15 $M_\odot$. The electron capture rate includes the effects of the inverse reaction and the neutron–proton mass difference. This decreases the electron density at the collapse stage, and led to rather large values of the maximum inward velocity and of the corresponding mass. For neutrino transport, they use a leakage model and an equilibrium diffusion model, respectively, for the thin and thick stages and a grey atmosphere model to assess the effect of neutrino precipitation on the collapse. The authors found this effect to be small, the energy precipitation to be not more than 10^{-5} the neutrino energy loss, and the momentum precipitation not more than 10^{-6} the gravitational acceleration.

065.102 Study of boundary effects of brightness variations of a spherical star with two hot precessing spots.
D. P. Kyurkchieva, V. G. Shkodrov.
Dokl. Bolg. Akad. Nauk, Vol. 37, No. 8, p. 995 – 998 (1984).

065.103 On the secular stability of rotating stars.
J. N. Imamura, J. L. Friedman, R. H. Durisen.
Birth and evolution of neutron stars: issues raised by millisecond pulsars, p. 191 – 197 (1984). – See Abstr. 012.115.

The authors calculate the secular stability limits of rotating polytropes to nonaxisymmetric perturbations of low m. They consider polytropic indices ranging fom 1 to 3 and several angular momentum distributions. Results are most conveniently presented in terms of the t–parameter, defined as the ratio of the rotational kinetic energy to the absolute value of the gravitational energy of the fluid.

065.104   **The birth and slow evolution of fizzlers.**
J. E. Tohline.
Birth and evolution of neutron stars: issues raised by millisecond pulsars, p. 198 – 204 (1984). – See Abstr. 012.115.

If the initial ratio of rotational to gravitational energy β_0 in the core of a highly evolved, massive star is $\beta_0 > "\beta_{min}"$, the core will not collapse all the way to neutron star densities. Centrifugal forces will stabilize the object at sub–nuclear densities and a fizzler – not a pulsar – will be formed. A simple model governing global energy balance in these objects is used to predict what the critical value β_{min} for stellar cores is.

065.105   **On the secular instability of axisymmetric rotating stars to gravitational radiation reaction.**
R. A. Managan.
Bull. Am. Astron. Soc., Vol. 17, No. 2, p. 580 (1985). Abstract. – See Abstr. 010.065.

065.106   **The synthesis of heavy elements by charged particle neutron producing processes.**
T. G. Harrison.
Proceedings of the Southwest Regional Conference for Astronomy and Astrophysics, Vol. 10, p. 29 – 32 (1985). – See Abstr. 012.125.

Over the past several years the author has proposed that heavy element nucleosynthesis occurs in low mass stars via photoneutron processes. He has not, however, been able to generate the necessary exposures to account for the buildup of elements heavier than Fe by employing this mechanism. As a direct result of this research the author has however been able to propose an entirely new idea, based upon epithermal proton and α–particle neutron producing reactions, which might not only explain the synthesis of the heavy elements in low mass red giants, but also a whole host of phenomenon seen in these stars.

TITAN: a computer program for calculating stellar evolution.
See Abstr. 021.013.

On the PSIVA (*Probing Stellar Interiors via Variability and Activity*) approach to stellar seismology and activity from space.
See Abstr. 051.010.

A survey for photometric variability from space.
See Abstr. 051.011.

The properties and effects on stellar burning of fractionally charged nuclei.
See Abstr. 061.017.

Unconventional ^{12}C production in Population III stars.
See Abstr. 061.021.

The production and survival of ^{205}Pb in stars, and the ^{205}Pb – ^{205}Tl s–process chronometry.
See Abstr. 061.026.

Helium detonation in pancake stars.
See Abstr. 061.027.

Anisotropic emission of neutrinos originating in beta–decay processes under the influence of a magnetic field of high intensity.
See Abstr. 061.030.

Stellar weak interaction rates for intermediate–mass nuclei. IV. Interpolation procedures for rapidly varying lepton capture rates using effective log (ft) – values.
See Abstr. 061.046.

Magnetic monopoles in stellar interiors.
See Abstr. 061.070.

The ^{13}N(p, γ)^{14}O reaction at stellar energies.
See Abstr. 061.092.

Status of helium burning of ^{12}C.
See Abstr. 061.106.

Indirect investigation of proton capture reactions at stellar energies.
See Abstr. 061.107.

The E2 contribution to the ^{12}C(α,γ)^{16}O reaction at stellar energies in a coupled channel approach.
See Abstr. 061.108.

Neutron capture reactions in astrophysics.
See Abstr. 061.109.

Dynamic stellar neutron capture nucleosynthesis: the need for more nuclear data for the s–process.
See Abstr. 061.111.

The late evolution of WC stars: weak s–process and neutron–rich nuclei in cosmic rays.
See Abstr. 061.132.

Helium plasma: pycnonuclear triple alpha reaction rates in stars.
See Abstr. 062.046.

Nonlinear dynamos: a complex generalization of the Lorenz equations.
See Abstr. 062.174.

Carbon–enriched stellar envelopes: nuclei of planetary nebulae and R Coronae Borealis stars.
See Abstr. 064.081.

The accuracy of the quadrupole approximation for the gravitational radiation from pulsating stars.
See Abstr. 066.043.

General relativistic, partially degenerate semirelativistic isothermal spheres of arbitrary temperature.
See Abstr. 066.045.

Compact objects in bimetric general relativity.
See Abstr. 066.082.

Nonstatic radiating spheres in general relativity.
See Abstr. 066.149.

Quark stars and abnormal neutron stars.
See Abstr. 067.127.

Nuclear matter at high density or high temperature in astrophysical phenomena.
See Abstr. 067.166.

Sub–saturation phases of nuclear matter.
See Abstr. 067.167.

Solar and stellar magnetic cycles: unity and multiplicity.
See Abstr. 080.073.

Comparing α Cen with the sun: chemical composition, mixing–length, age, and p–mode oscillation spectrum.
See Abstr. 080.105.

Solar neutrinos: prospects for detection and implications.
See Abstr. 080.118.

Interior structure and evolution of planets.
See Abstr. 091.013.

Clues on the origin of the solar system.
See Abstr. 107.006.

On the initial masses and evolutionary origins of Wolf–Rayet stars.
See Abstr. 115.001.

The evolutionary status of OB stars with peculiar nitrogen spectra.
See Abstr. 115.011.

The measure of the stars.
See Abstr. 115.023.

Observational comparisons for evolutionary tracks of massive stars.
See Abstr. 115.029.

Theory tested by means of the stars.
See Abstr. 116.014.

Instability in the red star of semi–detached binary systems.
See Abstr. 117.025.

The evolution of massive close binary systems with mass loss and overshooting. I. Mass transfer in systems with initial masses of $20 + 10 \, M_\odot$.
See Abstr. 117.033.

Secular evolution of magnetic cataclysmic variables.
See Abstr. 117.064.

Possible evolution of a triple system into Epsilon Aurigae.
See Abstr. 117.081.

Hydrodynamical modeling of mass transfer from cataclysmic variable secondaries.
See Abstr. 117.196.

Evolution of close binary systems.
See Abstr. 117.282.

Cataclysmic binaries and related objects in the context of stellar evolution.
See Abstr. 117.283.

The formation of massive white dwarfs in cataclysmic binaries.
See Abstr. 117.284.

Interacting binary stars: nomenclature – the stellar zoo.
See Abstr. 117.287.

X–ray binaries: end points of binary evolution.
See Abstr. 117.288.

Angular momentum loss and the evolution of binaries of extreme mass ratio.
See Abstr. 117.299.

The proportion of binaries having a degenerate companion: implications on the formation of barium stars.
See Abstr. 117.343.

Observational evidence for the evolution of contact binary stars.
See Abstr. 117.365.

Tidal interactions of close binary systems.
See Abstr. 117.366.

Wolf–Rayet stars and binarity.
See Abstr. 117.367.

The late stages of interactive stellar evolution.
See Abstr. 117.373.

The age and helium content of the eclipsing binary AI Phoenicis.
See Abstr. 119.041.

Binary systems among the peculiar cool stars.
See Abstr. 120.038.

Cepheid evolution.
See Abstr. 122.062.

Theory of Cepheid pulsation: excitation mechanisms.
See Abstr. 122.063.

Non–adiabatic effects on pulsation periods.
See Abstr. 122.064.

Nonadiabatic, nonlinear pulsations of bump Cepheids – a new approach.
See Abstr. 122.065.

Theoretical study of Cepheid light curves.
See Abstr. 122.066.

Theoretical models of W Virginis variables.
See Abstr. 122.077.

A new possible resonance for population II Cepheids.
See Abstr. 122.078.

Hydrodynamic models of Population II Cepheids.
See Abstr. 122.079.

Some masses for Population I and II Cepheids.
See Abstr. 122.080.

The effects of convection of RR Lyrae stars.
See Abstr. 122.085.

Maia variables and upper–main–sequence phenomena.
See Abstr. 122.098.

A comparison of theoretical and observed RR Lyrae light curves via Fourier decomposition.
See Abstr. 122.194.

On the discrepancy between the observed and theoretical light curves in RR Lyrae variables.
See Abstr. 122.195.

Hydrodynamic simulations of recurrent novae.
See Abstr. 124.001.

Eruptive Ereignisse als Kennzeichen später Sternentwicklung.
See Abstr. 124.012.

Hubble's constant and exploding carbon–oxygen white dwarf models for Type I supernovae.
See Abstr. 125.033.

Model atmospheres for Type I supernovae.
See Abstr. 125.118.

Physical models for Type I supernovae and the distance scale.
See Abstr. 125.120.

Velocity fields in binary protostellar clouds: an alternative to retrograde rotation.
See Abstr. 131.081.

Star formation in rotating, magnetized molecular disks.
See Abstr. 131.274.

Possible consequences of gas accretion for the initial mass function of star clusters.
See Abstr. 131.278.

Planetoidal hypothesis of CP (*chemically peculiar*) F, A, and B star formation: possibilities and prospects.
See Abstr. 131.369.

On the distances of planetary nebulae.
See Abstr. 134.033.

An analysis of the nonradial oscillations of the central star of the planetary nebula K1–16.
See Abstr. 134.048.

On the consequences of a new initial–final mass relation for low and intermediate mass stars and the birthrate of planetary nebulae.
See Abstr. 134.063.

Cosmic rays from Wolf–Rayet stars.
See Abstr. 144.332.

Dynamical consequences of star collisions for core–envelope structure in red giants.
See Abstr. 151.041.

Waves and gravitational instability in a rotating polytropic cylinder.
See Abstr. 151.162.

Time–dependent star formation in OB associations.
See Abstr. 152.003.

Evolution of low–mass stars in the Alpha Persei cluster.
See Abstr. 153.015.

The extended giant branches of intermediate age globular clusters in the Magellanic Clouds. IV.
See Abstr. 154.012.

Bimodal cyanogen distributions in moderately metal–poor globular clusters. II. Evidence for "mixing" and "saturation".
See Abstr. 154.090.

Carbon deflagrating supernovae and the chemical history of the solar neighbourhood.
See Abstr. 155.002.

The evolutionary connection between S and C stars: evidence from star clusters and the Magellanic Clouds.
See Abstr. 156.028.

066 Relativistic Astrophysics, Gravitation Theory

066.001   On the relativistic theory of astrometric observations. III. Radio interferometry of distant sources.
N. V. Pavlov.
Astron. Zh., Tom 62, Vyp. 1, p. 172–180 (1985). In Russian. English translation in Sov. Astron., Vol. 29, No. 1.

The relativistic VLBI equation and expressions for time delay and fringe frequency are deduced in the post–Newtonian approximation of general relativity. In the limits of the adopted accuracy these relations include all relativistic corrections due to the gravitational field of the sun and the ground–based observer's motion.

066.002   Gravitational field energy density for spheres and black holes.
D. Lynden–Bell, J. Katz.
Mon. Not. R. Astron. Soc., Vol. 213, No. 1, p. 21P–25P (1985).

The authors show by physical arguments that static spherical systems have a coordinate–independent field energy density. For Schwarzschild's spacetime it is $+(g^2/8\pi G)$ where $g = \{GM/[(1+m/2\bar{r})^3\bar{r}^2]\}$ and $\bar{r}$ is the isotropic coordinate. The total field energy outside $\bar{r}$ is $GM^2/2\bar{r}$. Schwarzschild's $r = 2m$ corresponds to $\bar{r} = GM/2c^2$ so the field energy outside a Schwarzschild black hole totals Mc^2. In this sense all the energy remains outside the hole.

066.003   Geometry of canonical variables in gravity theories.
M. Antonowicz, W. Szczyrba.
Lett. Math. Phys., Vol. 9, No. 1, p. 43–49 (1985).

The authors show how to pass from an SL(2, C) covariant to an SU(2) covariant formulation of the theories of gravity. The construction determines the canonical and gauge variables of the theory and establishes an appropriate framework for a Hamiltonian picture.

066.004   Computer–aided classification of geometries in general relativity; example: the Petrov type D vacuum metrics.
J. E. Åman.
Classical general relativity, p. 1–4 (1984). – See Abstr. 012.011.

A computer program performing a coordinate invariant classification of the curvature tensor and a number of its derivatives following the algorithm by Karlhede (1980) has been implemented. As an example a classification of the Petrov type D vacuum metrics is presented.

066.005   A singularity arising from the coherent generation of gravitational waves by electromagnetic waves.
H. Ardavan.
Classical general relativity, p. 5–14 (1984). – See Abstr. 012.011.

The gravitational field arising from a circularly moving electromagnetic wave packet is considered and it is shown that the linearized Einstein's equations, on a flat spacetime background, predict infinitely large values for the curvature invariants at any points within the electromagnetic wave packet where the phase speed of the wave packet equals the speed of light *in vacuo*.

066.006   Shear–free flows of a perfect fluid.
A. Barnes.
Classical general relativity, p. 15–23 (1984). – See Abstr. 012.011.

Recent results on shear–free fluid flows in general relativity are reviewed.

066.007   Chaotic behaviour and the Einstein equations.
J. D. Barrow.
Classical general relativity, p. 25–41 (1984). – See Abstr. 012.011.

An elementary introduction to the description of chaotic dynamical systems is given. These ideas are applied to the homogeneous Bianchi cosmologies and chaotic behaviour is shown to exist in types VIII and IX in vacuum. Various consequences of this for investigations of the general solution to the Einstein equations are discussed.

066.008   Cosmic no–hair theorems.
W. Boucher.
Classical general relativity, p. 43–52 (1984). – See Abstr. 012.011.

The role of the cosmological constant Λ in solutions of Einstein's field equations is examined and the prospects for obtaining field theories with $\Lambda = 0$ are discussed.

066.009   A generalization of the Lie derivative.
P. Dolan.
Classical general relativity, p. 53–62 (1984). – See Abstr. 012.011.

The theory of the Lie derivative for arbitrary contravariant tensor fields is presented. An application to the theory of Killing–Yano tensor fields on Riemannian manifolds is given.

066.010 Generalized Robertson–Walker metrics.
P. S. Florides.
Classical general relativity, p. 63 – 75 (1984). – See Abstr. 012.011.
This paper demonstrates that generalized Robertson–Walker (GRW) metrics are (1) expressible in time–independent form, (2) of constant curvature, (3) Einstein spaces. Furthermore there are only six such GRW metrics.

066.011 Absence of asymptotically flat solutions of Einstein's equations which are periodic and empty near infinity.
G. W. Gibbons, J. M. Stewart.
Classical general relativity, p. 77 – 94 (1984). – See Abstr. 012.011.
This paper presents a proof of the theorem that any asymptotically flat spacetime which is empty and periodic in time at infinity must in fact be stationary near infinity.

066.012 Algebraic coordinate conditions.
J. N. Goldberg.
Classical general relativity, p. 95 – 102 (1984). – See Abstr. 012.011.
The propagation of an algebraic coordinate condition on the 2 + 2 formalism of general relativity is considered for a null surface.

066.013 Curvature and metric in general relativity.
G. S. Hall.
Classical general relativity, p. 103 – 120 (1984). – See Abstr. 012.011.
This paper examines the question whether the space–time curvature in a smooth manifold determines the metric uniquely. It also investigates the relationship between the energy–momentum tensor and the Weyl tensor resulting from this metric.

066.014 Exact solutions for rotating charged dust.
J. N. Islam.
Classical general relativity, p. 121 – 130 (1984). – See Abstr. 012.011.
Earlier work by the author on rotating charged dust is summarized. A new global exact solution for cylindrically symmetric differentially rotating charged dust in Newton–Maxwell theory is presented. Finally, a new exact solution for cylindrically symmetric rigidly rotating charged dust in general relativity is given.

066.015 Asymptotically flat rotating solutions.
J. N. Islam.
Classical general relativity, p. 131 – 143 (1984). – See Abstr. 012.011.
The author reviews work on asymptotically flat axisymmetric stationary (rotating) solutions of Einstein's equations.

066.016 Algebraic computing in general relativity.
M. A. H. MacCallum.
Classical general relativity, p. 145 – 171 (1984). – See Abstr. 012.011.

066.017 The use of "REDUCE" in finding exact solutions of the Quadratic Poincaré Gauge field equations.
J. D. McCrea.
Classical general relativity, p. 173 – 182 (1984). – See Abstr. 012.011.
This paper describes the application of the REDUCE algebraic computing system to the problem of deriving solutions to the Quadratic Poincaré Gauge (QPG) field equations. An outline of the QPG theory is given and a specific stationary axially symmetric solution is derived.

066.018 Co–ordinates for expanding type N vacuum solutions.
C. B. G. McIntosh.
Classical general relativity, p. 183 – 192 (1984). – See Abstr. 012.011.
Null tetrads and co–ordinates used to describe expanding type N vacuum solutions of Einstein's field equations are given for two standard systems: the usual "real" one, and one based on a complex double Kerr–Schild form of the metric. The relationship between the two systems is discussed.

066.019 Cosmic censorship, persistent curvature and asymptotic causal pathology.
R. P. A. C. Newman.
Classical general relativity, p. 193 – 207 (1984). – See Abstr. 012.011.
Cosmic censorship is a concept which was introduced by Penrose (1969) to describe situations in general relativity wherein singularities resulting from collapse are hidden from view. This paper investigates conditions in space–time, which are sufficient to ensure that some form of censorship holds, and reviews the various types of singularities to which these censorship conditions may apply.

066.020 Gravitational radiation.
B. F. Schutz.
Classical general relativity, p. 209 – 230 (1984). – See Abstr. 012.011.
This paper reviews recent developments in the theory of gravitational radiation. It studies small–amplitude waves in vacuum, both in linearized theory and in perturbations of nonlinear solutions, and reviews work on large–amplitude waves. Further, it discusses the interaction of waves with matter, including the quadrupole formulas of linearized theory and a short review of current gravitational–wave detectors. The final section describes how to formulate the Newtonian limit and extend the quadrupole formulas to self–gravitating weak–field systems.

066.021 Numerical relativity.
J. M. Stewart.
Classical general relativity, p. 231 – 262 (1984). – See Abstr. 012.011.
This review of numerical methods in general relativity is mainly concerned with finite difference methods, the Cauchy problem, and the characteristic initial value problem.

066.022 Charged rotating dust in general relativity.
N. Van den Bergh, P. Wils.
Classical general relativity, p. 263 – 269 (1984). – See Abstr. 012.011.
The problem of an axially symmetric configuration of charged rotating dust is studied under the additional assumption that the Lorentz force on a typical particle of the fluid vanishes. It is proved that, for constant charge to mass ratio, this assumption leads to rigid rotation or cylindrical symmetry.

066.023 Scalar gravitation and cosmology.
J. A. Ferrari.
Nuovo Cimento B, Vol. 82B, Ser. 2, No. 2, p. 192 – 202 (1984). Abstr. in Phys. Abstr., Vol. 88, No. 1248, Entry 5066 (1985).

066.024 On the generalizations of Nariais GRT–solution in the Brans–Dicke theory.
D. Lorenz–Petzold.
Prog. Theor. Phys., Vol. 71, No. 6, p. 1426 – 1428 (1984). Abstr. in Phys. Abstr., Vol. 88, No. 1249, Entry 10191 (1985).

066.025 On the Nutku–Halil solution for colliding impulsive gravitational waves.
S. Chandrasekhar, V. Ferrari.
Proc. R. Soc. London, Ser. A, Vol. 396, No. 1810, p. 55 – 74 (1984). Abstr. in Phys. Abstr., Vol. 88, No. 1249, Entry 10196 (1985).

066.026 Inflation and generation of perturbations in broken–symmetric theory of gravity.
B. L. Spokoiny (*B. L. Spokoinij*).
Phys. Lett. B, Vol. 147B, No. 1 – 3, p. 39 – 43 (1984). Abstr. in Phys. Abstr., Vol. 88, No. 1249, Entry 10211 (1985).

066.027 Acoustic quality factor of an aluminium alloy for gravitational wave antennae below 1K.
E. Coccia, T. O. Ninikoski.
Lett. Nuovo Cimento, Vol. 41, Ser. 2, No. 7, p. 242 – 246 (1984). Abstr. in Phys. Abstr., Vol. 88, No. 1249, Entry 10245 (1985).

066.028 Gravitational radiation of rapidly rotating drop of homogeneous magnetized gravitating liquid near bifurcation point.
V. P. Tsvetkov.
Phys. Lett. A, Vol. 105A, No. 1 – 2, p. 34 – 35 (1984). Abstr. in Phys. Abstr., Vol. 88, No. 1249, Entry 14470 (1985).

066.029 de Sitter invariant gravity coupled with matter and its cosmological consequences.
T. Fukuyama.
Ann. Phys. (N.Y.), Vol. 157, No. 2, p. 321 – 341 (1984). Abstr. in Phys. Abstr., Vol. 88, No. 1250, Entry 15014 (1985).

066.030 Static space–time in general scalar tensor theory.
N. Banerjee, S. B. Duttachoudhury, A. Banerjee.
J. Math. Phys., Vol. 25, No. 11, p. 3299 – 3302 (1984). Abstr. in Phys. Abstr., Vol. 88, No. 1250, Entry 15017 (1985).

066.031 Sensitivity of a Weber–type resonant antenna to monochromatic gravitational waves.
G. V. Pallottino, G. Pizzella.
Nuovo Cimento C, Vol. 7C, Ser. 1, No. 2, p. 155 – 168 (1984). Abstr. in Phys. Abstr., Vol. 88, No. 1250, Entry 18659 (1985).

066.032 Non–static fluid spheres in general relativity.
A. Mészáros.
Astrophys. Space Sci., Vol. 108, No. 2, p. 415 – 417 (1985).
A few years ago, Banerjee and Banerji had given a class of exact interior solutions for a perfect spherically–symmetric fluid distribution with inhomogeneous density and pressure. This class serves as a simple model for gravitational collapse with inhomogeneous matter and, therefore, may be of an essential astrophysical importance. It is proved that these solutions are, in fact, identical with Faulkes' earlier solutions.

066.033 Comment on the BDT–Bianchi type–VI_0 stiff matter solution.
D. Lorenz–Petzold.
Astrophys. Space Sci., Vol. 108, No. 2, p. 419 – 421 (1985).
The author points out that the Brans–Dicke–Bianchi type VI_0 stiff matter solution recently given by Ram and Singh is not the most general solution of the corresponding field equations. Moreover, this solution has no vacuum limit and the general relativistic limit is obtained only after making an asymptotic expansion. In this paper the author rediscusses the entire problem in a different way. In the limit $\phi = $ const., $\omega \rightarrow \infty$, he obtains both the stiff matter and the vacuum relativistic limit first given by Ellis and MacCallum (1969) in analytic form.

066.034 Comment on the GRT–LRS Bianchi type V solutions.
D. Lorenz–Petzold.
Astrophys. Space Sci., Vol. 108, No. 2, p. 423 – 424 (1985).
The author points out that the locally rotationally symmetric (LRS) Bianchi type–V electromagnetic solutions recently given by Roy and Singh (1983) on the basis of the general theory of relativity (GRT) are wrong. The correct electromagnetic solutions are nothing but the vacuum and stiff matter solutions first given by Ftaclas (1978), which have been generalized by Jantzen (1980) and independently by Lorenz (1981).

066.035 Dynamical effects of radiation on the orbits of bodies in the Schwarzschild metric. I.
J. Buitrago, E. Mediavilla.
Astrophys. Space Sci., Vol. 109, No. 1, p. 77 – 86 (1985).
The authors present a generalization of the Poynting–Robertson effect to the general relativity, obtaining subsequently, in the Schwarzschild metric, the differential equations which describe the orbital evolution of a small spherical body under the combined action of radiation forces and gravitation.

066.036 Einstein–Kahler solutions in some cosmological space–times.
T. Dereli, M. Onder, R. W. Tucker.
Classical Quantum Gravity, Vol. 1, No. 6, p. L67 – L73 (1984). Abstr. in Phys. Abstr., Vol. 88, No. 1251, Entry 19164 (1985).

066.037 The use of generating techniques for space–times with two non–null commuting Killing vectors in vacuum and stiff perfect fluid cosmological models.
D. W. Kitchingham.
Classical Quantum Gravity, Vol. 1, No. 6, p. 677 – 694 (1984). Abstr. in Phys. Abstr., Vol. 88, No. 1251, Entry 19169 (1985).

066.038 Non–linear effects in the determination of the gravitational constant by means of the swing time method.
Y. T. Chen.
Phys. Lett. A, Vol. 106A, No. 1 – 2, p. 19 – 22 (1984). Abstr. in Phys. Abstr., Vol. 88, No. 1251, Entry 24281 (1985).

066.039 On static spherical symmetric solutions of the Bach–Einstein gravitational field equations.
H.–J. Schmidt.
Astron. Nachr., Vol. 306, No. 2, p. 67 – 70 (1985).
For field equations of 4th order, following from a Lagrangian "Ricci scalar plus Weyl scalar", it is shown (using methods of non–standard analysis) that in a neighbourhood of Minkowski space there do not exist regular static spherically symmetric solutions. With that (besides the known local expansions about $r = 0$ and $r = \infty$ resp.) for the first time a global statement on the existence of such solutions is given. Finally, this result will be discussed in connection with Einstein's particle programme.

066.040 Observations of the light deflection during the solar eclipse on 15 February 1961.
F. Schmeidler.
Astron. Nachr., Vol. 306, No. 2, p. 71 – 76 (1985). In German.
During the total eclipse, a plate of the stars near the sun was taken on the Monte Cónero near Ancona in Italy, on which 12 stars were measurable. Comparison with another plate of the same stars taken at night showed strong distortion on the eclipse plate which was very probably due to anomalous refraction. The differences of coordinates between both plates were represented by a general second–order expression. The value $L = 1\overset{..}{.}98 \pm 0\overset{..}{.}46$ was derived for the gravitational deflection on the solar limb.

066.041 Interpretation of observations of light deflection at the solar limb.
F. Schmeidler.
Astron. Nachr., Vol. 306, No. 2, p. 77 – 80 (1985). In German.
The agreement between radioastronomical measurements of the gravitational deflection of light at the solar limb and the theory of relativity is compatible with the excess found by most optical observers. The optical excess is caused by stars so near to the sun that radioastronomical observations are impossible. Optical and radioastronomical measurements both make it likely that there is an additional term inversely proportional to a power higher than the first of the distance from the sun. Results found by Bouet indicate that this additional term is of solar origin.

066.042 A new formulation of gravitational lens theory, time–delay, and Fermat's principle.
P. Schneider.
Astron. Astrophys., Vol. 143, No. 2, p. 413 – 420 (1985).
Recently it was shown that any gravitational lens produces at least one amplified image. For the proof of this theorem, a new formalism for dealing with light bending problems was introduced. After discussing the assumptions which lead to the lens equation in an expanding inhomogeneous universe (Friedmann–Robertson–Walker in the "mean"), the author examines this scalar formalism in greater detail. He shows that this scalar formalism not only simplifies light bending problems, but also gives new physical insights into the lens equation. The applicability of Fermat's principle to light bending problems in an expanding clumpy universe is investigated, and the possibility to obtain the

Hubble constant from measurements of the time delay is discussed.

066.043   The accuracy of the quadrupole approximation for the gravitational radiation from pulsating stars.
E. Balbinski, S. Detweiler, L. Lindblom, B. F. Schutz.
Mon. Not. R. Astron. Soc., Vol. 213, No. 3, p. 553 – 561 (1985).

Two methods for computing the damping of the non–radial oscillations of stars due to gravitational radiation emission are compared: (1) solving the complete linearized equations of general relativity and (2) using an approximate method based on the quadrupole formula. It is shown that the results of the fully relativistic calculation approach the results based on the quadrupole formula for sufficiently non–relativistic stars (i.e. $GM/c^2 R$ sufficiently small). The approximation is worse for larger polytropic indices; i.e. for greater central condensation.

066.044   Gravitational amplification of light by a non–static and thick lens.
F. Hammer.
Astron. Astrophys., Vol. 144, No. 2, p. 408 – 412 (1985).

The specific effects of a non–static and thick gravitational lens on the amplification of light are studied in the frame of a two–step vacuole model (which is made up of a high density Friedmann solution, separated from the background universe by an empty Schwarzschild solution). The author solves here the Optical Scalar Equation (Sachs, 1961) for a radial light beam, to a much higher order than the previous computations based on the thin lens and bounded and stationary mass hypotheses. A substantial correction with respect to the classical formula is obtained for large amplification, a situation which may actually be reached for astrophysical lenses like very rich clusters or superclusters of galaxies at relativistic distances.

066.045   General relativistic, partially degenerate semirelativistic isothermal spheres of arbitrary temperature.
T. W. Edwards.
Astrophys. J., Vol. 288, No. 2, p. 630 – 643 (1985).

The structure of general relativistic isothermal configurations is investigated for the case of a partially degenerate semirelativistic (PD–SR) gas including radiation pressure and energy effects but at an arbitrary constant temperature. This work extends previous investigations of Newtonian PD–SR isothermal configurations. For nonrotating configurations, a two–dimensional (central degeneracy and temperature) analytical continuum of configurations is shown to yield a number of limiting cases, some of which differ from their Newtonian counterparts. Numerical solutions have been obtained, and results are presented for a number of physically interesting quantities.

066.046   The graviton luminosity of the Sun and other stars.
R. J. Gould.
Astrophys. J., Vol. 288, No. 2, p. 789 – 794 (1985).

Graviton production in electron–electron (e–e) and electron–ion (e–z) scattering is evaluated in the Born approximation. The calculation is compared with that for photon production, that is, Coulomb quadrupole bremsstrahlung. Application is made to the Sun, and it is found that for the solar plasma the main contribution to the graviton luminosity comes from the central core at $r/R \approx 0.1$. The total luminosity (L_g) in gravitons is about 7.9×10^{14}ergs s^{-1}. The quantum–mechanical aspects of the solar L_g problem are discussed, and it is shown why a previous classical calculation overestimated L_g by about an order of magnitude. Production of gravitons in binary collisions in other types of stars is discussed briefly. White dwarfs have a typical graviton luminosity $L_{WD} \sim 10^{19}$ergs s^{-1}, while neutron stars have $L_{NS} \sim 10^{25}$ergs s^{-1}.

066.047   Minilensing of multiply imaged quasars: flux variations and vanishing of images.
K. Subramanian, S. M. Chitre, D. Narasimha.
Astrophys. J., Vol. 289, No. 1, p. 37 – 51 (1985).

The gravitational influence of minilenses which are located in lens galaxies and clusters and which intercept the light paths from multiply imaged quasars is investigated by developing an analyt-

ical framework. The mathematical formulation is adopted for deriving flux variations due to minilensing, and explicit light curves are computed numerically for a variety of minilens paths coming close to the images of the twin quasar 0957 + 561 A,B. It is demonstrated that under certain circumstances the operation of minilensing can lead to the vanishing of an image.

066.048   Numerical analysis of one–soliton solutions on a Bianchi type–II background.
A. Curir, M. Francaviglia.
Gen. Relativ. Gravitation, Vol. 17, No. 1, p. 1 – 14 (1985).

The asymptotic behavior of a solution of vacuum Einstein equations describing the propagation of a single "soliton wave" on a Bianchi type–II background is investigated numerically. In the framework of the oscillatory approach to a cosmological singularity, the "transitional behavior" corresponding to a critical value of a fundamental integration constant is analyzed.

066.049   Cauchy's problem and Huygens' principle for relativistic higher spin wave equations in an arbitrary curved space–time.
V. Wünsch.
Gen. Relativ. Gravitation, Vol. 17, No. 1, p. 15 – 38 (1985).

Relativistic spin s, nonzero mass equations are given which in an arbitrary curved space–time are internally consistent. By means of Riesz' integration method a representation theorem for the solution of Cauchy's problem is proved. A necessary and sufficient condition for the validity of Huygens' principle is stated from which follows that only in space–times of constant curvature do the field equations satisfy Huygens' principle.

066.050   On Møller's tetrad theory of gravitation. Gravitational radiation.
D. Sáez.
Gen. Relativ. Gravitation, Vol. 17, No. 1, p. 39 – 53 (1985).

In the background of Møller's theory the losses of energy and angular momentum of a nongravitationally bound system due to the radiation of gravitational waves are evaluated. The post–Newtonian limit is studied. When the modulus of the coupling constant of the theory is smaller than 10^3 and some reasonable conditions are assumed, the results here obtained appear to be identical to those of general relativity.

066.051   Spontaneously broken symmetries in gravitational fields.
V. N. Melnikov (*V. N. Mel'nikov*), V. M. Nikolaenko, S. V. Orlov.
Gen. Relativ. Gravitation, Vol. 17, No. 1, p. 63 – 88 (1985).

The authors study models with material and gravitational fields unified in a selfconsistent manner. They use a semiclassical approach, where the gravitational field is classical, but the other fields may be quantum ones. The mechanism of spontaneous symmetry breaking is due to the conformal generalization of the Higgs fields in a curved space–time. Gravitation is considered as a gauge field and then in the usual Einstein version.

066.052   Cylindrically symmetric Einstein–Yang–Mills–Higgs gauge configurations.
R. P. Mondaini.
Gen. Relativ. Gravitation, Vol. 17, No. 2, p. 101 – 109 (1985).

Two solutions are obtained for coupled Einstein–Yang–Mills–Higgs fields with cylindrical symmetry and rigid rotation. The Higgs fields are responsible for the creation of singularities and infinite energy densities at the cylinder's axis.

066.053   Complex relativity and real solutions. I: Introduction.
C. B. G. McIntosh, M. S. Hickman.
Gen. Relativ. Gravitation, Vol. 17, No. 2, p. 111 – 132 (1985).

This is the first of a series of papers on complex spaces and their use in complex relativity. The basic aim is to develop the theory of complex relativity insofar as it helps in obtaining, and understanding, real solutions of Einstein's vacuum equations as slices of complex solutions. The basic equations and key entities are presented. The basic relativistic language used is that of Newman and Penrose.

066.054 Nonlinear connections for curved twistor spaces.
A. D. Helfer.
Gen. Relativ. Gravitation, Vol. 17, No. 2, p. 133 – 147 (1985).

A holomorphic connection on (1, 0)–vector fields which is intrinsically defined on any curved twistor space is described. Although it is a local operator, it is given in terms of the nonlocal geometry of the twistor space corresponding to the local geometry of the space–time.

066.055 Space–times admitting Penrose–Floyd tensors.
P. Taxiarchis.
Gen. Relativ. Gravitation, Vol. 17, No. 2, p. 149 – 166 (1985).

The author presents the full integrability conditions for the Penrose–Floyd equation. Then he proceeds to the integration of the Einstein equations under the assumption of the existence of a Penrose–Floyd tensor and finds all space–times admitting such a tensor.

066.056 Gravity and grand unified theories.
L. Parker, D. J. Toms.
Gen. Relativ. Gravitation, Vol. 17, No. 2, p. 167 – 171 (1985).

It is currently believed that quantum effects play a dominant role in the evolution of the very early universe. The large curvature of space–time at early times gives rise to unexpected phenomena, such as the creation of particles by the expanding universe, which have strong dynamical consequences. This paper discusses some new consequences of grand unified theories that do depend in an essential way on curvature.

066.057 A new class of ideal clocks.
C. M. Will.
Gen. Relativ. Gravitation, Vol. 17, No. 2, p. 173 – 177 (1985).

The author has found a new class of ideal clocks within general relativity. They are self–gravitating systems such as rotating stars, rotating black holes, and binary star systems. The gravitational redshift of the observed period of rotation of such clocks in a given, weak external gravitational field is shown to be the same as that of an ideal atomic clock. This result is important for the binary pulsar PSR 1913 + 16, where the gravitational redshift of the pulsar's frequency caused by the companion's gravitational field is an observable effect.

066.058 The Schwarzschild H space.
J. R. Porter.
Gen. Relativ. Gravitation, Vol. 17, No. 2, p. 179 – 186 (1985).

The usual construction of an H space from asymptotic data on I^+ uses only the radiative component of the asymptotic gravitational field and ignores the longitudinal part of the field. The first step of giving a direct H space construction for the Schwarzschild asymptotic data is presented here.

066.059 On solutions of Einstein and Einstein–Yang–Mills equations with (maximal) conformal subsymmetries.
S. Sinzinkayo, J. Demaret.
Gen. Relativ. Gravitation, Vol. 17, No. 2, p. 187 – 201 (1985).

The authors consider the maximal subgroups of the conformal group (which have in common as a subgroup the group of pure spatial rotations) as isometry groups of conformally flat space–times. They identify the corresponding cosmological solutions of Einstein's field equations. For each of them, the authors investigate the possibility that it could be generated by an $SU(2)$ Yang–Mills field built from a scalar field identical with the square root of the conformal factor defining the space–time metric tensor.

066.060 Gravitational lensing and galaxy shape.
R. L. Webster.
Mon. Not. R. Astron. Soc., Vol. 213, No. 4, p. 871 – 888 (1985).

A sufficiently large mass inhomogeneity will gravitationally lense high–redshift galaxies. It is possible to calculate the size of this effect if assumptions are made about the distribution of mass in clusters. If most of the mass in the Universe is clumped on the scale of clusters, then the ellipticity distribution of distant sources could be used to discriminate between different values of total mass density. A population of intrinsically circular sources ob-

served at a redshift of one would set limits on the amount of matter in clusters. In addition, the effects of lensing by clusters on specific geometric measurements of a single object might be substantial.

066.061 The limit of mass for gravitational collapse and nuclear forces at short distance.
J. Buitrago, E. Mediavilla.
Astrophys. Space Sci., Vol. 109, No. 2, p. 407 – 409 (1985).

On the basis of the current knowledge about the size of the neutron, the nucleon–nucleon interaction, and the limit on the ratio $Gm/c^2R < 4/9$ imposed by General Relativity, the authors find an upper limit of about $2\,M_\odot$, for stars of uniform density in the neutron phase. The importance of an accurate determination of the neutron radius for neutron stars studies is pointed out.

066.062 Is gravitational energy real?
J. Maddox.
Nature, Vol. 314, No. 6007, p. 129 (1985).

The long–recognized problem of Einstein's theory of gravitation, that gravitational energy is not unambiguously defined, may be unavoidable.

066.063 The principle of equivalence at finite temperature.
J. F. Donoghue, B. R. Holstein, R. W. Robinett.
Gen. Relativ. Gravitation, Vol. 17, No. 3, p. 207 – 214 (1985).

The authors demonstrate that the equivalence principle is violated by radiative corrections to the gravitational and inertial masses at finite temperature. They argue that this result can be attributed to the Lorentz noninvariance of the finite temperature vacuum.

066.064 Planck length as the lower bound to all physical length scales.
T. Padmanabhan.
Gen. Relativ. Gravitation, Vol. 17, No. 3, p. 215 – 221 (1985).

The effect of quantum fluctuations of gravity on the measurement of proper distances is considered. It is shown that, when the length scales are of the order of Planck length the concept of a unique distance between points ceases to exist. It is also shown that the quantum expectation value of the proper length is bounded from below by Planck length in any space–time.

066.065 Exact solutions for nonstatic perfect fluid spheres with shear and an equation of state.
N. van den Bergh, P. Wils.
Gen. Relativ. Gravitation, Vol. 17, No. 3, p. 223 – 243 (1985).

It is shown that fairly general assumptions about the relevant physical and mathematical quantities in the study of nonstatic and spherically symmetric perfect fluid configurations usually single out the Friedmann–Lemaitre solutions as the only physically plausible ones. In addition three new classes of exact solutions having shear and acceleration are presented, as well as a generalization of Wesson's stiff fluid solution.

066.066 On a new interior Schwarzschild solution.
N. K. Kofinti.
Gen. Relativ. Gravitation, Vol. 17, No. 3, p. 245 – 249 (1985).

It is shown explicitly that a new interior Schwarzschild solution satisfies a set of necessary and sufficient conditions for a spherically symmetric metric to join smoothly onto the vacuum field at a nonnull boundary surface. Moreover, the conditions do not prevent the radius of a spherical distribution from assuming values arbitrarily close to the Schwarzschild radius.

066.067 Incompleteness of timelike submanifolds with nonvanishing second fundamental form.
J. K. Beem, P. E. Ehrlich.
Gen. Relativ. Gravitation, Vol. 17, No. 3, p. 293 – 300 (1985).

Let M be a properly immersed timelike hypersurface of Minkowski space and assume that M has a strictly positive second fundamental form. If each point of M is of diagonal type and dim $M \geqslant 3$, then the Ricci curvature of M is strictly positive on all (nonzero) nonspace–like vectors. Thus M satisfies both the generic and strong energy conditions and a singularity theorem for M may be established.

066.068   Graviton ionization of hydrogenlike ions.
A. S. Burundukoff (*A. S. Burundukov*).
Gen. Relativ. Gravitation, Vol. 17, No. 4, p. 311 – 318 (1985).
Calculations are made of the ionization process of hydrogen atoms or hydrogenlike ions by gravitons in both nonrelativistic and relativistic cases. The process of graviton deuteron fission is also examined.

066.069   Generalized Robertson–Walker metrics and some of their properties. II.
P. S. Florides.
Gen. Relativ. Gravitation, Vol. 17, No. 4, p. 319 – 341 (1985).
The Robertson–Walker (RW) metrics, of dimensionality four and signature –2, are generalized to metrics of dimensionality $(n+1)$ and of arbitrary signature, $n\,(>1)$ being an arbitrary integer. In canonical coordinates these generalized Robertson–Walker (GRW) metrics are functions of the coordinate t. The following statements are proved to be equivalent: The GRW metrics are (1) expressible in t–independent form, (2) of constant curvature, (3) Einstein spaces.

066.070   Conformal Einstein spaces.
C. N. Kozameh, E. T. Newman, K. P. Tod.
Gen. Relativ. Gravitation, Vol. 17, No. 4, p. 343 – 352 (1985).
The authors study conformal transformations in four-dimensional manifolds. In particular, they present a new set of two necessary and sufficient conditions for a space to be conformal to an Einstein space. The first condition defines the class of spaces conformal to C spaces, whereas the last one (the vanishing of the Bach tensor) gives the particular subclass of C spaces which are conformally related to Einstein spaces.

066.071   Scattering for the wave equation on the Schwarzschild metric.
J. Dimock.
Gen. Relativ. Gravitation, Vol. 17, No. 4, p. 353 – 369 (1985).
The author studies the wave equation for the Schwarzschild metric. Wave operators are constructed which yield solutions with given asymptotic behavior either at infinity or on the horizon. Asymptotic completeness is demonstrated for these wave operators.

066.072   A case of dual interpretation of Einstein–Maxwell fields.
P. Wils, N. van den Bergh.
Gen. Relativ. Gravitation, Vol. 17, No. 4, p. 381 – 385 (1985).
It is proved that the stationary semicylindrically symmetric Einstein–Maxwell solutions given in an earlier paper allow an interpretation as charge without mass fields. The system of equations is simplified and a previously unknown solution is given.

066.073   Gravitational–wave emission by gravitating systems in the post–Newtonian approximation.
V. P. Tsvetkov.
Sov. Astron., Vol. 28, No. 4, p. 394 – 395 (1984). English translation of 38.066.012.

066.074   Time and singularity.
J. Demaret, M. J. Gotay.
The Big Bang and Georges Lemaitre, p. 123 – 131 (1984). – See Abstr. 012.043.
The authors show that the occurrence of quantum gravitational collapse and, more generally, the validity of Wheeler's "rule of unanimity" are inextricably linked to the classical choice of time. The crucial distinction is between "fast" and "slow" times, that is, between times which give rise to complete or incomplete classical evolution respectively. The authors conjecture that unitary slow–time quantum dynamics is always non–singular, while unitary fast–time quantum dynamics inevitably leads to

collapse. These findings are illustrated by an analysis of the dust–filled Friedmann–Lemaitre–Robertson–Walker universes.

066.075   Algebraic programming in general relativity and cosmology.
Y. De Rop, A. Moussiaux, P. Tombal, A. Ronveaux, J. Demaret, J. L. Hanquin.
The Big Bang and Georges Lemaitre, p. 329 – 340 (1984). – See Abstr. 012.043.
Some algebraic programs, developed recently by a collaboration of physicists from Liège and Namur, are briefly described. They deal with topics in general relativity and cosmology, such as the field equations for Bianchi cosmological models, the general relativistic Hamiltonian formalism and its application to some vacuum inhomogeneous space–times and the search for metrics of stationary axisymmetric space–times.

066.076   Pioneer 10 search for gravitational waves – limits on a possible isotropic cosmic background of radiation in the microhertz region.
J. D. Anderson, B. Mashhoon.
Astrophys. J., Vol. 290, No. 2, p. 445 – 448 (1985).
The nature of the response of the Doppler tracking system to a stochastic background of gravitational radiation is discussed. Using data acquired in 1981 by the Deep Space Network with the Pioneer 10 spacecraft, the authors place upper limits on the energy density of the background in three frequency bands extending from 7×10^{-7} to 10^{-4} Hz, a region that has been inaccessible previously by any technique.

066.077   On negative time dilation, the yttrium effect and applications to chemistry, physics and cosmology.
F. S. Dunay.
Indian J. Theor. Phys., Vol. 30, No. 3, p. 255 – 265 (1982). Abstr. in Phys. Abstr., Vol. 88, No. 1252, Entry 24741 (1985).

066.078   Role of Maxwellian distribution in an expanding universe.
H. E. Kandrup.
Phys. Rev. D, Vol. 30, No. 10, p. 2067 – 2075 (1984). Abstr. in Phys. Abstr., Vol. 88, No. 1252, Entry 24805 (1985).

066.079   Gravitation and irreversibility.
A. Neacsu.
Int. J. Theor. Phys., Vol. 23, No. 10, p. 1009 – 1024 (1984). Abstr. in Phys. Abstr., Vol. 88, No. 1252, Entry 24808 (1985).

066.080   On the connection between the cosmological constant and the gravitational constant in a theory of induced gravity.
M. D. Pollock.
Phys. Lett. B, Vol. 148B, No. 4 – 5, p. 287 – 290 (1984). Abstr. in Phys. Abstr., Vol. 88, No. 1252, Entry 24816 (1985).

066.081   Gravitational forces.
R. A. Waldron.
Speculations Sci. Technol., Vol. 7, No. 3, p. 177 – 189 (1984). Abstr. in Phys. Abstr., Vol. 88, No. 1252, Entry 29834 (1985).

066.082   Compact objects in bimetric general relativity.
A. Harpaz, N. Rosen.
Astrophys. J., Vol. 291, No. 2, p. 417 – 421 (1985).
The field equations of the bimetric general relativity theory proposed by one of the authors (N. Rosen), in the static form, are solved in order to investigate the structure of a star. It is found that for an ordinary star the bimetric theory gives the same results as the Einstein general relativity theory. However, for a collapsed star the two theories give different results. In the bimetric theory a configuration in hydrostatic equilibrium exists for a collapsed star filling its Schwarzschild sphere. In general relativity no equilibrium configuration exists in this region, and the star shrinks to a point singularity to form a black hole.

066.083 The equations of motion of a test particle with spin and self–gravity.
T. W. Noonan.
Astrophys. J., Vol. 291, No. 2, p. 422 – 446 (1985).

Papapetrou's equations of motion of a spinning particle in general relativity are extended to the case of a nonsymmetric stress–energy tensor in order to include the test particle's self–gravity. The author derives the spin tensor and the momentum tensor and their equations of motion, including gravitational radiation reaction. The treatment of a test particle includes self–gravity, charge, and electromagnetic moment. It is concluded that the test particle's Newtonian gravitational potential energy contributes to the test particle's inertial mass (under electromagnetic forces) in the same way as do other forms of mass–energy (rest mass, internal energy, and kinetic energy).

066.084 Second–order coefficients for radiating fluids.
D. Jou, D. Pavón.
Astrophys. J., Vol. 291, No. 2, p. 447 – 449 (1985).

Based on the Ernst method of projection operators, the authors determine the linear coefficients of second–order transport equations for radiating fluids.

066.085 Reply to C. M. Will on the axially symmetric two–body problem in general relativity.
F. I. Cooperstock, P. H. Lim.
Astrophys. J., Vol. 291, No. 2, p. 460 – 463 (1985).

The authors consider the recent paper by Will which purports to demonstrate that the gravitational radiation which they had computed from their model two–body free–fall system is consistent with the so–called "quadrupole formula". The authors show that in fact the system presented by Will is different from their own and that the illegitimate application of the quadrupole formula to their system leads to a smaller flux than that which is correctly deduced using general relativity and a proper consideration of nonlinearities.

066.086 A note on the kinematics of the tachyon fluid.
T. Singh, G. Prasad.
Astrophys. Space Sci., Vol. 110, No. 2, p. 393 – 395 (1985).

It is shown that the kinematical parameters associated with the congruence formed by tachyonic motion can be defined in the manner of Greenberg, but not that of Ehlers. The space–like counterpart of Raychaudhuri's equation has also been obtained.

066.087 On the singularity in Rosen's bimetric theory of gravitation.
A. V. Sarkissian (*A. V. Sarkisyan*), R. M. Avakian (*R. M. Avakyan*), V. T. Karapetian (*V. T. Karapetyan*).
Astrophys. Space Sci., Vol. 111, No. 1, p. 197 – 201 (1985).

It is shown that Rosen's bimetric theory of gravitation also has a Schwarzschild–type singularity. The radius of the singularity – the gravitational radius – is defined by the secondary mass M^1. Simultaneously it is shown that only those configurations are stable, which are located in the range of central densities corresponding to the increasing branch of the $M^1 = M^1(\varrho_c)$ curve.

066.088 Newtonian limit in the Poincaré gauge field theory of gravitation.
Y. Zhang.
Sci. Sin., Ser. A, Vol. 28, No. 4, p. 399 – 404 (1985).

This paper discusses the Newtonian limit of five Lagrangians without ghosts and tachyons by means of the approximate solutions in the Poincaré gauge field theory of gravitation.

066.089 The charge, magnetic dipole and Lense–Thirring effect in general theory of gravity.
G. G. Arutyunyan, V. V. Papoyan.
Astrofizika, Tom 21, Vyp. 3, p. 587 – 594 (1984). In Russian. English translation in Astrophysics, Vol. 21, No. 3.

Three physical problems are solved in the framework of the general theory of gravity. The gravitational field of the point charged mass and the expression for the vector potential for the dipole magnetic field is found. The angular velocity of the in-volvement of the local frame into rotation of the central body is evaluated.

066.090 Signal shape recovery in gravitational–wave experiments.
S. I. Babichenko, V. N. Rudenko, A. G. Yagola.
Sov. Astron., Vol. 28, No. 5, p. 531 – 537 (1984). English translation of 38.066.176.

066.091 Development of the ideas on time.
D.–E. Liebscher, I. D. Novikov.
Priroda, No. 4, p. 15 – 20 (1985). In Russian.

066.092 On Kaluza–Klein theories.
N. Straumann.
MPA Rep., No. 177, 18 pp. (1985).

The Kaluza–Klein theory of gravity and electromagnetism and their extensions to non–Abelian gauge groups are generalized to fibre bundles, which are associated to principle fibre bundles and whose standard fibres are homogeneous spaces. A simple derivation of the generalized "Kaluza–Klein miracle" is given.

066.093 On the formulation of the positive energy theorem in Kaluza–Klein theories.
O. M. Moreschi, G. A. J. Sparling.
MPA Rep., No. 178, 24 pp. (1985).

The positive energy theorem is formulated in the context of Kaluza–Klein theories. Different cases are considered including the situation in which no symmetry is assumed. This work offers a new technique for stability considerations in Kaluza–Klein theories.

066.094 Absorption of gravitational energy by a viscous compressible fluid in a curved spacetime.
D. Papadopoulos, F. P. Esposito.
Astrophys. J., Vol. 292, No. 2, p. 330 – 338 (1985).

The authors derive the equations of motion for a viscous compressible fluid moving in a curved spacetime. The response of the fluid to an incident gravitational wave propagating in this spacetime is determined by solving the fluid equations. It is found that the response can be decomposed with a sound wave and a shear wave whose damping heats a surface layer.

066.095 The gravitational lens as an astronomical diagnostic.
W. C. Saslaw, D. Narasimha, S. M. Chitre.
Astrophys. J., Vol. 292, No. 2, p. 348 – 356 (1985).

The authors examine how a point gravitational lens images a diffuse background source, and show that under a wide range of circumstances the image contains a "gravity ring". The chief criterion for producing this phenomenon is that the angular scale of brightness variation in the source be smaller than the angular radius of the cone of inversion of the image. The observation of gravity rings, which should soon become resolvable with very long baseline radio interferometry, would give an independent (or unique) estimate of fundamental quantities such as the mass or distance of the deflector star and the distance of the source.

066.096 On the intrinsic entropy of the gravitational field.
L. Smolin.
Gen. Relativ. Gravitation, Vol. 17, No. 5, p. 417 – 437 (1985).

The author shows that in linearized general relativity it is impossible to construct a detector by the use of which the quantum state of the linearized gravitational field could be reliably determined. If this property is true of the full theory then one can conclude that a certain proportion of both the energy and information carried by a gravitational wave is irreversibly lost, and that there is a corresponding intrinsic entropy associated with any distribution of gravitational radiation.

066.097 Gravitational field strength and generalized Komar integral.
W. Simon.
Gen. Relativ. Gravitation, Vol. 17, No. 5, p. 439 – 459 (1985).

The author defines a "gravitational field strength" in theories of the Einstein–Cartan type admitting a Killing vector. This field strength is a second rank, antisymmetric, divergence–free tensor, whose ("Komar") integral over a closed 2–surface gives a physically meaningful quantity. The author finds conditions on the Lagrange density of the theory which ensure the existence of such a tensor, and shows that they are satisfied for $N = 2$–supergravity and for a special case of the bosonic sector of $N = 4$–supergravity.

066.098   On the possibility of a box for holding gravitational radiation in thermal equilibrium.
D. Garfinkle, R. M. Wald.
Gen. Relativ. Gravitation, Vol. 17, No. 5, p. 461 – 473 (1985).

The authors show that it is possible in principle to build a box which will hold gravitational radiation for a time long enough to thermalize it. The box is a thin spherical shell of charged matter with a large redshift at the surface of the shell. The radiation is kept in the box by the gravitational potential of the shell and is thermalized by the conversion between gravitational and electromagnetic radiation. The authors calculate the time for escape of the radiation and show that it is longer than the conversion time.

066.099   Complex relativity and real solutions II: classification of complex bivectors and metric classes.
G. S. Hall, M. S. Hickman, C. B. G. McIntosh.
Gen. Relativ. Gravitation, Vol. 17, No. 5, p. 475 – 491 (1985).

This paper continues the examination of real metrics and their properties from the viewpoint of complex relativity as initiated by McIntosh and Hickman. Tetrads of real metrics can be formally complexified by complex coordinate transformations and tetrad rotations and their properties investigated from the viewpoint of complex relativity.

066.100   A characterization of Robertson–Walker spaces by null sectional curvature.
S. G. Harris.
Gen. Relativ. Gravitation, Vol. 17, No. 5, p. 493 – 498 (1985).

066.101   Note on cosmic censorship.
F. J. Tipler.
Gen. Relativ. Gravitation, Vol. 17, No. 5, p. 499 – 507 (1985).

A number of recent theorems by Królak and Newman purport to prove cosmic censorship by showing that "strong curvature" singularities must be hidden behind horizons. The author proves that Newman's "null, strong curvature" condition, which he imposes on certain classes of null geodesics to restrict curvature growth in the space–time, does not hold in many physically realistic space–times: it is not satisfied by any null geodesic in the relevant class in any open Friedmann cosmological model, nor does it hold for any null geodesic in the relevant class in maximal Schwarzschild space.

066.102   The foundations of general relativity.
H. Bondi.
Q. J. R. Astron. Soc., Vol. 26, No. 2, p. 122 – 126 (1985). Invited lecture at the McVittie conference, Canterbury, 1984 June 2.

066.103   Understanding the solutions of Einstein's equations.
M. A. H. MacCallum.
Q. J. R. Astron. Soc., Vol. 26, No. 2, p. 127 – 136 (1985). Invited lecture at the McVittie conference, Canterbury, 1984 June 2.

The study of solutions of Einstein's equations of general relativity is of long standing, but has experienced a recent resurgence. New methods have been used, and new results obtained, for problems concerned with general classes of solutions, with approximate solutions and with exact solutions. Here a number of these developments are reviewed. They concern the identification and classification of solutions, the methods of finding new solutions, the properties of the space of solutions and general properties of classes of solutions, and the physical and astrophysical implications of particular solutions.

066.104   Thermal gravitational radiation from stellar objects and its possible detection.
C. Sivaram.
Bull. Astron. Soc. India, Vol. 12, No. 4, p. 350 – 356 (1984).

There has been a lot of current interest in attempting to detect or indirectly infer the presence of gravitational radiation especially from compact objects like pulsars. However, applications of general relativity lead to explore the possibility of compact stellar objects generating high frequency ($10^{16} - 10^{21}$Hz) thermal gravitational radiation which in the case of very young neutron stars could be rather high. Models of the earliest Planck phase of the universe would predict a thermal gravitational wave background whose detection (at frequencies $\sim 10^{11}$Hz) would enable to make a choice between various cosmological models and particle physics models that attempt to unify fundamental interactions.

066.105   Analytic result for the properties of gravitational waves emitted by a large class of model sources.
H. Beltrami, W. Y. Chau.
Astrophys. Space Sci., Vol. 111, No. 2, p. 335 – 341 (1985).

Expressions for the mass quadrupole moment tensor in the model for a wide variety of astrophysical objects are shown to be identical in form. This makes it possible to obtain analytical expressions for the gravitational radiation emitted by the sources, as well as the angular distribution, polarization dependence, and the wave forms of the radiation.

066.106   Einstein's gravity with massive gravitons.
A. Mészáros.
Astrophys. Space Sci., Vol. 111, No. 2, p. 399 – 405 (1985).

Graviton may, in principle, have a small non–zero mass. In this paper the relevant theory of the massive graviton with six polarisations is developed. The drastic impact of a non–zero mass of the graviton on cosmology is also illustrated.

066.107   Around–the–world relativistic Sagnac experiment.
D. W. Allan, M. A. Weiss, N. Ashby.
Science, Vol. 228, No. 4695, p. 69 – 70 (1985).

In 1971 Hafele and Keating carried portable atomic clocks east and then west around the world and verified the Sagnac effect, a special relativity effect attributable to the earth's rotation. In the study reported here observations of the effect were made by using electromagnetic signals instead of portable clocks to make clock comparisons. Global Positioning System satellites transmit signals that can be viewed simultaneously from remote stations on the earth; thus an around–the–world Sagnac experiment can be performed with electromagnetic signals. The effect is larger than that occurring when portable clocks are used. The average error over a 3–month experiment was only 5 nanoseconds.

066.108   Gravitational lens effects of neutrino celestial objects of uniform density.
C. Xu, X. Wu.
High energy astrophysics and cosmology, p. 481 – 488 (1983). – See Abstr. 012.068.

The purpose of this paper is to study the transparent gravitational lens effects of a neutrino celestial object. The solutions of the equations of null geodesics are presented.

066.109   A method of determining apparent horizons in three–dimensional numerical relativity.
T. Nakamura, Y. Kojima, K. Oohara.
Phys. Lett. A, Vol. 106A, No. 5 – 6, p. 235 – 238 (1984). Abstr. in Phys. Abstr., Vol. 88, No. 1253, Entry 30279 (1985).

066.110   Velocity dependent inertial induction: an extension of Mach's principle.
A. Ghosh.
Pramāna, Vol. 23, No. 5, p. L671 – L674 (1984). Abstr. in Phys. Abstr., Vol. 88, No. 1253, Entry 30281 (1985).

066.111   The Einstein–Rosen gravitational waves and cosmology.
M. Carmeli, C. Charach.
Found. Phys., Vol. 14, No. 10, p. 963 – 986 (1984). Abstr. in Phys. Abstr., Vol. 88, No. 1253, Entry 30282 (1985).

066.112   Does a cosmic censor exist?
W. Israel.
Found. Phys., Vol. 14, No. 11, p. 1049 – 1059 (1984). Abstr. in Phys. Abstr., Vol. 88, No. 1253, Entry 30291 (1985).

066.113   Gravitational waves in an expanding universe.
G. E. Tauber.
Found. Phys., Vol. 14, No. 12, p. 1169 – 1183 (1984). Abstr. in Phys. Abstr., Vol. 88, No. 1253, Entry 30293 (1985).

066.114   Generalized theory of gravitation.
J. W. Moffat.
Found. Phys., Vol. 14, No. 12, p. 1217 – 1252 (1984). Abstr. in Phys. Abstr., Vol. 88, No. 1253, Entry 30294 (1985).

066.115   The motion of a classical spinning particle in a Riemann–Cartan space–time.
R. Amorim.
Gen. Relativ. Gravitation, Vol. 17, No. 6, p. 525 – 533 (1985).

A consistent set of equations of motion for classical charged particles with spin and magnetic dipole moment in a Riemann–Cartan space–time is generated from a constrained Lagrangian formalism. The equations avoid the spurious free helicoidal solutions and at the same time conserve the canonical condition of normalization of the 4–velocity.

066.116   The gravitational contribution to the momentum of a medium in general relativity.
T. W. Noonan.
Gen. Relativ. Gravitation, Vol. 17, No. 6, p. 535 – 544 (1985).

The time–space components of the macroscopic stress–energy tensor of a medium in general relativity (derived in a previous paper) are examined for the momentum density and energy flux of the microscopic gravitational fields. Agreement with both the Newtonian approximation and Chandrasekhar's post–Newtonian approximation is noted.

066.117   Space as a "bucket of dust".
G. Nagels.
Gen. Relativ. Gravitation, Vol. 17, No. 6, p. 545 – 557 (1985).

A pregeometry of space is considered, in which the only structure imposed on individual points is a uniform probability of adjacency between any two arbitrarily chosen points. This probability is not specified (but only assumed small) and neither is the total number of points specified (but is assumed large). It is shown that the most likely distribution of points has similarities with a closed three–dimensional space of constant positive curvature.

066.118   A constraint on physical quantum states in cosmological space–times.
A. H. Najmi, A. C. Ottewill.
Gen. Relativ. Gravitation, Vol. 17, No. 6, p. 573 – 578 (1985).

The problem of constructing the Hilbert space of physical states for a free scalar quantum field propagating on a cosmological background is considered. The concept of energy-momentum for such a field is discussed.

066.119   On the scalar–tetradic theories of gravitation: cosmology and gravitational radiation.
D. Sáez.
Gen. Relativ. Gravitation, Vol. 17, No. 6, p. 579 – 593 (1985).

The so–called scalar–tetradic theories A and B generalize Møller's theory of gravitation. In this work, some results previously obtained in Møller's theory are extended to the theories A and B. When the parameters W and λ^* of the theory A are not too small (or large), the losses of energy and angular momentum of a nongravitationally bound system are proved to be the same as in general relativity. The basic equations of cosmology are derived and some topics are treated.

066.120   Can Einstein's definition of simultaneity be considered a convention?
R. de Ritis, S. Guccione.
Gen. Relativ. Gravitation, Vol. 17, No. 6, p. 595 – 598 (1985).

The authors give an easy way of falsifying the Reichenbach–Grünbaum conventionalist interpretation of the definition of simultaneity of distant events.

066.121   Propagation of gravitational waves in Robertson–Walker backgrounds.
A. I. Janis.
Gen. Relativ. Gravitation, Vol. 17, No. 6, p. 599 – 612 (1985).

The electric and magnetic parts of the linearized Weyl tensor, when the stress–energy tensor is that of a perfect fluid and the background is of Robertson–Walker type, are known to satisfy wave equations that differ by the presence of a source term for the electric part. It is shown here that all of the allowed solutions of the inhomogeneous equation can be obtained by applying a differential operator to the solutions of the homogeneous equation; consequently, electric–type and magnetic–type gravitational waves have the same propagation properties.

066.122   Uniqueness theorem for anti–de Sitter spacetime.
W. Boucher, G. W. Gibbons, G. T. Horowitz.
Phys. Rev. D, Vol. 30, No. 12, p. 2447 – 2451 (1984). Abstr. in Phys. Abstr., Vol. 88, No. 1254, Entry 36063 (1985).

066.123   Vacuum $\langle T_\mu{}^\nu \rangle$ in Schwarzschild spacetime.
K. W. Howard.
Phys. Rev. D, Vol. 30, No. 12, p. 2532 – 2547 (1984). Abstr. in Phys. Abstr., Vol. 88, No. 1254, Entry 36065 (1985).

066.124   The confrontation between general relativity and experiment: an update.
C. M. Will.
Phys. Rep., Vol. 113, No. 6, p. 345 – 422 (1984). Abstr. in Phys. Abstr., Vol. 88, No. 1254, Entry 36067 (1985).

066.125   Unorthodox curvature interactions, black holes, and inflation.
R. K. Unz.
Phys. Rev. D, Vol. 30, No. 12, p. 2521 – 2527 (1984). Abstr. in Phys. Abstr., Vol. 88, No. 1254, Entry 36077 (1985).

066.126   Severe limitations upon the ultracentrifuge gravitational redshift experiment.
F. N. Weber.
J. Appl. Phys., Vol. 56, No. 11, p. 3347 – 3349 (1984). Abstr. in Phys. Abstr., Vol. 88, No. 1254, Entry 40479 (1985).

066.127   Conformally invariant gravitation and electromagnetism – problems and prospects.
G. Papini.
Sir Arthur Eddington Centenary Symposium. Vol. 1: Relativistic astrophysics and cosmology, p. 89 – 110 (1984). – See Abstr. 012.080.

A conformally invariant, unified field theory of gravitation and electromagnetism is discussed. It is obtained from the theories of Weyl and Dirac by making the Dirac scalar field $\beta(x)$ complex and adding a constraint to the variational principle. The electromagnetic field is generated by topological singularities in the phase of $\beta(x)$, while $|\beta(x)|$ mediates the coupling between gravitation and electromagnetism.

066.128   Naked singularities and causality violation.
C. J. S. Clarke.
Sir Arthur Eddington Centenary Symposium. Vol. 1: Relativistic astrophysics and cosmology, p. 111 – 127 (1984). – See Abstr. 012.080.

A survey is given of some recent work on the cosmic censorship hypothesis by Newman, Krolak, de Felice and the author, which now gives a much clearer idea of when naked singularities are possible, and when not.

066.129   Equations of motion for radiationless particles in the general theory of relativity.
G. Bandyopadhyay.
Sir Arthur Eddington Centenary Symposium. Vol. 1: Relativistic astrophysics and cosmology, p. 128 – 136 (1984). – See Abstr. 012.080.

This paper gives a critical review of Einstein's ideas of the equations of motion which connects field with motion. It gives an example (so far unknown) of a paradox that may result in the absence of a similar connection in classical mechanics.

066.130   On quantum gravity with dynamical torsion.
Yu. N. Obukhov, E. A. Nazarovsky.
Sir Arthur Eddington Centenary Symposium. Vol. 1: Relativistic astrophysics and cosmology, p. 137 – 169 (1984). – See Abstr. 012.080.

The authors study the renormalizability properties of the quantum gravity model with propagating torsion. The structure of the one–loop effective action divergences is analysed. It is shown that the model is non–renormalizable since the cosmological term divergence never vanishes. The authors consider also the possibility of quantising only the Lorentz gauge field while treating the metric as a purely classical field. Some general problems of the quantum gauge theory of gravity are also discussed.

066.131   Gravitation and elementary particles.
D. Ivanenko.
Sir Arthur Eddington Centenary Symposium. Vol. 1: Relativistic astrophysics and cosmology, p. 170 – 172 (1984). – See Abstr. 012.080.

066.132   The induced torsional long range interaction in the Einstein–Cartan theory.
P. I. Pronin.
Sir Arthur Eddington Centenary Symposium. Vol. 1: Relativistic astrophysics and cosmology, p. 173 – 175 (1984). – See Abstr. 012.080.

This note briefly describes a gravitation theory with torsion which is based on an action functional constructed from the curvature scalar of the Riemannian–Cartan manifold (or U_4).

066.133   Unified field theories: from Eddington and Einstein up to now.
H. F. Goenner.
Sir Arthur Eddington Centenary Symposium. Vol. 1: Relativistic astrophysics and cosmology, p. 176 – 196 (1984). – See Abstr. 012.080.

This paper discusses the non–symmetric unified field theory (UFT) introduced, in the form of affine theories, first by Eddington and Einstein. It tries to elucidate the geometrical structure of UFT and to follow critically some of its developments until today.

066.134   Nonsymmetric gravitation theory and its experimental consequences.
J. W. Moffat.
Sir Arthur Eddington Centenary Symposium. Vol. 1: Relativistic astrophysics and cosmology, p. 197 – 223 (1984). – See Abstr. 012.080.

The nonsymmetric gravitation theory (NGT) based on a nonsymmetric hyperbolic Hermitian $g_{\mu\nu}$ is reviewed. The basic mathematical structure of the theory is formulated in terms of an eight–dimensional Riemannian geometry that is reduced to a four–dimensional nonsymmetric structure by an algebraic reduction, based on a hyperbolic complex extension (ring) of the tangent space. The experimental consequences of the theory for the solar system are reviewed and it is shown that the predictions of the theory are consistent with all current solar system tests. The prediction for the periastron shift of a binary system in NGT is compared with the more than 80 years of data for the eclipsing binary system DI Herculis.

066.135   General solutions to the linearized Einstein equations and initial data for three dimensional time evolution of pure gravitational waves.
T. Nakamura.
Prog. Theor. Phys., Vol. 72, No. 4, p. 746 – 760 (1984). Abstr. in Phys. Abstr., Vol. 88, No. 1255, Entry 40991 (1985).

066.136   Comments on the source of Gödel–type metrics.
E. P. Vasconcellos Vaidya, M. L. Bedran, M. M. Som.
Prog. Theor. Phys., Vol. 72, No. 4, p. 857 – 859 (1984). Abstr. in Phys. Abstr., Vol. 88, No. 1255, Entry 41002 (1985).

066.137   Gravitational collapse of a radiating shell of matter.
R. Pim, K. Lake.
Phys. Rev. D, Vol. 31, No. 2, p. 233 – 240 (1985). Abstr. in Phys. Abstr., Vol. 88, No. 1255, Entry 41003 (1985).

066.138   Gravitation: field theory par excellence Newton, Einstein, and beyond.
H. Yilmaz.
Hadronic J., Vol. 7, No. 5, p. 1164 – 1206 (1984). Abstr. in Phys. Abstr., Vol. 88, No. 1255, Entry 41009 (1985). – See Abstr. 012.083.

066.139   On a Lie–isotopic theory of gravity.
M. Gasperini.
Nuovo Cimento A, Vol. 83A, Ser. 11, No. 4, p. 309 – 326 (1984). Abstr. in Phys. Abstr., Vol. 88, No. 1255, Entry 41010 (1985).

066.140   Strings in a conformally invariant theory of gravitation and electromagnetism.
G. Papini.
Phys. Lett. A, Vol. 107A, No. 1, p. 26 – 28 (1985). Abstr. in Phys. Abstr., Vol. 88, No. 1255, Entry 41011 (1985).

066.141   Gravitation and Lie–isotopic algebra.
M. Gasperini.
Hadronic J., Vol. 7, No. 5, p. 971 – 1018 (1984). Abstr. in Phys. Abstr., Vol. 88, No. 1255, Entry 41027 (1985). – See Abstr. 012.083.

066.142   The search for gravitational waves.
G. Pizzella.
Phys. Bull., Vol. 35, No. 12, p. 508 – 510 (1984). Abstr. in Phys. Abstr., Vol. 88, No. 1255, Entry 41044 (1985).

066.143   The transition function for a "cool" gravitating medium.
Z. N. Chumak.
Tr. Astrofiz. Inst. Alma–Ata, Tom 43, p. 59 – 66 (1984). In Russian.

066.144   The generalized variational problem in conformal theories of gravitation. 1.
T. S. Kozhanov, Eh. G. Mychelkin.
Tr. Astrofiz. Inst. Alma–Ata, Tom 43, p. 92 – 107 (1984). In Russian.

066.145   Post–Newtonian equations of motion of a system of spin bodies. 1.
L. M. Chechin.
Tr. Astrofiz. Inst. Alma–Ata, Tom 43, p. 108 – 119 (1984). In Russian.

066.146   Equations of motion of the two–body problem taking into account the internal structure and intrinsic rotation in general relativity.
M. M. Abdil'din.
Tr. Astrofiz. Inst. Alma–Ata, Tom 43, p. 120 – 124 (1984). In Russian.

066.147   **Conformal–covariant formulation of scalar–tensor theories of gravitation.**
T. S. Kozhanov, Eh. G. Mychelkin.
Tr. Astrofiz. Inst. Alma–Ata, Tom 43, p. 125 – 135 (1984). In Russian.

066.148   **Upper bounds for the red–shift parameter in perfect–fluid spheres.**
M. R. Anderson.
Nuovo Cimento B, Vol. 84B, Ser. 11, No. 2, p. 160 – 166 (1984). Abstr. in Phys. Abstr., Vol. 88, No. 1257, Entry 52028 (1985).

066.149   **Nonstatic radiating spheres in general relativity.**
K. D. Krori, P. Borgohain, R. Sarma.
Phys. Rev. D, Vol. 31, No. 4, p. 734 – 741 (1985). Abstr. in Phys. Abstr., Vol. 88, No. 1257, Entry 52031 (1985).

066.150   **Some time–dependent axially symmetric metrics generalising the Weyl and the Einstein–Rosen line elements.**
M. A. J. Vandyck.
Classical Quantum Gravity, Vol. 2, No. 2, p. 241 – 252 (1985). Abstr. in Phys. Abstr., Vol. 88, No. 1257, Entry 52052 (1985).

066.151   **Negative energy fluxes and moving mirrors in curved space.**
W. R. Walker.
Classical Quantum Gravity, Vol. 2, No. 2, p. L37 – L40 (1985). Abstr. in Phys. Abstr., Vol. 88, No. 1257, Entry 52091 (1985).

066.152   **Gravitational lenses.**
K. Gorski.
Postepy Fiz., Vol. 35, No. 6, p. 611 – 613 (1984). In Polish. Abstr. in Phys. Abstr., Vol. 88, No. 1257, Entry 57607 (1985).

066.153   **Infinitesimal null isotropy and Robertson–Walker metrics.**
L. Koch–Sen.
J. Math. Phys., Vol. 26, No. 3, p. 407 – 410 (1985). Abstr. in Phys. Abstr., Vol. 88, No. 1258, Entry 58188 (1985).

066.154   **Some simple type D solutions to the Einstein equations with sources.**
G. F. Torres del Castillo, J. F. Plebanski.
J. Math. Phys., Vol. 26, No. 3, p. 477 – 481 (1985). Abstr. in Phys. Abstr., Vol. 88, No. 1258, Entry 58192 (1985).

066.155   **Casimir energy in quantum gravity in $R^1 \times T^3$ space–time.**
O. Abe.
Prog. Theor. Phys., Vol. 72, No. 6, p. 1225 – 1232 (1984). Abstr. in Phys. Abstr., Vol. 88, No. 1258, Entry 58226 (1985).

066.156   **The influence of the gravitational field on some electromagnetic phenomena.**
H. Schabe.
Exp. Tech. Phys., Vol. 33, No. 1, p. 7 – 14 (1985). In German. Abstr. in Phys. Abstr., Vol. 88, No. 1259, Entry 63781 (1985).

066.157   **Gravitation without black holes.**
A. G. Agnese, M. La Camera.
Phys. Rev. D, Vol. 31, No. 6, p. 1280 – 1286 (1985). Abstr. in Phys. Abstr., Vol. 88, No. 1259, Entry 63796 (1985).

066.158   **General relativity: progress, problems, and prospects.**
W. Israel.
Can. J. Phys., Vol. 63, No. 1, p. 34 – 43 (1985). Abstr. in Phys. Abstr., Vol. 88, No. 1260, Entry 69180 (1985).

066.159   **Special relativistic effects in rapidly moving plasmoids.**
R. Cowsik, P. P. Deo, N. Krishnan, M. M. Vasanthi.
18th International Cosmic Ray Conference, Vol. 2, p. 234 – 237 (1983). – See Abstr. 012.096.

The authors investigate the problem of synchrotron emission from relativistic plasmoids. A self consistent calculation of the equipartition magnetic field reduces its strength considerably from the value estimated without invoking relativistic motion. A dramatic increase in the cut–off frequency of synchrotron radiation occurs thereby alleviating the need for in–situ acceleration in these objects. These ideas are discussed with special reference to the extragalactic sources M87 and 3C 273.

066.160   **A fresh look at special relativity.**
A. Ramakrishnan.
18th International Cosmic Ray Conference, Vol. 5, p. 501 – 504 (1983). – See Abstr. 012.096.

066.161   **Some properties of an exact solution for a non–static uniform density sphere with singularity in general relativity.**
H. Knutsen.
Phys. Scr., Vol. 31, No. 1, p. 10 – 14 (1985).

A class of exact solutions for a time–dependent uniform density sphere is given. The model is non–regular since there exists an interior surface where the pressure is infinite. However, it is shown that this surface is always hidden by trapped surfaces. The equation for the zero pressure boundary of the fluid sphere is investigated, and it is found that this equation always has just one solution when integration constants are properly chosen.

066.162   **Evolution of spherical shells in general relativity.**
J. S. Høye, I. Linnerud, K. Olaussen, R. Sollie.
Phys. Scr., Vol. 31, No. 2, p. 97 – 102 (1985).

The authors discuss the evolution of thin spherical shells within general relativity. They start directly from the Einstein equations coupled to an arbitrary spherical matter distribution, and show that the resulting dynamics can be solved exactly in the limit when this distribution approaches an infinitesimally thin shell. The procedure can be generalized to systematically take into account finite thickness corrections.

066.163   **Refining the estimate of the gravitational constant.**
B. W. Rust, J. H. Dunn.
Bull. Am. Astron. Soc., Vol. 17, No. 2, p. 550 (1985). Abstract. – See Abstr. 010.065.

066.164   **The positive energy theorem and its extensions.**
G. T. Horowitz.
Asymptotic behavior of mass and spacetime geometry, p. 1 – 21 (1984). – See Abstr. 012.102.

Recent progress in understanding the concept of gravitational energy in general relativity is reviewed.

066.165   **Mass and angular momentum at the quasi–local level in general relativity.**
R. Penrose.
Asymptotic behavior of mass and spacetime geometry, p. 23 – 30 (1984). – See Abstr. 012.102.

A new expression for mass–energy in general relativity is derived. The expression refers to a spacelike 2–surface S, whose topology is that of 2–sphere, drawn arbitrarily in an arbitrary space–time M.

066.166   **The positive mass theorem and black holes.**
M. J. Perry.
Asymptotic behavior of mass and spacetime geometry, p. 31 – 40 (1984). – See Abstr. 012.102.

The positive energy theorem for the ADM and Bondi mass is reviewed. This is then extended to spacetimes which contain black holes. Various further generalizations are then briefly outlined.

066.167   **Exotic differentiable structures on Euclidean 4–space.**
C. H. Taubes.
Asymptotic behavior of mass and spacetime geometry, p. 41 – 43 (1984). – See Abstr. 012.102.

066.168   **A new approach to the vacuum Einstein equation.**
C. N. Kozameh, E. T. Newman.
Asymptotic behavior of mass and spacetime geometry, p. 45 – 55 (1984). – See Abstr. 012.102.

066.169   The existence of maximal surfaces in asymptotically flat spacetimes.
R. Bartnik.
Asymptotic behavior of mass and spacetime geometry, p. 57 – 60 (1984). – See Abstr. 012.102.

066.170   A gauge invariant index theorem for asymptotically flat manifolds.
C. H. Taubes.
Asymptotic behavior of mass and spacetime geometry, p. 85 – 94 (1984). – See Abstr. 012.102.

066.171   On the boundary conditions for gravitational and gauge fields at spatial infinity.
A. Ashtekar.
Asymptotic behavior of mass and spacetime geometry, p. 95 – 109 (1984). – See Abstr. 012.102.

066.172   The numerical characteristic initial value problem.
J. S. Welling.
Asymptotic behavior of mass and spacetime geometry, p. 123 – 142 (1984). – See Abstr. 012.102.

This article describes the solution of Einstein's equations on a single null hypersurface for an axisymmetric, non–rotating system. The gravitational radiation which reaches infinity along this hypersurface and the mass of the system as measured on this hypersurface is calculated. In particular, the effect of gravitational initial data on the flux of radiation is studied.

066.173   The gravitational Hamiltonian.
J. M. Nester.
Asymptotic behavior of mass and spacetime geometry, p. 155 – 163 (1984). – See Abstr. 012.102.

066.174   Twistors, asymptotic symmetries and conservation laws at null and spatial infinity.
W. T. Shaw.
Asymptotic behavior of mass and spacetime geometry, p. 165 – 176 (1984). – See Abstr. 012.102.

The relationship between Penrose's 2–surface twistors defined on infinite spheres and generators of asymptotic symmetry groups is elucidated, at both null and spatial infinity.

066.175   Two–surface twistors and conformal embedding.
B. Jeffryes.
Asymptotic behavior of mass and spacetime geometry, p. 177 – 184 (1984). – See Abstr. 012.102.

The existence of a constant norm on Penrose's two–surface twistor space is discussed and it is shown that this corresponds to whether the surface is embeddable in conformal real Minkowski space.

066.176   Remarks on the cosmic censorship conjecture.
V. Moncrief.
Asymptotic behavior of mass and spacetime geometry, p. 185 – 196 (1984). – See Abstr. 012.102.

066.177   On some (con–)formal properties of Einstein's field equations and their consequences.
H. Friedrich.
Asymptotic behavior of mass and spacetime geometry, p. 197 – 208 (1984). – See Abstr. 012.102.

066.178   Developments and predictions.
J. N. Goldberg.
Gravitation, geometry and relativistic physics, p. 1 – 17 (1984). – See Abstr. 012.103.

This paper reviews research in general relativity over the past 35 years. Its main topics are gravitational radiation, conservation laws, black holes, and quantum gravity. An outlook on future developments in the field is included.

066.179   Radiative gravitational fields and radiation reaction forces in general relativity.
L. Blanchet.
Gravitation, geometry and relativistic physics, p. 18 – 28 (1984). – See Abstr. 012.103.

The author defines a Post–Minkowskian iteration method for solving Einstein's vacuum equations and gives the general structure of the solution when $r \to 0$ or $c \to +\infty$. The method is then used to derive an expression for the radiation reaction force density in the case of a non–relativistic source.

066.180   Multipoles particles in general relativity: the Weyl and Kerr metrics.
J. Martín, E. Ruiz, M. J. Senosiaín.
Gravitation, geometry and relativistic physics, p. 29 – 39 (1984). – See Abstr. 012.103.

A method is described for obtaining approximate solutions of the Einstein equations relative to a multipole particle. The method is applied to the Kerr metric, and its reliability is checked in the reproduction and interpretation of the Weyl metric (static axisymmetric vacuum solutions).

066.181   Unconstrained degrees of freedom of gravitational field and the positivity of gravitational energy.
J. Kijowski.
Gravitation, geometry and relativistic physics, p. 40 – 50 (1984). – See Abstr. 012.103.

The space of Cauchy data for Einstein equations is effectively reduced with respect to Gauss–Codazzi constraints. The mixed initial value – boundary value problem is analysed. The role of boundary degrees of freedom is discussed. The energy–positivity is obtained as a simple consequence of the construction used.

066.182   A method for generating exact solutions of Einstein's field equations.
J. Hajj–Boutros.
Gravitation, geometry and relativistic physics, p. 51 – 53 (1984). – See Abstr. 012.103.

066.183   Causal relativistic thermodynamics of transitory processes in electromagnetic continuous media.
C. Barrabès.
Gravitation, geometry and relativistic physics, p. 54 – 56 (1984). – See Abstr. 012.103.

This note describes a study in general relativity of the constitutive equations of electromagnetic media, when submitted to transitory processes. The formalism is general enough to include electromagnetic deformable solids as well as electromagnetic fluids. Applications of this work may for instance concern the study of magnetospheres, pulsars, black–hole accretion rings and the early eras of the cosmological evolution.

066.184   La relativité générale: une théorie sans problème(s)?
J. Eisenstaedt.
Gravitation, geometry and relativistic physics, p. 57 – 76 (1984). – See Abstr. 012.103.

066.185   Theories of gravity and experimental tests in the post–Newtonian limit.
P. Teyssandier.
Gravitation, geometry and relativistic physics, p. 154 – 173 (1984). – See Abstr. 012.103.

The postulates upon which is founded general relativity (in particular the principle of equivalence) are in fact the underlying foundations of a large class of competing theories, the so–called metric theories of gravity. This review discusses experiments intended to distinguish between the various metric theories. To analyze the experiments in the slow–motion, weak–field limit, the most useful tool is the parametrized post–Newtonian formalism. After a brief review of the main features of this formalism, the paper presents the different experiments performed in order to determine the post–Newtonian parameters and it discusses the theoretical implications of the results already obtained.

066.186   Survey of relativistic effects in geodesy and fundamental astronomy.
C. Boucher, J. F. Lestrade.
Gravitation, geometry and relativistic physics, p. 174 – 186 (1984). – See Abstr. 012.103.

066.187   Close–up on gravitational lensing: the gravitational mirages.
C. Vanderriest.
Gravitation, geometry and relativistic physics, p. 265 – 280 (1984). – See Abstr. 012.103.

Among the various manifestations of gravitational lensing, this paper emphasizes the interest of gravitational mirages (defined as situations where several images of a single source are produced). It first recalls the principle of this phenomenon and the conditions for its occurrence. Then, a few possible uses of gravitational mirages, particularly in cosmology, are discussed. Finally, the properties of the presently known five cases of gravitational lensing of quasars are reviewed.

066.188   Amplification of light by gravitational lens: dynamics and thick lens effects.
F. Hammer.
Gravitation, geometry and relativistic physics, p. 281 – 285 (1984). – See Abstr. 012.103.

The specific effects of dynamics and extension of a gravitational lens on the amplification of light are studied in the frame of a two–step vacuole model.

066.189   Astronomical tests of general relativity.
J. Grygar.
Pokroky Mat., Fyz. Astron., Vol. 29, No. 5, p. 274 – 280 (1984). In Czech.

066.190   Lower dimensional gravity.
R. Jackiw.
Nucl. Phys. B, Part. Phys., Vol. B252, No. 1 – 2, p. 343 – 356 (1985). – See Abstr. 012.105.

Gravity theory on a line and in the plane is reviewed. The triviality of the planar Einstein model is avoided by adding sources and a topological mass term. A constant curvature model for two dimensional space–time, analogous to the theory in three dimensional space–time, is proposed.

066.191   The period of orbit as a test of the general theory of relativity.
H. G. Preston.
Diss. Abstr. Int., Sect. B, Vol. 45, No. 6, p. 1811 (1984). Thesis, University of California, 255 pp. (1984). Order No. DA8420270.

066.192   Physics of nonmagnetic relativistic thermal plasmas.
C. D. Dermer.
Diss. Abstr. Int., Sect. B, Vol. 45, No. 7, p. 2199 (1985). Thesis, University of California, 162 pp. (1984). Order No. DA8423914.

066.193   The determination of the post–Newtonian parameters in gravitational theory using laser ranging to the LAGEOS satellite.
M. A. Vincent.
Diss. Abstr. Int., Sect. B, Vol. 45, No. 7, p. 2201 (1985). Thesis, University of Texas, 226 pp. (1984). Order No. DA8421812.

066.194   Testing the nonsymmetric theory of gravitation.
J. C. McDow.
Diss. Abstr. Int., Sect. B, Vol. 45, No. 9, p. 2957 (1985). Thesis, University of Toronto (1984).

066.195   Relativistic astrophysics and gravitation theory.
Ya. B. Zel'dovich.
Main directions of astronomical investigations at the Moscow University, p. 31 – 38 (1985). In Russian. – See Abstr. 003.049.

066.196   Bäcklund transformations in general relativity.
D. Kramer, G. Neugebauer.
Solutions of Einstein's equations: techniques and results, p. 1 – 25 (1984). – See Abstr. 012.112.

066.197   Prolongation structures and differential forms.
B. K. Harrison.
Solutions of Einstein's equations: techniques and results, p. 26 – 54 (1984). – See Abstr. 012.112.

Prolongation structures, which are incomplete Lie algebras, are useful in finding Bäcklund transformations. Their application to general relativity is discussed here.

066.198   Vector Bäcklund transformations and associated superposition principle.
F. J. Chinea.
Solutions of Einstein's equations: techniques and results, p. 55 – 67 (1984). – See Abstr. 012.112.

In recent years, several different methods have been introduced in order to generate exact solutions of the vacuum Einstein equations having two commuting Killing fields. Bäcklund transformations were applied to general relativity for the first time. The transformations discussed here are a generalization of the vector–type Bäcklund transformations.

066.199   HKX transformations: an introduction.
C. Hoenselaers.
Solutions of Einstein's equations: techniques and results, p. 68 – 84 (1984). – See Abstr. 012.112.

The author gives an introduction into the technique of HKX transformations for generating new axisymmetric stationary solutions of Einstein's vacuum field equations from old ones. A review of the important definitions and theorems and a scetch of the proofs is given.

066.200   HKX transformations: some results.
W. Dietz.
Solutions of Einstein's equations: techniques and results, p. 85 – 112 (1984). – See Abstr. 012.112.

The author simplifies considerably the original representation of the HKX transformations by means of determinants. He presents some examples to illustrate the action of these transformations. The equivalence of HKX and Kramer–Neugebauer transformations is shown. Finally, the analysis of two balancing Kerr–NUT objects is outlined.

066.201   The Geroch group is a Banach Lie group.
B. G. Schmidt.
Solutions of Einstein's equations: techniques and results, p. 113 – 127 (1984). – See Abstr. 012.112.

The infinite parameter group acting on stationary, axisymmetric solutions of Einstein's vacuum field equations found by Geroch can be considered as a Banach Lie group.

066.202   On the homogeneous Hilbert problem for effecting Kinnersley–Chitre transformations.
I. Hauser.
Solutions of Einstein's equations: techniques and results, p. 128 – 175 (1984). – See Abstr. 012.112.

This paper presents a detailed account of the application of the homogeneous Hilbert problem for generating new stationary axisymmetric vacuum spacetimes by Kinnersley–Chitre transformations.

066.203   The homogeneous Hilbert problem: practical application.
F. J. Ernst.
Solutions of Einstein's equations: techniques and results, p. 176 – 185 (1984). – See Abstr. 012.112.

This paper describes the Hauser–Ernst formalism for generating solutions of the Einstein–Maxwell field equations.

066.204 Noniterative method for constructing exact solutions of Einstein equations.
D.-S. Guo.
Solutions of Einstein's equations: techniques and results, p. 186 – 198 (1984). – See Abstr. 012.112.
By applying the homogeneous Hilbert problem of the Hauser–Ernst formalism, the author developed a noniterative method for obtaining a quite general exact solution of the Einstein vacuum gravitational field equation.

066.205 Inverse scattering, differential geometry, Einstein–Maxwell solitons and one soliton Bäcklund transformations.
M. Gürses.
Solutions of Einstein's equations: techniques and results, p. 199 – 234 (1984). – See Abstr. 012.112.
A survey of the inverse scattering transform method is given and a differential geometric interpretation of the inverse scattering equations is presented. Einstein–Maxwell field equations for space–times admitting nunnull commuting two Killing vector fields are integrated by giving the $2N$–soliton construction. One soliton constructions of the gravitational field and of the self-dual Yang–Mills field equations are shown to be equivalent to the recently found Bäcklund transformations.

066.206 Classical aspects of Yang–Mills theories.
B. C. Xanthopoulos.
Solutions of Einstein's equations: techniques and results, p. 235 – 251 (1984). – See Abstr. 012.112.
The Yang–Mills equations, their symmetries, the physical quantities of their solutions, the characterization of different classes of solutions, and the similarities between the Yang–Mills and the Einstein equations are reviewed at an introductory level.

066.207 Exact solutions of $\mathbb{C}P^n$ models.
C. Reina.
Solutions of Einstein's equations: techniques and results, p. 252 – 275 (1984). – See Abstr. 012.112.
Harmonic maps theory include some examples of non linear field theories of physical interest, among which are stationary axisymmetric electro–vacuum Einstein equations. After a short review of such applications, this paper concentrates on the so called $\mathbb{C}P^n$ models. In particular it discusses the energy spectrum and the space of parameters of exact solutions of these models on the Riemann sphere.

066.208 Explicit and hidden symmetries of dimensionally reduced (super–) gravity theories.
P. Breitenlohner, D. Maison.
Solutions of Einstein's equations: techniques and results, p. 276 – 310 (1984). – See Abstr. 012.112.
This paper presents a systematic investigation of the explicit and hidden symmetries of dimensionally reduced 'gravity–type' theories. It also discusses the role played by the infinite-dimensional Geroch group of which various realizations are known: Bäcklund, Kinnersley–Chitre, Inverse Scattering and Riemann–Hilbert transformations.

066.209 N–Kerr particles.
M. Yamazaki.
Solutions of Einstein's equations: techniques and results, p. 311 – 320 (1984). – See Abstr. 012.112.
The metric of stationary axisymmetric, asymptotically flat exact fields of N collinear separated Kerr masses balanced under gravitation and rotational repulsion is described in a concise unified form in terms of six determinants of degree N. The conditions for regularity of the metric on the symmetric axis are given.

066.210 Algebraically special, shearfree, diverging, and twisting vacuum and Einstein–Maxwell fields.
H. Stephani.
Solutions of Einstein's equations: techniques and results, p. 321 – 333 (1984). – See Abstr. 012.112.
This paper gives a short review of the known classes of algebraically special, diverging, and twisting vacuum solutions and

indicates how and why it was possible to find them. It then concentrates on methods of constructing Einstein–Maxwell-fields from the vacuum solutions.

066.211 The Newtonian limit.
B. F. Schutz.
Solutions of Einstein's equations: techniques and results, p. 367 – 391 (1984). – See Abstr. 012.112.
The author discusses in detail the development of the Newtonian and post–Newtonian approximations to general relativity. By using an initial–value approach, he is able to show that the post-Newtonian hierarchy through gravitational–radiation-reaction order is an asymptotic approximation to general relativity, thereby verifying the validity of the quadrupole formula for radiation reaction. It is also shown with equal rigor that the radiation from nearly–Newtonian systems obeys the far–field quadrupole formula (Landau–Lifshitz formula). The author discusses also the relationships of observables to post–Newtonian quantities by the method of osculating Newtonian orbits.

066.212 Projective relativity and exact solutions.
E. Schmutzer.
Solutions of Einstein's equations: techniques and results, p. 392 – 402 (1984). – See Abstr. 012.112.
After some introductory historical annotations on projective relativity a new version of this type of theories, called Projective Unified Field Theory, and its physical interpretation are presented. Within this framework the situation with respect to exact solutions is sketched.

066.213 Exact Brans–Dicke–Bianchi solutions.
D. Lorenz–Petzold.
Solutions of Einstein's equations: techniques and results, p. 403 – 435 (1984). – See Abstr. 012.112.
Using a new method of reduction of the Brans–Dicke field equations the author derives various new vacuum solutions as well as some perfect fluid solutions for the class of spatially homogeneous space–times in a unified manner. Exact solutions are given for the isotropic Friedmann–Robertson–Walker models and for the anisotropic Bianchi types I – IX and for the related Kantowski–Sachs model.

066.214 The apparent velocity of an object moving in a Kerr space–time background.
J.-l. Zhang, S.-p. Xiang.
Acta Astron. Sin., Vol. 26, No. 1, p. 92 – 97 (1985). In Chinese.
The effects of strong gravitational fields on moving light sources have been studied, and general relativistic formulae of apparent velocities in Kerr fields have been given. Numerical calculations show that Kerr fields have amplifying effects on the apparent velocities.

066.215 Exact solutions of plane gravitational waves under harmonic conditions.
H. Liu, P. Zhou (*P.–y. Chou*).
Sci. Sin., Ser. A, Vol. 28, No. 6, p. 628 – 637 (1985).
Under harmonic conditions exact solutions of plane gravitational waves which are free from singularities and contain Einstein's solutions for weak fields as approximations are obtained. The physical interpretations of the solutions and the relationship with other two typical kinds of solutions are studied in detail.

General relativity.
See Abstr. 003.009.

Relativistic astronomy.
See Abstr. 003.119.

Spinors and space–time. Vol. 1: Two–spinor calculus and relativistic fields.
See Abstr. 003.151.

A first course in general relativity.
See Abstr. 003.165.

Constrained dynamics. With application to Yang–Mills theory, general relativity, classical spin, dual string model.
See Abstr. 003.175.

Theory and experiment in gravitational physics.
See Abstr. 003.186.

Supersymmetry and supergravity '82. Proceedings of the Trieste September 1982 School.
See Abstr. 012.118.

Fifteen years of Lunar Laser Ranging and future developments.
See Abstr. 013.021.

Space experiments in relativity and gravitation.
See Abstr. 013.079.

Spacecraft navigation and relativity.
See Abstr. 014.040.

Sensitivity analysis of a two–mode resonant gravitational–wave detector with arbitrary tuning.
See Abstr. 034.113.

Initial operation at liquid–helium temperature of the $M = 2270$ kg Al 5056 gravitational–wave antenna of the Rome Group.
See Abstr. 034.114.

The interferometric detection of gravitational waves.
See Abstr. 034.165.

The development of long baseline gravitational radiation detectors at Glasgow University.
See Abstr. 034.166.

Improved sensitivities in laser interferometers for the detection of gravitational waves.
See Abstr. 034.167.

Noise statistics of Pioneer 11 VLF gravitational wave search data.
See Abstr. 036.125.

Galaxies seen through gravitational lenses.
See Abstr. 036.206.

Formulae for calculating the corrections for the light deflection to apparent places.
See Abstr. 041.007.

Relativistic astrometry.
See Abstr. 041.015.

Radiointerferometric astrometry and general theory of relativity.
See Abstr. 041.016.

Universal time, lunar tidal deceleration and relativistic effects from observations of transits, eclipses and occultations in the XVIIIth – XXth centuries.
See Abstr. 042.009.

Virial oscillations of celestial bodies. IV. The Lyapunov stability of motion.
See Abstr. 042.036.

A note on upper limits for hypothetic variation in the Newtonian constant of gravitation.
See Abstr. 043.008.

Long–range antigravity.
See Abstr. 061.007.

Self–duality in Euclidean supergravity.
See Abstr. 061.018.

The importance of impatience.
See Abstr. 061.023.

The lifetime of an elementary particle in a field.
See Abstr. 061.052.

Supergravity and the unification of fundamental interactions.
See Abstr. 061.062.

Kaluza–Klein theories and the Dirac monopole.
See Abstr. 061.075.

Magnetic monopoles in grand unified and Kaluza–Klein theories.
See Abstr. 061.077.

Causality of classical supergravity.
See Abstr. 061.143.

Supergravities.
See Abstr. 061.144.

Prospects and problems of locally supersymmetric Kaluza–Klein theories.
See Abstr. 061.146.

Erratum: "Bremsstrahlung spectra at relativistic temperatures" [Acta Astron., Vol. 31, No. 4, p. 457 – 469 (1981)].
See Abstr. 062.075.

A viable mechanism to establish relativistic thermal particle distribution functions in cosmic sources.
See Abstr. 063.014.

Euler potential method in three–dimensional stellar wind problems.
See Abstr. 064.077.

On the nonradial pulsations of general relativistic stellar models.
See Abstr. 065.053.

Effects of general relativity on the mass ejection of neutrino–trapping supernovae.
See Abstr. 065.087.

On the secular instability of axisymmetric rotating stars to gravitational radiation reaction.
See Abstr. 065.105.

Extrinsic curvature for the two–black–hole problem.
See Abstr. 067.073.

Gravitational lensing and the regions surrounding neutron stars.
See Abstr. 067.107.

General relativistic effects on the pulse profile of fast pulsars.
See Abstr. 067.114.

Strong gravity effects in the temperature profile of accretion disks.
See Abstr. 067.116.

The maximum mass of a neutron star.
See Abstr. 067.117.

Gravitational radiation from a particle scattered by a Kerr black hole.
See Abstr. 067.152.

Rotational splitting of global solar oscillations and its relevance to tests of general relativity.
See Abstr. 080.019.

Newtonian N–body calculations of the advance of Mercury's perihelion.
See Abstr. 092.002.

Relativistic effects in the earth–moon dynamics.
See Abstr. 094.017.

Rigorous treatment of the heliocentric motion of stars.
See Abstr. 111.009.

Relativistic apsidal motion in the eclipsing binary systems V1143 Cygni and EK Cephei.
See Abstr. 119.010.

On the post–Newtonian effects in the millisecond pulsar 1937 + 214.
See Abstr. 126.021.

The binary pulsar – a test for general relativity?
See Abstr. 126.098.

Direct physical collisions and the evolution of dense stellar systems.
See Abstr. 151.042.

Equilibrium non–rotating magnetized disks as a plane analog to the n = 3 polytrope.
See Abstr. 151.082.

Collisional stellar dynamics in general relativity: an overview.
See Abstr. 151.117.

Testing theories of superluminal motion.
See Abstr. 158.351.

2237 + 0305: a new and unusual gravitational lens.
See Abstr. 159.051.

The gravitational lens effect and the surface density of quasars near foreground galaxies.
See Abstr. 159.067.

2237 + 0305: a new and unusual gravitational lens.
See Abstr. 159.088.

A test for a gravitational lens hypothesis.
See Abstr. 159.116.

The multiple images of the quasar 0957 + 561.
See Abstr. 159.122.

The gravitationally lensed quasar 0957 + 561: VLA observations and mass models.
See Abstr. 159.123.

Millisecond pulsars, gravitational waves and the inflationary universe.
See Abstr. 161.002.

A model for the early universe and the connection between gravitation and the quantum nature of matter.
See Abstr. 161.020.

The Kasner type Universe in new general relativity.
See Abstr. 161.024.

Cosmological solutions of Bi–metric theories of gravitation.
See Abstr. 161.029.

Inhomogeneous generalisation of Bianchi type–VI_0 cosmological model of perfect fluid distribution in general relativity.
See Abstr. 161.050.

Electromagnetic field in the structure of space–time.
See Abstr. 161.054.

Relations for a Kaluza–Klein cosmology with variable rest mass.
See Abstr. 161.056.

Gravitational lensing effects of vacuum strings: exact solutions.
See Abstr. 161.092.

On model–dependent bounds on H_0 from gravitational images: application to Q0957 + 561A,B.
See Abstr. 161.097.

Comment on the BDT–Bianchi type–I stiff matter solution.
See Abstr. 161.115.

Comment on the Brans–Dicke–Bianchi type–III vacuum solution.
See Abstr. 161.116.

Anisotropic cosmological model in a new scalar–tensor theory.
See Abstr. 161.118.

A higher limit to the mean galaxy mass allowed by gravitational lens image distortion.
See Abstr. 161.120.

On the use of measured time delays in gravitational lenses to determine the Hubble constant.
See Abstr. 161.162.

The specific physical character of the expansion of the Universe.
See Abstr. 161.163.

Comment on the Bianchi type–V vacuum model.
See Abstr. 161.181.

Spatially homogeneous dynamics: a unified picture.
See Abstr. 161.189.

Bianchi type–VI_0 perfect fluid cosmological models with magnetic field.
See Abstr. 161.212.

The statistics of gravitational lenses. II. Apparent evolution in the quasars' luminosity function.
See Abstr. 161.232.

Generalized cosmological models.
See Abstr. 161.255.

The de Sitter universe and "projective relativity".
See Abstr. 161.256.

Perturbative approach to self consistent cosmologies.
See Abstr. 161.257.

Radiation and five–dimensional cosmology.
See Abstr. 161.259.

A mechanism for reducing the value of the cosmological constant.
See Abstr. 161.260.

Mach's principle. I. Initial state of the Universe.
See Abstr. 161.281.

Mach's principle. II. Realization of Dirac's hypothesis in Brans–Dicke theory with cosmological term.
See Abstr. 161.282.

TCP, quantum gravity, the cosmological constant and all that. . .
See Abstr. 161.287.

Cosmologies with quasiregular singularities. I. Spacetimes and test waves.
See Abstr. 161.316.

Cosmologies with quasiregular singularities. II. Stability considerations.
See Abstr. 161.317.

The cosmic field tensor in bimetric general relativity.
See Abstr. 161.325.

Flux conservation by an isolated Schwarzschild gravitational lens.
See Abstr. 161.350.

Cosmic supergravity.
See Abstr. 161.373.

Kaluza–Klein cosmology and the inflationary universe.
See Abstr. 161.374.

The dimensional reduction transition.
See Abstr. 161.375.

Integral constraints on stress energy perturbations in general relativity and applications to cosmology.
See Abstr. 161.386.

Exact solutions in cosmology.
See Abstr. 161.398.

Quantum cosmogonic model.
See Abstr. 161.399.

067 Astrophysics of Compact Objects (Neutron Stars, Black Holes)

067.001   The number of critical points in polytropic accretion onto black holes.
K. M. Chang.
Astron. Astrophys., Vol. 142, No. 1, p. 212 – 213 (1985).

Ray (1980) pointed out that for accretion of polytropic gas by a Schwarzschild black hole, there exists a second critical point for $1 < \gamma < 5/3$. Here it is shown that the second critical point is not physically meaningful in that the temperature at that critical point is negative.

067.002   Plasma radiation during γ–ray bursts.
E. P. Liang.
Nature, Vol. 313, No. 5999, p. 202 – 204 (1985).

Most γ–ray bursts are thought to originate from or near the surface of neutron stars. The author proposes that γ–ray bursters may also strongly emit 0.05 – 1 keV XUV radiation simultaneously with the γ–rays, in the form of two narrow spectral lines, corresponding to the ω_e and $2\omega_e$ plasma radiation, where $\omega_e = (4\pi n_e e^2/m_e)^{1/2}$ is the electron plasma frequency of the γ–emitting (and pair annihilation) region of the burster. Since electron densities are typically $n_e \geqslant 10^{24} - 10^{26} \mathrm{cm}^{-3}$, it follows $\hbar\omega_e \geqslant 37 - 373$ eV.

067.003   Softening of neutron stars.
P. J. Siemens.
Nature, Vol. 313, No. 6002, p. 430 (1985).

067.004   Ultra–high energy γ rays and cosmic rays from accreting degenerate stars.
G. Chanmugam, K. Brecher.
Nature, Vol. 313, No. 6005, p. 767 – 768 (1985).

Ultra–high energy (UHE) γ–ray emission, with photon energies up to 2×10^{16}eV, has been detected from the galactic X–ray source Cygnus X–3, implying the acceleration of charged particles there to even greater energies. The authors present a model for acceleration of particles to such energies in binary–star systems containing accreting magnetized neutron stars.

067.005   Magnetic fluxes across black holes.
J. Bičák, V. Janiš.
Mon. Not. R. Astron. Soc., Vol. 212, No. 4, p. 899 – 915 (1985).

By using solutions of Maxwell's equations on the Kerr background the authors show that the flux of the magnetic fields, generated by stationary, axially symmetric sources, across one half of the surface of the horizon decreases as the angular momentum of the hole increases and becomes zero when the hole is an extreme Kerr black hole. The flux can, however, be large if the field is not axially symmetric. Starting from an exact solution of Einstein's equations representing a Schwarzschild black hole in a magnetic universe, the authors find that there exists a finite upper bound on the magnetic field strength for which the flux across the horizon has a maximum value.

067.006   2–dimensional low luminosity accretion onto neutron stars.
J. G. Kirk.
Astron. Astrophys., Vol. 142, No. 2, p. 430 – 436 (1985).

Two–dimensional analytic solutions are presented for accretion onto a magnetized neutron star. The flow is assumed to contain a shock front and immediately behind it there is a thin layer in which gas pressure is important. At all other points radiation pressure is assumed to dominate over gas pressure.

067.007   Theories of accreting X–ray pulsars.
P. Mészáros.
AIP Conf. Proc., No. 115, p. 165 – 178 (1984). – See Abstr. 012.005.

The author reviews current models of X–ray pulsars in the light of our observational knowledge. The physics of the accretion column and the polar cap are discussed, and the main physical processes described, emphasizing the peculiarities introduced by the strong magnetic field. Radiative transfer methods adapted to these extreme conditions are outlined. The author then discusses recent calculations on the self–consistent structure of the X–ray emission region, aimed at providing realistic theoretical spectra and pulse shapes.

067.008   Accretion by magnetic neutron stars.
F. K. Lamb.
AIP Conf. Proc., No. 115, p. 179 – 214 (1984). – See Abstr. 012.005.

Most of the bright, periodically pulsing X–ray sources in the Galaxy are strongly magnetic, rotating neutron stars accreting matter from a binary companion. This review first describes the qualitative features that are characteristic of all flows onto strongly magnetic neutron stars and then discusses accretion from a thin disk and spherically symmetric radial accretion in more detail, with emphasis on the scale and structure of the magnetosphere. Work on plasma flow inside the magnetosphere and the physics of accretion torques is described. Finally, some interesting unsolved problems in neutron star magnetospheric physics are indicated.

067.009   Accretion onto magnetized neutron stars: magnetospheric structure and stability.
J. Arons, D. J. Burnard, R. I. Klein, C. F. McKee, R. E. Pudritz, S. M. Lea.
AIP Conf. Proc., No. 115, p. 215 – 234 (1984). – See Abstr. 012.005.

A summary is given of the magnetospheric physics of accreting magnetized neutron stars. The role of hydromagnetic instabilities is described, with particular attention to the formation of polar caps in spherical and disk accretion, and to the origin of spin down torques in a centrifugally driven stellar wind, when the accretion is from a disk. The most elementary consequences for formation of the emergent spectrum are also discussed.

067.010 Two–dimensional gas dynamic models of polar cap accretion onto magnetized neutron stars.
R. I. Klein, J. Arons, S. M. Lea.
AIP Conf. Proc., No. 115, p. 235 – 242 (1984). – See Abstr. 012.005.
Preliminary results are presented from a two–dimensional radiation gas dynamics study of polar cap accretion. In particular, it is found that photon advection leads to fan beam formation when the accretion mass flux is independent of distance from the magnetic axis and that collisional excitation of cyclotron line photons is at least 10 times more important than bremsstrahlung as a source of photons. There is also evidence for the onset of a new form of overstable convection.

067.011 Spin–reversed accretion onto neutron stars and period fluctuation of X–ray pulsars.
H. Inoue.
AIP Conf. Proc., No. 115, p. 243 – 245 (1984). – See Abstr. 012.005.
In the canonical model of the period change for X–ray pulsars, two origins of the period fluctuation are considered. One is a fluctuation of the angular momentum of the accreted matter and the other is a redistribution of the angular momentum inside the neutron star. The author considers the possibilities of spin–reversed accretion in a binary system in terms of the first possibility.

067.012 Self–consistent description of axially symmetric magnetospheres of accreting neutron stars.
H. Herold, H. Ruder.
AIP Conf. Proc., No. 115, p. 246 – 248 (1984). – See Abstr. 012.005.
The authors reconsider the problem of spherical accretion onto strongly magnetized compact objects and derive, using ideal MHD, a second order differential equation for the selfconsistent magnetic field structure. Classifying this equation as partly hyperbolic and partly elliptic, the authors discuss the different regions of subsonic, supersonic, superalfvénic and subalfvénic accretion flow.

067.013 The thermonuclear flash model for X–ray bursts.
R. E. Taam.
AIP Conf. Proc., No. 115, p. 263 – 272 (1984). – See Abstr. 012.005.
The X–ray burst phenomenon is interpreted in terms of a nuclear shell flash on the surface of an accreting neutron star. The thermal and nuclear evolution of the envelope is described for the hydrogen–helium shell flash and the nearly pure helium shell flash cases. The differences between shell flash modes are discussed and compared with the average properties of X–ray bursts. Several outstanding problems that remain for the thermonuclear flash model are listed and possible resolutions are suggested.

067.014 Repeated thermonuclear flashes on an accreting neutron star.
S. E. Woosley, T. A. Weaver.
AIP Conf. Proc., No. 115, p. 273 – 297 (1984). – See Abstr. 012.005.
The thermonuclear activity accompanying the accretion of hydrogen–rich material onto the surface of a neutron star is numerically simulated. Several different values of (constant) accretion rate and metallicity are employed and models are propagated through multiple episodes of bursting activity. In one case, 12 consecutive X–ray bursts are modelled and the energy transport into the stellar core determined. Erratic bursting activity is exhibited by systems having relatively low values of accretion rate and metallicity. In extreme cases, hydrogen ignition may be forestalled until the density required for electron capture on free protons is attained. The observational consequence of such an extreme case is a rapid X–ray transient. A new mode of instability

is reported that may be related to "rapidly recurrent" X–ray bursters.

067.015 A theoretical calculation of rapid X–ray transients and radius expansion.
S. Starrfield, W. Sparks, J. Truran, S. Kenyon.
AIP Conf. Proc., No. 115, p. 298 – 301 (1984). – See Abstr. 012.005.
The authors present the results of a calculation of a thermonuclear runaway on a 10 km neutron star which produced a precursor, radius expansion, and after the envelope had begun to shrink, a second X–ray burst about 2500 seconds later. Although such an event has not yet been observed, decreasing the initial envelope mass should bring the calculations into better agreement with the observations.

067.016 Shell flashes interacting with the core of neutron stars.
M. Y. Fujimoto, T. Hanawa, I. Iben Jr., M. B. Richardson.
AIP Conf. Proc., No. 115, p. 302 – 305 (1984). – See Abstr. 012.005.

067.017 Nuclear physics problems for accreting neutron stars.
R. K. Wallace, S. E. Woosley.
AIP Conf. Proc., No. 115, p. 319 – 324 (1984). – See Abstr. 012.005.
The importance of $p(e^- \nu)n$ and of (p, γ) reactions on ^{56}Ni during a thermonuclear runaway on a neutron star surface is pointed out. A fast 16–isotope approximate nuclear reaction network is developed that is suitable for use in hydrodynamic calculations of such events.

067.018 Emission region of X–ray bursters.
R. Hoshi.
AIP Conf. Proc., No. 115, p. 325 – 329 (1984). – See Abstr. 012.005.
X–ray bursts are believed to be thermonuclear flashes in the surface layers of accreting neutron stars. The author shows that Comptonization and free–free absorption in the outermost surface layers are important for interpreting the observed parameters of X–ray bursters.

067.019 Quasi–static winds from neutron stars.
F. Melia, P. C. Joss.
AIP Conf. Proc., No. 115, p. 330 – 332 (1984). – See Abstr. 012.005.

067.020 The theory of gamma–ray bursts.
S. E. Woosley.
AIP Conf. Proc., No. 115, p. 485 – 511 (1984). – See Abstr. 012.005.
A γ-ray burst occurs when a strongly magnetic neutron star experiences either a thermonuclear explosion in degenerate material accumulated over a long period of time or the sudden, brief, greatly super–Eddington accretion of matter. Certain aspects of the thermonuclear model are briefly reviewed and the necessity that low–luminosity pulsars (e.g. X Per) occasionally undergo nuclear runaways stressed. Greatest attention is devoted to mechanisms for producing the hard spectrum of γ–ray bursts. A leading candidate utilizes the high temperature developed behind an accretion shock. A new model for the production of optical flashes accompanying γ–ray bursts is developed which relies upon the cyclotron emission of electrons at $\sim 10^8$cm (in a wind or accretion flux) pumped by Compton collisions with γ–rays from the burst.

067.021 Physics of gamma–ray burst spectra.
D. Q. Lamb.
AIP Conf. Proc., No. 115, p. 512 – 547 (1984). – See Abstr. 012.005.
It is now widely believed that gamma–ray bursts come from strongly magnetic neutron stars. This paper reviews the thermal and nonthermal radiation processes that can occur at high temperatures and in strong magnetic fields near neutron stars. The

luminosity and spectral properties of these processes are compared with observation.

067.022 Gamma burst emission from neutron star accretion.
S. A. Colgate, A. G. Petschek, R. Sarracino.
AIP Conf. Proc., No. 115, p. 548 – 554 (1984). – See Abstr. 012.005.
A model for emission of the hard photons of gamma bursts is presented. The model assumes accretion at nearly the Eddington limited rate onto a neutron star without a magnetic field. Initially soft photons are heated as they are compressed between the accreting matter and the star. A large electric field due to relatively small charge separation is required to drag electrons into the star with the nuclei against the flux of photons leaking out through the accreting matter.

067.023 The formation of planetesimals in highly compact binary stellar systems and the nature of cosmic gamma–ray bursts.
P. C. Joss, S. Rappaport.
AIP Conf. Proc., No. 115, p. 555 – 557 (1984). – See Abstr. 012.005.
Collisions between comets (or asteroids) and neutron stars were first proposed by Harwit and Salpeter as a mechanism for producing cosmic gamma–ray bursts. The authors present here a speculative scenario for strongly enhancing the collision rate of neutron stars with small bodies: the *in situ* formation of planetesimals within a disk surrounding a neutron star in a low–mass close–binary system. They find that this scenario may explain, in a natural way, several empirical characteristics of gamma–ray bursts.

067.024 Gamma–ray burst emission: a jet and fireball model.
R. E. Lingenfelter, G. J. Hueter.
AIP Conf. Proc., No. 115, p. 558 – 567 (1984). – See Abstr. 012.005.
The authors discuss some of the properties, problems and implications of the jet and fireball model of gamma–ray burst emission. This model was suggested by constraints on the opacity of $\sim$ MeV photons to photon–photon pair production.

067.025 Size–frequency distribution of gamma–ray bursts from thermonuclear runaway on neutron stars accreting interstellar gas.
J. C. Higdon, R. E. Lingenfelter.
AIP Conf. Proc., No. 115, p. 568 – 577 (1984). – See Abstr. 012.005.
The authors present some preliminary results of calculations indicating that runaway thermonuclear burning of interstellar gas accreted onto magnetic neutron stars can account for the observed size–frequency distribution of gamma–ray bursts.

067.026 Gamma–ray bursts – the roundabout way?
P. Kafka, F. Meyer.
AIP Conf. Proc., No. 115, p. 578 – 580 (1984). Abstract. – See Abstr. 012.005.

067.027 Inverse comptonization vs. thermal synchrotron.
E. E. Fenimore, R. W. Klebesadel, J. G. Laros.
AIP Conf. Proc., No. 115, p. 590 – 596 (1984). – See Abstr. 012.005.
There are currently two radiation mechanisms being considered for gamma–ray bursts: thermal synchrotron and inverse comptonization. They are mutually exclusive because thermal synchrotron requires a magnetic field of $\sim 10^{12}$ Gauss, whereas inverse comptonization cannot produce a monotonic spectrum if the field is larger than 10^{11} and is too inefficient relative to thermal synchrotron unless the field is less than 10^{9} Gauss. It is concluded that thermal synchrotron is more consistent with the observations if the sources are ~ 40 kpc away, whereas inverse comptonization is more consistent if they are ~ 300 pc away.

067.028 Physics of the synchrotron model of cosmic gamma–ray bursts.
E. P. Liang.
AIP Conf. Proc., No. 115, p. 597 – 604 (1984). – See Abstr. 012.005.
The author reviews the theoretical arguments leading to the thermal synchrotron model of cosmic gamma–ray bursts. He proposes a magnetic flare – resonant absorption mechanism for the production of the hot electrons that can account for most of the unusual properties of the emission layer.

067.029 Spectral models of bursting neutron stars.
R. Z. Yahel, A. Braun.
AIP Conf. Proc., No. 115, p. 605 – 610 (1984). – See Abstr. 012.005.
The authors calculate self–consistent model atmospheres of bursting neutron stars. It is found that for effective temperatures $\theta_e \equiv kT/mc^2 \sim 0.1$, the equilibrium density of $e^- - e^+$ pairs in the atmosphere is one to few hundred times larger than the proton density. However, the effect of photon production by pair annihilations on the emergent spectrum is only marginal. For $\theta_e \lesssim 0.1$, this spectrum is roughly described by a black body with reduced intensity.

067.030 Magnetohydrodynamical model of pulsar rotation. Exact periodic solutions.
O. I. Bogoyavlensky (*O. I. Bogoyavlenskij*).
Astron. Nachr., Vol. 306, No. 1, p. 35 – 41 (1985).
A rotating rigid body with ellipsoidal cavity filled with magnetic fluid is considered as a pulsar model. Dynamical equations for the pulsar model are derived and investigated, certain integrable cases are indicated. Three–parameter sets of periodic solutions integrable in terms of elliptic functions of the time variable are obtained. A formula is derived for the period of rotation and magneto–rotational oscillations of the pulsar.

067.031 Accretion onto neutron stars: the stopping of accreting matter in the neutron star atmosphere.
G. S. Miller.
News Lett. Astron. Soc. N.Y., Vol. 2, No. 7, p. 7 – 8 (1985). Abstract. – See Abstr. 010.241.

067.032 Neutron star seismology.
P. N. McDermott.
News Lett. Astron. Soc. N.Y., Vol. 2, No. 7, p. 19 – 20 (1985). Abstract. – See Abstr. 010.241.

067.033 Black hole thermodynamics from a possible model for internal structure.
P. Goswami, K. P. Sinha.
Pramāna, Vol. 23, No. 3, p. 381 – 384 (1984). Abstr. in Phys. Abstr., Vol. 88, No. 1248, Entry 9789 (1985).

067.034 Energy gaps in crust–phase of neutron stars.
T. Takatsuka.
Prog. Theor. Phys., Vol. 71, No. 6, p. 1432 – 1435 (1984). Abstr. in Phys. Abstr., Vol. 88, No. 1249, Entry 14660 (1985).

067.035 Vaidya spacetime as an evaporating black hole.
Y. Kuroda.
Prog. Theor. Phys., Vol. 71, No. 6, p. 1422 – 1425 (1984). Abstr. in Phys. Abstr., Vol. 88, No. 1249, Entry 14664 (1985).

067.036 Energy–extraction processes from a Kerr black hole immersed in a magnetic field. II. The formalism.
S. V. Dhurandhar, N. Dadhich.
Phys. Rev. D, Vol. 30, No. 8, p. 1625 – 1631 (1984). Abstr. in Phys. Abstr., Vol. 88, No. 1250, Entry 15012 (1985).

067.037 Phase transition of superdense neutron star matter.
Z.–W. Liu.
Phys. Lett. B, Vol. 147B, No. 1 – 3, p. 157 – 161 (1984). Abstr. in Phys. Abstr., Vol. 88, No. 1250, Entry 15390 (1985).

067.038　Equation of state and cooling rate of neutron stars with the charged pion condensate by a realistic calculation.
T. Tatsumi, R. Tamagaki.
Perspectives in nuclear physics at intermediate energies,, p. 518 – 524 (1984). Abstr. in Phys. Abstr., Vol. 88, No. 1250, Entry 18856 (1985). – See Abstr. 012.017.

067.039　The field of an accelerating black hole embedded in a magnetic Universe.
K. D. Krori, M. Barua.
Can. J. Phys., Vol. 62, No. 9, p. 889 – 897 (1984). Abstr. in Phys. Abstr., Vol. 88, No. 1250, Entry 18859 (1985).

067.040　Comments on "incomplete black–hole evaporation".
C. Massa.
Lett. Nuovo Cimento, Vol. 41, Ser. 2, No. 10, p. 351 (1984). Abstr. in Phys. Abstr., Vol. 88, No. 1250, Entry 18860 (1985).

067.041　Black hole in an asymmetric electromagnetic field: new ponderomotive effects – spin precession and drift.
D. V. Gal'tsov, V. I. Petukhov, A. N. Aliev.
Phys. Lett. A, Vol. 105A, No. 7, p. 346 – 350 (1984). Abstr. in Phys. Abstr., Vol. 88, No. 1250, Entry 18861 (1985).

067.042　Naked singularities in the Vaidya spacetime.
Y. Kuroda.
Prog. Theor. Phys., Vol. 72, No. 1, p. 63 – 72 (1984). Abstr. in Phys. Abstr., Vol. 88, No. 1250, Entry 18862 (1985).

067.043　Equilibrium of two rotating charged black holes and the Dirac string.
A. Tomimatsu.
Prog. Theor. Phys., Vol. 72, No. 1, p. 73 – 82 (1984). Abstr. in Phys. Abstr., Vol. 88, No. 1250, Entry 18863 (1985).

067.044　Thomas–Fermi study of the bubble phase in hot dense matter.
E. Suraud.
Astron. Astrophys., Vol. 143, No. 1, p. 108 – 115 (1985).
A Thomas–Fermi equation of state of hot dense matter, calculated without restricting the variational space to Fermi functions, is presented. Comparisons with Hartree–Fock and earlier Thomas–Fermi calculations are given. The author studies in detail the existence of the bubble configurations occurring just below nuclear matter density and discusses their influence on the equation of state.

067.045　Proton mixing in π^0–condensed phase of neutron star matter.
T. Takatsuka.
Prog. Theor. Phys., Vol. 72, No. 2, p. 252 – 265 (1984). Abstr. in Phys. Abstr., Vol. 88, No. 1251, Entry 19581 (1985).

067.046　The damping effects of the vibrations in the core of a neutron star.
Q. D. Wang, T. Lu.
Phys. Lett. B, Vol. 148B, No. 1 – 3, p. 211 – 214 (1984). Abstr. in Phys. Abstr., Vol. 88, No. 1251, Entry 24466 (1985).

067.047　Thermal structure of magnetized neutron–star envelopes.
L. Hernquist.
Mon. Not. R. Astron. Soc., Vol. 213, No. 2, p. 313 – 336 (1985).
The influence of a strong magnetic field ($B \sim 10^{10} - 10^{14}$G) on the thermal structure of neutron–star envelopes is investigated using the most recent calculations of radiative and electronic thermal conductivities. In particular, the relation between the core temperature and the heat flux is considered for effective surface temperatures in the range $T_s = 10^{5.5} - 10^{6.5}$K. For a purely vertical magnetic field it is found that quantum effects will enhance (relative to the zero–field case) the heat flux, for a fixed core temperature, by a factor $\lesssim 3$. It is further argued that the anisotropic nature of electron transport in a magnetic field will suppress the heat flux for a more realistic field geometry by a

factor $\lesssim 3$. Thus the magnetic field is expected to have only a minor effect on neutron–star cooling.

067.048　Twisted accretion discs: the Bardeen–Petterson effect.
S. Kumar, J. E. Pringle.
Mon. Not. R. Astron. Soc., Vol. 213, No. 2, p. 435 – 442 (1985).
The authors calculate the effects of Lense–Thirring precession on the inner regions of an accretion disc by solving the equations governing the tilt of a twisted, viscous disc derived by Papaloizou and Pringle. They confirm the earlier conclusion of Bardeen and Petterson that the rotation in the inner regions of the disc aligns with the spin of the hole. They find, however, that previous estimates of the radius at which this alignment occurs are incorrect.

067.049　Electrosphere of an aligned magnetized neutron star.
J. Krause–Polstorff, F. C. Michel.
Mon. Not. R. Astron. Soc., Vol. 213, No. 2, p. 43P – 49P (1985).
A fundamentally new self–consistent solution for the electrosphere of an aligned magnetized neutron star is presented. Unlike previous models the electrospheres are finite in extent. This avoids the light cylinder problem. The results may provide a basis for pulsar models.

067.050　Black holes or compact stars?
A. Harpaz, N. Rosen.
Astron. Astrophys., Vol. 143, No. 2, p. L5 – L6 (1985).
On the basis of the bimetric general relativity theory an investigation has been carried out on the structure of a star. It has been found that for ordinary stars the two theories give the same results. However, the bimetric theory permits the existence of a star in a collapsed state in which it fills its Schwarzschild sphere and is in hydrostatic equilibrium.

067.051　The initial mass limit for neutron star and black hole formation.
H. Schild, A. Maeder.
Astron. Astrophys., Vol. 143, No. 2, p. L7 – L10 (1985).
The membership of pulsars to young ($\lesssim 10^7$y) clusters and associations is analysed. As expected, the rate of pulsars associated to stellar groups is a strongly decreasing function of the pulsar ages. Stars with an initial mass on the zero age main sequence up to 50 $M_\odot$ terminate their evolution as neutron stars. This suggests that black holes, if any exist, are formed from stars initially more massive than 50 $M_\odot$. Among the consequences of black holes being formed only from large initial stellar masses (at least in Pop. I) the authors note: (1) It is not likely that the missing mass is locked in the form of black holes since their expected number is vey low. (2) The direct progenitors of black holes are Wolf–Rayet stars of late WN or early WC subtype.

067.052　Pulsar space charging.
J. Krause–Polstorff, F. C. Michel.
Astron. Astrophys., Vol. 144, No. 1, p. 72 – 80 (1985).
It is found that stable self–consistent static solutions for a pulsar magnetosphere can be constructed with the magnetosphere having vacuum gaps separating the positive and negative regions of space charge. The magnetosphere is confined well within the light cylinder thus avoiding the problems of the Goldreich and Julian model. The total system charge is a free parameter in the family of models.

067.053　The natural angular momentum distribution in the study of thick disks around black holes.
S. K. Chakrabarti.
Astrophys. J., Vol. 288, No. 1, p. 1 – 6 (1985).
The thick disk model is reviewed with an angular momentum distribution chosen to have a power law relation with the von Zeipel parameter $\lambda = (l/\Omega)^{1/2}$, where l denotes the specific angular momentum and Ω denotes the angular velocity. This choice enables the author to integrate the relativistic Euler equation independent of the background geometry. The consequences with regard to the Schwarzschild and Kerr black hole geometries are briefly investigated.

067.054 Analytic structure of cosmic radio jets: a preliminary investigation.
S. K. Chakrabarti.
Astrophys. J., Vol. 288, No. 1, p. 7 – 13 (1985).

The structure of the steady radio jet is discussed by analyzing the relativistic Euler equation in the case where surfaces of constant angular momentum coincide with surfaces of constant angular velocities. A closed analytic expression for the potential and other physical quantities is derived in a general stationary background geometry. The effect of the angular momentum component and the radial momentum component on the structure of the jet is studied in detail for Schwarzschild geometry. The possibility of obtaining arbitrarily well–collimated jets very near to a black hole is discussed.

067.055 Relativistic wind termination: jets and synchrotron nebulae.
F. C. Michel.
Astrophys. J., Vol. 288, No. 1, p. 138 – 141 (1985).

The author examines an idealized model describing the termination of a relativistic wind owing to its interaction with surrounding nonrelativistic matter, such as the interaction of a pulsar wind with the supernova remnant shell. It is assumed that the large–scale electric and magnetic fields out to the termination distance are controlled by the wind source, in contrast to previous work treating the wind as an isotropic MHD flow. Two effects are found: (1) forced synchrotron radiation from the bulk of the injected particles, and (2) formation of oppositely directed jets of ultrarelativistic particles.

067.056 Electromagnetic damping of neutron star oscillations.
P. N. McDermott, M. P. Savedoff, H. M. Van Horn.
News Lett. Astron. Soc. N.Y., Vol. 2, No. 4, p. 7 (1983). Abstract. – See Abstr. 010.242.

067.057 Radiative transfer in the accretion column of X–ray pulsars: effects from the hot spot.
M. Soffel, H. Herold, H. Ruder, J. Ventura.
Astron. Astrophys., Vol. 144, No. 2, p. 485 – 495 (1985).

The influence of the hot spot that very likely forms at the base of the accretion column of X–ray pulsars with X–ray luminosities not much larger than $\sim 10^{30}$W onto emitted spectra, angle dependent fluxes and pulse shapes is discussed. In the optically thin regime the hot spot is modelled by a thin slab and the radiative transfer through the atmosphere is treated in the zero scattering approximation. Special attention is paid to the influence of atmospheric bulk motion that leads to pulse–phase dependent cyclotron line centers. In the optically thick regime radiative transport is treated in the diffusion approximation, neglecting effects from Comptonization. A detailed mathematical two–dimensional photon–diffusion model is presented for a homogeneous and isothermal cylindrical accretion column.

067.058 Standing shocks in accretion flows onto black holes.
K. M. Chang, J. P. Ostriker.
Astrophys. J., Vol. 288, No. 2, p. 428 – 437 (1985).

Using standard radial, time–independent, fluid hydrodynamic equations with simplified heating and cooling functions, the authors find that solutions with standing shocks occur and are uniquely defined in a part of the luminosity–efficiency plane in which no stationary solutions (without shocks) otherwise exist. Necessary conditions for the existence of solutions with two sonic points are derived. Fifteen specific solutions are presented. Typically the outer (isothermal) sonic point is at $10^{7}r_{s}$, the standing shock at $10^{6}r_{s}$, and the inner sonic point at $10^{4}r_{s}$, (r_{s} Schwarzschild radius of the black hole).

067.059 On the existence of an exterior toroidal region in the nonaligned pulsar magnetosphere.
R. F. Martin Jr.
Astrophys. J., Vol. 288, No. 2, p. 665 – 671 (1985).

The magnetosphere of a pulsar with nonaligned rotation and magnetic axes is studied numerically by considering the motion of single charged particles in the radiation and static fields assumed to exist outside of the inner magnetosphere. Well–defined trapped orbits, localized somewhat beyond the light cylinder, are found for a physically interesting region of parameter space. These attracting orbits, centered on the rotational equator, have a position which is independent of initial conditions over a wide range. Properties of these orbits are used to develop a simple model of a toroidal electron plasma region outside the star plus inner magnetosphere.

067.060 On the relaxation times in the superfluid cores of neutron stars.
D. M. Sedrakyan, K. M. Shakhabasyan, A. G. Movsisyan.
Astrofizika, Tom 22, Vyp. 1, p. 137 – 144 (1985). In Russian. English translation in Astrophysics, Vol. 22, No. 1.

The velocity relaxation times of normal electrons on the proton flux lines occurring on account of the drag effect in the "npe"–phase of a neutron star are considered. It is shown that the "npe"–phase is rigidly connected with the star crust.

067.061 Wind interactions above accretion discs: a model for broad–line regions and collimated outflow.
M. D. Smith, D. J. Raine.
Mon. Not. R. Astron. Soc., Vol. 212, No. 2, p. 425 – 445 (1985).

The interaction of a wind from an active galactic nucleus with a Compton–heating–induced wind from an accretion disc is studied. The nuclear wind is taken as initially supersonic and spherically symmetric. The disc wind arises when the disc surface is exposed to a hard and powerful X–ray source. Three classes of interaction are identified in terms of the relation between the pressure on the disc surface and the corresponding thermal and ram pressures in the nuclear wind.

067.062 The electric field of a pulsar magnetosphere.
D. F. Smith, L. A. Muth, J. Arons.
Astrophys. J., Vol. 289, No. 1, p. 165 – 172 (1985).

Numerical solutions are found for the electric field of a simplified model of a radio pulsar's magnetosphere. The magnetic field assumed is that of a rotating point dipole with modulation in radius and azimuth chosen to emulate the effect of the field line sweepback expected in a full magnetospheric solution. The magnetic and rotation axes were assumed to be orthogonal. Nonneutral beams of particles were injected along the polar field at radii $r = 0.1\,R_{L}$, where R_{L} is the light cylinder distance. The properties of these beams were chosen according to an existing theory of low–altitude particle acceleration. The results show that the maximum noncorotation potential reached is 20% – 35% of the full vacuum potential drop which can occur over the polar caps when the neighboring closed zone is a good conductor.

067.063 Superconductivity and superfluidity in neutron stars and decay of pulsar magnetic fields.
A. G. Muslimov, A. I. Tsygan.
Pis'ma Astron. Zh., Tom 11, No. 3, p. 196 – 202 (1985). In Russian. English translation in Sov. Astron. Lett., Vol. 11.

It is suggested that in neutron stars with proton superconductivity the fluxoids are buoyant. This may lead to a rapid ($\lesssim 5 \times 10^{4}$years) expulsion of the magnetic flux from the superconducting core to the subcrustal region and subsequent decay within the outer crust. The considered effect may be the physical reason why the characteristic time of the decay of pulsar magnetic fields ($\sim 10^{6}$years) corresponds to the ohmic dissipation time within the neutron star crust.

067.064 Hydrogen–like atoms on the surface of neutron stars – intense magnetic field effects.
A. C. Williams, W. Darbro, M. C. Weisskopf, R. F. Elsner.
Astrophys. J., Vol. 289, No. 2, p. 782 – 791 (1985).

It is known that very strong ($> 10^{12}$gauss) magnetic fields exist on the surface of some neutron stars. Because of the mixing of spherical and cylindrical symmetries, the analytical solution to the problem of a hydrogen atom in a uniform magnetic field is impossible to obtain. In this work, a variational wave function is used to describe the properties of hydrogen–like atoms in intense magnetic fields including first–order relativistic effects. Special attention is given to the transition matrix elements for the $2p_{0} \rightarrow 1s_{0}$ transition in Fe XXVI.

067.065 Photodissociation of molecules in the magnetic field of a neutron star.
V. K. Khersonskij.
Astrofiz. Issled. Izv. Spets. Astrofiz. Obs., Tom 19, p. 54 – 59 (1985). In Russian. English translation in Bull. Spec. Astrophys. Obs. – North Caucasus.

The cross section of the photodissociation of the H_2^+ molecular ion in a magnetic field of $10^{12} - 10^{13}$ G is calculated. The author considered the case when the molecular axis is directed along the magnetic field and photodissociation goes from the vibrational levels of the low electron state to the continuum of this state. It is shown that the cross section increases when the vibrational level number υ grows and at $\upsilon \geqslant 16 - 18$ this cross section decreases when the magnetic field increases.

067.066 Revival of the Penrose process for astrophysical applications.
S. M. Wagh, S. V. Dhurandhar, N. Dadhich.
Astrophys. J., Vol. 290, No. 1, p. 12 – 14 (1985).

By utilizing the curious existence of negative energy orbits outside the event horizon of the Kerr black hole, Penrose proposed that rotational energy of the black hole can be extracted. It was shown by Bardeen et al. and Wald that this process cannot offer an astrophysically viable mechanism for high–energy jets, because for significant gain in energy the breakup of the incident particle must itself be relativistic. Here the authors show that the inconvenient requirement of relativistic splitting could be easily overcome for a black hole immersed in an electromagnetic field.

067.067 Interaction of $e^\pm$ with photons in the magnetospheres of neutron stars.
N. S. Kardashev, I. G. Mitrofanov, I. D. Novikov.
Inst. kosm. issled. Akad. Nauk SSSR. Prepr., No. 891, 30 pp. (1984). In Russian. Abstr. in Ref. Zh., 51. Astron., 1.51.852 (1985).

067.068 Isothermal neutron star cores.
M. C. Durgapal, R. S. Fuloria, K. Pande.
Astrophys. Space Sci., Vol. 109, No. 2, p. 241 – 248 (1985).

The authors have shown that some of the standard equations of state, when applied to NS cores, correspond to constancy of some adiabatic exponents. It has been shown that the equation of state, $P = KE$, corresponds to $\Gamma_1 = \Gamma_2 = \Gamma_3 = 1 + K$ and the equation of state, $dP/dE = K$, corresponds to $\Gamma_3 = 1 + K$. For isothermal NS, the local temperature T can be expressed in terms of pressure P, energy density E, and rest mass density ϱ. Equation of state corresponding to $\Gamma = \Gamma_2$ is obtained as: $P = E/\ln(K/E)$ and the equation corresponding to $\Gamma = \Gamma_3$ comes out as: $E = P \ln(K/P)$. When core equation corresponding to $\Gamma = \Gamma_2$ or $\Gamma = \Gamma_3$ is used in the core, the continuity of dP/dE at the core-envelope boundary can be ensured, along with the continuity of P, E, λ, and v. The parameters of isothermal NS cores corresponding to the cases $\Gamma = \Gamma_2$ and $\Gamma = \Gamma_3$, have been obtained. The maximum mass of these NS cores comes out to be $2.7 M_\odot$.

067.069 Cascade processes in the surface layers of pulsars.
I. L. Rozental (*I. L. Rozental'*), V. V. Usov.
Astrophys. Space Sci., Vol. 109, No. 2, p. 365 – 371 (1985).

The influence of the Landau–Pomeranchuk effect on the development of a shower generated by ultrarelativistic particles bombarding the surface of a pulsar is discussed. Because of this effect, the path length of the shower increases while low–energy photon generation is strongly suppressed. In view of this, the mechanism of pair production suggested by Cheng, Ruderman, and Jones for the pulsar magnetosphere, may be essential only for pulsars whose magnetic field intensity at the surface lies in a relatively narrow range of around $B \cong 10^{12}$ G.

067.070 The electric field in axisymmetric pulsar magnetospheres.
R. Burman.
Astrophys. Space Sci., Vol. 109, No. 2, p. 403 – 405 (1985).

A knowledge of the non–corotational electric potential required to support particles in corotation with the star is used to deduce the global qualitative forms that the potential can take in axisymmetric pulsar magnetospheres.

067.071 Gravitational radiation from a solid–crust neutron star.
M. A. Alpar, D. Pines.
Nature, Vol. 314, No. 6009, p. 334 – 336 (1985).

The role of gravitational radiation in the angular momentum loss of millisecond pulsars is investigated. Balancing gravitational radiation power against the rate of rotational energy loss, the authors find that if the effective triaxiality ε_{eff} exceeds $\varepsilon_{max} \sim 10^{-9}$, gravitational radiation alone would spin a millisecond pulsar down at a faster rate than the observed $\dot{P}$. It is shown that $\varepsilon_{eff} \ll \varepsilon_{max}$, so that gravitational radiation is not likely to play a major role in the evolution of millisecond pulsars even in the absence of damping. Damping times for precession of the solid crust are estimated to be $< 2 \times 10^3$ yr.

067.072 Conformal structure of a Schwarzschild black hole immersed in a Friedmann universe.
R. A. Sussman.
Gen. Relativ. Gravitation, Vol. 17, No. 3, p. 251 – 291 (1985).

The evolution of a Schwarzschild black hole in an expanding Friedmann universe is described using the same coordinate patch for both geometries. Comoving and extended Kruskal coordinates are considered and compared for the cases $k = 0$ and $k = 1$. The conformal structure and some global topological aspects of the Schwarzschild–Friedmann system are examined with the help of diagrams in comoving and extended Kruskal coordinates.

067.073 Extrinsic curvature for the two–black–hole problem.
A. D. Kulkarni.
Gen. Relativ. Gravitation, Vol. 17, No. 3, p. 301 – 310 (1985).

The solutions of the momentum constraints on the Einstein–Rosen manifold with two bridges, representing two black holes, are analyzed. These solutions are in the form of an infinite series. Their higher order terms are shown to fall off as r^{-6}. These terms add multipole moments and gravitational radiation to the initial data and do not contribute to the linear and (spin) angular momenta of the black holes.

067.074 Kerr black hole thermodynamical fluctuations.
D. Pavón, J. M. Rubí.
Gen. Relativ. Gravitation, Vol. 17, No. 4, p. 387 – 396 (1985).

The authors study, in the flat space–time limit approximation, the thermodynamical fluctuations of a massive rotating uncharged black hole. Energy and angular momentum correlations as well as other correlations are computed and interpreted.

067.075 Thermomagnetic phenomena and the cooling of neutron stars with a magnetic field.
V. A. Urpin.
Astron. Zh., Tom 62, Vyp. 2, p. 258 – 267 (1985). In Russian. English translation in Sov. Astron., Vol. 29, No. 2.

The stability of degenerate envelopes of neutron stars which possess a strong magnetic field ($\sim 10^{11} - 10^{12}$ G) is considered. It is shown that at early stages of evolution thermomagnetic instability may develop in the envelopes. The characteristic timescale and the conditions necessary for such an instability to develop are determined. Estimates of the small–scale magnetic fields and of the velocity of hydrodynamical motions generated by the instability are performed. Small–scale motions are able to transport the heat efficiently and to change significantly the cooling rate of neutron stars. The coefficient of turbulent "thermal conductivity" is evaluated.

067.076 New models of neutron stars.
V. M. Lipunov.
Zemlya Vselennaya, No. 2, p. 24 – 33 (1985). In Russian.

067.077 Stability of mass transfer in a neutron star plus degenerate dwarf binary.
P. F. J. Bonsema, E. P. J. van den Heuvel.
Astron. Astrophys., Vol. 146, No. 1, p. L3 – L5 (1985).

The authors consider mass transfer from a zero hydrogen abundance electron–degenerate secondary to a 1.4 $M_\odot$ neutron-star primary. They find that for secondary masses: (1) below $m_2 \sim 0.005\, M_\odot$, and (2) between $0.40\, M_\odot$ and $0.66\, M_\odot$, runaway mass transfer and disruption of the secondary may occur. For $m_2 \sim 0.005\, M_\odot$ the instability is reached within the age of the galaxy. For $m_2 > 0.66\, M_\odot$ the mass transfer is found to be always unstable, whereas for $0.005\, M_\odot \leqslant m_2 \leqslant 0.40\, M_\odot$ it is always stable. The amount of angular momentum available in the case of unstable mass transfer from a secondary with mass $\leqslant 0.005\, M_\odot$ is insufficient to produce a single radio pulsar with a period <3 milliseconds. Unstable mass transfer from a degenerate companion with $m_2 > 0.66\, M_\odot$ can, however, produce a 1.55 msec pulsar.

067.078 Accretion onto magnetized neutron stars.
J. G. Kirk.
Proc. Astron. Soc. Aust., Vol. 5, No. 4, p. 446 – 456 (1984).

Models of accretion onto neutron stars are discussed concentrating on the dynamics of the flow near the stellar surface. The discussion falls into two parts: models in which radiation pressure is assumed to play no role and such models in which radiation pressure dominates in at least some part of the flow.

067.079 The axisymmetric pulsar magnetosphere.
L. Mestel.
ESA Spec. Publ., ESA SP–207, p. 93 – 99 (1984). – See Abstr. 012.044.

The paper describes a possible model for the steady spin–down of an aligned rotating magnetic neutron star, within the framework of classical physics. The electron currents which exert the braking torque on the star flow from the polar caps out to and beyond the light–cylinder, where they emit gamma radiation; the associated loss of angular momentum enables the currents to cross field lines and so return to the star at lower latitudes. The high energies retained by some electrons returning to the strong magnetic regions near the star suggest that the model would spontaneously convert into a quantum magnetosphere, with pair production continually triggered by the circulating primary electrons.

067.080 Thermal convection and magnetic fields in accretion disks of black holes.
G. D. Chagelishvili, J. G. Lominadze.
ESA Spec. Publ., ESA SP–207, p. 131 – 138 (1984). – See Abstr. 012.044.

The processes preceding the formation of a hot optically thin corona in the inner region of an accretion disk (IRAD) of some black holes have been studied. In particular, thermal convection in IRAD has been studied in the presence of the differential rotation and toroidal magnetic field; it has been found that the magnetic field suppresses thermal convection only when its pressure becomes comparable with that of radiation. The equations of the turbulent dynamo are used to show that in IRAD, in the main, the toroidal component of a large–scale magnetic field increases, reaching its maximum value in a sufficiently short time. In addition, it is shown that a small–scale magnetic field, appearing as a result of the large–scale magnetic field entanglement by the convective turbulence, remains several times weaker than the toroidal component of the large–scale magnetic field.

067.081 About static and dynamic solutions of the aligned rotator model.
E. Asseo, D. Beaufils, R. Pellat.
ESA Spec. Publ., ESA SP–207, p. 247 – 250 (1984). – See Abstr. 012.044.

The authors study the possibility for a magnetospheric cold, force free, finite, non–corotating plasma to be separated from a corotating plasma (including the neutron star) by a vacuum gap limited by two force free surfaces. They conclude that axisymmetric vacuum bubbles cannot exist. The authors then start with beams of particles which propagate up and down, along the magnetic dipole field lines, between the stellar surface and a non force–free surface located at a finite distance, beyond which vacuum exists. They obtain that there is no vacuum potential that can be linked to such a plasma potential. An alternative configuration is studied.

067.082 Gamma–ray bursts from remnant neutron star disks.
F. C. Michel.
Astrophys. J., Vol. 290, No. 2, p. 721 – 727 (1985).

The author explores the consequences of a disk of matter orbiting an old neutron star. Such a disk will very slowly expand under internal viscous forces. When the inner edge approaches sufficiently close to the star, there should be a runaway ionization/accretion instability that will cause the inner part of the disk to be precipitated onto the neutron star surface. Estimates suggest that energies of 10^{39}ergs or more could be released with rise times as fast as 0.3 ms. The model seems broadly consistent with observed properties of γ–ray bursters and could accommodate a close (i.e., not in N49) origin for the 1979 March 5 event.

067.083 On the role of rotation and magnetic field in the structure and energetics of AGN.
G. Belvedere.
Mem. Soc. Astron. Ital., Vol. 55, No. 3, p. 619 – 623 (1984). – See Abstr. 012.050.

In the general framework of the mean field electrodynamics, two different approaches are outlined here: dynamo models of non–collapsed rotators and dynamo action in accretion disks onto collapsed objects.

067.084 Pair production in spherical accretion onto black holes.
A. L. Schultz, R. H. Price.
Astrophys. J., Vol. 291, No. 1, p. 1 – 7 (1985).

In models of active galactic nuclei involving the spherical accretion onto a black hole of plasma heated by the dissipation of magnetic fields, turbulence, etc., conditions in the plasma suggest the possibility that pair–production effects may be significant. The authors investigate numerically the extent and consequences of pair production in such situations using a Monte Carlo approach to Comptonization. It is found that the production of pairs is self–regulating, that pair density never greatly exceeds baryon density, and that the omission of pair production is not a serious flaw in these models.

067.085 Field theoretical model for nuclear and neutron matter. II. Neutron stars.
J. Diaz Alonso, J. M. Ibáñez Cabanell.
Astrophys. J., Vol. 291, No. 1, p. 308 – 318 (1985).

The authors analyze the consequences on relativistic stellar configurations of two new equations of state (EOS) for neutron matter. The EOS have been obtained by solving a general field (Lagrangian) theoretical model for nuclear interaction which contains many models studied in the literature as particular cases. The model describes an assembly of nucleons interacting via scalar (σ) mesons, pions (π), and vector (ω and ϱ) mesons; it is solved in the renormalized Hartree approximation, and relativistic effects are included. The general relativistic stellar structure equations are solved for both equations of state and the models are compared with present observational evidence.

067.086 Quantum effects near a rotating dyon black hole.
Z. Zhao, D. Zhang.
Kexue Tongbao (Beijing), Vol. 29, No. 10, p. 1303 – 1306 (1984). Abstr. in Phys. Abstr., Vol. 88, No. 1252, Entry 24804 (1985).

067.087 Superfluidity of neutron matter. I. Singlet pairing.
L. Amundsen, E. Ostgaard.
Ark. Fys. Semin. Trondheim, No. 12, p. 1 – 54 (1984). Abstr. in Phys. Abstr., Vol. 88, No. 1252, Entry 25245 (1985).

067.088   Superfluidity of neutron matter. II. Triplet pairing.
L. Amundsen, E. Ostgaard.
Ark. Fys. Semin. Trondheim, No. 16, p. 1 – 52 (1984). Abstr. in Phys. Abstr., Vol. 88, No. 1252, Entry 29982 (1985).

067.089   Gravitational stability of cold Bose star.
E. Takasugi, M. Yoshimura.
Z. Phys., C, Vol. 26, No. 2, p. 241 – 243 (1984). Abstr. in Phys. Abstr., Vol. 88, No. 1252, Entry 29946 (1985).

067.090   The role of pion excitations in a nucleon medium in the problem of luminosity of neutron stars.
D. N. Voskresenskij, A. V. Senatorov.
Pis'ma ZhEhTF, Tom 40, No. 9, p. 395 – 398 (1984). In Russian. Abstr. in Ref. Zh., 51. Astron., 4.51.392 (1985).

067.091   Feeding a gamma–ray burster.
R. I. Epstein.
Astrophys. J., Vol. 291, No. 2, p. 822 – 833 (1985).
The energetics of a γ–ray burster can be explained by the rapid accretion of $\gtrsim 10^{19}$g of matter onto a neutron star surface. It is proposed that this sudden infall of matter may be the result of a radiation–driven instability which occurs at the boundary between an accretion disk and the surface of a slowly rotating neutron star. Limits on the physical parameters of the accretion disk are derived from the observed properties of γ–ray bursts. The model yields estimates for the recurrence time for bursts, for the steady emission from the accretion disks, and for the flux of reprocessed radiation that is emitted during outbursts.

067.092   Black hole normal modes: a semianalytic approach.
B. F. Schutz, C. M. Will.
Astrophys. J., Lett. Ed., Vol. 291, No. 2, p. L33 – L36 (1985).
The authors present a new semianalytic technique for determining the complex normal mode frequencies of black holes. The method is based on the WKB approximation. It yields a simple analytic formula that gives the real and imaginary parts of the frequency in terms of the parameters of the black hole and of the field whose perturbation is under study, and in terms of the quantity $(n+1/2)$, where $n = 0, 1, 2,...$ and labels the fundamental mode, first overtone mode, and so on. For a Schwarzschild black hole good agreement is found between the WKB estimates and numerical results.

067.093   Envelopes of rapidly rotating neutron stars.
D. de Niem, U. Geppert, H.–J. Wiebicke.
Astrophys. Space Sci., Vol. 110, No. 2, p. 331 – 336 (1985).
Motivated by the discovery of the millisecond pulsars, the authors consider the effect of rapid rotation on the envelope of a neutron star. Solving the equation of hydrostatic equilibrium they find expressions for the density and oblateness as functions of radius and polar angle.

067.094   Gravitational properties of a neutron star with magnetic charge and magnetic moment.
Y. Wang, Q. Peng.
Sci. Sin., Ser. A, Vol. 28, No. 4, p. 422 – 431 (1985).
Taking the magnetic R–N metric as a space–time background, the authors derive a static magnetic field in this curved space–time. Substituting the derived static magnetic field as a source of gravitational field into the right–hand side of Einstein–Maxwell equation, they solve the Einstein–Maxwell equation with a step–by–step approximation method by the first degree, and derive a new metric. This metric describes the external field of a neutron star with magnetic charge and magnetic dipole moment.

067.095   Magnetic moments of neutron stars from real baryon gas.
D. M. Sedrakyan, K. M. Shakhabasyan, A. G. Movsisyan.
Astrofizika, Tom 21, Vyp. 3, p. 547 – 561 (1984). In Russian. English translation in Astrophysics, Vol. 21, No. 3.
The thermodynamics of the rotating superfluid solution in the "npe"–phase of a spherical neutron star is considered. The effect of the entrainment of superfluid protons by rotating superfluid neutrons leads to the occurrence of the array of rectilinear neutron vortices parallel to the axis of rotation. The values of integral characteristics of neutron star–mass, radius, magnetic moment as functions of the central density are obtained. The conditions of the appearance of neutron and proton vortex lines in a spherical star are also considered.

067.096   Temperature regime of neutron stars heated by nucleon decay.
I. D. Novikov, T. V. Perevodchikova.
Sov. Astron., Vol. 28, No. 5, p. 545 – 546 (1984). English translation of 38.067.072.

067.097   Two–phase accretion model for emission–line regions in quasars and active galactic nuclei.
A. Wandel, M. Milgrom, A. Yahil.
Astrophys. J., Vol. 292, No. 1, p. 206 – 216 (1985).
A model is constructed for the broad–line region of quasars and active galactic nuclei. A two–phase medium, made of cool photoionized filaments embedded in a hot gas, and assumed to be in a quasi–spherical infall, is producing the line emission, as well as supplying the mass needed to power the central continuum source. Taking into account conduction and radiative cooling in the filaments, an asymptotic solution is found, in which the ionization parameter in the clouds is almost independent of the distance from the central source. The observational data on the emission lines imply high central masses, $10^7 – 10^{10} M_{\odot}$; high accretion rates, $0.3 < \dot{M} < 10^3 M_{\odot}\mathrm{yr}^{-1}$; and low efficiencies, $e \lesssim 0.001$.

067.098   The triple alpha reaction at low temperatures in accreting white dwarfs and neutron stars.
K. Nomoto, F.–K. Thielemann, S. Miyaji.
MPA Rep., No. 171, 33 pp. (1984). Submitted to Astron. Astrophys.

067.099   Magnetic fluxes across black holes.
J. Bičák, V. Janiš.
MPA Rep., No. 172, 36 pp. (1984). To appear in Mon. Not. R. Astron. Soc.

067.100   On the influence of interaction and phase transition on the structure of a superdense configuration.
A. A. Mikhal'chuk.
Izv. vuzov. Fiz., Tomsk, 13 pp. (1984). In Russian. Abstr. in Ref. Zh., 51. Astron., 3.51.175 (1985).

067.101   The collapse of dense star clusters to supermassive black holes: the origin of quasars and AGNs.
S. L. Shapiro, S. A. Teukolsky.
Astrophys. J., Lett. Ed., Vol. 292, No. 2, p. L41 – L44 (1985). With plate L1.
Fully general relativistic calculations of the gravitational collapse of collisionless equilibrium systems have recently been carried out by Shapiro and Teukolsky. The authors apply the results of such calculations, together with recent Newtonian Fokker–Planck calculations of the "gravothermal catastrophe", to the dynamical evolution of a dense cluster of compact stars – neutron stars or stellar mass black holes. A plausible scenario for the formation of supermassive black holes via the collapse of such clusters embedded in evolved galactic nuclei is described. The process leads naturally to the birth of supermassive black holes of the "right size" to explain quasars and AGNs: $10^6 \lesssim M/M_{\odot} \lesssim 10^9$.

067.102   Generation of relativistic particles in pulsar magnetospheres.
H. Herold, T. Ertl, H. Ruder.
Mitt. Astron. Ges., Nr. 63, p. 174 – 176 (1985). – See Abstr. 012.063.
The authors studied the trajectories of individual charged particles in the electromagnetic vacuum fields of an aligned rotator.

067.103   The distribution of the projected density of stars near a Reissner–Nordström black hole in a globular cluster.
Y. Shen, Q. Zhang.
J. Nanjing Univ., Vol. 21, No. 1, p. 100–106 (1985). In Chinese.
The distribution of the projected density of stars near a Reissner–Nordström black hole in a globular cluster is discussed from the general theory of relativity.

067.104   Electrosphere of the aligned rotator pulsar model.
F. C. Michel, J. Krause–Polstorff.
Bull. Am. Astron. Soc., Vol. 16, No. 4, p. 943 (1984). Abstract. – See Abstr. 010.062.

067.105   Cyclotron lines in accreting magnetic white dwarfs with an application to VV Puppis.
P. Barrett, G. Chanmugam.
Bull. Am. Astron. Soc., Vol. 16, No. 4, p. 943 (1984). Abstract. – See Abstr. 010.062.

067.106   Neutron star oscillations in the presence of a vertical magnetic field.
B. W. Carroll, P. N. McDermott, M. P. Savedoff, J. H. Thomas, H. M. Van Horn, E. G. Zweibel, C. A. Morrow, C. J. Hansen.
Bull. Am. Astron. Soc., Vol. 16, No. 4, p. 943 (1984). Abstract. – See Abstr. 010.062.

067.107   Gravitational lensing and the regions surrounding neutron stars.
C. Ftaclas, M. Kearney, K. Pechenick.
Bull. Am. Astron. Soc., Vol. 16, No. 4, p. 944–945 (1984). Abstract. – See Abstr. 010.062.

067.108   X–ray induced stellar winds: a mass supply for QSO's?
G. M. Voit, J. M. Shull.
Bull. Am. Astron. Soc., Vol. 16, No. 4, p. 952 (1984). Abstract. – See Abstr. 010.062.

067.109   Gravitational waveforms generated by highly nonspherical core collapse.
C. R. Evans, L. L. Smarr, J. R. Wilson.
Bull. Am. Astron. Soc., Vol. 16, No. 4, p. 964 (1984). Abstract. – See Abstr. 010.062.

067.110   Can black hole formation accompanying massive star collapse involve significant conversion of rest mass into neutrinos?
A. Burrows.
Bull. Am. Astron. Soc., Vol. 16, No. 4, p. 964 (1984). Abstract. – See Abstr. 010.062.

067.111   The dynamic effects of angular momentum on accretion flows near black holes.
J. F. Hawley, L. L. Smarr.
Bull. Am. Astron. Soc., Vol. 16, No. 4, p. 964 (1984). Abstract. – See Abstr. 010.062.

067.112   Gamma ray burst model.
S. A. Colgate, A. G. Petschek.
Bull. Am. Astron. Soc., Vol. 16, No. 4, p. 1016 (1984). Abstract. – See Abstr. 010.062.

067.113   Accretion model of gamma ray bursts.
A. G. Petschek, S. A. Colgate.
Bull. Am. Astron. Soc., Vol. 16, No. 4, p. 1017 (1984). Abstract. – See Abstr. 010.062.

067.114   General relativistic effects on the pulse profile of fast pulsars.
B. Datta, R. C. Kapoor.
Nature, Vol. 315, No. 6020, p. 557–559 (1985).
The authors report here the implications of large space–time curvature and large rotation (specifically, the dragging of inertial frames introduced by rotation) on the trajectories of photons, and hence, on the pulse profile of fast pulsars. Space–time curva-
ture leads to substantial amounts of divergence in the pulse width and reduction in the pulse intensity whereas rotation produces a tilt of the pulse cone from its original direction of emission, and deforms the cone, introducing an asymmetry in the (flattened) pulse profile leading to a time delay in the arrival of photons emitted within the pulse cone.

067.115   Poloidal flow in axisymmetric pulsar magnetospheres.
R. R. Burman.
Aust. J. Phys., Vol. 38, No. 1, p. 97–112 (1985).
The author uses the integrals of the motion, including a complete integral for toroidal flow, to reorganize and simplify the fundamental electrodynamic and hydrodynamic equations for axisymmetric magnetospheres. The extension of the complete integral for purely toroidal flow to incorporate poloidal motion that is closely tied to the poloidal magnetic field lines shows that the inclusion of the poloidal flow in the analysis at least reduces the mismatch problem.

067.116   Strong gravity effects in the temperature profile of accretion disks.
H. G. Paul.
Astron. Nachr., Vol. 306, No. 3, p. 117–122 (1985).
The temperature profile of accretion disks moving axially symmetric in the static, spherically symmetric field of a compact, massive object is calculated without specifying the gravitational theory. Using the covariantly written viscous, fluid–dynamical equations of motion, the α viscosity law and free–free radiation of a cooling region is obtained under very general assumptions about the behaviour of the external gravitational field. The used gravitational theory is reflected by inner boundary processes taking place in the region of strong gravity; the cooling region could give a significant "bremsstrahlung" correction to the "cool" part of the spectrum radiated away from the outer, weak gravity region of the disk.

067.117   The maximum mass of a neutron star.
R. S. Sarracino, M. J. Eccles.
Astrophys. Space Sci., Vol. 111, No. 2, p. 375–381 (1985).
The authors consider the possibility that gravitational energy may play a local as well as a global role in the behavior of matter in strong gravitational fields. A particular idealized equation, suggested as representing uniform energy density in general relativity, is examined, and its stability with respect to oscillatory and convective perturbations shown to be consistent with general relativistic hydrodynamics, subject to a new physical effect predicted for the behavior of fluids moving in strong fields. The authors calculate from this idealized equation the mass of a non-rotating neutron star.

067.118   Effect of dynamical friction on the escape of a super-massive black hole from a galaxy.
R. C. Kapoor.
Astrophys. Space Sci., Vol. 112, No. 2, p. 347–359 (1985).
The autor has used the impulsive approximation technique to numerically estimate the effect of dynamical friction on the motion of a supermassive black hole (mass $\cong 10^9 M_\odot$) through a galaxy (mass $= 10^{11} M_\odot$) which has recoiled from the center of the latter as a result of anisotropic emission of gravitational radiation or asymmetric plasma emission.

067.119   Pulsar magnetospheres in binary systems.
A. I. Ershkovich, J. F. Dolan.
Astrophys. J., Vol. 293, No. 1, p. 25–30 (1985).
The criterion for stability of a tangential discontinuity interface in a magnetized, perfectly conducting inviscid plasma is investigated by deriving the dispersion equation including the effects of both gravitational and centrifugal acceleration. The results are applied to neutron star magnetospheres in X–ray binaries. The Kelvin–Helmholtz instability appears to be important in determining whether MHD waves of large amplitude generated by instability may intermix the plasma effectively, resulting in accretion onto the whole star as suggested by Arons and Lea and leading to no X–ray pulsar behavior.

067.120   Spectra of gamma–ray bursts.
J. M. Hameury, J. P. Lasota, S. Bonazzola,
J. Heyvaerts.
Astrophys. J., Vol. 293, No. 1, p. 56 – 68 (1985).

The authors show that γ–ray burst spectra are basically thermal synchrotron spectra emitted in a cold "photosphere" by electrons excited to high Landau levels by high–energy photons that are beamed along the magnetic field lines. This high–energy radiation is produced in a corona by the interaction of soft, thermal photons and synchrotron photons with one–dimensional, relativistic electrons in a strong magnetic field. These coronal electrons are accelerated by short–scale magnetic reconnection. Monte–Carlo simulations that include Compton and resonant scattering produce spectra in good agreement with the observations between 20 keV and 1 MeV.

067.121   Surface conditions in accreting neutron stars.
M. Y. Fujimoto, R. Hoshi.
Astrophys. J., Vol. 293, No. 1, p. 268 – 272 (1985).

The structure of the boundary layer where the accretion disk interacts with an accreting neutron star surface is studied using a one–zone approximation to observe the effects of the accretion on the thermal structure of the surface layers and on the progress of shell burning. The kinematic viscosity is parameterized by means of the critical Reynolds number R_{cr}. For large R_{cr} the boundary layer lies well inside the stellar photosphere, where the dissipation of kinetic energy plays a critical role in heating the envelope and in leading to the ignition of shell flashes. The implications of these results for models of X–ray burst sources are discussed.

067.122   The radius of a neutron star: an interpretation of absorption lines from X–ray burster X1636–536.
M. Y. Fujimoto.
Astrophys. J., Lett. Ed., Vol. 293, No. 1, p. L19 – L22 (1985).

An interpretation of absorption lines in X–ray burst spectra recently observed from X1636–536 is presented, taking into account the influence of angular momentum associated with accreted gas upon the surface structure. For large accretion rates as in the case of X–ray bursters, the surface layers are expected to rotate very rapidly with the accretion disk near the equator, which affects the formation of absorption lines. In particular, the transverse Doppler effect due to this rotation is shown to be important in evaluating the redshift as well as the general relativistic effect. The mass–radius relation of the underlying neutron star is derived with allowances made for both effects.

067.123   Preliminary investigations into a class of thick disk models.
R. E. Wilson, J. H. Hunter.
Bull. Am. Astron. Soc., Vol. 17, No. 1, p. 512 (1985). Abstract. – See Abstr. 010.064.

067.124   Magnetic buoyancy in QSO accretion disks: the underlying physics and analytic approximations.
P. J. Sakimoto, F. V. Coroniti.
Bull. Am. Astron. Soc., Vol. 17, No. 1, p. 518 (1985). Abstract. – See Abstr. 010.064.

067.125   Thermal X–ray emission from isolated older pulsars: a new heating mechanism.
J. Huang, R. E. Lingenfelter, Q. Peng, K. Huang.
High energy astrophysics and cosmology, p. 12 – 18 (1983). – See Abstr. 012.068.

A new pulsar heating mechanism, magnetic dipole radiation of superfluid neutrons, is proposed. It is also a new mechanism for pulsar spindown, $\dot{P} \propto P^2$.

067.126   The distribution of the magnetic inclination in a pulsar's polar cap model.
X. Wu, F. Wu, G. Qiao, J. Cheng, G. Deng.
High energy astrophysics and cosmology, p. 19 – 23 (1983). – See Abstr. 012.068.

Magnetic inclination φ is an important parameter of a pulsar. The investigation of the distribution and the evolution of φ is very

significant in determining the magnetosphere structure and emission mechanism. These problems are discussed in this paper.

067.127   Quark stars and abnormal neutron stars.
Q. Qu.
High energy astrophysics and cosmology, p. 45 – 63 (1983). – See Abstr. 012.068.

New theoretical results on the phase transition from neutron matter to quark matter in compact stars are reviewed.

067.128   X–ray observations by the Einstein satellite and neutron star cooling.
S. Tsuruta.
High energy astrophysics and cosmology, p. 64 – 76 (1983). – See Abstr. 012.068.

Einstein effective temperatures for X–ray pulsars are compared with heating and cooling theories of neutron stars.

067.129   The X–ray burst sources.
Q. Li.
High energy astrophysics and cosmology, p. 77 – 91 (1983). – See Abstr. 012.068.

Theoretical properties of thermonuclear flashes on the surfaces of accreting neutron stars are compared with the observed properties of X–ray bursters.

067.130   Physical processes in the strongly magnetized accretion plasma of a neutron star.
J. G. Kirk.
High energy astrophysics and cosmology, p. 92 – 111 (1983). – See Abstr. 012.068.

The theory of cyclotron and synchrotron radiation in the strong ($\sim 10^{12}$ gauss) magnetic field of a neutron star is briefly outlined. A comparison with observed spectra from gamma–ray bursts is presented.

067.131   Abnormal neutron stars with abnormal protons.
Q. Qu, Z. Wang, T. Lu, Y. Chu.
High energy astrophysics and cosmology, p. 112 – 119 (1983). – See Abstr. 012.068.

067.132   The quark–cluster phase of superdense matter and a possible mechanism of strong fields in neutron stars.
G. Yang, L. Luo.
High energy astrophysics and cosmology, p. 120 – 125 (1983). – See Abstr. 012.068.

067.133   Neutrino cyclotron radiation from superfluid vortexes in neutron stars: a new mechanism for pulsar spin down.
Q. Peng, K. Huang, J. Huang.
High energy astrophysics and cosmology, p. 126 – 128 (1983). – See Abstr. 012.068.

067.134   Accretion disks.
F. Meyer.
High energy astrophysics and cosmology, p. 283 – 293 (1983). – See Abstr. 012.068.

The basic properties of accretion disks around compact stellar objects are briefly reviewed.

067.135   Interaction of accretion disks and rotating magnetic neutron stars.
U. Anzer, G. Börner.
High energy astrophysics and cosmology, p. 294 – 315 (1983). – See Abstr. 012.068.

067.136   Further investigation of rapid X–ray bursts.
D. Wang.
High energy astrophysics and cosmology, p. 316 – 322 (1983). – See Abstr. 012.068.

A model is suggested for the rapid X–ray burster MXB 1730–335. Instabilities in the accretion disk around a magnetic neutron star give rise to the X–ray bursts. Calculation shows that bursts with energy less than 5×10^{38} erg are caused by the tearing mode instability; bursts with energy larger than

8×10^{38}erg are due to the tearing mode instability which in turn brings about a Kruskal–Schwarzschild instability at the inner boundary of the accretion disk.

067.137   Black holes and spherically symmetric accretion.
D. Freihoffer.
High energy astrophysics and cosmology, p. 327 – 342 (1983). – See Abstr. 012.068.

After reviewing present observational evidence for the existence of black holes, this paper discusses spherical accretion onto black holes and estimates the total luminosities generated by this process as a function of black hole mass and accretion rate.

067.138   The Hawking evaporation of Dirac particles in general Kerr–Newman black hole background.
C. Xu, Y. Shen.
High energy astrophysics and cosmology, p. 343 – 348 (1983). – See Abstr. 012.068.

067.139   The theory of an object with a compact core and its applications.
L. Fang, S. Xiang.
High energy astrophysics and cosmology, p. 394 – 400 (1983). – See Abstr. 012.068.

In this paper the authors propose a theory for two classes of compact objects with the core–halo structure, namely, the loaded polytropes and the loaded neutrino objects. The theory is applied to analyze some properties of galactic nuclei and quasars.

067.140   Neutron stars are giant hypernuclei?
N. K. Glendenning.
Astrophys. J., Vol. 293, No. 2, p. 470 – 493 (1985).

Neutron stars are studied in the framework of Lagrangian field theory of interacting nucleons, hyperons, and mesons, which is solved in the mean field approximation. The theory is constrained to account for the four bulk properties of nuclear matter; the saturation binding and density, compressibility, and charge symmetry energy. The cores of the heavier neutron stars are found to be dominated by hyperons, and the total hyperon population for such stars is 15% – 20%, depending on whether pions condense or not.

067.141   Numerical simulation of the growth of thick accretion disks.
D. Clarke, S. Karpik, R. N. Henriksen.
Astrophys. J., Suppl. Ser., Vol. 58, No. 1, p. 81 – 106 (1985).

The authors present extensive numerical simulations of the accretion of adiabatic and isothermal material with net angular momentum by a central gravitating point mass. The calculation is restricted to axial symmetry and to Newtonian gravity. The initial state is the appropriate Bondi flow, which is perturbed initially by adding abruptly a distribution of specific angular momentum. The evolution of the flow toward a thick equatorial disk with meridional internal circulation and an evacuated funnel near the axis is followed. An outer region of Bondi flow reforms after the passage of an expansion wave or shock wave.

067.142   Hydrodynamic simulations of a combined hydrogen, helium thermonuclear runaway on a 10 km neutron star.
S. Starrfield, S. Kenyon, J. W. Truran, W. M. Sparks.
Cataclysmic variables and low–mass X–ray binaries, p. 133 – 137 (1985). – See Abstr. 012.072.

The authors have used a Lagrangian, hydrodynamic stellar evolution computer code to evolve a thermonuclear runaway in the accreted hydrogen rich envelope of a 1.0 $M_\odot$, 10 km neutron star. The simulation produced an outburst which lasted about 2000 sec and peak effective temperature was 3 keV. The peak luminosity exceeded $2 \times 10^5 L_\odot$. A shock wave caused a precursor in the light curve which lasted 10^{-5}sec.

067.143   Stability of radiative shock waves.
J. N. Imamura, R. A. Chevalier, R. H. Durisen, M. T. Wolff.
Cataclysmic variables and low–mass X–ray binaries, p. 261 – 265 (1985). – See Abstr. 012.072.

Radially accreting degenerate dwarfs are thought to produce hard X–rays in radiative shocks formed when the accreting plasma strikes their surfaces. Recent nonlinear time–dependent numerical calculations of funneled accretion onto magnetic degenerate dwarfs found that such shocks are unstable under certain circumstances (Langer, Chanmugam, and Shaviv 1982, hereafter LCS). The authors review their own study of this instability. They first present results of their own linear and nonlinear analyses. They then compare their results to those of LCS and discuss the observational implications of their work. Lastly, the authors summarize their major conclusions.

067.144   Accretion discs.
J. I. Katz.
Cataclysmic variables and low–mass X–ray binaries, p. 359 – 377 (1985). – See Abstr. 012.072.

Accretion discs are believed to be important in a wide variety of astronomical objects, ranging from quasars to cataclysmic variables. Attempts to construct their theory from first principles founder on our ignorance of the processes of angular momentum transport. Phenomenological theories and empirical constraints, although model dependent, are useful. Empirical evidence and constraints are found from the study of systems as diverse as Her X–1, SS433, and the SU UMa stars.

067.145   Accretion disks.
W. M. Sparks, G. S. Kutter.
Cataclysmic variables and low–mass X–ray binaries, p. 429 – 433 (1985). – See Abstr. 012.072.

Derivations are made for the mass and the mass turnover time scale of an accretion disk as a function of the accretion rate, the observed disk radius, the non–viscous disk radius and two parameters. These parameters depend on the effectiveness of viscosity and tidal angular momentum loss. Application is made to DQ Herculis.

067.146   Nucleon superfluidity in neutron star matter with pion condensation.
T. Takatsuka, R. Tamagaki.
Nucleon–nucleon interaction and nuclear many–body problems, p. 726 – 728 (1984). Abstr. in Phys. Abstr., Vol. 88, No. 1253, Entry 30833 (1985). – See Abstr. 012.076.

067.147   Soliton interactions in pulsar magnetospheres.
A. D. Verga, C. Ferro Fontan.
Phys. Lett. A, Vol. 106A, No. 5 – 6, p. 281 – 284 (1984). Abstr. in Phys. Abstr., Vol. 88, No. 1253, Entry 35735 (1985).

067.148   Local black holes are type D on the horizon.
D. Papadopoulos, B. C. Xanthopoulos.
Nuovo Cimento B, Vol. 83B, Ser. 11, No. 2, p. 113 – 126 (1984). Abstr. in Phys. Abstr., Vol. 88, No. 1253, Entry 35738 (1985).

067.149   Static spin–3/2 perturbations of a two–black–hole system.
F. Embacher, P. C. Aichelburg.
Phys. Rev. D, Vol. 30, No. 12, p. 2457 – 2461 (1984). Abstr. in Phys. Abstr., Vol. 88, No. 1254, Entry 36068 (1985).

067.150   Pulsars: high magnetic field laboratories with 10^8T.
H. Ruder, H. Herold, W. Rosner, G. Wunner.
Physica B, C, Vol. 127B + C, No. 1 – 3, p. 11 – 25 (1984). Abstr. in Phys. Abstr., Vol. 88, No. 1254, Entry 40651 (1985). – See Abstr. 012.078.

067.151   Coulomb collisions in strong magnetic fields.
G. Miller, I. Wasserman.
Phys. Rev. A, Vol. 31, No. 1, p. 120 – 133 (1985). Abstr. in Phys. Abstr., Vol. 88, No. 1254, Entry 40653 (1985).

067.152 Gravitational radiation from a particle scattered by a Kerr black hole.
Y. Kojima, T. Nakamura.
Prog. Theor. Phys., Vol. 72, No. 3, p. 494 – 504 (1984). Abstr. in Phys. Abstr., Vol. 88, No. 1254, Entry 40655 (1985).

067.153 Nonthermal radiation from neutron stars.
J. I. Katz.
Astrophys. Lett., Vol. 24, No. 4, p. 183 – 187 (1985).
Many of the observed properties of gamma ray bursters and Geminga imply that these objects are single neutron stars with strong nonthermal radiation processes. The author suggests tests of this hypothesis, and considers the relation between these objects and pulsars, an apparently disjoint set of neutron stars which are similarly single (with few exceptions) and nonthermal radiators.

067.154 Remarks on the double–Kerr solution.
C. Hoenselaers.
Prog. Theor. Phys., Vol. 72, No. 4, p. 761 – 767 (1984). Abstr. in Phys. Abstr., Vol. 88, No. 1255, Entry 40992 (1985).

067.155 Stability of charged rotating black holes in the eikonal approximation.
B. Mashhoon.
Phys. Rev. D, Vol. 31, No. 2, p. 290 – 293 (1985). Abstr. in Phys. Abstr., Vol. 88, No. 1255, Entry 41004 (1985).

067.156 Soliton turbulence in a strongly magnetized plasma. Applications to the coherent radioemission of pulsars.
A. D. Verga, C. F. Fontan.
Plasma Phys. Controlled Fusion, Vol. 27, No. 1, p. 19 – 45 (1985). Abstr. in Phys. Abstr., Vol. 88, No. 1255, Entry 46108 (1985).

067.157 Ultracompact ($R < 3\,M$) objects in general relativity.
B. R. Iyer, C. V. Vishveshwara, S. V. Dhurandhar.
Classical Quantum Gravity, Vol. 2, No. 2, p. 219 – 228 (1985). Abstr. in Phys. Abstr., Vol. 88, No. 1257, Entry 52023 (1985).

067.158 Dense matter in stellar collapse and neutron stars.
C. J. Pethick.
Nucl. Phys. A, Vol. A434, p. 587 – 604 (1985). Abstr. in Phys. Abstr., Vol. 88, No. 1257, Entry 52871 (1985).

067.159 An aligned rotator model for the Crab pulsar.
J. E. Skjervold, E. Ostgaard.
Ark. Fys. Semin. Trondheim, No. 21, p. 1 – 34 (1984). Abstr. in Phys. Abstr., Vol. 88, No. 1257, Entry 57773 (1985).

067.160 Monopoles in pulsar PSR 1929 + 10.
K. Freese.
Monopole '83, p. 147 – 151 (1984). Abstr. in Phys. Abstr., Vol. 88, No. 1257, Entry 57777 (1985). – See Abstr. 012.088.

067.161 Neutron star physics and monopole flux limits.
J. Harvey.
Monopole '83, p. 137 – 145 (1984). Abstr. in Phys. Abstr., Vol. 88, No. 1257, Entry 57780 (1985). – See Abstr. 012.088.

067.162 The rotating Dyon black holes in a magnetic Universe.
Y.-J. Wang.
Acta Phys. Sin., Vol. 33, No. 12, p. 1728 – 1732 (1984). In Chinese. Abstr. in Phys. Abstr., Vol. 88, No. 1257, Entry 57781 (1985).

067.163 Black hole in thermal equilibrium with a scalar field: the back–reaction.
J. W. York Jr.
Phys. Rev. D, Vol. 31, No. 4, p. 775 – 784 (1985). Abstr. in Phys. Abstr., Vol. 88, No. 1257, Entry 57782 (1985).

067.164 Spherically symmetric systems of fields and black holes. IV. No room for black–hole evaporation in the reduced configuration space?
P. Hajicek.
Phys. Rev. D, Vol. 31, No. 4, p. 785 – 795 (1985). Abstr. in Phys. Abstr., Vol. 88, No. 1257, Entry 57783 (1985).

067.165 Black holes in compactified supergravity.
F. A. Bais.
Monopole '83, p. 51 (1984). Abstr. in Phys. Abstr., Vol. 88, No. 1257, Entry 57785 (1985). – See Abstr. 012.088.

067.166 Nuclear matter at high density or high temperature in astrophysical phenomena.
M. Soyeur.
Nucl. Sci. Appl., Sect. A, Vol. 2, No. 1, p. 21 – 50 (1984). Abstr. in Phys. Abstr., Vol. 88, No. 1258, Entry 58725 (1985).

067.167 Sub–saturation phases of nuclear matter.
R. D. Williams, S. E. Koonin.
Nucl. Phys. A, Vol. A435, No. 3 – 4, p. 844 – 858 (1985). Abstr. in Phys. Abstr., Vol. 88, No. 1258, Entry 58728 (1985).

067.168 Apparent horizons of time–symmetric initial value for three black holes.
K. Oohara, T. Nakamura, Y. Kojima.
Phys. Lett. A, Vol. 107A, No. 9, p. 452 – 455 (1985). Abstr. in Phys. Abstr., Vol. 88, No. 1258, Entry 63374 (1985).

067.169 Tidal effects: explosive disruption of stars by a giant black hole.
J.-P. Luminet.
Ann. Phys. (Paris), Vol. 10, No. 2, p. 101 – 200 (1985). In French. Abstr. in Phys. Abstr., Vol. 88, No. 1259, Entry 68845 (1985).

067.170 The space–time metric inside a charged black hole.
S. Wenda, Z. Shitong.
Nuovo Cimento B, Vol. 85B, Ser. 11, No. 2, p. 142 – 148 (1985). Abstr. in Phys. Abstr., Vol. 88, No. 1260, Entry 69188 (1985).

067.171 The gravitational shock wave of a massless particle.
T. Dray, G.'t Hooft.
Nucl. Phys. B, Part. Phys., Vol. B253, No. 1, p. 173 – 188 (1985). Abstr. in Phys. Abstr., Vol. 88, No. 1260, Entry 69196 (1985).

067.172 Superfluidity of neutron matter. I. Singlet pairing.
L. Amundsen, E. Ostgaard.
Nucl. Phys. A, Vol. A437, No. 2, p. 487 – 508 (1985). Abstr. in Phys. Abstr., Vol. 88, No. 1260, Entry 69706 (1985).

067.173 Phase transitions in dense matter and explosive neutron star phenomena.
B. Kämpfer.
Phys. Lett. B, Vol. 153B, No. 3, p. 121 – 123 (1985). Abstr. in Phys. Abstr., Vol. 88, No. 1260, Entry 69708 (1985).

067.174 Calculations of mass and moment of inertia for neutron stars.
T. Molnvik, E. Ostgaard.
Nucl. Phys. A, Vol. A437, No. 1, p. 239 – 252 (1985). Abstr. in Phys. Abstr., Vol. 88, No. 1260, Entry 74249 (1985).

067.175 Do black holes physically exist?
P. F. Gonzalez–Diaz.
Ann. Phys. (Leipzig), Vol. 41, No. 4 – 5, p. 353 – 356 (1984). Abstr. in Phys. Abstr., Vol. 88, No. 1260, Entry 74251 (1985).

067.176 Observational evidence for black holes.
J. B. Hutchings.
Am. Sci., Vol. 73, No. 1, p. 52 – 59 (1985). Abstr. in Phys. Abstr., Vol. 88, No. 1260, Entry 74252 (1985).

067.177 Thermonuclear runaway on neutron stars accreting interstellar gas as the origin of gamma ray bursts.
R. E. Lingenfelter, J. C. Higdon.
18th International Cosmic Ray Conference, Vol. 1, p. 62 – 65 (1983). – See Abstr. 012.096.

The calculated range of accretion rates of interstellar gas onto the polar caps of magnetic neutron stars is such that runaway thermonuclear burning of the accreted matter may occur, producing gamma ray bursts.

067.178 Tidal disruption of stars by a massive black hole as a possible source of the 0.511 MeV annihilation line.
M. Cassé, P. Durouchoux, J. P. Luminet.
18th International Cosmic Ray Conference, Vol. 1, p. 80 (1983). – See Abstr. 012.096.

067.179 Acceleration to ultra high energies in magnetospheres of young pulsars.
V. S. Berezinsky (*V. S. Berezinskij*).
18th International Cosmic Ray Conference, Vol. 2, p. 275 – 278 (1983). – See Abstr. 012.096.

The processes of acceleration and energy losses are considered for magnetospheres of young pulsars with vacuum gaps near the star's surface. Several models of this type are analyzed. For pulsar NP 0532 in Crab the maximum energy for Fe–nuclei is found to be around 10^{18}eV and the total number of these particles emitted during the lifetime of the pulsar is N $\gtrsim 10^{38}$. It can explain the observed cosmic ray intensity at E $\sim 10^{18}$eV.

067.180 Frictional acceleration of high–energy particles in pulsar vicinity.
E. G. Berezhko.
18th International Cosmic Ray Conference, Vol. 2, p. 279 – 282 (1983). – See Abstr. 012.096.

A process of frictional acceleration of high–energy charged particles caused by the presence of shearing flows in electron-positron plasma of a pulsar wind is considered. It is shown that this process can be the most natural explanation of high–energy part ($\gtrsim 10^{16}$Hz) of electromagnetic emission of Crab Nebula.

067.181 On possible evolution of pulsars.
Yu. I. Neshpor.
18th International Cosmic Ray Conference, Vol. 2, p. 402 – 405 (1983). – See Abstr. 012.096.

On the basis of published data analysis, the braking index n = 3.1 ± 0.1 was estimated for pulsars whose total luminosity exceeds 2×10^{32}erg s^{-1} and is less than 5×10^{35}erg s^{-1}.

067.182 Theoretical cyclotron and pulse profiles of X–ray pulsars.
P. Mészáros, W. Nagel.
Bull. Am. Astron. Soc., Vol. 17, No. 2, p. 555 (1985). Abstract. – See Abstr. 010.065.

067.183 A comparison of recent numerical calculations of stellar core collapse.
S. W. Bruenn.
Bull. Am. Astron. Soc., Vol. 17, No. 2, p. 566 (1985). Abstract. – See Abstr. 010.065.

067.184 A neutron star is born.
A. Burrows, J. M. Lattimer.
Bull. Am. Astron. Soc., Vol. 17, No. 2, p. 566 (1985). Abstract. – See Abstr. 010.065.

067.185 Do pulsars have elliptical beams?
J. J. Barnard.
Bull. Am. Astron. Soc., Vol. 17, No. 2, p. 567 (1985). Abstract. – See Abstr. 010.065.

067.186 Accretion disk model for quasar PG1211 + 143.
B. Czerny, M. Elvis, J. Bechtold.
Bull. Am. Astron. Soc., Vol. 17, No. 2, p. 575 – 576 (1985). Abstract. – See Abstr. 010.065.

067.187 Limb darkening and polarization of accretion disks.
K. Phillips, P. Mészáros.
Bull. Am. Astron. Soc., Vol. 17, No. 2, p. 576 – 577 (1985). Abstract. – See Abstr. 010.065.

067.188 Incompressible hydrodynamic and magnetohydrodynamic accretion flows in galactic nuclei.
H. A. Scott, R. V. E. Lovelace.
Bull. Am. Astron. Soc., Vol. 17, No. 2, p. 577 (1985). Abstract. – See Abstr. 010.065.

067.189 The evolution of the stellar and interstellar medium components of active galactic nuclei.
L. P. David, H. Cohn, R. H. Durisen.
Bull. Am. Astron. Soc., Vol. 17, No. 2, p. 577 – 578 (1985). Abstract. – See Abstr. 010.065.

067.190 Shock acceleration in active galactic nuclei and quasars.
D. Kazanas, D. Ellison.
Bull. Am. Astron. Soc., Vol. 17, No. 2, p. 587 (1985). Abstract. – See Abstr. 010.065.

067.191 Limits on gamma–ray burst modulation by neutron star rotation.
J. P. Norris, K. S. Wood, U. D. Desai.
Bull. Am. Astron. Soc., Vol. 17, No. 2, p. 604 (1985). Abstract. – See Abstr. 010.065.

067.192 Comptonization of relativistic electron beams in pulsar cascades.
J. K. Daugherty.
Bull. Am. Astron. Soc., Vol. 17, No. 2, p. 604 (1985). Abstract. – See Abstr. 010.065.

067.193 Consequences of neutrino production in the Cygnus X–3 system.
A. K. Harding, F. W. Stecker, J. J. Barnard.
Bull. Am. Astron. Soc., Vol. 17, No. 2, p. 604 (1985). Abstract. – See Abstr. 010.065.

067.194 Photon $-e^{\pm}$ interactions in neutron star magnetospheres.
N. S. Kardashev, I. G. Mitrofanov, I. D. Novikov.
Sov. Astron., Vol. 28, No. 6, p. 651 – 657 (1984). English translation of 38.067.135.

067.195 Durability of the accretion disk of millisecond pulsars.
F. C. Michel, A. J. Dessler.
Science, Vol. 228, No. 4702, p. 1015 – 1016 (1985).

Pulsars with pulsation periods in the millisecond range are thought to be neutron stars that have acquired an extraordinarily short spin period through the accretion of stellar material spiraling down onto the neutron star from a nearby companion. The end result of the accretion process is an object that looks much like a miniature (about 100 km), heavy version of Saturn: a central object (the neutron star) surrounded by a durable disk.

067.196 Metric fluctuations and the entropy of black holes.
J. W. York Jr.
Asymptotic behavior of mass and spacetime geometry, p. 111 – 122 (1984). – See Abstr. 012.102.

This paper reviews recent studies of the quantum zero–point fluctuations of a black hole metric and of the statistical entropy calculation for a black hole. In addition, it formulates an exact conservation law for the fluctuating metric that fully supports its interpretation as a non–classical equilibrium metric for a black hole.

067.197 Time–asymmetric initial data for N black holes.
A. D. Kulkarni.
Asymptotic behavior of mass and spacetime geometry, p. 144 – 153 (1984). – See Abstr. 012.102.

The time–asymmetric initial value problem for N–body systems consisting of N black holes is described. The spacelike initial hypersurface is taken to be the Einstein–Rosen manifold with N bridges. An explicit procedure using an infinite series is developed to construct solutions of the momentum constraints on the Einstein–Rosen manifold. A sufficient condition for the absolute convergence of a wide range of these solutions is presented.

067.198   Thermodynamical fluctuations of massive black holes.
D. Pavón, J. M. Rubí.
Gravitation, geometry and relativistic physics, p. 286 – 289 (1984). – See Abstr. 012.103.

067.199   Properties of bosonic black holes.
W. Thirring.
Nucl. Phys. B, Part. Phys., Vol. B252, No. 1 – 2, p. 357 – 361 (1985). – See Abstr. 012.105.

The gravitational collapse of bosonic matter leads to small black holes. Their properties and physical effects are discussed.

067.200   Can γ quanta really be captured by pulsar magnetic fields?
H. Herold, H. Ruder, G. Wunner.
Phys. Rev. Lett., Vol. 54, No. 13, p. 1452 – 1455 (1985).

Recently it was pointed out by Shabad and Usov that in the vicinity of pulsars high–energy photons can be captured by the strong magnetic field without creating pairs. The authors have reconsidered this process and find that the captured "photon" is transformed into bound positronium. This positronium can be ionized by the intense electric fields present in the polar cap region, or by thermal radiation from the neutron star surface, whereby the free electrons and positrons are regained that are urgently required for the models of pulsar radio emission.

067.201   Statistical mechanical origin of the entropy of a rotating, charged black hole.
W. H. Zurek, K. S. Thorne.
Phys. Rev. Lett., Vol. 54, No. 20, p. 2171 – 2175 (1985).

It is shown that the entropy of a rotating, charged black hole is (1) the logarithm of the number of quantum mechanically distinct ways that the hole could have been made, and (2) the logarithm of the number of configurations that the hole's "atmosphere", as measured by stationary observers, could assume in the presence of its background noise of acceleration radiation. In addition, a proof is given of the generalized second law of thermodynamics.

067.202   Is Bekenstein's conjecture true for charged black holes?
B. W. Schumacher.
Phys. Rev. Lett., Vol. 54, No. 24, p. 2643 – 2645 (1985).

It is known that the thermal radiation from an uncharged, nonrotating black hole into a vacuum carries an entropy about 4/3 times the entropy lost by the hole. For a black hole placed in a radiation bath with a temperature infinitesimally close to that of the hole, the entropy of the radiation is equal to the hole's entropy loss. This previously known result is extended to the case of a charged black hole, which gives further support to the statistical interpretation of black hole entropy.

067.203   Theoretical studies of the pulsar magnetosphere.
R. F. Martin Jr.
Diss. Abstr. Int., Sect. B, Vol. 45, No. 1, p. 230 (1984). Thesis, University of Illinois, 167 pp. (1983). Order No. DA8409994.

067.204   The vertical structure and stability of alpha model accretion disks.
J. K. Cannizzo.
Diss. Abstr. Int., Sect. B, Vol. 45, No. 7, p. 2199 (1985). Thesis, University of Texas, 191 pp. (1984). Order No. DA8421675.

067.205   Transport properties of the outer crust of magnetic neutron stars.
G. J. Hartke.
Diss. Abstr. Int., Sect. B, Vol. 45, No. 8, p. 2581 (1985). Thesis, University of Massachusetts, 185 pp. (1984). Order No. DA8424857.

067.206   A numerical study of nonspherical black hole accretion.
J. F. Hawley.
Diss. Abstr. Int., Sect. B, Vol. 45, No. 11, p. 3533 (1985). Thesis, University of Illinois, 277 pp. (1984). Order No. DA8502167.

067.207   Thermal and magnetic properties of neutron stars.
L. E. Hernquist.
Diss. Abstr. Int., Sect. B, Vol. 45, No. 11, p. 3533 – 3534 (1985). Thesis, California Institute of Technology, 312 pp. (1985). Order No. DA8501333.

067.208   Two temperature accretion disks in active galactic nuclei.
R. Matsumoto, S. Kato, J. Fukue.
Theoretical aspects on structure, activity, and evolution of galaxies: III, p. 102 – 106 (1985). – See Abstr. 012.110.

Accretion disk models around supermassive black holes are studied. Steady solutions satisfying inner– and outer boundary conditions are obtained by a relaxation method in the case of two temperature disks. The effect of viscosity on the structure near the inner edge of disks is clarified.

067.209   Vibrational damping by quark matter inside neutron stars.
Q.–d. Wang, T. Lu.
Chin. Astron. Astrophys., Vol. 9, No. 2, p. 159 – 163 (1985). English translation of Acta Astrophys. Sin., Vol. 5, No. 1, p. 59 – 66 (1985).

Free quark state may exist in the central region of massive neutron stars. The authors discuss the damping of the vibration of such neutron stars through the quark weak interaction. The damping time scale may be as short as tens of milliseconds. Damping with such short time constant may possibly be reflected in some γ–ray burst phenomena.

067.210   Energetics of the Kerr–Newman black hole by the Penrose process.
M. Bhat, S. Dhurandhar, N. Dadhich.
J. Astrophys. Astron., Vol. 6, No. 2, p. 85 – 100 (1985).

The authors have studied in detail the energetics of the Kerr–Newman black hole by the Penrose process using charged particles. It turns out that the presence of electromagnetic fields offers very favourable conditions for energy extraction by allowing for a region with enlarged negative energy states much beyond $r = 2M$, and higher negative values for energy. The efficiency of the process is examined. A specific example of over 100 per cent efficiency is given.

067.211   The estimation of some important parameters of pulsars and the comparison between the RS model and observations.
X.–j. Wu, G.–j. Qiao, X.–y. Xia.
Acta Astron. Sin., Vol. 26, No. 1, p. 69 – 75 (1985). In Chinese.

Two problems are discussed in this paper: the first, how to estimate the beamwidth of the emission cone, and the second, where does the line of sight sweep over the emission cone. According to the geometrical relationship of the polar cap model, the authors give formulae of these parameters from observational data of polarization and apparent beamwidth. The computed results have been compared with the RS model.

067.212   On the possibility of efficient electron–positron pair production in the vicinity of pulsars and accreting black holes.
B. E. Shtern.
Astron. Zh., Tom 62, Vyp. 3, p. 529 – 541 (1985). In Russian. English translation in Sov. Astron., Vol. 29, No. 3.

A Monte Carlo simulation of a synchrotron reactor has been made for the case when an intense e^+e^- pair production due to $\gamma\gamma$ interactions (an electron–positron reactor) takes place. A time–dependent solution has been calculated under the assumption of a homogeneous medium with fixed initial conditions. It turned out that for a rather wide variety of initial conditions 10^{-1} of the initial energy, injected into the reactor in form of hard particles, is converted into the e^+e^- rest mass, and a characteristic photon spectrum is established with a sharp high–energy cut–off near 1 MeV. In the framework of the model of an e^+e^- reactor the origin of the annihilation gamma–ray line, such as the one observed towards the galactic centre, seems quite natural. Thereby, the positron source may be an accreting black hole as well as a young pulsar ejecting not less than 10^{39} erg/s in form of a hard photon flow from the magnetic poles.

067.213 The origin of neutron stars.
R. A. Chevalier.
Birth and evolution of neutron stars: issues raised by millisecond pulsars, p. 73 – 85 (1984). – See Abstr. 012.115.

Neutron stars are the likely outcome of the electron capture induced collapse of the O–Ne–Mg cores of $8 - 10\ M_\odot$ stars and the photodisintegration induced collapse of the Fe–Ni cores of more massive stars. Stars with mass $> 100\ M_\odot$ do not leave neutron star remnants. Carbon ignition in the core of a $4 - 8\ M_\odot$ star is likely to lead to complete disruption, but collapse may be a possibility. The cores of stars with initial mass $\lesssim 10\ M_\odot$ can become white dwarfs in binary systems. The O–Ne–Mg white dwarfs probably form neutron stars. Observations which bear on these theoretical scenarios are reviewed.

067.214 Models for the formation of binary and millisecond radio pulsars.
E. P. J. van den Heuvel.
Birth and evolution of neutron stars: issues raised by millisecond pulsars, p. 86 – 106 (1984). – See Abstr. 012.115.

The two observed classes of binary radio pulsars (very close and very wide systems, respectively) are expected to have been formed by the later evolution of binaries consisting of a neutron star and a normal companion star, in which the companion was (considerably) more massive than the neutron star, or less massive than the neutron star, respectively. In the first case the companion of the neuton star in the final system will be a fairly massive white dwarf, in a circular orbit, or a neutron star in an eccentric orbit. In the second case the final companion to the neutron star will be a low–mass ($\sim 0.3\ M_\odot$) helium white dwarf in a wide and nearly circular orbit.

067.215 Neutron star seismology: understanding the oscillation modes.
P. N. McDermott, C. J. Hansen, R. Buland, H. M. Van Horn.
Birth and evolution of neutron stars: issues raised by millisecond pulsars, p. 173 – 181 (1984). – See Abstr. 012.115.

Spherical, non–rotating, non–magnetic neutron stars can sustain non–radial oscillations over periods ranging from tenths of milliseconds to tens of seconds. The authors are seeking an understanding of the nature of these modes in order to provide a foundation for investigations of the more complex situation that prevails in pulsars. The models that the authors have studied consist of three components: a fluid core, a solid crust, and at the surface a fluid ocean.

067.216 Gravitational radiation from a solid crust neutron star.
M. A. Alpar, D. Pines.
Birth and evolution of neutron stars: issues raised by millisecond pulsars, p. 182 – 190 (1984). – See Abstr. 012.115.

A rapidly rotating neutron star with a solid crust will have a rotationally induced oblateness that is constrained by the rigidity of the solid. It is shown that the effective triaxiality and the gravitational radiation output are small, in agreement with the very small P of PSR 1937 + 21.

067.217 Thermal origin of neutron star magnetic fields.
J. H. Applegate, R. D. Blandford, L. Hernquist.
Birth and evolution of neutron stars: issues raised by millisecond pulsars, p. 205 – 212 (1984). – See Abstr. 012.115.

Mechanisms by which magnetic fields can be generated on neutron stars are discussed. Applications to ordinary radio pulsars, millisecond pulsars, and rapidly spinning pulsars in plerionic supernova remnants are discussed.

067.218 The stability of magnetic fields of isolated and binary neutron stars.
G. Chanmugam.
Birth and evolution of neutron stars: issues raised by millisecond pulsars, p. 213 – 219 (1984). – See Abstr. 012.115.

It is suggested that convective instabilities in cooling neutron stars may lead to magnetic field decay. Since rotation may have a stabilizing influence, the rotational history of the star is more important, than the age of the star, in determining whether its magnetic field decays or not.

067.219 On the possibility of plasma oscillations of a pulsar's magnetosphere in the region of closed field lines.
M. A. Mamedov.
Astrofizika, Tom 22, Vyp. 3, p. 585 – 597 (1985). In Russian. English translation in Astrophysics, Vol. 22, No. 3.

The dispersion equation for the spectrum of plasma oscillations in the region of closed force lines of the magnetosphere of a pulsar is obtained in a uniformly rotating coordinate system with a neutron star. The dispersion relations for the oscillations propagating along and transversal to the closed force lines of the magnetosphere of the pulsar are analysed. It has been shown that in the pulsar plasma in the region of closed force lines high–frequency cyclotron radio waves can radiate.

General relativity.
See Abstr. 003.009.

Accretion power in astrophysics.
See Abstr. 003.012.

Beyond the black hole.
See Abstr. 003.065.

Physics of dense matter.
See Abstr. 003.196.

Les trous noirs.
See Abstr. 011.029.

Perspectives on space and astrophysical plasma physics.
See Abstr. 013.001.

Graphic displays of gravitational initial data.
See Abstr. 021.003.

Virial oscillations of celestial bodies. IV. The Lyapunov stability of motion.
See Abstr. 042.036.

Space Telescope observations of globular clusters.
See Abstr. 051.085.

Neutrons in science and technology.
See Abstr. 061.003.

Helium detonation in pancake stars.
See Abstr. 061.027.

Thermal cyclotron radiation in astrophysics.
See Abstr. 061.035.

Vacuum polarization by a magnetic field and its astrophysical manifestations.
See Abstr. 061.036.

Upper bound on entropy.
See Abstr. 061.050.

The magnetic monopole.
See Abstr. 061.055.

High energy neutrino astrophysics ($10^2 - 10^7$ GeV) with emphasis on Cyg X–3.
See Abstr. 061.061.

Monopolar contamination of normal sequence stars.
See Abstr. 061.094.

Some possible tests of the inapplicability of Pauli's exclusion principle.
See Abstr. 061.095.

Neutrino energy spectra from relativistic plasma.
See Abstr. 061.131.

Pair production, Comptonization and dynamics in astrophysical plasmas.
See Abstr. 062.002.

Formation, equilibrium and stability of jets.
See Abstr. 062.008.

Equilibria in stronlgy magnetized pair plasmas.
See Abstr. 062.041.

Surface emissivity of an optically thick, magnetized plasma and luminosities of accreting neutron stars.
See Abstr. 062.042.

Radiation from charges driven by large–amplitude longitudinal plasma waves.
See Abstr. 062.068.

Power–law X–ray and gamma–ray emission from relativistic thermal plasmas.
See Abstr. 062.069.

On the spectrum of longitudinal waves in the hot degenerate plasma.
See Abstr. 062.088.

Nonlinear waves in the magnetosphere of a pulsar.
See Abstr. 062.089.

Electron–positron pairs in a mildly relativistic plasma in active galactic nuclei.
See Abstr. 062.096.

Hydrodynamics of astrophysical jets.
See Abstr. 062.097.

An analytic model for hydromagnetospheres.
See Abstr. 062.099.

Yang–Mills magnetohydrodynamics: nonrelativistic theory.
See Abstr. 062.168.

Solitons in weakly nonlinear electron–positron plasmas and pulsar microstructures.
See Abstr. 062.173.

Dissipation by thermal forces in quantum plasmas.
See Abstr. 062.175.

Theory of optical flashes.
See Abstr. 063.004.

Non–thermal synchrotron radiation in a strong magnetic field.
See Abstr. 063.005.

Cyclotron line profiles and photon diffusion in a hot plasma.
See Abstr. 063.021.

Accretion disk electrodynamics.
See Abstr. 064.002.

Stellar winds driven by super–Eddington luminosities.
See Abstr. 064.035.

Two–photon cyclotron emission in accretion columns.
See Abstr. 064.046.

Mechanics of the affine star model.
See Abstr. 065.002.

Tidal squeezing of stars by Schwarzschild black holes.
See Abstr. 065.003.

Evolution of binary systems into transient sources.
See Abstr. 065.010.

The evolution of binaries of moderate primordial mass ($\lesssim 10\ M_\odot$) and the formation of transient and explosive objects due to accumulation instabilities in such binaries.
See Abstr. 065.011.

Pion stars.
See Abstr. 065.035.

Equilibrium models of differentially rotating, completely catalyzed, zero–temperature configurations with central densities intermediate to white dwarf and neutron star densities.
See Abstr. 065.061.

Gravitational collapse revisited.
See Abstr. 065.081.

Supernova theory.
See Abstr. 065.082.

Neutrinos from collapsing stars.
See Abstr. 065.086.

Spherically symmetric radiation in gravitational collapse.
See Abstr. 065.097.

On the secular stability of rotating stars.
See Abstr. 065.103.

Gravitational field energy density for spheres and black holes.
See Abstr. 066.002.

Gravitational radiation of rapidly rotating drop of homogeneous magnetized gravitating liquid near bifurcation point.
See Abstr. 066.028.

General relativistic, partially degenerate semirelativistic isothermal spheres of arbitrary temperature.
See Abstr. 066.045.

The graviton luminosity of the Sun and other stars.
See Abstr. 066.046.

A new class of ideal clocks.
See Abstr. 066.057.

The limit of mass for gravitational collapse and nuclear forces at short distance.
See Abstr. 066.061.

On a new interior Schwarzschild solution.
See Abstr. 066.066.

Scattering for the wave equation on the Schwarzschild metric.
See Abstr. 066.071.

Compact objects in bimetric general relativity.
See Abstr. 066.082.

Thermal gravitational radiation from stellar objects and its possible detection.
See Abstr. 066.104.

On the binary nature of cosmic γ–ray burst sources.
See Abstr. 143.031.

A model of the object Geminga: degenerate dwarf rotating around a black hole.
See Abstr. 143.037.

A model of object Geminga as a degenerate white dwarf orbiting around a black hole.
See Abstr. 143.055.

Evidence for 500 TeV gamma–ray emission from Hercules X–1.
See Abstr. 143.060.

Corequake and shock heating model of the 5 March 1979 γ–ray burst.
See Abstr. 143.067.

On the role of neutrons in the formation of radiation in γ–ray burst sources.
See Abstr. 143.068.

Gamma ray sources – black holes?
See Abstr. 143.082.

Much ado about Geminga.
See Abstr. 143.105.

Consequences of positron fluxes from densely shrouded sources.
See Abstr. 144.096.

Constraints on accretion powered positron sources.
See Abstr. 144.097.

Cosmic ray sources and propagation.
See Abstr. 144.364.

Black holes and the shapes of galaxies.
See Abstr. 151.072.

Triaxial galaxies containing massive black holes or central density cusps.
See Abstr. 151.074.

Can a moderately massive black hole reverse core collapse in globular clusters?
See Abstr. 151.090.

Direct Fokker–Planck calculations.
See Abstr. 151.102.

Close encounters.
See Abstr. 151.103.

Monte Carlo simulations of the $2+1$ dimensional Fokker–Planck equation: spherical star clusters containing massive, central black holes.
See Abstr. 151.125.

Can a moderately massive black hole reverse core collapse?
See Abstr. 151.126.

Dynamics of open star clusters.
See Abstr. 151.129.

Black hole remnants in globular clusters.
See Abstr. 154.067.

Ultra–high energy cosmic ray production by current disruption in active galactic nuclei.
See Abstr. 158.014.

Small scale structure and the spinar model for active galactic nuclei.
See Abstr. 158.121.

Galactic mergers, starburst galaxies, quasar activity and massive binary black holes.
See Abstr. 158.244.

Gamma ray observations of active galactic nuclei as a test of black hole accretion models.
See Abstr. 158.282.

Quasar and related problems – some recent developments.
See Abstr. 159.118.

Can primordial black holes solve the overproduction problem of monopoles?
See Abstr. 161.266.

Sun

071 Photosphere, Spectrum

071.001 Dependence of the properties of magnetic fluxtubes on area factor or amount of flux.
J. O. Stenflo, J. W. Harvey.
Sol. Phys., Vol. 95, No. 1, p. 99 – 118 (1985).

Stokes I and V line profiles with high signal–to–noise ratio of the 1 Fe I $\lambda\lambda5247.06$ and 5250.22 Å lines have been recorded in a number of regions with different amount of magnetic flux near disc center, from 'non–magnetic' regions to strong plages. The objective has been to study how the intrinsic fluxtube properties may depend on the amount of flux concentration, i.e., on the magnetic area factor. Indirectly, the area factor should be related to the average fluxtube diameter.

071.002 Apparature function and determination of the blanketing coefficient in the solar spectrum.
P. P. Kozak.
Probl. Kosm. Fiz., Vyp. 19, p. 45 – 47 (1984). In Russian. – See Abstr. 003.002.

071.003 Dynamics of the 5–min oscillation spectrum in the quiet photosphere.
S. A. Druzhinin, N. I. Kobanov, M. V. Nikonova.
Soln. Dannye, Byull., 1984, No. 10, p. 50 – 54 (1985). In Russian.

Time variations of the frequency spectrum of 5–minute oscillations in the quiet photosphere are analysed. Observational data obtained with respect to wavenumber K were used for the analysis. For $K_h = 24''$, $20''$, and $18''$ the frequency variations of the principal maximum were found to have a quasi–periodic character. The most typical time scale of these variations is about 60 min. Quasi–periods of 20 min, 30 to 40 min, and of 90 min are revealed as well. Some plausible suggestions as to how this effect is caused are made.

071.004 Center–to–limb darkening in the Fraunhofer line spectrum (6000 – 6600 Å).
V. I. Troyan.
Soln. Dannye, Byull., 1984, No. 10, p. 79 – 84 (1985). In Russian.

On the basis of center–to–limb observations of Fraunhofer lines in the region 6000 – 6600 Å the coefficient of the solar disc darkening in the absorption line profile is determined.

071.005 Asymmetry and shift of three Fe I photospheric lines in solar active regions.
F. Cavallini, G. Ceppatelli, A. Righini.
Astron. Astrophys., Vol. 143, No. 1, p. 116 – 121 (1985).

The asymmetry and the red–shift of three Fe I solar lines around 6300 Å have been measured in active regions with a Fabry–Perot spectrometer. At the disk center, the active region line bisectors are always displaced to the red and their shape is heavily modified. This red–shift is such to nearly compensate the convective blue–shift for strong magnetic fields, suggesting that the inhibition of convection in magnetic regions might be responsible for the red–shift instead of the often invoked "downdraft". Moving from the center to the limb, at about 0.56 solar radii, the red–shift changes to a slight blue–shift. This effect might explain why the center to limb red–shift of the line wavelengths is found to be smaller along the solar equator than along the meridian on large ensemble of data.

071.006 Population mechanisms for the He$^+$ $n = 3$ levels determined from measurements of the solar 1640 Å emission.
J. F. Seely, U. Feldman.
Mon. Not. R. Astron. Soc., Vol. 213, No. 2, p. 417 – 434 (1985).

The intensities of the $3s$–$2p$, $3p$–$2s$, and $3d$–$2p$ components of the He II 1640 Å line have been measured from solar data taken by the NRL slit spectrograph on Skylab. The relative and absolute population densities of the $3s$, $3p$, and $3d$ levels of He$^+$ are determined near the white–light limb in various solar regions (coronal hole, quiet Sun, active regions, and a prominence) and in a flare region on the solar disc. The population density measurements are interpreted using a rate equation model for the He$^+$ ion. It is found that self–absorption of the $3p$–$1s$ radiation contributes to the population of the $3p$ level in all the solar regions that are studied and is particularly important in the active regions, the prominence and the flare. Recombination contributes to the population of the $3d$ level in the active regions near the limb and in the prominence. In all the solar regions that are studied, the $3p$–$2s$ component of the 1640 Å line is more intense than the $3d$–$2p$ and $3s$–$2p$ components.

071.007 Radiative lifetimes for Nb II and the problem of the solar abundance of niobium.
P. Hannaford, R. M. Lowe, E. Biémont, N. Grevesse.
Astron. Astrophys., Vol. 143, No. 2, p. 447 – 450 (1985).

The radiative lifetimes of 27 levels in Nb II have been measured by the method of laser–induced fluorescence from sputtered metal vapour. These results have been combined with recent solar data to determine the solar abundance of niobium. The result, $A_{Nb} = 1.42 \pm 0.06$, in good agreement with meteorites, is a factor of 5 lower than that recently derived from Nb I lines by Kwiatkowski et al. (1982). Some possible explanations of the discrepancy are considered.

071.008 Solar line blocking for disk–center and disk–averaged radiation from 3300 to 6860 Å.
H. Neckel, D. Labs.
Sol. Phys., Vol. 95, No. 2, p. 229 – 238 (1985).

The line blocking is tabulated for 10 Å ($\lambda < 6300$ Å) or 20 Å ($\lambda > 6300$ Å) wide intervals. It follows from the spectral averages and the local continuum derived by Neckel and Labs from high–resolution Fourier transform spectra, which had been obtained by J. Brault at Kitt Peak. The internal accuracy (the scatter) is in the order of 0.1 %. Significant systematic errors arising from local distortions of the adopted continuum level can be excluded. Larger errors are to be expected only near the Balmer limit where the localization of the "continuum" is very ambiguous.

071.009 Water vapor and Fe 5250.2.
W. Livingston, L. Wallace.
Sol. Phys., Vol. 95, No. 2, p. 251 – 252 (1985).

The commonly used magnetograph line of Fe 5250.2 Å is found to be weakly blended by telluric water vapor. This circumstance could bias solar rotation measurements.

071.010 The blanketing effect from an analysis of the absorption profiles of the solar spectrum (6149 – 6189 Å). I. Calculation from residual intensities.
V. I. Troyan, V. K. Yarmolenko.
Soln. Dannye, Byull., 1984, No. 12, p. 75 – 79 (1985). In Russian.

The integrated flux for the sun as a star is calculated for five spectrum models in the field 6149 – 6189 Å. The blanketing effect is estimated. The error from the actually observed flow is estimated, and it does not exceed 1%.

071.011 Macro–microturbulence in the solar photosphere.
V. A. Sheminova.
Kinematika Fiz. Nebesn. Tel, Tom 1, No. 2, p. 50 – 52 (1985). In Russian.

Comparing the equivalent widths and the central depths of about 200 Fe I lines with calculated values the distribution of the large–scale motion velocities in the photosphere has been obtained. The motions are anisotropic. At the heights from 200 to 500 km the radial component of V_{mac} decreases from 2.0 to 1.3 km s^{-1}, and the tangential one decreases from 2.3 to 1.0 km s^{-1}.

071.012 Formation of weak Fraunhofer lines of the solar spectrum.
B. T. Babii (*B. T. Babij*), M. M. Koval'chuk.
Sov. Astron., Vol. 28, No. 4, p. 451 – 454 (1984). English translation of 38.071.005.

071.013 Observations of a solar latitude–dependent limb brightness variation.
J. R. Kuhn, K. G. Libbrecht, R. H. Dicke.
Astrophys. J., Vol. 290, No. 2, p. 758 – 764 (1985).

A small latitude–dependent photospheric excess brightness variation has been observed from 131 days of data obtained with the Princeton Solar Distortion Telescope. Using an analytic model to separate the influence of faculae from the brightness signal, the authors obtain, for the summer of 1983, a temperature difference of 0.6 ± 0.1K between the poles and the regions at $\pm 53°$ solar latitude, with the polar regions being hotter.

071.014 The photospheric temperature structure of magnetic fluxtubes.
S. K. Solanki.
ESA Spec. Publ., ESA SP–220, p. 63 – 66 (1984). – See Abstr. 012.045.

The temperature stratifications of plage and network fluxtubes are determined by comparing a large number of observed Fe I and Fe II lines with LTE model calculations. The data are also compared with profiles published in the literature. It is shown that in order to reproduce the data correctly, a sharp dip in the fluxtube temperature must occur in the photosphere, where the temperature of fluxtube and surroundings become similar (at equal optical depth). The effects of different pressure stratifications on the depths of the iron lines are briefly discussed and it is shown that the pressure stratification giving the best agreement with the data corresponds to a rapid expansion of the fluxtube near the height at which the temperature dip occurs.

071.015 The response of the line K I 7699 to the solar oscillations.
C. Marmolino, G. Roberti, M. Vazquez, G. Severino, H. Wöhl.
ESA Spec. Publ., ESA SP–220, 191 – 194 (1984). – See Abstr. 012.045.

The authors studied the time behaviour of the K I 7699 line profile in presence of acoustic–gravity waves with periods of 300 s, 180 s and 30 s.

071.016 A model for the dynamics of the deep photosphere.
A. Nesis.
ESA Spec. Publ., ESA SP–220, p. 203 – 204 (1984). – See Abstr. 012.045.

The small–scale ($< 3.''5$) vertical and horizontal velocities of different layers in the atmosphere were measured using spectrograms with absorption lines of different strength. The vertical and horizontal velocities of convective origin decrease up to a height of 150 km and 200 km respectively. The results infer the existence of a secondary velocity field at higher photospheric levels induced by convective motions.

071.017 Variability of the quiet photospheric network.
R. Muller, T. Roudier.
ESA Spec. Publ., ESA SP–220, p. 239 – 240 (1984). – See Abstr. 012.045.

High resolution photographs of the photospheric network taken in the Ca II K 3933 Å line and λ4308 Å are analysed in order to study the variation, in latitude and over the solar cycle, of its density.

071.018 On the difference of effective depths in formation of moderate and faint Fraunhofer lines.
Eh. A. Gurtovenko, V. A. Sheminova.
Soln. Dannye, Byull., 1985, No. 1, p. 70 – 74 (1985). In Russian.

It is shown that the tables by the authors (1978) of effective depths of faint Fraunhofer lines are correct for lines having central depths $d_c < 0.1$. A conclusion is made on the expediency of calculation of the depths of formation of moderate and strong Fraunhofer lines in the solar spectrum.

071.019 The blanketing effect from an analysis of absorption line profiles of the solar spectrum (6149 – 6189 Å). II. Correction for the effect using equivalent widths.
V. I. Troyan, V. K. Yarmolenko.
Soln. Dannye, Byull., 1985, No. 1, p. 74 – 78 (1985). In Russian.

The integral flux of the sun as a star is evaluated and the blanketing effect is estimated taking into account equivalent widths. The results differ from the value of the observed flux by less than 1%.

071.020 On the equivalent widths determination of some lines arising from the a^{5}D term of neutral iron in the solar spectrum.
A. G. Gasanalizade.
Soln. Dannye, Byull., 1985, No. 1, p. 78 – 83 (1985). In Russian.

Using the absolute oscillator strengths for Fe I lines by the Oxford scale and the absolute curve of growth of iron for two atmospheric models the author determines the equivalent widths of 21 solar lines of Fe I arising from the term a^{5}D.

071.021 On the importance of some parameters of the solar spectrum synthesis of neutral potassium under non–LTE conditions. III. Temperature.
N. G. Shchukina.
Soln. Dannye, Byull., 1985, No. 2, p. 84 – 87 (1985). In Russian.

The temperature effects on the profile of the resonance line λ7699 are discussed. The increase of temperature by 300K in the temperature minimum region is shown to change the central residual intensity of this line by 12%.

071.022 The solar O I λ7773 triplet. II. Analysis using line inversion techniques.
A. Kavetsky, B. J. O'Mara.
Sol. Phys., Vol. 96, No. 1, p. 1 – 10 (1985).

Profiles of the O I λ7773 triplet obtained at a spatial resolution $\sim 0.5''$ are analyzed using spectral line inversion techniques. Inferences are made about departures from LTE, convective velocity fields, and solar temperature fluctuations.

071.023 Technique of correction for the H$_2$O 10830.34 Å blend during the time of observations of the neutral helium line He I 10830.38 Å.
I. I. Sokolova, Eh. V. Kandrashov.
Astron. Tsirk., No. 1349, p. 5 – 7 (1984). In Russian.

071.024 Europium abundance in the solar photosphere.
A. V. Perekhod.
Kinematika Fiz. Nebesn. Tel, Tom 1, No. 3, p. 75 – 77 (1985). In Russian.

The relative abundance of europium has been determined in the undisturbed solar photosphere using five line profiles of Eu II in the spectral region $\lambda\lambda$600 – 740 nm, both from the equivalent widths (four lines) and by the synthetic spectrum method (line 617.304 nm). The Eu abundance in the sun is 0.71 ± 0.08 (in the scale $\lg A_H = 12$).

071.025 Determination of vanadium abundance in the solar photosphere accounting for the effect of hyperfine structure.
B. T. Babij, R. E. Rikalyuk.
Kinematika Fiz. Nebesn. Tel, Tom 1, No. 3, p. 78 – 82 (1985). In Russian.

From the analysis of 25 profiles of neutral vanadium lines without influence of hyperfine structure with 3 solar atmospheric models the abundance of vanadium is obtained within 4.08 to 3.91 dex as function of the adopted solar atmospheric model.

071.026 The shift of the solar 584–Å He I line, and evidence for interstellar wind from the interplanetary EUV background.
M. S. Burgin.
Sov. Astron., Vol. 28, No. 5, p. 510 – 514 (1984). English translation of 38.071.013.

071.027 Solar spectrum analysis through the curve of growth.
V. Lojen.
Publ. Astron. Opservatorije Beogr., No. 33, p. 45 – 49 (1985). – See Abstr. 012.061.

An example is presented of studying the solar spectrum by means of a curve of growth. Using spectrometric methods, profiles of five, and the equivalent widths of fourteen absorption lines of the solar spectrum were measured. An empirical curve of growth for Fe I has been established.

071.028 Differential photometry of very low contrast solar structures in white light.
E. J. Seykora.
Bull. Am. Astron. Soc., Vol. 16, No. 4, p. 991 – 992 (1984). Abstract. – See Abstr. 010.062.

071.029 Center to limb behavior of some strong Fraunhofer lines.
J. C. LoPresto, A. K. Pierce.
Bull. Am. Astron. Soc., Vol. 16, No. 4, p. 1001 (1984). Abstract. – See Abstr. 010.062.

071.030 Magnetic fields, downdrafts, and granulation in the solar photosphere.
A. M. Title, T. D. Tarbell.
Bull. Am. Astron. Soc., Vol. 16, No. 4, p. 1001 (1984). Abstract. – See Abstr. 010.062.

071.031 Doppler width of CN lines in the sun.
L. M. Punetha, G. C. Joshi.
Bull. Astron. Soc. India, Vol. 12, No. 4, p. 393 – 397 (1984).

The Doppler widths of 0–1, 1–2 and 2–3 vibrational bands of the violet system of CN molecules in the solar atmosphere increase linearly with equivalent width up to $\cong 20$ mÅ.

071.032 Morphological evolution of an emerging flux region.
J. J. Brants, J. C. M. Steenbeek.
Sol. Phys., Vol. 96, No. 2, p. 229 – 252 (1985).

The authors describe the morphological evolution of photospheric features in an emerging flux region (EFR), on the basis of high–resolution photographs taken with the Vacuum Tower Telescope at Sacramento Peak Observatory. Individual alignments of darkened intergranular lanes have a lifetime of only about 10 min; they may represent the tops of emerging flux loops. Roundish darkened patches within the intergranular lanes (protopores) may precede the birth of a pore, or may disappear again within a few hours. The birth of one pore coincides with the area of a conspicuous downflow observed in the spectrograms. The majority of the pores in the EFR grow in area and darken; their growth times vary between 1 and 6 hr. Various modes of growth are observed.

071.033 Identification of vibration–rotation lines of CH in the solar infrared spectrum.
A. J. Sauval, N. Grevesse.
Astron. Express, Vol. 1, Nos. 4 – 6, p. 153 – 158 (1985).

Transitions of the fundamental band of CH have recently been measured with very high accuracy by Lubic and Amano (1984). The lines are shown to be present in the solar infrared spectrum ($2580 - 2940$ cm^{-1}). The role these lines could play in solar and stellar spectroscopy is discussed.

071.034 Calculation of the profile of the inner wings of the K Ca II line under the condition of total frequency redistribution.
T. M. Minasyants, G. S. Minasyants.
Tr. Astrofiz. Inst. Alma–Ata, Tom 41, p. 114 – 119 (1983). In Russian.

071.035 Observations of long lived very low contrast solar structures in white light.
E. J. Seykora.
Bull. Am. Astron. Soc., Vol. 17, No. 2, p. 610 (1985). Abstract. – See Abstr. 010.065.

071.036 Filigree–granulation movie.
R. B. Dunn.
Bull. Am. Astron. Soc., Vol. 17, No. 2, p. 633 (1985). Abstract. – See Abstr. 010.067.

071.037 Full disk continuum photometry with the NSO/Tucson Vacuum Telescope.
D. G. Luttermoser, H. P. Jones.
Bull. Am. Astron. Soc., Vol. 17, No. 2, p. 639 – 640 (1985). Abstract. – See Abstr. 010.067.

071.038 Variation of the calcium K line profile over solar cycle 21.
O. R. White, W. C. Livingston.
Bull. Am. Astron. Soc., Vol. 17, No. 2, p. 640 (1985). Abstract. – See Abstr. 010.067.

071.039 Progress on solar opacity.
R. L. Kurucz.
Bull. Am. Astron. Soc., Vol. 17, No. 2, p. 640 – 641 (1985). Abstract. – See Abstr. 010.067.

071.040 Effects of magnetic fields on the asymmetry of photospheric line profiles.
S. L. Keil.
Bull. Am. Astron. Soc., Vol. 17, No. 2, p. 642 (1985). Abstract. – See Abstr. 010.067.

071.041 3–D behavior of buoyant magnetic flux tubes in granules and supergranules.
G. W. Simon, H. U. Schmidt, N. O. Weiss.
Bull. Am. Astron. Soc., Vol. 17, No. 2, p. 642 (1985). Abstract. – See Abstr. 010.067.

071.042 Activity cycle variations of photospheric lines in the sun's irradiance spectrum.
W. Livingston, L. Wallace.
Bull. Am. Astron. Soc., Vol. 17, No. 2, p. 644 (1985). Abstract. – See Abstr. 010.067.

071.043 Investigation of photospheric limb–darkening variation between 1980 and 1985.
L. D. Petro, P. V. Foukal, W. A. Rosen, A. K. Pierce, R. L. Kurucz.
Bull. Am. Astron. Soc., Vol. 17, No. 2, p. 644 (1985). Abstract. – See Abstr. 010.067.

071.044 Variations of the asymmetry of disk–integrated solar line profiles.
D. H. Bruning, B. LaBonte.
Sol. Phys., Vol. 97, No. 1, p. 1 – 7 (1985).

Mean line bisector positions were found for the neutral iron line at $\lambda5250.2$ using disk–integrated sunlight. After correction for the apparent time variation of the instrumental profile, it was found that the mean bisector position was constant during the period from May 1982 to February 1983. The correlation between the total magnetic flux as measured at Mount Wilson and the line asymmetry results of Livingston is not high. In particular, the magnetic flux dropped in 1982, suggesting a large line asymmetry that was not observed.

071.045   Drawings of the solar photosphere.
L. Křivský.
Říše hvězd, Vol. 66, No. 2, p. 34 – 35 (1985). In Czech.

071.046   On grey pores in the solar photosphere.
V. A. Krat, L. D. Parfinenko, M. N. Stoyanova.
Soln. Dannye, Byull., 1985, No. 3, p. 60 – 65 (1985). In Russian.
There are small–scale dark elements in the solar photosphere – grey pores that have a lower contrast (10 – 30%) than the proper pores. Their sudden disappearance and appearance can be explained as a plunge of their magnetic fields in the vertical direction followed by their rise into the outer region of the radiative equilibrium.

071.047   Central intensity, halfwidth and equivalent width of selected Fraunhofer lines of the solar spectrum for 1979 –
the last year of the 11–year series of observations.
E. K. Kokhan.
Soln. Dannye, Byull., 1985, No. 3, p. 83 – 88 (1985). In Russian.
Parameters of the central intensity, halfwidth and equivalent width of solar absorption lines for 1979 are given. The observations were made in nondisturbed solar regions. Information on the material obtained for the 11–year observing period is given.

071.048   Correction for the H_2O 10830.34 Å blended line and instrumental distortions in observations in the He I
10830.38 Å line in the spectrum of the sun as a star.
I. I. Sokolova.
Soln. Dannye, Byull., 1985, No. 4, p. 74 – 82 (1985). In Russian.
Some quantitative results of studies of the 10830.34 Å blended line distorting the He line 10830.38 Å is given using observations of the sun as a star in 1983 – 1984.

071.049   Pressure broadening and solar limb effect.
I. Vince, M. S. Dimitrijević, V. Kršljanin.
Progress in stellar spectral line formation theory, p. 373 – 380 (1985). – See Abstr. 012.108.
The pressure broadening contribution to the solar limb effect is discussed for the case of a spectral series using the Na I $3p$–ns and $4p$–ns series as an example. It is found that the influence of the pressure broadening on the solar limb effect is reduced due to different signs of hydrogen and electron impact shifts.

071.050   Results of the 11–year programme for spectrophotometry of selected Fraunhofer line profiles. Motions in the
solar atmosphere.
V. A. Krat, E. K. Kokhan.
Izv. Glav. Astron. Obs. Pulkovo, Astrometr. Astrofiz., No. 202, p. 49 – 70 (1984). In Russian.
The observational results of an 11–year spectrophotometric survey of selected Fraunhofer lines showing the independence of the line–profile parameters on the phase of the cycle were given earlier. In the course of the investigation the temperature sensitive line C I 5380.322 Å, the so called 7 "magnetic" lines and a faint, but free from blends, Fe II 6084.106 Å line were introduced in the observational programme. Here the authors give the results of the spectrophotometric study of these lines. Particular attention is paid to the asymmetry of the lines. The reduction of their profiles using the new method leads to a new picture of statistically large–scale moving photospheric formations. The statistical role of the oversheets in granules and chromospheric "rain" is evaluated. Some features of the model of the line formation of the temperature and magnetic sensitive lines are given.

071.051   Observations of selected Fraunhofer lines in 1979.
E. K. Kokhan, V. A. Krat, K. A. Kandaurova.
Izv. Glav. Astron. Obs. Pulkovo, Astrometr. Astrofiz., No. 201, p. 77 – 83 (1985). In Russian.

During 1969 – 1979 observations of the profiles of solar lines were made according to the program «Long–period variations of Fraunhofer lines». The observations were carried out with the Pulkovo double–pass monochromator. True profiles of the lines in digital form and their central intensities are given for the last year of the 11–year series of the observations.

Solar flux atlas from 296 to 1300 nm.
See Abstr. 002.011.

Ground–based optical observations during SMM.
See Abstr. 013.047.

Establishment of internally consistent solar scales of oscillator strengths and abundances of chemical elements. V. Scandium.
See Abstr. 022.010.

On the establishment of internally consistent solar scales of oscillator strengths and abundances of chemical elements. VI. Neutral vanadium.
See Abstr. 022.039.

Accurate experimental lifetimes of excited levels in Nd II.
See Abstr. 022.087.

Patchy reconnection and magnetic ropes in astrophysical plasmas.
See Abstr. 062.005.

Photospheric flux changes and the MHD approximation.
See Abstr. 062.086.

NLTE spectral line formation in a three–dimensional atmosphere with velocity fields.
See Abstr. 063.099.

Rapid non–LTE calculations of Balmer lines and hydrogen ionization: the solar case.
See Abstr. 064.025.

The Evershed flow as a steady–state homogeneous phenomenon.
See Abstr. 072.057.

Sunspot umbral oscillations in the photosphere and low chromosphere.
See Abstr. 072.095.

HRTS observations of spicular emission at transition region temperatures above the solar limb.
See Abstr. 073.061.

Ejection of chromospheric material associated with injection of electrons in the solar corona.
See Abstr. 073.081.

Calculation of the solar magnetic field from values observed in the photosphere.
See Abstr. 075.006.

Amplification and maintenance of thin magnetic flux tubes by compressible convection.
See Abstr. 075.018.

Nonlinear time–evolution of kink–unstable magnetic flux tubes in the convection zone of the Sun.
See Abstr. 075.019.

Expected intensities of solar neon–like ions.
See Abstr. 076.016.

Long and short term variation of the 10.7 cm solar flux. The photospheric granules and the Zürich numbers.
See Abstr. 077.026.

On the energetics of the solar supergranulation.
See Abstr. 080.024.

Report of IAU Commission 12: Radiation and structure of the solar atmosphere (*Radiation et structure de l'atmosphère solaire*).
See Abstr. 080.149.

072 Sunspots, Faculae, Activity Cycles, Solar Patrol

072.001 Velocity and asymmetry mapping of sunspots.
G. Küveler, E. Wiehr.
Astron. Astrophys., Vol. 142, No. 1, p. 205 – 211 (1985).

Using the two dimensional diode array system at the Locarno solar observatory velocity and asymmetry maps of several sunspots were measured in the photospheric non–split 4.2 eV line Fe 7090.4. It is found that the velocity and asymmetry contours of individual spots correspond very well. The Evershed velocity field is taken from Doppler displacements of the very line core, it does not show any discontinuity at the penumbra borders. From time series of identical sunspots a slow temporal development of the Evershed field is seen, indicating a time delay with respect to the development of the visible spot structure. The vertical upflow needed to feed the observed horizontal flux is estimated to be of the order of $20 - 100$ m s^{-1}.

072.002 Vortical distribution of sunspots.
S.-I. Akasofu.
Planet. Space Sci., Vol. 33, No. 3, p. 275 – 277 (1985).

A remarkable vortical distribution of sunspots, photographed at National Solar Observatory, Tucson, may be an important indication that a vortex flow is essential in the formation of sunspots. It is argued that the concept of vortex flow around a spot is not a heresy, since a generation mechanism of non–trivial force–free fields requires the presence of time varying vorticity in the photosphere.

072.003 Sunspot numbers: October 1984.
Sky Telesc., Vol. 69, No. 1, p. 87 (1985).

072.004 Sunspot numbers: November 1984.
Sky Telesc., Vol. 69, No. 2, p. 182 (1985).

072.005 High–resolution spectroscopy of active regions. 1. Observing procedures.
C. Zwaan, J. J. Brants, L. E. Cram.
Sol. Phys., Vol. 95, No. 1, p. 3 – 14 (1985).

The authors describe an observing program designed to obtain spectra of sunspots, pores, and other features in active regions using the Vacuum Tower Telescope and Echelle Spectrograph at Sacramento Peak Observatory. The spectral lines used in this study have been especially chosen to allow pointed studies of fine structure in the intensity distribution, and in the velocity and magnetic fields in the photospheric levels of active regions, and to relate this structure to chromospheric observations made in the Ca II H line.

072.006 High–resolution spectroscopy of active regions. 2. Line–profile interpretation, applied to an emerging flux region.
J. J. Brants.
Sol. Phys., Vol. 95, No. 1, p. 15 – 36 (1985).

This paper discusses the analysis of spectrograms of solar magnetic structures obtained with high spatial resolution in both directions of circular polarization, with application to observations of flux emergence. The author assumes each spatial resolution element to contain two atmospheric components: mean non–magnetic photosphere and unresolved magnetic structure. He first defines observable spectral–line parameters, and then calibrates the derivation of magnetic–structure parameters by computer simulations in which he varies the fractional contribution of the magnetic component, its field strength and its field inclination, and the line–of–sight velocity difference between the components. This grid of synthesized profiles then serves to define the uniqueness and margins of trial–and–error fits to the observed parameters. The author presents results for an emerging flux region.

072.007 Interpretation of oscillations in UV lines observed above sunspot umbrae.
J. Staude, Y. D. Žugžda (*Yu. D. Zhugzhda*), V. Locāns.
Sol. Phys., Vol. 95, No. 1, p. 37 – 44 (1985).

The authors' theory of a resonator for slow magneto–atmospheric waves in the chromosphere of a sunspot umbra has been used to check different models of the structure of the chromosphere and transition region. Oscillations of velocity and intensity in C IV, Si IV, and O IV lines observed by Gurman et al. (1982) on the SMM spacecraft have been compared with the calculated oscillations. The observed spectrum of resonant peaks could well be explained by a gradient model of the umbral chromosphere. There is strong evidence for a concentration of the observed oscillations in cold fine structure elements of the transition region. Isothermal rather than adiabatic oscillations in the cold elements should be assumed in order to explain the observed fluctuations of line intensity.

072.008 On the spatial structure of a penumbra.
R. B. Teplitskaya.
Sol. Phys., Vol. 95, No. 1, p. 45 – 49 (1985).

A series of spatial intensity profiles across the sunspot penumbra is obtained at different wavelengths within the Ca II K line. A number of photometric properties of the penumbra are outlined, which may be useful for constructing a relevant inhomogeneous model.

072.009 Peculiarities of the dynamics of activity of the July 1983 sunspot group.
A. V. Kryundal', L. D. Parfinenko.
Soln. Dannye, Byull., 1984, No. 10, p. 63 – 68 (1985). In Russian.

The magnetic field and radial velocities were of stationary character in the July 1983 group No 204. The eastern following sunspot was of low flare activity; however, there were instabilities (inhomogeneities) of the longitudinal magnetic field and radial velocities. Sudden rises with a velocity of $2 - 4$ km per sec of large areas of the sunspot, previously lowering, took place. The kinetic energy of such a phenomenon was 10^{31}erg per 10 min, which may cause "release" of energy, thus decreasing the flare activity.

072.010 Manifestation of supergranulation in sunspots.
V. I. Ivanchuk, O. I. Koval'chuk.
Probl. Kosm. Fiz., Vyp. 19, p. 41 – 44 (1984). In Russian. – See Abstr. 003.002.

072.011 Sonnenfleckenrelativzahlen des SONNE–Netzes und des S.I.D.C. für November 1984.
Sterne Weltraum, 24. Jahrg., Nr. 1, p. 31 (1985).

072.012 Sonnenfleckenrelativzahlen des SONNE–Netzes und des S.I.D.C. für Dezember 1984.
Sterne Weltraum, 24. Jahrg., Nr. 2, p. 92 (1985).

072.013 Sonnenfleckenrelativzahlen des SONNE–Netzes und des S.I.D.C. für Januar 1985.
Sterne Weltraum, 24. Jahrg., Nr. 3, p. 152 (1985).

072.014 Sunspot numbers: December 1984.
Sky Telesc., Vol. 69, No. 3, p. 284 (1985).

072.015 Sunspot numbers: January 1985.
Sky Telesc., Vol. 69, No. 4, p. 375 (1985).

072.016 Lohnt sich eine Reduktion der k–Faktoren des Relativzahlnetzes SONNE durch Seeing–Angaben nach der Kiepenheuer–Skala?
J. Dreyhsig.
Sonne, Jahrg. 9, Nr. 33, p. 24 – 28 (1985).

072.017 Statistical mechanics of velocity and magnetic fields in solar active regions.
V. Krishan.
Sol. Phys., Vol. 95, No. 2, p. 269 – 280 (1985).

A statistical mechanics of the velocity and magnetic fields is formulated for an active region plasma. The plasma subjected to the conservation laws emerges in a most probable state which is described by an equilibrium distribution function containing a Lagrange multiplier for every invariant of the system. The Lagrange multipliers are determined by demanding that the measured expectation values of the invariants be reproduced. For a numerical exercise, the author has assumed some probable values for these invariants. The total energy of a coronal loop is estimated from energy balance considerations. Doppler widths of the UV and EUV lines excited in the coronal loop plasma give a measure of the root–mean–square velocities.

072.018 Impulsive phenomena in a small active region.
G. L. Withbroe, S. R. Habbal, R. Ronan.
Sol. Phys., Vol. 95, No. 2, p. 297 – 310 (1985).

The temporal and spatial variations of EUV emission from a small growing active region were investigated. The EUV fluctuations in this small active region are similar to those observed in coronal bright points, suggesting that impulsive heating is an important, perhaps dominant form of heating the upper chromospheric and lower coronal plasmas in small magnetic bipolar regions. The responsible mechanism most likely involves the rapid release of magnetic energy, possibly associated with the emergence of magnetic flux from lower levels into the chromosphere and corona.

072.019 The internal magnetic field of the Sun and peculiarities of the solar activity cycles.
M. I. Pudovkin, E. E. Benevolenska (*E. E. Benevolenskaya*).
Sol. Phys., Vol. 95, No. 2, p. 381 – 390 (1985).

The existence of prolonged periods of abnormally low solar activity (such as the Maunder minimum) is explained within the framework of Leighton's model of a solar cycle with a hypothetical internal magnetic field of the Sun taken into account.

072.020 On the R_z–sunspot relative number variations.
M. R. Attolini, M. Galli, G. Cini Castagnoli.
Sol. Phys., Vol. 95, No. 2, p. 391 – 395 (1985).

The Zürich sunspot relative number R_z series has been analysed by the cyclogram method. The amplitude and the frequency variations of the Fourier 11 yr component between 1700 – 1983 A.D., were determined in a continuous way. Four distinct time intervals with significantly different characteristics of the periodicities are observed and discussed. Their second harmonics are also considered. The periodicity changes are contemporary to those of the 11 yr cycles. Around the year 1903 it seems that an important event has happened in the Sun.

072.021 A calculation of the multi–level configuration of an active region from measurements of the magnetic field with a magnetograph in several spectral lines.
V. A. Romanov.
Soln. Dannye, Byull., 1984, No. 11, p. 69 – 72 (1985). In Russian.

An empirical force model of the magnetic field of an active region is described. The magnetic and current fields in the model are continuous and strictly satisfy Maxwell's equations for stationary configurations. The calculated magnetic field at corresponding levels of the solar atmosphere agrees well with the multi–level magnetographic measurements of the total magnetic vector $\vec{H}$. The number of such levels is not limited.

072.022 A study of peculiarities of the latitudinal distribution of solar activity.
V. A. Magerramov.
Soln. Dannye, Byull., 1984, No. 11, p. 72 – 76 (1985). In Russian.

The results of an analysis of the latitudinal distribution of sunspot groups of the Zürich classes A + B + C and E + F for the northern and southern solar hemispheres by the method of polynomial approximation are given. Two–dimensional gradient–vector studies of the Maunder "butterfly diagrams" enabled the author to detect a number of properties of the latitudinal distribution of the solar activity unavailable for a one–dimensional analysis.

072.023 A study of time variations of the sunspot structure using the photographic differentiation method.
Kh. I. Abdusamatov.
Soln. Dannye, Byull., 1984, No. 12, p. 84 – 88 (1985). In Russian.

The method of photographic differentiation is used for a study of the development of the sunspot structure and proper motions of separate sunspots (structures) with respect to the whole group (neighbouring sunspots). Some results of the migration of pores to the leader of the group, dark elements of the penumbra towards the photosphere and others obtained by this method are given.

072.024 Sonnenfleckenrelativzahlen des SONNE–Netzes und des S.I.D.C. für Februar 1985.
Sterne Weltraum, 24. Jahrg., Nr. 4, p. 211 (1985).

072.025 Sunspot numbers: February 1985.
Sky Telesc., Vol. 69, No. 5, p. 476 (1985).

072.026 Quasi–biennial oscillation in sunspot activity.
E. M. Apostolov.
Bull. Astron. Inst. Czech., Vol. 36, No. 2, p. 97 – 102 (1985).

A 233–year series of monthly means of relative sunspot numbers for the period 1749 – 1981 is analysed. Different statistical methods are used: autocorrelation power spectrum and complex demodulation analysis. It is found that there exists a quasi–biennial oscillation with a basic period of 25.6 months. The lifetime of the quasi–biennial oscillation is one solar cycle and contains five cycles of oscillations and, for most solar cycles, the oscillation is formed at the beginning of the solar cycle and is destroyed at the end. The time variations of amplitude and oscillation period during the solar cycle are analysed as well.

072.027 A.L.P.O.S.S. solar observations for rotations 1732 – 1734.
R. E. Hill.
Strolling Astron., Vol. 31, Nos. 1 – 2, p. 16 – 21 (1985).

072.028 Solar cycle phase variations.
D. R. Whitehouse.
Astron. Astrophys., Vol. 145, No. 2, p. 451 – 453 (1985).

Two calculated solar cycles are compared with the real data to look for non–random phase variations, which are found. These are used to estimate the epochs of the next solar minimum and maximum as 1988.2 ± 0.2 and 1991.8 ± 0.2 respectively. It is also suggested that the 80–year radial oscillation may be weakly present in cycle length data.

072.029 An elementary energetic model of a sunspot.
A. A. Solov'ev.
Sov. Astron., Vol. 28, No. 4, p. 447 – 451 (1984). English translation of 38.072.002.

072.030 Modeling of the 22–year solar activity cycle within the framework of the dynamo theory with allowance for the primary field.
M. I. Pudovkin, E. E. Benevolenskaya.
Sov. Astron., Vol. 28, No. 4, p. 458 – 461 (1984). English translation of 38.072.003.

072.031 Oscillatory convection in sunspots.
F. Cattaneo.
ESA Spec. Publ., ESA SP–220, p. 47 – 50 (1984). – See Abstr. 012.045.

The structure of oscillatory convection in the layers below the umbral photosphere is discussed. A linear stability analysis of a simple model problem describing a polytropic layer with a vertical magnetic field shows that when the Alfvén speed and sound speed are comparable oscillatory convection has a mixed character exhibiting properties of both fast and slow magnetoacoustic waves. The analysis further reveals that in this regime overstability is possible even in convectively stable layers. The nature of the destabilizing mechanism is briefly discussed.

072.032   Variability of the structure of the granulation over the solar activity cycle.
R. Muller, T. Roudier.
ESA Spec. Publ., ESA SP–220, p. 51 – 54 (1984). – See Abstr. 012.045.

High resolution photographs of solar granulation, obtained between 1976 and 1983, have been computer processed, in order to study the structure of the granulation and its variation over the solar cycle.

072.033   Linear models of acoustic waves in sunspot umbrae.
J. B. Gurman, J. W. Leibacher.
ESA Spec. Publ., ESA SP–220, p. 205 – 206 (1984). – See Abstr. 012.045.

The authors interpret the 5.5 to 8.5 mHz oscillations observed in umbral chromospheres and transition regions as acoustic waves propagating parallel, or nearly parallel, to the temperature gradient.

072.034   Observational aspects of sunspot oscillations.
B. W. Lites.
ESA Spec. Publ., ESA SP–220, p. 207 – 209 (1984). – See Abstr. 012.045.

Several new observational results of oscillations in sunspots are presented. Observations of umbral oscillations in Ca II H and He I 10830 Å show that they have high amplitude in the upper chromosphere. The umbral oscillations have several co–spatial frequency modes.

072.035   The flux rope models checked by some observations.
J. I. Garcia de la Rosa.
ESA Spec. Publ., ESA SP–220, p. 229 – 230 (1984). – See Abstr. 012.045.

The high resolution observation of the birth of 73 active regions (AR's) allows checking the three main models of sunspots. It is concluded that the large AR's ($\phi \gtrsim 5 \times 10^{21}$ Mx) result from the emergence of a compact and twisted flux rope, while the small ones are produced by the clustering of flux tubes by the supergranular motions.

072.036   A consideration of the solar cycle in the light of the behaviour of the large active regions.
J. I. Garcia de la Rosa.
ESA Spec. Publ., ESA SP–220, p. 231 – 232 (1984). – See Abstr. 012.045.

072.037   Solar cycle variations of sunspot temperatures.
P. Maltby, S. B. Barth, P. B. Lilje, F. W. Vikanes.
ESA Spec. Publ., ESA SP–220, p. 233 – 234 (1984). – See Abstr. 012.045.

Observations obtained during 1983 strengthen the finding that the umbra/photosphere intensity ratio for large sunspots is a linear function of the phase in the solar cycle. Data for the period 1968 – 1983 are presented, thus covering most of the solar cycles 20 and 21.

072.038   Solar cycle variation of ephemeral active regions.
K. L. Harvey.
ESA Spec. Publ., ESA SP–220, p. 235 – 236 (1984). – See Abstr. 012.045.

Ephemeral active regions (ER) have been identified and counted for selected periods from 1970 to mid–1984 using the daily, full–disk photospheric magnetograms taken by the National Solar Observatory. The number of ER varies in phase with the solar cycle.

072.039   The first effect of the amplitude of a solar cycle.
P. A. Simon, J. P. Legrand.
ESA Spec. Publ., ESA SP–220, p. 237 – 238 (1984). – See Abstr. 012.045.

The authors discuss a "phase/intensity" relationship, discovered in a sample of 18 cycles, between the poloidal and the toroidal components of the solar magnetic field.

072.040   Observations concerning energy balance in solar magnetic regions.
G. A. Chapman, A. D. Herzog, J. K. Lawrence.
ESA Spec. Publ., ESA SP–220, p. 241 – 242 (1984). – See Abstr. 012.045.

Variations in the solar irradiance detected by the Solar Maximum Mission satellite have shown that sunspots alter the flow of heat near the photosphere. Analysis of these observations suggest (1) that there is storage of energy in active regions and (2) a significant fraction (over one–half) of this stored energy is radiated from magnetic elements (faculae) of the active region.

072.041   A siphon flow model for Evershed motion.
D. Dialetis, C. E. Alissandrakis.
ESA Spec. Publ., ESA SP–220, p. 245 – 246 (1984). – See Abstr. 012.045.

The authors have studied the motion of an inviscid isothermal plasma under the influence of gravity along the lines of force of an axisymmetric magnetic field. The general behaviour of the solutions is in good agreement with the observations.

072.042   On running penumbral waves.
L. M. Small, B. Roberts.
ESA Spec. Publ., ESA SP–220, p. 257 – 259 (1984). – See Abstr. 012.045.

Following the work of Nye & Thomas the authors model running penumbral waves as magnetoacoustic–gravity modes propagating along the lower boundary of the penumbral magnetic field. They examine the theoretical properties of surface modes on the boundary and show that it is not necessary to invoke trapping of magnetoacoustic–gravity modes to produce the confinement of the modes to the photospheric chromospheric region. Running penumbral waves are interpreted as fast magnetoacoustic surface modes.

072.043   About sudden disappearances and reappearances.
J. L. Ballester.
ESA Spec. Publ., ESA SP–220, p. 275 – 276 (1984). – See Abstr. 012.045.

An extensive study (1931 – 1981) of sudden disappearances (SD) and reappearances, and their relationships with active centers, maximum and minimum of solar cycle, latitude, coronal holes, flares, emerging flux and age of neutral lines is now in progress at the University of Palma de Mallorca, trying to obtain a picture of the external causes that might be involved in the production of SD and the circumstances that might affect reappearances. In this paper, some preliminary results are presented.

072.044   The rise and fall of sunspot group 18962: a case of magnetic submergence.
H. Zirin.
Astrophys. J., Vol. 291, No. 2, p. 858 – 859 (1985). With plates 20 – 21.

Normal sunspot groups emerge from below the surface as a big arch; the footpoints form a dipole that moves apart and the fields eventually diffuse and disappear. BBSO 18962 was a small dipole with elements which, after spreading apart in the first day of existence, drew back together and disappeared below the surface. It is proposed that the flux loop was pulled back down by magnetic tension and submerged. The circumstances suggest that this behavior may not be uncommon.

072.045   A force model of the magnetic field of a unipolar sunspot.
V. A. Romanov.
Soln. Dannye, Byull., 1985, No. 1, p. 66 – 70 (1985). In Russian.

The results of calculations of a unipolar sunspot using the force model of the magnetic field are given. It is shown that the distribution of current density and Lorentz force have complicated fine structure in the sunspot. An analysis of the obtained results is made.

072.046   On a connection between fluctuations of the emission of an active region in the radio and optical ranges.
L. P. Ipatova, S. P. Leonenko, M. B. Ogir',
B. A. Poperechenko, L. V. Yasnov.
Soln. Dannye, Byull., 1985, No. 1, p. 88 – 94 (1985). In Russian.
A detailed analysis of simultaneous observations of an active region in optical and radio ranges is given. The analysis showed that the events coincide in the radio and optical ranges to an accuracy of 1 – 1.5 min. The events were not detected through a summarized emission of all the flocculi, but through a dominating effect of the emission of some flocculi regions. The plotted power spectra and correlation functions have a number of maxima in the interval of periods from 3 to 8 minutes.

072.047   A study of the stability and comparison of different series of Wolf numbers.
M. N. Gnevyshev, Yu. A. Nagovitsyn, E. Yu. Nagovitsyna.
Soln. Dannye, Byull., 1985, No. 2, p. 72 – 79 (1985). In Russian.
A study has been made of the stability of 15 series of relative Wolf numbers for the period from 1957 to 1976. Estimates using 6 statistic criteria have been made.

072.048   Some photometric characteristics of the fine structure of the complex sunspot group No. 228–229 of July 1982.
A. V. Kryundal', V. B. Makulov, M. N. Stoyanova.
Soln. Dannye, Byull., 1985, No. 2, p. 79 – 84 (1985). In Russian.
A classification of light bridges in the complex sunspot group No. 228–229 of July 1982 is proposed. The classification is based on photometric measurements and structural peculiarities of light bridges. It is concluded that the bridges are stable as a whole during the evolution of a sunspot.

072.049   The interpretation of sunspot magnetic field observations.
M. G. Adam.
Sol. Phys., Vol. 96, No. 1, p. 27 – 34 (1985).
Magnetic field strengths and directions of the lines of force have been measured over two large sunspots in 1975 and 1976 using Treanor's method. Further refinements in observational technique reduce the effects of instrumental polarization to a small phase change. The new observations confirm the existence of differences between the polarization states of the red and violet Zeeman σ–components in some regions of the spots. These differences, which are especially associated with light bridges and streamers, are attributed to magneto–optical effects, coupled with Doppler shifts, in extraneous material lying over the spots.

072.050   On the change of the contrast of faculae over the solar disk.
M. M. Musaev.
Astron. Tsirk., No. 1344, p. 4 – 7 (1984). In Russian.

072.051   Sonnenfleckenrelativzahlen des SONNE–Netzes und des S.I.D.C. für März 1984.
Sterne Weltraum, 24. Jahrg., Nr. 5, p. 272 (1985).

072.052   Sunspot numbers: March 1985.
Sky Telesc., Vol. 69, No. 6, p. 577 (1985).

072.053   The atmosphere of a sunspot based on observations in the X–ray, extreme ultraviolet, optical, and radio ranges.
J. Staude, F. Fürstenberg, J. Hildebrandt, A. Krüger,
J. Jakimiec, V. N. Obridko, M. Siarkowski, B. Sylwester,
J. Sylwester.
Sov. Astron., Vol. 28, No. 5, p. 557 – 563 (1984). English translation of 38.072.030.

072.054   Comparative characteristics of stellar and sunspot spectra.
O. G. Badalyan, V. N. Obridko.
Sov. Astron., Vol. 28, No. 5, p. 564 – 568 (1984). English translation of 38.072.031.

072.055   On steadying the monthly fluctuations of heliophysical indices.
K. S. Voichishin (*K. S. Vojchishin*), M. I. Stodilka.
Sov. Astron., Vol. 28, No. 5, p. 569 – 573 (1984). English translation of 38.072.032.

072.056   Solar luminosity fluctuations during the disk transit of an active region.
J. K. Lawrence, G. A. Chapman, A. D. Herzog, J. C. Shelton.
Astrophys. J., Vol. 292, No. 1, p. 297 – 308 (1985).
The authors present monochromatic, photometric observations, obtained with a 512 element linear diode array, of the solar irradiance fluctuations caused by an active region during its entire disk transit in 1982 August. The maximum sunspot fluctuation, as a fraction of quiet sun irradiance, is about –800 parts per million (ppm). Faculae have a maximum irradiance fluctuation of about +200 ppm. By integrating over the viewing angle during disk transit the authors are able to determine that, for visible wavelengths, the facular luminosity excess is about 50% of the sunspot luminosity deficit.

072.057   The Evershed flow as a steady–state homogeneous phenomenon.
D. Dialetis, P. Mein, C. E. Alissandrakis.
Astron. Astrophys., Vol. 147, No. 1, p. 93 – 102 (1985).
The authors have studied the steady–state characteristics of the Evershed flow in the chromosphere and in the photosphere and have compared them with the topology of the magnetic field. They have used the Multichannel Subtractive Double Pass Spectrograph to obtain line of sight velocity maps in H_α and the Meudon magnetograph for mapping the photospheric velocity and magnetic field in sunspot regions.

072.058   Magnetic polarities and maximum field strengths of one selected sunspot group with time distances of about twenty minutes during the period 1983 June 22 – June 27.
H. Künzel.
HHI Suppl. Ser. Sol. Data, Vol. 5, No. 5, p. 101 – 123 (1983).

072.059   Observations solaires. Rotations 1730 – 1743, 23 décembre 1982 – 8 janvier 1984.
E. Tifrea, V. Dinulescu, A. Dimitriu, S. Dinulescu, G. Maris,
A. Oncică, C. Dumitrache.
Cent. Astron. Sci. Spat., Obs. Sol., 60 pp. Price Lei 6.25 (1984).

072.060   Planetary sunspot periods as a deception of solar activity. Sunspot relative numbers as an error.
K. Pilowski.
Univ. Hannover, Astron. Stn., Veröff., Nr. 16, p. 21 – 28 (1985).

072.061   Manila Observatory, Solar Division. Solar Maps and Activity. 1984 October – 1985 April.
F. J. Heyden.
Sol. Maps Act. (1984/85).

072.062   The determination of the reduction factors for sunspot observations at the Astronomical Observatory Zagreb.
G. Kren, B. Vršnak, V. Ruždjak.
Hvar Obs. Bull., Vol. 8, No. 1, p. 1 – 12 (1984).
Sunspot observations performed at the Astronomical Observatory Zagreb in the period May 1979 – June 1983 were investigated. The reduction factors of the observed Wolf numbers to the international Wolf numbers have been obtained. The dependence of the reduction factors on the seeing has been analysed.

072.063   Sunspots in collision.
V. Gaizauskas, K. L. Harvey.
Bull. Am. Astron. Soc., Vol. 16, No. 4, p. 928 – 929 (1984). Abstract. – See Abstr. 010.062.

072.064   Sunspot oscillations and the short–period cutoff for global p–mode oscillations.
R. L. Moore, D. M. Rabin.
Bull. Am. Astron. Soc., Vol. 16, No. 4, p. 978 (1984). Abstract. – See Abstr. 010.062.

072.065 Magnetic field inclination and the double return flux sunspot model.
J. K. Lawrence, M. Zhu.
Bull. Am. Astron. Soc., Vol. 16, No. 4, p. 979 (1984). Abstract. – See Abstr. 010.062.

072.066 Observations concerning the energy budget of a solar activity complex.
G. A. Chapman, A. D. Herzog, J. K. Lawrence, J. C. Shelton.
Bull. Am. Astron. Soc., Vol. 16, No. 4, p. 991 (1984). Abstract. – See Abstr. 010.062.

072.067 Bimodality of the solar cycle.
D. M. Rabin, R. L. Moore, R. M. Wilson.
Bull. Am. Astron. Soc., Vol. 16, No. 4, p. 993 (1984). Abstract. – See Abstr. 010.062.

072.068 Preliminary analysis of multi–color reticon data at the San Fernando Observatory.
A. D. Herzog, S. F. Mason, G. A. Chapman, J. K. Lawrence.
Bull. Am. Astron. Soc., Vol. 16, No. 4, p. 1001 (1984). Abstract. – See Abstr. 010.062.

072.069 X–ray, ultraviolet, optical and magnetic structure in and near an active region.
B. M. Haisch, T. D. Tarbell, M. E. Bruner, L. W. Acton, R. M. Bonnet, M. J. Hagyard.
Bull. Am. Astron. Soc., Vol. 16, No. 4, p. 1002 (1984). Abstract. – See Abstr. 010.062.

072.070 High resolution VLA observations of an active region during a partial solar eclipse.
D. E. Gary, G. J. Hurford.
Bull. Am. Astron. Soc., Vol. 16, No. 4, p. 1003 (1984). Abstract. – See Abstr. 010.062.

072.071 Il Sole. Il 21° ciclo di attività solare e le nuove acquisizioni sui rapporti tra il Sole e i pianeti. Il "caso" Geminga.
P. Stroppa.
Coelum, Vol. 53, N. 3, p. 139 – 152 (1985).

072.072 Variations in the rotational velocity of sunspot groups during their lifetimes.
M. H. Gokhale, K. M. Hiremath.
Bull. Astron. Soc. India, Vol. 12, No. 4, p. 398 – 403 (1984).

The authors determine the mean values, the r.m.s. deviations, and the standard errors of the mean, for differences between the "initial", "overall" and "final" angular velocities of the rotation of sunspot groups which occurred during the solar cycles 1923 – 1933, 1934 – 1944 and 1945 – 1954. They find a general trend that spot groups are accelerated during the early phase of their lifetime, and decelerated in the late phase, yielding a net deceleration over the whole lifetime.

072.073 On the nature of quasi–periodic oscillations of the magnetic field and of the velocity in a sunspot.
A. V. Baranov.
Investigation of magnetic fields and active formations on the sun, p. 26 – 32 (1984). In Russian. Abstr. in Ref. Zh., 51. Astron., 5.51.331 (1985). – See Abstr. 003.022.

072.074 Approximation of averaged characteristics of the magnetic field of a sunspot on the basis of a potential model.
A. V. Baranov.
Investigation of magnetic fields and active formations on the sun, p. 33 – 48 (1984). In Russian. Abstr. in Ref. Zh., 51. Astron., 5.51.332 (1985). – See Abstr. 003.022.

072.075 Connection of magnetic holes with development of sunspot groups.
L. F. Lazareva.
Investigation of magnetic fields and active formations on the sun, p. 63 – 70 (1984). In Russian. Abstr. in Ref. Zh., 51. Astron., 5.51.333 (1985). – See Abstr. 003.022.

072.076 On high–latitude spots outside the wings of "Maunder butterflys".
V. F. Chistyakov.
Investigation of magnetic fields and active formations on the sun, p. 10 – 26 (1984). In Russian. Abstr. in Ref. Zh., 51. Astron., 5.51.372 (1985). – See Abstr. 003.022.

072.077 Motion of active longitudes depending on the phase of the 11–year cycle.
A. P. Kramynin.
Investigation of magnetic fields and active formations on the sun, p. 61 – 63 (1984). In Russian. Abstr. in Ref. Zh., 51. Astron., 5.51.373 (1985). – See Abstr. 003.022.

072.078 Provisional sunspot numbers for December 1984 – May 1985.
Yamamoto Circ., Nos. 2033 – 2035, 2037, 2038, 2041 (1985). In Japanese.

072.079 Definitive sunspot numbers 1984.
Yamamoto Circ., No. 2036 (1985). In Japanese.

072.080 Solar activity in the Hilbert space.
L. V. Lukina, A. N. Lukin.
Izhevsk. mekh. inst. Ustinov, 8 pp. (1984). In Russian. Abstr. in Ref. Zh., 51. Astron., 6.51.324 (1985).

072.081 Some characteristics of spot groups and noise storms connected with them.
A. A. Gnezdilov.
Physics of solar activity, p. 118 – 128 (1983). In Russian. Abstr. in Ref. Zh., 51. Astron., 6.51.328 (1985). – See Abstr. 003.024.

072.082 Sonnenfleckenrelativzahlen des SONNE–Netzes und des S.I.D.C. für April 1985.
Sterne Weltraum, 24. Jahrg., Nr. 6, p. 335 (1985).

072.083 On the asymmetry of Wilson's sunspot effect.
L. Hejna, A. A. Solov'ev.
Bull. Astron. Inst. Czech., Vol. 36, No. 3, p. 183 – 186 (1985).

The validity of the interpretation of the asymmetric Wilson's effect, observed and interpreted by Obashev et al., is discussed. It is shown that this interpretation is contrary to reality and a new simple geometrical model of a sunspot with an asymmetric variation of the Wilson effect, which agrees well with theoretical conditions and observations is presented.

072.084 Long–time changes of solar and geomagnetic activity and cosmic ray density.
N. P. Chirkov.
Geomagn. Aehron., Tom 25, No. 3, p. 492 – 494 (1985). In Russian. English translation in Geomagn. Aeron.

072.085 Magnetographic observations of magnetic fields in spots from the Fe II line.
V. N. Milovanov.
Tr. Astrofiz. Inst. Alma–Ata, Tom 41, p. 3 – 6 (1983). In Russian.

072.086 Magnetographic observations of faculae.
V. N. Milovanov.
Tr. Astrofiz. Inst. Alma–Ata, Tom 41, p. 7 – 11 (1983). In Russian.

072.087 Quasi–periodic vibrations of signals of a magnetograph during the observation of sunspots.
N. N. Morozov.
Tr. Astrofiz. Inst. Alma–Ata, Tom 41, p. 19 – 29 (1983). In Russian.

072.088 Solar activity in the periods of anomalous cosmic ray variations (solar maxima of cycles 19 – 21).
M. V. Alania (*M. V. Alaniya*), R. T. Gushchina, L. I. Dorman.
18th International Cosmic Ray Conference, Vol. 10, p. 99 – 103 (1983). – See Abstr. 012.096.

072.089   Filaments and the magnetic field of an active region.
V. P. Maksimov, L. V. Ermakova.
Astron. Zh., Tom 62, Vyp. 3, p. 558 – 561 (1985). In Russian. English translation in Sov. Astron., Vol. 29, No. 3.
A combined study of magnetograms and filtergrams in the Hα and Hβ lines showed that the existence of quiescent filaments within two active regions studied is associated with specific values of the longitudinal magnetic field gradient near the polarity reversal line.

072.090   The Sun in 1984.
B. E. Stonehouse.
South. Stars, Vol. 31, No. 2, p. 153 – 155 (1985).

072.091   On the possibility of energy balance over the lifetime of a solar activity complex.
J. K. Lawrence, G. A. Chapman, A. D. Herzog.
Bull. Am. Astron. Soc., Vol. 17, No. 2, p. 610 – 611 (1985). Abstract. – See Abstr. 010.065.

072.092   Heat and mass flow around sunspots.
D. H. Bruning, B. J. LaBonte.
Bull. Am. Astron. Soc., Vol. 17, No. 2, p. 611 (1985). Abstract. – See Abstr. 010.065.

072.093   Frequent ultraviolet brightenings in solar active regions.
J. G. Porter, J. Toomre, K. B. Gebbie.
Bull. Am. Astron. Soc., Vol. 17, No. 2, p. 629 (1985). Abstract. – See Abstr. 010.067.

072.094   Umbral oscillations: correlation of amplitudes between two chromospheric heights.
B. W. Lites.
Bull. Am. Astron. Soc., Vol. 17, No. 2, p. 631 (1985). Abstract. – See Abstr. 010.067.

072.095   Sunspot umbral oscillations in the photosphere and low chromosphere.
J. H. Thomas, B. W. Lites.
Bull. Am. Astron. Soc., Vol. 17, No. 2, p. 631 (1985). Abstract. – See Abstr. 010.067.

072.096   On the broad–band circular polarization signature of sunspots.
A. Skumanich, B. W. Lites.
Bull. Am. Astron. Soc., Vol. 17, No. 2, p. 631 (1985). Abstract. – See Abstr. 010.067.

072.097   Sunspots in collision.
V. Gaizauskas, K. L. Harvey.
Bull. Am. Astron. Soc., Vol. 17, No. 2, p. 632 (1985). Abstract. – See Abstr. 010.067.

072.098   On the origin of sunspots and the descent of neutral hydrogen.
K. H. Schatten, H. G. Mayr.
Bull. Am. Astron. Soc., Vol. 17, No. 2, p. 633 (1985). Abstract. – See Abstr. 010.067.

072.099   Sunspots and the large–scale environment for the great active region of April, 1984.
P. S. McIntosh.
Bull. Am. Astron. Soc., Vol. 17, No. 2, p. 633 (1985). Abstract. – See Abstr. 010.067.

072.100   Preliminary results from the extreme limb photometer 1982 and 1983 observing seasons.
J. Oseas, G. A. Chapman.
Bull. Am. Astron. Soc., Vol. 17, No. 2, p. 639 (1985). Abstract. – See Abstr. 010.067.

072.101   Numerically simulated fields, currents and plasma properties in a solar active region due to photospheric shear.
Y. C. Xiao, S. T. Wu, J. J. Bao, J. B. Smith Jr.
Bull. Am. Astron. Soc., Vol. 17, No. 2, p. 641 (1985). Abstract. – See Abstr. 010.067.

072.102   Are there constant–alpha force–free magnetic fields in solar active regions?
M. J. Hagyard, E. A. West, M. O'Farrell.
Bull. Am. Astron. Soc., Vol. 17, No. 2, p. 641 (1985). Abstract. – See Abstr. 010.067.

072.103   Periodicities of 2.2 and 22 years in solar activity and solar – planetary relationships.
I. Predeanu.
Top. Astrophys. Astron. Space Sci., Vol. 1, p. 31 – 39 (1985).

072.104   The influence of faculae on sunspot heat blocking.
W.–H. Chiang, P. Foukal.
Sol. Phys., Vol. 97, No. 1, p. 9 – 20 (1985).
The authors study the influence of faculae on sunspot heat blockage using a thermal model based on eddy heat diffusion through the convection zone. The facula is represented as a localized area of excess emission surrounding the sunspot, which is represented as a thermal plug. The computations using a range of reasonable combinations of spot and facular depths show no significant influence of the facula on the long storage times of heat blocked by sunspots. However, the local cooling of surface layers produced by excess facular emission in this model propagates globally within the convection zone in a similar way to the heating produced by a spot.

072.105   Activity of sunspots and solar constant variations during 1980.
J. Pap.
Sol. Phys., Vol. 97, No. 1, p. 21 – 33 (1985).
The purpose of this paper is to investigate the connections between the variation of the solar constant and the age, magnetic structure and state of development of active regions.

072.106   Emerging flux in active regions.
M. Liggett, H. Zirin.
Sol. Phys., Vol. 97, No. 1, p. 51 – 58 (1985).
The authors have compared the rates at which flux emerges in active and quiet solar regions within the sunspot belts. The emerging flux regions (EFRs) were identified by the appearance of arch filament structures in Hα. All EFRs in high–resolution films of active regions made at Big Bear in 1978 were counted. The comparable rate of flux emergence in quiet regions was obtained from SGD (*Solar Geophysical Data*) data and independently from EFRs detected outside the active region perimeter on the same films. The rate of flux emergence is 10 times higher in active regions than in quiet regions. A sample of all active regions in 31 days of 1983 gave a ratio of 7.5.

072.107   A new model for flux emergence and the evolution of sunspots and the large–scale fields.
P. S. McIntosh, P. R. Wilson.
Sol. Phys., Vol. 97, No. 1, p. 59 – 79 (1985).
Existing models for the evolution of sunspots and sunspot groups, describing the subsurface structure of the magnetic fields and their interactions with the convective motions, are briefly reviewed. It is shown that they are generally unable to account for the most recent data concerning the relationship between the large–scale solar magnetic field structures and the magnetic fields of active regions. A new model is put forward based on the expulsion of toroidal magnetic flux by the dominant (i.e. giant) cells of the convection zone. The compatibility of the model with other observations is discussed and its implications for theories of the solar cycle are noted.

072.108   Report of IAU Commission 10: Solar activity (*Activité solaire*).
E. Tandberg–Hanssen.
Trans. IAU, Vol. XIXA, p. 57 – 96 (1985). – See Abstr. 003.046.

072.109 Visual observation of the Sun in the year 1983 in Czechoslovakia.
L. Schmied.
Říše hvězd, Vol. 65, No. 8, p. 164 – 169 (1984). In Czech.

072.110 11–year cycles of sunspots according to Greenwich photographic observation.
M. Kopecký.
Říše hvězd, Vol. 66, No. 2, p. 36 – 37 (1985). In Czech.

072.111 Which 11–year cycle of solar activity was the most important?
M. Kopecký.
Říše hvězd, Vol. 66, No. 6, p. 106 – 107 (1985). In Czech.

072.112 Daily maps of the sun and of magnetic fields of sunspots.
Soln. Dannye, Byull., 1984, 1985, Nos. 10 – 12, 1 – 4 (1985). In Russian.

072.113 On the possibility of numerical modelling of MHD–processes in solar active regions. I.
V. I. Dolgopolov.
Soln. Dannye, Byull., 1985, No. 3, p. 51 – 59 (1985). In Russian.
A method of numerical modelling of the magnetic field evolution and motions in solar active regions is analysed. For constructing a physical model a system of macroscopic magneto-hydrodynamics equations is used.

072.114 A study of helical oscillations in sunspots.
A. A. Pevtsov, I. S. Sattarov.
Soln. Dannye, Byull., 1985, No. 3, p. 65 – 71 (1985). In Russian.
On the basis of observations and published data on 17 sunspots helical oscillations of different types have been detected: close to harmonic, and increasing and decreasing with time.

072.115 Characteristics of fluctuations in solar activity for the years 1982 – 1984.
Yu. I. Vitinskij.
Soln. Dannye, Byull., 1985, No. 3, p. 72 – 74 (1985). In Russian.
Values of the fluctuation index and indices of "perturbations" for Wolf numbers and solar radio emission flux density at 2800 MHz during 1982 – 1984 are given. Also their positive and negative strong fluctuations for these years are listed. Some peculiarities of these characteristics are dicussed.

072.116 Twisting of the magnetic field and motions in complex sunspot groups. I.
R. N. Ikhsanov, G. P. Shchegoleva.
Soln. Dannye, Byull., 1985, No. 4, p. 51 – 57 (1985). In Russian.
The sunspot group No. 114(1960) with the large flares on March 29 and April 1, 1960 is considered. On the basis of the study of its structure, proper motions, magnetic field, flare activity and development of the flare magnetic configuration in the sunspot group are discussed. An exceptional anomaly of the group is pointed out. The nature and development of the magnetic configuration in sunspot groups are discussed.

072.117 On the possibility of numerical modelling of MHD–processes in solar active regions. II.
V. I. Dolgopolov.
Soln. Dannye, Byull., 1985, No. 4, p. 58 – 65 (1985). In Russian.
Problems connected with the quantitative prediction of magnetic and velocity fields in active regions on the sun are discussed on the basis of numerical integration of the MHD–model equations. By numerical experiments with the help of vector-magnetograph data the questions of initial fields agreement, filtration of errors, different boundary conditions are analysed. The magnetic field, obtained as a result of prediction 1 day in advance shows a satisfactory correlation (60%) with the observational data for a corresponding period of the active region evolution under research.

072.118 Numerical calculation of force–free field in solar active regions. II. An improved Chiu–Hilton method.
F. Wu, Z.–z. Wang.
Chin. Astron. Astrophys., Vol. 9, No. 1, p. 44 – 48 (1985). English translation of Acta Astrophys. Sin., Vol. 4, No. 4, p. 272 – 278 (1984).
A good method of finding the magnetic field above an active region from the normal field at the photosphere in the linear force–free field model is the one by Chiu and Hilton. In this paper the authors examine the effect of the eigenfunctions on the field configuration in the solution. They propose a method of estimating the eigenfunctions from Hα photographs and hence to get a more realistic solution.

072.119 The three–dimensional magnetic field structure of Hale Region 16747 (1980 April) and its evolution.
Y.–z. Lin, Z.–z. Wang, F. Wu, Y.–j. Ding, Q.–f. Hong.
Chin. Astron. Astrophys., Vol. 9, No. 2, p. 134 – 138 (1985). English translation of Acta Astrophys. Sin., Vol. 5, No. 1, p. 19 – 25 (1985).

072.120 A possible mechanism for the formation of faculae.
X.–b. Ni, Y.–t. Jiang, Z.–z. Chen, C. Fang.
Acta Astron. Sin., Vol. 26, No. 2, p. 107 – 113 (1985). In Chinese.
The authors propose a new facular heating mechanism: i.e., the formation of the faculae as a result of Joule dissipation of the Hall current which is produced by the interaction between the granule velocity field and the inter–granule magnetic field in the solar active region. The authors also suggest a possible explanation for the formation of the facular fine structure.

072.121 Umbra motions in two complex sunspot groups and their relationship with the magnetic field and flare activity.
R. N. Ikhsanov, G. P. Shchegoleva.
Izv. Glav. Astron. Obs. Pulkovo, Astrometr. Astrofiz., No. 201, p. 96 – 103 (1985). In Russian.
A measurement of proper motions of umbrae in the sunspot group N 379 (1959) with proton flares and in the sunspot group N 101 (1963) with a great number of large flares was carried out. A comparison of proper motions of sunspots with the development of the magnetic field and flare activity shows a close relationship between them.

072.122 L'activité solaire.
M.–J. Martres, G. Zlicaric.
Astronomie, Vol. 99, p. 48 – 49, 105, 160 – 161, 212 – 213, 264 – 265, 316 (1985).

Naked–eye observations of sunspots in Bohemia in the year 1139.
See Abstr. 004.025.

J. R. Wolf and the Zürich sunspot relative numbers.
See Abstr. 005.003.

The 1^1S–n^1P/1^1S–2^1P emission–line ratios in O VII as temperature diagnostics for solar flares and active regions.
See Abstr. 022.098.

A possible method for the determination of the magnitude and inclination of a sunspot magnetic field.
See Abstr. 036.215.

Initial results from the repaired Solar Maximum Mission and future prospects.
See Abstr. 051.009.

Waves in inhomogeneous media.
See Abstr. 062.102.

On the numerical computation of nonlinear force–free magnetic fields.
See Abstr. 062.142.

Cooling of magnetic flux tubes as a mechanism for suppression of magnetic buoyant escape of the flux tubes from the Sun and stars.
See Abstr. 064.056.

Asymmetry and shift of three Fe I photospheric lines in solar active regions.
See Abstr. 071.005.

Variability of the quiet photospheric network.
See Abstr. 071.017.

High–resolution photography of the solar chromosphere. XX. True widths of Hα active region loops.
See Abstr. 073.010.

Twisting motions in a disturbed solar filament.
See Abstr. 073.014.

The behaviour of prominence areas in the 11–year cycle and their relationship with other solar events.
See Abstr. 073.016.

Autowave structure of flare–active regions on the sun.
See Abstr. 073.021.

Consecutive homologous flares and their relation to sunspot motions.
See Abstr. 073.025.

Sunspot motions and magnetic shears as precursors of flares.
See Abstr. 073.033.

Magnetic field structure changes in the vicinity of solar flares.
See Abstr. 073.036.

Bidimensional spectroscopy of the solar chromosphere during the Maximum Year.
See Abstr. 073.054.

Chromospheric flare activity in solar cycle 20.
See Abstr. 073.063.

Singularity of solar rotation and flare productivity.
See Abstr. 073.074.

Rocket observation of the EUV images of a solar flare and active regions.
See Abstr. 073.079.

Peculiarities of the spectrum of a flare occurring above a sunspot umbra.
See Abstr. 073.086.

Studies on some aspects of the solar Hα flares of different visual features.
See Abstr. 073.095.

Spicules and surges.
See Abstr. 073.097.

Chromospheric umbral oscillations.
See Abstr. 073.105.

Multiwavelength observations of a preflare solar active region using the VLA.
See Abstr. 073.126.

The correlation of solar flare production with magnetic energy in active regions.
See Abstr. 073.172.

The nature of the major solar flare activity during solar cycles 19, 20 and 21.
See Abstr. 073.191.

Steady flows in active regions observed with the He I 10830 Å line.
See Abstr. 073.225.

North–south asymmetry of the solar activity and the chromospheric flares.
See Abstr. 073.231.

Heating of sunspot chromospheres by slow–mode acoustic shock waves.
See Abstr. 073.245.

Density and temperature structure of sunspot chromospheres by spectral synthesis.
See Abstr. 073.248.

Variations of the total brightness of the white–light corona with the phase of the solar cycle.
See Abstr. 074.052.

Coronal transient apparent morphology and the associated solar activity.
See Abstr. 074.098.

The polar crown during two solar cycles.
See Abstr. 074.118.

On the behaviour of the coronal emission line λ5303 Å above active regions.
See Abstr. 074.128.

Solar cycle variation of energy and photospheric magnetic flux from coronal holes.
See Abstr. 074.142.

MHD analysis of the evolution of solar magnetic fields and currents in an active region.
See Abstr. 075.001.

Les traceurs du champ magnétique en physique solaire.
See Abstr. 075.007.

Magnetic shear. I. Hale region 16918.
See Abstr. 075.011.

Harmonic analysis of the solar magnetic field.
See Abstr. 075.021.

Magnetic shear. II. Hale region 17244.
See Abstr. 075.022.

Height distribution of the magnetic field in active region No. 325 (HR 17751).
See Abstr. 075.027.

On the height gradient of a magnetic field in an active region.
See Abstr. 075.028.

On the character of decrease of the magnetic field with height above spots of the class J.
See Abstr. 075.029.

Numerical simulations of the mean solar magnetic field during the sunspot cycle.
See Abstr. 075.038.

A stochastic model for the mean solar magnetic field during the sunspot cycle.
See Abstr. 075.039.

Configurations of the magnetic field in sunspot groups and solar flares. II. Regularities of the development of flare magnetic configurations.
See Abstr. 075.043.

Results from the Ultraviolet Spectrometer and Polarimeter: non–flare investigations.
See Abstr. 076.012.

Solar microwave emission in active regions.
See Abstr. 077.002.

Spatio–temporal features of the development of microwave emission of active regions and flares.
See Abstr. 077.004.

A working model of the solar *S*–component radio emission.
See Abstr. 077.015.

Solar radio fluxes as indices of solar activity.
See Abstr. 077.021.

Long and short term variation of the 10.7 cm solar flux. The photospheric granules and the Zürich numbers.
See Abstr. 077.026.

Dual frequency observations of solar microwave bursts using the VLA.
See Abstr. 077.030.

Some results of model calculations of the solar S–component radio emission.
See Abstr. 077.033.

Long–period quasi–periodic variations of the shortwave radiation flux on the sun and flare activity of sunspot groups.
See Abstr. 077.040.

Solar noise storms and magnetic sector structures.
See Abstr. 077.043.

On heating of solar active regions by magnetic energy dissipation.
See Abstr. 080.003.

Non–linear dynamos.
See Abstr. 080.071.

The solar dynamo and the complex Lorenz equations.
See Abstr. 080.072.

Solar constant variabilities.
See Abstr. 080.109.

11–year variations of the ^{14}C content in the earth's atmosphere during the period 1850 – 1940.
See Abstr. 082.040.

Geomagnetic and solar data. September 1984.
See Abstr. 084.015.

Geomagnetic and solar data. October 1984.
See Abstr. 084.066.

Geomagnetic and solar data. November 1984.
See Abstr. 084.071.

Solar activity and variation of geomagnetic disturbance.
See Abstr. 084.083.

Geomagnetic and solar data. January 1985.
See Abstr. 084.113.

Geophysical records of a tree: new application for studying geomagnetic field and solar activity changes during the past 10^4 years.
See Abstr. 084.140.

Solar cycle dependence of the location of the Venus bow shock.
See Abstr. 093.048.

Peculiarities of development of the heads of comets at different heliocentric distances in the solar activity cycle.
See Abstr. 102.007.

The configuration of a heliospheric current sheet and some features of its evolution over the cycle of solar activity.
See Abstr. 106.071.

Solar and stellar activities.
See Abstr. 116.059.

The galactic cosmic ray intensity minimum in the inner and outer heliosphere in solar cycle 21.
See Abstr. 144.026.

The role of inhomogeneities of the magnetic field forming in the interplanetary medium in the 11–year cycle of cosmic rays.
See Abstr. 144.032.

Solar activity and modulation of the cosmic ray intensity.
See Abstr. 144.054.

On the degree of correlation between the 27–day variations in cosmic rays and in solar activity.
See Abstr. 144.219.

073 Chromosphere, Flares, Prominences

073.001   New identifications of Fe XVII spectral lines in solar flares.
U. Feldman, G. A. Doschek, J. F. Seely.
Mon. Not. R. Astron. Soc., Vol. 212, No. 3, p. 41P – 45P (1985).

The authors review the identifications of Fe XVII transitions between the $2s^2 2p^5 3s$, $3p$, and $3d$ configurations recently published by Jupén. The review is based on examining spectroheliograms of solar flares obtained by a Naval Research Laboratory instrument on Skylab (S082–A). The authors agree with the majority of identifications given by Jupén, but find different wavelengths for a few of the lines. The relative intensities of the lines are qualitatively in agreement with calculations. The authors also identify a broad line at 1153.20 Å with the $2p^5 3s^3 P_0 \rightarrow 2p^5 3s^3 P_1$ forbidden transition of Fe XVII. This line was found in flare spectra obtained by the NRL slit spectrograph on Skylab (S082–B).

073.002   Acceleration of runaway electrons and Joule heating in solar flares.
G. D. Holman.
Unstable current systems and plasma instabilities in astrophysics, p. 191 – 196 (1985). – See Abstr. 012.002 (IAU Symp. No. 107).

It is of interest to study the electric field acceleration of "runaway" electrons and the simultaneous Joule heating of the thermal plasma in light of results from flare observations, without recourse to a specific flare model. Some of the results of such a study are summarized here.

073.003   The coalescence instability in solar flares.
T. Tajima, F. Brunel, J.–I. Sakai, L. Vlahos, M. R. Kundu.
Unstable current systems and plasma instabilities in astrophysics, p. 197 – 210 (1985). – See Abstr. 012.002 (IAU Symp. No. 107).

The non–linear coalescence instability of current carrying solar loops can explain many of the characteristics of the solar flares. The plasma compressibility leads to the explosive phase of loop coalescence and its overshoot results in amplitude oscillations in temperatures by adiabatic compression and decompression. The presence of strong electric fields and super–Alfvénic flows plays an important role in the production of non–thermal particles. A qualitative explanation of the physical processes taking place during the non–linear stages of the instability is given.

073.004   Quasi–static evolution of force–free magnetic fields and a model for two–ribbon solar flares.
J. J. Aly.
Unstable current systems and plasma instabilities in astrophysics, p. 221 – 224 (1985). – See Abstr. 012.002 (IAU Symp. No. 107).

The author shows that a sheared 2–D force–free field can evolve in a quasi–static way towards an open configuration, and applies this result to a qualitative theory of two–ribbon solar flares.

073.005   Current sheets in solar flares.
E. R. Priest.
Unstable current systems and plasma instabilities in astrophysics, p. 233 – 244 (1985). – See Abstr. 012.002 (IAU Symp. No. 107).

Until recently magnetic reconnection in solar flares was discussed simplistically in terms of either a spontaneous tearing mode instability or a driven Petschek mode. Now the subtle relationship between these two extremes is much better understood. Current sheets may form and reconnection may be initiated in many different ways. There are also a variety of nonlinear pathways from a reconnection instability and several types of driven reconnection. In solar flares current sheets may be important as new flux emerges from below the photosphere and also as a magnetic arcade closes down after being blown open by an eruptive instability. Numerical simulations of these sheets are described.

073.006   A unified treatment of the filament and flare instabilities.
G. Van Hoven.
Unstable current systems and plasma instabilities in astrophysics, p. 263 – 271 (1985). – See Abstr. 012.002 (IAU Symp. No. 107).

Filaments and flares occur in sheared magnetic structures as a result of radiative cooling and resistive reconnection, respectively. A new integrated theory of these two unstable processes is described, which includes the relevant effects of magnetohydrodynamics and energy transport. The normally dissociated thermal and tearing phenomena are coupled together by a temperature–dependent Coulomb resistivity. As a result, the filamentation and flaring instabilities of a sheared field may coexist, as is familiar from the solar example. The growth rates and spatial structures of these two modes are detailed here.

073.007   Energy release in solar flares.
P. A. Sturrock, P. Kaufmann, D. F. Smith.
Unstable current systems and plasma instabilities in astrophysics, p. 293 – 298 (1985). – See Abstr. 012.002 (IAU Symp. No. 107).

The authors discuss the question how many phases of energy release are involved in flares and what are their characteristics.

073.008   Beam–return current systems in solar flares.
D. S. Spicer, R. N. Sudan.
Unstable current systems and plasma instabilities in astrophysics, p. 519 – 520 (1985). – See Abstr. 012.002 (IAU Symp. No. 107).

The importance of electron beams in solar flare dynamics is well known. In order to understand the dynamics produced by beams it is essential to have a clear understanding of the role beam driven return currents play and whether electrostatic or inductive electric fields maintain the return current. The authors show that inductive electric fields are responsible for driving return currents under solar conditions. The significant conclusions that follow from this result, as applicable to solar flares, are presented.

073.009   Collisionless effects on beam–return current systems in solar flares.
L. Vlahos, H. L. Rowland.
Unstable current systems and plasma instabilities in astrophysics, p. 521 – 525 (1985). – See Abstr. 012.002 (IAU Symp. No. 107).

A large fraction of the electrons which are accelerated during the impulsive phase of solar flares stream towards the chromosphere and are unstable to the growth of plasma waves. The linear and non–linear evolution of plasma waves as a function of time is analyzed with the use of a set of rate equations that follow in time the non–linearly coupled system of plasma waves–ion fluctuations. The authors' main conclusion is that the beam–return current system is interconnected and how the return current is carried is determined by the beam generated strong turbulence.

073.010   High–resolution photography of the solar chromosphere. XX. True widths of Hα active region loops.
R. J. Bray, R. E. Loughhead.
Astron. Astrophys., Vol. 142, No. 1, p. 199 – 204 (1985).

The true widths are measured of two very thin (sub–arcsec) loops observed on the disk at Hα + 1.0 Å at a height of several thousand kilometers. A realistic deconvolution procedure is applied to correct for the combined instrumental profile of telescope and atmosphere. This yields upper and lower limits to the true widths, corresponding to the possible range in the magnitude of the effective aperture of the telescope (Fried parameter) at the moment of observation.

073.011   Shock formation time and the viability of prompt MeV proton shock acceleration in solar flares.
D. F. Smith, S. H. Brecht.
J. Geophys. Res., Vol. 90, No. A1, p. 205 – 209 (1985). – See Abstr. 012.003.

A critical analysis is made of a proposed mechanism to explain the $\lesssim 2$– to 100–s delays observed in some flares between γ rays due to several MeV protons and hard X rays due to ~ 100–keV electrons.

073.012   A white light flare on 25 April 1984.
   T. Natori.
Astron. Her., Vol. 78, No. 2, p. 56 (1985). In Japanese.

073.013   Return currents in solar flares: collisionless effects.
   H. L. Rowland, L. Vlahos.
Astron. Astrophys., Vol. 142, No. 2, p. 219 – 224 (1985).

If the primary, precipitating electrons in a solar flare are unstable to beam plasma interactions, it is shown that strong Langmuir turbulence can seriously modify the way in which a return current is carried by the background plasma. In particular, the return (or reverse) current will not be carried by the bulk of the electrons, but by a small number of high velocity electrons. For beam/plasma densities $> 10^{-3}$, this can reduce the effects of collisions on the return current. For higher density beams where the return current could be unstable to current driven instabilities, the effects of strong turbulence anomalous resistivity is shown to prevent the appearance of such instabilities. Again in this regime, how the return current is carried is determined by the beam generated strong turbulence.

073.014   Twisting motions in a disturbed solar filament.
   B. Schmieder, M. A. Raadu, J. M. Malherbe.
Astron. Astrophys., Vol. 142, No. 2, p. 249 – 255 (1985).

Time sequences of the vertical velocity field in a disturbed solar active region filament (AR 2646) have been derived using the MSDP spectrograph operating in $H\alpha$ at the Meudon Solar Tower. Following a period of systematic upward velocities a pair of closely aligned elongated regions of oppositely directed velocities develops and persists over several minutes. These velocities are interpreted in terms of a twisted magnetic flux rope model for the filament. The initial upward motions indicate that the flux rope is rising. This should lead to an expansion since the surrounding pressure is decreasing with height. Conservation of the current and the magnetic flux along the filament then requires twisting motions as the flux rope adjusts to a new radial equilibrium when rising into the corona.

073.015   Propagation of energetic electron streams in solar flares.
   Y. Mok.
Sol. Phys., Vol. 95, No. 1, p. 181 – 188 (1985).

The microscopic stability of an electron stream flowing down to the photosphere from the corona is examined. It is found that, while a power–law distribution is stable in the low–density corona, it is unstable against the generation of magnetized electron plasma waves in the high–density photosphere. The scattering of these energetic electrons may alter their radiation signatures.

073.016   The behaviour of prominence areas in the 11–year cycle and their relationship with other solar events.
M. N. Gnevyshev, V. I. Makarov.
Sol. Phys., Vol. 95, No. 1, p. 189 – 192 (1985).

On the basis of the data on prominence areas for 1880 – 1976 and positions of the boundary background magnetic field for 1955 – 1982 it is shown that the maximum development of prominences and their poleward migration, accompanied by the magnetic field reversal, coincides with the first maximum of the 11–year solar cycle, which is characterized by an enhancement of solar activity at all latitudes. The second maximum is an increase of all features, including prominences but in the low latitudes only. That prominence zone migrates poleward in the following 11–year solar cycle.

073.017   Neutrons and gamma rays from solar flares.
   R. Ramaty, R. J. Murphy.
AIP Conf. Proc., No. 115, p. 628 – 640 (1984). – See Abstr. 012.005.

The theory of neutron and gamma–ray production in flares is reviewed and comparisons of the calculations with data are made. The principal conclusions pertain to the accelerated proton and electron numbers and spectra in flares and to the interaction site of these particles in the solar atmosphere. For the June 21, 1980 flare, from which high–energy neutrons and high–energy (>10 MeV) photons were seen, the electron–to–proton ratio is energy dependent and much smaller than unity at energies greater than 1 MeV. The interaction site of these particles appears to be the solar chromosphere.

073.018   Spectrophotometry of the bright quiet prominence of June 30, 1981.
M. N. Pasechnik, E. G. Rudnikova.
Probl. Kosm. Fiz., Vyp. 19, p. 22 – 25 (1984). In Russian. – See Abstr. 003.002.

073.019   Investigation of limb flares of July 25, 1959.
   V. A. Ostapenko, V. K. Yarmolenko, A. N. Yakobchuk.
Probl. Kosm. Fiz., Vyp. 19, p. 25 – 34 (1984). In Russian. – See Abstr. 003.002.

073.020   Thermal emission mechanism and physical conditions in chromospheric flares on the sun.
V. A. Ostapenko, P. N. Polupan.
Probl. Kosm. Fiz., Vyp. 19, p. 34 – 41 (1984). In Russian. – See Abstr. 003.002.

073.021   Autowave structure of flare–active regions on the sun.
   Z. B. Korobova, Eh. I. Mogilevskij.
Soln. Dannye, Byull., 1984, No. 10, p. 71 – 79 (1985). In Russian.

Some properties of helical structures in the spatial distribution of sunspots in sunspots in an active region, which are in agreement with the conception of the autowave nature of a flare–active region, are discussed. The latter follows from the synergetic idea of the evolution of solar active regions.

073.022   Chromospheric emission lines in the coronal spectrum of July 31, 1981.
R. A. Gulyaev, V. L. Merzlyakov.
Soln. Dannye, Byull., 1984, No. 10, p. 91 – 94 (1985). In Russian.

Faint emissions in H and K Ca II and $H\alpha$ are present in a coronal spectrum obtained during the total solar eclipse of July 31, 1981. It is shown that above emissions have not coronal origin, but they are conditioned by aureole scattering of prominence radiation.

073.023   Flare Build–up Study – homologous flares group – Part I.
M.-J. Martres, B. E. Woodgate, N. Mein, Z. Mouradian, J. Rayrole, B. Schmieder, G. Simon, I. Soru–Escaut.
Adv. Space Res., Vol. 4, No. 7, p. 5 – 10 (1984). – See Abstr. 012.009.

Analysis, importances, frequency, evolution, localisation, associated phenomena as well as productive areas of homologous flares are discussed.

073.024   Progress in the study of homologous flares on the sun – Part II.
B. E. Woodgate, M.-J. Martres, J. B. Smith Jr., K. T. Strong, M. K. McCabe, M. E. Machado, V. Gaisauskas, R. T. Stewart, P. A. Sturrock.
Adv. Space Res., Vol. 4, No. 7, p. 11 – 17 (1984). – See Abstr. 012.009.

Studies of groups of homologous flares in active regions in 1980 have been made using a variety of space and ground based instruments. Detailed properties of three of these groups have been studied, and are combined to form a possible sequence of events.

073.025   Consecutive homologous flares and their relation to sunspot motions.
L. Gesztelyi.
Adv. Space Res., Vol. 4, No. 7, p. 19 – 22 (1984). – See Abstr. 012.009.

Flares observed in the central part of the Hale region 17098 in September 1980 are reported.

073.026 Homologous flares and the evolution of NOAA Active Region 2372.
K. T. Strong, J. B. Smith Jr., M. K. McCabe, M. E. Machado, J. L. R. Saba, G. M. Simnett.
Adv. Space Res., Vol. 4, No. 7, p. 23 – 26 (1984). – See Abstr. 012.009.

A detailed record of the evolution of NOAA Active Region 2372 has been compiled by the FBS Homology Study Group. It was one of the most prolific flare–producing regions observed by SMM. The flares occurred in distinct stages which corresponded to particular evolutionary phases in the development of the active region magnetic field.

073.027 Recurrent mass ejections observed in Hα and C IV.
B. Schmieder, G. Simon, M.–J. Martres, P. Mein, N. Mein, E. Tandberg–Hanssen.
Adv. Space Res., Vol. 4, No. 7, p. 27 – 30 (1984). – See Abstr. 012.009.

Comparison of the temporal evolution of Hα and C IV brightnesses shows a weak phase lag between Hα and C IV maxima, in the case of homologous flares, with C IV brightness maxima preceding Hα maxima. The analysis of the variation of the ejection velocities is expected to lead to the determination of an energy balance.

073.028 The homologous flare sites and the general solar activity.
M.–J. Martres, Z. Mouradian, I. Soru–Escaut.
Adv. Space Res., Vol. 4, No. 7, p. 31 (1984). Abstract. – See Abstr. 012.009.

073.029 Activity in the homologous flare site.
N. Mein, M.–J. Martres, G. Simon, P. Mein, I. Soru–Escaut.
Adv. Space Res., Vol. 4, No. 7, p. 33 – 35 (1984). – See Abstr. 012.009.

The evolution of a site where homologous flares occurred on June 8, 1980 is analysed by using observations both in the photosphere and in the chromosphere. The homology is discussed through space, energy and dynamical aspects. The criteria are used in order to propose the definition of a coefficient of homology.

073.030 Role of newly emerging flux in the flare process.
E. R. Priest.
Adv. Space Res., Vol. 4, No. 7, p. 37 – 48 (1984). – See Abstr. 012.009.

New flux emerging from below the photosphere is believed to give rise to small flares and also to be capable of triggering large events when extra energy is stored in the overlying field. A summary is given of the observations of emerging flux, together with the current theoretical ideas on its behaviour.

073.031 The stability of magnetic fields relevant to two–ribbon flares.
A. W. Hood.
Adv. Space Res., Vol. 4, No. 7, p. 49 – 52 (1984). – See Abstr. 012.009.

The preflare structure, prior to two–ribbon flares, is thought to consist of magnetic field arcades. As a first approximation, the magnetic field is assumed to be invariant along the length of the arcade. The ideal MHD stability of such structures is studied using the energy method. The dense photosphere is simulated by line–typing the magnetic field and a discussion of boundary conditions is presented.

073.032 Numerical simulation of reconnection in an emerging magnetic flux region.
T. C. Forbes.
Adv. Space Res., Vol. 4, No. 7, p. 53 – 56 (1984). – See Abstr. 012.009.

The resistive MHD equations are numerically solved in two dimensions for an initial–boundary–value problem which simulates reconnection between an emerging magnetic flux region and an overlying coronal magnetic field. The solution involves both ideal–MHD and resistive–MHD processes, and the solution

shows an evolution which is remarkably suggestive of the preflare, impulsive, and main phases of the flare–cycle.

073.033 Sunspot motions and magnetic shears as precursors of flares.
L. Dezsö, G. Csepura, O. Gerlei, Á. Kovács, I. Nagy.
Adv. Space Res., Vol. 4, No. 7, p. 57 – 60 (1984). – See Abstr. 012.009.

Using full–disc white light photoheliograms, the authors have studied umbrae motion and variations in sunspot areas in a large activity complex over 4 solar rotations. On the basis of the observational data they illustrate with typical examples to what extent rapid spot motions are associated with flare occurrences.

073.034 Relationships of a growing magnetic flux region to flares.
S. F. Martin, R. D. Bentley, A. Schadee, A. Antalova, A. Kucera, L. Dezsö, L. Gesztelyi, K. L. Harvey, H. Jones, S. H. B. Livi, J. Wang.
Adv. Space Res., Vol. 4, No. 7, p. 61 – 70 (1984). – See Abstr. 012.009.

The authors have identified and analyzed the evolution of flare sites at the boundaries of a major new and growing magnetic flux region within a complex of active regions, Hale No. 16918 (1980).

073.035 The role of magnetic field shear in solar flares.
M. J. Hagyard, R. L. Moore, A. G. Emslie.
Adv. Space Res., Vol. 4, No. 7, p. 71 – 80 (1984). – See Abstr. 012.009.

The character of magnetic shear and its involvement in the buildup and release of flare energy are reviewed and illustrated.

073.036 Magnetic field structure changes in the vicinity of solar flares.
B. Kalman.
Adv. Space Res., Vol. 4, No. 7, p. 81 – 85 (1984). – See Abstr. 012.009.

Changes in the structure of the sunspot group and its magnetic field are studied in Hale Region 17644 (May 1981) in connection with the May 16 3B/X1 flare. The characteristic changes, also found in HR 16850 (May 1980) and HR 17098 (September 1980), are the following: Rapid motions of umbrae of opposite polarity in the vicinity of the magnetic zero line, parallel to this line, but in opposite direction. Appearance of new small spots before the flare, leading to a more complicated field structure. Simplification of the magnetic structure after the flare in some days, i.e. decrease of spot areas in the affected territory and the straightening of the magnetic zero line.

073.037 Spectroscopic study of plasma parameters for solar active regions and flares.
S. L. Mandelstam (*S. L. Mandel'shtam*), A. M. Urnov, I. A. Zhitnik.
Adv. Space Res., Vol. 4, No. 7, p. 87 – 89 (1984). – See Abstr. 012.009.

Results are given of the study of active regions and flares by a high resolution Mg XI ion spectra obtained aboard rockets and a satellite. It is shown that there is a noticable similarity in the physical conditions in the plasma of active regions and flares.

073.038 Study of combined soft and hard X–ray images of solar flares.
M. E. Machado, A. M. Hernández, M. G. Rovira, C. V. Sneibrun.
Adv. Space Res., Vol. 4, No. 7, p. 91 – 94 (1984). – See Abstr. 012.009.

The authors have studied soft and hard X–ray images of 13 solar flares from six active regions observed by the Hard X–ray Imaging Spectrometer. The results indicate the presence of pre-hard X–ray burst excesses in the 11.5 – 30.0 keV range, indicating a slow buildup of the acceleration process or a strong preheating. During the impulsive phase, all of the events show the simultaneous energization of neighboring field structures, which share about equal amounts of the released energy. This association seems to be indicative of strong acceleration and energy release triggered by the interaction between magnetic loops.

073.039 Flare precursors and onset.
G. Van Hoven, G. J. Hurford.
Adv. Space Res., Vol. 4, No. 7, p. 95 – 103 (1984). – See Abstr. 012.009.

The authors report on the progress of a search for precursors that have direct physical connections to the start of subsequent solar flares. The discussion includes recent results at radio, visible, ultraviolet, and X–ray wavelengths, which are relevant to the pre–impulsive (onset) phase. The authors also relate the aspects of a theoretical scenario, based on magnetic reconnection with transport–coefficient phase changes, for explaining flare onset. The pertinent time scales for pre–implusive temporal developments are discussed.

073.040 Origin and location of chromospheric evaporation in flares.
E. Antonucci, D. Marocchi, G. M. Simnett.
Adv. Space Res., Vol. 4, No. 7, p. 111 – 115 (1984). – See Abstr. 012.009.

Observation of two flares obtained with the Solar Maximum Mission spectrometers indicate that at flare onset the emission in soft (3.5 – 8 keV) and hard (16 – 30 keV) X–rays is predominant at the footpoints of the flaring loops. Since, at the same time, blue–shifts are observed in the soft X–ray spectra from the plasma at temperature of 10^7K, the authors infer that material is injected at high velocity into the coronal loops from the footpoints. These areas are also the sites of energy deposition, since their emission in hard X–rays is due to non–thermal electrons penetrating in the denser atmosphere. Hence, chromospheric evaporation occurs where energy is deposited. During the impulsive phase, the configuration of the flare region changes indicating that the flaring loop is progressively filled by hot plasma.

073.041 Identification of two X–ray miniflares with Hα–subflares.
A. Schadee, V. Gaizauskas.
Adv. Space Res., Vol. 4, No. 7, p. 117 – 120 (1984). – See Abstr. 012.009.

Active regions show many short–lived emissions in the 3.5 – 5.5 keV range that are 100 to 1000 times weaker than "normal" X–ray flares. The hypothesis that they may well be miniflares is supported by the simultaneous occurrence of 2 Hα–subflares at the site of weak X–ray sources.

073.042 Solar–flare neutrons and gamma rays.
R. J. Murphy, R. Ramaty.
Adv. Space Res., Vol. 4, No. 7, p. 127 – 136 (1984). – See Abstr. 012.009.

Calculations of neutron and gamma–ray production in solar flares are reviewed and compared with neutron and gamma–ray data from the 21 June 1980 and 3 June 1982 flares, as well as gamma–ray data from other flares. The implied charged–particle numbers and spectra are compared with interplanetary observations.

073.043 First–order Fermi shock acceleration in solar flares.
D. C. Ellison, R. Ramaty.
Adv. Space Res., Vol. 4, No. 7, p. 137 – 141 (1984). – See Abstr. 012.009.

First order Fermi shock acceleration of electrons, protons and alpha particles is compared to observations of energetic particle events. For each event, a unique shock compression ratio produces spectra in good agreement with observation. The simple model predicts that the acceleration time to a given energy will be approximately equal for electrons and protons and, for reasonable solar parameters, can be less than 1 second to $\sim$100 MeV.

073.044 High–temperature phenomena in flares.
T. Takakura, K. Tanaka, E. Hiei.
Adv. Space Res., Vol. 4, No. 7, p. 143 – 152 (1984). – See Abstr. 012.009.

High temperature phenomena occurring in solar flares are reviewed based on hard X–ray images and spectral analyses of highly ionized iron lines observed aboard the Hinotori spacecraft. Five basic flare components are proposed, i.e., impulsive, gradual hard, thermal, quasi thermal and hot thermal components. A flare shows some combination of the five components. Energy release and transport for each component would give a lot of variety to the hard X–ray image, spectrum and time history of X–rays.

073.045 On the correlation of microwave bursts with Hα flares.
S. Urpo, V. Ruzdjak, H. Teräsranta.
Adv. Space Res., Vol. 4, No. 7, p. 153 – 156 (1984). – See Abstr. 012.009.

Results of the direct comparison of solar microwave recordings at the Metsähovi Radio Research Station and Hα films at the Hvar Observatory are reported. This comparison reveals that the correlation of microwave gradual rise and fall (GRF) events (time scale tens of minutes) with Hα flares is practically 100 percent. On the other hand the correlation of the impulsive microwave bursts (time scale tens of seconds) with Hα flares is low. From this it can be concluded that the main contribution to the microwave GRF bursts comes from thermal radiation.

073.046 Hard X–ray and radio investigations prior to or during the impulsive phase of solar flares.
M. Pick, A. Raoult.
Adv. Space Res., Vol. 4, No. 7, p. 175 – 178 (1984). – See Abstr. 012.009.

This paper focusses on the activity prior and during the impulsive phase of solar flares. Observations give evidence for electron acceleration prior to the impulsive phase. The association between type III groups and hard X–ray bursts becomes closer with increasing starting frequency of the former observed during the impulsive phase. It is shown that pure type III burst groups, when they are X–ray associated, do not correspond to an intense X–ray emission. At the opposite, the type III/V events can be associated with strong X–ray emission. Radioheliograph observations bring constraints on the geometry of the injection/acceleration site.

073.047 Magnetic reconnection in a high–temperature plasma of solar flares.
B. V. Somov, V. S. Titov.
Adv. Space Res., Vol. 4, No. 7, p. 183 – 185 (1984). – See Abstr. 012.009.

073.048 Calculation of X–ray polarization during flares.
E. Haug, G. Elwert, R. R. Rausaria.
Adv. Space Res., Vol. 4, No. 7, p. 187 – 189 (1984). – See Abstr. 012.009.

The polarization of X–radiation emitted by electrons which are accelerated during flares is investigated in a non–thermal model.

073.049 Energy–transfer processes in flares.
D. M. Rust.
Adv. Space Res., Vol. 4, No. 7, p. 191 – 198 (1984). – See Abstr. 012.009.

073.050 Do all flares occur within a hierarchy of magnetic loops?
R. A. Harrison, G. M. Simnett.
Adv. Space Res., Vol. 4, No. 7, p. 199 – 202 (1984). – See Abstr. 012.009.

073.051 Differential emission measure analysis of hot–flare plasma from Solar–Maximum Mission X–ray data.
J. Jakimiec, J. Sylwester, J. R. Lemen, R. Mewe, R. D. Bentley, A. Fludra, J. Schrijver, B. Sylwester.
Adv. Space Res., Vol. 4, No. 7, p. 203 – 207 (1984). – See Abstr. 012.009.

The authors have investigated differential emission measure (DEM) distribution of hot flare plasma (T > 10MK) using SMM X–ray data from Bent Crystal Spectrometer and Hard X–ray Imaging Spectrometer. Typical examples of the DEM distribution are discussed.

073.052   The magnetohydrodynamic development of two–ribbon flares or a five–finger theory for solar flares.
J. S. Kaastra.
Adv. Space Res., Vol. 4, No. 7, p. 209 – 210 (1984). – See Abstr. 012.009.
A semi–analytical model for the electrodynamic development of two–ribbon flares is presented.

073.053   Analysis of the 1980 November 18 limb flare observed by the Hard X–ray Imaging Spectrometer (HXIS).
E. Haug, G. Elwert, P. Hoyng.
Adv. Space Res., Vol. 4, No. 7, p. 211 – 213 (1984). – See Abstr. 012.009.

073.054   Bidimensional spectroscopy of the solar chromosphere during the Maximum Year.
B. Caccin, A. Falchi, R. Falciani, G. Roberti, L. A. Smaldone.
Adv. Space Res., Vol. 4, No. 7, p. 215 – 219 (1984). – See Abstr. 012.009.
The bidimensional spectroscopy method has been tested and used for active regions and flare studies. The authors present the detailed photometric morphology of the observed active area and the longitudinal velocity field pattern.

073.055   Combined analysis of soft and hard X–ray spectra from flares.
A. H. Gabriel, E. Bely-Dubau, J. C. Sherman, L. E. Orwig, J. Schrijver.
Adv. Space Res., Vol. 4, No. 7, p. 221 – 225 (1984). – See Abstr. 012.009.
A method has been developed for interpreting the combined data set from the Bent Crystal Spectrometer, the Hard X–ray Imaging Spectrometer and the Hard X–Ray Burst Spectrometer on the Solar Maximum Mission covering both line and continuum X–ray emission from 4 kV up to 500 kV. The observations are fitted to a model including thermal and non–thermal electron components.

073.056   Study of energy release in flares.
J.–C. Hénoux.
Adv. Space Res., Vol. 4, No. 7, p. 227 – 237 (1984). – See Abstr. 012.009.
Recent advances in the study of energy release in flares are reviewed. Progress has been made in modelling coronal X–ray emission and the chromospheric response to energy input. These advances are based on theoretical studies and on the comparison of complementary data obtained from spacecraft and ground–based observatories. The implication of these results on the primary energy release process is discussed and prospects for new research are presented.

073.057   Interpretation of hard X–ray images during the impulsive phase of a limb flare.
M. E. Machado, G. Lerner.
Adv. Space Res., Vol. 4, No. 7, p. 239 – 241 (1984). – See Abstr. 012.009.
The authors show that the observations of a limb flare, in which a hard X–ray (16 – 30 keV) source is seen at the boundary between two interacting magnetic structures, indicate the presence of hot ($T \gtrsim 6 \times 10^7$K) plasma within the region. Non–thermal bremsstrahlung processes do not agree with these observations. Possible causes of the heating are discussed.

073.058   Observational evidence for chromospheric footpoint penetration of nonthermal electrons during two well–observed flares.
R. C. Canfield, T. A. Gunkler, A. L. Kiplinger.
Adv. Space Res., Vol. 4, No. 7, p. 255 – 258 (1984). – See Abstr. 012.009.
The authors study two events observed as part of a coordinated observing program between the Solar Maximum Mission and Sacramento Peak Observatory: the flares of 14.56 UT, 7 May 1980 and 15.22 UT, 24 June 1980. They can distinguish effects of intense nonthermal electron heating from those of high conduction and pressure from the overlying flare corona. Both flares show the signature of intense chromospheric heating by fast electrons, temporally correlated with X–ray light curves at $E > 27$ keV, and spatially associated with X–ray emission sites at $E > 16$ keV.

073.059   Eruption of huge magnetic systems from the Sun.
B. Rompolt.
Adv. Space Res., Vol. 4, No. 7, p. 357 – 361 (1984). – See Abstr. 012.009.
A number of evidences is presented supporting an idea that both the eruptive prominences as well as the associated white light transients are generated by eruptions of a huge magnetic field system from the solar surface. The presented opinion is based on a detailed comparison of the geometry, the characteristic features during the expansion and the location on the limb of the white light transients.

073.060   Increasing solar chromosphere line intensities with solar activity.
P. Lemaire.
Adv. Space Res., Vol. 4, No. 8, p. 29 – 35 (1984). – See Abstr. 012.010.
The OSO–8/L.P.S.P. multichannel experiment has recorded simultaneous chromospheric profiles in H and K Ca II, h and k Mg II, Lα and Lβ H I lines. These lines are formed from the temperature minimum (4200K) to the higher part of the chromosphere (30000K). A data set recorded over quiet sun, plage, penumbra and filament is presented. The observations, data reduction and calibration are described. Then a comparison between line intensities is made and results are discussed.

073.061   HRTS observations of spicular emission at transition region temperatures above the solar limb.
J. W. Cook, G. E. Brueckner, J.–D. F. Bartoe, D. G. Socker.
Adv. Space Res., Vol. 4, No. 8, p. 59 – 62 (1984). – See Abstr. 012.010.
Slit spectra and spectroheliograph observations were obtained during the fourth rocket flight of the High Resolution Telescope and Spectrograph (HRTS) on 7 March 1983. The C IV spectroheliograms show general spiked emission above the limb, and also several small loop– or prominence–like events. Slit spectra along the tops of several of these structures show tilted features which could be interpreted as rotational velocities of approximately 50 km s^{-1}.

073.062   The chromosphere–corona transition zone above an active region.
O. Kjeldseth Moe, Ø. Andreassen, P. Maltby, J.–D. F. Bartoe, G. E. Brueckner, K. R. Nicolas.
Adv. Space Res., Vol. 4, No. 8, p. 63 – 66 (1984). – See Abstr. 012.010.
Intensities and profiles of ion emission lines between 1170 Å and 1700 Å from an active region on the Sun are measured from spectra obtained with the Naval Research Laboratory's High Resolution Telescope and Spectrograph (1975). Fine structure variation of intensities and gas flow velocities in the temperature range 20,000 – 200,000K are determined and electron pressures are measured.

073.063   Chromospheric flare activity in solar cycle 20.
Š. Knoška, J. Petrášek.
Contrib. Astron. Obs. Skalnaté Pleso, Vol. 12, p. 165 – 260 (1984).
The paper gives the daily values of the flare index for solar cycle 20, i.e. the years 1966 – 1976. Separate indices were calculated for the north and south solar hemisphere and for longitudinal zones of 30° from the central meridian.

073.064   ³He–rich solar flares.
S. Ramadurai, M. N. Vahia, S. Biswas, K. Sukurai.
Pramāna, Vol. 23, No. 3, p. 305 – 311 (1984). Abstr. in Phys. Abstr., Vol. 88, No. 1248, Entry 9706 (1985).

073.065 Interpretation of millimeter and sub–millimeter observations of the solar chromosphere.
L. E. Cram.
Int. J. Infrared Millimeter Waves, Vol. 5, No. 8, p. 1165 – 1177 (1984). Abstr. in Phys. Abstr., Vol. 88, No. 1250, Entry 18745 (1985).

073.066 Quasi–static evolution of sheared force–free fields and the solar flare problem.
J. J. Aly.
Astron. Astrophys., Vol. 143, No. 1, p. 19 – 22 (1985).
The author reports some new results showing the possible evolution of a two–dimensional force–free field in the half–space $\{z > 0\}$ toward an open field. This evolution is driven by shearing motions applied to the feet of the field lines on the boundary $\{z = 0\}$. The author discusses the consequences of his results for a model of the two–ribbon solar flare.

073.067 Oscillations in the chromosphere and transition region above sunspot umbrae. A photospheric or a chromospheric resonator?
Y. D. Žugžda (*Yu. D. Zhugzhda*), V. Locāns, J. Staude.
Astron. Astrophys., Vol. 143, No. 1, p. 201 – 205 (1985).
The authors compare the basic features of the model for a chromospheric resonator for slow waves proposed by the present authors (ZLS) with those of the model for a photospheric resonator for fast waves proposed by Thomas and Scheuer (1982; TS). Moreover, the model predictions are tested for agreement with recent observations of umbral oscillations. Contrary to the claim by Thomas (1984), the observations in the umbral chromosphere and transition region are in agreement with the ZLS model but are difficult to reconcile with the TS model. Finally, the authors outline necessary improvements in the existing simplified theories.

073.068 Equilibre MHD et instabilités dans les protubérances ou filaments solaires.
J.–M. Malherbe.
Champs magnétiques stellaires, p. 383 – 412 (1984). – See Abstr. 012.022.

073.069 Heating in the solar mantle.
C. Chiuderi.
NASA Conf. Publ., NASA CP–2358, p. 101 – 105 (1985). Abstract. – See Abstr. 012.023.

073.070 Energetic electron heating and chromospheric evaporation during a well–observed compact flare.
R. C. Canfield, T. A. Gunkler.
Astrophys. J., Vol. 288, No. 1, p. 353 – 362 (1985).
From the Solar Maximum Mission and Sacramento Peak Observatory the authors observed the compact solar flare of 1980 May 7 with spatial, spectral, and temporal resolution in both X–rays and Hα profiles, throughout the impulsive phase. They compare the observed flare Hα profiles with theoretical Hα profiles based on physical models of chromospheric flare processes and model parameters inferred from the X–ray observations.

073.071 An impulsive solar burst observed in Hα, microwaves, and hard X–rays.
D. E. Gary, F. Tang.
Astrophys. J., Vol. 288, No. 1, p. 385 – 395 (1985).
The authors present Hα, 10.6 GHz microwave, and greater than 100 keV X–ray observations of a single impulsive spike flare that occurred on 1980 May 28, 1947 UT. The isolated high flux spike provides an excellent opportunity to infer the time behavior of the electron acceleration that is assumed to yield the microwave and hard X–ray flux profiles. It is found that, although the time profile consists of a simple, single peak, the spike is actually due to two acceleration episodes in separate sources. The authors adopt a source model for microwave and hard X–ray sources, in which the microwave source is due to electrons trapped at the top of a magnetic loop, and the hard X–ray source is due to thick–target emission at the footpoints.

073.072 Observational evidence for thermal wave fronts in solar flares.
D. M. Rust, G. M. Simnett, D. F. Smith.
Astrophys. J., Vol. 288, No. 1, p. 401 – 409 (1985). With plate 14.
Images in 3.5 – 30 keV X–rays obtained during the first few minutes of seven solar flares show rapid motions. In each case X–ray emission first appeared at one end of a magnetic field structure, and then propagated along the field at a velocity between 800 and 1700 km s^{-1}. The observed X–ray structures were 45,000 – 230,000 km long. The fast–moving fronts are interpreted as electron thermal conduction fronts, since their velocities are consistent with conduction at the observed temperatures of $1 - 3 \times 10^7$K. The inferred conductive heat flux of up to $\sim 10^{10}$ergs s^{-1}cm^{-2} accounts for most of the energy released in the flares, implying that the flares were primarily thermal phenomena.

073.073 Oscillations of the Sun's chromosphere. II. Hα line centre and wing filtergram time sequences.
F. Kneer, M. von Uexküll.
Astron. Astrophys., Vol. 144, No. 2, p. 443 – 451 (1985). = Mitt. Kiepenheuer–Inst., Nr. 229.
In order to investigate the dynamics of the solar chromosphere and to reveal candidates for chromospheric heating mechanisms the authors have performed a Fourier analysis of time sequences (total duration 128 min) of Hα photographic filtergrams taken simultaneously at disc centre in line centre and ± 0.5 Å from the line centre.

073.074 Singularity of solar rotation and flare productivity. I.
I. Soru–Escaut, M.–J. Martres, Z. Mouradian.
Astron. Astrophys., Vol. 145, No. 1, p. 19 – 24 (1985).
The authors study specific filament–flare situations in which the time sequence includes the five previous rotations up to the flare time. They derived a long–range estimate of the events from the "Cartes Synoptiques de la Chromosphère Solaire" published by Paris–Meudon Observatory. As for the spatial distribution of the flares, they have used the maps of the same publication obtained by the compilation of world data cinematographic survey. The paper is divided into two parts. In the first part, the authors discuss the relationship between the filaments and the global flaring emission of the active centres, then the evolution of this relation from 1974 to 1980, including a minimum and maximum of activity. The most eruptive areas are studied in the second part, which allows to offer a better definition of the spatial flare–filament association and to discuss the interpretation of the event in terms of differential solar rotation.

073.075 Gas dynamics in the impulsive phase of solar flares. II. The structure of the transition region – a diagnostic of energy transport processes.
A. G. Emslie, F. Nagai.
Astrophys. J., Vol. 288, No. 2, p. 779 – 788 (1985).
Optically thin line emission formed in the solar transition region provides sensitive diagnostics of the atmospheric structure at these levels. Observations of such emission lines during the impulsive phase of solar flares show that a good correlation exists between the energy input rate and the intensity I of a given line. This paper calculates and contrasts the predicted temporal behavior of the differential emission measure for two models of energy transport in a flaring loop. A model in which the energy is transported throughout the flaring loop principally by collisional degradation of a beam of accelerated suprathermal electrons does adequately reproduce observed behavior.

073.076 Electron acceleration in solar flares and the transition from nonthermal to thermal hard X–ray phases.
D. F. Smith.
Astrophys. J., Vol. 288, No. 2, p. 801 – 805 (1985).
Observations are reviewed which indicate that hard X–rays during the impulsive phase of a flare typically start with a primarily nonthermal phase which undergoes a transition to a primarily thermal phase as the flare progresses. Recent theoretical work on the modified two–stream instability as an efficient electron accelerator and modeling of thermal hard X–ray sources is considered.

A scenario which is termed the dissipative thermal model is proposed to explain the observations.

073.077   Microwave and X–ray observations of delayed brightenings at sites remote from the primary flare locations.
H. Nakajima, B. R. Dennis, P. Hoyng, G. Nelson, T. Kosugi, K. Kai.
Astrophys. J., Vol. 288, No. 2, p. 806 – 819 (1985).

The authors present five examples of solar flares observed with the 17 GHz interferometer at Nobeyama in which a secondary microwave burst occurred at a distance of $10^5 - 10^6$km from the primary flare site. The secondary microwave burst in all five cases had a similar time profile to the primary burst with a delay of 2 – 25 s. Two of the events were accompanied by meterwave type III/V bursts located high in the corona between the primary and secondary sites. For two of the other events, X–ray images of the secondary source were obtained with the SMM satellite. These observations strongly suggest in each of the five events that the distant microwave burst was produced by electrons with energies of 10 – 100 keV which were channeled along a huge loop from the main flare site to the remote location.

073.078   Remarks on the magnetic support of quiescent prominences.
U. Anzer, E. Priest.
Sol. Phys., Vol. 95, No. 2, p. 263 – 268 (1985).

The development of magnetic field structures which can lead to prominence configurations of the Kuperus–Raadu type is discussed. Starting from streamer type configurations and preserving the total current in the system the authors find that simple two–dimensional static configurations lead to prominences which in general lie systematically much lower than the heights found from observations. It is therefore concluded that either more complex field configurations are needed to explain the recent observations by Leroy et al. (1983) or the initial configurations must be very special.

073.079   Rocket observation of the EUV images of a solar flare and active regions.
T. Hirayama, K. Tanaka, T. Watanabe, K. Akita, T. Sakurai, K. Nishi.
Sol. Phys., Vol. 95, No. 2, p. 281 – 295 (1985).

Images of a flare and active regions were obtained in the extreme ultraviolet emission lines with a spatial resolution of less than ten seconds of arc together with one–dimensional scanning at 1650 Å. A microchannel plate was used as a detector. The relationship between the shape of the flare and the structure of the photospheric magnetic field is discussed. A map of the electron temperature distribution derived from the intensity ratio of the Lyman continua at 880 Å and 815 Å showed a lower temperature in regions of higher activity. A very small geometrical thickness of 50–500 m in the C III emitting region of the flare was found. And the layer emitting the continuum in 1650 Å is shown to be at a temperature of 5300K in the flare and 4700K in active regions.

073.080   Magnetic field structures of hard X–ray flares observed by HINOTORI spacecraft.
T. Sakurai.
Sol. Phys., Vol. 95, No. 2, p. 311 – 321 (1985).

The magnetic field structure of five flares observed by HINOTORI spacecraft is studied. The double source structure of impulsive flares seems to indicate hard X–ray emission from the two footpoints of a flaring loop, but the potential field computation does not reproduce a loop connecting the two sources. Therefore the magnetic field could be in a sheared configuration. On the other hand gradual flares are characterized by hard X–ray sources located in the corona, $2 - 4 \times 10^4$km above the photosphere. The potential field modeling is found to give a reasonable fitting in this type of flares, and the hard X–ray sources are located at the top of the magnetic loop or arcade. This configuration is consistent with the thick–target trap model of the hard X–ray bursts.

073.081   Ejection of chromospheric material associated with injection of electrons in the solar corona.
N. Mein, Y. Avignon.
Sol. Phys., Vol. 95, No. 2, p. 331 – 342 (1985).

Observations of a type III radio event and of concurrent Hα absorbing features are related. They were obtained with the Nançay Radioheliograph and the Multichannel Subtractive Double Pass spectrograph operating in Hα at Meudon. The authors are looking for the signature at chromosphere levels of the acceleration of the electron beams triggering the type III bursts. Some promising results are pointed out: the relationship between velocities in Hα and the occurrence of the type III bursts, the shape of Hα line which reveals turbulent motions, the presence of parasitic magnetic polarity and the probable existence of a shock wave. A schematic scenario of the phenomenon is proposed.

073.082   Dependence of the flare stream velocity on magnetic field orientation.
M. I. Pudovkin, S. A. Zaitseva (*S. A. Zajtseva*), S. P. Puchenkina.
Sol. Phys., Vol. 95, No. 2, p. 371 – 380 (1985).

The main parameters of a flare stream – its velocity, magnetic field intensity, plasma density, and temperature – are shown to depend on the mutual orientation of magnetic fields in the main body of the stream and within its compressed solar wind region. The mutual orientation also affects the characteristics of the flares and, in particular, the probability of solar type IV radio emission. The results obtained are explained within the framework of a model that takes into account magnetic field reconnection at the front surface of the stream body.

073.083   On a possibility of forecasting reliably some flare situations.
V. M. Efimenko, V. V. Tel'nyuk–Adamchuk.
Soln. Dannye, Byull., 1984, No. 11, p. 59 – 63 (1985). In Russian.

A reliable forecasting (100 per cent) of some separate flare situations (large flares, flare activity level) is discussed. The mathematical method of fictitious disturbances or expansion on the main components has been applied using some procedure of main components separation. The obtained relations for forecasting some flare events and their absence are given.

073.084   Character of brightness and area variations of flocculi in the Hα and K Ca II lines.
N. B. Ograpishvili.
Soln. Dannye, Byull., 1984, No. 11, p. 88 – 93 (1985). In Russian.

The Hα and K Ca II spectroheliograms obtained on July 14 – 23, August 12 – 16, September 2 – 8 and September 4 – 13, 1964 were reduced. The analysis of the data obtained shows that the development of the flocculi was not gradual, but had a pulsating character. The relative brightness of the flocculi in the K Ca II lines is much larger than that of the Hα line at the initial stage of the flocculi evolution.

073.085   On the solar faculae heating mechanism.
Eh. V. Kononovich, I. V. Mironova, B. E. Serebryakov.
Soln. Dannye, Byull., 1984, No. 12, p. 49 – 54 (1985). In Russian.

Solar faculae heating is suggested to be due to current Joule dissipation connected flux tubes. A corresponding model of the temperature–height dependence is calculated.

073.086   Peculiarities of the spectrum of a flare occurring above a sunspot umbra.
N. S. Shilova.
Soln. Dannye, Byull., 1984, No. 12, p. 62 – 68 (1985). In Russian.

The Balmer line profiles of H_8, NaD_1 and Fe I 4271 Å and widths of the lines $H_3 - H_8$ taken by Kubota et al. (1974) in a flare above a sunspot are considered.

073.087   On the structure of polar faculae.
V. I. Makarov, V. V. Makarova.
Soln. Dannye, Byull., 1984, No. 12, p. 88 – 94 (1985). In Russian.

Coordinates of polar faculae and charts of their spatial distribution are given using photoheliograms for 1970 – 1978. Four types of polar faculae are defined according to their expansion and compactness. On the basis of identification of polar faculae with magnetic field knots it is shown that polar faculae are classified into unipolar and bipolar faculae using magnetic characteristics as a criterion. The polarity of the magnetic field of bipolar faculae, originated after the general magnetic field reversal, is opposite to that of sunspots at that period and coincides with the polarity of the magnetic field of sunspots in the following cycle.

073.088   Interplanetary perturbation from a flare triplet in May 1981 according to measurements aboard Prognoz 8.
G. N. Zastenker, N. L. Borodkova, K. G. Ivanov, N. V. Mikerina.
Kosm. Issled., Tom 23, Vyp. 1, p. 134 – 142 (1985). In Russian. English translation in Cosm. Res.

073.089   Flare loop radiative hydrodynamics. V. Response to thick–target heating.
G. H. Fisher, R. C. Canfield, A. N. McClymont.
Astrophys. J., Vol. 289, No. 1, p. 414 – 424 (1985).
The authors have modeled the hydrodynamic and radiative response of a preflare loop atmosphere to short (5 s) bursts of energy in the form of energetic nonthermal electrons. Energy fluxes in the calculations range from $10^9 \mathrm{ergs\,cm^{-2}s^{-1}}$ to $10^{11}\mathrm{ergs\,cm^{-2}s^{-1}}$. The authors have improved on previous hydrodynamic flare calculations by taking into account optically thick losses in the flare chromosphere, by spatially resolving the flare transition region, and by self–consistently accounting for conductive flux saturation. Results of these calculations are shown by displaying the temperature, density, pressure, and velocity at selected times for a range of energy fluxes.

073.090   Flare loop radiative hydrodynamics. VI. Chromospheric evaporation due to heating by nonthermal electrons.
G. H. Fisher, R. C. Canfield, A. N. McClymont.
Astrophys. J., Vol. 289, No. 1, p. 425 – 433 (1985).
The response of the solar chromosphere to flare heating by nonthermal electrons is examined. A number of interesting phenomena appear in the numerical solutions of the equations of hydrodynamics and radiative transfer. Here the authors discuss one aspect of these results: the phenomenon of chromospheric evaporation. They present results for a range of heating fluxes and show that an energy flux threshold exists distinguishing "gentle" evaporation from "explosive" evaporation.

073.091   Flare loop radiative hydrodynamics. VII. Dynamics of the thick–target heated chromosphere.
G. H. Fisher, R. C. Canfield, A. N. McClymont.
Astrophys. J., Vol. 289, No. 1, p. 434 – 441 (1985).
It is shown that hydrodynamic phenomena in the chromospheric portion of the flaring solar atmosphere depend dramatically on whether chromospheric evaporation by thick–target fast–electron heating is "gentle" or "explosive". In the case of gentle evaporation, velocities in the upper chromosphere are upward. In the case of explosive evaporation, the overpressure of the evaporated material drives downward motion in the residual flare chromosphere. The plasma driven downward by explosive evaporation is cool and dense in comparison with the chromospheric material ahead of it. The authors review previous discussions of these "chromospheric condensations" and develop a simple model for the formation and propagation of chromospheric condensations.

073.092   Flare model and the energy balance in the upper chromosphere.
Eh. A. Baranovskij.
Izv. Krymskoj Astrofiz. Obs., Tom 69, p. 100 – 108 (1984). In Russian. English translation in Bull. Crimean Astrophys. Obs., Vol. 69.
A plane–parallel model atmosphere has been constructed for approximate simulation of observations of Lyman and Balmer lines and Lyman continuum ($\lambda 902$ Å) for flares of importance 2 – 3. It is concluded that in the region of Hα – Hγ lines formation the hydrogen atoms density is no less than $10^{15}\mathrm{cm^{-3}}$ and the temperature does not exceed 10000K.

073.093   Si II line ratios in the Sun.
P. L. Dufton, A. E. Kingston.
Astrophys. J., Vol. 289, No. 2, p. 844 – 848 (1985).
New atomic data for Si II have been used to predict level populations and emission–line intensity ratios for electron densities and temperatures appropriate to the solar atmosphere. The electron impact collision rates are significantly increased by the presence of complex resonance structure in the low–energy collision strength. Generally good agreement is found with observations obtained from the NRL slit spectrograph on Skylab. The effect of deviations from a Maxwellian electron energy distribution on the predicted line intensities is also briefly discussed.

073.094   Analysis of the flare of May 16th, 1981 with a complex space–time structure using opical, X–ray data and radio observations.
V. N. Ishkov, A. K. Markeev, V. V. Fomichev, G. P. Chernov, I. M. Chertok, O. B. Likin, N. F. Pisarenko, M. Karlicky, A. Tlamicha, F. Fárník, B. Valníček, B. Kalman.
Bull. Astron. Inst. Czech., Vol. 36, No. 2, p. 81 – 96 (1985). With plates 1 – 4.
Using heliograms, obtained with a tuneable Hα–filter, dynamic spectra of radio emissions in the 1000 – 45 MHz frequency band, burst records at a number of field frequencies and observations of X–ray emissions from the Prognoz–8 satellite, a comprehensive analysis has been carried out of the large solar flare of May 16th, 1981. The location and dynamics of Hα–knots and also the features of radio and X–ray bursts indicate that this particular event is an example of a flare with a complex space–time structure.

073.095   Studies on some aspects of the solar Hα flares of different visual features.
T. Chakravorti, T. K. Das, M. K. Das Gupta.
Bull. Astron. Inst. Czech., Vol. 36, No. 2, p. 122 – 127 (1985).
Solar Hα flares of different visual features have been analysed in respect of their general characteristics, association with radio bursts and sunspots.

073.096   Heating and acceleration processes in hot thermal and impulsive solar flares.
S. Tsuneta.
Astrophys. J., Vol. 290, No. 1, p. 353 – 358 (1985).
A simple model to describe the unusual characteristics of hot thermal flares and the overall hard X–ray time evolution of impulsive flares is presented. Under the assumption of a steady electric field applied by an external driver during the course of flare evolution, it is shown that a temporal variation of background plasma density due to chromospheric evaporation plays an essential role in the overall time evolution of impulsive bursts, and that high plasma density in a flaring loop is responsible for the occurrence of a hot thermal flare.

073.097   Spicules and surges.
M. L. Blake, P. A. Sturrock.
Astrophys. J., Vol. 290, No. 1, p. 359 – 368 (1985). With plate 4.
The authors adopt the position that spicules, macrospicules, and surges are manifestations of the same phenomenon occurring on different scales. They search for a mechanism that can be successfully applied to explain the phenomenon on all three scales. It is found that the Pikel'ner model, according to which gas is transported in a sequence of "magnetic sacks" which may, for instance, form as the result of reconnection, can reproduce the kinematic properties of spicules and surges. A modification of this mechanism, involving two different magnetic field configurations, can explain the collimation of spicules and surges. The implications of the Pikel'ner mechanism for the heating of the solar corona are examined.

073.098 Physical conditions in the cool parts of prominences. III. The Sr$^+$/Ba$^+$ resonance line ratios and the internal Lyman–alpha flux.
D. A. Landman.
Astrophys. J., Vol. 290, No. 1, p. 369 – 379 (1985).

This paper is devoted to the determination of the physical properties of the low–temperature material in prominences and spicules. It uses the differential effect of H Lyα on the ionization balance in Sr$^+$ and Ba$^+$ to provide a measure of the internal Lyα field in quiescent prominences. The statistical equilibria for these ions have been recomputed using expanded model atoms and new photoionization cross sections obtained according to a scaled Thomas–Fermi method with core polarization effects included. The calculated level population ratios are used to interpret Ba$^+$ and Sr$^+$ resonance line intensity measurements made in a large number of prominences at Haleakala.

073.099 Characteristics of the white–light source in the 1981 April 24 solar flare.
S. R. Kane, J. J. Love, D. F. Neidig, E. W. Cliver.
Astrophys. J., Lett. Ed., Vol. 290, No. 1, p. L45 – L48 (1985).

The large white–light flare on 1981 April 24 (∼ 1358 UT) was very well observed at the hard X–ray, optical, and radio wavelengths. Energetic particles escaping from the Sun were detected in the interplanetary space and in the vicinity of the Earth. The flare had distinct *impulsive* and *gradual* phases and provided the best available measurements of the optical continuum in a solar flare. In this letter the authors present these observations and discuss their interpretation in terms of the energetics of the flare and the role of energetic electrons in the production of optical continuum emission.

073.100 Gas dynamics in the impulsive phase of solar flares. III. Energy transport in a flaring loop.
F. Nagai, C. Jordan.
Astron. Astrophys., Vol. 146, No. 1, p. 25 – 37 (1985).

By using an electron beam bombardment model in the impulsive phase of solar flares, the authors investigate the relation between the dynamical behaviour of a flare plasma and the evolution of the spatial distribution of various energy terms. They evaluate the energy input (gain) or output (loss) rates integrated along the entire loop. Moreover, the authors show how the flare energy is distributed between various energy terms by integrating these rates over time. A one–fluid, two–temperature model is adopted and nonthermal electrons with power–law spectrum are assumed to be injected isotropically from the loop apex.

073.101 Energy and angular distributions of energetic flare electrons and their X–radiation. I. Initially monoenergetic electrons.
E. Haug, G. Elwert, R. R. Rausaria.
Astron. Astrophys., Vol. 146, No. 1, p. 159 – 167 (1985).

For initially monoenergetic flare electrons penetrating the solar atmosphere the evolution of the energy and angular distributions is calculated. The ambient atmosphere is assumed to be a fully ionized thermal plasma of protons and electrons. Analytically treated small–angle scattering processes are combined with Monte Carlo calculations of large–angle deflections by electron-electron and electron–proton collisions. Small–angle multiple scattering determines the mean energy loss, while single collisions with large deflections provide the main spread of the electrons' energy and angular distributions. With the aid of these distribution functions spectra of the total bremsstrahlung emitted in various layers of the atmosphere, represented by their column densities, are calculated for various relativistic as well as nonrelativistic primary electron energies. By summing over the layers the spectra of the bremsstrahlung emitted by the flare electrons in the whole plasma are determined.

073.102 Activation of solar flares.
P. J. Cargill, S. Migliuolo, A. W. Hood.
ESA Spec. Publ., ESA SP–207, p. 57 – 63 (1984). – See Abstr. 012.044.

The physics of the activation of two–ribbon solar flares via the MHD instability of coronal arcades is presented. The destabiliza-tion of a preflare magnetic field is necessary in order for a rapid energy release, characteristic of the impulsive phase of the flare, to occur. The authors examine the stability of a number of configurations and discuss the physical consequences and relative importance of varying pressure profiles and different sets of boundary conditions (involving field–line tying). Interchange modes, driven unstable by pressure gradients, are candidates for instability. Shearless vs. sheared equilibria are also discussed.

073.103 Neutron and gamma–ray signatures for particle acceleration in solar flares.
R. Ramaty, R. J. Murphy.
ESA Spec. Publ., ESA SP–207, p. 83 – 90 (1984). – See Abstr. 012.044.

Calculations of neutron and gamma–ray production in solar flares are reviewed and compared with neutron and gamma–ray data from the 21 June 1980 and 3 June 1982 flares, as well as with gamma–ray data from other flares. The implied charged–particle numbers and spectra are compared with interplanetary observations.

073.104 Signatures of the coalescence instability in solar flares.
H. Nakajima, T. Tajima, F. Brunel, J. Sakai.
ESA Spec. Publ., ESA SP–207, p. 193 – 196 (1984). – See Abstr. 012.044.

Double sub–peak structures in the quasi–periodic oscillations found in the time profiles of two solar flares on 1980 June 7 and 1982 November 26 are well explained in terms of the coalescence instability of two current loops. This interpretation is supported by the observations of two microwave sources and their interaction for the November 26 flare. The difference of both sub–peak structures and time scales between the two flares are discussed from the viewpoint of different plasma parameters in the authors' computer simulations.

073.105 Chromospheric umbral oscillations.
W. Mattig, M. von Uexküll, F. Kneer.
ESA Spec. Publ., ESA SP–220, p. 59 (1984). – See Abstr. 012.045.

073.106 A numerical simulation of the formation of solar prominences.
J. M. Malherbe, T. G. Forbes, E. R. Priest.
ESA Spec. Publ., ESA SP–220, p. 119 – 122 (1984). – See Abstr. 012.045.

The radiative–resistive MHD equations are numerically solved in two–dimensions for a magnetic field configuration that starts with a vertical current sheet which is line–tied at its base and is in mechanical, but not radiative, equilibrium. The aim of the present study is to determine whether this initial configuration can achieve a prominence–like equilibrium in the presence of magnetic reconnection and tearing in the current–sheet.

073.107 A study of the material motion into an eruptive prominence.
T. Prokakis, D. Dialetis, C. Macris.
ESA Spec. Publ., ESA SP–220, p. 243 – 244 (1984). – See Abstr. 012.045.

073.108 Magnetic instabilities in solar filaments: models of twisting motions and ejecta.
B. Schmieder, J. M. Malherbe, M. Raadu.
ESA Spec. Publ., ESA SP–220, p. 273 – 274 (1984). – See Abstr. 012.045.

073.109 Prominences, two–ribbon flares and coronal transients.
G. W. Pneuman.
ESA Spec. Publ., ESA SP–220, p. 277 – 278 (1984). – See Abstr. 012.045.

A scenario for prominence formation and eruption with an associated coronal transient and two–ribbon flare is suggested.

073.110   Solar Maximum Mission results on the energetics of the impulsive phase of solar flares.
E. Antonucci, A. H. Gabriel.
ESA Spec. Publ., ESA SP–220, p. 279 – 280 (1984). – See Abstr. 012.045.

The presence of chromospheric material evaporating in response to localized heating to coronal temperatures is inferrred from the observations of systematic plasma upflows during the rise of the soft X–ray emission in solar flares.

073.111   A study of the morphology of solar flares as observed by the hard X–ray imaging spectrometer.
R. A. Harrison, G. M. Simnett.
ESA Spec. Publ., ESA SP–220, p. 281 – 282 (1984). – See Abstr. 012.045.

The authors have conducted an analysis of solar flares in the 3.5 – 30 keV energy region. They show that the majority of flares appear to conform to the following scenario: in 3.5 – 30 keV X–rays, the impulsive burst defines a system of low lying coronal magnetic loops. The soft X–ray (< 10 keV) flare subsequently expands into a larger, well resolved loop system and frequently an even larger overlying structure is involved with footpoints separated by several hundred thousand km.

073.112   Magnetic field structures of hard X–ray flares observed by HINOTORI spacecraft.
T. Sakurai.
ESA Spec. Publ., ESA SP–220, p. 283 – 284 (1984). – See Abstr. 012.045.

073.113   Ultraviolet observations of solar flares from the Solar Maximum Misson.
C. –c. Cheng.
Mem. Soc. Astron. Ital., Vol. 55, No. 4, p. 663 – 672 (1984). – See Abstr. 003.017.

UV results are reviewed in three main topics: (1) plasma diagnostics in the flare transition zone plasmas, (2) spatial and temporal evolutions of the UV and hard X–ray bursts, and (3) energy release processes in the impulsive phase. The review is not meant to be exhaustive.

073.114   Soft X–ray spectral diagnostics from flares.
F. Bely–Dubau, A. H. Gabriel.
Mem. Soc. Astron. Ital., Vol. 55, No. 4, p. 685 – 697 (1984). – See Abstr. 003.017.

073.115   Line broadenings and shifts during the impulsive phase of flares.
E. Antonucci.
Mem. Soc. Astron. Ital., Vol. 55, No. 4, p. 699 – 712 (1984). – See Abstr. 003.017.

Two important results on the dynamics of the thermal plasma during the explosive phase of flares have been obtained from the high resolution SXR spectra by studying line profiles. Large non–thermal broadenings and blue–shifted components are observed in flare lines. The first spectral property is related to the existence of significant turbulent, or random, mass motions, which originate at the time of the formation of the thermal plasma. Blue–shifted components indicate that ascensional flows are present in the thermal plasma during the impulsive phase. These upflows indicate that high temperature plasma is transferred from the chromosphere into the active coronal loops. This process, known as chromospheric evaporation, occurs in response to localized heating of the chromosphere by electron beams or heat conduction during the impulsive energy release.

073.116   Magnetic reconnection in two–ribbon flares – theory and applications.
R. A. Kopp, G. Poletto.
Mem. Soc. Astron. Ital., Vol. 55, No. 4, p. 737 – 747 (1984). – See Abstr. 003.017.

Two–ribbon flares and their various accompaniments comprise perhaps the most spectacular and energetic of all solar transient phenomena. The authors apply the so–called magnetic reconnection theory of two–ribbon flares and flare loops to flares observed both during the SMM and earlier. A comparison of data for the two–ribbon flare of 21 May, 1980 with model predictions allows to conclude with a reasonable degree of certainty that magnetic reconnection has been directly witnessed in the corona above the flare site.

073.117   Solar flare hydrodynamics.
G. Peres, S. Serio.
Mem. Soc. Astron. Ital., Vol. 55, No. 4, p. 749 – 762 (1984). – See Abstr. 003.017.

The Palermo–Harvard numerical code for confined solar loop atmospheres is presented, evidencing general problems and advantages of the numerical modelling of flares, and discussing recent results of this code and their comparison with SMM observations. Future improvements and probable evolutions of this kind of models are discussed.

073.118   X–ray spectrum simulations for flaring loop models: transient ionization effects during the impulsive phase.
R. Mewe.
Mem. Soc. Astron. Ital., Vol. 55, No. 4, p. 763 – 772 (1984). – See Abstr. 003.017.

X–ray spectra were simulated for observations with the bent crystal spectrometer (BCS) on the Solar Maximum Mission spacecraft using results of a time–dependent one–dimensional numerical code for a dynamic flaring loop model. It turns out that in all cases considered during about the first minute of the flare a strong depletion (relative to the continuum) of high-ionization spectral lines takes place at electron densities below about $10^{11} \mathrm{cm}^{-3}$. A preliminary comparison to observations with the BCS shows some indications for such effects in a few strong flares, but clearly more sensitive instruments will be needed in future to exploit such transient ionization effects as valid density diagnostics for hot solar flares.

073.119   A study of flare buildup from simultaneous observations in microwave, Hα, and UV wavelengths.
M. R. Kundu, V. Gaizauskas, B. E. Woodgate, E. J. Schmahl, R. Shine, H. P. Jones.
Astrophys. J., Suppl. Ser., Vol. 57, No. 3, p. 621 – 630 (1985). With plates 7 – 12. With a correction in Vol. 58, No. 1, p. 195 (1985).

The authors have combined the simultaneous high resolution observations at 6 cm wavelength (VLA), in Hα (Ottawa River Solar Observatory), and in the ultraviolet (SMM satellite) of preflare activity on 1980 June 25. The authors have derived a number of important conclusions regarding flare buildup from the 6 cm intensity and polarization changes, the Hα transverse and Doppler motions, and the C IV brightenings and upflows around the site of the filament which disrupted before the flare.

073.120   Optical radiation of flares.
N. N. Stepanyan.
Complex investigations of the sun, p. 145 – 153 (1982). In Russian. Abstr. in Ref. Zh., 51. Astron., 4.51.346 (1985). – See Abstr. 012.046.

073.121   Some results of an investigation of the structure of flares in Hα and its comparison with observations at other wavelengths.
A. N. Babin.
Complex investigations of the sun, p. 154 – 159 (1982). In Russian. Abstr. in Ref. Zh., 51. Astron., 4.51.347 (1985). – See Abstr. 012.046.

073.122   Physical conditions and peculiarities of the subtelescopic structure of flares.
L. N. Kurochka.
Complex investigations of the sun, p. 160 – 175 (1982). In Russian. Abstr. in Ref. Zh., 51. Astron., 4.51.348 (1985). – See Abstr. 012.046.

073.123 Some results of polarization observations of flares in the visible region of the spectrum.
A. N. Koval'.
Complex investigations of the sun, p. 176 – 182 (1982). In Russian. Abstr. in Ref. Zh., 51. Astron., 4.51.349 (1985). – See Abstr. 012.046.

073.124 Some questions of the dynamics and radio radiation of flare loops.
V. V. Zajtsev.
Complex investigations of the sun, p. 105 – 118 (1982). In Russian. Abstr. in Ref. Zh., 51. Astron., 4.51.352 (1985). – See Abstr. 012.046.

073.125 Spectroscopic diagnostics of the plasma turbulence in solar flares: modern state and perspectives.
E. A. Oks.
Complex investigations of the sun, p. 183 – 191 (1982). In Russian. Abstr. in Ref. Zh., 51. Astron., 4.51.355 (1985). – See Abstr. 012.046.

073.126 Multiwavelength observations of a preflare solar active region using the VLA.
M. R. Kundu, R. K. Shevgaonkar.
Astrophys. J., Vol. 291, No. 2, p. 860 – 864 (1985).
A preflare active region was studied using the VLA at 2, 6, and 20 cm. At 2 cm the region is composed of two components located in regions of opposite polarity. Both components are preheated prior to the impulsive onset of a flare. However, one component develops new structures during preburst phase, and the burst occurs in this location. The authors believe that the new structures represent emerging flux regions which interact with an overlying loop to produce a neutral sheet, which ultimately is responsible for triggering the flare.

073.127 Rocket spectrogram of a solar flare in the 10 – 100 Å region.
L. W. Acton, M. E. Bruner, W. A. Brown, B. C. Fawcett, W. Schweizer, R. J. Speer.
Astrophys. J., Vol. 291, No. 2, p. 865 – 878 (1985).
The soft (10 – 100 Å) X–ray spectrum of an M–class solar flare was observed with a high–resolution (0.02 Å) rocket–borne spectrograph on 1982 July 13. Several hundred emission lines characteristic of temperatures from about 0.5 to 7×10^6K have been photographically recorded. Spectral lines from nickel, iron, chromium, calcium, sulphur, silicon, aluminium, magnesium, neon, oxygen, nitrogen, and carbon are tabulated and discussed with extensive references to earlier work. Absolute line intensities are given and the calibration of the telescope–spectrograph is discussed.

073.128 On the high–energy neutrons and gamma rays in solar flares.
G. E. Kocharov, N. Z. Mandzhavidze.
Sov. Astron. Lett., Vol. 10, No. 5, p. 321 – 322 (1984). English translation of 38.073.034.

073.129 Prediction of large solar flares from the level of EUV– and X–ray emission.
E. A. Bruevich, A. A. Nusinov.
Soln. Dannye, Byull., 1985, No. 2, p. 58 – 61 (1985). In Russian.
The probability of large Hα flares is shown to be a function of the ratio $R = I_{8-20}/I_{584}$, where I_{8-20} is the X–ray flux at 8 – 20 Å and I_{584} is the EUV flux at $\lambda = 584$ Å.

073.130 Flux relations between hard X–rays and microwaves for both impulsive and extended solar flares.
K. Kai, T. Kosugi, N. Nitta.
Publ. Astron. Soc. Jpn., Vol. 37, No. 1, p. 155 – 162 (1985).
The correlation of peak fluxes between hard X–rays and microwaves from solar flares was reexamined separately for impulsive and extended bursts using 61 events. (1) For impulsive bursts $F_R = 37.2F_X^{0.77}$ with a small scatter of 0.3 orders of magnitude (rms), where F_R is the 17 GHz peak flux in sfu and F_X is the hard X–ray peak flux integrated over 67 – 152 keV in

photons s^{-1}cm^{-2}. (2) Extended bursts deviate systematically above the regression line derived for impulsive bursts by a factor of 2 – 10 (excess of radio emission). The radio excess of extended bursts can be explained by a relatively abundant population of relativistic electrons trapped in large magnetic loops.

073.131 Evidence for systematic flows in the transition region around prominences.
O. Engvold, E. Tandberg–Hanssen, E. Reichmann.
Sol. Phys., Vol. 96, No. 1, p. 35 – 51 (1985).
The solar transition region in the neighbourhood of prominences has been studied from observations with the Ultraviolet Spectrometer and Polarimeter of SMM. Dopplergrams from observations of the transition–region lines C IV λ1548 Å and Si IV λ1393 Å give velocity amplitudes typically in the range ± 15 km s^{-1}. Prominences are found to be located very close to dividing lines between areas of up– and down–draughts in the transition–region. The observed pattern suggests that the 10^5K gas flows take place within arcades of magnetic loops. An additional band of blue–ward Doppler shifts is frequently seen close to quiescent prominences. This may be the source of outward flowing matter along the helmet streamers above filament channels.

073.132 Quiescent prominence thread models.
J. M. Fontenla, M. Rovira.
Sol. Phys., Vol. 96, No. 1, p. 53 – 92 (1985).
The authors have calculated prominence thread models for different values of the center temperature and pressure. They have simultaneously solved the radiative transfer, statistical equilibrium and ionization equilibrium equations assuming a three–level atom plus continuum. They have also computed the energy balance equation including the hydrogen radiative losses from their calculations, plus other radiative losses and heat conduction. Some models have been calculated assuming possible variations in thermal conductivity and heating terms.

073.133 The kinematic processes in solar prominences and flares and their spectral features.
S.–h. Ye, J.–h. Jin.
Sol. Phys., Vol. 96, No. 1, p. 113 – 128 (1985).
In this paper various models of mass motions in solar prominences and flares, such as expansion, contraction and rotation without or with depth gradients, are considered. The variation of the source function with depth is also taken into account. Under these conditions the profiles of the first Balmer lines are calculated. The various effects of mass motions on spectral lines are studied. The authors have established four methods for the derivation of the velocities of motions as well as their gradients from the corresponding spectral features. These methods have been preliminarily applied to observational data, mainly those of the spectra–spectroheliograph (SSHG) of the Yunnan Observatory.

073.134 Initial phase of chromospheric evaporation in a solar flare.
E. Antonucci, B. R. Dennis, A. H. Gabriel, G. M. Simnett.
Sol. Phys., Vol. 96, No. 1, p. 129 – 142 (1985).
The authors discuss the initial phase of chromospheric evaporation during a solar flare observed with instruments on the Solar Maximum Mission on May 21, 1980 at 20:53 UT. The observations provide further support for interpreting the plasma upflows as the mechanism responsible for the formation of the soft X–ray flare, identified with chromospheric evaporation. Moreover, it can be concluded that evaporation occurred in two regimes: an initial slow evaporation, observed as a motion of most of the thermal plasma, followed by a high–speed evaporation lasting as long as the soft X–ray emission of the flare was increasing, that is as long as plasma accumulation was observed in corona.

073.135 Coronal explosions.
C. de Jager.
Sol. Phys., Vol. 96, No. 1, p. 143 – 156 (1985).
The author searched for a new phenomenon, called "coronal explosions", in three solar flares, and found them in all three. A coronal explosion is the propagation of a density wave through

the flaring area. The wave emerges from one or two small areas (the "sources") which are close to, but not identical with the sources of hard X–ray burst emission. In all three cases the explosion starts at the end of the impulsive phase, during or after the last hard ($\gtrsim$ 20 keV) X–ray burst. The velocities of propagation range between 1800 and a few tens of km s^{-1}, and tend to decrease with time. It is suggested that the bursts are magnetohydrodynamical (shock) waves moving downward into denser regions.

073.136   High–resolution X–ray spectra of solar flares. VII. A long–duration X–ray flare associated with a coronal mass ejection.
R. W. Kreplin, G. A. Doschek, U. Feldman, N. R. Sheeley Jr., J. F. Seely.
Astrophys. J., Vol. 292, No. 1, p. 309 – 318 (1985).

X–ray spectra from the P78–1 spacecraft are discussed for a long–duration X–ray flare that occurred for at least 6 hr behind and above the west solar limb on 1980 November 14. The X–ray flare was associated with a large coronal mass ejection that was observed with the white–light coronagraph on P78–1. Emission measures for this flare vary between 10^{48} and 10^{49}cm^{-3} and are probably larger, since part of the emission may have been occulted by the solar limb. The combination of emission measure and source size implies the existence of small, high–density structures within a considerably larger volume.

073.137   Asymmetry of the spectral lines of solar prominences and flares and their velocity fields.
S.-h. Ye, J.-h. Jin.
Publ. Purple Mt. Obs., Vol. 3, No. 3, p. 1 – 17 (1984). In Chinese.

Taking into consideration the variations of velocity fields and source functions with depth, the authors have calculated the profiles of the first Balmer lines of a solar prominence and the corresponding filament.

073.138   A large flare of 3 – 4 September 1982.
H. Morishita.
Tokyo Astron. Bull., Second Ser., No. 272, p. 3123 – 3153 (1985).

Three flares successively occurred in two active regions and the last one showed one of the longest strands of the flaring region connecting the two active regions. The length of the strands reached 0.44 radius of the solar disk, which is 3.4×10^5km. The flares, their associated filaments and loop prominence system (post–flare loops) are phenomenologically described in photographs and drawings.

073.139   Eruption chromosphérique, October – December 1981.
Q. Bull. Sol. Act., Vol. 23, Part III, p. 121 – 158 (1981).

073.140   A possible explanation of spatial structure of X–rays $\gtrsim$ 100 keV in solar flares observed by the _PVO/ISEE 3_ spacecraft.
P. K. Koul, K. L. Moza, P. N. Khosa, R. R. Rausaria.
Astrophys. J., Vol. 292, No. 2, p. 725 – 732 (1985).

Evolution of electron energy and angular distributions in the solar atmosphere has been studied for large deflections by combining small–angle analytical treatment with Monte Carlo technique. The incident electron energies considered are 60, 100, and 300 keV having 0°, 30°, and 60° incidence angle in the beginning. The authors have computed the bremsstrahlung X–ray flux ratio due to 60° and 0° electron incidence angles for different photon energies with height and have compared it with observations. The calculated flux ratio corresponds well with the observations and reproduces well the rise time of X–ray flux of Kane and Anderson.

073.141   The helical prominence of May 26, 1982.
B. Vršnak.
Hvar Obs. Bull., Vol. 8, No. 1, p. 13 – 23 (1984).

The observation of a prominence with helical structure is described and interpreted. Dynamics, stability and the internal structure of the prominence are discussed.

073.142   The flares of May 14 and 16, 1981, August 19, 1981, October 14, 1983 and the associated radio events.
V. Ruždjak, B. Vršnak, P. Kotrč, C. Mercier, H. J. Schober, A. Schroll, S. Urpo, P. Zlobec.
Hvar Obs. Bull., Vol. 8, No. 1, p. 25 – 49 (1984).

Observational data for the flares of May 14 and 16, 1981, August 19, 1981 and October 14, 1983 in the optical and radio ranges are presented. It seems that the complexity of the photospheric–chromospheric aspect and its time evolution is reflected in the radio range.

073.143   Acceleration of ultrarelativistic electrons in solar flares.
S. V. Bulanov, L. V. Kurnosova, Ya. Yu. Ogul'chanskij, L. A. Razorenov, M. I. Fradkin.
Pis'ma Astron. Zh., Tom 11, No. 5, p. 383 – 389 (1985). In Russian. English translation in Sov. Astron. Lett., Vol. 11.

The role of synchrotron losses in the process of acceleration of electrons in the vicinities of magnetic–field zero points in solar flares is considered and typical energies of accelerated particles are estimated. Experimental data are obtained that evidence in favour of acceleration of ultrarelativistic electrons (up to 1 GeV) in the solar flare of April 29, 1973 and steepening of the energy spectrum of electrons. The synchrotron losses are suggested to be a possible cause of the spectrum steepening.

073.144   Energy release in solar flares.
P. A. Sturrock, P. Kaufmann, R. L. Moore, D. F. Smith.
Bull. Am. Astron. Soc., Vol. 16, No. 4, p. 890 (1984). Abstract. – See Abstr. 010.062.

073.145   Solar flare neutrons and gamma rays.
R. Ramaty, R. J. Murphy.
Bull. Am. Astron. Soc., Vol. 16, No. 4, p. 890 (1984). Abstract. – See Abstr. 010.062.

073.146   Gamma–ray line spectroscopy.
R. J. Murphy, R. Ramaty, D. J. Forrest.
Bull. Am. Astron. Soc., Vol. 16, No. 4, p. 890 – 891 (1984). Abstract. – See Abstr. 010.062.

073.147   Directionality of relativistic bremsstrahlung from solar flares.
C. Dermer, R. Ramaty.
Bull. Am. Astron. Soc., Vol. 16, No. 4, p. 891 (1984). Abstract. – See Abstr. 010.062.

073.148   On the generation of intense electron beams in solar flares by shock waves.
M. E. Pesses.
Bull. Am. Astron. Soc., Vol. 16, No. 4, p. 891 (1984). Abstract. – See Abstr. 010.062.

073.149   Detection of a 158 day periodicity in the solar hard X–ray flare rate.
A. L. Kiplinger, B. R. Dennis, L. E. Orwig.
Bull. Am. Astron. Soc., Vol. 16, No. 4, p. 891 (1984). Abstract. – See Abstr. 010.062.

073.150   Helium resonance lines in the solar flare of 15 June 1973.
J. G. Porter, K. B. Gebbie, L. J. November.
Bull. Am. Astron. Soc., Vol. 16, No. 4, p. 891 (1984). Abstract. – See Abstr. 010.062.

073.151   The Hα spectral counterpart of hard X–ray microflares.
R. C. Canfield, T. R. Metcalf.
Bull. Am. Astron. Soc., Vol. 16, No. 4, p. 891 – 892 (1984). Abstract. – See Abstr. 010.062.

073.152   Solar prominence model based on eigenvalue magnetohydrostatic solutions.
V. A. Osherovich.
Bull. Am. Astron. Soc., Vol. 16, No. 4, p. 928 (1984). Abstract. – See Abstr. 010.062.

073.153 Microwave and hard X–ray imaging of a solar limb flare.
E. J. Schmahl, M. R. Kundu.
Bull. Am. Astron. Soc., Vol. 16, No. 4, p. 1002 – 1003 (1984).
Abstract. – See Abstr. 010.062.

073.154 Simultaneous observations of solar flares at 6 and 20 cm wavelengths using the VLA.
M. Melozzi, R. K. Shevgaonkar, M. R. Kundu.
Bull. Am. Astron. Soc., Vol. 16, No. 4, p. 1003 (1984). Abstract. – See Abstr. 010.062.

073.155 The large flare November 7, 1980: a test of chromospheric evaporation theories?
J. T. Karpen, G. A. Doschek, U. Feldman.
Bull. Am. Astron. Soc., Vol. 16, No. 4, p. 1003 – 1004 (1984). Abstract. – See Abstr. 010.062.

073.156 On helium excitation in quiescent prominences.
D. A. Landman.
Bull. Am. Astron. Soc., Vol. 16, No. 4, p. 1004 (1984). Abstract. – See Abstr. 010.062.

073.157 The formation of prominences by condensation modes in magnetized cylindrical plasmas.
C.–H. An.
Bull. Am. Astron. Soc., Vol. 16, No. 4, p. 1004 (1984). Abstract. – See Abstr. 010.062.

073.158 Physical conditions in large–scale details of the chromosphere.
G. I. Kornienko.
Investigation of magnetic fields and active formations on the sun, p. 51 – 61 (1984). In Russian. Abstr. in Ref. Zh., 51. Astron., 5.51.317 (1985). – See Abstr. 003.022.

073.159 Comparative investigation of the chromospherical structure on filtergrams with low (4 – 5″) and high (1 – 2″) spatial resolution.
G. I. Kornienko.
Investigation of magnetic fields and active formations on the sun, p. 48 – 50 (1984). In Russian. Abstr. in Ref. Zh., 51. Astron., 5.51.318 (1985). – See Abstr. 003.022.

073.160 Correlation of X–ray and microwave radio radiation of solar flares from observations in March – April 1979.
G. E. Kocharov, G. A. Matveev, V. O. Najdenov,
V. F. Mel'nikov, T. S. Podstrigach, Yu. E. Charikov.
Radioizluch. Solntsa, Leningrad, No. 5, p. 45 – 59 (1984). In Russian. Abstr. in Ref. Zh., 51. Astron., 5.51.342 (1985).

073.161 On thermal escape of electrons in solar flares and polarization of hard X–ray emission.
Yu. I. Skrynnikov, B. V. Somov.
Complex investigations of the sun, p. 72 – 89 (1982). In Russian. Abstr. in Ref. Zh., 51. Astron., 5.51.345 (1985). – See Abstr. 012.046.

073.162 Ion acceleration in helium–3 rich solar flares.
L. G. Kocharov, Ya. V. Dvoryanchikov.
Fiz.–tekh. inst. Akad. Nauk SSSR. Prepr., No. 906, 25 pp. (1984). In Russian. Abstr. in Ref. Zh., 51. Astron., 5.51.361 (1985).

073.163 Hydrodynamical response of the chromosphere to heating by X–ray emission of a flare.
A. A. Kaltsenaus, B. V. Somov.
Complex investigations of the sun, p. 96 – 104 (1982). In Russian. Abstr. in Ref. Zh., 51. Astron., 5.51.362 (1985). – See Abstr. 012.046.

073.164 Energy accumulation and release in solar flares.
B. V. Somov.
Complex investigations of the sun, p. 6 – 49 (1982). In Russian. Abstr. in Ref. Zh., 51. Astron., 5.51.363 (1985). – See Abstr. 012.046.

073.165 On a possibility of routine forecast of flare activity.
V. A. Burov, N. I. Kantserovskaya,
R. Yu. Siromolot.
Radioizluch. Solntsa, Leningrad, No. 5, p. 150 – 153 (1984). In Russian. Abstr. in Ref. Zh., 51. Astron., 5.51.377 (1985).

073.166 On the evolution of activity of a quiet prominence.
N. S. Shilova, L. I. Starkova.
Physics of solar activity, p. 99 – 110 (1983). In Russian. Abstr. in Ref. Zh., 51. Astron., 6.51.292 (1985). – See Abstr. 003.024.

073.167 Relationship between γ–ray emission, radio bursts and proton fluxes from solar flares.
I. M. Chertok, V. V. Fomichev.
Inst. zem. magn., ionos., rasprostr. radiovoln Akad. Nauk SSSR. Prepr., No. 49a/523, 20 pp. (1984). Abstr. in Ref. Zh., 51. Astron., 6.51.311 (1985).

073.168 Model of a solar flare with subphotospheric energy source.
A. D. Chertkov, A. A. Polyakov, Yu. I. Dokuchaev.
Magnitosfer. issled., Moskva, No. 4, p. 73 – 87 (1984). In Russian. Abstr. in Ref. Zh., 51. Astron., 6.51.314 (1985).

073.169 On a two–component model of microwave radiation of solar flares.
V. A. Kovalev.
Physics of solar activity, p. 111 – 117 (1983). In Russian. Abstr. in Ref. Zh., 51. Astron., 6.51.316 (1985). – See Abstr. 003.024.

073.170 Methods of solar flare forecast.
A. A. Florinskij.
Magnitosfer. issled., Moskva, No. 4, p. 5 – 16 (1984). In Russian. Abstr. in Ref. Zh., 51. Astron., 6.51.332 (1985).

073.171 Vector magnetic fields in prominences. II. He I D_3 Stokes profiles analysis for two quiescent prominences.
C. W. Querfeld, R. N. Smartt, V. Bommier,
E. Landi Degl'Innocenti, L. L. House.
Sol. Phys., Vol. 96, No. 2, p. 277 – 292 (1985).
The Stokes components of He I D_3 emission in two quiescent prominences, using full spectral profile measurements, are analyzed to derive vector magnetic fields. Two independently developed schemes, based on the Hanle effect, are used for interpretation. Derived magnetic field vector solutions for each pair of linear polarization Stokes profiles corresponding to an observational point in the prominence are, intrinsically, not uniquely determined, and a set of possible solutions is usually obtained. However, mutual consistency of these solutions with those independently predicted by the form of the circular polarized component, allow, in almost all cases, rejection of all solutions of a set except one symmetrical pair. Of such a pair, a unique solution can be determined with a high confidence level by reference to independent potential field information. Field vectors are found usually to be close to horizontal and normal to the prominence surface.

073.172 The correlation of solar flare production with magnetic energy in active regions.
E. B. Mayfield, J. K. Lawrence.
Sol. Phys., Vol. 96, No. 2, p. 293 – 305 (1985).
An investigation of 531 active regions was made to determine the correlation between energy released by flares and the available energy in magnetic fields of the regions. Regions with magnetic flux greater than 10^{21} maxwell during the years 1967–1969, which included sunspot maximum, were selected for the investigation. A linear regression analysis of flare production on magnetic flux showed that the flare energy is correlated with magnetic energy.

073.173 An example for solar flares caused by magnetic field non–equilibrium.
N. Seehafer.
Sol. Phys., Vol. 96, No. 2, p. 307 – 316 (1985).

Using a photospheric magnetogram of the solar activity complex HR 16862, 16863, 16864 at the end of May 1980 as boundary data a sequence of three–dimensional force–free magnetic fields is calculated, including numerical field–line tracing. The field is assumed not to be frozen–in and $|\alpha|$, which is a measure of the magnetic free energy, is assumed to increase with time due to some dynamo mechanism. Variation of α, starting from $\alpha = 0$, produces a catastrophe–like change of the topology of a field–line system corresponding to an arch filament system. This topological change is interpreted as causing a series of large homologous flares with synchronous flaring in the two spots.

073.174 The interpretation of hard X–ray polarization measurements in solar flares.
J. Leach, A. G. Emslie, V. Petrosian.
Sol. Phys., Vol. 96, No. 2, p. 331 – 337 (1985).

The authors review recent observations of polarization of moderately hard X–rays in solar flares and compare them with the predictions of recent detailed modeling of hard X–ray bremsstrahlung production by non–thermal electrons. They find that the recent advances in the complexity of the modeling lead to substantially lower predicted polarizations than in earlier models and more fully highlight how various parameters play a role in determining the polarization of the radiation field. The new predicted polarizations are comparable to those predicted by thermal modeling of solar flare hard X–ray production, and both are in agreement with the observations.

073.175 Great microwave bursts and hard X–rays from solar flares.
H. J. Wiehl, D. A. Batchelor, C. J. Crannell,
B. R. Dennis, P. N. Price, A. Magun.
Sol. Phys., Vol. 96, No. 2, p. 339 – 356 (1985).

The microwave and hard X–ray characteristics of 13 solar flares that produced microwave fluxes greater than 500 solar flux units have been analyzed. These great microwave bursts were observed in the frequency range from 3 to 35 GHz at Bern, and simultaneous hard X–ray observations were made in the energy range from 30 to 500 keV with the Hard X–Ray Burst Spectrometer on the Solar Maximum Mission spacecraft. The principal aim of this analysis is to determine whether or not the same distribution of energetic electrons can explain both emissions. The temporal and spectral behaviors of the microwaves as a function of frequency and the X–rays as a function of energy were tested for correlations, with results suggesting that optically thick microwave emission, at a frequency near the peak frequency, originates in the same electron population that produces the hard X–rays.

073.176 Analysis of H I Balmer–alpha emission from an eruptive prominence above 3 solar radii.
R. M. E. Illing, R. G. Athay, A. J. Hundhausen.
Bull. Am. Astron. Soc., Vol. 17, No. 1, p. 514 (1985). Abstract. – See Abstr. 010.064.

073.177 Two classes of gamma–ray line flares: impulsive and gradual.
T. Bai, A. L. Kiplinger, B. R. Dennis.
Bull. Am. Astron. Soc., Vol. 17, No. 1, p. 519 (1985). Abstract. – See Abstr. 010.064.

073.178 Characteristic patterns of solar magnetic fields around the positions of flares with enhanced helium–3 abundance.
V. Bumba.
Bull. Astron. Inst. Czech., Vol. 36, No. 3, p. 177 – 182 (1985).

A comparison of the positions of flares with enhanced helium–3 abundance with the magnetic situation in which these flares occurred has been made. Concentration of these usually small events into "magnetic active longitudes" is striking. Practically all studied flares develop in a magnetically specific and well characterized situation. The pattern of the magnetic field lines in the discovered magnetic situation is indicated.

073.179 Acceleration of runaway electrons and Joule heating in solar flares.
G. D. Holman.
Astrophys. J., Vol. 293, No. 2, p. 584 – 594 (1985).

The electric–field acceleration of electrons out of a thermal plasma and the simultaneous Joule heating of the plasma are studied. Acceleration and heating time scales are derived and compared, and upper limits are obtained on the acceleration volume and the rate at which electrons can be accelerated. These upper limits, determined by the maximum magnetic–field strength observed in flaring regions, place stringent restrictions on the acceleration process. The implications of these results for the microwave and hard X–ray emission from solar flares are examined.

073.180 A new method for determining temperature and emission measure during solar flares from light curves of soft X–ray line fluxes.
P. L. Bornmann.
Astrophys. J., Vol. 293, No. 2, p. 595 – 608 (1985).

The resonance, intercombination, and forbidden lines of Ne IX, Mg XI, Si XIII, S XV, Ca XIX, and Fe XXV, as well as the Lyman–α line of O VIII and the resonance lines of Fe XIX, were observed with the SMM during the decay phase of the solar flare on November 5, 1980. The rates at which the observed line fluxes decayed were not constant. For all but the highest temperature lines observed, the rate changed abruptly, causing the fluxes to fall at a more rapid rate later in the flare decay. This behavior is proposed to be due to the decreasing temperature of the flare plasma tracking the rise and subsequent fall of each line emissivity function. This explanation is used to empirically model the observed light curves and to estimate the temperature and the change in emission measure of the plasma as a function of time during the decay phase.

073.181 Time histories of gamma– and hard X–ray emissions from solar flares.
M. Yoshimori.
J. Phys. Soc. Jpn., Vol. 53, No. 12, p. 4499 – 4506 (1984). Abstr. in Phys. Abstr., Vol. 88, No. 1256, Entry 51603 (1985).

073.182 Investigation of filament dynamics in the lower chromosphere from observations in the D lines of Na I.
G. K. Ajmanova.
Tr. Astrofiz. Inst. Alma–Ata, Tom 41, p. 109 – 113 (1983). In Russian.

073.183 Ratios of the fluence of 2.22 MeV gamma–ray line to the fluence of 4.44 MeV gamma–ray line in solar flares.
M. Yoshimori.
J. Phys. Soc. Jpn., Vol. 54, No. 2, p. 487 – 489 (1985). Abstr. in Phys. Abstr., Vol. 88, No. 1258, Entry 63282 (1985).

073.184 Microwave and hard X–ray emission during solar flare of 19 Oct. 1981.
V. K. Verma.
Indian J. Radio Space Phys., Vol. 13, No. 5, p. 159 – 163 (1984). Abstr. in Phys. Abstr., Vol. 88, No. 1259, Entry 68757 (1985).

073.185 Creation of current sheets in sheared force–free fields and two–ribbon solar flares.
J. J. Aly.
18th International Cosmic Ray Conference, Vol. 4, p. 2 – 5 (1983). – See Abstr. 012.096.

The author suggests a model for a two–ribbon flare. In this model, the coronal force–free field is brought by the photospheric motions into a quasi–open configuration, which then becomes unstable to the development of a tearing mode. A large scale reconnection process is then initiated, during which the stored magnetic free energy is released and converted into other forms.

073.186   Solar flare of April 14, 1982.
S. P. Bagare, K. R. Sivaraman.
18th International Cosmic Ray Conference, Vol. 4, p. 20 (1983).
Abstract. – See Abstr. 012.096.

073.187   Further isotopic studies of heavy nuclei in the 9/23/78 solar flare.
R. A. Mewaldt, J. D. Spalding, E. C. Stone.
18th International Cosmic Ray Conference, Vol. 4, p. 42 – 45 (1983). – See Abstr. 012.096.

The authors report measurements and interpretation of He to Mg (2 < Z < 12) isotopes in the 9/23/78 solar flare. In particular, they have obtained a more accurate ^{22}Ne/^{20}Ne ratio by extending the energy interval for isotope analysis. They continue to find a significant difference between the ^{22}Ne/^{20}Ne ratio in this flare and that for the solar wind.

073.188   A search for ^{2}H, ^{3}H, and ^{3}He in large solar flares.
R. A. Mewaldt, E. C. Stone.
18th International Cosmic Ray Conference, Vol. 4, p. 52 – 55 (1983). – See Abstr. 012.096.

The results of a new study of solar flare H and He isotopes imply that earlier observations have significantly overestimated the abundances of ^{2}H, ^{3}H, and ^{3}He in large solar flares. The authors find no evidence that solar flare nuclei have suffered any significant amount of fragmentation before escaping from the Sun.

073.189 ^{3}He rich solar flares.
S. Ramadurai, M. N. Vahia, S. Biswas.
18th International Cosmic Ray Conference, Vol. 4, p. 56 – 58 (1983). – See Abstr. 012.096.

A new subgroup of ^{3}He rich solar flares is found, having a constant ^{3}He/H ratio as a function of proton flux. The authors believe that the new group is indicative of a dominant role of nuclear spallation reactions in its origin.

073.190   Behavior of high–energy particles associated with a solar flare on April 27, 1981.
K. Sakurai.
18th International Cosmic Ray Conference, Vol. 4, p. 93 – 96 (1983). – See Abstr. 012.096.

In association with a solar flare on April 27, 1981, the emissons of several gamma ray lines and hard X–rays among others were observed. The observation on the gamma ray line emissions of energy 2.223 MeV indicates that an enormous number of neutrons were produced due to the interaction of high–energy nuclei accelerated in the flare with background atoms in the photosphere in the flare region or its neighboring region. However, it seems that most of these accelerated nuclei were never released into outer space.

073.191   The nature of the major solar flare activity during solar cycles 19, 20 and 21.
R. S. Yadav, Badruddin, N. R. Yadav, S. P. Agrawal.
18th International Cosmic Ray Conference, Vol. 4, p. 227 – 230 (1983). – See Abstr. 012.096.

Utilizing the major solar flare data for the period 1955 – 79, a comprehensive analysis of these flares is presented.

073.192   Solar flares with photon emission above 10 MeV. Measurements with the gamma ray experiment on board the SMM–satellite.
E. Rieger, C. Reppin, G. Kanbach, D. J. Forrest, E. L. Chupp, G. H. Share.
18th International Cosmic Ray Conference, Vol. 10, p. 338 – 341 (1983). – See Abstr. 012.096.

The authors list the flares with emission >10 MeV, show a time history of one of the events, give the time integrated fluxes in certain energy bands and discuss the origin of the photons above 10 MeV.

073.193   Microwave pulsations in a solar flare: magneto-hydrodynamic and plasma models.
V. V. Zajtsev, A. V. Stepanov, A. M. Sterlin.
Pis'ma Astron. Zh., Tom 11, No. 6, p. 463 – 468 (1985). In Russian. English translation in Sov. Astron. Lett., Vol. 11.

The second microwave pulsations in a solar flare are shown to be caused by MHD oscillations of a flaring loop resulting from rapid energy release. Millisecond pulsations arise due to a nonlinear oscillation regime of plasma wave loss–cone instability in the flaring loop.

073.194   Theory of solar flare produced by cascade shock waves.
W.–r. Hu.
Curr. Top. Chin. Sci., Sect. E: Astron., Vol. 2, p. 178 – 183 (1984). = Kexue Tongbao (Beijing), Vol. 27, No. 4, p. 405 – 410 (1982).

In this paper, the solar flare triggered by cascade shock waves is discussed theoretically. The wave energy excited in the convective region can be converted into transverse magnetic energy in the active region, and as a result of its twisted instability, it can be converted into the kinetic energy of the shock wave. The processes manifested as the optic phenomena, the particle acceleration and the plasma jet in the solar flare can be explained by the fact that the cascade shock is excited in the active region.

073.195   The motion of the flare loop prominence on April 27, 1981.
X.–m. Gu, S.–c. Li, Z. Li, B.–s. Li.
Curr. Top. Chin. Sci., Sect. E: Astron., Vol. 2, p. 184 – 187 (1984). = Kexue Tongbao (Beijing), Vol. 27, No. 5, p. 519 – 522 (1982).

The authors made Hα chromospheric photographic observations and Hα spectroheliographic observations of the flare loop prominence. Certain motion characteristics and dynamic phenomena of the flare loop prominence during its eruption and ascension are derived.

073.196   Filament evolution: energy build–up, eruption and oscillations.
P. C. H. Martens, N. P. M. Kuin.
Bull. Am. Astron. Soc., Vol. 17, No. 2, p. 592 (1985). Abstract. – See Abstr. 010.065.

073.197   Search for chromospheric velocity signatures of He I 10830 Å dark points.
R. Holt, A. Park, J. Archibald, D. Mullan.
Bull. Am. Astron. Soc., Vol. 17, No. 2, p. 593 (1985). Abstract. – See Abstr. 010.065.

073.198   Post–flare loop heating by trapped superthermal protons.
H. S. Hudson.
Bull. Am. Astron. Soc., Vol. 17, No. 2, p. 628 (1985). Abstract. – See Abstr. 010.067.

073.199   Images of a major compact flare in hard X–rays and H–alpha.
Z. Svestka, S. F. Martin.
Bull. Am. Astron. Soc., Vol. 17, No. 2, p. 628 (1985). Abstract. – See Abstr. 010.067.

073.200   Survey of solar X–ray flare dynamics.
D. A. Batchelor, D. M. Rust.
Bull. Am. Astron. Soc., Vol. 17, No. 2, p. 628 (1985). Abstract. – See Abstr. 010.067.

073.201   Observations of impulsive–phase chromospheric dynamics and energetic–electron–heated flare kernels.
R. C. Canfield, T. A. Gunkler, A. L. Kiplinger.
Bull. Am. Astron. Soc., Vol. 17, No. 2, p. 628 (1985). Abstract. – See Abstr. 010.067.

073.202 Non–thermal excitation of the white light source in the 24 April 1981 (~ 1358 UT) solar flare.
S. R. Kane, J. J. Love, D. F. Neidig, E. W. Cliver.
Bull. Am. Astron. Soc., Vol. 17, No. 2, p. 628 – 629 (1985). Abstract. – See Abstr. 010.067.

073.203 Impulsive and gradual gamma–ray/proton flares: their rates of occurrence in the same active region.
T. Bai.
Bull. Am. Astron. Soc., Vol. 17, No. 2, p. 629 (1985). Abstract. – See Abstr. 010.067.

073.204 Flare electron densities using X–ray line ratios.
W. A. Brown, M. E. Bruner, L. W. Acton, H. E. Mason.
Bull. Am. Astron. Soc., Vol. 17, No. 2, p. 629 (1985). Abstract. – See Abstr. 010.067.

073.205 Observations of the physical conditions in solar flare transition zone plasmas from SMM.
C.–C. Cheng, R. Pallavicini, E. Tandberg–Hanssen.
Bull. Am. Astron. Soc., Vol. 17, No. 2, p. 629 (1985). Abstract. – See Abstr. 010.067.

073.206 Element abundances from solar flare spectra.
G. A. Doschek, U. Feldman, J. F. Seely.
Bull. Am. Astron. Soc., Vol. 17, No. 2, p. 629 – 630 (1985). Abstract. – See Abstr. 010.067.

073.207 The structure of transition region loops.
S. K. Antiochos.
Bull. Am. Astron. Soc., Vol. 17, No. 2, p. 631 (1985). Abstract. – See Abstr. 010.067.

073.208 Modeling the solar chromosphere with submillimeter limb brightness profiles.
C. Lindsey, L. Hermans.
Bull. Am. Astron. Soc., Vol. 17, No. 2, p. 631 (1985). Abstract. – See Abstr. 010.067.

073.209 Spicule dynamics: long time behavior.
A. C. Sterling, J. V. Hollweg.
Bull. Am. Astron. Soc., Vol. 17, No. 2, p. 631 (1985). Abstract. – See Abstr. 010.067.

073.210 The K/Na abundance ratio in quiescent prominences.
E. Pilger, D. A. Landman.
Bull. Am. Astron. Soc., Vol. 17, No. 2, p. 633 (1985). Abstract. – See Abstr. 010.067.

073.211 Magnetic constriction and energy balance in the upper transition region.
J. F. Dowdy, A. G. Emslie, R. L. Moore.
Bull. Am. Astron. Soc., Vol. 17, No. 2, p. 633 (1985). Abstract. – See Abstr. 010.067.

073.212 Nonlocal heat transport in flaring solar loops.
D. F. Smith.
Bull. Am. Astron. Soc., Vol. 17, No. 2, p. 634 (1985). Abstract. – See Abstr. 010.067.

073.213 Radiative shocks in the solar flare chromosphere.
G. H. Fisher.
Bull. Am. Astron. Soc., Vol. 17, No. 2, p. 634 (1985). Abstract. – See Abstr. 010.067.

073.214 The importance of proton beam pressure in solar flares.
D. H. Tamres, R. C. Canfield, A. N. McClymont.
Bull. Am. Astron. Soc., Vol. 17, No. 2, p. 634 – 635 (1985). Abstract. – See Abstr. 010.067.

073.215 Coulomb and Ohmic flare heating by nonthermal electrons.
A. N. McClymont, R. C. Canfield, J. C. Brown.
Bull. Am. Astron. Soc., Vol. 17, No. 2, p. 635 (1985). Abstract. – See Abstr. 010.067.

073.216 Flare decay energy transport analysis utilizing X–ray temperature measurements.
H. A. Garcia.
Bull. Am. Astron. Soc., Vol. 17, No. 2, p. 635 (1985). Abstract. – See Abstr. 010.067.

073.217 Directivity of flare gamma–rays.
V. Petrosian.
Bull. Am. Astron. Soc., Vol. 17, No. 2, p. 635 (1985). Abstract. – See Abstr. 010.067.

073.218 The physics of filament formation in two dimensions.
L. Sparks, G. Van Hoven.
Bull. Am. Astron. Soc., Vol. 17, No. 2, p. 636 (1985). Abstract. – See Abstr. 010.067.

073.219 The MHD description of the dynamical formation of the prominence magnetic field configuration.
J. J. Bao, S. T. Wu, C.–H. An.
Bull. Am. Astron. Soc., Vol. 17, No. 2, p. 637 (1985). Abstract. – See Abstr. 010.067.

073.220 A 157–day periodicity of flare occurrence observed in microwave data.
R. S. Bogart, T. Bai.
Bull. Am. Astron. Soc., Vol. 17, No. 2, p. 644 (1985). Abstract. – See Abstr. 010.067.

073.221 Hα microflares at the limit of hard X–ray detectability.
T. R. Metcalf, R. C. Canfield.
Bull. Am. Astron. Soc., Vol. 17, No. 2, p. 644 (1985). Abstract. – See Abstr. 010.067.

073.222 Transport of relativistic electrons in the flare plasma.
J. M. McTiernan, V. Petrosian.
Bull. Am. Astron. Soc., Vol. 17, No. 2, p. 645 (1985). Abstract. – See Abstr. 010.067.

073.223 The structure of high temperature flare plasma in electron–heated flare models.
A. G. Emslie.
Bull. Am. Astron. Soc., Vol. 17, No. 2, p. 645 (1985). Abstract. – See Abstr. 010.067.

073.224 Solar flare heating by non–thermal electrons in a tapered magnetic loop.
C. Sankaranarayanan, A. G. Emslie.
Bull. Am. Astron. Soc., Vol. 17, No. 2, p. 645 – 646 (1985). Abstract. – See Abstr. 010.067.

073.225 Steady flows in active regions observed with the He I 10830 Å line.
B. W. Lites, S. L. Keil, G. B. Scharmer, A. A. Wyller.
Sol. Phys., Vol. 97, No. 1, p. 35 – 49 (1985).
The authors show that the He I 10830 Å line gives reliable Doppler shift measurements in the upper chromosphere above active regions. Persistent flow patterns in active regions observed near the solar limb show features previously noted in Dopplergrams using the C IV transition region ultraviolet emission line. Unlike the C IV measurements, however, the He I absorption shows a strong correlation with the line–of–sight velocity images in certain regions of some active regions.

073.226 Hard X–ray bremsstrahlung produced by electrons escaping a high–temperature thermal source in a solar flare.
L. Nocera, Yu. I. Skrynnikov, B. V. Somov.
Sol. Phys., Vol. 97, No. 1, p. 81 – 105 (1985).

The problem of production of flare hard X–rays by bremsstrahlung from hot thermal escaping electrons in a chromospheric plasma is studied. The Landau kinetic equation is solved near the thermal source of energized electrons in a homogeneous magnetic tube to compute the anisotropic inhomogeneous distribution of the thermal escaping electrons. The intensity and polarization of hard X–rays is also computed and a comparison of theoretical results with observational data is made.

073.227 Time delay between Hα and hard X–ray emissions during impulsive solar flares.
V. K. Verma, M. C. Pande.
Sol. Phys., Vol. 97, No. 1, p. 107 – 112 (1985).

This investigation shows that statistically there are significant time delays between Hα and hard X–ray (HXR) emissions during solar flares; most impulsive flares produce HXR emissions up to $\sim$ 1 min before and up to 2 min after the onset of Hα emission. HXR emissions are also found to be peaked up to 2 min before the Hα emissions.

073.228 On the role of energetic particles in solar flares.
R. Pérez–Enríquez.
Sol. Phys., Vol. 97, No. 1, p. 131 – 144 (1985).

The author shows that a model in which particles play the most important role in the production of a flare, with the magnetic field as an efficient storage, can explain in a natural way most features of this important transient phenomenon.

073.229 Sudden disappearance of quiescent prominences.
G. P. Apushkinskii (*G. P. Apushkinskij*),
N. A. Topchilo.
Sov. Astron., Vol. 28, No. 6, p. 672 – 677 (1984). English translation of 38.073.106.

073.230 Determination of the magnetic field in a quiescent solar prominence from the polarization characteristics of the emitted light.
M. B. Gornyi (*M. B. Gornyj*), D. V. Kupriyanov,
B. G. Matisov.
Sov. Astron., Vol. 28, No. 6, p. 677 – 682 (1984). English translation of 38.073.107.

073.231 North–south asymmetry of the solar activity and the chromospheric flares.
Z. Krušina.
Říše hvězd, Vol. 65, No. 7, p. 143 – 145 (1984). In Czech.

073.232 To the classification of solar flares.
L. Křivský.
Říše hvězd, Vol. 65, No. 9, p. 185 – 187 (1984). In Czech.

073.233 Observations and analysis of solar flares using Hα spectral profiles.
T. A. Gunkler.
Diss. Abstr. Int., Sect. B, Vol. 45, No. 1, p. 229 (1984). Thesis, University of California, 161 pp. (1984). Order No. DA8408721.

073.234 Energetic electrons in impulsive solar flares.
D. A. Batchelor.
Diss. Abstr. Int., Sect. B, Vol. 45, No. 4, p. 1217 (1984). Thesis, University of North Carolina, 175 pp. (1984). Order No. DA8415789.

073.235 The radiative hydrodynamics of flare loops heated by impulsive bursts of energetic electrons.
G. H. Fisher.
Diss. Abstr. Int., Sect. B, Vol. 45, No. 6, p. 1811 (1984). Thesis, University of California, 152 pp. (1984). Order No. DA8418283.

073.236 Ultraviolet spectral diagnostics of solar flares and heating events.
J. G. Porter.
Diss. Abstr. Int., Sect. B, Vol. 45, No. 9, p. 2957 (1985). Thesis, University of Colorado, 147 pp. (1984). Order No. DA8428675.

073.237 Partial redistribution interlocking in the solar chromosphere.
P. Heinzel, I. Hubený.
Progress in stellar spectral line formation theory, p. 137 – 142 (1985). – See Abstr. 012.108.

Starting with the model of a quiet solar chromosphere, the authors have calculated the relative probabilities of radiative and natural population of the second and third hydrogen levels, pertinent to various population processes. The analysis indicates that, while the Lα line is formed by resonance scattering between the first two levels, the third hydrogen level, from which Lβ and Hα are generated, is populated partly by direct photoexcitation $1 \rightarrow 3$ (about 55%), and partly by two–photon absorption $1 \rightarrow 2 \rightarrow 3$ (about 45%).

073.238 Vertical and horizontal heating in solar chromosphere.
M.–t. Song.
Chin. Astron. Astrophys., Vol. 9, No. 1, p. 8 – 14 (1985). English translation of 38.073.113.

073.239 Simultaneous monochromatic and spectral observations of two large loop prominence groups.
L.–s Cui, J. Hu, G.–p. Ji, X.–b. Ni, Y.–r. Huang, C. Fang.
Chin. Astron. Astrophys., Vol. 9, No. 1, p. 49 – 53 (1985). English translation of Acta Astrophys. Sin., Vol. 4, No. 4, p. 279 – 285 (1984).

From a combined analysis of the simultaneously obtained monochromatic photographs in Hα and spectral data in 3 wavelength intervals, the authors have derived the morphological structure of the groups as well as the law of motion, the kinetic temperature, density and turbulence velocity distribution of the matter in the loops.

073.240 Turbulence heating of plages.
Y. Tong, J.–s. Du, X.–j. Mao, S.–z. Lu.
Chin. Astron. Astrophys., Vol. 9, No. 2, p. 114 – 118 (1985). English translation of Acta Astron. Sin., Vol. 25, No. 4, p. 382 – 389 (1984).

The authors consider the heating of the solar plages to be due to the Alfvén waves having ponderomotive force. Assuming a certain field configuration, they derive a formula for the turbulence heating of plages by Alfvén waves. Their calculated plage temperature distribution is in agreement with the observed one.

073.241 Morphological characters of the two–ribbon flare of 1981 May 13 and their explanation.
Z.–y. Zhu, H.–a. Wu, T.–j. Cao.
Chin. Astron. Astrophys., Vol. 9, No. 2, p. 119 – 123 (1985). English translation of Acta Astron. Sin., Vol. 25, No. 4, p. 401 – 408 (1984).

The authors give a summary of the morphology of the two–ribbon flare of 1981 May 13. One striking feature is that the Hα flare began at about 03 38 UT and the double–ribbon structure was formed about 03 46, before the impulsive phase of the radio 3 cm burst at 04 11 UT. They explain this behavior in terms of Hyder's model. The authors give rough estimates of the energy density, the height of prominence and the infall matter at the different radio increments, and a qualitative explanation for the appearances of the single–peak structure in the radio burst.

073.242 On the microwave millisecond spike emission and its associated phenomena during the impulsive phase of large solar flare.
C.–s. Li, Q.–j. Fu, Y.–h. Yan, S.–y. Jiang, H.–w. Li.
Chin. Astron. Astrophys., Vol. 9, No. 2, p. 128 – 133 (1985). English translation of Acta Astrophys. Sin., Vol. 5, No. 1, p. 9 – 18 (1985).

The authors propose a model of two acceleration regions, which can explain, on the basis of microwave maser caused by a "hollow–beam" distribution of electrons, the presence of milli-second spikes in the event of 1981 May 16 and their absence in the event of 1981 October 12, and the enhanced continuous emission in the latter.

073.243   Analysis of the X–ray/Hα flare of 1980 July 14.
A.–a. Xu, Y.–h. Tang.
Chin. Astron. Astrophys., Vol. 9, No. 2, p. 164 – 169 (1985). English translation of Chin. J. Space Sci., Vol. 5, No. 1, p. 11 – 18 (1985).

Using the X–ray data from the SMM satellite and optical data from the Yunnan Observatory, the authors analysed the Class 3B flare of 1980 July 14. They obtain the time variation of the X–ray spectrum, calculated the total number of electrons at the time of the flare and their mean energy and measured and compared the positions of the Hα flare and the X–ray burst source. The results obtained support the newly emergent flux model of flares.

073.244   The action of a compressing and shearing non–force–free magnetic field on the build–up and triggering of a two–ribbon flare.
Q.–r. Su.
Chin. Astron. Astrophys., Vol. 9, No. 2, p. 170 – 174 (1985). English translation of Acta Astron. Sin., Vol. 25, No. 4, p. 376 – 381 (1985).

The corresponding field configurations of the solar isothermal atmosphere obtained from a series of non–constant shearing magnetic fields after considering both the gravity and gas pressure are examined and compared. When the shearing magnetic field increases, magnetic islands will appear, followed by shearing, quasi–three–dimensional field structures approaching the open state, and with almost anti–parallel field lines. It is possible that this is related to the initial phase of a two–ribbon flare.

073.245   Heating of sunspot chromospheres by slow–mode acoustic shock waves.
M. G. Lee, H. S. Yun, J. J. Hyun.
J. Korean Astron. Soc., Vol. 18, No. 1, p. 15 – 31 (1985).

The authors have calculated the dissipation rates of upward–travelling slow–mode acoustic shock waves in umbral chromospheres for two umbral chromosphere models. The results show that the slow–mode acoustic shock waves with a period of about 20 seconds can heat the low umbral chromospheres travelling with a mechanical energy flux of $2.6 \times 10^6 \mathrm{erg/cm^2 s}$ at a height of 300 – 400 km above the temperature minimum region.

073.246   A comprehensive analysis of the solar loop prominence of 27 April 1981.
T.–j. Cao, A.–a. Xu, Y.–j. Ding.
Acta Astron. Sin., Vol. 26, No. 1, p. 42 – 50 (1985). In Chinese.

From morphological features of Hα images it is suggested that the main loop of the event is possibly a projection of a post flare loop system on the sky instead of a single loop. The authors think it is possible that there was a two–ribbon flare which occurred behind the limb and the two–ribbon was extended to the line of sight. The occurrence of this kind of a two–ribbon flare in a sunspot group is also discussed in this paper.

073.247   The pulsation of filaments and a possible mechanism of homologous flares.
B.–r. Luo, A.–a. Xu.
Acta Astron. Sin., Vol. 26, No. 2, p. 101 – 106 (1985). In Chinese.

The close relation between the changes of filament morphology and repeated eruptions of homologous flares has been discovered by analysing the sequence of the homologous flares which erupted on July 12 in 1982: the recurrence of the homologous flares is modulated by the pulsation of the filament. It is proposed that new impulsively emerging magnetic flux may cause the dynamic pinch of the filament plasma, and it results in repeated eruptions of flares. The theoretical calculation is consistent with the observed result.

073.248   Density and temperature structure of sunspot chromospheres by spectral synthesis.
A. K. Dobson–Hockey, W. E. Baggett.
Proceedings of the Southwest Regional Conference for Astronomy and Astrophysics, Vol. 10, p. 11 – 21 (1985). – See Abstr. 012.125.

The authors have been engaged in producing a model of the chromosphere above a sunspot umbra which will simultaneously satisfy Ca II spectral line observations and transition zone densities. This work has led them to investigate more thoroughly the effects on the Ca II H and K line profiles caused by changes in the temperature and density structure of the chromosphere. To this end the authors have produced a grid of models with varying temperature structures and various column masses at the top of the chromosphere. They present the results of their investigations.

The night F region of the ionosphere in the period of flares on the sun.
See Abstr. 003.058.

Ground–based optical observations during SMM.
See Abstr. 013.047.

The 1^1S–$n^1P/1^1S$–2^1P emission–line ratios in O VII as temperature diagnostics for solar flares and active regions.
See Abstr. 022.098.

Method of mathematical reduction of results of observations of spicule occultations by the moon.
See Abstr. 036.160.

Analysis of a time–varying scene observed with a digital CCD camera.
See Abstr. 036.183.

The HESP/R satellite project.
See Abstr. 051.007.

Initial results from the repaired Solar Maximum Mission and future prospects.
See Abstr. 051.009.

Forced reconnection by nonlinear MHD waves and sequential triggering of solar flares.
See Abstr. 062.092.

Explosive transverse electric field and particle acceleration during magnetic collapse.
See Abstr. 062.093.

Energy balance in resistive instabilities.
See Abstr. 062.107.

Bremsstrahlung spectra from thick–target electron beams with noncollisional energy losses.
See Abstr. 063.058.

Radiative transfer diagnostics: Understanding multi–level transfer calculations.
See Abstr. 063.095.

Rapid non–LTE calculations of Balmer lines and hydrogen ionization: the solar case.
See Abstr. 064.025.

Interpretation of oscillations in UV lines observed above sunspot umbrae.
See Abstr. 072.007.

Impulsive phenomena in a small active region.
See Abstr. 072.018.

About sudden disappearances and reappearances.
See Abstr. 072.043.

Modelling of non thermal hard X ray emission observed during solar flares.
See Abstr. 076.034.

Observational evidence of continuous input of energetic electrons in the corona during large solar flares.
See Abstr. 076.035.

Solar hard X–ray bremsstrahlung production by proton beams?
See Abstr. 076.038.

The anisotropy of gamma–rays from solar flares.
See Abstr. 076.039.

Absolute wavelength measurements of solar UV emission lines.
See Abstr. 076.040.

Time series images of the UV chromosphere and transition zone.
See Abstr. 076.042.

Hard X–ray images of possible reconnection in the flare of 21 May, 1980.
See Abstr. 076.045.

The impulsive hard X–rays from solar flares.
See Abstr. 076.047.

Spatio–temporal features of the development of microwave emission of active regions and flares.
See Abstr. 077.004.

Solar burst with millimetre–wave emission at high frequency only.
See Abstr. 077.005.

Simultaneous dual wavelength observations of an impulsive microwave burst using the V.L.A.
See Abstr. 077.010.

Radio radiation as information source on proton fluxes from solar flares.
See Abstr. 077.024.

Some results of spectrographic investigations of the radio radiation of solar flares in the 8 – 12 GHz range.
See Abstr. 077.025.

Dual frequency observations of solar microwave bursts using the VLA.
See Abstr. 077.030.

On the radio radiation of solar flares and features of coronal magnetic fields.
See Abstr. 077.037.

Radio spikes and the fragmentation of flare energy release.
See Abstr. 077.041.

Solar noise storms and magnetic sector structures.
See Abstr. 077.043.

Spectral correlation between high energy solar flare particle events and associated radio bursts.
See Abstr. 077.045.

Correlation of solar decimetric radio bursts with X–ray flares.
See Abstr. 077.050.

A comparison of solar ³helium–rich events with type II bursts and coronal mass ejections.
See Abstr. 078.007.

Helios 1 energetic particle observations of the solar gamma–ray/ neutron flare events of 1982 June 3 and 1980 June 21.
See Abstr. 078.008.

High–energy neutrons and γ–quanta and characteristics of flare particles.
See Abstr. 078.010.

Characteristics of accelerated particles and regions of their generation for powerful solar flares.
See Abstr. 078.011.

An analysis of relations between the characteristics of solar cosmic rays and indices of solar activity.
See Abstr. 078.012.

Deep space observations of the east–west asymmetry of solar energetic storm particle events: Voyagers 1 and 2.
See Abstr. 078.013.

Solar ³He–rich events and nonrelativistic electron events: a new association.
See Abstr. 078.014.

Multispaceraft observations of the east–west asymmetry of solar energetic storm particle events.
See Abstr. 078.017.

Secular variations in solar flare proton fluxes.
See Abstr. 078.020.

Elemental composition of solar energetic particles.
See Abstr. 078.027.

The solar flare electrons with the energy $E_e > 100$ MeV.
See Abstr. 078.035.

The time history of 2.22 MeV line emission in solar flares.
See Abstr. 078.037.

Solar flare neutron fluxes derived from interplanetary charged particle measurements.
See Abstr. 078.038.

Generation of accelerated particles and hard radiation during solar flare.
See Abstr. 078.040.

Solar high energy particles and comprehensive solar flare index.
See Abstr. 078.060.

High–energy solar protons from a flare of April 4, 1981.
See Abstr. 078.065.

Proton acceleration in a solar flare on April 1, 1981.
See Abstr. 078.066.

Elemental composition of nuclei above silicon emitted in a large flare of August 4, 1972.
See Abstr. 078.069.

Solar neutrons from the impulsive flare on 1982 June 3 at 1143 UT.
See Abstr. 078.073.

Energetic particle and γ–ray generation in the solar matter.
See Abstr. 078.082.

Two classes of energetic particle events associated with impulsive and long duration flares.
See Abstr. 078.084.

Effects of CO molecules on the outer solar atmosphere: a time–dependent approach.
See Abstr. 080.001.

Radiative and reconnection instabilities: compressible and viscous effects.
See Abstr. 080.005.

Magnetic reconnection in a high–temperature plasma of solar flares. I. Effect of gradient instabilities.
See Abstr. 080.006.

The chromospheric brightness oscillation spectrum from filter observations in the Hβ–line.
See Abstr. 080.058.

Oscillations of the Sun's chromosphere. III. Simultaneous Hα observations from two sites.
See Abstr. 080.059.

Plasma and nuclear processes in solar matter.
See Abstr. 080.061.

Detection of solar flare neutrinos.
See Abstr. 080.126.

Variations in the solar rotation rate derived from Ca^+ K plage areas.
See Abstr. 080.147.

Report of IAU Commission 12: Radiation and structure of the solar atmosphere (*Radiation et structure de l'atmosphère solaire*).
See Abstr. 080.149.

On the intensification of [O I] 630 nm emission of the night sky luminescence during solar flares.
See Abstr. 082.070.

Study of large solar flares and their helio–longitudinal effects on geomagnetic storms during 1975 – 79.
See Abstr. 084.124.

Energy supply processes for magnetospheric substorms and solar flares: tippy bucket model or pitcher model?
See Abstr. 085.006.

Solar flare neon and solar cosmic ray fluxes in the past using gas–rich meteorites.
See Abstr. 105.133.

A simplified model for timing the arrival of solar flare–initiated shocks.
See Abstr. 106.008.

Distant heliospheric results on interplanetary shock propagation.
See Abstr. 106.010.

Efficiency of charged–particle acceleration by shocks in supernova remnants and in solar flares.
See Abstr. 125.036.

Tracking of flare–generated interplanetary shock waves, responsible for Forbush decreases, by observing type II radio bursts.
See Abstr. 144.011.

Study of solar flare associated cosmic ray decreases during 1975 – 79.
See Abstr. 144.240.

074 Corona, Solar Wind

074.001   The phase power spectrum of the solar wind measured by long–baseline interferometry at 81.5 MHz.
W. G. Rees, P. J. Duffett–Smith.
Mon. Not. R. Astron. Soc., Vol. 212, No. 2, p. 463 – 470 (1985).

The authors have measured the phase power spectrum of the solar wind at 81.5 MHz using an interferometer with a baseline of 130 km. A simple extrapolation of the power law observed by spacecraft at lower spatial frequencies is consistent with the authors' data, and the best–fitting power law has an index for the two–dimensional phase power spectrum of 3.3. While not ruling out the Gaussian model by Readhead, Kemp & Hewish (RKH) for spatial frequencies above $5 \times 10^{-3} km^{-1}$, the data suggest that the phase power in their model is too small. This is supported by the good agreement between the results and recent measurements by spacecraft. Unless the scattering power of the solar wind has changed since 1976, this implies that the interplanetary scintillation (IPS) compactness parameter, R, deduced from IPS observations using the RKH model, may be too large by up to 30 per cent.

074.002   Energy dissipation mechanisms in the solar corona.
J. Heyvaerts.
Unstable current systems and plasma instabilities in astrophysics, p. 95 – 111 (1985). – See Abstr. 012.002 (IAU Symp. No. 107).

Present ideas concerning the electric heating of the solar corona are reviewed. Some theories aiming at the evaluation of the net rate of energy dissipation are described. A short account is given of a recent analytical study based on a generalization of Taylor's hypothesis concerning the evolution of magnetic helicity in plasma with a large magnetic Reynolds number.

074.003   A unified theory of coronal heating.
J. A. Ionson.
Unstable current systems and plasma instabilities in astrophysics, p. 139 – 143 (1985). – See Abstr. 012.002 (IAU Symp. No. 107).

This presentation focuses upon the coronal heating problem and reports the results of Ionson's (1984) unified theory of electrodynamic heating. This generalized theory unveils a variety of new heating mechanisms and links together previously proposed processes.

074.004   Velocity shear instabilities in the anisotropic solar wind.
S. Migliuolo.
Unstable current systems and plasma instabilities in astrophysics, p. 371 – 374 (1985). – See Abstr. 012.002 (IAU Symp. No. 107).

The linear and quasilinear theory of perturbations in finite–β (β is the ratio of plasma pressure to magnetic energy density), collisionless plasmas, that have sheared (velocity) flows, is developed. A simple, one–dimensional magnetic field geometry is assumed to adequately represent solar wind conditions near the sun (i.e., at R $\cong$ 0.3 AU).

074.005   Coronal arcades in the sun and their hydromagnetic stability.
A. Ray.
Unstable current systems and plasma instabilities in astrophysics, p. 483 – 486 (1985). – See Abstr. 012.002 (IAU Symp. No. 107).

Two ribbon flares on the sun are sometimes preceded by luminous arcade like structures in the corona with filament activity. The arcades which are associated with the coronal magnetic field are sometimes without these filaments when they are found to be relatively long lived. The reasons underlying the stability and instability of these structures may be relevant in understanding the basic mechanism of two ribbon flares. Here the author presents results on the equilibrium structure of coronal arcades and their hydromagnetic stability.

074.006   Concerning the dynamics of energetic protons in coronal magnetic loops: dispersion effects of Alfvén waves.
V. A. Mazur, A. V. Stepanov.
Unstable current systems and plasma instabilities in astrophysics, p. 559 (1985). – See Abstr. 012.002 (IAU Symp. No. 107).

It is shown that the existence of plasma density inhomogeneities (ducts) elongated along the magnetic field in coronal loops, and of Alfvén wave dispersion, associated with considering the gyrotropy $U \equiv \omega/\omega_i \ll 1$ (Leonovich et al., 1983), leads to the possibility of a quasi–longitudinal propagation (wave guiding) of Alfvén waves. Here ω is the frequency of Alfvén waves, ω_i is the proton gyrofrequency.

074.007   On the distribution of θ_{Bn} for shocks in the solar wind.
J. K. Chao, Y. H. Chen.
J. Geophys. Res., Vol. 90, No. A1, p. 149 – 153 (1985). – See Abstr. 012.003.

The authors use the observed interplanetary magnetic fields and their associated fluctuations to model the distribution of the shock normal angles, θ_{Bn}, at various heliocentric distances. Comparison with available observations of θ_{Bn} are given.

074.008   Coronal mass ejections and interplanetary shocks.
N. R. Sheeley Jr., R. A. Howard, M. J. Koomen, D. J. Michels, R. Schwenn, K. H. Mühlhäuser, H. Rosenbauer.
J. Geophys. Res., Vol. 90, No. A1, p. 163 – 175 (1985). – See Abstr. 012.003.

A comparison between Solwind observations of coronal mass ejections (CME's) and Helios 1 observations of interplanetary shocks during 1979 – 1982 indicates that 72% of the shocks were associated with large, low–latitude mass ejections on the nearby limb. Most of the associated CME's had speeds in excess of 500 km/s, but some of them had speeds in the range 200 – 400 km/s. An additional 26% of the shocks may have been associated with CME's. Only 2% of the shocks clearly lacked CME's. As the average level of sunspot activity declined during 1982, the shock frequency also declined, but the observed shocks and some of their associated CME's had unusually high speeds well in excess of 1000 km/s.

074.009   Characteristics of coronal mass ejections associated with solar frontside and backside metric type II bursts.
S. W. Kahler, E. W. Cliver, N. R. Sheeley Jr., R. A. Howard, M. J. Koomen, D. J. Michels.
J. Geophys. Res., Vol. 90, No. A1, p. 177 – 182 (1985). – See Abstr. 012.003.

The authors compare fast ($v \geqslant 500$ km s^{-1}) coronal mass ejections (CME's) with reported metric type II bursts to study the properties of CME's associated with coronal shocks.

074.010   The formation of a standing shock in a polytropic solar wind model within $1 - 10\ R_s$.
S. R. Habbal.
J. Geophys. Res., Vol. 90, No. A1, p. 199 – 204 (1985). – See Abstr. 012.003.

The author shows how a one–fluid polytropic solar wind model exhibits properties similar to an isothermal wind when localized momentum addition and/or rapid area divergence produce multiple critical points in the flow. In particular, it is shown that when the sonic transition in the flow occurs closer to the coronal base, multiple steady solutions can exist. These multiple steady solutions consist of a continuous solution passing through the innermost critical point and other steady solutions involving a steady shock transition. By following the temporal evolution of the solar wind from a steady state with one critical point to a steady state with three critical points, the author shows that a standing shock solution is more likely to develop than a continuous solution when momentum deposition occurs close to the coronal base and the equation of motion admits multiple steady solutions. This result is particularly relevant to the solar wind when momentum deposition occurs as a result of a rapidly diverging coronal hole geometry.

074.011   Observation of a coronal transient from 1.2 to 6 solar radii.
R. M. E. Illing, A. J. Hundhausen.
J. Geophys. Res., Vol. 90, No. A1, p. 275 – 282 (1985).

The authors describe in detail the eruptive prominence associated coronal mass ejection of August 5, 1980, as seen in both the Solar Maximum Mission (SMM) coronagraph polarimeter and the Mauna Loa Observatory (MLO) K coronameter and prominence monitor. This event gives the first detailed look at the propagation of a "depletion" transient into the outer corona.

074.012   Nonlinear evolution of slow waves in the solar wind.
T. Hada, C. F. Kennel.
J. Geophys. Res., Vol. 90, No. A1, p. 531 – 535 (1985).

The authors show by numerical simulation using a hybrid code that comparison of the nonlinear steepening rate, calculated from fluid theory, with the linear collisionless damping rate, defines reasonably well the parameters for which fast and slow MHD waves should steepen. The results indicate that, whereas fast modes should ordinarily steepen, steepened slow waves should occur rarely in the solar wind near 1 AU.

074.013   Comments on "Density distribution in looplike coronal transients: a comparison of observations and a theoretical model" by D. G. Sime, R. M. MacQueen, and A. J. Hundhausen [J. Geophys. Res., Vol. 89, No. A4, p. 2113 – 2121 (1984)].
M. Dryer, S. T. Wu.
J. Geophys. Res., Vol. 90, No. A1, p. 559 – 561 (1985). See Abstr. 37.074.039.

074.014   Reply to the comments on "Density distribution in looplike coronal transients: a comparison of observations and a theoretical model" by M. Dryer and S. T. Wu [J. Geophys. Res., Vol. 90, No. A1, p. 559 – 561 (1985)].
D. G. Sime, R. M. MacQueen, A. J. Hundhausen.
J. Geophys. Res., Vol. 90, No. A1, p. 563 – 564 (1985). See Abstr. 074.013.

074.015   Alfvén wave dissipation in the solar wind.
M. Dobrowolny, G. Torricelli–Ciamponi.
Astron. Astrophys., Vol. 142, No. 2, p. 404 – 410 (1985).

The authors propose linear Landau damping on oblique Alfvén waves as a significant dissipation mechanism acting on Alfvénic turbulence during propagation in the solar wind. A corresponding calculation of the evolution of the turbulent spectrum is shown to indeed reproduce the main features of the HELIOS observations of magnetic turbulence. In particular, the characteristic behaviour of dissipation length versus frequency derived from the data is well reproduced.

074.016   Small–scale flux emergence and the evolution of equatorial coronal holes.
J. M. Davis.
Sol. Phys., Vol. 95, No. 1, p. 73 – 82 (1985).

To study the formation and development of coronal holes, their association with X–ray bright points has been investigated. The areal density of X–ray bright points was measured within the boundaries of coronal holes and was found to increase linearly with time for each of the three, long–lived, equatorial coronal holes of the Skylab era. Analysis of the data shows that the effect is not the result of global changes in bright point number and is therefore a property of the restricted longitude region which contains the coronal hole. The bright point density at the time of the hole's formation was also measured and, although the result is more uncertain, was found to be similar to the bright point number over the solar surface. No association was found between bright points and the rate of change of coronal hole area.

074.017   Differential rotation of coronal holes.
R. N. Shelke, M. C. Pande.
Sol. Phys., Vol. 95, No. 1, p. 193 – 197 (1985).

Using KPNO helium 10830 Å synoptic charts of Carrington rotations 1716 through 1739, and by assembling a time sequence representing single latitude zone, rotational properties of coronal holes for five zones of latitudes have been examined. It seems that the rotation period of coronal holes is a function of latitude, thus reflecting differential rotation of coronal holes.

074.018   The white emission of a hydrostatic corona.
O. G. Badalyan, M. A. Livshits.
Astron. Zh., Tom 62, Vyp. 1, p. 132 – 140 (1985). In Russian. English translation in Sov. Astron., Vol. 29, No. 1.

The analytical expression for the brightness distribution of the K corona as a function of the distance from the centre of the solar disk, $B(\varrho)$, is obtained under the assumption of the hydrostatic density distribution at $T = $ const. It is shown that the standard form of numerous observed dependences $\ln B(\varrho)$ is connected mainly with the validity of the above–mentioned simple assumptions in the layers above the inner corona which consists of closed loops. A convenient method is suggested for the determination of the temperature in the corona at the levels $(1.2 – 2.5)R_\odot$ from the centre; the values $T \approx 1.5 \times 10^6$K at equatorial latitudes and $T \approx 1.1 \times 10^6$K in polar regions of the "minimum" corona are obtained.

074.019   Motion and variation of structural formations in the solar corona (Review). II. Coronal transients–ejections.
N. I. Dzyubenko, V. I. Ivanchuk, O. S. Popov.
Probl. Kosm. Fiz., Vyp. 19, p. 3 – 17 (1984). In Russian. – See Abstr. 003.002.

074.020   Investigation of the structure of the solar corona from materials of the total solar eclipse of July 31, 1981.
G. P. Marchenko, L. A. Tsymbalyuk.
Probl. Kosm. Fiz., Vyp. 19, p. 18 – 22 (1984). In Russian. – See Abstr. 003.002.

074.021   Solar wind decrease at high heliographic latitudes detected from Prognoz interplanetary Lyman alpha mapping.
R. Lallement, J. L. Bertaux, V. G. Kurt.
J. Geophys. Res., Vol. 90, No. A2, p. 1413 – 1423 (1985).

New evidence for a latitudinal decrease of the solar wind mass flux is presented from observations of the interplanetary Lyman alpha emission collected in 1976 and 1977 with satellites Prognoz 5 and 6. The flow of interstellar hydrogen atoms in the solar system is ionized by EUV solar radiation and charge exchange with solar wind protons which accounts for about 80% of the total ionization rate. The resulting gradual decrease of the neutral H density from the upwind region down to the downwind region observed from Ly α intensity measurements allowed the determination of the absolute value of the total ionization rate β for one H atom at 1 AU against ionization. Collected in 1976 and 1977 at five places in the solar system, the measurements are compared to different models.

074.022   V.L.A. observations of flare build–up in coronal loops.
K. R. Lang, R. F. Willson.
Adv. Space Res., Vol. 4, No. 7, p. 105 – 110 (1984). – See Abstr. 012.009.

Very Large Array (V.L.A.) measurements at 20 cm wavelength map emission from coronal loops with second–of–arc angular resolution at time intervals as short as 3.3 seconds. The total intensity of the 20 cm emission describes the evolution and structure of the hot plasma that is detected by satellite X–ray observations of coronal loops. The circular polarization of the 20 cm emission describes the evolution, strength and structure of the coronal magnetic field.

074.023   Coronal mass–ejection events.
R. R. Fisher.
Adv. Space Res., Vol. 4, No. 7, p. 163 – 174 (1984). – See Abstr. 012.009.

Nearly fifteen years have passed since the discovery of coronal mass ejection events from the solar atmosphere. Progress in the interpretation of the observational results has led to a body of knowledge concerning the geometrical and evolutionary properties, physical characteristics, and the association of this type of event with other forms of solar activity. Recent interpretive results taken from the large body of observational data now available are discussed below in some detail. A classification system based on kinetic properties of these events is presented.

074.024   Post–flare thermal waves in the solar corona.
Z. Švestka.
Adv. Space Res., Vol. 4, No. 7, p. 179 – 182 (1984). – See Abstr. 012.009.

The author discusses coronal arches initiated by two–ribbon flares in the underlying active region. Temperature maps of the arches are also discussed. The observations show clearly that the post–flare X–ray arches must be natural components of two–ribbon flares, formed or revived during the flare formation.

074.025   Coronal interconnection of two active regions observed in 3.5 – 5.5 keV X–rays.
F. Fárnik, H. F. van Beek.
Adv. Space Res., Vol. 4, No. 7, p. 243 – 246 (1984). – See Abstr. 012.009.

Using HXIS data, the authors have studied further development of the coronal arch extending towards SE above the active region AR 17255 in November 1980. They estimate physical characteristics of the interconnection, with AR 17251. This interconnection may depend on flaring activity in active regions.

074.026   Dynamical behaviour of the corona in association with radio emissions.
G. Trottet, Y. Avignon, A. Kerdraon, N. Mein, M. Pick.
Adv. Space Res., Vol. 4, No. 7, p. 271 – 273 (1984). – See Abstr. 012.009.

Simultaneous observations of radio type III bursts and Hα mass ejecta are related. The authors are looking to the signature at low levels in the corona of the electron beam acceleration triggering type III bursts. The results deal with the relationship between the type III occurrence and optical features: the presence of velocities in Hα, the shape of Hα line which reveals turbulent motions and the probable existence of a shock wave.

074.027   The relationship between coronal mass ejections and solar flares.
G. M. Simnett, R. A. Harrison.
Adv. Space Res., Vol. 4, No. 7, p. 279 – 282 (1984). – See Abstr. 012.009.

X–ray observations show that at a time consistent with a coronal mass ejection onset there is a small, soft X–ray burst (precursor). Generally this is followed some 20 – 30 m later by a more significant flare. The authors present a model which accounts for the relationship between the coronal mass ejection and the precursor using $10^2 – 10^3$ keV protons as the energy transfer agent. High correlation between these events and a subsequent flare suggests that there may be a feedback mechanism operating from the coronal mass ejection.

074.028   A multiple type–II burst associated with a coronal transient and its MHD simulation.
T. E. Gergely, M. R. Kundu, S. T. Wu, M. Dryer, Z. Smith, R. T. Stewart.
Adv. Space Res., Vol. 4, No. 7, p. 283 – 286 (1984). – See Abstr. 012.009.

A large coronal transient took place on 8 May 1981. The transient was related to an M7.7/2B flare and was associated with at least two coronal type II bursts. The authors carry out two dimensional MHD simulations of the event, taking into account the observed velocity, position, and size of the type II bursts.

They simulate the multiple shocks observed during the event and their interaction, and discuss some results of the simulation.

074.029   Dynamical models of coronal transients and interplanetary disturbances.
M. Dryer, D. F. Smart.
Adv. Space Res., Vol. 4, No. 7, p. 291 – 301 (1984). – See Abstr. 012.009.

The authors review the status of the best "off–the–shelf" tool available for the study of dynamical behavior of coronal transients and traveling interplanetary disturbances. This tool involves numerical solution of the initial–boundary value problem of multi–dimensional time–dependent magnetohydrodynamics. Conclusions reached after a recent critique of the MHD paradigm's application to coronal transients are examined and found to have limited validity. Test cases for an MHD simulation are provided and described with some details.

074.030   Energy input in solar flares and coronal explosions.
C. de Jager.
Adv. Space Res., Vol. 4, No. 7, p. 303 – 306 (1984). – See Abstr. 012.009.

A coronal explosion is a density wave observed in X–ray images of solar flares. The wave occurs at the end of the impulsive phase, which is the time at which the flare's thermal energy content has reached its maximum value. It starts in a small area from where it spreads out, mainly into one hemisphere, with velocities that tend to rapidly decrease with time, and which are between $\sim 10^3$ and a few tens of km s^{-1}. The author interprets them as magnetohydrodynamic waves that (mainly) move downward from the low corona into denser regions.

074.031   The statistical properties of coronal mass ejections during 1979 – 1981.
R. A. Howard, N. R. Sheeley Jr., M. J. Koomen, D. J. Michels.
Adv. Space Res., Vol. 4, No. 7, p. 307 – 310 (1984). – See Abstr. 012.009.

The Solwind coronagraph on the P78–1 earth–orbiting satellite has been monitoring the Sun routinely at 10–minute intervals during the 5–year interval from April, 1979 to the present. In a statistical analysis of about 1000 mass ejections observed through the end of 1981, the authors find an average occurrence rate of 1.8 mass ejections per day. Histograms of speed, central latitude, angular span, brightness, and other parameters have been constructed, and properties such as shape classification have been tabulated.

074.032   Synoptic observations of coronal transients and their interplanetry consequences.
D. J. Michels, N. R. Sheeley Jr., R. A. Howard, M. J. Koomen, R. Schwenn, K. H. Mühlhäuser, H. Rosenbauer.
Adv. Space Res., Vol. 4, No. 7, p. 311 – 321 (1984). – See Abstr. 012.009.

About 50,000 coronal images have been examined, out of a five–year total of 68,000, and a standardized listing of more than 1,200 coronal transients for the period 1979 – 1982 has been prepared. The dynamical characteristics of the active corona, as they are beginning to emerge from the data, are presented: Coronal mass ejections exercise significant influence on the interplanetary solar wind. They are the source of disturbances that are frequent and energetic, that tend to be somewhat focussed, that often reach shock intensity, and that propagate to large heliocentric distances, sometimes causing major geomagnetic storms.

074.033   A catalogue of high–speed streams in the solar wind with preference for large–scale and long–term velocity structures.
L. Kulčár, V. Letfus, M. Litavský.
Contrib. Astron. Obs. Skalnaté Pleso, Vol. 12, p. 113 – 132 (1984).

The authors present a catalogue of 375 high–speed solar wind streams, which covers the period of 223 Bartels rotations (No. 1784 – 2006) from November 29, 1963 to May 23, 1980. The catalogue was compiled using the interplanetary data published by King (1977, 1979, 1983) and it reflects large–scale and long–term structures of enhanced velocities in the solar wind. The characteristic value, with respect to which the authors define a high–speed stream, is the median value of the daily average velocities for every Bartels rotation increased by an additional value of 80 km/sec.

074.034   Solar–stellar outer atmospheres and energetic particles, and galactic cosmic rays.
J.–P. Meyer.
Astrophys. J., Suppl. Ser., Vol. 57, No. 1, p. 173 – 204 (1985).

The heavy element compositions of the solar corona, solar wind (SW), solar energetic particles (SEP), and galactic cosmic-ray (GCR) sources are remarkably similar. They all show an underabundance of heavy elements with first ionization potential (FIP) $\gtrsim 9$ eV relative to elements with lower FIP, by a factor of $\sim 4 - 6$. This suggests that an ion–neutral separation takes place during the rise of matter from solar chromosphere to corona, and that the SEP and SW compositions primarily reflect that of the corona. As for GCRs, they should be mainly MeV–stellar energetic particles first "injected" by flares out of the coronae of later type F to M stars which were later on reaccelerated to GeV energies by strong interstellar shock waves (SNR; OB star wind).

074.035   A leaky waveguide model for MHD wave driven winds from coronal holes.
J. M. Davila.
NASA Conf. Publ., NASA CP–2358, p. 183 – 187 (1985). – See Abstr. 012.023.

High speed solar wind streams are now known to originate in discrete open field magnetic structures within the solar corona called coronal holes. The simple model presented in this paper demonstrates that coronal holes can act as waveguides for MHD waves. These waves serve as a source of additional acceleration for the solar wind.

074.036   Theoretical model of the solar corona during sunspot minimum. II. Dynamic approximation.
V. A. Osherovich, E. B. Gliner, I. Tzur.
Astrophys. J., Vol. 288, No. 1, p. 396 – 400 (1985).

The theoretical quasi–static model of the solar corona during sunspot minimum developed in Paper I is extended to include a quasi–radial outflow. The dynamic equation derived for the combination of a dipole–like and a radial field is used to calculate the electric current density around the Sun in the region $1.5\,R_\odot \leqslant R \leqslant 5\,R_\odot$. Comparison with the current density given by the quasi–static model shows that the outflow decreases the current density only slightly in the dynamic case. The dynamic coronal model with a magnetic quadrupole field is also considered, in relation to the north–south asymmetry in the solar corona.

074.037   Measurements of the solar wind velocity with EISCAT.
G. Bourgois, W. A. Coles, G. Daigne, J. Silen, T. Turunen, P. J. Williams.
Astron. Astrophys., Vol. 144, No. 2, p. 452 – 462 (1985).

The first observations of interplanetary scintillations with EISCAT facilities are reported. The results obtained are consistent with earlier work and show two particularly interesting features. The density fluctuations are anisotropic with an axial ratio of about 2. However the major axis can deviate as much as $20°$ from the radial direction. A two dimensional random velocity field is observed with an axial ratio of 1.3 to 3.0 aligned approximately radially. The analysis demonstrates that reliable solar wind measurements can be conducted with EISCAT antennas at solar elongation between 15 and 70 solar radii.

074.038   The fate of sunward streaming protons associated with coronal mass ejections.
G. M. Simnett.
Astron. Astrophys., Vol. 145, No. 1, p. 139 – 143 (1985).

It is suggested that the sunward streaming protons accelerated in the travelling interplanetary shock which accompanies a coronal mass ejection have sufficient energy to power the long duration solar X–ray event frequently correlated with the ejections. The energy flux of such protons over a 10^4s period is estimated to be comparable to the total energy content of the X–ray emitting plasma during the long decay of a typical event.

074.039   The structure and variability of the solar wind.
J. B. Zirker.
Effects of variable mass loss on the local stellar environment, p. 25 – 37 (1984). – See Abstr. 012.028.

The purpose of this paper is to focus on the question: Does the total mass loss of the solar wind vary in time; and if so, by what amount?

074.040   A coronal magnetic field model with volume and sheet currents.
R. Wolfson.
Astrophys. J., Vol. 288, No. 2, p. 769 – 778 (1985).

The model developed in this paper allows the description of a corona containing both volume and sheet currents. The model displays abrupt field reversals at large heliocentric distances, consistent with observations of sector boundary crossings. Unlike previous source surface and current sheet models, the present model allows for volume currents in the lower corona, while including a current sheet that forces field lines open. Comparison of solutions with and without volume currents suggests that the current sheet remains the dominant influence on the overall shape of the coronal magnetic field.

074.041   Clues to the mode of excitation of Fe X ions in the solar corona from the 1980 eclipse observations.
J. Singh.
Sol. Phys., Vol. 95, No. 2, p. 253 – 262 (1985).

The line and continuum intensities deduced from the multislit spectra of the (Fe X) coronal emission line taken at the 1980 eclipse are used to discuss the relative roles of radiative and collisional excitation mechanisms. It is shown that for $R/R_\odot < 1.2$, collisional excitation is the predominant mode. Collisional as well as radiative excitation is equally important for $1.2 < R/R_\odot < 1.4$, whereas beyond $1.4\,R_\odot$ radiative excitation becomes dominant. The line width measurements indicate that a large number of locations have half–widths around 1.3 Å. The maximum half–width is reached at $\sim 1.4\,R_\odot$ with an average value of 1.6 Å.

074.042   Coronal mass ejections: acceleration and surface associations.
R. M. MacQueen.
Sol. Phys., Vol. 95, No. 2, p. 359 – 361 (1985).

The fact that eruptive–prominence associated coronal mass ejection events may be accelerated over significant heights and times in the corona complicates the determination of possible surface or low coronal associations. A specific example of one such eruptive–prominence associated event, that observed in both the inner and outer solar corona on August 5, 1980, is used to illustrate the magnitude of the uncertainty of determining an onset time of the ejection. It is noted that such uncertainties may influence statistically–determined associations.

074.043   Helios observations of the earthward–directed mass ejection of 27 November, 1979.
B. V. Jackson.
Sol. Phys., Vol. 95, No. 2, p. 363 – 370 (1985).

The Helios spacecraft zodiacal light photometers are used to observe the earthward–directed solar mass ejection transient of 27 November, 1979 described by Howard et al. (1982). The brightness increase moved outward directly along the Sun–Earth line over a period of approximately 24 hr, indicating a strong collimation of the ejection. The outward motion and mass estimates of the ejected material from the photometers compared with near–Earth observations from IMP spacecraft show that at least a portion of the density increase observed at Earth on 29 and 30 November was associated with this ejection.

074.044   A study of the λ6374 Å Fe X and λ5303 Å Fe XIV lines from a slit spectrogram of the corona on July 31, 1981.
K. I. Nikol'skaya.
Soln. Dannye, Byull., 1984, No. 11, p. 81 – 87 (1985). In Russian.

The lines λ5303 Å Fe XIV and λ6374 Å Fe X are studied using echelle spectrograms of the solar corona at 3600 – 7000 Å taken during the total solar eclipse of July 31, 1981. Integral intensities and equivalent widths of the lines, half–widths of their profiles and radial velocities at various points of the corona along the spectrograph slit are measured. The obtained results are compared with other studies of these lines from eclipse and out of eclipse observations.

074.045   Numerical modelling of the interaction between the solar wind and cometary plasma.
A. S. Lipatov.
Kosm. Issled., Tom 23, Vyp. 1, p. 158 – 166 (1985). In Russian. English translation in Cosm. Res.

074.046   Coupling of newborn ions to the solar wind by electromagnetic instabilities and their interaction with the bow shock.
D. Winske, C. S. Wu, Y. Y. Li, Z. Z. Mou, S. Y. Guo.
J. Geophys. Res., Vol. 90, No. A3, p. 2713 – 2726 (1985).

The process by which the solar wind assimilates newly ionized atoms is important for understanding the presence of planetary or interstellar helium in the solar wind, the dynamics of the Active Magnetospheric Particle Tracer Explorers (AMPTE) lithium releases in front of the earth's bow shock, and the formation of cometary tails. In this paper the authors examine how newborn ions can be coupled to the solar wind in the direction parallel to the magnetic field by means of electromagnetic instabilities driven by the distribution of newborn ions. The linear properties of three instabilities are analyzed and compared with numerical solutions of the linear dispersion equation, while their nonlinear behavior is followed by means of computer simulation to obtain the characteristic time for the pickup process. With a primary emphasis on the AMPTE lithium releases, various degrees of realism are introduced into the calculations to model the upstream conditions and the intersection of the lithium with the bow shock.

074.047   Solar wind stagnation near comets.
A. A. Galeev, T. E. Cravens, T. I. Gombosi.
Astrophys. J., Vol. 289, No. 2, p. 807 – 819 (1985).

The nature of the solar wind flow near comets is examined analytically in this paper. In particular, typical values for the stagnation pressure and magnetic barrier strength are estimated, taking into account magnetic field line tension and change–exchange cooling of the mass–loaded solar wind. A knowledge of the strength of the magnetic barrier is required in order to determine the location of the ionopause surface which separates contaminated solar wind plasma from the outflowing plasma of the cometary ionosphere.

074.048   Thermal stability of coronal loops. I. The equilibrium structure and the stability equation.
A. N. McClymont, I. J. D. Craig.
Astrophys. J., Vol. 289, No. 2, p. 820 – 833 (1985).

The thermal stability of solar coronal loops is analyzed. It is shown that loop stability is sensitive to the nature of the heating mechanism, and that meaningful results can follow only from a consistent treatment of the structure and energy balance of the chromosphere. Realistic thermal coupling of the corona and chromosphere requires that the heating rate vary smoothly throughout the upper chromosphere. The stability equation of the equilibrium model is developed and its basic properties are investigated. It is concluded that, for the case of uniform heating, coronal loops are neutrally stable.

074.049 Thermal stability of coronal loops. II. Symmetric modes and constraints on the heating mechanism.
A. N. McClymont, I. J. D. Craig.
Astrophys. J., Vol. 289, No. 2, p. 834 – 843 (1985).

The nature of the coronal heating mechanism is examined in the light of the general analysis of thermal stability of coronal loops developed in Paper I. It is demonstrated that, if the coronal heating mechanism depends on temperature and pressure, a necessary condition on the pressure dependence of the heating mechanism must be satisfied for loop stability. Loosely speaking, the heating rate per atom must not increase with pressure faster than a power between zero and one, the precise value of which depends on the temperature dependence of the heating mechanism. The process of chromospheric evaporation is examined in terms of this condition.

074.050 The solar coronal X–ray spectrum from 5.5 to 12 Å.
D. L. McKenzie, P. B. Landecker, U. Feldman, G. A. Doschek.
Astrophys. J., Vol. 289, No. 2, p. 849 – 857 (1985).

Solar X–ray spectra in the wavelength range 5.5 – 12 Å have been measured by the SOLEX spectrometers aboard the USAF P78–1 satellite. The spectra were measured under a variety of flaring and nonflaring conditions. High sensitivity, attained by summing data from several successive spectral scans, enabled the detection of 85 lines, 22 of which remain unidentified, in this wavelength range. The lines of Fe XXII – XXIV are especially important in this wavelength range. For many of these lines, theoretical and observed line strengths were compared. Diagnostically valuable line ratios were evaluated for the helium–like species Mg XI, Al XII, and Si XIII.

074.051 The total brightness of the corona during the solar eclipse of February 16, 1980.
M. Rybanský, V. Rušin.
Bull. Astron. Inst. Czech., Vol. 36, No. 2, p. 73 – 77 (1985).

The total brightness of the white–light corona was determined for the total solar eclipse of February 16, 1980. Its value for the standard region $1.03 \leqslant \varrho \leqslant 6.00$ is $J_k = 1.29 \times 10^{-6}$ of the mean brightness of the solar disk at $\lambda_{eff} = 570$ nm. The values for solar eclipses of 1962, 1965, 1966, 1968 and 1970 were determined in a similar way. It was found that, in spite of the nearly spherically symmetrical appearance of the white–light corona in the solar activity maximum, the radiation maximum during this period is produced in the equatorial regions $(\pm 45°)$.

074.052 Variations of the total brightness of the white–light corona with the phase of the solar cycle.
V. Rušin, M. Rybanský.
Bull. Astron. Inst. Czech., Vol. 36, No. 2, p. 77 – 81 (1985).

The good agreement is demonstrated between the total brightness of the white–light corona in the standard region $(1.03 - 6.00\ R_\odot)$ in dependence on the phase of the solar cycle. Its observed values are within the interval of 0.48 (minimum) and 1.42 (maximum) $\times 10^{-6}$ of the mean brightness of the solar disk. Within the range of observational errors, no difference in the values of the total brightness of the white–light corona was observed between the individual cycles during this century. The dependence between ellipticity and the total brightness was determined in relation to the solar cycle phase.

074.053 Electron cyclotron maser instability in the solar corona: the role of superthermal tails.
L. Vlahos, R. R. Sharma.
Astrophys. J., Vol. 290, No. 1, p. 347 – 352 (1985).

The effect of a superthermal component of electrons on the loss–cone–driven electron cyclotron maser instability is analyzed. It is found that for a superthermal tail with temperature ~ 10 keV (1) the first harmonic (X– and O–mode) is suppressed for $n_t/n_r \approx 1$ (n_t and n_r are the densities of superthermal tail and loss–cone electrons) and (2) the second harmonic (X– and O– modes) is suppressed for $n_t/n_r \lesssim 10^{-1}$. A qualitative discussion on the formation of superthermal tails is presented.

074.054 Lower energy ions in co–rotating interaction regions at 1 AU: evidence for statistical ion acceleration.
I. G. Richardson.
Planet. Space Sci., Vol. 33, No. 5, p. 557 – 569 (1985).

The possibility of the statistical acceleration of solar wind ions to energies above 10 keV in the vicinity of co–rotating high speed solar wind streams by scattering from hydromagnetic waves is considered.

074.055 On viscous interaction of solar wind streams.
N. P. Korzhov, V. V. Mishin, V. M. Tomozov.
Astron. Zh., Tom 62, Vyp. 2, p. 371 – 376 (1985). In Russian. English translation in Sov. Astron., Vol. 29, No. 2.

Within the framework of a simple model the evolution of fronts between solar wind streams with different velocities, as they propagate from the sun to the earth's orbit, is considered. It is shown that the presence of stream velocity shears near the fronts leads to an excitation of the Kelvin–Helmholtz instability. Taking into account turbulent viscosity, brought about by the development of this instability makes it possible to explain the observed broadening of the fore–fronts of high–speed solar wind streams within the earth's orbit.

074.056 Alfvén waves in current–carrying solar magnetic flux tubes.
N. F. Cramer, I. J. Donnelly.
Proc. Astron. Soc. Aust., Vol. 5, No. 4, p. 481 – 483 (1984).

For a commonly employed model of a current–carrying solar magnetic flux tube, the authors have shown the existence of discrete global eigenmodes of the shear or torsional Alfvén wave. This wave does not suffer Alfvén resonance damping in an axially uniform magnetic field, but as the field along the coronal loop decreases, the wave eventually encounters the Alfvén resonance where its energy may be dumped.

074.057 Calculation of stationary magnetohydrodynamic flows in the solar corona.
Yu. V. Pisanko.
Geomagn. Aehron., No. 1, p. 17 – 22 (1985). In Russian. English translation in Geomagn. Aeron.

074.058 Discovery of the solar wind.
B. Rossi.
ESA Spec. Publ., ESA SP–207, p. 27 – 32 (1984). – See Abstr. 012.044.

The paper describes the satellite experiment which, in 1961, produced the first direct evidence of a solar wind, measured its density and its velocity, and established the existence of a region of space (the geomagnetic cavity) that surrounds the Earth and is shielded from the solar wind by the Earth's magnetic field.

074.059 Velocity distributions of solar wind ions and electrons.
E. Marsch.
ESA Spec. Publ., ESA SP–207, p. 33 – 40 (1984). – See Abstr. 012.044.

A survey of solar wind ion and electron three–dimensional velocity distributions is presented with emphasis on Helios measurements between 0.3 and 1 AU. The observed distributions strongly deviate from Maxwellians. Their detailed form distinctly correlates with the magnetic sector and solar wind stream structure. Among the most prominent non–thermal features are temperature anisotropies, heat fluxes, double ion streams, differential motion between protons and alpha particles, and the narrow field–aligned electron "strahl". Individual spectra are discussed and statistical results are presented.

074.060 The influence of the local solar wind structures on the propagation of low energy particles.
M. Amati, C. Paizis.
ESA Spec. Publ., ESA SP–207, p. 293 – 296 (1984). – See Abstr. 012.044.

The authors investigate the role that the high speed solar wind streams (HSS) may play in the confinement and propagation of low energy particles. They find that there exists a good correlation between the electron quiet time increases at 1 AU (QTI) and the HSS. The authors therefore propose that confinement and propagation in the local solar wind structures can in principle explain the morphology of the QTI without invoking source effects as current models do. Finally the authors test this interpretation by using data from the corotating nuclei events.

074.061   Coronal structures.
F. Chiuderi Drago.
ESA Spec. Publ., ESA SP–220, p. 109 – 114 (1984). – See Abstr. 012.045.
Outer and inner coronal structures are shortly reviewed.

074.062   Fe XIII 10747 Å and Fe XIV 5303 Å coronal emission lines polarization.
J. Arnaud.
ESA Spec. Publ., ESA SP–220, p. 115 – 118 (1984). – See Abstr. 012.045.
The polarization direction is found to be mainly radial. The average polarization rate is below standard theoretical models by a factor about 3. The existence of a large scale pattern of the polarization provides evidence for the existence of a large scale structure in the coronal magnetic field. The emission line polarization brightness provides a new method to study the heterogeneity of the plasma and abundance of Fe in the corona.

074.063   Mechanisms of magnetic heating of the solar corona.
J. Heyvaerts.
ESA Spec. Publ., ESA SP–220, p. 123 – 127 (1984). – See Abstr. 012.045.
Present ideas concerning the electric heating of the solar corona are briefly reviewed. The author considers mechanisms of MHD wave dissipation in strong horizontal gradients of the Alfvén velocity, as well as turbulence associated dissipation. Then he considers the evolution of D.C. coronal currents in the presence of active magnetic reconnection, and stresses the interest of considering the existence of approximate global invariants associated with such phenomena.

074.064   Heating of coronal loops by tearing turbulence.
P. K. Browning.
ESA Spec. Publ., ESA SP–220, p. 129 – 132 (1984). – See Abstr. 012.045.
The heating of a coronal loop by reconnection, in response to slow photospheric motions is discussed. It is shown that the method is invariant to the choice of gauge. General evolution equations of a coronal loop, modelled as a straight cylinder, are then derived. For solid body motions, it is shown that α is constant in time, and the heating is calculated.

074.065   Observational constraints on heating processes.
C. Jordan, B. Mendoza, R. S. Gill.
ESA Spec. Publ., ESA SP–220, p. 133 – 136 (1984). – See Abstr. 012.045.
A brief account is given of how spectroscopic measurements of emission line fluxes and widths relate to the plasma parameters relevant to wave heating processes, e.g. the magnetic field strength and its fluctuation, the damping length and plasma β. The need for further measurements of line profiles and of flux and profile periodicities in the inner corona is stressed and it is shown how systematic studies across coronal holes might elucidate the solar coronal heating mechanism.

074.066   Magnetic instabilities.
G. Einaudi.
ESA Spec. Publ., ESA SP–220, 147 – 153 (1984). – See Abstr. 012.045.
Some theoretical aspects are reviewed of both the ideal and resistive instabilities that are related to solar phenomena. The interest in the stability of solar coronal structures derives from their observed lifetime (much longer than the relevant hydromagnetic timescale) coupled with their active behaviour. This means that they must be generally stable with respect to fast, destructive instabilities, typically those governed by the ideal MHD equations, and, at the same time, they must allow some slow, non–disrupting, dissipative process to take place. The present review will concentrate on linear stability analysis, which provides essential information on the onset of the instabilities.

074.067   Axisymmetric MFS equilibria of solar coronal loops: an integral equation approach.
V. A. Delogu.
ESA Spec. Publ., ESA SP–220, 155 – 157 (1984). – See Abstr. 012.045.
The equilibrium magnetofluidostatic problem is formulated as an extremum problem for the energy integral. The resulting equations, obtained from general theorems of the calculus of variations, are transformed in integral equations, and these, together with conditions specifying the position of the feet of the field line, form a system of integral equations. Previously known configurations, like the pinch, are easily found with this method, which allows the analysis of two dimensional configurations of general aspect. Some comments on the intrinsical limits of this method and on the fundamental aspects of the MFS problem for coronal loops are also presented.

074.068   The role of differential rotation in magnetic energy storage.
R. A. Zappala, F. Zuccarello, M. Kuperus.
ESA Spec. Publ., ESA SP–220, 159 – 162 (1984). – See Abstr. 012.045.
The authors investigate a possible mechanism for magnetic energy storage in coronal loops without making assumptions about dissipative mechanisms.

074.069   Coronal heating by turbulent phase mixed MHD waves.
L. Nocera, E. R. Priest.
ESA Spec. Publ., ESA SP–220, p. 249 – 251 (1984). – See Abstr. 012.045.
The authors study the possibility of transferring mechanical and magnetic energy from MHD waves propagating in the solar corona to smaller and smaller scales via nonlinear effects, starting from a phase–mixed configuration. The efficiency of this turbulent cascade and the turbulent spectra (both in space and time) are worked out numerically. The time spectrum suggests the onset of chaotic behaviour. The resulting heating rate meets the energetic requirements to heat the corona.

074.070   The possible role of electrical currents in coronal phenomena.
E. Gliner, V. Osherovich, I. Tzur.
ESA Spec. Publ., ESA SP–220, p. 261 – 262 (1984). – See Abstr. 012.045.
On the basis of reviewing several models of solar coronal phenomena, the authors argue that the difference between the minimum and maximum types of the solar corona is due to electrical coronal currents.

074.071   Coronal structures observed at metric wavelengths with the Nançay radioheliograph.
C. E. Alissandrakis, P. Lantos, E. Nicolaidis.
ESA Spec. Publ., ESA SP–220, p. 263 – 264 (1984). – See Abstr. 012.045.
The authors present two–dimensional maps of the Sun at 169 MHz, obtained with the Nançay radioheliograph, with resolution of 1.5′ by 4.2′.

074.072   Empirical model of the solar corona using solar–cycle related parameters.
S. Koutchmy, M. Loucif.
ESA Spec. Publ., ESA SP–220, p. 265 – 267 (1984). – See Abstr. 012.045.
Using a set of eclipse pictures, the distribution of stream lines over the outer corona was analyzed.

074.073   On the variation of conditions of propagation of radio waves in the corona during the solar activity cycle.
A. I. Efimov, O. I. Yakovlev, V. K. Shtrykov, V. I. Rogal'skij.
Radiotekh. i ehlektron., Moskva, Tom 29, No. 7, p. 1277 – 1279 (1984). In Russian. Abstr. in Ref. Zh., 51. Astron., 3.51.413 (1985).

074.074 Numerical calculation of the solar wind in the interval of heliocentric distances (1 – 10) $R_\odot$.
Yu. V. Pisanko.
Simp. No. 4 KAPG po solnech. zem.–fiz., Sochi, noyab., 1984. Tez. dokl., Moskva, p. 50 (1984). In Russian. Abstr. in Ref. Zh., 62. Issled. Kosm. Prostranstva, 3.62.409 (1985).

074.075 Solar wind interaction with planetary objects.
V. Formisano.
Mem. Soc. Astron. Ital., Vol. 55, No. 3, p. 511 – 514 (1984). – See Abstr. 012.049.

074.076 Large post–flare arch–like structures in the solar corona.
Z. Svestka.
Mem. Soc. Astron. Ital., Vol. 55, No. 4, p. 725 – 736 (1984). – See Abstr. 003.017.
The Hard X–ray Imaging Spectrometer aboard the SMM detected very large arch–like structures in the solar corona, which are formed, or become enhanced, after two–ribbon flares. Flares of other kind do not affect them. These arches are seen in $3.5 - \lesssim 10\,\mathrm{keV}$ X–rays for 10 hours or more, and temperature in them may exceed $20 \times 10^6\,\mathrm{K}$. Some appear to be stationary, while others may be slowly expanding. Thermal disturbances propagate upwards through some (or all) of them with speeds of $\sim 10\ \mathrm{km/s}$.

074.077 Analysis of active region loop flows as observed from SMM.
G. Poletto, R. A. Kopp.
Mem. Soc. Astron. Ital., Vol. 55, No. 4, p. 773 – 785 (1984). – See Abstr. 003.017.
The authors discuss the indirect evidence for the presence of plasma motions in loops that derives either from observational data or from theoretical arguments. They review direct observations of plasma flows, mostly acquired by the SMM ultraviolet Spectrometer and Polarimeter experiment. The results of one recent attempt to compare theory and observations are presented and discussed.

074.078 Intercomparison of numerical flare–loop models during the NASA–SMM workshop series on solar flares.
R. A. Kopp.
Mem. Soc. Astron. Ital., Vol. 55, No. 4, p. 811 – 822 (1984). – See Abstr. 003.017.
The flare numerical modeling activity at the Solar Maximum Workshop represented the dedicated efforts of several individuals and substantial computer resources of their respective institutions. Whereas the original intent of this undertaking was to establish a benchmark reference calculation for a standard coronal loop configuration, in retrospect one can say that its true value was to demonstrate that none of the existing codes is without limitations. The author shows how difficult it is to intercompare the results obtained with diverse and highly complex computer codes.

074.079 Response of Earth and Venus ionospheres to corotating solar wind stream of 3 July 1979.
H. A. Taylor Jr., P. A. Cloutier, M. Dryer, S. T. Suess, A. Barnes, R. S. Wolff.
Earth, Moon, Planets, Vol. 32, No. 3, p. 275 – 290 (1985).
During May–July 1979 a sequence of distinct, recurrent coronal regions developed at the Sun. Analysis of these regions and the associated solar wind characteristics indicates a corresponding sequence of corotating streams, identifiable over wide distances. The time series of solar wind velocity variations observed at Earth, Venus, and the Helios–A positions during June–July attests to intervals of corotating stream propagation. The response to the intercepted stream is consistent with independent investigations which have shown that the variability of the solar wind momentum flux is an important factor in the solar wind–ionosphere interaction at both planets.

074.080 A leaky magnetohydrodynamic waveguide model for the acceleration of high–speed solar wind streams in coronal holes.
J. M. Davila.
Astrophys. J., Vol. 291, No. 1, p. 328 – 338 (1985).
It is well established observationally that high–speed solar wind streams originate in coronal hole regions in the solar corona. One suggested source for the additional momentum of these streams is "wave pressure" generated by magnetohydrodynamic (MHD) waves. In the paper the effect of coronal hole magnetic structure on the propagation of MHD waves of all periods is considered. It is found that the coronal hole structure acts as a "leaky" MHD waveguide, i.e., wave flux which enters at the base of the coronal hole is only weakly guided by the coronal hole structure. The force on the coronal hole plasma due to the propagation of the leaky wave modes is calculated. It is found that the net force consists of two terms: (1) magnetic wave pressure and (2) magnetic wave tension. The calculations imply that the coupling of the solar wind plasma to the turbulence in the photosphere may not be as efficient as previous theories of wave propagation in the corona have indicated.

074.081 The observational features and theoretical explanations of the coronal transients.
W.–r. Hu.
Chin. J. Space Sci., Vol. 4, No. 3, p. 232 – 239 (1984). In Chinese. Abstr. in Phys. Abstr., Vol. 88, No. 1252, Entry 29916 (1985).

074.082 A non–axisymmetric structure of coronal magnetic loop.
S. Wang.
Kexue Tongbao (Beijing), Vol. 29, No. 8, p. 1028 – 1032 (1984). Abstr. in Phys. Abstr., Vol. 88, No. 1252, Entry 29919 (1985).

074.083 Stereoscopic determination of the three–dimensional geometry of coronal magnetic loops.
R. Berton, T. Sakurai.
Sol. Phys., Vol. 96, No. 1, p. 93 – 111 (1985).
The three–dimensional shape of coronal magnetic loops is restored from extreme ultraviolet (XUV) images of the Sun (Skylab mission 3, 1973) by using the perspective effect due to the solar rotation. A method is developed which depends on the assumption only that the magnetic structures under consideration are (at least geometrically) stable within the time interval used for restoration. Large scale loops interconnecting different active regions are studied. They are found to lie approximately in planes inclined from the local vertical. Generally these loops are asymmetric. This tendency is also confirmed by the computation of coronal magnetic fields based on the photospheric magnetic data.

074.084 Coronal heating by resonant (A.C.) and nonresonant (D.C.) mechanisms.
J. A. Ionson.
Astron. Astrophys., Vol. 146, No. 2, p. 199 – 203 (1985).
Recent studies of solar coronal heating by Ionson (1984) and Heyvaerts and Priest (1984) have focused upon two categorically different mechanisms. These mechanisms involve the generation of A.C. currents (resonant mechanisms) and D.C. currents (nonresonant mechanisms). The major outcome of this investigation is that the efficiency of D.C. mechanisms is severely limited by the constraint that the dissipation time is comparable to the correlation time of the convective driver and smaller than the magnetic flux leakage time. For solar conditions, this suggests that D.C. processes play the dominant role in heating young active region loops ($l = 1.2 \times 10^9\,\mathrm{cm}$) while A.C. mechanisms dominate the heating of active region loops ($l = 6 \times 10^9\,\mathrm{cm}$) and large scale loops ($l = 1.6 \times 10^{10}\,\mathrm{cm}$).

074.085 New results on the Pioneer Venus Orbiter February 10 – 11, 1982 events: a solar wind disturbance, not a comet.
D. S. Intriligator.
Geophys. Res. Lett., Vol. 12, No. 4, p. 187 – 190 (1985).
The author studies the characteristics of a series of disturbances observed on February 10 – 11, 1982 by the plasma ana-

lyzer and the magnetometer on the Pioneer Venus Orbiter obtained in the solar wind upstream of Venus. The author concludes that the events were associated with the propagation of a solar wind disturbance of coronal origin and not with an encounter with a comet or other local outgassing object.

074.086   Persistence of shocks to large distances in the solar wind.
S. E. Kayser.
J. Geophys. Res., Vol. 90, No. A5, p. 3967 – 3972 (1985). – See Abstr. 012.003.

Present hydrodynamic models of solar wind streams predict that interactions will cause interplanetary shocks to decay and large–scale structures to coalesce and smooth out, with a decay length of 10 – 15 AU for moderate or small–amplitude shocks. The Pioneer 10 plasma data, extending 1 – 30 AU, are examined in the light of such predictions. It is found that coalescence of streams into a single stream per solar rotation does occur, in general, but that considerable structure remains by 30 AU. The leading edge of a stream often exhibits a velocity jump of > 20 km/s; many of these may be shocks. There is a characteristic velocity–density–temperature signature of these distant streams which differs from the close–in double–shock signature. A unique transient was seen in July 1982, at 28 AU, with a velocity jump of about 235 km/s.

074.087   Observations of He$^+$ ions in the solar wind.
N. F. Pissarenko (*N. F. Pisarenko*), E. M. Dubinin (*Eh. M. Dubinin*), A. V. Zakharov, E. Yu. Budnik, R. Lundin.
J. Geophys. Res., Vol. 90, No. A5, p. 4367 – 4372 (1985).

This paper describes results obtained from the Prognoz 8 ion composition experiment which indicates that He$^+$ ions may be more abundant in the solar wind than previous measurements have indicated. The experiment has a sufficiently high mass resolution to accurately discriminate between H$^+$, He^{++}, He$^+$, and O$^+$ ions over the energy range 0.35 – 5.2 keV/el. In a few cases, He$^+$/He^{++} ratios up to ≈ 0.05 have been found, but in most cases when He$^+$ was observed it was generally of the order 0.3 – 3% of the He^{++} content. The authors compare their results with previous measurements of He$^+$ ions in the solar wind and discuss also other constituents with $M/q = 4$ as alternative candidates. Finally, a possible explanation for the high He$^+$ abundance in the disturbed solar wind is given.

074.088   On western coronal condensations of July 31, 1981.
A. N. Ushakov, L. N. Yaroslavskij, A. B. Delone, E. A. Makarova, G. V. Yakunina, V. N. Charugin, M. M. Dagaev, V. S. Zhegulev.
Astron. Tsirk., No. 1340, p. 2 – 4 (1984). In Russian.

074.089   The microwave structure of hot coronal loops.
G. D. Holman, M. R. Kundu.
Astrophys. J., Vol. 292, No. 1, p. 291 – 296 (1985).

The thermal cyclotron emission from model dipole magnetic loops is computed. It is shown that a simple, isothermal dipole loop can show a great deal of spatial and polarization structure at microwave frequencies. Two qualitatively distinct microwave loop structures can be distinguished: (1) "thin loop", observed as a string of independent microwave peaks, corresponding to different harmonics of the local electron gyrofrequency; and (2) "thick loop", the harmonics are merged, so that a more continuous microwave structure is observed.

074.090   The F–corona and the circum–solar dust: evidences and properties.
S. Koutchmy, P. L. Lamy.
Inst. Astrophys. Paris, Pré–Publ., No. 83, 16 pp. (1984). To appear in IAU Colloq. No. 85 "Properties and interaction of interplanetary dust".

074.091   Coronal heating in closely packed flux tubes: a Taylor–Heyvaerts relaxation theory.
P. K. Browning, T. Sakurai, E. R. Priest.
MPA Rep., No. 181, 35 pp. (1985). Submitted to Astron. Astrophys.

074.092   Coronal holes: a driven MHD waveguide model.
J. M. Davila.
Bull. Am. Astron. Soc., Vol. 16, No. 4, p. 928 (1984). Abstract. – See Abstr. 010.062.

074.093   Flows in coronal loops: do syphon flows exist?
A. N. McClymont, I. J. D. Craig.
Bull. Am. Astron. Soc., Vol. 16, No. 4, p. 928 (1984). Abstract. – See Abstr. 010.062.

074.094   The differential emission measure in the lower transition region.
S. K. Antiochos.
Bull. Am. Astron. Soc., Vol. 16, No. 4, p. 928 (1984). Abstract. – See Abstr. 010.062.

074.095   Temporal and spatial variations of solar coronal bright points observed with the VLA.
S. R. Habbal, A. Cowell, R. Ronan, G. L. Withbroe, R. K. Shevgaonkar, M. R. Kundu.
Bull. Am. Astron. Soc., Vol. 16, No. 4, p. 929 (1984). Abstract. – See Abstr. 010.062.

074.096   VLA observations of thermal and non–thermal emission from coronal loops.
K. R. Lang, R. F. Willson.
Bull. Am. Astron. Soc., Vol. 16, No. 4, p. 929 (1984). Abstract. – See Abstr. 010.062.

074.097   On the distribution and variation of streamers in the lower solar corona.
R. Fisher, D. G. Sime.
Bull. Am. Astron. Soc., Vol. 16, No. 4, p. 929 (1984). Abstract. – See Abstr. 010.062.

074.098   Coronal transient apparent morphology and the associated solar activity.
D. G. Sime.
Bull. Am. Astron. Soc., Vol. 16, No. 4, p. 929 (1984). Abstract. – See Abstr. 010.062.

074.099   The onset of coronal mass ejections.
R. Wolfson, S. Gould.
Bull. Am. Astron. Soc., Vol. 16, No. 4, p. 929 – 930 (1984). Abstract. – See Abstr. 010.062.

074.100   Coronal mass ejections and sudden filament disappearances.
N. R. Sheeley Jr., R. A. Howard, K. L. Harvey.
Bull. Am. Astron. Soc., Vol. 16, No. 4, p. 930 (1984). Abstract. – See Abstr. 010.062.

074.101   Helios observations of the earthward–directed mass ejection of 27 November 1979.
B. V. Jackson.
Bull. Am. Astron. Soc., Vol. 16, No. 4, p. 930 (1984). Abstract. – See Abstr. 010.062.

074.102   Center–to–limb variation of transition region redshifts.
R. A. Shine, B. E. Woodgate, J. B. Gurman.
Bull. Am. Astron. Soc., Vol. 16, No. 4, p. 992 (1984). Abstract. – See Abstr. 010.062.

074.103   Coronal heating and the quasi–static evolution of magnetic fields.
A. A. van Ballegooijen.
Bull. Am. Astron. Soc., Vol. 16, No. 4, p. 1003 (1984). Abstract. – See Abstr. 010.062.

074.104   Coronal field changes associated with a disappearing sunspot.
G. J. Hurford.
Bull. Am. Astron. Soc., Vol. 16, No. 4, p. 1005 (1984). Abstract. – See Abstr. 010.062.

074.105   Association of coronal holes with large scale solar magnetic fields.
R. N. Shelke, M. C. Pande.
Bull. Astron. Soc. India, Vol. 12, No. 4, p. 357 – 363 (1984).

Association of coronal holes with large scale solar magnetic fields has been investigated. It is concluded that coronal holes always form in the large unipolar magnetic field patterns, and no specific polarity rule holds true for low–latitude coronal holes. The coronal holes show a strong tendency to bypass the opposite polarity regions in their location. Coronal holes get fragmented into pieces, if opposite polarity flux encroaches and/or erupts into their boundaries.

074.106   Coronal holes and long–lived unipolar magnetic regions.
R. N. Shelke, M. C. Pande.
Bull. Astron. Soc. India, Vol. 12, No. 4, p. 404 – 410 (1984).

Association of coronal holes with long–lived unipolar magnetic regions (LLUMRs) which survived for four solar rotations is examined. It is noticed that every coronal hole fits in well within the LLUMR. The coronal holes are also seen to dissolve as a consequence of the LLUMR boundary zone evolution and emergence of opposite polarity regions within the coronal hole boundary.

074.107   Electromagnetic lower hybrid instability in the solar wind.
G. S. Lakhina.
Astrophys. Space Sci., Vol. 111, No. 2, p. 325 – 334 (1985).

Low frequency electromagnetic lower hybrid waves (so–called hybrid whistlers) propagating nearly transverse to the magnetic field can be driven unstable by a resonant interaction with 'halo' electron distributions carrying solar wind heat flux. The observations of low frequency whistlers having high values of B/E ratios (B and E being the magnitude of the wave magnetic and electric field, respectively) and propagating at large oblique angles to the magnetic field B_0 behind interplanetary shocks, can be satisfactorily explained in terms of electromagnetic lower hybrid instability. The instability is also relevant to the generation mechanism of correlated whistler and electron plasma oscillation bursts detected on ISEE–3.

074.108   Thermal radio radiation of neutral current layers in the solar corona.
E. Ya. Zlotnik.
Complex investigations of the sun, p. 140 – 144 (1982). In Russian. Abstr. in Ref. Zh., 51. Astron., 5.51.323 (1985). – See Abstr. 012.046.

074.109   Dependence of the capacity of heating of the coronal plasma on its parameters for regions with open configuration of the magnetic field.
S. I. Molodykh.
Issled. Geomagn., Aehron., Fiz. Solntsa, Moskva, No. 69 (1984). In Russian. Abstr. in Ref. Zh., 51. Astron., 5.51.326 (1985).

074.110   Some peculiarities of magnetic reconnection in the solar corona.
B. V. Somov, V. S. Titov.
Complex investigations of the sun, p. 50 – 71 (1982). In Russian. Abstr. in Ref. Zh., 51. Astron., 5.51.351 (1985). – See Abstr. 012.046.

074.111   Diagnostics of solar wind velocity according to data of two diametrally located observatories.
A. D. Kosta, A. S. Potapov.
Issled. Geomagn., Aehron. Fiz. Solntsa, Moskva, No. 70, p. 169 – 171 (1984). In Russian. Abstr. in Ref. Zh., 62. Issled. Kosm. Prostranstva, 5.62.395 (1985).

074.112   Solar wind deceleration at the periphery of a cometary cloud.
I. S. Veselovskij.
Issled. Geomagn., Aehron. Fiz. Solntsa, Moskva, No. 69, p. 111 – 121 (1984). In Russian. Abstr. in Ref. Zh., 62. Issled. Kosm. Prostranstva, 5.62.400 (1985).

074.113   The magnetic field of solar wind near sector boundaries.
M. I. Pudovkin, D. I. Ponyavin, S. A. Zajtseva, V. V. Lebedeva.
Magnitosfer. issled., Moskva, No. 4, p. 19 – 23 (1984). In Russian. Abstr. in Ref. Zh., 51. Astron., 6.51.320 (1985).

074.114   Observations of steady anomalous magnetic heating in thin current sheets.
P. C. H. Martens, G. H. J. van den Oord, P. Hoyng.
Sol. Phys., Vol. 96, No. 2, p. 253 – 275 (1985).

A faint steadily emitting loop–like structure has been observed by HXIS (*Hard X–Ray Imaging Spectrometer*) in its low energy channels (3.5–8.0 keV) on November 5/6, 1980. These HXIS observations have permitted the authors to follow the thermal evolution of this loop for a period of about 15 hr and from this study they conclude that only a fraction of 0.1% of the volume of the loop is steadily heated at the rather large rate of 0.6 erg cm^{-3}s^{-1}. The authors identify the source of the X–ray emission in this paper with the Hα filament in the same region. The hot X–ray emitting plasma and the cool plasma radiating in Hα are thermally separated by the strong magnetic field. The main conclusion of the paper is that for the first time direct evidence is found for the steady dissipation of coronal magnetic fields via enhanced resistivity in thin current sheets.

074.115   Over–reflection of hydromagnetic planetary–gravity waves at the solar helmet streamers and magnetic sectors.
O. M. El Mekki.
Sol. Phys., Vol. 96, No. 2, p. 397 – 411 (1985).

Over–reflection of propagating hydromagnetic planetary-gravity waves incident on the current sheets of the helmet streamers and magnetic sectors of the solar corona is investigated. It is shown that over–reflection arises only if the wavenumbers, or the energy fluxes per unit mass, perpendicular to the current sheet of the incident and the transmitted waves in both cases are in opposite directions. The over–reflected waves then draw magnetic energy from the sun's field and communicate it to the interplanetary magnetic field.

074.116   Kinematics of coronal mass ejections associated with eruptive prominences.
W. J. Wagner.
Bull. Am. Astron. Soc., Vol. 17, No. 1, p. 514 (1985). Abstract. – See Abstr. 010.064.

074.117   A photometric comparison of the EUV– and K–coronas.
F. Q. Orrall, G. J. Rottman, R. R. Fisher, R. Munro.
Bull. Am. Astron. Soc., Vol. 17, No. 1, p. 514 (1985). Abstract. – See Abstr. 010.064.

074.118   The polar crown during two solar cycles.
P. S. McIntosh.
Bull. Am. Astron. Soc., Vol. 17, No. 1, p. 519 (1985). Abstract. – See Abstr. 010.064.

074.119   Coronal voids observed during the July 31, 1981 solar eclipse.
V. Rušin, M. Rybanský.
Bull. Astron. Inst. Czech., Vol. 36, No. 3, p. 175 – 176 (1985). With plates 1 – 2.

The authors report the existence of two coronal voids observed during the July 31, 1981 solar eclipse. They extended from the low corona to a height of 1.8 R$_\odot$ or 3.5 R$_\odot$. The authors assume that they could have fulfilled a function similar to that of a neutral sheet. No coronal prominence was observed to have condensed during eclipse totality.

074.120   An example of numerical calculation of the spatial structure of the solar wind.
Yu. V. Pisanko.
Geomagn. Aehron., Tom 25, No. 3, p. 371 – 376 (1985). In Russian. English translation in Geomagn. Aeron.

074.121   Solar wind.
J.–L. Steinberg, P. Couturier.
Recherche, Vol. 15, No. 161, p. 1494 – 1502 (1984). In French.
Abstr. in Phys. Abstr., Vol. 88, No. 1253, Entry 35632 (1985).

074.122   Helios spacecraft and earth perspective observations of three looplike solar mass ejection transients.
B. V. Jackson, R. A. Howard, N. R. Sheeley Jr., D. J. Michels,
M. J. Koomen, R. M. E. Illing.
J. Geophys. Res., Vol. 90, No. A6, p. 5075 – 5081 (1985).

Three looplike mass ejection transients observed from earth with the SOLWIND coronagraph and the solar maximum mission coronagraph are imaged by Helios spacecraft zodiacal light photometers. Because the Helios spacecraft are not earth orbiting, views of these ejections from the two perspectives allow conclusions to be drawn about their three–dimensional shapes. The mass ejection of May 24, 1979, in Helios data is concentrated in an outer structure followed by bright features separated by a region of depleted material. The ejections of June 18 and 29, 1980, appear restricted in position angle in Helios observations too less and the same, respectively, as in coronagraph observations. The observations imply that the ejections essentially retain their basic structure and speed out to heights (0.2 – 0.4 AU) observed by the Helios spacecraft.

074.123   Changes in the microturbulence spectrum of the solar wind during high–speed streams.
W. A. Coles, J. P. Filice.
J. Geophys. Res., Vol. 90, No. A6, p. 5082 – 5088 (1985).

Interplanetary scintillation (IPS) observations are used to investigate the distribution of microscale density fluctuations in high–speed solar wind streams. A new technique for estimating the electron density power spectrum which can be applied when the solar wind is grossly inhomogeneous is presented. This technique requires accurate knowledge of the radio source structure, in situ measurements of the solar wind velocity, and well-calibrated IPS observations. The estimation algorithm is applied to data obtained during 1973 and 1974 when the solar wind was dominated by recurrent fast streams.

074.124   Comment on "A reexamination of rotational and tangential discontinuities in the solar wind" by M. Neugebauer et al.
F. Mariani, B. Bavassano, U. Villante.
J. Geophys. Res., Vol. 90, No. A6, p. 5363 (1985). See Abstr. 38.074.005.

074.125   Reply to the comment on "A reexamination of rotational and tangential discontinuities in the solar wind".
M. Neugebauer.
J. Geophys. Res., Vol. 90, No. A6, p. 5364 (1985). See Abstr. 074.124.

074.126   The heating of the solar wind by Alfvénic fluctuations.
C.–y. Tu, H. Chen.
Chin. J. Space Sci., Vol. 4, No. 4, p. 277 – 284 (1984). In Chinese. Abstr. in Phys. Abstr., Vol. 88, No. 1255, Entry 45892 (1985).

074.127   Some characteristics of the solar wind in the ^{3}He–rich solar flares time.
M. Slivka, L. G. Kocharov, Ya. V. Dvoryanchikov.
Acta Phys. Slovaca, Vol. 34, No. 6, p. 363 – 368 (1984). Abstr. in Phys. Abstr., Vol. 88, No. 1255, Entry 46017 (1985).

074.128   On the behaviour of the coronal emission line λ5303 Å above active regions.
Eh. Ya. Vil'koviskij, T. M. Minasyants, G. S. Minasyants,
S. O. Obashev.
Tr. Astrofiz. Inst. Alma–Ata, Tom 41, p. 30 – 36 (1983). In Russian.

074.129   Interferometric observations of the F corona of the sun in the Mg I 5184 Å line during the eclipse of July 31, 1981 and preliminary results of a determination of radial velocities of circumsolar dust.
P. V. Shcheglov, L. I. Shestakova, A. K. Ajmanov.
Tr. Astrofiz. Inst. Alma–Ata, Tom 41, p. 70 – 82 (1983). In Russian.

074.130   Preferred solar longitude zones in affecting solar activity and geophysical parameters.
R. L. Singh, K. N. Mishra, D. P. Tiwari, R. P. Mishra,
P. L. Jain, A. K. Jain, A. P. Mishra.
18th International Cosmic Ray Conference, Vol. 3, p. 527 – 530 (1983). – See Abstr. 012.096.

It has been inferred that the sources of solar wind streams are clustered near preferred longitude zones on the sun, rotating rigidly with the rotation period of the sun. In the present study the authors have observed the effect of these preferred longitudes on the solar activity and geophysical parameters for the different phases of solar cycle 20.

074.131   The correlation of coronal mass ejections with energetic flare proton events.
S. W. Kahler, R. E. McGuire, D. V. Reames,
T. T. von Rosenvinge, N. R. Sheeley Jr., M. J. Koomen,
R. A. Howard, D. J. Michels.
18th International Cosmic Ray Conference, Vol. 4, p. 6 – 9 (1983). – See Abstr. 012.096.

E > 4 MeV proton events presumed due to solar flares are compared with coronal mass ejections observed with an orbiting coronagraph.

074.132   The effect of solar–wind convection on charged particle transport in interplanetary space.
J. A. Earl.
18th International Cosmic Ray Conference, Vol. 4, p. 119 – 122 (1983). – See Abstr. 012.096.

074.133   On the temperature of the active and quiet inner corona.
K. I. Nikol'skaya.
Astron. Zh., Tom 62, Vyp. 3, p. 562 – 568 (1985). In Russian. English translation in Sov. Astron., Vol. 29, No. 3.

In the framework of a model solar corona consisting in its quiet and active regions of the arches of different temperatures and equal densities the relative temperature composition of four coronal disturbances and of the quiet inner corona is studied for different epochs on the basis of eclipse observations of the corona's optical spectrum.

074.134   Interpretation of coronal synoptic observations.
R. H. Munro, R. R. Fisher.
Bull. Am. Astron. Soc., Vol. 17, No. 2, p. 632 (1985). Abstract. – See Abstr. 010.067.

074.135   Three–dimensional magnetostatic models of coronal structures.
T. J. Bogdan, B. C. Low.
Bull. Am. Astron. Soc., Vol. 17, No. 2, p. 632 (1985). Abstract. – See Abstr. 010.067.

074.136   A test for large–angle radio scattering in the solar corona.
D. G. Wentzel, P. Zlobec, M. Messerotti.
Bull. Am. Astron. Soc., Vol. 17, No. 2, p. 632 – 633 (1985). Abstract. – See Abstr. 010.067.

074.137   Wave speeds in the corona and the dynamics of mass ejections.
S. T. Suess, R. L. Moore.
Bull. Am. Astron. Soc., Vol. 17, No. 2, p. 636 (1985). Abstract. – See Abstr. 010.067.

074.138   The broadening of looplike solar coronal transients.
R. M. MacQueen, D. Cole.
Bull. Am. Astron. Soc., Vol. 17, No. 2, p. 636 (1985). Abstract. –
See Abstr. 010.067.

074.139   The mass distribution of coronal mass ejections.
B. V. Jackson, R. A. Howard, M. J. Koomen,
D. J. Michels, N. R. Sheeley Jr.
Bull. Am. Astron. Soc., Vol. 17, No. 2, p. 636 (1985). Abstract. –
See Abstr. 010.067.

074.140   White light and X–ray studies of the coronal mass ejection onset phase.
R. A. Harrison.
Bull. Am. Astron. Soc., Vol. 17, No. 2, p. 636 (1985). Abstract. –
See Abstr. 010.067.

074.141   On the formation of coronal cavity.
C.-H. An, S. T. Suess, E. Tandberg–Hanssen.
Bull. Am. Astron. Soc., Vol. 17, No. 2, p. 636 – 637 (1985). Abstract. – See Abstr. 010.067.

074.142   Solar cycle variation of energy and photospheric magnetic flux from coronal holes.
D. F. Webb, J. M. Davis.
Bull. Am. Astron. Soc., Vol. 17, No. 2, p. 637 (1985). Abstract. –
See Abstr. 010.067.

074.143   The transition region, corona, and solar wind in a coronal hole.
J. V. Hollweg, C. J. Pollock.
Bull. Am. Astron. Soc., Vol. 17, No. 2, p. 637 (1985). Abstract. –
See Abstr. 010.067.

074.144   MHD flux leakage from a turbulently driven coronal hole.
J. M. Davila.
Bull. Am. Astron. Soc., Vol. 17, No. 2, p. 637 (1985). Abstract. –
See Abstr. 010.067.

074.145   Results from the coronal photometry program at NSO, I: Three–line observations of the corona in 1984.
R. C. Altrock, R. R. Fisher, D. G. Sime.
Bull. Am. Astron. Soc., Vol. 17, No. 2, p. 637 – 638 (1985). Abstract. – See Abstr. 010.067.

074.146   Results from the coronal photometry program at NSO, II: Rotation of the green corona over the solar cycle.
D. G. Sime, R. R. Fisher, R. C. Altrock.
Bull. Am. Astron. Soc., Vol. 17, No. 2, p. 638 (1985). Abstract. –
See Abstr. 010.067.

074.147   Results from the coronal photometry program at NSO, III: The green line and white light corona compared.
R. R. Fisher, R. C. Altrock, D. G. Sime.
Bull. Am. Astron. Soc., Vol. 17, No. 2, p. 638 (1985). Abstract. –
See Abstr. 010.067.

074.148   Inferences concerning solar wind momentum and energy balance drawn from coronal density observations.
R. Lallement, T. E. Holzer, R. H. Munro.
Bull. Am. Astron. Soc., Vol. 17, No. 2, p. 638 (1985). Abstract. –
See Abstr. 010.067.

074.149   The generation of helical structure in coronal fields.
M. A. Berger.
Bull. Am. Astron. Soc., Vol. 17, No. 2, p. 638 – 639 (1985). Abstract. – See Abstr. 010.067.

074.150   Viscous normal modes on coronal inhomogeneities.
R. S. Steinolfson.
Bull. Am. Astron. Soc., Vol. 17, No. 2, p. 643 (1985). Abstract. –
See Abstr. 010.067.

074.151   Detailed morphology of flare–associated coronal loops.
R. N. Smartt, R. Jain.
Bull. Am. Astron. Soc., Vol. 17, No. 2, p. 645 (1985). Abstract. –
See Abstr. 010.067.

074.152   A gallery of disconnection transients.
R. M. E. Illing.
Bull. Am. Astron. Soc., Vol. 17, No. 2, p. 646 (1985). Abstract. –
See Abstr. 010.067.

074.153   Modelling disturbances due to a pressure perturbation at the coronal base.
E. Hildner, S. T. Wu.
Bull. Am. Astron. Soc., Vol. 17, No. 2, p. 646 (1985). Abstract. –
See Abstr. 010.067.

074.154   A positional comparison between coronal mass ejection events and solar type II bursts.
R. D. Robinson, R. T. Stewart.
Sol. Phys., Vol. 97, No. 1, p. 145 – 157 (1985).
From radio positions obtained with the Culgoora radio–heliograph the authors have confirmed the close association between type II and CME (*Coronal Mass Ejection*) events deduced by Sheeley et al. (1984). Using height–time plots and intensity maps the authors present evidence that type II events can occur at the leading edge of the mass ejection, within the region well behind the leading edge or in the absence of a CME. Type II events located at the CME leading edge may be generated by the CME and need not be flare–related, while those bursts not associated with the CME leading edge would most probably be generated by a blast wave created during the impulsive phase of the flare. The authors have examined current type II theories and conclude that the type II burst is most probably created during the interaction of a shock wave with the closed magnetic field regions within the transient.

074.155   Two–dimensional pressure structure of a coronal loop.
V. Krishan.
Sol. Phys., Vol. 97, No. 1, p. 183 – 189 (1985).
Representation of the coronal loop plasma in a state generated by the superposition of two Chandrasekhar–Kendall functions leads to a two–dimensional spatial profile of the plasma pressure. It is found that the radial variation of pressure corresponding to the larger spatial widths of the hotter lines does not exist all along the length of the loop. A twisted configuration of the plasma is obtained. The pressure or the temperature is still maximum at the top of the loop but only near the axis. For smaller spatial scales which are determined from the ratio of the toroidal to poloidal magnetic fluxes, the radial pressure variation exhibits oscillations.

074.156   White–light coronal transients observed from Skylab May 1973 to February 1974: a classification by apparent morphology.
R. H. Munro, D. G. Sime.
Sol. Phys., Vol. 97, No. 1, p. 191 – 201 (1985).
The apparent morphologies of the major coronal mass ejection transients observed during the Skylab mission with the High Altitude Observatory's white–light coronagraph are described and illustrated. The 77 major events are grouped into classes referred to as Loop, Filled Bottle, Material Injected into Streamer, Ray, Cloud, and Streamer Separation events, with 14 being Unclassifiable because of incomplete observations.

074.157   Collisionless deceleration of stabilized electron beams in the solar corona.
A. B. Eremin, V. V. Zaitsev (*V. V. Zajtsev*).
Sov. Astron., Vol. 28, No. 6, p. 668 – 672 (1984). English translation of 38.074.061.

074.158 Influence of nonuniformity of the solar corona on one–dimensional quasi–linear relaxation of an electron beam.
Yu. A. Sukovatov.
Soln. Dannye, Byull., 1985, No. 3, p. 80 – 83 (1985). In Russian.
The generation of noncollective plasma waves by an electron beam is investigated. Plasma waves become noncollective in a nonuniform plasma. The growth rate of instability decreases with time. The possibility of stabilization of a weak unbounded electron beam is shown.

074.159 Alfvén waves in a nonuniform layer in a gravity field.
Yu. A. Sukovatov.
Soln. Dannye, Byull., 1985, No. 4, p. 66 – 69 (1985). In Russian.
The influence of gravity on the surface of Alfvén waves and the Alfvén continuum is investigated. The Alfvén continuum is asymptotically stable. The instability of surface waves is similar to the case of sharp non–uniformity.

074.160 Studies of the microscale density fluctuations in the solar wind using interplanetary scintillations.
J. P. Filice.
Diss. Abstr. Int., Sect. B, Vol. 45, No. 2, p. 583 – 584 (1984). Thesis, University of California, 157 pp. (1984). Order No. DA8412474.

074.161 Magnetic helicity in the solar corona.
M. A. Berger.
Diss. Abstr. Int., Sect. B, Vol. 45, No. 6, p. 1810 (1984). Thesis, Harvard University, 139 pp. (1984). Order No. DA8419438.

074.162 Transfer of Lyman–α radiation in solar coronal loops.
P. Gouttebroze, J.–C. Vial, G. Tsiropoula.
Progress in stellar spectral line formation theory, p. 359 – 372 (1985). – See Abstr. 012.108.
The emission and scattering of Lyman–α radiation within the loop–like structures of the solar corona are investigated, for a large range of physical conditions within these objects. Results from partial and complete redistribution computations are compared. A series of predictions, concerning line profiles, integrated intensities, and directional diagrams are given for observation diagnosis.

074.163 Variation of Alfvén fluctuations across a slow shock near the Sun.
F.–s. Wei, C.–y. Tu.
Chin. Astron. Astrophys., Vol. 9, No. 1, p. 54 – 59 (1985). English translation of Acta Astrophys. Sin., Vol. 4, No. 4, p. 294 – 303 (1984).
The authors consider how the Alfvén fluctuation varies as it passes through a slow MHD shock, assuming the fluctuation to originate on the Sun and such slow shocks are present between 2 and 6 solar radii. The results are (1) low latitude region in the vicinity of the neutral sheet is where the fluctuation can propagate with no or little distortion; (2) high latitude region far from the neutral sheet is a region of small distortion only for low Alfvén–Mach numbers (about 0.3 or lower); (3) the observation of the Alfvén fluctuations in space is consistent with, and lends support to their originating in the Sun and the existence of slow shock around the Sun.

Report of ESA's topical team on solar and heliospheric physics.
See Abstr. 013.073.

Analyses of experimental observations of electron temperatures in the near wake of a model in a laboratory–simulated solar wind plasma.
See Abstr. 022.112.

Role of the gaunt factor in the derivation of dielectronic recombination coefficient.
See Abstr. 022.131.

Erratum: 'Measuring electron density in coronal active regions. II: A multichannel coronagraph with a photoelectric spectrograph and a reflex monitor at λ5303 Å' [Sol. Phys., Vol. 94, No. 1, p. 117 – 131 (1984)].
See Abstr. 032.002.

Solar coronal studies using normal–incidence X–ray optics.
See Abstr. 035.007.

Coronal–hole detectability on solar–type stars.
See Abstr. 036.133.

The Solar and Heliospheric Observatory, SOHO – a Phase–A project of the European Space Agency.
See Abstr. 051.008.

Coronal physics with the Solar and Heliospheric Observatory.
See Abstr. 051.048.

Thermal cyclotron radiation in astrophysics.
See Abstr. 061.035.

Reconnection in sheared magnetic fields in space and astrophysics.
See Abstr. 062.007.

Phase mixing of propagating Alfvén waves.
See Abstr. 062.021.

Bending waves on current sheets.
See Abstr. 062.031.

Alfvén surface waves along coronal streamers.
See Abstr. 062.079.

An analytic model for hydromagnetospheres.
See Abstr. 062.099.

Waves in inhomogeneous media.
See Abstr. 062.102.

Viscous dissipation in the tearing instability.
See Abstr. 062.188.

Overheated open coronal regions.
See Abstr. 064.014.

Three–dimensional structures of magnetostatic atmospheres. I. Theory.
See Abstr. 064.080.

Statistical mechanics of velocity and magnetic fields in solar active regions.
See Abstr. 072.017.

Impulsive phenomena in a small active region.
See Abstr. 072.018.

Report of IAU Commission 10: Solar activity (*Activité solaire*).
See Abstr. 072.108.

Chromospheric emission lines in the coronal spectrum of July 31, 1981.
See Abstr. 073.022.

Eruption of huge magnetic systems from the Sun.
See Abstr. 073.059.

Heating in the solar mantle.
See Abstr. 073.069.

Spicules and surges.
See Abstr. 073.097.

Prominences, two–ribbon flares and coronal transients.
See Abstr. 073.109.

Solar Maximum Mission results on the energetics of the impulsive phase of solar flares.
See Abstr. 073.110.

Initial phase of chromospheric evaporation in a solar flare.
See Abstr. 073.134.

Coronal explosions.
See Abstr. 073.135.

High–resolution X–ray spectra of solar flares. VII. A long–duration X–ray flare associated with a coronal mass ejection.
See Abstr. 073.136.

Filament evolution: energy build–up, eruption and oscillations.
See Abstr. 073.196.

The structure of transition region loops.
See Abstr. 073.207.

Magnetic constriction and energy balance in the upper transition region.
See Abstr. 073.211.

Using high energy cosmic particles to probe the coronal magnetic field.
See Abstr. 075.024.

Large–scale magnetic field on the sun and recurrent high–velocity streams of solar wind plasma.
See Abstr. 075.031.

Space–time relations of magnetic fields on the sun with the IMF, solar wind and geomagnetic activity in May –July 1979.
See Abstr. 075.032.

Imaging of reconnection processes in hard X–rays.
See Abstr. 076.007.

Results from the Hard X–ray Imaging Spectrometer.
See Abstr. 076.014.

The EUV structure of a solar active region.
See Abstr. 076.031.

Time series images of the UV chromosphere and transition zone.
See Abstr. 076.042.

Observational evidence for magnetic reconnection in microwave solar bursts.
See Abstr. 077.001.

Observations of solar radio bursts from meter to kilometer wavelengths.
See Abstr. 077.012.

Storms of U–bursts and the stability of coronal loops.
See Abstr. 077.016.

Polarization of solar noise storm continuum and plasma wave density in the corona.
See Abstr. 077.017.

Velocities of type II solar radio events.
See Abstr. 077.018.

VLA observations of narrow–band decimetric burst emission.
See Abstr. 077.028.

On the origin of continuum emission during decametric solar noise storms.
See Abstr. 077.042.

A comparison of solar ^{3}helium–rich events with type II bursts and coronal mass ejections.
See Abstr. 078.007.

Balloon observation of the 1983 solar eclipse in Indonesia.
See Abstr. 079.141.

Hydromagnetic buoyancy force in the solar atmosphere.
See Abstr. 080.004.

Observations of electron density in the solar corona during sunspot minimum and global electric currents around the sun.
See Abstr. 080.018.

Dynamics of the 1054 UT March 22, 1979, substorm event: CDAW 6.
See Abstr. 084.017.

Coupling between the solar wind and the magnetosphere: CDAW 6.
See Abstr. 084.018.

Solar wind control of magnetospheric pressure (CDAW 6).
See Abstr. 084.019.

Magnetotail energy storage and release during the CDAW 6 substorm analysis intervals.
See Abstr. 084.020.

Recurrent geomagnetic activity: evidence for long–lived stability in solar wind structure.
See Abstr. 084.022.

The application of dimensional analysis to the problem of solar wind–magnetosphere energy coupling.
See Abstr. 084.042.

Coupling of the solar wind to measures of magnetic activity.
See Abstr. 084.043.

Solar wind control of electric fields and currents in the magnetosphere and ionosphere.
See Abstr. 084.056.

The influence of the ionospheric conductivity on the potential difference of the electric field across a plasma layer.
See Abstr. 084.107.

Dependence of hydromagnetic energy spectra near $L = 2$ and $L = 3$ on upstream solar wind parameters.
See Abstr. 084.110.

Observations of polar–cap ionospheric signatures of solar wind/ magnetosphere coupling.
See Abstr. 084.134.

Geomagnetic effects at high latitudes connected with the dynamical shock of the solar wind.
See Abstr. 085.014.

Solar wind–Venus interaction affecting ionospheric electron density profile.
See Abstr. 093.041.

Saturn radio emission and the solar wind: Voyager–2 studies.
See Abstr. 100.039.

Plasmaphysikalische Prozesse bei der Wechselwirkung von Kometen mit dem Sonnenwind.
See Abstr. 102.052.

On the role of quasi–resonance recharging of cometary molecules of nitrogen and helium ions with solar wind particles in cometary tails of the $N_1{}^+$ infrared bands.
See Abstr. 102.058.

About the interaction between high speed wind streams from coronal holes and comets.
See Abstr. 102.063.

Formation of ion acoustic solitary waves upstream of the earth's bow shock.
See Abstr. 106.006.

Distant heliospheric results on interplanetary shock propagation.
See Abstr. 106.010.

A study of the formation, evolution, and decay of shocks in the heliosphere between 0.5 and 30.0 AU.
See Abstr. 106.011.

Some characteristics of propagation of flare– or CME (*coronal mass ejections*)–associated interplanetary shock waves.
See Abstr. 106.019.

Latitude variation of recurrent MeV–energy proton flux enhancements in the heliocentric radial range 11 to 20 AU and possible correlation with solar coronal hole dynamics.
See Abstr. 106.027.

Interplanetary field enhancements in the solar wind: evidence for cometesimals at 0.72 and 1.0 AU?
See Abstr. 106.029.

Origin of strong interplanetary shocks.
See Abstr. 106.033.

Sector structure of the interplanetary magnetic field and high–velocity solar wind fluxes.
See Abstr. 106.041.

Cosmic–ray picture of the heliosphere.
See Abstr. 106.050.

Evidence for long period Alfvén waves in the inner solar system.
See Abstr. 106.054.

Interplanetary magnetic field enhancements in the solar wind: statistical properties at 1 AU.
See Abstr. 106.062.

Numerically–simulated formation and propagation of interplanetary shocks.
See Abstr. 106.066.

The brightness of the interplanetary medium in Thomson scattering from 0.3 to 1.0 AU: comparison with a view from Helios B.
See Abstr. 106.083.

Report of IAU Commission 49: The interplanetary plasma and the heliosphere (*Plasma interplanétaire et l'héliosphère*).
See Abstr. 106.084.

A stability analysis of the spiral sector transition region in the interplanetary magnetic field.
See Abstr. 106.087.

Drag–like effects on heliospheric neutrals due to elastic collisions with solar wind ions.
See Abstr. 131.026.

Cosmic ray intensity variations and the solar wind velocity.
See Abstr. 144.006.

Variation of cosmic rays and solar wind properties with respect to the heliospheric current sheet. 1. Five–GeV protons and solar wind speed.
See Abstr. 144.029.

Cosmic ray acceleration by the solar wind.
See Abstr. 144.040.

North–south cosmic ray anisotropy in recurrent high–speed solar wind streams.
See Abstr. 144.060.

Peculiarities of galactic cosmic ray modulation in 1970 – 1974.
See Abstr. 144.194.

Cosmic ray intensity variations and solar wind velocity.
See Abstr. 144.239.

Cosmic ray decreases and cold plasma in the heliosphere.
See Abstr. 144.292.

Cosmic ray anisotropy relevant to high–velocity solar wind fluxes.
See Abstr. 144.413.

075 Magnetic Fields

075.001 MHD analysis of the evolution of solar magnetic fields and currents in an active region.
S. T. Wu, J. F. Wang, E. Tandberg-Hanssen.
Unstable current systems and plasma instabilities in astrophysics, p. 487 – 490 (1985). – See Abstr. 012.002 (IAU Symp. No. 107).
The authors have used a self-consistent magnetohydrodynamics model to study the evolution of solar magnetic fields in an active region. The problem has been cast as an initial boundary-value problem based on explicit mathematical formalism (i.e., method of projected characteristics), whereby a variety of horizontal photospheric motions can be treated. In this paper the authors deal specifically with photospheric shear motions in the active region.

075.002 Studies of the large–scale structure of the solar surface using the method of polynomial approximation.
V. A. Magerramov.
Soln. Dannye, Byull., 1984, No. 10, p. 54 – 63 (1985). In Russian.
Observational material of the magnetic field of an active region and the velocity field during a solar flare has been processed using the method of polynomial approximation. It is shown that the distributions of the vectors grad $|H|$ and grad $H_{\parallel}$ are similar, while the distribution of the vector grad $H_{\perp}$ differs significantly from them. Grad $|H|$, grad $H_{\parallel}$ and grad $H_{\perp}$ are found to be equal to $0.019 - 0.022$ Gauss/km, $0.024 - 0.163$ Gauss/km and $0.08 - 0.046$ Gauss/km, respectively. It is shown that the configuration of the velocity field during a solar flare restores with a period of 6 minutes and the energy release is of an impulse character.

075.003 Latitude zonal structures of the large–scale magnetic field of the sun during 1880 – 1935.
V. I. Makarov.
Soln. Dannye, Byull., 1984, No. 10, p. 68 – 71 (1985). In Russian.
On the basis of the reduction of the latitude distribution of prominence areas for 1880 – 1935 the existence of latitude zonal structures of the radial component of the solar magnetic field, obtained earlier from Hα charts for 1955 – 1982, is confirmed.

075.004 Fine–scale structure of solar magnetic fields.
J. O. Stenflo.
Adv. Space Res., Vol. 4, No. 8, p. 5 – 16 (1984). – See Abstr. 012.010.
Direct mapping of the circular polarization in spectral lines provides information on the morphology and evolution of the partially resolved magnetic structures. In reviewing recent results, special attention is payed to the question of flux disappearance, since it is fundamental for understanding the solar cycle, and depends on a knowledge of the fine–scale structures. The author uses the Fourier transform spectrometer data to illustrate the diagnostic contents of the line–ratio technique, and then indicates how a statistical approach with 400 Fe I lines has recently been applied. In particular he discusses the implications of the observed Stokes V asymmetries for flux tube dynamics.

075.005 The creation of fine structure by magnetic fields.
B. Roberts.
Adv. Space Res., Vol. 4, No. 8, p. 17 – 27 (1984). – See Abstr. 012.010.
The solar plasma is strongly structured by the presence of magnetic fields. This structuring is manifest in the photosphere in the form of flux tubes, from the readily visible sunspots to the sub–telescopic intense tubes, so that the atmosphere is divided into strong–field media or field–free media. In the corona, by contrast, the magnetic field permeates the whole of the atmosphere and structuring consists principally of density and temperature inhomogeneities. The author discusses some of the causes of magnetic structuring including kinematic concentration, convective collapse and magnetoconvection for photospheric tubes, spicules in the chromosphere, and thermal instability for coronal loops.

075.006 Calculation of the solar magnetic field from values observed in the photosphere.
L. Hannakam.
Arch. Elektrotech., Vol. 67, No. 6, p. 353 – 359 (1984). In German. Abstr. in Phys. Abstr., Vol. 88, No. 1248, Entry 9692 (1985).

075.007 Les traceurs du champ magnétique en physique solaire.
P. Mein.
Champs magnétiques stellaires, p. 341 – 368 (1984). – See Abstr. 012.022.

075.008 Etude de la rotation des traceurs magnétiques observés sur les spectrohéliogrammes K IV (Collection de Meudon) – Résultats préliminaires.
E. Ribes, P. Mein.
Champs magnétiques stellaires, p. 369 – 374 (1984). – See Abstr. 012.022.

075.009 Détermination des champs magnétiques dans les structures solaires non résolues.
M. Semel.
Champs magnétiques stellaires, p. 375 – 382 (1984). – See Abstr. 012.022.

075.010 Propriétés des écoulements stationnaires dans les structures magnétiques solaires. Application au champ photosphérique à petite échelle.
E. Ribes.
Champs magnétiques stellaires, p. 413 – 441 (1984). – See Abstr. 012.022.

075.011 Magnetic shear. I. Hale region 16918.
R. G. Athay, H. P. Jones, H. Zirin.
Astrophys. J., Vol. 288, No. 1, p. 363 – 372 (1985). With plates 7 – 13.
Material motion observed in spectral lines of C IV, C II, and Ca II formed in the chromosphere–corona transition region and upper chromosphere exhibits patterns that are closely identified with magnetic field structure at photospheric levels. Assuming that the fluid flow follows magnetic lines of force, the authors use chromospheric and transition region Dopplergrams to infer the broad features of the magnetic field geometry in these upper layers. For Hale region 16918 they find an area in the transition region and upper chromosphere, centered roughly over the photospheric magnetic neutral line, in which the lines of force show a strong tendency to parallel the photospheric neutral line. The authors interpret this as evidence for magnetic shear, which is pronounced in the upper layers of the atmosphere.

075.012 Theorie des solaren Magnetfeldes.
H. C. Spruit.
Sonne, Jahrg. 9, Nr. 33, p. 6 – 12 (1985).

075.013 Propagation of nonlinear, radiatively damped longitudinal waves along magnetic flux tubes in the solar atmosphere.
G. Herbold, P. Ulmschneider, H. C. Spruit, R. Rosner.
Astron. Astrophys., Vol. 145, No. 1, p. 157 – 169 (1985).
For solar magnetic flux tubes the authors compare three types of waves: longitudinal MHD tube waves, acoustic tube waves propagating in the same tube geometry but with rigid walls and ordinary acoustic waves in plane geometry. They find that the effect of distensibility of the tube is small and that longitudinal waves are essentially acoustic tube waves. Due to the tube geometry there is considerable difference between longitudinal waves or acoustic tube waves and ordinary acoustic waves. Longitudinal waves as well as acoustic tube waves show a smaller amplitude growth, larger shock formation heights, smaller mean chromospheric temperature but a steeper dependence of the temperature gradient on wave period.

075.014   Periodic variations of the solar magnetic field.
V. A. Kotov, L. S. Levitskij.
Izv. Krymskoj Astrofiz. Obs., Tom 69, p. 90 – 99 (1984). In Russian. English translation in Bull. Crimean Astrophys. Obs., Vol. 69.

Dominant recurrent periods – or rotation periods – of the mean solar magnetic field have been studied using the polarity of the interplanetary magnetic field (IMF, 1926 – 1981) as inferred from polar geomagnetic observations and direct measurements of the magnetic field of the sun seen as a star (1968 – 1981). The power spectra of these data reveal a set of discrete lines associated with 27 – 29 days (synodic) periods of the solar rotation. The most prominent peaks, in particular those at 26.95, 27.20, 27.38 and 28.21 days periods, exhibit remarkable phase–coherency over 56 years. It is suggested that these discrete lines (periods) belong to certain latitudinal zones on the sun where the magnetic field rotates coherently and almost independently (in regard to the phase) of the 11–years cycle and of the polar field reversals occurring near epochs of the solar activity maxima. In addition, the present analysis of 56 years of the IMF polarity shows the existence of a strong annual variation which appears to be, in the main part, a by–product of the well–known Rosenberg–Coleman's effect in the IMF predominant polarity.

075.015   Magnetic fields on the sun and the north–south component of transient variations of the interplanetary magnetic field at 1 AU.
F. Tang, S.–I. Akasofu, E. Smith, B. Tsurutani.
J. Geophys. Res., Vol. 90, No. A3, p. 2703 – 2712 (1985).

In order to study the relationship between solar magnetic fields and the transient variations of the north–south component, B_z, of the interplanetary magnetic field, IMF, at 1 AU, the authors collect (1) flares from unusual north–south oriented active regions; (2) large IMF B_z events; (3) large flares with comprehensive flare index > 12 and then investigate the associated IMF B_z changes or the magnetic field of the initiating flares, whichever the case may be.

075.016   Magnetic buoyancy instabilities in the convective overshoot zone.
D. W. Hughes.
ESA Spec. Publ., ESA SP–220, p. 55 – 58 (1984). – See Abstr. 012.045.

This paper contains the results of a linear stability analysis for a stratified magnetic layer, incorporating the effects of diffusion, wave motion and rotation and shows the occurrence of instabilities to be widespread.

075.017   On the structure of magnetic fields in the solar convection zone.
M. Schüssler.
ESA Spec. Publ., ESA SP–220, p. 67 – 76 (1984). – See Abstr. 012.045.

The following concepts are discussed in turn and tentatively put together in a picture of a boundary layer dynamo near the bottom of the convection zone: expulsion of magnetic flux and vorticity, fragmentation and accumulation of fields, the dominating forces that govern the evolution of a magnetic structure.

075.018   Amplification and maintenance of thin magnetic flux tubes by compressible convection.
M. R. E. Proctor, N. O. Weiss.
ESA Spec. Publ., ESA SP–220, p. 77 – 80 (1984). – See Abstr. 012.045.

The authors present a model that includes both the effects of diffusion and proper treatment of the Lorentz forces, for a thin tube whose depth is of the order of a scale height. The model includes both magnetic pressure (leading to evacuation of the tube) and curvature forces. It is found that while small tubes are limited principally by pressure effects, tubes with fluxes $\sim 10^{18}$ mx exert an important retarding force on the convection that causes them.

075.019   Nonlinear time–evolution of kink–unstable magnetic flux tubes in the convection zone of the Sun.
F. Moreno–Insertis.
ESA Spec. Publ., ESA SP–220, p. 81 – 84 (1984). – See Abstr. 012.045.

The nonlinear development of the kink instability of a magnetic flux tube initially lying in mechanical equilibrium in the deep convection zone is presented. A numerical code was developed; for the convection zone stratification the model by Spruit (1977) was used.

075.020   Convective collapse and overstable oscillations in solar flux tubes.
S. S. Hasan.
ESA Spec. Publ., ESA SP–220, p. 227 – 228 (1984). – See Abstr. 012.045.

The collapse of solar flux tubes by a convective instability is investigated. The final state consists of overstable oscillations with periods typically around 1000 s.

075.021   Harmonic analysis of the solar magnetic field.
J. T. Hoeksema, P. H. Scherrer.
ESA Spec. Publ., ESA SP–220, p. 269 – 270 (1984). – See Abstr. 012.045.

The spherical harmonics of the global solar magnetic field have been calculated using photospheric field measurements from the Stanford Solar Observatory from 1976 – 1983. The field evolution during the solar cycle is analyzed.

075.022   Magnetic shear. II. Hale region 17244.
R. G. Athay, H. P. Jones, H. Zirin.
Astrophys. J., Vol. 291, No. 1, p. 344 – 355 (1985).

A $B\gamma(\delta)$ sunspot group with growing δ–spots of trailing polarity shows evidence in Hα filament structure of a transition from a state of weak magnetic shear to a state of strong shear. The shear develops in the chromosphere and transition region to the corona overlying the photospheric magnetic neutral line separating the δ–spots from the leading polarity at a time when the δ–spots are undergoing rapid growth. Several major flares occur along the sheared portion of the neutral line following the shear development.

075.023   High resolution observations of magnetic features on the sun.
K. P. Topka, T. D. Tarbell.
Bull. Am. Astron. Soc., Vol. 16, No. 4, p. 991 (1984). Abstract. – See Abstr. 010.062.

075.024   Using high energy cosmic particles to probe the coronal magnetic field.
J. Linsley.
Bull. Am. Astron. Soc., Vol. 16, No. 4, p. 1004 – 1005 (1984). Abstract. – See Abstr. 010.062.

075.025   Estimate of magnetic fields in regions of generation of microwave radio bursts on the sun.
A. S. Grebinskij, A. P. Sedov.
Radioizluch. Solntsa, Leningrad, No. 5, p. 120 – 126 (1984). In Russian. Abstr. in Ref. Zh., 51. Astron., 5.51.339 (1985).

075.026   Synergetic conception of the evolution of structures of magnetic fields on the sun.
Eh. I. Mogilevskij.
Physics of solar activity, p. 56 – 62 (1983). In Russian. Abstr. in Ref. Zh., 51. Astron., 6.51.298 (1985). – See Abstr. 003.024.

075.027   Height distribution of the magnetic field in active region No. 325 (HR 17751).
B. A. Ioshpa, V. G. Utrobin.
Physics of solar activity, p. 79 – 85 (1983). In Russian. Abstr. in Ref. Zh., 51. Astron., 6.51.299 (1985). – See Abstr. 003.024.

075.028   On the height gradient of a magnetic field in an active region.
B. A. Ioshpa.
Physics of solar activity, p. 86 – 93 (1983). In Russian. Abstr. in Ref. Zh., 51. Astron., 6.51.300 (1985). – See Abstr. 003.024.

075.029   On the character of decrease of the magnetic field with height above spots of the class J.
F. A. Ermakov.
Physics of solar activity, p. 94 – 98 (1983). In Russian. Abstr. in Ref. Zh., 51. Astron., 6.51.301 (1985). – See Abstr. 003.024.

075.030   Evolution of the geoeffective region No. 1775 (SGD) on the sun in July 1981.
B. A. Ioshpa, Eh. I. Mogilevskij, L. I. Starkova, V. G. Utrobin, J. Sikora.
Physics of solar activity, p. 63 – 78 (1983). In Russian. Abstr. in Ref. Zh., 51. Astron., 6.51.302 (1985). – See Abstr. 003.024.

075.031   Large–scale magnetic field on the sun and recurrent high–velocity streams of solar wind plasma.
B. V. Zhiromskij.
Magnitosfer. issled., Moskva, No. 4, p. 17 – 18 (1984). In Russian. Abstr. in Ref. Zh., 51. Astron., 6.51.319 (1985).

075.032   Space–time relations of magnetic fields on the sun with the IMF, solar wind and geomagnetic activity in May –July 1979.
M. I. Pudovkin, D. I. Ponyavin.
Geomagn. Aehron., Tom 25, No. 3, p. 488 – 490 (1985). In Russian. English translation in Geomagn. Aeron.

075.033   Particle acceleration in local magnetic structures on the sun.
N. N. Kontor.
18th International Cosmic Ray Conference, Vol. 4, p. 12 – 15 (1983). – See Abstr. 012.096.

The dynamics of the production of local magnetic structures in the solar atmosphere, such as active regions, ephemeral regions, X–ray bright points, and others, is studied. The relationships of the local magnetic structures to the realization of various models for solar flares and processes of particle acceleration on the sun is discussed.

075.034   The association of He I λ10830 'dark points' and the evolution of the quiet sun magnetic fields.
K. L. Harvey, F. Tang, V. Gaizauskas.
Bull. Am. Astron. Soc., Vol. 17, No. 2, p. 632 (1985). Abstract. – See Abstr. 010.067.

075.035   Potential fields and magnetic canopies.
H. P. Jones.
Bull. Am. Astron. Soc., Vol. 17, No. 2, p. 633 – 634 (1985). Abstract. – See Abstr. 010.067.

075.036   A movie of the solar magnetic field 1974 – 1984.
J. W. Harvey, W. Ditsler.
Bull. Am. Astron. Soc., Vol. 17, No. 2, p. 634 (1985). Abstract. – See Abstr. 010.067.

075.037   The magnetic equilibrium of vertical, thick flux tubes near the solar surface.
V. J. Pizzo.
Bull. Am. Astron. Soc., Vol. 17, No. 2, p. 642 (1985). Abstract. – See Abstr. 010.067.

075.038   Numerical simulations of the mean solar magnetic field during the sunspot cycle.
N. R. Sheeley Jr., C. R. DeVore.
Bull. Am. Astron. Soc., Vol. 17, No. 2, p. 642 (1985). Abstract. – See Abstr. 010.067.

075.039   A stochastic model for the mean solar magnetic field during the sunspot cycle.
C. R. DeVore, N. R. Sheeley Jr.
Bull. Am. Astron. Soc., Vol. 17, No. 2, p. 642 (1985). Abstract. – See Abstr. 010.067.

075.040   Some peculiarities of large–scale solar magnetic fields.
V. S. Berdichevskaya.
Soln. Dannye, Byull., 1985, No. 3, p. 74 – 79 (1985). In Russian.

By using the Atlas of Solar Magnetic Fields (1967) for each solar Carrington rotation (NN 1417 – 1461, August 1959 – November 1962) at different latitudes the longitudinal sizes of unipolar regions of the S– and N– large–scale (background) magnetic field were measured. The similarity of the variation in these quantities shows the possible existence of some superlarge–scale (global) field.

075.041   Numerical simulations of stellar convective dynamos. III. At the base of the convection zone.
G. A. Glatzmaier.
Geophys. Astrophys. Fluid Dyn., Vol. 31, No. 1/2, p. 137 – 150 (1985).

The author describes numerical simulations of giant–cell solar convection and magnetic field generation. Nonlinear, three–dimensional, time–dependent solutions of the anelastic magnetohydrodynamic equations are presented for a stratified, rotating, spherical shell of ionized gas. The simulations suggest that the solar dynamo may be operating at the base of the convection zone in the transition region between the stable interior and the turbulent convective region.

075.042   Structure and evolution of the large scale solar and heliospheric magnetic fields.
J. T. Hoeksema.
Diss. Abstr. Int., Sect. B, Vol. 45, No. 6, p. 1811 (1984). Thesis, Stanford University, 233 pp. (1984). Order No. DA8420551.

075.043   Configurations of the magnetic field in sunspot groups and solar flares. II. Regularities of the development of flare magnetic configurations.
R. N. Ikhsanov.
Izv. Glav. Astron. Obs. Pulkovo, Astrometr. Astrofiz., No. 201, p. 84 – 95 (1985). In Russian.

Properties of flare magnetic configurations occurring during the interaction of two or more rising rope systems of the magnetic field are considered. The observational data were studied from the point of view of the structure of the magnetic field, the development of sunspots in a group and their proper motions and data on flares in Hα for individual sunspot groups with large flares. It is shown that the considered flare configurations of the magnetic field explain well many of the regularities that have been previously found.

A possible method for the determination of the magnitude and inclination of a sunspot magnetic field.
See Abstr. 036.215.

Magnetohydrodynamic stability of an axisymmetric, line–tied, diamagnetic plasmoid embedded in a uniform magnetic field.
See Abstr. 062.061.

Photospheric flux changes and the MHD approximation.
See Abstr. 062.086.

Nonlinear development of convective instability within slender flux tubes. II. The effect of radiative heat transport.
See Abstr. 062.143.

Spontaneous reconnection of magnetic field lines in a collisionless plasma.
See Abstr. 062.177.

A variational approach to the non–linear force–free magnetic fields.
See Abstr. 062.186.

Calculation on nonlinear force–free magnetic fields.
See Abstr. 062.187.

Dependence of the properties of magnetic fluxtubes on area factor or amount of flux.
See Abstr. 071.001.

Effects of magnetic fields on the asymmetry of photospheric line profiles.
See Abstr. 071.040.

3–D behavior of buoyant magnetic flux tubes in granules and supergranules.
See Abstr. 071.041.

High–resolution spectroscopy of active regions. 2. Line–profile interpretation, applied to an emerging flux region.
See Abstr. 072.006.

Statistical mechanics of velocity and magnetic fields in solar active regions.
See Abstr. 072.017.

The internal magnetic field of the Sun and peculiarities of the solar activity cycles.
See Abstr. 072.019.

A calculation of the multi–level configuration of an active region from measurements of the magnetic field with a magnetograph in several spectral lines.
See Abstr. 072.021.

Modeling of the 22–year solar activity cycle within the framework of the dynamo theory with allowance for the primary field.
See Abstr. 072.030.

Observations concerning energy balance in solar magnetic regions.
See Abstr. 072.040.

The rise and fall of sunspot group 18962: a case of magnetic submergence.
See Abstr. 072.044.

A force model of the magnetic field of a unipolar sunspot.
See Abstr. 072.045.

The interpretation of sunspot magnetic field observations.
See Abstr. 072.049.

X–ray, ultraviolet, optical and magnetic structure in and near an active region.
See Abstr. 072.069.

On the nature of quasi–periodic oscillations of the magnetic field and of the velocity in a sunspot.
See Abstr. 072.073.

Approximation of averaged characteristics of the magnetic field of a sunspot on the basis of a potential model.
See Abstr. 072.074.

Connection of magnetic holes with development of sunspot groups.
See Abstr. 072.075.

Magnetographic observations of magnetic fields in spots from the Fe II line.
See Abstr. 072.085.

Magnetographic observations of faculae.
See Abstr. 072.086.

Quasi–periodic vibrations of signals of a magnetograph during the observation of sunspots.
See Abstr. 072.087.

Filaments and the magnetic field of an active region.
See Abstr. 072.089.

Numerically simulated fields, currents and plasma properties in a solar active region due to photospheric shear.
See Abstr. 072.101.

Are there constant–alpha force–free magnetic fields in solar active regions?
See Abstr. 072.102.

A new model for flux emergence and the evolution of sunspots and the large–scale fields.
See Abstr. 072.107.

Twisting of the magnetic field and motions in complex sunspot groups. I.
See Abstr. 072.116.

Numerical calculation of force–free field in solar active regions. II. An improved Chiu–Hilton method.
See Abstr. 072.118.

Umbra motions in two complex sunspot groups and their relationship with the magnetic field and flare activity.
See Abstr. 072.121.

Equilibre MHD et instabilités dans les protubérances ou filaments solaires.
See Abstr. 073.068.

Remarks on the magnetic support of quiescent prominences.
See Abstr. 073.078.

Magnetic field structures of hard X–ray flares observed by HINOTORI spacecraft.
See Abstr. 073.080.

Dependence of the flare stream velocity on magnetic field orientation.
See Abstr. 073.082.

The correlation of solar flare production with magnetic energy in active regions.
See Abstr. 073.172.

An example for solar flares caused by magnetic field nonequilibrium.
See Abstr. 073.173.

Characteristic patterns of solar magnetic fields around the positions of flares with enhanced helium–3 abundance.
See Abstr. 073.178.

A coronal magnetic field model with volume and sheet currents.
See Abstr. 074.040.

Alfvén waves in current–carrying solar magnetic flux tubes.
See Abstr. 074.056.

A non–axisymmetric structure of coronal magnetic loop.
See Abstr. 074.082.

Coronal field changes associated with a disappearing sunspot.
See Abstr. 074.104.

Association of coronal holes with large scale solar magnetic fields.
See Abstr. 074.105.

Some peculiarities of magnetic reconnection in the solar corona.
See Abstr. 074.110.

The magnetic field of solar wind near sector boundaries.
See Abstr. 074.113.

Solar cycle variation of energy and photospheric magnetic flux from coronal holes.
See Abstr. 074.142.

A gallery of disconnection transients.
See Abstr. 074.152.

Modelling disturbances due to a pressure perturbation at the coronal base.
See Abstr. 074.153.

Observational evidence for magnetic reconnection in microwave solar bursts.
See Abstr. 077.001.

Creation of high–energy electron tails by the lower–hybrid waves and its relevance to Type II and III bursts.
See Abstr. 077.003.

A working model of the solar S–component radio emission.
See Abstr. 077.015.

On the radio radiation of solar flares and features of coronal magnetic fields.
See Abstr. 077.037.

Long nonlinear waves in a compressible magnetically structured atmosphere. I. Slow sausage waves in a magnetic slab.
See Abstr. 080.002.

Convective instability in a solar flux tube. II. Nonlinear calculations with horizontal radiative heat transport and finite viscosity.
See Abstr. 080.020.

The influence of convection zone flows on magnetic fields in the sun.
See Abstr. 080.130.

Report of IAU Commission 12: Radiation and structure of the solar atmosphere (*Radiation et structure de l'atmosphère solaire*).
See Abstr. 080.149.

076 UV, X, Gamma Radiation

076.001 Possible evidence for stochastic acceleration of electrons in solar hard X–ray bursts observed by *SMM*.
J. C. Brown, J. M. Loran.
Mon. Not. R. Astron. Soc., Vol. 212, No. 1, p. 245 – 255 (1985).

It is shown that the dynamic, hard X–ray spectra of the events of 1980 March 29 and June 7 observed by the Solar Maximum Mission (HXRBS) exhibit an anticorrelation of photon flux and spectral steepness. This is exhibited in terms of systematic loci followed by the event in the plane (flux I, spectral index γ). These observations are compared with a theoretical model, developed from Benz, involving injection of electrons into a thick target region from a fluctuating slab in which they are stochastically accelerated. The data are found to be in reasonable accord with the model predictions and are used to obtain constraints on plasma conditions in the acceleration site. Theoretical implications of this result are discussed, as are possible sources of deviation between the data and the theory.

076.002 An interpretation of the millisecond time variation in hard X–ray solar flares.
V. Krishan, M. R. Kundu.
Unstable current systems and plasma instabilities in astrophysics, p. 299 – 301 (1985). – See Abstr. 012.002 (IAU Symp. No. 107).

Recent observations of the fast time variability in the hard X–ray emission from solar flares are interpreted. The fast spikes are assumed to be superimposed on the thermal X–ray emission. The rise and fall of a spike are caused by disruptions in the plasma. The rise time represents the impulsive heating time and the decay or fall time represents a quick cooling of the plasma due to the accelerating growth rate of the m = 1 tearing mode. The estimated characteristic time durations of the spike are found to be in good agreement with the observed ones.

076.003 Current status of the dissipative thermal model for solar hard X–ray bursts.
D. F. Smith.
Unstable current systems and plasma instabilities in astrophysics, p. 509 – 512 (1985). – See Abstr. 012.002 (IAU Symp. No. 107).

Up until about five years ago all models for hard X–ray bursts consisted of streaming nonthermal electrons interacting with an ambient plasma (Brown 1975). Stimulated by observations of hard X rays with thermal spectra, analysis of a thermal model in which all the electrons in a given volume are heated to a temper-

ature $T_e \cong 10^8$K was begun. It was recognized from the beginning that some electrons in the tail of the distribution would escape through the conduction fronts formed and mimic nonthermal streaming electrons. This thermal model with loss of electrons or dissipation became known as the dissipative thermal model (Emslie and Vlahos 1980). If the escaping electrons are not replenished, they will cease to make a contribution after a fraction of a second and the source will become a pure thermal source. It is shown here that collisional replenishment (Smith and Brown 1980) is too slow.

076.004 Measurements of solar hard X–ray bursts with high spectral resolution and sensitivity.
R. P. Lin.
AIP Conf. Proc., No. 115, p. 619 – 627 (1984). – See Abstr. 012.005.

This paper reviews the first observations of solar hard X–ray bursts obtained with a combination of a high spectral resolution, $\lesssim 1$ keV, germanium spectrometer and a large area, low background, NaI/CsI phoswich scintillation detector. In observations of a large flare, the Ge spectrometer was able to identify for the first time a superhot, $\gtrsim 3 \times 10^7$K thermal component. With the high sensitivity of this detector complement, a total of ~ 25 solar hard X–ray bursts with peak fluxes $\sim 10 - 10^2$ times smaller than in solar flare hard X–ray bursts were observed in 141 minutes of observation. These hard X–ray microflares last a few to a few tens of seconds.

076.005 High energy transients from the Sun.
E. L. Chupp.
AIP Conf. Proc., No. 115, p. 641 (1984). Abstract. – See Abstr. 012.005.

076.006 Organisation of a unified system of energetic calibration of X–ray experiments.
B. Valníček, F. Fárník, B. Sylwester, J. Sylwester.
Adv. Space Res., Vol. 4, No. 7, p. 121 – 123 (1984). – See Abstr. 012.009.

X–ray data obtained by the Prognoz 5, 6, 7 and 8 hard X–ray photometers are compared with the measurements carried out by similar instruments aboard the Solrad 11, ISEE 3, SMM and Hinotori satellites. Using the method of relative amplitude analysis, the apparent disagreement in the energy discrimination level calibration between the instruments is pointed out. The results of the comparison and the possible sources of disagreement are given.

076.007 Imaging of reconnection processes in hard X–rays.
Z. Švestka, G. Poletto.
Adv. Space Res., Vol. 4, No. 7, p. 287 – 290 (1984). – See Abstr. 012.009.

The Hard X–ray Spectrometer aboard the SMM detected several events of energy release late in the development of two-ribbon flares. One such event, at 21:12 UT on 21 May, 1980 is studied in detail. The site of new brightening first became visible in hard X–rays (> 22 keV) and only afterwards showed up at lower energies. It was clearly located high in the corona so that one can identify it with energy release at the tops of newly formed post–flare loops.

076.008 Results from the transition region camera.
B. H. Foing, R. M. Bonnet.
Adv. Space Res., Vol. 4, No. 8, p. 43 – 53 (1984). – See Abstr. 012.010.

Three series of high resolution ultraviolet pictures of the Sun have been obtained during the three flights of rocket experiment T.R.C. (Transition Region Camera) which took place on 3 July 1979, 23 September 1980, and 13 July 1982. These pictures reveal many structures in Lα and ultraviolet continua at 160 nm and 220 nm. The scientific objectives, instrumentation, flight conditions and campaigns of simultaneous observations are described.

076.009 Ultraviolet spectroscopy of the chromosphere and transition zone at high spatial and temporal resolution.
K. P. Dere.
Adv. Space Res., Vol. 4, No. 8, p. 55 – 58 (1984). – See Abstr. 012.010.

076.010 Expressions to determine temperatures and emission measures for solar X–ray events from GOES measurements.
R. J. Thomas, R. Starr, C. J. Crannell.
Sol. Phys., Vol. 95, No. 2, p. 323 – 329 (1985).

The authors have developed expressions which give the effective color temperatures and corresponding emission measures for solar X–ray events observed with instruments onboard any of the GOES satellites. To simulate the solar X–ray input at a variety of plasma temperatures, they used theoretical spectra provided by D. L. McKenzie. These spectra were folded through the wavelength dependent transfer functions for the two GOES detectors as given by Donnelly et al. (1977). The resulting detector responses and their ratio as a function of plasma temperature were then fit with simple analytic curves. These fits reproduce the calculated color temperatures within 2% and the calculated emission measures within 5%. With the theoretical spectra provided by McKenzie, similar expressions can be determined for any pair of broadband X–ray detectors whose sensitivities are limited to wavelengths between 0.2 and 100 Å.

076.011 Spectral analysis of the preflare X–ray emission of the sun.
A. A. Zhdanov, Yu. E. Charikov.
Pis'ma Astron. Zh., Tom 11, No. 3, p. 216 – 221 (1985). In Russian. English translation in Sov. Astron. Lett., Vol. 11.

A spectral analysis of the data obtained from the Prognoz 5 – 7 stations in the course of the experiment on the measurement of solar soft X–rays before hard X–ray bursts has been carried out (Kocharov et al., 1983). Several quasi–periodic components (with periods ~ 1.5 hour, ~ 40, ~ 10, ~ 2 min) have been found, and time variations of their parameters while approaching the burst have been studied. These components may serve as one of the indications of a pre–flare situation.

076.012 Results from the Ultraviolet Spectrometer and Polarimeter: non–flare investigations.
E. Tandberg–Hanssen, W. Henze Jr.
Mem. Soc. Astron. Ital., Vol. 55, No. 4, p. 653 – 662 (1984). – See Abstr. 003.017.

The major topics described are sunspot research including magnetic field measurements, oscillations, and models; mass motions in quiet and active regions including steady flows and acoustic waves; and prominence research including physical conditions, dynamics, and mass motions around prominences. Also discussed are studies of UV bursts, the formation of the Cl I line at 1351 Å, ozone in the terrestrial atmosphere, and active regions using correlated observations from other instruments on the spacecraft or on the ground.

076.013 Results from the X–ray Polychromator on SMM.
J. L. Culhane, L. W. Acton, A. H. Gabriel.
Mem. Soc. Astron. Ital., Vol. 55, No. 4, p. 673 – 684 (1984). – See Abstr. 003.017.

The X–ray Polychromator on the SMM includes two high wavelength resolution crystal spectrometers of different design. The instruments operated successfully from launch until the failure of the spacecraft attitude control system in December 1980. Very many important results have been obtained in observations of the soft X–ray emitting plasma that is heated during solar flares. A number of the more interesting observations are presented.

076.014 Results from the Hard X–ray Imaging Spectrometer.
M. E. Machado.
Mem. Soc. Astron. Ital., Vol. 55, No. 4, p. 713 – 723 (1984). – See Abstr. 003.017.

This paper summarizes some of the results: (a) Observations of hard X–ray ($\geqslant 16$ keV) bright sources at the feet of coronal loops. (b) Gradual hard X–ray emission from high temperature ($\geqslant 4 \times 10^7$ K) discrete sources, located within the flare loops. (c) Evidence of flaring activity triggered by the interaction of magnetic structures. (d) The existence of gigantic coronal arches in soft X–rays. (e) The presence of hot plasma ($\geqslant 10^7$ K) in the active region corona, even in the absence of flares.

076.015 Short–period pulsations in solar hard X–ray bursts recorded by Venera 13, 14.
S. V. Bogovalov, A. F. Iyudin, Yu. D. Kotov, E. V. Shugal,
V. Sh. Dolidze, V. M. Zenchenko, G. Vedrenne, M. Niel,
C. Barat, G. Chambon, R. Talon.
Sov. Astron. Lett., Vol. 10, No. 5, p. 286 – 288 (1984). English translation of 38.076.007.

076.016 Expected intensities of solar neon–like ions.
A. K. Bhatia, S. O. Kastner.
Sol. Phys., Vol. 96, No. 1, p. 11 – 26 (1985).

A study of the expected intensities of the stronger solar neon–like ion emission lines, some not yet observed, is carried out to compare with the observational situation. The potential usefulness of the $2p^5 3s(^3P_2) - 2p^6$ forbidden line as a density diagnostic is discussed and new electric quadrupole lines in the soft X–ray range are noted. "Observability diagrams" are presented as a convenient overview of the known and unobserved lines. The S VII resonance lines appear to have anomalous intensities.

076.017 The solar X–ray flux from flares in the hv $\geqslant 2$ keV region.
M. A. Livshits.
Astron. Tsirk., No. 1341, p. 1 – 4 (1984). In Russian.

076.018 The shortest time scales present in solar hard X–ray bursts.
J. C. Brown, J. M. Loran, A. L. MacKinnon.
Astron. Astrophys., Vol. 147, No. 1, p. L10 – L12 (1985).

It is pointed out that isolated transient features in solar hard X–ray burst light curves, recently announced as having time scales of order 10 ms, have not yet been shown to be inconsistent with extreme statistical fluctuations in the Poisson counting noise. It is then shown that as far as persistent rapid time variations are concerned, the available data do not demand the existence of any intrinsic solar time scales shorter than about 100 ms. This conclusion is supported by the fact that the high frequency power level of the Fourier transformed data approaches the Poisson noise expectation value above about 10 Hz and that the correlation of time scales calculated in widely separated energy channels deteriorates as the integration time used approaches 100 ms.

076.019 Characteristics of gamma–ray line flares.
T. Bai, B. Dennis.
Astrophys. J., Vol. 292, No. 2, p. 699 – 715 (1985).

Observations of solar γ–rays by the SMM have demonstrated that energetic protons and ions are rapidly accelerated during the impulsive phase. The authors have studied the characteristics of the γ–ray line (GRL) flares observed by SMM. The main characteristics of GRL flares are as follows: (1) delay of high–energy hard X–rays; (2) flat hard X–ray spectra (average power–law index 3.4); (3) emission of type II and/or type IV radio bursts; (4) intense hard X–ray and microwave emission.

076.020 What the sun can tell us concerning phenomena and spectroscopy.
G. A. Doschek.
Bull. Am. Astron. Soc., Vol. 16, No. 4, p. 984 (1984). Abstract. – See Abstr. 010.062.

076.021 Solar hard X–ray and UV continuum bursts simultaneous to within 0.1 s.
L. E. Orwig, B. E. Woodgate, M. P. Nakada.
Bull. Am. Astron. Soc., Vol. 16, No. 4, p. 1004 (1984). Abstract. – See Abstr. 010.062.

076.022 Solar line profiles and line shifts in the EUV: 200 to 650 Å.
W. E. Behring, R. H. Cornett, L. E. Brotzman, M. R. Bracken.
Bull. Am. Astron. Soc., Vol. 16, No. 4, p. 1004 (1984). Abstract. – See Abstr. 010.062.

076.023 Method of forecast of a solar ultraviolet radiation stream at a fixed wavelength.
S. S. Fomin.
Magnitosfer. issled., Moskva, No. 3, p. 91 – 95 (1984). In Russian. Abstr. in Ref. Zh., 51. Astron., 5.51.378 (1985).

076.024 The relation between hard X–ray and transition–region line emission in solar flares.
J. T. Mariska, A. I. Poland.
Sol. Phys., Vol. 96, No. 2, p. 317 – 330 (1985).

Observational evidence suggests that both the hard X–ray and ultraviolet emission from the impulsive phase of flares result from an electron beam. The authors present the results of model calculations that are consistent with this theory. The impulsive phase is envisioned as occurring in many small magnetically confined loops, each of which maintains an electron beam for only a few seconds. This model successfully matches several observed aspects of the impulsive phase. The calculations indicate that UV emission lines formed below a temperature of about $10^5 K$ will arise predominantly from the chromospheric region heated by the electron beam to transition region temperatures. Emission lines formed at higher temperatures will be produced in the transition region.

076.025 The sun and nearby stars: microwave observations at high resolution.
M. R. Kundu, K. R. Lang.
Science, Vol. 228, No. 4695, p. 9 – 15 (1985).

High–resolution microwave observations are providing new insights into the nature of active regions and eruptions on the sun and nearby stars. The strength, evolution, and structure of magnetic fields in coronal loops can be determined by multiple–wavelength observations with the Very Large Array. Flare models can be tested with Very Large Array snapshot maps. Magnetic changes that precede solar eruptions on time scales of tens of minutes involve primarily emerging coronal loops and the interactions of two or more loops. Nearby main–sequence stars of late spectral type emit slowly varying microwave radiation and stellar microwave bursts that show striking similarities to those of the sun.

076.026 Plausible mechanisms for rapid acceleration of protons during solar flares.
T. Bai.
18th International Cosmic Ray Conference, Vol. 4, p. 26 – 29 (1983). – See Abstr. 012.096.

The author summarizes the characteristics of the SMM gamma–ray line flares, he reports a further study on the delay of high–energy hard X rays, and he discusses plausible mechanisms for proton acceleration in gamma–ray line flares.

076.027 Solar gamma–ray observations by the Hinotori satellite.
M. Yoshimori, K. Okudaira, Y. Hirasima, I. Kondo.
18th International Cosmic Ray Conference, Vol. 4, p. 85 – 88 (1983). – See Abstr. 012.096.

Gamma–ray data of time histories and energy spectra from four large solar flares are analysed. These four flares show apparent nuclear deexcitation lines and a neutron capture line. Problems relating to the particle acceleration and the gamma–ray emission are studied based on these data.

076.028 The limb darkening of neutron capture line at 2.22 MeV and gamma–ray line emission model in solar flares.
M. Yoshimori, K. Okudaira, Y. Hirasima, I. Kondo.
18th International Cosmic Ray Conference, Vol. 4, p. 89 – 92 (1983). – See Abstr. 012.096.

Ratios between the intensity of neutron capture line at 2.22 MeV and that of C deexcitation line at 4.44 MeV for eight large flares observed by the Hinotori satellite were analysed. The results show clear evidence of the limb darkening of 2.22 MeV line. From this results, the site of nuclear reactions and the interaction model in solar flares are studied.

076.029 X–ray precursors of solar flares.
Yu. E. Charikov, V. G. Pharaphonov.
18th International Cosmic Ray Conference, Vol. 4, p. 109 – 112 (1983). – See Abstr. 012.096.

Soft X–ray precursors of solar flares have been studied. From 300 observed flares in 1977 September to December, 286 X–ray precursors have been found. Physical parameters of these precursors are calculated and the theoretical model is advanced.

076.030 Multiple γ–quantum generation in nucleus–nucleus interactions as a source of quasi–continuous γ–rays of solar flares.
B. M. Kuzhevskij.
18th International Cosmic Ray Conference, Vol. 4, p. 113 – 116 (1983). – See Abstr. 012.096.

076.031 The EUV structure of a solar active region.
K. R. Sivaraman, P. K. Raju.
18th International Cosmic Ray Conference, Vol. 4, p. 222 (1983). Abstract. – See Abstr. 012.096.

076.032 Calculation of the density of the energy flux from solar flares by ionospheric data.
I. D. Kozin, B. T. Zhumabaev, B. M. Rubinstein (*B. M. Rubinshtejn*), B. A. Turkeeva.
18th International Cosmic Ray Conference, Vol. 10, p. 273 – 275 (1983). – See Abstr. 012.096.

The method of determination of the density of the energy flux of X–rays during the solar flares is given.

076.033 Nuclear gamma rays and interplanetary proton events.
E. W. Cliver, D. J. Forrest, R. E. McGuire, T. T. von Rosenvinge.
18th International Cosmic Ray Conference, Vol. 10, p. 342 – 345 (1983). – See Abstr. 012.096.

From 1980 – 1982, the Gamma Ray Spectrometer on the Solar Maximum Mission satellite observed the impulsive phases of sixteen western hemisphere flares that were associated with prompt solar proton events. The authors find a lack of correlation between the peak 10 MeV near–Earth proton fluxes and prompt gamma–ray–line fluences. The two largest proton events did not have detectable emission above 300 keV. For the 9 December 1981 event the authors obtain an upper limit for the density of the ion acceleration region of $\leqslant 8 \times 10^9 \mathrm{cm}^{-3}$ for an acceleration time constant of 1500 s.

076.034   Modelling of non thermal hard X ray emission observed during solar flares.
G. Trottet, N. Vilmer.
18th International Cosmic Ray Conference, Vol. 10, p. 346 – 349 (1983). – See Abstr. 012.096.

A model of hard X ray sources is proposed to deduce the characteristics of both the population of energetic electrons and of the region where they are injected or accelerated. In this contribution a hard X ray source is briefly presented and the diffusion of the electrons by Coulomb collisions and by waves is discussed.

076.035   Observational evidence of continuous input of energetic electrons in the corona during large solar flares.
G. Trottet, A. Kerdraon, L. Klein, P. Lantos, M. Pick, N. Vilmer.
18th International Cosmic Ray Conference, Vol. 10, p. 361 – 364 (1983). – See Abstr. 012.096.

Observations of hard and soft X–rays during the gradual and late phases of large solar flares are compared with metre–wave continua. Only a common, continuous/repetitive injection/acceleration of electrons in different coronal structures can account for the observations. There is no long lasting storage of electrons in the corona.

076.036   EUV solar line profiles: 200 to 650 Å.
W. E. Behring, R. H. Cornett, L. E. Brotzman, M. R. Bracken.
Bull. Am. Astron. Soc., Vol. 17, No. 2, p. 592 (1985). Abstract. – See Abstr. 010.065.

076.037   Solar hard X–ray and UV continuum bursts simultaneous to within 0.1 s.
L. E. Orwig, B. E. Woodgate, M. P. Nakada.
Bull. Am. Astron. Soc., Vol. 17, No. 2, p. 609 (1985). Abstract. – See Abstr. 010.065.

076.038   Solar hard X–ray bremsstrahlung production by proton beams?
A. G. Emslie, J. C. Brown.
Bull. Am. Astron. Soc., Vol. 17, No. 2, p. 609 – 610 (1985). Abstract. – See Abstr. 010.065.

076.039   The anisotropy of gamma–rays from solar flares.
W. T. Vestrand, D. J. Forrest, E. L. Chupp, E. Rieger, G. Share.
Bull. Am. Astron. Soc., Vol. 17, No. 2, p. 610 (1985). Abstract. – See Abstr. 010.065.

076.040   Absolute wavelength measurements of solar UV emission lines.
M. E. Bruner, R. A. Shine.
Bull. Am. Astron. Soc., Vol. 17, No. 2, p. 630 (1985). Abstract. – See Abstr. 010.067.

076.041   Relative contribution of lines and continua to solar far ultraviolet variability.
J. W. Cook, M. E. VanHoosier.
Bull. Am. Astron. Soc., Vol. 17, No. 2, p. 630 (1985). Abstract. – See Abstr. 010.067.

076.042   Time series images of the UV chromosphere and transition zone.
K. P. Dere, J.–D. F. Bartoe, G. E. Brueckner.
Bull. Am. Astron. Soc., Vol. 17, No. 2, p. 630 (1985). Abstract. – See Abstr. 010.067.

076.043   Solar UV and EUV temporal characteristics.
R. F. Donnelly.
Bull. Am. Astron. Soc., Vol. 17, No. 2, p. 640 (1985). Abstract. – See Abstr. 010.067.

076.044   The height distribution of solar hard X–rays in non–thermal models.
A. J. Fennelly, A. G. Emslie.
Bull. Am. Astron. Soc., Vol. 17, No. 2, p. 645 (1985). Abstract. – See Abstr. 010.067.

076.045   Hard X–ray images of possible reconnection in the flare of 21 May, 1980.
Z. Švestka, G. Poletto.
Sol. Phys., Vol. 97, No. 1, p. 113 – 129 (1985).

An analysis of the growth of X–ray loops in the flare of 21 May, 1980, observed by HXIS (*Hard X–Ray Imaging Spectrometer*) on board SMM spacecraft, has been carried out with high time resolution in six energy channels from 3.5 to 30 keV. This analysis has revealed that the tops of the loops stay for minutes at a given altitude before, quite abruptly, other loop tops begin to appear above them. One of the jumps in altitude, from ~ 27000 to ~ 45000 km if the loops extended radially, which occurred quite late in the flare development, is studied in detail. The fact that the tops of higher loops were first seen in the 22 – 30 keV energy channel, and only minutes later at lower energies, suggests a new release of energy in a very small volume high in the corona. A magnetic reconnection of previously distended field lines appears to be a likely candidate for the observed phenomenon.

076.046   The solar spectral irradiance between 150 and 200 nm.
J. E. Mentall, B. Guenther, D. Williams.
J. Geophys. Res., Vol. 90, No. D1, p. 2265 – 2271 (1985).

Three rocket measurements of the solar spectral irradiance have been obtained for the wavelength region 150 – 200 nm. These flights occurred on November 16, 1978; May 22, 1980; and October 16, 1981. Evidence was found for a variation in the solar irradiance with the solar cycle but smaller than that reported by Mount and Rottman (1981). There is good agreement between the October 16, 1981, flight and the last two flights of Mount and Rottman (1983, 1985).

076.047   The impulsive hard X–rays from solar flares.
J. Leach.
Diss. Abstr. Int., Sect. B, Vol. 45, No. 3, p. 901 – 902 (1984). Thesis, Stanford University, 295 pp. (1984). Order No. DA8412862.

076.048   High resolution hard X–ray spectra of solar and cosmic sources.
R. A. Schwartz.
Diss. Abstr. Int., Sect. B, Vol. 45, No. 9, p. 2958 (1985). Thesis, University of California, 115 pp. (1984). Order No. DA8427092.

Solar high resolution balloon spectra obtained in the 190 – 300 nm wavelength band.
See Abstr. 035.005.

X–ray polarimetry: the measurement of the polarization of solar flare X–rays and the design of a Compton polarimeter.
See Abstr. 035.138.

Solar and cosmic X– and gamma–ray studies with long duration balloon flights.
See Abstr. 051.039.

Bremsstrahlung spectra from thick–target electron beams with noncollisional energy losses.
See Abstr. 063.058.

X–ray, ultraviolet, optical and magnetic structure in and near an active region.
See Abstr. 072.069.

Frequent ultraviolet brightenings in solar active regions.
See Abstr. 072.093.

The Hα spectral counterpart of hard X–ray microflares.
See Abstr. 073.151.

Microwave and hard X–ray imaging of a solar limb flare.
See Abstr. 073.153.

Correlation of X–ray and microwave radio radiation of solar flares from observations in March – April 1979.
See Abstr. 073.160.

Hydrodynamical response of the chromosphere to heating by X–ray emission of a flare.
See Abstr. 073.163.

Relationship between γ–ray emission, radio bursts and proton fluxes from solar flares.
See Abstr. 073.167.

The interpretation of hard X–ray polarization measurements in solar flares.
See Abstr. 073.174.

Great microwave bursts and hard X–rays from solar flares.
See Abstr. 073.175.

Two classes of gamma–ray line flares: impulsive and gradual.
See Abstr. 073.177.

A new method for determining temperature and emission measure during solar flares from light curves of soft X–ray line fluxes.
See Abstr. 073.180.

Time histories of gamma– and hard X–ray emissions from solar flares.
See Abstr. 073.181.

Ratios of the fluence of 2.22 MeV gamma–ray line to the fluence of 4.44 MeV gamma–ray line in solar flares.
See Abstr. 073.183.

Microwave and hard X–ray emission during solar flare of 19 Oct. 1981.
See Abstr. 073.184.

Behavior of high–energy particles associated with a solar flare on April 27, 1981.
See Abstr. 073.190.

Images of a major compact flare in hard X–rays and H–alpha.
See Abstr. 073.199.

Survey of solar X–ray flare dynamics.
See Abstr. 073.200.

Non–thermal excitation of the white light source in the 24 April 1981 (~ 1358 UT) solar flare.
See Abstr. 073.202.

Impulsive and gradual gamma–ray/proton flares: their rates of occurrence in the same active region.
See Abstr. 073.203.

Flare electron densities using X–ray line ratios.
See Abstr. 073.204.

Element abundances from solar flare spectra.
See Abstr. 073.206.

Directivity of flare gamma–rays.
See Abstr. 073.217.

Hard X–ray bremsstrahlung produced by electrons escaping a high–temperature thermal source in a solar flare.
See Abstr. 073.226.

On the microwave millisecond spike emission and its associated phenomena during the impulsive phase of large solar flare.
See Abstr. 073.242.

Small–scale flux emergence and the evolution of equatorial coronal holes.
See Abstr. 074.016.

Post–flare thermal waves in the solar corona.
See Abstr. 074.024.

Energy input in solar flares and coronal explosions.
See Abstr. 074.030.

The fate of sunward streaming protons associated with coronal mass ejections.
See Abstr. 074.038.

The solar coronal X–ray spectrum from 5.5 to 12 Å.
See Abstr. 074.050.

Large post–flare arch–like structures in the solar corona.
See Abstr. 074.076.

Analysis of active region loop flows as observed from SMM.
See Abstr. 074.077.

Observations of steady anomalous magnetic heating in thin current sheets.
See Abstr. 074.114.

A photometric comparison of the EUV– and K–coronas.
See Abstr. 074.117.

Solar cycle variation of energy and photospheric magnetic flux from coronal holes.
See Abstr. 074.142.

Solar burst with millimetre–wave emission at high frequency only.
See Abstr. 077.005.

Relative positions of microwave and hard X–ray burst sources.
See Abstr. 077.009.

VLA observations of narrow–band decimetric burst emission.
See Abstr. 077.028.

Microbursts at meter–decameter wavelengths.
See Abstr. 077.031.

Coronal path of electron beams during the impulsive phase of flares.
See Abstr. 077.046.

Correlation of solar decimetric radio bursts with X–ray flares.
See Abstr. 077.050.

Helios 1 energetic particle observations of the solar gamma–ray/ neutron flare events of 1982 June 3 and 1980 June 21.
See Abstr. 078.008.

Solar flare neutrons and gamma ray lines.
See Abstr. 078.039.

Generation of accelerated particles and hard radiation during solar flare.
See Abstr. 078.040.

Estimation of the conditions in the charged–particle acceleration region in solar flares from the "Venera 11, 13, 14" data on energetic electrons and hard X–rays.
See Abstr. 078.051.

On the collective deceleration of particles during solar flares.
See Abstr. 078.074.

Energetic particle and γ–ray generation in the solar matter.
See Abstr. 078.082.

The role of non–classical transport in the formation of the Ly–α temperature plateau.
See Abstr. 080.098.

A kind of short wave band SID records and its application in the observation and study of solar X–ray flares.
See Abstr. 083.012.

The solar stellar connection in the far ultraviolet.
See Abstr. 114.006.

Gamma–ray burst emission above 1 MeV: SMM observations.
See Abstr. 143.046.

077 Radio, Infrared Radiation

077.001 Observational evidence for magnetic reconnection in microwave solar bursts.
M. R. Kundu.
Unstable current systems and plasma instabilities in astrophysics, p. 185 – 190 (1985). – See Abstr. 012.002 (IAU Symp. No. 107).

The author first discusses a set of 6 cm observations made with the NRAO Very Large Array (VLA) (spatial resolution $\sim 2''$) that pertain to changes in the coronal magnetic field configurations that took place before the onset of an impulsive burst observed on 14 May 1980. The author also discusses a second set of 6 cm VLA observations (spatial resolution $18''$ arc) where several interacting loops were involved in triggering the onset of an impulsive burst observed on June 24, 1980, 19:57:00 UT. Both sets of observations are examples of magnetic reconnection process being involved in accelerating microwave emitting electrons.

077.002 Solar microwave emission in active regions.
B. Lokanadham, P. K. Subramanian, M. S. Reddy, B. M. Reddy, R. Lakshmi.
Unstable current systems and plasma instabilities in astrophysics, p. 225 – 230 (1985). – See Abstr. 012.002 (IAU Symp. No. 107).

Multi–frequency observations of solar microwave bursts recorded during solar maximum period 1980 – 81 are analysed and compared with X–ray data for studying the nature of microwave emissions from active regions. Most of the microwave burst spectra showed that the spectral index below the peak frequency is always less than 2.

077.003 Creation of high–energy electron tails by the lower–hybrid waves and its relevance to Type II and III bursts.
M. Tanaka, K. Papadopoulos.
Unstable current systems and plasma instabilities in astrophysics, p. 505 – 508 (1985). – See Abstr. 012.002 (IAU Symp. No. 107).

According to the closed or open geometry of the magnetic field lines, two different processes involving the waves are possible for the electron acceleration. First, when the magnetic field lines are partly open such as in Type III bursts, the waves excited in the closed magnetic loop must propagate in space into the open field line region. Otherwise, the accelerated electrons never escape into the free space. Secondly, when the acceleration takes place only in the closed magnetic field, the excited waves act as catalyst and accelerate electrons in the source region (for example, Type II bursts).

077.004 Spatio–temporal features of the development of microwave emission of active regions and flares.
G. Ya. Smolkov (*G. Ya. Smol'kov*).
Unstable current systems and plasma instabilities in astrophysics, p. 555 – 558 (1985). – See Abstr. 012.002 (IAU Symp. No. 107).

At SibIZMIR a stepwise commissioning of the Siberian Solar Radio Telescope is being under way. One–dimensional radio brightness distributions obtained during phase adjustment of the operating model, an 8–element interferometer, and a stepwise commissioning of the W–beam enabled to refine some of the earlier understanding and to gain new insights into spatio–temporal features of development of microwave emission from active regions and flares.

077.005 Solar burst with millimetre–wave emission at high frequency only.
P. Kaufmann, E. Correia, J. E. R. Costa, A. M. Zodi Vaz, B. R. Dennis.
Nature, Vol. 313, No. 6001, p. 380 – 382 (1985).

The authors present the first high sensitivity and high time–resolution observations taken simultaneously at 90 GHz ($\lambda = 3.3$ mm) and at 30 GHz ($\lambda = 10$ mm). These have identified a unique impulsive burst on 21 May 1984 with fast pulsed emission that was considerably more intense at 90 GHz than at lower frequencies. Hard X–ray time structures at energies above 25 keV were almost identical to the 90 GHz structures to better than 1 s.

077.006 The simplest solar microbursts flux and circular polarization at 22 GHz.
P. Kaufmann, E. Correia, J. E. R. Costa, H. S. Sawant, A. M. Zodi Vaz.
Sol. Phys., Vol. 95, No. 1, p. 155 – 165 (1985).

The simplest solar microwave microbursts detected with high sensitivity may be the response to the simpler energetic burst injections. Seventeen events from this category were identified in a series of more than 150 bursts recorded in 21 – 26 November, 1982. This first systematic study suggests that microbursts e–folding rise times concentrate into two classes of time scales, $0.05\,s < t \ll 1$ s and $0.5\,s \lesssim t \lesssim 2$ s. Microbursts' circular polarization present a dominant steady or slowly varying component that sets in before maximum emission. In some cases a faster component of polarization was found superimposed, which is not always well correlated in time with flux.

077.007 Different time constants of solar decimetric bursts in the range 100 – 1000 MHz.
H. J. Wiehl, A. O. Benz, M. J. Aschwanden.
Sol. Phys., Vol. 95, No. 1, p. 167 – 179 (1985).

Between 1980, January 1 and 1981, December 31 a total of 664 'decimetric pulsation' events (DCIM) were observed with the Zürich spectrometers in the frequency range 100 to 1000 MHz. The class of DCIM bursts can be divided into two groups depending on their duration and thus reflecting different physical mechanisms. Each of the two groups can be further divided into small and large bandwidth subgroups. Short decimetric events ($\lesssim 1$ s) are most abundant in this frequency range. They may be caused by fast transients in the solar atmosphere. The half–power bandwidth of the shortest DCIM bursts, the millisecond spikes, were found to be 6 to 12 MHz. The long lasting DCIM bursts (5 s to 300 s) exhibit a gradual and smooth time profile. Such long lasting events indicate the presence of trapped particles in magnetic fields.

077.008 Peculiarities of the radio emission of solar type III bursts.
A. B. Eremin.
Pis'ma Astron. Zh., Tom 11, No. 2, p. 139 – 144 (1985). In Russian. English translation in Sov. Astron. Lett., Vol. 11.

From the viewpoint of a new scheme of the generation of the fundamental radiation type III solar radio bursts an interpretation is given for the experimental data on the ratio of frequencies and of frequency drift rates of harmonic and fundamental radiation. A conclusion is made on the fact that electron fluxes generating the burst undergo a noticeable deceleration (from 0.6 to 0.3 c) at the path 10^{11}cm. A possibility is shown to define the mean electron flux velocity from the value of the ratio of harmonic component frequencies in type III bursts.

077.009 Relative positions of microwave and hard X–ray burst sources.
M. R. Kundu.
Adv. Space Res., Vol. 4, No. 7, p. 157 – 162 (1984). – See Abstr. 012.009.

Simultaneous microwave and hard X–ray imaging observations of 12 bursts show that it is difficult to discern a general pattern between microwave and hard X–ray burst locations. In general, the microwave source is displaced from the hard X–ray source. The commonly believed behavior of the microwave source being located near the top and hard X–ray source near the footpoints of a loop appears to be true in some cases but not all.

077.010 Simultaneous dual wavelength observations of an impulsive microwave burst using the V.L.A.
R. K. Shevgaonkar, M. R. Kundu.
Adv. Space Res., Vol. 4, No. 7, p. 247 – 250 (1984). – See Abstr. 012.009.

Simultaneous observations of a microwave burst at 2 and 6 cm wavelengths were carried out with the Very Large Array (VLA). The 6 cm burst source is located close to a magnetic neutral line, presumably near the top of a flaring loop, while the 2 cm emission originates from the footpoints of the loop. It is concluded that the 6 cm emission is dominated by gyrosynchrotron radiation of the thermal electrons in the bulk heated plasma at a temperature of $\sim 4 \times 10^7$K, while the 2 cm emission is due to nonthermal particles released and accelerated during the flare process. A DC electric field flare model leads to the estimation of the strength of the electric field to be $0.2 - 4 \mu$ statvolt cm^{-1} in the flaring region.

077.011 Association of time structures of solar bursts at millimetric and at metric waves.
H. S. Sawant, P. Kaufmann, E. Correia, J. E. R. Costa, P. Zlobec, M. Messerotti, L. Fornasari.
Adv. Space Res., Vol. 4, No. 7, p. 251 – 254 (1984). – See Abstr. 012.009.

Preliminary results obtained by comparing mm–wave burst structures with 408, 327 and 237 MHz indicate that (1) for the majority of major time structures (time scales of the order of 1 sec) observed at 22 GHz bursts, corresponding type III bursts have been observed at 237 MHz, however (2) start times at mm–λ and m–λ are not often coincident at two wavelengths.

077.012 Observations of solar radio bursts from meter to kilometer wavelengths.
M. R. Kundu, R. G. Stone.
Adv. Space Res., Vol. 4, No. 7, p. 261 – 270 (1984). – See Abstr. 012.009.

The authors review some results obtained with the Clark Lake multifrequency radioheliograph at meter–decameter wavelengths and from satellite multifrequency directive observations at hectometer and kilometer wavelengths. They present evidence that type III electrons propagate in dense coronal streamers, and that frequently observed microbursts (presumably of type III) at meter–decameter wavelengths are due to plasma radiation. They discuss observations of hectometer and kilometer type III radio storms which reveal information about active region structures, interplanetery magnetic field configuration, and solar wind acceleration.

077.013 Origin of fine structures in solar radio bursts.
G. S. Lakhina, B. Buti.
Pramāna, Vol. 23, No. 3, p. 343 – 349 (1984). Abstr. in Phys. Abstr., Vol. 88, No. 1248, Entry 9707 (1985).

077.014 Quasi–linear propagation of a spatially inhomogeneous electron stream.
B. N. Levin.
Astron. Astrophys., Vol. 143, No. 1, p. 54 – 58 (1985).

The region of plasma turbulence associated with Type III radio burst lags behind the leading edge of the electron stream which was observed experimentally near the earth's orbit. It can be explained in terms of a model of quasi–linear relaxation.

077.015 A working model of the solar S–component radio emission.
A. Krüger, J. Hildebrandt, F. Fürstenberg.
Astron. Astrophys., Vol. 143, No. 1, p. 72 – 76 (1985).

Numerical results of an emission model for local sources of the S–component of solar radio radiation are compiled. Both thermal bremsstrahlung and gyromagnetic emission are considered. The model is distinguished by (1) the choice of continuous height distributions of the input plasma parameters (N_e, T), and (2) by detailed numerical work. The aim was to investigate the influence of the strength and the scale height of the magnetic fields, and to test various models of the chromosphere–corona transition region. The results are presented in the form of brightness distributions, spectra of flux density, and degree of polarization. The effective radii of the sources of gyromagnetic emission are also computed.

077.016 Storms of U–bursts and the stability of coronal loops.
Y. Leblanc, M. Hoyos.
Astron. Astrophys., Vol. 143, No. 2, p. 365 – 373 (1985).

U–burst activity as observed in the frequency range of 25 – 75 MHz with the large array and sweep frequency spectrograph of Nançay, is analysed by using 5 years of observations. The authors show evidence for a close relationship between Type II events and U–burst groups, and that several Type II bursts occurred during Type III–U storms. It is concluded that the existence of frequent U–burst storms implies large magnetic arches stable for as long as a few days and that these arches are very probably diverging to explain J and L–shaped bursts. The association of Type II events and U–burst activity is interpreted as a result of coronal mass ejections which are responsible for transient loops.

077.017 Polarization of solar noise storm continuum and plasma wave density in the corona.
A. O. Benz, P. Zolliker.
Astron. Astrophys., Vol. 144, No. 1, p. 227 – 231 (1985).

A statistical method for background subtraction of polarization observations has been developed. It separates the signal from a background with a much larger time constant. The method has been applied to the continuum component of a noise storm at various frequencies. The circular polarization was found to be constant in frequency within the statistical error. Its average value in time and frequency was $89.0 \pm 1.5\%$. Well accepted assumptions are used to determine the fraction of the radiation emitted at the harmonic of the plasma frequency from observations. The resulting flux of the harmonic is not significantly different from zero. The observed flux of the fundamental is used to derive the plasma wave density in the source of the noise storm and to predict the flux at the harmonic from theory.

077.018 Velocities of type II solar radio events.
R. D. Robinson.
Sol. Phys., Vol. 95, No. 2, p. 343 – 357 (1985).
Radial velocities for 144 simple but representative type II bursts were determined from measured frequency time histories.

The velocity distribution is peaked in the region between 500 and 700 km s^{-1} (with the exact value dependent upon the coronal density model assumed) and skewed towards the larger velocities. The measured velocity is dependent upon the properties of the flare event but does not appear to be related to other characteristics of the radio burst. Comparisons show that the group of type II events studied had a velocity distribution which was comparable with that for coronal mass ejection events seen in association with type II bursts.

077.019   Solar narrow–band radio bursts near 2.0 GHz.
O. G. Gontarev, A. P. Klassen.
Soln. Dannye, Byull., 1984, No. 12, p. 72 – 75 (1985). In Russian.
The results of observations of two solar narrow–band radio bursts near 2.0 GHz with band widths of 150 MHz and intensity 17 and 800 per cent of radio emission level before the bursts are given.

077.020   Oscillations of the degree of circular polarization of the proton region on the sun.
V. I. Abramenko, L. I. Tsvetkov.
Izv. Krymskoj Astrofiz. Obs., Tom 69, p. 123 – 130 (1984). In Russian. English translation in Bull. Crimean Astrophys. Obs., Vol. 69.
The oscillations of the circular polarization of the local radio source related to the spot group McMath region 13043 have been studied. The observations were carried out at the wavelengths 3.5, 2.5 and 1.9 cm in July 1 – 7, 1974. The oscillations were studied by means of the Fourier method and by Maximum Entropy Spectral Analysis (MESA). The fundamental formulas to calculate the power spectrum are presented.

077.021   Solar radio fluxes as indices of solar activity.
M. Nicolet, L. Bossy.
Planet. Space Sci., Vol. 33, No. 5, p. 507 – 555 (1985).
The daily solar radio flux values at 9400, 3750, 2000 and 1000 MHz and at 2800 MHz observed since 1957 at Toyokawa and Ottawa, respectively, have been used to provide new information on the solar radio fluxes as indices of solar activity. After an examination of the yearly mean values at each frequency, another investigation based on mean ratios during periods of 18 or 6 months indicates that a close connection is observed between the radio fluxes in the cm region and that anomalies related to calibration problems can be detected. The regression analysis of the daily values of the fluxes during at least 25 years and a special test on the sensitivity may provide final information on the stability of the data with respect to time and solar activity. The method is capable of detecting long–term trends corresponding to instrumental drifts.

077.022   Variations of the integral radio emission of the sun at 6.3 and 8.6 mm wavelengths.
S. A. Pelyushenko.
Astron. Zh., Tom 62, Vyp. 2, p. 377 – 379 (1985). In Russian. English translation in Sov. Astron., Vol. 29, No. 2.
On the basis of data of high–resolution measurements of the solar disk brightness distribution made synchronously at the wavelengths 6.3 and 8.6 mm the variations of the S–component of the solar integral radio emission during the period of maximum of the 11–year activity cycle of the sun are estimated.

077.023   Coordinated radio observations during SMM.
F. Chiuderi–Drago.
Mem. Soc. Astron. Ital., Vol. 55, No. 4, p. 801 – 810 (1984). – See Abstr. 003.017.
Coordinated radio–space observations performed during SMM are described. The attention is particularly focussed on centimetric, X–ray and ultraviolet observations of solar active regions. The main results are the following: (1) Very high values of the magnetic field strength (600 + 900 Gauss) at coronal level are required to reproduce the brightness temperature and the polarization pattern observed above the sunspots. (2) A bright, strongly polarized radio source corresponding to a moving spot eventually inflowing into a larger one is observed: its interpretation requires either the presence of a strong non–potential

field in the corona overlying the moving spot or a non–thermal mechanism such as synchrotron emission by mildly relativistic electrons in the local potential field. Some results on correlated coronograph and metric wavelengths observations of type II and type IV radio bursts detected during coronal transients are also presented.

077.024   Radio radiation as information source on proton fluxes from solar flares.
S. T. Akin'yan, V. V. Fomichev, I. M. Chertok.
Complex investigations of the sun, p. 119 – 130 (1982). In Russian. Abstr. in Ref. Zh., 51. Astron., 4.51.359 (1985). – See Abstr. 012.046.

077.025   Some results of spectrographic investigations of the radio radiation of solar flares in the 8 – 12 GHz range.
M. M. Kobrin, Yu. V. Tikhomirov, V. M. Fridman.
Complex investigations of the sun, p. 131 – 139 (1982). In Russian. Abstr. in Ref. Zh., 51. Astron., 4.51.362 (1985). – See Abstr. 012.046.

077.026   Long and short term variation of the 10.7 cm solar flux. The photospheric granules and the Zürich numbers.
J. Xanthakis, C. Poulakos.
Astrophys. Space Sci., Vol. 111, No. 1, p. 179 – 188 (1985).
Based on diurnal values of the total radio flux density at 10.7 cm as well as on corresponding Zürich numbers the relation giving the radio–flux as function of R is established. The behaviour of low and high values of R and F_{2800} during the time interval 1957 – 1976 is studied. Preliminary conclusions drawn by other investigators are confirmed. A prediction of the total radio–flux for the 22nd solar cycle is given.

077.027   Quenching of the beam–plasma instability by large–scale density fluctuations in 3 dimensions.
L. Muschietti, M. V. Goldman, D. Newman.
Sol. Phys., Vol. 96, No. 1, p. 181 – 198 (1985).
A model is presented to explain the highly variable yet low level of Langmuir waves measured in situ by spacecraft when electron beams associated with type III solar bursts are passing by; the low level of excited waves allows the propagation of such streams from the Sun to well past 1 AU without catastrophic energy losses. The model is based, first, on the existence of large–scale density fluctuations that are able to efficiently diffuse small–k beam–unstable Langmuir waves in phase space, and, second, on the presence of a significant isotropic non–thermal tail in the distribution function of the background electron population, which is capable of stabilizing larger k modes.

077.028   VLA observations of narrow–band decimetric burst emission.
R. F. Willson.
Sol. Phys., Vol. 96, No. 1, p. 199 – 207 (1985).
The Very Large Array was used to observe a multiply–impulsive solar radio burst at several wavelengths near 20 cm. The observations indicate that the impulsive emission was nearly 100% circularly polarized and originated in small regions of $\sim 10'' - 20''$ in size. For one of the impulsive spikes, the author finds evidence of narrow–band emission that could be attributed to an electron–cyclotron maser. The radio data are also compared with soft X–ray data and interpreted in light of a model in which the coronal plasma is heated by maser burst emission.

077.029   Solar radio emission, January – December 1980.
Q. Bull. Sol. Act., Vol. 22, Part V, p. 1 – 83 (1980).

077.030   Dual frequency observations of solar microwave bursts using the VLA.
R. K. Shevgaonkar, M. R. Kundu.
Astrophys. J., Vol. 292, No. 2, p. 733 – 751 (1985).
Simultaneous observations at 2 and 6 cm wavelengths of a solar active region and of microwave bursts were carried out with the VLA. The quiescent 6 cm emission is strongly associated with photospheric sunspots and is dominated by gyroresonance radia-

tion. The emission at 2 cm, on the other hand, is due to free–free mechanism, and it originates from the chromosphere–corona transition region. The bursts observed have multiple peaks. Two–dimensional snapshot maps have been produced at 6 and 2 cm. From the brightness temperature and the degree of circular polarization, the magnetic field in the microwave burst sources has been estimated. The generating mechanisms responsible for 6 and 2 cm radiation are discussed; it is concluded that the 6 cm radiation in the bursts studied here originates from the bulk heated plasma, whereas the 2 cm radiation is due to the nonthermal particles generated in the energy–release process.

077.031   Microbursts at meter–decameter wavelengths.
M. R. Kundu, T. E. Gergely, R. Loiacono.
Bull. Am. Astron. Soc., Vol. 16, No. 4, p. 892 (1984). Abstract. – See Abstr. 010.062.

077.032   Extreme limb profiles of the sun at far infrared and submillimeter wavelengths.
C. Lindsey, E. E. Becklin, F. Q. Orrall, M. W. Werner,
J. T. Jefferies, I. Gatley.
Bull. Am. Astron. Soc., Vol. 16, No. 4, p. 992 (1984). Abstract. – See Abstr. 010.062.

077.033   Some results of model calculations of the solar S–component radio emission.
A. Krüger, J. Hildebrandt.
Astron. Nachr., Vol. 306, No. 3, p. 157 – 166 (1985).
Numerical calculations of special characteristics of the solar S–component microwave radiation are presented on the basis of recent sunspot and plage models. Quantitative results are discussed and can be used for the plasma diagnostics of solar active regions by comparisons with observations with high spatial and spectral resolution. The possibility of generalized applications to magnetic stars and stellar activity is briefly noted.

077.034   On plasma mechanisms of radiation under conditions of the solar atmosphere.
L. P. Ipatova, L. V. Yasnov.
Radioizluch. Solntsa, Leningrad, No. 5, p. 128 – 137 (1984). In Russian. Abstr. in Ref. Zh., 51. Astron., 5.51.324 (1985).

077.035   On the determination of the spectral composition of solar radio radiation near $\lambda = 3$ cm with the method of spectral–temporal analysis.
A. R. Abbasov, Sh. Sh. Gusejnov, V. M. Somsikov.
Radioizluch. Solntsa, Leningrad, No. 5, p. 164 – 170 (1984). In Russian. Abstr. in Ref. Zh., 51. Astron., 5.51.325 (1985).

077.036   On fluctuations of radio radiation of local sources from observations with the RT–22.
I. I. Berulis, B. Ya. Losovskij, S. V. Makagonov,
N. G. Franchuk, L. V. Yasnov.
Radioizluch. Solntsa, Leningrad, No. 5, p. 59 – 78 (1984). In Russian. Abstr. in Ref. Zh., 51. Astron., 5.51.343 (1985).

077.037   On the radio radiation of solar flares and features of coronal magnetic fields.
V. V. Alekseev, T. V. Levashova, A. P. Molchanov,
I. E. Pogodin, A. G. Stupishin.
Radioizluch. Solntsa, Leningrad, No. 5, p. 78 – 101 (1984). In Russian. Abstr. in Ref. Zh., 51. Astron., 5.51.354 (1985).

077.038   Peculiarities of upwelling regions of solar radio radiation from RATAN–600 observations.
S. A. Andrianov.
Radioizluch. Solntsa, Leningrad, No. 5, p. 119 – 120 (1984). In Russian. Abstr. in Ref. Zh., 51. Astron., 5.51.365 (1985).

077.039   On an interrelation of limb ejections with type III radio bursts.
V. S. Loskutnikov.
Investigation of magnetic fields and active formations on the sun, p. 87 – 91 (1984). In Russian. Abstr. in Ref. Zh., 51. Astron., 5.51.374 (1985). – See Abstr. 003.022.

077.040   Long–period quasi–periodic variations of the shortwave radiation flux on the sun and flare activity of sunspot groups.
E. V. Ivanov.
Physics of solar activity, p. 160 – 168 (1983). In Russian. Abstr. in Ref. Zh., 51. Astron., 6.51.284 (1985). – See Abstr. 003.024.

077.041   Radio spikes and the fragmentation of flare energy release.
A. O. Benz.
Sol. Phys., Vol. 96, No. 2, p. 357 – 370 (1985).
Decimetric radio events with large numbers of spikes during the impulsive phase of flares have been selected. In the observing range of 100 to 1000 MHz some flares have of the order of 10000 spikes or more. The average half–power bandwidth of spikes has been measured to be only 1.5% of the spike frequency. Since the emission frequency is determined by some source parameter (such as plasma frequency or gyrofrequency) the source dimension must be a small fraction of the scale length. From the flare configuration a typical upper limit of the dimension of 200 km is found. Four events were analyzed in detail and compared to UV, SXR, and HXR data. The density of the loops where the SXR and HXR emission was observed has been measured before the flare. The plasma frequency well agrees with the observed frequency of spikes. The spikes thus originate close to or in the energy release region.

077.042   On the origin of continuum emission during decametric solar noise storms.
B. N. Levin, V. O. Rapoport.
Sol. Phys., Vol. 96, No. 2, p. 371 – 380 (1985).
The stationary current of diffusely–distributed super–thermal electrons along a weakly inhomogeneous coronal magnetic field is considered as a possible model of the noise storm continuum source in decametric wavelengths. It is shown that the realization of such a streaming leads to a considerably increased level of plasma noise in the diffuse component region and then to enhanced radio emission from this region.

077.043   Solar noise storms and magnetic sector structures.
R. T. Stewart.
Sol. Phys., Vol. 96, No. 2, p. 381 – 395 (1985).
A synoptic study of the occurrence and polarization of 160 MHz noise storms recorded at Culgoora during the current solar cycle shows that the storm sources occur in large unipolar cells extending $> 90°$ in solar longitude and $\leqslant 60°$ in latitude, with lifetimes of ~ 1 yr. From solar maximum onwards these large cells stretch across the solar equator to form a longitudinal sector pattern reminiscent of that observed in the interplanetary magnetic field. Comparisons with published heliospheric current sheet simulations support this conclusion. The noise storms occur in the strong magnetic fields above large, complex, flare–active sunspots. Unlike most active regions, those associated with noise storms do not always have dominant sunspots as leaders. Rather, about one–third have the dominant sunspot as a follower.

077.044   On peculiarities of a connection of proton stream intensity, the meter component of solar radio bursts and PCA–type absorption.
Yu. B. Vedeneev, M. A. Fasakhova.
Geomagn. Aehron., Tom 25, No. 3, p. 385 – 387 (1985). In Russian. English translation in Geomagn. Aeron.

077.045   Spectral correlation between high energy solar flare particle events and associated radio bursts.
A. P. Mishra, R. L. Singh.
18th International Cosmic Ray Conference, Vol. 4, p. 21 – 24 (1983). – See Abstr. 012.096.
The rigidity spectral exponent of high energy solar particle events ($E \gtrsim 1$ GeV) has been found to be correlated with the U–shaped peak flux density spectra of associated radio burst. The results are of significance in understanding the production and acceleration mechanism operating in source region during these events.

077.046 Coronal path of electron beams during the impulsive phase of flares.
G. Trottet, M. Pick, A. Raoult, P. Lantos.
18th International Cosmic Ray Conference, Vol. 10, p. 365 – 368 (1983). – See Abstr. 012.096.
Results of type III burst observations obtained at meter wavelength are compared with white light images of the corona, hard X ray bursts and 6 centimeter radio maps. The electron path through the corona and the structure of the acceleration site are dicussed.

077.047 Type III solar bursts from Voyager spacecraft.
C. Sawyer, J. W. Warwick.
Bull. Am. Astron. Soc., Vol. 17, No. 2, p. 630 (1985). Abstract. – See Abstr. 010.067.

077.048 Full disk maps of the sun at 1.4 GHz.
T. S. Bastian, G. A. Dulk.
Bull. Am. Astron. Soc., Vol. 17, No. 2, p. 632 (1985). Abstract. – See Abstr. 010.067.

077.049 Spatial structure of a Type III electron beam.
T. N. LaRosa.
Bull. Am. Astron. Soc., Vol. 17, No. 2, p. 645 (1985). Abstract. – See Abstr. 010.067.

077.050 Correlation of solar decimetric radio bursts with X–ray flares.
M. J. Aschwanden, H. J. Wiehl, A. O. Benz, S. R. Kane.
Sol. Phys., Vol. 97, No. 1, p. 159 – 172 (1985).
Several hundred radio bursts in the decimetric wavelength range (300 – 1000 MHz) have been compared with simultaneous soft and hard X-ray emission. Long lasting (type IV) radio events have been excluded. The association of decimetric emission with hard X–rays has been found to be surprisingly high (48%). The association rate increases with bandwidth, duration, number of structural elements, and maximum frequency. Type III–like bursts are observed up to the upper limit of the observed band. This demonstrates that the corona is transparent up to densities of about 10^{10}cm^{-3}, contrary to previous assumptions. This can only be explained in an inhomogenous corona with the radio source being located in a dense structure.

077.051 Rapid fluctuations in the position, size, and brightness of intense solar metre–wave radio sources.
R. A. Duncan.
Sol. Phys., Vol. 97, No. 1, p. 173 – 182 (1985).
A comparison of quiescent type I solar radio sources with concurrent intense impulsive type III, V, and type II sources shows that whereas the type I sources are usually small and stable the type III, V, and II sources are usually large and unstable. The author concludes that the large size and variability of the type III, V, and II radio sources cannot be attributed to instrumental error or ionospheric refraction but must instead reflect the size and variability of the coronal structures on which they arise.

077.052 Polarization bursts of solar microwave radio emission.
O. V. Epifanova.
Soln. Dannye, Byull., 1985, No. 3, p. 88 – 94 (1985). In Russian.
There are solar microwave bursts observable only in circular polarization. These bursts are analysed and compared with other active processes taking place simultaneously on the sun. A good time coincidence proved to be between the centimeter polarization bursts and the decimeter intensity bursts. A possible cause of these events is assumed to be a changing magnetic field. Possible causes of the magnetic field variation for each event are discussed.

077.053 On the internal regularities of the 11–year cycles of solar radio flux density at 2800 MHz.
Yu. I. Vitinskij, E. V. Miletskij.
Soln. Dannye, Byull., 1985, No. 4, p. 86 – 91 (1985). In Russian.
On the basis of the data of the solar radio flux density at 2800 MHz (Ottawa, 1954 – 1984) models for five phases of the 11–year solar cycles No. 19, 20, 21 are constructed.

077.054 A magnetic loop structure model of slowly varying radio sources observed in solar eclipse.
X.–h. Luo, D.–y. Yao, S.–c. Ji, R.–y. Zhao.
Acta Astron. Sin., Vol. 26, No. 1, p. 51 – 61 (1985). In Chinese.
Based on physical conditions above an active region of a bipolar sunspot group and the gyromagnetic radiation mechanism, the authors propose a magnetic loop structure model of slowly varying radio sources and obtain the brightness temperature distributions at wavelengths of 2, 3.2, 8.2 and 21 cm, respectively.

077.055 Pulsation mechanisms of solar microwave bursts.
J.–x. Yao.
Acta Astron. Sin., Vol. 26, No. 1, p. 62 – 68 (1985). In Chinese.
In this paper the author suggests that the pulsation phenomena of solar microwave bursts can be explained by the modulation mechanism of gyro–synchrotron radiation.

077.056 An analysis of the millisecond solar radio emission.
R.–y. Zhao, S.–z. Jin, Q.–j. Fu.
Acta Astron. Sin., Vol. 26, No. 2, p. 114 – 122 (1985). In Chinese.
The present paper gives a preliminary analysis for 128 activity events, observed from May 1981 to June 1983.

Introduction to solar radio astronomy and radio physics.
See Abstr. 003.122.

Forty years of solar radio astronomy – a history of major advances.
See Abstr. 004.192.

The radiometer and polarimeters at 80, 35, and 17 GHz for solar observations at Nobeyama.
See Abstr. 033.020.

Balloon–borne detection system for solar infrared brightness temperature.
See Abstr. 035.075.

Magnetic vortex tubes, jets and nonthermal sources.
See Abstr. 062.059.

The loss–cone driven instability for Langmuir waves in an unmagnetized plasma.
See Abstr. 062.127.

Induced emission of radiation near $2\omega_e$ by a synchrotron–maser instability.
See Abstr. 063.083.

High resolution VLA observations of an active region during a partial solar eclipse.
See Abstr. 072.070.

Some characteristics of spot groups and noise storms connected with them.
See Abstr. 072.081.

Report of IAU Commission 10: Solar activity (*Activité solaire*).
See Abstr. 072.108.

On the correlation of microwave bursts with Hα flares.
See Abstr. 073.045.

Hard X–ray and radio investigations prior to or during the impulsive phase of solar flares.
See Abstr. 073.046.

An impulsive solar burst observed in Hα, microwaves, and hard X–rays.
See Abstr. 073.071.

Microwave and X–ray observations of delayed brightenings at sites remote from the primary flare locations.
See Abstr. 073.077.

Ejection of chromospheric material associated with injection of electrons in the solar corona.
See Abstr. 073.081.

Dependence of the flare stream velocity on magnetic field orientation.
See Abstr. 073.082.

Analysis of the flare of May 16th, 1981 with a complex space–time structure using opical, X–ray data and radio observations.
See Abstr. 073.094.

Studies on some aspects of the solar Hα flares of different visual features.
See Abstr. 073.095.

A study of flare buildup from simultaneous observations in microwave, Hα, and UV wavelengths.
See Abstr. 073.119.

Some questions of the dynamics and radio radiation of flare loops.
See Abstr. 073.124.

Multiwavelength observations of a preflare solar active region using the VLA.
See Abstr. 073.126.

Flux relations between hard X–rays and microwaves for both impulsive and extended solar flares.
See Abstr. 073.130.

Microwave and hard X–ray imaging of a solar limb flare.
See Abstr. 073.153.

Simultaneous observations of solar flares at 6 and 20 cm wavelengths using the VLA.
See Abstr. 073.154.

Correlation of X–ray and microwave radio radiation of solar flares from observations in March – April 1979.
See Abstr. 073.160.

Relationship between γ–ray emission, radio bursts and proton fluxes from solar flares.
See Abstr. 073.167.

On a two–component model of microwave radiation of solar flares.
See Abstr. 073.169.

Great microwave bursts and hard X–rays from solar flares.
See Abstr. 073.175.

Microwave and hard X–ray emission during solar flare of 19 Oct. 1981.
See Abstr. 073.184.

Morphological characters of the two–ribbon flare of 1981 May 13 and their explanation.
See Abstr. 073.241.

On the microwave millisecond spike emission and its associated phenomena during the impulsive phase of large solar flare.
See Abstr. 073.242.

Characteristics of coronal mass ejections associated with solar frontside and backside metric type II bursts.
See Abstr. 074.009.

Dynamical behaviour of the corona in association with radio emissions.
See Abstr. 074.026.

A multiple type–II burst associated with a coronal transient and its MHD simulation.
See Abstr. 074.028.

Electron cyclotron maser instability in the solar corona: the role of superthermal tails.
See Abstr. 074.053.

Coronal structures observed at metric wavelengths with the Nançay radioheliograph.
See Abstr. 074.071.

The microwave structure of hot coronal loops.
See Abstr. 074.089.

A test for large–angle radio scattering in the solar corona.
See Abstr. 074.136.

A positional comparison between coronal mass ejection events and solar type II bursts.
See Abstr. 074.154.

Estimate of magnetic fields in regions of generation of microwave radio bursts on the sun.
See Abstr. 075.025.

Observational evidence of continuous input of energetic electrons in the corona during large solar flares.
See Abstr. 076.035.

A comparison of solar ^{3}helium–rich events with type II bursts and coronal mass ejections.
See Abstr. 078.007.

Radio radius of the sun toward the polar directions measured during the solar maximum at 2.25 cm wavelength.
See Abstr. 080.028.

Arguments for fundamental emission by the parametric process $L \rightarrow T + S$ in interplanetary type III bursts.
See Abstr. 106.039.

Tracking of flare–generated interplanetary shock waves, responsible for Forbush decreases, by observing type II radio bursts.
See Abstr. 144.011.

078 Cosmic Radiation

078.001 The baseline composition of solar energetic particles.
J.–P. Meyer.
Astrophys. J., Suppl. Ser., Vol. 57, No. 1, p. 151 – 171 (1985).

The author analyzes all existing spacecraft observations of the highly variable heavy element composition of solar energetic particles (SEP) during non–^{3}He–rich events. All data show the imprint of an ever–present basic composition pattern (dubbed "mass–unbiased baseline") that differs from the photospheric composition by a simple bias related to first ionization potential. In each particular observation, this mass–unbiased baseline composition is being distorted by an additional bias, which is always a monotonic function of mass (or Z). This latter bias varies in amplitude and even sign from observation to observation.

078.002 Dynamics and forecast of radiation characteristics of solar cosmic rays.
V. V. Bengin, L. I. Miroshnichenko, V. M. Petrov.
Kosm. Issled., Tom 23, Vyp. 1, p. 123 – 133 (1985). In Russian. English translation in Cosm. Res.

078.003 Statistic acceleration and stationary transport of solar cosmic rays in the interplanetary space.
V. V. Smirnova.
Geomagn. Aehron., Tom 25, No. 2, p. 177 – 183 (1985). In Russian. English translation in Geomagn. Aeron.

078.004 Characteristics of Forbush–type cosmic ray intensity decreases observed during solar cycle 21.
A. G. Fenton, K. B. Fenton, J. E. Humble.
Proc. Astron. Soc. Aust., Vol. 5, No. 4, p. 590 – 593 (1984).

The authors present results of a preliminary analysis of data from four sea–level neutron monitors during several of the large Forbush–type decreases which have occurred since the beginning of solar cycle 21 in June 1976.

078.005 The cosmic ray solar flare increase of 16 February 1984.
A. G. Fenton, K. B. Fenton, J. E. Humble.
Proc. Astron. Soc. Aust., Vol. 5, No. 4, p. 593 – 594 (1984).

078.006 Dynamics of the low–latitude boundary of penetration of low–energy solar protons into the magnetosphere.
T. A. Ivanova, S. N. Kuznetsov, Eh. N. Sosnovets, L. V. Tverskaya.
Geomagn. Aehron., No. 1, p. 7 – 12 (1985). In Russian. English translation in Geomagn. Aeron.

078.007 A comparison of solar ^{3}helium–rich events with type II bursts and coronal mass ejections.
S. Kahler, D. V. Reames, N. R. Sheeley Jr., R. A. Howard, M. J. Koomen, D. J. Michels.
Astrophys. J., Vol. 290, No. 2, p. 742 – 747 (1985).

The authors ask whether the energetic particles of ^{3}He–rich events are accelerated in the same process as that resulting in particles of normal–abundance events. They first present a list of 66 ^{3}He–rich events observed with the Goddard Space Flight Center particle detector on *ISEE 3*. It is then shown that these events are not statistically associated with either of the two common signatures of normal–abundance events, metric type II bursts and coronal mass ejections. This indicates that enhanced abundance events may be produced only in the impulsive phases of flares, while normal abundance events are produced in subsequent flare shock waves.

078.008 *Helios 1* energetic particle observations of the solar gamma–ray/neutron flare events of 1982 June 3 and 1980 June 21.
F. B. McDonald, M. A. I. Van Hollebeke.
Astrophys. J., Lett. Ed., Vol. 290, No. 2, p. L67 – L71 (1985).

The characteristics of the energetic particles associated with the solar γ–ray/neutron flare events of 1982 June 3 and 1980 June 21 observed by the Goddard cosmic–ray experiment on *Helios 1* (at heliocentric distances of 0.57 and 0.54 AU, respectively) differ in several important respects from typical solar particle increases. In particular, the 1982 June 3 event has a proton energy spectrum which fits a remarkable flat power law in kinetic energy with a spectral index of 1.2, an electron/proton ratio of 1 at 4 MeV, and a small but well–defined precursor event that began some 3 hr before the impulsive flare increase. The implications for the particle acceleration process in solar flares are discussed.

078.009 Isotope fractionation in solar plasma.
G. S. Anufriev, B. S. Boltenkov, I. N. Kapitonov, L. V. Usacheva.
Vses. simp. No. 10 po stabil. izotopam v geokhimii, Moskva, 3 – 5 dek., 1984. Tez. dokl. Moskva, p. 155 (1984). In Russian. Abstr. in Ref. Zh., 51. Astron., 4.51.212 (1985).

078.010 High–energy neutrons and γ–quanta and characteristics of flare particles.
G. E. Kocharov, N. Z. Mandzhavidze.
Izv. Akad. Nauk SSSR. Ser. fiz., Tom 48, No. 11, p. 2212 – 2216 (1984). In Russian. Abstr. in Ref. Zh., 51. Astron., 4.51.358 (1985).

078.011 Characteristics of accelerated particles and regions of their generation for powerful solar flares.
M. I. Savchenko.
Izv. Akad. Nauk SSSR. Ser. fiz., Tom 48, No. 11, p. 2217 – 2220 (1984). In Russian. Abstr. in Ref. Zh., 51. Astron., 4.51.366 (1985).

078.012 An analysis of relations between the characteristics of solar cosmic rays and indices of solar activity.
V. A. Bondarenko, M. V. Zil', A. V. Kolomenskij.
Soln. Dannye, Byull., 1985, No. 2, p. 52 – 57 (1985). In Russian.

A close statistical relationship is shown to exist between annual Wolf numbers and frequency of flares of importance 1, 2, 3 using the data of the 19th and 20th solar cycles. The distribution of enhancement of solar cosmic rays caused by solar flares of different importance is analysed. The enhancements of solar cosmic rays caused by flares of importance 1, 2 and 3 are statistically classified.

078.013 Deep space observations of the east–west asymmetry of solar energetic storm particle events: Voyagers 1 and 2.
E. T. Sarris, R. B. Decker, S. M. Krimigis.
J. Geophys. Res., Vol. 90, No. A5, p. 3961 – 3965 (1985). – See Abstr. 012.003.

Measurements of energetic proton ($E_p \geqslant 520$ keV) intensities by the Voyagers 1 and 2 deep space probes, during solar flare–induced shock waves, indicate that the formation of solar energetic storm particle (ESP) events depends critically on the heliolongitudes of different locations on the large–scale shock front with respect to the source flare site. It is shown that large ESP ion intensity enhancements are observed for solar flare sites to the east of the spacecraft meridian, whereas only weak ESP events are associated with shock crossings where the flare sites are west of the spacecraft meridian. The results are explained in terms of the average "interplanetary magnetic field–shock front" configurations encountered by the Voyagers at various heliolongitudes of the solar flare–generated shock wave.

078.014 Solar ^{3}He–rich events and nonrelativistic electron events: a new association.
D. V. Reames, T. T. von Rosenvinge, R. P. Lin.
Astrophys. J., Vol. 292, No. 2, p. 716 – 724 (1985).

In 15 months of observation by the *ISEE 3* spacecraft, the authors find that virtually all solar $\gtrsim 1.3$ MeV per nucleon ^{3}He–rich events are associated with impulsive 2 to $\sim 10^2$ keV electron events, although many electron events were not accompanied by detectable ^{3}He increases. The electron events and their related type III solar radio bursts provide, for the first time, identification of the flares which produce ^{3}He–rich events. ^{3}He appears to be accelerated at the flash phase of solar flares along with nonrelativistic electrons.

078.015   Spectral investigations of proton–active regions on the sun in the centimeter range.
O. V. Korobchuk, N. G. Peterova.
Radioizluch. Solntsa, Leningrad, No. 5, p. 102 – 114 (1984). In Russian. Abstr. in Ref. Zh., 51. Astron., 5.51.375 (1985).

078.016   Forecast of the development of proton fluxes of solar cosmic rays from initial data obtained with artificial earth satellites.
N. K. Pereyaslova, M. N. Nazarova, N. A. Mikirova.
Radioizluch. Solntsa, Leningrad, No. 5, p. 137 – 150 (1984). In Russian. Abstr. in Ref. Zh., 51. Astron., 5.51.376 (1985).

078.017   Multispacecraft observations of the east–west asymmetry of solar energetic storm particle events.
E. T. Sarris, S. M. Krimigis.
Sol. Phys., Vol. 96, No. 2, p. 413 – 421 (1985).
Multispacecraft observations of energetic protons ($E_p \geqslant 500$ keV) were obtained by the APL/JHU instruments on board the IMP–7 and 8 spacecraft and the Voyager–1 and 2 deep space probes, in order to study the generation of solar flare energetic storm particle (ESP) events at widely separated locations on the same shock front. These locations are presumably characterized, on the average, by different 'interplanetary magnetic field–shock front' configuration, i.e. quasi–perpendicular (quasi–parallel) shocks for eastern (western) solar flare sites. The multispacecraft energetic proton observations show that substantial differences in the ESP proton intensity enhancements are detected at these energies for locations on the shock front with wide heliolongitude separations.

078.018   Solar cosmic ray proton fluxes in the last hundred million years.
T. R. Venkatesan, M. N. Rao, C. M. Nautiyal, J. T. Padia.
18th International Cosmic Ray Conference, Vol. 2, p. 366 – 369 (1983). – See Abstr. 012.096.
Results of systematic study of selected minerals and size fractions of lunar soils obtained by resolving the solar cosmic ray proton produced neon isotopes, suggest that the solar flare proton fluxes in the last hundred million years are about five times higher than the values determined in the last couple of million years using lunar rocks and soils.

078.019   Long term solar flare fluxes based on aluminum–26 depth profile in lunar rocks.
P. N. Shukla, M. B. Potdar, N. Bhandari.
18th International Cosmic Ray Conference, Vol. 2, p. 370 – 372 (1983). – See Abstr. 012.096.
Solar proton fluxes have been deduced from Al–26 depth profile resulting mainly from their interactions with Mg, Al and Si in lunar rocks. The Al–26 profile is found to yield a flux which can be represented by $(J_s, R_o) = (125, 125)$. The data indicate that the solar proton flux has remained nearly constant during last few million years.

078.020   Secular variations in solar flare proton fluxes.
J. N. Goswami, R. Jha, D. Lal, R. C. Reedy, R. E. McGuire.
18th International Cosmic Ray Conference, Vol. 2, p. 373 – 376 (1983). – See Abstr. 012.096.
Proton fluences in contemporary solar flare events (1965 – 82) are analysed to obtain values of average flux and characteristic rigidity. Comparison of contemporary values with long–term averaged values, based on lunar sample data, indicate that the ancient solar flare proton spectra were harder.

078.021   Structural features of solar cosmic ray penetration into the high–latitude regions of the earth's magnetosphere inferred from Intercosmos–19 data.
Yu. V. Mineev, E. S. Spirkova, G. A. Glukhov, Yu. P. Kratenko.
18th International Cosmic Ray Conference, Vol. 3, p. 262 – 265 (1983). – See Abstr. 012.096.
A good correlation of the pulsation periods of the penetrating protons and the real geomagnetic fields is noted. Possible mechanisms of the observed features of penetration of solar cosmic ray protons are discussed.

078.022   The effect of geomagnetic disturbances on the response of a ground–based neutron monitor during an anisotropic solar cosmic ray event.
E. O. Flückiger, D. F. Smart, M. A. Shea.
18th International Cosmic Ray Conference, Vol. 3, p. 435 – 438 (1983). – See Abstr. 012.096.
The authors demonstrate the effect of geomagnetic disturbances on the response of a ground–based neutron monitor during an anisotropic solar cosmic ray event.

078.023   The SCR flux structure in the polar caps.
A. S. Biryukov, T. A. Ivanova, L. M. Kovrygina, S. N. Kuznetsov, E. N. Sosnovetz (*Eh. N. Sosnovets*), L. V. Tverskaya, I. Kimak, K. Kudela.
18th International Cosmic Ray Conference, Vol. 3, p. 461 – 464 (1983). – See Abstr. 012.096.
The electron and proton fluxes were studied in the polar caps aboard Intercosmos–17 and Cosmos–900 during the solar particle events of November 22, 1977 and February 13, 1978. A comparison with a theoretical polar cap structure is made.

078.024   Solar proton and alpha differential energy spectra.
R. E. McGuire, T. T. von Rosenvinge, F. B. McDonald.
18th International Cosmic Ray Conference, Vol. 4, p. 11 (1983). Abstract. – See Abstr. 012.096.

078.025   On the spectrum dynamics of solar cosmic rays.
L. I. Miroshnichenko.
18th International Cosmic Ray Conference, Vol. 4, p. 16 – 19 (1983). – See Abstr. 012.096.
The time evolution of solar cosmic ray (SCR) spectrum parameters is being analyzed. The form of SCR spectrum in the Earth's environment and in the source is being discussed and the absolute flux upper estimates for solar protons are obtained. Theoretical estimates are compared with observations for several solar proton events of the 19th – 21st solar cycles.

078.026   Temporal variations of solar flare nucleon abundances and their implications for acceleration and coronal propagation.
G. M. Mason, G. Gloeckler, D. Hovestadt.
18th International Cosmic Ray Conference, Vol. 4, p. 31 – 34 (1983). – See Abstr. 012.096.
The authors show that in poorly–connected solar particle events the ~ 1 MeV/nucleon He/H, O/He and Fe/He ratios and their random fluctuations are uncorrelated with flare site location.

078.027   Elemental composition of solar energetic particles.
H. Breneman, E. C. Stone, R. E. Vogt.
18th International Cosmic Ray Conference, Vol. 4, p. 35 (1983). Abstract. – See Abstr. 012.096.

078.028   On the mechanism of solar cosmic ray enrichment by heavy ions.
L. G. Kocharov, A. V. Orishchenko.
18th International Cosmic Ray Conference, Vol. 4, p. 37 – 40 (1983). – See Abstr. 012.096.
The authors investigate the mechanism of heavy ion enrichment of solar cosmic rays in detail.

078.029   Isotopic composition of solar energetic particles.
S. P. Christon, E. C. Stone, R. E. Vogt.
18th International Cosmic Ray Conference, Vol. 4, p. 46 (1983). Abstract. – See Abstr. 012.096.

078.030   Solar flare neon composition.
C. M. Nautiyal, J. T. Padia, M. N. Rao,
T. R. Venkatesan.
18th International Cosmic Ray Conference, Vol. 4, p. 47 (1983).
Abstract. – See Abstr. 012.096.

078.031   Solar ^{3}He–rich events observed on ISEE–3.
D. V. Reames, T. T. von Rosenvinge.
18th International Cosmic Ray Conference, Vol. 4, p. 48 – 51
(1983). – See Abstr. 012.096.
The authors describe some results of a survey of approximately
4 years of ISEE–3 data that seeks to characterize the ^{3}He–rich
solar particle events more completely.

078.032   On possibilities of explaining the dependence of ^{3}He/^{4}He ratio on energy.
L. G. Kocharov, Ya. V. Dvoryanchikov, M. Slivka.
18th International Cosmic Ray Conference, Vol. 4, p. 59 (1983).
Abstract. – See Abstr. 012.096.

078.033   Survey of He$^+$/He^{2+}–abundance ratios in energetic particle events.
D. Hovestadt, B. Klecker, G. Gloeckler, F. M. Ipavich,
M. Scholer.
18th International Cosmic Ray Conference, Vol. 4, p. 61 – 64
(1983). – See Abstr. 012.096.
The authors report first results of a systematic study of the
helium   charge   distribution   in   the   energy   range
$0.4 – 0.62$ MeV/nucleon in energetic particle events. They found
an average He$^+$/He^{2+} ratio of 0.12 ± 0.04. No obvious association with optical flare events could be observed for the events
richest in He$^+$. The authors did not find significant differences of
the proton energy spectra, and the abundance of helium relative
to   protons   and   heavy   ions   for   He$^+$–rich   events
(He$^+$/He^{2+} > 0.3) and for events with He$^+$/He^{2+} < 0.3, respectively.

078.034   Solar flare particle fluences during solar cycles 19, 20 and 21.
R. E. McGuire, J. N. Goswami, R. Jha, D. Lal, R. C. Reedy.
18th International Cosmic Ray Conference, Vol. 4, p. 66 – 69
(1983). – See Abstr. 012.096.
Satellite data for solar flare particle events during solar
cycle 21 (up to July 1982) have been analysed to obtain event–
integrated fluxes of energetic protons and alpha particles. Thirty
nine events with proton fluences (E > 10 MeV) > 10^7cm^{-2} occurred during 1976 – 1982. The fluxes are compared to those of
cycle 20 and 19. There is no definitive correlation between solar
cycle averaged proton fluxes and sunspot numbers.

078.035   The solar flare electrons with the energy E_e > 100 MeV.
L. V. Kurnosova, L. A. Razorenov, M. I. Fradkin.
18th International Cosmic Ray Conference, Vol. 4, p. 70 – 73
(1983). – See Abstr. 012.096.
Detailed analysis of cosmic ray composition during the solar
flare on the 29 of April 1973 led to the conclusion that there was
a flux of solar electrons with energy E_e > 100 MeV in the flare.
The energy spectrum of the electrons was obtained up to the
energy $E_e \sim 1$ GeV.

078.036   The solar cosmic ray neutron event on June 3, 1982.
H. Debrunner, E. O. Flückiger, E. L. Chupp,
D. J. Forrest.
18th International Cosmic Ray Conference, Vol. 4, p. 75 – 78
(1983). – See Abstr. 012.096.
The authors discuss the neutron monitor–measurements at
Jungfraujoch and estimate the intensity of the solar neutrons
above the earth's atmosphere.

078.037   The time history of 2.22 MeV line emission in solar flares.
T. A. Prince, D. J. Forrest, E. L. Chupp, G. Kanbach,
G. H. Share.
18th International Cosmic Ray Conference, Vol. 4, p. 79 – 82
(1983). – See Abstr. 012.096.

The authors examine the time dependence of 2.22 MeV emission using observations of the Solar Maximum Mission gamma–
ray spectrometer. They will determine the decay time constant for
the 2.22 MeV emission process and from this derive implications
concerning the density at which neutrons are captured and the
value of the ^{3}He/H ratio. They will also set upper limits on the
number of low energy neutrons produced in solar flares.

078.038   Solar flare neutron fluxes derived from interplanetary charged particle measurements.
P. Evenson, R. Kroeger, P. Meyer, D. Müller.
18th International Cosmic Ray Conference, Vol. 4, p. 97 – 100
(1983). – See Abstr. 012.096.
Observations of interplanetary protons of high energy produced by the decay of solar neutrons are made. The authors
determine the spectrum of the decay protons for the June 21,
1980 neutron event. The measurements suggest that neutron
emission from solar flares is isotropic and that different flares
emit neutrons with similar spectra.

078.039   Solar flare neutrons and gamma ray lines.
R. E. Lingenfelter, R. Ramaty, R. J. Murphy,
B. Kozlovsky.
18th International Cosmic Ray Conference, Vol. 4, p. 101 – 104
(1983). – See Abstr. 012.096.
The authors have derived the energy spectrum of accelerated
protons and nuclei at the site of the June 21, 1980 limb flare by
a new technique. This energy spectrum is very similar to the
energy spectra of 7 disk flares for which the accelerated particle
spectra have been previously derived.

078.040   Generation of accelerated particles and hard radiation during solar flare.
G. E. Kocharov, G. A. Kovaltsov (*G. A. Koval'tsov*),
L. G. Kocharov.
18th International Cosmic Ray Conference, Vol. 4, p. 105 – 108
(1983). – See Abstr. 012.096.
Features of particle acceleration from X– and γ–observations
are considered.

078.041   The influence of the ionic charge composition on heavy ion abundance variations during solar energetic particle events.
B. Klecker, M. Scholer.
18th International Cosmic Ray Conference, Vol. 4, p. 123 – 126
(1983). – See Abstr. 012.096.
The authors investigate the influence of the actual ionic charge
distribution of He, C, and O on compositional variations during
the onset phase of solar energetic particle events, using a numerical model for the interplanetary propagation.

078.042   Is quasilinear description applicable to the propagation of low–energy ($10^5 – 10^6$eV nucl^{-1}) flare accelerated protons and alpha particles?
E. I. Daibog (*E. I. Dajbog*), V. G. Kurt, V. G. Stolpovskii
(*V. G. Stolpovskij*), G. P. Skrebtsov, A. A. Kolchin,
M. I. Verigin.
18th International Cosmic Ray Conference, Vol. 4, p. 127 – 130
(1983). – See Abstr. 012.096.
The time profiles of the protons and alpha particles of
$0.1 – 100$ MeV nucl^{-1} detected in energetic solar particle events
1977 are considered.

078.043   The effect of prolonged injection of flare–generated particles on the parameters of their propagation in inter-planetary space.
G. A. Trebukhovskaya.
18th International Cosmic Ray Conference, Vol. 4, p. 131 – 134
(1983). – See Abstr. 012.096.
The variations of the propagation parameters of flare–
generated particles in interplanetary space are treated as functions of particle injection duration. The temporal behaviour of
the electron flux with energies 30 keV anisotropy and intensity at
1 a.u. are calculated for two types of injection function.

078.044   The three–dimensional diffusion of high–energy solar cosmic rays.
Y.–n. Huang.
18th International Cosmic Ray Conference, Vol. 4, p. 135 – 138 (1983). – See Abstr. 012.096.

Basic features of solar cosmic ray events are discussed by using the solution of the three–dimensional diffusion equation of solar cosmic ray propagation, which is derived on the basis of a new "spiral coordinate system" for the large–scale interplanetary spiral magnetic field. The solution depends on the solar magnetic longitude of the flare producing cosmic rays and agrees well with the observations.

078.045   Study of anisotropic distribution of relativistic solar cosmic ray flux.
R. L. Singh, A. P. Mishra, S. K. Nigam, M. Shrivastava.
18th International Cosmic Ray Conference, Vol. 4, p. 139 – 142 (1983). – See Abstr. 012.096.

An extensive analysis has been performed to study the pitch angle anisotropy of recently observed relativistic solar cosmic ray events.

078.046   Comparison of the energy spectra and pitch angle distributions for solar particle events.
H. Debrunner, E. O. Flückiger, H. Neuenschwander, M. Schubnell, J. A. Lockwood.
18th International Cosmic Ray Conference, Vol. 4, p. 144 – 147 (1983). – See Abstr. 012.096.

The authors have analyzed solar particle events to determine the energy spectra and pitch angle distributions of the solar protons near earth as a function of time. These data form the basis for discussing the coronal and interplanetary propagation of the solar particles.

078.047   Rigidity dependence of the interplanetary scattering mean free path determined from rise–phase nucleonic abundance variations.
G. M. Mason, G. Gloeckler, D. Hovestadt.
18th International Cosmic Ray Conference, Vol. 4, p. 149 – 152 (1983). – See Abstr. 012.096.

The authors conclude that in the range $\sim 80 - 400$ MV the interplanetary scattering mean free path increases with rigidity.

078.048   Time–dependent rigidity spectrum and spatial anisotropy of the ground level event on May 7, 1978.
M. Lumme, M. Nieminen, J. Peltonen, J. J. Torsti, E. Vainikka, E. Valtonen.
18th International Cosmic Ray Conference, Vol. 4, p. 157 – 160 (1983). – See Abstr. 012.096.

Data from the world–wide network of neutron monitors has been used in determining the rigidity spectrum and spatial anisotropy of the ground level event on May 7, 1978.

078.049   Variations of solar proton anisotropy on November 10 – 11, 1981.
S. I. Ermakov, N. N. Kontor, G. P. Lyubimov, N. V. Pereslegina, V. I. Tulupov, E. A. Chuchkov.
18th International Cosmic Ray Conference, Vol. 4, p. 162 – 164 (1983). – See Abstr. 012.096.

The Venera–13, 14 data are used to study the reasons for observing the sign–alternating anisotropy of 1.5 – 3.5 MeV solar proton flux. The time variations of the flux are analyzed. The observed effect is shown to be due to high solar activity and to charged–particle motion along the loop magnetic fields in the interplanetary medium.

078.050   Quasistationary injection of solar protons with energies above 30 MeV.
G. P. Lyubimov, N. V. Pereslegina.
18th International Cosmic Ray Conference, Vol. 4, p. 165 – 168 (1983). – See Abstr. 012.096.

078.051   Estimation of the conditions in the charged–particle acceleration region in solar flares from the "Venera 11, 13, 14" data on energetic electrons and hard X–rays.
E. I. Daibog (E. I. Dajbog), V. G. Kurt, Yu. I. Logachev, V. G. Stolpovskii (V. G. Stolpovskij), V. Sh. Dolidze, V. M. Zenchenko, M. Niel, R. Talon.
18th International Cosmic Ray Conference, Vol. 4, p. 169 – 172 (1983). – See Abstr. 012.096.

078.052   Proton effectiveness of active regions throughout solar activity cycle.
S. I. Ermakov, N. N. Kontor, G. P. Lyubimov, T. I. Morozova, N. N. Pavlov, T. G. Khotilovskaya.
18th International Cosmic Ray Conference, Vol. 4, p. 173 – 175 (1983). – See Abstr. 012.096.

Long–term prediction of low–energy (1 – 5 MeV) proton fluxes in interplanetary space obtained using the model of active processes on the Sun and in interplanetary space is compared with spacecraft data for 1965 – 1982. The dependence of low–energy solar proton fluxes on the period and number of solar activity cycle is noted and analyzed.

078.053   Differences in the rising portion of relativistic solar protons and solar electrons at the earth.
M. A. Shea, D. F. Smart.
18th International Cosmic Ray Conference, Vol. 4, p. 177 – 180 (1983). – See Abstr. 012.096.

The onset profiles of relativistic particle events are examined. The authors found either a delay in the release of the relativistic electrons into the interplanetary medium, or different propagation characteristics while enroute to their detection point.

078.054   The spectra of solar proton ground level events recorded at Sanae.
P. H. Stoker, P. A. Louw.
18th International Cosmic Ray Conference, Vol. 4, p. 181 – 184 (1983). – See Abstr. 012.096.

The time history of solar proton spectra is investigated.

078.055   Spatial distribution of the solar particles during the 7 May 1978 flare.
J. A. Lockwood, H. Debrunner.
18th International Cosmic Ray Conference, Vol. 4, p. 185 – 188 (1983). – See Abstr. 012.096.

An extended study has been made of the satellite and ground-based cosmic–ray data for the nearly scatter–free solar particle event on 7 May 1978.

078.056   Cosmic ray flare increases.
K. B. Fenton, A. G. Fenton, J. E. Humble.
18th International Cosmic Ray Conference, Vol. 4, p. 189 (1983). Abstract. – See Abstr. 012.096.

078.057   Intercomparison of isotropic and anisotropic diffusion models in relation to high energy solar particle events.
A. P. Mishra, K. N. Mishra, M. Shrivastava, S. K. Nigam, R. L. Singh.
18th International Cosmic Ray Conference, Vol. 4, p. 190 – 193 (1983). – See Abstr. 012.096.

The isotropic and anisotropic diffusion models suggested earlier, have been compared and an attempt has been made to test the appropriateness of the model in explaining the time history of recently observed ground level events.

078.058   Energy spectra of solar cosmic rays according to balloon and neutron monitor data.
G. A. Bazilevskaya, V. S. Makhmutov, M. S. Grigoryan, T. N. Charakhch'yan.
18th International Cosmic Ray Conference, Vol. 4, p. 194 – 197 (1983). – See Abstr. 012.096.

The energy spectra of solar protons were obtained in the range 100 MeV – several GeV on the basis of joint analysis of stratospheric and neutron monitor measurements. The most powerful events of 1976 – 1982 were found occurring during increasing and declining phases of solar activity cycle.

078.059 The results of stratospheric measurements of solar cosmic rays with energy more than 100 MeV in October 1981.
G. A. Bazilevskaya, Yu. I. Stozhkov, V. S. Makhmutov,
T. N. Charakhch'yan, L. P. Borovkov, E. V. Vashenyuk,
L. L. Lazutin, D. Yu. Bimenov.
18th International Cosmic Ray Conference, Vol. 4, p. 198 – 201 (1983). – See Abstr. 012.096.
On October 8 – 15, 1981, the solar cosmic protons with the energy more than 100 MeV were observed in the stratosphere. The data on the time profile and energy spectra are reported. An unusual north–south anisotropy was measured in the event of October 12.

078.060 Solar high energy particles and comprehensive solar flare index.
B. Y. Zhu, M. Wada.
18th International Cosmic Ray Conference, Vol. 4, p. 202 – 205 (1983). – See Abstr. 012.096.
The solar high energy particle events are investigated in relation to the experimental comprehensive solar flare indices.

078.061 Anisotropic propagation stage of solar cosmic rays.
L. I. Dorman, Yu. I. Fedorov, M. E. Katz
(*M. E. Kats*), S. F. Nosov, B. A. Shakhov.
18th International Cosmic Ray Conference, Vol. 10, p. 63 – 66 (1983). – See Abstr. 012.096.
An equation of cosmic ray transfer describing the propagation of high–energy charged particles in interplanetary space and including rapid variations of cosmic ray anisotropy is derived.

078.062 Fluctuations of low–energy solar cosmic rays as inferred from observation data for September – December 1977.
I. Ya. Libin, G. M. Blokh, B. M. Kuzhevsky
(*B. M. Kuzhevskij*).
18th International Cosmic Ray Conference, Vol. 10, p. 78 – 81 (1983). – See Abstr. 012.096.
Data obtained by studying the 30 keV electron and 30 MeV proton flux intensities are studied. Power spectra of intensity fluctuations in quiet, preflare and flare–time periods are calculated. The solar origin of the fluctuations is suggested.

078.063 Dynamics of the boundaries of solar cosmic ray penetration into the Earth's magnetosphere during a strong magnetic storm.
L. A. Darchieva, A. V. Dronov, T. A. Ivanova,
L. M. Kovrygina, Yu. V. Mineev, Eh. N. Sosnovets,
E. S. Spirkova, L. V. Tverskaya.
18th International Cosmic Ray Conference, Vol. 10, p. 233 – 236 (1983). – See Abstr. 012.096.
The data of simultaneous measurements on board three polar satellites (Cosmos–900 and 1067, Intercosmos–19) are used to study the features of penetrations of solar >1 MeV protons and >30 keV electrons in the Earth's magnetosphere during the strong magnetic storm on April 3, 1979.

078.064 On the solar neutrons observation on high mountain neutron monitor.
Yu. E. Efimov, G. E. Kocharov, K. Kudela.
18th International Cosmic Ray Conference, Vol. 10, p. 276 – 278 (1983). – See Abstr. 012.096.
Analysis of the enhancement in the Lomnicky Stit neutron monitor during the intense flare on 3 June 1982 has been made. It is shown that the obtained increase can be explained as the direct registration of high energy solar neutrons.

078.065 High–energy solar protons from a flare of April 4, 1981.
N. N. Volodichev, N. A. Mamontova,
O. Yu. Nechaev, I. A. Savenko.
18th International Cosmic Ray Conference, Vol. 10, p. 299 – 302 (1983). – See Abstr. 012.096.
Possible acceleration and propagation in interplanetary space of high–energy protons from this flare are discussed on the basis of a comparison of solar proton generation with the optical, radio, X–ray, and gamma–ray emissions.

078.066 Proton acceleration in a solar flare on April 1, 1981.
N. N. Volodichev, N. A. Mamontova,
O. Yu. Nechaev, I. A. Savenko.
18th International Cosmic Ray Conference, Vol. 10, p. 303 – 306 (1983). – See Abstr. 012.096.
Possible acceleration of particles up to high energies in this flare is discussed.

078.067 On the determination of the total energy of particles accelerated in solar flares.
L. G. Kocharov.
18th International Cosmic Ray Conference, Vol. 10, p. 311 – 313 (1983). – See Abstr. 012.096.
The possibility is proposed to find the minimum (or characteristic) energy of proton spectra at the sun and accordingly the total energetics of accelerated particles.

078.068 The September 1979 solar cosmic ray event.
D. C. Hamilton, G. Gloeckler, B. Klecker.
18th International Cosmic Ray Conference, Vol. 10, p. 314 – 317 (1983). – See Abstr. 012.096.
The authors present observations from Voyager 2 at 5.6 AU and from ISEE–3 and IMP–8 near Earth of a long–lived solar particle event in September 1979. This event illustrates the importance of knowing the solar connection longitude of the observer in interpreting intensity–time profiles, the success of the standard spherically symmetric interplanetary transport model in predicting the profiles of particles $\gtrsim 10$ MeV/nuc, and the apparent importance of interplanetary acceleration of particles.

078.069 Elemental composition of nuclei above silicon emitted in a large flare of August 4, 1972.
N. Durgaprasad, M. N. Vahia, S. Biswas.
18th International Cosmic Ray Conference, Vol. 10, p. 318 – 321 (1983). – See Abstr. 012.096.
Elemental abundance of even charged nuclei between $Z = 16$ and 26 in the energies of 10 to 40 MeV/amu was measured. The relative abundances of these elements with reference to oxygen is presented.

078.070 Elemental and isotopic composition of solar energetic particles: preliminary results from the Phoenix I telescope.
J. A. Simpson, J. P. Wefel, R. Zamow.
18th International Cosmic Ray Conference, Vol. 10, p. 322 – 325 (1983). – See Abstr. 012.096.
The Phoenix I instrument on the S81–1 mission was designed to resolve isotopes in the charge range He to Ni. Analysis of four periods of moderate flare activity shows no strong evidence for mass dependent acceleration in the range C to Si; indeed, the mass abundance ratios are consistent with "solar system" composition.

078.071 Ionization states of helium in ^{3}He–rich solar energetic particle events.
B. Klecker, D. Hovestadt, G. Gloeckler, E. Möbius,
F. M. Ipavich, M. Scholer.
18th International Cosmic Ray Conference, Vol. 10, p. 326 – 329 (1983). – See Abstr. 012.096.
The authors report results of a systematic study of the ionic charge state of helium in the energy range 0.6 – 1.0 MeV/nucleon for ^{3}He–rich solar energetic particle events during the time period August 1978 to October 1979. ^{3}He–rich solar energetic particle events do not show significant abundances of ^{3}He^{+}.

078.072 Direct determination of the ionic charge distribution of heavy ions in Fe–rich solar energetic particle events.
B. Klecker, D. Hovestadt, G. Gloeckler, E. Möbius,
F. M. Ipavich, M. Scholer.
18th International Cosmic Ray Conference, Vol. 10, p. 330 – 333 (1983). – See Abstr. 012.096.
The authors report results of a systematic study of elemental and isotopic abundances and heavy ion charge states in the energy range 0.3 – 1.8 MeV/nucleon in energetic particle events. The high mean charge of iron as observed during Fe–rich and

³He–Fe–rich solar energetic particle events is compatible with the resonant heating mechanism proposed by Fisk (1978).

078.073   Solar neutrons from the impulsive flare on 1982 June 3 at 1143 UT.
E. L. Chupp, D. J. Forrest, G. H. Share, G. Kanbach, H. Debrunner, E. O. Flueckiger.
18th International Cosmic Ray Conference, Vol. 10, p. 334 – 337 (1983). – See Abstr. 012.096.

A transient flux of high energy solar neutrons from 50 MeV to ~1 GeV has been detected by the Gamma Ray Spectrometer on the Solar Maximum Mission satellite following an intense burst of high energy photons (< 100 MeV). The authors summarize the observations, compare them with the Jungfraujoch neutron monitor data, and estimate both the time dependent neutron flux at the Earth and the neutron emisson spectrum at the sun.

078.074   On the collective deceleration of particles during solar flares.
L. G. Kocharov.
18th International Cosmic Ray Conference, Vol. 10, p. 350 – 352 (1983). – See Abstr. 012.096.

The deceleration of energetic electrons by the direct electric field while penetrating the upper chromosphere is possible if the flux density is high. This collective deceleration has influence on the predicted intensity and spectrum of the hard X–ray burst generated during the solar flare.

078.075   A multi–spacecraft study of the coronal and interplanetary transport of solar cosmic rays. I.
R. E. McGuire, M. A. I. van Hollebeke, N. Lal.
18th International Cosmic Ray Conference, Vol. 10, p. 353 – 356 (1983). – See Abstr. 012.096.

Multiple spacecraft observations of a collection of flare events are analyzed to separate and determine the longitudinal and radial dependence of the times of maximum flux, the relative maximum fluxes and the decay time constants. The intensity study gives new direct information on the spatial distribution of flare particles.

078.076   A multi–spacecraft study of the coronal and interplanetary transport of solar cosmic rays. II.
R. E. McGuire, N. Lal, M. A. I. van Hollebeke.
18th International Cosmic Ray Conference, Vol. 10, p. 357 – 360 (1983). – See Abstr. 012.096.

This paper is concerned with the modelling and interpretation of the observations.

078.077   The study of propagation of SCR (*solar cosmic rays*) from flares in the terms of diffusion approximation taking into account their acceleration by turbulence.
E. V. Kolomeets, V. N. Sevast'yanov.
18th International Cosmic Ray Conference, Vol. 10, p. 369 – 372 (1983). – See Abstr. 012.096.

078.078   The delayed energetic particle event of June 6 – 10, 1979.
T. T. von Rosenvinge, D. V. Reames.
18th International Cosmic Ray Conference, Vol. 10, p. 373 – 376 (1983). – See Abstr. 012.096.

The authors report on observations from the ISEE–3 spacecraft of an unusual enegetic particle event associated with the passage of an interplanetary shock. They present the energetic particle data, investigate the absence at ISEE–3 of prompt particles from the parent flare, and demonstrate that the shock was strongly accelerating particles at and beyond 1 AU.

078.079   Dependence of energy spectrum of solar cosmic rays and its variation on electromagnetic conditions in interplanetary space.
D. Yu. Bimenov, B. N. Dyusembaev, E. V. Kolomeets, V. N. Sevast'yanov.
18th International Cosmic Ray Conference, Vol. 10, p. 377 – 380 (1983). – See Abstr. 012.096.

From balloon data and the data of Mirny neutron monitor and of "Meteor" satellite the energy spectra of solar cosmic rays, their temporal transformation and their propagation parameters were determined.

078.080   Energetic solar cosmic rays in the outer heliosphere.
R. Goeman, W. R. Webber.
18th International Cosmic Ray Conference, Vol. 10, p. 385 – 388 (1983). – See Abstr. 012.096.

The authors study intermediate energy (3.5 – 56 MeV) protons in the outer heliosphere using Pioneer 10 and 11 data. They investigate the properties of solar flare associated flux increases for radial distances between 5.8 and 28.5 AU.

078.081   Understanding the heliosphere and its energetic particles.
L. F. Burlaga.
18th International Cosmic Ray Conference, Vol. 12, p. 21 – 60 (1983). – See Abstr. 012.096.

Contents: Corotating flows, corotating Forbush decreases, and 27–day variations. Transients and Forbush decreases. Systems and transients and long–lasting Forbush decreases. Eleven–year variations.

078.082   Energetic particle and γ–ray generation in the solar matter.
G. E. Kocharov.
18th International Cosmic Ray Conference, Vol. 12, p. 235 – 256 (1983). – See Abstr. 012.096.

Contents: Some questions of acceleration and physics of pre–flare condition. Common characteristics of the Sun and solar cosmic rays. Characteristics of solar cosmic rays: energy spectra, elemental and isotopic composition, ³He–rich flares and ionization states of SCR. Solar γ–rays and neutrons: flare high energy γ–rays and neutrons, 2.2 MeV line and low energy solar neutrons, on the two groups of solar γ–ray flares, on the energy spectra of particles in situ, on the direct connection between the proton and electron acceleration processes.

078.083   Coronal and interplanetary propagation of solar cosmic rays.
A. G. Fenton.
18th International Cosmic Ray Conference, Vol. 12, p. 257 – 278 (1983). – See Abstr. 012.096.

Contents: The solar wind and interplanetary magnetic field. Coronal propagation of solar cosmic rays. Interplanetary propagation, theory and observations. Solar neutrinos and other topics.

078.084   Two classes of energetic particle events associated with impulsive and long duration flares.
H. V. Cane, R. E. McGuire, T. T. von Rosenvinge.
Bull. Am. Astron. Soc., Vol. 17, No. 2, p. 610 (1985). Abstract. – See Abstr. 010.065.

078.085   Statistical properties of solar proton events.
F.–m. Hu.
Acta Astron. Sin., Vol. 25, No. 4, p. 390 – 400 (1984). In Chinese.

Sounding system for determination of the intensity of solar cosmic radiation.
See Abstr. 034.037.

X–ray experiment aboard the artificial earth satellite Prognoz 9.
See Abstr. 035.032.

The monitoring and prediction of solar particle events – an experience report.
See Abstr. 051.029.

Neutrons and gamma rays from solar flares.
See Abstr. 073.017.

Neutron and gamma–ray signatures for particle acceleration in solar flares.
See Abstr. 073.103.

Solar flare neutrons and gamma rays.
See Abstr. 073.145.

Relationship between γ–ray emission, radio bursts and proton fluxes from solar flares.
See Abstr. 073.167.

Further isotopic studies of heavy nuclei in the 9/23/78 solar flare.
See Abstr. 073.187.

A search for ^{2}H, ^{3}H, and ^{3}He in large solar flares.
See Abstr. 073.188.

^{3}He rich solar flares.
See Abstr. 073.189.

Behavior of high–energy particles associated with a solar flare on April 27, 1981.
See Abstr. 073.190.

Solar flares with photon emission above 10 MeV. Measurements with the gamma ray experiment on board the SMM–satellite.
See Abstr. 073.192.

Post–flare loop heating by trapped superthermal protons.
See Abstr. 073.198.

On the role of energetic particles in solar flares.
See Abstr. 073.228.

Solar–stellar outer atmospheres and energetic particles, and galactic cosmic rays.
See Abstr. 074.034.

Plausible mechanisms for rapid acceleration of protons during solar flares.
See Abstr. 076.026.

Nuclear gamma rays and interplanetary proton events.
See Abstr. 076.033.

Radio radiation as information source on proton fluxes from solar flares.
See Abstr. 077.024.

On peculiarities of a connection of proton stream intensity, the meter component of solar radio bursts and PCA–type absorption.
See Abstr. 077.044.

Spectral correlation between high energy solar flare particle events and associated radio bursts.
See Abstr. 077.045.

Solar particle corotating events from shocked neutral sheets.
See Abstr. 080.128.

Entry of solar energetic protons into the distant plasma sheet – dependence on distance from the neutral sheet.
See Abstr. 084.044.

Solar flare neon and solar cosmic ray fluxes in the past using gas–rich meteorites.
See Abstr. 105.133.

The features of solar proton propagation in interplanetary space.
See Abstr. 106.077.

Absolute value and functional dependence of interplanetary pitch angle scattering derived from Helios observations at 0.5 AU.
See Abstr. 106.080.

Recurrent double flares of pitch–angle anisotropy of cosmic rays and their connection with solar proton events.
See Abstr. 144.048.

Rapporteur paper for sessions MG1, MG3, and MG4 (*of the conference*): modulation theory, interplanetary propagation, and interplanetary acceleration.
See Abstr. 144.453.

Cosmic rays as probes of heliosphere and magnetospheres: rapporteur talk on MG–2 and –5 (*of the conference*).
See Abstr. 144.454.

079 Solar Eclipses

079.001   Sonnenfinsternisübersicht 1984 – 2084.
E. Riedel.
Sonne, Jahrg. 9, Nr. 33, p. 18 – 20 (1985).

079.002   On precalculation of the visibility of solar eclipses by means of computers.
D. D. Polozhentsev.
Astron. Geod., Tomsk, No. 12, p. 123 – 127, 146 (1984). In Russian. Abstr. in Ref. Zh., 51. Astron., 2.51.85 (1985).

079.003   The scintillation theory of eclipse shadow bands.
J. L. Codona.
Bull. Am. Astron. Soc., Vol. 17, No. 2, p. 640 (1985). Abstract. – See Abstr. 010.067.

Solar eclipse 1984 May 30

079.101   Ringförmige Sonnenfinsternis 30.5.1984 in Marokko.
A. Sudy.
Sonne, Jahrg. 9, Nr. 33, p. 16 – 17 (1985).

Solar eclipse 1981 July 31

079.111   The 1981 solar total eclipse corona: II. Global absolute photometrical analysis.
C. Lebecq, S. Koutchmy, G. Stellmacher.
Inst. Astrophys. Paris, Pré–Publ., No. 99, 24 pp. (1985). To appear in Astron. Astrophys.

Solar eclipse 1945 July 9

079.121   The total solar eclipse of July 9, 1945.
K. Bracher.
Publ. Astron. Soc. Pac., Vol. 97, No. 588, p. 199 (1985). Abstract. – See Abstr. 010.281.

Solar eclipse 1983 June 11

079.141   Balloon observation of the 1983 solar eclipse in Indonesia.
H. Tanabe, S. Isobe, H. Akiyama, Y. Koma, T. Okabe,
J. Nishimura, T. Maihara, K. Mizutani, J. Soegijo,

T. E. Hariadi, S. Indrawan, S. Slamet, P. Anondo, T. Tatang,
S. Agus, W. Mulyana, V. R. Suroto.
Adv. Space Res., Vol. 5, No. 1, p. 69 – 72 (1985). – See Abstr.
012.038.

A balloon observation of the total solar eclipse on 11 June
1983 was carried out as a cooperative work between Japanese and
Indonesian teams. The observation was a photo–polarimetry of
the F corona in both visual and near–infrared regions. As a preliminary result, an excess in infrared brightness has been found
near the position of 3.8 $R_\odot$ west from the sun, which may be due
to thermal emission from a high–temperature dust cloud located
around the sun.

Solar eclipse 1982 December 15

079.161 **Observation of the partial solar eclipse of December 15,
1982.**
I. Bosnić, V. Jamičić, D. Šimunić.
Publ. Astron. Opservatorije Beogr., No. 33, p. 92 – 93 (1985). –
See Abstr. 012.061.

Solar eclipse 1984 November 22 – 23

079.181 **Observing the 22/23 Nov. 1984 solar total eclipse in
New Caledonia.**
S. Koutchmy, C. Nitschelm.
Inst. Astrophys. Paris, Pré–Publ., No. 84, 16 pp. (1984). To be
published in Sky Telesc.

079.182 **Observing the Papua New Guinea solar eclipse 1984
November 23.**
B. W. Soulsby, D. Miller, T. B. Tregaskis.
Aust. J. Astron., Vol. 1, No. 1, p. 10 – 26 (1985).

Canon der Sonnenfinsternisse –2003 bis + 2526.
See Abstr. 002.128.

Uses for ancient eclipse records.
See Abstr. 004.001.

**Clues to the mode of excitation of Fe X ions in the solar corona
from the 1980 eclipse observations.**
See Abstr. 074.041.

**The total brightness of the corona during the solar eclipse of Febru-
ary 16, 1980.**
See Abstr. 074.051.

Coronal voids observed during the July 31, 1981 solar eclipse.
See Abstr. 074.119.

**Atmospheric hydroxyl response to the partial solar eclipse of
May 30, 1984.**
See Abstr. 082.067.

080 Atmosphere, Figure, Internal Constitution, Neutrinos, Rotation, etc.

080.001 **Effects of CO molecules on the outer solar atmosphere:
a time–dependent approach.**
D. Muchmore, P. Ulmschneider.
Astron. Astrophys., Vol. 142, No. 2, p. 393 – 400 (1985).

CO can be an important source of cooling near the solar temperature minimum. The authors investigate the influence of
radiation in the 4.6 µm fundamental vibration–rotation band on
the structure of the Sun's outer atmosphere. The model calculations employ a 1–D hydrodynamical code in conjunction with a
two–frequency treatment of radiative transfer. Resulting models
have a bistable character. Radiative equilibrium atmospheres for
$T_{eff} < 5800$K are dominated in outer layers by CO cooling which
draws the temperature down to $T \lesssim 3000$K. For hotter models,
CO plays no important role. The chromosphere adjusts from one
state to another in a time of the order of one hour. Acoustic
heating destroys CO in the outermost parts of a model but for
low and moderately high acoustic fluxes CO continues to dominate the region of the temperature minimum.

080.002 **Long nonlinear waves in a compressible magnetically
structured atmosphere. I. Slow sausage waves in a magnetic slab.**
E. G. Merzljakov, M. S. Ruderman.
Sol. Phys., Vol. 95, No. 1, p. 51 – 68 (1985).

The Benjamin–Ono equation is derived for long slow sausage
waves propagating in a vertical magnetic slab embedded into a
stratified atmosphere, provided that the slab thickness is much
smaller than the scale height of the atmosphere. The soliton
propagation in a nonstratified atmosphere is discussed. The approximate formulas describing the slow evolution of the amplitude and the length of a soliton propagating in a very weakly
stratified atmosphere are obtained. The exact soliton–like solu-

tion for an atmosphere with a linearly growing temperature is
found.

080.003 **On heating of solar active regions by magnetic energy
dissipation.**
M. Kumar, U. Narain.
Sol. Phys., Vol. 95, No. 1, p. 69 – 72 (1985).

The Tucker's proposal of heating of active regions by magnetic
energy dissipation has been found to be unsatisfactory in its
present form because it predicts unacceptably high values of the
twisting velocity for the magnetic field.

080.004 **Hydromagnetic buoyancy force in the solar atmosphere.**
T. Yeh.
Sol. Phys., Vol. 95, No. 1, p. 83 – 97 (1985).

The author tries to explain and ascertain the required distribution of hydromagnetic stress that gives rise to an essentially uniform force density, just like the gravitational force density, in
view that the plasma in an immersed body behaves like a compressible body. In the solar atmosphere hydromagnetic buoyancy
force has an obliquely upward direction, with a component in the
direction opposite to the downward gravity. It provides an upward force to counterbalance or even to exceed the downward
gravitational force. Such an upward force is the dynamic cause
for the stationary equilibrium of quiescent prominences and outward motion of coronal transients.

080.005 **Radiative and reconnection instabilities: compressible
and viscous effects.**
T. Tachi, R. S. Steinolfson, G. Van Hoven.
Sol. Phys., Vol. 95, No. 1, p. 119 – 140 (1985).

Filaments and flares are prominent indicators of the magnetic
fields of solar activity. Their (tearing or reconnection) mecha-

nisms are resistively coupled in sheared magnetic fields of the kind existing in active regions. The present paper includes the effects of compressibility and viscosity, which are most prominent at short wavelengths. The results show that compressibility affects the radiative mode, including a modest increase of its growth rate, and that viscosity modifies the tearing mode, partially through a decrease of its growth rate. A comprehensive discussion of the mode structures and flows is presented. The strongest effect found is a reversal, at very long wavelengths, of the radiative cooling of the resistive interior layer of the tearing mode, caused by compressional heating.

080.006   Magnetic reconnection in a high–temperature plasma of solar flares. I. Effect of gradient instabilities.
B. V. Somov, V. S. Titov.
Sol. Phys., Vol. 95, No. 1, p. 141 – 153 (1985).
A simple self–consistent model of a high–temperature turbulent current sheet (HTCS) is considered. The anomalous character of plasma conductivity in a sheet is assumed to be due to gradient instabilities. The possibility of a low threshold of their excitation is demonstrated by an example of temperature–drift instability. Application of the HTCS model of the 'hot' or 'main' phase of a solar flare is discussed. The model consistently explains many observed properties of this phase.

080.007   Fine structure of Fraunhofer lines and the structure of the solar atmosphere.
R. I. Kostyk.
Astron. Zh., Tom 62, Vyp. 1, p. 112 – 123 (1985). In Russian. English translation in Sov. Astron., Vol. 29, No. 1.
A semi–empirical, inhomogeneous model of the solar atmosphere is suggested using the fine structure (asymmetry) of weak Fraunhofer lines observed in the center of the solar disk. The model differs from previous ones in one principal aspect: there is no correlation between the granulation intensity and velocity fluctuations.

080.008   Velocity field at the temperature minimum of the solar atmosphere.
Eh. A. Gurtovenko, V. A. Sheminova, R. J. Rutten.
Astron. Zh., Tom 62, Vyp. 1, p. 124 – 131 (1985). In Russian. English translation in Sov. Astron., Vol. 29, No. 1.
The weak Fraunhofer lines in the near wings of H, K Ca II lines have been analysed to study the velocity amplitude of the general velocity field in the middle and outer photospheric layers. The results confirm the basic well–known data on the velocity amplitude in the middle photospheric layers. Besides, it is shown that the radial and tangential components of the velocity amplitude continue to decrease with height also in the outer photosphere.

080.009   Determination of iron abundance in the solar atmosphere by profiles of ion absorption lines.
B. T. Babij, R. E. Rikalyuk.
Kinematika Fiz. Nebesn. Tel, Tom 1, No. 1, p. 61 – 65 (1985). In Russian.
Using 27 Fe II absorption line profiles and Vernazza et al., HSRA, and Holweger–Müller model atmospheres the abundance of iron in the solar atmosphere is obtained: 7.56 ± 0.06 in the Blackwell oscillator strength system. The effect of the model atmosphere, damping constant, velocity field, and the oscillator strength system of the abundance derived is studied.

080.010   Models of the solar convective zone.
A. S. Gadun.
Kinematika Fiz. Nebesn. Tel, Tom 1, No. 1, p. 66 – 74 (1985). In Russian.
Models of the solar convective zone consistent with Abraham's and Eben's model of the internal solar structure on the inner boundary, and HSRA, HOLMU, VAL 80 semi–empirical models and Nelson's (1978) theoretical two–dimensional model on the outer boundary were obtained on the basis of the mixing length theory. Calculated velocities of rising convective elements were compared with the observed ones.

080.011   On the heating mechanism of magnetic flux loops in the solar atmosphere.
M. T. Song, S. T. Wu.
Adv. Space Res., Vol. 4, No. 7, p. 275 – 278 (1984). – See Abstr. 012.009.
The authors have investigated physical heating mechanisms due to ponderomotive forces exerted by turbulent waves along curved magnetic flux loops. Results show that the temperature difference between the inside and outside of the flux loop can be classified into three parts. Solar atmospheric heating is illustrated via an example leading to the formulation of plages.

080.012   Immediate and long–term prospects for helioseismology.
D. O. Gough.
Adv. Space Res., Vol. 4, No. 8, p. 85 – 102 (1984). – See Abstr. 012.010.
Recent extensive measurements of frequencies of free oscillation of the Sun have permitted a first direct esimate of the variation of sound speed and angular velocity throughout the Sun. The results hint that the answers to some tantalizing questions concerning the Sun's interior structure and its history are almost within grasp. Optimists believe that in a few years a world-wide network of ground–based observing stations will give us important clues. However, it may be necessary to make observations from space before we can be sure of the answers.

080.013   Solar total irradiance and sunspot area in 1981.
H. S. Hudson.
Adv. Space Res., Vol. 4, No. 8, p. 113 – 116 (1984). – See Abstr. 012.010.
The author shows that a small correlation exists between the solar irradiance variations and active–regions time–scales.

080.014   Solar variability for periods of days to months.
C. Fröhlich.
Adv. Space Res., Vol. 4, No. 8, p. 117 – 120 (1984). – See Abstr. 012.010.
The time series of total solar irradiance determinations from ACRIM on the Solar Maximum Mission satellite (SMM) of 270 days and from the ERB experiment on NIMBUS 7 of 1445 days are analysed for periods greater than a few days. Comparison of the spectra of both with the spectrum of projected sunspot area over the corresponding time periods show high coherence for periods of 7 to about 25 days and for periods longer than about 30 to 35 days. In the vicinity and at the 27–day rotational period of the Sun, however, the coherence between sunspot areas and irradiance is small. This means that there is a signal in the irradiance which cannot be due to the sunspot area.

080.015   Observed relation between solar luminosity and radius.
C. Fröhlich, J. A. Eddy.
Adv. Space Res., Vol. 4, No. 8, p. 121 – 124 (1984). – See Abstr. 012.010.
The authors find an apparent increase in solar diameter of about 0.03″ per year during the period of increasing solar irradiance of about 0.025% per year and a weaker suggestion of a similar, subsequent decline while the solar irradiance decreases.

080.016   North–south asymmetries of the solar magnetic field in conjunction with the location of the heliospheric current sheet on both sides of the solar equator.
V. P. Tritakis.
Adv. Space Res., Vol. 4, No. 8, p. 125 – 128 (1984). – See Abstr. 012.010.
Reported heliospheric current–sheet displacements from the equatorial plane in the period 1947 – 1977 have been found to be in agreement with north–south asymmetries of the solar magnetic field. The author concludes that the heliospheric current sheet has its origin on the solar surface while its location with respect to the solar equator appears to be affected by the variability of the lower layers of the solar interior.

080.017   A review of the present state of gravity modes oscillations of the Sun.
A. B. Severny (*A. B. Severnyj*), V. A. Kotov.
Adv. Space Res., Vol. 4, No. 8, p. 129 – 131 (1984). – See Abstr. 012.010.

Short review of ground–based observations of the Sun and of solar gravity modes detection is presented with the emphasis on 160^m oscillation as predominant, long–time phase–coherent oscillation. The limitations imposed by the influence of Earth's atmosphere and geoseismic waves on observations of solar oscillations are pointed out together with a necessity of the observations of solar oscillations with the aid of space–probes.

080.018   Observations of electron density in the solar corona during sunspot minimum and global electric currents around the sun.
V. Osherovich, E. B. Gliner, I. Tzur.
Adv. Space Res., Vol. 4, No. 8, p. 133 – 139 (1984). – See Abstr. 012.010.

The observed difference in electron density between the equatorial plane and the polar direction is compared for three empirical distributions: the Allen distribution, the Saito distribution, and the Saito–Munro–Jackson distribution. It is shown that from $1.5\,R_\odot$ to $5\,R_\odot$ the observed difference in electron density is sufficient to result in a global azimuthal electric current flow around the sun. The dependence of the calculated current density on the radial distance and solar latitude is discussed.

080.019   Rotational splitting of global solar oscillations and its relevance to tests of general relativity.
H. A. Hill.
Int. J. Theor. Phys., Vol. 23, No. 8, p. 683 – 699 (1984). Abstr. in Phys. Abstr., Vol. 88, No. 1250, Entry 18744 (1985). – See Abstr. 012.015.

080.020   Convective instability in a solar flux tube. II. Nonlinear calculations with horizontal radiative heat transport and finite viscosity.
S. S. Hasan.
Astron. Astrophys., Vol. 143, No. 1, p. 39 – 45 (1985). = Mitt. Kiepenheuer–Inst. Sonnenphys., Nr. 245.

Convective instability in a thin flux tube is examined in the presence of horizontal radiative heat transport and finite viscosity. The temporal behaviour of flux tubes initially in hydrostatic and thermal equilibrium is studied by solving the nonlinear time dependent equations for a thin flux tube. An important result of the investigation is the demonstration of overstable oscillations with periods typically about 1000 s in intense flux tubes on the Sun. Detailed calculations are presented for a broad range of parameters which characterize the strength of the magnetic field and the tube radius in the initial state.

080.021   Present status of the solar neutrino problem.
N. Itoh.
Grand unified theories and cosmology, p. 250 (1984). Abstr. in Phys. Abstr., Vol. 88, No. 1251, Entry 24412 (1985). – See Abstr. 012.020.

080.022   Nuclear physics in the solar neutrino problem.
K. Nagatani.
Grand unified theories and cosmology, p. 251 – 274 (1984). Abstr. in Phys. Abstr., Vol. 88, No. 1251, Entry 24413 (1985). – See Abstr. 012.020.

080.023   Trillingen op de zon.
W. van Tend.
Zenit, 12. Jaarg., No. 2, p. 64 – 67 (1985).

080.024   On the energetics of the solar supergranulation.
P. J. Gierasch.
Astrophys. J., Vol. 288, No. 2, p. 795 – 800 (1985).

Thermodynamic arguments demonstrate that kinetic energy dissipation must occur in the convection zone. The dissipation rate depends on the temperature gradient, which is at a minimum in the hydrogen ionization zone. As a result, the kinetic energy dissipation rate per density scale height shows a maximum just below the photosphere and another maximum beneath the ionization zone. The author proposes that intensified convection at these two levels is the explanation for the two distinct phenomena of granulation and supergranulation.

080.025   Torsional oscillations of low mode.
H. B. Snodgrass, R. Howard.
Sol. Phys., Vol. 95, No. 2, p. 221 – 228 (1985).

Standing wave torsional oscillations of wavenumber 1/2 and 1 hemisphere^{-1} are studied using an improved fit to Mount Wilson magnetograph data. These oscillations are seen to be in phase with each other and with the magnetic activity cycle, and seem best represented as a flexing of the differential rotation curve. Superposing them gives a differential rotation which at solar minimum is slightly flattened at the equator but considerably ($\sim 5\%$) steepened at the poles, and also tends to produce a travelling wave with wavenumber 1 hemisphere^{-1} that moves from pole to equator as the cycle progresses.

080.026   Differential rotation of the solar atmosphere determined from "radiotracers" at 8.2 mm wavelength.
Yu. A. Solonskij, E. D. Khilov.
Soln. Dannye, Byull., 1984, No. 11, p. 64 – 68 (1985). In Russian.

Radio scans of the sun obtained at $\lambda = 8.2$ mm in July 1980 were used to determine the rotation rate of different features in the solar atmosphere. The migration of identified radio and Hα filaments of the solar disk was followed for 15 days.

080.027   Line–of–sight velocity differential signal monitoring at two closely–spaced levels of the solar atmosphere.
M. L. Demidov, N. I. Kobanov.
Soln. Dannye, Byull., 1984, No. 12, p. 79 – 83 (1985). In Russian.

A brief account of a method of measurement of the line–of–sight velocity at two levels of the solar atmosphere is given. The method enables a reliable recording of the signal of the five minute oscillations for a height difference of about 25 km. Under the assumption of a vertically propagating wave an approximate estimate of the relevant wavelength and phase velocity is obtained.

080.028   Radio radius of the sun toward the polar directions measured during the solar maximum at 2.25 cm wavelength.
A. F. Bachurin.
Izv. Krymskoj Astrofiz. Obs., Tom 69, p. 130 – 133 (1984). In Russian. English translation in Bull. Crimean Astrophys. Obs., Vol. 69.

In May 1981 the solar radio radius was measured toward the polar directions. The radio radius value at wavelength 2.25 cm is equal to $(1.035 \pm 0.001)\,R_\odot$. The value increase of the radio radius in polar regions at solar activity maximum in comparison with that at minimum is confirmed.

080.029   Erratum: "Measurements of solar total irradiance and its variability" [Space Sci. Rev., Vol. 38, Nos. 3/4, p. 203 – 242 (1984)].
R. C. Willson.
Space Sci. Rev., Vol. 39, Nos. 3/4, p. 381 – 382 (1984). See Abstr. 38.080.004.

080.030   Solar models with differential rotation and toroidal magnetic fields.
C. Talmadge, S. Richter, E. Fischbach.
Astrophys. J., Vol. 290, No. 1, p. 337 – 343 (1985).

Solar models which could possibly solve the solar neutrino problem without incorporating neutrino oscillations are discussed. In particular it is found that it is possible to have solar models with rapid differential rotation and low surface oblateness, if one allows for the presence of strong toroidal magnetic fields with the proper configuration.

080.031   A search for nuclear–burning instabilities in the Sun.
R. L. Gilliland.
Astrophys. J., Vol. 290, No. 1, p. 344 – 346 (1985).

A search for coupled thermal–nuclear oscillations in the solar core is made using a stellar evolution code with a full nuclear reaction network. The time scales of ^{3}He burning and thermal relaxation suggest that any such oscillations would occur with periods of $10^5 - 10^7$yr. The evolutionary computations consistently show strong damping of abundance perturbations, thus lending support to the conclusions based on linear analyses that the Sun does not have thermal instabilities.

080.032   Turbulence in the solar atmosphere and in the interplanetary plasma.
I. V. Chashej, V. I. Shishov.
Fiz. inst. Akad. Nauk SSSR. Prepr., No. 282, 20 pp. (1983). In Russian. Abstr. in Ref. Zh., 51. Astron., 1.51.529 (1985).

080.033   Magnetohydrodynamical model of some instationary phenomena on the sun.
L. B. Tsirul'nik.
Dokl. Akad. Nauk AzSSR, Tom 40, No. 3, p. 15 – 19 (1984). In Russian. Abstr. in Ref. Zh., 51. Astron., 2.51.323 (1985).

080.034   Overview of solar seismology: oscillations as probes of internal structure and dynamics in the Sun.
J. Toomre.
Solar seismology from space, p. 7 – 39 (1984). – See Abstr. 012.041.

The physical nature of solar oscillations is reviewed. The nomenclature of the subject and the techniques used to interpret the oscillations are discussed. Many of the acoustic and gravity waves that can be observed in the atmosphere of the Sun are actually resonant or standing modes of the interior; precise measurements of the frequencies of such modes allow deductions of the internal structure and dynamics of this star. The scientific objectives of such studies of solar seismic disturbances, or of solar seismology, will be outlined. The reasons for why it would be very beneficial to carry out further observations of solar oscillations both from ground–based networks and from space will be discussed.

080.035   What would a dynamo theorist like to know about the dynamics of the solar convection zone?
P. A. Gilman.
Solar seismology from space, p. 41 – 48 (1984). – See Abstr. 012.041.

Current issues concerning solar dynamo theory, recent results from global compressible convection models, and what we need to know about differential rotation and convection are discussed.

080.036   Solar inverse theory.
D. Gough.
Solar seismology from space, p. 49 – 78 (1984). – See Abstr. 012.041.

Helioseismological inversion, as with the inversion of any other data, is divided into three phases. The first is the solution of the so–called forward problem: namely, the calculation of the eigenfrequencies of a theoretical equilibrium state. The second is an attempt to understand the results, either empirically or analytically. The rather more abstract third phase is to pose and solve an inverse problem, which seeks to find a plausible equilibrium model of the Sun whose eigenfrequencies are consistent with observation. The three phases are briefly discussed in this review, and the third, which is not yet widely used in helioseismology, is illustrated with some selected inversions of artificial solar data.

080.037   Rotational inversion from global solar oscillations.
J. Christensen–Dalsgaard, D. Gough.
Solar seismology from space, p. 79 – 93 (1984). – See Abstr. 012.041.

The authors investigate the degree to which various sets of solar oscillations can resolve the solar internal rotation. Genuine observations were simulated by the following procedure: First an artificial angular velocity was invented by one of the authors, and from it the rotational splitting of a set of normal modes was calculated; to that was added some random noise. The result was treated as artificial data by the other author, acting as an observer, who attempted to recover the rotation law by using the Backus–Gilbert optimal averaging procedure. The observer knew neither the original rotation law nor the amount of noise that had been added. Finally his conclusion was compared with the actual artificial angular velocity.

080.038   Sensitivity of inferred subphotospheric velocity field to mode selection, analysis technique and noise.
F. Hill, D. Gough, J. Toomre.
Solar seismology from space, p. 95 – 111 (1984). – See Abstr. 012.041.

The horizontal velocity immediately below the photosphere can be inferred from observations of high–degree solar oscillations by an optimal–averaging inversion technique. The authors investigate the sensitivity of the results to various details of both the inversion and the determination of the frequencies. The results are shown to be quite stable to the choice of most parameters, suggesting that this procedure produces reliable estimates of the subsurface velocity.

080.039   Observations of intermediate–degree solar oscillations.
J. W. Harvey, T. L. Duvall Jr.
Solar seismology from space, p. 165 – 172 (1984). – See Abstr. 012.041.

A progress report on observations of intermediate degree oscillations is presented. The authors list frequencies of zonal p–mode oscillations with amplitudes in excess of ~ 2 cm s^{-1}. These frequencies show systematic disagreement with recent theoretical calculations. The frequencies are compared with asymptotic formula estimates. Small scatter is obtained for low degree modes but large scatter at large degree. A first look at sectoral harmonic observations shows that magnetic active regions provide a major signal at low frequencies.

080.040   Detection of solar gravity mode oscillations.
P. H. Scherrer.
Solar seismology from space, p. 173 – 182 (1984). – See Abstr. 012.041.

An analysis of solar velocity data obtained at the Stanford Solar Observatory has shown the existence of solar global oscillations. The oscillations are in the range 45 to 105 µHz (160 to 370 minutes) and are interpreted as internal gravity modes of degree $l = 1$ and $l = 2$.

080.041   Solar gravity modes from ACRIM/SMM irradiance data.
C. Fröhlich, P. Delache.
Solar seismology from space, p. 183 – 193 (1984). – See Abstr. 012.041.

The record of 280 days of continuous data of the ACRIM radiometer on board the Solar Maximum Mission satellite is analysed in the frequency range from 10 to 80 µHz. Gravity modes of degree one and two with orders from about 10 to several hundreds can be localized. A statistical method to determine the fundamental period T_0 and the rate of rotation v_R as seen by rotational splitting is described and the results for $33.5 < T_0 < 45.5$ minutes and $0.4 < v_R < 2.0$ µHz presented. They indicate a rather high T_0 and it cannot be excluded that it is above the upper limit analysed.

080.042   Observations of low–degree modes from the Solar Maximum Mission.
M. Woodard.
Solar seismology from space, p. 195 – 197 (1984). Extended Abstract. – See Abstr. 012.041.

080.043   Implications of observed frequencies of solar p modes.
J. Christensen–Dalsgaard, D. Gough.
Solar seismology from space, p. 199 – 204 (1984). – See Abstr. 012.041.

The authors present a preliminary comparison of the observed frequencies of 5–min modes reported by Duvall & Harvey in these proceedings with theoretical frequencies for a traditional solar model. The differences between observations and theory can be understood qualitatively in terms of two separate sources of error in the frequency calculation, one near the solar surface and the other at the base of the convection zone. There is no indication of errors in the deep interior of the model.

080.044   Observed spatial properties of the solar eigenfunctions and the implications for the existence of resolved multiplets.
T. P. Caudell, H. A. Hill, R. J. Bos.
Solar seismology from space, p. 207 – 217 (1984). – See Abstr. 012.041.

080.045   Optimized response functions for two–dimensional observations of solar oscillations.
J. Christensen–Dalsgaard.
Solar seismology from space, p. 219 – 253 (1984). – See Abstr. 012.041.

Reasonably realistic response functions are calculated for solar oscillation observations made in Doppler velocity which is averaged over a grid of $30'' \times 30''$ pixels. In a simulation of an analysis scheme proposed for the HAO/SPO Fourier Tachometer the responses for the individual pixels are combined using Chebychev weighting functions. It is shown how a technique developed for geophysical inversion may be used to find linear combinations, concentrated within a fairly narrow range of $l-$ and $m-$values, of these transformed responses. The extension of this analysis to bidirectional observations is discussed.

080.046   The effects of image motion on the l–v diagram.
F. Hill.
Solar seismology from space, p. 255 – 262 (1984). – See Abstr. 012.041.

The author reports the results of a preliminary study of the effect of image motion produced by seeing in the Earth's atmosphere on the l–v diagram of high degree solar oscillations.

080.047   The effects of seeing on noise.
R. K. Ulrich, E. J. Rhodes Jr., A. Cacciani, S. Tomczyk.
Solar seismology from space, p. 263 – 270 (1984). – See Abstr. 012.041.

The authors discuss the effect of the supergranulation velocity field combined with seeing smearing of the solar image on the measurement of solar oscillations. Depending on the nature of the observational velocity determination scheme, the image motions can shift the background velocity pattern and produce a source of noise that reduces the quality of the observations. The authors give a rough estimate for the magnitude of this effect and present observational results which are consistent with this estimate.

080.048   The effects of a nearly 100% duty cycle on observations of solar oscillations.
F. Hill.
Solar seismology from space, p. 271 – 277 (1984). – See Abstr. 012.041.

Power spectra of window functions with duty cycles between 80% and 99% and with randomly spaced gaps are computed and their effect on observations of solar oscillations are discussed. It is found that for all the cases considered, observations of solar oscillations would not be severely impacted as long as the gap structure is random rather than periodic.

080.049   Analysis and interpretation of synthetic time strings of oscillation data.
B. W. Mihalas, J. Christensen–Dalsgaard, T. M. Brown.
Solar seismology from space, p. 279 – 292 (1984). – See Abstr. 012.041.

Artificial strings of solar oscillation data with gaps and noise, corresponding to the output of different spatial filter functions, are analyzed.

080.050   Numerical simulations of convectively excited gravity waves.
G. A. Glatzmaier.
Solar seismology from space, p. 315 – 324 (1984). – See Abstr. 012.041.

Magneto–convection and gravity waves are numerically simulated with a nonlinear, three–dimensional, time–dependent model of a stratified, rotating, spherical fluid shell heated from below.

080.051   On the influence of turbulent motions on non–radial oscillations.
B. R. Durney.
Solar seismology from space, p. 325 – 334 (1984). – See Abstr. 012.041.

The effect of turbulent motions on oscillations is studied, considering only the coupling between turbulent and oscillatory velocities. In this case, the turbulence affects the oscillations through the Reynolds stresses in the momentum equation for the pulsations. A simple model of turbulence is adopted to evaluate these Reynolds stresses and the perturbed eigenfrequencies are expressed as a function of certain averages of the turbulent velocities.

080.052   The stability of the low degree five minute solar oscillations.
R. B. Kidman, A. N. Cox.
Solar seismology from space, p. 335 – 343 (1984). – See Abstr. 012.041.

In this paper the authors discuss the decay rate for many of the low degree p modes observed as 5 minute oscillations of the sun.

080.053   An excitation mechanism for solar five–minute oscillations of intermediate and high degree.
H. M. Antia, S. M. Chitre, D. Narasimha.
Solar seismology from space, p. 345 – 348 (1984). – See Abstr. 012.041.

The overstability of acoustic modes trapped in the solar convection zone is studied with mechanical and thermal effects of turbulence included, in an approximate manner, through the eddy transport coefficients. The numerical results turn out to be in reasonable accord with the observed power–spectrum of the five–minute oscillations of intermediate and high degree.

080.054   The resonant count diagram and solar g mode oscillations.
D. B. Guenther, P. Demarque.
Solar seismology from space, p. 349 – 350 (1984). – See Abstr. 012.041.

080.055   Fine structure of solar acoustic oscillations due to rotation.
P. R. Goode, W. Dziembowski.
Solar seismology from space, p. 351 – 353 (1984). – See Abstr. 012.041.

080.056   Ionization equilibrium and equation of state in the solar interior.
F. J. Rogers.
Solar seismology from space, p. 357 – 370 (1984). – See Abstr. 012.041.

Many–body formulations of the equations of state are restated as a set of Saha–like equations. It is shown that the resulting equations are unique and convergent. These equations are similar to the usual Saha equations to the order of the Debye–Hückel theory. Higher order corrections, however, require a more general formulation. It is demonstrated that the positive free energy resulting from the interaction of unscreened particles in high orbits depletes the occupation of these states, without the introduction of shifted energy levels.

080.057   The sensitivity of solar eigenfrequencies to the treatment of the equation of state.
R. K. Ulrich, E. J. Rhodes Jr.
Solar seismology from space, p. 371 – 377 (1984). – See Abstr. 012.041.

The authors have examined the sensitivity of solar eigenfrequencies to uncertainties in the equation of state. The principal uncertainties in the equation of state involve the treatment of pressure ionization, the Debye–Hückel coulomb corrections and the treatment of many–particle interaction effects.

080.058   The chromospheric brightness oscillation spectrum from filter observations in the Hβ–line.
V. E. Merkulenko, M. N. Mishina.
Astron. Astrophys., Vol. 146, No. 1, p. L9 – L10 (1985).

Using filtergrams taken in the wing of the Hβ–line the authors have detected a spectrum of brightness oscillations in the chromosphere near the solar limb. The spectrum clearly shows the following periods: 77, 45, 32, 26, 22, 15, 9.2, and 3.0 minutes. The sequence of these periods is in best agreement with the p–modes of radial pulsation of the Sun for the 1968 non–standard solar model by Rouse (1977).

080.059   Oscillations of the Sun's chromosphere. III. Simultaneous Hα observations from two sites.
M. von Uexküll, K. Kneer, W. Mattig, A. Nesis, W. Schmidt.
Astron. Astrophys., Vol 146, No. 1, p. 192   194 (1985). = Mitt. Kiepenheuer–Inst. Nr. 244.

The authors analyze time sequences of Hα filtergrams taken simultaneously from two distant observatories, Capri and Izaña. By means of a coherence analysis the authors discriminate between instrumental effects including seeing and truly solar intensity fluctuations. Waves with periods as short as 60 s are present in the solar chromosphere; the lower limit is set by the time resolution of the observations.

080.060   A multimode investigation of granular and supergranular motions. I. Boussinesq model.
R. Van der Borght, P. Fox.
The Big Bang and Georges Lemaitre, p. 269 – 276 (1984). – See Abstr. 012.043.

Using a Boussinesq model this paper studies the interacting convective motions in the solar atmosphere, which give rise to granulation and supergranulation.

080.061   Plasma and nuclear processes in solar matter.
G. E. Kocharov.
ESA Spec. Publ., ESA SP–207, p. 65 – 81 (1984). – See Abstr. 012.044.

A brief overview is given on some plasma and nuclear processes in the interior of the Sun and solar atmosphere. A serious discrepancy between the standard theory and experiment on detecting neutrinos, born deep in the interior of the Sun, is discussed mainly from plasma and nuclear processes point of view. The problem of particle and photon generation is considered based on the experimental results of simultaneous registration of different particles and radiations. The main attention is paid to the questions of greatest interest: where the particle acceleration occurs; what are the main characteristics of the solar flare accelerator; what are the physical conditions in the solar plasma before the flare, during and after it.

080.062   Changes of solar luminosity and radius following secular perturbations in the convective envelope.
A. S. Endal, S. Sofia, L. W. Twigg.
Astrophys. J., Vol. 290, No. 2, p. 748 – 757 (1985).

The authors have computed the nonadiabatic response of the Sun to a variety of spherically symmetric, secular perturbations, acting at various depths in the convective envelope. In the superadiabatic layers near the surface, small changes in the efficiency of convective energy transport produce substantial luminosity changes, but negligible changes of the total radius. Below the superadiabatic layer, perturbations of the (mixing length) convection equations have little effect on the models. The effects of perturbing the total pressure and the internal energy at various

depths and the influence of magnetic fields concentrated near the interface between the radiative core and the convective envelope are also calculated.

080.063   Detection and classification of resolved multiplet members of the solar 5 minute oscillations through solar diameter–type observations.
H. A. Hill.
Astrophys. J., Vol. 290, No. 2, p. 765 – 781 (1985).

Individual modes of low–degree 5 minute oscillations have been identified in solar diameter observations. These modes have n–values in the range $12 \leqslant n \leqslant 27$ and l–values in the range $0 \leqslant l \leqslant 6$, where n and l represent the radial order and the spherical harmonic degree of the eigenfunction respectively. In total, 184 modes belonging to 83 multiplets have been resolved and classified. The study of these modes has been accomplished without superposed–frequency analysis. The first–order rotational splitting in m is derived. The observed width of the individual modes is ≈ 1 µHz.

080.064   Solar oscillations.
J. Christensen–Dalsgaard.
ESA Spec. Publ., ESA SP–220, p. 3 – 12 (1984). – See Abstr. 012.045.

080.065   Solar oscillations from two widely separated stations.
G. J. Bamford, H. B. van der Raay, P. L. Palle, T. Roca Cortes.
ESA Spec. Publ., ESA SP–220, p. 13 – 16 (1984). – See Abstr. 012.045.

The analysis of data obtained from Hawaii and Tenerife yield details of the low l 5 minute modes. Not only is almost 24 hour coverage possible but also the correlation of the two independent data sets is invaluable in establishing the reality of any observed phenomenon. Possible evidence for the coherence, decay, excitation and splitting of the prominent lines is found. Preliminary experiments on the selection of sectoral modes by spatial filters are discussed.

080.066   Global solar oscillations in irradiance and velocity: a comparison.
C. Fröhlich, H. B. van der Raay.
ESA Spec. Publ., ESA SP–220, p. 17 – 20 (1984). – See Abstr. 012.045.

The results of 7 hours of solar spectral irradiance measurements from a stratospheric balloon (1983) are analysed for frequencies between 1 ad 5 mHz. Several significant peaks of up to 6 ppm in the red (778 nm) and of up to 15 ppm in the blue (380 nm) can be identified as 5–minute p–mode oscillations of the Sun. From bivariate time series analysis a more or less constant relative phase close to zero and an amplitude ratio between 1.5 and 3.2 with a maximum around 3 mHz is found. By comparing the spectra with velocity measurements a phase difference between the velocity and irradiance signals of 120 to 180° is found and amplitude ratios of 15 and 50 ppm/m s^{-1} for the red and blue respectively.

080.067   Mixing in the solar interior.
H. C. Spruit.
ESA Spec. Publ., ESA SP–220, p. 21 – 27 (1984). – See Abstr. 012.045.

It seems appropriate to study in detail which mixing processes can occur in a star, and why the effective diffusivity resulting from these processes should be a weakly varying multiple of the kinematic viscosity. The author reviews the extent to which these questions can be answered at present.

080.068   On the determination of the solar rotation and indications of the solar differential rotation from an analysis of solar integrated light.
T. Drescher, H. Wöhl, G. Küveler.
ESA Spec. Publ., ESA SP–220, p. 29 – 32 (1984). – See Abstr. 012.045.

The rotation modulation of the emission measure of the Ca$^+$K line within the integrated solar light is used to determine the solar rotation period. Attempts are made to determine changes of this period due to the differential rotation and the different latitude regions of the activity.

080.069   Models of solar differential rotation.
R. M. Pidatella, M. Stix, G. Belvedere, L. Paternó.
ESA Spec. Publ., ESA SP–220, p. 33 – 35 (1984). – See Abstr. 012.045.

The models presented give the dependence of the angular velocity at the Sun's equator on the depth of the convection zone. In order to drive differential rotation two different mechanisms are considered: latitude dependence of the mixing length and the anisotropy of the turbulent viscosity. The influence of boundary conditions at the bottom of the convection zone on the models are also studied.

080.070   Magnetoconvection: the interaction of convection and small scale magnetic fields.
Å. Nordlund.
ESA Spec. Publ., ESA SP–220, p. 37 – 46 (1984). – See Abstr. 012.045.

Present consensus about the interaction of convection and magnetic fields is reviewed, with special emphasis on the interaction of convection in the solar convection zone with magnetic fields on the scale of granulation and supergranulation. Computer simulations of Boussinesq systems are reviewed, and additional effects due to stratification and radiation are discussed. The particular mechanisms that determine the degree of evacuation of magnetic fields in the solar photosphere are discussed, and some limitations of the "flux tube" concept are pointed out. Finally, some recent results from numerical simulations of convection in a stratified medium with magnetic fields are discussed, and possible consequences for activity in the chromosphere and corona are mentioned.

080.071   Non–linear dynamos.
A. A. Ruzmaikin (*A. A. Ruzmajkin*).
ESA Spec. Publ., ESA SP–220, p. 85 – 89 (1984). – See Abstr. 012.045.

The basic large–scale features of solar activity are explained by dynamo theory in terms of mean–field magnetohydrodynamics. The helicity dependence on the magnetic field results in stabilization of the basic 22–year oscillations. The excitation of two modes (of dipole and quadrupole symmetry) can create non–linear beats which may explain the secular (60 – 80 years) modulation. The observed torsional waves can be interpreted as a back action of the dynamo waves on the differential rotation. The actual variations in solar activity are non–periodic. There are random recurrent episodes of reduced activity like the Maunder Minimum. The non–linear dynamo models demonstrating the oscillatory–chaotic behaviour are discussed.

080.072   The solar dynamo and the complex Lorenz equations.
C. A. Jones.
ESA Spec. Publ., ESA SP–220, p. 91 – 95 (1984). – See Abstr. 012.045.

Nonlinear mechanisms in the dynamos associated with the sun and other lower main sequence stars are discussed in the light of truncated model dynamo equations. The predicted relationships between rotation rate, field strength and cycle frequency are examined; the time–series derived from these model equations is examined, and compared with time–series coming from stochastic models.

080.073   Solar and stellar magnetic cycles: unity and multiplicity.
G. Belvedere.
ESA Spec. Publ., ESA SP–220, p. 101 – 105 (1984). – See Abstr. 012.045.

The analysis of the characteristics of stellar cycles and their dependence on stellar parameters as the mass, the convection zone depth, the rotation rate and the age should highlight the dynamo operation modes and, in particular, the large and small time scale dependence of the solar cycle, in the framework of the mutual interaction between the linear and non–linear approachs.

080.074   The internal rotation of the Sun inferred from the rotational splitting of acoustic and gravity mode multiplets.
H. A. Hill, R. D. Rosenwald, D. S. Yakowitz, W. Campbell.
ESA Spec. Publ., ESA SP–220, 187 – 188 (1984). – See Abstr. 012.045.

Measures of the equatorial surface frequency and measures of rotational splitting from 100 nonradial multiplets of acoustic modes and internal gravity modes are used to infer the internal rotation of the sun. It was found that the Ω rises $\sim 30\%$ just below the surface and is proportional to $r^{-1.5}$ in the convection zone. The Ω then makes a sharp transition to $\sim 3 \,\mu Hz$ at the bottom of the convection zone and decreases by $\sim 8\%$ in going from the convection zone to the core.

080.075   Ten years observations of long–period oscillations of the Sun.
V. A. Kotov, A. B. Severny (*A. B. Severnyj*), T. T. Tsap.
ESA Spec. Publ., ESA SP–220, 189 – 190 (1984). – See Abstr. 012.045.

The measurements of Doppler shift of a Fraunhofer spectral line were performed over the last 10 years. The power spectrum of the data shows a major peak at 160^m period which is interpreted as multiple resonance of gravity modes of the Sun. This 160^m oscillation reveals remarkable phase stability during the entire decade.

080.076   Conditions of the small oscillations excitation in the different solar models with the mixed core.
W. G. Gavryusev (*V. G. Gavryusev*), E. A. Gavryuseva.
ESA Spec. Publ., ESA SP–220, p. 197 – 198 (1984). – See Abstr. 012.045.

The growth rates of dipole and quadrupole oscillations have been calculated in quasi–adiabatic approximation in different solar models with mixed core. The oscillation is great inside the core when the node of the radial displacement is on the mixed core boundary. This can help to understand the peculiarity of the 160–min oscillation in a solar model with mixed core.

080.077   Cyclic flows in the solar convection zone due to the dynamo reaction.
I. Tuominen, H. Virtanen, F. Krause, G. Rüdiger.
ESA Spec. Publ., ESA SP–220, p. 225 – 226 (1984). – See Abstr. 012.045.

080.078   Effects of atmospheric stratification on dissipative acoustic and Alfvén waves.
L. M. B. C. Campos.
ESA Spec. Publ., ESA SP–220, p. 253 – 255 (1984). – See Abstr. 012.045.

The author considers the two simplest wave modes in a compressible atmosphere, in the presence of a magnetic field, namely, acoustic– and Alfvén–gravity waves, with dissipation respectively by fluid viscosity and electrical resistance.

080.079   Solar magnetohydrodynamics; Symposium review.
C. de Jager.
ESA Spec. Publ., ESA SP–220, p. 287 – 290 (1984). – See Abstr. 012.045.

080.080   Global oscillations of the sun.
L. Paternó.
Mem. Soc. Astron. Ital., Vol. 55, No. 3, p. 537 – 542 (1984). – See Abstr. 012.050.

080.081   Torsional 11–year oscillations of the sun.
A. Z. Dolginov, A. G. Muslimov.
Pis'ma Astron. Zh., Tom 11, No. 4, p. 278 – 285 (1985). In Russian. English translation in Sov. Astron. Lett., Vol. 11.

The possibility is considered that the observed 11–year torsional oscillations of the sun may be primary eigen–mode oscillations of the surface layers excited by convection. Estimates of both frequency and amplitude of the torsional wave are obtained. The magnetic manifestation of the solar cycle probably results from the modulation (by torsional oscillations) of the field generation process.

080.082 Solar torsional oscillations: a net pattern with wavenumber 2 as artifact.
H. B. Snodgrass.
Astrophys. J., Vol. 291, No. 1, p. 339 – 343 (1985).
A net solar torsional oscillation pattern is uncovered through a new analysis of Mount Wilson Doppler data. This pattern, found from zonal fits, without subtraction of a global fit, consists of a relative polar spin–up around solar maximum, alternating with a single traveling wave that runs from mid latitude to low latitude during the rest of the cycle. It is suggested that these are separate phenomena, and thus that the previously inferred pole-to–equator traveling pattern with wavenumber 2 per hemisphere may be a mathematical artifact.

080.083 Completely convective model of the sun.
A. D. Chertkov.
Magnitosfer. issled., Moskva, No. 3, p. 98 – 114 (1984). In Russian. Abstr. in Ref. Zh., 51. Astron., 4.51.383 (1985).

080.084 Study of rapidly growing overstability which can be excited within the solar core.
Yu. V. Vandakurov.
Soln. Dannye, Byull., 1985, No. 1, p. 83 – 87 (1985). In Russian.
Preliminary data are given of a numerical study of nonadiabatic stability of a solar model with a marginally stable core. In the case where the superadiabaticity in the core is sufficiently large, sudden excitation of a rapidly growing overstability is possible.

080.085 The stability of solar gravity–mode oscillations and the structure of the sun.
A. G. Kosovichev, A. B. Severnyj.
Sov. Astron. Lett., Vol. 10, No. 5, p. 284 – 286 (1984). English translation of 38.080.059.

080.086 On the reliability of observed oscillations of the solar diameter.
V. F. Chistyakov.
Soln. Dannye, Byull., 1985, No. 2, p. 65 – 71 (1985). In Russian.
The following criteria for the reliability of long–term time series of the solar diameter as obtained from meridian circle observations are given: 1) the solar diameter varies constantly, while the diameter of Venus stays constant for the same time period, 2) similarity of the run of long–term series of the solar diameter variations from observations of a number of astronomers simultaneously with one and the same telescope, vertical and horizontal diameter variations from observations with one and the same telescope, simultaneous observations of the diameter of the sun at various observatories, 3) correlation of oscillations of the solar diameter and solar activity.

080.087 Can mirage phenomena be traced on the Sun?
Y. Öhman.
Sol. Phys., Vol. 96, No. 1, p. 209 – 212 (1985).
It is suggested, that effects of total reflection and refraction may appear sometimes in various objects on the Sun in connection with grazing incidence towards layers formed by magnetic fields and perhaps also by electrostatic double layers or shock waves. It seems possible that such effects may influence sometimes the contrast of 'bright' structures as well as 'dark' structures at selective wavelengths as well as in white light.

080.088 Lithium depletion during the contraction phase of the Sun.
N. Kiziloğlu, D. Eryurt–Ezer.
Astron. Astrophys., Vol. 146, No. 2, p. 384 – 386 (1985).
Convective overshooting, which can extend the mixing region in proportion to the depth of the convective zone during evolution, can explain the observed depletion of lithium in the surface layers of the Sun. However, it can be inferred from Li observations (Skumanich, 1972) that strong overshooting does not exist in stars of age 10^8 to 10^9yr.

080.089 Sur la variabilité de la constante solaire.
C. de Charentenay, S. Koutchmy.
Inst. Astrophys. Paris, Pré–Publ., No. 92, 12 pp. (1985). To appear in C. R. Acad. Sci., Sér. II.

080.090 A search for long–lived velocity fields at the solar poles.
B. R. Durney, L. E. Cram, D. B. Guenther, S. L. Keil, D. M. Lytle.
Astrophys. J., Vol. 292, No. 2, p. 752 – 762 (1985).
A search has been made in the polar regions of the Sun for large–scale (50 – 200 Mm) velocity fields with lifetimes of the order of the solar rotation period ($\gtrsim 30$ days). The observations show that any such large–scale, long–lived velocity patterns in the polar regions must have an amplitude less than 5 m s^{-1}. Marginally significant detections (at the $2 - 3$ σ level) were made of two kinds of structures with amplitudes of order 3 m s^{-1}. One has a rotation period ~ 38 days and a scale ~ 150 Mm; the other has a period ~ 24 days and a scale ~ 100 Mm.

080.091 Chlorine and gallium solar neutrino experiments.
J. N. Bahcall, B. T. Cleveland, R. Davis Jr., J. K. Rowley.
Astrophys. J., Lett. Ed., Vol. 292, No. 2, p. L79 – L82 (1985).
The authors reevaluate the expected capture rates and their uncertainties for the chlorine and gallium solar neutrino experiments using improved laboratory data and new theoretical calculations. They also derive a minimum value for the flux of solar neutrinos that is expected provided only (1) that the sun is currently producing energy by fusing light nuclei at the rate that it is emitting energy in the form of photons from its surface and (2) that nothing happens to solar neutrinos on their way to earth. These results are used – together with Monte Carlo simulations – to determine how much gallium is required for a solar neutrino experiment.

080.092 Solar oscillations, sunspot cycles, and climatic change.
T. Landscheidt.
Weather and climate responses to solar variations, p. 293 – 308 (1983). – See Abstr. 012.086.

080.093 Solar irradiance and g–modes.
C. L. Wolff, J. R. Hickey.
Bull. Am. Astron. Soc., Vol. 16, No. 4, p. 978 (1984). Abstract. – See Abstr. 010.062.

080.094 Solar oscillations observations with the Fourier tachometer II.
T. M. Brown.
Bull. Am. Astron. Soc., Vol. 16, No. 4, p. 978 (1984). Abstract. – See Abstr. 010.062.

080.095 Absolute torsional oscillations of the sun.
H. B. Snodgrass.
Bull. Am. Astron. Soc., Vol. 16, No. 4, p. 978 (1984). Abstract. – See Abstr. 010.062.

080.096 Does the sun really have small, high–Z core?
C. A. Rouse.
Bull. Am. Astron. Soc., Vol. 16, No. 4, p. 978 – 979 (1984). Abstract. – See Abstr. 010.062.

080.097 Observations of solar velocity fields with large–format CCD cameras at the Mount Wilson Observatory.
E. J. Rhodes Jr., A. Cacciani, S. Tomczyk, R. K. Ulrich, P. Dumont, R. F. Howard.
Bull. Am. Astron. Soc., Vol. 16, No. 4, p. 979 (1984). Abstract. – See Abstr. 010.062.

080.098 The role of non–classical transport in the formation of the Ly–α temperature plateau.
S. P. Owocki, R. C. Canfield, A. N. McClymont.
Bull. Am. Astron. Soc., Vol. 16, No. 4, p. 992 (1984). Abstract. – See Abstr. 010.062.

080.099   The scale height of the latitude dependent term of the sun's internal rotation.
R. D. Rosenwald, H. A. Hill, D. S. Yakowitz.
Bull. Am. Astron. Soc., Vol. 16, No. 4, p. 999 – 1000 (1984). Abstract. – See Abstr. 010.062.

080.100   Solar diameter–type observations of the 160 min solar oscillation: detection and interpretation.
H. A. Hill, J. Tash.
Bull. Am. Astron. Soc., Vol. 16, No. 4, p. 1000 (1984). Abstract. – See Abstr. 010.062.

080.101   Episodic mixing and solar oscillations.
R. B. Kidman, A. N. Cox.
Bull. Am. Astron. Soc., Vol. 16, No. 4, p. 1000 (1984). Abstract. – See Abstr. 010.062.

080.102   Inertial oscillations in the solar convection zone. Part I.
D. B. Guenther, P. A. Gilman.
Bull. Am. Astron. Soc., Vol. 16, No. 4, p. 1000 (1984). Abstract. – See Abstr. 010.062.

080.103   Inertial oscillations in the solar convection zone. Part II: A cylindrical model for equatorial regions.
P. A. Gilman, D. B. Guenther.
Bull. Am. Astron. Soc., Vol. 16, No. 4, p. 1000 – 1001 (1984). Abstract. – See Abstr. 010.062.

080.104   New observations of solar oscillations at SCLERA.
W. M. Czarnowski, L. Yi, B. J. Beardsley, H. A. Hill, T. P. Caudell.
Bull. Am. Astron. Soc., Vol. 16, No. 4, p. 1001 (1984). Abstract. – See Abstr. 010.062.

080.105   Comparing α Cen with the sun: chemical composition, mixing–length, age, and p–mode oscillation spectrum.
P. Demarque, D. B. Guenther.
Bull. Am. Astron. Soc., Vol. 16, No. 4, p. 1001 (1984). Abstract. – See Abstr. 010.062.

080.106   The role of diamagnetic material in the solar atmosphere.
P. J. Cargill, K. B. MacGregor, G. W. Pneuman.
Bull. Am. Astron. Soc., Vol. 16, No. 4, p. 1005 (1984). Abstract. – See Abstr. 010.062.

080.107   New insights into the solar–stellar connection.
T. R. Ayres.
Bull. Am. Astron. Soc., Vol. 16, No. 4, p. 1011 (1984). Abstract. – See Abstr. 010.062.

080.108   Speed of sound in the solar interior.
J. Christensen–Dalsgaard, T. L. Duvall Jr., D. O. Gough, J. W. Harvey, E. J. Rhodes Jr.
Nature, Vol. 315, No. 6018, p. 378 – 382 (1985).
Frequencies of solar 5–min oscillations can be used to determine directly the sound speed of the solar interior. The determination described here does not depend on a solar model, but relies only on a simple asymptotic description of the oscillations in terms of trapped acoustic waves.

080.109   Solar constant variabilities.
C. de Charentenay, S. Koutchmy.
C. R. Acad. Sci., Sér. II, Tome 301, No. 3, p. 151 – 156 (1985). In French.
The solar constant variabilities, as observed aboard SMM with the ACRIM experiment, are analyzed using a weekly average. To reproduce the measured variations, both radiative deficit produced by sunspots and radiative excess produced by faculae are introduced on a weekly basis. International sunspot numbers and Meudon Observatory synoptic maps are used. For long term variation the proposed model predicts a relative decrease of the solar constant at the time of the sunspot activity minimum.

080.110   Time variations of the solar neutrino flux.
H. J. Haubold, E. Gerth.
Astrophys. Space Sci., Vol. 112, No. 2, p. 397 – 405 (1985).
The Fourier analysis of the argon–37 production rate for runs 18 – 80 observed in Davis's well–known solar neutrino experiment is presented. The method of Fourier analysis with the unequally–spaced data of Davis and associates is described and the discovered periods are compared with recently published results for the analysis of the data of runs 18 – 69. The harmonic analysis of the data of runs 18 – 80 shows time variations of the solar neutrino flux with periods $\pi = 8.33, 5.26, 2.13, 1.56, 0.83, 0.64, 0.54,$ and 0.50 yr, respectively, which confirms earlier computations.

080.111   On a correlation of fast oscillations of rotation of the northern and southern hemispheres of the sun.
L. M. Kornienko, V. F. Chistyakov.
Investigation of magnetic fields and active formations on the sun, p. 3 – 9 (1984). In Russian. Abstr. in Ref. Zh., 51. Astron., 5.51.313 (1985). – See Abstr. 003.022.

080.112   The problem of the backward current in heating of the solar atmosphere by accelerated electrons.
V. A. Sermulyn'sh, B. V. Somov.
Complex investigations of the sun, p. 90 – 95 (1982). In Russian. Abstr. in Ref. Zh., 51. Astron., 5.51.327 (1985). – See Abstr. 012.046.

080.113   Neutrino activity of the sun.
Yu. S. Kopysov.
Chastitsy i kosmologiya. Chast' 1. Moskva, p. 82 – 101 (1984). In Russian. Abstr. in Ref. Zh., 51. Astron., 6.51.331 (1985).

080.114   Solar p–mode eigenfrequencies are decreased by turbulent convection.
T. M. Brown.
Science, Vol. 226, No. 4675, p. 687 – 689 (1984).
Average solar p–mode eigenfrequencies are decreased by large fluctuating velocity fields in the upper convection zone. This effect is greatest for modes with large horizontal wave numbers and frequencies. It is large enough to affect estimates of the depth of the convection zone and may carry useful information about the structure of solar convective turbulence.

080.115   Excitation of rapidly growing overstability in the solar core.
Yu. V. Vandakurov.
Sov. Astron. Lett., Vol. 10, No. 6, p. 365 – 369 (1984). English translation of 38.080.075.

080.116   The nature of the 11–year solar torsional oscillations.
N. I. Kliorin, A. A. Ruzmajkin.
Sov. Astron. Lett., Vol. 10, No. 6, p. 390 – 392 (1984). English translation of 38.080.118.

080.117   What accelerator mass spectrometry can do for solar physics.
G. Newkirk Jr.
Nucl. Instrum. Methods Phys. Res., Sect. B, Vol. 233, No. 2, p. 404 – 410 (1984). Abstr. in Phys. Abstr., Vol. 88, No. 1253, Entry 35556 (1985). – See Abstr. 012.077.

080.118   Solar neutrinos: prospects for detection and implications.
W. C. Haxton.
Proceedings of the XIth International Conference on Neutrino Physics and Astrophysics, p. 214 – 228 (1984). – See Abstr. 012.081.
The nuclear physics, particle physics, and astrophysics motivations for completing a program of solar neutrino spectroscopy are reviewed.

080.119 Is iron soluble in the Sun?
G. Dávid, G. Marx, I. Ruff.
Proceedings of the XIth International Conference on Neutrino Physics and Astrophysics, p. 254 – 255 (1984). – See Abstr. 012.081.

In the solar interior the hydrogen and iron ions carry very different charges what results in separation of a hydrogen rich and an iron rich phase. The low concentration of iron in the hydrogen rich phase produces small opacity and low central temperature. This thermodynamical conclusion offers a straightforward explanation for the absence of boron neutrinos at the chlorine experiment and predicts a positive result for the gallium experiment.

080.120 Solar neutrino spectroscopy.
W. C. Haxton.
AIP Conf. Proc., No. 123, p. 1026 – 1036 (1984). Abstr. in Phys. Abstr., Vol. 88, No. 1256, Entry 46784 (1985). – See Abstr. 012.082.

080.121 Report on solar neutrino experiments.
R. Davis Jr, B. T. Cleveland, J. K. Rowley.
AIP Conf. Proc., No. 123, p. 1037 – 1050 (1984). Abstr. in Phys. Abstr., Vol. 88, No. 1256, Entry 46785 (1985). – See Abstr. 012.082.

080.122 Solar evolution and solar neutrinos.
N. Kiziloğlu, D. Eryurt–Ezer.
METU (Middle East Technical University, Ankara, Turkey) Journal of Pure and Applied Sciences, Vol. 16, No. 2/3, p. 217 – 231 (1983).

The evolutionary scenario of the sun is followed up to its present age by constructing standard and contaminated solar models. Both evolutionary sequences start from the point where collapsing protosun becomes stable against further dynamical collapse. For the contaminated solar models, the surface is assumed to accrete material from the interstellar medium during the evolution at a rate to increase the heavy element abundance to its present observed value. The predicted total solar neutrino fluxes are 3.48 SNU and 4.68 SNU for accreted and standard solar models, respectively.

080.123 A possible periodic change of the neutrino flux from the sun.
K. Sakurai.
18th International Cosmic Ray Conference, Vol. 4, p. 210 – 213 (1983). – See Abstr. 012.096.

Neutrinos from the ^{8}B–decay process have been observed since 1970 by Davis et al. Examining the observed results on their flux at the earth, it is shown that this flux is varying with a period of about twenty–six months. This periodic change of the flux may reflect upon the quasi–biennial change in some physical processes inside the sun.

080.124 p–p chain non–resonant reactions with a quasi non–degenerate distribution.
S. A. S. Ahmed, H. D. Goswami.
18th International Cosmic Ray Conference, Vol. 4, p. 214 – 217 (1983). – See Abstr. 012.096.

A quasi non–degenerate distribution is designed primarily to account for ^{37}Cl neutrino counting rate. The authors investigate some p–p chain reactions using the new distribution and compare them with the Maxwell–Boltzmann results.

080.125 Cosmic ray and neutrino flux in Davis' experiment.
G. A. Bazilevskaya, S. I. Nikolskii (*S. I. Nikol'skij*), Yu. I. Stozhkov, T. N. Charakhch'yan, A. M.–A. Mukhamedzhanov.
18th International Cosmic Ray Conference, Vol. 4, p. 218 – 221 (1983). – See Abstr. 012.096.

An important correlation is found between the neutrino flux ν in Davis' installation and galactic cosmic rays. The ν flux seems also to be generated during the large solar cosmic ray events.

080.126 Detection of solar flare neutrinos.
I. N. Erofeeva, S. I. Lyutov, V. S. Murzin, E. V. Kolomeets, J. Albers, P. Kotzer.
18th International Cosmic Ray Conference, Vol. 7, p. 104 – 107 (1983). – See Abstr. 012.096.

The aim of this work is to estimate possible neutrino fluxes from solar flares on the opposite side of the Sun.

080.127 Solar neutrinos, solar activity cycle and quasibiennial variation.
P. Raychaudhuri.
18th International Cosmic Ray Conference, Vol. 7, p. 108 – 111 (1983). – See Abstr. 012.096.

From the observed solar neutrino flux the author has shown that an annual, 11 year quasibiennial variation (nonstationary) of solar neutrino flux exists. It is pointed out that the appearance of quasibiennial variation is due to the interaction of solar core to the solar surface layers.

080.128 Solar particle corotating events from shocked neutral sheets.
D. J. Mullan, J. Pérez–Peraza, M. Gálvez, M. Alvarez–Madrigal.
18th International Cosmic Ray Conference, Vol. 10, p. 307 – 310 (1983). – See Abstr. 012.096.

The authors study the effect of a travelling shock wave on a magnetic neutral sheet. Specifically, they consider the interaction between a flare shock and a large coronal neutral sheet, such as that which is associated with a helmet–streamer.

080.129 Short–time variations of the solar neutrino luminosity.
H. J. Haubold, E. Gerth.
18th International Cosmic Ray Conference, Vol. 10, p. 389 – 392 (1983). – See Abstr. 012.096.

The authors confirm results obtained by Sakurai (1979) that the observed solar neutrino flux has a tendency to vary with quasi–biennial periodicity.

080.130 The influence of convection zone flows on magnetic fields in the sun.
K. H. Schatten, H. G. Mayr, I. Harris.
Bull. Am. Astron. Soc., Vol. 17, No. 2, p. 610 (1985). Abstract. – See Abstr. 010.065.

080.131 Viscosity losses in the solar convective zone.
J. B. Blizard.
Bull. Am. Astron. Soc., Vol. 17, No. 2, p. 610 (1985). Abstract. – See Abstr. 010.065.

080.132 Solar radius variations determined from eight solar eclipses, 1715 – 1984.
A. D. Fiala, D. W. Dunham, J. B. Dunham, S. Sofia.
Bull. Am. Astron. Soc., Vol. 17, No. 2, p. 624 (1985). Abstract. – See Abstr. 010.066.

080.133 Nonlocal thermal transport in the solar atmosphere.
J. T. Karpen, C. R. DeVore.
Bull. Am. Astron. Soc., Vol. 17, No. 2, p. 630 (1985). Abstract. – See Abstr. 010.067.

080.134 Solar meridional flow during 1982 – 1984.
J. B. Zirker, R. F. Howard.
Bull. Am. Astron. Soc., Vol. 17, No. 2, p. 634 (1985). Abstract. – See Abstr. 010.067.

080.135 Dynamical linear instability of solar mass ejection.
M. T. Song, S. T. Wu.
Bull. Am. Astron. Soc., Vol. 17, No. 2, p. 635 (1985). Abstract. – See Abstr. 010.067.

080.136 The detection of global (giant cell) convective wave flows on the sun.
H. Yoshimura, P. H. Scherrer, R. S. Bogart, J. T. Hoeksema.
Bull. Am. Astron. Soc., Vol. 17, No. 2, p. 639 (1985). Abstract. – See Abstr. 010.067.

080.137   Observations of low–degree p–mode oscillations in 1984.
H. M. Henning, P. H. Scherrer.
Bull. Am. Astron. Soc., Vol. 17, No. 2, p. 639 (1985). Abstract. –
See Abstr. 010.067.

080.138   Comparison of the differential Doppler shift and differential radius observations of long period solar oscillations.
H. A. Hill.
Bull. Am. Astron. Soc., Vol. 17, No. 2, p. 639 (1985). Abstract. –
See Abstr. 010.067.

080.139   Inversion techniques for helioseismology.
W. Jeffrey, R. Rosner.
Bull. Am. Astron. Soc., Vol. 17, No. 2, p. 639 (1985). Abstract. –
See Abstr. 010.067.

080.140   Solar irradiance variations derived from Mt. Wilson Observatory daily magnetograms.
G. A. Chapman, J. E. Boyden.
Bull. Am. Astron. Soc., Vol. 17, No. 2, p. 640 (1985). Abstract. –
See Abstr. 010.067.

080.141   Alfvénic pulses in the solar atmosphere.
J. T. Mariska, J. V. Hollweg.
Bull. Am. Astron. Soc., Vol. 17, No. 2, p. 643 (1985). Abstract. –
See Abstr. 010.067.

080.142   Changes in subsurface horizontal velocities inferred from observations of high degree 5–minute solar oscillations.
F. Hill, J. Toomre, D. O. Gough.
Bull. Am. Astron. Soc., Vol. 17, No. 2, p. 643 (1985). Abstract. –
See Abstr. 010.067.

080.143   Deconvolution methods for analysis of solar oscillation observations.
P. H. Scherrer.
Bull. Am. Astron. Soc., Vol. 17, No. 2, p. 643 (1985). Abstract. –
See Abstr. 010.067.

080.144   Amplitude ratio of solar p–mode intensity and Doppler oscillations.
T. L. Duvall Jr., J. W. Harvey, M. A. Pomerantz.
Bull. Am. Astron. Soc., Vol. 17, No. 2, p. 643 (1985). Abstract. –
See Abstr. 010.067.

080.145   An upper limit on the size of giant cells.
A. A. van Ballegooijen.
Bull. Am. Astron. Soc., Vol. 17, No. 2, p. 643 – 644 (1985). Abstract. – See Abstr. 010.067.

080.146   On the generalization of the mixing length theory to rotating convection zones.
B. R. Durney.
Bull. Am. Astron. Soc., Vol. 17, No. 2, p. 644 (1985). Abstract. –
See Abstr. 010.067.

080.147   Variations in the solar rotation rate derived from Ca⁺ K plage areas.
J. Singh, T. P. Prabhu.
Sol. Phys., Vol. 97, No. 1, p. 203 – 212 (1985).
Daily calcium plage areas for the period 1951 – 1981 (which include the solar cycle 19 and 20) have been used to derive the rotation period of the Sun at latitude belts $10 - 15°N$, $15 - 20°N$, $10 - 15°S$, and $15 - 20°S$ and also for the entire visible solar disk. The mean rotation periods derived from $10 - 20°S$ and N, total active area and sunspot numbers were 27.5, 27.9, and 27.8 days (synodic), respectively. A power spectral analysis of the derived rotation rate as a function of time indicates that the rotation rate in each latitude belt varies over time scales ranging from the solar activity cycle, down to about 2 years. The rotation rates derived from sunspot numbers also behave similarly though the dependence over the solar cycle is not very apparent.

080.148   Torsional oscillations of the sun.
H. B. Snodgrass, R. Howard.
Science, Vol. 228, No. 4702, p. 945 – 952 (1985).
The sun's differential rotation has a cyclic pattern of change that is tightly correlated with the sunspot, or magnetic activity, cycle. This pattern can be described as a torsional oscillation, in which the solar rotation is periodically sped up or slowed down in certain zones of latitude while elsewhere the rotation remains essentially steady. The zones of anomalous rotation move on the sun in wavelike fashion, keeping pace with and flanking the zones of magnetic activity. It is uncertain whether this torsional oscillation is a globally coherent ringing of the sun of whether it is a local pattern caused by and causing local changes in the magnetic fields. In either case, it may be an important link in the connection between the rotation and the cycle that is widely believed to exist but is not yet understood.

080.149   Report of IAU Commission 12: Radiation and structure of the solar atmosphere (*Radiation et structure de l'atmosphère solaire*).
R. W. Noyes.
Trans. IAU, Vol. XIXA, p. 97 – 119 (1985). – See Abstr. 003.046.

080.150   Predicted solar neutrino rates in the Q–nuclear solar model.
B. Sur, R. N. Boyd.
Phys. Rev. Lett., Vol. 54, No. 5, p. 485 – 487 (1985).
It has been shown previously that the existence of a tiny abundance of nuclei which have an additional embedded hadronic particle, Q–nuclei, can solve the solar–neutrino problem. The authors present detailed predictions of solar–neutrino detection rates for detectors of ^{37}Cl, ^{71}Ga, and ^{115}In. It is found that, while Q–nuclei could reduce the ^{37}Cl detection rate from that of the standard model to the experimental value, they would also produce a dramatic increase in the rates for the ^{71}Ga and ^{115}In detectors.

080.151   The shape of the sun.
K. G. Libbrecht.
Diss. Abstr. Int., Sect. B, Vol. 45, No. 1, p. 230 (1984). Thesis, Princeton University, 143 pp. (1984). Order No. DA8409631.

080.152   Resonant three–wave interactions and an application to the 160 minute solar oscillation.
D. B. Guenther.
Diss. Abstr. Int., Sect. B, Vol. 45, No. 3, p. 900 – 901 (1984). Thesis, Yale University, 136 pp. (1983). Order No. DA8413087.

080.153   Nonlocal and nonlinear effects on solar oscillations.
J. D. Logan.
Diss. Abstr. Int., Sect. B, Vol. 45, No. 8, p. 2581 (1985). Thesis, University of Arizona, 456 pp. (1984). Order No. DA8424907.

080.154   Short–period oscillations in the total solar irradiance.
M. F. Woodard.
Diss. Abstr. Int., Sect. B, Vol. 45, No. 8, p. 2582 (1985). Thesis, University of California (1984).

The summer 1984 solar oscillation program of the Mount Wilson 60–foot solar telescope.
See Abstr. 013.067.

Report of ESA's topical team on solar and heliospheric physics.
See Abstr. 013.073.

A mean spherical approximation of the solubility of iron in the internal solar plasma.
See Abstr. 022.037.

Influence of different line broadening mechanisms on the limb–effect within Na I (4s²S–np²P⁰) series.
See Abstr. 022.118.

[205]Pb: accelerator mass spectrometry of a very heavy radioisotope and the solar neutrino problem.
See Abstr. 022.141.

A two kiloton liquid argon detector for solar neutrinos and proton decay.
See Abstr. 034.008.

The scanning heliometer.
See Abstr. 034.045.

Solar disk sextant at GSFC.
See Abstr. 034.087.

Solar neutrino detectors based on indium neutrino reaction.
See Abstr. 034.106.

The gallium solar neutrino detector and neutrino oscillations.
See Abstr. 034.107.

Neutral current detector for neutrino physics, astronomy, and geology.
See Abstr. 034.108.

A Fabry–Perot etalon for differential spectral imaging.
See Abstr. 034.161.

A compact Dopplergraph/magnetograph suitable for space–based measurements of solar oscillations and magnetic fields.
See Abstr. 035.008.

Narrowband spatial filtering in differential measurements of the line–of–sight velocity of the solar atmosphere.
See Abstr. 036.032.

On the expected performance of a solar oscillation network.
See Abstr. 036.057.

The effect of spatial smearing on solar Doppler measurements. I. Mathematical formulation and application to measurements of solar rotation.
See Abstr. 036.058.

The basic equations for scanning heliometer measurement of solar diameters.
See Abstr. 036.148.

The solar radial velocity and implications for spectroscopic planetary detection.
See Abstr. 036.208.

A determination of precise heliographic coordinates at the Kislovodsk Station of the Pulkovo Observatory. I.
See Abstr. 041.008.

A determination of precise heliographic coordinates at the Kislovodsk Mountain Station of the Pulkovo Observatory. II.
See Abstr. 041.009.

Scientific observing plans for the SOUP (*Solar Optical Universal Polarimeter*) instrument on Spacelab 2 in July, 1985.
See Abstr. 051.107.

The properties and effects on stellar burning of fractionally charged nuclei.
See Abstr. 061.017.

Magnetic monopoles in stellar interiors.
See Abstr. 061.070.

Neutrino oscillations, parity violation and solar neutrinos.
See Abstr. 061.128.

Solar abundances and the role of nucleogenesis in low–to–medium mass stars in the Galaxy.
See Abstr. 061.148.

Kinematic dynamo theory for an arbitrary mean flow.
See Abstr. 062.101.

Cooling of magnetic flux tubes as a mechanism for suppression of magnetic buoyant escape of the flux tubes from the Sun and stars.
See Abstr. 064.056.

Numerical simulations of stellar convective dynamos. II Field propagation in the convection zone.
See Abstr. 065.045.

The graviton luminosity of the Sun and other stars.
See Abstr. 066.046.

Solar line blocking for disk–center and disk–averaged radiation from 3300 to 6860 Å.
See Abstr. 071.008.

The response of the line K I 7699 to the solar oscillations.
See Abstr. 071.015.

Results of the 11–year programme for spectrophotometry of selected Fraunhofer line profiles. Motions in the solar atmosphere.
See Abstr. 071.050.

Oscillatory convection in sunspots.
See Abstr. 072.031.

Solar luminosity fluctuations during the disk transit of an active region.
See Abstr. 072.056.

Sunspot oscillations and the short–period cutoff for global p–mode oscillations.
See Abstr. 072.064.

Solar activity in the Hilbert space.
See Abstr. 072.080.

Activity of sunspots and solar constant variations during 1980.
See Abstr. 072.105.

Report of IAU Commission 10: Solar activity (*Activité solaire*).
See Abstr. 072.108.

Heating in the solar mantle.
See Abstr. 073.069.

Oscillations of the Sun's chromosphere. II. Hα line centre and wing filtergram time sequences.
See Abstr. 073.073.

Singularity of solar rotation and flare productivity.
See Abstr. 073.074.

Magnetic instabilities.
See Abstr. 074.066.

Studies of the large–scale structure of the solar surface using the method of polynomial approximation.
See Abstr. 075.002.

The creation of fine structure by magnetic fields.
See Abstr. 075.005.

Etude de la rotation des traceurs magnétiques observés sur les spectrohéliogrammes K IV (Collection de Meudon) – Résultats préliminaires.
See Abstr. 075.008.

On the structure of magnetic fields in the solar convection zone.
See Abstr. 075.017.

Convective collapse and overstable oscillations in solar flux tubes.
See Abstr. 075.020.

Magnetic flux ropes of Venus: a paradigm for helical magnetic structures in astrophysical systems.
See Abstr. 093.001.

An upper limit for the solar acceleration.
See Abstr. 111.006.

Did the young sun rotate at 100 km/s?
See Abstr. 116.035.

On the determination of stellar rotation and differential rotation from chromospheric activity data.
See Abstr. 116.039.

Much ado about Geminga.
See Abstr. 143.105.

Earth

081 Structure, Figure, Gravity, Orbit, etc.

081.001 Internal oscillations in the earth's fluid core.
S. Friedlander.
Geophys. J. R. Astron. Soc., Vol. 80, No. 2, p. 345 – 361 (1985).
A formulation is given for the eigenvalue problem describing internal oscillations in the earth's fluid core. The description is appropriate for either the Boussinesq or the subseismic approximations to the full system of governing equations for a rotating, compressible, inhomogeneous, self–gravitating thick shell of fluid.

081.002 Geopotential harmonics of order 15 and 30, from analysis of the orbit of satellite 1971–10B.
D. M. C. Walker.
Planet. Space Sci., Vol. 33, No. 1, p. 97 – 107 (1985).
The satellite 1971–10B passed through exact 15th–order resonance on 30 March 1981 and orbital parameters have been determined at 52 epochs from some 3500 observations using the RAE orbit refinement program, PROP, between September 1980 and October 1981. The variations in inclination and eccentricity during this time have been analysed, and six lumped 15th–order harmonic coefficients and two 30th–order coefficients have been evaluated. The 15th–order coefficients are the best yet obtained for an orbital inclination near 65°; and previously there were no 30th–order coefficients available at this inclination. The lumped coefficients have been used to test the Goddard Earth Model GEM 10B: there is good agreement for seven of the eight coefficients.

081.003 Individual geopotential coefficients of order 15 and 30, from resonant satellite orbits.
D. G. King–Hele, D. M. C. Walker.
Planet. Space Sci., Vol. 33, No. 2, p. 223 – 238 (1985).
The analysis of variations in satellite orbits when they pass through 15th–order resonance (15 revolutions per day) yields values of lumped geopotential harmonics of order 15, and sometimes of order 30. The 15th–order lumped harmonics obtained from 24 such analyses over a wide range of orbital inclinations are used here to determine individual harmonic coefficients of order 15 and degree 15, 16,...35; and the 30th–order lumped harmonics (from eight of the analyses) are used to evaluate individual coefficients of order 30 and degree 30, 32,...40. The accuracy of the 15th–order coefficients of degree 15, 16,...23 is equivalent to 1 cm in geoid height, while the 30th–order coefficients of degree 30, 32 and 34 are determined with an accuracy which is equivalent to better than 2 cm in geoid height. The results are used to assess the accuracy of the Goddard Earth Model 10B.

081.004 Lower mantle heterogeneity, dynamic topography and the geoid.
B. H. Hager, R. W. Clayton, M. A. Richards, R. P. Comer, A. M. Dziewonski.
Nature, Vol. 313, No. 6003, p. 541 – 545 (1985). With a correction in Vol. 314, No. 6013, p. 752 (1985).
Density contrasts in the lower mantle, inferred using seismic tomography, drive viscous flow; this results in kilometres of dynamically maintained topography at the core–mantle boundary and at the Earth's surface. The total gravity field due to interior density contrasts and dynamic boundary topography predicts the longest–wavelength components of the geoid remarkably well.

081.005 Multistage accretion and core formation of the earth.
S.–s. Sun.
Nature, Vol. 313, No. 6004, p. 628 – 629 (1985).
This note discusses recent hypotheses to explain the earth's accretion from the solar nebula and the subsequent formation of its metal core within a period of 100 million years, about 4.6 thousand million years ago.

081.006 When does small–scale convection begin beneath oceanic lithosphere?
W. R. Buck.
Nature, Vol. 313, No. 6005, p. 775 – 777 (1985).
Convection on a scale smaller than the horizontal dimensions of lithospheric plates can be produced by instabilities in the upper or lower boundary layer of the larger mantle flow. The author describes here a numerical model of small–scale convection in a fluid of variable viscosity. The results indicate that recently observed gravity anomalies showing a pattern of highs and lows aligned in the direction of oceanic plate motion may be the result of small–scale mantle flow.

081.007 Variations of the Earth's albedo deduced from the ashen light of the Moon.
H. Hilbrecht, G. Küveler.
Earth, Moon, Planets, Vol. 32, No. 1, p. 1 – 7 (1985).
An analysis of 1210 visual brightness estimations of the Moon's ashen light is presented, performed by a working group of amateur astronomers from June 1972 to December 1973. In the Moon phase interval $0.1 \leqslant T_b \leqslant 0.7$ the brightness expressed in a semi–empirical scale, S_G, is found to be linearly related to the phase. Monthly deviations from the mean brightness show well defined winter maxima (January) and summer minima (July). Within the referenced period the brightness of the ashen light tends to increase, whereas the solar magnetic activity decreased. In addition, minor correlations and, respectively, anti–correlations are found at stratospheric temperature and, respectively density. On account of the nature of the ashen light its variations are regarded as fluctuations of the Earth's albedo.

081.008 Equatorial flattening and principal moments of inertia of the Earth.
M. Burša, Z. Šíma.
Stud. Geophys. Geod., Vol. 28, No. 1, p. 9 – 10 (1984).
The best fitting parameters of the tri–axial Earth's ellipsoid have been computed on the basis of recent satellite data, as well as, the corresponding principal moments of inertia.

081.009 Thermal regime of the continental lithosphere.
P. Morgan, J. H. Sass.
J. Geodyn., Vol. 1, No. 2, p. 143 – 166 (1984). Abstr. in Phys. Abstr., Vol. 88, No. 1248, Entry 9166 (1985).

081.010 The types of the Earth's crust.
V. V. Beloussov, N. I. Pavlenkova.
J. Geodyn., Vol. 1, No. 2, p. 167 – 183 (1984). Abstr. in Phys. Abstr., Vol. 88, No. 1248, Entry 9167 (1985).

081.011 A universal thermal equation–of–state.
O. L. Anderson.
J. Geodyn., Vol. 1, No. 2, p. 185 – 214 (1984). Abstr. in Phys. Abstr., Vol. 88, No. 1248, Entry 9168 (1985).

081.012 Earth.
M. H. Carr.
NASA Spec. Publ., NASA SP–469, p. 78 – 105 (1984). – See Abstr. 003.004.
Contents: Introduction. General properties. Plate tectonics. The interior of Earth. Formation of Earth. The Archean. Evolution of the atmosphere and hydrosphere.

081.013 Representation of the Earth's gravity potential by means of a sum of potentials of two single layers.
G. A. Meshcheryakov.
Publ. Astron. Inst. Czech. Acad. Sci., No. 58, p. 199 – 209 (1984). In Russian. – See Abstr. 012.018.
New approach to the construction of multipoint models of the geopotential is discussed. The Earth's potential is represented, as usual, by means of the main and residual parts. The first is interpreted as potential of an ellipsoid with ellipsoidal–layers structure. This potential is replaced by the potential of a planar layer – the focal disk. The formulae for computation of the density of such layers are presented.

081.014 The use of geosynchronous satellites for determining parameters of the geopotential.
A. S. Sochilina.
Publ. Astron. Inst. Czech. Acad. Sci., No. 58, p. 211 – 231 (1984). – See Abstr. 012.018.

081.015 Geopotential representation by means of a point mass system with complex parameters.
K. V. Kholshevnikov, S. M. Poleshchikov.
Publ. Astron. Inst. Czech. Acad. Sci., No. 58, p. 243 – 248 (1984). In Russian. – See Abstr. 012.018.
Representation of the Earth's gravity field by means of a system of attracting point masses is considered. Models with complex parameters appear to be perspective in spite of their more complicated character. A criterion is established for the potential model to be a single–valued function.

081.016 Accuracy of Earth gravity field models as a function of order of harmonic coefficients and a draft of a laser geodynamic twin–satellite.
J. Klokočník.
Publ. Astron. Inst. Czech. Acad. Sci., No. 58, p. 309 (1984). Abstract. – See Abstr. 012.018.

081.017 "Frequency windows" from Earth gravity models and resonant solutions – test of accuracy.
Z. Šíma, J. Klokočník.
Publ. Astron. Inst. Czech. Acad. Sci., No. 58, p. 311 (1984). Abstract. – See Abstr. 012.018.

081.018 Possibility of a thermal catastrophe in the earliest period of the Earth as revealed by the volatile content of igneous crustal and mantle rocks.
A. Szalay.
Acta Phys. Hung., Vol. 55, No. 1 – 4, p. 427 – 442 (1984). Abstr. in Phys. Abstr., Vol. 88, No. 1251, Entry 23698 (1985).

081.019 Density anomalies, geoid shape and stresses.
B. Gadomska, R. Teisseyre.
Acta Geophys. Pol., Vol. 32, No. 1, p. 1 – 24 (1984). Abstr. in Phys. Abstr., Vol. 88, No. 1251, Entry 23710 (1985).

081.020 A cometary impact model for the source of Libyan Desert glass.
W. R. Seebaugh, A. M. Strauss.
J. Non–Cryst. Solids, Vol. 67, No. 1 – 3, p. 511 – 519 (1984). Abstr. in Phys. Abstr., Vol. 88, No. 1251, Entry 23767 (1985). – See Abstr. 012.016.

081.021 Zwaartekracht meten in gewichtloosheid.
W. van Tend.
Zenit, 12. Jaarg., No. 3, p. 106 – 107 (1985).

081.022 Further comparisons of Earth gravity models by means of lumped coefficients.
J. Klokočník.
Bull. Astron. Inst. Czech., Vol. 36, No. 1, p. 27 – 43 (1985). With plates 1 – 4.
The paper further develops the comparison of lumped coefficients, observed from resonant perturbations of satellite orbits, or reconstituted from various resonant solutions for the individual harmonic coefficients, or from harmonic coefficients in the Earth gravity field models (EMs). If these types of information are mutually independent, the accuracy of the EMs (for a particular order of harmonics) can be tested. The tests can be prepared for arbitrary orbital inclination I and selected order m, using "fictitious" satellite orbits. Numerical results are now available for $30 \leqslant I \leqslant 140°$, $2 \leqslant m \leqslant 15$ plus $m = 30$. They are valid for circular ($q = 0$) as well as eccentric ($q = 1$) orbits.

081.023 The origin and earliest state of the earth's hydrosphere.
J. G. Cogley, A. Henderson–Sellers.
Rev. Geophys. Space Phys., Vol. 22, No. 2, p. 131 – 175 (1984).
Contents: Introduction. The astrophysical environment. Constraints from the rock record. Climatology of the early earth. The early hydrosphere and the origin of life.

081.024 How many minerals were found on the earth and on the moon?
D. Yu. Pushcharovskij.
Priroda, No. 1, p. 117 – 118 (1985). In Russian.

081.025 Tidal and rotational perturbations of Earth's gravity gradients.
F. Bocchio.
Geophys. J. R. Astron. Soc., Vol. 81, No. 2, p. 463 – 467 (1985).
In view of present attempts to build orbiting gravity gradiometers with an accuracy of about 10^{-4}E, perturbing gravitational gradients arising from tide–generating bodies and from rotational deformations induced by polar motion are evaluated.

081.026 Studies on the earth tidal observations in China.
P. Melchior, B. Ducarme, M. Van Ruymbeke, C. Poitevin, T. Fang, H.–t. Hsu, R.–h. Li, D.–s. Chen.
Acta Geophys. Sin., Vol. 28, No. 2, p. 142 – 154 (1985). In Chinese.

081.027 Isotopic relations and rare gas abundance in the accretion period of the earth.
G. S. Anufriev, A. Ya. Krylov.
Cosmochemistry and meteoritics, p. 54 – 64 (1984). In Russian. Abstr. in Ref. Zh., 51. Astron., 2.51.157 (1985). – See Abstr. 012.037.

081.028 Interpretations of mass extinction.
M. J. Benton.
Nature, Vol. 314, No. 6011, p. 496 – 497 (1985).

081.029 Terrestrial impactors at geological boundary events: comets or asteroids?
P. R. Weissman.
Nature, Vol. 314, No. 6011, p. 517 – 518 (1985).
Evidence has been presented for a 26 – 28–Myr periodicity in both the terrestrial extinction record and the age of large well-dated impact craters on the Earth. A cometary source controlled by perturbations associated with either the Solar System's galactic z–motion, which has a 33–Myr half–period, or an unseen solar companion star in a 26 – 28–Myr orbit has been suggested for the impacts. Identifications of meteoritic material in the impact melts of the craters used to demonstrate the periodicity, however, are indicative of highly differentiated impactors, which are more representative of asteroids than of comets. Clustering of crater ages at some dates may be indicative of random comet showers, some associated with extinctions and some not.

081.030   Interval of formation of the earth.
L. K. Levskij.
Geokhimiya, No. 11, p. 1667 – 1673 (1984). In Russian. Abstr. in
Ref. Zh., 51. Astron., 3.51.196 (1985).

**081.031   Isotopically–geochemical calculation of the balance in a
chondrite model of the earth.**
Yu. D. Pushkarev, L. A. Nejmark.
Vses. simp. No. 10 po stabil. izotopam v geokhimii, Moskva,
3 – 5 dek., 1984. Tez. dokl. Moskva, p. 154 (1984). In Russian.
Abstr. in Ref. Zh., 51. Astron., 4.51.171 (1985).

**081.032   Local approximation of the quasi–geoid from astro-
geodetic data.**
W. Keller.
Mitt. Lohrmann–Obs. Tech. Univ. Dresden, Nr. 51, p. 84 – 85
(1984). In German. – See Abstr. 012.058.

**081.033   On the motion of the principal axes of inertia within the
elastic tri–axial earth.**
J. Vondrák.
Mitt. Lohrmann–Obs. Tech. Univ. Dresden, Nr. 51, p. 86 (1984).
Abstract. See 37.081.013. – See Abstr. 012.058.

**081.034   Darstellung des Gravitationspotentials der Erde in
kartesischen Basisfunktionen.**
M. Jarosch.
Dtsch. Geod. Komm. Bayer. Akad. Wiss., Reihe A, Heft
Nr. 100, 81 pp. (1984).

**081.035   Zur Definition eines körperfesten Bezugssystems für
eine deformierbare Erde und seine analytische Bewe-
gungsform unter dem Einfluß externer Kräfte.**
P. Georgiadou.
Dtsch. Geod. Komm. Bayer. Akad. Wiss., Reihe C, Heft Nr. 297,
81 pp. (1984).

081.036   Expandiert die Erde?
K. Bretterbauer.
Österr. Z. Vermessungswes. Photogramm., 72. Jahrg., Heft 3,
p. 81 – 93 (1984).
    Geodesy as a science aims beyond measuring the figure of the
earth, and tries an interpretation of its present state. Therefore,
a simplified presentation of the complex problem of earth expansion is given.

**081.037   Corrections to the earth's ellipticity for astronomical
refraction.**
N. A. Vasilenko.
Kinematika Fiz. Nebesn. Tel, Tom 1, No. 3, p. 71 – 74 (1985). In
Russian.
    Two types of corrections for the earth's ellipticity for astro-
nomical refraction are given. The first type is due to the fact that
the rays from the observed object pass through the atmospheric
layers at some different angles than in the case of a spherically
stratified medium. The second one is specified by variations of
atmospheric density due to variations of gravity on the earth's
surface. The dependence of these corrections on the latitude,
azimuth and zenith distance is analysed.

081.038   Comparative planetology and early history of the earth.
V. L. Barsukov.
Zemlya Vselennaya, No. 3, p. 8 – 16 (1985). In Russian.

**081.039   Terrestrial xenon isotope constraints on the early history
of the Earth.**
M. Ozima, F. A. Podosek, G. Igarashi.
Nature, Vol. 315, No. 6019, p. 471 – 474 (1985).
    Comparison between ^{129}I–radiogenic ^{129}Xe and ^{244}Pu–
fissiogenic ^{136}Xe components in terrestrial xenon suggests that
the Earth's inner region accreted a few tens of millions of years
earlier than the outer region from which the atmosphere evolved.
The results also indicate that there has been no substantial mixing
of the two regions since the Earth's accretion.

**081.040   Patterns of family extinction depend on definition and
geological timescale.**
A. Hoffman.
Nature, Vol. 315, No. 6021, p. 659 – 662 (1985).
    The recent claim of an approximate 26–Myr periodicity in the
pattern of mass species extinction over the past 250 Myr has
already stimulated much astrophysical speculation and debate on
causal mechanisms. However, as shown here, the evidence for
that claim is strongly contingent on arbitrary decisions concerning the absolute dating of stratigraphical boundaries, the culling
of the database and the definition of what is mass extinction as
opposed to background extinction. This evidence becomes insufficient under other plausible geological timescales and other acceptable definitions of mass extinction.

**081.041   Sobre las fuerzas móviles principales del desarrollo
geologico de la Tierra.**
G. P. Tamrazya.
Bol. Acad. Cienc. Fis. Mat. Nat., Tomo 43, Nos. 131 – 132,
p. 229 – 279 (1983).

081.042   Geological rhythms and cometary impacts.
M. R. Rampino, R. B. Stothers.
Science, Vol. 226, No. 4681, p. 1427 – 1431 (1984).
    Time–series analysis reveals two dominant, stable long–term
periodicities approximately equal to 33 ± 3 and 260 ± 25 million
years in the known series of geological and biological upheavals
during the Phanerozoic Eon. Because the cycles of these episodes
agree in period and phase with the cycles of impact cratering on
Earth, these results suggest that periodic comet impacts strongly
influence global tectonism and biological evolution. These two
periodicities could arise from interactions of the solar system with
interstellar clouds as the solar system moves cyclically through
the Galaxy.

081.043   The geological significance of the geoid.
C. G. Chase.
Annu. Rev. Earth Planet. Sci., Vol. 13, p. 97 – 117 (1985). – See
Abstr. 003.032.
    The paper focuses on the interpretation of global gravity in the
form of the geoid and concentrates on what has already been
learned from it about the structure of the Earth at long–to–
intermediate wavelengths.

081.044   Seismic tomography of the Earth's interior.
A. M. Dziewonski, D. L. Anderson.
Am. Sci., Vol. 72, No. 5, p. 483 – 494 (1984). Abstr. in Phys.
Abstr., Vol. 88, No. 1253, Entry 35181 (1985).

081.045   On the expanding Earth hypothesis.
M. Burša.
Stud. Geophys. Geod., Vol. 28, No. 3, p. 215 – 223 (1984).
    The hypothesis of an expanding Earth is discussed on the basis
of lunar laser ranging. It is proved that the given data do not
indicate a secular increase in the principal moment of the Earth's
inertia which would have to occur if the Earth were really expanding.

081.046   The Earth is expanding and we don't know why.
H. Owen.
New Sci., Vol. 104, No. 1431, p. 27 – 30 (1984). Abstr. in Phys.
Abstr., Vol. 88, No. 1256, Entry 51042 (1985).

081.047   Geodynamic thermal runaway with melting.
J. R. Ockendon, A. B. Tayler, S. H. Emerman,
D. L. Turcotte.
J. Fluid Mech., Vol. 152, p. 301 – 314 (1985). Abstr. in Phys.
Abstr., Vol. 88, No. 1259, Entry 68326 (1985).

081.048 Attraction potential of the rotating earth in the outer space.
A. I. Rybakov.
Tr. Gos. Astron. Inst. Shternberg, Tom 57, p. 5 – 19 (1985). In Russian.

Expansions of the gravitational potential in harmonic polynomials up to the fourth degree inclusive are given for both the fixed and rotating earth.

081.049 A simple model of whole–mantle convection.
D. E. Loper.
J. Geophys. Res., Vol. 90, No. B2, p. 1809 – 1836 (1985).

081.050 Thermal evolution models for the earth.
U. R. Christensen.
J. Geophys. Res., Vol. 90, No. B4, p. 2995 – 3007 (1985).

Thermal evolution models for the earth which are based on a parameterization of the convective heat transport are critically reexamined.

081.051 Large body impacts through geologic time.
E. M. Shoemaker.
Patterns of change in Earth evolution, p. 15 – 40 (1984). – See Abstr. 012.106.

In this paper the evidence concerning the present and past rates of collision of asteroids and comets with the Earth is briefly summarized. The sources of asteroids and comets are then considered and the probable variations in the collision rate of large bodies with the Earth during the last 4 Gyr are estimated.

081.052 Geochemical markers of impacts and of their effects on environments.
K. J. Hsü.
Patterns of change in Earth evolution, p. 63 – 74 (1984). – See Abstr. 012.106.

Geochemical markers include anomalies in trace element abundances, in stable isotope ratios, and abrupt changes in mineralogic and chemical composition. Enrichment of siderophile and depletion of rare–earth elements indicate extraterrestrial impact. The worldwide geochemical records across Cretaceous/Tertiary boundary sections all point to mass mortality caused by a large body impact which also triggered environmental catastrophes and mass extinction.

081.053 Time and space scales of mantle convection.
F. M. Richter.
Patterns of change in Earth evolution, p. 271 – 289 (1984). – See Abstr. 012.106.

081.054 Isotopic evolution of the crust and mantle.
R. K. O'Nions.
Patterns of change in Earth evolution, p. 291 – 302 (1984). – See Abstr. 012.106.

The age and chemical evolution of the continents together with the nature and locale of their depleted mantle complement have been investigated using naturally occurring isotope tracers. This review paper summarizes briefly major results from these studies and presents an outline of future research directions.

081.055 Secular rotational motions and the mechanical structure of a dynamical viscoelastic Earth.
D. A. Yuen, R. Sabadini.
Phys. Earth Planet. Inter., Vol. 36, Nos. 3 – 4, p. 391 – 412 (1984).

081.056 On present–day gravimetry and some of its problems.
V. V. Brovar, V. L. Panteleev, M. U. Sagitov.
Main directions of astronomical investigations at the Moscow University, p. 157 – 170 (1985). In Russian. – See Abstr. 003.049.

081.057 Planetary rotation and invertebrate skeletal patterns: prospects for extant taxa.
W. W. Hughes.
Geophys. Surv., Vol. 7, No. 2, p. 169 – 183 (1985). – See Abstr. 012.111.

081.058 Growth rhythms, evolution of the Earth's interior, and origin of the Metazoa.
G. D. Rosenberg.
Geophys. Surv., Vol. 7, No. 2, p. 185 – 199 (1985). – See Abstr. 012.111.

The purpose of this paper is to suggest that, not only do fossil growth patterns imply that the Earth's moment of inertia changed most rapidly long before the end of the Precambrian, but that the consequent establishment of relatively oxygen–rich outer layers (atmosphere, hydrosphere, and crust) and growing length of day were requirements for the origin of the Metazoa with accretionary skeletons.

Intrinsic geodesy.
See Abstr. 003.043.

The story of the earth.
See Abstr. 003.073.

Genesis on planet earth.
See Abstr. 003.080.

History of the Earth's crust.
See Abstr. 003.084.

Evolution of matter and energy on a cosmic and planetary scale.
See Abstr. 003.194.

150 years gravimetry at the Sternberg Astronomical Institute.
See Abstr. 009.041.

Cratering theories bombarded.
See Abstr. 011.008.

Problems of data bases in geophysics.
See Abstr. 013.015.

Comments on the summations of spherical harmonics in the geopotential evaluation theories of Deprit and others.
See Abstr. 021.022.

Balloon–borne, high–altitude gravimetry.
See Abstr. 035.027.

Microgravity experiment system utilizing a balloon.
See Abstr. 035.028.

Precession–nutation torque in terms of the Stokes constants.
See Abstr. 044.007.

Influence of time variation in the second zonal harmonic on polar motion.
See Abstr. 044.012.

The secular acceleration of the Earth's spin.
See Abstr. 044.093.

Altimetry–gravimetry functional boundary value problem.
See Abstr. 045.002.

Astronomy, geodesy and geophysics.
See Abstr. 045.028.

Geopotential Research Mission (GRM).
See Abstr. 051.019.

Model of conjugated point masses of a smoothed field for LAGEOS orbit determination.
See Abstr. 052.003.

On orbit determination with lumped coefficients.
See Abstr. 052.012.

Improvement of the shadow function and its influence on the orbital elements of satellites.
See Abstr. 052.014.

Altimetry, orbits and tides.
See Abstr. 052.055.

Radial variations of a satellite orbit due to gravitational errors: implications for satellite altimetry.
See Abstr. 052.071.

Magnetic reversals and mass extinctions.
See Abstr. 084.072.

A point–mass method for modeling a planet's gravitational field.
See Abstr. 091.036.

K–U–Th systematics of matter of planetary bodies of the solar system.
See Abstr. 091.048.

Relativistic effects in the earth–moon dynamics.
See Abstr. 094.017.

Mars–Earth geographical comparisons: a pictorial view.
See Abstr. 097.001.

Periodic comet showers and planet X.
See Abstr. 101.002.

Structure of associations of leading chemical elements in olivine pallasites, lunar and terrestrial rocks.
See Abstr. 105.034.

Possibility of collisions of comets and minor planets with the Earth.
See Abstr. 105.253.

Cometary impacts, molecular clouds, and the motion of the Sun perpendicular to the galactic plane.
See Abstr. 107.013.

082 Atmosphere (Refraction, Scintillation, Extinction, Airglow, Site Testing)

082.001 Brightness of the O_2 atmospheric bands in the daytime thermosphere.
W. R. Skinner, P. B. Hays.
Planet. Space Sci., Vol. 33, No. 1, p. 17 – 22 (1985).

The Fabry–Perot interferometer on Dynamics Explorer 2 was used as a low sensitivity photometer to study the O_2 atmospheric A band during the daytime. A study of the brightness of the emission showed that the assumed source of $O_2(b^1\Sigma_g{}^+)$ in the thermosphere, $O(^1D)$, can account for the observed intensity up to about 250 km but with a significantly different scale height. This combined with an enhanced brightness above this altitude suggests an additional source for this emission.

082.002 Spectroscopic measurements of atmospheric HCN at northern and southern latitudes.
M. A. H. Smith, C. P. Rinsland.
Geophys. Res. Lett., Vol. 12, No. 1, p. 5 – 8 (1985).

A number of lines belonging to the v_3 fundamental vibration–rotation band of hydrogen cyanide (HCN) have been identified in atmospheric infrared absorption spectra obtained on high-altitude balloon flights at 32°N in 1976 and at 30°S in 1977. Quantitative analysis of these spectral features has resulted in the first determination of atmospheric HCN concentrations in the southern hemisphere, as well as an independent estimate of stratospheric HCN abundance in the northern hemisphere.

082.003 Retrieval of non–LTE vertical structure from a spectrally resolved infrared limb radiance profile.
A. S. Zachor, R. D. Sharma.
J. Geophys. Res., Vol. 90, No. A1, p. 467 – 475 (1985).

The authors describe a method for recovering information about the upper mesosphere and thermosphere from a spectrally resolved measurement of the IR limb radiance profile.

082.004 Intensity variations of geocoronal Balmer alpha emission. 1. Observational results.
P. Shih, F. L. Roesler, F. Scherb.
J. Geophys. Res., Vol. 90, No. A1, p. 477 – 490 (1985).

Ground–based observations of geocoronal Balmer α (Hα) emission were performed with a large–aperture, dual–etalon Fabry–Perot spectrometer at a resolving power of about 25,000, which clearly isolates the geocoronal emission from galactic and zodiacal light backgrounds.

082.005 Downward flux of atmospheric 63–μm emission from atomic oxygen at balloon altitudes.
T. A. Clark, D. A. Naylor, R. T. Boreiko, J. M. Hoogerdijk, B. Fitton, M. F. Kessler, R. J. Emery.
Nature, Vol. 313, No. 5999, p. 206 – 207 (1985).

The 63–μm emission line from the ground electronic state fine-structure transition $(^3P_1 - {}^3P_2)$ of atomic oxygen, first suggested as a major source of thermospheric cooling by Bates, has been measured over a range of thermospheric altitudes. The authors report here high spectral resolution measurements of O I emission at 30 km. The downward O I flux is measured to be $(2.4 \pm 0.5) \times 10^{-5} W\,m^{-2}sr^{-1}$, somewhat larger than expected on the basis of previous rocket measurements or theoretical predictions. Furthermore, this value is found to be independent of zenith angle.

082.006 Mesospheric ozone variations caused by gravity waves.
G. Brasseur.
Nature, Vol. 313, No. 6000, p. 270 (1985).

This note comments briefly on new ozone observations obtained with the Solar Mesosphere Explorer satellite.

082.007 Global and seasonal variability of the temperature and composition of the middle atmosphere.
J. J. Barnett, M. Corney, A. K. Murphy, R. L. Jones, C. D. Rodgers, F. W. Taylor, E. J. Williamson, N. M. Vyas.
Nature, Vol. 313, No. 6002, p. 439 – 443 (1985)

The Stratospheric and Mesospheric Sounder on board the experimental meteorological satellite Nimbus 7, launched in 1978, was designed to study the global structure, radiative properties, chemistry and dynamics of the stratosphere and mesosphere. The instrument obtained data over a four and a half year period, until spring 1984. This review considers the results obtained.

082.008 Ein Rückgang der Zahl der klaren Nachtstunden?
R. Ziener, K.-H. Mau.
Astron. Raumfahrt, 23. Jahrg., Heft 1, p. 6 – 8 (1985).

082.009   Monte Carlo simulation of atmospheric gamma–ray scattering.
D. J. Morris.
AIP Conf. Proc., No. 115, p. 665 – 668 (1984). – See Abstr. 012.005.
The scattering of γ–rays (energy greater than 20 keV) from an astronomical point source in the earth's atmosphere is simulated using Monte Carlo methods. The intensity of scattered γ–rays is estimated as a function of energy, direction, altitude, and geographic location. A few examples of the calculated flux are presented.

082.010   Anmerkungen zum Thema "Atmosphärische Extinktion" (I).
B.–C. Kämper.
BAV Rundbrief, 34. Jahrg., Nr. 1, p. 12 – 23 (1985).

082.011   A simple model of the transient response of the thermosphere to impulsive forcing.
R. L. Walterscheid, D. J. Boucher Jr.
J. Atmos. Sci., Vol. 41, No. 6, p. 1062 – 1072 (1984). Abstr. in Phys. Abstr., Vol. 88, No. 1248, Entry 9399 (1985).

082.012   Theory of the two wavelength dispersion method for determination of refractive index for laser observations.
E. Pedrero Gonzales.
Publ. Astron. Inst. Czech. Acad. Sci., No. 58, p. 163 – 174 (1984). In Russian. – See Abstr. 012.018.
The two wavelength dispersion method for determination of the refractive index for laser observations is discussed. The difference between terrestrial and satellite observations is analyzed. Finally, parameters of a satellite laser radar are proposed.

082.013   Density scale height determined from the motion of Dash 2 satellite.
Y. E. Helali, M. Y. Tawadraus.
Publ. Astron. Inst. Czech. Acad. Sci., No. 58, p. 315 – 324 (1984). – See Abstr. 012.018.
The density scale height H is determined using the equation of change rate of the orbital period of Dash 2 satellite. The computed values of H are compared with values obtained by Jacchia (1970). Jacchia's values of H are given up to an altitude of 2500 km. The authors extend Jacchia's values of H up to an altitude of 3110.7 km. Their values of H are much greater than Jacchia's values for higher altitudes, while at smaller altitudes the values are approximately the same as Jacchia's values.

082.014   Importance of wave dissipation in the behaviour of equatorial and tropical thermosphere.
C. Berger, M. Ill, F. Barlier.
Publ. Astron. Inst. Czech. Acad. Sci., No. 58, p. 325 – 332 (1984). – See Abstr. 012.018.
From accelerometer total density data collected during the last minimum of the solar cycle, density scale heights have been extensively computed. Effects of waves dissipation are evidenced on this data set which reflects temperature behaviour. It appears that tides of high orders (or planetary waves) are quite significant and variable.

082.015   Is there any "after effect" in density variations of the neutral atmosphere?
E. Illés–Almár, P. Bencze, F. Märcz.
Publ. Astron. Inst. Czech. Acad. Sci., No. 58, p. 333 – 337 (1984). – See Abstr. 012.018.
It is demonstrated that simultaneously with variations of ionospheric absorption there are unmodeled increases of neutral density 6 – 7 days after geomagnetic storms, i.e. there is an "after effect" in the neutral atmosphere as well.

082.016   Models of the thermosphere for usage in satellite dynamics.
L. Sehnal, I. V. Tupikova.
Publ. Astron. Inst. Czech. Acad. Sci., No. 58, p. 339 – 343 (1984). – See Abstr. 012.018.
Models of the variations and distribution of the upper atmosphere density are constructed, using modelled as well as observed data. Theoretical expression of the model is used in the equations of motion to study the drag effects of the atmosphere.

082.017   Magnetospheric effects on the upper atmosphere.
G. C. Reid.
ESA Spec. Publ., ESA SP–217, p. 495 – 502 (1984). – See Abstr. 012.021.
This paper reviews recent developments in the understanding of the influence of the magnetosphere on both the dynamics and the chemical composition of the atmosphere.

082.018   Atmospheric extinction in relation to stellar photometry.
A. W. J. Cousins.
Mon. Notes Astron. Soc. S. Afr., Vol. 44, Nos. 1 – 2, p. 10 – 17 (1985).
All ground–based stellar photometry is affected by atmospheric extinction. The effects are considered both for a laminar atmosphere and one that is mottled. Current methods for correcting observations are acceptable in most cases. Some improvement is possible if the observations are made in duplicate or with a simultaneous multi–channel photometer.

082.019   Survey of planetary–scale traveling waves: the state of theory and observations.
M. L. Salby.
Rev. Geophys. Space Phys., Vol. 22, No. 2, p. 209 – 236 (1984).
Contents: Introduction. Historical overview. Motivation: normal modes of an unbounded isothermal atmosphere in uniform rotation. Normal modes in realistic atmospheric conditions. Observational support. Realistic considerations. Concluding remarks.

082.020   Atmospheric wave propagations in the mesopause region observed by the OH(8,3) band, NaD, O_2A (8645 Å) band and O I 5577 Å nightglow emissions.
H. Takahashi, P. P. Batista, Y. Sahai, B. R. Clemesha.
Planet. Space Sci., Vol. 33, No. 4, p. 381 – 384 (1985).

082.021   Atmospheric spectroscopy in relation to atmospheric chemistry.
B. A. Thrush.
J. Quant. Spectrosc. Radiat. Transfer, Vol. 32, No. 5/6, p. 373 – 379 (1984). – See Abstr. 012.026.
The review discusses spectroscopic measurements in the Earth's atmosphere in relation to laboratory studies and shows how measurements, particularly of the ratios of trace species, can be used to check the understanding of particular parts of the very complex chemistry and physics of atmospheric processes.

082.022   Atmospheric transmission in the $750 – 2000$ cm^{-1} region.
D. G. Murcray.
J. Quant. Spectrosc. Radiat. Transfer, Vol. 32, No. 5/6, p. 381 – 396 (1984). – See Abstr. 012.026.
Infrared solar spectra have been obtained during a series of flights with a balloon borne Fourier Transform Spectrometer system. Major emphasis during these flights was placed on obtaining data during sunset. The spectra thus contain information on the atmospheric transmission over long stratospheric paths. These spectra have been analyzed to obtain information on the constituents responsible for the observed absorptions and the distribution with altitude of several of these constituents of interest in stratospheric chemistry. These spectra have also been used to determine the spectroscopic parameters for several compounds such as O_3 which are difficult to study in the laboratory. A description of the instrumentation used to obtain the data, samples of the spectra obtained, and details of the analysis are given.

082.023   The high resolution submillimetre spectrum of the stratosphere.
B. Carli.
J. Quant. Spectrosc. Radiat. Transfer, Vol. 32, No. 5/6, p. 397 – 405 (1984). – See Abstr. 012.026.
The spectrum of the stratosphere in the $7 – 90$ cm^{-1} spectral interval has been measured with a resolution of 0.0033 cm^{-1}

unapodized. The structure of this rich spectrum is described and the assignments that have been made up to date are revised.

082.024   On the effect of mean inclinations of equal–density atmospheric layers on the astronomical refraction.
A. F. Kantorov.
Kinematika Fiz. Nebesn. Tel, Tom 1, No. 2, p. 78 – 84 (1985). In Russian.

Using the data from the aeroclimatic reference book the maps of equal–density surfaces are compiled for a free atmosphere over the USSR territory and the inclination of surfaces is determined. The effect of "mean" inclinations of surfaces on astronomical refraction is estimated.

082.025   Atmospheric effects and selection of places for astrometric stations.
G. Teleki.
Kinematika Fiz. Nebesn. Tel, Tom 1, No. 2, p. 85 – 91 (1985). In Russian.

The paper deals with the earth's atmosphere effect on astrometric instruments and on star seeing. The problems of protection against the atmospheric effect as well as the criteria for selection of places for astrometric stations are discussed.

082.026   Comparison of visibility of stars at day time for three stations.
N. F. Minyajlo.
Kinematika Fiz. Nebesn. Tel, Tom 1, No. 2, p. 92 – 93 (1985). In Russian.

The visibility of stars at day time is compared for three stations (Goloseevo, Kislovodsk, Majdanak). The stars observed at the mountain stations are shown to be fainter by one – two magnitudes.

082.027   HIRIS rocketborne spectra of infrared fluorescence in the $O_3(v_3)$ band near 100 km.
W. T. Rawlins, G. E. Caledonia, J. J. Gibson, A. T. Stair Jr.
J. Geophys. Res., Vol. 90, No. A3, p. 2896 – 2904 (1985).

High–resolution $O_3(v_3)$ spectral emission data, obtained between 80 and 125 km during an aurora by the rocketborne HIRIS cryogenic interferometer spectrometer, have been analyzed by a spectral simulation/least squares technique. Data obtained below 110 km exhibit behavior independent of local auroral strength and thus appear to represent quiescent nightglow. The derived vibrational state population distributions for levels $v_3' \geqslant 1$ can be expressed as vibrational "temperatures" of 400 – 600K. This result is interpreted in terms of formation of $O_3(v_3')$ by three–body recombination of O with O_2 as well as other radiative and collisional processes. Additional data, obtained near 120 km and sampling intense auroral arcs, indicate a hitherto unexpected auroral enhancement of $O_3(v_3)$ fluorescence intensity and vibrational state distribution.

082.028   Singlet oxygen nightglow in the atmospheres of Earth, Venus and Mars.
R. G. H. Greer, D. P. Murtagh.
Ir. Astron. J., Vol. 17, No. 1, p. 25 – 30 (1985). – See Abstr. 012.034.

082.029   Auroral zone effects on hydrogen geocorona structure and variability.
T. E. Moore, A. P. Biddle, J. H. Waite Jr., T. L. Killeen.
Planet. Space Sci., Vol. 33, No. 5, p. 499 – 505 (1985).

082.030   Upper limits to |O| in the lower thermosphere from airglow.
W. E. Sharp.
Planet. Space Sci., Vol. 33, No. 5, p. 571 – 575 (1985).

082.031   Methods for cloud cover estimation.
D. L. Glackin, J. R. Huning, J. H. Smith, T. L. Logan.
Solar seismology from space, p. 305 – 312 (1984). – See Abstr. 012.041.

Several methods for cloud cover estimation are described relevant to assessing the performance of a ground–based network of solar observatories. These methods rely on ground and satellite data sources and provide meteorological or climatological information. Related studies completed and now in progress at the Jet Propulsion Laboratory are discussed.

082.032   Analysis of planetary waves in the upper atmosphere calculated on the basis of rocket and satellite data and comparison with model calculations.
N. Griger, G. Shmits, I. V. Bugaeva, G. R. Zakharov, L. A. Ryazanov, D. A. Tarasenko.
Issled. verkhn. atmos. Zemli. Tr. 3 Mezhdunar. simp. po kosm. meteorol., Moskva, 7 – 10 apr. 1981, Leningrad, p. 951 – 959 (1984). In Russian. Abstr. in Ref. Zh., 62. Issled. Kosm. Prostranstva, 2.62.518 (1985).

082.033   Middle atmospheric NO and NO_2 observed by the Spacelab grille spectrometer.
J. Laurent, M.–P. Lemaitre, J. Besson, A. Girard, C. Lippens, C. Muller, J. Vercheval, M. Ackerman.
Nature, Vol. 315, No. 6015, p. 126 – 127 (1985).

The authors report here the first observation of NO from the low thermosphere down to the low stratosphere by instrumentation similar to that used previously on board balloon gondolas, but on this occasion the observation platform was Spacelab I, which gave access to higher altitudes. NO and NO_2 were observed simultaneously.

082.034   Seasonal variation of turbulence in the lower thermosphere.
E. A. Zhadin.
Geomagn. Aehron., Tom 25, No. 2, p. 239 – 242 (1985). In Russian. English translation in Geomagn. Aeron.

082.035   Tables of astronomical refraction for the Observatory on Mt. Majdanak.
V. S. Borovskikh, A. I. Nefed'eva.
Kazan. inzh.–stroit. inst. Kazan', 7 pp. (1984). In Russian. Abstr. in Ref. Zh., 51. Astron., 3.51.139 (1985).

082.036   Some results of an investigation of the optical parameters of the atmosphere.
M. R. Fedyanin, A. M. Morozov.
Astron. geod., Tomsk, No. 10, p. 178 – 186 (1984). In Russian. Abstr. in Ref. Zh., 51. Astron., 3.51.945 (1985).

082.037   Where is the Earth's missing xenon?
J. F. Wacker, E. Anders.
Meteoritics, Vol. 19, No. 4, p. 327 – 328 (1984). Abstract. – See Abstr. 010.641.

082.038   Empirical adjustment for atmospheric transmittances based on calculated and measured radiances.
V. Parikh.
J. Quant. Spectrosc. Radiat. Transfer, Vol. 33, No. 3, p. 227 – 236 (1985).

The study summarizes important aspects of transmittance correction for the high–resolution infrared detector aboard the TIROS–N satellite.

082.039   On a connection of the tropospheric angular refraction in different regions of electromagnetic waves.
M. N. Kolomejtseva, A. S. Medovikov, V. V. Vinogradov, R. M. Nigamat'yanov.
Vladim. politekh. inst. Vladimir, 11 pp. (1984). In Russian. Abstr. in Ref. Zh., 51. Astron., 4.51.109 (1985).

082.040   11–year variations of the ^{14}C content in the earth's atmosphere during the period 1850 – 1940.
G. E. Kocharov, R. Ya. Metskhvarishvili, V. M. Ostryakov, S. L. Tsereteli.
Soln. Dannye, Byull., 1985, No. 2, p. 61 – 64 (1985). In Russian.

The presence of an inverse correlation of solar activity and the radiocarbon abundance in the earth's atmosphere during the period 1850 – 1940 is experimentally established. Such a correlation is due to modulation of galactic cosmic rays by the sun.

082.041 Shorbulak in the Eastern Pamirs: a promising site for astronomical observations.
I. I. Kanaev, G. B. Sholomitskij, I. A. Maslov,
V. M. Grozdilov.
Astrophys. Space Phys. Rev., Vol. 3, p. 329 – 349 (1984). – See Abstr. 003.019. Revised and extended English translation of 34.009.013.
This paper discusses the results of astronomical site testing on Shorbulak mountain (4350 m) in the Eastern Pamirs. Prospects for submillimeter photometry of extended sources are examined.

082.042 Infrared measurements of atmospheric ethane (C_2H_6) from aircraft and ground–based solar absorption spectra in the 3000 cm^{-1} region.
M. T. Coffey, W. G. Mankin, A. Goldman, C. P. Rinsland,
G. A. Harvey, V. Malathy Devi, G. M. Stokes.
Geophys. Res. Lett., Vol. 12, No. 4, p. 199 – 202 (1985).
A number of prominent Q–branches of the v_7 band of C_2H_6 have been identified near 3000 cm^{-1} in aircraft and ground–based infrared solar absorption spectra. The aircraft spectra provide the column amount above 12 km at various altitudes. The column amount is strongly correlated with tropopause height and can be described by a constant mixing ratio of 0.46 ppbv in the upper troposphere and a mixing ratio scale height of 3.9 km above the tropopause. The ground–based spectra yield a column of 9.0×10^{15} molecules cm^{-2} above 2.1 km; combining these results implies a tropospheric mixing ratio of approximately 0.63 ppbv.

082.043 A spectral search for Lyman–Birge–Hopfield band nightglow from Spacelab 1.
M. R. Torr, D. G. Torr, J. W. Eun.
J. Geophys. Res., Vol. 90, No. A5, p. 4427 – 4433 (1985).
The Lyman–Birge–Hopfield (LBH) system of N_2 requires 8.5 eV for its excitation. Measurements were reported several years ago of significant intensities ($\lesssim 1$ kR) of this system at mid–latitudes at night. It thus becomes important to establish whether such LBH nightglow is generally present in the nightglow in order to identify the source mechanism. In this paper the authors examine the ultraviolet spectra gathered at night from Spacelab 1 for further information on this phenomenon.

082.044 Characteristics of the astroclimate of the two high–altitude expeditions of the Sternberg Institute at Tien Shan and Mt. Majdanak.
Eh. V. Kononovich, E. A. Makarova, I. V. Mironova,
O. B. Smirnova.
Astron. Tsirk., No. 1345, p. 3 – 7 (1984). In Russian.

082.045 Refraction Tables of Pulkovo Observatory.
V. K. Abalakin (Editor).
Fifth edition. Central Astronomical Observatory, Academy of Sciences of the USSR. Nauka Publishing House, Leningrad Section, Leningrad, USSR. 49 pp. Price 40 Kop. (1985).
The four previous editions of the Refraction Tables of Pulkovo Observatory (1870, 1905, 1930, 1956) were based on the classical theory developed by the Pulkovo astronomer Hugo Gyldén, and no essential alterations were introduced since then into every new edition. Unlike the previous editions the present tables have been compiled on a completely new basis by using recent results of study of physical and optical properties of the Earth's atmosphere.

082.046 Seasonal variations of the Hvar extinction coefficients.
J. Arsenijević.
Publ. Astron. Opservatorije Beogr., No. 33, p. 36 – 40 (1985). – See Abstr. 012.061.
Mean monthly values of the extinction coefficients, measured at the Hvar Observatory in the period 1972 – 78, and the gas density of some atmospheric layers at +45° latitude are discussed. High anticorrelation of these quantities over the atmospheric layers up to 7.5 km is stated.

082.047 Preliminary investigation of the image motion at the Belgrade Observatory.
O. Atanacković.
Publ. Astron. Opservatorije Beogr., No. 33, p. 83 – 87 (1985). – See Abstr. 012.061.
Image motion at different zenith distances has been investigated on the basis of star trails recorded photographically. Correlation between the image motion and meteorological parameters has been established. Correlation proved rather close with the wind velocity and with relative humidity

082.048 Satellite–earth range measurements. I. Correction of the excess path length due to atmospheric water vapour by ground based microwave radiometry.
G. Elgered, B. Rönnäng, E. Winberg, J. Askne.
Res. Lab. Electron. Onsala Space Obs., Res. Rep., No. 147, 6 + 112 pp. (1985).
This report deals with the excess delay of radio waves penetrating the neutral atmosphere of the earth, especially the delay due to water vapour, and reviews the state of the art of sounding techniques and instrumentation, in particular microwave radiometry. The physical background of the water vapour radiometer (WVR) method is presented and seven different types of WVR are described in detail.

082.049 Computations of refraction errors based on star programs in VZT observations.
N. Kikuchi.
Proc. Int. Latitude Obs. Mizusawa, No. 23, p. 9 – 16 (1984). In Japanese.

082.050 On the sensitive measurement of horizontal temperature gradients of air near an astrometric instrument for correcting anomalous refraction.
N. Hu, Z. Wang, X. Jiang.
Publ. Int. Latitude Obs. Mizusawa, Vol. 17, No. 2, p. 35 – 46 (1984).
Anomalous refraction is believed to be the main error source for classical astrometry. This paper suggests that by measuring the small difference of two average temperature values for two long air columns, which are close to the star light beam, then the anomalous refraction taking place between these two air columns can be obtained in real–time. Suitable measuring equipment with a sensitivity of 0.003°C in measuring the temperature difference of air columns corresponding to a sensitivity of 0.″008 in determining the anomalous refraction are under development.

082.051 Die Bestimmung des Refraktionskoeffizientmittelwerts durch Radiosondenmessungen.
K. Horváth.
Geod. Geophys. Veröff., Reihe III, Heft 51, p. 10 – 21 (1984).

082.052 Die atmosphärische Refraktion – berechnet, gemessen oder eine unbekannte Größe.
C. Gergov.
Geod. Geophys. Veröff., Reihe III, Heft 51, p. 44 – 51 (1984). In Russian.

082.053 Mt. Graham information.
N. J. Woolf.
Prepr. Steward Obs., No. 564, 96 pp. (1985).
In 1980, a search started for a new site for an astronomical observatory for the University of Arizona. The search covered that part of the continental U.S. shown to be most cloud free, and resulted in picking two sites most appropriate for further study, Mt. Graham Arizona, and Telescope Peak California. For reasons of ease of access and logistics the authors' attention has been concentrated exclusively on the former. This article reports on the information obtained to date, together with a report on progress in acquiring the site for astronomical use, and plans for facilities on the site.

082.054 Seeing measurements on Mt. Graham.
B. L. Ulich, W. B. Davison.
Prepr. Steward Obs., No. 568, 19 pp. (1985).

A transportable 30 cm aperture telescope was constructed and used with a CCD camera to measure stellar image motion as seen from the summit of Mt. Graham in southeastern Arizona. No large differences were found among three different sites or between different heights of the telescope above the ground. Direct measurements of thermal turbulence in the first few tens of meters above ground indicate that telescopes may be located within a few meters off the ground and well below treetop height without significant seeing degradation.

082.055 Model for artificial night–sky illumination.
R. H. Garstang.
Bull. Am. Astron. Soc., Vol. 16, No. 4, p. 904 – 905 (1984). Abstract. – See Abstr. 010.062.

082.056 Is Mt. Wilson the best interferometric site in the world?
P. Nisenson, R. W. Noyes, M. Shao.
Bull. Am. Astron. Soc., Vol. 16, No. 4, p. 908 (1984). Abstract. – See Abstr. 010.062.

082.057 Status report on Mt. Graham site testing.
N. J. Woolf, J. R. P. Angel, B. L. Ulich, W. B. Davison, P. A. Strittmatter, J. T. Williams.
Bull. Am. Astron. Soc., Vol. 16, No. 4, p. 923 (1984). Abstract. – See Abstr. 010.062.

082.058 Lower stratosphere trace gas detection using aircraft–borne active chemical ionization mass spectrometry.
F. Arnold, G. Hauck.
Nature, Vol. 315, No. 6017, p. 307 – 309 (1985).

Acetaldehyde and acetone are expected to be formed as trace gases in the free troposphere from photochemical conversion of non–methane hydrocarbons and to represent key precursors of peroxyacetyl nitrate. The authors report on a search for these trace gases and other species with large proton affinities in the tropopause using aircraft–borne active chemical ionization mass spectrometry (ACIMS), a novel method for atmospheric trace gas detection.

082.059 Measurements of stellar scintillation using photon counting statistics.
B. Stecklum.
Astron. Nachr., Vol. 306, No. 3, p. 145 – 156 (1985).

The properties of the stellar scintillation were investigated using high speed photometry. The zenithal dependence of the total scintillation power turned out to be weaker than predicted by scintillation theory. The dependence of the total scintillation power on wavelength and the distribution of the intensity fluctuations according to a log–normal distribution agree with the theoretical predictions. The temporal behaviour of the stellar scintillation shows correlation time scales of less than 10 ms and some ten seconds which indicate the presence of the inner and outer scale of the atmospheric turbulence. The influence of other processes which cause intensity variations on the measurement of the total scintillation power was analyzed.

082.060 Observations of Talcott refraction pairs with a Zeiss zenith telescope.
F. I. Podshchipkova, V. K. Budz'ko.
Vrashchenie i priliv. deformatsii Zemli, Kiev, No. 16, p. 42 – 47 (1984). In Russian. Abstr. in Ref. Zh., 51. Astron., 5.51.105 (1985).

082.061 Determination of astronomical refraction anomalies at large zenith distances in an off–shore region.
V. A. Kovalenko, D. I. Maslich, M. I. Rusin, I. S. Sidorov.
Vrashchenie i priliv. deformatsii Zemli, Kiev, No. 16, p. 47 – 53 (1984). In Russian. Abstr. in Ref. Zh., 51. Astron., 5.51.106 (1985).

082.062 Increasing astronomical observing time by means of cloud dispersal.
I. V. Litvinov, R. G. Tsverava.
Astrofiz. Issled. Izv. Spets. Astrofiz. Obs., Tom 19, p. 115 – 117 (1985). In Russian. English translation in Bull. Spec. Astrophys. Obs. – North Caucasus.

082.063 Intensity profile of the 22° halo.
D. K. Lynch, P. Schwartz.
J. Opt. Soc. Am. A, Vol. 2, No. 4, p. 584 – 589 (1985).

The authors report the first relative intensity measurements of the 22° halo, obtained from photographic photometry of a halo of exceptional brightness and uniformity. The maximum brightness occurs at 22°.8, and a relative minimum occurs at 19°.7. The full width at one half the maximum intensity is 3°.4. A low–intensity tail reaches from 29° to 39°.

082.064 Atmospheric IR propagation.
D. H. Höhn, W. Steffens, A. Kohnle.
Infrared Phys., Vol. 25, No. 1/2, p. 445 – 456 (1985). – See Abstr. 012.065.

The state–of–the–art of IR atmospheric propagation modelling – important to both scientific and practical applications of IR radiation – is reviewed.

082.065 Observations of 63 µm atomic oxygen emission in the earth's atmosphere from balloon altitudes: astronomical implications.
D. A. Naylor, J. M. Hoogerdijk, R. T. Boreiko, T. A. Clark, B. Fitton, M. F. Kessler, R. J. Emery.
Infrared Phys., Vol. 25, No. 1/2, p. 485 – 489 (1985). – See Abstr. 012.065.

Measurements of $^3P_1 \rightarrow {}^3P_2$ 63 µm atomic oxygen (O I) emission in the earth's atmosphere obtained with a balloon–borne telescope and high–resolution Fourier transform spectrometer are presented. Three results emerge from analysis of this data: (1) the frequency of the O I $^3P_1 \rightarrow {}^3P_2$ transition was determined to be 158.2693 ± 0.003 cm^{-1}; (2) the integrated line intensity of the atmospheric O I emission was determined to be $2.4 \pm 0.5 \times 10^{-5}$ Wm^{-2}sr^{-1}; and (3) the integrated line intensity of the atmospheric O I emission was found to be constant over a range of zenith angles corresponding to air–mass values between 1.27 and > 20. The implications of these results on astronomical observations of O I emission are discussed.

082.066 First sodium nightglow results for Natal.
V. W. J. H. Kirchhoff, H. Takahashi.
Planet. Space Sci., Vol. 33, No. 6, p. 757 – 760 (1985).

The first year of sodium nightglow observations from Natal (6°S, 35°W) are examined. Time variations appear to follow a pattern of their own, different from low latitude results. The major seasonal peak occurs in September–October and the average variation during the night decreases from dusk to dawn. Statistics on cloud coverage show that Natal has roughly only about 3 clear hours per night. The best observing period is April with an average of 5 clear hours per night.

082.067 Atmospheric hydroxyl response to the partial solar eclipse of May 30, 1984.
C. R. Burnett, E. B. Burnett.
Geophys. Res. Lett., Vol. 12, No. 5, p. 263 – 266 (1985).

A large amplitude oscillation in the vertical column abundance of atmospheric OH has been observed in ground–based spectroscopic absorption measurements from Fritz Peak Observatory, Colorado during and after the partial solar eclipse of May 30, 1984. An initial OH reduction during the eclipse was followed by an underdamped oscillation having a period of about one hour; the OH abundances returned to normal values two hours after the eclipse termination. This is believed to be the first observation of a "ringing" response of any atmospheric constituent to a solar eclipse.

082.068 On the structure and dynamics of the thermosphere.
  H. G. Mayr, I. Harris, F. Varosi, F. A. Herrero,
H. Volland, N. W. Spencer, A. E. Hedin, R. E. Hartle,
H. A. Taylor Jr., L. E. Wharton, G. R. Carignan.
Adv. Space Res., Vol. 5, No. 4, p. 283 – 288 (1985). – See Abstr.
012.066.
  Thermospheric temperature, composition and wind measurement from the Dynamics Explorer satellite (DE-2) are interpreted using a three dimensional, multiconstituent spectral
model. The analysis accounts for tides driven by the absorbed
solar radiation as well as energy and momentum coupling involving the magnetosphere and lower atmosphere. The authors discuss phenomena associated with the annual tide, polar circulation, magnetic storms and substorms.

082.069 A guide to extinction and differential refraction.
  D. G. Schleicher, R. M. Wagner.
IHW Newsl., No. 6, p. 29 – 30 (1985).

**082.070 On the intensification of [O I] 630 nm emission of the
  night sky luminescence during solar flares.**
S. V. Avakyan, G. S. Kudryashev, L. M. Fishkova.
Geomagn. Aehron., Tom 25, No. 3, p. 415 – 419 (1985). In Russian. English translation in Geomagn. Aeron.

**082.071 Determination of the relation between the refraction an-
  gle and the perigee of the line of sight with methods of
photographic astrometry from results of experiments aboard the
Salyut 6 station.**
G. M. Grechko, T. P. Kiseleva, V. K. Nikolaeva, Eh. M. Saar.
Kosm. Issled., Tom 23, Vyp. 3, p. 471 – 476 (1985). In Russian.
English translation in Cosm. Res.

082.072 Noctilucent clouds over western Europe during 1983.
  D. M. Gavine.
Meteorol. Mag., Vol. 113, No. 1346, p. 272 – 277 (1984). Abstr.
in Phys. Abstr., Vol. 88, No. 1254, Entry 40404 (1985).

**082.073 Geocoronal structure: the effects of solar radiation
  pressure and the plasmasphere interaction.**
J. Bishop.
J. Geophys. Res., Vol. 90, No. A6, p. 5235 – 5245 (1985).
  The theory of planetary exospheres is extended to incorporate
solar radiation pressure in a rigorous manner, and an evaporative
geocoronal prototype (classical, motionless exobase) is constructed using Liouville's theorem.

**082.074 Wavelength effects of soil–derived aerosols on atmo-
  spheric scattering coefficient.**
N. S. Kopeika, J. Budner.
Opt. Acta, Vol. 31, No. 11, p. 1197 – 1201 (1984). Abstr. in Phys.
Abstr., Vol. 88, No. 1256, Entry 51267 (1985).

**082.075 The transmittance of the atmosphere in the region of the
  High–altitude Station of the Astrophysical Institute.**
Yu. M. Zavarzin.
Tr. Astrofiz. Inst. Alma–Ata, Tom 42, p. 106 – 110 (1983). In
Russian.

**082.076 Production of the upper atmospheric sodium from im-
  pinging meteors.**
N. Maruyama.
J. Geomagn. Geoelectr., Vol. 36, No. 8, p. 305 – 316 (1984).
Abstr. in Phys. Abstr., Vol. 88, No. 1257, Entry 57489 (1985).

**082.077 Spectral thermal infrared emission of the terrestrial
  atmosphere.**
G. Finger, F. K. Kneubühl.
Infrared and millimeter waves. Vol. 12. Electromagnetic waves in
matter, Part II, p. 145 – 193 (1984). Abstr. in Phys. Abstr.,
Vol. 88, No. 1259, Entry 68530 (1985). See Abstr. 003.034.

082.078 A catalogue of atmospheric gamma ray lines.
  C. A. Ayre, P. N. Bhat, Y.-q. Ma, R. H. Myers,
M. G. Thompson.
18th International Cosmic Ray Conference, Vol. 1, p. 81 – 84
(1983). – See Abstr. 012.096.
  During the balloon flight of a solid state detector, Hp(Ge),
actively shielded by a minimum thickness of 12.5 cm of Na I,
some sixty background lines have been observed in the energy
range 50 keV to 4 MeV. The origin of ~ 50 of these lines has been
established. A catalogue of the lines is presented and the presence
of members of the three naturally occurring radioactive series is
demonstrated.

082.079 ^{81}Kr production rate in the atmosphere.
  V. V. Kuzminov, A. A. Pomansky (*A. A. Pomanskij*).
18th International Cosmic Ray Conference, Vol. 2, p. 357 – 360
(1983). – See Abstr. 012.096.
  A comparison of the calculated results and the measurements
of the ^{81}Kr activity in the atmospheric krypton indicates that the
average cosmic ray intensity in the atmosphere in the last
hundred thousand years was about equal to the present one.

**082.080 Characteristics of atmospheric gamma–rays flux with
  energy above 30 MeV in near–earth cosmic space.**
A. M. Gal'per, V. M. Gratchev (*V. M. Grachev*),
V. V. Dmitrenko, V. B. Komarov, S. E. Ulin.
18th International Cosmic Ray Conference, Vol. 3, p. 465 – 468
(1983). – See Abstr. 012.096.

**082.081 Estimation of the vertical profile of atmospheric
  temperature from cosmic ray components II.**
A. Inoue, M. Kodama.
18th International Cosmic Ray Conference, Vol. 10, p. 225 – 228
(1983). – See Abstr. 012.096.
  Miyazaki and Wada studied the possibility to estimate atmospheric temperature from observation of muon components. The
authors reexamine this method by using the partial temperature
coefficients calculated by Sagisaka through the equations of hadronic cascades in atmosphere.

082.082 Atmospheric distortion and blurring.
  L. J. November, R. B. Dunn.
Bull. Am. Astron. Soc., Vol. 17, No. 2, p. 640 (1985). Abstract. –
See Abstr. 010.067.

082.083 A mechanism of corpuscular–atmospheric connections.
  É. R. Mustel' (Eh. R. Mustel').
Sov. Astron., Vol. 28, No. 6, p. 689 – 691 (1984). English transla-
tion of 38.082.090.

082.084 Overview of geodetic refraction studies.
  F. K. Brunner.
Geodetic refraction, p. 1 – 6 (1984). – See Abstr. 003.044.

082.085 Two wavelength angular refraction measurement.
  D. C. Williams, H. Kahmen.
Geodetic refraction, p. 7 – 31 (1984). – See Abstr. 003.044.

082.086 Water vapor radiometry in geodetic applications.
  G. M. Resch.
Geodetic refraction, p. 53 – 84 (1984). – See Abstr. 003.044.

**082.087 Temperature and humidity structure in the lower atmo-
  sphere.**
E. K. Webb.
Geodetic refraction, p. 85 – 141 (1984). – See Abstr. 003.044.

**082.088 Atmospheric refraction effects in time and latitude
  observations using classical techniques.**
I. Naito, C. Sugawa.
Geodetic refraction, p. 181 – 187 (1984). – See Abstr. 003.044.
  This paper addresses some of the roots of the actual atmospheric refraction errors in time and latitude observations
through the classical techniques such as the VZT, the PZT and
the astrolabe. The basis of the observing system for removing the

refraction effects is discussed and the meteorological point of view is considered. These instruments account for about thirty percent of all instruments employed now by the IPMS.

082.089 The equations of electromagnetic wave propagation in a refractive medium corotating with the Earth.
E. W. Grafarend.
Geodetic refraction, p. 189 – 208 (1984). – See Abstr. 003.044.

082.090 Rocket–borne measurements of atmospheric infrared fluxes.
J. C. Ulwick, K. D. Baker, A. T. Stair Jr., W. Frings, R. Hennig, K. U. Grossmann, E. R. Hegblom.
J. Atmos. Terr. Phys., Vol. 47, No. 1 – 3, p. 123 – 131 (1985).

During the Energy Budget Campaign two rockets were simultaneously launched as part of salvo B, one from Andøya Rocket Range, Norway, and one from Esrange, Sweden, and each carried a liquid helium cooled infrared spectrometer (not the same design) as the principal experiment. The 15 μm CO_2 results from the Esrange rocket are included in the paper with the 15 μm CO_2, 9.6 μm O_3 and 5.4 μm NO Andøya results. The paper is concerned with these principal IR emitters, comparisons to model calculations and previous auroral results and the effect of Joule heating on these infrared atmospheric emitters.

082.091 Ground–based atmospheric infrared and visible emission measurements.
D. J. Baker, A. J. Steed, G. A. Ware, D. Offermann, G. Lange, H. Lauche.
J. Atmos. Terr. Phys., Vol. 47, No. 1 – 3, p. 133 – 145 (1985).

Ground–based measurements of night–sky near–infrared and visible emissions were made at the Andenes, Norway, and Kiruna, Sweden, rocket launch sites during the Energy Budget Campaign of 1980. Optical measurements were made using visible and infrared photometers, a Michelson interferometer and a grating spectrometer. The spectral range of the spectrometers was λ0.85 through 1.7 μm. Photometer wavelength coverages were at λ1.7 μm, 1.27 μm, 6300 Å, 5577 Å, 5525 Å, and 3914 Å. Emission intensities of OH, O_2, N_2^+ and O airglow and auroral species are presented. Correlations were observed between the $O_2(a^1\Delta_g)$ emission at λ1.27 μm and the OH(8,5) emission at λ1.3 μm. Values of the rotational temperatures of the OH airglow were derived. Comparisons were made of airglow intensities and temperatures with indices of auroral zone geomagnetic activity and with key atmospheric measurements made by other investigators who participated in the campaign.

082.092 Concentrations of H_2O and NO in the mesosphere and the lower thermosphere at high latitudes.
K. U. Grossmann, W. G. Frings, D. Offermann, L. André, E. Kopp, D. Krankowsky.
J. Atmos. Terr. Phys., Vol. 47, No. 1 – 3, p. 291 – 300 (1985).

Water vapour and nitric oxide concentrations in the mesosphere and lower thermosphere were derived from infrared emission and positive ion composition measurements above northern Europe during the Energy Budget Campaign 1980.

082.093 Measurements of the angular distribution of auroral–zone bremsstrahlung in the middle atmosphere.
J. R. Barcus, R. W. Fathauer, R. A. Goldberg.
J. Atmos. Terr. Phys., Vol. 47, No. 4, p. 401 – 409 (1985).

Measurements of auroral–zone X–rays during rocket flights over Alaska in March 1978 have been analyzed to obtain angular distributions of electron bremsstrahlung in the atmosphere at altitudes of 45 – 65 km.

082.094 The 2.7 and 6.3 μm H_2O band emissions in the middle atmosphere.
R. O. Manuilova, G. M. Shved.
J. Atmos. Terr. Phys., Vol. 47, No. 5, p. 413 – 422 (1985).

Radiation transfer in the 6.3 μm H_2O band in the mesosphere and lower thermosphere has been considered. For the day and night and for a set of vertical [H_2O] profiles, the populations of the H_2O vibrational states have been estimated. The populations have been used to estimate the zenith and limb total band radiance intensities for the 2.7 and 6.3 μm H_2O bands in the altitude range 40 – 110 km. The calculated radiance intensities have been used for the analysis of rocket and satellite measurements of the emissions.

082.095 Satellite observations of the nitric oxide dayglow: implications for the behavior of mesospheric and lower–thermospheric odd nitrogen.
J. E. Frederick, G. N. Serafino.
J. Geophys. Res., Vol. 90, No. D2, p. 3821 – 3830 (1985).

082.096 Report of IAU Commission 21: Light of the night sky (*Lumière du ciel nocturne*).
R. H. Giese.
Trans. IAU, Vol. XIXA, p. 227 – 234 (1985). – See Abstr. 003.046.

082.097 Report of IAU Commission 50: Identification and protection of existing and potential observatory sites (*Protection des sites d'observatoires existants et potentiels*).
A. Hoag.
Trans. IAU, Vol. XIXA, p. 707 – 712 (1985). – See Abstr. 003.046.

082.098 Sudden changes in atmospheric composition and climate.
O. B. Toon.
Patterns of change in Earth evolution, p. 41 – 61 (1984). – See Abstr. 012.106.

Volcanic eruptions, collisions with galactic dust lanes, and asteroid impacts could all alter the composition of the atmosphere in significant ways. Asteroid impacts may be the major cause of large, sudden changes in atmospheric composition and climate by injection of dust and nitrogen oxides. Volcanic eruptions produce much less dust than large asteroid impacts but may still cause significant climatic changes. Galactic dust lane collisions probably do not affect the Earth's climate unless very dense dust lanes are encountered.

082.099 Investigation of the spectral transmittance of the atmosphere in Lesniki with the help of the A'Hearn interference filters.
V. V. Kleshchenok, K. I. Churyumov.
Komet. Tsirk., No. 340 (1985). In Russian.

082.100 Site choosing of Qing–Hai mm–wave radioastronomy station of Purple Mountain Observatory.
P. Han.
Acta Astron. Sin., Vol. 26, No. 2, p. 187 – 196 (1985). In Chinese.

082.101 An evaluation of the chromatic refraction in visual solar observations.
A. V. Devyatkin, E. G. Zhilinskij.
Izv. Glav. Astron. Obs. Pulkovo, Astrometr. Astrofiz., No. 202, p. 32 – 36 (1984). In Russian.

An evaluation of the influence of chromatic refraction on visual solar observations is carried out in accordance with the theory by O. A. Melnikov and G. G. Lengauer. For large zenith distances the correction is 0″.4. The chromatic effect "Sun minus star" is shown to reach also 0″.4 for some stars.

082.102 On the problem of anomalous refraction.
P. N. Fedorov, A. V. Shul'ga.
Izv. Glav. Astron. Obs. Pulkovo, Astrometr. Astrofiz., No. 202, p. 37 – 39 (1984). In Russian.

Formulas for correction of day light observations for the anomalous vertical and lateral refraction due to the different temperature inside and outside a pavilion are derived.

082.103 On the accuracy of measurements of spectral coefficients of extinction.
G. A. Alekseeva, V. V. Novikov, V. B. Novopashennyj, D. E. Shchegolev.
Izv. Glav. Astron. Obs. Pulkovo, Astrometr. Astrofiz., No. 201, p. 118 – 121 (1985). In Russian.

Simultaneous spectrophotometric observations of stars with two closely situated telescopes of essentially different diameters

(D = 200, 600 mm) did not show systematic differences in the determination of the spectral coefficients of extinction and enabled a numerical estimation of the mean error of one measurement of the extinction coefficient. Recommendations are given for the organization of spectrophotometric observations.

Wind as a geological process on Earth, Mars, Venus and Titan.
See Abstr. 003.001.

The global climate.
See Abstr. 003.003.

Chemistry of atmospheres. An introduction to the atmospheres of Earth, the planets, and their satellites.
See Abstr. 003.027.

Aeronomy of the middle atmosphere.
See Abstr. 003.068.

Satellite sensing of a cloudy atmosphere: observing the third planet.
See Abstr. 003.107.

Intercomparison of stratospheric/mesospheric data. Proceedings of the Topical Meeting of the COSPAR Interdisciplinary Scientific Commission A (Meeting A1) of the COSPAR Twenty–fifth Plenary Meeting held in Graz, Austria, 25th June – 7th July 1984.
See Abstr. 012.006.

Laboratory measurement of the O I 1173/989 Å branching ratio.
See Abstr. 022.001.

The dissociative recombination of N_2^+ ($v = 0, 1$) as a source of metastable atoms in planetary atmospheres.
See Abstr. 022.003.

Intensity measurements and self–broadening coefficients in the γ band of O_2 at 628 nm using intracavity laser–absorption spectroscopy (ICLAS).
See Abstr. 022.106.

Pure rotational transitions of H_2O molecules in the $8 - 14\,\mu m$ atmospheric window.
See Abstr. 022.129.

The role of intersystem collisional transfer of excitation in the determination of N_2 vibronic level populations. Application to $B'^3\Sigma_u^- - B^3\pi_g$ band intensity measurements.
See Abstr. 022.149.

N_2^+ Meinel band quenching.
See Abstr. 022.158.

Measurement of atmospheric seeing with Ronchi telescopes.
See Abstr. 032.028.

A balloon–borne microwave limb sounder for stratospheric measurements.
See Abstr. 035.015.

A balloon–borne sun–tracking multichannel photometer for atmospheric aerosol measurements.
See Abstr. 035.026.

Astrophysical and geophysical observations with PIRAMIG/ Salyut 7 experiment.
See Abstr. 035.049.

Estimation of phase errors on laser observations of artificial earth satellites.
See Abstr. 036.021.

Precise computation of the refractive index correction for laser observations.
See Abstr. 036.023.

Atmospheric turbulence influence on the range error and maximum range.
See Abstr. 036.025.

Passive microwave remote sensing in meteorology and atmospheric physics.
See Abstr. 036.055.

Visualization of atmospheric inhomogeneities during solar observations.
See Abstr. 036.071.

Iterative method of the atmospheric extinction reduction in multicolor broadband stellar photometry.
See Abstr. 036.083.

Effects of finite spectral bandwidth and focusing error on the transfer function in stellar speckle interferometry.
See Abstr. 036.145.

The University of Wyoming's small scientific balloon program.
See Abstr. 051.035.

Space Telescope observations of the Earth's upper atmosphere by stellar occultations.
See Abstr. 051.065.

Use of nonstationary models of the atmosphere for the description of motion of the IKB 1300 artificial earth satellite.
See Abstr. 052.053.

Multimode radiative transfer in finite optical media. I. Fundamentals.
See Abstr. 063.006.

Multimode radiative transfer in finite optical media. II. Solutions.
See Abstr. 063.007.

Water vapor and Fe 5250.2.
See Abstr. 071.009.

The scintillation theory of eclipse shadow bands.
See Abstr. 079.003.

The effects of image motion on the l–v diagram.
See Abstr. 080.046.

The effects of seeing on noise.
See Abstr. 080.047.

Earth.
See Abstr. 081.012.

Corrections to the earth's ellipticity for astronomical refraction.
See Abstr. 081.037.

Images of the earth's aurora and geocorona from the Dynamics Explorer Mission.
See Abstr. 084.097.

Some results of scanner spectrophotometric measurements of aurora and airglow from IK–Bulgaria–1300 satellite and prospects of similar measurements.
See Abstr. 084.135.

Rocket–borne EUV–visible emission measurements.
See Abstr. 084.136.

Evidence of atmospheric gravity waves produced during the 11 June 1983 total solar eclipse.
See Abstr. 085.002.

Plasma–gas interactions in planetary atmospheres and their relevance for the terrestrial hydrogen budget.
See Abstr. 091.052.

The impact of spectroscopic parameters on the composition of the Jovian atmosphere discussed in connection with recent laboratory, Earth and planetary observation programs.
See Abstr. 099.022.

Residual mass from atmospheric ablation of small meteoroids.
See Abstr. 104.001.

Meteoroid influx and ionization irregularities.
See Abstr. 104.010.

The contribution of sporadic meteors to the ionization of the upper atmosphere.
See Abstr. 104.016.

The role of thermal motions of air particles for fireballs.
See Abstr. 104.030.

Interaction of large meteoritic bodies with the earth atmosphere.
See Abstr. 105.233.

The young Sun and the atmospheres of Earth.
See Abstr. 107.007.

On the role of meteoritic impacts in the formation of organic molecules.
See Abstr. 107.009.

On the dependence of atmospheric extinction coefficients of the Strömgren system on spectral type, luminosity, and interstellar reddening.
See Abstr. 113.026.

On the atmospheric extinction coefficients for heterochromatic photometry of stars.
See Abstr. 113.034.

Pulsations of cosmic rays and atmospheric pressure near periods of 160 and 80 min.
See Abstr. 144.062.

083 Ionosphere

083.001   **Global large scale structures in the F region.**
S. H. Gross.
J. Geophys. Res., Vol. 90, No. A1, p. 553 – 558 (1985).

083.002   **Mean auroral E–region plasma convection patterns measured by SABRE.**
J. A. Waldock, T. B. Jones, E. Nielsen.
Nature, Vol. 313, No. 5999, p. 204 – 206 (1985).
Since April 1982, a new radar auroral backscatter system, SABRE (Sweden and Britain Radar auroral Experiment) has been in operation in northern Europe, which allows estimates to be made of plasma convection in the auroral E region over a large area ($\sim 200,000$ km^2) with high spatial and temporal resolution. Here the authors present the flow patterns averaged over this period of radar operation for different levels of magnetic activity, revealing a definite rotation of the whole convection pattern towards earlier local times with increasing activity.

083.003   **Low–frequency fluctuations of the electric field in the equatorial ionosphere.**
V. L. Patel, P. Lagos.
Nature, Vol. 313, No. 6003, p. 559 – 560 (1985).
High time–resolution radar measurements of ionospheric electric fields have recently permitted studies of low–frequency hydromagnetic waves. The purpose of this study is to examine the fluctuations in the electric field on a time scale longer than ion cyclotron periods, $T_{ci} \sim 0.6$ s, which have not been studied at low latitudes. The authors report here observations of these waves with periods larger than 45 s from the high time–resolution data taken at Jicamarca, Peru.

083.004   **Contour dynamics – an interface method for studying the evolution of large density gradient ionospheric plasma clouds.**
E. A. Overman II, N. J. Zabusky.
Physica D, Vol. 12D, No. 1 – 3, p. 145 – 153 (1984). Abstr. in Phys. Abstr., Vol. 88, No. 1248, Entry 9437 (1985). – See Abstr. 012.012.

083.005   **Ionosphere–magnetosphere coupling and convection.**
R. A. Wolf, R. W. Spiro.
ESA Spec. Publ., ESA SP–217, p. 417 – 426 (1984). – See Abstr. 012.021.
Some IMS–associated attempts at quantitative modeling of specific observed ionosphere–magnetosphere events are reviewed, including a theoretical model of convection, algorithms for deducing ionospheric current and electric–field patterns from sets of ground magnetograms and appropriate ionospheric conductivity information, and empirical models of ionospheric conductances, polar–cap potential drops, etc. A few topics in the active research area of magnetic–field–aligned electric fields are reviewed very briefly, particularly magnetic–mirror effects and double layers.

083.006   **Estimation of ionospheric electric fields and currents from a regional magnetometer array.**
M. Murison, A. D. Richmond, S. Matsushita, W. Baumjohann.
J. Geophys. Res., Vol. 90, No. A4, p. 3525 – 3530 (1985).

083.007   **Local auroral disturbance in the morning sector of the upper ionosphere as a standing electromagnetic wave.**
Eh. M. Dubinin, P. L. Izrajlevich, I. Kutiev, N. S. Nikolaeva, I. M. Podgornyj.
Inst. kosm. issled. Akad. Nauk SSSR Prepr., No. 926, 29 pp. (1984). In Russian. Abstr. in Ref. Zh., 62. Issled. Kosm. Prostranstva, 2.62.444 (1985).

083.008   **Predicting the ionosphere.**
D. G. Cole.
Nature, Vol. 315, No. 6016, p. 182 (1985).

083.009   **Annual variations of the parameters of the day–time F2 layer in the region of the geomagnetic equator.**
T. Yu. Leshchinskaya, A. V. Mikhajlov.
Geomagn. Aehron., No. 1, p. 42 – 46 (1985). In Russian. English translation in Geomagn. Aeron.

083.010   **Connection of absolute and relative ion concentrations in the lower part of the F region.**
L. A. Antonova, G. S. Ivanov–Kholodnyj.
Geomagn. Aehron., No. 1, p. 47 – 50 (1985). In Russian. English translation in Geomagn. Aeron.

083.011   **On meridional winds in the ionosphere.**
A. I. Gvelesiani.
Geomagn. Aehron., No. 1, p. 58 – 62 (1985). In Russian. English translation in Geomagn. Aeron.

083.012 A kind of short wave band SID records and its application in the observation and study of solar X–ray flares.
Z. Xiao, S.–l. Zhang, J.–c. Zuo.
Chin. J. Space Sci., Vol. 4, No. 3, p. 191 – 197 (1984). In Chinese.
Abstr. in Phys. Abstr., Vol. 88, No. 1252, Entry 29777 (1985).

083.013 Zur Berechnung ionosphärischer Refraktionskorrekturen für VLBI–Beobachtungen aus simultanen Dopplermessungen nach Satelliten.
F. J. Lohmar.
Mitt. Geod. Inst. Rheinischen Friedrich–Wilhelms–Univ. Bonn, Nr. 67, 111 pp. (1985).

083.014 The glow of the night ionosphere due to energetic ion beams.
T. G. Adeishvili (*T. G. Adejshvili*), T. I. Gagua, G. G. Managadze.
Adv. Space Res., Vol. 5, No. 4, p. 251 – 254 (1985). – See Abstr. 012.066.
This paper discusses photometric measurements made of the ionospheric excitation of the line $\lambda 5577$ Å at the time of electron beam injection from a rocket into the earth ionosphere.

083.015 Plasma motion along the lines of the magnetic field in the ionosphere.
A. V. Pavlova.
Izv. Vyssh. Uchebn. Zaved., Radiofiz., Tom 27, No. 11, p. 1474 – 1477 (1984). In Russian. Abstr. in Ref. Zh., 62. Issled. Kosm. Prostranstva, 6.62.416 (1985).

083.016 Modelling of the daily development of stratifications of the F 2 region of the equatorial ionosphere. I. The aeronomic aspect.
V. A. Surotkin, A. A. Namgaladze, O. P. Kolomijtsev.
Geomagn. Aehron., Tom 25, No. 3, p. 394 – 399 (1985). In Russian. English translation in Geomagn. Aeron.

083.017 Distribution of the ion concentration according to height in the outer ionosphere.
G. S. Ivanov–Kholodnyj, Yu. K. Kalinin.
Geomagn. Aehron., Tom 25, No. 3, p. 400 – 405 (1985). In Russian. English translation in Geomagn. Aeron.

083.018 Structure of the subauroral ionosphere during a magnetospheric storm from data of the Intercosmos 19 AES.
M. G. Deminov, A. T. Karpachev, Yu. V. Kushnerevskij, J. Šmilauer.
Geomagn. Aehron., Tom 25, No. 3, p. 406 – 410 (1985). In Russian. English translation in Geomagn. Aeron.

083.019 Intensive localized disturbances of the auroral ionosphere.
Eh. M. Dubinin, I. M. Podgornyj, V. M. Balebanov, L. Bankov, N. Bankov, G. L. Gdalevich, Ts. Dachev, L. I. Zhuzgov, I. Kutiev, V. I. Lazarev, N. S. Nikolaeva, K. Serafimov, G. Stanev, D. Teodos'ev.
Kosm. Issled., Tom 23, Vyp. 3, p. 449 – 465 (1985). In Russian. English translation in Cosm. Res.

083.020 A theoretical study of the global F region for June solstice, solar maximum, and low magnetic activity.
J. J. Sojka, R. W. Schunk.
J. Geophys. Res., Vol. 90, No. A6, p. 5285 – 5298, 5311 – 5313 (1985).
The authors constructed a time–dependent, three–dimensional, multi–ion numerical model of the global ionosphere at F region altitudes. The model takes account of all the processes included in the existing regional models of the ionosphere.

083.021 The latitudinal dependence of the corpuscular ionization of the ionospheric F–region during the Forbush–decrease of cosmic rays intensity.
I. D. Kozin, B. T. Zhumabaev, B. M. Rubinstein (*B. M. Rubinshtejn*).
18th International Cosmic Ray Conference, Vol. 3, p. 520 – 522 (1983). – See Abstr. 012.096.

083.022 The reaction of F–region of the ionosphere on cosmic rays.
I. D. Kozin, B. T. Zhumabaev, B. M. Rubinstein (*B. M. Rubinshtejn*).
18th International Cosmic Ray Conference, Vol. 10, p. 245 – 247 (1983). – See Abstr. 012.096.
The change of electron concentration in the maximum of F–2 region during the cosmic ray flares was considered.

083.023 Determination of the ionization by cosmic rays during daytime.
I. D. Kozin, B. T. Zhumabaev, B. M. Rubinstein (*B. M. Rubinshtejn*).
18th International Cosmic Ray Conference, Vol. 10, p. 270 – 272 (1983). – See Abstr. 012.096.
It is supposed that two sources determine the ionization atmosphere velocity in the levels of D–region: one of them is the electromagnetic radiation of the sun and the other is particle radiation that is the proton flux.

083.024 Contribution on "Intercosmos–Bulgaria–1300 satellite": large–scale structure and dynamics of the polar ionosphere during geomagnetic storms.
K. Serafimov, I. Kutiev.
Results of the ARCAD 3 project and of the recent programmes in magnetospheric and ionospheric physics, p. 735 – 745 (1985). – See Abstr. 012.097.
To analyze the auroral F–region dynamics during the course of a geomagnetic storm, data from simultaneous measurements of electron precipitation fluxes in the energy range 0.2 – 15 keV, emission intensity of 6300 Å, 5577 Å and 4278 Å, field aligned currents, ion drift and density and temperature of the cold plasma around the midnight sector of the auroral oval are used, as well as data from ground based magnetometers. It is shown that the main structure characteristics have different inertia to the changes of the intensity of the auroral electrojet.

083.025 Comparative magnetospheric/ionospheric studies using Dynamics Explorer spacecraft and ground–based radars.
J. L. Green, M. O. Chandler, C. R. Chappell.
Results of the ARCAD 3 project and of the recent programmes in magnetospheric and ionospheric physics, p. 885 – 894 (1985). – See Abstr. 012.097.
Coincident observations between the Dynamics Explorer 1 and 2 spacecraft with Chatanika and Arecibo ground–based radar are briefly reviewed. Ionospheric–magnetospheric coupling processes in the plasmasphere, main trough, auroral zone, and polar cap are inferred from the density, temperature, composition, and angular distributions of the low–energy plasma observed from the E–region ionosphere out into the magnetosphere to an altitude of 2.5 earth radii.

083.026 D–region variability.
B. S. N. Prasad, S. Chandramma.
Gerlands Beitr. Geophys., Band 94, Heft 3, p. 177 – 186 (1985).
Rocket data on positive ion and electron densities have been analysed using a simplified D–region model. Mesospheric nitric oxide concentration, electron loss coefficient, and thermal structure are derived for different D–region conditions.

083.027 Simultaneous measurements of the ionospheric electric field by probes and radar methods.
R. Grabowski, C. Hanuise, E. Nielsen, J. P. Villain, H. Wolf.
J. Atmos. Terr. Phys., Vol. 47, No. 1 – 3, p. 41 – 48 (1985).
Ionospheric electric field values are presented, obtained simultaneously by the double probe technique on board a rocket and by two incoherent backscatter radar installations. The measurements were performed during auroral activity over northern Scandinavia. The spatial distribution of the field reveals pronounced local variations.

083.028 **A method for determination of the complete $N_e(h)$–profile by combined satellite and ground–based optical and plasma measurements.**
K. B. Serafimov.
Dokl. Bolg. Akad. Nauk, Tome 37, No. 12, p. 1633 – 1636 (1984).

The author deals with combinations of satellite measurements of the local electron (or ion) concentration, of the red O line ($\lambda6300$ Å) and the u.v. O line ($\lambda1356$ Å) with ground–based vertical sounding with an ionosonde. He shows that in this case it is possible to build a complete profile $N_e(h)$ in the entire ionosphere from the lowest sounding level up to 1000 km above the earth.

083.029 **Dayside high–latitude ionospheric current systems.**
W. Baumjohann, E. Friis–Christensen.
The polar cusp. J. A. Holtet, A. Egeland (Editors). D. Reidel Publishing Company, Dordrecht – Boston – Lancaster (1985), p. 223 – 234.

This paper reviews the present understanding of the current flow in the dayside high–latitude ionosphere. The authors describe the morphology of this current circuit (on the basis of ground–based and low–altitude satellite magnetic measurements) and then point out its essential elements and their relation to the dynamics of the interplanetary magnetic field. Subsequently, they discuss possible generation mechanisms for these currents, especially merging between interplanetary and magnetospheric field lines.

The night F region of the ionosphere in the period of flares on the sun.
See Abstr. 003.058.

Wave phenomena in the ionosphere and in cosmic plasma.
See Abstr. 003.096.

Intercomparison of stratospheric/mesospheric data. Proceedings of the Topical Meeting of the COSPAR Interdisciplinary Scientific Commission A (Meeting A1) of the COSPAR Twenty–fifth Plenary Meeting held in Graz, Austria, 25th June – 7th July 1984.
See Abstr. 012.006.

Response of Earth and Venus ionospheres to corotating solar wind stream of 3 July 1979.
See Abstr. 074.079.

Ground–based and near–earth observations during the IMS.
See Abstr. 084.045.

IMS advances in studies of field–aligned currents and their related electrodynamics.
See Abstr. 084.048.

Electrodynamics of magnetosphere–ionosphere coupling.
See Abstr. 084.049.

Comparison of magnetospheric and ionospheric electric fields using different empirical mapping factors.
See Abstr. 084.051.

Magnetosphere–ionosphere interaction through hydromagnetic waves.
See Abstr. 084.057.

Consequences of imperfect magnetosphere–ionosphere coupling.
See Abstr. 084.131.

Magnetospheric–ionospheric coupling deduced from measurements above a discrete auroral arc.
See Abstr. 084.132.

Change of the critical frequencies of the E layer with solar activity at various phases of the solar cycle.
See Abstr. 085.009.

Influence of the state of the ionosphere on physical processes in meteor trains.
See Abstr. 104.003.

Corotating interplanetary streams and associated ionospheric disturbances at Venus and Earth.
See Abstr. 106.021.

On the influence of the sector structure of the interplanetary magnetic field on the upper ionosphere of the earth.
See Abstr. 106.035.

Ionospheric detection of X–ray pulsars.
See Abstr. 142.047.

084 Aurorae, Geomagnetic Field, Magnetosphere

084.001 Computer simulation of reconnection in planetary magnetospheres.
J. Birn.
Unstable current systems and plasma instabilities in astrophysics, p. 167 – 184 (1985). – See Abstr. 012.002 (IAU Symp. No. 107).

The earth's magnetosphere provides an ideal opportunity to model reconnection in well known geometries that are close enough to the idealized analytic models to make a comparison of the computer models with analytic theory meaningful. In addition more detailed, even three–dimensional, models can be used for a comparison with extended data from in situ observations. The computer studies have basically confirmed the reconnection picture that was based on two–dimensional steady state models and linear analytic theory. The three–dimensional models in particular have also added a lot more information on the reconnection process and the structure of flow, magnetic fields, and currents including many features that are consistent with observations and empirical models of geomagnetic substorms.

084.002 Observations of the Earth's cross–tail current sheet and their implications.
A. T. Y. Lui.
Unstable current systems and plasma instabilities in astrophysics, p. 303 – 307 (1985). – See Abstr. 012.002 (IAU Symp. No. 107).

Observations of the neutral sheet in the Earth's magnetotail are presented to show different magnetic signatures of the neutral sheet which have been used to infer (1) wave profiles on the neutral sheet surface, (2) magnetic islands embedded in the neutral sheet, and (3) localized turbulent magnetic field regions. The occurrence of these features even at magnetospheric quiet conditions suggests that the above features are intrinsic to the current sheet and may possibly play a role in its stability. There are indications that these features are common to other current sheets in space.

084.003 Non–stochastic acceleration of protons in the magnetic neutral sheet.
J.–i. Sakai, R. Sugihara.
Unstable current systems and plasma instabilities in astrophysics, p. 513 – 518 (1985). – See Abstr. 012.002 (IAU Symp. No. 107).

A rapid non–stochastic proton acceleration mechanism by electrostatic waves during the substorm activity in the magnetospheric tail is presented to explain the origin of energetic protons (up to MeV). The protons are accelerated normal to the neutral sheet. Near a reconnection point, however, the protons are also accelerated along the sheet by a second process.

084.004 Geomagnetic field analysis – III. Magnetic fields on the core–mantle boundary.
D. Gubbins, J. Bloxham.
Geophys. J. R. Astron. Soc., Vol. 80, No. 3, p. 695 – 713 (1985).

084.005 Analytical two–dimensional model of a quadrupole magnetosphere.
H. K. Biernat, N. I. Kömle, H. I. M. Lichtenegger.
Planet. Space Sci., Vol. 33, No. 1, p. 45 – 52 (1985).

A model of a quadrupole magnetosphere is developed with the aid of conformal mappings. Such a type of magnetosphere might have occurred at the Earth during times of polarity reversals of the dipole field, when the residual field essentially may have had a quadrupole structure. The magnetospheric field–line configuration is calculated for an axial quadrupole directed perpendicular to the solar–wind flow. The model is two–dimensional and leads to analytical formulae.

084.006 Ground–based observations of O_2 ($b^1\Sigma_g{}^+-X^3\Sigma_g{}^-$) atmospheric bands in high latitude auroras.
K. Henriksen, G. G. Sivjee, C. S. Deehr, H. K. Myrabø.
Planet. Space Sci., Vol. 33, No. 1, p. 119 – 125 (1985).

084.007 Some auroral properties from far ultraviolet observations.
L. Monchick, M. J. Linevsky, C. I. Meng, S. Favin, S. Chakrabarti, F. Paresce.
Planet. Space Sci., Vol. 33, No. 2, p. 175 – 181 (1985).

084.008 Projection of auroral intensity contours into the magnetosphere.
M. M. Shepherd, G. G. Shepherd.
Planet. Space Sci., Vol. 33, No. 2, p. 183 – 189 (1985).

Isointensity contours of 630 nm auroral emission are traced into the magnetosphere, using two different empirical magnetic field models, the Mead–Fairfield model, and the Hedgecock–Thomas model.

084.009 The height, spectrum and mechanism of type–B red aurora and its bearing on the excitation of $O(^1S)$ in aurora.
R. L. Gattinger, F. R. Harris, A. Vallance Jones.
Planet. Space Sci., Vol. 33, No. 2, p. 207 – 221 (1985). = Natl. Res. Council Can., pap. No. 23862.

084.010 Solar wind variations and geomagnetic storms: a study of individual storms based on high time resolution ISEE 3 data.
S.–I. Akasofu, C. Olmsted, E. J. Smith, B. Tsurutani, R. Okida, D. N. Baker.
J. Geophys. Res., Vol. 90, No. A1, p. 325 – 340 (1985).

084.011 Radar and photometric measurements of an intense type A red aurora.
R. M. Robinson, S. B. Mende, R. R. Vondrak, J. U. Kozyra, A. F. Nagy.
J. Geophys. Res., Vol. 90, No. A1, p. 457 – 466 (1985).

On the evening of March 5, 1981, an intense, type A red aurora appeared over southern Alaska. Radar and photometric measurements were made of the aurora from the Chatanika radar site. The line of sight intensity of the 630.0–nm emissions exceeded 150 kR and was accompanied by enhanced emissions at 486.1 and 427.8 nm. The Chatanika radar measured electron densities of $10^6 cm^{-3}$ and electron temperatures of 6000 K at an altitude of 400 km and an invariant latitude of 59° in association with the aurora. Comparison of optical and radar measurements indicated that the 630.0–nm emissions were produced to a large degree by thermal excitation of $O(^1D)$ in the region of high electron temperatures and densities. Model calculations are discussed.

084.012 Near equality of ion phase space densities at Earth, Jupiter, and Saturn.
A. F. Cheng, S. M. Krimigis, T. P. Armstrong.
J. Geophys. Res., Vol. 90, No. A1, p. 526 – 530 (1985).

Energetic ion phase space densities are nearly equal in the Earth's plasmasphere, the Io plasma torus/inner Jovian plasma sheet, and the heavy ion plasma torus of Saturn. The thermal ion phase space densities are likewise nearly the same in all three cases. This near equality of ion phase space densities suggests that common physical processes govern the ion transport, sources, and losses in these regions.

084.013 Comment on "The visual aurora as a predictor of solar activity" by S. M. Silverman [J. Geophys. Res., Vol. 88, No. A10, p. 8123 – 8128 (1983)].
J. P. Legrand, P. A. Simon.
J. Geophys. Res., Vol. 90, No. A1, p. 565 – 567 (1985). See Abstr. 34.084.033.

084.014 Reply to the comment on "The visual aurora as a predictor of solar activity" by J. P. Legrand and P. A. Simon [J. Geophys. Res., Vol. 90, No. A1, p. 565 – 567 (1985)].
S. M. Silverman.
J. Geophys. Res., Vol. 90, No. A1, p. 569 – 572 (1985). See Abstr. 084.013.

084.015 Geomagnetic and solar data. September 1984.
H. E. Coffey.
J. Geophys. Res., Vol. 90, No. A1, p. 574 (1985).

084.016 Statistical structure of geomagnetic reversals.
D. V. Kent.
Nature, Vol. 313, No. 5997, p. 15 (1985).

084.017 Dynamics of the 1054 UT March 22, 1979, substorm event: CDAW 6.
R. L. McPherron, R. H. Manka.
J. Geophys. Res., Vol. 90, No. A2, p. 1175 – 1190 (1985). – See Abstr. 012.008.
The physical processes involved in the transfer of energy from the solar wind to the magnetosphere, and release associated with substorms, have been examined in a sequence of Coordinated Data Analysis Workshops (CDAW 6). Magnetic storms of March 22 and 31, 1979, were chosen to study the problem, using a data base from 13 spacecraft and about 130 ground–based magnetometers. This paper describes the March 22 storm, in particular the large, isolated substorm at 1054 UT which followed an interval of magnetic calm. The authors summarize the observations in the solar wind, in various regions of the magnetosphere, and at the ground, synthesizing these observations into a description of the substorm development. They then give their interpretation of these observations and test their consistency with the reconnection model.

084.018 Coupling between the solar wind and the magnetosphere: CDAW 6.
B. T. Tsurutani, J. A. Slavin, Y. Kamide, R. D. Zwickl, J. H. King, C. T. Russell.
J. Geophys. Res., Vol. 90, No. A2, p. 1191 – 1199 (1985). – See Abstr. 012.008.
An extensive study of the causes and manifestations of geomagnetic activity has been carried out as part of the sixth Coordinated Data Analysis Workshop, CDAW 6. It is the purpose of this paper to determine the coupling between the solar wind and the magnetosphere for the two selected analysis intervals by using the interplanetary field and plasma observations from ISEE 3 and IMP 8 and the geomagnetic activity indicators developed by CDAW 6 participants. From the field and plasma data, interplanetary indices were formed and compared to geomagnetic activity via correlation analyses.

084.019 Solar wind control of magnetospheric pressure (CDAW 6).
D. H. Fairfield.
J. Geophys. Res., Vol. 90, No. A2, p. 1201 – 1204 (1985). – See Abstr. 012.008.
The CDAW 6 data base is used to compare solar wind and magnetospheric pressures. The flaring angle of the tail magnetopause is determined by assuming that the component of solar wind pressure normal to the tail boundary is equal to the total pressure within the tail. Results indicate an increase in the tail flaring angle from 18° to 32° prior to the 1055 substorm onset and a decrease to 25° after the onset. This behavior supports the concept of tail energy storage before the substorm and subsequent release after the onset.

084.020 Magnetotail energy storage and release during the CDAW 6 substorm analysis intervals.
D. N. Baker, T. A. Fritz, R. L. McPherron, D. H. Fairfield, Y. Kamide, W. Baumjohann.
J. Geophys. Res., Vol. 90, No. A2, p. 1205 – 1216 (1985). – See Abstr. 012.008.
Using solar wind, magnetotail, and geostationary particle and field data, as well as ground–based information, the authors present the evidence for enhanced solar wind–magnetosphere coupling and concomitant increases of stored magnetotail energy. Clear examples of energy storage are found prior to the 1055 and 1435 UT substorms of March 22, as well as the 0250 UT and 2250 UT substorms on April 1 and March 31, respectively. In these cases the authors estimate the total energy increase in the tail prior to the substorm onsets, and they estimate the dissipation rate of this energy during the substorms themselves.

084.021 Magnetotail plasma observations during the 1054 UT substorm on March 22, 1979 (CDAW 6).
G. Paschmann, N. Sckopke, E. W. Hones Jr.
J. Geophys. Res., Vol. 90, No. A2, p. 1217 – 1229 (1985). – See Abstr. 012.008.
It is the purpose of this paper to report plasma observations on ISEE 1 and 2 during the 1054 substorm on March 22, 1979, and discuss their implications for the substorm process.

084.022 Recurrent geomagnetic activity: evidence for long–lived stability in solar wind structure.
H. H. Sargent III.
J. Geophys. Res., Vol. 90, No. A2, p. 1425 – 1428 (1985).
The 27–day recurrence index is presented as a diagnostic for long–lived stable structures in the solar wind, in the ecliptic plane at 1 AU. The index may also be of use as a tool for inferring coronal hole evolution over the past 116 years. A 22–year periodic asymmetry, which must certainly have its source at the sun, is clearly evident in the analysis.

084.023 The aurora.
R. J. Livesey.
J. Br. Astron. Assoc., Vol. 95, No. 2, p. 67 – 69 (1985).

084.024 Auroral images during the substorm of 27 July 1979.
M. W. J. Scourfield, P. A. Wakerley, P. R. Sutcliffe, D. P. Smits.
S.Afr. J. Phys., Vol. 7, No. 1, p. 15 – 19 (1984). Abstr. in Phys. Abstr., Vol. 88, No. 1248, Entry 9407 (1985).

084.025 Proton aurora during the magnetospheric substorm of 27 July 1979.
P. R. Sutcliffe, D. P. Smits.
S.Afr. J. Phys., Vol. 7, No. 1, p. 20 – 23 (1984). Abstr. in Phys. Abstr., Vol. 88, No. 1248, Entry 9408 (1985).

084.026 Photometric observations of the aurora during the substorm of 27 July 1979.
I. S. Dore.
S.Afr. J. Phys., Vol. 7, No. 1, p. 24 – 27 (1984). Abstr. in Phys. Abstr., Vol. 88, No. 1248, Entry 9409 (1985).

084.027 Supply of energy to the magnetosphere during the main phase of magnetic storms.
P. Ochabova.
Stud. Geophys. Geod., Vol. 28, No. 1, p. 82 – 89 (1984). Abstr. in Phys. Abstr., Vol. 88, No. 1248, Entry 9444 (1985).

084.028 N_2^+ emission in sunlit cusp and night–side aurora.
K. Henriksen.
Ann. Geophys., Vol. 2, No. 4, p. 457 – 462 (1984). Abstr. in Phys. Abstr., Vol. 88, No. 1251, Entry 24051 (1985).

084.029 O I 7774 Å and 8446 Å emissions from night–side and midday cusp auroras.
G. G. Sivjee, A. B. Christensen, K. Henriksen, A. E. Belon.
Ann. Geophys., Vol. 2, No. 4, p. 463 – 466 (1984). Abstr. in Phys. Abstr., Vol. 88, No. 1251, Entry 24052 (1985).

084.030 The relationship between the Northern Lights and energy exchange from interplanetary space to the upper atmosphere.
P. E. Sandholt, A. Egeland.
Fra Fys. Verden, Vol. 46, No. 3, p. 56 – 60 (1984). In Norwegian. Abstr. in Phys. Abstr., Vol. 88, No. 1251, Entry 24091 (1985).

084.031   Results from the GEOS–1 and GEOS–2 spacecraft and their contributions to the IMS.
K. Knott.
ESA Spec. Publ., ESA SP–217, p. 9 – 23 (1984). – See Abstr. 012.021.
Contents: Introduction. The environment at the geostationary orbit. Wave–particle interactions. Multi–satellite observations. Studies of magnetic conjugate phenomena. The GEOS equilibrium potential. Conclusion.

084.032   Some contributions to knowledge of the magnetospheric plasma by ISEE–1 investigators.
K. W. Ogilvie.
ESA Spec. Publ., ESA SP–217, p. 25 – 38 (1984). – See Abstr. 012.021.
The ISEE project has made substantial contributions to our knowledge of the magnetosphere during the period of the IMS, especially in the discipline of space plasma physics. This paper reviews results obtained during approximately the first two years of the operation of ISEE–1 and –2, and touches on relevant results of ISEE–3.

084.033   Observations of auroras and related plasma phenomena by the Kyokko and Jikiken satellites in IMS.
H. Oya.
ESA Spec. Publ., ESA SP–217, p. 39 – 48 (1984). – See Abstr. 012.021.
Contents: Introduction. Mission summary and objectives. Scientific achievements by UV aurora and energy particle detectors. Auroral kilometric radiation. Plasmapause and plasma convections. Wave particle interactions. Concluding remarks.

084.034   The measurements on Prognoz–4, 5, 6, 7 during IMS.
A. A. Galeev.
ESA Spec. Publ., ESA SP–217, p. 49 – 52 (1984). – See Abstr. 012.021.
During the International Magnetospheric Study (IMS) from 1975 to 1978 four satellites of the Prognoz series were launched, equipped with scientific instruments for measuring plasma and magnetic fields in the near–earth and interplanetary space as well as solar electromagnetic and corpuscular radiation. The paper describes the modes of plasma and field measurements of these satellites. The paper gives also a brief review of original results on plasmasphere heating, ion composition of the plasma mantle and penetration of solar wind plasma into the magnetosphere.

084.035   The earth's magnetopause.
G. Paschmann.
ESA Spec. Publ., ESA SP–217, p. 53 – 64 (1984). – See Abstr. 012.021.
Owing to the IMS, great advances have been made in recent years in our understanding of the magnetopause boundary region. This paper is intended to provide a brief review of the major achievements.

084.036   The structure and dynamics of the magnetosphere: progress in the IMS.
C. T. Russell.
ESA Spec. Publ., ESA SP–217, p. 67 – 76 (1984). – See Abstr. 012.021.
While many of our ideas about the structure and dynamics of the magnetosphere had their origins prior to the IMS, many of these concepts crystalized and many existing theories were confirmed during the IMS. This paper reviews the state of our knowledge at the beginning of the IMS, how it matured during the IMS and where we stand today.

084.037   Observations of the magnetospheric boundary layers.
T. E. Eastman.
ESA Spec. Publ., ESA SP–217, p. 77 – 83 (1984). – See Abstr. 012.021.
Recent results on magnetospheric boundary layers are reviewed with an emphasis on their dynamical importance based on hot plasma observations, energetic particle signatures, heavy ion contributions and the effects of waveparticle interactions.

084.038   Hot plasma and energetic particles in the earth's outer magnetosphere: new understandings during the IMS.
D. N. Baker, T. A. Fritz.
ESA Spec. Publ., ESA SP–217, p. 85 – 96 (1984). – See Abstr. 012.021.
The authors review the major accomplishments made during the IMS period in clarifying magnetospheric particle variations in the region from roughly geostationary orbit altitudes into the deep magnetotail. The authors divide their review into three topic areas: (1) acceleration processes; (2) transport processes; and (3) loss processes.

084.039   Magnetic field in the earth's outer magnetosphere: Prognoz–6,7 data.
A. E. Antonova, E. G. Eroshenko, V. A. Styazhkin.
ESA Spec. Publ., ESA SP–217, p. 117 – 120 (1984). – See Abstr. 012.021.
The Prognoz–6,7 measurements of magnetic field at 8–25 R_E in the outer magnetosphere and in the adjacent regions from 1977 to 1979 carried out within the IMS program are presented. The experimental data are compared with the magnetospheric model predictions.

084.040   Satellite–borne electric field measurements with spherical double probes.
C.–G. Fälthammar, P.–A. Lindqvist, C. Cattell, F. Mozer, V. Formisano, A. Pedersen.
ESA Spec. Publ., ESA SP–217, p. 125 – 129 (1984). – See Abstr. 012.021.
The paper briefly summarizes the experimentally determined characteristics of the electric field in the following boundary regions: (1) bow shock, (2) dayside magnetopause, (3) flanks of the magnetosphere, (4) high latitude plasmasheet boundary and (5) neutral sheet.

084.041   Response of magnetospheric substorm activity to changes in the interplanetary magnetized plasma.
G. Rostoker.
ESA Spec. Publ., ESA SP–217, p. 153 – 156 (1984). – See Abstr. 012.021.

084.042   The application of dimensional analysis to the problem of solar wind–magnetosphere energy coupling.
L. F. Bargatze, R. L. McPherron, D. N. Baker, E. W. Hones Jr.
ESA Spec. Publ., ESA SP–217, p. 157 – 160 (1984). – See Abstr. 012.021.
The analyses reported in this paper assume that only magnetohydrodynamic processes are important in controlling the rate of energy transfer. The study utilizes ISEE–3 solar wind observations, the AE index, and U_T from three 10–day intervals during the IMS. Simple linear regression and histogram techniques are used to find the value of the MHD coupling exponent, α, which is consistent with observations of magnetospheric response. The form of the solar wind energy transfer rate is obtained by substitution into an equation of the interplanetary variables whose exponents depend upon α.

084.043   Coupling of the solar wind to measures of magnetic activity.
R. L. McPherron, R. A. Fay, C. R. Garrity, L. F. Bargatze, C. R. Clauer, D. N. Baker, C. Searls.
ESA Spec. Publ., ESA SP–217, p. 161 – 166 (1984). – See Abstr. 012.021.
The technique of linear prediction filtering has been used to generate empirical response functions relating the solar wind electric field to the most frequently used magnetic indices, AL, AU, Dst and ASYM.

084.044   Entry of solar energetic protons into the distant plasma sheet – dependence on distance from the neutral sheet.
P. Poutlis, E. T. Sarris.
ESA Spec. Publ., ESA SP–217, p. 171 – 175 (1984). – See Abstr. 012.021.
The access of solar energetic protons at different parts of the distant plasma sheet is studied by simultaneous observations of

step discontinuities in the solar proton intensities both in the interplanetary space and inside the plasma sheet regime. It is determined that the time delay for access of the solar energetic protons to the position of the spacecraft inside the plasma sheet increases as function of its distance from the neutral sheet.

084.045   Ground–based and near–earth observations during the IMS.
W. Baumjohann.
ESA Spec. Publ., ESA SP–217, p. 215 – 220 (1984). – See Abstr. 012.021.
For the IMS numerous ground–based networks of magnetometers, riometers, all–sky cameras as well as coherent and incoherent backscatter radar facilities were installed or upgraded. Together with rocket– and balloon–borne observations and measurements made by low–altitude satellites this comprehensive set of data provided a unique opportunity for studying the physics of the auroral ionosphere and the magnetosphere–ionosphere coupling.

084.046   Particle sources, transport, storage and precipitation.
K. Schindler.
ESA Spec. Publ., ESA SP–217, p. 221 – 229 (1984). – See Abstr. 012.021.
This paper discusses theoretical and observational aspects of the energy flow through the magnetosphere with emphasis on progress made during the IMS–period.

084.047   Ring current dynamics and plasma sheet sources.
L. R. Lyons.
ESA Spec. Publ., ESA SP–217, p. 233 – 242 (1984). – See Abstr. 012.021.
The two main questions concerning the storm time ring current are what is the source of the energized plasma that forms the ring current, and how does the ring current decay. Progress on these questions is reviewed in this paper.

084.048   IMS advances in studies of field–aligned currents and their related electrodynamics.
Y. Kamide.
ESA Spec. Publ., ESA SP–217, p. 243 – 256 (1984). – See Abstr. 012.021.
The main purpose of this paper is to attempt to evaluate the IMS progress in studies of field–aglined currents and their related electrodynamic processes in the magnetosphere–ionosphere system.

084.049   Electrodynamics of magnetosphere–ionosphere coupling.
J. R. Kan.
ESA Spec. Publ., ESA SP–217, p. 257 – 266 (1984). – See Abstr. 012.021.
The main purpose of this review is to bring forth a coherent picture of physical processes fundamental to the magnetosphere–ionosphere coupling. Emphasis is placed on the theoretical aspects of the coupling.

084.050   Auroral X–rays and equatorial electrons in the morning sector.
K. M. Torkar, W. Riedler, G. Kremser, A. Korth, S. Ullaland, J. Stadsnes, J. Bjordal, L. P. Block, I. B. Iversen, J. Kangas, P. Tanskanen.
ESA Spec. Publ., ESA SP–217, p. 319 – 324 (1984). – See Abstr. 012.021.

084.051   Comparison of magnetospheric and ionospheric electric fields using different empirical mapping factors.
R. Schmidt, E. Nielsen, A. Pedersen.
ESA Spec. Publ., ESA SP–217, p. 355 – 358 (1984). – See Abstr. 012.021.

084.052   The relationship between auroral electrojet and visual aurora.
M. Ayukawa, K. Makita.
ESA Spec. Publ., ESA SP–217, p. 379 – 382 (1984). – See Abstr. 012.021.
On the basis of ground magnetic field and auroral observations in Antarctica during the IMS period, the authors examined fluctuation of small scale auroral electrojet current regions and compared them to spatial and temporal variation of visual auroras by using three ground stations data.

084.053   On–off characteristics of auroral pulsations.
T. Yamamoto.
ESA Spec. Publ., ESA SP–217, p. 387 – 390 (1984). – See Abstr. 012.021.

084.054   Multiple correlation between auroral and magnetic pulsations.
T. Oguti, J. H. Meek, K. Hayashi.
ESA Spec. Publ., ESA SP–217, p. 391 – 394 (1984). – See Abstr. 012.021.

084.055   Multiple correlation between auroral and magnetic pulsations. 2. Determination of electric currents and electric fields around a pulsating auroral patch.
T. Oguti, K. Hayashi.
ESA Spec. Publ., ESA SP–217, p. 395 – 398 (1984). – See Abstr. 012.021.

084.056   Solar wind control of electric fields and currents in the magnetosphere and ionosphere.
O. A. Troshichev.
ESA Spec. Publ., ESA SP–217, p. 407 – 416 (1984). – See Abstr. 012.021.
This review is concerned with the experimental evidence for the direct relationships between the solar wind parameters and the electric fields and currents in the magnetosphere, and with possible mechanisms responsible for these relationships.

084.057   Magnetosphere–ionosphere interaction through hydromagnetic waves.
T. Tamao.
ESA Spec. Publ., ESA SP–217, p. 427 – 435 (1984). – See Abstr. 012.021.
Interaction processes between the magnetosphere and ionosphere are reviewed in two different regimes. In the first there is no significant kinetic energy flux of energetic particles along field lines, and the classical shear Alfvén wave with its field–aligned currents and Poynting flux plays an essential role in the localized interaction processes. In the second regime where the field–aligned kinetic energy flux of particles can not be neglected, several possibilities including wave dispersion and particle acceleration are discussed. Finally a few quantitative numerical simulations for the coupling interaction are briefly surveyed.

084.058   Solar wind control of magnetospheric convection.
S. W. H. Cowley.
ESA Spec. Publ., ESA SP–217, p. 483 – 494 (1984). – See Abstr. 012.021.

084.059   A structural classification of geomagnetic pulsations.
F. W. Menk, K. D. Cole.
ESA Spec. Publ., ESA SP–217, p. 709 – 713 (1984). – See Abstr. 012.021.
A structural classification of mid–latitude geomagnetic pulsation activity is presented and its application to Pc–type pulsations described. The existence of physically discrete classes of regular pulsations is illustrated in this way. The use of such a classification scheme in obtaining a better understanding of generation processes is also indicated.

084.060   Advances in magnetospheric plasma–wave research during the IMS.
R. R. Anderson.
ESA Spec. Publ., ESA SP–217, p. 751 – 760 (1984). – See Abstr. 012.021.

084.061   How strong was the earth's magnetic field?
J. Shaw.
Geophys. J. R. Astron. Soc., Vol. 81, No. 1, p. 309 (1985). Abstract. – See Abstr. 012.025.

084.062   Geomagnetic field modelling using cubic splines.
R. A. Langel, D. R. Barraclough, D. J. Kerridge.
Geophys. J. R. Astron. Soc., Vol. 81, No. 1, p. 326 (1985). Abstract. – See Abstr. 012.025.

084.063   The relationship between the IMF magnitude and the frequency of Pc 3, 4 pulsations on the ground: simultaneous events.
T. J. Odera, W. F. Stuart.
Planet. Space Sci., Vol. 33, No. 4, p. 387 – 393 (1985).

Several attempts have been made to predict the strength of the interplanetary magnetic field (IMF) from the frequency of Pc 3, 4 pulsations measured on the ground. In this paper the authors show the correlation between the IMF magnitude and the frequency of coincident pulsation events in a network of five stations in the IGS magnetometer array.

084.064   Energetic distribution of protons with $0.05 \leqslant E \leqslant 50$ MeV in the radiation belts of the earth.
M. I. Panasyuk, Eh. N. Sosnovets.
Kosm. Issled., Tom 23, Vyp. 1, p. 106 – 112 (1985). In Russian. English translation in Cosm. Res.

084.065   The aurora, 1983.
R. J. Livesey.
J. Br. Astron. Assoc., Vol. 95, No. 3, p. 100 – 105 (1985).

This report summarizes observations of the polar aurora and magnetic field disturbances collected in 1983 by the Aurora Section and its Magnetometry Group.

084.066   Geomagnetic and solar data. October 1984.
H. E. Coffey.
J. Geophys. Res., Vol. 90, No. A2, p. 1786 – 1787 (1985).

084.067   A note on the nature of the distant geomagnetic tail magnetopause and boundary layer.
J. T. Gosling, M. F. Thomsen, D. W. Swift, L. C. Lee.
Geophys. Res. Lett., Vol. 12, No. 3, p. 153 – 154 (1985).

Recently reported plasma and magnetic field measurements of the distant geomagnetic tail magnetopause and boundary layer (i.e. mantle) are compared with numerical simulation results for an "open" boundary. Most aspects of the observations are consistent with the simulation results provided that the normal magnetic field component at the magnetopause is generally small.

084.068   The mechanism of MHD waves for magnetospheric substorms.
H. Long, N.-h. Xu, W.-r. Hu.
Acta Geophys. Sin., Vol. 28, No. 2, p. 127 – 132 (1985). In Chinese.

The mechanism of a magnetospheric substorm which is produced by the energy transportation from the fluctuation kinetic energy of solar wind is extended to the case of collisionless plasma process. The fluctuations of solar wind quantities at the magnetopause excite the compressed Alfvén waves, which propagate through the collisionless plasma in the tail of the magnetosphere. It is also shown that the energetic particles in the plasma sheet may be produced by the dissipation of energy from the solar wind and are not necessarily originated from the solar wind directly.

084.069   A semiannual geomagnetic variation that is not dependent on the heliographic latitude effect.
P. R. Robinson.
Planet. Space Sci., Vol. 33, No. 5, p. 577 – 580 (1985).

Hourly means of the geomagnetic elements recorded at Lerwick have been analysed to determine the effect of monthly sunspot number on the solar and lunar daily variations. The diurnal term of the solar variation in declination is found to have a distinct semiannual component that is independent of sunspot number. Thus this semiannual variation is not generated by the heliographic latitude or axial process proposed by Cortie (1912).

084.070   Computer modeling of magnetotail convection.
J. Birn, K. Schindler.
J. Geophys. Res., Vol. 90, No. A4, p. 3441 – 3447 (1985).

The authors studied the dynamic evolution of the geomagnetic tail in response to a dawn–to–dusk electric field E_y applied at the high–latitude boundary by means of a two–dimensional nonlinear MHD code.

084.071   Geomagnetic and solar data. November 1984.
H. E. Coffey.
J. Geophys. Res., Vol. 90, No. A4, p. 3547 – 3548 (1985).

084.072   Magnetic reversals and mass extinctions.
D. M. Raup.
Nature, Vol. 314, No. 6009, p. 341 – 343 (1985).

Previous analyses of the time distribution of reversals of the earth's magnetic field have yielded mixed results. There have been repeated suggestions that field reversal is linked to biological extinction. The author presents here the results of a study of the reversal record of the past 165 Myr. A stationary periodicity of 30 Myr emerges (superimposed on the non–stationarities already established by others), which predicts pulses of increased reversal activity centred at 10, 40, 70,...Myr BP.

084.073   Spatial distribution of protons at large and small heights in the radiation belts. Comparison between theory and experiment.
M. I. Panasyuk, S. Ya. Rejzman, Eh. N. Sosnovets.
NII yader. fiz. MGU. Moskva, 29 pp. (1984). In Russian. Abstr. in Ref. Zh., 62. Issled. Kosm. Prostranstva, 2.62.417 (1985).

084.074   Connection between variations of the geomagnetic activity index and parameters of interplanetary scintillations.
V. I. Vlasov, V. I. Shishov, T. D. Shishova.
Geomagn. Aehron., Tom 25, No. 2, p. 254 – 258 (1985). In Russian. English translation in Geomagn. Aeron.

084.075   VLF waves of anomalous polarization in the earth's magnetosphere.
V. P. Voevudskij, B. V. Lundin.
Geomagn. Aehron., Tom 25, No. 2, p. 339 – 340 (1985). In Russian. English translation in Geomagn. Aeron.

084.076   A comparison of three satellite models of the main geomagnetic field.
N. P. Ben'kova, G. I. Kolomijtseva.
Geomagn. Aehron., Tom 25, No. 2, p. 343 – 345 (1985). In Russian. English translation in Geomagn. Aeron.

084.077   On a connection of the radial variation of rigidity and of the form of the proton spectrum in the earth's radiation belts.
A. S. Kovtyukh.
Geomagn. Aehron., No. 1, p. 23 – 28 (1985). In Russian. English translation in Geomagn. Aeron.

084.078   Atomic nuclei in the composition of the inner radiation belt of the earth.
R. A. Nymmik.
Geomagn. Aehron., No. 1, p. 29 – 34 (1985). In Russian. English translation in Geomagn. Aeron.

084.079 Rocket measurements of the atmospheric band $O_2(B'\Sigma g^+ - X^3\Sigma g^-)$ (0 – 0) in aurorae.
O. I. Yagodkina, V. G. Vorob'ev, T. M. Tarasova.
Geomagn. Aehron., No. 1, p. 73 – 78 (1985). In Russian. English translation in Geomagn. Aeron.

084.080 Emissions of O I and N_2^+ in aurorae at day time.
V. S. Davydov, L. S. Evlashin, M. V. Orlova, G. F. Tulinov, T. A. Khviyuzova, V. N. Khokhlov.
Geomagn. Aehron., No. 1, p. 79 – 82 (1985). In Russian. English translation in Geomagn. Aeron.

084.081 Particle transport in planetary magnetospheres.
T. J. Birmingham.
ESA Spec. Publ., ESA SP–207, p. 49 – 56 (1984). – See Abstr. 012.044.
Particle energization in two contrasting planetary magnetospheres, Earth's and Jupiter's, is the theme of the paper. Current understanding of the large scale magnetic and electric fields in which charged particles move is first presented. Orbit theory in the adiabatic approximation is then sketched. General conditions for adiabatic breakdown at each of three levels of periodicity are presented. High energy losses and lower energy sources argue for the existence of magnetospheric accelerations. Non–adiabatic acceleration processes are mentioned briefly. Greatest attention is paid to slow diffusive energization by particle interactions with electromagnetic fluctuations. This mechanism seems adequate at Earth but, operating alone, is in trouble at Jupiter.

084.082 On the problem of sudden commencements of geomagnetic storms as manifestations of shock waves.
J. Halenka.
Stud. Geophys. Geod., Vol. 28, No. 2, p. 164 – 172 (1984). Abstr. in Phys. Abstr., Vol. 88, No. 1252, Entry 29797 (1985).

084.083 Solar activity and variation of geomagnetic disturbance.
K. Sun, X.–h. Cheng, Y.–h. Ming, J.–l. Jing, X. Wang, Z.–p. Li, H. Li.
Chin. J. Space Sci., Vol. 4, No. 3, p. 216 – 221 (1984). In Chinese. Abstr. in Phys. Abstr., Vol. 88, No. 1252, Entry 29799 (1985).

084.084 Magnetic field change across the earth's bow shock: comparison between observations and theory.
D. Winterhalter, M. G. Kivelson, R. J. Walker, C. T. Russell.
J. Geophys. Res., Vol. 90, No. A5, p. 3925 – 3933 (1985). – See Abstr. 012.003.
The authors have examined 204 bow shock crossings observed in the magnetic field and plasma data on ISEE 1. Using the measured upstream field and plasma data, they calculated the downstream field magnitudes by using the single–fluid MHD Rankine–Hugoniot jump conditions, and they compared the results with the observed downstream field magnitudes.

084.085 Numerical studies on magnetotail formation and driven reconnection.
K. Min, H. Okuda, T. Sato.
J. Geophys. Res., Vol. 90, No. A5, p. 4035 – 4045 (1985).
The formation of the magnetotail configuration under the influence of the solar wind plasma flow entering the magnetosphere and the associated driven reconnection in the magnetotail region are studied numerically by means of a two–dimensional time–dependent nonlinear resistive MHD code.

084.086 On the configuration of the polar cusps in earth's magnetosphere.
G.–H. Voigt, R. A. Wolf.
J. Geophys. Res., Vol. 90, No. A5, p. 4046 – 4054 (1985).
The authors discuss the physical conditions that determine the configuration of the polar cusps in earth's magnetosphere.

084.087 Poynting vector as a diagnostic of hydromagnetic wave structure.
H. Junginger.
J. Geophys. Res., Vol. 90, No. A5, p. 4155 – 4163 (1985).
The Poynting vector of resonant shear Alfvén waves which are generated by surface waves in a cold, inhomogeneous plasma is investigated in a simple model of the magnetosphere with ionospheres of finite conductivity. The results are applied to solar–wind–driven magnetospheric pulsations with periods between 100 and 600 s.

084.088 A theoretical model of polar cap auroral arcs.
J. R. Kan, W. J. Burke.
J. Geophys. Res., Vol. 90, No. A5, p. 4171 – 4177 (1985).
A theory of the polar cap auroral arcs is proposed under the assumption that the magnetic field reconnection occurs in the cusp region on tail field lines during northward interplanetary magnetic field (IMF) conditions. Requirements of a convection model during northward IMF are enumerated based on observations and fundamental theoretical considerations. The theta aurora can be expected to occur on the closed field lines convecting sunward in the central polar cap, while the less intense regular polar cap arcs can occur either on closed or open field lines. The dynamo region for the polar cap arcs is required to be on closed field lines convecting tailward in the plasma sheet which is magnetically connected to the sunward convection in the central polar cap.

084.089 Auroral electrodynamics with current and voltage generators.
R. L. Lysak.
J. Geophys. Res., Vol. 90, No. A5, p. 4178 – 4190 (1985).
The electrodynamic structure of auroral currents is studied in the steady state by coupling Maxwell's equations with effective Ohm's laws for ionospheric Pedersen currents, field–aligned currents and currents in the generator region.

084.090 Extreme ultraviolet emissions for monitoring auroras in dark and daylight hemispheres.
C.–I. Meng, S. Chakrabarti.
J. Geophys. Res., Vol. 90, No. A5, p. 4261 – 4268 (1985).
Observations of auroral extreme ultraviolet (EUV) emissions from the noon–midnight polar–orbiting Space Test Program (STP) 78–1 satellite are used to investigate the spectral features of the dayside auroral oval, nightside auroral oval, and the dayglow. The purpose of this study is to evaluate the possibility of using EUV atmospheric emissions for global auroral imaging and to study the temporal variations of the auroral EUV intensity with geomagnetic activity.

084.091 EUV (300–900 Å) spectrum of polar cap and cusp emissions near local noon.
S. Chakrabarti.
J. Geophys. Res., Vol. 90, No. A5, p. 4421 – 4426 (1985).
Emissions in the 300–to–900–Å wavelength range near the dayside oval were observed on April 5, 1979, over the south pole by EUV spectrometer on board the STP78–1 satellite. The spectra, obtained from 600 km, near local noon, are dominated by spectral lines of singly ionized atomic oxygen. Most prominent features in this passband are O II 538–to–539, 617–, and O II 834–Å emission lines, all of which have intensities exceeding 100 R. Several weak O II and N II features are present in the spectrum with intensities ranging from 10 to 60 R. The brightest spectral feature is O II 834 Å, which was recorded with an intensity exceeding 1 kR.

084.092 Zur Dynamotheorie des Erdmagnetfeldes.
M. Stix.
Phys. Bl., 41. Jahrg., Heft 6, p. 152 – 155 (1985).

084.093 Beam driven Alfvénic turbulence in the earth's bow shock.
S. Ghosh, K. Papadopoulos.
Bull. Am. Astron. Soc., Vol. 16, No. 4, p. 922 (1984). Abstract. – See Abstr. 010.062.

084.094 On the trail of Earth's tail.
S. W. H. Cowley.
Nature, Vol. 315, No. 6017, p. 281 – 282 (1985).
This note discusses some new results on the geomagnetic tail obtained with the ISEE–3 spacecraft.

084.095   Evidence for an increase in cosmogenic ^{10}Be during a geomagnetic reversal.
G. M. Raisbeck, F. Yiou, D. Bourles, D. V. Kent.
Nature, Vol. 315, No. 6017, p. 315–317 (1985).

The authors report here evidence in marine sediments for an increase in cosmogenic ^{10}Be production in the Earth's atmosphere during the Brunhes–Matuyama geomagnetic reversal 730,000 yr ago. In addition to confirming an increase in cosmogenic isotope production, the results provide information on the magnitude and duration of the geomagnetic intensity decrease during such an event, and the depth at which remanent magnetism is acquired in marine sediments.

084.096   Plasma convection model and generation of longitudinal currents in the earth's magnetosphere.
G. Krymsky (*G. Krymskij*), P. Krymsky (*P. Krymskij*), Yu. Romaschenko (*Yu. Romashchenko*).
Adv. Space Res., Vol. 5, No. 4, p. 15–18 (1985). – See Abstr. 012.066.

The aim of the paper is to investigate processes in the magnetosphere and in particular the problems of the interaction of the solar wind with the earth's magnetic field to produce large–scale convection, electric fields and longitudinal currents in the magnetosphere. The investigation is carried out in the frame of magnetic hydrodynamics.

084.097   Images of the earth's aurora and geocorona from the Dynamics Explorer Mission.
L. A. Frank, J. D. Craven, R. L. Rairden.
Adv. Space Res., Vol. 5, No. 4, p. 53–68 (1985). – See Abstr. 012.066.

Several results from analyses of auroral and geocoronal images from the Dynamcis Explorer Mission are summarized. (1). The motion of the transpolar arc of a theta aurora is found to be correlated with the y–component of the interplanetary magnetic field. The arc motion is in the general direction of the y–component. (2) A sequence of global images of a small auroral substorm shows the initial development of intense luminosities in a relatively small spatial region, or "bright spot", in the pre-midnight sector of the auroral oval and a subsequent appearance of an expanding area of lesser intensities at lower latitudes and contiguous to the midnight boundary of the bright spot. This evolution of auroral luminosities is discussed in detail. (3) A series of images of the Earth's geocorona in scattered solar Ly α emissions is used to obtain a best–fit spherical model of atomic hydrogen densities in the earth's exosphere.

084.098   On the formation of auroral arcs.
K. Stasiewicz.
Adv. Space Res., Vol. 5, No. 4, p. 83–86 (1985). – See Abstr. 012.066.

A new mechanism for auroral arc formation is presented. The characteristic linear shape of auroral arcs is determined by magnetically connected plasma clouds in the distant equatorial magnetosphere. These clouds originate as high speed plasma beams in the magnetotail and in the solar wind. It is found that the free energy for driving an auroral arc can be provided by the difference of pressure between the cloud and the ambient plasma.

084.099   Passive mode locking in masers with non–equidistant spectra.
P. A. Bespalov.
Zh. Ehksp. Teor. Fiz., Tom 87, No. 6, p. 1894–1905 (1984). In Russian. Abstr. in Ref. Zh., 62. Issled. Kosm. Prostranstva, 6.62.455 (1985).

084.100   MHD modelling of the earth's magnetosphere.
C. C. Wu.
Computer simulation of space plasmas, p. 155–177 (1985). – See Abstr. 003.031.

A global MHD model of the earth's magnetosphere is defined. Some numerical aspects of the model which include shock capturing techique, nonuniform grid system and multiple time scale problem are discussed. Also presented are some recent results.

084.101   Particle behavior in the magnetosphere.
R. A. Wolf, R. W. Spiro.
Computer simulation of space plasmas, p. 227–254 (1985). – See Abstr. 003.031.

The Rice Convection Model deals with large–scale processes in the Earth's inner and middle magnetosphere, including coupling to the ionosphere. This paper reviews work on the model, with emphasis on the assumptions made, the basic equations, and the numerical methods. The theoretical basis of the model is compared and contrasted with standard magnetohydrodynamics. The limitations imposed by the major assumptions are discussed. Model inputs and boundary conditions are listed, and the methods of specifying them discussed. Some physical conclusions and insights that have been gained from the model are listed and described very briefly.

084.102   Hybrid simulation techniques applied to the earth's bow shock.
D. Winske, M. M. Leroy.
Computer simulation of space plasmas, p. 255–278 (1985). – See Abstr. 003.031.

The application of a hybrid simulation model, in which the ions are treated as discrete particles and the electrons as a massless charge–neutralizing fluid, to the study of the earths bow shock is discussed. The essentials of the numerical methods are described in detail: movement of the ions, solution of the electromagnetic fields and electron fluid equations, and imposition of appropriate boundary and initial conditions. Examples of results of calculations for perpendicular shocks are presented. Results for oblique shocks are also presented to show how the magnetic field and ion motion differ from the perpendicular case.

084.103   A connection of geomagnetic pulsations appearance in the range of 0.1–1.0 Hz with the B_z–component of the interplanetary magnetic field.
Eh. T. Matveeva, V. A. Troitskaya, F. Z. Fejgin.
Geomagn. Aehron., Tom 25, No. 3, p. 428–431 (1985). In Russian. English translation in Geomagn. Aeron.

084.104   Convection of magnetospheric plasma on open field lines.
I. I. Alekseev, E. S. Belen'kaya.
Geomagn. Aehron., Tom 25, No. 3, p. 450–457 (1985). In Russian. English translation in Geomagn. Aeron.

084.105   Secular geomagnetic variation in North–West Africa from archeomagnetic and modelling data.
M. Kovacheva, N. P. Ben'kova.
Geomagn. Aehron., Tom 25, No. 3, p. 523–525 (1985). In Russian. English translation in Geomagn. Aeron.

084.106   Macroscopic parameters of electron streams of a plasma layer from measurements in geostationary orbits.
V. I. Dovgij, I. V. Inzhelevskaya, V. N. Kanaleev, V. I. Lazarev, N. K. Osipov, M. V. Tel'tsov, A. G. Kozlov.
Kosm. Issled., Tom 23, Vyp. 3, p. 481–484 (1985). In Russian. English translation in Cosm. Res.

084.107   The influence of the ionospheric conductivity on the potential difference of the electric field across a plasma layer.
V. G. Pivovarov.
Kosm. Issled., Tom 23, Vyp. 3, p. 484–486 (1985). In Russian. English translation in Cosm. Res.

084.108   Double layers above the aurora.
M. Temerin, F. S. Mozer.
Second symposium of plasma double layers and related topics, p. 119–127 (1984). Abstr. in Phys. Abstr., Vol. 88, No. 1253, Entry 35485 (1985). – See Abstr. 012.075.

084.109   The northern light: from mystery to modern space science.
A. Egeland, A. Brekke.
Endeavour New Ser., Vol. 8, No. 4, p. 188–193 (1984). Abstr. in Phys. Abstr., Vol. 88, No. 1254, Entry 40408 (1985).

084.110 Dependence of hydromagnetic energy spectra near $L = 2$ and $L = 3$ on upstream solar wind parameters.
A. Wolfe, A. Meloni, L. J. Lanzerotti, C. G. Maclennan,
J. Bamber, D. Venkatesan.
J. Geophys. Res., Vol. 90, No. A6, p. 5117 – 5131, 5307 – 5310 (1985).
The paper reports the results of correlative studies of solar wind and magnetospheric phenomena to elucidate the transport of hydromagnetic energy in the magnetosphere.

084.111 Model of oval and polar cap arc configurations.
Y. T. Chiu, N. U. Crooker, D. J. Gorney.
J. Geophys. Res., Vol. 90, No. A6, p. 5153 – 5157 (1985).
A model of oval and polar cap arc configurations has been formulated by combining the theory of antiparallel magnetic merging and theories of the evening discrete arc and of the morningside arc systems.

084.112 The poleward leap of the auroral electrojet as seen in auroral images.
E. W. Hones Jr.
J. Geophys. Res., Vol. 90, No. A6, p. 5333 – 5337 (1985).

084.113 Geomagnetic and solar data. January 1985.
H. E. Coffey.
J. Geophys. Res., Vol. 90, No. A6, p. 5365 – 5368 (1985).

084.114 Photometric and interferometric observations of recent SAR arc events.
T. Watanabe, J. S. Kim.
J. Geomagn. Geoelectr., Vol. 36, No. 1, p. 1 – 10 (1984). Abstr. in Phys. Abstr., Vol. 88, No. 1255, Entry 45856 (1985).

084.115 Particle simulation in a dipole field.
B. L. Smith, H. Okuda.
Second symposium on plasma double layers and related topics, p. 321 – 326 (1984). Abstr. in Phys. Abstr., Vol. 88, No. 1255, Entry 45857 (1985). – See Abstr. 012.075.

084.116 Evolution of ion holes and large potential structures in simulations of bounded current–driven magnetized plasma systems.
N. F. Otani.
Second symposium on plasma double layers and related topics, p. 296 – 301 (1984). Abstr. in Phys. Abstr., Vol. 88, No. 1255, Entry 45876 (1985). – See Abstr. 012.075.

084.117 Can Buneman double layers be driven in auroral plasmas?
N. Singh, R. W. Schunk.
Second symposium on plasma double layers and related topics, p. 364 – 369 (1984). Abstr. in Phys. Abstr., Vol. 88, No. 1255, Entry 45877 (1985). – See Abstr. 012.075.

084.118 Additional drifts in the magnetospheric double current layer.
D. K. Callebaut, A. M. van den Buys.
Second symposium on plasma double layers and related topics, p. 370 – 376 (1984). Abstr. in Phys. Abstr., Vol. 88, No. 1255, Entry 45878 (1985). – See Abstr. 012.075.

084.119 Effect of geometry in magnetospheric double current layer.
D. K. Callebaut, A. M. van den Buys.
Second symposium on plasma double layers and related topics, p. 377 – 386 (1984). Abstr. in Phys. Abstr., Vol. 88, No. 1255, Entry 45879 (1985). – See Abstr. 012.075.

084.120 Spatial–temporal characteristics of flickering spots in flickering auroras.
M. Kunitake, T. Oguti.
J. Geomagn. Geoelectr., Vol. 36, No. 4, p. 121 – 138 (1984). Abstr. in Phys. Abstr., Vol. 88, No. 1256, Entry 51394 (1985).

084.121 Determination of a quadrupole magnetopause configuration in two dimensions.
H. K. Biernat, N. I. Komle, H. I. M. Lichtenegger.
Sitzungsber., Österr. Akad. Wiss., Math.–Naturwiss. Kl. Abt. II, Band 192, Heft 8 – 10, p. 319 – 328 (1983). Abstr. in Phys. Abstr., Vol. 88, No. 1257, Entry 57550 (1985).

084.122 A comparison between GEOS 1 magnetic–field measurements and some models of the geomagnetic field.
E. Amata, M. Candidi, R. Orfei, C. Signorini.
Nuovo Cimento C, Vol. 7C, Ser. 1, No. 4, p. 397 – 412 (1984). Abstr. in Phys. Abstr., Vol. 88, No. 1259, Entry 68630 (1985).

084.123 Influence of interplanetary magnetic field and solar wind parameters on main phase of geomagnetic storms.
J. P. Shukla, R. P. Mishra, U. Singh, R. P. Mishra, R. L. Singh.
18th International Cosmic Ray Conference, Vol. 3, p. 274 – 277 (1983). – See Abstr. 012.096.

084.124 Study of large solar flares and their helio–longitudinal effects on geomagnetic storms during 1975 – 79.
S. K. Garde, A. K. Jain, P. K. Pandey, P. K. Shrivastava, S. M. Sharma.
18th International Cosmic Ray Conference, Vol. 3, p. 278 – 281 (1983). – See Abstr. 012.096.

084.125 Fluxes of excess high–energy charged particles under and inside radiation belts of the earth.
S. I. Prokopyev (*S. I. Prokop'ev*), Yu. G. Shafer.
18th International Cosmic Ray Conference, Vol. 3, p. 469 – 472 (1983). – See Abstr. 012.096.
The experimental results of intensity angular distribution of charged particles with energy more than tens of MeV obtained on geophysical rockets are presented.

084.126 Influence of interplanetary magnetic field sector structure on geomagnetic field variations.
R. P. Mishra, J. P. Shukla, R. L. Singh, R. P. Mishra.
18th International Cosmic Ray Conference, Vol. 3, p. 476 – 479 (1983). – See Abstr. 012.096.
The influence of the interplanetary magnetic field sector structure and its boundary crossings on geomagnetic field variations have been observed for the recent solar cycle. It is inferred that on or near the sector boundary passage date geomagnetic field is largely influenced. The cause of these variations is discussed.

084.127 Modulation diffusion and stochastic instability of high–energy protons from the inner radiation belt at small altitudes.
I. V. Amirkhanov, A. A. Gusev, V. D. Iliyin (*V. D. Il'in*), A. N. Iliyina (*A. N. Il'ina*), G. I. Pugacheva.
18th International Cosmic Ray Conference, Vol. 3, p. 480 – 482 (1983). – See Abstr. 012.096.

084.128 Effects of sudden commencements on particle intensities at synchronous orbit.
T. Suda, T. Kohno, M. Wada.
18th International Cosmic Ray Conference, Vol. 10, p. 124 – 127 (1983). – See Abstr. 012.096.
The particle increase events at SSC (sudden commencement of geomagnetic storm) observed by Japanese Geostationary Meteorological Satellite (GMS–1) are analyzed.

084.129 Registration of high energy electrons ($F_e > 100$ MeV) in the equatorial region.
L. Just, K. Kudela, A. A. Gusev, G. I. Pugacheva.
18th International Cosmic Ray Conference, Vol. 10, p. 279 (1983). Abstract. – See Abstr. 012.096.

084.130 Imaging aurorae under full sunlight by vacuum ultraviolet observation.
R. E. Huffman, C.–I. Meng.
Results of the ARCAD 3 project and of the recent programmes in magnetospheric and ionospheric physics, p. 35 (1985). Abstract. – See Abstr. 012.097.

084.131 Consequences of imperfect magnetosphere–ionosphere coupling.
J. R. Kan.
Results of the ARCAD 3 project and of the recent programmes in magnetospheric and ionospheric physics, p. 123 – 142 (1985).
– See Abstr. 012.097.

084.132 Magnetospheric–ionospheric coupling deduced from measurements above a discrete auroral arc.
K. Brüning, C. K. Goertz.
Results of the ARCAD 3 project and of the recent programmes in magnetospheric and ionospheric physics, p. 145 – 151 (1985).
– See Abstr. 012.097.

084.133 The hot ion composition in the magnetosphere.
B. Hultqvist.
Results of the ARCAD 3 project and of the recent programmes in magnetospheric and ionospheric physics, p. 177 – 217 (1985).
– See Abstr. 012.097.

A summary of some recent progress based on hot ion composition measurements on board S3-3, GEOS–1, 2, ISEE–1, PROGNOZ–7, SCATHA and Dynamics Explorer–1 is presented, starting with the low and high latitude boundary layers and continuing inward with the lobes, plasma sheet and ring current regions. The first observations of individual flow vectors of different ion species in the LLBL are described and their interpretation is discussed. The PROGNOZ–7 measurements in the mantle are briefly reviewed. Some major results of the DE–1 observations in the lobes and of the ISEE–1 data taken in the plasma sheet and in the ring current region are presented. The GEOS–1, 2 and PROGNOZ–7 measurements of ring current ions and some consequences are finally summarized.

084.134 Observations of polar–cap ionospheric signatures of solar wind/magnetosphere coupling.
O. de la Beaujardière, V. B. Wickwar, J. D. Kelly.
Results of the ARCAD 3 project and of the recent programmes in magnetospheric and ionospheric physics, p. 353 – 365 (1985).
– See Abstr. 012.097.

084.135 Some results of scanner spectrophotometric measurements of aurora and airglow from IK–Bulgaria–1300 satellite and prospects of similar measurements.
V. M. Balebanov, A. K. Kuzmin (*A. K. Kuz'min*),
M. M. Gogoshev, Ts. N. Gogosheva, N. P. Petkov, K. Kinev.
Results of the ARCAD 3 project and of the recent programmes in magnetospheric and ionospheric physics, p. 747 – 764 (1985).
– See Abstr. 012.097.

084.136 Rocket–borne EUV–visible emission measurements.
G. Schmidtke, G. Stasek, C. Wita, P. Seidl,
K. D. Baker.
J. Atmos. Terr. Phys., Vol. 47, No. 1 – 3, p. 147 – 158 (1985).

The results from auroral measurements with spectrometers and photometers are reported for three successful rocket flights with the spectra (50 – 640 nm) and height profiles presented. The ratios of 391.4/557.7 nm show characteristic differences for diffuse, pulsating and stable arc conditions met. The total energy content in the electromagnetic radiation measured is $0.61 \, \mathrm{mWm}^{-2}$ for the diffuse aurora (1980) and $0.67 \, \mathrm{mWm}^{-2}$ for the stable arc (1981). The spatial distribution of selected emissions exhibit distinct differences for the auroral events observed.

084.137 The zonal harmonic model of polarity transitions: a test using successive reversals.
F. Theyer, E. Herrero–Bervera, V. Hsu, S. R. Hammond.
J. Geophys. Res., Vol. 90, No. B2, p. 1963 – 1982 (1985).

A recently developed zonal model for the last geomagnetic field reversal, which describes time– and latitude–dependent transitional behavior of intensity and inclination in terms of dominance of low–order field harmonics, is tested using a latitudinal and chronological succession of transition records.

084.138 The near–earth magnetic field at 1980 determined from Magsat data.
R. A. Langel, R. H. Estes.
J. Geophys. Res., Vol. 90, No. B3, p. 2495 – 2509 (1985).

Data from the Magsat spacecraft for November 1979 through April 1980 and from 91 magnetic observatories for 1978 through 1982 are used to derive a spherical harmonic model of the earth's main magnetic field and its secular variation at epoch 1980.0.

084.139 A comparison of satellite and observatory estimates of geomagnetic secular variation.
D. R. Barraclough.
J. Geophys. Res., Vol. 90, No. B3, p. 2523 – 2526 (1985).

A comparison is made between secular variation values computed at the positions of 168 magnetic observatories from two spherical harmonic models based almost entirely on the 7–month spread of Magsat data and observed secular variation values derived from the observatory annual mean values. It appears that a time span of more than 7 months is needed to produce accurate models of the secular variation from satellite data.

084.140 Geophysical records of a tree: new application for studying geomagnetic field and solar activity changes during the past 10^4 years.
D. Lal, J. R. Arnold, K. Nishiizumi.
Meteoritics, Vol. 20, No. 2, Part 2, p. 403 – 414 (1985). – See Abstr. 003.048.

Suess showed that tree rings contain a true record of fluctuations in atmospheric $^{14}C/^{12}C$ ratios, that the fluctuations were outside the errors of measurements and, therefore, the tree rings provide a record of the variations in the cosmic ray production rate and changes in the geochemical distribution of radiocarbon. In this paper the authors discuss briefly the geophysical records of a tree, and suggest that the ^{10}Be content of tree samples may also permit measuring changes in the cosmic ray flux reaching them during the past $\leqslant 10^4$ years. This should allow one to deduce maximum changes in the geomagnetic dipole field and solar activity during this period.

084.141 An assessment of the near–surface accuracy of the International Geomagnetic Reference Field 1980 model of the main geomagnetic field.
N. W. Peddie, A. K. Zunde.
Phys. Earth Planet. Inter., Vol. 37, No. 1, p. 1 – 4 (1985).

The new International Geomagnetic Reference Field (IGRF) model of the main geomagnetic field for 1980 is based heavily on measurements from the MAGSAT satellite survey. Assessment of the accuracy of the new model, as a description of the main field near the Earth's surface, is important because the accuracy of models derived from satellite data can be adversely affected by the magnetic field of electric currents in the ionosphere and the auroral zones. The authors assess the near–surface accuracy of the new model by comparing it with values for 1980 derived from annual means from 69 magnetic observatories, and by comparing it with WC 80, a model derived from near–surface data. The comparison with observatory–derived data shows that the new model describes the field at the 69 observatories about as accurately as would a model derived solely from near–surface data.

084.142 Sensitivity of cosmic ray trajectory calculations to geomagnetic field model representations.
J. E. Humble, M. A. Shea, D. F. Smart.
Phys. Earth Planet. Inter., Vol. 37, No. 1, p. 12 – 19 (1985).

084.143 Perpendicular error effect in the DGRF model proposals.
F. J. Lowes.
Phys. Earth Planet. Inter., Vol. 37, No. 1, p. 25 – 34 (1985).

084.144 The effects of the non–uniform distribution of magnetic observatory data on secular variation models.
F. S. Barker, D. R. Barraclough.
Phys. Earth Planet. Inter., Vol. 37, No. 1, p. 65 – 73 (1985).

The effect of the poor spatial distribution of magnetic observatory data on spherical harmonic models of the geomagnetic secular variation is investigated. Values of the secular variation of the north, east and vertical components of the geomagnetic field were synthesized from the 1980 IGRF at the positions of 184 currently operating magnetic observatories. The differences between the results of these analyses and the IGRF are discussed and compared with the input data distributions.

084.145   Review of lake sediment palaeomagnetic data. Part 1.
K. M. Creer.
Geophys. Surv., Vol. 7, No. 2, p. 125 – 160 (1985). – See Abstr. 012.111.
This first part of a two part review is essentially descriptive: palaeomagnetic data sets used to compile the published reference curves of geomagnetic secular variations through Holocene time are critically examined, particularly with respect to coring techniques used for sampling, reliability of the measured directions of remanent magnetization and transformation to timescales.

084.146   Plasma and field observations in the exterior cusp, entry layer, and plasma mantle.
N. Sckopke.
The polar cusp. J. A. Holtet, A. Egeland (Editors). D. Reidel Publishing Company, Dordrecht – Boston – Lancaster (1985), p. 1 – 7.
This report briefly summarises the basic properties of the magnetopause boundary layer, in particular those of its high–latitude portions. Emphasis is placed on the results of the Heos 2 studies.

Quantitative aspects of magnetospheric physics.
See Abstr. 003.128.

Das Phänomen des Polarlichtes.
See Abstr. 003.195.

Waves in space plasmas: highlights of a conference held in Hawaii, 7 – 11 February 1983.
See Abstr. 011.007.

Perspectives on space and astrophysical plasma physics.
See Abstr. 013.001.

Report of ESA's topical team on space plasma physics.
See Abstr. 013.074.

Vibrational temperature and excess vibrational energy of molecular nitrogen in the ground state derived from N_2^+ emission bands in aurora.
See Abstr. 022.002.

A method to detect ultra high energy electrons using Earth's magnetic field as a radiator.
See Abstr. 036.193.

The flow at the core mantle boundary as a possible source of excitation of the Chandler wobble.
See Abstr. 044.061.

The role of scientific ballooning for exploration of the magnetosphere.
See Abstr. 051.040.

Balloon measurements in future exploration of auroral zone phenomena.
See Abstr. 051.041.

Magnetic field reconnection in cosmic plasmas.
See Abstr. 062.004.

Patchy reconnection and magnetic ropes in astrophysical plasmas.
See Abstr. 062.005.

Introduction to particle simulation models and their application to electrostatic plasma waves in space.
See Abstr. 062.157.

Particle simulation of electromagnetic waves and its application to space plasmas.
See Abstr. 062.158.

Relativistic code applied to radiation generation.
See Abstr. 062.159.

Principles of magnetohydrodynamic simulation in space plasmas.
See Abstr. 062.161.

Anomalous transport by Kelvin–Helmholtz instabilities.
See Abstr. 062.162.

Vlasov simulations of ion acoustic double layers.
See Abstr. 062.163.

Simulation models for space plasmas and boundary conditions as a key to their design and analysis.
See Abstr. 062.164.

A parametric survey of the first critical Mach number for a fast MHD shock.
See Abstr. 062.176.

Spontaneous reconnection of magnetic field lines in a collisionless plasma.
See Abstr. 062.177.

Longitudinal friction and intermediate Mach number collisionless transverse magnetosonic shocks.
See Abstr. 062.178.

Almost parallel electromagnetic wave propagation in a hot anisotropic plasma.
See Abstr. 062.189.

Long–time changes of solar and geomagnetic activity and cosmic ray density.
See Abstr. 072.084.

Coupling of newborn ions to the solar wind by electromagnetic instabilities and their interaction with the bow shock.
See Abstr. 074.046.

Space–time relations of magnetic fields on the sun with the IMF, solar wind and geomagnetic activity in May –July 1979.
See Abstr. 075.032.

Structural features of solar cosmic ray penetration into the high–latitude regions of the earth's magnetosphere inferred from Intercosmos–19 data.
See Abstr. 078.021.

Dynamics of the boundaries of solar cosmic ray penetration into the Earth's magnetosphere during a strong magnetic storm.
See Abstr. 078.063.

Is there any "after effect" in density variations of the neutral atmosphere?
See Abstr. 082.015.

Magnetospheric effects on the upper atmosphere.
See Abstr. 082.017.

HIRIS rocketborne spectra of infrared fluorescence in the $O_3(\nu_3)$ band near 100 km.
See Abstr. 082.027.

Mean auroral E–region plasma convection patterns measured by SABRE.
See Abstr. 083.002.

Ionosphere–magnetosphere coupling and convection.
See Abstr. 083.005.

Local auroral disturbance in the morning sector of the upper iono-sphere as a standing electromagnetic wave.
See Abstr. 083.007.

Comparative magnetospheric/ionospheric studies using Dynamics Explorer spacecraft and ground–based radars.
See Abstr. 083.025.

Energy supply processes for magnetospheric substorms and solar flares: tippy bucket model or pitcher model?
See Abstr. 085.006.

Geomagnetic effects at high latitudes connected with the dynamical shock of the solar wind.
See Abstr. 085.014.

Mass loading of planetary atmospheres by rocky satellites – II: Transport and enhanced lifetimes of satellite ejecta in planetary magnetospheres.
See Abstr. 091.021.

The transit effect of comet Halley and the magnetic field in its plasma tail.
See Abstr. 103.958.

IMF B_y–dependent plasma flow and Birkeland currents in the day-side magnetosphere. 1. Dynamics Explorer observations.
See Abstr. 106.016.

IMF B_y–dependent plasma flow and Birkeland currents in the day-side magnetosphere. 2. A global model for northward and south-ward IMF.
See Abstr. 106.017.

A stability analysis of the spiral sector transition region in the interplanetary magnetic field.
See Abstr. 106.087.

Geomagnetic influence on the muon charge ratio of cosmic rays.
See Abstr. 144.030.

Influence of the quiet asymmetrical magnetosphere on asymptotic directions of cosmic rays.
See Abstr. 144.050.

Acceleration of particles in the magnetotail.
See Abstr. 144.216.

Changes in asymptotic directions for various geomagnetic storm conditions.
See Abstr. 144.279.

Procedure for estimating the change in asymptotic directions at a mid–latitude cosmic ray station resulting from a local change in cosmic ray cutoff rigidity.
See Abstr. 144.280.

Suprathermal H^+ and He^{++} in the distant geomagnetic tail: ISEE–3 observations.
See Abstr. 144.380.

Energetic He^+ and He^{++} in the earth's outer radiation belt.
See Abstr. 144.382.

085 Solar-terrestrial Relations

085.001   Aeronomical aspects of mesospheric photodissociation: processes resulting from the solar H Lyman–alpha line.
M. Nicolet.
Planet. Space Sci., Vol. 33, No. 1, p. 69 – 80 (1985).

085.002   Evidence of atmospheric gravity waves produced during the 11 June 1983 total solar eclipse.
E. J. Seykora, A. Bhatnagar, R. M. Jain, J. L. Streete.
Nature, Vol. 313, No. 5998, p. 124 – 125 (1985).
During a solar eclipse the Moon's shadow moves at supersonic speed through the Earth's atmosphere. Chimonas and Hines suggested that the resultant cooling of the atmosphere would generate a bow wave of atmospheric gravity waves. The authors report the detection of a ground–level pressure wave detected at three stations in India and one station in Java, Indonesia. The microbarometer recordings indicate that a wave disturbance was recorded at each station with a quasi–period of ~ 4 h and a wave velocity of ~ 320 m s^{-1}.

085.003   Separation of geomagnetic disturbances caused by flares, coronal holes and filament cavities.
V. P. Mikhajlutsa, M. N. Gnevyshev.
Soln. Dannye, Byull., 1984, No. 10, p. 85 – 90 (1985). In Russian.
It is shown that flares of importance 1 and more in the west solar hemisphere, coronal holes and filament cavities cause geomagnetic disturbances of high, moderate and low amplitude, respectively, measured by a diurnal planetary index A_p. The number of disturbances of each source and the corresponding range of A_p are determined using 1977 – 1983 data.

085.004   Solar–terrestrial electromagnetic phenomena; the role of the breach in the Sun's toroid.
K. L. McDonald.
Bull. Univ. Utah, No. 145, p. 1 – 133 (1984). Abstr. in Phys. Abstr., Vol. 88, No. 1249, Entry 14549 (1985).

085.005   Solar cycles in the last centuries in ^{10}Be and δ^{18}O in polar ice and in thermoluminescence signals of a sea sediment.
G. C. Castagnoli, G. Bonino, M. R. Attolini, M. Galli, J. Beer.
Nuovo Cimento C, Vol. 7C, Ser. 1, No. 2, p. 235 – 244 (1984). Abstr. in Phys. Abstr., Vol. 88, No. 1250, Entry 18612 (1985).

085.006   Energy supply processes for magnetospheric sub-storms and solar flares: tippy bucket model or pitcher model?
S.–I. Akasofu.
Astrophys. Space Sci., Vol. 108, No. 1, p. 81 – 93 (1985).
In the past, both magnetospheric substorms and solar flares have almost exclusively been discussed in terms of explosive magnetic reconnection. Such a model may conceptually be illustrated by the so–called 'tippy–bucket model' , which causes a sudden conversion of stored magnetic energy. However, recent observations indicate that magnetospheric substorms can be understood as a result of a directly driven process which can conceptually be illustrated by the 'pitcher model' in which the output rate varies in harmony with the input rate. It is also possible that solar flare phenomena are directly driven by a photospheric dynamo.

085.007 22–year cycle in air temperature of the northern hemisphere and in tree rings (Northern Kazakstan).
A. I. Ol'.
Soln. Dannye, Byull., 1984, No. 12, p. 69 – 72 (1985). In Russian.
  Examples of global and regional manifestations of the 22–year solar activity cycle in the climate of the earth are given.

085.008 Cyclic variations of heliogeomagnetic activity manifested in variations of the zonal pressure gradient in the troposphere.
R. V. Smirnov, É. V. Kononovich (*Eh. V. Kononovich*),
S. V. Startsev.
Sov. Astron., Vol. 28, No. 4, p. 455 – 457 (1984). English translation of 38.085.002.

085.009 Change of the critical frequencies of the E layer with solar activity at various phases of the solar cycle.
A. A. Nusinov.
Geomagn. Aehron., No. 1, p. 133 – 135 (1985). In Russian. English translation in Geomagn. Aeron.

085.010 Energetics of solar–terrestrial relations.
Eh. I. Mogilevskij.
Simp. No. 4 KAPG po solnech.–zem. fiz., Sochi, noyab., 1984.
Tez. dokl. Moskva, p. 159 (1984). In Russian. Abstr. in Ref. Zh., 51. Astron., 3.51.375 (1985).

085.011 Connection between the variations of solar activity and the concentration of radio carbon in the earth's atmosphere.
V. A. Dergachev, G. E. Kocharov, S. Kh. Tleugaliev.
Complex investigations of the sun, p. 192 – 202 (1982). In Russian. Abstr. in Ref. Zh., 51. Astron., 4.51.374 (1985). – See Abstr. 012.046.

085.012 Polar rain: solar coronal electrons in the earth's magnetosphere.
D. H. Fairfield, J. D. Scudder.
J. Geophys. Res., Vol. 90, No. A5, p. 4055 – 4068 (1985).

085.013 An apparent solar magnetic effect on atmospheric zonal index in the Australia/New Zealand region.
S. W. Goulter, G. F. A. Ward.
J. Geophys. Res., Vol. 90, No. A5, p. 4347 – 4353 (1985).
  An apparent response of atmospheric zonal index to the passage past the earth of solar magnetic sector boundaries is detected in the Australia/New Zealand/ Southwest Pacific region during the period 1957 – 1978. The apparent response, revealed by the method of superposed epoch analysis, is highly nonrandom, the statistical significance of which is assessed in a number of different ways. All approaches concur in showing that it is extremely unlikely that the apparent response could be produced by a chance mechanism.

085.014 Geomagnetic effects at high latitudes connected with the dynamical shock of the solar wind.
V. M. Mishin, S. B. Lunyushkin, T. I. Sajfudinova,
V. V. Shelomentsev, D. Sh. Shirapov.
Issled. Geomagn., Aehron. Fiz. Solntsa, Moskva, No. 70, p. 196 – 203 (1984). In Russian. Abstr. in Ref. Zh., 62. Issled. Kosm. Prostranstva, 5.62.417 (1985).

085.015 Theoretical aspects of ionosphere–magnetosphere–solar wind coupling.
D. J. Southwood.
Adv. Space Res., Vol. 5, No. 4, p. 7 – 14 (1985). – See Abstr. 012.066.
  The author initially reviews magnetospheric plasma distributions emphasizing the explanative power of the open model of the magnetosphere. Next he turns attention to the role of MHD wave phenomena both as transients and as standing structures in the coupled flow system. Finally the physics of the magnetopause flux transfer events is discussed. The logic of the existing interpretation suggests a miniature twin vortex convection system should be created in part of the polar cap ionosphere magnetically connecting to the magnetopause flux tube.

085.016 Sunspot cycle and thermal structure of equatorial middle atmosphere.
S. Devanarayanan, K. Mohanakumar.
J. Geophys. Res., Vol. 90, No. A6, p. 5357 – 5362 (1985).
  Correlation and linear regression analyses of the variations in the mean stratospheric (15 – 50 km) and the mean mesospheric (50 – 80 km) temperatures with sunspot number variations during a complete solar cycle (1971 – 1982) have been made over the equatorial rocket launching station, Thumba (8°N, 77°E), India.

085.017 On the possibility of long–range forecasts of geomagnetic activity.
J. Halenka.
Stud. Geophys. Geod., Vol. 28, No. 3, p. 294 – 305 (1984). Abstr. in Phys. Abstr., Vol. 88, No. 1256, Entry 50932 (1985).

085.018 Solar cycle variation of thermospheric nitric oxide.
T. Ogawa, N. Iwagami, Y. Kondo.
J. Geomagn. Geoelectr., Vol. 36, No. 8, p. 317 – 340 (1984). Abstr. in Phys. Abstr., Vol. 88, No. 1257, Entry 57490 (1985).

085.019 On the low energy limit of the polar cap absorption events.
G.–l. Zhang.
Curr. Top. Chin. Sci., Sect. E: Astron., Vol. 2, p. 175 – 177 (1984). = Kexue Tongbao (Beijing), Vol. 27, No. 1, p. 52 – 54 (1982).

085.020 Spatial regularities of the manifestation of solar activity in the troposphere.
R. V. Smirnov.
Sov. Astron., Vol. 28, No. 6, p. 683 – 689 (1984). English translation of 38.085.037.

085.021 An assessment of thermal, wind, and planetary wave changes in the middle and lower atmosphere due to 11–year UV flux variations.
L. B. Callis, J. C. Alpert, M. A. Geller.
J. Geophys. Res., Vol. 90, No. D1, p. 2273 – 2282 (1985).
  A two–dimensional radiative equilibrium calculation with fixed dynamic heating is used to calculate variations in atmospheric temperature due to solar UV flux variations which are assumed to be associated with the 11–year solar cycle.

085.022 The manifestation of solar cyclicity in the midlatitude characteristics of the troposphere.
Eh. V. Kononovich, V. V. Mikhnevich, R. V. Smirnov.
Soln. Dannye, Byull., 1985, No. 4, p. 69 – 74 (1985). In Russian.
  The frequency characteristics of the connection between troposphere temperature anomalies horizontal gradients and solar-geomagnetic activity are considered. Manifestiation of the 22–year and 11–year cyclicity at high latitudes is shown.

Introduction élémentaire à la physique cosmique et à la physique des relations Soleil–Terre.
See Abstr. 003.125.

The IVth symposium on solar–terrestrial physics, held in Sochi in November 1984.
See Abstr. 012.048.

Weather and climate responses to solar variations. Second International Symposium on Solar Terrestrial Influences on Weather and Climate, held at Boulder, Colo, USA, 2 – 6 August 1982.
See Abstr. 012.086.

Solar–terrestrial research opportunities – a look to the future.
See Abstr. 013.006.

G–EXEC and the World Data Centre.
See Abstr. 013.014.

Soho and Cluster: Europe's possible contribution to the ISTP (*International Solar Terrestrial Physics Programme*).
See Abstr. 051.021.

The first effect of the amplitude of a solar cycle.
See Abstr. 072.039.

Periodicities of 2.2 and 22 years in solar activity and solar – planetary relationships.
See Abstr. 072.103.

Solar oscillations, sunspot cycles, and climatic change.
See Abstr. 080.092.

Planetary System

091 Physics and Dynamics of the Planetary System

091.001 Solar system studies with the IUE: 1982 – 1984.
J. Caldwell.
NASA Conf. Publ., NASA CP–2349, p. 27 – 31 (1984). – See Abstr. 012.001.
Planetary observations with the IUE in the last two years are summarized. These include work on the atmospheres of comets and the giant planets; aurorae; rings and circumplanetary environment. Planetary requirements for future Earth–orbital astronomical observatories are also described.

091.002 Collisions in the solar system. – I. Impacts of the Apollo–Amor–Aten asteroids upon the terrestrial planets.
D. I. Steel, W. J. Baggaley.
Mon. Not. R. Astron. Soc., Vol. 212, No. 4, p. 817 – 836 (1985).
The collision probability between each of the presently–known population of four Aten, 34 Apollo and 38 Amor asteroids and each of the terrestrial planets is determined by a new technique. The influx to the Earth is found to be one impact per 160000 yr, but this figure is biased by the inclusion of four recently–discovered low–inclination Apollos. Excluding these four the rate would be one per 250000 yr, in line with previous estimates. The impact rate is highest for the Earth, being around twice that of Venus. The rates for Mercury and Mars using the present sample are about one per 5 Myr and one per 1.5 Myr respectively.

091.003 Between fire and ice: the planets.
R. Gore.
Natl. Geogr., Vol. 167, No. 1, p. 4 – 51 (1985).

091.004 Ringed planets: still mysterious – II.
J. N. Cuzzi.
Sky Telesc., Vol. 69, No. 1, p. 19 – 23 (1985).

091.005 Local harmonic analysis of planetary Doppler gravity data.
A. W. G. Kunze.
Earth, Moon, Planets, Vol. 32, No. 2, p. 173 – 181 (1985).
Planetary gravity fields represented in terms of spherical harmonics or surface mass distributions do not have the necessary resolution to permit gravity analysis of local features. Doppler gravity maps representing residual line–of–sight acceleration have much greater resolution. These gravity data may be converted to vertical gravity anomalies by expressing the anomalous local gravitational potential over small rectangular areas in terms of a modified double Fourier series constrained by local Doppler gravity data. The vertical derivative of the resulting potential yields the vertical gravity components at desired altitudes. The resolution of the resulting normalized free air anomaly maps is limited only by that of the original Doppler gravity data. Extended gravity maps may be constructed this way using a moving window approach. It is anticipated that much of the lunar frontside can be mapped at resolutions ranging from 1 to 4 deg of arc.

091.006 Eccentricities and inclinations of orbits of growing planets.
I. N. Ziglina.
Astron. Zh., Tom 62, Vyp. 1, p. 141 – 152 (1985). In Russian. English translation in Sov. Astron., Vol. 29, No. 1.
The evolution of eccentricities e and inclinations i of orbits of the terrestrial planets as a result of collisions and encounters with preplanetary bodies during the accumulation process is considered. The encounters were more numerous and more important in comparison with the collisions. The differential equations describing the variation of the expected average values of e^2 and i^2 during the planet's growth are derived.

091.007 Planetary size comparisons: a photographic study.
S. P. Meszaros.
NASA Tech. Memo., NASA TM–85017, 6 + 94 pp. (1985).
Over the past two decades NASA spacecraft have visited many of the planets and moons of the solar system. During these missions many detailed photographs were taken. This publication utilizes some of the best of the photos to show size comparisons – to scale – of these planets and moons.

091.008 A geographic comparison of selected large–scale planetary surface features.
S. P. Meszaros.
NASA Tech. Memo., NASA TM–86147, 6 + 63 pp. (1985).
This publication is a compilation of photographic and cartographic comparisons of large, well–known geographical features on planets and moons in the solar system. Included are structures caused by impacts, volcanism, tectonics, and other natural forces. Each feature is discussed individually and then those of similar origin are compared at the same scale.

091.009 Resonances in the solar system.
Eh. I. Nesmyanovich.
Probl. Kosm. Fiz., Vyp. 19, p. 84 – 93 (1984). In Russian. – See Abstr. 003.002.

091.010 Planetary cartography in the next decade (1984 – 1994).
R. G. Strom, R. M. Batson, J. M. Boyce, M. E. Davies, F. J. Doyle, T. C. Duxbury, M. P. Golombek, H. Masursky, R. S. Saunders, W. L. Sjogren, J. R. Underwood.
NASA Spec. Publ., NASA SP–475, 6 + 71 pp. (1984).
Contents: Introduction. Summary of recommendations. Utilization of planetary cartography. Cartographic data base. Current status of planetary cartography. Planetary cartography ten-year plan.

091.011 La vie dans les planètes telluriques?
J. Blamont.
Astronomie, Vol. 99, p. 55 – 65 (1985).

091.012 Problèmes actuels de planétologie.
R. Dejaiffe.
Rev. Quest. Sci., Vol. 155, No. 3, p. 279 – 300 (1984).

091.013 Interior structure and evolution of planets.
W. B. Hubbard.
Planets, their origin, interior and atmosphere, p. 61 – 162 (1984). – See Abstr. 012.019.
Contents: (1) Basic concepts: presumed initial composition, comparison with stellar structure. (2) Structure of cold bodies: $T = 0$ thermodynamics, experimental data at high pressure, the radius–mass diagram. (3) Heat flow: review of data, the Kelvin mechanism, differentiation, radioactivity, tidal heating. (4) Other diagnostics of interior structure: response to rotation, tidal response, magnetic field. (5) The terrestrial planets. (6) The Jovian

planets: evolution, thermal structure, heat flow. (7) Jovian planet satellites.

091.014   An introduction to the evolution of planetary atmospheres.
D. Gautier.
Planets, their origin, interior and atmosphere, p. 163 – 287 (1984). – See Abstr. 012.019.
Contents: An overview of planetary atmospheres of the solar system. The structure of planetary atmospheres. Mechanisms of long term change in planetary atmospheres. Sources of atmospheric gases. The giant planets. Formation of terrestrial atmospheres. Evolution of atmospheres of terrestrial planets. The atmospheres of Titan, Triton, and Pluto. The future of solar system exploration.

091.015   Extraterrestrial ice.
J. Klinger.
Recherche, Vol. 15, No. 158, p. 1060 – 1070 (1984). In French. Abstr. in Phys. Abstr., Vol. 88, No. 1251, Entry 24306 (1985).

091.016   Formation of rings in the equatorial plane of gaseous stars (*and planets*).
Y. Aihara.
Acta Astronaut., Vol. 11, No. 6, p. 313 – 317 (1984). Abstr. in Phys. Abstr., Vol. 88, No. 1251, Entry 24320 (1985).

091.017   The influences of planetary environments on the eruption styles of volcanoes.
L. Wilson.
Vistas Astron., Vol. 27, Part 4, p. 333 – 360 (1984).
Volcanism appears to have been a major process in the formation of surface materials on each of the silicate–dominated planets and satellites in the solar system at some stage in its development. Observations of the types and extents (in space and time) of volcanic activity on a planetary body can provide important information on the interior structure and chemistry, and on the near–surface environmental conditions.

091.018   Ultraviolet spectroscopy of the outer planets: current knowledge and unresolved problems.
G. E. Hunt, V. Moore.
J. Quant. Spectrosc. Radiat. Transfer, Vol. 32, No. 5/6, p. 439 – 451 (1984). – See Abstr. 012.026.
The paper summarises the UV observations for the outer planets and the interpretations of these results; then discusses the unresolved problems including the need for further laboratory studies and simultaneous solar data.

091.019   Patterns of fracture and tidal stresses due to nonsynchronous rotation: implications for fracturing on Europa.
P. Helfenstein, E. M. Parmentier.
Icarus, Vol. 61, No. 2, p. 175 – 184 (1985).
This study considers the global patterns of fracture that would result from nonsynchronous rotation of a tidally distorted planetary body. The incremental horizontal stresses in a thin elastic or viscous shell due to a small displacement of the axis of maximum tidal elongation are derived, and the resulting stress distributions are applied to interpret the observed pattern of fracture lineaments on Europa.

091.020   Mass loading of planetary magnetospheres by rocky satellites – I: Production of rocky ejecta.
M. Alexander, P. Anz, D. Lyons, W. Tanner, Y.–L. Chen, J. A. M. McDonnell.
Adv. Space Res., Vol. 4, No. 9, p. 23 – 26 (1984). – See Abstr. 012.029.
Recent hypervelocity studies have been conducted which simulate the collision of interplanetary dust with rocky planetary satellite surfaces. Preliminary flux–mass distributions of micron and submicron ejecta from these hypervelocity impact studies have been determined. Several models of the flux–mass distribution of primary interplanetary dust are used to determine ratios of satellite surface ejecta and primary meteoroid flux–mass distri-

butions. The results are used in a second model to determine the ejecta spatial mass densities near the surface of the satellite.

091.021   Mass loading of planetary atmospheres by rocky satellites – II: Transport and enhanced lifetimes of satellite ejecta in planetary magnetospheres.
M. Alexander, P. Anz, T. Hyde, A. Hargrave, L. Lodhi, S. Lodhi, W. Tanner.
Adv. Space Res., Vol. 4, No. 9, p. 27 – 30 (1984). – See Abstr. 012.029.
The mass–flux distributions of lunar ejecta at the surface of the magnetopause for a complete lunar orbit are presented. Spatial mass densities of lunar ejecta in specific zones of the magnetosphere provide a means to compare sporadic interplanetary dust spatial mass densities in the same zones.

091.022   Modification of planetary atmospheres by material from the rings.
S. K. Atreya.
Adv. Space Res., Vol. 4, No. 9, p. 31 – 40 (1984). – See Abstr. 012.029.
The atmospheres and ionospheres of ringed planets can be, and perhaps are, modified by the injection of gaseous neutral and ionized species, and dust of ring origin. Although no direct evidence for such interaction exists, many of the unresolved characteristics of planetary composition, thermal structure and ionosphere would be understood if the rings supplied certain materials to their parent planets.

091.023   Effects of electrostatic forces on the vertical structure of planetary rings.
O. Havnes, G. E. Morfill.
Adv. Space Res., Vol. 4, No. 9, p. 85 – 90 (1984). – See Abstr. 012.029.
The authors have calculated the vertical structure of planetary dust rings as it results from a balance between an electrostatic force on the dust grains and the vertical component of the gravitational force from the central planet.

091.024   Entry of dust particles into planetary magnetospheres.
D. A. Mendis.
Adv. Space Res., Vol. 4, No. 9, p. 111 – 120 (1984). – See Abstr. 012.029.
The author reviews the sources of dust in planetary magnetospheres and discusses their physics and their dynamics under the combined action of both planetary gravitational and magnetospheric electromagnetic forces.

091.025   Planetary rings.
J. A. Burns.
Adv. Space Res., Vol. 4, No. 9, p. 121 – 134 (1984). – See Abstr. 012.029.
The paper begins with a précis of the primary processes that govern the form of planetary rings. It then summarizes the ring systems of Jupiter, Saturn, and Uranus with modest emphasis on structures where dust interacts with magnetospheres.

091.026   Exploration of planetary atmospheres: current knowledge, future opportunities and the possible role of Europe.
G. E. Hunt.
J. Br. Interplanet. Soc., Vol. 38, No. 5, p. 217 – 221 (1985).
In the past 15 years there has been a rapid advance in our understanding of planetary atmospheres through the wealth of data provided by missions that have observed Venus to Saturn, the continued improvements in ground–based and Earth orbiting telescopic studies and supporting numerical studies. The author briefly reviews the current understanding of the planetary atmospheres where processes may now be quantitatively compared with terrestrial phenomena.

091.027   Zur Stabilität der Bahnen natürlicher Monde.
N. Giesinger.
Sternenbote, 28. Jahrg., Nr. 3, p. 38 – 46 (1985).

091.028   What color is the solar system?
A. T. Young.
Sky Telesc., Vol. 69, No. 5, p. 399 – 403 (1985).

091.029   Chemical evolution on the giant planets and Titan.
J. Caldwell, T. Owen.
Adv. Space Res., Vol. 4, No. 12, p. 51 – 58 (1984). – See Abstr.
012.036.

The authors summarize the current status of atmospheric chemistry in the atmospheres of the outer solar system with special emphasis on the question of HCN formation on Jupiter, differences between polar and equatorial compositions on Jupiter, the coloration of the Great Red Spot, and the unique environment of Titan.

091.030   Organic syntheses in gas phase and chemical evolution in planetary atmospheres.
F. Raulin, A. Bossard.
Adv. Space Res., Vol. 4, No. 12, p. 75 – 82 (1984). – See Abstr.
012.036.

Atmospheric chemistry may be one of the important pathways to the synthesis of organic compounds in a planetary periphery. Depending on the nature of the carbon source (CH_4, CO or CO_2), the main composition of the atmosphere, and the respective roles of the various energy sources, is it possible, and to what extent, to produce organics? What kind of gaseous mixture is the most favourable to prebiotic organic syntheses? How far can the results of laboratory works be extrapolated to the case of planetary atmospheres? These questions are discussed, on the basis of several available laboratory data, and by considering the main atmospheric composition of the planets of the solar system, and the list of organic compounds which have already been detected in their atmospheres.

091.031   On volatile components and highland material of planets.
K. P. Florenskij, O. V. Nikolaeva.
Geokhimiya, No. 9, p. 1251 – 1267 (1984). In Russian. Abstr. in Ref. Zh., 51. Astron., 1.51.265 (1985).

091.032   Addition from external sources and dissipation of rare gases in the atmospheres of terrestrial planets.
A. K. Pavlov.
Cosmochemistry and meteoritics, p. 131 – 135 (1984). In Russian. Abstr. in Ref. Zh., 51. Astron., 1.51.270 (1985). – See Abstr. 012.037.

091.033   A theory of satellite sweeping.
M. Paonessa, A. F. Cheng.
J. Geophys. Res., Vol. 90, No. A4, p. 3428 – 3434 (1985).

A theory of energetic charged particle absorption by insulating moons is presented that includes gyrophase dependence of the absorption probability when the gyroradius is not small compared to the satellite radius. This effect increases the average lifetime against absorption by factors of approximately 2 to 6 above previous estimates at Saturn. The sweeping lifetime then becomes significantly greater than the strong diffusion lifetime. A rigorous expression for the average lifetime against absorption, appropriate for a drift phase averaged radial diffusion equation, is derived assuming (1) a uniform distribution of gyrocenter equatorial plane crossing points in the sweeping corridor and (2) randomization of gyrophase. The pitch angle and energy dependences of the sweeping lifetime are found to be significantly different from the predictions of previous analytical estimates.

091.034   Rare gases in the atmospheres of terrestrial planets and early stages of evolution of the solar system.
M. N. Izakov.
Inst. kosm. issled. Akad. Nauk SSSR, Prepr., No. 904, 51 pp. (1984). In Russian. Abstr. in Ref. Zh., 51. Astron., 2.51.147 (1985).

091.035   The free oscillations of the giant planets: phase–transition boundary conditions.
S. V. Vorontsov.
Sov. Astron., Vol. 28, No. 4, p. 410 – 414 (1984). English translation of 38.091.003.

091.036   A point–mass method for modeling a planet's gravitational field.
A. L. Tserklevich, S. D. Volzhanin.
Sov. Astron. Lett., Vol. 10, No. 4, p. 229 – 231 (1984). English translation of 38.091.001.

091.037   Structural and thermal icy satellite models: preliminary results.
P. J. Thomas.
Proc. Astron. Soc. Aust., Vol. 5, No. 4, p. 487 – 489 (1984).

091.038   Non–gravitational forces in the evolution of the solar system.
J. Kovalevsky.
The Big Bang and Georges Lemaitre, p. 181 – 194 (1984). – See Abstr. 012.043.

Among the numerous cases of resonances encountered in the solar system, many of the satellite to satellite and spin–orbit couplings can be explained as the result of an evolution driven by tidal forces. Models of such evolution are described and it is shown that they all lead to a similar type of reduced equations. It is proposed to solve these equations in form of perturbation of a simpler "restricted tidal problem". Such an approach gives a simple picture of the capture as well as conditions for a permanent capture into a resonant situation, as it is the case of the major satellite resonances and the Mercury rotation.

091.039   On the stability of the Solar System as hierarchical dynamical system.
A. Milani, A. M. Nobili.
The Big Bang and Georges Lemaitre, p. 219 – 230 (1984). – See Abstr. 012.043.

The stability of the Solar System as hierarchical dynamical system is investigated with analytical methods (1) by using the topological stability criterion of the general 3–body problem for 3–body subsystems, (2) by exploiting the hierarchical arrangement of the whole system to develop a new perturbation theory containing as smallness parameters both mass and distance ratios. At the first order in these parameters one gets for the inner Solar System a minimum lifetime of $\sim 10^8$y.

091.040   Magnetospheres in the solar system.
R. L. McNutt Jr.
ESA Spec. Publ., ESA SP–207, p. 41 – 48 (1984). – See Abstr. 012.044.

Magnetospheres have been found at all of the planets of the solar system visited by spacecraft. A moon of at least one planet has a magnetosphere and comets may also exhibit such a structure. Magnetospheres can be broadly divided into two categories: intrinsic and induced, depending upon whether the central object possesses its own magnetic field or not. The author considers general concepts which can be applied to the magnetospheres in the solar system and also discusses some significant differences between them.

091.041   The stability of the solar system: an investigation based on new methods.
A. Milani, A. M. Nobili.
Mem. Soc. Astron. Ital., Vol. 55, No. 3, p. 485 – 488 (1984). – See Abstr. 012.049.

091.042   Cosmochemistry and structure of the giant planets and their satellites.
D. J. Stevenson.
Icarus, Vol. 62, No. 1, p. 4 – 15 (1985). = Contrib. Div. Geol. Planet. Sci., Calif. Inst. Technol., Pasadena, No. 4169.

091.043   Programs for determination of the coordinates of planetary surface features from photos made with space instrumentation.
S. G. Valeev, M. G. Shamarin.
Astron. Geod., Tomsk, No. 12, p. 115 – 122, 146 (1984). In Russian. Abstr. in Ref. Zh., 52. Geod. Aehrosemka, 2.52.126 (1985).

091.044   New insights into solar system objects.
H. W. Moos.
Bull. Am. Astron. Soc., Vol. 16, No. 4, p. 1011 – 1012 (1984). Abstract. – See Abstr. 010.062.

091.045   D/H ratio in the outer solar system.
W. Hayden Smith.
Bull. Am. Astron. Soc., Vol. 16, No. 4, p. 1027 (1984). Abstract. – See Abstr. 010.063.

091.046   The solar system in modern astronomy.
S. K. Runcorn.
Q. J. R. Astron. Soc., Vol. 26, No. 2, p. 137 – 146 (1985). Invited lecture at the McVittie conference, Canterbury, 1984 June 2.

091.047   Resonances in the solar system.
A. M. Molchanov, G. A. Shishlovskaya.
IXth Mezhdunar. konf. po nelinejn. kolebaniyam, Kiev, 30 avg. – 6 sent., 1981. Tom 3. Kiev, p. 183 – 184 (1984). In Russian. Abstr. in Ref. Zh., 51. Astron., 6.51.70 (1985).

091.048   K–U–Th systematics of matter of planetary bodies of the solar system.
A. T. Bazilevskij.
Geokhimiya, No. 2, p. 131 – 141 (1985). In Russian. Abstr. in Ref. Zh., 62. Issled. Kosm. Prostranstva, 6.62.312 (1985).

091.049   Energy conversion processes in the outer planets.
P. J. Gierasch, B. J. Conrath.
Recent advances in planetary meteorology, p. 121 – 146 (1985). – See Abstr. 012.071.
   Energy conversion processes which are potentially important in the outer planets at pressures greater than ∼0.1 bar are reviewed. Generation of buoyancy contrasts by condensation of various constituents is discussed with emphasis on the possible significance of phase changes in substances such as Si and Mg compounds at deep levels. It is demonstrated that, in the absence of non–equilibrium thermodynamic processes, strong kinetic energy generation must accompany the transport of heat out of the high temperature planetary interiors. The possibly dominant role of lagged para hydrogen conversion in the convective transport of heat at levels where T < 300K is discussed. Measurements which may ultimately contribute to a better understanding of energy conversion processes are summarized.

091.050   Planetary atmospheres as laboratories of geophysical fluid dynamics.
C. B. Leovy, G. E. Hunt.
Recent advances in planetary meteorology, p. 147 – 156 (1985). – See Abstr. 012.071.
   The authors briefly review some of the major observational results and theoretical problems of planetary atmospheres, and point out some connections with important problems of geophysical fluid dynamics.

091.051   Condensed matter physics of planets: puzzles, progress and predictions.
D. J. Stevenson.
High pressure in science and technology, Part 3, p. 375 – 368 (1984). Abstr. in Phys. Abstr., Vol. 88, No. 1254, Entry 40511 (1985). – See Abstr. 012.079.

091.052   Plasma–gas interactions in planetary atmospheres and their relevance for the terrestrial hydrogen budget.
H. U. Nass, H. J. Fahr.
J. Geophys., Vol. 56, No. 1, p. 34 – 46 (1984). Abstr. in Phys. Abstr., Vol. 88, No. 1255, Entry 45855 (1985).

091.053   Properties of the planetary materials He, SiO_2, and N_2 at dynamic high pressures and temperatures.
W. J. Nellis, N. C. Holmes, A. C. Mitchell, M. Van Thiel, H. B. Radousky, D. C. Hamilton, S. Henning.
J. Phys. Colloq., Vol. 45, No. C–8, p. 105 – 107 (1984). Abstr. in Phys. Abstr., Vol. 88, No. 1256, Entry 51491 (1985). – See Abstr. 012.085.

091.054   High pressure physics and chemistry in giant planets and their satellites.
D. J. Stevenson.
J. Phys. Colloq., Vol. 45, No. C–8, p. 97 – 103 (1984). Abstr. in Phys. Abstr., Vol. 88, No. 1256, Entry 51532 (1985). – See Abstr. 012.085.

091.055   External control of planetary radio emission.
H. O. Rucker, M. D. Desch.
Sitzungsber., Österr. Akad. Wiss., Math.–Naturwiss. Kl. Abt. II, Band 192, Heft 8 – 10, p. 433 – 441 (1983).
   Nonthermal radio emission is an indicator of magnetospheric activity and therefore provides a valuable tool for remote sensing of planetary magnetospheres. Using Voyager data from the Plasma Science experiment, the Magnetometer experiment, and the Planetary Radio Astronomy experiment, studies on the correlation between solar wind parameters and the Saturn kilometric radiation (SKR) showed that the solar wind ram pressure is the only significant correlator with SKR intensity variations. Recent results on the external control of other planetary radio emissions, i.e. the earth's kilometric radiation and the Jovian non–Io–related emission are examined and discussed.

091.056   Analytical description of profiles through planetary atmospheres.
K. Rawer.
Acta Astronaut., Vol. 11, No. 9, p. 607 – 608 (1984). Abstr. in Phys. Abstr., Vol. 88, No. 1258, Entry 63239 (1985).

091.057   On the dynamics of rotating fluids and planetary atmospheres: a summary of some recent work.
R. Hide.
Pure Appl. Geophys., Vol. 121, No. 3, p. 365 – 374 (1983). Abstr. in Phys. Abstr., Vol. 88, No. 1258, Entry 63240 (1985). – See Abstr. 012.091.

091.058   Convection–driven zonal flows in the major planets.
F. H. Busse.
Pure Appl. Geophys., Vol. 121, No. 3, p. 375 – 390 (1983). Abstr. in Phys. Abstr., Vol. 88, No. 1258, Entry 63241 (1985). – See Abstr. 012.091.

091.059   A demonstration of the Sun's motion around the barycenter of the Solar System.
J. Hardorp.
Bull. Am. Astron. Soc., Vol. 17, No. 2, p. 592 (1985). Abstract. – See Abstract 010.065.

091.060   Fault type predictions from stress distributions on planetary surfaces: importance of fault initiation depth.
M. P. Golombek.
J. Geophys. Res., Vol. 90, No. B4, p. 3065 – 3074 (1985).
   The purpose of the paper is (1) to review briefly the generally accepted and most commonly employed criterion for fault prediction from calculated stresses on planetary surfaces, (2) to point out problems in neglecting the effects of nonisotropic overburden stresses and the use of realistic material constants, (3) to show a well–known example (mascon loading on the moon) where neglecting this effect predicts a different type of fault than is actually observed, and (4) to suggest an explanation for the absence of strike–slip faults on planet and satellite surfaces.

091.061   Lightning on other planets.
K. Rinnert.
J. Geophys. Res., Vol. 90, No. D4, p. 6225 – 6237 (1985).
   The necessary conditions for producing lightning discharges, the experimental possibilities for investigating extraterrestrial

lightning, and the scientific objectives of planetary lightning research are briefly discussed. Present knowledge on the composition, structure, and dynamics of the atmospheres for the extraterrestrial planets and the satellites Io and Titan are reviewed in terms of their importance for the production of lightning. From the knowledge on planetary atmospheres, intensive lightning activity can be expected to exist in the Jupiter and Saturn cloud systems; this is less conclusive for Venus and unlikely for the other bodies. Electrical activity is not completely ruled out for Martian dust storms. Optical and RF wave measurements from spacecraft have yielded evidence of possible lightning activity on Venus, Jupiter, and Saturn. These observations are reviewed and discussed.

091.062 Report of IAU Commission 16: Physical study of planets and satellites (*Etude physique des planètes et des satellites*).
V. G. Tejfel'.
Trans. IAU, Vol. XIXA, p. 189 – 192 (1985). – See Abstr. 003.046.

091.063 Report of the IAU working group for planetary system nomenclature (*Nomenclature du système planétaire*).
H. Masursky.
Trans. IAU, Vol. XIXA, p. 725 – 726 (1985). – See Abstr. 003.046.

091.064 Rings of planets.
V. Pohánka.
Kozmos, Vol. 15, No. 5, p. 152 – 158 (1984). In Slovak.

091.065 A simple model of convection in the terrestrial planets.
A. C. Fowler.
Geophys. Astrophys. Fluid Dyn., Vol. 31, No. 3/4, p. 283 – 309 (1985).
The author studies analytically the simplest fluid mechanical model which can mimic the convective behavior which is thought to occur in the solid mantles of the terrestrial planets. The convecting materials are polycrystalline rocks, whose creep behavior depends very strongly on temperature and probably also on pressure.

091.066 On computing the coordinates of the inner planets and the Sun with a given limited accuracy.
N. I. Glebova.
Byull. Inst. Teor. Astron., Tom 15, No. 7 (170), p. 360 – 374 (1985). In Russian.
The paper deals with description of a technique for constructing simplified theories of motion of the Sun, Mercury, Venus, and Mars on the basis of Newcomb's complete theories for these bodies' motions. The expressions for the orbital elements and the tables representing the perturbations in longitude, latitude, and logarithm of the radius–vector are given. These are transformed to more convenient form with argument reckoned from the initial date 1985, January 0.0. The algorithm is described to compute the planetary coordinates, and estimates are made for the time span 1983 – 2000 with respect to the number of terms retained in expansions needed for calculation of various coordinates to various limited accuracy ranging from 1' to 1".

091.067 The observations of major planets and their use for ephemeris astronomy problems.
M. L. Sveshnikov.
Byull. Inst. Teor. Astron., Tom 15, No. 7 (170), p. 375 – 382 (1985). In Russian.
The value of planetary observations for modern astrometry and celestial mechanics is presented. The difficulties are discussed which must be overcome for solution of problems using planetary observational data.

091.068 Millisecond pulsars and the location of the solar system barycenter.
M. Prószyński.
Birth and evolution of neutron stars: issues raised by millisecond pulsars, p. 287 – 293 (1984). – See Abstr. 012.115.
The possibility of using millisecond pulsars to more accurately determine the location of the solar system barycenter is discussed. Two sets of the light–travel time corrections (necessary to refer pulse arrival–time data to the solar system barycenter) are obtained; one from the old MIT PEP311 ephemerides and the second from the most recent JPL DE200/LE200 ephemerides. The comparison of these two sets gives a rough estimate of the possible uncertainties in the location of the solar system barycenter and in pulsar parameters that are derived from pulse arrival–time data.

091.069 The collisional dynamics of particulate disks.
F. H. Shu, G. R. Stewart.
Icarus, Vol. 62, No. 3, p. 360 – 383 (1985).
The authors use a Krook equation, modified to allow collisions to be inelastic, to describe the dynamics of a particulate disk. By a simple heuristic argument, they compute the effective collision rate in a disk of spherical particles with a power–law distribution of sizes. For Saturn's rings, the effective collision rate for momentum transport is substantially lower than that conventionally estimated on the basis of an observed optical depth at visual wavelengths. The authors generalize the formation to include the effects of gravitational scattering. This may be important for the discussion of gas clouds in the disk of a spiral galaxy, and it is probably central to the accumulation of planets from smaller bodies in the primitive solar nebula.

Bibliography of Earth, Moon, and Planets.
See Abstr. 002.006.

Bibliography of Earth, Moon, and Planets.
See Abstr. 002.031.

Wind as a geological process on Earth, Mars, Venus and Titan.
See Abstr. 003.001.

Planetary geology in the 1980s.
See Abstr. 003.014.

Chemistry of atmospheres. An introduction to the atmospheres of Earth, the planets, and their satellites.
See Abstr. 003.027.

Rings: discoveries from Galileo to Voyager.
See Abstr. 003.086.

Le système solaire et son exploration.
See Abstr. 003.191.

Atlas des Sonnensystems.
See Abstr. 003.197.

Early observations of thermal planetary radio emission.
See Abstr. 004.195.

The letters between Titius and Bonnet and the Titius–Bode law of planetary distances.
See Abstr. 005.027.

Names of landscapes of other worlds.
See Abstr. 011.010.

Conference of the working group on the nomenclature of the planetary system.
See Abstr. 011.027.

Perspectives on space and astrophysical plasma physics.
See Abstr. 013.001.

Solar system science – Survey Committee report.
See Abstr. 013.072.

Report of ESA's topical team on planetary science.
See Abstr. 013.075.

Charges on dust particles.
See Abstr. 022.032.

On the importance of fluff on the electric behaviour of cosmic grains.
See Abstr. 022.033.

Organic chemistry by irradiation in space.
See Abstr. 022.048.

Prebiotic syntheses of purines and pyrimidines.
See Abstr. 022.050.

An apparently first–order transition between two amorphous phases of ice induced by pressure.
See Abstr. 022.055.

High–energy neutron induced prompt gamma–rays: chemical remote sensing of planetary surfaces.
See Abstr. 022.066.

Solar wind and cosmic ray irradiation of grains and ices – application to erosion and synthesis of organic compounds in the solar system.
See Abstr. 022.067.

Determination of the products state in dissociative recombination of molecular ions: N_2^+ and O_2^+ ions.
See Abstr. 022.096.

Impact and explosion crater ejecta, fragment size, and velocity.
See Abstr. 022.126.

Impact cratering mechanics: relationship between the shock wave and excavation flow.
See Abstr. 022.127.

Reaction of O^+, CO^+, H^+, and CH_5^+ ions with atomic hydrogen.
See Abstr. 022.140.

Experimental electron scattering spectrum of atomic oxygen.
See Abstr. 022.156.

Clustered impacts: experiments and implications.
See Abstr. 022.182.

Molecular spectroscopy and planetary atmospheres.
See Abstr. 036.016.

Declinations of the sun and major planets obtained from observations with the Wanschaff vertical circle at the Goloseevo Observatory in 1977 – 1979.
See Abstr. 041.014.

Accurate methods in general planetary theory.
See Abstr. 042.006.

The stability of our solar system.
See Abstr. 042.020.

Evolution of the rotation of a viscoelastic planet on a circular orbit in a central force field.
See Abstr. 042.112.

Space Telescope observations of aurorae on the giant planets.
See Abstr. 051.064.

Through the system of Saturn to the history of the matter of planets.
See Abstr. 051.067.

Bending waves on current sheets.
See Abstr. 062.031.

Review of the critical ionization velocity effect in space.
See Abstr. 062.172.

Intensity of radiation reflected by a vertically inhomogeneous medium through multiple scattering.
See Abstr. 063.029.

Polarized light in planetary atmospheres for perpendicular directions.
See Abstr. 063.037.

Orbit perturbation: evolution of a Keplerian disk.
See Abstr. 064.033.

Solar wind interaction with planetary objects.
See Abstr. 074.075.

Comparative planetology and early history of the earth.
See Abstr. 081.038.

Computer simulation of reconnection in planetary magnetospheres.
See Abstr. 084.001.

Report of IAU Commission 20: Positions and motions of minor planets, comets and satellites (*Positions et mouvements des petites planètes, des comètes et des satellites*).
See Abstr. 098.098.

The soliton model for the Great Red Spot and other large vortices in planetary atmospheres.
See Abstr. 099.039.

Some morphologic systematics of complex impact structures.
See Abstr. 105.215.

Energetic particle impacts on frozen surfaces: laboratory simulation and solar system applications.
See Abstr. 106.047.

Occurrence of giant impacts during the growth of the terrestrial planets.
See Abstr. 107.031.

A new empirical formula for satellite distances and its explanation.
See Abstr. 107.037.

092 Mercury

092.001 Mercury.
R. G. Strom.
NASA Spec. Publ., NASA SP–469, p. 12 – 55 (1984). – See Abstr. 003.004.
Contents: Introduction. General surface characteristics. Craters and basins. Major surface units. Tectonics. Surface history and thermal evolution.

092.002 Newtonian N–body calculations of the advance of Mercury's perihelion.
J. V. Narlikar, N. C. Rana.
Mon. Not. R. Astron. Soc., Vol. 213, No. 3, p. 657 – 663 (1985).

The rate of advance of the perihelion of planet Mercury is calculated by a numerical integration of the Newtonian equations of motion and gravitation. It is found that the rate fluctuates but has a steady trend of ~ 528.95 arcsec cy^{-1} over a long–term period of about five centuries. Taking into account the observed precession the authors therefore find that the general relativistic correction explains all but about 2.3 arcsec cy^{-1} of the discrepancy. It may be necessary to invoke another cause like solar oblateness to account for the residual effect.

092.003 Mercury: the world closest to the Sun.
B. M. Cordell.
Mercury, Vol. 13, No. 5, p. 136 – 146 (1985).

092.004 Mercury: thermal emission at radio wavelengths. I. 3.3– and 28–mm observed brightness temperatures and periodicities therein.
E. E. Epstein, B. H. Andrew.
Icarus, Vol. 62, No. 3, p. 448 – 457 (1985).

The authors have made extensive Mercury observations at 3.3 mm and 28 mm to search for variations of the disk–average brightness temperature as a function of longitude, to search for other periodicities, and to constrain the range of values of thermophysical parameters by comparing the data with models. This paper deals with the observations and the first two objectives.

Discovery of Mercury's rotation.
See Abstr. 004.196.

093 Venus

093.001 Magnetic flux ropes of Venus: a paradigm for helical magnetic structures in astrophysical systems.
R. C. Elphic.
Unstable current systems and plasma instabilities in astrophysics, p. 43 – 46 (1985). – See Abstr. 012.002 (IAU Symp. No. 107).

The magnetic flux ropes of Venus are small scale (ion gyroradius) cylindrically symmetric structures observed in situ by the Pioneer Venus orbiter in the largely magnetic field–free ionosphere of the planet. The magnetic structure of flux ropes indicates that they are helical kink unstable for $\lambda > 100$ km. The altitude distribution of the fractional volume occupied by ropes can be explained by invoking the helical kink instability as ropes convect to lower altitudes. Furthermore, the low altitude orientations of ropes can also be explained by helical kink instability. It may be that this instability is at work in other astrophysical phenomena. For example, it may cause filaments of sub–photospheric magnetic flux to protrude into the solar atmosphere.

093.002 Chemical composition of Venus clouds.
V. A. Krasnopolsky (*V. A. Krasnopol'skij*).
Planet. Space Sci., Vol. 33, No. 1, p. 109 – 117 (1985).

From estimates of drying effect in the cloud layer, data of the Venera 14 X–ray fluorescent spectroscopy, and evaluation of photochemical production of sulfuric acid, it follows that sulfuric acid and/or products of its further conversion should constitute not only the Mode 2 particles but most of the Mode 3 particles as well. The eddy mixing coefficient equals 2×10^4 cm^{2}s^{-1} in the cloud layer. The presence of ferric chloride in the cloud layer is indicated by the Venus u.v. absorption spectrum in the range of 3200 – 5000 Å, by the Venera 12 X–ray fluorescent spectrum, by the coincidence of the calculated $FeCl_3$ condensate density profile and that of the Mode 1 in the middle and lower cloud layer, as well as by the upward flux of $FeCl_3$ from the middle cloud layer which provides the necessary concentration of $FeCl_3$ in H_2SO_4 solution. Other possible absorbers such as sulfur, ammonium pyrosulfite, nitrosylsulfuric acid, etc. are discussed.

093.003 Transport of venusian rolling "stones" by wind?
R. Greeley, J. R. Marshall.
Nature, Vol. 313, No. 6005, p. 771 – 773 (1985).

Speculation that aeolian processes operate on Venus has been confirmed tentatively by analyses of results from Venera spacecraft. The authors describe simulations of venusian wind processes, which show that particles are moved by "rolling" at wind speeds as much as 30% lower than those required for saltation threshold. The formation of small sand ridges and grooves oriented parallel to the wind direction is associated with the rolling of grains in venusian simulations and these structures may be unique aeolian features on Venus.

093.004 Periodicity of the polarization of the equatorial region of Venus.
O. M. Starodubtseva.
Pis'ma Astron. Zh., Tom 11, No. 2, p. 156 – 160 (1985). In Russian. English translation in Sov. Astron. Lett., Vol. 11.

Polarization characteristics of ultraviolet cloud formations were studied in five spectral bands. An inverse correlation between the intensity and the degree of polarization of these formations is found. Periodic temporal variability of the polarization degree is revealed. The period is 4.55 ± 0.05 days on the average for all spectral bands.

093.005 Magnetic field in the wake of Venus and the formation of ionospheric holes.
K. Marubashi, J. M. Grebowsky, H. A. Taylor Jr., J. G. Luhmann, C. T. Russell, A. Barnes.
J. Geophys. Res., Vol. 90, No. A2, p. 1385 – 1398 (1985).

The authors study the structure of the Venusian nightside ionosphere using data from the Pioneer Venus orbiter. The object of the investigation is to analyze the dynamics in the interaction region between the magnetosheath and the ionosphere, and to gain an insight into their effects on the formation of ionospheric holes. It is shown that the detailed characteristics of the hole morphology suggest that the ionospheric hole and magnetosheath–ionosphere interaction regions are inseparably related to each other.

093.006 Plasma measurements of the Pioneer Venus Orbiter in the Venus ionosheath: evidence for plasma heating near the ionopause.
H. Pérez–de–Tejada, D. S. Intriligator, F. L. Scarf.
J. Geophys. Res., Vol. 90, No. A2, p. 1759 – 1764 (1985).

A study of the angular spread of plasma fluxes detected in the Venus ionosheath is presented. It is shown that in the inner regions of the ionosheath, where the flux intensity is severely decreased with respect to solar wind values, the width of the azimuthal distribution of the local plasma is comparable to, or even larger than, that of the stronger fluxes measured in the outer ionosheath. The observed variation of the angular width suggests the existence of a source of heating near the ionopause but is not consistent with the overall cooling that would be expected if mass loading and charge exchange collisions were solely responsible for the interaction process at that boundary. Dissipative phenomena associated with local plasma turbulent processes seem to be required to account for the broad angular distributions seen near the ionopause.

093.007 Venusbeobachtung.
D. Niechoy.
Orion, 43. Jahrg., Nr. 206, p. 4 – 6 (1985).

093.008 The mapping of Venus.
P. Moore.
J. Br. Astron. Assoc., Vol. 95, No. 2, p. 50 – 61 (1985).

093.009 Venus.
R. S. Saunders, M. H. Carr.
NASA Spec. Publ., NASA SP–469, p. 56 – 77 (1984). – See Abstr. 003.004.

Contents: Introduction. Early telescopic observations. Orbital and rotational motions. Earth–based radar observations of the surface. Spacecraft observations. Constraints on the composition of Venus. Venera Lander results. Eolian erosion and transport. Chemical weathering. Global topography and surface roughness. Surface roughness. Bright radar rings. Gravity. Plate tectonics on Venus.

093.010 Zonal winds in the middle atmosphere of Venus from Pioneer Venus radio occultation data.
M. Newman, G. Schubert, A. J. Kliore, I. R. Patel.
J. Atmos. Sci., Vol. 41, No. 12, p. 1901 – 1913 (1984). Abstr. in Phys. Abstr., Vol. 88, No. 1251, Entry 24309 (1985).

093.011 The volcanoes and clouds of Venus.
R. G. Prinn.
Sci. Am., Vol. 252, No. 3, p. 36 – 43 (1985).

Radar maps of Venus and chemical analysis of its atmosphere and crust imply the existence of active volcanoes. The sulfur gases they release form a global cover of sulfuric acid clouds.

093.012 Venussichtbarkeit 1983.
E. Freydank.
Sterne Weltraum, 24. Jahrg., Nr. 2, p. 94 (1985).

093.013 Analysis of the errors of results of radio occultations of the Venus dayside ionosphere due to its nonsphericity.
A. L. Gavrik, L. N. Samoznaev.
Kosm. Issled., Tom 23, Vyp. 1, p. 148 – 157 (1985). In Russian. English translation in Cosm. Res.

093.014 Pioneer Venus suprathermal electron flux measurements in the Venus umbra.
W. C. Knudsen, K. L. Miller.
J. Geophys. Res., Vol. 90, No. A3, p. 2695 – 2702 (1985).

The authors provide in this paper a more definitive description of the suprathermal electron flux field observed in the Venus umbra by the Lockheed–IPW RPA than was presented previously (Spenner et al., 1981). The present study confirms the main properties attributed to the limited suprathermal electron flux measurements presented in the previous study and yields additional information about the altitude and solar zenith angle variations of the flux. The authors compare, to the extent possible, the P–V suprathermal fluxes with those measured by the Venera spacecraft.

093.015 On the variations in Venus' rotation.
M. Burša.
Bull. Astron. Inst. Czech., Vol. 36, No. 2, p. 71 – 73 (1985).

The variations in Venus' rotation, generated by the torques due to the non–zonal terms in the aphroditopotential exerted on Venus by the Earth, have been computed. The torques are compared with the tidal torque due to the Sun; their magnitudes do not support the hypothesis that the Earth controls the spin of Venus significantly.

093.016 Volcanism on Venus: a connecting link?
L. V. Ksanfomaliti.
Sov. Astron. Lett., Vol. 10, No. 4, p. 257 – 261 (1984). English translation of 38.093.008.

093.017 Photographic position observations of Venus at the Main Astronomical Observatory of the Academy of Sciences of the Ukrainian SSR in 1980.
E. M. Sereda.
Glav. astron. obs. Akad. Nauk SSSR. Kiev, 22 pp. (1984). In Russian. Abstr. in Ref. Zh., 51. Astron., 4.51.89 (1985).

093.018 Indirect evidence of volcanism on Venus.
L. V. Ksanfomaliti.
Astron. Vestn., Tom 18, No. 4, p. 310 – 321 (1984). In Russian. Abstr. in Ref. Zh., 51. Astron., 4.51.161 (1985).

093.019 On the height extend of the night ionosphere of Venus.
I. K. Osmolovskij, N. A. Savich, L. N. Samoziev.
Radiotekh. ehlektron., Moskva, Tom 29, No. 12, p. 2302 – 2306 (1984). In Russian. Abstr. in Ref. Zh., 51. Astron., 4.51.168 (1985).

093.020 First results of a geological–morphological analysis of radar pictures of the Venus surface obtained with the automatic interplanetary stations Venera 15 and Venera 16.
V. L. Barsukov, A. T. Bazilevskij, A. A. Pronin, R. O. Kuz'min, V. P. Kryuchkov, O. V. Nikolaeva, I. M. Chernaya, G. A. Burba, N. N. Bobina, V. P. Shashkina, V. A. Kotel'nikov, O. N. Rzhiga, G. M. Petrov, Yu. N. Aleksandrov, A. I. Sidorenko, V. M. Kovtunenko, R. S. Kremnev, A. F. Bogomolov, M. N. Meshkov, N. V. Zherikhin, Yu. S. Tyuflin, Eh. L. Akim, M. S. Markov, A. L. Sukhanov.
Dokl. Akad. Nauk SSSR. Ser. Mat. Fiz., Tom 279, No. 4, p. 946 – 950 (1984). In Russian. Abstr. in Ref. Zh., 62. Issled. Kosm. Prostranstva, 4.62.350 (1985).

093.021 Net thermal radiation in the atmosphere of Venus.
H. E. Revercomb, L. A. Sromovsky, V. E. Suomi, R. W. Boese.
Icarus, Vol. 61, No. 3, p. 521 – 538 (1985).

The four entry probes of the Pioneer Venus mission measured the radiative net flux in the atmosphere of Venus at latitudes of 60°N, 31°S, 27°S, and 4°N. The net flux measurements from two of the three Pioneer Venus small probes contained large errors below the clouds, and the measurements from the large probe were affected by an error which became significant a short distance below the cloud deck and grew to an extremely large value at the surface. Plausible explanations for the errors have been found and the corrected flux profiles are presented. They are compared to flux profiles from radiative transfer calculations. The implication for the global distribution of water vapor and radiative cooling in the lower atmosphere, is discussed.

093.022 Venus gravity west of Beta Regio.
Z. M. Goldberg, R. D. Reasenberg.
Icarus, Vol. 62, No. 1, p. 129 – 142 (1985).

Doppler tracking data from the Pioneer Venus Orbiter (PVO) have been used to estimate the anomalous gravity field in the region of Venus west of Beta Regio. The topographic map for this region, derived from the PVO radar, has been filtered to have the same distortions and degree of smoothing as the gravity map. The undulations of the gravity are about 0.2 times as large as expected from the topography on the assumption that the latter is uncompensated. A comparison of the gravity and topography by means of the spectral admittance is consistent with Airy compensation at a depth of 50 km if the surface material has a density of 2.6 g/cm^3. However, this is not a unique interpretation.

093.023 A mechanism for the low nightside temperatures in the Venus thermosphere.
B. F. Gordiets, Yu. N. Kulikov.
Sov. Astron. Lett., Vol. 10, No. 5, p. 291 – 293 (1984). English translation of 38.093.016.

093.024 Sky polarization on Venus.
V. M. Loskutov, I. N. Minin, K. I. Selyakov.
Sov. Astron. Lett., Vol. 10, No. 5, p. 325 – 326 (1984). English translation of 38.093.019.

093.025 Reduction of data of a survey of the Venus surface transmitted by the stations Venera 15 and Venera 16.
A. F. Bogomolov, G. I. Skrypnik, I. M. Bokshtejn,
M. A. Kronrod, P. A. Chochia, M. Yu. Bergman,
L. V. Kudrin, A. V. Bashnin.
Kosm. Issled., Tom 23, Vyp. 2, p. 179 – 190 (1985). In Russian. English translation in Cosm. Res.

093.026 Infrared experiment aboard the automatic interplanetary stations Venera 15 and Venera 16. I. Methods and first results.
D. Oertel, V. I. Moroz, J. Nopirakowsky, V. M. Linkin,
H. Jahn, R. S. Kremnev, H. Becker–Ross, W. Stadthaus,
V. V. Kerzhanovich, I. A. Matsygorin, A. V. D'yachkov,
L. I. Khlyustova, W. Berwald, M. Ulich, H. Drischer,
W. Skrbek, H. Studemund, R. Schuster, G. Kaiser,
S. P. Ignatova, I. A. Zelenov, I. D. Tserenin, D. Spänkuch,
W. Döhler, K. Schäfer, L. V. Zasova, E. A. Ustinov,
G. Fellberg, A. N. Lipatov, A. A. Shurupov, N. G. Havenson.
Kosm. Issled., Tom 23, Vyp. 2, p. 191 – 205 (1985). In Russian. English translation in Cosm. Res.

093.027 Infrared experiment aboard the automatic interplanetary stations Venera 15 and Venera 16. II. Preliminary results of restoration of temperature profiles.
D. Spänkuch, L. V. Zasova, K. Schäfer, E. A. Ustinov,
Yu. Gyul'dner, V. I. Moroz, W. Döhler, V. M. Linkin,
R. Dubois, A. V. D'yachkov, H. Becker–Ross, A. N. Lipatov,
W. Stadthaus, I. A. Matsygorin, D. Oertel,
V. V. Kerzhanovich, J. Nopirakowsky, H. Jahn, G. Fellberg,
R. Schuster, A. A. Shurupov.
Kosm. Issled., Tom 23, Vyp. 2, p. 206 – 220 (1985). In Russian. English translation in Cosm. Res.

093.028 Infrared experiment aboard the automatic interplanetary stations Venera 15 and Venera 16. III. Some conclusions on the structure of clouds based on an analysis of spectra.
L. V. Zasova, D. Spänkuch, V. I. Moroz, K. Schäfer,
E. A. Ustinov, W. Döhler, V. M. Linkin, D. Oertel,
A. V. D'yachkov, H. Becker–Ross, I. A. Matsygorin,
J. Nopirakowsky, V. V. Kerzhanovich, R. Dubois,
A. N. Lipatov, W. Stadthaus, A. A. Shurupov.
Kosm. Issled., Tom 23, Vyp. 2, p. 221 – 235 (1985). In Russian. English translation in Cosm. Res.

093.029 Infrared experiment aboard the automatic interplanetary stations Venera 15 and Venera 16. IV. Preliminary results of an analysis of spectra in the region of the H$_2$O and SO$_2$ absorption bands.
V. I. Moroz, W. Döhler, E. A. Ustinov, K. Schäfer,
L. V. Zasova, D. Spänkuch, A. V. D'yachkov, R. Dubois,
V. M. Linkin, D. Oertel, V. V. Kerzhanovich,

J. Nopirakowsky, I. A. Matsygorin, H. Becker–Ross,
A. A. Shurupov, W. Stadthaus, A. N. Lipatov.
Kosm. Issled., Tom 23, Vyp. 2, p. 236 – 247 (1985). In Russian. English translation in Cosm. Res.

093.030 Infrared experiment aboard the automatic interplanetary stations Venera 15 and Venera 16. V. Preliminary results of an analysis of fields of brightness temperatures and thermal streams.
V. M. Linkin, K. Schäfer, I. A. Matsygorin, W. Döhler,
V. I. Moroz, R. Dubois, A. V. D'yachkov, D. Spänkuch,
V. V. Kerzhanovich, D. Oertel, L. V. Zasova, J. Nopirakowsky,
E. A. Ustinov, H. Becker–Ross, A. N. Lipatov, W. Stadthaus,
A. A. Shurupov.
Kosm. Issled., Tom 23, Vyp. 2, p. 248 – 258 (1985). In Russian. English translation in Cosm. Res.

093.031 Infrared radiation of Venus: approximate methods of calculation of the spectrum in the absorption bands of atmospheric gases.
V. I. Moroz, L. V. Zasova.
Kosm. Issled., Tom 23, Vyp. 2, p. 259 – 267 (1985). In Russian. English translation in Cosm. Res.

093.032 Exogeneous processes and roughness of the Venus surface according to radar observations.
V. P. Kryuchkov, A. A. Pronin.
Kosm. Issled., Tom 23, Vyp. 2, p. 268 – 275 (1985). In Russian. English translation in Cosm. Res.

093.033 Venus: eastern elongation 1979 – 1980.
J. McCue, J. Nichol.
J. Br. Astron. Assoc., Vol. 95, No. 4, p. 157 – 161 (1985).
Visual and photographic studies of Venus during its eastern elongation 1979 – 80 are described and analyzed.

093.034 A radar tour of Venus.
J. K. Beatty.
Sky Telesc., Vol. 69, No. 6, p. 507 – 510 (1985).

093.035 First map of Venus.
O. N. Rzhiga.
Zemlya Vselennaya, No. 3, p. 2 – 7 (1985). In Russian.

093.036 Total injection of water vapor into the Venus atmosphere.
V. A. Krasnopolsky (*V. A. Krasnopol'skij*).
Icarus, Vol. 62, No. 2, p. 221 – 229 (1985).
The atomic hydrogen mixing ratio at the homopause and the deuterium fractionation factor equal 0.5 ppm and 2.2%, respectively. The function which defines hydrogen amounts in the lower thermosphere relative to the lower atmospheric water vapor mixing ratio has been calculated for the sulfuric acid trap. The author considers the effects of the enhanced solar EUV radiation in the past as well as the supersaturation of water vapor at the cold trap on the nonthermal hydrogen escape. If the present amount of carbon dioxide on Venus were present in the atmosphere throughout the entire history of the planet, then an initial water amount equal to a full terrestrial ocean is consistent with the present deuterium–to–hydrogen ratio.

093.037 Modelling of streams of thermal radiation and of water content in the subcloud atmosphere of Venus.
V. P. Shari.
Inst. prikl. mat. Akad. Nauk SSSR. Prepr., No. 166, 23 pp. (1984). In Russian. Abstr. in Ref. Zh., 51. Astron., 6.51.140 (1985).

093.038 Investigation of the thermal structure of the upper atmosphere of Venus by 10 µm heterodyne spectroscopy.
H. U. Käufl, H. Rothermel, S. Drapatz.
Infrared Phys., Vol. 25, No. 1/2, p. 505 – 512 (1985). – See Abstr. 012.065.
The Venusian atmosphere has been investigated using a 10 µm heterodyne spectrometer (spectral resolution $\lambda/\Delta\lambda = 3 \times 10^6$)

during upper conjunction and during elongation in 1983. In both cases a variety of spectra from the subsolar region were taken with high spatial resolution showing fully–resolved lines of the 10 µm laser transition of CO_2. From these measurements kinetic and rotational temperatures are derived ($T \sim 200K$). These spectroscopically–derived temperatures represent the atmospheric temperature for extended areas (1000 – 6000 km dia) at an altitude of 100 – 120 km. They are in agreement with on–the–spot temperature measurements indirectly obtained by space probes during descent. Multiple–line structures have been observed in the 10 µm spectra, which indicate a wave–like perturbation on the vertical temperature profile above a cloud layer at ~ 20 mb atmospheric pressure.

093.039   Evidence for mass–loading of the Venus magnetosheath.
    J. G. Luhmann, C. T. Russell, J. R. Spreiter,
S. S. Stahara.
Adv. Space Res., Vol. 5, No. 4, p. 307 – 311 (1985). – See Abstr. 012.066.
    The observed magnetic field configuration in the Venus magnetosheath contains infomation about the solar wind mass–loading processes occurring as a result of the extension of the neutral atmosphere into the magnetosheath. In this paper, magnetic field signatures of various mass–loading processes are discussed and experimental results from the Pioneer Venus Orbiter magnetometer experiment are examined for evidence of these signatures.

093.040   Electron densities and temperatures in the Venus ionosphere: effects of solar EUV, solar wind pressure and magnetic field.
R. C. Elphic, L. H. Brace, C. T. Russell.
Adv. Space Res., Vol. 5, No. 4, p. 313 – 316 (1985). – See Abstr. 012.066.
    The authors investigate the dependence of day and nightside Venus ionospheric conditions on solar EUV, solar wind conditions and the associated magnetic fields in the ionosphere. They use electron density and temperature normalized by empirical models of Theis et al. (1984) to investigate variations about mean conditions.

093.041   Solar wind–Venus interaction affecting ionospheric electron density profile.
R. N. Singh, R. Prasad.
Adv. Space Res., Vol. 5, No. 4, p. 317 – 320 (1985). – See Abstr. 012.066.
    The equilibrium electron density profile has been computed and compared with measured profiles by Venera 9 and Mariner 5 and 10. The computed electron density profile is seen to show discrepancies with the measured data. The contribution of solar wind interaction induced convection to the equilibrium electron density profile has been estimated. It is found that the convective processes are less important at lower altitudes, whereas at higher altitudes their contribution becomes dominant. The night side Venus ionosphere is formed due to the transport of O^+ and impact ionization of neutral gases by suprathermal electrons. The discrepancies in theoretical and measured electron density profiles provide a clear indication of an additional energy source of solar wind origin.

093.042   Venus: volcanism and rift formation in Beta Regio.
    D. B. Campbell, J. W. Head, J. K. Harmon,
A. A. Hine.
Science, Vol. 226, No. 4671, p. 167 – 170 (1984).
    A new high–resolution radar image of Beta Regio, a Venus highland area, confirms the presence of a major tectonic rift system and associated volcanic activity. The lack of identifiable impact craters, together with the apparent superposition of the Theia Mons volcanic structure on the rift system, suggest that at least some of the volcanic activity occurred in relatively recent geologic time. The presence of topographically similar highland areas elsewhere on Venus (Aphrodite Terra, Dali Chasma, and Diana Chasma) suggests that rifting and volcanism are significant processes on Venus.

093.043   I've looked at clouds from both sides now.
    D. A. Allen.
Aust. J. Astron., Vol. 1, No. 1, p. 1 – 9 (1985).
    This is a personal account of the events that led up to the discovery of the cloud structure on the dark side of the planet Venus. The discovery was made less than two years ago, and no further observations have yet been possible. The significance of the discovery, and the possibilities for future work are reviewed.

093.044   The Maxwell Montes region, surveyed by the Venera 15, Venera 16 orbiters.
V. A. Kotel'nikov, É. L. Akim (*Eh. L. Akim*),
Yu. N. Aleksandrov, N. A. Armand, A. T. Bazilevskij,
A. F. Bogomolov, V. M. Dubrovin, B. Ya. Fel'dman,
V. I. Kaevitser, V. M. Kovtunenko, R. S. Kremnev,
A. P. Krivtsov, G. A. Krylov, A. A. Krymov,
I. L. Kucheryavenkova, E. P. Molotov, G. M. Petrov,
O. N. Rzhiga, A. S. Selivanov, A. M. Shakhovskoj,
V. A. Shubin, A. I. Sidorenko, V. P. Sinilo, A. V. Sknarya,
G. A. Sokolov, V. P. Sorokin, K. G. Sukhanov,
V. F. Tikhonov, Yu. S. Tyuflin, A. S. Vyshlov, A. I. Zakharov,
N. V. Zherikhin, V. E. Zimov.
Sov. Astron. Lett., Vol. 10, No. 6, p. 369 – 373 (1984). English translation of 38.093.067.

093.045   Equatorial flattenings of planets: Venus.
    M. Burša, Z. Šíma.
Bull. Astron. Inst. Czech., Vol. 36, No. 3, p. 129 – 138 (1985).
    The best–fitting tri–axial parameters of Venus have been determined. It has been demonstrated that the figure of Venus is strange, differing from all terrestrial bodies in the solar system: in the equatorial zone the aphroditoid surface is situated above the best–fitting tri–axial Venus ellipsoid. Deflections of the vertical at Venus' surface have been computed; their global interpretation has been given.

093.046   The sulfur cycle and clouds of Venus.
    R. G. Prinn.
Recent advances in planetary meteorology, p. 1 – 15 (1985). – See Abstr. 012.071.
    In this paper the overall chemical cycle responsible for the maintenance of significant amounts of the lithophilic element sulfur in the atmosphere of Venus and the photochemical and thermochemical reactions involved in formation and destruction of the cloud particles themselves are elucidated. The author emphasizes the susceptibility of these processes to change (e.g., due to episodic volcanism) with concomitant feedbacks to the Venusian climate.

093.047   Pioneer Venus orbiter ultraviolet spectrometer limb observations: analysis and interpretation of the 166– and 156–nm data.
L. J. Paxton.
J. Geophys. Res., Vol. 90, No. A6, p. 5089 – 5096 (1985).
    Pioneer Venus orbiter ultraviolet spectrometer (PVOUVS) limb observations at 166 and 156 nm have been analyzed, and emission mechanisms at these wavelengths have been modeled. The observations prove to be diagnostic of both the thermal carbon density and the O_2 to CO_2 mixing ratio. The photons seen by the PVOUVS at 156 and 166 nm arise from a variety of sources: photoelectron impact dissociative excitation of CO and CO_2, photodissociative excitation of CO and CO_2, photoelectron impact excitation of C, solar resonance scattering by C, and the underlying CO fourth–positive bands (CO4PG). Solar resonant scattering by atomic carbon is an important contribution to the observed intensity and, furthermore, a diagnostic of the O_2 to CO_2 mixing ratio. An O_2 to CO_2 mixing ratio of 0.3% yielded the best agreement between the modeled and observed limb intensities for both the 156– and 166–nm data sets. For that O_2 to CO_2 mixing ratio, the atomic carbon density reaches a maximum at 155 km of 5×10^6 cm^{-3}.

093.048   Solar cycle dependence of the location of the Venus bow shock.
C. J. Alexander, C. T. Russell.
Geophys. Res. Lett., Vol. 12, No. 6, p. 369 – 371 (1985).

Since Pioneer Venus has now been in orbit over six years, well over half a solar cycle, the authors are provided with an excellent opportunity to test the dependence of the bow shock location on the solar cycle. They show that there is a good correlation betweeen the fall off in solar productivity and the retreat in bow shock position, that indeed the long–term variability in bow shock location is responsive to the EUV output of the sun.

093.049   Solar control of the Venus ionosphere.
S. J. Bauer.
Sitzungsber., Österr. Akad. Wiss., Math.–Naturwiss. Kl. Abt. II, Band 192, Heft 8 – 10, p. 309 – 317 (1983).

This paper discusses, in general terms, the analytical relationships resulting from the solar control of the Venus ionosphere useful for reference models such as the Venus International Reference Atmosphere.

093.050   Baroclinic instability in the Venus atmosphere.
R. E. Young, H. Houben, L. Pfister.
J. Atmos. Sci., Vol. 41, No. 15, p. 2310 – 2333 (1984). Abstr. in Phys. Abstr., Vol. 88, No. 1257, Entry 57636 (1985).

093.051   Wave propagation through the Venusian atmosphere.
R. Misra.
Indian J. Theor. Phys., Vol. 30, No. 4, p. 297 – 301 (1982). Abstr. in Phys. Abstr., Vol. 88, No. 1260, Entry 74096 (1985).

093.052   Dimer spectra of carbon dioxide in the atmospheres of Venus and Mars.
K. Fox, S. Kim.
Bull. Am. Astron. Soc., Vol. 17, No. 2, p. 591 – 592 (1985). Abstract. – See Abstr. 010.065.

093.053   Photographic position observations of Venus in 1980.
B. S. Vozdvizhenskij.
Tr. Gos. Astron. Inst. Shternberg, Tom 57, p. 218 – 242 (1985). In Russian.

A catalogue of 801 Venus positions is published.

093.054   Venus topography: a harmonic analysis.
B. G. Bills, M. Kobrick.
J. Geophys. Res., Vol. 90, No. B1, p. 827 – 836 (1985).

A model of Venusian global topography has been obtained by fitting an eighteenth–degree harmonic series to Pioneer Venus orbiter radar altimeter data. The mean radius is (6051.45 ± 0.04) km. The corresponding mean density is (5244.8 ± 0.5) kg m^{-3}. The center of figure is displaced from the center of mass by (0.339 ± 0.088) km towards $(6.6 \pm 10.1)°$N, $(148.8 \pm 7.7)°$E. The figure of Venus is distinctly triaxial, but the orientation and magnitudes of the principal topographic axes correlate rather poorly with the gravitational principal axes. However, the higher–degree harmonics of topography and gravity are significantly correlated. The topographic variance spectrum of Venus is very similar in form to those of the moon, Mars, and especially Earth.

093.055   Radar evidence for cratering on Venus.
B. A. Burns, D. B. Campbell.
J. Geophys. Res., Vol. 90, No. B4, p. 3037 – 3047 (1985).

To date many craterlike features have been detected in earth–based radar and Pioneer Venus imagery of the surface of Venus, but whether they are of impact or endogenic origin is still in question. The morphologic and polarization characteristics of these features, as observed from earth–based radar data obtained at the Arecibo Observatory, provide additional information that may be useful in their identification. This paper examines the extent to which this information can be used, in conjunction with our knowledge of the morphologic and polarization properties of impact craters and volcanic calderas on other terrestrial bodies, to constrain inferences about the origin of the craterlike features on Venus.

093.056   Analysis of regional slope characteristics on Venus and earth.
V. L. Sharpton, J. W. Head III.
J. Geophys. Res., Vol. 90, No. B5, p. 3733 – 3740 (1985).

Regional slope values for earth and Venus are calculated for each $3° \times 3°$ region of topography, and the global characteristics of these magnitudes are examined and compared.

093.057   Venus and Mars at the 15th COSPAR General Assembly.
M. Burša.
Říše hvězd, Vol. 65, No. 12, p. 245 – 249, 254 – 255 (1984). In Czech.

093.058   Atmosphere and surface of Venus.
P. Lála.
Vesmír, Vol. 63, No. 6, p. 167 – 170 (1984). In Czech.

093.059   Analysis and interpretation of observations of airglow at 297 nm in the Venus thermosphere.
M. A. LeCompte.
Diss. Abstr. Int., Sect. B, Vol. 45, No. 9, p. 2957 (1985). Thesis, University of Colorado, 241 pp. (1984). Order No. DA8428664.

093.060   Photographic position observations of Venus at Pulkovo in 1980.
N. M. Bronnikova.
Izv. Glav. Astron. Obs. Pulkovo, Astrometr. Astrofiz., No. 201, p. 64 – 67 (1985). In Russian.

The results of photographic observations of Venus made with the Pulkovo normal astrograph during 1980 are given. The mean square errors of one position of the planet are as follows: $\sigma_\alpha = \pm 0.''20$, $\sigma_\delta = \pm 0.''33$ (winter–spring); $\sigma_\alpha = \pm 0.''10$, $\sigma_\delta = 0.''23$ (autumn). The reference stars were taken from the AGK3.

093.061   Temperature structure in the lower atmosphere of Venus: new results derived from Pioneer Venus entry probe measurements.
L. A. Sromovsky, H. E. Revercomb, V. E. Suomi.
Icarus, Vol. 62, No. 3, p. 458 – 493 (1985).

The interpretation of unexpected characteristics of Pioneer Venus temperature measurements, and of the large difference between these and the Venera results, is aided by new Venus temperature profiles derived from engineering measurements of the Pioneer Venus Small–Probe Net Flux Radiometer instruments. To facilitate correction of a temperature–dependent radiometric response, these instruments monitored the temperatures of their deployed radiation detectors. The accurate calibration of the temperature sensors, and their strong thermal coupling to the atmosphere, make it possible to deduce atmospheric temperatures within 2K (at most altitudes) using a simple two–component thermal model to account for lag effects.

093.062   Deuterium on Venus: model comparisons with Pioneer Venus observations of the predawn bulge ionosphere.
S. Kumar, H. A. Taylor Jr.
Icarus, Vol. 62, No. 3, p. 494 – 504 (1985).

A model of the predawn bulge ionosphere composition and structure is constructed and compared with the ion mass spectrometer measurements from the Pioneer Venus Orbiter during orbits 117 and 120. Particular emphasis is given to the identification of the mass–2 ion which the authors find unequivocally due to D^+ (and not H_2^+). The atmospheric D/H ratio of 1.4% and 2.5% is obtained at the homopause (~ 130 km) for the two orbits. The H_2^+ contribution to the mass–2 ion density is less than 10%, and the H_2 mixing ratio must be < 0.1 ppm at 130 km altitude. The He^+ data require a downward He^+ flux of $\sim 2 \times 10^7 cm^{-2} sec^{-1}$ in the predawn region which suggests that the light ions also flow across the terminator from day to night along with the observed O^+ ion flow.

Mean collision frequencies of O^+–He and H^+–He for ionospheric investigations.
See Abstr. 022.035.

Analyses of experimental observations of electron temperatures in the near wake of a model in a laboratory–simulated solar wind plasma.
See Abstr. 022.112.

Infrared–Fourier–spectrometers in the Venusian orbit.
See Abstr. 035.071.

Construction of the optical block of the infrared–Fourier–spectrometer FS 1/4.
See Abstr. 035.072.

Thermal regime of the optical block of the infrared–Fourier–spectrometer FS 1/4.
See Abstr. 035.073.

Physical–technical conception of the Infrared–Fourier–Spectrometers for 'Venera 15' and 'Venera–16'.
See Abstr. 035.082.

USSR leads the way to Halley's comet.
See Abstr. 051.094.

Venus Radar Mapper attitude control system.
See Abstr. 052.065.

Patchy reconnection and magnetic ropes in astrophysical plasmas.
See Abstr. 062.005.

Response of Earth and Venus ionospheres to corotating solar wind stream of 3 July 1979.
See Abstr. 074.079.

Zwaartekracht meten in gewichtloosheid.
See Abstr. 081.021.

Singlet oxygen nightglow in the atmospheres of Earth, Venus and Mars.
See Abstr. 082.028.

Natural 10 μm band CO_2 laser in the atmospheres of Mars and Venus.
See Abstr. 097.011.

Corotating interplanetary streams and associated ionospheric disturbances at Venus and Earth.
See Abstr. 106.021.

094 Moon

094.001   Megaregolith thickness, heat flow, and the bulk composition of the Moon.
K. L. Rasmussen, P. H. Warren.
Nature, Vol. 313, No. 5998, p. 121 – 124 (1985).

Lunar heat flow data have been interpreted to imply that the Moon's uranium content is about 46 ng g^{-1}, about twice that of the Earth's mantle. Based on a new model that takes into account the considerable, but variable, thickness of porous, low-conductivity megaregolith, the thickness of the lunar lithosphere, and the nonrepresentative composition of the crust at one of the two sites where heat flow was measured, the authors estimate here that the Moon's uranium content is roughly 19 ng g^{-1}. The Moon's bulk composition appears far less exotic than generally assumed.

094.002   Relative and absolute ages of individual craters and the rate of infalls on the Moon in the post–Imbrium period.
R. B. Baldwin.
Icarus, Vol. 61, No. 1, p. 63 – 91 (1985).

This paper contains a reasonably successful attempt to determine relative ages and then absolute ages of individual craters younger than Imbrium, and the rate of infalls onto the Moon as a function of time. After the tail of the massive premare bombardment became depleted before 3×10^9 years ago, there was a period of minimal numbers of infalls. The rate of infalls increased rather steadily from this minimum to the present.

094.003   Numerical models for the study of motion of lunar satellites.
M. Gousidou–Koutita.
Earth, Moon, Planets, Vol. 32, No. 1, p. 21 – 45 (1985).

The present study deals with numerical modeling of the elliptic restricted three–body problem as well as of the perturbed elliptic restricted three–body (Earth–Moon–Satellite) problem by a fourth body (Sun). Two numerical algorithms are established and investigated. The first is based on the method of the series solution of the differential equations and the second is based on a 5th–order Runge–Kutta method. The applications concern the solution of the equations and integrals of motion of the circular and elliptical restricted three–body problem as well as the search for periodic orbits.

094.004   Numerical investigation of collision orbits of lunar satellites.
M. Gousidou–Koutita.
Earth, Moon, Planets, Vol. 32, No. 2, p. 135 – 163 (1985).

In the present study an investigation of the collision orbits of natural satellites of the Moon (considered to be of finite dimensions) is developed, and the tendency of natural satellites of the Moon to collide on the visible or the far side of the Moon is studied. The collision course of the satellite is studied up to its impact on the lunar surface for perturbations of its initial orbit arbitrarily induced, for example, by the explosion of a meteorite.

094.005   Topographic mapping of the Moon.
S. S. C. Wu.
Earth, Moon, Planets, Vol. 32, No. 2, p. 165 – 172 (1985).

Contour maps of the Moon have been compiled by photogrammetric methods that use stereoscopic combinations of all available metric photographs from the Apollo 15, 16, and 17 missions. The maps utilize the same format as the existing NASA shaded–relief Lunar Planning Charts (LOC–1, –2, –3, and –4), which have a scale of 1:2750000. The map contour interval is 500 m. A control net derived from Apollo photographs by Doyle and others was used for the compilation. Contour lines and elevations are referred to the new topographic datum of the Moon, which is defined in terms of spherical harmonics from the lunar gravity field. Compilation of all four LOC charts was completed on analytical plotters from 566 stereo models of Apollo metric photographs that cover approximately 20% of the Moon. This is the first step toward compiling a global topographic map of the Moon at a scale of 1:5000000.

094.006   The lunar ellipsoid of inertia.
V. S. Kislyuk.
Kinematika Fiz. Nebesn. Tel, Tom 1, No. 1, p. 41 – 48 (1985). In Russian.

The problem is discussed for establishing a system of dynamical parameters of the moon which characterize the lunar ellipsoid of inertia and its orientation. It is shown that the modern models of the gravitational field of the moon give a more or less reliable value of a latitude shift of the earth–ward principal moment of inertia axis from the mean sub–earth point. An analogous shift in longitude is obtained with most uncertainty because of inaccurate knowledge of the harmonic S_{33}.

094.007 Profile of the moon from data of an Apollo 11 spacecraft survey.
A. S. Duma, V. S. Kislyuk.
Kinematika Fiz. Nebesn. Tel, Tom 1, No. 1, p. 49 – 54 (1985). In Russian.

Absolutisation of the moon's profile obtained by spacecraft Apollo 11 is carried out by means of bench mark points the heights of which were determined from the selenodetic data of the nearside lunar hemisphere and the results of Apollo 15 and Apollo 16 laser altimetry. The first information about the relief of some regions was obtained for the farside of the moon.

094.008 Chronology and petrogenesis of a 1.8 g lunar granitic clast: 14321,1062.
C.-Y. Shih, L. E. Nyquist, D. D. Bogard, J. L. Wooden, B. M. Bansal, H. Wiesmann.
Geochim. Cosmochim. Acta, Vol. 49, No. 2, p. 411 – 426 (1985).

A study was undertaken to determine the chronology of a pristine granite clast (1062) from Apollo 14 breccia 14321 using Rb–Sr, Sm–Nd and ^{39}Ar–^{40}Ar methods. The genesis of the granite as constrained by the isotopic results and trace element characteristics is discussed.

094.009 Moon.
D. E. Wilhelms.
NASA Spec. Publ., NASA SP–469, p. 106 – 205 (1984). – See Abstr. 003.004.

Contents: Introduction. Craters. Basins. Terra breccias. Maria. Mare basalts. Tectonism. Geologic history.

094.010 Quench media effects on iron partitioning and ordering in a lunar glass.
M. D. Dyar, D. P. Birnie III.
J. Non–Cryst. Solids, Vol. 67, No. 1 – 3, p. 397 – 412 (1984). Abstr. in Phys. Abstr., Vol. 88, No. 1251, Entry 24302 (1985). – See Abstr. 012.016.

094.011 Orientation of the moon's inertia ellipsoid from measurements of the position angles of lunar craters.
V. S. Kislyuk.
Pis'ma Astron. Zh., Tom 11, No. 3, p. 222 – 226 (1985). In Russian. English translation in Sov. Astron. Lett., Vol. 11.

Using the earth–based astrometric measurements of position angles of the directions "crater Mösting A – marginal craters", the earthward axis of the principal moment of inertia is found to be displaced $289'' \pm 25''$ east and $80'' \pm 12''$ south from the "prime radius" of the moon.

094.012 The topography of the Phocylides and Nasmyth area of the Moon.
D. G. Buczynski, P. Wade.
J. Br. Astron. Assoc., Vol. 95, No. 3, p. 106 – 109 (1985).

Methods of deriving the relative heights of features in the Phocylides/Nasmyth area from Earth–based drawings are outlined and discussed. Profiles across the craters in an east/west direction and of their east walls (IAU sense, as throughout) are presented and compared with explanations for dawn appearance of the craters proposed by BAA Lunar Section members in the 1950s.

094.013 Megarelief and figure of the moon from a harmonic analysis of heights of the lunar surface.
I. V. Gavrilov, V. S. Kislyuk, V. I. Belan.
Kinematika Fiz. Nebesn. Tel, Tom 1, No. 2, p. 24 – 29 (1985). In Russian.

Earth–based hypsometric data as well as the data obtained from spacecraft Zond 8 and Apollo 15, 16, 17 are used to find the coefficients of the expansion of the lunar topography in series of spherical harmonics up to the 6th order and the 4th degree. It is shown that the correlation of the moon's megarelief with the selenoid figure is very small.

094.014 Planetary perturbations on the libration of the Moon.
M. Moons.
Celest. Mech., Vol. 34, Nos. 1 – 4, p. 263 – 273 (1984). – See Abstr. 012.030.

A theory of the libration of the Moon, completely analytical with respect to the harmonic coefficients of the lunar gravity field, was recently built (Moons, 1982). In complement to this theory, the author has now computed the planetary effects on the libration. For the main problem, as well as for the planetary perturbations, the motion of the center of mass of the Moon is described by the ELP 2000 solution (Chapront and Chapront–Touze, 1983).

094.015 An analysis of lunar occultations in the years 1955 – 1980 using the new lunar ephemeris ELP 2000.
M. Sôma.
Celest. Mech., Vol. 35, No. 1, p. 45 – 88 (1985).

About 60000 observations of lunar occultations made during 1955 – 1980 are analysed using recently–developed semi-analytical solution ELP 2000–82 for the Moon's position in order to determine the constants in the lunar theory and to investigate the tidal term in the Moon's mean longitude and the motions of the perigee and node of the lunar orbit. The equinox correction and systematic correction to the fundamental star catalogue and the correction to the datum of the lunar–profile in Watt's charts are also investigated.

094.016 Anomalous distribution of large, fresh lunar craters.
M. T. Kitt.
Strolling Astron., Vol. 31, Nos. 1 – 2, p. 22 – 26 (1985).

094.017 Relativistic effects in the earth–moon dynamics.
V. A. Brumberg, T. V. Ivanova.
Tr. Inst. Teor. Astron., Leningrad, Vyp. 19, p. 3 – 30 (1985). In Russian.

Equations of motion of the earth and moon convenient for numerical integration have been derived in the framework of the mass–point PPN formalism taking into account β, γ parameters and the coordinate parameter α. The Lagrangian of the geocentric lunar motion is expanded up to the first degree terms in eccentricity of the orbit of the sun and the solar parallax. The first and the most essential second degree terms in lunar motion have been found by iterations with the aid of the Poisson processor up to order m^7. The variational terms and mean motions of the perigee and node have been obtained up to order m^8. These expressions may be used to complete the analytical theories of the Newtonian motion of the moon elaborated recently by different authors. In conclusion the relativistic reduction of lunar observations is outlined in relation to the problem of determining the lunar orbital constants.

094.018 Genetic types of lunar silicate globules.
N. A. Ashikhmina, O. A. Bogatikov, O. D. Rodeh, D. I. Frikh–Khar, A. Tsimbal'nikova, V. Tsilek, B. Kolman.
Dokl. Akad. Nauk SSSR. Ser. Mat. Fiz., Tom 277, No. 2, p. 434 – 437 (1984). In Russian. Abstr. in Ref. Zh., 51. Astron., 1.51.322; 62. Issled. Kosm. Prostranstva, 1.62.394 (1985).

094.019 The heliometer as a tool for the measure of the Moon.
M. Moutsoulas.
Astrophys. Space Sci., Vol. 110, No. 1, p. 197 – 201 (1985). – See Abstr. 012.039.

The heliometer has been the only instrument for the measurement of the lunar physical libration for more than a century. Bessel (1839), who introduced the use of the heliometer for the systematic measurement of the relative positions of craters on the lunar disc, has also developed the necessary formulation for the calculation of the lunar physical libration from the heliometric measurements. That methodology is presented, and results obtained by Bessel's students and other investigators who followed Bessel's method, are discussed.

094.020 Theoretical interpretations of the empirically obtained laws in the motion of the moon.
A. A. Aryutkina.
Arzamas. gos. ped. inst. Arzamas, 44 pp. (1984). In Russian. Abstr. in Ref. Zh., 51. Astron., 3.51.102 (1985).

094.021 X–ray and electron–microprobe investigation of peculiarities of disintegration of lunar pyroxene.
N. I. Organova, A. I. Gorshkov, N. A. Ashikhmina, I. M. Marsij, A. V. Mokhov.
Izv. Akad. Nauk SSSR. Ser. geol., No. 11, p. 86 – 92 (1984). In Russian. Abstr. in Ref. Zh., 51. Astron., 3.51.249; 62. Issled. Kosm. Prostranstva, 3.62.328 (1985).

094.022 Spectral reflectance studies of the Orientale region of the Moon.
B. R. Hawke, P. Lucey, J. F. Bell, P. D. Spudis.
Meteoritics, Vol. 19, No. 4, p. 235 – 236 (1984). Abstract. – See Abstr. 010.641.

094.023 Lunar palaeomagnetism, polar wandering and the existence of primeval lunar satellites.
S. K. Runcorn.
Meteoritics, Vol. 19, No. 4, p. 304 – 305 (1984). Abstract. – See Abstr. 010.641.

094.024 The origin of selected lunar geochemical anomalies: implications for early volcanism and the formation of light plains.
B. R. Hawke, P. D. Spudis, P. E. Clark.
Earth, Moon, Planets, Vol. 32, No. 3, p. 257 – 273 (1985).
The Apollo orbital geochemistry, photogeologic, and other remote sensing data sets were used to identify and characterize geochemical anomalies on the eastern limb and farside of the Moon and to investigate the processes responsible for their formation. The anomalies are commonly associated with Imbrian- or Nectarian–aged light plains units which exhibit dark–haloed impact craters. The results strongly indicate that those geochemical anomalies associated with light plains deposits which display dark–haloed impact craters result from the presence of basaltic units that are either covered by varying thickness of highland debris or have a surface contaminated with significant amounts of highlands material. Basaltic volcanism on the eastern limb and farside of the Moon was more extensive in both space and time than has been accepted.

094.025 Geological map and optical characteristics of the lunar surface.
A. L. Sukhanov.
Comparative planetology, p. 17 – 22 (1984). In Russian. Abstr. in Ref. Zh., 52. Geod. Aehrosemka, 4.52.244 (1985). – See Abstr. 012.057.

094.026 Remote sensing of lunar pyroclastic mantling deposits.
L. R. Gaddis, C. M. Pieters, B. R. Hawke.
Icarus, Vol. 61, No. 3, p. 461 – 489 (1985).
Mantling deposits on the Moon are considered to be pyroclastic units emplaced on the lunar surface as a result of explosive fire fountaining. These pyroclastic units are characterized as having low albedos, having smooth fine–textured surfaces, and consisting in part of homogeneous, Fe–bearing volcanic glass and partially crystallized spheres. Mantling units exhibit low returns on depolarized 3.8–cm radar maps, indicating an absence of surface scatterers in the 1– to 50–cm–size range. A number of reflectance spectra from several regional pyroclastic deposits are presented. The Rima Bode region is discussed as an example. On the basis of the remote sensing data summarized and presented, five new areas have been identified which may represent higher–albedo regional pyroclastic deposits.

094.027 How the lunar craters weren't formed.
P. Moore.
J. Br. Astron. Assoc., Vol. 95, No. 4, p. 154 – 156 (1985).
A review of eccentric theories to explain the formation of the craters on the Moon.

094.028 The influence of systematic corrections to the limb–profile heights in the marginal zone charts upon the derivation of zero points of star catalogues, of the parameters of motion and rotation of the moon.
L. N. Kizyun.
Kinematika Fiz. Nebesn. Tel, Tom 1, No. 3, p. 33 – 41 (1985). In Russian.
A procedure for determining the systematic corrections to the limb–profile heights in marginal zone charts of the moon is suggested. The dependence of the apparent radius on the libration is studied by comparing the eccentricity and inclination corrections of the lunar orbit for different limbs. On the basis of reduction of Washington meridian observations the position of the centre of reference data on Watts' charts is obtained with respect to the ephemeris centre of mass. The expression is obtained for systematic corrections to the limb–profile heights in Watts' charts.

094.029 Determination of the orientation angles of a selenodetic coordinate system based on photographic position observations of the moon (November 1979 – May 1981).
N. A. Vasilenko, V. S. Kislyuk, R. L. Semerenko.
Kinematika Fiz. Nebesn. Tel, Tom 1, No. 3, p. 42 – 45 (1985). In Russian.
A new series of photographic observations of the moon against the stellar background were used for determining the orientation of the selenodetic system realized by the Consolidated Catalogue of 4900 basic points on the lunar surface with respect to the selenodetic coordinate system, the third axis of which is connected with the mean direction to the earth.

094.030 Numerical modeling of the physical libration of the moon.
A. A. Shiryaev.
Sov. Astron., Vol. 28, No. 5, p. 599 – 601 (1984). English translation of 38.094.026.

094.031 Untersuchungen zur Verbesserung Brownscher Mondentfernungen aufgrund von Variationen der Integrationskonstanten und anderer Parameter.
K.–P. Stoffels.
Mitt. Geod. Inst. Rheinischen Friedrich–Wilhelms–Univ. Bonn, Nr. 66, 132 pp. (1985).

094.032 The parameters of the physical libration of the moon obtained from the second series of I. V. Belkovich.
Yu. A. Nefed'ev.
Kazan. univ. Kazan', 12 pp. (1985). In Russian. Abstr. in Ref. Zh., 51. Astron., 5.51.71 (1985).

094.033 Investigation of the second series of A. A. Nefedev using Morrison's corrections.
Yu. A. Nefed'ev, G. M. Stolyarov.
Kazan. univ. Kazan', 6 pp. (1985). In Russian. Abstr. in Ref. Zh., 51. Astron., 5.51.72 (1985).

094.034 Petrological characteristics of similarity between anorthosite gabbro–dolerites of trap intrusions and lunar anorthosites.
M. D. Tomshin.
Geokhimiya i mineral. bazitov i ul'trabazitov Sib. platformy. Yakutsk, p. 28 – 43 (1984). In Russian. Abstr. in Ref. Zh., 51. Astron., 5.51.193; 62. Issled. Kosm. Prostranstva, 5.62.323 (1985).

094.035 Absolute coordinates of lunar craters according to heliometric measurements at the Engelhardt Astronomical Observatory.
Yu. A. Nefed'ev.
Kazan. univ. Kazan', 21 pp. (1985). In Russian. Abstr. in Ref. Zh., 62. Issled. Kosm. Prostranstva, 5.62.321 (1985).

094.036   Lunar sample 14425: characterization and resemblance to high–magnesium microtektites.
J. A. O'Keefe, B. P. Glass.
Science, Vol. 227, No. 4686, p. 515 – 516 (1985).
Measurements by energy–dispersive X–ray analysis of the surface of lunar sample 14425, a large glass bead, yield a noritic composition enriched in aluminum and magnesium and, as compared with other norites, depleted in iron and especially calcium. The sample is close in composition to the most basic microtektites.

094.037   The magma ocean concept and lunar evolution.
P. H. Warren.
Annu. Rev. Earth Planet. Sci., Vol. 13, p. 201 – 240 (1985). – See Abstr. 003.032.

094.038   Radiation history of meteoritic and lunar material by track data.
L. L. Kashkarov, L. I. Genaeva, A. K. Lavrukhina.
18th International Cosmic Ray Conference, Vol. 8, p. 128 – 131 (1983). – See Abstr. 012.096.
Under study were tracks formed by the iron group nuclei of solar cosmic rays observed in microcrystals of olivine and pyroxene from the lithic chondrules of the brecciated chondrite Weston as well as from the lunar regolith samples and microbreccias of "Luna–16", "Luna–20" and "Luna–24" sites. The obtained results indicate on that the relict irradiation of initial meteorite material took place at the early stage of formation of the solar system bodies and formation of the lithic chondrules occurred in the processes similar to formation of lunar microbreccias without a significant heating.

094.039   Estructura y las propiedades de la Luna pronosticadas a punta de pluma y demostradas después por el experimento "Apolo".
G. P. Tamrazián.
Bol. Acad. Cienc. Fis. Mat. Nat., Tomo 43, Nos. 133 – 134, p. 194 – 203 (1983).

094.040   An error analysis for lunar trajectories.
C. M. de la Barre.
Bull. Am. Astron. Soc., Vol. 17, No. 2, p. 622 (1985). Abstract. – See Abstr. 010.066.

094.041   Elevation profiles of the lunar surface from Apollo 15, 16, 17 laser altimetry.
V. A. Nikonov.
Tr. Gos. Astron. Inst. Shternberg, Tom 57, p. 194 – 217 (1985). In Russian.
Eleven Apollo tracks were calculated from several independent sources. 2120 elevations of lunar surface points were obtained. The horizontal accuracy of points is $\pm 0°4$ of longitude and latitude, the vertical accuracy is ± 0.4 km. Elevations are given with respect to a 1738.0 km sphere about the center of mass of the moon.

094.042   Analisi preliminare dei dati fotometrici visuali relativi ai crateri Tycho e Posidonius.
F. Foresta Martin, M. Missori, M. Pigliucci, D. Ricci, P. Tosi.
Astronomia, N. 2, p. 17 – 22 (1985).
This research is devoted to the study of the lunation curves of the craters Tycho and Posidonius, through a series of visual estimates of the albedo of selected points. The two curves collected in the period February 1983 – July 1984 show a similar behaviour, with two peaks in almost identical conditions of lunar surface illumination.

094.043   The method of approximate cluster analysis and the three–dimensional diagram of optical characteristics of the lunar surface.
N. N. Evsyukov.
Sov. Astron., Vol. 28, No. 6, p. 692 – 697 (1984). English translation of 38.094.067.

094.044   The deep structure of lunar basins: implications for basin formation and modification.
S. R. Bratt, S. C. Solomon, J. W. Head, C. H. Thurber.
J. Geophys. Res., Vol. 90, No. B4, p. 3049 – 3064 (1985).
The authors present models for the structure of the crust and upper mantle beneath lunar impact basins from an inversion of gravity and topographic data from the nearside of the moon.

094.045   On constructing a model of the moon's physical libration based on numerical integration of equations of lunar rotational and orbital motion.
A. A. Shiryaev.
Byull. Inst. Teor. Astron., Tom 15, No. 7 (170), p. 396 – 405 (1985). In Russian.
The paper deals with problems concerned with modelling numerically the moon's physical libration under the gravitational influence of the earth and the sun, these bodies being assumed to be mass points, and the moon's gravitational potential being developed to the fourth degree spherical harmonics. The results of the numerical integration of simultaneous equations of the moon's physical libration and of orbital motions of the moon and perturbing bodies obtained on the time–span of 3100 days have been compared with Eckhardt's (1981) analytical theory.

094.046   On the solution of a secular system of equations of motion of the moon in trigonometric form.
V. A. Brumberg, T. V. Ivanova.
Byull. Inst. Teor. Astron., Tom 15, No. 8 (171), p. 424 – 439 (1985). In Russian.
The trigonometric theory of the secular perturbations in the motion of the major planets enables to construct a trigonometric solution of the secular system for the motion of the moon. This solution has been built up to the third degree terms inclusively with respect to the eccentricities and inclinations of all bodies. The secular system for the lunar motion has been obtained on the basis of the analytical solution of the main problem taking into account the indirect secular planetary inequalities in semi-analytical form.

094.047   Determination of the tidal acceleration of the moon from eclipses and occultations during the 18th – 20th century.
E. Yu. Saramonova.
Byull. Inst. Teor. Astron., Tom 15, No. 8 (171), p. 449 – 456 (1985). In Russian.
The value of the tidal acceleration of the moon has been determined from analysis of solar eclipses, occultations of stars and planets by the moon during 18th – 20th century. Occultations of planets are first used for this purpose. The correction taken from analysis of the transits of Mercury and Venus are used to pass from universal to ephemeris time. The obtained value of the tidal acceleration is $(-22.9 \pm 0.8)''\text{century}^{-2}$.

094.048   Computation of the moon's coordinates with not high accuracy.
M. A. Fursenko.
Byull. Inst. Teor. Astron., Tom 15, No. 8 (171), p. 473 – 477 (1985). In Russian.
A simplified theory of the moon's motion which permits computing the moon's position with given accuracy has been developed. To evaluate the accuracy of coordinates computed by the theory a comparison with the precise lunar ephemeris over the interval 1983 – 2000 has been performed.

Bibliography of Earth, Moon, and Planets.
See Abstr. 002.006.

Bibliography of Earth, Moon, and Planets.
See Abstr. 002.031.

Erdmond: Vorderseite–Rückseite. Kosmos Handkarte 1:12000000.
See Abstr. 002.094.

The moon's acceleration and its physical origins. Vol. 2: As deduced from general lunar observations.
See Abstr. 003.145.

Bessel and librations of the Moon.
See Abstr. 004.033.

Newton's lunar mass error.
See Abstr. 004.053.

Investigations of the moon at the Sternberg Astronomical Institute.
See Abstr. 009.036.

Making the moon from a big splash.
See Abstr. 011.028.

Fifteen years of Lunar Laser Ranging and future developments.
See Abstr. 013.021.

Mondbeobachtung nach dem "Apollo–Schock".
See Abstr. 036.041.

Comparison of brightnesses of lunar features by three methods of observations.
See Abstr. 036.069.

Regular motions of a satellite and some small effects in the motion of the moon and Phobos.
See Abstr. 042.005.

Universal time, lunar tidal deceleration and relativistic effects from observations of transits, eclipses and occultations in the XVIIIth – XXth centuries.
See Abstr. 042.009.

Stability of L_4 and L_5 against radiation pressure.
See Abstr. 042.025.

Solar cosmic ray proton fluxes in the last hundred million years.
See Abstr. 078.018.

Long term solar flare fluxes based on aluminum–26 depth profile in lunar rocks.
See Abstr. 078.019.

How many minerals were found on the earth and on the moon?
See Abstr. 081.024.

Mass loading of planetary magnetospheres by rocky satellites – I: Production of rocky ejecta.
See Abstr. 091.020.

Mass loading of planetary atmospheres by rocky satellites – II: Transport and enhanced lifetimes of satellite ejecta in planetary magnetospheres.
See Abstr. 091.021.

K–U–Th systematics of matter of planetary bodies of the solar system.
See Abstr. 091.048.

Asteroid–meteorite connection: regolith effects implied by lunar reflectance spectra.
See Abstr. 098.038.

A comparison between terrestrial impact glasses and lunar volcanic glasses: the case of fluorine.
See Abstr. 105.012.

Structure of associations of leading chemical elements in olivine pallasites, lunar and terrestrial rocks.
See Abstr. 105.034.

The influence of gravitational body force in meteoritic chrondrule and lunar glass formation.
See Abstr. 105.060.

Lunar meteorites in Japanese collection of the Yamato meteorites.
See Abstr. 105.189.

Collisional balance of the meteoritic complex.
See Abstr. 106.063.

Formation of the prelunar accretion disk.
See Abstr. 107.030.

On the problem of weak modulation of GCR few million years ago.
See Abstr. 144.169.

^{22}Na and ^{26}Al in Luna 24 samples.
See Abstr. 144.170.

095 Lunar Eclipses

095.001   **A propos de l'éclipse totale de Lune du 4 mai 1985.**
J. Meeus.
Astronomie, Vol. 99, p. 140 – 141 (1985).

095.002   **Die totale Mondfinsternis vom 4. Mai 1985.**
H. Mucke.
Sternenbote, 28. Jahrg., Nr. 4, p. 62 – 74 (1985).

095.003   **Die photometrische Bestimmung der Schattendichte während der Halbschattenfinsternis am 8. November 1984.**
D. Böhme.
Orion, 43. Jahrg., Nr. 208, p. 105 – 106 (1985).

Canon of lunar eclipses, –2002 to + 2526.
See Abstr. 002.127.

Uses for ancient eclipse records.
See Abstr. 004.001.

096 Lunar and Planetary Occultations

096.001 **Planetary occultations of stars in 1985.**
D. W. Dunham.
Sky Telesc., Vol. 69, No. 1, p. 56 – 57 (1985).

096.002 **Lunar occultation highlights for 1985.**
D. W. Dunham.
Sky Telesc., Vol. 69, No. 1, p. 58 – 59 (1985).

096.003 **Occultation d'une étoile par la planète Mars.**
J. Meeus.
Astronomie, Vol. 99, p. 91 – 92 (1985).

096.004 **Grazing occultations.**
D. W. Dunham.
Occultation Newsl., Vol. 3, No. 11, p. 224 – 227 (1985).

096.005 **Grazing occultations.**
D. Stockbauer.
Occultation Newsl., Vol. 3, No. 11, p. 230 – 231 (1985).

096.006 **Sternbedeckungen durch Kleinplaneten im Jahre 1985.**
W. Palzer.
Sterne Weltraum, 24. Jahrg., Nr. 1, p. 34 – 35 (1985).

096.007 **The April 22 occultation of Hya $-20°51699$ by Uranus and its rings.**
F. J. Jablonski, J. Barroso Jr.
Astron. Astrophys., Vol. 144, No. 1, p. 249 – 250 (1985).
In spite of the presence of cirrus over the observatory the authors obtained several timings for the occultation of the star Hya $-20°51699$ by the main Uranus rings in April 22, 1982, using a pulse counting photometer with 100 ms integration times through a combination of an RCA C31034 photomultiplier and 1 mm of UG 6 filter attached to the Brazilian 1.60 m telescope at Brasópolis ($\lambda = 45°34'57''6$, $\phi = -22°32'04''0$, $h = 1870$ m).

096.008 **Occultations of stars and radio sources by comets: predictions and observing prospects.**
E. Bowell, L. H. Wasserman, W. A. Baum, R. L. Millis, K. Lumme.
Cometary astrometry, p. 105 – 122 (1984). – See Abstr. 012.024.

096.009 **Photoelectric observations of lunar occultations of stars. Angular diameter of the star 61 δ^1 Tauri.**
V. G. Kornilov, A. V. Mironov, E. M. Trunkovskii (*E. M. Trunkovskij*), Kh. F. Khaliullin, A. M. Cherepashchuk.
Sov. Astron., Vol. 28, No. 4, p. 431 – 437 (1984). English translation of 38.096.002.

096.010 **Une occultation d'étoile par les anneaux d'Uranus en 1985.**
G. E. Taylor.
Astronomie, Vol. 99, p. 239 – 241 (1985).

096.011 **Photographic observation of the occultation of Jupiter on March 6, 1983.**
N. Čabrić, A. Tomić, V. Čelebonović.
Publ. Astron. Opservatorije Beogr., No. 33, p. 75 – 79 (1985). – See Abstr. 012.061.

On the basis of 47 photographs of the occultation of Jupiter the times of contacts, the topocentric conjunction and the position angles are determined.

096.012 **Occultations d'étoiles par la Lune, observées à l'équatorial de 45 cm à l'aide d'un équipement de télévision en 1980 et 1981.**
J. Dommanget.
Bull. Astron., Vol. 9, No. 6, p. 320 – 321 (1984).

096.013 **Lunar occultation observations of M8E–IR.**
M. Simon, D. Peterson, A. Longmore, J. Storey, A. Tokunaga.
Bull. Am. Astron. Soc., Vol. 16, No. 4, p. 938 (1984). Abstract. – See Abstr. 010.062.

096.014 **Observations of occultations of stars by the moon in Poltava in 1982.**
B. F. Sincheskul.
Vrashchenie i priliv. deformatsii Zemli, Kiev, No. 16, p. 62 – 69 (1984). In Russian. Abstr. in Ref. Zh., 51. Astron., 5.51.91 (1985).

096.015 **Sternbedeckungen durch den Mond und ihre astrophysikalische Bedeutung.**
B. Stecklum.
Sterne, 61. Band, Heft 2, p. 70 – 82 (1985).

Universal time, lunar tidal deceleration and relativistic effects from observations of transits, eclipses and occultations in the XVIIIth – XXth centuries.
See Abstr. 042.009.

An analysis of lunar occultations in the years 1955 – 1980 using the new lunar ephemeris ELP 2000.
See Abstr. 094.015.

Report of IAU Commission 20: Positions and motions of minor planets, comets and satellites (*Positions et mouvements des petites planètes, des comètes et des satellites*).
See Abstr. 098.098.

Results from observations of the 15 June 1983 occultation by the Neptune system.
See Abstr. 101.008.

Predictions of occultations of stars by comet Giacobini–Zinner and prediction updates for comet Halley.
See Abstr. 103.047.

Updated predictions of occultations of stars by comet Halley.
See Abstr. 103.943.

Photometry of occultation candidate stars. I. Uranus 1985 and Saturn 1985 – 1991.
See Abstr. 113.021.

Brightness distribution over the disk of μ Geminorum, derived from lunar occultations.
See Abstr. 113.035.

097 Mars, Mars Satellites

097.001 Mars–Earth geographical comparisons: a pictorial view.
S. P. Meszaros.
NASA Tech. Memo., NASA TM–86166, 5+35 pp. (1985).

This publication is a collection of pictorial comparisons of prominent physiographic features found on Mars and Earth, consisting of equal–scale side–by–side pairs or cartographic overlays. Martian features compared with terrestrial ones include Valles Marineris, the Tharsis bulge, Olympus Mons, the Hellas and Argyre basins, areas of catastrophic flooding, and the polar regions. The illustrations are accompanied by a brief descriptive text and bibliography.

097.002 Mars.
M. H. Carr.
NASA Spec. Publ., NASA SP–469, p. 206 – 263 (1984). – See Abstr. 003.004.

Contents: Introduction. General properties of the surface. The view from the Viking Landers. The atmosphere and surface volatiles. Craters and crater ages. Densely cratered terrain. Sparsely cratered plains. Volcanoes. The Tharsis bulge. Canyons. Channels and valleys. Wind. The poles.

097.003 Martian bore waves of the Tharsis Region: a comparison with Australian atmospheric waves of elevation.
A. O. Pickersgill.
J. Atmos. Sci., Vol. 41, No. 8, p. 1461 – 1473 (1984). Abstr. in Phys. Abstr., Vol. 88, No. 1250, Entry 18687 (1985).

097.004 Linear baroclinic instability in the Martian atmosphere.
J. R. Barnes.
J. Atmos. Sci., Vol. 41, No. 9, p. 1536 – 1550 (1984). Abstr. in Phys. Abstr., Vol. 88, No. 1251, Entry 24311 (1985).

097.005 Satellite orbits and ephemerides.
S. Ferraz–Mello.
Celest. Mech., Vol. 34, Nos. 1 – 4, p. 223 – 241 (1984). – See Abstr. 012.030.

This review considers theoretical work leading to the actual knowledge of the motions of all planetary satellites, except the Moon. It covers recent calculations of inequalities of satellites' motions and the determination of their orbits on the basis of existing observational data. The results are presented and some of them discussed in views of possible future developments.

097.006 Investigations of the chemical composition of planetary soil with automatic means.
V. I. Chesnokov.
Cosmochemistry and meteoritics, p. 92 – 100 (1984). In Russian. Abstr. in Ref. Zh., 62. Issled. Kosm. Prostranstva, 1.62.405 (1985).

097.007 Martian atmospheric photochemistry and composition during periods of low obliquity.
B. L. Lindner, B. M. Jakosky.
J. Geophys. Res., Vol. 90, No. A4, p. 3435 – 3440 (1985).

During periods of low obliquity, previous work has shown that martian CO_2 partial pressures decreased to 0.1 mbar; CO_2 partial pressure decreased to 0.02 mbar prior to the formation of the Tharsis bulge. The permanent polar caps act as a cold trap and projected global average water vapor abundances drop to possibly as low as 10^{-5} precipitable μm. As a result, the odd hydrogen catalytic cycle would not be effective at recombining CO and O back into CO_2, and as much as 0.12 mbar of CO and 0.06 mbar of O_2 could exist. These increased abundances would radically affect surface oxidation, change the lower atmospheric thermal structure, and completely alter the upper atmosphere.

097.008 Geomorphology of Mars.
N. V. Makarova, Ya. G. Kats, V. V. Kozlov, E. D. Sulidi–Kondrat'ev.
Comparative planetology, p. 44 – 50 (1984). In Russian. Abstr. in Ref. Zh., 52. Geod. Aehrosemka, 4.52.246 (1985). – See Abstr. 012.057.

097.009 Mars: dual–polarization radar observations with extended coverage.
J. K. Harmon, S. J. Ostro.
Icarus, Vol. 62, No. 1, p. 110 – 128 (1985).

Thirteen–centimeter–wavelength radar observations of Mars made in 1982 at Arecibo Observatory yield accurate measurements of the full backscatter spectrum in two orthogonal polarizations. The data, which were obtained for several widely separated subradar longitudes at 24°N latitude, provide the first global view of the distribution of small–scale surface roughness on Mars. The diffuse component of the echo exhibits strong spatial variations. Areas of maximum depolarization correlate well with volcanic regions (Tharsis and Elysium), while the heavily cratered upland terrain yields relatively low depolarization. On the average, the northern Martian tropics yield higher diffuse radar cross sections ($\sigma^D = 0.05 - 0.12$) and a higher degree of disk–integrated depolarization ($\mu_c = 0.1 - 0.4$) than is found for the Moon, Mercury, and Venus.

097.010 Model of the composition of the Martian ionosphere in the region of the photochemical equilibrium.
A. V. Pavlov.
Kosm. Issled., Tom 23, Vyp. 2, p. 276 – 282 (1985). In Russian. English translation in Cosm. Res.

097.011 Natural 10 μm band CO_2 laser in the atmospheres of Mars and Venus.
G. I. Stepanova, G. M. Shved.
Pis'ma Astron. Zh., Tom 11, No. 5, p. 390 – 394 (1985). In Russian. English translation in Sov. Astron. Lett., Vol. 11.

It is shown that in the daytime atmosphere of Mars a natural laser exists permanently, whereas in the daytime atmosphere of Venus it appears to arise only when internal gravity waves propagate through the atmosphere. The volume laser amplification coefficient and the amplification for a single passage are estimated.

097.012 Possible precipitation of ice at low latitudes of Mars during periods of high obliquity.
B. M. Jakosky, M. H. Carr.
Nature, Vol. 315, No. 6020, p. 559 – 561 (1985).

Most of the old cratered highlands of Mars are dissected by branching river valleys that appear to have been cut by running water yet liquid water is unstable everywhere on the martian surface. The authors suggest that during periods of very high obliquities, ice could accumulate at low latitudes as a result of sustained sublimation of ice from the poles and transport of the water vapour equatorwards. At low latitudes, the water vapour would saturate the atmosphere and condense onto the surface where it would accumulate until lower obliquities prevailed. Partial melting of the ice could have provided runoff to form the channels or replenish the groundwater system.

097.013 The evolution of CO_2 on Mars.
R. Kahn.
Icarus, Vol. 62, No. 2, p. 175 – 190 (1985).

At an average location on the surface of Mars, the pressure of CO_2 varies seasonally between about 6 and 8 mbar. Outgassing models suggest that at least 140 mbar, and possibly as much as 3000 mbar, of CO_2 have been placed in the atmosphere over geologic time. Neither the polar caps nor the regolith alone appears to be an adequate repository for the CO_2. It is argued that carbonate rock is the most reasonable reservoir for the excess

CO_2 and that the rock formation process can explain the current CO_2 pressure.

097.014   Ice–lubricated gravity spreading of the Olympus Mons aureole deposits.
K. L. Tanaka.
Icarus, Vol. 62, No. 2, p. 191 – 206 (1985).

The huge aureole deposits and perimeter scarp of Olympus Mons in the Tharsis region of Mars have puzzled planetary geologists since they were first observed by Mariner 9. The model presented in the paper proposes that moderate amounts of ground ice in the regolith induced shear failure in a preaureole outer base of Olympus Mons; hence upper shield deposits decoupled from and slid over underlying materials. This supposition is shown to be compatible with (1) the stress conditions of the aureole deposits, (2) the structure of the aureoles, (3) the possible presence of ice in the regolith, and (4) the formation of the basal scarp surrounding Olympus Mons.

097.015   Interannual variability of Martian weather.
C. B. Leovy, J. E. Tillman, W. R. Guest, J. Barnes.
Recent advances in planetary meteorology, p. 69 – 84 (1985). – See Abstr. 012.071.

Pressure, temperature, imaging, and wind data from the Mutch Memorial Station, the Viking lander located in Mars' subtropics, are used to demonstrate the existence of two distinct regimes of northern hemisphere winter weather on Mars.

097.016   Martian local dust storms.
P. B. James.
Recent advances in planetary meteorology, p. 85 – 99 (1985). – See Abstr. 012.071.

Eolian transport appears to be the major constructional and erosional mechanism on Mars; the movement of dust by winds is responsible for the changes in coloration and albedo on the martian surface which were once attributed to vegetation and liquid water. As in the arid and semi–arid regions of Earth, dust storms are among the most significant martian meteorological phenomena. It is with these dust storms that this paper is concerned.

097.017   The effects of Martian topography upon atmospheric motions and a possible mechanism that produces dust storms on Mars.
M.–c. Li, Y.–q. Xie.
Chin. J. Space Sci., Vol. 4, No. 4, p. 314 – 323 (1984). In Chinese. Abstr. in Phys. Abstr., Vol. 88, No. 1255, Entry 45971 (1985).

097.018   Time–dependent model of the Martian atmosphere for use in orbit lifetime and sustenance studies.
R. D. Culp, A. I. Stewart.
J. Astronaut. Sci., Vol. 32, No. 3, p. 329 – 341 (1984). Abstr. in Phys. Abstr., Vol. 88, No. 1256, Entry 51536 (1985).

097.019   Mars: thickness of the lithosphere from the tectonic response to volcanic loads.
R. P. Comer, S. C. Solomon, J. W. Head.
Rev. Geophys., Vol. 23, No. 1, p. 61 – 92 (1985).

The authors estimate the thickness of the elastic lithosphere of Mars in the vicinity of several major surface loads. The primary observational constraints are the geometry of each load and the locations of extensional tectonic features which satisfy a set of criteria for origin by lithospheric flexure. The authors relate the tectonic features to lithospheric stresses principally through the theory for flexure of a thin elastic shell; viscoelastic and thick plate models are also examined.

097.020   Annual heat balance of Martian polar caps: Viking observations.
D. A. Paige, A. P. Ingersoll.
Science, Vol. 228, No. 4704, p. 1160 – 1168 (1985).

The authors present an extensive compilation of Viking Infrared Thermal Mapper solar reflectance and infrared emission observations of the Martian north and south polar regions. The observations span an entire Mars year, and are used to determine annual radiation budgets and heat budgets for the core regions of the north and south permanent or residual polar caps. The results define the current behavior of CO_2 frost at the Martian poles and provide new clues to the properties and processes that may be responsible for this behavior.

097.021   Global map of eolian features on Mars.
A. W. Ward, K. B. Doyle, P. J. Helm, M. K. Weisman, N. E. Witbeck.
J. Geophys. Res., Vol. 90, No. B2, p. 2038 – 2056 (1985).

The authors examine a variety of eolian features and discuss their regional and global distributions and orientations. They discuss both streak and dune patterns (some of which have been reported in other studies), plus the patterns of large erosion features such as yardangs, which might record long–term winds perhaps having directions different from those that formed the ephemeral features.

097.022   Gravity and lithospheric stress on the terrestrial planets with reference to the Tharsis region of Mars.
N. H. Sleep, R. J. Phillips.
J. Geophys. Res., Vol. 90, No. B6, p. 4469 – 4489 (1985).

An analytical theory for isostatic compensation in one–plate planets, including membrane stresses in the lithosphere, self–gravitation, and rotational ellipticity, is applied to Mars.

097.023   Big impact on the northern hemisphere of Mars.
Z. Urban.
Říše hvězd, Vol. 66, No. 4, p. 73 – 75 (1985). In Czech.

097.024   Mars, a volatile–rich planet.
G. Dreibus, H. Wänke.
Meteoritics, Vol. 20, No. 2, Part 2, p. 367 – 381 (1985). – See Abstr. 003.048.

The detection of a trapped Martian atmosphere–like component in the shergottite EETA 79001 provides the most conclusive evidence that SNC–meteorites are rocks from Mars. If we assume that the parent body of the SNC–meteorites is indeed Mars, these meteorites can be used to estimate the abundance of volatile elements on Mars. It is found that Mars contains a number of volatile elements in concentrations exceeding those of the Earth. The low abundance of primordial rare gases on Mars is explained by drastic depletion during the escape of the early Martian atmosphere.

097.025   A model for the climatic behavior of water on Mars.
S. M. Clifford.
Diss. Abstr. Int., Sect. B, Vol. 45, No. 8, p. 2580 – 2581 (1985). Thesis, University of Massachusetts, 302 pp. (1984). Order No. DA8424858.

097.026   Positions of Mars from photographic observations in the opposition with the Pulkovo normal astrograph during 1979 – 1980.
L. V. Zhukov.
Izv. Glav. Astron. Obs. Pulkovo, Astrometr. Astrofiz., No. 201, p. 68 – 70 (1985). In Russian.

Twenty–nine positions of Mars determined from photographic observations with the Pulkovo normal astrograph in the opposition during 1979 – 1980 are given. Mean square errors of one position are equal to $\sigma_\alpha = \pm 0\overset{s}{.}016$ and $\sigma_\delta = \pm 0\overset{''}{.}30$.

Conclusion of Viking Lander imaging investigation. Picture catalog of experiment data record.
See Abstr. 002.017.

Mars: Westliche und östliche Hemisphäre. Kosmos Handkarte 1:23500000.
See Abstr. 002.093.

On Mars: exploration of the red planet 1958 – 1978.
See Abstr. 003.087.

More news from Mars.
See Abstr. 011.025.

The dissociative recombination of N_2^+ ($v = 0, 1$) as a source of metastable atoms in planetary atmospheres.
See Abstr. 022.003.

The use of proportional counters for X–ray radiometric analysis of the composition of planetary surfaces.
See Abstr. 035.031.

Regular motions of a satellite and some small effects in the motion of the moon and Phobos.
See Abstr. 042.005.

The Mars dual orbiter.
See Abstr. 051.097.

Singlet oxygen nightglow in the atmospheres of Earth, Venus and Mars.
See Abstr. 082.028.

Dimer spectra of carbon dioxide in the atmospheres of Venus and Mars.
See Abstr. 093.052.

Venus and Mars at the 15th COSPAR General Assembly.
See Abstr. 093.057.

Martian atmospheric carbon dioxide and weathering products in SNC meteorites.
See Abstr. 105.042.

Martian atmospheric CO_2 in an Antarctic meteorite?
See Abstr. 105.064.

Balloon–borne far–infrared spectrophotometry of the galactic center region.
See Abstr. 155.107.

098 Minor Planets

098.001 Positions of asteroids obtained during 1982 with the GPO telescope at ESO, Chile.
H. Debehogne, G. Hahn, C.–I. Lagerkvist.
Astron. Astrophys., Suppl. Ser., Vol. 59, No. 1, p. 87 – 94 (1985).
The authors present 239 positions of 11 numbered and 3 unnumbered asteroids obtained during March and April 1982 and 126 positions of 21 numbered and 19 unnumbered asteroids during September 1982 with the GPO telescope at ESO, La Silla.

098.002 Positions of asteroids observed at La Silla–GPO, in September 1983. 3 discoveries.
H. Debehogne.
Astron. Astrophys., Suppl. Ser., Vol. 59, No. 1, p. 95 – 97 (1985).
A sample of asteroids was observed at ESO, La Silla, in September 1983 with the GPO. The results presented here were obtained with the measuring machine OPTRONICS at ESO in Garching. The discovery of three new asteroids is reported.

098.003 Positions of asteroids (1982).
H. Debehogne, G. De Sanctis, V. Zappalà.
Astron. Astrophys., Suppl. Ser., Vol. 59, No. 1, p. 99 – 101 (1985).
318 positions of 35 asteroids were obtained from plates taken in 1982 by means of the GPO of ESO, La Silla. 5 new asteroids were also discovered. The reductions were made by the dependence method using 6 or 7 reference stars.

098.004 Minor planets' positions and discoveries obtained in February 1983 at the European Southern Observatory, La Silla.
H. Debehogne, R. R. de Freitas Mourão.
Astron. Astrophys., Suppl. Ser., Vol. 59, No. 1, p. 103 – 116 (1985).
In February 1983, the authors have observed minor planets at ESO, La Silla. The instrument GPO was used. The measurements and reductions were performed at the Royal Observatory of Belgium (Uccle Obs.) with the Ascorecord Zeiss measuring machine and by using five reference stars from the SAO Catalogue.

098.005 Minor planets discoveries at the GPO, ESO–La Silla in September 1983. Dependences of stars for catalogue improvement and future perturbation studies.
H. Debehogne.
Astron. Astrophys., Suppl. Ser., Vol. 59, No. 1, p. 117 – 135 (1985).
In August and September 1983, the author has observed with the GPO instrument at ESO, La Silla, some minor planets. Some of the data were reduced at Uppsala, Rio de Janeiro and at ESO, Garching. A total of 583 positions has been obtained by using reduction techniques available at Uccle, Belgium.

098.006 Positions of asteroids obtained with the GPO telescope at ESO, Chile and with the Kvistaberg Schmidt telescope.
C.–I. Lagerkvist, K. Olofsson, A. From, G. Hammarbäck, P. Magnusson, O. Morell.
Astron. Astrophys., Suppl. Ser., Vol. 59, No. 1, p. 137 – 138 (1985).
The authors present 101 positions of asteroids obtained during August 1982 with the GPO astrograph at ESO, Chile and with the Kvistaberg Schmidt telescope during September 1979 and February 1981.

098.007 Positions of selected minor planets (1981 – 1983).
G. De Sanctis, W. Ferreri, V. Zappalà.
Astron. Astrophys., Suppl. Ser., Vol. 59, No. 1, p. 139 – 143 (1985).
525 accurate positions are presented for 52 asteroids observed from September 1981 to July 1983 at the Observatory of Torino. Reductions were made using the dependence method with 6 reference stars taken from the AGK3 and SAO catalogues.

098.008 Minor planets at unusually favorable oppositions in 1985.
F. Pilcher.
Minor Planet Bull., Vol. 12, No. 1, p. 1 – 5 (1985).

098.009 Photoelectric program for small telescopes in 1985.
M. Di Martino, V. Zappalà.
Minor Planet Bull., Vol. 12, No. 1, p. 6 (1985).

098.010 Photoelectric photometry opportunities February – March.
A. W. Harris, V. Zappalà.
Minor Planet Bull., Vol. 12, No. 1, p. 6 – 7 (1985).

098.011 Low phase angle asteroids.
D. J. Tholen.
Minor Planet Bull., Vol. 12, No. 1, p. 8 (1985).

098.012 Minor planet cluster passages in 1985.
D. L. Welch, M. Kaitting, C. Cunningham.
Minor Planet Bull., Vol. 12, No. 1, p. 8 – 9 (1985).

098.013 Rotations of 1168 Brandia and 1219 Britta.
F. Pilcher, R. P. Binzel, D. J. Tholen.
Minor Planet Bull., Vol. 12, No. 1, p. 10 (1985).

Visual and photoelectric lightcurves of minor planet 1168 Brandia indicate a probable synodic rotation period of 11.444 ± 0.002 hours. Visual and photoelectric lightcurves of 1219 Britta disclose a definitive synodic rotation period of 5.575 ± 0.001 hours.

098.014 The occultation of AG +29°398 by 93 Minerva.
R. L. Millis, L. H. Wasserman, E. Bowell,
O. G. Franz, R. Nye, W. Osborn, A. Klemola.
Icarus, Vol. 61, No. 1, p. 124 – 131 (1985).

The occultation of AG +29°398 by 93 Minerva on 22 November 1982 was successfully observed at 10 sites. The data are best fitted by a circular limb profile having a diameter of 170.8 ± 1.4 km, a value that agrees well with the published radiometric diameter for this asteroid. However, evidence of significant departure from a spherical shape is found in the occultation observations and in photometric measurements on Minerva made at Lowell Observatory over several months. Additional observations are needed to specify definitively the three-dimensional figure of Minerva.

098.015 Speckle interferometry of asteroids. I. 433 Eros.
J. D. Drummond, W. J. Cocke, E. K. Hege,
P. A. Strittmatter, J. V. Lambert.
Icarus, Vol. 61, No. 1, p. 132 – 151 (1985).

Analytic expressions for the semimajor and semiminor axes and an orientation angle of the ellipse projected by a triaxial ellipsoid (an asteroid) and of the ellipse segment cast by a terminator across the ellipsoid as functions of the dimensions and pole of the body and the asterocenteric position of the Earth and Sun are derived. Applying these formulae to observations of the Earth–approaching asteroid 433 Eros obtained with the speckle interferometry system of Steward Observatory on December 17 – 18, 1981, and January 17 – 18, 1982, the following dimensions are derived: $(40.5 \pm 3.1 \text{ km}) \times (14.5 \pm 2.3 \text{ km}) \times (14.1 \pm 2.4 \text{ km})$. Eros' north pole is found to lie within $14°$ of RA = 0^h16^m Dec. = $+43°$ (ecliptic longitude $23°$, latitude $+37°$). These dimensions, together with a lightcurve from December 18, 1981, lead to a geometric albedo of 0.156 ± 0.010. A series of two-dimensional power spectra and autocorrelation functions of the resolved asteroid clearly show it spinning in space.

098.016 Rotational properties of ten main belt asteroids: analysis of the results obtained by photoelectric photometry.
M. A. Barucci, M. Fulchignoni, R. Burchi, V. D'Ambrosio.
Icarus, Vol. 61, No. 1, p. 152 – 162 (1985). With a correction in Vol. 62, No. 1, p. 173 (1985).

V photoelectric lightcurves of ten main belt asteroids (11 Partenope, 20 Massalia, 31 Euphrosyne, 41 Daphne, 55 Pandora, 71 Niobe, 79 Eurynome, 129 Antigone, 344 Desiderata, and 387 Aquitania), obtained during the 1981 – 1983 oppositions, are reported. The rotation period of 11 Partenope is $P = 7.83$ hr and that of 344 Desiderata $P = 10.53$ hr. The shape and the pole coordinates of 20 Massalia, 31 Euphrosyne, and 129 Antigone were also derived and those of 41 Daphne confirmed. The lightcurves of the remaining objects are presented: a preliminary discussion of their possible rotational properties and their morphological features is given.

098.017 Thermal modelling of asteroids and its application to IRAS data.
S. F. Green.
Observatory, Vol. 105, No. 1064, p. 4 (1985). Abstract. – See Abstr. 010.721.

098.018 Nouveaux noms de petites planètes.
J. Meeus.
Astronomie, Vol. 99, p. 141 – 142 (1985).

098.019 Observations of asteroidal appulses and occultations.
J. Stamm.
Occultation Newsl., Vol. 3, No. 11, p. 225 – 230 (1985).

098.020 Een periodieke buur van de aarde.
G. W. E. Beekman.
Zenit, 12. Jaarg., No. 4, p. 131 (1985).

098.021 The depletion of the outer asteroid belt.
A. Milani, A. M. Nobili.
Astron. Astrophys., Vol. 144, No. 2, p. 261 – 274 (1985).

The outstanding problem in the outer asteroid belt ($a > 3.2$ AU) is the evident depletion and extremely uneven distribution of the objects. The authors investigate the gravitational hypothesis with numerical experiments based on different methods. First, regions of ordered and chaotic motion are established in the framework of the restricted 3–body problem. Then the eccentricity of Jupiter is introduced and escapers are searched for in the planar elliptic restricted 3–body model. They are found from well inside the stability boundary of the Hill's criterion in the circular problem. According to these results, most of the outer asteroid belt has been depleted by ejection due to the eccentricity of Jupiter's orbit (in turn perturbed by Saturn) over short timescales of the order of 10^5 yr. If this is true, collisions cannot have played a significant role and big asteroids rather than families of small objects should be expected in the outer belt. The observational data seem to corroborate this conclusion.

098.022 Do we observe light curves of binary asteroids?
A. Cellino, R. Pannunzio, V. Zappalà, P. Farinella,
P. Paolicchi.
Astron. Astrophys., Vol. 144, No. 2, p. 355 – 362 (1985).

Among the asteroids of intermediate size, whose shape is probably controlled by self–gravitation, the authors have selected a sample of 10 objects for which some "anomalous" light curve features indicate a possible binary nature. Applying the procedure described by Leone et al. (1984), they have derived from the rotational properties of these hypothetical binary systems the values of the geometrical and physical parameters yielding a satisfactory agreement between the light curves computed from the models and the observed ones. In particular they have obtained shapes, densities and mass ratios of the suspected binary asteroids; only the latter parameter is found to be remarkably sensitive to a possible contribution of surface scattering effects to the light curve amplitude.

098.023 Rotational properties of asteroids 2, 12, 80, 145 and 354 obtained by photoelectric photometry.
R. Burchi, V. D'Ambrosio, P. Tempesti, N. Lanciano.
Astron. Astrophys., Suppl. Ser., Vol. 60, No. 1, p. 9 – 15 (1985).

Photoelectric observations of the asteroids 2 Pallas, 12 Victoria, 80 Sappho, 145 Adeona and 354 Eleonora have been carried out at the Teramo Astronomical Observatory during the 1982 and 1983 oppositions and also in 1971 in the case of 12 Victoria. The observations of 2 Pallas confirmed its period of rotation and provided some indications of its shape; its pole coordinates were recalculated. For 80 Sappho a probable period of about 14 hours is suggested. A discussion on the uncertainty of the rotation period of 145 Adeona is given on the basis of the new lightcurves obtained. The rotation pole of 354 Eleonora was determined.

098.024 A possible satellite of (146) Lucina.
J. E. Arlot, J. Lecacheux, C. Richardson, W. Thuillot.
Icarus, Vol. 61, No. 2, p. 224 – 231 (1985).

This paper reports on an observation of the appulse of (146) Lucina with AGK3 +17°1309 made on April 18, 1982, at the Meudon Observatory in France. During this observation a secondary event occurred and was recorded. A Nocticon camera mounted on the 1–m telescope and video equipment was used. The Meudon observation is described. Observational data from other sites have been collected. A possible interpretation is the existence of a faint satellite in the neighborhood of (146) Lucina. The observation leads to a diameter of at least 5.7 km and a projected distance of 1600 km from the primary.

098.025 Speckle interferometry of asteroids. II. 532 Herculina.
J. D. Drummond, E. K. Hege, W. J. Cocke,
J. D. Freeman, J. C. Christou, R. P. Binzel.
Icarus, Vol. 61, No. 2, p. 232 – 240 (1985).

Speckle interferometry of 532 Herculina performed on January 17 and 18, 1982, yields triaxial ellipsoid dimensions of $(263 \pm 14) \times (218 \pm 12) \times (215 \pm 12)$ km, and a north pole for the asteroid within 7° of RA = 7^h47^m and DEC = $-39°$. In addition, a "spot" some 75% brighter than the rest of the asteroid is inferred from both speckle observations and Herculina's lightcurve history. This bright complex extends over a diameter of 55° (115 km) of the asteroid's surface. No evidence for a satellite is found from the speckle observations.

098.026 Pole orientation of 16 Psyche by two independent methods.
E. F. Tedesco, R. C. Taylor.
Icarus, Vol. 61, No. 2, p. 241 – 251 (1985).

Nineteen new lightcurves of 16 Psyche are presented along with a pole orientation derived using two independent methods, namely, photometric astrometry (PA) and magnitude–amplitude–shape–aspect (MASA). The pole orientations found using these two methods agree to within 4°. The results from applying PA were prograde rotation, a sidereal period of $0^d1748143 \pm 0^d0000003$, and a pole at longitude 223° and latitude $+37°$, with an uncertainty of 10°; and, from applying MASA a pole at $220 \pm 1°$, $+40 \pm 4°$, and a modeled triaxial ellipsoid shape with $a/b = 1.33 \pm 0.02$ and $b/c = 1.33 \pm 0.07$.

098.027 Observations of minor planets with the Very Large Array.
P. K. Seidelmann, G. H. Kaplan, K. J. Johnston, C. M. Wade.
Celest. Mech., Vol. 34, Nos. 1 – 4, p. 39 – 48 (1984). – See Abstr. 012.030.

The weak thermal emission from the largest minor planets can be detected and measured at all points around their orbits at microwave frequencies using the Very Large Array. Position determinations of astrometric quality have been obtained and flux measurements have provided size estimates. When enough precise positional observations have been accumulated, the orbits of the minor planets and the Earth can be determined. This will allow the equinox to be located within the radio reference frame, providing a truly fundamental coordinate system for radio source positions. It will also provide a means of relating the optical and radio (quasar) coordinate systems.

098.028 Formation of the Kirkwood gaps in the asteroid belt.
A. Lemaitre.
Celest. Mech., Vol. 34, Nos. 1 – 4, p. 329 – 341 (1984). – See Abstr. 012.030.

A possible mechanism to explain the depletion of the Kirkwood gaps in the asteroid belt would be the slow dissipation of the solar nebula at the origin of the Solar System. The effects of this dissipation on a uniform distribution of asteroids are explored by means of the adiabatic invariant theory for the 2/1, 3/1 and 5/2 resonance cases. The framework is the restricted, circular and planar three body problem.

098.029 Resonant structure of the outer asteroid belt.
A. Milani, A. M. Nobili.
Celest. Mech., Vol. 34, Nos. 1 – 4, p. 343 – 355 (1984). – See Abstr. 012.030.

By using the planar elliptic restricted 3–body model the authors have investigated the motion of outer belt asteroids which had not been suspected to librate. They find 3 cases of $\tilde{\omega}$ libration and 11 cases of e, $\tilde{\omega}$ coupling that can be explained within the theory of secular resonances. It is thus established that in the outer belt only resonant and dynamically protected asteroids can have lifetimes of the same order as the age of the Solar System.

098.030 Trojan orbits in secular resonances.
R. Bien, J. Schubart.
Celest. Mech., Vol. 34, Nos. 1 – 4, p. 425 – 434 (1984). – See Abstr. 012.030.

A near equality between the nodal rates of suitably defined Trojan orbits and Jupiter represents an important type of a secular resonance. This case is realized by the model Sun–Jupiter–Saturn–Trojan, referred to the invariable plane. A second theoretical example is based on the elliptic three–body problem Sun–Jupiter–Trojan, where the vanishing nodal rate of a special Trojan orbit and the vanishing rate of Jupiter's longitude of perihelion define a secular resonance. The authors investigate the perturbations in the asteroidal inclinations and the nodes and consider the possibility of a libration.

098.031 Critical inclination of Trojan asteroids.
B. Érdi.
Celest. Mech., Vol. 34, Nos. 1 – 4, p. 435 – 441 (1984). – See Abstr. 012.030.

The author's earlier solution for Trojan asteroids is developed further. It is shown that depending on the amplitude of libration around the Lagrangian point L_4, there is a critical inclination which determines the sign of the variation of the ascending node. If the orbital inclination of a Trojan is smaller than the critical one, then the ascending node decreases and otherwise it increases. The variation of the eccentricity and of the longitude of the perihelion has also a dependence on the critical inclination.

098.032 An application of Labrouste's method to quasi–periodic asteroidal motion.
J. Schubart, R. Bien.
Celest. Mech., Vol. 34, Nos. 1 – 4, p. 443 – 452 (1984). – See Abstr. 012.030.

In an earlier paper the authors have applied Labrouste's method to single orbital elements in order to isolate periodicities. However, in practice the investigation of two–dimensional vectors, where the components are combinations of orbital elements, can be useful. In the present paper the authors apply Labrouste's method to vectorial components of this type and represent the results by two–dimensional graphs. Examples refer to the Trojan case of asteroidal motion.

098.033 Secular perturbations of asteroids with commensurable mean motions.
Y. Kozai.
The Big Bang and Georges Lemaitre, p. 207 – 215 (1984). – See Abstr. 012.043.

Secular perturbations of asteroids are derived for mean motion resonance cases under the assumptions that the disturbing planets are moving along circular orbits on the same plane and that critical arguments are fixed at stable equilibrium points. Under these assumptions the equations of motion are reduced to those of one degree of freedom with the energy integral. The same method is also applied to Pluto–Neptune system.

098.034 A mechanism of depletion for the Kirkwood's gaps.
J. Henrard, A. Lemaitre.
The Big Bang and Georges Lemaitre, p. 217 – 218 (1984). Abstract. – See Abstr. 012.043.

098.035 Numerical theory of motion of the minor planet Icarus.
V. A. Shefer.
Astron. geod., Tomsk, No. 10, p. 57 – 71 (1984). In Russian. Abstr. in Ref. Zh., 51. Astron., 3.51.100 (1985).

098.036 Determination of the orientation of the rotational axis taking asteroid (433) Eros as an example.
N. I. Koshkin.
Mater. konf. mol. uchenykh Odes. astron. nauchn.–proizv. akad.–univ. kompleksa, Odessa, 15 fevr., 1983. Odessa, p. 80 – 84 (1983). In Russian. Abstr. in Ref. Zh., 51. Astron., 3.51.219 (1985).

098.037 Radar investigation of asteroids.
S. J. Ostro.
Meteoritics, Vol. 19, No. 4, p. 286 – 287 (1984). Abstract. – See Abstr. 010.641.

098.038 Asteroid–meteorite connection: regolith effects implied by lunar reflectance spectra.
C. M. Pieters.
Meteoritics, Vol. 19, No. 4, p. 290 – 291 (1984). Abstract. – See Abstr. 010.641.

098.039 On the potential importance of carbon during metal core segregation in planetoids.
F. Ulff–Møller.
Meteoritics, Vol. 19, No. 4, p. 326 (1984). Abstract. – See Abstr. 010.641.

098.040 Collisional evolution in the asteroid system.
P. Farinella, P. Paolicchi, V. Zappalà.
Mem. Soc. Astron. Ital., Vol. 55, No. 3, p. 489 – 492 (1984). – See Abstr. 012.049.

098.041 The study of the asteroid's lightcurves on the basis of laboratory simulations.
M. A. Barucci, M. Fulchignoni, A. Di Paoloantonio, C. Giuliani.
Mem. Soc. Astron. Ital., Vol. 55, No. 3, p. 493 – 497 (1984). – See Abstr. 012.049.
The aim of the experiments is to determine the parameters of the asteroids that influence the lightcurves: the orientation parameters, the shape, the surface morphology and the coating material.

098.042 Photographic observation of a possible occultation between minor planets 1981 EM$_4$ and 1983 TC.
E. Colombini.
Mem. Soc. Astron. Ital., Vol. 55, No. 3, p. 499 – 501 (1984). – See Abstr. 012.049.

098.043 The asteroids: the major scientific issues.
C. R. Chapman.
Adv. Space Res., Vol. 5, No. 2, p. 121 – 122 (1985). – See Abstr. 012.051.
The author summarizes some of the themes that have been developed in the last few years that will be particularly interesting to follow up once spacecraft missions to representative asteroids become a reality.

098.044 Review of asteroidal dynamics.
H. Scholl.
Adv. Space Res., Vol. 5, No. 2, p. 123 – 132 (1985). – See Abstr. 012.051.
It is a major goal of research in the field of asteroidal dynamics to relate the distribution of orbital elements to the evolutionary history of the asteroidal belt. The distribution of orbital elements shows characteristics which are presumably of cosmogonic significance. Gaps are associated with resonance phenomena. The clustering of asteroidal semi–major axes is associated with resonances or like in the case of Hirayama families with an early break–up of a parent body. Another dynamical problem present the planet–crossers, in particular the Earth–crossers since their low collisional lifetimes require a source. Presumably, the Earth–crossers are a mixture of asteroidal collisional fragments and of extinct cometary nuclei. For this reason, Earth–crossers are of high interest as targets for spaceflights.

098.045 Five–color polarimetry of the asteroid 16 Psyche.
I. N. Bel'skaya, Yu. S. Efimov, D. F. Lupishko, N. M. Shakhovskoj.
Pis'ma Astron. Zh., Tom 11, No. 4, p. 286 – 291 (1985). In Russian. English translation in Sov. Astron. Lett., Vol. 11.
UBVRI polarimetry of the asteroid 16 Psyche was carried out on June 4 and 5, 1983 in order to verify the presence of preferable orientation of the surface particles. Analysis of the observational data didn't show significant periodic variations of the Stokes parameters q and u with the asteroid's rotation cycle. This evidences the absence of any preferable orientation of surface particles, not coinciding with the direction of the axis of rotation.

098.046 The eight–color asteroid survey: results for 589 minor planets.
B. Zellner, D. J. Tholen, E. F. Tedesco.
Icarus, Vol. 61, No. 3, p. 355 – 416 (1985).
Results are presented from reflection spectrophotometry of 589 minor planets in a photometric system using eight filter passbands ranging from 0.34– to 1.04–μm wavelength. The sampling completeness approaches or exceeds 50% of the numbered asteroids for the near–Earth objects, the Hungarias, the Nysa family, the Cybeles, the Hildas, and the Trojans. The general evolution of predominant compositional type from S to C to D with increasing heliocentric distance is evident, as is the spectral homogeneity of the Eos, Koronis, Nysa, and Themis families.

098.047 Asteroids in cometary orbits.
G. Hahn, H. Rickman.
Icarus, Vol. 61, No. 3, p. 417 – 442 (1985).
Orbital integrations are presented for a total of 14 asteroids with perihelia inside 1.7 AU and with aphelion distances in excess of 4 AU, 10 of which were discovered in 1979 – 1984. The integrations were normally extended over approximately ± 1000 years in a three–body model (Sun–Jupiter–asteroid). A wide variety of orbital evolutions is found, and some of them evidently belong to the cometary, chaotic type. Three such cases are identified with certainty (1983 SA, 1983 XF, and 1984 BC). An asteroidal motion is found for the well–observed object 1979 VA. A stable libration around the 2/1 resonance is found for 1981 FD. A long–lasting libration around the 5/3 resonance performed by 1982 YA is probably unstable. Temporary librations are also found for 1983 SA (4/3 resonance) and 1983 XF (2/1 resonance).

098.048 Visible and near–infrared lightcurves of eight asteroids.
R. S. McCheyne, N. Eaton, A. J. Meadows.
Icarus, Vol. 61, No. 3, p. 443 – 460 (1985).
B lightcurves are presented for seven asteroids (4, 20, 29, 31, 39, 115, and 349) together with visual and infrared color curves (B–V, B–K, V–J, and J–K). A V broadband lightcurve for 40 Harmonia is also included. Color variations are observed for three asteroids (4, 31, and 115). The variation in the colors of 4 Vesta is discussed in terms of differences in surface composition. Pole positions and shapes are estimated for 20 Massalia, 29 Amphitrite, 31 Euphrosyne, and 39 Laetitia. Various UBVRIJHK colors are listed for these asteroids as well as for 5, 19, 44, 52, 83, 145, 386, and 471.

098.049 The pole orientation of asteroid 433 Eros determined by photometric astrometry.
R. C. Taylor.
Icarus, Vol. 61, No. 3, p. 490 – 496 (1985).
Previous photometric astrometry poles are reviewed. The results for asteroid 433 Eros are: prograde rotation; a sidereal period of 0.219588 ± 0.000005 day; and a north pole at 22° longitude, +9° latitude. The uncertainty of the pole is 10°.

098.050 Erratum: "Physical study of asteroids: lightcurves and rotational periods of six asteroids" [Icarus, Vol. 60, No. 3, p. 541 – 546 (1984)].
M. Di Martino.
Icarus, Vol. 61, No. 3, p. 539 (1985). See Abstr. 38.098.084.

098.051 Collisional history of asteroids: evidence from Vesta and the Hirayama families.
D. R. Davis, C. R. Chapman, S. J. Weidenschilling, R. Greenberg.
Icarus, Vol. 62, No. 1, p. 30 – 53 (1985).
Collisional evolution studies of asteroids indicate that the initial asteroid population at the time mean collisional velocities were pumped up to ~ 5 km/sec was only modestly larger than it is today; i.e., the asteroid belt was already depleted relative to the mean surface density elsewhere in the planetary region. Numerical simulations of the collisional evolution of hypothetical initial asteroid populations have been run. A model is presented for calculating the fragmental size distribution for the disruption of

large, gravitationally bound bodies in which the material strength is increased by hydrostatic self–compression.

098.052   Finalmente un asteroide da vicino.
V. Zappalà.
Orione, Vol. 5, N. 2, p. 36 – 37 (1985).

098.053   Infrared observations of the extinct cometary candidate minor planet (3200) 1983 TB.
S. F. Green, A. J. Meadows, J. K. Davies.
Mon. Not. R. Astron. Soc., Vol. 214, No. 3, p. 29P – 36P (1985).

The discovery of Apollo asteroid (3200) 1983 TB in an orbit virtually coincident with the Geminid meteor stream has raised the possibility that it is an extinct cometary nucleus. The authors present infrared observations between 1 and 20 µm from which an albedo of 0.11 ± 0.02 and diameter of (4.70 ± 0.5) km are derived. Both the thermal spectrum and the reflected colours, disagree with the expected properties of an extinct cometary nucleus. However the blue *JHK* colours and moderate albedo, when combined with optical data, show 1983 TB to be a unique object among the known Apollo asteroids.

098.054   Asteroids and amateur astronomers.
J. U. Gunter.
Mercury, Vol. 14, No. 1, p. 9 – 13, 30 (1985).

098.055   New cases of ambiguity among large asteroids' spin rates.
V. Zappalà, M. Di Martino, A. Hanslmeier, H. J. Schober.
Astron. Astrophys., Vol. 147, No. 1, p. 35 – 38 (1985).

New observational evidence supports the importance of the ambiguity problem about determination of the spin rate of asteroids. In the range of objects larger than about 150 km, where equilibrium figures are thought to exist, and therefore where irregular fragments should be absent, four cases of periods shorter by a factor two than the previously adopted values, believed unambiguous, where recently found. This paper presents results for two of them (409 Aspasia and 423 Diotima) attempting to give plausible physical interpretations, but admitting that a quantitative explanation cannot be reached with the present knowledge of the asteroids' surface morphology.

098.056   Orbital elements of numbered minor planets.
Minor Planet Circ., Nos. 9315 – 9716 (1985).
The minor planets are listed according to their definitive number. Newly numbered objects are indicated by an asterisk. The names of the orbit computers are given behind the respective M.P.C. numbers:
(26) 9581, (29) 9350, (50) 9581, (143) 9458, (157), (160) 9581, (167), (262), 9582, (263) 9458, (278) 9459, (457), (649) 9350, (802) 9459, (1038) 9350, (1116), (1125) 9459, (1161) 9350, (1162) 9414, (1253) 9459, (1297) 9414, (1525) 9459, (1544), (1669), (1670), (1697), (1801), (1830) 9460, (1854), (1890) 9461, (2201) 9678, (2224), (2330), (2764) 9461 W. Landgraf; (3167)*–(3169)* 9351–9352 B. G. Marsden; (3170)*–(3174)* 9353–9354 C. M. Bardwell; (3175)*–(3176)* 9357–9358 K. Hurukawa; (3177)*–(3178)* 9358–9359, (3179)*–(3180)* 9415–9416 S. Nakano; (3181)* 9417–9418 K. Hurukawa; (3182)* 9419 T. Urata; (3183)*–(3194)* 9419–9423 C. M. Bardwell; (3195)*–(3201)* 9426–9428 B. G. Marsden; (3202)*–(3210)* 9461–9464 C. M. Bardwell; (3211)*–(3218)* 9466–9468 B. G. Marsden; (3219)*–(3222)* 9470–9471 S. Nakano; (3223)*–(3226)* 9474–9475 K. Hurukawa; (3227)*–(3228)* 9582–9583 S. Nakano;(3229)*–(3237)* 9585–9588   C. M. Bardwell;   (3238)*–(3244)*   9590–9593 B. G. Marsden; (3245)*–(3246)* 9594 D. W. E. Green; (3247)*–(3248)* 9679 T. Urata; (3249)*–(3251)* 9680 K. Hurukawa; (3252)*–(3254)* 9685–9686 B. G. Marsden; (3255)*–(3259)* 9688–9689 C. M. Bardwell.

098.057   Orbital elements of unnumbered minor planets.
Minor Planet Circ., Nos. 9315 – 9716 (1985).
The unnumbered minor planets are sorted by their provisionary designation. The names of the orbit computers are given behind the respective M.P.C. numbers:

[A915 TE] 9469 B. G. Marsden; [A919 SD] 9583 S. Nakano; [1928 SL] 9687 B. G. Marsden; [1929 TD₁], [1930 VD] 9684, [1931 TE₄] 9471 S. Nakano; [1938 DB₁] 9588 C. M. Bardwell; [1938 DN₁], [1940 ED] 9684 S. Nakano; [1941 WA] 9464 C. M. Bardwell; [1948 RD] 9583, [1948 WF] 9685, [1949 PQ] 9583, [1949 QC₁] 9583–9584, [1951 AB], [1952 JH] 9359, [1953 PR] 9360, [1958 GQ] 9416 S. Nakano; [1964 UC] 9588 C. M. Bardwell; [1966 BO] 9471, [1967 JP] 9416 S. Nakano; [1969 TB₂] 9476 K. Hurukawa; [1971 QN] 9472 S. Nakano; [1971 QP₁] 9469 B. G. Marsden; [1971 SN₂] 9472 S. Nakano; [1971 UD₁] 9465 C. M. Bardwell; [1973 DS] 9472 S. Nakano; [1973 QB₂] 9476 K. Hurukawa; [1974 SB₁] 9472, [1974 SU₁] 9473 S. Nakano; [1974 VG] 9354 C. M. Bardwell; [1975 ES] 9473 S. Nakano;   [1975 VA₉]   9477,   [1975 VG₉] 9584   T. Urata; [1976 GO₈] 9593 B. G. Marsden; [1976 SE₁] 9416 S. Nakano; [1976 SP₄] 9595 H. Oishi; [1976 YP₂] 9423, [1977 DD₃] 9465 C. M. Bardwell; [1977 EN₁] 9593 B. G. Marsden; [1977 EO₁], [1977 PE₁] 9476 K. Hurukawa; [1977 QC₄] 9584 S. Nakano; [1977 QG₄] 9465, [1977 QA₅] 9355, [1978 NE] 9423–9424 C. M. Bardwell; [1978 NT₁] 9595 H. Oishi; [1978 OJ], [1978 PR₄] 9424 C. M. Bardwell; [1978 PS₄] 9473 S. Nakano; [1978 QQ₂] 9682 H. Oishi; [1978 TO₇] 9355 C. M. Bardwell; [1978 UF₂] 9352 B. G. Marsden; [1979 QP₈] 9681 K. Hurukawa; [1979 QA₁₀] 9473, [1979 SK₁₁], [1979 SL₁₁] 9417 S. Nakano; [1979 SM₁₁] 9418 K. Hurukawa; [1979 WX₃] 9682 H. Oishi; [1980 FV] 9465, [1980 FE₁₂] 9589 C. M. Bardwell; [1980 OA] 9594 B. G. Marsden; [1980 OD] 9681 K. Hurukawa; [1980 PF] 9469 B. G. Marsden; [1980 TG₅], [1981 EJ₅] 9683 H. Oishi; [1981 EY₈] 9424, [1981 EY₁₇] 9690, [1981 ED₂₁], [1981 EE₂₇] 9589 C. M. Bardwell; [1981 FB] 9595 D. W. E. Green; [1981 FD], [1981 GD₁] 9687 B. G. Marsden; [1981 JH] 9683 H. Oishi; [1981 JZ] 9353 B. G. Marsden; [1981 JY₁] 9683 H. Oishi; [1981 VA] 9687 B. G. Marsden; [1981 XA] 9466, [1981 YC], [1982 CD] 9355 C. M. Bardwell; [1982 DV] 9428–9429 B. G. Marsden; [1982 KG₁] 9466 C. M. Bardwell; [1982 UJ₈] 9358 K. Hurukawa; [1982 VZ] 9360 S. Nakano; [1983 CB₃] 9356 C. M. Bardwell; [1983 EA] 9469, [1983 QD] 9469–9470, [1983 SA] 9429, [1983 VA] 9430, [1983 WF₁] 9687–9688, [1983 XF], [1984 BC] 9430 B. G. Marsden; [1984 DV] 9360, [1984 DF₁] 9474 S. Nakano; [1984 HA₁] 9690, [1984 QO] 9424, [1984 QE₁] 9590 C. M. Bardwell;   [1984 SR]   9584   S. Nakano;   [1984 SV] 9414–9415, [1984 SU₃] 9415 W. Landgraf; [1984 SW₃] 9356 C. M. Bardwell; [1984 SM₄] 9415 W. Landgraf; [1984 SQ₅] 9682 K. Hurukawa; [1984 UT] 9590 C. M. Bardwell; [1984 UW] 9418 K. Hurukawa; [1984 UA₂], [1984 UC₂] 9356, [1984 UL₂] 9357 C. M. Bardwell; [1984 VA] 9361 S. Nakano; [1984 WB] 9590 C. M. Bardwell; [1984 WK] 9418 K. Hurukawa; [1984 YV] 9690 C. M. Bardwell; [1985 AF] 9680 T. Urata; [1985 DQ] 9678–9679 W. Landgraf.

098.058   Observations of minor planets.
Minor Planet Circ., Nos. 9320 – 9349, 9394 – 9413, 9442 – 9457, 9516 – 9579, 9605 – 9676 (1985).

Observations made at the following stations are published: Brescia, Brorfelde, Bucharest, Bulgarian Natl. Obs., Burlington, N.J., Caussols, Cavriana, Centro Astron. Yebes, El Leoncito, ESO, Fabra Obs., Geisei, Goethe Link Obs., Haute Provence, Hoher List, Karasuyama, Kitt Peak, Kleť, Lick Obs., Lincoln Lab., Lowell Obs., Lowell Obs. Anderson Mesa Stn., Mauna Kea, Mt. John Obs., Mt. Palomar, Oak Ridge Obs., Oss. Chaonis, Perth Obs., Pino Torinese, Purple Mountain Obs., Reintal, S. Vittore (Bologna), SAAO Cape Town, Seewalchen, Shizuoka, Siding Spring, Tautenburg, Toyota, Turku, Victoria, Zimmerwald.

098.059   Identifications and identification changes of minor planets.
Minor Planet Circ., Nos. 9389, 9437, 9514, 9601 (1985).

098.060   Ephemerides of minor planets and comets.
Minor Planet Circ., Nos. 9361 – 9388, 9430 – 9436, 9479 – 9512, 9596 – 9600, 9690 – 9716 (1985).

098.061 Orbital elements of one–opposition minor planets.
Minor Planet Circ., Nos. 9349 – 9350, 9413 – 9414, 9457 – 9458, 9579 – 9581, 9676 – 9678 (1985).

098.062 New names of minor planets.
Minor Planet Circ., Nos. 9477 – 9479 (1985).

098.063 Photoelektrische Photometrie der Kleinplaneten (54) Alexandra und (372) Palma.
H. Haupt, A. Hanslmeier.
Anz. Österr. Akad. Wiss., Math.–Naturwiss. Kl., Band 121, p. 69 – 74 (1984). = Mitt. Universitätssternw. Graz, Nr. 107.

The asteroids (54) Alexandra and (372) Palma have been observed photoelectrically with the 60 cm telescope of the Leopold–Figl–Observatory. For (54) an amplitude of $0^{m}14$ and a $V_{max} = 10^{m}28$ has been found. The period was not observed in full but seems to agree with the previously published one ($P = 7^{h}04$). A new period of $6^{h}58$ has been derived for (372) that seems not be in contradiction with earlier observations. The amplitude was $0^{h}11$ and the $V_{max} = 11^{m}54$.

098.064 On the diameter distribution of some asteroid families.
X.–h. Zhou, Z.–x. Wu, Y.–y. Zhang.
Publ. Purple Mt. Obs., Vol. 3, No. 3, p. 22 – 36 (1984). In Chinese.

Based on the TRIAD data of magnitudes and diameters, a statistic of diameter distribution has been made for 10 asteroid families. The relation between diameter (D) and accumulated number (N) for Themis, Eos and Koronis families follows a single power law of $N = \alpha D^{-\beta}$, the values of β are in accordance with those of collision fragments in the laboratory. The distribution for the Flora family must be represented by two values of β, and for the other families by one or two values of β.

098.065 Determination of the pole orientation of an asteroid: the amplitude–aspect relation revisited.
A. Pospieszalska–Surdej, J. Surdej.
ESO Sci. Prepr., No. 369, 30 pp. (1985). Submitted to Astron. Astrophys.

098.066 Speckle interferometry of asteroids. III. 511 Davida.
J. D. Drummond, E. K. Hege.
Prepr. Steward Obs., No. 578, 42 pp. (1985).

511 Davida was observed with the technique of speckle interferometry at Steward Observatory's 2.3 m telescope on May 3, 1982. Based on 5 ten minute observations, its dimensions were found to be $(465 \pm 33) \times (358 \pm 39) \times (258 \pm 52)$ km. Such a shape falls close to an equilibrium figure of a "rubble pile", suggesting a density of 1.4 ± 0.4 g/cm³. The authors derive and apply to Davida a new simultaneous amplitude magnitude aspect method for finding, from photometric data only, axial ratios and rotational pole coordinates.

098.067 Observations photographiques de petites planètes effectuées à l'astrographe double de 40 cm au cours de l'année 1982.
H. Debehogne.
Bull. Astron., Vol. 9, No. 6, p. 283 – 288 (1984).

098.068 Observations photographiques de petites planètes effectuées en 1982 à l'équatorial GPO de 40 cm de l'Observatoire Austral Européen (ESO) à La Silla (Chili).
H. Debehogne.
Bull. Astron., Vol. 9, No. 6, p. 289 – 298 (1984).

098.069 Surface properties of asteroids.
H. J. Schober.
Hvar Obs. Bull., Vol. 8, No. 1, p. 51 – 56 (1984).

The physical properties of the surfaces of asteroids can be studied by various methods, such as photometry and/or spectrophotometry as well as polarimetry. A survey is given of the methods used and of the properties derived form the observations, and of possible classification systems.

098.070 Speckle interferometry results for 12 Victoria and 4 Vesta.
J. Drummond, E. K. Hege.
Bull. Am. Astron. Soc., Vol. 16, No. 4, p. 922 (1984). Abstract. – See Abstr. 010.062.

098.071 The puzzling oscillation of Ceres' period.
V. Zappalà, P. Farinella, P. Paolicchi.
Bull. Am. Astron. Soc., Vol. 16, No. 4, p. 1026 (1984). Abstract. – See Abstr. 010.063.

098.072 Do we observe lightcurves of binary asteroids?
A. Cellino, R. Pannunzio, V. Zappalà, P. Farinella, P. Paolicchi.
Bull. Am. Astron. Soc., Vol. 16, No. 4, p. 1027 (1984). Abstract. – See Abstr. 010.063.

098.073 The occultation diameter of 47 Aglaja.
R. L. Millis, L. H. Wasserman, R. M. Williamon, D. W. Dunham, P. L. Manly, R. W. Olson, W. E. Baggett, P. D. Maley, K. W. Ziegler, A. R. Klemola.
Bull. Am. Astron. Soc., Vol. 16, No. 4, p. 1027 (1984). Abstract. – See Abstr. 010.063.

098.074 8– to 13–µm spectra of asteroids.
S. F. Green, N. Eaton, D. K. Aitken, P. F. Roche, A. J. Meadows.
Icarus, Vol. 62, No. 2, p. 282 – 288 (1985).

The authors present 8– to 13–µm spectra of the main–belt asteroids 1, 2, 6, 7, 10, 16, 45, 51, 56, 65, 78, and 451. None of these exhibit pronounced emission features, and all can be fitted reasonably well using a standard thermal model. They also present wide– and narrowband photometry of (19) Fortuna, which has previously been reported to show an emission feature in this region.

098.075 Variation of the UBV colors of S–class asteroids with semimajor axis and diameter.
S. F. Dermott, J. Gradie, C. D. Murray.
Icarus, Vol. 62, No. 2, p. 289 – 297 (1985).

The mean UBV color of S–class asteroids varies markedly with distance from the Sun and may also vary with diameter, implying either that the surface properties of the asteroids have been modified by space weathering or, as suggested in the paper, that there are at least two major subclasses of S–class asteroids with different mean locations.

098.076 The rotation period and pole orientation of asteroid 4 Vesta.
R. C. Taylor, S. Tapia, E. F. Tedesco.
Icarus, Vol. 62, No. 2, p. 298 – 304 (1985).

The authors present arguments to show that both the visual lightcurves and polarimetric curves of Vesta are consistent with either the 5– or 10–hr period. It appears that the simpler model of a near spherical body with either one surface feature, or one high albedo hemisphere, is more plausible than a symmetric ellipsoid with both of the two broad sides having identical high albedos. For that reason, the authors favor the 5 hr 20 min period.

098.077 Ultraviolet reflectance properties of asteroids.
P. S. Butterworth, A. J. Meadows.
Icarus, Vol. 62, No. 2, p. 305 – 318 (1985).

An analysis of the UV spectra of 28 asteroids obtained with the International Ultraviolet Explorer satellite is presented. The spectra lie within the range 2100 – 3200 Å. The results are examined in terms of both asteroid classification and of current ideas concerning the surface mineralogy of asteroids. For all the asteroids examined, UV reflectivity declines approximately linearly toward shorter wavelengths. In general, the same taxonomic groups are seen in the UV as in the visible and IR.

098.078   Observations of minor planets. IV.
F. Börngen, K. Kirsch.
Astron. Nachr., Vol. 306, No. 3, p. 167 – 169 (1985). In German.
A summary is given of the minor planet survey performed in 1983 on Tautenburg Schmidt plates. The authors discovered 147 planets and calculated 272 positions for them. Tautenburg observations could give a tribute to fourteen planets numbered in the period of this report. The planet 1981 VL$_2$ discovered in Tautenburg has received the permanent number (2861). In honor of Prof. Hermann Lambrecht who died 4.6.1983 this planet obtained the name Lambrecht.

098.079   Visual magnitudes of 1983 TB and comets
P/Schaumasse (1984m) and Levy–Rudenko (1984t).
Yamamoto Circ., No. 2034 (1985). In Japanese.

098.080   1985 JA.
Yamamoto Circ., Nos. 2039, 2040 (1985). In Japanese.

098.081   1983 TB and comet Shoemaker (1984s).
IAU Circ., No. 4029 (1985).

098.082   (3200) 1983 TB.
IAU Circ., No. 4034 (1985).

098.083   1981 VA.
IAU Circ., No. 4056 (1985).

098.084   1982 RB.
IAU Circ., No. 4062 (1985).

098.085   1985 JA.
IAU Circ., Nos. 4063, 4064, 4079 (1985).

098.086   Could an asteroid be a comet in disguise?
R. A. Kerr.
Science, Vol. 227, No. 4689, p. 930 – 931 (1985).
Two asteroids of the inner solar system are strong candidates for once–active comets that now masquerade as inert hunks of rock.

098.087   Wavelength dependence of polarization of asteroids.
R. V. Myers, B. A. Whitney, K. H. Nordsieck.
Bull. Am. Astron. Soc., Vol. 17, No. 1, p. 515 (1985). Abstract. – See Abstr. 010.064.

098.088   Numerically–analytical method of investigation of the
evolution of asteroid orbits.
M. A. Vashkov'yak.
Kosm. Issled., Tom 23, Vyp. 3, p. 335 – 346 (1985). In Russian. English translation in Cosm. Res.

098.089   Asteroids – the comet connection.
J. Davies.
New Sci., Vol. 104, No. 1435 – 1436, p. 46 – 48 (1984). Abstr. in Phys. Abstr., Vol. 88, No. 1255, Entry 45974 (1985).

098.090   The modulation curve of a minor planet.
B. Morando.
Processing of scientific data from the ESA astrometry satellite HIPPARCOS, p. 125 – 132 (1985). – See Abstr. 012.087.
Taking for the surface of a minor planet a model used for Mars by the astrolabe observers, the modulation coefficients and phases of both harmonics have been calculated for different values of the phase angle and the apparent diameter. The phases are found to correspond to the position of the photocentre and the greater the apparent diameter of the planet the smaller are the modulation coefficients.

098.091   Can comets become asteroids?
J. Davies.
Astronomy, Vol. 13, No. 1, p. 66 – 70 (1985). Abstr. in Phys. Abstr., Vol. 88, No. 1257, Entry 57641 (1985).

098.092   Die Entdeckung des kleinen Planeten (2861) Lambrecht.
F. Börngen.
Sterne, 61. Band, Heft 2, p. 67 – 69 (1985).

098.093   Minor planets: close encounters of the apparent kind.
D. A. Pierce, A. L. Whipple.
Bull. Am. Astron. Soc., Vol. 17, No. 2, p. 622 (1985). Abstract. – See Abstr. 010.066.

098.094   The large–scale structure of the asteroid belt.
B. Zellner, A. Thirunagari, D. Bender.
Icarus, Vol. 62, No. 3, p. 501 – 511 (1985).
The authors examine the distributions of 2888 numbered minor planets over orbital inclination, eccentricity, and semimajor axis, and define 19 zones which they believe adequately to isolate the selection biases in survey programs of the physical properties of minor planets. Six numbered asteroids have exceptional orbits and fall into no zone. The authors also call attention to rather sharp upper limits, which become increasingly stringent at larger heliocentric distances, on orbital inclinations and eccentricity.

098.095   Photoelectric   photometry   of   asteroids   9 Metis,
18 Melpomene, 60 Echo, 116 Sirona, 230 Athamantis,
694 Ekard, and 1984 KD.
K. W. Zeigler, W. B. Florence.
Icarus, Vol. 62, No. 3, p. 512 – 517 (1985).
Photoelectric observations of seven asteroids were made from Gila Observatory between October 14, 1983, and June 21, 1984. The following synodic rotational periods and amplitudes are reported: 9 Metis, $P = 5.04$ hr, $\Delta M = 0.05$; 18 Melpomene, $P = 11.570$ hr, $\Delta M = 0.22$; 60 Echo, $P = 25.208$ hr, $\Delta M = 0.22$; 116 Sirona, $P = 12.028$ hr, $\Delta M = 0.42$; 230 Athamantis, $P = 23.99$ hr, $\Delta M > 0.20$; 694 Ekard, $P = 5.925$ hr, $\Delta M = 0.50$; 1984 KD, $P = 1.97$ hr, $\Delta M = 0.26$.

098.096   Ephemerides of minor planets for 1986.
Yu. V. Batrakov (Editor).
Published by Institut Teoreticheskoj Astronomii Akademii Nauk SSSR. Izdatel'stvo Nauka, Leningrad–skoe Otdelenie, Leningrad. 329 pp. Price 4 Rbl. (1985). ISSN 0201-7806. In Russian and English.
Contents: Introduction, information on new elements, elements of planets (1) – (3143), lost objects, opposition dates, ephemerides of bright planets, ephemerides of some unusual planets, critical list.

098.097   Theory of the motion of (126) Velleda.
G. T. Arazov, S. A. Gabibov.
Sov. Astron., Vol. 28, No. 6, p. 705 – 708 (1984). English translation of 38.098.112.

098.098   Report of IAU Commission 20: Positions and motions of
minor planets, comets and satellites (*Positions et mouvements des petites planètes, des comètes et des satellites*).
E. Roemer.
Trans. IAU, Vol. XIXA, p. 207 – 226 (1985). – See Abstr. 003.046.

098.099   1984 HA$_1$ – new Trojan discovered?
J. Bouška.
Říše hvězd, Vol. 65, No. 10, p. 204 – 208 (1984). In Czech.

098.100   Asteroid taxonomy from cluster analysis of photometry.
D. J. Tholen.
Diss. Abstr. Int., Sect. B, Vol. 45, No. 7, p. 2201 (1985). Thesis, University of Arizona, 162 pp. (1984). Order No. DA8421985.

098.101   Observations of minor planets at the Crimean Astro-
physical Observatory.
T. M. Smirnova.
Komet. Tsirk., No. 334 (1984). With a correction in No. 338 (1985). In Russian.

098.102 Stellar occultations by minor planets in 1985.
I. S. Balinskaya.
Komet. Tsirk., No. 334 (1984). In Russian.

098.103 Observations of the minor planet (4) Vesta in Tartu.
H. Raudsaar.
Komet. Tsirk., No. 341 (1985). In Russian.

098.104 Observations of minor planets in Saratov.
L. N. Berdnikov, K. N. Grankin, I. A. Rakitin.
Byull. Inst. Teor. Astron., Tom 15, No. 7 (170), p. 406 (1985).
Concerning minor planets (2) Pallas, (6) Hebe, (7) Iris, (27) Euterpe, and (216) Kleopatra.

098.105 Determination of maximum likelihood orbital parameters using parameters otbained for separate apparitions.
Yu. V. Batrakov.
Byull. Inst. Teor. Astron., Tom 15, No. 8 (171), p. 419 – 423 (1985). In Russian.
A method has been developed for obtaining the maximum likelihood orbital parameters (Keplerian elements or equivalents) based on observations covering several apparitions (oppositions) of a celestial body – comet, minor planet. The orbital parameters determined at every apparition independently of other apparitions are used as basic data to determine the final orbit by this method. These intermediate orbital parameters serve as a kind of synthetic measurements or normal places.

098.106 Precise positions of minor planets from observations with the Pulkovo normal astrograph in 1975 – 1980.
L. S. Koroleva, O. N. Orlova, L. A. Firsova.
Izv. Glav. Astron. Obs. Pulkovo, Astrometr. Astrofiz., No. 202, p. 40 – 45 (1984). In Russian.
154 precise positions of 7 minor planets are given and compared with their ephemerides published by the Institute for Theoretical Astronomy of the Academy of Sciences of the USSR.

098.107 Precise positions of the minor planet Parthenope from observations with the Pulkovo normal astrograph in 1974 – 1980.
L. S. Koroleva.
Izv. Glav. Astron. Obs. Pulkovo, Astrometr. Astrofiz., No. 202, p. 46 – 48 (1984). In Russian.
Precise positions of the minor planet Parthenope in opposition are given for 1974 – 1980.

098.108 Binary precessing asteroids.
A. W. Harris, R. P. Binzel.
Bull. Am. Astron. Soc., Vol. 17, No. 2, p. 624 (1985). Abstract. – See Abstr. 010.066.

Bibliography of Earth, Moon, and Planets.
See Abstr. 002.007.

Asteroids, comets, meteors II.
See Abstr. 011.023.

IRAS asteroid workshop number 4: report and recommendations. Workshop held at the Jet Propulsion Laboratory, California Institute of Technology, Pasadena, Calif., 12 – 14 February 1985.
See Abstr. 012.042.

IRAS moving–object search.
See Abstr. 013.009.

Laboratory photometry of asteroids and atmosphereless bodies. I. The S.A.M. apparatus.
See Abstr. 022.025.

A portable digital data acquisition system for asteroid occultation photometry.
See Abstr. 034.057.

A portable digital data acquisition system for asteroid occultation photometry.
See Abstr. 034.073.

Methods of remote surface chemical analysis for asteroid missions.
See Abstr. 036.015.

Methods of remote surface chemical analysis for asteroid missions.
See Abstr. 036.087.

Observations of minor planets at the Dresden Lohrmann Observatory.
See Abstr. 041.019.

Stability of binary asteroids.
See Abstr. 042.038.

Studies on orbital resonance. I.
See Abstr. 042.062.

Studies on orbital resonance. II.
See Abstr. 042.063.

Initial orbit determination: final word.
See Abstr. 042.069.

On the last Brown conjecture.
See Abstr. 042.120.

Minor planet observations and the fundamental reference system.
See Abstr. 043.003.

An asteroid for the asking.
See Abstr. 051.003.

The J.W.G. (*Joint Working Group*) asteroid mission.
See Abstr. 051.050.

Fly–by missions to near–earth asteroids.
See Abstr. 051.051.

Large body impacts through geologic time.
See Abstr. 081.051.

Collisions in the solar system. – I. Impacts of the Apollo–Amor–Aten asteroids upon the terrestrial planets.
See Abstr. 091.002.

Sternbedeckungen durch Kleinplaneten im Jahre 1985.
See Abstr. 096.006.

Report of IAU Commission 15: Physical study of comets, minor planets, and meteorites (*L'étude physique des comètes, des petites planètes et des météorites*).
See Abstr. 102.053.

Comet formation in molecular clouds.
See Abstr. 102.064.

Spectroscopic identification of probable pallasite parent bodies.
See Abstr. 105.050.

Strain analysis of the Leoville chondrite and conditions in asteroidal interiors.
See Abstr. 105.063.

The asteroidal source region of ordinary chondrites.
See Abstr. 105.182.

Unmelted meteoritic debris in the Late Pliocene iridium anomaly: evidence for the ocean impact of a nonchondritic asteroid.
See Abstr. 105.202.

Earthward bound from chaotic regions of the asteroid belt.
See Abstr. 105.209.

Meteorites may follow a chaotic route to Earth.
See Abstr. 105.210.

Asteroidal source of ordinary chondrites.
See Abstr. 105.212.

Interplanetary field enhancements in the solar wind: evidence for cometesimals at 0.72 and 1.0 AU?
See Abstr. 106.029.

The association of interplanetary field enhancements with the asteroid 2201 Oljato and the comparison with simulations.
See Abstr. 106.061.

Interplanetary magnetic field enhancements and their association with the asteroid 2201 Oljato.
See Abstr. 106.064.

Asteroids and comets.
See Abstr. 107.002.

Asteroids, comets, and planet formation.
See Abstr. 107.005.

099 Jupiter, Jupiter Satellites

099.001   The spatial dependence of the Jovian auroral emissions.
T. E. Skinner, H. W. Moos, G. E. Ballester.
NASA Conf. Publ., NASA CP–2349, p. 497 – 498 (1984). – See Abstr. 012.001.

099.002   The Jovian stratosphere in the ultraviolet.
R. Wagener, J. Caldwell.
NASA Conf. Publ., NASA CP–2349, p. 499 (1984). Abstract. – See Abstr. 012.001.

099.003   Long term variability of the Io torus.
H. W. Moos, T. E. Skinner, P. D. Feldman, S. T. Durrance, J. L. Bertaux, M. C. Festou.
NASA Conf. Publ., NASA CP–2349, p. 500 (1984). Abstract. – See Abstr. 012.001.

099.004   Opening of the magnetic field lines in a fast rotating magnetosphere, with an application to Jupiter.
J. J. Aly.
Unstable current systems and plasma instabilities in astrophysics, p. 217 – 220 (1985). – See Abstr. 012.002 (IAU Symp. No. 107).

The author considers a simple model of magnetosphere around a fast rotating Jupiter–like object possessing a spin–aligned dipolar moment μ. In this model, low–energy plasma released by inner sources located beyond the corotation radius diffuses outward through closed lines, forming a thin equatorial disk in which there is a quasi–static balance between the centrifugal force and the magnetic tension. At some critical radius r_0, however, the magnetic field is no longer strong enough to hold the matter, and blows open, the matter escaping freely (this model has been introduced by Hill and Carbary, 1978).

099.005   The global resonance of the Jovian radiation belts.
P. A. Bespalov.
Pis'ma Astron. Zh., Tom 11, No. 1, p. 72 – 77 (1985). In Russian. English translation in Sov. Astron. Lett., Vol. 11.

Oscillation regimes of cyclotron instability in the Jovian radiation belts are investigated. Essentially the eigenfrequencies of these oscillations in the external Jovian magnetosphere do not depend on the magnetic shell and coincide with the rotation velocity of the planet. This gives an account of the ten–hour modulation of energetic electron fluxes which was observed experimentally.

099.006   Large–amplitude MHD waves upstream of the Jovian bow shock: reinterpretation.
M. L. Goldstein, H. K. Wong, A. F. Viñas, C. W. Smith.
J. Geophys. Res., Vol. 90, No. A1, p. 302 – 310 (1985).

Observations of large–amplitude MHD waves upstream of the Jovian bow shock have previously been interpreted as arising from a resonant electromagnetic ion beam instability (Goldstein et al., 1983). That interpretation was based on the conclusion that the observed fluctuations were predominantly right elliptically polarized in the solar wind rest frame. Because it has been noted by the authors that the fluctuations are, in fact, left elliptically polarized, a reanalysis of the observations is necessary. In this paper the authors investigate several mechanisms for producing left–hand–polarized MHD waves in the observed frequency range.

099.007   Plasma conditions inside Io's orbit: Voyager measurements.
F. Bagenal.
J. Geophys. Res., Vol. 90, No. A1, p. 311 – 324 (1985).

The Voyager 1 ion data that were obtained inside the orbit of Io allow accurate determination of convective velocity, temperature, and density of the major ionic species (S^+, O^+, S^{2+} and O^{2+} ions). The irregular radial profiles of ion temperature and flux tube content are not consistent with simple models of radial transport of plasma from a source near Io. The evidence of a source of ions well inside Io's orbit is provided by the detection of molecular (SO_2^+) ions at 5.3 R_J, the prevalence of non–Maxwellian tails to the ion distribution functions, the persistent presence of oxygen ions throughout the inner torus, and a 1 – 3% lag behind corotation outside 5.4 R_J.

099.008   Voyager observations of ion phase space densities in the Jovian magnetosphere.
M. Paonessa.
J. Geophys. Res., Vol. 90, No. A1, p. 521 – 525 (1985).

Data from the Voyager low–energy charged particle experiment were used to calculate ion phase space densities at constant first and second invariants in the inner Jovian magnetosphere.

099.009   Correction to "Sulfur and oxygen escape from Io and a lower limit to atmospheric SO_2 at Voyager 1 encounter" [J. Geophys. Res., Vol. 89, No. A9, p. 7399 – 7406 (1984)].
S. Kumar.
J. Geophys. Res., Vol. 90, No. A1, p. 573 (1985). See Abstr. 38.099.025.

099.010   Volcanic sulphur flows on Io.
M. H. Carr.
Nature, Vol. 313, No. 6005, p. 735 – 736 (1985).

099.011   Upcoming mutual events of Jupiter's moons.
K. Aksnes, F. Franklin.
Sky Telesc., Vol. 69, No. 2, p. 116 – 118 (1985).

099.012   Models, figures, and gravitational moments of the Galilean satellites of Jupiter and icy satellites of Saturn.
V. N. Zharkov, V. V. Leontjev (*V. V. Leont'ev*),
A. V. Kozenko.
Icarus, Vol. 61, No. 1, p. 92 – 100 (1985).
A general theory for the figures of satellites, which are synchronously rotating in the gravitational field of a planet, is developed to the first approximation. Love numbers, figure parameters, and gravitational moments for two– and three–layer models of the Galilean satellites, Titan, and Saturn's icy satellites are calculated.

099.013   The SO_2 atmosphere and ionosphere of Io: ion chemistry, atmospheric escape, and models corresponding to the Pioneer 10 radio occultation measurements.
S. Kumar.
Icarus, Vol. 61, No. 1, p. 101 – 123 (1985).
Models of Io's ionosphere at the time of the Pioneer 10 encounter are constructed in the presence of an SO_2–Na atmosphere on Io. The formation of the observed ionosphere on the downstream side requires precipitation of electrons; solar EUV alone is inadequate. Electron impact in the range 500 – 800 eV on an SO_2 atmosphere with a surface density of $4 \times 10^{10} cm^{-3}$ provides the best fit to the Pioneer 10 radio occultation entry data.

099.014   Bursts of type N in Jupiter's decametric radio spectra.
J. J. Riihimaa.
Earth, Moon, Planets, Vol. 32, No. 1, p. 9 – 19 (1985).
Dynamic spectra of Jupiter's decametric emission often display narrow–band features, referred to as events of type N (Carr et al., 1983). The average bandwidth of these emissions is in the vicinity of 200 kHz, their durations are typically in the decasecond range, and their $f–t$ slopes are small and random. Although the N–bursts can be described as narrow–band L–bursts, it seems that they are related to S–bursts in their area of occurrence in the Io–B region, the durations of the emission envelopes, and their bandwidths. Possible implications are discussed.

099.015   Conjecture about a hurricane system in the Jovian atmosphere.
H. G. Mayr, K. Maeda, I. Harris.
Earth, Moon, Planets, Vol. 32, No. 2, p. 183 – 192 (1985).
Kuiper (1972) had suggested that the Great Red Spot (GRS) of Jupiter is a giant hurricane. The authors present further arguments in support of this idea and propose that it may also apply to the smaller vortices such as the white and brown ovals (barges). The GRS with its large size and long life time (indicating that it is very deep) is unique, and the authors suggest that it may have been induced by meteor impact.

099.016   Revised ion temperatures for Voyager plasma measurements in the Io plasma torus.
F. Bagenal, R. L. McNutt Jr., J. W. Belcher, H. S. Bridge,
J. D. Sullivan.
J. Geophys. Res., Vol. 90, No. A2, p. 1755 – 1757 (1985).
The authors have discovered that an error was made in computing ion temperatures for some of the results derived from the positive ion data from the plasma science experiment on the Voyager spacecraft. The reported results directly affected by this error are the ion temperatures in the inner magnetosphere of Jupiter given by Bagenal et al. (1980) and Bagenal and Sullivan (1981) plus a single ion temperature calculated from Voyager 1 data obtained in Saturn's magnetosphere and quoted by Bridge et al. (1981).

099.017   Gravity field of the Jovian system from Pioneer and Voyager tracking data.
J. K. Campbell, S. P. Synnott.
Astron. J., Vol. 90, No. 2, p. 364 – 372 (1985).
Analysis of the Doppler–tracking data and star–satellite imaging from the Voyager 1 and Voyager 2 spacecraft, combined with a reanalysis of the Pioneer 10 and Pioneer 11 Doppler tracking has yielded improved values for the Galilean satellite masses, and the rotational pole, mass, and harmonic coefficients of Jupiter. Mass results are expressed as the product GM, in units of $(km^3 s^{-2})$. The Galilean satellite masses are (5961 ± 10) for Io, (3201 ± 10) for Europa, (9887 ± 3) for Ganymede, and (7181 ± 3) for Callisto. The mass of the Jupiter system is (126712767 ± 100). A mass for Amalthea could not be reliably estimated. The right ascension and declination of the pole of Jupiter relative to the mean Earth equator and equinox of 1950.0 is $268°.001 \pm 0°.005$ and $64°.504 \pm 0°.001$, respectively.

099.018   Glass on the surfaces of Io and Amalthea.
J. Gradie, S. J. Ostro, P. C. Thomas, J. Veverka.
J. Non–Cryst. Solids, Vol. 67, No. 1 – 3, p. 421 – 432 (1984).
Abstr. in Phys. Abstr., Vol. 88, No. 1251, Entry 24318 (1985). –
See Abstr. 012.016.

099.019   The enigma called Io.
D. Morrison.
Sky Telesc., Vol. 69, No. 3, p. 198 – 205 (1985).

099.020   What colors the surface of Io?
J. Gradie.
News Lett. Astron. Soc. N.Y., Vol. 2, No. 4, p. 11 – 12 (1983).
Abstract. – See Abstr. 010.242.

099.021   Les astronomes amateurs et la campagne "PHEMU85".
J.–E. Arlot, A. Figer, W. Thuillot.
Astronomie, Vol. 99, p. 179 – 185 (1985).

099.022   The impact of spectroscopic parameters on the composition of the Jovian atmosphere discussed in connection with recent laboratory, Earth and planetary observation programs.
A. Chedin, N. A. Scott.
J. Quant. Spectrosc. Radiat. Transfer, Vol. 32, No. 5/6, p. 453 – 461 (1984). – See Abstr. 012.026.
The paper discusses problems of spectroscopic character encountered when analysing the infrared spectra recorded by the recent mission to Jupiter and Saturn by the Voyager IRIS experiment. These problems are discussed in conjunction with similar problems occurring in the Earth's atmosphere and extensively studied from operational satellite platforms. Preliminary conclusions issuing from inter–comparisons of transmittance codes are presented and their quantitative impact on the interpretation of planetary spectra are also reported.

099.023   Near–infrared spectroscopy of the atmosphere of Jupiter.
F. W. Taylor, S. B. Calcutt.
J. Quant. Spectrosc. Radiat. Transfer, Vol. 32, No. 5/6, p. 463 – 477 (1984). – See Abstr. 012.026.
The spectroscopic properties of the atmosphere of Jupiter are reviewed and the uses of spectroscopy for remote sensing of the Jovian composition, thermal structure and cloud properties are discussed. The limitations of earth–based and existing spacecraft spectroscopy and the prospects for the Near Infrared Mapping Spectrometer on the forthcoming Galileo orbiter are considered.

099.024   Europan surface phenomena.
A. Eviatar, A. Bar-Nun, M. Podolak.
Icarus, Vol. 61, No. 2, p. 185 – 191 (1985).
The interaction of corotating iogenic plasma with the surface of Europa in light of recent ice sputtering, experimental results, and published Voyager data has been examined. It has been found that the residual atmosphere of Europa is made up of sputtered molecular oxygen and is exospheric from the surface outwards. It was also found that if sputtering, redistribution, and escape are considered and the sulfur dioxide/water mixing ratio is held constant over a UV observing depth, the observed sulfur dioxide density on the trailing hemisphere lends support to the hypothesis that liquid water from the interior of Europa is boiling out and being deposited as a frost layer on the surface at the rate of about 0.04 µm/year.

099.025   Voyager 1 imaging and IRIS observations of Jovian methane absorption and thermal emission: implications for cloud structure.
R. A. West, P. Kupferman, H. Hart.
Icarus, Vol. 61, No. 2, p. 311 – 342 (1985).

Images from three filters of the Voyager 1 wide–angle camera were used to measure the continuum reflectivity and spectral gradient near 6000 Å and the 6190–Å band methane/continuum ratio for a variety of cloud features in Jupiter's atmosphere.

099.026   Dust motion in Jupiter's tilted magnetic field.
L. Schaffer, J. A. Burns.
Adv. Space Res., Vol. 4, No. 9, p. 107 – 110 (1984). – See Abstr. 012.029.

The authors outline an analytical method for studying the motion of charged dust particles that orbit an oblate planet having a tilted, offset, dipolar magnetic field. The computed trajectories closely mimic previous numerical results; equilibrium dust potentials must be less than 10 volts or the Jovian ring would be thicker than observed. The authors identify several Lorentz resonances, where the periods of components of the Lorentz force, as seen by a reference particle moving in the equatorial plane, match the particle's orbital period; several seem to be near observed features of the Jovian ring system.

099.027   Formation of Jupiter's rings.
E. Grün.
Adv. Space Res., Vol. 4, No. 9, p. 135 – 136 (1984). – See Abstr. 012.029.

The author summarizes the basic ideas and open questions concerning the physical phenomena governing Jupiter's rings.

099.028   Voyager observations of lower hybrid noise in the Io plasma torus and anomalous plasma heating rates.
D. D. Barbosa, F. V. Coroniti, W. S. Kurth, F. L. Scarf.
Astrophys. J., Vol. 289, No. 1, p. 392 – 408 (1985).

A study of Voyager 1 electric field measurements obtained by the plasma wave instrument in the Io plasma torus has been carried out. The data reveal the presence of persistent peaks in electric field spectra in the frequency range 100 – 600 Hz consistent with their identification as lower hybrid noise for a heavy-ion plasma of sulfur and oxygen. Typical wave intensities are $0.1\ mV\ m^{-1}$. A theoretical analysis of lower hybrid wave generation by a bump–on–tail ring distribution of ions is given. The model is appropriate for plasmas with a superthermal pickup ion population present.

099.029   On the theory of S–bursts in the decameter radio emission of Jupiter.
V. V. Zajtsev, E. Ya. Zlotnik, V. E. Shaposhnikov.
Pis'ma Astron. Zh., Tom 11, No. 3, p. 208 – 215 (1985). In Russian. English translation in Sov. Astron. Lett., Vol. 11.

A mechanism of generation of S–bursts in decameter Jovian radio emission is proposed. The mechanism is based on the excitation of plasma waves at frequencies close to the electron gyrofrequency by nonequilibrium electrons in the Io flux tube and on the conversion of these waves into a fast extraordinary mode due to induced scattering by nonequilibrium ions.

099.030   Coupled low–energy – ring current plasma diffusion in the Jovian magnetosphere.
D. Summers, G. L. Siscoe.
J. Geophys. Res., Vol. 90, No. A3, p. 2665 – 2671 (1985).

The authors derive a functional form of the diffusion coefficient appropriate to fully developed, steady state, centrifugally driven turbulence in Jupiter's magnetosphere, incorporating in the diffusion coefficient the effects of the pressure gradient of the ring current. They set up a steady state, coupled diffusion model to describe the simultaneous outwardly diffusing low–energy (Iogenic) plasma and the inwardly diffusing ring current plasma.

099.031   An explanation for the H Ly α longitudinal asymmetry in the equatorial spectrum of Jupiter: an outcrop of paradoxical energy deposition in the exosphere.
D. E. Shemansky.
J. Geophys. Res., Vol. 90, No. A3, p. 2673 – 2694 (1985).

An analysis of the Voyager EUV spectra of the Jupiter sunlit equatorial emissions shows no evidence for a substantial dependence of atomic hydrogen abundance on magnetic longitude, required by earlier theories of the H Ly α longitudinal asymmetry. An explanation for the H Ly α bulge phenomenon is advanced in this work that conforms to the observations and does not require a strong asymmetry in atomic hydrogen abundance.

099.032   The effect of the dynamical parameters in the motion of the Galilean satellites of Jupiter.
W. Thuillot.
Celest. Mech., Vol. 34, Nos. 1 – 4, p. 245 – 253 (1984). – See Abstr. 012.030.

The construction of an analytical theory of the motion of the Galilean satellites of Jupiter requires to keep track of the dynamical parameters, that is, the masses of the satellites, and the harmonic coefficients of the potential of the planet J_2 and J_4. This is realized here. But as in other theories the solution becomes partly numerical from the resolution of an autonomous system. The aim of this paper is to present a method to obtain developed solutions of this autonomous system.

099.033   Libration of Laplace's argument in the Galilean satellites theory.
J. Henrard.
Celest. Mech., Vol. 34, Nos. 1 – 4, p. 255 – 262 (1984). – See Abstr. 012.030.

In the Galilean satellites motion, the Laplace argument $\lambda_1 - 3\lambda_2 + 2\lambda_3$ librates around the value π. The amplitude of libration is very small so that the classical theories have not been set up to take into account large librations. On the other hand large librations have to be considered in case of a description of possible scenarios of capture into resonance by tidal effects. The aim of this paper is to present a new way of applying Hamiltonian perturbation methods to the problem of the Galilean satellites in such a way that the theory is valid for large librations. Preliminary results from such a theory are discussed.

099.034   Synoptic report for the 1980 – 81 apparition of the planet Jupiter.
P. K. Mackal.
Strolling Astron., Vol. 31, Nos. 1 – 2, p. 2 – 10 (1985).

099.035   The 1985 – 86 mutual events of the Galilean satellites.
J. E. Westfall.
Strolling Astron., Vol. 31, Nos. 1 – 2, p. 10 – 16 (1985).

099.036   Galilean satellite eclipse ephemeris for 1985.
J. E. Westfall.
Strolling Astron., Vol. 31, Nos. 1 – 2, p. 21 – 24 (1985).

099.037   Evolution of the Galilean moons of Jupiter.
Yu. A. Khodak.
Cosmochemistry and meteoritics, p. 170 – 180 (1984). In Russian. Abstr. in Ref. Zh., 51. Astron., 2.51.187 (1985). – See Abstr. 012.037.

099.038   Optical parameters of the atmosphere in the region of the equatorial belt of Jupiter based on the results of observations in 1980.
A. N. Aksenov, V. D. Vdovichenko, K. Yu. Ibragimov.
Sov. Astron., Vol. 28, No. 4, p. 441 – 446 (1984). English translation of 38.099.009.

099.039   The soliton model for the Great Red Spot and other large vortices in planetary atmospheres.
M. V. Nezlin.
Sov. Astron. Lett., Vol. 10, No. 4, p. 221 – 226 (1984). English translation of 38.099.001.

099.040 Forgotten observations of Io?
A. V. Arkhipov.
Zemlya Vselennaya, No. 1, p. 45 – 46 (1985). In Russian.

099.041 The $^{14}N/^{15}N$ ratio in the Jovian atmosphere.
P. Drossart, T. Encrenaz, M. Combes.
Astron. Astrophys., Vol. 146, No. 1, p. 181 – 184 (1985).
Spectroscopic data of Jupiter between 850 and 915 cm^{-1} have been recorded and analyzed, as well as previous published data, for a determination of the Jovian $^{14}N/^{15}N$ ratio. It is shown that the main uncertainty in this determination comes from the choice of the atmospheric model used for the synthetic spectra, and in particular the modelization of the Jovian cloud structure. The result is $^{14}N/^{15}N = 125\,(+145,-75)$ which does not completely exclude the terrestrial value $^{14}N/^{15}N = 270$ but still suggests a significant ^{15}N enrichment.

099.042 Implications of the Galilean satellites ice envelope explosions. I. The motion of fragments inside and beyond Jupiter's sphere of action.
I. I. Agafonova, E. M. Drobyshevski (*Eh. M. Drobyshevskij*).
Earth, Moon, Planets, Vol. 32, No. 3, p. 241 – 255 (1985).
Explosions of the electrolyzed ice envelopes of the Galilean satellites resulted in the appearance of a large number of ice fragments deep inside Jupiter's sphere of action. Gravitational perturbations by the Galilean satellites transferred these fragments from satellite orbits into the periphery of the sphere of action and beyond it. The fragments move initially in the direction of a satellite's motion tangentially to its orbit. If ejected with a sufficient velocity, the fragments can leave Jupiter's sphere of action going both inside and outside its orbit. The results obtained may be used to shed light on the origin of the irregular satellites and Trojans.

099.043 On the nature of plasma waves in the Io torus.
A. A. Galeev, I. Kh. Khabibrakhmanov.
Pis'ma Astron. Zh., Tom 11, No. 4, p. 292 – 297 (1985). In Russian. English translation in Sov. Astron. Lett., Vol. 11.
The electric field oscillations with frequencies higher than the low–hybrid resonance frequency and below the electron cyclotron frequency detected in the Io plasma torus by Voyager 1 are considered to be related to generation of superthermal electrons due to "loss cone" instability of plasma containing a small number of hot ionized atoms of the Io atmosphere. The spectral energy density of the electric field and the distribution of superthermal electrons are calculated.

099.044 [S II] images of the Io torus.
C. B. Pilcher, J. H. Fertel, J. S. Morgan.
Astrophys. J., Vol. 291, No. 1, p. 377 – 393 (1985).
The authors report the results of a study of the Io plasma torus in which ~ 90 images in the [S II] $\lambda6731$ line were acquired during two four–night intervals in 1981 March and April. The authors show that, in general, the brightness variations with longitude can be accounted for by the presence of two localized plasma source regions. They find support for the earlier conclusion drawn from ground–based data that the red [S II] emission is often brightest approximately in the active sector, and show that the region of strongest S II intensity modulation with magnetic longitude is essentially coincident with the region of brightest S III EUV emission. The possible effects of subcorotational velocities on ion density distributions in the torus are also investigated. Evidence is presented for east–west variations in the torus geometry. Finally, the pertinent geometrical aspects of the data are discussed.

099.045 Meteorology of Jupiter and Saturn.
R. Bibe.
Astron. Vestn., Tom 18, No. 4, p. 263 – 271 (1984). In Russian. Abstr. in Ref. Zh., 51. Astron., 4.51.185 (1985).

099.046 Geodetic reference networks on the Jupiter and Saturn satellites.
M. Davis.
Astron. Vestn., Tom 18, No. 4, p. 272 – 280 (1984). In Russian. Abstr. in Ref. Zh., 51. Astron., 4.51.186 (1985).

099.047 A summary of whistlers observed by Voyager 1 at Jupiter.
W. S. Kurth, B. D. Strayer, D. A. Gurnett, F. L. Scarf.
Icarus, Vol. 61, No. 3, p. 497 – 507 (1985).
The Voyager 1 observations of whistlers at Jupiter are summarized in order to provide a basis for further analyses of the density profile of the Io plasma torus as well as to support studies of atmospheric lightning at Jupiter. All the whistlers detected by Voyager 1 fell into three general regions in the torus at radial distances ranging between 5 and 6 R_J. The whistler dispersions are presented in statistical form for each of the three groups of events and analyzed in view of the structure of the Io plasma torus as determined by plasma measurements.

099.048 Models of the millimeter–centimeter spectra of the giant planets.
I. de Pater, S. T. Massie.
Icarus, Vol. 62, No. 1, p. 143 – 171 (1985).
A comparison of various model atmospheric calculations with data for all four giant planets is shown. It appeared that ammonia must be depleted in the upper atmospheres of all four planets by a factor of 4–5 with respect to the solar abundance for Jupiter (and Saturn) and by a factor of 100–200 for Uranus and Neptune. At deeper layers the optical depth is larger, due either to a larger abundance of ammonia or to absorption by the presence of water. Given the vertical ammonia distribution in the atmospheres as derived from the centimeter data, a best fit to the millimeter spectra of all four planets was found by changing the high frequency tail of the ammonium lineshape profile. It was found that a line profile which at millimeter wavelengths more closely resembles a Van Vleck–Weisskopf lineshape than the usually adopted Ben Reuven profile gives a rather satisfactory fit to the data of all four gaseous planets.

099.049 Le groupement des satellites de Jupiter du 6 juillet 1984.
L. Rébuffat.
Astronomie, Vol. 99, p. 257 (1985).

099.050 Le nubi di Giove.
M. Arpino.
Orione, Vol. 5, N. 2, p. 21 – 25 (1985).

099.051 Seasonal activity on Jupiter?
A. P. Vid'machenko, A. F. Steklov, N. F. Minyajlo.
Sov. Astron. Lett., Vol. 10, No. 5, p. 289 – 290 (1984). English translation of 38.099.029.

099.052 Jovian cloud stratification.
A. V. Morozhenko.
Sov. Astron. Lett., Vol. 10, No. 5, p. 323 – 325 (1984). English translation of 38.099.031.

099.053 Triggered Jovian radio emissions.
W. Calvert.
Geophys. Res. Lett., Vol. 12, No. 4, p. 179 – 182 (1985).
Certain Jovian radio emissions seem to be triggered from outside, by much weaker radio waves from the sun. Recently found in the Voyager observations near Jupiter, such triggering occurs at hectometric wavelengths during the arrival of solar radio bursts, with the triggered emissions lasting sometimes more than an hour as they slowly drifted toward higher frequencies. Like the previous discovery of similar triggered emissions at the earth, this suggests that Jupiter's emissions might also originate from natural radio lasers.

099.054   High time resolution plasma wave and magnetic field observations of the Jovian bow shock.
S. L. Moses, F. V. Coroniti, C. F. Kennel, F. L. Scarf,
E. W. Greenstadt, W. S. Kurth, R. P. Lepping.
Geophys. Res. Lett., Vol. 12, No. 4, p. 183 – 186 (1985).

High time resolution (60 ms) Voyager magnetometer and plasma wave measurements of a strong (fast Mach number 16), quasi–perpendicular Jovian bow shock reveal an abrupt change in the plasma wave spectrum at the leading edge of the shock foot. Upstream electron plasma waves terminate at the leading edge, and are replaced by a lower–frequency broadband spectrum of ion–acoustic–like waves, which terminates at the main shock ramp. The clear association with the foot region of the lower frequency component suggests that it is generated by reflected ions. If the upstream plasma waves are generated by an escaping electron heat flux, their termination at the leading edge suggests that electrons are heated by the low–frequency waves in the shock foot.

099.055   Energetic ions upstream of Jupiter's bow shock.
S. M. Krimigis, R. D. Zwickl, D. N. Baker.
J. Geophys. Res., Vol. 90, No. A5, p. 3947 – 3960 (1985). – See Abstr. 012.003.

Comprehensive measurements of ion spectra, composition, and anisotropies over the energy range ~ 30 keV to ~ 5 MeV in Jupiter's foreshock, magnetosheath, and magnetopause were obtained by the low–energy charged particle instrument during the Voyager 1 and 2 encounters with the planet in 1979. Detailed analyses of the spectral and compositional signatures are presented.

099.056   Electrodynamic interaction of Ganymede with the Jovian magnetosphere and the radial spread of wake–associated disturbances.
G. F. Tariq, T. P. Armstrong, J. W. Lowry.
J. Geophys. Res., Vol. 90, No. A5, p. 3995 – 4009 (1985).

To understand and explain the Voyager 2 observations of Ganymede's wake–associated disturbances, an investigation of the electrodynamic interaction of the Jovian magnetospheric plasma with the satellite has been carried out.

099.057   Occultations of the Galilean satellites in 1985 and 1986.
K. Aksnes, F. Franklin.
I.A.P.P.P. Commun., No. 19, p. 23 – 25 (1985).

099.058   The geology and geochemistry of the Galilean satellites of Jupiter.
R. J. Stevenson.
South. Stars, Vol. 31, No. 1, p. 85 – 94 (1984).

Jupiter's Galilean satellites constitute a "mini solar system" with increasing density, rock content and geological activity from the outermost, Callisto, with corresponding decrease in ice content and surficial age inward toward Io. Block models and a segmental comparative diagram are shown to compare and elucidate these interesting and diverse worlds.

099.059   Some morphological features of Ganymede craters.
M. A. Kazennov.
Astron. Tsirk., No. 1354, p. 3 – 4 (1984). In Russian.

099.060   Long term stability of the Io high temperature plasma torus.
H. W. Moos, T. E. Skinner, S. T. Durrance, P. D. Feldman,
M. C. Festou, J. L. Bertaux.
Inst. Astrophys. Paris, Pré–Publ., No. 94, 51 pp. (1985). To appear in Astrophys. J.

099.061   Observations photographiques de grosses planètes et de leurs satellites effectuées en 1983 à l'équatorial GPO de 40 cm de l'Observatoire Austral Européen (ESO) à La Silla (Chili).
H. Debehogne.
Bull. Astron., Vol. 9, No. 6, p. 307 – 308 (1984).

099.062   Infrared spectro–spatial mapping of Jupiter between 8 and 14 μm.
J. H. Goebel, F. C. Witteborn, J. D. Bregman,
C. R. McCreight, W. Wisniewski.
Bull. Am. Astron. Soc., Vol. 16, No. 4, p. 922 (1984). Abstract. – See Abstr. 010.062.

099.063   Deuterium in the outer solar system.
W. H. Smith.
Bull. Am. Astron. Soc., Vol. 16, No. 4, p. 922 (1984). Abstract. – See Abstr. 010.062.

099.064   Recent determinations of $^{12}C/^{13}C$ and $^{14}N/^{15}N$ on Jupiter.
T. Encrenaz, P. Drossart, M. Combes, J. Lacy, E. Serabyn,
A. Tokunaga.
Bull. Am. Astron. Soc., Vol. 16, No. 4, p. 922 (1984). Abstract. – See Abstr. 010.062.

099.065   UV diagnostic of the Io torus.
P. Bruston.
Bull. Am. Astron. Soc., Vol. 16, No. 4, p. 1027 (1984). Abstract. – See Abstr. 010.063.

099.066   Explanation of the inward displacement of Io's hot plasma torus and consequences for sputtering sources.
J. A. Linker, M. G. Kivelson, M. A. Moreno, R. J. Walker.
Nature, Vol. 315, No. 6018, p. 373 – 378 (1985).

Radio profiles of the ion density and flux–tube content in the Io torus have peak values inside Io's orbit, even though Io is the effective source of these ions. Formation of an inward peak constrains either the velocity distributions or source regions of sputtered neutrals. A further constraint is that the ionization of neutrals on trapped trajectories that return to Io's surface must be limited. A dominant sulphur source is most easily reconciled with these constraints.

099.067   Refraction scattering as origin of the anomalous radar returns of Jupiter's satellites.
T. Hagfors, T. Gold, H. M. Ierkic.
Nature, Vol. 315, No. 6021, p. 637 – 640 (1985).

Three satellites of Jupiter, Callisto, Ganymede and Europa, give anomalous radar returns. The authors show here that this behaviour would arise as a consequence of "refraction scattering" rather than the familiar "reflection scattering" and discuss mechanisms that lead to the predominance of such scattering on these satellites.

099.068   Polar frost formation on Ganymede.
R. E. Johnson.
Icarus, Vol. 62, No. 2, p. 344 – 347 (1985).

The suggested models of polar frost formation on Ganymede are reviewed. A model in which plasma bombardment changes the reflectance characteristics of the icy surface is proposed.

099.069   Blowoff of an atmosphere and possible application to Io.
D. M. Hunten.
Geophys. Res. Lett., Vol. 12, No. 5, p. 271 – 273 (1985).

Diffusion equations for two components are set up with mutual drag terms included. It is found that these terms become important when the flux of an escaping gas is of the order of its diffusion limit. In this situation, other gases can be dragged along (blowoff) and both scale heights tend towards equality. It appears that this may occur on Io, giving a strong tendency to equalize the scale heights of O, S, and SO_2 without any appeal to eddy diffusion.

099.070   Differential rotation of the magnetic fields of gaseous planets.
A. J. Dessler.
Geophys. Res. Lett., Vol. 12, No. 5, p. 299 – 302 (1985).

The author argues that, with regard to the spin rate of magnetic field structure as a function of latitude, the behavior of the magnetic fields of gaseous planets is more analogous to the Sun than the Earth. Certain Jovian magnetospheric phenomena differ in repetition period by 3%. In order to explain Jupiter's two distinct periodicities, it is hypothesized that the spin period of the planet's magnetic features is a function of both latitude and the size of the feature, with smaller high–latitude features rotating slower than either low–latitude features or the dominant dipole moment. Similarly, the low–latitude planetary spin period of Saturn is shorter than the presently accepted single value because the present value is based on a high–latitude magnetic phenomenon.

099.071   Magnetic longitude variations in the Io torus.
C. B. Pilcher, J. S. Morgan.
Adv. Space Res., Vol. 5, No. 4, p. 337 – 345 (1985). – See Abstr. 012.066.

The authors review the evidence for both systematic variations in the optical properties of the Io torus with magnetic longitude and for plasma corotational lag. They summarize the results of a recent paper in which they showed that observed longitudinal variations in [S II] emission and contemporaneous measurements indicating little longitudinal variation in total charge density can be explained in terms of local (i.e., longitudinally confined) plasma sources and effects of corotational lag. The authors present new measurements of both [S II] and [S III] emissions as a function of magnetic longitude that indicate local plasma production, but no significant departure of the bulk plasma from corotation. The data suggest, however, that the longitudes of plasma formation drift in the sense of subcorotation.

099.072   The Jovian nebula: a post–Voyager perspective.
J. T. Trauger.
Science, Vol. 226, No. 4672, p. 337 – 341 (1984).

Voyager 1 carried a diverse collection of magnetospheric probes through the inner Jovian magnetosphere in March 1979. The ensuing data analysis and theoretical investigation provided a comprehensive description of the Jovian nebula, a luminous torus populated with newly released heavy ions drawn from Io's surface. Recent refinements in Earth–based imaging instrumentation are used to extend the Voyager in situ picture in temporal and spatial coverage.

099.073   Io's sodium cloud.
B. A. Goldberg, G. W. Garneau, S. K. LaVoie.
Science, Vol. 226, No. 4674, p. 512 – 516 (1984).

The first two–dimensional images of the source region of Io's neutral sodium cloud have been acquired by ground–based observation. Observed asymmetries in its spatial brightness distribution provide new evidence that the cloud is supplied by sodium that is ejected nonisotropically from Io or its atmosphere. The data demonstrate that the cloud exhibits a persistent systematic behavior coupled with Io's orbital position, a distinct "east–west orbital asymmetry", a variety of spatial morphologies, and true temporal changes. The geometric stability of the sodium source is also indicated.

099.074   Photochemistry and clouds of Jupiter, Saturn and Uranus.
S. K. Atreya, P. N. Romani.
Recent advances in planetary meteorology, p. 17 – 68 (1985). – See Abstr. 012.071.

Photochemistry of ammonia, methane, phosphine, hydrogen sulfide, methylamine, hydrogen cyanide and carbon monoxide in the atmospheres of Jupiter, Saturn and Uranus is discussed. Condensation of ammonia, ammonium hydrosulfide, water, methane, ethane and acetylene below and near the tropopause of these planets is formulated. Whenever necessary, new calculations are included. Candidates for the upper atmospheric hazes, and the reddish–brown chromophore in the clouds of Jupiter and Saturn are discussed.

099.075   The meteorology of Jupiter and Saturn.
G. E. Hunt, D. Godfrey, K. Haines, V. Moore.
Recent advances in planetary meteorology, p. 101 – 120 (1985). – See Abstr. 012.071.

There has been a rapid increase in the knowledge of the atmospheres of Jupiter and Saturn from the recent spacecraft missions, and the corresponding and complementary ground–based telescopic observations. The authors summarize their knowledge of these planetary atmospheres and review the current understanding of their meteorologies.

099.076   Nonisotropic coronal atmosphere on Io.
E. M. Sieveka, R. E. Johnson.
J. Geophys. Res., Vol. 90, No. A6, p. 5327 – 5331 (1985).

A model is presented for calculating nonisotropic coronal atmospheres. This is tested by comparison with the analytic results for an isotropic atmosphere. It is then used to consider differences between sublimated and sputtered corona on Io with reference to the ion and electron bombardment of such coronae when the exobase is at or near the surface.

099.077   Propagation of Jovian electrons in the interplanetary space.
Y.–n. Huang.
Chin. J. Space Sci., Vol. 4, No. 4, p. 269 – 276 (1984). In Chinese. Abstr. in Phys. Abstr., Vol. 88, No. 1255, Entry 45976 (1985).

099.078   Five–year (1977–82) observations of Jovian decametric radiation by the 1 km baseline interferometer at Mt. Zao Observatory.
H. Oya, A. Morioka, M. Kondo, T. Kondo, S. Miura, K. Miyashita, M. Tokumaru, K. Okutani, T. Nagai.
J. Geomagn. Geoelectr., Vol. 36, No. 1, p. 11 – 31 (1984). Abstr. in Phys. Abstr., Vol. 88, No. 1255, Entry 45978 (1985).

099.079   Large–scale turbulence in the Jovian atmosphere.
J. L. Mitchell, T. Maxworthy.
Turbulence and chaotic phenomena in fluids, p. 543 – 547 (1984). Abstr. in Phys. Abstr., Vol. 88, No. 1258, Entry 63253 (1985). – See Abstr. 012.093.

099.080   Jovian electron events in the orbit of the earth.
Y.–n. Huang.
18th International Cosmic Ray Conference, Vol. 3, p. 123 – 126 (1983). – See Abstr. 012.096.

The convection–diffusion equation of Jovian electron transport is derived based on a new "spiral coordinate system". An approximate analytical solution is obtained under certain assumptions and a new model on Jovian electron transport in interplanetary space and the modulation of Jovian electrons in the orbit of the earth is presented.

099.081   Probable contribution to cosmic rays by Jupiter.
B. Mitra, S. K. Bose, S. R. Ganguly.
18th International Cosmic Ray Conference, Vol. 3, p. 182 – 185 (1983). – See Abstr. 012.096.

Neutron–monitor data for 1959 – 1965 of Mt. Norikura and for 1972 – 1977 of Tokyo have been analysed with a view to investigate the possible contribution to cosmic rays by Jupiter. The results indicate Jupiter as a source of cosmic rays.

099.082   The secondary production of gamma rays in the Jovian magnetosphere.
J. B. Blake, S. H. Margolis.
18th International Cosmic Ray Conference, Vol. 9, p. 33 – 36 (1983). – See Abstr. 012.096.

The authors discuss the results of some calculations of secondary gamma rays using as input data the fluxes of electrons and protons measured by the Pioneer and Voyager spacecraft.

099.083   Magnetospheric plasma interaction with Io's atmosphere.
M. A. McGrath Kinnally, E. M. Sieveka, R. E. Johnson.
Bull. Am. Astron. Soc., Vol. 17, No. 2, p. 591 (1985). Abstract. – See Abstr. 010.065.

099.084   Spectral analysis of groove spacing on Ganymede.
R. E. Grimm, S. W. Squyres.
J. Geophys. Res., Vol. 90, No. B2, p. 2013 – 2021 (1985).

Grooves on Ganymede are interpreted as extensional tectonic features whose regular spacing is an indicator of the thickness of a brittle surface layer, or "lithosphere", at the time of deformation. The authors have performed a statistical analysis of groove spacing on Ganymede, taking Fourier transforms of 157 photometric profiles across sets of grooves and examining the resultant power spectra for peaks representing the dominant topographic periodicities.

099.085   Volcanic activity on the satellite Io.
M. Eliáš.
Říše hvězd, Vol. 66, No. 2, p. 28 – 31 (1985). In Czech.

099.086   Charged particle erosion of icy satellites: coronal atmospheres and surface redistribution.
E. M. Sieveka.
Diss. Abstr. Int., Sect. B, Vol. 45, No. 5, p. 1502 (1984). Thesis, University of Virginia, 165 pp. (1983). Order No. DA8416400.

099.087   Temporal and spatial variations in the intensity of ultraviolet emissions from Jupiter and the Io torus.
T. E. Skinner.
Diss. Abstr. Int., Sect. B, Vol. 45, No. 5, p. 1502 (1984). Thesis, Johns Hopkins University, 104 pp. (1984). Order No. DA8414299.

099.088   Analysis of motions of Jupiter's Great Red Spot and white ovals.
H. Guitar.
Diss. Abstr. Int., Sect. B, Vol. 45, No. 11, p. 3533 (1985). Thesis, New Mexico State University, 160 pp. (1985). Order No. DA8501892.

099.089   Photographic observations of the Galilean satellites of Jupiter at Pulkovo in 1904 – 1910. A comparison with theory and an analysis.
T. P. Kiseleva.
Izv. Glav. Astron. Obs. Pulkovo, Astrometr. Astrofiz., No. 201, p. 71 – 76 (1985). In Russian.
Observations of the Galilean satellites made by S. K. Kostinsky with the Pulkovo normal astrograph in 1904 – 1910 are analysed. These observations are compared with Sampson's theory and with modern observations made with the same instrument in 1977 – 1978.

099.090   Temporal and spatial variations in the Io torus.
J. S. Morgan.
Icarus, Vol. 62, No. 3, p. 389 – 414 (1985).
Spectrographic data on the Io torus from 15 nights of observations spread over a 4–month period in 1981 are presented. The [S II] $\lambda\lambda$6716, 6731; [S II] $\lambda\lambda$4069, 4076; [O II] $\lambda\lambda$3726, 3729, and [S III] λ3722 lines were simultaneously measured on each spectrogram. An east–west asymmetry was observed in the optical emissions, showing larger western intensities and a more diffuse and radially extensive nebula to the east. The [S II] line ratios indicate that very–high–density regions are present in the torus.

099.091   Laboratory simulations of PH_3 photolysis in the atmospheres of Jupiter and Saturn.
J. P. Ferris, H. Khwaja.
Icarus, Vol. 62, No. 3, p. 415 – 424 (1985).
The authors report studies on the effect of pressure, temperature, light wavelength and intensity, and components of the atmospheres of the Jovian planets on the photolysis of PH_3.

Jupiter und Saturn. Ergebnisse der Planetenforschung.
See Abstr. 003.036.

Nomenclature of the details of the relief of the Galilean satellites of Jupiter.
See Abstr. 003.070.

Galileo, planetary atmospheres, and prograde revolution.
See Abstr. 004.101.

The Great Inequality of Jupiter and Saturn: from Kepler to Laplace.
See Abstr. 004.104.

The discovery of Jupiter bursts.
See Abstr. 004.193.

Discovery of the Jupiter radiation belts.
See Abstr. 004.194.

Use of microwave detected microwave–optical double resonance to assign the 6450 Å band of NH_3.
See Abstr. 022.019.

Statistical mechanics of light elements at high pressure. VII. A perturbative free energy for arbitrary mixtures of H and He.
See Abstr. 022.062.

Hydrogen broadening of vibrational–rotational transitions of ammonia lying near 6450 Å.
See Abstr. 022.070.

On the $S_0(0)$ and $S_1(0)$ spectra of the H_2–H_2 dimer.
See Abstr. 022.088.

High resolution emission spectrum of H_2 between 78 and 118 nm.
See Abstr. 022.153.

Comparative performances of an objective grating concept in UV spectroscopic observations of solar system objects. I. Imaging capabilities.
See Abstr. 034.001.

Infrared speckle imaging.
See Abstr. 036.149.

Stability of L_4 and L_5 against radiation pressure.
See Abstr. 042.025.

Resonant structure of the outer solar system.
See Abstr. 042.075.

Periodic orbits of the general three–body problem for the Sun–Jupiter–Saturn system.
See Abstr. 042.076.

Posizioni dei satelliti di Giove per l'anno 1985.
See Abstr. 046.005.

Jupiter's satellites.
See Abstr. 046.011.

Ephémérides des satellites de Jupiter, Saturne et Uranus pour 1986. (Ephemerides of the satellites of Jupiter, Saturn and Uranus for 1986).
See Abstr. 046.034.

Space Telescope studies of Jupiter and Saturn simulated by Voyager observations.
See Abstr. 051.063.

Planning a probing voyage to Jupiter.
See Abstr. 051.074.

The Galileo scan platform pointing control system – a modern control theoretic viewpoint.
See Abstr. 052.066.

A perturbation theory for the free oscillations of a self–gravitating sphere with phase boundaries.
See Abstr. 062.130.

Near equality of ion phase space densities at Earth, Jupiter, and Saturn.
See Abstr. 084.012.

Particle transport in planetary magnetospheres.
See Abstr. 084.081.

Passive mode locking in masers with non–equidistant spectra.
See Abstr. 084.099.

Patterns of fracture and tidal stresses due to nonsynchronous rotation: implications for fracturing on Europa.
See Abstr. 091.019.

Chemical evolution on the giant planets and Titan.
See Abstr. 091.029.

The free oscillations of the giant planets: phase–transition boundary conditions.
See Abstr. 091.035.

Satellite orbits and ephemerides.
See Abstr. 097.005.

Resonant structure of the outer asteroid belt.
See Abstr. 098.029.

The composition and origin of the Iapetus dark material.
See Abstr. 100.009.

Application of a radiative transfer model to bright icy satellites.
See Abstr. 100.010.

On detection of radio bursts associated with Jovian and Saturnian lightning.
See Abstr. 100.022.

Motions of the perihelions of Neptune and Pluto.
See Abstr. 101.007.

No sign of cosmic–ray generation from Jupiter, detectable at ground–based stations.
See Abstr. 144.073.

100 Saturn, Saturn Satellites

100.001   Observations astrométriques et comparaison à la théorie des satellites de Saturne effectuées à l'Observatoire de Bordeaux sur le réfracteur de 38 cm durant les oppositions de 1981 et 1982.
G. Dourneau, M. R. Dulou, J. F. Le Campion.
Astron. Astrophys., Vol. 142, No. 1, p. 91 – 99 (1985).

Astrometric positions of Saturn satellites have been obtained during 1981 and 1982 oppositions as part of a continuing program at the Bordeaux Observatory, France. About 170 plate exposures have been taken with the 14 inches (38 cm) refractor which has a focal length of approximately 7 m and gives a plate scale of 30.120″/mm. The observations have been reduced by means of AGK3 or SAO reference field stars.

100.002   The restitution coefficient of colliding ice particles of the Saturn rings.
N. N. Gor'kavyj.
Pis'ma Astron. Zh., Tom 11, No. 1, p. 66 – 71 (1985). In Russian.
English translation in Sov. Astron. Lett., Vol. 11.

It is shown that mutual collisions of particles of Saturn's rings result in covering the particles by a few–millimeter friable ice dust layer which protects the ice from further destruction. Analytical dependences of the restitution coefficient on the collision velocities are derived for a simple impact model in the cases of smooth particles and particles covered by a friable layer of various thickness.

100.003   Ring structure in Saturn's outer magnetosphere.
H. S. Mahra, A. K. Pandey, V. Mohan, B. B. Sanwal.
Nature, Vol. 313, No. 5997, p. 38 – 39 (1985).

The presence of a particulate or gaseous ring in Saturn's outer magnetosphere at distances of 14 and 19 radii from Saturn was predicted by Lazarus et al. based on the analysis of ion density measurements by Voyagers 1 and 2 and Pioneer 11. The ring structure at 14 radii from Saturn was detected by Vasundhara et al. from occultation observations of the star SAO 158913 on 24 and 25 March 1984. The authors report the detection of a ring system at 19 radii using photoelectric observations of the occultation of the star SAO 158763 on 12 May 1984.

100.004   Mögliche Ursachen für die Bildung der Saturn-ringstrukturen.
G. Konrad.
Astron. Raumfahrt, 23. Jahrg., Heft 1, p. 2 – 6 (1985).

100.005   Erratum: "Saturn's nonaxisymmetric ring edges at 1.95 R_s and 2.27 R_s." [Icarus, Vol. 60, No. 1, p. 17 – 28 (1984)].
C. Porco, G. E. Danielson, P. Goldreich, J. B. Holberg, A. L. Lane.
Icarus, Vol. 61, No. 1, p. 173 (1985). See Abstr. 38.100.024.

100.006   On the problem of evolution of the resonance motion of the Mimas–Tethys system.
I. G. Chugunov, G. V. Stolyarov, V. V. Stolyarov.
Astron. Zh., Tom 62, Vyp. 1, p. 181 – 182 (1985). In Russian.
English translation in Sov. Astron., Vol. 29, No. 1.

In the classical statement of the problem of Saturn satellites resonance motion the question of the character of the solutions of the resonance equation over a long interval of time is investigated, taking into account the variations of orbital elements.

100.007   The stability of the oscillation motion of charged grains in the Saturnian ring system.
R.–l. Xu, H. L. F. Houpis.
J. Geophys. Res., Vol. 90, No. A2, p. 1375 – 1384 (1985).

The stability of charged grains given a random initial velocity is determined by a perturbation approach for the gravito-electrodynamic forces encountered in the corotating plasma environment of Saturn. Saturn's magnetic field is assumed to be a dipole with a small northward offset. If the equatorial component of the grain's initial velocity is given a specific value V_{s0}, the general results of two previous theories (Northrop and Hill, 1982; Mendis et al., 1982) are reproduced, including the critical radius for z instability at 1.616 R_s. However, for initial velocities other than V_{s0} a large spectrum of new results occur. The authors discuss the implications of these results for the formation of the Saturnian ring system.

100.008   Far–infrared rotational transitions of PH_3 in Saturn.
P. J. Viscuso, G. J. Stacey, N. T. Kurtz, C. Fuller, M. Harwit.
News Lett. Astron. Soc. N.Y., Vol. 2, No. 4, p. 13 (1983). Abstract. – See Abstr. 010.242.

100.009   The composition and origin of the Iapetus dark material.
J. F. Bell, D. P. Cruikshank, M. J. Gaffey.
Icarus, Vol. 61, No. 2, p. 192 – 207 (1985).

Telescopic observations, laboratory simulations, and photogeological studies have been conducted to investigate the composition of the dark material on Iapetus and the reasons for its peculiar distribution. The results of the authors' earlier studies (D. P. Cruikshank et al., 1983) are confirmed and extended. An Iapetus–like bombardment regime may be the cause of the large hemispheric asymmetry in regolith properties on Callisto.

100.010   Application of a radiative transfer model to bright icy satellites.
B. J. Buratti.
Icarus, Vol. 61, No. 2, p. 208 – 217 (1985).

A radiative transfer model, derived largely from the work of B. W. Hapke (1981) and J. D. Goguen (1981) is fit to Voyager imaging observations of Europa, Mimas, Enceladus, and Rhea. It is possible to place constraints on the single–scattering albedo, the porosity of the optically active upper regolith, the single–particle phase function, and, in the cases of Europa and Mimas, the mean slope angle of macroscopic surface features. The texture of the surfaces of the Saturnian satellites appears to be similar to the Earth's moon. However, Europa is found to have a distinctly more compact regolith and a more forward–scattering single–particle phase function.

100.011   A seasonal climate model of the atmospheres of the giant planets at the Voyager encounter time. I. Saturn's stratosphere.
B. Bézard, D. Gautier.
Icarus, Vol. 61, No. 2, p. 296 – 310 (1985).

A radiative seasonal model which incorporates a multilayer radiative transfer treatment at wavelengths longward of 7 µm is presented and applied to Saturn's stratosphere. Opacities due to H_2–He, CH_4, C_2H_2, and C_2H_6 are included. Season–dependent insolation is shown to produce a strong hemispheric asymmetry decreasing with depth at the Voyager encounter times, and seasonal amplitudes of 30K at the poles are predicted in the high stratosphere. The ring–modulated dependence of the insolation and the orbital eccentricity are shown to have a significant effect. The possible role of aerosols in the stratospheric heating is analyzed.

100.012   Particle erosion mechanisms and mass redistribution in Saturn's rings.
R. H. Durisen.
Adv. Space Res., Vol. 4, No. 9, p. 13 – 21 (1984). – See Abstr. 012.029.

A variety of physical processes can erode the surfaces of planetary ring particles. According to current estimates, the most efficient of these over the bulk of Saturn's rings is hypervelocity impact by 100 micron to one centimeter radius meteoroids. The atoms, molecules, and fragments ejected from ring particles by erosion arc across the rings along elliptical orbits to produce a tenuous halo of solid ejecta and an extensive gaseous atmosphere. Continuous exchange of ejecta between different ring regions can lead to net radial transport of mass and angular momentum. The equations governing this ballistic transport process are presented and discussed.

100.013   Stability and mass flow in Saturn's rings.
W.–H. Ip.
Adv. Space Res., Vol. 4, No. 9, p. 53 – 62 (1984). – See Abstr. 012.029.

Besides gravitational effects, interesting electrodynamical processes could also take place in the vicinity of the rings of Saturn. The author describes several of the electrodynamical mechanisms (with emphasis on their corresponding electric fields and current systems) which have been postulated to be of importance in determining the mass transport of the ring system. Further points are made that, besides mass exchange between the rings and the planetary atmosphere, the mass injection from the rings could also have significant effect on the mass and energy budget of the magnetosphere, maintenance of the E ring, the Titan hydrogen torus as well as aeronomic process in the upper atmosphere of Titan.

100.014   Formation of Saturn's spokes.
C. K. Goertz.
Adv. Space Res., Vol. 4, No. 9, p. 137 – 141 (1984). – See Abstr. 012.029.

The author analyzes in more detail than before the requirements and predictions of the model about dust particle size and optical depth of the spokes in Saturn's rings and the radial speed of spoke evolution.

100.015   Kinematics of Saturn's spokes.
E. Grün, G. W. Garneau, R. J. Terrile, T. V. Johnson, G. E. Morfill.
Adv. Space Res., Vol. 4, No. 9, p. 143 – 148 (1984). – See Abstr. 012.029.

Voyager 2 images of Saturn's rings have been analyzed for spoke activity. More than 80 and 40 different spokes have been measured at the morning and at the evening ansa, respectively.

100.016   Saturn's E ring.
T. W. Hill.
Adv. Space Res., Vol. 4, No. 9, p. 149 – 157 (1984). – See Abstr. 012.029.

The author reviews the evidence bearing on the following questions: What is known of the composition and structure of the E ring? What is the life expectancy of E–ring particles against various known destruction and transport mechanisms? What is the likelihood that known source mechanisms can maintain a permanent E–ring structure in the presence of such losses?

100.017   A comparison of the latitudes of Saturn's belts derived from Voyager imagery and Earth–based observation.
A. J. Hollis.
J. Br. Astron. Assoc., Vol. 95, No. 3, p. 110 – 112 (1985).

Mean belt latitudes for Saturn have been derived from measures of Voyager photographs and also from drawings of the planet made by members of the BAA Saturn Section. A comparison between the results from each source is made and a possible explanation of the discrepancies is made.

100.018   Dynamics of ring–satellite systems around Saturn and Uranus.
N. Borderies.
Celest. Mech., Vol. 34, Nos. 1 – 4, p. 297 – 327 (1984). – See Abstr. 012.030.

The author first recalls the observations concerning the opaque rings of Saturn and Uranus. Then a model is described which represents the kinematics of these rings. Finally collisions, self–gravity and satellite perturbations are treated.

100.019   The organic aerosols of Titan.
B. N. Khare, C. Sagan, W. R. Thompson, E. T. Arakawa, F. Suits, T. A. Callcott, M. W. Williams, S. Shrader, H. Ogino, T. O. Willingham, B. Nagy.
Adv. Space Res., Vol. 4, No. 12, p. 59 – 68 (1984). – See Abstr. 012.036.

A dark reddish organic solid, called tholin, is synthesized from simulated Titanian atmospheres by irradiation with high energy electrons in a plasma discharge. The visible reflection spectrum of this tholin is found to be similar to that of high altitude aerosols responsible for the albedo and reddish color of Titan. The real (n) and imaginary (k) parts of the complex refractive index of thin films of Titan tholin prepared by continuous D.C. discharge through a $0.9 N_2/0.1 CH_4$ gas mixture at 0.2 mb is determined from X–ray to microwave frequencies. Values of n ($\cong 1.65$) and k ($\cong 0.004$ to 0.08) in the visible are consistent with deductions made by ground–based and spaceborne observations of Titan. Molecular analysis of the volatile component of this tholin was performed by sequential and non–sequential pyrolytic gas chromatography/mass spectrometry. More than one hundred organic compounds are released. Many of these molecules are implicated in the origin of life on Earth, suggesting Titan as a contemporary laboratory environment for prebiological organic chemistry on a planetary scale.

100.020 On the resonance motion of the Saturn satellites.
G. V. Stolyarov.
Upr., nadezhnost', navigatsiya. Saransk, p. 140 – 145 (1984). In Russian. Abstr. in Ref. Zh., 51. Astron., 1.51.111 (1985).

100.021 The secondary radiation under Saturn's A–B–C rings produced by cosmic ray interactions.
J. F. Cooper, J. H. Eraker, J. A. Simpson.
J. Geophys. Res., Vol. 90, No. A4, p. 3415 – 3427 (1985).
From measurements on the Pioneer 11 spacecraft, electrons and protons under Saturn's A–B–C rings were reported by Simpson et al. (1980) and Chenette et al. (1980) and identified as the secondary products of cosmic ray nucleons with energies >10 GeV interacting with ring matter. The authors have extended their analysis and have also found both secondary gamma ray fluxes with energies >15 MeV and fragmentation nuclei (H and heavier nuclei) with energies <10 MeV/nucleon, which are consistent with this interpretation. A comparison of the measured low–energy proton and gamma ray fluxes with fluxes from numerical simulations of the secondary particle production and propagation in material slabs (protons) and the Earth's atmosphere (gamma rays) results in an A–B–C ring mean surface density. Direct measurement of the A–B–C ring surface density in radial intervals of ~300 – 1000 km is obtained from the measured higher–energy secondary particles (electrons >25 MeV and protons >67 MeV). The surface density measurements are shown to be consistent with results derived from Voyager ring radio/optical opacity measurements and the observed damping of density waves in the ring.

100.022 On detection of radio bursts associated with Jovian and Saturnian lightning.
P. Zarka.
Astron. Astrophys., Vol. 146, No. 1, p. L15 – L18 (1985).
During both Voyager–Saturn encounters, the planetary radio astronomy instrument recorded intense short–lived radio bursts. These bursts originate from an extended long–lived storm system located in Saturn's equatorial atmosphere. However, no comparable radio emissions were detected during Voyager–Jupiter encounters, although lightning activity is known to exist on this planet. The author shows that this non–detection is due to strong absorption suffered by the radio waves propagating through Jupiter's lower ionospheric layers. Consequently, from the detection of Saturnian radio bursts, the author infers a stringent upper limit to the electron density in Saturn's lower ionosphere, where no observational result is available. These results are compared with the case of the Earth.

100.023 Construction of a numerical theory of motion of Phoebe, the IXth satellite of Saturn.
L. E. Bykova, V. V. Shikhalev, V. A. Yurga.
Astron. geod., Tomsk, No. 10, p. 104 – 113 (1984). In Russian. Abstr. in Ref. Zh., 51. Astron., 3.51.103 (1985).

100.024 The atmosphere of Titan.
T. Owen.
Astron. Vestn., Tom 18, No. 4, p. 281 – 292 (1984). In Russian. Abstr. in Ref. Zh., 51. Astron., 4.51.194 (1985).

100.025 Directivity of Saturn electrostatic discharges and ionospheric implications.
P. Zarka.
Icarus, Vol. 61, No. 3, p. 508 – 520 (1985).
From a detailed analysis of the intensity distribution of Saturn electrostatic discharges (SED) as a function of time during the Voyager 1 encounter with Saturn, the total beaming pattern of the SED source, which is found to be isotropic, is determined. This result allows for the derivation of the diurnal variations of the peak electron density over the dayside equatorial ionosphere of Saturn, and thus explains the puzzling features observed during the Voyager 1 encounter with Saturn – the longer extent of the SED visibility after closest approach and the existence of double–humped SED episodes before the encounter – by ionospheric absorption of the SED radio emission.

100.026 Comments on "The Z_3 model of Saturn's magnetic field and the Pioneer 11 vector helium magnetometer observations" by J. E. P. Connerney, M. H. Acuña, and N. F. Ness [J. Geophys. Res., Vol. 89, No. A9, p. 7541 – 7544 (1984)].
L. Davis Jr., E. J. Smith.
J. Geophys. Res., Vol. 90, No. A5, p. 4461 – 4464 (1985). See Abstr. 38.100.012.

100.027 Reply to the comments on "The Z_3 model of Saturn's magnetic field and the Pioneer 11 vector helium magnetometer observations".
J. E. P. Connerney, M. H. Acuña, N. F. Ness.
J. Geophys. Res., Vol. 90, No. A5, p. 4466 (1985). See Abstr. 100.026.

100.028 Saturn, 1981 – 1982.
A. W. Heath.
J. Br. Astron. Assoc., Vol. 95, No. 4, p. 143 – 150 (1985).

100.029 The critical ionization velocity as a mechanism for producing Titan's plasma tail.
A. A. Galeev, I. Kh. Khabibrakhmanov.
Sov. Astron., Vol. 28, No. 5, p. 520 – 523 (1984). English translation of 38.100.020.

100.030 Collisional dynamics of particles in Saturn's rings.
I. G. Shukhman.
Sov. Astron., Vol. 28, No. 5, p. 574 – 585 (1984). English translation of 38.100.021.

100.031 Wavy edges suggest moonlet in Encke's gap.
J. N. Cuzzi, J. D. Scargle.
Astrophys. J., Vol. 292, No. 1, p. 276 – 290 (1985).
Voyager images have revealed radial undulations of the inner and outer edges of the 325 km wide Encke gap in Saturn's A ring. These waves are present at some, but not all, longitudes. Their locations and wavelengths provide strong indirect evidence for the presence of at least one dominant moonlet of about 10 km radius orbiting near the center of the gap. Implications for "shepherding" theory are discussed.

100.032 Observations photographiques des satellites de Saturne effectuées en 1982 à l'équatorial GPO de 40 cm de l'Observatoire Austral Européen (ESO) à La Silla (Chili).
H. Debehogne.
Bull. Astron., Vol. 9, No. 6, p. 299 – 306 (1984).

100.033 Damping of nonlinear density waves in Saturn's rings.
L. Dones, F. H. Shu, J. J. Lissauer, J. N. Cuzzi, C. Yuan.
Bull. Am. Astron. Soc., Vol. 16, No. 4, p. 922 – 923 (1984). Abstract. – See Abstr. 010.062.

100.034 Spectra of N_2–Ar dimers in the atmosphere of Titan.
K. Fox, S. Kim.
Bull. Am. Astron. Soc., Vol. 16, No. 4, p. 1002 (1984). Abstract. – See Abstr. 010.062.

100.035 The evolution of spokes in Saturn's ring B.
A. F. Cook, R. T. E. Barrey, G. E. Hunt.
Bull. Am. Astron. Soc., Vol. 16, No. 4, p. 1026 (1984). Abstract. – See Abstr. 010.063.

100.036 The atmosphere of Saturn: an analysis of the Voyager radio occultation measurements.
G. F. Lindal, D. N. Sweetnam, V. R. Eshleman.
Astron. J., Vol. 90, No. 6, p. 1136 – 1146 (1985).
During occultations of Voyager 1 and 2 by Saturn, the radio links from the spacecraft were utilized to probe the atmosphere of the planet at latitudes ranging from 36.7°N to 79.7°S. The measurements, which were conducted at wavelengths of 13 cm (S band) and 3.6 cm (X band), have yielded profiles in height of the electron number density in the ionosphere and the gas refractivity, number density, pressure, temperature, and ammonia abundance in the troposphere and stratosphere. From the ver-

tical pressure profiles obtained at different latitudes, the size and shape of Saturn's isobaric surfaces are determined.

100.037   Analysis of the orbits of Titan, Hyperion, and Iapetus by numerical integration and by analytical theories.
A. T. Sinclair, D. B. Taylor.
Astron. Astrophys., Vol. 147, No. 2, p. 241 – 246 (1985).

The orbits of Titan, Hyperion, and Iapetus have been generated by both numerical integration and analytical theories, and fitted to astrometric observations made in the period 1967 – 1982. For Titan and Iapetus the two fits to the observations were of similar accuracy, but for Hyperion the numerical integration was significantly better. The numerical integration gave good determinations of the mass ratio Titan/Saturn, the dynamical flatting J_2 of Saturn and the mass ratio Saturn/Sun.

100.038   Interaction of Titan's atmosphere with Saturn's magnetosphere.
R. E. Hartle.
Adv. Space Res., Vol. 5, No. 4, p. 321 – 332 (1985). – See Abstr. 012.066.

The author concentrates on the interaction of Titan's atmosphere with Saturn's magnetosphere and compares it with the similar one at Venus, where a wealth of knowledge has been gathered for more than a decade through the Venera, Mariner and Pioneer missions at Venus.

100.039   Saturn radio emission and the solar wind: Voyager–2 studies.
M. D. Desch, H. O. Rucker.
Adv. Space Res., Vol. 5, No. 4, p. 333 – 336 (1985). – See Abstr. 012.066.

Voyager 2 data from the Plasma Science experiment, the Magnetometer experiment and the Planetary Radio Astronomy experiment were used to analyze the relationship between parameters of the solar wind/interplanetary medium and the nonthermal Saturn radiation. Solar wind and interplanetary magnetic field properties were combined to form quantities known to be important in controlling terrestrial magnetospheric processes.

100.040   Saturne, après Voyager, avant le vaisseau spatial Cassini.
C. Muller.
Ciel Terre, Vol. 101, No. 3, p. 71 – 76 (1985).

A description of the Saturn system is made after five years of interpretation of the Voyager results. The specificities of the planet, its rings and its satellite Titan, lead to the conception of Cassini: a new ESA–NASA space mission to Saturn of which the first preliminary designs are introduced.

100.041   Electrostatic waves in the Saturn rings.
P. V. Bliokh, V. V. Yaroshenko.
Astron. Zh., Tom 62, Vyp. 3, p. 569 – 579 (1985). In Russian. English translation in Sov. Astron., Vol. 29, No. 3.

Azimuth–dependent disturbances of the space charge density arising on the background of a complex radial structure of the rings are considered. The wave modes and possible instability mechanisms are analysed.

100.042   On the stability of Saturn's rings.
N. N. Gor'kavyj.
Pis'ma Astron. Zh., Tom 11, No. 6, p. 469 – 474 (1985). In Russian. English translation in Sov. Astron. Lett., Vol. 11.

The stability of Saturn's rings is investigated taking into account mutual gravitational scattering of the particles.

100.043   Some dynamical models for the Saturn co–orbital satellites.
R. A. Broucke, A. Konopliv.
Bull. Am. Astron. Soc., Vol. 17, No. 2, p. 626 (1985). Abstract. – See Abstr. 010.066.

100.044   Osservazioni visuali e fotografiche di Saturno nel 1983.
G. Adamoli.
Astronomia, N. 2, p. 6 – 10 (1985).

100.045   Impact cratering history of the Saturnian satellites.
J. B. Plescia, J. M. Boyce.
J. Geophys. Res., Vol. 90, No. B2, p. 2029 – 2037 (1985).

In order to model the flux history at the Saturnian satellites, observational data regarding cratering on the satellites have been combined with theory to provide an estimate of the absolute ages of the satellite surfaces.

100.046   Resonances of Saturn's rings.
K. Pintová.
Říše hvězd, Vol. 66, No. 3, p. 41 – 45 (1985). In Czech.

100.047   Volatiles in the outer solar system: I. Thermodynamics of clathrate hydrates. II. Ethane ocean on Titan. III. Evolution of primordial Titan atmosphere.
J. I. Lunine.
Diss. Abstr. Int., Sect. B, Vol. 45, No. 12, p. 3847 – 3848 (1985). Thesis, California Institute of Technology, 346 pp. (1985). Order No. DA8501341.

100.048   On the time scale of fragment ejection from the action sphere of Saturn by gravitational perturbations of Titan.
I. I. Agafonova.
Komet. Tsirk., No. 334 (1984). In Russian.

100.049   Bending waves and the structure of Saturn's rings.
J. J. Lissauer.
Icarus, Vol. 62, No. 3, p. 433 – 447 (1985).

The surface mass density profiles at four locations within Saturn's rings are calculated using Voyager spacecraft images of spiral bending waves. Bending waves are vertical corrugations in Saturn's rings which are excited at vertical resonances of a moon, e.g., Mimas. Observations of bending waves can be used to determine the surface density in regions of Saturn's rings near vertical resonances. The average surface density of the outer B ring near Mimas' 4:2 inner vertical resonance is 54 ± 10 g cm^{-2}. Surface densities ranging from 24 g cm^{-2} to 45 g cm^{-2} are found in the A ring.

Jupiter und Saturn. Ergebnisse der Planetenforschung.
See Abstr. 003.036.

The Great Inequality of Jupiter and Saturn: from Kepler to Laplace.
See Abstr. 004.104.

The nature of Saturn's rings: James E. Keeler's "Prettiest application of Doppler's principle".
See Abstr. 004.105.

U.A.I. Saturn section.
See Abstr. 013.090.

Use of microwave detected microwave–optical double resonance to assign the 6450 Å band of NH₃.
See Abstr. 022.019.

Sputtering processes: erosion and chemical change.
See Abstr. 022.031.

Hydrogen broadening of vibrational–rotational transitions of ammonia lying near 6450 Å.
See Abstr. 022.070.

On the $S_0(0)$ and $S_1(0)$ spectra of the H_2–H_2 dimer.
See Abstr. 022.088.

Infrared spectra of gaseous mononitriles: application to the atmosphere of Titan.
See Abstr. 022.125.

Parametric excitation of density waves in circular streams of charged particles (model of the Saturn rings).
See Abstr. 022.132.

High resolution emission spectrum of H_2 between 78 and 118 nm.
See Abstr. 022.153.

Stability of L_4 and L_5 against radiation pressure.
See Abstr. 042.025.

Resonant structure of the outer solar system.
See Abstr. 042.075.

Periodic orbits of the general three–body problem for the Sun–Jupiter–Saturn system.
See Abstr. 042.076.

Ephémérides des satellites de Jupiter, Saturne et Uranus pour 1986. (Ephemerides of the satellites of Jupiter, Saturn and Uranus for 1986).
See Abstr. 046.034.

Cassini – a concept for a Titan probe.
See Abstr. 051.020.

Space Telescope studies of Jupiter and Saturn simulated by Voyager observations.
See Abstr. 051.063.

Dust in magnetised plasmas: basic theory and some applications.
See Abstr. 062.062.

Nonlinear spiral density waves: an inviscid theory.
See Abstr. 062.112.

Near equality of ion phase space densities at Earth, Jupiter, and Saturn.
See Abstr. 084.012.

Chemical evolution on the giant planets and Titan.
See Abstr. 091.029.

The collisional dynamics of particulate disks.
See Abstr. 091.069.

Satellite orbits and ephemerides.
See Abstr. 097.005.

Models, figures, and gravitational moments of the Galilean satellites of Jupiter and icy satellites of Saturn.
See Abstr. 099.012.

Meteorology of Jupiter and Saturn.
See Abstr. 099.045.

Geodetic reference networks on the Jupiter and Saturn satellites.
See Abstr. 099.046.

Models of the millimeter–centimeter spectra of the giant planets.
See Abstr. 099.048.

Differential rotation of the magnetic fields of gaseous planets.
See Abstr. 099.070.

Photochemistry and clouds of Jupiter, Saturn and Uranus.
See Abstr. 099.074.

The meteorology of Jupiter and Saturn.
See Abstr. 099.075.

Charged particle erosion of icy satellites: coronal atmospheres and surface redistribution.
See Abstr. 099.086.

Laboratory simulations of PH_3 photolysis in the atmospheres of Jupiter and Saturn.
See Abstr. 099.091.

Motions of the perihelions of Neptune and Pluto.
See Abstr. 101.007.

Bolometric albedos of Titan, Uranus, and Neptune.
See Abstr. 101.022.

Titan and the dispersal of the proto–Saturnian nebula.
See Abstr. 107.015.

101 Uranus, Neptune, Pluto, Transplutonian Planets

101.001   Ultraviolet observations of Uranus and Neptune below 3000 Å.
J. Caldwell, R. Wagener, T. Owen, M. Combes, T. Encrenaz.
NASA Conf. Publ., NASA CP-2349, p. 501 (1984). Abstract. –
See Abstr. 012.001.

101.002   Periodic comet showers and planet X.
D. P. Whitmire, J. J. Matese.
Nature, Vol. 313, No. 5997, p. 36 – 38 (1985).

The discovery that Pluto's mass is insufficient to explain the discrepancies in the motions of the outer planets has led to the prediction of a tenth planet (planet X) beyond the orbit of Pluto (50 – 100 AU). Further, the existence of a belt or disk of comets beyond the orbit of Neptune ($\gtrsim 35$ AU) has been proposed in connection with some theories of the origin of the Solar System as a possible source of short–period comets. The authors point out that the existence of both planet X and the comet disk at their expected distances may explain not only the observed planetary motions and the origin of comets, but also the recently reported 28–Myr periodicity in terrestrial cratering and in mass extinctions. The cratering period is associated with the precession of the perihelion of planet X caused by the perturbations of the outer planets.

101.003   Planeten–Porträts: Neue Aufnahmen von Uranus, Neptun und Pluto.
W. Engelhardt.
Orion, 43. Jahrg., Nr. 206, p. 12 – 14 (1985).

101.004   In search of Uranian decametric emission.
G. R. Lebo, L. T. Roth.
Publ. Astron. Soc. Pac., Vol. 97, No. 588, p. 177 – 179 (1985). =
Contrib. Dep. Astron. Univ. Florida, No. 80.

Recent evidence suggests that the planet Uranus possesses a strong magnetic field, raising the possibility that Uranus may, like Jupiter, emit low–frequency radio radiation. Uranus was monitored some 106 hours in the spring of 1983 using the 26.3 MHz array at the University of Florida Dixie County Radio Observatory. No radiation was discerned as Uranian yielding an upper flux density limit of 400 Jy (1 Jy = 10^{-26} W M^{-2}Hz^{-1}).

101.005 The enigma of the Uranian satellites' orbital eccentricities.
S. W. Squyres, R. T. Reynolds, J. J. Lissauer.
Icarus, Vol. 61, No. 2, p. 218 – 223 (1985).

Using recently published determinations of the diameters and orbital elements of the Uranian satellites and assuming reasonable dissipation functions and rigidities for icy satellites, the eccentricity decay times for the satellites were calculated. For the inner three, decay times are on the order of $10^7 - 10^8$ years, making it difficult to understand why these satellites still have their observed eccentricities. Several possible explanations are discussed.

101.006 Approximation methods in celestial mechanics. Application to Pluto's motion.
J. Chapront.
Celest. Mech., Vol. 34, Nos. 1 – 4, p. 165 – 184 (1984). – See Abstr. 012.030.

Various methods of approximation for the computation of planetary perturbations are investigated: Fourier, Fourier-Chebyschev and Legendre. Application is made to Pluto's motion. Based upon JPL's ephemeris DE200, Pluto's mean elements are provided and also a compact ephemeris covering 2 centuries.

101.007 Motions of the perihelions of Neptune and Pluto.
H. Kinoshita, H. Nakai.
Celest. Mech., Vol. 34, Nos. 1 – 4, p. 203 – 217 (1984). – See Abstr. 012.030.

Five outer planets are numerically integrated over five million years in the Newtonian frame. The argument of Pluto's perihelion librates about 90 degrees with an amplitude of about 23 degrees. The period of the libration depends on the mass of Pluto. The motion of Neptune's perihelion is more sensitive to the mass of Pluto. With the initial conditions which do not lie in the resonance region between Neptune and Pluto, a close approach between them takes place frequently and the orbit of Pluto becomes unstable and irregular.

101.008 Results from observations of the 15 June 1983 occultation by the Neptune system.
W. B. Hubbard, J. E. Frecker, J.-A. Gehrels, T. Gehrels,
D. M. Hunten, L. A. Lebofsky, B. A. Smith, D. J. Tholen,
F. Vilas, B. Zellner, H. P. Avey, K. Mottram, T. Murphy,
B. Varnes, B. Carter, A. Nielsen, A. A. Page, H. H. Fu,
H. H. Wu, H. D. Kennedy, M. D. Waterworth, H. J. Reitsema.
Astron. J., Vol. 90, No. 4, p. 655 – 667 (1985).

The authors obtained eight observations of Neptune occultations from six sites in the southwestern Pacific on 15 June 1983. They used the data to search for evidence of rings around Neptune down to a distance of 0.03 Neptune radii from the planetary surface, but the results were negative. An astrometric analysis of the timings yields a solution for the equatorial radius α_0 of Neptune at $1\,\mu$ bar pressure and the oblateness e of this level. The results are $\alpha_0 = 25295 \pm 50$ km, $e = 0.022 \pm 0.004$, from which the authors derive α_1 (equatorial radius at 1 bar pressure) $= 24830 \pm 100$ km and a rotation period $P = 15^h\,(+3^h, -2^h)$. These results are consistent with the hypothesis that Neptune and Uranus have homologous mass distributions. The determinations of Neptune atmospheric temperatures at 1 μbar pressure are consistent with earlier results and when combined with all previous data give an average value of 156 ± 10K. There is only slight evidence for any latitude dependence in this quantity. Profiles with a high signal–to–noise ratio suggest the possible presence of an absorbing layer at altitudes higher than the 1–μbar level.

101.009 Neptune revealed.
Space Educ., Vol. 1, No. 9, p. 386 – 389 (1985).

101.010 A photometric study of Pluto near perihelion. II. Rotation period and color indices.
V. M. Lyutyj, V. P. Tarashchuk.
Sov. Astron. Lett., Vol. 10, No. 4, p. 226 – 229 (1984). English translation of 38.101.001.

101.011 Models of Uranus and Neptune.
T. V. Gudkova, V. N. Zharkov.
Astron. Vestn., Tom 18, No. 4, p. 293 – 309 (1984). In Russian. Abstr. in Ref. Zh., 51. Astron., 4.51.196 (1985).

101.012 Auch Planet Neptun besitzt einen Ring.
M. J. Schmidt.
Orion, 43. Jahrg., Nr. 208, p. 85 – 86 (1985).

101.013 Pluto and Charon: the dance begins.
J. K. Beatty.
Sky Telesc., Vol. 69, No. 6, p. 501 – 502 (1985).

101.014 Atmospheric temperature profiles of Uranus and Neptune.
H. Moseley, B. Conrath, R. F. Silverberg.
Astrophys. J., Lett. Ed., Vol. 292, No. 2, p. L83 – L86 (1985).

The authors present far–infrared spectrophotometry of Uranus and Neptune in the 30 – 55 μm spectral range. The measurements in six independent spectral bands are used to derive atmospheric temperature profiles for these planets. Both planets are found to have tropopause temperatures near 53K, with Neptune having a stronger stratospheric temperature inversion than Uranus. Effective temperatures of 57.7 ± 1.8K and 58.2 ± 1.9K are obtained for Uranus and Neptune, respectively, confirming the large internal heat source in Neptune.

101.015 Photometry of the Pluto/Charon system.
D. J. Tholen, E. F. Tedesco.
Bull. Am. Astron. Soc., Vol. 16, No. 4, p. 923 (1984). Abstract. – See Abstr. 010.062.

101.016 Possible detection of a 10–kilometer sized object around Neptune.
F. Roques, B. Sicardy, P. Bouchet, A. Brahic, R. Häfner,
J. Lecacheux, J. Manfroid.
Bull. Am. Astron. Soc., Vol. 16, No. 4, p. 1027 – 1028 (1984). Abstract. – See Abstr. 010.063.

101.017 Occultation and transient phenomena of Pluto and its satellite.
IAU Circ., No. 4040 (1985).

101.018 Remote sensing of the magnetic moment of Uranus: predictions for Voyager.
T. W. Hill, A. J. Dessler.
Science, Vol. 227, No. 4693, p. 1466 – 1469 (1985).

Power is supplied to a planet's magnetosphere from the kinetic energy of planetary spin and the energy flux of the impinging solar wind. A fraction of this power is available to drive numerous observable phenomena, such as polar auroras and planetary radio emissions. In this report the present understanding of these power transfer mechanisms is applied to Uranus to make specific predictions of the detectability of radio and auroral emissions by the planetary radio astronomy and ultraviolet spectrometer instruments aboard the Voyager spacecraft before its encounter with Uranus at the end of January 1986.

101.019 The search for eclipses in the Pluto–Charon system.
R. P. Binzel, J. D. Mulholland.
Bull. Am. Astron. Soc., Vol. 17, No. 1, p. 514 – 515 (1985). Abstract. – See Abstr. 010.064.

101.020 Astrometric CCD observations of the Uranian moons.
D. Pascu, P. K. Seidelmann, R. E. Schmidt,
E. J. Santoro, J. L. Hershey.
Bull. Am. Astron. Soc., Vol. 17, No. 2, p. 622 (1985). Abstract. – See Abstr. 010.066.

101.021 The detection of eclipses in the Pluto–Charon system.
R. P. Binzel, D. J. Tholen, E. F. Tedesco,
B. J. Buratti, R. M. Nelson.
Science, Vol. 228, No. 4704, p. 1193 – 1195 (1985).

The first eclipses between Pluto and its satellite ("Charon") were detected in January and February 1985, confirming the satellite's existence. Eclipses lasting a few hours will now occur at 3.2–day intervals for the next 5 to 6 years and then will cease for about 120 years. Careful observations of these eclipses will allow greatly improved determinations to be made of several physical parameters for the Pluto–Charon system: the diameters of the planet and satellite, the surface albedo distribution on one hemisphere of the planet, the orbit of the satellite, and the mass of the planet and hence its density. Knowledge of the density will provide a constraint on models of Pluto's bulk composition.

101.022   Bolometric albedos of Titan, Uranus, and Neptune.
J. S. Neff, T. A. Ellis, J. Apt, J. T. Bergstralh.
Icarus, Vol. 62, No. 3, p. 425 – 432 (1985).

The geometric albedos of Titan, Uranus, and Neptune have been measured from 2000 to 3175 Å by Caldwell et al. (1981), from 3500 to 10,500 Å by Neff et al. (1984) and from 6440 to 25,360 Å by Apt, Singer and Clark. The integrated solar flux in this spectral interval amounts to 97% of the solar constant. These data sets were combined and integrated to find the bolometric geometric albedo of each object. Preliminary determinations of the phase funtions were used to compute the Bond albedos and effective temperatures. The effective temperatures are compared with bolometric temperatures determined from brightness temperatures in the 10–μm to 5–mm region of the spectrum.

The observations of Neptune by Galileo.
See Abstr. 004.026.

Sputtering processes: erosion and chemical change.
See Abstr. 022.031.

Resonant structure of the outer solar system.
See Abstr. 042.075.

Ephémérides des satellites de Jupiter, Saturne et Uranus pour 1986. (Ephemerides of the satellites of Jupiter, Saturn and Uranus for 1986).
See Abstr. 046.034.

The January 1986 Voyager–Uranus encounter.
See Abstr. 051.121.

The April 22 occultation of Hya −20°51699 by Uranus and its rings.
See Abstr. 096.007.

Une occultation d'étoile par les anneaux d'Uranus en 1985.
See Abstr. 096.010.

Satellite orbits and ephemerides.
See Abstr. 097.005.

Secular perturbations of asteroids with commensurable mean motions.
See Abstr. 098.033.

Models of the millimeter–centimeter spectra of the giant planets.
See Abstr. 099.048.

Observations photographiques de grosses planètes et de leurs satellites effectuées en 1983 à l'équatorial GPO de 40 cm de l'Observatoire Austral Européen (ESO) à La Silla (Chili).
See Abstr. 099.061.

Deuterium in the outer solar system.
See Abstr. 099.063.

Photochemistry and clouds of Jupiter, Saturn and Uranus.
See Abstr. 099.074.

The composition and origin of the Iapetus dark material.
See Abstr. 100.009.

Dynamics of ring–satellite systems around Saturn and Uranus.
See Abstr. 100.018.

On a cometary family of Uranus.
See Abstr. 102.049.

102 Comets (Origin, Structure, Atmospheres, Dynamics)

102.001   Implications of the presence of S_2 in cometary ice.
P. D. Feldman, M. F. A'Hearn.
NASA Conf. Publ., NASA CP–2349, p. 502 – 503 (1984). – See Abstr. 012.001.

102.002   Formation environment of cometary nuclei in the primordial solar nebula.
T. Yamamoto.
Astron. Astrophys., Vol. 142, No. 1, p. 31 – 36 (1985).

The author investigates the formation environment of comets in the primordial solar nebula. He presents a sublimation sequence for various species of possible constituents of the nuclear ice, which would have condensed on the grain surface in the parent interstellar cloud, by calculating the temperature of grains in the solar nebula. He obtains an allowed range of the nebular temperature in the formation region of cometary nuclei from a condition for retention of the ices of the nuclear composition. Combining this result with models of the solar nebula, he discusses the region for the formation of cometary nuclei in the solar nebula. It is shown that cometary nuclei formed at least beyond the region between the formation regions of Saturn and Uranus. Finally, the author estimates an upper limit of the grain temperature in the region of comet formation at an earlier stage of the solar nebula.

102.003   Prediction of deuterium abundance in comets.
V. Vanýsek, P. Vanýsek.
Icarus, Vol. 61, No. 1, p. 57 – 59 (1985). – See Abstr. 012.004.

Using a chemical scheme based on ion–molecule reactions in cool interstellar clouds, the possibility of a deuterium enrichment of volatile material in comets is discussed. It is assumed that the hydrogen–containing molecules are deuterated before accretion and condensation on the core-mantle dust particles from which the cometesimals are formed. The D/H ratio in comets may be enhanced in respect to the average value by a factor of 10^2. Therefore comets are not promising objects for testing the primordial deuterium abundance.

102.004   Implications of the observed distributions of very long period comet orbits.
R. S. Harrington.
Icarus, Vol. 61, No. 1, p. 60 – 62 (1985).

With allowance for galactic perturbations and observational error, the observed distributions of sizes and orientations of very long period comets are consistent with a uniform distribution of comets within the Oort Cloud.

102.005 Physical characteristics of comets of the years 1976 – 1977.
D. A. Andrienko, V. N. Vashchenko, S. K. Vsekhsvyatskij,
A. V. Karpenko, L. M. Kostenko.
Probl. Kosm. Fiz., Vyp. 19, p. 47 – 58 (1984). In Russian. – See Abstr. 003.002.

102.006 Transplutonian family of long–period comets.
S. K. Vsekhsvyatskij, A. A. Demenko.
Probl. Kosm. Fiz., Vyp. 19, p. 59 – 61 (1984). In Russian. – See Abstr. 003.002.

102.007 Peculiarities of development of the heads of comets at different heliocentric distances in the solar activity cycle.
D. A. Andrienko, V. N. Vashchenko, A. I. Loza.
Probl. Kosm. Fiz., Vyp. 19, p. 61 – 68 (1984). In Russian. – See Abstr. 003.002.

102.008 Role of electric fields in the cometary environment.
T. E. Cravens, T. I. Gombosi, B. E. Gribov,
M. Horanyi, K. Kecskemety, A. Korosmezey, M. L. Marconi,
D. A. Mendis, A. F. Nagy, R. Z. Sagdeev, V. I. Shevchenko,
V. D. Shapiro, K. Szego.
Report KFKI–1984–41, Hungarian Acad. Sci., Budapest, Hungary, 47 pp. (1984). Abstr. in Phys. Abstr., Vol. 88, No. 1248, Entry 9681 (1985).

102.009 The LeBlanc bands of the CN radical in the spectra of comets.
P. D. Singh, A. M. Gomez Balboa, J. A. De Freitas Pacheco.
Astrophys. Space Sci., Vol. 108, No. 1, p. 153 – 160 (1985).

Spontaneous emission rates and absorption oscillator strengths for prominent $\Delta V \leqslant 4$ sequence bands of the $B^2\Sigma^+ - A^2\Pi_i$ transition of the CN molecule are estimated. The wavelengths of some lines observed in the coma spectrum of the comet Bradfield 1980t as well as in several comets coincide with these $\Delta V \leqslant 4$ sequence LeBlanc bands of the CN radical. Formation and destruction of the CN radical in the coma of a comet are discussed in the framework of gas phase reactions.

102.010 La nube di Oort.
P. Tempesti.
Coelum, Vol. 53, N. 1, p. 29 – 33 (1985).

102.011 Physical properties of comets.
P. S. Butterworth.
Vistas Astron., Vol. 27, Part 4, p. 361 – 419 (1984).

In preparation for the imminent arrival at perihelion of Comet Halley in February 1986, many groups have been perfecting their instruments and plans for Halley by observing other periodic comets. Since 1980 this has produced a lot of new information, which it is the principle aim of this paper to review.

102.012 Current ideas on the nature of comets.
J. Rahe.
Cometary astrometry, p. 7 – 13 (1984). – See Abstr. 012.024.

102.013 Comets and nongravitational forces.
B. G. Marsden.
Cometary astrometry, p. 163 – 166 (1984). – See Abstr. 012.024.

A summary is given of the procedure now widely used for the allowance of nongravitational effects in the motions of comets. Some results are mentioned, and an innovative variation of the procedure is briefly discussed.

102.014 On the orbit improvement results obtained by referring old cometary observations to the SAO Star Catalog.
M. Bielicki, G. Sitarski, K. Ziołkowski.
Cometary astrometry, p. 203 – 206 (1984). – See Abstr. 012.024.

An algorithm is presented for the automatic reduction to the SAO Star Catalog of old micrometer cometary observations of the type "comet minus star". Results of a numerical test are presented showing the distinct influence of such a reduction on orbit determination results.

102.015 Numerical simulations of comet nuclei. I. Water–ice comets.
G. Herman, M. Podolak.
Icarus, Vol. 61, No. 2, p. 252 – 266 (1985).

A one–dimensional simulation of pure water–ice cometary nuclei is presented, and the effect of the nucleus as a heat reservoir is considered. The phase transition from amorphous to crystalline ice is studied for two cases: (1) where the released latent heat goes entirely into heating adjacent layers and (2) where the released latent heat goes entirely into sublimation.

102.016 Numerical simulations of comet nuclei. II. The effect of the dust mantle.
M. Podolak, G. Herman.
Icarus, Vol. 61, No. 2, p. 267 – 277 (1985).

The insulating effect of an evolving dust mantle is examined. The role of this mantle in determining the surface temperature of the ice core is studied as a function of the mass fraction of the dust in the ice–dust mixture and the thermal conductivity of the nucleus. Using the so–called "loose lattice" model of D. A. Mendis and G. D. Brin (1977) it was found that both high dust to ice ratios and high core conductivities inhibit mantle blowoff. Indeed, it is often possible to build an essentially permanent dust mantle around an ice nucleus, so that the nucleus will take on an asteroidal appearance.

102.017 A model for the hydrogen coma of a comet.
Y. Kitamura, O. Ashihara, T. Yamamoto.
Icarus, Vol. 61, No. 2, p. 278 – 295 (1985).

A model for the hydrogen coma of a comet on the basis of the Monte Carlo method is presented. In this model isotropic ejections of H atoms produced by photodissociation of H_2O and OH, thermalization of the H atoms due to collisions with ambient H_2O molecules, and the solar radiation pressure have been taken into account. Velocity distribution functions of the H atoms at various positions in the coma, as well as their density and outflow velocity profiles, have been calculated. Lyman α isophotes and its line profiles in the optically thin region are computed by using the velocity distribution function.

102.018 Aeronomical processes in cometary atmospheres: the carbon compounds' puzzle.
M. C. Festou.
Adv. Space Res., Vol. 4, No. 9, p. 165 – 175 (1984). – See Abstr. 012.029.

The deduction of coma abundances, ultimately leading to the abundance of parent molecules escaping the nucleus of a comet, requires to study a very large number of physical phenomena. Those include excitation processes of the emission of the light, production and loss processes of coma species, interaction with the solar wind... etc. In this paper, as an example, the author reviews the knowledge on carbon bearing species and attempts to estimate the C/OH abundance ratio in various comets. He reaches the interesting conclusion that this parameter greatly varies from comet to comet.

102.019 Heterogeneous grain model in comets.
T. Mukai, S. Mukai.
Adv. Space Res., Vol. 4, No. 9, p. 207 – 210 (1984). – See Abstr. 012.029.

Based on the Maxwell–Garnet expression for the optical constant of heterogeneous material, the temperature of grain consisting of homogeneous matrix and small sized impurity can be examined. It is found that the heterogeneous grain model can explain the evidence observed in the comets, i.e. (1) higher production rate of water molecules at large solar distance due to sublimation from water–ice with magnetite inclusions, and (2) elevated color temperature, which frequently coexists with a 10 μm silicate peak, as the thermal emission of obsidian contaminated by small magnetite inclusions.

102.020 A fine mist of very small comet dust particles.
J. M. Greenberg.
Adv. Space Res., Vol. 4, No. 9, p. 211 – 212 (1984). – See Abstr. 012.029.

Very little consideration has been given to the possibility of cometary particles as small as 0.01 µ, radius. However, if comets are aggregated interstellar dust there is good reason to expect that such particles should provide a very significant contribution to the cometary dust spectrum.

102.021 Heterogeneous grain morphologies and acceleration mechanisms in cometary coma dust dynamics: mass envelope dispersion.
W. C. Carey, J. A. M. McDonnell, C. S. Welch, J. C. Zarnecki.
Adv. Space Res., Vol. 4, No. 9, p. 217 – 220 (1984). – See Abstr. 012.029.

102.022 Thermal model and thermo–mechanical stresses in cometary nuclei.
E. Kührt, D. Möhlmann.
Adv. Space Res., Vol. 4, No. 9, p. 225 – 228 (1984). – See Abstr. 012.029.

Spherically symmetric radial temperature profiles of cometary nuclei have been determined numerically (and for simplified models analytically) in dependence on the orbital position of the periodic comet Halley. These temperature fields in the nucleus are connected with thermal stress fields which have been calculated with the assumption of elastic properties of cometary matter. The remarkable result is the possible existence of stresses, strong enough to cause internal cracking of the nucleus and break–ups of the cometary surface.

102.023 Plasma processes in cometary atmospheres.
A. A. Galeev, A. S. Lipatov.
Adv. Space Res., Vol. 4, No. 9, p. 229 – 237 (1984). – See Abstr. 012.029.

The paper reviews the studies of collective plasma processes responsible for anomalously fast exchange of energy between the plasma components, for a structure of the boundaries of the characteristic regions in the zone of the comet interaction with the solar wind, and for the anomalous transport across these boundaries. The position and structure of the cometary bow shock are obtained by numerical simulation using the particles in cells method. To study different plasma instabilities in the cometary atmospheres the simple analytic model of the cometary plasma environment is used that is based on the semikinetic description of the solar wind loading by cometary ions. The influence of a self–consistent electric field on the cometary plasma outflow is briefly discussed.

102.024 Comet–solar wind interactions: a dusty point of view.
W.–H. Ip.
Adv. Space Res., Vol. 4, No. 9, p. 239 – 247 (1984). – See Abstr. 012.029.

Anticipating the new results from the space missions to Comet Halley and Comet Giacobini–Zinner, the author makes a brief review of recent theoretical and observational studies of dust–plasma environment. In order to relate different disciplines in cometary research in the context of comet–solar wind interaction, two separate issues: (1) surface processes and (2) plasma processes are considered to indicate how various kinds of observations of cometary dust comas and tails may be used to infer the conditions of solar wind – comet interaction and the corresponding plasma processes in the cometary ionospheres and ion tails (and vice–versa).

102.025 Energetic positive ions in the cometary "foreshock" region.
V. Formisano, E. Amata.
Adv. Space Res., Vol. 4, No. 9, p. 253 – 254 (1984). – See Abstr. 012.029.

102.026 Searching for cometary parent molecules at radio wavelengths.
J. Crovisier.
Astron. J., Vol. 90, No. 4, p. 670 – 674 (1985).

Most cometary parent molecules can only be studied through their rotational lines in the microwave or submillimeter range, or through their vibrational bands in the medium infrared. Recent radio observations of these molecules are reviewed. Their results can only be interpreted if detailed excitation models are worked out. The prevailing excitation mechanisms are thermal excitation by collisions in the inner coma, and infrared excitation of the fundamental bands of vibration followed by fluorescence in the outer coma. The dynamical evolution of molecular excitation must be followed when the molecules expand from the nucleus into the coma. The cases for H_2O, CO, HCN, and NH_3 molecules are reviewed in more detail and future prospects for radio spectroscopy of comets are indicated.

102.027 The aphelion anisotropy of long–period comets.
T. P. Ray.
Ir. Astron. J., Vol. 17, No. 1, p. 58 – 60 (1985). – See Abstr. 012.035.

102.028 Statistical catalogue of the parameters of orbits of nearly parabolic comets in Laplacian coordinates.
V. V. Radzievskij, V. P. Tomanov.
Vologod. gos. ped. inst. Vologda, 169 pp. (1984). In Russian. Abstr. in Ref. Zh., 51. Astron., 1.51.118 (1985).

102.029 Secular brightness decrease and structure of the nuclei of periodic comets.
O. V. Dobrovol'skij, Kh. I. Ibadinov, S. I. Gerasimenko.
Dokl. Akad. Nauk TadzhSSR, Tom 27, No. 4, p. 189 – 200 (1984). In Russian. Abstr. in Ref. Zh., 51. Astron., 2.51.219 (1985).

102.030 On the radiation of comets in the X–ray emission of the sun.
S. Ibadov.
Dokl. Akad. Nauk TadzhSSR, Tom 27, No. 5, p. 258 – 261 (1984). In Russian. Abstr. in Ref. Zh., 51. Astron., 2.51.220 (1985).

102.031 Explanation of the distributions of short–period comets with respect to angular orbital elements.
A. M. Kazantsev, L. M. Sherbaum.
Sov. Astron., Vol. 28, No. 4, p. 438 – 441 (1984). English translation of 38.102.001.

102.032 Magnetic field penetration into a cometary ionosphere.
Z. M. Ioffe.
Sov. Astron., Vol. 28, No. 4, p. 474 – 475 (1984). English translation of 38.102.002.

102.033 Dynamics of comets.
A. Manara.
Mem. Soc. Astron. Ital., Vol. 55, No. 3, p. 503 – 506 (1984). – See Abstr. 012.049.

Local work review and work programs at the Milano Observatory as well as up–to–date theories on the origin of comets are reported.

102.034 The Tisserand constants of short–period comets.
V. V. Radzievskij, V. P. Tomanov.
Gor'k. gos. ped. inst. Gor'kij, 12 pp. (1984). In Russian. Abstr. in Ref. Zh., 51. Astron., 4.51.221 (1985).

102.035 On the vectorial model of cometary atmospheres.
V. A. Dranevich.
Sov. Astron. Lett., Vol. 10, No. 5, p. 297 – 299 (1984). English translation of 38.102.020.

102.036 Wind instability and the helical comet–tail structures.
S. G. Gestrin, V. M. Kontorovich.
Sov. Astron. Lett., Vol. 10, No. 5, p. 329 – 331 (1984). English translation of 38.102.021.

102.037 Astrophysical applications of a laboratory investigation of solidity of matrices.
I. S. Lizunkova.
Astron. Tsirk., No. 1343, p. 3 – 6 (1984). In Russian.
This paper presents an application of a laboratory investigation on cometary dust.

102.038 Kometen – eine Übersicht.
H.–M. Hahn.
Orion, 43. Jahrg., Nr. 208, p. 76 – 81 (1985).

102.039 Comets and their origin.
R. D. Chapman, J. C. Brandt.
Mercury, Vol. 14, No. 1, p. 2 – 8, 30 (1985).

102.040 On the generation of hot plasma and short–wave radiation in comets.
S. Ibadov.
Astron. Tsirk., No. 1353, p. 1 – 2 (1984). In Russian.

102.041 On a version of the hypothesis of interstellar origin of comets.
A. S. Guliev.
Kinematika Fiz. Nebesn. Tel, Tom 1, No. 3, p. 7 – 12 (1985). In Russian.
The modification by V. V. Radzievskij and V. P. Tomanov of the Laplace hypothesis on the origin of long–period comets is criticized. Some arguments in favour of this hypothesis are shown to be the results of insufficiently strict statistical approach to observational data.

102.042 On diffusion of comets by the Tisserand constant.
Yu. G. Babenko.
Kinematika Fiz. Nebesn. Tel, Tom 1, No. 3, p. 13 – 16 (1985). In Russian.
The existence of diffusion of comets by the Tisserand constant is shown from the study of the orbital evolution of both real and hypothetic comets. It is noted that the diffusion is significant only for comets with Tisserand constants exceeding 0.450, but is negligible in other cases.

102.043 The fluorescence spectrum of the CN radical in comets.
J. M. Zucconi, M. C. Festou.
Inst. Astrophys. Paris, Pré–Publ., No. 96, 56 pp. (1985). To appear in Astron. Astrophys.

102.044 The surface brightness of Na D emission in comets.
S. Wyckoff, S. Konno.
Bull. Am. Astron. Soc., Vol. 16, No. 4, p. 923 (1984). Abstract. – See Abstr. 010.062.

102.045 Dynamical influence of molecular cloud encounters on the Oort cloud of comets.
M. V. Torbett, R. Smoluchowski, K. D. Borne.
Bull. Am. Astron. Soc., Vol. 16, No. 4, p. 923 (1984). Abstract. – See Abstr. 010.062.

102.046 The magnetic field in outer parts of cometary atmospheres.
I. S. Veselovskij.
Issled. Geomagn., Aehron. Fiz. Solntsa, Moskva, No. 69, p. 122 – 126 (1984). In Russian. Abstr. in Ref. Zh., 62. Issled. Kosm. Prostranstva, 5.62.401 (1985).

102.047 Time–dependent dusty gasdynamical flow near cometary nuclei.
T. I. Gombosi, T. E. Cravens, A. F. Nagy.
Astrophys. J., Vol. 293, No. 1, p. 328 – 341 (1985).
This paper presents time–dependent solutions to the coupled dusty hydrodynamics equations describing the spherically symmetric expansion of cometary neutral gas in the vicinity of a cometary nucleus. The sublimation process is represented by gas outflow from a dust–covered reservoir containing stationary gas whose pressure and density values are determined by the sublimating and surface temperatures. The time evolution of a cometary outburst is modeled. It is found that, as a result of the strong gas–dust interaction in the inner coma region, a "slow" disturbance in both the dust and gas parameters will be created in addition to the familiar gas blast–wave solution.

102.048 On the structure of the cometary nucleus.
B. Donn, P. A. Daniels, D. W. Hughes.
Bull. Am. Astron. Soc., Vol. 17, No. 1, p. 520 (1985). Abstract. – See Abstr. 010.064.

102.049 On a cometary family of Uranus.
V. P. Tomanov.
Tr. Astrofiz. Inst. Alma–Ata, Tom 40, p. 98 – 103 (1983). In Russian.

102.050 The interval of observations of comets as a criterion of probability of their discovery and modern cometary statistics.
V. V. Radzievskij, M. A. Mamedov, M. L. Ivanov.
Astron. Zh., Tom 62, Vyp. 3, p. 580 – 589 (1985). In Russian. English translation in Sov. Astron., Vol. 29, No. 3.
It is shown that the interval of observation of comets is the best criterion of their discovery; this criterion having a number of advantages over Everhart's one. Taking into account the probability of discovery of comets, a considerable prevalence of long–period comets with retrograde motion is established. Three possible interpretations of this phenomenon are suggested. The intervals between observations are used for a statistical analysis of the comets' distribution over the longitude of perihelion and the perihelion distance.

102.051 The icy halo of a comet and the temperature of the inner coma.
D. V. Bisikalo, V. S. Strel'nitskij.
Pis'ma Astron. Zh., Tom 11, No. 6, p. 475 – 480 (1985). In Russian. English translation in Sov. Astron. Lett., Vol. 11.
A spherically–symmetric hydrodynamical model of the inner cometary coma consisting of H_2O vapour and H_2O icy grains ejected from the nucleus is calculated for a comet of medium gas production rate (like the Halley comet) at the heliocentric distance of 1 AU.

102.052 Plasmaphysikalische Prozesse bei der Wechselwirkung von Kometen mit dem Sonnenwind.
T. Roatsch, M. Danz, K. Sauer.
Sterne, 61. Band, Heft 2, p. 83 – 93 (1985).

102.053 Report of IAU Commission 15: Physical study of comets, minor planets, and meteorites (L'étude physique des comètes, des petites planètes et des météorites).
C. R. Chapman.
Trans. IAU, Vol. XIXA, p. 167 – 187 (1985). – See Abstr. 003.046.

102.054 Dust properties determined from backscattering in the interplanetary and interstellar medium.
R. V. Myers.
Diss. Abstr. Int., Sect. B, Vol. 45, No. 5, p. 1501 (1984). Thesis, University of Wisconsin, 135 pp. (1984). Order No. DA8413266.

102.055 The fluorescence of cometary OH and CN.
D. G. Schleicher.
Diss. Abstr. Int., Sect. B, Vol. 45, No. 7, p. 2200 (1985). Thesis, University of Maryland, 205 pp. (1983). Order No. DA8421259.

102.056 Quasi–resonance recharging of negative hydrogen ions and positive sodium ions in cometary atmospheres.
V. I. Cherednichenko.
Komet. Tsirk., No. 337 (1985). In Russian.

102.057 Rendezvous effects in cometary statistics.
V. V. Radzievskij.
Komet. Tsirk., No. 339 (1985). In Russian.

102.058 On the role of quasi–resonance recharging of cometary molecules of nitrogen and helium ions with solar wind particles in cometary tails of the N_1^+ infrared bands.
V. I. Cherednichenko.
Komet. Tsirk., No. 340 (1985). In Russian.

102.059 Method for the determination of the type of molecule source in cometary atmospheres.
V. A. Dranevich.
Komet. Tsirk., No. 340 (1985). In Russian.

102.060 On the nature of condensations of cometary nuclei.
I. S. Lizunkova, E. A. Kajmakov.
Komet. Tsirk., No. 341 (1985). In Russian.

102.061 Calculation of the intensity of dust release from the nucleus of a comet.
M. Z. Markovich.
Komet. Tsirk., No. 342 (1985). In Russian.

102.062 Models of CN and C_2 comae for comet Tago–Sato–Kosaka and comet Bennett.
Z.–w. Hu, S.–d. Chen, J.–n. Zhao.
Chin. Astron. Astrophys., Vol. 9, No. 1, p. 86 – 90 (1985). English translation of Chin. J. Space Sci., Vol. 4, Nos. 3/4, p. 331 – 337 (1984).

Based on the CN and C_2 comae isophotes for the two comets given by Rahe et al. and the relevant theory of physical chemistry, the authors have deduced the distributions of the CN and C_2 molecules in the coma, their scale heights and mean lifetimes. The results favor the viewpoint that HCN is the parent of CN, and that C_2H_2 is the parent of C_2.

102.063 About the interaction between high speed wind streams from coronal holes and comets.
A. Sanchez.
IHW Newsl., No. 7, p. 2 – 4 (1985).

This paper tries to present some considerations about a possible observational study of solar wind streams and comets with a specific model for the next approach of Halley's Comet.

102.064 Comet formation in molecular clouds.
S. V. M. Clube, W. M. Napier.
Icarus, Vol. 62, No. 3, p. 384 – 388 (1985).

The observed properties of the long–period comet system, and its periodic disturbance by galactic forces manifesting as terrestrial impact episodes, may be indicative of a comet capture/escape cycle as the Solar System orbits the Galaxy. A mean number density of comets in molecular clouds of $\sim 10^{-1\pm1} AU^{-3}$ is implied. This is sufficient to deplete metals from the gaseous component of the interstellar medium, as observed, but leads to the problem of how stars are formed nevertheless with solar metal abundances. Formation of comets *prior* to stars in dense systems of near–zero energy may be indicated, and isotope signatures in cometary particles may be diagnostic of conditions in young spiral arm material.

Bibliography of Earth, Moon, and Planets.
See Abstr. 002.007.

Bibliography of Earth, Moon, and Planets.
See Abstr. 002.031.

Kometen und Biosphäre bei Isaac Newton.
See Abstr. 004.202.

Asteroids, comets, meteors II.
See Abstr. 011.023.

A potential extraterrestrial species: the SH^+ molecular ion.
See Abstr. 022.009.

Radiation chemical experiments relevant to studies of cometary nuclei: remarks on working conditions.
See Abstr. 022.049.

Simulation of cosmic dust spectra.
See Abstr. 022.124.

The triplet–singlet $\tilde{a}^3 A''-\tilde{X}^1 A'$ emission spectra of hydroxy-benzaldehydes and a comparison with CO^+ comet tail spectra.
See Abstr. 022.139.

Reaction of O^+, CO^+, H^+, and CH_5^+ ions with atomic hydrogen.
See Abstr. 022.140.

Construction of periodic orbits and problems of capture.
See Abstr. 042.054.

Numerical model of time–dependent hydrodynamic flows with mass loading.
See Abstr. 062.063.

Numerical modelling of the interaction between the solar wind and cometary plasma.
See Abstr. 074.045.

Solar wind stagnation near comets.
See Abstr. 074.047.

Solar wind deceleration at the periphery of a cometary cloud.
See Abstr. 074.112.

Large body impacts through geologic time.
See Abstr. 081.051.

Extraterrestrial ice.
See Abstr. 091.015.

Infrared observations of the extinct cometary candidate minor planet (3200) 1983 TB.
See Abstr. 098.053.

Could an asteroid be a comet in disguise?
See Abstr. 098.086.

Asteroids – the comet connection.
See Abstr. 098.089.

Can comets become asteroids?
See Abstr. 098.091.

Report of IAU Commission 20: Positions and motions of minor planets, comets and satellites (*Positions et mouvements des petites planètes, des comètes et des satellites*).
See Abstr. 098.098.

Determination of maximum likelihood orbital parameters using parameters otbained for separate apparitions.
See Abstr. 098.105.

Periodic comet showers and planet X.
See Abstr. 101.002.

C_2 photolytic processes in cometary comae.
See Abstr. 103.020.

Models of cometary emission in the 18–cm OH transitions: the predicted behavior of comet Halley.
See Abstr. 103.936.

Chondrites from comets?
See Abstr. 105.194.

Interplanetary field enhancements in the solar wind: evidence for cometesimals at 0.72 and 1.0 AU?
See Abstr. 106.029.

Asteroids and comets.
See Abstr. 107.002.

Asteroids, comets, and planet formation.
See Abstr. 107.005.

A lower limit on cosmic rays flux from complex molecules in cometary comae.
See Abstr. 144.042.

103 Comets (Individual Objects)

103.001 Comets in 1976.
B. G. Marsden, D. W. E. Green, E. Roemer.
Q. J. R. Astron. Soc., Vol. 26, No. 1, p. 68 – 80 (1985).

103.002 Comets in 1977.
B. G. Marsden, D. W. E. Green.
Q. J. R. Astron. Soc., Vol. 26, No. 1, p. 81 – 91 (1985).

103.003 Comets in 1978.
B. G. Marsden, D. W. E. Green.
Q. J. R. Astron. Soc., Vol. 26, No. 1, p. 92 – 105 (1985).

103.004 Comets in 1979.
B. G. Marsden.
Q. J. R. Astron. Soc., Vol. 26, No. 1, p. 106 – 114 (1985).

103.005 Comet Digest.
J. E. Bortle.
Sky Telesc., Vol. 69, No. 1, p. 88 (1985).
Concerning comets: 1984m P/Schaumasse, 1984t Levy-Rudenko.

103.006 Comet Digest.
J. E. Bortle.
Sky Telesc., Vol. 69, No. 2, p. 187 (1985).
Concerning comets: 1982i P/Halley, 1984i Austin, 1984k P/Arend-Rigaux, 1984m P/Schaumasse, 1984q P/Shoemaker, 1984s Shoemaker, 1984t Levy-Rudenko.

103.007 Photometric observations of long–period comets at large heliocentric distances in the years 1927 to 1955.
J. Svoreň.
Contrib. Astron. Obs. Skalnaté Pleso, Vol. 12, p. 7 – 44 (1984).
This paper is a continuation of the published list of photometric observations of long–period comets (see 37.103.010) and contains a list of photometric observations of 22 long–period comets which passed through perihelion in the years 1927 to 1955. 1176 estimates and brightness measurements of comets are given together with time, data on the magnitude type, type, diameter of objective and light–gathering power of the instrument used, references to literature and calculated values of heliocentric and geocentric distances, as well as the phase angles for the dates of observations.

103.008 Comet observations made at the Skalnaté Pleso Observatory in the years 1972 – 1975.
M. Antal, J. Svoreň, E. M. Pittich.
Contrib. Astron. Obs. Skalnaté Pleso, Vol. 12, p. 75 – 98 (1984).
The paper presents the results of position photographing of comets carried out at the Skalnaté Pleso Observatory in the years 1972 – 1975. 252 observations of 14 comets are given together with the list of reference stars and dependences.

103.009 Determination of photometric parameters of long–period comets at large heliocentric distances. I. Comets observed in the years 1861 – 1946.
J. Svoreň.
Contrib. Astron. Obs. Skalnaté Pleso, Vol. 12, p. 133 – 164 (1984).
Using compiled photometric observations of long–period comets, their photometric parameters at large distances from the Sun were determined.

103.010 Tabulation of comet observations.
Int. Comet Q., Vol. 7, No. 1, p. 17 – 23 (1985).
Concerning comets: 1983n P/Crommelin, 1983v P/Hartley-IRAS, 1983w P/Clark, 1984c P/Neujmin 1, 1984g P/Wolf-Harrington, 1984h P/Faye, 1984i Austin, 1984j P/Takamizawa, 1984k P/Arend-Rigaux, 1984m P/Schaumasse, 1984o Meier, 1984q P/Shoemaker 1, 1984r Shoemaker, 1984s Shoemaker, 1984t Levy-Rudenko.

103.011 Recent news and research concerning comets.
D. W. E. Green.
Int. Comet Q., Vol. 7, No. 1, p. 23 – 25 (1985).
Concerning comets: 1982i P/Halley, 1984i Austin, 1984q P/Shoemaker 1, 1984r Shoemaker, 1984s Shoemaker, 1984t Levy-Rudenko, 1984u P/Shoemaker 2, 1984v Hartley.

103.012 Comets for the visual observer in 1985.
A. Hale.
Int. Comet Q., Vol. 7, No. 1, p. 25 – 28 (1985).
Concerning comets: 1974 II P/Schwassmann-Wachmann 1, 1974 XVI P/Honda-Mrkos-Pajdusakova, 1975 I P/Boethin, 1978 XIV P/Ashbrook-Jackson, 1978 XX P/Haneda-Campos, 1982i P/Halley, 1984e P/Giacobini-Zinner, 1984f Shoemaker, 1984g P/Wolf-Harrington, 1984k P/Arend-Rigaux, 1984m P/Schaumasse, 1984p P/Tsuchinshan 1.

103.013 Comet Digest.
J. E. Bortle.
Sky Telesc., Vol. 69, No. 3, p. 285 (1985).
Concerning comets: 1982i P/Halley, 1984t Levy-Rudenko.

103.014 Comet Digest.
J. E. Bortle.
Sky Telesc., Vol. 69, No. 4, p. 376 (1985).
Concerning comets: 1982i P/Halley, 1984k P/Arend-Rigaux, 1984m P/Schaumasse, 1984p P/Tsuchinshan 1, 1984t Levy-Rudenko.

103.015 Ultraviolet spectroscopy of cometary comae: an update.
P. D. Feldman.
Adv. Space Res., Vol. 4, No. 9, p. 177 – 184 (1984). – See Abstr. 012.029.
Since its launch in 1978 the International Ultraviolet Explorer satellite observatory has been used to record ultraviolet spectra of nearly two dozen comets. These observations have been applied principally to studies of the composition, chemistry and evolution of the gaseous coma and more recently, with the substantially increased data base, to comparative analyses.

103.016 A new calibration of the semi–empirical photometric theory for Halley and other comets.
R. L. Newburn Jr.
Adv. Space Res., Vol. 4, No. 9, p. 185 – 188 (1984). – See Abstr. 012.029.
A massive new body of data, comprised of spectrophotometry of 17 comets by Newburn and Spinrad, has now become available for calibration of the basic theory, the semi–empirical photometric theory, used for modelling of Comet Halley. A redetermination of the constant R and the function δ has been made, and no change is needed. A new light curve for Halley and a lower dust to gas ratio prove to make roughly compensating changes in dust densities.

103.017 A comparison of the dust properties in recent periodic comets.
M. S. Hanner.
Adv. Space Res., Vol. 4, No. 9, p. 189 – 196 (1984). – See Abstr. 012.029.
Measurements of the thermal emission from the cometary dust coma can be used to derive the rate of dust production from the nucleus as well as the size distribution of absorbing grains. More than ten short–period comets have now been observed in the infrared over a wide range in heliocentric distance. Dust production rates are derived for these comets based on theoretical models of the thermal emission from small absorbing grains and calculations of dust grain velocities. The mean size and albedo of the dust grains is similar in these comets, with the exception of Comet Crommelin, which seems to have had larger, darker grains.

103.018 IRAS observations of cometary dust.
R. G. Walker, H. H. Aumann.
Adv. Space Res., Vol. 4, No. 9, p. 197 – 201 (1984). – See Abstr. 012.029.

The large beam size of the Infrared Astronomy Satellite (IRAS) focal plane detector array is well suited to measuring the low level thermal emission from cometary dust. Eight comets discovered in 1983 and nine previously known periodic comets were observed by IRAS during its ten month active lifetime. Dust production rates are derived for a wide range of heliocentric distances. Grain properties are inferrred from application of simple models to the long wavelength spectral energy distribution.

103.019 Recent infrared observations of comets with UKIRT and IRAS.
J. C. Zarnecki, N. Eaton, J. A. M. McDonnell, A. J. Meadows, W. C. Carey, G. H. MacDonald.
Adv. Space Res., Vol. 4, No. 9, p. 203 – 206 (1984). – See Abstr. 012.029.

A preliminary analysis of infrared observations of comets P/Crommelin and P/Tempel 1 is presented. Comet P/Crommelin was observed from UKIRT over the range 1 – 20 micron, using standard filters. From the shape of the thermal emission spectrum, the temperature of the dust grains is estimated and also the dust production rate. Comet P/Tempel 1 was observed with the Infrared Astronomical Satellite (IRAS). The emission is found to be considerably extended and there is also evidence for temperature variation of the dust grains as indicated by the 12 to 25 micron flux ratio.

103.020 C_2 photolytic processes in cometary comae.
A. L. Cochran.
Astrophys. J., Vol. 289, No. 1, p. 388 – 391 (1985).

The authors present observations of C_2 column densities of comets Tuttle 1980h and Meier 1980q. Nonequilibrium chemical models are fitted to the data. It is shown that new rates for the photolytic reactions affecting the creation and destruction of C_2 are necessary. These new rates are: $[C_2H + h\nu \rightarrow C_2 + H]$ $5 \times 10^{-4} s^{-1}$ for creation and $[C_2 + h\nu \rightarrow C + C]$ $4 \times 10^{-6} s^{-1}$ for destruction. Other possible methods for altering the photometric profiles of C_2 are discussed.

103.021 Tabulation of comet observations.
Int. Comet Q., Vol. 7, No. 2, p. 48 – 66 (1985).

Concerning comets: 1961 VIII Seki, 1962 III Seki–Lines, 1963 I Ikeya, 1963 III Alcock, 1964 VIII Ikeya, 1965 VIII Ikeya–Seki, 1966 V Kilston, 1967 II Rudnicki, 1968 I Ikeya–Seki, 1968 VI Honda, 1969 IX Tago–Sato–Kosaka, 1970 II Bennett, 1971 V Toba, 1973 XII Kohoutek, 1974 II P/Schwassmann–Wachmann 1, 1974 III Bradfield, 1975 IX Kobayashi–Berger–Milon, 1975 X Suzuki–Saigusa–Mori, 1975 XII Mori–Sato–Fujikawa, 1976 VI West, 1976 XI P/d'Arrest, 1979 X Bradfield, 1980 XI P/Encke, 1980 XIII P/Tuttle, 1980 XV Bradfield, 1981 II Panther, 1982i P/Halley, 1983 XIII P/Kopff, 1983n P/Crommelin, 1983v P/Hartley–IRAS, 1983w P/Clark, 1984c P/Neujmin, 1984f Shoemaker, 1984g P/Wolf–Harrington, 1984h P/Faye, 1984i Austin, 1984j P/Takamizawa, 1984k P/Arend–Rigaux, 1984m P/Schaumasse, 1984p Tsuchinshan 1, 1984q P/Shoemaker 1, 1984s Shoemaker, 1984t Levy–Rudenko.

103.022 Recent news and research concerning comets.
D. W. E. Green.
Int. Comet Q., Vol. 7, No. 2, p. 67 (1985).

Concerning comets: 1982i P/Halley, 1984k P/Arend–Rigaux, 1984m P/Schaumasse, 1984p P/Tsuchinshan 1, 1984s Shoemaker, 1984t Levy–Rudenko.

103.023 Comet Digest.
J. E. Bortle.
Sky Telesc., Vol. 69, No. 5, p. 473 (1985).

Concerning comets: 1982i P/Halley, 1984e P/Giacobini–Zinner, 1984f Shoemaker.

103.024 Comets for 1985.
J. V. Scotti.
Strolling Astron., Vol. 31, Nos. 1 – 2, p. 26 – 28 (1985).

103.025 Photographic observations of comets.
G. De Sanctis, W. Ferreri, V. Zappalà.
Acta Astron., Vol. 34, No. 3, p. 395 – 396 (1984).

66 positions of four comets (1983d IRAS–Araki–Alcock, 1982j Tempel 1, 1982k Kopff, 1982 VIII Churyumov–Gerasimenko) are given as obtained from photographic observations made at the Observatory of Torino from October 1982 to June 1983.

103.026 Die 1983 und 1984 entdeckten Kometen.
R. Lukas.
Sterne Weltraum, 24. Jahrg., Nr. 5, p. 273 (1985).

103.027 Comet Digest.
J. E. Bortle.
Sky Telesc., Vol. 69, No. 6, p. 578 (1985).

Concerning comets: 1984e P/Giacobini–Zinner, 1984t Levy–Rudenko.

103.028 Roman numeral designations of comets in 1983.
Minor Planet Circ., Nos. 9389 – 9390 (1985).

103.029 Observations of comets.
Minor Planet Circ., Nos. 9316 – 9319, 9390 – 9393, 9438 – 9442, 9514 – 9516, 9602 – 9605 (1985).

Observations made at the following stations are published: Belgrade, Brescia, Bulgarian Natl. Obs., Burlington, N.J., Calar Alto, Caussols, Centro Astron. Yebes, Chamberlin Obs., Colchester, Conder Brow, Crimean Astrophys. Obs., Eastfield, Engelhardt Obs. Zelenchukskaya Stn., ESO, Geisei, Heidelberg–Königstuhl, Hemingford Abbots, Kambah (Canberra), Kitt Peak, Kleť, Lick Obs., Lowell Obs. Anderson Mesa Stn., McDonald Obs., Mt. John Obs., Mt. Palomar, Oak Ridge Obs., Oss. Chaonis, Perth, Purple Mountain Obs., S. Vittore (Bologna), Sanglok, Scheuren Obs. (Leverkusen), Sendai Obs. Ayashi Stn., Shokin Majdanak, South Wonston, Stakenbridge, Tautenburg, Tokyo Obs. Kiso Stn., Victoria, Woolston Obs., Yunnan, Zimmerwald.

Observations of the following comets are published: 1975 VIII Lovas, 1975 XII Mori–Sato–Fujikawa, 1976 IV Bradfield, 1976 V Bradfield, 1977 VII P/Gehrels 3, 1977 IX West, 1977 XIV Kohler, 1978 VII Bradfield, 1978 XVIII Bradfield, 1978 XX P/Haneda–Campos, 1979 X Bradfield, 1981 II Panther, 1981 XIX P/Swift–Gehrels, 1982 I Bowell, 1982 VIII P/Churyumov–Gerasimenko, 1982 X P/Gunn, 1982i P/Halley, 1983 II P/Bowell–Skiff, 1983 VII IRAS–Araki–Alcock, 1983 X P/Tempel 2, 1983 XI P/Tempel 1, 1983 XII Černis, 1983 XIII P/Kopff, 1983 XIV P/IRAS, 1983 XV Shoemaker, 1983 XVI IRAS, 1983h P/Johnson, 1983m P/Wolf, 1983n P/Crommelin, 1983s P/Wild 2, 1983w P/Clark, 1984c P/Neujmin 1, 1984e P/Giacobini–Zinner, 1984f Shoemaker, 1984g P/Wolf–Harrington, 1984h P/Faye, 1984i Austin, 1984j P/Takamizawa, 1984k P/Arend–Rigaux, 1984m P/Schaumasse, 1984p P/Tsuchinshan 1, 1984q P/Shoemaker 1, 1984r Shoemaker, 1984s Shoemaker, 1984t Levy–Rudenko, 1984u P/Shoemaker 2, 1984v Hartley, 1985a P/Ashbrook–Jackson, 1985b P/Russell 1.

103.030 Orbital elements of comets.
Minor Planet Circ., Nos. 9315 – 9716 (1985).

The comets are listed according to their Roman numeral designation or preliminary designation. The names of the authors are given behind the respective M.P.C. numbers.
1982 I Bowell 9425 B. G. Marsden; 1984e P/Giacobini–Zinner 9351 D. K. Yeomans; 1984f Shoemaker 9426, 1984i Austin 9425, 1984q P/Shoemaker 1 9425, 1984r Shoemaker 9425, 9685, 1984s Shoemaker 9425, 1984t Levy–Rudenko 9351, 9425, 9685, 1984u P/Shoemaker 2 9351, 1984v Hartley 9351, 9426, 9685 B. G. Marsden.

103.031 Ephemerides of comets.
Minor Planet Circ., Nos. 9361 – 9363, 9431 – 9432, 9596 – 9598, 9691 – 9695 (1985).
Concerning ephemerides of the following comets:
1975 I P/Boethin, 1978 I P/Schuster, 1978 VIII P/Whipple, 1978 XII P/Daniel, 1978 XXII P/Giclas, 1979 IV P/Holmes, 1979 V P/Russell 1, 1980 XI P/Encke, 1982 I Bowell, 1982 X P/Gunn, 1982i P/Halley, 1983s P/Wild 2, 1984c P/Neujmin 1, 1984r Shoemaker, 1984s Shoemaker, 1984t Levy–Rudenko, 1984u P/Shoemaker 2, 1984v Hartley.

103.032 Comets in 1980.
B. G. Marsden.
Q. J. R. Astron. Soc., Vol. 26, No. 2, p. 156 – 167 (1985).

103.033 Recoveries of periodic comets.
Br. Astron. Assoc. Circ., No. 648 (1985).
Concerning comets: 1985a P/Ashbrook–Jackson, 1985b P/Russell 1, 1985c P/Honda–Mrkos–Pajdušaková, 1985d P/Tsuchinshan 2.

103.034 Magnitudes of comets.
Yamamoto Circ., No. 2035 (1985). In Japanese.
Concerning comets: 1984s Shoemaker, 1984f Shoemaker, 1984p P/Tsuchinshan 1, 1984g P/Wolf–Harrington.

103.035 On predicting and analyzing comet light curves.
M. Festou.
IHW Newsl., No. 6, p. 21 – 26 (1985).

103.036 Conversion of data on cometary tail lengths from old Chinese records to angular measure.
R. Podstanická.
Bull. Astron. Inst. Czech., Vol. 36, No. 3, p. 186 – 188 (1985).
The meaning of the units in which the apparent angular lengths of cometary tails are expressed in the ancient and mediaeval Chinese records, is discussed. Conversion factors to the present angular measure are estimated by comparing the distribution of tail lengths in the 1st to 16th century Chinese records with that of the 19th century European naked–eye data.

103.037 Osservazioni delle comete Austin 1982g e IRAS–Araki– Alcock 1983d.
A. Milani.
Astronomia, N. 2, p. 11 – 16 (1985).

103.038 Observations of comets in Zelenchukskaya.
Komet. Tsirk., No. 334 (1984). In Russian.
Concerning the comets: 1984i Austin and 1984g Wolf– Harrington.

103.039 Observations of comets in Kleť.
Komet. Tsirk., No. 334 (1984). In Russian.
Concerning the comets: 1984f Shoemaker and 1984q Shoemaker.

103.040 Observations of comets in Zelenchukskaya.
Komet. Tsirk., No. 336 (1985). In Russian.
Concerning the comets: P/Shoemaker 1 (1984q) and P/Schaumasse (1984m).

103.041 Observations of comets in Zelenchukskaya.
Komet. Tsirk., Nos. 337, 341 (1985). In Russian.
Concerning the comets: Faye (1984h), Arend–Rigaux (1984k), and Schaumasse (1984m).

103.042 Observations of comets in Kiev.
Komet. Tsirk., No. 341 (1985). In Russian.
Concerning the comets: 1984i Austin and 1984f Shoemaker.

103.043 IHW: astrometry.
D. K. Yeomans, R. M. West, R. S. Harrington, B. G. Marsden.
IHW Newsl., No. 7, p. 18 – 19 (1985).
Contents: Recent Halley and Giacobini–Zinner observations received. Old Halley data being re–reduced. Halley and Giacobini–Zinner orbit and ephemeris updates. Astrometric observations of comet Giacobini–Zinner requested.

103.044 IHW: infrared studies.
R. Knacke.
IHW Newsl., No. 7, p. 19 – 21 (1985).

103.045 IHW: near–nucleus studies.
S. Larson, Z. Sekanina, J. Rahe.
IHW Newsl., No. 7, p. 21 – 23 (1985).

103.046 IHW: spectroscopy and spectrophotometry.
S. Wyckoff, P. A. Wehinger, B. Boothman.
IHW Newsl., No. 7, p. 24 – 29 (1985).
Contents: Update on comet P/Halley (1982i). Update on comet P/Giacobini–Zinner (1984e).

103.047 Predictions of occultations of stars by comet Giacobini– Zinner and prediction updates for comet Halley.
E. Bowell, L. H. Wasserman.
IHW Newsl., No. 7, p. 30 – 37 (1985).

Periodic comet Crommelin

103.101 The orbit of periodic comet Crommelin between the years 1000 and 2100.
M. C. Festou, B. Morando, P. Rocher.
Astron. Astrophys., Vol. 142, No. 2, p. 421 – 429 (1985).
The evolution of the orbit of comet P/Crommelin is investigated over a period of 11 centuries. It is shown that the present orbit is very stable and must have existed for still longer periods of time. A capture by a giant planet of a comet coming from the Oort cloud seems an unlikely mechanism for explaining the presently observed orbit of the comet. It is assumed that the comet was once very active, very bright, and that the action of non-gravitational forces on its nucleus decreased its aphelion distance from a very large value to its present value of about 18 AU. During the course of this long evolutionary process, the comet lost considerable amounts of dust and volatile material. The nucleus of comet Crommelin has probably today a radius of at least 1.5 km and only a small fraction of its surface is ice/snow covered.

103.102 A search for 18–cm OH emission from comet Crommelin.
W. T. S. Deich, J. M. Cordes, Y. Terzian.
Astron. J., Vol. 90, No. 2, p. 373 – 374 (1985).
Observations were made at 18 cm of the periodic comet Crommelin on January 30 – February 1, and February 9 – 12, 1984 with the 305–m Arecibo antenna (beam width ~ 3.2 arcmin). No detection of the OH 1665– and 1667–MHz lines was made above a 3σ limit of 9 mJy. The authors also did not detect any continuum emission at 18 cm above a 3σ level of 5 mJy.

103.103 Spectrophotometric observations of comet P/Crommelin (1983n).
M. C. Festou, M. Dennefeld, E. Maurice, J. Bouvier.
Astron. Astrophys., Vol. 144, No. 1, p. L5 – L8 (1985).
Observations of comet P/Crommelin were conducted from February 28 till April 4, 1984 and permitted the evaluation of the gas production rates of the main gaseous species emitting in the optical window over a five week period. The mean variation of the gaseous output of the comet with the heliocentric distance (post perihelion branch of the orbit) varied approximately as $r^{-4.5}$, a value often found in comets having accomplished a large number of revolutions around the sun.

103.104 La comète P/Crommelin (1983n).
M. Festou.
Astronomie, Vol. 99, p. 117 – 123 (1985).

103.105 Spectrophotometric observations of comet P/Crommelin (1983n).
M. C. Festou, M. Dennefeld, E. Maurice, J. Bouvier.
Inst. Astrophys. Paris, Pré–Publ., No. 87, 7 pp. (1984). To appear in Astron. Astrophys.

103.106 IUE observations of comet P/Crommelin (1983n).
M. C. Festou, W. C. Carey, A. Evans, M. K. Wallis, H. U. Keller.
Inst. Astrophys. Paris, Pré–Publ., No. 101, 15 pp. (1985). To appear in Astron. Astrophys.

103.107 The gaseous activity of comet Crommelin.
M. F. A'Hearn, P. V. Birch, R. L. Millis, D. G. Schleicher.
IHW Newsl., No. 6, p. 19 – 21 (1985).

Comet 1983 VII IRAS–Araki–Alcock

103.121 The interpretation of the radio continuum emission from comet IRAS–Araki–Alcock (1983d).
C. M. Walmsley.
Astron. Astrophys., Vol. 142, No. 2, p. 437 – 440 (1985).

The author discusses recent radio continuum observations of comet 1983d (IRAS–Araki–Alcock). The available evidence suggests that the size of the source of continuum emission is approximately 100 km. Models invoking emission due to mm–sized grains have difficulty in accounting for the observed flux because such particles either evaporate or are swept out of the beam on a time scale of approximately a day. An alternative is that the emission is due to a large number of roughly meter–sized boulders distributed in a "halo" approximately 100 km in size around the cometary nucleus. These boulders could also contribute to the observed radar echo.

103.122 Infrared spectrophotometry of comet IRAS–Araki–Alcock (1983d): a bare nucleus revealed?
M. S. Hanner, D. K. Aitken, R. Knacke, S. McCorkle, P. F. Roche, A. T. Tokunaga.
Icarus, Vol. 62, No. 1, p. 97 – 109 (1985).

Spectra of the central core and surrounding coma of comet IRAS–Araki–Alcock (1983d) were obtained at $8 - 13\,\mu m$ on 11 May and $2 - 4\,\mu m$ on 12 May 1983. Spatially resolved measurements at $10\,\mu m$ with a 4–arcsec beam showed that the central core was more than 100 times brighter than the inner coma only 8 arcsec away; for radially outflowing dust, the brightness ratio would be a factor of 8. The observations of the central core are consistent with direct detection of a nucleus having a radius of approximately 5 km. The temperature of the sunlit hemisphere was $>300K$. Spectra of the core are featureless, while spectra of the coma suggest weak silicate emission. The spectra show no evidence for icy grains. The dust production rate on 11.4 May was $\sim 10^5 g/sec$, assuming that the gas flux from the dust–producing areas on the nucleus was $\sim 10^{-5} g/cm^2/sec$.

103.123 Temperature of comet IRAS–Araki–Alcock (1983d).
R. H. Brown, D. P. Cruikshank, D. Griep.
Icarus, Vol. 62, No. 2, p. 273 – 281 (1985).

Infrared observations of comet IRAS–Araki–Alcock indicate that the thermal spectrum of its nuclear condensation was approximately blackbody. There is evidence of thermal radiation in the 1.5– to 2.6–μm spectrophotometry that has a color temperature of $\approx 330K$, in agreement with the longer wavelength observations. Spatial scans of the nuclear condensation at 11.6 and 20 μm show that most of the thermal radiation came from an unresolved area at the center of the nuclear condensation.

103.124 Seeing comet IRAS with a radar eye.
R. F. Jurgens.
Spaceflight, Vol. 27, No. 5, p. 221 – 224 (1985).

103.125 C_2 imagery of the inner coma of comet IRAS–Araki–Alcock.
R. J. Oliversen, J. M. Hollis, L. W. Brown.
Bull. Am. Astron. Soc., Vol. 17, No. 2, p. 591 (1985). Abstract. – See Abstr. 010.065.

103.126 Photographic observations of comet IRAS–Araki–Alcock (1983d) in Odessa.
Komet. Tsirk., No. 338 (1985). In Russian.

103.127 The nondetection of continuum radiation from Comet IRAS–Araki–Alcock (1983d) at 2– to 6–cm wavelengths and its implication on the icy–grain halo theory.
I. De Pater, C. M. Wade, H. L. F. Houpis, P. Palmer.
Icarus, Vol. 62, No. 3, p. 349 – 359 (1985).

Observations of Comet IRAS–Araki–Alcock have been made with the VLA at 6 and 2 cm, when the comet was at geocentric distances of 0.08 and 0.035 AU, respectively. The authors show that the "conventional" icy–grain halo theory is not adequate to explain the data.

Periodic comet Churyumov–Gerasimenko

103.141 Spectral and polarization peculiarities of comet Churyumov–Gerasimenko (1982f).
K. I. Churyumov, D. I. Gorodetskij, F. K. Rspaev, M. K. Wallis.
Probl. Kosm. Fiz., Vyp. 19, p. 68 – 74 (1984). In Russian. – See Abstr. 003.002.

Comet 1984t Levy–Rudenko

103.161 Drawings of comet Levy–Rudenko 1984t.
S. J. O'Meara.
Int. Comet Q., Vol. 7, No. 1, p. 11 – 13 (1985).

103.162 Comet Levy–Rudenko 1984t.
Br. Astron. Assoc. Circ., No. 647 (1985).
Ephemeris 1985 February 19 – March 31.

103.163 Comet Levy–Rudenko (1984t).
Yamamoto Circ., Nos. 2035, 2037 (1985). In Japanese.

103.164 Comet Levy–Rudenko (1984t).
IAU Circ., Nos. 4032, 4035, 4037, 4045, 4057 (1985).

103.165 New comet Levy–Rudenko (1984t).
Komet. Tsirk., Nos. 336, 337, 339, 340, 342 (1985). In Russian.

Periodic comet Giacobini–Zinner

103.181 The comet Giacobini–Zinner handbook. An observer's guide to the first comet to be explored by a spacecraft.
D. K. Yeomans, J. C. Brandt.
Jet Propulsion Laboratory, California Institute of Technology, Pasadena, Calif., USA, 69 pp. (1985).

With contributions from R. Farquhar, M. Niedner, N. Divine, C. Morris, G. Holman, S. Maran, T. von Rosenvinge, and Z. Sekanina.

103.182 Precession model for the nucleus of periodic comet Giacobini–Zinner.
Z. Sekanina.
Astron. J., Vol. 90, No. 5, p. 827 – 845 (1985).

A model has been developed for the nucleus of P/Giacobini–Zinner, one of the "erratic" short–period comets whose orbital motions exhibit discontinuities in the nongravitational perturbations due to outgassing. The present approach is based on the precession model, applied to P/Encke by Whipple and Sekanina and to P/Kopff by Sekanina, but employs a new feature, a concept of interdependence between the precession and outgassing patterns. It is shown that the variations with time in the nongravitational parameter A_2, determined by Yeomans from his orbital solutions, can be fitted satisfactorily only with the precession models that imply a highly flattened nucleus, a very rapid rotation, and a pattern of surface outgassing that is virtually symmetrical with respect to the subsolar meridian. The adopted model is discussed in detail.

103.183   Periodic comet Giacobini–Zinner (1984e).
Yamamoto Circ., Nos. 2038, 2039 (1985). In Japanese.

103.184   Occultation of BD + 45°1922 by periodic comet Giacobini–Zinner.
Yamamoto Circ., No. 2041 (1985). In Japanese.

103.185   Periodic comet Giacobini–Zinner (1984e).
IAU Circ., Nos. 4053, 4061, 4070, 4075 (1985).

103.186   Komet P/Giacobini–Zinner (1984e).
K. Güssow, J. Linder.
Sterne Weltraum, 24. Jahrg., Nr. 6, p. 334 – 335 (1985).

103.187   MHD simulations of comet P/Giacobini–Zinner.
J. L. Giuliani Jr., J. A. Fedder, J. G. Lyon, M. B. Niedner Jr.
Bull. Am. Astron. Soc., Vol. 17, No. 2, p. 591 (1985). Abstract. – See Abstr. 010.065.

103.188   Comet Giacobini–Zinner.
Z. Ceplecha.
Vesmír, Vol. 64, No. 5, p. 295 (1985). In Czech.

103.189   Short–period comet Giacobini–Zinner (1984e).
Komet. Tsirk., No. 338 (1985). In Russian.

103.190   Comet Giacobini–Zinner and the Draconids.
Yu. V. Evdokimov.
Komet. Tsirk., No. 340 (1985). In Russian.

103.191   Photograph of comet P/Giacobini–Zinner 1984e.
T. Gehrels, J. Scotti.
IHW Newsl., No. 7, p. 1 – 2 (1985).

103.192   Ephemeris (with perturbations) for comet Giacobini–Zinner.
D. K. Yeomans.
IHW Newsl., No. 7, p. 38 – 42 (1985).

Comet 1982 VI Austin

103.201   An estimate of the magnetic field strength in the OH coma of comet Austin (1982 VI).
E. Gérard.
Astron. Astrophys., Vol. 146, No. 1, p. 1 – 10 (1985).

Left and right–circular polarization spectra of comet Austin (1982 VI) were taken between 3 and 12 August 1982 at 1667 MHz and 1665 MHz, the main Λ doubling transitions of the OH molecule within the ground state $^2\Pi_{3/2} J = 3/2$. A significant frequency shift was observed between 6 and 12 August that the author attributes to a Zeeman displacement corresponding to an average line–of–sight magnetic field of $+50 \pm 21$ γ. The sign is compatible with the polarity of the interplanetary magnetic field (IMF) at the comet, inferred from existing data. The amplitude requires an amplification of the IMF which is typically 15 γ at 0.75 AU, the average heliocentric distance of C/Austin. The similarity of the 1667 MHz and 1665 MHz total power spectra excludes a magnetic field strength larger than 300 γ, whatever the field inclination to the line–of–sight.

103.202   Complex observations of comet Austin 1982g. I. On the systematic differences of positions from observations in two–colour systems.
G. R. Kastel', V. K. Rozenbush, N. V. Kharchenko.
Kinematika Fiz. Nebesn. Tel, Tom 1, No. 3, p. 3 – 6 (1985). In Russian.

The systematic differences of positions of the brightness centres of comet Austin 1982g have been obtained in photographic and photovisual spectral regions from simultaneous observations in two–colour systems. With allowance for the phase angle this difference is $1''.24 \pm 0''.22$; that corresponds to the linear distance 595 ± 106 km. The physical reasons of the effect are discussed.

Peridic comet Encke

103.221   Surface photometry of comet P/Encke.
S. Djorgovski, H. Spinrad.
Astron. J., Vol. 90, No. 5, p. 869 – 876 (1985).

The authors have developed a scheme to clean cometary digital images from offending background–star trails, and applied this technique to a pair of deep Kitt Peak 4–m plates of comet P/Encke, taken in October 1980. Simultaneous and subsequent digital spectra have been obtained at Lick Observatory. The non–polluted coma images show a strong asymmetric sunward–oriented fan/jet, and an extended and rounder (mostly gaseous) main coma, out to $\sim 10^5$ km radius. The stellar–trail point–spread function has a narrow width ($\sigma \sim 0.6$ arcsec), so that spatial resolution better than ~ 300 km is achieved at the comet. The photometric gradient near the nucleus is very steep, strongly suggesting an icy–grain component which evaporates quickly (at radii $\leqslant 500$ km) in the sunlight. Further from the nucleus, the profile becomes shallower, bluer, and more gas dominated. The effect of solar radiation pressure on C_2, CN, and other molecules is probably responsible for the rounding of the outer, faint isophotes. The source of the molecules is likely to be larger than the nucleus itself, and a substantial fraction may originate in the jet.

Comet 1976 VI West

103.241   Radius and albedo of the nucleus of comet West 1976 VI.
I. N. Potapov.
Astron. geod., Tomsk, No. 10, p. 175 – 177 (1984). In Russian. Abstr. in Ref. Zh., 51. Astron., 3.51.265 (1985).

Periodic comet Stephan–Oterma

103.261   Spatially resolved spectrophotometry of comet P/Stephan–Oterma.
A. L. Cochran, E. S. Barker.
Icarus, Vol. 62, No. 1, p. 72 – 81 (1985).

Observations of comet P/Stephan–Oterma were made with an Intensified Dissector Scanner spectrograph on the McDonald Observatory 2.7–m telescope during the period from July 1980 to February 1981. These spectra cover a range of heliocentric distances from 2.3 AU preperihelion to 1.8 AU postperihelion. A small aperture was used to map the spatial distributions of the gases in the coma. Column densities of the observed cometary emissions (CN, C_3, CH, and C_2) were calculated and it is shown that Stephan–Oterma appeared nearly spherically symmetric.

103.262   Nonequilibrium chemical analysis of the coma of comet P/Stephan–Oterma.
A. L. Cochran.
Icarus, Vol. 62, No. 1, p. 82 – 96 (1985).

A computer code to calculate the time–dependent nonequilibrium chemistry taking place within the coma of a comet has been developed. This code incorporates 1249 chemical reactions involving 128 species. It was shown that (1) HCN is the parent for CN; (2) C_2H_2 is a parent for C_2; (3) pure gas–phase chemistry with known species cannot adequately reproduce the observed C_3 but a single step process can; and (4) at least prior to perihelion, the vaporization rate seems to have been controlled by water vaporization.

Comet 1977 XIV Kohler

103.281   **Photometric observation and nuclear diameter of comet Kohler.**
S.-c. Wang, Y.-z. Wu.
Publ. Purple Mt. Obs., Vol. 3, No. 3, p. 37 – 44 (1984). In Chinese.

The results of photographic photometry of comet Kohler are presented. On the basis of the magnitudes of the cometary head and diameters of the coma the nuclear diameter is computed to be (1.27 ± 0.05) km. The relation between cometary photometry and nuclear diameter is discussed.

Comet 1983 XII Černis

103.301   **Observations photographiques de la comète Černis (1983*l*) effectuées en 1983 à l'équatorial GPO de 40 cm de l'Observatoire Austral Européen (ESO) à La Silla (Chili).**
H. Debehogne.
Bull. Astron., Vol. 9, No. 6, p. 309 – 310 (1984).

Comet 1979 X Bradfield

103.321   **Observation of rapidly–growing helical waves in the plasma tail of comet Bradfield 1979 X.**
M. B. Niedner, A. I. Ershkovich, J. C. Brandt.
Bull. Am. Astron. Soc., Vol. 16, No. 4, p. 923 (1984). Abstract. – See Abstr. 010.062.

Periodic comet Neujmin 1

103.341   **The nucleus of comet P/Neujmin 1.**
M. A'Hearn, H. Campins, L. McFadden, R. Millis.
Bull. Am. Astron. Soc., Vol. 16, No. 4, p. 1026 (1984). Abstract. – See Abstr. 010.063.

103.342   **The nucleus of comet P/Neujmin 1.**
M. A'Hearn, H. Campins, L. McFadden, R. Millis.
Bull. Am. Astron. Soc., Vol. 16, No. 4, p. 1027 (1984). Abstract. – See Abstr. 010.063.

Comet 1984s Shoemaker

103.361   **Comet Shoemaker 1984s.**
Br. Astron. Assoc. Circ., No. 647 (1985).
Ephemeris 1985 February 14 – March 21.

103.362   **Comet Shoemaker (1984s).**
IAU Circ., Nos. 4045, 4049 (1985).

103.363   **New comet Shoemaker 1984s.**
Komet. Tsirk., Nos. 335 – 337, 342 (1985). In Russian.

Periodic comet Schaumasse

103.381   **Comet P/Schaumasse 1984m.**
Br. Astron. Assoc. Circ., No. 647 (1985).
Ephemeris 1985 February 4 – March 16.

103.382   **Periodic comet Schaumasse (1984m).**
IAU Circ., Nos. 4030, 4057 (1985).

103.383   **Photographic observations of comet Schaumasse at the Southern Station of the Sternberg Institute.**
N. V. Metlova, V. G. Metlov.
Komet. Tsirk., No. 341 (1985). In Russian.

Comet 1985e Machholz

103.401   **New comet Machholz 1985e.**
Br. Astron. Assoc. Circ., No. 648 (1985).
Preliminary parabolic elements and ephemeris 1985 June 8 – 18.

103.402   **Comet Machholz (1985e).**
Yamamoto Circ., Nos. 2039 – 2041 (1985). In Japanese.

103.403   **Comet Machholz (1985e).**
IAU Circ., Nos. 4067, 4069, 4071, 4072, 4074, 4078 (1985).

Periodic comet Ashbrook–Jackson

103.421   **P/Ashbrook–Jackson (1985a).**
Yamamoto Circ., No. 2036 (1985). In Japanese.

103.422   **Periodic comet Ashbrook–Jackson (1985a).**
IAU Circ., No. 4048 (1985).

103.423   **Short–period comet Ashbrook–Jackson (1948 IX).**
Komet. Tsirk., Nos. 338, 341 (1985). In Russian.

Comet 1984f Shoemaker

103.441   **Comet Shoemaker (1984f).**
Yamamoto Circ., Nos. 2037, 2040 (1985). In Japanese.

103.442   **Comet Shoemaker (1984f).**
IAU Circ., Nos. 4029, 4052, 4066 (1985).

Periodic comet Russell 1

103.461   **Periodic comet Russell 1 (1985b).**
Yamamoto Circ., No. 2038 (1985). In Japanese.

103.462   **Periodic comet Russell 1 (1985b).**
IAU Circ., No. 4053 (1985).

Periodic comet Honda–Mrkos–Pajdušáková

103.481   **Periodic comet Honda–Mrkos–Pajdušáková (1985c).**
Yamamoto Circ., No. 2038 (1985). In Japanese.

103.482   **Periodic comet Honda–Mrkos–Pajdušáková (1985c).**
IAU Circ., No. 4055 (1985).

Periodic comet Tsuchinshan 2

103.501   **Periodic comet Tsuchinshan 2 (1985d).**
Yamamoto Circ., No. 2039 (1985). In Japanese.

103.502   **Periodic comet Tsuchinshan 2 (1985d).**
IAU Circ., No. 4063 (1985).

Periodic comet Tsuchinshan 1

103.521   **Periodic comet Tsuchinshan 1 (1984p).**
Yamamoto Circ., No. 2040 (1985). In Japanese.

103.522   **Periodic comet Tsuchinshan 1 (1984p).**
IAU Circ., Nos. 4030, 4066 (1985).

Periodic comet Wolf–Harrington

103.541   Periodic comet Wolf–Harrington (1984g).
IAU Circ., No. 4031 (1985).

Periodic comet Arend–Rigaux

103.561   Periodic comet Arend–Rigaux (1984k).
IAU Circ., Nos. 4038, 4041 (1985).

103.562   Observations of comet Arend–Rigaux (1984k) in Zelen-chukskaya.
Komet. Tsirk., No. 341 (1985). In Russian.

Comet 1985f Hartley

103.581   Comet Hartley (1985f).
IAU Circ., Nos. 4077, 4079 (1985).

Periodic comet Tuttle

103.601   CCD observations of comet Tuttle 1980 XIII: the H_2O^+ ionosphere.
W.-H. Ip, U. Fink, J. R. Johnson.
Astrophys. J., Vol. 293, No. 2, p. 609 – 615 (1985).
A CCD spectrum of comet Tuttle 1980h has been analyzed with emphasis on the emission of H_2O^+ ions. The fine angular resolution (1″5) and the capability of absolute brightness calibration of the CCD instrument enabled the authors to determine the spatial concentration of the H_2O^+ ions of this faint comet and the total number of these ions confined within a spherical region. Solar photoionization of the H_2O atmosphere can account for the production of the H_2O^+ ions observed in the confined region. The dimension of the H_2O^+ ionosphere at the time of observation was found to be comparatively small.

Comet 1984r Shoemaker

103.621   New comet Shoemaker (1984r).
Komet. Tsirk., Nos. 334, 336 (1984/1985). In Russian.

Periodic comet Shoemaker 1

103.641   Periodic comet Shoemaker 1984q.
Komet. Tsirk., Nos. 334, 336 (1984/1985). In Russian.

Periodic comet Takamizawa

103.661   Observations of comet Takamizawa (1984j).
Komet. Tsirk., Nos. 335, 336 (1985). In Russian.

Periodic comet Shoemaker 2

103.681   New comet Shoemaker (1984u).
Komet. Tsirk., No. 337 (1985). In Russian.

Comet 1984i Austin

103.701   Observations of comet Austin (1984i).
Komet. Tsirk., Nos. 337, 342 (1985). In Russian.

Comet 1970 II Bennett

103.721   Relative abundances of H, C, N, O, Na and Si in the nucleus of comet Bennett 1970 II.
O. V. Dobrovol'skij, I. N. Matveev.
Komet. Tsirk., No. 337 (1985). In Russian.

Comet 1984v Hartley

103.741   New comet Hartley (1984v).
Komet. Tsirk., No. 338 (1985). In Russian.

Periodic comet Wolf

103.761   Observations of comet Wolf (1983m) and their comparison with calculations.
E. I. Kazimirchak–Polonskaya.
Komet. Tsirk., No. 338 (1985). In Russian.

Periodic comet Gehrels 3

103.781   Encounter of comet Gehrels 3 with Jupiter.
N. Yu. Emel'yanenko.
Komet. Tsirk., No. 341 (1985). In Russian.

Periodic comet Hartley–IRAS

103.801   Positions of comet Hartley–IRAS (1983v).
Komet. Tsirk., No. 342 (1985). In Russian.

Periodic comet Faye

103.821   Observations of periodic comet Faye (1984h).
Komet. Tsirk., No. 342 (1985). In Russian.

Periodic comet Halley

103.901   What is at the centre of Halley's comet and how fast does it spin?
D. W. Hughes.
Nature, Vol. 313, No. 5999, p. 178 (1985).

103.902   The International Halley Watch.
S. J. Edberg.
Spaceflight, Vol. 27, No. 4, p. 171 – 173 (1985).

103.903   La comète de Halley à son retour de 1910.
R. Boyer, É. Neyvoz, E. Stram.
Astronomie, Vol. 99, p. 15 – 31 (1985).

103.904   La comète de Halley à son retour de 1910.
R. Boyer, É. Neyvoz, E. Stram.
Astronomie, Vol. 99, p. 67 – 81 (1985).

103.905   The size, mass, mass loss and age of Halley's comet.
D. W. Hughes.
Mon. Not. R. Astron. Soc., Vol. 213, No. 1, p. 103 – 109 (1985).
Halley's comet (1910 II, 1982i) has a D^2p_v value of 5.90 ± 0.33 km^2, where D km is the diameter of the nucleus and p_v is the geometric albedo. The mass of the nucleus is $7.5 \times 10^{15} \varrho p_v^{-1.5}$g, where ϱ is the density. Reasonable assumptions yield diameter and mass values of 9.4 km and 2.2×10^{17}g. In the 1910 apparition the comet lost a mass of 2.8×10^{14}g which is equivalent to an absolute magnitude change of 9×10^{-4} per apparition. The mass of the meteor stream, produced by the decay of P/Halley, is consistent with the statement that the comet has had 2300 previous close passages of the Sun.

103.906   La cometa di Halley.
P. Tempesti.
Coelum, Vol. 53, N. 1, p. 1 – 18 (1985).

103.907   La grande cometa del 1910.
G. Di Giovanni.
Coelum, Vol. 53, N. 1, p. 19 – 28 (1985).

103.908 L'avvistamento della cometa di Halley all'osservatorio di Asiago.
C. Bonoli, F. Bortoletto, M. D'Alessandro.
Coelum, Vol. 53, N. 1, p. 35 – 37 (1985).

103.909 Komeet Halley in de computer.
K. Velt, G. Schilling.
Zenit, 12. Jaarg., No. 4, p. 132 – 138 (1985).

103.910 Numerical methods and regularized transformations in the problem of forecasting the motion of comet Halley.
V. A. Shefer.
Astron. Geod., Tomsk, No. 12, p. 37 – 42, 142 (1984). In Russian. Abstr. in Ref. Zh., 51. Astron., 2.51.64 (1985).

103.911 Near–nucleus studies of comet Halley.
S. Larson, Z. Sekanina.
Cometary astrometry, p. 14 – 20 (1984). – See Abstr. 012.024.

Quantitative studies of the near–nucleus dust coma structure of comet Halley are possible using modern digital image processing techniques on photographs taken in 1910. The authors review recent investigations carried out in conjunction with the Near–Nucleus Studies Net of the IHW to better understand the characteristics and behavior of comet Halley. A new image processing algorithm developed to enhance coma feature boundaries permits their evolution over as many as three days to be followed. The features can be modelled to derive information on the nucleus spin vector, particle sizes, ejection velocities and distribution of emission areas on the nucleus.

103.912 The orbital motion of comet Halley.
V. V. Savchenko.
Cometary astrometry, p. 176 – 187 (1984). – See Abstr. 012.024.

This paper briefly presents the methodology of constructing an exact theory, analyzes the effect of all kinds of errors in building a mathematical model of motion and the effect of a shift of Halley's center–of–mass relative to the optical center. The theory is based on the optical, angular observations of 1911 back to 1759 reduced to the mean equator and epoch of 1950.0 as well as the relative observations of 1682. Deviations of calculated values from measured values do not exceed 1".5 which confirms the good degree of accuracy of this theory.

103.913 The work on comet Halley's orbit at ESOC.
T. Morley, F. Hechler.
Cometary astrometry, p. 188 – 202 (1984). – See Abstr. 012.024.

The presently foreseen target for the GIOTTO spacecraft is a fly–by past the sunlit side of the nucleus of comet Halley with a distance at closest approach of 500 km. The predominant error source in the final navigation is due to uncertainties in the comet's ephemeris. Efforts are being undertaken at ESOC to improve the prediction of the comet's state at the time of encounter by GIOTTO. Some of the aspects involved in the problem are discussed. Also addressed are the difficulties in the mathematical modelling of the non–gravitational forces and possible offsets of the comet's centre–of–light from its centre–of–mass.

103.914 Dust environment models for comet P/Halley: support for targeting of the GIOTTO S/C.
J. Fertig, G. H. Schwehm.
Adv. Space Res., Vol. 4, No. 9, p. 213 – 216 (1984). – See Abstr. 012.029.

In March 1986 ESA's GIOTTO spacecraft will fly by P/Halley's nucleus at a distance of a few hundred kilometres. The near nucleus dust environment the probe will traverse poses a hazard with respect to physical damage as well as to attitude disturbance with the possible loss of ground station contact. To predict S/C survivability and dust impact rates for the experiments, a model of the spatial distribution of the dust in the nucleus' vicinity has been constructed.

103.915 Thermal modeling of Halley's comet.
P. R. Weissman, H. H. Kieffer.
Adv. Space Res., Vol. 4, No. 9, p. 221 – 224 (1984). – See Abstr. 012.029.

The comet thermal model of Weissman and Kieffer is used to calculate gas production rates and other parameters for the 1986 perihelion passage of Halley's comet.

103.916 Space missions to Halley's comet.
Adv. Space Res., Vol. 4, No. 9, p. 263 – 264 (1984). – See Abstr. 012.029.

103.917 Comet Halley observations to date.
J. Rahe.
Adv. Space Res., Vol. 4, No. 9, p. 265 – 272 (1984). – See Abstr. 012.029.

Following Comet Halley's recovery on 16 October, 1982, several astrometric, photometric, and spectroscopic observations have been obtained. The results derived from these observations are presented.

103.918 Radio occultation experiments with Comet Halley.
M. K. Bird, P. Edenhofer, H. Porsche, H. Volland.
Adv. Space Res., Vol. 4, No. 9, p. 303 – 306 (1984). – See Abstr. 012.029.

Previous radio occultation investigations on cometary comae and tails have included refraction measurements and intensity scintillations of natural radio sources used to derive the density and structure of the cometary plasma. Significant improvements in the coverage and sensitivity of these measurements will be achieved during the present apparition of Comet Halley. The comet missions GIOTTO and VEGA will also feature passive radio science experiments designed to measure comet–induced Doppler shifts of the dual–frequency spacecraft signals during Halley flyby. A brief survey of these radio occultation measurement techniques and their application in the specific case of Comet Halley are presented.

103.919 EUV observations of comet Halley.
H. W. Ripken, H. J. Fahr, G. Lay.
Adv. Space Res., Vol. 4, No. 9, p. 307 – 310 (1984). – See Abstr. 012.029.

The observation of EUV emissions of comet Halley and its plasma–gas environment by means of rocket– or satellite–borne resonance absorption cell spectrophotometer devices is planned. The technical outlay of the payload, the estimated EUV intensities, and the scientific objectives of this mission are presented.

103.920 Proposed optical observations of comet Halley.
S. Rath, D. Jadhav, P. Choudhary, A. Tillu.
Adv. Space Res., Vol. 4, No. 9, p. 311 – 313 (1984). – See Abstr. 012.029.

A comprehensive programme is proposed for optical observations of comet Halley based on the wide range of available facilities such as photometers, monochromators, interferometers and a polarimeter. Feasibility study of the proposed investigations with reference to existing facilities is carried out for each technique and needs for additional instrumentation are established. Conclusions for optimum utilization of existing facilities are presented.

103.921 Far Eastern observations of Halley's comet: 240 BC to AD 1368.
F. R. Stephenson, K. K. C. Yau.
J. Br. Interplanet. Soc., Vol. 38, No. 5, p. 195 – 216 (1985).

Halley's comet has been observed at every return since 12 BC and may possibly be traced as far back as 240 BC. The observations by the ancient astronomers of China, Japan and Korea are a valuable contribution to the understanding of the past history of this famous comet. A collection of oriental records of the comet from earliest times down to the period when detailed European observations became available is presented here. A number of comparisons between the dates of perihelion passage derived from observations and those calculated from various theories are made. The apparent path of Halley's comet changes considerably from one apparition to the next. Diagrams showing the computed path at all of the returns discussed here are presented.

103.922 The astrometry network of the International Halley Watch.
D. K. Yeomans.
Int. Comet Q., Vol. 7, No. 2, p. 31 – 34 (1985).

103.923 Una nueva visita del cometa Halley.
J. C. Cersosimo.
Rev. Astron., Tomo 56, No. 230, p. 5 – 9 (1984).

103.924 Comet notes: III.
D. H. Levy.
Strolling Astron., Vol. 31, Nos. 1 – 2, p. 28 – 30 (1985).

103.925 Colour, albedo and nucleus size of Halley's comet.
D. P. Cruikshank, W. K. Hartmann, D. J. Tholen.
Nature, Vol. 315, No. 6015, p. 122 – 124 (1985). With a correction in Nature, Vol. 315, No. 6021, p. 690 (1985).
Halley's comet (1982i) is now bright enough for broadband photometry by large infrared telescopes. The authors report here on photometry of the comet in $BVJK$ broadband filters during a time when the coma was very weak and presumed to contribute negligibly to the broadband photometry. The V–J and J–K colours suggest that the colour of the nucleus of Halley's comet is similar to that of the D–type asteroids, which in turn suggests that the surface of the nucleus is of albedo <0.1.

103.926 Meeting comet Halley.
V. M. Balebanov.
Zemlya Vselennaya, No. 1, p. 25 – 33 (1985). In Russian.

103.927 Coma morphology and dust–emission pattern of periodic comet Halley. III. Additional high–resolution images taken in 1910.
S. M. Larson, Z. Sekanina.
Astron. J., Vol. 90, No. 5, p. 823 – 826 (1985). With plates 67 – 73.
The image–processing algorithm introduced by the authors in Paper I (Larson and Sekanina, 1984) is applied to high–resolution images of periodic comet Halley taken at the Lick, Helwan, Lowell, and Vienna Observatories during the period of May 12 – June 2, 1910 to close the gap between the May and June sequences of the Mount Wilson plates already processed in Paper I and subsequently analyzed in Paper II (Sekanina and Larson, 1984). With the exception of five days around the time of P/Halley's transit across the Sun's disk, the combined set of the Mount Wilson, Lick, Helwan, Lowell, and Vienna images presents a daily, and sometimes twice daily, record of coma features between May 5 and June 6, 1910. General characteristics of the coma features in this augmenting sample of images are discussed.

103.928 L'esplorazione ravvicinata della cometa di Halley.
F. F. Martin.
Coelum, Vol. 53, N. 2, p. 85 – 98 (1985).

103.929 La cometa di Halley e la missione spaziale Giotto.
C. B. Cosmovici.
Orione, Vol. 5, N. 2, p. 11 – 16 (1985).

103.930 Previsione di visibilità della cometa P/Halley.
A. Maitan.
G. A.A.B., Anno 20, N. 78, p. 8 – 10 (1985).

103.931 Year posts round the orbit of Halley's comet.
D. W. Hughes.
J. Br. Astron. Assoc., Vol. 95, No. 4, p. 162 – 163 (1985).
A note on plotting to scale the orbit of Halley's comet and its changing velocity.

103.932 Die Erscheinung des Halleyschen Kometen in den Jahren 1835 und 1836.
S. Röser.
Sterne Weltraum, 24. Jahrg., Nr. 5, p. 250 – 256 (1985).

103.933 Beobachtungsmöglichkeiten und Positionsberechnungen des Halleyschen Kometen 1985/86.
M. Belter.
Sterne Weltraum, 24. Jahrg., Nr. 5, p. 284 – 285 (1985).

103.934 More sites for observing Halley's comet.
E. M. Brooks.
Sky Telesc., Vol. 69, No. 6, p. 485 – 486 (1985).

103.935 Première détection amateur de la comète de Halley en Europe.
P. Martinez.
Astronomie, Vol. 99, p. 281 – 287 (1985).

103.936 Models of cometary emission in the 18–cm OH transitions: the predicted behavior of comet Halley.
F. P. Schloerb, E. Gerard.
Astron. J., Vol. 90, No. 6, p. 1117 – 1135 (1985). = Five Coll. Radio Astron. Dep., Contrib. No. 593.
The importance of observations of the OH radical in the 18 cm wavelength Λ–doublet during the return of Halley's comet in 1985 – 1986 is emphasized. This paper presents models of the OH emission from the comae of comets and predictions of the brightness and spectra of Halley's comet during the coming apparition as an aid to prospective observers.

103.937 Periodic comet Halley (1982i).
Yamamoto Circ., Nos. 2033, 2035 (1985). In Japanese.

103.938 Periodic comet Halley (1982i).
IAU Circ., Nos. 4025, 4029, 4034, 4041 (1985).

103.939 IHW: astrometry.
D. K. Yeomans, R. M. West, R. S. Harrington, B. G. Marsden.
IHW Newsl., No. 6, p. 31 – 32 (1985).

103.940 IHW: large–scale phenomena.
J. C. Brandt, M. B. Niedner, J. Rahe.
IHW Newsl., No. 6, p. 32 – 33 (1985).

103.941 IHW: spectroscopy and spectrophotometry.
S. Wyckoff, P. Wehinger.
IHW Newsl., No. 6, p. 33 – 41 (1985).

103.942 IHW: orbit updates – Halley orbit.
D. K. Yeomans.
IHW Newsl., No. 6, p. 41 – 42 (1985).

103.943 Updated predictions of occultations of stars by comet Halley.
E. Bowell, L. H. Wasserman.
IHW Newsl., No. 6, p. 43 – 46 (1985).

103.944 Brightness variations in comet P/Halley determined by the least scatter algorithm.
C. L. Morbey.
Astron. Express, Vol. 1, Nos. 4 – 6, p. 133 – 136 (1985).
A simple period finding technique shows that the available photometric magnitudes of comet P/Halley can be assembled on periods of 2.005 and 0.684 day with scatter 0.22 and 0.24 magnitude respectively.

103.945 Observing Halley's comet.
D. W. Hughes.
Eur. J. Phys., Vol. 6, No. 1, p. 1 – 12 (1985). Abstr. in Phys. Abstr., Vol. 88, No. 1255, Entry 40857 (1985).

103.946 On the photometric parameters of comet Halley.
I. Toth.
Report KFKI–1984–112, Hungarian Acad. Sci., Budapest, Hungary, 14 pp. (1984). Abstr. in Phys. Abstr., Vol. 88, No. 1257, Entry 57654 (1985).

103.947 Computing the ephemeris of comet Halley.
R. Browne.
Astronomy, Vol. 13, No. 2, p. 75 – 78 (1985). Abstr. in Phys. Abstr., Vol. 88, No. 1259, Entry 68729 (1985).

103.948 The comet Halley ephemeris development effort.
D. K. Yeomans.
Bull. Am. Astron. Soc., Vol. 17, No. 2, p. 622 (1985). Abstract. – See Abstr. 010.066.

103.949 A two–body ephemeris generation program for observing comet Halley.
M. S. Keesey.
Bull. Am. Astron. Soc., Vol. 17, No. 2, p. 622 (1985). Abstract. – See Abstr. 010.066.

103.950 The International Halley Watch.
S. J. Edberg.
Astronomia, Suppl. al N. 2, p. 19 – 20 (1985).

103.951 A comparison of the new and old computations of the long–term motion of Halley's comet.
K. Ziolkowski.
Astronomia, Suppl. al N. 2, p. 25 – 28 (1985).

103.952 Amateur studies of Halley's comet.
S. J. Edberg.
Astronomia, Suppl. al N. 2, p. 29 – 31 (1985).

103.953 International Halley Watch.
J. Svoreň.
Kozmos, Vol. 15, No. 5, p. 148 – 151 (1984). In Slovak.

103.954 The history of comet Halley.
Ľ. Kresák.
Kozmos, Vol. 16, Nos. 2 – 3, p. 48 – 51, 94 – 97 (1985). In Slovak.

103.955 How old is the comet Halley?
V. Vanýsek.
Vesmír, Vol. 64, No. 7, p. 373 – 376 (1985). In Czech.

103.956 Observations of comet Halley.
Komet. Tsirk., Nos. 335, 336, 338 – 342 (1985). In Russian.

103.957 Observations of comet Halley in 1835 – 1836.
N. A. Belyaev, Yu. D. Medvedev, Yu. A. Chernetenko.
Byull. Inst. Teor. Astron., Tom 15, No. 7 (170), p. 357 – 359 (1985). In Russian.
All circumstances of comet Halley's apparition in 1835 – 1836 have been carefully studied. The problem concerning the more accurate definition of the reference star positions and their employment in this apparition has been considered. The results of the analysis of the comet Halley observations in 1835 – 1836 are presented.

103.958 The transit effect of comet Halley and the magnetic field in its plasma tail.
Z.-y. Li.
Chin. Astron. Astrophys., Vol. 9, No. 2, p. 124 – 127 (1985). English translation of Acta Astrophys. Sin., Vol. 5, No. 1, p. 1 – 8 (1985).
When the plasma tail of a comet sweeps past the Earth, it may cause geomagnetic storms. In this paper, in addition to analysis of this phenomenon, a specific theoretical investigation of the comet tail as the storm source is made. It is found that the tail is capable of storing a large amount of magnetic energy and that the interaction between the tail and the Earth's magnetosphere can lead to magnetic storms. Finally, a simple discussion is made on other possible mechanisms.

103.959 Observations of comet Halley 1982i at the six–meter telescope.
G. K. Nazarchuk.
IHW Newsl., No. 7, p. 5 – 7 (1985).

103.960 Photometric parameters of Halley's comet (1982i) based on B, V, R photometry.
S. I. Gerasimenko, N. N. Kiselev, G. P. Chernova.
IHW Newsl., No. 7, p. 7 – 10 (1985).
Electrophotometry of comet Halley was made in February – April 1985.

103.961 Ephemeris (with perturbations) for comet Halley.
D. K. Yeomans.
IHW Newsl., No. 7, p. 43 – 49 (1985).

103.962 Predictions of the hydrogen Lyman α coma of Comet Halley.
R. R. Meier, H. U. Keller.
Icarus, Vol. 62, No. 3, p. 521 – 537 (1985).
Theoretical predictions of the expected Lyman α emission from the hydrogen coma of Comet Halley, as seen from Earth and Venus orbits, are presented for selected dates during 1985 – 1986 when the comet is within 2 AU of the Sun. Recommendations for scaling to Lyman β and Balmer α are given.

103.963 Comet Halley news.
S. Wyckoff, P. Wehinger.
Mercury, Vol. 14, No. 3, p. 74 – 75 (1985).

Desired characteristics of catalogs for cometary astrometry.
See Abstr. 002.014.

The IHW reference star catalogs for comets Halley and Giacobini–Zinner.
See Abstr. 002.015.

Halley's comet. A bibliography.
See Abstr. 002.018.

Halley's Comet, 1755 – 1984. A bibliography.
See Abstr. 002.090.

Manuale IHW (International Halley Watch). Guida allo studio scientifico della cometa per astronomi non professionisti.
See Abstr. 002.091.

Halley's comet: a bibliography of Canadian newspaper sources, 1835 – 36 and 1910.
See Abstr. 002.092.

Comets: a descriptive catalog.
See Abstr. 002.125.

Halley's Comet.
See Abstr. 003.008.

The comet book: a guide for the return of Halley's comet.
See Abstr. 003.074.

Der Halleysche Komet.
See Abstr. 003.093.

Edmond Halley: the man and his comet.
See Abstr. 003.104.

La cometa di Halley.
See Abstr. 003.129.

Le comete.
See Abstr. 003.159.

The science digest book of Halley's comet.
See Abstr. 003.180.

Mr. Halley's comet.
See Abstr. 003.192.

Asimov's guide to Halley's comet.
See Abstr. 003.193.

The glory of gravity – Halley's comet 1759.
See Abstr. 004.011.

Halley's comet in Babylonia.
See Abstr. 004.037.

Records of Halley's comet on Babylonian tablets.
See Abstr. 004.038.

The first predicted return of comet Halley.
See Abstr. 004.043.

Ancient apparitions of Halley's comet.
See Abstr. 004.073.

The First International Halley Watch: worldwide anticipation and
observation of comet Halley, 1755 – 1759.
See Abstr. 004.102.

IRAS moving–object search.
See Abstr. 013.009.

Recent discoveries of comets with the Palomar 46–cm Schmidt
camera.
See Abstr. 013.010.

The Association's plans for observing Halley's Comet.
See Abstr. 013.011.

US experiments and expertise contribute to Soviet Halley probe.
See Abstr. 013.024.

Astrometry at La Silla.
See Abstr. 013.027.

The astrometry network of observers in China.
See Abstr. 013.029.

The astrometry network of observers in Japan.
See Abstr. 013.030.

The astrometry network of observers in U.S.S.R.
See Abstr. 013.031.

The Astrometry Network of the IHW and the P/Crommelin trial
run results.
See Abstr. 013.036.

The International Halley Watch from Australia.
See Abstr. 013.040.

Halley–specialist Donald K. Yeomans.
See Abstr. 013.053.

Halley's comet.
See Abstr. 014.026.

Measurement of charged secondary particle emission under ion and
neutral impact.
See Abstr. 022.075.

Contactless determination of the conductivity of the white paint
PCB–Z.
See Abstr. 022.076.

Measurements of integral yields of charged secondary particles
using neutral beams simulating a cometary fly–by.
See Abstr. 022.077.

Giotto residual ionization.
See Abstr. 022.078.

Impact ionization from gold, aluminum and PCB–Z.
See Abstr. 022.079.

Experimental investigations on ion emission with dust impact on
solid surfaces.
See Abstr. 022.080.

Ion emission from solid surfaces: comparison of dust impact with
other excitations.
See Abstr. 022.081.

Electrostatic charging of the Giotto spacecraft due to neutral gas
impact.
See Abstr. 022.082.

A model of plasma cloud expansion generated by a dust particle
impact.
See Abstr. 022.083.

Giotto plasmasheath and wake charging near Halley's comet:
revised parameters.
See Abstr. 022.084.

Simulations of satellite–plasma interaction, including electron
space charges.
See Abstr. 022.085.

Laboratory measurement on impact ionization by neutrals and
floating potential of a spacecraft during encounter with Halley's
comet.
See Abstr. 022.130.

Halley's comet exploration and Usuda's large antenna.
See Abstr. 033.056.

Astrometry of comets using hypersensitized type 2415 film.
See Abstr. 034.026.

The instrument IKS and its calibration.
See Abstr. 035.016.

The Infrared Imaging Channel of the IKS Instrument for detection
of Comet Halley nucleus.
See Abstr. 035.017.

Oriented platform for the Vega–probe to the Halley comet, position detection, working program.
See Abstr. 035.018.

A spectral photopolarimeter for Giotto: Halley Optical Probe Experiment.
See Abstr. 035.019.

First calibration measurements with the dust impact detector
DIDSY–IPM.
See Abstr. 035.020.

The Giotto spacecraft configuration and its achievement.
See Abstr. 035.021.

Anticipated impact plasma problems for the Copernic–RPA experiment in the cometary environment.
See Abstr. 035.037.

An impact plasma monitor for cometary missions.
See Abstr. 035.038.

Plasma environment effects on the Giotto ion mass spectrometer.
See Abstr. 035.039.

Thermal design of the Giotto spacecraft.
See Abstr. 035.040.

Cometary astrometry with Schmidt cameras.
See Abstr. 036.047.

Kitt Peak measurements of P/Halley positions.
See Abstr. 036.049.

Problems and procedures of cometary astrometry.
See Abstr. 036.050.

Obtaining accurate comet positions despite some inaccurate catalog stars.
See Abstr. 036.051.

Manuale di osservazione delle comete.
See Abstr. 036.097.

Remarks on cometary astrometry.
See Abstr. 041.006.

The orbits of comets Halley and Giacobini–Zinner.
See Abstr. 042.008.

Uniform approximation in problems of celestial mechanics and numerical integration of equations of motion of comet Halley.
See Abstr. 042.051.

Giotto's enkele reis naar komeet Halley.
See Abstr. 051.002.

Wetenschappelijke ruimtevloot zet koers naar komeet Halley.
See Abstr. 051.022.

Comet coma sample return via Giotto II.
See Abstr. 051.023.

The International Cometary Explorer Mission to P/Giacobini–Zinner and P/Halley.
See Abstr. 051.024.

Astrometric needs for the ISEE–3/ICE mission to comet Giacobini–Zinner.
See Abstr. 051.032.

Prospects for Giotto and Halley.
See Abstr. 051.043.

Sample return from a comet flyby.
See Abstr. 051.044.

The Giotto mission to Halley's comet.
See Abstr. 051.045.

A proposed GIOTTO/comet coma sample return mission.
See Abstr. 051.046.

Spacecraft environment during the Giotto–Halley encounter: a summary.
See Abstr. 051.052.

Japanse vluchten naar komeet Halley.
See Abstr. 051.054.

Some questions of planning flights to comets.
See Abstr. 051.056.

A Giacobini–Zinner Watch (GZW).
See Abstr. 051.082.

The Spartan Halley Mission.
See Abstr. 051.083.

Space missions to comet Halley: Giotto, Vega, and Planet–A.
See Abstr. 051.084.

USSR leads the way to Halley's comet.
See Abstr. 051.094.

VEGA – the international space experiment to observe Halley's comet.
See Abstr. 051.095.

Occultations of stars and radio sources by comets: predictions and observing prospects.
See Abstr. 096.008.

Ephemerides of minor planets and comets.
See Abstr. 098.060.

Visual magnitudes of 1983 TB and comets P/Schaumasse (1984m) and Levy–Rudenko (1984t).
See Abstr. 098.079.

1983 TB and comet Shoemaker (1984s).
See Abstr. 098.081.

Implications of the presence of S_2 in cometary ice.
See Abstr. 102.001.

Physical characteristics of comets of the years 1976 – 1977.
See Abstr. 102.005.

The LeBlanc bands of the CN radical in the spectra of comets.
See Abstr. 102.009.

A model for the hydrogen coma of a comet.
See Abstr. 102.017.

Models of CN and C_2 comae for comet Tago–Sato–Kosaka and comet Bennett.
See Abstr. 102.062.

The Perseids and comet Swift–Tuttle 1862 III.
See Abstr. 104.023.

The flux of meteoroids in the vicinity of the orbit of comet Halley.
See Abstr. 104.049.

Meteoren van komeet 1983d?
See Abstr. 104.052.

Meteor showers of comet Halley.
See Abstr. 104.058.

Possible disturbance of the interplanetary magnetic field by the passage of a comet as observed by three spacecraft.
See Abstr. 106.060.

A catalog of radio sources to be occulted by comets P/Halley and P/Giacobini–Zinner.
See Abstr. 141.022.

104 Meteors, Meteor Streams

104.001 Residual mass from atmospheric ablation of small meteoroids.
E. J. Nicol, J. MacFarlane, R. L. Hawkes.
Planet. Space Sci., Vol. 33, No. 3, p. 315 – 320 (1985).

Numerical solutions of the equations of meteor ablation in the Earth's atmosphere have been obtained using a variable step size Runge–Kutta technique in order to determine the size of the residual mass resulting from atmospheric flight. The equations used include effects of meteoroid heat capacity and thermal radiation, and a realistic atmospheric density profile. Results were obtained for initial masses in the range $10^{-7} - 10^{-2}$g, and for initial velocities less than 24 km s^{-1} (results indicated no appreciable residual mass for meteors with velocities above 24 km s^{-1} in this mass range). When the results are expressed in terms of the size of the residual mass following atmospheric ablation as a function of the initial mass and velocity, it is found that the final residual mass is almost independent of the original mass of the meteoroid, but very strongly dependent on the original velocity. This strong velocity dependence coupled with the weak dependence on the original mass has important consequences for the sampling of ablation product micrometeorites.

104.002 Structure of meteor streams. III. The Quadrantids.
A. A. Tkachuk.
Probl. Kosm. Fiz., Vyp. 19, p. 74 – 79 (1984). In Russian. – See Abstr. 003.002.

104.003 Influence of the state of the ionosphere on physical processes in meteor trains.
S. M. Levitskij.
Probl. Kosm. Fiz., Vyp. 19, p. 80 – 84 (1984). In Russian. – See Abstr. 003.002.

104.004 Changes in the activity of the Perseid meteor shower 1944 – 1953.
J. Zvolánková.
Contrib. Astron. Obs. Skalnaté Pleso, Vol. 12, p. 45 – 63 (1984).

On the basis of more than 17000 visual records of Perseids obtained at the Skalnaté Pleso Observatory over a period of 9 years the activity of the Perseid meteor shower in the years 1944 – 1953 (with the exception of 1945) is compared. A considerable difference is found between the individual years as can be seen from the annual coefficients which cover the interval (0.73; 1.87). It appears that the activity of the Perseids before and after their maximum is asymmetrical.

104.005 Nongravitational effects affecting small meteoroids in interplanetary space.
I. Kapišinský.
Contrib. Astron. Obs. Skalnaté Pleso, Vol. 12, p. 99 – 111 (1984).

The set of twenty best–known nongravitational effects affecting meteoroids in interplanetary space has been divided, using a suitably chosen criterion, into a group of four destructive, seven disruptive and ten disturbing effects. Each of the effects is briefly characterized, giving the appropriate literature. The suggested task demonstrates the possibility of using and the usefulness of the presented classification of nongravitational effects.

104.006 Activity and flux of the Orionid meteor shower from visual observations.
V. Porubčan, J. Zvolánková.
Contrib. Astron. Obs. Skalnaté Pleso, Vol. 12, p. 279 – 285 (1984).

Visual observations of the Orionid meteor shower carried out at the Skalnaté Pleso Observatory in 1945 – 1950, are analysed from the viewpoint of activity and flux determination. The maximum of activity appears at the solar longitude of 207.8°, with the particle flux of about 2×10^{-3}km^{-2}h^{-1} down to the visual magnitude $+4$. The total mass influx of the Orionids is estimated at 360 kg per apparition for the whole Earth, down to the same magnitude limit.

104.007 The dispersion of the Geminid stream by planetary perturbations.
J. Jones, K. R. Wheaton.
Observatory, Vol. 105, No. 1065, p. 34 – 36 (1985).

104.008 Double–station observations of 454TV meteors. I. Trajectories.
T. Sarma, J. Jones.
Bull. Astron. Inst. Czech., Vol. 36, No. 1, p. 9 – 24 (1985).

Double–station meteor observations made with two different low–light–level television systems are described which together span the magnitude range 0.5 to 8.5 absolute magnitudes. Besides being more accurate than previous television meteor surveys the present data set constitutes a much larger sample than was previously available. The atmospheric trajectories, velocities, brightnesses and mass are presented together with the beginning and ending heights and the heights of maximum luminosity.

104.009 The relation between meteor optical brightness and properties of the ionized trail. III. Double station observations (results of the Ondřejov Meteor Expeditions in 1972 and 1973).
V. Znojil, J. Hollan, M. Šimek.
Bull. Astron. Inst. Czech., Vol. 36, No. 1, p. 44 – 56 (1985).

The results of telescopic meteor observation from two stations are presented. A more accurate relation between the electron line density α (in m^{-1} units) and the absolute visual meteor magnitude is given: $M = (40.9 \pm 0.3)$–2.5 log α. The results of the observation justify some conclusions mentioned in Znojil et al. (1981). Furthermore, a significant number of non–specular echoes has been ascertained, the reflection point of which is far from the regular one. An analysis of radiant distribution and heights of telescopic meteors is also presented. The paper contains a list of meteors, recorded both by telescopes from two stations and by radar.

104.010 Meteoroid influx and ionization irregularities.
W. J. Baggaley.
Bull. Astron. Inst. Czech., Vol. 36, No. 1, p. 56 – 59 (1985).

The seasonal variation in the daily dawn maximum influx of radar meteors is compared with the dawn occurrence of ionized irregularities in the earth's upper atmosphere. Results of a long series of data covering 37 years of ionospheric soundings and 4 years of meteor influx are used in seeking a relation between the two phenomena.

104.011 Double–station observations of 454 TV meteors. II. Orbits.
J. Jones, T. Sarma.
Bull. Astron. Inst. Czech., Vol. 36, No. 2, p. 103 – 115 (1985).

The orbital elements together with their probable errors for 454 faint meteors spanning the range 0.5 to 8.5 in absolute magnitude are presented. The observations were made with two very sensitive television systems between May 1981 and August 1982. Both the cometary and asteroidal groups of meteors discovered in photographic meteor surveys are also present in the TV data but the cometary group with randomly inclined orbits appears to be more diffuse than for the photographic meteors. Many TV meteors have retrograde orbits with aphelia > 3 A.U. and also a greater proportion of orbits have very small perihelia when compared with the Super–Schmidt meteors. These features seem to suggest that the meteoroid population is in a state of dynamic equilibrium with a source of TV meteoroids replenishing those lost by Jovian perturbations and solar heating.

104.012 Double–station observations of 454 TV meteors. III. Populations.
J. Jones, T. Sarma, Z. Ceplecha.
Bull. Astron. Inst. Czech., Vol. 36, No. 2, p. 116 – 122 (1985).

This paper describes the results of a population analysis based on the beginning heights of 454 doubly observed TV meteors.

Both the A and C levels previously seen in the Super–Schmidt data are present and the existence of the C3 group previously reported in a small TV meteor survey is confirmed. For the C or higher of the two levels radiation cooling appears to be the dominant heat loss mechanism while for the lower level thermal conduction into the interior of the meteoroid seem to be more important. The material properties of the particles in the upper level are difficult to determine. Comparison of theory and observation seems to demand that the A meteoroids be of a dense material with a fairly high thermal conductivity such as stone. The origin of the C3 group is discussed.

104.013 Observing meteors, III.
D. H. Levy.
Strolling Astron., Vol. 31, Nos. 1 – 2, p. 32 – 33 (1985).

104.014 Meteor astronomy. Comets, minor planets and meteor streams.
R. A. Mackenzie.
Space Educ., Vol. 1, No. 9, p. 411 – 414 (1985).

104.015 Ionization efficiency in meteor phenomena.
K. Kh. Saidov.
Dokl. Akad. Nauk TadzhSSR, Tom 27, No. 2, p. 73 – 75 (1984). In Russian. Abstr. in Ref. Zh., 51. Astron., 2.51.234 (1985).

104.016 The contribution of sporadic meteors to the ionization of the upper atmosphere.
Z. M. Ioffe, N. P. Marchenko, L. N. Rubtsov.
Dokl. Akad. Nauk TadzhSSR, Tom 27, No. 3, p. 137 – 140 (1984). In Russian. Abstr. in Ref. Zh., 51. Astron., 2.51.235 (1985).

104.017 Formation of meteor plasma taking fragmentation into account.
G. G. Novikov, A. V. Blokhin.
Fiz.-tekh. inst. Akad. Nauk SSSR, Prepr., No. 901, 33 pp. (1984). In Russian. Abstr. in Ref. Zh., 51. Astron., 2.51.236 (1985).

104.018 Application of the Gauss–Halphen–Goryachev method to investigate the orbital evolution of meteor streams.
A. A. Sukhotin.
Astron. geod., Tomsk, No. 10, p. 121 – 124 (1984). In Russian. Abstr. in Ref. Zh., 51. Astron., 3.51.99 (1985).

104.019 Modelling of the distribution of meteor matter in the solar system.
V. S. Zabolotnikov.
Kazan. inzh.-stroit. inst. Kazan', 20 pp. (1984). In Russian. Abstr. in Ref. Zh., 51. Astron., 3.51.289 (1985).

104.020 Evolution of the orbits and conditions of the encounter of the Geminid and Quadrantid meteor streams with the earth.
P. B. Babadzhanov, Yu. V. Obrubov.
Astron. geod., Tomsk, No. 10, p. 125 – 130 (1984). In Russian. Abstr. in Ref. Zh., 51. Astron., 3.51.292 (1985).

104.021 On a method for determining the sensitivity of a radar station during meteor observations.
G. V. Andreev, G. O. Ryabova.
Astron. geod., Tomsk, No. 10, p. 131 – 136 (1984). In Russian. Abstr. in Ref. Zh., 51. Astron., 3.51.293 (1985).

104.022 Variations of ^{210}Pb fallout on the surface of the earth.
G. N. Bondarenko, T. I. Koromyslichenko.
Cosmochemistry and meteoritics, p. 164 – 169 (1984). In Russian. Abstr. in Ref. Zh., 51. Astron., 3.51.296 (1985). – See Abstr. 012.037.

104.023 The Perseids and comet Swift–Tuttle 1862 III.
J. A. Russell.
Meteoritics, Vol. 19, No. 4, p. 305 – 306 (1984). Abstract. – See Abstr. 010.641.

104.024 The fragmentation of meteoroids. II. Quasicontinuous fragmentation parameters derived from meteor observations.
G. G. Novikov, V. N. Lebedinets, A. V. Blokhin.
Sov. Astron. Lett., Vol. 10, No. 5, p. 327 – 328 (1984). English translation of 38.104.013.

104.025 Electrophonic meteor fireballs.
C. Keay.
South. Stars, Vol. 31, No. 1, p. 11 – 16 (1984).

104.026 Peculiarities of the evolution of the Geminid and Quadrantid meteor swarms.
P. B. Babadzhanov, Yu. V. Obrubov.
Sov. Astron., Vol. 28, No. 5, p. 585 – 590 (1984). English translation of 38.104.012.

104.027 Bright bolide in South Siberia.
D. F. Anfinogenov, V. G. Fast.
Zemlya Vselennaya, No. 3, p. 72 – 75 (1985). In Russian.

104.028 Activity of meteor streams in the years 1983 – 1984.
V. V. Martynenko, A. S. Levina.
Zemlya Vselennaya, No. 3, p. 89 – 93 (1985). In Russian.

104.029 Fictitious daily meteor radiants.
A. Tomić, N. Čabrić.
Publ. Astron. Opservatorije Beogr., No. 33, p. 80 – 82 (1985). – See Abstr. 012.061.

104.030 The role of thermal motions of air particles for fireballs.
V. Padevĕt.
Publ. Astron. Inst. Czech. Acad. Sci., No. 59, p. 1 – 50 (1985).

104.031 The role of thermal motions of particles of ablated meteoric material for fireballs.
V. Padevĕt.
Publ. Astron. Inst. Czech. Acad. Sci., No. 59, p. 51 – 87 (1985).

104.032 Observations of Geminids and Quadrantids.
J. Rendtel, I. Rendtel, F. Otto, R. Koschack.
Astron. Nachr., Vol. 306, No. 3, p. 171 – 176 (1985). In German.
Geminids and Quadrantids are probably observed for only about 100 years. Due to perturbations of their orbits rapid variations occur. A change in the Geminid rate profile can not be found from observations. Observations of Geminids and Quadrantids from Arbeitskreis Meteore are compared with Perseid results. The authors find an increasing number of bright meteors around their activity peaks. At the same time the population index r decreases for Perseids and Quadrantids, while it increases for Geminids. From this behaviour the authors estimate the increase of the number density to fainter magnitude classes for the three showers.

104.033 Peculiarities of radio observations of meteors in two opposite directions.
B. V. Kal'chenko, B. L. Kashcheev, A. N. Olejnikov, T. B. Tomashevskaya.
Meteor. issled., Moskva, No. 11, p. 5 – 14 (1984). In Russian. Abstr. in Ref. Zh., 51. Astron., 5.51.248 (1985).

104.034 On a possibility of determining the physical parameters of transition–type meteor trains taking into account polarization effects.
O. A. Solyanik, V. N. Olejnikov.
Meteor. issled., Moskva, No. 11, p. 43 – 47 (1984). In Russian. Abstr. in Ref. Zh., 51. Astron., 5.51.249 (1985).

104.035 Determination of the coordinates of the radiant and complex reflection coefficients of transition–type meteor trains in the presence of polarization effects.
O. A. Solyanik.
Meteor. issled., Moskva, No. 11, p. 48 – 57 (1984). In Russian. Abstr. in Ref. Zh., 51. Astron., 5.51.250 (1985).

104.036   On the ionization coefficient in atomic collisions in a meteor train.
K. Kh. Saidov.
Dokl. Akad. Nauk TadzhSSR, Tom 27, No. 7, p. 372 – 375 (1984). In Russian. Abstr. in Ref. Zh., 51. Astron., 5.51.251 (1985).

104.037   Meteor swarms and jet streams.
E. N. Kramer, I. S. Shestaka.
Meteor. issled., Moskva, No. 11, p. 72 – 80 (1984). In Russian. Abstr. in Ref. Zh., 51. Astron., 5.51.252 (1985).

104.038   Statistics of perihelion directions of nearly parabolic and hyperbolic orbits of meteor bodies.
B. L. Kashcheev, S. V. Kolomiets.
Meteor. issled., Moskva, No. 11, p. 81 – 88 (1984). In Russian. Abstr. in Ref. Zh., 51. Astron., 5.51.253 (1985).

104.039   Meteor matter inflow during the activity epoch of the Geminid stream.
O. I. Bel'kovich, N. I. Sulejmanov.
Meteor. issled., Moskva, No. 11, p. 89 – 94 (1984). In Russian. Abstr. in Ref. Zh., 51. Astron., 5.51.254 (1985).

104.040   Method of statistical testing and the problem of the origin of meteor swarms.
L. A. Katasev, N. V. Kulikova.
Meteor. issled., Moskva, No. 11, p. 95 – 109 (1984). In Russian. Abstr. in Ref. Zh., 51. Astron., 5.51.255 (1985).

104.041   Theory of quasi–continuous fragmentation of meteor bodies taking deceleration into account.
P. B. Babadzhanov, G. G. Novikov, V. N. Lebedinets, A. V. Blokhin.
Fiz.–tekh. inst. Akad. Nauk SSSR. Prepr., No. 919, 35 pp. (1984). In Russian. Abstr. in Ref. Zh., 51. Astron., 5.51.256 (1985).

104.042   Maps of density distribution of the radiants of sporadic meteors obtained with account for the physical factor.
Yu. A. Pupyshev, T. K. Filimonova, N. V. Nikiforova.
Meteor. rasprostr. radiovoln, Kazan', No. 19, p. 22 – 42 (1984). In Russian. Abstr. in Ref. Zh., 51. Astron., 6.51.194 (1985).

104.043   Distribution of the number of meteor radio reflections.
A. A. Gajdaev, A. N. Kamalov.
Meteor. rasprostr. radiovoln, Kazan', No. 19, p. 43 – 48 (1984). In Russian. Abstr. in Ref. Zh., 51. Astron., 6.51.195 (1985).

104.044   Boundaries of the active region of a radiant in meteor scattering of radio waves.
G. I. Lyutershtejn, I. V. Ermolina.
Meteor. rasprostr. radiovoln, Kazan', No. 19, p. 48 – 55 (1984). In Russian. Abstr. in Ref. Zh., 51. Astron., 6.51.196 (1985).

104.045   Polarization characteristics of radio waves reflected from a meteor trail.
R. G. Khuzyashev.
Meteor. rasprostr. radiovoln, Kazan', No. 19, p. 59 – 62 (1984). In Russian. Abstr. in Ref. Zh., 51. Astron., 6.51.197 (1985).

104.046   Analysis of the duration of meteor reflections in the numerical solution of the diffraction problem.
R. G. Khuzyashev.
Meteor. rasprostr. radiovoln, Kazan', No. 19, p. 62 – 68 (1984). In Russian. Abstr. in Ref. Zh., 51. Astron., 6.51.198 (1985).

104.047   Diffraction of a spherical wave on an infinite meteor trail.
Yu. I. Orlov, R. G. Khuzyashev.
Meteor. rasprostr. radiovoln, Kazan', No. 19, p. 69 – 73 (1984). In Russian. Abstr. in Ref. Zh., 51. Astron., 6.51.199 (1985).

104.048   Use of the numerical solution of the problem of diffraction of radio waves on a meteor trail for the determination of the parameters of the trail.
N. S. Andrianov, R. A. Kurganov, R. G. Khuzyashev.
Meteor. rasprostr. radiovoln, Kazan', No. 19, p. 80 – 86 (1984). In Russian. Abstr. in Ref. Zh., 51. Astron., 6.51.200 (1985).

104.049   The flux of meteoroids in the vicinity of the orbit of comet Halley.
G. Cevolani, A. Hajduk.
Nuovo Cimento C, Vol. 7C, Ser. 1, No. 4, p. 447 – 457 (1984). Abstr. in Phys. Abstr., Vol. 88, No. 1259, Entry 68738 (1985).

104.050   Orioniden en Tauriden. Najaarszwermen 1984 goed waargenomen.
R. Veltman.
Radiant, Jaarg. 7, Nr. 1, p. 11 – 15 (1985).

104.051   Waarnemingen winter 1984 – 1985.
R. Veltman.
Radiant, Jaarg. 7, Nr. 2, p. 22 – 26 (1985).
Three meteor showers were covered from November 23 – December 27: Geminids, Ursids and Taurids. During 15 different nights, 29 observers reported 2862 meteors, both shower and sporadic within 272.45 hours of net observing time.

104.052   Meteoren van komeet 1983d?
P. Jenniskens.
Radiant, Jaarg. 7, Nr. 2, p. 31 – 33 (1985).

104.053   Lyridenaktie 1985: Lyriden te Bussloo.
H. Betlem.
Radiant, Jaarg. 7, Nr. 3, p. 40 (1985).

104.054   Geminiden 1984. A telescopic ZHR (*Zenit Hourly Rate*).
P. Jenniskens, F. Witte.
Radiant, Jaarg. 7, Nr. 3, p. 44 – 55, 56 (1985).
From telescopic meteor observations the authors will derive the number of meteors which can be seen if our eyes would have the limiting magnitude of a telescope and the radiant of a stream is in the zenith.

104.055   Report of IAU Commission 22: Meteors and interplanetary dust (*Météorites et la poussière interplanétaire*).
O. I. Bel'kovich.
Trans. IAU, Vol. XIXA, p. 235 – 252 (1985). – See Abstr. 003.046.

104.056   The fireball Valeč of 3 August 1984.
Z. Ceplecha.
Říše hvězd, Vol. 66, No. 1, p. 3 – 5 (1985). In Czech.

104.057   Activity of the radar Orionids 1984.
J. Váňa.
Říše hvězd, Vol. 66, No. 3, p. 58 (1985). In Czech.

104.058   Meteor showers of comet Halley.
A. Hajduk.
Kozmos, Vol. 16, No. 1, p. 4 – 6 (1985). In Slovak.

104.059   The bodies producing fireballs.
Z. Ceplecha.
Kozmos, Vol. 16, No. 2, p. 40 – 43 (1985). In Czech.

104.060   Harderwijk herfstakties.
K. Miskotte.
Werkgroepnieuws, Vol. 13, Nr. 1, p. 4 – 5 (1985).
In the period of 21 October to 4 November 1984 the activity of the Orionids and Taurids was examined by a team of observers from Harderwijk (O.S.M.), The Netherlands.

104.061 **De Lyriden 1984.**
P. Roggemans.
Werkgroepnieuws, Vol. 13, Nr. 1, p. 6 (1985).

104.062 **Perseiden 1984.**
P. Roggemans.
Werkgroepnieuws, Vol. 13, Nr. 1, p. 8 – 13 (1985).

104.063 **Orioniden & Tauriden 1984.**
P. Roggemans.
Werkgroepnieuws, Vol. 13, Nr. 1, p. 13 – 15 (1985).

104.064 **Harderwijk: Geminiden.**
K. Miskotte.
Werkgroepnieuws, Vol. 13, Nr. 1, p. 15 – 17 (1985).

104.065 **Radiowaarnemingen.**
C. Steyaert.
Werkgroepnieuws, Vol. 13, Nr. 1, p. 22 – 24 (1985).

104.066 **Canada: Perseids.**
B. Katz.
Werkgroepnieuws, Vol. 13, Nr. 1, p. 29 – 32 (1985).

104.067 **The η–Aquariids 1984.**
J. Wood.
Werkgroepnieuws, Vol. 13, Nr. 1, p. 33 – 34 (1985).

104.068 **The Lyrids 1984.**
J. Wood.
Werkgroepnieuws, Vol. 13, Nr. 1, p. 34 (1985).

104.069 **Italian fireball.**
M. Eltri, E. Stomeo.
Werkgroepnieuws, Vol. 13, Nr. 1, p. 35 (1985).

104.070 **Harderwijk: Geminiden.**
K. Miskotte.
Werkgroepnieuws, Vol. 13, Nr. 2, p. 49 – 50 (1985).

104.071 **Meteoren–Observatorium Cyclops: Lyriden.**
K. Jobse.
Werkgroepnieuws, Vol. 13, Nr. 3, p. 76 – 77 (1985).

104.072 **Radiowerk: Lyriden.**
C. Steyaert.
Werkgroepnieuws, Vol. 13, Nr. 3, p. 77 – 78 (1985).

104.073 **Japan: orbital data.**
T. Ochiai.
Werkgroepnieuws, Vol. 13, Nr. 3, p. 88 – 94 (1985).

104.074 **Denmark: the Orionids.**
P. Aldrich.
Werkgroepnieuws, Vol. 13, Nr. 3, p. 98 – 99 (1985).

104.075 **Observations of a bright meteor.**
Yu. V. Dubrovskij.
Komet. Tsirk., No. 339 (1985). In Russian.

104.076 **The age of the Geminid meteor stream.**
E. N. Kramer, I. S. Shestaka.
Komet. Tsirk., No. 342 (1985). In Russian.

104.077 **Meteor studies, a new IHW discipline.**
P. B. Babadzhanov, A. Hajduk, B. A. Linblad,
B. A. McIntosh.
IHW Newsl., No. 7, p. 12 – 17 (1985).
The new discipline is responsible for the scientific program and coordination of meteor studies in the IHW. The main observing techniques will utilize meteor radars, photoelectric devices, photographic methods (direct and spectroscopic) and visual recording (naked–eye and telescopic). The new program will not interfere with processes and procedures already established for meteor observations. The IHW program is an opportunity to coordinate and combine many different types of observations relating to the Halley meteor showers.

Asteroids, comets, meteors II.
See Abstr. 011.023.

Meteor bodies in the interplanetary space and in the earth's atmosphere. All–Union conference held in Suzdal, 17 – 22 September 1984.
See Abstr. 012.047.

Development of meteor investigations in Tomsk.
See Abstr. 013.042.

Production of the upper atmospheric sodium from impinging meteors.
See Abstr. 082.076.

Infrared observations of the extinct cometary candidate minor planet (3200) 1983 TB.
See Abstr. 098.053.

Comet Giacobini–Zinner and the Draconids.
See Abstr. 103.190.

The size, mass, mass loss and age of Halley's comet.
See Abstr. 103.905.

Collisional balance of the meteoritic complex.
See Abstr. 106.063.

105 Meteorites, Meteorite Craters

105.001   Cathodoluminescence zoning and minor elements in forsterites from the Murchison (C2) and Allende (C3V) carbonaceous chondrites.
I. M. Steele, J. V. Smith, C. Skirius.
Nature, Vol. 313, No. 6000, p. 294 – 297 (1985).
Cathodoluminescence photographs of forsterite grains in Murchison (C2) and Allende (C3) meteorites reveal a blue core (inclusion–free) with planar boundaries to a red or dark rim (inclusions of glass, metal and silicate). The authors have also performed high–precision electron microprobe analyses revealing in these forsterites unusually large amounts of the "minor" elements Al, Ti and Ca in the blue cores (and sharp chemical changes at the blue/red boundary), suggesting formation by crystallization at high temperatures from a source rich in these metals.

105.002   Minor–element signature of relic olivine grains in deep–sea particles – a match with forsterites from C2 meteorites.
I. M. Steele, J. V. Smith, D. E. Brownlee.
Nature, Vol. 313, No. 6000, p. 297 – 299 (1985).
Deep–sea particles (DSP) extracted magnetically from oceanic sediments are the products of aerodynamic heating and cooling of extraterrestrial materials. The authors describe the results of high–precision electron microprobe analyses on relic forsterites (26 spots) in 13 different DSP, comparing their "minor"–element (Mn, Cr, Ca, Ti, Al) signatures with those of olivines found in several types of meteorites.

105.003   Deep–sea stony spherules and the primordial nebula.
D. W. G. Sears.
Nature, Vol. 313, No. 6003, p. 528 (1985).

105.004   The stable carbon isotopes in enstatite chondrites and Cumberland Falls.
P. Deines, F. E. Wickman.
Geochim. Cosmochim. Acta, Vol. 49, No. 1, p. 89 – 95 (1985).
The carbon isotopic composition of the total carbon in the enstatite chondrites Indarch, Abee, St. Marks, Pillistfer, Hvittis and Daniel's Kuil and the enstatite achondrite Cumberland Falls has been measured. The empirical relationship between carbon isotopic composition and total carbon content is distinct from that of carbonaceous and ordinary chondrites.

105.005   H–chondrites: trace element clues to their origin.
J. W. Morgan, M.–J. Janssens, H. Takahashi, J. Hertogen, E. Anders.
Geochim. Cosmochim. Acta, Vol. 49, No. 1, p. 247 – 259 (1985).
The authors have analyzed 10 H–chondrites for 20 trace elements, using RNAA. The data show that H–chondrites are not isochemical. The abundance pattern of siderophiles varies systematically with petrologic type.

105.006   The compositional classification of chondrites: IV. Ungrouped chondritic meteorites and clasts.
G. W. Kallemeyn, J. T. Wasson.
Geochim. Cosmochim. Acta, Vol. 49, No. 1, p. 261 – 270 (1985).
Six specimens of unusual chondritic materials were analyzed by neutron activation for 30 elements in order to assess their degree of chondritic compositional pristinity and to search for evidence of genetic links to other chondrites. Five have highly recrystallized textures; the other, the Cumberland Falls chondrite, has suffered minor metamorphic recrystallization. The elemental composition of the Cumberland Falls chondrite is virtually identical to that of LL chondrites, and its O–isotope composition is closely similar to those of some unequilibrated ordinary chondrites including LL Semarkona.

105.007   A major revision of iron meteorite cooling rates – an experimental study of the growth of the Widmanstätten pattern.
C. Narayan, J. I. Goldstein.
Geochim. Cosmochim. Acta, Vol. 49, No. 2, p. 397 – 410 (1985).
AEM (*analytical electron microscopy*) techniques were employed to study the growth of intragranular kamacite in Fe–Ni–P alloys containing between 5% and 10% Ni and 0 and 1.0 wt% P. The experimental evidence was used to develop a numerical model to simulate kamacite growth and hence predict cooling rates of meteorites from observed Widmanstätten patterns.

105.008   Origin of howardites, diogenites and eucrites: a mass balance constraint.
P. H. Warren.
Geochim. Cosmochim. Acta, Vol. 49, No. 2, p. 577 – 586 (1985).
A mass balance constraint indicates that more than half of the basaltic components in howardites formed as residual liquids from fractional crystallization of melts that had earlier produced the diogenite–like pyroxene cumulate components.

105.009   Examination of meteorites.
A. Christiansen, L. Larsen, H. Roy–Poulsen.
Fys. Tidsskr., Vol. 82, No. 1, p. 12 – 43 (1984). In Danish. Abstr. in Phys. Abstr., Vol. 88, No. 1248, Entry 9684 (1985).

105.010   On identification of ultraheavy cosmic nuclei in extraterrestrial olivines.
V. P. Perelyghin (*V. P. Perelygin*), S. G. Stetsenko, O. O. Otgonsuren (*O. Otgonsurehn*), R. Ignatova, G. Ya. Starodub, M. Haiduc, D. Hasegan, M. Bytyci, B. Jacubi, R. Antanasievici, J. Todorovici, P. Vater, R. Brandt, R. Schpor.
Rev. Roum. Phys., Vol. 29, No. 7, p. 615 – 621 (1984). Abstr. in Phys. Abstr., Vol. 88, No. 1250, Entry 18613 (1985).

105.011   Glasses formed by hypervelocity impact.
D. Stöffler.
J. Non–Cryst. Solids, Vol. 67, No. 1 – 3, p. 465 – 502 (1984). Abstr. in Phys. Abstr., Vol. 88, No. 1251, Entry 23765 (1985). – See Abstr. 012.016.

105.012   A comparison between terrestrial impact glasses and lunar volcanic glasses: the case of fluorine.
C. Koeberl, W. Kiesl, F. Kluger, H. H. Weinke.
J. Non–Cryst. Solids, Vol. 67, No. 1 – 3, p. 637 – 648 (1984). Abstr. in Phys. Abstr., Vol. 88, No. 1251, Entry 23773 (1985). – See Abstr. 012.016.

105.013   Tektites and terrestrial meteorite craters: possible associations.
R. S. Dietz.
J. Non–Cryst. Solids, Vol. 67, No. 1 – 3, p. 649 – 655 (1984). Abstr. in Phys. Abstr., Vol. 88, No. 1251, Entry 23774 (1985). – See Abstr. 012.016.

105.014   Aspects of a glassy meteorite from the moon bearing on some problems in extraterrestrial glass–making.
R. F. Fudali, M. Kreutzberger, G. Kurat, F. Brandstätter.
J. Non–Cryst. Solids, Vol. 67, No. 1 – 3, p. 383 – 396 (1984). Abstr. in Phys. Abstr., Vol. 88, No. 1251, Entry 24356 (1985). – See Abstr. 012.016.

105.015   The formation of chondrule–containing extraterrestrial materials: evidence for rapid solidification under microgravity conditions.
P. Z. Budka.
J. Non–Cryst. Solids, Vol. 67, No. 1 – 3, p. 413 – 419 (1984). Abstr. in Phys. Abstr., Vol. 88, No. 1251, Entry 24357 (1985). – See Abstr. 012.016.

105.016 Absolute isotopic abundances of Ti in meteorites.
F. R. Niederer, D. A. Papanastassiou,
G. J. Wasserburg.
Geochim. Cosmochim. Acta, Vol. 49, No. 3, p. 835 – 851 (1985).

The absolute isotope abundance of Ti has been determined in Ca–Al–rich inclusions from the Allende and Leoville meteorites and in samples of whole meteorites. The absolute Ti isotope abundances differ by a significant mass dependent isotope fractionation transformation from the previously reported abundances, which were normalized for fractionation using $^{46}Ti/^{48}Ti$. Therefore, the absolute compositions define distinct nucleosynthetic components from those previously identified or reflect the existence of significant mass dependent isotope fractionation in nature. The authors provide a general formalism for determining the possible isotope compositions of the exotic Ti from the measured composition, for different values of isotope fractionation in nature and for different mixing ratios of the exotic and normal components.

105.017 Meteor crater: what's in a name?
W. G. Hoyt.
Publ. Astron. Soc. Pac., Vol. 97, No. 588, p. 200 (1985). Abstract. – See Abstr. 010.281.

105.018 A search for presolar organic matter in meteorites.
J. Yang, S. Epstein.
Geophys. Res. Lett., Vol. 12, No. 2, p. 73 – 76 (1985).

The D/H ratios and the $^{13}C/^{12}C$ ratios of acid–insoluble organic matter of 4 meteorites, Ochansk (H4), Plainview (H5), Gladstone (H6) and Odessa (IA), were measured. δD values for hydrogen extracted by stepwise combustion were negative, down to -28%. $\delta^{13}C$ values were also negative except in the case of the carbon coming off at the highest temperature steps for Plainview and Odessa meteorites. The concentrations of ^{13}C–rich carbon were 3 – 5 orders of magnitude smaller than those found in Murchison meteorite, suggesting that relic grains of stellar condensates were mostly destroyed in the meteorites examined.

105.019 Superheavy elements in meteorites: expectations and disappointments.
Yu. A. Shukolyukov.
Priroda, No. 1, p. 20 – 29 (1985). In Russian.

105.020 Das Tunguska–Objekt war kein Komet.
A. Dill.
Astron. Raumfahrt, 22. Jahrg., Heft 2, p. 32 – 36 (1985).

105.021 The carbon and nitrogen isotopic composition of ureilites: implications for their genesis.
M. M. Grady, I. P. Wright, P. K. Swart, C. T. Pillinger.
Geochim. Cosmochim. Acta, Vol. 49, No. 4, p. 903 – 915 (1985).

Stable carbon isotope analysis of a suite of ureilite achondrites has established the existence of a variation in isotopic composition of the main carbonaceous phases (graphite/diamond) amongst the samples; that the diamond has been derived from the graphite without significant fractionation is apparent. The distribution of isotopic compositions of these phases does not follow a regular pattern, rather the samples appear to scatter into two discrete populations.

105.022 The origin and history of the metal and sulfide components of chondrules.
J. N. Grossman, J. T. Wasson.
Geochim. Cosmochim. Acta, Vol. 49, No. 4, p. 925 – 939 (1985).

The chief goal of this paper is to assess the relative contributions of the various inferred processes to the origin of chondrules, and to the solar nebula from which they formed.

105.023 Impact melting of the Cachari eucrite 3.0 Gy ago.
D. D. Bogard, G. J. Taylor, K. Keil, M. R. Smith,
R. A. Schmitt.
Geochim. Cosmochim. Acta, Vol. 49, No. 4, p. 941 – 946 (1985).

To increase the data base on specific impact events in the histories of meteorite parent bodies, the authors have made ^{39}Ar–^{40}Ar age measurements on both the glass and host rock in Cachari. They also made additional microprobe measurements of the glass and determined the bulk composition of glass and host by instrumental neutron activation analysis. The results indicate that the glass formed by impact melting on the Cachari parent body 3.0 Gy ago (1 Gy = 10^9 years).

105.024 Willy: a prize noble Ur–Fremdling – its history and implications for the formation of Fremdlinge and CAI.
J. T. Armstrong, A. El Goresy, G. J. Wasserburg.
Geochim. Cosmochim. Acta, Vol. 49, No. 4, p. 1001 – 1022 (1985).

A detailed mineralogic and chemical study of Willy, a very large (150 μm diameter) Fremdling from the Allende CAI 5241, was performed and compared to other Fremdlinge from Allende CAI's 5241 and TS–34 in an attempt to understand the nature and mode of formation of these exotic and complex objects. From the observed chemistry and texture, a multistage sequence of formation of Willy, possibly occurring in the solar nebula can be deduced.

105.025 The chemical conditions on the parent body of the Murchison meteorite: some conclusions based on amino, hydroxy and dicarboxylic acids.
E. T. Peltzer, J. L. Bada, G. Schlesinger, S. L. Miller.
Adv. Space Res., Vol. 4, No. 12, p. 69 – 74 (1984). – See Abstr. 012.036.

105.026 The role of kinetics in thermal transformations of carbonaceous chondrites.
T. V. Malysheva.
Cosmochemistry and meteoritics, p. 72 – 81 (1984). In Russian. Abstr. in Ref. Zh., 51. Astron., 1.51.376 (1985). – See Abstr. 012.037.

105.027 Formation of chondrules in space.
Eh. V. Sobotovich, V. P. Semenenko.
Cosmochemistry and meteoritics, p. 3 – 33 (1984). In Russian. Abstr. in Ref. Zh., 51. Astron., 1.51.378 (1985). – See Abstr. 012.037.

105.028 On the nature of the anomalous components of Xe in the carbonaceous chondrite Kainsaz CO.
A. V. Fisenko, Dang Vu Minh, L. F. Semenova,
A. K. Lavrukhina, Yu. A. Shchukolyukov.
Dokl. Akad. Nauk SSSR. Ser. Mat. Fiz., Tom 274, No. 3, p. 698 – 702 (1984). In Russian. Abstr. in Ref. Zh., 51. Astron., 1.51.381 (1985).

105.029 On the nature of isotope anomalies in meteorites.
A. K. Lavrukhina.
Cosmochemistry and meteoritics, p. 33 – 53 (1984). In Russian. Abstr. in Ref. Zh., 51. Astron., 1.51.382 (1985). – See Abstr. 012.037.

105.030 Abundance of magnesium isotopes in meteorites and terrestrial rocks.
S. P. Ol'shtynskij, L. V. Kononenko.
Cosmochemistry and meteoritics, p. 122 – 130 (1984). In Russian. Abstr. in Ref. Zh., 51. Astron., 1.51.383 (1985). – See Abstr. 012.037.

105.031 Radioactivity of the Malakal and Kutais chondrites.
V. D. Gorin, G. K. Ustinova.
Cosmochemistry and meteoritics, p. 140 – 146 (1984). In Russian. Abstr. in Ref. Zh., 51. Astron., 1.51.384 (1985). – See Abstr. 012.037.

105.032 Model experiments on the investigation of metal–troilite interactions in normal chondrites.
A. V. Fisenko, Z. A. Lavrent'eva, A. K. Lavrukhina.
Cosmochemistry and meteoritics, p. 100 – 104 (1984). In Russian. Abstr. in Ref. Zh., 51. Astron., 1.51.430 (1985). – See Abstr. 012.037.

105.033 Secondary minerals and the structure of the Kaali, Severnyj Kolchim and Erofeevka meteorites.
I. A. Yudin, V. N. Loginov, A. M. Dymkin, V. A. Vilisov, V. G. Gmyra, N. F. Obotnin.
Cosmochemistry and meteoritics, p. 105 – 117 (1984). In Russian. Abstr. in Ref. Zh., 51. Astron., 1.51.431 (1985). – See Abstr. 012.037.

105.034 Structure of associations of leading chemical elements in olivine pallasites, lunar and terrestrial rocks.
Yu. K. Burkov, V. D. Kolomenskij, V. S. Pevzner, V. A. Gubanov.
Cosmochemistry and meteoritics, p. 135 – 139 (1984). In Russian. Abstr. in Ref. Zh., 51. Astron., 1.51.432 (1985). – See Abstr. 012.037.

105.035 The cosmogenic factor in geological processes.
B. A. Troshechev.
Cosmochemistry and meteoritics, p. 197 – 204 (1984). In Russian. Abstr. in Ref. Zh., 51. Astron., 1.51.437 (1985). – See Abstr. 012.037.

105.036 The Loganchia impact crater.
S. A. Vishnevskij, Yu. A. Dolgov.
Dokl. Akad. Nauk SSSR. Ser. Mat. Fiz., Tom 275, No. 3, p. 684 – 688 (1984). In Russian. Abstr. in Ref. Zh., 51. Astron., 1.51.442 (1985).

105.037 Meteoritic matter in impactites.
V. I. Fel'dman, I. G. Kapustkina, L. B. Granovskij, L. V. Sazonova.
Cosmochemistry and meteoritics, p. 147 – 151 (1984). In Russian. Abstr. in Ref. Zh., 51. Astron., 1.51.444 (1985). – See Abstr. 012.037.

105.038 Comparative analysis of the microcomponent composition of the rocks of the Boltysh meteorite crater.
A. A. Val'ter, G. K. Eremenko, V. E. Tepikin, L. K. Magur, I. A. Lejko.
Cosmochemistry and meteoritics, p. 151 – 164 (1984). In Russian. Abstr. in Ref. Zh., 51. Astron., 1.51.445 (1985). – See Abstr. 012.037.

105.039 Alkali elements in the impactites of the Elgygytgyn and Boltysh meteorite craters.
E. P. Gurov, E. P. Gurova.
Cosmochemistry and meteoritics, p. 205 – 211 (1984). In Russian. Abstr. in Ref. Zh., 51. Astron., 1.51.446 (1985). – See Abstr. 012.037.

105.040 Experimental results of selective condensation of alkali-rich vapour and ultra–alkali impactites.
O. V. Parfenova, O. I. Yakovlev.
Cosmochemistry and meteoritics, p. 117 – 122 (1984). In Russian. Abstr. in Ref. Zh., 51. Astron., 1.51.451 (1985). – See Abstr. 012.037.

105.041 Criteria of ablation of meteoritic bodies.
A. I. Vislyj.
Aehrodinam. vkhoda tel v atmos. planet. Moskva, p. 58 – 66 (1983). In Russian. Abstr. in Ref. Zh., 51. Astron., 2.51.244 (1985).

105.042 Martian atmospheric carbon dioxide and weathering products in SNC meteorites.
R. H. Carr, M. M. Grady, I. P. Wright, C. T. Pillinger.
Nature, Vol. 314, No. 6008, p. 248 – 250 (1985).
Analysis of the $^{12}C/^{13}C$ isotopic ratio in the SNC meteorite EETA 79001 supports the hypothesis of a Martian origin of these meteorites. Another carbon–containing species, believed to be carbonate has been found in the SNC meteorite Nakhla and may be a product of atmospheric weathering on Mars.

105.043 Meteorites of Chukotski.
G. F. Pavlov, A. A. Plyashkevich, N. E. Savva.
Geol. i polez. iskopaemye. Sev.–Vost. Azii. Vladivostok, p. 190 – 201 (1984). In Russian. Abstr. in Ref. Zh., 51. Astron., 3.51.297 (1985).

105.044 Impact rocks of the Arizona crater (USA).
O. A. Bogatikov, R. V. Boyarskaya, S. V. Soboleva, D. I. Frikh–Khar.
Izv. Akad. Nauk SSSR. Ser. geol., No. 11, p. 19 – 25 (1984). In Russian. Abstr. in Ref. Zh., 51. Astron., 3.51.332 (1985).

105.045 The Loganchinsk astrobleme in the traps of the Tunguska syneclise.
S. A. Vishnevskij.
Inst. geol. i geofiz. SO Akad. Nauk SSSR., Prepr., No. 1, 28 pp. (1984). In Russian. Abstr. in Ref. Zh., 51. Astron., 3.51.333 (1985).

105.046 Study of the problem of the Tunguska meteorite at Tomsk University and at the Tomsk Department of VAGO.
N. V. Vasil'ev.
Astrón. geod., Tomsk, No. 10, p. 40 – 48 (1984). In Russian. Abstr. in Ref. Zh., 51. Astron., 3.51.345 (1985).

105.047 Chondrule rims and interchondrule matrix in U.O.C's.
C. Alexander, D. Barber, R. Hutchison.
Meteoritics, Vol. 19, No. 4, p. 184 – 185 (1984). Abstract. – See Abstr. 010.641.

105.048 Meteorite concentrations in Antarctica – how complete is the picture?
J. O. Annexstad.
Meteoritics, Vol. 19, No. 4, p. 185 – 186 (1984). Abstract. – See Abstr. 010.641.

105.049 Fremdlinge in Leoville and Allende CAI: clues to post–formation cooling and alteration.
J. T. Armstrong, I. D. Hutcheon, G. J. Wasserburg.
Meteoritics, Vol. 19, No. 4, p. 186 – 187 (1984). Abstract. – See Abstr. 010.641.

105.050 Spectroscopic identification of probable pallasite parent bodies.
J. F. Bell, M. J. Gaffey, B. R. Hawke.
Meteoritics, Vol. 19, No. 4, p. 187 – 188 (1984). Abstract. – See Abstr. 010.641.

105.051 ^{244}Pu fission track and metallographic cooling rates of Toluca and Copiapo iron (IA) meteorites.
Y. Bekheiri.
Meteoritics, Vol. 19, No. 4, p. 188 (1984). Abstract. – See Abstr. 010.641.

105.052 Noble gas chronology of LL chondrites.
T. Bernatowicz, M. Honda, F. Podosek.
Meteoritics, Vol. 19, No. 4, p. 188 – 189 (1984). Abstract. – See Abstr. 010.641.

105.053 Raman microscopy of the Lodran meteorite.
R. W. Bild, D. R. Tallant.
Meteoritics, Vol. 19, No. 4, p. 190 (1984). Abstract. – See Abstr. 010.641.

105.054 Anomalous isotopic composition of chromium in Allende inclusions.
J. L. Birck, C. J. Allègre.
Meteoritics, Vol. 19, No. 4, p. 190 – 192 (1984). Abstract. – See Abstr. 010.641.

105.055 **Bulk compositions of Al–rich chondrules in ordinary and carbonaceous chondrites: variations and similarities.**
A. Bischoff.
Meteoritics, Vol. 19, No. 4, p. 192 (1984). Abstract. – See Abstr. 010.641.

105.056 **Perovskite–hibonite–spinel–bearing, refractory inclusions and Ca–Al–rich chondrules in enstatite chondrites.**
A. Bischoff, K. Keil, D. Stöffler.
Meteoritics, Vol. 19, No. 4, p. 193 – 194 (1984). Abstract. – See Abstr. 010.641.

105.057 **On the origin of excess ^{40}Ar in the four shergottite–achondrites.**
D. Bogard.
Meteoritics, Vol. 19, No. 4, p. 195 (1984). Abstract. – See Abstr. 010.641.

105.058 **Trace element abundances in rim layers of an Allende Type A coarse–grained, Ca–Al–rich inclusion.**
W. V. Boynton, D. A. Wark.
Meteoritics, Vol. 19, No. 4, p. 195 – 197 (1984). Abstract. – See Abstr. 010.641.

105.059 **Mg isotopic measurements in fine–grained Ca–Al–rich inclusions.**
C. A. Brigham, D. A. Papanastassiou, G. J. Wasserburg.
Meteoritics, Vol. 19, No. 4, p. 198 – 199 (1984). Abstract. – See Abstr. 010.641.

105.060 **The influence of gravitational body force in meteoritic chrondrule and lunar glass formation.**
P. Z. Budka.
Meteoritics, Vol. 19, No. 4, p. 201 (1984). Abstract. – See Abstr. 010.641.

105.061 **Speculations on the formation of metallic meteorite phases.**
P. Z. Budka, F. F. Milillo.
Meteoritics, Vol. 19, No. 4, p. 201 – 202 (1984). Abstract. – See Abstr. 010.641.

105.062 **Confirmation of cosmogenic neon from precompaction irradiation of Kapoeta and Murchison.**
M. W. Caffee, C. M. Hohenberg, T. D. Swindle, J. N. Goswami.
Meteoritics, Vol. 19, No. 4, p. 203 (1984). Abstract. – See Abstr. 010.641.

105.063 **Strain analysis of the Leoville chondrite and conditions in asteroidal interiors.**
P. M. Cain, H. Y. McSween Jr.
Meteoritics, Vol. 19, No. 4, p. 203 – 204 (1984). Abstract. – See Abstr. 010.641.

105.064 **Martian atmospheric CO_2 in an Antarctic meteorite?**
R. H. Carr, I. P. Wright, C. T. Pillinger.
Meteoritics, Vol. 19, No. 4, p. 204 – 205 (1984). Abstract. – See Abstr. 010.641.

105.065 **High temperature phase equilibria in the enstatite chondrite system.**
W. A. Cassidy.
Meteoritics, Vol. 19, No. 4, p. 206 (1984). Abstract. – See Abstr. 010.641.

105.066 **Anomalous silver in sulfide nodules.**
J. H. Chen, G. J. Wasserburg.
Meteoritics, Vol. 19, No. 4, p. 206 – 207 (1984). Abstract. – See Abstr. 010.641.

105.067 **Structural development in the Santa Catharina meteorite.**
R. S. Clarke Jr.
Meteoritics, Vol. 19, No. 4, p. 207 – 208 (1984). Abstract. – See Abstr. 010.641.

105.068 **Massive schreibersite in the Bellsbank and Santa Luzia meteorites.**
R. S. Clarke Jr., R. Sallamuthu, J. I. Goldstein.
Meteoritics, Vol. 19, No. 4, p. 208 (1984). Abstract. – See Abstr. 010.641.

105.069 **s–process Nd in Allende residues.**
D. D. Clayton.
Meteoritics, Vol. 19, No. 4, p. 208 – 209 (1984). Abstract. – See Abstr. 010.641.

105.070 **Ion probe determinations of the REE contents of individual meteoritic phosphate grains.**
G. Crozaz, E. Zinner.
Meteoritics, Vol. 19, No. 4, p. 209 – 210 (1984). Abstract. – See Abstr. 010.641.

105.071 **Boron cosmochemistry.**
D. B. Curtis, E. S. Gladney.
Meteoritics, Vol. 19, No. 4, p. 211 (1984). Abstract. – See Abstr. 010.641.

105.072 **Fractionation of siderophiles in cosmic spherules and its implications.**
J. Czajkowski, P. Englert.
Meteoritics, Vol. 19, No. 4, p. 212 – 213 (1984). Abstract. – See Abstr. 010.641.

105.073 **A scandalously refractory inclusion in Ornans.**
A. M. Davis.
Meteoritics, Vol. 19, No. 4, p. 214 (1984). Abstract. – See Abstr. 010.641.

105.074 **Low temperature diffusion coefficients in the Fe–Ni and FeNiP systems – application to meteorite cooling rates.**
D. C. Dean, J. I. Goldstein.
Meteoritics, Vol. 19, No. 4, p. 214 – 215 (1984). Abstract. – See Abstr. 010.641.

105.075 **Plutonium and uranium in individual crystals of merrilite and apatite of St. Severin.**
S. De Chazal, G. Crozaz, M. Bourot–Denise, P. Pellas.
Meteoritics, Vol. 19, No. 4, p. 216 – 217 (1984). Abstract. – See Abstr. 010.641.

105.076 **The significance of two pyroxene mafic clasts in basaltic achondrites.**
J. S. Delaney.
Meteoritics, Vol. 19, No. 4, p. 218 (1984). Abstract. – See Abstr. 010.641.

105.077 **Trace elements in Antarctic H5 chondrites: weathering effects and comparison with non–Antarctic falls.**
J. E. Dennison, D. W. Lingner, M. E. Lipschutz.
Meteoritics, Vol. 19, No. 4, p. 219 (1984). Abstract. – See Abstr. 010.641.

105.078 **Petrology and shock age of the Palo Blanco Creek eucrite.**
T. Dickinson, K. Keil, L. LaPaz, D. Bogard, R. A. Schmitt, M. R. Smith, M. Rhodes.
Meteoritics, Vol. 19, No. 4, p. 219 – 220 (1984). Abstract. – See Abstr. 010.641.

105.079 **Shatter cones: definitive criteria for meteorite impact.**
R. S. Dietz, J. F. McHone.
Meteoritics, Vol. 19, No. 4, p. 221 (1984). Abstract. – See Abstr. 010.641.

105.080   The Kendleton L4 fragmental breccia: parent body surface history.
A. J. Ehlmann, K. Keil, E. R. D. Scott, H. W. Weber,
L. Schultz, T. K. Mayeda, R. N. Clayton.
Meteoritics, Vol. 19, No. 4, p. 221 – 222 (1984). Abstract. – See Abstr. 010.641.

105.081   Trace elements in high–temperature inclusions from Murchison.
V. Ekambaram, S. M. Sluk, L. Grossman, A. M. Davis.
Meteoritics, Vol. 19, No. 4, p. 222 – 223 (1984). Abstract. – See Abstr. 010.641.

105.082 ^{53}Mn in main fragments of the Norton County meteorite.
P. Englert, R. Sarafin.
Meteoritics, Vol. 19, No. 4, p. 223 – 224 (1984). Abstract. – See Abstr. 010.641.

105.083   The chemistry and origin of refractory metal particles from Ca–Al–rich inclusions in carbonaceous chondrites.
B. Fegley, H. Palme.
Meteoritics, Vol. 19, No. 4, p. 225 (1984). Abstract. – See Abstr. 010.641.

105.084   A new H–3 chondrite from Study Butte, Texas.
K. Fredriksson, R. S. Clarke Jr., R. Pugh.
Meteoritics, Vol. 19, No. 4, p. 225 – 226 (1984). Abstract. – See Abstr. 010.641.

105.085   Comparison of chemical compositions of Yamato polymict eucrites.
T. Fukuoka.
Meteoritics, Vol. 19, No. 4, p. 227 (1984). Abstract. – See Abstr. 010.641.

105.086   North American tektites and microtektites from Barbados, West Indies.
B. P. Glass, C. A. Burns, D. H. Lerner, A. Sanfilippo.
Meteoritics, Vol. 19, No. 4, p. 228 (1984). Abstract. – See Abstr. 010.641.

105.087   Aqueous alteration on meteorite parent bodies: possible role of "unfrozen" water and the Antarctic meteorite analogy.
J. L. Gooding.
Meteoritics, Vol. 19, No. 4, p. 228 – 229 (1984). Abstract. – See Abstr. 010.641.

105.088   Ureilite petrogenesis: clues from a graphite and metal–bearing intrusive complex, Disko Island, Greenland.
C. A. Goodrich.
Meteoritics, Vol. 19, No. 4, p. 230 – 231 (1984). Abstract. – See Abstr. 010.641.

105.089   Size dependence of chondrule textural types.
J. N. Goswami.
Meteoritics, Vol. 19, No. 4, p. 231 (1984). Abstract. – See Abstr. 010.641.

105.090   Exposure history and compaction age of Banten CM chondrite.
J. N. Goswami, K. Nishiizumi.
Meteoritics, Vol. 19, No. 4, p. 231 – 232 (1984). Abstract. – See Abstr. 010.641.

105.091   Cosmogenic records in Antarctic meteorites – II.
J. N. Goswami, K. Nishiizumi, J. R. Arnold.
Meteoritics, Vol. 19, No. 4, p. 232 (1984). Abstract. – See Abstr. 010.641.

105.092   Chemical variations among chondrules from Quingzhen (EH3).
J. N. Grossman, E. R. Rambaldi, R. S. Rajan, A. E. Rubin,
J. T. Wasson.
Meteoritics, Vol. 19, No. 4, p. 232 – 233 (1984). Abstract. – See Abstr. 010.641.

105.093   Thermoluminescence as a paleothermometer?
R. K. Guimon, K. S. Weeks, B. D. Keck,
D. W. G. Sears.
Meteoritics, Vol. 19, No. 4, p. 233 – 234 (1984). Abstract. – See Abstr. 010.641.

105.094   Refractory inclusions in amoeboid olivine aggregates in Allende.
A. Hashimoto, L. Grossman.
Meteoritics, Vol. 19, No. 4, p. 234 – 235 (1984). Abstract. – See Abstr. 010.641.

105.095 ^{22}Na, ^{60}Co and long–lived cosmogenic radionuclides in meteorite falls.
U. Herpers, P. Englert.
Meteoritics, Vol. 19, No. 4, p. 236 – 237 (1984). Abstract. – See Abstr. 010.641.

105.096 ^{26}Al, ^{60}Co, ^{53}Mn and ^{22}Na profiles in the Jilin drill cores.
G. Heusser, Z. Ouyang, W. Yi, D. Kaether, E. Pernicka,
W. Hampel, T. Kirsten.
Meteoritics, Vol. 19, No. 4, p. 237 – 238 (1984). Abstract. – See Abstr. 010.641.

105.097   Pairing in Antarctic mesosiderites.
R. H. Hewins.
Meteoritics, Vol. 19, No. 4, p. 238 (1984). Abstract. – See Abstr. 010.641.

105.098   X–radiography of slices of the Allende meteorite.
D. Heymann, M. J. Smith, J. B. Anderson.
Meteoritics, Vol. 19, No. 4, p. 238 – 239 (1984). Abstract. – See Abstr. 010.641.

105.099   Magnesium and calcium isotopes in hibonite–bearing CAIs.
R. W. Hinton, L. Grossman, G. J. MacPherson.
Meteoritics, Vol. 19, No. 4, p. 240 – 241 (1984). Abstract. – See Abstr. 010.641.

105.100   An SEM study of preterrestrial alteration effects in ALHA 77003.
R. M. Housley.
Meteoritics, Vol. 19, No. 4, p. 242 – 243 (1984). Abstract. – See Abstr. 010.641.

105.101   Excess ^{41}K in Allende CAI: a hint re–examined.
I. E. Hutcheon, J. T. Armstrong, G. J. Wasserburg.
Meteoritics, Vol. 19, No. 4, p. 243 – 244 (1984). Abstract. – See Abstr. 010.641.

105.102   Mg isotopic studies of Leoville "compact" type A CAI.
I. E. Hutcheon, J. T. Armstrong, G. J. Wasserburg.
Meteoritics, Vol. 19, No. 4, p. 244 – 245 (1984). Abstract. – See Abstr. 010.641.

105.103   Chondrules in the Bishunpur LL3 chondrite.
R. Hutchison, C. Alexander, D. J. Barber.
Meteoritics, Vol. 19, No. 4, p. 245 – 246 (1984). Abstract. – See Abstr. 010.641.

105.104   Major element chemical compositions of chondrules in unequilibrated chondrites.
Y. Ikeda.
Meteoritics, Vol. 19, No. 4, p. 246 (1984). Abstract. – See Abstr. 010.641.

105.105 **Atmospheric heating of meteorites: results from nuclear track studies.**
R. Jha.
Meteoritics, Vol. 19, No. 4, p. 246 – 247 (1984). Abstract. – See Abstr. 010.641.

105.106 **Extreme incompatibility of Pb during the crystallization of magmatic iron meteorites.**
J. H. Jones, S. R. Hart.
Meteoritics, Vol. 19, No. 4, p. 248 (1984). Abstract. – See Abstr. 010.641.

105.107 **Meteorite Hg diffusion studies.**
S. Jovanovic, G. W. Reed Jr.
Meteoritics, Vol. 19, No. 4, p. 248 – 250 (1984). Abstract. – See Abstr. 010.641.

105.108 **The composition of enstatite chondrites.**
G. W. Kallemeyn, J. T. Wasson.
Meteoritics, Vol. 19, No. 4, p. 250 (1984). Abstract. – See Abstr. 010.641.

105.109 **Evolutionary history of CI and CM chondrites.**
J. F. Kerridge, J. D. Macdougall.
Meteoritics, Vol. 19, No. 4, p. 250 – 251 (1984). Abstract. – See Abstr. 010.641.

105.110 **Spectroscopic evidence of regolith maturation.**
T. V. V. King, M. J. Gaffey, L. A. McFadden.
Meteoritics, Vol. 19, No. 4, p. 251 – 252 (1984). Abstract. – See Abstr. 010.641.

105.111 **Geochemistry of Muong–Nong type tektites V: unusual ferric/ferrous ratios.**
C. Koeberl, F. Kluger, W. Kiesl.
Meteoritics, Vol. 19, No. 4, p. 253 – 254 (1984). Abstract. – See Abstr. 010.641.

105.112 **The identification of group II inclusions by electron microprobe analysis of perovskite.**
A. S. Kornacki, J. A. Wood.
Meteoritics, Vol. 19, No. 4, p. 254 – 256 (1984). Abstract. – See Abstr. 010.641.

105.113 **Inclusions in IAB irons: is the partial differentiation model compatible with barometry and chronometry?**
A. Kracher.
Meteoritics, Vol. 19, No. 4, p. 256 – 257 (1984). Abstract. – See Abstr. 010.641.

105.114 **Investigations of taenite from iron meteorites and chondrites.**
L. Larsen, H. Roy–Poulsen, N. O. Roy–Poulsen, L. Vistisen, G. B. Jensen, J. M. Knudsen.
Meteoritics, Vol. 19, No. 4, p. 258 – 259 (1984). Abstract. – See Abstr. 010.641.

105.115 **Barntrup: LL–4 chondrite.**
G. R. Levi–Donati, F. Brandstätter, G. Kurat.
Meteoritics, Vol. 19, No. 4, p. 260 (1984). Abstract. – See Abstr. 010.641.

105.116 **Interstellar carbon grains from the Murchison meteorite: electron microscopy.**
R. S. Lewis, M. Ohtsuki.
Meteoritics, Vol. 19, No. 4, p. 260 – 261 (1984). Abstract. – See Abstr. 010.641.

105.117 **Mobile trace elements and thermal histories of H4–6 chondrites: comparison with L4–6 chondrites.**
D. W. Lingner, T. J. Huston, M. E. Lipschutz.
Meteoritics, Vol. 19, No. 4, p. 261 – 262 (1984). Abstract. – See Abstr. 010.641.

105.118 **Refractory inclusions in the Mighei C2 meteorite.**
G. J. MacPherson.
Meteoritics, Vol. 19, No. 4, p. 262 – 263 (1984). Abstract. – See Abstr. 010.641.

105.119 **Investigations concerning the magmatic iron meteorites: static vs. dynamic experiments.**
D. J. Malvin, J. H. Jones, M. J. Drake.
Meteoritics, Vol. 19, No. 4, p. 263 – 264 (1984). Abstract. – See Abstr. 010.641.

105.120 **Heavy noble gases associated with neon–E.**
O. K. Manuel, R. Ramaduri, S. Ramaduri.
Meteoritics, Vol. 19, No. 4, p. 264 – 265 (1984). Abstract. – See Abstr. 010.641. With a correction in Vol. 20, No. 1, p. 173 (1985).

105.121 **Meteorite distributions on the main icefield of the Allan Hills region, Antarctica.**
U. B. Marvin.
Meteoritics, Vol. 19, No. 4, p. 265 – 266 (1984). Abstract. – See Abstr. 010.641.

105.122 **Oxygen isotopes in deep sea spherules.**
T. K. Mayeda, R. N. Clayton, D. E. Brownlee.
Meteoritics, Vol. 19, No. 4, p. 266 – 267 (1984). Abstract. – See Abstr. 010.641.

105.123 **Thermal history of the Peace River L6 chondrite based on ^{40}Ar–^{39}Ar measurements.**
P. McConville, G. Turner.
Meteoritics, Vol. 19, No. 4, p. 267 – 268 (1984). Abstract. – See Abstr. 010.641.

105.124 **Oxygen isotopes and ferromagnesian chondrule populations in carbonaceous chondrites.**
H. Y. McSween Jr.
Meteoritics, Vol. 19, No. 4, p. 271 (1984). Abstract. – See Abstr. 010.641.

105.125 **Neutron activation analysis of stoney spherules from a marine sediment sample.**
H. T. Millard Jr., P. Englert.
Meteoritics, Vol. 19, No. 4, p. 271 – 272 (1984). Abstract. – See Abstr. 010.641.

105.126 **Chemical zoning and homogenization of olivines in ordinary chondrites.**
M. Miyamoto, D. S. McKay, G. A. McKay, M. B. Duke.
Meteoritics, Vol. 19, No. 4, p. 272 – 273 (1984). Abstract. – See Abstr. 010.641.

105.127 **Correlated isotopic anomalies in the elements silicon and magnesium from Allende inclusions.**
C. Molini–Velsko, T. K. Mayeda, R. N. Clayton.
Meteoritics, Vol. 19, No. 4, p. 273 – 274 (1984). Abstract. – See Abstr. 010.641.

105.128 **Shock deformation texture of olivine crystals of the EETA 79001 shergottite.**
H. Mori, H. Takeda.
Meteoritics, Vol. 19, No. 4, p. 275 (1984). Abstract. – See Abstr. 010.641.

105.129 **Actinide chemistry of Allende components.**
M. T. Murrell, D. S. Burnett.
Meteoritics, Vol. 19, No. 4, p. 275 – 276 (1984). Abstract. – See Abstr. 010.641.

105.130 **The low–energy secondary cosmic ray flux: detectors, in iron meteorites.**
S. V. S. Murty, K. Marti.
Meteoritics, Vol. 19, No. 4, p. 276 – 277 (1984). Abstract. – See Abstr. 010.641.

105.131   Silica–niningerite–enstatite clasts in the primitive enstatite chondrites.
H. Nagahara.
Meteoritics, Vol. 19, No. 4, p. 277 – 278 (1984). Abstract. – See Abstr. 010.641.

105.132   Unique clasts with V–shaped REE pattern in L6 chondrites.
N. Nakamura, K. Yanai, Y. Matsumoto.
Meteoritics, Vol. 19, No. 4, p. 278 – 279 (1984). Abstract. – See Abstr. 010.641.

105.133   Solar flare neon and solar cosmic ray fluxes in the past using gas–rich meteorites.
C. M. Nautiyal, M. N. Rao.
Meteoritics, Vol. 19, No. 4, p. 279 (1984). Abstract. – See Abstr. 010.641.

105.134   Evidence for the formation and the size of a metal core in the eucrite parent body.
H. E. Newsom.
Meteoritics, Vol. 19, No. 4, p. 279 – 280 (1984). Abstract. – See Abstr. 010.641.

105.135   Hydrothermal alteration of suevite impact ejecta at the Ries meteorite crater, F.R. Germany.
H. E. Newsom, G. Graup.
Meteoritics, Vol. 19, No. 4, p. 280 – 282 (1984). Abstract. – See Abstr. 010.641.

105.136   Bishopville: the importance of a field search.
H. H. Nininger.
Meteoritics, Vol. 19, No. 4, p. 282 (1984). Abstract. – See Abstr. 010.641.

105.137   Cosmogenic nuclides in peculiar meteorites.
K. Nishiizumi, J. R. Arnold, D. Elmore.
Meteoritics, Vol. 19, No. 4, p. 283 (1984). Abstract. – See Abstr. 010.641.

105.138   Sr and Nd isotopic systematics of EETA 79001.
L. Nyquist, J. Wooden, B. Bansal, H. Wiesmann, C.-Y. Shih.
Meteoritics, Vol. 19, No. 4, p. 284 (1984). Abstract. – See Abstr. 010.641.

105.139   Schöllhornite, $Na_{0.3}(H_2O)_1[CrS_2]$, a new mineral in the Norton County enstatite achondrite.
A. Okada, K. Keil, B. F. Leonard, I. D. Hutcheon.
Meteoritics, Vol. 19, No. 4, p. 284 – 285 (1984). Abstract. – See Abstr. 010.641.

105.140   The chondrite Higashi–Koen.
A. Okada, M. Shima, S. Murayama, N. Takaoka.
Meteoritics, Vol. 19, No. 4, p. 285 – 286 (1984). Abstract. – See Abstr. 010.641.

105.141   Ureilites: the case of missing diamonds and a new neon component.
U. Ott, H. P. Löhr, F. Begemann.
Meteoritics, Vol. 19, No. 4, p. 287 – 288 (1984). Abstract. – See Abstr. 010.641.

105.142   The irradiation history of Dhurmsala meteorite.
J. T. Padia, M. N. Rao, J. N. Goswami, P. Englert, U. Herpers.
Meteoritics, Vol. 19, No. 4, p. 288 – 289 (1984). Abstract. – See Abstr. 010.641.

105.143   Absence of excess ^{26}Mg in anorthite from the Vaca Muerta mesosiderite.
D. A. Papanastassiou, G. J. Wasserburg, U. B. Marvin.
Meteoritics, Vol. 19, No. 4, p. 289 – 290 (1984). Abstract. – See Abstr. 010.641.

105.144   Silicate inclusions in IVA iron meteorites.
M. Prinz, C. E. Nehru, J. S. Delaney, K. Fredriksson, H. Palme.
Meteoritics, Vol. 19, No. 4, p. 291 – 292 (1984). Abstract. – See Abstr. 010.641.

105.145   H chondritic clasts in a Yamato L6 chondrite: implications for metamorphism.
M. Prinz, C. E. Nehru, M. K. Weisberg, J. S. Delaney, K. Yanai, H. Kojima.
Meteoritics, Vol. 19, No. 4, p. 292 – 293 (1984). Abstract. – See Abstr. 010.641.

105.146   Salem meteorite.
R. N. Pugh.
Meteoritics, Vol. 19, No. 4, p. 293 (1984). Abstract. – See Abstr. 010.641.

105.147   Possible neutron effects in the Elephant Morraine shergottite.
R. S. Rajan, E. Rambaldi, A. S. Tamhane, G. Poupeau.
Meteoritics, Vol. 19, No. 4, p. 293 – 294 (1984). Abstract. – See Abstr. 010.641.

105.148   Coexisting chalcophile and lithophile uranium in Qingzhen (EH3) chondrite.
E. R. Rambaldi, R. S. Rajan, M. T. Murrell, D. S. Burnett.
Meteoritics, Vol. 19, No. 4, p. 295 (1984). Abstract. – See Abstr. 010.641.

105.149   Fine–grained millimeter–sized objects in type 3 ordinary chondrites and their relation to chondrules and matrix.
S. I. Recca, E. R. D. Scott, G. J. Taylor, K. Keil.
Meteoritics, Vol. 19, No. 4, p. 296 – 297 (1984). Abstract. – See Abstr. 010.641.

105.150   Cosmogenic radionuclides and exposure histories of SNC meteorites.
R. C. Reedy.
Meteoritics, Vol. 19, No. 4, p. 297 – 298 (1984). Abstract. – See Abstr. 010.641.

105.151   On the record of galactic cosmic ray flux and exposure histories of iron meteorites.
S. Regnier, B. Lavielle, K. Marti, G. N. Simonoff.
Meteoritics, Vol. 19, No. 4, p. 298 – 299 (1984). Abstract. – See Abstr. 010.641.

105.152   Compositions of 7 Allan Hills polymict eucrites and one diogenite.
A. M. Reid, A. P. le Roex.
Meteoritics, Vol. 19, No. 4, p. 299 (1984). Abstract. – See Abstr. 010.641.

105.153   A study of tetrataenite.
K. B. Reuter, J. I. Goldstein, D. B. Williams, E. P. Butler.
Meteoritics, Vol. 19, No. 4, p. 299 – 300 (1984). Abstract. – See Abstr. 010.641.

105.154   Magnetite in the Essebi and Haripura CM chondrites.
M. W. Rowe, M. Hyman, E. B. Ledger.
Meteoritics, Vol. 19, No. 4, p. 302 – 303 (1984). Abstract. – See Abstr. 010.641.

105.155   Mössbauer spectroscopy and X–ray diffraction of samples from the Santa Catharina iron meteorite.
H. Roy–Poulsen, L. Larsen, N. O. Roy–Poulsen, L. Vistisen, R. S. Clarke Jr., G. B. Jensen, J. M. Knudsen.
Meteoritics, Vol. 19, No. 4, p. 303 (1984). Abstract. – See Abstr. 010.641.

105.156 The Colony meteorite and the possible existence of a new chemical subgroup of CO3 chondrites.
A. E. Rubin, J. A. James, B. D. Keck, K. S. Weeks, D. W. G. Sears, E. Jarosewich.
Meteoritics, Vol. 19, No. 4, p. 303 – 304 (1984). Abstract. – See Abstr. 010.641.

105.157 Cosmogenic nuclides in Antarctic achondrites and chondrites.
R. Sarafin, U. Herpers, R. Wieler, P. Signer.
Meteoritics, Vol. 19, No. 4, p. 307 – 308 (1984). Abstract. – See Abstr. 010.641.

105.158 The fractionation of siderophile and volatile elements on the eucrite parent body (EPB).
W. Schmitt, G. Weckwerth, H. Wänke.
Meteoritics, Vol. 19, No. 4, p. 308 – 309 (1984). Abstract. – See Abstr. 010.641.

105.159 Terrestrial ages of Antarctic meteorites.
L. Schultz, M. Freundel.
Meteoritics, Vol. 19, No. 4, p. 310 (1984). Abstract. – See Abstr. 010.641.

105.160 Metamorphism of type 3 carbonaceous and ordinary chondrites.
E. R. D. Scott, G. J. Taylor.
Meteoritics, Vol. 19, No. 4, p. 310 – 311 (1984). Abstract. – See Abstr. 010.641.

105.161 Experimental and theoretical study of the formation of the IIAB, IIIAB and IVA iron meteorite chemical groups from the parent liquid.
R. Sallamuthu, J. I. Goldstein.
Meteoritics, Vol. 19, No. 4, p. 311 – 312 (1984). Abstract. – See Abstr. 010.641.

105.162 Crater ages, comet showers, and the putative "death star".
E. M. Shoemaker, R. F. Wolfe.
Meteoritics, Vol. 19, No. 4, p. 313 (1984). Abstract. – See Abstr. 010.641.

105.163 Three new chondrite finds from Roosevelt County, New Mexico.
P. P. Sipiera.
Meteoritics, Vol. 19, No. 4, p. 313 – 314 (1984). Abstract. – See Abstr. 010.641.

105.164 Minor elements in relict olivine grains of deep–sea spheres: match with Mg–rich olivines from C2 meteorites.
J. V. Smith, I. M. Steele, D. E. Brownlee.
Meteoritics, Vol. 19, No. 4, p. 314 – 315 (1984). Abstract. – See Abstr. 010.641.

105.165 Minor elements and cathodoluminescence of Mg–rich olivines from Murchison and Allende carbonaceous meteorites.
I. M. Steele, C. M. Skirius, J. V. Smith.
Meteoritics, Vol. 19, No. 4, p. 315 – 316 (1984). Abstract. – See Abstr. 010.641.

105.166 ^{40}Ar–^{39}Ar evidence for two discrete tektite–forming events in the Australian–southeast Asian area.
D. Storzer, E. K. Jessberger, N. Klay, G. A. Wagner.
Meteoritics, Vol. 19, No. 4, p. 317 (1984). Abstract. – See Abstr. 010.641.

105.167 Thermoluminescence age of Meteor Crater, Arizona.
S. R. Sutton.
Meteoritics, Vol. 19, No. 4, p. 317 – 318 (1984). Abstract. – See Abstr. 010.641.

105.168 Noble gases in SNC meteorites.
T. D. Swindle, M. W. Caffee, C. M. Hohenberg, G. B. Hudson, R. S. Rajan.
Meteoritics, Vol. 19, No. 4, p. 318 – 319 (1984). Abstract. – See Abstr. 010.641.

105.169 Ordinary eucrites with slowly cooled textures and their crystallization history.
H. Takeda, H. Mori, Y. Ikeda.
Meteoritics, Vol. 19, No. 4, p. 319 – 320 (1984). Abstract. – See Abstr. 010.641.

105.170 A quantitative look at chondrite metamorphism.
G. J. Taylor, E. R. D. Scott.
Meteoritics, Vol. 19, No. 4, p. 320 – 321 (1984). Abstract. – See Abstr. 010.641.

105.171 Peculiar spherules from Antarctica and their origin.
Y. Tazawa, Y. Fujii.
Meteoritics, Vol. 19, No. 4, p. 321 (1984). Abstract. – See Abstr. 010.641.

105.172 Petrography of Cape York and Grant: irons with simple Pd–Ag systematics.
J. M. Teshima, G. J. Wasserburg, A. El Goresy.
Meteoritics, Vol. 19, No. 4, p. 322 (1984). Abstract. – See Abstr. 010.641.

105.173 Polymict eucrite ALHA81011: equilibrated clasts in a glassy matrix.
A. H. Treiman.
Meteoritics, Vol. 19, No. 4, p. 323 – 324 (1984). Abstract. – See Abstr. 010.641.

105.174 Core formation in the shergottite parent body (SPB).
A. H. Treiman, M. J. Drake.
Meteoritics, Vol. 19, No. 4, p. 324 – 325 (1984). Abstract. – See Abstr. 010.641.

105.175 Noble gases in the unshocked ureilite Allan Hills 78019.
J. F. Wacker.
Meteoritics, Vol. 19, No. 4, p. 326 – 327 (1984). Abstract. – See Abstr. 010.641.

105.176 Unexplained Fe, Ni and S anomalies in CV chondrite components.
D. A. Wark.
Meteoritics, Vol. 19, No. 4, p. 329 – 330 (1984). Abstract. – See Abstr. 010.641.

105.177 Origin of howardites, diogenites and eucrites: a mass balance constraint.
P. H. Warren.
Meteoritics, Vol. 19, No. 4, p. 330 – 331 (1984). Abstract. – See Abstr. 010.641.

105.178 Cumberland Falls (chondrite), Suwahib (Buwah) and other ordinary chondrites showing evidence of postaccretional reduction.
J. T. Wasson, G. W. Kallemeyn.
Meteoritics, Vol. 19, No. 4, p. 331 (1984). Abstract. – See Abstr. 010.641.

105.179 Chemical relationships among shergottites, nakhlites, and Chassigny.
G. Weckwerth, H. Wänke.
Meteoritics, Vol. 19, No. 4, p. 331 – 332 (1984). Abstract. – See Abstr. 010.641.

105.180 A new class of enstatite chondrite?
K. S. Weeks, D. W. G. Sears.
Meteoritics, Vol. 19, No. 4, p. 332 – 333 (1984). Abstract. – See Abstr. 010.641.

105.181   **Geochemistry of Muong–Nong type tektites VI: major element determinations and inhomogeneities.**
H. H. Weinke, C. Koeberl.
Meteoritics, Vol. 19, No. 4, p. 333 – 335 (1984). Abstract. – See Abstr. 010.641.

105.182   **The asteroidal source region of ordinary chondrites.**
G. W. Wetherill.
Meteoritics, Vol. 19, No. 4, p. 335 (1984). Abstract. – See Abstr. 010.641.

105.183   **Cosmogenic nuclides in a cross section of the 300 kg Knyahinya chondrite.**
R. Wieler, P. Signer, U. Herpers, R. Sarafin, G. Bonani, H. J. Hofmann, E. Morenzoni, M. Nessi, M. Suter.
Meteoritics, Vol. 19, No. 4, p. 335 – 336 (1984). Abstract. – See Abstr. 010.641.

105.184   **Remeasurement of nitrogen in EETA 79001 glass.**
R. C. Wiens, R. H. Becker, R. O. Pepin.
Meteoritics, Vol. 19, No. 4, p. 336 – 337 (1984). Abstract. – See Abstr. 010.641.

105.185   **Clastic texture of meteorites.**
L. L. Wilkening, S. E. Jensen, K. Schaudt.
Meteoritics, Vol. 19, No. 4, p. 337 – 338 (1984). Abstract. – See Abstr. 010.641.

105.186   **Petrology of some ordinary chondrite regolith breccias: implications for parent body history.**
C. V. Williams, K. Keil, A. E. Rubin, A. S. Miguel.
Meteoritics, Vol. 19, No. 4, p. 338 (1984). Abstract. – See Abstr. 010.641.

105.187   **Searching for impact craters using Space Shuttle photography.**
C. A. Wood, C. Dailey, W. Daley, G. Wells.
Meteoritics, Vol. 19, No. 4, p. 338 – 339 (1984). Abstract. – See Abstr. 010.641.

105.188   **Laser Raman Microprobe study of mineral phases in meteorites.**
B. Wopenka, S. A. Sandford.
Meteoritics, Vol. 19, No. 4, p. 340 – 341 (1984). Abstract. – See Abstr. 010.641.

105.189   **Lunar meteorites in Japanese collection of the Yamato meteorites.**
K. Yanai, H. Kojima, T. Katsushima.
Meteoritics, Vol. 19, No. 4, p. 342 – 343 (1984). Abstract. – See Abstr. 010.641.

105.190   **Search for isotopic anomalies in Odessa (IA), Ochansk (H4), Plainview (H5), and Gladstone (H6).**
J. Yang, S. Epstein.
Meteoritics, Vol. 19, No. 4, p. 343 – 344 (1984). Abstract. – See Abstr. 010.641.

105.191   **Noble gases in the Bells (C2) and Sharps (H3) meteorites.**
M. G. Zadnik.
Meteoritics, Vol. 19, No. 4, p. 344 – 345 (1984). Abstract. – See Abstr. 010.641.

105.192   **Hydrothermal alteration of CM carbonaceous chondrites: implications of the identification of tochilinite as one type of meteoritic PCP.**
M. E. Zolensky.
Meteoritics, Vol. 19, No. 4, p. 346 – 347 (1984). Abstract. – See Abstr. 010.641.

105.193   **Guin: an ungrouped iron with silicate inclusions.**
P. Zong, A. E. Rubin, J. T. Wasson, J. Westcott.
Meteoritics, Vol. 19, No. 4, p. 347 – 348 (1984). Abstract. – See Abstr. 010.641.

105.194   **Chondrites from comets?**
R. L. Stratford.
Speculations Sci. Technol., Vol. 7, No. 3, p. 149 – 153 (1984). Abstr. in Phys. Abstr., Vol. 88, No. 1252, Entry 29899 (1985).

105.195   **^{21}Ne–^{60}Co classification of Jilin samples and their relative positions in the parent body.**
X.-x. Zhou, B. Li, Z.-y. Ouyang.
Kexue Tongbao (Beijing), Vol. 29, No. 10, p. 1362 – 1365 (1984). Abstr. in Phys. Abstr., Vol. 88, No. 1252, Entry 29900 (1985).

105.196   **Isotopic composition of hydrogen in carbonaceous chondrites: are carbonaceous chondrites an analog to the protomatter of the earth?**
A. L. Devirts, E. P. Lagutina, A. A. Ul'yanov.
Vses. simp. No. 10 po stabil. izotopam v geokhimii, Moskva, 3 – 5 dek., 1984. Tez. dokl. Moskva, p. 145 (1984). In Russian. Abstr. in Ref. Zh., 51. Astron., 4.51.245 (1985).

105.197   **Isotopic composition of hydrogen of the carbonaceous chondrite Kaidun 1.**
A. L. Devirts, A. V. Ivanov, E. P. Lagutina, A. A. Ul'yanov, Yu. A. Shukolyukov.
Vses. simp. No. 10 po stabil. izotopam v geokhimii, Moskva, 3 – 5 dek., 1984. Tez. dokl. Moskva, p. 144 (1984). In Russian. Abstr. in Ref. Zh., 51. Astron., 4.51.246 (1985).

105.198   **Anomalous isotopic composition of oxygen in the refractory inclusions of the carbonaceous chondrite Efremovka.**
V. I. Ustinov, A. A. Ul'yanov, E. A. Gavrilov, Yu. A. Shukolyukov.
Vses. simp. No. 10 po stabil. izotopam v geokhimii, Moskva, 3 – 5 dek., 1984. Tez. dokl. Moskva, p. 142 (1984). In Russian. Abstr. in Ref. Zh., 51. Astron., 4.51.248 (1985).

105.199   **Isotopic composition of oxygen in the new carbonaceous chondrite Kaidun.**
A. V. Ivanov, A. A. Ul'yanov, V. P. Strizhov, Yu. A. Shukolyukov.
Vses. simp. No. 10 po stabil. izotopam v geokhimii, Moskva, 3 – 5 dek., 1984. Tez. dokl. Moskva, p. 241 (1984). In Russian. Abstr. in Ref. Zh., 51. Astron., 4.51.249 (1985).

105.200   **Isotopic composition of Xe in the acid–insoluble radicals of the chondrite Efremovka C 3 and their chemical composition.**
A. V. Fisenko, L. F. Semenova, Dang Vu Minh, G. M. Kolesov, K. I. Ignatenko, A. K. Lavrukhina, Yu. A. Shukolyukov.
Vses. simp. No. 10 po stabil. izotopam v geokhimii, Moskva, 3 – 5 dek., 1984. Tez. dokl. Moskva, p. 143 (1984). In Russian. Abstr. in Ref. Zh., 51. Astron., 4.51.253 (1985).

105.201   **Astroblemes and endogeneous ring structures.**
V. I. Fel'dman.
Probl. geol. i razvedki mestorozhd. polez. iskopaemykh Sibiri. Tez. dokl. konf., Tomsk, 2 – 4 fevr., 1983. Tomsk, p. 32 (1983). In Russian. Abstr. in Ref. Zh., 51. Astron., 4.51.271 (1985).

105.202   **Unmelted meteoritic debris in the Late Pliocene iridium anomaly: evidence for the ocean impact of a nonchondritic asteroid.**
F. T. Kyte, D. E. Brownlee.
Geochim. Cosmochim. Acta, Vol. 49, No. 5, p. 1095 – 1108 (1985).
The authors' observations imply that an asteroid impacted into the Antarctic Basin in the southeastern Pacific at ~2.3 Ma resulting in deposition of a monolayer of mm–sized particles in the vicinity of Eltanin core E13–3. The projectile was composed of howardite and siderophile–rich (metal?) components and may have had a diameter of a few hundred meters. An oceanic impact is suggested by the large amount of siderophile–rich molten debris and the incorporation of seawater produced abundant vesicles and added Na, K, Cl, and radiogenic Sr to the debris.

105.203 **Mineral chemistry and origin of spinel–rich inclusions in the Allende CV3 chondrite.**
A. S. Kornacki, J. A. Wood.
Geochim. Cosmochim. Acta, Vol. 49, No. 5, p. 1219 – 1237 (1985).

The authors report the results of detailed mineralogical study of 36 spinel–bearing Allende inclusions and discuss selected aspects of their mineral chemistry and petrology.

105.204 **Unequilibrated ordinary chondrites: a tentative subclassification based on volatile–element content.**
E. Anders, M. G. Zadnik.
Geochim. Cosmochim. Acta, Vol. 49, No. 5, p. 1281 – 1291 (1985).

For unequilibrated ordinary chondrites, two measures of primitiveness are available: volatile content, in principle reflecting accretion conditions from the solar nebula, and metamorphism, reflecting reheating in the parent bodies. These two measures do not always correlate, and the authors have therefore developed a tentative classification scheme based on volatile content that complements the Sears et al. (1980) scheme based on metamorphism.

105.205 **Erratum: "Chemical compositions of refractory inclusions in the Murchison C2 chondrite" [Geochim. Cosmochim. Acta, Vol. 48, No. 10, p. 2089 – 2105 (1984)].**
R. W. Hinton.
Geochim. Cosmochim. Acta, Vol. 49, No. 5, p. 1293 (1985). See Abstr. 38.105.070.

105.206 **Laboratory infrared transmission spectra of individual interplanetary dust particles from 2.5 to 25 microns.**
S. A. Sandford, R. M. Walker.
Astrophys. J., Vol. 291, No. 2, p. 838 – 851 (1985).

Dust particles collected in the stratosphere that have chondritic elemental abundances provide a new form of extraterrestrial matter for laboratory study. Spectral transmission measurements from 2.5 to 25 µm of 26 such particles show that almost all have a dominant 10 µm silicate absorption feature. Twenty–two of the particles can be grouped into one of three spectral classes referred to as olivines, layer–lattice silicates, and pyroxenes after the minerals that provide the best match to the dust spectra. Measurements of large D/H enrichments and solar flare nuclear tracks confirm that at least some particles are interplanetary dust particles. The layer–lattice silicates have additional bands at 3.0, 6.0, 6.8, and 11.4 µm. A comparison of these spectral features with the infrared spectra of comets and interstellar grains is presented.

105.207 **Negative polarization does not imply a dusty asteroid surface.**
Yu. G. Shkuratov, L. A. Akimov, V. P. Tishkovets.
Sov. Astron. Lett., Vol. 10, No. 5, p. 331 – 332 (1984). English translation of 38.105.055.

105.208 **Constraints on dust formation in stellar outflows and destruction in the ISM from the meteoritic record.**
J. A. Nuth.
Bull. Am. Astron. Soc., Vol. 16, No. 4, p. 921 (1984). Abstract. – See Abstr. 010.062.

105.209 **Earthward bound from chaotic regions of the asteroid belt.**
C. D. Murray.
Nature, Vol. 315, No. 6022, p. 712 (1985).

This note comments on recent research supporting the hypothesis that meteorites are asteroidal material delivered to the Earth. Calculations by J. Wisdom show that orbits of asteroids near the 3/1 resonance of Jupiter could evolve from a typical low-eccentricity orbit to the high eccentricity necessary for Earth-crossing orbits (see also Abstr. 105.210).

105.210 **Meteorites may follow a chaotic route to Earth.**
J. Wisdom.
Nature, Vol. 315, No. 6022, p. 731 – 733 (1985).

It is widely believed that meteorites originate in the asteroid belt, but the precise dynamical mechanism whereby material is transported to Earth has eluded discovery. Recently the author has shown that there is a large chaotic zone in the phase space near the 3/1 mean motion commensurability with Jupiter and that the chaotic trajectories within this zone have particularly large variations in orbital eccentricity. Here he presents a numerical integration which demonstrates that at least some of these chaotic trajectories do have the properties required to transport meteoritic material from the asteroid belt to Earth. Combined with the Monte Carlo calculations which show that the resulting meteorites are consistent with all the observational constraints, the case for this chaotic route to Earth is fairly strong.

105.211 **Poorly graphitized carbon as a new cosmothermometer for primitive extraterrestrial materials.**
F. J. M. Rietmeijer, I. D. R. Mackinnon.
Nature, Vol. 315, No. 6022, p. 733 – 736 (1985).

A recent addition to the suite of primitive extraterrestrial materials available for study by cosmochemists includes particles collected from the stratosphere called chondritic porous (CP) aggregates. The authors describe here the nature of the most abundant carbon phase in a carbon–rich CP aggregate (sample no. W7029*A) collected from the stratosphere as part of the Johnson Space Center Cosmic Dust Program. By comparison with experimental and terrestrial studies of poorly graphitized carbon (PGC), it is shown that the graphitization temperature, or the degree of ordering in the PGC, may provide a useful cosmothermometer for primitive extraterrestrial materials.

105.212 **Asteroidal source of ordinary chondrites.**
G. W. Wetherill.
Meteoritics, Vol. 20, No. 1, p. 1 – 22 (1985).

The orbital evolution of asteroidal fragments with diameters ranging from 10 cm to 20 km, injected into the 3:1 Kirkwood gap at 2.50 A.U., has been investigated using Monte Carlo techniques. It is assumed that this material can become Earth-crossing on a time scale of 10^6 years, as a result of a chaotic zone discovered by Wisdom, associated with the 3:1 resonance. This phenomenon, as well as close encounter planetary perturbations, the secular resonance, and the ablative effects of the Earth's atmosphere are included in the determination of the orbital characteristics of meteorites impacting the Earth derived by fragmentation of this asteroidal material. It is found that the predicted meteorite orbits closely match those found for observed ordinary chondrites, and the total flux is in approximate agreement with the observed fall rate of ordinary chondrites.

105.213 **Chemical variations among L–chondrites. IV. Analyses, with petrographic notes, of 13 L–group and 3 LL–group chondrites.**
E. Jarosewich, R. T. Dodd.
Meteoritics, Vol. 20, No. 1, p. 23 – 36 (1985).

The authors review procedures for selecting, preparing and analyzing meteorite samples, present new analyses of 16 ordinary chondrites, and discuss variations of Fe, S and Si in the L–group.

105.214 **Uranium distribution and "excessive" U–He ages in iron meteoritic troilite.**
D. E. Fisher.
Meteoritics, Vol. 20, No. 1, p. 37 – 41 (1985).

Uranium was measured by fission track analysis in meteoritic troilite and graphite. The distribution is extremely heterogeneous, with a few high–U grains dominating the total abundances. U/He ages cannot be estimated from such a distribution pattern, and therefore previously reported excessive ages are not valid.

105.215 **Some morphologic systematics of complex impact structures.**
R. J. Pike.
Meteoritics, Vol. 20, No. 1, p. 49 – 68 (1985).

Similarities among impact structures on different planets and satellites suggest that the cratering process transcends variations in both target and impactor. In particular, impact may control the spacing of concentric rings, if not their actual emplacement.

In at least four respects the scaled horizontal dimensions of complex meteorite–impact structures on Earth resemble those of multi–ring basins and large craters on the Moon, Mars, Mercury, and some outer satellites. Two minor differences in morphology suggest that uniquely terrestrial conditions may control some horizontal dimensions of meteorite craters.

105.216 Noble gases and the classification of Brachina.
U. Ott, H.–P. Löhr, F. Begemann.
Meteoritics, Vol. 20, No. 1, p. 69 – 78 (1985).

Noble–gas systematics show that Brachina is not a member of the SNC–group of meteorites. The same evidence argues strongly against any simple genetic relationship with eucrites. The noble–gas abundance pattern resembles closely that in silicate inclusions from the iron meteorites Campo del Cielo and Udei Station. Abundances of cosmic–ray produced ^{3}He and ^{21}Ne indicate an exposure age of ~ 2.4 Ma.

105.217 Mineralogy, petrology and chemistry of the Messina (Italy) chondrite.
B. Baldanza, M. Triscari.
Meteoritics, Vol. 20, No. 1, p. 79 – 87 (1985).

The meteorite which fell near Messina, Italy, on 16 July 1955 is a typical olivine–hypersthene (L–group) chondrite. The stone contains abundant chondrules. The matrix consists chiefly of broken chondrules with tiny fragments of crystals and rare amorphous material. Chondrules form more than 42% of the meteorite by volume.

105.218 E–chondrites: significance of the partition of elements between "silicate" and "sulphide".
A. J. Easton.
Meteoritics, Vol. 20, No. 1, p. 89 – 101 (1985).

It is shown that there are chemical differences in the silicates and sulphides within the E–chondrites that preclude these meteorites from belonging to an isochemical evolutionary series derived from a common parent material.

105.219 Mineralogy of the Bocaiuva iron meteorite: a preliminary study.
C. Desnoyers, M. Christophe–Michel–Levy, I. S. Azevedo, R. B. Scorzelli, J. Danon, E. Galvão da Silva.
Meteoritics, Vol. 20, No. 1, p. 113 – 124 (1985).

This preliminary study underlines the unusual nature of this iron meteorite and raises several questions concerning the genetic relations between silicates, sulfide and metal, and the thermal history of the whole material.

105.220 Lukanga Swamp: probably astrobleme.
S. Vrána.
Meteoritics, Vol. 20, No. 1, p. 125 – 139 (1985).

This article summarizes information relevant to the interpretation of the Lukanga Swamp – a depression some 52 km in diameter in central Zambia – as a considerably eroded astrobleme. The results of reconnaissance geological mapping, an airborne magnetic survey and petrological study of a small set of samples of impact rocks are presented. Provisional correlation of the brecciation event with the local geological record suggests a Paleozoic age of the astrobleme.

105.221 Rare earth element abundances in separated phases of Mayo Belwa, an enstatite achondrite.
A. L. Graham, P. Henderson.
Meteoritics, Vol. 20, No. 1, p. 141 – 149 (1985).

Separated fractions from the Mayo Belwa aubrite have been analysed for the rare earth elements. The data show no ytterbium anomalies but the feldspar and magnetic fractions show complementary europium anomalies.

105.222 Catalog of the Collection of Meteorites at the University of California, Los Angeles.
J. T. Wasson, A. E. Rubin, E. Widom.
Meteoritics, Vol. 20, No. 1, p. 151 – 170 (1985).

105.223 Nd and Sr isotopic compositions of tektite material from Barbados and their relationship to North American tektites.
H. H. Ngo, G. J. Wasserburg, B. P. Glass.
Geochim. Cosmochim. Acta, Vol. 49, No. 6, p. 1479 – 1485 (1985).

The purpose of this study is to determine the Nd and Sr isotopic composition of individual microtektites and tektite fragments from Barbados in order to establish the possible sources of these microtektites and their relationship to known tektite strewn fields.

105.224 Isotopic composition of hydrogen in some L– and LL– chondrites.
A. L. Devirts, E. P. Lagutina, Yu. A. Shukolyukov.
Geokhimiya, No. 1, p. 35 – 40 (1985). In Russian. Abstr. in Ref. Zh., 51. Astron., 5.51.269 (1985).

105.225 Variation of the isotopic composition of xenon in meteoritic matter under the action of shock waves.
Z. K. Mil'nikova, Dang Vu Minh, Yu. A. Shukolyukov, A. K. Lavrukhina, S. V. Pershin.
Vses. simp. No. 10 po stabil. izotopam v geokhimii. Moskva, 3 – 5 dek., 1984. Tez. dokl. Moskva, p. 125 (1984). In Russian. Abstr. in Ref. Zh., 51. Astron., 5.51.275 (1985).

105.226 Composition and content of normal and isoprenoid hydrocarbons in the Pavel meteorite.
Ch. P. Ivanov, R. Zh. Stoyanova, G. Stoev.
Dokl. Bolg. Akad. Nauk, Vol. 37, No. 9, p. 1227 – 1230 (1984). Abstr. in Ref. Zh., 51. Astron., 5.51.291 (1985).

105.227 Pyroxenes of normal chondrites of the H– and L–groups.
G. V. Baryshnikova, S. A. Stakheeva, K. I. Ignatenko, A. K. Lavrukhina.
Geokhimiya, No. 1, p. 20 – 34 (1985). In Russian. Abstr. in Ref. Zh., 51. Astron., 5.51.292 (1985).

105.228 New data on the ontogenesis of rhabdites in the Sikhote–Alin meteorite.
D. P. Grigor'eva, Yu. L. Kretser.
Dokl. Akad. Nauk SSSR. Ser. Mat. Fiz., Tom 279, No. 5, p. 1215 – 1217 (1984). In Russian. Abstr. in Ref. Zh., 51. Astron., 5.51.293 (1985).

105.229 The carbonaceous chondrite Kaidun – a new type of meteoritic breccia.
A. V. Ivanov, A. A. Ul'yanov, A. Ya. Skripnik, M. A. Nazarov, N. I. Kononkova.
Dokl. Akad. Nauk SSSR. Ser. Mat. Fiz., Tom 280, No. 2, p. 473 – 475 (1985). In Russian. Abstr. in Ref. Zh., 51. Astron., 6.51.219 (1985).

105.230 History of the study of the Tunguska meteorite problem in the years after the war (1958 – 1969).
N. V. Vasil'ev.
Meteoritic investigations in Siberia, p. 3 – 22 (1984). In Russian. Abstr. in Ref. Zh., 51. Astron., 6.51.245 (1985). – See Abstr. 003.023.

105.231 Heliophysical hypothesis of the nature of the Tunguska phenomenon.
V. K. Zhuravlev, A. N. Dmitriev.
Meteoritic investigations in Siberia, p. 128 – 141 (1984). In Russian. Abstr. in Ref. Zh., 51. Astron., 6.51.246 (1985). – See Abstr. 003.023.

105.232 Some peculiarities of the local structure of traces of the 1908 Tunguska catastrophe.
S. P. Golenetskij, V. V. Stepanok.
Meteoritic investigations in Siberia, p. 63 – 67 (1984). In Russian. Abstr. in Ref. Zh., 51. Astron., 6.51.247 (1985). – See Abstr. 003.023.

105.233 Interaction of large meteoritic bodies with the earth atmosphere.
V. P. Korobejnikov, P. I. Chushkin, L. V. Shurshalov.
Meteoritic investigations in Siberia, p. 99 – 117 (1984). In Russian. Abstr. in Ref. Zh., 51. Astron., 6.51.248 (1985). – See Abstr. 003.023.

105.234 Carbon in the composition of the Tunguska meteorite.
Yu. A. L'vov.
Meteoritic investigations in Siberia, p. 83 – 88 (1984). In Russian. Abstr. in Ref. Zh., 51. Astron., 6.51.251 (1985). – See Abstr. 003.023.

105.235 On the existence of astroblemes in the Kuznetzk Alatau.
O. G. Konovalova.
Kompleks. aehrokosm. issled. Sibiri, Novosibirsk, p. 57 – 61 (1984). In Russian. Abstr. in Ref. Zh., 51. Astron., 6.51.255 (1985).

105.236 Information aspect of investigations of the 1908 Tunguska phenomenon.
D. V. Demin, A. N. Dmitriev, V. K. Zhuravlev.
Meteoritic investigations in Siberia, p. 30 – 49 (1984). In Russian. Abstr. in Ref. Zh., 51. Astron., 6.51.256 (1985). – See Abstr. 003.023.

105.237 Fission–track annealing characteristics of meteoritic phosphates.
P. Mold, R. K. Bull, S. A. Durrani.
Nucl. Tracks Radiat. Meas., Vol. 9, No. 2, p. 119 – 128 (1984). Abstr. in Phys. Abstr., Vol. 88, No. 1253, Entry 35597 (1985).

105.238 Spallogenic nuclides in meteorites by conventional and accelerator mass spectrometry.
R. Sarafin, G. Bonani, U. Herpers, P. Signer, H. Hofmann, M. Nessi, E. Morenzoni, M. Suter, R. Wieler, W. Wolfli.
Nucl. Instrum. Methods Phys. Res., Sect. B, Vol. 233, No. 2, p. 411 – 414 (1984). Abstr. in Phys. Abstr., Vol. 88, No. 1253, Entry 35598 (1985). – See Abstr. 012.077.

105.239 Meteorites found in Antarctic ice.
B. Cervelle.
Recherche, Vol. 15, No. 160, p. 1454 – 1456 (1984). In French. Abstr. in Phys. Abstr., Vol. 88, No. 1253, Entry 35599 (1985).

105.240 Accelerator mass spectrometry for heavy isotopes at Oxford (OSIRIS).
S. H. Chew, T. J. L. Greenway, K. W. Allen.
Nucl. Instrum. Methods Phys. Res., Sect. B, Vol. 233, No. 2, p. 179 – 184 (1984). Abstr. in Phys. Abstr., Vol. 88, No. 1254, Entry 40557 (1985). – See Abstr. 012.077.

105.241 Shock compression of Jilin meteorite and pyrolite model.
W.–z. Lin.
Chin. J. Space Sci., Vol. 4, No. 4, p. 338 – 346 (1984). In Chinese. Abstr. in Phys. Abstr., Vol. 88, No. 1255, Entry 46002 (1985).

105.242 A transmission electron microscopic study of the Bethany iron meteorite.
F. Hasan, H. J. Axon.
J. Mater. Sci., Vol. 20, No. 2, p. 590 – 596 (1985). Abstr. in Phys. Abstr., Vol. 88, No. 1256, Entry 51585 (1985).

105.243 Comments on the existence of superheavy element tracks in meteoritic crystals.
J. S. Yadav, A. P. Sharma, G. N. Flerov, V. P. Perelygin, S. G. Stetsenko.
18th International Cosmic Ray Conference, Vol. 2, p. 38 – 41 (1983). – See Abstr. 012.096.
Measurements of ultraheavy and superheavy nuclei tracks are made in olivine crystals of Eagle Station and Marjalahti meteorites using the partial annealing technique.

105.244 Cosmic ray induced thermoluminescence in meteorites.
D. Sengupta, N. Bhandari, A. K. Singhvi.
18th International Cosmic Ray Conference, Vol. 2, p. 350 – 353 (1983). – See Abstr. 012.096.
The authors have carried out a study on TL levels on a suite of eight chondrites with effective preatmospheric radius in the range of 6 to 29 cm. A comparative study of the depth profile of TL with those of fossil track density, ^{26}Al and ^{53}Mn allows to determine the cosmic ray energy spectrum in the eV – MeV range.

105.245 Isotope production in meteorites and cosmic ray variation.
N. Bhandari.
18th International Cosmic Ray Conference, Vol. 2, p. 354 – 356 (1983). – See Abstr. 012.096.
The depth profiles of cosmic ray tracks and radio–isotopes such as ^{53}Mn, ^{26}Al, ^{60}Co etc. measured in some chondrites indicate that the isotope production rates are a sensitive function of the size of the meteoroid and depth within it.

105.246 On the Fe–group nuclei track density and track length distribution in different meteorites.
A. P. Sharma, J. S. Yadav, G. N. Flerov, V. P. Perelygin, S. G. Stetsenko.
18th International Cosmic Ray Conference, Vol. 2, p. 377 – 380 (1983). – See Abstr. 012.096.
The results of investigations of Fe–group nuclei track density and track length distribution in olivine and pyroxene crystals picked up from various locations of seven different meteorites are presented.

105.247 Hatte der Tunguska–Meteorit einen Nachfolger?
E. Hantzsche, J. Classen.
Sterne, 61. Band, Heft 2, p. 94 – 95 (1985).

105.248 Chaotic zone yields meteorites.
R. A. Kerr.
Science, Vol. 228, No. 4704, p. 1186 (1985).

105.249 Penecontemporaneous metamorphism, fragmentation, and reassembly of ordinary chondrite parent bodies.
R. E. Grimm.
J. Geophys. Res., Vol. 90, No. B2, p. 2022 – 2028 (1985).
The author reviews the thermal histories of ordinary chondrites and the canonical internal heating, or "onion–shell" models. He then analyzes the thermal and accretional requirements of the metamorphosed planetesimal model and suggests an alternative, consistent with the metallographic cooling rate constraints, in which ordinary chondrite parent bodies are collisionally fragmented and then rapidly reassembled before metamorphic heat has been dissipated.

105.250 Thermoluminescence measurements on shock–metamorphosed sandstone and dolomite from Meteor Crater, Arizona. 1. Shock dependence of thermoluminescence properties.
S. R. Sutton.
J. Geophys. Res., Vol. 90, No. B5, p. 3683 – 3689 (1985).

105.251 Thermoluminescence measurements on shock–metamorphosed sandstone and dolomite from Meteor Crater, Arizona. 2. Thermoluminescence age of Meteor Crater.
S. R. Sutton.
J. Geophys. Res., Vol. 90, No. B5, p. 3690 – 3700 (1985).

105.252 Interstellar dust in meteorites.
M. Bukovanská.
Kozmos, Vol. 16, No. 1, p. 8 – 11 (1985). In Slovak.

105.253 Possibility of collisions of comets and minor planets with the Earth.
Z. Ceplecha.
Vesmír, Vol. 64, No. 2, p. 91 – 92 (1985). In Czech.

105.254 The Colony meteorite and variations in CO3 chondrite properties.
A. E. Rubin, J. A. James, B. D. Keck, K. S. Weeks,
D. W. G. Sears, E. Jarosewich.
Meteoritics, Vol. 20, No. 2, Part 1, p. 175 – 196 (1985).

The Colony meteorite is an accretionary breccia containing several millimeter– to centimeter–size chondritic clasts embedded in a chondritic host. Colony is one of the least equilibrated CO3 chondrites; it has an unrecrystallized texture and contains compositionally heterogeneous olivine and low–Ca pyroxene, kamacite with low Ni and Co and high Cr, amoeboid inclusions with low FeO and MnO, a fine–grained silicate matrix with very high FeO, and numerous small chondrules with clear pink glass. However, Colony differs from normal CO chondrites in several respects: although Al, Sc, V, Cr, Ir, Fe, Au and Ga abundances are consistent with a CO chondrite classification, certain lithophiles (Mg and Mn), siderophiles (Ni and Co) and chalcophiles (Se and Zn) are depleted by factors of 10–40%. The shape of Colony's thermoluminescence glow curve is similar to that of Allan Hills A77307 (another unequilibrated chondrite with CO3 petrological characteristics) and different from those of normal CO chondrites.

105.255 Rare earth and other elements in components of the Abee enstatite chondrite.
R. M. Frazier, W. V. Boynton.
Meteoritics, Vol. 20, No. 2, Part 1, p. 197 – 218 (1985).

Abee clast samples, a matrix sample, a dark inclusion, magnetic and non–magnetic samples, and bulk samples were analyzed by neutron activation analysis (NAA). The REE were determined by radiochemical NAA. Na, K, Sc, Cr, Mn, Fe, Co, Ni, Zn, As, Se, Sm, Ir, Au were determined by instrumental NAA.

105.256 Classification of eight ordinary chondrites from Texas.
A. J. Ehlmann, K. Keil.
Meteoritics, Vol. 20, No. 2, Part 1, p. 219 – 227 (1985).

Based on optical microscopy and electron microprobe analyses, eight previously undescribed or poorly known chondrites were classified into compositional groups, petrologic types, and degree of shock alteration.

105.257 Amphibole and hercynite spinel in Shergotty and Zagami: magmatic water, depth of crystallization, and metasomatism.
A. H. Treiman.
Meteoritics, Vol. 20, No. 2, Part 1, p. 229 – 243 (1985).

Amphibole and spinel occur in the Shergotty and Zagami meteorites only in magmatic inclusions in pigeonite. The trapped magma is essentially identical to the parental magmas for Shergotty and Zagami. The amphibole is a kaersutite with minimal halogen content; by inference, it must have been hydrous. If so, the Shergotty and Zagami melts contained at least 0.2 wt % H_2O and were probably H_2O–undersaturated. Pervasive replacement of magnetite through reduction reactions suggests that Shergotty and Zagami interacted with hydrogen–rich fluids during their cooling.

105.258 Noble gases in the Bells (C2) and Sharps (H3) chondrites.
M. G. Zadnik.
Meteoritics, Vol. 20, No. 2, Part 1, p. 245 – 257 (1985).

This paper reports elemental and isotopic measurements of the noble gases from Bells and Sharps together with quantities derived from them, such as cosmogenic exposure and gas retention ages, as well as primordial gas contents.

105.259 Bocaiuva – a silicate–inclusion bearing iron meteorite related to the Eagle–Station pallasites.
D. J. Malvin, J. T. Wasson, R. N. Clayton, T. K. Mayeda,
W. da Silva Curvello.
Meteoritics, Vol. 20, No. 2, Part 1, p. 249 – 273 (1985).

Bocaiuva is a unique meteorite consisting of major metal having a high Ge/Ga ratio and minor (~ 50 mg/g) silicates. The silicates are generally chondritic and consist of major olivine (Fa7.7) and orthopyroxene (Fs7.6) and minor plagioclase (Ab49,

An49) and clinopyroxene (Fs4.5, Wo42). The low alkali content of the silicates is the only property inconsistent with a chondritic composition. Based on metal composition Bocaiuva seems distantly related to certain iron meteorites having similar Ge contents and similar Ge/Ga ratios, but detailed comparison with six such irons shows none to be closely related to Bocaiuva.

105.260 The Meteoritical Bulletin, No. 63.
A. L. Graham (Editor).
Meteoritics, Vol. 20, No. 2, Part 1, p. 275 – 283 (1985).

Place of fall/find, date of fall/find, class and type, number of individual specimens, total weight and circumstances of fall/find of some meteorites are given.

105.261 Petrology of the Cangas de Onis and Nulles regolith breccias: implications for parent body history.
C. V. Williams, A. E. Rubin, K. Keil, A. San Miguel.
Meteoritics, Vol. 20, No. 2, Part 2, p. 331 – 345 (1985). – See Abstr. 003.048.

Cangas de Onis and Nulles are H chondrite regolith breccias from northern Spain. Uniform mineral compositions in both Cangas de Onis and Nulles indicate that their matrices consist almost entirely of comminuted equilibrated clasts. If these meteorites are representative samples of the regoliths in which they resided, the regoliths were compositionally homogeneous at the time of breccia consolidation. Zoned taenites within the clastic matrix of Cangas de Onis scatter widely on composition-dimension plots, indicating that these taenites cooled at different rates (about 1 – 1000 K/m.y.) at various depths (1 – 150 km). This suggests that the H chondrite parent body was disrupted and reassembled.

105.262 Some chemical and isotopic observations in chondrules.
K. Fredriksson, S. V. S. Murty, K. Marti.
Meteoritics, Vol. 20, No. 2, Part 2, p. 347 – 357 (1985). – See Abstr. 003.048.

The authors report chemical, nitrogen and noble gas isotopic observations in individual chondrules of the Bjurböle (L4) and Dhajala (H3,4) chondrites. The $\delta^{15}N$ is $+9.4$ for the Bjurböle "matrix", ranges from $+6.4$ to -20.0 in chondrules, and appears to be related to the chemical composition. Nitrogen isotope systematics seem to require the presence of more than one component of distinct signatures. Although the authors cannot rule out complex processes of chondrule and matrix formation involving element and isotope fractionations, the oxygen isotope systematics also seem to require distinct reservoirs. The authors discuss the results in terms of the lack of equilibrations among chondrule minerals, as well as with the host meteorite matter, and also the paradox of the constant Fe/Mg ratio of olivines in Bjurböle and of the majority of ordinary chondrites.

105.263 Shock heating of chondrules.
W. B. Thompson.
Meteoritics, Vol. 20, No. 2, Part 2, p. 359 – 365 (1985). – See Abstr. 003.048.

The possibility that chondrules have been heated by shock waves in the early solar nebula is explored. Suitable heating cycles would be produced in the wakes of cometary bodies in eccentric orbits; however, since these are brought to rest with respect to the nebula in a fairly short time, the total mass of the exciting solid bodies would have been about 10% of that of the heated gas.

105.264 The identification of Group II inclusions in carbonaceous chondrites by electron probe microanalysis of perovskite.
A. S. Kornacki, J. A. Wood.
Earth Planet. Sci. Lett., Vol. 72, No. 1, p. 74 – 86 (1985).

105.265 A transmission electron microscope study of pyroxene chondrules in equilibrated L–group chondrites.
S. Watanabe, M. Kitamura, N. Morimoto.
Earth Planet. Sci. Lett., Vol. 72, No. 1, p. 87 – 98 (1985).

105.266   Noble gases and the history of Jilin meteorite.
F. Begemann, Z. Li, S. Schmitt–Strecker,
H. W. Weber, Z. Xu.
Earth Planet. Sci. Lett., Vol. 72, No. 2/3, p. 247 – 262 (1985).

105.267   Conditions of the cosmic ray exposure of the Jilin chondrite.
G. Heusser, Z. Ouyang, T. Kirsten, U. Herpers, P. Englert.
Earth Planet. Sci. Lett., Vol. 72, No. 2/3, p. 263 – 272 (1985).

The pre–atmospheric size of Jilin was reconstructed. Jilin was almost spherical and had a pre–atmospheric radius of about 85 cm. The cosmogenic nuclides in Jilin can be explained by a two–stage exposure history. The first stage occurred near the surface of the parent body in 2π geometry and lasted about 10 Myr. Jilin was then expelled from its parent body and exposed to cosmic rays in a second (4π) stage lasting approximately 0.4 Myr. The measured concentration profiles of spallogenic nuclides are compared to calculated production rates.

105.268   Spallogenic ^{10}Be in the Jilin chondrite.
D. K. Pal, R. K. Moniot, T. H. Kruse, C. Tuniz,
G. F. Herzog.
Earth Planet. Sci. Lett., Vol. 72, No. 2/3, p. 273 – 275 (1985).

105.269   Dating Jilin and constraints on its temperature history.
N. Müller, E. K. Jessberger.
Earth Planet. Sci. Lett., Vol. 72, No. 2/3, p. 276 – 285 (1985).

The surface of the meteoroid was heated probably 400,000 years ago at the separation from its parent body. Model calculations suggest that the heating event was moderate (initial surface temperature 500 – 600°C) and brief (surface cooling rate 1 – 2°C/h). Heating of meteoroids at separation from their parent bodies might be a common phenomenon. It could explain "late" disturbances in some isotope systems and could be the driving force for hydrothermal alterations.

105.270   Nuclear track study of Jilin chondrite.
P. Pellas, M. Bourot–Denise.
Earth Planet. Sci. Lett., Vol. 72, No. 2/3, p. 286 – 298 (1985).

105.271   Mobile trace element contents in Jilin chondrite.
Y. Sakuragi, M. E. Lipschutz.
Earth Planet. Sci. Lett., Vol. 72, No. 2/3, p. 299 – 303 (1985).

The authors determined ppm–ppt levels of Co, Se, Ga, Rb, Cs, Te, Bi, Ag, In, Tl, Zn and Cd (arranged empirically in order of increasing mobility at 1000°C) by radiochemical neutron activation analysis in eight Jilin samples. All elements but Cd are present at levels similar to those in H4–6 chondrites with long K/Ar age (i.e., presumably mildly shocked). These levels are low relative to those in analogous L4–6 chondrites suggesting that H chondrites formed and/or evolved under higher pre–shock temperatures than did mildly shocked L chondrites. Time–temperature conditions during shock–loading of Jilin parent material were mild relative to those in strongly shocked L chondrites, being sufficient at most only to mobilize Cd.

105.272   Thermoluminescence studies on Jilin meteorite.
G. A. Wagner.
Earth Planet. Sci. Lett., Vol. 72, No. 2/3, p. 304 – 306 (1985).

105.273   Thorium and uranium abundances in the Jilin H5 chondrite.
E. Pernicka.
Earth Planet. Sci. Lett., Vol. 72, No. 2/3, p. 307 – 310 (1985).

Thorium and uranium abundances have been measured in the Jilin H5 chondrite by radiochemical neutron activation analysis. Although the abundances of (35.8 ± 1.5) ppb Th and (14.9 ± 0.9) ppb U are within the range of literature values, the ratio Th/U is about 25% lower than the average value for H chondrites. Terrestrial addition of uranium appears to be the most likely explanation.

105.274   Evidence for oxidizing conditions in the solar nebula from Mo and W depletions in refractory inclusions in carbonaceous chondrites.
B. Fegley Jr., H. Palme.
Earth Planet. Sci. Lett., Vol. 72, No. 4, p. 311 – 326 (1985).

105.275   Self–shielding in O_2 – a possible explanation for oxygen isotopic anomalies in meteorites?
O. Navon, G. J. Wasserburg.
Earth Planet. Sci. Lett., Vol. 73, No. 1, p. 1 – 16 (1985).

105.276   Transmission electron microscopy of L–group chondrites. 1. Natural shock effects.
J. R. Ashworth.
Earth Planet. Sci. Lett., Vol. 73, No. 1, p. 17 – 32 (1985).

105.277   Transmission electron microscopy of L–group chondrites. 2. Experimentally annealed Kyushu.
J. R. Ashworth, L. G. Mallinson.
Earth Planet. Sci. Lett., Vol. 73, No. 1, p. 33 – 40 (1985).

105.278   Ion probe determinations of the rare earth concentrations of individual meteoritic phosphate grains.
G. Crozaz, E. Zinner.
Earth Planet. Sci. Lett., Vol. 73, No. 1, p. 41 – 52 (1985).

105.279   Cosmic ray effects in the Antarctic meteorite Allan Hills A 78084.
R. Sarafin, M. Bourot–Denise, G. Crozaz, U. Herpers,
P. Pellas, L. Schultz, H. W. Weber.
Earth Planet. Sci. Lett., Vol. 73, No. 2/4, p. 171 – 182 (1985).

105.280   The submicroscopic structure of the unequilibrated ordinary chondrites Chainpur, Mezö–Madaras and Tieschitz: a transmission electron–microscopic study.
J. Töpel–Schadt, W. F. Müller.
Earth Planet. Sci. Lett., Vol. 74, No. 1, p. 1 – 12 (1985).

Impact and explosion crater ejecta, fragment size, and velocity.
See Abstr. 022.126.

Production of ^{7}Be, ^{22}Na, ^{24}Na and ^{10}Be from Al in a 4π–irradiated meteorite model.
See Abstr. 022.142.

Chondrule volume percents: statistical behavior and visual–estimation guides produced by numerical simulation.
See Abstr. 036.135.

The production and survival of ^{205}Pb in stars, and the ^{205}Pb – ^{205}Tl s–process chronometry.
See Abstr. 061.026.

Stellar neutron capture rates for ^{46}Ca and ^{48}Ca.
See Abstr. 061.031.

Neutron capture and total cross sections for ^{48}Ca: astrophysical implications.
See Abstr. 061.101.

Solar abundances and the role of nucleogenesis in low–to–medium mass stars in the Galaxy.
See Abstr. 061.148.

A cometary impact model for the source of Libyan Desert glass.
See Abstr. 081.020.

Radiation history of meteoritic and lunar material by track data.
See Abstr. 094.038.

Mars, a volatile–rich planet.
See Abstr. 097.024.

Asteroid–meteorite connection: regolith effects implied by lunar reflectance spectra.
See Abstr. 098.038.

Report of IAU Commission 15: Physical study of comets, minor planets, and meteorites (*L'étude physique des comètes, des petites planètes et des météorites*).
See Abstr. 102.053.

Report of IAU Commission 22: Meteors and interplanetary dust (*Météorites et la poussière interplanétaire*).
See Abstr. 104.055.

Hydrated interplanetary dust particle linked with carbonaceous chondrites?
See Abstr. 106.034.

Diagenesis in interplanetary dust: chondritic porous aggregate W7029*A.
See Abstr. 106.044.

A hydrated interplanetary dust particle containing calcium– and aluminum–rich pyroxene: possible relations to carbonaceous chondrites.
See Abstr. 106.045.

Discovery of nuclear tracks in interplanetary dust.
See Abstr. 106.065.

Asteroids, comets, and planet formation.
See Abstr. 107.005.

A possible ^{126}Sn chronometer for the early solar system.
See Abstr. 107.018.

Meteoritic constraints on processes in the solar nebula.
See Abstr. 107.021.

Aluminium clues to the formation of the Solar System.
See Abstr. 107.029.

The optical properties of dust in the mid–IR silicate bands.
See Abstr. 112.030.

Analysis of cosmic rays with $Z \geqslant 50$ registered in meteorite minerals.
See Abstr. 144.070.

The time– and space–averaged cosmic ray flux experienced by meteorites.
See Abstr. 144.166.

Can we get the cosmic ray actinide abundance from the study of tracks in meteorites?
See Abstr. 144.325.

106 Interplanetary Matter, Interplanetary Magnetic Field, Zodiacal Light

106.001 Zodiacal light gathered along the line of sight. Retrieval of the local scattering coefficient from photometric surveys of the ecliptic plane.
R. Dumont, A. C. Levasseur–Regourd.
Planet. Space Sci., Vol. 33, No. 1, p. 1 – 9 (1985).

Associating to each photometric observation in the ecliptic the second intersection of the line of sight with the terrestrial orbit offers a clue for retrieving considerable information about the local quantities (elemental brightness, or scattering coefficient) all along that line. At one peculiar point outside the orbit ("martian node") and at one inside the orbit ("quasi–radial node"), the scattering coefficient is remarkably insensitive to the mathematical model, if assumed sufficiently smooth, as strongly suggested by the smoothness of the brightness vs the geometric parameters. The locations of these "nodes of lesser uncertainty" in the ecliptic plane are fairly favourable in order to deconvolve the heliocentric (r) from the angular (θ) dependences of the scattering coefficient, D. Various photometric results in the range 0.5 – 1.5 a.u. can be derived about these two functions, without recourse to the classical but suspicious assumption of a uniform composition of the interplanetary matter.

106.002 Anisotropy of > 35 keV ions in corotating particle events at 1 AU.
I. G. Richardson.
Planet. Space Sci., Vol. 33, No. 2, p. 147 – 157 (1985).

The anisotropy of 35 – 1000 keV ions in two corotating particle events associated with high–speed solar wind streams at 1 AU is examined in terms of the diffusion–convection propagation model using data from the Energetic Proton Anisotropy Spectrometer on *ISEE*–3.

106.003 Acceleration of > 47 keV ions and > 2 keV electrons by interplanetary shocks at 1 AU.
B. T. Tsurutani, R. P. Lin.
J. Geophys. Res., Vol. 90, No. A1, p. 1 – 11 (1985). – See Abstr. 012.003.

The authors present initial results from a survey of the effects of interplanetary shocks on energetic $\geqslant 2$ keV electrons and $\geqslant 47$ keV ions, as observed by the field, plasma, and energetic particle experiments on the ISEE 3 spacecraft. Shock normals, velocities, Mach numbers, and upstream and downstream plasma parameters were determined for 37 forward shocks out of a total of 55 shocks observed between August 1978 and November 1979.

106.004 Characteristics of energetic particle events associated with interplanetary shocks.
K.–P. Wenzel, R. Reinhard, T. R. Sanderson, E. T. Sarris.
J. Geophys. Res., Vol. 90, No. A1, p. 12 – 18 (1985). – See Abstr. 012.003.

The authors present observations of the 35– to 1600–keV proton intensity–time profiles and the three–dimensional 35– to 56–keV anisotropy distributions recorded during two interplanetary shock events on ISEE 3, and discuss these in the light of current particle acceleration models.

106.005 Observations of three–dimensional anisotropies of 35– to 1000–keV protons associated with interplanetary shocks.
T. R. Sanderson, R. Reinhard, P. van Nes, K.–P. Wenzel.
J. Geophys. Res., Vol. 90, No. A1, p. 19 – 27 (1985). – See Abstr. 012.003.

The authors present results of a detailed analysis of three–dimensional anisotropies of protons in the energy range 35 – 1000 keV observed in association with interplanetary shocks on ISEE 3. They compare observations of high time resolution anisotropies made close to the shock in seven energy channels with theoretical predictions for Fermi acceleration and shock drift acceleration, and find good evidence for both types of acceleration.

106.006 Formation of ion acoustic solitary waves upstream of the earth's bow shock.
M. J. Pangia, N. C. Lee, G. K. Parks.
J. Geophys. Res., Vol. 90, No. A1, p. 95 – 98 (1985). – See Abstr. 012.003.

The turbulent plasma development of Lee and Parks is applied to the solar wind approaching the earth's bow shock region. The ponderomotive force contribution is due to ion acoustic waves propagating in the direction of the ambient magnetic field. In this case, the envelope of the ion acoustic wave is shown to satisfy the cubic Schroedinger equation. Modulational instabilities exist for waves in the solar wind, thereby predicting the generation of solitary waves. This analysis further identifies that the ion acoustic waves which exhibit this instability have short wavelengths.

106.007 Doppler scintillation observations of interplanetary shocks within 0.3 AU.
R. Woo, J. W. Armstrong, N. R. Sheeley Jr., R. A. Howard, M. J. Koomen, D. J. Michels, R. Schwenn.
J. Geophys. Res., Vol. 90, No. A1, p. 154 – 162 (1985). – See Abstr. 012.003.

Near–sun spacecraft Doppler scintillation observations have been combined with Solwind coronagraph and Helios 1 plasma measurements to provide more definitive measurements of the evolution and propagation of interplanetary shock waves between the sun and earth orbit than have been available from previous observations.

106.008 A simplified model for timing the arrival of solar flare–initiated shocks.
D. F. Smart, M. A. Shea.
J. Geophys. Res., Vol. 90, No. A1, p. 183 – 190 (1985). – See Abstr. 012.003.

106.009 The evolution of interplanetary shocks.
H. V. Cane.
J. Geophys. Res., Vol. 90, No. A1, p. 191 – 197 (1985). – See Abstr. 012.003.

The behavior of interplanetary (IP) shocks is investigated by making intercomparisons of a number of properties of 48 shocks which generated radio emission in the IP medium, i.e., IP type II events.

106.010 Distant heliospheric results on interplanetary shock propagation.
J. D. Mihalov.
J. Geophys. Res., Vol. 90, No. A1, p. 210 – 216 (1985). – See Abstr. 012.003.

Solar wind plasma data from Pioneers 10 and 11 and the Pioneer Venus Orbiter for 1978 to mid–1981 have been examined for forward shock signatures, and associations have been made with solar flares. (IMP 8 shocks from early 1978 have similarly been included for completeness.) The numbers of flare–associated shocks observed at the various heliocentric distances of the spacecraft (ranging from 0.7 to 24 AU) are deduced, and the shock strengths and energies are given. Significantly fewer shocks associated with flares are observed at the greater distances, presumably because more cases have decreased below the threshold for inclusion. Three cases in which the same shock was probably detected by spacecraft at different heliocentric distances indicate deceleration and weakening at the greater distances.

106.011 A study of the formation, evolution, and decay of shocks in the heliosphere between 0.5 and 30.0 AU.
Z. K. Smith, M. Dryer, R. S. Steinolfson.
J. Geophys. Res., Vol. 90, No. A1, p. 217 – 220 (1985). – See Abstr. 012.003.

The spatial and temporal evolution of corotating interaction regions is examined between 0.5 and 30 AU using a solar wind simulation with a continuous sinusoidally varying velocity, density, and temperature input pulse train. The authors show (1) the well–known formation of forward–reverse MHD shock ensembles, followed by (2) their interaction that results in a highly nonlinear "pressure" wave that eventually reforms into a new but more irregular set of forward–reverse shock pairs. As a result of these compound interactions, the overall temperature decays much more slowly than the classical steady state adiabatic radial dependency of $R^{-4/3}$.

106.012 Coalescence of two pressure waves associated with stream interactions.
Y. C. Whang, L. F. Burlaga.
J. Geophys. Res., Vol. 90, No. A1, p. 221 – 232 (1985). – See Abstr. 012.003.

An MHD unsteady one–dimensional model is used to simulate the interaction and coalescence of two pressure waves in the outer heliosphere.

106.013 Determination of local temperatures and of albedo gradient in zodiacal cloud from IRAS radiometric data.
A.–C. Levasseur–Regourd, R. Dumont.
C. R. Acad. Sci., Sér. II, Tome 300, No. 3, p. 109 – 112 (1985). In French.

106.014 The infrared sky: zodiacal dust bands and infrared cirrus.
H. L. Shipman.
Phys. Today, Vol. 38, No. 1, p. S12 (1985). Short review paper. – See Abstr. 013.003.

106.015 Predicted interplanetary distribution of Lyman–α intensity and polarization.
J. M. Ajello, G. E. Thomas.
Icarus, Vol. 61, No. 1, p. 163 – 170 (1985).

The authors investigate how the uncertainties in current model calculations could be reduced through the use of polarization measurements made from interplanetary spacecraft. In particular they inquire how a mapping of the degree of linear polarization made from a spacecraft at various locations in the Solar System can improve our knowledge of the interstellar wind parameters, number density, temperature, and velocity. The sky distribution of intensity and polarization has been calculated using a variety of models for the neutral hydrogen. It is found that the polarization distribution over the sky is quite different from that of the intensity distribution. It is also shown that the maximum degree of polarization of the Lyman–α line increases with heliocentric distance of the spacecraft, varying from 0 up to ~18% at 20 AU.

106.016 IMF B_y–dependent plasma flow and Birkeland currents in the dayside magnetosphere. 1. Dynamics Explorer observations.
J. L. Burch, P. H. Reiff, J. D. Menietti, R. A. Heelis,
W. B. Hanson, S. D. Shawhan, E. G. Shelley, M. Sugiura,
D. R. Weimer, J. D. Winningham.
J. Geophys. Res., Vol. 90, No. A2, p. 1577 – 1593, 1631 – 1638 (1985).

Plasma, magnetic–field and dc electric–field observations from Dynamics Explorers 1 and 2 are used to investigate the morphology of solar–wind ion injection, Birkeland currents, and plasma convection in the morning sector for both positive and negative interplanetary magnetic field (IMF) B_y components. The results of the study are used to construct a B_y–dependent global convection model for southward IMF.

106.017 IMF B_y–dependent plasma flow and Birkeland currents in the dayside magnetosphere. 2. A global model for northward and southward IMF.
P. H. Reiff, J. L. Burch.
J. Geophys. Res., Vol. 90, No. A2, p. 1595 – 1609 (1985).

This paper extends to all interplanetary magnetic field (IMF) orientations the qualitative convection pattern presented by Burch et al. (see Abstr. 106.016).

106.018 Low–energy proton interaction with interplanetary magnetohydrodynamic discontinuities.
V. P. Tritakis.
Adv. Space Res., Vol. 4, No. 7, p. 323 – 326 (1984). – See Abstr. 012.009.

Low energy proton measurements associated with interplanetary MHD discontinuities were detected by the DFH instrument on ISEE–3 spacecraft during the solar maximum period, September 1978 – March 1980. The observations confirm that local magnetohydrodynamic conditions, especially discontinuities, significantly affect the propagation of low energy particles in the interplanetary medium.

106.019 Some characteristics of propagation of flare– or CME (*coronal mass ejections*)–associated interplanetary shock waves.
J. K. Chao.
Adv. Space Res., Vol. 4, No. 7, p. 327 – 330 (1984). – See Abstr. 012.009.

A physical mechanism is described to calculate the probability that a weak shock which enters a turbulent solar wind region will degenerate into a MHD wave. That is, the shock would disappear as an entropy–generate entity. This model also suggests that most interplanetary shock waves cannot propagate continuously with a smooth shock surface. It is suggested that the surface of an interplanetary shock will be highly distorted and that parts of the shock surface can degenerate into MHD waves or even disappear during its global propagation through interplanetary space. This model of shock propagation also applies to corotating shocks.

106.020 Radio–scintillation observations of interplanetary disturbances.
T. Watanabe, T. Kakinuma.
Adv. Space Res., Vol. 4, No. 7, p. 331 – 341 (1984). – See Abstr. 012.009.

The authors briefly review current interplanetary disturbances by scintillation techniques, then discuss several examples of interplanetary disturbances observed in 1978 – 1981 around the maximum of the Solar Cycle No. 21.

106.021 Corotating interplanetary streams and associated ionospheric disturbances at Venus and Earth.
H. Taylor, P. Cloutier, M. Dryer, S. Suess, A. Barnes,
R. Wolff, A. Stern.
Adv. Space Res., Vol. 4, No. 7, p. 343 – 346 (1984). – See Abstr. 012.009.

During the summer of 1979, solar coronal structure was such that a sequence of recurrent regions produced a corresponding sequence of corotating solar wind streams, with pronounced downstream signatures. One of these stream events passed Earth on July 3, and was observed later at Venus late on July 11th, with similar characteristics. Corresponding in–situ measurements at Earth from the Atmospheric Explorer–E satellite and at Venus from the Pioneer Venus Orbiter are examined for evidence of comparable perturbations of the planetary ionospheres. The ionosphere disturbances appear to be closely associated with large variations in the solar wind momentum flux.

106.022 Some features of the interplanetary disturbances in the post–Solar Maximum Year period.
G. N. Zastenker, N. L. Borodkova.
Adv. Space Res., Vol. 4, No. 7, p. 347 – 352 (1984). – See Abstr. 012.009.

Features of strong interplanetary disturbances (including 14 shock waves) are considered by the solar wind plasma measurements onboard the PROGNOZ–8 satellite. Examination of large–scale structure of the plasma fluxes enabled the authors to discover extreme values of proton temperature ($\sim 10^6$K) and density ($\sim 10^2 \text{cm}^{-3}$) in some cases. The energy transferred by the interplanetary shock waves ($10^{31} - 10^{32}$erg) and their deceleration are estimated. Determination of the plasma parameter jumps for protons and α–particles at the shock front made it possible to estimate the potential barrier (40 – 400V) depending on magnetosonic Mach number.

106.023   The inner heliosphere from solar minimum to solar maximum: short– and long–term variations in the energetic particle population.
R. Müller–Mellin, G. Wibberenz.
Adv. Space Res., Vol. 4, No. 7, p. 353 – 356 (1984). – See Abstr. 012.009.

Between 1975 and 1983 HELIOS 1 scanned the interplanetary medium between 0.3 and 1 AU 31 times. The observed variations in the differential and integral flux of protons and helium nuclei in the energy range from 4 to >50 MeV/n are characterized by large temporal changes in the intensities, moderate changes in the energy spectrum and changes in the gradient below the detection level (60%). The onset of solar activity near the end of 1977 is accompanied by a monotonous decrease of galactic cosmic radiation. The successive reduction of the cosmic ray intensity to the level of solar maximum is discussed in view of the role of large transient disturbances as compared to processes as diffusion, convection, adiabatic energy losses and drifts.

106.024   Infrared measurements of zodiacal light.
T. L. Murdock, S. D. Price.
Astron. J., Vol. 90, No. 2, p. 375 – 386 (1985).

The infrared zodiacal emission at wavelengths between 2 and 30 μm has been measured with a cryogenically cooled absolute radiometer from a rocket platform. Solar elongation angles between $22° \leqslant \varepsilon \leqslant 180°$ and elevation angles between the North Ecliptic Pole and 60° south ecliptic latitude were recovered. At a spectral resolution of ~ 2 μm the emission shows no strong spectral features and the results are consistent with previous rocket measurements. Results of detailed modeling of the interplanetary dust cloud are presented.

106.025   The plane of symmetry of the zodiacal light and the structure of the Gegenschein.
C. Winkler, T. Schmidt–Kaler, W. Schlosser.
Astron. Astrophys., Vol. 143, No. 1, p. 194 – 200 (1985). In German.

Super wide–angle photographs obtained by means of the Bochum spherical mirror camera with a FOV of 145° are used to study the plane of symmetry of the zodiacal light and the structure of the Gegenschein.

106.026   A theory of energization of solar wind electrons by the Earth's bow shock.
M. M. Leroy, A. Mangeney.
Ann. Geophys., Vol. 2, No. 4, p. 449 – 456 (1984). Abstr. in Phys. Abstr., Vol. 88, No. 1251, Entry 24257 (1985).

106.027   Latitude variation of recurrent MeV–energy proton flux enhancements in the heliocentric radial range 11 to 20 AU and possible correlation with solar coronal hole dynamics.
S. P. Christon, E. C. Stone.
Geophys. Res. Lett., Vol. 12, No. 2, p. 109 – 112 (1985).

Recurrent low energy ($\gtrsim 0.5$ MeV) proton flux enhancements, reliable indicators of corotating plasma interaction regions in interplanetary space, have been observed on the Voyager 1 and 2 and Pioneer 11 spacecraft in the heliographic latitude range 2°S to 23°N and the heliocentric radial range 11 to 20 AU. After a period of rather high correlation between fluxes at different latitudes in early 1983, distinct differences develop. The evolution of the fluxes appears to be related to the temporal and latitudinal dynamics of solar coronal holes, suggesting that information about the latitudinal structure of solar wind stream sources propagates to these distances.

106.028   The interplanetary dust environment beyond 1 AU and in the vicinity of the ringed planets.
H. Fechtig.
Adv. Space Res., Vol. 4, No. 9, p. 5 – 11 (1984). – See Abstr. 012.029.

Two Pioneer 10/11 dust experiments observed the dust distribution beyond 1 AU in the 10 to 100 μm diameter size range for the first time directly with contradicting results. The penetration experiment saw a constant flux out to 20 AU while the optical experiment observed a decrease of the dust number densities until 3.3 AU, but no scattered light was recorded further out. An attempt is made to explain these observations on the basis of the socalled ''Greenberg''–particles: cometary core/mantle grains with organic mantle material. The observed enhancement of the dust flux by 1 or 2 orders of magnitudes near Jupiter and Saturn are interpreted as being caused by gravitational focussing, ejecta from Jovian/Saturnian satellites and electrostatic fragmentation products.

106.029   Interplanetary field enhancements in the solar wind: evidence for cometesimals at 0.72 and 1.0 AU?
M. R. Arghavani, C. T. Russell, J. G. Luhmann.
Adv. Space Res., Vol. 4, No. 9, p. 255 – 259 (1984). – See Abstr. 012.029.

A new class of interplanetary magnetic disturbance has been identified which consists of a nearly symmetric rise and fall of the magnetic field surrounding a cusp–shaped maximum. These disturbances have been hypothesized to be caused by the mass-loading of the solar wind by small outgassing bodies. The clustering of these events in space suggests that not all the events are independent. Clustering is greatest at 0.72 AU because of one very strong family of events associated with the orbit of the asteroid 2201 Oljato. The events are larger at 0.72 AU than at 1 AU. The timing of the disturbances at both 1 AU and 0.72 AU relative to Oljato suggests the presence of outgassing debris both in front of and behind the asteroid.

106.030   The evolution of the IMF sector structure for the period of 1926 – 1980.
V. N. Obridko.
Soln. Dannye, Byull., 1984, No. 11, p. 54 – 58 (1985). In Russian.

The evolution of the IMF sector structure has been studied using an objective method with application of autocorrelation functions. It is shown that the main state of the IMF is the two–sector structure. The four–sector structure appears at the descending branch of the solar cycle and has a short life time.

106.031   MHD processes in the outer heliosphere.
L. F. Burlaga.
Space Sci. Rev., Vol. 39, Nos. 3/4, p. 255 – 316 (1984).

Contents: (1) Introduction. (2) Basic equations (dynamics, thermodynamics, electrodynamics, coordinates). (3) Large–scale variations (magnetic field strength, magnetic field direction, components of **B**, sectors, large–scale field strength fluctuations, shocks and interfaces). (4) Corotating configurations (corotating streams, filtering, pressure waves without fast streams, large–scale patterns). (5) Magnetic clouds (transients, nature of magnetic clouds, size, magnetic field configuration). (6) Stream interactions (classification, entrainment, twin streams, recurrent streams). (7) Flow systems and large–scale fluctuations (scales, corotating systems, transient and mixed systems, shells, stationary and homogeneity. (8) Radial evolution of large–scale fluctuations (velocity fluctuations, magnetic field strength fluctuations, filtering and entrainment, turbulence). (9) Discussion.

106.032   Large–scale electrochemistry in ice lunar–like bodies and the nature of the small bodies of the solar system.
Eh. M. Drobyshevskij.
Fiz.–tekh. inst. Akad. Nauk SSSR, Prepr., No. 897, 22 pp. (1984). In Russian. Abstr. in Ref. Zh., 51. Astron., 1.51.259 (1985).

106.033 Origin of strong interplanetary shocks.
A. Hewish, S. J. Tappin, G. R. Gapper.
Nature, Vol. 314, No. 6007, p. 137 – 140 (1985).
Observations of 900 radio sources have been used to study the shape and motion of large–scale interplanetary transients associated with shock disturbances at 1 AU. The variations of plasma density and speed and the zones in the solar atmosphere from which the transients originate suggest an origin related to intermittent flows of enhanced speed from coronal holes at mid-latitudes on the Sun.

106.034 Hydrated interplanetary dust particle linked with carbonaceous chondrites?
K. Tomeoka, P. R. Buseck.
Nature, Vol. 314, No. 6009, p. 338 – 340 (1985).
The authors report here the results of transmission electron microscope observations of a hydrated interplanetary dust particle (IDP) containing Fe–, Mg–rich smectite or mica as a major phase. Fassaite, a Ca, Al clinopyroxene, also occurs in this particle, and one of the crystals exhibits solar–flare tracks. Fassaite is a major constituent of the Ca–, Al–rich refractory inclusions found in the carbonaceous chondrites. Its presence in this particle, therefore, suggests that there may be a link between hydrated IDPs and carbonaceous chondrites.

106.035 On the influence of the sector structure of the interplanetary magnetic field on the upper ionosphere of the earth.
O. P. Kolomijtsev, M. A. Livshits, T. N. Soboleva.
Geomagn. Aehron., Tom 25, No. 2, p. 206 – 210 (1985). In Russian. English translation in Geomagn. Aeron.

106.036 Henyey–Greenstein representation of the mean volume scattering phase function for zodiacal dust.
S. S. Hong.
Astron. Astrophys., Vol. 146, No. 1, p. 67 – 75 (1985).
To determine the scattering characteristics of interplanetary particles, the author substitutes a linear combination of three Henyey–Greenstein functions for the mean volume scattering phase function in the zodiacal light brightness integral. Residuals are obtained by comparing the observed zodiacal light in the ecliptic with results of the integral. Minimization of the residuals leads to a scattering function which has a strong peak in the forward direction, an isotropic part at intermediate scattering angles and a mild enhancement in the backward direction. The same method is employed to analyze the polarized components of the zodiacal light.

106.037 Spectra of MHD–turbulence of the interplanetary plasma with account for nonlinear absorption.
I. V. Chashej, V. I. Shishov.
Geomagn. Aehron., No. 1, p. 1 – 6 (1985). In Russian. English translation in Geomagn. Aeron.

106.038 Annual variations of the correlation of the B_x and B_z components of the interplanetary magnetic field in the GSEQ, GSE and GSM coordinate systems.
D. I. Ponyavin, A. V. Usmanov.
Geomagn. Aehron., No. 1, p. 128 – 129 (1985). In Russian. English translation in Geomagn. Aeron.

106.039 Arguments for fundamental emission by the parametric process L → T + S in interplanetary type III bursts.
I. H. Cairns.
ESA Spec. Publ., ESA SP–207, p. 281 – 284 (1984). – See Abstr. 012.044.
Recent observations of low frequency ion acoustic–like waves associated with Langmuir waves present during interplanetary type III bursts lead the author to consider plasma emission mechanisms and wave processes involving ion acoustic waves. He concludes that the parametric process L → T + S may be the important emission process for the fundamental radiation of interplanetary type III bursts.

106.040 A new radio emission at 3 kHz in the outer heliosphere.
W. S. Kurth, D. A. Gurnett, F. L. Scarf, R. L. Poynter.
ESA Spec. Publ., ESA SP–207, p. 285 – 288 (1984). – See Abstr. 012.044.
Evidence of a radio source in the outer heliosphere is given based on observations made by the plasma wave receivers on Voyagers 1 and 2 at heliocentric radial distances ranging from 13 to 20 AU. The radio emission is observed in the frequency range of 2 to 3 kHz, and is above the local electron plasma frequency. The maximum spectral density of the emission recorded to date is about $10^{-14} V^2 m^{-2} Hz^{-1}$. The bandwidth of the radio noise is about 1 kHz. One of several possible sources considered for the emission is radiation at the second harmonic of the plasma frequency at the heliopause. Should this interpretation be correct, the data reported represent the first remote observations of the heliopause.

106.041 Sector structure of the interplanetary magnetic field and high–velocity solar wind fluxes.
B. V. Zhiromskij.
Magnitosfer. issled., Moskva, No. 3, p. 25 – 29 (1984). In Russian. Abstr. in Ref. Zh., 62. Issled. Kosm. Prostranstva, 3.62.410 (1985).

106.042 Discovery of nuclear tracks in interplanetary dust.
J. P. Bradley, D. E. Brownlee.
Meteoritics, Vol. 19, No. 4, p. 197 – 198 (1984). Abstract. – See Abstr. 010.641.

106.043 D/H ratios in interplanetary dust and their relationship to IR, Raman, and EDX observations.
K. D. McKeegan, S. A. Sandford, R. M. Walker, B. Wopenka, E. Zinner.
Meteoritics, Vol. 19, No. 4, p. 269 – 270 (1984). Abstract. – See Abstr. 010.641.

106.044 Diagenesis in interplanetary dust: chondritic porous aggregate W7029*A.
F. J. M. Rietmeijer, I. D. R. Mackinnon.
Meteoritics, Vol. 19, No. 4, p. 301 (1984). Abstract. – See Abstr. 010.641.

106.045 A hydrated interplanetary dust particle containing calcium– and aluminum–rich pyroxene: possible relations to carbonaceous chondrites.
K. Tomeoka, P. R. Buseck.
Meteoritics, Vol. 19, No. 4, p. 322 – 323 (1984). Abstract. – See Abstr. 010.641.

106.046 Magnesium and silicon isotopic composition on interplanetary dust particles.
E. Zinner, A. Fahey, K. D. McKeegan.
Meteoritics, Vol. 19, No. 4, p. 345 – 346 (1984). Abstract. – See Abstr. 010.641.

106.047 Energetic particle impacts on frozen surfaces: laboratory simulation and solar system applications.
V. Pirronello.
Mem. Soc. Astron. Ital., Vol. 55, No. 3, p. 561 – 567 (1984). – See Abstr. 012.050.
Recent work on the interaction of fast particles with ices is reviewed. Emphasis is placed on both the experimental results of laboratory simulations and the relevant astrophysical applications relative to the solar system.

106.048 Space Shuttle microabrasion foil experiment (MFE): implications for aluminium oxide sphere contamination of near–earth space.
W. C. Carey, D. G. Dixon, J. A. M. McDonnell.
Adv. Space Res., Vol. 5, No. 2, p. 87 – 90 (1985). – See Abstr. 012.051.

106.049 Photometric properties of zodiacal light particles.
K. Lumme, E. Bowell.
Icarus, Vol. 62, No. 1, p. 54 – 71 (1985).

The authors derive empirically an expression for the mean single–particle phase function of zodiacal cloud particles. The dependence of brightness on ecliptic latitude is investigated. By incorporating polarization data for both the zodiacal light and atmosphereless bodies, the authors derive the single–scattering albedo, the number density, and particle size, and determine their dependences on heliocentric distance. They calculate the total mass and extinction of the zodiacal cloud.

106.050 Cosmic–ray picture of the heliosphere.
D. Venkatesan.
Johns Hopkins APL Tech. Dig., Vol. 6, No. 1, p. 4 – 19 (1985).

Cosmic rays were discovered about 75 years ago. During the last quarter of a century, the study of the time variations of cosmic rays has progressed considerably and has been transformed into an investigation of cosmic rays over time and space. The study has served as a useful tool for probing the interplanetary medium, a dynamic and complex region reborn with a new name – the heliosphere. In situ observations by satellites and spacecraft have enlarged and enhanced our understanding of the heliosphere, so a clearer picture of this region of solar influence is gradually emerging.

106.051 Diagnostics of the radial component of the interplanetary magnetic field.
V. M. Litinskij, P. V. Sumaruk.
Soln. Dannye, Byull., 1985, No. 2, p. 91 – 93 (1985). In Russian.

It is shown that it is possible to detect the hourly value and sign of the radial component of the interplanetary magnetic field using the known values of its azimuthal component.

106.052 Observations of 35– to 1600–keV protons and low–frequency waves upstream of interplanetary shocks.
T. R. Sanderson, R. Reinhard, P. van Nes, K.–P. Wenzel, E. J. Smith, B. T. Tsurutani.
J. Geophys. Res., Vol. 90, No. A5, p. 3973 – 3980 (1985). – See Abstr. 012.003.

The authors compare measurements of energetic protons in the range 35 – 1600 keV and low–frequency waves (periods of ~ 6 s) on ISEE 3 associated with the passage of the large oblique shock of April 5, 1979, which exhibits an extended foreshock, and they attempt to identify the energy of the particles which are responsible for the waves. By comparing intensity profiles of both waves and particles as a function of upstream distance, and taking account of the relation between the energy of the particles and the period of the waves, the authors are able to identify protons with energies of a few hundred keV as being responsible for the waves in the extended foreshock.

106.053 A major shock–associated energetic storm particle event wherein the shock plays a minor role.
P. van Nes, E. C. Roelof, R. Reinhard, T. R. Sanderson, K.–P. Wenzel.
J. Geophys. Res., Vol. 90, No. A5, p. 3981 – 3994 (1985). – See Abstr. 012.003.

106.054 Evidence for long period Alfvén waves in the inner solar system.
R. Bruno, B. Bavassano, U. Villante.
J. Geophys. Res., Vol. 90, No. A5, p. 4373 – 4377 (1985).

Magnetic field and plasma data of Helios 1 and 2 between 0.3 and 1 AU have been used to investigate the Alfvénic character of the solar wind fluctuations with periods above 1 hour. Clear evidence for the existence of very long period Alfvén waves, up to ~ 15 hours (in the spacecraft frame) close to the sun, has been obtained. The observations at different heliocentric distances suggest that the longest wavelengths are removed from the Alfvénic regime leaving the sun.

106.055 Theoretical interpretation of the observed interplanetary magnetic field radial variation in the outer solar system.
S. T. Suess, B. T. Thomas, S. F. Nerney.
J. Geophys. Res., Vol. 90, No. A5, p. 4378 – 4382 (1985).

Analysis of interplanetary magnetic field (IMF) data from the outer solar system has shown that the azimuthal component of the IMF falls off more rapidly with radius between 1 and 10 AU than as r^{-1}, where an r^{-1} dependence is predicted in classical theory. This result is consistent with meridional flow transporting magnetic flux away from the heliographic equator. An identical effect has been predicted to result from the force due to the meridional gradient of the azimuthal component of the IMF. Here, the data are evaluated in the context of the theory and it is shown that the observed flux deficits generally agree with predictions by the MHD model.

106.056 Photographic observations of the inner zodiacal light aboard Saliout 7.
G. Nikolsky (*G. M. Nikol'skij*), S. Koutchmy, P. L. Lamy, I. A. Nesmjanovich (*I. A. Nesmyanovich*).
Inst. Astrophys. Paris, Pré–Publ., No. 91, 4 pp. (1985). To appear in IAU Colloq. No. 85 "Properties and interaction of interplanetary dust".

106.057 Das elektrostatische Potential und die Fragmentation von Fluffy–Teilchen im interplanetaren Raum, in Kometennähe und in der Erdmagnetosphäre.
H. Böhnhardt, H. Fechtig, J. Kissel.
Mitt. Astron. Ges., Nr. 63, p. 191 – 193 (1985). – See Abstr. 012.063.

106.058 IRAS observations of the zodiacal background emission.
J. C. Good, T. N. Gautier III, M. G. Hauser.
Bull. Am. Astron. Soc., Vol. 16, No. 4, p. 921 – 922 (1984). Abstract. – See Abstr. 010.062.

106.059 Small scale structures in the zodiacal light.
S. S. Hong, N. Y. Misconi, M. H. H. van Dijk, J. L. Weinberg, G. N. Toller.
Bull. Am. Astron. Soc., Vol. 16, No. 4, p. 1002 (1984). Abstract. – See Abstr. 010.062.

106.060 Possible disturbance of the interplanetary magnetic field by the passage of a comet as observed by three spacecraft.
C. T. Russell, J. L. Phillips, M. R. Arghavani, J. G. Luhmann, K. Schwingenschuh.
Bull. Am. Astron. Soc., Vol. 16, No. 4, p. 1028 (1984). Abstract. – See Abstr. 010.063.

106.061 The association of interplanetary field enhancements with the asteroid 2201 Oljato and the comparison with simulations.
J. G. Luhmann, J. L. Phillips, M. R. Arghavani, C. T. Russell, J. Fedder.
Bull. Am. Astron. Soc., Vol. 16, No. 4, p. 1028 (1984). Abstract. – See Abstr. 010.063.

106.062 Interplanetary magnetic field enhancements in the solar wind: statistical properties at 1 AU.
M. R. Arghavani, C. T. Russell, J. G. Luhmann, R. C. Elphic.
Icarus, Vol. 62, No. 2, p. 230 – 243 (1985).

A new class of magnetic field signatures in the solar wind has been studied using interplanetary data obtained in the vicinity of Earth. Typical behavior of these signatures is a slow rise of the total magnetic field to a sharp peak and then a gradual and almost symmetric decrease to the background interplanetary magnetic field. A total of 45 events have thus been found in a survey of the magnetometer data of two spacecraft (ISEE–3 and IMP–8), corresponding to a total of more than 6 years of available survey data. The peak magnetic pressure of these events is always less than the solar wind dynamic pressure by a factor of 10 or more. These signatures are similar to those observed at Venus by the Pioneer Venus Orbiter. Based on these observations and the previously discovered signatures at Venus, the cause of these events is postulated to be small outgassing bodies which have passed by the spacecraft.

106.063   Collisional balance of the meteoritic complex.
    E. Grün, H. A. Zook, H. Fechtig, R. H. Giese.
Icarus, Vol. 62, No. 2, p. 244 – 272 (1985).

Taking into account meteoroid measurements by *in situ* experiments, zodiacal light observations, and oblique angle hypervelocity impact studies, it is found that the observed size distributions of lunar microcraters usually do not represent the interplanetary meteoroid flux for particles with masses $\lesssim 10^{-10}$g. From the steepest observed lunar crater size distribution a "lunar flux" is derived which is up to 2 orders of magnitude higher than the interplanetary flux at the smallest particle masses. New models of the "lunar" and "interplanetary" meteoroid fluxes are presented. A detailed analysis of the effects of mutual collisions (i.e., destruction of meteoroids and production of fragment particles) and of radiation pressure has been performed which yielded a new picture of the balance of the meteoritic complex.

106.064   Interplanetary magnetic field enhancements and their association with the asteroid 2201 Oljato.
C. T. Russell, R. Aroian, M. Arghavani, K. Nock.
Science, Vol. 226, No. 4670, p. 43 – 45 (1984).

Comparison of the times of occurrence of a newly discovered type of disturbance in the interplanetary magnetic field at 0.72 astronomical unit with the passage of the small Venus–crossing asteroid, 2201 Oljato, reveals a possible association, but the source of these disturbances appears to be associated with outgassing material in the Oljato orbit some distance behind the asteroid and not with the asteroid itself.

106.065   Discovery of nuclear tracks in interplanetary dust.
    J. P. Bradley, D. E. Brownlee, P. Fraundorf.
Science, Vol. 226, No. 4681, p. 1432 – 1434 (1984).

Nuclear tracks have been identified in interplanetary dust particles (IDP's) collected from the stratosphere. The presence of tracks unambiguously confirms the extraterrestrial nature of IDP's, and the high track densities (10^{10} to 10^{11} per square centimeter) suggest an exposure age of approximately 10^4years within the inner solar system.

106.066   Numerically–simulated formation and propagation of interplanetary shocks.
S. T. Wu.
Computer simulation of space plasmas, p. 179 – 201 (1985). – See Abstr. 003.031.

With recent advancements in the space program a large data base of in–situ measurements of solar wind plasma parameters has been accumulated by the observers and deposited in the World Data Centers. From these fruitful data, the understanding of interplanetary shock wave physics has been advanced significantly. The numerical simulation method is contributing to this understanding. In this paper, the author presents several numerical methods for treating the shock propagation problem and uses a specific example to demonstrate the capability of the simulation method.

106.067   Interplanetary spheromacs.
    K. G. Ivanov, A. F. Kharshiladze.
Geomagn. Aehron., Tom 25, No. 3, p. 377 – 380 (1985). In Russian. English translation in Geomagn. Aeron.

106.068   Cosmogenic isotopes and the velocity of cosmic matter capture by the earth.
A. V. Blinov, A. N. Konstantinov, V. I. Mlotehk.
Cosmochemistry and meteoritics, p. 64 – 72 (1984). In Russian. Abstr. in Ref. Zh., 51. Astron., 2.51.240 (1985). – See Abstr. 012.037.

106.069   Cosmic dust: collection and research.
    D. E. Brownlee.
Annu. Rev. Earth Planet. Sci., Vol. 13, p. 147 – 173 (1985). – See Abstr. 003.032.

The paper reviews the aspects of cosmic dust that are important to its collection and utilization as a resource of extraterrestrial material. Most of the collected dust particles appear to be primitive and sometimes unique solar system materials that contain important clues to early solar system processes and environments. Collected dust can also be studied to investigate selected properties of the interplanetary medium and the terrestrial environment. The treatment of the origin and evolution of dust in the solar system is focused on aspects important to the interpretation of sample results. The bulk of the paper reviews the results of laboratory studies of dust samples collected in the stratosphere and deep–sea sediments.

106.070   The dust that lights up the Zodiac.
    D. Dixon, T. McDonnell, B. Carey.
New Sci., Vol. 105, No. 1438, p. 26 – 29 (1985). Abstr. in Phys. Abstr., Vol. 88, No. 1255, Entry 45983 (1985).

106.071   The configuration of a heliospheric current sheet and some features of its evolution over the cycle of solar activity.
N. P. Korzhov.
18th International Cosmic Ray Conference, Vol. 3, p. 106 – 109 (1983). – See Abstr. 012.096.

The author has compiled a catalog of maps of polarities of the global magnetic field of the Sun and of configurations of the interplanetary current sheet in 1971 – 1978. Some main properties and dynamics of the heliospheric current sheet configuration are revealed.

106.072   The interplanetary shock event of November 11/12, 1978 – a comprehensive test of acceleration theory.
K.–P. Wenzel, T. R. Sanderson, P. van Nes, C. F. Kennel, F. L. Scarf, F. V. Coroniti, C. T. Russell, G. K. Parks, E. J. Smith, W. C. Feldman.
18th International Cosmic Ray Conference, Vol. 3, p. 131 – 134 (1983). – See Abstr. 012.096.

A comprehensive study of the November 11/12, 1978 shock event based on energetic particle, solar wind, magnetic field and wave data from the ISEE–3, –1 and –2 spacecraft has been undertaken both from the energetic particles and the collisionless shock point of view. The energy density of 10 – 50 keV protons accelerated by the shock is found to be equivalent to the upstream magnetic field energy density. The observations are in quantitative agreement with Lee's (1983) self–consistent theory for the excitation of hydromagnetic waves and the acceleration of ions upstream of interplanetary shocks.

106.073   Acceleration and drift in latitude at the solar–wind termination shock.
J. R. Jokipii.
18th International Cosmic Ray Conference, Vol. 3, p. 154 (1983). Abstract. – See Abstr. 012.096.

106.074   Coupled hydromagnetic wave excitation and ion acceleration at interplanetary shocks.
M. A. Lee.
18th International Cosmic Ray Conference, Vol. 3, p. 155 (1983). Abstract. – See Abstr. 012.096.

106.075   Association of interplanetary particles with radio S.A. (shock acceleration) events.
T. T. von Rosenvinge, D. V. Reames, H. V. Cane.
18th International Cosmic Ray Conference, Vol. 4, p. 10 (1983). Abstract. – See Abstr. 012.096.

106.076   The interplanetary mean free path of low energy ions in front of interplanetary shock waves.
M. Scholer, D. Hovestadt, F. M. Ipavich.
18th International Cosmic Ray Conference, Vol. 4, p. 143 (1983). Abstract. – See Abstr. 012.096.

106.077 The features of solar proton propagation in interplanetary space.
S. I. Ermakov, N. N. Kontor, G. P. Lyubimov, T. I. Morozova, N. N. Pavlov, N. V. Pereslegina, T. G. Khotilovskaya.
18th International Cosmic Ray Conference, Vol. 4, p. 153 – 156 (1983). – See Abstr. 012.096.
The authors report observations made by the Venera–11 space probe in 1979, July – August between 1 AU and 0.85 AU.

106.078 Origin of cosmic dust and cosmogenic nuclides.
H. Hasegawa.
18th International Cosmic Ray Conference, Vol. 9, p. 325 (1983). Abstract. – See Abstr. 012.096.

106.079 Acceleration and spatial diffusion of energetic particles obtained with Helios data.
J. F. Valdés–Galicia, X. Mousas, J. J. Quenby, F. Neubauer, R. Schwenn.
18th International Cosmic Ray Conference, Vol. 10, p. 104 – 107 (1983). – See Abstr. 012.096.
The authors have performed numerical simulations using Helios 1/2 IMF data at distances from 0.3 to 0.67 AU from the Sun. Space diffusion and statistical acceleration of particles are described.

106.080 Absolute value and functional dependence of interplanetary pitch angle scattering derived from Helios observations at 0.5 AU.
C. K. Ng, G. Wibberenz, G. Green, H. Kunow.
18th International Cosmic Ray Conference, Vol. 10, p. 381 – 384 (1983). – See Abstr. 012.096.
Further analysis of the 28 March 1976 event has been carried out using the focused transport model, with a Parker spiral interplanetary magnetic field.

106.081 A physical model for the interplanetary dust emission.
L. J. Rickard, E. Dwek, R. A. White, M. G. Hauser.
Bull. Am. Astron. Soc., Vol. 17, No. 2, p. 591 (1985). Abstract. – See Abstr. 010.065.

106.082 Induced shape on a cloud of interplanetary dust.
B. Å. S. Gustafson.
Bull. Am. Astron. Soc., Vol. 17, No. 2, p. 623 (1985). Abstract. – See Abstr. 010.066.

106.083 The brightness of the interplanetary medium in Thomson scattering from 0.3 to 1.0 AU: comparison with a view from Helios B.
A. Venkataraman, H. S. Hudson, B. V. Jackson.
Bull. Am. Astron. Soc., Vol. 17, No. 2, p. 638 (1985). Abstract. – See Abstr. 010.067.

106.084 Report of IAU Commission 49: The interplanetary plasma and the heliosphere (*Plasma interplanétaire et l'héliosphère*).
I. W. Roxburgh.
Trans. IAU, Vol. XIXA, p. 697 – 705 (1985). – See Abstr. 003.046.

106.085 A comprehensive investigation of the dynamics of micron and submicron lunar ejecta in heliocentric space.
A. D. Hargrave.
Diss. Abstr. Int., Sect. B, Vol. 45, No. 12, p. 3847 (1985). Thesis, Baylor University, 124 pp. (1984). Order No. DA8503687.

106.086 Dissipation mechanism of interplanetary Alfvénic fluctuation.
C.-y. Tu, L. Dong.
Chin. Astron. Astrophys., Vol. 9, No. 1, p. 60 – 65 (1985). English translation of Acta Astrophys. Sin., Vol. 4, No. 4, p. 304 – 312 (1984).
Observations on board Helios 1 and 2 have shown that, in 0.3 – 0.9 AU, the decay of the Alfvénic fluctuations is greater than is calculated by the WBK solution of the propagation of Alfvén waves. The additional decay can be expressed by an exponential factor with a "dissipation length" λ, which varies with the frequency but is approximately constant for frequencies higher than 0.01 Hz. In this paper, the authors deduce the analytical expression for the dissipation length in both the wave energy cascade model and the viscous decay model.

106.087 A stability analysis of the spiral sector transition region in the interplanetary magnetic field.
S. Wang, L. Zhu.
Chin. Astron. Astrophys., Vol. 9, No. 1, p. 80 – 85 (1985). English translation of Chin. J. Space Sci., Vol. 4, Nos. 3/4, p. 261 – 268 (1984).
The authors use a four–layer model in a stability analysis of the ME type spiral sector transition in the interplanetary magnetic field. They deduce that waves in the spiral sector transition region may be a source that triggers off the Kelvin–Helmholtz instability of the magnetopause.

Bibliography of Earth, Moon, and Planets.
See Abstr. 002.031.

Links between astronomical observations of protostellar clouds and laboratory measurements of interplanetary dust: the 6.8 µm carbonate band.
See Abstr. 022.068.

Two–stage acoustic penetration and impact detector for micrometeoroid and space debris application.
See Abstr. 035.036.

Astrophysical and geophysical observations with PIRAMIG/ Salyut 7 experiment.
See Abstr. 035.049.

Shock drift acceleration in the presence of waves.
See Abstr. 062.033.

Electromagnetic ion beam instabilities: hot beams at interplanetary shocks.
See Abstr. 062.058.

The relationship between the electric fields associated with plasma expansion and double layers.
See Abstr. 062.169.

Shock formation and decay in hydrodynamic flows with mass loading.
See Abstr. 062.171.

A parametric survey of the first critical Mach number for a fast MHD shock.
See Abstr. 062.176.

Moment method for multiple scattering of solar Lα radiation in the nearby interstellar medium.
See Abstr. 063.070.

Interplanetary perturbation from a flare triplet in May 1981 according to measurements aboard Prognoz 8.
See Abstr. 073.088.

The phase power spectrum of the solar wind measured by long–baseline interferometry at 81.5 MHz.
See Abstr. 074.001.

On the distribution of θ_{Bn} for shocks in the solar wind.
See Abstr. 074.007.

Coronal mass ejections and interplanetary shocks.
See Abstr. 074.008.

Solar wind decrease at high heliographic latitudes detected from Prognoz interplanetary Lyman alpha mapping.
See Abstr. 074.021.

Dynamical models of coronal transients and interplanetary disturbances.
See Abstr. 074.029.

Synoptic observations of coronal transients and their interplanetry consequences.
See Abstr. 074.032.

Measurements of the solar wind velocity with EISCAT.
See Abstr. 074.037.

The fate of sunward streaming protons associated with coronal mass ejections.
See Abstr. 074.038.

Response of Earth and Venus ionospheres to corotating solar wind stream of 3 July 1979.
See Abstr. 074.079.

Persistence of shocks to large distances in the solar wind.
See Abstr. 074.086.

Over–reflection of hydromagnetic planetary–gravity waves at the solar helmet streamers and magnetic sectors.
See Abstr. 074.115.

The effect of solar–wind convection on charged particle transport in interplanetary space.
See Abstr. 074.132.

Studies of the microscale density fluctuations in the solar wind using interplanetary scintillations.
See Abstr. 074.160.

Variation of Alfvén fluctuations across a slow shock near the Sun.
See Abstr. 074.163.

Magnetic fields on the sun and the north–south component of transient variations of the interplanetary magnetic field at 1 AU.
See Abstr. 075.015.

Space–time relations of magnetic fields on the sun with the IMF, solar wind and geomagnetic activity in May –July 1979.
See Abstr. 075.032.

Observations of solar radio bursts from meter to kilometer wavelengths.
See Abstr. 077.012.

Solar noise storms and magnetic sector structures.
See Abstr. 077.043.

Statistic acceleration and stationary transport of solar cosmic rays in the interplanetary space.
See Abstr. 078.003.

Deep space observations of the east–west asymmetry of solar energetic storm particle events: Voyagers 1 and 2.
See Abstr. 078.013.

Rigidity dependence of the interplanetary scattering mean free path determined from rise–phase nucleonic abundance variations.
See Abstr. 078.047.

Variations of solar proton anisotropy on November 10 – 11, 1981.
See Abstr. 078.049.

The delayed energetic particle event of June 6 – 10, 1979.
See Abstr. 078.078.

Turbulence in the solar atmosphere and in the interplanetary plasma.
See Abstr. 080.032.

Report of IAU Commission 21: Light of the night sky (*Lumière du ciel nocturne*).
See Abstr. 082.096.

Coupling between the solar wind and the magnetosphere: CDAW 6.
See Abstr. 084.018.

Solar wind control of electric fields and currents in the magnetosphere and ionosphere.
See Abstr. 084.056.

The relationship between the IMF magnitude and the frequency of Pc 3, 4 pulsations on the ground: simultaneous events.
See Abstr. 084.063.

Connection between variations of the geomagnetic activity index and parameters of interplanetary scintillations.
See Abstr. 084.074.

A connection of geomagnetic pulsations appearance in the range of 0.1 – 1.0 Hz with the B_z–component of the interplanetary magnetic field.
See Abstr. 084.103.

The influence of the ionospheric conductivity on the potential difference of the electric field across a plasma layer.
See Abstr. 084.107.

Influence of interplanetary magnetic field sector structure on geomagnetic field variations.
See Abstr. 084.126.

Mass loading of planetary magnetospheres by rocky satellites – I: Production of rocky ejecta.
See Abstr. 091.020.

Mass loading of planetary atmospheres by rocky satellites – II: Transport and enhanced lifetimes of satellite ejecta in planetary magnetospheres.
See Abstr. 091.021.

Propagation of Jovian electrons in the interplanetary space.
See Abstr. 099.077.

Jovian electron events in the orbit of the earth.
See Abstr. 099.080.

Dust properties determined from backscattering in the interplanetary and interstellar medium.
See Abstr. 102.054.

Nongravitational effects affecting small meteoroids in interplanetary space.
See Abstr. 104.005.

Report of IAU Commission 22: Meteors and interplanetary dust (*Météorites et la poussière interplanétaire*).
See Abstr. 104.055.

Laboratory infrared transmission spectra of individual interplanetary dust particles from 2.5 to 25 microns.
See Abstr. 105.206.

Drag–like effects on heliospheric neutrals due to elastic collisions with solar wind ions.
See Abstr. 131.026.

Results of an investigation of interstellar neutral hydrogen and helium in the solar system.
See Abstr. 131.117.

On removing the zodiacal emission from IRAS all–sky data.
See Abstr. 133.017.

Voyager observations of the far UV background at the North Galactic Pole.
See Abstr. 142.044.

Interplanetary cosmic ray intensity: 1972 – 1984 and out to 32 AU.
See Abstr. 144.005.

Variation of cosmic rays and solar wind properties with respect to the heliospheric current sheet. 1. Five–GeV protons and solar wind speed.
See Abstr. 144.029.

The role of inhomogeneities of the magnetic field forming in the interplanetary medium in the 11–year cycle of cosmic rays.
See Abstr. 144.032.

A study of solar modulation of medium energy primary cosmic ray nuclei.
See Abstr. 144.057.

Some questions of particle acceleration between approaching fronts of shock waves in interplanetary medium.
See Abstr. 144.209.

Anisotropies of 35 – 56 keV ions associated with interplanetary shocks.
See Abstr. 144.211.

A statistical study of interplanetary shock associated proton intensity increases.
See Abstr. 144.212.

Energetic particle spectra upstream and downstream of interplanetary shock waves.
See Abstr. 144.213.

Analysis of cosmic ray pitch–angle anisotropy during the Forbush–effect in June 1972 by the method of spectrographic global survey.
See Abstr. 144.235.

Cosmic ray decreases and cold plasma in the heliosphere.
See Abstr. 144.292.

Diffusive first–order Fermi acceleration at quasi–parallel interplanetary shocks: injection of thermal ions.
See Abstr. 144.397.

Characteristics of particle events associated with interplanetary shocks.
See Abstr. 144.398.

ISEE–3 observations of 35 – 1600 keV protons and low frequency waves upstream of an interplanetary shock.
See Abstr. 144.399.

Rapporteur paper for sessions MG1, MG3, and MG4 (*of the conference*): modulation theory, interplanetary propagation, and interplanetary acceleration.
See Abstr. 144.453.

107 Cosmogony

107.001   A model for diagenesis in proto–planetary bodies.
F. J. M. Rietmeijer.
Nature, Vol. 313, No. 6000, p. 293 – 294 (1985).

In the earliest stage of planetary–body formation, a proto–planet presumably consists of loosely–accumulated solid grains, or clusters of grains, with ice mantles and water–ice grains. Physical and chemical reactions, similar to diagenesis in the terrestrial permafrost, may occur and may be enhanced by an interfacial water layer between solid grains and water–ice at temperatures below the melting point of water–ice. The model proposed here eases the thermal constraints imposed by earlier models of low–temperature aqueous alteration in primitive extraterrestrial materials in which formation of an aqueous medium promoting these alterations required complete melting of the water–ice.

107.002   Asteroids and comets.
T. Gehrels.
Phys. Today, Vol. 38, No. 2, p. 32 – 35, 37, 39 – 41 (1985).

The formation of the sun and planets left remnants, such as the million objects larger than 1 km between the orbits of Mars and Jupiter, that give us a look at the original building blocks of the solar system.

107.003   Three–dimensional calculations of the formation of the presolar nebula from a slowly rotating cloud.
A. P. Boss.
Icarus, Vol. 61, No. 1, p. 3 – 9 (1985). – See Abstr. 012.004.

The presolar nebula may have formed from the collapse of a very slowly rotating interstellar cloud. The first three–dimensional, hydrodynamical calculations of the collapse of such clouds are presented. The models include radiative transfer in the Eddington approximation, as well as detailed equations of state appropriate for the nonisothermal regime of protostellar evolution.

107.004   Modifications of grains by particle bombardment in the early solar system.
G. Strazzulla.
Icarus, Vol. 61, No. 1, p. 48 – 56 (1985). – See Abstr. 012.004.

Much observational and theoretical evidence supports the idea that the young Sun passed through a high–activity or T Tau phase. Among the various phenomena of activity a high flux of energetic ($\sim$ MeV) protons is expected. Some solid–state effects induced by ion fluence on solid surfaces are discussed on the basis of recent experimental results. In particular it is shown that erosion by particles is a very important destruction mechanism for ice grains in a large range of grain dimensions and distances from the Sun. This should significantly contribute to sweep out material from the solar system in its early evolutive stage. The release of new molecules in the gas phase and the synthesis of new organic materials from simple gases condensed on solid substrates are discussed. The laboratory infrared features of the new polymer–like materials are compared with spectra observed in organic compounds extracted from some meteorites.

107.005   Asteroids, comets, and planet formation.
M. H. Carr.
NASA Spec. Publ., NASA SP–469, p. 5 – 11 (1984). – See Abstr. 003.004.

107.006   Clues on the origin of the solar system.
H. Reeves.
Planets, their origin, interior and atmosphere, p. 1 – 60 (1984). – See Abstr. 012.019.

Contents: (1) Introduction: the protosolar nebula, a scenario in need of improvements. (2) Solar system observations: fossil radioactivities, isotopic anomalies in graphite grains (s–process, r–process), the oxygen anomaly, the puzzle of Ne–22, the origin of stardust in meteorites, deuterated molecules. (3) Astronomical observations: non–thermal processes in the Orion infrared cluster, jets from young stars, X–rays and γ–rays from young stars, atoms of ^{26}Al in space. (4) Clues from the physics of the solar system: the thermal barrier, the magnetic barrier, the angular momentum barrier, gravitational collapse of grains, gas–grains separation during the formation of planetary bodies. (5) Appendix: history of nucleosynthesis.

107.007   The young Sun and the atmospheres of Earth.
K. Rynefors, G. F. Gahm.
Observatory, Vol. 105, No. 1065, p. 36 – 39 (1985).

107.008   The formation of the planetary system.
W. M. Tscharnuter.
Celest. Mech., Vol. 34, Nos. 1 – 4, p. 289 – 296 (1984). – See Abstr. 012.030.

The basic ideas concerning solar system formation were developed by Kant (1755) and Laplace (1796) whose starting point was the so–called nebular hypothesis. The great advantage of the nebular hypothesis is that many regularities, e.g. prograde motions of all planets and asteroids in almost coplanar orbits, can be explained. Observations in the radio and infrared region strongly support the nebular hypothesis provided that the angular momentum problem can be solved in some way. Three possibilities are listed: (1) magnetic fields via Alfvén waves which can transport angular momentum from the contracting cloud fragment into the external medium, (2) turbulent friction, (3) gravitational torques exerted by high amplitude spiral or bar–like density waves in the nebula.

107.009   On the role of meteoritic impacts in the formation of organic molecules.
P. Cerroni, G. Martelli.
Adv. Space Res., Vol. 4, No. 12, p. 97 – 102 (1984). – See Abstr. 012.036.

It is suggested that the UV radiation, and shock and plasma phenomena which accompanied the hypervelocity impacts of solid bodies (meteorites and comets) onto the surface of the young Earth may have contributed to the synthesis of prebiotic organic molecules in the primitive atmosphere in a larger amount than was thought previously. The mechanisms responsible for this synthesis are discussed using information obtained from recent experimental and theoretical work on macroscopic hypervelocity impacts.

107.010   Sticking experiments and non–gravitational component of the mechanism of the growth of planets.
J. Leliwa–Kopystynski.
J. Phys. Colloq., Vol. 45, No. C–8, p. 109 – 112 (1984). Abstr. in Phys. Abstr., Vol. 88, No. 1256, Entry 51522 (1985). – See Abstr. 012.085.

107.011   Hydrodynamical models of the evolution of planets.
A. N. Karapats.
Dag. univ. Makhachkala, 30 pp. (1984). In Russian. Abstr. in Ref. Zh., 51. Astron., 2.51.146 (1985).

107.012   Characteristics of hydrides and their connection with the behaviour of elements in protoplanetary matter.
N. P. Semenenko.
Cosmochemistry and meteoritics, p. 81 – 92 (1984). In Russian. Abstr. in Ref. Zh., 51. Astron., 2.51.158 (1985). – See Abstr. 012.037.

107.013   Cometary impacts, molecular clouds, and the motion of the Sun perpendicular to the galactic plane.
P. Thaddeus, G. A. Chanan.
Nature, Vol. 314, No. 6006, p. 73 – 75 (1985).

Raup and Sepkoski have presented evidence from marine fossils for a 26–Myr periodicity in the occurrence of mass extinctions. Using the same data Rampino and Stothers obtained a different period, 30 ± 1 Myr, which agrees with the 33 ± 3–Myr half–period for the vertical oscillation of the Solar System about the plane of the Galaxy. To explain this agreement they suggest that encounters with molecular clouds perturb the Sun's family of comets, causing many to enter the inner Solar System where one or more collide with the Earth. It is shown here that the molecular cloud layer near the Sun is too extended and, as a consequence, the modulation of cloud encounters is too weak for a statistically significant period to be extracted from the nine extinctions analysed by Rampino and Stothers.

107.014   Formation of the planets and asteroids: some difficulties of accretion theories.
M. M. Komesaroff.
Proc. Astron. Soc. Aust., Vol. 5, No. 4, p. 457 – 459 (1984).

107.015   Titan and the dispersal of the proto–Saturnian nebula.
K. Hourigan.
Proc. Astron. Soc. Aust., Vol. 5, No. 4, p. 459 – 461 (1984).

The author examines some theories of the proto–Saturnian nebula. He is interested in exploring whether the presence of Titan in a viscously evolving nebula leads to any restrictions being placed on the timescales of nebula dispersal and satellite accretion.

107.016   Nebula tides and gap formation.
K. Hourigan, M. P. Schwarz.
Proc. Astron. Soc. Aust., Vol. 5, No. 4, p. 461 – 464 (1984).

The role of density waves in transferring angular momentum from the orbit of a planetoid to a gaseous nebula, and vice versa, is outlined. Two mechanisms proposed for wave damping, viz. turbulent viscosity and non–linear wave development, may possibly be effective for density waves generated in the proto–solar nebula by bodies of Jovian–size.

107.017   Dynamic thermal episodes in the protosolar nebula: development of models from observations on CAI's.
T. E. Bunch, S. Chang, P. Cassen, D. Hollenbach.
Meteoritics, Vol. 19, No. 4, p. 202 (1984). Abstract. – See Abstr. 010.641.

107.018   A possible ^{126}Sn chronometer for the early solar system.
J. R. De Laeter, K. J. R. Rosman.
Meteoritics, Vol. 19, No. 4, p. 217 (1984). Abstract. – See Abstr. 010.641.

107.019   The chemistry of rare earth elements in the solar nebula.
J. W. Larimer, H. A. Bartholomay, B. Fegley.
Meteoritics, Vol. 19, No. 4, p. 258 (1984). Abstract. – See Abstr. 010.641.

107.020   Chemical evolution of the solar nebula: a new model.
B. M. P. Trivedi.
Meteoritics, Vol. 19, No. 4, p. 325 – 326 (1984). Abstract. – See Abstr. 010.641.

107.021   Meteoritic constraints on processes in the solar nebula.
J. A. Wood.
Meteoritics, Vol. 19, No. 4, p. 339 – 340 (1984). Abstract. – See Abstr. 010.641.

107.022   The solar system during the T–Tau phase.
G. Strazzulla.
Mem. Soc. Astron. Ital., Vol. 55, No. 3, p. 481 – 484 (1984). – See Abstr. 012.049.

107.023   Modern problems of the cosmogony of the solar system.
V. S. Safronov.
Astron. Vestn., Tom 18, No. 4, p. 322 – 341 (1984). In Russian. Abstr. in Ref. Zh., 51. Astron., 4.51.149 (1985).

107.024   Orbital resonances in the solar nebula: implications for planetary accretion.
S. J. Weidenschilling, D. R. Davis.
Icarus, Vol. 62, No. 1, p. 16 – 29 (1985).

The influence of gas drag and gravitational perturbations by a planetary embryo on the orbit of a planetesimal in the solar nebula was examined. Non–Keplerian rotation of the gas causes secular decay of the orbit. If the planetesimal's orbit is exterior to the perturber's, resonant perturbations oppose this drag and can cause it to be trapped in a stable orbit at a commensurability of order $j/(j+1)$, where j is an integer. Numerical and analytical demonstrations show that resonant trapping occurs for wide ranges of perturbing mass, planetesimal size, and j. Fragments smaller than a critical size can pass through resonances under the influence of drag and be accreted by the embryo. This effect speeds accretion and tends to prevent dynamical isolation of planetary embryos, making gas–rich scenarios for planetary formation more plausible.

107.025   Protosatellite swarm dynamics.
G. V. Pechernikova, S. V. Maeva, A. V. Vityazev.
Sov. Astron. Lett., Vol. 10, No. 5, p. 293 – 297 (1984). English translation of 38.107.008.

107.026   Gas accretion of the protoplanetary nebula by growing terrestrial planets and early stages of the evolution of the solar system.
M. N. Izakov.
Kosm. Issled., Tom 23, Vyp. 2, p. 283 – 295 (1985). In Russian. English translation in Cosm. Res.

107.027   Origin of the solar system.
V. S. Safronov, A. V. Vityazev.
Astrophys. Space Phys. Rev., Vol. 4, p. 1 – 98 (1985). – See Abstr. 003.020. Revised and extended English translation of 34.107.012.
This paper is a review of work on the origin of the solar system. New results obtained by Soviet researchers are presented in greater detail. Primary attention is devoted to the following problems: evolution of the solar nebula, formation of the preplanetary disk, physicochemical processes in the disk, and the accumulation of the terrestrial planets. The critical importance of research at the interface between planetary cosmogony and adjacent sciences is noted.

107.028   Preferential perihelion and aphelion distances, and planetary formation.
N. A. Barricelli.
Theor. Pap., Vol. 3, Nr. 6, p. 76 – 122 (1985).
An interpretation of a series of precise relationships between planetary perihelion and aphelion distances, leading to the definition of preferential perihelion and aphelion distances, designated as meeting distances is outlined. The interpretation is based on Alfvén's jet stream theory and the hypothesis of harmonic resonance between intersecting jet streams.

107.029   Aluminium clues to the formation of the Solar System.
D. D. Clayton.
Nature, Vol. 315, No. 6021, p. 633 – 634 (1985).

107.030   Formation of the prelunar accretion disk.
A. G. W. Cameron.
Icarus, Vol. 62, No. 2, p. 319 – 327 (1985).
The present paper is concerned with the formation of the prelunar accretion disk. The process of expansion of the rock vapors from the site of the impact has been studied by a particle–in–cell representation of the hydrodynamics in which several parameters have been varied. The objective has been to see whether significantly more than a lunar mass of material can be put into orbit by the collision and to see whether this material possesses significantly more than enough angular momentum to assemble a lunar mass at the Roche limit at about 3 Earth radii. It has been found that reasonable conditions exist in which these requirements are satisfied.

107.031   Occurrence of giant impacts during the growth of the terrestrial planets.
G. W. Wetherill.
Science, Vol. 228, No. 4701, p. 877 – 879 (1985).
Three–dimensional Monte Carlo simulations of the accumulation of the terrestrial planets in the absence of gas drag produced results that are in general agreement with the number and distribution of the present planets. The accumulation process appears to be characterized by impact of bodies as large as three times the mass of Mars at velocities of about 9 kilometers per second.

107.032   The stellar upper mass limit in an OB association and implications for cosmic rays and cosmic abundances.
J. B. Blake, D. S. P. Dearborn.
18th International Cosmic Ray Conference, Vol. 9, p. 293 – 296 (1983). – See Abstr. 012.096.
Some implications are considered of the possibility that collapse of the solar nebula was induced by the supernova of an extremely massive star or, at least, such a star was in the vicinity of the protosolar nebula prior to its collapse to form the solar system.

107.033   Mechanical models of close approaches and collisions of large protoplanets.
W. M. Kaula, A. E. Beachey.
Bull. Am. Astron. Soc., Vol. 17, No. 2, p. 623 – 624 (1985). Abstract. – See Abstr. 010.066.

107.034   Origin of the solar system.
Z. Pokorný.
Pokroky Mat., Fyz. Astron., Vol. 29, No. 2, p. 80 – 88 (1984). In Czech.

107.035   Extinct nuclides – "Much ado about nothing".
G. J. Wasserburg.
Meteoritics, Vol. 20, No. 2, Part 2, p. 295 – 310 (1985). – See Abstr. 003.048.
A concise review of the status of research on short–lived nuclei is presented. The importance of these nuclei is very great in spite of the fact that they are essentially absent today (except for cosmic–ray products). The significance of these nuclei for understanding broader cosmic problems is outlined and it is shown that they are a key to the earliest processes in solar system formation and possibly provide a link with presolar processes in the interstellar medium or of intense activity of the early sun.

107.036   History and current understanding of the Suess abundance curve.
K. Marti, H. D. Zeh.
Meteoritics, Vol. 20, No. 2, Part 2, p. 311 – 320 (1985). – See Abstr. 003.048.
Brief accounts are presented on the study of nuclide abundances and the data base for theories of element synthesis. The most recent empirical solar system abundances confirm the Suess curve remarkably well. The poorly understood structures in the abundance curve are discussed in the light of recent models on nuclear structure and stability. Some prospects for future experimental research are outlined, as well as tests regarding the evolution of the "cosmic" abundances.

107.037   A new empirical formula for satellite distances and its explanation.
Z.–w. Hu, Z.–x. Chen.
Chin. Astron. Astrophys., Vol. 9, No. 1, p. 91 – 94 (1985). English translation of Chin. J. Space Sci., Vol. 4, Nos. 3/4, p. 240 – 244 (1984).
The authors found a new empirical formula for the distance of the n–th satellite in the Jovian, Saturnian and Uranian systems, $a_n = B_1 \times B^n$, with just two constants B_1 and B for each system. The difference between the observed distances and the values calculated according to this formula is generally less than 10%. This theoretical analysis shows that one type of radial perturbation in the planetesimal disk will lead to instability and hence the formation of gaseous rings with enhanced density. Within these rings, the planetesimals stick together to form the satellites, and it is the form of the distribution of the rings that leads to the distance law.

The cosmic history of the biogenic elements and compounds.
See Abstr. 003.050.

Cratering theories bombarded.
See Abstr. 011.008.

Making the moon from a big splash.
See Abstr. 011.028.

Homogeneous condensation of gaseous mixtures of Si, Fe, O, N, and C in relative cosmic abundance and implications for astronomical condensation.
See Abstr. 022.069.

Impact and explosion crater ejecta, fragment size, and velocity.
See Abstr. 022.126.

Laboratory simulation of planetesimal collision. 2. Ejecta velocity distribution.
See Abstr. 022.181.

The orbital evolution of Pluto–like objects.
See Abstr. 042.007.

Space research and the new approach to the mechanics of fluid media in cosmos.
See Abstr. 062.087.

Orbit perturbation: evolution of a Keplerian disk.
See Abstr. 064.033.

Multistage accretion and core formation of the earth.
See Abstr. 081.005.

Earth.
See Abstr. 081.012.

The origin and earliest state of the earth's hydrosphere.
See Abstr. 081.023.

Isotopic relations and rare gas abundance in the accretion period of the earth.
See Abstr. 081.027.

Interpretations of mass extinction.
See Abstr. 081.028.

Terrestrial impactors at geological boundary events: comets or asteroids?
See Abstr. 081.029.

Interval of formation of the earth.
See Abstr. 081.030.

Terrestrial xenon isotope constraints on the early history of the Earth.
See Abstr. 081.039.

Patterns of family extinction depend on definition and geological timescale.
See Abstr. 081.040.

Geological rhythms and cometary impacts.
See Abstr. 081.042.

Problèmes actuels de planétologie.
See Abstr. 091.012.

Rare gases in the atmospheres of terrestrial planets and early stages of evolution of the solar system.
See Abstr. 091.034.

Non–gravitational forces in the evolution of the solar system.
See Abstr. 091.038.

On the stability of the Solar System as hierarchical dynamical system.
See Abstr. 091.039.

The collisional dynamics of particulate disks.
See Abstr. 091.069.

Formation of the Kirkwood gaps in the asteroid belt.
See Abstr. 098.028.

Collisional history of asteroids: evidence from Vesta and the Hirayama families.
See Abstr. 098.051.

Evolution of the Galilean moons of Jupiter.
See Abstr. 099.037.

Implications of the Galilean satellites ice envelope explosions. I. The motion of fragments inside and beyond Jupiter's sphere of action.
See Abstr. 099.042.

Periodic comet showers and planet X.
See Abstr. 101.002.

Formation environment of cometary nuclei in the primordial solar nebula.
See Abstr. 102.002.

Prediction of deuterium abundance in comets.
See Abstr. 102.003.

Deep–sea stony spherules and the primordial nebula.
See Abstr. 105.003.

Penecontemporaneous metamorphism, fragmentation, and reassembly of ordinary chondrite parent bodies.
See Abstr. 105.249.

Shock heating of chondrules.
See Abstr. 105.263.

Evidence for oxidizing conditions in the solar nebula from Mo and W depletions in refractory inclusions in carbonaceous chondrites.
See Abstr. 105.274.

The origin of planetary systems in the scenario of star formation.
See Abstr. 118.029.

Planetoidal hypothesis of CP (*chemically peculiar*) F, A, and B star formation: possibilities and prospects.
See Abstr. 131.369.

Are cosmic ray isotopic "anomalies" due to the solar system being anomalous?
See Abstr. 144.167.

The parameters of irradiation by accelerated particles during the different stages of the solar system matter evolution.
See Abstr. 144.168.

Stars

111 Parallaxes, Proper Motions, Radial Velocities, Space Motions, Distances

111.001 Radial velocities of blue horizontal branch field stars.
J. Sommer–Larsen, P. R. Christensen.
Mon. Not. R. Astron. Soc., Vol. 212, No. 4, p. 851 – 856 (1985).
Radial velocities of blue horizontal branch field (bhbf) stars have been measured. The solar motion solution is determined for a sample of bhbf stars. It is found to be 172 ± 40 km s^{-1}. The Z–dispersion of the local velocity ellipsoid of bhbf stars is determined. It is found to be 114 ± 28 km s^{-1}.

111.002 Radial velocities of southern stars obtained with the photoelectric scanner CORAVEL. III. 790 late–type bright stars.
J. Andersen, B. Nordström, A. Ardeberg, M. Benz, M. Imbert, H. Lindgren, N. Martin, E. Maurice, M. Mayor, L. Prévot.
Astron. Astrophys., Suppl. Ser., Vol. 59, No. 1, p. 15 – 36 (1985).
This paper presents 1595 photoelectric radial velocity observations for 790 bright southern stars of spectral type F5 and later. It includes all such stars in the Bright Star Catalogue without previous radial velocity data. Based on two CORAVEL observations per star, the mean velocities are accurate to about 0.15 km s^{-1} r.m.s. for sharp–lined constant stars, errors increasing somewhat for stars with significant rotation. 46 probable and 127 certain velocity variables have been found, including 10 double–lined spectroscopic binaries. The present data complete and supersede the preliminary data for ~ 500 stars supplied by the authors in advance of publication for inclusion in the fourth edition of the Bright Star Catalogue (Hoffleit, 1982).

111.003 Investigation of the proper motions of reference stars at galactic latitudes $|b| > 45°$.
S. P. Rybka.
Kinematika Fiz. Nebesn. Tel, Tom 1, No. 1, p. 88 – 93 (1985). In Russian.
Reference stars at high galactic latitudes with relatively small proper motions are recommended to be used for decreasing the cosmic error of stellar proper motions obtained with the Goloseevo long–focus astrograph.

111.004 Radial velocity and four–color measures of halo A–type stars.
A. G. D. Philip.
News Lett. Astron. Soc. N.Y., Vol. 2, No. 7, p. 17 – 18 (1985).
Abstract. – See Abstr. 010.241.

111.005 U.S. Naval Observatory parallaxes of faint stars. List VII.
R. S. Harrington, C. C. Dahn, V. V. Kallarakal, B. Y. Riepe, J. W. Christy, H. H. Guetter, H. D. Ables, A. V. Hewitt, F. J. Vrba, R. L. Walker.
Astron. J., Vol. 90, No. 1, p. 123 – 129 (1985).
Trigonometric parallaxes, relative proper motions, and photoelectric photometry are presented for 106 stars in 97 astrometric plate series.

111.006 An upper limit for the solar acceleration.
J. Thornburg.
Mon. Not. R. Astron. Soc., Vol. 213, No. 1, p. 27P – 28P (1985).
It has been suggested that the Sun may be a member of a binary star system. In this case, the barycentre of the (known) Solar System would be accelerated towards the companion object. Consideration of the binary pulsar PSR 1913+16's orbital motion gives an upper limit of $|a| \lesssim 10^{-9}$m s^{-2} for the component of any Solar System acceleration in its direction.

111.007 Determination of absolute proper motions in a KSZ field with two astrographs.
W. R. Dick.
Astron. Nachr., Vol. 306, No. 2, p. 91 – 96 (1985). In German.
In the field No. 153 of the KSZ program relative proper motions of 144 stars up to $15\overset{m}{.}0$ are presented as well as parameters for the reduction to absolute proper motions with reference to 6 galaxies. The external error of one proper motion is 0.014″/yr. It is shown that in a small field the statistical method for the estimation of the magnitude error is not able to give significant results due to the variance of the proper motions. It is concluded that in practice the long focus and the wide angle astrographs of the Main Observatory of the Ukrainian Academy of Science in Kiev can be combined for the derivation of proper motions.

111.008 IUE data analysis: radial velocities from line coincidence statistics.
D. J. Bord, J. P. Davidson.
Astron. Astrophys., Vol. 143, No. 2, p. 461 – 465 (1985).
The authors have employed the method of wavelength coincidence statistics (WCS) to determine radial velocities from International Ultraviolet Explorer (IUE) satellite data for several chemically peculiar stars. By repeatedly carrying out the entire WCS procedure with a sequence of radial velocities, the authors demonstrate that significance–weighted mean velocities can be obtained which are accurate to within the uncertainties imposed by the stellar wavelengths, viz. $\leqslant 5$ km s^{-1} for sharp–lined stars. The use of this approach to derive velocity information from IUE spectra containing wavelength scale errors and to identify stars with composite spectra is discussed.

111.009 Rigorous treatment of the heliocentric motion of stars.
P. Stumpff.
Astron. Astrophys., Vol. 144, No. 1, p. 232 – 240 (1985).
The author presents a theory of the apparent heliocentric motion of stars which handles the perspective effects and the effects due to special relativity rigorously. The observational equivalents of the true (inertial) proper motion and radial velocity are derived. The deviations of the conventional astrometric methods from the rigorous theory are studied theoretically on the example of a fictitious star. Numerical examples for a few nearby stars are given. Although the deviations are very small, they may become detectable in the future. The theory is also applied to the case of very large space velocities, and it is shown that it produces the same effects which have been proposed as one possible explanation for the superluminal velocities observed in the nuclei of active galaxies.

111.010 The zero–point of the ”Fick” radial–velocity stars.
C. Neese, H. L. Detweiler, K. M. Yoss.
Publ. Astron. Soc. Pac., Vol. 97, No. 587, p. 78 – 84 (1985).
Observations obtained over seven observing runs with the KPNO coudé–feed spectrograph indicate no detectable zero–

point error in the "Fick" stellar radial–velocity system. The coudé–feed zero point was established through observations of solar–system objects for which radial velocities were derived from their orbital elements.

111.011   The existence of wide pairs among late–type stars.
A. R. Upgren.
News Lett. Astron. Soc. N.Y., Vol. 2, No. 5, p. 22 (1984). Abstract. – See Abstr. 010.243.

111.012   The velocity dispersion of carbon stars at the north galactic pole.
J. R. Mould, D. P. Schneider, G. A. Gordon, M. Aaronson, J. W. Liebert.
Publ. Astron. Soc. Pac., Vol. 97, No. 588, p. 130 – 137 (1985).

Nine carbon stars at high galactic latitude, including the first eight carbon stars from the Case survey have a corrected line–of–sight velocity dispersion of 101 km s^{-1}. If they are similar to the carbon stars found in dwarf spheroidal galaxies, they range in distance from 5 to 50 kpc. Their velocity dispersion is higher than the prediction of the cylindrical model for the galactic velocity ellipsoid by Ratnatunga and Freeman, but equal to that of galactic globular clusters. One of these stars may have an age of less than six billion years.

111.013   The proper motions of LS V +46°21 and AS 84, two "central" star candidates for S216.
K. Cudworth, R. J. Reynolds.
Publ. Astron. Soc. Pac., Vol. 97, No. 588, p. 175 – 176 (1985).

The 12th magnitude blue star LS V +46°21 has a proper motion of about $2\overset{''}{.}5\pm0\overset{''}{.}6$ century^{-1} primarily to the east. This motion suggests that LS V +46°21 may be the "central" star of S216.

111.014   Velocity ellipses in the galactic plane: regional variations and spectral distributions.
S. M. Freitas.
Astron. Astrophys., Suppl. Ser., Vol. 60, No. 1, p. 63 – 69 (1985).

Considering the behaviour of the velocity components of the nearby stars studied in two previous papers, and using the same population and stratifications, the author studies the orientations of the velocity ellipses in the galactic plane. The variations found seem to be due to local effects rather than to peculiarities related to age or spectral type.

111.015   Magnitude equation of the Pulkovo catalogue of stellar proper motions in areas with galaxies.
N. V. Kharchenko, S. S. Prilepina.
Kinematika Fiz. Nebesn. Tel, Tom 1, No. 2, p. 43 – 49 (1985). In Russian.

The magnitude equation of stellar proper motions in 85 areas with galaxies of the Pulkovo catalogue is investigated. The method of correction for the magnitude equation is applied by using the same catalogue data. On the basis of the corrected proper motions correction values of precession constants are determined: $\Delta k = +8\pm14$, $\Delta n = 33\pm14$ (in units $0\overset{''}{.}0001$/year).

111.016   Parallaxes and proper motions. XVII.
A. R. Upgren, E. W. Weis, H. L. Nations, J. T. Lee.
Astron. J., Vol. 90, No. 4, p. 652 – 654 (1985).

Parallaxes and proper motions are presented for 23 stars in 19 star fields. Seven of the target stars and one other star have no previous parallax determinations. New photometry is also presented for two of the stars for which previous data had not been published. Two of the target stars are resolved binaries for which both components were measured separately; one of the two is not a physical pair. This list is the first in the series which makes use of the PDS microdensitometer of the Yale Observatory with five of the 19 star fields being measured there. This is also the first list to report parallaxes of subdwarf candidates.

111.017   Proper motions of a group of R Coronae Borealis type stars.
G. Torres, L. A. Milone, M. M. Villada de Arnedo.
Astron. J., Vol. 90, No. 4, p. 680 – 684 (1985).

The authors present proper motions and positions on the FK4 system for eight variable stars of the R CrB type. Values for UW Cen, Y Mus, S Aps, RS Tel, and RY Sgr were derived by making use of all available astrometric data (photographic and meridian) and newly obtained plates; XX Cam, R CrB, and U Aqr were also included for completeness, with p.m. derived from literature values. Results indicate considerably small motions for these stars; a comparison with other published values is made when possible.

111.018   On the observation of double stars for the determination of their proper motions.
M. S. Zverev.
Astrophys. Space Sci., Vol. 110, No. 1, p. 117 – 118 (1985). – See Abstr. 012.039.

The state of meridian observations of the DS (double stars) program is reported. The program contains close double stars whose precise positions can be determined by the visual method only. In some declination zones these stars have been observed at a few observatories of the USSR and in Strasbourg, Santiago de Chile and Beograd. The importance of further DS observations, particularly in the southern sky, is stressed.

111.019   A search for CPM (*common proper motion*) stars.
J. L. Halbwachs.
Astrophys. Space Sci., Vol. 110, No. 1, p. 159 (1985). Abstract. – See Abstr. 012.039.

111.020   Potential of astrographic catalog star first–epoch positions for proper–motion determination.
A. Fresneau.
Astron. J., Vol. 90, No. 5, p. 892 – 895 (1985).

An experiment is described in which recent Palomar Schmidt plate observations are combined with astrographic catalog positions of one century ago, in order to derive relative proper motions for stars in the apparent visual–magnitude range from 9.0 to 11.5. Stars with tangential velocities greater than 0.06 arcsec/yr extend the list of stars detected with tangential velocities greater than 0.18 arcsec/yr.

111.021   Proper motions of 4949 geodetic stars with declinations from +90° to –90°.
E. V. Khrutskaya.
Glav. astron. obs. Akad. Nauk SSSR. Leningrad, 19 pp. (1984). In Russian. Abstr. in Ref. Zh., 52. Geod. Aehrosemka, 4.52.85 (1985).

111.022   Radial velocities for 28 southern young open clusters.
J. Hron, H. M. Maitzen, A. F. J. Moffat, T. Schmidt–Kaler, N. Vogt.
Astron. Astrophys., Suppl. Ser., Vol. 60, No. 3, p. 355 – 364 (1985).

The authors present radial velocities for 83 OB–stars in 28 southern young open clusters. The internal and external accuracy of the data is in good agreement with previous results obtained with the same equipment. From a comparison with radial velocities existing in the literature no systematic radial velocity differences for stars in common are found.

111.023   Radial velocities from S.I.M.B.A.D.
F. Ochsenbein.
Bull. Inf. Cent. Données Stellaires, No. 28, p. 17 – 18 (1985).

The paper provides figures concerning the radial velocity data obtained with classical methods.

111.024   Radial velocities of nearby stars.
H. Jahreiss, W. Gliese.
Bull. Inf. Cent. Données Stellaires, No. 28, p. 19 – 23 (1985).

Selon les prévisions actuelles, le troisième catalogue des étoiles proches du soleil fournira les mesures des vitesses radiales d'environ 65% des étoiles. Les vitesses radiales de la plupart des étoiles faibles resteront inconnues.

111.025 Proper motion survey with the 48–inch Schmidt telescope. LXV. Proper motions for 6,056 stars in the Pleiades–Hyades region.
W. J. Luyten, S. Morris, G. Hill, H. S. Hughes.
Separate Print, University of Minnesota, Minneapolis, Minn. 55455, USA. 104 pp. (1985).

111.026 Recent results from precision radial velocity measurements.
R. D. McClure.
Bull. Am. Astron. Soc., Vol. 16, No. 4, p. 874 – 875 (1984). Abstract. – See Abstr. 010.062.

111.027 An updated trigonometric parallax for LHS 2924.
D. G. Monet, C. C. Dahn.
Bull. Am. Astron. Soc., Vol. 16, No. 4, p. 1014 (1984). Abstract. – See Abstr. 010.062.

111.028 New determination of proper motions of faint fundamental stars with declinations from +90° to –20°.
Ya. S. Yatskiv, A. D. Polozhentsev.
Inst. teor. fiz. Akad. Nauk USSR. Prepr., No. 151 R, 20 pp. (1984). In Russian. Abstr. in Ref. Zh., 51. Astron., 6.51.93 (1985).

111.029 Investigation of the method of construction of proper motions from the declination of stars of the N30 catalogue.
V. V. Vityazev, I. L. Vygovskaya.
Vestn. Leningr. Univ., Mat. Mekh. Astron., No. 1, p. 84 – 89 (1985). In Russian. Abstr. in Ref. Zh., 51. Astron., 6.51.94 (1985).

111.030 Radial velocities and kinematics of barium stars.
P. K. Lu, J. Miller.
Bull. Am. Astron. Soc., Vol. 17, No. 1, p. 512 (1985). Abstract. – See Abstr. 010.064.

111.031 Short–interval estimations of trigonometric parallaxes.
G. Gatewood, J. Stein, C. Difatta, J. Kiewiet de Jonge, J. Prosser, T. Reiland.
Publ. Astron. Soc. Pac., Vol. 97, No. 590, p. 345 – 347 (1985).
The high precision of a new astrometric detector allows the estimation of trigonometric parallaxes in a fraction of the time required to complete the full study.

111.032 Radial velocities of late–type Population II stars.
A. Ardeberg, H. Lindgren.
Stellar radial velocities, p. 151 – 170 (1985). – See Abstr. 012.095 (IAU Colloq. No. 88).
The authors describe their survey of cool Population II stars. With *uvby* photometry they have covered a total of about 5000 stars in an attempt to derive an unbiased sample of late–type stars suitable for a study of galactic distribution and kinematical properties as a function of evolutionary phase. Especially for the metal–poor stars of the sample the authors have made high-accuracy CORAVEL scanner measurements of radial velocities. So far, around 1700 stars have been observed for radial velocities, most of them more than once. A comparison between the radial-velocity data and preliminary metallicity indices confirms the early evolutionary phase reached in the survey.

111.033 A search for high velocity stars.
A. Florsch.
Stellar radial velocities, p. 183 – 186 (1985). – See Abstr. 012.095 (IAU Colloq. No. 88).
Preliminary results of an objective–prism search for high-velocity stars in the galactic halo and in the Small Magellanic Cloud are presented.

111.034 Radial velocity measurements with objective prisms for the HIPPARCOS program.
C. Fehrenbach.
Stellar radial velocities, p. 189 – 198 (1985). In French. With a summary English translation. – See Abstr. 012.095 (IAU Colloq. No. 88).

111.035 Radial velocities for the HIPPARCOS program.
M. Duflot.
Stellar radial velocities, p. 199 – 206 (1985). In French. With a summary English translation. – See Abstr. 012.095 (IAU Colloq. No. 88).

111.036 Image tube radial velocities of southern stars.
P. K. Lu.
Stellar radial velocities, p. 207 – 212 (1985). – See Abstr. 012.095 (IAU Colloq. No. 88).
Radial velocities have been determined for 243 southern barium and high proper motion stars based on 346 image–tube (IT) spectra. These spectra were obtained using the Yale 1–m telescope at CTIO with a dispersion of 43 Å/mm near H–gamma. Some preliminary results of this survey are presented.

111.037 A critical examination of the Stock Velocity Survey.
W. Osborn, D. J. MacConnell.
Stellar radial velocities, p. 231 – 239 (1985). – See Abstr. 012.095 (IAU Colloq. No. 88).
Recently Stock has carried out a large scale radial velocity survey of faint southern stars using objective–prism plates obtained with the Curtis Schmidt. The goals of this survey were (1) to identify new high–velocity objects and (2) to provide an unbiased radial velocity sample suitable for statistical studies of stellar kinematics. This paper presents the results of an investigation of how well the Stock Velocity Survey meets these goals.

111.038 Radial velocity variations in cool supergiants.
H. C. Harris.
Stellar radial velocities, p. 283 – 288 (1985). – See Abstr. 012.095 (IAU Colloq. No. 88).
A survey of F, G, and K supergiants has been carried out with the DAO radial velocity spectrometer, an efficient instrument for detecting low–amplitude velocity variations in cool stars. Observations of 78 stars over five seasons show generally good agreement with CORAVEL results for spectroscopic binaries. The majority of supergiants show low–amplitude variability, with amplitudes typically 1 to 2 km s^{-1}. The width of the cross–correlation profile has been measured for 58 supergiants. It reveals 14 stars with unusually broad lines, indicative of rotation velocities of 15 to 35 km s^{-1}. Several have short–period binary companions and may be in synchronous rotation.

111.039 Bright giants with small amplitude long period velocity variability.
W. I. Beavers.
Stellar radial velocities, p. 289 – 297 (1985). – See Abstr. 012.095 (IAU Colloq. No. 88).
Observations during the first eight years of operation of the Fick Observatory photoelectric radial velocity spectrometer have led to the development of a list of approximately seventy bright late giant stars with suspected small amplitude long period velocity variations. Preliminary SB1 orbits are reported for eight of the list members which have been observed through at least one complete cycle. Estimates of the period and velocity semiamplitude are made for other confirmed variables.

111.040 CORAVEL measurements of IAU and southern potential radial–velocity standard stars. A zero point discussion.
M. Mayor, E. Maurice.
Stellar radial velocities, p. 299 – 310 (1985). – See Abstr. 012.095 (IAU Colloq. No. 88).
Radial velocity measurements have been carried out since 1981 with the spectrometer CORAVEL at ESO. Almost one thousand measurements of IAU radial velocity standard stars and of potential southern standard stars have been acquired (mean preci-

sion per measurement 0.2 km/s). In the present paper the authors correct the radial velocities of the bright IAU standard stars so that they now belong to the same system as the faint ones. After elimination of variable velocity stars and stars showing large differences between IAU values and recent radial–velocity determinations, an homogeneous list of 34 IAU standard stars is obtained. These stars are distributed between the declinations $\delta = -82°$ and $\delta = +28°$.

111.041   Radial velocity variations in CN–peculiar and normal K giant stars.
G. C. L. Aikman, R. D. McClure.
Stellar radial velocities, p. 321 – 324 (1985). – See Abstr. 012.095 (IAU Colloq. No. 88).
A sample of 54 CN–enriched G and K giant stars monitored for radial velocity variations for the last four years seems to have a binary frequency comparable to a similar sample of normal giant stars.

111.042   Radial–velocity standards.
A. H. Batten.
Stellar radial velocities, p. 325 – 334 (1985). – See Abstr. 012.095 (IAU Colloq. No. 88).
The author reviews the presently available lists of standard–velocity stars approved by IAU Commission 30 and suggests some for major revisions.

111.043   A search for constant velocity B and A stars.
F. C. Fekel.
Stellar radial velocities, p. 335 – 340 (1985). – See Abstr. 012.095 (IAU Colloq. No. 88).
A search for constant velocity stars of B and A spectral type was begun recently at Dyer Observatory. Included in the survey are 6 early B stars, 8 late B stars and 15 early A stars. All have $v \sin i \leqslant 55 \, \mathrm{km \, s^{-1}}$ and about two–thirds have $v \sin i \leqslant 25 \, \mathrm{km \, s^{-1}}$.

111.044   DDO standard velocity star measures 1968 – 1984.
K. W. Kamper.
Stellar radial velocities, p. 341 – 344 (1985). – See Abstr. 012.095 (IAU Colloq. No. 88).
Results of radial velocity measurements of 35 IAU standard stars at David Dunlop Observatory are briefly summarized.

111.045   A new radial velocity survey at the NGP.
A. J. Adamson, R. W. Hilditch, G. Hill, W. A. Fisher.
Stellar radial velocities, p. 355 – 357 (1985). – See Abstr. 012.095 (IAU Colloq. No. 88).
Preliminary results from a radial velocity survey of all known O to F8 stars to the 11th magnitude, within 15° of the North Galactic Pole, are presented.

111.046   Radial velocity and four–color measures of halo A–type stars.
A. G. D. Philip.
Stellar radial velocities, p. 359 – 362 (1985). – See Abstr. 012.095 (IAU Colloq. No. 88).

111.047   Kinematical properties of high velocity B–stars.
P. K. Lu.
Bull. Am. Astron. Soc., Vol. 17, No. 2, p. 547 (1985). Abstract. – See Abstr. 010.065.

111.048   The Barnard's star perturbation.
L. W. Fredrick, P. A. Ianna.
Bull. Am. Astron. Soc., Vol. 17, No. 2, p. 551 (1985). Abstract. – See Abstr. 010.065.

111.049   A photometric search for nearby stars in the NLTT catalogue.
E. W. Weis.
Bull. Am. Astron. Soc., Vol. 17, No. 2, p. 553 (1985). Abstract. – See Abstr. 010.065.

111.050   Recent parallax results from the McCormick southern hemisphere program at the Mt. Stromlo Observatory.
P. A. Ianna.
Bull. Am. Astron. Soc., Vol. 17, No. 2, p. 553 (1985). Abstract. – See Abstr. 010.065.

111.051   Astrometric efforts on low luminosity stars.
C. C. Dahn.
Bull. Am. Astron. Soc., Vol. 17, No. 2, p. 558 (1985). Abstract. – See Abstr. 010.065.

111.052   Defining a bias–free sample of nearby stars.
A. R. Upgren.
Bull. Am. Astron. Soc., Vol. 17, No. 2, p. 624 – 625 (1985). Abstract. – See Abstr. 010.066.

111.053   Dynamical data from Schmidt plate surveys.
J. L. Russell, B. J. McLean.
Bull. Am. Astron. Soc., Vol. 17, No. 2, p. 625 (1985). Abstract. – See Abstr. 010.066.

111.054   Report of IAU Commission 30: Radial velocities (*Vitesses radiales*).
A. G. D. Philip.
Trans. IAU, Vol. XIXA, p. 375 – 382 (1985). – See Abstr. 003.046.

111.055   Determination of proper motions of stars with respect to galaxies in twenty selected areas of the Pulkovo plan.
N. A. Shakht.
Izv. Glav. Astron. Obs. Pulkovo, Astrometr. Astrofiz., No. 201, p. 39 – 43 (1985). In Russian.
The results of the determination of proper motions of stars relative to galaxies in twenty areas of the Pulkovo zone are given. The reductions of proper motions to absolute ones are calculated using galaxies, the AGK3 stars and the statistical method. Secular parallaxes have been obtained for the $13^{m}.8$ and $15^{m}.7$ stars.

Improving the nearby star data base.
See Abstr. 002.008.

A corrected and uniformly formatted edition of the Abt and Biggs (1972) *Bibliography of Stellar Radial Velocities*.
See Abstr. 002.077.

Catalogue of proper motions of stars relative to galaxies in nine selected areas of southern declination.
See Abstr. 002.086.

Catalogue of proper motions of stars relative to galaxies in nine selected sky areas.
See Abstr. 002.087.

Catalogue of proper motions of stars relative to galaxies in nine southern sky areas.
See Abstr. 002.088.

Catalogue of proper motions of 4423 stars relative to galaxies in 41 areas of the sky.
See Abstr. 002.089.

Mean stellar radial velocities catalogue.
See Abstr. 002.098.

Bibliographic catalogue (*stellar radial velocities*).
See Abstr. 002.099.

Bibliography of radial velocity papers (1981 – 1984).
See Abstr. 002.100.

Systematic differences and average errors in the new edition of the Yale Parallax Catalogue.
See Abstr. 002.104.

Compte rendu du Colloque UAI 88 – vitesses radiales stellaires, Schenectady, 24 – 27 octobre 1985.
See Abstr. 011.022.

The DAO radial velocity spectrometer and recent results.
See Abstr. 034.125.

The determination of absolute proper motions on Schmidt plates at Tautenburg.
See Abstr. 036.092.

Detectability of extrasolar planetary transits.
See Abstr. 036.098.

Logiciel de pointage des raies spectrales et mesures des vitesses radiales.
See Abstr. 036.117.

Le programme de vitesses radiales de Strasbourg au prisme–objectif de Fehrenbach.
See Abstr. 036.118.

Stellar radial velocities of high precision: techniques and results.
See Abstr. 036.184.

Digital stellar speedometry.
See Abstr. 036.185.

Radial velocity information in solar–type spectra.
See Abstr. 036.187.

Radial velocities from CCD detectors.
See Abstr. 036.188.

Absolute astronomical accelerometry.
See Abstr. 036.191.

Das Radialgeschwindigkeitssystem der Fundamentalsterne.
See Abstr. 041.022.

Report of IAU Commission 24: Photographic astrometry (*Astrométrie photographique*).
See Abstr. 041.051.

Stellar lineshifts induced by photospheric convection.
See Abstr. 064.088.

The reddening (and distance?) of P Cygni.
See Abstr. 113.010.

A photometric survey of the stars of largest H.
See Abstr. 113.066.

Spectrophotometric studies of faint Luyten proper motion stars.
See Abstr. 114.118.

Luminosity and kinematics of barium stars.
See Abstr. 115.026.

Accuracy of close binary mass determinations from parallaxes.
See Abstr. 117.129.

The binary frequency of extreme subdwarfs revisited.
See Abstr. 117.157.

Metallic–line binaries.
See Abstr. 120.034.

High–velocity halo binary systems.
See Abstr. 120.035.

A search for halo binaries: the first results.
See Abstr. 120.036.

A search for radial velocity variations in the central stars of southern planetary nebulae and planetary–like objects.
See Abstr. 134.012.

Limits on parallax and proper motion of an optical counterpart of Geminga.
See Abstr. 143.023.

Radial velocities of F and G dwarfs in the Orion cluster region.
See Abstr. 152.008.

Evolution of low–mass stars in the Alpha Persei cluster.
See Abstr. 153.015.

Auswahl von Mitgliedern offener Sternhaufen zur Messung der Eigenbewegung der Haufen mit dem Astrometriesatelliten HIPPARCOS.
See Abstr. 153.033.

A proper motion study of the Pleiades cluster.
See Abstr. 153.047.

Proper motion studies of stars in and around open clusters.
See Abstr. 153.048.

The Hyades: membership and convergent point from radial velocities.
See Abstr. 153.051.

Photometry, proper motions, and membership in the globular cluster M71.
See Abstr. 154.003.

Radial velocities of stars in globular clusters: a look into ω Cen and 47 Tuc.
See Abstr. 154.030.

Radial velocities and proper motions of globular cluster stars.
See Abstr. 154.051.

Heavy remnants in globular cluster cores.
See Abstr. 154.078.

Galaxy population structure from proper motions.
See Abstr. 155.023.

The velocity–age relation for B and A stars.
See Abstr. 155.060.

Solar motion and galactic rotation obtained from the proper motions of faint stars in areas with galaxies.
See Abstr. 155.082.

Kinematics of K giants in the outer galactic halo.
See Abstr. 155.100.

Some results of a stellar–statistical investigation on the basis of proper motions of bright stars.
See Abstr. 155.109.

A constraint on the past star formation rate from the kinematics of nearby stars.
See Abstr. 155.144.

A new halo survey.
See Abstr. 155.152.

Current programs of local galactic structure and evolution.
See Abstr. 155.153.

Kinematics of halo red giants.
See Abstr. 155.154.

The effect of incompleteness among nearby stars.
See Abstr. 155.169.

CORAVEL radial velocities of supergiants in the Small Magellanic Cloud.
See Abstr. 156.026.

112 Stellar Environments (Chromospheres, Coronae, Stellar Winds, Shells, Masers, etc.)

112.001   Hot stars: six years of progress.
C. D. Garmany.
NASA Conf. Publ., NASA CP–2349, p. 17 – 26 (1984). – See Abstr. 012.001.

Mass loss is known to take place across the H–R diagram, but nowhere is it more pronounced than in the upper left hand corner of that figure. Here, winds from luminous stars peel away a solar mass in a million years or less. The winds of the Of, OB supergiants, and W–R stars have received the most attention because they are so readily studied. Their effect on stellar evolution is significant also. On the other hand, it is the B stars whose variability has been most thoroughly studied with IUE, undoubtedly because historically these stars have been known to vary. This review concentrates on the current state of the observational understanding of stellar winds, with particular emphasis on how the observations fit various theoretical predictions.

112.002   Outer atmospheres of cool stars observed with IUE.
S. L. Baliunas.
NASA Conf. Publ., NASA CP–2349, p. 64 – 79 (1984). – See Abstr. 012.001.

The author seeks answers to questions such as: Are the mechanisms exciting the outer solar atmosphere present on other stars? What influences and controls these processes, and in particular, what is the role of the magnetic fields? Ultraviolet observations of cool stars in conjunction with coronal X–ray or ground–based chromospheric measurements, or both, have served both to refine and to define the understanding of stellar activity.

112.003   Coronal effects on the winds of early–type stars.
W. L. Waldron.
NASA Conf. Publ., NASA CP–2349, p. 215 – 218 (1984). – See Abstr. 012.001.

The X–ray emission from early–type stars can be a dominant factor in determining the ionization structure of a stellar wind. Using a base coronal model, the UV resonance lines of N V, Si IV, and C IV are calculated by subjecting them to changes in the coronal emission measure, coronal temperature, and mass loss rate. The calculations predict that a unique behavior is present in the variations of these line profiles. This suggests that it would be possible to distinguish which quantity was responsible for the variation. A preliminary search for this predicted variability, using the high–resolution SWP data from the IUE archives, has been conducted for several stars.

112.004   Superionized species and winds in low luminosity B and Be stars.
P. K. Barker, J. M. Marlborough, J. D. Landstreet.
NASA Conf. Publ., NASA CP–2349, p. 219 – 222 (1984). – See Abstr. 012.001.

High dispersion IUE spectra have been obtained of 72 luminosity class III–V Be stars and 82 luminosity class IV–V normal B stars. For normal B stars, the C IV $\lambda\lambda1548,50$ resonance doublet is observed only at type B2 or earlier, whereas C IV can occur as late as B9 among the Be stars. In both groups of stars the average C IV equivalent width is greatest at the earliest spectral types, and decreases rapidly in strength toward later types; however, at each spectral subtype a large range of C IV equivalent widths exists. For the Be stars there is no correlation between the equivalent widths of C IV absorption and Hα emission. Neither normal B nor Be stars show any clear correlation between C IV equivalent width and projected rotational velocity. C IV emission is definitely present in only two Be stars.

112.005   Stellar winds in the young galactic cluster NGC 6530.
E. Böhm–Vitense, P. Hodge.
NASA Conf. Publ., NASA CP–2349, p. 223 – 226 (1984). – See Abstr. 012.001.

The authors have studied line profiles for O stars on or close to the main sequence in the young galactic cluster NGC 6530.

P Cygni profiles are seen for $T_{eff} > 36000$K. Different stars show, however, different lines and different outflow velocities. The degree of ionization in the wind appears to depend on T_{eff}.

112.006   Main sequence B stars with strong winds.
D. Massa, B. D. Savage.
NASA Conf. Publ., NASA CP–2349, p. 227 – 230 (1984). – See Abstr. 012.001.

The authors obtained low resolution IUE observations of B0 – 2 V stars in several young open clusters as part of a project to derive the UV extinction to the clusters. The aim of this study was to use the extinction found from the B stars to deredden the cluster O stars. In the process, it became obvious that the B0 – 2 V stars in the nuclear cluster of Sco OB1, NGC 6231, had very unusual UV spectra. This discovery prompted the authors to obtain a high dispersion IUE spectrum of one of the peculiar NGC 6231 B stars, BD –41°7719.

112.007   IUE and ESO observations of the P Cyg star AG Carinae and its ring nebula.
R. Viotti, A. Altamore, M. Barylak, A. Cassatella,
R. Gilmozzi, C. Rossi.
NASA Conf. Publ., NASA CP–2349, p. 231 – 234 (1984). – See Abstr. 012.001.

From the study of the optical and ultraviolet spectrum of the P Cyg variable AG Car, the authors conclude that this high luminosity ($M_{bol} \cong -8.3$), high mass loss star is a massive object evolving towards the WR region after a cool supergiant phase. Like in other superluminous stars, the large photometric and spectroscopic variations are attributed to flux redistribution of the stellar radiation due to structure variations of the expanding atmosphere.

112.008   Coordinated ultraviolet and visual observations of ω Orionis 1982 – 1983.
C. A. Grady, G. Sonneborn, C.–C. Wu, D. P. Hayes,
E. F. Guinan, P. K. Barker, H. F. Henrichs.
NASA Conf. Publ., NASA CP–2349, p. 235 – 238 (1984). – See Abstr. 012.001.

Coordinated IUE high dispersion and visual observations of ω Orionis in 1982 – 1983 covering periods of both low and high B–band linear polarization have shown that line profile variability seen in the ultraviolet resonance lines of Si IV and C IV is not correlated with the visual continuum polarization state of the star. Variations in the visual colors and continuum fluxes are correlated with the amount of linear polarization. Hα emission strength variations are only weakly correlated with the polarization activity. The variation seen in C IV and Si IV is consistent with density perturbations propagating through the wind acceleration zone. The observed data are interpreted as indicating that the stellar wind is latitudinally separated from the continuum scattering envelope.

112.009   The spectroscopic antics of S 18/SMC.
S. N. Shore, N. Sanduleak, D. A. Allen.
NASA Conf. Publ., NASA CP–2349, p. 239 – 242 (1984). – See Abstr. 012.001.

The authors present optical (3600 – 10000 Å) and IUE (1200 – 3200 Å) observations of the peculiar, luminous early–type supergiant star S 18/SMC which cover the period 1981 – 1983. This is supplemented by objective prism and high dispersion studies from 1967 – 1980. There is no evidence for a red companion. The UV spectrum and derived N enhancement is quite similar to Sanduleak's star in the LMC. The authors present evidence for a nitrogen overabundance of a factor of the order of 70, and discuss a preliminary stellar wind model for the star.

112.010 An episodic red–wing structure of Si IV, λ1394 Å, in γ Cas.
V. Doazan, G. Sedmak, R. Stalio, R. N. Thomas, A. J. Willis.
NASA Conf. Publ., NASA CP–2349, p. 243 – 248 (1984). – See Abstr. 012.001.

IUE observations of γ Cas obtained in January 1983 show the conspicuous presence of a red–wing structure in the Si IV, λ1394 Å, resonance line in γ Cas. Combining long–term variations of γ Cas in the visual, summarized in two preceding papers, with published far UV data, the authors note that this red feature seems to occur preferentially at epochs where the V/R ratio of the violet and red emission peaks at Hα is less than one. It is suggested that these two characteristics, visual and far UV, are linked to the flow–deceleration in the outer–atmosphere.

112.011 Chromospheric activity in M giants.
T. Y. Steiman–Cameron, H. R. Johnson, R. K. Honeycutt.
NASA Conf. Publ., NASA CP–2349, p. 441 – 444 (1984). – See Abstr. 012.001.

Low–resolution IUE spectra have been obtained with the LWR camera for fifteen cool giant stars ranging in spectral type from K4.8 through M5.9. These spectra have been used to examine chromospheric activity in late type giants and to evaluate the extent to which nonradiative heating affects the upper levels of cool giant photospheres. The program stars are from the Ridgway et al. (1980) sample. In this paper the authors present the preliminary analysis of the observations.

112.012 Chromospheric emission lines in high–resolution LWR spectra (2200 – 3000 Å) of Gamma Cru (M3 III) and Alpha Ori (M2 Iab).
K. G. Carpenter.
NASA Conf. Publ., NASA CP–2349, p. 450 – 453 (1984). – See Abstr. 012.001.

The identity and characteristics of the chromospheric emission features in the 2200 – 3000 Å region of high–resolution spectra of γ Cru and α Ori are summarized. The velocities, fluxes, and asymmetries of a set of Fe II lines are discussed and the information gained from flux measurements of the C II (UV 0.01) lines is presented. An analysis of the Fe II lines in the α Ori spectra indicates the general shape of the velocity versus radius relation in its wind. The C II (UV 0.01) data are combined with measures of the C II (UV 1) flux to estimate the electron density and temperature in the wind and the geometric extent of the C II emitting region in both stars.

112.013 Active late–type stars and the applicability of coronal loop models.
M. S. Giampapa, L. Golub, G. Peres, S. Serio, G. S. Vaiana.
NASA Conf. Publ., NASA CP–2349, p. 454 – 457 (1984). – See Abstr. 012.001.

The authors combine far ultraviolet IUE observations with existing soft X–ray measurements obtained by Einstein (HEAO–B) satellite observatory for a sample of solar–type stars. They utilize the resulting data–set and a new coronal loop model numerical code developed at the Harvard–Smithsonian Center for Astrophysics to perform a preliminary investigation of the applicability of coronal loop models to solar–type stars. They demonstrate that semi–empirical, coronal loop models can be applied to account for observed stellar transition region and coronal emission.

112.014 Co–rotating interaction regions in stellar winds.
D. J. Mullan.
NASA Conf. Publ., NASA CP–2349, p. 458 – 461 (1984). – See Abstr. 012.001.

A co–rotating interaction region (CIR) forms in a stellar wind when a fast stream from a rotating star overtakes a slow stream. CIR's have been studied in detail in the solar wind over the past decade. Here, the author points out their usefulness in interpreting several spectroscopic features in stars of various types, including "hybrid" stars, OB stars, and cool supergiants.

112.015 Betelgeuse at maximum luminosity.
A. K. Dupree, G. Sonneborn, S. L. Baliunas, E. F. Guinan, L. Hartmann, D. P. Hayes.
NASA Conf. Publ., NASA CP–2349, p. 462 – 467 (1984). – See Abstr. 012.001.

Betelgeuse (Alpha Ori; M2 Iab) was extremely bright at optical wavelengths and in the Mg II resonance lines during January and February 1984 when an intrinsic brightening occurred in the photosphere and chromosphere. Linear polarization in the B–band at this time was not anomalous when compared to earlier epochs. The core of the Hα line was redshifted by about 10 km/s with respect to the photospheric lines during January/February as compared to measurements made five months previously. There may be periodic variations in the chromospheric flux.

112.016 Ultraviolet, radio and X–ray observations of hybrid stars.
S. A. Drake, A. Brown, J. L. Linsky.
NASA Conf. Publ., NASA CP–2349, p. 472 – 475 (1984). – See Abstr. 012.001.

In order to understand the nature of the circumstellar regions in the so–called hybrid (–chromosphere) stars, the authors have analyzed existing long wavelength IUE data of these stars, obtained new 6 cm radio observations with the VLA, and compiled all available X–ray observations. They conclude that the low–velocity absorption components seen in the Mg II h and k lines of hybrids are almost certainly interstellar and that only the high–velocity components are indicative of the stellar wind speeds. The mass loss rates of ionized material obtained from the radio data are $\lesssim 2 - 4 \times 10^{-9} M_\odot \mathrm{yr}^{-1}$ for the three hybrids observed to date.

112.017 Rotational modulation of chromospheric emission in cool giants and "hybrid" stars.
J. W. Brosius, D. J. Mullan, R. E. Stencel.
NASA Conf. Publ., NASA CP–2349, p. 476 – 479 (1984). – See Abstr. 012.001.

The authors have used IUE archival data to study temporal variations of the Mg II h and k emission lines in 8 late–type giants. They present evidence that the variations are periodic in nature. They argue that the periodicities can be interpreted in terms of rotation and find that the four fastest rotators in their sample are "hybrid" stars.

112.018 Ultraviolet and optical spectroscopy and polarimetry of the helium weak star HD 21699: evidence for a magnetically controlled stellar wind.
D. N. Brown, S. N. Shore, C. T. Bolton, S. J. Hulbert, G. Sonneborn.
NASA Conf. Publ., NASA CP–2349, p. 483 – 486 (1984). – See Abstr. 012.001.

The authors have obtained high dispersion SWP spectra with IUE, and contemporaneous optical Zeeman polarimetry and spectroscopy, of the helium weak star HD 21699 = HR 1063. They discuss this set of observations in the context of a magnetic star which is losing mass, having constrained that outflow to corotate with the stellar surface. If the magnetic period is correct, the greatest mass outflow seems to be from above the polar regions. The comparison between optical and UV spectrum phenomenology is also briefly discussed.

112.019 Magnetospheres and winds in the helium weak stars: observations of C IV in upper main sequence CP stars.
D. N. Brown, S. N. Shore, P. K. Barker, G. Sonneborn.
NASA Conf. Publ., NASA CP–2349, p. 487 – 490 (1984). – See Abstr. 012.001.

The authors present a sample of eleven helium weak and Si stars, observed at high dispersion. The stars span the range B2.5 to B6 and include both magnetic and nonmagnetic as well as rapid and slow rotators. In the two stars designated "sn" by Abt, the C IV profile is strong while all of the other stars can be explained by numerous blended lines of Fe III and similar ions. Most of the stars have been observed several times, and only HD 21699 appears to show large amplitude spectral variations.

112.020   The circumstellar envelopes around OH/IR stars.
P. J. Diamond, R. P. Norris, P. R. Rowland,
R. S. Booth, L.-A. Nyman.
Mon. Not. R. Astron. Soc., Vol. 212, No. 1, p. 1 – 21 (1985).

The authors have observed the 1612–MHz OH maser emission from five OH/IR stars using MERLIN in its spectral–line mode. The observations yield the angular radii of the circumstellar envelopes. These can be combined with the linear radii obtained from phase–lag techniques to calculate source distances. The distances obtained differ significantly from the kinematic estimates and imply peculiar velocities of between 10 and 30 km s^{-1}. The authors have also determined the absolute positions of nine OH/IR objects, and find that they are coincident with the optical positions of the stars, where these positions are available. The authors have collated all published 1612–MHz OH maser maps of OH/IR stars. They find that the observations are generally consistent with the expanding–shell model, although significant asymmetries are present.

112.021   The unusual OH envelope of U Orionis.
J. M. Chapman, R. J. Cohen.
Mon. Not. R. Astron. Soc., Vol. 212, No. 2, p. 375 – 384 (1985).

MERLIN maps of the OH maser emission from U Orionis reveal an incomplete ring of 1665–MHz masers of radius 10^{15}cm and two compact clusters of 1612–MHz masers. The 1612–MHz masers are separated along the same position angle at which the break occurs in the ring of 1665–MHz masers. The ring distribution of 1665–MHz masers can be explained in terms of a thin–shell model for the masers. The model implies a stellar velocity of -39.5 ± 1.0 km s^{-1}. The model also makes a number of predictions which it may be possible to test as the maser emission from U Ori continues to evolve. The 1665–MHz masers show a regular velocity–gradient around the ring. No simple explanation for this is completely satisfactory, although it seems that rotation of the circumstellar envelope probably contributes to the gradient observed. The stellar magnetic field may provide the necessary coupling to transfer angular momentum from the star to the circumstellar gas.

112.022   First spectroscopic observations with the 14–m radio telescope at the Centro Astronomico de Yebes: SiO masers in evolved stars.
A. Barcia, V. Bujarrabal, J. Gómez–González,
J. Martín–Pintado, P. Planesas.
Astron. Astrophys., Vol. 142, No. 1, p. L9 – L12 (1985).

The authors present observations of the SiO ($v = 1$ and 2, $J = 1$–0) transitions at 43 GHz towards 40 objects, as a first step in an extensive observing program of SiO maser emission from evolved stars that they are carrying out at the Centro Astronómico de Yebes. The detection of 12 new SiO stars and the confirmation of one previous tentative detection are reported. A careful choice of objects with respect to infrared flux and stellar phase has yielded a high detection rate for oxygen–rich Mira variables, in spite of the moderate sensitivity. These are the first spectroscopic results obtained with the 14–m mm–wave radio telescope at Yebes, the system is briefly described in the text.

112.023   Spektroskopische und photometrische Untersuchungen der B–Überriesen mit zirkumstellarem Staub in den Magellanschen Wolken.
F.-J. Zickgraf.
Diss. Naturwiss.-Math. Gesamtfak. Ruprecht-Karls-Univ., Heidelberg, F.R. Germany, 4 + 97 pp. (1985).

Spektroskopische und photometrische Beobachtungen vom Satelliten–UV bis in das Infrarotgebiet der acht B[e]–Sterne der Magellanschen Wolken (R 4, R 50, R 66, R 82, R 126, Hen S12, Hen S22 und Hen S134), die thermische Strahlung von zirkumstellarem Staub zeigen, werden beschrieben.

112.024   Detection of OH radicals from IRAS sources.
B. M. Lewis, J. Eder, Y. Terzian.
Nature, Vol. 313, No. 5999, p. 200 – 202 (1985).

An efficient method for detecting new OH/infrared stars is to begin with IRAS source positions, selected for appropriate infrared colours, and using radio–line observations to confirm the OH

properties. The authors demonstrate the validity of this approach here, using the Arecibo 305 m radio–telescope to confirm the 1,612 MHz line observations of sources in IRAS Circulars 8 and 9; the present observations identify 21 new OH/infrared stars. The new sources have weaker 1,612 MHz fluxes, bluer (60 – 25) μm colours and a smaller mean separation between the principal emission peaks than previous samples.

112.025   The peculiar X–ray and radio star AS431.
J.-P. Caillault, G. A. Chanan, D. J. Helfand,
J. Patterson, J. A. Nousek, L. O. Takalo, G. D. Bothun,
R. H. Becker.
Nature, Vol. 313, No. 6001, p. 376 – 378 (1985).

During a systematic survey of X–ray flux–limited late–type stars, the authors have rediscovered a highly reddened emission–line star, previously listed as AS431. They report here Einstein observations revealing that AS431 has a highly absorbed X–ray spectrum and a relatively strong intrinsic flux of $\geqslant 5 \times 10^{-12}$erg cm^{-2}s^{-1}. Observations at 20 cm and 6 cm with the VLA show that it is also a moderately strong radio source (~ 35 mJy). These data, together with optical and infrared observations, suggest a model in which both the radio and X–ray emissions arise in a chaotic stellar wind emerging from a single luminous Wolf–Rayet star.

112.026   OH emission from Mira variables, infrared stars and molecular clouds.
A. Slootmaker, J. Herman, H. J. Habing.
Astron. Astrophys., Suppl. Ser., Vol. 59, No. 3, p. 465 – 483 (1985).

A sample of 79 stellar sources has been searched for the $\lambda = 18$ cm OH lines. Seven of the eighteen newly detected OH sources have a stellar origin and the others probably originate from molecular clouds. Five sources were detected in other lines than previously known, and for a number of already known (stellar) OH masers new features were found. The 1612 MHz line of the Mira variable UX Cyg was found to be linearly polarized.

112.027   Time variations and shell sizes of OH masers in late–type stars.
J. Herman, H. J. Habing.
Astron. Astrophys., Suppl. Ser., Vol. 59, No. 3, p. 523 – 555 (1985).

The authors present the results of a program in which the flux density of OH masers in late–type stars was measured to determine the variability, its period and amplitude. Normal Mira variables have "radio" periods that are equal to their optical periods (~ 400 days). Most of the related, but optically unidentified OH/IR stars have much longer periods, up to ~ 2000 days. A significant fraction (25%) of the OH/IR stars shows small amplitude, or no variations. The light curves of different peaks in the line profile of one star have the same periodicity and shape, but differ slightly in phase. From these phase differences linear dimensions of the OH shells can be derived; typical values for the radii are $\sim 8 \times 10^{15}$cm for Mira variables and $\sim 5 \times 10^{16}$cm for OH/IR stars.

112.028   IRAS findings of possible solid material orbiting nearby stars.
H. L. Shipman.
Phys. Today, Vol. 38, No. 1, p. S8 – S9 (1985). Short review paper. – See Abstr. 013.003.

112.029   Comment on the IRAS infrared spectrum of Zeta Puppis (O4 If).
M. G. Wolfire, W. L. Waldron, J. P. Cassinelli.
Astron. Astrophys., Vol. 142, No. 2, p. L25 – L28 (1985).

The infrared spectrum of ζ Pup that was obtained with the IRAS satellite has recently been compared by Lamers, Waters, and Wesseleus (1984) with flux distributions from models with coronal zones. Here the authors reconsider the thick corona and warm wind models used by Lamers et al. to interpret the IRAS data and find them unable to explain the X–ray observations. The thin coronal models of Waldron, however, are found to be capable of fitting the IRAS data. The question of the presence of base coronal regions must be considered still open.

112.030 The optical properties of dust in the mid–IR silicate bands.
B. Pégourié, R. Papoular.
Astron. Astrophys., Vol. 142, No. 2, p. 451 – 460 (1985).

IR photometry and spectrophotometry is presented for two G and two M supergiants and one M giant, all exhibiting a narrow 10 µm silicate feature. These spectra, together with six other, previously available ones, form a data base from which the authors deduce the optical efficiency of the circumstellar dust material over the range 8 – 30 µm. This efficiency peaks at 10 µm and is similar to that of amorphous Mg_2SiO_4. Similarly, using 13 available spectra of protostellar objects, it is shown that the dust in molecular clouds and in the ISM is composed of small (<1 µm) grains with $Q_{ext}(\lambda)/a$ practically identical to that of amorphous $MgSiO_3$, and peaking at 9.5 µm. The two kinds of dust are clearly different. The results of meteoritic chemical analysis and of the theory of composite gas condensation are used to evaluate the likelihood of amorphous Mg_2SiO_4 and $MgSiO_3$ being present in circumstellar and interstellar dust respectively.

112.031 On the nature of OH/IR stars.
D. Engels.
The Milky Way galaxy, p. 131 (1985). Abstract. – See Abstr. 012.007 (IAU Symp. No. 106).

112.032 Circumstellar dust distributions around variable stars.
A. J. Mayes.
Thesis, Univ. Keele, England (1983). Abstr. in Phys. Abstr., Vol. 88, No. 1249, Entry 10034 (1985).

112.033 Spectrophotometric investigation of Be stars.
P. S. Goraya, M. Singh.
Astrophys. Space Sci., Vol. 108, No. 1, p. 161 – 173 (1985).

The continuum energy distribution data of seven Be and five normal B stars have been presented in the wavelength range $\lambda\lambda 3200 – 8000$ Å. Empirical effective temperatures of these stars have been derived by comparing the observed continuum energy distributions with the computed energy distributions given by Kurucz (1979). The effective temperatures of all Be stars observed here except KX And are in fair agreement with those of normal B stars. The variable nature of the Be stars has been discussed. No excess or deficiency in the mean flux of normal B stars was detected.

112.034 VLA line observations of OH/IR stars.
J. Herman, B. Baud, H. J. Habing, A. Winnberg.
Astron. Astrophys., Vol. 143, No. 1, p. 122 – 135 (1985).

The authors used the VLA in a spectral line mode at 1612 MHz to investigate the structure of eleven OH/IR stars. Ten objects have large radial velocities and are expected to be close to their tangential points. For six resolved sources very accurate distances could be derived by combining the angular radii of the OH shell with availabe phase lag diameters. A comparison of these geometric distances with kinematic distances gives a determination of the distance to the Galactic Centre of 9.2 ± 1.2 kpc. The OH shells appear to be fairly symmetric, with deviations fom spherical symmetry of less than 20%, and rather thin, the thickness being at most 20% of the radius. The OH density falls off as r^{-2}, or slightly steeper.

112.035 Infrarotenergieverteilung von OB–Sternen mit Massenverlust.
C. Leitherer.
Diss. Naturwiss.–Math. Gesamtfak. Ruprecht–Karls–Univ., Heidelberg, F.R. Germany, 5+111 pp. (1985).

The author computed the theoretical infrared flux distribution of an early–type star surrounded by a stellar wind. The influence of the wind's density and temperature–distribution on the emergent infrared flux is investigated. A comparison with observed infrared fluxes allows to determine the velocity field of stars with known mass–loss rates.

112.036 The hybrid spectrum of the LMC hypergiant R126.
F.–J. Zickgraf, B. Wolf, O. Stahl, C. Leitherer, G. Klare.
Astron. Astrophys., Vol. 143, No. 2, p. 421 – 430 (1985).

An extensive spectroscopic (1150 – 6700 Å) and photometric ($UBVRIJHKLMN$) study of the peculiar LMC hypergiant R126 has been carried out. The authors derive stellar parameters $T_{eff} = 22,500$K, $R = 72\,R_\odot$ and $M_{bol} = -10.5$. The corresponding spectral type is B0.5 Ia$^+$. An estimate of the initial and present mass leads to $70 – 80\,M_\odot$ and $\sim 40\,M_\odot$, respectively. Thermal radiation of a cool dust component with $T_{bb} = 900$K is observed in the infrared. The hybrid character of the line spectrum, i.e., broad UV resonance absorption lines contrasting sharp emission lines, leads to the conclusion that R126 has a two component stellar wind leading to a disk–like outer configuration. The disk is supposed to be formed by a slow (~ 40 km s^{-1}), cool, dense wind. A stellar rotation close to the break–up velocity is regarded as the reason for this two–component structure.

112.037 Narrow absorption components in Be star winds.
C. A. Grady.
NASA Conf. Publ., NASA CP–2358, p. 57 – 61 (1985). – See Abstr. 012.023.

This paper discusses briefly IUE observations of narrow absorption lines in the spectra of ω Ori (B2 IIIe), 66 Oph (B2 IVe), and 59 Cyg (B1.5 IVe). Models of the formation of these lines in the stellar winds from the Be stars are outlined.

112.038 Ultraviolet spectral morphology of O–type stellar winds.
N. R. Walborn.
NASA Conf. Publ., NASA CP–2358, p. 66 – 69 (1985). – See Abstr. 012.023.

A compilation of IUE spectra of about 120 stars is used to demonstrate that the prominent stellar wind profiles in ultraviolet O–type spectra display strong systematic trends, and a high degree of correlation with the optical spectral types, among the majority of normal stars.

112.039 Coronal temperatures.
J. H. Swank.
NASA Conf. Publ., NASA CP–2358, p. 86 – 90 (1985). – See Abstr. 012.023.

This paper presents a brief summary of the main results from a survey of X–ray spectra of stellar coronae obtained with the Einstein Observatory. A sample of 19 stars, including close binaries and both late type and early type stars, is considered.

112.040 Stellar winds, observational evidence for a hot–cool star connection.
W. L. Waldron.
NASA Conf. Publ., NASA CP–2358, p. 95 – 100 (1985). – See Abstr. 012.023.

Stellar wind data has been collected for a total of 272 stars representing all spectral types including Wolf–Rayet stars. Two significant correlations are found relating the wind luminosity (L_w) to the bolometric luminosity and the terminal velocity (v_∞) of the stellar wind to the stellar effective temperature. Least–squared fits to the data suggest that $L_W \sim L_{Bol}{}^2$ and $v_\infty \sim T_{eff}{}^{1.8}$.

112.041 Is the ratio of observed X–ray luminosity to bolometric luminosity in early–type stars really a constant?
W. L. Waldron.
NASA Conf. Publ., NASA CP–2358, p. 159 – 163 (1985). – See Abstr. 012.023.

The observed X–ray emission from early–type stars can be explained by the recombination stellar wind model. The model predicts that the true X–ray luminosity from the base coronal zone can be $10^2 – 10^4$ times greater than the observed X–ray luminosity. From the models, scaling laws have been found for the true and observed X–ray luminosities. It is shown that these scaling laws are very successful in predicting the general observed X–ray characteristics of O and B stars.

112.042 On the role played by lines in radiatively driven stellar winds depending on the position of the stars in the HR diagram.
M. C. Migozzi, J.–P. J. Lafon.
NASA Conf. Publ., NASA CP–2358, p. 236 – 240 (1985). – See Abstr. 012.023.

This paper searches for correlations between the mass loss rates, terminal velocities, luminosities and other stellar parameters for hot stars of various spectral types and compares the observations with theoretical predictions from models of purely radiation–driven stellar winds.

112.043 Summary of the origin of nonradiative heating/ momentum in hot stars.
A. B. Underhill.
NASA Conf. Publ., NASA CP–2358, p. 243 – 254 (1985). – See Abstr. 012.023.

112.044 Early–type stars in OB associations in the infrared. II. A discussion of density and temperature distributions in stellar winds.
C. Bertout, C. Leitherer, O. Stahl, B. Wolf.
Astron. Astrophys., Vol. 144, No. 1, p. 87 – 97 (1985). With a correction in Vol. 147, No. 1, p. 186 (1985).

The authors present numerical computations of the infrared flux distribution in early–type stars with mass loss. After assuming that infrared radiation is due both to the stellar photosphere and to free–free and free–bound transitions in the ionized stellar wind, they study the influence of mass–loss rate and wind velocity–law on the infrared flux. By matching computed and observed flux distributions, they investigate the velocity law of a sample of OB stars with spectral types ranging from O4 to B2. While the velocity law derived from the infrared flux of O stars is consistent with that expected from a radiatively–driven wind, early B stars show on the average a smoother velocity gradient.

112.045 Rotational modulation of chromospheric emission in cool giants and "hybrid" stars.
J. W. Brosius, D. J. Mullan, R. E. Stencel.
Astrophys. J., Vol. 288, No. 1, p. 310 – 328 (1985).

The authors have used archival data from the IUE to study temporal variations of the Mg II h and k emission lines in eight late–type giants. They present evidence that the variations are periodic in nature and argue that the periodicities can be interpreted in terms of rotation. The four fastest rotators in the sample appear to be "hybrid" stars.

112.046 Carbon IV absorption troughs in the ultraviolet spectra of Be stars: gone with the wind?
P. K. Barker, J. M. Marlborough.
Astrophys. J., Vol. 288, No. 1, p. 329 – 337 (1985).

IUE observations of five Be stars reveal striking and previously unrecognized forms of stellar wind variability. Broad C IV $\lambda\lambda$1548, 1550 absorption troughs, which underlie shortward–shifted narrow absorption components, may disappear or reappear. The C IV equivalent width can vary by a factor $\gtrsim$ 5 on a time scale of $\sim$3 months, implying considerable changes in the mass loss rate or source of superionization in the wind or both. The shifted narrow absorption components may themselves disappear or reappear, and are not necessarily superposed on a broad absorption trough.

112.047 The IUE Mg II chromospheric emission feature of barium stars.
P. K. Lu.
News Lett. Astron. Soc. N.Y., Vol. 2, No. 5, p. 20 – 21 (1984). Abstract. – See Abstr. 010.243.

112.048 VLA 6 cm continuum observations of OH/IR stars.
J. Herman, B. Baud, H. J. Habing.
Astron. Astrophys., Vol. 144, No. 2, p. 514 – 515 (1985).

A number of OH/IR stars with accurately known positions were searched for radio continuum emission at a wavelength of 6 cm (5 GHz). None were detected at a level of $\sim$0.1 mJy. The lack of detections in this and previous searches implies either that the transition time between the tip of the asymptotic giant branch, the locus of the OH/IR stars, and the planetary nebula stage is much shorter than the duration of the "superwind" phase, or that there is a noticeable difference between the OH/IR stars on the one hand and Vy2–2, the only known example of an object in the transition stage, on the other.

112.049 The Be stars.
V. Doazan.
Effects of variable mass loss on the local stellar environment, p. 115 – 124 (1984). – See Abstr. 012.028.

The author summarizes the different kinds of variability observed in Be stars. She presents recent and new observations of θ CrB in the far–UV. The author reviews the different interpretations proposed to explain Be stars atmospheric structure. She stresses the importance of the role of variable mass loss in the production of the Be phenomenon and suggests, in the light of the θ CrB observations, that a variable non–radiative flux might also be characteristic of Be stars atmospheres.

112.050 Winds in red giants.
D. Reimers.
Effects of variable mass loss on the local stellar environment, p. 149 – 162 (1984). – See Abstr. 012.028.

Main emphasis in this review is on more recent optical, and in particular ultraviolet, observations of winds in red giants.

112.051 A multitransitional study of linear polarization in SiO maser emission.
R. Barvainis, C. R. Predmore.
Astrophys. J., Vol. 288, No. 2, p. 694 – 702 (1985).

Linear polarization has been measured in up to four SiO maser transitions for seven sources, including the Orion maser source associated with IRc2. Polarization variability characteristics are reported for some sources which were observed several times over a 6 month period in the $v = 1$, $J = 2 - 1$ transition. Stars with low envelope expansion velocities appear to have more highly polarized masers than those with high expansion velocities. Comparison of polarization of individual maser features in different transitions indicates that emission from the same rotational transition in different vibrational states often arises in the same volume of gas, but that masers of different rotational transitions within the same vibrational state arise in different regions.

112.052 The X–ray corona of Procyon.
J. H. M. M. Schmitt, F. R. Harnden Jr., G. Peres, R. Rosner, S. Serio.
Astrophys. J., Vol. 288, No. 2, p. 751 – 755 (1985).

The authors have detected X–ray emission from the nearby system Procyon A/B (F5 IV + DF) using the IPC on board the Einstein Observatory. Analysis of the X–ray pulse height spectrum suggests that the observed X–ray emission originates in Procyon A rather than in the white dwarf companion Procyon B, since the derived X–ray temperature log T = 6.2, agrees well with temperatures found for quiescent solar X–ray emission. The technique of loop modeling is applied to Procyon's corona.

112.053 X–ray observations of Wolf–Rayet stars.
W. T. Sanders, J. P. Cassinelli, R. V. Myers, K. A. van der Hucht.
Astrophys. J., Vol. 288, No. 2, p. 756 – 763 (1985).

X–ray observations have been made of a very luminous O8f star, four late WN stars, and eight O VI Wolf–Rayet stars. The X–ray pulse height distributions of the O8f star is analyzed to obtain information on source temperature, emission measure and location relative to the stellar wind. Only upper limits are obtained for the X–rays from the Wolf–Rayet stars.

112.054 Closed coronal structures. VI. Far–ultraviolet and X–ray emission from active late–type stars and the applicability of coronal loop models.
M. S. Giampapa, L. Golub, G. Peres, S. Serio, G. S. Vaiana.
Astrophys. J., Vol. 289, No. 1, p. 203 – 212 (1985).

The authors present far–ultraviolet line fluxes of prominent transition region emission lines, as obtained with the IUE satellite, for a sample of solar–type stars. The ultraviolet observations are combined with existing soft X–ray measurements obtained by the Einstein Observatory. The authors utilize the resulting data set and a new coronal loop model numerical code to perform a preliminary investigation of the applicability of coronal loop models to solar–type stars. In general, it is found that the addition of non–simultaneous ultraviolet observations to a previously acquired soft X–ray data set does not provide a sufficient constraint on the range of possible loop filling factors and pressures for loop model atmospheres that may be producing the observed X–ray and transition region emissions.

112.055 Chromospheric Hα emission in F8 – G3 dwarfs, and its connection with the T Tauri stars.
G. H. Herbig.
Astrophys. J., Vol. 289, No. 1, p. 269 – 278 (1985). = Lick Obs. Bull., No. 992.

A chromospheric component in the center of Hα, analogous to the emission cores in Ca II H, K, has been detected in high signal–to–noise scans of about 40 F8 – G3 V stars by substraction of a low–activity standard, β CVn. The strength of this emission spike is about 100 mÅ in the most active stars. Similar emission cores are present in the infrared Ca II lines. The decay of Hα surface fluxes with age on the main sequence can be represented fairly well by a power law with exponent near –0.4, but equally well by exponential decay with an e–folding time of $3 – 5 \times 10^9$yr. These results for main–sequence stars are compared with observations of chromospheric activity in T Tauri stars and it is concluded that different mechanisms are responsible for active phenomena in the two types of stars.

112.056 Circumstellar shells in the Large Magellanic Cloud.
O. Stahl.
Messenger, No. 39, p. 13 – 15 (1985).

112.057 The Local Stellar Environment (LSE) – the B emission–line stars.
V. Doazan.
Messenger, No. 39, p. 17 – 20 (1985).

112.058 On the problem of the luminous emission line stars.
R. Viotti.
Messenger, No. 39, p. 30 – 33 (1985).

112.059 Characteristics and interpretation of the photometric variability of Eta Carinae and its nebula.
A. M. van Genderen, P. S. Thé.
Space Sci. Rev., Vol. 39, Nos. 3/4, p. 317 – 373 (1984).

A short review is given on the history of the peculiar variable object η Car and on a number of relevant references describing and discussing its physical characteristics and behaviour, based on different types of observational techniques.

112.060 Far–infrared spectrum of IRC + 10216.
T. N. Rengarajan, G. G. Fazio, C. W. Maxson, B. McBreen, S. Serio, S. Sciortino.
Astrophys. J., Vol. 289, No. 2, p. 630 – 633 (1985).

The authors present far–infrared observations of IRC + 10216 in the wavelength range 20 – 160 μm. The observations were performed simultaneously in four bands with effective wavelengths of 21, 42, 73, and 135 μm, respectively. The scan profile across the source at 21 μm is found to be wider than that for a point source. Interpreting the far–infrared spectrum on the basis of a model in which circumstellar grains are heated by a central source, the radial density dependence is r^{-2}, and the emissivity dependence is λ^{-}n, it is found that $n = 1 – 1.2$.

112.061 The geometric extent of C II (UV 0.01) emitting regions around luminous, late–type stars.
K. G. Carpenter, A. Brown, R. E. Stencel.
Astrophys. J., Vol. 289, No. 2, p. 676 – 680 (1985).

The authors present a method by which the geometric extent of the chromospheres around late–type stars can be estimated from measurements of the total emission–line flux and line ratios within the C II (UV 0.01) multiplet. Application of this technique to a sample of 15 late–type stars indicates a clear difference in the radial extent of the chromospheres around coronal and noncoronal stars. The former stars appear to have very thin chromospheres ($\leqslant 0.1\%$ of the photospheric radius), while the latter stars have chromospheres extending, on average, out to 2.5 photospheric radii.

112.062 Variable mass loss in P Cygni.
D. P. Hayes.
Astrophys. J., Vol. 289, No. 2, p. 726 – 731 (1985).

The optical continuum (B band) linear polarization of the star P Cygni was intensively monitored over an interval spanning 15 months. The detected polarization variations indicate anisotropic mass flows which are both spatially and temporally varying. Lower limit estimates have been made on the amounts of mass deposited in the circumstellar envelope during various activity episodes. Dynamical stellar instabilities are likely to account for the mass flow variations.

112.063 Polarization properties and time variations of the SiO maser emission of W Hydrae.
F. O. Clark, T. H. Troland, J. S. Miller.
Astrophys. J., Vol. 289, No. 2, p. 756 – 764 (1985).

The authors have measured the polarization properties of the $\upsilon = 1, J = 2–1$ SiO circumstellar maser emission from W Hya over a period of 3.5 yr. They present data concerning Stokes parameter I and linear polarization. The W Hya line profiles exhibit peaks which are well represented by a superposition of a small number of narrow Gaussian features, each having a well–defined position angle of maximum linear polarization. The data offer the first evidence for a small velocity pulsation in the SiO maser emission features which may be related to the envelope pulsation observed in the infrared. It is argued that the SiO maser is located extremely close to the star.

112.064 Ultraviolet radiation from stellar flares and its relation to the coronal X–ray emission for dMe stars.
J. G. Doyle, C. J. Butler.
Ir. Astron. J., Vol. 17, No. 1, p. 19 (1985). Abstract. – See Abstr. 012.034.

112.065 Infrared spectra and interstellar reddening of anonymous type II OH/IR stars.
R. D. Gehrz, S. G. Kleinmann, S. Mason, J. A. Hackwell, G. L. Grasdalen.
Astrophys. J., Vol. 290, No. 1, p. 296 – 306 (1985).

The authors report infrared positions and multicolor infrared photometry for a sample of type II OH/IR stars. The infrared colors and 11.4 μm silicate optical depths of the confirmed sources in this group increase as a function of distance, suggesting that interstellar reddening must be taken into account in assessing their infrared energy distributions and physical characteristics.

112.066 Interpretation of the spectrum of Gamma Cassiopeiae from 1 to 1.7 microns.
R. P. Lowe, J. M. Moorhead, W. H. Wehlau, P. K. Barker, J. M. Marlborough.
Astrophys. J., Vol. 290, No. 1, p. 325 – 336 (1985).

For the Be star γ Cas the authors present observations in the 1.0 – 1.7 μm spectral region of both the infrared continuum and the high–n Brackett, Paβ, Paγ, and He I λ10830 emission–line profiles. The authors also present conventional observations of Balmer lines made at the same time. A comparison is made with a theoretical model of γ Cas developed in detail by Poeckert and Marlborough. The model Brackett 14 profile is in reasonable agreement with the observations, but for Paβ a poor fit is obtained. Prospects for adjusting the model parameters to achieve better representation of the observations are explored briefly.

112.067   Observations of the SiC₂ radical toward IRC + 10216 at 1.27 centimeters.

L. E. Snyder, C. Henkel, J. M. Hollis, F. J. Lovas.

Astrophys. J., Lett. Ed., Vol. 290, No. 1, p. L29 – L33 (1985).

The first centimeter–wave transition of the recently identified SiC_2 radical has been observed in the envelope of the evolved carbon star IRC +10216. The excellent agreement between the measured astronomical rest frequency and the predicted frequency, and the measured line intensity support the SiC_2 identification. The high–resolution line profile and mapping data are used to estimate the size of the IRC +10216 SiC_2 envelope and the abundance of SiC_2 relative to H_2.

112.068   MgS grain component in circumstellar shells.

J. H. Goebel, S. H. Moseley.

Astrophys. J., Lett. Ed., Vol. 290, No. 1, p. L35 – L39 (1985).

This letter presents far–infrared spectrophotometry (30 – 55 μm) and photometry (53 – 200 μm) which define, for the first time, the long wavelength limit of the previously unidentified 30 μm emission feature found in certain extreme carbon star spectra. The spectral similarities are sufficiently similar to those of solid MgS that MgS is proposed to be the band carrier. This is interpreted as the first direct evidence that chemical surface reactions occur on dust grains in circumstellar environments.

112.069   The circumstellar envelope of S 106–IRS 4.

M. Felli, M. Simon, J. Fischer, F. Hamann.

Astron. Astrophys., Vol. 145, No. 2, p. 305 – 310 (1985).

The authors present new observations that help set the parameters of the ionized circumstellar envelope of S 106–IRS 4. The part of the envelope that is optically thick at 1.35 cm wavelength is smaller than $0''.15$ diameter which corresponds to 90 AU at 600 pc distance. The profiles of the Brackett–α and $-\gamma$ lines are somewhat different with half power widths of 121 ± 10 and 181 ± 15 km s^{-1}, respectively. The He I (2^1P–2^1S) line is detected at the S 106 nebula but not at IRS 4. The He I line emission of the nebula indicates that the central star of IRS 4 must have an effective temperature of about 35,000K. Comparison of the wind model scenario presented by Felli et al. (1984) with the present data and the Paschen line and Paschen edge data of McGregor et al. (1984) shows that the model encounters difficulties when observables that require details of the velocity field and of the innermost regions of the flow are considered.

112.070   Multiplex imagery of the infrared core of Eta Carinae.

S. Bensammar, N. Letourneur, F. Perrier, M. Friedjung, R. Viotti.

Astron. Astrophys., Vol. 146, No. 1, p. L1 – L2 (1985).

Infrared observation of Eta Car at the 3.6 m ESO telescope with the Multiplex coding technique, have revealed the presence of a compact core with a full width at 1/e intensity of $\sim 0''.13$ at 1.25 to 3.7 μm, and $0''.23$ at 4.8 μm. This is most likely the inner layer of the dust envelope where the grains start to condense from the intense stellar wind.

112.071   Cocoon stars in M17.

R. Chini, E. Krügel.

Astron. Astrophys., Vol. 146, No. 1, p. 175 – 180 (1985).

Photometric data between 2 and 20 μm for eight very young stars within M17 are presented. Combining these observations with previous *UBVRIJH* measurements the authors obtain energy distributions that exhibit the largest IR excesses known for optically visible stars today. From spectroscopy between 6800 and 9400 Å they derive spectral types that are very uniform, either B1 V or B2 V. Detailed radiative transfer models for the emission from these stars are constructed. Assuming that matter is distributed spherically around the stars all spectra could be fitted by a superposition of three distinct components. It is not possible to fit the data for a matter density around the stars that is constant or displays a power law distribution. The authors interpret the cocoons as remnants from the protostellar clouds and argue that the outer ones are formed by the effects of radiation pressure on the dust.

112.072   HD 160538 and HD 185510: two active–chromosphere stars with hot companions.

F. C. Fekel, T. Simon.

Astron. J., Vol. 90, No. 5, p. 812 – 816 (1985).

Short–wavelength ultraviolet spectra of two active–chromosphere giants, HD 160538 and HD 185510, show hot stellar continua and Lα absorption features. The hot companion of HD 160538 is a white dwarf whose ultraviolet spectrum is matched best to the energy distribution of a $T_{\text{eff}} = 30000$K and $\log g = 8$ model atmosphere. The ultraviolet spectrum of HD 185510 shows numerous absorption features and is identified as a B subdwarf. Since the giant star of each of these binary systems has no known abundance anomalies, these systems appear to be consistent with the mass–transfer scenario of barium–star formation.

112.073   Spectroscopy of the winds from Hubble–Sandage stars in M31 and M33.

S. J. Kenyon, J. S. Gallagher III.

Astrophys. J., Vol. 290, No. 2, p. 542 – 550 (1985).

The authors have obtained intermediate resolution (1 Å) spectra of five Hubble–Sandage (H–S) stars in M31 (AE And, AF And, and Variable A–1) and M33 (Variables B and C). P Cygni profiles are present on all strong permitted lines, indicating photospheric outflow velocities of $\sim 100 - 300$ km s^{-1} and mass–loss rates of $1 - 5 \times 10^{-5} M_\odot$yr^{-1}. These dense, low-velocity winds are characteristic of optically thick flows driven by continuum radiation pressure, and can be produced, in principle, by either (1) very massive single stars evolving off the main sequence, or (2) massive binary stars undergoing an epoch of rapid mass exchange.

112.074   The circumstellar H₂O maser emission associated with four late–type stars.

K. J. Johnston, J. H. Spencer, P. F. Bowers.

Astrophys. J., Vol. 290, No. 2, p. 660 – 670 (1985).

The positions and structure of H_2O masers associated with four long–period variable stars (RX Boo, R Aql, RR Aql, and NML Cyg) have been measured, using the VLA with a spatial resolution of $0''.07$. The H_2O features appear to be unresolved knots ($< 0''.07$) distributed over not more than $\sim 0''.4$. For the Mira variables R Aql and RR Aql the velocity and spatial structure change considerably with time. The estimated sizes of the H_2O maser regions are $\sim 8 \times 10^{14}$cm for RX Boo, R Aql, and RR Aql. The supergiant star NML Cyg has the largest maser region (10^{16}cm). The positional accuracy for individual maser features ranges between $0''.03$ and $0''.09$ and is referenced to the position of extragalactic radio sources.

112.075   The 2200 Å circumstellar dust absorption feature in the spectra of three bright RV Tauri stars.

S. R. Baird, J. A. Cardelli.

Astrophys. J., Vol. 290, No. 2, p. 689 – 695 (1985).

IUE observations of the 2200 Å absorption feature of circumstellar dust were made of the three RV Tauri stars AC Her, U Mon, and R Sct. U Mon was found to have a dust shell with a much lower column density at 2200 Å than R Sct, requiring U Mon to have a much more extended dust shell than R Sct to account for its greater infrared excess. AC Her and R Sct were observed twice, at different phases of their cycles, and no major change in the 2200 Å absorption feature with phase was detected.

112.076   Winds in central stars of planetary nebulae.

M. Cerruti–Sola, M. Perinotto.

Astrophys. J., Vol. 291, No. 1, p. 237 – 246 (1985).

Properties of winds have been studied with low–resolution *IUE* spectra of 60 central stars of planetary nebulae. Twenty–two stars out of 42 with a measurable stellar continuum are seen to display P Cygni profiles in the UV lines. The presence of a wind appears to be correlated with the stellar temperature and radius. It is argued that when the gravity is smaller than $\log g = 5.2$, a wind is always present in a central star of a planetary nebula. When it is larger, the presence is less frequent. The observed terminal velocities range between 1400 and 5000 km s^{-1}. An attempt to determine mass loss rates leads to values between 3×10^{-10} and $10^{-7} M_\odot \mathrm{yr}^{-1}$.

112.077 The near–infrared spectrum of Eta Carinae.
D. A. Allen, T. J. Jones, A. R. Hyland.
Astrophys. J., Vol. 291, No. 1, p. 280 – 290 (1985).

Spectroscopy of the luminous star η Carinae has been performed in the atmospheric windows between 1.0 and 2.3 µm. The spectral resolution employed, up to about 1200, is the highest yet used at these wavelengths on this object. The authors have recorded more than 80 emission lines, and offer identifications for 90% of them, mostly with species also seen at optical wavelengths. In comparison with other wave bands, the emission lines are little affected by either circumstellar reddening or optical depth effects. There is evidence that the central star is hotter than 40,000K. η Car is clearly overabundant in helium, but probably underabundant in iron in the gaseous state.

112.078 *IUE* observations of Beta Pictoris: an *IRAS* candidate for a proto–planetary system.
Y. Kondo, F. C. Bruhweiler.
Astrophys. J., Lett. Ed., Vol. 291, No. 1, p. L1 – L5 (1985).

The authors have obtained new *IUE* observations of β Pic, which was identified from *IRAS* observations as one of the several candidate stars associated with proto–planetary systems (reported by Aumann and colleagues) and whose near–infrared imagery by Smith and Terrile showed an edge–on disk surrounding the star. New high–resolution spectra, when compared with previously acquired data by Slettebak and Carpenter, revealed sharp, variable absorption in the resonance lines as well as metastable transitions of Fe II at the velocity of the photosphere of β Pic. These features must arise either in an extended gaseous envelope around the star or a circumstellar nebula. If the variable Fe II absorption originates in a circumstellar disk, it implies that the disk is possibly clumpy at distances less than 1 – 2 AU from the star.

112.079 Chromospheric activity and TiO bands in M giants.
T. Y. Steiman–Cameron, H. R. Johnson,
R. K. Honeycutt.
Astrophys. J., Lett. Ed., Vol. 291, No. 2, p. L51 – L54 (1985).

Low–resolution *IUE* spectra of 23 cool giant stars ranging from K3 through M6 have been used to examine chromospheric activity in late–type giants. The decrease in the fractional flux of the Mg II resonance lines (an indicator of chromospheric activity) with effective temperature is confirmed. A strong correlation is found between relative TiO band strengths, as measured by the Wing TiO index, and the level of chromospheric activity, as measured by Mg II fluxes. Cool giants which have weak TiO bands relative to the mean TiO strengths for stars of similar color also have Mg II lines that are weaker than the mean.

112.080 Spectrophotometric study of X Persei.
B. B. Sanwal, B. S. Rautela, S. C. Joshi.
Astrophys. Space Sci., Vol. 110, No. 2, p. 301 – 309 (1985).

The continuum energy distribution of the emission line star X Per (O9.5 III–V) has been obtained in the wavelength range 340 – 710 nm and has been compared with the energy distribution of α Cam in the same wavelength range. The continuum of the star is found to be modified by the circumstellar envelope. A number density of the order of 10^{11} in the envelope has been obtained from the observations of Hα in emission.

112.081 Comet – star close encounters and transient emission line activity of stars.
H. C. Bhatt.
Astron. Astrophys., Vol. 146, No. 2, p. 363 – 365 (1985).

Close encounters between cometary bodies and stars are considered. It is shown that during such encounters disintegration of the comet provides substantial amount of circumstellar matter which can cause observable spectroscopic and photometric changes in the star.

112.082 The hydrogen emission lines in the spectrum of γ Cassiopeiae in 1979 – 1980.
Z. K. Ivanova, T. S. Galkina.
Izv. Krymskoj Astrofiz. Obs., Tom 69, p. 34 – 39 (1984). In Russian. English translation in Bull. Crimean Astrophys. Obs., Vol. 69.

The profiles of Hα, Hβ and Hγ lines have been constructed according to 1979 – 1980 observations. The structures of these emission line profiles were similar to each other in the period of these observations – well noticeable asymmetry with prevailing intensity in the violet edge, in some cases a double–component emission with ratio V/R > 1 was clearly manifested. Spectral characteristics of the Hα and Hβ emission line profiles are defined.

112.083 Patrol electrophotometric observations of the Hα emission in the spectra of γ Cas.
N. I. Bondar', N. I. Shakhovskaya.
Izv. Krymskoj Astrofiz. Obs., Tom 69, p. 40 – 42 (1984). In Russian. English translation in Bull. Crimean Astrophys. Obs., Vol. 69.

During 28.10.80 – 25.12.80 narrow–band electrophotometric observations of the Hα emission intensity of the Be star γ Cas were obtained. Variations of intensity markedly exceeding the level of atmospheric and instrumental noise adopted by the observation of the comparison star were not detected.

112.084 Dynamical activity in V1016 Cygni.
W. A. Feibelman, R. P. Fahey.
Astrophys. J., Lett. Ed., Vol. 292, No. 1, p. L15 – L18 (1985).

New high–dispersion *IUE* observations show the C IV emission lines in V1016 Cyg to be split, suggesting evidence for dynamical activity. The [Mg V] line shows structure nearly identical to the C IV lines when plotted in velocity space.

112.085 Molecular lines in IRC + 10°216 and CIT 6.
C. Henkel, H. E. Matthews, M. Morris, S. Terebey,
M. Fich.
Astron. Astrophys., Vol. 147, No. 1, p. 143 – 154 (1985).

Observations of molecular lines between 18 and 150 GHz are reported for the carbon stars IRC +10°216 and CIT 6, and in addition, at 18 GHz, for a number of other stars.

112.086 The ultraviolet spectrum of the Be star HD 50138.
D. Hutsemékers.
Astron. Astrophys., Suppl. Ser., Vol. 60, No. 3, p. 373 – 388 (1985).

The ultraviolet spectrum of the B8 Ve star HD 50138, recorded during the Dec. 1978 – Jan. 1982 period with IUE, is analysed in the present paper. The spectrum is found to be crowded by absorption lines, mostly those of singly ionized iron peak elements. In addition there are broad lines essentially due to superionized species (Si IV, C IV) and a P Cygni profile for the Mg II doublet. A few Fe II emission lines are also detected. The slowly expanding envelope surrounding HD 50138 appears to be decelerated outward with a decreasing ionization. The physical state of this envelope seems to suffer strong variations.

112.087 Near infrared spectroscopy of γ Cas: constraints on the velocity field in the envelope.
A. A. Chalabaev, J. P. Maillard.
ESO Sci. Prepr., No. 365, 27 pp. (1985). To appear in Astrophys. J.

112.088 Time variations of the SiO ($v = 0$, $J = 2$–1) emission from circumstellar shells.
L.–Å. Nyman, H. Olofsson.
Onsala Space Obs., Prepr., No. 85:1, 25 pp. (1985). To appear in Astron. Astrophys.

112.089 The peculiar SiO ($v = 2$, $J = 2$–1) maser in χ Cygni.
H. Olofsson, O. E. H. Rydbeck, L.–Å. Nyman.
Onsala Space Obs., Prepr., No. 85:12, 21 + 11 pp. (1985). To appear in Astron. Astrophys.

112.090 The chromospheric He I D$_3$ line in main–sequence stars.
A. C. Danks, D. L. Lambert.
ESO Sci. Prepr., No. 370, 25 pp. (1985). To appear in Astron. Astrophys.

112.091 Condensation onto grains in the outflows from mass–losing red giants.
M. Jura, M. Morris.
Astrophys. J., Vol. 292, No. 2, p. 487 – 493 (1985).

The authors describe the conditions under which molecules condense onto already formed grains in the outer circumstellar envelopes of mass–losing red giants. Although there are a number of uncertain parameters, it is clear that condensation is particularly effective for species which can bind most tightly to the grains and in envelopes with large amounts of dust. For example, the authors can explain the presence of both ice and water vapor in the unusual circumstellar envelope of the mass–losing, oxygen–rich star, OH 231.8 + 4.2. In most stars with cool outflows, SiO is probably depleted onto grains, while condensation onto grains is not important for such well–studied species as CO, HCN, and H$_2$.

112.092 Radio and infrared observations of OH/IR stars at the tangential point and near the galactic center.
B. Baud, A. I. Sargent, M. W. Werner, A. F. Bentley.
Astrophys. J., Vol. 292, No. 2, p. 628 – 639 (1985).

The authors have determined accurate radio positions and subsequently made simultaneous infrared and OH radio measurements of a sample of OH/IR stars near the tangential point and in the galactic center. Intrinsic physical parameters of the stars and their dust shells, as well as the maser luminosities, have been determined. They are consistent with the stars being at the top of the asymptotic giant branch; the dense circumstellar shells suggest that they are evolving rapidly through a phase of high mass loss. It is shown that the OH luminosity is strongly dependent on the degree of reddening of the circumstellar shell, i.e., on the stellar mass–loss rate. In addition, a quantitative relation between the OH luminosity and the optical depth in the 10 µm silicate feature has been determined.

112.093 Mass loss from evolved stars. III. Mass loss rates for fifty stars from CO $J = 1$–0 observations.
G. R. Knapp, M. Morris.
Astrophys. J., Vol. 292, No. 2, p. 640 – 669 (1985).

The authors present observations at 2.6 mm wavelength of the CO $J = 1$–0 line made with the Bell Laboratories 7 m antenna in the directions of 105 cool, evolved stars having envelopes produced by mass loss. Emission was detected from 50 stars, of which about half are M–type Mira variables, half are carbon or S stars, and two are supergiants. Twenty of these envelopes are detected for the first time in the CO (1–0) line. The data are used to derive values for the stellar systemic velocity, the terminal outflow velocity of the wind, and the mass loss rate $\dot{M}$ for the envelopes. The values of $\dot{M}$ derived from CO observations cover a range of $\sim 10^{-7}$ to a few $\times 10^{-4} M_\odot \mathrm{yr}^{-1}$. A preliminary estimate of the total galactic rate of mass return to the interstellar medium by such stars gives $\dot{M}_T \sim 0.3\, M_\odot \mathrm{yr}^{-1}$.

112.094 Circumstellar material and the light variations of RV Tauri stars.
T. Lloyd Evans.
Bull. Etoiles Tardives Spectre Particulier, No. 2, p. 3 – 4 (1985). Abstract.

112.095 The outer atmosphere of the carbon star TX Piscium.
K. Eriksson, B. Gustafsson, H. R. Johnson, F. Querci, M. Querci, J. H. Baumert, M. Carlsson, H. Olofsson.
Bull. Etoiles Tardives Spectre Particulier, No. 2, p. 5 (1985). Abstract. – To appear in Astron. Astrophys.

112.096 Temporal variations in UV spectra of the red giant C star, TW Hor.
M. Querci, F. Querci.
Bull. Etoiles Tardives Spectre Particulier, No. 2, p. 6 (1985). Abstract. – To appear in Astron. Astrophys.

112.097 Properties and nature of Be stars. 12. The UV line spectrum of KX And – is there a hot primary?
S. Stefl.
Bull. Inf. Etoiles Be, No. 11, p. 7 (1985). Abstract. – Submitted to Bull. Astron. Inst. Czech.

112.098 How strong is the evidence of superionization and large mass outflows in B/Be stars?
I. Hubeny, S. Stefl, P. Harmanec.
Bull. Inf. Etoiles Be, No. 11, p. 9 – 12 (1985). Abstract. – Submitted to Bull. Astron. Inst. Czech.

112.099 Suggested new observational criteria and a new working hypothesis for the structural modeling of Be star envelopes.
D. Baade.
Bull. Inf. Etoiles Be, No. 11, p. 13 – 14 (1985). Summary. – To be published in Astron. Astrophys.

112.100 Rapid spectroscopic and photometric variations in o Cas.
J. Horn, P. Koubský, H. Božic, K. Pavlovski.
Inf. Bull. Variable Stars, No. 2659, 6 pp. (1985).

112.101 X–ray observations of late–type stars.
R. A. Stern.
Bull. Am. Astron. Soc., Vol. 16, No. 4, p. 880 (1984). Abstract. – See Abstr. 010.062.

112.102 Speckle interferometry of OH/IR stars at 5 and 10 microns.
M. L. Cobb, J. D. Fix.
Bull. Am. Astron. Soc., Vol. 16, No. 4, p. 894 (1984). Abstract. – See Abstr. 010.062.

112.103 IUE and IRAS observations of luminous M stars with varying gas–to–dust ratios.
W. Hagen, K. G. Carpenter, R. E. Stencel.
Bull. Am. Astron. Soc., Vol. 16, No. 4, p. 895 (1984). Abstract. – See Abstr. 010.062.

112.104 Chromospheric activity and TiO in M giants.
T. Y. Steiman–Cameron, H. R. Johnson, R. H. Honeycutt.
Bull. Am. Astron. Soc., Vol. 16, No. 4, p. 895 (1984). Abstract. – See Abstr. 010.062.

112.105 Chromospheric expansion velocities in late K and M giants.
S. A. Drake, J. L. Linsky.
Bull. Am. Astron. Soc., Vol. 16, No. 4, p. 895 – 896 (1984). Abstract. – See Abstr. 010.062.

112.106 Molecular ions in the circumstellar envelope of IRC + 10216.
A. E. Glassgold, R. Lucas, A. Omont.
Bull. Am. Astron. Soc., Vol. 16, No. 4, p. 937 (1984). Abstract. – See Abstr. 010.062.

112.107 Radio continuum from the surroundings of V645 Cyg and MWC 1080.
L. F. Rodríguez, S. Curiel, J. Cantó, J. M. Torrelles.
Bull. Am. Astron. Soc., Vol. 16, No. 4, p. 938 (1984). Abstract. – See Abstr. 010.062.

112.108 Three–component analysis of ultraviolet emission lines of solar–type stars.
T. Simon.
Bull. Am. Astron. Soc., Vol. 16, No. 4, p. 939 (1984). Abstract. – See Abstr. 010.062.

112.109   The surface morphology of solar–type Hyades stars.
R. R. Radick, D. K. Duncan, G. W. Lockwood.
Bull. Am. Astron. Soc., Vol. 16, No. 4, p. 939 (1984). Abstract. –
See Abstr. 010.062.

112.110   On the behavior of excess chromospheric Hα emission in late–type stars and its correlation with coronal X–ray emission.
A. Skumanich, A. Young, J. Stauffer, B. W. Bopp.
Bull. Am. Astron. Soc., Vol. 16, No. 4, p. 940 (1984). Abstract. –
See Abstr. 010.062.

112.111   He I λ5876 as a chromospheric activity indicator.
J. R. Varsik, S. C. Wolff, J. N. Heasley.
Bull. Am. Astron. Soc., Vol. 16, No. 4, p. 940 (1984). Abstract. –
See Abstr. 010.062.

112.112   Na D lines in late–type stars: observational results and semiempirical model chromospheres.
D. C. Boice.
Bull. Am. Astron. Soc., Vol. 16, No. 4, p. 940 (1984). Abstract. –
See Abstr. 010.062.

112.113   The filling factor of active regions of non–dMe stars.
M. S. Giampapa.
Bull. Am. Astron. Soc., Vol. 16, No. 4, p. 940 – 941 (1984). Abstract. – See Abstr. 010.062.

112.114   Water masers in late–type stars.
M. Elitzur, B. Cooke.
Bull. Am. Astron. Soc., Vol. 16, No. 4, p. 941 (1984). Abstract. –
See Abstr. 010.062.

112.115   Spectroscopy of the winds from Hubble–Sandage stars in M31 and M33.
S. J. Kenyon, J. S. Gallagher.
Bull. Am. Astron. Soc., Vol. 16, No. 4, p. 970 – 971 (1984). Abstract. – See Abstr. 010.062.

112.116   Voyager FUV observations of Be stars.
G. J. Peters, R. S. Polidan, R. Stalio.
Bull. Am. Astron. Soc., Vol. 16, No. 4, p. 974 (1984). Abstract. –
See Abstr. 010.062.

112.117   Voyager far–UV observations of the variable star ζ Tau.
T. E. Carone, R. S. Polidan.
Bull. Am. Astron. Soc., Vol. 16, No. 4, p. 974 (1984). Abstract. –
See Abstr. 010.062.

112.118   New 43 GHz SiO observations with the MPIfR 100 m telescope.
P. R. Jewell, C. M. Walmsley, T. L. Wilson, L. E. Snyder, D. Engels.
Bull. Am. Astron. Soc., Vol. 16, No. 4, p. 1013 (1984). Abstract. – See Abstr. 010.062.

112.119   Magnetic structure in cool stars. VIII. The Mg II h and k surface fluxes in relation to the Mt. Wilson photometric Ca II H and K measurements.
B. J. Oranje, C. Zwaan.
Astron. Astrophys., Vol. 147, No. 2, p. 265 – 272 (1985).

The authors determine relations between the Mg II h + k line core flux, at the stellar surface and in absolute units, and the Ca II H + K surface flux as derived from the photometric fluxes measured at Mt. Wilson. Finally they compare the chromospheric properties of stars of various colour and luminosity by means of the combined Mg II h + k and Ca II H + K data.

112.120   Time variations of the SiO ($v = 0$, $J = 2$–1) emission from circumstellar shells.
L.-Å. Nyman, H. Olofsson.
Astron. Astrophys., Vol. 147, No. 2, p. 309 – 316 (1985).

The authors present observations of the SiO ($v = 0$, $J = 2$–1) line towards eight late type stars, some of them first detections. Four sources have been observed on several occasions during a

four year period and temporal variations in the line shapes are apparent in two cases. The authors have also obtained CO ($J = 1$–0) spectra and upper limits on the isotopic species ^{29}SiO towards some of these sources.

112.121   Spectrophotometry of bright Be stars.
P. S. Goraya.
Astrophys. Space Sci., Vol. 112, No. 1, p. 1 – 11 (1985).

The author presents new measurements of the distribution of energy in the continuum for eight Be stars in the optical region (λλ3200 – 7600 Å). The effective temperatures of these stars have been estimated from their observed fluxes. It is found that, in general, 'pole–on' stars show near–infrared excess emission. It is interesting to note that the Balmer jumps for stars having an infrared excess are systematically smaller than for those lacking the infrared excess. Variability of ultraviolet and infrared excess emissions in these stars has been discussed.

112.122   On a possible contribution of molecules to the process of dust formation in the atmospheres of some carbon stars.
D. N. Dojkov.
Mater. Konf. mol. uchenykh Odess. astron. nauchn.–proizv. akad.–univ. kompleksa, Odessa, 15 fev. 1983. Odessa, p. 55 – 56 (1983). In Russian. Abstr. in Ref. Zh., 51. Astron., 5.51.431 (1985).

112.123   UBVRI photometry of the southern chromospherically active star – HD 139084.
A. Udalski, E. H. Geyer.
Inf. Bull. Variable Stars, No. 2692, 4 pp. (1985).

112.124   HR 1362: a chromospherically active variable with a 5–month period.
L. J. Boyd, R. M. Genet, D. S. Hall, W. S. Barksdale, R. E. Fried, G. W. Henry, J. E. Pearsall, N. F. Wasson.
Inf. Bull. Variable Stars, No. 2696, 4 pp. (1985).

112.125   A new shell star: HD 50845.
J. Sahade, A. E. Ringuelet.
Inf. Bull. Variable Stars, No. 2710, 3 pp. (1985).

112.126   HD 217188: a long–period chromospherically active variable.
L. J. Boyd, R. M. Genet, D. S. Hall, G. W. Henry.
Inf. Bull. Variable Stars, No. 2727, 3 pp. (1985).

112.127   Mass loss from evolved stars. IV. The dust–to–gas ratio in the envelopes of Mira variables and carbon stars.
G. R. Knapp.
Astrophys. J., Vol. 293, No. 1, p. 273 – 280 (1985).

The gas and grain loss rates have been calculated for a sample of 40 evolved giant stars, consisting of 18 carbon stars and 22 oxygen–rich (OH/IR and Mira) stars, for which good observations exist. The analysis shows that the gas–to–dust ratio is roughly constant in circumstellar envelopes, with values of ∼160 by mass for oxygen–rich stars and ∼400 by mass for carbon–rich stars, over a wide range of mass–loss rates. The gas–to–dust ratio in the Galaxy thus appears to be determined by the available heavy element abundances.

112.128   Mass loss from evolved stars. V. Observations of the ^{12}CO and ^{13}CO $J = 1$ 0 lines in Mira variables and carbon stars.
G. R. Knapp, K. M. Chang.
Astrophys. J., Vol. 293, No. 1, p. 281 – 287 (1985).

Emission in the ^{13}CO $J = 1$–0 line of CO was detected from 15 circumstellar envelopes (three tentatively) produced by mass loss from evolved stars. The intensity ratios for the ^{13}CO:^{12}CO lines are on the average higher for stars with [O] > [C] than for carbon stars. While individual values are not well determined, the observations suggest that Mira stars have ^{12}C/^{13}C in the range 5 – 20, while for carbon stars the range is 30 – 100.

112.129   The gaseous component of the disk around Beta Pictoris.
L. M. Hobbs, A. Vidal–Madjar, R. Ferlet,
C. E. Albert, C. Gry.
Astrophys. J., Lett. Ed., Vol. 293, No. 1, p. L29 – L33 (1985).

Optical spectra of α Lyr, α PsA, and β Pic have been obtained at a velocity resolution of 3 km s^{-1}. No circumstellar absorption lines of Ca II or Na I are detected toward α Lyr or α PsA at sensitive limits. In the favorable case of β Pic, where the circumstellar disk imaged by Smith and Terrile is seen nearly edge-on, a strong, narrow, circumstellar Ca II K absorption line previously reported by Slettebak and weaker, still narrower circumstellar Na I D lines are detected. The total gaseous column density of hydrogen along a radius of the circumstellar disk is $1 \times 10^{18} \lesssim N(\mathrm{H}) \lesssim 4 \times 10^{20}\mathrm{cm}^{-2}$. A simplified model of the density distribution in the gaseous disk yields a characteristic total density $n_\mathrm{H} \approx 10^5 \mathrm{cm}^{-3}$.

112.130   OH maser emission from infrared stars.
V. V. Burdyuzha.
Inst. kosm. issled. Akad. Nauk SSSR. Prepr., No. 948, 25 pp. (1984). In Russian. Abstr. in Ref. Zh., 51. Astron., 6.51.398 (1985).

112.131 μ Centauri.
IAU Circ., Nos. 4051, 4057 (1985).

112.132   HD 193793 = WR 140.
IAU Circ., No. 4056 (1985).

112.133   V1302 Aquilae.
IAU Circ., No. 4072 (1985).

112.134   A circumstellar disk around β Pictoris.
B. A. Smith, R. J. Terrile.
Science, Vol. 226, No. 4681, p. 1421 – 1424 (1984).

A circumstellar disk has been observed optically around the fourth–magnitude star β Pictoris. First detected in the infrared by the Infrared Astronomy Satellite last year, the disk is seen to extend to more than 400 astronomical units from the star, or more than twice the distance measured in the infrared by the Infrared Astronomy Satellite. Because the circumstellar material is in the form of a highly flattened disk rather than a spherical shell, it is presumed to be associated with planet formation. It seems likely that the system is relatively young and that planet formation either is occurring now around β Pictoris or has recently been completed.

112.135   Quasi–periodic modulation in the stellar wind of ξ Per.
H. F. Henrichs, R. K. Prinja, I. D. Howarth,
C. C. Wu, G. Sonneborn, C.   . Grady.
Bull. Am. Astron. Soc., Vol. 17, No. 1, p. 513 (1985). Abstract. –
See Abstr. 010.064.

112.136   Ultraviolet observations of Beta Pictoris: a possible protoplanetary system?
F. C. Bruhweiler, Y. Kondo.
Bull. Am. Astron. Soc., Vol. 17, No. 1, p. 520 (1985). Abstract. –
See Abstr. 010.064.

112.137   Stellar winds from hot stars in the Magellanic Clouds.
C. D. Garmany, P. S. Conti.
Astrophys. J., Vol. 293, No. 2, p. 407 – 413 (1985).

The authors have observed O and B stars in the LMC and SMC with *IUE* in order to compare their winds and mass–loss rates with those of their galactic counterparts. It is found that while the wind terminal velocities are generally about 20% lower than in galactic stars of the same spectral type, it is not clear that the mass–loss rates are significantly lower. This conclusion differs from the conclusions of other studies, and the authors discuss the reasons: most importantly that they are examining a less luminous set of stars with new spectral types determined from slit spectra.

112.138   Bipolar molecular outflows: T Tauri stars and Herbig–Haro objects.
S.–U. Choe.
Diss. Abstr. Int., Sect. B, Vol. 45, No. 5, p. 1501 (1984). Thesis, University of Minnesota, 125 pp. (1984). Order No. DA8418459.

112.139   Hydrogen spectral line and radio continuum emission from bipolar stellar winds.
F. Hamann, M. Simon.
Bull. Am. Astron. Soc., Vol. 17, No. 2, p. 557 (1985). Abstract. –
See Abstr. 010.065.

112.140   Detection of weak 1665 MHz emission in Type II OH/IR stars.
B. M. Lewis.
Bull. Am. Astron. Soc., Vol. 17, No. 2, p. 560 (1985). Abstract. –
See Abstr. 010.065.

112.141   Three dimension mapping of the atmosphere of an active K–dwarf star.
E. F. Guinan, S. W. Wacker, S. L. Baliunas, J. G. Loeser.
Bull. Am. Astron. Soc., Vol. 17, No. 2, p. 569 (1985). Abstract. –
See Abstr. 010.065.

112.142   Study of velocity structure in IRC + 10216 using acetylene line profiles.
T. Kostiuk, F. Espenak, D. Deming, M. J. Mumma, D. Zipoy, J. Keady.
Bull. Am. Astron. Soc., Vol. 17, No. 2, p. 570 (1985). Abstract. –
See Abstr. 010.065.

112.143   Short–timescale UV line profile variability in three Be stars.
C. A. Grady, G. Sonneborn, C.–C. Wu, H. F. Henrichs,
R. Prinja, I. Howarth.
Bull. Am. Astron. Soc., Vol. 17, No. 2, p. 599 (1985). Abstract. –
See Abstr. 010.065.

112.144   Recent observations of the violet opacity of N–type carbon stars.
D. R. Faulkner, H. R. Johnson, R. K. Honeycutt.
Bull. Am. Astron. Soc., Vol. 17, No. 2, p. 599 (1985). Abstract. –
See Abstr. 010.065.

112.145   Optical and infrared polarization as a probe of matter near stars.
M. G. Lacasse.
Diss. Abstr. Int., Sect. B, Vol. 45, No. 1, p. 229 – 230 (1984). Thesis, University of Rochester, 204 pp. (1984). Order No. DA8409148.

112.146   Theoretical models for the polarization of astronomical masers.
L. R. Western.
Diss. Abstr. Int., Sect. B, Vol. 45, No. 1, p. 231 (1984). Thesis, University of Illinois, 191 pp. (1983). Order No. DA8410071.

112.147   Satellite observations of the extreme ultraviolet and far ultraviolet radiation fields.
R. A. Kimble.
Diss. Abstr. Int., Sect. B, Vol. 45, No. 3, p. 901 (1984). Thesis, University of California, 141 pp. (1983). Order No. DA8413454.

112.148   Stellar winds and their dynamical influence on the interstellar medium.
D. Van Buren.
Diss. Abstr. Int., Sect. B, Vol. 45, No. 3, p. 902 (1984). Thesis, University of California, 158 pp. (1983). Order No. DA8413615.

Photometric observing campaign of Be stars. Progress Report No. 8.
See Abstr. 013.058.

Spectroscopic observing campaign of Be stars.
See Abstr. 013.059.

Laboratory identification of the emission features near 3.5 μm in the pre–main–sequence star HD 97048.
See Abstr. 022.026.

Laboratory infrared spectra of predicted condensates in carbon–rich stars.
See Abstr. 022.054.

Coronal–hole detectability on solar–type stars.
See Abstr. 036.133.

Image restoration by the shift–and–add algorithm.
See Abstr. 036.142.

Detection of nebulosity around stars using a Fabry–Perot spectrometer: a new technique.
See Abstr. 036.146.

X–ray astronomy and plasma astrophysics.
See Abstr. 062.090.

The Cerenkov microwave line emission of the hydroxyl radical.
See Abstr. 063.075.

Molecular emission from expanding circumstellar envelopes: polarization and profile asymmetries.
See Abstr. 064.004.

The scaling of coronal models from one star to another.
See Abstr. 064.013.

Overheated open coronal regions.
See Abstr. 064.014.

Co–rotating interaction regions in stellar winds: particle acceleration and non–thermal radio emission in hot stars.
See Abstr. 064.015.

Synchrotron emission from chaotic stellar winds.
See Abstr. 064.016.

Winds from rotating, magnetic, hot stars.
See Abstr. 064.017.

Non–radiative energy from differential rotation in hot stars.
See Abstr. 064.018.

Radiatively driven winds and what they imply: a review of radiative driven instabilities and their importance in hot stars.
See Abstr. 064.019.

Magnetic effects in the heating and modification of flows in the outer stellar atmospheres with application to early type stars.
See Abstr. 064.020.

Effect of scattering on instabilities in line–driven stellar winds.
See Abstr. 064.021.

New instabilities in line driven winds.
See Abstr. 064.022.

Radiative amplification of acoustic waves in hot stars.
See Abstr. 064.023.

Does nucleation theory apply to the formation of refractory circumstellar grains?
See Abstr. 064.026.

Empirical–theoretical structural patterns for stellar atmospheres and their local environment relative to variable mass–loss.
See Abstr. 064.030.

The interaction of the stellar wind with the local environment.
See Abstr. 064.031.

Multiline transfer and the dynamics of stellar winds.
See Abstr. 064.032.

Synchrotron emission from chaotic stellar winds.
See Abstr. 064.037.

Extreme–ultraviolet emission from cool star outer atmospheres.
See Abstr. 064.038.

Mass loss rates of Wolf–Rayet stars and the velocity structure of their winds.
See Abstr. 064.044.

Molecular structure in the envelopes of metal–rich red giants.
See Abstr. 064.052.

The gravity dependence of the Hα width in late–type stars.
See Abstr. 064.053.

Two–dimensional models of stellar wind bubbles. V. The efficiency of momentum transfer between winds and dense clumps in free wind regions.
See Abstr. 064.062.

Infrared excesses in Be stars: an axisymmetric model.
See Abstr. 064.063.

Formation of bipolar flow by the stellar wind embedded in a molecular disk.
See Abstr. 064.076.

Amplification and propagation of sound waves in early–type stars.
See Abstr. 064.082.

Effects of multiquantum transitions on molecular populations in grain–forming circumstellar environments
See Abstr. 064.085.

Radio spectra of collimated, ionized stellar winds.
See Abstr. 064.095.

Chromospheric dust formation and mass loss.
See Abstr. 064.096.

Report of IAU Commission 36: Theory of stellar atmospheres (*Théories des atmosphères stellaires*).
See Abstr. 064.100.

The effect of abundance values on partial redistribution line computations.
See Abstr. 064.102.

Observational problems in spectral line formation.
See Abstr. 064.103.

Current problems of line formation in early–type stars.
See Abstr. 064.104.

Stellar surface inhomogeneities and the interpretation of stellar spectra.
See Abstr. 064.105.

Computed He II spectra for Wolf–Rayet stars.
See Abstr. 064.107.

Partial redistribution in the winds of red giants.
See Abstr. 064.108.

Modeling lines formed in the expanding chromospheres of red giants.
See Abstr. 064.109.

Mass distribution and evolutionary scheme for central stars of planetary nebulae.
See Abstr. 065.001.

Theoretical and observational studies of stellar activity.
See Abstr. 065.099.

The white emission of a hydrostatic corona.
See Abstr. 074.018.

Solar–stellar outer atmospheres and energetic particles, and galactic cosmic rays.
See Abstr. 074.034.

The reddening (and distance?) of P Cygni.
See Abstr. 113.010.

Three–colour UBV photometry of the Be–shell star EW Lac.
See Abstr. 113.031.

Some infrared photometric characteristics of Be stars.
See Abstr. 113.051.

Photoelectric photometry of the stars HD 29647 and HDE 283809 in the Taurus dark cloud.
See Abstr. 113.052.

The central star of NGC 40.
See Abstr. 114.001.

Multi–year & possibly periodic variations in the UV spectrum of 56 Pegasi.
See Abstr. 114.005.

A progress report on the analysis of long exposure SWP high resolution spectra of cool stars.
See Abstr. 114.007.

Precise measurements of radial velocities of emission lines in the far–ultraviolet spectra of late–type stars.
See Abstr. 114.008.

Departures from LTE populations vs. anomalous abundances; the effects of heating.
See Abstr. 114.030.

An approach to ultraviolet spectral classification of OB stars with *IUE* data.
See Abstr. 114.055.

Ultraviolet spectra and chromospheres of R stars.
See Abstr. 114.063.

Recent changes in the spectrum of the Be star 48 Persei.
See Abstr. 114.071.

Alpha Trianguli Australis (K2 II – III): hybrid or composite?
See Abstr. 114.087.

Ultraviolet spectral morphology of the O stars. III. The ON and OC stars.
See Abstr. 114.088.

Spectrophotometric study of Be stars.
See Abstr. 114.120.

IUE spectra of G0 V – G5 V solar–type stars.
See Abstr. 114.130.

The infrared spectra of red variables: II. The SC and CS stars.
See Abstr. 114.142.

Report of IAU Commission 29: Stellar spectra (*Spectres stellaires*).
See Abstr. 114.165.

Ultraviolet radiation from stellar flares and the coronal X–ray emission for dwarf–Me stars.
See Abstr. 116.004.

Evidence for non–radiative activity in hot stars.
See Abstr. 116.010.

Evidence for non–radiative activity in stars with $T_{eff} < 10{,}000K$.
See Abstr. 116.011.

Observations of nonthermal radio emission from early–type stars.
See Abstr. 116.012.

Theory tested by means of the stars.
See Abstr. 116.014.

Short time scale periodicity in Hα emission from the main–sequence star H II 1883.
See Abstr. 116.015.

Polarization measurements of late–type stars at Hβ and Ca II K.
See Abstr. 116.018.

A search for magnetic fields in Be stars.
See Abstr. 116.019.

An Einstein Observatory X–ray survey of main–sequence stars with shallow convection zones.
See Abstr. 116.028.

Excess polarization of starlight and neutral absorption.
See Abstr. 116.034.

Detailed photometric and polarimetric coverage of a major 1984 mass loss episode in the Be star ω Ori.
See Abstr. 116.044.

The evolution of chromospheric activity and the spin–down of solar–type stars.
See Abstr. 116.058.

Strange F + Be binary star systems.
See Abstr. 117.011.

Linear polarization of symbiotic stars.
See Abstr. 117.044.

The first UV studies of the optical candidate for the X–ray source 1118–61.
See Abstr. 117.068.

Ultraviolet spectroscopic observations of HD 77581 (Vela X–1 = 4U 0900–40).
See Abstr. 117.082.

IUE spectroscopy, visible–band photometry, and polarimetry of HD 47732 (V641 Monocerotis).
See Abstr. 117.096.

High spectral resolution observations of the coronal X–ray emission from the RS CVn binary Sigma Corona Borealis.
See Abstr. 117.111.

Expanding envelopes of binary stars. I. The cone model.
See Abstr. 117.124.

A search for OH emission from symbiotic stars.
See Abstr. 117.140.

High–velocity winds in close binaries with accretion disks. II. The view along the plane of the disk.
See Abstr. 117.145.

UV and optical observations of variability in the WR + compact candidate HD 96548.
See Abstr. 117.165.

The binary system MWC 349.
See Abstr. 117.183.

Absorption spectra from massive X–ray binaries.
See Abstr. 117.245.

Activity of contact binary systems.
See Abstr. 117.364.

High–resolution photo of protoplanetary disk orbiting star.
See Abstr. 118.003.

A simultaneous X–ray and radio observation of a flare from Algol.
See Abstr. 119.048.

A search for X–ray emitting coronal structures in Algol.
See Abstr. 119.049.

A study of ultraviolet spectra of ζ Aurigae/VV Cephei systems. VII. Chromospheric density distribution and wind acceleration region.
See Abstr. 119.053.

Observations of the hydrogen deficient binary Upsilon Sagittarii.
See Abstr. 119.107.

UV observations and results on pre–main sequence stars.
See Abstr. 121.001.

A search for young stellar chromospheres in NGC 2264.
See Abstr. 121.005.

Atmospheric properties of RU Lupi derived from high– and low–resolution IUE spectra.
See Abstr. 121.007.

Upper limits to coronal line emission from X–ray detected T Tauri stars.
See Abstr. 121.008.

FU Orionis star winds.
See Abstr. 121.015.

Photoelektrische UBV–Beobachtungen des jungen Sternes V1685 Cygni.
See Abstr. 121.041.

Jets from young stars – R Monocerotis.
See Abstr. 121.051.

Jets from young stars.
See Abstr. 121.052.

Molecular–line observations of energetic outflows around R CrA and AFGL 2591.
See Abstr. 121.056.

Collimated mass outflows associated with Herbig Ae– and Be–stars.
See Abstr. 121.061.

Simultaneous observations of Ca II K and Mg II k in T Tauri stars.
See Abstr. 121.062.

Young stars and dense cores: 1 – 20 µm observations of circumstellar dust emission.
See Abstr. 121.069.

Nonradial pulsation and mass loss in early B stars.
See Abstr. 122.026.

Investigation of the variability of V351 Ori. II. Analysis of the basic peculiarities of light and color variations.
See Abstr. 122.132.

Temporal variations in UV spectra of the red giant C star, TW Hor.
See Abstr. 122.134.

Peculiar emission lines in the spectrum of the flare star YZ Canis Minoris.
See Abstr. 122.185.

V605 Aql – Schlüssel zur späten Sternentwicklung?
See Abstr. 124.011.

Eruptive Ereignisse als Kennzeichen später Sternentwicklung.
See Abstr. 124.012.

The destruction of dust grains in collisions of a supernova with a circumstellar medium.
See Abstr. 125.001.

Infra–red astronomy of supernovae.
See Abstr. 125.011.

The distribution of the local interstellar medium derived from Mg II column densities towards seven cool stars.
See Abstr. 131.068.

On the source of the ^{26}Al observed in the interstellar medium.
See Abstr. 131.080.

Optical polarisation of R CrA and T CrA – NGC 6729.
See Abstr. 131.301.

Galactic OH and H_2O masers.
See Abstr. 131.371.

Bipolar ejection of matter from hot stars.
See Abstr. 132.015.

OH masers associated with compact H II regions in the G45.1 + 0.1 complex.
See Abstr. 132.022.

An infrared study of the giant H II region NGC 3603.
See Abstr. 132.024.

The spectra of Orion A and IRC + 10216 between 72.2 and 91.1 GHz.
See Abstr. 132.026.

Speckle observations of the central source in the bipolar nebula NGC 2346.
See Abstr. 134.010.

Central stars of planetary nebulae.
See Abstr. 134.023.

The wind from the nucleus of the planetary nebula M4–18.
See Abstr. 134.046.

The Einstein soft X–ray survey of the Pleiades.
See Abstr. 153.016.

Ultraviolet and X–ray observations of NGC 2264.
See Abstr. 153.050.

OH/IR stars within 50 parsecs of the galactic center.
See Abstr. 155.102.

Extreme Be stars in the Small Magellanic Cloud.
See Abstr. 156.018.

113 Photometric Properties

113.001   *uvby* Hβ photometry of UV–bright stars.
B. R. Wade, L. F. Smith.
Mon. Not. R. Astron. Soc., Vol. 212, No. 1, p. 77 – 91 (1985).
uvby Hβ photometry is presented for 90 stars taken from an early version of the Carnochan & Wilson catalogue of stars that have very negative UV colours. Two have definite UV excesses, (HD 36629, and HD 81307). Four early–B stars have UV colours too positive for their visible classification, and β–indices that indicate higher luminosities than appear possible on galactic distribution grounds. Six late–B stars appear to have discordant flux distributions for which there are no obvious explanations. The authors suggest that the high population of subluminous stars derived by Carnochan & Wilson is the product of the statistical treatment used and the extreme patchiness in the interstellar absorption which gives rise to large numbers of little–reddened stars.

113.002   A search for low–harmonic pulsation in the Ap star CS Vir.
T. J. Kreidl.
Mon. Not. R. Astron. Soc., Vol. 212, No. 2, p. 337 – 342 (1985).
Analysis of 9.2 hr of differential photoelectric photometry of the Ap star CS Vir (HD 125248) obtained at Lowell Observatory resulted in an upper limit of approximately 0.004 mag on the peak–to–peak amplitude of any brightness variation having a period between 10 and 90 min. Consequently, the possible quasi-periodic variability reported by Maitzen & Moffat is not verified. A slight possibility that CS Vir is undergoing low–amplitude, aperiodic variations cannot be dismissed.

113.003   A search for rapid oscillations in HD 92664.
D. W. Kurtz.
Astron. Astrophys., Vol. 142, No. 1, p. 89 – 90 (1985).
Scatter in the rotational light curves of HD 92664 (HR 4185) has led Mégessier and Hensberge (1983) to suggest that this A0pSi magnetic variable may be a rapidly oscillating Ap star. If this were true, it would be of considerable importance. High speed photometric observations of HD 92664 obtained on 3 nights, including the phase of (negative) magnetic maximum, give no indication of oscillations.

113.004   *RGU* three–colour photometry of a starfield in anticentre–direction (A6).
R. P. Fenkart, L. Topaktas, G. Kandemir, S. Boydağ.
Astron. Astrophys., Suppl. Ser., Vol. 59, No. 1, p. 73 – 81 (1985).
For 1362 stars down to $G_{lim} = 18^m$, situated in a starfield of 0.217 square degrees in anticentre–direction and measured on *RGU* plates taken at the 48″ Palomar Schmidt telescope, the reddening, the density and the luminosity functions have been determined and compared with the corresponding results in another anticentre–field (A5) which lies only ∼7° apart.

113.005   Photoelectric Hβ photometry for A and F stars brighter than $V = 14^m$ in four areas in directions towards the South Galactic Pole.
T. B. Andersen, K. S. Jensen.
Astron. Astrophys., Suppl. Ser., Vol. 59, No. 2, p. 361 – 366 (1985).
Photoelectric Hβ photometry is presented for 145 stars brighter than $V \cong 14^m$ in four regions covering about 60 sq. degrees at galactic latitude $b^{II} \cong -60°$. The stars are for the majority of spectral classes A and F, and the Hβ photometry is part of a major programme searching for Population II A – F stars brighter than $V = 15.5$. Also, the *b*–magnitudes of the Strömgren *uvby* photometric system are given as derived from the counts in the Hβ–wide channel.

113.006   *UBVRI* photoelectric photometry of nearby stars.
G. Rosselló, R. Calafat, F. Figueras, C. Jordi, J. Núñez, J. M. Paredes, F. Sala, J. Torra.
Astron. Astrophys., Suppl. Ser., Vol. 59, No. 3, p. 399 – 401 (1985).
UBVRI photometry results are presented for nearby stars of Gliese's Catalogue (1969) and its supplements.

113.007   *UBV* photometry of stars whose positions are accurately known. II.
T. Oja.
Astron. Astrophys., Suppl. Ser., Vol. 59, No. 3, p. 461 – 464 (1985).
UBV photometry is presented for stars of the AGK3R and NPZT catalogues between BD declinations 50° and 55°.

113.008   Photoelectric observations of stars in the WBVR system.
N. I. Dorokhov, A. V. Dragunova, N. N. Zakozhurnikova, N. S. Komarov, T. N. Korotkikh, V. D. Motrich.
Probl. Kosm. Fiz., Vyp. 19, p. 102 – 114 (1984). In Russian. – See Abstr. 003.002.

113.009   Photometry of the extreme helium star BD + 1°4381.
C. S. Jeffery, R. A. Malaney.
Mon. Not. R. Astron. Soc., Vol. 213, No. 2, p. 61P – 65P (1985).
The authors present photometry of the extreme helium star BD + 1°4381 which shows light variations with an amplitude of 0.04 mag on a time–scale of ∼20 day. The variations are consistent with radial pulsations.

113.010   The reddening (and distance?) of P Cygni.
D. G. Turner.
Astron. Astrophys., Vol. 144, No. 1, p. 241 – 244 (1985).
A space reddening of $E_{B-V} = 0.53 \pm 0.02$ is derived for P Cyg using *UBV* data for an early–type star in its immediate vicinity. This star may be a member of a previously–undetected open cluster which contains P Cyg. If P Cyg is physically associated with the star, its absolute magnitude would be $M_V \leqslant -8.41 \pm 0.25$. The unreddened $(B-V)$ and $(U-B)$ colours of P Cyg are demonstrated to be in reasonably good agreement with those predicted for a star of its spectral classification (B2 0e), particularly if account is taken of the contribution to the stellar continuum arising from circumstellar emission associated with its strong stellar wind. The reddening for P Cyg recently estimated by Lamers et al. (1983) is too large by $0^m.10$ in E_{B-V}.

113.011   Secondary standard stars in the four–color system for larger telescopes.
A. G. D. Philip.
News Lett. Astron. Soc. N.Y., Vol. 2, No. 4, p. 17 – 18 (1983). Abstract. – See Abstr. 010.242.

113.012   Field giants: selection efficiency from photometry and metallicity gradient in the galactic bulge.
A. M. Spaenhauer, F. Thevenin.
Astron. Astrophys., Vol. 144, No. 2, p. 327 – 334 (1985).
Spectroscopic observations of 29 field giant stars covering a wide range of metal abundance are presented ($-2.5 \leqslant [M/H]_\odot^* \leqslant 0.1$). Using these stars for the physical calibration of the *RGU* broadband colours, the authors derive the metallicity gradient in a field of the inner bulge (SA 133, $l = 6°5$, $b = 10°3$) with the help of photographic *RGUBV* photometry of 150 K–type giants. Furthermore they discuss briefly the role of broadband magnitude colour surveys for the search of late type field giants.

113.013   Photometry of stars in the *uvgr* system.
S. M. Kent.
Publ. Astron. Soc. Pac., Vol. 97, No. 588, p. 165 – 174 (1985).
Photoelectric photometry is presented for over 400 stars using the *uvgr* system of Thuan and Gunn. Stars were selected to cover

a wide range of spectral type, luminosity class, and metallicity. A mean main sequence is derived along with reddening curves and approximate transformations to the $UBVR$ system. The calibration of the standard–star sequence is significantly improved.

113.014   On the shape of the _uvby_ lightcurves of CP stars.
G. Mathys, J. Manfroid.
Astron. Astrophys., Suppl. Ser., Vol. 60, No. 1, p. 17 – 42 (1985).

The authors present the lightcurves in the Strömgren system of 56 CP stars. The observational data have been fitted by a sine wave and its first harmonic. The least–square parameters are tabulated and it is shown that such a fit describes very well the CP variations in most cases, the only exceptions being due to observational uncertainties.

113.015   The relationship between the MK system and photometric classification.
D. L. Crawford.
The MK process and stellar classification, p. 191 – 212 (1984). – See Abstr. 012.033.

113.016   Chemically peculiar A to F stars: a photometric approach.
B. Hauck.
The MK process and stellar classification, p. 213 – 221 (1984). – See Abstr. 012.033.

The Geneva photometric system is very suitable for the study of chemically peculiar A to F stars. It permits the use of quantitative parameters for the CP1 to CP4 stars and metal–weak stars. However, the metallicity index might be enhanced for some other kinds of stars, especially δ Scuti stars and giant A to F stars.

113.017   "Predicting" spectral classification from photometric data?
A. Heck.
The MK process and stellar classification, p. 222 – 229 (1984). – See Abstr. 012.033.

This paper confronts spectroscopic and photometric data through modern statistical algorithms. Performances of predicting MK spectral classifications by various methods are discussed. Recommendations for terminology are also presented.

113.018   A comparison between the MK classification in the Michigan Spectral Catalogue and uvbyβ photometry.
H. E. P. Nielsen.
The MK process and stellar classification, p. 230 – 237 (1984). – See Abstr. 012.033.

113.019   Possible contributions of the Geneva photometric boxes to the MK classification.
B. Nicolet.
The MK process and stellar classification, p. 238 – 242 (1984). – See Abstr. 012.033.

113.020   Early–type MK–standard stars in the Strömgren four-color and Geneva photometric systems.
A. G. D. Philip, B. Hauck.
The MK process and stellar classification, p. 243 – 260 (1984). – See Abstr. 012.033.

113.021   Photometry of occultation candidate stars. I. Uranus 1985 and Saturn 1985 – 1991.
L. M. French, G. Morales, A. S. Dalton, J. J. Klavetter, S. R. Conner.
Astron. J., Vol. 90, No. 4, p. 668 – 669 (1985).

$V–I$ photometry is presented for five stars to be occulted by the Uranian rings during 1985, and four stars to be occulted by Saturn or its rings during the period 1985 – 1991. $V–I$ colors have been used to predict approximate K magnitudes.

113.022   Some basics of JHKLM photometry.
I. S. Glass.
Ir. Astron. J., Vol. 17, No. 1, p. 1 – 10 (1985).

Systematic factors affecting the accuracy of JHKLM photometry and its absolute calibration are discussed. A comparison is made between the systems of standards currently available.

113.023   H beta line variability in magnetic Ap stars. I.
J. Madej, K. Jahn, K. Stepień.
Acta Astron., Vol. 34, No. 4, p. 419 – 432 (1984).

Preliminary results of photometric measurements of H beta in several Ap stars are presented. Periodic variations are found certainly in θ Aur and α²CVn, and possibly in φ Dra. For the other stars upper limits for variations of H beta are determined. Observed amplitudes are transformed into variations of equivalent width assuming specific profile variations. The results show that variations of equivalent width of H beta in investigated stars are of the order of 10% or less.

113.024   Photographic B, V photometry of stars in the region of the open cluster NGC 6910.
Eh. A. Gerts.
Glav. Astron. Obs. Akad. Nauk USSR. Kiev, 15 pp. (1984). In Russian. Abstr. in Ref. Zh., 51. Astron., 2.51.437 (1985).

113.025 $(U–B)_0$, $(B–V)_0$ diagram for stars of the luminosity class V with solar chemical composition.
I. M. Kopylov.
Astron. Zh., Tom 62, Vyp. 2, p. 348 – 355 (1985). In Russian. English translation in Sov. Astron., Vol. 29, No. 2.

A line of intrinsic colours, that is a $(U–B)_0$, $(B–V)_0$ diagram, is constructed for stars O5 – K5 of the luminosity class V with solar chemical composition ([Fe/H] = 0).

113.026   On the dependence of atmospheric extinction coefficients of the Strömgren system on spectral type, luminosity, and interstellar reddening.
G. Tautvaišiene, V. Straižys.
Astron. Zh., Tom 62, Vyp. 2, p. 404 – 407 (1985). In Russian. English translation in Sov. Astron., Vol. 29, No. 2.

The atmospheric extinction coefficients of the photometric system $uvby$ are analysed by the numerical integration method. The extinction coefficients for the color indices $u–y$ and index differences c_1 are shown to depend in a complicated manner on spectral types and luminosities of stars and on their interstellar reddening. The effect can be accounted for by using interstellar reddening–free atmospheric extinction coefficients which are calculated from the interstellar reddening–free photometric Q-parameters.

113.027   Absolute calibration of photometry at 1 through 5 μm.
H. Campins, G. H. Rieke, M. J. Lebofsky.
Astron. J., Vol. 90, No. 5, p. 896 – 899 (1985).

The solar–analog method is used to determine the absolute calibration of photometry in the J, H, K, L, and M bands to an accuracy of 3%, (5% at M). This calibration agrees well with the direct calibration obtained by Blackwell et al. (1983); this agreement gives confidence that the calibration is known to an accuracy of 2% to 3%. The agreement also implies that the behavior of solar type stars is well understood in the infrared and that such stars make reliable standards for calibration of nonstandard photometric bands and for planetary reflectance studies. Both methods indicate that Vega is slightly brighter than indicated by atmospheric models.

113.028   An absolute photometric system at 10 and 20 μm.
G. H. Rieke, M. J. Lebofsky, F. J. Low.
Astron. J., Vol. 90, No. 5, p. 900 – 906 (1985).

Two new direct calibrations at 10 and 20 μm are presented, in which terrestrial flux standards are referred to infrared standard stars. These measurements give both good agreement and higher accuracy when compared with previous direct calibrations. As a result, the absolute calibrations at 10 and 20 μm have now been determined with accuracies of 3% and 8%, respectively. A variety of absolute calibrations based on extrapolation of stellar spectra from the visible to 10 μm are reviewed. Current atmospheric models of A–type stars underestimate their fluxes by about 10% at 10 μm, whereas models of solar–type stars agree well with the direct calibrations. The photometric system at 10 and 20 μm is updated to reflect the new absolute calibration, to base its zero point directly on the colors of A0 stars, and to improve the accuracy in the comparison of the standard stars.

113.029 Limb darkening laws in photometric systems.
A. A. Rubashevskij.
Inst. probl. materialoved. Akad. Nauk USSR., Prepr., No. 20, 65 pp. (1984). In Russian. Abstr. in Ref. Zh., 51. Astron., 3.51.155 (1985).

113.030 *UBV* photometry of new Hα stars in the association Orion OB1d.
K. G. Gasparyan.
Astrofizika, Tom 22, Vyp. 2, p. 325 – 333 (1985). In Russian. English translation in Astrophysics, Vol. 22, No. 2.
The results of *UBV* photometry of new Hα stars in the Orion OB1d association are presented. Photometric *UBV* data for 92 emission stars are obtained. The observed stars are located above the main sequence on the *U–B, B–V* diagram. This indicates the existence of ultraviolet excess in their spectra.

113.031 Three–colour UBV photometry of the Be–shell star EW Lac.
L. H. Iliev (*L. Kh. Iliev*), B. Z. Kovachev.
Dokl. Bolg. Akad. Nauk, Vol. 37, No. 8, p. 983 – 985 (1984). Abstr. in Ref. Zh., 51. Astron., 4.51.470 (1985).

113.032 System of highly accurate photoelectric standards in WBVR bands.
A. V. Mironov, V. G. Moshkalev, Kh. F. Khaliullin.
Astron. Tsirk., No. 1351, p. 1 – 4 (1984). In Russian.

113.033 Infrared observations of low mass objects.
F. J. Low.
Bull. Am. Astron. Soc., Vol. 17, No. 2, p. 545 (1985). Abstract. – See Abstr. 010.065.

113.034 On the atmospheric extinction coefficients for heterochromatic photometry of stars.
T. V. Vybornaya, G. A. Terez.
Kinematika Fiz. Nebesn. Tel, Tom 1, No. 3, p. 17 – 24 (1985). In Russian.
With the aim of studying the bandwidth effect the atmospheric extinction coefficients are calculated for the photometric system UBV and the system of the photometer. 35 curves of energy distribution in the spectra of stars of different spectral classes and luminosities and 5 spectral curves of the terrestrial atmosphere transparency characterising the observing sites with excellent, mean and low transparency and assumed as initial data.

113.035 Brightness distribution over the disk of μ Geminorum, derived from lunar occultations.
M. B. Bogdanov, A. M. Cherepashchuk.
Sov. Astron., Vol. 28, No. 5, p. 549 – 556 (1984). English translation of 38.113.008.

113.036 *VBLUW*–photometry of stars in small fields around 13 planetary nebulae.
R. Gathier.
Astron. Astrophys., Suppl. Ser., Vol. 60, No. 3, p. 399 – 423 (1985).
Photometry in the Walraven *VBLUW* system is presented for stars in small fields around 13 planetary nebulae. The fields have an angular radius of ~0°.3. Finding charts are given for all stars measured: most stars are brighter than $m_v = 14^m$. Except for stars that were observed with short test–integrations and for stars fainter than $m_v \approx 14^m$, the photometry usually has an accuracy of better than 1%.

113.037 Four–colour and Hβ photometry of blue stars selected from a balloon–ultraviolet survey and other sources.
W. Tobin.
Astron. Astrophys., Suppl. Ser., Vol. 60, No. 3, p. 459 – 470 (1985).
As part of a programme to search for B stars apparently far form the Galactic plane, new *uvby* and/or Hβ photometry has been obtained at Chiran for 105 stars mostly at intermediate or high galactic latitudes. Twenty–eight of the fainter stars were selected from a comparison of the Laboratoire d'Astronomie Spatiale balloon–ultraviolet survey at λ = 200 nm with the corresponding blue and red prints of the Palomar Observatory Sky Survey. The remaining 77 stars were selected from a variety of other, less systematic sources. They are mostly B stars.

113.038 Galactic structure photoelectric standards.
R. S. Stobie, G. Gilmore, N. Reid.
Astron. Astrophys., Suppl. Ser., Vol. 60, No. 3, p. 495 – 502 (1985).
Photoelectric *UBVRI* magnitudes are given for standard stars in eight fields, which are suitable for studying galactic structure. There are 20 – 30 standards in each field covering a magnitude range $8 < V < 16$ and area of sky typically 1 degree square.

113.039 CCD stellar sequences in galactic structure fields.
R. S. Stobie, R. Sagar, G. Gilmore.
Astron. Astrophys., Suppl. Ser., Vol. 60, No. 3, p. 503 – 515 (1985).
Deep *BVI* photometric sequences in the magnitude range $12 < V < 22$ have been obtained with a CCD camera for the purpose of calibrating stellar magnitudes in selected galactic structure fields.

113.040 UBVRI photometry of red stars.
L. Celis S.
Astrophys. Prepr. Ser., No. 10, 34 pp. (1985). To appear in Astrophys. J.

113.041 The absolute flux of six hot stars in the ultraviolet (912 – 1600 Å).
T. N. Woods, P. D. Feldman, G. H. Bruner.
Astrophys. J., Vol. 292, No. 2, p. 676 – 686 (1985).
Six hot stars were observed on 1984 March 2 from a sounding rocket. The absolute fluxes from γ²Vel, ζ Pup, α CMa, γ Ori, β Tau, and ε Per were measured in the spectral region between 912 and 1600 Å at 10 Å resolution. Comparisons with revised Voyager 1 and Voyager 2 Ultraviolet Spectrometer data and previous sounding rocket data are evaluated.

113.042 Photometry of HD 23838.
P. Vivekananda Rao, M. B. K. Sarma, P. C. Agrawal, M. V. K. Apparao.
Inf. Bull. Variable Stars, No. 2654, 2 pp. (1985).

113.043 Recent photometry of the helium–weak star V396 Persei.
H. J. Landis, H. Louth, D. S. Hall.
Inf. Bull. Variable Stars, No. 2662, 3 pp. (1985).

113.044 Photometric variations of the supergiant B3 Ia: HD 178129.
C. Megessier.
Inf. Bull. Variable Stars, No. 2689, 4 pp. (1985).

113.045 HD 102077 – a new BY Draconis star?
A. Udalski, E. H. Geyer.
Inf. Bull. Variable Stars, No. 2691, 4 pp. (1985).

113.046 The intrinsic UV colors of O stars.
D. Massa, B. D. Savage.
Bull. Am. Astron. Soc., Vol. 16, No. 4, p. 893 (1984). Abstract. – See Abstr. 010.062.

113.047 Mauna Kea L′ and M infrared standards.
W. M. Sinton
Bull. Am. Astron. Soc., Vol. 16, No. 4, p. 901 (1984). Abstract. – See Abstr. 010.062.

113.048 Absolute calibration of photometry at 1 through 5 μm.
H. Campins, G. H. Rieke, M. J. Lebofsky.
Bull. Am. Astron. Soc., Vol. 16, No. 4, p. 941 (1984). Abstract. – See Abstr. 010.062.

113.049   Eight–color observations of suspected M supergiants.
D. J. MacConnell, R. F. Wing.
Bull. Am. Astron. Soc., Vol. 16, No. 4, p. 970 (1984). Abstract. –
See Abstr. 010.062.

113.050   A new photometric _WBVR_ system.
Kh. Khaliullin (_Kh. F. Khaliullin_), A. V. Mironov,
V. G. Moshkalyov (_V. G. Moshkalev_).
Astrophys. Space Sci., Vol. 111, No. 2, p. 291 – 323 (1985).

A new photometric system _WBVR_ is described. It is close to _UBVR_ but has been defined by strictly fixed response curves and the secondary standards distributed uniformly in the northern sky have been constructed. The magnitudes of these standards have been coordinated to the self–consistent system. All standards have been examined for a possible brightness variability.

113.051   Some infrared photometric characteristics of Be stars.
P. S. Goraya.
Astrophys. Space Sci., Vol. 112, No. 1, p. 203 – 208 (1985).

A photometric study of a large sample of Be stars is reported. Infrared homogeneous observational data in the _JHKL_ system are used to derive some photometric characteristics of Be stars, as a class. New infrared observations of 34 Be stars are included in the present paper. Infrared two–colour diagrams are used to investigate the presence of infrared emission in Be stars. The origin of infrared excess in relation to Balmer line emission in Be stars is discussed.

113.052   Photoelectric photometry of the stars HD 29647 and HDE 283809 in the Taurus dark cloud.
V. Straižys, K. Černis, D. S. Hayes.
Astrophys. Space Sci., Vol. 112, No. 2, p. 251 – 257 (1985).

Photoelectric Vilnius photometry of the B–type stars HD 29647 and HDE 283809 in the direction of the Taurus molecular cloud indicates their brightness and energy distribution to be constant within 1 – 2%. The interstellar extinction law is determined for the star HDE 283809 from the photometry data in the Vilnius and _UBVRJHKL_ systems. The interstellar extinction law for the two stars is found to be the same in the infrared, however, it is very different in the near ultraviolet. The new spectra of HDE 283809 confirm the earlier classification and indicate an absence of emission in the hydrogen lines. The observed peculiarities of the energy distribution in the spectrum of HDE 283809 apparently originate in interstellar or circumstellar dust, not in the star itself.

113.053   JHK photometry of carbon stars and their effective temperature.
H. Gao, P.–s. Chen, Y. Zhang.
Chin. Astron. Astrophys., Vol. 9, No. 2, p. 150 – 153 (1985). English translation of Acta Astrophys. Sin., Vol. 5, No. 1, p. 44 – 49 (1985).

Infrared photometry in the J, H and K bands of 24 carbon stars is presented. Their distribution in the (J–H) – (H–K) two–color diagram is described and their bolometric corrections and bolometric magnitudes are given. The effective temperature of the stars was derived in two different ways. The authors suggested that, pending accurate measurement of its angular diameter, the effective temperature of a carbon star may be evaluated by using $T_e = 7070/[(J-K)_J + 0.88]$.

113.054   UBV photometry of HR 8752.
E. Zsoldos, K. Oláh.
Inf. Bull. Variable Stars, No. 2715, 3 pp. (1985).

113.055   CD –30°5135.
G. Welin.
Inf. Bull. Variable Stars, No. 2717, 1 p. (1985).

113.056   Photometric behavior of HR 8752 = V509 Cas.
E. M. Halbedel.
Inf. Bull. Variable Stars, No. 2718, 2 pp. (1985).

113.057   No Cepheid–like variability in SAO 096478 = NSV 03374.
E. Poretti, J. F. Le Borgne.
Inf. Bull. Variable Stars, No. 2719, 2 pp. (1985).

113.058   The detection of rapid variability in HR 5156.
J. M. Matthews, W. H. Wehlau.
Inf. Bull. Variable Stars, No. 2725, 4 pp. (1985).

113.059   Photometric methods of stellar abundance determinations.
R. Buser.
La composition chimique des étoiles dans le voisinage solaire, p. 71 – 79 (1985). – See Abstr. 012.067.

A brief review is given of the major photometric systems used in determining stellar metallicity. Recent applications are described to illustrate the importance of photometric methods in the context of galactic structure and evolution. It is stressed that synthetic photometry and the Centre de Données Stellaires should continue their roles as efficient tools for the comparative study of different data sets.

113.060   Metallicity determination of F and G stars using the Walraven photometric system.
C. F. Trefzger.
La composition chimique des étoiles dans le voisinage solaire, p. 81 (1985). – See Abstr. 012.067.

113.061   Video camera/CCD standard stars (KPNO video camera/CCD standards consortium).
C. A. Christian, M. Adams, J. V. Barnes, H. Butcher,
D. S. Hayes, J. R. Mould, M. Siegel.
Publ. Astron. Soc. Pac., Vol. 97, No. 590, p. 363 – 372 (1985).

Photoelectric _BVRI_ photometry of several fields is presented. These data are intended to be used as "standard fields" for calibration of area detectors, specifically CCDs. The fields are located at intervals of 4 – 6 hours and are accessible from Northern Hemisphere observatories. The photometry has been converted to the "standard" system of Landolt (1983).

113.062   Photometric properties of peculiar red giants.
R. F. Wing.
Cool stars with excesses of heavy elements, p. 61 – 85 (1985). – See Abstr. 012.101.

Four kinds of information are commonly obtained through photometric observations: (1) magnitudes and colors on various systems; (2) energy distributions, leading to estimates of effective temperature, tests of model atmospheres, and the detection of companions or circumstellar shells; (3) spectroscopic data from narrow–band photometry of strong spectral features; and (4) light curves of variable stars. The photometric properties of the peculiar red giants – the cool carbon and S–type stars, and the warmer R–type carbon stars, CH stars, barium stars, and related objects – are reviewed with emphasis on what can be learned from each of these kinds of photometric data.

113.063   Absolute spectrophotometry of F–, G–, K–, and M–type stars.
R. Kiehling, J. Dachs.
Cool stars with excesses of heavy elements, p. 87 – 91 (1985). – See Abstr. 012.101.

Absolute spectral energy distributions were measured photoelectrically from 3600 Å to 8600 Å with a resolution of 10 Å for 62 bright southern and equatorial stars of intermediate and late spectral types of all luminosity classes in order to provide flux calibration and a grid of secondary spectrophotometric standards for the MK system. Most of the stars are MK standards, several are slightly peculiar (CN, Ba). Quantitative classificiation criteria are derived from measured line and band strengths.

113.064   A new photometric system for monitoring carbon star variability.
F. Querci, M. Querci.
Cool stars with excesses of heavy elements, p. 99 – 104 (1985). – See Abstr. 012.101.

113.065   Bright star photometry and the APTS.
D. C. Crawford.
Bull. Am. Astron. Soc., Vol. 17, No. 2, p. 549 (1985). Abstract. –
See Abstr. 010.065.

113.066   A photometric survey of the stars of largest H.
R. G. Probst.
Bull. Am. Astron. Soc., Vol. 17, No. 2, p. 553 – 554 (1985). Abstract. – See Abstr. 010.065.

113.067   Photometric standard stars for the far UV.
J. B. Holberg.
Bull. Am. Astron. Soc., Vol. 17, No. 2, p. 554 (1985). Abstract. –
See Abstr. 010.065.

113.068   Deep 4m CCD survey: stars in the $\bar{J}\,\bar{R}\,\bar{I}$ system.
P. C. Boeshaar, J. A. Tyson, P. O. Seitzer.
Bull. Am. Astron. Soc., Vol. 17, No. 2, p. 558 (1985). Abstract. –
See Abstr. 010.065.

113.069   IRAS observations of anomalous IR emission in G–type stars.
S. F. Odenwald.
Bull. Am. Astron. Soc., Vol. 17, No. 2, p. 569 (1985). Abstract. –
See Abstr. 010.065.

113.070   Report of IAU Commission 25: Stellar photometry and polarimetry (*Photométrie et polarimétrie stellaires*).
J. Tinbergen.
Trans. IAU, Vol. XIXA, p. 259 – 267 (1985). – See Abstr. 003.046.

113.071   Three–colour UBV photometry of the Be–shell star EW Lac.
L. H. Iliev, B. Z. Kovachev, N. A. Tomov.
Dokl. Bolg. Akad. Nauk, Tome 37, No. 8, p. 983 – 985 (1984).

The authors discuss the results obtained for EW Lac by using electrophotometric observations carried out on six nights: Sept. 27 and 28, 1982, Sept. 16 and 30 and Oct. 16 and 17, 1983. The observing runs had a duration of about 5 hours with a time interval of 4 to 6 minutes between individual measurings.

Photometric catalogue for stars in Selected Areas and other fields in the RGU–system (X). Photometry of fields in and near to the Milky Way: Anticenter 5, Carina (IC 2581), Centaurus III, Aquila II, Cassiopeia (NGC 7654).
See Abstr. 002.010.

***uvbyβ* photoelectric photometric catalogue (magnetic tape).**
See Abstr. 002.016.

Photometric observing campaign of Be stars. Progress Report No. 8.
See Abstr. 013.058.

A telescope for automatic photometry.
See Abstr. 032.034.

The status of absolute flux calibration in the 1050 – 3200 Å spectral region.
See Abstr. 035.134.

Ein photometrisches Zwei–Blenden–Verfahren.
See Abstr. 036.040.

Iterative method of the atmospheric extinction reduction in multi-color broadband stellar photometry.
See Abstr. 036.083.

Precision and method for HIPPARCOS photometry.
See Abstr. 051.090.

About the accuracy in magnitude determination.
See Abstr. 051.091.

Measurements of stellar scintillation using photon counting statistics.
See Abstr. 082.059.

Radial velocity and four–color measures of halo A–type stars.
See Abstr. 111.004.

Radial velocities of late–type Population II stars.
See Abstr. 111.032.

Radial velocity and four–color measures of halo A–type stars.
See Abstr. 111.046.

Spektroskopische und photometrische Untersuchungen der B–Überriesen mit zirkumstellarem Staub in den Magellanschen Wolken.
See Abstr. 112.023.

Spectrophotometric investigation of Be stars.
See Abstr. 112.033.

Infrarotenergieverteilung von OB–Sternen mit Massenverlust.
See Abstr. 112.035.

The hybrid spectrum of the LMC hypergiant R126.
See Abstr. 112.036.

Characteristics and interpretation of the photometric variability of Eta Carinae and its nebula.
See Abstr. 112.059.

Spectrophotometric study of X Persei.
See Abstr. 112.080.

Magnetic structure in cool stars. VIII. The Mg II h and k surface fluxes in relation to the Mt. Wilson photometric Ca II H and K measurements.
See Abstr. 112.119.

Spectrophotometry of bright Be stars.
See Abstr. 112.121.

IUE spectrophotometry of helium weak and silicon stars.
See Abstr. 114.009.

IUE observations of faint standard stars for calibration of the Space Telescope.
See Abstr. 114.013.

Spectrophotometry of peculiar B and A stars. XVII. 63 Andromedae, HD 34452, Epsilon Ursae Majoris, CQ Ursae Majoris, CU Virginis, CS Virginis and Beta Coronae Borealis.
See Abstr. 114.019.

Effective temperature of Ap stars: a comparison between several photometric estimators.
See Abstr. 114.031.

Abundances in field dwarf stars. I. Atmospheric parameters.
See Abstr. 114.032.

Spectrophotometric study of Be stars.
See Abstr. 114.120.

Spectrophotometric characteristics of barium stars.
See Abstr. 114.143.

Calibrations of photometric indices on new temperature scales and problem of spectral classification in M and C stars.
See Abstr. 114.148.

Report of IAU Commission 45: Stellar classification (*Classification stellaire*).
See Abstr. 114.166.

M supergiants in the Milky Way and the Magellanic Clouds: colors, spectral types, and luminosities.
See Abstr. 115.004.

Measurement of stellar integrated flux in the wavelength range 370 nm – 950 nm.
See Abstr. 115.012.

Differences between the absolute magnitudes of main–sequence stars determined from colours $B–V$ and those determined from red colours $R–I$ or $V–I$ or $V–R$.
See Abstr. 115.021.

A search for short periodic variations of the Ap Si star HD 92664.
See Abstr. 116.005.

Detailed photometric and polarimetric coverage of a major 1984 mass loss episode in the Be star ω Ori.
See Abstr. 116.044.

Photometric variations of HD 25267.
See Abstr. 117.080.

Infrarot–Photometrie der novaähnlichen Systeme RW Sex und CPD –48°1577.
See Abstr. 117.210.

Photometric variability of stars belonging to the Trapezium–type systems.
See Abstr. 118.007.

New period determinations for variable CP stars.
See Abstr. 122.016.

Some properties of S Mira variables.
See Abstr. 122.189.

UBVRI observations of Magellanic supergiants and Cepheids.
See Abstr. 122.209.

CPD –59°2857: a new red variable star.
See Abstr. 123.002.

The galactic reddening law: the evidence from $uvby\beta$ photometry of B stars.
See Abstr. 131.019.

Bandwidth effects on interstellar reddening and on the R–ratio.
See Abstr. 131.270.

Ultraviolet studies of interstellar matter.
See Abstr. 131.358.

JHKL observations of *IRAS* sources – II.
See Abstr. 133.002.

Multicolour UBVRi photographic and photoelectric photometry of members and possible members of T–associations T1 and T3 Tauri.
See Abstr. 152.002.

uvby β photometry of southern clusters – VI. NGC 3766.
See Abstr. 153.002.

New photometric data for the red giants in the open cluster NGC 5822: membership, chemical composition and physical properties.
See Abstr. 153.021.

A *UBVRI* photoelectric sequence in 47 Tuc.
See Abstr. 154.025.

Deep photometry of globular clusters. V. Age derivations and their implications for galactic evolution.
See Abstr. 154.027.

Stellar chemical–abundance gradient in the direction of the South Galactic Pole – preliminary results.
See Abstr. 155.019.

Studies of O – F5 stars at the galactic poles.
See Abstr. 155.022.

RGU–photometry of a starfield in the direction to Scutum (S1').
See Abstr. 155.073.

The application of three–colour photometry on galactic starfields with strong interstellar extinction (Aquila II).
See Abstr. 155.075.

New limits on the surface density of M dwarfs. I. Photographic survey and preliminary CCD data.
See Abstr. 155.096.

A photometric survey of two high galactic latitude fields.
See Abstr. 155.170.

Infrared photometry and the comparative stellar content of dwarf spheroidals in the galactic halo.
See Abstr. 157.089.

Investigations of radio objects with continuous optical spectra. The results of four–colour electrophotometric observations.
See Abstr. 158.286.

114 Spectra, Temperatures, Chemical Composition, etc.

114.001   The central star of NGC 40.
L. Bianchi, M. Grewing.
NASA Conf. Publ., NASA CP–2349, p. 262 – 265 (1984). – See Abstr. 012.001.

The authors present high and low resolution *IUE* observations of the central star of the planetary nebula NGC 40. From consideration of the 220.0 nm interstellar extinction feature they determine the reddening towards NGC 40 to be E(B–V) = 0.50. The continuum distribution is best fitted with a blackbody emission with temperature between 80000 and 100000K. Comparison with existing model atmospheres for WR stars is also discussed. The line spectrum is dominated by strong emission lines from different ionization stages of C. Almost all the emission line flux is of stellar origin, as all the lines are extremely broad. P Cygni profiles of the C IV and Si IV resonance lines and of He II λ1640 indicate an expanding wind with a terminal velocity of V_∞ = –1800 km/s.

114.002   A search for white–dwarf companions of subgiant CH stars.
H. E. Bond.
NASA Conf. Publ., NASA CP–2349, p. 289 – 292 (1984). – See Abstr. 012.001.

The binary hypothesis for the origin of subgiant CH stars implies that they should have white–dwarf companions (the remnants of the former red–giant primary stars). To test this hypothesis, the author obtained low–dispersion *IUE* spectra for a total of 19 subgiant CH stars (most of which had been found on objective–prism plates by the writer during the past several years). The *IUE* archive also contains short–wavelength spectra (obtained by E. Böhm–Vitense) for two further subgiant CH stars. None of these 21 subgiant CH stars has shown any evidence for white–dwarf companions.

114.003   Temporal UV emission from the peculiar star RX Puppis.
M. Kafatos, A. G. Michalitsianos, J. Brugioni.
NASA Conf. Publ., NASA CP–2349, p. 326 – 329 (1984). – See Abstr. 012.001.

The authors have monitored the peculiar emission star RX Puppis with the IUE in low and high dispersion. The ultraviolet spectrum of RX Puppis is characterized by strong permitted and intercombination emission lines similar to that observed in slow novae and symbiotic stars. The authors present a description of their IUE data and suggest that the temporal nature of the line profile structure is consistent with an ejection event having occurred.

114.004   Ultraviolet analysis of the peculiar F supergiant HD 112374.
E. Böhm–Vitense, C. Proffitt.
NASA Conf. Publ., NASA CP–2349, p. 352 – 354 (1984). – See Abstr. 012.001.

The authors have studied the ultraviolet energy distribution of the metal–poor supergiant HD 112347. They need a temperature T_{eff} = 5500 ± 100K, log g = –0.3 ± 0.3 and a metal deficiency of log $Z/Z_\odot$ = –0.7 in order to find agreement between theoretical and observed ultraviolet energy distributions with a reddening of E(B–V) ⩽ 0.1 consistent with its galactic latitude of + 36°.

114.005   Multi–year & possibly periodic variations in the UV spectrum of 56 Pegasi.
R. E. Stencel, J. E. Neff, R. D. McClure.
NASA Conf. Publ., NASA CP–2349, p. 400 – 403 (1984). – See Abstr. 012.001.

Radical variations in the Mg II emission profile of the late type supergiant 56 Peg have been observed to occur during the course of five years of IUE operations. Pronounced and possibly periodic changes in asymmetry, emission relative velocity and photospheric radial velocity are reported. The data can be viewed as either reflecting binary motion, or a strong enhancement (flare?) in the Mg II forming region, on a 4 to 5 year timescale.

114.006   The solar stellar connection in the far ultraviolet.
J. O. Bennett, T. R. Ayres, G. J. Rottman.
NASA Conf. Publ., NASA CP–2349, p. 437 – 440 (1984). – See Abstr. 012.001.

The authors compare the far–ultraviolet "activity" of solar-type stars, as measured with the IUE, to that of the Sun, as measured by the Solar Irradiance Monitor of the Solar Mesosphere Explorer. The goal of the study is to explore the relationships between the "ultraviolet activity" at different levels of the atmospheres of solar type stars. Secondary goals are to establish the strength of the solar ultraviolet activity within the class of solar–type stars, and to examine the amplitudes of rotational modulations of the solar emission lines during the declining portion of the current sunspot cycle. A unique aspect of the study is that the spectral resolution of the SME instrument (7.5 Å) compares very favorably with that of the low dispersion model of the IUE (5 Å).

114.007   A progress report on the analysis of long exposure SWP high resolution spectra of cool stars.
J. L. Linsky, T. R. Ayres, A. Brown, K. Carpenter, C. Jordan, P. Judge, B. Gustafsson, K. Eriksson, M. Saxner, O. Engvold, E. Jensen, O. K. Moe, T. Simon.
NASA Conf. Publ., NASA CP–2349, p. 445 – 449 (1984). – See Abstr. 012.001.

During the last few years very long exposure, high–dispersion SWP spectra of many stars located throughout the cool half of the HR diagram have been obtained. Included are dwarf stars of spectral type G0 V – M2 V, G9.5 III – M5 II giants, G2 Ib – M2 Iab supergiants, a number of RS CVn–type systems, and barium stars. The authors present a nearly complete summary of the data set, deleting only the very short exposures and observations of pre–main–sequence stars. Given the importance of this data set and the many questions that it can answer with appropriate data reduction and extensive modeling efforts, the authors summarize briefly what has been and is being done with these data.

114.008   Precise measurements of radial velocities of emission lines in the far–ultraviolet spectra of late–type stars.
T. R. Ayres, O. Engvold, O. K. Moe, T. Simon, C. Jordan, P. Judge, A. Brown, J. L. Linsky.
NASA Conf. Publ., NASA CP–2349, p. 468 – 471 (1984). – See Abstr. 012.001.

The authors have measured the radial velocities of emission lines in deep SWP echelle exposures of several late–type dwarf and giant stars, taken with special observing precautions to ensure the assignment of precise wavelength scales. The goal was to search for absolute and differential Doppler shifts of emission lines formed at different temperatures in the stellar outer atmospheres.

114.009   IUE spectrophotometry of helium weak and silicon stars.
S. N. Shore, D. N. Brown.
NASA Conf. Publ., NASA CP–2349, p. 491 – 494 (1984). – See Abstr. 012.001.

The authors have surveyed a sample of helium weak stars using the low dispersion SWP and LWR cameras of IUE during 1983. The aim is to study systematics of the broad continuum feature at 1400 Å as a function of effective temperature, spectral anomaly and age. For the Orion association, the authors have five stars; for Sco–Cen they have two and for Alpha Per they have one. The range from B2 to B6 is covered. Only HD 34452, the most extreme silicon star in the sample, has a very strong 1400 feature.

114.010   IUE low–dispersion reference atlas.
A. Heck, D. Egret, M. Jaschek, C. Jaschek.
NASA Conf. Publ., NASA CP–2349, p. 507 – 510 (1984). – See Abstr. 012.001.

This atlas, published by ESA and essentially devoted to normal stars, presents 229 graphic spectra together with the corresponding fluxes and an ultraviolet spectral type. The preparation of this publication has indeed confirmed that MK classifications cannot simply be transferred to the ultraviolet range. A set of transparencies are illustrating the reference sequences constructed from the ultraviolet data. A magnetic–type copy of all the spectra pertaining to this atlas is available from the Stellar Data Center in Strasbourg.

114.011   An IUE high–resolution atlas of O–type spectra.
N. R. Walborn, R. J. Panek, J. N. Heckathorn.
NASA Conf. Publ., NASA CP–2349, p. 511 (1984). – See Abstr. 012.001.

114.012   Statistical classification of IUE low–dispersion stellar spectra: progress report.
D. Egret, A. Heck, P. Nobelis, J.–C. Turlot.
NASA Conf. Publ., NASA CP–2349, p. 512 – 515 (1984). – See Abstr. 012.001.
Statistical data analysis methods have been applied to a set of stellar low–dispersion IUE spectra. From distances between flux vectors in an adequate multivariate space, these algorithms build statistical classification schemes in the UV which can be confronted with classical morphological approaches. A progress report of the study is given.

114.013   IUE observations of faint standard stars for calibration of the Space Telescope.
A. W. Harris, C. Gry, P. Benvenuti, A. Cassatella, R. Gilmozzi, B. J. M. Hassall, A. Talavera, W. Wamsteker.
NASA Conf. Publ., NASA CP–2349, p. 516 – 518 (1984). – See Abstr. 012.001.
Many scientific programmes envisaged for the ST will require spectrophotometric observations with precise determination of absolute flux as a function of wavelength. The aim of the authors' programme, which is being carried out in collaboration with the ST calibration team, is to provide accurate, absolute IUE spectrophotometric data for a number of faint standard stars with magnitudes near the limit of IUE's observing capability. The scope of the programme and the precautions taken to maximise the accuracy of the IUE data are discussed.

114.014   Stellar ultraviolet flux distributions in the 912 – 1200 Å wavelength range.
G. R. Carruthers, H. M. Heckathorn, C. B. Opal.
NASA Conf. Publ., NASA CP–2349, p. 534 – 537 (1984). – See Abstr. 012.001.
The authors have conducted a series of three sounding rocket experiments aimed at measurement of stellar ultraviolet flux distributions in the wavelength range below 1200 Å, the effective limit of previous calibrations such as those of OAO–2, TD–1, and IUE. Previous sounding rocket and Voyager UVS measurements in this wavelength range were discordant with each other and with model atmosphere predictions. The authors describe the calibration and data reduction procedures used for their investigations, as well as preliminary results.

114.015   A deep optical survey of a small region in Aquarius – II. Stellar spectroscopy.
D. C. Morton, P. A. Krug, K. P. Tritton.
Mon. Not. R. Astron. Soc., Vol. 212, No. 2, p. 325 – 335 (1985).
This paper describes the objective prism study of 753 stellar objects (606 with $B \leqslant 20.0$) in a survey field of 0.31 degree^2 at $l = 36°5$, $b = -51°1$. Slit spectra were obtained for 103 of the objects, including 33 classified as possibly unusual from the prism images. All the slit spectra revealed normal stars except 7 of the latter group, 3 of which turned out to be white dwarfs, and 4 to be QSOs. The authors found no evidence for other objects with noticeable emission lines or an unusual energy distribution between 3200 and 5000 Å. The number of white dwarfs is consistent with the model of the galactic disc adopted by Bahcall & Soneira, but favours the higher density luminosity function for white dwarfs of Green and Liebert, Dahn & Sion. One of the white dwarfs belongs to the rare class of DF or DZ. The authors also present finding charts for all objects in the survey.

114.016   The surface gravity of Arcturus from MgH lines, strong metal lines and the ionization equilibrium of iron.
R. A. Bell, B. Edvardsson, B. Gustafsson.
Mon. Not. R. Astron. Soc., Vol. 212, No. 3, p. 497 – 515 (1985).
The surface gravity of Arcturus is estimated from the strengths of MgH features, from strong metal lines, and from the Fe I – Fe II ionization equilibrium. The MgH lines give $\log g = 1.8$ (cgs units), for the effective temperature of 4375K found by Frisk et al. This value of $\log g$ is consistent with the gravity derived from a sample of strong pressure–broadened lines of Fe I, Ca I and Na I and from the ionization equilibrium of Fe. The mass of Arcturus is found to be 0.95 $M_\odot$ when the gravity determination from MgH is used. The effective temperature is uncertain by about 50K; a reduction of the temperature by this amount reduces the logarithmic gravity found from MgH to 1.6 and the mass to 0.7 $M_\odot$.

114.017   The infrared spectra of red variables – I. Super–lithium–rich stars.
P. A. Whitelock, R. M. Catchpole.
Mon. Not. R. Astron. Soc., Vol. 212, No. 4, p. 873 – 878 (1985).
Infrared spectra of the super–lithium–rich stars: RZ Sgr (Se), T Sgr (Se), Henize 166 (SC) and VX Aql (CS) show particularly strong absorption in the first overtone vibration–rotation band of CO at 2.3 μm; possible reasons for this are discussed. An unidentified highly variable absorption feature at 1.26 μm has been observed. It is particularly strong in the spectrum of T Sgr.

114.018   Abundance analysis from the UV spectrum of the He–w star HR 6000.
F. Castelli, M. Cornachin, M. Hack, C. Morossi.
Astron. Astrophys., Suppl. Ser., Vol. 59, No. 1, p. 1 – 14 (1985).
A high resolution spectrum of the He–w star HR 6000, covering the SW and LW regions, has been analysed. A fully line-blanketed model atmosphere with parameters $T_e = 14000$K, $\log g = 4$, microturbulence $\xi = 2\,\mathrm{km\,s^{-1}}$ has been adopted. This model is in agreement with the photometric data and yields the same abundance for the Fe II and Fe III ions. By using the synthetic spectrum method abundances for a large number of elements were determined.

114.019   Spectrophotometry of peculiar B and A stars. XVII. 63 Andromedae, HD 34452, Epsilon Ursae Majoris, CQ Ursae Majoris, CU Virginis, CS Virginis and Beta Coronae Borealis.
D. M. Pyper, S. J. Adelman.
Astron. Astrophys., Suppl. Ser., Vol. 59, No. 3, p. 369 – 397 (1985).
Optical region spectrophotometry results are presented for seven Ap stars. Additional data for two silicon stars, HD 34452 and CU Vir, show that both have energy distributions that can be fit by a single global model atmosphere, except for the λ4200 and λ5200 broad absorption features. The energy distributions of the five cooler Ap stars cannot be fit by a single model atmosphere. CQ UMa, CS Vir and β CrB all show complicated variations with phase at different wavelengths. Their energy distributions show combined effects of flux redistribution and differential line blanketing.

114.020   A model atmosphere analysis of the F5 IV – V subgiant Procyon.
M. Steffen.
Astron. Astrophys., Suppl. Ser., Vol. 59, No. 3, p. 403 – 427 (1985).
The spectrum of the main component of the Procyon system is analyzed on the basis of the high–quality Procyon Atlas (Griffin, 1979). Basic atmospheric parameters of this prominent object appear to be well–determined. The surface gravity is known with unusual accuracy, independent of spectroscopic analysis, and has the value $\log g = 4.0 \pm 0.1$. Inconsistencies between this value and spectroscopic determinations of the surface gravity are investigated. It is shown that no simple model can convincingly explain these inconsistencies which are likely to occur in F type stars in general. Departures from LTE can account for at least part of the discrepancies. The author argues that fairly reliable

abundances can be obtained using preferably weak lines and a model with a formal effective temperature of 6750K. Abundances of 28 elements are derived. The envelope of Procyon has essentially solar composition; substantial contamination with processed matter of the white dwarf progenitor can be excluded.

114.021 A high dispersion analysis of a giant star in 47 Tucanae.
S. D'Odorico, R. G. Gratton, D. Ponz.
Astron. Astrophys., Vol. 142, No. 2, p. 232 – 236 (1985).

A high dispersion analysis of a 47 Tucanae giant of 13.5 m_v is presented, together with analogous material for Arcturus. Although the star is fainter than any former observed at high dispersion in 47 Tucanae, its higher temperature reduces the continuum tracing problems. The iron abundance obtained is [Fe/H] = -0.8 ± 0.2, in good agreement with the most recent low dispersion and photometric estimates.

114.022 SIT Vidicon and IDS spectra of central stars of planetary nebulae.
R. H. Méndez, R. P. Kudritzki, K. P. Simon.
Astron. Astrophys., Vol. 142, No. 2, p. 289 – 296 (1985).

The authors present monochromatic blue magnitudes, spectral descriptions and non–LTE model atmosphere analyses of several central stars of planetary nebulae (CSPN). The analyses yield the T_{eff}, log g and He abundances in the atmospheres of these stars. The authors have found new examples of H–deficient central stars: K1–27, Longmore 3 and Longmore 4. The observed positions of 12 CSPN on the log g – log T_{eff} diagram are compared with theoretical post–AGB evolutionary tracks.

114.023 UV–bright stars in galactic globular clusters, their far–UV spectra and their contribution to the globular cluster luminosity.
K. S. de Boer.
Astron. Astrophys., Vol. 142, No. 2, p. 321 – 332 (1985).

Ultraviolet–bright stars in 7 galactic globular clusters have been observed with 4 – 7 Å spectral resolution between 1150 Å and 3200 Å. The UV–bright stars, which are in the post–AGB (PAGB) phase of evolution, have effective temperatures between 19,000 and 35,000K, and have luminosities of about $10^{3.3} L_\odot$. They generate on average about 0.3% of the total cluster luminosity. The ionizing photons produced by UV–bright stars may be an essential factor for the ionization structure of gas in the Milky Way halo. The PAGB star spectra were dereddened with laws taking into account the regional differences in the Milky Way. Doing so, the earlier reported division of the UV–bluest clusters into an EB and B category is less pronounced.

114.024 Optical spectrum of HDE 245770 and X–ray flares of the transient source A 0535 + 26.
O. Eh Aab.
Astron. Zh., Tom 62, Vyp. 1, p. 70 – 80 (1985). In Russian. English translation in Sov. Astron., Vol. 29, No. 1.

Investigation results are presented for the optical component of the transient X–ray source A 0535 + 26, the Be star HDE 245770. The investigation is based on spectrograms obtained during three years. A list of lines and their parameters within the range $\lambda\lambda 3450 - 7000$ Å is given. Variability of the spectrum for the three years and a violent change in the spectrum in January 1981 are noted.

114.025 Peculiar stars in the Pleiades group.
V. G. Klochkova, I. M. Kopylov.
Astron. Zh., Tom 62, Vyp. 1, p. 87 – 93 (1985). In Russian. English translation in Sov. Astron., Vol. 29, No. 1.

For 22 peculiar stars from the Pleiades group the quantitative spectral types, the type and degree of peculiarity of the spectra, the effective temperatures, the surface gravity and the velocities of rotation are determined. The age of the stars of the group $t = 3 \times 10^7$ years is estimated from the $[M_v, (U-B)_0]$ diagram.

114.026 Determination of abundances of chemical elements in the atmospheres of K giants.
N. S. Komarov, T. V. Mishenina.
Kinematika Fiz. Nebesn. Tel, Tom 1, No. 1, p. 77 – 79 (1985). In Russian.

Twelve K giants are investigated by the differential curve–of–growth method. The atmospheric parameters and abundances of chemical elements are determined relative to Arcturus. Most stars studied have solar abundances.

114.027 Objective–prism discoveries in the northern sky. II.
W. P. Bidelman.
Astron. J., Vol. 90, No. 2, p. 341 – 345 (1985).

This paper lists 244 newly recognized peculiar or otherwise interesting northern–hemisphere stars noted on recent moderate–dispersion objective–prism plates taken with the Burrell Schmidt at its new location on Kitt Peak.

114.028 Transition regions of solar type stars observed by the International Ultraviolet Explorer satellite.
E. De Castro, M. J. Fernandez Figueroa, M. Rego.
An. Fis., Ser. B, Vol. 80, No. 2, p. 132 – 138 (1984). In Spanish. Abstr. in Phys. Abstr., Vol. 88, No. 1248, Entry 9739 (1985).

114.029 An ultraviolet line list for O star spectra.
C. A. Dean, F. C. Bruhweiler.
Astrophys. J., Suppl. Ser., Vol. 57, No. 1, p. 133 – 143 (1985).

A composite ultraviolet spectral line identification list, applicable to the O stars, is presented. Archival high–dispersion ultraviolet spectra obtained by the IUE and recent laboratory data for the iron peak ions reveal more than 550 identifiable spectral features within the far–ultraviolet region ($\lambda\lambda 1150 - 2000$). The spectra analyzed are for the O subdwarfs BD +75°325 and BD +28°4211 and the sharp–lined main–sequence star HD 46202. The wide range in effective temperatures for the selected stars renders this line list suitable for use with all stars of spectral type O.

114.030 Departures from LTE populations vs. anomalous abundances; the effects of heating.
A. B. Underhill.
NASA Conf. Publ., NASA CP–2358, p. 148 – 158 (1985). – See Abstr. 012.023.

It is demonstrated how the deposit of nonradiative heat and momentum in the mantle of a hot star affects the interpretation of the spectrum of the star. Some examples are presented for H and He showing how changes in the electron temperature affect the solution of the equations of statistical equilibrium. The observed spectra of the Wolf–Rayet stars HD 191765, HD 192103, and HD 192163 are compatible with a normal H/He abundance ratio.

114.031 Effective temperature of Ap stars: a comparison between several photometric estimators.
T. Lanz.
Astron. Astrophys., Vol. 144, No. 1, p. 191 – 198 (1985).

Effective temperatures are inferred for 28 Bp stars with the Blackwell–Shallis method. They are compared to photometric estimations, which are made with Δm_{2100}, $(B-V)_J$, $(B2-G)$, $(R-I)_c$, and $(J-K)$. It is shown that $(B2-G)$ and $(R-I)_c$ are the most interesting temperature estimators for peculiar stars. A new calibration of $(B2-G)_0$ is then provided for Si and He weak stars. Linear radii of Bp stars are found to be similar to those of main sequence late B–type stars.

114.032 Abundances in field dwarf stars. I. Atmospheric parameters.
J. B. Laird.
Astrophys. J., Suppl. Ser., Vol. 57, No. 2, p. 389 – 404 (1985).

Atmospheric parameters have been determined for 116 field dwarf stars, plus 10 faint field giants and three Hyades dwarfs. Effective temperatures were found from new $R-I$ photometry, plus published $R-I$, $b-y$, and $V-K$ colors. Surface gravities and metallicities were derived from Strömgren photometry from the literature and intermediate dispersion spectra, calibrated using published high dispersion results and parallax data. Estimated random errors in the final parameters are ± 70K in T_{eff}, ± 0.15 in $\log g$, and ± 0.10 in [Fe/H].

114.033 Line identifications, line strengths, and continuum flux measurements in the ultraviolet spectrum of Arcturus.
K. G. Carpenter, R. F. Wing, R. E. Stencel.
Astrophys. J., Suppl. Ser., Vol. 57, No. 2, p. 405 – 422 (1985).

The ultraviolet spectrum of Arcturus has been observed at high resolution (~ 0.2 Å) with the *IUE* satellite. Line identifications, mean absolute "continuum" flux measurements, integrated absolute emission–line fluxes, and measurements of selected absorption line strengths are presented for the 2250 – 2930 Å region. In the 1150 – 2000 Å region, identifications are given primarily on the basis of low–resolution spectra. Chromospheric emission lines have been identified with low–excitation species. Iron lines strongly dominate the identifications in the 2250 – 2930 Å region, Fe II accounting for $\sim 86\%$ of the emission features and Fe I for 43% of the identified absorption features.

114.034 Lithium abundances of southern F, G, and K dwarfs.
D. R. Soderblom.
Publ. Astron. Soc. Pac., Vol. 97, No. 587, p. 54 – 56 (1985).

Observations are reported of the lithium feature (6708 Å) in some bright southern stars, most of which are F, G, or K dwarfs. Three of these stars have been suggested as belonging to the Ursa Major Group. Two of these three have the large Li abundance and strong Ca II H and K emission expected of such young stars. The third potential Ursa Major Group member has little Li, but is also not a true kinematic member. No stars were found with abnormal ^{6}Li/^{7}Li ratios.

114.035 On H–alpha variability in Alpha Lyrae.
J. C. Charlton, B. S. Meyer.
Publ. Astron. Soc. Pac., Vol. 97, No. 587, p. 60 – 61 (1985).

Observations were made of α Lyr in an attempt to confirm an earlier report of Hα variability in that star. These observations showed no variability during 17.5 hours of monitoring distributed over six nights in a 43–day interval.

114.036 Analysis of three field halo stars and the chemical evolution of the Galaxy.
B. Barbuy, F. Spite, M. Spite.
Astron. Astrophys., Vol. 144, No. 2, p. 343 – 354 (1985).

High resolution electronographic plates were used to study three extreme Population II stars. Abundances of [M/H] = -2.2, -2.5, and -3.0 are found. Abundance determinations are presented for the following elements: the light elements, C, N; the carbon–burning products: Mg, Al; the silicon–burning products: Ca, Ti, Cr, Fe, Ni and the s and r–process elements: Y, Ba, Eu. The nitrogen–to–iron ratio is always above solar. The abundance of the light metal Al varies from star to star. There seems to be a trend for Ca to be overabundant. The s–process elements are always overdeficient whereas the europium–to–iron ratio seems to be solar at all metallicities.

114.037 An extremely metal–poor star with r–process overabundances.
C. Sneden, C. A. Pilachowski.
Astrophys. J., Lett. Ed., Vol. 288, No. 2, p. L55 – L58 (1985).

A high–resolution spectroscopic study of the very metal-deficient giant HD 110184 confirms the presence of strong rare earth features. An abundance analysis relative to the classic metal–poor giant HD 122563 yields heavy element abundance enhancements of roughly a factor of 2 for iron–peak elements, but factor of 4–10 for the heavy elements. The heavy element abundance pattern in HD 110184 is consistent only with r–process neutron synthesis reactions in preceding stellar generation(s).

114.038 The thorium abundance in the atmospheres of Am stars.
L. S. Lyubimkov, I. S. Savanov.
Astrofizika, Tom 22, Vyp. 1, p. 63 – 74 (1985). In Russian. English translation in Astrophysics, Vol. 22, No. 1.

It has been shown that in the spectra of relatively cool Am stars quite a strong line Th II $\lambda 4019.13$ must be present, if thorium is overabundant in their atmospheres over ten times. While investigating four Am stars in the spectra of three of them – 15 Vul, 16 Ori and 63 Tau – the authors found the sharp line Th II $\lambda 4019.13$, but in the spectrum of 81 Tau there were no obvious signs of this line. Comparing synthetic spectra with the observed ones the authors found the abundance of thorium to be equal to lg ε(Th) = $1.5 - 1.8$ for 15 Vul, 16 Ori and 63 Tau and lg ε(Th) $\leqslant 0.4$ for 81 Tau. Thus three of the four Am stars are considerably overabundant in thorium in relation to the solar value lg ε(Th)$_\odot$ = 0.2.

114.039 An improvement of the MK classification using continuous spectra and the average energy distributions of O9 – A0 stars.
E. V. Ruban.
Astrofizika, Tom 22, Vyp. 1, p. 75 – 86 (1985). In Russian. English translation in Astrophysics, Vol. 22, No. 1.

From O9 – A0 stars the most representative ones of spectral subclasses were selected. This enabled the author to deduce average distributions in the spectra using the uniform data obtained by the southern expedition of the Academy of Sciences of the USSR in 1971 – 1973.

114.040 Observations of an sdO star in the globular cluster M22.
J. W. Glaspey, S. Demers, A. F. J. Moffat, M. Shara.
Astrophys. J., Vol. 289, No. 1, p. 326 – 330 (1985).

Spectra of the hot, UV–bright star II–81 in the globular cluster M22 show absorption lines of H I, He I, and He II, and have been used to estimate a spectral type of O6. The measured radial velocity of -140 km s^{-1} agrees well with the published cluster velocity of -152 km s^{-1} and supports the probable membership of this star in the cluster. The profile of Hδ implies a value of $\log g = 4.2$. Using $M_V = 0.5$ for this star, a mass of 0.44 $M_\odot$ is derived.

114.041 Relative isotopic abundances of zirconium in R Cygni and V Cancri.
A. C. Zook.
Astrophys. J., Vol. 289, No. 1, p. 356 – 362 (1985).

The relative abundances of the isotopes of zirconium, determined from the isotopic splitting of the $^1\Pi - ^1\Sigma$ (0,1) band head of ZrO, have been found for the two S stars R Cygni and V Cancri. The nonzero abundance of ^{93}Zr indicates that nucleosynthesis has taken place in these two stars within the recent past. The abundances are consistent with all of the most common scenarios of s–process nucleosynthesis.

114.042 Extremeley metal–deficient red giants. III. Chemical abundance patterns in field halo giants.
R. E. Luck, H. E. Bond.
Space Telesc. Sci. Inst., Prepr. Ser., No. 36, 41 pp. (1985). To appear in Astrophys. J.

114.043 High–resolution near–infrared spectra of normal and super–metal–rich K giants.
H. E. Bond, D. Burstein, S. M. Faber, R. E. Luck.
Space Telesc. Sci. Inst., Prepr. Ser., No. 42, 19 pp. (1985). To appear in Astron. J.

114.044 The atmosphere of the Am star 15 Vul.
L. S. Lyubimkov, I. S. Savanov.
Izv. Krymskoj Astrofiz. Obs., Tom 69, p. 50 – 58 (1984). In Russian. English translation in Bull. Crimean Astrophys. Obs., Vol. 69.

On the basis of high–dispersion spectra a model atmosphere analysis of the Am star 15 Vul has been fulfilled. The following values of effective temperature, surface gravity and microturbulent velocity were obtained: T_{eff} = 8100$\pm$200K, $\log g = 3.5 \pm 0.2$, $\xi_t = 4.8 \pm 0.5$ km/s. It is shown that Fe II and Ti II lines lead to somewhat higher ξ_t values in comparison with Fe I lines. The abundances of 25 elements were found. Mass, radius, luminosity and age of 15 Vul were found using evolutionary calculations. These parameters are essentially close for the Am stars 15 Vul and 68 Tau, but the chemical compositions of their atmospheres are different.

114.045   Abundances in field dwarf stars. II. Carbon and nitrogen abundances.
J. B. Laird.
Astrophys. J., Vol. 289, No. 2, p. 556 – 569 (1985).

Intermediate–dispersion spectra of 116 field dwarf stars, plus 10 faint field giants and 3 Hyades dwarfs, have been used to derive carbon and nitrogen abundances relative to iron. The program sample includes both disk and halo stars, spanning a range in [Fe/H] of $+0.50$ to -2.45. Synthetic spectra of CH and NH bands have been used to determine carbon and nitrogen abundances. The results are discussed in connection with the chemical evolution of the Galaxy and the sites of C, N, and Fe nucleosynthesis.

114.046   The MK system and the MK process.
W. W. Morgan.
The MK process and stellar classification, p. 18 – 25 (1984). – See Abstr. 012.033.

A methodology is described that permits the development of new spectral classification systems that would supplement the MK system by making possible accurate classification of Population II stars, and other eccentric categories of stellar spectra.

114.047   What is wrong with the MK system?
P. C. Keenan.
The MK process and stellar classification, p. 29 – 42 (1984). – See Abstr. 012.033.

114.048   Some problems of composite spectra.
W. P. Bidelman.
The MK process and stellar classification, p. 45 – 54 (1984). – See Abstr. 012.033.

The classification of stars showing composite spectra is discussed, particularly those cases in which both stars possess essentially normal spectra but are too close ($\varrho \leqslant 2$ arcseconds) to be studied separately in the optical region. Several examples of dubious or incorrect classifications are given. A list of 178 close binaries from *The Bright Star Catalogue* having substantially evolved primaries is given. A number of these do not exhibit composite spectra, but their secondaries can be detected in the far ultraviolet.

114.049   Natural groups: the gateways for the MK classification structure.
M. F. McCarthy.
The MK process and stellar classification, p. 55 – 59 (1984). – See Abstr. 012.033.

114.050   The meaning of spectral type for O and WR stars.
A. B. Underhill.
The MK process and stellar classification, p. 60 – 67 (1984). – See Abstr. 012.033.

Factors affecting the estimation of relative abundances from O and WR spectra are reviewed, and it is noted that the apparent strengths of many absorption and emission lines are influenced strongly by the amount of nonradiative energy and momentum deposited in the mantle of the star.

114.051   On the spectral classification of WN stars.
A. A. Nikitin, A. A. Sapar, T. H. Feklistova (*T. Kh. Feklistova*), A. F. Kholtygin.
The MK process and stellar classification, p. 68 – 71 (1984). – See Abstr. 012.033.

114.052   The paradoxical behavior of the Si IV lines at spectral type O3.
A. B. Underhill.
The MK process and stellar classification, p. 72 – 76 (1984). – See Abstr. 012.033.

The paradoxical behavior of the Si IV lines $\lambda\lambda 4088$, 4116 and $\lambda\lambda 1393$, 1402 is pointed out, and it is noted that similar behavior is shown by the Mg II lines $\lambda\lambda 9217$, 9243 and $\lambda\lambda 2795$, 2803. The Mg^+ and Si^{+3} ions are isoelectronic and the observed behavior

can be understood when doubly excited states are added to the model atom used to explain the observations.

114.053   MK criteria applied to a scanner spectral atlas of the cooler stars.
D. E. Turnshek, P. C. Boeshaar.
The MK process and stellar classification, p. 77 – 82 (1984). – See Abstr. 012.033.

Spectral scans of 74 G, K, M, C, and S stars, many from the *Atlas of Spectra of the Cooler Stars* (Keenan and McNeil, 1976), have been made with the Steward Observatory Red Reticon system to form an atlas of red spectral standard stars (Turnshek et al., 1983). Here the authors review the atlas by presenting plots of relative flux versus wavelength, covering the wavelength range of 4300 to 7900 Å for 4 representative MK–standard stars. Several applications of the atlas are noted.

114.054   Classification of ultraviolet spectra.
M. Jaschek, C. Jaschek.
The MK process and stellar classification, p. 290 – 304 (1984). – See Abstr. 012.033.

A sample of 375 spectra taken with *IUE* is analyzed. It is shown that spectral types and luminosities can be assigned from UV criteria alone. The agreement with MK spectral classification is good. The procedure followed is illustrated for O– and B–type stars.

114.055   An approach to ultraviolet spectral classification of OB stars with *IUE* data.
N. R. Walborn, R. J. Panek.
The MK process and stellar classification, p. 305 – 320 (1984). – See Abstr. 012.033.

A systematic survey of O to B0 spectra between 1200 and 1900 Å has been undertaken, by means of *IUE* high–dispersion data rebinned to a resolution of 0.25 Å. Examples of spectral type and luminosity effects are shown, as well as pronounced anomalies in the ultraviolet C– and N–features of two ON stars. The consistent behavior of both photospheric and stellar–wind features in the ultraviolet, and their correlation with the optical spectral classifications, are emphasized. The remarkable luminosity effect in the Si IV resonance doublet is an outstanding problem for astrophysical interpretation.

114.056 T_{eff} determination from UV and visual spectrophotometry and comparison with MK classification.
M. L. Malagnini, C. Morossi, R. Faraggiana.
The MK process and stellar classification, p. 321 – 333 (1984). – See Abstr. 012.033.

A sample of normal MK B5 – F7 non–supergiant stars has been analyzed. The quantity defined as $R = \log(F_{1965}/F_{5445})$ has been computed for 162 stars whose spectral distribution in the optical and/or in the UV region is available. In general, the mean values of the index R correlate univocally with the MK spectral types, thus supporting the congruence of the information derived from different regions of the spectrum. Moreover, the R values appear to correlate closely with the effective temperature as derived by comparing the observed spectral energy distributions with Kurucz's models. The derived $T_{eff}(R)$ calibration would permit the determination of T_{eff} for a very large sample of stars, for which the flux in the 1965 Å band has been observed, thus providing one of the desirable links between observations and theoretical predictions.

114.057   Discovery of λ Bootis stars.
H. A. Abt.
The MK process and stellar classification, p. 340 – 345 (1984). – See Abstr. 012.033.

114.058   An unstable Ofpe star in the LMC.
N. R. Walborn.
The MK process and stellar classification, p. 346 – 347 (1984). – See Abstr. 012.033.

114.059 Spectral classification of individual stars in nearby galaxies.
R. M. Humphreys.
The MK process and stellar classification, p. 374 – 384 (1984). – See Abstr. 012.033.

The author first summarizes previous spectral classification work in other galaxies and then mentions some of the problems of classification of faint absorption–line stars and what she foresees as the needs for the future.

114.060 A distant carbon star: an accident at the edge of the Galaxy.
B. Margon.
Mercury, Vol. 13, No. 5, p. 148 – 149 (1985).

114.061 A quest for H–alpha variability of BD + 60°2522.
U. Hänni.
Tartu Astrofüüs. Obs. Teated, Nr. 74, p. 25 – 32 (1985).

The author has examined the spectral variability of BD + 60°2522, the central star of NGC 7635 ("Bubble Nebula"), observing its Hα line contour during 5 months. No inference of periodic activity can be drawn on the basis of these, data, but small–amplitude spectral variations in the observed time–scales cannot be ruled out.

114.062 The parameters of the atmosphere and metallicity of the star θ Leo (A2 V).
V. G. Klochkova, V. E. Panchuk, V. V. Tsymbal.
Astrofiz. Issled. Izv. Spets. Astrofiz. Obs., Tom 19, p. 22 – 27 (1985). In Russian. English translation in Bull. Spec. Astrophys. Obs. – North Caucasus.

Using the method of model atmospheres for the star θ Leo T_e = 9300K, $\lg g$ = 3.4, $\lg \varepsilon$(Fe) = –4.46, $\lg \varepsilon$(Cr) = –6.28, $\lg \varepsilon$(Ti) = –6.86 are determined. The errors of the method are estimated.

114.063 Ultraviolet spectra and chromospheres of R stars.
J. A. Eaton, H. R. Johnson, G. T. O'Brien, J. H. Baumert.
Astrophys. J., Vol. 290, No. 1, p. 276 – 283 (1985).

The authors have obtained long–wavelength *IUE* spectra of 13 normal R stars and two hydrogen–deficient R0 supergiants. The ultraviolet spectra of these stars are unremarkable at the 6 Å resolution of *IUE* and are very similar to the UV spectra of G and K giants. Both the R and N stars are only very weak sources of chromospheric emission when compared with the Sun and with K and M giants. The low rate of chromospheric emission of R stars relative to the bulk of late–type giants is consistent with the interpretation of the early R stars as products of mixing during the helium flash.

114.064 The lithium isotope ratio in five F or G dwarfs.
L. M. Hobbs.
Astrophys. J., Vol. 290, No. 1, p. 284 – 288 (1985).

Observations of the Li I λ6707 doublet obtained at high resolution and photometric precision are reported for two F dwarfs and three G dwarfs known to have narrow lines and strong λ6707 absorption. Upper limits $R \equiv {}^6Li/{}^7Li < 0.1$ are obtained for four of the stars, while $R \lesssim 0.1$ for ξ UMa A. The total Li/H abundances range from 3×10^{-10} to 10×10^{-10}, with uncertainties probably not exceeding a factor of 2.

114.065 Light–element abundances in 20 F and G dwarfs.
J. Tomkin, D. L. Lambert, S. Balachandran.
Astrophys. J., Vol. 290, No. 1, p. 289 – 295 (1985).

High–resolution red and near–infrared Reticon spectra of atomic lines are used to determine Na, Mg, Al, Si, Ca, and Sc abundances for 20 F and G dwarfs of the disk. The results, combined with those of recent investigations of light elemets in halo dwarfs, show that the abundances of these elements relative to Fe are distinctly different in the disk and halo.

114.066 Mean energy distributions in the spectra of O9 – A0 stars.
E. I. Hagen–Thorn, E. V. Ruban.
Glav. astron. obs. Akad. Nauk SSSR, Leningrad, 42 pp. (1984). In Russian. Abstr. in Ref. Zh., 51. Astron., 1.51.633 (1985).

114.067 Spectrophotometric analysis of the atmosphere of the star γ UMi.
V. V. Leushin, M. Yu. Nevskij, O. A. Praskova.
Rostov n/D univ. Rostov n/D, 29 pp. (1983). In Russian. Abstr. in Ref. Zh., 51. Astron., 1.51.641 (1985).

114.068 H–gamma line variability in φ Draconis.
B. Musielok.
Acta Astron., Vol. 34, No. 3, p. 387 – 394 (1984).

Variations of the profile of Hγ line in the silicon Ap star φ Dra were measured. The width of the line is variable whereas the central depth is constant during the rotational period. The variations of the Hγ line are correlated with variations of the effective magnetic field but not with the light curve or variations of the Si II lines. The hypothesis of electric currents flowing in the atmosphere of an Ap star proposed by Stepień (1978) can account for the measured variations.

114.069 Line identification in the ultraviolet spectrum of the Ap star ε Ursae Majoris.
M. Jasiński, A. Woszczyk.
Acta Astron., Vol. 34, No. 4, p. 455 – 462 (1984).

The authors present a detailed identification of the absorption lines in the ultraviolet spectrum of ε UMa. The list contains 770 lines. The identified lines belong mainly to iron and chromium.

114.070 An ultraviolet identification list for the Ap star Iota Cassiopeiae.
M. Jasiński, A. Woszczyk.
Acta Astron., Vol. 34, No. 4, p. 463 – 468 (1984).

The authors present a detailed identification of the absorption lines in the ultraviolet spectrum of ι Cas. The list contains 412 lines. The identified lines belong mainly to iron and chromium. Also many lines of manganese are present.

114.071 Recent changes in the spectrum of the Be star 48 Persei.
P. S. Goraya.
Astrophys. Space Sci., Vol. 109, No. 2, p. 373 – 380 (1985).

A search of rapid and slow spectral variations of Balmer lines with time resolution from seconds, minutes, months, and years is carried out for 48 Per. In total, 40 spectral scans in the Hα and 13 spectral scans in the λλ3500 – 5300 Å region were secured during 6 nights. The results of this study show that, in general, there are rapid variations of Hα at the limit of the noise level. Large changes with time scales of months and years in Hα and higher Balmer lines have been investigated for the first time in 48 Per.

114.072 Rapid spectral variability of stars.
S. V. Marchenko.
Inst. teor. fiz. Akad. Nauk USSR. Prepr., No. 80R, 27 pp. (1984). In Russian. Abstr. in Ref. Zh., 51. Astron., 2.51.555 (1985).

114.073 Discovery of carbon stars in the galactic bulge.
M. Azzopardi, J. Lequeux, E. Rebeirot.
Astron. Astrophys., Vol. 145, No. 2, p. L4 – L6 (1985).

The authors report the discovery of a total of 15 carbon star candidates in the galactic centre windows. These objects were detected by their strong Swan bands of C_2 in a green Grens survey with the CFH 3.6–m telescope. Three of them have been confirmed as carbon stars by IDS observations at the ESO 3.6–m telescope. These carbon stars are relatively blue and of low V luminosity, and exhibit a strong stellar Na I doublet. They do not fit into present schemes for stellar populations in the galactic centre and are either low–metallicity, relatively young objects or members of binary systems.

114.074   On the nature of carbon stars.
A. N. Zaritovskij, M. A. Marsagishvili,
Yu. K. Melik–Alaverdyan.
Soobshch. Byurakan. Obs., Vyp. 54, p. 55 – 59 (1983). In Russian.

A statistical analysis of some parameters of carbon stars is presented. The assumption is made that carbon stars are young objects with a form of activity typical for early stars of stellar evolution.

114.075   Is HD 23878 a mild CP star?
G. Mathys, J. Manfroid, A. Heck.
Inf. Bull. Variable Stars, No. 2738, 2 pp. (1985).

114.076   Long–term and orbital variability of the optical spectrum of HDE 245770 = A 0535 + 26.
O. Eh. Aab.
Astron. Zh., Tom 62, Vyp. 2, p. 339 – 347 (1985). In Russian. English translation in Sov. Astron., Vol. 29, No. 2.

Further results of investigation of the Be star HDE 245770 are presented. Strengthening of absorption and emission lines by 40 – 50 per cent in January 1981 is noted. Connection of this with the X–ray flare is discussed. The presence of periodicity with $P = 34\overset{d}{.}5$ in the intensity of the emission line Hβ is shown. The spectral class Sp = O9.90±0.10 of the star and the absolute magnitude $M_v = 5\overset{m}{.}8$ are determined. The dimensions of the envelope in the lines Hβ ($r = 1.5\ R_*$) and Hα ($r = 5\ R_*$) are estimated.

114.077   The Bp and Ap stars in the Scorpius–Centaurus moving cluster. II.
V. G. Klochkova, I. M. Kopylov.
Sov. Astron. Lett., Vol. 10, No. 4, p. 212 – 215 (1984). English translation of 38.114.001.

114.078   Spectroscopic analysis of extreme metal–poor "dwarfs". II. Improved model atmospheres and detailed abundances.
P. Magain.
Astron. Astrophys., Vol. 146, No. 1, p. 95 – 112 (1985).

The author presents the results of a spectroscopic analysis of the classical "subdwarfs" HD 19445 and HD 140283 on the basis of new observational material. This analysis makes use of empirical model atmospheres designed to reproduce the observed continuous flux and the Fe I excitation equilibrium. It is shown that these models are also able to reproduce the strong line profiles. The difference between the empirical and theoretical models for these extreme metal–poor dwarfs is tentatively attributed to non–local convection.

114.079   Spectrum analysis of the giant Am star HR 178.
C. van't Veer–Menneret, M. F. Coupry, C. Burkhart.
Astron. Astrophys., Vol. 146, No. 1, p. 139 – 148 (1985).

A fine abundance analysis of the Am star HR 178 is performed in the visual region, absolute and relative to 15 Vulpeculae, with a line–blanketed convective model atmosphere. This evolved non–pulsating star shows the typical abundance anomalies of the M.S. classical Am stars. HR 178 cannot be regarded as a δ Del star because of the Ca and Sc deficiency, but is similar to other Am stars in the blue side of the Cepheid instability strip with Sc and Ca anomalies not decreasing with advancing evolution. The authors notice a discrepancy between the very strong metallicity indices from visual and UV photometries and the moderate over-abundance of Fe for this star, which shows exceptional depth of the strong lines and a high plateau of the curve of growth.

114.080   Synthetic OH band spectra in G and K stars.
M. S. Bessell, S. M. G. Hughes, P. L. Cottrell.
Proc. Astron. Soc. Aust., Vol. 5, No. 4, p. 547 – 552 (1984).

114.081   New carbon stars found in a hemispheric survey.
C. B. Stephenson.
Astron. J., Vol. 90, No. 5, p. 784 – 786 (1985).

105 new, cool galactic carbon stars are listed, found in 11 years of objective–prism surveys covering upwards of 25000 deg^2 (al-most all) north of declination –25°. The stars are mostly brighter than visual magnitude 13.5.

114.082   Red horizontal–branch stars in the Galactic disk.
J. A. Rose.
Astron. J., Vol. 90, No. 5, p. 787 – 802 (1985).

A class of red horizontal–branch (RHB) stars, similar to those in the "metal–rich" globular cluster M71, has been identified in the Galactic disk, using a quantitative three–dimensional spectral classification system developed earlier (Rose 1984) that uses 2.5–Å resolution spectra in the blue. A prototype for this class is the G5 III star HD 79452, which has been found by Helfer and Wallerstein (1968) to have [Fe/H] = –0.85 and $M_v = +1$. The RHB stars are shown to be evolved stars on the basis of the strength of their Sr II λ4077 line, and are distinguished from post–main–sequence stars evolving through the same region of the HR diagram because of the unique appearance of their CN λ3883 and λ4216 bands. A preliminary estimate has been made of their space density, scale height perpendicular to the Galactic plane, and kinematics by surveying G5 – G7 stars in the Upgren (1962) North Galactic Pole survey.

114.083   Strong–lined G dwarfs in the Galactic disk.
J. A. Rose.
Astron. J., Vol. 90, No. 5, p. 803 – 811 (1985).

A class of very strong–lined (VSL) main–sequence stars in the Galactic disk has been found among a sample of G5–G8 stars in the Upgren (1962) North Galactic Pole survey. These stars were not noted previously as VSL stars because the great strength of their CN bands led to their misclassification as giants. The above stars show VSL characteristics in the CN λ3883 band strength, in an index that measures the overall strength of unresolved (at 2.5–Å resolution) Fe I lines, and in photometric broadband colors that presumably measure the overall line blanketing. These VSL dwarf stars constitute about 15% of the Galactic disk, a result that agrees with the fraction of disk giants that are VSL, but that is substantially higher than the fraction of VSL dwarfs reported by Taylor (1970). The velocity dispersion of the VSL stars reported in the paper is consistent with disk kinematics.

114.084   On the ultraviolet iron spectrum of pre–white dwarfs.
D. Schönberner, J. S. Drilling.
Astrophys. J., Lett. Ed., Vol. 290, No. 2, p. L49 – L53 (1985). = Contrib. Louisiana State Univ. Obs., No. 190.

The authors have detected numerous lines of Fe V, Fe VI, and Fe VII in high–resolution *IUE* spectra of two central stars and a number of very hot subdwarfs. The strengths of these lines vary widely from star to star, suggesting that the abundance of iron in the atmospheres of the immediate progenitors of white dwarfs in strongly variable. A possible explanation is elemental separation by gravity and radiation pressure.

114.085   A method for spectral classification of faint stars from their spectral photographs in the ultraviolet.
G. A. Gurzadyan.
Dokl. Akad. Nauk SSSR. Ser. Mat. Fiz., Tom 278, No. 4, p. 839 – 842 (1984). In Russian. Abstr. in Ref. Zh., 51. Astron., 3.51.463 (1985).

114.086   New Hα emission–line stars in the region of the Khavtasi 193 dark nebula.
M. Tsvetkov, E. Semkov.
Astrofizika, Tom 22, Vyp. 2, p. 421 – 423 (1985). In Russian. English translation in Astrophysics, Vol. 22, No. 2.

On a plate obtained on the 40″–Schmidt telescope of the Byurakan Astrophysical Observatory with 4°–objective–prism 6 new and 3 known Hα emission stars as well as the probable variability of 33 stars from Kun's list of the Hα stars are detected.

114.087   Alpha Trianguli Australis (K2 II – III): hybrid or composite?
T. R. Ayres.
Astrophys. J., Lett. Ed., Vol. 291, No. 1, p. L7 – L10 (1985).

The prototype "hybrid–spectrum" giant, α Trianguli Austra-lis, exhibits a far–ultraviolet continuum which is considerably bluer than would be expected of a star of its optical colors, suggesting the presence of a previously unrecognized companion. If the K type primary is as luminous as indicated by the widths of its Ca II and Hα lines, the companion could be an early F type dwarf that only recently has arrived on the main sequence. Indeed, the flux of C IV from α TrA – an important measure of "hybridness" – would not be inconsistent with that expected from a very young, chromospherically active F star.

114.088   Ultraviolet spectral morphology of the O stars. III. The ON and OC stars.
N. R. Walborn, R. J. Panek.
Astrophys. J., Vol. 291, No. 2, p. 806 – 811 (1985).
Key regions in the 1200 – 1900 Å spectra of two ON dwarfs and four ON/OC supergiants are illustrated and discussed, by means of high–resolution data from the IUE archives. Marked anomalies are displayed by both stellar wind and photospheric C and N features, relative to the consistent framework provided by normal ultraviolet O–type spectra. The opposite departures from normal behavior of the N versus C features strongly suggest an origin in abundance effects.

114.089   Empirical effective temperatures of late O, B, A and early F stars.
E. Theodossiou.
Mon. Not. R. Astron. Soc., Vol. 214, No. 3, p. 327 – 335 (1985).
Empirical effective temperatures of 99 late O, B, A and early F stars have been derived by combining space observations in the ultraviolet with ground–based observations and by fitting the computed fluxes to the observed fluxes in the visible and ultraviolet spectral ranges. Effective temperatures found here are in good agreement with those derived by previous workers.

114.090   Observed and computed spectral flux distribution of non–supergiant O9 – G8 stars. III. Determination of T_{eff} for the stars in the Breger Catalogue.
C. Morossi, M. L. Malagnini.
Astron. Astrophys., Suppl. Ser., Vol. 60, No. 3, p. 365 – 372 (1985).
The effective temperatures and angular diameters of non-supergiant O9 – G8 stars are determined from visible spectrophotometry. The results, which refer to 302 stars included in the Breger Catalogue, are derived from the comparison between observed flux distributions and the predictions of Kurucz' models. The uncertainties to be expected in individual results are discussed; their sizes are of the order of 5% in effective temperature and 10% in angular diameter.

114.091   Spectrum analysis of the giant Am star HR 178.
C. Van't Veer–Menneret, M. F. Coupry, C. Burkhart.
Inst. Astrophys. Paris, Pré–Publ., No. 89, 31 pp. (1985). To appear in Astron. Astrophys.

114.092   Extremely metal–deficient red giants. III. Chemical abundance patterns in field halo giants.
R. E. Luck, H. E. Bond.
Astrophys. J., Vol. 292, No. 2, p. 559 – 577 (1985).
The authors have determined chemical abundances in 36 metal–poor field red giants, using model–stellar–atmosphere analysis techniques. The observational data consist of image-tube echelle spectrograms obtained with the 4 m reflectors at Kitt Peak and Cerro Tololo. The abundance analyses confirm the well–known overall enhancements of the light metals (Na through Ti, except for Sc), and deficiencies of the heavy "s–process" elements (Sr, Y, Zr, and Ba), in iron–poor stars. The elements heavier than Ba are not systematically underabundant relative to iron but do appear to show a significant star–to–star scatter. The implications of the results for the chemical evolution and the sites of nucleosynthesis in the galactic halo are discussed.

114.093   The determination of the helium abundance in main–sequence B stars.
S. C. Wolff, J. N. Heasley.
Astrophys. J., Vol. 292, No. 2, p. 589 – 600 (1985).
Measurements of the strengths of the lines He I λλ4387 and 4026 in main–sequence B–type stars in the Orion, I Lac, and Sco–Cen associations, the α Per cluster, and the field form the basis of a detailed analysis of the accuracy of helium abundances derived from photospheric lines. The helium abundances of all the groups of stars in this sample are the same to within approximately 10%, and the authors estimate that relative abundances can be derived with sufficient accuracy to detect with confidence abundance differences as small as 20%. The accuracy of the absolute abundance of helium, which is found to be 0.085 from λ4026 alone, is no better than about 15% because of possible systematic errors in placement of the continuum, in the modeling of non–LTE effects, and in the calculations of line broadening.

114.094   Oxygen isotopic abundances in evolved stars. I. Six barium stars.
M. J. Harris, D. L. Lambert, V. V. Smith.
Astrophys. J., Vol. 292, No. 2, p. 620 – 627 (1985).
Oxygen isotope ratios have been measured in four Ba II stars. The abundance determinations were made by analysis of high-resolution 2.3 μm spectra. In this spectral region the first overtone vibration rotation bands of the CO molecule's ground electronic state give rise to a large number of absorption lines, including lines due to the isotopically substituted species $^{13}C^{16}O$, $^{12}C^{17}O$, and $^{12}C^{18}O$. The stars examined were the Ba II stars HD 178717, HD 121447, HD 101013, and HR 774, and the mild barium stars o Vir and 16 Ser. The implications of the observed oxygen isotopic abundances for the evolutionary state of these stars are discussed.

114.095   Old stellar populations. II. An analysis of K–giant spectra.
S. M. Faber, E. D. Friel, D. Burstein, C. M. Gaskell.
Astrophys. J., Suppl. Ser., Vol. 57, No. 4, p. 711 – 741 (1985). = Lick Obs. Bull., No. 999.
Strong absorption features are calibrated in 110 field and globular cluster K giants versus T_e, [Fe/H], and gravity. The calibrations are intended for stellar population models of early–type galaxies. Line strengths are based on 9 Å resolution IDS spectra and include CN, Mg b, Na D, and two iron lines. An iron index, ⟨ΔFe⟩, is shown to be a good metallicity indicator and insensitive to surface gravity. Super metal–rich giants show enhancements in ⟨ΔFe⟩ and in many other lines and blends.

114.096   Barium stars indicators.
M. Rego, M. Cornide.
Bull. Etoiles Tardives Spectre Particulier, No. 2, p. 1 – 2 (1985).

114.097   An R Star in the central bulge of the Galaxy.
T. Lloyd Evans.
Bull. Etoiles Tardives Spectre Particulier, No. 2, p. 4 (1985). Abstract. – Submitted to Mon. Not. R. Astron. Soc.

114.098   The abundances of carbon, nitrogen, oxygen and their isotopes in the atmospheres of four SC stars.
J. F. Dominy, G. Wallerstein, N. B. Suntzeff.
Bull. Etoiles Tardives Spectre Particulier, No. 2, p. 7 (1985). Abstract. – Submitted to Astrophys. J.

114.099   Further radial velocity measurements of HD 36705.
J. L. Innis, G. J. Nelson, D. W. Coates, K. Thompson.
Inf. Bull. Variable Stars, No. 2667, 2 pp. (1985).

114.100   UV–Spektroskopie von 3 Hyperriesen.
H. Zekl.
Mitt. Astron. Ges., Nr. 63, p. 179 – 180 (1985). – See Abstr. 012.063.

114.101   Line profile variations in α Lyr and α CrB.
K. E. Sedwick, E. L. Wright.
Bull. Am. Astron. Soc., Vol. 16, No. 4, p. 892 (1984). Abstract. –
See Abstr. 010.062.

114.102   Ultraviolet variability and flux redistribution in the Ap Si star 56 Arietis.
G. Sonneborn, R. J. Panek.
Bull. Am. Astron. Soc., Vol. 16, No. 4, p. 893 (1984). Abstract. –
See Abstr. 010.062.

114.103   Abundances of carbon, nitrogen, and oxygen in metallic–line A stars.
S. W. Roby.
Bull. Am. Astron. Soc., Vol. 16, No. 4, p. 893 – 894 (1984). Abstract. – See Abstr. 010.062.

114.104   Combined low resolution ultraviolet and optical spectrophotometry.
M. H. Slovak, A. D. Code, M. R. Meade.
Bull. Am. Astron. Soc., Vol. 16, No. 4, p. 894 (1984). Abstract. –
See Abstr. 010.062.

114.105   Optical studies of Wolf–Rayet (carbon and oxygen) stars.
A. V. Torres.
Bull. Am. Astron. Soc., Vol. 16, No. 4, p. 895 (1984). Abstract. –
See Abstr. 010.062.

114.106   High–resolution spectra of southern carbon stars.
J. H. Black, E. F. van Dishoeck.
Bull. Am. Astron. Soc., Vol. 16, No. 4, p. 895 (1984). Abstract. –
See Abstr. 010.062.

114.107   Discovery of carbon stars in the direction of the galactic center.
M. Azzopardi, J. Lequeux, E. Rebeirot.
Bull. Am. Astron. Soc., Vol. 16, No. 4, p. 911 (1984). Abstract. –
See Abstr. 010.062.

114.108   Discovery of two unique, subluminous, hot stars.
D. Crampton, A. P. Cowley, E. M. Sion.
Bull. Am. Astron. Soc., Vol. 16, No. 4, p. 966 (1984). Abstract. –
See Abstr. 010.062.

114.109   Moderate–resolution ground–based spectrophotometry of hot, evolved stars detected in the S201 far–UV survey.
H. M. Heckathorn, C. B. Opal.
Bull. Am. Astron. Soc., Vol. 16, No. 4, p. 966 (1984). Abstract. –
See Abstr. 010.062.

114.110   Spectroscopic and photometric results from a complete magnitude limited survey of UVX and faint blue stars.
S. B. Howell, K. J. Mitchell.
Bull. Am. Astron. Soc., Vol. 16, No. 4, p. 966 (1984). Abstract. –
See Abstr. 010.062.

114.111   A southern survey for extremely metal–poor stars.
T. C. Beers, S. A. Shectman, G. W. Preston.
Bull. Am. Astron. Soc., Vol. 16, No. 4, p. 970 (1984). Abstract. –
See Abstr. 010.062.

114.112   High–dispersion spectroscopy of barium–poor stars.
J. B. Laird, D. J. Bord.
Bull. Am. Astron. Soc., Vol. 16, No. 4, p. 973 (1984). Abstract. –
See Abstr. 010.062.

114.113   A possible explanation for the Bond–Neff depression in barium stars.
A. McWilliam, V. V. Smith.
Bull. Am. Astron. Soc., Vol. 16, No. 4, p. 973 (1984). Abstract. –
See Abstr. 010.062.

114.114   Calcium chloride in the IUE spectra of carbon stars.
P. D. Bennett, H. R. Johnson.
Bull. Am. Astron. Soc., Vol. 16, No. 4, p. 973 (1984). Abstract. –
See Abstr. 010.062.

114.115   Elemetal abundances in early–type stars.
D. S. Leckrone, S. J. Adelman.
Bull. Am. Astron. Soc., Vol. 16, No. 4, p. 973 – 974 (1984). Abstract. – See Abstr. 010.062.

114.116   Ultraviolet spectra of field horizontal–branch A–type stars.
A. G. D. Philip, D. P. Huenemoerder, D. S. Hayes.
Bull. Am. Astron. Soc., Vol. 16, No. 4, p. 974 (1984). Abstract. –
See Abstr. 010.062.

114.117   Heavy element abundance variations in Population II stars.
C. Sneden, C. A. Pilachowski.
Bull. Am. Astron. Soc., Vol. 16, No. 4, p. 999 (1984). Abstract. –
See Abstr. 010.062.

114.118   Spectrophotometric studies of faint Luyten proper motion stars.
C. C. Dahn, J. W. Liebert, P. C. Boeshaar, R. G. Probst.
Bull. Am. Astron. Soc., Vol. 16, No. 4, p. 1014 (1984). Abstract. – See Abstr. 010.062.

114.119   Remote OB stars in Puppis.
P. B. Stetson, M. P. FitzGerald.
Astron. J., Vol. 90, No. 6, p. 1060 – 1075 (1985). = Contrib. Univ. Waterloo Obs., No. 106.
Spectroscopic observations are presented for 54 remote OB stars in Puppis, for which *UBV* photometry has been published by Reed and FitzGerald. The combination of MK spectral classification and the photometric data suggests that the overwhelming majority of these stars lie at a common distance ~6 kpc from the Sun, and may form a physical association.

114.120   Spectrophotometric study of Be stars.
P. S. Goraya.
Astrophys. Space Sci., Vol. 112, No. 2, p. 325 – 336 (1985).
A large sample of Be stars has been studied spectrophotometrically in the visible region. The continuum energy distribution data for 23 Be stars included in the list of Harmanec et al. (1983) are presented and discussed in the wavelength range λλ3200 Å – 8000 Å. For 15 Be stars the observations reported in the present work are new. By comparing the observed continua with models, the effective temperatures of these stars have been estimated. It is found that, in general Be stars have lower effective temperature than the corresponding normal B stars.

114.121   Comparison of observations of some stars in the infrared region with catalogue data.
V. V. Dragomiretskij.
Mater. Konf. mol. uchenykh Odess. astron. nauchn.–proizv. akad.–univ. kompleksa, Odessa, 15 fev. 1983. Odessa, p. 42 – 47 (1983). In Russian. Abstr. in Ref. Zh., 51. Astron., 5.51.412 (1985).

114.122   Investigation of the star HD 19845 with the curve–of–growth method.
T. V. Mishenina.
Mater. Konf. mol. uchenykh Odess. astron. nauchn.–proizv. akad.–univ. kompleksa, Odessa, 15 fev. 1983. Odessa, p. 48 – 51 (1983). In Russian. Abstr. in Ref. Zh., 51. Astron., 5.51.428 (1985).

114.123   Rapid spectral variability of stars.
S. V. Marchenko.
Inst. teor. fiz. Akad. Nauk USSR. Prepr., No. 80 R, 27 pp. (1984). In Russian. Abstr. in Ref. Zh., 51. Astron., 5.51.500 (1985).

114.124   Taxonomie.
   C. Jaschek.
La composition chimique des étoiles dans le voisinage solaire,
p. 5 – 12 (1985). – See Abstr. 012.067.

114.125   Etoiles riches en métaux.
   G. Cayrel de Strobel.
La composition chimique des étoiles dans le voisinage solaire,
p. 13 – 26 (1985). – See Abstr. 012.067.
   With the help of the [Fe/H] Catalogue the author has discussed
two groups of metal–rich stars. The first one is composed by G
and K dwarfs and subgiants, the second one by bright yellow
luminosity class III, II and I giants. If the metal enrichment of the
first group is certainly real, that of the second group is doubtful
and to a great extent probably due to weaknesses in the theoretical treatment of their model atmospheres.

**114.126   Abondances NLTE de Ca, Mg, C, Si dans quelques
            étoiles A.**
R. Freire Ferrero.
La composition chimique des étoiles dans le voisinage solaire,
p. 27 – 47 (1985). – See Abstr. 012.067.
   Some results on Ca, Mg, C and Si abundances are given for
some A stars of the solar neighbourhood. The abundances are
determined via the comparison of computed and observed line
resonance profiles, in the framework of the NLTE theory, with
atom models reduced to a few atomic levels. The obtained NLTE
abundance values are then compared to the LTE ones obtained
from the literature. Finally, the possibility to enlarge this method
for abundance determinations and also its limitations and constraints are discussed.

114.127   Les étoiles pauvres en métaux.
   M. Spite.
La composition chimique des étoiles dans le voisinage solaire,
p. 49 – 60 (1985). – See Abstr. 012.067.
   The metal poor stars are old stars which were formed a long
time ago when the matter in the Galaxy was poor in heavy
elements. The determination of the chemical composition of the
stars as a function of their age helps us to understand how the
different elements which exist today in the Universe were built.
There has been rapid progress recently in this area and the author
tries to give an idea of the present state of the problem.

**114.128   Mesure des abondances relatives en métaux légers dans
            les étoiles de population II.**
P. François.
La composition chimique des étoiles dans le voisinage solaire,
p. 61 – 69 (1985). – See Abstr. 012.067.
   Abundance of light metals (Al–Na–Mg–Si) in population II
stars enables to test the different scenarios of carbon burning.
Dwarf stars are particularly interesting because their atmospheres have not yet been contaminated by the yields of their own
nucleosynthesis. The author presents a study bearing on 32 dwarf
stars observed at the 3.6 m CFH telescope and the 1.4 m CAT
telescope of ESO.

114.129   Technetium abundances in red giant atmospheres.
   T. A. Kipper, M. A. Kipper.
Sov. Astron. Lett., Vol. 10, No. 6, p. 363 – 365 (1984). English
translation of 38.114.065.

114.130   IUE spectra of G0 V – G5 V solar–type stars.
   B. M. Haisch, G. Basri.
Astrophys. J., Suppl. Ser., Vol. 58, No. 1, p. 179 – 193 (1985).
   One approach to the study of stellar activity is to investigate
the properties of solar twins. The authors thus present an atlas of
IUE short–wavelength spectra for a set of bright G0 V – G5 V
stars. These are shown to manifest a range of qualitatively different chromospheric and transition region spectra; in particular
significant differences are found in radiative fluxes originating at
the temperature minimum. A comprehensive survey of observational data and physical parameters has been compiled and tabulated as a reference compendium.

114.131   Ultraviolet line identifications for Tau Scorpii.
   J. B. Rogerson Jr., M. W. Ewell Jr.
Astrophys. J., Suppl. Ser., Vol. 58, No. 2, p. 265 – 287 (1985).
   Identifications are presented for absorption features found in
the ultraviolet spectrum of the B0 V star, τ Scorpii. The spectrum, produced by the Copernicus satellite, covers the wavelengths from 949 to 1560 Å. A correction to the first–order wavelength scale of the τ Scorpii atlas is derived.

114.132   He2–442: two objects under the same name.
   V. P. Arhipova (*V. P. Arkhipova*), V. F. Esipov,
B. F. Yudin.
Astrophys. Lett., Vol. 24, No. 4, p. 205 – 209 (1985).
   Spectral observations of the peculiar object He2–442, carried
out in 1982, revealed that it represents essentially two stellar
sources – He2–442A and He2–442B. They are located at an angular distance about 6 arcsec with a north–east – south–west orientation. In the region of 4000 to 6800 Å, both sources produce
emission spectra similar to those of planetary nebulae of excitation class 3 – 4. The most prominent feature of spectrum of
He2–442B, in distinction from that of He2–442A, is the presence
of He II 4686 line in it. Photometric observations in the UBV
system have shown that He2–442A is a variable source.

114.133   Spectrophotometric study of the Be star Maia.
   L. M. Sapargalieva.
Tr. Astrofiz. Inst. Alma–Ata, Tom 42, p. 73 – 80 (1983). In Russian.
   Spectroscopic variations of the Be star Maia during January
1971 – February 1972 are described. The variations of electron
density, amount of hydrogen and helium atoms above 1 cm^2 of
the photosphere and the Balmer jump are investigated. Equivalent widths of hydrogen, helium and metal lines are calculated.

**114.134   Variation of absolute energy distribution in the spectrum
            of Pleione in the period of shell outburst.**
A. Kh. Mamatkazina, Z. A. Kudryavtseva.
Tr. Astrofiz. Inst. Alma–Ata, Tom 42, p. 81 – 105 (1983). In Russian.
   The absolute energy distributions in the spectrum of Pleione
were determined on 43 registrograms obtained during 15 nights
in 1969 – 1975. A sharp variation in the spectrum has been re-
marked between 1970 and 1972. Variations in the spectrum from
night to night with a typical time scale of about ten minutes have
been observed.

114.135   Spectrophotometric investigation of the Be star Alcyone.
   L. M. Sapargalieva.
Tr. Astrofiz. Inst. Alma–Ata, Tom 44, p. 49 – 61 (1984). In Rus-
sian.

**114.136   Identification of the telluric H$_2$O lines in the photo-
            graphic infrared spectral region of carbon stars.**
Y. Fujita.
Proc. Jpn. Acad., Ser. B, Vol. 60, No. 7, p. 217 – 221 (1984).
From Phys. Abstr., Vol. 88, No. 1257, Entry 57730 (1985).

**114.137   The abundances of hydrogen, helium and heavy elements
            in Procyon.**
V. V. Leushin.
Astron. Zh., Tom 62, Vyp. 3, p. 602 – 604 (1985). In Russian.
English translation in Sov. Astron., Vol. 29, No. 3.
   The abundances of hydrogen, helium and heavy elements
($X = 0.77, Z = 0.035, Y = 0.195$) and the evolutionary status of
Procyon are determined by means of model atmosphere analysis
and evolutionary tracks.

**114.138   Ultraviolet   spectrum   of   the   Wolf–Rayet   star
            HD 192163: identification of a great number of blended
emission lines of Fe V and Fe VI.**
T. Nugis, A. Sapar.
Pis'ma Astron. Zh., Tom 11, No. 6, p. 455 – 457 (1985). In Rus-
sian. English translation in Sov. Astron. Lett., Vol. 11.
   From the study of a high–resolution spectrum of the WN6 star
HD 192163 observed by means of the IUE satellite it has been

concluded that most of the previously unidentified spectral features in the region 1245 – 1720 Å are due to the overlapping of pure or of P Cygni–type emission lines of Fe V and Fe VI. The estimated Fe/He ratio is found to be close to the solar value.

114.139 Praktische Leuchtkraftklassifikation für Amateur-astronomen.
D. Böhme, E. Pollmann.
Sterne, 61. Band, Heft 2, p. 108 – 116 (1985).

114.140 Taxonomy of late–type giants.
C. Jaschek.
Cool stars with excesses of heavy elements, p. 3 – 14 (1985). – See Abstr. 012.101.

The giants of spectral types G, K, and M constitute the taxonomically most complex region of the HR diagram. This paper reviews spectral classification systems for normal giants, for carbon stars, S stars, barium stars, CH stars, Li stars, weak–G–band stars and CN stars.

114.141 Spectral classification and the relations between peculiar giants.
S. B. Yorka, P. C. Keenan.
Cool stars with excesses of heavy elements, p. 15 – 18 (1985). – See Abstr. 012.101.

The spectral classification of giant barium stars is briefly discussed.

114.142 The infrared spectra of red variables: II. The SC and CS stars.
R. M. Catchpole, P. A. Whitelock.
Cool stars with excesses of heavy elements, p. 19 – 23 (1985). – See Abstr. 012.101.

Representative 1 – 4 μm spectra of cool stars across the transition 1 > C/O > 1 are shown. The 3.1 μm absorption feature is seen to be a characteristic of CS as well as C stars, though not of SC or S stars and is therefore a sensitive indicator of C/O variation when C/O ≅ 1.0. The depths of the various absorption features within the spectra of CS stars differ considerably from one star to the next while the SC stars all show very similar spectra.

114.143 Spectrophotometric characteristics of barium stars.
P. K. Lu, A. R. Upgren.
Cool stars with excesses of heavy elements, p. 25 – 29 (1985). – See Abstr. 012.101.

Image–tube spectra with dispersions of 43 Å/mm at H–gamma and intermediate–band photometry on the DDO system were obtained for about 300 barium stars. A statistical study of color indices and various abundance ratios for this sample is given.

114.144 Identification of photographic infrared spectral region in carbon stars.
Y. Fujita.
Cool stars with excesses of heavy elements, p. 31 – 35 (1985). – See Abstr. 012.101.

Microphotometric tracings of the spectrograms of carbon stars with a dispersion of 8 Å/mm from the spectral region λ7800 to λ8800 are discussed in connection with the identification of ^{12}CN, ^{13}CN, HCN, [C I] λ8727, and telluric H_2O lines.

114.145 Peculiar cool stars in planetary nebulae – the spectrum of FG Sge.
A. Acker.
Cool stars with excesses of heavy elements, p. 37 – 41 (1985). – See Abstr. 012.101.

A list of 11 central stars of planetary nebulae with peculiar cool spectra is presented and the spectrum of the most remarkable of them, FG Sge is briefly discussed.

114.146 The discovery and frequency of barium stars.
W. P. Bidelman.
Cool stars with excesses of heavy elements, p. 43 – 46 (1985). – See Abstr. 012.101.

The frequency of barium stars as compared to normal giants and intermediate supergiants of classes G and K is assessed

through a study of the characteristics of HD barium–star suspects south of δ = –26° found on moderate–dispersion objective–prism plates. For stars of the type that can be found from such plates, the frequency appears to lie between one–half and one percent.

114.147 Ultraviolet spectra of N, R, and S stars.
H. R. Johnson, T. B. Ake, J. A. Eaton.
Cool stars with excesses of heavy elements, p. 53 – 57 (1985). – See Abstr. 012.101.

IUE spectra of N, R, and S stars are displayed and discussed. Carbon stars of type N have IUE spectra similar to M stars. Spectra of early R stars (R0 – R3) closely resemble those of late G or early K. Late R stars (R5 – R8) show spectra like late K or early M stars. Ultraviolet spectra of S stars show a wide variation in the strength of Mg II lines.

114.148 Calibrations of photometric indices on new temperature scales and problem of spectral classification in M and C stars.
T. Tsuji.
Cool stars with excesses of heavy elements, p. 93 – 97 (1985). – See Abstr. 012.101.

In carbon stars heavy line–blanketing effect by strong molecular absorption changes the atmospheric structure in such a way that even the most fundamental spectral features used for spectral classification are seriously disturbed. For this reason, some considerations on atmospheric structure are necessary before one can establish the physical basis for the spectral classification of carbon stars.

114.149 The chemical composition of cool stars: I – The barium stars.
D. L. Lambert.
Cool stars with excesses of heavy elements, p. 191 – 223 (1985). – See Abstr. 012.101.

The chemical composition of barium stars and their putative relatives is reviewed. Main sequence stars showing heavy element excesses, subgiant CH stars, classical and mild barium giants, Sr–Ba supergiants, and early R stars are reviewed. A cursory interpretation of the abundance anomalies in terms of the key processes of nucleosynthesis is presented. Mass transfer across a binary system is identified as a plausible explanation for composition of barium stars.

114.150 Rare earth patterns in red giants and blue dwarfs.
C. R. Cowley, R. C. Dempsey.
Cool stars with excesses of heavy elements, p. 225 – 229 (1985). – See Abstr. 012.101.

Lanthanides are intercompared in normal and chemically peculiar red giants and A dwarfs. There is a surprising similarity, much of which may be accounted for in terms of the observational techniques, the properties of the atomic spectra, an the column densities above optical depth unity. There are important differences between the CP lanthanide anomalies and those of the red giants, especially for the elements praseodymium and europium.

114.151 Statistical equilibrium of Fe I/Fe II in cool stars.
W. Steenbock.
Cool stars with excesses of heavy elements, p. 231 – 235 (1985). – See Abstr. 012.101.

The author investigates the statistical equilibrium of Fe I/Fe II in the sun and in the red giant Pollux (K0 III). The solar iron abundance is hardly affected by departures from LTE. In Pollux, non–LTE abundance corrections are 0.1 to 0.2 dex and depend on equivalent width and excitation potential.

114.152 Chemical composition of barium stars.
N. Kovacs.
Cool stars with excesses of heavy elements, p. 237 – 241 (1985). – See Abstr. 012.101.

A model atmosphere analysis of seven barium stars has been carried out, based on high–dispersion spectra. The s–process elements are found to be enhanced relative to iron by factors between 2 (HD 139195) and 30 (HD 92626). HD 65699 shows a special abundance pattern insofar as out of the s–process elements only strontium and barium are enhanced. A "mild" s–process is proposed in order to explain this result.

114.153   Abundance analysis of cool carbon stars.
K. Utsumi.
Cool stars with excesses of heavy elements, p. 243 – 247 (1985). – See Abstr. 012.101.

Curve–of–growth analyses for thirty cool carbon stars have been made, and physical parameters and relative abundances for fourteen elements have been determined. In J–type cool carbon stars, abundances of s–process elements with respect to Fe are nearly normal. In normal carbon stars of C5 to C8, heavy metals are overabundant by factors of 10 to 100, and rare–earth elements are overabundant by a factor of about 10.

114.154   Calcium chloride in the IUE spectra of carbon stars.
P. D. Bennett, H. R. Johnson.
Cool stars with excesses of heavy elements, p. 249 – 254 (1985). – See Abstr. 012.101.

An identification of CaCl in the IUE spectra of N–type carbon stars is proposed. This is based primarily on the coincidence of several strong and previously unidentified stellar features with the strongest band heads of the D–X system of CaCl. It is possible that these lines carry information about the upper photosphere and perhaps even the temperature minimum.

114.155   Niobium in R And (S6, 6e) and HR 1105 (S5, 3).
D. N. Davis.
Cool stars with excesses of heavy elements, p. 261 – 266 (1985). – See Abstr. 012.101.

Lines of the first multiplet of niobium are strong in R And and HR 1105. These lines are also present in other S stars: HR 8714, R Cam, V Cnc, R CMi, and T Sgr. They are also visible in the M stars, β Peg and μ UMa. An approximation to the abundance ratio, Nb/Fe, has been deduced from pairs of lines having nearly equal intensity. In R And the ratio is about 200 times the solar value.

114.156   C, N, O, and their isotope abundances in coolest stars of the red giant branch.
T. Tsuji.
Cool stars with excesses of heavy elements, p. 295 – 299 (1985). – See Abstr. 012.101.

A quantitative analysis based on model atmospheres is applied to a high resolution IR spectrum of α Herculis (M5 Ib–II). Abundances of CO, OH, and CN are derived and the isotopic ratios $^{12}C/^{13}C$ and $^{16}O/^{17}O$ are determined for this oxygen–rich star.

114.157   An objective prism survey of the equatorial Selected Areas in the red spectral region.
T. H. Robertson, T. M. Jordan.
Bull. Am. Astron. Soc., Vol. 17, No. 2, p. 554 (1985). Abstract. – See Abstr. 010.065.

114.158   Oxygen in intermediate–mass supergiants.
R. E. Luck, D. L. Lambert.
Bull. Am. Astron. Soc., Vol. 17, No. 2, p. 560 (1985). Abstract. – See Abstr. 010.065.

114.159   Spectroscopic comparison of "sn" stars.
L. Sun, D. N. Brown.
Bull. Am. Astron. Soc., Vol. 17, No. 2, p. 570 (1985). Abstract. – See Abstr. 010.065.

114.160   CCD spectroscopy and Wing 8–color photometry of new galactic M supergiants.
D. J. MacConnell, R. F. Wing, E. Costa.
Bull. Am. Astron. Soc., Vol. 17, No. 2, p. 595 (1985). Abstract. – See Abstr. 010.065.

114.161   Unwidened thin prism spectra of high galactic latitude fields.
A. G. D. Philip.
Bull. Am. Astron. Soc., Vol. 17, No. 2, p. 595 (1985). Abstract. – See Abstr. 010.065.

114.162   Spectroscopic properties of 89 members of the Kiso ultraviolet survey.
R. McMahan, G. Wegner.
Bull. Am. Astron. Soc., Vol. 17, No. 2, p. 598 (1985). Abstract. – See Abstr. 010.065.

114.163   Spectral classifications for late M stars in the Dearborn and IRC catalogues.
R. F. Wing, P. G. Kakaletris.
Bull. Am. Astron. Soc., Vol. 17, No. 2, p. 599 (1985). Abstract. – See Abstr. 010.065.

114.164   Ultraviolet spectra of CH stars.
H. R. Johnson, J. H. Baumert, J. A. Eaton.
Bull. Am. Astron. Soc., Vol. 17, No. 2, p. 600 (1985). Abstract. – See Abstr. 010.065.

114.165   Report of IAU Commission 29: Stellar spectra (*Spectres stellaires*).
J. Jugaku.
Trans. IAU, Vol. XIXA, p. 353 – 374 (1985). – See Abstr. 003.046.

114.166   Report of IAU Commission 45: Stellar classification (*Classification stellaire*).
V. Straižys.
Trans. IAU, Vol. XIXA, p. 645 – 652 (1985). – See Abstr. 003.046.

114.167   On the spectral classification of faint stars.
G. A. Gurzadyan.
Astrofizika, Tom 22, Vyp. 3, p. 515 – 529 (1985). In Russian. English translation in Astrophysics, Vol. 22, No. 3.

The perspectives of spectral classification of faint stars using space observations are examined. The impossibility of a simple spectral classification of faint reddened stars is shown on the basis of colorimetric data in ultraviolet (2000 – 3000 Å), as well as with the combination of ground–based observations.

114.168   Line studies of Wolf–Rayet stars: the WN subclass.
D. N. Perry.
Diss. Abstr. Int., Sect. B, Vol. 45, No. 1, p. 230 (1984). Thesis, University of Colorado, 316 pp. Order No. DA8408064.

114.169   A spectrophotometric study of temporal variations in the optical spectrum of SS433.
R. M. Wagner.
Diss. Abstr. Int., Sect. B, Vol. 45, No. 1, p. 231 (1984). Thesis, Ohio State University, 127 pp. (1984). Order No. DA8410439.

114.170   Carbon and nitrogen abundances in field dwarf stars.
J. B. Laird.
Diss. Abstr. Int., Sect. B, Vol. 45, No. 3, p. 901 (1984). Thesis, Yale University, 112 pp. (1983). Order No. DA8413097.

114.171   Heavy–element abundances in two subgiant CH stars.
K. Krishnaswamy, C. Sneden.
Publ. Astron. Soc. Pac., Vol. 97, No. 591, p. 407 – 417 (1985).

New high–resolution, high signal–to–noise–ratio spectra have been gathered for subgiant CH stars HD 4395 and HD 216219. A detailed model–atmosphere abundance analysis has been carried out, with emphasis on the determination of abundances for elements heavier than iron. The heavy–element abundance patterns in the two stars are quite similar, and appear to have been produced by neutron exposures ($\tau = 0.2$) which are less than those in the typical classical barium star. The evolutionary states of these stars are rediscussed in light of these new results.

114.172   Energy distribution in the spectra of bright stars in the near IR–range.
N. L. Alekseev, G. A. Alekseeva, A. A. Arkharov,
Yu. A. Belyaev, N. V. Bogoroditskaya, E. I. Hagen–Thorn,
V. D. Galkin, L. N. Zhukova, L. A. Kamionko, V. V. Novikov,
V. B. Novopashennyj, V. P. Pakhomov, T. A. Polozhentseva,
E. V. Ruban, Yu. N. Chistyakov.
Izv. Glav. Astron. Obs. Pulkovo, Astrometr. Astrofiz., No. 202,
p. 71 – 82 (1984). In Russian.
The energy distribution is given for 15 southern and northern
bright stars in the range 5000 – 11100 Å. A photoelectric spec-
trum scanner for measurements of stellar spectra, methods of
correction for atmospheric extinction and absolute calibration
are described. The final results listed in a table are discussed and
compared with other results.

114.173   Na D lines in late–type stars: empirical results.
D. C. Boice.
Proceedings of the Southwest Regional Conference for Astron-
omy and Astrophysics, Vol. 10, p. 23 (1985). Abstract. – See
Abstr. 012.125.

A catalogue of [Fe/H] determinations, 1984 edition.
See Abstr. 002.001.

**100,000 MK types: a mid–course look at the HD reclassification
project.**
See Abstr. 002.020.

Spectroscopic data bases.
See Abstr. 002.021.

Retrieving data for faint stars.
See Abstr. 002.022.

**A finding list for the multiplet tables of *NSRDS–NBS3*, Sections
1 – 10.**
See Abstr. 002.025.

A general catalogue of galactic S stars, second edition.
See Abstr. 002.053.

The Copernicus ultraviolet spectral atlas of Gamma Pegasi.
See Abstr. 002.059.

The IUE Low–Dispersion Spectra Reference Atlas.
See Abstr. 002.101.

A catalog of G5 – M stars in a region at the South Galactic Pole.
See Abstr. 002.105.

MK Spectral Classifications. Sixth General Catalogue.
See Abstr. 002.108.

**Photometric observing campaign of Be stars. Progress Report
No. 8.**
See Abstr. 013.058.

Spectroscopic observing campaign of Be stars.
See Abstr. 013.059.

Progress report on NASA IUE Regional Data Analysis Facilities.
See Abstr. 013.066.

**Laboratory identification of the emission features near 3.5 μm in
the pre–main–sequence star HD 97048.**
See Abstr. 022.026.

Electronic–detector arrays for spectral classification.
See Abstr. 034.033.

Spectral classification of OB stars with digital detectors.
See Abstr. 034.034.

Electronic detectors and computer–assisted spectral classification.
See Abstr. 034.035.

**Application of an intensified silicon vidicon to astronomical spec-
trophotometry.**
See Abstr. 034.171.

Calibrations of wavelengths in SWP echelle spectra.
See Abstr. 036.001.

Interactive reduction of echelle spectrograms.
See Abstr. 036.030.

An analysis of scattered light in low dispersion IUE spectra.
See Abstr. 036.044.

Progress in automation techniques for MK classification.
See Abstr. 036.063.

A program for two–dimensional classification.
See Abstr. 036.064.

**High dispersion spectroscopy trials using an echelle spectrograph
with CCD camera.**
See Abstr. 036.078.

**Investigation of random and fixed pattern noise in high dispersion
IUE spectra.**
See Abstr. 036.127.

IUE autographic spectral features – setting the limits.
See Abstr. 036.128.

Taming the Medusa: data reduction.
See Abstr. 036.129.

Spectrophotometry from uncalibrated Schmidt plates.
See Abstr. 036.154.

Some methods for searching for weak spectra.
See Abstr. 036.198.

Astrophysical significance of spectral line shape investigations.
See Abstr. 063.061.

**On the relevance of the MK system and process to the theory of
stellar atmospheres.**
See Abstr. 064.039.

Computed He II spectra for Wolf–Rayet stars.
See Abstr. 064.045.

Hydrogen–deficient atmospheres for cool carbon stars.
See Abstr. 064.059.

Stellar lineshifts induced by photospheric convection.
See Abstr. 064.088.

Model atmospheres for peculiar red giant stars.
See Abstr. 064.090.

**Model atmospheres for M (super–) giants with different abun-
dances of the heavy metals and the CNO group.**
See Abstr. 064.091.

Observational problems in spectral line formation.
See Abstr. 064.103.

Current problems of line formation in early–type stars.
See Abstr. 064.104.

**Stellar surface inhomogeneities and the interpretation of stellar
spectra.**
See Abstr. 064.105.

Analysis of Zr and Tc abundances from S–stars using the *s*–process with an exponential distribution of neutron exposures.
See Abstr. 065.092.

Comparative characteristics of stellar and sunspot spectra.
See Abstr. 072.054.

Superionized species and winds in low luminosity B and Be stars.
See Abstr. 112.004.

Coordinated ultraviolet and visual observations of ω Orionis 1982 – 1983.
See Abstr. 112.008.

The spectroscopic antics of S 18/SMC.
See Abstr. 112.009.

An episodic red–wing structure of Si IV, λ1394 Å, in γ Cas.
See Abstr. 112.010.

Chromospheric activity in M giants.
See Abstr. 112.011.

Chromospheric emission lines in high–resolution LWR spectra (2200 – 3000 Å) of Gamma Cru (M3 III) and Alpha Ori (M2 Iab).
See Abstr. 112.012.

Ultraviolet and optical spectroscopy and polarimetry of the helium weak star HD 21699: evidence for a magnetically controlled stellar wind.
See Abstr. 112.018.

Magnetospheres and winds in the helium weak stars: observations of C IV in upper main sequence CP stars.
See Abstr. 112.019.

Spektroskopische und photometrische Untersuchungen der B–Überriesen mit zirkumstellarem Staub in den Magellanschen Wolken.
See Abstr. 112.023.

Comment on the IRAS infrared spectrum of Zeta Puppis (O4 If).
See Abstr. 112.029.

Spectrophotometric investigation of Be stars.
See Abstr. 112.033.

The hybrid spectrum of the LMC hypergiant R126.
See Abstr. 112.036.

Ultraviolet spectral morphology of O–type stellar winds.
See Abstr. 112.038.

The IUE Mg II chromospheric emission feature of barium stars.
See Abstr. 112.047.

The Be stars.
See Abstr. 112.049.

On the problem of the luminous emission line stars.
See Abstr. 112.058.

Interpretation of the spectrum of Gamma Cassiopeiae from 1 to 1.7 microns.
See Abstr. 112.066.

The near–infrared spectrum of Eta Carinae.
See Abstr. 112.077.

IUE observations of Beta Pictoris: an *IRAS* candidate for a proto–planetary system.
See Abstr. 112.078.

Spectrophotometric study of X Persei.
See Abstr. 112.080.

Comet – star close encounters and transient emission line activity of stars.
See Abstr. 112.081.

Molecular lines in IRC + 10°216 and CIT 6.
See Abstr. 112.085.

The ultraviolet spectrum of the Be star HD 50138.
See Abstr. 112.086.

Near infrared spectroscopy of γ Cas: constraints on the velocity field in the envelope.
See Abstr. 112.087.

The outer atmosphere of the carbon star TX Piscium.
See Abstr. 112.095.

Temporal variations in UV spectra of the red giant C star, TW Hor.
See Abstr. 112.096.

Properties and nature of Be stars. 12. The UV line spectrum of KX And – is there a hot primary?
See Abstr. 112.097.

Three–component analysis of ultraviolet emission lines of solar–type stars.
See Abstr. 112.108.

Spectrophotometry of bright Be stars.
See Abstr. 112.121.

Field giants: selection efficiency from photometry and metallicity gradient in the galactic bulge.
See Abstr. 113.012.

The relationship between the MK system and photometric classification.
See Abstr. 113.015.

Chemically peculiar A to F stars: a photometric approach.
See Abstr. 113.016.

"Predicting" spectral classification from photometric data?
See Abstr. 113.017.

A comparison between the MK classification in the Michigan Spectral Catalogue and uvbyβ photometry.
See Abstr. 113.018.

Possible contributions of the Geneva photometric boxes to the MK classification.
See Abstr. 113.019.

Early–type MK–standard stars in the Strömgren four–color and Geneva photometric systems.
See Abstr. 113.020.

H beta line variability in magnetic Ap stars. I.
See Abstr. 113.023.

The absolute flux of six hot stars in the ultraviolet (912 – 1600 Å).
See Abstr. 113.041.

The intrinsic UV colors of O stars.
See Abstr. 113.046.

JHK photometry of carbon stars and their effective temperature.
See Abstr. 113.053.

Photometric methods of stellar abundance determinations.
See Abstr. 113.059.

Metallicity determination of F and G stars using the Walraven photometric system.
See Abstr. 113.060.

Photometric properties of peculiar red giants.
See Abstr. 113.062.

Absolute spectrophotometry of F–, G–, K–, and M–type stars.
See Abstr. 113.063.

A new photometric system for monitoring carbon star variability.
See Abstr. 113.064.

Effective temperatures, angular diameters and radii of Ap stars.
See Abstr. 115.002.

An empirical Hγ luminosity calibration for class V – III stars.
See Abstr. 115.003.

M supergiants in the Milky Way and the Magellanic Clouds: colors, spectral types, and luminosities.
See Abstr. 115.004.

A convincing M_v–W(Hγ) calibration for A and B supergiants.
See Abstr. 115.006.

A new Hγ–absolute magnitude calibration.
See Abstr. 115.007.

Which map of absolute magnitudes: Keenan or Schmidt–Kaler?
See Abstr. 115.008.

The evolutionary status of OB stars with peculiar nitrogen spectra.
See Abstr. 115.011.

Measurement of stellar integrated flux in the wavelength range 370 nm – 950 nm.
See Abstr. 115.012.

Spectral types and luminosities of supergiants in open clusters.
See Abstr. 115.022.

On the rarity of FK Com stars.
See Abstr. 116.007.

Magnetic Ap stars.
See Abstr. 116.041.

The first UV studies of the optical candidate for the X–ray source 1118–61.
See Abstr. 117.068.

Spectral observations of the composite system HD 45166 (B8 V + hot component + gas stream).
See Abstr. 117.113.

The dependence of He I 10830 absorption on X–ray luminosity in RS CVn binaries and very active F and G main–sequence stars.
See Abstr. 117.163.

The spectrum of SS433 in the stage of eruptive activity (July, 1980).
See Abstr. 117.193.

The spectroscopic study of astrometric binaries.
See Abstr. 118.028.

Spectral types of companions to variable visual double stars.
See Abstr. 118.060.

Variations of the H and K emission lines of singly ionized calcium in the eclipsing binary star system AR Lacertae.
See Abstr. 119.105.

HR 5637: the unveiling of a binary.
See Abstr. 120.003.

Binary systems among the peculiar cool stars.
See Abstr. 120.038.

A search for companions to irregular variables of type S.
See Abstr. 120.040.

Variability of ultraviolet emission lines in the carbon star TX Psc.
See Abstr. 122.001.

The fading of R Coronae Borealis.
See Abstr. 122.002.

Flares and Hα in emission stars in the region of the Orion nebula.
See Abstr. 122.097.

Observations of Hβ and He II λ4686 lines in the spectra of UV Ceti–type stars during flares.
See Abstr. 122.100.

Temporal variations in UV spectra of the red giant C star, TW Hor.
See Abstr. 122.134.

Discovery of flare activity in G 119–62.
See Abstr. 122.172.

Some properties of S Mira variables.
See Abstr. 122.189.

Carbon isotope ratios in oxygen rich Mira and SRa variables.
See Abstr. 122.190.

Excitation of some La II, Gd II and V I lines by the fluorescence mechanism in the spectra of the long–period variable o Ceti.
See Abstr. 122.191.

CPD –59°2857: a new red variable star.
See Abstr. 123.002.

A possible observational constraint on the production of lithium by galactic novae.
See Abstr. 124.015.

IUE and Voyager observations of very hot O–type subdwarfs.
See Abstr. 126.002.

Analysis of high–dispersion IUE spectra of subdwarf B stars.
See Abstr. 126.003.

IUE observations of interstellar lines.
See Abstr. 131.003.

The distribution of the local interstellar medium derived from Mg II column densities towards seven cool stars.
See Abstr. 131.068.

Ice in the Taurus molecular cloud: modelling of the 3–μm profile.
See Abstr. 131.132.

The abundance of oxygen in the interstellar medium.
See Abstr. 131.185.

Interstellare C$_2$–Linien in frühen Überriesen.
See Abstr. 131.207.

Planetoidal hypothesis of CP (*chemically peculiar*) F, A, and B star formation: possibilities and prospects.
See Abstr. 131.369.

Ultraviolet studies of the young populous cluster NGC 2100 in the LMC.
See Abstr. 153.001.

Mg II emission of late main sequence stars in open clusters.
See Abstr. 153.022.

On the helium abundance in globular cluster stars.
See Abstr. 154.032.

Chemical abundances of field horizontal–branch stars of globular clusters.
See Abstr. 154.079.

Bimodal cyanogen distributions in moderately metal–poor globular clusters. I. Carbon and nitrogen abundances for six M5 giants.
See Abstr. 154.089.

Bimodal cyanogen distributions in moderately metal–poor globular clusters. II. Evidence for "mixing" and "saturation".
See Abstr. 154.090.

A new halo survey.
See Abstr. 155.152.

Haystacks, needles and carbon stars.
See Abstr. 155.172.

Highly ionized gas associated with the Small Magellanic Cloud.
See Abstr. 156.002.

O–type stars in the Magellanic Clouds.
See Abstr. 156.021.

Wolf–Rayet stars in "lazy" galaxies: a statistical approach.
See Abstr. 157.011.

Discovery of the first S star in NGC 6822.
See Abstr. 157.123.

Discovery of the first S star in NGC 6822.
See Abstr. 157.201.

115 Luminosities, Masses, Diameters, HR and other Diagrams

115.001 On the initial masses and evolutionary origins of Wolf–Rayet stars.
R. M. Humphreys, M. Nichols, P. Massey.
Astron. J., Vol. 90, No. 1, p. 101 – 108 (1985).

Based on a study of the HR diagrams of the stellar associations and clusters with definite and probable Wolf–Rayet star members, the authors find that the lower limit to the initial masses of the Wolf–Rayet stars is greater than 30 $M_\odot$, and that 80% of them have initial masses greater than 50 $M_\odot$. No significant difference is found between the WN and WC stars for the lower limits to their initial masses. It has been suggested that some Wolf–Rayet stars have previously passed through an M supergiant stage. However, most red supergiants have evolved from stars with initial masses between 15 and 30 $M_\odot$; and therefore have too low an initial mass to evolve to the Wolf–Rayet stage. The authors also discuss the evidence for a gradient in the M supergiant to Wolf–Rayet star ratio in our galaxy and M33.

115.002 Effective temperatures, angular diameters and radii of Ap stars.
M. J. Shallis, J. E. F. Baruch, A. J. Booth, M. J. Selby.
Mon. Not. R. Astron. Soc., Vol. 213, No. 2, p. 307 – 312 (1985).

The Infrared Flux Method is used to obtain effective temperatures and angular diameters of 7 Ap stars, to an accuracy of around 4 and 8 per cent respectively. The temperatures obtained are lower than normally quoted for Ap stars but in agreement with recent results from other methods. Radii obtained for six of these stars using parallax values, combined with three from a previous study, do not confirm the expectation that Ap radii are in excess of those for normal A–type stars.

115.003 An empirical Hγ luminosity calibration for class V – III stars.
C. G. Millward, G. A. H. Walker.
Astrophys. J., Suppl. Ser., Vol. 57, No. 1, p. 63 – 76 (1985).

High signal–to–noise Reticon spectra for 87 members of eight open clusters and associations, together with 37 stars having reliable parallaxes, have been used to calibrate the $W(H\gamma) - M_v$ relation for spectral types O to early A of luminosity classes V – III. The new calibration has a mean probable dispersion of ± 0.28 mag. The distance modulus of the Pleiades, the calibration's anchor cluster, is 5.54 ± 0.06 mag. The calibration is compared to other early–type star calibrations.

115.004 M supergiants in the Milky Way and the Magellanic Clouds: colors, spectral types, and luminosities.
J. H. Elias, J. A. Frogel, R. M. Humphreys.
Astrophys. J., Suppl. Ser., Vol. 57, No. 1, p. 91 – 131 (1985).

Spectral classifications and visual and infrared photometry for over 200 red supergiants in the Magellanic Clouds are presented. Intrinsic colors for M supergiants in the Milky Way, the Large Magellanic Cloud, and the Small Magellanic Cloud are derived. The Magellanic Cloud stars are generally bluer and earlier in spectral type than Milky Way red supergiants because of lower metal abundances in the Magellanic Clouds. Luminosity effects on the intrinsic colors are also discussed.

115.005 Remarks concerning the masses of horizontal–branch A–type stars.
A. G. D. Philip.
News Lett. Astron. Soc. N.Y., Vol. 2, No. 5, p. 17 – 19 (1984). – See Abstr. 010.243.

115.006 A convincing M_v–W(Hγ) calibration for A and B supergiants.
G. A. H. Walker, C. G. Millward.
Astrophys. J., Vol. 289, No. 2, p. 669 – 675 (1985).

From Reticon spectra of 31 B0 – A5 supergiants in the direction of h and χ Persei the authors find a very close correlation between the equivalent width of Hγ and V_0, after correcting for a well–defined, linear dependence of $V_0/W(H\gamma)$ on spectral type (temperature) and of $V_0/W(H\gamma)$ on $W(H\gamma)$ within a spectral type (degree of ionization). The authors have established a tentative, absolute magnitude calibration by adopting a distance modulus of 11.8 and taking V_0 values from Wildey. Apart from the uncertainty in zero point, the probable error in M_v, from a single supergiant spectrum, is 0.18 mag which is remarkably small since many of the stars are probably not members of the double cluster.

115.007 A new Hγ–absolute magnitude calibration.
C. G. Millward, G. A. H. Walker.
The MK process and stellar classification, p. 261 – 276 (1984). – See Abstr. 012.033.

A new Hγ–absolute magnitude calibration is presented for luminosity classes III to V based on high signal–to–noise (120 to 1500) Reticon spectra. The new calibration has a mean probable dispersion of ± 0.28 magnitudes and spectral–type corrections are not necessary. Comparisons are made with calibrations in the literature. The authors derive a distance modulus of 11.11 for NGC 2244. Preliminary attempts to predict spectral types using line–depth ratios from the Reticon spectra are given.

115.008   Which map of absolute magnitudes: Keenan or Schmidt–Kaler?
C. J. Corbally, R. F. Garrison.
The MK process and stellar classification, p. 277 – 289 (1984). – See Abstr. 012.033.

Visual binaries, with new MK classifications and area–scanner photometry, have been used to decide between maps of MK classes to absolute magnitudes. The Keenan map, especially with an improved giant calibration (Egret et al., 1982), agreed with the photometry better than the Schmidt–Kaler (1965) map. A preliminary Garrison (1978) map, as yet only for main–sequence stars, was also found to be good.

115.009   Absolute–magnitude calibration of normal stars (V).
S. Grenier, A. E. Gómez, C. Jaschek, M. Jaschek, A. Heck.
Astron. Astrophys., Vol. 145, No. 2, p. 331 – 338 (1985).

A statistical–parallax algorithm based on the principle of maximum likelihood has been used to determine the absolute magnitudes of field stars belonging to the luminosity classes V and III and to the spectral range B5 – F5. The final results are gathered for apparent–magnitude and distance–limited samples, respectively. They represent a significant improvement compared to previous absolute–magnitude calibrations. Moreover, they agree with modern stellar evolution theories. The authors also rediscuss their previous results on the location of various peculiar–star groups in the HR diagram. Maximum ages for the different groups are also derived from stellar–evolution theory.

115.010   Metallicities and distances of galactic clusters as determined from UBV–data. I. The effects of metallicity and reddening on the colors of main–sequence stars.
L. M. Cameron.
Astron. Astrophys., Vol. 146, No. 1, p. 59 – 66 (1985).

Three effects of the ultraviolet excesses of stars in the $(U–B)$, $(B–V)$–plane and the M_V, $(B–V)$–plane are investigated: (1) Blanketing lines in the $(U–B)$, $(B–V)$–plane. (2) The influence of the metallicity on the absolute magnitude. (3) The effects of the interstellar reddening on a star in the $(U–B)$, $(B–V)$–plane. These three aspects are essential for any analysis of the metallicities and distances of open clusters involving UBV data.

115.011   The evolutionary status of OB stars with peculiar nitrogen spectra.
H. Schild.
Astron. Astrophys., Vol. 146, No. 1, p. 113 – 118 (1985).

More than 60% of OB stars with peculiar nitrogen spectra are members of stellar associations. This relationship and the luminosities of these stars are used to study their evolutionary status.

115.012   Measurement of stellar integrated flux in the wavelength range 370 nm – 950 nm.
A. D. Petford, S. K. Leggett, D. E. Blackwell, A. J. Booth, C. M. Mountain, M. J. Selby.
Astron. Astrophys., Vol. 146, No. 1, p. 195 – 198 (1985).

A technique is described for the measurement of stellar integrated flux over the range 370 nm to 950 nm using a spectrometer with a Reticon detector for comparison with a standard star. Data from Johnson and Mitchell 13–colour photometry are included in the analysis of observations. Absolute flux measurements are presented for 30 stars, together with values corrected for interstellar extinction.

115.013   The Malmquist correction.
C. Jaschek, A. E. Gómez.
Astron. Astrophys., Vol. 146, No. 2, p. 387 – 388 (1985).

The Malmquist correction is derived using different shapes of the luminosity function. It is shown that the value of the correction depends considerably on the exact shape adopted.

115.014   The bolometric luminosities of type II OH/IR sources.
M. W. Feast.
Observatory, Vol. 105, No. 1066, p. 85 – 89 (1985).

OH/IR variables, of known period and with absolute bolometric luminosities derived from kinematic distances, lie in the mean on an extrapolation of the Mira period–luminosity relationship to longer periods. The scatter about the P–L relationship can be accounted for by errors in the kinematic distances resulting from a velocity dispersion of 32 km/s.

115.015   Masse–Radius–Diagramm der Sterne zur Deutung und zur Realität.
K. Pilowski.
Univ. Hannover, Astron. Stn., Veröff., Nr. 16, p. 15 – 20 (1985).

115.016   R 136a and the central object in the giant H II region NGC 3603 resolved by holographic speckle interferometry.
G. Weigelt, G. Baier, R. Ladebeck.
Messenger, No. 40, p. 4 – 7 (1985).

115.017   Precision and accuracy of published stellar occultation angular diameters.
N. M. White, B. Feierman.
Bull. Am. Astron. Soc., Vol. 16, No. 4, p. 939 (1984). Abstract. – See Abstr. 010.062.

115.018   Interpreting Alpha Orionis speckle interferometry.
A. Y. S. Cheng, E. K. Hege, P. A. Strittmatter, L. Goldberg.
Bull. Am. Astron. Soc., Vol. 16, No. 4, p. 939 (1984). Abstract. – See Abstr. 010.062.

115.019   A and B supergiants as standard candles.
G. A. H. Walker, C. G. Millward.
Bull. Am. Astron. Soc., Vol. 16, No. 4, p. 947 (1984). Abstract. – See Abstr. 010.062.

115.020   The revised Yale isochrones and luminosity functions.
E. M. Green, P. Demarque, C. R. King.
Bull. Am. Astron. Soc., Vol. 16, No. 4, p. 997 (1984). Abstract. – See Abstr. 010.062.

115.021   Differences between the absolute magnitudes of main–sequence stars determined from colours $B–V$ and those determined from red colours $R–I$ or $V–I$ or $V–R$.
W. Gliese.
La composition chimique des étoiles dans le voisinage solaire, p. 85 – 87 (1985). – See Abstr. 012.067.

115.022   Spectral types and luminosities of supergiants in open clusters.
P. C. Keenan, R. E. Pitts.
Publ. Astron. Soc. Pac., Vol. 97, No. 590, p. 297 – 299 (1985).

Revised MK spectral types of supergiants and bright giants in open clusters are combined with published cluster moduli to calibrate the luminosity classes in terms of visual absolute magnitudes.

115.023   The measure of the stars.
N. Henbest.
New Sci., Vol. 104, No. 1434, p. 33 – 36 (1984). Abstr. in Phys. Abstr., Vol. 88, No. 1259, Entry 68772 (1985).

115.024   The absolute magnitude of barium stars.
C. Jaschek, M. Jaschek, S. Grenier, A. Gomez, A. Heck.
Cool stars with excesses of heavy elements, p. 185 – 188 (1985). – See Abstr. 012.101.

The authors have derived the mean absolute magnitude of a sample of "certain" and of "marginal" Ba stars. The certain Ba stars behave as normal giants, both in luminosity and kinematical properties.

115.025   **The luminosity functions of the main and degenerate sequences.**
J. W. Liebert.
Bull. Am. Astron. Soc., Vol. 17, No. 2, p. 545 (1985). Abstract. –
See Abstr. 010.065.

115.026   **Luminosity and kinematics of barium stars.**
J. Miller, P. K. Lu.
Bull. Am. Astron. Soc., Vol. 17, No. 2, p. 553 (1985). Abstract. –
See Abstr. 010.065.

115.027   **Far–UV flux variability in OB stars.**
R. S. Polidan, T. E. Carone, C. Campbell.
Bull. Am. Astron. Soc., Vol. 17, No. 2, p. 554 (1985). Abstract. –
See Abstr. 010.065.

115.028   **The luminosity function of the halo main sequence in the solar neighbourhood.**
P. C. Dawson.
Bull. Am. Astron. Soc., Vol. 17, No. 2, p. 560 (1985). Abstract. –
See Abstr. 010.065.

115.029   **Observational comparisons for evolutionary tracks of massive stars.**
W. M. Brunish.
Bull. Am. Astron. Soc., Vol. 17, No. 2, p. 603 (1985). Abstract. –
See Abstr. 010.065.

The brightest stars.
See Abstr. 003.081.

An astronomical application of deconvolution by the maximum entropy method.
See Abstr. 036.179.

Stellar evolution, nucleosynthesis and dredge–up in cool giants.
See Abstr. 065.091.

Photoelectric observations of lunar occultations of stars. Angular diameter of the star 61 δ^1Tauri.
See Abstr. 096.009.

Speckle interferometry of OH/IR stars at 5 and 10 microns.
See Abstr. 112.102.

Photometry of stars in the *uvgr* system.
See Abstr. 113.013.

The surface gravity of Arcturus from MgH lines, strong metal lines and the ionization equilibrium of iron.
See Abstr. 114.016.

The atmosphere of the Am star 15 Vul.
See Abstr. 114.044.

Observed and computed spectral flux distribution of non–supergiant O9 – G8 stars. III. Determination of T_{eff} for the stars in the Breger Catalogue.
See Abstr. 114.090.

The chemical composition of cool stars: I – The barium stars.
See Abstr. 114.149.

Nonthermal radio emission and the HR diagram.
See Abstr. 116.013.

Wolf–Rayet stars in the Magellanic Clouds. III. The WO4 + O4 V binary Sk 188 in the SMC.
See Abstr. 117.195.

Wolf–Rayet stars and binarity.
See Abstr. 117.367.

Radius determination for nine short–period Cepheids.
See Abstr. 122.040.

Surface brightness radii and distances of Cepheids and the period–radius relationship.
See Abstr. 122.041.

A generalization of the CORS method to determine Cepheid radii: theory and application.
See Abstr. 122.042.

Baade–Wesselink radii of long–period Cepheids: new observational results.
See Abstr. 122.043.

Cepheid radii from infrared photometry.
See Abstr. 122.044.

Distances and radii of classical Cepheids.
See Abstr. 122.045.

The zero–point of the Cepheid luminosity scale from a calibration of the luminosities of early–type stars.
See Abstr. 122.070.

Some masses for Population I and II Cepheids.
See Abstr. 122.080.

Ultraviolet studies of the young populous cluster NGC 2100 in the LMC.
See Abstr. 153.001.

IUE observations of blue horizontal branch globular clusters and UV–bright stars.
See Abstr. 154.001.

A deep luminosity function for 47 Tucanae.
See Abstr. 154.058.

Wolf–Rayet stars in nearby galaxies: tracers of the most massive stars.
See Abstr. 157.051.

The brightest stars in nearby galaxies. V. Cepheids and the brightest stars in the dwarf galaxy Sextans B compared with those in Sextans A.
See Abstr. 157.198.

116 Rotation, Magnetic Fields, Activity, Polarization, Radio Radiation

116.001   Narrow–band polarization measurements of late–type stars.
D. Clarke, A. Brooks.
Mon. Not. R. Astron. Soc., Vol. 212, No. 1, p. 211 – 218 (1985).

Narrow–band polarization measurements at 4870 Å and across the Ca II K line are reported for eight middle to late–type stars. For three stars (α CMi, α Boo and α^1Cen) which are regularly used as unpolarized broad–band standards, measurements were made at $\pm 10^{-4}$ accuracy or better. At 4870 Å, null results were obtained and with this passband (10 Å), these stars may be considered as narrow–band zero polarization standards. However, around the Ca II K region, the situation is different, there being a strong suspicion that α CMi exhibits a weak fluorescence polarization. A polarization dip at the K line in α Sco has been detected with 95 per cent confidence.

116.002   An evaluation of magnetic fields of some Ae/Be Herbig stars.
Yu. N. Gnedin, M. A. Pogodin.
Pis'ma Astron. Zh., Tom 11, No. 1, p. 37 – 43 (1985). In Russian. English translation in Sov. Astron. Lett., Vol. 11.

A possibility of the interpretation of observed polarization of some Ae/Be Herbig stars by scattering in the circumstellar electron envelope with magnetic field is considered. The magnetic fields have been evaluated to be $10^2 - 10^3$ gauss for various models of a gaseous envelope. An explanation of the variability of the polarization of HD 200775 by variations of the magnetic field parameters is proposed.

116.003   Development and investigation of an algorithm of determination of Ap stars magnetic field geometry by means of solving the inverse problem.
N. E. Piskunov.
Pis'ma Astron. Zh., Tom 11, No. 1, p. 44 – 50 (1985). In Russian. English translation in Sov. Astron. Lett., Vol. 11.

A mathematical formulation of the inverse problem of determination of magnetic field geometry from the polarization profiles of spectral lines is given. A solving algorithm is proposed. A set of model calculations has shown the effectiveness of the algorithm, the high precision of magnetic star model parameters obtained and also the advantages of the inverse problem method over the commonly used method of interpretation of effective field curves.

116.004   Ultraviolet radiation from stellar flares and the coronal X–ray emission for dwarf–Me stars.
J. G. Doyle, C. J. Butler.
Nature, Vol. 313, No. 6001, p. 378 – 380 (1985).

The authors correlate Einstein observations of the X–ray flux of quiescent dMe stars with the time–averaged energy emitted by flares in the Johnson–U band, showing that the X–ray energy emitted by the coronae of these stars is about an order of magnitude greater than the U–band flare energy. From the estimate of the ratio of the total radiation emitted to the U–band flux, it is possible that, if a similar amount of energy were dissipated in the stellar atmosphere, then the observed flare events could heat the coronae of these stars.

116.005   A search for short periodic variations of the Ap Si star HD 92664.
C. Mégessier, P. North, M. Burnet, A. Duquennoy.
Astron. Astrophys., Suppl. Ser., Vol. 59, No. 3, p. 485 – 490 (1985).

The Si star HD 92664 has been continuously monitored in Geneva photometry during one night around the phase of magnetic maximum in order to detect the high frequency oscillations suggested by Mégessier and Hensberge (1983). No variation with periods shorter than 50 min was detected, but a much longer, 3^h16 period might be present.

116.006   On the magnetic field estimation of the HgMn star α And.
Yu. V. Glagolevskij, I. I. Romanyuk, V. D. Bychkov, I. D. Najdenov.
Pis'ma Astron. Zh., Tom 11, No. 2, p. 107 – 111 (1985). In Russian. English translation in Sov. Astron. Lett., Vol. 11.

Six measurements of the magnetic field of the HgMn star α And have been carried out. On the basis of these data along with 5 measurements made by Borra and Landstreet (1980) the root–mean–square value of the field $\langle B_e \rangle = (33 \pm 19)$ Gauss was determined.

116.007   On the rarity of FK Com stars.
W. Hagen, R. E. Stencel.
Astron. J., Vol. 90, No. 1, p. 120 – 122 (1985).

Very high–dispersion spectra (2.5 Å mm^{-1}) were obtained of 31 southern late–type stars, predominantly early G giants, in an effort to find new rapidly rotating, active stars which would be FK Com–like. Measurements of linewidths and the strength of chromospheric Ca II K–line emission are presented, but no new star could be added to the class of "rapid rotators". Space densities and evolutionary lifetimes for FK Com stars are discussed.

116.008   Synchrotron model for a radio burst of HR 1099.
S. Borghi, F. Chiuderi–Drago.
Astron. Astrophys., Vol. 143, No. 1, p. 226 – 230 (1985).

The radio spectrum of a long–lasting burst observed in the flaring star HR 1099, in the frequency range $1.4\,\mathrm{GHz} \leqslant \nu \leqslant 15.5$ GHz, is interpreted in terms of gyrosynchrotron radiation by mildly relativistic electrons. The observed flux at lower frequencies puts a firm lower limit on the dimension of the source, $d \sim 4\text{–}5\,R_*$, making unlikely the hypothesis that the burst, at least at decimetric wavelengths, involves only a small portion of the stellar atmosphere. An electron density $N_R \propto R^{-\alpha}$ in a uniform field $B \sim 100$ G perfectly fits the observed points if $\alpha = 5$.

116.009   Activité stellaire.
A. Mangeney.
Champs magnétiques stellaires, p. 467 – 485 (1984). – See Abstr. 012.022.

116.010   Evidence for non–radiative activity in hot stars.
J. P. Cassinelli.
NASA Conf. Publ., NASA CP–2358, p. 2 – 23 (1985). – See Abstr. 012.023.

Evidence for non–radiative activity in early–type stars is reviewed by examining observations from the X–ray region to the radio domain. In particular, the paper discusses data which indicate (1) the existence of high temperature regions in stellar atmospheres, which are too hot to be heated by radiative processes; (2) the existence of gas flows from isolated regions of a stellar surface ("plumes"); (3) the existence of strong magnetic fields in stellar atmospheres.

116.011   Evidence for non–radiative activity in stars with $T_{\mathrm{eff}} < 10,000$K.
J. L. Linsky.
NASA Conf. Publ., NASA CP–2358, p. 24 – 46 (1985). – See Abstr. 012.023.

Major advances in the acquisition of evidence for and the understanding of nonradiative heating and other activity in stars cooler than $T_{\mathrm{eff}} = 10,000$K has occurred in the last few years primarily as a result of the IUE and Einstein spacecraft and the VLA microwave facility. The author critically reviews this evidence and comments on the trends that are now becoming apparent. The existence for nonradiatively heated outer atmospheric layers (chromospheres, transition regions, and coronae) in dwarf stars cooler than spectral type A7, in F and G giants, pre–main sequence stars, and close binary systems is unambiguous, and chromospheres exist in the K and M giants and supergiants. The existence of nonradiative heating in the outer layers of the A stars remains undetermined despite repeated searches at all wavelengths.

116.012 Observations of nonthermal radio emission from early–type stars.
D. C. Abbott, J. H. Bieging, E. Churchwell.
NASA Conf. Publ., NASA CP–2358, p. 47 – 52 (1985). – See Abstr. 012.023.

As a part of a wider survey of radio emission from O, B, and Wolf–Rayet (WR) stars, the authors discovered five new stars whose radio emission is dominated by a nonthermal mechanism of unknown origin. The characteristics of this new class of nonthermal radio emitters are (1) a high radio luminosity, typically 10^{19}ergs s^{-1}Hz^{-1}, (2) a spectral index α ranging from 0.0 to –0.7, (3) lack of pronounced variability, and (4) no measurable polarization.

116.013 Nonthermal radio emission and the HR diagram.
D. M. Gibson.
NASA Conf. Publ., NASA CP–2358, p. 70 – 74 (1985). – See Abstr. 012.023.

Perhaps the most reliable indicator of non–radiative heating/momentum in a stellar atmosphere is the presence of nonthermal radio emission. To date, 77 normal stellar objects have been detected and identified as nonthermal sources. This paper presents a list of the objects and discusses the HR diagram of normal nonthermal radio stars.

116.014 Theory tested by means of the stars.
L. Golub.
NASA Conf. Publ., NASA CP–2358, p. 106 – 120 (1985). – See Abstr. 012.023.

Recent X–ray observations of M dwarfs and Pleiades stars are used to discuss the relation between surface magnetic fields and the heating of stellar coronae. The viability of the α–ω dynamo model for stellar magnetic fields is then examined.

116.015 Short time scale periodicity in Hα emission from the main–sequence star H II 1883.
G. W. Marcy, D. K. Duncan, R. D. Cohen.
Astrophys. J., Vol. 288, No. 1, p. 259 – 265 (1985).

Hα emission having a full width at its base of 700 km s^{-1} is observed in the main–sequence K star, H II 1883. Both the centroid and the width of the Hα emission vary synchronously with the 0.24 day rotation period of this Pleiades star. Based on the ratio Hα/Hβ, the absence of forbidden lines, and the inferred gas kinematics, this star must have an extended emission region of order one stellar radius in size. The periodic Doppler shifts of Hα probably arise from a wind with oppositely directed streams.

116.016 The photometric effect of rotation in the A stars.
G. W. Collins II, R. C. Smith.
Mon. Not. R. Astron. Soc., Vol. 213, No. 3, p. 519 – 552 (1985).

The authors have used detailed stellar atmosphere models to compute for the first time the photometric effects of differential as well as rigid rotation; results are given for both the Johnson and the Strömgren systems for late B, A and early F stars; they have obtained statistical results for an ensemble of stars in that spectral range, for comparison with the observed properties of galactic clusters. The well–known qualitative result is confirmed that differential rotation produces a larger scatter in photometric diagrams than does rigid rotation, but the authors find a number of quantitative subtleties that significantly complicate any comparison with observations. Photometry alone can only put rather weak constraints on the degree of differential rotation within stars.

116.017 Determining the spatial orientation of stellar rotation axes.
D. R. Soderblom.
Publ. Astron. Soc. Pac., Vol. 97, No. 587, p. 57 – 59 (1985).

Recent proposals for determining the spatial orientation of stellar rotation axes are discussed and are shown to be impractical.

116.018 Polarization measurements of late–type stars at Hβ and Ca II K.
D. Clarke, H. E. Schwarz, B. G. Stewart.
Astron. Astrophys., Vol. 145, No. 1, p. 232 – 240 (1985).

Polarization measures at the Hβ and Ca II K lines of 25 late–type stars (mainly K spectral type) are assessed for intrinsic polarizations and polarimetric changes across their spectral features. The detectivity of normalised Stokes parameters ranges from about ± 0.0002 to ± 0.0020 for Hβ and from ± 0.0005 to ± 0.0050 for Ca II K.

116.019 A search for magnetic fields in Be stars.
P. K. Barker, J. D. Landstreet, J. M. Marlborough, I. B. Thompson.
Astrophys. J., Vol. 288, No. 2, p. 741 – 745 (1985).

No mean longitudinal magnetic fields have been detected in a sample of 15 Be and related stars, with $\sigma \approx 70 - 250$ gauss. These objects are thus magnetically indistinguishable from normal upper main–sequence stars. However, large surface fields could escape detection, depending upon field geometry and orientation. Models for Be envelopes which require predominantly toroidal fields are not seriously constrained by these observed limits on effective field strength.

116.020 *BVR* observations of polarimetric data of stars in reflection nebulae.
L. A. Pavlova, F. K. Rspaev.
Astrofizika, Tom 22, Vyp. 1, p. 145 – 152 (1985). In Russian. English translation in Astrophysics, Vol. 22, No. 1.

Observational polarimetric data for the nucleus of reflection nebulae and field stars obtained with B,V,R filters are presented. A connetion is found between the direction of the polarization vector and the structure of the reflection nebula and the shape of the cloud in which the nebula is situated. The direction of the plane of polarization for the stars behind the filamentary structure of the nebula coincides with the direction of the filaments.

116.021 The rotation of horizontal–branch stars. III. Members of the globular cluster M4.
R. C. Peterson.
Astrophys. J., Vol. 289, No. 1, p. 320 – 325 (1985).

Radial velocities and line breadths are measured for nine blue horizontal–branch stars in the globular cluster M4. The mean velocity is 72.3 ± 0.9 km s^{-1}, and the intrinsic velocity dispersion is 1.9 ± 0.3 km s^{-1}. No line broadening was detected, which indicates the complete absence of rotation as large as $v \sin i = 15$ km s^{-1} in all of the nine stars. This establishes a level of rotation in M4 lower than that found either among blue–horizontal–branch stars of the field or among those in the globular clusters M3, M5, and M13.

116.022 γ Cyg – a magnetic periodic star?
S. I. Plachinda, E. S. Dmitrienko, A. B. Severnyj.
Izv. Krymskoj Astrofiz. Obs., Tom 69, p. 42 – 50 (1984). In Russian. English translation in Bull. Crimean Astrophys. Obs., Vol. 69.

Effective longitudinal magnetic fields of the supergiant γ Cyg were measured in 1977 and 1981. The star β CrB with known variable magnetic field was observed simultaneously to control the measurements. Periodic, with period $8\overset{d}{.}2594$, variations of the longitudinal magnetic field strength in the limits –40 to +40 Gs were suspected for γ Cyg.

116.023 Coordinated multiband observations of stellar flares.
M. Rodonò, B. H. Foing, J. L. Linsky, J. C. Butler, B. M. Haisch, D. E. Gary, D. M. Gibson.
Messenger, No. 39, p. 9 – 10 (1985).

116.024   An investigation of the mean surface magnetic fields of Ap stars.
Yu. V. Glagolevskij, V. D. Bychkov, I. I. Romanyuk, N. M. Chunakova.
Astrofiz. Issled. Izv. Spets. Astrofiz. Obs., Tom 19, p. 28 – 36 (1985). In Russian. English translation in Bull. Spec. Astrophys. Obs. – North Caucasus.

The measurements of stellar magnetic fields were continued for which the strongest mean surface fields H_s have been predicted on the basis of the Geneva photometric system (the list of Cramer and Maeder). The strongest surface fields were discovered in the stars HD 147010, HD 111133 and HD 27309.

116.025   A search for the magnetic fields in stars with anomalous helium lines.
Yu. V. Glagolevskij, N. M. Chunakova.
Astrofiz. Issled. Izv. Spets. Astrofiz. Obs., Tom 19, p. 37 – 40 (1985). In Russian. English translation in Bull. Spec. Astrophys. Obs. – North Caucasus.

The results of measuring the magnetic fields in 5 helium–weak stars and one helium–rich star are presented. Fields have been detected in 2 stars of the Si spectroscopic subclass (HD 183339 and HD 217833) and confirmed in one helium–rich star (HD 184927).

116.026   Erratum: "Evidence of decay of the magnetic fields of Ap stars" [Astron. Astrophys., Suppl. Ser., Vol. 58, No. 2, p. 387 – 403 (1984)].
P. North, N. Cramer.
Astron. Astrophys., Suppl. Ser., Vol. 60, No. 2, p. 353 (1985). See Abstr. 38.116.027.

116.027   Determination of the velocity and inclination angle of rotating stars from IUE data.
M. Ruusalepp, A. Sapar, L. Sapar.
Tartu Astrofüüs. Obs. Publ., Tom 50, p. 152 – 169 (1984).

The results obtained on the rotation velocity and inclination angle of 7 rapidly rotating early–type B stars are summarized in a table.

116.028   An Einstein Observatory X–ray survey of main–sequence stars with shallow convection zones.
J. H. M. M. Schmitt, L. Golub, F. R. Harnden Jr., C. W. Maxson, R. Rosner, G. S. Vaiana.
Astrophys. J., Vol. 290, No. 1, p. 307 – 320 (1985).

The authors report the results of an X–ray survey of bright late A and early F stars on the main sequence. Only stars without any spectral peculiarities, listed in the Yale Bright Star Catalogue and observed by the Einstein Observatory, were included in the sample. The authors find significantly larger X–ray luminosities for the sample binaries than for the single stars. The nature of the X–ray emission from binary as well as from single stars is discussed and the correlations between X–ray luminosities and rotation are studied. It is argued that the observed X–ray emission in the sample stars originates from coronae, produced by magnetic dynamos in the convection zones of these stars.

116.029   Light variations of a spherical star with two hot precessing spots.
D. P. Kyurkchieva, V. G. Shkodrov.
Dokl. Bolg. Akad. Nauk, Vol. 37, No. 5, p. 553 – 555 (1984). In Russian. Abstr. in Ref. Zh., 51. Astron., 1.51.832 (1985).

116.030   Beginnings of asteroseismology.
D. Gough.
Nature, Vol. 314, No. 6006, p. 14 – 15 (1985).

This note discusses the prospects for the science of determining the internal structure of stars from the properties of dynamical oscillations, known as asteroseismology.

116.031   A possible eruptive event on M–supergiant μ Cephei.
J. Arsenijević.
Astron. Astrophys., Vol. 145, No. 2, p. 430 – 432 (1985).

A sudden fast brightening of μ Cep amounting to 0.034 magnitude in V spectral region has been observed on August 3, 1981 at 0^h31^m9 UT. The observation was done with the Belgrade polarimeter using 4^s integration intervals. Standard deviation of the measurements was ± 0.011 mag. The shape and duration of the disturbed portion of the light curve of the observed event evokes those of an ordinary M–dwarf flare.

116.032   Coronal heating and stellar flares.
D. R. Whitehouse.
Astron. Astrophys., Vol. 145, No. 2, p. 449 – 450 (1985).

A relationship between mechanical energy input into dwarf star coronae and the amount of energy released as flares is demonstrated. Since flares are generally held to be the result of magnetic field annihilation in loop structures this relationship strengthens the view that the mechanical energy into such coronae is also via magnetic loops.

116.033   On the character of light polarization of red supergiants.
R. A. Vardanyan.
Soobshch. Byurakan. Obs., Vyp. 54, p. 27 – 41 (1983). In Russian.

A detailed analysis of the nature of light polarization of red supergiants is given. It is shown that compared with the apparently faint red supergiants the apparently brighter ones have an excess of light polarization in case of the same interstellar absorption.

116.034   Excess polarization of starlight and neutral absorption.
R. A. Vardanyan.
Soobshch. Byurakan. Obs., Vyp. 54, p. 42 – 54 (1983). In Russian.

It is shown that the polarization of the light of M, O, B type stars is partly due to the neutral absorbing circumstellar clouds and clouds situated within stellar associations. In the directions of the centres of stellar associations the neutral absorption is about 1^m3 in the mean.

116.035   Did the young sun rotate at 100 km/s?
D. K. Duncan, G. W. Marcy, R. D. Cohen.
ESA Spec. Publ., ESA SP–220, p. 213 – 215 (1984). – See Abstr. 012.045.

In order to better understand the early history of the sun, the authors decided to study stars of similar mass but very young age. A useful sample is found in the Pleiades. They investigated in detail the most rapid rotator, HII 1883.

116.036   A search for differential rotation in cool stars using Ca II H and K emission.
A. C. Porter, S. L. Baliunas, J. H. Horne, R. W. Noyes, D. K. Duncan, J. Frazer, H. Lanning, A. Misch, J. Mueller, D. Soyumer, A. H. Vaughan, L. Woodard.
Bull. Am. Astron. Soc., Vol. 17, No. 1, p. 512 (1985). Abstract. – See Abstr. 010.064.

116.037   Dependence of the polarization degree of the light of cool supergiants on their I–K colour.
R. A. Vardanyan.
Astrofizika, Tom 22, Vyp. 2, p. 335 – 338 (1985). In Russian. English translation in Astrophysics, Vol. 22, No. 2.

The dependence of the polarization degree of the light of cool supergiants on their I–K colours is given. For some of these stars the mean value of the degree of interstellar polarization is estimated.

116.038   An FK Comae–like star in NGC 188.
H. C. Harris, R. D. McClure.
Publ. Astron. Soc. Pac., Vol. 97, No. 589, p. 261 – 263 (1985).

The giant–branch star I–1 in NGC 188 has broad lines, indicative of rotation with $v \sin i = 24$ km s^{-1}. During four years of observation, it has had a constant radial velocity equal to the cluster velocity. In addition, it has weak and possibly variable emission in the Ca H and K and Hα lines. In these respects, I–1 is similar to the FK Comae stars, although it exhibits slower rotation and much milder activity than the known FK Comae stars. It is likely to be a single star that has already evolved through the FK Comae state, having recently merged with a close binary companion.

116.039 On the determination of stellar rotation and differential rotation from chromospheric activity data.
R. L. Gilliland, R. Fisher.
Publ. Astron. Soc. Pac., Vol. 97, No. 589, p. 285 – 293 (1985).

Methods of determining stellar rotation and differential rotation from data sets of chromospheric activity indicators are discussed from the point of view of establishing required properties of the time series. Problems inherent in determining differential rotation and partial solutions to these difficulties are addressed. A solar data set of plage areas from the summer of 1982 is used to illustrate the technique.

116.040 Observations of the emission from some stars at millimeter wavelengths.
N. S. Nesterov.
Izv. Krymskoj Astrofiz. Obs., Tom 69, p. 71 – 78 (1984). In Russian. English translation in Bull. Crimean Astrophys. Obs., Vol. 69.

The results of observations of 5 α^2CVn–type stars, 5 emission-line stars, the object SS433 and 4 possibly related objects at 13.5 and 8.15 mm are presented.

116.041 Magnetic Ap stars.
V. L. Khokhlova.
Astrophys. Space Phys. Rev., Vol. 4, p. 99 – 159 (1985). – See Abstr. 003.020. Revised and extended English translation of 34.116.027.

Contents: (1). Introduction. (2). General properties of magnetic Ap stars. (3) Chemical inhomogeneity of the surface of magnetic Ap stars and methods for studying the chemical composition of their atmospheres. (4). Mapping of chemical elements on the surfaces of magnetic Ap stars. (5). Data on the surface structure of the magnetic field. (6). The relationship between the magnetic field and chemical anomalies. (7). Interpretation of the observed properties of magnetic Ap stars.

116.042 The Mg II emission in BD –0°4234.
S. M. Rucinski.
Observatory, Vol. 105, No. 1066, p. 77 – 80 (1985).

The strength of the Mg II emission in this metal–poor, high-velocity binary consisting of synchronized K–type dwarfs follows the same extension of the dependence on the inverse Rossby number which holds for rapidly–rotating Population I stars. This result indicates that the efficiency of the dynamo mechanism might be independent of metal abundance.

116.043 Narrow band polarimetry of a sample of cool giants and supergiants.
H. E. Schwarz.
Astron. Astrophys., Vol. 147, No. 1, p. 111 – 114 (1985).

Polarimetric observations of a sample of cool giant and supergiant stars are presented. The measurements were made with the Glasgow polarimeter using narrow wavelength bands (10 Å FWHM) centred on Ca II K and Hβ. In the data reduction the biasing effects have been taken into account using the techniques described by Clarke et al. (1983). The contribution made by the interstellar polarization (ISP) has been calculated and removed from the data. The resulting polarization values show a correlation with spectral type, the cooler types being more highly polarized. This behaviour of the degree of polarization is in qualitative agreement with the model predictions of Schwarz and Clarke (1984).

116.044 Detailed photometric and polarimetric coverage of a major 1984 mass loss episode in the Be star ω Ori.
E. F. Guinan, D. P. Hayes.
Bull. Inf. Etoiles Be, No. 11, p. 3 – 5 (1985).

116.045 Investigation of the Ap star α^2CVn magnetic field geometry from circular polarization profiles of metallic lines.
Yu. V. Glagolevskij, N. E. Piskunov, V. L. Khokhlova.
Pis'ma Astron. Zh., Tom 11, No. 5, p. 371 – 377 (1985). In Russian. English translation in Sov. Astron. Lett., Vol. 11.

The Ap star α^2CVn magnetic field model was determined from circular polarization line profiles of iron and titanium. Solution of the inverse problem was used to obtain parameters of the model, taking into account chemical inhomogeneities of the stellar surface.

116.046 Variation of magnetic field strength of V474 Mon.
Yu. S. Romanov, S. N. Udovichenko, M. S. Frolov.
Pis'ma Astron. Zh., Tom 11, No. 5, p. 378 – 382 (1985). In Russian. English translation in Sov. Astron. Lett., Vol. 11.

The observations of the δ Sct variable star V474 Mon made during three nights of September 1982 have shown the variability of the magnetic field strength occurring with the fundamental pulsation period and the amplitude of approximately 1500 Gauss.

116.047 Microwave emission from late type dwarf stars UV Ceti and YZ CMi.
M. R. Kundu, R. K. Shevgaonkar.
Bull. Am. Astron. Soc., Vol. 16, No. 4, p. 892 (1984). Abstract. – See Abstr. 010.062.

116.048 On the polarization of Alpha Orionis.
L. R. Doherty, J. P. Cassinelli.
Bull. Am. Astron. Soc., Vol. 16, No. 4, p. 892 – 893 (1984). Abstract. – See Abstr. 010.062.

116.049 A survey of rotation and chromospheric emission among solar–type stars of the solar neighborhood.
D. R. Soderblom.
Bull. Am. Astron. Soc., Vol. 16, No. 4, p. 894 (1984). Abstract. – See Abstr. 010.062.

116.050 Large and small amplitude UV activity in cool giants.
I. Oznovich, D. M. Gibson.
Bull. Am. Astron. Soc., Vol. 16, No. 4, p. 896 (1984). Abstract. – See Abstr. 010.062.

116.051 Activity cycles of lower main–sequence stars: eighteen years of research.
S. L. Baliunas, R. A. Donahue, J. H. Horne, R. W. Noyes, A. Porter, R. L. Gilliland, D. K. Duncan, J. Frazer, H. Lanning, A. Misch, J. Mueller, D. Soyumer, A. H. Vaughan, O. C. Wilson, L. A. Woodard.
Bull. Am. Astron. Soc., Vol. 16, No. 4, p. 899 (1984). Abstract. – See Abstr. 010.062.

116.052 Objective characterization of stellar cycles.
R. L. Gilliland, S. L. Baliunas.
Bull. Am. Astron. Soc., Vol. 16, No. 4, p. 899 (1984). Abstract. – See Abstr. 010.062.

116.053 VLA observations of rotationally modulated coronal emission from AU Mic.
J. J. Cox, D. M. Gibson.
Bull. Am. Astron. Soc., Vol. 16, No. 4, p. 900 (1984). Abstract. – See Abstr. 010.062.

116.054 An X–ray survey of solar–type stars.
J. A. Bookbinder, P. Majer, L. Golub, R. Rosner.
Bull. Am. Astron. Soc., Vol. 16, No. 4, p. 940 (1984). Abstract. – See Abstr. 010.062.

116.055 Red giant rotation.
D. K. Duncan.
Bull. Am. Astron. Soc., Vol. 16, No. 4, p. 941 (1984). Abstract. – See Abstr. 010.062.

116.056   Magnetic jets, helium weakness and the *sn* stars.
S. N. Shore, D. N. Brown.
Bull. Am. Astron. Soc., Vol. 16, No. 4, p. 974 (1984). Abstract. –
See Abstr. 010.062.

**116.057   The extremely rapidly rotating M–dwarf star
Gliese 890.**
A. Young, A. Skumanich, K. B. MacGregor, S. Temple.
Bull. Am. Astron. Soc., Vol. 16, No. 4, p. 1014 (1984). Ab-
stract. – See Abstr. 010.062.

**116.058   The evolution of chromospheric activity and the spin-
down of solar–type stars.**
T. Simon, G. Herbig, A. M. Boesgaard.
Astrophys. J., Vol. 293, No. 2, p. 551 – 574 (1985).

The authors report on observations made with the IUE satellite of nearly two dozen F7 – G2 dwarfs, dated on the basis of
their surface Li content in order to examine the age dependence
of ultraviolet chromospheric and transition region emission lines.
The IUE data for solar–type stars are compared with observations of T Tauri stars (both published and newly reported here)
in order to determine whether the pattern of main–sequence
chromospheric decay shown by stars older than $\sim 10^8$yr extends
back to ages of 10^6yr appropriate for T Tauri stars. The analysis
of the time decay of stellar ultraviolet and X–ray emission establishes a relationship between emission level and axial rotation,
which is expressed in terms of the Rossby parameter (here taken
to be the stellar rotation period divided by a characteristic turnover time for subphotospheric convection). It is suggested that
the evolution of main– sequence activity reflects a diminishing
coverage of stellar active regions with age.

116.059   Solar and stellar activities.
V. Bumba.
Stud. Cercet. Fiz., Vol. 36, No. 5, p. 438 – 447 (1984). In Rumanian. Abstr. in Phys. Abstr., Vol. 88, No. 1258, Entry 63315
(1985).

**116.060   The age and evolutionary status of the peculiar G–giant
FK Comae.**
C. R. Robinson, E. F. Guinan.
Bull. Am. Astron. Soc., Vol. 17, No. 2, p. 551 (1985). Abstract. –
See Abstr. 010.065.

**116.061   On the dependence of the magnetic field of chemically
peculiar stars on the rotation period.**
Yu. V. Glagolevskij.
Astrofizika, Tom 22, Vyp. 3, p. 545 – 550 (1985). In Russian. English translation in Astrophysics, Vol. 22, No. 3.

The dependence of the main surface magnetic field B_s of chemically peculiar stars on the period of rotation P was revealed. This
dependence in the range $P < 8^d$ has direct correlation while in
the range $P > 8^d$ it is reverse.

**116.062   Observations of flares at 6 and 20 cm wavelengths from
late type dwarf stars UV Ceti and AT Mic.**
M. R. Kundu, P. D. Jackson, M. Melozzi.
Bull. Am. Astron. Soc., Vol. 17, No. 2, p. 589 – 590 (1985). Abstract. – See Abstr. 010.065.

**116.063   X–ray variability in K and M dwarfs observed by *Ein-
stein*.**
C. Ambruster, S. Sciortino, L. Golub.
Bull. Am. Astron. Soc., Vol. 17, No. 2, p. 598 (1985). Abstract. –
See Abstr. 010.065.

**Point and interval estimation of the true unbiased degree of linear
polarization in the presence of low signal–to–noise ratios.**
See Abstr. 036.004.

Mesure des champs magnétiques à grande échelle.
See Abstr. 036.037.

Cross–correlation spectroscopy using CORAVEL.
See Abstr. 036.186.

Magnetic vortex tubes, jets and nonthermal sources.
See Abstr. 062.059.

**Co–rotating interaction regions in stellar winds: particle accelera-
tion and non–thermal radio emission in hot stars.**
See Abstr. 064.015.

Synchrotron emission from chaotic stellar winds.
See Abstr. 064.016.

Winds from rotating, magnetic, hot stars.
See Abstr. 064.017.

Non–radiative energy from differential rotation in hot stars.
See Abstr. 064.018.

**Magnetic effects in the heating and modification of flows in the
outer stellar atmospheres with application to early type stars.**
See Abstr. 064.020.

Synchrotron emission from chaotic stellar winds.
See Abstr. 064.037.

Extreme–ultraviolet emission from cool star outer atmospheres.
See Abstr. 064.038.

The effect of gravity on the X–ray and UV emission of cool stars.
See Abstr. 064.098.

**Stellar surface inhomogeneities and the interpretation of stellar
spectra.**
See Abstr. 064.105.

**Champ magnétique stellaire à grande échelle. Le modèle du ro-
tateur oblique.**
See Abstr. 065.022.

**Study of boundary effects of brightness variations of a spherical
star with two hot precessing spots.**
See Abstr. 065.102.

**The sun and nearby stars: microwave observations at high resolu-
tion.**
See Abstr. 076.025.

**Some results of model calculations of the solar S–component radio
emission.**
See Abstr. 077.033.

**Coordinated ultraviolet and visual observations of ω Orionis
1982 – 1983.**
See Abstr. 112.008.

Chromospheric activity in M giants.
See Abstr. 112.011.

Active late–type stars and the applicability of coronal loop models.
See Abstr. 112.013.

Co–rotating interaction regions in stellar winds.
See Abstr. 112.014.

**Rotational modulation of chromospheric emission in cool giants
and ”hybrid” stars.**
See Abstr. 112.017.

**Ultraviolet and optical spectroscopy and polarimetry of the helium
weak star HD 21699: evidence for a magnetically controlled stellar
wind.**
See Abstr. 112.018.

Magnetospheres and winds in the helium weak stars: observations of C IV in upper main sequence CP stars.
See Abstr. 112.019.

The peculiar X–ray and radio star AS431.
See Abstr. 112.025.

Is the ratio of observed X–ray luminosity to bolometric luminosity in early–type stars really a constant?
See Abstr. 112.041.

Summary of the origin of nonradiative heating/momentum in hot stars.
See Abstr. 112.043.

Rotational modulation of chromospheric emission in cool giants and "hybrid" stars.
See Abstr. 112.045.

Closed coronal structures. VI. Far–ultraviolet and X–ray emission from active late–type stars and the applicability of coronal loop models.
See Abstr. 112.054.

Chromospheric Hα emission in F8 – G3 dwarfs, and its connection with the T Tauri stars.
See Abstr. 112.055.

HD 160538 and HD 185510: two active–chromosphere stars with hot companions.
See Abstr. 112.072.

Molecular lines in IRC + 10°216 and CIT 6.
See Abstr. 112.085.

Chromospheric activity and TiO in M giants.
See Abstr. 112.104.

Magnetic structure in cool stars. VIII. The Mg II h and k surface fluxes in relation to the Mt. Wilson photometric Ca II H and K measurements.
See Abstr. 112.119.

UBVRI photometry of the southern chromospherically active star – HD 139084.
See Abstr. 112.123.

HR 1362: a chromospherically active variable with a 5–month period.
See Abstr. 112.124.

HD 217188: a long–period chromospherically active variable.
See Abstr. 112.126.

μ Centauri.
See Abstr. 112.131.

The solar stellar connection in the far ultraviolet.
See Abstr. 114.006.

Departures from LTE populations vs. anomalous abundances; the effects of heating.
See Abstr. 114.030.

Discovery of λ Bootis stars.
See Abstr. 114.057.

IUE spectra of G0 V – G5 V solar–type stars.
See Abstr. 114.130.

Study of the magnetic field of the β Lyrae system.
See Abstr. 117.035.

Synchronization in early–type close binaries.
See Abstr. 117.042.

Time–variable, excess radio emission from Cyg OB2 No. 5.
See Abstr. 117.043.

Linear polarization of symbiotic stars.
See Abstr. 117.044.

Photometric variations of HD 25267.
See Abstr. 117.080.

IUE observations of the RS CVn systems Z Her, TY Pyx and HD 155555.
See Abstr. 117.094.

Dual polarization VLBI observations of stellar binary systems at 5 GHz.
See Abstr. 117.099.

Starspot areas and temperatures in nine binary systems with late–type components.
See Abstr. 117.106.

Spectral and temporal studies of various late–type stars.
See Abstr. 117.170.

Atmospheric structures in AR Lac. I. Mapping quiescent features by occultations and Doppler imaging.
See Abstr. 117.220.

Atmospheric structures in Ar Lac. II. A spatially resolved chromospheric active region.
See Abstr. 117.221.

Atmospheric structures in AR Lac. III. VLA observations of the radio eclipses during the 1984 campaign.
See Abstr. 117.222.

Starspots and the Mg II emission of VW Cephei.
See Abstr. 117.257.

Activity of contact binary systems.
See Abstr. 117.364.

Radio emission from RS CVn binary systems.
See Abstr. 117.381.

The ellipticity effect and a migrating wave in the chromospherically active triple system V772 Her.
See Abstr. 118.057.

The rotation of the primary of Algol.
See Abstr. 119.019.

RZ Scuti as a double contact binary.
See Abstr. 119.028.

A simultaneous X–ray and radio observation of a flare from Algol.
See Abstr. 119.048.

Photometry of active Algols.
See Abstr. 119.103.

AY Ceti: a flaring, spotted star with a hot companion.
See Abstr. 120.045.

Rotation and activity of T Tauri stars.
See Abstr. 121.023.

Simultaneous observations of Ca II K and Mg II k in T Tauri stars.
See Abstr. 121.062.

The rotational velocity of the rapidly oscillating Ap star HD 83368.
See Abstr. 122.009.

Photoelectric measurements of BY Draconis carried out at the Catania Astrophysical Observatory from 1967 to 1970.
See Abstr. 122.015.

The phase variations of intrinsic polarization of ten cepheids.
See Abstr. 122.017.

The extraordinary magnetic variation of the helium–strong star HD 37776: a quadrupole field configuration.
See Abstr. 122.099.

Optical photometry and spectroscopy of the flare star Gliese 229 (= HD 42581).
See Abstr. 122.107.

Discovery of flare activity in G 119–62.
See Abstr. 122.172.

Hydrogen spectrum in magnetic white dwarfs: $H\alpha$, $H\beta$, and $H\gamma$ transitions.
See Abstr. 126.103.

Polarization measurements of the Bok globule B361.
See Abstr. 131.010.

Search for radio–continuum emission from galactic O–type and Wolf–Rayet stars off the plane.
See Abstr. 132.010.

A search for atomic hydrogen associated with planetary nebulae and evolved stars.
See Abstr. 134.066.

Flux density observations of natural radio sources for radio position astronomy.
See Abstr. 141.023.

Correlation between X–ray luminosity and the deviation from the main sequence colour diagram for stellar X–ray sources.
See Abstr. 142.064.

Radial velocities of F and G dwarfs in the Orion cluster region.
See Abstr. 152.008.

Evolution of low–mass stars in the Alpha Persei cluster.
See Abstr. 153.015.

The Einstein soft X–ray survey of the Pleiades.
See Abstr. 153.016.

Einstein X–ray survey of the Pleiades: the dependence of X–ray emission on stellar age.
See Abstr. 153.028.

Ultraviolet and X–ray observations of NGC 2264.
See Abstr. 153.050.

A deep Westerbork survey of areas with multicolor Mayall 4 m plates. III. Photometry and spectroscopy of faint source identifications.
See Abstr. 158.112.

117 Close Binaries (Observations, Theory)

117.001   Interacting binary systems.
J. Rahe.
NASA Conf. Publ., NASA CP–2349, p. 51 – 63 (1984). – See Abstr. 012.001.

The present paper summarizes the kind of information that has been derived from ultraviolet observations of interacting binary stars and points out problems of current interest.

117.002   IUE observations of the "jet" emission feature in R Aquarii.
A. G. Michalitsianos, J. M. Hollis, M. Kafatos.
NASA Conf. Publ., NASA CP–2349, p. 163 – 166 (1984). – See Abstr. 012.001.

IUE low dispersion observations of the "jet" emission feature in the symbiotic variable R Aquarii were obtained over the course of two years. A comparison SWP $\lambda\lambda1200 - 2000$ spectra obtained of both this feature and the central UV star indicates significant differences exist between these emission regions; Si III] $\lambda1893$ which is prominent in the central star is virtually absent in the "jet". Based upon analyses of UV and optical emission line spectra, the spectral properties of the feature suggest it is a highly excited tenuous region $\sim 10^4 cm^{-3}$ characterized by prominent forbidden nebular line emission.

117.003   The outburst and long term behavior of the triple period cataclysmic variable TV Col.
P. Szkody, M. Mateo.
NASA Conf. Publ., NASA CP–2349, p. 297 – 300 (1984). – See Abstr. 012.001.

The authors have analyzed 48 of their IUE spectra and 16 archival spectra of the cataclysmic variable TV Col obtained over a 4 year period in order to study variability associated with the orbital period of 5.49^h, the photometric period of 5.19^h and a longer beat period of 4.024 days. The most extreme event occurred in November, 1982 when TV Col underwent an unprecedented 2 magnitude continuum outburst in the UV and optical with large increases in the high excitation lines of He II 1640 and N V 1240. The only repetition timescale apparent from all the data appears to be the 4 days beat period. The flux distributions from UV through 2.2 μ can usually be represented by a power law with $F_\lambda \propto \lambda^{-2.3}$ and a black body component with T near 10000K.

117.004   P Cygni profiles in HL CMa.
J. C. Raymond.
NASA Conf. Publ., NASA CP–2349, p. 301 – 304 (1984). – See Abstr. 012.001.

The C IV doublet in spectra of the dwarf nova HL CMa shows strong P Cygni profiles during the late rise, peak and decline from outburst maximum. Very high velocity emission wings are seen in some spectra. The N V, Si III and Si IV lines generally show blueshifted absorption with very weak emission components. Models of the ionization state of the wind provide a basis for discussion of the wind geometry.

117.005   New results on PU Vul.
M. Friedjung, M. Ferrari–Toniolo, P. Persi, A. Altamore, A. Cassatella, R. Viotti.
NASA Conf. Publ., NASA CP–2349, p. 305 – 308 (1984). – See Abstr. 012.001.

The authors analyze coordinated UV and IR observations of the nova–like variable PU Vul made during the 1983 September active phase. A reddening of E(B–V) = 0.49 is obtained, while the 1980 fading is attributed to absorption by dust formed in the interactive binary system, containing an active warm component and an M giant.

117.006 Blue companions of Cepheids.
E. Böhm–Vitense, C. Proffitt.
NASA Conf. Publ., NASA CP–2349, p. 344 – 347 (1984). – See Abstr. 012.001.

Nineteen Cepheids, known or suspected to have blue companions, have been observed. For 11 of them, blue companions were indeed seen, though many of them were fainter in the UV than suspected. For four Population I Cepheids the suspected companions were not seen. For none of the Population II Cepheids could a companion be detected. For the observed companions the authors have determined T_{eff}, luminosities and masses from their position in the HR diagram.

117.007 Observations of cataclysmic variable star winds.
F. A. Córdova, K. O. Mason.
NASA Conf. Publ., NASA CP–2349, p. 377 – 381 (1984). – See Abstr. 012.001.

High velocity winds are common among cataclysmic variable stars (CVs) with luminous accretion disks. Here the authors summarize what has been learned about these winds from IUE spectrophotometry.

117.008 Ultraviolet spectroscopy of the interacting binary U Sagittae.
G. E. McCluskey Jr., Y. Kondo.
NASA Conf. Publ., NASA CP–2349, p. 382 – 386 (1984). – See Abstr. 012.001.

The interacting Algol–type binary U Sagittae has been observed with the IUE. Eleven high resolution spectra in the far–ultraviolet and nine high resolution spectra in the mid–ultraviolet were obtained during one orbital cycle in June 1983. The presence of Si IV and C IV indicates the existence of regions considerably hotter than a normal B8 V photosphere. The existence and behavior of these features suggests that a hot "pseudo-photosphere" is created around the B8 star by matter flowing from the G–type companion.

117.009 Ultraviolet observations of circumstellar matter in Algol type binary systems.
G. J. Peters, R. S. Polidan.
NASA Conf. Publ., NASA CP–2349, p. 387 – 390 (1984). – See Abstr. 012.001.

Properties of four components to the circumstellar matter in Algol type interacting binary systems which have been investigated with the aid of timed high resolution IUE observations are discussed. Included are the high temperature accretion region, gas stream, restricted domains of material outflow, and the general wind. In particular, the authors comment on the geometrical distribution in the system, physical conditions, flow patterns, and stability. CX Dra serves as an illustrative example and a model for the circumstellar matter in this system is presented.

117.010 The essence of binary star coalescence – a coordinated ground–based photometric and IUE ultraviolet study of FK Comae.
J. D. Dorren, E. F. Guinan.
NASA Conf. Publ., NASA CP–2349, p. 391 – 395 (1984). – See Abstr. 012.001.

It is concluded for FK Com that the light variations, the strong chromospheric and transition region line emissions, and the apparently frequent and random flare–like events that occur in the UV and visible regions are manifestations of extreme surface activity. The level of the surface activity is consistent with that expected from the period–activity relations found for the chromospherically active RS CVn variables (Bopp and Stencel 1981). The authors speculate that there may be significant non–uniform mass outflows from the star and conclude that this stage of binary–star evolution may be very short, making these kinds of objects extremely rare.

117.011 Strange F + Be binary star systems.
S. B. Parsons, B. W. Bopp, Y. Kondo.
NASA Conf. Publ., NASA CP–2349, p. 396 – 399 (1984). – See Abstr. 012.001.

The authors discuss IUE spectra of 5 stars with F or F + B optical spectra and strong Be–star like H alpha emission.

117.012 Three short–period binaries seen at high–dispersion: UX Ari, Iota Tri and HR 5110.
I. R. Little–Marenin, T. R. Ayres, T. Simon.
NASA Conf. Publ., NASA CP–2349, p. 404 – 407 (1984). – See Abstr. 012.001.

The three short–period binary systems UX Ari (P = 6.43791 days), Iota Tri (P = 14.732 days) and HR 5110 (P = 2.61328 days) represent three different evolutionary stages. The first two systems are classical RS CVn binaries whereas HR 5110 has the characteristics of an Algol system. The authors analyzed high–dispersion short–wavelength as well as long–wavelength IUE spectra of all three systems.

117.013 Physical models for the UV continua of symbiotic stars.
S. J. Kenyon.
NASA Conf. Publ., NASA CP–2349, p. 408 – 411 (1984). – See Abstr. 012.001.

Low resolution IUE spectra represent a unique opportunity to examine the hot component in a symbiotic binary. Reddening–free color indices have been developed which serve to identify the nature of the hot components in these interesting systems. The hot components are divided into two categories: accreting main sequence stars and hot stellar sources. Symbiotic stars are therefore not a homogeneous group of variables, and at least two formation mechanisms are needed to produce the observed sample.

117.014 The eclipsing binary U Sagittae: evidence for CNO processing and mass exchange in the past.
J. J. Dobias, M. J. Plavec.
NASA Conf. Publ., NASA CP–2349, p. 412 – 415 (1984). – See Abstr. 012.001.

The authors have determined the system parameters of U Sagittae. The components are of spectral types B7.5 V and G4 IV–III, and the system is at a distance of approximately 290 pc. The authors confirm the conclusion by Tomkin that the masses are about 5.5 and 1.9 solar masses, respectively. The system is a typical semidetached system of the Algol type. The profiles of the absorption lines on high–dispersion IUE spectra can be matched if solar abundances are assumed except for carbon and nitrogen. The abundance of carbon is only 9% of its solar abundance, and N is overabundant by about a factor of 4.5. It is concluded that the system underwent large mass transfer.

117.015 IUE and optical observations of LMC X–ray binaries.
L. Bianchi, M. Pakull.
NASA Conf. Publ., NASA CP–2349, p. 416 – 419 (1984). – See Abstr. 012.001.

The first IUE observations of two LMC X–ray binary candidates, LMC X–1 (star 32) and 1E0501.8 (Sk-7036) are presented. The authors derive the foreground reddening to be E(B–V) = 0.05 and the additional reddening in the LMC to be E(B–V) = 0.11 for Sk-7036 and E(B–V) = 0.32 for LMC X–1. The spectra of Sk-7036 show a significant variation of the continuum, the flux at 1300 Å being stronger at phase 0.2 by 40% compared to the flux observed at a minimum of the light curve. The latter can be represented by a model atmosphere with T_{eff} = 25000K, log g = 3.0, in agreement with the spectral type derived from the optical colors. The UV continuum of star 32 can be modeled with T_{eff} = 35000K, log g = 3.5, which is appropriate for an O7 – 9 giant.

117.016 Circumstellar material in Algols and Serpentids.
M. J. Plavec, J. J. Dobias, P. B. Etzel, J. L. Weiland.
NASA Conf. Publ., NASA CP–2349, p. 420 – 423 (1984). – See Abstr. 012.001.

The observations of the emission lines of C IV, N V, Si IV, etc. in interacting binaries of the Algol and W Serpentis types are reviewed. So far, seven "Serpentids" are known, and seven bona fide Algol–type semidetached systems have been found to display the same kind of emission lines, only on a weaker scale. Interesting differences are observed in both the absolute and relative strengths of the emission lines.

117.017 Mu Sagittarii and Beta Lyrae: combined IUE/Voyager observations.
M. J. Plavec, R. S. Polidan.
NASA Conf. Publ., NASA CP–2349, p. 424 – 427 (1984). – See Abstr. 012.001.

A hot companion to the B8 Ia supergiant μ Sagittarii was discovered by observations with three satellites: Copernicus, IUE, and Voyager. The companion is probably a B2 V star, but could be as hot as B0. Voyager observations in the region 900 – 1700 Å have been started also for β Lyrae, and show that the disk-shaped secondary is the dominating source in the EUV.

117.018 Tasks for observations of close binary systems in the extreme ultraviolet.
Z. Kopal, J. Rahe.
NASA Conf. Publ., NASA CP–2349, p. 432 – 434 (1984). – See Abstr. 012.001.

Close binary systems, which by virtue of the inclination of their orbits to the celestial sphere happen to be eclipsing variables, have in the past century contributed greatly to our knowledge of the physics of their constituent stars; and much more can be expected to be learned from them in the future because of new observations which can be carried out from spacecraft operating well outside the terrestrial atmosphere. The aims of the communication presented here are some more specific contributions which can be confidently expected from them once the sensitivity of their detectors extend beyond the Lyman limit of the hydrogen atom.

117.019 On the evolution of accretion disc flow in cataclysmic variables – III. Outburst properties of constant and uniform–α model discs.
D. N. C. Lin, J. Papaloizou, J. Faulkner.
Mon. Not. R. Astron. Soc., Vol. 212, No. 1, p. 105 – 149 (1985). = Contrib. Lick Obs., No. 437.

The authors examine the stability and evolution of some simple accretion disc models in which the viscosity is prescribed by an ad hoc uniform–α model. They are primarily concerned with systems in which the mass–input rate from the secondary to the disc around the primary, $\dot{M}_*$, is assumed to be constant. However, initial calculations with variable mass–input rates are also performed. The time–dependent visual magnitude light–curves are constructed for cataclysmic binaries with a range of disc size, primary mass, mass–input rate, and magnitude of viscosity (i.e. value of α). Comparisons are made between these theoretical results and the observed properties of various subclasses of cataclysmic variables. These results indicate that the observational differences between novae and dwarf novae may be due to differences in the mass–input rate. The authors' models can reproduce the gross observational features of U Gem–type dwarf–novae outbursts. For SS Cyg and Z Cam systems, modestly variable mass–input rates may also be required.

117.020 Dwarf novae in outburst: simultaneous ultraviolet and optical observations of RU Pegasi and TZ Persei.
C. la Dous, F. Verbunt, R. Schoembs, R. W. Argyle,
D. H. P. Jones, A. Schwarzenberg–Czerny, B. J. M. Hassall,
J. E. Pringle, R. A. Wade.
Mon. Not. R. Astron. Soc., Vol. 212, No. 1, p. 231 – 243 (1985).

The authors present ultraviolet and optical observations of the dwarf novae RU Peg and TZ Per during decline from outburst. In both systems the flux drops simultaneously at all wavelengths. In TZ Per the flux distributions during standstill and during the decline from different outbursts are similar. The authors discuss briefly the implications of these results for theoretical models.

117.021 Rapid oscillations in CPD –48°1577.
B. Warner, D. O'Donoghue, S. Allen.
Mon. Not. R. Astron. Soc., Vol. 212, No. 1, p. 9P – 13P (1985).

High–speed photometry of the recently discovered bright cataclysmic variable star CPD –48°1577 shows the occasional presence of coherent periodic oscillations of brightness with periods in the range 24 – 29 s and amplitudes ∼0.001 mag. The authors discuss the observed phase and amplitude variations of these oscillations.

117.022 The magnetic field of AM Herculis.
D. T. Wickramasinghe, B. Martin.
Mon. Not. R. Astron. Soc., Vol. 212, No. 2, p. 353 – 358 (1985).

The authors show that spectroscopic observations of the magnetic white dwarf in AM Herculis obtained in 1979 when the system was in a low state indicate that the magnetic field geometry is different from that of a centred dipole. A model based on an offset dipole field geometry gives reasonable agreement with observations and suggests that the magnetic field at the accreting pole may be significantly lower ($\sim 1.4 \times 10^7$G) than has previously been assumed.

117.023 EXOSAT soft X–ray observations of EX Hydrae.
F. A. Cordova, K. O. Mason, S. M. Kahn.
Mon. Not. R. Astron. Soc., Vol. 212, No. 2, p. 447 – 461 (1985).

During a continuous 26 hour X–ray observation of the 98–min binary EX Hydrae using the CMA detector on EXOSAT, a 67–min modulation was detected with a pulse–fraction of about 30 per cent. The phase of the modulation coincides with that of the 67–min optical modulation which has a pulse–fraction of 15 per cent. The X–ray and optical pulse shapes are similar. A modulation with a period of 98 min is also found in a power spectral analysis of the EXOSAT data. This variation has an amplitude half that of the 67–min X–ray modulation. An analysis of extensive optical data reveals that, in addition to the 67–min variation, there are two optical modulations with pulsed fractions of 5 per cent: one at half the orbital period, or 49.1 min, and one at 46.4 min. A moderate resolution ($\lambda/\Delta\lambda \sim 10 – 50$) soft X–ray spectrum of EX Hydrae in the energy range 0.04 – 2 keV was also obtained during the observation through the use of an objective transmission grating.

117.024 The emission line profiles of 2A 0311–227.
K. Mukai, P. Charles.
Mon. Not. R. Astron. Soc., Vol. 212, No. 3, p. 609 – 621 (1985).

The authors have obtained time resolved spectrophotometry of the AM Her–type system 2A 0311–227 (= EF Eri) in 1983 February. The velocities of the complex emission–line profiles are deduced from multicomponent fits and three distinct emitting regions are identified, although their location within the system is unclear. The overall behaviour of 2A 0311–227 is compared with the observations of Bailey and Ward in 1979 indicating that much of the structure is highly stable and therefore a permanent feature of the mass–transfer process.

117.025 Instability in the red star of semi–detached binary systems.
D. A. Edwards.
Mon. Not. R. Astron. Soc., Vol. 212, No. 3, p. 623 – 630 (1985).

The solution of the linear adiabatic wave equation has been found for a range of evolutionary models in both single–star and binary (line–of–centres) gravitational potentials, with Eulerian and Lagrangian surface pressure boundary conditions. The eigenvalues determining the stability of the envelope from different boundary conditions converge onto the same dynamically unstable value as the stellar surface approaches the L_1 point. The properties of this instability are shown to be well correlated with the position and existence of the hydrogen/helium I ionization zone. Off–line–of–centres calculations show that the instability patch about the L_1 point is approximately 15 scale–heights in radius for a typical 1 $M_\odot$ main–sequence star.

117.026 Dwarf novae in outburst: simultaneous ultraviolet and optical observations of VW Hydri.
A. Schwarzenberg–Czerny, M. Ward, D. A. Hanes,
D. H. P. Jones, J. E. Pringle, F. Verbunt, R. A. Wade.
Mon. Not. R. Astron. Soc., Vol. 212, No. 3, p. 645 – 655 (1985).

The authors present simultaneous spectrophotometry of the dwarf nova VW Hydri in the range 1200 – 7000 Å. The main set of observations cover one complete outburst, including the rise and the decline. Comparing these data with data from other outbursts of VW Hyi it is found that all the data can be interleaved. This underlines the similarity in the behaviour of the continuum flux distribution from outburst to outburst. In particular the authors confirm the discovery by Hassall et al. that the outburst starts at optical wavelengths and spreads later to the ultraviolet.

117.027 Simultaneous linear and circular polarimetry of EF Eri.
M. Cropper.
Mon. Not. R. Astron. Soc., Vol. 212, No. 3, p. 709 – 721 (1985).

The design and construction of a polarimeter for simultaneous high–speed linear and circular polarimetry are described. Observations are presented for the AM Her–type binary EF Eri. Modelling of the average behaviour is carried out to obtain better geometrical parameters for the inclination and magnetic colatitude of the accretion column. Linear and circular polarization curves are synthesized from the position–angle data and calculations of cyclotron beaming effects. The validity of the simple centred dipole model for the primary magnetic field is discussed in terms of the variations seen from cycle to cycle.

117.028 Erratum: "Flux density and structure variations in SS 433" [Mon. Not. R. Astron. Soc., Vol. 209, No. 4, p. 869 – 879 (1984)].
R. E. Spencer.
Mon. Not. R. Astron. Soc., Vol. 212, No. 3, p. 751 (1985). See Abstr. 38.117.025.

117.029 Magnetodynamical processes in interacting magnetospheres of RS CVn binaries.
Y. Uchida, T. Sakurai.
Unstable current systems and plasma instabilities in astrophysics, p. 281 – 286 (1985). – See Abstr. 012.002 (IAU Symp. No. 107).

Magnetodynamical processes in RS CVn binaries are discussed in the scheme of Active–Longitude–Belt picture (Uchida and Sakurai, 1983) in which the "photometric wave" is due to a number of spot pairs which emerge, drift across, and submerge in the "active longitude belt" on the K star. The formation of the corona and the origin of flares in these close binary systems having starspots are interpreted in terms of the reconnections of the magnetic flux tubes of the companion star with the emerging and submerging pairs of spots on the K star. The injection of the hot plasma into the large scale pole–to–spot connections is required to explain the extended corona with large emission measure.

117.030 Discovery of a cataclysmic variable type companion of the M3 III giant 4 Dra with IUE.
D. Reimers.
Astron. Astrophys., Vol. 142, No. 1, p. L16 – L18 (1985).

IUE observations show that the bright ($m_v \cong 5$) normal M3 III giant 4 Dra has a blue companion. The spectrum of the companion of 4 Dra bears similarity to UV spectra of magnetic white dwarfs like AM Her and cataclysmic variables or old novae with some additional low excitation lines possibly formed in the radiatively excited wind of the M giant. The spectrum of the 4 Dra companion is different from those of all other known WD companions of red giants. 4 Dra is apparently a member of a new subclass of interacting binary systems – a normal M giant plus an AM Her or cataclysmic variable type star.

117.031 HM Sagittae and V1016 Cygni in 1982 – 83: unforeseen variations in the IR spectrum of 1016 Cyg.
A. P. Ipatov, O. G. Taranova, B. F. Yudin.
Astron. Astrophys., Vol. 142, No. 1, p. 85 – 88 (1985).

During the period from 1982 to 1983, systematic photometric observations, in the $UBVRJHKLMN$ system, and spectrophotometric ones, in the range of 0.33 to 0.75 µm with $\Delta\lambda = 50\text{Å}$, of V1016 Cyg and HM Sge were carried out. They have shown that no significant variations in the photometric behavior of HM Sge have taken place over the past two years. At the same time, the ratio of the fluxes in the Hα and Hβ lines as well as in the lines of [Ne III] 3869 and [O III] 5007 was found to decrease. The latter change must be due to the decreasing density in the circumstellar envelope of HM Sge. The observations of V1016 Cyg in the summer of 1983 indicate that the energy distribution pattern in the infrared range has changed drastically. No unusual variations in the visible brightness of V1016 Cyg were noticed. The observa-

tions of V1016 Cyg are indicative of an additional thick ($\tau_J \approx 1$) dust envelope.

117.032 Theoretical interpretation of the nutational effects in SS433.
R. Cogotti, R. Ruffini, Song Doo Jong.
Astron. Astrophys., Vol. 142, No. 1, p. 124 – 130 (1985).

The nutational effects induced by a companion star on the emitting regions of SS433 are analysed within the framework of the ring model. Analytic expressions are derived for the variations of the shifted lines due to the nutation of the disk; the amplitudes and periodicities of these variations are given. Constraints on the inclination angles and masses of the system are derived. These results are compared and contrasted with those of the "twin jet" model.

117.033 The evolution of massive close binary systems with mass loss and overshooting. I. Mass transfer in systems with initial masses of $20 + 10\,M_\odot$.
C. H. B. Sybesma.
Astron. Astrophys., Vol. 142, No. 1, p. 171 – 180 (1985).

A study is made of the different mass transfer phases encountered in the evolution of a system of a $20\,M_\odot$ primary with an initial mass ratio of 0.5. Evolutionary sequences are calculated, including the effects of mass loss and overshooting from the convective core for different initial periods, all of them resulting in a case A of mass transfer. All the sequences evolve through both a case A and a case B of mass transfer and end up as Wolf–Rayet binaries of varying periods and mass ratios. It is found that initial periods smaller than 3 days result in a contact system being formed during the initial mass transfer phase. It is shown that short period WR binaries with large mass ratios can be formed without having to assume that mass is lost from the system. Semiconvection is found to be of no importance in the evolution.

117.034 SS433 – a new candidate for a black hole.
Eh. A. Antokhina, A. M. Cherepashchuk.
Pis'ma Astron. Zh., Tom 11, No. 1, p. 10 – 16 (1985). In Russian. English translation in Sov. Astron. Lett., Vol. 11.

The inverse problem in interpretation of the optical light curves of SS433 is solved in the framework of the model of a geometrically thick accretion disk around a relativistic object. It is shown that the relativistic object in the binary system SS433 is most probably a black hole.

117.035 Study of the magnetic field of the β Lyrae system.
M. Yu. Skul'skij.
Pis'ma Astron. Zh., Tom 11, No. 1, p. 51 – 54 (1985). In Russian. English translation in Sov. Astron. Lett., Vol. 11.

In addition to previous data about 110 new Zeeman spectrograms of β Lyr have been obtained. From all available data the orbital phase dependence of the effective magnetic field turns out to be quasi-sinusoidal with amplitude ± 600 gauss and the mean value 1400 gauss. A depression of the continuum at λ5200 has been found.

117.036 The symbiotic star AS 296: optical and infrared photometry in 1982 – 1983.
O. G. Taranova, B. F. Yudin.
Pis'ma Astron. Zh., Tom 11, No. 1, p. 55 – 59 (1985). In Russian. English translation in Sov. Astron. Lett., Vol. 11.

Photometric $UBVRJHK$ observations of the symbiotic star AS 296 are presented. No light variations exceeding 0^m2 are found. The cool star is classified as M5 III. The color excess is $E(B-V) = 0^m43 \pm 0^m04$. The luminosity of the hot component amounts to $\sim 200\,L_\odot$ if provided its distance to be 2.2 kpc.

117.037 An extended X–ray low state from Hercules X–1.
A. N. Parmar, W. Pietsch, S. McKechnie, N. E. White, J. Trümper, W. Voges, P. Barr.
Nature, Vol. 313, No. 5998, p. 119 – 121 (1985).

Hercules X–1 exhibits a 35–day cycle in its X–ray intensity in addition to its pulsar rotational and orbital periodicities of 1.24 s and 1.7 days respectively. The authors report here observations made with the EXOSAT Observatory between 1983 June and August that failed to detect the expected 35–day variation in X–ray intensity, although low–level extended X–ray emission was seen. The observations suggest that a temporary change in the disk structure may have occurred such that the disk was in the line of sight throughout.

117.038   An X–ray corona in SS Cygni?
A. R. King, M. G. Watson, J. Heise.
Nature, Vol. 313, No. 6000, p. 290 – 291 (1985).

A recent model for the X–ray emission of non–magnetic cataclysmic variables suggests that most of the observed hard X–ray flux originates in a radiatively–cooling quasi–hydrostatic corona surrounding the white dwarf as it accretes from its companion. The authors present here some results from a ~ 7.5–h EXOSAT observation of the dwarf nova SS Cygni in its quiescent state, which confirm many of the predictions of this model.

117.039   The origin of soft X–ray pulsations in dwarf novae at outburst and the DQ Herculis phenomenon.
A. R. King.
Nature, Vol. 313, No. 6000, p. 291 – 292 (1985).

A recent study of X–ray emission from non–magnetic cataclysmic variables concludes that most of the hard ($\gtrsim 2$ keV) flux is produced in a corona formed by the accreting material around the white dwarf. The model suggests that soft X–ray production is controlled largely by transport processes into the white–dwarf body. The author points out that, because even weak magnetic fields can exert a marked influence on transport processes, there is a natural explanation for the quasi–periodic oscillations seen in soft X–ray observations of dwarf novae at outburst.

117.040   Transient quasi–periodic oscillations in the X–ray flux of Cygnus X–3.
M. van der Klis, F. A. Jansen.
Nature, Vol. 313, No. 6005, p. 768 – 771 (1985).

The authors describe X–ray data on Cyg X–3, obtained with EXOSAT, showing transient quasi–periodic oscillations with amplitudes between 5 and 20% of the $1 - 10$ keV flux and periods in the range $50 - 1,500$ s. The oscillations persisted typically for $5 - 40$ (quasi–oscillation) cycles, occurred exclusively in the phase interval $0.0 - 0.75$, were seen during the high state and probably also during the low state of the source, and, at least once, were observed with two different periods simultaneously. It is suggested that the quasi–periodic flux variations are caused by quasi–periodic phenomena in an accretion disk which partially occults the X–ray source.

117.041   On the ultraviolet spectrum of AG Peg.
M. V. Penston, D. A. Allen.
Mon. Not. R. Astron. Soc., Vol. 212, No. 4, p. 939 – 954 (1985).

AG Peg has displayed the slowest known nova outburst. Since the system contains an M giant, it can be classified as a symbiotic star. The authors present ultraviolet spectra of AG Peg taken with the *IUE* satellite. They find a variety of regimes having characteristically different spectral features. These include a high–velocity wind in which high–ionization P Cygni lines are seen, and a nebular emission zone of lower velocity but very high electron density. These data are interpreted in terms of a thermonuclear event on the surface of a white dwarf which has gradually accreted hydrogen–rich gas from the wind of the M giant.

117.042   Synchronization in early–type close binaries.
G. Giuricin, F. Mardirossian, M. Mezzetti.
Astron. Astrophys., Suppl. Ser., Vol. 59, No. 1, p. 37 – 42 (1985).

The authors have rediscussed the synchronism between rotation and revolution in early–type (from O to F5) close binaries. Confirming that a strong tendency to synchronism extends to pairs wider than usually thought, the study further specifies that this tendency is stronger in the late A and especially in the F–type stars than in the hotter ones; but in contrast with previous contentions, it does not appear to be stronger in stars evolved from the main sequence (with respect to dwarfs) or in Am stars (with respect to normal stars).

117.043   Time–variable, excess radio emission from Cyg OB2 No. 5.
P. Persi, M. Ferrari–Toniolo, M. Tapia, M. Roth, L. F. Rodriguez.
Astron. Astrophys., Vol. 142, No. 2, p. 263 – 267 (1985).

VLA measurements of the 6 cm flux of Cyg OB2 No. 5 reveal that it has increased by a factor of 3 between 1980 and 1983. The spectral index of this excess radio emission is 0.2 ± 0.1. The radio variability of this and other O–type stars recently reported appears to have time scales of months. The evidence of a 6.6 day periodicity found at 2.2 μm indicates that the IR variability observed in the contact binary system Cyg OB2 No. 5 originates from the eclipsing stars instead of from the expanding envelope. This result places important constraints on models where the mass loss rate of the star suffers changes. It is yet unclear if the radio emission has a thermal or a nonthermal nature. A model in terms of the ejection of a dense shell by the O–type stars is discussed.

117.044   Linear polarization of symbiotic stars.
R. Schulte–Ladbeck.
Astron. Astrophys., Vol. 142, No. 2, p. 333 – 340 (1985).

Multifilter linear polarization observations of 16 northern symbiotic stars are presented. Four stars, namely AS 338, HM Sge, V1016 Cyg, R Aqr are shown to possess intrinsic polarization. In the remaining program stars, an intrinsic component could not be distinguished from the larger interstellar polarization. Intrinsic linear polarization is now detected in 8 out of 18 symbiotic stars, for which polarization measurements are published in the literature. The author suggests that the major polarization mechanism in symbiotic systems might be scattering from dust grains, located in the extended atmospheres of the late–type primaries and the circumstellar nebulae. A gradual transition from S–type (less circumstellar dust) to D–type (more circumstellar dust) symbiotic stars is proposed.

117.045   Variations in the variability of RR Tel.
A. Heck, J. Manfroid.
Astron. Astrophys., Vol. 142, No. 2, p. 341 – 345 (1985).

uvby photometric data, IUE FES measurements and AAVSO visual estimates of the very slow nova RR Tel show that this object has gone through periodic variations in visible light since it started fading in 1949 after its outburst in 1944. The period however is not constant and varies between 350 and 410 days. Besides a general increase in magnitude, the visible lightcurve is also marked by two events corresponding to sudden decreases of brightness in 1949 ($\approx 1^m$) and 1962/3 ($\approx 0.5^m$). There are also strong suspicions of occasional presence of fast variations of y.

117.046   Are galactic bulge X–ray burst sources leftovers of disrupted globular clusters?
J. van Paradijs, W. H. G. Lewin.
Astron. Astrophys., Vol. 142, No. 2, p. 361 – 362 (1985).

It has been suggested by Grindlay and Hertz (1983), and Grindlay (1984) that X–ray bulge sources outside globular clusters have been formed inside globular clusters that have subsequently been tidally disrupted. They argue that the observed proximity – in four of twelve cases – of a normal star and an X–ray burst source is observational evidence for their suggestion. In this paper the authors show that Grindlay and Hertz made an error of a factor ~ 10 in their statistical analysis, and that the alignments of the normal stars and the X–ray sources are probably chance coincidences.

117.047   The initial parameters of the semidetached binary TV Cas.
J. P. De Greve, A. C. de Landtsheer, W. Packet.
Astron. Astrophys., Vol. 142, No. 2, p. 367 – 374 (1985).

Using recently determined parameters of the Algol system TV Cas the origin of the system is discussed from a conservative as well as from a nonconservative point of view concerning mass transfer. Comparison with evolutionary computations points towards an initial mass of the loser of 3 $M_\odot$ evolving through a nonconservative early case B. The initial mass of the gainer was 2.4 $M_\odot$ indicating that 25% of the total initial mass together with

40% of the total angular momentum is lost from the system. This gainer is overluminous with respect to its mass, but does not attain the observed luminosity. The age of the system is 3.24×10^8yr.

117.048   A comparison of soft X–ray transients and dwarf novae.
J. van Paradijs, F. Verbunt.
AIP Conf. Proc., No. 115, p. 49 – 62 (1984). – See Abstr. 012.005.

The authors review some basic characteristics of soft X–ray transients (rise, decay and recurrence times, maximum luminosity and total outburst energies). Based on a literature search they compare the optical spectra of low–mass X–ray binaries with those of cataclysmic variables.

117.049   On the structure, stability and evolution of accretion disks in soft X–ray transient sources.
D. N. C. Lin, R. E. Taam.
AIP Conf. Proc., No. 115, p. 83 – 102 (1984). – See Abstr. 012.005.

Recurrent soft X–ray transient sources and dwarf novae have many remarkably similar observational properties. Both classes of cataclysmic variables are thought to be composed of close binaries with one component filling its Roche lobe. These systems undergo considerable luminosity variation on the timescale ranging from weeks to months. The outburst properties in both classes can be adequately described in terms of a model in which the outbursts are regulated by a thermal instability associated with the partially ionized regions of an accretion disk around a compact accreting component of the close binary system. The authors show how observational properties obtained in these two classes of cataclysmic variables may be combined to provide an understanding of the nature of the outbursts.

117.050   Binary period decrease by unstable orbit.
N. Shibazaki, K. Makishima, H. Inoue, T. Murakami.
AIP Conf. Proc., No. 115, p. 143 – 145 (1984). – See Abstr. 012.005.

The orbital period change due to an unstable binary orbit has been examined for 6 binary X–ray pulsars. It is shown that the orbital period decrease in Cen X–3 can be interpreted by this mechanism. SMC X–1 is also a candidate for the unstable orbit phenomenon.

117.051   Investigation of SU Ursae Majoris.
N. F. Vojkhanskaya, I. I. Nazarenko.
Astron. Zh., Tom 62, Vyp. 1, p. 81 – 86 (1985). In Russian. English translation in Sov. Astron., Vol. 29, No. 1.

A spectral investigation is made on the descending branch of a typical outburst and in the brightness minimum. The equivalent widths of all the emission lines and the amplitude of the Balmer jump vary randomly. The radial velocities vary with the period $3^h5^m \pm 10^m$ and with a period of the order of 2^h. Narrow deep minima appear sometimes on the radial velocity curve.

117.052   Synthesis of light curves of X–ray binary systems.
T. S. Khruzina.
Astron. Zh., Tom 62, Vyp. 1, p. 94 – 102 (1985). In Russian. English translation in Sov. Astron., Vol. 29, No. 1.

An algorithm of calculation of theoretic light curves of an X–ray binary system with eccentric orbit is presented. The modification of the light curve connected with the change of the orbit's orientation in space, its inclination, and also of some parameters of the accretion disc around the point source is considered.

117.053   Cool spots on the surface of the components of V523 Cas and CC Com binary systems.
G. V. Zhukov.
Pis'ma Astron. Zh., Tom 11, No. 2, p. 101 – 106 (1985). In Russian. English translation in Sov. Astron. Lett., Vol. 11.

Broadband photometry of the W UMa–type systems V523 Cas and CC Com has been analysed. It is shown that the observed changes of the light curves and B–V colour can be explained by solar–like cool spots on the surface of the primary component.

117.054   Evolution of low–mass close binaries: minimum orbital periods of dwarf binaries.
A. V. Tutukov, A. V. Fedorova, E. V. Ergma, L. R. Yungel'son.
Pis'ma Astron. Zh., Tom 11, No. 2, p. 123 – 132 (1985). In Russian. English translation in Sov. Astron. Lett., Vol. 11.

The evolution of low–mass close binaries with compact primaries and slightly evolved secondaries has been studied numerically taking into account the angular momentum loss due to a magnetic stellar wind and radiation of gravitational waves. The evolution of such systems leads to the formation of close binaries with orbital periods $\sim$20m – 80m, the mass exchange rate up to $\sim 10^{-9} M_\odot$/yr, and to the formation of a low–mass helium dwarf. Thus ultrashort–period cataclysmic and X–ray systems with periods lower than $\sim$80m can be formed.

117.055   On the dependence of accretion velocity on the dipole orientation in magnetic close binary systems.
I. L. Andronov.
Probl. Kosm. Fiz., Vyp. 19, p. 114 – 118 (1984). In Russian. – See Abstr. 003.002.

117.056   Photoelectric observations of the symbiotic stars CH Cyg, AG Dra and the peculiar object PU Vul during the maximum of their activity in the years 1981 – 1982.
D. Chochol, L. Hric, A. Skopal, J. Papoušek.
Contrib. Astron. Obs. Skalnaté Pleso, Vol. 12, p. 261 – 278 (1984).

The photoelectric U, B, V observations of the symbiotic stars CH Cyg, AG Dra and the peculiar object PU Vul, performed during maximum of activity of all three objects in the years 1981 – 1982 are published. The observations show different kinds of short and long–term photometric variations.

117.057   Does the gamma–ray radiation from SS433 originate from streaming dust particles?
H. L. Helfer, M. P. Savedoff.
News Lett. Astron. Soc. N.Y., Vol. 2, No. 7, p. 23 – 24 (1985). Abstract. – See Abstr. 010.241.

117.058   V Sagittae: Beobachtungen 1983/1984.
H. Grzelczyk.
BAV Rundbrief, 34. Jahrg., Nr. 1, p. 5 – 6 (1985).

At about JD 2445925 V Sge started another active period, which is still going on. A set of 112 out of 199 visual estimates covering the years 1983/84 is presented.

117.059   Two almost–contact semidetached systems: ZZ Aurigae and AX Virginis.
K.–C. Leung, D.–s. Zhai, Q.–y. Liu, Y.–l. Yang.
Astron. J., Vol. 90, No. 1, p. 115 – 119 (1985).

Photoelectric observations of ZZ Aur and AX Vir secured from Yunnan and Beijing observatories were analyzed by the Wilson and Devinney method. Appropriate values of mass ratios of these systems are found. The systems have a semidetached configuration where the lower–mass components are in contact with their respective Roche surfaces. Both of the higher–mass components very nearly fill their Roche lobes. They are very close to a critical contact configuration. The evolutionary state of this type of system is considered.

117.060   A radial–velocity study of V523 Cassiopeiae.
E. F. Milone, B. J. Hrivnak, W. A. Fisher.
Astron. J., Vol. 90, No. 2, p. 354 – 357 (1985). = Publ. Rothney Astrophys. Obs., No. 26.

Spectra of the short–period contact system V523 Cas have been obtained beginning in August 1981. The spectral type of the system is about K5. Velocities of the two components were obtained by cross correlating the spectra of V523 Cas with those of velocity standards of similar spectral type. A mass ratio of 0.42 ± 0.02 and velocity semiamplitudes of 96 ± 5 and 231 ± 4 km s^{-1} were derived, neglecting tidal and eclipse effects. Ca II H and K emission was observed in the system. It is associated with the more massive component.

117.061   The massive near–contact binary system V348 Carinae (HD 90707) in IC 2581.
R. W. Hilditch, T. Lloyd Evans.
Mon. Not. R. Astron. Soc., Vol. 213, No. 1, p. 75 – 83 (1985).

Light curves on the *UBV* system are presented for the first time for this B1 III + B binary system which has an orbital period of 5.6 day and is an established member of the open cluster IC 2581. Using previously published spectroscopic data, the three light curves are analysed and it is concluded that both components lie close to their respective limiting Roche surfaces. The total mass of the system is $\sim 65\,M_\odot$ and its age of $\sim 7 \times 10^6$yr, derived from evolutionary models for massive single stars, is in good agreement with the age of the cluster derived from isochrones matched to the colour–magnitude diagram.

117.062   On the tidal evolution of massive X–ray binaries: the tidal evolution time–scales for very long orbital periods.
J. C. B. Papaloizou, G. J. Savonije.
Mon. Not. R. Astron. Soc., Vol. 213, No. 1, p. 85 – 96 (1985).

The authors extend their previous calculations of tidal evolution rates for massive X–ray binaries to systems with very long orbital periods. They use a simple model in which internal gravity waves generated in the massive star by the time–dependent tidal potential propagate towards the surface layers where they are absorbed. The consequent energy dissipation and angular momentum transfer are responsible for the tidal evolution. It is found that for moderately low orbital frequencies, due to the interference of waves generated at different locations, the tidal angular momentum transfer is an oscillating function of orbital frequency.

117.063   Images of accretion discs – I. The eclipse mapping method.
K. Horne.
Mon. Not. R. Astron. Soc., Vol. 213, No. 1, p. 129 – 141 (1985).

A method of mapping the surface brightness distributions of accretion discs in eclipsing cataclysmic binaries is described and tested with synthetic eclipse data. Accurate synthetic light curves are computed by numerical simulation of the accretion disc eclipse, and images of the disc are reconstructed by maximum entropy methods. This eclipse mapping method permits powerful tests of accretion disc theory by deriving the spatial structure of discs from observational data with a minimum of model-dependent assumptions. Mass transfer rates for eclipsing cataclysmic variables can be measured by this method with an uncertainty of about a factor of two.

117.064   Secular evolution of magnetic cataclysmic variables.
A. R. King, J. Frank, H. Ritter.
Mon. Not. R. Astron. Soc., Vol. 213, No. 1, p. 181 – 189 (1985).

The authors show that the observed period distribution of magnetic cataclysmic variables can be easily understood on the basis of current theories of the secular evolution of these systems, and the formation of the period gap in particular, provided that the white dwarfs have magnetic moments μ *either* in the range $10^{33} \lesssim \mu \lesssim 10^{34}$G cm^3, *or* $\mu < 10^{30}$G cm^3, roughly as observed for isolated white dwarfs. The absence of intermediate polars with orbital periods <2 hr rules out the possibility that their fields are systematically weaker than those of the AM Her systems.

117.065   Investigation of a wind model for cataclysmic variable ultraviolet resonance line emission.
J. Drew, F. Verbunt.
Mon. Not. R. Astron. Soc., Vol. 213, No. 1, p. 191 – 213 (1985).

The authors investigate the ionization and thermal structure of a spherically symmetric, radiatively driven wind emanating from the vicinity of the white dwarf in a cataclysmic variable. It is found for plausible wind and radiation field parameters that the wind ionization is always high. Furthermore, adiabatic cooling ensures that the electron temperature declines to $T_e \lesssim 10^4$K beyond ~ 5 white–dwarf radii. Theoretically derived line fluxes are compared with the results of *IUE* observations of the eclipsing nova–like variables, UX UMa and RW Tri. The observed ratios between the N V λ1240, Si IV λ1397 and C IV λ1549 emis-sion line strengths cannot be matched. It is concluded that the ultraviolet resonance lines observed in the spectra of UX UMa and RW Tri, at least, cannot form in a single volume.

117.066   Starspot modelling of the RS CVn binary HK Lacertae.
K. Olah, J. A. Eaton, D. S. Hall, G. W. Henry, E. W. Burke Jr., C. R. Chambliss, R. E. Fried, H. J. Landis, H. Louth, T. R. Renner, H. J. Stelzer, R. P. Wasatonic.
Astrophys. Space Sci., Vol. 108, No. 1, p. 137 – 152 (1985).

Investigating more than 270 nightly mean magnitudes of the long–period RS CVn binary HK Lac, the authors can draw some conclusions about the nature of its complicated light variations. Analysis with a starspot model indicates two comparably large spots which appear to have maintained their separate identities for the last 15 yr and drifted in longitude separation from each other smoothly by only about 45°. The phase of the two spots indicates both are rotating very nearly synchronously with the orbital motion. The latitudes of the two spots are consistent with solar–type differential rotation.

117.067   The short time–scale light variability of HD 153919 revisited.
E. Gosset.
Astrophys. Space Sci., Vol. 108, No. 2, p. 323 – 335 (1985).

The author has analysed all the available sets of visual photometric data concerning the X–ray binary HD 153919. This was made in order to clear up the controversy on the possible existence or absence of the small amplitude light variability ($T \sim 90$ min). The main conclusion is that at the 99.4% confidence level, a periodicity ($\nu = 15.5$ day^{-1}) is detected in one data set. In this case, it can be safely concluded that the short time–scale variability does exist. No other self–conclusive detection can be definitely reported; nevertheless, at the 90% confidence level, the star has a tendency to vary with characteristic frequencies of $\nu \sim 15$ day^{-1}, $\nu \sim 30$ day^{-1} and possibly higher multiples.

117.068   The first UV studies of the optical candidate for the X–ray source 1118–61.
M. J. Coe, B. J. Payne.
Astrophys. Space Sci., Vol. 109, No. 1, p. 175 – 181 (1985).

The X–ray pulsator, 1118–61, discovered in 1975 by the Ariel–5 satellite, has been observed for the first time at UV wavelengths using IUE. Its UV spectrum clearly identifies the optical counterpart as a Be star similar to that seen in other X–ray pulsators. From the 2200 Å absorption feature a distance estimate of ~ 4 kpc is obtained and the carbon–IV profile reveals the mass loss rate from the primary. The discrepancy seen in other similar systems between the optical/UV mass loss rate and that needed to power the X–ray source is also observed in this source. Possible models are reviewed.

117.069   The Lyman alpha emission in W Ursae Majoris.
S. M. Rucinski, O. Vilhu, J. A. J. Whelan.
Astron. Astrophys., Vol. 143, No. 1, p. 153 – 159 (1985).

The authors present results of reductions of six low resolution IUE–SWP spectra properly exposed at Lyα during one orbital revolution of W UMa, the prototype for late–type contact binaries. Two different methods have been used to eliminate the geocoronal contribution to the emission line. The effect of the interstellar H I absorption has been modelled for rotational profiles at different orbital phases. The stellar emission seems to follow variations of the bolometric brightness so that the absorption–corrected average over the orbit is $f(\mathrm{Ly}\alpha)/f(\mathrm{bol}) = (4.2 \pm 0.6)10^{-5}$, a value about 9 times larger than for the Sun. This result indicates a fairly uniform chromosphere around the common surface of the system.

117.070 H_α observations of the enigmatic Of binary BD $+40°4220$ = Cyg OB2 No. 5 = V729 Cyg.
J. M. Vreux.
Astron. Astrophys., Vol. 143, No. 1, p. 209 – 215 (1985).

The structure of the H_α emission profile of the spectrum of BD $+40°4220$ has always been described as complex. Using high quality observations the author shows that the variations of the brightest part of the profile are periodic. Nevertheless, they cannot be accounted for by a mere redshift of the radial velocity curve of the absorption lines of the secondary. The author proposes a model in which this component of the emission profile is produced by material accreting on a limited area of the surface of the secondary.

117.071 The dynamical behaviour of SS433.
G. W. Collins II.
Mon. Not. R. Astron. Soc., Vol. 213, No. 2, p. 279 – 293 (1985). With microfiche MN 213/1.

The author develops a dynamical formalism for describing the motion of a distorted object moving under the influence of the gravitational fields present in a close binary system. The object is assumed to be centred about one of the components and to have a form or shape sustained by axial forces such as rotation in addition to tides generated by the companion. Although such a shape will not be temporally constant, the moments of inertia are expressible in terms of the canonical coordinates of the problem and vary in a periodic manner. The dynamical behaviour of this "elastic" body can be expressed in terms of periodic motion analogous to that found in a rigid body.

117.072 The continuum variability of the puzzling X–ray three–period cataclysmic variable 2A 0526–328 (TV Col).
J. M. Bonnet Bidaud, C. Motch, M. Mouchet.
Astron. Astrophys., Vol. 143, No. 2, p. 313 – 320 (1985).

A study of the continuum variability of the cataclysmic variable 2A 0526–328 (TV Col) from UV to optical is presented. The ultraviolet flux is found to be modulated with the 5.5 h spectroscopic period but not with the two other periods present in the system (5.2 h and 4 d). A 5.5 h modulation is also observed in the optical at 3650 Å with an amplitude similar to the one observed at the 5.2 h period. The energy distribution from UV to optical of the modulated flux at the 5.2 h and 5.5 h periods is well described in terms of two blackbody components with temperatures of 16,000K and less than 11,000K respectively. Infrared measurements indicate a secondary star with a temperature of 4300K and a radius of 0.75 $R_\odot$ at 500 pc compatible with an orbital period of 5.5 h. The nature of the three periods is discussed.

117.073 Radio emission from RS CVn binaries. I. VLA survey and period–radio luminosity relationship.
R. L. Mutel, J. F. Lestrade.
Astron. J., Vol. 90, No. 3, 493 – 498 (1985).

A VLA survey of radio emission from 36 close binary stellar systems with RS CVn properties is reported. Eight new sources were detected. A summary of all published reports of radio emission from RS CVn systems is presented. There appears to be a correlation between maximum radio luminosity and rotational period, with a tentative functional form $L_R \propto P^{-0.7}$. Rapid rotators (periods $\cong$ 2 days) may be underluminous compared with the extrapolated trend from longer–period systems. The luminosity–period correlation probably results from a dynamo mechanism which produces strong magnetic fields and, in turn, enhances the nonthermal radio emission. The decrease in radio luminosity at short periods may be caused by a saturation of energy deposition in the chromosphere, possibly because the surface of the active star has become covered with spotted regions.

117.074 Hydrogen alpha observations of RS Canum Venaticorum stars. IV. Gas streams in RT Lacertae.
D. P. Huenemoerder.
Astron. J., Vol. 90, No. 3, 499 – 503 (1985).

Hα spectroscopy of RT Lac with extensive phase and time coverage presents evidence for intermittent gas flow in this semi-detached RS CVn system. Observed line profiles have been compared to synthetic profiles formed from standard–star spectra. Differences between the observed and synthetic line profiles show both excess emission and excess absorption. The observations suggest mass loss through the outer Lagrangian point (L2). Radii determined from $v \cdot \sin i$ and the assumption of synchronous rotation are consistent with previously determined photometric values.

117.075 Two very similar late–type contact systems: BX and BB Pegasi.
K.–c. Leung, D.–s. Zhai, Y.–x. Zhang.
Astron. J., Vol. 90, No. 3, 515 – 521 (1985).

The two W–type systems BX and BB Peg are very similar in many respects: period of $0\overset{d}{.}280$ and $0\overset{d}{.}361$, mass ratio of 2.80, 2.81, differential temperature of 227 and 335K, overcontact of 10% and 12%, and the spectral type of the brighter component (Star 2) of G4.5 and F8. No unusual period change has been reported for the systems. With all these similarities, one may expect that the systems are almost identical. However, the light curves of BB Peg indicate that there are considerable short–time-scale variations. On the contrary, BX Peg is fairly stable.

117.076 Superausbrüche und Superhumps bei SU UMa–Sternen.
Y. Osaki.
Sterne Weltraum, 24. Jahrg., Nr. 2, p. 85 – 86 (1985).

117.077 Zwergnovae: Beobachtungsergebnisse und ihre Interpretation. V. Über die Mitarbeit von Amateurastronomen.
N. Vogt.
Sterne Weltraum, 24. Jahrg., Nr. 3, p. 143 – 146 (1985).

117.078 The eclipsing dwarf nova OY Carinae. III. Photometry during the superoutburst of January 1980.
W. Krzeminski, N. Vogt.
Astron. Astrophys., Vol. 144, No. 1, p. 124 – 132 (1985).

Extended photoelectric photometry of OY Car in super-outburst is presented (*UBVRI*, high speed *B* and infrared *H* data). A superhump period $P_s = 0\overset{d}{.}064631$ is derived which exceeds the orbital period P_0 by 2.4%. The variations of eclipse parameters and their wavelength dependence are discussed in terms of the eccentric disc model suggested previously.

117.079 Recurrence behaviour of dwarf novae: the Kukarkin–Parenago relation.
J. van Paradijs.
Astron. Astrophys., Vol. 144, No. 1, p. 199 – 201 (1985).

The amplitudes and average recurrence times of dwarf novae are correlated. The slope of the Kukarkin–Parenago relation suggests an underlying regularity in the average dwarf–nova outburst behaviour which may imply that on average the amount of matter transferred during quiescence has an approximately constant value for all dwarf novae.

117.080 Photometric variations of HD 25267.
J. Manfroid, G. Mathys, A. Heck.
Astron. Astrophys., Vol. 144, No. 1, p. 251 – 253 (1985).

Two periodicities are found in the photometric variations of HD 25267. One of the periods (1.2 d) seems to be the same as the magnetic period and is thus attributed to a CP component. The origin of the variations with the second period (3.8 d) remains uncertain.

117.081 Possible evolution of a triple system into Epsilon Aurigae.
P. P. Eggleton, J. E. Pringle.
Astrophys. J., Vol. 288, No. 1, p. 275 – 276 (1985).

If the companion of a Roche–lobe–filling red supergiant is itself a binary (of short period), the process of accretion onto the companion is likely to be different from the usual situation where the gainer is single. The authors suggest that the evolution of a triple system like HD 157978/9 – (A V + A V, $3\overset{d}{.}86$) + G III, 1170^d – may lead it to a state where it could resemble quite closely the enigmatic eclipsing system ε Aur (? + F0 Iap, 9890^d).

117.082 Ultraviolet spectroscopic observations of HD 77581 (Vela X–1 = 4U 0900–40).
K. Sadakane, R. Hirata, J. Jugaku, Y. Kondo, M. Matsuoka, Y. Tanaka, G. Hammerschlag–Hensberge.
Astrophys. J., Vol. 288, No. 1, p. 284 – 291 (1985).

Ultraviolet spectra of HD 77581 obtained with the *IUE* satellite in 1982 December and 1983 January are analyzed. The effective temperature is found to be $25,000 \pm 1000$K. The distance of

1.9 ± 0.2 kpc is derived from its luminosity $\log L/L_\odot = +5.53$, which is obtained from the effective temperature and from its radius. Circumstellar matter is detected not only in the C IV and Si IV lines but also in Al III, Fe III, C II, and Mg II lines. Profiles of these lines are analyzed to study the physical state and the motion of the circumstellar matter.

117.083 Ultraviolet, optical, and infrared observations of the intermediate polar TV Columbae.
M. Mateo, P. Szkody, J. Hutchings.
Astrophys. J., Vol. 288, No. 1, p. 292 – 304 (1985). With a correction in Vol. 292, No. 2, p. 763 (1985).

The authors analyze 43 *IUE* spectra of the X–ray discovered, triply periodic cataclysmic variable, TV Col. This data set covers the interval between 1980 April and 1983 September, excluding observations during and immediately after the large flare of 1982 November 22. High–speed photometric data taken simultaneously with some of the ultraviolet observations as well as non-simultaneous infrared photometry are also presented. The results show that the UV flux varies with the 4 day period discovered by Motch in 1981. It is inferred that this modulation corresponds to the periodic heating of a normally 9000K source due to reprocessing of beamed X–ray and (possibly) EUV radiation from the vicinity of the degenerate star. Models, where this source is located either at the disk hot spot or at the secondary star, are discussed.

117.084 Discovery of a faint U Geminorum star on a U.K. Schmidt objective–prism plate.
A. V. Filippenko, W. L. W. Sargent, C. Hazard.
Publ. Astron. Soc. Pac., Vol. 97, No. 587, p. 41 – 44 (1985).

The authors have discovered a faint ($m_v \approx 18^m5$) emission–line star, 1502 + 09, during an objective–prism survey conducted with the U.K. Schmidt telescope. It is probably a cataclysmic variable of the U Geminorum (dwarf nova) type. The equivalent width of Hβ emission is used to derive a very approximate distance of 350 pc. At $b \approx 54°$, the corresponding height above the Galactic plane (~ 280 pc) is considerably larger than the scale height generally associated with such stars.

117.085 Infrared photometry of cataclysmic variables I. Discovery of ellipsoidal variations in TW Virginis.
M. Mateo, P. Szkody, M. Bolte.
Publ. Astron. Soc. Pac., Vol. 97, No. 587, p. 45 – 50 (1985).

The authors present infrared (*JHK*) photometry of the cataclysmic variables TW Vir, AF Cam, and SW UMa. The data on TW Vir include time series observations at H over an entire orbital period which can be fit by a sine curve (constrained to have $P = 0.5 \, P_{orb}$) with a semiamplitude of 0.063 ± 0.017 mag. This is interpreted as due to the ellipsoidal variations of the tidally distorted M3 (± 1)V secondary which contributes $\geq 71\%$ of the total K flux of the system. Using the calibration of Bailey (1981) on these results sets a lower limit of 340 pc on the distance to TW Vir. Comparing these results to those derived from the ellipsoidal variations of U Gem, the authors find that the secondary of TW Vir very nearly or completely fills its Roche lobe and that the upper limit on its distance is 610 pc.

117.086 Photometric observations and analysis of the close binary system DV Aquarii.
A. Okazaki, A. Yamasaki, C. Nurwendaya, H. L. Malasan.
Publ. Astron. Soc. Pac., Vol. 97, No. 587, p. 62 – 66 (1985).

Photoelectric *BV* light curves of the close binary system DV Aqr are presented. These light curves are analyzed with a synthesis method to determine the photometric elements. Physical properties of the system are briefly discussed. It is found that DV Aqr is a detached system consisting of a late–A and a K–subgiant star.

117.087 The enigmatic star EZ Pegasi – a mystery solved?
S. B. Howell, B. W. Bopp.
Publ. Astron. Soc. Pac., Vol. 97, No. 587, p. 72 – 77 (1985).

EZ Peg, a ninth–magnitude G star that has been classified by various authors as an irregular variable, a U Gem system, and a contact binary is shown to have all the spectroscopic and photo-metric characteristics of an active–chromosphere RS CVn binary. The authors suggest that the reported "outburst" of 1943, when the spectrum appeared to be that of a B star, never occurred. The strong Ca II H and K reversals, viewed with low spectral resolution, caused the photospheric Ca II absorption to appear abnormally weak, mimicking a much earlier spectral type.

117.088 The binary X–ray pulsar 1E 2259 + 59 – a descendant of an AM Her type system?
V. M. Lipunov, K. A. Postnov.
Astron. Astrophys., Vol. 144, No. 2, p. L13 – L14 (1985).

A possible evolutionary scenario for the binary X–ray pulsar 1E 2259 + 59 is suggested. In the framework of this scenario the observational properties of the binary, the supernova remnant and the neutron star are explained in a self–consistent manner. AM Her–like systems (polars) are suggested to be possible progenitors of this class of X–ray pulsars.

117.089 The RS CVn–type binary SV Camelopardalis: evidence of dark spots from *UBV* observations and IR fluxes.
A. Cellino, F. Scaltriti, M. Busso.
Astron. Astrophys., Vol. 144, No. 2, p. 315 – 320 (1985).

New *UBV* light curves and some infrared *JHK* observations of the RS CVn–type binary SV Cam are presented. New determinations of primary minimum epochs confirm the presence of a light–time effect, with a period $U = 74.7$ yr. The overall shape of the light curve appears to vary in time by several hundredths of magnitude, due to a distortion wave whose fast migration causes an inversion in the levels of the maxima in a few months. IR excesses strongly suggest the presence of cool regions, in agreement with the common hypothesis that the light variations are due to starspots. By the method of Vogt (1981) spot temperatures turn out to be near 3800K, about 1500K cooler than the quiet photosphere. The spot hypothesis is also consistent with the changes in luminosity observed in the interval 1969 – 1984, which are shown to be tightly correlated with the cycles of stellar activity, as derived by the model of Busso et al. (1984).

117.090 Irradiation–induced mass–overflow instability as a possible cause of superoutbursts in SU UMa stars.
Y. Osaki.
Astron. Astrophys., Vol. 144, No. 2, p. 369 – 380 (1985).

The heating of the secondary's atmosphere in cataclysmic variables by strong far UV and soft X–ray radiation due to accretion and the resulting increase in mass–overflow rate are shown to lead to a positive feed–back instability between accretion and mass–overflow in certain circumstances. It is suggested that this irradiation–induced mass–overflow instability is a possible cause of superoutbursts in SU UMa stars. Examination of irradiation effects and of mass–overflow rates indicates that this instability operates only in those cataclysmic variables with very short periods and with low mass–transfer rates, a condition which SU UMa stars seem to satisfy. A new superhump model is proposed along this line.

117.091 The dissipation factor in contact binaries.
J. Hazlehurst.
Astron. Astrophys., Vol. 145, No. 1, p. 25 – 40 (1985).

It is argued that a large part of the light emitted from the secondary components of W UMa stars could have its origin in the dissipation of mechanical energy. This possibility could indicate the way towards a resolution of the so–called "light curve paradox" which arises from attempts to represent the W UMa stars by theoretical models.

117.092 The ultraviolet spectrum of the symbiotic star HM Sge.
B. E. A. Mueller, H. Nussbaumer.
Astron. Astrophys., Vol. 145, No. 1, p. 144 – 156 (1985).

The authors present IUE fluxes of the symbiotic star HM Sge, extracted from IUE observations made during the years 1978 – 1983. They argue in favour of an accretion disc model. They suggest that the sudden flux increase in He II and the more highly ionized atoms were due to a decrease from a high mass accretion rate, leading to a luminosity above the Eddington limit, to a lower rate leading to a luminosity below that limit. A thermonuclear runaway model might, however, present an alternative. For elemental abundances it is found that by number He/H = 0.14, that within the uncertainty limits, the oxygen and neon abundances are essentially solar, with a slight overabundance of carbon and silicon, and nitrogen is enhanced by approximately a factor 8. Contrary to an occasionally stated opinion the authors do not believe that HM Sge should be associated with a Wolf–Rayet star.

117.093 Magnetic braking and the period gap of cataclysmic binaries.
H. Ritter.
Astron. Astrophys., Vol. 145, No. 1, p. 227 – 231 (1985).

It is shown that, if the period gap of cataclysmic binaries is to be understood according to the model of Spruit and Ritter (1983) and Rappaport et al. (1983), the secondary of systems with an orbital period near 3 hours must suffer mass loss on a time scale which is shorter by about a factor of 10 than the Kelvin–Helmholtz time of the secondary in thermal equilibrium. The corresponding mass loss rate must be of order $(1-2) \times 10^{-9} M_\odot \mathrm{yr}^{-1}$. The author finds that magnetic braking according to Verbunt and Zwaan (1981) can account for such a high mass transfer rate. The braking law derived by Patterson (1984), on the other hand, results in mass transfer rates for systems near the upper edge of the gap which are too low to account for the observed width of the gap. Furthermore, this braking law results in a period distribution of nova–like systems and dwarf novae which is inconsistent with observations.

117.094 IUE observations of the RS CVn systems Z Her, TY Pyx and HD 155555.
M. J. Fernández–Figueroa, E. de Castro, A. Giménez.
Astron. Astrophys., Suppl. Ser., Vol. 60, No. 1, p. 5 – 8 (1985).

Short wavelength ($\lambda\lambda 1150 - 1950$ Å) low resolution ultraviolet spectra of three RS CVn regular–period binary systems, obtained with the IUE satellite, have been examined. Newly observed spectra were compared with those retrieved from the IUE data bank. The surface fluxes of chromospheric and transition region lines were obtained and are briefly discussed. The enhancements relative to the quiet sun were derived; they increase with the line formation temperature and in all cases are higher than those found among single active stars.

117.095 Variability in symbiotic stars and related objects.
R. Viotti.
Effects of variable mass loss on the local stellar environment, p. 79 – 89 (1984). – See Abstr. 012.028.

117.096 IUE spectroscopy, visible–band photometry, and polarimetry of HD 47732 (V641 Monocerotis).
R. H. Koch, B. J. Hrivnak, D. H. Bradstreet, W. Blitzstein, R. J. Pfeiffer, P. M. Perry.
Astrophys. J., Vol. 288, No. 2, p. 731 – 740 (1985).

Several high– and low–dispersion IUE spectra are described for the hot, massive close binary HD 47732. A weak stellar wind is measurable with an expansion velocity of ~ 500 km s^{-1}. There is difficulty in fitting a model atmosphere to all of the IUE continuum and visible band and IR measures. After alternative explanations are presented and rejected, these difficulties are resolved by appealing to a 22,500K model atmosphere and a distinctive interstellar extinction law. Numerous V band measures show the polarization to be variable. Models of the light curves assume a sporadic development of a systemic shell and consider light variations due to the elliptical shape of the component stars.

117.097 On the light variability of V1016 Cygni after outburst.
V. P. Arkhipova.
Perem. Zvezdy, Tom 22, No. 1, p. 25 – 30 (1983). In Russian.

The UBV light curves of V1016 Cyg from 1971 to 1982 were derived. During 10 years the B brightness declined by 0ᵐ.3, that of U increased by Oᵐ.3. The U–B index became bluer from –0ᵐ.6 to –1ᵐ.1, mainly due to strengthening of the Balmer continuum.

117.098 Disruption of light He companions in accreting neutron star binaries.
M. A. Ruderman, J. Shaham.
Astrophys. J., Vol. 289, No. 1, p. 244 – 246 (1985).

An old neutron star, being spun up to become a radio pulsar by accretion from a very low–mass He secondary, will ultimately tidally disrupt the secondary before the latter's mass reaches $4 \times 10^{-3} M_\odot$. Even if angular momentum loss from the binary is carried away only by gravitational radiation, the formation of an isolated, rapidly spinning pulsar in this way will take less than 10^{10}yr.

117.099 Dual polarization VLBI observations of stellar binary systems at 5 GHz.
R. L. Mutel, J. F. Lestrade, R. A. Preston, R. B. Phillips.
Astrophys. J., Vol. 289, No. 1, p. 262 – 268 (1985).

The milli–arcsecond radio structures of seven binary stellar systems (σ CrB, SZ Psc, HR 1099, UX Ari, HR 5110, Algol, and Cyg X–1) were determined using an intercontinental VLBI array at 5 GHz. Two sources (HR 5110, UX Ari) were undergoing intense radio outbursts during the observations. The sources UX Ari and Algol had a core–halo structure. In each case the core size was smaller than an individual stellar diameter, and the halo was comparable in size to the binary system. A simple expanding coronal loop model is proposed in which a flare event originates in a compact, optically thick source on the surface of the spotted star and radiates by gyrosynchrotron radiation. One or more coronal loops, whose feet are tied to the active region, expand into the outer corona, eventually becoming about as large as the binary system.

117.100 Time–resolved spectroscopy of long–period DQ Herculis stars.
W. R. Penning.
Astrophys. J., Vol. 289, No. 1, p. 300 – 309 (1985).

The author presents high–dispersion time–resolved spectroscopic observations of four long–period DQ Herculis stars (intermediate polars); V1223 Sgr, H2215–086, 3A 0729 + 103, and AO Psc. In two of these objects he was able to clearly detect orbital modulations, permitting an investigation of dynamic parameters. In all four objects short–period coherent variations in wavelength were discovered. The periods in each case turned out to be close to the published photometric white dwarf coherent periods. This lends the strongest kinematic evidence to date in support of an oblique magnetic rotator model.

117.101 Electrophotometric and spectroscopic observations of the peculiar star CH Cygni during 1980 – 1981.
T. S. Galkina, N. I. Shakhovskaya, N. I. Bondar'.
Izv. Krymskoj Astrofiz. Obs., Tom 69, p. 19 – 30 (1984). In Russian. English translation in Bull. Crimean Astrophys. Obs., Vol. 69.

The results of the spectroscopic and electrophotometric observations of the peculiar star CH Cygni obtained from July 1980 till November 1981 are reported. During the whole period of observations the brightness of CH Cygni was continuously increasing within the spectral region from $\lambda 3500$ to $\lambda 8000$ Å with increase of the amplitude toward the short–wave region. From July 1980 till November 1981 the brightness of the star in the visual spectral region increased up to ~ 5ᵐ6. The emission H I, Fe II, [Fe II], [O I] lines as well as the molecular absorption bands of TiO were observed. The observational results qualitatively confirm the model proposed by Luud. According to this model CH Cygni is supposed to be a double star which consists of a semiregular giant M6 III and a white dwarf with an enveloping transient accretion disk formed during the higher maxima of the semiregular star.

117.102 Symbiotic eclipse–binary stars CI Cyg and V1329 Cyg. Light curves comparison.
V. P. Arkhipova, T. S. Belyakina.
Izv. Krymskoj Astrofiz. Obs., Tom 69, p. 30 – 34 (1984). In Russian. English translation in Bull. Crimean Astrophys. Obs., Vol. 69.

The light curves of the symbiotic eclipse–binary stars CI Cyg ($P = 855^d$) and V1329 Cyg ($P = 950^d$) have been compared.

117.103   On the luminosity dependence of the light curve of AM Herculis.
I. L. Andronov.
Pis'ma Astron. Zh., Tom 11, No. 3, p. 203 – 207 (1985). In Russian. English translation in Sov. Astron. Lett., Vol. 11.

The luminosity dependence of the orbital light curves of AM Herculis in yellow light was investigated for the "active" state of the system. In the range of mean brightness $12^m4 – 13^m1$ the amplitude of flux variations remains nearly constant. As the system becomes brighter the main maximum is getting more pronounced and the light curve resembles that of U Geminorum outside outburst.

117.104   AS 338 in outburst, or how I found my "pet symbiotic".
R. Schulte–Ladbeck.
Messenger, No. 39, p. 3 – 6 (1985).

117.105   W Serpentis stars – a new class of interacting binaries.
W. Strupat, H. Drechsel, J. Rahe.
Messenger, No. 39, p. 40 – 42 (1985).

117.106   Starspot areas and temperatures in nine binary systems with late–type components.
C. H. Poe, J. A. Eaton.
Astrophys. J., Vol. 289, No. 2, p. 644 – 659 (1985).

A computer program that generates light curves for rotating spotted stars has been used to solve multicolor light curves of BY Dra, II Peg, HK Lac, HR 7275, IM Peg, λ And, UX Ari, σ Gem, and LX Per. The flux ratio of spot to photosphere in the photometric bands was determined from a revision of the Barnes–Evans relations. One or two circular spots could accurately reproduce the photometric behavior for most of these systems. The wavelength dependence of limb darkening was found to contribute a surprisingly large amount of the variation in the visible colors. The changes in the spot activity on II Peg were followed from 1976 through 1982. Definite spot temperatures were derived for most of the systems. The temperature of the spots ranges from 4050 to 3330K, while the temperature of the unspotted photospheres of the stars varied by only about 400K. The data suggest that the spots consist primarily of material at a single temperature, and that they are not divided into roughly equal amounts of umbral and penumbral material.

117.107   The geometry of the AM Herculis variable E1405–451.
I. R. Tuohy, N. Visvanathan, D. T. Wickramasinghe.
Astrophys. J., Vol. 289, No. 2, p. 721 – 725 (1985).

The authors present linear polarization observations and photometry of the AM Herculis variable E1405–451, obtained at the AAT. A pronounced, but transient, linear polarization pulse was detected at photometric maximum. The authors also find significant residual polarization which shows a position angle variation of $\Delta\theta \sim 70°$ with binary phase. This variation can be modeled with a system geometry having an orbital inclination of $\sim 60°$, and a colatitude of $\sim 20°$ for the active magnetic pole of the white dwarf.

117.108   The R Aquarii system at optical and radio wavelengths.
J. M. Hollis, M. Kafatos, A. G. Michalitsianos, H. A. McAlister.
Astrophys. J., Vol. 289, No. 2, p. 765 – 773 (1985).

Observations of the symbiotic binary R Aquarii environment were obtained with the VLA at 2, 6, and 20 cm at the same epoch. It is demonstrated that the compact double radio source is not associated with R Aquarii but is an extragalactic background object. The spectral index of the compact nebula surrounding R Aquarii indicates that the emission is thermal and the nebula is ionized by an unseen, hot companion to the Mira–like variable R Aquarii. The spectral index and polarization observations of the extended jet $\sim 6''$ away from R Aquarii indicate that this amorphous source is definitely thermal and optically thin in nature. It is suggested that the jet can be best explained by enhanced mass exchange occurring periodically in the symbiotic system.

117.109   Spectral peculiarities in the recent spectra of PU Vul (Nova Vul 1979) and CH Cyg.
Y. Yamashita, Y. Norimoto, K. H. Yoo.
The MK process and stellar classification, p. 337 – 339 (1984). – See Abstr. 012.033.

117.110   Spectroscopic orbits for the cataclysmic binaries CM Delphini, V380 Ophiuchi, and VW Vulpeculae.
A. W. Shafter.
Astron. J., Vol. 90, No. 4, p. 643 – 646 (1985).

Spectroscopic observations are presented for CM Del, V380 Oph, and VW Vul which reveal their orbital periods to be 0.162, 0.16, and 0.073 days, respectively. The spectra of CM Del and V380 Oph appear to be characteristic of novalike systems. In particular, the spectra suggest relatively high mass–transfer rates in these two systems. Such high mass–transfer rates are typical of novalike variables in particular, and of cataclysmic variables with orbital periods between 0.13 and 0.17 days in general. The ultrashort period system VW Vul has a spectrum which suggests a lower mass–transfer rate than in either CM Del or V380 Oph. The spectrum of VW Vul is typical of a dwarf nova.

117.111   High spectral resolution observations of the coronal X–ray emission from the RS CVn binary Sigma Corona Borealis.
P. C. Agrawal, T. H. Markert, G. R. Riegler.
Mon. Not. R. Astron. Soc., Vol. 213, No. 4, p. 761 – 771 (1985).

Results of the high spectral resolution observation of the RS CVn binary σ CrB, made with the Focal Plane Crystal Spectrometer (FPCS) on the Einstein Observatory, are reported. A spectral scan in 800 – 840 eV interval shows clear presence of X–ray emission line at 826 eV identified with $^1S_0 – ^1P_1$ transition of Fe XVII while a prominent peak at 1007 eV in the scan band 986 – 1014 eV is attributed to a blend of lines due to highly ionized iron. Using the observed fluxes of the lines and the Raymond–Smith model, best–fit values of corona temperature and volume emission measures are derived to be (6.92 with errors $+0.84$ and $-0.75) \times 10^6$K and (1.74 with errors $+0.55$ and $-0.69) \times 10^{53}$cm^{-3} respectively. Implications of these results are briefly discussed.

117.112   X–ray binaries and high–energy gamma rays.
D. J. Fegan.
Ir. Astron. J., Vol. 17, No. 1, p. 16 – 19 (1985). – See Abstr. 012.034.

117.113   Spectral observations of the composite system HD 45166 (B8 V + hot component + gas stream).
T. Nugis, M. Ruusalepp.
Tartu Astrofüüs. Obs. Teated, Nr. 74, p. 3 – 6 (1985).

The system HD 45166 was observed by the authors during the years 1978 – 1980. It was found that the spectrum remained steady for long periods of time, but at some periods variations in emission–line intensities took place. Maximum changes of the equivalent widths of some emission lines reached two times. The composite system HD 45166 consists of a B8 V star and of a hot component. The system seems to be situated pole–on and there probably exists a stream of matter linking the hot and B8 V components.

117.114   Spectroscopic observations of the Wolf–Rayet binary HD 193077.
K. Annuk.
Tartu Astrofüüs. Obs. Teated, Nr. 74, p. 7 – 19 (1985).

On the basis of 26 spectrograms the equivalent widths of the emission and absorption lines of HD 193077 have been measured. It has been found that the strengths of the emission lines in 1981 were about twice greater than in 1982. Comparing the equivalent widths of HD 193077 with the equivalent widths of single WN5 and WN6 stars, the author concludes that the O–type component is three times brighter than the WR component of HD 193077.

117.115 AG Pegasi in the period of 1982 – 83.
M. Ilmas.
Tartu Astrofüüs. Obs. Teated, Nr. 74, p. 20 – 24 (1985).

Observations of AG Peg are presented. The emission lines of AG Peg have been found to vary depending on phase. The broad–band emission components of H and He II lines originate in the same region near the rotating hot component of AG Peg. Assuming that an expanding envelope surrounds the binary system one can explain the He I line intensities using the Sobolev Escape Probability method.

117.116 Ultraviolet spectrum of the close binary system AO Cassiopeiae obtained with the IUE.
V. Harvig, A. Sapar, L. Sapar.
Tartu Astrofüüs. Obs. Teated, Nr. 74, p. 33 – 48 (1985).

The results of an analysis of the ultraviolet spectrum obtained for the close binary system AO Cas with the aid of the IUE satellite are presented. A number of the photospheric lines of the primary or/and of the secondary star have been identified and their equivalent widths have been found.

117.117 On the nature of envelopes surrounding W UMa–type stars.
I. Pustyl'nik.
Tartu Astrofüüs. Obs. Publ., Tom 50, p. 191 – 220 (1984).

A review of the observed features of W UMa–type stars is given with special emphasis on the peculiarities seen in the infra-red and ultraviolet regions. The role of possible circumstellar dust and its influence upon colors and the light curves are discussed. Necessary formulae are derived to calculate the light curve of a binary consisting of two spherical stars embedded in a spherical homogeneous dust envelope.

117.118 The minimum period of hydrogen–deficient cataclysmic binaries.
R. Sienkiewicz.
Acta Astron., Vol. 34, No. 3, p. 325 – 330 (1984).

The dependence of the minimum orbital period of cataclysmic binaries on the hydrogen content is investigated. A standard model of such systems composed of a low–mass secondary star transferring mass to a compact primary star due to the decay of the binary orbit resulting from gravitational radiation, is adopted. The secondary is assumed to be chemically homogeneous with the hydrogen mass fraction ranging from 0.5 to 0.01. Such binaries can have orbital periods as short as 40 minutes. The relevance of these models to the four observed binaries having periods shorter than 80 minutes is discussed.

117.119 RR Pictoris – an intermediate polar?
M. Kubiak.
Acta Astron., Vol. 34, No. 3, p. 331 – 343 (1984).

Five nights of photoelectric observations in the UBV system with 5 second time–resolution give evidences of an eclipse of the hot spot in the system – particularly in band U. Fourier analysis of the data reveals the presence of coherent brightness modulation in all bands with a period of about 15 min. In many respects RR Pic is similar to intermediate polars and very probably is a member of this group of objects.

117.120 IUE observations of W UMa systems AE Phoenicis and TY Mensae.
S. M. Ruciński.
Acta Astron., Vol. 34, No. 3, p. 345 – 352 (1984).

SWP spectra of AE Phe and TY Men are presented and discussed. The spectrum of AE Phe is typical in showing strong emission lines which originate in the chromosphere and transition region. On the basis of the strength of the He II emission line the author predicts that AE Phe should be a moderately strong X–ray source. The spectrum of TY Men is too weakly exposed for a full discussion. It is pointed out that this star is one of the most important for understanding of the activity in W UMa systems.

117.121 Ca II H and K emissions in binaries. III. G and K type dwarfs and subgiants.
R. Głebocki, A. Stawikowski.
Acta Astron., Vol. 34, No. 3, p. 365 – 375 (1984).

Ca II K line emission intensities, I_K, are estimated for 30 main sequence and RS CVn binary stars. A decrease in emission intensity with increasing period is found. For all late type binaries a good correlation exists between I_K and the rotational velocity. This is interpreted as indication of increased magnetic activity due to the dynamo action forced by tidal synchronization of rotational and orbital periods. Analysis of K line intensities, X–ray fluxes and mass loss rates indicate that in main sequence and slightly evolved binaries the increase of chromosphere–corona activity is not anticorrelated with strong mass loss.

117.122 UBV photometry of H 2252–035 (AO Psc).
M. Kubiak.
Acta Astron., Vol. 34, No. 4, p. 397 – 407 (1984).

Presented are results of UBV fast photometry of AO Psc made during five nights in October 1982. The shapes of light variations corresponding to three periodicities observed in the system are determined. Similar spectral characteristics of the light modulated with the orbital and 859 s periods suggest that both components have common origin, being probably a result of reprocessing of the original X–rays in the disk and in the disk–bulge, respectively. Determination of the moments of maxima, together with previous observations, gives upper limits of period variations equal to 3×10^{-9}s/s and 9×10^{-12}s/s for the orbital and 859 s periods, respectively.

117.123 Photometric observations of W UMa–type system SS Ari.
J. Kałużny, G. Pojmański.
Acta Astron., Vol. 34, No. 4, p. 445 – 453 (1984).

UBV photometry of SS Ari obtained in 1982 is given. The analysis of the period variations indicates that the $O–C$ diagram has a sine–like shape. However, on the basis of available times of minima, the authors cannot conclude if the period variations are due to continuous or random, abrupt changes. Light curve synthesis solutions are calculated for both possible cases (A–type and W–type light curves).

117.124 Expanding envelopes of binary stars. I. The cone model.
W. Neutsch, H. Schmidt.
Astrophys. Space Sci., Vol. 109, No. 2, p. 249 – 257 (1985).

HD 152270 is a binary system containing an O5 star and a Wolf–Rayet companion of spectral type WC 6–7. The authors developed two simple descriptions for the envelope. It is the aim of this paper to discuss the validity of one of them, called the 'cone model'. The cone model is compared with a more physical one which is based on a Monte–Carlo simulation of the stellar wind flow and is believed to be a much better approximation to reality. Furthermore, the authors are able to test the assumptions of the cone model by quantitatively reducing spectral observations of HD 152270.

117.125 UBV photometric analysis of V758 Centauri.
S. L. Lipari, R. F. Sisteró.
Astrophys. Space Sci., Vol. 109, No. 2, p. 271 – 276 (1985).

Classical R–M and synthetic W–D analysis of V758 Centauri are presented. Two solutions (semi–detached and contact) were found from differential corrections approach. The semi–detached model is physically acceptable since the system is thermally decoupled. The solution for this case and the photometric data are consistent with a B9 primary and A9 secondary components having parameters close to main–sequence values. It is suggested that V758 Centauri is a B–type W UMa system at the broken–contact phase predicted by the Thermal Relaxation Oscilllations theory.

**117.126   De media aequinoctiorum praecessione atque de nuta-
tione stellae SS433.**
R. Ruffini, S. D. Jong.
Astrophys. Space Sci., Vol. 110, No. 1, p. 89 – 94 (1985). – See
Abstr. 012.039.

On the basis of Bessel's classical papers De Media Aequinoctiorum Praecessione and De Nutatione compact formulae are
given to interpret the observed variations in the shifted lines of
SS433 in terms of nutational and precessional effects of a disk in
a binary system.

117.127   The frequency of binaries with a degenerate component.
J. L. Halbwachs.
Astrophys. Space Sci., Vol. 110, No. 1, p. 149 – 152 (1985). – See
Abstr. 012.039.

The frequency of binaries with degenerate secondary components was evaluated according to the spectral types of the primaries. It appears that this proportion is 25% for binaries with giant
primary components, and less than about 17% for dwarfs.

117.128   G 82–23: a new subdwarf–white dwarf binary.
I. Bues, G. Rupprecht.
Astrophys. Space Sci., Vol. 110, No. 1, p. 163 – 168 (1985). – See
Abstr. 012.039.

Photometry and spectrophotometry of the proper motion star
G 82–23 are presented. A comparison with subdwarfs and white
dwarfs in the same range of temperature shows only partial
agreement. If the parallax is taken into account, the best explanation of this object seems to be a binary structure with a K–
subdwarf and a DC–white dwarf.

**117.129   Accuracy of close binary mass determinations from par-
allaxes.**
W. van Hamme, R. E. Wilson.
Astrophys. Space Sci., Vol. 110, No. 1, p. 169 – 175 (1985). – See
Abstr. 012.039.

Absolute masses for W Ursae Majoris and Algol–type close
binaries can be determined from their parallax, if observed, and
the relative sizes of the stars and their mass ratio, obtained from
a light curve solution. An error propagation study compares the
typical order of magnitude of the various terms involved, and
shows how accurate parallaxes have to be in order to make the
procedure work, i.e., making the parallax term not larger than the
combined non–parallax terms, and producing reasonably low
mass errors. Some comments are made on the possibilities with
respect to the HIPPARCOS program.

**117.130   The 2–second optical variability in E1405–451: origin
and coherency.**
S. Larsson.
Astron. Astrophys., Vol. 145, No. 2, p. L1 – L3 (1985).

Earlier optical observations of the AM Her–object E1405–451
have shown variability on timescales of about 2 seconds in this
object. The author's observations show this variability to be due
to quasi–periodic oscillations with a coherence time on the order
of 1 minute. The broad frequency distribution of the oscillations
in time averaged power spectra is caused both by frequency drift
and by the simultaneous presence of more than one oscillation
frequency. The color of the oscillating radiation component allows the author to identify it with the cyclotron radiation and
therefore localize the oscillations to the accretion column.

**117.131   Magnetohydrostatics in the polar caps of accreting
magnetized white dwarfs.**
J. M. Hameury, J. P. Lasota.
Astron. Astrophys., Vol. 145, No. 2, p. L10 – L12 (1985).

The authors discuss the recently proposed mechanism (Livio,
1984) by which quasi–periodic luminosity variations could be
obtained in accreting magnetized white dwarfs. They show by
order of magnitude and numerical calculations that breaking of
the magnetic field (if any) cannot give the required periods. This
particular mechanism has therefore to be excluded as a possible
explanation of quasi–periodicities in cataclysmic variables.

117.132   EX Hya: the slowest DQ Her star?
F. Jablonski, I. C. Busko.
Mon. Not. R. Astron. Soc., Vol. 214, No. 2, p. 219 – 228 (1985).

The authors report the results of a photometric programme on
the cataclysmic variable EX Hya. New eclipse timings, as well as
new timings for the 67–min photometric modulation are presented and discussed. The authors suggest that EX Hya is a
member of the DQ Her subclass of cataclysmic variables, in
which the optical modulation is thought to be associated to the
non–synchronous rotation of a white dwarf in the binary system.

117.133   *UBV* images of the Z Cha accretion disc in outburst.
K. Horne, M. C. Cook.
Mon. Not. R. Astron. Soc., Vol. 214, No. 2, p. 307 – 317 (1985).

Images of the accretion disc in the dwarf nova Z Cha are
derived from *UBV* photometry of two eclipses obtained in the
early stage of the 1983 March outburst. Two–colour and colour–
magnitude diagrams for the accretion disc demonstrate that its
surface consists of opaque thermal radiators. A distance of
105 ± 20 ($K_R/400$ km s^{-1}) pc is derived, where K_R is the assumed
radial velocity semi–amplitude of the red dwarf companion star.
Temperatures in the disc range from ~ 7000K at its outer rim to
~ 40000K near the white dwarf at its centre. The radial temperature profile is consistent with steady mass transfer at the rate
$10^{-8.9 \pm 0.3} M_\odot \text{yr}^{-1}$.

117.134   Tidal flows in eccentric binary systems.
A. Z. Dolginov, E. V. Smel'chakova.
Astron. Zh., Tom 62, Vyp. 2, p. 301 – 305 (1985). In Russian.
English translation in Sov. Astron., Vol. 29, No. 2.

Expressions for the tidal flow velocities in close eccentric binaries are obtained. The third spherical harmonic is taken into
account in the multipole expansion of the tidal potential. The
solutions are applied to some concrete systems.

**117.135   Eclipse duration of the point source in X–ray binary
systems with eccentric orbits.**
T. S. Khruzina.
Astron. Zh., Tom 62, Vyp. 2, p. 356 – 364 (1985). In Russian.
English translation in Sov. Astron., Vol. 29, No. 2.

An algorithm of the calculation of the eclipse duration of the
compact source in close binary systems with eccentric orbits and
the tables of dependence between the duration angle and inclination of the orbital plane for systems with different mass ratio,
orbit eccentricity, and orbit orientation in space are presented.
The influence of rotation asynchronism of the optical star on the
eclipse duration is analysed.

**117.136   Model light curves for close binaries with a common
scattering envelope.**
I. B. Pustyl'nik, L. Einasto.
Sov. Astron. Lett., Vol. 10, No. 4, p. 215 – 217 (1984). English
translation of 38.117.002.

117.137   Massive close binary systems.
A. M. Cherepashchuk.
Zemlya Vselennaya, No. 1, p. 16 – 24 (1985). In Russian.

117.138   Further observations of EX Hydrae.
K. M. Hill, R. D. Watson.
Proc. Astron. Soc. Aust., Vol. 5, No. 4, p. 532 – 535 (1984).

The authors summarise the constraints on the source of the
67 min flux modulation in EX Hya.

117.139   The RS CVn type star PZ Tel.
J. L. Innis, D. W. Coates, K. Thompson.
Proc. Astron. Soc. Aust., Vol. 5, No. 4, p. 540 – 543 (1984).

PZ Tel is a double lined binary whose components are of approximately equal luminosities, but this is yet to be confirmed.
The authors suggest that the photometric variations are due to
the presence of large cooler starspots on the photosphere of one
or both components, as seems to be the case for related systems.
The rapid changes in the observed light curve imply equally rapid
changes in the distribution of the starspots, and make this an
interesting object for further study.

117.140 A search for OH emission from symbiotic stars.
R. P. Norris, D. A. Allen, R. F. Haynes, A. E. Wright.
Proc. Astron. Soc. Aust., Vol. 5, No. 4, p. 562 – 565 (1984).

A sensitive search has been made for OH maser emission from a sample of 16 symbiotic stars. This sample has been selected on the basis of infrared optical depth and variability, so that the stars within it have circumstellar shells similar to those seen in the well–known OH/IR and OH/Mira stars. There were no significant detections, except for one unassociated background source, and the authors conclude that the presence of a hot binary companion inhibits any possible OH maser action.

117.141 Light–curve solution and interpretation for CG Cygni.
S. A. Naftilan, E. F. Milone.
Astron. J., Vol. 90, No. 5, p. 761 – 766 (1985).

Combining spectroscopic radial–velocity measures and photometric light–curve solutions has allowed the authors to model the short–period RS CVn–like system CG Cygni. The mass ratio is 1.0 and the system is detached. Long–term brightening is best explained by a steady increase in the temperature of the hotter star. The photometric distortion "wave" is rapidly variable and has a complex shape. Standard starspot models fail to reproduce the observed light curves.

117.142 The mass of AW Ursae Majoris.
M. J. Rensing, S. W. Mochnacki, C. T. Bolton.
Astron. J., Vol. 90, No. 5, p. 767 – 772 (1985).

The data from 22 spectra of AW Ursae Majoris obtained at the David Dunlap Observatory have been analyzed. Radial velocities of the primary component were measured using Griffin–type cross–correlation techniques. Line profiles were modeled to estimate the distortion of the measured radial–velocity curve due to proximity effects. The radial–velocity amplitude of the primary was found to be 22.2 ± 0.9 km s^{-1}. The mass of AW Ursae Majoris was found to be $1.3 \pm 0.2\, M_\odot$, if the mass ratio q is 0.079, and $1.7 \pm 0.3\, M_\odot$ if $q = 0.070$. Such masses are much smaller than the values found in previous studies. The new result agrees well with evolutionary theories of contact binaries if AW UMa is a terminal–age main–sequence system.

117.143 Formation of optical–line emitting regions in the jets of SS433.
G. Bodo.
ESA Spec. Publ., ESA SP–207, p. 311 – 314 (1984). – See Abstr. 012.044.

117.144 On fast X–ray rotators with long–term periodicities.
S. Naranan, R. F. Elsner, W. Darbro, P. E. Hardee, B. D. Ramsey, D. A. Leahy, M. C. Weisskopf, A. C. Williams, P. G. Sutherland, J. E. Grindlay.
Astrophys. J., Vol. 290, No. 2, p. 487 – 495 (1985).

Using results from SAS 3 and the Einstein Observatory, the authors show that the pulse–period history of the 13.5 s pulsing X–ray source LMC X–4 is consistent with standard accretion and torque models only if LMC X–4 is a fast rotator for which the accretion torques nearly cancel. This result leads to an estimate for the neutron star's magnetic field strength $\sim 1.2 \times 10^{13}$G. The authors then review the evidence that Her X–1 and SMC X–1 are fast and intermediate–to–fast rotators respectively, and examine the properties of the 1 – 2 month X–ray intensity variations. They discuss the possibility that tilted precessing accretion disks about these three neutron stars are slaved to the precession of the companion stars.

117.145 High–velocity winds in close binaries with accretion disks. II. The view along the plane of the disk.
F. A. Córdova, K. O. Mason.
Astrophys. J., Vol. 290, No. 2, p. 671 – 682 (1985).

The authors present phase–resolved, ultraviolet spectra of the eclipsing cataclysmic binaries RW Trianguli and DQ Herculis. Both stars display N V λ1240, Si IV λ1400, C IV λ1549, and He II λ1640 in emission. The strongest spectral feature in both stars is C IV λ1549. In RW Tri the blue wing of this line extends to velocities of -3000 km s^{-1}. In both stars there is a deep eclipse of the UV continuum, representing an occultation of the interior regions of an accretion disk surrounding the degenerate dwarf. None of the spectral lines in either binary system are eclipsed to the same degree as the UV continuum, indicating that the lines are formed in an extended region above and below the disk. The distinctive morphology of RW Tri's lines suggests an origin in an accelerating wind.

117.146 A photometric study and analysis of XY Leonis.
B. J. Hrivnak.
Astrophys. J., Vol. 290, No. 2, p. 696 – 706 (1985).

Photoelectric observations of the short–period contact system XY Leo have been obtained over the past six seasons, and include three seasonal light curves. Times of minimum light indicate a series of period changes, or perhaps the effects of a third body. Season–to–season changes are seen in the light curves. The light curves are analyzed, using the new spectroscopic mass ratio of Hrivnak et al., and absolute parameters are determined. An examination of the available photometric and spectroscopic data indicates that the intrinsic light–curve variability is consistent with starspot activity on one or both components.

117.147 TT Arietis: the low state.
A. W. Shafter, P. Szkody, J. Liebert, W. R. Penning, H. E. Bond, A. D. Grauer.
Astrophys. J., Vol. 290, No. 2, p. 707 – 720 (1985).

The authors present photometric and spectroscopic observations extending from the ultraviolet through the infrared of the novalike variable TT Ari, which were obtained during its recent (1982) low state. The observations indicate that mass transfer did not cease completely, but appeared to be occurring sporadically during the low state. The spectroscopic observations reveal strong, narrow Balmer and He I emission superposed on a blue continuum when mass transfer is diminished. The blue and UV continuum is most likely produced by the photosphere of the white dwarf with $T_{\text{eff}} \gtrsim 50000$K. It is shown that the low–state emission probably arises in the chromosphere of the secondary star.

117.148 Investigations of the early spectral class close binary system XZ Cep with the curve of growth method.
L. V. Glazunova.
Mater. konf. mol. uchenykh Odes. astron., nauchn.–proivz. akad.–univ. kompleksa, Odessa, 15 fevr., 1983. Odessa, p. 19 – 25 (1983). In Russian. Abstr. in Ref. Zh., 51. Astron., 3.51.487 (1985).

117.149 On emission line profiles in the spectra of magnetic close binary systems.
I. L. Andronov.
Mater. konf. mol. uchenykh Odes. astron. nauchn.–proizv. akad.–univ. kompleksa, Odessa, 15 fevr., 1983. Odessa, p. 6 – 13 (1983). In Russian. Abstr. in Ref. Zh., 51. Astron., 3.51.618 (1985).

117.150 Precession and nutation in SS 433.
F. Ciatti.
Mem. Soc. Astron. Ital., Vol. 55, No. 3, p. 591 – 595 (1984). – See Abstr. 012.050.

117.151 *UBVRI* photometry of W Ursae Majoris.
A. P. Linnell.
Astrophys. J., Suppl. Ser., Vol. 57, No. 3, p. 611 – 619 (1985).

High–speed *UBVRI* photometry of W Ursae Majoris produced 15,930 observations covering one complete cycle on 1984 March 5 – 6. Primary minimum is flat during totality, in contrast to a substantial tilt during annular phases of secondary minimum. The tilt is qualitatively consistent with a starspot origin. There is an observed greater color change at primary minimum than secondary, and this effect is unexplained by the known color changes arising from wavelength dependence of limb darkening. There is qualitative consistency with a Ruciński hot secondary model.

117.152   The low–mass binary X–ray sources observed from TENMA.
K. Mitsuda.
Astron. Her., Vol. 78, No. 4, p. 98 – 102 (1985). In Japanese.

117.153   Search for gamma–ray line emission from SS 433.
C. J. MacCallum, A. F. Huters, P. D. Stang, M. Leventhal.
Astrophys. J., Vol. 291, No. 2, p. 486 – 491 (1985).

A balloon–borne germanium γ–ray telescope was flown over Palestine, Texas, to search for Doppler–shifted nuclear lines recently reported from SS 433 by the *HEAO 3* workers. No line activity was detected. If present at the reported time–averaged intensities, the reported lines would have been detected at the $2 - 3\,\sigma$ level.

117.154   A search for relativistic companions of runaway OB stars.
A. A. Aslanov, L. N. Kornilova, A. M. Cherepashchuk.
Sov. Astron. Lett., Vol. 10, No. 5, p. 278 – 281 (1984). English translation of 38.117.103.

117.155   The flickering of CH Cygni.
V. P. Reshetnikov, T. N. Khudyakova.
Sov. Astron. Lett., Vol. 10, No. 5, p. 281 – 283 (1984). English translation of 38.117.104.

117.156   Further evidence for precession of the optical star in the Cygnus X–1 system.
I. M. Kopylov, V. V. Sokolov.
Sov. Astron. Lett., Vol. 10, No. 5, p. 315 – 318 (1984). English translation of 38.117.100.

117.157   The binary frequency of extreme subdwarfs revisited.
L. L. Stryker, J. E. Hesser, G. Hill, G. S. Garlick, L. M. O'Keefe.
Publ. Astron. Soc. Pac., Vol. 97, No. 589, p. 247 – 260 (1985).

New spectroscopic data for a large sample of subdwarfs, taken over a four–year period, are presented and analyzed using statistical tests, including the F–ratio test (the ratio of variances of external to internal errors). The authors find a lower limit of 20% to 30% for the binary frequency for a sample of 46 subdwarfs. The semiamplitude limit is $\sim 14\,\mathrm{km\,s^{-1}}$. A subset of twelve subdwarfs with the lowest metallicities ([Fe/H] < -1.5) has a similar frequency, $\sim 30\%$. Data from Abt and Levy and from Crampton and Hartwick are also reanalyzed by these methods. Candidate binary stars are listed for further study.

117.158   The 1984 eclipse of the symbiotic binary SY Muscae.
S. J. Kenyon, A. G. Michalitsianos, J. H. Lutz, M. Kafatos.
Publ. Astron. Soc. Pac., Vol. 97, No. 589, p. 268 – 271 (1985).

The authors present ultraviolet and optical observations of the 1984 eclipse of the symbiotic binary star SY Mus. The optical light curve shows a 627–day variation which is reflected in the intensity of the far–UV continuum ($\lambda < 2000$ Å) and in the intensities of all strong, permitted UV emission lines. This contrasts sharply with other eclipsing systems, in which some high ionization permitted lines show little evidence for large–scale variability. The behavior of the emission lines and the UV continuum is most naturally understood if the hot stellar source and a surrounding ionized nebula in the SY Mus binary are eclipsed by a red–giant companion every 627 days. The depth of the eclipse in the He II $\lambda 1640$ emission line allows estimating the radius of the partially eclipsed He$^+$ region ($75\,R_\odot$) and that of the cool giant ($60\,R_\odot$), for a distance of 1.3 kpc.

117.159   Disk–instability model for outbursts of dwarf novae. II. Full–disk calculations.
S. Mineshige, Y. Osaki.
Publ. Astron. Soc. Jpn., Vol. 37, No. 1, p. 1 – 18 (1985).

The authors have calculated the time evolution of unstable accretion disks by solving full–disk equations in the nonthermal equilibrium situation. By assigning high α (where α is the stan-

dard viscosity parameter) in hot state and low α in cool state, one can reproduce the basic feature of dwarf–nova outbursts by the disk–instability model. Numerical results are presented for two typical cases of outbursts. The local variations in the accretion disks and the global propagation of the transition waves are discussed in detail. The propagation of transition waves is fairly complicated particularly in the case when a heating wave propagating outward is reflected in the middle of the disk as a cooling wave.

117.160   Optical observations of X 0331 + 53.
K. Kodaira, S. Nishimura, M. Kondo, S. Kikuchi, H. Ando, S. Isobe, Y. Mikami, Y.–i. Takeda, J. Jugaku, S. Okamura, K.–i. Ishida, H. Maehara, M. Shimizu, E. Watanabe, Y. Norimoto, K. Okida, M. Yutani, K. Takagishi, T. Omodaka, S. Sato, M. Nishida, K. Ogura, A. Yamasaki, N. Kaneko, M. Nishimura, A. Okazaki.
Publ. Astron. Soc. Jpn., Vol. 37, No. 1, p. 97 – 106 (1985).

The initial follow–up observations of a candidate for the optical counterpart of X 0331 + 53 are summarized. The spectroscopic and photometric data suggest that this object of $V = 15.2$ mag may be a heavily interstellar reddened early–type star ($M \gtrsim 10\,M_\odot, A_v \sim 7.4$ mag) located at a distance larger than ~ 1 kpc.

117.161   Expanding envelopes of binary stars. II. Motion in extremely hot envelopes.
W. Neutsch, H. Schmidt.
Astrophys. Space Sci., Vol. 110, No. 2, p. 257 – 276 (1985).

In this paper, which is a continuation of Neutsch and Schmidt (1985), the authors investigate the structure of binary envelopes under the hypothesis that at least one of the companions produces a very intense radiation field. The limiting case in which the forces due to the system's rotation (i. e., Coriolis and centrifugal forces) as well as gas pressure can be neglected is solved analytically using a classical result of Euler. Beyond this the velocity and density distributions in the envelope are determined.

117.162   A model for symbiotic star Z Andromedae.
B. F. Yudin.
Astrophys. Space Sci., Vol. 110, No. 2, p. 277 – 291 (1985).

An analysis of the results of observations of the symbiotic star Z And has shown that no definite model can be derived at present on their basis. If the hot component is essentially an accreting white dwarf with a hydrogen–burning shell source, then the gas envelope must be optically thin for Lc–emission and its T_e must be in the neighborhood of 2.6×10^4K. And if the hot component is a main sequence star with an accretion disk around it, then it is classified with red dwarfs. The electron temperature of the gas envelope must be 1.5×10^4K. Substitution of a solar–type star for the first–named component in the binary 'red dwarf + red giant' system will lead to a significant decrease in the excitation of the combination spectrum.

117.163   The dependence of He I 10830 absorption on X–ray luminosity in RS CVn binaries and very active F and G main–sequence stars.
M. V. Mekkaden.
Astrophys. Space Sci., Vol. 110, No. 2, p. 413 – 416 (1985).

A study of the relationship between soft X–ray luminosity and He I 10830 equivalent width in RS CVn binaries and very active late–type main–sequence stars is made. Long–period RS CVns and very active F and G main–sequence stars show strong dependence on their X–ray luminosities. This is attributed to the dominance of coronal excitation of the line in these stars. Short–period RS CVn binaries have lower He I 10830 absorption compared to their X–ray emission. This phenomenon is explained by the presence of He I emission region also in their chromospheres.

117.164   The first IUE observations of LMC X–1 (star 32).
L. Bianchi, M. Pakull.
Astron. Astrophys., Vol. 146, No. 2, p. 242 – 248 (1985).

The first far UV observations of LMC X–1 (star 32) are presented. Form the UV continuum distribution the authors derive a value for the reddening towards star 32 of $E(B-V) = 0.05$

(galactic foreground extinction) plus 0.32 (extinction in the LMC). The dereddened stellar continuum can be reproduced by a model atmosphere with $T_{eff} = 35,000K$ (log $g = 3.5$), confirming the spectral type O7–9 III derived from the optical spectrum. Continuum emission is significantly detected from the surrounding H II region N 159, within a region of $\sim 5''$ radius around the star, suggesting that stellar light is scattered by dust particles in the nebula.

117.165 UV and optical observations of variability in the WR + compact candidate HD 96548.

L. J. Smith, C. Lloyd, E. N. Walker.
Astron. Astrophys., Vol. 146, No. 2, p. 307 – 316 (1985).

IUE SWP high resolution spectra of the WR + compact candidate HD 96548 are presented which show considerable variability in the P Cygni and emission line profiles. Previously published visible photometry is re-analysed with new data; it is found that all the photometry is definitely periodic, with a period of 5.879 d or its one day aliases at 0.855 and 1.204 d. The two most likely interpretations of the UV profile changes are discussed: either modification of the WR wind by X–rays from a compact companion or intrinsic stellar wind variations.

117.166 Accretion disks heated by luminous central stars.

M. Friedjung.
Astron. Astrophys., Vol. 146, No. 2, p. 366 – 368 (1985).

In certain cases heating of an accretion disk by radiation from a bright central star, can be more important than heating by dissipation of gravitational energy. Simple calculations have been performed, which indicate that the spectrum of radiation re-emitted by a disk heated in this way, can be similar to that of the classical Lynden–Bell law. The excessively high accretion rates deduced for some objects (some old novae, symbiotic stars, the luminous star S 22, etc...) assuming heating by gravitational dissipation, may be explained by the presence of excess heating due to the presence of a bright central star, which in the case of an old nova may not have returned to a quiescent state.

117.167 Infrared observations of variable stars. 2. Spectrophotometry of CH Cygni.

O. G. Taranova, V. I. Shenavrin.
Perem. Zvezdy, Tom 21, No. 6, p. 817 – 818 (1983). In Russian.

Results of spectral observations of CH Cyg in the 0.6 – 2.5 μm region are presented. Continuous radiation of CH Cyg in the 0.6 – 1.1 μm region in quiet and active state is belonging to a giant of the spectral class M6. The distance to CH Cyg is 360 pc estimated by means of the λ2.2 μm radiation. The radius of the giant is 450 $R_\odot$.

117.168 Symbiotic stars.

A. A. Boyarchuk.
Astrophys. Space Phys. Rev., Vol. 3, p. 123 – 155 (1984). – See Abstr. 003.019. Revised and extended English translation of 34.117.138.

Contents: (1). Introduction. (2). Basic observations. (3). Physical conditions in symbiotic stars. (4). Models of symbiotic stars.

117.169 The role of the magnetic stellar wind in low–mass close binary evolution.

A. V. Tutukov.
Astrofizika, Tom 21, Vyp. 3, p. 573 – 586 (1984). In Russian. English translation in Astrophysics, Vol. 21, No. 3.

Magnetic stellar wind (MSW) can play a decisive role in the origin and evolution of low–mass binaries with semimajor axis smaller than $\sim 10 - 12\ R_\odot$, if at least one of the components has a convective envelope. The MSW can also explain the origin and, possibly, evolution of W UMa, cataclysmic binary stars and the origin of a part of "blue stragglers".

117.170 Spectral and temporal studies of various late–type stars.

A. C. Brinkman, E. H. B. M. Gronenschild, R. Mewe, I. McHardy, J. P. Pye.
Adv. Space Res., Vol. 5, No. 3, p. 65 – 68 (1985). – See Abstr. 012.059.

The RS CVn stars Capella and σ^2CrB have been measured with EXOSAT in soft and medium X–rays for about 24 hours each and the less active late–type star Procyon for about 6.5 hours. In addition, the RS CVn star λ And was twice observed about one month apart for a total of about 7 hours, with the ME and the LE in the photometer mode only. All three RS CVn stars were detected with the ME–detector. A preliminary discussion is given by comparing the various spectra.

117.171 The soft X–ray superoutburst of VW Hydri: 14 second periodicity.

H. van der Woerd, J. Heise, F. Paerels.
Adv. Space Res., Vol. 5, No. 3, p. 77 – 80 (1985). – See Abstr. 012.059.

EXOSAT observations of the dwarf nova VW Hydri reveal a strong soft X–ray flux during optical superoutburst. The onset of the X–ray outburst was delayed by 2.5 days compared to the optical outburst. A modulation of the extreme soft X–ray flux was detected, consistent with a coherent ($Q > 10^7$) pulsation with a period of 14.07 seconds, probably reflecting the rotation period of the white dwarf. If this is indeed the case, VW Hydri is the fastest rotating white dwarf detected so far.

117.172 Energy spectra of non–pulsating low mass X–ray binaries.

M. Matsuoka, K. Mitsuda.
Adv. Space Res., Vol. 5, No. 3, p. 101 – 104 (1985). – See Abstr. 012.059.

Detailed analyses of data obtained with the gas scintillation proportional counter on Tenma have been done for four low mass X–ray binaries (LMXB's); that is, Sco X–1, 4U1608–52, GX 349+2, and GX 5–1. The paper presents a new scenario concerning the spectra of LMXB's based on the observational fact from Tenma. The energy spectra of these sources can be expressed by a sum of two spectral components; that is, a hard component with a blackbody of temperature kT $\cong$ 2 keV and soft component with a multi–color blackbody of maximum temperature kT $\cong$ 1.4 keV which is expected from the optically thick accretion disk.

117.173 Is Cygnus X–3 a low–mass X–ray binary?

M. van der Klis, F. Jansen.
Adv. Space Res., Vol. 5, No. 3, p. 109 – 112 (1985). – See Abstr. 012.059.

The authors present examples of the quasi–periodic variations in the X–ray flux of Cyg X–3 which they have recently found during observations of this source with EXOSAT. Amplitudes and periods of the variations range from 5% to 20% of the total flux and from 50 to 1500 s, respectively. The tentative interpretation of these quasi–periodicities, the occurrence of quasi–periodic phenomena in an accretion disk which partially occults the X–ray source, points towards an analogy of Cyg X–3 with certain "dipping" low–mass X–ray binaries such as 4U 1822–37. The authors point out, however, that there are also fundamental differences between Cyg X–3 and this type of low–mass X–ray binary.

117.174 Exosat observations of the galactic bulge X–ray source GX 17 + 2.

M. Sztajno, J. Trümper, H. U. Zimmermann, A. Langmeier.
Adv. Space Res., Vol. 5, No. 3, p. 113 – 116 (1985). – See Abstr. 012.059.

The authors have measured the X–ray flux of the bright galactic bulge source GX 17+2 in the energy range 1 – 20 keV using the Exosat ME experiment. During 8 hours of continuous observation an X–ray flare was observed (lasting ~ 1 hr) followed by an intensity increase. The data show intensity dips with a quasi-period of ~ 1.4 hours and quasi–periodic oscillations on time scale of 200 – 500 sec, which are possibly connected with oscillations of an accretion disc. The spectrum can be fitted by two blackbody spectra with $kT_1 \sim 1$ keV, and $kT_2 \sim 2$ keV, respectively, and an iron line at 6.3 ± 0.3 keV having 130 eV equivalent width.

117.175   Optical and X–ray observations of 4U2129 + 47/V1727 Cyg in a quiescent state.
W. Pietsch, H. Steinle, M. Gottwald.
Adv. Space Res., Vol. 5, No. 3, p. 117 – 120 (1985). – See Abstr. 012.059.

The authors observed the 5.2 h X–ray binary 4U2129 + 47 for more than one orbital cycle on 29 September and 4 October 1983 using the LE, ME and GSPC detectors of Exosat. In neither detector an X–ray flux from the source could be detected. Quasi-simultaneous UBV observations failed to detect the large amplitude light curve reported in earlier observations but show the optical companion in a low intensity state. The large amplitude light curve has been interpreted as due to X–ray heating of the optical star by the X–ray source similar to the system Her X–1/HZ Her. The optical observations indicate that the heating X–ray source has been shut off and nicely explain that Exosat failed to detect the source.

117.176   Search for millisecond rotational periods in some low–mass X–ray binaries observed by Exosat.
A. Langmeier, M. Sztajno, J. Trümper.
Adv. Space Res., Vol. 5, No. 3, p. 121 – 123 (1985). – See Abstr. 012.059.

The authors present the results of a search for millisecond rotational pulsations in the X–ray emission of five low–mass X–ray binaries: MXB 1728–34, 4U 1702–42, 4U 1705–44, 4U 1755–33 and GX 17 + 2. The data were obtained by the Exosat Medium Energy Experiment Argon Counters (1 – 20 keV) with a time resolution of 7.8125 milliseconds. The authors searched for periods in the range from 16 msec to 8 sec. Upper limits of 1.5 – 5% were found at a 3 sigma level.

117.177   Hard X–ray observation of galactic X–ray sources.
F. Frontera, D. Dal Fiume, W. Dusi, E. Morelli, G. Spada, G. Ventura.
Adv. Space Res., Vol. 5, No. 3, p. 125 – 128 (1985). – See Abstr. 012.059.

A large (1455 cm^2) hard X–ray telescope was successfully launched aboard a stratospheric balloon on October 4, 1980. During this flight four galactic X–ray sources were observed, namely the transient recurrent X–ray pulsar A0535 + 26, the Crab Nebula, Cygnus X–1 and X Persei. The authors report the results on the latter two sources.

117.178   Absolute photometry of HM Sagittae in 1983.
O. D. Dokuchaeva, E. B. Kostyakova, A. Yu. Shchelkanova.
Astron. Tsirk., No. 1345, p. 1 – 3 (1984). In Russian.

117.179   New WR–type star in CMa.
A. L. Gyul'budagyan, K. G. Gasparyan, R. Sh. Natsvlishvili.
Astron. Tsirk., No. 1348, p. 7 – 8 (1984). In Russian.

117.180   Cygnus X–3: cosmic–ray powerhouse.
D. H. Smith.
Sky Telesc., Vol. 69, No. 6, p. 497 – 500 (1985).

117.181   Photoelectric photometry of VW Hydri during the 1983 November superoutburst.
B. F. Marino, W. S. G. Walker, G. C. D. Herdman, W. H. Allen.
South. Stars, Vol. 31, No. 1, p. 61 – 71 (1984).

Photoelectric observations made at the Auckland Observatory and at Blenheim during the 1983 November superoutburst of VW Hydri are summarised. These show a two event structure in the light and colour curves with superhumps first appearing on the third night of the outburst.

117.182   New photoelectric observations of the Wolf–Rayet star HD 197406 with probable relativistic companion.
I. I. Antokhin.
Astron. Tsirk., No. 1350, p. 1 – 5 (1984). In Russian.

New photoelectric observations of the Wolf–Rayet star HD 197406 with probable relativistic companion were obtained during 1981 – 1983. The period of photometric variability of this star is stable over about 1500 periods. The stability of period of HD 197406 gives strong argument for the double nature of this star.

117.183   The binary system MWC 349.
M. Cohen, J. H. Bieging, J. W. Dreher, W. J. Welch.
Astrophys. J., Vol. 292, No. 1, p. 249 – 256 (1985).

Using the VLA the authors have resolved the ionized zone surrounding MWC 349. They see definite evidence for an interaction between a wind from MWC 349B and that from MWC 349A, in the form of an arc of enhanced radio emission. This interaction shows that the two stars form a binary system. From the 5 GHz data a wind temperature of 8700 ± 300K and a mass loss rate of $(1.2 \pm 0.05) \times 10^{-5} M_\odot \text{yr}^{-1}$ are derived for MWC 349A. The authors present an optical spectrum of the B component of this binary system that enables a direct spectral classification of MWC 349B as a B0 III star, at a distance of 1.2 kpc. The unusual geometry of the radio emission, the low infrared luminosity, and this extra extinction could all be explained by an edge–on, circumstellar, dusty disk close to the orbital plane of the binary.

117.184   X–ray emission from Be star binaries.
K. M. V. Apparao.
Astrophys. J., Vol. 292, No. 1, p. 257 – 259 (1985).

A model in which a neutron star in an eccentric inclined orbit around a Be star passes through a ring of matter around the Be star has been used to explain the recurrent flares of X–ray emission from these systems. The optical emission during X–ray flares is due to partial absorption and reprocessing of the X–rays in the ring.

117.185   Optical and infrared pulsations from the HZ Herculis binary system during the 1983 prolonged X–ray low state.
J. Middleditch, R. C. Puetter, C. R. Pennypacker.
Astrophys. J., Vol. 292, No. 1, p. 267 – 275 (1985).

Pulsed infrared and optical flux from HZ Her has been detected during the 1983 prolonged X–ray low state by simultaneous observations with the Lick Observatory Shane 3.1 m and Crossley 91 cm telescopes. The pulsed fluxes in the 1.0 – 2.5 μm bandpass and the 3200 – 7500 Å bandpass agree in pulse frequency and phase and were measured to be 31 μJy and 13 μJy, respectively. These pulsed fluxes indicate that the reprocessed pulsation spectrum is consistent with optically thin thermal bremsstrahlung radiation with a characteristic temperature of 25,000K which is modulated in intensity.

117.186   XB 1905 + 000: the most distant optically identified X–ray burster?
C. Chevalier, S. A. Ilovaisky, P. A. Charles.
Astron. Astrophys., Vol. 147, No. 1, p. L3 – L5 (1985).

The authors have identified the optical counterpart of the X–ray burst source XB 1905 + 000 using multicolor CCD images obtained with the ESO 2.2 m telescope in May 1984. This faint star (V = 20.5) has a large UV excess (U–B = –0.5, B–V = 0.5), is found within the combined SAS–3/HEAO A–3 error box, and is furthermore consistent with the reprocessed Einstein HRI source position. Its X–ray to optical luminosity ratio is roughly 400, within the range for other low–mass binaries. If its absolute visual magnitude is close to the average value for other bursters, it may then be the most distant (20 kpc) optically identified object of this class.

117.187   Wolf–Rayet stars in the Magellanic Clouds. III. The WO4 + O4 V binary Sk 188 in the SMC.
A. F. J. Moffat, J. Breysacher, W. Seggewiss.
ESO Sci. Prepr., No. 356, 22 pp. (1984). Submitted to Astrophys. J.

117.188   On the number of W–R stars in the Large Magellanic Cloud.
M. Azzopardi, J. Breysacher.
ESO Sci. Prepr., No. 366, 16 pp. (1985). To appear in Astron. Astrophys.

117.189   Untersuchungen der Zyklotron–Resonanzeffekte im Röntgenspektrum von Her X–1.
W. Voges.
MPE Rep., No. 191, 123 pp. (1985).

117.190   Photoelectric observations of CH Cygni. III.
J. Vennik, T. Tuvikene, L. Luud.
Tartu Astrofüüs. Obs. Publ., Tom 50, p. 170 – 174 (1984). In Russian.

The results from UBVIJHK photometry of CH Cygni during the Julian days 2444215 – 2445009 and the light curve for 1967 – 1982 have been given. Rapid brightness variations exceeding $0^{\rm m}\!.1$ have been detected during the observations. In 1981 the brightness of CH Cygni was nearly $0^{\rm m}\!.5$ greater than has been observed ever before.

117.191   Photoelectric observations of V711 Tau = HR 1099.
M. B. K. Sarma, B. D. Ausekar.
Contrib. Nizamiah Japal–Rangapur Obs., No. 15, 33 pp. (1981).

117.192   Accretion disks heated by luminous central stars.
M. Friedjung.
Inst. Astrophys. Paris, Pré–Publ., No. 93, 11 pp. (1985). To appear in Astron. Astrophys.

117.193   The spectrum of SS433 in the stage of eruptive activity (July, 1980).
I. M. Kopylov, R. N. Kumajgorodskaya, T. A. Somova.
Astron. Zh., Tom 62, Vyp. 2, p. 323 – 338 (1985). In Russian. English translation in Sov. Astron., Vol. 29, No. 2.

The behaviour of W_λ, R_c, $\Delta\lambda\,(1/2)$ and v_r parameters of "stationary" and "relativistic" emission lines and absorption ("blue" and "red") components of "stationary" H and He I lines in the spectrum of SS433 during the large optical flare ($\Delta m \sim 0^{\rm m}\!.8$) from July 13 to 20 was studied. It is shown that the optical flare was accompanied by considerable changes of intensities, widths, depths and radial velocities of all the lines in the spectrum of SS433. Essential differences in the character of the radial velocity, profile and intensity variations of the positive (H^+, He I$^+$) and negative (H^-, He I$^-$) systems of "relativistic" lines are noted. Evidences are given in favour of the fact that the explosion occurred in the central parts of the disk near the compact component of the system and was accompanied by a disk envelope collapse and by an increase of relativistic jets activity.

117.194   The optical continua of magnetic variables.
G. D. Schmidt, H. S. Stockman, S. A. Grandi.
Prepr. Steward Obs., No. 579, 53 pp. (1985).

Optical circular spectropolarimetry and spectroscopy with full phase coverage are presented for four well–known AM Herculis magnetic variables. The authors propose a refinement to the "standard" accretion model for magnetic variables which appears to improve the agreement with observed optical continua. The new picture suggests an accretion profile which is not perfectly collimated, where the optical/IR continuum can be dominated by a low–density halo around the compact hard X–ray shock.

117.195   Wolf–Rayet stars in the Magellanic Clouds. III. The WO4 + O4 V binary Sk 188 in the SMC.
A. F. J. Moffat, J. Breysacher, W. Seggewiss.
Astrophys. J., Vol. 292, No. 2, p. 511 – 516 (1985).

Emission and antiphased absorption line orbits ($P = 16.644\,d$, $e = 0$) show that this massive system contains a Wolf–Rayet (W–R) star with relatively low mass ratio ($M_{\rm W-R}/M_\odot \approx 0.27$). It is interpreted to be the most evolved W–R star known in a close, massive binary and fits the scenario of Barlow and Hummer that oxygen–rich WO stars are in a late–WC stage as a result of advanced nuclear burning.

117.196   Hydrodynamical modeling of mass transfer from cataclysmic variable secondaries.
R. L. Gilliland.
Astrophys. J., Vol. 292, No. 2, p. 522 – 534 (1985).

Results of detailed computer simulations of mass loss from a cataclysmic variable star secondary are presented. The calculations involve solution of the nonlinear hydrodynamical equations of stellar structure under varying degrees of approximation in an attempt to determine stability of the mass loss process. Dynamical sequences treating only the region near the Lagrangian point were constructed in a manner consistent with the assumption of Roche geometry. The effects of mass flow nonorthogonal to the Roche equipotential surfaces were treated in a very simple way. The finding of stable mass transfer implies that instability of the secondary star is not the mechanism leading to cataclysmic variable outbursts.

117.197   X–ray emission from cataclysmic variables with accretion disks. I. Hard X–rays.
J. Patterson, J. C. Raymond.
Astrophys. J., Vol. 292, No. 2, p. 535 – 549 (1985).

The authors discuss how the currently available X–ray and optical data can be used to constrain boundary–layer models for the origin of the hard X–ray emission in cataclysmic variables. They compare the observed and predicted dependences of the X–ray luminosity and X–ray–to–visual flux ratio F_x/F_v on accretion rate. This comparison reveals that systems with a low accretion rate are invariably strong X–ray emitters, with $F_x/F_v \sim 3$. The X–ray fluxes of stars with a high accretion rate are not consistent with any of the published models, but could be explained by invoking a density gradient in the boundary layer. The authors also discuss the effects of reprocessing boundary–layer radiation on the white dwarf surfaces.

117.198   X–ray emission from cataclysmic variables with accretion disks. II. EUV/soft X–ray radiation.
J. Patterson, J. C. Raymond.
Astrophys. J., Vol. 292, No. 2, p. 550 – 558 (1985).

About half of the gravitational luminosity released by gas accreting onto a white dwarf through a disk should emerge from the star/disk *boundary layer*. For the accretion rates present in many cataclysmic variables, theory predicts that this luminosity should be in the form of an optically thick EUV/soft X–ray component, with $T_e \approx (1-3)\times10^5$K. The authors compare the theoretical predictions with presently available soft X–ray observations and find satisfactory agreement. They also attempt to constrain the boundary layer radiation by comparing observed and predicted strengths of the He II $\lambda1640$ and $\lambda4686$ emission lines, assuming that these are produced by photoionization in the upper layers of the disk.

117.199   On the nature of the UX Ursae Majoris–type nova–like variables: CPD –48°1577.
E. M. Sion.
Astrophys. J., Vol. 292, No. 2, p. 601 – 605 (1985).

A series of low–dispersion spectra and one high–dispersion spectrum of CPD –48°1577 were obtained with the IUE during 1983 April 19 – 20 and 1984 February 27 – 28. The ultraviolet spectra reveal broad, highly ionized, violet–displaced, asymmetric absorption features with C IV showing a P Cygni emission component as well. The inferred velocity of the outflowing wind is ~ 3000 km s^{-1}. An energy distribution fit covering the far–ultraviolet to infrared yields an accretion rate $\dot{M}_{\rm acc} \approx 5\times10^{-9}M_\odot{\rm yr}^{-1}$. The derived parameters establish the system as a member of the UX Ursae Majoris subclass of nova–like variables.

117.200 35 day spectroscopic effects in HZ Herculis.
J. B. Hutchings, E. M. Gibson, D. Crampton,
W. A. Fisher.
Astrophys. J., Vol. 292, No. 2, p. 670 – 675 (1985).
New spectroscopic data are presented of HZ Her, which have
~2 Å resolution, and cover both 1.7 day orbital and 35 day (precessional?) periods. Data include measures of principal
absorption–line velocities and asymmetries, and emission–line
velocities and intensities. Orbital parameters derived from
Balmer and He I velocities are found to be modulated over half
the 35 day period of the system. This can be generally understood
in terms of a variable X–ray heating, modulated by a thick precessing disk about the secondary. Models for such a variable
heating fit the principal effects and yield a primary star orbital
velocity of 83 ± 3 km s^{-1}.

**117.201 Additional identifications of high ionization stages of
iron and nickel in the ultraviolet spectrum of the slow
nova RR Telescopii.**
A. J. J. Raassen.
Astrophys. J., Vol. 292, No. 2, p. 696 – 698 (1985).
The wavelengths of forbidden transitions within the ground
configuration for around five times ionized iron and nickel have
been calculated from level values, established from new, partly
unpublished laboratory analyses. The predicted wavelengths
have been compared with recently published results of *IUE* observations of the slow nova RR Tel. Several unclassified or partly
classified lines are assigned to transitions belonging to the multiplet $^5D - ^3D$ in the $3d^4$ configuration of Fe V, while four unclassified lines in the line list of RR Tel with wavelengths below
3000 Å are identified as forbidden Fe VII, Ni VI, and Ni VII
transitions.

**117.202 Nova–like objects and dwarf novae during outburst – a
comparative study.**
W. F. Wargau.
Messenger, No. 40, p. 7 – 11 (1985).

**117.203 The dwarf nova SVS 2549, a shortperiodic eclipsing sys-
tem.**
V. P. Goranskij, S. Yu. Shugarov, E. I. Orlowsky,
V. Yu. Rahimov.
Inf. Bull. Variable Stars, No. 2653, 4 pp. (1985).

**117.204 Hα emission in RS CVn stars: BD + 61°1211 and
HD 37847.**
H.–s. Tan, X.–f. Liu.
Inf. Bull. Variable Stars, No. 2669, 4 pp. (1985).

**117.205 Improved period for the cataclysmic variable
V795 Her = PG 1711 + 336: is it an outstanding object?**
A. V. Baidak, N. A. Lipunova, S. Yu. Shugarov,
V. G. Moshkalev, I. M. Volkov.
Inf. Bull. Variable Stars, No. 2676, 2 pp. (1985).

117.206 AK Her – new V light curve and period change.
Z. Głownia.
Inf. Bull. Variable Stars, No. 2677, 3 pp. (1985).

117.207 Light curve of BV Dra.
M. Hamdy, N. S. Awadalla, A. B. Morcos.
Inf. Bull. Variable Stars, No. 2682, 2 pp. (1985).

117.208 Light variation of BW Dra.
N. S. Awadalla, M. Hamdy, A. B. Morcos.
Inf. Bull. Variable Stars, No. 2683, 2 pp. (1985).

117.209 HR 6384: a probable interacting binary.
T. B. Ake, S. B. Parsons.
Inf. Bull. Variable Stars, No. 2686, 2 pp. (1985).

**117.210 Infrarot–Photometrie der novaähnlichen Systeme
RW Sex und CPD –48°1577.**
K. Haug, H. Drechsel.
Mitt. Astron. Ges., Nr. 63, p. 193 – 194 (1985). – See Abstr.
012.063.

**117.211 Objective grating X–ray spectroscopy of compact
sources.**
S. M. Kahn.
Bull. Am. Astron. Soc., Vol. 16, No. 4, p. 880 (1984). Abstract. –
See Abstr. 010.062.

**117.212 VW Cephei color curves and the W–subclass W Ursae
Majoris phenomenon.**
A. P. Linnell.
Bull. Am. Astron. Soc., Vol. 16, No. 4, p. 883 (1984). Abstract. –
See Abstr. 010.062.

117.213 Photometric analyses of S Cancri.
P. B. Etzel, E. C. Olson.
Bull. Am. Astron. Soc., Vol. 16, No. 4, p. 883 (1984). Abstract. –
See Abstr. 010.062.

117.214 Recent photometric observations of CG Cygni.
S. A. Naftilan, G. Trager.
Bull. Am. Astron. Soc., Vol. 16, No. 4, p. 883 (1984). Abstract. –
See Abstr. 010.062.

117.215 Evidence for a starspot cycle of V711 Tauri.
E. F. Guinan, S. W. Wacker.
Bull. Am. Astron. Soc., Vol. 16, No. 4, p. 883 – 884 (1984). Ab-
stract. – See Abstr. 010.062.

**117.216 Color snapshot of a dwarf nova accretion disk in out-
burst.**
K. Horne, M. C. Cook.
Bull. Am. Astron. Soc., Vol. 16, No. 4, p. 884 (1984). Abstract. –
See Abstr. 010.062.

**117.217 The evolution of the UV line spectrum in the dwarf nova
SS Cyg through outburst.**
R. S. Polidan, J. B. Holberg, A. V. Holm, T. E. Carone.
Bull. Am. Astron. Soc., Vol. 16, No. 4, p. 884 (1984). Abstract. –
See Abstr. 010.062.

117.218 The many faces of HR 1099.
T. R. Ayres, J. O. Bennett, J. L. Linsky, T. Simon.
Bull. Am. Astron. Soc., Vol. 16, No. 4, p. 893 (1984). Abstract. –
See Abstr. 010.062.

**117.219 An Hα survey of short period RS CVn and W UMa
binaries: general characteristics.**
S. C. Barden.
Bull. Am. Astron. Soc., Vol. 16, No. 4, p. 893 (1984). Abstract. –
See Abstr. 010.062.

**117.220 Atmospheric structures in AR Lac. I. Mapping quiescent
features by occultations and Doppler imaging.**
F. M. Walter, D. M. Gibson, A. Brown, K. G. Carpenter,
J. L. Linsky, M. Rodono, C. Eyles.
Bull. Am. Astron. Soc., Vol. 16, No. 4, p. 896 (1984). Abstract. –
See Abstr. 010.062.

**117.221 Atmospheric structures in Ar Lac. II. A spatially re-
solved chromospheric active region.**
J. E. Neff, F. M. Walter, J. L. Linsky, D. M. Gibson,
M. Rodono.
Bull. Am. Astron. Soc., Vol. 16, No. 4, p. 896 (1984). Abstract. –
See Abstr. 010.062.

**117.222 Atmospheric structures in AR Lac. III. VLA observa-
tions of the radio eclipses during the 1984 campaign.**
D. M. Gibson.
Bull. Am. Astron. Soc., Vol. 16, No. 4, p. 897 (1984). Abstract. –
See Abstr. 010.062.

117.223 **Infrared spectroscopy of the symbiotic star CH Cygni.**
K. H. Hinkle, W. W. G. Scharlach,
A. D. Shaw–Hanson.
Bull. Am. Astron. Soc., Vol. 16, No. 4, p. 897 (1984). Abstract. –
See Abstr. 010.062.

117.224 **Simultaneous Voyager, IUE, and ground based observations of SS Cyg.**
R. S. Polidan, J. B. Holberg, A. V. Holm, J. A. Mattei,
T. E. Carone.
Bull. Am. Astron. Soc., Vol. 16, No. 4, p. 898 (1984). Abstract. –
See Abstr. 010.062.

117.225 **Balmer decrements in active chromosphere stars.**
D. P. Huenemoerder, L. W. Ramsey.
Bull. Am. Astron. Soc., Vol. 16, No. 4, p. 912 (1984). Abstract. –
See Abstr. 010.062.

117.226 **Photoelectric photometry of the W UMa system HI Pup.**
J. R. Kern, B. B. Bookmyer.
Bull. Am. Astron. Soc., Vol. 16, No. 4, p. 912 (1984). Abstract. –
See Abstr. 010.062.

117.227 **IUE and optical observations of the symbiotic stars He2–38 and He2–106.**
J. H. Lutz.
Bull. Am. Astron. Soc., Vol. 16, No. 4, p. 913 (1984). Abstract. –
See Abstr. 010.062.

117.228 **On fast X–ray rotators with long–term periodicities.**
R. F. Elsner, W. Darbro, B. D. Ramsey,
M. C. Weisskopf, A. C. Williams, S. Naranan, P. E. Hardee,
D. Leahy, P. G. Sutherland, J. E. Grindlay.
Bull. Am. Astron. Soc., Vol. 16, No. 4, p. 913 (1984). Abstract. –
See Abstr. 010.062.

117.229 **Coherent wavelength variations in DQ Herculis stars.**
W. R. Penning.
Bull. Am. Astron. Soc., Vol. 16, No. 4, p. 913 (1984). Abstract. –
See Abstr. 010.062.

117.230 **Surprisingly long orbital periods for two cataclysmic binary systems in the Palomar–Green sample.**
J. R. Thorstensen.
Bull. Am. Astron. Soc., Vol. 16, No. 4, p. 914 (1984). Abstract. –
See Abstr. 010.062.

117.231 **The 1983 September radio outburst of Cyg X–3.**
K. J. Johnston, J. Spencer, R. S. Simon, E. Waltman,
R. E. Spencer, R. Swinney, G. Pooley, P. Angerhofer,
D. Florkowski, D. McCarthy, D. Matsakis, R. Hjellming.
Bull. Am. Astron. Soc., Vol. 16, No. 4, p. 914 (1984). Abstract. –
See Abstr. 010.062.

117.232 **Cyg X–3 not seen in the COS–B > 100 MeV gamma rays.**
P. A. Caraveo, G. F. Bignami.
Bull. Am. Astron. Soc., Vol. 16, No. 4, p. 933 (1984). Abstract. –
See Abstr. 010.062.

117.233 **Ultra high energy gamma–rays and cosmic rays from accreting degenerate stars.**
K. Brecher, G. Chanmugam.
Bull. Am. Astron. Soc., Vol. 16, No. 4, p. 934 (1984). Abstract. –
See Abstr. 010.062.

117.234 **Evidence for a cyclotron feature from Cygnus X–3.**
J. C. Ling, W. A. Mahoney, W. A. Wheaton,
A. S. Jacobson.
Bull. Am. Astron. Soc., Vol. 16, No. 4, p. 934 – 935 (1984). Abstract. – See Abstr. 010.062.

117.235 **The spectra of four symbiotic SS stars in Sagittarius.**
B. M. Lasker, T. D. Kinman, M. F. McCarthy.
Bull. Am. Astron. Soc., Vol. 16, No. 4, p. 939 (1984). Abstract. –
See Abstr. 010.062.

117.236 **A wind model for ultraviolet resonance lines in cataclysmic variables.**
F. Verbunt, J. E. Drew.
Bull. Am. Astron. Soc., Vol. 16, No. 4, p. 943 (1984). Abstract. –
See Abstr. 010.062.

117.237 **Search for gamma ray lines from SS433.**
B. J. Geldzahler, G. H. Share, R. L. Kinzer,
D. J. Forrest, E. Reiger, E. L. Chupp.
Bull. Am. Astron. Soc., Vol. 16, No. 4, p. 963 (1984). Abstract. –
See Abstr. 010.062.

117.238 **The magnetic accretion disk in cataclysmic variables.**
D. H. Ferguson, D. S. Dearborn.
Bull. Am. Astron. Soc., Vol. 16, No. 4, p. 971 (1984). Abstract. –
See Abstr. 010.062.

117.239 **Hercules X–1/HZ Herculis: yet another periodicity?**
S. D. Vrtilek.
Bull. Am. Astron. Soc., Vol. 16, No. 4, p. 982 (1984). Abstract. –
See Abstr. 010.062.

117.240 **Slow variability of three X–ray pulsars.**
D. E. Gruber.
Bull. Am. Astron. Soc., Vol. 16, No. 4, p. 982 (1984). Abstract. –
See Abstr. 010.062.

117.241 **Pseudo white dwarfs.**
J. Patterson.
Bull. Am. Astron. Soc., Vol. 16, No. 4, p. 982 (1984). Abstract. –
See Abstr. 010.062.

117.242 **Implications of observed pulse frequency variations for the structure of the neutron in Vela X–1.**
F. K. Lamb, G. J. Zylstra, P. E. Boynton, J. E. Deeter.
Bull. Am. Astron. Soc., Vol. 16, No. 4, p. 983 (1984). Abstract. –
See Abstr. 010.062.

117.243 **Implication of observed pulse frequency variations for the accretion flow in Vela X–1.**
N. Shibazaki, F. K. Lamb, G. J. Zylstra, P. E. Boynton,
J. E. Deeter.
Bull. Am. Astron. Soc., Vol. 16, No. 4, p. 983 (1984). Abstract. –
See Abstr. 010.062.

117.244 **Cygnus X–1: the 294–day period, a likely disk precession.**
J. C. Kemp, G. D. Henson, D. J. Kraus.
Bull. Am. Astron. Soc., Vol. 16, No. 4, p. 983 – 984 (1984). Abstract. – See Abstr. 010.062.

117.245 **Absorption spectra from massive X–ray binaries.**
T. Kallman, J. Swank.
Bull. Am. Astron. Soc., Vol. 16, No. 4, p. 984 (1984). Abstract. –
See Abstr. 010.062.

117.246 **New insights into interacting stellar systems.**
Y. Kondo.
Bull. Am. Astron. Soc., Vol. 16, No. 4, p. 1011 (1984). Abstract. – See Abstr. 010.062.

117.247 **Spectral imagery of R Aquarii in O III 5007 Å.**
W. V. Schempp.
Bull. Am. Astron. Soc., Vol. 16, No. 4, p. 1012 (1984). Abstract. – See Abstr. 010.062.

117.248 Accretion characteristics of the AM Her stars CW 1103+254, E 1114+182 and PG 1550+191.
P. Szkody, J. W. Liebert, R. J. Panek.
Bull. Am. Astron. Soc., Vol. 16, No. 4, p. 1014 (1984). Abstract. – See Abstr. 010.062.

117.249 Photometry of EX Hya.
B. S. Shylaja.
Astrophys. Space Sci., Vol. 111, No. 2, p. 407 – 411 (1985).
   Photometric observations of EX Hya in B and V filters are reported. The 67 min modulation of the light curve is also found to be in good agreement with the results of earlier studies. The $(B-V)$ colour variation with respect to the 67 min variation is found to be opposite to those of typical colour variation during hump/superhump activity in other dwarf novae. The model of an intermediate polar is discussed.

117.250 Photoelectric light curves of RT Coronae Borealis and their analysis.
C. Ibanoğlu, A. Y. Ertan, O. Tümer, S. Evren, Z. Tunca.
Astrophys. Space Sci., Vol. 112, No. 1, p. 133 – 143 (1985).
   The first photoelectric light curves of the RS CVn–type eclipsing binary RT CrB are presented. There is no indication about the existence of wave like distortion on its light curve. The light curves were analyzed by using Wood's and Nelson, Davis, and Etzel's methods. The radii of the components are nearly identical. The solutions indicate that the cooler component has a later spectral type between K0 and K2.

117.251 The short–period eclipsing binary V1010 Ophiuchi does it have an eccentric orbit?
K.-c. Leung, G.-j. Qiao.
Astrophys. Space Sci., Vol. 112, No. 2, p. 267 – 272 (1985).
   V1010 Oph is an unusually complicated close binary. Leung and Wilson (1977) found the system had an overcontact configuration. Subsequently, Margoni et al. (1981) claimed that the system had an eccentric orbit with $e = 0.25$ based upon their spectroscopic study. A theoretical radial velocity curve based on the photometric parameters (e.g., $e = 0$) of Leung and Wilson has been computed. The computed curve fits the Asiago and Kitt Peak data very well. It is suggested that the asymmetry in the observed radial velocity arose from tidal, reflection, and eclipse effects rather than from orbital eccentricity. Other arguments against an eccentric orbit for V1010 Oph are also discussed. It is concluded that the eccentricity derived from the spectroscopic study may be spurious.

117.252 The symbiotic star AG Pegasi: UBV photometry during 1962 – 1984.
T. S. Belyakina.
Inf. Bull. Variable Stars, No. 2697, 2 pp. (1985).

117.253 The symbiotic star Z Andromedae: UBV photometry during 1974 – 1984.
T. S. Belyakina.
Inf. Bull. Variable Stars, No. 2698, 2 pp. (1985).

117.254 NSV 6044: a W UMa star.
B. E. Helt.
Inf. Bull. Variable Stars, No. 2699, 4 pp. (1985).

117.255 A flare on the contact binary CN And.
Y.-l. Yang, Q.-y. Liu.
Inf. Bull. Variable Stars, No. 2705, 3 pp. (1985).

117.256 New times of minima, and a recent period increase, in U Cephei.
E. C. Olson, J. P. Hickey, L. Humes, V. Paylor.
Inf. Bull. Variable Stars, No. 2707, 3 pp. (1985).

117.257 Starspots and the Mg II emission of VW Cephei.
J. A. Eaton.
Inf. Bull. Variable Stars, No. 2711, 3 pp. (1985).

117.258 Limits on the variability of Epsilon Eridani and Delta Eridani.
J. A. Eaton, C. H. Poe.
Inf. Bull. Variable Stars, No. 2712, 2 pp. (1985).

117.259 On the cycle–length of RZ Leonis.
G. A. Richter.
Inf. Bull. Variable Stars, No. 2714, 2 pp. (1985).

117.260 FS Lup – variable star type EA or EB?
A. Terzan, P. Didelon.
Inf. Bull. Variable Stars, No. 2716, 3 pp. (1985).

117.261 Photometric observations of the RS CVn binary σ CrB.
P. Vivekananda Rao, M. B. K. Sarma, P. C. Agrawal, M. V. K. Apparao.
Inf. Bull. Variable Stars, No. 2721, 3 pp. (1985).

117.262 Analysis of the non–variability in the 78–day binary HR 503.
J. A. Eaton, L. J. Boyd, R. M. Genet, D. S. Hall.
Inf. Bull. Variable Stars, No. 2722, 3 pp. (1985).

117.263 Spectra of 10 symbiotic stars.
M. T. Martel, R. Gravina.
Inf. Bull. Variable Stars, No. 2724, 6 pp. (1985).

117.264 Behaviour of the X–ray binary V1727 Cygni = 4U 2129+47 in 1984.
W. Götz.
Inf. Bull. Variable Stars, No. 2732, 2 pp. (1985).

117.265 Observations of AT Cancri in the season 1984/85.
W. Götz.
Inf. Bull. Variable Stars, No. 2734, 3 pp. (1985).

117.266 Optical behaviour of the polar ST Leonis Minoris = CW 1103+254.
W. Götz.
Inf. Bull. Variable Stars, No. 2735, 3 pp. (1985).

117.267 HT Cassiopeiae (unusual dwarf nova).
Yamamoto Circ., No. 2033 (1985). In Japanese.

117.268 AG Draconis.
Yamamoto Circ., No. 2036 (1985). In Japanese.

117.269 The new eclipsing magnetic binary system E1114+182.
P. Biermann, G. D. Schmidt, J. Liebert, H. S. Stockman, S. Tapia, H. Kühr, P. A. Strittmatter, S. West, D. Q. Lamb.
Astrophys. J., Vol. 293, No. 1, p. 303 – 320 (1985).
   The authors present a comprehensive analysis of E1114+182, the first eclipsing AM Herculis binary system and the shortest period eclipsing cataclysmic variable known. This paper reports (1) the time–resolved X–ray observations which led to its recognition as an AM Her system with a ~90 minute orbital period, (2) the current optical photometric and polarimetric ephemeris and a description of its phase–modulated properties, (3) the detailed photometric eclipse profile, and (4) the highly variable spectroscopic behavior. This information is used to determine systematic parameters and glean new information on the line emission regions. The data put severe constraints on current torque models for keeping the binary and white dwarf rotation in phase.

117.270 IUE results on the AM Herculis stars CW 1103, E1114, and PG 1550.
P. Szkody, J. Liebert, R. J. Panek.
Astrophys. J., Vol. 293, No. 1, p. 321 – 327 (1985).
   The authors present IUE data on three AM Her stars (CW 1103+254), E1114+182, and PG 1550+191) which are used in conjunction with optical and IR fluxes to study the accretion characteristics of these systems in relation to other polars. The time–resolved IUE spectra of CW 1103 show that the col-

umn contributes little to the UV, while the white dwarf, with a temperature of $\sim 13{,}000$K and a distance of ~ 140 pc, is the dominant source of light.

117.271 RZ Leonis.
IAU Circ., Nos. 4026, 4027, 4036 (1985).

117.272 HT Cassiopeiae.
IAU Circ., Nos. 4027, 4037 (1985).

117.273 IR Geminorum.
IAU Circ., No. 4030 (1985).

117.274 AG Draconis.
IAU Circ., Nos. 4038, 4045, 4054, 4073 (1985).

117.275 PG 0834 + 488.
IAU Circ., No. 4038 (1985).

117.276 CH Cygni.
IAU Circ., No. 4055 (1985).

117.277 AM Herculis.
IAU Circ., No. 4075 (1985).

117.278 High–resolution coudé observations of the cataclysmic variable SS Cygni.
F. V. Hessman.
Bull. Am. Astron. Soc., Vol. 17, No. 1, p. 512 (1985). Abstract. – See Abstr. 010.064.

117.279 Orbital radial velocity curves for symbiotic stars.
M. Garcia.
Bull. Am. Astron. Soc., Vol. 17, No. 1, p. 519 (1985). Abstract. – See Abstr. 010.064.

117.280 Proximity effects and radial velocity solutions for W UMa contact binaries.
W. Van Hamme, R. E. Wilson.
Bull. Am. Astron. Soc., Vol. 17, No. 1, p. 519 (1985). Abstract. – See Abstr. 010.064.

117.281 Evidence for periodic 10^{15}eV gamma–ray emission from Hercules X–1.
J. W. Elbert, R. M. Baltrusaitis, G. L. Cassiday, R. Cooper, P. R. Gerhardy, E. C. Loh, Y. Mizumoto, P. Sokolsky, P. Sommers, D. Steck, S. Wasserbaech.
Bull. Am. Astron. Soc., Vol. 17, No. 1, p. 522 (1985). Abstract. – See Abstr. 010.064.

117.282 Evolution of close binary systems.
H.–C. Thomas.
High energy astrophysics and cosmology, p. 190 – 206 (1983). – See Abstr. 012.068.
This paper reviews stellar evolution calculations for close binaries. Emphasis is given to the crucial role of mass transfer when total mass and angular momentum are conserved.

117.283 Cataclysmic binaries and related objects in the context of stellar evolution.
H. Ritter.
High energy astrophysics and cosmology, p. 207 – 269 (1983). – See Abstr. 012.068.
This paper reviews theoretical models of the formation and evolution of cataclysmic binaries. These systems contain white dwarf primaries accreting material from low–mass companions.

117.284 The formation of massive white dwarfs in cataclysmic binaries.
W.–Y. Law, H. Ritter.
High energy astrophysics and cosmology, p. 275 – 278 (1983). – See Abstr. 012.068.
White dwarfs (WD's) in cataclysmic binaries (CB's) contrast with single white dwarfs in that, while the masses of most single WD's fall into a narrow range of $\sim 0.12\, M_\odot$ around the mean

mass of $\sim 0.58\, M_\odot$, the masses of WD's in CB's are distributed quite uniformly over almost the entire permitted mass range. This makes a discussion of the formation of higher mass ($\gtrsim 1\, M_\odot$) WD's of particular interest.

117.285 Spectacular changes in the 1985 UV spectrum of CH Cyg.
P. L. Selvelli, M. Hack.
Astron. Express, Vol. 1, Nos. 4 – 6, p. 115 – 126 (1985).
The ultraviolet spectrum of the symbiotic star CH Cygni observed on January 23, 1985, has radically changed from the authors' previous observations. The far ultraviolet continuum is completely flat and a large number of emission lines have appeared: a very broad Lyα, previously completely absent, C IV, N V, Si IV resonance doublets, previously in absorption (C IV and Si IV) or absent (N V), practically all the Fe II lines, previously in absorption, besides the emissions of O I at λλ1304, 1641, Fe II (multiplet 191), C III] 1909 and Si III] 1892, which were always present. Radical variations are also observed in the visual spectrum obtained on January 29, 1985, suggesting that the outburst is near to the end.

117.286 Interacting binary stars: introduction.
J. E. Pringle.
Interacting binary stars, p. 1 – 20 (1985). – See Abstr. 003.030.
Contents: The shapes of stars in close binary systems. Transfer of mass. The effect of mass transfer and mass loss on orbital parameters. Eccentric orbits and non–synchronism.

117.287 Interacting binary stars: nomenclature – the stellar zoo.
P. P. Eggleton.
Interacting binary stars, p. 21 – 37 (1985). – See Abstr. 003.030.
Contents: The evolutionary expansion of stars. Observational evidence for Roche lobe overflow. Observational evidence for "weak interaction" in binaries. Observational evidence for "formerly interacting" binaries.

117.288 X–ray binaries: end points of binary evolution.
A. C. Fabian.
Interacting binary stars, p. 71 – 84 (1985). – See Abstr. 003.030.
Contents: General considerations. The observed X–ray binaries – some particular examples.

117.289 Contact binaries: observational properties.
S. M. Rucinski.
Interacting binary stars, p. 85 – 112 (1985). – See Abstr. 003.030.
Contents: Contact systems as binary stars. Period–colour relation. The light curve synthesis. Spectroscopic observations. Mass ratios. The degree of contact. The internal mass–luminosity relation. A and W types. Chromospheric and coronal activity.

117.290 Contact binaries: theory.
S. M. Rucinski.
Interacting binary stars, p. 113 – 127 (1985). – See Abstr. 003.030.
Contents: Basic problems. The Lucy models. Models with temperature differences. The contact discontinuity models. Thermal–relaxation–oscillation models. Angular momentum loss. Towards the final model. Unsolved problems.

117.291 Cataclysmic variables: observational overview.
R. A. Wade, M. J. Ward.
Interacting binary stars, p. 129 – 176 (1985). – See Abstr. 003.030.
Contents: Introduction. Long–term behaviour of cataclysmic variable stars. Orbital phenomena in disc–type cataclysmic variables. Rapid variability of cataclysmic binary stars. Magnetic cataclysmic variable stars. The recurrent novae.

117.292 Cataclysmic variables: theoretical overview.
G. T. Bath, J. E. Pringle.
Interacting binary stars, p. 177 – 195 (1985). – See Abstr. 003.030.
The authors consider in detail the theoretical structure of cataclysmic variable stars. They concentrate their attention on the

non–magnetic or disc stars in which the transfer flow is not disrupted by a strong magnetic field attached to the white dwarf.

117.293   Interacting binary stars: appendix.
R. Pennington.
Interacting binary stars, p. 197 – 199 (1985). – See Abstr. 003.030.
Various properties of the Roche lobes are tabulated.

117.294   On the binary model for CH Cygni.
L. S. Luud, T. Tomov.
Sov. Astron. Lett., Vol. 10, No. 6, p. 360 – 363 (1984). English translation of 38.117.123.

117.295   On the ratio of the X–ray and gamma–ray emission of SS433.
S. V. Bogovalov, Yu. D. Kotov.
Sov. Astron. Lett., Vol. 10, No. 6, p. 380 – 382 (1984). English translation of 38.117.246.

117.296   The orbit of the HDE 245770 system.
O. É. Aab (*O. Eh. Aab*).
Sov. Astron. Lett., Vol. 10, No. 6, p. 386 – 389 (1984). English translation of 38.117.247.

117.297   The influence of an optically thick disc on the light curve of SX Cas.
K. Pavlovski, S. Kříž.
Bull. Astron. Inst. Czech., Vol. 36, No. 3, p. 153 – 159 (1985).
The influence of an optically thick disc on the light–curves of eclipsing binaries is studied for the case of SX Cas. This system belongs to the strongly interacting binaries of the W Ser type. The authors have assumed it to be a semidetached system, with a hot gainer surrounded by an opaque disc. A special computer code was developed to take into account the disc influence. By trial and error the disc parameters have been derived by fitting the theoretical and observed light curves.

117.298   Photoelectric observations of CN Andromedae: an unusual overcontact binary.
J. B. Rafert, N. L. Markworth, E. J. Michaels.
Publ. Astron. Soc. Pac., Vol. 97, No. 590, p. 310 – 321 (1985).
Recent observations of the W Ursae Majoris System CN And reveal a large temperature difference (≈ 1500K) between substantially overcontact components. The light curves were used to improve the ephemeris and to provide photometric solutions of this system by the computer model of Wilson and Devinney. CN And is probably an example of an extreme overcontact configuration for a binary undergoing thermal relaxation oscillations, with components in poor thermal contact. This contention is supported by anomalous photometric activity roughly centered upon secondary eclipse.

117.299   Angular momentum loss and the evolution of binaries of extreme mass ratio.
R. E. Taam, R. A. Wade.
Astrophys. J., Vol. 293, No. 2, p. 504 – 507 (1985).
Results are presented for 10 evolutionary sequences of close binary systems of extreme mass ratio. Angular momentum loss associated with material outflow from the system, a consequence of the ineffectiveness of tidal torques, as well as gravitational radiation is included in the calculations. The application of the results to the millisecond pulsar PSR 1937 + 214 suggests that its origin does not involve the continuous accretion of matter in a low mass X–ray binary system.

117.300   Evolution of cataclysmic binaries.
B. Paczynski.
Cataclysmic variables and low–mass X–ray binaries, p. 1 – 14 (1985). – See Abstr. 012.072.
A standard scenario for the evolutionary origin of cataclysmic binaries is presented. In this scenario there should be a lot of short period binaries made of pairs of white dwarfs, which were produced by the systems with initial mass ratios close to unity. About 10% of all "single" white dwarfs could be short period binaries, or could have passed through such a phase some time in the past. Such binaries should evolve due to angular momentum losses caused by gravitational radiation and could produce R CrB stars, AM CVn cataclysmic binaries, DB white dwarfs, and Type I Supernovae. Inconsistencies in the standard evolutionary scenario are discussed.

117.301   Pre–cataclysmic binaries.
H. E. Bond.
Cataclysmic variables and low–mass X–ray binaries, p. 15 – 27 (1985). – See Abstr. 012.072.
The author refers to binary nuclei of planetary nebulae, and detached close binaries containing hot subdwarfs or white dwarfs, as "pre–cataclysmic" binaries.

117.302   Abell 41, a cataclysmic variable progenitor.
A. D. Grauer.
Cataclysmic variables and low–mass X–ray binaries, p. 29 – 33 (1985). – See Abstr. 012.072.
The central star of the planetary nebula Abell 41 has been found to be a detached binary star system with a period of 2 hours 43 minutes. The data are consistent with a model consisting of a sdO primary of about 0.6 $M_\odot$ and a dM companion of 0.1 to 0.3 $M_\odot$. The existence of this star provides direct evidence for theoretical work suggesting that binary stars which have suffered catastrophic angular momentum loss through the ejection of a planetary nebula are the progenitors of the cataclysmic variables.

117.303   The breakdown of nuclear quasi–equilibrium in highly compact binaries.
P. C. Joss, S. Rappaport.
Cataclysmic variables and low–mass X–ray binaries, p. 35 – 38 (1985). – See Abstr. 012.072.

117.304   A systematic study of magnetic braking in low–mass binaries.
F. Verbunt, S. Rappaport, P. C. Joss.
Cataclysmic variables and low–mass X–ray binaries, p. 39 – 43 (1985). – See Abstr. 012.072.

117.305   Magnetic braking and the origin and evolution of close low–mass binaries.
A. Tutukov.
Cataclysmic variables and low–mass X–ray binaries, p. 45 – 53 (1985). – See Abstr. 012.072.
The author discusses some aspects of the origin and evolution of cataclysmic binaries and W UMa stars driven by an assumed magnetic stellar wind.

117.306   On the evolutionary status of bright, low–mass X–ray sources.
R. F. Webbink, S. Rappaport, G. J. Savonije.
Cataclysmic variables and low–mass X–ray binaries, p. 55 – 59 (1985). – See Abstr. 012.072.

117.307   Low–mass X–ray binaries.
J. E. McClintock, S. Rappaport.
Cataclysmic variables and low–mass X–ray binaries, p. 61 – 77 (1985). – See Abstr. 012.072.

117.308   Galactic bulge X–ray burst sources from disrupted globular clusters?
J. E. Grindlay, P. Hertz.
Cataclysmic variables and low–mass X–ray binaries, p. 79 – 91 (1985). – See Abstr. 012.072.
The origin of the bright galactic bulge X–ray sources, or GX sources, is unclear despite intensive study for the past 15 years. The authors suggest that the fact that many (or most) of the GX sources are X–ray burst sources (GXRBS) and are otherwise apparently identical to the luminous X–ray sources found in globular cluster cores implies that they too may have a globular cluster origin. The possibility that the compact X–ray binaries found in globulars are ejected is constrained by observations of

CVs in and out of clusters. Implications for globular cluster evolution and the GXRBS themselves are discussed.

117.309 X–ray luminous white dwarf binaries in globular clusters.
P. Hertz.
Cataclysmic variables and low–mass X–ray binaries, p. 99 – 102 (1985). – See Abstr. 012.072.

An X–ray survey of galactic globular clusters has been conducted. An apparently distinct class of X–ray sources has been discovered. They are identified as accreting white dwarfs in close binaries. Implications concern the binary formation rate in globular clusters and the number of compact objects present.

117.310 Three–body interactions and cataclysmic binaries in globular clusters.
P. Hut, F. Verbunt.
Cataclysmic variables and low–mass X–ray binaries, p. 103 – 106 (1985). – See Abstr. 012.072.

117.311 The evolution of highly compact binaries in globular clusters.
J. H. Krolik, A. Meiksin, P. C. Joss.
Cataclysmic variables and low–mass X–ray binaries, p. 107 – 116 (1985). – See Abstr. 012.072.

The authors report on detailed calculations of the secular evolution, observational appearence, and numbers of highly compact binaries in globular clusters. They present examples of the sorts of histories that are possible for highly compact binaries in globular clusters, and calculate the rate at which mass transfer is induced by gravitational encounters. The authors then describe the observational appearence of these systems, and report on calculations of their expected numbers. They also briefly discuss the relationship of highly compact binaries in globular clusters to the bright X–ray sources of the Galactic bulge.

117.312 The mass transfer rate in X1916–053: is it driven by gravitational radiation?
J. H. Swank, R. E. Taam, N. E. White.
Cataclysmic variables and low–mass X–ray binaries, p. 143 – 146 (1985). – See Abstr. 012.072.

A 50 minute period for a binary system harboring an X–ray burster would allow several alternatives for the mass–giving secondary, including a H shell burning plus He degenerate core composite model. The authors use the burst properties of X1916–053 (4U1915–05/MXB1916–053) to argue against the He degenerate as well as the He main sequence solutions and to estimate whether for any of the other solutions the mass transfer rate could be consistent with that expected from gravitational radiation (GR). Within uncertainty of a factor of 2 the transfer rate for the composite model solution is consistent with GR, but enhancement by other mechanisms should be investigated.

117.313 COS–B X–ray observations of Cygnus X–3.
J. M. Bonnet–Bidaud, M. van der Klis.
Cataclysmic variables and low–mass X–ray binaries, p. 147 – 150 (1985). – See Abstr. 012.072.

The X–ray source Cygnus X–3 was further observed by the X–ray detector onboard the COS–B satellite from November 1981 to February 1982. This brings to 220 days the total duration of observation of this source by COS–B. Preliminary results from the last observations are presented.

117.314 The AM Herculis magnetic variables.
J. Liebert, H. S. Stockman.
Cataclysmic variables and low–mass X–ray binaries, p. 151 – 177 (1985). – See Abstr. 012.072.

The observational properties of the ten known AM Her systems are reviewed. A multitude of components of continuum and line radiation from these objects are outlined. The authors discuss the important physical processes involved in the accretion flow and funnel shock, with emphasis on the properties of the emission line regions. Other topics include the properties of the white dwarf primaries and M dwarf secondaries, the maintenance of synchronous rotation, and the implications of "clumpings" in

the known orbital periods (80 – 115 minutes) and magnetic field strengths (20 – 35 megagauss).

117.315 Recent developments in the theory of AM Her and DQ Her stars.
D. Q. Lamb.
Cataclysmic variables and low–mass X–ray binaries, p. 179 – 218 (1985). – See Abstr. 012.072.

The author describes a number of advances in the theory of accreting magnetic white dwarfs, and discusses several unresolved issues concerning these stars.

117.316 CW1103+254: the AM Her object that has everything.
G. D. Schmidt.
Cataclysmic variables and low–mass X–ray binaries, p. 219 – 223 (1985). – See Abstr. 012.072.

With a dozen AM Her systems now known, each new object becomes less important in its own right, but each holds promise for presenting unusual variations or new perspectives on the phenomenon. The recently discovered binary CW1103+254 is particularly simple in its observed behavior, yet is proving to be a great aid in refining our understanding of accretion in these CVs with magnetic primaries.

117.317 AM Herculis: an outburst at 4.9 GHz.
T. S. Bastian, G. A. Dulk, G. Chanmugam.
Cataclysmic variables and low–mass X–ray binaries, p. 231 – 235 (1985). – See Abstr. 012.072.

The authors report the results of radio observations of AM Her with the Very Large Array (VLA). The quiescent emission first discovered by Chanmugam and Dulk (1982) at 4.9 GHz is confirmed and upper limits to the flux density at 1.5 GHz and 15 GHz obtained. The authors also report the discovery of a remarkable outburst at 4.9 GHz which was essentially 100% circularly polarized. The outburst is probably due to an electron–cyclotron maser which operates near the red–dwarf companion in a region where the magnetic field is ≈ 1000 gauss.

117.318 Synchronization of magnetic white dwarfs in close binary systems.
F. K. Lamb, J.–J. Aly, M. C. Cook, D. Q. Lamb.
Cataclysmic variables and low–mass X–ray binaries, p. 237 – 245 (1985). – See Abstr. 012.072.

Asynchronous rotation of strongly magnetic white dwarfs in close binary systems drives substantial field–aligned electrical currents between the magnetic star and its companion. The resulting magnetohydrodynamic torque is able to account for the heretofore unexplained synchronous rotation of the strongly magnetic degenerate dwarf component in systems like AM Her, VV Pup, AN UMa, and EF Eri. The electric fields produced by even a small asynchronism are large enough to accelerate electrons to high energies, and may lead to radio emission. The total energy dissipation rate in systems with white dwarf spin periods as short as 1^m may reach 10^{33}ergs s^{-1}. Total luminosities of this order may be a characteristic feature of such systems.

117.319 Orientations of AM Her stars from their polarization properties: the case of the missing AM Her stars.
J. J. Brainerd, D. Q. Lamb.
Cataclysmic variables and low–mass X–ray binaries, p. 247 – 256 (1985). – See Abstr. 012.072.

The authors use observation of the polarization properties of the AM Her stars to determine their orientations. They find that the AM Her stars are randomly distributed with respect to their inclination angle i but not with respect to their magnetic colatitude δ. The stars are concentrated along the semicircle $\cos^2\delta + \cos^2 i = 1$, where δ is the angle between the magnetic moment and the spin axis of the degenerate dwarf. This result implies that the discovery of AM Her stars is strongly affected by observational selection effects. The authors suggest that these effects are a result of the way in which the AM Her stars have been identified observationally, and of the angular properties of high harmonic cyclotron emission. The authors estimate that there are about three times as many AM Her stars as have been found so far.

117.320   Time dependent accretion flows in the AM Her systems.
S. H. Langer.
Cataclysmic variables and low–mass X–ray binaries, p. 257 – 260 (1985). – See Abstr. 012.072.

The author considers time dependent accretion onto white dwarfs with strong magnetic fields. For certain conditions this flow is unstable and leads to large periodic variations in the hard X–ray luminosity. If the magnetic field is large enough, electron cyclotron radiation will stabilize the flow. The author also discusses the probable observational appearance of the oscillations found in his calculations.

117.321   Intermediate polars.
B. Warner.
Cataclysmic variables and low–mass X–ray binaries, p. 269 – 279 (1985). – See Abstr. 012.072.

117.322   A comparison of H2252–035/AO Psc and 4U1849–31/V1223 Sgr.
J. van Paradijs, S. van Amerongen, M. de Kool, M. Pakull, H. van der Woerd.
Cataclysmic variables and low–mass X–ray binaries, p. 283 – 286 (1985). – See Abstr. 012.072.

From five–colour photometry of the triple–periodic systems H2252–035 and V1223 Sgr, which have very similar orbital and pulsation periods, the authors find a significant difference in the wavelength dependence of their short period pulsations. The results indicate that in H2252–035 this pulse originates in a hot region, probably on the white dwarf, whereas in V1223 Sgr both pulses are the result of reprocessing of X–rays (in the far side of the accretion disk, and in the companion star, respectively).

117.323   Ultra–violet radiation from dwarf nova discs.
B. J. M. Hassall.
Cataclysmic variables and low–mass X–ray binaries, p. 287 – 297 (1985). – See Abstr. 012.072.

The discs in dwarf novae are the major source of ultra–violet light. The UV observations are reviewed and the continuum is compared with several steady disc models, at quiescence and outburst. The results are discussed in terms of the mass accretion rates, and the time–dependent behaviour of the continuum during the rise is presented.

117.324   On the nature of dwarf novae.
J. Smak.
Cataclysmic variables and low–mass X–ray binaries, p. 299 (1985). Abstract. – See Abstr. 012.072.

117.325   Dwarf novae eruptions.
V. J. Mantle.
Cataclysmic variables and low–mass X–ray binaries, p. 301 – 305 (1985). – See Abstr. 012.072.

The author shows how observed features of dwarf novae eruptions, in particular the eruption decay time, spectral evolution and stream impact bright–spot behaviour are accounted for by the bursting mass transfer model.

117.326   Accretion instability models for dwarf novae and X–ray transients.
J. K. Cannizzo, J. C. Wheeler, P. Ghosh.
Cataclysmic variables and low–mass X–ray binaries, p. 307 – 313 (1985). – See Abstr. 012.072.

Steady state and time dependent calculations are presented of a model for accretion instability in which matter accumulates in a cold torus and then undergoes a thermal instability causing the matter to heat and flow down onto the central star. This model is compared to the observations of dwarf novae and certain examples of both hard and soft X–ray transients.

117.327   Dwarf novae – a hot dam instability.
J. Faulkner, D. N. C. Lin, J. C. B. Papaloizou.
Cataclysmic variables and low–mass X–ray binaries, p. 315 – 322 (1985). – See Abstr. 012.072.

When mass input rates are low enough, accretion disks cease to be fully ionized. The strong and oppositely directed temperature dependences of opacity (e.g. in hydrogen ionization regions) drive a thermal instability which modulates mass transfer throughout the disk. The conventional thin–disk treatment breaks down in cool regions; instead, the surface boundary condition dominates the structure of these largely radiative zones. Convection, when it occurs, is inefficient and essentially irrelevant. The authors modify the conventional treatment to produce useful approximations to local vertical energy losses. The authors' local cooling modification is then incorporated into global disk computations.

117.328   First detection of radio emission from a dwarf nova.
A. O. Benz, E. Fürst, A. L. Kiplinger.
Cataclysmic variables and low–mass X–ray binaries, p. 331 – 335 (1985). Reprinted from Nature. – See Abstr. 012.072.

The authors describe a search for radio emission at 4.75 GHz from dwarf novae that has been carried out with the 100–m telescope at Effelsberg, F.R. Germany. They have searched for radio emission from six dwarf novae and a source was discovered at the position of SU UMa. The source could only be detected during optical outburst and was below the threshold during quiescence. The authors suggest that the radio emission arises from a non–thermal process.

117.329   Photoionization models for the wind from TW Vir.
T. Kallman.
Cataclysmic variables and low–mass X–ray binaries, p. 337 – 342 (1985). – See Abstr. 012.072.

117.330   On measuring the radial velocity of white dwarfs in cataclysmic binaries.
A. W. Shafter.
Cataclysmic variables and low–mass X–ray binaries, p. 355 – 358 (1985). – See Abstr. 012.072.

117.331   The cataclysmic variable GD552.
R. J. Stover.
Cataclysmic variables and low–mass X–ray binaries, p. 379 – 384 (1985). – See Abstr. 012.072.

Time–resolved spectroscopic observations of the proper motion star GD552 were made. Radial velocity variations were measured for the Hβ emission line. No velocity variations were detected, with an upper limit of 35 kilometers per second on the semi–amplitude of the K–velocity of the accretion disk. A strong "S–wave" component was observed in the emission lines, and, as a result, the orbital period could be determined. The most likely orbital period is 1 hour 36 minutes, with a night–to–night cycle count uncertainty of ±1.

117.332   A summary of the UV, optical and IR properties of disks in CV's.
P. Szkody.
Cataclysmic variables and low–mass X–ray binaries, p. 385 – 401 (1985). – See Abstr. 012.072.

The available observations of dwarf novae and novalike systems that are relevant to the study of accretion in cataclysmic variables are reviewed. This includes UV data from IUE and optical spectroscopy and infrared photometry. The flux distributions are compared to available steady state models to obtain an estimate of the mass accretion and the range of this parameter among the dwarf novae (U Gem, Z Cam and SU UMa types) and the novalikes (UX UMa, AM Her, DQ Her types). The normal and abnormal range of variation in the continuum and line fluxes is discussed.

117.333   White dwarf/accretion disk interactions: instabilities of infinite, differentially rotating cylinders.
B. W. Carroll, H. M. Van Horn.
Cataclysmic variables and low–mass X–ray binaries, p. 403 – 405 (1985). – See Abstr. 012.072.

117.334 Boundary layers and coronae in cataclysmic variables.
K. A. Jensen.
Cataclysmic variables and low–mass X–ray binaries, p. 407 – 416 (1985). – See Abstr. 012.072.

X–ray observations of novae, nova–like variables, and dwarf novae in outburst show them to be weaker X–ray sources than predicted by boundary layer models. The discrepancy can be resolved if these systems have optically thick boundary layers with temperatures less than 30 eV, provided the neutral hydrogen column densities of circumstellar gas are greater than $10^{21} cm^{-2}$. The absorption of greater than 99 percent of the flux from the boundary layer must occur in a region nearer the white dwarf than the region in which spectral lines are formed. The weak hard X–ray fluxes observed from these systems probably do not come from the boundary layer but may be produced in an accretion disk corona.

117.335 Accretion disks in symbiotic stars and their relationship to cataclysmic binaries.
S. J. Kenyon.
Cataclysmic variables and low–mass X–ray binaries, p. 417 – 427 (1985). – See Abstr. 012.072.

117.336 Distances, accretion rates, and space densities of cataclysmic variables: piece o'cake, and other heresies.
J. Patterson.
Cataclysmic variables and low–mass X–ray binaries, p. 435 – 444 (1985). – See Abstr. 012.072.

117.337 Polarimetric observations of Iota and Theta 2 Orionis.
H. G. Luna.
Astrophys. Lett., Vol. 24, No. 4, p. 211 – 216 (1985).

UBV polarimetric observations of ι Ori and θ^2Ori have been obtained in order to detect variable linear polarization. While ι Ori has not shown significant variations, θ^2Ori presents phase variations of low amplitude. In addition, this latter system shows a clear wavelength dependence of the polarization angle, which can be explained if the intrinsic component of the observed polarization is due to Thomson scattering in a circumstellar envelope.

117.338 Short–term variability of hard X–rays from Cyg X–1.
M. Nakagawa, H. Sakurai, M. Uchida.
18th International Cosmic Ray Conference, Vol. 1, p. 1 (1983). Abstract. – See Abstr. 012.096.

117.339 Photometric observations of the X–ray dwarf nova HL Canis Majoris.
S. Seetha, T. M. K. Marar, D. P. Sharma, K. Kasturirangan, U. R. Rao, J. C. Bhattacharyya.
18th International Cosmic Ray Conference, Vol. 9, p. 5 – 8 (1983). – See Abstr. 012.096.

High speed optical photometric observations of the X–ray dwarf nova HL Canis Majoris (1E 0643–1648), discovered with the Einstein Observatory, are presented. Results of long term monitoring and attempts to detect photometric eclipses at the reported orbital periods of 0.22 d or 0.18 d of the binary system are presented.

117.340 A tentative explanation for the energy variation of the cyclotron feature in Her X–1.
N. Prantzos, P. Durouchoux.
18th International Cosmic Ray Conference, Vol. 9, p. 9 – 12 (1983). – See Abstr. 012.096.

The observed energy variation in the Her X–1 cyclotron line seems to be correlated to the line's intensity and the hard X–ray continuum luminosity of the source. A tentative interpretation of such a correlation in terms of the Bonazzola et al. (1979) model is made.

117.341 High speed photometry of dwarf nova EX Hydrae.
D. P. Sharma, T. M. K. Marar, S. Seetha, K. Kasturirangan, U. R. Rao, J. C. Bhattacharyya.
18th International Cosmic Ray Conference, Vol. 9, p. 330 – 333 (1983). – See Abstr. 012.096.

High speed optical photometric observations of the hard X–ray source EX Hya, carried out with the reflector of the Kavalur observatory are presented. A new ephemeris for the photometric period confirms the trend towards decreasing period on a time scale of $\sim 10^7$ years. Evidence for the presence of interpulses in the light curves between successive photometric maxima is shown.

117.342 The Algol scenario.
E. Budding.
South. Stars, Vol. 31, No. 2, p. 125 – 141 (1985).

The paper presents a general review of current data and ideas on Algol type binaries. The central "Roche lobe overflow" problem is discussed in both its hydrodynamic and evolutionary contexts. A final resume presents some outstanding unsolved issues.

117.343 The proportion of binaries having a degenerate companion: implications on the formation of barium stars.
J. L. Halbwachs.
Cool stars with excesses of heavy elements, p. 337 – 343 (1985). – See Abstr. 012.101.

The theoretical proportion of barium stars was derived from the frequency of binaries with degenerate secondary components. It appears that the proportion of Ba II stars should be of about 0.5% among dwarfs and more than 1% among giants. The frequency of dwarf stars among Ba II stars later than F6 should be about 5%.

117.344 Mira's role in RR Tel revisited.
A. Heck, J. Manfroid.
Cool stars with excesses of heavy elements, p. 349 – 353 (1985). – See Abstr. 012.101.

Strömgren photometry, IUE measurements and AAVSO visual estimates of the very slow nova RR Tel show that the object has gone through periodic variations in visual light since it started fading in 1949 after the outburst in 1944. The period is however not constant and varies between 350 and 410 days. Synchronized variations of the Mira component in the infrared and of the blue component indicate probably that the Mira mass outflow varies in phase with its pulsations. Large variations occur in visual light on a time scale of a few weeks.

117.345 Comparative time–resolved spectrophotometry of cataclysmic variables.
E. M. Schlegel, R. K. Honeycutt, R. H. Kaitchuck.
Bull. Am. Astron. Soc., Vol. 17, No. 2, p. 550 (1985). Abstract. – See Abstr. 010.065.

117.346 A possible explanation for the unusual light curves of V444 Cygni.
A. B. Underhill, R. P. Fahey.
Bull. Am. Astron. Soc., Vol. 17, No. 2, p. 552 (1985). Abstract. – See Abstr. 010.065.

117.347 The binary nature of the symbiotic star EG Andromedae.
N. A. Oliversen, C. M. Anderson, M. Slovak, R. E. Stencel.
Bull. Am. Astron. Soc., Vol. 17, No. 2, p. 552 (1985). Abstract. – See Abstr. 010.065.

117.348 An analysis of UV spectra of Algol–type binary stars.
H. Cugier, J. Hardorp.
Bull. Am. Astron. Soc., Vol. 17, No. 2, p. 553 (1985). Abstract. – See Abstr. 010.065.

117.349 Eclipse by X–ray induced shock in GX301–2.
D. A. Leahy, N. Kawai, K. Koyama, F. Makino,
M. Matsuoka.
Bull. Am. Astron. Soc., Vol. 17, No. 2, p. 555 (1985). Abstract. –
See Abstr. 010.065.

117.350 Neutrino binaries.
K. Brecher, G. Chanmugam.
Bull. Am. Astron. Soc., Vol. 17, No. 2, p. 579 (1985). Abstract. –
See Abstr. 010.065.

117.351 A variable gas–stream model for U Herculis.
G. J. Qiao, K. C. Leung, Y. F. Li.
Bull. Am. Astron. Soc., Vol. 17, No. 2, p. 583 – 584 (1985). Abstract. – See Abstr. 010.065.

117.352 The early–type semidetached system LY Aurigae.
Y. F. Li, K. C. Leung.
Bull. Am. Astron. Soc., Vol. 17, No. 2, p. 584 (1985). Abstract. –
See Abstr. 010.065.

117.353 Ultraviolet observations of the early type contact binary SX Aur.
G. J. Peters, R. S. Polidan, A. P. Linnell.
Bull. Am. Astron. Soc., Vol. 17, No. 2, p. 584 (1985). Abstract. –
See Abstr. 010.065.

117.354 A spectroscopic study of VZ Psc.
B. J. Hrivnak, E. F. Milone.
Bull. Am. Astron. Soc., Vol. 17, No. 2, p. 584 (1985). Abstract. –
See Abstr. 010.065.

117.355 Spectral variations in Sigma Geminorum.
Z. Eker.
Bull. Am. Astron. Soc., Vol. 17, No. 2, p. 588 (1985). Abstract. –
See Abstr. 010.065.

117.356 H–alpha emission and polarization of RS CVn stars.
X. F. Liu, H. S. Tan.
Bull. Am. Astron. Soc., Vol. 17, No. 2, p. 588 (1985). Abstract. –
See Abstr. 010.065.

117.357 PG 1012–029 and winds in high–inclination cataclysmics.
R. K. Honeycutt, E. M. Schlegel.
Bull. Am. Astron. Soc., Vol. 17, No. 2, p. 588 (1985). Abstract. –
See Abstr. 010.065.

117.358 Radio emission from DQ Her stars.
J. A. Bookbinder, D. Q. Lamb.
Bull. Am. Astron. Soc., Vol. 17, No. 2, p. 589 (1985). Abstract. –
See Abstr. 010.065.

117.359 Optical observations of new X–ray emitting cataclysmic variables.
D. A. H. Buckley, I. R. Tuohy, R. Remillard, H. Bradt,
D. A. Schwartz.
Bull. Am. Astron. Soc., Vol. 17, No. 2, p. 589 (1985). Abstract. –
See Abstr. 010.065.

117.360 Time resolved spectrophotometry of the AM Her system E2003+225.
P. J. McCarthy, J. T. Clarke, S. Bowyer.
Bull. Am. Astron. Soc., Vol. 17, No. 2, p. 589 (1985). Abstract. –
See Abstr. 010.065.

117.361 Hα activity and the light curve of the RS Canum Venaticorum star DH Leo (HD 86590).
S. C. Barden, L. W. Ramsey, R. E. Fried.
Bull. Am. Astron. Soc., Vol. 17, No. 2, p. 597 – 598 (1985). Abstract. – See Abstr. 010.065.

117.362 Photoelectric photometry of cataclysmic variable star candidates.
J. W. Wilson, H. R. Miller, J. L. Africano, R. J. Quigley.
Bull. Am. Astron. Soc., Vol. 17, No. 2, p. 598 (1985). Abstract. –
See Abstr. 010.065.

117.363 Something strange from Cygnus X–3.
M. M. Waldrop.
Science, Vol. 228, No. 4705, p. 1298 (1985).

117.364 Activity of contact binary systems.
S. M. Rucinski.
Interacting binaries, p. 13 – 49 (1985). – See Abstr. 012.104.

This review discusses various indicators of stellar activity in contact binaries of the W Ursae Majoris type. It considers evidence for activity in the photospheres, chromospheres, transition regions, and coronae of these late–type systems. Observational data on flare activity and on period changes are also briefly reviewed.

117.365 Observational evidence for the evolution of contact binary stars.
S. W. Mochnacki.
Interacting binaries, p. 51 – 82 (1985). – See Abstr. 012.104.

This paper reviews data relevant to the evolution of contact binary stars. Statistics, photometric surveys, period changes and the results of light curve fitting are reviewed. These data are then compared with evolutionary models. A new collation of data for 55 W Ursae Majoris systems is presented. Simple evolutionary models including magnetic braking have been calculated, and show good agreement with observations if contact lifetimes are long.

117.366 Tidal interactions of close binary systems.
G. J. Savonije, J. C. B. Papaloizou.
Interacting binaries, p. 83 – 102 (1985). – See Abstr. 012.104.

Contents: (1) Introduction. (2) The perturbing potential. (3) The hydrostatic response; the weak friction model. (4) Equations for tidal evolution. (5) Tidal dissipation in stars with convective envelopes. (6) Tidal dissipation in stars with radiative envelopes. (7) Concluding remarks.

117.367 Wolf–Rayet stars and binarity.
A. J. Willis.
Interacting binaries, p. 103 – 126 (1985). – See Abstr. 012.104.

After a brief review of current ideas concerning the physical and chemical nature of Wolf–Rayet (WR) stars, this paper discusses the detection methods for WR binaries and their statistics in the Galaxy and Magellanic Clouds. It then summarises current knowledge of WR masses and orbital parameters with inferences for evolutionary mechanisms involved in producing a WR star. Finally the question of whether or not WR + collapsar systems exist is addressed.

117.368 From Algol to Beta Lyrae.
M. J. Plavec.
Interacting binaries, p. 155 – 178 (1985). – See Abstr. 012.104.

The structure and evolutionary status of β Lyrae systems is reviewed. The best model available at present considers β Lyrae as an Algol–type semi–detached system in a phase of fairly rapid mass transfer, i.e. younger than typical Algols. The most important conclusion drawn in this article is that β Lyrae is not so unique as it is usually believed. Some characteristics in β Lyrae actually appear already in such classical Algols like U Cephei and RW Tauri. The mass–accreting components in these systems are surrounded by a hot, turbulent layer which probably expands and which is the seat of emission lines of fairly high ionization discovered in the far ultraviolet. In β Lyrae and the so–called W Serpentis stars, the circumstellar hot turbulent shell is much more extensive and probably also denser, and may be forming a kind of a "superchromosphere".

117.369   Symbiotic stars.
S. J. Kenyon.
Interacting binaries, p. 179 – 203 (1985). – See Abstr. 012.104.

Possible binary models for the small class of eruptive variables known as symbiotic stars are outlined. A discussion of the observable properties of such models leads to the conclusion that symbiotic stars are likely to be main sequence or white dwarf stars accreting material from a red (super)giant primary. Recent observations of symbiotic stars confirm this notion, and three systems (CI Cyg – a red giant with an accreting main sequence star companion, V1016 Cyg – a Mira variable wih a hot white dwarf companion, and V443 Her – a red giant with a hot white dwarf companion) have been chosen to illustrate the impact modern observations have had on understanding the symbiotic phenomenon.

117.370   Optical characteristics of X–ray binaries.
S. A. Ilovaisky.
Interacting binaries, p. 205 – 248 (1985). – See Abstr. 012.104.

This review highlights the important role played by accretion disks in massive X–ray binaries and emphasizes the increasing amount of circumstantial evidence for the existence of massive collapsed objects (”black holes”) in Cyg X–1, LMC X–3 and LMC X–1. The existence of ”tilted”, apparently precessing disks may be more widespread than previously believed (LMC X–4, Cyg X–1, 1907+09) and still poses basic theoretical problems. The X–ray heated accretion disk is the dominant source of light in low–mass X–ray binaries and explains the optical activity of these systems (pulses, bursts, flickering) in terms of reprocessed X–rays. The absence of X–ray eclipses in these systems has been interpreted in terms of thick disks which shield the companion star. In some transient sources long–period modulations may be due to ”precessing” disks. The nature of the compact source in some well–studied low–mass X–ray binaries is discussed.

117.371   Massive X–ray binaries.
N. E. White.
Interacting binaries, p. 249 – 287 (1985). – See Abstr. 012.104.

An OB star loses mass via a stellar wind and any collapsed binary companion to such a star will capture a fraction of that material, releasing gravitational potential energy predominantly as X–ray emission. If in addition the OB star fills its critical potential lobe, then material will also spill onto the compact star via an accretion disk. About one quarter of the 100 or so known galactic X–ray sources with luminosities greater than 10^{32} erg s^{-1} have been identified with such systems. This review discusses the accretion process, the photoabsorption of X–rays by the stellar wind, and the evolutionary status of these systems.

117.372   Cataclysmic variables as interacting binary stars.
R. A. Wade.
Interacting binaries, p. 289 – 326 (1985). – See Abstr. 012.104.

After a brief description of the visual and ultraviolet properties of cataclysmic variables (CVs), this review concentrates on two areas: (1) the basic binary star model of CVs and the evidence supporting it, and (2) ”standard” accretion disks as the means of mass exchange between the component stars. The emphasis is on observations, particularly on assessing how reliably the physical quantities of interest can be or have been determined.

117.373   The late stages of interactive stellar evolution.
R. E. Nather.
Interacting binaries, p. 349 – 366 (1985). – See Abstr. 012.104.

The final stages in the evolution of cataclysmic binary systems are discussed. Possible descendents of these systems may be the small handful of known interacting objects with orbital periods so short that only two collapsed objects can be involved. This paper concentrates on these interacting collapsed objects already known, and speculates on those not yet known observationally but which ought to exist in any reasonably designed universe.

117.374   Stars that go hump in the night: the SU UMa stars.
B. Warner.
Interacting binaries, p. 367 – 392 (1985). – See Abstr. 012.104.

This paper reviews the ”superhump” phenomenon occurring during the outburst phases of dwarf novae of the SU Ursae Majoris type. After summarizing current observational data, the paper outlines the intermediate polar model for the superhumps and presents a detailed comparison of its theoretical features with the observations.

117.375   Interacting binaries – summing up.
V. Trimble.
Interacting binaries, p. 393 – 410 (1985). – See Abstr. 012.104.

A summary of major recent achievements and outstanding problems in research on interacting binaries (contact binaries, cataclysmic binaries, symbiotic stars, X–ray binaries) is presented.

117.376   Report of IAU Commission 42: Close binary stars (*Etoiles binaires serrées*).
A. H. Batten.
Trans. IAU, Vol. XIXA, p. 583 – 606 (1985). – See Abstr. 003.046.

117.377   Combined observation of binary stars.
Z. Komárek.
Říše hvězd, Vol. 66, No. 1, p. 15 – 17 (1985). In Czech.

117.378   On the dynamics of the common gas envelope of a binary stellar system.
V. G. Shkodrov, T. R. Bonev.
Dokl. Bolg. Akad. Nauk, Tome 37, No. 6, p. 707 – 710 (1984).

117.379   A note on two RS Canum Venaticorum stars: LX Persei and SZ Piscium.
J. R. Percy.
J. R. Astron. Soc. Can., Vol. 79, No. 3, p. 113 – 118 (1985).

New and previously–published photometric observations have been used to derive more precise values for the periods of migration of the photometric distortion waves in LX Persei and SZ Piscium. They are 479 days and 1574 days (4.31 years), respectively. Both are retrograde and both are constant over the interval of the observations used. The amplitudes of the waves in both stars vary by a factor of four or more, on time scales which are similar to (but not necessarily equal to) the periods of migration.

117.380   The physical nature of the symbiotic stars.
S. J. Kenyon.
Diss. Abstr. Int., Sect. B, Vol. 45, No. 1, p. 229 (1984). Thesis, University of Illinois, 357 pp. (1983). Order No. DA8409968.

117.381   Radio emission from RS CVn binary systems.
D. J. Doiron.
Diss. Abstr. Int., Sect. B, Vol. 45, No. 7, p. 2199 – 2200 (1985). Thesis, University of Iowa, 131 pp. (1984). Order No. DA8423552.

117.382   A model for eclipsing cataclysmic variables.
J. A. Magnuson.
Diss. Abstr. Int., Sect. B, Vol. 45, No. 9, p. 2957 (1985). Thesis, Dartmouth College, 133 pp. (1984). Order No. DA8426776.

117.383   Stellar surface phenomena: a spectroscopic study of the RS Canum Venaticorum binaries.
S. E. Smith.
Diss. Abstr. Int., Sect. B, Vol. 45, No. 10, p. 3257 (1985). Thesis, University of Toledo, 312 pp. (1984). Order No. DA8429741.

117.384   The optical orbit of the X–ray pulsar binary 0535–668 (= A0538–66).
J. B. Hutchings, D. Crampton, A. P. Cowley, E. Olszewski,
I. B. Thompson, N. Suntzeff.
Publ. Astron. Soc. Pac., Vol. 97, No. 591, p. 418 – 422 (1985).

Spectroscopic data are presented from the optical primary of the LMC X–ray source 0535–668 during its optical low state. From these data the star appears as a normal B1 star, slightly evolved off the main sequence. Adopting the X–ray outburst period of 16.65 days, radial velocities indicate an orbit with a high eccentricity, and periastron passage very close to the X–ray flux peak. Probable masses are normal for the primary and the pulsar. Orbital parameters and implied quantities are discussed.

117.385   Continuous time–resolved spectroscopy of the AM Herculis binary 0139–68.
J. B. Hutchings, A. P. Cowley, D. Crampton.
Publ. Astron. Soc. Pac., Vol. 97, No. 591, p. 423 – 427 (1985).

The authors present results of single–trailed spectrograms of the magnetic accreting white–dwarf binary 0139–68. The spectral resolution is ~ 2.5 Å and the time resolution is 45 seconds. Line–profile changes and radial velocities are derived and described. A discussion of possible orbital velocities indicates probable masses of $0.2 - 0.6\ M_\odot$ for the white–dwarf primary and $0.3\ M_\odot$ for the M–star secondary. The position of the active pole is suggested.

117.386   The two periods of TT Arietis.
J. R. Thorstensen, J. Smak, F. V. Hessman.
Publ. Astron. Soc. Pac., Vol. 97, No. 591, p. 437 – 445 (1985).

The authors obtained velocities of the cataclysmic variable TT Ari in its high and intermediate photometric states, with the aim of clarifying the spectroscopic period. It is demonstrated that the spectroscopic and photometric periods are different. The authors find that no single period fits *all* the available data well. They conclude that the velocities *change phase* between the high and the intermediate states. Low–state data communicated by Shafter et al. (1984) are just one–half cycle out of phase with the present high–state ephemeris. The intermediate–state velocities lag the high–state velocities by 0.28 of a cycle; if the high–state velocities reflect the motion of the white dwarf in the system, the emission in the intermediate state may originate near the hot spot and gas stream in the system.

117.387   A new photometric study of the binary U Pegasi.
D.–s. Zhai, K. C. Leung, R.–x. Zhang.
Chin. Astron. Astrophys., Vol. 9, No. 2, p. 98 – 105 (1985). English translation of Acta Astron. Sin., Vol. 25, No. 4, p. 336 – 347 (1984).

A period study of the times of minima from 1896 to 1980 of U Peg was performed. The system was found to have a secular period decrease, $\Delta p/p$ of -1.32×10^{-10} or -4.16×10^{-3} sec/yr, as well as a short term (17 years) sinusoidal oscillation with a semiamplitude of 0.00323 day. It is suggested that oscillating term is caused by the light–time effect of an unseen third body. The third body may be a M6 main sequence star with a mass of $0.16\ M_\odot$.

117.388   Two similar detached binary systems: AT and AZ Camelopardalis.
D.–s. Zhai, K.–C. Leung, R.–x. Zhang, J.–t. Zhang.
Chin. Astron. Astrophys., Vol. 9, No. 2, p. 139 – 144 (1985). English translation of Acta Astrophys. Sin., Vol. 5, No. 1, p. 26 – 35 (1985).

Photoelectric BV observations of AT Cam are presented. The O–C diagram of the times of light minimum shows a probable long–term variation in the period. Photometric solutions using the Wilson–Devinney method were made for both AT Cam and AZ Cam. Both are found to be detached systems, but the components are close to the critical surfaces. A brief discussion is given on their evolutionary stages, based on the fractional radii and the mass–ratios obtained.

117.389   Spectral observations of Beta Lyrae in the visual region during June 1976. II. The emission spectrum.
A. N. Dadaev.
Izv. Glav. Astron. Obs. Pulkovo, Astrometr. Astrofiz., No. 201, p. 110 – 117 (1985). In Russian.

This discussion is based on the observational data used in a previous paper (1982). When studied in the visual region the emission spectrum confirms the properties of the binary shown earlier: (1) the emission is produced by circumstellar matter which first and foremost is a system of gaseous streams flowing around both stars; (2) the emission can be best detected on the outer limbs of the stars, being most intensive during eclipses due to a peculiar distribution of gaseous masses within the system; (3) data on relative velocities determined from shifts of emission peaks permit to construct a scheme of streams and to make an identification of chemical elements which are present in the circumstellar gas.

117.390   The period behaviour of AU Serpentis.
H. D. Kennedy.
Inf. Bull. Variable Stars, No. 2742, 4 pp. (1985).

117.391   BV photometry and period variation of GO Cygni.
C. Sezer, Ö. Gülmen, N. Güdür.
Inf. Bull. Variable Stars, No. 2743, 5 pp. (1985).

117.392   Photoelectric $H_\beta W$ and $H_\beta N$ observations of W UMa (BD + 56°1400).
B. M. Davan.
Inf. Bull. Variable Stars, No. 2745, 2 pp. (1985).

117.393   UBV observations of symbiotic stars in July and October 1982.
M. T. Martel, R. Gravina.
Inf. Bull. Variable Stars, No. 2750, 2 pp. (1985).

117.394   An interesting episode in the history of CH Cygni.
M. H. Rodriguez.
Inf. Bull. Variable Stars, No. 2753, 2 pp. (1985).

117.395   Optical and infrared pulsations from the HZ Herculis binary system during the 1983 prolonged X–ray low state.
J. Middleditch, R. C. Puetter, C. R. Pennypacker.
Proceedings of the Southwest Regional Conference for Astronomy and Astrophysics, Vol. 10, p. 33 – 35 (1985). Abstract. – See Abstr. 012.125.

General Catalogue of Variable Stars.
See Abstr. 002.027.

Bibliography and program notes on close binaries, No. 40, (Material received by March 15, 1984).
See Abstr. 002.034.

Bibliography and program notes on close binaries, No. 41. (Material received by September 15, 1984).
See Abstr. 002.035.

A data base for RS CVn binary systems.
See Abstr. 002.130.

Accretion power in astrophysics.
See Abstr. 003.012.

Small–telescope photoelectric data acquisition at Pace University.
See Abstr. 013.081.

The Universe, bit by bit.
See Abstr. 014.042.

Fairborn Observatory automatic photoelectric telescopes.
See Abstr. 032.035.

A digital technique for the separation of the eclipses of a white dwarf and an accretion disc.
See Abstr. 036.100.

Maximum entropy reconstruction of accretion disk images from eclipse data.
See Abstr. 036.212.

Fission sequences of self–gravitating and rotating fluid with internal motion.
See Abstr. 042.001.

The Roche limit based on vibrational stability.
See Abstr. 042.117.

The Tenma mission.
See Abstr. 051.058.

High energy neutrino astrophysics ($10^2 - 10^7$ GeV) with emphasis on Cyg X–3.
See Abstr. 061.061.

Quasi–two–dimensional cosmic jets.
See Abstr. 062.032.

Cosmic jets.
See Abstr. 062.134.

Theory of optical flashes.
See Abstr. 063.004.

Polarization of intrinsic radiation of tidally distorted stars with electron–scattering atmospheres. I. The case of conservative scattering.
See Abstr. 063.008.

Polarization of intrinsic radiation of tidally distorted stars with electron–scattering atmospheres. II. An account of true absorption.
See Abstr. 063.012.

Bowen fluorescence mechanism in X–ray binaries.
See Abstr. 063.046.

The polarization properties of magnetic accretion columns. III. A grid of uniform temperature and shock front models.
See Abstr. 063.078.

Magnetic field reconnection in differentially rotating accretion disks.
See Abstr. 064.003.

Magnetic effects in the heating and modification of flows in the outer stellar atmospheres with application to early type stars.
See Abstr. 064.020.

Model atmosphere for the bright component of the binary system υ Sgr.
See Abstr. 064.034.

Optically thick radial accretion onto main sequence stars and degenerate dwarfs.
See Abstr. 064.071.

The reflection effect in model stellar atmospheres. I. Grey atmospheres with convection.
See Abstr. 064.075.

Radiation transfer in accretion disk coronae.
See Abstr. 064.086.

The life and times of an intermediate mass star – in isolation/in a close binary.
See Abstr. 065.005.

Evolution of binary systems into transient sources.
See Abstr. 065.010.

The evolution of binaries of moderate primordial mass ($\lesssim 10\ M_\odot$) and the formation of transient and explosive objects due to accumulation instabilities in such binaries.
See Abstr. 065.011.

Collapse anisotropy for massive stars.
See Abstr. 065.036.

Recurrent novae as a consequence of the accretion of solar material onto a $1.38\ M_\odot$ white dwarf.
See Abstr. 065.043.

A simple model for binary star evolution.
See Abstr. 065.049.

On the observed properties and long term structure and evolution of white dwarfs in cataclysmic variables.
See Abstr. 065.063.

Stellar evolution and binaries.
See Abstr. 065.083.

On initial masses of stars that become Wolf–Rayet stars.
See Abstr. 065.089.

Formation of helium stars with extended envelopes in binaries.
See Abstr. 065.090.

Ultra–high energy γ rays and cosmic rays from accreting degenerate stars.
See Abstr. 067.004.

Theories of accreting X–ray pulsars.
See Abstr. 067.007.

Accretion by magnetic neutron stars.
See Abstr. 067.008.

Accretion onto magnetized neutron stars: magnetospheric structure and stability.
See Abstr. 067.009.

Spin–reversed accretion onto neutron stars and period fluctuation of X–ray pulsars.
See Abstr. 067.011.

The formation of planetesimals in highly compact binary stellar systems and the nature of cosmic gamma–ray bursts.
See Abstr. 067.023.

Radiative transfer in the accretion column of X–ray pulsars: effects from the hot spot.
See Abstr. 067.057.

Stability of mass transfer in a neutron star plus degenerate dwarf binary.
See Abstr. 067.077.

Pulsar magnetospheres in binary systems.
See Abstr. 067.119.

X–ray observations by the Einstein satellite and neutron star cooling.
See Abstr. 067.128.

Stability of radiative shock waves.
See Abstr. 067.143.

Accretion discs.
See Abstr. 067.144.

Accretion disks.
See Abstr. 067.145.

Theoretical cyclotron and pulse profiles of X–ray pulsars.
See Abstr. 067.182.

HD 160538 and HD 185510: two active–chromosphere stars with hot companions.
See Abstr. 112.072.

A search for white–dwarf companions of subgiant CH stars.
See Abstr. 114.002.

Multi–year & possibly periodic variations in the UV spectrum of 56 Pegasi.
See Abstr. 114.005.

Some problems of composite spectra.
See Abstr. 114.048.

Alpha Trianguli Australis (K2 II – III): hybrid or composite?
See Abstr. 114.087.

Peculiar cool stars in planetary nebulae – the spectrum of FG Sge.
See Abstr. 114.145.

Report of IAU Commission 29: Stellar spectra (*Spectres stellaires*).
See Abstr. 114.165.

A spectrophotometric study of temporal variations in the optical spectrum of SS433.
See Abstr. 114.169.

An Einstein Observatory X–ray survey of main–sequence stars with shallow convection zones.
See Abstr. 116.028.

Observations of the emission from some stars at millimeter wavelengths.
See Abstr. 116.040.

The Mg II emission in BD $-0°4234$.
See Abstr. 116.042.

A W UMa variable in the visual binary system ADS 9019: a *UBV* photoelectric investigation.
See Abstr. 118.006.

High resolution problems in astrometry and spectroscopy of binaries and their significance for the progress of stellar physics.
See Abstr. 118.020.

A spectroscopic study of 44i Bootis.
See Abstr. 118.061.

Accretion in the Zeta Aurigae and 32 Cygni shock cones.
See Abstr. 119.001.

Eclipse observations of Epsilon Aurigae.
See Abstr. 119.002.

Egress spectra of the new eclipsing supergiant 22 Vulpeculae.
See Abstr. 119.004.

RW Comae Berenices. II. Spectroscopy.
See Abstr. 119.009.

Photoelectric photometry of the eclipsing binary MM Herculis: preliminary results.
See Abstr. 119.011.

Photoelectric photometry of the peculiar system RT Lacertae.
See Abstr. 119.013.

RZ Scuti as a double contact binary.
See Abstr. 119.028.

EG Cep – an almost contact binary.
See Abstr. 119.033.

Epsilon Aurigae.
See Abstr. 119.034.

A redetermination of the luminosity, distance, and reddening of Beta Lyrae.
See Abstr. 119.038.

Differential *UBV* photometry of the eclipsing binary SVS 2086.
See Abstr. 119.043.

Photometry of active Algols.
See Abstr. 119.103.

Photoelectric radial velocities. Paper XII. The orbits of another four spectroscopic binaries in the Clube Selected Areas.
See Abstr. 120.002.

Binary systems among the peculiar cool stars.
See Abstr. 120.038.

The optical morphology of the R Aqr jet.
See Abstr. 122.011.

Photographic photometry of V426 Ophiuchi.
See Abstr. 122.093.

Emission–line variability in WR stars: short periods and nonradial pulsations?
See Abstr. 122.123.

A search for southern hemisphere long–period variable stars with close companions.
See Abstr. 123.006.

Classification of variable stars in the light of current conceptions of their evolution.
See Abstr. 123.007.

The old nova GK Per: discovery of the X–ray pulse period.
See Abstr. 124.201.

The 1983 outburst of GK Persei.
See Abstr. 124.202.

The old nova V603 Aql: an intermediate polar?
See Abstr. 124.381.

Pre–existing dust shells around type–I supernovae.
See Abstr. 125.077.

White dwarf stars at ultraviolet wavelengths.
See Abstr. 126.001.

The white dwarf companion of the mild Ba star ξ^1Cet.
See Abstr. 126.009.

Episodic accretion events on to a white dwarf.
See Abstr. 126.023.

Temperatures and luminosities of white dwarfs in dwarf novae.
See Abstr. 126.044.

Pairs of pulsars possibly joined in the past in binary systems.
See Abstr. 126.061.

The binary frequency of extreme subdwarfs revisited.
See Abstr. 126.086.

Detection of a compact companion of the mild barium star ξ^1Ceti.
See Abstr. 126.091.

A search for non–interacting, close binary white dwarfs.
See Abstr. 126.093.

The binary pulsar – a test for general relativity?
See Abstr. 126.098.

R–S model and radio binary pulsars.
See Abstr. 126.099.

On the masses of white dwarfs in cataclysmic binaries.
See Abstr. 126.121.

High–velocity interstellar gas in the line–of–sight to HD 50896.
See Abstr. 131.006.

Cataclysmic variables as probes of X–ray properties of interstellar grains.
See Abstr. 131.297.

Planetoidal hypothesis of CP (*chemically peculiar*) F, A, and B star formation: possibilities and prospects.
See Abstr. 131.369.

Speckle observations of the central source in the bipolar nebula NGC 2346.
See Abstr. 134.010.

Origin of the new axisymmetric radio sources.
See Abstr. 141.002.

A new non–thermal galactic radio source with a possible binary system.
See Abstr. 141.016.

Optical spectroscopy and photometry of the periodic X–ray transient A0538–66 (X0535–668) during an outburst and an OFF state.
See Abstr. 142.002.

The spectra of X–ray transients.
See Abstr. 142.004.

X–ray properties of the galactic bulge sources.
See Abstr. 142.005.

Hard X–ray recurrent transients.
See Abstr. 142.006.

Erratic variability in MXB 1916–05.
See Abstr. 142.009.

Recent observations of transient and variable X–ray sources with satellites "Hakucho" and "Tenma".
See Abstr. 142.010.

X–ray pulsars observed with Hakucho and Tenma.
See Abstr. 142.012.

Discovery of eclipses from the X–ray burst source MXB 1659–29.
See Abstr. 142.013.

Joint timing analysis of A 0535+26 at 1980 outburst.
See Abstr. 142.014.

SMC X–1: another tilted precessing accretion disk?
See Abstr. 142.015.

The occasional disappearance of GX 301–2 pulses.
See Abstr. 142.016.

Steady emission from recurrent transient pulsar A 0535+26.
See Abstr. 142.017.

Globular cluster origin of X–ray bursters.
See Abstr. 142.020.

The discovery of 4.4 second X–ray pulsations from the rapidly variable X–ray transient V0332+53.
See Abstr. 142.027.

The nature of the low–luminosity globular cluster X–ray sources.
See Abstr. 142.028.

Optische Beobachtungen der Röntgenquelle E2003+225.
See Abstr. 142.035.

4U 1624–49.
See Abstr. 142.055.

High–resolution spectroscopic observation of Vela X–1 in the hard X–ray energy range.
See Abstr. 142.060.

Spectrophotometry of the probable optical counterpart of the transient X–ray source V0332+53.
See Abstr. 142.062.

Long term variability of pulse profile of Her X–1.
See Abstr. 142.077.

Circinus X–1: a laboratory for studying the accretion phenomenon in compact binary X–ray sources.
See Abstr. 142.086.

Ultrahigh–energy gamma–ray astronomy.
See Abstr. 143.004.

Variability of high–energy gamma–ray flux of Cyg X–3.
See Abstr. 143.027.

On the binary nature of cosmic γ–ray burst sources.
See Abstr. 143.031.

First observation of ultra–high–energy γ rays from LMC X–4.
See Abstr. 143.032.

Line features from Cygnus X–1 and the Crab nebula in the energy range 30 – 270 keV.
See Abstr. 143.039.

Is Cygnus X–3 a source of gamma rays or of new particles?
See Abstr. 143.054.

A model of object Geminga as a degenerate white dwarf orbiting around a black hole.
See Abstr. 143.055.

Evidence for 500 TeV gamma–ray emission from Hercules X–1.
See Abstr. 143.060.

Very high energy gamma rays from Cygnus X3?
See Abstr. 143.077.

The 4.8 hour period of gamma–rays from Cygnus X–3 at energies $E \geqslant 2 \times 10^{15}$eV.
See Abstr. 143.079.

Observations of high energy gamma rays from Cygnus X–3 and other sources with the Whipple Observatory Large Aperture Camera.
See Abstr. 143.091.

Ultra high energy cosmic rays and the relevance of Cygnus X–3.
See Abstr. 144.056.

Accreting binaries, diffuse radio emission, and galactic cosmic rays.
See Abstr. 144.460.

Collisional coalescence of binaries in globular clusters.
See Abstr. 151.089.

Evolution of stellar systems including tidally captured binaries.
See Abstr. 151.091.

The role of hard binaries in cluster evolution.
See Abstr. 151.116.

Numerical simulations of encounters of hard binaries.
See Abstr. 151.120.

Physical interactions between stars.
See Abstr. 151.123.

Monte–Carlo calculations.
See Abstr. 151.124.

Two body capture in large N body systems.
See Abstr. 151.127.

Evolution of a massive binary in a star field.
See Abstr. 151.173.

The structure and internal kinematics of open clusters.
See Abstr. 153.045.

X–raying the dynamics of globular clusters.
See Abstr. 154.053.

A search for cataclysmic binaries in the globular cluster M3.
See Abstr. 154.063.

A survey for galactic plane ultraviolet–excess objects: space densities for white dwarfs, subdwarfs, and cataclysmic variables.
See Abstr. 155.194.

Fluorescent excitation of Fe II, Mn II, Ti II, N I lines by C IV, N V, O VI,... emission lines in the spectra of symbiotic stars & Seyfert galaxies.
See Abstr. 158.007.

118 Visual Binaries, Multiple Stars, Astrometric Binaries

118.001   Visual measurements of southern double stars.
G. Torres.
Astron. Astrophys., Suppl. Ser., Vol. 59, No. 3, p. 449 – 454 (1985).

A double–star observing program is described which has been started at Córdoba with a 30 cm refractor. Usually neglected IDS pairs south of –60° are selected for observation. 174 micrometric observations of 78 wide pairs are presented together with a few measurements of 9 double stars not found in the IDS. Also reported is a list of 135 additional measurements of 26 test–stars.

118.002   Revised orbital elements of the visual binary stars ADS 2616 AB and ADS 4229 AB.
M. Scardia.
Astron. Astrophys., Suppl. Ser., Vol. 59, No. 3, p. 455 – 460 (1985). In French.

Revised orbital elements of the visual binary stars ADS 2616 AB and ADS 4229 AB are given. The dynamical parallaxes and total masses of the systems have been calculated.

118.003   High–resolution photo of protoplanetary disk orbiting star.
B. M. Schwarzschild.
Phys. Today, Vol. 38, No. 2, p. 19 – 21 (1985).

118.004   VB 8B: brown dwarf or planet?
R. A. Schorn.
Sky Telesc., Vol. 69, No. 2, p. 126 (1985).

118.005   Digital speckle interferometry of 72 binary stars.
Yu. Yu. Balega, I. I. Balega.
Pis'ma Astron. Zh., Tom 11, No. 2, p. 112 – 122 (1985). In Russian. English translation in Sov. Astron. Lett., Vol. 11.

Measurements of position angles and separations of 72 binary stars observed by means of the digital speckle interferometer with the 6–m telescope are presented. The internal error of measurements is close to 0″.003 in both coordinates.

118.006   A W UMa variable in the visual binary system ADS 9019: a UBV photoelectric investigation.
R. L. Walker, C. R. Chambliss.
Astron. J., Vol. 90, No. 2, p. 346 – 353 (1985).

ADS 9019 is a close visual binary with a period of approximately 300 yr and a semimajor axis of 1.2 arcsec. In 1979, the light of the system was discovered to be varying with a period of about 0.4 days. Visual observations indicate that the B component is the variable in this system. Light curves in UBV were obtained that show the system contains an eclipsing binary of the W UMa type. Geometric and photometric elements are derived from the observations using the Wood model, and these indicate that the components of the eclipsing pair are identical to each other and are in a contact configuration.

118.007   Photometric variability of stars belonging to the Trapezium–type systems.
J. J. Claria, E. Lapasset, H. Levato, S. Malaroda.
Astrophys. Space Sci., Vol. 108, No. 1, p. 95 – 100 (1985).

The authors report the variability of 21 stars belonging to 17 systems considered to be Trapezium–type systems by Allen et al. (1977). In order to avoid serious problems due to possible light contamination by neighbouring stars, only systems with angular separations between components equal or greater than five seconds of arc have been considered. Among the new variables here reported there are five having ΔV variations greater than $0^{m}\!.3$. For six systems the authors have spectrograms of the individual components. Three of the six systems are probably optical configurations and only two of the variables in the remaining three systems may be physical companions.

118.008   On the inclination of extra–solar planetary orbits.
B. Campbell, R. F. Garrison.
Publ. Astron. Soc. Pac., Vol. 97, No. 588, p. 180 – 182 (1985).

The authors show that the inclination of a planetary orbital axis can be determined, on the assumption that it is the same as the inclination of the stellar rotation axis. The latter may be derived from the apparent stellar rotation speed ($v \sin i$) and the rotation period determined from the modulation or strength of Ca II emission.

118.009   On the kinematics of trapezium–type multiple systems.
G. N. Salukvadze.
Astrofizika, Tom 22, Vyp. 1, p. 97 – 104 (1985). In Russian. English translation in Astrophysics, Vol. 22, No. 1.

On the basis of up–to–date observational material including the photographic observations performed by the author the kinematics of selected O – B2 spectral class trapezium–type multiple stars has been studied. Out of 15 trapezia investigated 14 show expansion. Among them 11 trapezia expand totally and in three trapezia only one component shows recession.

118.010 Physical parameters of visual binaries with computed orbits.
Z. T. Krajcheva, E. I. Popova, A. V. Tutukov, L. R. Yungel'son.
Astrofizika, Tom 22, Vyp. 1, p. 105 – 120 (1985). In Russian. English translation in Astrophysics, Vol. 22, No. 1.

The main selection effects that define the sample of visual binaries with computed orbits are analysed. The main physical parameters of 473 visual binaries are determined. The distribution of these binaries over the major semi–axes of orbits after correction for selection effects corresponds to the law $dN \propto d\lg a$. The mass function of primaries is the Salpeter function.

118.011 The MK classification of close visual binaries.
C. J. Corbally, R. F. Garrison.
The MK process and stellar classification, p. 367 – 368 (1984). – See Abstr. 012.033.

118.012 High–resolution imaging from Mauna Kea: the astrometric binary Mu Cassiopeiae.
M. J. Pierce, R. J. Lavery.
Astron. J., Vol. 90, No. 4, p. 647 – 651 (1985).

The authors have obtained high–resolution CCD images, at a wavelength of 1 µm, of the Population II binary µ Cassiopeiae and have detected the secondary. By averaging four short–exposure images, a composite image of FWHM = 0.43″ was obtained. With subsequent model fitting, the authors find a separation and position angle of $0.93'' \pm 0.06''$ and $224.3° \pm 2.6°$, respectively. These data, combined with recent astrometric orbits, yield masses of $M(A) = 0.61 \pm 0.30\,M_{\odot}$, $M(B) = 0.15 \pm 0.07\,M_{\odot}$, and a helium abundance of $Y = 0.42 \pm 0.30$.

118.013 A study of visual double stars with early type primaries. IV. Astrophysical data.
K. P. Lindroos.
Astron. Astrophys., Suppl. Ser., Vol. 60, No. 2, p. 183 – 211 (1985).

Astrophysical parameters (MK class, colour excess, absolute magnitude, distance, effective temperature, mass and age) are derived from calibrations of the $uvby\beta$ indices for the members of 253 double stars with O or B type primaries and faint secondaries. The photometric spectral classification is compared to the MK classes and the agreement is very good. The derived data together with spectroscopic and $JHKL$ data are used for deciding which pairs are likely to be physical and which are optical and it is shown that 98 (34%) of the secondaries are likely to be members of physical systems. For 90% of the physical pairs the projected separations between the components is less than 25000 AU. A majority of the physical secondaries are late type stars and 50% of them are contracting towards the zero–age main–sequence. Also presented are new $uvby\beta$ data for 43 secondaries and a computer programme for determining astropysical parameters from $uvby\beta$ data.

118.014 New double stars (19th series) discovered at Nice.
P. Couteau.
Astron. Astrophys., Suppl. Ser., Vol. 60, No. 2, p. 237 – 239 (1985). In French.

The author gives a list of 100 close visual binaries discovered at the 50 cm refractor.

118.015 Measurements of double stars at Nice with 74 and 50 cm refractors.
P. Couteau.
Astron. Astrophys., Suppl. Ser., Vol. 60, No. 2, p. 241 – 259 (1985). In French.

The author gives 717 measures of 287 binaries made at the 74 and 50 cm refractors. These are mostly very close binaries discovered recently, for which a rapid orbital motion is evident for about thirty per cent.

118.016 Photographic observations of visual double stars.
E. van Albada–van Dien.
Astron. Astrophys., Suppl. Ser., Vol. 60, No. 2, p. 315 – 324 (1985).

The author presents the results of 770 plates on 238 pairs of visual double stars taken in 1976 – 1979 with the 60 cm visual refractor of the Bosscha Observatory at Lembang, Java.

118.017 Orbital elements for fifteen visual binaries.
P. Baize.
Astron. Astrophys., Suppl. Ser., Vol. 60, No. 2, p. 333 – 337 (1985). In French.

The orbital elements for fifteen pairs are given: 6 revised (Kpr 7 and 108, ADS 8032, 11080, 13135 and 16448) 9 new (B2530, ADS 7674, 8198, 8347, 8831, 9545, 9688, 12741 and 14648). For each pair one finds the measures and their discussion, the residuals O–C, the dynamical parallax and the ephemeris.

118.018 Infrared detection of a close cool companion to Van Biesbroeck 8.
D. W. McCarthy Jr., R. G. Probst, F. J. Low.
Astrophys. J., Lett. Ed., Vol. 290, No. 1, p. L9 – L13 (1985).

The authors have detected, via infrared speckle interferometry, a cool object 1″ from the very low–luminosity star VB 8, 3 mag fainter than VB 8 at 2.2 µm. Measurements at 1.6 and 2.2 µm give $T_e = 1360K$ and (assuming a physical association) $R = 0.09\,R_{\odot}$, $L = 3 \times 10^{-5}L_{\odot}$, consistent with a substellar brown dwarf. These observations may constitute the first direct detection of an extrasolar planet.

118.019 HIPPARCOS astrometric binaries.
J. Dommanget.
Astrophys. Space Sci., Vol. 110, No. 1, p. 47 – 63 (1985). – See Abstr. 012.039.

The Input Catalogue of some 100,000 stars that is presently prepared for observation by the astrometric satellite HIPPARCOS, will contain many double multiple systems. Because of the HIPPARCOS observation technique, these systems have to be divided in a few particular categories that are described and discussed. Each of them leads to specific considerations concerning the contribution of the HIPPARCOS observations. The category of very close pairs to which HIPPARCOS will certainly add many systems newly discovered during the mission, is compared to that of the few astrometric pairs that have been discovered by groundbased techniques.

118.020 High resolution problems in astrometry and spectroscopy of binaries and their significance for the progress of stellar physics.
F. van't Veer.
Astrophys. Space Sci., Vol. 110, No. 1, p. 65 – 74 (1985). – See Abstr. 012.039.

Two centuries were necessary for binaries to become, after having been unknown and ignored, one of the outstanding subjects of stellar research. This remarkable evolution has resulted now in the reversed situation with the question: do single stars really exist? It is recognized by everybody, that many essential properties of stars can only be obtained from observations of close and wide binaries. It is the aim of this contribution to see how binaries can help to better understand binaries and single stars, even if the existence of the latter is not yet definitely confirmed.

118.021 Statistical models for HIPPARCOS binaries.
S. Söderhjelm.
Astrophys. Space Sci., Vol. 110, No. 1, p. 77 – 87 (1985). – See Abstr. 012.039.

A statistical model for the binaries in the solar neighbourhood is constructed, using schematic but plausible distribution functions for the semi–major axes, the mass–ratios, and the eccentricities. The model is 'calibrated' to give correct numbers of observed visual binaries and is then used to study the (closer) binaries of relevance for the HIPPARCOS mission. One main purpose is to estimate the influence of astrometric binaries observed by their photocentres.

118.022 On the determination of double star orbits.
H. Eichhorn.
Astrophys. Space Sci., Vol. 110, No. 1, p. 119 – 124 (1985). =
Commun. Dep. Astron., Univ. Florida, Gainesville, No. 77. –
See Abstr. 012.039.

It is proposed to improve the convergence of the determination
of the elements of the orbits of visual binaries by using not only
first, but second–order derivatives in the development of the
appropriate equations of condition. Also, some improvements of
the Kowalsky–Seeliger method are suggested which improve the
accuracy of the orbit used as first approximation.

**118.023 Distribution of astrometric binaries in various galactic
latitudes.**
A. N. Goyal.
Astrophys. Space Sci., Vol. 110, No. 1, p. 125 – 131 (1985). – See
Abstr. 012.039.

The conclusions of the present paper broadly are: (1) The galactic concentration of doubles by comparing the distributions in
galactic latitudes $0° < \beta \leqslant 20°$ and $\beta > 40°$ is nearly twice as
large as the galactic concentration of stars in general. (2) The
astrographic catalogues are not complete in the fainter magnitudes. (3) The large value of the ratio $T{:}O_k$ (observed to optical
number of pairs) from Kreiken's formula shows that almost
all stars in the group $0'' < d \leqslant 5''$ and quite a few in the other
two groups, viz., $5'' < d \leqslant 10''$ and $10'' < d \leqslant 15''$ might be
shown true binaries. The doubles are probably stars of
Population I. (4) The logarithm to the base 10 of the cumulative
counts can be represented by an empirical relation
$A + B(\Delta m{-}1.5) + C(\Delta m{-}1.5)^2$.

**118.024 Astrometric binaries and their role in double star astron-
omy.**
T. Herczeg.
Astrophys. Space Sci., Vol. 110, No. 1, p. 133 – 134 (1985). – See
Abstr. 012.039.

118.025 Close triple star systems.
F. B. Wood.
Astrophys. Space Sci., Vol. 110, No. 1, p. 137 – 141 (1985). – See
Abstr. 012.039.

Triple star systems, especially those in which one star has very
small mass may be more common than has been generally considered. Here is summarized some of the recent evidence supporting
this possibility.

118.026 Luminosity distribution function for visual binary stars.
K. D. Rakos.
Astrophys. Space Sci., Vol. 110, No. 1, p. 143 – 148 (1985). – See
Abstr. 012.039.

The 'cosmic scatter' of 147 of the best known visual binary
stars on the Main Sequence is discussed and a new estimation of
the luminosity distribution function for multiple star systems is
presented. As long as the mass ratio q of a close binary is not
smaller than 0.5, the distribution of close binary components is
identical to the van Rhijn luminosity–function. For smaller mass
ratios ($q < 0.5$) the number of close companions decreases
rapidly. It appears that less than 13% of visual binaries in the
author's sample are simple binary systems.

118.027 Trapezium–type multiple systems.
L. V. Mirzoyan, G. N. Salukvadze.
Astrophys. Space Sci., Vol. 110, No. 1, p. 153 – 158 (1985). – See
Abstr. 012.039.

The non–stable nature of stellar motions in Trapezium–type
multiple systems is discussed. The disintegration time of such
systems is several orders of magnitude less than the life–time of
their components, and considerably less than the disintegration
time of stellar associations. The importance of ground–based and
orbital observations of Trapezium–type systems is stressed.

118.028 The spectroscopic study of astrometric binaries.
G. Ishida.
Astrophys. Space Sci., Vol. 110, No. 1, p. 161 – 162 (1985). – See
Abstr. 012.039.

The spectroscopic observations of several astrometric binaries,
obtained by the author during the past 20 years, are described.
Main emphasis is placed on the determination of orbital elements
of visual binaries, and the detection of unseen companions.

**118.029 The origin of planetary systems in the scenario of star
formation.**
P. Paolicchi.
Mem. Soc. Astron. Ital., Vol. 55, No. 3, p. 469 – 479 (1984). – See
Abstr. 012.049.

**118.030 New Trapezium–type systems in Taurus Dark Clouds
(TDC).**
A. S. Khodzhaev.
Astrofizika, Tom 22, Vyp. 2, p. 425 – 428 (1985). In Russian. En-
glish translation in Astrophysics, Vol. 22, No. 2.

Two new probable Trapezium–type triples in TDC have been
discovered, the components of both systems being variable stars.

118.031 Stellar companion appears to be giant planet.
D. Helfand.
Phys. Today, Vol. 38, No. 4, p. 19 – 21 (1985).

**118.032 Photographic measurements of ADS 11632, a binary
with a possible unseen companion.**
N. A. Shakht.
Sov. Astron. Lett., Vol. 10, No. 5, p. 319 – 321 (1984). English
translation of 38.118.016.

**118.033 Van Biesbroeck 8 and van Biesbroeck 10: dark compan-
ions of nearby stars.**
R. S. Harrington, B. J. Harrington.
Mercury, Vol. 14, No. 1, p. 14 – 15 (1985).

118.034 A photoelectric study of PZ Cassiopeiae.
L. S. Kudashkina.
Astron. Tsirk., No. 1351, p. 4 – 6 (1984). In Russian.

118.035 Micrometric observations of visual double stars.
G. Massone.
Astron. Astrophys., Suppl. Ser., Vol. 60, No. 3, p. 485 – 494
(1985).

This paper reports visual micrometric measurements of 149
double stars obtained at the Torino Astronomical Observatory.

118.036 The IDS triple star systems.
G. M. Popović.
Publ. Astron. Opservatorije Beogr., No. 33, p. 88 – 91 (1985). –
See Abstr. 012.061.

It was found that the IDS Catalogue incorporated 4140 triple
star systems out of which 506 displayed features of clear or very
probable physical connection between their components. The
analysis showed that most of the triples consisted of close A and
B components and a distant C component, and a lesser number
of close B and C and a distant A. There is not a distinct discontinuity between the ε Lyr and T Orionis types.

**118.037 Observations photographiques d'étoiles doubles et
d'étoiles multiples effectuées en 1984 à l'équatorial Gautier (Carte du Ciel).**
S. M. Kazeza.
Bull. Astron., Vol. 9, No. 6, p. 311 – 319 (1984).

**118.038 Tendance évolutive au rapport des masses dans les binai-
res visuelles.**
A. Jorissen.
Bull. Astron., Vol. 9, No. 6, p. 325 – 329 (1984).

From orbital data concerning 811 visual binaries, a diagram
has been constructed in order to investigate an eventual secular
evolution of their mass ratio. After indicating that selection effects are not able to explain entirely the global shape of the

diagram, the author shows that, for a certain range of the parameters describing secular mass loss, this diagram is in good agreement with theoretical predictions.

118.039   Discovery of eclipses and long–period variability in the triple system HR 6469.
L. J. Boyd, R. M. Genet, D. S. Hall, W. T. Persinger,
R. E. Fried, N. F. Wasson, H. J. Stelzer, R. D. Lines,
P. A. Brooks, D. Hoff.
Inf. Bull. Variable Stars, No. 2675, 7 pp. (1985).

118.040   An IUE study of the RS CVn binary DH Leo (HD 86590).
J. S. Shaw, E. F. Guinan, D. A. Fraquelli.
Bull. Am. Astron. Soc., Vol. 16, No. 4, p. 914 (1984). Abstract. –
See Abstr. 010.062.

118.041   Detection of an infrared source near VB 8: the first extra–solar planet?
D. W. McCarthy Jr., R. G. Probst.
Bull. Am. Astron. Soc., Vol. 16, No. 4, p. 965 (1984). Abstract. –
See Abstr. 010.062.

118.042   A nearby newly resolved low mass astrometric binary.
P. A. Ianna, J. R. Rohde, D. W. McCarthy Jr.
Bull. Am. Astron. Soc., Vol. 16, No. 4, p. 965 (1984). Abstract. –
See Abstr. 010.062.

118.043   High resolution X–ray observations of nearby binary systems: flaring and evidence for unseen companions.
D. E. Harris, H. M. Johnson.
Bull. Am. Astron. Soc., Vol. 16, No. 4, p. 983 (1984). Abstract. –
See Abstr. 010.062.

118.044   Micrometer measures of 711 double stars.
R. L. Walker Jr.
Publ. U.S. Nav. Obs., Second Ser., Vol. 25, Part II, 73 pp. (1985).

The paper presents 3053 micrometer measures of 711 double stars and is part of a continuing series of measures. The observations were obtained using the 40–inch Ritchey–Chrétien and the 61–inch astrometric reflectors at the Flagstaff Station of the U.S. Naval Observatory and the 36–inch refractor at the Lick Observatory in the interval between 1972 and 1979.

118.045   Gamma Cae: a binary with a suspected variable secondary.
R. B. Culver, P. A. Ianna.
Inf. Bull. Variable Stars, No. 2736, 2 pp. (1985).

118.046   Orbites nouvelles.
Circ. Inf., Nos. 95, 96 (1985).

118.047   Etoiles doubles nouvelles.
W. D. Heintz, P. Couteau.
Circ. Inf., Nos. 95, 96 (1985).

118.048   More than a planet, almost a star.
M. M. Waldrop.
Science, Vol. 227, No. 4682, p. 44 (1985).

The faint companion of van Biesbroeck 8 is the first known "brown dwarf" – a cosmic ember not quite massive enough to ignite.

118.049   Detection of an X–ray flare in the RS CVn binary Sigma Coronae Borealis.
P. C. Agrawal, A. R. Rao, G. R. Riegler, R. Stern.
18th International Cosmic Ray Conference, Vol. 1, p. 7 – 10 (1983). – See Abstr. 012.096.

The source spectrum during the flare was found to be considerably harder compared to that observed in the quiescent state. Details of the various characteristics of the X–ray flare are described.

118.050   The possibility of exploration of stellar systems by radar.
O. N. Rzhiga.
Astron. Zh., Tom 62, Vyp. 3, p. 500 – 505 (1985). In Russian.
English translation in Sov. Astron., Vol. 29, No. 3.

The tasks which can be solved by radar investigations of stellar systems, as well as the means required for this are considered.

118.051   Interferometric measurements of binary stars with the GSU ICCD speckle camera.
H. A. McAlister, W. I. Hartkopf, D. J. Hutter, O. G. Franz.
Bull. Am. Astron. Soc., Vol. 17, No. 2, p. 551 (1985). Abstract. –
See Abstr. 010.065.

118.052   On the number of comets around other single stars.
C. C. Fristrom, C. Alcock.
Bull. Am. Astron. Soc., Vol. 17, No. 2, p. 558 (1985). Abstract. –
See Abstr. 010.065.

118.053   A search for dark companions around M dwarfs.
G. W. Marcy, V. Lindsay.
Bull. Am. Astron. Soc., Vol. 17, No. 2, p. 558 (1985). Abstract. –
See Abstr. 010.065.

118.054   On a possible close companion to α Ori.
M. Karovska, R. W. Noyes, F. Roddier, P. Nisenson,
R. V. Stachnik.
Bull. Am. Astron. Soc., Vol. 17, No. 2, p. 598 – 599 (1985). Abstract. – See Abstr. 010.065.

118.055   The companion to VB8.
R. S. Harrington.
Bull. Am. Astron. Soc., Vol. 17, No. 2, p. 624 (1985). Abstract. –
See Abstr. 010.066.

118.056   Orbits of 11 visual binaries.
G. A. Starikova.
Tr. Gos. Astron. Inst. Shternberg, Tom 57, p. 243 – 244 (1985).
In Russian.

From available published data and from new observations revised orbits of 11 visual binary stars are obtained. The new orbits are in better agreement with observation.

118.057   The ellipticity effect and a migrating wave in the chromospherically active triple system V772 Her.
L. J. Boyd, R. M. Genet, D. S. Hall, W. T. Persinger.
Inf. Bull. Variable Stars, No. 2747, 6 pp. (1985).

118.058   Report of IAU Commission 26: Double and multiple stars (*Etoiles doubles et multiples*).
M. G. Fracastoro.
Trans. IAU, Vol. XIXA, p. 269 – 275 (1985). – See Abstr. 003.046.

118.059   The quadruple system HR 3337.
G. A. Bakos.
J. R. Astron. Soc. Can., Vol. 79, No. 3, p. 119 – 129 (1985).

From spectrograms with a dispersion of 12 Åmm^{-1} the orbital elements of both components of this visual–spectroscopic binary have been derived. The companion having a period of 2.5 days was found to be an eclipsing binary.

118.060   Spectral types of companions to variable visual double stars.
E. M. Halbedel.
Publ. Astron. Soc. Pac., Vol. 97, No. 591, p. 434 – 436 (1985).

Spectral types are given for 45 companions to 42 variable visual doubles.

118.061   A spectroscopic study of 44i Bootis.
X.–f. Liu, S.–q. Peng, G.–w. Cha.
Chin. Astron. Astrophys., Vol. 9, No. 2, p. 145 – 149 (1985). English translation of Acta Astrophys. Sin., Vol. 5, No. 1, p. 36 – 43 (1985).

The authors observed the spectrum of 44i Boo in 1981 March and in 1983 March. They found that the equivalent widths of the

Ca II K and Ca I 4227 Å lines showed periodic variations with different phases, and that the profile of the Hα line also varied in step. A preliminary analysis is made and the recent IUE data and a simple model atmosphere with stellar activity of 44i Boo (B) is proposed.

118.062   On the accuracy of determination of the parameters of the apparent motion using photographic observations of a short arc of the orbit of a visual double star.
O. V. Kiyaeva.
Izv. Glav. Astron. Obs. Pulkovo, Astrometr. Astrofiz., No. 201, p. 44 – 51 (1985). In Russian.
  A new method for the determination of the orbits of a visual double star requires parameters of the apparent motion in the middle of the observed arc with a very high accuracy. The actual accuracy of determination of the parameters of the apparent motion in dependence on observational errors is considered and how long the observed series should be. Special attention is paid to the determination of the radius of curvature. Two methods for the determination of this parameter are given and the errors of the methods are studied.

118.063   Le stelle doppie astrometriche e interferometriche.
B. Cester.
Orione, Vol. 5, N. 1, p. 11 – 16 (1985).

Sky catalogue 2000.0. Volume 2: Double stars, variable stars and nonstellar objects.
See Abstr. 002.019.

Fourth catalog of orbits of visual binary stars.
See Abstr. 002.069.

The Washington Visual Double Star Catalog, 1984.0.
See Abstr. 002.070.

Current work on binary and multiple stars.
See Abstr. 013.037.

Work on astrometric binaries in China.
See Abstr. 013.038.

Peter van de Kamp en zijn "lieve Barnard's Ster".
See Abstr. 013.052.

Candidates for astrometric or direct–imaging detection of extra–solar planets, and comments on the photometric method.
See Abstr. 036.056.

Modern methods in the astrometry of double stars.
See Abstr. 036.076.

Detectability of extrasolar planetary transits.
See Abstr. 036.098.

On a possibility of observations of variable stars in visual binary systems with the area scanning technique.
See Abstr. 036.105.

Algorithme pour analyse des images d'étoiles doubles visuelles fournies par un équipement de télévision.
See Abstr. 036.119.

Double star analysis strategy.
See Abstr. 036.162.

Double star detection.
See Abstr. 036.163.

Estimation of double star parameters.
See Abstr. 036.164.

Imaging approach to multiple star recognition.
See Abstr. 036.165.

Accurate differential magnitudes of binary star components as obtained using the SAA algorithm.
See Abstr. 036.199.

Reduction and analysis techniques developed for the GSU ICCD speckle camera.
See Abstr. 036.200.

The solar radial velocity and implications for spectroscopic planetary detection.
See Abstr. 036.208.

Weighting of phases and grid coordinates.
See Abstr. 041.030.

Numerical experiments on planetary orbits in double stars.
See Abstr. 042.027.

The existence of wide pairs among late–type stars.
See Abstr. 111.011.

On the observation of double stars for the determination of their proper motions.
See Abstr. 111.018.

A search for CPM (*common proper motion*) stars.
See Abstr. 111.019.

The X–ray corona of Procyon.
See Abstr. 112.052.

Ultraviolet observations of Beta Pictoris: a possible protoplanetary system?
See Abstr. 112.136.

Infrared observations of low mass objects.
See Abstr. 113.033.

Which map of absolute magnitudes: Keenan or Schmidt–Kaler?
See Abstr. 115.008.

Possible evolution of a triple system into Epsilon Aurigae.
See Abstr. 117.081.

Absolute dimensions of eclipsing binaries. VII. V1647 Sagittarii.
See Abstr. 119.022.

Analysis of light curves of the eclipsing binary u Herculis.
See Abstr. 119.024.

Photoelectric radial velocities. Paper XII. The orbits of another four spectroscopic binaries in the Clube Selected Areas.
See Abstr. 120.002.

Contribution to the study of F, G, K, M binaries III. Orbital elements of the long period spectroscopic binaries HD 102928 and HD 145206.
See Abstr. 120.007.

Speckle image reconstruction of a northern optical companion to T Tau.
See Abstr. 121.066.

On the problem of duplicity of flare stars.
See Abstr. 122.216.

The occurrence of peculiar stars in clusters and visual systems.
See Abstr. 153.017.

Alpha Tauri CD: another Hyades binary.
See Abstr. 153.049.

119 Eclipsing Binaries

119.001 **Accretion in the Zeta Aurigae and 32 Cygni shock cones.**
I. A. Ahmad, R. D. Chapman, R. Stencel, Y. Kondo.
NASA Conf. Publ., NASA CP–2349, p. 357 – 360 (1984). – See Abstr. 012.001.

Zeta Aurigae and 32 Cygni are binary stars consisting of a cool supergiant primary and a hot dwarf secondary. The authors have observed variations in the Mg II 2800 Å and C IV 1550 Å doublets of these stars near the time of secondary minimum. Longward–shifted absorption is seen in the Mg II lines of both stars which may be due to material accreting onto the B star behind the shock front. Parameters for the interaction dynamics in the case of Zeta Aur confirms the general conclusions of Ahmad, Chapman and Kondo (1983). In the case of 32 Cygni the observed aberration angle is similar to that of Zeta Aur, but the system seems more difficult to model in detail. For Zeta Aur the angle of aberration of the accreting material from the radial direction is about 38° and the half width of the shock cone is about 11°. For 32 Cygni the aberration angle is about 44°.

119.002 **Eclipse observations of Epsilon Aurigae.**
T. B. Ake, T. Simon.
NASA Conf. Publ., NASA CP–2349, p. 361 – 364 (1984). – See Abstr. 012.001.

The authors summarize the major characteristics in the UV of the 1982–84 eclipse of ε Aur, from first through third contact. The only strong emission lines in the UV are O I 1304 and Mg II 2800. The Mg II emission has developed a broad, blue–shifted component of –300 km s^{-1}, which the authors speculate may be a signature of accretion by the secondary of gas ejected by the primary.

119.003 **UV observations of Epsilon Aurigae during ingress and totality.**
B. M. Altner, R. D. Chapman, Y. Kondo, R. E. Stencel.
NASA Conf. Publ., NASA CP–2349, p. 365 – 368 (1984). – See Abstr. 012.001.

Analysis of SWP and LWR spectra of Epsilon Aurigae taken during the pre–eclipse, ingress and total phases of the present eclipse has provided further constraints on models of this enigmatic system. Both high and low dispersion spectra indicate that the eclipse starts earlier, ends later and is deeper in the UV compared to visual wavelengths. The authors confirm their earlier observation that the eclipse depth is wavelength dependent shortward of 2400 Å. Abrupt changes in the light curve appear at all wavelengths, suggestive of discontinuities in the opacity of a ring of material surrounding the secondary object.

119.004 **Egress spectra of the new eclipsing supergiant 22 Vulpeculae.**
T. B. Ake III, S. B. Parsons, Y. Kondo.
NASA Conf. Publ., NASA CP–2349, p. 369 – 372 (1984). – See Abstr. 012.001.

The G–type primary in the single–lined spectroscopic binary 22 Vul was found with IUE in April 1983 to eclipse its recently identified B9 main sequence companion. The system thus joins the few known Zeta Aurigae systems in which a hot dwarf can be used to probe the extended atmosphere of a cool supergiant. Outside of eclipse, a wind from the system with terminal velocity of 300 km s^{-1} is conspicuous in the Mg II lines. Enhanced Fe II absorption seen well away from geometrical eclipse implies an extensive envelope around the G star. Some of the 22 Vul spectra closely resemble those of other interacting binaries, suggesting common properties of relatively dense, warm plasmas in these systems.

119.005 **The stellar wind in the B0 V–type eclipsing binary Y Cygni**
N. D. Morrison, C. D. Garmany.
NASA Conf. Publ., NASA CP–2349, p. 373 – 376 (1984). – See Abstr. 012.001.

In several high–dispersion SWP images taken near eclipse and at quadrature, the authors studied the profiles of the resonance lines of Si IV, C IV, and N V. In general, the spectrum is similar to that of the single B0 V–type star τ Scorpii. In the spectrum obtained at quadrature, the C IV profile is shifted in accordance with the orbital motion of the hotter, less massive star. Therefore, it is likely that the wind–borne C IV absorption originates in this star. By fitting the C IV and N V lines with synthetic profiles computed from the program written by Olson (1982 and references therein), the authors obtained a terminal velocity of 1500 km s^{-1}. Application of the technique of Garmany et al. (1981) yields $\dot{M} = 10^{-8.4} M_\odot yr^{-1}$.

119.006 **Four–colour photometry of eclipsing binaries. XXIII. Light curves of GZ Canis Majoris.**
J. Andersen, J. V. Clausen, B. Nordström.
Astron. Astrophys., Suppl. Ser., Vol. 59, No. 2, p. 349 – 355 (1985).

Complete *uvby* light curves of the southern detached double–lined Am eclipsing binary GZ CMa (8$^{\rm m}$0) are presented.

119.007 **Photoelectric radial velocities of eclipsing binaries. I. Orbital elements of BF Dra.**
M. Imbert.
Astron. Astrophys., Suppl. Ser., Vol. 59, No. 2, p. 357 – 360 (1985). In French.

A spectroscopic orbit has been determined for the double–lined eclipsing binary BF Dra. The radial velocities were obtained with the Coravel radial velocity spectrometer. The observations show that the photometric period must be doubled.

119.008 **The light variations of the triple system HD 165590.**
G. A. Bakos, J. Tremko.
Contrib. Astron. Obs. Skalnaté Pleso, Vol. 12, p. 65 – 74 (1984).

Photometric observations of HD 165590 extending over a period of six years have shown that the eclipsing system undergoes systematic light variations in the form of flares and short–time– and long–time–scale changes. An attempt has been made to analyse and to interpret these changes.

119.009 **RW Comae Berenices. II. Spectroscopy.**
E. F. Milone, B. J. Hrivnak, G. Hill, W. A. Fisher.
Astron. J., Vol. 90, No. 1, p. 109 – 114 (1985). Publ. Rothney Astrophys. Obs., No. 23.

Since 1977, 36 spectroscopic exposures of the 0$^{\rm d}$24 W UMa eclipsing binary RW Com have been obtained, permitting the classification and the velocities and emission behavior of the components to be studied. The radial velocities were determined using the cross–correlation method. The best determination of the mass ratio was $m_2/m_1 = 0.34 \pm 0.02$, where star 2 is the star eclipsed at primary minimum, and is therefore the less massive component.

119.010 **Relativistic apsidal motion in the eclipsing binary systems V1143 Cygni and EK Cephei.**
A. Giménez, T. E. Margrave.
Astron. J., Vol. 90, No. 2, p. 358 – 363 (1985).

A detailed analysis is presented in this paper of the period variations exhibited by two detached eclipsing binary systems: V1143 Cygni and EK Cephei, and new photoelectric times of minimum light are given. The observed period changes are explained in terms of a secular displacement of the periastron, and it is shown in both cases that the general relativistic contribution to the observed rates of apsidal motion is important. The authors find an apsidal–motion period of $(1.07 \pm 0.06) \times 10^4$yr for V1143 Cyg with a relativistic contribution of 43%, while for EK Cep they find a period of $(0.41 \pm 0.12) \times 10^4$yr with a relativistic contribution of 46%.

119.011   Photoelectric photometry of the eclipsing binary MM Herculis: preliminary results.
S. Evren.
Astrophys. Space Sci., Vol. 108, No. 1, p. 113 – 123 (1985).

The RS CVn–type eclipsing binary star MM Her has been observed in two colours, B and V, in 1979, 1980, and 1983. Several minima times were obtained during the observations and new light elements calculated. The light curves of the system obtained in blue and yellow lights show a significant wave–like distortion which migrates towards the decreasing orbital phases. Its migration period was estimated to be about 3.5 yr. The amplitudes of the wave–like distortion in B and V appear to change each year. The primary minimum of the system is a total eclipse with a duration of $0^{d}_{.}08$.

119.012   *UBV* photometry and light–curve analysis of Algol.
H. M. K. Al–Naimiy, A. A. A. Mutter, H. A. Flaih.
Astrophys. Space Sci., Vol. 108, No. 2, p. 227 – 236 (1985).

Photoelectric observations of the eclipsing variable β Per, were obtained in *UBV* standard system, and new elements for the primary minimum were determined as J.D. = 2445641.5135, $O-C = 0^{d}_{.}0.009$. The light curves of the system were analysed using Fourier techniques in the frequency–domain. The fractional radii of both components are $r_1 = 0.217 \pm 0.002$, $r_2 = 0.233 \pm 0.002$ and $i = 85.5 \pm 0.5$. Absolute elements were derived and the effective temperatures are $T_1 = 11800K$, $T_2 = 5140K$.

119.013   Photoelectric photometry of the peculiar system RT Lacertae.
S. Evren, Z. Tunca, C. Ibanoğlu, O. Tümer.
Astrophys. Space Sci., Vol. 108, No. 2, p. 383 – 392 (1985).

The observations of the peculiar eclipsing binary RT Lac were continued during the observing seasons of 1982 and 1983. The shallowest primary eclipse was observed in 1983. When the light curves, obtained in six years successively, are interpreted together it is seen that the variations of the brightness at mid–primary and second maximum are similar to each other. The colour variations at mid–primary also follow the variations in the brightness. These circumstances indicate that the starspot hypothesis is insufficient alone to explain all the phenomena observed. The earlier spectral type G9 companion may be intrinsically variable and the unusual distortion on the light curves following the secondary eclipse may arise from the gas–stream from larger K star to the G9 star.

119.014   The variability of the light curve of the WR–eclipsing binary CV Serpentis.
N. A. Lipunova.
Astrophys. Space Sci., Vol. 109, No. 1, p. 57 – 62 (1985).

Two sets of observations of the WR–eclipsing binary CV Ser were carried out in 1982 and 1983 in the standard photometric system *UBV*. Since 1982, the depth of the atmospheric eclipse has changed and does not show in the V data obtained in 1983. There is significant intrinsic variability ($\gtrsim 0^{m}_{.}05$) in the light curves of CV Ser over both sets of observations and it is suggested that this is due to the changing mass–loss rate from the envelope of the WC8 star on time–scales from days to months. The observed mass–loss rate can change the mass by some $10^{-5} M_\odot y^{-1}$. An orbital inclination $i = 72 \pm 2°$ is obtained.

119.015   UBV photometry of ε Aurigae during the 1982 – 84 eclipse.
T. Ôki, H. Yoshinari, I. Sekiya, K. Hirayama.
Sci. Rep. Fukushima Univ., No. 34, p. 19 – 23 (1984).

The observations of ε Aur using the three–color photometric system were made during the last eclipse. The data were obtained from observations for 72 nights in the period from October 1982 to April 1984. The second contact seemed to occur 5 days earlier than the date predicted by Gyldenkerne, while the third contact appeared to be appreciably delayed (by about 20 days). The color did not change noticeably during the eclipse except in the active phase. Near the end of the totality anomalous brightening occurred for over 40 days. It was conspicuous in the ultraviolet region. The gradient of the light curve in the egress phase was steeper than that of the ingress phase.

119.016   Photometric analyses of S Cancri.
P. B. Etzel, E. C. Olson.
Astron. J., Vol. 90, No. 3, 504 – 514 (1985).

Photometric solutions are made on several photoelectric light curves of S Cancri previously published by Crawford and Olson (1980). Two widely independent methods are employed to give adopted equivalent volume radii of $r_A = 0.0860$ and $r_B = 0.1880$, and an orbital inclination of $i = 85.31$ deg. Spectrophotometric observations are presented and analyzed to give revised spectral classifications of B9.5 V and G8 to K0 III and temperatures of $T_A = 10500K$ and $T_B = 4750$ to $4500K$ for the components. The cool Roche–lobe–filling secondary exhibits changes in color with phase due to oblateness and reflection. Anomalous photometric behavior is correlated with changes in a nonstellar Ca II K–line absorption feature that suggests the presence of an extended atmosphere around the cool mass–losing component. This is a phenomenon never before noted in Algol–type binaries.

119.017   Das Amateurprojekt "Epsilon Aurigae".
D. Böhme.
Sterne Weltraum, 24. Jahrg., Nr. 1, p. 38 (1985).

119.018   Beobachtungen an Epsilon Aurigae.
K.–P. Timm.
Sterne Weltraum, 24. Jahrg., Nr. 3, p. 153 (1985).

119.019   The rotation of the primary of Algol.
J. Tomkin, H.–s. Tan.
Publ. Astron. Soc. Pac., Vol. 97, No. 587, p. 51 – 53 (1985).

The authors have used high signal–to–noise ratio Reticon observations of the 5875 Å He I line to estimate an equatorial velocity $V = 49 \pm 3$ km s^{-1} for the rotation of the primary. This is not significantly different from the synchronous velocity $V = 52 \pm 3$ km s^{-1} and confirms the conclusion of previous studies that the rotation of the primary is synchronous.

119.020   High–dispersion spectroscopy of the eclipse of Epsilon Aurigae at visible and ultraviolet wavelengths.
S. Ferluga, M. Hack.
Astron. Astrophys., Vol. 144, No. 2, p. 395 – 402 (1985).

Visual observations: The authors report some preliminary results of the study of three photographic spectrograms obtained during the ingress, totality and egress phases and compare them with one spectrogram taken before the start of the eclipse. They describe the characteristics of the "shell" spectrum which are similar to those observed during previous eclipses. Ultraviolet observations: Comparing a set of high–resolution IUE spectra, taken at different epochs during the eclipse, the authors derive the following preliminary results. The pre–eclipse far–UV activity of March 1982 is accompanied in the mid–UV by a general rising of the continuum. The depth of the eclipse deviates very slightly from greyness. Additional deepening during the eclipse is found over the absorption wings of the Mg II λ2800 doublet. The mid–eclipse brightening seems to come from a rarefaction of the material, responsible for the continuous (grey) absorption factor.

119.021   *IUE* and optical spectral scans of U Sagittae: an analysis and comparison with U Cephei.
J. J. Dobias, M. J. Plavec.
Publ. Astron. Soc. Pac., Vol. 97, No. 588, p. 138 – 150 (1985).

The authors have determined the effective temperatures and surface gravities of the components of U Sge, using a combination of optical spectral scans and *IUE* spectra. The primary component is found to be a B7.5 V star with $T_{eff} = 12,250 (\pm 250)K$ and $\log g = 3.9 \pm 0.1$. The secondary component is of spectral type G4 III–IV. The best value for distance is 295 ± 20 pc, and the color excess is $E(B-V) = 0^{m}_{.}06$. The mass ratio is very close to 3. The authors discovered emission lines of the W Serpentis type in an *IUE* spectrum taken in the total primary eclipse. Although the observed fluxes of these lines are very low, the powers emitted are half as strong as those seen in U Cep. This is rather surprising, since no other evidence of circumstellar matter or mass transfer activity is found.

119.022 Absolute dimensions of eclipsing binaries. VII. V1647 Sagittarii.
J. Andersen, A. Giménez.
Astron. Astrophys., Vol. 145, No. 1, p. 206 – 214 (1985).

V1647 Sgr is an early A–type double–lined eclipsing binary with a highly eccentric orbit of rather short period ($e = 0.41$, $P = 3^{d}28$). An accurate photometric analysis and first determination of the apsidal motion was published by Clausen et al. (1977). This paper presents new, accurate spectroscopic orbits of the system for determination of the absolute dimensions of the components, and new times of eclipse which allow the determination of the apsidal motion period with much improved accuracy. The F-type visual companion h 5000 B is physically bound to V1647 Sgr. Its $uvby\beta$ indices indicate that it, too, is unevolved and of approximately Hyades chemical composition.

119.023 The eclipsing variable HR 2800.
C. Sterken, H. W. Duerbeck, H. Hensberge, J. Manfroid, O. Stahl, D. vander Linden.
Astron. Astrophys., Suppl. Ser., Vol. 60, No. 1, p. 1 – 4 (1985).

HR 2800 is a bright B star exhibiting eclipses with a depth of 0.25 magnitudes in V and with a period of 24.6 days.

119.024 Analysis of light curves of the eclipsing binary u Herculis.
H. Rovithis–Livaniou, P. E. Rovithis.
Astron. Astrophys., Suppl. Ser., Vol. 60, No. 1, p. 71 – 74 (1985).

B and V light curves of u Her are analysed using Frequency Domain techniques. New elements for the system are given.

119.025 Narrow–band observations in UV continuum of V444 Cygni.
V. G. Kornilov.
Perem. Zvezdy, Tom 21, No. 6, p. 835 – 839 (1983). In Russian.

The results of narrow–band observations in the UV continuum (3100 Å, 3520 Å, 3780 Å) for the binary system V444 Cyg containing a Wolf–Rayet star are given. The light curve solution by Cherepashchuk's method argues the predominating role of electron scattering in the Wolf–Rayet envelope.

119.026 Photoelectric observations of IS Cassiopeiae, an eclipsing variable in SA 19.
M. M. Zakirov.
Perem. Zvezdy, Tom 21, No. 6, p. 841 – 849 (1983). In Russian.

Results of photoelectric UBVR observations of the eclipsing binary IS Cas are given in a table. In some phases out of eclipse the light curve of IS Cas is distorted by oscillations. The light curves of IS Cas in BVR were derived by method Russell–Merrill with limb darkening $x_1 = x_2 = 0.6$. The absolute elements of the binary system IS Cas were evaluated under the assumption that the primary is a normal star A3 V. IS Cas seems to be a system of a normal star A3 V and a subgiant G7 IV.

119.027 Revision of the photometric elements of the eclipsing variables CW Cep, V539 Ara, AG Per, AR Aur, RS Cha, ZZ Boo.
R. A. Botsula.
Perem. Zvezdy, Tom 21, No. 6, p. 851 – 859 (1983). In Russian.

The revision was carried out using Lavrov's method. Photometric elements were obtained for AR Aur and RS Cha with O'Connell's and Clausen's light curves. Coefficients of limb darkening close to theoretical ones were determined for the AR Aur and ZZ Boo components. Elements consistent with each other for two minima and all colours were obtained for RS Cha and AG Per. Accuracy of elements was estimated for the nonstationary systems CW Cep, V539 Ara, AG Per.

119.028 RZ Scuti as a double contact binary.
R. E. Wilson, W. Van Hamme, L. E. Pettera.
Astrophys. J., Vol. 289, No. 2, p. 748 – 755 (1985). = Dep. Astron., Univ. Florida, Contrib. No. 76.

Among Algol binaries with rapidly rotating primary stars there are a few which might be rotating at the centrifugal limit. Such systems would be double–contact binaries, a new morphological type in which both components accurately fill their limiting lobes.

Double contact has physical significance only if at least one component rotates faster than synchronously. Here the authors check on the possibility that RZ Scuti is in double contact, and find that it is, using a model which treats nonsynchronism and other physical effects in some detail. The primary's rotation rate is among the parameters found by least squares adjustment of the light curves.

119.029 AL Sculptoris – a bright eclipsing binary, discovered by accident.
M. de Groot.
Ir. Astron. J., Vol. 17, No. 1, p. 56 – 57 (1985). – See Abstr. 012.035.

119.030 CCD based B and V lightcurves for the eclipsing binary NJL 5 in Omega Centauri.
K. S. Jensen, H. E. Jørgensen.
Astron. Astrophys., Suppl. Ser., Vol. 60, No. 2, p. 229 – 236 (1985).

Using differential small–aperture photometry on CCD frames accurate B and V lightcurves for the 16th magnitude eclipsing binary NJL 5 in the Omega Centauri field (Niss et al., 1978) have been obtained. The adopted reduction method is discussed in some detail and an evaluation of errors is carried out. An unexpected source of systematic colour dependent errors is pointed out. If this pitfall is avoided internal errors of around 0.01 mag are obtained. External errors are estimated to be 0.02 mag for NJL 5 due to the severe crowding of stars around the variable.

119.031 UBV photometric observations of TT Aurigae.
A. A. Wachmann.
Astron. Astrophys., Suppl. Ser., Vol. 60, No. 2, p. 349 – 352 (1985).

UBV photometric observations of the $1^{d}33$ early B type eclipsing binary, TT Aur, carried out at the Bergedorf Observatory primarily in 1967 – 1968 are presented.

119.032 The secondary component of AA Doradus.
K. Włodarczyk.
Acta Astron., Vol. 34, No. 3, p. 381 – 385 (1984).

A new photometric solution of the light curve of AA Dor is obtained by application of the Wilson–Devinney procedure. The secondary component is found to be a cool red dwarf or a substellar object with temperature below 4500K and luminosity below $0.003 L_{\odot}$.

119.033 EG Cep – an almost contact binary.
J. Kałużny, I. Semeniuk.
Acta Astron., Vol. 34, No. 4, p. 433 – 444 (1984).

552 V and 565 B photoelectric observations of short period eclipsing binary EG Cep have been obtained and analysed with the Wilson and Devinney method. The photometric analysis suggests that EG Cep is most probably a semidetached system in the slow stage of case A mass exchange evolving to a contact configuration. The authors discuss a possibility of getting unique mass ratio for some systems by using light curve synthesis technique, even if the systems have deep minima and total eclipses.

119.034 Epsilon Aurigae.
R. D. Chapman.
Astrophys. Space Sci., Vol. 110, No. 1, p. 177 – 182 (1985). – See Abstr. 012.039.

In April 1984, fourth contact ended the two year long eclipse of Epsilon Aurigae. An astrometric study of the system was carried out by van de Kamp (1978) leading to the conclusion that the orbit is seen very close to edge on. The eclipse was monitored by a number of groups from the ground and from spacecraft such as the IUE. Ultraviolet observations of the system from IUE have thrown new light on the nature of the system. The secondary object is probably a cold, dusty accretion disk surrounding a star that is completely hidden inside the disk.

119.035 Interpretation of new narrow band light curves of the eclipsing binary u Her.
W. E. C. J. van der Veen.
Astron. Astrophys., Vol. 145, No. 2, p. 380 – 386 (1985).

New narrow band light curves of the eclipsing variable u Her, obtained in the Utrecht Photometric System, are interpreted by means of the computational model of R. E. Wilson and E. J. Devinney. The results show that u Her is a semi–detached system in which the secondary fills its Roche lobe. Probably the system is at the end of its mass transfer. Because the observations were carried out in wavelengths up to the near infrared the properties of the secondary could be determined more accurately than has been done before. Especially the results derived from the light curves at 672, 782, and 871 nm show good agreement. The derived masses, radii and inclination are practically independent of the temperature of the primary component, whereas the temperature of the secondary component depends linearly on it.

119.036 Helium in the atmospheres of binary stars.
V. V. Leushin.
Sov. Astron., Vol. 28, No. 4, p. 427 – 430 (1984). English translation of 38.119.014.

119.037 Period lengthening in the V444 Cygni system and the mass loss by the Wolf–Rayet component.
Kh. F. Khaliullin, A. I. Khaliullina, A. M. Cherepashchuk.
Sov. Astron. Lett., Vol. 10, No. 4, p. 250 – 251 (1984). English translation of 38.119.022.

119.038 A redetermination of the luminosity, distance, and reddening of Beta Lyrae.
J. J. Dobias, M. J. Plavec.
Astron. J., Vol. 90, No. 5, p. 773 – 783 (1985).

The authors have redetermined the distance to β Lyrae and found that it probably lies between 345 and 400 pc, with the most likely value being 370 pc. With the corresponding true distance modulus of 7.8 mag, they find that the eclipsing system of β Lyrae has a maximum absolute visual magnitude of –4.7 mag. Using Wilson's model, the authors conclude that the average absolute visual magnitude of the primary component is –4.1 mag, so that the star is best classified as B8.5 or B9 II–Ib. The visual absolute magnitude of the secondary component is –3.3 mag. The color excess is small, $E(B–V) = 0.04$ mag. These data are based on optical and IUE scans of the brightest visual companion to β Lyrae, HD 174664, and on an analysis of the hydrogen line profiles in its spectrum. The authors find that the star is mildly evolved within the main–sequence band. Its effective temperature (14250K) and surface gravity ($\log g \cong 4.0$) correspond most closely to those of stars classified as B6 V. This conclusion creates a certain evolutionary dilemma, since the age of HD 174664 should not exceed 20 – 30 million years if the basic model of β Lyrae as an Algol–type binary is correct while the authors' result is 48×10^6yr.

119.039 Calculation of the photometric elements of a binary system with fixed inclination.
I. Kudzej.
Mater. konf. mol. uchenykh Odes. astron. nauchn.–proizv. akad.–univ. kompleksa, Odessa, 15 fevr., 1983. Odessa, p. 38 – 41 (1983). In Russian. Abstr. in Ref. Zh., 51. Astron., 3.51.598 (1985).

119.040 Variation of the Hγ line profile in the spectrum of the eclipsing binary V367 Cyg.
E. V. Menchenkova.
Mater. konf. mol. uchenykh Odes. astron. nauchn.–proizv. akad.–univ. kompleksa, Odessa, 15 fevr., 1983. Odessa, p. 14 – 18 (1983). In Russian. Abstr. in Ref. Zh., 51. Astron., 3.51.605 (1985).

119.041 The age and helium content of the eclipsing binary AI Phoenicis.
D. A. VandenBerg, B. J. Hrivnak.
Astrophys. J., Vol. 291, No. 1, p. 270 – 279 (1985).

Comparisons of new theoretical isochrones for heavy–element abundances $Z = 0.0169$ (solar) and $Z = 0.04$ with recently published parameters of AI Phoenicis suggest that this system has an age of approximately $(3.6 \pm 0.7) \times 10^9$yr and a helium content of $Y = 0.38 \pm 0.05$. The indicated uncertainty is largely due to the lack of precise knowledge about the metallicity of the binary, since the fits to the data by both sets of isochrones are exceedingly good.

119.042 Radial velocity measurements of the spectra of Zeta Aurigae outside eclipses.
R. Koppmann.
Astrophys. Space Sci., Vol. 110, No. 2, p. 321 – 330 (1985).

Radial velocities of both components of Zeta Aurigae have been measured on 39 grating spectra obtained in the interval February 1970 – November 1981. The evaluation of the orbital elements of the primary component confirmed the elements observed so far. The masses of the components were found to be $M_K\sin^3 i = 6.4 \pm 1.7$ and $M_B\sin^3 i = 4.5 \pm 0.9$ solar masses. With the elements obtained a radial velocity curve of the B–star has been calculated. Comparison of the radial velocities derived from the hydrogen lines of the B–star with the calculated radial velocity curve shows systematic deviations which indicate that these lines originate partly in an expanding circumstellar envelope of the system.

119.043 Differential UBV photometry of the eclipsing binary SVS 2086.
G. Ciardo, L. del Giudice, D. Nappi, L. Santillo.
Astrophys. Space Sci., Vol. 111, No. 1, p. 123 – 129 (1985).

Photoelectric UBV differential observations of the eclipsing binary SVS 2086 are presented. The system shows a W UMa–type light curve, with an amplitude of 0.5 in V light; the period seems to be longer than previously believed.

119.044 108th list of minima of eclipsing binaries.
BBSAG Bull., No. 75, p. 1 – 3 (1985).

119.045 NSV 3772 Monocerotis: detection of the period.
K. Locher.
BBSAG Bull., No. 75, p. 4 – 5 (1985).

119.046 109th list of minima of eclipsing binaries.
BBSAG Bull., No. 76, p. 1 – 5 (1985).

119.047 BD + 37°2641 Bootis: a large amplitude EA binary.
K. Locher.
BBSAG Bull., No. 76, p. 5 (1985).

119.048 A simultaneous X–ray and radio observation of a flare from Algol.
A. N. Parmar, J. L. Culhane, N. E. White, G. H. J. van den Oord.
Adv. Space Res., Vol. 5, No. 3, p. 69 – 72 (1985). – See Abstr. 012.059.

An X–ray flare was observed from Algol using the low and medium energy detectors on the EXOSAT observatory. Spectra obtained during the flare are well fitted by thermal continua while an Fe XXV emission feature was also detected. The strength of this feature indicates a cosmic abundance for iron. The data indicate that the flare occurred in a loop of height approximately 0.25 of the K star radius and with a magnetic field > 300 Gauss.

119.049 A search for X–ray emitting coronal structures in Algol.
J. L. Culhane, N. E. White, S. Kahn, A. N. Parmar, R. J. Blissett, B. Kellett.
Adv. Space Res., Vol. 5, No. 3, p. 73 – 76 (1985). – See Abstr. 012.059.

Algol was observed with the low energy imaging X–ray telescope and the medium energy detectors on the ESA EXOSAT spacecraft during the time of secondary optical eclipse when the B star passes in front of its K type companion. An examination of the X–ray light curves allows the authors to set preliminary lower limits to the size of an X–ray emitting corona associated with the K star. The medium energy detector indicates a continuum temperature of 24.3×10^6K. The detection of Fe XVII and

Fe XVIII emission lines in an objective grating spectrum of the source positively indicates the presence of hot coronal material.

119.050   U Ophiuchi: the third body in the system?
N. N. Samus', M. P. Galkina.
Astron. Tsirk., No. 1340, p. 1 – 2 (1984). In Russian.

119.051   IRAS observations of ε Aurigae during the 1983 eclipse.
D. J. Stickland.
Observatory, Vol. 105, No. 1066, p. 90 – 93 (1985).
The recent eclipse of the primary star in the ε Aurigae system coincided with the operational lifetime of the Infrared Astronomical Satellite. Observations made with the survey instrument enable the energy distribution of the eclipsing body to be defined more precisely.

119.052   Spotted stars in Algol–type eclipsing binaries?
E. C. Olson.
I.A.P.P.P. Commun., No. 19, p. 6 – 13 (1985).

119.053   A study of ultraviolet spectra of ζ Aurigae/VV Cephei systems. VII. Chromospheric density distribution and wind acceleration region.
K.-P. Schröder.
Astron. Astrophys., Vol. 147, No. 1, p. 103 – 110 (1985).
Results from 54 chromospheric eclipse spectra of three ζ Aurigae type binary systems taken with IUE are presented. Curves of growth have been constructed at 20 phases. In order to fit average density models to the chromospheres, 3 samples of column densities (for 31 Cygni, 32 Cygni and ζ Aurigae K giants) have been used. Assuming continuous outflow of matter and knowledge of $\dot{M}$ (Che et al, 1983) wind velocities can be derived. While ζ Aurigae's wind acceleration region shows temporarily a deficiency in density and mass loss at the 1979 eclipse, density gradients and wind acceleration of 31 Cygni and 32 Cygni are very similar in the corresponding eclipses in 1981 and 1982.

119.054   Two–colour photometry and analysis of the eclipsing binary IM Aurigae.
Ö. Gülmen, C. Sezer, N. Güdür.
Astron. Astrophys., Suppl. Ser., Vol. 60, No. 3, p. 389 – 397 (1985).
Photoelectric observations of the eclipsing binary IM Aur have been carried out in B and V colours at the Ege University Observatory. The light curves were analyzed by means of Wood's model simulation program WINK and the Wilson–Devinney computer code. The system was found to be a detached binary with an undersized subgiant secondary nearly filling its Roche limit. The absolute elements of the system are also deduced.

119.055   Photoelectric UBV photometry of Epsilon Aurigae (preliminary results).
M. Muminović, M. Stupar.
Publ. Astron. Opservatorije Beogr., No. 33, p. 41 – 43 (1985). – See Abstr. 012.061.

119.056   Observations of eclipsing binaries in the years 1981 – 1983.
Z. Mikulášek.
Contrib. Nicholas Copernicus Obs. Planetarium Brno, No. 26, p. 4 – 41 (1985).
The paper contains determination of 1197 minima of 154 eclipsing binaries derived from visual and photographic observations performed at Czechoslovak public observatories and in astronomical clubs in the years 1981 – 1983.

119.057   The eclipsing binary DP Cephei.
J. Borovička, V. Wagner.
Contrib. Nicholas Copernicus Obs. Planetarium Brno, No. 26, p. 42 – 45 (1985).

119.058   The improvement of AA UMa period.
J. Borovička.
Contrib. Nicholas Copernicus Obs. Planetarium Brno, No. 26, p. 45 – 49 (1985).

119.059   Photoelectric observations of R Canis Majoris.
K. R. Radhakrishnan, M. B. K. Sarma.
Contrib. Nizamiah Japal–Rangapur Obs., No. 16, 48 pp. (1982).

119.060   Untersuchung des Bedeckungsveränderlichen AK Canis Minoris auf Sonneberger Platten.
B. Fuhrmann.
Mitt. Veränderliche Sterne, Band 10, Heft 4, p. 96 – 97 (1984).

119.061   Photoelektrische Beobachtung des Bedeckungsminimums von Epsilon Aurigae (1982 bis 84).
D. Böhme.
Mitt. Veränderliche Sterne, Band 10, Heft 5, p. 119 (1985).
Photoelectric UBV observations of the recent eclipsing minimum of ε Aurigae are presented.

119.062   Photoelectric observations of V375 Cas.
R.-x. Zhang, J.-t. Zhang, C.-s. Li, D.-s. Zhai.
Inf. Bull. Variable Stars, No. 2651, 3 pp. (1985).

119.063   Photoelectric observations of V541 Cas and its period.
J.-t. Zhang, R.-x. Zhang, C.-s. Li, D. s. Zhai.
Inf. Bull. Variable Stars, No. 2652, 4 pp. (1985).

119.064   The light changes of RT Lacertae.
C. Ibanoğlu, S. Evren, Z. Tunca, O. Tümer.
Inf. Bull. Variable Stars, No. 2655, 3 pp. (1985).

119.065   Two–colour photoelectric light curves of the eclipsing binary V367 Cyg.
M. Can Akan.
Inf. Bull. Variable Stars, No. 2656, 3 pp. (1985).

119.066   Recent data on 22 Vulpeculae.
J. D. Fernie, R. Lyons.
Inf. Bull. Variable Stars, No. 2658, 3 pp. (1985).

119.067   HD 200776: a new eclipsing binary in Cygnus.
E. M. Halbedel.
Inf. Bull. Variable Stars, No. 2663, 3 pp. (1985).

119.068   Photometry of the 1984 eclipse of 22 Vulpeculae.
C. D. Scarfe, R. M. Robb.
Inf. Bull. Variable Stars, No. 2668, 3 pp. (1985).

119.069   Period change in the eclipsing binary system V471 Tauri.
T. E. Beach.
Inf. Bull. Variable Stars, No. 2670, 3 pp. (1985).

119.070   A new photoelectric time of minimum for VZ Psc.
E. Poretti.
Inf. Bull. Variable Stars, No. 2671, 1 p. (1985).

119.071   Photoelectric photometry of Epsilon Aurigae.
P. Flin, M. Winiarski, S. Zoła.
Inf. Bull. Variable Stars, No. 2678, 8 pp. (1985).

119.072   BD + 37°2641, a possible eclipsing binary system.
R. Peniche, S. F. González, J. H. Peña.
Inf. Bull. Variable Stars, No. 2690, 4 pp. (1985).

119.073   Variationen der Linienprofile im optischen Spektrum von W Ser.
W. Strupat.
Mitt. Astron. Ges., Nr. 63, p. 194 – 196 (1985). – See Abstr. 012.063.

119.074   UV–Spektroskopie von 31 Cygni.
E. Rädlein.
Mitt. Astron. Ges., Nr. 63, p. 198 – 200 (1985). – See Abstr. 012.063.

119.075   Radiative properties of eclipsing binaries.
D. M. Popper.
Bull. Am. Astron. Soc., Vol. 16, No. 4, p. 882 – 883 (1984). Abstract. – See Abstr. 010.062.

119.076   Photoelectric light curves of UZ Puppis.
B. B. Bookmyer, J. R. Kern.
Bull. Am. Astron. Soc., Vol. 16, No. 4, p. 883 (1984). Abstract. – See Abstr. 010.062.

119.077   Accretion columns in eclipsing binary stars.
I. A. Ahmad, S. B. Parsons.
Bull. Am. Astron. Soc., Vol. 16, No. 4, p. 884 (1984). Abstract. – See Abstr. 010.062.

119.078   Epsilon Aurigae 1982 – 1984 eclipse campaign report.
J. L. Hopkins, R. E. Stencel.
Bull. Am. Astron. Soc., Vol. 16, No. 4, p. 910 (1984). Abstract. – See Abstr. 010.062.

119.079   Results from the August 1984 eclipse of 22 Vulpeculae.
S. B. Parsons, T. B. Ake.
Bull. Am. Astron. Soc., Vol. 16, No. 4, p. 913 (1984). Abstract. – See Abstr. 010.062.

119.080   Light variations of HD 207739.
R. H. Bloomer Jr.
Bull. Am. Astron. Soc., Vol. 16, No. 4, p. 913 – 914 (1984). Abstract. – See Abstr. 010.062.

119.081   Recent BVRI light curves of the spotted eclipsing binary RS Canum Venaticorum.
J. A. Eaton, C. H. Poe.
Bull. Am. Astron. Soc., Vol. 16, No. 4, p. 914 (1984). Abstract. – See Abstr. 010.062.

119.082   Polarimetry of Epsilon Aurigae: 1982 – 1984 eclipse.
G. D. Henson, J. C. Kemp, D. J. Kraus.
Bull. Am. Astron. Soc., Vol. 16, No. 4, p. 1013 (1984). Abstract. – See Abstr. 010.062.

119.083   Modelling ε Aurigae without solid particles.
A. Y. S. Cheng, N. J. Woolf.
Bull. Am. Astron. Soc., Vol. 16, No. 4, p. 1013 (1984). Abstract. – See Abstr. 010.062.

119.084   *UBV* photometry of DX Aquarii.
R. K. Srivastava, B. K. Sinha.
Astrophys. Space Sci., Vol. 111, No. 2, p. 225 – 236 (1985).

U, B, V photometry of the eclipsing binary DX Aqr has been presented for the first time. The period of the system comes out to be nearly half the period given by Strohmeier (1966) and a value of $0\overset{d}{.}472502$ has been determined. The colour of the system has been discussed. The primary eclipse is a total occultation. The orbit seems to be eccentric. The system belongs to Algol type. DX Aqr may be a very complicated system. The present results are considerably different from those of earlier investigations.

119.085   Photoelectric photometry and photometric elements of LX Persei.
O. Tümer, C. Ibanoğlu, S. Evren, Z. Tunca.
Astrophys. Space Sci., Vol. 112, No. 2, p. 273 – 286 (1985).

The authors present observations of the eclipsing binary LX Per, which were obtained in 1983. There is a wave–like distortion at outside eclipses with an amplitude ranging from 0.03 to 0.08 mag. The wave–like distortion was removed with a new approach. Then, the light curves were analyzed by the methods of Wood and Nelson, Davis, and Etzel. The absolute parameters of the components were also calculated. The physical parameters of the components indicate that the cooler star has separated or has been departing from the main sequence.

119.086   Photoelectric times of minimum light of HI Puppis.
J. R. Kern, B. B. Bookmyer.
Inf. Bull. Variable Stars, No. 2693, 1 p. (1985).

119.087   BD + 61°277 confirmed as an eclipsing binary star.
D. R. Faulkner.
Inf. Bull. Variable Stars, No. 2702, 2 pp. (1985).

119.088   Confirmation of the regular intrinsic variability of AU Monocerotis.
L. Lorenzi.
Inf. Bull. Variable Stars, No. 2704, 5 pp. (1985).

119.089   HD 167971 – an Of–type eclipsing binary.
O. Stahl, D. Forbes, G. Klare, C. Leitherer, B. Wolf, F.–J. Zickgraf.
Inf. Bull. Variable Stars, No. 2726, 4 pp. (1985).

119.090   Spectrophotometric analysis of the system V367 Cygni.
O. Yu. Zozulina, L. V. Karasinskaya, V. V. Leushin.
Rostov n/D. univ. Rostov n/D, 34 pp. (1984). In Russian. Abstr. in Ref. Zh., 51. Astron., 6.51.606 (1985).

119.091   Light curve variations using Roche geometry.
T. A. Nagy.
Bull. Am. Astron. Soc., Vol. 17, No. 1, p. 513 (1985). Abstract. – See Abstr. 010.064.

119.092   On the detection of a common envelope in early type eclipsing binaries.
A. Theokas.
Bull. Am. Astron. Soc., Vol. 17, No. 1, p. 520 (1985). Abstract. – See Abstr. 010.064.

119.093   Time–resolved UV spectroscopy of BE Ursae Majoris.
J. B. Hutchings, A. P. Cowley.
Publ. Astron. Soc. Pac., Vol. 97, No. 590, p. 328 – 332 (1985).

Far–ultraviolet spectra have been obtained of the eclipsing, hot subdwarf binary, BE UMa, with phase coverage in the 2.3–day orbital period concentrated near minimum light. No clearly phase–related spectral changes were found, but the UV continuum appeared to be in a highly variable state, with changes of a factor of two occurring on a time scale of hours. The hot star in this system may be one of the few known PG 1159–035 variables, which show nonradial pulsations. If so, this is the first such star known in a binary system and thus should be important for determining the physical properties of these stars.

119.094   *B* and *V* light curves of UZ Puppis.
B. B. Bookmyer.
Publ. Astron. Soc. Pac., Vol. 97, No. 590, p. 349 – 354 (1985).

The eclipsing binary system UZ Pup was observed on five consecutive nights. The observations covering the eclipse portions of the curves yielded two epochs of minimum light. A study of the period of the system indicates that it has probably remained constant for more than 50 years. The light curves, defined by 431 observations with the B filter and 426 with the V filter, are β Lyrae–type in appearance, and they are asymmetric. The maximum following primary conjunction is brighter than that following secondary conjunction. The system becomes redder during primary eclipse, which is a transit, and there appears to be a slight reddening before and after secondary minimum.

119.095   On the solution of elements of eclipsing variable stars.
O. Demircan.
METU (Middle East Technical University, Ankara, Turkey) Journal of Pure and Applied Sciences, Vol. 16, No. 1, p. 157 – 161 (1983).

Errors in a solution of the elements of an eclipsing variable star are caused not only by the uncertainties in the observational quantities but also by the geometrical behaviour of the functions used in the modelling. In the present paper the determination of the elements has been reexamined in some detail and the indetermined case has been illustrated by a simple example on a theoretical model.

119.096   Spectral investigation of the eclipsing binary V367 Cyg.
V. G. Karetnikov, E. V. Menchenkova.
Astron. Zh., Tom 62, Vyp. 3, p. 542 – 551 (1985). In Russian.
English translation in Sov. Astron., Vol. 29, No. 3.

From 15 diffraction spectrograms spectral investigations of the eclipsing binary V367 Cyg have been carried out. The star's spectrum is mainly typical for an envelope with emissions in the lines $H_\alpha - H_\gamma$. The line Mg II λ4481 Å and the wings of hydrogen lines belong to the primary star of the system. The lines of the secondary component have not been found. Structure and velocity of motion of gaseous media in the system have been studied from the emission components of the H_γ line. The envelope around the main star rotates with the velocity of 119 ± 20 km/s. The gas flow from the secondary component passing through the Lagrange point L_1 is observed.

119.097   Discovery of N V absorption in Zeta Aur.
I. A. Ahmad.
Bull. Am. Astron. Soc., Vol. 17, No. 2, p. 552 (1985). Abstract. – See Abstr. 010.065.

119.098   The solar–type eclipsing binary FL Lyrae.
C. H. Lacy, D. M. Popper, M. L. Frueh.
Bull. Am. Astron. Soc., Vol. 17, No. 2, p. 583 (1985). Abstract. – See Abstr. 010.065.

119.099   IUE high dispersion spectra of interacting binary Mu Sagittarii.
Y. Kondo, F. C. Bruhweiler, G. E. McCluskey.
Bull. Am. Astron. Soc., Vol. 17, No. 2, p. 584 (1985). Abstract. – See Abstr. 010.065.

119.100   RX Cas, observations 1982 – 1984.
M. Fernandes.
BAV Mitt., 34. Jahrg., Nr. 2, p. 49 – 69 (1985). In German.

RX Cas was observed photoelectrically in B and V. 128 observations were obtained in both colours between 3D 2445196 and ...670. The existence of different stages in maximum brightness is confirmed. The change between these stages was not continuous, within a few days the star shifts into another photometric behaviour. It remains then closely to a mean light curve ($\pm 0^m.05$), interrupted by outbursts and the reverse of it lasting up to ten days.

119.101   Der langperiodische Bedeckungsveränderliche V371 Cygni.
H. Grzelczyk.
BAV Mitt., 34. Jahrg., Nr. 2, p. 70 – 72 (1985).

119.102   Spectroscopy of the novalike binary RW Trianguli.
N. F. Vojkhanskaya.
Sov. Astron., Vol. 28, No. 6, p. 665 – 667 (1984). English translation of 38.119.094.

119.103   Photometry of active Algols.
E. C. Olson.
Interacting binaries, p. 127 – 154 (1985). – See Abstr. 012.104.

While light curves of several short period Algol binaries have shown transient instabilities from the ultraviolet to the near infrared, only in RW Tau and most notably in U Cep have these disturbances been large enough to allow quantitative analysis. From a study of the spectral distribution of light excesses and losses in U Cep, the author derives properties of "equatorial bulges", hot spot, and stream during periods of activity presumably associated with transient, enhanced, mass flows. He also discusses the nature of large light losses that appeared outside eclipse and in primary eclipse egress during active intervals. It is shown that small surges in cool star brightness preceded and accompanied intervals of activity. Apart from such surges, the cool star brightness has varied over the last eight years in a way somewhat suggestive of a magnetic activity cycle.

119.104   Unique eclipsing variable IU Aurigae.
P. Mayer.
Říše hvězd, Vol. 65, No. 10, p. 201 – 204 (1984). In Czech.

119.105   Variations of the H and K emission lines of singly ionized calcium in the eclipsing binary star system AR Lacertae.
S. W. Hoffman.
Diss. Abstr. Int., Sect. B, Vol. 45, No. 4, p. 1217 – 1218 (1984). Thesis, University of Florida, 217 pp. (1983). Order No. DA8415127.

119.106   Short time–scale variations of Epsilon Aurigae in the H–alpha region.
H.–s. Tan.
Chin. Astron. Astrophys., Vol. 9, No. 1, p. 39 – 43 (1985). English translation of Acta Astrophys. Sin., Vol. 4, No. 4, p. 264 – 271 (1984).

During the total eclipse of ε Aur, 18 spectra in the Hα region were taken. Variations on short time–scales were observed in the radial velocities of the metallic lines and Hα. The radial velocities of the emission features are about –60, +12 and +57 km/s; that of the absorption component is +20 km/s. The radial velocity of the strongest emission feature also varies, but by much smaller amounts than the metallic lines and the variation appears to be periodic. A model in which the secondary is a B star surrounded by hot, active clouds is proposed.

119.107   Observations of the hydrogen deficient binary Upsilon Sagittarii.
N. Kameswara Rao, V. R. Venugopal.
J. Astrophys. Astron., Vol. 6, No. 2, p. 101 – 112 (1985).

The absolute magnitude M_v of the hydrogen deficient binary υ Sgr has been estimated as -4.8 ± 1.0 from the distribution of the interstellar reddening, polarization and interstellar lines of the surrounding stars. From the ANS observations obtained at the time of the secondary eclipse, it appears that the hotter secondary is surrounded by a disc with colours of a B8 – B9 star. The λ1550 C IV absorption line arising in the stellar wind does not show any change in strength during the secondary minimum. The upper limit to the mass–loss rate from the high temperature wind is estimated as $\leqslant 5 \times 10^{-7} M_\odot yr^{-1}$ from the 2 cm and 6 cm radio observations.

119.108   Researches on the photoelectric photometry of the eclipsing binary WX Cnc.
Y.–l. Yang, Q.–y. Liu, L. Lu.
Acta Astron. Sin., Vol. 25, No. 4, p. 325 – 335 (1984). In Chinese.

A total of 672 yellow and 674 blue observations were obtained from 20 to 27 Jan., 1982, using the 100–cm telescope and its photometer at Yunnan Observatory. All measurements are corrected for differential extinctions and transformed to the standard UBV system. Two times of primary minima and two times of secondary minima are obtained, respectively. The eclipse type is that of partial eclipse and transit at the primary eclipse. The hotter component is of type F5 V and the colder component is of type F8 V, corresponding to the color indices. From the mass–luminosity relation, the same mass ratio, 0.71 is obtained. The results show that WX Cnc is a main sequence detached binary star.

119.109   Photoelectric photometry of SZ Lyn and its period changes.
S.–y. Jiang, Y.–w. Huang, Z.–h. Guo.
Acta Astron. Sin., Vol. 26, No. 2, p. 135 – 143 (1985). In Chinese.

The authors have made photoelectric photometry of SZ Lyn in 15 nights, and have recorded 19 maxima. Using a second order least squares solution, they found that the period of SZ Lyn is increasing with a rate of 0.165 seconds per 100 years, or $\dot{P} = (6.32 \pm 0.62) \times 10^{-12}$ d/period.

119.110   A new exploration of ε Aur during ingress and totality of the 1982 – 1984 eclipse.
X.-f. Liu, S.-q. Peng, G.-w. Cha, X.-y. Li.
Acta Astron. Sin., Vol. 26, No. 2, p. 144 – 151 (1985). In Chinese.
Spectrograms of ε Aur were obtained using the 102 cm telescope of Yunnan Observatory and the 90 cm telescope of Beijing Observatory during the period August 1982 to March 1984. Variations of Hα profiles and radial velocities of atoms with phase during ingress and totality were obtained. Curves-of-growth analysis of ε Aur is made in order to determine the turbulent velocity of this F0 Ia supergiant.

119.111   Period change in the eclipsing binary GK Cephei.
J. E. Isles.
Inf. Bull. Variable Stars, No. 2741, 1 p. (1985).

119.112   Contact times for the 1982–4 eclipse of Epsilon Aurigae.
P. C. Schmidtke, S. I. Ingvarsson, J. L. Hopkins, R. E. Stencel.
Inf. Bull. Variable Stars, No. 2748, 2 pp. (1985).

General Catalogue of Variable Stars.
See Abstr. 002.027.

Modern methods in the astrometry of double stars.
See Abstr. 036.076.

Maximum entropy reconstruction of accretion disk images from eclipse data.
See Abstr. 036.212.

Three dimension mapping of the atmosphere of an active K–dwarf star.
See Abstr. 112.141.

Ultraviolet spectroscopy of the interacting binary U Sagittae.
See Abstr. 117.008.

Ultraviolet observations of circumstellar matter in Algol type binary systems.
See Abstr. 117.009.

The eclipsing binary U Sagittae: evidence for CNO processing and mass exchange in the past.
See Abstr. 117.014.

Tasks for observations of close binary systems in the extreme ultraviolet.
See Abstr. 117.018.

V Sagittae: Beobachtungen 1983/1984.
See Abstr. 117.058.

Two almost–contact semidetached systems: ZZ Aurigae and AX Virginis.
See Abstr. 117.059.

Two very similar late–type contact systems: BX and BB Pegasi.
See Abstr. 117.075.

The eclipsing dwarf nova OY Carinae. III. Photometry during the superoutburst of January 1980.
See Abstr. 117.078.

Possible evolution of a triple system into Epsilon Aurigae.
See Abstr. 117.081.

The RS CVn–type binary SV Camelopardalis: evidence of dark spots from UBV observations and IR fluxes.
See Abstr. 117.089.

Dual polarization VLBI observations of stellar binary systems at 5 GHz.
See Abstr. 117.099.

UBV photometric analysis of V758 Centauri.
See Abstr. 117.125.

Accuracy of close binary mass determinations from parallaxes.
See Abstr. 117.129.

UBV images of the Z Cha accretion disc in outburst.
See Abstr. 117.133.

Further observations of EX Hydrae.
See Abstr. 117.138.

The mass of AW Ursae Majoris.
See Abstr. 117.142.

A photometric study and analysis of XY Leonis.
See Abstr. 117.146.

UBVRI photometry of W Ursae Majoris.
See Abstr. 117.151.

The 1984 eclipse of the symbiotic binary SY Muscae.
See Abstr. 117.158.

The dwarf nova SVS 2549, a shortperiodic eclipsing system.
See Abstr. 117.203.

Atmospheric structures in AR Lac. I. Mapping quiescent features by occultations and Doppler imaging.
See Abstr. 117.220.

Atmospheric structures in Ar Lac. II. A spatially resolved chromospheric active region.
See Abstr. 117.221.

Atmospheric structures in AR Lac. III. VLA observations of the radio eclipses during the 1984 campaign.
See Abstr. 117.222.

Photoelectric light curves of RT Coronae Borealis and their analysis.
See Abstr. 117.250.

The short–period eclipsing binary V1010 Ophiuchi does it have an eccentric orbit?
See Abstr. 117.251.

The new eclipsing magnetic binary system E1114 + 182.
See Abstr. 117.269.

Contact binaries: observational properties.
See Abstr. 117.289.

The influence of an optically thick disc on the light curve of SX Cas.
See Abstr. 117.297.

Photoelectric observations of CN Andromedae: an unusual over-contact binary.
See Abstr. 117.298.

The Algol scenario.
See Abstr. 117.342.

From Algol to Beta Lyrae.
See Abstr. 117.368.

A model for eclipsing cataclysmic variables.
See Abstr. 117.382.

Two similar detached binary systems: AT and AZ Camelopardalis.
See Abstr. 117.388.

BV photometry and period variation of GO Cygni.
See Abstr. 117.391.

A W UMa variable in the visual binary system ADS 9019: a *UBV* photoelectric investigation.
See Abstr. 118.006.

The quadruple system HR 3337.
See Abstr. 118.059.

Spectroscopic orbit of R CMa.
See Abstr. 120.024.

Comparison stars for some eclipsing binaries.
See Abstr. 123.027.

On detection of astrophysical chaos phenomena.
See Abstr. 142.030.

120 Spectroscopic Binaries

120.001   A search for the spectroscopic binary companion of Polaris using IUE spectra.
N. R. Evans.
NASA Conf. Publ., NASA CP–2349, p. 428 – 431 (1984). – See Abstr. 012.001.
Polaris has a spectroscopic orbit from an extensive series of observations as well as more uncertain astrometric elements. Spectra exposed down to 3200 Å have previously been searched for light from a companion, presumably a hotter main sequence star, but none has been found. In this study IUE low dispersion spectra have been searched for any light from a companion by forming ratios with observations of stars with spectral types similar to Polaris. There is no sign of a companion as early as A7 V. Mass limits are discussed.

120.002   Photoelectric radial velocities. Paper XII. The orbits of another four spectroscopic binaries in the Clube Selected Areas.
R. F. Griffin.
Mon. Not. R. Astron. Soc., Vol. 212, No. 3, p. 663 – 669 (1985).
Orbits are presented for four more ninth–magnitude stars of HD type K0. HD 65195, a system containing a giant star, has the unusually short period of 38 days. HD 68874 has a remarkably large mass function; but no secondary star is seen in the radial-velocity traces or in a classification–dispersion spectrogram, although some abnormality exists in the object's narrow–band photometric properties. A possible resolution of the difficulty is that the secondary may itself be a close binary, so HD 68874 may be a system analogous to HR 6497 and d Serpentis.

120.003   HR 5637: the unveiling of a binary.
B. Noordanus, P. L. Cottrell, T. B. Ake III.
Mon. Not. R. Astron. Soc., Vol. 212, No. 4, p. 931 – 938 (1985).
HR 5637, previously classified as G1/3 Ib/II but considered composite by Bidelman & MacConnell has been examined both photometrically and spectroscopically. The spectroscopic observations spanned the wavelength interval 1000 – 7000 Å. Using these data in various combinations, the star is shown to be a binary with components of G2–5 Ib and B8 V, with a magnitude difference $\Delta V \sim 2.5 \pm 0.3$ mag. These results are confirmed by ultraviolet (*IUE*) observations.

120.004   Spectroscopic binary orbits from photoelectric radial velocities. Paper 60: 10 Leonis.
R. F. Griffin.
Observatory, Vol. 105, No. 1064, p. 7 – 12 (1985).

120.005   Has α Vir ever stopped pulsating?
E. Chapellier, J. C. Valtier, J. P. Sareyan, J. M. Le Contel, D. Ducatel, P. J. Morel.
Astron. Astrophys., Vol. 143, No. 2, p. 466 – 468 (1985).
Photometric observations of α Vir have been obtained in 1983. On one excellent night, light variations have been detected. The observed maximum is in phase with the maxima which had been observed before the damping of the pulsation some ten years ago. The value of the period has been redetermined. This result confirms that the pulsation has never actually stopped.

120.006   Spectroscopic binary orbits from photoelectric radial velocities. Paper 61: HD 25099.
R. F. Griffin.
Observatory, Vol. 105, No. 1065, p. 29 – 31 (1985).

120.007   Contribution to the study of F, G, K, M binaries III. Orbital elements of the long period spectroscopic binaries HD 102928 and HD 145206.
N. Ginestet, J. M. Carquillat, A. Pédoussaut, R. Nadal.
Astron. Astrophys., Vol. 144, No. 2, p. 403 – 407 (1985). In French.
The spectroscopic binaries HD 102928 and HD 145206 were observed at the Observatoire de Haute–Provence in order to determine their spectrographic orbits. The orbital elements have been obtained. The authors show that the interferometric companion found by McAlister et al. (1983) at 0″.170 of HD 102928 is not the spectroscopic secondary but, probably, a third body in the system.

120.008   Spectroscopic orbits for 16 more binaries in the Hyades field.
R. F. Griffin, J. E. Gunn, B. A. Zimmerman, R. E. M. Griffin.
Astron. J., Vol. 90, No. 4, p. 609 – 642 (1985).
A photoelectric survey of the radial velocities of Hyades candidates has provided data for 16 more spectroscopic orbits in addition to the 11 the authors have already published.

120.009   The ellipsoidal variable ψ Orionis and the β Cephei instability region.
M. Jerzykiewicz.
Acta Astron., Vol. 34, No. 4, p. 409 – 417 (1984).
Photoelectric observations of this well–known spectroscopic binary and β Cephei suspect are presented. No short–period variations with an amplitude exceeding $0^{m}005$ are found, but the ellipsoidal character of the light variation is confirmed. Moreover, it is shown that in the colour–magnitude plane the primary component of the system falls below the low–luminosity boundary of the β Cephei instability region. The use of ψ Ori by Waelkens and Rufener (1983) for testing their hypothesis that tidal interactions have a damping effect on the β Cephei–type pulsations is thus found to be questionable.

120.010   Erratum: "Iterative methods for determination of parameters of spectroscopic binaries" [Acta Astron., Vol. 33, No. 3 – 4, p. 431 – 439 (1983)].
T. Z. Dworak.
Acta Astron., Vol. 34, No. 4, p. 494 (1984). See Abstr. 37.120.001.

120.011 75 Pegasi, a very–short–period extreme mass ratio binary.
D. P. Hube, A. F. Gulliver.
Publ. Astron. Soc. Pac., Vol. 97, No. 589, p. 280 – 284 (1985).
  The bright star 75 Peg has long been known to be a spectroscopic binary. The orbital parameters are presented here for the first time, and include the very short period of $0^d5021035$. The system has one of the most extreme values of mass ratio known. The most direct interpretation of the available spectroscopic observations leads to the conclusion that the primary fills, or is very close to filling, its critical Roche lobe, and will begin to lose mass to its companion before leaving the main sequence.

120.012 Light curve of the RS Canum Venaticorum system V711 Tauri = HR 1099 during 1981 – 1982.
M. B. K. Sarma, B. D. Ausekar, B. V. N. S. Prakasa Rao,
J. Y. Oh, I.–S. Nha.
Publ. Astron. Soc. Jpn., Vol. 37, No. 1, p. 107 – 122 (1985).
  Observations of the RS CVn–type spectroscopic binary HR 1099 (V711 Tau) in V and B passbands for the year 1981 – 82 season are reported. It is noticed that HR 1099 has passed through an unusually active phase during January – February 1982.

120.013 Frequency of Bp – Ap stars among spectroscopic binaries.
M. Gerbaldi, M. Floquet, B. Hauck.
Astron. Astrophys., Vol. 146, No. 2, p. 341 – 351 (1985).
  Improving previous studies with more numerous published values, the authors have reexamined the binary frequency for Bp – Ap stars. They have analysed the period and the eccentricity distributions for Bp – Ap stars, compared to normal stars of various spectral types. Remarkably this analysis reveals a great deficiency among low eccentricity systems for all the peculiar stars, except the Hg–Mn ones. The authors discuss also the synchronism for the systems for which the photometric period is known. Finally, they compare the values of the parameter $\Delta(V1 – G)$ of the Geneva photometry, which is a measurement of the $\lambda5200$ depression, for different binary systems.

120.014 On regular brightness increases of the Wolf–Rayet star HD 186943.
N. A. Lipunova.
Astron. Tsirk., No. 1346, p. 3 – 5 (1984). In Russian.

120.015 Spectroscopic binary orbits from photoelectric radial velocities. Paper 62: EZ Pegasi.
R. F. Griffin.
Observatory, Vol. 105, No. 1066, p. 81 – 85 (1985).
  EZ Peg, which has had a confused history in the astronomical literature, is shown to be a double–lined spectroscopic binary with slightly unequal components (both late–G dwarfs) in a circular orbit of period 11.66 days.

120.016 Photometry needed for the new Zeta–Aurigae system 22 Vul.
S. B. Parsons, T. B. Ake.
I.A.P.P.P. Commun., No. 19, p. 26 – 27 (1985).

120.017 Frequency of Bp, Ap stars among the spectroscopic binaries.
M. Gerbaldi, M. Floquet, B. Hauck.
Inst. Astrophys. Paris, Pré–Publ., No. 82, 47 pp. (1984). To be published in Astron. Astrophys.

120.018 Binaires spectroscopiques dans le voisinage solaire.
J. L. Halbwachs.
Bull. Inf. Cent. Données Stellaires, No. 28, p. 25 – 27 (1985).
  A radial velocity survey of nearby stars was undertaken with the Coravel instrument of the Geneva Observatory at the Observatoire de Haute–Provence. Almost all program stars have at least two measures, and some spectroscopic binaries (SB) were already detected. The proportion of SB's detected among K dwarfs is now 15%, 1/4 of them being SBII. The proportion of SBII among all SB is however less important, when only the nearby stars are considered. This is due to the incompleteness of the sample for parallaxes smaller than $0''049$.

120.019 Additional double–lined eclipsing binaries observed with CCD detectors.
C. H. Lacy.
Inf. Bull. Variable Stars, No. 2685, 2 pp. (1985).

120.020 Observations of the X–ray luminous SB1 system HD 149162.
H. M. Johnson, M. Mayor.
Bull. Am. Astron. Soc., Vol. 16, No. 4, p. 884 (1984). Abstract. – See Abstr. 010.062.

120.021 HR 6697 revisited.
R. B. Culver, P. A. Ianna, H. A. McAlister.
Bull. Am. Astron. Soc., Vol. 16, No. 4, p. 884 (1984). Abstract. – See Abstr. 010.062.

120.022 New classifications for spectrum binaries.
P. C. Schmidtke.
Bull. Am. Astron. Soc., Vol. 16, No. 4, p. 912 (1984). Abstract. – See Abstr. 010.062.

120.023 IUE observations of the RS CVn binary – TZ Tri.
I. R. Little–Marenin, T. R. Ayres, A. Young.
Bull. Am. Astron. Soc., Vol. 16, No. 4, p. 912 (1984). Abstract. – See Abstr. 010.062.

120.024 Spectroscopic orbit of R CMa.
K. R. Radhakrishnan, K. D. Abhyankar,
M. B. K. Sarma.
Bull. Astron. Soc. India, Vol. 12, No. 4, p. 411 – 416 (1984).
  Spectroscopic observations of the Algol–type binary R CMa during 1980 – 81 are presented and definitive orbital elements obtained by combining all available measurements of radial velocity. Near circularity of the orbit and low mass function are confirmed.

120.025 Spectroscopic binaries near the North Galactic Pole. Paper 11: HD 105982.
R. F. Griffin.
J. Astrophys. Astron., Vol. 6, No. 1, p. 71 – 73 (1985).
  Photoelectric radial–velocity measurements show that HD 105982 is a spectroscopic binary with a period of 3.7 years.

120.026 HD 22403: a new variable with Ca II H and K emission.
A. V. Raveendran, S. Mohin, M. V. Mekkaden.
Inf. Bull. Variable Stars, No. 2694, 2 pp. (1985).

120.027 HD 115781: a large–amplitude ellipsoidal variable.
R. D. Lines, W. S. Barksdale, H. J. Stelzer,
D. S. Hall.
Inf. Bull. Variable Stars, No. 2728, 4 pp. (1985).

120.028 Three modes of variability in the bright star HR 7428.
W. S. Barksdale, L. J. Boyd, R. M. Genet,
R. E. Fried, D. S. Hall, W. T. Persinger, D. B. Hoff,
S. I. Ingvarsson, P. Nielsen, H. J. Stelzer, N. F. Wasson.
Inf. Bull. Variable Stars, No. 2737, 4 pp. (1985).

120.029 Short term variability of γ^2 Velorum – evidence for a compact companion?
T. Stiff, S. Jeffers, W. G. Weller.
Bull. Am. Astron. Soc., Vol. 17, No. 1, p. 511 – 512 (1985). Abstract. – See Abstr. 010.064.

120.030 Preliminary orbital elements for the Am binary HD 434.
D. P. Hube, A. F. Gulliver.
J. R. Astron. Soc. Can., Vol. 79, No. 2, p. 49 – 53 (1985).
  Preliminary spectroscopic orbital elements are presented for the metallic–lined A–star HD 434. Most of the orbital elements are based only on data obtained during the past eight years, but a few radial velocities measured more than 4 decades earlier have contributed to the reliable determination of an orbital period of

$34\overset{d}{.}26014$. The system has the moderately large mass function of $0.122\,M_{\odot}$.

120.031   A spectroscopic orbit of the late B star 27 Leo.
P. Harmanec, M. Ouhrabka, F. Žd'árský.
Bull. Astron. Inst. Czech., Vol. 36, No. 3, p. 160 – 172 (1985).
With plates 3 – 4.

The orbital period, 137.3 days and the first orbital elements of the B 9 IV star 27 Leo were derived on the basis of 26 Ondřejov coudé spectrograms, 33 older radial–velocity measurements found in the astronomical literature, and 6 new Victoria spectrograms kindly made available by D. P. Hube. The orbit of the star is highly eccentric and needs further confirmation using more data. Certain suspected spectral peculiarities of the object are briefly discussed. Photometric eclipses of the binary components are not excluded.

120.032   Radial–velocity orbits for six single–line spectroscopic binaries.
W. I. Beavers, J. J. Salzer.
Publ. Astron. Soc. Pac., Vol. 97, No. 590, p. 355 – 362 (1985).

Orbital elements are reported for three single–line spectroscopic binaries, HR 2692, HR 3725, and HR 5911, none of which have previously determined orbits. In addition, improved orbital elements are reported for three stars with previously determined orbits of low quality.

120.033   The discovery of a spectroscopic binary in M3.
D. W. Latham, M. L. Hazen–Liller, C. P. Pryor.
Stellar radial velocities, p. 269 – 273 (1985). – See Abstr. 012.095 (IAU Colloq. No. 88).

To extend the search for spectroscopic binaries in the globular cluster M3, the authors have obtained more than 300 new radial velocities for the 111 giants previously observed by Gunn and Griffin (1979). For one of the stars, von Zeipel 164, four observations spanning ten years show a velocity variation with an amplitude of at least 18 km s^{-1} and a period of perhaps a few years. The authors believe this to be a strong candidate for the first spectroscopic binary to be found in a globular cluster.

120.034   Metallic–line binaries.
H. A. Abt.
Stellar radial velocities, p. 275 – 281 (1985). – See Abstr. 012.095 (IAU Colloq. No. 88).

A new radial–velocity study of 55 Am stars reveals 16 SB2s, 20 SB1s, and 22 visual or occultation companions for a total of more than one companion per primary. Only about 75% of the primaries are in binaries with periods less than 100 days, even if one allows for unobservable small inclinations and low–mass secondaries. Therefore it may not be true that tidal interactions in close binaries is the only mechanism necessary to produce the low rotational velocities that allow diffusion to act, unless small reductions (factor of 1.4) in rotational velocities occur in binaries of longer periods.

120.035   High–velocity halo binary systems.
A. Ardeberg, H. Lindgren.
Stellar radial velocities, p. 371 – 374 (1985). – See Abstr. 012.095 (IAU Colloq. No. 88).

Preliminary results from a search for high–velocity binaries in a sample of halo stars observed with CORAVEL are reported.

120.036   A search for halo binaries: the first results.
D. W. Latham, R. P. Stefanik, B. W. Carney.
Stellar radial velocities, p. 381 – 384 (1985). – See Abstr. 012.095 (IAU Colloq. No. 88).

The authors report the first results from a program to monitor the stars in the Carney and Latham (1985) halo survey for velocity variations due to orbital motion. From the sample of 300 stars with high space velocities, 24 spectroscopic binary candidates have been identified.

120.037   A search with Reticon and CCD detectors for new spectroscopic components of binary and multiple stars.
F. C. Fekel.
Stellar radial velocities, p. 389 – 392 (1985). – See Abstr. 012.095 (IAU Colloq. No. 88).

Some preliminary results of a search for the secondaries of single–lined solar–type binaries are reported.

120.038   Binary systems among the peculiar cool stars.
R. D. McClure.
Cool stars with excesses of heavy elements, p. 315 – 329 (1985). – See Abstr. 012.101.

Radial velocity observations with an accuracy of about 0.5 km s^{-1} have been obtained over the last five years for samples of strong CN, Ba II, CH, sgCH, and R stars. These data indicate that those stars that exhibit excesses of s process elements as well as carbon, the Ba II and CH stars, are probably all binary systems. Those stars, on the other hand, that show no excess of s process elements, the strong CN and R stars, show the normal binary frequency for K giants. These results are discussed in terms of possible mixing scenarios.

120.039   The first binary orbit for an S–type star – HR 1105.
R. F. Griffin.
Cool stars with excesses of heavy elements, p. 331 (1985). Abstract. – See Abstr. 012.101.

120.040   A search for companions to irregular variables of type S.
B. F. Peery Jr.
Cool stars with excesses of heavy elements, p. 333 – 336 (1985). – See Abstr. 012.101.

IUE archival spectra of S–type stars that may be cool analogues of Ba II stars have been inspected for evidence of hot companions. Although the currently accessible data for 9 stars are of marginal quality, spectra of the brightest candidate, HR 1105, show strong evidence for the presence of a hot companion of low luminosity.

120.041   LSS 2018: a double–lined spectroscopic binary central star with an extremely large reflection effect.
A. U. Landolt, J. S. Drilling.
Bull. Am. Astron. Soc., Vol. 17, No. 2, p. 594 (1985). Abstract. – See Abstr. 010.065.

120.042   The distributions of periods and amplitudes of late–type spectroscopic binaries.
R. F. Griffin.
Interacting binaries, p. 1 – 12 (1985). – See Abstr. 012.104.

Photoelectric radial–velocity measurements have permitted the discovery of groups of spectroscopic binaries in which the usual observational discrimination against long periods and small amplitudes is much reduced. The distributions of periods and amplitudes in these groups is quite different from those of binaries with published orbits. In particular, the $N - \log P$ distribution appears to be a monotonically rising function for at least the range $3 < P < 3000$ days.

120.043   The double–lined spectroscopic binary Psi Orionis.
W. Lu.
Publ. Astron. Soc. Pac., Vol. 97, No. 591, p. 428 – 433 (1985).

From high–dispersion spectrograms measured by cross–correlation techniques, the spectroscopic orbits for both components of ψ Ori have been determined. From spectrophotometric measurements, the spectral types are found to be B1 III (primary) and B2 V (secondary), while the projected rotational velocities ($v \sin i$) are 95 km s^{-1} and 75 km s^{-1}, respectively. The spectroscopic parallax corresponds to a distance of 640 parsecs.

120.044   Spectroscopic binaries near the North Galactic Pole. Paper 12: 6 Boötis.
R. F. Griffin.
J. Astrophys. Astron., Vol. 6, No. 2, p. 77 – 83 (1985).

Photoelectric radial–velocity measurements show that 6 Boötis undergoes small periodic variations of velocity. An orbit with a 2.6 year period and a semi–amplitude of little more than

1 km s^{-1} is derived. The amplitude is smaller by a factor of two than that of any plausible orbit previously derived from radial velocities.

120.045   AY Ceti: a flaring, spotted star with a hot companion.
D. M. Gibson, T. Simon, F. C. Fekel Jr.
Proceedings of the Southwest Regional Conference for Astronomy and Astrophysics, Vol. 10, p. 25 (1985). Abstract. – See Abstr. 012.125.

Sky catalogue 2000.0. Volume 2: Double stars, variable stars and nonstellar objects.
See Abstr. 002.019.

General Catalogue of Variable Stars.
See Abstr. 002.027.

Errata in catalogues of spectroscopic binary orbits.
See Abstr. 002.071.

Programmes d'observation des binaires spectroscopiques et informations du 14ème catalogue complémentaire.
See Abstr. 013.060.

Cross–correlation spectroscopy using CORAVEL.
See Abstr. 036.186.

Radial velocities of southern stars obtained with the photoelectric scanner CORAVEL. III. 790 late–type bright stars.
See Abstr. 111.002.

Radial velocity variations in cool supergiants.
See Abstr. 111.038.

Bright giants with small amplitude long period velocity variability.
See Abstr. 111.039.

Radial velocity variations in CN–peculiar and normal K giant stars.
See Abstr. 111.041.

The quadruple system HR 3337.
See Abstr. 118.059.

Egress spectra of the new eclipsing supergiant 22 Vulpeculae.
See Abstr. 119.004.

Detection of a large, very faint emission nebula surrounding α Virginis.
See Abstr. 132.008.

1983 observations of the eclipsing nucleus of the bipolar planetary nebula NGC 2346.
See Abstr. 134.002.

Ultraviolet observations of the binary nucleus of the planetary nebula NGC 1360.
See Abstr. 134.056.

The binary population of the Hyades.
See Abstr. 153.053.

The spatial distribution of spectroscopic binaries and blue stragglers in the open cluster M67.
See Abstr. 153.059.

An extension of the search for spectroscopic binaries in M3.
See Abstr. 154.062.

121 Early-stage Stars (T Tauri Stars, Herbig-Haro Objects, etc.)

121.001   UV observations and results on pre–main sequence stars.
C. L. Imhoff.
NASA Conf. Publ., NASA CP-2349, p. 81 – 92 (1984). – See Abstr. 012.001.
IUE has given the first good look at the ultraviolet spectra of pre–main sequence objects. This review concentrates on the observations and results obtained for the T Tauri stars and related objects. The T Tauri stars represent a normal phase of pre–main sequence evolution for solar mass stars. As such they provide an invaluable probe of the physical mechanisms involved in star formation and the genesis of stellar magnetic fields and surface activity.

121.002   The changing ultraviolet spectrum of Herbig–Haro object No. 1.
K. H. Böhm, E. Böhm-Vitense, E. W. Brugel.
NASA Conf. Publ., NASA CP–2349, p. 167 – 170 (1984). – See Abstr. 012.001.
IUE spectra of H–H1 taken in March 1982 and December 1983 show steady qualitative changes of the short wavelength emission line spectrum. Earlier IUE studies (in 1979 and 1980) had shown the typical "high excitation object" ultraviolet spectrum with the C IV 1550 and the C III] 1909 lines being very strong. These line fluxes have steadily decreased and are presently not detectable on a 4–1/2 hour exposure. While the spectrum looked very similar to that of H–H2 in 1980, it now gives more the impression of the spectrum of a low excitation H–H object with a possible presence of fluorescent lines from the H$_2$ Lyman bands.

121.003   Short wavelength observations of HH–2H and HH–24A.
E. W. Brugel, J. M. Shull.
NASA Conf. Publ., NASA CP-2349, p. 171 – 174 (1984). – See Abstr. 012.001.
There still remains some debate about the exact origin of the very blue optical and ultraviolet continuum in Herbig–Haro objects. The authors believe that there is reasonably good evidence for a two–photon continuum in HH–2H. It has recently been suggested that there is a significant contribution to the UV continuum from scattered light in the general Orion region. The authors present evidence from spatially resolved IUE spectra which indicate the continuum and emission line fluxes have the same spatial distribution. A very low ultraviolet extinction is indicated based upon an enhanced two–photon origin for the UV continuum. A recent observation of HH–24A shows an interesting absence of high excitation emission lines and a weak continuum.

121.004   Dust in the region of Herbig–Haro objects: the case of NGC 1999.
J. A. Cardelli, K. H. Böhm.
NASA Conf. Publ., NASA CP-2349, p. 175 – 178 (1984). – See Abstr. 012.001.
The authors have used observations of the reflection nebula NGC 1999 and its illuminating star V380 Ori to examine the properties of dust in the region of Herbig–Haro objects H–H 1 and H–H 2. The ratio of nebular intensity to stellar flux is fairly flat in the UV. If reflected light represents a strong contribution to the continua of H–H objects, then the results seem to rule out

geometries in which this reflection arises from forward scattering of light from imbedded T Tauri stars. In addition, the authors have examined the extinction to V380 Ori.

121.005   A search for young stellar chromospheres in NGC 2264.
T. Simon.
NASA Conf. Publ., NASA CP-2349, p. 183 – 186 (1984). – See Abstr. 012.001.

Pre–main–sequence stars in the galactic cluster NGC 2264 were observed with IUE to search for ultraviolet chromospheric and transition region (TR) emission lines. Chromospheric emission was detected only in the faint Hα emission–line star NX Mon (W 79); its Mg II surface flux is comparable to that observed for very active T Tauri stars. Upper limits on chromospheric and TR line emission of the remaining observed stars are an order of magnitude below the surface fluxes and normalized fluxes of T Tau stars and of the chromospherically–active Ae star HR 5999. These upper limits are comparable to the emission–line strengths of solar–type stars in the Hyades and UMa clusters and of the youngest field stars.

121.006   The UV extinction of HR 5999.
J. H. Hecht, A. V. Holm, T. B. Ake III, C. L. Imhoff, N. A. Oliversen, G. Sonneborn.
NASA Conf. Publ., NASA CP-2349, p. 318 – 321 (1984). – See Abstr. 012.001.

The irregular Ae variable HR 5999 has been observed by IUE during a period of minimum light ($7.4 \leqslant M_v \leqslant 7.7$). By comparing these spectra with spectra of the star taken five years previously near maximum light ($M_v \sim 6.9$) the authors have obtained a UV extinction curve. The shape of the curve as a function of wavelength does not resemble other extinction curves. A possible cause of the extinction could be 110 nm graphite particles in the star's atmosphere along the line of sight. It is proposed that this star is intermediate as a dust producer between other Ae and T Tauri stars, whose variability appears to be due to spotting, and R CrB variables where dust production is the cause of the light extinction.

121.007   Atmospheric properties of RU Lupi derived from high– and low–resolution IUE spectra.
A. Brown, M. V. Penston, R. Johnstone, C. Jordan, N. P. M. Kuin, M. T. V. T. Lago, B. Gross, J. L. Linsky.
NASA Conf. Publ., NASA CP-2349, p. 338 – 341 (1984). – See Abstr. 012.001.

High– and low–dispersion IUE spectra of the pre–main sequence star, RU Lupi, have been obtained using both the SWP and LWR cameras. Strong P Cygni line profiles are seen in Mg II and Fe II emission lines, indicating that the lines are formed in the stellar wind of RU Lupi. An increase in transition region line widths is seen with increasing temperature, indicating that kinematic broadening mechanisms are dominant. The transition region density is $\sim 3 \times 10^{10} \mathrm{cm}^{-3}$ derived from the Si III λ1892/C III λ1909 line ratio. The status of the atmospheric modeling of RU Lupi is discussed.

121.008   Upper limits to coronal line emission from X–ray detected T Tauri stars.
M. T. V. T. Lago, M. V. Penston, R. M. Johnstone.
Mon. Not. R. Astron. Soc., Vol. 212, No. 1, p. 151 – 162 (1985).

High dispersion AAT spectrograms of the two T Tauri stars, GW Orionis and T Tauri, are used to set upper limits to the intensities of the coronal emission line of [Fe XIV] at λ5303 Å. Comparison made with the intensities of chromospheric and the transition–region emission lines and of the X–ray flux shows that in GW Ori the [Fe XIV] is fainter than might have been expected by comparison with the Sun. Overall differences in temperature structure in the wind regions of the different stars are apparent. A crude model is fitted to the emission measures giving the overall behaviour of density and temperature with height in the chromospheres and transition regions. Three possible explanations for the weakness of the coronal lines are presented of which variability is perhaps the most likely.

121.009   Infrared photometry of young (Orion–type) stars.
H. J. Walker.
Observatory, Vol. 105, No. 1064, p. 5 (1985). Abstract. – See Abstr. 010.721.

121.010   Some systematic trends in the color variations of T Tauri stars at visible wavelengths.
F. J. Vrba, A. E. Rydgren, D. S. Zak, J. T. Schmelz.
Astron. J., Vol. 90, No. 2, p. 326 – 332 (1985).

The dependence of $U-B$, $B-V$, and $V-I$ color on V magnitude is examined for ten well–observed T Tauri stars, based on photometry from numerous sources. It is found that the "color slopes" $d(B-V)/dV$ and $d(V-I)/dV$, due to variability, differ significantly between stars and tend to be larger for T Tauri stars of later spectral type. Furthermore, when the color slopes are small, the $B-V$ color slope is significantly less than the $V-I$ color slope. These results are in accord with the hypothesis that the primary source of large–amplitude brightness variations in T Tauri stars is a changing mix of photospheric regions, hot plage regions, and dark spots on the stellar surface.

121.011   Herbig–Haro objects.
M. Rozyczka.
Postepy Astron., Vol. 31, No. 4, p. 279 – 302 (1983). In Polish. Abstr. in Phys. Abstr., Vol. 88, No. 1250, Entry 18903 (1985).

121.012   Herbig–Haro objects and FU Orionis eruptions. The case of HH 57.
B. Reipurth.
Astron. Astrophys., Vol. 143, No. 2, p. 435 – 442 (1985).

Optical spectroscopy, CCD images and infrared observations of a newly brightened star next to HH 57 show that the star is a fifth member of the FU Ori class. Similar observations of the star associated with HH 101 indicate that this star may also once have undergone an FU Ori eruption. Both HH 57 and HH 101 are situated at the front of elongated bubble structures emanating from their associated stars. A small nebula extends from the opposite side of the HH 57 star, suggesting a bipolar outflow. Arguments are presented in support of speculations that HH objects may be powered by FU Ori eruptions.

121.013   X–ray activity in pre–main sequence stars.
E. D. Feigelson.
NASA Conf. Publ., NASA CP-2358, p. 75 – 80 (1985). – See Abstr. 012.023.

This paper summarizes new X–ray observations of low mass pre–main sequence (PMS) stars and discusses their interpretation with respect to winds and surface activity. Some recent optical and radio observations supporting the X–ray evidence for giant flares on PMS stars are also considered.

121.014   Active phenomena in the pre–main sequence star AB Aur.
F. Praderie, T. Simon, A. M. Boesgaard, P. Felenbok, C. Catala, J. Czarny, A. Talavera.
NASA Conf. Publ., NASA CP-2358, p. 81 – 85 (1985). – See Abstr. 012.023.

The Herbig Ae star AB Aur presents short time scale variability in the Mg II and Ca II resonance lines. A qualitative model of the expanding envelope, involving fast and slow streams in a corotating structure, is proposed to explain the Mg II spectral variability.

121.015   FU Orionis star winds.
U. Bastian, R. Mundt.
Astron. Astrophys., Vol. 144, No. 1, p. 57 – 63 (1985).

The authors present high–quality H_α and NaD line profiles (resolution about 13 km s^{-1}) of four FU Orionis objects. The profiles indicate strong and cool stellar winds, accelerated very close to the stellar surfaces to velocities of several hundred km s^{-1}. An outward increasing temperature in the wind acceleration region is indicated. The derived wind properties are similar to those derived for other young stellar objects, suggesting a common wind acceleration mechanism.

121.016   The line profiles generated in the bow shocks of a Herbig–Haro object.
S.-U. Choe, K.-H. Böhm, J. Solf.
Astrophys. J., Vol. 288, No. 1, p. 338 – 341 (1985).

In order to interpret their recent observations of spatially resolved emission–line profiles in Herbig–Haro object HH 1, the authors have calculated the position–dependent $H\beta$ profile for a model of a bow shock around a spherical body using an approximate theory which leads to a hyperbolic bow shock. They derive a theoretical "position–velocity" map with a shock velocity of 350 km s^{-1} for $H\beta$. The observations show that HH 1 shows a "position–velocity" map of a simple bow shock directed away from the Cohen–Schwartz star, as would be expected for an "interstellar bullet" model.

121.017   The T Tauri phenomenon.
A. E. Rydgren.
News Lett. Astron. Soc. N.Y., Vol. 2, No. 5, p. 5 – 6 (1984). Abstract. – See Abstr. 010.243.

121.018   The T Tauri stars.
M. Cohen.
Effects of variable mass loss on the local stellar environment, p. 93 – 114 (1984). – See Abstr. 012.028.

121.019   Optical spectroscopy of the outflow source in L1551.
M. Sarcander, T. Neckel, H. Elsässer.
Astrophys. J., Lett. Ed., Vol. 288, No. 2, p. L51 – L54 (1985).

Cassegrain spectra of the jet and the nebulosities associated with the molecular outflow region in L1551 have been used to derive radial velocities, electron densities, and extinction. The observed emission–line spectra show the characteristic features of shock–excited gas. In the jet emanating from IRS 5, radial velocities up to -210 km s^{-1} and a mean electron density of 10^3cm^{-3} are found. The extinction deduced from the $H\alpha/H\beta$ ratio is rather low with values of $A_V < 3$ mag. On the other hand, an obscuration of $A_V \gtrsim 20$ mag is derived for IRS 5, which appears to be a very young pre–main–sequence star of about 1 solar mass.

121.020   On the variability of hydrogen and calcium emission in the spectra of SU Aur.
V. P. Grinin, A. S. Mitskevich, L. V. Timoshenko.
Astrofizika, Tom 22, Vyp. 1, p. 43 – 49 (1985). In Russian. English translation in Astrophysics, Vol. 22, No. 1.

The patrol spectroscopic observations of the T Tauri star SU Aur revealed the following features of variability of emission spectra: (1) appearance in single nights of the redshifted absorption components in the $H\alpha$ profile, (2) $H\alpha$ flare in the night 11/12 October 1982, which was accompanied by an outburst of matter from the star, (3) variability of emission in H and K Ca II lines with a characteristic time about 1^h. The origin of the variability is briefly discussed.

121.021   An FU Orionis star associated with Herbig–Haro object 57.
J. A. Graham, J. A. Frogel.
Astrophys. J., Vol. 289, No. 1, p. 331 – 341 (1985).

The authors describe a variable star of the FU Orionis type which has appeared on the edge of Herbig–Haro object 57. The star was discovered in 1983, although it had evidently begun to flare optically about 3 yr before. Past and present photographs of the region are reproduced. Photometry out to 20 μm is presented. The photometry shows a strong infrared excess and is consistent with an F star surrounded by a two–component dust shell with temperatures of 2000K and 325K. Spectroscopic observations show that the star is ejecting an optically thin shell which is producing a strong and broad P Cygni absorption profile at $H\alpha$ with a violet absorption edge at -440 km s^{-1}. The $H\alpha$ emission is sharp and at near zero velocity. In the $\lambda 8500$ region, the Ca II triplet is observed in emission, also at near zero velocity.

121.022   Multifrequency radio images of L1551 IRS 5.
J. H. Bieging, M. Cohen.
Astrophys. J., Lett. Ed., Vol. 289, No. 1, p. L5 – L8 (1985).

L1551 IRS 5 is detected and resolved at 1.5, 5, and 15 GHz with the VLA. At 15 GHz, the central core is well resolved into two pointlike sources, surrounded by weaker extended emission, on a line at p.a. 191°. The northern (brighter) source is coincident with the peak at 5 GHz. The jet seen at 5 GHz is barely detected at 15 GHz. The jet has $\alpha \approx 0.3$, while the two pointlike sources have $\alpha \approx 1$. It is suggested that L1551 IRS 5 is a binary system whose orbital plane coincides with the CS molecular toroid.

121.023   Rotation and activity of T Tauri stars.
J. Bouvier, C. Bertout.
Messenger, No. 39, p. 33 – 38 (1985).

121.024   An extremely variable radio star in the Rho Ophiuchi cloud.
E. D. Feigelson, T. Montmerle.
Astrophys. J., Lett. Ed., Vol. 289, No. 2, p. L19 – L23 (1985). With plates L1 – L2.

During a multi–epoch radio continuum survey of the ϱ Ophiuchi star formation cloud perfomed with the VLA, rapid radio variability was seen in the young star DoAr 21. Typically present at levels of $2 – 5$ mJy, it rose up to 48 mJy at 5 GHz on time scales of hours on 1983 February 18, during which the spectrum was steeply inverted. Interpretation in terms of mass loss, which is usually invoked to explain T Tauri star radio emission, encounters difficulties. (Gyro)synchrotron radiation from energetic electrons in a magnetic loop several times larger than the star provides a better model.

121.025   Herbig–Haro objects 1 and 2: another infrared candidate for their energy source.
L. F. Rodríguez, M. Roth, M. Tapia.
Mon. Not. R. Astron. Soc., Vol. 214, No. 1, p. 9P – 13P (1985).

A possible source for the excitation energy of Herbig–Haro objects 1 and 2 has been detected at wavelengths of $1 – 4$ μm. Its position coincides with a recent detection of a 6–cm radio continuum source and the centre of an NH_3 torus orientated perpendicularly to the line joining H–H 1 and 2. Its infrared $(1 – 200$ μm) luminosity is approximately $10 \, L_\odot$, consistent with a reddened $(A_V \cong 12)$ T Tauri–type star. From upper limits in visual and near–infrared plates and from infrared photometry, a reddening law which resembles that of the Orion Belt and Sword regions has been calculated.

121.026   Coronal lines and the structure of the upper atmospheres of T Tauri stars.
S. A. Lamzin.
Astron. Zh., Tom 62, Vyp. 2, p. 306 – 313 (1985). In Russian. English translation in Sov. Astron., Vol. 29, No. 2.

The relations are obtained allowing to determine the parameters of stellar coronae from the intensities of the strongest forbidden coronal lines. A joint analysis of the UV and X–ray observations together with the results of searches for coronal lines in the case of T Tau stars is performed. It is shown that thermal conductivity plays a negligible role in the energy balance in the atmospheres of these stars at $T \gtrsim 2 \times 10^5 K$. The prospects for the search for coronal lines in the spectra of T Tau stars are discussed.

121.027   The magnetic field of T Tauri.
Yu. N. Gnedin, N. P. Red'kina.
Sov. Astron. Lett., Vol. 10, No. 4, p. 255 – 257 (1984). English translation of 38.121.009.

121.028   Plasma processes in Herbig–Haro objects.
I. Axnäs, N. Brenning, G. F. Gahm.
ESA Spec. Publ., ESA SP–207, p. 229 – 233 (1984). – See Abstr. 012.044.

The UV excess from highly ionized elements in the high–excitation Herbig–Haro objects is discussed. It is proposed that it can be explained if a sheath is formed in the object, with heating in excess of ordinary shock wave calculations. Such a sheath could be the result of dynamo interaction between a Herbig–Haro cloudlet and a magnetized stellar wind, with a transverse magnetic field of the order of 0.1 mGauss.

121.029 Mechanical energy release by protostars to molecular clouds.
G. Silvestro, M. Robberto.
ESA Spec. Publ., ESA SP–207, p. 235 – 236 (1984). – See Abstr. 012.044.

The mass–loss rates of low–mass stars powering high–velocity molecular outflows cannot be accounted for by steady stellar wind. The authors argue that energetic outbursts of the FU Orionis type are likely to give an appreciable contribution to the observed flows. They estimate local birthrate and rate of energy input to molecular clouds by these high–velocity outflow sources.

121.030 Considerable brightening of the variable RY Tauri in 1983 – 1984: spectroscopic and photometric observations.
G. V. Zajtseva, E. A. Kolotilov, P. P. Petrov, A. E. Tarasov, V. I. Shenavrin, A. G. Shcherbakov.
Pis'ma Astron. Zh., Tom 11, No. 4, p. 271 – 277 (1985). In Russian. English translation in Sov. Astron. Lett., Vol. 11.

Results of *UBVJHK* photometry and spectroscopy ($\lambda\lambda 3600 - 5000$ and Hα region) of the T Tauri type star RY Tau made during September 1983 – April 1984 are given. Considerable increase in brightness of the star from $V \approx 10\overset{m}{.}8$ to $V \approx 9\overset{m}{.}5$ without noticeable changes in colours U–B and B–V is discovered. The profile of the Hα emission line varies significantly from night to night.

121.031 Two new Herbig–Haro objects.
T. Yu. Magakyan.
Sov. Astron. Lett., Vol. 10, No. 5, p. 276 – 278 (1984). English translation of 38.121.019.

121.032 New objects found at a scanning of the Palomar Sky Survey.
A. L. Gyul'budagyan.
Astron. Tsirk., No. 1342, p. 7 – 8 (1984). In Russian.

121.033 Linear polarization of T Tauri at maximum brightness.
N. P. Kiselev, N. P. Red'kina, K. V. Tarasov, G. P. Chernova.
Astron. Tsirk., No. 1353, p. 5 – 8 (1984). In Russian.

121.034 Das photometrische Verhalten des T–Tauri–Sternes DI Cephei.
S. Rößiger.
Mitt. Veränderliche Sterne, Band 10, Heft 4, p. 86 – 89 (1984).

Photoelectric and photographic observations of the T Tauri star DI Cephei over a range of some decades are presented. It is possible to investigate especially the long–time–scale variability of this object.

121.035 Rotation and X–ray activity in T Tauri stars.
J. Bouvier, C. Bertout, W. Benz, M. Mayor.
Inst. Astrophys. Paris, Pré–Publ., No. 85, 5 pp. (1984). To appear in Proceedings of the Eighth European Regional Meeting on "Nearby molecular clouds", Toulouse, 17 – 21 September 1984.

121.036 The stellar content of nearby clouds – T Tauri stars.
C. Bertout.
Inst. Astrophys. Paris, Pré–Publ., No. 86, 15 pp. (1984). To appear in Proceedings of the Eighth European Regional Meeting on "Nearby molecular clouds", Toulouse, 17 – 21 September 1984.

121.037 Cold outflows, energetic winds, and enigmatic jets around young stellar objects.
C. J. Lada.
Prepr. Steward Obs., No. 571, 92 pp. (1985).

The author summarizes the current status of our observational knowledge concerning energetic outflows associated with young stellar objects. Particular emphasis is placed on results derived from millimeter–wave, molecular–line observations which so far have provided some of the most dramatic and illuminating information pertaining to the energetic outflow phenomena. The author stars his review with a brief discussion of the increasing evidence for an outflow phase of early stellar evolution provided by observations of T Tauri stars, FU Ori outbursts, Herbig–Haro objects, water masers, and compact infrared emission line sources. This is followed by a detailed review of our knowledge of energetic outflows of cold molecular material around young stellar objects.

121.038 Erratum: "Mass loss in T Tauri stars: observational studies of the cool parts of their stellar winds and expanding shells" [Astrophys. J., Vol. 280, No. 2, p. 749 – 770 (1984)].
R. Mundt.
Astrophys. J., Vol. 292, No. 2, p. 763 (1985). See Abstr. 37.121.045.

121.039 The unexpected ultraviolet variability of Herbig–Haro object 1.
E. W. Brugel, K. H. Böhm, J. M. Shull, E. Böhm–Vitense.
Astrophys. J., Lett. Ed., Vol. 292, No. 2, p. L75 – L78 (1985).

Between 1979 and 1983 the line fluxes of the C IV 1550 and C III] 1909 emission lines in HH 1 have decreased monotonically by factors of at least 4 – 6, while no indications of drastic changes in the optical range have been found. This result is based on four IUE spectra obtained by three different groups of observers. These relatively rapid changes can be used to estimate the thickness of the shocked layers and preshock density.

121.040 Broad band spectral energy distributions of T Tauri stars in the Taurus–Auriga region.
A. E. Rydgren, J. T. Schmelz, D. S. Zak, F. J. Vrba.
Publ. U.S. Nav. Obs., Second Ser., Vol. 25, Part 1, 114 pp. (1984).

The authors have compiled a database of spectral classifications, photoelectric UBVRI photometry and infrared photometric observations for 61 T Tauri and related stars in the Taurus–Auriga dark cloud complex. These data are taken primarily from published sources. This information is presented in the form of a catalog and as the plotted spectral energy distribution for each star. The collected data and plots should be helpful in the study of T Tauri variability and T Tauri spectral energy distributions. Finding charts for these stars are also provided.

121.041 Photoelektrische UBV–Beobachtungen des jungen Sternes V1685 Cygni.
S. Rößiger.
Mitt. Veränderliche Sterne, Band 10, Heft 5, p. 111 – 112 (1985).

Photoelectric UBV observations of the pre–main sequence star BD +40°4124 show only a small amplitude in the light variation. There has not been found any dependence of the (B–V) colour index upon the variation in V. From this the author concludes that the light variation is produced primarily by clouds of circumstellar condensation products.

121.042 Spektroskopische Beobachtungen von PV Cephei, der Zentralquelle eines bipolaren CO–Ausflusses.
M. Sarcander, T. Neckel.
Mitt. Astron. Ges., Nr. 63, p. 128 – 129 (1985). – See Abstr. 012.063.

PV Cep ist ein T Tauri ähnlicher Stern, der in einer ca. 3.5 großen Dunkelwolke liegt. Er ist die zentrale Quelle eines bipolaren CO–Ausflusses, der sich von NW nach SE (Pos. Winkel etwa 330°) erstreckt.

121.043 H₂–emission from T Tauri stars and related objects.
H. Zinnecker, R. Mundt, P. M. Williams, W. J. Zealey.
Mitt. Astron. Ges., Nr. 63, p. 234 (1985). Abstract. – See Abstr. 012.063.

121.044   Energetics of low luminosity young stellar objects.
R. I. Thompson.
Bull. Am. Astron. Soc., Vol. 16, No. 4, p. 920 (1984). Abstract. –
See Abstr. 010.062.

121.045   A transition from chromosphere to wind for Mg II in T Tauri stars.
G. Basri, N. Calvet, C. L. Imhoff, M. S. Giampapa.
Bull. Am. Astron. Soc., Vol. 16, No. 4, p. 938 (1984). Abstract. –
See Abstr. 010.062.

121.046   A search for extended radio trails from Herbig–Haro objects.
A. P. Lane, S. P. Reynolds.
Bull. Am. Astron. Soc., Vol. 16, No. 4, p. 971 – 972 (1984). Abstract. – See Abstr. 010.062.

121.047   Proper motions of Herbig–Haro objects in the Orion Nebula.
B. F. Jones, M. Walker.
Bull. Am. Astron. Soc., Vol. 16, No. 4, p. 972 (1984). Abstract. –
See Abstr. 010.062.

121.048   Detection of radio continuum emission from HH1 and HH2 from their central exciting source.
S. H. Pravdo, R. H. Becker, K. S. Sellgren, S. Curiel, J. Cantó, L. F. Rodríguez, J. M. Torrelles.
Bull. Am. Astron. Soc., Vol. 16, No. 4, p. 972 (1984). Abstract. –
See Abstr. 010.062.

121.049   High resolution spectra of Herbig–Haro object H–H 32.
K. H. Böhm, J. Solf, A. Raga.
Bull. Am. Astron. Soc., Vol. 16, No. 4, p. 972 (1984). Abstract. –
See Abstr. 010.062.

121.050   The H^0 two–photon continuous spectrum of HH 43 and dark cloud extinction.
R. D. Schwartz, M. A. Dopita.
Bull. Am. Astron. Soc., Vol. 16, No. 4, p. 972 (1984). Abstract. –
See Abstr. 010.062.

121.051   Jets from young stars – R Monocerotis.
E. W. Brugel, T. Bührke, R. Mundt.
Bull. Am. Astron. Soc., Vol. 16, No. 4, p. 972 (1984). Abstract. –
See Abstr. 010.062.

121.052   Jets from young stars.
R. Mundt, T. Bührke, E. W. Brugel.
Bull. Am. Astron. Soc., Vol. 16, No. 4, p. 973 (1984). Abstract. –
See Abstr. 010.062.

121.053   Ultraviolet continuum and Mg II emission rotational modulation in three pre–main sequence stars.
C. L. Imhoff, M. S. Giampapa.
Bull. Am. Astron. Soc., Vol. 16, No. 4, p. 997 (1984). Abstract. –
See Abstr. 010.062.

121.054   On the nature of FU Orionis stars.
L. Hartmann, S. J. Kenyon.
Bull. Am. Astron. Soc., Vol. 16, No. 4, p. 998 (1984). Abstract. –
See Abstr. 010.062.

121.055   CCD imagery, high resolution spectroscopy, and polarization measurements of HH objects in Cepheus A.
P. Hartigan, C. J. Lada, J. T. Stocke, S. Tapia.
Bull. Am. Astron. Soc., Vol. 16, No. 4, p. 998 (1984). Abstract. –
See Abstr. 010.062.

121.056   Molecular–line observations of energetic outflows around R CrA and AFGL 2591.
C. K. Walker, C. J. Lada, P. Hartigan.
Bull. Am. Astron. Soc., Vol. 16, No. 4, p. 998 (1984). Abstract. –
See Abstr. 010.062.

121.057   UBVRI photometric monitoring of the low–mass T Tauri stars BP Tau, DN Tau, GK Tau, and GI Tau.
F. J. Vrba, A. E. Rydgren, D. S. Zak, P. F. Chugainov (*P. F. Chugajnov*), N. I. Shakhovskaya.
Bull. Am. Astron. Soc., Vol. 16, No. 4, p. 998 (1984). Abstract. –
See Abstr. 010.062.

121.058   Photometric behaviour of DR Tauri in the season 1984/85.
W. Götz.
Inf. Bull. Variable Stars, No. 2731, 1 p. (1985).

121.059   Detection of radio continuum emission from Herbig–Haro objects 1 and 2 and from their central exciting source.
S. H. Pravdo, L. F. Rodríguez, S. Curiel, J. Cantó, J. M. Torrelles, R. H. Becker, K. Sellgren.
Astrophys. J., Lett. Ed., Vol. 293, No. 1, p. L35 – L38 (1985). With plate L1.

The region in Orion containing HH 1 and HH 2 was observed with the VLA at 20, 6, and 2 cm on several occasions from 1981 to 1984. At lower resolution, four continuum sources were detected. Two of these sources coincide positionally with HH 1 and HH 2. At 6 cm and higher resolution, HH 1 is resolved into at least two components. The emission is probably bremsstrahlung originating in the same region where the visible line emission is produced. At a position closely centered between HH 1 and HH 2, the authors detected an object that they interpret at the energy source of the system.

121.060   T Tauri.
IAU Circ., No. 4039 (1985).

121.061   Collimated mass outflows associated with Herbig Ae– and Be–stars.
S. E. Strom, K. M. Strom, S. C. Wolff, J. Morgan.
Bull. Am. Astron. Soc., Vol. 17, No. 1, p. 514 (1985). Abstract. –
See Abstr. 010.064.

121.062   Simultaneous observations of Ca II K and Mg II k in T Tauri stars.
N. Calvet, G. Basri, C. L. Imhoff, M. S. Giampapa.
Astrophys. J., Vol. 293, No. 2, p. 575 – 583 (1985).

The authors present the first simultaneous, calibrated observations of the Ca II K and Mg II k resonance lines in T Tauri stars. It is found that for T Tauri stars with $M > 1.5\,M_\odot$, which have radiative cores and tend to be fast rotators, the k line seems to arise in an extended region (probably also responsible for the Hα emission), whereas the K line apparently originates closer to the highly inhomogeneous stellar surface. The lower mass stars, which are fully convective and tend to be slow rotators, are more easily described by a largely chromospheric model, consistent with main–sequence activity structures but at greater values of the nonradiative flux.

121.063   The T Tauri stars.
M. Cohen.
Phys. Rep., Vol. 116, No. 4, p. 173 – 249 (1984). Abstr. in Phys. Abstr., Vol. 88, No. 1256, Entry 51635 (1985).

121.064   The Van Vleck Observatory T Tauri monitoring program: season four.
W. Herbst, J. F. Booth.
Bull. Am. Astron. Soc., Vol. 17, No. 2, p. 556 (1985). Abstract. –
See Abstr. 010.065.

121.065   An observational test of pre–main sequence evolutionary tracks.
M. Wenz, K. M. Strom, S. E. Strom, S. C. Wolff.
Bull. Am. Astron. Soc., Vol. 17, No. 2, p. 556 (1985). Abstract. –
See Abstr. 010.065.

121.066 Speckle image reconstruction of a northern optical companion to T Tau.
P. Nisenson, M. Karovska, R. Stachnik, R. Noyes.
Bull. Am. Astron. Soc., Vol. 17, No. 2, p. 556 (1985). Abstract. –
See Abstr. 010.065.

121.067 UBVRI photometric monitoring of the late–type PMS stars AA Tau, DH Tau, DI Tau, GG Tau, and HP Tau/G2.
A. E. Rydgren, F. J. Vrba, P. F. Chugainov (*P. F. Chugajnov*), N. I. Shakhovskaya.
Bull. Am. Astron. Soc., Vol. 17, No. 2, p. 556 (1985). Abstract. –
See Abstr. 010.065.

121.068 Emission line variability of RY Tau, DR Tau and SU Aur.
A. Brown, F. M. Walter, K. G. Carpenter, C. Jordan, P. Judge.
Bull. Am. Astron. Soc., Vol. 17, No. 2, p. 556 (1985). Abstract. –
See Abstr. 010.065.

121.069 Young stars and dense cores: $1 - 20\ \mu m$ observations of circumstellar dust emission.
P. C. Myers, R. D. Mathieu, P. J. Benson, C. A. Beichman, G. Fuller.
Bull. Am. Astron. Soc., Vol. 17, No. 2, p. 557 (1985). Abstract. –
See Abstr. 010.065.

121.070 Photoelectric photometry of DI Cephei during 1981 – 1984.
J. Kelemen.
Inf. Bull. Variable Stars, No. 2744, 4 pp. (1985).

121.071 A Fuor–like variable star in Orion.
E. Chavira, M. Peimbert, G. Haro.
Inf. Bull. Variable Stars, No. 2746, 2 pp. (1985).

121.072 Atmosferas de estrellas T Tauri.
N. Calvet.
Acta Cient. Venezolana, Vol. 34, p. 1 – 15 (1983).
The most recent observations concerning T Tauri stars are reviewed and interpreted in terms of a model based on the solar analogy. This model is able to give a self–consistent description for the atmospheres of T Tauri stars, although some problems remain. Both observational and theoretical work is suggested which will improve our knowledge of the atmospheres of these stars.

General Catalogue of Variable Stars.
See Abstr. 002.027.

Stellar winds and the interstellar medium.
See Abstr. 064.001.

Optically thick radial accretion onto main sequence stars and degenerate dwarfs.
See Abstr. 064.071.

Two–dimensional models of stellar wind bubbles. II. Variable mass–loss rates and the possibility of outer–shell fragmentation in relation to the origin of interstellar bullets.
See Abstr. 064.072.

Two–dimensional models of stellar wind bubbles. III. Self–confining flows in media with strong density gradients.
See Abstr. 064.073.

Two–dimensional models of stellar wind bubbles. IV. Properties of bow shocks around dense clumps in free wind regions.
See Abstr. 064.074.

Lunar occultation observations of M8E–IR.
See Abstr. 096.013.

Chromospheric Hα emission in F8 – G3 dwarfs, and its connection with the T Tauri stars.
See Abstr. 112.055.

Bipolar molecular outflows: T Tauri stars and Herbig–Haro objects.
See Abstr. 112.138.

Report of IAU Commission 29: Stellar spectra (*Spectres stellaires*).
See Abstr. 114.165.

An evaluation of magnetic fields of some Ae/Be Herbig stars.
See Abstr. 116.002.

The evolution of chromospheric activity and the spin–down of solar–type stars.
See Abstr. 116.058.

Interacting binary stars: nomenclature – the stellar zoo.
See Abstr. 117.287.

Report of IAU Commission 27: Variable stars (*Etoiles variables*).
See Abstr. 122.206.

The source of illumination of the bipolar nebula near NGC 7129.
See Abstr. 131.012.

Bipolar flows and X–ray emission from young stellar objects.
See Abstr. 131.065.

OH observations of cloud complexes in Taurus.
See Abstr. 131.094.

VLA observations of ammonia and continuum in regions with high–velocity gaseous outflows.
See Abstr. 131.095.

GSS 30: an infrared reflection nebula in the Ophiuchus dark cloud.
See Abstr. 131.122.

Radio and optical observations of the jets from L1551 IRS 5.
See Abstr. 131.142.

CO $J = 3$–2 observations of molecular line sources having high–velocity wings.
See Abstr. 131.145.

Polarimetry of infrared sources in bipolar CO flows.
See Abstr. 131.157.

Infrared spectroscopy of carbon monoxide in GL 2591 and OMC–1:IRc2.
See Abstr. 131.162.

Macroscopic turbulence in molecular clouds.
See Abstr. 131.198.

Low mass pre–main sequence stars in the Serpens molecular cloud.
See Abstr. 131.230.

High–velocity gas flows associated with H_2 emission regions: how are they related and what powers them?
See Abstr. 131.293.

Predicted long–slit, high–resolution emission–line profiles from interstellar bow shocks.
See Abstr. 131.299.

Optical polarisation of R CrA and T CrA – NGC 6729.
See Abstr. 131.301.

Study of the bipolar nebula associated with LkHα 208.
See Abstr. 131.309.

Gas jets associated with star formation.
See Abstr. 131.355.

Photoionization model of the emission and abundance of the T Tauri nebula.
See Abstr. 132.038.

Infrared observations of a compact H II region in Monoceros (GGD 12–15).
See Abstr. 132.043.

GL 961–W: a pre–main–sequence object.
See Abstr. 133.011.

X–ray observations of the runaway stars HD 206327 and 26 Cephei and of the λ¹Orionis region.
See Abstr. 142.029.

Ultraviolet and X–ray observations of NGC 2264.
See Abstr. 153.050.

122 Intrinsic Variables (Pulsating Variables, Spectrum Variables, etc.)

122.001   Variability of ultraviolet emission lines in the carbon star TX Psc.
J. H. Baumert, H. R. Johnson.
NASA Conf. Publ., NASA CP–2349, p. 322 – 325 (1984). – See Abstr. 012.001.

During the course of their study on the R and N stars the authors have obtained multiple observations of the carbon star TX Psc. They report here the probable time variations of the Mg II 2800 Å line and possible variation of the C II 2325 Å line in the spectra of this object.

122.002   The fading of R Coronae Borealis.
A. V. Holm, J. Hecht, C.–C. Wu, B. Donn.
NASA Conf. Publ., NASA CP–2349, p. 330 – 333 (1984). – See Abstr. 012.001.

The authors obtained IUE spectra of R CrB near the beginning of one of its infrequent minima. These spectra show major changes in the apparent spectrum. The authors attempt to identify new features and to interpret the overall change in the continuum in terms of an obscuring cloud model.

122.003   Observations of RR Lyrae and X Arietis with the IUE satellite.
J. T. Bonnell, R. A. Bell.
NASA Conf. Publ., NASA CP–2349, p. 334 – 337 (1984). – See Abstr. 012.001.

Seven low–dispersion spectra of RR Lyr and 11 of X Ari were obtained with the LWR camera of IUE. RR Lyr was also observed in the low–dispersion mode with the SWP camera (9 spectra) and in the high–dispersion mode with the LWR camera (3 spectra). The ability of the model atmospheres and synthetic spectra to match ultraviolet observations of these stars indicates that the parameters of these stars derived by Manduca et al. (1981) from the interpretation of VR photometry using spectrum synthesis techniques are valid. Difficulties encountered in the analysis of the IUE observations of RR Lyr are believed to be due to the star's secondary cycle.

122.004   Mg II emission of Pop II long–period Cepheids.
M. Parthasarathy, S. B. Parsons.
NASA Conf. Publ., NASA CP–2349, p. 342 – 343 (1984). – See Abstr. 012.001.

The Population II Cepheid pulsational variables AL Vir (P = 10.3 days) and W Vir (P = 17.3 days) were observed with IUE at two phases each. Only low–dispersion mid–UV exposures were feasible, but were sufficient to show a great range in Mg II emission behavior. The relationship to phase variation of emission in classical (Pop I) Cepheids of long period is unclear.

122.005   Ultraviolet observations of Population II Cepheids.
E. Böhm–Vitense, C. Proffitt, G. Wallerstein.
NASA Conf. Publ., NASA CP–2349, p. 348 – 351 (1984). – See Abstr. 012.001.

The two Population II Cepheids, ST Pup and W Vir, have nearly the same length of period and B–V colors, yet spectral types are very different, indicating large differences in metal abundances. The authors have determined metal abundances, T_{eff} and the color excess from the observed discontinuities at 1700 Å, at 2600 Å, the ultraviolet to visual colors, and the 2400 Å absorption band. The observations for ST Pup were made shortly after maximum light. For ST Pup the authors found E(B–V) = 0.2, T_{eff} = 6600±100K at 0.5 days after maximum. A metal abundance of [A/H] = –1.7±0.2 was determined. For W Vir a metal abundance of [A/H] ∼ –0.9 is suggested.

122.006   CCD observations of the RR Lyrae stars in NGC 2210 and the distance to the Large Magellanic Cloud.
A. R. Walker.
Mon. Not. R. Astron. Soc., Vol. 212, No. 2, p. 343 – 352 (1985).

CCD photometry of 9 RR Lyrae stars in the LMC globular cluster NGC 2210 suggests that the true distance modulus of the LMC is 18.42±0.10 mag, assuming $<M_V>_{RR}$ = 0.60 and E(B–V) = 0.060 for NGC 2210. The implications of this result on the Cepheid distance scale are discussed. The results demonstrate that high–accuracy photometry on stars as faint as 20 mag is possible with a combination consisting of a CCD camera and a 1–m telescope.

122.007   Period change in Magellanic Cloud Cepheids.
H. P. Deasy, P. A. Wayman.
Mon. Not. R. Astron. Soc., Vol. 212, No. 2, p. 395 – 411 (1985).

The data for period change in some 115 Cepheid variable stars in the Magellanic Clouds are presented. The accuracy of each determination is estimated and for all but three of the stars it is possible to estimate whether the rate of change of period prior to 1966/67 is maintained subsequently. About 40 per cent of the stars show period variations, similar to the percentage estimated for galactic Cepheids by Hoffmeister (1967). About half of these variations cannot be regarded as being constant in rate. Probably the interpretation of the data lies either in small atmospheric changes that produce phase–discontinuities that accumulate or in small luminosity changes that produce random period changes.

122.008 Line doubling in the 272–day long–period variable V Cancri.
J. F. Dominy, G. Wallerstein, N. B. Suntzeff.
Mon. Not. R. Astron. Soc., Vol. 212, No. 3, p. 671 – 675 (1985).

From Fourier Transform Spectra and high–dispersion optical spectra, the authors have measured doubled absorption lines of the CO molecule and various atomic species soon after three maxima of V Cnc. The velocity difference between doubled atomic features in the optical is 17.5 km s^{-1}, while for CO at 2.3 µm the difference is 21 km s^{-1}. The authors have assembled and compared the data for line doubling in long–period variables over the interval in period of 150 to 450 days. There is no correlation of the amplitude of doubling with period, but there is a correlation of doubling amplitude with the light amplitude as measured at 10400 Å. The relationships between these parameters and the theory of shock waves in long–period variables is discussed.

122.009 The rotational velocity of the rapidly oscillating Ap star HD 83368.
B. W. Carney, R. C. Peterson.
Mon. Not. R. Astron. Soc., Vol. 212, No. 3, p. 33P – 35P (1985).

A rotational velocity $v_{rot}\sin i = 33 \pm 3$ km s^{-1} has been measured for the Ap star HD 83368, confirming the prediction of 32 km s^{-1} (Kurtz) from an 'oblique pulsator' model.

122.010 Energy dissipation mechanisms in flare stars.
D. J. Mullan.
Unstable current systems and plasma instabilities in astrophysics, p. 245 – 262 (1985). – See Abstr. 012.002 (IAU Symp. No. 107).

The author summarizes the relevant physical conditions in the atmosphere of a flare star. He finds that even outside flares, the dissipation of mechanical energy in the atmospheres of flare stars is in a certain sense qualitatively different from what is observed in the sun. Further, he discusses energy dissipation in flares and finds that, at least in the very coolest flare stars, the distinction between the flaring and the non–flaring states of the star becomes difficult to define.

122.011 The optical morphology of the R Aqr jet.
N. Mauron, J. L. Nieto, J. P. Picat, G. Lelievre, H. Sol.
Astron. Astrophys., Vol. 142, No. 1, p. L13 – L15 (1985).

An optical map of the inner nebulosity of the active symbiotic star R Aqr, taken in near UV light with a 1–arcsec resolution, is presented and compared with previous observations, especially a radio VLA 6 cm map. It confirms the existence of a counterjet feature and brings new evidence for bipolar ejection. An upper limit to the velocity of the knots in the jet is obtained. As a result, the recent appearance of this bright jet, at least of its external condensation B, is probably due to some other mechanism than actual, recent matter ejection, perhaps sudden ionization by the variable hot central component.

122.012 Infrared variability of the reflection nebulosity around RS Puppis.
A. J. Mayes, A. Evans, M. F. Bode.
Astron. Astrophys., Vol. 142, No. 1, p. 48 – 54 (1985).

The authors consider the "infrared echo" of the optical variability of the cepheid variable RS Puppis, that arises from the variable heating of the dust grains in its reflection nebulosity. Detailed far infrared observations of the nebulosity can provide a valuable determination of the distance of RS Puppis, independent of interstellar extinction and of the scattering function for the grains. Near infrared observations of RS Puppis, obtained at the South African Astronomical Observatory, are presented in an appendix.

122.013 Peculiarities in the period distribution for Mira variables.
N. V. Kharchenko.
Pis'ma Astron. Zh., Tom 11, No. 1, p. 60 – 65 (1985). In Russian. English translation in Sov. Astron. Lett., Vol. 11.

The distribution of the periods for Mira variables was found to be nonmonotonous. The period distribution function has some gaps. Possible reasons of this phenomenon are discussed.

122.014 A period–luminosity relation for Mira variables in globular clusters and its impact on the distance scale.
J. W. Menzies, P. A. Whitelock.
Mon. Not. R. Astron. Soc., Vol. 212, No. 4, p. 783 – 797 (1985).

JHKL photometry is presented for 31 red variables in 15 galactic globular clusters. The photometry of the Mira variables is used to find absolute bolometric magnitudes and an M_{bol}–log P relation which differs from the one found for LMC Miras. This can be understood only if there is some systematic error in the globular cluster and/or LMC distance scales or if there is some fundamental difference between the cluster Miras and those in the LMC.

122.015 Photoelectric measurements of BY Draconis carried out at the Catania Astrophysical Observatory from 1967 to 1970.
S. Cristaldi.
Astron. Astrophys., Suppl. Ser., Vol. 59, No. 2, p. 187 – 193 (1985).

The photoelectric measurements of BY Draconis carried out at the Catania Astrophysical Observatory from 1967 to 1970, together with some comments are reported.

122.016 New period determinations for variable CP stars.
J. Manfroid, G. Mathys.
Astron. Astrophys., Suppl. Ser., Vol. 59, No. 3, p. 429 – 432 (1985).

New or improved photometric periods are presented for 29 CP stars.

122.017 The phase variations of intrinsic polarization of ten cepheids.
T. A. Polyakova.
Pis'ma Astron. Zh., Tom 11, No. 2, p. 133 – 138 (1985). In Russian. English translation in Sov. Astron. Lett., Vol. 11.

The consideration of variations of intrinsic polarization parameters with the light curve phase for ten cepheids shows that the polarization degree slightly increases twice during the period, and the plane of polarization turns considerably also twice with one of the turnings being more sharp than the other.

122.018 Variations of the period of 35 variables in the globular cluster M15.
A. F. Gordenko, A. V. Klabukova, N. N. Fashchevskij.
Probl. Kosm. Fiz., Vyp. 19, p. 93 – 102 (1984). In Russian. – See Abstr. 003.002.

122.019 Sur l'origine des ondes de choc dans l'atmosphère des étoiles Mira.
D. Gillet.
Bull. Assoc. Fr. Obs. Etoiles Variables, No. 31, p. 3 – 7 (1985).

122.020 Neue instantane Elemente für RR Lyr.
P. Ringe.
BAV Rundbrief, 34. Jahrg., Nr. 1, p. 1 – 4 (1985).

New elements for RR Lyr, based on about thirty observations, are Max = JD $2442995.446 + 0.566839 \cdot E$.

122.021 Metal abundances of RR Lyrae variables in selected galactic star fields. IV. The Lick Astrograph field RR I (MWF 361) in Serpens and Ophiuchus.
T. D. Kinman, R. P. Kraft, E. Friel, N. B. Suntzeff.
Astron. J., Vol. 90, No. 1, p. 95 – 100 (1985). = Lick Obs. Bull., No. 993.

Preston Δs values are given for 36 RR Lyrae variables in the Lick Astrograph field RR I ($l = 11°$, $b = +29°$). Thirty–four of these variables are of type a and account for over 80% of the variables of this type in this field. Their $\langle \Delta s \rangle = 5.80 \pm 0.39$ and differs significantly from the $\langle \Delta s \rangle = 7.65 \pm 0.22$ of the 52 RR Lyraes which have previously been observed in the north galactic pole and anticenter regions of the halo. It is shown that this difference is not produced by the differing z distributions of the two samples, but is related to their differing galactocentric distances. Δs shows a continuous monotonic increase with the galactocentric distance.

122.022 Double–mode RR Lyrae stars in the Draco dwarf galaxy.
J. M. Nemec.
Astron. J., Vol. 90, No. 2, p. 204 – 239 (1985).

From analysis of the photometry of Baade and Swope (1961), ten double–mode RR Lyrae (RRd) stars in the Draco dwarf galaxy are identified. The physical characteristics of the stars are derived and compared with those of RRd stars in the Galaxy. Their periods and period ratios are consistent with simultaneous radial pulsation in the fundamental and the first–overtone modes. For each star, the amplitudes of the first–overtone pulsations of the stars are larger than the amplitudes of the fundamental pulsations, as seen in galactic RRd stars. Average beat masses are determined for the sample. Metal abundances for the stars, estimated from their positions in the period–amplitude diagram, are presented. A distance of 84 ± 12 kpc is derived for the stars. Correlations involving the entire system of Draco RR Lyrae stars are investigated.

122.023 Double–mode RR Lyrae stars in M15: reanalysis, and experiments with simulated photometry.
J. M. Nemec.
Astron. J., Vol. 90, No. 2, p. 240 – 253 (1985).

Periods, amplitudes, and beat masses for 12 double–mode RR Lyrae (RRd) stars in the galactic globular cluster NGC 7078 (M15) are derived from the photometry of Sandage, Katem, and Sandage (1981). The same methods used to study the RRd stars in the Draco dwarf galaxy have been used to study the M15 stars. Uncertainties in the periods are discussed and light curves are presented for all the stars. Experiments are described that show the effects of random scatter on the derived periods of the RRd stars in M15 and Draco. They provide evidence that the detected secondary oscillations probably are real.

122.024 The effect of synchronization on the modal selection in classical cepheids.
M. Takeuti.
Astrophys. Space Sci., Vol. 109, No. 1, p. 99 – 109 (1985).

The coupled self–exciting oscillator model is investigated in the non–resonant case and applied to classical cepheids. The modal selection in these models is explained as the result of the synchronization caused by the mutual interaction between different modes. By using linear adiabatic coupling coefficients, it is shown that the fundametal mode suppresses the first overtone mode marginally in short–period classical cepheids. The double periodicity of several cepheids is expected as the result of a small change of physical state in the outer envelopes of these stars.

122.025 Non–adiabatic effects on the pulsation periods.
T. Aikawa.
Astrophys. Space Sci., Vol. 109, No. 1, p. 183 – 189 (1985).

Weight functions for the non–adiabatic radial pulsations are introduced. It is shown from behavior of these functions that the pulsation periods in classical Cepheids are determined essentially in the adiabatic region of stellar envelopes and, on the other hand, those of low surface–gravity models are strongly affected in the region where the acoustic waves are strongly coupled with the radiation fields. The fact is important for understanding basic difference of the pulsation properties between classical Cepheids and low surface–gravity models.

122.026 Nonradial pulsation and mass loss in early B stars.
G. D. Penrod, M. Smith.
NASA Conf. Publ., NASA CP-2358, p. 53 – 56 (1985). – See Abstr. 012.023.

This paper summarizes recent observational evidence for non-radial pulsations in nearly all sharp–lined B stars near the main sequence. Furthermore, it reviews evidence that recurrent temporary ejection of circumstellar shells in these stars may be triggered by nonradial stellar pulsations. As examples, recent observations of ϱ Leo and λ Eri are presented.

122.027 Light variations of the B–type star HD 160202.
G. A. Bakos.
NASA Conf. Publ., NASA CP-2358, p. 62 – 65 (1985). – See Abstr. 012.023.

Light variations of the B–type star HD 160202 in the galactic cluster M6 are presented. A flare type increase in brightness by about three magnitudes in a time interval of 40 minutes has been recorded.

122.028 Delta–Scuti–Sterne.
M. Breger.
Sterne Weltraum, 24. Jahrg., Nr. 1, p. 21 – 23 (1985).

122.029 Light variations of 28 Andromedae.
R. Garrido, S. F. Gonzalez, A. Rolland, M. A. Hobart, P. Lopez de Coca, J. H. Peña.
Astron. Astrophys., Vol. 144, No. 1, p. 211 – 214 (1985).

Analysis of existing and new photometric observations of the δ Scuti type star 28 Andromedae indicates that it is a mono-periodic variable star with a period of $0\overset{d}{.}0693$ with notable amplitude variations from season to season. Evidence presented suggests that 28 And is a second overtone pulsator.

122.030 Distances to Magellanic Clouds from observations of Cepheids at 1.05 microns.
N. Visvanathan.
Astrophys. J., Vol. 288, No. 1, p. 182 – 186 (1985).

Observations in the IV (1.05 µm) and V (5500 Å) wave bands are presented for 13 Cepheids in the SMC, 14 in the LMC, and nine in groups and clusters in our Galaxy having periods between 2 and 87 days. The $<IV>$ magnitudes of Cepheids are used in conjunction with periods to construct a period–luminosity $(P–L)$ relation in the SMC, the LMC, and the Galaxy. Through the absolutely calibrated relation in our Galaxy, the author obtains distance moduli of 19.11 ± 0.07 and 18.82 ± 0.07 for the SMC and the LMC, respectively, based on a distance modulus of 3.29 for the Hyades.

122.031 Pulsational mode typing in line–profile variables. VI. Nonradial modes in the remarkable B star Epsilon Persei.
M. A. Smith.
Astrophys. J., Vol. 288, No. 1, p. 266 – 274 (1985).

Epsilon Persei is a B0.7 III star with no radial velocity or photometric variations at optical wavelengths, but it shows pronounced variations in line shape. The author has obtained nearly 200 high–quality CCD observations of the Si III $\lambda4567$ triplet on two pairs of nights separated by a month. It is possible to fit a representative sample of 40 of them with three prograde sectorial nonradial pulsation modes with degrees $l = 6, 4$, and 1 or 2. The observed frequencies appear to be in the ratio of 6:4:1 with $\sigma_6 \approx 10.9$ cycles day^{-1}. The periods in ε Per demonstrate that both p– and g–type modes can coexist in a B star. Similarities and possible physical links between this type of rotating variable and the 53 Persei and β Cephei variables are suggested.

122.032 The RR Lyrae stars in and around the LMC globular cluster NGC 2257.
J. M. Nemec, J. E. Hesser, P. Ugarte P.
Astrophys. J., Suppl. Ser., Vol. 57, No. 2, p. 287 – 328 (1985). With plates 1 – 6.

Forty–one RR Lyrae stars in the LMC halo globular cluster NGC 2257, 16 of which are newly identified, are studied. Within 25' of NGC 2257, 47 new field variable stars are also identified and studied using new photographic plates taken at CTIO from 1971 to 1982. The estimated completeness of the discoveries is $\sim 82\%$ in the cluster and $\sim 53\%$ in the field. Periods, light curves, epochs of maximum light, rise times, and mean and median magnitudes are derived and are compared with those of galactic RR Lyrae stars. The field variables appear to show a wider metallicity range than do the cluster variables.

122.033 The secular period behavior of 38 RR Lyrae stars in the LMC globular cluster NGC 2257.
J. M. Nemec, M. L. Hazen–Liller, J. E. Hesser.
Astrophys. J., Suppl. Ser., Vol. 57, No. 2, p. 329 – 348 (1985).

The secular period behavior of 38 RR Lyrae stars in the Large Magellanic Cloud globular cluster NGC 2257 is studied using new PDS and visual photometry of plates taken during the years 1951 – 1982. Mean period change rates are derived under various assumptions. Assuming uniformly changing periods, it is of interest that the mean period change rates are significantly different for the ab– and the c–type variables. The period change rates show no correlation with position in the C–M diagram, as is found for variables in M15 by Smith and Sandage.

122.034 A classification of Miras from their visual and near–infrared light curves: an attempt to correlate them with their evolution.
M. O. Mennessier.
Astron. Astrophys., Vol. 144, No. 2, p. 463 – 470 (1985).

Using cluster analysis methods, Miras are classified from the global features of their visual and near–infrared light curves. Groups of stars, stable against various clustering analyses, are defined. Some correlations between the membership to one cluster and physical properties of the Miras are searched. The authors do not find a break between the properties of the clusters but rather a continuous variation from one to another. The classification of Miras which were not included in stars for clustering confirms the established correlations. The minimum luminosity of a Mira likely appears as a more stable state than its maximum.

122.035 Photometric properties of HD 200925.
M. D. Joner, S. B. Johnson.
Publ. Astron. Soc. Pac., Vol. 97, No. 588, p. 153 – 157 (1985).

Photometry of HD 200925 in the $uvby$ and β systems has been secured and analyzed. The data strongly indicate that the variable is multiperiodic. A reddening value, $E(b-y) = 0\overset{m}{.}040$, has been derived from the variable– and comparison–star observations. Analysis of the intrinsic $(b-y)$ and c_1 values for one primary pulsation cycle are used to determine a mean effective temperature $\langle T_{eff}\rangle = 7470K$, and a mean surface gravity $\langle\log g\rangle = 3.87$. HD 200925 is found to have a somewhat higher abundance, [Fe/H] = +0.33, than stars in the Hyades. These values do not conform to the presently accepted properties for dwarf Cepheid variables.

122.036 Fundamental parameters of Cepheids.
J. W. Pel.
Cepheids: theory and observations, p. 1 – 16 (1985). – See Abstr. 012.027. (IAU Colloq. No. 82).

The author reviews some of the present knowledge of "fundamental parameters" of Cepheids. He discusses only the "classical" type of Cepheids, and says nothing about the W Virginis stars. The author concentrates on observational results, and on the constraints that can be derived from them on some of the fundamental parameters, giving most attention to the calibration of L and T_{eff}.

122.037 The double–mode Cepheids.
L. A. Balona.
Cepheids: theory and observations, p. 17 – 29 (1985). – See Abstr. 012.027. (IAU Colloq. No. 82).

Recent observations of double–mode Cepheids and proposed candidates are reviewed. It appears that the change of modal content reported in some stars may not be real. Observations show that the double–mode Cepheids are indistinguishable from normal Cepheids of similar period. This poses a severe problem for pulsation theory which predicts very low masses from the ratio of first overtone to fundamental period. Attempts to resolve this problem and the problem of modal selection are discussed.

122.038 Cepheid temperatures derived from energy distributions.
T. J. Teays, E. G. Schmidt.
Cepheids: theory and observations, p. 30 – 31 (1985). – See Abstr. 012.027. (IAU Colloq. No. 82).

122.039 Radial velocities of classical Cepheids.
T. G. Barnes III, T. J. Moffett.
Cepheids: theory and observations, p. 32 – 33 (1985). – See Abstr. 012.027. (IAU Colloq. No. 82).

122.040 Radius determination for nine short–period Cepheids.
G. Burki.
Cepheids: theory and observations, p. 34 – 37 (1985). – See Abstr. 012.027. (IAU Colloq. No. 82).

122.041 Surface brightness radii and distances of Cepheids and the period–radius relationship.
W. P. Gieren.
Cepheids: theory and observations, p. 38 – 42 (1985). – See Abstr. 012.027. (IAU Colloq. No. 82).

122.042 A generalization of the CORS method to determine Cepheid radii: theory and application.
B. Caccin, B. Buonaura, A. Onnembo, G. Russo, A. M. Sambuco, C. Sollazzo.
Cepheids: theory and observations, p. 43 – 47 (1985). – See Abstr. 012.027. (IAU Colloq. No. 82).

The CORS method for the empirical determination of the radii of pulsating variables (Caccin et al., 1981; Sollazzo et al., 1981) is discussed in the framework of the quasistatic approximation to the variations of the atmospheric parameters (Unno, 1965) and reformulated in a way that does not make direct use of theoretical calibrations of the photometric system in terms of model atmospheres.

122.043 Baade–Wesselink radii of long–period Cepheids: new observational results.
I. M. Coulson, J. A. R. Caldwell, W. Gieren.
Cepheids: theory and observations, p. 48 – 50 (1985). – See Abstr. 012.027. (IAU Colloq. No. 82).

The radii of Galactic Cepheids as determined from a version of the Baade–Wesselink technique are shown to depend upon the colour index used to define the temperature scale. The (V–I) radii are systematically larger than the commonly used (B–V) radii and are probably better estimators of the true radii. There still remains a problem, however, in reconciling the period–radius relation with the period–luminosity–colour relation.

122.044 Cepheid radii from infrared photometry.
D. L. Welch, N. R. Evans, G. Drukier.
Cepheids: theory and observations, p. 51 – 52 (1985). – See Abstr. 012.027. (IAU Colloq. No. 82).

The authors have determined radii for Cepheids using near–infrared K(2.2 μm) photometry and Balona's approach to the Baade–Wesselink method of radius determination.

122.045 Distances and radii of classical Cepheids.
T. G. Barnes III, T. J. Moffett.
Cepheids: theory and observations, p. 53 – 55 (1985). – See Abstr. 012.027. (IAU Colloq. No. 82).

The authors have used new BVRI photometry and radial velocities of a selection of bright classical Cepheids to determine their distances and radii through the surface brightness method.

122.046 Light–curve parameters of northern galactic Cepheids.
T. J. Moffett, T. G. Barnes III.
Cepheids: theory and observations, p. 56 – 57 (1985). – See Abstr. 012.027. (IAU Colloq. No. 82).

The BVRI light–curve parameters of 112 galactic Cepheids are determined by Fourier analysis using over 4,000 differential photoelectric observations. This catalog (see Abstr. 38.122.013) is similar to Schaltenbrand and Tammann's except that it is based on a homogeneous data set and the Fourier coefficients are given.

122.047 Infrared observations of galactic Cepheids.
J. A. Fernley, R. F. Jameson, M. R. Sherrington.
Cepheids: theory and observations, p. 58 – 62 (1985). – See Abstr. 012.027. (IAU Colloq. No. 82).

The authors present BVJHK observations of the galactic Cepheids T Vul, U Vul and T Mon. Using this data they find reasonable agreement with current versions of the period–luminosity relations, in both the optical and infrared. From these relations and existing optical and infrared data for LMC Cepheids the authors find for the LMC (1) $A_V = 0.35$ and (2) a distance modulus of 18.50. In addition they present the P–L–C relation.

122.048   New evidence for mass loss in classical Cepheids.
M. J. Stift.
Cepheids: theory and observations, p. 63 – 66 (1985). – See Abstr. 012.027. (IAU Colloq. No. 82).

122.049   Classical Cepheids: period changes and mass loss.
H. Deasy.
Cepheids: theory and observations, p. 67 – 70 (1985). – See Abstr. 012.027. (IAU Colloq. No. 82).

122.050   Duplicity, mass loss and the Cepheid mass anomaly.
G. Burki.
Cepheids: theory and observations, p. 71 – 74 (1985). – See Abstr. 012.027. (IAU Colloq. No. 82).

122.051   Duplicity among the Cepheids in the northern hemisphere.
L. Szabados.
Cepheids: theory and observations, p. 75 – 78 (1985). – See Abstr. 012.027. (IAU Colloq. No. 82).

The aim of the paper is not only to give an estimation of the percentage of binaries in a previously unused sample but also to recommend for further analysis more than twenty Cepheids suspected of having a companion.

122.052   A search for Cepheid binaries using the Ca II H and K lines.
N. R. Evans.
Cepheids: theory and observations, p. 79 – 80 (1985). – See Abstr. 012.027. (IAU Colloq. No. 82).

A survey of 24 classical Cepheids has been made to search for blue companions using the Ca II H and K lines. It is shown that this technique can detect an early A companion for a typical Cepheid. A blue companion of SU Cas was discovered and upper limits for the companions for a number of previously suspected binaries were established.

122.053   On the binary nature of 89 Herculis.
A. Arellano Ferro.
Cepheids: theory and observations, p. 81 – 82 (1985). – See Abstr. 012.027. (IAU Colloq. No. 82).

The radial velocities of 89 Herculis between 1977 and 1981 show a clear periodicity of about 285 days. This periodic variation is interpreted as the orbital variation of 89 Her around an unseen companion. From the orbital elements no support is found for a low mass ($M \sim 2\,M_\odot$) for 89 Her, but rather a high mass ($13 < M/M_\odot < 24$) is preferred.

122.054   Non–Cepheids in the instability strip.
W. P. Bidelman.
Cepheids: theory and observations, p. 83 – 84 (1985). – See Abstr. 012.027. (IAU Colloq. No. 82).

122.055   Cepheid–like supergiants in the halo.
D. D. Sasselov.
Cepheids: theory and observations, p. 85 – 88 (1985). – See Abstr. 012.027. (IAU Colloq. No. 82).

The newly proposed UU Her–type stars are discussed; their main features being long–period variability and high galactic latitude. It is unlikely that the UU Her stars originated in the plane of the galaxy as they are now located more than a kiloparsec above it.

122.056   The quasi–Cepheid nature of Rho Cassiopeiae.
J. R. Percy, D. Keith.
Cepheids: theory and observations, p. 89 – 90 (1985). – See Abstr. 012.027. (IAU Colloq. No. 82).

122.057   On the brightness and pulsational properties of yellow supergiants.
A. Arellano Ferro.
Cepheids: theory and observations, p. 91 – 92 (1985). – See Abstr. 012.027. (IAU Colloq. No. 82).

Most of the existing data in the literature is used to search for the periodicity of five yellow supergiant variables. The pulsational mode and Q–values are discussed, being the luminosity the key parameter for their accurate determination. An attempt to find the spherical harmonic l from the light and colour variation phase shift is carried out.

122.058   Some remarks on the unique Cepheid HR 7308.
N. R. Simon.
Cepheids: theory and observations, p. 93 – 94 (1985). – See Abstr. 012.027. (IAU Colloq. No. 82).

122.059   Kinematic evidence for fundamental mode pulsation in the short–period classical Cepheid SU Cassiopeiae.
D. G. Turner, D. W. Forbes, R. W. Lyons, R. J. Havlen.
Cepheids: theory and observations, p. 95 – 97 (1985). – See Abstr. 012.027. (IAU Colloq. No. 82).

122.060   A study of the short–period Cepheid EU Tauri.
W. P. Gieren.
Cepheids: theory and observations, p. 98 – 99 (1985). – See Abstr. 012.027. (IAU Colloq. No. 82).

122.061   Abundance analysis of Alpha Ursae Minoris.
S. Giridhar.
Cepheids: theory and observations, p. 100 – 103 (1985). – See Abstr. 012.027. (IAU Colloq. No. 82).

The author has derived atmospheric abundances of the bright Cepheid α UMi in order to study the abundance anomalies in different elements. The evolutionary status of α UMi is discussed in the light of the derived abundances.

122.062   Cepheid evolution.
S. A. Becker.
Cepheids: theory and observations, p. 104 – 125 (1985). – See Abstr. 012.027. (IAU Colloq. No. 82).

A review of the phases of stellar evolution relevant to Cepheid variables of both Types I and II is presented. Despite great progress in the past two decades, uncertainties still remain in such areas as how to best model convective overshoot, semiconvection, stellar atmospheres, rotation, and binary evolution as well as uncertainties in important physical parameters such as the nuclear reaction rates, opacity, and mass loss rates. The potential effect of these uncertainties on stellar evolution models is discussed. Finally, comparisons between theoretical predictions and observations of Cepheid variables are presented for a number of cases.

122.063   Theory of Cepheid pulsation: excitation mechanisms.
J. P. Cox.
Cepheids: theory and observations, p. 126 – 146 (1985). – See Abstr. 012.027. (IAU Colloq. No. 82).

The various excitation mechanisms (eight in all) that have been proposed to account for the vibrational instability of variable stars, are surveyed. The most widely applied one is perhaps the "envelope ionization mechanism". This can account for most of the essential characteristics of the "instability strip". A simple explanation of the period–luminosity relation of classical Cepheids is given. A few outstanding problems in pulsation theory are also listed.

122.064   Non–adiabatic effects on pulsation periods.
T. Aikawa.
Cepheids: theory and observations, p. 147 – 148 (1985). – See Abstr. 012.027. (IAU Colloq. No. 82).

It is well known that the difference between the adiabatic pulsation periods and the corresponding non–adiabatic periods is small in classical Cepheids. However, the difference becomes significantly large in low surface–gravity models. In this paper the author discusses the origins of the difference between the two pulsation periods. The weight functions for non–adiabatic pulsation periods are given.

122.065   Nonadiabatic, nonlinear pulsations of bump Cepheids – a new approach.
J. R. Buchler, M. J. Goupil, J. Klapp.
Cepheids: theory and observations, p. 149 – 152 (1985). – See Abstr. 012.027. (IAU Colloq. No. 82).
Finite amplitude pulsations of realistic stellar models are analyzed within the framework of a promising asymptotic perturbation approach and the results are compared with those of numerical hydrodynamic studies.

122.066   Theoretical study of Cepheid light curves.
C. G. Davis.
Cepheids: theory and observations, p. 153 – 156 (1985). – See Abstr. 012.027. (IAU Colloq. No. 82).
The author describes some recent model results for long period Cepheids in an attempt to understand the "dips" and possible get another handle on Cepheid masses.

122.067   Magellanic Cloud and other extragalactic Cepheids: some current topics.
M. W. Feast.
Cepheids: theory and observations, p. 157 – 165 (1985). – See Abstr. 012.027. (IAU Colloq. No. 82).
Contents: Introduction. Chemical abundances of extragalactic Cepheids. The significance of accurate interstellar reddenings for Magellanic Cloud Cepheids. The P–L–C relation and the structure of the Magellanic Clouds. Period–colour relation. Radii and surface brightness. Infrared work on Magellanic Cloud Cepheids. Zero point of the P–L–C relation.

122.068   Cepheid variables as extra–galactic distance indicators.
B. F. Madore.
Cepheids: theory and observations, p. 166 – 198 (1985). – See Abstr. 012.027. (IAU Colloq. No. 82).
The role of Cepheids variables in establishing the inner distance scale to nearby galaxies is discussed. Emphasis is placed on the necessity for broad wavelength coverage in attempting to account for metallicity differences and reddening internal to the parent galaxies. In addition linear detectors are essential in minimizing the effects of any unresolved background contribution to the photometry. Recent infrared observations of Cepheids in Local Group galaxies are surveyed and all published data on extragalactic Cepheids are presented for convenient access.

122.069   The Cepheid luminosity scale.
E. G. Schmidt.
Cepheids: theory and observations, p. 199 – 200 (1985). – See Abstr. 012.027. (IAU Colloq. No. 82).

122.070   The zero–point of the Cepheid luminosity scale from a calibration of the luminosities of early–type stars.
L. A. Balona, R. R. Shobbrook.
Cepheids: theory and observations, p. 201 – 202 (1985). – See Abstr. 012.027. (IAU Colloq. No. 82).
A new calibration of the absolute magnitudes of early–type stars in terms of the (β, c_0) photometric system is used to establish the distance moduli of clusters containing Cepheids. The zero points of the period–luminosity and period–luminosity–colour relations are calculated and compared to previous determinations.

122.071   Some remarks concerning the accuracy of photometric reddenings for classical Cepheids.
D. G. Turner.
Cepheids: theory and observations, p. 205 – 208 (1985). – See Abstr. 012.027. (IAU Colloq. No. 82).

122.072   Are the Cepheids in cluster nuclei a rare breed?
D. G. Turner.
Cepheids: theory and observations, p. 209 – 211 (1985). – See Abstr. 012.027. (IAU Colloq. No. 82).

122.073   A search for long–period Cepheids in associations.
S. van den Bergh.
Cepheids: theory and observations, p. 212 – 214 (1985). – See Abstr. 012.027. (IAU Colloq. No. 82).
Skeleton photoelectric sequences and plates in U, B and V have been obtained for fields surrounding all known Cepheids with $P \gtrsim 12$ days in the Southern Milky Way. The study of 14 of these fields has been completed and is discussed in this paper.

122.074   Leavitt variables: the brightest Cepheid variables and their implications for the distance scale.
G. R. Grieve, B. F. Madore, D. L. Welch.
Cepheids: theory and observations, p. 215 – 218 (1985). – See Abstr. 012.027. (IAU Colloq. No. 82).
Two low–amplitude variable supergiants in the Large Magellanic Cloud, S65–08 and S65–48 are each found to have periods of approximately 250 days. The optical data suggest that these stars are high–luminosity Cepheid variables falling more than one magnitude brighter than any other known Cepheids in the LMC. To honour the astronomer who discovered the first of these highest–luminosity Cepheids, the authors have sub–classified the variables with $\log P > 1.8$ as being "Leavitt variables". As soon as these long–period variables are discovered in other external galaxies, reliable distances should be possible out to $(m–M) \sim 30$.

122.075   The infrared distance scale: the Galaxy and the Magellanic Clouds.
D. L. Welch, C. W. McAlary, R. A. McLaren, B. F. Madore.
Cepheids: theory and observations, p. 219 – 222 (1985). – See Abstr. 012.027. (IAU Colloq. No. 82).
The advantages of using near–infrared photometry of Cepheids to determine distances to nearby galaxies are now well known. The authors summarize the current state of the infrared period–luminosity (P–L) relations for the Galaxy, the LMC, and the SMC and present composite P–L relations derived from all available photometry. The authors give distance moduli for the LMC and SMC based on these data and briefly report on the status of other work in progress.

122.076   Population II Cepheids.
H. C. Harris.
Cepheids: theory and observations, p. 232 – 245 (1985). – See Abstr. 012.027. (IAU Colloq. No. 82).
A review of Cepheids in globular clusters finds 46 stars that are highly probable members of 21 clusters, all of intermediate to low metallicity with blue horizontal branches. In order to separate Type II Cepheids in the field from classical Cepheids, an analysis is made of the distribution of all Cepheids: 144 are sufficiently far from the galactic plane to be considered Type II. Their properties are compared with the cluster Cepheids. Some recent studies are reviewed that are improving the understanding of low–mass Cepheids.

122.077   Theoretical models of W Virginis variables.
A. Bridger.
Cepheids: theory and observations, p. 246 – 249 (1985). – See Abstr. 012.027. (IAU Colloq. No. 82).
W Virginis variables are the population II counterparts of the classical Cepheids, although they do not show quite the same trends as are seen in the latter. The paper summarises the results of a series of hydrodynamic (nonlinear) models constructed in an attempt to reproduce the observed characteristics of these stars.

122.078   A new possible resonance for population II Cepheids.
A. N. Cox, R. B. Kidman.
Cepheids: theory and observations, p. 250 – 253 (1985). – See Abstr. 012.027. (IAU Colloq. No. 82).

122.079   Hydrodynamic models of Population II Cepheids.
Yu. A. Fadeyev (*Yu. A. Fadeev*), A. B. Fokin.
Cepheids: theory and observations, p. 254 – 255 (1985). – See Abstr. 012.027. (IAU Colloq. No. 82).
A short report is given on hydrodynamic modelling for BL Her and W Vir pulsating variables.

122.080   Some masses for Population I and II Cepheids.
R. B. Kidman, A. N. Cox.
Cepheids: theory and observations, p. 256 – 259 (1985). – See Abstr. 012.027. (IAU Colloq. No. 82).

122.081   The two Pop II Cepheids in the globular cluster Messier 10.
C. Clement, H. Sawyer Hogg, K. Lake.
Cepheids: theory and observations, p. 260 – 261 (1985). – See Abstr. 012.027. (IAU Colloq. No. 82).

The authors examine the variations in the periods of the two Cepheids of M10 over the interval 1912 to 1983 and 1931 to 1983, respectively.

122.082   Eight Pop II Cepheids recently identified in globular clusters.
C. Clement, H. Sawyer Hogg, T. Wells.
Cepheids: theory and observations, p. 262 – 263 (1985). – See Abstr. 012.027. (IAU Colloq. No. 82).

122.083   Fourier decomposition parameters for the halo RR Lyrae variables U and V Caeli.
L. Hansen, J. O. Petersen.
Cepheids: theory and observations, p. 272 – 275 (1985). – See Abstr. 012.027. (IAU Colloq. No. 82).

UBVRI light curves are obtained for the two halo RR Lyrae variables U Caeli with period 0.420 days and V Caeli with period 0.571 days. It is shown that their light curve characteristics are very similar to those of field RR Lyrae stars. Fourier decompositions are studied for all five magnitudes and the resulting amplitude ratios and phase differences are discussed.

122.084   Comparison of the pulsation properties of the RR Lyrae stars in ω Centauri with those of classical Cepheids.
J. O. Petersen.
Cepheids: theory and observations, p. 276 – 279 (1985). – See Abstr. 012.027. (IAU Colloq. No. 82).

The author analyses 130 photographic mean light curves of RR Lyrae variables in ω Centauri taken from Martin (1938). He (1) compares the Fourier decomposition parameters of the ω Cen RR Lyrae stars with those of the field variables as studied by Simon and Teays, (2) discusses the evidence for progression sequences among the ω Cen variables and (3) compares the basic pulsation properties of the RRab variables in ω Cen with those of classical Cepheids.

122.085   The effects of convection of RR Lyrae stars.
R. F. Stellingwerf.
Cepheids: theory and observations, p. 280 – 283 (1985). – See Abstr. 012.027. (IAU Colloq. No. 82).

The effect of convection in RR Lyrae stars has been investigated using nonlinear models that include the effects of time dependence, turbulent pressure, convective overshooting, and convection–pulsation interaction.

122.086   The long term behaviour of two variables in the globular cluster M56.
A. Wehlau, P. Rice, M. Wehlau, H. Sawyer Hogg.
Cepheids: theory and observations, p. 284 – 287 (1985). – See Abstr. 012.027. (IAU Colloq. No. 82).

122.087   Periodical features in the Lc variable star BU Geminorum during the years 1972 – 1977.
A. Buzzoni.
GEOS Circ., SR 6, 9 pp. (1985).

The results of the observations of BU Gem, made by the GEOS observers during the years 1972 to 1977 are presented. Starting from the total of 1034 visual magnitude estimates a mean light curve has been calculated via a computerized method. A weak semi–regular variability of the star is evident with a period of about 240 days and a more marked irregular fluctuation with a time–scale of a thousand days. There is no evidence of any eclipses caused by suggested duplicity of the star, the red supergiant being the main responsible for the light variations. On the basis of the photometric behaviour, a classification as an Lc is suggested for BU Gem.

122.088   Parameters of the light curve of CE Cassiopeiae.
L. N. Berdnikov.
Perem. Zvezdy, Tom 21, No. 6, p. 819 – 822 (1983). In Russian.

The light curve parameters of the cepheids CE Cas a and CE Cas b corrected by published photographic observations are given.

122.089   New electrophotometric UBV observations of the variable DH Pegasi.
I. F. Alaniya, O. P. Abuladze.
Perem. Zvezdy, Tom 21, No. 6, p. 823 – 826 (1983). In Russian.

112 photoelectric UBV observations of the RR Lyrae variable DH Peg were performed in September and October of 1980. A strong variation of the light curve for a month is pointed out.

122.090   CF Persei is a semiregular variable.
G. K. Filip'ev.
Perem. Zvezdy, Tom 21, No. 6, p. 831 – 833 (1983). In Russian.

The results of photoelectric BVR observations of the variable CF Per are presented. It is shown that CF Per is a pulsating giant of SR type with a period equal to approximately 60^d.

122.091   A catalogue of distances and light absorption for cepheids.
G. R. Ivanov, Yu. N. Efremov, N. S. Nikolov.
Perem. Zvezdy, Tom 21, No. 6, p. 861 – 871 (1983). In Russian.

Data on the space positions of 409 classical cepheids of our Galaxy are given. New distances and colour excesses are obtained for 309 cepheids with photoelectric data. This number exceeds those in the preceding catalogue by Fernie and Hube (1968) by 40 %. Colour excesses are based on the multicolour photometry by Dean et al. (1978) and luminosities are obtained from van den Bergh's (1976) data about cluster cepheids.

122.092   Photoelectric photometry of WW Vulpeculae in 1967 – 1982.
G. V. Zajtseva.
Perem. Zvezdy, Tom 22, No. 1, p. 1 – 8 (1983). In Russian.

Results are presented of photoelectric photometry of the irregular variable WW Vul carried out in the UBV system in 1967 – 1982. Several light weakenings with amplitude more than 2^m were observed. A non–heterogenuous relation between the V–brightness and (B–V), (U–B) colour indices is discovered. A quasi–periodicity with period of 404 days is found in light weakenings of WW Vul.

122.093   Photographic photometry of V426 Ophiuchi.
S. Yu. Shugarov.
Perem. Zvezdy, Tom 22, No. 1, p. 31 – 36 (1983). In Russian.

An analysis is carried out of 242 photographic observations of the variable (JD 2414868–45232). The object displays outbursts with an amplitude of $1^m – 1{.}^m5$, the outbursts in 1976 – 77 being recurrent with a cycle of $32^d{.}3$. An analogy with Q Cyg (Nova Cygni 1876) and V794 Aql, a cataclysmic variable and an X–ray source, is pointed out. The photographic observations are given in a table.

122.094   Photographic observations of the semiregular variable star Z Sagittae (V1, NGC 6838).
A. Yu. Pogosyants.
Perem. Zvezdy, Tom 22, No. 1, p. 85 – 91 (1983). In Russian.

The variability of Z Sge is investigated on the basis of approximately 300 plates (1960 – 1981). At first the period of the star was $175^d{.}3$. Near JD 2438700 the period changed by jump and became equal to $190^d{.}8$. The amplitude of the star's light variations reaches 2^m in B. The star's location in the colour–magnitude diagram is shown.

122.095   Photographic U, B, V photometry of the components of the double cepheid CE Cassiopeiae.
P. N. Kholopov, Yu. N. Efremov.
Perem. Zvezdy, Tom 22, No. 1, p. 93 – 100 (1983). In Russian.

Results of photographic measurements of brightness of each component of CE Cas with the aid of an iris astrophotometer are given. Mean light curves of each component in the B, V systems are obtained. The results are based on 73 plates obtained during 1963 – 1965.

122.096   Spectral study of the irregular variables SV Cep, UX Ori and DD Ser.
L. V. Timoshenko.
Astrofizika, Tom 22, Vyp. 1, p. 51 – 61 (1985). In Russian. English translation in Astrophysics, Vol. 22, No. 1.

A two–dimensional quantitative classification is carried out for three variables with nonperiodic light minima SV Cep, UX Ori and DD Ser. Mean spectral classes are determined for each star according to some classification criteria based on metal lines: A0 for SV Cep, A3 for UX Ori, and A5 for DD Ser. The spectral class of SV Cep is found to be changed from A0 to A3 from night to night. The mean absolute magnitudes over SV Cep, UX Ori and DD Ser spectra are $M_V = -0^m5$, $M_V = 0^m6$ and $M_V = 0^m2$, respectively. The depths of hydrogen lines and K Ca II are well above the depths in standard star spectra belonging to the same spectral and luminosity classes. The study of DD Ser photographic observations available in literature and covering the period of 25 years permits to suspect a minimum cycle of 4 years.

122.097   Flares and Hα in emission stars in the region of the Orion nebula.
Eh. S. Parsamyan.
Astrofizika, Tom 22, Vyp. 1, p. 87 – 96 (1985). In Russian. English translation in Astrophysics, Vol. 22, No. 1.

An analysis of flare activity of RW Ori variables has shown that the percentage of flare stars among variables with A $\geqslant 1^m$ is about 40 and among A $\geqslant 2^m$ about 60; therefore the higher the RW Ori activity, the higher is the percentage of flare stars among them. It has been shown that the average flare frequency of stars with RW Ori activity does not differ noticeably from the flare frequency of the rest of the flare stars. Among flare stars about 10% have RW Ori activity. The shapes of distributions of flare and Hα stars by magnitude are similar. The luminosity function of Hα stars increases up to m = 18.

122.098   Maia variables and upper–main–sequence phenomena.
B. J. McNamara.
Astrophys. J., Vol. 289, No. 1, p. 213 – 219 (1985).

A sample of four stars, including Maia itself, which are located within the Maia instability strip are investigated for photometric variability. The data consist of over 1600 differential Strömgren y–magnitudes collected during 11 nights of observation. No evidence is found for variability over the 0.1 – 0.3 day period range suggested for Maia stars. Two stars, Merope and Atlas, do seem to show variability, but the lengths of their periods imply a closer relation to the 53 Persei stars. It is suggested that the Maia stars do not exist as a separate class of variable stars but are an extension of the 53 Persei phenomenon to cooler temperatures. Convective core overshooting is hypothesized to be the mechanism responsible for the variability of these stars.

122.099   The extraordinary magnetic variation of the helium–strong star HD 37776: a quadrupole field configuration.
I. B. Thompson, J. D. Landstreet.
Astrophys. J., Lett. Ed., Vol. 289, No. 1, p. L9 – L13 (1985).

Extensive longitudinal magnetic field measurements of the helium–strong B star HD 37776 have been obtained. It is found that when these data are phased with a period of 1.538 days derived from helium line variations, an extraordinary double–wave magnetic curve results. It is argued that this curve is in fact the real magnetic curve of the star and that HD 37776 is therefore the first known star in which a quadrupole–like field geometry dominates the usual dipole found in other magnetic stars.

122.100   Observations of Hβ and He II λ4686 lines in the spectra of UV Ceti–type stars during flares.
P. P. Petrov, P. F. Chugajnov, A. G. Shcherbakov.
Izv. Krymskoj Astrofiz. Obs., Tom 69, p. 3 – 18 (1984). In Russian. English translation in Bull. Crimean Astrophys. Obs., Vol. 69.

Spectroscopic observations and photoelectric B system observations of the flare stars AD Leo, DT Vir, YZ CMi and UV Cet are reported. Three flares of AD Leo and three flares of YZ CMi were recorded. In two flares of AD Leo and two flares of YZ CMi the increase of the central intensity of Hβ was observed 10 – 20 minutes before the flare maxima. The emission line He II λ4686 was found neither in the quiet state of the stars nor during the flares. The following conclusions are drawn: 1) preflares are characterized by a prevailing increase of the line emission; 2) the emission wings of Hβ occur during the flare maxima owing to the Stark effect and mass motions; 3) only a very weak He II λ4686 emission may appear during the flare maxima due to cascade recombinations of He III caused by the increase of the X–ray flux.

122.101   Double emission and line absorption doubling in Mira stars: a new approach.
D. Gillet, P. Bouchet, R. Ferlet, E. Maurice.
Messenger, No. 39, p. 38 – 40 (1985).

122.102   Investigation of the variability of V351 Ori. I. Observations.
G. U. Koval'chuk.
Kinematika Fiz. Nebesn. Tel, Tom 1, No. 2, p. 30 – 36 (1985). In Russian.

Photoelectric estimates of brightness obtained during 103 nights showed periods of quiet and active stages in light variations. At normal state the variable shows light variations with an amplitude up to 0.3^m (V). Weakenings of the variable's light by 2^m (V) are noted. They are accompanied by noticeable colour changes. A brief history of the variable star is presented.

122.103   Spectral classification of southern–hemisphere Mira variables.
R. Crowe.
The MK process and stellar classification, p. 348 – 352 (1984). – See Abstr. 012.033.

122.104   The identification of the double lines in the photographic–infrared spectrum of the Mira–type carbon star U Cygni near maximum light.
M. Hirai.
The MK process and stellar classification, p. 353 – 361 (1984). – See Abstr. 012.033.

The author reports the results of further spectroscopic analysis of the double lines which were suggested by Hirai and Nagasawa (1980) in the photographic–infrared spectrum of U Cygni near maximum light. Each line has two components, and the excitation temperatures derived from the short– and long–wavelength components are about 3,000K and 2,520K, respectively. The turbulent velocity of the s–component is 1.26 times larger than of the l–component. Other physical parameters related to the cool atmosphere are obtained using both the synthetic–spectrum and curve–of–growth methods. Possible models to explain the double–line phenomena in the Mira–type carbon star U Cygni are discussed.

122.105   MK classification of Cepheids.
R. P. Gauthier, R. F. Garrison.
The MK process and stellar classification, p. 362 – 366 (1984). – See Abstr. 012.033.

A multiphase classification study of a sample of 26 Cepheids from the Southern Hemisphere at the relatively high dispersion of 67 Å/mm has been undertaken and complemented with the recent high–quality photometric data of Pel (1976) in order to observe the results of increasing period and amplitude of pulsation on the line spectrum.

122.106   New radial pulsation constants for the β Cephei variables.
R. R. Shobbrook.
Mon. Not. R. Astron. Soc., Vol. 214, No. 1, p. 33 – 44 (1985).

Recent new calibrations of luminosities, temperatures and bolometric corrections for B stars in terms of the β index and the Strömgren parameter c_0 have necessitated the recalculation of the radial pulsation constants, Q, for the β Cephei (of β Canis Majoris) variable stars. Corrections for the effect of binaries on the absolute magnitudes, derived both from the luminosity calibration and from the mean distance moduli of those variables in clusters, are calculated in an Appendix. The mean value of Q, although determined from absolute magnitudes which are about 0.4 mag fainter than those from previous calibrations of the β index, still suggests that the majority of the variables are pulsating in the first overtone radial mode, as have most investigations in recent years.

122.107   Optical photometry and spectroscopy of the flare star Gliese 229 (= HD 42581).
P. B. Byrne, J. G. Doyle, J. W. Menzies.
Mon. Not. R. Astron. Soc., Vol. 214, No. 1, p. 119 – 130 (1985).

The authors present optical flare photometry and a search for spotted variations on the star Gl 229. These results rule out the presence of a large–scale asymmetric spot distribution and indicate a low level of flare activity. This is in agreement with other indicators of stellar activity including coronal X–ray emission and radiative losses from the lower chromosphere as gauged from the strength of Ca H and K and the Balmer lines.

122.108   The shock–induced variability of the Hα emission profile in Mira. II.
D. Gillet, R. Ferlet, E. Maurice, P. Bouchet.
ESO Sci. Prepr., No. 368, 31 pp. (1985). To appear in Astron. Astrophys.

122.109   Photographic photometry of variable stars based on the Tartu photoplate collection. III. Area of μ Persei: FM and FO Per.
L. Kalv, A. Maripuu.
Tartu Astrofüüs. Obs. Publ., Tom 50, p. 175 – 177 (1984).

122.110   Photographic photometry of variable stars in the α Sagittae region. TZ Sge and eleven more pulsating stars.
L. Kalv, P. Kalv, V. Harvig, L. Leis.
Tartu Astrofüüs. Obs. Publ., Tom 50, p. 178 – 190 (1984).

The authors continue publishing the results of the investigations of variable stars from the plate collection of Tartu Astrophysical Observatory.

122.111   Photoelectric photometry of the red variable HD 157010.
M. Jerzykiewicz.
Acta Astron., Vol. 34, No. 3, p. 353 – 364 (1984).

UBV observations of HD 157010 are presented and discussed. It is found that the star exhibits SRb type light and colour variations. These variations can be understood in terms of temperature sensitivity of the continuous spectrum and TiO blanketing. The surface area effect turns out to be very small, perhaps absent altogether. Hence, the star's UBV magnitudes become effective temperature parameters. These results are then extended to other low–range red giant variables and shown to be consistent with observations. In addition, the $0^{m}_{.}02$ upper limit for the interstellar reddening of M92 is confirmed and a small correction to the intrinsic B–V colours of red giants is derived.

122.112   Photoelectric observations of BS Aqr.
D. Kozerska, K. Stepień.
Acta Astron., Vol. 34, No. 3, p. 377 – 380 (1984).

UBV observations of BS Aqr were obtained. Because the comparison star turned out to be variable its variability was taken into account when reducing the observations of BS Aqr. BS Aqr

is a typical large–amplitude δ Scuti–type variable and its period steadily decreases.

122.113   EW Scuti: a double–mode cepheid.
J. Cuypers.
Astron. Astrophys., Vol. 145, No. 2, p. 283 – 285 (1985).

Period analyses of observations of the cepheid EW Scuti published by Bakos (1950) and Eggen (1973) reveal the existence of two periods in the light variations $P_0 = 5^{d}_{.}8195$ and $P_1 = 4^{d}_{.}0646$. The period ratio $P_1/P_0 = 0.6984$ characterizes the star as a double–mode cepheid. A study of the confidence intervals derived for periods found with a phase dispersion minimization analysis establishes the stability of the periods to the limits of detection.

122.114   U and Hα electrophotometry of flares of EV Lac.
A. S. Melkonyan.
Soobshch. Byurakan. Obs., Vyp. 54, p. 15 – 23 (1983). In Russian.

The results of simultaneous U and Hα electrophotometry of flares of EV Lac are presented. It is shown that the detection of flares in the U band is much easier than in the Hα line when an Hα–filter of FWHM = 12 Å is used.

122.115   Polarimetric and photometric observations of five flares of the star EV Lac.
M. A. Eritsyan.
Soobshch. Byurakan. Obs., Vyp. 54, p. 24 – 26 (1983). In Russian.

The results of polarimetric and photometric observations of the flare star EV Lac are given. It is shown that during flares as well as in the normal state of the star no polarization exceeding the observational errors has been observed.

122.116   Study of the variability of the δ Scuti stars. VIII. HR 4684 and the resonance mechanism in dwarf pulsators.
E. Antonello, G. Guerrero, L. Mantegazza, M. Scardia.
Astron. Astrophys., Vol. 146, No. 1, p. 11 – 16 (1985).

The light curve analysis of HR 4684 provides an observational indication of the presence in δ Scuti stars of the parametric resonance mechanism suggested by Dziembowski. Probably, this process plays a general role in the pulsation of dwarf stars, by exciting low order modes which, in a linear approach, should be stable, and limiting the pulsational amplitudes. Physical features and pulsational modalities, both in HR 4684 and in another δ Scuti star (KW 207), agree with the assumption of the presence of parametric resonance phenomena in normal δ Scuti stars.

122.117   A catalogue of field Type II Cepheids.
H. C. Harris.
Astron. J., Vol. 90, No. 5, p. 756 – 760 (1985).

A catalogue of field Type II Cepheids is presented. The primary list consists of 152 Cepheids sufficiently far from the galactic plane to be very probably Type II stars. A second list contains 56 additional stars that are likely, but less certain, Type II Cepheids, including both stars estimated to be at large distances from the galactic plane but with uncertain distances and stars close to the galactic plane believed to be Type II for independent reasons.

122.118   The unusual pulsating variable XZ Ceti.
T. J. Teays, N. R. Simon.
Astrophys. J., Vol. 290, No. 2, p. 683 – 688 (1985).

The peculiar pulsating variable XZ Ceti has a period of 0.8231 day, suggesting that it is a Bailey type ab RR Lyrae star, but it has a light curve shape and amplitude which is like that of the Bailey type c variables. The authors have obtained a new photometric light curve which verifies this long period. Energy distributions were obtained from spectrum scans, which are used to derive a temperature and approximate surface gravity of this star. Pulsation models suggest the possibility that XZ Ceti is an anomalous Cepheid or, perhaps, an overtone BL Her star.

122.119 A photometric and radial velocity study of six southern Cepheids. I. The data.
I. M. Coulson, J. A. R. Caldwell, W. P. Gieren.
Astrophys. J., Suppl. Ser., Vol. 57, No. 3, p. 595 – 609 (1985).

New observations of the galactic Cepheids AQ Car (period = $9{.}^{\rm d}77$), XX Cen ($10{.}^{\rm d}95$), XY Car ($12{.}^{\rm d}44$), TT Aql ($13{.}^{\rm d}75$), XX Car ($15{.}^{\rm d}71$), and XZ Car ($16{.}^{\rm d}65$) are presented. 365 photometric measures in the $UBVRI_c$ system and 297 radial velocities are combined with previous data to produce light, color, and velocity curves of almost total phase coverage.

122.120 New flare stars in the Mon I association.
Eh. S. Parsamyan, L. Rosino, O. S. Chavushyan.
Astrofizika, Tom 22, Vyp. 2, p. 315 – 324 (1985). In Russian. English translation in Astrophysics, Vol. 22, No. 2.

A search for flare stars in the region of the Mon I association was carried out by one of the authors (L. R.) during the years 1974 – 75 and 1981 at the Asiago Astrophysical Observatory. During 55 hours of effective time 150 plates were obtained. As a result of the examination of these plates which cover a region of about 25 square degrees centered in the cluster NGC 2264 29 new flare stars have been found, 17 of which are probably members of the NGC 2264 cluster.

122.121 Interstellar polarization from cepheid polarimetry.
T. A. Polyakova.
Sov. Astron. Lett., Vol. 10, No. 5, p. 312 – 315 (1984). English translation of 38.122.091.

122.122 Observations of RR Lyrae and X Arietis with the IUE satellite.
J. T. Bonnell, R. A. Bell.
Publ. Astron. Soc. Pac., Vol. 97, No. 589, p. 236 – 246 (1985).

The observed fluxes have been compared with fluxes calculated using synthetic spectra and angular diameters determined by Manduca et al. (1981) from photometry at longer wavelengths. The observed fluxes for RR Lyr were found to be significantly fainter than the computed ones. RR Lyr was also found to be 0.4 mag fainter at 2200 Å and 2400 Å during minimum light than it was during previous ultraviolet observations with the ANS satellite. One of the high–dispersion exposures made near maximum light was deep enough to reveal the cores of the Mg II h and k lines. No emission was observed and a strong interstellar absorption component was present. An excellent agreement between observed and calculated fluxes was found for X Ari, an extremely metal–poor RR Lyrae star which has no observed secondary cycle.

122.123 Emission–line variability in WR stars: short periods and nonradial pulsations?
J.–M. Vreux.
Publ. Astron. Soc. Pac., Vol. 97, No. 589, p. 274 – 279 (1985).

It is shown that nearly all the published periods in the emission–line profiles for the so–called "WR + compact–companion" systems can be easily related to each other. It is suggested that the emission–line variability is due to single–star nonradial pulsations instead of spiralling–in neutron star scenarios.

122.124 A survey for red variables in the LMC. I.
I. S. Glass, N. Reid.
Mon. Not. R. Astron. Soc., Vol. 214, No. 3, p. 405 – 418 (1985).

The first results of a UK Schmidt–based V– and I–band photographic survey for large–amplitude red variable stars in a field of the LMC are presented. 40 red variables were found and periods were determined for 24. Infrared JHK photometry is given for all 40. Eight of the sample constitute a subgroup of relatively luminous M–type stars having periods, where determined, in excess of 555 day. Thirteen are probably carbon Mira variables including nine extremely red ones similar to the galactic example R For. The remainder are red–giant variables of which nine are probably M–type Miras. Further evidence for the period–luminosity relationship obeyed by all LMC M– and C–type Mira variables is also presented.

122.125 Mass loss by R Coronae Borealis–type stars.
A. Eh. Rozenbush.
Astron. Tsirk., No. 1343, p. 6 – 8 (1984). In Russian.

122.126 Investigation of the G star V460 Cyg with scanning spectra.
N. S. Komarov, A. G. Cherkass.
Astron. Tsirk., No. 1346, p. 5 – 7 (1984). In Russian.

122.127 V421 Herculis.
I. M. Volkov.
Astron. Tsirk., No. 1347, p. 7 – 8 (1984). In Russian.

122.128 Small amplitude red variables: a brief introduction.
J. R. Percy.
I.A.P.P.P. Commun., No. 19, p. 14 – 19 (1985).

122.129 Catch the minima: a real challenge on V348 Sgr.
A. Heck, J. Manfroid.
I.A.P.P.P. Commun., No. 20, p. 31 – 36 (1985).

122.130 Further observations of the Mira star CR Muscae.
B. F. Marino, W. S. G. Walker.
South. Stars, Vol. 31, No. 1, p. 72 – 76 (1984).

Earlier observations of the Mira star CR Muscae are reviewed and the data are updated to include more recent photoelectric measurements made up to the end of 1983. A revised period of 201.6 days is determined.

122.131 On the light variability of PG 0134 + 70.
S. Yu. Shugarov.
Astron. Tsirk., No. 1350, p. 5 – 7 (1984). In Russian.

122.132 Investigation of the variability of V351 Ori. II. Analysis of the basic peculiarities of light and color variations.
G. U. Koval'chuk.
Kinematika Fiz. Nebesn. Tel, Tom 1, No. 3, p. 25 – 32 (1985). In Russian.

The basic peculiarities of light and color variations of the irregular variable V351 Ori are analysed. The global decreasing of brightness up to $\Delta V = 2^{\rm m}$ is the main process of variability. It is accompanied by noticeable color variations, namely, by the reddening of the star on the initial stages of light decrease and by its further "blueing" when the star goes to the minimum. On the basis of color–magnitude analysis the optical properties of the circumstellar matter are obtained.

122.133 An improved calibration of the near–infrared period–luminosity relations for Cepheids.
D. L. Welch, C. W. McAlary, B. F. Madore, R. A. McLaren, G. Neugebauer.
Astrophys. J., Vol. 292, No. 1, p. 217 – 221 (1985).

Near–infrared photometry of galactic cluster Cepheids is used to establish the J, H, and K period–luminosity relations at mean light. The relations are derived by adopting the distance moduli and reddenings of Caldwell for EV Sct, CF Cas, CV Mon, V Cen, V367 Sct, U Sgr, DL Cas, and S Nor, and those of Pel and Havlen for RS Pup.

122.134 Temporal variations in UV spectra of the red giant C star, TW Hor.
M. Querci, F. Querci.
Astron. Astrophys., Vol. 147, No. 1, p. 121 – 126 (1985).

Time behaviour in IUE spectra of the semi–regular cool carbon star, TW Hor (N0; C7,2) is reported. The excitation of the emission lines and their temporal variability (over month, day and possibly hour time–scales) are discussed through acoustic wave heating mechanisms.

122.135 Period changes in β CMa stars.
E. Chapellier.
Astron. Astrophys., Vol. 147, No. 1, p. 135 – 142 (1985).

New observations and a careful analysis of published data of β CMa stars for which secular variations of pulsational periods have been claimed, lead to the conclusion that probably no such variations exist. For the best observed stars (BW Vul, β Cep, σ Sco, δ Cet, 12 Lac, β CMa, and α Vir) a better fit of the data is obtained assuming that positive and negative period jumps occur. Comparison with other types of variable stars shows that this phenomenon is a general characteristic of pulsating stars. That means that the variations in the period are produced by an unexplained physical phenomenon. The lack of secular variations of the period confirms that the β CMa stars are in the end of the core hydrogen burning phase.

122.136   Ca II H and K line variability in the Ap star HD 43819.
S. J. Adelman.
Inf. Bull. Variable Stars, No. 2701, 2 pp. (1985).

122.137   The shock–induced variability of emission profiles in S Car.
D. Gillet, E. Maurice, P. Bouchet, R. Ferlet.
ESO Sci. Prepr., No. 362, 42 pp. (1985). To appear in Astron. Astrophys.

122.138   Untersuchungen zum Lichtwechsel des doppeltperiodischen δ–Cephei–Sterns CO Aurigae.
B. Fuhrmann, R. Luthardt, R. H. Schult.
Mitt. Veränderliche Sterne, Band 10, Heft 4, p. 79 – 85 (1984).
   Conventional Argelander estimations of eight hundred photographic plates of Sonneberg Observatory are made. The light curve elements of the double period cepheid are improved and found to be stable within the observational accuracy.

122.139   Elemente des Mirasterns NSV 14466.
H. Geßner.
Mitt. Veränderliche Sterne, Band 10, Heft 4, p. 94 – 95 (1984).

122.140   Beobachtung des RRa–Veränderlichen 0940 + 10 im Sternbild Leo auf Sonneberger Platten.
B. Fuhrmann.
Mitt. Veränderliche Sterne, Band 10, Heft 4, p. 101 – 102 (1984).

122.141   Visuelle Beobachtungen von R Coronae Borealis.
Mitt. Veränderliche Sterne, Band 10, Heft 4, p. 109 – 110 (1984).

122.142   Spectral and luminous variation of long period red variable stars.
L. Celis S.
Astrophys. Prepr. Ser., No. 9, 28 pp. (1985). Submitted to Astron. J.

122.143   The R CrB star RY Sagittarii as a pulsating star.
T. Lloyd Evans.
Bull. Etoiles Tardives Spectre Particulier, No. 2, p. 3 (1985). Abstract. – Submitted to Mon. Not. R. Astron. Soc.

122.144   Visible and infrared study of L.P.V. (*Long–period variables*) with the 1 m telescope.
P. Bouchet, T. Le Bertre.
Messenger, No. 40, p. 17 – 19 (1985).

122.145   Beobachtung des halbregelmäßigen Veränderlichen 2114 + 46 auf Sonneberger Platten.
B. Fuhrmann.
Mitt. Veränderliche Sterne, Band 10, Heft 5, p. 113 – 115 (1985).

122.146   Maxima von SU Camelopardalis und RR Cancri.
P. Kroll, B. Tietze.
Mitt. Veränderliche Sterne, Band 10, Heft 5, p. 115 – 116 (1985).

122.147   Beobachtung der Mira–Sterne DR Cygni und GO Aurigae auf Sonneberger Platten.
F. Vohla.
Mitt. Veränderliche Sterne, Band 10, Heft 5, p. 116 – 117 (1985).

122.148   Visuelle Lichtkurve von Alpha Herculis.
D. Böhme.
Mitt. Veränderliche Sterne, Band 10, Heft 5, p. 118 (1985).

122.149   Visuelle Beobachtungen des roten Kohlenstoffsterns VY Ursae Maioris 1980 bis 1984 Juni.
M. Rätz, K. Rätz.
Mitt. Veränderliche Sterne, Band 10, Heft 5, p. 120 – 121 (1985).

122.150   Periodenänderung des RR–Lyrae–Sterns IR Monocerotis.
B. Calvo, S. Rößiger.
Mitt. Veränderliche Sterne, Band 10, Heft 5, p. 121 – 122 (1985).
   New elements for the RR Lyrae star IR Mon are presented. The authors' findings suggest that there was an alteration of period in the early fifties.

122.151   A new period for AX UMa.
R. Boninsegna.
GEOS Circ., RR 8, 4 pp. (1985).

122.152   Spectra of flare stars.
B. R. Pettersen, M. K. Tsvetkov.
Inf. Bull. Variable Stars, No. 2660, 2 pp. (1985).

122.153   Correlation between the mean amplitude of the flares and the luminosity of the flare stars.
N. D. Melikian (*N. D. Melikyan*).
Inf. Bull. Variable Stars, No. 2661, 4 pp. (1985).

122.154   The behaviour of FG Vul in 1984.
W. Götz, R. Luthardt.
Inf. Bull. Variable Stars, No. 2664, 2 pp. (1985).

122.155   A possible new flare star in the Pleiades (M45) region.
M. Santangelo.
Inf. Bull. Variable Stars, No. 2665, 1 p. (1985).

122.156   On the type of variability of the star H 42 in the Andromeda galaxy.
G. R. Ivanov, R. Kourtev.
Inf. Bull. Variable Stars, No. 2666, 4 pp. (1985).

122.157   Times of maximum light of the Delta Scuti star CY Aqr.
J. H. Peña, R. Peniche, M. A. Hobart.
Inf. Bull. Variable Stars, No. 2672, 2 pp. (1985).

122.158   Verification of Cepheids in open clusters and associations as distance indicators.
A. Opolski.
Inf. Bull. Variable Stars, No. 2688, 3 pp. (1985).

122.159   Ultraviolet spectroscopy of R Coronae Borealis.
A. V. Holm, L. R. Doherty.
Bull. Am. Astron. Soc., Vol. 16, No. 4, p. 897 (1984). Abstract. – See Abstr. 010.062.

122.160   The recent deep minimum of VX Sagittarii.
R. F. Wing, G. M. Wahlgren.
Bull. Am. Astron. Soc., Vol. 16, No. 4, p. 897 (1984). Abstract. – See Abstr. 010.062.

122.161   Rectification of nine epochs of photometry for the LMC Cepheid HV 5541.
L. P. Connolly.
Bull. Am. Astron. Soc., Vol. 16, No. 4, p. 897 (1984). Abstract. – See Abstr. 010.062.

122.162   Observations of TiO absorption and Balmer–line emission in RV Tauri variables.
G. M. Wahlgren, R. F. Wing, N. M. White.
Bull. Am. Astron. Soc., Vol. 16, No. 4, p. 897 – 898 (1984). Abstract. – See Abstr. 010.062.

122.163 Synoptic observations of BW Vulpeculae.
D. Barry, J. B. Holberg, N. Schneider,
D. Rautenkranz, R. S. Polidan, I. Furenlid, T. Margrave,
M. Alvarez, R. Michel, R. Joyce.
Bull. Am. Astron. Soc., Vol. 16, No. 4, p. 898 (1984). Abstract. –
See Abstr. 010.062.

122.164 Observational studies of Cepheids. III. Catalog of light curve parameters.
T. J. Moffett, T. G. Barnes III.
Bull. Am. Astron. Soc., Vol. 16, No. 4, p. 899 – 900 (1984). Abstract. – See Abstr. 010.062.

122.165 The determination of the Cepheid period–luminosity–color relation: the effect of random photometric errors.
T. E. Lutz.
Bull. Am. Astron. Soc., Vol. 16, No. 4, p. 900 (1984). Abstract. –
See Abstr. 010.062.

122.166 A maximum likelihood investigation of RR Lyrae kinematics and absolute magnitudes.
S. L. Hawley, T. G. Barnes III, W. H. Jefferys.
Bull. Am. Astron. Soc., Vol. 16, No. 4, p. 966 (1984). Abstract. –
See Abstr. 010.062.

122.167 Fourier decompositions for short–period Type II Cepheids.
N. R. Simon.
Bull. Am. Astron. Soc., Vol. 16, No. 4, p. 970 (1984). Abstract. –
See Abstr. 010.062.

122.168 Pulsation masses of globular cluster RR Lyrae variables.
A. N. Cox, R. B. Kidman, S. W. Hodson.
Bull. Am. Astron. Soc., Vol. 16, No. 4, p. 1012 (1984). Abstract. – See Abstr. 010.062.

122.169 Time dependent turbulence in RR Lyrae.
R. F. Stellingwerf, W. Benz.
Bull. Am. Astron. Soc., Vol. 16, No. 4, p. 1012 (1984). Abstract. – See Abstr. 010.062.

122.170 The two micron spectrum of R Scuti.
D. Mozurkewich, R. D. Gehrz, K. H. Hinkle,
D. L. Lambert.
Bull. Am. Astron. Soc., Vol. 16, No. 4, p. 1013 (1984). Abstract. – See Abstr. 010.062.

122.171 The search, discovery, and use of Cepheid variables as distance indicators: past procedures and future strategies.
B. F. Madore, W. L. Freedman.
Astron. J., Vol. 90, No. 6, p. 1104 – 1112 (1985).
  Studies of the periods and magnitudes of Cepheids in external galaxies are critically reviewed in the context of distance determinations. An important property of Cepheid light curves is identified, namely, that despite differing amplitudes and periods, the decline rate of each Cepheid as a function of phase is approximately constant. Making use of this property, and the tight correlation of color and luminosity with phase, a strategy for future distance–scale determinations is suggested.

122.172 Discovery of flare activity in G 119–62.
B. R. Pettersen.
Astron. Astrophys., Vol. 147, No. 2, p. 328 – 330 (1985).
  Seven flares were recorded in 3h of photometric monitoring in the U-filter, of the star G 119–62. Significant variability in the strength of hydrogen Balmer lines was also detected on a timescale of one day. Photometric and spectroscopic criteria indicate that G 119–62 has a spectral type close to dM4.5e.

122.173 Investigation of the pulsation instability of the star AR Her.
A. N. Rudenko, B. N. Firmanyuk.
Mater. Konf. mol. uchenykh Odess. astron. nauchn.–proizv.
akad.–univ. kompleksa, Odessa, 15 fev. 1983. Odessa, p. 57 – 62

(1983). In Russian. Abstr. in Ref. Zh., 51. Astron., 5.51.459
(1985).

122.174 Variation of the Hα line profile in the spectrum of DH Pegasi.
G. A. Garbuzov.
Mater. Konf. mol. uchenykh Odess. astron. nauchn.–proizv.
akad.–univ. kompleksa, Odessa, 15 fev. 1983. Odessa, p. 2 – 5
(1983). In Russian. Abstr. in Ref. Zh., 51. Astron., 5.51.460
(1985).

122.175 Dramatic changes in the polarization of AR Pup.
A. V. Raveendran, N. Kameswara Rao,
M. R. Deshpande, U. C. Joshi, A. K. Kuleshrestha.
Inf. Bull. Variable Stars, No. 2713, 2 pp. (1985).

122.176 A Cepheid–like variable in the Andromeda galaxy with an extremely long period.
G. R. Ivanov.
Inf. Bull. Variable Stars, No. 2729, 3 pp. (1985).

122.177 Flare stars in the Pleiades.
K. P. Tsvetkova, A. G. Tsvetkova, M. K. Tsvetkov.
Inf. Bull. Variable Stars, No. 2730, 2 pp. (1985).

122.178 Behaviour of KR Aur in the season 1984/85.
W. Götz.
Inf. Bull. Variable Stars, No. 2733, 2 pp. (1985).

122.179 Equivalent widths of absorption lines in the spectrum of the yellow semiregular variable star VW Dra. I. The $\lambda\lambda 4000 – 5000$ Å region.
S. M. Andrievskij, E. N. Makarenko.
Odes. univ. Odessa, 36 pp. (1984). In Russian. Abstr. in Ref. Zh.,
51. Astron., 6.51.429 (1985).

122.180 U Aquarii.
IAU Circ., No. 4073 (1985).

122.181 TT Arietis.
IAU Circ., No. 4077 (1985).

122.182 SiO observations of long–term Mira variables.
J. D. White, A. Leong.
Bull. Am. Astron. Soc., Vol. 17, No. 1, p. 511 (1985). Abstract. –
See Abstr. 010.064.

122.183 Photometry and frequency analysis of line profile variables.
L. A. Balona, C. A. Engelbrecht.
Mon. Not. R. Astron. Soc., Vol. 214, No. 4, p. 559 – 574 (1985).
  The authors have monitored three line profile variables or 53 Per stars and one broad line star for light variations. The observations comprise about 750 measurements of each star and its comparison spread over 28 nights in two seasons. From these results the authors detect light variability in all programme stars with semi–amplitudes ranging from 0.8 to 8.0 millimag. The coherent periods detected do not agree with those found from the line profile variability. The authors discuss possible reasons for this discrepancy.

122.184 A photometric and spectrographic study of the variable star HD 94033.
D. H. McNamara, K. G. Budge.
Publ. Astron. Soc. Pac., Vol. 97, No. 590, p. 322 – 327 (1985).
  New photometric $(uvby\beta)$ and spectrographic observations of the dwarf Cepheid HD 94033 are described. The light amplitude, $0\overset{m}{.}80$ in V, and radial–velocity amplitude, $2K = 52$ km s^{-1}, are unusually large. The mean effective temperature and surface gravity are $\langle T_{eff}\rangle = 7650$K and $\langle \log g\rangle = 4.14$. The mean radial velocity, $+268$ km s^{-1}, and low–metal abundance, [Fe/H] $\sim$ -2.4, indicate that the star is a very extreme Population II star. Standard evolutionary theory indicates the mass is 0.9 $M_{\odot}$ and the age is 6.5×10^9 years. In many respects the star is similar to GD 428.

122.185   Peculiar emission lines in the spectrum of the flare star YZ Canis Minoris.
B. M. Haisch, M. S. Giampapa.
Publ. Astron. Soc. Pac., Vol. 97, No. 590, p. 340 – 344 (1985).

During a series of spectral scans in search of flares on the dMe star YZ CMi the authors observed prominent chromospheric emission lines of Hγ λ4340, Hδ λ4100, Ca II H λ3968 blended with Hϵ λ3970, Ca II K λ3934, and Hζ λ3889. During one five-minute scan the authors recorded the following peculiar transient phenomena: (1) the appearance of an exceptionally strong line at $\sim$ λ4007; (2) the simultaneous appearance of another prominent feature at $\sim$ λ4276; and (3) a change in the ratio of (Ca II H + Hϵ)/Ca II K. The authors discuss the possible origin of this unusual transient spectrum.

122.186   Near–infrared photometry of unidentified IRC stars. III. The Mira variables of spectral type M10.
G. W. Lockwood.
Astrophys. J., Suppl. Ser., Vol. 58, No. 1, p. 167 – 177 (1985).

Near–infrared magnitudes and spectral types are given for 38 very red sources from the *Two–Micron Sky Survey*, all of which are Mira variables that attain spectral types later than M9.5 at minimum light. These stars on the average exhibit significantly larger amplitudes at 1 μm, longer periods, later spectral types, and redder near–infrared colors than the few well–known visually bright M10 Mira variables. Photometry is also tabulated for 107 additional IRC sources earlier than M9.5.

122.187   CORAVEL radial–velocity curves for Cepheids in the Large Magellanic Cloud.
M. Imbert, J. Andersen, A. Ardeberg, C. Bardin, W. Benz, H. Lindgren, N. Martin, E. Maurice, M. Mayor, B. Nordström, L. Prevot.
Stellar radial velocities, p. 375 – 380 (1985). – See Abstr. 012.095 (IAU Colloq. No. 88).

122.188   Measures of RR Lyrae stars.
G. Burki, G. Meylan, M. Mayor.
Stellar radial velocities, p. 399 – 403 (1985). – See Abstr. 012.095 (IAU Colloq. No. 88).

Radial velocity curves and Geneva photometry are used to derive mean stellar radii for the RR Lyrae stars RR Cet, DX Del, BS Aqr, and DY Peg.

122.189   Some properties of S Mira variables.
M. S. Vardya.
Cool stars with excesses of heavy elements, p. 105 – 109 (1985). – See Abstr. 012.101.

Some correlations between period of pulsation, spectral type at maximum, mean difference in visual magnitude between maxima and minima, intensity of Tc I line, ratio of the number of days from minimum to the next following maximum to the period, and infrared colour index [11 μ]–[3.5 μ] are described for S Mira variables.

122.190   Carbon isotope ratios in oxygen rich Mira and SRa variables.
K. H. Hinkle, W. W. G. Scharlach.
Cool stars with excesses of heavy elements, p. 255 – 259 (1985). – See Abstr. 012.101.

Carbon isotope ratios have been measured in oxygen rich Mira and SRa variables brighter than $K = +0.6$. Results are now available for 34 stars (25 Miras and 9 SRa variables). The longer period Miras have a ^{12}C/^{13}C ratio about three times larger than the shorter period Miras. For Miras, period, Tc abundance and ^{12}C/^{13}C are shown to be functionally related.

122.191   Excitation of some La II, Gd II and V I lines by the fluorescence mechanism in the spectra of the long–period variable o Ceti.
S. Grudzińska.
Cool stars with excesses of heavy elements, p. 267 – 268 (1985). – See Abstr. 012.101.

In the near infrared spectrum of the star o Ceti some emission lines of La II and Gd II were found. It is possible that some lines of La II are excited by Hβ and some of Gd II by Hϵ. The lines of V I from the upper level $z^4 P^0_{1/2}$ are excited by the line 4412.25 of Cr I.

122.192   Cepheid masses and cepheid binaries.
E. Böhm–Vitense.
Bull. Am. Astron. Soc., Vol. 17, No. 2, p. 559 (1985). Abstract. – See Abstr. 010.065.

122.193   IUE spectra of three dwarf cepheids.
C. R. Sturch, E. Nelan, C.–C. Wu.
Bull. Am. Astron. Soc., Vol. 17, No. 2, p. 559 (1985). Abstract. – See Abstr. 010.065.

122.194   A comparison of theoretical and observed RR Lyrae light curves via Fourier decomposition.
N. R. Simon.
Bull. Am. Astron. Soc., Vol. 17, No. 2, p. 559 (1985). Abstract. – See Abstr. 010.065.

122.195   On the discrepancy between the observed and theoretical light curves in RR Lyrae variables.
T. Aikawa, N. R. Simon.
Bull. Am. Astron. Soc., Vol. 17, No. 2, p. 559 (1985). Abstract. – See Abstr. 010.065.

122.196   Ultraviolet observations of the system containing the Cepheid SU Cyg.
N. R. Evans, E. Böhm–Vitense, C. T. Bolton.
Bull. Am. Astron. Soc., Vol. 17, No. 2, p. 559 – 560 (1985). Abstract. – See Abstr. 010.065.

122.197   Period–luminosity–abundance relations for galactic Cepheids.
O. J. Eggen.
Bull. Am. Astron. Soc., Vol. 17, No. 2, p. 585 (1985). Abstract. – See Abstr. 010.065.

122.198   Spectroscopy of anomalous Cepheids in the Sculptor dwarf galaxy.
H. A. Smith.
Bull. Am. Astron. Soc., Vol. 17, No. 2, p. 597 (1985). Abstract. – See Abstr. 010.065.

122.199   RR Lyrae stars of long period and high amplitude.
E. P. Belserene.
Bull. Am. Astron. Soc., Vol. 17, No. 2, p. 597 (1985). Abstract. – See Abstr. 010.065.

122.200   Peculiar transients in the light curve of the dwarf Cepheid XX Cygni.
A. C. Sadun, M. Ressler.
Bull. Am. Astron. Soc., Vol. 17, No. 2, p. 597 (1985). Abstract. – See Abstr. 010.065.

122.201   IRAS views of R Coronae Borealis stars.
B. E. Schaefer.
Bull. Am. Astron. Soc., Vol. 17, No. 2, p. 598 (1985). Abstract. – See Abstr. 010.065.

122.202   The peculiar spectrum of HQ Mon.
G. M. Wahlgren, R. F. Wing, R. H. Kaitchuck, S. R. Baird, D. J. MacConnell, D. W. Dawson.
Bull. Am. Astron. Soc., Vol. 17, No. 2, p. 599 (1985). Abstract. – See Abstr. 010.065.

122.203   Chemical abundances in Cepheid stars.
M. D. Suran.
Top. Astrophys. Astron. Space Sci., Vol. 1, p. 1 – 24 (1985).

A new method for the determination of mass and chemical composition of Cepheids is presented. Part of the mass discrepancy is shown to be removed by the change of chemical compositions of Cepheids. A general relation between abundances and galactocentric radius is obtained.

122.204   Das Minimum 1983/84 von R Coronae Borealis.
H. Grzelczyk.
BAV Mitt., 34. Jahrg., Nr. 2, p. 73 – 75 (1985).

122.205   Beobachtungen an δ–Cephei Sternen 1984.
J. Schmidt.
BAV Mitt., 34. Jahrg., Nr. 2, p. 75 – 79 (1985).

122.206   Report of IAU Commission 27: Variable stars (*Etoiles variables*).
N. H. Baker.
Trans. IAU, Vol. XIXA, p. 277 – 311 (1985). – See Abstr. 003.046.

122.207   Two big flares of UV Cet on November 9, 1982.
K. P. Panov.
Dokl. Bolg. Akad. Nauk, Tome 37, No. 5, p. 557 – 559 (1984).

122.208   Spectroscopic studies of classical Cepheids.
R. P. Gauthier.
Diss. Abstr. Int., Sect. B, Vol. 45, No. 2, p. 584 (1984). Thesis, University of Toronto (1983).

122.209   UBVRI observations of Magellanic supergiants and Cepheids.
G. R. Grieve.
Diss. Abstr. Int., Sect. B, Vol. 45, No. 2, p. 584 (1984). Thesis, University of Toronto (1983).

122.210   Studies of extragalactic RR Lyrae stars.
J. M. Nemec.
Diss. Abstr. Int., Sect. B, Vol. 45, No. 5, p. 1501 (1984). Thesis, University of Washington, 330 pp. (1984). Order No. DA8419179.

122.211   The spectra of Mira variable stars.
R. A. Crowe.
Diss. Abstr. Int., Sect. B, Vol. 45, No. 9, p. 2955 (1985). Thesis, University of Toronto (1984).

122.212   The color excesses of classical Cepheids and the structure of our Galaxy.
C. Kim.
Diss. Abstr. Int., Sect. B, Vol. 45, No. 9, p. 2956 (1985). Thesis, Brigham Young University, 114 pp. (1984). Order No. DA8427784.

122.213   Radius and absolute magnitude of a Cepheid variable.
A. J. Wesselink.
Supernovae as distance indicators, p. 166 – 170 (1985). – See Abstr. 012.107.
A brief summary of our present knowledge of the diameters and luminosities of the Cepheid variables is presented.

122.214   Photometric observations of BW Vul.
J. H. Jung, S.–W. Lee.
J. Korean Astron. Soc., Vol. 18, No. 1, p. 1 – 13 (1985).
The authors present the data of photoelectric photometric observations of BW Vul carried out for four nights during the period of 1982 – 1984. Using all the published data, a new ephemeris of maximum time is derived, in which the period of light variation is $P = 0^{d}.20102977$ and its increasing rate is 2.2 sec/century.

122.215   The energy spectrum of flares of UV Cet–type stars and physical meaning of several statistical characteristics of these stars.
R. E. Gershberg.
Astrofizika, Tom 22, Vyp. 3, p. 531 – 544 (1985). In Russian. English translation in Astrophysics, Vol. 22, No. 3.
Accounting the observed power character of the energy spectrum of flares of UV Cet–type stars, several statistical characteristics of these stars are considered. It is shown that the mean amplitude of flares is mainly determined with the amplitude of the faintest flare that can be registered at the star under consider-
ation and therefore – contrary to tradition – the mean flare amplitude cannot be used as a measure of flare activity of the star. On the basis of the Catalogue of flare stars in the Pleiades by Haro, Chavira and Gonzalez (1982), the luminosity function of these stars is constructed. Using this function and the revealed dependence of flare mean frequencies on stellar absolute magnitudes, the distribution of flare stars in the Pleiades along flare mean frequencies is constructed.

122.216   On the problem of duplicity of flare stars.
Eh. S. Parsamyan.
Astrofizika, Tom 22, Vyp. 3, p. 633 – 635 (1985). In Russian. English translation in Astrophysics, Vol. 22, No. 3.
It has been shown that the percentage of visual binary stars among flare stars is more than 3 times higher than in randomly selected areas.

122.217   The discovery of rapid oscillations in the Ap star HD 134214.
T. J. Kreidl.
Inf. Bull. Variable Stars, No. 2739, 2 pp. (1985).

122.218   New flare star SVS 2559 Persei.
W. Wenzel.
Inf. Bull. Variable Stars, No. 2740, 1 p. (1985).

122.219   Photoelectric observations of the flare star AD Leo.
V. Reglero, J. J. Fuensalida, M. J. Arevalo, J. L. Ballester.
Inf. Bull. Variable Stars, No. 2752, 3 pp. (1985).

122.220   Maia variables and upper–main–sequence phenomena.
B. J. McNamara.
Proceedings of the Southwest Regional Conference for Astronomy and Astrophysics, Vol. 10, p. 63 (1985). Abstract. – See Abstr. 012.125.

Sky catalogue 2000.0. Volume 2: Double stars, variable stars and nonstellar objects.
See Abstr. 002.019.

General Catalogue of Variable Stars.
See Abstr. 002.027.

Eddington's contribution to astrophysics and pulsation theory.
See Abstr. 004.109.

Activité de l'A.F.O.E.V. en 1984.
See Abstr. 010.102.

On a possibility of observations of variable stars in visual binary systems with the area scanning technique.
See Abstr. 036.105.

The Feinheit method: a phase–independent scheme for calibrating the period–luminosity relation of classical Cepheids.
See Abstr. 036.134.

Spectrophotometry from uncalibrated Schmidt plates.
See Abstr. 036.154.

Cross–correlation spectroscopy using CORAVEL.
See Abstr. 036.186.

Statistical properties of phase dispersion minimization period finding techniques.
See Abstr. 036.209.

Stochastic electron acceleration in stellar coronae.
See Abstr. 064.009.

A periodic shock wave model for Mira variable atmospheres.
See Abstr. 064.064.

A dynamical instability as a driving mechanism for stellar oscillations.
See Abstr. 065.008.

The rapidly oscillating Ap stars.
See Abstr. 065.018.

Current problems in horizontal–branch theory: some implications for RR Lyrae variables.
See Abstr. 065.027.

Observable quantities of nonradial pulsations in the presence of slow rotation.
See Abstr. 065.055.

Horizontal–branch sequences for globular cluster RR Lyrae stars.
See Abstr. 065.069.

The effect of opacity on theoretical models of RR Lyrae stars.
See Abstr. 065.071.

Hydrodynamic models for population–II cepheids.
See Abstr. 065.075.

Radial pulsation of less–massive yellow supergiants. I. 1 $M_\odot$, 3200 $L_\odot$ models.
See Abstr. 065.076.

Stellar evolution and binaries.
See Abstr. 065.083.

The s–process within the R Coronae Borealis star U Aquarii.
See Abstr. 065.093.

Report of IAU Commission 35: Stellar constitution (*Constitution des étoiles*).
See Abstr. 065.096.

Thermal effects in stellar pulsations.
See Abstr. 065.098.

Proper motions of a group of R Coronae Borealis type stars.
See Abstr. 111.017.

Betelgeuse at maximum luminosity.
See Abstr. 112.015.

First spectroscopic observations with the 14–m radio telescope at the Centro Astronomico de Yebes: SiO masers in evolved stars.
See Abstr. 112.022.

OH emission from Mira variables, infrared stars and molecular clouds.
See Abstr. 112.026.

Time variations and shell sizes of OH masers in late–type stars.
See Abstr. 112.027.

Circumstellar dust distributions around variable stars.
See Abstr. 112.032.

A multitransitional study of linear polarization in SiO maser emission.
See Abstr. 112.051.

Characteristics and interpretation of the photometric variability of Eta Carinae and its nebula.
See Abstr. 112.059.

Variable mass loss in P Cygni.
See Abstr. 112.062.

Polarization properties and time variations of the SiO maser emission of W Hydrae.
See Abstr. 112.063.

Spectroscopy of the winds from Hubble–Sandage stars in M31 and M33.
See Abstr. 112.073.

The circumstellar H_2O maser emission associated with four late–type stars.
See Abstr. 112.074.

The 2200 Å circumstellar dust absorption feature in the spectra of three bright RV Tauri stars.
See Abstr. 112.075.

Mass loss from evolved stars. III. Mass loss rates for fifty stars from CO $J = 1$–0 observations.
See Abstr. 112.093.

Circumstellar material and the light variations of RV Tauri stars.
See Abstr. 112.094.

Mass loss from evolved stars. IV. The dust–to–gas ratio in the envelopes of Mira variables and carbon stars.
See Abstr. 112.127.

Mass loss from evolved stars. V. Observations of the ^{12}CO and ^{13}CO $J = 1$–0 lines in Mira variables and carbon stars.
See Abstr. 112.128.

A search for low–harmonic pulsation in the Ap star CS Vir.
See Abstr. 113.002.

A search for rapid oscillations in HD 92664.
See Abstr. 113.003.

Photometry of the extreme helium star BD + 1°4381.
See Abstr. 113.009.

Photometric properties of peculiar red giants.
See Abstr. 113.062.

A new photometric system for monitoring carbon star variability.
See Abstr. 113.064.

The infrared spectra of red variables: II. The SC and CS stars.
See Abstr. 114.142.

Niobium in R And (S6, 6e) and HR 1105 (S5, 3).
See Abstr. 114.155.

Oxygen in intermediate–mass supergiants.
See Abstr. 114.158.

The bolometric luminosities of type II OH/IR sources.
See Abstr. 115.014.

Coronal heating and stellar flares.
See Abstr. 116.032.

Variation of magnetic field strength of V474 Mon.
See Abstr. 116.046.

Blue companions of Cepheids.
See Abstr. 117.006.

The essence of binary star coalescence – a coordinated ground–based photometric and IUE ultraviolet study of FK Comae.
See Abstr. 117.010.

HM Sagittae and V1016 Cygni in 1982 – 83: unforeseen variations in the IR spectrum of 1016 Cyg.
See Abstr. 117.031.

Variations in the variability of RR Tel.
See Abstr. 117.045.

Electrophotometric and spectroscopic observations of the peculiar star CH Cygni during 1980 – 1981.
See Abstr. 117.101.

The R Aquarii system at optical and radio wavelengths.
See Abstr. 117.108.

Infrared observations of variable stars. 2. Spectrophotometry of CH Cygni.
See Abstr. 117.167.

Photometric variability of stars belonging to the Trapezium–type systems.
See Abstr. 118.007.

Has α Vir ever stopped pulsating?
See Abstr. 120.005.

The ellipsoidal variable ψ Orionis and the β Cephei instability region.
See Abstr. 120.009.

A search for companions to irregular variables of type S.
See Abstr. 120.040.

The UV extinction of HR 5999.
See Abstr. 121.006.

Identification charts and secondary UBVR standards to observe antiflare stars.
See Abstr. 123.003.

Classification of variable stars in the light of current conceptions of their evolution.
See Abstr. 123.007.

Relationship of classical novae to other eruptive variables.
See Abstr. 124.010.

Pre–existing dust shells around type–I supernovae.
See Abstr. 125.077.

The effective temperature of Wolf 485A and the statistics of ZZ Ceti stars.
See Abstr. 126.033.

Two–phase ultraviolet spectrophotometry of the pulsating white dwarf ZZ Piscium.
See Abstr. 126.036.

Discovery of oxygen in the PG 1159 degenerate stars: a direct evolutionary link to O VI planetary nebula nuclei and confirmation of pulsation theory.
See Abstr. 126.067.

A measurement of secular evolution in the pre–white dwarf star PG 1159–035.
See Abstr. 126.069.

Discovery of oxygen in the PG1159 degenerate stars: a direct evolutionary link to O VI planetary nebula nuclei and confirmation of pulsation theory.
See Abstr. 126.082.

Strömgren photometry of ZZ Ceti and other DA white dwarfs.
See Abstr. 126.089.

Optical absorption from the high velocity neutral hydrogen complex C in the spectrum of the RR Lyrae star BT Draconis.
See Abstr. 131.273.

The A–1 survey of fast X–ray transients.
See Abstr. 142.007.

A search for OB associations near long–period Cepheids. III. U Carinae, XZ Carinae, QY Centauri, VX Crucis, and AA Normae.
See Abstr. 152.004.

Mapping of the Beta Cepheid instability strip: the cluster NGC 6231.
See Abstr. 153.003.

A systematic search for members of the Hyades Supercluster. V. The red giants.
See Abstr. 153.009.

On the distance to the open cluster Lyngå 6.
See Abstr. 153.014.

CCD photometry of Galactic clusters containing Cepheid variables – I. Lyngå 6.
See Abstr. 153.018.

CCD photometry of galactic clusters containing Cepheid variables. II. NGC 6067.
See Abstr. 153.019.

A systematic search for members of the Hyades Supercluster. IV. The metallic–line stars and ultrashort–period Cepheids.
See Abstr. 153.039.

Reddening, distance modulus and age of the globular cluster NGC 6121 (M4) from the properties of RR Lyrae variables.
See Abstr. 154.004.

Two mid–giant branch variable stars im M15.
See Abstr. 154.010.

Variable stars in the globular cluster M13. III. Short–period variables.
See Abstr. 154.014.

Period variations of RR Lyrae–type variable stars in the globular cluster M22.
See Abstr. 154.015.

Observations of variable stars in the globular cluster M15.
See Abstr. 154.016.

The variable stars in the globular cluster NGC 6569.
See Abstr. 154.040.

Globular clusters and RR Lyrae variables.
See Abstr. 154.092.

RR Lyrae stars and the distant galactic halo: distribution, chemical composition, kinematics, and dynamics.
See Abstr. 155.078.

BVI reddenings of Magellanic Cloud Cepheids.
See Abstr. 156.003.

Bright red variables of large amplitude in the Magellanic Clouds.
See Abstr. 156.004.

New distances to the Magellanic Clouds.
See Abstr. 156.009.

A sample of long–period variables in the bar of the Large Magellanic Cloud and evidence for a recent burst of star formation.
See Abstr. 156.014.

The distance to M33 based on BVRI CCD observations of Cepheids.
See Abstr. 157.061.

The distances to nearby galaxies from near–infrared photometry of Cepheids.
See Abstr. 157.062.

Cepheids in the dwarf galaxies Sextans A, Sextans B, and WLM.
See Abstr. 157.151.

The distance to M31 from H(1.6 μm) photometry of its Cepheids.
See Abstr. 157.175.

The brightest stars in nearby galaxies. V. Cepheids and the brightest stars in the dwarf galaxy Sextans B compared with those in Sextans A.
See Abstr. 157.198.

On cepheids in the galaxy M31.
See Abstr. 157.250.

123 Variable Stars (Surveys, Lists of Observations, Charts, etc.)

123.001   **Tableaux des observations reçues à l'AFOEV pendant les mois d'octobre, novembre et décembre 1984.**
Bull. Assoc. Fr. Obs. Etoiles Variables, No. 31, p. 12 – 49 (1985).

123.002   **CPD –59°2857: a new red variable star.**
E. Gosset, D. Hutsemekers, J. Surdej.
Publ. Astron. Soc. Pac., Vol. 97, No. 587, p. 67 – 71 (1985).
Two (C3 and C4) of the four comparison stars monitored during the photometric study of the peculiar emission–line star GG Car have been found to be variable. Available photometric and spectroscopic data allow to classify C4 $\equiv$ CPD –59°2857 as a late–type giant (M3 – 4 III) belonging to the red variable stars of the semiregular or irregular type.

123.003   **Identification charts and secondary UBVR standards to observe antiflare stars.**
A. F. Pugach, G. U. Koval'chuk.
Perem. Zvezdy, Tom 22, No. 1, p. 9 – 23 (1983). In Russian.
A photoelectric UBVR magnitudes catalogue and charts of 30 check stars for 15 antiflare stars are given. A procedure of transformation of instrumental magnitudes to standard ones is briefly described.

123.004   **New variable stars in the region of the association T4 Cygni.**
V. Satyvoldiev.
Perem. Zvezdy, Tom 22, No. 1, p. 37 – 38 (1983). In Russian.
Data on 18 new variable stars discovered by the author are reported as a supplement to a previously published list of new variable stars by Satyvoldiev, 1979. Approximate coordinates and magnitudes are given in a table. Charts are presented on a figure.

123.005   **Variable stars.**
S. Dunlop.
J. Br. Astron. Assoc., Vol. 95, No. 3, p. 118 – 121 (1985).

123.006   **A search for southern hemisphere long–period variable stars with close companions.**
J. E. Meyer, S. J. Shawl.
Publ. Astron. Soc. Pac., Vol. 97, No. 589, p. 272 – 273 (1985).
A null photographic search for hot companions to 632 southern long–period variable stars is reported.

123.007   **Classification of variable stars in the light of current conceptions of their evolution.**
P. N. Kholopov.
Astrophys. Space Phys. Rev., Vol. 3, p. 97 – 121 (1984). – See Abstr. 003.019. Revised and extended English translation of 34.122.131.
Contents: (1). Introduction. Reasons for light variability. (2). Young variable stars. (3). Variable stars of intermediate age. (4). Old variable stars. (5). Close eclipsing binary systems. (6). Explosive and nova–like close binary systems. (7). Close binary X–ray systems. (8). Conclusions.

123.008   **Information on photoelectric observations of variable stars deposited at the Odessa Astronomical Observatory.**
E. N. Makarenko.
Astron. Tsirk., No. 1340, p. 7 – 8 (1984). In Russian.

123.009   **Forty suspected variable stars.**
F. C. Fekel, D. S. Hall.
I.A.P.P.P. Commun., No. 20, p. 37 – 40 (1985).

123.010   **List of variable stars discovered in the USSR and preliminary SVS designations.**
Astron. Tsirk., No. 1350, p. 7 – 8 (1984). In Russian.

123.011   **1985 annual predictions of maxima and minima of long period variables.**
J. A. Mattei.
Am. Assoc. Variable Star Obs. Bull., No. 48, 14 pp. (1985).
Predicted dates for 1985 of maxima and minima of 564 long period and semiregular variables are given in this Bulletin.

123.012   **Mehrfarben–Beobachtungen von V18, V19, V20, V21, V22, V23 im Kugelhaufen M3.**
I. Meinunger.
Mitt. Veränderliche Sterne, Band 10, Heft 4, p. 89 – 93 (1984).

123.013   **Bearbeitung von 50 Veränderlichen (Feld γ Aquilae, Teil II).**
H. Geßner.
Mitt. Veränderliche Sterne, Band 10, Heft 4, p. 103 (1984).

123.014   **Beobachtungsergebnisse des Arbeitskreises "Veränderliche Sterne" im Kulturbund der DDR (Teil XI).**
Mitt. Veränderliche Sterne, Band 10, Heft 4, p. 104 – 109 (1984).

123.015   **Some new spectral classifications for named and suspected variable stars.**
C. B. Stephenson.
Inf. Bull. Variable Stars, No. 2657, 2 pp. (1985).

123.016   **Is CD Herculis = BD + 46°2308 variable?**
R. Luthardt.
Inf. Bull. Variable Stars, No. 2674, 2 pp. (1985).

123.017   **Search for peculiar stars in the region of the association Cassiopeia OB14.**
T. Radoslavova.
Inf. Bull. Variable Stars, No. 2679, 2 pp. (1985).

123.018 **Automatic photoelectric telescope: second and third quarter 1984 observations.**
L. J. Boyd, R. M. Genet, D. S. Hall.
Inf. Bull. Variable Stars, No. 2680, 2 pp. (1985).

123.019 **The 67ᵗʰ name–list of variable stars.**
P. N. Kholopov, N. N. Samus', E. V. Kazarovets, N. B. Perova.
Inf. Bull. Variable Stars, No. 2681, 32 pp. (1985).

123.020 **Discovery of variability in HR 9024.**
J. L. Hopkins, L. J. Boyd, R. M. Genet, D. S. Hall.
Inf. Bull. Variable Stars, No. 2684, 4 pp. (1985).

123.021 **Confirmation of the 525 day period variability for HD 2453.**
F. A. Catalano, S. Vaccari.
Inf. Bull. Variable Stars, No. 2687, 2 pp. (1985).

123.022 **HD 212936 – eine spannende Veränderlichenbeobachtung.**
M. Pfleiderer, S. Marx.
Mitt. Astron. Ges., Nr. 63, p. 201 – 202 (1985). – See Abstr. 012.063.

123.023 **On the nature of M28 V7.**
S. F. Anderson, B. Margon.
Bull. Am. Astron. Soc., Vol. 16, No. 4, p. 898 (1984). Abstract. – See Abstr. 010.062.

123.024 **The Maia variables – a true class of variable stars?**
B. J. McNamara.
Bull. Am. Astron. Soc., Vol. 16, No. 4, p. 970 (1984). Abstract. – See Abstr. 010.062.

123.025 **Tableaux des observations reçues à l'AFOEV en janvier, février et mars 1985.**
Bull. Assoc. Fr. Obs. Etoiles Variables, No. 32, p. 14 – 43 (1985).

123.026 **Note about the variability of HD 23838.**
D. Hoffleit.
Inf. Bull. Variable Stars, No. 2695, 1 p. (1985).

123.027 **Comparison stars for some eclipsing binaries.**
E. C. Olson.
Inf. Bull. Variable Stars, No. 2706, 3 pp. (1985).

123.028 **Étoiles à helium faible photométriquement variables.**
P. Renson.
Inf. Bull. Variable Stars, No. 2708, 4 pp. (1985).

123.029 **Photoelectric observations of γ Cas, X Per, and BU Tau.**
D. Böhme.
Inf. Bull. Variable Stars, No. 2723, 2 pp. (1985).

123.030 **Variable star in Orion.**
IAU Circ., No. 4035 (1985).

123.031 **New variable stars found in southern open clusters.**
E. Lapasset, J. J. Clariá.
Inf. Bull. Variable Stars, No. 2749, 5 pp. (1985).

I.A.U. archives of unpublished observations of variable stars: 1981 – 84 data.
See Abstr. 002.013.

General Catalogue of Variable Stars. The Fourth Edition containing information on variable stars discovered and designated till 1982. Volume 1. Constellations Andromeda – Crux.
See Abstr. 002.023.

General Catalogue of Variable Stars.
See Abstr. 002.027.

La quatrième édition du G.C.V.S.
See Abstr. 002.085.

Activité de l'A.F.O.E.V. en 1984.
See Abstr. 010.102.

Spektroskopische Veränderlichenbeobachtung.
See Abstr. 036.150.

108th list of minima of eclipsing binaries.
See Abstr. 119.044.

109th list of minima of eclipsing binaries.
See Abstr. 119.046.

RX Cas, observations 1982 – 1984.
See Abstr. 119.100.

Variable stars in the globular cluster M92.
See Abstr. 154.013.

124 Novae

124.001   Hydrodynamic simulations of recurrent novae.
S. Starrfield, W. M. Sparks, J. W. Truran, E. M. Sion.
NASA Conf. Publ., NASA CP–2349, p. 313 – 317 (1984). – See Abstr. 012.001.

The authors have continued their simulations of the 1979 outburst of the recurrent nova U Scorpii. They have used a Lagrangian, hydrodynamic computer code which now incorporates accretion in the evolution to the outburst. They computed three evolutionary sequences in an attempt to understand the very rapid outburst and short recurrence time of this most unusual nova. It is now possible to reproduce the CNO composition of the ejected material, the light curve, the amount of ejected material, and the kinetic energy of the ejecta.

124.002   IRAS observations of novae.
A. Evans.
Observatory, Vol. 105, No. 1064, p. 6 – 7 (1985). Abstract. – See Abstr. 010.721.

124.003   Les résidus d'explosions de novae.
Y. Andrillat.
Astronomie, Vol. 99, p. 33 – 35 (1985).

124.004   The recurrent nova U Scorpii in post–outburst quiescence.
D. A. Hanes.
Mon. Not. R. Astron. Soc., Vol. 213, No. 2, p. 443 – 449 (1985).

A low–dispersion spectrum is presented for the recurrent nova U Scorpii after its return to post–outburst quiescence. The measured brightness and observed spectral features are the same as when it was observed serendipitously before the 1979 outburst. The spectrum shows He II emission lines on a nearly featureless continuum; as reported earlier by Williams et al. (1981), hydrogen is seen neither in emission nor in absorption. A spectral type of G0 ($\pm$5) V is inferred from identified absorption features (Ca H and K lines, the Mg b triplet) and JHK infrared photometry; however, it is not possible to rule out a giant star as the companion object. If it is indeed a dwarf, U Sco lies at a distance of $\sim 3.5 \pm 1.5$ kpc.

124.005   Die Helligkeitsentwicklung der Novae 1980 bis 1984.
R. Lukas.
Sterne Weltraum, 24. Jahrg., Nr. 1, p. 37 (1985).

124.006   Ergebnisse der Novasuche auf La Silla.
R. Lucas.
Sterne Weltraum, 24. Jahrg., Nr. 2, p. 105 (1985).

124.007   Outer atmospheric structures of classical novae during and after outburst.
R. Stalio.
Effects of variable mass loss on the local stellar environment, p. 59 – 77 (1984). – See Abstr. 012.028.

In an earlier article (Stalio, 1983) the velocity structure of the novae atmospheres during the outburst phase (from when the nova is detected to when it comes back to its pre–nova state) has been described in some detail. Here the author goes a step further: he discusses the kind of time dependent mass ejection that can produce the observed outburst and how a sudden enhancement of the mass flux determines the structure of the extended envelope during the whole episode.

124.008   Observations of two novae in M87.
C. Pritchet, S. van den Bergh.
Astrophys. J., Lett. Ed., Vol. 288, No. 2, p. L41 – L43 (1985). With plate L2.

A $2\rlap{.}'2 \times 3\rlap{.}'3$ field in the eastern part of the Virgo Cluster giant elliptical M87 = NGC 4486 was observed with a CCD detector for 60 minutes on each of five nights. The limiting magnitude of these observations was $B \approx 25.5$. Two novae were discovered: the first of these reached $B(\max) \lesssim 24.9$ on, or before, 1984 February 24, whereas the second attained $B(\max) \lesssim 24.1$ on, or slightly after, 1984 February 28. The present observations indicate that the distance modulus of M87 lies in the range $30.4 < (m-M)_B < 32.4$.

124.009   Nova shells. II. Calibration of the distance scale using novae.
J. G. Cohen.
Astrophys. J., Vol. 292, No. 1, p. 90 – 103 (1985).

Eight new spatially resolved nova shells have been found by imaging with a digital detector through a narrow Hα filter, and two old novae have been recovered. The 11 novae with the best determined maximum luminosities at outburst of the sample of 21 novae with reliable distances are used to derive a $M_v(\max)$ – rate of decline relationship. The author finds $M_v(\max, \text{corr}) = -10.70(\pm 0.30) + 2.41(\pm 0.23)\log(t_2)$, where t_2 is the time in days to decline 2 mag below maximum light. Previously published surveys of novae in M31 are used to redetermine the distance to that galaxy.

124.010   Relationship of classical novae to other eruptive variables.
N. Vogt.
Astrophys. Prepr. Ser., No. 11, 57 pp. (1985). Chapter 10 in "Classical novae", M. F. Bode, A. Evans (Editors), John Wiley & Sons Ltd., Chichester, England.

124.011   V605 Aql – Schlüssel zur späten Sternentwicklung?
W. C. Seitter.
Mitt. Astron. Ges., Nr. 63, p. 181 – 186 (1985). – See Abstr. 012.063.

Die Autorin untersucht, ob es sich bei dem Auftreten einer ungewöhnlichen Nova nahe dem Zentrum eines sehr ausgedehnten und lichtschwachen planetarischen Nebels um ein Objekt handelt, das Aufschluß über die Entwicklungsgeschichte von Kernen planetarischer Nebel und Novae geben kann.

124.012   Eruptive Ereignisse als Kennzeichen später Sternentwicklung.
W. C. Seitter.
Mitt. Astron. Ges., Nr. 63, p. 187 – 189 (1985). – See Abstr. 012.063.

Die Autorin hat mit V605 Aql ein vermutliches Bindeglied zwischen planetarischen Nebeln und Novae gefunden. Sie sucht nun nach beobachtbaren Gemeinsamkeiten der beiden Objektklassen und weiterer eruptiver Sterne.

124.013   Novae and galactic chemical evolution.
M. Peimbert, A. Sarmiento.
Bull. Am. Astron. Soc., Vol. 16, No. 4, p. 971 (1984). Abstract. – See Abstr. 010.062.

124.014   RS Ophiuchi – first radio detection of a recurrent nova outburst.
S. Padin, R. J. Davis, M. F. Bode.
Nature, Vol. 315, No. 6017, p. 306 (1985).

The authors report the first radio detection of an outburst from a recurrent nova (RS Ophiuchi). The observations begin only 18 days after the optical maximum and show a radio "light" curve which is different from that of classical novae. The brightness temperature of $\sim 10^7$K, calculated for this event, greatly exceeds the 10^4K seen in classical novae, and suggests a non–thermal origin for the radio emission from RS Ophiuchi.

124.015   A possible observational constraint on the production of lithium by galactic novae.
F. D'Antona.
Astrophys. Space Sci., Vol. 112, No. 2, p. 361 – 365 (1985).

Recent computations (D'Antona and Mazzitelli, 1982), together with the general scheme of evolution of cataclysmic binaries (CBs), lead to a conclusion that the secondaries in those CBs having periods shorter than ~ 4.5 hr have a large ^{3}He content in the envelope if the nova systems have an age of some billion years. The consequence on the frequency of novae outbursts is shortly examined. If lithium is produced by galactic novae, the ^{7}Li content of old disk stars should be very close to the Population II content.

124.016   RS Ophiuchi.
Br. Astron. Assoc. Circ., No. 647 (1985).

124.017   Recurrent nova RZ Leonis.
Yamamoto Circ., Nos. 2033, 2034 (1985). In Japanese.

124.018   Recurrent nova RS Ophiuchi.
Yamamoto Circ., Nos. 2034, 2036 (1985). In Japanese.

124.019   RS Ophiuchi.
IAU Circ., Nos. 4030, 4031, 4036, 4048, 4049, 4050, 4066, 4067, 4074 (1985).

124.020   NSV 7429.
IAU Circ., No. 4075 (1985).

124.021   Orbital periods of novae before eruption.
B. E. Schaefer, J. Patterson.
Cataclysmic variables and low–mass X–ray binaries, p. 343 – 346 (1985). – See Abstr. 012.072.

Nova Coronae Austrinae 1981

124.101   The abundances of the elements – carbon to silicon – in Nova Coronae Austrinae 1981.
S. Starrfield, R. E. Williams, W. M. Sparks, S. Wyckoff, J. W. Truran, E. P. Ney.
NASA Conf. Publ., NASA CP–2349, p. 309 – 312 (1984). – See Abstr. 012.001.

The authors observed Nova Coronae Austrinae 1981 systematically over a period of seven months with the IUE satellite. These spectra cover the range 1150 – 3200 Å, and are mostly at low dispersion. Several months after maximum visual light, high ionization forbidden lines of Na, Mg and Al appeared and then declined over a period of two months. The abundances deduced from these lines imply that N, O, Ne, Na, Mg and Al are all enhanced with respect to He while carbon is not. The observations of this nova, plus the observations of Nova Cygni 1975 and Nova Aquilae 1982, imply that some fraction of novae contain white dwarfs that have undergone non–degenerate carbon burning prior to entering a close binary–nova phase of evolution.

124.102   Ultraviolet spectral evolution and heavy element abundances in Nova Coronae Austrinae 1981.
R. E. Williams, E. P. Ney, W. M. Sparks, S. G. Starrfield, S. Wyckoff, J. W. Truran.
Mon. Not. R. Astron. Soc., Vol. 212, No. 4, p. 753 – 766 (1985).

Nova CrA 1981, a moderately fast galactic nova, was systematically observed over a period of seven months with the IUE satellite. The initial evolution of the UV spectrum was similar to that of previously observed novae; however, several months after maximum, weak emission due to very high ionization forbidden lines of Na, Mg, Al, and Si appeared, eventually decreasing in strength over the next two months. The spectral development of the lines is consistent with the photoionization of an expanding shell; a high–temperature 'coronal' origin is ruled out. Abundances deduced from the spectra show that most of the observed heavy elements in the ejected material are enhanced with respect to helium, with neon the most abundant of these.

Nova Vulpeculae 1979 = PU Vulpeculae

124.121   Optical variability of PU Vulpeculae, the novalike object Kuwano, in 1983.
E. A. Kolotilov.
Sov. Astron. Lett., Vol. 10, No. 4, p. 253 – 254 (1984). English translation of 38.124.103.

124.122   Peculiar object Kuwano–Honda (PU Vul).
D. Chochol.
Kozmos, Vol. 16, No. 3, p. 82 – 83 (1985). In Slovak.

Nova Vulpeculae 1670 = CK Vulpeculae

124.141   Unravelling the oldest and faintest recovered nova: CK Vulpeculae (1670).
M. M. Shara, A. F. J. Moffat, R. F. Webbink.
Space Telesc. Sci. Inst., Prepr. Ser., No. 39, 67 pp. (1985). To appear in Astrophys. J.

Nova Puppis 1942 = CP Puppis

124.161   Probable detection of a very short orbital period for the old classical nova CP Puppis (1942).
A. Bianchini, M. Friedjung, F. Sabbadin.
Inf. Bull. Variable Stars, No. 2650, 2 pp. (1985).

Nova Delphini 1967 = HR Delphini

124.181   On the variable structure of absorption lines in the spectrum of nova HR Del.
M. D. Dabaev.
Pis'ma Astron. Zh., Tom 11, No. 5, p. 362 – 370 (1985). In Russian. English translation in Sov. Astron. Lett., Vol. 11.

Variations in the structure of absorption lines in the course of the evolution of the spectrum of nova HR Del are analyzed using spectrograms taken with the 2–m telescope. All lines have variable multicomponent structure. Radial velocities and profiles of hydrogen and metal lines do not essentially differ. A group of absorption lines corresponding to the third outburst of HR Del is found.

Nova Persei 1901 = GK Persei

124.201   The old nova GK Per: discovery of the X–ray pulse period.
M. G. Watson, A. R. King, J. Osborne.
Mon. Not. R. Astron. Soc., Vol. 212, No. 4, p. 917 – 930 (1985).

The authors report EXOSAT observations of GK Per during an optical outburst. The hard X–ray emission (> 2 keV) shows a strong, coherent modulation at a period of 351 s, identifying GK Per as the latest member of the 'intermediate polar' subclass of magnetic cataclysmic variables. The X–ray spectrum is complex with evidence for two components above 2 keV, both of which show large low–energy cut–offs, and a third, much softer component below ~ 0.1 keV. These data are discussed in the context of current ideas on accretion on to magnetic white dwarfs.

124.202   The 1983 outburst of GK Persei.
P. Szkody, J. A. Mattei, M. Mateo.
Publ. Astron. Soc. Pac., Vol. 97, No. 589, p. 264 – 267 (1985).

The authors present the optical light curve of the 1983 minor outburst of GK Per obtained by the AAVSO in the context of the last few outbursts which, in general, follow a rising trend in brightness. An optical spectrum obtained at the peak of outburst showing enhanced He II, C III + N III emission is compared to the outburst spectra of other cataclysmic variables.

124.203   The old–nova GK Per (1901): evidence for a time–delay between its X–ray and optical outbursts.
A. Bianchini, F. Sabbadin.
Inf. Bull. Variable Stars, No. 2751, 2 pp. (1985).

Nova Vulpeculae 1984 No. 1 = PW Vulpeculae

124.221   Nova Vulpeculae 1984, No. 1 – courbe de lumière.
Bull. Assoc. Fr. Obs. Etoiles Variables, No. 31, p. 8 (1985).

124.222   Nova Vulpeculae 1984 No. 1.
H. Grzelczyk.
BAV Rundbrief, 34. Jahrg., Nr. 1, p. 7 – 11 (1985).

124.223   Nova Vulpeculae 1.
G. Comello, H. Feijth.
Zenit, 12. Jaarg., No. 4, p. 148 – 150 (1985).

124.224   Nova Vulpeculae (1) 1984.
H. Grzelzyk.
Sterne Weltraum, 24. Jahrg., Nr. 4, p. 213 – 214 (1985).

124.225   New star in the Vulpecula constellation.
I. B. Voloshina.
Zemlya Vselennaya, No. 2, p. 33 (1985). In Russian.

124.226   Behaviour of nova Vulpeculae 1984 near the maximum.
R. I. Noskova.
Astron. Tsirk., No. 1348, p. 1 – 3 (1984). In Russian.

124.227   Photometry of the nova Vulpeculae 1984 from August to September 1984.
O. G. Taranova, V. I. Shenavrin, A. Eh. Nadzhip.
Astron. Tsirk., No. 1348, p. 3 – 4 (1984). In Russian.

124.228   Évolution spectrale de la nova Vulpeculae 1984 I au voisinage du maximum de lumière.
Y. Andrillat, L. Houziaux.
Bull. Inf. Cent. Données Stellaires, No. 28, p. 111 – 124 (1985).

124.229   Spectroscopic observations of nova Vulpeculae 1984 No. 1.
J. R. Sowell, J. B. Laird, J. R. Thorstensen.
Inf. Bull. Variable Stars, No. 2700, 3 pp. (1985).

124.230   Photoelectric observations of PW Vul.
J. Papoušek.
Inf. Bull. Variable Stars, No. 2720, 3 pp. (1985).

124.231   Nova Vulpeculae 1984 No. 1 (PW Vulpeculae).
Yamamoto Circ., Nos. 2033, 2034, 2036, 2039 (1985). In Japanese.

124.232   Nova Vulpeculae 1984 No. 1 (PW Vulpeculae).
IAU Circ., Nos. 4028, 4046, 4060, 4065 (1985).

Nova Vulpeculae 1984 No. 2

124.241   La nova Vulpeculae 1984, No. 2.
E. Schweitzer.
Bull. Assoc. Fr. Obs. Etoiles Variables, No. 31, p. 9 – 10 (1985).

124.242   Nova Vulpeculae 1984 No. 2.
Yamamoto Circ., Nos. 2033, 2034, 2039 (1985). In Japanese.

124.243   Observation des raies métalliques dans Nova Vul 1984 (2).
Y. Andrillat.
La composition chimique des étoiles dans le voisinage solaire, p. 89 – 98 (1985). – See Abstr. 012.067.
Spectra of Nova Vul 1984 (2) have been obtained 9 days after the maximum in the blue and in the near infrared region. They show numerous emissions due principally to H I, He I, C I, N I, O I, Mg I, Ca II, Fe II, Ti II, Sr II, Si II, Mg II. These emissions are double and flanked by 2 absorption components blueshifted to −650 and −1337 km/sec. Mg II (7877 – 7896 Å) is abnormally strong.

124.244   Nova Vulpeculae 1984 No. 2.
IAU Circ., Nos. 4025, 4026, 4028, 4033, 4059, 4065 (1985).

124.245   A second nova in Vulpecula during 1984.
Astronomer, Vol. 21, No. 249, p. 143 (1985). Abstr. in Phys. Abstr., Vol. 88, No. 1258, Entry 63345 (1985).

Nova Serpentis 1909 = RT Serpentis

124.261   A spectrum of the symbiotic Nova RT Serpentis.
S. W. Mochnacki, G. Starkman.
Publ. Astron. Soc. Pac., Vol. 97, No. 588, p. 151 – 152 (1985).
The spectrum of RT Ser (Nova Serpentis 1909) was observed in 1979 in the range 6300 Å to 6750 Å, at a resolution of 0.6 Å. In addition to the dominant Hα line, weak emission lines of [O I] $\lambda6363$ and He I $\lambda6678$ were detected. These lines have radial velocities of 64, 82, and 77 km s^{-1}, respectively. Emitting regions with different physical conditions around RT Ser appear to have different radial velocities.

Nova Cygni 1876 = Q Cygni

124.281   Photometry of Nova Q Cygni in minimum light.
S. Yu. Shugarov.
Perem. Zvezdy, Tom 21, No. 6, p. 807 – 816 (1983). In Russian.
50 photoelectric UBV measures of the old Nova Q Cygni (1876) were obtained during 1980 – 1981. Its brightness was also estimated on 544 plates (1907 – 1982). The majority of bursts occur cyclically, the cycle being either 54^{d}99 or 64^{d}61. The available observations do not permit to distinguish between the two values. The position of the variable in the two–colour diagram, sections of the light curve, mean light curves and the O–C deviations for the above periods are presented. The photoelectric and photographic observations are given in a table.

Nova Herculis 1934 = DQ Herculis

124.301   Five–color photometry of DQ Herculis (N Her 1934). I.
E. S. Dmitrienko, Yu. S. Efimov, N. M. Shakhovskoj.
Astrofizika, Tom 22, Vyp. 1, p. 31 – 42 (1985). In Russian. English translation in Astrophysics, Vol. 22, No. 1.
Five–color light curves of N Her 1934 with almost complete phase coverage have been obtained for the first time. The observations were performed in 1982 – 1983. There are day–to–day and month–to–month light variations of the system as observed in 1954 – 1956 and in 1975. The out–of–eclipse light of DQ Her itself became brighter as compared with that in 1978 and was in 1982 – 1983 at the light level of 1954. The mid–eclipse light level of DQ Her changed insignificantly.

Nova Muscae 1983

124.321   Nova Muscae 1983: coordinated observations from X–rays to the infrared regime.
J. Krautter, K. Beuermann, H. Ögelman.
Messenger, No. 39, p. 25 – 30 (1985).

124.322   The chemical composition of the Nova Muscae 1983 ejecta.
J. A. de Freitas Pacheco, S. J. Codina.
Mon. Not. R. Astron. Soc., Vol. 214, No. 4, p. 481 – 490 (1985).
The authors present spectroscopic observations of Nova Muscae 1983 obtained during the nebular phase. The line widths indicate an expansion velocity of about 1080 km s^{-1} and the Balmer decrement indicates an interstellar reddening $E(B-V)$ of about 0.43. The derived ionic ratios suggest that helium, oxygen, nitrogen and iron are overabundant in the nova ejecta. Comparison with other novae shows that the brightest objects have an enhanced oxygen (CNO) abundance.

Nova Sagittarii 1983

124.341   Discovery of nova Sgr 1983.
H. Kosai.
Astron. Her., Vol. 78, No. 3, p. 84 – 85 (1985).

Nova Cygni 1978 = V1668 Cygni

124.361   Optical variability of V1668 Cygni = Nova Cygni 1978 during the nova and post–nova phase.
A. Piccioni, A. Guarnieri, C. Bartolini, F. Giovannelli.
Acta Astron., Vol. 34, No. 4, p. 473 – 485 (1984).
Photoelectric observations of Nova Cygni 1978, obtained during 18 nights in 1978 and one night in 1981 at the Bologna Observatory with the 152–cm telescope equipped with a fast double *UBV* photometer are presented and analysed. Photographic observations from July 1983 show that the magnitude of the nova was greater than 19th.

124.362   An optical spectrum of Nova Cygni 1978.
F. Giovannelli, C. Bartolini, A. Guarnieri,
A. Piccioni.
Acta Astron., Vol. 34, No. 4, p. 487 – 493 (1984).
The authors present a spectrum of Nova Cyg 1978 in the range 3850 – 5050 Å taken at the Cassegrain focus of the 152–cm telescope of the Bologna Astronomical Observatory using a Boller and Chivens 26767 spectrograph. The most important features are strong Balmer emission lines and Fe II emission lines, which show P Cygni profiles. The most important group of absorption lines is that of Ti II. Several other lines are recognized.

Nova Aquilae 1918 = V603 Aquilae

124.381   The old nova V603 Aql: an intermediate polar?
R. Haefner, K. Metz.
Astron. Astrophys., Vol. 145, No. 2, p. 311 – 320 (1985).
Extensive photometric and polarimetric observations of the old nova V603 Aql are presented. The light curve is characterized by the appearance of a hump structure repeating with a period 3^h28^m6, clearly different from the known orbital (spectroscopic) period of 3^h18^m9. Both, linear and circular polarization are modulated with a period of 2^h48^m0. A magnetic field in the order of 10^6Gauss can be derived from the degree of the circular polarization. The authors propose an intermediate polar model combined with a temporarily existing eccentric disc which can account for most of the observed properties of this system.

124.382   Far–UV observations of the old nova V603 Aql.
T. E. Carone, R. S. Polidan.
Bull. Am. Astron. Soc., Vol. 17, No. 2, p. 589 (1985). Abstract. – See Abstr. 010.065.

Nova Cygni 1975 = V1500 Cygni

124.401   Investigation of the strongest lines in the spectrum of Nova Cygni 1975.
M. P. Yasinskaya.
Mater. Konf. mol. uchenykh Odess. astron. nauchn.–proizv. akad.–univ. kompleksa, Odessa, 15 fev. 1983. Odessa, p. 26 – 37 (1983). In Russian. Abstr. in Ref. Zh., 51. Astron., 5.51.527 (1985).

Nova Cephei 1971

124.421   Photoelectric observations of nova Cephei 1971.
R. Burchi, V. D'Ambrosio.
Inf. Bull. Variable Stars, No. 2703, 2 pp. (1985).

Nova Normae 1985

124.441   Nova Normae 1985.
Yamamoto Circ., No. 2034 (1985). In Japanese.

124.442   Nova Normae 1985.
IAU Circ., Nos. 4030, 4031, 4033, 4035, 4037 (1985).

The discovery of radio novae.
See Abstr. 004.198.

Recurrent novae as a consequence of the accretion of solar material onto a 1.38 $M_\odot$ white dwarf.
See Abstr. 065.043.

CNO abundances resulting from diffusion in accreting nova progenitors.
See Abstr. 065.047.

On the ultraviolet spectrum of AG Peg.
See Abstr. 117.041.

Variations in the variability of RR Tel.
See Abstr. 117.045.

Photoelectric observations of the symbiotic stars CH Cyg, AG Dra and the peculiar object PU Vul during the maximum of their activity in the years 1981 – 1982.
See Abstr. 117.056.

Accretion disks heated by luminous central stars.
See Abstr. 117.166.

Accretion disks heated by luminous central stars.
See Abstr. 117.192.

Cataclysmic variables: observational overview.
See Abstr. 117.291.

Mira's role in RR Tel revisited.
See Abstr. 117.344.

Report of IAU Commission 42: Close binary stars (*Etoiles binaires serrées*).
See Abstr. 117.376.

E2000 + 223: a newly discovered old nova.
See Abstr. 142.039.

Novae in the galactic halo?
See Abstr. 155.122.

Hα observations of novae in M31.
See Abstr. 157.189.

Observations of two novae in M87.
See Abstr. 157.192.

125 Supernovae, Supernova Remnants

125.001   The destruction of dust grains in collisions of a supernova with a circumstellar medium.
H. Itoh.
Mon. Not. R. Astron. Soc., Vol. 212, No. 2, p. 309 – 323 (1985).
Dust grains exist in systems composed of a Type II supernova and a circumstellar medium (CSM). The grains may subsequently suffer intense sputtering by X–ray emitting plasmas during the collision of the supernova ejecta with the CSM. The author obtains an approximate expression for the amount of grain erosion in both the blast–shocked CSM and the reverse–shocked ejecta, assuming their self–similar interaction. The distortion of the grain size distribution is also calculated. In the remnant of SN 1979c, the population of the grains of radius 0.1 µm, for example, will be diminished substantially in both the ejecta and the CSM, while in the remnant of SN 1980k, it will be affected considerably only in the ejecta. The infrared fluxes expected from the blast–shocked CSM are roughly estimated. The present destruction mechanism may be responsible for the severe underabundance of dust in the Cassiopeia A supernova remnant.

125.002   Structure of the Crab Nebula: intensity and polarization at 20 cm.
T. Velusamy.
Mon. Not. R. Astron. Soc., Vol. 212, No. 2, p. 359 – 365 (1985).
VLA maps of the Crab Nebula at 20 cm with high dynamic range and resolution 15×15 arcsec^2 are presented. These maps show enhanced radio emission associated with optical filaments and an excellent correlation between depolarization at 20 cm and the bright optical filaments on the near side of the nebula. The overall structure of the radio emission is discussed, in particular the possibility that a circum–nebular remnant shell coincides with the outer boundary of the Crab Nebula.

125.003   A recurrence of the Cassiopeia A flux anomaly.
L. T. Walczowski, K. L. Smith.
Mon. Not. R. Astron. Soc., Vol. 212, No. 2, p. 27P – 31P (1985).
In the late 1970s, an anomalous increase in the flux density of Cassiopeia A at 38 MHz was noted. The authors present a series of new observations which indicate that the flux density at this frequency is still abnormally high and that whatever mechanism caused the supernova remnant to flare is still active.

125.004   Two remarkable bright supernova remnants.
P. A. Shaver, C. J. Salter, A. R. Patnaik,
J. H. van Gorkom, G. C. Hunt.
Nature, Vol. 313, No. 5998, p. 113 – 115 (1985).
Between July 1981 and August 1984, the authors used the Very Large Array in its B, C and D configurations to clarify the morphology of several high–brightness sources from the Milne supernova remnant catalogue. Of these, the two sources G349.7+0.2 and G357.7–0.1 were found to be most unusual.

125.005   A shadowed flow in the stem of the Crab nebula?
P. Morrison, D. Roberts.
Nature, Vol. 313, No. 6004, p. 661 – 662 (1985).
The faint radio and emission line 'jet' outward from the northern boundary of the Crab nebula (here called the 'stem') appears as a neat right cylinder which is lengthening and expanding. The authors model the glowing cylinder as the convected margin of a gas cloud that accidentally cast its shadow across the nearly ballistic flow of the stellar envelope ejected in the supernova explosion.

125.006   X–ray morphology of the Crab nebula.
W. Brinkmann, B. Aschenbach, A. Langmeier.
Nature, Vol. 313, No. 6004, p. 662 – 664 (1985).
The Crab nebula is so far the only object where the direct interaction between a pulsar and the surrounding remnants of the supernova can be studied. The authors present an X–ray image of the Crab nebula taken with the Einstein Observatory High Resolution Imager and deconvolved using Fourier techniques.

The picture uncovers for the first time spatial structures that further refine the ideas about the energy injection mechanism and the electron propagation process in the nebula.

125.007   A model for SNR evolution in a cloudy medium and its application to N49.
P. Shull Jr., J. E. Dyson, F. D. Kahn, K. A. West.
Mon. Not. R. Astron. Soc., Vol. 212, No. 4, p. 799 – 808 (1985).
The authors model the evolution of a supernova remnant in a cloudy interstellar medium (ISM). They first examine how the radiation fields and the winds of an early–type progenitor star alter its surroundings, and then study the effects of the supernova's blast wave as it propagates through the modified ISM. The properties of the well–observed supernova remnant N49 are in accord with this model, even though it contains several approximations which can be considerably improved.

125.008   Supernovae: mileposts of the universe.
D. H. Smith.
Sky Telesc., Vol. 69, No. 1, p. 18 (1985).

125.009   Infrared observations of the Crab Nebula.
P. L. Marsden.
Observatory, Vol. 105, No. 1064, p. 7 (1985). Abstract. – See Abstr. 010.721.

125.010   Einstein X–ray observations of the Monoceros supernova remnant.
D. A. Leahy, S. Naranan, K. P. Singh.
Mon. Not. R. Astron. Soc., Vol. 213, No. 1, p. 15P – 19P (1985).
The Monoceros nebula is seen in the optical and radio wavelengths as a ring of 3.5° angular diameter and is believed to be an old supernova remnant (SNR). The authors report the first positive detection of X–rays from the Monoceros nebula, confirming its supernova remnant nature. Reprocessed Einstein imaging proportional counter observations of soft X–rays (0.2 – 5 keV) were used to produce a contour map of the surface brightness distribution.

125.011   Infra–red astronomy of supernovae.
G. Pearce.
Thesis, Univ. Keele, England (1983). Abstr. in Phys. Abstr., Vol. 88, No. 1249, Entry 14647 (1985).

125.012   On the nature of the supernova remnant 0540–69.3 in the Large Magellanic Cloud.
G. Srinivasan, D. Bhattacharya.
Curr. Sci., Vol. 53, No. 10, p. 513 – 516 (1984). Abstr. in Phys. Abstr., Vol. 88, No. 1250, Entry 18902 (1985).

125.013   Neutral hydrogen in the vicinity of galactic radio sources. III. Evolutionary characteristics of old supernova remnants.
I. V. Gosachinskij, V. K. Khersonskij.
Astrophys. Space Sci., Vol. 108, No. 2, p. 303 – 317 (1985).
Observations of neutral gas in the vicinity of 24 SNRs were performed on the RATAN–600 radio telescope with resolution of $2' \times 130' \times 6.3$ km s^{-1}. The observations demonstrate that kinematics and spatial structure of the gas in the vicinity of 12 SNRs is typical of the expanding envelopes. This paper considers in detail only those enveloping the SNRs. The dimensions and expanding velocities of the envelopes are defined and the evolutionary parameters of the SNRs estimated.

125.014   VLA observations of three extragalactic SNR at 20 and 6 cm.
J. R. Dickel, S. D'Odorico, A. Silverman.
Astron. J., Vol. 90, No. 3, 414 – 417, 557 – 559 (1985).
A radio search has been conducted for optical SNR candidates in the external galaxies NGC 185, IC 1613, and NGC 6822. A faint, nonthermal radio source has been found at the expected

position in IC 1613. Upper limits are placed on the objects in NGC 185 and NGC 6822. All three appear fainter than typical SNR of the same size in the Milky Way.

125.015 Der Krebs–Nebel – ein Prüfstein für Supernova–Modellvorstellungen.
W. Brinkmann.
Sterne Weltraum, 24. Jahrg., Nr. 2, p. 76 – 79 (1985).

125.016 Imaging X–ray spectrophotometric observation of SN 1006.
M. H. Vartanian, K. S. K. Lum, W. H.–M. Ku.
Astrophys. J., Lett. Ed., Vol. 288, No. 1, p. L5 – L9 (1985).

An imaging gas scintillation proportional counter at the focal plane of a grazing incidence telescope was carried aloft by a sounding rocket and used to observe soft X–ray emission from the supernova remnant SN 1006. The instrument obtained the first unambiguous detection of strong $(8 \pm 1 \times 10^{-2}$ photons cm^{-2}s^{-1}) oxygen Heα and Lyα emisson at 0.59 ± 0.02 keV from SN 1006. This line emission is consistent with emission from a 1.8×10^{6}K plasma, although hard X–ray emission from an additional power law or a higher temperature component was also observed.

125.017 An X–ray study of the remnant of SN185 AD.
R. Pisarski, D. Helfand.
News Lett. Astron. Soc. N.Y., Vol. 2, No. 4, p. 15 (1983). Abstract. – See Abstr. 010.242.

125.018 Non–spherical supernova remnants. II. The interaction of remnants with molecular clouds.
G. Tenorio–Tagle, P. Bodenheimer, H. W. Yorke.
Astron. Astrophys., Vol. 145, No. 1, p. 70 – 80 (1985).

Two–dimensional numerical hydrodynamical calculations are presented which describe the evolution of a remnant resulting from supernova explosions of approximately 10^{51}erg in or near molecular clouds. Five evolutionary sequences are calculated for which the explosion site within the cloud is placed at different distances from the edge. An additional sequence is calculated in which two explosions occur inside the cloud at different locations and different times. Finally, if a supernova explodes outside a cloud in a medium with much lower density than that of the cloud, there is little disruption of the cloud. Several calculations were repeated under the assumption of a non–radiative gas. The morphological appearance and evolution of these remnants are compared and contrasted with the solutions which include cooling.

125.019 G27.4 + 0.0: a galactic supernova remnant with a central compact source.
G. A. Kriss, R. H. Becker, D. J. Helfand, C. R. Canizares.
Astrophys. J., Vol. 288, No. 2, p. 703 – 706 (1985).

The authors present high–resolution X–ray and radio images of the galactic supernova remnant G27.4 + 0.0. The X–ray image reveals a centrally located compact source of emission which accounts for 25% ± 2% of the emission. The rest of the emission is found in diffuse patches defining a circle of diameter 4′. The radio image is an imcomplete shell of comparable diameter with no apparent counterpart to the compact X–ray source. The distance and the X–ray luminosity of the supernova remnant are discussed and it is suggested that the central source is most likely an accretion–powered X–ray binary.

125.020 The evolution of Tycho's supernova remnant.
J. R. Dickel, E. M. Jones.
Astrophys. J., Vol. 288, No. 2, p. 707 – 717 (1985). With plate 19.

The evolution of Tycho's supernova remnant has been modeled with a numerical calculation based on a similarity solution developed by Chevalier in 1982. The new model has one important addition: a shell of matter, presumably a presupernova planetary nebula, is needed around the exploding white dwarf in order to reproduce some features in the observed morphology of the SNR. In particular, a very thin "rim" surrounding the main shell of the SNR cannot be reproduced without the planetary nebula. The evolution has been followed from a time of 10^{7}s to the present epoch $(1.3 \times 10^{10}$s).

125.021 Observations of 14 supernovae.
D. Yu. Tsvetkov.
Perem. Zvezdy, Tom 22, No. 1, p. 39 – 48 (1983). In Russian.

Photographic observations of 14 supernovae obtained at Sternberg Astronomical Institute from 1960 to 1980 are reported. Observations of 6 supernovae were not elaborated earlier, for other supernovae new magnitudes were obtained based on new determination of magnitudes of comparison stars. For 11 supernovae the light curves are presented and their parameters are derived.

125.022 Formation of filaments in supernova remnants.
A. A. Rumyantsev.
Astrofizika, Tom 22, Vyp. 1, p. 157 – 162 (1985). In Russian. English translation in Astrophysics, Vol. 22, No. 1.

A new mechanism of plasma condensation in the filaments due to shock waves collisions in the expanding shells is considered. It has been found that the electron gas losing energy due to ion excitation and emission behind the refracted fronts is being compressed. The ion–electron energy exchange is investigated. The formula for the top gas compression in the formed filaments is found.

125.023 Predicted continuum spectra of Type II supernovae: LTE results.
G. Shaviv, R. Wehrse, R. V. Wagoner.
Astrophys. J., Vol. 289, No. 1, p. 198 – 202 (1985).

The continuum spectral energy distribution of the flux emerging from Type II supernovae is calculated from quasi–static radiative transfer through a power–law density gradient, assuming radiative equilibrium and LTE. It is found that the Balmer jump disappears at high effective temperatures and low densities, while the spectrum resembles that of a dilute blackbody but is flatter with a sharper cutoff at the short–wavelength end. A significant UV–excess is found in all models calculated.

125.024 Improved optical spectrophotometry of supernova remnants in M33.
W. P. Blair, R. P. Kirshner.
Astrophys. J., Vol. 289, No. 2, p. 582 – 597 (1985).

The authors present spectrophotometry of 12 supernova remnants in the galaxy M33. The spectra generally cover the wavelength range 4750 – 7000 Å at a resolution of 7 Å. These data are analyzed in the light of recent shock model calculations and are used to investigate abundance gradients in M33. Gradients in nitrogen and sulfur abundances are clearly present and are similar to those found in our Galaxy. The results are compared with the abundance gradients from H II regions in M33. The authors investigate evolutionary models for the M33 remnants and find that the distribution of the number of remnants with diameter is consistent with free expansion of the remnants to diameters of ~ 26 pc.

125.025 Interaction of supernovae with the interstellar medium.
R. A. Chevalier.
Gas in the interstellar medium, p. 128 – 138 (1984). – See Abstr. 012.032.

The review concentrates on two aspects of supernova remnants which are of particular current interest. The first concerns the possibility of circumstellar rather than interstellar interaction. There is growing evidence for the occurrence of this interaction. The second is the question of what component of the interstellar medium fills most of the volume. This basic question is still controversial.

125.026 The production of X–rays in supernova remnants.
F. D. Kahn.
Gas in the interstellar medium, p. 139 – 143 (1984). – See Abstr. 012.032.

125.027   Radio observations of supernova remnants.
A. J. B. Downes, D. A. Green, S. F. Gull.
Gas in the interstellar medium, p. 144 – 168 (1984). – See Abstr. 012.032.

The authors review the current state of statistical work on supernova remnants (SNRs), describe the "typical" structure of both young and old SNRs, and attempt to divide these objects into different evolutionary categories. The authors also investigate some of the selection effects, and comment on the possibility that there are large numbers of very young ($\lesssim 2000$ yr old) SNRs in the Galaxy which have so far escaped detection.

125.028   Supernova remnants as probes of the interstellar medium.
D. H. Clark.
Gas in the interstellar medium, p. 169 – 171 (1984). – See Abstr. 012.032.

It is an interesting question as to whether the remnants of supernovae, observed at wavelengths from the radio to X–rays, can be used to probe the interstellar medium (ISM) into which they are expanding. Certainly attempts have been made to use observational data from supernova remnants (SNRs) to investigate the composition, density, temperature, and morphology of the interstellar medium. The purpose of this paper is to urge caution in such endeavours.

125.029   Exploding stars, superbubbles, and the HEAO observations.
W. H. Tucker.
Mercury, Vol. 13, No. 5, p. 130 – 135, 152 (1985).

125.030   The type I supernovae absolute magnitude brightness decline rate relation and the Hubble constant.
M. Jõeveer.
Tartu Astrofüüs. Obs. Publ., Tom 50, p. 327 – 334 (1984). In Russian.

The absolute magnitude M_{pg} brightness decline rate β relation $M_{pg} = -20^{m}41 + 0.17(\beta - 10)$ is derived on the basis of supernovae in the Local Supercluster. For the Local Group infall velocity and the global value of the Hubble constant the estimates $\Delta V = 144 \pm 103$ km/s and $H_0 = 45 \pm 7$ km/s/Mpc are obtained.

125.031   Radio lines of heavy elements from hot gas in the remnants of supernova explosions and clusters of galaxies.
R. A. Syunyaev, E. M. Churazov.
Inst. kosm. issled. Akad. Nauk SSSR, Prepr., No. 910, 22 pp. (1984). In Russian. Abstr. in Ref. Zh., 51. Astron., 2.51.106 (1985).

125.032   Supernovae and the distance scale.
S. van den Bergh.
Nature, Vol. 314, No. 6009, p. 320 (1985).

125.033   Hubble's constant and exploding carbon–oxygen white dwarf models for Type I supernovae.
W. D. Arnett, D. Branch, J. C. Wheeler.
Nature, Vol. 314, No. 6009, p. 337 – 338 (1985).

The immediate progenitor of a Type I supernova (SN I) is thought to be a mass–accreting carbon–oxygen (C–O) white dwarf in a binary system. In the supernova explosion the C–O fuel is partly incinerated to radioactive ^{56}Ni. The optical luminosity results from the trapping and thermalization of the γ rays and positrons emitted by the decay of ^{56}Ni through ^{56}Co to ^{56}Fe. The authors estimate the amount of ^{56}Ni synthesized and the corresponding peak luminosity of a SN I from the model. This estimate is combined with the observed Hubble diagram of SN I to derive the value of the Hubble constant H_0. It is found that H_0 is in the range $39 – 73$ km s^{-1}Mpc^{-1} with a best estimate of 59 km s^{-1}Mpc^{-1}.

125.034   Observations of supernovae in NGC 4753 and NGC 4051.
D. Yu. Tsvetkov.
Astron. Zh., Tom 62, Vyp. 2, p. 365 – 370 (1985). In Russian. English translation in Sov. Astron., Vol. 29, No. 2.

Photographic observations for two supernovae discovered in 1983 in NGC 4753 and NGC 4051 and revised results for the supernova 1965i in NGC 4753 are reported. Parameters of light curves and colour excesses are derived. Distances and absolute magnitudes are discussed, comparison is made with type I supernovae observed in the clusters of galaxies to which NGC 4753 and NGC 4051 belong.

125.035   Millimeter–wavelength lines of heavy elements predicted from the hot gas in supernova remnants and galaxy clusters.
R. A. Syunyaev, E. M. Churazov.
Sov. Astron. Lett., Vol. 10, No. 4, p. 201 – 205 (1984). English translation of 38.125.001.

125.036   Efficiency of charged–particle acceleration by shocks in supernova remnants and in solar flares.
S. V. Bulanov, I. V. Sokolov.
Sov. Astron. Lett., Vol. 10, No. 4, p. 247 – 249 (1984). English translation of 38.125.014.

125.037   Supernova remnant searches at Molonglo.
A. J. Turtle, B. Y. Mills.
Proc. Astron. Soc. Aust., Vol. 5, No. 4, p. 537 – 540 (1984).

125.038   Clark Lake observations of IC 443 and Puppis A.
W. C. Erickson, M. J. Mahoney.
Astrophys. J., Vol. 290, No. 2, p. 596 – 601 (1985).

The supernova remnant IC 443 has been observed at 73.8, 57.1, 38.5, and 30.9 MHz with the Clark Lake TPT telescope. Puppis A has been observed at 57.5 MHz. In each case, the observed brightness distributions are essentially identical to those found at higher frequencies. This shows that free–free absorption is negligible in these remnants and that the spectrum of the radio emission is constant throughout each remnant The spectrum of IC 443 is shown to be a power law from 20 MHz to 11 GHz with a spectral index of -0.36 ± 0.02.

125.039   Supernovae 1985B, 1985C, 1985D.
Yamamoto Circ., No. 2035 (1985). In Japanese.

125.040   Double radio features of supernova remnant maps.
S. F. Pimenov.
Pis'ma Astron. Zh., Tom 11, No. 4, p. 265 – 270 (1985). In Russian. English translation in Sov. Astron. Lett., Vol. 11.

Downstream two–component magnetic field perturbations are shown to cause "strong–strong" and "strong–weak" double radio features of supernova remnant maps. Magnetic field estimation is derived from the distance between feature components.

125.041   An evolutionary history for the Crablike, pulsar–powered supernova remnant 0540–69.3.
S. P. Reynolds.
Astrophys. J., Vol. 291, No. 1, p. 152 – 155 (1985).

The author reviews the radio, optical, and X–ray observations of the Crablike supernova remnant in the Large Magellanic Cloud, 0540–69.3, and its newly discovered 50 ms X–ray pulsar. Simple evolutionary considerations restrict the age of the object to the range 800 – 1100 yr. The appearance of the object in all frequency bands can be understood if it is a combination of emission resulting from several solar masses of fast–moving material interacting with the surrounding medium, producing the radio emission, and 0.2 – 2 solar masses of processed material which has been swept into a shell and accelerated by the pulsar. The original rotational energy is inferred to be $1.5 – 4.2 \times 10^{49}$ ergs, depending only insensitively on model details; this implies an initial period for the pulsar of 30 ± 8 ms.

125.042 Erratum: "The evolution of nonthermal supernova remnants. II. Can radio supernovae become plerions?"
[Astrophys. J., Vol. 285, No. 1, p. 134 – 140 (1984)].
R. Bandiera, F. Pacini, M. Salvati.
Astrophys. J., Vol. 291, No. 1, p. 394 (1985). See Abstr. 38.125.045.

125.043 Results of absolute measurements of the intensity of Cassiopeiae A, Cygnus A and the Crab nebula at 100 – 150 cm wavelengths and their spectra in the 60 – 150 cm range.
M. E. Miller.
Izv. Vyssh. Uchebn. Zaved., Radiofiz., Tom 27, No. 11, p. 1468 – 1471 (1984). In Russian. Abstr. in Ref. Zh., 51. Astron., 4.51.603 (1985).

125.044 How a supernova explodes.
H. A. Bethe, G. Brown.
Sci. Am., Vol. 252, No. 5, p. 40 – 48 (1985).
When a large star runs out of nuclear fuel, the core collapses in milliseconds. The subsequent "bounce" of the core generates a shock wave so intense that it blows off most of the star's mass.

125.045 Similarity solutions for the structure of supernova blast waves driven by clumped ejecta. I. Undecelerated clumps.
A. J. S. Hamilton.
Astrophys. J., Vol. 291, No. 2, p. 523 – 543 (1985).
New similarity solutions are presented for the structure of supernova blast waves driven by clumped supernova ejecta. The solutions are obtained using a spherically symmetric two–fluid hydrodynamic model in which clumps of ejecta stream through diffuse background gas. The clumps couple to the diffuse gas through mass ablation and the associated momentum and energy transfer. The most striking difference between these solutions and previous solutions is that clumps may move ahead of the shock front in the ambient medium, leading a precursor which heats the medium in advance of the main shock. The results are applied to the supernova remnants Cas A and Tycho.

125.046 Self–consistent models for the X–ray emission from supernova remnants: an application to Kepler's remnant.
J. P. Hughes, D. J. Helfand.
Astrophys. J., Vol. 291, No. 2, p. 544 – 560 (1985).
The authors have developed a novel solution to the problem of time–dependent ionization in a shock–heated plasma and have incorporated it into a standard, spherically symmetric hydrodynamic shock code to study the evolution of supernova remnants (SNRs). The ionization calculation uses the eigenvalue method for the matrix formed from the coupled system of rate equations expressing the time development of the ionization structure. Combined with the numerical shock code which treats emission both from the ambient interstellar gas and from the supernova ejecta, this model serves as a powerful tool in understanding the X–ray emission from young SNRs. As a first application, the authors have fitted all the available observations of Kepler's SNR obtained with the imaging and spectral instruments of the Einstein Observatory.

125.047 Why are there no type II supernovae in irregular galaxies?
I. S. Shklovskij.
Sov. Astron. Lett., Vol. 10, No. 5, p. 302 – 303 (1984). English translation of 38.125.030.

125.048 Radio observations of the December 1982 lunar occultation of the Crab Nebula.
M. I. Agafonov, A. M. Aslanyan, A. P. Barabanov, I. T. Bubukin, A. G. Gulyan, V. P. Ivanov, I. A. Malyshev, R. M. Martirosyan, K. S. Stankevich, S. P. Stolyarov.
Sov. Astron. Lett., Vol. 10, No. 5, p. 305 – 307 (1984). English translation of 38.125.031.

125.049 Adiabatic deceleration of secondary antiprotons in the envelopes of supernova exploding in dense clouds.
S. A. Stephens, B. G. Mauger.
Astrophys. Space Sci., Vol. 110, No. 2, p. 337 – 350 (1985).
Antiprotons ($\bar{p}$) in cosmic rays are believed to be secondary particles created by the collisions of cosmic rays with interstellar gas. On the basis of the well–established models of cosmic–ray propagation, the expected flux of $\bar{p}$ in the few hundred MeV region is very small compared to the observed flux. In order to explain the observational data, the authors have examined the possibility of $\bar{p}$ being produced in supernova envelopes during the expansion phase and undergoing the consequent adiabatic deceleration. It is shown that the $\bar{p}$ data can be well explained by the authors' model.

125.050 Supernova remnants: observational data. Evolution in the interstellar medium.
T. A. Lozinskaya.
Astrophys. Space Phys. Rev., Vol. 3, p. 35 – 96 (1984). – See Abstr. 003.019. Revised and extended English translation of 34.125.104.
The results of optical, X–ray, and radio observations of supernova remnants are presented. Different aspects of the interaction of supernova shells with the interstellar gas are examined: acceleration of clouds in the expanding shell, evolution of remnants in a medium with different density, and the filamentary structure of old remnants. The $\Sigma(D)$ dependence, the nature of the radio emission of shell–like remnants, plerions, and large–scale loops are analyzed. Estimates of the frequency of supernova explosions in the Galaxy and the mass of presupernovae are presented.

125.051 X–ray spectra of supernova remnants.
A. E. Szymkowiak.
NASA Tech. Memo., NASA TM–86169, 4 + 113 pp. (1985).
X–ray spectra were obtained from fields in three supernova remnants with the Solid State Spectrometer of the HEAO–2 satellite. These spectra, which contain lines from K–shell transitions of several abundant elements with atomic numbers between 10 and 22, were compared with various models, including some of spectra that would be produced by adiabatic phase remnants when the time–dependence of the ionization is considered.

125.052 The X–ray structure of the Crab nebula.
W. Brinkmann, B. Aschenbach, A. Langmeier, G. Hasinger, T. Bork.
Adv. Space Res., Vol. 5, No. 3, p. 45 – 47 (1985). – See Abstr. 012.059.

125.053 Exosat observations of the supernova remnant Cas A.
F. A. Jansen, S. P. McKechnie, P. A. J. de Korte, J. A. M. Bleeker, E. H. B. M. Gronenschild, A. Peacock, G. Manzo, G. Branduardi–Raymont, B. Kellett.
Adv. Space Res., Vol. 5, No. 3, p. 49 – 52 (1985). – See Abstr. 012.059.
The young supernova remnant Cas A has been observed with the imaging proportional counter of the Exosat observatory. High quality spatially resolved, spectral data allow for the first time the determination of the temperature structure of the remnant. Preliminary results on the distribution of temperature and emission measure over the remnant are presented.

125.054 X–ray observations of Vela–X.
A. Smith, H. U. Zimmermann.
Adv. Space Res., Vol. 5, No. 3, p. 53 – 56 (1985). – See Abstr. 012.059.
Exosat ME observations of the Vela–X region are presented. It is found that the 2 – 10 keV emission is divided into two components. One is associated with the pulsar but is probably extended by 10 – 20 arc minutes, the other is associated with the Vela–X radio nebula and is probably more extended ($\gtrsim 1$ degree). The Vela–X component is softer than the pulsar component. No pulsed emission is observed.

125.055   An [Fe X] λ6374 image of part of the Cygnus Loop.
R. G. Teske, R. P. Kirshner.
Astrophys. J., Vol. 292, No. 1, p. 22 – 28 (1985).
The authors have obtained an image of part of the Cygnus Loop in the [Fe X] λ6374 line at a resolution of $2\rlap{.}''5$, using a CCD preceded by an interference filter with a 9.5 Å bandpass. The data show a smooth intensity distribution associated with the supernova blast wave; superposed on this is a cloudlike surface-brightness pattern loosely associated with the optical nebulosity. The structures have properties consistent with evaporation of cloudlets into the blast wave interior.

125.056   Optical emission–line properties of evolved galactic supernova remnants.
R. A. Fesen, W. P. Blair, R. P. Kirshner.
Astrophys. J., Vol. 292, No. 1, p. 29 – 48 (1985).
New optical spectrophotometric data are presented for the supernova remnants CTB 1, OA 184, VRO 42.05.01, S 147, the Monoceros Loop, G 206.9 + 2.3, and G 65.3 + 5.7. These data are combined with published spectral data to study some of the general properties of evolved galactic supernova remnants. It is found that [O I] and [O II] line strengths, when used in conjunction with the usual Hα/[S II] ratio test, provide an excellent additional diagnostic for discriminating remnants from H II regions. The authors derive a galactic nitrogen abundance gradient of $d\log(\mathrm{N/H})/dR = -0.088$ dex kpc^{-1}, which is in agreement with that derived from H II regions. However, no abundance gradients for oxygen or sulfur are indicated from the remnant data.

125.057   Photometric classification and basic parameters of type I supernovae.
Yu. P. Pskovskij.
Sov. Astron., Vol. 28, No. 6, p. 658 – 664 (1984). English translation of 38.125.077.

125.058   Problems of supernova remnants.
L. Woltjer.
ESO Sci. Prepr., No. 358, 13 pp. (1985). To appear in Proceedings of the workshop on supernovae and their remnants, Bangalore, India.

125.059   Optical observations of the jet of the Crab nebula.
M.–P. Véron–Cetty, P. Véron, L. Woltjer.
ESO Sci. Prepr., No. 360, 14 pp. (1985). Submitted to Astron. Astrophys.

125.060   Optical spectrophotometric study of SNRs in the LMC.
I. J. Danziger, E. M. Leibowitz.
ESO Sci. Prepr., No. 367, 47 pp. (1985). To appear in Mon. Not. R. Astron. Soc.

125.061   Einige neue Supernovaüberreste mit zum Teil ungewöhnlichen Eigenschaften.
E. Fürst, W. Reich, P. Reich, Y. Sofue.
Mitt. Astron. Ges., Nr. 63, p. 147 – 148 (1985). – See Abstr. 012.063.
Vierzehn Supernovaüberreste wurden gefunden, deren Emission polarisiert ist und/oder einen nichtthermischen Spektralindex aufweist.

125.062   Zwiebelschalenmodell für Supernova–Überreste.
R. Beck, L. O'C. Drury, H. J. Völk.
Mitt. Astron. Ges., Nr. 63, p. 149 – 150 (1985). – See Abstr. 012.063.

125.063   Determination of the integral parameters of type II supernovae.
I. Yu. Litvinova, D. K. Nadezhin.
Pis'ma Astron. Zh., Tom 11, No. 5, p. 351 – 356 (1985). In Russian. English translation in Sov. Astron. Lett., Vol. 11.
Theoretical relations are derived between the properties of the light–curve plateau observed for type II supernovae and the explosion energy E, the mass of the ejected envelope M, the presupernova radius R. It is found that $E \approx 6 \times 10^{50}$ erg, $M \approx 6\,M_{\odot}$ and $R \approx 1000\,R_{\odot}$ for an "average" type II supernova, whereas for SN 1969l these parameters are $\approx 10^{51}$ erg, $\approx 10\,M_{\odot}$ and $\approx 500\,R_{\odot}$ respectively.

125.064   Permitted oxygen lines in supernova remnant spectra.
P. F. Winkler, R. P. Kirshner.
Bull. Am. Astron. Soc., Vol. 16, No. 4, p. 925 (1984). Abstract. – See Abstr. 010.062.

125.065   Optical and UV spectra of supernova remnants in M33.
W. P. Blair, R. P. Kirshner, J. C. Raymond.
Bull. Am. Astron. Soc., Vol. 16, No. 4, p. 925 – 926 (1984). Abstract. – See Abstr. 010.062.

125.066   EXOSAT observations of the X–ray SNR 1E 1149.6 + 6209, first seen with the Einstein IPC.
G. F. Bignami, P. A. Caraveo, S. Mereghetti, G. G. C. Palumbo.
Bull. Am. Astron. Soc., Vol. 16, No. 4, p. 926 (1984). Abstract. – See Abstr. 010.062.

125.067   X–ray halos around supernova remnants.
C. Mauche, P. Gorenstein.
Bull. Am. Astron. Soc., Vol. 16, No. 4, p. 926 (1984). Abstract. – See Abstr. 010.062.

125.068   Structure of one million degree gas in the Cygnus Loop.
B. E. Woodgate, L. W. Brown.
Bull. Am. Astron. Soc., Vol. 16, No. 4, p. 926 (1984). Abstract. – See Abstr. 010.062.

125.069   Are supernova remnant filaments really filaments?
J. J. Hester.
Bull. Am. Astron. Soc., Vol. 16, No. 4, p. 926 (1984). Abstract. – See Abstr. 010.062.

125.070   A new model of SNR evolution in a cloudy medium.
P. Shull Jr., J. E. Dyson, F. D. Kahn, K. A. West.
Bull. Am. Astron. Soc., Vol. 16, No. 4, p. 926 (1984). Abstract. – See Abstr. 010.062.

125.071   Clark Lake observations of the supernova remnant HB21.
M. J. Mahoney, N. Kassim.
Bull. Am. Astron. Soc., Vol. 16, No. 4, p. 995 – 996 (1984). Abstract. – See Abstr. 010.062.

125.072   A two–dimensional spectrum of a nonradiative shock filament in the Cygnus Loop.
R. A. Fesen, H. Itoh.
Bull. Am. Astron. Soc., Vol. 16, No. 4, p. 996 (1984). Abstract. – See Abstr. 010.062.

125.073   Calculated X–ray emission from models of Tycho's supernova remnant.
B. W. Smith, E. M. Jones, J. C. Raymond.
Bull. Am. Astron. Soc., Vol. 16, No. 4, p. 996 (1984). Abstract. – See Abstr. 010.062.

125.074   Changes in structure of the Crab Nebula's jet.
T. R. Gull, R. A. Fesen.
Bull. Am. Astron. Soc., Vol. 16, No. 4, p. 996 (1984). Abstract. – See Abstr. 010.062.

125.075   Is G 192.8–1.1 a discrete supernova remnant, or part of the Origem loop?
J. L. Caswell.
Astron. J., Vol. 90, No. 6, p. 1076 – 1081 (1985).
A region of nonthermal radio emission near the boundary of the Orion and Gemini constellations has long been regarded as an old supernova remnant but there is uncertainty as to whether the emission corresponds to a quite small remnant (G 192.8–1.1) with diameter ~1.5 deg or a much larger remnant (the Origem loop) with diameter ~5 deg. A new map of the region at 1.4 GHz has now been made with the Penticton synthesis telescope, covering a 2–deg field with good resolution (several arcmin) and high

sensitivity. The structure revealed by the new data supports the small–diameter remnant interpretation.

125.076 High–resolution radio observations of the supernova remnants CTB 104A (G 93.7–0.3) and 3C 434.1 (G 94.0 + 1.0).
T. L. Landecker, L. A. Higgs, R. S. Roger.
Astron. J., Vol. 90, No. 6, p. 1082 – 1093 (1985).
Radio emission from the SNRs CTB 104A (G 94.3–0.3) and 3C 434.1 (G 94.0 + 1.0) has been mapped with the DRAO Synthesis Telescope with 1′ resolution at 21–cm wavelength. Both remnants show features characteristic of shell SNRs and are best classified as such. CTB 104A has a thick shell and considerable central emission. 3C 434.1 has a steep boundary and mostly circular outline. Its strongest emission is within the shell, but is probably a combination of projected shell emission and a superimposed extragalactic source. A list of small–diameter sources in the vicinity of these remnants is given.

125.077 Pre–existing dust shells around type–I supernovae.
G. Pearce, A. J. Mayes.
Astrophys. Space Sci., Vol. 111, No. 2, p. 419 – 424 (1985).
The possibility of the existence, around type–I supernovae, of dust shells which existed before a supernova outburst is considered. None have so far been detected observationally; and any dust around the progenitor radiating in the near infrared would evaporate at outburst. Far infrared observations of the two possible types of progenitor, R Coronae Borealis–type stars and dwarf novae, would be useful to indicate whether there is any dust around them which would survive a supernova outburst.

125.078 Supernovae 1985E and 1985B.
Yamamoto Circ., No. 2036 (1985). In Japanese.

125.079 Supernovae 1985I and 1985J.
Yamamoto Circ., No. 2038 (1985). In Japanese.

125.080 Supernovae 1985B, 1985C, 1985D.
IAU Circ., No. 4038 (1985).

125.081 Supernovae (1985E and 1985B).
IAU Circ., No. 4046 (1985).

125.082 Supernovae (1985I, 1985H and 1985A).
IAU Circ., No. 4058 (1985).

125.083 Supernovae (1985J and 1985I).
IAU Circ., No. 4059 (1985).

125.084 Supernovae (1985L and 1985M).
IAU Circ., No. 4077 (1985).

125.085 Optical studies of Cassiopeia A. VII. Recent observations of the structure and evolution of the nebulosity.
S. van den Bergh, K. Kamper.
Astrophys. J., Vol. 293, No. 2, p. 537 – 541 (1985). With plates 24 – 34.
Palomar observations obtained between 1951 and 1983 are used to study the evolution of Cas A. Most quasi–stationary flocculi (QSFs) are found to have lifetimes $\gtrsim 25$ yr. Those QSFs outside the main radio shell appear to have below–average lifetimes. It is suggested that tadpole–like QSFs may be older than more compact flocculi situated in regions that presently contain fast–moving knots. High excitation fast–moving knots, which radiate in [O III], seem to be somewhat more ephemeral than knots that radiate in [S II]. New observations of quasi–stationary flocculi yield an expansion age of $11,000 \pm 2000$ yr.

125.086 The decrease with time of the radio flux of the Crab Nebula.
H. D. Aller, S. P. Reynolds.
Astrophys. J., Lett. Ed., Vol. 293, No. 2, p. L73 – L75 (1985).
Observations of the Crab Nebula at 8.0 GHz over the period 1968 – 1984 show a secular decrease in the flux density at a rate of $-0.167\% \pm 0.015\%$ yr^{-1}. This observed value is in good quantitative agreement with the rate predicted by a theoretical model for the evolution of the Crab Nebula. The authors discuss inferences that can be made about the characteristic pulsar slowdown time scales in Crab–like remnants.

125.087 Supernova remnants in the Magellanic Clouds. III.
D. S. Mathewson, V. L. Ford, I. R. Tuohy, B. Y. Mills, A. J. Turtle, D. J. Helfand.
Astrophys. J., Suppl. Ser., Vol. 58, No. 2, p. 197 – 200 (1985). With plates 4 – 17.
The authors have obtained narrow–band optical images of seven SNR candidates in the LMC selected on the basis of radio and/or X–ray observations. Four of the candidates are confirmed as new SNRs. The object 0536–692 appears to be a "superbubble" resulting from one or more supernovae and the stellar winds from the large OB stellar association, NGC 2044, within its interior. These results bring the total number of SNRs with optical identifications in the Large Magellanic Cloud to 32.

125.088 Possible optical counterparts to the X–ray point source in the supernova remnant CTB 80.
W. P. Blair, R. E. Schild.
Astrophys. Lett., Vol. 24, No. 4, p. 189 – 195 (1985).
The authors present a three color CCD image of the central region of the supernova remnant CTB 80 along with astrometry and photometry of many stars in the field. The color image does not show evidence of heavy or variable dust absorption in the surrounding region. Using an Einstein HRI position for the central X–ray point source, the authors have identified two possible optical counterparts at magnitudes $V = 19.9$ and $V = 20.9$. Comparison of the intrinsic colors and magnitudes of these candidates are made to the optical properties of Crab and Vela pulsars, and they are found to be viable candidates.

125.089 The optical structure of the central core in the peculiar supernova remant CTB 80.
R. A. Fesen, T. R. Gull.
Astrophys. Lett., Vol. 24, No. 4, p. 197 – 204 (1985).
Deep and high resolution Hα + [N II], [O III], and red continuum interference–filter images of the bright central radio core of the supernova remnant CTB 80 are presented. These photographs reveal a surprisingly filamentary structure not visible from previous imaging and indicate considerable morphological differences between [O III] and Hα + [N II] emission regions. The Hα + [N II] filaments appear elongated in the E–W direction like that of the radio emission plateau surrounding the central core. The [O III] emission is much smaller in angular size but matches well the position, shape and size of the core's 6 cm radio emission. The authors also detect the two possible optical counterparts reported by Blair and Schild to the remnant's X–ray point source and note that both may have coincident Hα + [N II] emission.

125.090 Upper limits to supernova prompt emission in gamma rays from the Franco–Soviet SIGNE experiment.
C. Barat, R. Hayles, K. Hurley, M. Niel, G. Vedrenne, I. V. Estulin (*I. V. Ehstulin*), A. V. Kuznetsov, V. M. Zenchenko.
18th International Cosmic Ray Conference, Vol. 1, p. 49 (1983). Abstract. – See Abstr. 012.096.

125.091 On correlated observations of SN'e and their young shells.
V. S. Berezinsky (*V. S. Berezinskij*), V. L. Ginzburg, O. F. Prilutsky (*O. F. Prilutskij*).
18th International Cosmic Ray Conference, Vol. 2, p. 310 – 313 (1983). – See Abstr. 012.096.
The radiations produced by SN and its young shell, the possibility of regular search for SN explosions in optically invisible parts of the Galaxy and the program of urgent observations in the case of unexpected discovery of SN are discussed.

125.092 Detection of supernovae explosions by thermoluminescence.
G. C. Castagnoli, G. Bonino, S. Miono.
18th International Cosmic Ray Conference, Vol. 2, p. 389 – 391 (1983). – See Abstr. 012.096.

A new method for the detection in terrestrial materials of the traces of the galactic supernovae explosions has been proposed. Sharp peaks in the thermoluminescence intensity profile of a recent sea sediment have been found in coincidence with the historically recorded supernovae events during the last 15 centuries. The detection of some TL peaks not correlated with visually observed SN explosions is discussed.

125.093 Observations of the Cygnus Loop at 40 MHz with the Clark Lake Teepee Tee.
P. D. Jackson, N. E. Kassim, M. R. Kundu.
Bull. Am. Astron. Soc., Vol. 17, No. 2, p. 545 – 546 (1985). Abstract. – See Abstr. 010.065.

125.094 High resolution X–ray spectroscopy of the Cygnus Loop.
P. W. Vedder, C. R. Canizares, T. H. Markert,
T. Pfafman, A. Pradhan.
Bull. Am. Astron. Soc., Vol. 17, No. 2, p. 546 (1985). Abstract. – See Abstr. 010.065.

125.095 Multiwavelength investigation of the adolescent supernova remnant IC 443.
M. L. McCollough, S. L. Mufson, J. Dickel, R. White,
R. Petrie, R. Chevalier.
Bull. Am. Astron. Soc., Vol. 17, No. 2, p. 546 (1985). Abstract. – See Abstr. 010.065.

125.096 N63A – an unusual supernova remnant in the LMC.
W. P. Blair, R. P. Kirshner, J. C. Raymond,
P. F. Winkler.
Bull. Am. Astron. Soc., Vol. 17, No. 2, p. 546 (1985). Abstract. – See Abstr. 010.065.

125.097 X–ray studies of the supernova remnant N132D.
J. P. Hughes.
Bull. Am. Astron. Soc., Vol. 17, No. 2, p. 546 (1985). Abstract. – See Abstr. 010.065.

125.098 Cassiopeia A at 86 GHz: spectral and rotation measure differences.
J. Kenney, W. A. Dent.
Bull. Am. Astron. Soc., Vol. 17, No. 2, p. 546 – 547 (1985). Abstract. – See Abstr. 010.065.

125.099 Peculiar Type I supernovas.
A. Uomoto, R. P. Kirshner.
Bull. Am. Astron. Soc., Vol. 17, No. 2, p. 565 (1985). Abstract. – See Abstr. 010.065.

125.100 IRAS observations of the supernova remnant Cassiopeia A.
E. Dwek, H. L. Dinerstein, F. C. Gillett, M. G. Hauser,
W. L. Rice.
Bull. Am. Astron. Soc., Vol. 17, No. 2, p. 596 (1985). Abstract. – See Abstr. 010.065.

125.101 The supernova remnant OA 184 (G 166.2 + 2.5) and associated H I.
D. Routledge, T. L. Landecker, J. F. Vaneldik.
Bull. Am. Astron. Soc., Vol. 17, No. 2, p. 596 (1985). Abstract. – See Abstr. 010.065.

125.102 CCD images of the Puppis A supernova remnant in [Fe X] λ6374 and [Fe XIV] λ5303.
R. G. Teske.
Bull. Am. Astron. Soc., Vol. 17, No. 2, p. 596 (1985). Abstract. – See Abstr. 010.065.

125.103 A soft X–ray survey of galactic supernova remnants with the Einstein Observatory.
R. L. Pisarski.
Diss. Abstr. Int., Sect. B, Vol. 45, No. 10, p. 3257 (1985). Thesis, Columbia University, 130 pp. (1984). Order No. DA8427449.

125.104 A rocket–borne X–ray observation of SN 1006 with the Imaging Gas Scintillation Proportional Counter.
M. H. Vartanian.
Diss. Abstr. Int., Sect. B, Vol. 45, No. 10, p. 3258 (1985). Thesis, Columbia University, 212 pp. (1984). Order No. DA8427491.

125.105 Multifrequency observations of recent supernovae.
N. Panagia.
Supernovae as distance indicators, p. 14 – 33 (1985). – See Abstr. 012.107.

This review discusses recent progress in determining the physical parameters of supernovae. Emphasis is put on extensive UV observations with IUE and on infrared photometry. The implications of these new data on the absolute luminosities of both Type I and Type II supernovae are analyzed.

125.106 Supernova observations at McDonald Observatory.
J. C. Wheeler.
Supernovae as distance indicators, p. 34 – 47 (1985). – See Abstr. 012.107.

The programs to obtain high quality spectra and photometry of supernovae at McDonald Observatory are reviewed. Spectra of recent Type I supernovae in NGC 3327, NGC 3625, and NGC 4419 are compared with those of SN 1981b in NGC 4536 to quantitatively illustrate both the homogeneity of Type I spectra at similar epochs and the differences in detail. Spectra of the recent supernova in NGC 991 give for the first time quantitative confirmation of a spectrally homogeneous, but distinct subclass of Type I supernovae which appear to be less luminous and to have lower excitation at maximum light than classical Type I supernovae.

125.107 Spectropolarimetry of supernovae.
M. L. McCall.
Supernovae as distance indicators, p. 48 – 61 (1985). – See Abstr. 012.107.

A new technique is introduced for testing the validity of the spherical symmetry approximation of the Baade method for determining distances to supernovae. Lines with P Cygni profiles produced by resonance scattering in an expanding atmosphere with non–circular isophotes should show a linear polarization in excess of that in the continuum. Thus, spectropolarimetry offers a means of measuring the roundness of a supernova envelope and assessing the reliability of Baade method distances. The method is applied to two Type I supernovae.

125.108 Radio emission from Type II supernovae.
K. W. Weiler.
Supernovae as distance indicators, p. 65 – 74 (1985). – See Abstr. 012.107.

Over the past five years the study of radio emission from supernovae has developed from a few weak detections of SN 1970g into a field with detailed, multifrequency radio "light curves" on several objects, with extensive spectral information, and with complex models for the radio supernova (RSN) phenomenon. This work is continuing and will provide within the next few years detailed studies of individual supernovae as well as statistics concerning the properties of the radio emission from different classes of supernovae.

125.109 Radio observations of historical, extragalactic supernovae.
J. J. Cowan, D. Branch.
Supernovae as distance indicators, p. 75 – 87 (1985). – See Abstr. 012.107.

VLA maps of the galaxy M83 (NGC 5236) at 6 and 20 cm reveal the presence of both non–thermal and thermal sources, lying predominantly along the inner edges of the optical spiral arms. A radio source coincident with the optical position of supernova 1957d is found to have a non–thermal spectrum; thus SN 1957d is confirmed as the first supernova of intermediate–age ($\sim 10 - 300$ years) to be detected at any wavelength. A second non–thermal source is tentatively identified with supernova 1950b, pending the measurement of a precise optical position of the supernova. A composite "radio light curve" for supernovae and young supernova remnants of known age is presented and discussed.

125.110 Discovery of an entire population of radio supernova candidates in the nucleus of Messier 82.
P. P. Kronberg.
Supernovae as distance indicators, p. 88 – 99 (1985). – See Abstr. 012.107.

Observations with the VLA by Kronberg, Biermann and Schwab have revealed about 40 discrete radio sources in the inner, visually obscured nucleus of the enigmatic galaxy M82. These radio sources, more luminous than any comparable objects in our Galaxy, are comparable to those of the half dozen radio supernovae found to date in other nearby galaxies. Repeated monitoring reveals that most of the brightest of these sources are declining in intensity on the scale of months to years. Both their radio luminosity and variability strongly suggest that M82 contains an entire population of radio supernovae. Some physical characteristics of this population are outlined.

125.111 Detecting supernova remnants in external galaxies.
J. R. Dickel, S. D'Odorico.
Supernovae as distance indicators, p. 100 – 106 (1985). – See Abstr. 012.107.

Reliable identifications of supernova remnants in external galaxies are slowly becoming available through the combined use of radio, optical, and X–ray surveys. The host galaxies represent a variety of Hubble types. Although the number of SNR detected in each galaxy is currently small, new telescopes and detectors in all three wavelength regimes should allow us to obtain larger samples in the near future.

125.112 Angular diameter determinations of radio supernovae and the distance scale.
N. Bartel.
Supernovae as distance indicators, p. 107 – 122 (1985). – See Abstr. 012.107.

A new method of determining extragalactic distances and inferring H_0 is presented. The method combines the determinations of radial expansion velocities of supernovae via optical spectroscopy and the determinations of angular expansion velocities via very long baseline interferometry. The data of two recent supernovae, SN 1979c in the galaxy M100 and SN 1980k in the galaxy NGC 6946, have been analyzed.

125.113 Supernova interaction with a circumstellar wind and the distance to SN 1979c.
R. A. Chevalier, C. Fransson.
Supernovae as distance indicators, p. 123 – 129 (1985). – See Abstr. 012.107.

There is observational evidence at a number of wavelengths for the interaction of the supernovae SN 1979c and SN 1980k with a circumstellar medium created by presupernova mass loss. Distance independent quantities can be used to develop a model for the interaction region. When these results are combined with a distance dependent quantity, the distance to the supernova is determined. The most useful distance dependent quantities are the radio angular diameter (from VLBI observations) and the X–ray flux. This method for distance determination is applied to the supernova SN 1979c.

125.114 Model and geometry dependence of radio distance determinations of extragalactic supernovae.
A. P. Marscher.
Supernovae as distance indicators, p. 130 – 137 (1985). – See Abstr. 012.107.

This paper discusses models for the radio radiation from extragalactic supernovae. It concentrates on the uncertainties involved in extracting angular diameters from VLBI data.

125.115 Optical supernovae and the Hubble constant.
D. Branch.
Supernovae as distance indicators, p. 138 – 150 (1985). – See Abstr. 012.107.

Three ways to use optical supernovae to estimate the value of the Hubble constant are reviewed: (1) Type I supernovae as standard candles, (2) the Baade–Wesselink method applied to Type II and Type I supernovae, and (3) the ^{56}Ni–radioactivity method for SNe I. It is likely that $H_0 \approx 50 - 60$ km s^{-1}Mpc^{-1}. If "ordinary" Type I supernovae are completely–disrupting carbon–oxygen white dwarfs near the Chandrasekhar mass, the ^{56}Ni–method provides a rigorous lower limit, $H_0 \gtrsim 40$, and a firm upper limit, $H_0 \lesssim 70$ km s^{-1}Mpc^{-1}.

125.116 Type I supernovae as standard candles.
R. Cadonau, A. Sandage, G. A. Tammann.
Supernovae as distance indicators, p. 151 – 165 (1985). – See Abstr. 012.107.

Observations are compiled to construct a mean light curve of SNe I in different colors. Individual SNe I show generally no systematic deviations from these templet light curves. The peak luminosity of absorption–free SNe I is uniform with an intrinsic rms scatter of <0.3 mag. The calibration of the peak luminosity yields $M_B(\text{max}) = -20.0 \pm 0.4$, which requires $H_0 = 43 \pm 10$ km s^{-1}Mpc^{-1}. SNe I are probably among the best standard candles known.

125.117 Astrophysical distances to Type II supernovae.
R. P. Kirshner.
Supernovae as distance indicators, p. 171 – 182 (1985). – See Abstr. 012.107.

This review discusses recent developments in methods to determine distances to extragalactic Type II supernovae by studying their expanding atmospheres from optical continuum and line observations.

125.118 Model atmospheres for Type I supernovae.
R. Harkness.
Supernovae as distance indicators, p. 183 – 191 (1985). – See Abstr. 012.107.

A thorough understanding of supernova atmospheres is needed now to determine the corrections which may be necessary before one can place much confidence in the Baade–Wesselink distance estimates. This paper discusses models of Type I supernovae and in particular the consequences of the detonating or deflagrating white dwarf mechanism.

125.119 Accuracy of model parameters from spectroscopic fine analyses of supernovae.
K. Hempe.
Supernovae as distance indicators, p. 192 – 199 (1985). – See Abstr. 012.107.

A model grid of hydrogen Balmer lines has been calculated for supernova atmospheres by the comoving frame technique. The theoretical line profiles have been used for an analysis of SN 1983 in NGC 1448. The comparison of theoretical and observed equivalent widths and line strengths gives information on the accuracy of the model parameters. It is found that temperature, density, and velocity structure can be well determined, while the photospheric radius is uncertain by a factor of 2.

125.120 Physical models for Type I supernovae and the distance scale.
J. C. Wheeler, P. G. Sutherland.
Supernovae as distance indicators, p. 200 – 208 (1985). – See Abstr. 012.107.

The degenerate carbon deflagration models for Type I supernova are consistent with a variety of observations of light curves and spectra. Progenitor systems consisting of carbon/oxygen white dwarfs with normal companions or of binary white dwarfs are discussed and limits are established on the amount of ^{56}Ni which can be ejected in models which match the observations. The luminosity at maximum light in these models is proportional to the amount of nickel ejected, and hence limits can be set on the distance scale and H_0. If the carbon deflagration model is correct, then $0.4 < H_0/(100 \text{ km/s/Mpc}) < 0.7$.

125.121 A preliminary discussion of Bose–Einstein diffusion in supernovae.
A. J. Fu, W. D. Arnett.
Supernovae as distance indicators, p. 209 – 221 (1985). – See Abstr. 012.107.

A new new method for the treatment of radiation dynamics in supernovae is discussed. The radiation field is characterized by a Bose–Einstein distribution with nonzero chemical potential. Radiative transport is treated by a diffusion technique. A bootstrap procedure is invoked to argue that the physical conditions thought to prevail during the explosion favor a Bose–Einstein radiation field for a significant portion of the evolution. A primary consequence of the theory may be that supernovae are dimmer than they are now thought to be and that distances are therefore smaller; values of H_0 inferred from supernovae would then be larger. As example, a calculation for SN 1983k in NGC 4699 is presented.

125.122 An optical synchrotron nebula around the X–ray pulsar 0540–693 in the Large Magellanic Cloud.
G. A. Chanan, D. J. Helfand, S. P. Reynolds.
Birth and evolution of neutron stars: issues raised by millisecond pulsars, p. 40 – 47 (1984). – See Abstr. 012.115.

The authors report the discovery of extended optical continuum emission around the recently discovered 50 ms X–ray pulsar in the supernova remnant 0540–693. Exposures in blue and red broadband filters made with the CTIO 4 m telescope and prime focus CCD show a center–brightened but clearly extended nebula about 4″ in diameter (FWHM), while an image in an [O III] filter shows an 8″ diameter shell (as reported earlier) which encloses the continuum source. 0540–693 is a system very similar to the Crab nebula and represents the second detection of optical synchrotron radiation in a supernova remnant.

125.123 On the nature of the Crablike, pulsar–powered supernova remnant 0540–693: the pulsar's initial period.
S. P. Reynolds.
Birth and evolution of neutron stars: issues raised by millisecond pulsars, p. 121 – 129 (1984). – See Abstr. 012.115.

The author uses radio, optical and X–ray observations of the Crablike supernova remnant in the Large Magellanic Cloud, 0540–693, to deduce the initial properties of its recently discovered 50 ms X–ray pulsar. Simple evolutionary considerations give the age of the object as 700 – 1000 years. This pulsar, the second for which initial parameters can be inferred, joins the Crab in having a relatively slow rotation at birth. The author discusses the significance of this finding for theories of pulsars and their supernova remnants.

125.124 Pulsar powered radio supernovae and the early evolution of plerions.
R. Bandiera, F. Pacini, M. Salvati.
Birth and evolution of neutron stars: issues raised by millisecond pulsars, p. 324 – 329 (1984). – See Abstr. 012.115.

The authors discuss the nature of radio supernovae assuming that their activity is due to the presence of a very fast internal pulsar. They then try to establish a direct evolutionary link between radio supernovae and plerions using the theory of pulsar powered supernova remnants.

125.125 A powerful young supernova remnant in another galaxy.
W. P. Blair, R. P. Kirshner, P. F. Winkler Jr.
Mercury, Vol. 14, No. 3, p. 80 – 82 (1985).

Supernova in NGC 7038

125.201 The spectrum of a Type I supernova in NGC 7038.
H. Schild.
Astron. Astrophys., Vol. 142, No. 2, p. 401 – 403 (1985).

An intensity calibrated spectrum in the range 4200 Å to 7100 Å of a Type I supernova about 7 weeks after maximum is presented.

Strong broad emission lines are observed but no P Cygni profiles or conspicuous absorption lines. The emission lines can be attributed to transitions of ions of helium and oxygen. Traces of hydrogen are possibly present but no carbon and nitrogen lines are detectable.

Supernova in NGC 224 (1885a)

125.221 Dunér's observations of SN 1885 (S Andromedae) in M31.
G. de Vaucouleurs, N. Hansson, G. Lyngå.
Publ. Astron. Soc. Pac., Vol. 97, No. 587, p. 30 – 31 (1985).

The light curve of SN 1885 in M31 from 1885 September 15 to November 15 is derived from ten visual observations made by N. C. Dunér with the 24–cm refractor of Lund Observatory.

125.222 The supernova of 1885 in M31.
G. de Vaucouleurs.
Bull. Am. Astron. Soc., Vol. 17, No. 2, p. 565 (1985). Abstract. – See Abstr. 010.065.

Supernova in NGC 4699 (1983k)

125.241 The supernova 1983k in NGC 4699: clues to the nature of Type II progenitors.
V. S. Niemela, M. T. Ruíz, M. M. Phillips.
Astrophys. J., Vol. 289, No. 1, p. 52 – 57 (1985). With plate 1.

Optical spectrographic and photometric observations of the Type II supernova 1983k in NGC 4699 are presented. For the first time, high quality spectra have been obtained of a Type II supernova before it reached maximum light. The first of these, taken nearly 10 days before maximum, showed high–ionization N III and He II emission lines atop a strong blue continuum. Near maximum, the emission lines disappeared, leaving weak H I, He I, and Ca II absorption lines. The light curve and the absorption lines seen at maximum provide independent evidence that the progenitor of this supernova had an extensive, preexisting circumstellar shell. The observed characteristics are consistent with the progenitor having been an exploding Wolf–Rayet star or a red supergiant.

Supernova in NGC 5236 (1983n)

125.261 The synchrotron selfabsorption of radio emission of the supernova 1983.51.
I. S. Shklovskij.
Pis'ma Astron. Zh., Tom 11, No. 4, p. 261 – 264 (1985). In Russian. English translation in Sov. Astron. Lett., Vol. 11.

The radio luminosity curve of the type I supernova 1983.51 is explained by synchrotron selfabsorption. This allows to determine a lower limit of the angular size of the radio source. It is concluded that the radius of the radio source is 3 times that of the supernova photosphere and the expansion velocity of the radio emitting envelope is > 60000 km/s. The nature of the supernova radio emission is discussed.

125.262 Radio emission from a Type I supernova.
R. A. Sramek.
Supernovae as distance indicators, p. 62 – 64 (1985). – See Abstr. 012.107.

VLB monitoring of the supernova SN 1983n in M83 resulted in a detailed light curve at $\lambda 6$ cm. Some data at $\lambda 20$ cm were also obtained.

Supernova in NGC 4753 (1983g)

125.281 Supernova 1983g and the distance to NGC 4753.
R. J. Buta, H. G. Corwin Jr., C. B. Opal.
Publ. Astron. Soc. Pac., Vol. 97, No. 589, p. 229 – 235 (1985).

Photoelectric and photographic UBV photometry of SN 1983g in NGC 4753 is presented. The light curves and color evolution show the typical forms for a type I supernova and indicate that maximum light occurred on JD2445430 when the magnitude was B(max) $\approx$ 12.9. The color evolution indicates a substantial reddening, $E(B-V) \approx 0^m.45$, for this SN, mostly due to internal dust. Using SN 1983g as a distance indicator, the authors estimate the distance modulus of NGC 4753 to be 29.7 ± 0.4. Implications of this distance for the properties of NGC 4753, an I0 galaxy, are briefly discussed.

Supernova in NGC 4490 (1982f)

125.301 **Observations of supernova 1982f in NGC 4490.**
D. Yu. Tsvetkov.
Astron. Tsirk., No. 1346, p. 1 – 3 (1984). In Russian.

Supernova in NGC 4321 (1979c)

125.321 **The ejected mass of the type II supernova SN 1979c.**
N. N. Chugaj.
Pis'ma Astron. Zh., Tom 11, No. 5, p. 357 – 361 (1985). In Russian. English translation in Sov. Astron. Lett., Vol. 11.

Interpretation of the profile and intensity of the Hα emission in the spectrum of the type II supernova SN 1979c on 72 days after the light maximum results in two independent estimates of the lower limit of the mass of the envelope ($M > 0.6\,M_\odot$). An analysis of the light curve on the basis of an approximate theory gives the value of the envelope mass of SN 1979c $M \approx 1\,M_\odot$, which is consistent with the lower limit obtained from Hα.

Supernova in NGC 4045 (1985B)

125.341 **Supernova in NGC 4045 (1985B).**
Yamamoto Circ., Nos. 2034 – 2036 (1985). In Japanese.

125.342 **Supernova 1985B in NGC 4045.**
IAU Circ., No. 4035 (1985).

Supernova in NGC 4451 (1985G)

125.361 **Supernova in NGC 4451 (1985G).**
Yamamoto Circ., Nos. 2036, 2037 (1985). In Japanese.

125.362 **Supernova 1985G.**
IAU Circ., Nos. 4049, 4050, 4052 (1985).

Supernova in NGC 4618 (1985F)

125.381 **Supernova 1985F (unusual object in NGC 4618).**
Yamamoto Circ., Nos. 2036, 2037 (1985). In Japanese.

125.382 **Supernova 1985F in NGC 4618.**
IAU Circ., Nos. 4042, 4048, 4049 (1985).

Supernova in NGC 3359 (1985H)

125.401 **Supernova 1985H in NGC 3359.**
Yamamoto Circ., No. 2037 (1985). In Japanese.

125.402 **Supernova 1985H in NGC 3359.**
IAU Circ., Nos. 4050, 4052, 4053 (1985).

Supernova in NGC 1023

125.421 **Supernova in NGC 1023.**
IAU Circ., Nos. 4025, 4028, 4031 (1985).

Supernova in NGC 2748 (1985A)

125.441 **Supernova 1985A in NGC 2748.**
IAU Circ., No. 4031 (1985).

Supernova in ESO 510–G48 (1985E)

125.461 **Supernova 1985E in ESO 510–G48.**
IAU Circ., No. 4041 (1985).

Supernova 1985K

125.481 **Supernova 1985K.**
IAU Circ., No. 4062 (1985).

Supernova 1985J

125.501 **Supernova 1985J.**
IAU Circ., Nos. 4066, 4070 (1985).

Supernova in NGC 3227 (1983u)

125.521 **Spectrophotometry of the supernova 1983u in NGC 3227.**
M. M. De Robertis, P. A. Pinto.
Astrophys. J., Lett. Ed., Vol. 293, No. 2, p. L77 – L81 (1985). = Lick Obs. Bull., No. 1007.

High–quality CCD and IDS spectra of the Type I supernova in NGC 3227 are presented. A broad–band light curve is derived from which the date and absolute magnitude at maximum light are estimated. While qualitative differences between other recent Type I's and SN 1983u are noted, it is the overall similarity of Type I spectra that remains most striking. Various features and interpretations of these spectra are discussed.

Supernova in NGC 1058

125.541 **Probable detection of radio emission from the peculiar supernova 1961v in NGC 1058.**
J. J. Cowan, D. Branch.
Bull. Am. Astron. Soc., Vol. 17, No. 2, p. 565 – 566 (1985). Abstract. – See Abstr. 010.065.

Supernova in NGC 991 (1984L)

125.561 **A Type I radio supernova in NGC 991.**
R. A. Sramek, K. W. Weiler, J. M. van der Hulst, N. Panagia.
Bull. Am. Astron. Soc., Vol. 17, No. 2, p. 566 (1985). Abstract. – See Abstr. 010.065.

Archeoastrophysics of supernovae and neutron stars.
See Abstr. 004.017.

A possible Roman record of the supernova of A.D. 185.
See Abstr. 004.124.

Defining the Crab nebula.
See Abstr. 011.002.

Supernovae up close (an after–dinner talk).
See Abstr. 015.043.

The CCD/transit instrument – scientific programs.
See Abstr. 032.030.

Software for an automated supernova search.
See Abstr. 032.033.

Techniques for mapping galactic SNRs with the MOST (*Molonglo Observatory Synthesis Telescope*).
See Abstr. 036.084.

Evaluation of Compton contamination events in an actively shielded Ge gamma–ray detector.
See Abstr. 036.138.

Isovector vibrational modes in heavy nuclei.
See Abstr. 061.069.

Stratification of radiation behind interstellar shock waves.
See Abstr. 062.073.

X–ray astronomy and plasma astrophysics.
See Abstr. 062.090.

The back action of accelerated particles on shock–front structure.
See Abstr. 062.131.

The hydrodynamics of clouds overtaken by supernova remnants. I. Cloud crushing phenomena.
See Abstr. 062.133.

Current–carrying jets.
See Abstr. 062.200.

Continuum spectral flux from Type II supernova model atmospheres.
See Abstr. 064.068.

Shock propagation in supernovae: concept of net ram pressure.
See Abstr. 065.016.

Type II supernova energetics.
See Abstr. 065.028.

Deflagration of white dwarfs as a model for type–I supernovae.
See Abstr. 065.032.

An expanding vortex site for the *r*–process in rotating stellar collapse.
See Abstr. 065.046.

Presupernova core structure and explosive nucleosynthesis.
See Abstr. 065.068.

Supernova theory.
See Abstr. 065.082.

Neutrinos from collapsing stars.
See Abstr. 065.086.

Effects of general relativity on the mass ejection of neutrino–trapping supernovae.
See Abstr. 065.087.

A numerical model for type II supernovae – the collapse stage.
See Abstr. 065.101.

Thomas–Fermi study of the bubble phase in hot dense matter.
See Abstr. 067.044.

Relativistic wind termination: jets and synchrotron nebulae.
See Abstr. 067.055.

A comparison of recent numerical calculations of stellar core collapse.
See Abstr. 067.183.

The origin of neutron stars.
See Abstr. 067.213.

The binary X–ray pulsar 1E 2259+59 – a descendant of an AM Her type system?
See Abstr. 117.088.

Optical pulsations in the Large Magellanic Cloud remnant 0540–69.3.
See Abstr. 126.013.

Beobachtung des Crab im harten Röntgenlicht.
See Abstr. 126.066.

Particle acceleration near pulsars: constraints from synchrotron nebulae.
See Abstr. 126.072.

A search for non–interacting, close binary white dwarfs.
See Abstr. 126.093.

IUE absorption line studies of highly ionized interstellar gas.
See Abstr. 131.001.

High–velocity interstellar gas in the line–of–sight to HD 50896.
See Abstr. 131.006.

Structure and origin of the Cygnus superbubble.
See Abstr. 131.056.

An upper limit on the proper motion of optical filaments near the Cetus Arc (Loop 2).
See Abstr. 131.083.

Star–forming regions near the supernova remnant IC 443.
See Abstr. 131.194.

Observations of radio recombination lines with principal quantum number $456 < \bar{n} < 634$ towards Cas–A.
See Abstr. 131.242.

Report of IAU Commission 34: Interstellar matter (*Matière interstellaire*).
See Abstr. 131.356.

Structure and evolution of superbubbles in the stratified gas distribution.
See Abstr. 131.364.

Observations of the radio source G 6.6–0.1 positionally coincident with the W28 SNR.
See Abstr. 132.009.

The planetary nebula surrounding Tycho's supernova remnant.
See Abstr. 134.045.

A new class of nonthermal radio sources.
See Abstr. 141.001.

G2.4+1.4, a smothered supernova?
See Abstr. 141.006.

Further observations of radio sources from the BG survey. III. CTB 80.
See Abstr. 141.010.

Absolute measurements of the radio radiation of Cassiopeia A, Cygnus A and the Crab nebula in the 60 – 100 cm range.
See Abstr. 141.012.

Radio studies of galactic nonthermal sources.
See Abstr. 141.037.

The compact radio source near G357.7–0.1.
See Abstr. 141.040.

On the astronomical nature of the sources of gamma–ray bursts.
See Abstr. 143.001.

Line features from Cygnus X–1 and the Crab nebula in the energy range 30 – 270 keV.
See Abstr. 143.039.

Possible detection of gamma–ray lines from the Crab Nebula.
See Abstr. 143.069.

Acceleration of cosmic rays in the Loop I "supernova remnant"?
See Abstr. 144.031.

The isotropy of ultra high energy cosmic rays and multiple supernova I galactic source.
See Abstr. 144.136.

Cosmic ray source spectra produced in supernova blast waves.
See Abstr. 144.138.

Shock–wave acceleration out of the hot interstellar medium: effect of charge states on composition.
See Abstr. 144.139.

Are young supernova remnants the acceleration sites of cosmic ray electrons?
See Abstr. 144.150.

Cosmic ray acceleration efficiency by supernova remnants.
See Abstr. 144.151.

Onion–shell model of cosmic ray acceleration in supernova remnants.
See Abstr. 144.152.

Acceleration of electrons in supernova remnants.
See Abstr. 144.153.

Adiabatic expansion of cosmic ray sources and the consequences for secondary antiprotons.
See Abstr. 144.335.

The energy spectrum of cosmic ray electrons in radio loop III.
See Abstr. 144.350.

Cosmic rays acceleration by large scale compressible and incompressible motions of plasma.
See Abstr. 144.368.

Possible characteristics of supernova induced enrichment of globular clusters.
See Abstr. 154.064.

Carbon deflagrating supernovae and the chemical history of the solar neighbourhood.
See Abstr. 155.002.

Loop I (the North Polar Spur) region – a quasi radio halo.
See Abstr. 155.041.

A 10–GHz radio–continuum survey of the galactic–plane region at the Nobeyama Radio Observatory – a complex region at $l = 22° – 25°$.
See Abstr. 155.050.

Nonequilibrium models for LMC supernova remnants.
See Abstr. 156.007.

A statistical study of the supernova remnants in the Large Magellanic Cloud.
See Abstr. 156.020.

Supernova remnants and the carbon abundance in M33.
See Abstr. 157.001.

Radio detection of historical supernovae and H II regions in M83.
See Abstr. 157.208.

Shock formation of the broad emission–line regions in QSOs and active galactic nuclei.
See Abstr. 158.092.

Similarity between the extended components of radio galaxies and plerion–type supernova remnants.
See Abstr. 158.130.

The nucleus of M82 at radio and X–ray bands: discovery of a new radio population of supernova candidates.
See Abstr. 158.142.

Detection of a supernova in the host galaxy of the QSO 1059 + 730.
See Abstr. 159.058.

Tycho's supernova and the Hubble constant.
See Abstr. 161.095.

Deuterium and ^{3}He formation by supernovae of the first generation.
See Abstr. 161.176.

Determining H_0 and q_0 from supernova spectra.
See Abstr. 161.204.

126 Degenerate Stars, White Dwarfs, Pulsars

126.001   White dwarf stars at ultraviolet wavelengths.
J. Liebert.
NASA Conf. Publ., NASA CP–2349, p. 93 – 100 (1984). – See Abstr. 012.001.

Several areas of white dwarf research which have been advanced by IUE observations are briefly reviewed.

126.002   IUE and Voyager observations of very hot O–type subdwarfs.
J. S. Drilling, J. B. Holberg, D. Schönberner.
NASA Conf. Publ., NASA CP–2349, p. 249 – 253 (1984). – See Abstr. 012.001.

The authors have observed 12 newly–discovered subdwarf O stars with IUE and have obtained complementary observations of two of these with Voyager 1. They find that these stars all have effective temperatures of at least 60,000K and that they occupy a region in the H–R diagram between the hottest white dwarfs and the central stars of planetary nebulae. Evidence is presented that the abundances of H, He, C, N, and possibly Fe vary widely from star to star. It is believed that these stars represent the state of evolution immediately preceding the white dwarf state, and that the abundance differences are due to diffusion and convective mixing in surface layers.

126.003   Analysis of high–dispersion IUE spectra of subdwarf B stars.
R. Lamontagne, F. Wesemael, G. Fontaine, E. M. Sion.
NASA Conf. Publ., NASA CP–2349, p. 254 – 257 (1984). – See Abstr. 012.001.

High–dispersion IUE spectra of the subdwarf B stars UV 1758 + 36 and Ton S–227 are presented. These spectra are characterized by the presence of a large number of photospheric low– and medium–excitation lines from numerous ions. The lines of C, N, and Si are used in conjunction with LTE metal line calculations to derive preliminary abundances for these elements. These results are compared and contrasted with those obtained for the hotter OB and O subdwarfs.

126.004   Further investigations of mass loss phenomena in hot white dwarfs and O subdwarfs.
F. C. Bruhweiler.
NASA Conf. Publ., NASA CP–2349, p. 269 – 272 (1984). – See Abstr. 012.001.

Previous *IUE* studies have found sharp, shortward–displaced absorption features in two DA white dwarfs (Bruhweiler and Kondo 1981, 1983) and in three O subdwarfs (Bruhweiler and Dean 1983) indicative of mass loss. Continued analysis of new and archival data has added one DA white dwarf and possibly two O subdwarfs to the growing list of subluminous stars displaying these features. One, possibly both, of these new O subdwarfs are of extreme low luminosity, intermediate between normal O subdwarfs and the hot white dwarfs. These observations provide evidence that a single mechanism is responsible for the unusual features seen in O subdwarfs and DA white dwarfs.

126.005   *IUE* spectrophotometry of the hot helium–rich PG1159 DO degenerates.
E. M. Sion, J. Liebert, S. Starrfield, F. Wesemael.
NASA Conf. Publ., NASA CP–2349, p. 273 – 276 (1984). – See Abstr. 012.001.

The authors have obtained low resolution IUE spectra of four PG1159 stars, PG1151–029, PG1424 + 535, PG1520 + 525 and an additional image of PG1159–035 as well as an optical ultraviolet spectrum of PG2131 + 066. The IUE (SWP) spectra suggest the presence of numerous metallic absorption features of C III, N III, Si IV, S III, C IV, N V and a few unidentified features. The metal absorption lines and He II have equivalent widths of a few angstroms. IUE/optical energy distributions are considered. Tentative identifications of C IV absorptions and, possibly, weak O VI features in the authors' optical ultraviolet reticon spectra

suggest a probable link to the subluminous Wolf–Rayet O VI planetary nuclei. The PG1159 DO degenerates are the hottest known ($T_e \gtrsim 10^5$K), high gravity (log g $\gtrsim$ 7) objects.

126.006   Self–consistent recalibration of IUE and determination of hot DA white dwarf effective temperatures.
D. S. Finley, G. Basri, S. Bowyer.
NASA Conf. Publ., NASA CP–2349, p. 277 – 280 (1984). – See Abstr. 012.001.

The authors have analyzed the spectra of over twenty stars identified as DA white dwarfs, with temperatures in the range 20,000 to 70,000K. In addition to the IUE data, they have collected all other available data on these stars from which temperature estimates may be made and compared the observed IUE fluxes with the FUV fluxes predicted by using the observed V magnitudes and temperatures obtained from non–IUE data. These comparisons indicate a need to revise the overall IUE calibration by + 10% in F_λ. The authors computed the appropriate correction as a function of wavelength, and applied this correction to the measured IUE fluxes. They then obtained temperatures with significantly more accuracy than is achievable with optical photometry.

126.007   Carbon, silicon, and oxygen in the white dwarf star GD 40.
H. L. Shipman.
NASA Conf. Publ., NASA CP–2349, p. 281 – 284 (1984). – See Abstr. 012.001.

An IUE spectrum of the metal–rich DB white dwarf star GD 40 shows no unequivocal spectral features in the range of the SWP camera (1200 – 2000 Å). This wavelength region contains resonance lines of many common elements: H I, C I, C II, Si II, and O I. Upper limits to the abundances of H, C, Si, and O are derived from model atmosphere calculations. These limits, combined with the detection of Ca II, Mg II, and Fe II, conflict with the predictions of diffusion theories, which have been widely invoked to explain the presence of metals in the spectra of the He–rich white dwarf stars.

126.008   Analysis of white dwarf Lyman α profiles.
J. B. Holberg, F. Wesemael.
NASA Conf. Publ., NASA CP–2349, p. 285 – 288 (1984). – See Abstr. 012.001.

Through the use of a doubly exposed large and small aperture SWP image it is possible to obtain Lyman α profiles for white dwarfs which are relatively free from geocoronal Lyman α contamination. For DA white dwarfs the analysis of both wings of these broad profiles yields well–determined temperatures and useful constraints upon surface gravity. Two examples of such an analysis are discussed, and the results compared with temperatures determined from Voyager far UV (912 – 1150 Å) fluxes.

126.009   The white dwarf companion of the mild Ba star ξ¹Cet.
E. Böhm–Vitense, C. Proffitt, H. Johnson.
NASA Conf. Publ., NASA CP–2349, p. 293 – 294 (1984). – See Abstr. 012.001.

The mild Ba star ξ^1Cet was found to have a hot companion. The absolute intensities and the relative energy distribution shows that it is a DA white dwarf with broad absorption bands around 1400 and 1650 Å. The temperature is determined to be $T_{eff} = 13000 \pm 1000$K.

126.010   Unpulsed radio emission from pulsars.
T. E. Perry, A. G. Lyne.
Mon. Not. R. Astron. Soc., Vol. 212, No. 2, p. 489 – 496 (1985).

The authors have searched for unpulsed radio emission from 25 pulsars. Four of these pulsars were found to exhibit a significant unpulsed flux, the ratio of the peak flux to unpulsed flux (beaming ratio) lying between 55 and 2300; for the others the authors find lower limits for the beaming ratio ranging from 40 to 16000. For the pulsar PSR 1929 + 10 it is also demonstrated that the unpulsed flux comes from a region within about 10^5km of the neutron star. The authors also report two previously undetected pulsed features: one situated at a longitude of 120° relative to the main pulse in the integrated profile of PSR 1929+10, and the other at a longitude of 225° in the integrated profile of PSR 1818–04.

126.011 The explanation of the 1400 and 1600 Å features in DA white dwarfs.
D. Koester, V. Weidemann, E.-M. Zeidler-K. T., G. Vauclair.
Astron. Astrophys., Vol. 142, No. 1, p. L5 – L8 (1985).

The authors present new IUE observations of 4 cool DA white dwarfs together with additional data for 8 objects from the literature. All objects show absorption features near 1400 and/or 1600 Å, which were already noted by other authors. These features are explained as due to quasi–molecular absorption of the H_2 (1600 Å) resp. H_2^+ (1400 Å) molecules.

126.012 A second measurement of a pulsar braking index.
R. N. Manchester, J. M. Durdin, L. M. Newton.
Nature, Vol. 313, No. 6001, p. 374 – 376 (1985).

The braking index, n, of a pulsar, which describes the dependence of the braking torque on rotation frequency, is a fundamental parameter of pulsar electrodynamics. The authors present radio timing observations of the X–ray and radio pulsar, PSR 1509–58, made over an approximately 2–year interval, which yield a second significant measurement of $n = 2.83 \pm 0.03$, more than that for the Crab pulsar, but less than the canonical value of 3.

126.013 Optical pulsations in the Large Magellanic Cloud remnant 0540–69.3.
J. Middleditch, C. Pennypacker.
Nature, Vol. 313, No. 6004, p. 659 – 661 (1985).

The X–ray pulsar PSR 0540–693 was discovered in the Large Magellanic Cloud supernova remnant, 0540–69.3, by Seward, Harnden and Helfand, as a pulse, with repetition period ~ 50 ms, in Einstein Observatory data. Previously, Clark et al. had noted that this remnant resembles the Crab Nebula because of the X–ray power law spectrum and suggested that the nebular emission was synchrotron radiation powered by a central pulsar. The authors have now detected pulsed optical emission for the X–ray pulsar, having a time–averaged magnitude of ~ 22.7.

126.014 A search for short–period pulsars.
R. N. Manchester, N. D'Amico, I. R. Tuohy.
Mon. Not. R. Astron. Soc., Vol. 212, No. 4, p. 975 – 986 (1985).

A search for short–period pulsars using the Parkes radio telescope at a frequency of 1.4 GHz has yielded four new radio pulsars. The search had a limiting sensitivity of ~ 1 mJy for pulsars of period >10 ms and dispersion measures <1600 cm^{-3}pc. It was directed at objects in the Galaxy and Magellanic Clouds likely to contain young pulsars such as supernova remnants and γ–ray sources. Only one of the four pulsars detected, PSR 1509–58, which was previously known as an X–ray pulsar, is apparently associated with the target object although one other pulsar–supernova remnant association is possible.

126.015 The influence of the twist effect of magnetic field lines on the polarization of pulsar radio emission.
Yu. P. Shitov.
Astron. Zh., Tom 62, Vyp. 1, p. 54 – 65 (1985). In Russian. English translation in Sov. Astron., Vol. 29, No. 1.

The twist phenomenon of pulsar magnetic field lines in the direction opposite to its rotation is caused by the braking torque of electromagnetic nature which leads to the slow–down of the pulsar's rotation. The expected character of the deformation of the open magnetic–field line cone (as compared to the undisturbed dipole field) is shown which is "perceived" by the relativistic particles moving along these lines with the assumption that the braking torque is caused mainly by the magnetic–dipole radiation reaction.

126.016 Superdispersion pulse delay of the pulsar PSR 0809+74 at metre wavelengths.
Yu. P. Shitov, V. M. Malofeev.
Pis'ma Astron. Zh., Tom 11, No. 2, p. 94 – 100 (1985). In Russian. English translation in Sov. Astron. Lett., Vol. 11.

A superdispersion time delay of the integrated pulse profile is discovered at metre wavelengths for the pulsar PSR 0809+74. After excluding interstellar dispersion this extra delay increases abruptly as the wavelength increases achieving ≈ 150 ms at 30 MHz. The superdispersion delay is interpreted as the radio emission cone distortion reflecting the twist effect of the pulsar's magnetic field lines due to electromagnetic braking of the pulsar's rotation.

126.017 Non–radial oscillations of white dwarfs and neutron stars.
B. W. Carroll, H. M. Van Horn.
News Lett. Astron. Soc. N.Y., Vol. 2, No. 7, p. 5 – 6 (1985). Abstract. – See Abstr. 010.241.

126.018 Magnetic field decay in white dwarfs.
C. Wendell.
News Lett. Astron. Soc. N.Y., Vol. 2, No. 7, p. 11 (1985). Abstract. – See Abstr. 010.241.

126.019 Le pulsar binaire. Un nouveau laboratoire où étudier la gravitation.
N. Deruelle.
Astronomie, Vol. 99, p. 111 – 116 (1985).

126.020 Magnetic–field decay and pulsar cooling.
J. F. Lodenquai.
Lett. Nuovo Cimento, Vol. 41, Ser. 2, No. 4, p. 119 – 122 (1984). Abstr. in Phys. Abstr., Vol. 88, No. 1249, Entry 14650 (1985).

126.021 On the post–Newtonian effects in the millisecond pulsar 1937+214.
R. K. Kochhar, C. Sivaram.
Curr. Sci., Vol. 53, No. 8, p. 420 – 421 (1984). Abstr. in Phys. Abstr., Vol. 88, No. 1250, Entry 18852 (1985).

126.022 Pulsars and interstellar medium: multiple regression analysis of related parameters.
E. Antonello, M. Fracassini.
Astrophys. Space Sci., Vol. 108, No. 1, p. 187 – 193 (1985).

An empirical relation which relates the 408 MHz galactic continuum background temperature (408GCBT) to dispersion measures, position and radio–luminosity of 325 pulsars is obtained. This relation shows that pulsars may be considered as galactic probes for the distribution of 408GCBT and interstellar electron density in interstellar medium.

126.023 Episodic accretion events on to a white dwarf.
D. Prialnik, M. Livio.
Mon. Not. R. Astron. Soc., Vol. 213, No. 2, p. 407 – 415 (1985).

The authors investigate by means of a hydrodynamic Lagrangian implicit code the collapse of a spherically symmetric gas cloud (of $5 \times 10^{-5} M_\odot$ or $10^{-7} M_\odot$) on to a 1 $M_\odot$ white dwarf. The gas cloud, extending from the surface of the white dwarf to several solar radii, is initially at rest, in thermal equilibrium and optically thick. Two models are constructed: (A) The accretion shock reaches the white–dwarf's surface, and (B) only ~ 70 per cent of the accretion shock reaches the surface. The relevance of the latter model to recurrent and dwarf novae is discussed.

126.024 A model of radio emission from the millisecond pulsar PSR 1937+214.
J. Gil.
Astron. Astrophys., Vol. 143, No. 2, p. 443 – 446 (1985).

A model of radio emission from the millisecond pulsar PSR 1937+214 is proposed. The emission region is expected to lie very close to the polar cap surface, where the contribution of the quadrupole component to the actual magnetic field may be significant. It is argued that the interpulse as well as the main pulse are emitted from the same magnetic pole.

126.025 Radio sources near the millisecond pulsar PSR 1937+214.
J. L. Caswell, T. L. Landecker, P. A. Feldman.
Astron. J., Vol. 90, No. 3, 488 – 492 (1985).

A 2–deg field surrounding the millisecond pulsar PSR 1937+214 has been mapped in the continuum at 1.4 GHz with a beam of 1 × 3 arcmin. The survey, made with the synthesis telescope at the Dominion Radio Astrophysical Observatory, provides a good combination of sky coverage, sensitivity, and resolution to search for emission from any associated supernova remnant of quite large angular size. Several extended emission regions lie in the field, but show no clear link with the pulsar. Observations of 21 small–diameter radio sources in the field are also summarized, but there is no evidence for any of these sources being associated with the pulsar.

126.026 Spectrum synthesis study of selected ultraviolet metal lines in hot DA white dwarf stars.
R. B. C. Henry, H. L. Shipman, F. Wesemael.
Astrophys. J., Suppl. Ser., Vol. 57, No. 1, p. 145 – 150 (1985).

Using pure hydrogen atmospheres of $\log g = 8$, the authors have calculated equivalent widths as a function of temperature and abundance for C II λ1335, C III λ977 and λ1176, C IV λ1549, N II λ1085, N V λ1240, Mg II λ2800, Si II λ1263, and Si IV λ1395. The calculations were carried out over a temperature range of $15{,}000 \leqslant T_{\mathrm{eff}} \leqslant 100{,}000$K and an abundance range of $-3 \geqslant \log[N(\mathrm{Z})/N(\mathrm{H})] \geqslant -7$, where Z refers to the relevant element.

126.027 Vortex creep and the internal temperature of neutron stars: the Crab pulsar and PSR 0525+21.
M. A. Alpar, R. Nandkumar, D. Pines.
Astrophys. J., Vol. 288, No. 1, p. 191 – 195 (1985).

The behavior of the Crab pulsar following the 1969 and 1975 glitches and the postglitch behavior of the old pulsar PSR 0525+21 are explained as resulting from internal torques due to the thermal creep of vortex lines in pinned superfluid regions. It is found that the postglitch behavior of both pulsars can be fitted with the hypothesis that the internal structure and vortex pinning sites in these pulsars are essentially the same as those hypothesized previously for the Vela pulsar, so that the only significant differences between these three pulsars reside in the spin down rates and internal temperatures. The temperature obtained for PSR 0525+21 supports the hypothesis that dissipation due to vortex creep is the dominant heat source in old pulsars.

126.028 The galactic population of pulsars.
A. G. Lyne, R. N. Manchester, J. H. Taylor.
Mon. Not. R. Astron. Soc., Vol. 213, No. 3, p. 613 – 639 (1985).

In order to draw statistical conclusions about the overall population of pulsars in the Galaxy, the authors have analysed a sample of 316 pulsars detected in surveys carried out at Jodrell Bank, Arecibo, Molonglo, and Green Bank. They quantify the important selection effects of each survey, and describe a statistically reliable pulsar distance scale. The authors conclude that the Galaxy contains approximately 70000 potentially observable pulsars with luminosities above 0.3 mJy kpc^2. The distribution in galactocentric radius rises smoothly through the solar location towards the galactic centre and peaks inside 6 kpc. The galactic z distribution has a scale height of about 400 pc, much larger than that of the OB stars from which pulsars form. The authors then consider the period and luminosity evolution of pulsars.

126.029 Discovery of a millisecond binary pulsar: P 1953+29.
V. Boriakoff, R. Buccheri, F. Fauci.
News Lett. Astron. Soc. N.Y., Vol. 2, No. 4, p. 32 (1983). Abstract. – See Abstr. 010.242.

126.030 Pulse nulling in pulsar radio emission.
W. Deich, J. Cordes.
News Lett. Astron. Soc. N.Y., Vol. 2, No. 5, p. 23 (1984). Abstract. – See Abstr. 010.243.

126.031 Two new color–selected magnetic DA white dwarfs.
J. Liebert, G. D. Schmidt, E. M. Sion, S. G. Starrfield, R. F. Green, T. A. Boroson.
Publ. Astron. Soc. Pac., Vol. 97, No. 588, p. 158 – 164 (1985).

The authors report the discovery of two fairly hot magnetic white dwarfs showing spectra with hydrogen Zeeman components. Dipolar field models show that PG 1533–057 has a polar field (B_p) strength of 31 megagauss (MG) while K813–14 has $B_p = 29$ MG. A third object, previously discovered by Downes and Margon, fits $B_p = 18$ MG. All have temperatures in the 11,000K – 20,000K range. These three DA stars have surface field strengths similar to the primaries of AM Herculis binary systems. The authors discuss the possibility that a significant fraction of the isolated magnetic degenerates could be progeny of magnetic accreting binary systems.

126.032 Atmospheric analysis of the carbon white dwarf G227–5.
G. Wegner, D. Koester.
Astrophys. J., Vol. 288, No. 2, p. 746 – 750 (1985).

Observations and a model atmosphere analysis of the spectrum of the white dwarf G227–5 ($= 1728+56$) are reported. This star's spectrum is dominated by lines of C I, but weaker features attributed to H I and He I are also found. With the gravity fixed at $\log g = 8$, the following atmospheric parameters are derived: $T_{\mathrm{eff}} = 12{,}500 + 500$K, and the number abundances C:He $= (3\pm1)\times 10^{-3}$ and H:He $= (2\pm1)\times 10^{-4}$. The evolutionary state of G227–5 is also examined.

126.033 The effective temperature of Wolf 485A and the statistics of ZZ Ceti stars.
F. Wesemael, G. Fontaine.
Astrophys. J., Vol. 288, No. 2, p. 764 – 768 (1985).

High–speed photometry shows that the bright DA white dwarf Wolf 485A is not a variable star, with an upper limit on the semi–amplitudes of luminosity variations of 0.0005 mag in the period range characteristic of the ZZ Ceti stars. However, its time–averaged multichannel color index ($G–R$) places it in the instability strip. New photometric and spectrophotometric observations conclusively show, however, that the effective temperature of Wolf 485A is around 15,000K, substantially above the blue edge of the instability strip ($T_e \approx 13{,}000$K). The source of this discrepancy most probably lies in erroneous multichannel colors.

126.034 On the theory of micropulses of the radio emission of pulsars.
A. B. Mikhajlovskij, O. G. Onishchenko, A. I. Smolyakov.
Pis'ma Astron. Zh., Tom 11, No. 3, p. 190 – 195 (1985). In Russian. English translation in Sov. Astron. Lett., Vol. 11.

It is proposed that the microstructure of pulsar radio emission may be explained as a result of propagation of nonlinear electromagnetic wave packets in pulsar plasma.

126.035 Emission lines in the magnetic white dwarf GD 356.
J. L. Greenstein, J. K. McCarthy.
Astrophys. J., Vol. 289, No. 2, p. 732 – 747 (1985).

GD 356 (1639+53, Gr 329) is a unique white dwarf showing the resolved Zeeman triplets of Hα and Hβ as emission lines. Multichannel spectrophotometry suggests $M_V = +13.3$. Measurements made at high signal–to–noise ratio with the double–spectrograph and CCD detectors at the Hale reflector are presented and analyzed. Multichannel information gives the temperature as 7500K; any red companion must be fainter than $M_V = +15$. The Balmer decrement is flat, with Hβ apparently stronger than Hα. The flat decrement suggests that the emission–line region has high density ($N_e \approx 10^{14}\mathrm{cm}^{-3}$) and is optically thick in Lyα. Its radial extent must be small; i.e., it is a chromosphere rather than a corona. Possible explanations for the existence of the emission–line region in GD 356 are examined.

126.036   Two–phase ultraviolet spectrophotometry of the pulsating white dwarf ZZ Piscium.
A. V. Holm, R. J. Panek, F. H. Schiffer III, H. E. Bond,
E. Kemper, A. D. Grauer.
Astrophys. J., Vol. 289, No. 2, p. 774 – 781 (1985).

The authors obtained spectra of the pulsating white dwarf ZZ Psc (= G 29–38) using the IUE. By using a multiple–exposure technique in conjunction with simultaneous ground–based exposure–metering photometry, they were able to obtain mean on–pulse and off–pulse spectra in the 1950 – 1310 Å wavelength range. The ratio of the time–averaged on–pulse to off–pulse spectra is best fitted by a temperature variation that is in phase with the optical light variation. This result is consistent with the hypothesis that the observed variation is due to a high–order non-radial pulsation.

126.037   Further identifications of hydrogen in Grw + 70°8247.
J. L. Greenstein, R. J. W. Henry, R. F. O'Connell.
Astrophys. J., Lett. Ed., Vol. 289, No. 2, p. L25 – L29 (1985).

Numerous spectra of the magnetic white dwarf Grw + 70°8247 were obtained with the double CCD spectrograph at Palomar with signal–to–noise ratio above 100. Five weak absorptions in the blue and four in the red are produced by hydrogen in a strong magnetic field. The authors compute new, precise, multiparameter, variational energy levels yielding intensity and wavelength, from 100 to 600 MG. They search for those Zeeman components with wavelengths nearly constant with field. Two such, which are relatively sharp and strong π transitions explain the Minkowski band, 4137 Å, as the $2s0$–$4f0$ and the 5855 Å band as the $2s0$–$3p0$.

126.038   Identification of the 1400 and 1600 Å features observed in the ultraviolet spectra of DA white dwarfs.
E. P. Nelan, G. Wegner.
Astrophys. J., Lett. Ed., Vol. 289, No. 2, p. L31 – L33 (1985).

The $\lambda1600$ absorption feature in the spectra of cool DA white dwarf stars has been identified as a resonance broadening of Lyα due to the hydrogen quasi molecule. Likewise, the $\lambda1400$ absorption feature in the spectra of cool and moderately warm DA white dwarf stars appears to be due to a Lyα satellite line arising from the hydrogen ion quasi molecule. The strength of both features is gravity sensitive and therefore promises to be an excellent indicator of surface gravity.

126.039   A search for millisecond pulsars in globular clusters.
T. T. Hamilton, D. J. Helfand, R. H. Becker.
Astron. J., Vol. 90, No. 4, p. 606 – 608 (1985).

Stimulated by the suggestion that the progenitors of millisecond pulsars are low–mass X–ray binary systems, the authors have surveyed a dozen relatively nearby globular clusters with the VLA in a search for fast pulsar candidates. In general, the number of detected radio sources is consistent with that expected from background source counts; the one source found within 1 cluster core radius does, however, have an anomalously steep radio spectrum and deserves follow–up observation. The implications of the results for millisecond pulsar formation scenarios are discussed.

126.040   Frequency drift in pulsar scintillation.
F. G. Smith, N. C. Wright.
Mon. Not. R. Astron. Soc., Vol. 214, No. 1, p. 97 – 107 (1985).

The dynamic spectra of interstellar scintillation at 408 MHz in 32 pulsars give values for bandwidth and time–scale, and for the characteristic rate of frequency drifting in spectral features. Frequency drift in scintillation patterns is related to large–scale structure in the electron density variations in the medium, while the bandwidth and time–scale of scintillation depend on smaller scale structure. A comparison of these two types of pattern characteristics show conformity with a Kolmogorov spectrum of irregularities extending over a range of scales from 10^9 to 10^{12}m.

126.041   Changing parameters along the path to the Vela pulsar.
P. A. Hamilton, P. J. Hall, M. E. Costa.
Mon. Not. R. Astron. Soc., Vol. 214, No. 1, p. 5P – 8P (1985).

Recent observations of the Vela pulsar, PSR 0833–45, confirm that the rotation measure of the source is increasing, and show that at the same time the dispersion measure is decreasing. The observations are interpreted in terms of a magnetized cloud moving out of the line–of–sight to the source.

126.042   Dependence of pulsewidth on the age of pulsars – a comment on Candy and Blair's analysis.
S. Krishnamohan.
Mon. Not. R. Astron. Soc., Vol. 214, No. 1, p. 15P – 17P (1985).

Candy and Blair have found that pulsewidth depends on the characteristic age of a pulsar and suggest that the emission cone aligns with the spin axis on a time–scale of $\sim 10^7$yr. A re-examination of the data used by them reveals no evidence for the reported dependence.

126.043   KPD 0005 + 5106: a post–PG 1159 type object?
R. A. Downes, J. Liebert, B. Margon.
Astrophys. J., Vol. 290, No. 1, p. 321 – 324 (1985).

In a survey for galactic plane UV–excess objects, the authors have discovered a very hot ($T \approx 10^5$K) helium–rich white dwarf. The object is spectroscopically similar to the pulsating PG 1159 class of stars, with narrow emission lines of He II, probably C IV, and possibly C III/N III, although there is no trace of carbon absorptions. This object appears to be an old PG 1159 star, having ceased pulsations during evolution toward a more typical DO white dwarf.

126.044   Temperatures and luminosities of white dwarfs in dwarf novae.
J. Smak.
Acta Astron., Vol. 34, No. 3, p. 317 – 323 (1984).

Far ultraviolet radiation observed in dwarf novae at minimum can only be attributed to their white dwarfs. In three systems white dwarfs are detected directly through their eclipses. These data are used to determine the effective temperatures and luminosities of white dwarfs. The resulting temperatures range from about $\log T_e = 4.1$ to about 4.9. The luminosities range from about $\log L_1 = 31.0$ to about 33.5. Radiation from white dwarfs is likely to be the source of excitation of the emission lines from disks. It is also argued that the heating by the white dwarfs can significantly modify the structure of the innermost parts of the disk and inhibit the incidence of thermal instability in that region.

126.045   A geometrical depolarization of pulsar radiation.
J. Gil, W. Rudnicki.
Astrophys. Space Sci., Vol. 109, No. 2, p. 381 – 386 (1985).

It is assumed that pulsar radiation originates in a polar cap region and that the emission mechanism is curvature radiation. It is further assumed that radiation reaching an observer at any time may represent contributions from several particle bunches moving relativistically along different magnetic field lines and radiating mutually incoherently. These assumptions are used to explain the minimum of linear polarization appearing near the profile centre of some pulsars.

126.046   Periodicity in the pulse microstructure of the radio emission of the pulsar PSR 1133+16 from results of three–frequency observations at meter wavelengths.
O. A. Kuz'min.
Astron. Zh., Tom 62, Vyp. 2, p. 234 – 239 (1985). In Russian. English translation in Sov. Astron., Vol. 29, No. 2.

Single pulses of PSR 1133+16 were observed simultaneously at the frequencies 102.5, 78.9 and 67.5 MHz with 10 µs time resolution. An analysis of the temporal fine structure with particular reference to the periodicity in the pulse microstructure was performed. The appearance of the periodicity in the pulse microstructure simultaneously at different frequencies is not fully cor-related.

126.047 On the orientation of radiation of pulsars.
  O. Kh. Gusejnov, I. M. Yusifov.
Astron. Zh., Tom 62, Vyp. 2, p. 240 – 251 (1985). In Russian.
English translation in Sov. Astron., Vol. 29, No. 2.

One of the important problems in the pulsar investigation is to
determine the intrinsic width of the radiation beam (2Θ), the
angle α between the magnetic axis and the rotation axis, and their
time evolution. The authors made an attempt to solve this problem concerning not every concrete pulsar, but PSRs as a whole.
They selected the pulsars with well–determined estimates of the
maximum velocity of the change in the polarization plane. The
angle α and the intrinsic width of the radiation diagram (2Θ)
turned out to undergo considerable changes in time by a factor
of ~ 5. This naturally leads to a great difference in the probability
of detecting young and old pulsars. The probability of detecting
PSRs with interpulses is also considered.

126.048 On two types of pulsars.
  I. F. Malov.
Astron. Zh., Tom 62, Vyp. 2, p. 252 – 257 (1985). In Russian.
English translation in Sov. Astron., Vol. 29, No. 2.

The arguments favouring the subdivision of pulsars into two
classes are considered: 1) short–period pulsars described by
Smith's model and 2) long–period pulsars for which the hollow–
cone model is valid. These arguments are: the difference in the
period dependence of the pulse width and of the total change of
the position angle of linear polarization through the integrated
profile; only type 1 pulsars have an interpulse; the difference in
the rms deviations in the times of arrival of pulses and in the
shapes of the direction pattern of the radio emission; the difference of the slowing–down mechanisms. The data for the recently
discovered PSR $1937+21$ ($P = 1.56$ ms) are shown to be in good
agreement with the above conception, this pulsar being a typical
representative of the first group of pulsars.

126.049 On the problem of evolution of pulsars.
  Yu. I. Neshpor.
Astron. Zh., Tom 62, Vyp. 2, p. 408 – 410 (1985). In Russian.
English translation in Sov. Astron., Vol. 29, No. 2.

On the basis of the analysis of published data the braking index
$n = 3.1 \pm 0.2$ was estimated for pulsars whose total luminosity
exceeds 2×10^{32} erg/s and is less than 5×10^{35} erg/s. Such an estimate of the braking index allows the suggestion that the evolu-
tion of the pulsars in question seems to proceed in accordance
with the law describing the energy losses by a rotating magnetic
dipole.

126.050 Spatial distribution of pulsars.
  O. Kh. Guseinov (*O. Kh. Gusejnov*), I. M. Yusifov.
Sov. Astron., Vol. 28, No. 4, p. 415 – 423 (1984). English transla-
tion of 38.126.009.

**126.051 Observations of interstellar scattering of radio emission
  of pulsar PSR 0329 + 54 at 102.5 MHz.**
M. V. Popov, V. A. Soglasnov.
Sov. Astron., Vol. 28, No. 4, p. 424 – 427 (1984). English translation of 38.126.010.

126.052 Pulsar space velocities.
  A. V. Tutukov, N. N. Chugaj, L. R. Yungel'son.
Sov. Astron. Lett., Vol. 10, No. 4, p. 244 – 247 (1984). English
translation of 38.126.018.

126.053 Electron outflow in pulsar magnetospheres.
  R. R. Burman.
Proc. Astron. Soc. Aust., Vol. 5, No. 4, p. 467 – 469 (1984).

The author uses the Mestel, Wang and Westfold
(1984)–formalism to see what types of moderately accelerated
flow from pulsar surfaces are predicted by their dissipation–free
flow theory incorporating a significant emission speed.

126.054 The luminosity distribution of pulsars.
  S. Pineault.
ESA Spec. Publ., ESA SP–207, p. 307 – 309 (1984). – See Abstr.
012.044.

The luminosity distribution of pulsars is investigated. A gap in
the middle of the observed distribution at $\log L_{gap} = 27.5$
(cgs units) is found to be statistically significant. The implications
are discussed in terms of the possible existence of two distinct
pulsar luminosity classes, considering in particular the standard
magnetic dipole model, the disk model and a recently proposed
model in which superfluid neutron vortexes provide the dominant torque. An alternative explanation for the gap is that pulsars
either evolve very rapidly or temporarily cease to radiate as they
reach the critical luminosity L_{gap}.

**126.055 Discriminant analysis of pulsar groups in the diagram
  $\dot{P}$ vs P.**
M. Fracassini, L. E. Pasinetti, G. Raffaelli.
ESA Spec. Publ., ESA SP–207, p. 315 – 317 (1984). – See Abstr.
012.044.

The physical meaning of groups of pulsars (and gaps) in the
diagram $\dot{P}$ vs P is analysed by means of the discriminant analysis.
The use of the true age of pulsars allows to obtain two orthogonal
discriminant functions, F_1 and F_2, which give a better statistical
estimate of the main physical parameters leading the distribution
of pulsars in the diagram F_1 vs F_2 (transformed of the diagram
$\dot{P}$ vs P), according to the evolutionary models proposed by some
authors.

126.056 Pulsars and interstellar medium.
  M. Fracassini, L. E. Pasinetti, E. Antonello.
ESA Spec. Publ., ESA SP–207, p. 319 – 321 (1984). – See Abstr.
012.044.

A statistical relation between 408 MHz galactic continuum
background temperature (T_{408}) and dispersion measure, position, radio–luminosity of 325 pulsars (PSR) was obtained by
means of multiple stepwise regression analysis. This relation
shows that PSR may be considered as probes of the galactic
places where the observed T_{408} and the electron density (n_e) of
the interstellar medium (ISM), derived by large–scale models, are
correlated. The n_e of this relation is that of all the electrons of the
ISM which contribute to T_{408} with thermal and non–thermal
emissions. Also a contribution from the intrinsic luminosity of
PSR to T_{408} cannot be excluded.

126.057 Brief search for low–mass objects.
  C. K. Kumar.
Publ. Astron. Soc. Pac., Vol. 97, No. 589, p. 294 – 296 (1985).

The L (3.4 µ) band fluxes of two nearby white dwarfs,
van Maanen 2 and EG 290, were measured to look for low–mass
companions. None were detected.

126.058 The drifting subpulses of PSR 0148–06.
  J. D. Biggs, P. A. Hamilton, P. M. McCulloch,
R. N. Manchester.
Mon. Not. R. Astron. Soc., Vol. 214, No. 3, p. 47P – 52P (1985).

Drifting subpulses have been detected at 645 MHz in observations of individual pulses from PSR 0148–06. The integrated
pulse profile has two well–resolved components. Distinct drifting
subpulses are visible in the earliest component whilst, in the later
component, drifting is weak. Both components have a strong
periodic amplitude modulation associated with the drifting subpulses.

**126.059 Fluctuations of pulsar emission with sub–microsecond
  time–scales.**
J. Gil.
Astrophys. Space Sci., Vol. 110, No. 2, p. 293 – 296 (1985).

A model of pulsar emission fluctuations down to sub–
microsecond time–scales is presented. The model is consistent
with the empirical amplitude modulated noise model. It is proposed that the time–scale of the shortest observed micropulses is
determined by the coherence of single–particle curvature radiation.

126.060 On the progenitors of white dwarfs.
J. S. Drilling, D. Schönberner.
Astron. Astrophys., Vol. 146, No. 2, p. L23 – L24 (1985). =
Contrib. Louisiana State Univ. Obs., No. 191.

Direct observational evidence is presented which indicates that
the immediate progenitors of white dwarfs are the central stars of
planetary nebulae (∼80%), other post–AGB objects (∼20%),
and post–HB objects not massive enough to climb the AGB
(∼0.7%). The combined birth rate for these objects
$(2 - 3 \times 10^{-12} \mathrm{pc}^{-3}\mathrm{yr}^{-1})$ is in satisfactory agreement with the
death rate of main–sequence stars and the birth rate of white
dwarfs.

126.061 Pairs of pulsars possibly joined in the past in binary
systems.
B. M. Vladimirskij.
Astrofizika, Tom 21, Vyp. 3, p. 535 – 545 (1984). In Russian. English translation in Astrophysics, Vol. 21, No. 3.

A method to recognize pairs of pulsars joined in the past in
binary systems is presented. This method uses essentially some
regularities derived from statistical analysis. A catalogue of pairs
of possible components of disrupted binary systems includes 27
pairs. A correlation was revealed between the distance of the
components of the pairs and probable time after disruption of the
system. It was found that the disruption of the system was probably isotropic in respect to the observer, but the axes of rotation
were aligned. The mean expanding velocity is 260 ± 150 km s^{-1}.
The z–component of the velocity of the pair after the first SN
explosion is about 30 km s^{-1}. Some plerions and γ–ray sources
also might be generated after the disruption of binary systems.

126.062 Soft X–ray characteristics of white dwarfs observed by
Exosat.
J. Heise, J. A. M. Bleeker, A. C. Brinkman,
E. H. B. M. Gronenschild, F. Paerels, M. Grewing,
C. Wulf–Mathies, K. Beuermann.
Adv. Space Res., Vol. 5, No. 3, p. 61 – 64 (1985). – See Abstr.
012.059.

Exosat has observed 19 hot white dwarfs with alleged strong
soft X–ray emission. Positive detection of a large fraction of this
sample was obtained, among these practically all hot DA dwarfs.
High–resolution spectral data indicates no traces of He in the
atmosphere of HZ43, i.e. n(He)/n(H) ⩽ 10^{-5} at a photospheric
temperature of 60000K. In contrast, the hot DA1 dwarf Feige 24
shows the presence of an appreciable He–abundance
(n(He)/n(H) ⩾ 10^{-3}); however no simple homogeneously mixed
H/He atmosphere can explain the observed spectral shape.

126.063 The optical spectrum of hydrogen at 160 – 350 million
Gauss in the white dwarf Grw + 70°8247.
J. R. P. Angel, J. Liebert, H. S. Stockman.
Astrophys. J., Vol. 292, No. 1, p. 260 – 266 (1985).

Spectra of the magnetic white dwarf Grw + 70°8247 with high
signal–to–noise ratio are compared with the hydrogen spectrum
up to 10^{9}G, derived from the recent work of Rösner et al. Nearly
all the definite absorption features coincide with transitions
whose energy is stationary or nearly stationary with field strength
in the range $(1.6 - 3.5) \times 10^8$G. Pi transitions show up clearly in
the spectrum of linear polarization. The complex spectrum of
circular polarization is generally consistent with dipolar field
geometry.

126.064 Depolarization of emission of some pulsars.
J. Gil, W. Rudnicki.
Astron. Astrophys., Vol. 147, No. 1, p. 184 – 185 (1985).

It is assumed that pulsar radiation originates in a polar cap
region and that the emission mechanism is curvature radiation. It
is further assumed that the radiation reaching an observer at any
one time may represent contributions from several particle
bunches moving relativistically along different magnetic field
lines and radiating incoherently. These assumptions are used to
explain the minimum of linear polarization appearing near the
profile centre of some pulsars.

126.065 New extragalactic pulsar discovered.
L. I. Gurvits.
Priroda, No. 5, p. 104 (1985). In Russian.

126.066 Beobachtung des Crab im harten Röntgenlicht.
G. Hasinger.
MPE Rep., No. 186, 103 pp. (1984).

126.067 Discovery of oxygen in the PG 1159 degenerate stars: a
direct evolutionary link to O VI planetary nebula nuclei
and confirmation of pulsation theory.
E. M. Sion, J. Liebert, S. G. Starrfield.
Astrophys. J., Vol. 292, No. 2, p. 471 – 476 (1985).

The authors report the discovery of strong O VI absorption/
emission lines in five members of the group of very hot degenerate stars whose prototype is the pulsating star PG 1159–35. The
optical spectra are dominated by C IV, O VI, and He II, but with
little evidence for nitrogen features. It is confirmed that the central star of the planetary nebula K1–16 is a member of the spectroscopic group. The results imply that significant amounts of
oxygen are present in the surface layers of these pulsating stars,
which supports the predicted driving mechanism of Starrfield
et al. Moreover, the isolated PG 1159 stars show similar spectral
features and have similar temperatures and luminosities to the
so–called O VI central stars of planetary nebulae.

126.068 Detection and analysis of photospheric CNO features in
the ultraviolet spectrum of the hot DO white dwarf
PG 1034 + 001.
E. M. Sion, J. Liebert, F. Wesemael.
Astrophys. J., Vol. 292, No. 2, p. 477 – 483 (1985).

High–resolution ultraviolet and optical spectra of the hot DO
white dwarf PG 1034 + 001 are presented. The ultraviolet observations revealed the presence of absorption features due to highly
ionized species, N V, C IV, and O V, at large redshifts
(> 35 km s^{-1}) with respect to the interstellar lines. On the basis
of these shifts, the authors identify these features as being of
photospheric origin. Preliminary abundances for C, N and O
relative to He are derived. An interpretation of the observed
abundances is made in terms of selective radiative forces acting
in a high–temperature photosphere. Implications for the evolutionary state of PG 1034 + 001 are briefly outlined.

126.069 A measurement of secular evolution in the pre–white
dwarf star PG 1159–035.
D. E. Winget, S. O. Kepler, E. L. Robinson, R. E. Nather,
D. O'Donoghue.
Astrophys. J., Vol. 292, No. 2, p. 606 – 613 (1985).

The authors have analyzed 96 hours of high–speed photometry
of the pulsating variable star PG 1159–035, obtaind from
1979 through 1984. They find that the pulsation period with the
largest amplitude, at 516 s, is changing at a rate $dP/dt =
(-1.2 \pm 0.1) \times 10^{-11}$s s^{-1}, corresponding to an evolutionary time
scale of $\tau \sim (1.4 \pm 0.1) \times 10^6$yr. This period change is consistent
with theoretical descriptions of a gravitationally contracting pre–
white dwarf object undergoing nonradial g–mode pulsations.

126.070 Die Emissionskomponenten der Balmerlinien im Spek-
trum des weißen Zwergsterns GD 1401.
I. Bues.
Mitt. Astron. Ges., Nr. 63, p. 196 – 198 (1985). – See Abstr.
012.063.

126.071 Identification of the 1400 and 1600 Å features observed
in the ultraviolet spectra of DA white dwarfs.
G. Wegner, E. P. Nelan.
Bull. Am. Astron. Soc., Vol. 16, No. 4, p. 894 – 895 (1984). Abstract. – See Abstr. 010.062.

126.072 Particle acceleration near pulsars: constraints from
synchrotron nebulae.
S. P. Reynolds.
Bull. Am. Astron. Soc., Vol. 16, No. 4, p. 926 – 927 (1984). Abstract. – See Abstr. 010.062.

126.073 **Gamma rays of 0.3 – 30 MeV from PSR 0531 + 21.**
R. S. White, W. Sweeney, O. T. Tümer, A. Zych.
Bull. Am. Astron. Soc., Vol. 16, No. 4, p. 935 (1984). Abstract. –
See Abstr. 010.062.

126.074 **Search for very high energy gamma rays from the Crab pulsar.**
O. T. Tümer, W. A. Wheaton, C. P. Godfrey, R. C. Lamb.
Bull. Am. Astron. Soc., Vol. 16, No. 4, p. 936 (1984). Abstract. –
See Abstr. 010.062.

126.075 **Astrometry of millisecond pulsar 1937 + 214 and binary pulsar 1913 + 160.**
D. C. Backer.
Bull. Am. Astron. Soc., Vol. 16, No. 4, p. 944 (1984). Abstract. –
See Abstr. 010.062.

126.076 **Optical pulsar in LMC remnant 0540–69.3.**
J. Middleditch, C. R. Pennypacker.
Bull. Am. Astron. Soc., Vol. 16, No. 4, p. 944 (1984). Abstract. –
See Abstr. 010.062.

126.077 **Survey for new pulsars by the HEAO A–1 experiment.**
J. P. Norris, K. S. Wood.
Bull. Am. Astron. Soc., Vol. 16, No. 4, p. 944 (1984). Abstract. –
See Abstr. 010.062.

126.078 **Circular dichroism in PG 1031 + 23.**
S. C. West, G. D. Schmidt.
Bull. Am. Astron. Soc., Vol. 16, No. 4, p. 945 (1984). Abstract. –
See Abstr. 010.062.

126.079 **The luminosity function of DA white dwarfs.**
T. A. Fleming, J. W. Liebert, R. F. Green.
Bull. Am. Astron. Soc., Vol. 16, No. 4, p. 945 (1984). Abstract. –
See Abstr. 010.062.

126.080 **Temperatures of DA white dwarfs determined from Lyman alpha profile analysis.**
J. B. Holberg, F. Wesemael, J. Basile.
Bull. Am. Astron. Soc., Vol. 16, No. 4, p. 965 (1984). Abstract. –
See Abstr. 010.062.

126.081 **The kinematics of cool white dwarfs with metals and the local interstellar medium.**
P. A. Aannestad, E. M. Sion.
Bull. Am. Astron. Soc., Vol. 16, No. 4, p. 965 (1984). Abstract. –
See Abstr. 010.062.

126.082 **Discovery of oxygen in the PG1159 degenerate stars: a direct evolutionary link to O VI planetary nebula nuclei and confirmation of pulsation theory.**
E. M. Sion, J. W. Liebert, S. G. Starrfield.
Bull. Am. Astron. Soc., Vol. 16, No. 4, p. 965 (1984). Abstract. –
See Abstr. 010.062.

126.083 **Hard X–ray time variability of the mean pulsar profile of the Crab pulsar.**
R. B. Wilson, G. J. Fishman, C. A. Meegan.
Bull. Am. Astron. Soc., Vol. 16, No. 4, p. 982 (1984). Abstract. –
See Abstr. 010.062.

126.084 **Refractive scintillation: a cause of compact source variability?**
D. R. Stinebring, J. J. Condon.
Bull. Am. Astron. Soc., Vol. 16, No. 4, p. 1010 – 1011 (1984).
Abstract. – See Abstr. 010.062.

126.085 **New insights into degenerate stars and interacting stellar systems.**
J. C. Raymond.
Bull. Am. Astron. Soc., Vol. 16, No. 4, p. 1011 (1984). Abstract. – See Abstr. 010.062.

126.086 **The binary frequency of extreme subdwarfs revisited.**
L. L. Stryker, J. E. Hesser, G. Hill, G. S. Garlick, L. M. O'Keefe.
Bull. Am. Astron. Soc., Vol. 16, No. 4, p. 1013 (1984). Abstract. – See Abstr. 010.062.

126.087 **Probing the early universe with the millisecond pulsar.**
B. Carr.
Nature, Vol. 315, No. 6020, p. 540 (1985).
This note comments on recent suggestions that the millisecond pulsar PSR 1937 + 21 may provide information about processes that occurred within the first millisecond of the Big Bang.

126.088 **High–precision timing observations of the millisecond pulsar PSR 1937 + 21.**
M. M. Davis, J. H. Taylor, J. M. Weisberg, D. C. Backer.
Nature, Vol. 315, No. 6020, p. 547 – 550 (1985).
Long–term studies of pulse arrival times for the pulsar PSR 1937 + 21 yield estimates of the pulsar's period, spin–down rate and celestial coordinates with unprecedented accuracies. Over periods exceeding six months, this pulsar is at least comparable in stability to the best man–made atomic clocks. The results place a limit on the energy density of low–frequency gravitational waves in the Universe.

126.089 **Strömgren photometry of ZZ Ceti and other DA white dwarfs.**
G. Fontaine, P. Bergeron, P. Lacombe, R. Lamontagne, A. Talon.
Astron. J., Vol. 90, No. 6, p. 1094 – 1103 (1985).
Strömgren colors for a sample of 71 stars classified as DA white dwarfs are presented. Comparison with the recent model atmospheres of the Kiel group indicates that the average gravity of 63 stars of the sample is $\log g = 7.98 \pm 0.31$, with no indication of a dependence on the effective temperature. This is consistent with the theoretical expectation that white dwarfs evolve at constant gravity. Correlations between Strömgren colors and Greenstein multichannel colors are also given, using 52 stars that have been observed in common in the two systems. Finally, observations of 11 pulsating objects reveal the existence of a narrow instability strip in the range $13000K \gtrsim T_e \gtrsim 11000K$ in a $[(u-b), (b-y)]$ two–color diagram.

126.090 **General relativistic effects of rotation on the structure and surface redshift of fast pulsars.**
B. Datta, R. C. Kapoor, A. Ray.
Bull. Astron. Soc. India, Vol. 12, No. 4, p. 343 – 349 (1984).
General relativistic effects of rotation on the structure and surface emission of the fast pulsar PSR 1937 + 214 are illustrated using a rotationally perturbed interior spherical metric. The results are found to differ markedly from those derived on the basis of simple spherical models, and are expected to be generally valid for the class of fast pulsars.

126.091 **Detection of a compact companion of the mild barium star ξ^1 Ceti.**
E. Böhm–Vitense, H. R. Johnson.
Astrophys. J., Vol. 293, No. 1, p. 288 – 293 (1985).
The authors observed with IUE a white dwarf companion for the mild Ba star ξ^1 Ceti. The relative energy distribution of the companion indicates a temperature of $13000 \pm 1000K$. Mass and radius depend on the distance of the system, but the mass, although probably less than 1 solar mass, seems quite large. The DA white dwarf shows several wide absorption bands. Some weak emission lines may be present, possibly from the Ba star.

126.092 **An analysis of the bright white dwarf CD –38°10980.**
J. B. Holberg, F. Wesemael, G. Wegner, F. C. Bruhweiler.
Astrophys. J., Vol. 293, No. 1, p. 294 – 302 (1985).
The bright DA white dwarf CD –38°10980 is comprehensively analyzed using all available data. Detailed comparisons of the overall energy distribution and Lyman–α and Balmer profiles are made with theoretical predictions based on a homogeneous grid of model atmospheres. These comparisons yield a consistent,

well–defined effective temperature and surface gravity for this star of $24,700 \pm 250$K and $\log g = 7.95 \pm 0.15$, respectively. A new gravitational redshift of 28.4 ± 4.8 km s^{-1} is derived from an *IUE* high–dispersion spectrum.

126.093   A search for non–interacting, close binary white dwarfs.
A. W. Shafter, E. L. Robinson.
Bull. Am. Astron. Soc., Vol. 17, No. 1, p. 513 (1985). Abstract. – See Abstr. 010.064.

126.094   Mass spectrum of pulsars.
C. Y. Fan, H. Hang.
High energy astrophysics and cosmology, p. 5 – 11 (1983). – See Abstr. 012.068.

Observational evidence for a scaling law between Jupiter and pulsars with respect to radiative energy generation by a rotating magnetic body is presented. This relation is then applied to derive a mass spectrum for pulsars.

126.095   The distributions of apparent beamwidth for the sheet – shape radiation beam model of pulsars.
G. Qiao, X. Wu.
High energy astrophysics and cosmology, p. 24 – 29 (1983). – See Abstr. 012.068.

One model for pulsar radio emission assumes that the radiation comes from the regions close to the magnetic equator and that the radiation beam is in the shape of a sheet of ring structure (sheet–shape radiation beam model). In this paper, the formula of the distributions of apparent beamwidth for the model is derived and the theoretical distributions are calculated by numerical integration. A comparison with observations is presented.

126.096   A statistical study of pulsar rotational braking.
G. Qiao, X. Wu.
High energy astrophysics and cosmology, p. 30 – 35 (1983). – See Abstr. 012.068.

This paper presents a statistical study of pulsar spindown rates is an attempt to distinguish between the braking mechanisms of magnetic dipole radiation and of unipolar induction.

126.097   An external factor affecting pulsar's $\dot{P}$ – the differential rotation of the Galaxy.
J. Rong, Y. Xia.
High energy astrophysics and cosmology, p. 36 – 40 (1983). – See Abstr. 012.068.

In this paper the effect of the differential rotation of the Galaxy on a pulsar's $\dot{P}$ is analysed.

126.098   The binary pulsar – a test for general relativity?
G. Börner.
High energy astrophysics and cosmology, p. 175 – 189 (1983). – See Abstr. 012.068.

The binary system of PSR 1913+16 contains the pulsar and an, as yet unknown, companion. If this star is a compact object too, then the data can be interpreted in terms of general relativistic effects. This leads to the conclusion that the decay of the orbit must be due to the emission of gravitational waves. The nature of the unseen companion is discussed in detail.

126.099   R–S model and radio binary pulsars.
Z. Wang.
High energy astrophysics and cosmology, p. 270 – 274 (1983). – See Abstr. 012.068.

A model is presented which attempts to explain the small number of observed binary radio pulsars. It is assumed that both components of these systems are neutron stars.

126.100   Neutron stars and the discovery of pulsars. Part 1.
G. Greenstein.
Mercury, Vol. 14, No. 2, p. 34 – 39, 62 (1985).

126.101   On the expected number of white dwarfs in globular clusters.
A. Renzini.
Astron. Express, Vol. 1, Nos. 4 – 6, p. 127 – 132 (1985).

Some methods for predicting the number of observable white dwarfs in globular star clusters are discussed and compared to each other. It is shown that one such method gives fallacious indications, which may mislead planning for future attempts at detecting white dwarfs in globulars. The number of detectable white dwarfs in some nearby clusters is then evaluated and discussed.

126.102   The evolving orientation of the magnetic axis of pulsars.
A. D. Kuz'min, I. M. Dagkesamanskaya, V. D. Pugachev.
Sov. Astron. Lett., Vol. 10, No. 6, p. 357 – 359 (1984). English translation of 38.126.052.

126.103   Hydrogen spectrum in magnetic white dwarfs: Hα, Hβ, and Hγ transitions.
R. J. W. Henry, R. F. O'Connell.
Publ. Astron. Soc. Pac., Vol. 97, No. 590, p. 333 – 339 (1985).

Using the results of an accurate variational calculation, the authors present a graphical display of the wavelengths of the Hα, Hβ and Hγ lines of hydrogen, for magnetic–field values ranging from 0 to 560 megagauss, which is believed to cover the range of fields found in magnetic white dwarfs. This is the first complete detailed compilation of such results.

126.104   Superfast pulsars.
E. P. J. van den Heuvel.
Ned. Tijdschr. Natuurkd. A, Vol. A51, No. 1, p. 30 – 35 (1985). In Dutch. Abstr. in Phys. Abstr., Vol. 88, No. 1258, Entry 63363 (1985).

126.105   Space distribution of the pulsars in the Galaxy and the nature of the γ–ray sources.
B. M. Vladimirsky (*B. M. Vladimirskij*).
18th International Cosmic Ray Conference, Vol. 1, p. 114 – 117 (1983). – See Abstr. 012.096.

126.106   Possible sporadicity in emission of very high energy gamma rays from Crab pulsar.
P. N. Bhat, S. K. Gupta, P. V. Ramana Murthy, P. R. Vishwanath.
18th International Cosmic Ray Conference, Vol. 1, p. 139 (1983). Abstract. – See Abstr. 012.096.

126.107   The energy spectrum of very high energy gamma rays from the Crab pulsar.
P. N. Bhat, S. K. Gupta, P. V. Ramana Murthy, S. C. Tonwar, P. R. Vishwanath.
18th International Cosmic Ray Conference, Vol. 1, p. 141 (1983). Abstract. – See Abstr. 012.096.

126.108   Possible transient emission of very high energy gamma rays from Vela pulsar.
P. N. Bhat, S. K. Gupta, P. V. Ramana Murthy, B. V. Sreekantan, P. R. Vishwanath.
18th International Cosmic Ray Conference, Vol. 1, p. 143 (1983). Abstract. – See Abstr. 012.096.

126.109   Very high energy gamma ray emission from Vela pulsar and PSR 0950.
P. N. Bhat, S. K. Gupta, P. V. Ramana Murthy, S. C. Tonwar, P. R. Vishwanath.
18th International Cosmic Ray Conference, Vol. 1, p. 144 (1983). Abstract. – See Abstr. 012.096.

126.110 Gamma rays of 1 – 30 MeV from the Vela pulsar PSR 0833–45.
O. T. Tümer, B. Dayton, J. L. Long, T. O'Neill, A. D. Zych, R. S. White.
18th International Cosmic Ray Conference, Vol. 9, p. 37 – 40 (1983). – See Abstr. 012.096.

Results are reported for observations of gamma rays of 1 – 30 MeV from the Vela pulsar PSR 0833–45 carried out with the UCR double scatter gamma ray telescope on a balloon launched from Alice Springs, Australia on November 10, 1981. Flux and energy distributions of the pulsed and non–pulsed gamma rays are presented.

126.111 Gamma rays from the Crab pulsar 0531 + 21.
J. L. Long, O. T. Tümer, A. D. Zych, R. S. White.
18th International Cosmic Ray Conference, Vol. 9, p. 41 – 44 (1983). – See Abstr. 012.096.

Results are reported for the observations of PSR 0531 + 21 carried out with the University of California, Riverside (UCR) gamma ray double Compton scatter telescope launched on a balloon from Palestine, Texas at 4.5 GV, on September 29, 1978.

126.112 COS–B upper limits on gamma–ray emission from radio pulsars.
R. Buccheri, K. Bennett, G. F. Bignami, J. B. G. M. Bloemen, V. Boriakoff, P. A. Caraveo, W. Hermsen, G. Kanbach, R. N. Manchester, J. L. Masnou, H. A. Mayer–Hasselwander, M. E. Özel, J. A. Paul, B. Sacco, L. Scarsi, A. W. Strong.
18th International Cosmic Ray Conference, Vol. 9, p. 49 – 52 (1983). – See Abstr. 012.096.

The authors report on the results of a search for pulsed gamma–ray emission from old radio pulsars in the COS–B data in the energy range 50 MeV to 2 GeV. The analysis has been done by using either pulsar parameters obtained in simultaneous radio and gamma–ray observations or parameters which could be reliably derived by interpolation from radio observations spanning the COS–B observations.

126.113 A search for high energy gamma rays from PSR 0531 and PSR 1953 using a large aperture camera.
M. Cawley, J. Clear, D. J. Fegan, K. Gibbs, P. Gorham, R. C. Lamb, I. MacRae, P. K. MacKeown, N. A. Porter, V. J. Stenger, K. E. Turver, T. C. Weekes.
18th International Cosmic Ray Conference, Vol. 9, p. 61 – 64 (1983). – See Abstr. 012.096.

A method of event selection based on pattern recognition is used to analyse high energy gamma ray observations of pulsars with periods 6 and 33 msec.

126.114 Study of the pulsations of the ZZ Ceti stars.
K. De S. Oliveira Filho.
Diss. Abstr. Int., Sect. B, Vol. 45, No. 7, p. 2200 (1985). Thesis, University of Texas, 185 pp. (1984). Order No. DA8421775.

126.115 Spectra of short–period pulsars in the light of the hypothesis of the two types of pulsars.
I. F. Malov.
Astron. Zh., Tom 62, Vyp. 3, p. 608 – 609 (1985). In Russian. English translation in Sov. Astron., Vol. 29, No. 3.

The lack of low–frequency turnovers in the spectra of PSR 0531 + 21 and 1937 + 21 may be explained if the generation of radio emission in these pulsars occurs near the light cylinder. Differences of high frequency cut–offs and spectral indices for long–period pulsars and short–period ones are discussed.

126.116 Frequency dependence of pulsar profiles.
T. H. Hankins, B. J. Rickett.
Bull. Am. Astron. Soc., Vol. 17, No. 2, p. 555 (1985). Abstract. – See Abstr. 010.065.

126.117 Sharp–lined features in the UV spectra of hot DA white dwarfs.
F. Bruhweiler.
Bull. Am. Astron. Soc., Vol. 17, No. 2, p. 559 (1985). Abstract. – See Abstr. 010.065.

126.118 The decametric wavelength spectra of fast pulsars.
W. C. Erickson, N. E. Kassim.
Bull. Am. Astron. Soc., Vol. 17, No. 2, p. 566 (1985). Abstract. – See Abstr. 010.065.

126.119 Statistical properties of pulsars.
R. T. Emmering, R. A. Chevalier.
Bull. Am. Astron. Soc., Vol. 17, No. 2, p. 566 – 567 (1985). Abstract. – See Abstr. 010.065.

126.120 Radiative forces, mass loss, and metal abundances in sdB stars.
F. Wesemael, G. Michaud, P. Bergeron, G. Fontaine.
Bull. Am. Astron. Soc., Vol. 17, No. 2, p. 599 (1985). Abstract. – See Abstr. 010.065.

126.121 On the masses of white dwarfs in cataclysmic binaries.
A. W. Shafter.
Diss. Abstr. Int., Sect. B, Vol. 45, No. 1, p. 230 – 231 (1984). Thesis, University of California, 230 pp. (1983). Order No. DA8409238.

126.122 A search for low luminosity pulsars and its implications for the pulsar luminosity function.
R. J. Dewey.
Diss. Abstr. Int., Sect. B, Vol. 45, No. 4, p. 1217 (1984). Thesis, Princeton University, 154 pp. (1984). Order No. DA8415021.

126.123 Measurement of pulsar parallax by very long baseline interferometry.
C. R. Gwinn.
Diss. Abstr. Int., Sect. B, Vol. 45, No. 5, p. 1501 (1984). Thesis, Princeton University, 148 pp. (1984). Order No. DA8417421.

126.124 Neutron stars and the discovery of pulsars. Part 2.
G. Greenstein.
Mercury, Vol. 14, No. 3, p. 66 – 73 (1985).

126.125 An introduction to millisecond pulsars.
D. C. Backer.
Birth and evolution of neutron stars: issues raised by millisecond pulsars, p. 3 – 10 (1984). – See Abstr. 012.115.

The author discusses pulsar 1937 + 214 which has the fastest known rotation period, 1.558 milliseconds.

126.126 Arrival observations of the millisecond pulsar 1937 + 21.
M. M. Davis, J. H. Taylor, J. M. Weisberg, D. C. Backer.
Birth and evolution of neutron stars: issues raised by millisecond pulsars, p. 12 – 23 (1984). – See Abstr. 012.115.

The period, period derivative and celestial coordinates of the 1.6 millisecond pulsar have been derived from eighteen months of Arecibo timing observations. The pulsar has proven to be an extremely accurate and stable clock. Residuals are presently about one microsecond, but improvements in the timing technique may provide an order of magnitude improvement.

126.127 The 6.1 millisecond binary pulsar.
V. Boriakoff, R. Buccheri, F. Fauci, K. Turner, M. M. Davis.
Birth and evolution of neutron stars: issues raised by millisecond pulsars, p. 24 – 31 (1984). – See Abstr. 012.115.

The 6.1 millisecond binary pulsar P1953 + 29 has been studied for approximately one year. This represents slightly over three cycles of the 117.3 day orbital period. Results of timing and other experiments in progress are presented. The results of some optical counterpart searches are also presented.

126.128   Polarimetry of the two fastest pulsars.
D. R. Stinebring, V. Boriakoff, J. M. Cordes,
W. Deich, A. Wolszczan.
Birth and evolution of neutron stars: issues raised by millisecond
pulsars, p. 32 – 39 (1984). – See Abstr. 012.115.

The authors report recent polarization observations of the
millisecond pulsars PSR 1937+214 and PSR 1953+290. The
PSR 1937+214 data, obtained at 431 and 1384 MHz, are fully
dedispersed and have a time resolution of 4 microseconds. The
preliminary data for PSR 1953+290 show little linear polarization at either 430 or 1400 MHz.

**126.129   Optical observations of the millisecond pulsars
PSR 1937+214 and PSR 1953+29.**
T. J. Loredo, G. R. Ricker, S. A. Rappaport, J. Middleditch.
Birth and evolution of neutron stars: issues raised by millisecond
pulsars, p. 48 – 58 (1984). – See Abstr. 012.115.

The authors report here the results of photometric and time–
resolved optical observations of the fields of the recently discovered ultra–fast radio pulsars, PSR 1937+214 and PSR 1953+29.
Deep images of the fields reveal no conspicuous optical counterpart for PSR 1937+214, though a star with $m_v \sim 20.5$ is coincident with the position of PSR 1953+29. Time–resolved images
of the fields of both objects, obtained with a new stroboscopic
technique, have yielded upper limits on the pulsed near–infrared
emission from each object. The authors briefly discuss constraints on the physical models for these systems derived from
their optical studies.

126.130   No detectable millisecond pulsar in the 1913+16 system.
S. R. Kulkarni, U. C. Berkeley, M. M. Davis,
J. H. Taylor.
Birth and evolution of neutron stars: issues raised by millisecond
pulsars, p. 59 – 62 (1984). – See Abstr. 012.115.

The observed absence of eclipses and the rate of advance of
periastron in the 1913+16 system constrain the companion of
PSR 1913+16 to be a compact object, such as a helium main-
sequence star, or a rapidly rotating white dwarf. On evolutionary
grounds it has been argued that the companion must be another
neutron star. In an attempt to find whether the companion was
a fast pulsar that went undetected in previous slow–period
searches, the authors fast–sampled an 8 MHz dedispersed signal
from the 1913+16 system and looked for pulses. They report the
absence of any fast pulsar with a flux density greater than 1/3 that
of the binary pulsar and a rotating rate smaller than 2500 Hz.

**126.131   A single pulse study of the millisecond pulsar
PSR 1937+214.**
A. Wolszczan, J. M. Cordes, D. R. Stinebring.
Birth and evolution of neutron stars: issues raised by millisecond
pulsars, p. 63 – 70 (1984). – See Abstr. 012.115.

Preliminary results of the search for strong pulses from the
millisecond pulsar PSR 1937+214, based on the Arecibo observations of this object at 431 MHz and 1400 MHz are presented.
The authors show that its single pulse behaviour does not differ
from that of the "slow" pulsars: detectable single pulse emission
is present in nearly every period and strong pulses occur approximately every 400,000 periods. Results of the correlation analysis
of the data indicate that pulse intensities are correlated over 3 – 4
periods and that there may be a significant correlation between
main pulse and interpulse emission.

**126.132   Isolated and binary millisecond pulsars and accretion
spun–up neutron stars.**
J. Shaham.
Birth and evolution of neutron stars: issues raised by millisecond
pulsars, p. 107 – 112 (1984). – See Abstr. 012.115.

The author concentrates here on three selected remarks regarding (1) the thermal origin of neutron stellar magnetic fields and
its relation to 1937+214, (2) the P – Ṗ relationship, and (3) the
companion of 1937+214.

126.133   The period distribution of fast pulsars.
A. K. Harding.
Birth and evolution of neutron stars: issues raised by millisecond
pulsars, p. 113 – 120 (1984). – See Abstr. 012.115.

Several proposed models for the origin of fast pulsars predict
a correlation between period and magnetic field strength, where
the shortest period pulsars should have the lowest fields, and
therefore the smallest period derivatives. Such a correlation, coupled with a lower limit on pulsar period from rotational stability
arguments, could produce a significant clumping in the period
distribution near the minimum stable period. The shape of the
distribution will depend on the initial magnetic field distribution
of fast pulsars and on the strength of the P, Ṗ correlation.

126.134   The origin of pulsar velocities.
V. Radhakrishnan.
Birth and evolution of neutron stars: issues raised by millisecond
pulsars, p. 130 – 137 (1984). – See Abstr. 012.115.

Ever since pulsars were found to have significant proper motions, the origin of the velocities has been an intriguing question.
The more recent finding that the velocities display a significant
correlation with the derived magnetic moments of the pulsars has
made the origin of the velocities appear even more mysterious.
The author advances arguments to show that the above correlation is not "causal", but "accidental". Pulsar velocities are determined by their binary histories and not governed in any way by
their magnetic fields.

126.135   Pulsar space velocities from interstellar scintillations.
J. M. Cordes, J. M. Weisberg.
Birth and evolution of neutron stars: issues raised by millisecond
pulsars, p. 138 – 143 (1984). – See Abstr. 012.115.

Interstellar scintillation observations have been used to determine transverse velocities of the underlying diffraction patterns
which, in turn, seem to be determined mostly by the pulsar space
velocities. The distribution of transverse speeds for 70 objects
shows large departures from a Maxwellian distribution both at
large and at small speeds, in agreement with interferometer results on 26 objects. Velocities show a correlation with magnetic
moment (or with period derivative or with spindown age) that
suggests a common link between space velocity and surface magnetic field.

126.136   The orbital inclination of PSR 0655+64.
A. G. Lyne.
Birth and evolution of neutron stars: issues raised by millisecond
pulsars, p. 144 – 145 (1984). – See Abstr. 012.115.

**126.137   A model of radio emission of the millisecond pulsar
PSR 1937+214.**
J. A. Gil.
Birth and evolution of neutron stars: issues raised by millisecond
pulsars, p. 146 – 150 (1984). – See Abstr. 012.115.

A model of radio emission of millisecond pulsar
PSR 1937+214 is proposed. The emission region is expected to
lie very close to the polar cap surface, where the contribution of
the quadrupole component to the actual magnetic field may be
significant. It is argued that the interpulse as well as the main
pulse are emitted from the same magnetic pole.

126.138   Pulsar statistics: a study of pulsar luminosities.
M. Prószyński, D. Przybycień.
Birth and evolution of neutron stars: issues raised by millisecond
pulsars, p. 151 – 157 (1984). – See Abstr. 012.115.

The authors find a statistically significant correlation between
pulsar luminosity at 400 MHz and both pulsar period and period
derivative. Fitting a phenomenological power–law model
$L_{model}(P,\dot{P}) \sim P^{\alpha}\dot{P}^{\beta}$ (where P is pulsar period, $\dot{P}$ period derivative and L radio luminosity) to the pulsar luminosity data, they
obtain $\alpha = -1.04 \pm 0.15$ and $\beta = 0.35 \pm 0.06$. The above values
suggest that pulsar radio luminosity varies roughly as the cube
root of the total loss of rotational energy.

126.139   Superfluidity in millisecond pulsars.
D. Pines, M. A. Alpar.
Birth and evolution of neutron stars: issues raised by millisecond pulsars, p. 161 – 172 (1984). – See Abstr. 012.115.

The authors review the evidence for superfluidity in the Vela pulsar, the Crab pulsar and PSR 0525+21, and examine the prospects for observing similar consequences of superfluidity in the already–discovered millisec pulsars. They consider, inter alia, the likelihood of observing glitches, the expected post–glitch behavior, and pulsar heating by energy dissipation due to the creep of neutron vortex lines in pinned superfluid regions of the crust.

126.140   Searches for short period pulsars at Jodrell Bank.
A. G. Lyne.
Birth and evolution of neutron stars: issues raised by millisecond pulsars, p. 223 – 226 (1984). – See Abstr. 012.115.

Two separate approaches are presently being taken at Jodrell Bank in the quest for short period pulsars. One experiment consists of a high frequency survey of the galactic plane for highly dispersed and/or short period pulsars which would not have been visible in the major surveys carried out at around 400 MHz. The second experiment is designed to detect millisecond pulsars using MERLIN as a filter for small diameter, linearly polarized sources, followed by a periodicity search on likely candidates using the 76 m MKIA telescope alone.

126.141   Southern hemisphere searches for short period pulsars.
R. N. Manchester.
Birth and evolution of neutron stars: issues raised by millisecond pulsars, p. 227 – 233 (1984). – See Abstr. 012.115.

Two searches of the southern sky for short period pulsars are briefly described. The first, made using the 64–m telescope at Parkes, is sensitive to pulsars with periods greater than about 10 ms and the second, made using the Molonglo radio telescope, has sensitivity down to periods of about 1.5 ms. Four pulsars were found in the Parkes survey and none in the Molonglo survey, although analysis of the latter is as yet incomplete.

126.142   The period distribution of pulsars.
R. J. Dewey, G. H. Stokes, D. J. Segelstein,
J. H. Taylor, J. M. Weisberg.
Birth and evolution of neutron stars: issues raised by millisecond pulsars, p. 234 – 241 (1984). – See Abstr. 012.115.

The searching methods used to discover most of the known pulsars have had sharply reduced sensitivities to periods less than a few tenths of a second, and virtually no sensitivity to periods < 30 ms. This selection effect produces a bias in the observed period distribution, and has probably prevented the detection of a number of interesting objects. The authors discuss the period bias quantitatively and estimate its effect on present knowledge of the pulsar period distribution. In the period range 1 – 100 ms, the facts remain highly uncertain. In an effort to improve the situation observationally, the authors have begun surveys at both Green Bank and Arecibo designed specifically to detect fast pulsars.

126.143   Two low–frequency searches for fast pulsars.
M. A. Stevens, C. E. Heiles, D. C. Backer,
W. M. Goss, U. J. Schwarz, J. S. Albinson, A. Purvis.
Birth and evolution of neutron stars: issues raised by millisecond pulsars, p. 242 – 244 (1984). – See Abstr. 012.115.

Two low–frequency surveys of the galactic plane have recently been completed. One survey used the NRAO 300–ft telescope at Green Bank looking for the frequency structure induced by interstellar scintillation. It covered the plane from $l = 15$ to 235 degrees and $|b| \leqslant 5$ degrees as well as selected fields where previous observations indicate the possible existence of pulsars. The other survey was conducted using the Westerbork Synthesis Radio Telescope looking for highly polarized compact sources. This covered the plane from $l = 0$ to 90 degrees and $|b| \leqslant 1.3$ degrees.

126.144   A Cambridge–VLA continuum search for fast pulsar candidates.
A. Purvis, J. E. Baldwin, P. J. Warner, W. M. Goss,
J. H. van Gorkom, S. R. Kulkarni, C. Heiles.
Birth and evolution of neutron stars: issues raised by millisecond pulsars, p. 252 – 264 (1984). – See Abstr. 012.115.

The millisecond pulsar was found because of its compactness and steep radio spectrum. If more rapid pulsars are in our Galaxy then they might be discovered by looking first for similar anomalous properties. This paper describes such a search. A new interplanetary scintillation survey has just been completed northward of –10° declination. The authors describe how a sample of candidate fast pulsars has been assembled by (1) selecting the smallest IPS sources at low galactic latitudes, (2) determining a list of candidate source positions from existing catalogues and new data from the Cambridge Low Frequency Synthesis Telescope and (3) eliminating those with double radio morphology using the VLA.

126.145   Arecibo search of pulsar candidates for pulses and interstellar scintillation.
C. Heiles, S. R. Kulkarni, A. Purvis, W. M. Goss,
J. H. van Gorkom.
Birth and evolution of neutron stars: issues raised by millisecond pulsars, p. 265 – 270 (1984). – See Abstr. 012.115.

The candidate sources were mainly obtained from the Cambridge Interplanetary Scintillation/VLA survey, described in the previous paper (Purvis et al, 1984). A small collection of sources from the Ooty interplanetary scintillation survey (Anantakrishnan and Rao, 1983), filled supernova remnants, X–ray binaries, and other exotic sources were also searched. All sources searched have flux densities at 1400 MHz > 10 mJy and declinations between 0° and 39°.

126.146   Arecibo pulsar searches in connection with gamma–ray sources. Status of the experiment.
V. Boriakoff, R. Buccheri, F. Fauci.
Birth and evolution of neutron stars: issues raised by millisecond pulsars, p. 271 – 276 (1984). – See Abstr. 012.115.

The authors report on the status of an experiment set up at the Arecibo Observatory for the search of fast pulsars in spatial coincidence with some of the gamma–ray sources of the 2CG catalogue. As a first result two new pulsars have been discovered: PSR 1953+29 (a binary pulsar with 6.1 ms pulsation period and 117.3 days orbital period) and PSR 1848+04 (a low–magnetic pulsar with 0.284 s pulsation period).

126.147   Elongated beams and millisecond pulsars.
R. Narayan.
Birth and evolution of neutron stars: issues raised by millisecond pulsars, p. 279 – 286 (1984). – See Abstr. 012.115.

It is argued on the basis of a variety of observations that pulsars have long fan–beams. The elongation is extreme in short period pulsars and decreases at longer periods. As a consequence fast pulsars have interpulses and relatively small variation of polarization position angle. In PSR 1937+21 the magnetic dipole axis and the line of sight appear to be tilted from the rotation axis by $\sim 80°$ and $\sim 40°$ respectively.

126.148   X–ray emission from fast pulsars.
A. F. Cheng.
Birth and evolution of neutron stars: issues raised by millisecond pulsars, p. 294 – 302 (1984). – See Abstr. 012.115.

Mechanisms for soft X–ray emission from fast pulsars are considered. Synchrotron radiation from plasma winds powered by a radio pulsar can yield observable fluxes at kiloparsec distances, for pulsar spindown energy loss exceeding $\sim 10^{34}$ erg s^{-1}. Unpulsed X–ray emission which is observed to be spatially extended must arise far outside the light cylinder. Such emission can be explained by synchrotron radiation from a pulsar wind which is confined by an external medium. A theory of ram pressure confinement is reviewed and applied to several objects including 1937+214. If unpulsed point source emission is observed, the emission may also arise much closer to the pulsar. The models for soft X–ray emission from pulsar winds appear to be consistent with observations of unpulsed X–rays from fast pulsars.

126.149 Gamma–ray emission from fast pulsars.
K. Brecher.
Birth and evolution of neutron stars: issues raised by millisecond pulsars, p. 303 – 309 (1984). – See Abstr. 012.115.

It is suggested that many, perhaps all, of the heretofore un-identified COS–B gamma–ray sources are fast pulsars. If the incoherent synchrotron emission mechanism at the speed–of–light cylinder is effective up to gamma–ray energies ($\cong 100$ MeV), then a population of old ($> 10^6$years), weak magnetic field ($B < 10^{10}$gauss), high angular velocity ($\omega > 10^3$sec^{-1}) pulsars may account for a number of properties of these sources.

126.150 Low frequency variability of pulsars – implications for timing of millisecond pulsars.
R. Blandford, R. Narayan.
Birth and evolution of neutron stars: issues raised by millisecond pulsars, p. 310 – 316 (1984). – See Abstr. 012.115.

Rickett, Coles and Bourgois (1984) have argued that long–term (months to years) variation in pulsar flux is caused by fluctuations in the interstellar electron density on length scales $\sim 10^{13-16}$cm. In this paper the authors show that there should then be correlated fluctuations in the pulse arrival time, pulse width, and angular size. PSR 1937+21 is suitable for detecting some of the new effects. The timing noise and pulse width variation in this pulsar is estimated assuming a power–law spectrum for the electron density fluctuations, normalized using scintillation data.

126.151 Further observations of the eight hour binary pulsar PSR 1913+16.
J. M. Weisberg, J. H. Taylor.
Birth and evolution of neutron stars: issues raised by millisecond pulsars, p. 317 – 323 (1984). – See Abstr. 012.115.

Timing measurements of the binary pulsar PSR 1913+16 now permit determination of the system's orbital elements and component masses to high precision. The pulsar and companion masses are 1.42 ± 0.03 and 1.40 ± 0.03 solar masses, respectively. The measured orbital period derivative is 1.00 ± 0.04 times the rate of orbital decay expected from emission of gravitational radiation. This agreement requires the equivalent energy density of a cosmic background of gravitational waves, with periods in the range ~ 10 to 10^4y, to be less than half that required to close the universe.

126.152 PSR 1953+29 – possible explanation of the large period derivative.
L. A. Nowakowski.
Birth and evolution of neutron stars: issues raised by millisecond pulsars, p. 330 – 334 (1984). – See Abstr. 012.115.

This paper attempts to explain larger $\dot{P}$ observed in PSR 1953+29 than expected from the spin–up theory in terms of the simultaneous magnetic field decay and counter–alignment, occuring on different time scales. The author also shows that this pulsar may evolve to period and period derivative similar to the majority of other known pulsars if its initial P and $\dot{P}$ were similar to present ones and counter–alignment is assumed.

126.153 Pulsar multi–component profiles: a phenomenological model.
M. Prószyński.
Birth and evolution of neutron stars: issues raised by millisecond pulsars, p. 335 – 336 (1984). – See Abstr. 012.115.

Pulsar radio emission is suggested to be confined to a number of narrow beams irregularly placed in the vicinity of the pulsar magnetic axis. The beams are assumed to originate from different streams of plasma that flow away from the region close to the polar cap. Finally, it is suggested that narrow, widely spaced components may result from a non–dipolar structure of the magnetic field near the star's surface.

The discovery of pulsars.
See Abstr. 004.188.

First observations with a new pulsar signal averager.
See Abstr. 033.057.

Proposed U. C. Berkeley fast pulsar search machine.
See Abstr. 033.059.

Peculiarities of reduction of pulse radio signals distorted when propagating in the interstellar medium.
See Abstr. 036.141.

Pulsar astrometry via VLBI.
See Abstr. 041.005.

Nonlinear wave conversion in electron–positron plasmas.
See Abstr. 062.047.

A dynamical instability as a driving mechanism for stellar oscillations.
See Abstr. 065.008.

Deflagration of white dwarfs as a model for type–I supernovae.
See Abstr. 065.032.

Detonation–wave structure and nucleosynthesis in exploding carbon–oxygen cores and white dwarfs.
See Abstr. 065.037.

Recurrent novae as a consequence of the accretion of solar material onto a 1.38 $M_\odot$ white dwarf.
See Abstr. 065.043.

CNO abundances resulting from diffusion in accreting nova progenitors.
See Abstr. 065.047.

Equilibrium models of differentially rotating, completely catalyzed, zero–temperature configurations with central densities intermediate to white dwarf and neutron star densities.
See Abstr. 065.061.

On the observed properties and long term structure and evolution of white dwarfs in cataclysmic variables.
See Abstr. 065.063.

Evolution of the nonradial pulsation properties of hot pre–white dwarf stars.
See Abstr. 065.064.

An analysis of nonradial pulsations of the central star of the planetary nebula K1–16.
See Abstr. 065.078.

Stellar evolution and binaries.
See Abstr. 065.083.

Report of IAU Commission 35: Stellar constitution (*Constitution des étoiles*).
See Abstr. 065.096.

Thermal gravitational radiation from stellar objects and its possible detection.
See Abstr. 066.104.

Magnetohydrodynamical model of pulsar rotation. Exact periodic solutions.
See Abstr. 067.030.

Pulsar space charging.
See Abstr. 067.052.

On the existence of an exterior toroidal region in the nonaligned pulsar magnetosphere.
See Abstr. 067.059.

The electric field of a pulsar magnetosphere.
See Abstr. 067.062.

Superconductivity and superfluidity in neutron stars and decay of pulsar magnetic fields.
See Abstr. 067.063.

Cascade processes in the surface layers of pulsars.
See Abstr. 067.069.

The electric field in axisymmetric pulsar magnetospheres.
See Abstr. 067.070.

Gravitational radiation from a solid–crust neutron star.
See Abstr. 067.071.

Envelopes of rapidly rotating neutron stars.
See Abstr. 067.093.

The triple alpha reaction at low temperatures in accreting white dwarfs and neutron stars.
See Abstr. 067.098.

Generation of relativistic particles in pulsar magnetospheres.
See Abstr. 067.102.

Cyclotron lines in accreting magnetic white dwarfs with an application to VV Puppis.
See Abstr. 067.105.

General relativistic effects on the pulse profile of fast pulsars.
See Abstr. 067.114.

Poloidal flow in axisymmetric pulsar magnetospheres.
See Abstr. 067.115.

The maximum mass of a neutron star.
See Abstr. 067.117.

Thermal X–ray emission from isolated older pulsars: a new heating mechanism.
See Abstr. 067.125.

The distribution of the magnetic inclination in a pulsar's polar cap model.
See Abstr. 067.126.

X–ray observations by the Einstein satellite and neutron star cooling.
See Abstr. 067.128.

Stability of radiative shock waves.
See Abstr. 067.143.

Soliton interactions in pulsar magnetospheres.
See Abstr. 067.147.

Pulsars: high magnetic field laboratories with 10^8T.
See Abstr. 067.150.

An aligned rotator model for the Crab pulsar.
See Abstr. 067.159.

Monopoles in pulsar PSR 1929 + 10.
See Abstr. 067.160.

Acceleration to ultra high energies in magnetospheres of young pulsars.
See Abstr. 067.179.

Frictional acceleration of high–energy particles in pulsar vicinity.
See Abstr. 067.180.

On possible evolution of pulsars.
See Abstr. 067.181.

Can γ quanta really be captured by pulsar magnetic fields?
See Abstr. 067.200.

Theoretical studies of the pulsar magnetosphere.
See Abstr. 067.203.

The estimation of some important parameters of pulsars and the comparison between the RS model and observations.
See Abstr. 067.211.

On the possibility of efficient electron–positron pair production in the vicinity of pulsars and accreting black holes.
See Abstr. 067.212.

The origin of neutron stars.
See Abstr. 067.213.

Models for the formation of binary and millisecond radio pulsars.
See Abstr. 067.214.

Neutron star seismology: understanding the oscillation modes.
See Abstr. 067.215.

Gravitational radiation from a solid crust neutron star.
See Abstr. 067.216.

Thermal origin of neutron star magnetic fields.
See Abstr. 067.217.

Millisecond pulsars and the location of the solar system barycenter.
See Abstr. 091.068.

A deep optical survey of a small region in Aquarius – II. Stellar spectroscopy.
See Abstr. 114.015.

On the ultraviolet iron spectrum of pre–white dwarfs.
See Abstr. 114.084.

Spectroscopic properties of 89 members of the Kiso ultraviolet survey.
See Abstr. 114.162.

The magnetic field of AM Herculis.
See Abstr. 117.022.

An X–ray corona in SS Cygni?
See Abstr. 117.038.

The origin of soft X–ray pulsations in dwarf novae at outburst and the DQ Herculis phenomenon.
See Abstr. 117.039.

Secular evolution of magnetic cataclysmic variables.
See Abstr. 117.064.

Investigation of a wind model for cataclysmic variable ultraviolet resonance line emission.
See Abstr. 117.065.

Disruption of light He companions in accreting neutron star binaries.
See Abstr. 117.098.

Time–resolved spectroscopy of long–period DQ Herculis stars.
See Abstr. 117.100.

UBV photometry of H 2252–035 (AO Psc).
See Abstr. 117.122.

The frequency of binaries with a degenerate component.
See Abstr. 117.127.

G 82–23: a new subdwarf–white dwarf binary.
See Abstr. 117.128.

Magnetohydrostatics in the polar caps of accreting magnetized white dwarfs.
See Abstr. 117.131.

The binary frequency of extreme subdwarfs revisited.
See Abstr. 117.157.

Pseudo white dwarfs.
See Abstr. 117.241.

Cataclysmic binaries and related objects in the context of stellar evolution.
See Abstr. 117.283.

The formation of massive white dwarfs in cataclysmic binaries.
See Abstr. 117.284.

Evolution of cataclysmic binaries.
See Abstr. 117.300.

Pre–cataclysmic binaries.
See Abstr. 117.301.

X–ray luminous white dwarf binaries in globular clusters.
See Abstr. 117.309.

Recent developments in the theory of AM Her and DQ Her stars.
See Abstr. 117.315.

Synchronization of magnetic white dwarfs in close binary systems.
See Abstr. 117.318.

Time dependent accretion flows in the AM Her systems.
See Abstr. 117.320.

On measuring the radial velocity of white dwarfs in cataclysmic binaries.
See Abstr. 117.330.

The late stages of interactive stellar evolution.
See Abstr. 117.373.

Continuous time–resolved spectroscopy of the AM Herculis binary 0139–68.
See Abstr. 117.385.

A shadowed flow in the stem of the Crab nebula?
See Abstr. 125.005.

X–ray morphology of the Crab nebula.
See Abstr. 125.006.

An evolutionary history for the Crablike, pulsar–powered supernova remnant 0540–69.3.
See Abstr. 125.041.

X–ray observations of Vela–X.
See Abstr. 125.054.

The decrease with time of the radio flux of the Crab Nebula.
See Abstr. 125.086.

On the nature of the Crablike, pulsar–powered supernova remnant 0540–693: the pulsar's initial period.
See Abstr. 125.123.

Pulsar powered radio supernovae and the early evolution of plerions.
See Abstr. 125.124.

Quasi–periodic scintillation patterns of the pulsars PSR 1133+16 and PSR 1642–03.
See Abstr. 131.054.

Small–scale electron density turbulence in the interstellar medium.
See Abstr. 131.077.

Low–frequency variability of pulsars.
See Abstr. 131.082.

Interstellar scintillation measurements of pulsars at 326.5 MHz.
See Abstr. 131.265.

Slow pulsar scintillation and the spectrum of interstellar electron density fluctuations.
See Abstr. 131.292.

The planetary nebula around the sdO star RWT 152.
See Abstr. 134.055.

The Galactic population and birthrate of X–ray pulsars.
See Abstr. 142.001.

Recent results on galactic X–ray sources.
See Abstr. 142.071.

The existence of a shower excess from the direction of the Crab Nebula.
See Abstr. 144.117.

Propagation and origin of energetic cosmic rays, $10^{14} - 10^{19}$ eV.
See Abstr. 144.147.

Scaling from Jupiter to pulsars and the acceleration of cosmic ray particles by pulsars. II.
See Abstr. 144.148.

Observations of the Crab pulsar near $10^{15} - 10^{16}$ eV.
See Abstr. 144.316.

A search for millisecond pulsar candidates in globular clusters.
See Abstr. 154.006.

A survey for galactic plane ultraviolet–excess objects: space densities for white dwarfs, subdwarfs, and cataclysmic variables.
See Abstr. 155.194.

Millisecond pulsars, gravitational waves and the inflationary universe.
See Abstr. 161.002.

Interstellar Matter, Nebulae

131 Interstellar Matter (Molecular Clouds, Reflection Nebulae, etc.), Star Formation

131.001 IUE absorption line studies of highly ionized interstellar gas.
B. D. Savage.
NASA Conf. Publ., NASA CP–2349, p. 3 – 16 (1984). – See Abstr. 012.001.

IUE high dispersion echelle spectra of galactic and extragalactic sources have generally revealed the presence of absorption by interstellar Si IV and C IV. Occasionally N V has been detected. The observational results relating to these species will be reviewed for H II regions, supernova remnants, galactic disk gas, galactic halo gas, and extragalactic gas. For the various regions, the likely origin of the ionization is considered.

131.002 The UV extinction laws in two very young clusters, NGC 6530 in the Galaxy and NGC 2100 in the LMC.
E. Böhm–Vitense, P. Hodge, D. Boggs.
NASA Conf. Publ., NASA CP–2349, p. 187 – 190 (1984). – See Abstr. 012.001.

The authors have studied the UV extinction for a number of O and B stars in the galactic cluster NGC 6530. They have also studied the extinction law for several stars in the LMC cluster NGC 2100. They find distinct differences for different stars in the same cluster. In NGC 6530 a correlation of the extinction law with T_{eff} or with position in the cluster is seen. In NGC 2100 the 2200 Å absorption appears to be much stronger in the center of the cluster. The authors have also studied the interstellar gas line absorption for the stars in NGC 6530.

131.003 IUE observations of interstellar lines.
D. P. Gilra.
NASA Conf. Publ., NASA CP–2349, p. 199 (1984). – See Abstr. 012.001.

For the study of interstellar lines about 350 spectra of about 40 stars were obtained from the IUE Vilspa data bank and analysed. The preliminary results are presented.

131.004 Observations of Mg I and Mg II in the local ISM.
F. Bruhweiler, W. Oegerle, E. Weiler, R. Stencel, Y. Kondo.
NASA Conf. Publ., NASA CP–2349, p. 200 – 203 (1984). – See Abstr. 012.001.

The authors have used high quality IUE data combined with that acquired by Copernicus to study the Mg II/Mg I ionization balance in the local interstellar medium within 50 pc of the Sun. The results are in agreement with the local cloud model as presented previously by Bruhweiler. The Mg I and Mg II column densities are used to place constraints on the physical conditions of the interstellar gas near the Sun.

131.005 The dependence of interstellar element depletions on mean space density.
A. W. Harris, G. E. Bromage, C. Gry.
NASA Conf. Publ., NASA CP–2349, p. 204 – 206 (1984). – See Abstr. 012.001.

Correlations of interstellar gas–phase depletions with hydrogen column density N(H), and with mean line–of–sight space density $\bar{n}(H) = N(H)/r$, are investigated for 14 elements. For the 6 most depleted elements, correlations with $\bar{n}(H)$ are clearly stronger than with N(H). In general, the survey supports the proposal that depletion is mainly governed by density–dependent processes. It also suggests that depletion occurs in relatively low density gas as well as in dense clouds.

131.006 High–velocity interstellar gas in the line–of–sight to HD 50896.
J. N. Heckathorn, R. A. Fesen.
NASA Conf. Publ., NASA CP–2349, p. 207 – 210 (1984). – See Abstr. 012.001.

A large interstellar structure at least 20 parsecs in radius and expanding at 15 – 20 km/sec has been detected in the line–of–sight of the Wolf–Rayet star HD 50896 by means of interstellar absorption lines in the IUE spectra of field stars. This structure may be the remnant left by the supernova explosion caused by the companion to HD 50896. If this scenario is correct, HD 50896 would then represent the second Wolf–Rayet stage of the evolution of a massive OB binary.

131.007 The variation of galactic interstellar extinction in the ultraviolet.
A. N. Witt, R. C. Bohlin, T. P. Stecher.
NASA Conf. Publ., NASA CP–2349, p. 211 (1984). Abstract. – See Abstr. 012.001.

131.008 Interstellar extinction in the nucleus of h Per.
A. Magazzu, R. Stalio.
NASA Conf. Publ., NASA CP–2349, p. 212 (1984). Abstract. – See Abstr. 012.001.

131.009 Time–dependent chemistry – II. Dependence of the chemistry on the initial [C]/[O] abundance ratio.
G. D. Watt.
Mon. Not. R. Astron. Soc., Vol. 212, No. 1, p. 93 – 103 (1985).

A time–dependent treatment of chemical modelling in dense interstellar clouds has been used to investigate the effects of different initial values for the carbon:oxygen ([C]/[O]) abundance ratio. The chemical network used involves 30 simple species containing only C, H and O components and a range of cloud densities from $n(H_2) = 10^3 cm^{-3}$ to $10^6 cm^{-3}$.

131.010 Polarization measurements of the Bok globule B361.
I. P. Williams, K. Vedi, W. K. Griffiths, H. C. Bhatt, P. V. Kulkarni, N. M. Ashok, R. E. Wallis.
Mon. Not. R. Astron. Soc., Vol. 212, No. 1, p. 181 – 187 (1985).

The results of the first measurements of the polarization of light from background stars passing through B361 are described. Nearly all the stars show that the direction of the polarized light is approximately north–south. If the polarization is caused by aligned grains within the globule then a magnetic field of the order of 50 – 100 μG is required. Both polarimetry and photometry confirm that two of the stars studied are very distant background stars while three of these stars were found to be foreground stars. The analysis indicates that the globule is not further away than 650 pc, but can only establish an approximate upper limit. This uncertainty in the data similarly affects all the other conclusions.

131.011 A polarization study of the cometary nebula Parsamyan 21.
P. W. Draper, R. F. Warren–Smith, S. M. Scarrott.
Mon. Not. R. Astron. Soc., Vol. 212, No. 1, p. 1P – 4P (1985).

Polarization data are presented for the cometary nebula Parsamyan 21. The nebula is a relatively highly polarized and illuminated by the apical star within it. Across the head of the nebula and including the illuminating star is a band of polarization which may be due to aligned grains in a circumstellar torus. Parsamyan 21 is very similar in polarization structure to the well–studied cometary nebulae NGC 2261 and 6729, illuminated by R Mon and R CrA respectively.

131.012 The source of illumination of the bipolar nebula near NGC 7129.
P. W. Draper, R. F. Warren–Smith, S. M. Scarrott.
Mon. Not. R. Astron. Soc., Vol. 212, No. 1, p. 5P – 8P (1985).

Polarization data are presented for a small nebula located in the southern outskirts of the reflection nebula NGC 7129. The nebula is bipolar, illuminated from within by a totally obscured star. The position of the star is $RA = 21^h41^m48\overset{s}{.}4 \pm 0\overset{s}{.}7$, $Dec = 65°50'39'' \pm 4''$ (1950).

131.013 The morphology of NGC 6334 IRS V–1.
T. Simon, H. M. Dyck, R. D. Wolstencroft,
R. R. Joyce, P. E. Johnson, I. S. McLean.
Mon. Not. R. Astron. Soc., Vol. 212, No. 1, p. 21P – 25P (1985).

High spatial resolution mapping of the region around the near–infrared source NGC 6334 IRS V–1 shows it to consist of three discrete sources, whose linear arrangement resembles a bipolar nebula. None of these sources corresponds to an object visible on the SERC *I* plate of this region. The high infrared polarization of the easternmost source IRS V–1 suggests that it may be a reflection nebula illuminated by a nearby hidden source, perhaps IRS V–2 or V–3.

131.014 Disc–like magneto–gravitational equilibria.
L. Mestel, T. P. Ray.
Mon. Not. R. Astron. Soc., Vol. 212, No. 2, p. 275 – 300 (1985).

Models are constructed of cool, disc–like clouds in magneto–gravitational equilibrium, immersed in a warm, low–density inter–cloud medium, and with the magnetic field a locally distorted part of the local galactic field. In each case, an area density distribution is adopted, and the detailed flux–to–mass distribution constructed from the electromagnetic and dynamical equations. As expected from simple virial theorem arguments, for each density distribution the non–dimensional total flux/mass parameter f cannot be less than a critical value f_c if mechanical equilibrium is to be maintained. Each such limiting model (with $f = f_c$) has a magnetic neutral point at the edge. It is found that in the various limiting models the flux–to–mass distribution has a nearly constant form (with f_c correspondingly nearly constant). The consequences of plasma–plus–field diffusion are briefly discussed.

131.015 A new search for and discovery of deuterated ammonia in three molecular clouds.
M. Olberg, M. Bester, G. Rau, T. Pauls, G. Winnewisser,
L. E. B. Johansson, Å. Hjalmarson.
Astron. Astrophys., Vol. 142, No. 1, p. L1 – L4 (1985).

An interstellar search for the two rotation–inversion components of the $1_{11} - 1_{01}$ transition of NH_2D was performed towards four selected molecular clouds DR21(OH), S140, L183 and TMC1. Whereas the 85.9 GHz transition was detected in DR21(OH), S140 and L183, the 110 GHz component is considerably weaker and only marginally detected in DR21(OH). High resolution observations of the 85.9 GHz transition allow unambiguous identification of NH_2D, and determination of optical depths and thus column densities by fitting LTE–profiles to the resolved hyperfine spectra. The observed intensity difference between the two rotation–inversion transitions in favour of the 85.9 GHz line is in agreement with the expected nuclear spin statistical weights of the levels involved, whereas strong chemical fractionation effects could cause the negative result in TMC1.

131.016 Star formation and the galactic magnetic field. The B–ϱ relation, flux freezing, and magnetic braking.
T. C. Mouschovias.
Astron. Astrophys., Vol. 142, No. 1, p. 41 – 47 (1985).

A recent paper by Mestel and Paris (1984) claims that the above author has misused the relation $B \propto \varrho^{1/2}$ between the magnetic field and the gas density in a contracting interstellar cloud, in that he has used an inappropriate proportionality constant. According to Mestel and Paris, this led to an incorrect time scale for the magnetic braking of an aligned rotator (Mouschovias 1979). The author documents here that, since the invention of that B–ϱ relation (Mouschovias 1976), he has used it consistently and with full awareness of its validity conditions. The apparent discrepancy concerning the braking time scale is due to (1) the different geometry considered by Mestel and Paris, and (2) their applying Mouschovias' expression to a situation that violates an important validity condition.

131.017 Observations of selected sources in the 2 cm line of H_2CO.
J. Martín–Pintado, T. L. Wilson, F. F. Gardner, C. Henkel.
Astron. Astrophys., Vol. 142, No. 1, p. 131 – 142 (1985).

One arcminute resolution measurements of the $2_{11} - 2_{12}$ line of H_2CO, at 2 cm, are presented. The observations were made toward nineteen OH and/or H_2O maser centers and compact H II regions. In addition, limited maps were made toward eight H II regions; these are shown for 3 sources, W31, W43, and W51, which are extended relative to the telescope beam. The absorption line optical depths were combined with appropriate 6 cm line data to estimate H_2 densities. These agree with previous estimates, to a factor of 2 – 3.

131.018 A model of clumped molecular clouds. I. Hydrostatic structure of dense cores.
E. Falgarone, J. L. Puget.
Astron. Astrophys., Vol. 142, No. 1, p. 157 – 170 (1985).

The authors have computed the equation of state and density structure of thermally supported self–gravitating condensations of interstellar gas mixed with dust. This structure depends only on two physical parameters: the dust temperature in the central core and the external ultra–violet field.

131.019 The galactic reddening law: the evidence from $uvby\beta$ photometry of B stars.
W. Tobin.
Astron. Astrophys., Vol. 142, No. 1, p. 189 – 198 (1985).

Values of interstellar reddening derived from $uvby$ photometry of intermediate and high latitude B stars are used to test the conflicting ideas of total galactic reddening expounded by Burstein and Heiles (1982) and de Vaucouleurs and Buta (1983). The Burstein and Heiles maps are used to determine the intrinsic colours of some slightly–reddened B stars. B stars with projected rotational velocities of $250 - 300$ km s^{-1} do not appear to be significantly redder than the Crawford (1978) standard relation.

131.020 Associations between neutral and ionized gas in Sgr A.
H. S. Liszt, W. B. Burton, J. M. van der Hulst.
Astron. Astrophys., Vol. 142, No. 2, p. 237 – 244 (1985).

The authors consider the morphology of the radio continuum radiation from Sgr A and that of several kinematic components of the neutral molecular and atomic gas, most notably the 50 km s^{-1} cloud. On a smaller scale, it is shown that Sgr A* and Sgr A (West) are encircled by a ring of gas, visible in H I and ^{12}CO, with radius $\cong 3$ pc and rotation velocity $\cong 110$ km s^{-1}. It is this ring, not the 50 km s^{-1} cloud, which provides $40 - 60$ km s^{-1} H I or H_2CO absorption toward Sgr A (West). The ring has a peculiar orientation with respect to the galactic equator but this geometry is also seen in the radio and infrared continuum radiation, and in the kinematics of ^{12}CO, over some 50 pc near Sgr A.

131.021 A high resolution H I absorption spectrum of Sgr A*.
H. S. Liszt, W. B. Burton, J. M. van der Hulst.
Astron. Astrophys., Vol. 142, No. 2, p. 245 – 248 (1985).

The authors have used the VLA to obtain an H I absorption spectrum of the compact source Sgr A* in the galactic nucleus at an angular resolution of 1″.5. Absorption at $40 - 60$ km s^{-1} is present at an optical depth $\cong 0.25$, as in the earlier, lower–resolution H I data (Liszt et al., 1983). The absence of such gas in the recent H_2CO absorption synthesis of Whiteoak et al. (1983) at 3″.5 resolution must therefore be ascribed to peculiarities in the absorbing material and perhaps to the $\cong 130$ times larger solid angle of Sgr A* at $\lambda 21$ cm. It cannot be due only to a background source geometry in which Sgr A* is physically removed from the thermally–emitting material which appears to surround it.

131.022 Oscillations in star formation and contents of a molecular cloud complex.
G. Bodifée, C. de Loore.
Astron. Astrophys., Vol. 142, No. 2, p. 297 – 315 (1985).

The authors present a model of a star–forming region with negative and positive feedback mechanisms, that should reflect more of the complex ecology of the galactic system. The model includes three components: cool H I gas, dusty molecular gas, and young stars with their associated H II regions, that interact on different time scales. Only local transformations, described as mass exchanges between the system components are considered in this study. No bulk transport phenomena are included since the model aims to describe mass transformations in a galactic star–forming subsystem of constant mass in a co–moving frame of reference.

131.023 CO ($J = 4 - 3$) spectroscopy in M17 SW and OMC1.
A. Schulz, A. R. Gillespie, E. Krügel.
Astron. Astrophys., Vol. 142, No. 2, p. 363 – 366 (1985).

The authors present spectra of the $J = 4 - 3$ transition of CO in the Kleinmann–Low object of Orion and in IRc1 in the molecular cloud M17 SW. OMC1 was observed for calibration. IRc1 in M17 is discussed in detail. The spectrum has the same T_A^* as the lower transitions but, with a halfwidth of 12.6 km s^{-1}, it is broader. The CO data are explained by a model for the surroundings of IRc1. In this model the gas density peaks towards IRc1, but the heating is dominated by the H II region of M17. Only turbulent and no systematic motions are present.

131.024 HCO$^+$ observations of the molecular cloud HH24–27.
L. T. Little, S. R. Davies, D. R. Vizard, B. N. Ellison, P. W. Riley, N. Matthews, G. H. Macdonald, L.–A. Nyman.
Astron. Astrophys., Vol. 142, No. 2, p. 378 – 380 (1985).

Observations of the $J = 1 - 0$ and $J = 3 - 2$ transitions of HCO$^+$ in the molecular cloud HH24–26 are consistent with a morphology of a disk structure with bipolar outflow normal to the disk. However, estimates of the molecular hydrogen density $\sim 10^6$cm^{-3} for the disk deduced from HCO$^+$ are not consistent with the estimates $\sim 10^4$cm^{-3} deduced from ammonia observations, although the distribution of emission in NH_3 and HCO$^+$ is similar. Reconciliation of the differences seems to require the presence of many small clumps within the source. As the gas pressure for a particle density $\sim 10^4$cm^{-3} does not appear capable of confining the observed bipolar outflows it is suggested that such clumps, elongated in a similar manner to the overall distribution, may be responsible for such channelling.

131.025 Temperature determinations in molecular clouds of the galactic center.
R. Güsten, C. M. Walmsley, H. Ungerechts, E. Churchwell.
Astron. Astrophys., Vol. 142, No. 2, p. 381 – 387 (1985).

Temperature measurements of the Sgr A and other galactic center molecular clouds are reported. These have been carried out using the (1, 1), (2, 2), (4, 4), and (5, 5) transitions of ammonia as well as the $J = 6 \rightarrow 5$ and $J = 5 \rightarrow 4$ transitions of the symmetric top molecules CH_3CN and CH_3CCH. In the Sgr A cloud, ammonia rotation temperatures are in the range $60 - 120$K and this is consistent with the observations of CH_3CN and CH_3CCH. The authors confirm the conclusion of Morris et al. (1983) that high

kinetic temperatures are characteristic of the entire galactic center region. Some possible heating mechanisms are discussed.

131.026 Drag–like effects on heliospheric neutrals due to elastic collisions with solar wind ions.
H. J. Fahr, H. U. Nass, D. Rucinski.
Astron. Astrophys., Vol. 142, No. 2, p. 476 – 486 (1985).

The authors show that neutral interstellar particles penetrating the heliosphere, besides being subject to specific loss processes, suffer elastic collisions with keV–solar wind ion species and thereby experience a loss of angular momentum and a reduction of the effective gravity. Thus their trajectories are changed into non–Keplerians which result in density and temperature distributions differing from those conventionally calculated on the basis of conserved angular momentum.

131.027 Structure and dynamics of the Bok globule B335.
M. A. Frerking, W. D. Langer, R. W. Wilson.
Icarus, Vol. 61, No. 1, p. 22 – 26 (1985). – See Abstr. 012.004.

CO maps of the Bok globule B335 are presented and used to derive its density profile, mass distribution, and rotational velocity structure. It is found that the cloud is in nearly hydrostatic equilibrium with a density profile that varies roughly as r^{-1} in the core and r^{-3} in the envelope. The observed rotation is unimportant in the force balance at the present stage of evolution.

131.028 A model for the C I (609 μm) emission of Orion.
A. G. G. M. Tielens, D. Hollenbach.
Icarus, Vol. 61, No. 1, p. 40 – 47 (1985). – See Abstr. 012.004.

A model for the energy balance and chemical equilibrium of the gas in photodissociation regions at the edge of molecular clouds, which are illuminated by strong FUV fields (6 eV $\leqslant h\nu \leqslant 13.6$ eV), has been developed. This model is used to calculate the emergent intensities in the fine structure lines of O I (63 μm, 145 μm), C I (609 μm, 370 μm), and C II (158 μm) and in the low–lying rotational transitions of CO. The numerical results show that column densities in the range 2×10^{17} to 2×10^{18}cm^2 can be expected from the C^+/C/CO transition region at the edge of molecular clouds. A detailed model is constructed for the Orion photodissociation region.

131.029 Nonlinear effects of thermal instability. Cooling of the interstellar medium.
A. G. Kritsuk.
Astron. Zh., Tom 62, Vyp. 1, p. 66 – 69 (1985). In Russian. English translation in Sov. Astron., Vol. 29, No. 1.

Early nonlinear stages of thermal instability are described by means of solutions of hydrodynamic equations in the short–wave approximation. It is shown that the nonlinear growth of small density perturbations leads to effective cooling of the thermally unstable gas. The possibility of cloud formation as a result of compression of unstable regions in the interstellar medium is considered.

131.030 On the abundance of deuterated molecules in the Galaxy.
Yu. A. Shchekinov.
Astron. Zh., Tom 62, Vyp. 1, p. 182 – 184 (1985). In Russian. English translation in Sov. Astron., Vol. 29, No. 1.

It is shown that the enrichment of interstellar molecular ions HCO$^+$ by deuterium is determined by metal abundances in the gas phase of the interstellar medium. The positive gradient of [DCO$^+$]/[HCO$^+$] observed in the Galaxy conforms to decreasing of interstellar deuterium abundance and increasing of the interstellar metal abundance in the central parts of the Galaxy by a factor of ten in comparison with the sun.

131.031 Investigation of diffuse matter distribution within a star formation region in Taurus.
N. F. Levina.
Kinematika Fiz. Nebesn. Tel, Tom 1, No. 1, p. 80 – 87 (1985). In Russian.

The distribution of dust material in the region centered at $\alpha_{1950} = 4^h34^m$, $\delta_{1950} = +26°$ is investigated. It is shown that the density maximum of absorbing material is placed at a distance from 50 pc to 200 pc. The dark cloud is located in the central part of the star–formation region at a distance of $\sim$125 pc. The cloud is very patchy and appears to be a complex of dark nebulae. The lower limit of the total mass of the dark cloud material is about 2000 $M_\odot$. A correlation was found between the location of the dust cloud and the most dense part of ^{12}CO emission.

131.032 Small–scale structure and motions in the interstellar gas.
J. M. Dickey.
The Milky Way galaxy, p. 311 – 317 (1985). – See Abstr. 012.007 (IAU Symp. No. 106).

The interstellar medium shows turbulence whose velocity and density spectra are surprisingly like a Kolmogoroff law, in spite of the fact that the turbulence is generated mainly by microscopic explosions, rather than a cascade of energy from larger to smaller scales.

131.033 Atomic hydrogen towards 3C 10.
J. S. Albinson.
The Milky Way galaxy, p. 319 – 320 (1985). – See Abstr. 012.007 (IAU Symp. No. 106).

131.034 Interferometric observations of the small–scale structure of galactic neutral hydrogen.
J. Crovisier, J. M. Dickey.
The Milky Way galaxy, p. 321 – 322 (1985). – See Abstr. 012.007 (IAU Symp. No. 106).

131.035 Interstellar sodium and galactic structure. A high-resolution survey.
E. Maurice, A. Ardeberg, H. Lindgren.
The Milky Way galaxy, p. 325 – 327 (1985). – See Abstr. 012.007 (IAU Symp. No. 106).

The authors have made a study of absorption lines of interstellar sodium covering a substantial part of the Galaxy at extremely high spectral resolution.

131.036 Molecular–cloud clusters and chains.
D. B. Sanders, D. P. Clemens, N. Z. Scoville, P. M. Solomon.
The Milky Way galaxy, p. 3 – 5, 329 – 330 (1985). – See Abstr. 012.007 (IAU Symp. No. 106).

The authors report some preliminary results from the Massachusetts – Stony Brook CO survey of the first galactic quadrant using the 14–meter millimeterwave telescope of the Five College Radio Astronomy Observatory. The survey contains approximately 50,000 observations spaced every 3 arcminutes in l and b between longitudes 0° and 90° and latitudes –1° and 1°.

131.037 Three large molecular complexes in Norma.
L. Bronfman, R. S. Cohen, P. Thaddeus, H. Alvarez.
The Milky Way galaxy, p. 331 – 332 (1985). – See Abstr. 012.007 (IAU Symp. No. 106).

A segment of the Milky Way in Norma, from $l = 327°$ to 335°, $|b| \leqslant 1°$, has been studied as part of the Columbia CO survey of the fourth galactic quadrant.

131.038 Star formation in the Orion Arm.
A. Blaauw.
The Milky Way galaxy, p. 335 – 342 (1985). – See Abstr. 012.007 (IAU Symp. No. 106).

The author assembles principal constituents of the morphology of the Orion Arm within 1500 pc as a starting point for the study of progression of star formation.

131.039 A survey of molecular clouds associated with young open star clusters.
D. Leisawitz, F. Bash.
The Milky Way galaxy, p. 343 – 344 (1985). – See Abstr. 012.007 (IAU Symp. No. 106).

131.040 A new study of the neutral hydrogen in Gould's Belt.
T. J. Sodroski, F. J. Kerr, R. P. Sinha.
The Milky Way galaxy, p. 345 – 346 (1985). – See Abstr. 012.007 (IAU Symp. No. 106).

The structure and kinematics of the neutral hydrogen associated with Gould's Belt have been studied using data of high velocity resolution and large latitude extent covering $l = 10° – 350°$.

131.041 Fine structure of molecular clouds within 1 minute of arc of the galactic center.
N. Kaifu, J. Inatani, T. Hasegawa, M. Morimoto.
The Milky Way galaxy, p. 367 – 369 (1985). – See Abstr. 012.007 (IAU Symp. No. 106).

The authors have observed the HCN($J = 1$–0) line in the vicinity of the galactic center with the 18″ beam of the Nobeyama 45 m telescope. Profiles were taken at 53 points within 1 arcmin from the galactic nucleus with a 10″ grid.

131.042 Highlights of high–velocity clouds.
H. van Woerden, U. J. Schwarz, A. N. M. Hulsbosch.
The Milky Way galaxy, p. 6 – 7, 387 – 408 (1985). – See Abstr. 012.007 (IAU Symp. No. 106).

Twenty years after their first discovery, the nature and origin of high–velocity clouds remain enigmatic. Yet, much important progress has been made in the study of their properties, and prospects are brightening that the problem of their distances, which holds the key to their understanding, may soon be solved. The present paper reviews recent results, partly unpublished.

131.043 A deep survey for high–velocity clouds.
A. N. M. Hulsbosch.
The Milky Way galaxy, p. 409 – 410 (1985). – See Abstr. 012.007 (IAU Symp. No. 106).

The deep Dwingeloo survey for high–velocity clouds is now nearing completion. The preliminary results now available for $0° \leqslant l \leqslant 200°$, $-70° \leqslant b + 70°$ are presented.

131.044 A survey for high–velocity clouds in the inner galaxy.
I. F. Mirabel, R. Morras.
The Milky Way galaxy, p. 411 – 412 (1985). – See Abstr. 012.007 (IAU Symp. No. 106).

The results from a search for high–velocity hydrogen around the direction to the galactic center are presented. About 2000 positions were surveyed with the 43–m radiotelescope of NRAO with an rms of 0.03K on a velocity interval of –1000 to +1000 km s^{-1}.

131.045 Observations of high–velocity clouds colliding in the anticenter.
R. Morras, I. F. Mirabel.
The Milky Way galaxy, p. 413 – 414 (1985). – See Abstr. 012.007 (IAU Symp. No. 106).

131.046 Star formation in molecular clouds.
F. H. Shu.
The Milky Way galaxy, p. 561 – 566 (1985). – See Abstr. 012.007 (IAU Symp. No. 106).

131.047 Chemical composition of interstellar material.
S. R. Pottasch.
The Milky Way galaxy, p. 575 – 584 (1985). – See Abstr. 012.007 (IAU Symp. No. 106).

Abundances in interstellar clouds, as determined from interstellar absorption lines, are discussed first, including abundances in "abnormal" (high–velocity) clouds. H II–region abundances are then discussed and compared to results from the interstellar clouds. The present status of an abundance gradient as determined from H II regions is given. Abundances in planetary nebulae are then given for various categories of nebulae, and compared to H II regions. Finally a short status report on abundances near the galactic center is given.

131.048   Interstellar nitrogen isotope ratios.
R. Güsten, H. Ungerechts.
The Milky Way galaxy, p. 585 – 586 (1985). – See Abstr. 012.007 (IAU Symp. No. 106).

The authors report first results of a systematic study of the galactic interstellar nitrogen isotope ratios, based upon measurements of the $(J,K) = (1,1)$ and $(2,2)$ lines of the $^{14}NH_3$ and $^{15}NH_3$ molecules. They find significant deviations of the $[^{14}N]/[^{15}N]$ abundance ratio from the solar–system value (~ 270), with the most extreme enrichment in the galactic–center region, where $[^{14}N]/[^{15}N] \sim 1500$. This value is consistent with a nitrogen nuclear–synthesis scheme in which ^{14}N is the main product of normal secondary CNO–burning, whereas ^{15}N is depleted by the same process.

131.049   IRAS observations of high–velocity flows.
J. P. Emerson.
Observatory, Vol. 105, No. 1064, p. 4 (1985). Abstract. – See Abstr. 010.721.

131.050   Synthesis of complex molecules via gas–phase chemistry in interstellar molecular clouds.
C. M. Leung.
News Lett. Astron. Soc. N.Y., Vol. 2, No. 7, p. 25 – 26 (1985). Abstract. – See Abstr. 010.241.

131.051   Optical spiral structure at $l = 30° - 70°$. II. The distribution of interstellar extinction.
D. Forbes.
Astron. J., Vol. 90, No. 2, p. 301 – 307 (1985).

New and existing observations of 300 OB stars are used to examine the distribution of interstellar extinction within the highly obscured region of the Milky Way between longitudes 30° and 70°. The region has been divided into 16 zones, each of relatively uniform extinction. The total visual absorption A_v as a function of distance for each zone is shown in a series of diagrams. In general, extinction throughout the region is confined to clouds within 500 – 1000 pc of the Sun. Beyond this distance, there is often little or no additional extinction out to distances of 4 – 5 kpc and sometimes farther.

131.052   The distribution of extinction in the dark cloud complex L1295 as derived from general star counts.
G. S. Rossano.
Astron. J., Vol. 90, No. 2, p. 308 – 309 (1985).

The distribution of extinction in the dark cloud complex L1295 has been derived from general star counts covering an area of 12.5 deg^2. The complex appears to have a generally filamentary structure. Several clouds are observed to be embedded along two ridge–like structures. A total mass of 9300 $M_\odot$ is observed in the complex. Two small, very dense clouds, approximately 1 pc in diameter, and with densities greater than 5600 H I/cm^3, have been detected.

131.053   Time–dependent chemistry – III. The effects of sulphur and oxygen depletion on the interstellar abundances of sulphur–bearing molecules.
G. D. Watt, S. B. Charnley.
Mon. Not. R. Astron. Soc., Vol. 213, No. 1, p. 157 – 166 (1985).

A time–dependent treatment of chemical modelling in dense interstellar clouds has been used to investigate the effects of different initial values for the carbon:oxygen ($[C]/[O]$) abundance ratio. The chemical network used involves 44 species (nitrogen chemistry omitted) and a range of cloud densities from $n(H_2) = 10^3$ to $10^6 cm^{-3}$. If excess carbon remains after CO formation, which dominates all usage of C and O, this leads to enhanced quantities of CS and H_2CS. For a residual excess of oxygen the enhanced sulphur species are OCS and SO_2 with SO only peaking for $[C]/[O]$ near to the solar ratio. The chemistry is not driven by collisions with H_2 molecules, and does not produce strong molecular enhancements with increasing density.

131.054   Quasi–periodic scintillation patterns of the pulsars PSR 1133 + 16 and PSR 1642–03.
A. Hewish, A. Wolszczan, D. A. Graham.
Mon. Not. R. Astron. Soc., Vol. 213, No. 1, p. 167 – 179 (1985).

A new series of interstellar scintillation measurements of the pulsars PSR 1133 + 16 and PSR 1642–03 at 408 and 1426 MHz made in 1982 and 1983 is presented. Quasi–periodic patterns are seen in the dynamic spectra of pulsar scintillation which change on time–scales ranging from one day to several months, but the coherence of these periodic modulations is maintained only over small portions of the time–frequency plane. These effects can be explained in terms of a theoretical model combining random diffraction from a number of small–scale electron density fluctuations in the interstellar medium with systematic refraction caused by larger irregularities.

131.055   Protonated interstellar ions of diatomic molecules: quantum–mechanical predictions.
G. Berthier.
Ann. Phys. (Paris), Vol. 9, No. 4, p. 617 – 620 (1984). Abstr. in Phys. Abstr., Vol. 88, No. 1249, Entry 14729 (1985).

131.056   Structure and origin of the Cygnus superbubble.
N. G. Bochkarev, T. G. Sitnik.
Astrophys. Space Sci., Vol. 108, No. 2, p. 237 – 302 (1985).

This paper summarizes and analyzes the results of radio, optical, infrared, and X–ray observations of a large sector of the sky in the constellation Cygnus ($\alpha \approx 19^h20^m - 22^h$, $\delta = 30 - 50°$; $l^{II} = 65 - 90°$, $|b^{II}| \leqslant 10°$). This region is associated with an extended X–ray source referred to as the Cygnus superbubble. About a quarter of the superbubble region is occupied by the extensively investigated multicomponent thermal radio source Cyg X. The region contains eight OB–associations. Between 50 and 75% of the Cygnus superbubble's X–ray emission can be ascribed to discrete sources, the rest being probably due to regions of coronal gas about 100 pc in diameter, created by stellar winds and, possibly, supernova explosions in individual associations. The objects that produce the X–ray and optical radiation of the presumed superbubble are located at distances from 0.5 to 2.5 kpc from the Sun in the Carina–Cygnus spiral arm.

131.057   Properties of dust in the Orion nebula.
P. Patriarchi, M. Perinotto.
Astron. Astrophys., Vol. 143, No. 1, p. 35 – 38 (1985).

New observations of the Orion nebula very close to the Trapezium stars have been taken with the IUE satellite in order to improve our knowledge of the dust properties in the nebula. The new data have been interpreted together with previous observations using multiscattering models with various distributions of the dust–to–gas ratio. The main result is that a single kind of dust cannot explain the radial behaviour of the scattered light. A possibility is that of a relatively large dust "cavity", such that the grains closer to the illuminating stars scatter in the UV more forward than the more distant ones. The albedo remains the same for all the grains, and moreover approximately constant with the wavelength, from the visual to the UV ($\tilde\omega \cong 0.45$).

131.058 $[^{12}C]/[^{13}C]$ ratios from formaldehyde in the inner galactic disk.
C. Henkel, R. Güsten, F. F. Gardner.
Astron. Astrophys., Vol. 143, No. 1, p. 148 – 152 (1985).

New formaldehyde measurements toward 6 sources in the inner galactic disk (5 kpc $\leqslant R_{GC} \leqslant$ 8 kpc) are reported. From measurements of the $1_{10} - 1_{11}$ line of $H_2^{12}C^{16}O$ and $H_2^{13}C^{16}O$ and of the $2_{11} - 2_{12}$ line of $H_2^{12}C^{16}O$, $[H_2^{12}C^{16}O]/[H_2^{13}C^{16}O]$ ratios are derived. The average corrected ratio is ~ 50 for the inner galactic disk. A decrease in the $[^{12}C]/[^{13}C]$ ratio by a factor of ~ 2 compared with the local interstellar medium is qualitatively consistent with predictions from galactic evolution models and the higher metallicity of the gas in the inner galactic disk.

131.059 Galactic dust: infrared spectrum and temperature distribution of the different components.
M. de Muizon, D. Rouan.
Astron. Astrophys., Vol. 143, No. 1, p. 160 – 175 (1985).

The authors examine four components of the dust, namely those in: the diffuse medium, the molecular clouds, the low density H II regions associated with OB clusters and the regions of active star formation. In each case, the large scale temperature distribution, resulting from the statistical dispersion of the physical parameters, (distance to heating sources, opacity, size, etc.) is evaluated from an appropriate model, and the corresponding emission spectrum computed. The homogeneity of several parameters (grain properties, radiation field etc.) allows a direct comparison of the results, which are given for the solar neighbourhood and the 5 kpc arm region.

131.060 Erratum: "One–millimeter continuum observations of IRAS and FIRSSE sources" [Astron. Astrophys., Vol. 135, No. 2, p. L14 – L17 (1984)].
R. Chini, P. G. Mezger, E. Kreysa, H.–P. Gemünd.
Astron. Astrophys., Vol. 143, No. 1, p. 244 (1985). See Abstr. 37.131.253.

131.061 Nonlinear dispersive waves in interstellar medium.
K. K. Ghosh, S. N. Ghosh, G. C. Pramanik.
Proc. Indian Natl. Sci. Acad., Part A, Vol. 49, No. 6, p. 579 – 585 (1983). Abstr. in Phys. Abstr., Vol. 88, No. 1251, Entry 24510 (1985).

131.062 Sticking coefficients for atoms and molecules at the surfaces of interstellar dust grains.
M. A. Leitch–Devlin, D. A. Williams.
Mon. Not. R. Astron. Soc., Vol. 213, No. 2, p. 295 – 306 (1985). With microfiche MN 213/1.

A fully quantum mechanical model of the particle–surface sticking process is developed, and calculations are performed for a variety of atoms, molecules and solids of relevance to the interstellar medium. The sticking coefficient is a sensitive function of the nature of the particle–surface interaction and the lattice phonon spectrum. Soft lattices and/or chemical adsorption ensure high sticking coefficients. For hard lattices and physical adsorption, sticking is unlikely to occur. The consequences for surface reactions in the interstellar medium are briefly discussed.

131.063 Erratum: "A semi–classical description of radiation transport in astrophysical masers" [Mon. Not. R. Astron. Soc., Vol. 211, No. 4, p. 799 – 811 (1984)].
D. Field, I. M. Richardson.
Mon. Not. R. Astron. Soc., Vol. 213, No. 2, p. 495 (1985). See Abstr. 38.131.192.

131.064 High–resolution surveys of the Sagittarius A molecular cloud complex in ammonia, carbon monoxide, and iso-cyanic acid.
J. T. Armstrong, A. H. Barrett.
Astrophys. J., Suppl. Ser., Vol. 57, No. 3, p. 535 – 570 (1985).

The Sgr A molecular cloud complex has individual cloud masses of $1 - 5 \times 10^5 M_\odot$. The rotation temperature $T_{rot}[NH_3(2-1)]$ is $\sim 35K$, and $T_{rot}[NH_3(6-3)]$ is $\sim 80K$, while CO and HNCO excitation temperatures are between 10K and 30K throughout the complex. The authors report results and data from surveys of five transitions of NH_3, HNCO and CO and supplemental observations of five other lines.

131.065 Bipolar flows and X–ray emission from young stellar objects.
Y. Uchida, K. Shibata.
NASA Conf. Publ., NASA CP–2358, p. 169 – 176 (1985). – See Abstr. 012.023.

The production of both large scale CO bipolar flows and small scale optical bipolar jets from the star–forming regions is interpreted in terms of a magnetic mechanism operating in an accretion model. It is shown by an axisymmetric 2.5–dimensional simulation that a large scale cold bipolar flow may be produced in the relaxation of the magnetic twist which is created by the rotational winding–up of the magnetic field in the contracting disk. In contrast, small scale warm bipolar jets may be driven by the recoiling shocks produced at the stellar surface by the infalling material, which is released from the inner edge of the disk through magnetic reconnections.

131.066 Warm dust in the neutral interstellar medium.
F. Boulanger, B. Baud, G. D. van Albada.
Astron. Astrophys., Vol. 144, No. 1, p. L9 – L12 (1985).

A comparative study of H I and IRAS maps of a $20° \times 18°$ field at high galactic latitude is presented. The galactic infrared emission is shown to be well correlated with the H I distribution. A quantitative relation between H I column density and far–infrared surface brightness is derived. From this correlation the possible use of IRAS observations to map high galactic latitude extinction is discussed. The infrared spectrum of one diffuse H I cloud, from 12 to 100 μm is presented. Its emission in the mid infrared (12 and 25 μm) is several orders of magnitude larger than predicted by standard dust models. The implications of the results on interstellar dust properties are discussed.

131.067 High resolution mapping of galactic H I in the direction of 3C147.
P. M. W. Kalberla, U. J. Schwarz, W. M. Goss.
Astron. Astrophys., Vol. 144, No. 1, p. 27 – 36 (1985).

High resolution H I emission and absorption data in the field of 3C147 have been obtained by combining observations from both the Westerbork and the 100 m Effelsberg telescopes. The H I brightness temperature distribution is clumpy. The absorption features can be identified with small scale (3′ to 6′) features having brightness temperatures in the range 4 to 10K. These clumps have excitation temperatures between 34 and 74K. Their sizes are 0.5 to 1 pc, densities 20 to 50 cm^{-3} and the H I masses 0.03 to 0.5 $M_\odot$, assuming a distance of 500 pc. The clumps of absorbing H I gas are associated with extended warm halos which dominate the observed H I emission. The fact that the absorbing gas is concentrated in small scale clumps provides severe constraints for the interpretation of previously determined apparent spin temperatures.

131.068 The distribution of the local interstellar medium derived from Mg II column densities towards seven cool stars.
G. Vladilo, J. E. Beckman, L. Crivellari, M. L. Franco, P. Molaro.
Astron. Astrophys., Vol. 144, No. 1, p. 81 – 86 (1985).

High resolution spectra containing the Mg II h and k resonance doublet have been obtained using the long wavelength spectrometer of IUE, for 7 late–type stars (F0 to G8). In one of these stars, β Hyi, two sharp absorption features were detected against the chromospheric emission cores. Using the study of Crutcher (1982) the authors can employ the velocity of one of these features to identify it as arising from the local interstellar medium (LISM), while the other feature appears due to chromospheric self–absorption. Although the other stars do not show separate intrinsic and LISM features it is possible to use evidence both from wavelength shifts and line half–widths to identify the interstellar absorption components. This gives the Mg II column densities along the lines of sight to five of these stars and upper limits towards the other two.

131.069 IDS and CCD study of the LMC excited blobs.
M. Heydari–Malayeri, G. Testor.
Astron. Astrophys., Vol. 144, No. 1, p. 98 – 108 (1985).

IDS spectrophotometry of emission lines in the 3650 – 7000 Å range and CCD narrow–band imagery are presented for the newly detected Large Magellanic Cloud compact nebulae N159 Blob and N11A. For comparison purposes, the LMC H II region N159A is studied as well. The chemical abundance of He, N, O, and S relative to H are derived and discussed. The extinction is found to be extraordinarily high especially for N159 Blob, while the extinction derived for N11A represents a rather "normal" value for the LMC. A remarkable aspect of the blobs is their strong continuum at Hβ. These objects may be excited by a ZAMS star of $T_{eff} \sim 44,000K$ ($\sim 40\ M_\odot$) along with a cluster of B and A stars playing a minor part in the excitation. The masses derived for N159 Blob and N11A are ~ 50 and 250 $M_\odot$ respectively.

131.070   Desorption from interstellar grains.
A. Léger, M. Jura, A. Omont.
Astron. Astrophys., Vol. 144, No. 1, p. 147 – 160 (1985).

Different desorption mechanisms from interstellar grains are considered to resolve the conflict between the observations of the presence of gaseous species in molecular clouds and their expected depletion onto grains. The authors concentrate on impulsive heating by X–rays and cosmic rays, and use recent measurements of the fluxes of these particles. The temperature increase of the grains following impulsive heating is estimated using new laboratory specific heat measurements in both the classical thermal evaporation picture and the chemical explosion model of Greenberg and d'Hendecourt. Both whole grain heating and local hot spot desorption are considered. The effects of the relatively rare heavy cosmic rays are found to dominate, except for the smallest grains.

131.071   The physical structure of the dark cloud B5.
L. G. Stenholm.
Astron. Astrophys., Vol. 144, No. 1, p. 179 – 185 (1985).

The author has derived a consistent physical model, for the dark dust cloud B5, from recent line observations of the 1 – 0 and 2 – 1 lines of ^{12}CO, ^{13}CO, and C^{18}O. He found a cloud mass of 622 $M_\odot$. The cloud seems to be in hydrostatic equilibrium and its bulk is heated by cosmic rays. In the region near the edge, the temperature increases strongly and the abundances decrease by nearly three orders of magnitude. This indicates that the interstellar radiation field plays an important role in the outer regions of the cloud. The turbulent velocity is supersonic. The simplest explanation is that the turbulence is due to non–dissipative Alfvén waves. The implied field strength is consistent with the non–detection of magnetic fields in dark clouds.

131.072   Interstellar cloud phase transitions: effects of metal abundances, grains and X–rays.
J. M. Shull, D. T. Woods.
Astrophys. J., Vol. 288, No. 1, p. 50 – 57 (1985).

The authors study the cloud–intercloud phase transition of interstellar gas in pressure equilibrium, including effects of metal abundance variations, grain heating, and X–ray fluxes from the galactic halo. Equilibrium solutions for the gas pressure P and hydrogen density n_H obey fairly universal scaling laws. Both grain and X–ray heating enhance the susceptibility of intercloud gas to thermal instability, possibly resulting in the conversion of intercloud gas to clouds in the inner disks of spiral galaxies. X–ray ionization may be responsible for a class of C IV – Si IV absorption lines seen above the galactic disk and in quasar absorption systems.

131.073   Thermal phases of interstellar and quasar gas.
S. Lepp, R. McCray, J. M. Shull, D. T. Woods, T. Kallman.
Astrophys. J., Vol. 288, No. 1, p. 58 – 64 (1985).

Interstellar gas may be in a variety of thermal phases, depending on how it is heated and ionized. The authors present a unified picture of the equation of state of interstellar and quasar gas for a variety of such mechanisms over a broad range of temperatures, densities, and column densities of absorbing matter. It is found that for select ranges of gas pressure, photoionizing flux, and heating, three thermally stable phases are allowed: coronal gas ($T > 10^5$K); warm gas ($T \approx 10^4$K); and cold gas ($T < 10^2$K). With attenuation of ultraviolet and X–ray radiation, the cold phase may undergo a transition to molecules. The authors discuss the underlying atomic physics behind each of these phase transitions, and describe their relevance to interstellar matter and quasars.

131.074   The effect of a weak shock on interstellar gas toward the ϱ Ophiuchi cloud.
K. A. Meyers, T. P. Snow, S. R. Federman, M. Breger.
Astrophys. J., Vol. 288, No. 1, p. 148 – 158 (1985).

The authors have studied the effect of a low–velocity shock on depletion in interstellar clouds. High–resolution Copernicus observations of interstellar absorption lines toward four stars in the ϱ Ophiuchi cloud complex were used to measure differential depletion in interstellar clouds separated by velocities of 10 – 15 km s^{-1}. The authors infer the presence of a shock wave with a velocity of about 10 km s^{-1} propagating away from the Sun in the direction of the Sco OB2 association. In comparing the preshock and postshock gas, it is found that the postshock gas is more depleted than the preshock gas and contains a large molecular component. It is concluded that a low–velocity shock, such as the one observed here, enhances grain growth in the shocked interstellar gas.

131.075   VLA observations of smooth, rapidly rotating NH₃ in the Sagittarius A "15 km s^{-1} cloud".
J. T. Armstrong, P. T. P. Ho, A. H. Barrett.
Astrophys. J., Vol. 288, No. 1, p. 159 – 163 (1985).

The authors observed NH$_3$(3,3) emission from the Sgr A "15 km s^{-1}" cloud, M –0.13–0.08, with the VLA. The emission in the VLA maps is in structures of $\sim 15''$ to $40''$ (~ 0.7 to 2 pc) size. Longitude–velocity plots show a velocity gradient of ~ 20 km s^{-1}arcmin^{-1} (~ 8 km s^{-1}pc^{-1}). If this represents gravitationally bound rotation, it implies a mass of $1.2 \times 10^5 M_\odot$ within a 2 pc radius.

131.076   The bright–rimmed molecular cloud around S140 IRS. I. CS ($J = 1 - 0$) observations.
M. Hayashi, T. Omodaka, T. Hasegawa, S. Suzuki.
Astrophys. J., Vol. 288, No. 1, p. 170 – 174 (1985).

The bright–rimmed molecular cloud around S140 IRS has been mapped in the CS ($J = 1 - 0$) emission with an angular resolution of $33''$. The relation between the CS emission and the distance from the ionization front is discussed, based on the position of the optical rim determined from an Hα photograph. The distribution of the CS emission in the vicinity of S140 IRS shows a barlike east–west elongation which is prominent at $V_{LSR} = -6$ km s^{-1}, and an arclike structure clearly seen at $V_{LSR} = -8$ km s^{-1}.

131.077   Small–scale electron density turbulence in the interstellar medium.
J. M. Cordes, J. M. Weisberg, V. Boriakoff.
Astrophys. J., Vol. 288, No. 1, p. 221 – 247 (1985).

Diffraction phenomena associated with the scattering of pulsar radiation are used to constrain the power spectrum and galactic distribution of electron density fluctuations. The authors combine new measurements of interstellar scintillation bandwidths and pulse broadening times with those from the literature to yield information on 76 lines of sight. For individual objects, the scaling of scintillation bandwidth with frequency is in agreement with a Kolmogorov spectrum. The study of all objects, however, indicates strong disagreement with a model in which turbulence is uniformly distributed. The data suggest a two–component model for electron density turbulence: (1) a clumped medium with scale height $H \lesssim 100$ pc; (2) a nearly uniform medium with $H \gtrsim 0.5$ kpc. Scattering material is ubiquitous but is preferentially located near regions of large energy input into the interstellar medium.

131.078   A study of depletions within the Rho Ophiuchi cloud based on IUE observations of HD 147889.
T. P. Snow, C. L. Joseph.
Astrophys. J., Vol. 288, No. 1, p. 277 – 283 (1985).

Archival IUE data on the deeply embedded ϱ Oph cloud star HD 147889 have been used to derive column densities for several species. Absolute depletions have been estimated, based on an assumed value of the total hydrogen column density, but they are uncertain due to uncertainty in N_H. The data appear to indicate that depletions are generally very large in this line of sight. There is also some indication that the relative depletions are enhanced in the densest portions of the cloud.

131.079 New maser lines of methanol.
M. Morimoto, M. Ohishi, T. Kanzawa.
Astrophys. J., Lett. Ed., Vol. 288, No. 1, p. L11 – L15 (1985).

The authors found that the $4_{-1} - 3_0 \, E$ line (36.169 GHz) of methanol (CH_3OH) is a maser in Sgr B2, with an antenna temperature of up to 60K and a velocity width of ~ 10 km s^{-1}. This maser is spatially extended with angular size of up to 90", is distributed along the edge of H II regions, and apparently is not directly attached to star forming regions. The authors also discovered a new maser line of CH_3OH, the $7_0 - 6_1 \, A$ line (44.069 GHz) in Sgr B2, W51, and two other galactic sources.

131.080 On the source of the ^{26}Al observed in the interstellar medium.
D. S. P. Dearborn, J. B. Blake.
Astrophys. J., Lett. Ed., Vol. 288, No. 1, p. L21 – L24 (1985).

Recent HEAO 3 observations have been interpreted by Mahoney and colleagues as requiring approximately 3 $M_\odot$ of ^{26}Al alive in the interstellar medium. Calculations briefly discussed here indicate that there is substantial production and dispersal of ^{26}Al in the stellar winds of O and W–R stars and suggest that the stellar winds of very massive stars are a significant source of ^{26}Al.

131.081 Velocity fields in binary protostellar clouds: an alternative to retrograde rotation.
A. P. Boss.
Astrophys. J., Lett. Ed., Vol. 288, No. 1, p. L25 – L28 (1985).

Observations of the emission from optically thin molecular species in several dense interstellar clouds have been interpreted as indicating rotation of cloud envelopes in one direction and of cloud cores in the opposite direction (retrograde rotation). This has been taken as evidence for the presence of magnetic fields sufficiently strong to have caused the retrograde rotation. However, it is shown that the velocity fields that are produced when a nonmagnetic interstellar cloud collapses to form a binary protostellar system yield spatial velocity maps that appear to be at least qualitatively consistent with the ^{13}CO observations. The model calculations are applied to the protostellar core of the cloud Barnard 5.

131.082 Low–frequency variability of pulsars.
R. Blandford, R. Narayan.
Mon. Not. R. Astron. Soc., Vol. 213, No. 3, p. 591 – 611 (1985).

Several recent observational and theoretical investigations have suggested that long–wavelength ($\leqslant 10^{14}$cm) density fluctuations in the interstellar medium may have a major influence on observations of pulsars and extragalactic radio sources. These long–wavelength fluctuations are responsible for refractive focusing, and may cause the monthly to annual variations in pulsar flux densities observed at low frequency. Fluctuations in the angular sizes, pulse arrival times and pulse widths of pulsars should also be observable and should be cross–correlated with the flux density variations. The magnitude of these correlations, and their dependence upon the time lag and the observing wavelength, for different power–law spectra of density fluctuations are estimated. Sensitive observations on selected pulsars will be able to confirm the importance of refractive effects in the interstellar medium and also determine the slope of the density fluctuation spectrum.

131.083 An upper limit on the proper motion of optical filaments near the Cetus Arc (Loop 2).
A. W. Jones, J. Meaburn.
Mon. Not. R. Astron. Soc., Vol. 213, No. 3, p. 711 – 716 (1985). With a correction in Vol. 215, No. 1, p. 159 (1985).

The authors have used a cross–correlation method to investigate the proper motion of optical filaments outside the radio loop known as the Cetus Arc (Loop 2). No motion was detected to a 3σ upper limit of 0.15 arcsec yr^{-1}. If the filaments are the result of a supernova explosion at the centre of the Cetus Arc, then the age is $T > 1.3 \times 10^6$yr, the optical radius is $R = 43 - 160$ pc, the expansion velocity is $v < 37$ km s^{-1} and the explosion energy is $(0.1 - 35) \times 10^{44}$J.

131.084 New results for ion–molecule reactions of HC_3N in dense interstellar clouds.
D. K. Bohme, A. B. Raksit.
Mon. Not. R. Astron. Soc., Vol. 213, No. 3, p. 717 – 720 (1985).

A selected–ion flow tube apparatus has been employed in the measurement of rate constants and product distributions at 296 ± 2K for the reactions of C^+, CH_3^+ and $C_2H_2^+$ with cyanoacetylene. The results are in disaccord with those obtained previously with the flowing afterglow technique by Freeman et al. and, in contrast to the conclusions drawn by these authors, establish a pivotal role for cyanoacetylene in the chemical evolution of interstellar clouds by ion–molecular reactions.

131.085 How stars form: a synthesis of modern ideas.
B. E. Turner.
Vistas Astron., Vol. 27, Part 4, p. 303 – 332 (1984).

In this review the author assembles different observational data into a global picture of star formation which traces how molecular clouds, both giant and small, produce stars of varying masses at varying times and with varying spatial distributions. He discusses the central role of turbulence (either energy cascade or torque–driven) in initially supporting molecular clouds and ultimately in determining the mass spectrum of the formed stars. Special emphasis is given to the formation of massive stars, which requires that several physical processes operate, as distinct from low–mass stars which are more easily formed.

131.086 Interstellar beryllium.
A. M. Boesgaard.
Publ. Astron. Soc. Pac., Vol. 97, No. 587, p. 37 – 40 (1985).

A new search for interstellar Be has been conducted which takes advantage of the high ultraviolet transparency of Mauna Kea and the photometrically accurate Reticon detector and supercoated (UV) optics of the Canada–France–Hawaii 3.6–m telescope. Spectra were obtained at 1.6 Å mm^{-1} of four early B stars on each of two nights and combined for each star to give signal–to–noise ratios of 65 to 300. The two stars with the highest quality data, ζ Per and δ Sco, yield upper limits on the Be II resonance line $\lambda 3130$ of 0.23 mÅ and 0.36 mÅ, respectively, smaller than any previously reported limits. The corresponding Be/H abundances are $\leqslant 4.8 \times 10^{-12}$ and $\leqslant 8.4 \times 10^{-12}$ or [Be/H] = -0.43 and -0.22, respectively, with solar–stellar Be/H = 1.3×10^{-11}.

131.087 Detectioan of high velocity emission in $NH_3(1,1)$ and (2,2) spectra towards NGC 2071.
T. Takano, J. Stutzki, G. Winnewisser, Y. Fukui.
Astron. Astrophys., Vol. 144, No. 2, p. L20 – L22 (1985).

High signal–to–noise NH_3 (J,K) = (1,1) and (2,2) spectra observed with the 100 m radiotelescope towards the central position of the reflection nebula NGC 2071 reveal high velocity emission of $\Delta v \sim 11$ km/s width superimposed on a spike ($\Delta v = 0.7$ km/s) and a narrow ($\Delta v = 1.8$ km/s) molecular emission line. The rotational temperature of the high velocity outflowing gas is determined to be close to 35K, suggesting a somewhat higher kinetic temperature ~ 45K. The small beam filling factor $\sim 1\%$ shows that the NH_3 originates from a gas with clumpy or filamentary structure on the scale of 10^{-2}pc.

131.088 Kinematics of the extended CO envelope of Heiles' Cloud 2.
T. Takano, Y. Fukui, H. Ogawa.
Astron. Astrophys., Vol. 144, No. 2, p. 363 – 368 (1985).

Heiles' Cloud 2 was observed over a $\sim 1°5 \times 1°5$ area using the $^{12}CO \, J = 1-0$ line with a high velocity–resolution of 0.1 km s^{-1}. The CO emission extends more widely than the H_2CO cloud and the region of the heavy visual extinction. The profiles show heavy saturation of the line all over the area except in the peripheral part, and the kinetic temperature of the cloud was estimated to be $(6 - 9) \pm 1$K. An integrated intensity map of CO emission shows a ridge of matter on the NE edge of the cloud. The ridge lies on the major axis of TMC 1 and might be formed through cloud–cloud collision between two components of Heiles' Cloud 2.

131.089   Gamma–ray emission from atomic and molecular gas at intermediate galactic latitudes.
A. W. Strong.
Astron. Astrophys., Vol. 145, No. 1, p. 81 – 89 (1985).

Recent developments in the study of the correlations between gamma–rays and local gas tracers are described. COS–B data in the range $10° < |b| < 20°$ are analysed using simultaneously 21–cm and galaxy count data in order to separate the roles of atomic and molecular gas. The new galaxy–count calibration of Strong (1983) is used and the results compared with those for the calibration of Strong and Lebrun (1982). The contribution from inverse Compton scattering is modelled and its absolute intensity derived by fitting to the data is in accord with that expected from known interstellar radiation fields. The resulting total predicted intensities from atomic and molecular gas and inverse Compton scattering are in better accord with the data than in earlier work (Strong et al., 1982). The results in the three energy ranges considered indicate a flatter spectrum for the gas emissivity than previously found.

131.090   A search for H_2 emission in bipolar nebulae and regions of interstellar shock.
J. P. Phillips, G. J. White, R. Harten.
Astron. Astrophys., Vol. 145, No. 1, p. 118 – 126 (1985).

The authors report a H_2 emission survey of five bipolar outflow sources (NGC 1333, M2–9, AS 353, S 106, V645 Cyg), and one region of shock interaction between an H II region and molecular cloud (NGC 281). Two of the sources (M2–9, NGC 1333) were detected in the $v = 1$–0S(1), and Q branch transitions of H_2, and a detailed analysis and modelling for these cases is provided.

131.091   Anomalous motions of H I clouds.
F. N. Bash, D. Leisawitz.
Astron. Astrophys., Vol. 145, No. 1, p. 127 – 134 (1985).

The authors have examined a set of H I clouds which have been found to be moving with anomalous velocities and find that the peculiar velocities associated with the Galaxy's spiral arms can explain their motions. Specifically, the authors' Ballistic Particle Model, which describes the motions of giant molecular clouds (GMCs) born in the spiral arms, produces GMCs having velocities similar to observed values in regions of the Galaxy where the anomalous H I clouds are found. In this explanation, the anomalous H I clouds are actually the residual H I in giant molecular clouds. It is shown that the expected molecular emission lines are found.

131.092   Rapid outbursts in the water maser Cepheus A.
K. Mattila, N. Holsti, M. Toriseva, R. Anttila, L. Malkamäki.
Astron. Astrophys., Vol. 145, No. 1, p. 192 – 200 (1985).

The authors have monitored the 22.2 GHz emission of the water maser in Cepheus A between October 1980 – October 1983. Several rapid bursts were observed with typical rise and fall times ranging from 3 to 30 d. Detailed observations are presented for one such burst which cover the event in April – May 1983 with time intervals of one or a few days. To model the intensity curve of the outburst the authors assume an instant release of radiation energy from a central (proto)star, leading to heating of the surrounding gas cloud where the water maser is located. A collisional pump model for the maser gives a good agreement with the observed intensity curve of the burst. This is in accordance with an equally good agreement obtained by Burke et al. (1978) between this model and the W3(OH) outburst.

131.093   Constraints on the sites of nitrogen nucleosynthesis from $^{15}NH_3$–observations.
R. Güsten, H. Ungerechts.
Astron. Astrophys., Vol. 145, No. 1, p. 241 – 250 (1985).

The authors present results on the interstellar nitrogen isotope abundance ratio towards molecular clouds in the local disk (S140, DR21(OH), W3(OH)) and the galactic center region, based upon observations of the $^{14,15}NH_3$ inversion transitions in the $(J, K) = (1,1)$, (2,2) rotational levels. They find that the local interstellar matter has been only moderately enriched,

$[^{14}N/^{15}N] \sim 400$, since the time the solar system was formed 4.5 Gyrs ago. From the weak $^{15}NH_3$ lines in the galactic center clouds an extreme enrichment by a factor of ~ 4 over the solar system ratio (270) is inferred. The authors analyze their results on the nitrogen enrichment, together with data on C, O nuclei in terms of a chemical evolution model.

131.094   OH observations of cloud complexes in Taurus.
J. G. A. Wouterloot, H. J. Habing.
Astron. Astrophys., Suppl. Ser., Vol. 60, No. 1, p. 43 – 59 (1985).

The authors present groundstate OH line observations of large molecular clouds in the Taurus region. Two cloud complexes can be distinguished on the basis of the radial velocities. Complex A, at 110 pc distance, has a size of 30×25 pc^2 and a mass of about $1.5 \times 10^4 M_\odot$. The excitation temperatures of the main lines are almost equal, at least at positions with a high apparent optical depth. The mean densities in the clouds are a few times 10^2 H_2 molecules per cm^3. The velocities vary irregularly within complex A. Complex B (distance 300 pc) with a size of 70×12 pc^2 and a mass of $4 \times 10^4 M_\odot$ shows a large velocity gradient in the direction of its longest dimension. There is a weak correlation between the distribution of the T Tauri stars and some properties (OH optical depth, CO temperature) of the molecular clouds.

131.095   VLA observations of ammonia and continuum in regions with high–velocity gaseous outflows.
J. M. Torrelles, P. T. P. Ho, L. F. Rodríguez, J. Cantó.
Astrophys. J., Vol. 288, No. 2, p. 595 – 603 (1985).

The authors present VLA observations of the (1, 1) inversion transition of NH_3 in seven regions with high–velocity gaseous outflows. Small ($\lesssim 0.1$ pc) condensations were detected in L1551, S140, and Cepheus A. The condensation in L1551 is displaced by $\sim 28''$ from IRS 5, the suspected source of energy of the region. Three individual condensations were detected in Cepheus A, one of them with a remarkably elongated shape. These condensations may play an important role in focusing the bipolar outflow in the region.

131.096   Optical and radio study of the Taurus molecular cloud toward HD 29647.
R. M. Crutcher.
Astrophys. J., Vol. 288, No. 2, p. 604 – 617 (1985).

Combined optical and radio wavelength observations of a dark molecular cloud are described. The line of sight to HD 29647, a B6 – 7 IV Hg–Mn star which is behind the outer envelope of Taurus Molecular Cloud 1, has been studied. Photometry yields $A_V = 3.7$ mag, $R \approx 3.6$, enhanced far–ultraviolet extinction, and an almost absent 2200 Å extinction feature. Optical interstellar lines of CN and C_2 are very strong, while lines of CH and K I are only moderate in strength. Millimeter lines of CO, ^{13}CO, $C^{18}O$, HCN, HCO$^+$, and CN were detected, while HC_3N, C_2H, and DCO$^+$ were not. Physical conditions are an H_2 density of 800 cm^{-3}, a kinetic temperature of 10K, and a fractional electron abundance of 10^{-6}.

131.097   The interstellar extinction law from 1 to 13 microns.
G. H. Rieke, M. J. Lebofsky.
Astrophys. J., Vol. 288, No. 2, p. 618 – 621 (1985).

Measurements from 1 to 13 μm are reported for o Sco and for stars in the galactic center. The interstellar extinction law toward these sources and toward VI Cyg No. 12 is the same from 1 to 13 μm. An improved estimate of the extinction law beyond 3 μm is presented, including an improved ratio of total to selective extinction, $R \approx 3.09 \pm 0.03$, and an improved ratio of total extinction to optical depth in the 10 μm silicate absorption, $A_v/\tau_{Si} = 16.6 \pm 2.1$.

131.098   An infrared study of the bipolar outflow region GGD 12–15.
P. M. Harvey, B. A. Wilking, M. Joy, D. F. Lester.
Astrophys. J., Vol. 288, No. 2, p. 725 – 730 (1985).

Infrared observations from 1 to 100 μm are presented for the region associated with a bipolar CO outflow source near the nebulous objects GGD 12 – 15. The authors find a luminous far–

infrared source associated with a radio continuum source in the area. This object appears to be a compact H II region around a nearly main–sequence B0 star. A faint 20 μm source has been found at the position of an H_2O maser 30″ northwest of the H II region. This object appears to be associated with but not coincident with a 2 μm reflection nebula. This structure is interpreted as evidence for a nonspherically symmetric, possibly disklike dust distribution around the exciting star for the maser.

131.099   Study of carbon monoxide formation in interstellar clouds.
L. N. Arshutkin.
Astrofizika, Tom 22, Vyp. 1, p. 163 – 176 (1985). In Russian. English translation in Astrophysics, Vol. 22, No. 1.
The abundance of atoms and molecules which are important in the thermal balance of interstellar clouds was calculated as a function of optical depth over a wide range of densities $(10 – 10^4 cm^{-3})$ and temperatures (10 – 100K). Prime attention was given to variations of C^+, C and CO concentrations.

131.100   Chemistry in dynamically evolving clouds.
S. P. Tarafdar, S. S. Prasad, W. T. Huntress Jr., K. R. Villere, D. C. Black.
Astrophys. J., Vol. 289, No. 1, p. 220 – 237 (1985).
The authors present a unified model of chemical and dynamical evolution of isolated, initially diffuse and quiescent interstellar clouds. This model uses a semiempirically derived dependence of the observed cloud temperatures on the visual extinction and density. Even low–mass, low–density, diffuse clouds can collapse in this model, because the inward pressure gradient force assists gravitational contraction. Theoretically predicted dependences of the column densities of various atoms and molecules, such as C and CO, on visual extinction in diffuse clouds are in accord with observations. Similarly, our predicted dependences of the fractional abundances of various chemical species on the total hydrogen density in the core of the dense clouds also agree with observations.

131.101   The star forming regions in the Monoceros R2 molecular cloud.
V. A. Hughes, J. G. N. Baines.
Astrophys. J., Vol. 289, No. 1, p. 238 – 243 (1985).
In order to determine if newly formed early–type stars are present in the large molecular cloud associated with the Mon R2 cluster of reflection nebulae, the whole cloud has been scanned at 3.2 and 10.55 GHz using the 46 m radio telescope of the Algonquin Radio Observatory. Only three H II regions were definitely detected, all of which are located in the central core region, and each has been previously observed. It is noted that all young stellar objects are located in the inner core of the cloud. The majority of all these objects lies close together along a line which is almost perpendicular to the galactic plane, implying that the formation of stars with spectral types ranging from B1 to B9 has only occurred in an annulus, plane, or line through the center of the cloud and nowhere else in the cloud.

131.102   Generation of turbulence by shock waves and cosmic ray diffusion in the interstellar medium.
A. M. Bykov, I. N. Toptygin.
Pis'ma Astron. Zh., Tom 11, No. 3, p. 184 – 189 (1985). In Russian. English translation in Sov. Astron. Lett., Vol. 11.
The interstellar shock waves distribution function is derived taking into account supernovae shock interactions with interstellar clouds. The turbulent fluctuation spectrum connected with shock waves as well as cosmic ray diffusion coefficients are obtained.

131.103   Moleküle im interstellaren Raum. I. Struktur und physikalische Eigenschaften der Molekülwolken.
T. Henning, J. Gürtler.
Sterne, 61. Band, Heft 1, p. 3 – 10 (1985).

131.104   Correlation of infrared dust emission, galaxy counts, and hydrogen column density.
G. Dall'Oglio, P. de Bernardis, S. Masi, F. Melchiorri, G. Moreno, R. Trabalza.
Astrophys. J., Vol. 289, No. 2, p. 609 – 612 (1985).
The authors analyze the correlations of far–infrared emission of interstellar dust, galaxy counts, and 21 cm atomic hydrogen emission. They conclude that the IR signal is the best available dust indicator. The correlations also provide infomation on the inhomogeneity of the galaxy distribution and on the molecular hydrogen column density.

131.105   Search for molecular oxygen in dense interstellar clouds.
P. F. Goldsmith, R. L. Snell, N. R. Erickson, R. L. Dickman, F. P. Schloerb, W. M. Irvine.
Astrophys. J., Vol. 289, No. 2, p. 613 – 617 (1985).
The authors have carried out a search for the 234 GHz $N = 2 \rightarrow 0$, $J = 1 \rightarrow 1$ transition of $^{16}O^{18}O$ using the 13.7 m FCRAO radio telescope. No emission was detected toward six giant molecular clouds. Observations of the 220 GHz $J = 2 \rightarrow 1$ transition of $C^{18}O$ yield column densities for this species $1 – 3 \times 10^{16} cm^{-2}$; the resulting limits on the $[O_2]/[CO]$ ratio lie between <0.5 and <4. The implications of these limits for the cloud chemistry are discussed.

131.106   The abundance of interstellar OD.
K. Croswell, A. Dalgarno.
Astrophys. J., Vol. 289, No. 2, p. 618 – 620 (1985).
The authors present the chemistry of OD and estimate the abundance of OD in diffuse and dense molecular clouds. They predict substantial enhancements in the abundance of OD relative to OH and argue that density ratios of 0.01 or greater are likely in many clouds. A search could add OD to the list of known interstellar molecules.

131.107   Molecular chemistry and the chemical evolution of interstellar clouds.
T. J. Millar.
Gas in the interstellar medium, p. 6 – 38 (1984). – See Abstr. 012.032.

131.108   Chemical enrichment and evolution of the interstellar medium.
M. Peimbert.
Gas in the interstellar medium, p. 55 – 65 (1984). – See Abstr. 012.032.
There are four problems related to the chemical evolution of the interstellar medium that are briefly discussed: (1) the $\Delta Y/\Delta Z$ ratio, (2) the relation between the yield and the heavy element abundances, (3) N production and the N/O versus O/H diagram, and (4) the relative importance of planetary nebulae and supernovae in the C enrichment of the interstellar medium.

131.109   Abundances of elements in the gas–phase of the interstellar medium.
P. M. Gondhalekar.
Gas in the interstellar medium, p. 66 – 83 (1984). – See Abstr. 012.032.
The abundances of elements in the interstellar medium (ISM) determined from absorption line studies, are considered. The abundances of elements are correlated with the local gas density in the ISM. This is true for gas in the Galactic plane and in the Galactic halo and is also true for shocked gas in the ISM. The data suggest that the elements are depleted during grain formation in the atmospheres of stars, there is no grain formation in the ISM. In the ISM grains are sputtered or destroyed returning elements to the gas–phase. The degree of sputtering or destructin depends on the shielding of grains. In high density clouds the shielding is higher and, therefore, abundances are lower.

131.110 Long–chain hydrocarbon molecules in the interstellar medium: search for 1–cyanobut–3–ene–1–yne, $CH_2 = CH–C \equiv C–C \equiv N$.
H. W. Kroto, D. McNaughton, L. T. Little, N. Matthews.
Mon. Not. R. Astron. Soc., Vol. 213, No. 4, p. 753 – 759 (1985).

In order to understand the implications of the apparently high abundance of long polyynes in the ISM, some molecules which are closely related chemically are being studied. A radio search for 1–cyanobut–3–ene–1–yne, $CH_2 = CH–C \equiv C–C \equiv N$, which is very closely related to $HC \equiv C–C \equiv C–C \equiv N$, the first polyyne to be detected, has yielded an upper limit to the abundance. The data suggest that the chemical processes which give rise to interstellar molecules are biased towards the production of highly unsaturated organic species. The implication of this and other results is discussed with regard to the chemistry of the ISM and future investigations of these important processes.

131.111 Erratum: "On the ortho–H_2/para–H_2 ratio in molecular clouds" [Mon. Not. R. Astron. Soc., Vol. 209, No. 1, p. 25 – 31 (1984)].
D. R. Flower, G. D. Watt.
Mon. Not. R. Astron. Soc., Vol. 213, No. 4, p. 991 (1985). See Abstr. 38.131.003.

131.112 Self–similar condensation of spherically symmetric self–gravitating isothermal gas clouds.
A. Whitworth, D. Summers.
Mon. Not. R. Astron. Soc., Vol. 214, No. 1, p. 1 – 25 (1985).

The authors show that similarity solutions for the condensation of spherically symmetric self–gravitating isothermal gas clouds are much more abundant than was hitherto believed; they have discovered that each of the discrete solutions found by previous authors must be replaced by a bounded two–parameter continuum of such solutions. Moreover, the two parameters (z_0, w_0) admit a clear and useful physical interpretation. z_0 characterizes the asymptotic form of a solution as $t \rightarrow -\infty$ (i.e. at very early times), and reflects how intrinsically unstable against contraction the cloud interior is from the outset. w_0 characterizes the asymptotic form of a solution as $t \rightarrow +\infty$ (i.e. at very late times) and reflects how important external pressure is in driving compression. The evolution of the flow is presented and discussed.

131.113 The interstellar abundance of nitrogen.
F. P. Keenan, A. Hibbert, P. L. Dufton.
Ir. Astron. J., Vol. 17, No. 1, p. 20 – 24 (1985). – See Abstr. 012.034.

131.114 Chemical and biological evolution in space.
J. M. Greenberg, P. Weber, W. Schutte.
Adv. Space Res., Vol. 4, No. 12, p. 41 – 49 (1984). – See Abstr. 012.036.

The authors focus their attention on the solid particles in space, otherwise known as interstellar grains or dust. The evolution of these particles which results from interactions with the gaseous atoms and molecules, with ultraviolet radiation, with each other and perhaps also with cosmic rays leads to the production of extremely complex organic molecules within the dust. Identification has been made of some smaller molecules in the grains but the precise identification of the complex molecules by purely astronomical methods will perhaps never be possible. However, current infrared observations combined with laboratory and theoretical studies give credence to their presence – indeed in enormous quantities – in space. How these molecules are formed and their quantities estimated is outlined in this paper.

131.115 On the determination of the physical parameters of the interstellar medium.
I. Pustyl'nik.
Tartu Astrofüüs. Obs. Publ., Tom 50, p. 221 – 279 (1984).

The well–known procedure of fitting the empirical curves of growth to the family of model curves is reversed. A new approach based on the least–squares criterion is formulated which fits rather theoretical curves of growth to the observational points and yields the self–consistent values of column densities and Doppler widths. The essential element of the proposed method is the so–called logarithmic line width, its definition following from the condition of the least squares. The relation between this quantity and the equivalent width is studied in detail, useful asymptotics are established. The revised curve–of–growth method has been used to redetermine column densities and Doppler widths for various species along the lines of sight toward ζ Oph, ζ Pup, γ Ara, ζ Per and o Per. The results obtained demonstrate the credibility of the proposed method, confirm basically the earlier determinations, some previous estimates are improved or revised.

131.116 Distances of high–velocity H I clouds in the galactic corona.
E. Saar, J. Jaaniste.
Tartu Astrofüüs. Obs. Publ., Tom 50, p. 280 – 295 (1984). In Russian.

Galactic or extragalactic H I clouds, if not too massive, must be in pressure equilibrium with the galactic coronal gas. Given the model of the corona and observed parameters of an H I cloud, one can find now the (two) possible distances of the cloud. Three examples are given, the solutions being either smaller than one kpc or about 30 – 60 kpc. Limitations and error bounds of the method are discussed.

131.117 Results of an investigation of interstellar neutral hydrogen and helium in the solar system.
E. N. Mironova.
Issled. soln. aktiv. i kosm. sistema Prognoz 1975 – 1978. Sbornik dokl. 5 Obedin. nauchn. chteniya po kosmonavt. Moskva, p. 91 – 97 (1984). In Russian. Abstr. in Ref. Zh., 62. Issled. Kosm. Prostranstva, 1.62.381 (1985).

131.118 Infrared extinction and polarization due to partially aligned spheroidal grains: models for the dust toward the BN object.
H. M. Lee, B. T. Draine.
Astrophys. J., Vol. 290, No. 1, p. 211 – 228 (1985).

The authors describe methods for computing the extinction and polarization due to partially aligned, precessing, spheroidal grains in the dipole approximation regime. These techniques are applied to develop models for the dust on the line of sight to the Becklin–Neugebauer (BN) object. The models, utilizing a mixture of graphite and silicate particles, some coated with ice mantles, are found to be capable of reproducing all important features of spectroscopic and polarimetric observations in the spectral range $2 \lesssim \lambda \lesssim 13 \, \mu m$.

131.119 High–latitude H I structure and the soft X–ray background.
K. Jahoda, D. McCammon, J. M. Dickey, F. J. Lockman.
Astrophys. J., Vol. 290, No. 1, p. 229 – 237 (1985).

The authors present 21 cm observations made with the NRAO 43 m telescope of 20 randomly selected intermediate and high galactic latitude regions. The data are examined for evidence of the neutral gas clumping required by models in which a substantial fraction of the diffuse soft X–ray background originates outside the galactic disk and is absorbed by interstellar gas. The authors find no such evidence and conclude that the degree of clumping required by such models must, if it exists, have characteristic angular scales less than 14′. It is therefore unlikely that a significant fraction of the X–ray flux originates in a galactic corona.

131.120 The optical interstellar–line spectrum of HD 147889.
R. M. Crutcher, Y.–H. Chu.
Astrophys. J., Vol. 290, No. 1, p. 251 – 255 (1985).

Observations of optical interstellar lines at blue wavelengths have been made of the very heavily reddened star HD 147889, which is behind the outer envelope of the ϱ Ophiuchi molecular cloud. The lines $\lambda3875$ of CN, $\lambda3933$ of Ca II, $\lambda4232$ of CH^+, $\lambda4300$ of CH, and possibly $\lambda4226$ of Ca I have been detected and measured. Using these and additional published data, the authors have analyzed the abundances and physical conditions of the ϱ Oph cloud toward HD 147889.

131.121 The molecular core associated with DR 21.
H. R. Dickel, P. T. P. Ho, M. C. H. Wright.
Astrophys. J., Vol. 290, No. 1, p. 256 – 260 (1985).

DR 21 has been observed in the $J = 1–0$ line of HCN with a resolution of $12.''5$ by $15.''5$ using the mm–wave interferometer at the Hat Creek Observatory. The HCN line is seen in both emission and absorption as an isolated clump immediately adjacent to the compact H II regions of DR 21. The full extent of the clump is 0.3 by 0.6 pc for an assumed distance of 2 kpc. The mass of the clump is estimated to be $\sim 200\ M_\odot$. Two distinct velocity features at -5.0 km s^{-1} and -2.4 km s^{-1} are present.

131.122 GSS 30: an infrared reflection nebula in the Ophiuchus dark cloud.
M. W. Castelaz, J. A. Hackwell, G. L. Grasdalen, R. D. Gehrz, C. Gullixson.
Astrophys. J., Vol. 290, No. 1, p. 261 – 272 (1985).

Surface brightness maps at Γ, H, K, L, and N, and polarimetric images at K and H of the object GSS 30 in the Ophiuchus dark cloud were made with the 2.3 m Wyoming infrared telescope. The images show that GSS 30 is a bi–lobed asymmetric reflection nebula detectable at H, K, and L. At H the nebulosity extends $22''$ along the major axis, and the illuminating star has a magnitude of 11.17; at K it extends $45''$ along the major axis, and the illuminating star has a magnitude of 8.88. A model is presented in which an accretion disk is seen nearly edge–on obscuring a star and polar reflection lobes. Density profiles and grain characteristics of the reflection nebula are briefly discussed.

131.123 Average density along interstellar lines of sight.
L. Spitzer Jr.
Astrophys. J., Lett. Ed., Vol. 290, No. 1, p. L21 – L24 (1985).

Recent studies indicate that the depletion of certain elements in the interstellar gas is well correlated with $<n_H>$, the mean particle density of hydrogen along each line of sight. It is shown that these results can be understood in terms of a simple theoretical model, based on random distributions of the two types of cold clouds which fit the observed distribution of color excess, E_{B-V}, together with a more uniform warm H I gas of lower density.

131.124 Volume extinction factors of interstellar grains.
D. B. Vaidya.
Astrophys. Space Sci., Vol. 109, No. 2, p. 393 – 394 (1985).

Volume extinction factors V_c have been evaluated for spherical grains at several refractive indices ($m = 1.6, m = 1.5, m = 1.33, m = 1.2, m = 1.16, m = 1.1$). From these results it is found that grains with higher–refractive index are more efficient for visual extinction than grains with low–refractive index.

131.125 More on the λ2800 Å 'interstellar extinction' feature.
A. McLachlan, K. Nandy.
Astrophys. Space Sci., Vol. 109, No. 2, p. 399 – 402 (1985).

The authors respond to comments made in a recent letter by Karim et al. (1984), and show that the examples of interstellar absorption at λ2800 Å that they have presented so far can all be attributed to overexposure of the IUE detectors.

131.126 Comparison of optical appearance and infrared emission of some high latitude extended dust clouds.
C. P. de Vries, R. S. Le Poole.
Astron. Astrophys., Vol. 145, No. 2, p. L7 – L9 (1985).

The extended 100 micron emission discovered by IRAS and called "interstellar cirrus" is also visible on modern sky survey Schmidt plates as faint, blue emission. In this paper the infrared emission and optical appearance for two "cirrus" clouds are compared. A dust temperature of about 20K is derived and a luminosity at 100 micron of about 9 MJy/Sr per magnitude extinction.

131.127 Westerbork observations of 6 cm H₂CO in W51A.
E. M. Arnal, W. M. Goss.
Astron. Astrophys., Vol. 145, No. 2, p. 369 – 376 (1985).

Using the Westerbork Synthesis Radio Telescope, the authors have observed the 6 cm H_2CO ($1_{11} \to 1_{10}$) absorption line against the continuum sources G 49.5–0.4 and G 49.4–0.3 (W51A). The observed profiles show a marked variation in both shape and optical depth. Conspicuous opacity enhancements on a linear scale size of 0.1 to 0.3 pc are observed in the area. The opacity enhancements are interpreted as being caused by clumps in the molecular material overlying the continuum sources. Evidence for an interaction between the ionized gas of component b(G 49.5–0.4) and the nearby high velocity stream has been found. Based on the observations a schematic representation of the spatial distribution of the different molecular complexes, and their associated H II regions, is proposed.

131.128 The interstellar radiation field and the production of inverse–Compton gamma rays in the Galaxy.
J. B. G. M. Bloemen.
Astron. Astrophys., Vol. 145, No. 2, p. 391 – 404 (1985).

The production of diffuse galactic gamma radiation above 1 MeV from the interaction of cosmic–ray electrons with the interstellar photon field, via the inverse–Compton process, is examined. Empirical modelling of the interstellar photon field throughout the Galaxy is performed, based on the emission functions of various stellar components for near–UV upto near–infrared wavelengths, on reemission from dust grains for the far infrared, and on the universal 2.7K blackbody emission. The dust absorption is taken into account. The resulting galactic distribution of the energy densities of the interstellar radiation field is given for selected wavelength intervals. The predicted Compton radiation is compared to the high–energy ($E > 70$ MeV) gamma–ray observations from the COS–B satellite.

131.129 CS 5–4 observations of OMC1: evidence for external heating of the quiescent gas.
R. Padman, P. F. Scott, D. R. Vizard, A. S. Webster.
Mon. Not. R. Astron. Soc., Vol. 214, No. 2, p. 251 – 270 (1985).

The authors have observed the $J = 5–4$ and $2–1$ transitions of $^{12}C^{32}S$ and $^{12}C^{34}S$ in the direction of OMC1 and the results provide strong support for a clumpy (or turbulent) model of the high–velocity doughnut source. By using the high–velocity emission as a quasi–continuum background point source the authors estimate the excitation temperature of the ridge gas as $15 - 35$K. They suggest that the ridge clouds are heated on the near side by radiation from M42 and the Trapezium stars, creating a temperature gradient in the cloud. CO emission then arises in a thin skin on the hot face of the cloud. The kinetic temperature of the ridge gas, and hence the Jeans mass and cooling function, are therefore similar to those in other warm molecular clouds. A simple model of this system is described which qualitatively reproduces many of the features observed.

131.130 Rate coefficients for the rotational excitation of CO by ortho– and para–H₂.
D. R. Flower, J. M. Launay.
Mon. Not. R. Astron. Soc., Vol. 214, No. 2, p. 271 – 277 (1985).

Rate coefficients for the rotational excitation of $^{12}C^{16}O$ by ortho– and para–H_2 are given for the low–temperature regime of interstellar molecular clouds. The rate coefficients derive from quantum mechanical coupled channels calculations of the relevant cross–sections and are intended for the analysis of the intensities of CO microwave and submillimetre lines.

131.131 An investigation of diffuse cloud chemistry. II. Inferred interstellar parameters.
A. P. C. Mann, D. A. Williams.
Mon. Not. R. Astron. Soc., Vol. 214, No. 2, p. 279 – 287 (1985).

The chemistry in diffuse clouds is discussed in terms of a low–density equilibrium model, and attention is given principally to those molecules which are formed predominantly in the gas phase. The model indicates that substantial variations in the cosmic ray flux and interstellar radiation field occur from cloud to cloud. If these variations are included, then, for the limited observational data set available, the low–density equilibrium model provides a satisfactory though not unique description of individual molecular clouds, apart from the case of CH^+. In particular, a high–density core in diffuse clouds is shown not to be obligatory.

131.132 Ice in the Taurus molecular cloud: modelling of the 3–μm profile.
C. E. P. M. van de Bult, J. M. Greenberg, D. C. B. Whittet.
Mon. Not. R. Astron. Soc., Vol. 214, No. 2, p. 289 – 305 (1985).

Detailed calculations of the absorption by interstellar core–mantle particles with mantles of different compositions are compared with observations of the 3 μm ice band in the Taurus molecular cloud. The strength and shape of the 3–μm band is shown to be a remarkably good diagnostic of the physical state and evolution of the dust in molecular clouds. The strength of the band is consistent with large fractional H_2O mantle concentrations, in the range 60 – 70 per cent, as predicted by theoretical studies of cloud chemistry and as expected from the high oxygen abundance in pre–molecular clouds.

131.133 Polycyclic aromatic hydrocarbons and the diffuse interstellar bands.
G. P. van der Zwet, L. J. Allamandola.
Astron. Astrophys., Vol. 146, No. 1, p. 76 – 80 (1985).

The authors discuss some of the thermodynamic and spectroscopic properties of polycyclic aromatic hydrocarbons which make them attractive candidates as carriers of the diffuse interstellar bands (DIB's). They point out that, in the diffuse medium, many of these species will be partially hydrogenated and positively charged and will absorb in the visible. The observed line–shapes and –widths of the DIB's can be explained by rovibronic band contours and linebroadening due to internal conversion. Since little information is available concerning the spectroscopic properties of such species in the gas phase, a considerable amount of laboratory and theoretical work is needed.

131.134 Are polycyclic aromatic hydrocarbons the carriers of the diffuse interstellar bands in the visible?
A. Léger, L. d'Hendecourt.
Astron. Astrophys., Vol. 146, No. 1, p. 81 – 85 (1985).

Large polycyclic aromatic hydrocarbons are the likely origin of the formerly called "Unidentified IR Emission Features". The authors show that these molecules or their ions are also attractive candidates for the carriers of the diffuse interstellar bands in the visible (DIBs). In particular, they are better candidates than the long carbon chains that had been proposed previously. A laboratory effort is undertaken to search for a spectroscopic support to this point. The authors predict the absence of the DIBs in special astrophysical environments such as those where all the carbon is locked up in CO. There are some indications that this is the actual situation but further observations are suggested.

131.135 Formaldehyde observations of OMC–1.
P. Bastien, W. Batrla, C. Henkel, T. Pauls,
C. M. Walmsley, T. L. Wilson.
Astron. Astrophys., Vol. 146, No. 1, p. 86 – 94 (1985).

Maps of the emission from the $2_{11} - 2_{12}$ K–doublet of $H_2{}^{12}C^{16}O$ at 2 cm (angular resolution of 60″) and the $3_{12} - 3_{13}$ line of $H_2{}^{12}C^{16}O$ at 1 cm (angular resolution 35″) have been made toward OMC–1. Four clouds are found: two adjacent to the BN/KL region with $v_{LSR} = 8$ and 10 km s^{-1}, another 90″ to the south with 6.5 km s^{-1} and one $\sim 80″$ to the north–east at a radial velocity of 10 km s^{-1}. The authors have also detected emission from the $2_{11} - 2_{12}$ line of $H_2{}^{13}CO$ at 2.2 cm. This result is combined with model calculations to estimate optical depths and H_2 densities. The dynamics and stability of the K–doublet emission line region are investigated. Comparisons of the spatial distribution of $2_{11} - 2_{12}$ H_2CO emission with that of O I and CO lines and the far infrared continuum are made.

131.136 The Taurus dark cloud around the position of 3C 111: an optical and $C^{18}O$ line study.
V. Ungerer, N. Mauron, J. Brillet, Nguyen–Quang–Rieu.
Astron. Astrophys., Vol. 146, No. 1, p. 123 – 133 (1985).

The authors present optical and radio observations of the region of the Taurus complex located around the position of the radiosource 3C 111. They performed prism–objective classification and UBV photometry of field stars and found that the visual extinction, in a line of sight up to 800 pc, is due to two absorption layers whose distance was estimated. They identified the more distant layer as the molecular cloud studied in the paper. By means of star counts and millimeter observations, the authors mapped the visual extinction and the $C^{18}O$ emission. They discuss the large scale density and velocity structure of this region and the correlation between $C^{18}O$ column density and visual extinction.

131.137 $^{15}NH_3$ in Orion–KL: the hot core is not so hot.
W. Hermsen, T. L. Wilson, C. M. Walmsley,
W. Batrla.
Astron. Astrophys., Vol. 146, No. 1, p. 134 – 138 (1985).

The six metastable inversion lines $(J,K) = (1,1)$ to $(6,6)$ of the rare isotopic species of ammonia, $^{15}NH_3$, have been measured toward Orion–KL. The radial velocities of the $^{15}NH_3$ lines agree with those of the Hot Component of $^{14}NH_3$, at 6 km s^{-1}. Assuming that the $^{15}NH_3$ is optically thin and all lines fill the same volume, the authors derive rotational temperatures $T_R = 110 \pm 10$K from the four lines of para and $T_R = 115 \pm 15$K from the two lines of ortho $^{15}NH_3$. Model calculations for H_2 densities of 10^7cm^{-3} show that this should be the kinetic gas temperature. Using these rotational temperatures, the authors calculate an ortho/para ratio for $^{15}NH_3$ of 0.7 (+0.5,–0.3); the LTE value is one. If both the $^{14}NH_3$ and $^{15}NH_3$ fill the same volume the authors derive a $^{14}N/^{15}N$ ratio of 170 (+140,–80).

131.138 Newly discovered sources of non–metastable ammonia.
R. Mauersberger, T. L. Wilson, W. Batrla,
C. M. Walmsley, C. Henkel.
Astron. Astrophys., Vol. 146, No. 1, p. 168 – 174 (1985).

Detections of the $(J,K) = (2,1)$ non–metastable inversion line of ammonia in eight sources near compact H II regions or OH and H_2O masers are reported. The authors report a first detection of the (5,3) line and a second detection of the (4,2) line, both found in W51e. If the non–metastable levels are collisionally excited, the H_2 densities should be greater than 10^5cm^{-3}, and hence non–metastable ammonia lines are expected to be indicators for high density regions. The measurements of various metastable NH_3 lines and model calculations allow estimates of the rotational, T_r, and kinetic temperatures, T_k, in these regions. The authors have estimated source sizes, virial masses, and average H_2 densities. These densities are compared with those obtained from statistical equilibrium calculations. The H_2 densities and source sizes are used to obtain column densities of H_2 and the abundance ratios of NH_3 relative to H_2. Comparisons with values of T_k and $n(H_2)$ from studies of other molecules are made.

131.139 Cloud–cloud collisions.
J. C. Lattanzio, J. J. Monaghan, H. Pongracic,
M. P. Schwarz.
Proc. Astron. Soc. Aust., Vol. 5, No. 4, p. 495 – 498 (1984).

The classical model of the interstellar medium consists of cool clouds moving through a warmer interstellar medium. More recent models have included the coronal gas (10^6K) as part of this medium. The authors consider the collisions of these clouds in order to determine whether these collisions initiate star formation and/or change the state of the interstellar medium.

131.140 Observations of 115 GHz CO emission towards the Carina nebula.
J. B. Whiteoak, R. E. Otrupcek.
Proc. Astron. Soc. Aust., Vol. 5, No. 4, p. 552 – 557 (1984).

The authors have carried out a survey of the 115 GHz CO emission (J = 1→0 transition) towards the Carina nebula (NGC 3372) using the Epping 4 m radio telescope of the CSIRO Division of Radiophysics. In many respects the Carina nebula is one of the outstanding objects of the Southern Milky Way, an extremely interesting volatile region of recent star formation.

131.141 The detection of shocked CO emission from G 333.6–0.2.
J. W. V. Storey.
Proc. Astron. Soc. Aust., Vol. 5, No. 4, p. 566 – 569 (1984).

This paper is published as a 38 stanza poem. The main results are:

In 1980, from the plane, he
found it in Orion,
but no more CO could he find
despite long hours of flyin'.
And so, he searched for southern sources
of this shocked H_2
and found it, in G 333
point six, minus, nought point two.
Intensity is really weak:
it's two point nought by ten
to the minus eighteenth power
(in watts per square cm).
That's thirty times as weak as we
detected in Orion.
No wonder it took several years
of concentrated tryin'.

131.142 Radio and optical observations of the jets from L1551 IRS 5.
R. L. Snell, J. Bally, S. E. Strom, K. M. Strom.
Astrophys. J., Vol. 290, No. 2, p. 587 – 595 (1985).

Radio continuum observations obtained with the VLA and optical CCD images are presented of the jets associated with the infrared source IRS 5 located in the core of the dark cloud L1551. The radio observations show that there are two jets aligned with the axis of the bipolar molecular outflow which extend for more than $10''$ from IRS 5. CCD images reveal an optical jet emanating from the southwest of IRS 5 coincident with the radio jet but no optical emission associated with the northeast radio jet. The absence of optical emission from the northeast jet is best explained by the presence of a dense disk of gas perpendicular to the outflow axis which obscures the northeast jet and may also be responsible for collimating the outflow into two initially opposed jets.

131.143 The detection of acetaldehyde in cold dust clouds.
H. E. Matthews, P. Friberg, W. M. Irvine.
Astrophys. J., Vol. 290, No. 2, p. 609 – 614 (1985).

Observations of the $1_{01}\to0_{00}$ rotational transitions of A and E state acetaldehyde are reported. The transitions were detected, for the first time in interstellar space, in the cold dust clouds TMC–1 and L134N, and in Sgr B2. The authors find a column density of $6\times10^{12}\,cm^{-2}$ in TMC–1. In the direction of Sgr B2, the CH_3CHO profile appears to consist of broad emission features from the hot molecular cloud core, together with absorption features resulting from intervening colder material. The possible detection of HC_9N toward IRC +10216 is also reported.

131.144 Shielding of CO from dissociating radiation in interstellar clouds.
A. E. Glassgold, P. J. Huggins, W. D. Langer.
Astrophys. J., Vol. 290, No. 2, p. 615 – 626 (1985).

The authors investigate the photodissociation of CO in interstellar clouds in the light of recent laboratory studies which suggest that line rather than continuum processes dominate its dissociation by ultraviolet radiation. Using a simple radiative transfer model, the authors estimate the shielding of representative dissociating bands including self–shielding, mutual shielding between different isotopes, and near coincidences with strong lines of H_2. The results are combined with an appropriate gas phase chemical model to determine how the abundances of the CO isotopes vary with depth into a cloud. Self–shielding and mutual shielding cause significant variations in isotopic ratios.

131.145 CO J = 3–2 observations of molecular line sources having high–velocity wings.
K. J. Richardson, G. J. White, L. W. Avery, J. C. G. Lesurf, R. H. Harten.
Astrophys. J., Vol. 290, No. 2, p. 637 – 652 (1985).

Twelve molecular line sources which exhibit high–velocity wings in the CO J = 1–0 rotational transition have been observed in the CO J = 3–2 line. In all cases the authors have detected high–velocity wings in this transition. Large velocity gradient modeling of the relative line intensities of the lowest three transitions indicates, for some of the sources, that densities

are present which are considerably higher than previously quoted estimates of average hydrogen density determined only from CO J = 1–0 data. The data are consistent with a model in which an expanding but decelerating shell of molecular material is swept up by a pre–main–sequence stellar wind or a magnetically driven shock.

131.146 On the detection of rubidium in diffuse interstellar clouds.
S. R. Federman, C. Sneden, W. V. Schempp, W. H. Smith.
Astrophys. J., Lett. Ed., Vol. 290, No. 2, p. L55 – L58 (1985).

A search for absorption from neutral rubidium at 7800 Å was conducted. No evidence for absorption to a 3 σ limit of $\leqslant1.5$ mÅ was seen in the diffuse interstellar gas toward the stars o Persei, ζ Persei, and ζ Ophiuchi. These results do not confirm the detection by Jura and Smith toward ζ Oph. A possible reason for the discrepancy is presented.

131.147 Detection of the 370 micron $^3P_2 - ^3P_1$ fine–structure line of [C I].
D. T. Jaffe, A. I. Harris, M. Silber, R. Genzel, A. L. Betz.
Astrophys. J., Lett. Ed., Vol. 290, No. 2, p. L59 – L62 (1985).

The authors report here the first detection of the 370 µm (809 GHz) $^3P_2 - ^3P_1$ fine–structure transition of neutral carbon. The detection of this line confirms the identification of the line seen at 492 GHz by Phillips et al. as the $^3P_1 - ^3P_0$ line of [C I]. The observations toward the core of the Orion molecular cloud show a line brightness temperature 1.7 ± 0.5 times that of the previously detected $^3P_1 - ^3P_0$ line. If both transitions arise in a single emission region that fills the beams used in the observations, an optically thin region with a temperature $\geqslant300K$ is the best explanation of the observed line ratio. The most likely location for a warm optically thin region is in the interface between the Orion H II region and the massive molecular cloud that lies behind it. The possibility that the [C I] emission may arise from an optically thick $\sim25K$ region is also discussed.

131.148 On the evidence for methane in Orion KL: a search for the 4.6 gigahertz line.
T. L. Wilson, L. E. Snyder.
Astrophys. J., Lett. Ed., Vol. 290, No. 2, p. L63 – L64 (1985).

A sensitive search for the J = 11 $E(2)$–$E(1)$ transition of interstellar methane (CH_4) has resulted in a peak upper limit which is much less than the value reported by Fox and Jennings in 1978. When combined with the negative results reported by Elldér et al. in 1980, these data rule out the detection of CH_4 in Orion KL previously claimed by Fox and Jennings.

131.149 Interstellar matter and grains.
G. Strazzulla.
Mem. Soc. Astron. Ital., Vol. 55, No. 3, p. 585 – 589 (1984). – See Abstr. 012.050.

Some recent observational, experimental and interpretative aspects concerning the grain properties and their influence on the general interstellar medium are briefly discussed.

131.150 The space distribution and internal structure of the Local dust arm in Cygnus.
T. A. Uranova.
Pis'ma Astron. Zh., Tom 11, No. 4, p. 251 – 260 (1985). In Russian. English translation in Sov. Astron. Lett., Vol. 11.

About two hundred values of dust density in the direction of the Local Cygnus arm are determined on the basis of photoelectric absorption curves, containing a thousand stars. The form and location of the dust arm axis are determined up to the distance of 4 kpc from the sun. It coincides with the neutral hydrogen filament and with the location of faint OB stars.

131.151 New nebulous objects.
V. M. Petrosyan.
Astrofizika, Tom 22, Vyp. 2, p. 423 – 425 (1985). In Russian. English translation in Astrophysics, Vol. 22, No. 2.

A list and photographs of 18 new nebulous objects connected with stars found on the Palomar Survey prints are presented.

131.152   Upper limits on the O_2/CO ratio in two dense interstellar clouds.
H. S. Liszt, P. A. Vanden Bout.
Astrophys. J., Vol. 291, No. 1, p. 178 – 182 (1985).

The authors have searched for a magnetic–dipole rotation transition of the isotope $^{16}O^{18}O$ toward ϱ Oph and Orion A at a wavelength of 1.3 mm. Upper limits on the intensity of this line imply that $N(O_2)/N(CO) \leqslant 0.13$ and 0.67 in the two sources, respectively, so that neither harbors a hitherto–undetected reservoir of oxygen in molecular form.

131.153   On the conversion of carbon monoxide intensities to molecular hydrogen abundances.
M. L. Kutner, C. M. Leung.
Astrophys. J., Vol. 291, No. 1, p. 188 – 201 (1985).

The authors present results of theoretical models (static spherical clouds with a microturbulent velocity field) to study the conversion of carbon monoxide (CO) line parameters into molecular hydrogen (H_2) column densities, $N(H_2)$. The three potential H_2 tracers investigated are the integrated ^{12}CO and ^{13}CO intensities, I_{12} and I_{13}, and the ^{13}CO LTE column density, N_{13}*. The usefulness of these tracers depending on various CO abundance, cloud size, gas density, and temperature is investigated. In particular, the analysis suggests that the conversion factor $N(H_2)/I_{12}$ for giant molecular clouds in the molecular ring of our galaxy may be a factor of 2 lower than the average used by many observers.

131.154   Millimetre and sub–millimetre astronomy.
A. S. Webster, M. S. Longair.
Contemp. Phys., Vol. 25, No. 6, p. 519 – 534 (1984). Abstr. in Phys. Abstr., Vol. 88, No. 1252, Entry 30024 (1985).

131.155   Le nouveau visage du milieu interstellaire.
A. Acker.
Astronomie, Vol. 99, p. 219 – 230 (1985).

131.156   Classical thermal evaporation of clouds: an electrostatic analogy.
S. A. Balbus.
Astrophys. J., Vol. 291, No. 2, p. 518 – 522 (1985).

The dynamics of the thermal evaporation of an ensemble of clouds in the classical regime is considered. It is shown that all previously known solutions are special cases of the more general class of irrotational flows. For this class of solutions the mass–loss rate is simply related to the electrostatic capacitance of a system of electrical conductors having identical geometry. Any individual cloud or ensemble geometry has a unique flow solution belonging to this class. Explicit analytic mass loss formulae for a variety of nonspherical and two cloud systems are given.

131.157   Polarimetry of infrared sources in bipolar CO flows.
S. Sato, T. Nagata, T. Nakajima, M. Nishida, M. Tanaka, T. Yamashita.
Astrophys. J., Vol. 291, No. 2, p. 708 – 715 (1985).

The results of polarimetry at 2.3 µm for nine infrared sources associated with high–velocity molecular outflow are reported. All the sources show large linear polarization ranging from 4% to more than 20%. The position angles of infrared polarization are perpendicular to the direction of bipolar molecular outflow and are parallel to the major axis of the associated molecular disk/torus. The infrared polarizations have a tendency to be orthogonal to the optical interstellar polarizations in the vicinity of the sources.

131.158   Measurement of spin temperatures in a rapidly moving H I shell.
S. R. Kulkarni, J. M. Dickey, C. Heiles.
Astrophys. J., Vol. 291, No. 2, p. 716 – 721 (1985).

The authors have obtained high quality H I absorption spectra toward 3C 123 and 3C 147. For the normal–velocity H I they report two upper limits to magnetic field strengths and an upper limit to the spin temperature of the intercloud medium. Anomalous–velocity H I is present along both lines of sight. This H I arises in a rapidly moving, large–angular–diameter H I structure. Emission and absorption properties of this H I are also anomalous. There is evidence suggesting that the H I structure is the result of collision of high–velocity clouds with the H I disk.

131.159   Photodissociation regions. I. Basic model.
A. G. G. M. Tielens, D. Hollenbach.
Astrophys. J., Vol. 291, No. 2, p. 722 – 746 (1985).

The authors have theoretically modeled the chemistry and heat balance of dense ($10^3 cm^{-3} < n_0 < 10^6 cm^{-3}$) neutral gas illuminated by far–ultraviolet (FUV) ($6 eV < h\nu < 13.6 eV$) fluxes $10^3 - 10^6$ times more intense than the ambient interstellar field. The one–dimensional models extend $A_v \sim 10$ into the neutral gas, and are primarily intended to study the FUV illuminated neutral gas (photodissociation regions) between molecular clouds and H II regions. The models relate observed line and continuum emission from these regions to physical parameters such as the gas density and temperature, the incident FUV flux, the elemental abundances, and the grain properties.

131.160   Photodissociation regions. II. A model for the Orion photodissociation region.
A. G. G. M. Tielens, D. Hollenbach.
Astrophys. J., Vol. 291, No. 2, p. 747 – 754 (1985).

The authors have constructed a model for the photodissociation region in Orion. This model matches the observed emission in the fine–structure lines of O I 63 µm and 145 µm, C I 609 µm, and C II 158 µm and the low–lying rotational lines of CO. It also explains the extended H_2 1–0 S(1) emission, the carbon recombination lines, and the C I 9850 Å line. The model fit implies that a fairly uniform density ($\sim 10^5 cm^{-3}$) photodissociation region lies behind the Trapezium stars. The peak temperature in this region is about 500K, and the intensity of the incident far–ultraviolet (FUV) field is approximately 10^5 times as intense as the ambient interstellar field.

131.161   The magnetic flux problem and ambipolar diffusion during star formation: one–dimensional collapse. II. Results.
T. C. Mouschovias, E. V. Paleologou, R. A. Fiedler.
Astrophys. J., Vol. 291, No. 2, p. 772 – 797 (1985).

The magnetic flux problem lies in the observation that fluxes of typical stars are between two and five orders of magnitude smaller than fluxes of corresponding masses at interstellar densities. The detailed calculations of collapsing model clouds described in this paper show that the flux–to–mass ratio in cloud cores can decrease by two to more than four orders of magnitude in less than 10^6 yr at neutral densities $n_n < 10^9 cm^{-3}$ due to ambipolar diffusion alone. The qualitative features of the evolution (e.g., infall of the core and, often, formation of a nonmagnetic shock; initial expansion of the envelope and formation of hydromagnetic shocks; dynamical decoupling of the field from the neutral matter and reduction of the flux–to–mass ratio in the core) are common to virtually all cases which develop a rapid enough ambipolar diffusion. Clouds are studied varying in temperature from 0 to 50K; in initial central density from 10^3 to $10^6 cm^{-3}$; in initial central magnetic field from 28.9 µG to 1.7 mG; in initial half–thickness from 0.26 to 11 pc; in initial central and surface degree of ionization from 10^{-7} to 10^{-9} and from 10^{-3} to 3×10^{-5}, respectively; in mass from 57 to 80,000 $M_\odot$.

131.162   Infrared spectroscopy of carbon monoxide in GL 2591 and OMC–1:IRc2.
T. R. Geballe, R. Wade.
Astrophys. J., Lett. Ed., Vol. 291, No. 2, p. L55 – L58 (1985).

Spectra of 4.7 µm fundamental band absorption lines of carbon monoxide in GL 2591 and OMC–1:IRc2 are presented. In GL 2591 the observed lines have total velocity widths of about 100 km s^{-1}, far larger than that in the 2.6 mm CO line seen toward this object. The spectra indicate ejection of material at ~ 100 km s^{-1} at a distance of $\sim 10^{15}$ cm from the central object, followed by deceleration. The single CO line profile observed in IRc2 appears similar to those seen previously in BN except for the possible presence of a faint blue wing.

131.163 Detection of the CO ($J = 7 \rightarrow 6$) rotational transition at $\lambda = 0.37$ millimeters toward Orion.
G. V. Schultz, E. J. Durwen, H. P. Röser, W. A. Sherwood, R. Wattenbach.
Astrophys. J., Lett. Ed., Vol. 291, No. 2, p. L59 – L61 (1985).

The $J = 7 \rightarrow 6$ rotational transition of carbon monoxide has been detected toward the Kleinmann–Low nebula. The peak antenna temperature is $110 \pm 35K$ centered at $V_{LSR} = +5 \text{ km s}^{-1}$; the FWHM line width is $35 \pm 5 \text{ km s}^{-1}$, and its optical depth $\leqslant 0.3$. The integrated intensity of the line between -20 and $+30 \text{ km s}^{-1}$ leads to a total flux density of $(2.5 \pm 0.5) \times 10^{-14} \text{W m}^{-2}$ in the beam of $35'' \pm 5''$.

131.164 Detection of $HC_{11}N$ in the cold dust cloud TMC–1.
M. B. Bell, H. E. Matthews.
Astrophys. J., Lett. Ed., Vol. 291, No. 2, p. L63 – L65 (1985).

The authors report the detection of the $J = 41 \rightarrow 40$ rotational transition of $HC_{11}N$ in the direction of the Taurus molecular cloud complex. The line strength is in good agreement with predicted values, and the rotational parameters, although determined more accurately with these results, remain unchanged from those obtained previously using observations of IRC $+10°216$ alone.

131.165 Observations of the peculiar OH maser G24.3 + 0.1.
M. I. Pashchenko.
Sov. Astron. Lett., Vol. 10, No. 5, p. 303 – 304 (1984). English translation of 38.131.115.

131.166 Water maser flares in W75N.
E. E. Lekht, R. L. Sorochenko.
Sov. Astron. Lett., Vol. 10, No. 5, p. 307 – 309 (1984). English translation of 38.131.116.

131.167 Space distribution of the interstellar clouds toward the Rosette Nebula.
N. G. Guseva, I. G. Kolesnik, S. G. Kravchuk.
Sov. Astron. Lett., Vol. 10, No. 5, p. 309 – 312 (1984). English translation of 38.131.117.

131.168 High–resolution spectroscopy of molecular hydrogen in OMC–1.
T. Matsumoto, T. Moritsugu, K. Uyama.
Publ. Astron. Soc. Jpn., Vol. 37, No. 1, p. 129 – 132 (1985).

H_2 line emission from the $S(1)$, $\upsilon = 1 \rightarrow 0$ (2.1218 μm) transition at peak 1 in OMC–1 was observed by using a Fabry–Perot interferometer with 10 km s^{-1} velocity resolution. The emission peak occurs as a velocity, V_{LSR}, of $5 \pm 1.4 \text{ km s}^{-1}$. The line profile observed with a $10''$ beam appears to consist of two components; one is a narrow peak component which is not well resolved yet, and the other is a wing component spread over a wide velocity range. The result suggests the presence of a fragmental thin shock front and emission from a bipolar outflowing gas.

131.169 Cloud fragmentation and stellar masses.
R. B. Larson.
Mon. Not. R. Astron. Soc., Vol. 214, No. 3, p. 379 – 398 (1985).

Theoretical arguments and numerical simulations both suggest that a star–forming cloud generally collapses to a flattened or filamentary configuration before fragmenting, and that fragmentation occurs as a result of the gravitational instability of the resulting layer or filament. A number of existing and new results for the stability of polytropic sheets, discs, and filaments are collected in this paper, and critical lengths and masses are derived for a variety of values of the polytropic exponent. The predictions of these stability analyses agree satisfactorily with the results of numerical simulations of the fragmentation of discs and filaments. The predicted critical masses in some well–studied regions of star formation also agree with the typical masses of the observed dense cloud clumps, and, in order of magnitude, with the masses of the young stars.

131.170 Observation of neutral hydrogen 21–cm absorption toward double sources.
J. Crovisier, J. M. Dickey, I. Kazès.
Astron. Astrophys., Vol. 146, No. 2, p. 223 – 234 (1985).

In order to study the small–scale structure of galactic neutral hydrogen, the authors have measured 21–cm absorption spectra toward background double sources. They have detected typical absorption variations of about 0.1 in optical depth over scalelengths less than 1 pc, and 0.2 to 0.3 over lengths greater than 1 pc. No structure is tractable below 0.2 pc, suggesting the existence of a threshold. This is consistent with a minimum diameter of about 1.5 pc for cold H I clouds. A decomposition of the profiles into Gaussian components suggests that the absorption variations are due to velocity variations rather than cloudlet fluctuations. These velocity variations would be due to ordered motions or random turbulent motions. In the latter hypothesis, $\Delta v \propto \lambda^{1/3}$ mimicking Kolmogoroff's law for incompressible turbulence.

131.171 Observations of ortho– and para–thioformaldehyde.
F. F. Gardner, B. Höglund, C. Shukre, A. A. Stark, T. L. Wilson.
Astron. Astrophys., Vol. 146, No. 2, p. 303 – 306 (1985).

Observations of the $3_{12}-2_{11}$ and $3_{13}-2_{12}$ transitions of ortho and the $3_{03}-2_{02}$ transitions of para H_2CS made with the Onsala and AT&T Bell Laboratories telescopes are reported and discussed.

131.172 A multiline study of a typical giant molecular cloud: S 147/S 153.
C. Kahane, S. Guilloteau, R. Lucas.
Astron. Astrophys., Vol. 146, No. 2, p. 325 – 336 (1985).

The molecular cloud S 147/S 153, taken as an example of typical giant molecular cloud with a moderate star formation activity, has been mapped in several molecular lines [$^{12}CO(1-0)$, $^{13}CO(1-0)$, $HCO^+(1-0)$, $^{12}CO(2-1)$, $^{13}CO(2-1)$, $NH_3(1,1)$, and $NH_3(2,2)$] with different spatial resolutions ($\sim 5'$, ~ 1.4, and ~ 0.7). The diffuse envelope shows a very smooth density structure and a velocity gradient strongly suggestive of an overall rotation of the cloud, even at the resolution of 1.4 towards the cloud core seen in ^{13}CO. On the other hand, the NH_3 emission, although mapped with a resolution only twice better, presents at least 3 different condensations in the cloud core, with a velocity pattern unrelated to the large scale gradient in the cloud. A detailed analysis of this NH_3 emission suggests that the condensations are composed of many small turbulent clumps of mass $\sim 1 \, M_\odot$ and size ~ 0.1 pc.

131.173 Ammonia observations of L 1551.
K. M. Menten, C. M. Walmsley.
Astron. Astrophys., Vol. 146, No. 2, p. 369 – 374 (1985).

The authors have used the Effelsberg 100–m Telescope to map the (1,1) and (2,2) transitions of NH_3 in the molecular cloud surrounding the infrared source L 1551–IRS 5. A search for water maser emission in the area was also carried out. The ammonia brightness distribution shows no obvious evidence for a rotating disk surrounding the infrared source. There is a structure approximately $30''$ south–west of IRS 5 which appears to be oriented roughly perpendicular to the bipolar outflow seen in CO observations. However, this is likely to be a background object and it is found that the high density gas seen in NH_3 shows little evidence of interaction with the flow. Temperatures are $9 - 12K$ throughout the cloud as expected for dark cloud gas heated by cosmic rays.

131.174 A plan view of the bipolar molecular outflow source G 35.2 N.
W. R. F. Dent, L. T. Little, N. Kaifu, M. Ohishi, S. Suzuki.
Astron. Astrophys., Vol. 146, No. 2, p. 375 – 380 (1985).

The authors present 15 arcsec resolution CO ($J = 1 \rightarrow 0$) and CS ($J = 2 \rightarrow 1$) observations of the molecular cloud associated with the star forming region G 35.2 N. The results suggest that the central object is undergoing well–collimated bipolar, high–velocity mass loss. Assuming an abundance ratio $[CO/H_2] = 5 \times 10^{-5}$ then the mass of outflowing gas is $\sim 60 \, M_\odot$.

Comparison with HCO^+ observations (Matthews et al., 1984) suggests that the $[HCO^+/^{12}CO]$ abundance ratio must be enhanced in this gas by a factor of $\gtrsim 20$ compared with that expected for warm, dense clouds. The CS data may point to the existence of a 0.2 pc disc perpendicular to the outflow.

131.175   On the cometary nebula Parsamian 21.
V. M. Petrosyan.
Astrofizika, Tom 21, Vyp. 3, p. 523 – 533 (1984). In Russian. English translation in Astrophysics, Vol. 21, No. 3.

The results of isodensitometric and spectrophotometric investigations of the cometary nebula P 21 and the coherent nucleus are presented. The shape of isodensities is different from that of normal stars and resembles R Mon. The spectral investigation of the nucleus of P 21 (1981 – 1982) shows that it has spectral type F2 – F5 V and an envelope expanding with V_r about 120 km/s.

131.176   Infrared spectra from interstellar dust grains.
J. M. Greenberg, F. Baas.
Adv. Space Res., Vol. 5, No. 3, p. 19 – 26 (1985). – See Abstr. 012.059.

131.177   Measurements of X–ray scattering from interstellar dust.
C. Mauche, P. Gorenstein, D. Fabricant.
Adv. Space Res., Vol. 5, No. 3, p. 141 – 144 (1985). – See Abstr. 012.059.

The authors have examined the surface brightness profiles of four of the brightest compact galactic X-ray sources observed with the Imaging Proportional Counter aboard the Einstein Observatory for the existence of halos produced by the scattering of X-rays from interstellar dust. The sources are Cyg X–3, 4U 1658–48, GX 13 + 1, and 4U 1254–69. The halos are apparent when a comparison is made between each source's measured surface brightness profile and a model profile based upon a point response function for each source.

131.178   Star formation in the turbulent protogalaxy.
V. G. Surdin.
Astron. Tsirk., No. 1353, p. 2 – 4 (1984). In Russian.

131.179   Structure of the W3/W4 star formation region. II. A study of interstellar absorption.
L. N. Kolesnik, M. D. Metreveli.
Kinematika Fiz. Nebesn. Tel, Tom 1, No. 3, p. 53 – 63 (1985). In Russian.

V and B magnitudes and spectral types have been obtained for 1150 stars of spectral classes O to A9 in an 18 square degree region at $l = 134°7$, $b = +0°92$. The interstellar absorption has been investigated by means of colour excesses. Analysis of these data indicates an absorption complex between 70 and 900 pc from the sun. The dust concentrations coincide with the local spiral arm as it is defined by associations and young open clusters. Most of the reddening occurs on the inner side of the local spiral arm.

131.180   Star formation: phase transition, not Jeans instability.
J. E. Tohline.
Astrophys. J., Vol. 292, No. 1, p. 181 – 187 (1985).

It is customarily thought that the Jeans instability is the primary mode by which interstellar gas transforms itself into stars. When rotational energy is included in the determination of the virial equilibrium structure of isothermal gases, however, every equilibrium sequence that terminates via the Jeans instability at mass M_{max} also has an allowed, first–order phase transition to a compact state at a mass $M_{pt} < M_{max}$. Therefore, diffuse gas clouds that are stable according to the Jeans criterion can fragment into high–density clumps in a non–dynamic manner analogous to the manner in which a gas–to–liquid phase transition occurs in an ordinary Van der Waals gas. This transition may help explain why molecular clouds are so clumpy in our Galaxy.

131.181   Far–infrared line intensities of H_2O and CO from warm molecular clouds.
T. Takahashi, D. J. Hollenbach, J. Silk.
Astrophys. J., Vol. 292, No. 1, p. 192 – 199 (1985).

The effects of cloud density, column density, dust temperature, water abundance, and velocity gradient on the emergent lines and profiles of ortho–H_2O and CO from a linearly expanding (or contracting) spherical cloud have been studied. An extended form of the escape probability approximation for the large velocity–gradient case is used to include warm ($T_d \approx 35 – 70K$) dust emission in an optically thick ($A_v \approx 10 – 100$) molecular cloud. The intensities of the strongest H_2O lines are found to depend mainly on the intensity of dust radiation and are good indicators of cloud optical depth and dust temperature. The intensities of the CO lines depend mainly on the collisional rates and measure cloud density and gas temperature.

131.182   Formation of OB clusters: CO, NH_3, and H_2O observations of the distant H II region complex in S128.
A. D. Haschick, P. T. P. Ho.
Astrophys. J., Vol. 292, No. 1, p. 200 – 205 (1985).

The spectral line emission from the ^{12}CO, ^{13}CO, $C^{18}O$, and NH_3 molecular species was mapped for the molecular cloud surrounding the S128 H II region. The ^{12}CO map shows an elongated cloud ($\sim 9' \times 3'$) having at least two cloud components at –74 and –72.8 km s^{-1}. A further H_2O maser was detected close to the compact H II region S128N. All the signposts of star formation, the H II regions, and the H_2O maser sources are found to lie along the transition between the –74 km s^{-1} and –72.8 km s^{-1} clouds, suggesting that the interaction of these two clouds may have initiated the observed star formation.

131.183   Detection of interstellar rotationally excited CH.
L. M. Ziurys, B. E. Turner.
Astrophys. J., Lett. Ed., Vol. 292, No. 1, p. L25 – L29 (1985).

Rotationally excited CH has been detected for the first time in the ISM. The first excited level (F_1, $N = 1$, $J = 3/2$) has been observed via its Λ–doubling transitions at 700 MHz toward W51 – the lowest frequency molecular lines detected to date. The two main hyperfine components are seen in absorption. The implications for CH excitation and abundance are discussed. It is found that CH is present with considerable abundance at high densities in W51, contrary to predictions of chemical models.

131.184   The discovery of a new masering transition of interstellar methanol.
T. L. Wilson, C. M. Walmsley, K. M. Menten, W. Hermsen.
Astron. Astrophys., Vol. 147, No. 1, p. L19 – L22 (1985).

A spectral line of strength ~ 50 Jy has been detected towards the compact H II region W3(OH). From the agreement with the laboratory line frequency, the authors identify this line with the $J_k = 2_1–3_0$ transition of E–type methanol. If, as with the $9_2–10_1$ A^+ line, the emission region is less than $1''$ in diameter, the line brightness temperature is $> 1.5 \times 10^5$K. The phenomenon appears to be relatively unusual however since a short survey of galactic sources in the $2_1–3_0$ E line revealed no other sources of this kind.

131.185   The abundance of oxygen in the interstellar medium.
F. P. Keenan, A. Hibbert, P. L. Dufton.
Astron. Astrophys., Vol. 147, No. 1, p. 89 – 92 (1985).

O I equivalent widths observed by the Copernicus satellite and Doppler widths determined from curves–of–growth for weak N I lines are used to rederive column densities towards 26 early–type stars. Oxygen is found to have a mean abundance of approximately 50% of its solar value towards both reddened and unreddened stars. In addition, there appears to be no correlation between depletion and total hydrogen column density $N(H_{TOT})$, in contrast to the results of York et al. who found a greater depletion for sightlines with log $N(H_{TOT}) < 20.5$.

131.186 On cloud shadows in wide beam scans of the diffuse soft X–ray sky.
J. Knude.
Astron. Astrophys., Vol. 147, No. 1, p. 155 – 160 (1985).

From the observed smoothness of the diffuse soft X–ray background the local ISM has been suggested to be either an extinction free, expanding emitting volume or an emission volume with very few clouds containing all interstellar mass. The background observations have furthermore been shown to have a very low probability, $\sim 1\%$, to be consistent with the cloud distribution predicted in the McKee and Ostriker model. This paper investigates the effect the observed distribution of clouds has on emission from regions more distant than ~ 100 pc as observed in a 0.01 sr beam. The expected intensity is computed from the spatial frequency of the clouds' solid angles and optical depths. The near identity of the model intensity variation to the observed variation and the rather low probability for consistency suggest the possibility of a non–local origin for some fraction of the observed background.

131.187 The nature of interstellar grains.
F. Hoyle.
Astrophys. Relativ., Prepr. Ser., No. 117, 39 pp. (1985).

131.188 An investigation of the interstellar extinction. II. Towards the mid–infrared sources in the galactic centre.
P. F. Roche, D. K. Aitken.
Anglo–Aust. Obs., Prepr., No. 201, 23 pp. (1985). Submitted to Mon. Not. R. Astron. Soc.

131.189 Far–UV to infrared photometric star formation tracers.
B. Rocca–Volmerange.
Inst. Astrophys. Paris, Pré–Publ., No. 88, 12 pp. (1985). To be published in "Proceedings of new aspects of photometry of galaxies", Toulouse, September 1984.

131.190 High resolution observations of interstellar Na I absorption towards Large Magellanic Cloud stars: I. R 116, R 127 and R 128 with the ESO coudé échelle spectrometer.
R. Ferlet, M. Dennefeld, E. Maurice.
Inst. Astrophys. Paris, Pré–Publ., No. 98, 20 pp. (1985). To appear in Astron. Astrophys.

131.191 Na I as a tracer of H I in the diffuse interstellar medium.
R. Ferlet, A. Vidal–Madjar, C. Gry.
Inst. Astrophys. Paris, Pré–Publ., No. 100, 28 pp. (1985). To appear in Astrophys. J.

131.192 The W3 molecular cloud.
H. A. Thronson Jr., C. J. Lada, T. Hewagama.
Prepr. Steward Obs., No. 573, 55 pp. (1985).

Extensive $J = 1 \rightarrow 0$ $^{12}C^{16}O$ and $^{13}C^{16}O$ observations of the W3 molecular cloud are presented and discussed. Four luminous star–forming regions, W3A, W3(OH), W3 North, and AFGL 333 are investigated. Both the large–scale (sizes up to ~ 20 pc) and small–scale (sizes down to ~ 1 pc) structures are considered, from which the authors find that molecular gas velocities are best explained as the result of material being swept up via the expansion of the neighboring giant W4 H II region. More significantly, the analysis of the motions within the W3A and W3(OH) clouds shows that, as predicted by the theory of externally–induced star formation, the young embedded stars are moving with the swept–up material, not with the undisturbed gas.

131.193 On the abundance of metals and the ionization state in absorbing clouds toward QSOs.
F. H. Chaffee Jr., C. B. Foltz, J. Bechtold, R. J. Weymann.
Prepr. Steward Obs., No. 580, 31 pp. (1985).

The authors present photoionization models for clouds of material illuminated by the integrated radiation of QSOs over a range of epochs and predict the column densities of the most abundant ions for clouds with $N(H\ I) = 5 \times 10^{16} cm^{-2}$. For a wide density range, C III is predicted to have the highest column density of all metallic ions, and the authors search for C III in the

$z = 3.321$ absorption system toward S5 0014+81. Their observations set an upper limit of the metal–to–hydrogen ratio of $10^{-3.5}$ compared to the solar value, the lowest value yet inferred for any extragalactic material.

131.194 Star–forming regions near the supernova remnant IC 443.
S. F. Odenwald, K. Shivanandan.
Astrophys. J., Vol. 292, No. 2, p. 460 – 463 (1985). With plate 5.

The supernova remnant (SNR) IC 443 (3C 157) represents a growing number of examples where SNRs appear to have collided with nearby molecular clouds or are surrounded by dense shells of neutral hydrogen. In the current study, the northern component of the IC 443 molecular cloud has been detected at 20, 27, and 93 μm. The bolometric luminosity of this far–infrared source is $\sim 130\ L_\odot$, consistent with the expected luminosity of a B7 V star. The authors discuss these findings in the context of theoretical models where SNRs trigger star–forming activity in nearby molecular clouds.

131.195 The rate of the radiative association reaction between CH_3^+ and NH_3 and its implications for interstellar chemistry.
E. Herbst.
Astrophys. J., Vol. 292, No. 2, p. 484 – 486 (1985).

A theory of radiative association reaction rates for systems in which one or more competitive exothermic binary channels exist has been developed. The theory is utilized to calculate the rate of the radiative association reaction between CH_3^+ and NH_3, which is important in the synthesis of CH_3NH_2 in dense interstellar clouds. This reaction is calculated to have a large rate coefficient despite the existence of two competitive exothermic channels.

131.196 Temperature fluctuations and infrared emission from interstellar grains.
B. T. Draine, N. Anderson.
Astrophys. J., Vol. 292, No. 2, p. 494 – 499 (1985).

Infrared emission spectra are computed for various mixtures of interstellar graphite and silicate grains, heated by ambient starlight. The effects of temperature fluctuations (due to the discrete nature of the heating) are included. It is found that temperature fluctuations in small graphite and silicate grains can result in substantial amounts of continuum emission at wavelengths $\lambda \lesssim 50$ μm. The graphite–silicate grain model, exposed to a "standard" interstellar radiation field, can account for the 60 μm and 100 μm emission observed by *IRAS* from several diffuse clouds, provided that the grain size distribution extends to very small sizes ($a_{min} \approx 3$ Å).

131.197 Spectroscopy of the 3 micron emission features.
T. R. Geballe, J. H. Lacy, S. E. Persson, P. J. McGregor, B. T. Soifer.
Astrophys. J., Vol. 292, No. 2, p. 500 – 505 (1985).

Spectroscopic observations at a resolving power of ~ 400 between 3.1 and 3.6 μm are presented of the planetary nebulae NGC 7027, BD +30°3639, and IC 418; the H II region S106; and the red rectangle HD 44179. The well–known emission features at 3.3 and 3.4 μm are resolved. The 3.4 μm feature is found to consist of two components: a sharp feature at 3.40 μm and a broad plateau centered approximately at 3.45 μm. The relative stengths of the 3.3, 3.40, and 3.45 μm features vary from source to source.

131.198 Macroscopic turbulence in molecular clouds.
J. Silk.
Astrophys. J., Lett. Ed., Vol. 292, No. 2, p. L71 – L74 (1985).

The hypothesis that outflows driven by pre–main–sequence stellar winds are driving shells that fragment and energize molecular clouds, sustaining them so that they do not undergo free–fall collapse, leads to predicted correlations between line width, hydrogen density, and clump scale. Both the correlation scaling and normalization are in good accord with the correlations observed over a wide range of cloud scales. Implications are drawn for the efficiency of star formation and for cloud lifetimes.

131.199   Galactic fountains, planetary nebulae and warm H I.
E. E. Salpeter.
Mitt. Astron. Ges., Nr. 63, p. 11 – 20 (1985). – See Abstr. 012.063.
An appreciable fraction of the H I gas observed in our galactic disk is quite warm (> 1000K) and has a moderately large velocity dispersion. The X–ray flux in the Galaxy is not sufficient to warm so much material. The author proposes an interplay between a galactic fountain of hot gas ($\sim 5 \times 10^5$K) and planetary nebulae produced above the galactic disk. Fossil planetary nebulae provide "nucleation seeds" on which some of the hot gas condenses, leading to a very inhomogeneous cloud complex with multiple cores. Flashes of ionizing radiation from other, newer planetary nebulae then warm some of the H I cores.

131.200   Astrophysik interstellarer Moleküle.
G. Winnewisser.
Mitt. Astron. Ges., Nr. 63, p. 35 – 48 (1985). – See Abstr. 012.063.

131.201   Die Entstehung des interstellaren Staubs.
E. Sedlmayr.
Mitt. Astron. Ges., Nr. 63, p. 75 – 88 (1985). – See Abstr. 012.063.
The physical and chemical "boundary conditions" for interstellar dust formation and their consequences for the formation of the original dust components are discussed. By a consistent calculation of circumstellar soot formation, the important thermodynamic and hydrodynamic implications for a dust forming shell are demonstrated. It is shown in particular that both the optical depth of the shell and the observed velocity structure can be explained quantitatively by dust formation.

131.202   H II regions and star formation – an overview of recent theoretical developments.
H. W. Yorke.
Mitt. Astron. Ges., Nr. 63, p. 89 – 103 (1985). – See Abstr. 012.063.
The evolutionary phases of star formation from the molecular cloud state to the evolved H II region surrounding an OB association are described. Recent theoretical developments pertaining to H II region evolution and star formation are critically examined. Current unresolved problems are discussed and suggestions for possible further studies are made.

131.203   Spektroskopische Beobachtungen des Boomerang–Nebels.
T. Neckel, H. J. Staude, M. Sarcander.
Mitt. Astron. Ges., Nr. 63, p. 126 – 127 (1985). – See Abstr. 012.063.

131.204 $^{15}NH_3$ im Orion–KL Nebel: der "heiße Kern" ist gar nicht so heiss.
W. Hermsen, T. L. Wilson, C. M. Walmsley, W. Batrla.
Mitt. Astron. Ges., Nr. 63, p. 132 (1985). – See Abstr. 012.063.

131.205   Methanolmaser in Sternentstehungsgebieten.
K. M. Menten, T. L. Wilson, C. M. Walmsley, R. Mauersberger, C. Henkel, K. J. Johnston.
Mitt. Astron. Ges., Nr. 63, p. 133 (1985). – See Abstr. 012.063.

131.206   The detection of the (7,7) line of NH_3.
R. Mauersberger, T. L. Wilson, C. Henkel, C. M. Walmsley, W. Hermsen.
Mitt. Astron. Ges., Nr. 63, p. 134 – 138 (1985). – See Abstr. 012.063.
New detections of the $(J,K) = (7,7)$ inversion line of ammonia in 15 sources and a map of Sgr B2 are presented.

131.207   Interstellare C_2–Linien in frühen Überriesen.
R. Gredel.
Mitt. Astron. Ges., Nr. 63, p. 152 – 154 (1985). – See Abstr. 012.063.

131.208   Interstellare Absorptionslinien zwischen 2000 Å and 3000 Å aus Spektren der BUSS–Datenbasis – das Mg I/Na I–Verhältnis als Strukturindikator des LISM.
H. Lenhart, K. S. de Boer.
Mitt. Astron. Ges., Nr. 63, p. 155 – 156 (1985). – See Abstr. 012.063.

131.209   Die interstellare Extinktionskurve bei O– und B–Sternen im Rosettennebel.
W. E. Celnik, T. Schmidt–Kaler.
Mitt. Astron. Ges., Nr. 63, p. 157 (1985). Abstract. – See Abstr. 012.063.

131.210   Optical and infrared observations of thin dust clouds.
C. P. de Vries.
Mitt. Astron. Ges., Nr. 63, p. 158 – 159 (1985). – See Abstr. 012.063.
Two of the so called "interstellar cirrus" clouds, in different parts of the sky, are compared for their extinction, surface brightness and infrared fluxes.

131.211   Correlation of IRAS spline maps with molecular clouds in the Carina spiral arm.
H. J. Walker, E. R. Deul, H. M. Butner, W. B. Burton.
Mitt. Astron. Ges., Nr. 63, p. 160 – 162 (1985). – See Abstr. 012.063.
In a recent paper Cohen et al. (1985) reported the identification of 37 molecular clouds in the Carina region, based on CO ($1\rightarrow0$) observations. The authors examined the IRAS spline maps at 100 µm to see if peaks in the diffuse dust emission could be associated with these positions. Many of the molecular clouds occur at the peaks in the 100 µm band maps, but there was a tighter correlation with the point sources from the IRAS catalog.

131.212   Gravitative Wechselwirkung zwischen dem System der Sterne und dem interstellaren Gas unter Berücksichtigung von Heiz– und Kühlprozessen.
B. M. Deiss, W. H. Kegel.
Mitt. Astron. Ges., Nr. 63, p. 170 (1985). – See Abstr. 012.063.

131.213   Can highly excited states of atoms serve as probes for external fields in the interstellar medium?
G. Wunner, H. Ruder, H. Herold.
Mitt. Astron. Ges., Nr. 63, p. 173 – 174 (1985). – See Abstr. 012.063.

131.214   Far–infrared and submillimeter astronomical spectroscopy.
R. Genzel, G. J. Stacey.
Mitt. Astron. Ges., Nr. 63, p. 215 – 233 (1985). – See Abstr. 012.063.
The authors review some of the highlights of the last ten years' far–infrared (FIR)/submillimeter spectroscopy (excluding planetary work), then show recent infrared and submillimeter measurements of the rotation curve and mass distribution in the central 10 pc of our Galaxy, and finally discuss in more detail the warm and dense neutral interstellar medium from the viewpoint of recent work in FIR/submillimeter spectroscopy.

131.215   Confirmation of the existence of two new interstellar molecules: C_3H and C_3O.
W. M. Irvine, P. Friberg, Å. Hjalmarson, L. E. B. Johansson, P. Thaddeus, R. D. Brown, P. D. Godfrey.
Bull. Am. Astron. Soc., Vol. 16, No. 4, p. 877 (1984). Abstract. – See Abstr. 010.062.

131.216   Star formation in rotating, magnetized molecular disks.
R. E. Pudritz.
Bull. Am. Astron. Soc., Vol. 16, No. 4, p. 877 (1984). Abstract. – See Abstr. 010.062.

131.217 On fragmentation and turbulence in collapsing gas clouds.
W. Benz.
Bull. Am. Astron. Soc., Vol. 16, No. 4, p. 878 (1984). Abstract. – See Abstr. 010.062.

131.218 The $^{12}C/^{13}C$ isotope ratio toward ζ Oph.
I. Hawkins, M. Jura, D. M. Meyer.
Bull. Am. Astron. Soc., Vol. 16, No. 4, p. 878 (1984). Abstract. – See Abstr. 010.062.

131.219 1.3 mm continuum mapping of Orion KL: evidence for a disk.
F. P. Schloerb, R. L. Snell, P. R. Schwartz.
Bull. Am. Astron. Soc., Vol. 16, No. 4, p. 878 (1984). Abstract. – See Abstr. 010.062.

131.220 Carbon monoxide as an interstellar mass tracer at high visual extinctions.
R. L. Dickman, W. Herbst, D. Chapman Taylor.
Bull. Am. Astron. Soc., Vol. 16, No. 4, p. 878 (1984). Abstract. – See Abstr. 010.062.

131.221 A new type of methanol maser.
C. Henkel, T. L. Wilson, C. M. Walmsley, P. R. Jewell, L. E. Snyder, R. Mauersberger.
Bull. Am. Astron. Soc., Vol. 16, No. 4, p. 879 (1984). Abstract. – See Abstr. 010.062.

131.222 Determination of the masses and energetics of six molecular outflows.
M. Margulis, C. J. Lada.
Bull. Am. Astron. Soc., Vol. 16, No. 4, p. 879 (1984). Abstract. – See Abstr. 010.062.

131.223 Resolution of the magnetic flux problem during star formation by ambipolar diffusion.
T. C. Mouschovias, E. Paleologou, R. A. Fiedler.
Bull. Am. Astron. Soc., Vol. 16, No. 4, p. 879 (1984). Abstract. – See Abstr. 010.062.

131.224 Gamma–ray observations of recently synthesized interstellar ^{26}Al.
W. A. Mahoney.
Bull. Am. Astron. Soc., Vol. 16, No. 4, p. 879 – 880 (1984). Abstract. – See Abstr. 010.062.

131.225 Echelle–CCD observations of interstellar lithium.
R. E. White.
Bull. Am. Astron. Soc., Vol. 16, No. 4, p. 910 (1984). Abstract. – See Abstr. 010.062.

131.226 Molecular spectroscopy with the new Cologne University 3 m radio telescope.
T. Pauls, G. Winnewisser.
Bull. Am. Astron. Soc., Vol. 16, No. 4, p. 915 (1984). Abstract. – See Abstr. 010.062.

131.227 An unusual giant molecular cloud in Monoceros.
R. J. Maddalena, P. Thaddeus.
Bull. Am. Astron. Soc., Vol. 16, No. 4, p. 919 (1984). Abstract. – See Abstr. 010.062.

131.228 Far–infrared H_2O intensities from warm molecular clouds.
T. Takahashi, D. Hollenbach, J. Silk.
Bull. Am. Astron. Soc., Vol. 16, No. 4, p. 920 (1984). Abstract. – See Abstr. 010.062.

131.229 Unresolved structure in an unbiased sample of molecular clouds.
H. M. Martin, D. B. Sanders, F. O. Clark.
Bull. Am. Astron. Soc., Vol. 16, No. 4, p. 920 (1984). Abstract. – See Abstr. 010.062.

131.230 Low mass pre–main sequence stars in the Serpens molecular cloud.
E. Churchwell, J. Koornneef.
Bull. Am. Astron. Soc., Vol. 16, No. 4, p. 920 (1984). Abstract. – See Abstr. 010.062.

131.231 Temperature fluctuations and infrared emission from interstellar grains.
B. T. Draine, N. Anderson.
Bull. Am. Astron. Soc., Vol. 16, No. 4, p. 920 (1984). Abstract. – See Abstr. 010.062.

131.232 High resolution IRAS observations of the core of the Ophiuchi dark cloud.
E. T. Young, C. J. Lada, B. A. Wilking, C. A. Beichman.
Bull. Am. Astron. Soc., Vol. 16, No. 4, p. 921 (1984). Abstract. – See Abstr. 010.062.

131.233 Rotation in dense star–forming clouds.
S. N. Vogel.
Bull. Am. Astron. Soc., Vol. 16, No. 4, p. 921 (1984). Abstract. – See Abstr. 010.062.

131.234 The Lynds 204 complex – magnetic field controlled evolution?
W. H. McCutcheon, F. J. Vrba, R. L. Dickman, D. P. Clemens.
Bull. Am. Astron. Soc., Vol. 16, No. 4, p. 921 (1984). Abstract. – See Abstr. 010.062.

131.235 Emission from interstellar shocks with grain destruction.
C. G. Seab, J. M. Shull, C. F. McKee.
Bull. Am. Astron. Soc., Vol. 16, No. 4, p. 927 (1984). Abstract. – See Abstr. 010.062.

131.236 A comparison of [S II] and Hα radial velocities, intensities, and line widths in the galactic background.
R. J. Reynolds.
Bull. Am. Astron. Soc., Vol. 16, No. 4, p. 927 (1984). Abstract. – See Abstr. 010.062.

131.237 Radio observations of bright rims.
P. R. Schwartz.
Bull. Am. Astron. Soc., Vol. 16, No. 4, p. 927 (1984). Abstract. – See Abstr. 010.062.

131.238 Statistical clustering properties of giant molecular cloud cores.
A. R. Rivolo, P. M. Solomon, D. B. Sanders.
Bull. Am. Astron. Soc., Vol. 16, No. 4, p. 937 (1984). Abstract. – See Abstr. 010.062.

131.239 Molecular structure of L134N.
D. A. Swade, F. P. Schloerb, R. L. Snell.
Bull. Am. Astron. Soc., Vol. 16, No. 4, p. 937 (1984). Abstract. – See Abstr. 010.062.

131.240 Evidence for a disk around Orion–IRc2.
R. L. Plambeck, S. N. Vogel, M. C. H. Wright.
Bull. Am. Astron. Soc., Vol. 16, No. 4, p. 937 (1984). Abstract. – See Abstr. 010.062.

131.241 Observation of the interstellar red band of CN.
R. M. Crutcher, B. L. Lutz.
Bull. Am. Astron. Soc., Vol. 16, No. 4, p. 937 – 938 (1984). Abstract. – See Abstr. 010.062.

131.242 Observations of radio recombination lines with principal quantum number $456 < \bar{n} < 634$ towards Cas–A.
W. C. Erickson, K. R. Anantharamaiah.
Bull. Am. Astron. Soc., Vol. 16, No. 4, p. 938 (1984). Abstract. – See Abstr. 010.062.

131.243 Observations of H_2 emission in DR 21.
D. Nadeau, S. Béland, R. Garden, T. R. Geballe.
Bull. Am. Astron. Soc., Vol. 16, No. 4, p. 939 (1984). Abstract. – See Abstr. 010.062.

131.244 New interstellar absorption lines of C_2 and CN.
E. F. van Dishoeck, J. H. Black.
Bull. Am. Astron. Soc., Vol. 16, No. 4, p. 958 (1984). Abstract. – See Abstr. 010.062.

131.245 Diffuse interstellar features at 5780 Å and 5797 Å.
M. E. Kaiser.
Bull. Am. Astron. Soc., Vol. 16, No. 4, p. 958 (1984). Abstract. – See Abstr. 010.062.

131.246 Infrared reflection nebulae in the Cha I, ρ Oph, and R CrA dark clouds.
M. W. Castelaz, J. A. Hackwell.
Bull. Am. Astron. Soc., Vol. 16, No. 4, p. 958 (1984). Abstract. – See Abstr. 010.062.

131.247 Interstellar absorption features in the direction of the compact infrared source W33A.
H. P. Larson, D. S. Davis, J. H. Black, U. Fink.
Bull. Am. Astron. Soc., Vol. 16, No. 4, p. 959 (1984). Abstract. – See Abstr. 010.062.

131.248 A triggering mechanism for star formation in W5A/S201.
P. Pismis, I. Hasse, M. Moreno.
Bull. Am. Astron. Soc., Vol. 16, No. 4, p. 959 (1984). Abstract. – See Abstr. 010.062.

131.249 High spatial resolution radio observations of the star formation region G351.8–0.5.
R. A. Gaume, R. L. Mutel, J. D. Fix.
Bull. Am. Astron. Soc., Vol. 16, No. 4, p. 959 (1984). Abstract. – See Abstr. 010.062.

131.250 Detection of interstellar methylcyanodiacetylene in the dark dust cloud TMC 1.
L. E. Snyder, T. L. Wilson, C. Henkel, P. R. Jewell, C. M. Walmsley.
Bull. Am. Astron. Soc., Vol. 16, No. 4, p. 959 (1984). Abstract. – See Abstr. 010.062.

131.251 Observations of several new transitions of interstellar HCO.
M. S. Schenewerk, L. E. Snyder, J. M. Hollis.
Bull. Am. Astron. Soc., Vol. 16, No. 4, p. 959 – 960 (1984). Abstract. – See Abstr. 010.062.

131.252 Spiral gravitational potentials and the mass growth of molecular clouds. II.
F. Valdes, J. Kwan.
Bull. Am. Astron. Soc., Vol. 16, No. 4, p. 960 (1984). Abstract. – See Abstr. 010.062.

131.253 The origin of the infrared luminosity in violent star formation regions.
A. W. Campbell, R. J. Terlevich.
Bull. Am. Astron. Soc., Vol. 16, No. 4, p. 960 (1984). Abstract. – See Abstr. 010.062.

131.254 Drag forces on ablating clumps in molecular clouds.
D. L. Gilden.
Bull. Am. Astron. Soc., Vol. 16, No. 4, p. 960 (1984). Abstract. – See Abstr. 010.062.

131.255 IRAS observations of extended, warm interstellar dust.
T. N. Gautier III, C. A. Beichman.
Bull. Am. Astron. Soc., Vol. 16, No. 4, p. 968 – 969 (1984). Abstract. – See Abstr. 010.062.

131.256 Infrared emission from protostars.
F. H. Shu, F. Adams.
Bull. Am. Astron. Soc., Vol. 16, No. 4, p. 973 (1984). Abstract. – See Abstr. 010.062.

131.257 IUE observations of interstellar hydrogen and deuterium toward Alpha Centauri B.
J. Murthy, R. C. Henry, H. W. Moos, W. B. Landsman, J. L. Linsky, J. L. Russell.
Bull. Am. Astron. Soc., Vol. 16, No. 4, p. 980 (1984). Abstract. – See Abstr. 010.062.

131.258 Measurement of X–ray scattering from interstellar grains.
P. Gorenstein, C. Mauche, D. Fabricant.
Bull. Am. Astron. Soc., Vol. 16, No. 4, p. 981 (1984). Abstract. – See Abstr. 010.062.

131.259 New insights into the hot component of the interstellar medium.
R. McCray.
Bull. Am. Astron. Soc., Vol. 16, No. 4, p. 984 (1984). Abstract. – See Abstr. 010.062.

131.260 The arcminute scale radio structure of W50.
S. A. Baum, R. Elston.
Bull. Am. Astron. Soc., Vol. 16, No. 4, p. 995 (1984). Abstract. – See Abstr. 010.062.

131.261 Numerical modelling of interstellar bubbles in an inhomogeneous medium.
M. T. Wolff, R. H. Durisen.
Bull. Am. Astron. Soc., Vol. 16, No. 4, p. 995 (1984). Abstract. – See Abstr. 010.062.

131.262 Star formation in dark clouds associated with the Gum nebula.
J. A. Graham.
Bull. Am. Astron. Soc., Vol. 16, No. 4, p. 998 (1984). Abstract. – See Abstr. 010.062.

131.263 Interstellar space contains the largest encountered atoms.
W. D. Watson.
Nature, Vol. 315, No. 6021, p. 630 – 631 (1985).
This note comments on the detection of extremely high Rydberg states in diffuse interstellar clouds (see also Abstr. 131.264).

131.264 Observation of highly excited radio recombination lines towards Cassiopeia A.
K. R. Anantharamaiah, W. C. Erickson, V. Radhakrishnan.
Nature, Vol. 315, No. 6021, p. 647 – 649 (1985).
The absorption line at 26 MHz observed by Konovalenko and Sodin towards Cas A was interpreted by Blake et al. as the 631α recombination line due to carbon or heavier elements. This was confirmed subsequently by Konovalenko and Sodin, who detected the 640α line at the same optical depth as the 631α line. The authors report new observations of these lines carried out near 26, 38, 52 and 68 MHz using the NRAO 91–m telescope. They detected absorption lines at all these frequencies, with a striking fourfold increase in width towards low frequencies due to collisional or radiation broadening. Assuming that these are recombination lines of carbon, the observed radial velocity corresponds well with that of the H I, OH, NH_3 and H_2CO absorption features that originate in the Perseus arm in the same direction.

131.265 Interstellar scintillation measurements of pulsars at 326.5 MHz.
V. Balasubramanian, S. Krishnamohan.
J. Astrophys. Astron., Vol. 6, No. 1, p. 35 – 47 (1985).
The authors have measured the decorrelation frequency (f_v) and decorrelation time (t_v) for 15 pulsars and combined their data with those of others. The relations obtained are broadly in agreement with those expected from a homogeneous interstellar medium and are in disagreement with earlier conclusions by others

that these relations steepen even for low–DM (dispersion measure) pulsars. The agreement suggests that the local interstellar medium is homogeneous at least up to a distance of about 2 kpc.

131.266 Detection of interstellar CCD.
F. Combes, F. Boulanger, P. J. Encrenaz, M. Gerin, M. Bogey, C. Demuynck, J. L. Destombes.
Astron. Astrophys., Vol. 147, No. 2, p. L25 – L26 (1985).

The first detection of interstellar CCD has been obtained in the $N=3–2$, $J=7/2–5/2$ line at 216.3732 GHz towards the Kleinman–Low nebula. The authors found an abundance ratio CCD/CCH of 4.5×10^{-2}, indicating in this molecule a deuterium enhancement similar to that found for HNC, but an order of magnitude higher than that for HCN. Negative results towards DR21(OH), NGC 2264, L134N, TMC–1, ϱ Oph B2 and IRC 10216 rule out a stronger D enhancement for CCD than for other deuterated molecules in these sources. Predictions of currently developed ion–molecule reaction schemes are consistent with the CCD line detected and negative results.

131.267 Interstellar reddening law towards the nucleus of h Per.
M. L. Franco, A. Magazzù, R. Stalio.
Astron. Astrophys., Vol. 147, No. 2, p. 191 – 196 (1985).

The authors construct the extinction curves in the direction of four early–B main sequence stars located in the nucleus of h Per: (BD $+56°501$, BD $+56°510$, BD $+56°516$ and BD $+56°517$) and compare them to one another and to the extinction curves derived from eight stars located in the same general area but lying at different distances. Data from *UBV* photometry and low dispersion ultraviolet spectroscopy were used to derive the curves. The interstellar reddening law along the line of sight of the four h Per stars is constant. The interstellar medium in the h Per direction has properties similar to those of the general h and χ Per direction. There is rather smooth decrease in the extinction per unit distance with increasing height from the galactic plane. There is no indication of a variation of the ratio bump strength to far–UV extinction with distance.

131.268 Radio recombination lines from the galactic plane in Cygnus.
A. Barcia, J. Gómez–González, F. J. Lockman, P. Planesas.
Astron. Astrophys., Vol. 147, No. 2, p. 237 – 240 (1985).

Radio recombination lines were observed from the galactic plane over $75° \leqslant l \leqslant 86°$ at two fequencies, 1.4 and 2.3 GHz, using different telescopes to obtain comparable angular resolution at the two transitions. The emission seems to come from gas that is optically very thin. Estimates of the emission measure derived from the recombination lines, when combined with Hα observations, imply visual extinctions of $>6^m$ in some directions, and a distance of >1 kpc to most of the nebulae.

131.269 Turbulent motions in molecular clouds.
G. A. Pellegatti Franco, R. D. Tarsia, R. J. Quiroga.
Astrophys. Space Sci., Vol. 111, No. 2, p. 343 – 354 (1985).

The authors have studied the behavior of the inner motions of OH, H_2CO, and CO molecular clouds. This study shows the existence of two main components of these clouds: the narrow one, associated to dense small clouds and a wide one 'representing' the large diffuse clouds seen in neutral hydrogen, the large clouds are the 'vortex' and intermediate state between turbulent and hydrodynamic motions in the Galaxy. For the dense clouds with sizes $d < 10$ pc the authors have found a relationship $\sigma \propto d^{0.38}$ consistent with the Kolmogorov law of turbulence; the densities and sizes of these clouds behave as $n \propto d^{-1}$. Also, the effects of the inner magnetic field in these clouds are discussed.

131.270 Bandwidth effects on interstellar reddening and on the R–ratio.
B. Cester, C. Marsi.
Astrophys. Space Sci., Vol. 112, No. 2, p. 311 – 320 (1985).

The energy distributions of 85 stars were combined with the transmittance of B and V filters and with an interstellar reddening law in order to investigate the effects of the bandwidth on the colour excess and on the ratio $R = A_V/E(B-V)$. Since these depend both on the stellar spectral type and on the quantity of absorbing matter, corrective formulae are proposed to allow for these effects.

131.271 Elemental depletions and the 217.5 nm interstellar feature.
W. W. Duley.
Astrophys. Space Sci., Vol. 112, No. 2, p. 321 – 324 (1985).

The strength of the 217.5 nm interstellar feature per unit volume of dust is strongly correlated with the presence of silicon and magnesium in dust but not with the presence of carbon.

131.272 Distribution of electron density and scattering inhomogeneities over the Galaxy.
N. Ya. Shapirovskaya, A. A. Bocharov.
Inst. kosm. issled. Akad. Nauk SSSR. Prepr., No. 936, 41 pp. (1984). In Russian. Abstr. in Ref. Zh., 51. Astron., 5.51.576 (1985).

131.273 Optical absorption from the high velocity neutral hydrogen complex C in the spectrum of the RR Lyrae star BT Draconis.
A. Songaila, D. G. York, L. L. Cowie, J. C. Blades.
Space Telesc. Sci. Inst., Prepr. Ser., No. 43, 16 pp. (1985). To appear in Astrophys. J., Lett. Ed.

131.274 Star formation in rotating, magnetized molecular disks.
R. E. Pudritz.
Astrophys. J., Vol. 293, No. 1, p. 216 – 229 (1985).

This paper presents a theory for star formation in rotating molecular disks in which accretion onto the protostellar core produces FUV radiation which heats the disk surfaces out to large radii. A hydromagnetic wind results in which heated gas is driven out along field lines which thread the disk and are aligned with the disk rotation axis. The centrifugally driven wind removes angular momentum from the disk at rates high enough to brake it down to protostellar specific values in 10^5yr. The wind drives an accretion rate through the disk at rates which are consistent with the accretion luminosity. This global analysis of star formation in a rotating, magnetized disk offers a unifying scheme for understanding both star formation and bipolar outflows. The disks in which massive stars form are predicted to be dense ($10^8 cm^{-3}$) and have rotation speeds of 4 km s^{-1}, scales of order 5×10^{16}cm, masses of order $10^2 M_\odot$, and axial ratios of 0.2.

131.275 Depletion of elements in the interstellar medium.
P. M. Gondhalekar.
Astrophys. J., Vol. 293, No. 1, p. 230 – 235 (1985).

The abundances of both refractory and volatile elements in the interstellar medium (ISM) are correlated with the density of neutral hydrogen in the clouds in the ISM. The density of neutral hydrogen in the clouds is determined from the population of excited fine–structure levels of neutral carbon. The correlation between abundances and density of neutral hydrogen in the clouds and the lack of correlation between depletion and dust density suggests that, in the gas phase of the ISM, the abundances of elements are determined by the shielding of grains in the interstellar clouds. There is little or no grain formation in the ISM.

131.276 Theoretical investigation of the interstellar CH_3NC/CH_3CN ratio.
D. J. DeFrees, A. D. McLean, E. Herbst.
Astrophys. J., Vol. 293, No. 1, p. 236 – 242 (1985).

Calculations have been performed to determine the abundance ratio of the metastable isomer CH_3NC to the stable isomer CH_3CN in dense interstellar clouds. According to gas phase, ion–molecule treatments, these molecules are both synthesized via protonated ion precursors. The calculations, which involve both ab initio quantum chemistry and equilibrium determinations, lead to a predicted CH_3NCH^+/CH_3CNH^+ formation rate ratio between 0.1 and 0.4. If this ratio is maintained in the neutral species formed from the precursor ions, theory predicts a sizable abundance for methyl isocyanide and lends credence to its tentative observation.

131.277 Absorption and recombination of hydrogen atoms on a model graphite surface.
S. Aronowitz, S. Chang.
Astrophys. J., Vol. 293, No. 1, p. 243 – 250 (1985).

The adsorption and recombination of atomic hydrogen on a model graphite grain have been examined in a series of calculations in which a modified, iterative, extended Hückel program was used. Results for the chemisorption of H atoms, their tunneling to adjacent sites, and their recombination to H_2 molecules on the grain surface are presented.

131.278 Possible consequences of gas accretion for the initial mass function of star clusters.
G. H. Smith.
Astrophys. J., Vol. 293, No. 1, p. 251 – 257 (1985).

The influence of gas accretion in determining the shape of the initial stellar mass function of a star cluster is investigated. Small condensations are assumed to form within gaseous protocluster clouds, and then to grow in mass by the accretion of gas from the environment. This accretion process is considered to continue up until such a time as the protocloud is ionized by the turning on of luminous O and B stars, or is dispersed by stellar winds and supernova explosions. Analytical expressions are derived for the stellar mass spectrum upon the assumption that the accretion rate onto a condensation is either a constant, or is proportional to the instantaneous mass of that condensation. In both cases the resultant stellar mass spectrum shows a turnover, i.e., below a certain mass, the luminosity function decreases with decreasing luminosity.

131.279 Optical absorption from the high–velocity neutral hydrogen complex C in the spectrum of the RR Lyrae star BT Draconis.
A. Songaila, D. G. York, L. L. Cowie, J. C. Blades.
Astrophys. J., Lett. Ed., Vol. 293, No. 1, p. L15 – L18 (1985).

The authors report the detection of interstellar Na D absorption lines at a LSR velocity of -85 km s^{-1}, and tentative detections of further Na D lines at -133 km s^{-1} and -110 km s^{-1}, in the spectrum of the RR Lyrae star BT Draconis which lies toward the high–velocity neutral hydrogen complex C (LSR velocity around -115 km s^{-1}). This suggests an upper limit of 2.1 kpc on the distance to this high–velocity complex and indicates that the complex has near cosmic metallicity.

131.280 The Orion B molecular jet.
D. B. Sanders, S. P. Willner.
Astrophys. J., Lett. Ed., Vol. 293, No. 1, p. L39 – L43 (1985).

CO maps of Orion B made with the FCRAO 14 m telescope have revealed a bipolar molecular jet centered within the region of maximum molecular column density approximately 0.5 south of the ionization front. The driving source has not yet been discovered, but the jet evidently marks a new area of star formation activity that is younger than the more evolved area to the north. The jet is the most highly collimated one known, and its energy suggests that the source is at least as luminous as an early B star.

131.281 Ionized polycyclic aromatic hydrocarbons and the diffuse interstellar bands.
M. K. Crawford, A. G. G. M. Tielens, L. J. Allamandola.
Astrophys. J., Lett. Ed., Vol. 293, No. 1, p. L45 – L48 (1985).

A good case has recently been made that the unidentified infrared emission features arise from positively charged, partially hydrogenated polycyclic aromatic hydrocarbons (PAHs). The authors suggest that these exceedingly stable ions are also the carriers of the diffuse interstellar bands. While neutral PAHs do not absorb in the visible, their ionized counterparts do. Because of their low ionization potential, a substantial fraction of the interstellar PAHs will be ionized. Visible spectra of the most stable PAH cations isolated in glasses are compared directly to the interstellar band spectra.

131.282 The molecular cloud size spectrum in the outer Galaxy.
S. Terebey, M. Fich, L. Blitz.
Bull. Am. Astron. Soc., Vol. 17, No. 1, p. 512 – 513 (1985). Abstract. – See Abstr. 010.064.

131.283 370 µm CO 7 → 6 observations of molecular cloud cores.
D. T. Jaffe, A. Harris, M. Silber, R. Genzel.
Bull. Am. Astron. Soc., Vol. 17, No. 1, p. 520 (1985). Abstract. – See Abstr. 010.064.

131.284 Refractive interstellar scintillation and radiosource variations.
B. J. Rickett.
Bull. Am. Astron. Soc., Vol. 17, No. 1, p. 521 (1985). Abstract. – See Abstr. 010.064.

131.285 A comparison of the effects of several models of turbulent viscosity on the early stages of star formation.
T. C. Vanajakshi, A. W. Jenkins Jr.
Bull. Am. Astron. Soc., Vol. 17, No. 1, p. 521 (1985). Abstract. – See Abstr. 010.064.

131.286 The CO cloud in the region of GL490.
Y. Fukui.
Bull. Am. Astron. Soc., Vol. 17, No. 1, p. 521 (1985). Abstract. – See Abstr. 010.064.

131.287 Sputtering of interstellar grains in shock processes.
G. Strazzulla, G. A. Baratta, A. Magazzù.
Astron. Express, Vol. 1, Nos. 4 – 6, p. 143 – 152 (1985).

Using experimental results and an empirical formula describing the sputtering yield of low energy light ions, the authors have calculated the erosion rates of non–thermal sputtering induced by supernova shock waves eroding graphite or silicate grains. The authors find that graphite and silicate grains are almost equally destroyed at shock velocities greater than about 70 km/sec. At lower velocities graphite is preferentially eroded.

131.288 Low–frequency (42, 57, 84 MHz) excited carbon lines toward Cassiopeia A.
A. A. Ershov, Yu. P. Ilyasov, E. E. Lekht, G. T. Smirnov,
V. T. Solodkov, R. L. Sorochenko.
Sov. Astron. Lett., Vol. 10, No. 6, p. 348 – 353 (1984). English translation of 38.131.141.

131.289 Decameter–wavelength carbon recombination lines toward Cassiopeia A.
A. A. Konovalenko.
Sov. Astron. Lett., Vol. 10, No. 6, p. 353 – 356 (1984). English translation of 38.131.142.

131.290 Density–wave interpretation of the Mars 7 and Venera 9 interstellar gas velocity measurements.
E. M. Grivnev.
Sov. Astron. Lett., Vol. 10, No. 6, p. 382 – 383 (1984). English translation of 38.131.226.

131.291 Decameter excited carbon lines in certain galactic objects.
A. A. Konovalenko.
Sov. Astron. Lett., Vol. 10, No. 6, p. 384 – 386 (1984). English translation of 38.131.227.

131.292 Slow pulsar scintillation and the spectrum of interstellar electron density fluctuations.
J. Goodman, R. Narayan.
Mon. Not. R. Astron. Soc., Vol. 214, No. 4, p. 519 – 537 (1985).

Intensity fluctuations in pulsars are commonly ascribed to scintillations produced by electron density inhomogeneities in the interstellar medium. The power spectrum of the inhomogeneities is believed to be a power law, $q^{-\beta}$, with $\beta = 11/3$ (Kolmogorov spectrum). The spectrum of intensity fluctuations produced by a thin phase–changing screen has already been worked out for the case $2 < \beta < 4$ and shows that two length–scales exist – a fast variation due to diffractive scintillation and a slow variation due to refractive effects. In this paper a theory is worked out for $4 < \beta < 6$ and asymptotic expressions are developed for the intensity fluctuation spectrum in various regimes.

131.293 High–velocity gas flows associated with H_2 emission regions: how are they related and what powers them?
J. Fischer, D. B. Sanders, M. Simon, P. M. Solomon.
Astrophys. J., Vol. 293, No. 2, p. 508 – 521 (1985).

The authors report on an investigation of high–velocity molecular material near the vibrationally excited H_2 emission regions in the objects DR 21, W75 N, NGC 7538, OMC-2, NGC 6334/Peak V, and the Herbig–Haro Object 2. The ^{12}CO ($J = 1$–0) and ^{13}CO ($J = 1$–0) transitions were used as tracers of the high–velocity molecular flows. All the H_2 sources are associated with bipolar CO high–velocity flows. A new high–spatial-resolution map ($11''$) of the DR 21 H_2 region is also presented. The ^{12}CO and ^{13}CO profiles are used to determine the spatial extent, optical depth, column density, momentum, and energy of the outflows. The possible role of radiation pressure and the H II expansion as driving mechanisms for the outflows is discussed.

131.294 Molecular clouds associated with compact H II regions. II. The rapidly rotating condensation associated with ON1.
X. W. Zheng, P. T. P. Ho, M. J. Reid, M. H. Schneps.
Astrophys. J., Vol. 293, No. 2, p. 522 – 529 (1985).

VLA observations of the ON1 region in the $(J,K) = (1,1)$ line of NH_3 and the H76α recombination line are reported. Dense and cool NH_3 condensations with size scales of $0.2 - 0.3$ pc are found near the ultracompact H II region, ON1. From the systematic shift in position with velocity, the authors interpret some of the condensations as parts of a rapidly rotating structure with a large velocity gradient of 11 ± 2 km s^{-1}pc^{-1}. The H76α recombination line has a peak velocity of 5.1 ± 2.5 km s^{-1} and is blueshifted with respect to both the NH_3 emission and the OH maser emission.

131.295 VLA observations of the 9_2–10_1 A^+ methanol masers toward W3(OH).
K. M. Menten, K. J. Johnston, T. L. Wilson, C. M. Walmsley, R. Mauersberger, C. Henkel.
Astrophys. J., Lett. Ed., Vol. 293, No. 2, p. L83 – L85 (1985).

Measurements at $2''$ resolution of the 9_2–10_1 A^+ emission line of methanol toward W3(OH) are presented. The frequency resolution was 12.2 kHz. The two velocity components seen in single-dish spectra are separated by $\sim 1''$, but each is unresolved. This implies intrinsic source sizes of $\leqslant 1''$. For the masing spike feature at -43.3 km s^{-1}, the line brightness temperature is $\geqslant 2 \times 10^4$K. For the broader emission at -43.9 km s^{-1}, it is $\geqslant 5000$K, showing that this emission is also caused by masing. The spike is $0\rlap{.}''5$ southwest and the broader feature is $1\rlap{.}''5$ northwest of the continuum peak.

131.296 A new general survey of high–velocity neutral hydrogen in the southern hemisphere.
E. Bajaja, C. E. Cappa de Nicolau, J. C. Cersosimo, N. Loiseau, M. C. Martín, R. Morras, C. A. Olano, W. G. L. Pöppel.
Astrophys. J., Suppl. Ser., Vol. 58, No. 1, p. 143 – 165 (1985).

A southern sky survey in the H I 21 cm line, at declinations between $-90°$ and $-15°$, was done at the Instituto Argentino de Radioastronomia. The observations were made on a $2°/\cos \delta \times 2°$ grid, covering a velocity range between -650 and $+650$ km s^{-1}, with an angular resolution of $34'$ (HPBW) and a velocity resolution of 16 km s^{-1}. The rms noise on the spectra is 0.025K. The results are presented in a table and maps. A whole–sky map containing all known data on high–velocity H I is also included.

131.297 Cataclysmic variables as probes of X–ray properties of interstellar grains.
M. F. Bode, A. Evans, G. A. Norwell.
Cataclysmic variables and low–mass X–ray binaries, p. 347 – 353 (1985). – See Abstr. 012.072.

Interstellar grain properties have previously been probed at wavelengths ranging from the infrared to the ultraviolet. Recent work by other authors has shown that one may also observe the effects of scattering by such grains at X–ray wavelengths. In this paper the authors suggest that investigations of the X–ray properties of interstellar grains may profitably be conducted in sight lines to variable sources. Particular emphasis is given in this context to cataclysmic variables and related objects.

131.298 Interstellar molecular synthesis via protonated formic acid.
H. Villinger, A. Saxer, W. Lindinger.
Seventh European Sectional Conference on the Atomic and Molecular Physics of Ionized Gases, p. 3A – 4 (1984). Abstr. in Phys. Abstr., Vol. 88, No. 1253, Entry 35811 (1985). – See Abstr. 012.056.

131.299 Predicted long–slit, high–resolution emission–line profiles from interstellar bow shocks.
A. C. Raga, K.-H. Böhm.
Astrophys. J., Suppl. Ser., Vol. 58, No. 2, p. 201 – 224 (1985).

The authors have computed the position–dependent emission-line profiles (called "position–velocity diagrams") for the lines Hβ, [N II] λ6583, [S II] λ6731, [O I] λ6300, and [O III] λ5007 which are formed in a model of a radiating interstellar bow shock of high Mach number. Such models have been suggested as an explanation of the emission–line spectra of Herbig–Haro objects in connection with the "interstellar bullet model".

131.300 An observational trend for large ($100 - 300$ pc) interstellar magnetic ($1 - 10$ µgauss) bubbles.
N. W. Broten, J. M. MacLeod, J. P. Vallée.
Astrophys. Lett., Vol. 24, No. 4, p. 165 – 171 (1985).

The observational properties of a sample comprising all large ($100 - 300$ pc) interstellar magnetic ($1 - 10$ µgauss) bubbles, for which the data have been published elsewhere, are summarized and discussed. An observational trend is found (for the first time) between B, the magnetic field strength in the shell, and K, the degree of compression defined as the cube of the ratio of the inner to the outer shell radii.

131.301 Optical polarisation of R CrA and T CrA – NGC 6729.
D. Ward–Thompson.
Thesis, Univ. Durham, England (1984). Abstr. in Phys. Abstr., Vol. 88, No. 1256, Entry 51747 (1985).

131.302 On a model construction of interstellar scattering processes with dust formation in the Galaxy.
D. A. Rozhkovskij.
Tr. Astrofiz. Inst. Alma–Ata, Tom 42, p. 3 – 17 (1983). In Russian.

Simple expressions and formulas are given for the approximation of the properties of the scattering function of interstellar dust particles and for calculation of the flux of light scattering by different spherical models of interstellar dust clouds.

131.303 On the nature of dust particles in reflection nebulae.
L. A. Pavlova.
Tr. Astrofiz. Inst. Alma–Ata, Tom 42, p. 18 – 27 (1983). In Russian.

Possible dust particle models are discussed for explanation of observed spectra of nuclei of reflection nebulae. The effects of different dust particle sizes on the spectral curve of interstellar extinction are considered.

131.304 Polarimetric and colorimetric observations of the nebula NGC 6913.
D. I. Gorodetskij, Eh. S. Eroshevich.
Tr. Astrofiz. Inst. Alma–Ata, Tom 42, p. 57 – 66 (1983). In Russian.

Colorimetric and polarimetric results across the surface of the diffuse nebula NGC 6913 are presented; the color indices $(B-V)_N$–$(B-V)_S$, $(U-B)_N$–$(U-B)_S$ as a function of distance from the luminous star are given. In accordance with the silicate model the concentration of dust in the nebula is 6×10^{-12}cm^{-3}.

131.305   Molecules in interstellar clouds.
E. F. van Dishoeck.
Ned. Tijdschr. Natuurkd. A, Vol. A51, No. 1, p. 38 – 40 (1985).
In Dutch. Abstr. in Phys. Abstr., Vol. 88, No. 1258, Entry 63450
(1985).

131.306   Characteristics of collapse of rotating isothermal clouds.
S. Narita, C. Hayashi, S. M. Miyama.
Prog. Theor. Phys., Vol. 72, No. 6, p. 1118 – 1136 (1984). Abstr.
in Phys. Abstr., Vol. 88, No. 1258, Entry 63451 (1985).

131.307   Interstellar chemistry.
J. Lequeux.
Recherche, Vol. 16, No. 164, p. 330 – 338 (1985). In French.
Abstr. in Phys. Abstr., Vol. 88, No. 1259, Entry 68914 (1985).

131.308   Microscopic dust in the infrared sky.
A. Leene, P. Wesselius.
Recherche, Vol. 16, No. 164, p. 402 – 405 (1985). In French.
Abstr. in Phys. Abstr., Vol. 88, No. 1259, Entry 68915 (1985).

131.309   Study of the bipolar nebula associated with LkHα 208.
J. V. Shirt.
Thesis, Univ. Durham, England (1984). Abstr. in Phys. Abstr.,
Vol. 88, No. 1260, Entry 74270 (1985).

**131.310   Cloud accretion and evaporation, and their consequences
for the diffusive galactic γ–ray emission.**
H. J. Völk.
18th International Cosmic Ray Conference, Vol. 1, p. 173 (1983).
– See Abstr. 012.096.

**131.311   Correlation of γ–ray intensities with column densities of
gas in the I.S.M.**
B. P. Houston, T.–p. Li, P. A. Riley, C.–x. Xu,
A. W. Wolfendale.
18th International Cosmic Ray Conference, Vol. 1, p. 174 – 177
(1983). – See Abstr. 012.096.
New data on the distribution of H_2 in the 4th quadrant have
been used, together with H I information, to determine the extent
of correlation with γ–ray intensities.

**131.312   The far infrared radiation field in Sgr B2 inferred from
the excitation of HNCO.**
E. Churchwell, P. Myers, R. V. Myers, D. Wood.
Bull. Am. Astron. Soc., Vol. 17, No. 2, p. 557 (1985). Abstract. –
See Abstr. 010.065.

131.313   Kinematic distances to cool H I clouds.
T. M. Bania.
Bull. Am. Astron. Soc., Vol. 17, No. 2, p. 561 – 562 (1985). Ab-
stract. – See Abstr. 010.065.

**131.314   Effect of compressible turbulence on ring formation in
rotating protostellar collapse.**
T. C. Vanajakshi.
Bull. Am. Astron. Soc., Vol. 17, No. 2, p. 562 (1985). Abstract. –
See Abstr. 010.065.

**131.315   New detections of interstellar SiO: a probe for shock
chemistry?**
L. M. Ziurys, P. Friberg, W. M. Irvine.
Bull. Am. Astron. Soc., Vol. 17, No. 2, p. 563 (1985). Abstract. –
See Abstr. 010.065.

**131.316   A high–resolution map of dust continuum emission from
OMC1.**
C. R. Masson, M. J. Claussen, K. Y. Lo, T. G. Phillips,
A. I. Sargent, N. Z. Scoville.
Bull. Am. Astron. Soc., Vol. 17, No. 2, p. 563 (1985). Abstract. –
See Abstr. 010.065.

**131.317   Interferometer maps of the CS J = 2–1 emission around
Orion IRc 2.**
L. G. Mundy, N. Z. Scoville, L. Baath, C. R. Masson,
D. P. Woody.
Bull. Am. Astron. Soc., Vol. 17, No. 2, p. 563 (1985). Abstract. –
See Abstr. 010.065.

131.318   Dense molecular cores and IRAS point sources.
F. O. Clark.
Bull. Am. Astron. Soc., Vol. 17, No. 2, p. 563 (1985). Abstract. –
See Abstr. 010.065.

131.319   Molecular emission in the core of L134N.
D. A. Swade, F. P. Schloerb, W. M. Irvine,
R. L. Snell.
Bull. Am. Astron. Soc., Vol. 17, No. 2, p. 563 (1985). Abstract. –
See Abstr. 010.065.

**131.320   Induced star formation in the molecular cloud associated
with Sharpless 254 – 258.**
J. A. Morgan, R. L. Snell, F. P. Schloerb.
Bull. Am. Astron. Soc., Vol. 17, No. 2, p. 563 (1985). Abstract. –
See Abstr. 010.065.

**131.321   Interferometer   mapping   of   the   high   velocity
CO emission in NGC 2071, W49, and NGC 7538.**
A. I. Sargent, M. J. Claussen, K. Y. Lo, C. Masson,
T. G. Phillips, D. B. Sanders.
Bull. Am. Astron. Soc., Vol. 17, No. 2, p. 564 (1985). Abstract. –
See Abstr. 010.065.

**131.322   CO (J = 1–0) observations of NGC 2071 and GL 490
with the 45–m telescope.**
Y. Fukui, T. Iwata, T. Takano.
Bull. Am. Astron. Soc., Vol. 17, No. 2, p. 564 (1985). Abstract. –
See Abstr. 010.065.

131.323   IRAS photometry of bipolar molecular flows.
D. Mozurkewich.
Bull. Am. Astron. Soc., Vol. 17, No. 2, p. 564 (1985). Abstract. –
See Abstr. 010.065.

**131.324   High latitude molecular clouds: star counts and dis-
tances.**
L. Magnani, C. DeVries, L. Blitz.
Bull. Am. Astron. Soc., Vol. 17, No. 2, p. 567 (1985). Abstract. –
See Abstr. 010.065.

**131.325   IRAS cirrus associated with high galactic latitude CO
clouds.**
M. G. Hauser, J. Weiland, L. Magnani, L. Blitz, R. White,
L. J. Rickard.
Bull. Am. Astron. Soc., Vol. 17, No. 2, p. 567 (1985). Abstract. –
See Abstr. 010.065.

131.326   A CO–survey of the high–latitude infrared cirrus.
H. W. de Vries, P. Thaddeus.
Bull. Am. Astron. Soc., Vol. 17, No. 2, p. 567 (1985). Abstract. –
See Abstr. 010.065.

131.327   Galactic H I in directions of low total column density.
F. J. Lockman, K. Jahoda, D. McCammon.
Bull. Am. Astron. Soc., Vol. 17, No. 2, p. 567 (1985). Abstract. –
See Abstr. 010.065.

131.328   Interstellar clouds near the sun.
P. C. Frisch, D. G. York.
Bull. Am. Astron. Soc., Vol. 17, No. 2, p. 568 (1985). Abstract. –
See Abstr. 010.065.

131.329   The hydrocarbon ring C_3H_2 is ubiquitous in the Galaxy.
H. E. Matthews, W. M. Irvine, S. C. Madden,
D. A. Swade.
Bull. Am. Astron. Soc., Vol. 17, No. 2, p. 568 (1985). Abstract. –
See Abstr. 010.065.

131.330 A trace of the chemistry of DCN.
M. R. Greason, H. A. Wootten.
Bull. Am. Astron. Soc., Vol. 17, No. 2, p. 568 (1985). Abstract. –
See Abstr. 010.065.

131.331 Observations of additional unidentified emission features.
J. D. Bregman, L. Allamandola, M. Cohen, J. Simpson,
A. Tielens, F. Witteborn, D. Wooden, D. Rank.
Bull. Am. Astron. Soc., Vol. 17, No. 2, p. 568 – 569 (1985). Abstract. – See Abstr. 010.065.

131.332 IR recombination lines from L 1536 and other low luminosity core or broad wing CO sources.
H. A. Smith, J. Fischer, P. R. Schwartz, T. R. Geballe.
Bull. Am. Astron. Soc., Vol. 17, No. 2, p. 569 (1985). Abstract. –
See Abstr. 010.065.

131.333 Far–infrared observations of OH and CO in the Orion Nebula.
P. J. Viscuso, G. J. Stacey, C. E. Fuller, N. T. Kurtz,
M. O. Harwit.
Bull. Am. Astron. Soc., Vol. 17, No. 2, p. 570 (1985). Abstract. –
See Abstr. 010.065.

131.334 Interstellar ammonia maser detected in star formation regions.
S. C. Madden, W. M. Irvine, H. E. Matthews, R. D. Brown,
P. D. Godfrey.
Bull. Am. Astron. Soc., Vol. 17, No. 2, p. 570 (1985). Abstract. –
See Abstr. 010.065.

131.335 A cylindrical magnetic field envelops the Orion elongated molecular cloud.
C. E. Heiles.
Bull. Am. Astron. Soc., Vol. 17, No. 2, p. 570 (1985). Abstract. –
See Abstr. 010.065.

131.336 UV colors of reflection nebulae and the scattering phase function of interstellar grains.
A. N. Witt.
Bull. Am. Astron. Soc., Vol. 17, No. 2, p. 571 (1985). Abstract. –
See Abstr. 010.065.

131.337 The ionization of the diffuse galactic gas.
J. S. Mathis.
Bull. Am. Astron. Soc., Vol. 17, No. 2, p. 596 (1985). Abstract. –
See Abstr. 010.065.

131.338 On the structure and stability of radiative shock waves.
E. Bertschinger.
Bull. Am. Astron. Soc., Vol. 17, No. 2, p. 597 (1985). Abstract. –
See Abstr. 010.065.

131.339 GMC's associated with H II regions in the inner Galaxy.
N. Z. Scoville, D. P. Clemens, D. B. Sanders,
P. M. Solomon, W. H. Waller.
Bull. Am. Astron. Soc., Vol. 17, No. 2, p. 606 (1985). Abstract. –
See Abstr. 010.065.

131.340 The largest molecular clouds in the galactic plane.
D. B. Sanders, N. Z. Scoville, D. P. Clemens,
P. M. Solomon.
Bull. Am. Astron. Soc., Vol. 17, No. 2, p. 606 (1985). Abstract. –
See Abstr. 010.065.

131.341 A study of CO emission from molecular clouds in the northern Milky Way.
T. M. Dame, I. A. Grenier, P. Thaddeus.
Bull. Am. Astron. Soc., Vol. 17, No. 2, p. 606 – 607 (1985). Abstract. – See Abstr. 010.065.

131.342 A CO survey of molecular clouds in Taurus, Auriga, and Perseus.
H. Ungerechts, P. Thaddeus.
Bull. Am. Astron. Soc., Vol. 17, No. 2, p. 607 (1985). Abstract. –
See Abstr. 010.065.

131.343 Viscous origin of the velocity dispersion of the giant molecular clouds.
C. J. Jog, J. P. Ostriker.
Bull. Am. Astron. Soc., Vol. 17, No. 2, p. 614 (1985). Abstract. –
See Abstr. 010.065.

131.344 The morphology and energetics of spatially extended molecular flows.
G. A. Wolf, C. J. Lada, J. Bally.
Bull. Am. Astron. Soc., Vol. 17, No. 2, p. 614 (1985). Abstract. –
See Abstr. 010.065.

131.345 The structure of the Taurus molecular cloud.
M. C. Heyer, F. P. Schloerb, R. L. Snell,
P. F. Goldsmith.
Bull. Am. Astron. Soc., Vol. 17, No. 2, p. 614 (1985). Abstract. –
See Abstr. 010.065.

131.346 CO mapping of the Orion molecular cloud.
F. P. Schloerb, R. L. Snell, P. F. Goldsmith.
Bull. Am. Astron. Soc., Vol. 17, No. 2, p. 614 (1985). Abstract. –
See Abstr. 010.065.

131.347 Extended molecular clouds around the CO outflow sources observed with the Nagoya 4–m telescope.
H. Ogawa, Y. Fukui, T. Takano.
Bull. Am. Astron. Soc., Vol. 17, No. 2, p. 614 – 615 (1985). Abstract. – See Abstr. 010.065.

131.348 Molecular line survey of Orion A from 215 to 247 GHz.
E. C. Sutton, G. A. Blake, C. R. Masson,
T. G. Phillips.
Bull. Am. Astron. Soc., Vol. 17, No. 2, p. 615 (1985). Abstract. –
See Abstr. 010.065.

131.349 Generic high velocity clouds do not exist.
G. l. Verschuur.
Bull. Am. Astron. Soc., Vol. 17, No. 2, p. 615 (1985). Abstract. –
See Abstr. 010.065.

131.350 Formaldehyde and ammonia absorption toward galactic and extragalactic sources.
A. G. Nash.
Bull. Am. Astron. Soc., Vol. 17, No. 2, p. 615 (1985). Abstract. –
See Abstr. 010.065.

131.351 Cosmic ray ionization in cold clouds.
S. Lepp, A. Dalgarno.
Bull. Am. Astron. Soc., Vol. 17, No. 2, p. 615 – 616 (1985). Abstract. – See Abstr. 010.065.

131.352 Detection of HCO$^+$ (J = 3–2) in the 2–pc ring surrounding Sgr A (West) and Sgr A*.
Å. Sandqvist, A. Wootten, R. B. Loren.
Bull. Am. Astron. Soc., Vol. 17, No. 2, p. 616 (1985). Abstract. –
See Abstr. 010.065.

131.353 Spatial–velocity maps, retrograde rotation, and binary protostellar clouds.
A. P. Boss.
Bull. Am. Astron. Soc., Vol. 17, No. 2, p. 616 (1985). Abstract. –
See Abstr. 010.065.

131.354 A reexamination of the kinematics of B163 and B163SW.
R. Arquilla.
Bull. Am. Astron. Soc., Vol. 17, No. 2, p. 616 (1985). Abstract. –
See Abstr. 010.065.

131.355   Gas jets associated with star formation.
W. J. Welch, S. N. Vogel, R. L. Plambeck,
M. C. H. Wright, J. H. Bieging.
Science, Vol. 228, No. 4706, p. 1389 – 1395 (1985).

Young stellar objects of both high and low luminosity emit energetic jets or winds of material that are often highly collimated and often bipolar. Near the stars, turbulent swept–up gas is observed in the emission of interstellar molecules such as carbon monoxide, and small, bright regions of water maser emission and the nebulous bright patches known as Herbig–Haro objects appear to be participating in the outflows. There are striking changes in chemical abundances associated with the attendant shocks. Probably every star goes through this phase, which may mark the end of its period of accretion.

131.356   Report of IAU Commission 34: Interstellar matter (*Matière interstellaire*).
M. Peimbert.
Trans. IAU, Vol. XIXA, p. 437 – 478 (1985). – See Abstr. 003.046.

131.357   Ultraviolet interstellar absorption toward stars in the Small Magellanic Cloud.
E. L. Fitzpatrick.
Diss. Abstr. Int., Sect. B, Vol. 45, No. 2, p. 584 (1984). Thesis, University of Wisconsin, 206 pp. (1984). Order No. DA8405422.

131.358   Ultraviolet studies of interstellar matter.
W. B. Landsman.
Diss. Abstr. Int., Sect. B, Vol. 45, No. 3, p. 901 (1984). Thesis, Johns Hopkins University, 142 pp. (1984). Order No. DA8414292.

131.359   The stellar distribution at $l = 245°$ $b = 0°$ in Puppis.
B. C. Reed.
Diss. Abstr. Int., Sect. B, Vol. 45, No. 5, p. 1501 – 1502 (1984). Thesis, University of Waterloo (1984).

131.360   Studies of galactic H I in 21–centimeter absorption.
S. R. Kulkarni.
Diss. Abstr. Int., Sect. B, Vol. 45, No. 9, p. 2956 (1985). Thesis, University of California, 232 pp. (1984). Order No. DA8427022.

131.361   The structure and angular momentum content of dark clouds.
R. Arquilla.
Diss. Abstr. Int., Sect. B, Vol. 45, No. 10, p. 3256 (1985). Thesis, University of Massachusetts, 335 pp. (1984). Order No. DA8500052.

131.362   Numerical study of a two–fluid hydrodynamical model of interstellar medium and population I stars.
W.–h. Chiang.
Diss. Abstr. Int., Sect. B, Vol. 45, No. 10, p. 3256 (1985). Thesis, Columbia University, 139 pp. (1984). Order No. DA8427366.

131.363   A review of line formation in molecular clouds.
J. E. Beckman.
Progress in stellar spectral line formation theory, p. 389 – 405 (1985). – See Abstr. 012.108.

A summary of the physical considerations governing line formation in molecular clouds with kinetic temperatures in the range $5K – 100K$, and hydrogen number densities in the range $10^2 – 10^6 cm^{-3}$ is presented. An exposition of the classical Sobolev formulation due to Goldreich and Kwan (1973) is followed by results of a simple microturbulent model with partial redistribution due to Deguchi and Kwan. The results illustrate that line formation theory in molecular clouds is still in a somewhat unsatisfactory state.

131.364   Structure and evolution of superbubbles in the stratified gas distribution.
K. Tomisaka.
Theoretical aspects on structure, activity, and evolution of galaxies: III, p. 25 – 32 (1985). – See Abstr. 012.110.

The structure and evolution of superbubbles occurring in the stratified gas disk are studied. When the density of interstellar matter at the midplace of the disk, n_0, is as 1.0 cm^{-3}, the superbubble becomes egg–shaped and its diameter reaches 500 pc, which is similar to the diameter of Cygnus superbubble. On the other hand, when n_0 is low as ~ 0.1 cm^{-3}, the shape becomes funnel–like and hot matter in the superbubble is pushed up about ~ 1 kpc.

131.365   Periodic star formation activity and color change of galaxies.
S. Ikeuchi, Y. Yoshii.
Theoretical aspects on structure, activity, and evolution of galaxies: III, p. 41 – 45 (1985). – See Abstr. 012.110.

Several proposals of periodic star formation activity are critically reviewed and a simple nonlinear model is presented. The behavior in the two–color diagram is examined. A possible star formation episode of dwarf galaxies is presented.

131.366   Behavior of interstellar gas clouds in galaxy encounters.
M. Noguchi, S. Ishibashi.
Theoretical aspects on structure, activity, and evolution of galaxies: III, p. 79 – 88 (1985). – See Abstr. 012.110.

The authors investigate the behavior of the system of interstellar gas clouds in a close encounter between two galaxies. Star formation process is included in the numerical code explicitly. Spiral structures are formed in the distribution of the clouds by the passage of the perturbing galaxy. These structures do not, however, last long and turn into a ring in a few dynamical times as a result of the cloud–cloud collisions. A burst of star formation is induced as the spiral structures develop after the perigalactic passage of the perturbing galaxy.

131.367   Determination of the physical parameters of NGC 2023 by its NH_3 rotation–inversion superfine spectra.
J. Xing, L.–p. Xu.
Acta Astron. Sin., Vol. 26, No. 2, p. 123 – 134 (1985). In Chinese.

131.368   On the problem of the shape of expanding supershells of neutral hydrogen.
S. A. Silich.
Astrofizika, Tom 22, Vyp. 3, p. 563 – 570 (1985). In Russian. English translation in Astrophysics, Vol. 22, No. 3.

The dependence of the shape of supershells on its generating mechanisms is discussed. It is shown that in the case of a supernova cascade propagating along the galactic disk all the supershells having radii greater than a critical value Z_c must be stretched along the galactic plane. The origin of spherical shells or shells stretching along normal to the galactic plane direction with sizes exceeding Z_c must be connected with another mechanism.

131.369   Planetoidal hypothesis of CP (*chemically peculiar*) F, A, and B star formation: possibilities and prospects.
E. M. Drobyshevski (*Eh. M. Drobyshevskij*).
A. F. Ioffe Physical–Technical Institute, Academy of Sciences of the USSR, Leningrad, Prepr., No. 942, 27 pp. (1985).

A possibility is analyzed of explaining the chemical anomalies of chemically peculiar (CP) F–A–B stars basing on the assumption of the formation of a large number of moonlike planetoids both in the course of separation of the components and in late stages of close binary evolution.

131.370   The relevance of Quasat to the star formation problem.
D. Downes, D. T. Emerson.
ESA Spec. Publ., ESA SP–213, p. 165 – 170 (1984). – See Abstr. 012.124.

The role of space VLBI observations (Quasat) of masers and molecular flows for progress in our understanding of the star formation process is indicated.

131.371 Galactic OH and H_2O masers.
R. S. Booth.
ESA Spec. Publ., ESA SP–213, p. 171 – 178 (1984). – See Abstr. 012.124.

The general properties of galactic OH and H_2O masers are discussed. The two main classes of masers, those in regions of star formation and those associated with evolved stars, are described. For both types it is clear that resolution and mapping capabilities of the Quasat mission can be used to good effect. Finally the problem of interstellar scattering is discussed.

131.372 H_2O maser outburst.
L. I. Matveyenko (*L. I. Matveenko*).
ESA Spec. Publ., ESA SP–213, p. 179 – 180 (1984). – See Abstr. 012.124.

H_2O maser outbursts have complex spatial structure. The brightness temperature of individual components is equal to $T_b \cong 10^{17} K$. An angular resolution of $\sim 50 \mu as$ may allow a detailed study of these components.

131.373 H_2O masers and distance measurements: the impact of Quasat.
M. J. Reid.
ESA Spec. Publ., ESA SP–213, p. 181 – 184 (1984). – See Abstr. 012.124.

Recently VLBI observations of the proper motions of H_2O masers have been used to determine distances to the sources. The increased angular resolution of Quasat could be useful to increase the acccuracy of the measurements and possibly to allow direct distance measurements to other galaxies.

Catalog of CO observations of galaxies.
See Abstr. 002.012.

Sky catalogue 2000.0. Volume 2: Double stars, variable stars and nonstellar objects.
See Abstr. 002.019.

Catalogue of dark nebulae and globules for galactic longitudes 240 to 360 degrees.
See Abstr. 002.068.

A galactic center molecular cloud atlas.
See Abstr. 002.106.

The cosmic history of the biogenic elements and compounds.
See Abstr. 003.050.

The beginnings of molecular radio astronomy.
See Abstr. 004.197.

Millimeter–Astronomie von Köln aus.
See Abstr. 013.025.

On the effects of irregularities of internal structure in determining the ultraviolet properties of interstellar grains.
See Abstr. 015.036.

Diatoms on Earth, comets, Europa and in interstellar space.
See Abstr. 015.037.

Adaptive mesh techniques for fronts in star formation.
See Abstr. 021.002.

Einstein A–coefficients for pure rotational transitions in the $H_2^{18}O$–molecule.
See Abstr. 022.005.

Contribution of large polycyclic aromatic molecules to the infrared emission of the interstellar medium.
See Abstr. 022.007.

Amorphous carbon grains: laboratory measurements in the 2000 Å – 40 μm range.
See Abstr. 022.008.

A potential extraterrestrial species: the SH^+ molecular ion.
See Abstr. 022.009.

A theoretical study of the rotation–vibration energy levels and dipole moment functions of CCN^+, CNC^+, and C_3.
See Abstr. 022.015.

The first experimental observation of electronic transitions in C_2^+ and C_2D^+.
See Abstr. 022.016.

Energetics of the protonation of CO: implications for the observation of HOC^+ in dense interstellar clouds.
See Abstr. 022.017.

Optical spectra of amorphous carbon grains.
See Abstr. 022.018.

Oscillator strengths for transitions in N I and the interstellar abundance of nitrogen.
See Abstr. 022.023.

Hot atoms in cosmic chemistry.
See Abstr. 022.047.

Organic chemistry by irradiation in space.
See Abstr. 022.048.

Prebiotic syntheses of purines and pyrimidines.
See Abstr. 022.050.

Polycyclic aromatic hydrocarbons and the unidentified infrared emission bands: auto exhaust along the Milky Way!
See Abstr. 022.053.

Laboratory infrared spectra of predicted condensates in carbon–rich stars.
See Abstr. 022.054.

Millimeter–wave spectrum of the CCO radical.
See Abstr. 022.064.

Links between astronomical observations of protostellar clouds and laboratory measurements of interplanetary dust: the 6.8 μm carbonate band.
See Abstr. 022.068.

Tabulated optical properties of graphite and silicate grains.
See Abstr. 022.086.

Photodissociation rates of OH, OD, and CN by the interstellar radiation field.
See Abstr. 022.089.

An update of and suggested increase in calculated radiative association rate coefficients.
See Abstr. 022.091.

Laboratory measurement of the $S(9)$ pure rotation frequency in H_2.
See Abstr. 022.092.

Direct measurement of the fundamental rotational transitions of the OH radical by laser sideband spectroscopy.
See Abstr. 022.093.

Millimeter wave spectrum of HCS.
See Abstr. 022.094.

Rotational excitation of HCO^+ by collisions with H_2.
See Abstr. 022.099.

Population ratios for the fine structure ground state of Si II applicable to the interstellar medium.
See Abstr. 022.100.

The case for interstellar micro–organisms.
See Abstr. 022.101.

The infrared and ultraviolet absorptions of micro–organisms and their relation to the Hoyle–Wickramasinghe hypothesis.
See Abstr. 022.102.

The ultraviolet absorbance of presumably interstellar bacteria and related matters.
See Abstr. 022.103.

The microwave and far–infrared spectra of the SiH radical.
See Abstr. 022.114.

Total synthesis of interstellar chemical compounds by high energy molecular beam bombardment on pure graphite.
See Abstr. 022.115.

CRESU study of the reaction $N^+ + H_2 \rightarrow NH^+ + H$ between 8 and 70K and interstellar chemistry implications.
See Abstr. 022.116.

The spectrum of magnesium hydride.
See Abstr. 022.119.

NLTE–Rechnungen zur Bildung interstellarer Moleküllinien in einem turbulenten Medium.
See Abstr. 022.120.

Laboratory detection of the C_3H radical.
See Abstr. 022.121.

Polarization and hyperfine splitting of water maser emission.
See Abstr. 022.122.

Simulation of cosmic dust spectra.
See Abstr. 022.124.

The rotational spectra of $HOCO^+$, $HOCS^+$, $HSCO^+$, and $HSCS^+$.
See Abstr. 022.128.

Reaction of O^+, CO^+, H^+, and CH_5^+ ions with atomic hydrogen.
See Abstr. 022.140.

The structure of linear SiCC: an ab initio SCF CI study including vibrational effects.
See Abstr. 022.144.

The millimetre wave spectrum of the $^{13}C^{14}N$ radical in its ground state.
See Abstr. 022.151.

Difference frequency laser spectroscopy of the v_1 band of $HOCO^+$.
See Abstr. 022.154.

Laboratory and astronomical identification of C_3H_2.
See Abstr. 022.174.

Laboratory and astronomical detection of the deuterated ethynyl radical CCD.
See Abstr. 022.175.

Point and interval estimation of the true unbiased degree of linear polarization in the presence of low signal–to–noise ratios.
See Abstr. 036.004.

Prospects for determining the cosmological helium–3 to helium–4 ratio via absorption–line studies of the local interstellar medium.
See Abstr. 036.099.

Photometric calibration and first results obtained with the UFT experiment on board the ASTRON satellite.
See Abstr. 051.066.

Solar abundances and the role of nucleogenesis in low–to–medium mass stars in the Galaxy.
See Abstr. 061.148.

Molecular hydrogen and thermal phases in astrophysics.
See Abstr. 061.149.

Quasi–two–dimensional cosmic jets.
See Abstr. 062.032.

Fragmentation of gas clouds.
See Abstr. 062.085.

Magnetic flux loss in gaseous layers.
See Abstr. 062.118.

Hyperfine selective collisional excitation of interstellar molecules.
See Abstr. 063.018.

On the interpretation of hyperfine–structure intensity anomalies in the NH_3 $(J, K) = (1, 1)$ inversion transition.
See Abstr. 063.019.

Circular polarization of interstellar absorption lines at radio frequencies.
See Abstr. 063.024.

Intensity of radiation reflected by a vertically inhomogeneous medium through multiple scattering.
See Abstr. 063.029.

A treatment of opacity suitable for media of low density and temperature. I.
See Abstr. 063.030.

Excitation of the hyperfine transitions of atomic hydrogen, deuterium, and ionized helium 3 by Lyman–alpha radiation.
See Abstr. 063.040.

Resonance lines in dusty gaseous nebulae.
See Abstr. 063.059.

Effects of source geometry on continuum radiation transport.
See Abstr. 063.064.

Moment method for multiple scattering of solar $L\alpha$ radiation in the nearby interstellar medium.
See Abstr. 063.070.

The Cerenkov microwave line emission of the hydroxyl radical.
See Abstr. 063.075.

One–dimensional scattering as a possible model for interpretation of emission of globules and dust nebulae. I.
See Abstr. 063.082.

Stellar winds and the interstellar medium.
See Abstr. 064.001.

Two–dimensional models of stellar wind bubbles. I. Numerical methods and their application to the investigation of outer shell instabilities.
See Abstr. 064.010.

Does nucleation theory apply to the formation of refractory circumstellar grains?
See Abstr. 064.026.

The interaction of the stellar wind with the local environment.
See Abstr. 064.031.

Formation of bipolar flow by the stellar wind embedded in a molecular disk.
See Abstr. 064.076.

Radio spectra of collimated, ionized stellar winds.
See Abstr. 064.095.

Collapse and formation of stars.
See Abstr. 065.007.

Do we understand rotating isothermal collapses yet?
See Abstr. 065.009.

Conditions for very massive protostellar accretion.
See Abstr. 065.067.

Anomalous magnetic field diffusion during star formation.
See Abstr. 065.073.

The shift of the solar 584–Å He I line, and evidence for interstellar wind from the interplanetary EUV background.
See Abstr. 071.026.

Extraterrestrial ice.
See Abstr. 091.015.

The collisional dynamics of particulate disks.
See Abstr. 091.069.

Dynamical influence of molecular cloud encounters on the Oort cloud of comets.
See Abstr. 102.045.

Dust properties determined from backscattering in the interplanetary and interstellar medium.
See Abstr. 102.054.

Comet formation in molecular clouds.
See Abstr. 102.064.

Laboratory infrared transmission spectra of individual interplanetary dust particles from 2.5 to 25 microns.
See Abstr. 105.206.

Constraints on dust formation in stellar outflows and destruction in the ISM from the meteoritic record.
See Abstr. 105.208.

Cometary impacts, molecular clouds, and the motion of the Sun perpendicular to the galactic plane.
See Abstr. 107.013.

Ultraviolet, radio and X–ray observations of hybrid stars.
See Abstr. 112.016.

OH emission from Mira variables, infrared stars and molecular clouds.
See Abstr. 112.026.

The optical properties of dust in the mid–IR silicate bands.
See Abstr. 112.030.

Spectrophotometric investigation of Be stars.
See Abstr. 112.033.

A multitransitional study of linear polarization in SiO maser emission.
See Abstr. 112.051.

Infrared spectra and interstellar reddening of anonymous type II OH/IR stars.
See Abstr. 112.065.

Observations of the SiC_2 radical toward IRC + 10216 at 1.27 centimeters.
See Abstr. 112.067.

MgS grain component in circumstellar shells.
See Abstr. 112.068.

The circumstellar envelope of S 106–IRS 4.
See Abstr. 112.069.

Cocoon stars in M17.
See Abstr. 112.071.

Condensation onto grains in the outflows from mass–losing red giants.
See Abstr. 112.091.

Mass loss from evolved stars. III. Mass loss rates for fifty stars from CO $J = 1$–0 observations.
See Abstr. 112.093.

Mass loss from evolved stars. IV. The dust–to–gas ratio in the envelopes of Mira variables and carbon stars.
See Abstr. 112.127.

Ultraviolet observations of Beta Pictoris: a possible protoplanetary system?
See Abstr. 112.136.

Stellar winds and their dynamical influence on the interstellar medium.
See Abstr. 112.148.

$uvby$ Hβ photometry of UV–bright stars.
See Abstr. 113.001.

RGU three–colour photometry of a starfield in anticentre–direction (A6).
See Abstr. 113.004.

The reddening (and distance?) of P Cygni.
See Abstr. 113.010.

Photoelectric photometry of the stars HD 29647 and HDE 283809 in the Taurus dark cloud.
See Abstr. 113.052.

BVR observations of polarimetric data of stars in reflection nebulae.
See Abstr. 116.020.

On the character of light polarization of red supergiants.
See Abstr. 116.033.

Excess polarization of starlight and neutral absorption.
See Abstr. 116.034.

Dependence of the polarization degree of the light of cool supergiants on their I–K colour.
See Abstr. 116.037.

Dust in the region of Herbig–Haro objects: the case of NGC 1999.
See Abstr. 121.004.

Optical spectroscopy of the outflow source in L1551.
See Abstr. 121.019.

Multifrequency radio images of L1551 IRS 5.
See Abstr. 121.022.

Mechanical energy release by protostars to molecular clouds.
See Abstr. 121.029.

Cold outflows, energetic winds, and enigmatic jets around young stellar objects.
See Abstr. 121.037.

Spektroskopische Beobachtungen von PV Cephei, der Zentralquelle eines bipolaren CO–Ausflusses.
See Abstr. 121.042.

The H^0 two–photon continuous spectrum of HH 43 and dark cloud extinction.
See Abstr. 121.050.

Molecular–line observations of energetic outflows around R CrA and AFGL 2591.
See Abstr. 121.056.

Photoelectric photometry of the red variable HD 157010.
See Abstr. 122.111.

Interstellar polarization from cepheid polarimetry.
See Abstr. 122.121.

A model for SNR evolution in a cloudy medium and its application to N49.
See Abstr. 125.007.

Non–spherical supernova remnants. II. The interaction of remnants with molecular clouds.
See Abstr. 125.018.

Interaction of supernovae with the interstellar medium.
See Abstr. 125.025.

Supernova remnants as probes of the interstellar medium.
See Abstr. 125.028.

Exploding stars, superbubbles, and the HEAO observations.
See Abstr. 125.029.

Supernova remnants: observational data. Evolution in the interstellar medium.
See Abstr. 125.050.

Pulsars and interstellar medium: multiple regression analysis of related parameters.
See Abstr. 126.022.

Changing parameters along the path to the Vela pulsar.
See Abstr. 126.041.

Pulsars and interstellar medium.
See Abstr. 126.056.

The kinematics of cool white dwarfs with metals and the local interstellar medium.
See Abstr. 126.081.

Far–infrared observations of two southern H II regions: RCW122 and G 351.6–1.3.
See Abstr. 132.007.

Optical measurements of the Trifid dust.
See Abstr. 132.018.

The form of the initial mass function in an H II complex in NGC 6946.
See Abstr. 132.019.

Compact H II regions: hydrogen recombination and OH maser lines.
See Abstr. 132.028.

Observations of radio recombination lines in the millimeter–wave spectrum of Orion A.
See Abstr. 132.034.

The distribution of H_2CO absorption towards W33.
See Abstr. 132.035.

Detection of molecular hydrogen in the Small Magellanic Cloud H II region N81.
See Abstr. 132.039.

Infrared sources and excitation of the W40 complex.
See Abstr. 132.041.

Infrared observations of a compact H II region in Monoceros (GGD 12–15).
See Abstr. 132.043.

Spectroscopic observations of the Horsehead Nebula.
See Abstr. 132.045.

Submillimeter wavelength observations of CS towards the core of M17.
See Abstr. 132.056.

H76α emission from the Sgr A "15 km/s cloud".
See Abstr. 132.057.

The correlation between excitation parameter and nebular luminosity for galactic H II regions, with application to regions of star formation in M83.
See Abstr. 132.067.

Near infrared sources in the molecular cloud G35.2–0.74.
See Abstr. 133.005.

New infrared absorption features due to solid phase molecules containing sulfur in W33A.
See Abstr. 133.006.

Infrared sources in the ϱ Ophiuchi dark cloud region.
See Abstr. 133.007.

Near–infrared sources in the molecular cloud G 35.2–0.74.
See Abstr. 133.008.

Preliminary observations of the luminosity function of IRAS AO sources in the Rho Ophiuchi cloud core.
See Abstr. 133.009.

CVF observations of several Orion–KL sites: the spectra of scattered light.
See Abstr. 133.016.

CCD observations of bipolar nebulae. II. LkHα 233.
See Abstr. 134.016.

Spectral characteristics of dust in carbon–rich objects.
See Abstr. 134.068.

On the interpretation of the ultra–soft X–ray background: the effects of an "embedded" cloud geometry.
See Abstr. 142.075.

COS–B gamma–ray sources and interstellar gas in the first galactic quadrant.
See Abstr. 143.040.

Gamma rays from the Orion molecular clouds.
See Abstr. 143.061.

Search for gamma ray lines in molecular clouds.
See Abstr. 143.070.

Interstellar absorption of very high energy gamma rays.
See Abstr. 143.093.

Cosmic rays and the Parker instability.
See Abstr. 144.002.

Cosmic ray propagation in the local superbubble.
See Abstr. 144.023.

Acceleration of cosmic rays in the Loop I "supernova remnant"?
See Abstr. 144.031.

Particle acceleration in giant molecular clouds.
See Abstr. 144.041.

Coupled hydromagnetic wave excitation and cosmic ray acceleration at interstellar shocks.
See Abstr. 144.137.

Can cosmic rays be accelerated in collapsing molecular clouds?
See Abstr. 144.149.

Magnetohydrodynamic turbulence in the cosmic ray confinement regions of the Galaxy.
See Abstr. 144.354.

The propagation of ultraheavy cosmic rays.
See Abstr. 144.462.

Stochastic star formation and spiral structure.
See Abstr. 151.029.

Spiral structure in galaxies: large–scale stochastic self–organization of interstellar matter and young stars.
See Abstr. 151.030.

Star formation in a density–wave–dominated, cloudy interstellar medium.
See Abstr. 151.031.

Spiral tracers and prestellar incubation periods in a cloudy interstellar medium.
See Abstr. 151.032.

Approximate rotation curve solutions for the evolution of a viscous protogalactic disk.
See Abstr. 151.067.

Die Verteilung von Blasen in Galaxien.
See Abstr. 151.078.

Stern–Gas–Reibung in Galaxienkernen. Eine numerische Methode zur Berechnung gekoppelter Gas– und Stellardynamik.
See Abstr. 151.079.

New stochastic star formation models of disk galaxy evolution.
See Abstr. 151.085.

Gravitational instabilities and the formation of molecular cloud complexes in spiral arms.
See Abstr. 151.088.

Dynamics of open star clusters.
See Abstr. 151.129.

The dynamical evolution of young open clusters.
See Abstr. 151.130.

Are cloud–cloud collisions necessary for global spiral structure?
See Abstr. 151.147.

Hydrodynamic theory of galactic shocks in a clumpy ISM.
See Abstr. 151.148.

Dissipative structures in galaxies – effects of advection.
See Abstr. 151.166.

N–body simulation of giant molecular clouds in a galaxy.
See Abstr. 151.167.

Time–dependent star formation in OB associations.
See Abstr. 152.003.

UBV three–colour photometric parameters of four galactic clusters near Cassiopeia.
See Abstr. 153.004.

The young open cluster Stock 16: an example of star formation in an elephant trunk?
See Abstr. 153.027.

Active star formation in NGC 2264.
See Abstr. 153.029.

Molekülwolken und die Lebensdauer offener Sternhaufen.
See Abstr. 153.032.

Far–infrared observations of young clusters embedded in the R Coronae Australis and Rho Ophiuchi dark clouds.
See Abstr. 153.042.

The star–formation history of very young clusters.
See Abstr. 153.043.

Galactic absorption lines measured from the IUE low dispersion spectra of active galaxies.
See Abstr. 155.001.

Survey of galactic H I emission at $|b| \leqslant 20°$.
See Abstr. 155.024.

The vertical distribution of galactic H I: the Arecibo–Green Bank survey.
See Abstr. 155.025.

H I at the outer edge of the Galaxy and its implications for galactic rotation.
See Abstr. 155.027.

CO survey of the southern Milky Way.
See Abstr. 155.029.

Distribution of CO in the southern Milky Way and large–scale structure in the Galaxy.
See Abstr. 155.030.

A CO (2–1) survey of the southern Milky Way.
See Abstr. 155.031.

The carbon monoxide distribution in the inner Galaxy.
See Abstr. 155.032.

Outer–Galaxy molecular clouds.
See Abstr. 155.033.

CH in the Galaxy.
See Abstr. 155.034.

High–energy gamma rays and the large–scale distribution of gas and cosmic rays.
See Abstr. 155.035.

High–energy galactic phenomena and the interstellar medium.
See Abstr. 155.037.

Giant clouds and star–forming regions as spiral–arm tracers.
See Abstr. 155.047.

The largest molecular complexes in the first galactic quadrant.
See Abstr. 155.048.

The global properties of the Galaxy. III. Maps of the ^{12}CO emission in the first quadrant of the Galaxy.
See Abstr. 155.058.

A UV survey of the galactic plane.
See Abstr. 155.059.

Parametrization of the flow of halo high–velocity clouds.
See Abstr. 155.064.

A model of the radio continuum filaments in the galactic center.
See Abstr. 155.065.

Wide–field kinematic Hα observations of the Milky Way with a scanning and imaging Fabry–Perot: preliminary results.
See Abstr. 155.072.

The local system of early type stars. Spatial extent and kinematics.
See Abstr. 155.074.

The application of three–colour photometry on galactic starfields with strong interstellar extinction (Aquila II).
See Abstr. 155.075.

Galactic chemical evolution and nucleocosmochronology: analytic quadratic models.
See Abstr. 155.076.

Interactions between the continuum sources in the galactic center and their immediate molecular environment.
See Abstr. 155.077.

Giant molecular clouds in the Galaxy. II. Characteristics of discrete features.
See Abstr. 155.079.

Observed properties of the "central parsec" of the Galaxy as manifestation of a permanent star formation process.
See Abstr. 155.081.

Annihilation radiation from the galactic center: positrons in dust?
See Abstr. 155.083.

The fraction of high velocity dispersion H I in the Galaxy.
See Abstr. 155.084.

The 157 micron [C II] luminosity of the Galaxy. II. The presence of knotlike features in the [C II] emission.
See Abstr. 155.085.

Cosmic gamma rays and the mass of molecular hydrogen in the Galaxy.
See Abstr. 155.086.

Molecular clouds in the Carina arm.
See Abstr. 155.089.

How much H_2 in the Milky Way?
See Abstr. 155.090.

Cosmic γ rays and the mass of gas in the Galaxy.
See Abstr. 155.091.

On the structure and mass of the galactic halo.
See Abstr. 155.104.

The Massachusetts–Stony Brook galactic plane CO survey: disk and spiral arm molecular cloud populations.
See Abstr. 155.112.

Detection of galactic [26]Al gamma radiation by the SMM spectrometer.
See Abstr. 155.115.

A possible galactic positron annihilation medium: neutral hydrogen.
See Abstr. 155.116.

The gaseous galactic halo.
See Abstr. 155.117.

The chemical evolution of the Milky Way.
See Abstr. 155.118.

Morphology of galactic dust and gas based on IRAS and other data.
See Abstr. 155.119.

Ultraviolet absorption by highly ionized halo gas near the galactic center.
See Abstr. 155.124.

The Massachusetts–Stony Brook CO survey of the galactic plane: disk and spiral arm molecular cloud populations.
See Abstr. 155.128.

Observations of the CO J = 7→6 line at 370 μm toward the galactic center.
See Abstr. 155.131.

Further observations of H and He in the very center of the Galaxy.
See Abstr. 155.139.

Mapping molecular hydrogen emission from the galactic center.
See Abstr. 155.140.

Molecular clouds in the Carina arm.
See Abstr. 155.141.

The thickness of the galactic H I layer.
See Abstr. 155.142.

Mass distribution in the galactic centre.
See Abstr. 155.143.

A constraint on the past star formation rate from the kinematics of nearby stars.
See Abstr. 155.144.

The interstellar medium in the outer Galaxy.
See Abstr. 155.147.

Giant molecular clouds in the Galaxy: the Massachusetts – Stony Brook CO galactic plane survey.
See Abstr. 155.181.

The large–scale distribution of molecules in the northern galactic disk.
See Abstr. 155.182.

A carbon monoxide survey of the galactic center.
See Abstr. 155.183.

The scale height of molecular clouds in the Carina arm.
See Abstr. 155.184.

A deep CO survey of molecular clouds in the fourth galactic quadrant.
See Abstr. 155.185.

The Bell Laboratories CO survey.
See Abstr. 155.186.

Spiral arm and interarm regions in the Bell Laboratories CO survey.
See Abstr. 155.187.

The Massachusetts–Stony Brook CO survey of the inner galaxy.
See Abstr. 155.188.

"Windows" in the galactic plane.
See Abstr. 155.189.

Massachusetts–Stony Brook CO galactic plane survey: large scale distribution of molecular clouds in the Galaxy.
See Abstr. 155.190.

CO(1–0) survey at high galactic latitudes.
See Abstr. 155.191.

Highly ionized gas associated with the Small Magellanic Cloud.
See Abstr. 156.002.

Interstellar dust in the Large Magellanic Cloud.
See Abstr. 156.010.

The diffuse interstellar medium in the Magellanic Clouds.
See Abstr. 156.012.

Age and processes of star formation in very young LMC associations.
See Abstr. 156.017.

Regional ultraviolet extinction variations in the Large Magellanic Cloud.
See Abstr. 156.019.

Interstellar dust in the Large Magellanic Cloud.
See Abstr. 156.032.

Molecular clouds in external galaxies.
See Abstr. 157.014.

Comments on the distribution of molecules in spiral galaxies.
See Abstr. 157.015.

Distribution and motions of CO in M51.
See Abstr. 157.016.

CO (2–1) observations of Maffei 2.
See Abstr. 157.017.

Spiral structure and kinematics of H I and H II in external galaxies.
See Abstr. 157.019.

Distribution and motion of CO in M31.
See Abstr. 157.025.

H I observations of galaxies in nearby groups.
See Abstr. 157.034.

The statistical distribution of the neutral–hydrogen content of elliptical galaxies.
See Abstr. 157.037.

Reddening of globular clusters in M31.
See Abstr. 157.057.

A very bright water vapour maser source in the galaxy NGC 3079.
See Abstr. 157.093.

Star formation in early–type galaxies.
See Abstr. 157.095.

Powerful extragalactic masers.
See Abstr. 157.101.

Star–formation bursts in the central regions of Sc galaxies.
See Abstr. 157.103.

CO abundances and star formation in the three irregular galaxies NGC 4449, NGC 4214, and NGC 3738.
See Abstr. 157.112.

Far–infrared spectroscopy of galaxies: the 158 micron C^+ line and the energy balance of molecular clouds.
See Abstr. 157.122.

Carbon monoxide isotope ratios in galactic centers and disks.
See Abstr. 157.149.

CO emission from IRAS galaxies.
See Abstr. 157.162.

Ultraviolet observations of galaxies with recent star formation.
See Abstr. 157.167.

Star–formation rates of interacting galaxies.
See Abstr. 157.168.

Star formation in grand design and flocculent spiral galaxies.
See Abstr. 157.169.

Multicolor near–infrared mapping of star–forming regions in interacting galaxies.
See Abstr. 157.181.

Peculiar motions and star formation in the interacting galaxy complex Mk 171 = NGC 3690 + IC 694.
See Abstr. 157.200.

Hydroxyl absorption in NGC 520, NGC 2623, and NGC 6240.
See Abstr. 157.207.

Giant molecular clouds in M31.
See Abstr. 157.234.

Star formation rates in disk galaxies: comparison of far–infrared and H–alpha emission.
See Abstr. 157.239.

The detailed gas distribution of NGC 6946.
See Abstr. 157.240.

The correlation of CO and Hα in M51.
See Abstr. 157.241.

A detailed study of dust in NGC 205 and NGC 185.
See Abstr. 157.242.

A comparative study of dust and young stars in three small galaxies.
See Abstr. 157.245.

Problems of galactic and extragalactic CO photometry.
See Abstr. 157.258.

A study of star formation in Shapley–Ames spiral galaxies based on IRAS far–infrared observations.
See Abstr. 157.259.

Recent star formation in interacting galaxies. II. Super starbursts in merging galaxies.
See Abstr. 158.090.

Shock formation of the broad emission–line regions in QSOs and active galactic nuclei.
See Abstr. 158.092.

Low frequency variability and interstellar focusing.
See Abstr. 158.107.

Star formation in the cooling flows of M87 and NGC 1275.
See Abstr. 158.184.

A model for starbursts in galaxies.
See Abstr. 158.203.

Water masers as starburst tracers in external galaxies.
See Abstr. 158.214.

OH maser in Markarian 273.
See Abstr. 158.262.

Star formation in galaxies: the active dwarfs.
See Abstr. 158.294.

H I deficiency and photometric evolution of spiral galaxies in the Virgo cluster.
See Abstr. 160.114.

A near–infrared survey of bright Virgo spiral galaxies.
See Abstr. 160.120.

Molecular clouds in spiral galaxies in the Virgo cluster.
See Abstr. 160.122.

Molecular clouds in H I–deficient Virgo spirals.
See Abstr. 160.123.

Search for young stars in Virgo dwarf ellipticals.
See Abstr. 160.129.

132 H II Regions, Emission Nebulae

132.001   Comments on the origins and variations of carbon and nitrogen in the ISM of galaxies as evident from IUE spectroscopy of H II regions.
R. J. Dufour.
NASA Conf. Publ., NASA CP–2349, p. 107 – 110 (1984). – See Abstr. 012.001.

During the last several years IUE observations of UV spectra of galactic and extragalactic H II regions have provided the first good quantitative measurements of gaseous–phase C abundances in the ISM of several galaxies over a broad range of "metallicity". This paper illustrates the observed relative variations of C, N, and O based on IUE and ground–based observations of H II regions and discusses the trends evident using current concepts regarding stellar nucleosynthesis of CNO element group and simple models of galactic chemical evolution.

132.002   IUE spectroscopy of extragalactic H II regions.
R. J. Dufour, F. H. Schiffer III, G. A. Shields.
NASA Conf. Publ., NASA CP–2349, p. 111 – 114 (1984). – See Abstr. 012.001.

This report summarizes progress made in studying the UV spectra of several prominent extragalactic H II regions with the IUE during the last two years. High dispersion spectra have been obtained of N81 and N66A in the SMC, the 30 Doradus Nebula in the LMC, NGC 2363 in the SBm galaxy NGC 2366, NGC 5471 in M101, and NGC 604 in M33, that are superior in quality to low dispersion spectra in the statistical accuracy and resolution of the nebular emission lines of ions such as C III], Si III], N III], and O III]. These data are combined with ground–based spectroscopy to diagnose physical conditions and abundances in over twenty extragalactic H II regions derived from recent IUE low dispersion observations and/or archival spectra are also presented.

132.003   Measurement of electron density in the H II regions DR21 and W3 from Stark broadening of radio recombination lines.
G. T. Smirnov.
Pis'ma Astron. Zh., Tom 11, No. 1, p. 17 – 26 (1985). In Russian. English translation in Sov. Astron. Lett., Vol. 11.

21 recombination line profiles ($56 \leqslant n \leqslant 156$, $\Delta n \geqslant 1$) were measured for the H II regions DR21 and W3. Stark broadening of the high n line profiles was extracted. Observed properties of the profiles were explained by two–component models with essentially different densities of the components. The electron densities of compact features of DR21 and shell–like structures of W3A and W3B were estimated.

132.004   On the nature of ring nebulae around WR and Of stars: prospects of X–ray observations.
N. G. Bochkarev, T. A. Lozinskaya.
Astron. Zh., Tom 62, Vyp. 1, p. 103 – 111 (1985). In Russian. English translation in Sov. Astron., Vol. 29, No. 1.

The expected X–ray emission of wind–blown bubbles is evaluated in terms of the model of Weaver et al. (1977). The results are presented for 8 ring nebulae around WR and Of stars. Their X–ray observations can be used to test whether the layer of hot wind exists inside a bubble or the hot wind dissipates quickly through the inhomogeneous shell. The X–ray observations of the nebulae NGC 6888 and Sh 119 are shown to be the most promising ones. Their expected spectra in the energy range 0.1 – 4 keV are presented.

132.005   Studies of small galactic nebulae.
K. C. Turner, Y. Terzian.
Astron. J., Vol. 90, No. 1, p. 59 – 64 (1985). With plates 5 – 7.

High–resolution VLA radio observations are reported for several small H II regions, including the cluster emission nebulae S258, S255, S257, and S256. The region between S255 and S257 at $\lambda 2$ cm reveals significant fine structure close to OH, H_2O, and infrared emitting sources. Observations of the cometary nebula PP59(S269) also reveal similar characteristics. These regions are considered to be active star–forming clouds. The cometary nebula PP40 was also observed and it is suggested that it might be a planetary nebula.

132.006   Structural similarities among nearby giant H II regions.
D. A. Hunter, J. S. Gallagher III.
Astron. J., Vol. 90, No. 1, p. 80 – 87 (1985). With plates 9 – 13.

The authors discuss the similarities in morphology and stellar content of a number of giant H II regions in several nearby late–type spiral and irregular galaxies. These are suggestive of parallel evolutionary developments in very large star–forming complexes which are located in gas–rich galaxies. The similarities between regions in galaxies with different global galactic properties, e.g., the presence of spiral arms, implies that star formation is primarily a local process which is independent of the mechanism responsible for initiating the star–formation process. The characteristics of individual giant H II complexes are found to be most consistent with sustained formation of relatively normal OB stars over time periods of at least several million years.

132.007   Far–infrared observations of two southern H II regions: RCW122 and G 351.6–1.3.
B. McBreen, G. G. Fazio, L. Loughran, T. N. Rengarajan.
Astron. J., Vol. 90, No. 1, p. 88 – 91 (1985).

The authors present a 1′ resolution far–infrared (40 – 250 μm) map of two southern H II regions, RCW122 and G 351.6–1.3. Both sources have luminosities in excess of $10^6 L_\odot$, requiring the presence of OB star clusters. Combining the near–infrared, far–infrared, and CO emission properties for RCW122, the authors find that the dust grains responsible for the emission are like amorphous silicates and that the luminosity per unit mass of the associated cloud is $\sim 6\, L_\odot/M_\odot$. The source, G 351.6–1.3 has very similar properties.

132.008   Detection of a large, very faint emission nebula surrounding α Virginis.
R. J. Reynolds.
Astron. J., Vol. 90, No. 1, p. 92 – 94 (1985).

Twenty–nine Hα scans were obtained with the Wisconsin 15–cm Fabry–Perot spectrometer in a 32° diameter region surrounding the high galactic latitude B star α Vir. The scans reveal an H II region $14° \times 18°$ in extent with an emission measure of 8 cm^{-6}pc, a gas density of about 0.6 cm^{-3}, and a radial velocity of -6 ± 1 km s^{-1} with respect to the local standard of rest. The total hydrogen ionization rate within the H II region is estimated to be $1.9 \pm 0.6 \times 10^{46} s^{-1}$.

132.009   Observations of the radio source G 6.6–0.1 positionally coincident with the W28 SNR.
M. D. Andrews, J. P. Basart, R. C. Lamb.
Astron. J., Vol. 90, No. 2, p. 310 – 314 (1985). With plates 16 – 17.

This paper reports scaled–array continuum and H76α line observations of the radio source G 6.6–0.1 which is positioned at the center of the W28 SNR. The source exhibits a bright core surrounded by extended emission which appears organized into arcs and wisps. G 6.6–0.1 is interpreted as a compact H II region with a unique morphology.

132.010   Search for radio–continuum emission from galactic O type and Wolf–Rayet stars off the plane.
J. P. Vallée, A. F. J. Moffat.
Astron. J., Vol. 90, No. 2, p. 315 – 317 (1985).

A study was made at radio wavelengths to detect continuum emission from several O–type and Wolf–Rayet type stars and of any associated nebula, located off the galactic plane. Within a radius of 10 arcmin in each field of view, the authors found typically only one *unrelated* background radio source with a continuum flux density stronger than 10 mJy at λ20.5 cm.

132.011   On recombination lines of excited hydrogen: 1. High emission measure H II regions. 2. Radio galaxies and quasars.
Z. V. Dravskikh, A. F. Dravskikh.
Spets. astrofiz. obs. Akad. Nauk SSSR. Prepr., No. 131, 12 pp. (1984). Abstr. in Ref. Zh., 51. Astron., 5.51.581 (1985).

132.012   Optical polarization in the bipolar nebula NGC 6302.
D. J. King, S. M. Scarrott, J. V. Shirt.
Mon. Not. R. Astron. Soc., Vol. 213, No. 1, p. 11P – 14P (1985).

Optical linear polarization observations are presented for the peculiar bipolar nebula NGC 6302. The data show the presence of dust throughout the nebula; the source of illumination is in the bright central optical condensation and coincides in position with the compact thermal radio and infrared object. The illuminating source is presumably responsible for exciting the nebula. The level of interstellar polarization towards NGC 6302 suggests a maximum distance of 1.5 kpc.

132.013   Determination of physical parameters in H II regions. Electron densities.
J. Zamorano, M. Rego.
An. Fis., Ser. B, Vol. 80, No. 2, p. 139 – 147 (1984). In Spanish. Abstr. in Phys. Abstr., Vol. 88, No. 1248, Entry 9858 (1985).

132.014   Gas and star content and spatial distribution in the giant extragalactic H II region Tol 89.
F. Durret, J. Bergeron, A. Boksenberg.
Astron. Astrophys., Vol. 143, No. 2, p. 347 – 354 (1985).

Tol 89 is a giant H II region 2.5 kpc in diameter, located in one of the arms of the spiral galaxy NGC 5398. The authors have observed this object in UV and at optical wavelengths. The UV spectrum reveals P Cygni profiles. The optical spectrum is typically that of an H II region, with a very blue continuum and strong emission lines, among which is a broad emission band at λ4650 confirming the presence of WR stars. Star formation in Tol 89 has occurred in bursts as suggested by the UV and optical energy distribution and the large WR to O star ratio, roughly

equal to 0.35. The recent burst of star formation is a few 10^6yr old.

132.015   Bipolar ejection of matter from hot stars.
P. Pismis.
NASA Conf. Publ., NASA CP-2358, p. 91 – 94 (1985). – See Abstr. 012.023.

Some results from a program to study the internal velocities in H II regions are presented. The velocity data for three H II regions and one planetary nebula combined with morphological properties strongly suggest that the nebulae were formed essentially by matter ejected from the central star and that ejection has occurred preferentially from diametrically opposite regions on the star, that is, in a bi–polar fashion.

132.016   The Rosette nebula. II. Radio continuum and recombination line observations.
W. E. Celnik.
Astron. Astrophys., Vol. 144, No. 1, p. 171 – 178 (1985).

The electron temperature and the abundance of He^+ within the Rosette nebula are determined from radio recombination line measurements on three positions in the H II region using the 100 m telescope at Effelsberg. The ratio $N(He^+)/N(H^+)$ could be determined to 0.12 ± 0.03. The averaged NLTE electron temperature of 5800 ± 700K turned out to be only 1100K larger than the corresponding LTE temperature. The distribution of the spectral index across the nebula is obtained from continuum intensity maps at the frequencies 1410 and 4750 MHz, which have angular resolutions of 9.'2 and 2.'4, respectively. For the H II region a shell type density model is suggested. Mean electron densities of various shell models range from 5 to 15 cm^{-3}. The observed continuum flux density corresponds to the UV luminosity of the O type stars embedded in the nebula.

132.017   Structure and origin of velocity fluctuations in the H II region Sharpless 142.
J.–R. Roy, G. Joncas.
Astrophys. J., Vol. 288, No. 1, p. 142 – 147 (1985).

Close to 41,000 Hα radial velocities have been measured across most of the evolved H II region S 142 ($\theta \approx 25'$), allowing a systematic study of velocity fluctuations. Using grids of different mesh size to subdivide the H II region, the mean velocity dispersion, σ, is found to be dependent on mesh size L. A well–defined correlation in the form of a power law $\sigma \approx L^{0.30}$ is found. Because of the ambiguous interpretation of this result in terms of turbulence, the structure function, B, which tests for velocity correlation at all scales has also been calculated. A brief qualitative analysis is attempted in terms of a turbulent energy cascade in a supersonic and compressible fluid. Kelvin–Helmholtz instabilities are suggested for the generation and maintenance of nebular turbulence.

132.018   Optical measurements of the Trifid dust.
B. T. Lynds, B. J. Canzian, E. J. O'Neil Jr.
Astrophys. J., Vol. 288, No. 1, p. 164 – 169 (1985). With plates 3 – 4.

CCD frames of the Trifid nebula have been obtained with interference filters isolating the Hα and Hβ emission lines and the contiuum near Hβ. The Balmer fluxes are compared with 4.875 GHz data, and a model of the nebula is discussed. By comparing Mathis's models of reflection nebulae with measurements of the intensity of the scattered continuum light, the authors estimate that the 4693 Å optical depth of M20 is ∼1.5, and that the grains have a high albedo and a strongly forward–throwing phase function.

132.019   The form of the initial mass function in an H II complex in NGC 6946.
K. DeGioia–Eastwood.
Astrophys. J., Vol. 288, No. 1, p. 175 – 181 (1985).

Evidence is beginning to accumulate that the initial mass function (IMF) is not the same everywhere but suffers at least spatial variations. The author has constructed a simple model aimed at determining the slope of the IMF in a cluster of young stars. She presents narrow–band photometry of H II complex C in the Sc I galaxy NGC 6946 and compares the observations with the results of the model. It is found that a precise value for the IMF slope is indeterminable. The star formation rate and efficiency of massive star formation are calculated using a Salpeter IMF and compared to the average values over the entire galaxy.

132.020   G 34.3 + 0.2: a "cometary" H II region.
M. J. Reid, P. T. P. Ho.
Astrophys. J., Lett. Ed., Vol. 288, No. 1, p. L17 – L19 (1985).

Radio–frequency observations of the H II complex G 34.3 + 0.2 made at 1″ resolution with the VLA indicate a component whose appearance resembles that of a comet, containing an ultracompact H II "head" and a diffuse "tail". This "cometary" H II region points (from tail through head) in the direction of the supernova remnant W44 which lies at a projected separation of 40 pc. The morphology of the cometary source might be explained as owing to the interaction of a newly formed O type star with a large (40 pc radius) shell.

132.021   Models of H II regions: heavy element opacity, variation of temperature.
R. H. Rubin.
Astrophys. J., Suppl. Ser., Vol. 57, No. 2, p. 349 – 387 (1985).

A homogeneous grid of 280 H II region models is presented. This is used to demonstrate that heavy element opacity is important over a large region of parameter space, dramatically shifting ionization equilibria to lower states compared with previous models. It is demonstrated that metallicity is by far the most sensitive factor affecting $<T_e>$. This work corroborates the interpretation that the observed rise in T_e with increasing galactocentric distance is due to a corresponding decrease in the relative heavy element abundances.

132.022   OH masers associated with compact H II regions in the G45.1 + 0.1 complex.
E. E. Baart, R. J. Cohen.
Mon. Not. R. Astron. Soc., Vol. 213, No. 3, p. 641 – 655 (1985).

The mainline OH masers in the G45.1 + 0.1 complex have been mapped at high resolution using MERLIN. Relative positions of the masers have been measured to an accuracy of ~ 10 milliarcsec. The maser emission is found to come from two sources OH45.07 + 0.13 and OH45.12 + 0.13 separated by 3 arcmin, each of which is associated with a compact H II region. The masers occur near the edges of the H II regions. It is suggested that the masers are in expanding envelopes of gas and dust which are being driven outwards at a velocity of ~ 6 km s^{-1}. The magnetic field direction determined from Zeeman splitting shows a surprising reversal between the two sources. The association between type I OH masers and compact H II regions is briefly reviewed.

132.023   VLA continuum observations of H II regions in M8 and W31.
C. Woodward, H. L. Helfer, J. Pipher.
News Lett. Astron. Soc. N.Y., Vol. 2, No. 4, p. 33 (1983). Abstract. – See Abstr. 010.242.

132.024   An infrared study of the giant H II region NGC 3603.
P. Persi, M. Tapia, M. Roth, M. Ferrari–Toniolo.
Astron. Astrophys., Vol. 144, No. 2, p. 275 – 281 (1985).

New infrared observations of several sources in the complex giant H II region NGC 3603 are presented. Three sources, Irs 1, Irs 2, and Irs 9, embedded in the optically visible nebula, have been analyzed in detail by means of CVF spectra between 2 and 4 µm and a 2.2 µm map. HD 97950, responsible for the ionization of the giant H II region, is shown to have an excess emission longward of ~ 2 µm, which, if interpreted as originating in an ionized stellar wind, implies a mass–loss rate of $2.5 \times 10^{-5} M_{\odot} \mathrm{yr}^{-1}$, typical for a Wolf–Rayet star. Finally, the source Irs 16, located NE of the bright nebula, was classified from the strong CO and H$_2$O absorption bands as late M; it shows a long–term variability in the K–band of about 6 mag.

132.025   Discovery of a very bright compact object N88A in the Small Magellanic Cloud.
G. Testor, M. Pakull.
Astron. Astrophys., Vol. 145, No. 1, p. 170 – 178 (1985).

The authors present new imagery and spectrophotometric results for the N88 H II region in the Small Magellanic Cloud. These observations using CCD and IDS detectors allowed to detect a very complex morphology. In contrast to the SRC(J) plates this region appears to be composed of several hot stars embedded in faint H II regions of about 8″ diameter, except for the central object N88A, which is very bright $\log I(\mathrm{H}\beta) = -10.62$, highly excited [O III] $\lambda\lambda 5007 + 4959/\mathrm{H}\beta > 10.5$, and highly reddened $E(B-V) = 0.6$. It is about 150 times more intense than the brighter surrounding H II region in Hβ. Thus N88A is probably the brightest, most absorbed, and one of the densest H II regions discovered in the SMC up to now.

132.026   The spectra of Orion A and IRC + 10216 between 72.2 and 91.1 GHz.
L. E. B. Johansson, C. Andersson, J. Elldér, P. Friberg, Å. Hjalmarson, B. Höglund, W. M. Irvine, H. Olofsson, G. Rydbeck.
Astron. Astrophys., Suppl. Ser., Vol. 60, No. 1, p. 135 – 168 (1985).

The complete spectra of Orion A and IRC + 10216 in the range 72.2 to 91.1 GHz, obtained with the 20 m mm–wave telescope at the Onsala Space Observatory, are presented.

132.027   Fabry–Perot observations of the unusual emission–line nebula S216.
R. J. Reynolds.
Astrophys. J., Vol. 288, No. 2, p. 622 – 629 (1985).

High spectral resolution scans of S216 in Hα, [N II] $\lambda 6583$, [S II] $\lambda 6716$, [O III] $\lambda 5007$, and Hβ were obtained with the University of Wisconsin large–aperture Fabry–Perot spectrometer. These observations indicate a mean nebular expansion velocity < 4 km s^{-1} and a gas temperature of 9400 ± 1100K. It is shown that interpretations of S216 as supernova remnant, as typical planetary nebula or as an O or B star H II region probably can be ruled out. The available data suggest that S216 is an extremely old ($\gtrsim 300{,}000$ yr) planetary nebula with an unusually low–expansion velocity. This interpretation predicts that S216 has a distance $\lesssim 80$ pc and a radius $\lesssim 1.1$ pc.

132.028   Compact H II regions: hydrogen recombination and OH maser lines.
G. Garay, M. J. Reid, J. M. Moran.
Astrophys. J., Vol. 289, No. 2, p. 681 – 697 (1985).

Hydrogen recombination lines at 15 and 22 GHz and hydroxyl (OH) maser emission at 1.6 GHz have been mapped toward nine compact H II regions using the VLA. The authors find that all OH masers have radial velocities which are redshifted from the velocities of the H II regions as deduced from their hydrogen recombination lines. The majority of the maser spots appear projected onto the most compact H II region in a complex. These observations are inconsistent with models in which the masers form in an expanding, shocked shell, and they are explained most simply if the OH maser sources are part of a remnant envelope which is still collapsing toward the newly formed star.

132.029   H II regions in the Galaxy and Magellanic Clouds.
M. G. Edmunds.
Gas in the interstellar medium, p. 84 – 99 (1984). – See Abstr. 012.032.

132.030   The halo of the 30 Doradus nebula.
J. Meaburn.
Gas in the interstellar medium, p. 114 – 127 (1984). – See Abstr. 012.032.

132.031 A long filament in the faint galactic Hα background.
P. M. Ogden, R. J. Reynolds.
Astrophys. J., Vol. 290, No. 1, p. 238 – 243 (1985).

Additional observations have been made of a long, narrow emission feature discovered on a map of the faint, galactic Hα background. The feature is a continuous filament of Hα emission about $0°\!.8$ wide, which appears to extend at least 15°. It has an apparent emission measure of $2.6\ \text{cm}^{-6}\text{pc}$ and a radial velocity of $-65\ \text{km s}^{-1}$ with respect to the local standard of rest. Measurements have also been made of the intensities and line profiles of [S II] λ6716 and [N II] λ6583. The results are consistent with the intensities predicted by models of a moderate velocity $(50-90\ \text{km s}^{-1})$ shock propagating through a low-density $(\sim 0.3\ \text{cm}^{-3})$ ambient medium.

132.032 Ionization equilibrium in isolated H II regions.
D. C. Eder.
Astrophys. J., Vol. 290, No. 1, p. 244 – 250 (1985).

Theoretical ionization equilibrium models are combined with Copernicus ultraviolet observations to study a particular H II region absorption–line component in the spectra of λ Sco and υ Sco. The λ Sco observations imply that the region is characterized by a hydrogen density of approximately $0.2\ \text{cm}^{-3}$ and an extent along the line of sight of approximately 16 pc. Three choices for the relative placement of λ Sco and υ Sco are considered. Conclusions are that λ Sco is the primary source of ionization and that there is a separation of approximately 5 pc between the H II region and λ Sco.

132.033 [O III] and [N III] far–infrared line emission in dusty H II blisters.
P. A. Aannestad, R. J. Emery.
Astron. Astrophys., Vol. 145, No. 2, p. 347 – 352 (1985).

The ionization structure in H, He, O, and N is calculated for a dusty "blister" assuming an exponential density gradient and a model–atmosphere ionizing spectrum with an effective temperature of 50,000K as given by Kurucz (1979). Charge exchange reactions are taken into account as are recent dielectronic recombination rates for the O and N ions. The 52 μm and 88 μm line intensities of [O III] and the 57 μm line intensity of [N III] are calculated for viewing angles from 0° to 90°. The far–infrared line intensities are found to closely co–vary with the radio free–free emission, although they peak at a slightly closer distance to the ionizing source (for a viewing angle $\neq 0°$). The electron density as derived from the 52 μm/88 μm line ratio is found to be up to a factor of three higher than the rms density derived from a radio measurement.

132.034 Observations of radio recombination lines in the millimeter–wave spectrum of Orion A.
D. Hoang–Binh, P. Encrenaz, R. A. Linke.
Astron. Astrophys., Vol. 146, No. 1, p. L19 – L21 (1985).

The authors have observed the H 42α, 40α, 39α, 52β, 57γ, and 60γ lines in the direction of the molecular cloud in Orion A. From the alpha line data, the derived values of the electron temperature, assuming local thermodynamic equilibrium (LTE), are close to 10000K, significantly higher than values $\cong 8500$K, inferred from optical studies. The authors interpret this as an evidence for a significant deviation from LTE. This is confirmed by the analysis of the higher order lines, and is consistent with the recent results of Wilson and Pauls (1984). The importance of observations in the millimeter–wave spectrum is discussed.

132.035 The distribution of H_2CO absorption towards W33.
F. F. Gardner, J. B. Whiteoak, R. E. Otrupcek.
Proc. Astron. Soc. Aust., Vol. 5, No. 4, p. 557 – 560 (1984).

132.036 Spatial variations in the physical conditions in the giant extragalactic H II region NGC 5471.
E. D. Skillman.
Astrophys. J., Vol. 290, No. 2, p. 449 – 461 (1985).

The giant extragalactic H II region NGC 5471 in M101 has been observed with high spatial resolution at both optical and radio wavelengths. Variations are found in reddening, gas temperature, ionic abundances, and stellar populations. Evidence for a supernova remnant has been discovered. An upper limit on the presence of Wolf–Rayet stars in NGC 5471 implies that the winds of Wolf–Rayet stars are not solely responsible for the large velocity dispersions in the ionized gas.

132.037 Determination of nebular density and temperature from radio recombination lines.
N. Odegard.
Astrophys. J., Suppl. Ser., Vol. 57, No. 3, p. 571 – 585 (1985).

A new method is described for determining mean values of nebular electron temperature and local electron density from observations of two Hnα lines, together with radio continuum observations. If the distance to the nebula is known, the filling factor and mass of the ionized gas can also be obtained. The analysis is applied to well–studied nebulae, and the results are in good agreement with results obtained by other methods.

132.038 Photoionization model of the emission and abundance of the T Tauri nebula.
V. V. Golovatyj, B. S. Novosyadlyj.
Astrofizika, Tom 22, Vyp. 2, p. 357 – 368 (1985). In Russian. English translation in Astrophysics, Vol. 22, No. 2.

The small emission nebula connected with the T Tau variable has been investigated. Ionization of atoms H, He, O, N, Ne, S and the emission–line spectrum for models of the T Tau nebula were calculated. Determined abundances in the nebula: $\text{He/H} \cong 0.07$; $\text{N/H} = 1.4 \times 10^{-5}$; $\text{O/H} = 6.5 \times 10^{-5}$; $\text{Ne/H} = 4.2 \times 10^{-5}$; $\text{S/H} = 2.1 \times 10^{-5}$.

132.039 Detection of molecular hydrogen in the Small Magellanic Cloud H II region N81.
J. Koornneef, F. P. Israel.
Astrophys. J., Vol. 291, No. 1, p. 156 – 159 (1985).

Two micron near–infrared spectrophotometry of the compact H II region N81 in the Small Magellanic Cloud is presented. The resulting spectrum is similar to that of the bright galactic H II region G333.6–0.2. The authors report the detection of molecular hydrogen emission, consistent with the presence of a mild shock moving into a dense ambient medium, which is due to the expansion of the H II region into an associated molecular cloud. Although the nebular core of N81 is virtually dust–free, a 3.76 μm excess suggests the presence of hot dust behind the visible nebula.

132.040 Ionization correction factors for low–excitation gaseous nebulae.
J. S. Mathis.
Astrophys. J., Vol. 291, No. 1, p. 247 – 259 (1985).

A large set of model H II regions has been computed with several stellar atmospheres, as well as widely varying nebular conditions and abundances. The fractions of various elements (N, S, Cl, Ar, C, and Ne) which are present in the observable ions $(N^+, Cl^{+2}, Ar^{+2}, C^{+2}$ and Ne^{+2}, respectively) are plotted against observed ionic ratios, O^+/O and S^+/S^{+2}. The ratio of O^0/O^+ can be used to estimate the average fraction of H^+ at the edge of the nebula. The role of intranebular dust is found to be unimportant. The results of applying the results to various regions in the well–observed Orion Nebula are encouraging.

132.041 Infrared sources and excitation of the W40 complex.
J. Smith, A. Bentley, M. Castelaz, R. D. Gehrz, G. L. Grasdalen, J. A. Hackwell.
Astrophys. J., Vol. 291, No. 2, p. 571 – 580 (1985).

Infrared images and photometry are presented for the W40 molecular cloud and H II region complex. The images, sampled through a 5″ aperture and observed at effective wavelengths of 2.3, 3.6, 4.9, 10.0, and 19.5 μm, reveal a cluster of seven compact infrared sources in W40, all but one of which can be associated with optically identified stars. Being heavily obscured and centered on the H II region, the three brightest sources (IRS 1a, 2a, 3a) are reasonable candidates for exciting both the dust and the H II plasma. The circumstellar regions observed for the brightest five sources have little impact on the overall energetics of the W40 dust. Instead, the most important mode of excitation appears to involve the diffuse clouds of dust.

132.042 A 6.5–mm continuum map of Orion A.
K. Akabane, Y. Sofue, H. Hirabayashi, M. Inoue, N. Nakai, T. Handa.
Publ. Astron. Soc. Jpn., Vol. 37, No. 1, p. 123 – 127 (1985).

A 46 GHz (6.5–mm wavelength) continuum map of Ori A has been made with an angular resolution of 36″. All the features in a region of 12′ × 12′ were mapped. The continuum peak position coincides well with those of the Trapezium star θ^1Orionis. A sharp bar–like ridge is clearly seen in the south–east corner of the region. The central part (core) of the region has a quasi-rectangular edge which is quite similar to that of 5 GHz (6 cm) results by previous workers.

132.043 Infrared observations of a compact H II region in Monoceros (GGD 12–15).
G. Olofsson, J. Koornneef.
Astron. Astrophys., Vol. 146, No. 2, p. 337 – 340 (1985).

The properties of a compact H II region situated in an area currently referred to as GGD 12–15, have been investigated by means of a CCD image at 0.9 μm, CVF spectrophotometry in the range 1.6 – 2.3 μm and photometry in the $K, L, N,$ and Q bands (2.2 – 20 μm). It is found that the classical picture is applicable; a recently formed early–type star, deeply embedded in a dark cloud ($A_V \sim 40$), provides the ionization rate for the compact H II region and the luminosity needed to explain the dust emission. Well outside the ionized region the authors observe a relatively bright nebulosity at 2.2 μm, which is possibly caused by the same (unknown) mechanism as that recently observed in reflection nebulae (Sellgren et al., 1983).

132.044 Spectrophotometry and chemical composition of the 30 Doradus nebula.
J. S. Mathis, Y.–H. Chu, D. E. Peterson.
Astrophys. J., Vol. 292, No. 1, p. 155 – 163 (1985).

Spectra of the 30 Doradus nebula in the Large Magellanic Cloud in the range λλ3727 – 7140 were obtained at 22 locations in the nebula. It is argued on several grounds that the dust which reddens the nebula is not well mixed with the nebular gas but is rather in front of it. Electron temperatures from the [O III] lines were used to obtain ionic abundances. Total elemental abundances were obtained from nebular models. The abundance ratios are similar to those in the Orion Nebula and the Sun for Ne/O, S/O, Ar/O, and Cl/O, although O/H in 30 Doradus is about half that in the Orion Nebula and 0.3 of that in the Sun.

132.045 Spectroscopic observations of the Horsehead Nebula.
T. Neckel, M. Sarcander.
Astron. Astrophys., Vol. 147, No. 1, p. L1 – L2 (1985).

Cassegrain spectra of the ”Horsehead Jet” have been obtained in order to study its emission line spectrum. The authors have found no evidence for shock excitation of the ”jet”. The emission line ratios in the ”jet” and in the adjacent H II region IC 434 are identical as well as the radial velocities. It is concluded that the Horsehead Jaw is part of IC 434 and not a shock excited collimated flow.

132.046 VLA continuum observations of compact H II regions.
C. E. Woodward, H. L. Helfer, J. L. Pipher.
Astron. Astrophys., Vol. 147, No. 1, p. 84 – 88 (1985).

Continuum observations at 5 GHz of the galactic H II regions W48, G 25.4–0.2S, G 8.1+0.2, and G 89.1+1.3 at moderately high spatial resolution ($<6″$) have been obtained with the VLA interferometer. Maps of the sources are presented and physical parameters of the H II regions are derived. Tentative models for the G 25.4–0.2S H II region utilizing the VLA radio interferometric map and available infrared data are discussed.

132.047 MWC 349: a bipolar nebula with a very hot central star.
R. L. White, R. H. Becker.
Space Telesc. Sci. Inst., Prepr. Ser., No. 46, 20 pp. (1985). To appear in Astrophys. J.

132.048 The spectra of Orion A and IRC + 10216 between 72.2 and 91.1 GHz.
L. E. B. Johansson, C. Andersson, J. Ellder, P. Friberg, Å. Hjalmarson, B. Höglund, W. M. Irvine, H. Olofsson, G. Rydbeck.
Onsala Space Obs., Prepr., No. 85:2, 69 pp. (1985). To appear in Astron. Astrophys., Suppl. Ser.

132.049 Internal dust in H II–regions.
G. Münch.
Mitt. Astron. Ges., Nr. 63, p. 65 – 74 (1985). – See Abstr. 012.063.

A review is given of the observable effects that scattering dust internal to H II–regions may have on their characteristic visual spectrum. The intrinsic color of optically thin lines is discussed, with special reference to the Balmer decrement. The effects of scattering on the transmission of line radiation formed in the deep interior of H II–regions is considered. Recent observations of He I λλ10829–30 emission line profiles in M42 are discussed, together with their interpretation through radiative transfer calculations. Absorption line observations of He I λλ10829–30 on θ^1Ori C and θ^2Ori A are presented and discussed in relation to the corresponding emission lines.

132.050 Die Kinematik der H II–Regionen M8 und M42.
A. Hänel.
Mitt. Astron. Ges., Nr. 63, p. 139 – 140 (1985). – See Abstr. 012.063.

Detailed radial velocity fields of [O III], Hα, [N II] and [S II] have been measured for the H II regions M8 and M42/43.

132.051 H64α–Beobachtungen von M8.
A. Hänel, R. Güsten.
Mitt. Astron. Ges., Nr. 63, p. 140 – 141 (1985). – See Abstr. 012.063.

132.052 Zum Verhältnis der Infrarot– zu Radiokontinuumstrahlung in H II–Gebieten und Galaxien.
G. F. O. Schnur, T. Schmidt–Kaler.
Mitt. Astron. Ges., Nr. 63, p. 167 (1985). – See Abstr. 012.063.

132.053 Near infrared mapping of W51.
S. J. Little, J. A. Hackwell, R. D. Gehrz, G. L. Grasdalen, J. Locke.
Bull. Am. Astron. Soc., Vol. 16, No. 4, p. 914 – 915 (1984). Abstract. – See Abstr. 010.062.

132.054 Kinematic properties of three giant H II regions.
L. Danly, Y.–H. Chu, K. DeGioia–Eastwood.
Bull. Am. Astron. Soc., Vol. 16, No. 4, p. 915 (1984). Abstract. – See Abstr. 010.062.

132.055 Far infrared surveys of sources in the Cygnus–X region.
S. F. Odenwald, K. Shivanandan, G. G. Fazio, C. Maxson, M. Campbell, T. N. Rengarajan.
Bull. Am. Astron. Soc., Vol. 16, No. 4, p. 927 (1984). Abstract. – See Abstr. 010.062.

132.056 Submillimeter wavelength observations of CS towards the core of M17.
R. L. Snell, N. R. Erickson, P. F. Goldsmith, B. L. Ulich, C. J. Lada, J. H. Black, A. Schulz, R. N. Martin.
Bull. Am. Astron. Soc., Vol. 16, No. 4, p. 937 (1984). Abstract. – See Abstr. 010.062.

132.057 H76α emission from the Sgr A ”15 km/s cloud”.
J. T. Armstrong, P. T. P. Ho, J. M. Jackson, A. H. Barrett.
Bull. Am. Astron. Soc., Vol. 16, No. 4, p. 959 (1984). Abstract. – See Abstr. 010.062.

132.058 Spatially resolved infrared observations of the Red Rectangle.
J. C. Dainty, J. L. Pipher, M. G. Lacasse, S. T. Ridgway.
Astrophys. J., Vol. 293, No. 2, p. 530 – 536 (1985).

Infrared one–dimensional speckle interferometric observations centered on HD 44179 (the "Red Rectangle") reveal spatially resolved extended emission at K, L, and M, of $1/e$ diameter $1\rlap{.}''05$ N–S and $0\rlap{.}''4$ E–W, as well as a central source of $1/e$ diameter $0\rlap{.}''2$. Polarimetry of K with a beam size encompassing both sources was also obtained; the low resultant polarization suggests that scattering off grains associated with the bipolar flow is not a plausible source of the infrared emission. The spatially resolved extended emission is elongated in the direction of the bipolar flow and is interpreted as thermal dust emission.

132.059 Theoretical models for H II regions. I. Diagnostic diagrams.
I. N. Evans, M. A. Dopita.
Astrophys. J., Suppl. Ser., Vol. 58, No. 1, p. 125 – 142 (1985).

Using an extensive grid of theoretical photoionization models with conditions appropriate to observed H II regions, the authors have developed a comprehensive set of diagnostic diagrams employing ratios of prominent emission lines which may be used to determine the ionization parameter in the nebula and the mean ionization temperature of the exciting star(s), provided that adequate spectrophotometric data are available. It is also possible to estimate the element–averaged metallicity in the nebula and star(s) from these diagnostics.

132.060 Polarization of light in the peculiar nebula NGC 7538.
Eh. S. Eroshevich, D. I. Gorodetskij.
Tr. Astrofiz. Inst. Alma–Ata, Tom 42, p. 67 – 72 (1983). In Russian.

The literature data on the peculiar nebula NGC 7538 are reviewed. Polarimetric observation results are given.

132.061 Spectrophotometric investigation of the "Trifid" nebula (NGC 6514, M20).
Yu. I. Glushkov, Z. V. Karyagina.
Tr. Astrofiz. Inst. Alma–Ata, Tom 44, p. 43 – 48 (1984). In Russian.

132.062 Physical parameter determination in extragalactic H II regions. Electron temperature.
J. Zamorano, M. Rego.
An. Fis., Ser. A, Vol. 80, No. 3, p. 196 – 209 (1984). In Spanish. Abstr. in Phys. Abstr., Vol. 88, No. 1259, Entry 68680 (1985).

132.063 High–resolution observations of compact H II regions.
K. D. Aliakberov, A. B. Berlin, V. Ya. Gol'nev, N. M. Lipovka, M. G. Mingaliev, N. A. Nizhel'skij, E. E. Spangenberg, L. M. Sharipova.
Astron. Zh., Tom 62, Vyp. 3, p. 482 – 490 (1985). In Russian. English translation in Sov. Astron., Vol. 29, No. 3.

The results of observations of 98 sources carried out with the RATAN–600 radio telescope at the wavelength 7.6 cm with the resolution as high as $18''$ and the sensitivity 0.5 Jy are presented. For many sources a complex structure and a core–halo structure have been detected. Twenty–five per cent of the sources turned out to be compact or supercompact H II regions.

132.064 Radio observations of the H II region S 121 and its surroundings.
L. A. Higgs, J. P. Vallée.
Bull. Am. Astron. Soc., Vol. 17, No. 2, p. 595 (1985). Abstract. – See Abstr. 010.065.

132.065 CCD observations of NGC 7538.
B. T. Lynds, E. J. O'Neil Jr.
Bull. Am. Astron. Soc., Vol. 17, No. 2, p. 596 (1985). Abstract. – See Abstr. 010.065.

132.066 H II regions in the outer Galaxy.
M. Fich.
Diss. Abstr. Int., Sect. B, Vol. 45, No. 3, p. 900 (1984). Thesis, University of California, 85 pp. (1983). Order No. DA8413377.

132.067 The correlation between excitation parameter and nebular luminosity for galactic H II regions, with application to regions of star formation in M83.
K. S. Rumstay.
Diss. Abstr. Int., Sect. B, Vol. 45, No. 6, p. 1812 (1984). Thesis, Ohio State University, 170 pp. (1984). Order No. DA8419006.

132.068 Nuclear H II regions in galaxies with emission lines.
G. T. Petrov, I. M. Yankoulova (*I. M. Yankulova*), V. K. Golev.
Dokl. Bolg. Akad. Nauk, Tome 37, No. 4, p. 411 – 414 (1984).

Sky catalogue 2000.0. Volume 2: Double stars, variable stars and nonstellar objects.
See Abstr. 002.019.

Partition functions and dissociation constants for HeH$^+$.
See Abstr. 022.074.

Population ratios for the fine structure ground state of Si II applicable to the interstellar medium.
See Abstr. 022.100.

Photographie extrem schwacher H II–Regionen. Teil 2.
See Abstr. 034.019.

Photometric calibration and first results obtained with the UFT experiment on board the ASTRON satellite.
See Abstr. 051.066.

Excitation of the hyperfine transitions of atomic hydrogen, deuterium, and ionized helium 3 by Lyman–alpha radiation.
See Abstr. 063.040.

Line fluorescence in astrophysics.
See Abstr. 063.045.

Characteristics and interpretation of the photometric variability of Eta Carinae and its nebula.
See Abstr. 112.059.

Cocoon stars in M17.
See Abstr. 112.071.

The near–infrared spectrum of Eta Carinae.
See Abstr. 112.077.

R 136a and the central object in the giant H II region NGC 3603 resolved by holographic speckle interferometry.
See Abstr. 115.016.

The unexpected ultraviolet variability of Herbig–Haro object 1.
See Abstr. 121.039.

Optical emission–line properties of evolved galactic supernova remnants.
See Abstr. 125.056.

IUE absorption line studies of highly ionized interstellar gas.
See Abstr. 131.001.

Observations of selected sources in the 2 cm line of H$_2$CO.
See Abstr. 131.017.

CO ($J = 4 - 3$) spectroscopy in M17 SW and OMC1.
See Abstr. 131.023.

Chemical composition of interstellar material.
See Abstr. 131.047.

Structure and origin of the Cygnus superbubble.
See Abstr. 131.056.

Properties of dust in the Orion nebula.
See Abstr. 131.057.

The bright–rimmed molecular cloud around S140 IRS. I. CS ($J = 1 - 0$) observations.
See Abstr. 131.076.

A search for H_2 emission in bipolar nebulae and regions of interstellar shock.
See Abstr. 131.090.

VLA observations of ammonia and continuum in regions with high–velocity gaseous outflows.
See Abstr. 131.095.

An infrared study of the bipolar outflow region GGD 12–15.
See Abstr. 131.098.

The star forming regions in the Monoceros R2 molecular cloud.
See Abstr. 131.101.

The molecular core associated with DR 21.
See Abstr. 131.121.

Newly discovered sources of non–metastable ammonia.
See Abstr. 131.138.

Observations of 115 GHz CO emission towards the Carina nebula.
See Abstr. 131.140.

Detection of the 370 micron $^3P_2 - {}^3P_1$ fine–structure line of [C I].
See Abstr. 131.147.

On the evidence for methane in Orion KL: a search for the 4.6 gigahertz line.
See Abstr. 131.148.

Le nouveau visage du milieu interstellaire.
See Abstr. 131.155.

Photodissociation regions. I. Basic model.
See Abstr. 131.159.

Photodissociation regions. II. A model for the Orion photodissociation region.
See Abstr. 131.160.

Detection of the CO ($J = 7 \to 6$) rotational transition at $\lambda = 0.37$ millimeters toward Orion.
See Abstr. 131.163.

Formation of OB clusters: CO, NH_3, and H_2O observations of the distant H II region complex in S128.
See Abstr. 131.182.

Detection of interstellar rotationally excited CH.
See Abstr. 131.183.

The discovery of a new masering transition of interstellar methanol.
See Abstr. 131.184.

The W3 molecular cloud.
See Abstr. 131.192.

Spectroscopy of the 3 micron emission features.
See Abstr. 131.197.

H II regions and star formation – an overview of recent theoretical developments.
See Abstr. 131.202.

Die interstellare Extinktionskurve bei O– und B–Sternen im Rosettennebel.
See Abstr. 131.209.

Far–infrared and submillimeter astronomical spectroscopy.
See Abstr. 131.214.

The Orion B molecular jet.
See Abstr. 131.280.

High–velocity gas flows associated with H_2 emission regions: how are they related and what powers them?
See Abstr. 131.293.

Molecular clouds associated with compact H II regions. II. The rapidly rotating condensation associated with ON1.
See Abstr. 131.294.

VLA observations of the $9_2-10_1\, A^+$ methanol masers toward W3(OH).
See Abstr. 131.295.

Induced star formation in the molecular cloud associated with Sharpless 254 – 258.
See Abstr. 131.320.

Interstellar ammonia maser detected in star formation regions.
See Abstr. 131.334.

GMC's associated with H II regions in the inner Galaxy.
See Abstr. 131.339.

Report of IAU Commission 34: Interstellar matter (*Matière interstellaire*).
See Abstr. 131.356.

Near infrared sources in the molecular cloud G35.2–0.74.
See Abstr. 133.005.

GL 961–W: a pre–main–sequence object.
See Abstr. 133.011.

CCD observations of bipolar nebulae. II. LkHα 233.
See Abstr. 134.016.

Results of observations made at RATAN–600 with a high–sensitivity radiometer at 7.6 cm wavelength of the region of the Galaxy in the interval $\alpha_{1950.0} = 17^h30^m$ to 19^h30^m and $\delta_{1950.0} = -25°20'$ to $-22°20'$.
See Abstr. 141.019.

The compact radio source near G357.7–0.1.
See Abstr. 141.040.

The young open cluster Stock 16: an example of star formation in an elephant trunk?
See Abstr. 153.027.

The dense stellar cores of giant H II regions.
See Abstr. 153.046.

A CO (2–1) survey of the southern Milky Way.
See Abstr. 155.031.

A 10–GHz radio–continuum survey of the galactic–plane region at the Nobeyama Radio Observatory – a complex region at $l = 22° - 25°$.
See Abstr. 155.050.

Interactions between the continuum sources in the galactic center and their immediate molecular environment.
See Abstr. 155.077.

Mass distribution in the galactic centre.
See Abstr. 155.143.

The chemistry of galaxies. I. The nature of giant extragalactic H II regions.
See Abstr. 157.039.

Separations of H II regions in galaxies.
See Abstr. 157.045.

A grism search of Messier 33 for emission–line objects.
See Abstr. 157.078.

Radio detection of historical supernovae and H II regions in M83.
See Abstr. 157.208.

Warmers: the missing link between Starburst and Seyfert galaxies.
See Abstr. 158.085.

H II regions and the value of H_0.
See Abstr. 161.075.

133 Infrared Sources

133.001 Results from the IRAS survey.
M. Rowan–Robinson.
Observatory, Vol. 105, No. 1064, p. 1 – 2 (1985). Abstract. – See Abstr. 010.721.

133.002 *JHKL* observations of *IRAS* sources – II.
P. Whitelock.
Mon. Not. R. Astron. Soc., Vol. 213, No. 2, p. 51P – 59P (1985).
JHKL photometry is given for 13 *IRAS* galaxies which are shown to have near–infrared colours similar to those of H II region galaxies. Sixteen *IRAS* stars with *JHKL* colours like those of blackbodies are discussed. Fourteen of these are proposed as likely OH/IR star candidates on the basis of their *JHKL* and *IRAS* colours, their variability, and their near–infrared spectra. Many of these stars are at high galactic latitude and the implications of finding such objects considerably above the galactic plane are discussed.

133.003 Far–infrared survey of the galactic disc. II. The sources.
E. Caux, J. L. Puget, G. Serra, R. Gispert, C. Ryter.
Astron. Astrophys., Vol. 144, No. 1, p. 37 – 48 (1985).
The authors present an almost complete survey of the galactic disc in the far–infrared obtained with a balloon–borne instrument. This paper follows an analysis of part of the data dealing with the large scale galactic structure (Caux et al., 1984). Here the complete data set is presented in the form of brightness contour maps. A new procedure has been adapted to separate the sources from the "unresolved galactic component". The authors present a catalogue of the brightest extended sources, and comparisons with the radio continuum at 5 GHz are made. A more detailed discussion of some sources is given and some general properties of extended far–infrared sources are deduced. The properties of the unresolved component show some differences with the corresponding average properties of the sources.

133.004 A near infrared survey of the southern galactic plane.
N. Epchtein.
Messenger, No. 39, p. 6 – 8 (1985).

133.005 Near infrared sources in the molecular cloud G35.2–0.74.
M. Tapia, M. Roth, P. Persi, M. Ferrari–Toniolo.
Mon. Not. R. Astron. Soc., Vol. 213, No. 4, p. 833 – 840 (1985).
Near–infrared (1 – 4 μm) observations of the molecular cloud G35.2–0.74 reveal the presence of four infrared sources in the vicinity of two previously reported centres of recent star formation. The northern part of G35.2–0.74 contains three point sources which are interpreted as highly obscured stars. Irs 1 coincides with H_2O and OH maser sources and seems to be a very young early–type star reddened by $A_V \cong 54$. This star is probably responsible for providing the pumping energy to the masers, the ionizing energy to the H II region and for the outflow that leads to the observed bipolarity. The southern part of G35.2–0.74 shows a diffuse 2.2–μm source with a flux distribution in the short–wavelength region compatible with free–free emission and a large excess at $\lambda \gtrsim 3$ μm attributed to warm ($T \gtrsim 1000K$) dust mixed with the gas. These data are consistent with a fully developed H II region. The evolutionary aspects of the region are discussed in terms of the available observations.

133.006 New infrared absorption features due to solid phase molecules containing sulfur in W33A.
T. R. Geballe, F. Baas, J. M. Greenberg, W. Schutte.
Astron. Astrophys., Vol. 146, No. 1, p. L6 – L8 (1985).
Two new absorption features are reported in the highly obscured infrared source W33A. The stronger of these occurs at 4.9 μm; based on laboratory experiments it is due to an as yet unidentified, sulfur–containing molecule on the mantles of dust grains. The weaker feature, at 3.9 μm, is identified as solid H_2S.

133.007 Infrared sources in the ϱ Ophiuchi dark cloud region.
T. Lanz.
Bull. Inf. Cent. Données Stellaires, No. 28, p. 125 – 129 (1985).
Dans la région obscurcie proche de ϱ Ophiuchi (HD 147933/4), la découverte de nombreuses sources infrarouges a été annoncée par divers auteurs. Une liste unique, contenant 168 sources avec leurs coordonnées 1950, ainsi que leurs identifications dans des listes antérieures, est formée afin de faciliter la compilation, puis l'usage, des données astronomiques s'y rapportant.

133.008 Near–infrared sources in the molecular cloud G 35.2–0.74.
M. Roth, M. Tapia, P. Persi, M. Ferrari–Toniolo.
Bull. Am. Astron. Soc., Vol. 16, No. 4, p. 919 (1984). Abstract. – See Abstr. 010.062.

133.009 Preliminary observations of the luminosity function of IRAS AO sources in the Rho Ophiuchi cloud core.
B. A. Wilking, E. T. Young, C. J. Lada.
Bull. Am. Astron. Soc., Vol. 16, No. 4, p. 958 (1984). Abstract. – See Abstr. 010.062.

133.010 IRAS 0423 + 536P03: a remarkable galactic infrared source.
E. E. Becklin, J. N. Heasley, D. L. DePoy, G. J. Hill, C. G. Wynn–Williams.
Bull. Am. Astron. Soc., Vol. 16, No. 4, p. 976 (1984). Abstract. – See Abstr. 010.062.

133.011   GL 961–W: a pre–main–sequence object.
M. W. Castelaz, G. L. Grasdalen, J. A. Hackwell,
R. W. Capp, D. Thompson.
Astron. J., Vol. 90, No. 6, p. 1113 – 1116 (1985). With plate 85.

High–spatial–resolution surface–brightness images at J (1.25 µm), H (1.6 µm), K (2.2 µm), and L' (3.8 µm) of the infrared source GL 961 are presented. The images show that GL 961 is an infrared double system with the components separated by 5.8 arcsec at a position angle of 254°. Extinction and polarization measurements show that both objects are embedded in the same dark cloud at a distance between 1000 and 1600 pc. It is shown that the two sources are associated with each other and that GL 961–E is at the position of the known compact H II region. The western source, GL 961–W, is shown to lie above the main sequence, and it is suggested that it is a very young $(1 - 2 \times 10^4\,\mathrm{yr})$ pre–main–sequence object.

133.012   Spectral types of a few "unidentified" Equatorial Infra-
red Catalogue 1 (EIC–1) sources.
T. P. Prabhu, K. V. K. Iyengar.
Bull. Astron. Soc. India, Vol. 12, No. 4, p. 417 – 422 (1984).

Spectral types are assigned to 13 "unidentified" Equatorial Infrared Catalogue 1 (EIC–1) sources from a study of their spectra in the wavelength interval 5200 – 8700 Å. Their spectral types range from M4.5 to M6.5. The authors infer from a combination of the data on their $[J–K]$ colours and optical spectra that these sources are stars of luminosity class III.

133.013   Optical identification of a serendipitous infrared source.
E. R. Craine, R. B. Culver, T. A. Fleming.
Publ. Astron. Soc. Pac., Vol. 97, No. 590, p. 303 – 306 (1985).

A 30 by 45–arc–min area in the region of an anonymous infra-red source discovered near IRAS 0453 + 444P03 has been exam-ined on near infrared photographs in order to identify very red stars in the vicinity which may be optical counterparts to infrared sources in this area. A list of 15 stars in the region is presented; the list contains the optical counterpart of the anonymous bright, serendipitous infrared source.

133.014   Properties of silicate particles in the infrared region.
L. A. Pavlova.
Tr. Astrofiz. Inst. Alma–Ata, Tom 44, p. 20 – 29 (1984). In Rus-sian.

133.015   Structure and evolution of compact infrared sources.
I. G. Kolesnik, S. G. Kravchuk.
Astron. Zh., Tom 62, Vyp. 3, p. 518 – 528 (1985). In Russian. English translation in Sov. Astron., Vol. 29, No. 3.

Results of an analysis of protostellar envelopes' evolution dur-ing massive star formation are used to derive characteristics of infrared sources detected in regions of star formation. Observa-tional criteria for the determination of the structure of the sources are presented and an evolutionary classification of compact infra-red sources is proposed.

133.016   CVF observations of several Orion–KL sites: the spectra
of scattered light.
S. R. McCorkle, R. F. Knacke.
Bull. Am. Astron. Soc., Vol. 17, No. 2, p. 562 (1985). Abstract. – See Abstr. 010.065.

133.017   On removing the zodiacal emission from IRAS all–sky
data.
R. A. White, J. C. Good, L. J. Rickard, J. Wieland,
M. G. Hauser.
Bull. Am. Astron. Soc., Vol. 17, No. 2, p. 591 (1985). Abstract. – See Abstr. 010.065.

Polycyclic aromatic hydrocarbons and the unidentified infrared emission bands: auto exhaust along the Milky Way!
See Abstr. 022.053.

Space astronomy: technical comments – infrared, submillimetre and radio astronomy.
See Abstr. 051.077.

Results obtained from the IRAS satellite.
See Abstr. 051.093.

Detection of OH radicals from IRAS sources.
See Abstr. 112.024.

OH emission from Mira variables, infrared stars and molecular clouds.
See Abstr. 112.026.

Speckle interferometry of OH/IR stars at 5 and 10 microns.
See Abstr. 112.102.

Molecular ions in the circumstellar envelope of IRC + 10216.
See Abstr. 112.106.

Multifrequency radio images of L1551 IRS 5.
See Abstr. 121.022.

The morphology of NGC 6334 IRS V–1.
See Abstr. 131.013.

CO ($J = 4 - 3$) spectroscopy in M17 SW and OMC1.
See Abstr. 131.023.

An infrared study of the bipolar outflow region GGD 12–15.
See Abstr. 131.098.

Radio and optical observations of the jets from L1551 IRS 5.
See Abstr. 131.142.

Polarimetry of infrared sources in bipolar CO flows.
See Abstr. 131.157.

Infrared spectroscopy of carbon monoxide in GL 2591 and OMC–1:IRc2.
See Abstr. 131.162.

Ammonia observations of L 1551.
See Abstr. 131.173.

An investigation of the interstellar extinction. II. Towards the mid–infrared sources in the galactic centre.
See Abstr. 131.188.

Spectroscopy of the 3 micron emission features.
See Abstr. 131.197.

Evidence for a disk around Orion–IRc2.
See Abstr. 131.240.

Interstellar absorption features in the direction of the compact infrared source W33A.
See Abstr. 131.247.

Microscopic dust in the infrared sky.
See Abstr. 131.308.

Interferometer maps of the CS $J = 2–1$ emission around Orion IRc 2.
See Abstr. 131.317.

Dense molecular cores and IRAS point sources.
See Abstr. 131.318.

An infrared study of the giant H II region NGC 3603.
See Abstr. 132.024.

Infrared sources and excitation of the W40 complex.
See Abstr. 132.041.

Spatially resolved infrared observations of the Red Rectangle.
See Abstr. 132.058.

The compact radio source near G357.7–0.1.
See Abstr. 141.040.

IRAS observations of M22.
See Abstr. 154.036.

Unidentified *IRAS* sources: ultrahigh–luminosity galaxies.
See Abstr. 157.090.

Infrared Seyferts: a new population of active galaxies?
See Abstr. 158.103.

134 Planetary Nebulae

134.001   IUE observations of planetary nebulae in the Magellanic Clouds.
S. P. Maran, L. H. Aller, T. R. Gull, C. D. Keyes,
T. P. Stecher.
NASA Conf. Publ., NASA CP–2349, p. 155 – 158 (1984). – See Abstr. 012.001.

Four more PN in the Clouds have been observed with IUE (LMC P2, P9, P33; SMC N43). Reliable C abundances were computed for the first time for each PN, and other abundances were calculated from ionization models taking the IUE and earlier groundbased spectra into consideration. The authors also reobserved three PN (LMC P40; SMC N2, N5) that were studied in 1981, to search for time variations that might be anticipated from the relatively high luminosities and masses that were derived for the central stars of these young, compact objects.

134.002   1983 observations of the eclipsing nucleus of the bipolar planetary nebula NGC 2346.
W. A. Feibelman, L. H. Aller.
NASA Conf. Publ., NASA CP–2349, p. 159 – 162 (1984). – See Abstr. 012.001.

The authors present IUE data of the eclipsing nucleus of NGC 2346 obtained at 3 phases of the 1983 observing season. At $\phi = 0.02$ a highly asymmetric C IV emission line is seen, but the He II line has a gaussian profile with a slight suggestion of an inverse P Cygni character. The stellar continuum in the 1200 – 1950 Å region and high excitation emission lines argue for a hot subdwarf component in the binary nucleus.

134.003   Shapes of planetary nebulae.
F. D. Kahn, K. A. West.
Mon. Not. R. Astron. Soc., Vol. 212, No. 4, p. 837 – 850 (1985).

The formation of planetary nebulae can be discussed in terms of an interacting stellar winds model, in which the slow wind from the progenitor star is later overtaken by the fast wind from the central star. As a result of the interaction the inner region of the slow wind is compressed into a dense shell. The authors investigate the consequences for the model of a red–giant wind in which the mass–loss rate is enhanced towards the equator. The resulting shell is elongated in the direction of the poles.

134.004   Numerical models of planetary nebulae evolution.
V. A. Okorokov, B. M. Shustov, A. V. Tutukov,
H. W. Yorke.
Astron. Astrophys., Vol. 142, No. 2, p. 441 – 450 (1985).

Evolutionary sequences of planetary nebulae (PN) were calculated numerically beginning from the moment of PN envelope detachment to its dissipation. Gas and dust were treated as two separate hydrodynamical components and the equations for their motion were solved simultaneously with the frequency and angle dependent equations of radiation transfer in spherical symmetry. Radiation pressure on the dust component, the motion of the ionization front and the effect of a fast stellar wind were also included in the numerical calculations. Spectra of the outgoing radiation are presented for the infrared and radio frequency ranges. The general outline of PN evolution is discussed and comparisons with some observations are made.

134.005   Models for the planetary nebulae NGC 4361 and NGC 1535: influence of the stellar wind on the nebular ionization.
J. Adam, J. Köppen.
Astron. Astrophys., Vol. 142, No. 2, p. 461 – 475 (1985).

Model atmospheres fitted to line profiles of the central stars (Méndez et al., 1981) are used to construct nebular ionization models for the planetary nebulae NGC 4361 and 1535. For NGC 4361 the optical and IUE nebular emission lines are reproduced satisfactorily well. For NGC 1535 the model atmospheres only emit less than 10^{-3} of the number of He$^+$ ionizing photons which are necessary to explain the observed nebular He II emission. The IUE spectrum of NGC 1535 shows P Cyg profiles of N V and O V and, in contrast, NGC 4361 does not. The authors find that a stellar wind with electron temperature $T_e = 3 \times 10^5$K and mass–loss–rate $\dot{M} = 10^{-7} M_\odot/a$ explains the P Cyg profiles and provides the required number of He$^+$ ionizing photons by thermal emission. Consequently, the He II Zanstra temperature of NGC 1535 does not represent the effective temperature of the central star.

134.006   Fine–structure lines.
S. R. Pottasch.
Observatory, Vol. 105, No. 1064, p. 5 – 6 (1985). Abstract. – See Abstr. 010.721.

134.007   L'évolution de la courbe de lumière de l'étoile centrale de la nébuleuse planétaire NGC 2346.
A. Acker.
Bull. Assoc. Fr. Obs. Etoiles Variables, No. 31, p. 1 – 2 (1985).

134.008   High–resolution radio observations of compact planetary nebulae.
S. Kwok.
Astron. J., Vol. 90, No. 1, p. 49 – 58 (1985). = Publ. Rothney Astrophys. Obs., No. 25.

The author reports the radio observations of ten compact planetary nebulae with angular resolutions of $\sim 0\rlap{.}''4$. With the exception of two objects which are optically thick at 5 GHz, all show definite shell structures characteristic of planetary nebulae. The observed physical parameters of the objects suggest that they are likely to be very young planetary nebulae.

134.009   *JHK* photometry of planetary nebulae.
P. A. Whitelock.
Mon. Not. R. Astron. Soc., Vol. 213, No. 1, p. 59 – 69 (1985).

Infrared (*JHK*) photometry is presented for 80 planetary nebulae. Sixty–two of these have infrared colours which are dominated by hydrogen and helium plasma emission, while only 12 show the effects of thermal dust emission to a greater or lesser extent. All of the nebulae are classified according to whether the principal source of infrared emission is nebular, stellar, dust or some combination of these three. A correlation is found between

$(J-H)_0$ and He^+/H. This arises because the J band includes the strong He I triplet line at 1.083 μm.

134.010 Speckle observations of the central source in the bipolar nebula NGC 2346.
J. Meaburn, J. R. Walsh, B. L. Morgan, J. C. Hebden, H. Vine, C. Standley.
Mon. Not. R. Astron. Soc., Vol. 213, No. 1, p. 35P – 38P (1985).

The core of the bipolar nebula NGC 2346 has been observed with the Imperial College Speckle Interferometer combined with the 3.9–m Anglo–Australian telescope. A disc of light 0.54 arcsec in diameter surrounds an inner, bright, unresolved source. It is suggested that this extended region may be light from the central A5 V star scattered off a cloud of dust $\cong 420$ AU in diameter.

134.011 The central star of NGC 2346: a dephasing light curve.
A. Acker, G. Jasniewicz.
Astron. Astrophys., Vol. 143, No. 1, p. L1 – L3 (1985).

New photometric observations of the central star of the bipolar planetary nebula NGC 2346 have been made in September and October 1984. The periodic light curve ($P \sim 16.0$ d) is shown to be totally dephased with that obtained by Kohoutek in January 1982. To take into account this new result, the eclipse models of Mendez et al. (1984) and Roth et al. (1984) are somewhat modified in the letter. A three component elongated toroidal cloud of cold dust is involved; this cloud is assumed to be rotating around the center of the nebula with a velocity near to 40 $R_\odot$/yr.

134.012 A search for radial velocity variations in the central stars of southern planetary nebulae and planetary–like objects.
H. J. Augensen.
Mon. Not. R. Astron. Soc., Vol. 213, No. 2, p. 399 – 405 (1985).

Results are presented of a radial velocity analysis of moderate–dispersion spectra of the central stars of the southern planetary nebulae NGC 2346, 3132, and He2–36, as well as of the questionable or misclassified planetaries NGC 6164–5, PK 289–0°1, CN1–1, and VV1–7. With the exception of NGC 2346, a confirmed binary, and possibly NGC 3132 and He2–36, no compelling evidence could be found for large–amplitude radial velocity variations on a time–scale of ~ 1 to ~ 10 day. The averaged heliocentric velocities for each star are listed and compared with any extant values of the radial velocity of the star or nebula that could be found in the literature.

134.013 New planetary nebulae in the Small Magellanic Cloud.
D. H. Morgan, A. R. Good.
Mon. Not. R. Astron. Soc., Vol. 213, No. 2, p. 491 – 494 (1985). With 13 plates.

A search of a deep objective prism plate of the SMC taken with the UK 1.2–m Schmidt telescope using a medium–dispersion objective prism has revealed 10 additional planetary nebula candidates. This raises the number of possible planetary nebulae in the SMC seen on objective prism plates to 44.

134.014 Highly ionized neon in the planetary nebula NGC 6302.
S. R. Pottasch, A. Preite–Martinez, F. M. Olnon, E. Raimond, D. A. Beintema, H. J. Habing.
Astron. Astrophys., Vol. 143, No. 2, p. L11 – L13 (1985).

The (low resolution) spectrum of the planetary nebula NGC 6302 in the wavelength range from 7.5 μm to 22 μm is presented. The strongest feature in the spectrum is due to a line at 7.7 μm, which is identified as Ne VI seen for the first time. The Ne V line at 14.3 μm is also seen for the first time. The weakness of this line relative to Ne V lines in the visible spectrum indicates that it is formed at a density of $3 \times 10^5 cm^{-3}$ which is considerably higher of $10^4 cm^{-3}$ at which the S II, O II and C III lines are formed. The Ne II and Ne III lines are also observed, making it possible to accurately derive the total neon abundance.

134.015 Ultraviolet observations of the central star in the planetary nebula 136+5°1.
J. N. Heckathorn, R. A. Fesen.
Astron. Astrophys., Vol. 143, No. 2, p. 475 – 477 (1985).

Ultraviolet spectra obtained with IUE are presented which confirm the Heckathorn et al. (1982) stellar candidate as the central star of the recently discovered planetary nebula 136+5°1. Spectra of the star covering the 1200 – 2000 Å range show a featureless continuum indicative of a black body temperature between 50,000 and 60,000K. Anomalously low flux values shortward of λ1500 are discusssed.

134.016 CCD observations of bipolar nebulae. II. LkHα 233.
C. Aspin, I. S. McLean, M. J. McCaughrean.
Astron. Astrophys., Vol. 144, No. 1, p. 220 – 226 (1985).

Imaging polarimetry and photometry in three colours (B, V, and R) are presented of the emission star/nebula Lick Hα 233. These new observations were made with the Royal Observatory, Edinburgh CCD Imaging Spectropolarimeter and have a pixel resolution of $\sim 0\rlap{.}''5$. The polarization maps clearly reveal a large disk or torus surrounding the central star and show that the associated faint nebulosity (or lobes) is part of the system. These "bipolar" type lobes are highly polarized indicative of an optically–thin reflection nebula. Polarization mapping is shown to be a powerful tool in detecting disks and bipolar lobes. The authors also discuss the origin of the large visual extinction towards LkHα 233 and conclude that an intrinsic component of extinction must be present.

134.017 Wind distances for planetary nebulae.
J. B. Kaler, J.–E. Mo, S. R. Pottasch.
Astrophys. J., Vol. 288, No. 1, p. 305 – 309 (1985).

A new method for the determination of distances to planetary nebulae is presented. If a star has a strong wind that produces a P Cygni profile in its spectrum, one may infer an escape velocity from the terminal velocity, which gives a relation involving stellar temperature, luminosity, and mass. The theory of stellar evolution provides a second relation among these three variables, so that once the star's temperature is known, one can simultaneously derive its mass and luminosity, and hence its distance. The method is applied to several test cases that illustrate its viability and current limitations.

134.018 Expansion velocities for selected planetary nebulae.
F. Sabbadin, S. Ortolani, A. Bianchini.
Mon. Not. R. Astron. Soc., Vol. 213, No. 3, p. 563 – 573 (1985).

The authors have derived the [O III] expansion velocity in 22 nebulae of the Catalogue of Galactic Planetary Nebulae by Perek & Kohoutek from high–dispersion spectra taken at the Astrophysical Observatory of Asiago (Italy). The sample contains planetaries in very different evolutionary phases. The analysis of all the existing data on expansion in planetary nebulae suggests that nebulae of the two sub–classes B and C of Grieg possess different expansion velocity versus nebular radius relationships. These relationships can be understood in terms of the evolutionary model suggested by Sabbadin et al. assuming final masses of the exciting stars in the range from 0.56 to about $0.80\ M_\odot$.

134.019 Radio synthesis observations of M2–9, the Butterfly nebula.
S. Kwok, C. R. Purton, H. E. Matthews, T. A. T. Spoelstra.
Astron. Astrophys., Vol. 144, No. 2, p. 321 – 326 (1985). = Publ. Rothney Astrophys. Obs., No. 29.

Observations of the bipolar planetary nebulae M2–9 have been made at five different radio wavelengths using the VLA and WSRT. The measurements allow separate determinations of the flux density spectra of the core and the nebular "wings" of the object. The core spectrum indicates mass loss from the central star. Unresolved ($< 1\rlap{.}''3$) knots of radio emission are observed in the nebula, which correspond to high–density condensations observed at optical wavelengths. A comparison of the radio emission measures of these condensations with the electron densities determined from optical measurements suggests a distance of ~ 1 kpc for M2–9. The authors propose that M2–9 is descended from a rotating red–giant progenitor and that the bipolar morphology is the results of a rapid expansion of the ionization front in the N–S directions.

134.020 New planetary nebulae of low surface brightness detected on UK–Schmidt plates.
H. Hartl, S. B. Tritton.
Astron. Astrophys., Vol. 145, No. 1, p. 41 – 44 (1985).

Fourteen new candidates for planetary nebulae were discovered on deep R and J plates taken at the UK 1.2 m Schmidt telescope. Photographs, positions and descriptions are presented. For objects visible on Palomar Observatory Sky Survey prints, brightnesses and distances calculated by use of Abell's (1966) formula are given.

134.021 Spectroscopic verification of suspected planetary nebulae. I.
L. Kohoutek, R. Pauls.
Astron. Astrophys., Suppl. Ser., Vol. 60, No. 1, p. 87 – 90 (1985).

Classification of eight objects as a planetary nebulae was confirmed according to the authors' image tube spectra. M4–1 is an emission line (Seyfert 2) galaxy, K2–13 is probably a plate fault, the status of He2–111 remains uncertain. Line identifications, estimates of line–strengths and mean radial velocities of all emission objects are given.

134.022 Spectroscopic verification of suspected planetary nebulae. II.
R. M. West, L. Kohoutek.
Astron. Astrophys., Suppl. Ser., Vol. 60, No. 1, p. 91 – 97 (1985).

Classification of seven objects as planetary nebulae was confirmed by means of image tube spectra. Eight further objects turned out not to be planetaries; the status of four remaining nebulae is still uncertain. The authors present mean radial velocities of those objects having emission lines.

134.023 Central stars of planetary nebulae.
M. Perinotto, M. Cerruti–Sola.
Effects of variable mass loss on the local stellar environment, p. 39 – 50 (1984). – See Abstr. 012.028.

First, the authors deal with the fundamental parameters of central stars of planetary nebulae. Then they discuss the evidence for fast winds in nuclei of planetary nebulae (PNN), consider a few detailed studies of these winds and compare their results with empirical parametrizations and theoretical predictions for hot stars. Furthermore the authors look at the importance of PNN winds for the nebulae and for the evolution of central stars. Finally they report about work in progress in several PNN on the behaviour of v_∞ and its relationship to some parameters.

134.024 Low–excitation planetary nebulae.
L. N. Kondrat'eva.
Astrofizika, Tom 22, Vyp. 1, p. 153 – 156 (1985). In Russian. English translation in Astrophysics, Vol. 22, No. 1.

The spectrophotometric results for three objects: M1–46, PC 12, M1–39 are presented. The relative emission–line intensities and electron densities are determined. The effective temperatures of the central stars are evaluated.

134.025 M4–18: a young, cool planetary nebula.
R. W. Goodrich, O. Dahari.
Astrophys. J., Vol. 289, No. 1, p. 342 – 355 (1985). = Lick Obs. Bull., No. 991.

The unresolved planetary nebula M4–18 has been studied spectroscopically from the ultraviolet to the near–infrared. The nebula is found to be of very low excitation, with electron density $N_e = (1.1 \pm 0.1) \times 10^4 \mathrm{cm}^{-3}$ and electron temperature $T_e = 5600 \pm 1300\mathrm{K}$, and a high O abundance. The central star is found to be among the coolest central stars known, with temperature $(22 \pm 2) \times 10^3\mathrm{K}$. M4–18 appears to be a very young planetary nebula, possibly only 4000 years old. It is roughly 3.5 kpc distant and is highly reddened.

134.026 A model of the planetary nebula NGC 2392 determined from velocity observations.
C. R. O'Dell, M. E. Ball.
Astrophys. J., Vol. 289, No. 2, p. 526 – 534 (1985).

High–resolution slit spectra are combined with published photometric and velocity data to develop a model of the double–ring planetary nebula NGC 2392. The inner ring is due to an plete, prolate spheroid shell with most of the material trated near the minor axis circumference. The spheroid is seen almost pole–on, open at the sharp ends, and expands with 93 km s^{-1}. The outer ring is a nearly round spheroid of ~ 16 km s^{-1} expansion velocity and a different tilt angle than the inner shell. A high–velocity stream of material is formed at the open ends of the inner shell and is accelerated by the central star's stellar wind to velocities of at least ± 190 km s^{-1}.

134.027 Chemical abundances in planetary nebulae in the Galaxy and in the Magellanic Clouds.
D. R. Flower.
Gas in the interstellar medium, p. 100 – 106 (1984). – See Abstr. 012.032.

134.028 IRAS observations of Sand. 3 and M1–67: two new planetary nebulae with Wolf–Rayet nuclei.
K. A. van der Hucht, T. A. Jurriens, F. M. Olnon, P. S. Thé, P. R. Wesselius, P. M. Williams.
Astron. Astrophys., Vol. 145, No. 2, p. L13 – L16 (1985).

IRAS observations of the objects Sand. 3 and 209 BAC, the latter in the nebula M1–67, which were listed as WR72 and WR124 in the 6–th Catalogue of Galactic Population I Wolf–Rayet stars, show IR emission of circumstellar dust with respectively $T_c \approx 85K$ and $T_c \approx 100K$. Such dust envelopes and the parameters the authors derive from them, are characteristic for planetary nebulae, confirming the planetary nucleus status of Sand. 3, [WO] and proving that of 209 BAC, [WN8].

134.029 Spectroscopy and photometry of the peculiar objects Henize 2–442A, B in 1983.
V. P. Arkhipova, V. F. Esipov, B. F. Yudin.
Sov. Astron. Lett., Vol. 10, No. 4, p. 252 – 253 (1984). English translation of 38.134.012.

134.030 Carbon abundances in planetary nebulae in the galactic bulge.
B. L. Webster.
Proc. Astron. Soc. Aust., Vol. 5, No. 4, p. 535 – 537 (1984).

Planetary nebulae in the galactic bulge come from a population of stars that are predominantly of solar oxygen abundance and with masses less than 3 solar masses. There is a small proportion of oxygen deficient (halo) and super–oxygen–rich objects and also a small fraction of more massive objects. In a few nebulae the carbon is significantly greater than the average value in disk planetaries and these objects would warrant further study because of their relevance to theories of carbon enrichment on the asymptotic giant branch.

134.031 Evolutionary calculations for planetary nebula nuclei with continuing mass loss and realistic starting conditions.
D. J. Faulkner, P. R. Wood.
Proc. Astron. Soc. Aust., Vol. 5, No. 4, p. 543 – 546 (1984).

This paper describes evolutionary calculations for nuclei of planetary nebulae (NPN) which have been made using a variety of assumptions regarding (1) mass of the NPN, (2) phase in the He shell flash cycle at which the NPN leaves the asymptotic giant branch, and (3) time variation of the mass loss rate. It is hoped that the comparison of these calculations with sufficiently precise observations, when these are available, will resolve some of the present uncertainties regarding planetary shell ejection.

134.032 Spectrophotometry of 12 planetary nebulae.
J. B. Kaler.
Astrophys. J., Vol. 290, No. 2, p. 531 – 541 (1985).

A dozen planetary nebulae have been observed between Hδ and [O II] $\lambda7325$ with the Red Reticon at the 2.3 m telescope of Steward Observatory, with the most extensive spectrophotometry done between Hγ and [Ar III] $\lambda7135$. The objects are a mixture of types, all but two drawn from wide–aperture surveys. For most of the nebulae [O III] and/or [N II] electron temperatures, [S II] electron densities, and He/H, O/H, and N/O ratios are derived. An exhaustive discussion of errors is included.

134.033 On the distances of planetary nebulae.
S. Kwok.
Astrophys. J., Vol. 290, No. 2, p. 568 – 577 (1985). = Publ. Rothney Astrophys. Obs., No. 27.

The evolution of the ionization structure of planetary nebulae is calculated from the central–star evolutionary models of Schönberner. The predictions of models in the $0.57 - 0.64\,M_\odot$ mass range are consistent with observed nebular parameters of planetary nebulae with known distances. The author also derives a semiempirical mass–radius relationship of the form $M \propto R^\beta$, where $\beta = 2.25 - 2.50$. This relationship suggests that the dependence of the radio flux density on angular size as the result of the expansion of an ionization front has approximately the same form as the flux–angular size dependence for planetary nebulae at different distances. Consequently, distances of ionization–bounded planetary nebulae cannot be reliably determined from the measurement of radio flux (or $H\beta$) and angular size alone.

134.034 Influence of the duplicity of stars on the forms of planetary nebulae.
I. G. Kolesnik, L. S. Pilyugin.
Inst. teor. fiz. Akad. Nauk SSSR. Prepr., No. 120R, 22 pp. (1984). In Russian. Abstr. in Ref. Zh., 51. Astron., 3.51.629 (1985).

134.035 Far–infrared line observations of planetary nebulae. I. The [O III] spectrum.
H. L. Dinerstein, D. F. Lester, M. W. Werner.
Astrophys. J., Vol. 291, No. 2, p. 561 – 570 (1985).

Observations of the far–infrared fine–structure lines of [O III] have been obtained for six planetary nebulae. The infrared measurements are combined with optical [O III] line fluxes to probe physical conditions in the gas. From the observed line intensity ratios, the authors obtain a simultaneous solution for electron temperature and density, as well as a means of evaluating the importance of inhomogeneities. Densities determined from the far–infrared [O III] lines agree well with density diagnostics from other ions, indicating a fairly homogeneous density in the emitting gas. Evidence for the existence of temperature variations in these nebulae is presented.

134.036 X–ray emission from the planetary nebula NGC 1360.
P. A. J. de Korte, J. J. Claas, F. A. Jansen, S. P. McKechnie.
Adv. Space Res., Vol. 5, No. 3, p. 57 – 59 (1985). – See Abstr. 012.059.

The Exosat observatory has detected the nucleus of NGC 1360 in four photometric energy bands. The data rules out that the emission is from blackbody origin. Initial fits made with LTE model atmosphere spectra require the presence of highly ionized oxygen and neon in the stellar atmosphere.

134.037 On the spectral variability of the planetary nebula IC 4997.
E. B. Kostyakova.
Astron. Tsirk., No. 1341, p. 4 – 6 (1984). In Russian.

134.038 Spectral observations of three planetary nebulae.
L. N. Kondrat'eva.
Astron. Tsirk., No. 1354, p. 1 – 2 (1984). In Russian.

134.039 High–resolution CO observations of NGC 7027.
C. R. Masson, K. W. Cheung, G. L. Berge, M. J. Claussen, G. M. Heiligman, R. B. Leighton, K. Y. Lo, A. T. Moffet, T. G. Phillips, A. I. Sargent, S. L. Scott, D. P. Woody.
Astrophys. J., Vol. 292, No. 2, p. 464 – 470 (1985).

The authors have made interferometer maps of CO(1–0) emission from NGC 7027. These maps show that the neutral gas has a flattened distribution, which is constraining the development of the ionized nebula contained within it. For the first time anomalous high–velocity CO emission has been detected. This emission originates in a layer of shocked, neutral gas surrounding the expanding H II region.

134.040 Radial velocities of non stellar objects and the case of planetary nebulae.
A. Acker.
Bull. Inf. Cent. Données Stellaires, No. 28, p. 5 – 8 (1985).

134.041 Halos von Planetarischen Nebeln.
H. Hippelein, M. Baessgen, M. Grewing.
Mitt. Astron. Ges., Nr. 63, p. 130 (1985). – See Abstr. 012.063.

134.042 Expansion alter Planetarischer Nebel.
F. Gieseking, H. Hippelein, R. Weinberger.
Mitt. Astron. Ges., Nr. 63, p. 131 (1985). – See Abstr. 012.063.

134.043 Narrow–band observations of planetary nebulae with a photon–counting imaging detector.
J. Barnstedt, M. Gutekunst, L. Bianchi, M. Grewing.
Mitt. Astron. Ges., Nr. 63, p. 212 – 214 (1985). – See Abstr. 012.063.

A newly developed photon–counting imaging detector has been used to obtain narrow–band (interference filter) observations of planetary nebulae. Results are presented for NGC 40 and NGC 6781.

134.044 Morphology of planetary nebulae from A to Z.
L. H. Aller, B. Zuckerman.
Bull. Am. Astron. Soc., Vol. 16, No. 4, p. 974 – 975 (1984). Abstract. – See Abstr. 010.062.

134.045 The planetary nebula surrounding Tycho's supernova remnant.
E. M. Jones, J. R. Dickel.
Bull. Am. Astron. Soc., Vol. 16, No. 4, p. 975 (1984). Abstract. – See Abstr. 010.062.

134.046 The wind from the nucleus of the planetary nebula M4–18.
R. A. Shaw.
Bull. Am. Astron. Soc., Vol. 16, No. 4, p. 975 (1984). Abstract. – See Abstr. 010.062.

134.047 Observations of variations in the Balmer decrement in NGC 3242.
R. W. Olson.
Bull. Am. Astron. Soc., Vol. 16, No. 4, p. 975 (1984). Abstract. – See Abstr. 010.062.

134.048 An analysis of the nonradial oscillations of the central star of the planetary nebula K1–16.
S. G. Starrfield, A. N. Cox, R. B. Kidman, W. D. Pesnell.
Bull. Am. Astron. Soc., Vol. 16, No. 4, p. 975 (1984). Abstract. – See Abstr. 010.062.

134.049 An ionization and dust model for the planetary nebula NGC 3918.
J. P. Harrington, R. E. S. Clegg, D. J. Monk.
Bull. Am. Astron. Soc., Vol. 16, No. 4, p. 976 (1984). Abstract. – See Abstr. 010.062.

134.050 Dynamical evolution of planetary nebulae.
S. Kwok, K. M. Volk.
Bull. Am. Astron. Soc., Vol. 16, No. 4, p. 976 (1984). Abstract. – See Abstr. 010.062.

134.051 The electron temperature and emission measure distributions in the planetary nebula NGC 7027.
C. T. Daub, J. P. Basart.
Bull. Am. Astron. Soc., Vol. 16, No. 4, p. 993 (1984). Abstract. – See Abstr. 010.062.

134.052 High resolution CO observations of NGC 7027.
C. R. Masson, K. W. Cheung, G. L. Berge,
M. J. Claussen, G. M. Heiligman, R. B. Leighton, K.-Y. Lo,
A. T. Moffet, T. G. Phillips, A. I. Sargent, S. L. Scott,
D. P. Woody.
Bull. Am. Astron. Soc., Vol. 16, No. 4, p. 993 (1984). Abstract. –
See Abstr. 010.062.

134.053 The location of molecular hydrogen in the planetary nebula NGC 7027.
H. A. Thronson Jr., J. A. Hackwell, G. L. Grasdalen,
R. D. Gehrz.
Bull. Am. Astron. Soc., Vol. 16, No. 4, p. 994 (1984). Abstract. –
See Abstr. 010.062.

134.054 Infrared excess in the nuclei of planetary nebulae.
A. F. Bentley, M. W. Castelaz, J. A. Hackwell.
Bull. Am. Astron. Soc., Vol. 16, No. 4, p. 994 (1984). Abstract. –
See Abstr. 010.062.

134.055 The planetary nebula around the sdO star RWT 152.
J. M. Pasachoff, K. B. Kwitter, P. Massey.
Bull. Am. Astron. Soc., Vol. 16, No. 4, p. 994 (1984). Abstract. –
See Abstr. 010.062.

134.056 Ultraviolet observations of the binary nucleus of the planetary nebula NGC 1360.
E. F. Guinan, K. Garlow, A. Theokas.
Bull. Am. Astron. Soc., Vol. 16, No. 4, p. 994 (1984). Abstract. –
See Abstr. 010.062.

134.057 Two–dimensional CCD spectrophotometry of the planetary nebulae NGC 40 and NGC 6826.
R. Quigley, G. H. Jacoby, J. L. Africano.
Bull. Am. Astron. Soc., Vol. 16, No. 4, p. 994 (1984). Abstract. –
See Abstr. 010.062.

134.058 The Bowen fluorescent mechanism in planetary nebulae.
L. Jones Likkel, L. H. Aller.
Bull. Am. Astron. Soc., Vol. 16, No. 4, p. 994 – 995 (1984). Abstract. – See Abstr. 010.062.

134.059 The internal motions in multiple shell planetary nebulae.
Y.-H. Chu, K. B. Kwitter, G. H. Jacoby, J. Kaler.
Bull. Am. Astron. Soc., Vol. 16, No. 4, p. 995 (1984). Abstract. –
See Abstr. 010.062.

134.060 Numerical hydrodynamic models of planetary nebulae.
M. Bobrowsky.
Bull. Am. Astron. Soc., Vol. 16, No. 4, p. 995 (1984). Abstract. –
See Abstr. 010.062.

134.061 On the existence and role of magnetic fields in planetary nebulae.
G. Pascoli.
Astron. Astrophys., Vol. 147, No. 2, p. 257 – 264 (1985). In French.
The author considers various aspects of magnetic fields in planetary nebulae. He discusses, in a simplified model, the existence of an internal magnetic field and its role in the morphology of planetary nebulae. The interstellar magnetic action is also considered.

134.062 V651 Monocerotis (central star in NGC 2346).
IAU Circ., Nos. 4036, 4047, 4064 (1985).

134.063 On the consequences of a new initial–final mass relation for low and intermediate mass stars and the birthrate of planetary nebulae.
D. C. V. Mallik.
Astrophys. Lett., Vol. 24, No. 4, p. 173 – 182 (1985).
Recent work on the initial–final mass relation for low–and intermediate–mass stars and the mass distribution of the nuclei of planetary nebulae sets a lower limit on the initial mass of progenitors of PN which is substantially higher than 1 $M_\odot$. With such a lower limit the theoretical estimates of the deathrate of main-sequence stars would be consistent with the observed PN birthrate only if the distance scale due to Cudworth (1974) were used. A raised lower limit on the progenitor mass also helps explain the low mean distances of planetary nebulae from the galactic plane and the paucity of PN in Population II systems.

134.064 Chemical composition of planetary nebulae.
L. N. Kondrat'eva.
Tr. Astrofiz. Inst. Alma–Ata, Tom 42, p. 39 – 56 (1983). In Russian.
Helium abundances are calculated for 28 planetary nebulae and neon, oxygen, nitrogen and sulfur abundances for 19 objects are presented.

134.065 Central stars of some planetary nebulae.
L. N. Kondrat'eva.
Tr. Astrofiz. Inst. Alma–Ata, Tom 44, p. 30 – 42 (1984). In Russian.

134.066 A search for atomic hydrogen associated with planetary nebulae and evolved stars.
P. R. Silverglate, S. E. Schneider, D. A. Altschuler.
Bull. Am. Astron. Soc., Vol. 17, No. 2, p. 547 (1985). Abstract. –
See Abstr. 010.065.

134.067 Evolution of model planetary nebulae.
M. Bobrowsky.
Bull. Am. Astron. Soc., Vol. 17, No. 2, p. 547 (1985). Abstract. –
See Abstr. 010.065.

134.068 Spectral characteristics of dust in carbon–rich objects.
R. F. Silverberg, H. Moseley, W. Glaccum.
Bull. Am. Astron. Soc., Vol. 17, No. 2, p. 594 (1985). Abstract. –
See Abstr. 010.065.

134.069 Theoretical models as tools for abundance determinations in high–excitation planetaries.
L. H. Aller, C. D. Keyes, W. A. Feibelman.
Bull. Am. Astron. Soc., Vol. 17, No. 2, p. 594 (1985). Abstract. –
See Abstr. 010.065.

134.070 Zanstra temperatures of planetary nebula central stars.
H. L. Shipman, R. B. C. Henry.
Bull. Am. Astron. Soc., Vol. 17, No. 2, p. 594 (1985). Abstract. –
See Abstr. 010.065.

134.071 Masses for the central stars of three planetary nebulae in the Magellanic Clouds.
H. J. Augensen, S. R. Heap.
Bull. Am. Astron. Soc., Vol. 17, No. 2, p. 594 – 595 (1985). Abstract. – See Abstr. 010.065.

134.072 Origin and evolution of planetary nebulae.
J. Podolský.
Říše hvězd, Vol. 66, No. 6, p. 114 – 115 (1985). In Czech.

134.073 Studies of southern planetary nebulae. I. Fluxes from bright planetary nebulae.
A. Gutiérrez–Moreno, H. Moreno, G. Cortés.
Publ. Astron. Soc. Pac., Vol. 97, No. 591, p. 397 – 403 (1985).
Observed Hβ fluxes, reddenings, and intrinsic forbidden line intensities relative to Hβ are given for 14 bright southern planetary nebulae, for several of which there were no previous data.

134.074 The ultraviolet spectrum of the planetary nebula M3–27.
W. A. Feibelman.
Publ. Astron. Soc. Pac., Vol. 97, No. 591, p. 404 – 406 (1985).
The planetary nebula M3–27 was observed in the low–dispersion mode of the *IUE* satellite in the region 1150 Å – 3200 Å. A fairly strong continuum is seen to extend down to 1150 Å, superposed with a relatively small number of emission lines. A high electron density is inferred from [O III] but regions of lower density also exist. The ultraviolet characteristics

of M3–27 are compared to those of M1–2. A carbon abundance of $\log\{C\} = 7.72$ is estimated from the C III] nebular line for an adopted $T_e = 15,000K$.

Sky catalogue 2000.0. Volume 2: Double stars, variable stars and nonstellar objects.
See Abstr. 002.019.

Berichte aus den Workshops.
See Abstr. 011.024.

Partition functions and dissociation constants for HeH^+.
See Abstr. 022.074.

The splitting of the $2s^2 2p^3\,^2P$ term in O II.
See Abstr. 022.138.

The lines of carbon, nitrogen and oxygen in the spectra of planetary nebulae. I. Transition probabilities and oscillator strengths.
See Abstr. 022.187.

The study of astrophysical interactions with the ESA Photon Counting Detector.
See Abstr. 035.052.

Solar abundances and the role of nucleogenesis in low–to–medium mass stars in the Galaxy.
See Abstr. 061.148.

Mass distribution and evolutionary scheme for central stars of planetary nebulae.
See Abstr. 065.001.

The life and times of an intermediate mass star – in isolation/in a close binary.
See Abstr. 065.005.

An analysis of nonradial pulsations of the central star of the planetary nebula K1–16.
See Abstr. 065.078.

The proper motions of LS V + 46°21 and AS 84, two "central" star candidates for S216.
See Abstr. 111.013.

Winds in central stars of planetary nebulae.
See Abstr. 112.076.

VBLUW–photometry of stars in small fields around 13 planetary nebulae.
See Abstr. 113.036.

The central star of NGC 40.
See Abstr. 114.001.

SIT Vidicon and IDS spectra of central stars of planetary nebulae.
See Abstr. 114.022.

He2–442: two objects under the same name.
See Abstr. 114.132.

Peculiar cool stars in planetary nebulae – the spectrum of FG Sge.
See Abstr. 114.145.

IUE and optical observations of the symbiotic stars He2–38 and He2–106.
See Abstr. 117.227.

Pre–cataclysmic binaries.
See Abstr. 117.301.

Abell 41, a cataclysmic variable progenitor.
See Abstr. 117.302.

V605 Aql – Schlüssel zur späten Sternentwicklung?
See Abstr. 124.011.

Eruptive Ereignisse als Kennzeichen später Sternentwicklung.
See Abstr. 124.012.

On the progenitors of white dwarfs.
See Abstr. 126.060.

Discovery of oxygen in the PG 1159 degenerate stars: a direct evolutionary link to O VI planetary nebula nuclei and confirmation of pulsation theory.
See Abstr. 126.067.

Discovery of oxygen in the PG1159 degenerate stars: a direct evolutionary link to O VI planetary nebula nuclei and confirmation of pulsation theory.
See Abstr. 126.082.

Chemical composition of interstellar material.
See Abstr. 131.047.

Spectroscopy of the 3 micron emission features.
See Abstr. 131.197.

Galactic fountains, planetary nebulae and warm H I.
See Abstr. 131.199.

Report of IAU Commission 34: Interstellar matter (_Matière interstellaire_).
See Abstr. 131.356.

Bipolar ejection of matter from hot stars.
See Abstr. 132.015.

Fabry–Perot observations of the unusual emission–line nebula S216.
See Abstr. 132.027.

Determination of nebular density and temperature from radio recombination lines.
See Abstr. 132.037.

Spatially resolved infrared observations of the Red Rectangle.
See Abstr. 132.058.

IRAS 0423 + 536P03: a remarkable galactic infrared source.
See Abstr. 133.010.

Radio Sources, X-ray Sources, Cosmic Rays

141 Radio Sources (Surveys, etc.)

141.001 A new class of nonthermal radio sources.
R. H. Becker, D. J. Helfand.
Nature, Vol. 313, No. 5998, p. 115 – 118 (1985).

Current radio catalogues list some 145 sources as galactic supernova remnants. Detailed maps including spatially–resolved spectral information and polarimetric data have been obtained for about half of the catalogued objects. While mapping many of the remnants not well studied to date, the authors have discovered at least two nonthermal extended objects which defy classification as supernova remnants. The two sources exhibit a marked axial symmetry with evidence for a compact component at one extreme of the axis. Their diffuse emission includes bright filamentary structures superimposed on a lower surface brightness component which gradually fades, becoming undetectable at the end of the axis opposite the compact source. The authors propose that these objects are representative of a new class of nonthermal radio sources.

141.002 Origin of the new axisymmetric radio sources.
D. J. Helfand, R. H. Becker.
Nature, Vol. 313, No. 5998, p. 118 – 119 (1985).

In an earlier letter, the authors have reported the discovery of two nonthermal radio sources lying near the galactic plane, which have surface brightness distributions radically different from those of the supernova remnants with which they had been formerly classified. Their marked axial symmetry, associated compact components, steep surface–brightness gradients and polarization structure are unlike those seen in any previously–identified class of radio emitters. Here, the authors discuss some of their physical properties (distances, ages, energetics) and describe briefly a scenario in which the observed radiation is produced by electrons generated in a relativistic outflow from high–velocity accreting binary systems.

141.003 An new Bologna sky survey at 408 MHz.
A. Ficarra, G. Grueff, G. Tomassetti.
Astron. Astrophys., Suppl. Ser., Vol. 59, No. 2, p. 255 – 347 (1985).

The authors present the first section of a new sky survey performed at 408 MHz with the "Northern Cross" Radiotelescope. A catalogue of 13354 radiosources is given, complete to 0.1 Jy in an area of 0.78 steradian included between $+37°15'$ and $47°37'$ declination. The instrument in its new version and the data reduction procedure are fully described, and the experimental errors are extensively discussed.

141.004 Observations of the source 4C 21.53 in the 10 – 25 MHz frequency range.
S. Ya. Braude, S. M. Zakharenko, K. P. Sokolov.
Astron. Zh., Tom 62, Vyp. 1, p. 34 – 37 (1985). In Russian. English translation in Sov. Astron., Vol. 29, No. 1.

The results of the measurements of the total flux density of the source 4C 21.53 at the frequencies 10, 12.6, 14.7, 16.7, 20, and 25 MHz are presented. It is shown that the spectrum of the compact components of the source remains linear (in the $\log S - \log v$ scale) with decreasing observational frequency down to 10 MHz.

141.005 Determination of flux densities of radio sources using broad–band radiometers of the radio telescope RATAN–600.
K. D. Aliakberov, M. G. Mingaliev, M. N. Naugol'naya, S. A. Trushkin, L. M. Sharipova, S. N. Yusupova.
Astrofiz. Issled. Izv. Spets. Astrofiz. Obs., Tom 19, p. 60 – 65 (1985). In Russian. English translation in Bull. Spec. Astrophys. Obs. – North Caucasus.

The list of 11 commonly used reference sources and their flux densities at cm–wavelengths are given. These sources are used for the effective area measurements. The experimental diagrams of the A_{ef} of the north sector and the south sector with periscope antennas are given. Estimates of errors in the flux–density measurements were made.

141.006 G2.4 + 1.4, a smothered supernova?
J. R. Graham.
Observatory, Vol. 105, No. 1064, p. 7 (1985). Abstract. – See Abstr. 010.721.

141.007 Structural details of the Sgr A complex.
F. Yusef–Zadeh, M. Morris.
News Lett. Astron. Soc. N.Y., Vol. 2, No. 7, p. 27 – 29 (1985). Abstract. – See Abstr. 010.241.

141.008 Frequency–dependence of the statistics of radio sources.
V. R. Amirkhanyan.
Astrophys. Space Sci., Vol. 108, No. 1, p. 125 – 136 (1985).

Without assuming the existence of two populations of radio sources, the counts vs frequency, spectral index vs flux density, and counts vs spectral index dependences have been derived. An assumption – with arguments to support it – is made that most extragalactic radio sources belong to the class of doubles whose visible parameters depend on their spatial orientation being relative to the observer.

141.009 OSRT observations of the Sco X–1 region at 327 MHz.
T. Velusamy, A. Pramesh Rao, S. Sukumar.
Mon. Not. R. Astron. Soc., Vol. 213, No. 3, p. 735 – 741 (1985).

An OSRT (Ooty Synthesis Radio Telescope) map of a $2° \times 1°$ field around Sco X–1 with resolution of 3×3 arcmin^2 at 327 MHz is presented. There is no evidence of any extended emission around Sco X–1 brighter than 50 mJy beam^{-1}. Two sources in the field, 1617–153 and 1616–160, lie at distances of 15 and 31 arcmin from Sco X–1 and are collinear with its triple source structure. High–resolution (100×50 arcsec^2) OSRT maps of these and Sco X–1 are presented. The spectra of the NE and central components in Sco X–1 show low–frequency turnovers around 1 GHz, while the SW components have a power–law non–thermal spectrum.

141.010 Further observations of radio sources from the BG survey. III. CTB 80.
F. Mantovani, W. Reich, C. J. Salter, P. Tomasi.
Astron. Astrophys., Vol. 145, No. 1, p. 50 – 58 (1985).

The authors present new observations of CTB 80 at 408, 1410, 1720, 2695, and 4750 MHz. These consist of both total power and linear polarization measurements. The morphology of the source and its spectral and polarization properties are described. The implications of the new data for existing theories of this peculiar object are considered. The possibility that the source may involve the interaction between two shell–type supernova remnants is also discussed. None of the theories is found to be entirely satisfactory at present.

141.011   A survey of the radio background at 38 MHz and the galactic spectral index.
J. Milogradov–Turin.
Publ. Astron. Opservatorije Beogr., No. 33, p. 28 – 30 (1985). Abstract of a thesis. – See Abstr. 012.061.

141.012   Absolute measurements of the radio radiation of Cassiopeia A, Cygnus A and the Crab nebula in the 60 – 100 cm range.
M. E. Miller.
Radiofizika, Tom 27, No. 7, p. 934 – 938 (1984). In Russian. Abstr. in Ref. Zh., 51. Astron., 1.51.1007 (1985).

141.013   A very deep Westerbork survey of a field previously observed with the VLA.
M. J. A. Oort, R. A. Windhorst.
Astron. Astrophys., Vol. 145, No. 2, p. 405 – 424 (1985).
The authors present a deep 21 cm radio survey of the Lynx 2 field selected from the Leiden–Berkeley Deep Survey areas. In the main part of the paper the authors give a review of the data reduction techniques that were used and present a complete radio sample which they used to determine the 1412 MHz source counts and the median angular size at the mJy level. A comparison of this survey with a survey of the same field with the Very Large Array is performed. A systematic position difference as a function of distance to the field center is found. The absolute uncertainties in the coordinate frame and flux scale are determined, as well as the signal–to–noise dependent error distribution. Corrections for systematic influence of the flux determining algorithm are discussed. Finally, a search was made for variable sources in the maps.

141.014   Observations of the Ooty radio sources at 102 MHz.
V. S. Artyukh, M. A. Ogannisyan, V. G. Panadzhyan.
Soobshch. Byurakan. Obs., Vyp. 54, p. 3 – 9 (1983). In Russian.
The results of observations of 52 Ooty radio sources are presented. The flux densities of the detected radio sources as well as the diameters and the flux densities of their compact components are determined.

141.015   Observations of Ooty radio sources at RATAN–600.
G. A. Oganyan, V. G. Panadzhyan.
Soobshch. Byurakan. Obs., Vyp. 54, p. 10 – 14 (1983). In Russian.
The results of observations of 20 Ooty radio sources at 968 and 3660 MHz are presented. The flux densities at 968 and 3660 MHz and spectral indices in the ranges 327 – 968 and 968 – 3660 MHz are determined.

141.016   A new non–thermal galactic radio source with a possible binary system.
E. Fürst, W. Reich, P. Reich, Y. Sofue, T. Handa.
Nature, Vol. 314, No. 6013, p. 720 – 721 (1985).
Helfand and Becker recently discussed the unusual filamentary appearance of the galactic radio sources G 357.7–0.1 and G 5.3–1.0, both of which had originally been classified as supernova remnants and proposed that these sources belong to a new class of non–thermal radio sources that originate in accreting binary systems containing a neutron star or a black hole. Here the authors present observations of the non–thermal galactic object G 18.95–1.1 demonstrating that this source is another candidate for that class of objects.

141.017   Survey of the galactic plane in the region of the associations Canis Majoris OB1, R1 carried out at 7.6 cm wavelength at the RATAN–600 radio telescope.
T. B. Pyatunina.
Astron. Zh., Tom 62, Vyp. 2, p. 218 – 225 (1985). In Russian. English translation in Sov. Astron., Vol. 29, No. 2.
The results of the galactic plane survey in the region of the associations Canis Majoris OB1, R1 made at the wavelength

7.6 cm are given. The total number of radio sources is 89. The survey is statistically complete down to 40 mJy.

141.018   On the nature of some galactic radio sources with large angular dimensions.
I. V. Gosachinskij.
Astron. Zh., Tom 62, Vyp. 2, p. 226 – 228 (1985). In Russian. English translation in Sov. Astron., Vol. 29, No. 2.
The spectral indices of 14 extended components in some well–known galactic objects were computed on the basis of the results of observations at the frequencies of 1.42 GHz, 0.408 GHz, and 4.875 GHz. It is found that 8 of them have nonthermal spectra and therefore may be supernova remnants. Some thermal sources with low angular dimensions are observed on the background of 5 of these SNRs.

141.019   Results of observations made at RATAN–600 with a high–sensitivity radiometer at 7.6 cm wavelength of the region of the Galaxy in the interval $\alpha_{1950.0} = 17^h30^m$ to 19^h30^m and $\delta_{1950.0} = -25°20'$ to $-22°20'$.
A. B. Berlin, V. Ya. Gol'nev, N. M. Lipovka, N. A. Nizhel'skij, E. E. Spangenberg.
Astron. Zh., Tom 62, Vyp. 2, p. 229 – 233 (1985). In Russian. English translation in Sov. Astron., Vol. 29, No. 2.
The results of observations of a giant H II region situated at the longitudes $4° - 10°$ at the wavelength 7.6 cm are presented. A catalogue has been compiled containing 63 sources for which the positions, flux densities, and angular dimensions have been determined. The radio brightness distribution of the background component of the galactic radio emission has been obtained. A weak background extended by more than $2°$ in galactic latitude has been registered. It is shown that the detected weak objects represent a more high–latitude population than the strong galactic sources, and are probably associated with the weak extended background.

141.020   On the radio source scintillation due to postshock plasma irregularities.
S. F. Pimenov.
Sov. Astron. Lett., Vol. 10, No. 4, p. 218 – 220 (1984). English translation of 38.141.001.

141.021   On the variability of some Parkes radio sources.
A. E. Wright.
Proc. Astron. Soc. Aust., Vol. 5, No. 4, p. 510 – 513 (1984).
The purpose of this paper is to give statistics of the variability of 68 pointing sources at 2700 and/or 5000 MHz.

141.022   A catalog of radio sources to be occulted by comets P/Halley and P/Giacobini–Zinner.
I. de Pater, F. P. Schloerb, A. H. Johnson.
Astron. J., Vol. 90, No. 5, p. 846 – 868 (1985).
A catalog of radio sources which lie along the paths of comets P/Halley and P/Giacobini–Zinner is presented to facilitate planning of radio–source occultation experiments. The sources were observed at 20 cm wavelength, using the VLA in the B configuration to provide accurate positions and flux densities for occultation predictions. Since the majority of these sources had not been previously mapped, the catalog and source maps may be of more general use than solely for occultation experiments.

141.023   Flux density observations of natural radio sources for radio position astronomy.
M. Fujishita, K. Nakajima, T. Yoshino, K. Koike, H. Kiuchi.
Proc. Int. Latitude Obs. Mizusawa, No. 23, p. 17 – 22 (1984). In Japanese.
Flux densities of some POLARIS and FK4 stars were observed on March 14, 1983 (UTC) using the 45 m antenna of Nobeyama Radio Observatory at 10 GHz. Observed POLARIS stars are 0106+013, 0235+164, 0355+508, 0552+398, 0851+202, 0923+392, 1226+023, 1404+286, 1641+399, 1642+690, 2200+420 and 2251+158. Observed flux densities in Jy are 4.5, 2.8, 7.2, 4.8, 7.4, 5.4, 46., 2.4, 15., 1.5, 3.1 and 14., respectively. Observed FK4 stars are FK4–111, FK4–224, FK4–472,

FK4–502, FK4–616 and FK4–705. Observed flux densities in Jy are 0.05, 0.01, 0.01, 0.04, 0.01 and 0.02, respectively. No big flare is observed. No flux density variation is observed in 0851+202 during the observation.

141.024   Radio surveys of the sky.
R. Wielebinski.
Bull. Inf. Cent. Données Stellaires, No. 28, p. 41 – 48 (1985). – See Abstr. 012.062.

In this review various sky surveys are discussed. No claim is made on completeness of information. The limiting parameters of the various surveys are given. The likely future survey developments are presented.

141.025   The 408 MHz all–sky continuum survey.
C. G. T. Haslam.
Bull. Inf. Cent. Données Stellaires, No. 28, p. 49 – 52 (1985). – See Abstr. 012.062.

The 408 MHz all–sky survey was begun in 1965 with the objective of providing an accurate wide dynamic range survey of the whole sky. The 408 MHz survey has been compiled from four sets of observations made using the 100 meter Effelsberg, 76 meter Jodrell Bank and 64 meter Parkes telescopes. The edited and corrected scans were compiled into survey maps using the NOD2 software library (Haslam, 1974) and published in Astronomy and Astrophysics Supplements (Haslam et al., 1982) as an atlas of maps in both α, δ and l, b coordinate systems. The data is also available on 9–track tape in NOD2 format.

141.026   The 2.3 GHz Rhodes survey of the southern sky.
A. Greybe.
Bull. Inf. Cent. Données Stellaires, No. 28, p. 53 – 56 (1985). – See Abstr. 012.062.

A southern sky 2.3 GHz survey is presented. The surveying techniques employed are described and it is shown how a good choice of these can determine the amount of work necessary to correct observations for unwanted effects. Some results are presented and relationships which exist with other surveys are commented on.

141.027   The first identifications of objects found in the 2.695 GHz galactic plane survey.
E. Fürst, W. Reich, P. Reich, Y. Sofue, T. Handa.
Bull. Inf. Cent. Données Stellaires, No. 28, p. 57 – 58 (1985). – See Abstr. 012.062.

Nous avons trouvé des sources nombreuses dans le "survey" radioélectrique du plan galactique à la fréquence de 2.695 GHz. L'observation supplémentaire à 1.42 GHz, 4.75 GHz et 10 GHz permet de déterminer la nature des 10 sources par le spectre et la polarisation: 3 restes de supernovae, 5 régions–H II, un reste de supernovae couvert par la région–H II RCW 164, et une radiogalaxie.

141.028   Absolute calibration of the 1420 MHz Stockert survey.
P. Reich, W. Reich.
Bull. Inf. Cent. Données Stellaires, No. 28, p. 59 – 62 (1985). – See Abstr. 012.062.

141.029   Spectral indices.
W. Reich, P. Reich.
Bull. Inf. Cent. Données Stellaires, No. 28, p. 63 – 64 (1985). – See Abstr. 012.062.

141.030   The Stockert 2.7 GHz radio continuum survey of the northern galactic plane.
K. Reif.
Bull. Inf. Cent. Données Stellaires, No. 28, p. 65 – 66 (1985). – See Abstr. 012.062.

The author reports on a large scale 11 cm radio continuum and linear polarization survey of the northern galactic plane (above –28 degrees declination) and the North Polar Spur region. The angular resolution is 0.3 degrees and the final r.m.s. noise will be below 20mK in brightness temperature.

141.031   Radio through optical studies of sources from the Green Bank–Bonn 5 GHz S–surveys.
H. Kühr.
Bull. Inf. Cent. Données Stellaires, No. 28, p. 67 – 68 (1985). – See Abstr. 012.062.

141.032   9 mm–Radiokontinuumsmessungen von SNRs am Effelsberger 100 Meter–Spiegel.
H. W. Morsi.
Mitt. Astron. Ges., Nr. 63, p. 142 – 146 (1985). – See Abstr. 012.063.

Im mm–Wellenlängenbereich gibt es bisher kaum Kartierungen des Radiokontinuums, so daß hier kurz auf die instrumentellen Gegebenheiten eingegangen wird.

141.033   5 GHz source counts from the MG survey.
C. L. Bennett, C. R. Lawrence, B. F. Burke.
Bull. Am. Astron. Soc., Vol. 16, No. 4, p. 1015 (1984). Abstract. – See Abstr. 010.062.

141.034   Decametric survey of discrete sources in the northern sky. X. Spectra of some radio sources in the frequency range 12.6 – 1415 MHz for declinations 52° to 60°.
S. Ya. Braude, N. K. Sharykin, K. P. Sokolov, S. M. Zakcharenko.
Astrophys. Space Sci., Vol. 111, No. 2, p. 237 – 252 (1985).

The data of the latest decametric band survey performed with the UTR–2 radio telescope are used along with other results obtained at higher frequencies (below 1415 MHz) for plotting spectra of 114 radio sources located in a sky strip between declinations 52° and 60°. Some parameters of the source spectra in the frequency range 12.6 – 1415 MHz are presented.

141.035   Survey of radio sources at 7.7 GHz with the RATAN–600 radio telescope.
V. V. Vitkovskij, A. G. Gorshkov, A. V. Ipatov, D. V. Korol'kov, V. V. Mardyshkin.
Spets. astrofiz. obs. Akad. Nauk SSSR. Prepr., No. 12 L, 29 pp. (1984). In Russian. Abstr. in Ref. Zh., 51. Astron., 5.51.413 (1985).

141.036   Radio sources in the deep sky survey with RATAN–600: cooperative studies of the 13th hour objects in Effelsberg, Tonanzintla and with RATAN–600.
R. Wielebinski, Yu. N. Parijskij, V. V. Vitkovskij, Z. V. Dravskikh, U. Klein, W. Reich, A. V. Temirova.
Pis'ma Astron. Zh., Tom 11, No. 6, p. 403 – 414 (1985). In Russian. English translation in Sov. Astron. Lett., Vol. 11.

Preliminary results of a cooperative research of radio sources found earlier with the RATAN–600 deep sky survey at 7.6 cm are reported. With the aid of the Effelsberg 100–m paraboloid flux densities from 19 radio sources at 6 and/or 11 cm are obtained and declination data for 13 objects are improved. For some radio sources brighter than 8 mJy three–frequency spectra are obtained. Optical identifications are performed.

141.037   Radio studies of galactic nonthermal sources.
R. H. Becker, D. J. Helfand.
Bull. Am. Astron. Soc., Vol. 17, No. 2, p. 595 (1985). Abstract. – See Abstr. 010.065.

141.038   Low radio frequency extragalactic variable sources.
D. B. Garrett.
Diss. Abstr. Int., Sect. B, Vol. 45, No. 3, p. 900 (1984). Thesis, University of Texas, 147 pp. (1983). Order No. DA8414373.

141.039   Report of IAU Commission 40: Radio astronomy (_Radio astronomie_).
K. I. Kellermann.
Trans. IAU, Vol. XIXA, p. 549 – 580 (1985). – See Abstr. 003.046.

141.040   The compact radio source near G357.7–0.1.
P. A. Shaver, S. R. Pottasch, C. J. Salter,
A. R. Patnaik, J. H. van Gorkom, G. C. Hunt.
Astron. Astrophys., Vol. 147, No. 1, p. L23 – L24 (1985).
The compact radio source near the supernova remnant G357.7–0.1 is a prominent infrared source in the IRAS catalogue, and its properties are typical for a normal H II region.

Decametric survey of discrete sources in the northern sky. IX. Source catalogue in the declination range 52° to 60°.
See Abstr. 002.033.

Corrections for biases in slope estimation.
See Abstr. 021.008.

A simple maximum entropy deconvolution algorithm.
See Abstr. 036.031.

Space astronomy: technical comments – infrared, submillimetre and radio astronomy.
See Abstr. 051.077.

Space astronomy: technical comments – interferometry missions.
See Abstr. 051.078.

VLA observations of three extragalactic SNR at 20 and 6 cm.
See Abstr. 125.014.

Radio sources near the millisecond pulsar PSR 1937 + 214.
See Abstr. 126.025.

Structure and origin of the Cygnus superbubble.
See Abstr. 131.056.

Refractive interstellar scintillation and radiosource variations.
See Abstr. 131.284.

The nonthermal radio emission of the galactic center.
See Abstr. 155.099.

A low frequency high resolution galactic plane survey.
See Abstr. 155.125.

An Arecibo survey for extragalactic hydroxyl.
See Abstr. 157.222.

Low frequency variability and interstellar focusing.
See Abstr. 158.107.

A deep Westerbork survey of areas with multicolor Mayall 4 m plates. III. Photometry and spectroscopy of faint source identifications.
See Abstr. 158.112.

The global radio continuum of nearby galaxies.
See Abstr. 158.284.

142 UV Sources, X-ray Sources, X-ray Background

142.001   The Galactic population and birthrate of X–ray pulsars.
D. G. Blair, B. N. Candy.
Mon. Not. R. Astron. Soc., Vol. 212, No. 1, p. 219 – 229 (1985).
In this paper, the authors use the surprising similarity of the observed distribution of X–ray and radio pulsars to show that the total Galactic population of X–ray pulsars is of the order of 10^4. Using lifetime estimates, it is shown that the birthrate of X–ray pulsars may be comparable to the radio pulsar birthrate, and could be as frequent as one per year.

142.002   Optical spectroscopy and photometry of the periodic X–ray transient A0538–66 (X0535–668) during an outburst and an OFF state.
R. H. D. Corbet, K. O. Mason, F. A. Córdova,
G. Branduardi–Raymont, A. N. Parmar.
Mon. Not. R. Astron. Soc., Vol. 212, No. 3, p. 565 – 590 (1985).
The authors present optical spectroscopy and photometry of the periodic X–ray transient A0538–66. Extensive spectroscopy performed in 1982 March follows the evolution of the multi-component line profiles through an outburst. Particularly distinctive are the broad, asymmetric emission lines which are interpreted as arising in an extensive stellar wind. Photometric measurements during an X–ray OFF state in 1983 May reveal V–band variability with an amplitude of 0.2 mag on a time–scale of days. The authors obtain a radial velocity curve that appears to confirm the model for A0538–66 in which a neutron star is in an eccentric orbit about the primary and generates X–ray and optical flares at periastron passage. The long–term evolution of the optical colours of A0538–66 during quiescence reflects the changing nature of the circumstellar matter that envelops the primary star. No evidence of the star's 69–ms X–ray periodicity is found in optical data taken during an outburst in 1982 May.

142.003   A new fast X–ray pulsar.
F. D. Seward.
Phys. Today, Vol. 38, No. 1, p. S8 (1985). Short review paper. – See Abstr. 013.003.

142.004   The spectra of X–ray transients.
N. E. White, J. L. Kaluzienski, J. H. Swank.
AIP Conf. Proc., No. 115, p. 31 – 48 (1984). – See Abstr. 012.005.
The properties of the persistent X–ray sources (such as X–ray bursts) are also seen from the transient X–ray source suggesting that they are related binary systems containing accreting compact objects. The spectral properties of the X–ray transients are considered and compared with those of the persistent sources. A comparative study of the spectra of the transients with those of the persistent sources suggests that two thirds of the transients seen to date result from episodic accretion onto neutron stars with either low magnetic fields (the soft bulge–like transients) or high magnetic fields (the hard pulsing transients). The remaining transients have unusually soft (ultra–soft) spectra similar to those of the black hole candidates in their high state.

142.005   X–ray properties of the galactic bulge sources.
N. Shibazaki, K. Mitsuda.
AIP Conf. Proc., No. 115, p. 63 – 71 (1984). – See Abstr. 012.005.
Results are presented on observations by HAKUCHO and TENMA satellites on the galactic bulge X–ray sources. Most of the sources observed showed a positive correlation between intensity and spectral hardness. GX 5–1 has been found to have two characteristic states. Quasi–periodic pulsations with 106 to 205 seconds have been observed from GX 349 + 2. Observational results are interpreted in terms of mass accretion onto an unmagnetized neutron star.

142.006   Hard X–ray recurrent transients.
N. Shibazaki.
AIP Conf. Proc., No. 115, p. 72 (1984). Abstract. – See Abstr. 012.005.

142.007 The A–1 survey of fast X–ray transients.
C. Ambruster, K. S. Wood.
AIP Conf. Proc., No. 115, p. 73 – 75 (1984). – See Abstr. 012.005.

A systematic search of the HEAO A–1 Sky Survey Experiment data base for fast (< 3 hour) transients is described. Seven new transient or flaring sources were found, all relatively faint. dMe flare stars are present in two of the error boxes and NGC 1841, a globular cluster in the LMC, lies just outside a third.

142.008 Ten years of Vela X–ray observations.
J. Terrell, W. C. Priedhorsky.
AIP Conf. Proc., No. 115, p. 76 – 79 (1984). – See Abstr. 012.005.

The Vela spacecraft, particularly Vela 5B, produced all–sky X–ray data of unprecedented length and completeness. Recent re–analysis has put the data in the form of 10–day skymaps covering a 7–year period, which have led to the discovery or confirmation of a number of long–term periodicities, and have made possible a time–lapse movie of the X–ray sky.

142.009 Erratic variability in MXB 1916–05.
P. Hertz, K. S. Wood.
AIP Conf. Proc., No. 115, p. 80 – 82 (1984). – See Abstr. 012.005.

Five day light curves of MXB 1916–05 obtained 6 months apart reveal two modes of behavior. In the Fall 1977 the source had several extended periods of low flux as well as generally erratic variability. In the Spring 1978 the flux steadily increased over several days. The 50 minute period was not detected. The erratic variability is interpreted as an instability in the accretion disk.

142.010 Recent observations of transient and variable X–ray sources with satellites "Hakucho" and "Tenma".
M. Oda.
AIP Conf. Proc., No. 115, p. 103 – 120 (1984). – See Abstr. 012.005.

Highlights of observational results obtained by "Hakucho" and "Tenma" are reviewed. Emphasis is given to the secular variations of X–ray pulsars and the time–behaviour of X–ray bursters.

142.011 Early results from the X–ray astronomy satellite "Tenma".
H. Inoue.
AIP Conf. Proc., No. 115, p. 121 – 130 (1984). – See Abstr. 012.005.

Preliminary results from observations of X–ray pulsars and X–ray bursters with the Japanese X–ray satellite "Tenma" are presented.

142.012 X–ray pulsars observed with Hakucho and Tenma.
F. Nagase, N. Sato, K. Makishima, N. Kawai, K. Mitani.
AIP Conf. Proc., No. 115, p. 131 – 138 (1984). – See Abstr. 012.005.

The pulse periods of eight pulsars Her X–1, Cen X–3, Vela X–1, A 0535+26, GX 301–2, 4U 1626–67, 4U 1538–52, and OAO 1657–415, have been measured with the Hakucho and Tenma satellites between March 1979 and July 1983. Not only the wind–fed X–ray pulsars such as Vela X–1, GX 301–2 and A 0535–26, but also Her X–1 and Cen X–3, which are considered to be disk–fed pulsars, exhibit a remarkable change in the spin–up rate.

142.013 Discovery of eclipses from the X–ray burst source MXB 1659–29.
L. R. Cominsky, K. S. Wood.
AIP Conf. Proc., No. 115, p. 139 – 142 (1984). – See Abstr. 012.005.

The authors have discovered a 7.1 h period and eclipses in the X–ray emission from MXB 1659–29. MXB 1659–29 is the first burst source to show persistent, regular eclipses. Irregular variability is observed during approximately 25% of the cycle (∼1.5 hours) immediately preceding the eclipse, which is only 15 minutes in duration. The authors derive limits to the mass of the binary companion to the X–ray emitting neutron star, and argue that the companion is less massive than the canonical prediction for Roche lobe filling main sequence stars.

142.014 Joint timing analysis of A 0535+26 at 1980 outburst.
N. Kawai, F. Nagase.
AIP Conf. Proc., No. 115, p. 146 – 149 (1984). – See Abstr. 012.005.

The pulse–timing data of A 0535+26, obtained by the Hakucho satellite in October 1980 are analyzed taking into consideration other observations. It is shown that an accretion torque model can explain the period change during the 1980 outburst.

142.015 SMC X–1: another tilted precessing accretion disk?
D. E. Gruber.
AIP Conf. Proc., No. 115, p. 150 – 153 (1984). – See Abstr. 012.005.

The 13 – 70 keV emission from the 0.7 s X–ray pulsar SMC X–1 was monitored by the UCSD/MIT instrument aboard HEAO–1 during three ∼80–day epochs in 1977 and 1978. The X–ray flux wandered slowly between a maximum value and unobservably small levels with a rough time scale of 60 days. This pattern resembles, but is less regular than, the ∼1 month cycles of Her X–1 and LMC X–4, which are attributed to regular occultation by a tilted precessing accretion disk. There is no evidence for cyclotron lines in the spectrum.

142.016 The occasional disappearance of GX 301–2 pulses.
H. Inoue.
AIP Conf. Proc., No. 115, p. 154 – 157 (1984). – See Abstr. 012.005.

The X–ray satellite Hakucho observed the binary X–ray pulsar GX 301–2 (4U 1223–62) from April 6 through 20 and May 16 through 21, 1982. The overall light curve shows two large flares peaked on April 8 – 9 and May 19 – 20, respectively. The system exhibits large pulse amplitude variations on a time scale of a few hours. A possible mechanism for these variations is proposed.

142.017 Steady emission from recurrent transient pulsar A 0535+26.
R. K. Manchanda, A. Bazzano, V. F. Polcaro, C. D. La Padula, P. Ubertini.
AIP Conf. Proc., No. 115, p. 158 – 164 (1984). – See Abstr. 012.005.

A steady hard X–ray emission between 20 – 100 keV was observed from the 104 sec pulsar A 0535+26 during the quiescent phase of transient activity. The present observations correspond to the binary phase of 0.7 taking 110 d as the binary period. The observed flux was comparable to ∼20 milli–crab and a power law spectrum with spectral index α ∼ 1.2 fits the data.

142.018 Some remarks about X–ray burst sources.
W. H. G. Lewin.
AIP Conf. Proc., No. 115, p. 249 – 256 (1984). – See Abstr. 012.005.

142.019 A precursor in a large X–ray burst and the expanding envelope of the neutron star.
Y. Tawara, T. Kii, S. Hayakawa.
AIP Conf. Proc., No. 115, p. 257 – 262 (1984). – See Abstr. 012.005.

A very long X–ray burst with precursor from XB 1715–321 was observed on July 20, 1982 by the Hakucho satellite. There is evidence for luminosity saturation and for very soft X–ray spectra for both the decline of the precursor and the rise of the main burst. The data are interpreted by the thermonuclear flash model for an accreting neutron star and by assuming a rapid expansion of the photospheric envelope of the neutron star during the flash.

142.020 Globular cluster origin of X–ray bursters.
J. E. Grindlay.
AIP Conf. Proc., No. 115, p. 306 – 318 (1984). – See Abstr. 012.005.

X–ray bursters and galactic bulge X–ray sources are reasonably well constrained in their basic nature but not in their origin. The author has suggested they may all have been produced by tidal capture in high density cores of globular clusters, which have now largely been disrupted by tidal stripping and shocking in the galactic plane. General arguments are presented for cluster disruption by the possible ring of giant molecular clouds in the Galaxy. Tests of the cluster disruption hypothesis are in progress and preliminary results are summarized here. The G–K star "companions" previously noted for at least 4 bursters have spectra (in the two cases observed) consistent with metal rich cluster giants.

142.021   EINSTEIN observations of the soft X–ray background close to the galactic plane.
J.–P. Caillault, S. M. Kahn.
News Lett. Astron. Soc. N.Y., Vol. 2, No. 7, p. 9 – 10 (1985).
Abstract. – See Abstr. 010.241.

142.022   Further spectroscopy of LMC X–ray candidates.
D. Crampton, A. P. Cowley, I. B. Thompson, J. B. Hutchings.
Astron. J., Vol. 90, No. 1, p. 43 – 48 (1985). With plates 3 – 4.

This paper presents optical spectroscopic observations, mostly obtained with the Shectograph on the du Pont telescope of the Las Campanas Observatory, of candidates for 12 X–ray sources in the LMC. Most objects were observed on ten consecutive nights. Five of the sources may be associated with Be stars in the LMC. Some previously proposed candidates have been eliminated as probable counterparts for the X–ray sources on the basis of the new data. Nine of these sources have only Einstein Observatory IPC positions with uncertainties of $\pm 30''$, so the identifications are tentative.

142.023   Comptonization–softening and the hard X–ray spectrum of Cyg X–1.
Z.–g. Ye, J.–h. You.
Astrophys. Space Sci., Vol. 109, No. 1, p. 155 – 173 (1985).

The Comptonization–softening of very hard X–ray photons with $E \lesssim m_0 c^2$ in the "cold" electron gas is discussed. The frequency diffusion equation for Comptonization of hard X–rays has been derived to the zero–temperature approximation. By use of this equation, and under the assumption of pair–annihilation origin of hard X–rays, the authors have calculated the energy spectrum with $E > 80$ keV, for Cyg X–1, which is in good agreement with the observation. The high–energy edge ~ 400 keV of the observed spectrum and the small bump in the range 100 – 200 keV can also be explained in this way.

142.024   High energy X–ray observations of COS–B gamma–ray sources from OSO–8.
J. F. Dolan, P. A. Caraveo, C. J. Crannell, B. R. Dennis, K. J. Frost, L. E. Orwig.
Astron. Astrophys., Vol. 143, No. 1, p. 1 – 7 (1985).

During the three years between satellite launch in June 1975 and turn–off in October 1978, the high energy X–ray spectrometer on board OSO–8 observed nearly all of the COS–B gamma–ray source positions given in the 2CG catalog. An X–ray source was detected at energies above 20 keV at the 6σ level of significance in the γ–ray error box containing 2CG342–02 and at the 3σ level of significance in the error boxes containing 2CG065 + 00, 2CG195 + 04, and 2CG311–01. No definite association between the X–ray and gamma–ray sources can be made from the authors' data alone. Upper limits are given for the 2CG sources from which no X–ray flux was detected above 20 keV.

142.025   The peculiar X–ray source AS 431.
J.–P. Caillault, G. A. Chanan, D. J. Helfand, J. Patterson, J. Nousek, L. Takala, G. Bothun.
News Lett. Astron. Soc. N.Y., Vol. 2, No. 5, p. 26 – 27 (1984).
Abstract. – See Abstr. 010.243.

142.026   High–resolution spectroscopy of the optical candidate for the X–ray transient X0331 + 53.
J. Stocke, D. Silva, J. H. Black, K. Kodaira.
Publ. Astron. Soc. Pac., Vol. 97, No. 588, p. 126 – 129 (1985).

High–resolution spectral scans (0.1 Å – 1 Å) of the optical candidate for the X–ray transient X0331 + 53 in January 1984 show broad hydrogen and helium emission (FWZI $= 335$ km s^{-1}) and strong interstellar absorption corresponding to approximately 7 magnitudes of visual extinction. No photospheric absorption and no radial–velocity variations in the emission lines greater than 0.5 km s^{-1} over a three–day period were found. This leads to the conclusion that the optical continuum does not arise at or near the primary star.

142.027   The discovery of 4.4 second X–ray pulsations from the rapidly variable X–ray transient V0332 + 53.
L. Stella, N. E. White, J. Davelaar, A. N. Parmar, R. J. Blissett, M. van der Klis.
Astrophys. J., Lett. Ed., Vol. 288, No. 2, p. L45 – L49 (1985).

Using the EXOSAT Observatory the authors have observed three outbursts from the transient X–ray source V0332 + 53 between 1983 November and 1984 January. They find that in addition to the rapid Cyg X–1–like variability previously reported by Tanaka et al. this source also displays stable pulsations with a period of 4.4 s. Doppler variations in the pulse period indicate that the pulsar is in a 34.25 day binary orbit with an eccentricity of 0.31. The times of periastron passage are close to those of the X–ray outbursts.

142.028   The nature of the low–luminosity globular cluster X–ray sources.
P. Hertz, K. S. Wood.
Astrophys. J., Vol. 290, No. 1, p. 171 – 184 (1985).

An X–ray survey of 134 galactic globular clusters has been conducted with *HEAO 1*. No new globular cluster X–ray sources were discovered, and improved upper limits were obtained on X–ray emission from 67 globular clusters. The lack of new sources confirms the gap in the globular cluster luminosity function noted by Hertz and Grindlay. A simple analytical calculation of the luminosity function has been carried out under the assumption that low–luminosity globular cluster X–ray sources are tidally captured white dwarfs accreting matter from a Roche lobe filling, low–mass companion. The calculated luminosity function of low–luminosity globular cluster X–ray sources agrees with that observed in the X–ray surveys.

142.029   X–ray observations of the runaway stars HD 206327 and 26 Cephei and of the λ^1Orionis region.
R. C. Stone, R. E. Taam.
Astrophys. J., Vol. 291, No. 1, p. 183 – 187 (1985). With plate 5.

Observations made with the IPC on the Einstein Observatory are reported for the runaway OB–type stars HD 206327 and 26 Cephei and also of the λ^1Orionis region. Only upper limits for the X–ray emissions from the two runaway stars were determined. Implications for possible compact companions are considered. A single IPC observation of the region centered on λ^1Orionis has found five serendipitous X–ray sources. An astrometric and photometric study of this region provides evidence that these sources are probable members of the association. Two of the sources are identified with normal OB stars, and the others are most likely pre–main–sequence stars that have evolved from T–Tauri stars.

142.030   On detection of astrophysical chaos phenomena.
W. Unno, K. Urata.
Astron. Her., Vol. 78, No. 4, p. 92 – 96 (1985). In Japanese.

142.031   Discovery of absorption lines in X–ray burst spectra from X1636–536.
H. Inoue, K. Koyama.
Adv. Space Res., Vol. 5, No. 3, p. 91 – 94 (1985). – See Abstr. 012.059.

Absorption features in the X–ray burst spectra from X1636–536 were observed with gas scintillation proportional counters on board the X–ray astronomy satellite Tenma. These are most probably due to the gravitationally redshifted iron K–absorption line near the surface of a neutron star.

142.032 **X–ray pulsars observed from Tenma.**
F. Nagase.
Adv. Space Res., Vol. 5, No. 3, p. 95 – 99 (1985). – See Abstr. 012.059.

Eleven X–ray pulsars have been observed by Tenma between March 1983 and April 1984. Detailed structures in the energy spectra of these X–ray pulsars were obtained using a set of gas scintillation proportional counters aboard Tenma together with pulse periods and pulse profiles. Most of the X–ray pulsars exhibit iron emission lines with the center energy at 6.4 keV.

142.033 **The rapid burster activity observed from Tenma.**
H. Kunieda.
Adv. Space Res., Vol. 5, No. 3, p. 105 – 108 (1985). – See Abstr. 012.059.

The activity of the rapid burster MXB 1730–335 has been recorded about twice a year from 1971 to April 1979. In August 1979, the activity was observed from the Japanese X–ray astronomy satellite Hakucho. After a long quiescence of four years, in August 1983 the activity was observed from the two Japanese X–ray astronomy satellites Hakucho and Tenma. In this paper, the authors present the results of both and implications on the properties of the bursts.

142.034 **Recent optical identifications of HEAO–1 X–ray sources.**
D. A. Schwartz, H. Bradt, D. Buckley, J. Patterson, R. Remillard, W. Roberts, I. Tuohy.
Adv. Space Res., Vol. 5, No. 3, p. 137 – 140 (1985). – See Abstr. 012.059.

The HEAO–1 satellite has produced the most complete, all-sky, X–ray survey. Approximately 350 out of 840 sources have certain or highly probable optical counterparts. The identifications are necessary for estimating distances to the objects, so that one measures properties intrinsic to the source, and for establishing the classifications of the astronomical systems. The authors review their methods and discuss the new active galactic nuclei X–ray counterparts.

142.035 **Optische Beobachtungen der Röntgenquelle E2003 + 225.**
B. Fuhrmann.
Mitt. Veränderliche Sterne, Band 10, Heft 4, p. 97 – 101 (1984).

Observations on 413 Sonneberg plates are discussed. The period, which has been refined slightly, and the mean brightness proved to be stable over 50 years. However, the shape of the light curve underwent strong variations.

142.036 **Optical studies of the X–ray transient EXO 0748–676.**
R. A. Wade, H. Quintana, K. Horne, T. R. Marsh.
Prepr. Steward Obs., No. 583, 11 pp. (1985). Submitted to Publ. Astron. Soc. Pac.

142.037 **Endpoints of stellar evolution: X–ray surveys of the Local Group.**
D. J. Helfand.
Bull. Am. Astron. Soc., Vol. 16, No. 4, p. 874 (1984). Abstract. – See Abstr. 010.062.

142.038 **The Einstein Observatory Medium Sensitivity Survey: present results and future prospects.**
T. Maccacaro.
Bull. Am. Astron. Soc., Vol. 16, No. 4, p. 880 (1984). Abstract. – See Abstr. 010.062.

142.039 **E2000 + 223: a newly discovered old nova.**
L. O. Takalo, J. A. Nousek.
Bull. Am. Astron. Soc., Vol. 16, No. 4, p. 898 (1984). Abstract. – See Abstr. 010.062.

142.040 **Ultrasoft X–ray background observations towards the north galactic pole.**
J. J. Bloch, K. Jahoda, M. Juda, D. McCammon, W. T. Sanders, S. L. Snowden.
Bull. Am. Astron. Soc., Vol. 16, No. 4, p. 910 (1984). Abstract. – See Abstr. 010.062.

142.041 **Long term X–ray observations of galactic bulge sources.**
W. C. Priedhorsky.
Bull. Am. Astron. Soc., Vol. 16, No. 4, p. 933 – 934 (1984). Abstract. – See Abstr. 010.062.

142.042 **Variability of galactic X–ray sources from HEAO A–1.**
P. Hertz, K. S. Wood.
Bull. Am. Astron. Soc., Vol. 16, No. 4, p. 944 (1984). Abstract. – See Abstr. 010.062.

142.043 **The CFA Einstein observatory high–sensitivity X–ray survey: imaging proportional counter results.**
F. A. Primini, R. I. Burg, J. Huchra, R. Schild, S. S. Murray.
Bull. Am. Astron. Soc., Vol. 16, No. 4, p. 964 (1984). Abstract. – See Abstr. 010.062.

142.044 **Voyager observations of the far UV background at the North Galactic Pole.**
J. B. Holberg.
Bull. Am. Astron. Soc., Vol. 16, No. 4, p. 981 (1984). Abstract. – See Abstr. 010.062.

142.045 **Einstein observations of the soft X–ray background close to the galactic plane.**
J.–P. Caillault, S. M. Kahn.
Bull. Am. Astron. Soc., Vol. 16, No. 4, p. 981 (1984). Abstract. – See Abstr. 010.062.

142.046 **Spectra and positions of galactic center high–energy X–ray sources.**
F. K. Knight, J. L. Matteson, G. V. Jung, R. E. Rothschild.
Bull. Am. Astron. Soc., Vol. 16, No. 4, p. 981 (1984). Abstract. – See Abstr. 010.062.

142.047 **Ionospheric detection of X–ray pulsars.**
H. S. Hudson, R. Te Kolste.
Bull. Am. Astron. Soc., Vol. 16, No. 4, p. 982 (1984). Abstract. – See Abstr. 010.062.

142.048 **A six–second periodic X–ray source in Carina.**
F. D. Seward, P. A. Charles.
Bull. Am. Astron. Soc., Vol. 16, No. 4, p. 983 (1984). Abstract. – See Abstr. 010.062.

142.049 **NGC 6212: an elliptical galaxy with a highly active nucleus.**
P. Biermann, R. Strom, N. Bartel.
Astron. Astrophys., Vol. 147, No. 2, p. L27 – L28 (1985).

The authors have identified a strong variable X–ray source with the elliptical galaxy NGC 6212. Westerbork observations at 327, 609 and 1412 MHz show the presence of an unresolved (< 12″) radio source with a nonthermal spectrum centered on the galaxy. The X–ray luminosity of $\sim 10^{43}$ erg sec^{-1}, variability and presence of a radio component suggest an active nucleus in NGC 6212, possibly a new BL Lac or OVV. The object deserves the attention of optical spectroscopists in particular.

142.050 **Erratum: ”A tentative interpretation of the cyclotron line energy variation in Her X–1” [Astron. Astrophys., Vol. 136, No. 2, p. 363 – 367 (1984)].**
N. Prantzos, P. Durouchoux.
Astron. Astrophys., Vol. 147, No. 2, p. 335 (1985). See Abstr. 38.142.004.

142.051 **EXO 0748–676.**
IAU Circ., Nos. 4039, 4043, 4047, 4054, 4057, 4076 (1985).

142.052 **GX 5–1 (4U 1758–250).**
IAU Circ., Nos. 4043, 4046 (1985).

142.053 **4U 1323–62.**
IAU Circ., No. 4044 (1985).

142.054 **EXO 1846–031.**
IAU Circ., Nos. 4051, 4059 (1985).

142.055 **4U 1624–49.**
IAU Circ., No. 4051 (1985).

142.056 **EXO 1747+214.**
IAU Circ., No. 4058 (1985).

142.057 **Scorpius X–1 = V818 Scorpii.**
IAU Circ., Nos. 4060, 4068 (1985).

142.058 **EXO 2030+375.**
IAU Circ., Nos. 4066, 4073 (1985).

142.059 **V1341 Cygni.**
IAU Circ., No. 4070 (1985).

142.060 **High–resolution spectroscopic observation of Vela X–1 in the hard X–ray energy range.**
J. Tueller, T. L. Cline, B. J. Teegarden, P. Durouchoux, N. Prantzos.
Bull. Am. Astron. Soc., Vol. 17, No. 2, p. 552 (1985). Abstract. – See Abstr. 010.065.

142.061 **X–ray sources in globular clusters and in the plane of the Galaxy.**
P. L. Hertz.
Mercury, Vol. 14, No. 2, p. 42 – 43 (1985).

142.062 **Spectrophotometry of the probable optical counterpart of the transient X–ray source V0332+53.**
R. K. Honeycutt, E. M. Schlegel.
Publ. Astron. Soc. Pac., Vol. 97, No. 590, p. 300 – 302 (1985).
Spectrophotometry shows that the probable optical counterpart of V0332+53 is an OB star with $E_{B-V} = 1.9$.

142.063 **Quasi–periodic oscillations in galactic bulge sources observed by Hakucho.**
M. Matsuoka.
Cataclysmic variables and low–mass X–ray binaries, p. 139 – 142 (1985). – See Abstr. 012.072.
The first object of the paper is to present the longer periodicities of two galactic bulge sources (GX 349+2 and 4U/MXB 1636–53) suggested by the Hakucho data. The second is to present a quasi–periodicity of GX 349+2 between 100 – 200 sec.

142.064 **Correlation between X–ray luminosity and the deviation from the main sequence colour diagram for stellar X–ray sources.**
P. K. Kunte, A. R. Rao, M. N. Vahia.
18th International Cosmic Ray Conference, Vol. 1, p. 13 – 15 (1983). – See Abstr. 012.096.
The authors have investigated the variations in the X–ray luminosity with deviation from main sequence colour diagram for stellar X–ray sources. It is observed that the early and late type stars follow two different patterns. They believe that the anomalies seen in this pattern are related to the interaction of the stars with their neighbourhood and propose that such regions should contribute to the composition of interstellar particles and to low energy cosmic rays.

142.065 **Decay of massive photinos and gravitinos and keV X–ray background.**
C. Sivaram.
18th International Cosmic Ray Conference, Vol. 1, p. 20 – 22 (1983). – See Abstr. 012.096.

142.066 **X rays from cosmic ray interactions.**
A. W. Wolfendale, C.–x. Xu, M. Sadzinska, J. Wdowczyk.
18th International Cosmic Ray Conference, Vol. 1, p. 23 – 26 (1983). – See Abstr. 012.096.
New data on diffuse X rays in the range 2 – 60 keV from HEAO–1 have been analysed and the dependence of emissivity on Galactic position has been determined. The extent to which the X rays can be understood in terms of cosmic ray interactions is assessed.

142.067 **The anisotropy of the diffuse cosmic X–ray background.**
M. R. Issa, M. Sadzinska, J. Wdowczyk, A. W. Wolfendale, C. X. Xu.
18th International Cosmic Ray Conference, Vol. 1, p. 27 – 30 (1983). – See Abstr. 012.096.
The recent data of Iwan et al. (1982) for diffuse X–rays in the range 2 – 60 keV have been analysed from the standpoint of the anisotropy in the extragalactic flux. It is concluded that there is no inconsistency with the well known anisotropy in the microwave background although the X–ray data are still not sufficiently accurate to enable a statement of clear consistency to be made.

142.068 **Discussion on zero–point field acceleration and applicability to cosmic rays, X–ray background and energetics of the intergalactic medium.**
A. Rueda.
18th International Cosmic Ray Conference, Vol. 1, p. 31 – 34 (1983). – See Abstr. 012.096.
As intergalactic space (IGS) occupies over 99% of the available space in the Universe and as the density is so low $(10^{-5} – 10^{-6} cm^{-3})$, the cumulative action of a real (as opposed to virtual) zero–point field on IGS particles should energize the intergalactic medium and contribute to the isotropic cosmic X and γ–ray backgrounds.

142.069 **X–ray emission associated with gamma–ray bursts.**
J. Nishimura, M. Fujii, T. Yamagami.
18th International Cosmic Ray Conference, Vol. 1, p. 50 – 53 (1983). – See Abstr. 012.096.
A mechanism for producing X rays at the time of a gamma–ray burst is investigated. The authors assume that X rays are emitted from the surface of a neutron star which is heated by the irradiation of the burst gamma rays. A large flux of X rays is expected in this model, and the astrophysical significance of this result is discussed in relation to the existing X–ray observations of the gamma–ray bursts.

142.070 **Results from the HXR 81 M hard X–rays sky survey experiment.**
P. Ubertini, A. Bazzano, C. D. La Padula, V. F. Polcaro, G. Zambon, R. K. Manchanda.
18th International Cosmic Ray Conference, Vol. 9, p. 1 – 4 (1983). – See Abstr. 012.096.
A deep survey between R.A. = 04^h00^m to 17^h20^m and dec = $23°$ to $53°$ was performed during a single long duration (17h) transmediterranean balloon flight (HXR 81 M) with a 10800 cm^2 geometric area payload. The flux levels and some of the spectral characteristics of 13 hard X–ray sources, determined during the experiment, are reported.

142.071 **Recent results on galactic X–ray sources.**
Y. Tanaka.
18th International Cosmic Ray Conference, Vol. 12, p. 91 – 108 (1983). – See Abstr. 012.096.
The author presents recent progress on binary X–ray sources. He attempts to point out current problems and puzzles that are yet to be solved. Contents: Classification of compact–binary X–ray sources. X–ray pulsars: pulse profiles and related problems, iron emission line and the question: fan beams or pencil beams?, pulse period variations. X–ray bursts: burst activity and thermonuclear flash model, burst peak luminosity, energy spectrum of burst and possible gravitational redshift, rapid burster. Black–hole candidates.

142.072 Einstein observations of Aql X–1.
M. Czerny, B. Czerny, J. E. Grindlay.
Bull. Am. Astron. Soc., Vol. 17, No. 2, p. 554 (1985). Abstract. –
See Abstr. 010.065.

142.073 A search for 0.2 – 2.0 keV millisecond pulsations in galactic bulge X–ray sources.
S. Mereghetti, J. E. Grindlay.
Bull. Am. Astron. Soc., Vol. 17, No. 2, p. 554 – 555 (1985). Abstract. – See Abstr. 010.065.

142.074 Discovery of the first X–ray burst from 4U2129 + 470.
M. R. Garcia, J. E. Grindlay.
Bull. Am. Astron. Soc., Vol. 17, No. 2, p. 565 (1985). Abstract. –
See Abstr. 010.065.

142.075 On the interpretation of the ultra–soft X–ray background: the effects of an "embedded" cloud geometry.
S. M. Kahn, P. Jakobsen.
Bull. Am. Astron. Soc., Vol. 17, No. 2, p. 567 – 568 (1985). Abstract. – See Abstr. 010.065.

142.076 Fluctuation analysis of high sensitivity IPC images.
R. I. Burg, R. Shafer, A. Fabian, S. S. Murray,
F. A. Primini.
Bull. Am. Astron. Soc., Vol. 17, No. 2, p. 578 (1985). Abstract.
See Abstr. 010.065.

142.077 Long term variability of pulse profile of Her X–1.
Y. Soong, D. E. Gruber, R. E. Rothschild.
Bull. Am. Astron. Soc., Vol. 17, No. 2, p. 589 (1985). Abstract. –
See Abstr. 010.065.

142.078 High–resolution X–ray spectroscopy of three galactic bulge sources.
S. D. Vrtilek, G. A. Chanan, D. J. Helfand, S. M. Kahn,
F. D. Seward.
Bull. Am. Astron. Soc., Vol. 17, No. 2, p. 590 (1985). Abstract. –
See Abstr. 010.065.

142.079 EXOSAT observations of the transient EXO 0748–676.
M. Gottwald, A. Parmar, P. Giommi, F. Haberl,
N. E. White.
Bull. Am. Astron. Soc., Vol. 17, No. 2, p. 590 (1985). Abstract. –
See Abstr. 010.065.

142.080 An X–ray survey of the North Ecliptic Pole region.
F. E. Marshall, G. A. Reichert, R. A. White.
Bull. Am. Astron. Soc., Vol. 17, No. 2, p. 596 – 597 (1985). Abstract. – See Abstr. 010.065.

142.081 The bright 6–Nov–77 fast transient: constraints on a quiescent counterpart.
A. Connors, L. Takala.
Bull. Am. Astron. Soc., Vol. 17, No. 2, p. 598 (1985). Abstract. –
See Abstr. 010.065.

142.082 Soft diffuse X–ray absorption and low and high velocity gas.
S. L. Snowden.
Bull. Am. Astron. Soc., Vol. 17, No. 2, p. 611 (1985). Abstract. –
See Abstr. 010.065.

142.083 Report of IAU Commission 48: High energy astrophysics (*Astrophysique de grande énergie*).
R. Giacconi.
Trans. IAU, Vol. XIXA, p. 695 (1985). – See Abstr. 003.046.

142.084 Discovery of intensity–dependent quasi–periodic oscillations in the X–ray flux of GX 5–1.
M. van der Klis, F. Jansen, J. van Paradijs, W. H. G. Lewin,
E. P. J. van den Heuvel, J. E. Trümper, M. Sztajno.
ESLAB 85/57. Space Science Department, European Space Research and Technology Centre (ESTEC), Keplerlaan 1,
Postbus 299, 2200 AG Noordwijk, The Netherlands, 28 pp.
(1985). Accepted by Nature.

142.085 Calculation of neutrino flux from Cygnus X–3.
T. K. Gaisser, T. Stanev.
Phys. Rev. Lett., Vol. 54, No. 20, p. 2265 – 2268 (1985).
The authors estimate the flux of neutrinos produced by protons accelerated by the compact partner in Cygnus X–3. Such neutrinos are produced in collisions with the atmosphere of the companion, and there can be substantial modulation due to absorption in the companion. The resulting flux of upward muons induced by neutrinos in rock below an underground detector is at the level of 1 per 1000 m^2 per year.

142.086 Circinus X–1: a laboratory for studying the accretion phenomenon in compact binary X–ray sources.
J. L. Robinson Saba.
Diss. Abstr. Int., Sect. B, Vol. 45, No. 3, p. 902 (1984). Thesis, University of Maryland (1983).

142.087 A survey of fast X–ray transients using the HEAO A–1 Sky Survey experiment.
C. W. Ambruster.
Diss. Abstr. Int., Sect. B, Vol. 45, No. 7, p. 2199 (1985). Thesis, University of Pennsylvania, 281 pp. (1984). Order No. DA8422885.

142.088 The X–ray sky, 1969 – 1976, and a 1973 transient.
J. Terrell.
Proceedings of the Southwest Regional Conference for Astronomy and Astrophysics, Vol. 10, p. 37 (1985). – See Abstr. 012.125.

Sky catalogue 2000.0. Volume 2: Double stars, variable stars and nonstellar objects.
See Abstr. 002.019.

Documentation for the machine–readable version of the *Revised S201 Catalog of Far–Ultraviolet Objects* (Page, Carruthers and Heckathorn 1982).
See Abstr. 002.036.

The sky in X–ray emission.
See Abstr. 003.055.

Corrections for biases in slope estimation.
See Abstr. 021.008.

All sky high resolution cameras for hard and soft X–rays.
See Abstr. 035.004.

EXOSAT – present status and future prospects.
See Abstr. 035.024.

An imaging optical/UV monitor for X–ray astronomy observatories.
See Abstr. 035.064.

The X–ray Timing Explorer.
See Abstr. 051.004.

The X–ray view of the universe.
See Abstr. 051.031.

Hard X–ray astronomy from balloons.
See Abstr. 051.038.

The Exosat mission.
See Abstr. 051.057.

The Tenma mission.
See Abstr. 051.058.

X–ray sky surveys and the ROSAT mission.
See Abstr. 051.068.

Detection of X–ray sources with ROSAT.
See Abstr. 051.069.

Space astronomy: technical comments – X– and gamma–ray astronomy.
See Abstr. 051.076.

EXOSAT, European X–Ray Astronomy Satellite.
See Abstr. 051.122.

X–ray astronomy and plasma astrophysics.
See Abstr. 062.090.

The X–radiation of optically thick plasma: calculation of the comptonization contribution.
See Abstr. 062.179.

Current–carrying jets.
See Abstr. 062.200.

Theory of optical flashes.
See Abstr. 063.004.

Two–photon cyclotron emission in accretion columns.
See Abstr. 064.046.

Radiation transfer in accretion disk coronae.
See Abstr. 064.086.

Evolution of binary systems into transient sources.
See Abstr. 065.010.

The evolution of binaries of moderate primordial mass ($\lesssim 10\ M_\odot$) and the formation of transient and explosive objects due to accumulation instabilities in such binaries.
See Abstr. 065.011.

Theories of accreting X–ray pulsars.
See Abstr. 067.007.

Accretion by magnetic neutron stars.
See Abstr. 067.008.

Accretion onto magnetized neutron stars: magnetospheric structure and stability.
See Abstr. 067.009.

Spin–reversed accretion onto neutron stars and period fluctuation of X–ray pulsars.
See Abstr. 067.011.

The thermonuclear flash model for X–ray bursts.
See Abstr. 067.013.

Repeated thermonuclear flashes on an accreting neutron star.
See Abstr. 067.014.

A theoretical calculation of rapid X–ray transients and radius expansion.
See Abstr. 067.015.

Shell flashes interacting with the core of neutron stars.
See Abstr. 067.016.

Emission region of X–ray bursters.
See Abstr. 067.018.

Radiative transfer in the accretion column of X–ray pulsars: effects from the hot spot.
See Abstr. 067.057.

The radius of a neutron star: an interpretation of absorption lines from X–ray burster X1636–536.
See Abstr. 067.122.

X–ray observations by the Einstein satellite and neutron star cooling.
See Abstr. 067.128.

The X–ray burst sources.
See Abstr. 067.129.

Further investigation of rapid X–ray bursts.
See Abstr. 067.136.

Hydrodynamic simulations of a combined hydrogen, helium thermonuclear runaway on a 10 km neutron star.
See Abstr. 067.142.

Theoretical cyclotron and pulse profiles of X–ray pulsars.
See Abstr. 067.182.

The origin of neutron stars.
See Abstr. 067.213.

High resolution hard X–ray spectra of solar and cosmic sources.
See Abstr. 076.048.

The peculiar X–ray and radio star AS431.
See Abstr. 112.025.

The X–ray corona of Procyon.
See Abstr. 112.052.

X–ray observations of Wolf–Rayet stars.
See Abstr. 112.053.

X–ray observations of late–type stars.
See Abstr. 112.101.

Optical spectrum of HDE 245770 and X–ray flares of the transient source A 0535 + 26.
See Abstr. 114.024.

Long–term and orbital variability of the optical spectrum of HDE 245770 = A 0535 + 26.
See Abstr. 114.076.

An Einstein Observatory X–ray survey of main–sequence stars with shallow convection zones.
See Abstr. 116.028.

IUE and optical observations of LMC X–ray binaries.
See Abstr. 117.015.

EXOSAT soft X–ray observations of EX Hydrae.
See Abstr. 117.023.

An extended X–ray low state from Hercules X–1.
See Abstr. 117.037.

An X–ray corona in SS Cygni?
See Abstr. 117.038.

The origin of soft X–ray pulsations in dwarf novae at outburst and the DQ Herculis phenomenon.
See Abstr. 117.039.

Transient quasi–periodic oscillations in the X–ray flux of Cygnus X–3.
See Abstr. 117.040.

Are galactic bulge X–ray burst sources leftovers of disrupted globular clusters?
See Abstr. 117.046.

A comparison of soft X–ray transients and dwarf novae.
See Abstr. 117.048.

On the structure, stability and evolution of accretion disks in soft X–ray transient sources.
See Abstr. 117.049.

Binary period decrease by unstable orbit.
See Abstr. 117.050.

The short time–scale light variability of HD 153919 revisited.
See Abstr. 117.067.

The first UV studies of the optical candidate for the X–ray source 1118–61.
See Abstr. 117.068.

The binary X–ray pulsar 1E 2259 + 59 – a descendant of an AM Her type system?
See Abstr. 117.088.

High spectral resolution observations of the coronal X–ray emission from the RS CVn binary Sigma Corona Borealis.
See Abstr. 117.111.

X–ray binaries and high energy gamma rays.
See Abstr. 117.112.

The minimum period of hydrogen–deficient cataclysmic binaries.
See Abstr. 117.118.

IUE observations of W UMa systems AE Phoenicis and TY Mensae.
See Abstr. 117.120.

UBV photometry of H 2252–035 (AO Psc).
See Abstr. 117.122.

Eclipse duration of the point source in X–ray binary systems with eccentric orbits.
See Abstr. 117.135.

On fast X–ray rotators with long–term periodicities.
See Abstr. 117.144.

The low–mass binary X–ray sources observed from TENMA.
See Abstr. 117.152.

The dependence of He I 10830 absorption on X–ray luminosity in RS CVn binaries and very active F and G main–sequence stars.
See Abstr. 117.163.

The first IUE observations of LMC X–1 (star 32).
See Abstr. 117.164.

UV and optical observations of variability in the WR + compact candidate HD 96548.
See Abstr. 117.165.

Spectral and temporal studies of various late–type stars.
See Abstr. 117.170.

Energy spectra of non–pulsating low mass X–ray binaries.
See Abstr. 117.172.

Is Cygnus X–3 a low–mass X–ray binary?
See Abstr. 117.173.

Exosat observations of the galactic bulge X–ray source GX 17 + 2.
See Abstr. 117.174.

Optical and X–ray observations of 4U2129 + 47/V1727 Cyg in a quiescent state.
See Abstr. 117.175.

Search for millisecond rotational periods in some low–mass X–ray binaries observed by Exosat.
See Abstr. 117.176.

Hard X–ray observation of galactic X–ray sources.
See Abstr. 117.177.

Cygnus X–3: cosmic–ray powerhouse.
See Abstr. 117.180.

Optical and infrared pulsations from the HZ Herculis binary system during the 1983 prolonged X–ray low state.
See Abstr. 117.185.

XB 1905 + 000: the most distant optically identified X–ray burster?
See Abstr. 117.186.

Untersuchungen der Zyklotron–Resonanzeffekte im Röntgenspektrum von Her X–1.
See Abstr. 117.189.

X–ray emission from cataclysmic variables with accretion disks. I. Hard X–rays.
See Abstr. 117.197.

X–ray emission from cataclysmic variables with accretion disks. II. EUV/soft X–ray radiation.
See Abstr. 117.198.

35 day spectroscopic effects in HZ Herculis.
See Abstr. 117.200.

Objective grating X–ray spectroscopy of compact sources.
See Abstr. 117.211.

Evidence for a cyclotron feature from Cygnus X–3.
See Abstr. 117.234.

Hercules X–1/HZ Herculis: yet another periodicity?
See Abstr. 117.239.

Slow variability of three X–ray pulsars.
See Abstr. 117.240.

Implications of observed pulse frequency variations for the structure of the neutron in Vela X–1.
See Abstr. 117.242.

Implication of observed pulse frequency variations for the accretion flow in Vela X–1.
See Abstr. 117.243.

Absorption spectra from massive X–ray binaries.
See Abstr. 117.245.

The new eclipsing magnetic binary system E1114 + 182.
See Abstr. 117.269.

X–ray binaries: end points of binary evolution.
See Abstr. 117.288.

On the evolutionary status of bright, low–mass X–ray sources.
See Abstr. 117.306.

Galactic bulge X–ray burst sources from disrupted globular clusters?
See Abstr. 117.308.

COS–B X–ray observations of Cygnus X–3.
See Abstr. 117.313.

Short–term variability of hard X–rays from Cyg X–1.
See Abstr. 117.338.

Photometric observations of the X–ray dwarf nova HL Canis Majoris.
See Abstr. 117.339.

A tentative explanation for the energy variation of the cyclotron feature in Her X–1.
See Abstr. 117.340.

High speed photometry of dwarf nova EX Hydrae.
See Abstr. 117.341.

Optical observations of new X–ray emitting cataclysmic variables.
See Abstr. 117.359.

Optical characteristics of X–ray binaries.
See Abstr. 117.370.

Massive X–ray binaries.
See Abstr. 117.371.

Report of IAU Commission 42: Close binary stars (*Etoiles binaires serrées*).
See Abstr. 117.376.

The optical orbit of the X–ray pulsar binary 0535–668 (= A0538–66).
See Abstr. 117.384.

High resolution X–ray observations of nearby binary systems: flaring and evidence for unseen companions.
See Abstr. 118.043.

Detection of an X–ray flare in the RS CVn binary Sigma Coronae Borealis.
See Abstr. 118.049.

A simultaneous X–ray and radio observation of a flare from Algol.
See Abstr. 119.048.

A search for X–ray emitting coronal structures in Algol.
See Abstr. 119.049.

Observations of the X–ray luminous SB1 system HD 149162.
See Abstr. 120.020.

The old nova GK Per: discovery of the X–ray pulse period.
See Abstr. 124.201.

X–ray morphology of the Crab nebula.
See Abstr. 125.006.

Einstein X–ray observations of the Monoceros supernova remnant.
See Abstr. 125.010.

On the nature of the supernova remnant 0540–69.3 in the Large Magellanic Cloud.
See Abstr. 125.012.

Imaging X–ray spectrophotometric observation of SN 1006.
See Abstr. 125.016.

Exploding stars, superbubbles, and the HEAO observations.
See Abstr. 125.029.

Self–consistent models for the X–ray emission from supernova remnants: an application to Kepler's remnant.
See Abstr. 125.046.

X–ray spectra of supernova remnants.
See Abstr. 125.051.

The X–ray structure of the Crab nebula.
See Abstr. 125.052.

Exosat observations of the supernova remnant Cas A.
See Abstr. 125.053.

X–ray observations of Vela–X.
See Abstr. 125.054.

EXOSAT observations of the X–ray SNR 1E 1149.6+6209, first seen with the Einstein IPC.
See Abstr. 125.066.

X–ray halos around supernova remnants.
See Abstr. 125.067.

Possible optical counterparts to the X–ray point source in the supernova remnant CTB 80.
See Abstr. 125.088.

The optical structure of the central core in the peculiar supernova remant CTB 80.
See Abstr. 125.089.

An optical synchrotron nebula around the X–ray pulsar 0540–693 in the Large Magellanic Cloud.
See Abstr. 125.122.

On the nature of the Crablike, pulsar–powered supernova remnant 0540–693: the pulsar's initial period.
See Abstr. 125.123.

Optical pulsations in the Large Magellanic Cloud remnant 0540–69.3.
See Abstr. 126.013.

Soft X–ray characteristics of white dwarfs observed by Exosat.
See Abstr. 126.062.

Survey for new pulsars by the HEAO A–1 experiment.
See Abstr. 126.077.

Hard X–ray time variability of the mean pulsar profile of the Crab pulsar.
See Abstr. 126.083.

X–ray emission from fast pulsars.
See Abstr. 126.148.

Structure and origin of the Cygnus superbubble.
See Abstr. 131.056.

High–latitude H I structure and the soft X–ray background.
See Abstr. 131.119.

Measurements of X–ray scattering from interstellar dust.
See Abstr. 131.177.

On cloud shadows in wide beam scans of the diffuse soft X–ray sky.
See Abstr. 131.186.

Measurement of X–ray scattering from interstellar grains.
See Abstr. 131.258.

OSRT observations of the Sco X–1 region at 327 MHz.
See Abstr. 141.009.

Limits on parallax and proper motion of an optical counterpart of Geminga.
See Abstr. 143.023.

First observation of ultra–high–energy γ rays from LMC X–4.
See Abstr. 143.032.

Ultra–high energy γ–ray astronomy using an EAS (*extensive air shower*) array: search for emission from 100 MeV sources.
See Abstr. 143.033.

Line features from Cygnus X–1 and the Crab nebula in the energy range 30 – 270 keV.
See Abstr. 143.039.

Einstein and Exosat observations of Geminga (1E 0630 + 178). A summary of the short– and medium–term variability data.
See Abstr. 143.042.

X–ray spectra of two cosmic gamma–ray bursts.
See Abstr. 143.051.

5 – 100 keV observations of GB790107, a γ–ray burst without γ–rays.
See Abstr. 143.059.

Very high energy gamma rays from Cygnus X3?
See Abstr. 143.077.

The 4.8 hour period of gamma–rays from Cygnus X–3 at energies $E \geqslant 2 \times 10^{15}$eV.
See Abstr. 143.079.

Observation of Oct. 16 1981 cosmic gamma burst by Hakucho satellite.
See Abstr. 143.088.

Low energy gamma ray observation of the Crab and Cygnus region with the "MISO" telescope.
See Abstr. 143.089.

Observations of high energy gamma rays from Cygnus X–3 and other sources with the Whipple Observatory Large Aperture Camera.
See Abstr. 143.091.

Observation of γ–rays $> 10^{15}$eV from Cygnus X–3.
See Abstr. 143.092.

New developments in high energy astrophysics.
See Abstr. 143.100.

Rapporteur paper on high energy gamma–ray astronomy, sessions XG4 and XG5 (*of the conference*).
See Abstr. 143.101.

Rapid X–ray burster as a source of cosmic rays.
See Abstr. 144.161.

The Einstein soft X–ray survey of the Pleiades.
See Abstr. 153.016.

Einstein X–ray survey of the Pleiades: the dependence of X–ray emission on stellar age.
See Abstr. 153.028.

X–raying the dynamics of globular clusters.
See Abstr. 154.002.

X–raying the dynamics of globular clusters.
See Abstr. 154.053.

On the ultraviolet background radiation of the Galaxy.
See Abstr. 155.039.

High–energy X–ray observations of the galactic center region.
See Abstr. 155.098.

The X–ray source populations of the Magellanic Clouds.
See Abstr. 156.008.

Detailed X–ray observations of M83.
See Abstr. 157.086.

Hot gas and massive halos around early type galaxies.
See Abstr. 157.226.

Galaxy photometry and X–ray astronomy.
See Abstr. 157.257.

X–ray emission from E and S0 galaxies with compact nuclear radio sources.
See Abstr. 158.140.

The nucleus of M82 at radio and X–ray bands: discovery of a new radio population of supernova candidates.
See Abstr. 158.142.

Exosat observations of active galactic nuclei.
See Abstr. 158.156.

X–ray line emission from 3C 120.
See Abstr. 158.198.

EXOSAT observations of the radio galaxy 3C 390.3.
See Abstr. 158.199.

Cen A X–ray activity, 1969 – 1979.
See Abstr. 158.201.

Long–term X–ray monitoring of 4 active galactic nuclei.
See Abstr. 158.202.

The wide angle tailed radio source NGC 2329 in the cluster A569.
See Abstr. 158.247.

1E 1402.3 + 0416.
See Abstr. 158.254.

NGC 3031.
See Abstr. 158.257.

H 0323 + 022.
See Abstr. 158.259.

NGC 6212.
See Abstr. 158.260.

X–ray variability of BL Lac objects: H2155–304 and PKS 0548–322.
See Abstr. 158.281.

Newly discovered BL Lacertae objects identified as bright X–ray source counterparts by the HEAO–1 scanning modulation collimator.
See Abstr. 158.305.

Rapid X–ray variability of active galaxies.
See Abstr. 158.316.

The radio jets of 3C 449.
See Abstr. 158.327.

Pavo XD–10, an X–ray QSO with extended optical structure.
See Abstr. 159.008.

The diversity of soft X–ray spectra in quasars.
See Abstr. 159.071.

New observations of the X–ray discovered QSO–galaxy pair 1E 0104.2 + 3153.
See Abstr. 159.082.

QSO Lyman continuum radiation and the ultraviolet extragalactic background.
See Abstr. 159.099.

A statistical X–ray QSOs classification.
See Abstr. 159.126.

The high energy X–ray spectrum of 3C 273.
See Abstr. 159.127.

The evolution of X–ray radiation of quasars.
See Abstr. 159.128.

X–ray emission possibly coincident with the radio tail of PKS 0301–123.
See Abstr. 160.062.

An X–ray study of the Centaurus cluster of galaxies using Einstein.
See Abstr. 160.063.

Exosat observations of the Perseus cluster.
See Abstr. 160.067.

X–ray survey of the Virgo cluster and comparison to field galaxies.
See Abstr. 160.131.

EXOSAT observations of the Virgo cluster.
See Abstr. 160.132.

A comparative study of cosmological evolution in the radio, optical and X–ray bands.
See Abstr. 161.074.

143 Gamma-ray Sources, Gamma-ray Background

143.001   On the astronomical nature of the sources of gamma–ray bursts.
I. S. Shklovskii (*I. S. Shklovskij*), I. G. Mitrofanov.
Mon. Not. R. Astron. Soc., Vol. 212, No. 3, p. 545 – 551 (1985).

The isotropy and $\log N/\log S$ distribution of localized gamma–ray bursts are shown to indicate that their sources belong to the extended galactic corona. Bursts are supposed to be generated by old neutron stars with age $<10^9$yr excited by strong starquakes. The radiated energy is estimated to be about 10^{42}erg per event. The nearest sources are at a distance of about 5 kpc. Practically all outbursts emitted inside the corona are detectable with available instruments. The anisotropy of the distribution of gamma–ray bursts on the sky may be well–pronounced only for the most powerful events. The identification of GRB 790305 with the SN remnant N49 in the Large Magellanic Cloud strongly confirms the authors' conclusion.

143.002   The galactic plane: a source of 1000 GeV gamma rays.
J. C. Dowthwaite, A. B. Harrison, I. W. Kirkman, H. J. Macrae, K. J. Orford, K. E. Turver, M. Walmsley.
Astron. Astrophys., Vol. 142, No. 1, p. 55 – 58 (1985).

A recent measurement using the atmospheric Cerenkov technique has strengthened previous indications that certain regions of the galactic plane are emitters of very high energy γ–rays. The flux of γ–rays of energy in excess of 1000 GeV has been measured and is 3×10^{-7}photon cm^{-2}s^{-1}ster^{-1} from parts of the Cygnus region. This flux is higher than that extrapolated from earlier measurements at an energy of 150 GeV in which the results were averaged over broader bands of longitude during shorter scans across the plane. The authors find no correlation between the very high energy γ–ray emission and the total hydrogen column density, but a possible correlation with the stellar density extends to an observed lack of emission near the galactic equator which was noted in previous VHE measurements at these longitudes.

143.003   Gamma–ray bursters.
B. E. Schaefer.
Sci. Am., Vol. 252, No. 2, p. 40 – 46 (1985).

Intense flashes of high–energy radiation appear unpredictably in the sky. Limited observational data have frustrated attempts to determine their cause, but likely mechanisms have been proposed.

143.004   Ultrahigh–energy gamma–ray astronomy.
K. Brecher.
Phys. Today, Vol. 38, No. 1, p. S11 – S12 (1985). Short review paper. – See Abstr. 013.003.

143.005   Gamma–ray bursts: a 1983 overview.
T. L. Cline.
AIP Conf. Proc., No. 115, p. 333 – 342 (1984). – See Abstr. 012.005.

Gamma–ray burst observations are reviewed with mention of new gamma–ray and optical transient measurements and with discussions of the controversial, contradictory and unresolved issues that have recently emerged: burst spectra appear to fluctuate in time as rapidly as they are measured, implying that any one spectrum may be incorrect; energy spectra can be obligingly fitted to practically any desired shape, implying, in effect, that no objective spectral resolution exists at all; burst fluxes and temporal quantities, including the total event energy, are characterized very differently with differing instruments; finally, the $\log N - \log S$ determinations are deficient in the weak bursts.

143.006   Gamma–ray bursts: recent Soviet and Franco–Soviet results, and observational overview.
K. Hurley.
AIP Conf. Proc., No. 115, p. 343 – 351 (1984). – See Abstr. 012.005.

New results on the time histories and energy spectra of gamma–ray bursts are presented from the SIGNE experiments. The KONUS instruments continue to detect low and high energy features in the energy spectra of bursts, and show evidence for a luminosity–temperature correlation in burst time histories and spectra. A summary is presented of some relevant observational facts concerning the time histories, energy spectra, $\log N - \log S$ relation, and positions and distances of bursters.

143.007   The spectral properties of gamma–ray bursts: a review of recent developments.
B. J. Teegarden.
AIP Conf. Proc., No. 115, p. 352 – 366 (1984). – See Abstr. 012.005.

Recent developments in the spectroscopy of gamma–ray bursts (GRB) are reviewed. The general question of the validity of the spectral results, particularly with regard to features in the spectrum, is discussed. Confirmations of these spectral features are summarized. Recent results from the KONUS experiments on Venera 13 and 14 are reviewed. The status of models of the continuum spectrum is summarized.

143.008   Frequency of fast, narrow gamma–ray bursts and burst classification.
J. P. Norris, T. L. Cline, U. D. Desai, B. J. Teegarden.
AIP Conf. Proc., No. 115, p. 367 – 372 (1984). – See Abstr. 012.005.

Evidence from the Vela satellites that very brief, ~ 0.1 s, gamma–ray bursts constitute a class distinct from the longer, highly structured bursts has been strengthened by the results of the Venera 11 and 12 KONUS experiments. The Goddard ISEE–3 Gamma–Ray Burst Spectrometer, utilizing a trigger criterion which is more likely to be independent of duration than previous experiments, has detected a sample of events which enhances this bimodal distribution.

143.009   Observation of an absorption feature in a gamma–ray burst spectrum.
G. J. Hueter.
AIP Conf. Proc., No. 115, p. 373 – 377 (1984). – See Abstr. 012.005.

A gamma–ray burst was detected on March 25, 1978 by the High Energy X–Ray and Low Energy Gamma–Ray Experiment on HEAO–1. The burst spectrum shows an absorption feature at 55 ± 5 keV with an equivalent width of 13 ± 3 keV, confirming the observation of similar features by the KONUS experiment. The burst spectrum also is characterized by a hard component extending from ~ 0.25 to 6 MeV.

143.010   3 – 10 keV and 0.1 to 2 MeV observations of four gamma–ray bursts.
J. G. Laros, W. D. Evans, E. E. Fenimore, R. W. Klebesadel, S. Shulman, G. Fritz.
AIP Conf. Proc., No. 115, p. 378 – 389 (1984). – See Abstr. 012.005.

Four catalogued γ–ray bursts that occurred between 79/3/7 and 79/7/31 have been observed over the 3 – 10 keV range by a joint NRL/Los Alamos experiment on the Air Force P78–1 satellite. The bursts were also well observed by members of the interplanetary network. In this paper the authors present hardness ratios, X–ray/γ–ray luminosity ratios, and time histories.

143.011   Observation of a cosmic gamma–ray burst on Hakucho.
M. Katoh, T. Murakami, J. Nishimura, T. Yamagami, M. Fujii, M. Itoh.
AIP Conf. Proc., No. 115, p. 390 – 398 (1984). – See Abstr. 012.005.

Hakucho, the X–ray observation satellite launched in Feb. 1979, has detected several gamma–ray bursts. Detailed X–ray spectra from the burst occurring 1981 Oct. 16 at 23:53 UT were obtained. The event has also been observed by PVO and SMM. Combining the Hakucho and PVO data has yielded a detailed time profile of the event, energy spectra at 4 s intervals covering the energy range of 1 – 55 keV and 120 – 2000 keV, and a source location. The total energy of the event beyond 1 keV is $(5 – 10) \times 10^{-4}$ erg/cm^2.

143.012   High–energy emission from gamma–ray bursts.
P. L. Nolan, G. H. Share, S. Matz, E. L. Chupp, D. J. Forrest, E. Rieger.
AIP Conf. Proc., No. 115, p. 399 – 402 (1984). – See Abstr. 012.005.

The authors discuss broad–band continuum spectroscopy of 17 gamma–ray bursts above 0.3 MeV. The spectra were fitted by 3 trial functions, none of which provided an adequate fit to all the spectra. Most were too hard for a thermal bremsstrahlung function. Harder functional forms, such as thermal synchrotron or power–law, provide better fits for most of the spectra.

143.013   Are there nuclear contributions to gamma–ray burst spectra?
S. M. Matz, E. L. Chupp, D. J. Forrest, G. H. Share, P. L. Nolan, E. Rieger.
AIP Conf. Proc., No. 115, p. 403 – 405 (1984). – See Abstr. 012.005.

The authors have examined the spectra of 38 γ–ray bursts observed by the Gamma Ray Spectrometer on the Solar Maximum Mission satellite for evidence of a nuclear contribution to the high energy flux. A sum of spectra from the nine bursts with detectable flux >4 MeV suggests but does not require a drop–off above 7 MeV. A cutoff between 7 and 8 MeV is consistent with a high energy spectrum dominated by nuclear lines.

143.014   Implications of the three GRB optical flashes.
B. E. Schaefer.
AIP Conf. Proc., No. 115, p. 406 – 408 (1984). – See Abstr. 012.005.

The current state of these optical observations is briefly summarized in this paper.

143.015   Periodicities in gamma–ray bursts.
K. S. Wood.
AIP Conf. Proc., No. 115, p. 409 – 411 (1984). – See Abstr. 012.005.

Gamma–ray burst models based on magnetic neutron stars face a problem of accounting for the scarcity of observed periods. Both this scarcity and the typical period found when any is detected are explained if the neutron stars are accreting in binary systems.

143.016   The gamma–ray burst spatial distribution log $N(>S)$ vs log S and $N(>S, l^{II}, b^{II})$ vs S.
M. C. Jennings.
AIP Conf. Proc., No. 115, p. 412 – 421 (1984). – See Abstr. 012.005.

Gamma–ray burst distance scales are reviewed. Most observations are inconsistent with metagalactic ($D > 30$ Mpc) or extragalactic (100 kpc $< D <$ 30 Mpc) distances. For galactic distances ($D < 100$ kpc) disks or a conventional halo are improbable. The possibility that bursts occupy the Galaxy's massive extended halo is examined. Such distributions provide the best fit to the data. The total galactic burst rate is ~ 400 yr^{-1} and the nearest burst source should lie between 5 and 1.4 kpc for burst repetitions of 1 to 50 yrs.

143.017   Measurement of the rate of weak gamma–ray bursts.
C. A. Meegan, G. J. Fishman, R. B. Wilson.
AIP Conf. Proc., No. 115, p. 422 – 425 (1984). – See Abstr. 012.005.

A sensitive search for weak gamma–ray bursts was conducted using data from two balloon flights. One burst was observed in 64 hours of observation. A 95% confidence upper limit to the burst rate is 2600 bursts/year with fluence greater than 4.4×10^{-7} ergs/cm^2.

143.018   The intrinsic gamma–ray burst luminosity function from observational data analysis.
B. M. Belli.
AIP Conf. Proc., No. 115, p. 426 – 428 (1984). – See Abstr. 012.005.

The spatial distribution of the gamma–ray bursts, selected according to their spectral hardness, appear spherical with divergence from an isotropic behavior depending on the spectral parameter kT. This can be explained by an intrinsic luminosity function and a disk or halo galactic distribution. If one assumes a galactic disk distribution an average peak burst luminosity of 10^{39} erg/sec is derived.

143.019   log N – log S is inconclusive.
R. W. Klebesadel, E. E. Fenimore, J. G. Laros.
AIP Conf. Proc., No. 115, p. 429 – 433 (1984). – See Abstr. 012.005.

The log N – log S data acquired by the Pioneer Venus Orbiter Gamma Burst Detector (PVO) are presented and compared to similar data from the Soviet KONUS experiment. Although the PVO data are consistent with and suggestive of a –3/2 power law distribution, the results are not adequate at this state of observations to differentiate between a –3/2 and a –1 power law slope.

143.020   The detection efficiency of cosmic gamma–ray sources and limits on the discrete source flux.
B. P. Houston, A. W. Wolfendale.
J. Phys. G, Vol. 10, No. 11, p. 1587 – 1598 (1984). Abstr. in Phys. Abstr., Vol. 88, No. 1250, Entry 18945 (1985).

143.021   A point source of very high energy gamma rays.
T. Kifune.
Grand unified theories and cosmology, p. 201 – 210 (1984). Abstr. in Phys. Abstr., Vol. 88, No. 1251, Entry 24561 (1985). – See Abstr. 012.020.

143.022   The gamma–ray deficit toward the galactic center.
L. Blitz, J. B. G. M. Bloemen, W. Hermsen, T. M. Bania.
Astron. Astrophys., Vol. 143, No. 2, p. 267 – 273 (1985).

The gamma–ray flux ($E > 300$ MeV) from the central few hundred parsecs of the Milky Way is shown to be nearly an order of magnitude smaller than the value expected from the H_2 masses generally estimated to be present in the center and from the average gamma–ray emissivity measured for the disk. This result implies that in the galactic center the gamma–ray emissivity is anomalously low, or that molecular hydrogen is nearly an order of magnitude less abundant than estimates made from CO observations.

143.023   Limits on parallax and proper motion of an optical counterpart of Geminga.
H. Sol, M. Tarenghi, C. Vanderriest, L. Vigroux, G. Lelièvre.
Astron. Astrophys., Vol. 144, No. 1, p. 109 – 114 (1985).

A sequence of several images of the recently proposed optical counterpart of the X–ray source 1E 0630+178, possibly associated with the bright γ–ray source Geminga, is presented and discussed. From these data covering 410 days, an upper limit of $0\overset{''}{.}1$ on its parallax can be derived, which places the object definitely further than 10 pc from the solar system. No evidence of proper motion has been detected up to $0\overset{''}{.}2$/yr, which again places it further than 10 pc for a tangential velocity of 10 km s^{-1} relatively to the Sun. UBV photometry is compatible with the object being a G star, without measurable long–term variability. Another possible identification involves a very faint ($B \cong 24.3$) object.

143.024   The optical content of the 1979 April 6 gamma ray burst error box.
C. Motch, H. Pedersen, S. A. Ilovaisky, C. Chevalier, K. Hurley, G. Pizzichini.
Astron. Astrophys., Vol. 145, No. 1, p. 201 – 205 (1985).

The authors present a complete analysis of the deepest available CCD composite images of the 1979 April 6 gamma–ray burst error box in red ($R < 24.1$) and white light (equivalent to $R < 25.0$). Of the nine objects contained in the error box, one is definitely stellar, a fainter one may also be stellar. Six others appear extended and are faint galaxies, while a seventh may also be extended. None of the objects within or near the error box is found to be variable on time scales from hours to months. If the 1979 April 6 quiescent optical counterpart is similar to that found by Pedersen et al. for the 1978 November 19 event, comparison of the gamma–ray characteristics of both events suggests the 1979 April 6 object could be fainter than the present optical detection threshold.

143.025   Erratum: "Time history, energy spectrum, and localization of an unusual gamma–ray burst" [Astrophys. J., Vol. 280, No. 1, p. 150 – 153 (1984)].
C. Barat, K. Hurley, M. Niel, G. Vedrenne, W. D. Evans, E. E. Fenimore, R. W. Klebesadel, J. G. Laros, T. L. Cline, I. V. Estulin (*I. V. Ehstulin*), V. M. Zenchenko, V. G. Kurt.
Astrophys. J., Vol. 288, No. 2, p. 833 (1985). See Abstr. 37.143.041.

143.026   High–energy emission in gamma–ray bursts.
S. M. Matz, D. J. Forrest, W. T. Vestrand, E. L. Chupp, G. H. Share, E. Rieger.
Astrophys. J., Lett. Ed., Vol. 288, No. 2, p. L37 – L40 (1985).

Between 1980 February and 1983 August the Gamma–Ray Spectrometer on the SMM satellite detected 72 events identified as being of cosmic origin. These events are an essentially unbiased subset of all γ–ray bursts. The measured spectra of these events show that high–energy (>1 MeV) emission is a common and energetically important feature. There is no evidence for a general high–energy cut–off or a distribution of cut–offs below ~ 6 MeV.

143.027   Variability of high–energy gamma–ray flux of Cyg X–3.
Yu. I. Neshpor, Yu. L. Zyskin, B. M. Vladimirskij, A. A. Stepanyan, V. P. Fomin.
Izv. Krymskoj Astrofiz. Obs., Tom 69, p. 59 – 63 (1984). In Russian. English translation in Bull. Crimean Astrophys. Obs., Vol. 69.

Observational data of high–energy γ rays with energy $E \geqslant 2 \times 10^{12}$eV obtained in 1972 – 1980 were analysed. It is shown that the γ–ray flux varies with the period $P = 34.1$ day. The authors suspect that there might be some correlation between the intensities in radio, X–ray and γ–ray ranges for the 34–d periodicity.

143.028   The periodicity of gamma–ray emission from the source 2CG 195+04.
Yu. L. Zyskin, D. B. Mukanov.
Izv. Krymskoj Astrofiz. Obs., Tom 69, p. 67 – 70 (1984). In Russian. English translation in Bull. Crimean Astrophys. Obs., Vol. 69.

Observations of the gamma–ray source 2CG 195+04 have been carried out in 1979 and 1981 in the energy range $>10^{12}$eV. The observational results show evidence for existing variability of the gamma–ray emission with the period ~ 59 s. It is suggested that the derivative of the period varies with time.

143.029   Gamma–ray burster recurrence time scales.
B. E. Schaefer, T. L. Cline.
Astrophys. J., Vol. 289, No. 2, p. 490 – 493 (1985).

Three optical transients have now been found which are associated with gamma–ray bursters (GRBs). The deduced recurrence time scale for these optical transients (τ_{opt}) will depend on the minimum brightness for which a flash would be detected. The authors present a detailed analysis using all available data of τ_{opt} as a function of E_γ/E_{opt}. For flashes similar to those found in the Harvard archives, the best estimate of τ_{opt} is 0.74 yr, with a 99% confidence interval from 0.20 to 0.65 yr. It is currently unclear whether the optical transients from GRBs also give rise to gamma–ray events.

143.030   One–photon and two–photon annihilation lines in gamma–bursts. I.
V. V. Zheleznyakov, A. A. Litvinchuk.
Astrophys. Space Sci., Vol. 109, No. 2, p. 293 – 307 (1985).

This paper and subsequent Paper II are an investigation of the annihilation line formation in gamma–ray bursts based on the assumption of positron production in a strong magnetic field. The authors discuss a two–photon annihilation line in this paper. It is shown that if the star magnetic field is greater than 3×10^{12}G, the relative flux in the line depends solely on the hardness of the continuum and is, as a rule, less than or about 10 – 20% of the total flux. This is consistent with the spectral data recorded by Venera–11 and Venera–12 space probes. The annihilation region formation above the hot polar spot is discussed, and positron density and annihilation region dimensions are estimated.

143.031   On the binary nature of cosmic γ–ray burst sources.
S. A. Rappaport, P. C. Joss.
Nature, Vol. 314, No. 6008, p. 242 – 245 (1985).

Optical flashes on archival plates have been detected from the error boxes of three gamma–ray burst sources. This letter shows that these flashes could be emitted by a collapsed star in a close binary system. Under the assumption that the optical flashes were produced by γ–ray bursts of about the same intensity as those observed, the authors find that nearby ($\lesssim 100$ pc) binary systems with secondaries whose masses are less than $\sim 0.06\, M_\odot$ can fit all the observational constraints for the three optical/γ–ray pair events.

143.032 First observation of ultra–high–energy γ rays from LMC X–4.
R. J. Protheroe, R. W. Clay.
Nature, Vol. 315, No. 6016, p. 205 – 207 (1985).

Ultra–high–energy (UHE) γ rays ($>10^{15}$eV) have recently been observed from Cygnus X–3 and from Vela X–1. Data from the University of Adelaide air–shower array at Buckland Park taken over a 3–year period have been analysed to search for evidence of UHE γ–ray emission from additional neutron–star binary X–ray sources having known orbital periods. The authors present here the result of this search and report the detection of UHE γ rays from LMC X–4 above 10^{16}eV with an integral flux of $(4.6\pm1.7)\times10^{-11}$ photons m^{-2}s^{-1}.

143.033 Ultra–high energy γ–ray astronomy using an EAS (*extensive air shower*) array: search for emission from 100 MeV sources.
R. J. Protheroe, R. W. Clay.
Proc. Astron. Soc. Aust., Vol. 5, No. 4, p. 586 – 589 (1984).

The authors have searched for evidence of γ–ray emission above 3×10^{15}eV from the 14 sources in the second COS–B catalogue (Swanenburg et al. 1981) which have declinations south of $\delta = 0°$.

143.034 Search for light flashes from a gamma ray burst source.
K. B. Fenton.
Proc. Astron. Soc. Aust., Vol. 5, No. 4, p. 594 – 596 (1984).

With the discovery of evidence for recurrent bursts from the direction of the supernova remnant N49 in the LMC (Golenetskii et al. 1984), it became evident that it may be worthwhile having a monitor trained on N49. This paper describes a suitable telescope and camera which has now been in operation since the beginning of 1984 in the grounds of Canopus Observatory.

143.035 Locations and time histories of five 1979 gamma–ray bursts.
J. G. Laros, W. D. Evans, E. E. Fenimore, R. W. Klebesadel,
J. Middleditch, C. Barat, K. Hurley, M. Niel, G. Vedrenne,
G. H. Nakano, W. L. Imhof, T. L. Cline, U. D. Desai,
B. E. Schaefer, B. J. Teegarden, I. V. Estulin (*I. V. Ehstulin*),
V. G. Kurt, G. A. Mersov, V. M. Zenchenko.
Astrophys. J., Vol. 290, No. 2, p. 728 – 734 (1985).

The authors have studied the locations and time histories of five γ–ray bursts that occurred between 1979 March 7 and March 31. The error box for GB 790325 has a typical dimension of ~1.5. The other localizations, while not precise enough for thorough optical examination, contribute to distribution studies and allow radio and X–ray observations, catalog searches, and other archival work. A search through selected catalogs did reveal one object, the star FY Aql (cataloged as a Mira–type variable but probably a dwarf nova) inside one of the γ–ray burst boxes.

143.036 Comparison of results of observations of the source Cas γ–1 in two energetic regions: 10^{12} and 10^8eV.
B. M. Vladimirskij, Yu. L. Zyskin, Yu. I. Neshpor,
A. A. Stepanyan, V. P. Fomin.
Izv. Akad. Nauk SSSR. Ser. fiz., Tom 48, No. 11, p. 2078 – 2079 (1984). In Russian. Abstr. in Ref. Zh., 51. Astron., 3.51.590 (1985).

143.037 A model of the object Geminga: degenerate dwarf rotating around a black hole.
G. S. Bisnovatyj-Kogan.
Astrofizika, Tom 22, Vyp. 2, p. 369 – 378 (1985). In Russian. English translation in Astrophysics, Vol. 22, No. 2.

The periodic flux variations of the object Geminga in γ and X–ray regions are interpreted as orbital rotation of a degenerate dwarf with $M_2 = 0.6\,M_\odot$ around a black hole with the total mass of the system $M \cong 5\,M_\odot$. The dwarf is supposed to fill the Roche lobe. Its orbital period coincides with the observed $P = 1$ min, and the gravitational radiation together with the mass transfer explain the increase of the period.

143.038 The frequency of weak gamma–ray bursts.
C. A. Meegan, G. J. Fishman, R. B. Wilson.
Astrophys. J., Vol. 291, No. 2, p. 479 – 485 (1985).

A search for weak γ–ray bursts was conducted using a sensitive balloon–borne detector. One burst was detected in 64 hr of observation. The upper limit to the burst rate is 2300 bursts yr^{-1} above 6×10^{-7}ergs cm^{-2} for simple spatial distribution models. Comparison with satellite results indicates that the slope of the log N–log S curve can be no steeper than –1 between 10^{-4} and 10^{-6}ergs cm^{-2}. A detailed procedure for calculating detector sensitivity to bursts is provided.

143.039 Line features from Cygnus X–1 and the Crab nebula in the energy range 30 – 270 keV.
H. Watanabe.
Astrophys. Space Sci., Vol. 111, No. 1, p. 157 – 169 (1985).

A balloon–borne germanium spectrometer was flown in an attempt to detect line–emission from Cyg X–1 and the Crab nebula in the energy range 30 – 270 keV. The experiment was carried out on 29 – 30 September, 1982. A line feature at 145 keV was observed from Cyg X–1. The intensity is $(1.34\pm0.31)\times10^{-2}$ photons cm^{-2}s^{-1} and the width is 14.3 keV FWHM. From the Crab nebula, a weak line feature with 1.8 σ excess was found around 78 keV.

143.040 COS–B gamma–ray sources and interstellar gas in the first galactic quadrant.
A. M. T. Pollock, K. Bennett, G. F. Bignami,
J. B. G. M. Bloemen, R. Buccheri, P. A. Caraveo, W. Hermsen,
G. Kanbach, F. Lebrun, H. A. Mayer–Hasselwander,
A. W. Strong.
Astron. Astrophys., Vol. 146, No. 2, p. 352 – 362 (1985).

The consequences are considered for the high–energy gamma–ray sources of the recently established correlation between the intensity above 300 MeV and tracers of atomic and molecular gas in the first quadrant of the galaxy. Using a likelihood method on the complete set of COS–B data it is shown that there are point–like gamma–ray sources which do not have counterparts in the gas data. They may be due to localised enhancements of the cosmic–ray density or, alternatively, may be quite independent of the gas. There are two sources near the edges of the Cygnus X complex and one near M17, all of which correspond to COS–B sources in the 2CG catalogue. Three other 2CG sources appear to be explained above 300 MeV by the gas alone. Two transient or variable sources have also been observed.

143.041 Very high–energy gamma–ray astronomy.
A. A. Stepanian (*A. A. Stepanyan*).
Astrophys. Space Phys. Rev., Vol. 4, p. 257 – 286 (1985). – See Abstr. 003.020. Revised and extended English translation of 34.143.036.

Contents: (1). The subject of gamma–ray astronomy. (2). Basic characteristics of the interaction of very high–energy gamma–rays with the Earth's atmosphere and the parameters of the Čerenkov light flash. (3). Basic characteristics of Čerenkov flash detectors. (4). Description of the instrument at the Crimean Astrophysical Observatory. (5). Brief historical account of the development of the apparatus used to search for discrete gamma–ray sources. (6). Basic results of observations of discrete fluxes of very high–energy gamma–rays. (7). Prospects for further development of techniques for detecting very high–energy gamma–rays.

143.042 Einstein and Exosat observations of Geminga (1E 0630 + 178). A summary of the short– and medium–term variability data.
G. F. Bignami, P. A. Caraveo, S. Mereghetti, L. Salotti.
Adv. Space Res., Vol. 5, No. 3, p. 145 – 148 (1985). – See Abstr. 012.059.

143.043 Rapid spectral variability of cosmic gamma–ray bursts.
I. G. Mitrofanov, V. Sh. Dolidze, C. Barat,
G. Vedrenne, M. Niel, K. Hurley.
Sov. Astron., Vol. 28, No. 5, p. 547 – 549 (1984). English translation of 38.143.014.

143.044 Gamma–ray source Geminga: a white dwarf rotating around a black hole?
G. S. Bisnovatyj–Kogan.
Priroda, No. 5, p. 86 – 87 (1985). In Russian.

143.045 COS–B gamma–ray surveys in the 70 MeV – 5000 GeV energy range.
H. A. Mayer–Hasselwander.
Bull. Inf. Cent. Données Stellaires, No. 28, p. 73 (1985). Abstract. – See Abstr. 012.062.

143.046 Gamma–ray burst emission above 1 MeV: SMM observations.
S. M. Matz.
Bull. Am. Astron. Soc., Vol. 16, No. 4, p. 918 – 919 (1984). Abstract. – See Abstr. 010.062.

143.047 Search for extragalactic sources of very high energy gamma rays.
P. Gorham, V. J. Stenger, R. C. Lamb, D. F. Liebing, K. Gibbs, T. C. Weekes, M. F. Cawley, D. J. Fegan, N. A. Porter, K. E. Turver.
Bull. Am. Astron. Soc., Vol. 16, No. 4, p. 934 (1984). Abstract. – See Abstr. 010.062.

143.048 A search for very high energy gamma–ray emission from CG 135 + 1 and CG 195 + 4.
D. F. Liebing, R. C. Lamb, K. Gibbs, T. C. Weekes, M. F. Cawley, D. J. Fegan, N. A. Porter, P. N. Gorham, V. J. Stenger, K. E. Turver.
Bull. Am. Astron. Soc., Vol. 16, No. 4, p. 934 (1984). Abstract. – See Abstr. 010.062.

143.049 Is there a separate population of short gamma–ray bursts?
M. C. Jennings.
Bull. Am. Astron. Soc., Vol. 16, No. 4, p. 935 (1984). Abstract. – See Abstr. 010.062.

143.050 The frequency of weak gamma–ray bursts using satellite cross–correlation.
M. M. Fikani, J. G. Laros.
Bull. Am. Astron. Soc., Vol. 16, No. 4, p. 935 – 936 (1984). Abstract. – See Abstr. 010.062.

143.051 X–ray spectra of two cosmic gamma–ray bursts.
M. Katoh, E. E. Fenimore, R. W. Klebesadel, J. G. Laros, M. Fujii, M. Itoh, T. Murakami, J. Nishimura, T. Yamagami.
Bull. Am. Astron. Soc., Vol. 16, No. 4, p. 936 (1984). Abstract. – See Abstr. 010.062.

143.052 The unusual gamma–ray burst of March 4, 1984.
R. W. Klebesadel, J. G. Laros, E. E. Fenimore.
Bull. Am. Astron. Soc., Vol. 16, No. 4, p. 1016 (1984). Abstract. – See Abstr. 010.062.

143.053 Spectral characteristics of the March 4, 1984 cosmic gamma–ray burst.
S. R. Kane, E. E. Fenimore, R. W. Klebesadel, J. G. Laros.
Bull. Am. Astron. Soc., Vol. 16, No. 4, p. 1016 (1984). Abstract. – See Abstr. 010.062.

143.054 Is Cygnus X–3 a source of gamma rays or of new particles?
A. Watson.
Nature, Vol. 315, No. 6019, p. 454 – 455 (1985).

143.055 A model of object Geminga as a degenerate white dwarf orbiting around a black hole.
G. S. Bisnovatyi–Kogan (G. S. Bisnovatyj–Kogan).
Nature, Vol. 315, No. 6020, p. 555 – 557 (1985).
Geminga (2CG 195 + 04) is a very bright γ–ray source. It has been recently identified with X–ray and optical sources. The observed periodicity is $P \approx 1$ min and it changes at a rate corre-sponding to a doubling time of about several hundred years. The author proposes here a model which consists of a degenerate dwarf ($M_d = 0.6\ M_\odot$) filling its Roche lobe and orbiting a black hole ($M_{bh} = 4.4\ M_\odot$) with a period $P = 1$ min. Gravitational radiation dissipates the angular momentum, which leads to mass transfer with a current rate $\sim 10^{-3} M_\odot \mathrm{yr}^{-1}$ and a period change $\dot{P}$ which agrees with the observed value.

143.056 Optical bursts – where are they?
K. Hurley.
Nature, Vol. 315, No. 6022, p. 715 – 716 (1985).
This note comments on recent attempts to observe simultaneous optical bursts from gamma–ray bursters.

143.057 One–photon and two–photon annihilation lines in gamma–bursts. II.
V. V. Zheleznyakov, A. A. Litvinchuk.
Astrophys. Space Sci., Vol. 112, No. 1, p. 25 – 49 (1985).
The paper studies the formation of a one–quantum annihilation line in the spectrum of gamma–ray bursts. The radiative transfer equation together with the positron density balance relation is solved, and the expected photon fluxes in the one– and two–quantum annihilation lines are calculated. The fractional luminosities of these lines versus source parameters are investigated.

143.058 2CG 195 + 04 (Geminga).
IAU Circ., No. 4058 (1985).

143.059 5 – 100 keV observations of GB790107, a γ–ray burst without γ–rays.
J. G. Laros, E. E. Fenimore, R. W. Klebesadel, S. R. Kane.
Bull. Am. Astron. Soc., Vol. 17, No. 1, p. 520 – 521 (1985). Abstract. – See Abstr. 010.064.

143.060 Evidence for 500 TeV gamma–ray emission from Hercules X–1.
R. M. Baltrusaitis, G. L. Cassiday, R. Cooper, J. W. Elbert, P. R. Gerhardy, E. C. Loh, Y. Mizumoto, P. Sokolsky, P. Sommers, D. Steck.
Astrophys. J., Lett. Ed., Vol. 293, No. 2, p. L69 – L72 (1985).
A signal (chance probability $= 2 \times 10^{-4}$) with a 1.24 s period has been observed from the direction of Hercules X–1. The signal's relatively long period and high shower energy conflict with some popular models of particle acceleration by pulsars. Optical and X–ray data support a picture in which energetic particles produce multi–TeV γ–rays by collisions with Hercules X–1's accretion disk.

143.061 Gamma rays from the Orion molecular clouds.
B. P. Houston, A. W. Wolfendale.
J. Phys. G, Vol. 11, No. 3, p. 407 – 420 (1985). Abstr. in Phys. Abstr., Vol. 88, No. 1257, Entry 57843 (1985).

143.062 Cosmic gamma–ray bursts observed by the Hinotori satellite.
M. Yoshimori, K. Okudaira, Y. Hirasima, I. Kondo.
18th International Cosmic Ray Conference, Vol. 1, p. 39 – 42 (1983). – See Abstr. 012.096.
Hinotori observed four cosmic gamma–ray bursts on Feb. 28, July 21, 1981 and Feb. 26 and March 13, 1982 with the gamma- and hard X–ray spectrometers. These burst data were studied for time histories and energy spectra.

143.063 Gamma–ray burst high time–resolution spectral observations made with the Solar Maximum Mission.
U. D. Desai, J. P. Norris, T. L. Cline, B. R. Dennis, K. J. Frost.
18th International Cosmic Ray Conference, Vol. 1, p. 43 – 46 (1983). – See Abstr. 012.096.
The authors confirm the existence of deficiencies of low energy photons. Significant spectral variations, primarily at the lower energies, are also seen on time scales $\approx .25$ s.

143.064 Archival searches for optical counterparts to gamma ray bursters.
J.-L. Ateia, C. Barat, K. Hurley, M. Niel, G. Vedrenne,
W. Wenzel, I. V. Estulin (*I. V. Ehstulin*), A. V. Kuznetsov,
V. M. Zenchenko, W. D. Evans, E. Fenimore, R. Klebesadel,
J. Laros, T. L. Cline, U. D. Desai.
18th International Cosmic Ray Conference, Vol. 1, p. 47 (1983).
Abstract. – See Abstr. 012.096.

143.065 Searches for optical emission coincident with gamma ray bursts.
J.-L. Ateia, C. Barat, K. Hurley, M. Niel, G. Vedrenne,
Z. Ceplecha, R. Hudec, I. V. Estulin (*I. V. Ehstulin*),
A. V. Kuznetsov, V. M. Zenchenko, T. L. Cline, U. D. Desai,
W. D. Evans, E. Fenimore, R. Klebesadel, J. Laros.
18th International Cosmic Ray Conference, Vol. 1, p. 48 (1983).
Abstract. – See Abstr. 012.096.

143.066 A distance model for gamma–ray bursts.
R. E. Lingenfelter, G. J. Hueter.
18th International Cosmic Ray Conference, Vol. 1, p. 54 – 57
(1983). – See Abstr. 012.096.

The authors discuss the apparent contradiction between the isotropic latitude distribution and the flat size–frequency distribution of gamma–ray bursts. They show that the data are not inconsistent but rather can be easily explained by a non–singular distribution of burst energies, such as that implied by a fireball model for gamma–ray bursts.

143.067 Corequake and shock heating model of the 5 March 1979 γ–ray burst.
D. C. Ellison, D. Kazanas.
18th International Cosmic Ray Conference, Vol. 1, p. 58 – 61
(1983). – See Abstr. 012.096.

Ramaty et al. (1980) have proposed a model of a neutron star corequake and subsequent shock heating of the atmosphere. The authors examine the overall energetics and characteristics of the radiation dominated, gas shocks, under the assumption of thermodynamic equilibrium, taking into account the e^+–e^- pair production behind the shock. Using values for the density typical to those expected for neutron star crusts ($\varrho \cong 10^2 – 10^4 \mathrm{g\,cm}^{-3}$), shock luminosities are obtained comparable with those required if the burst originated in the LMC.

143.068 On the role of neutrons in the formation of radiation in γ–ray burst sources.
F. A. Aharonian (*F. A. Agaronyan*).
18th International Cosmic Ray Conference, Vol. 1, p. 66 – 69
(1983). – See Abstr. 012.096.

The presence of redshifted emission lines in spectra of γ–ray bursts testifies that they are formed near the neutron stars. The spectral feature at energy 0.74 MeV in the spectrum of GB 781119 is of rather nuclear origin, being due to the radiation of the excited iron ^{56}Fe. An alternative hypothesis on the formation of the 0.84 MeV line in bombarding the iron crust of the neutron star by neutrons ejected from the star deeper layers has been discussed.

143.069 Possible detection of gamma–ray lines from the Crab Nebula.
C. A. Ayre, P. N. Bhat, Y.-q. Ma, R. M. Myers,
M. G. Thompson.
18th International Cosmic Ray Conference, Vol. 1, p. 70 – 73
(1983). – See Abstr. 012.096.

Results are presented of an experiment to search for γ–ray lines from the Crab Nebula region in the energy range 50 keV to 2000 keV using a cooled intrinsic germanium detector in a sodium iodide anti–coincidence shield. Some evidence for line features is observed at 404.4 keV and 1049 keV and a line at 78.9 keV has also been detected which shows some variability with time.

143.070 Search for gamma ray lines in molecular clouds.
P. Durouchoux, T. Montmerle, A. Jacobson, J. Ling,
W. Mahoney, G. R. Riegler, W. Wheaton.
18th International Cosmic Ray Conference, Vol. 1, p. 74 (1983).
Abstract. – See Abstr. 012.096.

143.071 Statistical reliability of gamma–ray line observations.
T.-p. Li, Y.-q. Ma.
18th International Cosmic Ray Conference, Vol. 1, p. 76 – 79
(1983). – See Abstr. 012.096.

A procedure has been proposed to evaluate the statistical significance and the confidence level of a positive source or line observation in gamma–ray astronomy. New estimates for the statistical reliability of reported gamma–ray lines with application of this method are presented.

143.072 Observation of an absorption feature in a gamma ray burst spectrum.
G. J. Hueter, H. S. Hudson, J. L. Matteson, R. E. Rothschild,
L. E. Peterson.
18th International Cosmic Ray Conference, Vol. 1, p. 95 – 98
(1983). – See Abstr. 012.096.

A gamma ray burst was detected on March 25, 1978. The burst spectrum shows an absorption feature at 55 ± 5 keV with an equivalent width of 13 ± 3 keV and a hard component extending from $\sim 0.25 – 6$ MeV. This component can be interpreted in terms of a fireball model for gamma ray bursts. The burst source has been localized to within a degree of RA = 237.5° and Dec = 76.2°.

143.073 The observation of discrete gamma–ray sources with $E_\gamma \geqslant 5$ MeV from Galactic anticenter region.
A. F. Iyudin, V. G. Kirillov–Ugryumov, Yu. D. Kotov,
A. P. Klarin, Yu. V. Smirnov, V. G. Tyshkevich, V. N. Yurov,
M. I. Fradkin, L. V. Kurnosova, S. V. Damle, G. S. Gokhale,
P. K. Kunte, B. V. Sreekantan.
18th International Cosmic Ray Conference, Vol. 1, p. 102 – 105
(1983). – See Abstr. 012.096.

Differential fluxes of the total Crab emission in the energy range 5 – 100 MeV, integral flux from the Seyfert galaxy 3C 120 and the upper limits of some discrete sources in Galactic anticenter region were determined from the Nov. 6, 1980 balloon flight of "Natalya–1" gamma–telescope.

143.074 Source catalogue of γ–rays from SASII data.
B. P. Houston, A. W. Wolfendale.
18th International Cosmic Ray Conference, Vol. 1, p. 106 – 109
(1983). – See Abstr. 012.096.

Many of the sources coincide in position and flux with those from the COS B experiment. Monte Carlo analyses have also been performed and it is concluded that many of the sources can be understood in terms of cosmic ray interactions with the lumpy interstellar gas.

143.075 Are there gamma–ray loud quasars?
R. Schlickeiser, W. Sieber.
18th International Cosmic Ray Conference, Vol. 1, p. 110 – 113
(1983). – See Abstr. 012.096.

143.076 On the periodicity of gamma–ray flux with the energy $> 10^{12}$eV from 2CG 195 + 4.
Yu. L. Zyskin, D. B. Mukanov.
18th International Cosmic Ray Conference, Vol. 1, p. 122 – 125
(1983). – See Abstr. 012.096.

143.077 Very high energy gamma rays from Cygnus X3?
C. Morello, G. Navarra, S. Vernetto.
18th International Cosmic Ray Conference, Vol. 1, p. 127 – 130
(1983). – See Abstr. 012.096.

The high energy gamma ray emission from Cygnus X3 has been studied by measuring the arrival directions of small air showers. The analysis of 1.8×10^5 showers within 5 degrees from the source indicates a periodic component (P ~ 4.8 h) at a 99% confidence level. A peak (2.8 s.d.) observed at phase 0.60–0.65 coincides with the maximum of the X–ray emission, and the corresponding flux is s $= 4.2 \pm 1.5 \times 10^{-12}$ph cm^{-2}sec^{-1}. A preliminary analysis indicates a 34.1 d phase and amplitude modulation.

143.078   Search for gamma–rays of energies E > 10^{15}eV from discrete sources.
W. Stamm, M. Samorski.
18th International Cosmic Ray Conference, Vol. 1, p. 131 – 134 (1983). – See Abstr. 012.096.

The data of the extensive air shower experiment at Kiel have been systematically scanned for possible gamma–ray sources in the energy range E > 10^{15}eV and declination range $\delta = 25° - 75°$. Photon fluxes for the six statistically most significant source candidates and 3 σ upper limit photon fluxes for COS B gamma–ray sources are presented.

143.079   The 4.8 hour period of gamma–rays from Cygnus X–3 at energies E $\geqslant$ 2 × 10^{15}eV.
M. Samorski, W. Stamm.
18th International Cosmic Ray Conference, Vol. 1, p. 135 – 138 (1983). – See Abstr. 012.096.

The 4.8 hour modulation of E $\geqslant 2 \times 10^{15}$eV gamma–rays from Cygnus X–3 has been analysed in detail on the basis of experimental data of the extensive air shower experiment at Kiel. A period $p_0 = 0.199,681,6 \pm 0.000,000,2$ d is found. For the period derivate an upper limit of $\dot{p} \lesssim 2 \times 10^{-9}$ is derived. The light curve has a significant narrow peak at phase 0.35 and there are indications for a second smaller peak at phase 0.65.

143.080   Steady emission of very high energy gamma rays from interesting celestial sources.
P. N. Bhat, S. K. Gupta, P. V. Ramana Murthy,
B. V. Sreekantan, S. C. Tonwar, P. R. Vishwanath.
18th International Cosmic Ray Conference, Vol. 1, p. 140 (1983). Abstract. – See Abstr. 012.096.

143.081   Search for excess showers from gamma ray sources.
T. Hara, N. Hayashida, M. Honda, K. Kamata,
T. Kifune, M. Nagano, G. Tanahashi, M. Teshima.
18th International Cosmic Ray Conference, Vol. 1, p. 145 (1983). Abstract. – See Abstr. 012.096.

143.082   Gamma ray sources – black holes?
S. Karakula, W. Tkaczyk, F. Giovannelli.
18th International Cosmic Ray Conference, Vol. 1, p. 148 – 151 (1983). – See Abstr. 012.096.

The authors determined the γ–ray production spectrum in the comoving plasma system for different temperatures and expected γ–ray spectrum for the case of spherically symmetric accretion of matter onto a black hole. The luminosity and emissivity of a typical black hole of 10 M$_\odot$ and mass accretion rate of 10^{-8}M$_\odot$/y were also determined. The predictions of the model gauged to 3C 273 are presented. A mass accretion rate M $\geqslant$ 400 M$_\odot$/y needs to fit the model with the experimental data for a total mass of the black hole M $\geqslant 2.5 \times 5 \times 10^9M_\odot$.

143.083   Diffuse component of gamma–ray radiation with $E_\gamma \geqslant$ 5 MeV measured by high altitude balloons.
A. F. Iyudin, V. G. Kirillov–Ugryumov, Yu. D. Kotov,
A. P. Klarin, Yu. V. Smirnov, V. G. Tyshkevich, V. N. Yurov,
M. I. Fradkin, L. V. Kurnosova, S. V. Damle, G. S. Gokhale,
P. K. Kunte, B. V. Sreekantan.
18th International Cosmic Ray Conference, Vol. 1, p. 157 – 160 (1983). – See Abstr. 012.096.

The gamma–radiation detected by the balloon–borne instruments can be divided into three components $I_\gamma = I_{\gamma cos} + I_{\gamma atm} + I_{\gamma loc}$ where $I_{\gamma cos}$ is the primary radiation, $I_{\gamma atm}$ is the secondary radiation in the atmosphere due to cosmic rays, and $I_{\gamma loc}$ is the secondary radiation in the material of the balloon and gamma–telescope caused by cosmic rays.

143.084   The bremsstrahlung component of the diffuse galactic gamma–ray emission at MeV energies.
V. Schönfelder, W. Sacher.
18th International Cosmic Ray Conference, Vol. 1, p. 161 – 164 (1983). – See Abstr. 012.096.

Under the assumption that most of the observed flux below 30 MeV is produced by electron bremsstrahlung, the energy spectrum of cosmic ray electrons in interstellar space below 100 MeV is derived.

143.085   Diffuse galactic emission of high energy gamma rays.
D. A. Kniffen, C. E. Fichtel, R. C. Hartman.
18th International Cosmic Ray Conference, Vol. 1, p. 165 – 168 (1983). – See Abstr. 012.096.

143.086   Origin of diffuse gamma–ray background.
R. J. Protheroe, D. Kazanas.
18th International Cosmic Ray Conference, Vol. 1, p. 169 – 172 (1983). – See Abstr. 012.096.

A new model for the origin of relativistic particles and gamma–rays in active galactic nuclei and quasars, together with recent HEAO–1 observations of the spectra of active galaxies from 2 to 165 keV, provide the basis for a re–examination of the nature of the extragalactic gamma–ray background.

143.087   On the spatial distribution of gamma–ray burst sources.
W. K. H. Schmidt.
18th International Cosmic Ray Conference, Vol. 9, p. 18 – 21 (1983). – See Abstr. 012.096.

Evidence for a spatially structured burst source distribution is searched for in the KONUS and VELA data. It is found that there is no conclusive evidence in favour of a spatially limited burst source distribution.

143.088   Observation of Oct. 16 1981 cosmic gamma burst by Hakucho satellite.
M. Katoh, T. Murakami, J. Nishimura, T. Yamagami,
M. Fujii, M. Itoh, I. Kondo.
18th International Cosmic Ray Conference, Vol. 9, p. 22 – 25 (1983). – See Abstr. 012.096.

Combining the Hakucho and PVO data, analysis has yielded a detailed time profile of the event, energy spectra at 4 s interval covering the energy range of 1 – 55 keV and 120 – 2000 keV, and a source location.

143.089   Low energy gamma ray observation of the Crab and Cygnus region with the "MISO" telescope.
F. Perotti, A. Della Ventura, G. Villa, L. Bassani, R. C. Butler,
G. Di Cocco, R. E. Baker, J. N. Carter, A. J. Dean.
18th International Cosmic Ray Conference, Vol. 9, p. 26 – 29 (1983). – See Abstr. 012.096.

During two balloon flights of the MISO telescope in 1979 and 1980, the Crab Nebula and Cygnus X–1 were studied in the hard X ray – soft γ ray energy band. The Crab Nebula was observed in 1980 and its spectrum was measured up to 2 MeV. Cygnus X–1 was observed on both occasions and was found to be in different hard X–ray states: in a superlow state in 1979 and in a low state 1980. No γ ray excess from this source was measured above few hundred keV.

143.090   Search for gamma ray lines from the Seyfert galaxy NGC 1275 and the COS B source 2CG 195 + 04.
C. A. Ayre, P. N. Bhat, Y. Q. Ma, R. M. Myers,
M. G. Thompson.
18th International Cosmic Ray Conference, Vol. 9, p. 30 – 32 (1983). – See Abstr. 012.096.

NGC 1275 and 2CG 195 + 04 have been observed during flight over Palestine (Texas, USA). The payload consisted of HP(Ge) detector actively shielded by 12.5 cm thick sodium iodide crystals. A total exposure factor on NGC 1275 of 5.4 m^{2}s has been obtained but considerably less on 2CG 195 + 04.

143.091 Observations of high energy gamma rays from Cygnus X–3 and other sources with the Whipple Observatory Large Aperture Camera.
J. Clear, M. F. Cawley, D. J. Fegan, K. Gibbs, R. C. Lamb,
P. K. MacKeown, N. A. Porter, T. C. Weekes.
18th International Cosmic Ray Conference, Vol. 9, p. 53 – 56 (1983). – See Abstr. 012.096.

Preliminary observations with the 19 element camera on the Whipple Observatory's 10 m Optical Reflector are reported. Two methods for selecting candidate gamma ray events are described. A small effect at phase 0.55 – 0.70 is seen from Cygnus X–3.

143.092 Observation of γ–rays $> 10^{15}$eV from Cygnus X–3.
J. Lloyd–Evans, R. N. Coy, A. Lambert, J. Lapikens,
M. Patel, R. J. O. Reid, A. A. Watson.
18th International Cosmic Ray Conference, Vol. 9, p. 65 – 68 (1983). – See Abstr. 012.096.

The existence of pulsed emission of ultra–high energy γ–rays ($> 10^{15}$eV) from Cyg X–3 with the characteristic period of 4.8^h, discovered by Samorski and Stamm (1983), is confirmed using data from three arrays at Haverah Park. The authors also provide strong evidence that the pulsed γ–ray spectrum steepens above 10^{16}eV.

143.093 Interstellar absorption of very high energy gamma rays.
M. F. Cawley, K. Gibbs, T. C. Weekes.
18th International Cosmic Ray Conference, Vol. 9, p. 69 – 72 (1983). – See Abstr. 012.096.

Recent reports of the detection of very high energy gamma rays ($10^{12} – 10^{16}$eV) from sources within the Galaxy prompt a reexamination of the absorption expected from photon–photon pair production. For a source such as Cygnus X–3 it is shown that the absorption may be used to limit the distance to the source.

143.094 The radial distribution of gamma rays and cosmic rays in the outer Galaxy.
J. B. G. M. Bloemen, K. Bennett, G. F. Bignami, L. Blitz,
P. A. Caraveo, M. Gottwald, W. Hermsen, F. Lebrun,
H. A. Mayer–Hasselwander, A. W. Strong.
18th International Cosmic Ray Conference, Vol. 9, p. 73 – 76 (1983). – See Abstr. 012.096.

The gamma–ray emissivity spectrum is derived for three distance ranges in the second and third galactic quadrants and is discussed in terms of the contribution from bremsstrahlung and π^0 decay and of the distribution of cosmic rays.

143.095 The production of galactic gamma–rays above 35 MeV.
K. H. Chik, E. C. M. Young.
18th International Cosmic Ray Conference, Vol. 9, p. 77 – 80 (1983). – See Abstr. 012.096.

Assuming that cosmic rays are coupled to galactic matter on a large scale, the authors work out the longitudinal distributions of γ–ray flux in two energy ranges, namely 35 – 100 MeV and > 100 MeV. The results suggest a strong coupling between the cosmic rays and the matter, supporting the supernova hypothesis for the origin of cosmic rays. In addition, interstellar electron spectra with intensities substantially greater than those currently used are required to reproduce the observed γ–ray flux in the 35 – 100 MeV range.

143.096 Gamma–rays from the inverse Compton scattering process.
C. B. Poon, E. C. M. Young.
18th International Cosmic Ray Conference, Vol. 9, p. 81 84 (1983). – See Abstr. 012.096.

The contribution to the galactic gamma–rays from the inverse Compton scattering process (ICS) has been studied in detail. The source functions at solar vicinity are derived and evaluated for the three important ICS components: 2.7K background radiation, starlight and far infrared.

143.097 γ rays measurements in E $>$ 4 MeV energy range.
J. M. Lavigne.
18th International Cosmic Ray Conference, Vol. 9, p. 89 (1983). Abstract. – See Abstr. 012.096.

143.098 COS–B measurements of high–latitude gamma rays and evidence for a component not related to the gas.
A. W. Strong, K. Bennett, G. F. Bignami, J. B. G. M. Bloemen,
R. Buccheri, P. A. Caraveo, W. Hermsen, F. Lebrun,
H. A. Mayer–Hasselwander, M. E. Özel.
18th International Cosmic Ray Conference, Vol. 9, p. 90 – 93 (1983). – See Abstr. 012.096.

The gamma ray latitude distribution for $10° < |b| < 90°$ measured by COS–B is compared with the expected emission from cosmic–ray/gas interactions. An additional component is found which is unrelated to the gas, and which may be interpreted as a gamma–ray 'halo'. The possibility of an inverse–Compton origin is discussed using diffuse X–ray data.

143.099 Upper limits to ultra–high energy ($> 10^{15}$eV) γ–ray fluxes from regions within the Galaxy.
A. Lambert, J. Lloyd–Evans, A. A. Watson.
18th International Cosmic Ray Conference, Vol. 9, p. 219 – 222 (1983). – See Abstr. 012.096.

The extensive air shower data recorded at Haverah Park during the period 1978 to 1982, for primary energies $> 10^{15}$eV, have been analysed to provide upper limits to the γ–ray fluxes from several candidate sources and for a narrow band along the galactic equator.

143.100 New developments in high energy astrophysics.
P. C. Agrawal.
18th International Cosmic Ray Conference, Vol. 12, p. 357 – 377 (1983). – See Abstr. 012.096.

This rapporteur paper presents a brief review of the present state of our knowledge and a summary of the new results presented in the various papers in the areas of X–ray and gamma–ray astronomy.

143.101 Rapporteur paper on high energy gamma–ray astronomy, sessions XG4 and XG5 (*of the conference*).
N. A. Porter.
18th International Cosmic Ray Conference, Vol. 12, p. 435 – 447 (1983). – See Abstr. 012.096.

High energy sources, such as Cyg X–3, Crab– and Vela pulsar, are treated. Then papers on diffuse high–energy gamma–rays are reported.

143.102 On the optical and radio radiation accompanying cosmic gamma–ray bursts.
N. N. Vzorov, L. P. Gorbachev, A. Yu. Matronchik,
K. S. Mozgov.
Pis'ma Astron. Zh., Tom 11, No. 6, p. 444 –.447 (1985). In Russian. English translation in Sov. Astron. Lett., Vol. 11.

The possibility of ground–based detection of electromagnetic radiation excited by gamma–ray bursts in the earth's atmosphere is considered.

143.103 Flux and distance limits for gamma ray bursters.
B. J. Carrigan, J. I. Katz.
Bull. Am. Astron. Soc., Vol. 17, No. 2, p. 580 (1985). Abstract. – See Abstr. 010.065.

143.104 Gamma–ray burst localizations.
R. A. Schwartz, J. C. Ling, W. A. Mahoney,
W. A. Wheaton, A. S. Jacobson.
Bull. Am. Astron. Soc., Vol. 17, No. 2, p. 604 (1985). Abstract. – See Abstr. 010.065.

143.105 Much ado about Geminga.
N. Deruelle.
Gravitation, geometry and relativistic physics, p. 238 – 248 (1984). In French. – See Abstr. 012.103.

A review of the observational properties of the strong gamma–ray source Geminga is presented. In particular, evidence for a 160 min periodicity in the source is examined and the reputed excitation of 160 min solar oscillations by gravitational waves from Geminga is critically discussed.

143.106 **Gamma–ray astronomy.**
J. Grygar.
Vesmír, Vol. 64, No. 1, p. 10 – 12 (1985). In Czech.

143.107 **A study of the temporal and spectral characteristics of gamma–ray bursts.**
J. P. Norris.
Diss. Abstr. Int., Sect. B, Vol. 45, No. 2, p. 584 – 585 (1984). Thesis, University of Maryland, 279 pp. (1983). Order No. DA8412036.

Catalog and bibliography of gamma–ray bursts.
See Abstr. 002.005.

Gamma–ray astronomy and the spirit of Cos–B.
See Abstr. 013.064.

The Explosive Transient Camera (ETC): an instrument for the detection of gamma–ray burst optical counterparts.
See Abstr. 032.004.

The rapidly moving telescope: an instrument for the precise study of optical transients.
See Abstr. 032.005.

An observatory to study 10^{10} to 10^{16}eV gamma rays.
See Abstr. 034.080.

The application of imaging to very high energy gamma ray astronomy.
See Abstr. 034.081.

Application of two dimensional imaging of atmospheric Cerenkov light to very high energy gamma ray astronomy.
See Abstr. 034.128.

An experiment to detect UHE γ–ray sources.
See Abstr. 034.134.

BATSE/GRO observational capabilities.
See Abstr. 035.003.

All sky high resolution cameras for hard and soft X–rays.
See Abstr. 035.004.

A balloon–borne instrumentation for cosmic gamma–ray burst detection and measurement.
See Abstr. 035.044.

Prospects for developing satellite–borne gamma–ray telescopes.
See Abstr. 035.047.

The observation program of COMPTEL.
See Abstr. 035.051.

Concept and design studies for the exploration of high energy gamma–ray astronomy on the Gamma–Ray Observatory.
See Abstr. 035.137.

Evaluation of Compton contamination events in an actively shielded Ge gamma–ray detector.
See Abstr. 036.138.

Gamma–ray astronomy and the study of electrons $> 10^{13}$eV and bursts using an array of γ–ray detectors in space platform.
See Abstr. 036.139.

A new statistic for the analysis of circular data with applications in ultra–high energy gamma–ray astronomy.
See Abstr. 036.151.

The High Energy Transient Explorer (HETE).
See Abstr. 051.005.

Solar and cosmic X– and gamma–ray studies with long duration balloon flights.
See Abstr. 051.039.

Space astronomy: technical comments – X– and gamma–ray astronomy.
See Abstr. 051.076.

Neutrino production from discrete high–energy gamma–ray sources.
See Abstr. 061.020.

Diffusive shock acceleration in modified shocks.
See Abstr. 062.001.

Theory of optical flashes.
See Abstr. 063.004.

The evolution of binaries of moderate primordial mass ($\lesssim 10\ M_\odot$) and the formation of transient and explosive objects due to accumulation instabilities in such binaries.
See Abstr. 065.011.

Plasma radiation during γ–ray bursts.
See Abstr. 067.002.

Ultra–high energy γ rays and cosmic rays from accreting degenerate stars.
See Abstr. 067.004.

The theory of gamma–ray bursts.
See Abstr. 067.020.

Physics of gamma–ray burst spectra.
See Abstr. 067.021.

Gamma burst emission from neutron star accretion.
See Abstr. 067.022.

The formation of planetesimals in highly compact binary stellar systems and the nature of cosmic gamma–ray bursts.
See Abstr. 067.023.

Gamma–ray burst emission: a jet and fireball model.
See Abstr. 067.024.

Size–frequency distribution of gamma–ray bursts from thermonuclear runaway on neutron stars accreting interstellar gas.
See Abstr. 067.025.

Gamma–ray bursts – the roundabout way?
See Abstr. 067.026.

Inverse comptonization vs. thermal synchrotron.
See Abstr. 067.027.

Physics of the synchrotron model of cosmic gamma–ray bursts.
See Abstr. 067.028.

Gamma–ray bursts from remnant neutron star disks.
See Abstr. 067.082.

Feeding a gamma–ray burster.
See Abstr. 067.091.

Gamma ray burst model.
See Abstr. 067.112.

Accretion model of gamma ray bursts.
See Abstr. 067.113.

Spectra of gamma–ray bursts.
See Abstr. 067.120.

Nonthermal radiation from neutron stars.
See Abstr. 067.153.

Thermonuclear runaway on neutron stars accreting interstellar gas as the origin of gamma ray bursts.
See Abstr. 067.177.

Tidal disruption of stars by a massive black hole as a possible source of the 0.511 MeV annihilation line.
See Abstr. 067.178.

Limits on gamma–ray burst modulation by neutron star rotation.
See Abstr. 067.191.

Comptonization of relativistic electron beams in pulsar cascades.
See Abstr. 067.192.

Il Sole. Il 21° ciclo di attività solare e le nuove acquisizioni sui rapporti tra il Sole e i pianeti. Il "caso" Geminga.
See Abstr. 072.071.

Solar–stellar outer atmospheres and energetic particles, and galactic cosmic rays.
See Abstr. 074.034.

Monte Carlo simulation of atmospheric gamma–ray scattering.
See Abstr. 082.009.

A catalogue of atmospheric gamma ray lines.
See Abstr. 082.078.

X–ray binaries and high–energy gamma rays.
See Abstr. 117.112.

Search for gamma–ray line emission from SS 433.
See Abstr. 117.153.

Cyg X–3 not seen in the COS–B > 100 MeV gamma rays.
See Abstr. 117.232.

Ultra–high–energy gamma–rays and cosmic rays from accreting degenerate stars.
See Abstr. 117.233.

Search for gamma ray lines from SS433.
See Abstr. 117.237.

Evidence for periodic 10^{15}eV gamma–ray emission from Hercules X–1.
See Abstr. 117.281.

Upper limits to supernova prompt emission in gamma rays from the Franco–Soviet SIGNE experiment.
See Abstr. 125.090.

Gamma rays of 0.3 – 30 MeV from PSR 0531 + 21.
See Abstr. 126.073.

Search for very high energy gamma rays from the Crab pulsar.
See Abstr. 126.074.

Space distribution of the pulsars in the Galaxy and the nature of the γ–ray sources.
See Abstr. 126.105.

Possible sporadicity in emission of very high energy gamma rays from Crab pulsar.
See Abstr. 126.106.

The energy spectrum of very high energy gamma rays from the Crab pulsar.
See Abstr. 126.107.

Possible transient emission of very high energy gamma rays from Vela pulsar.
See Abstr. 126.108.

Very high energy gamma ray emission from Vela pulsar and PSR 0950.
See Abstr. 126.109.

Gamma rays of 1 – 30 MeV from the Vela pulsar PSR 0833–45.
See Abstr. 126.110.

Gamma rays from the Crab pulsar 0531 + 21.
See Abstr. 126.111.

COS–B upper limits on gamma–ray emission from radio pulsars.
See Abstr. 126.112.

A search for high energy gamma rays from PSR 0531 and PSR 1953 using a large aperture camera.
See Abstr. 126.113.

Arecibo pulsar searches in connection with gamma–ray sources. Status of the experiment.
See Abstr. 126.146.

Gamma–ray emission from fast pulsars.
See Abstr. 126.149.

Gamma–ray emission from atomic and molecular gas at intermediate galactic latitudes.
See Abstr. 131.089.

The interstellar radiation field and the production of inverse–Compton gamma rays in the Galaxy.
See Abstr. 131.128.

Gamma–ray observations of recently synthesized interstellar ^{26}Al.
See Abstr. 131.224.

Cloud accretion and evaporation, and their consequences for the diffusive galactic γ–ray emission.
See Abstr. 131.310.

Correlation of γ–ray intensities with column densities of gas in the I.S.M.
See Abstr. 131.311.

High energy X–ray observations of COS–B gamma–ray sources from OSO–8.
See Abstr. 142.024.

X–ray emission associated with gamma–ray bursts.
See Abstr. 142.069.

Report of IAU Commission 48: High energy astrophysics (*Astrophysique de grande énergie*).
See Abstr. 142.083.

Acceleration of cosmic rays in the Loop I "supernova remnant"?
See Abstr. 144.031.

Antiprotons from thick cosmic–ray sources.
See Abstr. 144.055.

Possible contribution of primary gamma–rays to the observed cosmic–ray anisotropy.
See Abstr. 144.075.

Small scale anisotropies in cosmic rays near 10^{16}eV.
See Abstr. 144.120.

Time–dependent energetics of OB associations and gamma–ray sources.
See Abstr. 152.009.

High–energy gamma rays and the large–scale distribution of gas and cosmic rays.
See Abstr. 155.035.

High–energy galactic phenomena and the interstellar medium.
See Abstr. 155.037.

The high–latitude distribution of galactic gamma rays and possible evidence for a gamma–ray halo.
See Abstr. 155.038.

Annihilation radiation from the galactic center: positrons in dust?
See Abstr. 155.083.

Cosmic γ rays and the mass of gas in the Galaxy.
See Abstr. 155.091.

High–energy X–ray observations of the galactic center region.
See Abstr. 155.098.

The galactic gamma–ray distribution: implications for galactic structure and the radial cosmic–ray gradient.
See Abstr. 155.101.

Detection of galactic ^{26}Al gamma radiation by the SMM spectrometer.
See Abstr. 155.115.

A possible galactic positron annihilation medium: neutral hydrogen.
See Abstr. 155.116.

Gamma ray imaging of the Vela region and galactic plane at 2 – 20 MeV.
See Abstr. 155.129.

Diffusion characteristics of electrons in the Galaxy.
See Abstr. 155.157.

Gamma rays of 1 – 30 MeV from the galactic center region.
See Abstr. 155.159.

Galactic hydrogen distribution and the production of galactic gamma–rays.
See Abstr. 155.160.

The Large Magellanic Cloud: a test for gamma ray and cosmic ray origin theories.
See Abstr. 156.027.

Soft gamma–ray production in active galactic nuclei.
See Abstr. 158.011.

Gamma ray observations of active galactic nuclei as a test of black hole accretion models.
See Abstr. 158.282.

The development of the electromagnetic cascade showers initiated by high–energy electrons and gammay–rays in the hot photon gas.
See Abstr. 158.283.

144 Cosmic Rays

144.001 Self–confined cosmic rays.
D. G. Wentzel.
Unstable current systems and plasma instabilities in astrophysics, p. 341 – 354 (1985). – See Abstr. 012.002 (IAU Symp. No. 107).
The theory of self–confinement explains the isotropy of the bulk of the cosmic rays but not of cosmic rays above 10^3GeV; it has been a stimulus to the theory for cosmic–ray acceleration at supernova shocks; and, on inclusion of diffusion in a galactic wind, it may explain the uniform cosmic–ray density out to 18 kpc in our galaxy. Rapidly streaming electrons in clusters of galaxies, in supernova remnants, and near solar flares are accommodated by the theory when it is expanded to include the effects of hot plasmas and other wave modes. A "resonance gap" may prevent the turning backwards of streaming particles and thus allow streaming near the particle speed.

144.002 Cosmic rays and the Parker instability.
A. H. Nelson.
Unstable current systems and plasma instabilities in astrophysics, p. 361 – 363 (1985). – See Abstr. 012.002 (IAU Symp. No. 107).
It seems likely that the effect of cosmic rays helps to stabilize the galactic gas layer and makes the Parker instability (Parker, 1966, 1969, 1979) less plausible as a mechanism for forming molecular clouds.

144.003 Intensity variation of cosmic rays near the heliospheric current sheet.
Badruddin, R. S. Yadav, N. R. Yadav.
Planet. Space Sci., Vol. 33, No. 2, p. 191 – 201 (1985).
Cosmic ray intensity variations near the heliospheric current sheet – both above and below it – have been studied during 1964 – 76. Superposed epoch analysis of the cosmic ray neutron monitor data with respect to sector boundaries (i.e., heliospheric current sheet crossings) has been performed. The authors have used the data from neutron monitors well distributed in latitude over the Earth's surface. The results have been discussed in the light of theoretical and observational evidences.

144.004 Large–scale solar modulation of $\gtrsim 500$ MeV/nucleon galactic cosmic rays seen from 1 to 30 AU.
W. Fillius, I. Axford.
J. Geophys. Res., Vol. 90, No. A1, p. 517 – 520 (1985).
Using measurements of $\gtrsim 500$ MeV/nucleon cosmic rays obtained by Cerenkov counters on Pioneer 10 and Pioneer 11 and neutron monitor data from earth, the authors can observe the spatial and temporal development of cosmic ray modulation during the last solar maximum. These observations add new spatial dimensions to their knowledge and enable the authors to test and improve theories of the cosmic ray modulation process.

144.005 Interplanetary cosmic ray intensity: 1972 – 1984 and out to 32 AU.
J. A. Van Allen, B. A. Randall.
J. Geophys. Res., Vol. 90, No. A2, p. 1399 – 1412 (1985).
The paper gives a progress report on the authors' long–term study of the interplanetary intensity of the cosmic radiation out to great distances from the sun, using a system of detectors on the spacecraft Pioneer 10 and Pioneer 11.

144.006 Cosmic ray intensity variations and the solar wind velocity.
K. Fujimoto, H. Kojima.
Proc. Cosmic–Ray Res. Lab. Nagoya Univ., Vol. 27, No. 1, p. 51 – 56 (1984). In Japanese. Abstr. in Phys. Abstr., Vol. 88, No. 1248, Entry 9499 (1985).

144.007   Galactic cosmic rays above 10^{18}eV.
J. Wdowczyk, A. W. Wolfendale.
J. Phys. G, Vol. 10, No. 10, p. 1453 – 1463 (1984). Abstr. in Phys. Abstr., Vol. 88, No. 1249, Entry 14351 (1985).

144.008   Shape of energy spectrum of hadrons at sea level.
H. Arvela, M. Lumme, M. Mieminen, J. J. Torsti, E. Vainikka, E. Valtonen.
J. Phys. G, Vol. 10, No. 10, p. 1465 – 1471 (1984). Abstr. in Phys. Abstr., Vol. 88, No. 1249, Entry 14363 (1985).

144.009   Size dependence of frequency attenuation length of extensive air showers observed at Akeno.
M. Nagano, Y. Tan, S. Kawaguchi, T. Hara, Y. Hatano, N. Hayashida, K. Kamata, T. Kifune, G. Tanahashi.
J. Phys. G, Vol. 10, No. 10, p. L235 – L239 (1984). Abstr. in Phys. Abstr., Vol. 88, No. 1249, Entry 14390 (1985).

144.010   The anisotropy of cosmic rays below 10^{18}eV.
J. Wdowczyk, A. W. Wolfendale.
J. Phys. G, Vol. 10, No. 11, p. 1599 – 1608 (1984). Abstr. in Phys. Abstr., Vol. 88, No. 1250, Entry 18487 (1985).

144.011   Tracking of flare–generated interplanetary shock waves, responsible for Forbush decreases, by observing type II radio bursts.
S. Pinter.
Nuovo Cimento C, Vol. 7C, Ser. 1, No. 2, p. 223 – 234 (1984). Abstr. in Phys. Abstr., Vol. 88, No. 1250, Entry 18524 (1985).

144.012   An underground search for anomalous slow penetrating particles.
K. Kawagoe, T. Mashimo, S. Nakamura, M. Nozaki, S. Orito.
Lett. Nuovo Cimento, Vol. 41, Ser. 2, No. 9, p. 315 – 319 (1984). Abstr. in Phys. Abstr., Vol. 88, No. 1250, Entry 18557 (1985).

144.013   An accurate measurement of the sea–level muon spectrum within the range 4 to 3000 GeV/c.
B. C. Rastin.
J. Phys. G, Vol. 10, No. 11, p. 1609 – 1628 (1984). Abstr. in Phys. Abstr., Vol. 88, No. 1250, Entry 18608 (1985).

144.014   A study of the muon charge ratio at sea level within the momentum range 4 to 2000 GeV/c.
B. C. Rastin.
J. Phys. G, Vol. 10, No. 11, p. 1629 – 1638 (1984). Abstr. in Phys. Abstr., Vol. 88, No. 1250, Entry 18609 (1985).

144.015   New interpretation of the charge ratio of cosmic–ray muons.
Y. Minorikawa, K. Mitsui.
Lett. Nuovo Cimento, Vol. 41, Ser. 2, No. 10, p. 333 – 338 (1984). Abstr. in Phys. Abstr., Vol. 88, No. 1250, Entry 18610 (1985).

144.016   Diffusion and convection of relativistic electrons in non–thermal sources: steady–state solutions.
E. Massaro.
Astrophys. Space Sci., Vol. 108, No. 2, p. 369 – 381 (1985).

The author studies the propagation of relativistic electrons in stationary spherically–symmetric models of non–thermal sources. Electrons are injected by a collapsed object in the centre and then diffuse away in the presence of a flux of accreting matter. The electron density is calculated taking into account both the outward diffusion and the inward convection: it is assumed that the diffusion coefficient and the velocity field have a power dependence on the radial coordinate. The transport equation is solved for the particular case where the electron energy is constant; the case in which the particles suffer synchrotron or inverse–Compton losses is treated. The effects due to the convection are then discussed in comparison with the purely diffusive case.

144.017   Diffusion of the galactic cosmic rays in the vicinity of the solar system.
L. I. Dorman, A. Ghosh, V. S. Ptuskin.
Astrophys. Space Sci., Vol. 109, No. 1, p. 87 – 97 (1985).

Cosmic–ray propagation in the vicinity of 1 kpc from the Sun is considered. The data on the $10^{12} - 10^{15}$eV particle anisotropy, on 10^{12}eV electron spectrum, and on temporal cosmic–ray variations are analyzed. The diffusion coefficient $D(10^{12} - 10^{13}$eV$) = 10^{29} - 10^{30}$cm^{2}s^{-1} inferred from the analysis coincides with its standard value in the large–halo model with $h = 15$ kpc. The total power of cosmic–ray generation, about 3×10^{49}erg per SN in the proton component and about 10^{48}erg per SN in the electron component, typical of the galactic diffusion model is in agreement with the obtained parameters of local sources.

144.018   The flux of secondary anti–deuterons and antihelium produced in the interstellar medium.
O. C. Allkofer, D. Brockhausen.
Astrophys. Space Sci., Vol. 109, No. 1, p. 145 – 148 (1985).

Several measurements have been performed to find antiprotons in the primary cosmic radiation. Because it is difficult to get completely separated secondary produced antiprotons from primary ones, calculations based on accelerator results have been performed for the flux of secondary produced anti–deuterons and antihelium.

144.019   A plan of a monopole search experiment using the 100 m^2 calorimeter of the Akeno air shower array.
T. Hara, M. Honda, Y. Ohno, Y. Totsuka, M. Kobayashi, T. Kondo.
Grand unified theories and cosmology, p. 95 – 109 (1984). Abstr. in Phys. Abstr., Vol. 88, No. 1251, Entry 20062 (1985). – See Abstr. 012.020.

144.020   Some topics in cosmic rays.
J. Arafune.
Grand unified theories and cosmology, p. 169 – 176 (1984). Abstr. in Phys. Abstr., Vol. 88, No. 1251, Entry 24188 (1985). – See Abstr. 012.020.

144.021   Observation of ultra high energy cosmic rays ($\geqslant 10^{10}$GeV).
M. Nagano.
Grand unified theories and cosmology, p. 177 – 200 (1984). Abstr. in Phys. Abstr., Vol. 88, No. 1251, Entry 24189 (1985). – See Abstr. 012.020.

144.022   A very long baseline search for correlated air showers form relativistic dust grains.
C. L. Bhat, H. S. Rawat, H. Razdan, M. L. Sapru, S. C. Tonwar.
J. Phys. G, Vol. 10, No. 12, p. 1771 – 1776 (1984). Abstr. in Phys. Abstr., Vol. 88, No. 1251, Entry 24190 (1985).

144.023   Cosmic ray propagation in the local superbubble.
R. E. Streitmatter, V. K. Balasubrahmanyan, R. J. Protheroe, J. F. Ormes.
Astron. Astrophys., Vol. 143, No. 2, p. 249 – 255 (1985).

The authors suggest that Feature A, a ring of H I gas lying in the galactic plane, is part of a supershell which formed some 3×10^7yr ago. They examine the consequences of a closed magnetic supershell for cosmic ray propagation and conclude that there is no evidence which precludes the production and trapping of cosmic rays in such a region.

144.024   Ionic models of cosmic ray sources.
M. Arnaud, M. Cassé.
Astron. Astrophys., Vol. 144, No. 1, p. 64 – 71 (1985).

The authors present an update version of the Cassé and Goret (1978) ionization model of cosmic ray sources, with emphasis on the relevant atomic parameters (rates of ionization, recombination, and charge exchange). The cosmic ray source/local galactic (CRS/LG) abundance ratio is identified for each element of interest with the ionic fraction of this element, averaged over a temperature distribution. The authors try to find temperature conditions such that the CRS/LG ratios are reasonably reproduced from hydrogen to nickel. An exponential temperature distribution of the form $\exp(-T/7000K)$ with $5000K \leqslant T \leqslant 100,000K$ leads to a satisfactory fit for most of the elements.

144.025   The insensitivity of the cosmic ray galactic anisotropy to heliomagnetic polarity reversals.
K. Nagashima, S. Sakakibara, A. G. Fenton, J. E. Humble.
Planet. Space Sci., Vol. 33, No. 4, p. 395 – 405 (1985).

Using the cosmic ray sidereal and anti–sidereal diurnal variations observed underground in London and Hobart during the period 1958 – 1983, it is demonstrated that: (1) the phase changes of the apparent sidereal diurnal variation observed only in the Northern Hemisphere cannot be attributed to the change of the heliomagnetospheric modulation of galactic cosmic ray anisotropy caused by the polarity reversal of the solar magnetic field, but that they are due to the fluctuation of the spurious sidereal variation produced from the anisotropy responsible for the solar semi–diurnal variation; (2) the spurious sidereal variation can be eliminated from the apparent variation by using the observed anti–sidereal diurnal variation; and (3) after the elimination, the sidereal diurnal variations in the Northern and Southern Hemispheres almost coincide with each other and are stationary throughout the period, regardlesss of the polarity reversal of the heliomagnetosphere. The origin of the corrected sidereal variation is discussed.

144.026   The galactic cosmic ray intensity minimum in the inner and outer heliosphere in solar cycle 21.
D. Venkatesan, R. B. Decker, S. M. Krimigis, J. A. Van Allen.
J. Geophys. Res., Vol. 90, No. A3, p. 2905 – 2909 (1985).

The paper investigates the occurrence of the cosmic ray intensity minimum during solar cycle 21 at widely separated locations in the heliosphere. The data include in situ measurements by cosmic ray detectors on the earth–orbiting satellite IMP 8 ($E \gtrsim 35$ MeV), on the Voyagers 1 and 2 spacecraft ($E \gtrsim 70$ MeV), and on the Pioneers 10 and 11 spacecraft ($E \gtrsim 80$ MeV), located at distances from 1 to 27 AU.

144.027   Changes in radial gradients of low–energy cosmic rays between solar minimum and maximum: observations from 1 to 31 AU.
R. B. McKibben, K. R. Pyle, J. A. Simpson.
Astrophys. J., Lett. Ed., Vol. 289, No. 2, p. L35 – L39 (1985).

Observations of the fluxes and radial gradients of protons and helium ($10 < E < 70$ MeV n^{-1}) between the period of minimum solar modulation ending in 1977 and the period of maximum modulation in 1981 – 1983 show (1) that the variation in modulation with the phase of the solar cycle is much stronger at radii of 20 – 30 AU than at 1 AU, and (2) that at solar maximum, more than 99% of the total modulation takes place beyond 31 AU.

144.028   On the origin of high–energy cosmic rays.
J. R. Jokipii, G. E. Morfill.
Astrophys. J., Lett. Ed., Vol. 290, No. 1, p. L1 – L4 (1985).

The authors suggest that cosmic rays, up to the highest energies observed, originate in the Galaxy and are accelerated in astrophysical shock waves. If there is a galactic wind, in analogy with the solar wind, one expects a hierarchy of shocks ranging from supernova shocks to the galactic wind termination shock. This leads to a consistent model in which most cosmic rays, up to perhaps 10^{14}eV energy, are accelerated by supernova shocks, but that particles with energies of 10^{15}eV and higher are accelerated at the termination shock of the galactic wind.

144.029   Variation of cosmic rays and solar wind properties with respect to the heliospheric current sheet. 1. Five–GeV protons and solar wind speed.
G. Newkirk Jr., L. A. Fisk.
J. Geophys. Res., Vol. 90, No. A4, p. 3391 – 3414 (1985).

The central objectives of this paper are to determine the gradient of 5–GeV cosmic rays with respect to the heliospheric current sheet at 1 AU; to determine what, if any, effect the solar cycle has upon such gradients; and to evaluate the ability of several theoretical models to explain these gradients. Since the spatial variation of the velocity of the solar wind plays an important role both in determining the structure of the heliospheric magnetic field and in propagating cosmic rays, the authors investigate the corresponding profiles of solar wind speed.

144.030   Geomagnetic influence on the muon charge ratio of cosmic rays.
O. C. Allkofer, W. D. Dau, H. Jokisch, G. Klemke, R. C. Uhr, G. Bella, Y. Oren.
J. Geophys. Res., Vol. 90, No. A4, p. 3537 – 3539 (1985).

The charge ratio of muons from cosmic radiation is strongly modified by the earth's magnetic field in the low–energy region. With a magnetic spectrometer (DEIS project) erected at Tel–Aviv this influence was studied in the momentum range from 10 to 200 GeV.

144.031   Acceleration of cosmic rays in the Loop I "supernova remnant"?
C. L. Bhat, M. R. Issa, C. J. Mayer, A. W. Wolfendale.
Nature, Vol. 314, No. 6011, p. 515 – 517 (1985).

Attempts are being made to establish the origin of cosmic rays using the γ–ray method, that is, by dividing γ–ray intensity by gas–column density to chart the cosmic–ray intensity. Here the authors present evidence from satellite data on γ rays and from new results on the distribution of gas in the interstellar medium for an enhancement of electron and proton intensity in the Loop I supernova remnant. There follows from this a straightforward explanation for the origin of the bulk of the low–energy cosmic radiation.

144.032   The role of inhomogeneities of the magnetic field forming in the interplanetary medium in the 11–year cycle of cosmic rays.
V. P. Okhlopkov, L. S. Okhlopkova, T. N. Charakhch'yan.
NII yader. fiz. MGU. Moskva, 19 pp. (1984). In Russian. Abstr. in Ref. Zh., 62. Issled. Kosm. Prostranstva, 2.62.388 (1985).

144.033   Gradients of cosmic rays perpendicular to the ecliptic plane.
A. G. Zusmanovich, L. A. Mirkin.
Geomagn. Aehron., Tom 25, No. 2, p. 184 – 188 (1985). In Russian. English translation in Geomagn. Aeron.

144.034   Peculiarities of cosmic ray intensity variations and anisotropy during recurrent Forbush effects.
N. A. Nachkebiya, L. Kh. Shatashvili.
Geomagn. Aehron., Tom 25, No. 2, p. 189 – 192 (1985). In Russian. English translation in Geomagn. Aeron.

144.035   Cosmic ray intensity waves and the north–south anisotropy.
R. M. Jacklyn, M. L. Duldig, M. A. Pomerantz.
Proc. Astron. Soc. Aust., Vol. 5, No. 4, p. 581 – 586 (1984).

144.036   On time delay of modulated effects for cosmic rays of various energies.
A. G. Zusmanovich, O. N. Kryakunova.
Geomagn. Aehron., No. 1, p. 126 – 128 (1985). In Russian. English translation in Geomagn. Aeron.

144.037   Cosmic ray sources and confinement in the Galaxy.
L. Koch–Miramond.
The Big Bang and Georges Lemaitre, p. 315 – 325 (1984). – See Abstr. 012.043.

Recently the Saclay–Copenhagen spectrometer on board the satellite HEAO 3 has provided extremely accurate data on the elemental composition and energy spectra of cosmic ray nuclei from Be to Zn in the energy range 0.7 to 25 GeV/n. These data have been interpreted in the framework of the galactic diffusion model, either homogeneous or with a halo. The data tend to favor a rather flat halo, ~ 1 kpc thick.

144.038 Cosmic ray generation by plasma turbulence – development of a new paradigm in physics.
V. N. Tsytovich.
ESA Spec. Publ., ESA SP–207, p. 141 – 152 (1984). – See Abstr. 012.044.

A review of the development of a new concept (a new paradigm) for cosmic ray generation by plasma turbulence is given. Effectiveness of magnetohydrodynamic turbulence is compared with that of electrostatic plasma turbulence. A new mechanism of cosmic ray generation due to radiative processes is described. It is shown that the main properties of cosmic rays can be explained by this mechanism.

144.039 Cosmic ray composition and origin.
P. Király.
ESA Spec. Publ., ESA SP–207, p. 153 – 157 (1984). – See Abstr. 012.044.

Information on the composition of cosmic rays has considerably increased during the past few years, mainly due to the success of the HEAO–3, Ariel, and JACEE missions. Some minor components like antiprotons and ultrahigh energy gamma–rays have also been detected, and are of high diagnostic value. Although the predominant process of cosmic ray origin is still much debated, the steady progress of observation techniques gives hope to answer most of the outstanding questions in the foreseeable future.

144.040 Cosmic ray acceleration by the solar wind.
S. I. Petukhov, A. A. Turpanov, V. S. Nikolaev.
Izv. Akad. Nauk SSSR. Ser. fiz., Tom 48, No. 11, p. 2066 – 2069 (1984). In Russian. Abstr. in Ref. Zh., 51. Astron., 3.51.727 (1985).

144.041 Particle acceleration in giant molecular clouds.
V. A. Dogel', G. S. Sharov.
Izv. Akad. Nauk SSSR. Ser. fiz., Tom 48, No. 11, p. 2070 – 2073 (1984). In Russian. Abstr. in Ref. Zh., 51. Astron., 3.51.728 (1985).

144.042 A lower limit on cosmic rays flux from complex molecules in cometary comae.
V. Pirronello.
Mem. Soc. Astron. Ital., Vol. 55, No. 3, p. 507 – 510 (1984). – See Abstr. 012.049.

A method is suggested to obtain lower limits on the flux of cosmic rays pervading the Galaxy through observations of complex molecules in the comae of comets entering the solar system.

144.043 The isotopic composition of helium in the cosmic radiation above 11 gigavolts.
S. P. Jordan.
Astrophys. J., Vol. 291, No. 1, p. 207 – 218 (1985).

The author has measured the ^{3}He/^{4}He abundance ratio above 11 GV with a balloon–borne experiment that employs the geomagnetic method of isotope separation near Earth's equator. Above a kinetic energy of approximately 6 GeV per nucleon, the ^{3}He/^{4}He ratio is 0.24 ± 0.05 under the assumption that the helium spectrum at Earth is a power law in rigidity of the form $R^{-2.65}$. The implied interstellar mean escape pathlength of ~ 15 g cm^{-2}, with this assumption, is considerably greater than values determined from observations of heavier elements.

144.044 The influence of momentum distribution of source particles on Fermi acceleration in shock waves.
D.-y. Mao, D.-y. Wang.
Chin. J. Space Sci., Vol. 4, No. 3, p. 184 – 190 (1984). In Chinese. Abstr. in Phys. Abstr., Vol. 88, No. 1252, Entry 29812 (1985).

144.045 The mass composition of the primary cosmic rays in the low–TeV range derived from ground–level muon measurements.
O. C. Allkofer, W. D. Dau, H. Jokisch, G. Klemke, R. C. Hur, G. Bella, Y. Ohren.
Lett. Nuovo Cimento, Vol. 41, Ser. 2, No. 11, p. 373 – 376 (1984). Abstr. in Phys. Abstr., Vol. 88, No. 1252, Entry 29813 (1985).

144.046 Stationary and nonstationary modulation of cosmic rays.
M. V. Alaniya, R. G. Aslamazashvili, D. P. Bochikashvili, T. B. Bochorishvili, T. V. Dzhapiashvili, V. S. Tkemaladze.
Izv. Akad. Nauk SSSR. Ser. fiz., Tom 48, No. 11, p. 2103 – 2105 (1984). In Russian. Abstr. in Ref. Zh., 51. Astron., 4.51.368 (1985).

144.047 Nonstationary effects in the 11–year intensity variations of galactic cosmic rays.
A. G. Zusmanovich, O. N. Kryakunova, L. F. Churunova.
Izv. Akad. Nauk SSSR. Ser. fiz., Tom 48, No. 11, p. 2106 – 2109 (1984). In Russian. Abstr. in Ref. Zh., 51. Astron., 4.51.369 (1985).

144.048 Recurrent double flares of pitch–angle anisotropy of cosmic rays and their connection with solar proton events.
V. M. Dvornikov, V. E. Sdobnov, A. V. Sergeev.
Izv. Akad. Nauk SSSR. Ser. fiz., Tom 48, No. 11, p. 2140 – 2142 (1984). In Russian. Abstr. in Ref. Zh., 51. Astron., 4.51.370 (1985).

144.049 Fluctuation phenomena in cosmic rays observed on the earth, in the stratosphere and cosmic space.
L. I. Dorman, I. Ya. Libin.
Izv. Akad. Nauk SSSR. Ser. fiz., Tom 48, No. 11, p. 2143 – 2145 (1984). In Russian. Abstr. in Ref. Zh., 51. Astron., 4.51.371 (1985).

144.050 Influence of the quiet asymmetrical magnetosphere on asymptotic directions of cosmic rays.
O. A. Danilova, M. I. Tyasto.
Izv. Akad. Nauk SSSR. Ser. fiz., Tom 48, No. 11, p. 2243 – 2245 (1984). In Russian. Abstr. in Ref. Zh., 51. Astron., 4.51.372; 62. Issled. Kosm. Prostranstva, 4.62.384 (1985).

144.051 Cosmic ray flares and cosmogenic isotopes.
G. E. Kocharov.
Complex investigations of the sun, p. 203 – 207 (1982). In Russian. Abstr. in Ref. Zh., 51. Astron., 4.51.375 (1985). – See Abstr. 012.046.

144.052 Searches for narrow–angle anisotropies in the primary energy range 0.1 – 10 TeV.
O. C. Allkofer, W. D. Dau, H. Jokisch, G. Klemke, R. C. Uhr, G. Bella, Y. Oren.
Astrophys. J., Vol. 291, No. 2, p. 468 – 470 (1985).

The arrival directions and energy of 3×10^6 individual muons have been measured with the DEIS spectrometer at Tel Aviv. After appropriate corrections, the muons were sorted into four bins with respect to their production energy ($E > 300$ GeV, 300 GeV $> E > 125$ GeV, 125 GeV $> E > 50$ GeV, 50 GeV $> E > 23$ GeV). They cover the declination band from $-15°$ to $+21°$. Each group was investigated for narrow–angle anisotropies. The result is directional isotropy.

144.053 The origin of cosmic rays.
V. L. Ginzburg, V. S. Ptuskin.
Astrophys. Space Phys. Rev., Vol. 4, p. 161 – 256 (1985). – See Abstr. 003.020. Revised and extended English translation of 34.144.091.

Contents: (1). A general review. (2). Some data on cosmic rays. (3). The proton–nucleus component of cosmic rays in the Galaxy. (4) The electron component of cosmic rays and radio emission in the Galaxy. (5). Cosmic ray propagation in the Galaxy. (6). Closing remarks.

144.054 Solar activity and modulation of the cosmic ray intensity.
S.-I. Akasofu, C. Olmsted, J. A. Lockwood.
J. Geophys. Res., Vol. 90, No. A5, p. 4439 – 4447 (1985).

A transfer function F between a measure of the solar activity S and the 11–year cosmic ray intensity modulation I_0–I is determined for the two sunspot cycles 19 and 20. For S the authors used the monthly occurrence of major flares, all flares, sunspots and ssc's. They find a well–behaved transfer function for both major flares and all flares. It is a damped pulse with duration of 16 and 13 months for the major flares and all flares, respectively. These figures are significantly longer than that of 10 months obtained by Bowe and Hatton (1982).

144.055   Antiprotons from thick cosmic–ray sources.
P. O. Lagage, C. J. Cesarsky.
Astron. Astrophys., Vol. 147, No. 1, p. 127 – 134 (1985).

The authors examine the possibility of explaining the cosmic-ray antiproton flux observed by adding thick cosmic–ray sources to the thin sources of the standard leaky–box model. They show that in this model, if all cosmic rays issue from these sources, the mean thick–source grammage needed to account for the observed antiproton flux is only $\sim 7.5\ \mathrm{g\ cm^{-2}}$. However, thick sources will overproduce secondary cosmic–ray nuclei such as lithium, beryllium, boron if their grammage is less than $\sim 30\ \mathrm{g\ cm^{-2}}$. With this grammage, the fraction of cosmic rays coming from thick sources is $\sim 25\%$. The amounts of other thick–source products predicted by the model are also evaluated: deuterons, helium 3, interstellar lithium, beryllium and boron, positrons, gamma–rays and neutrinos.

144.056   Ultra high energy cosmic rays and the relevance of Cygnus X–3.
A. M. Hillas.
Bull. Am. Astron. Soc., Vol. 16, No. 4, p. 918 (1984). Abstract. – See Abstr. 010.062.

144.057   A study of solar modulation of medium energy primary cosmic ray nuclei.
P. K. Das, T. D. Goswami.
J. Astrophys. Astron., Vol. 6, No. 1, p. 57 – 60 (1985).

The energy spectra of primary cosmic rays were studied in the energy interval 150 to 450 MeV/nucl by using a balloon–borne cellulose–nitrate solid–state plastic detector. Effects of solar modulation were studied using the theoretical spectrum of H_1 nuclei near the solar minimum in 1964 as the demodulated spectrum. The "force–field" potential which fit the experimental results was estimated to be 270 MeV/nucl.

144.058   Cosmic ray antiprotons: experimental perspectives.
J. F. Ormes.
Bull. Am. Astron. Soc., Vol. 17, No. 1, p. 517 (1985). Abstract. – See Abstr. 010.064.

144.059   Cosmic–ray propagation in the solar neighborhood.
A. Ghosh, L. I. Dorman, V. S. Ptuskin.
Sov. Astron. Lett., Vol. 10, No. 6, p. 345 – 347 (1984). English translation of 38.144.029.

144.060   North–south cosmic ray anisotropy in recurrent high-speed solar wind streams.
N. V. Mymrina, L. I. Dorman, N. S. Kaminer, A. E. Kuz'micheva.
Geomagn. Aehron., Tom 25, No. 3, p. 381 – 384 (1985). In Russian. English translation in Geomagn. Aeron.

144.061   Multiple scattering of cosmic rays on moving irregularities of a magnetic field.
L. I. Dorman, V. Kh. Shogenov.
Geomagn. Aehron., Tom 25, No. 3, p. 491 – 492 (1985). In Russian. English translation in Geomagn. Aeron.

144.062   Pulsations of cosmic rays and atmospheric pressure near periods of 160 and 80 min.
A. M. Novikov, Yu. F. Rudnev, N. G. Skryabin, V. A. Starina, V. P. Ul'yanov.
Geomagn. Aehron., Tom 25, No. 3, p. 494 – 495 (1985). In Russian. English translation in Geomagn. Aeron.

144.063   High–energy electrons in the nonuniform galactic disk model.
L. C. Tan.
Astrophys. J., Vol. 293, No. 2, p. 414 – 423 (1985).

The nonuniform galactic disk model, which the author previously proposed to explain the unexpectedly large flux of interstellar antiprotons, is also able to account in a natural way for the spectral shape variation of high–energy cosmic–ray electrons. A comparison of model prediction with the measured electron data suggests that only about 5% of the observed protons are of local origin and that the electron intensity inside a spiral arm is much lower than that outside it.

144.064   Cosmic ray heavy ions at and above 40000 feet.
C. H. Tsao, R. Silberberg, J. R. Letaw.
IEEE Trans. Nucl. Sci., Vol. NS–31, No. 6, p. 1066 – 1068 (1984). Abstr. in Phys. Abstr., Vol. 88, No. 1253, Entry 35495 (1985). – See Abstr. 012.073.

144.065   Cosmic ray intensity distribution perpendicular to the solar equatorial plane.
T. Nosaka, S. Mori, S. Sagisaka.
J. Geomagn. Geoelectr., Vol. 36, No. 2, p. 73 – 81 (1984). Abstr. in Phys. Abstr., Vol. 88, No. 1255, Entry 45884 (1985).

144.066   Cosmic ray heavy–ion interactions: recent JACEE results.
T. H. Burnett, S. Dake, M. Fuki, J. C. Gregory, T. Hayashi, R. Holynski, J. Iwai, W. V. Jones, A. Jurak, J. J. Lord, O. Miyamura, H. Oda, T. Ogata, T. A. Parnell, T. Saito, T. Tabuki, Y. Takahashi, T. Tominaga, J. W. Watts, B. Wilczynska, R. J. Wilkes, W. Wolter, B. Wosiek.
AIP Conf. Proc., No. 123, p. 723 – 734 (1984). Abstr. in Phys. Abstr., Vol. 88, No. 1256, Entry 51442 (1985). – See Abstr. 012.082.

144.067   On the origin of cosmic rays and the value of fundamental length.
K. Namsrai.
Found. Phys., Vol. 15, No. 2, p. 129 – 143 (1985). Abstr. in Phys. Abstr., Vol. 88, No. 1257, Entry 57563 (1985).

144.068   Composition and spectra of cosmic–ray hadrons at sea level.
M. Nieminen, J. J. Torsti, E. Valtonen, H. Arvela, M. Lumme, J. Peltonen, E. Vainikka.
J. Phys. G, Vol. 11, No. 3, p. 421 – 437 (1985). Abstr. in Phys. Abstr., Vol. 88, No. 1257, Entry 57567 (1985).

144.069   Ultrahigh–energy cosmic–ray spectrum.
C. T. Hill, D. N. Schramm.
Phys. Rev. D, Vol. 31, No. 3, p. 564 – 580 (1985). Abstr. in Phys. Abstr., Vol. 88, No. 1257, Entry 57568 (1985).

144.070   Analysis of cosmic rays with $Z \geqslant 50$ registered in meteorite minerals.
B. Jakupi, M. Bytyci, V. P. Perelygin, S. G. Stetsenko, R. Antanasijevic, Z. Todorovic, A. B. Akopova, L. Y. Melkumyan.
Fizika, Vol. 16, No. 4, p. 377 – 384 (1984). Abstr. in Phys. Abstr., Vol. 88, No. 1257, Entry 57570 (1985).

144.071   Primary mass composition and its determination in the energy range between 10^5 and 10^8 GeV.
P. K. F. Grieder.
Nuovo Cimento A, Vol. 84A, Ser. 11, No. 4, p. 285 – 298 (1984). Abstr. in Phys. Abstr., Vol. 88, No. 1258, Entry 63181 (1985).

144.072   The power spectrum of cosmic rays. The low–frequency range $(1 \times 10^{-11} – 1 \times 10^{-5})$ Hz.
M. R. Attolini, S. Cecchini, M. Galli.
Nuovo Cimento C, Vol. 7C, Ser. 1, No. 4, p. 413 – 426 (1984). Abstr. in Phys. Abstr., Vol. 88, No. 1259, Entry 68633 (1985).

144.073   No sign of cosmic–ray generation from Jupiter, detectable at ground–based stations.
K. Nagashima, R. Tatsuoka.
Nuovo Cimento C, Vol. 7C, Ser. 1, No. 4, p. 379 – 390 (1984).
Abstr. in Phys. Abstr., Vol. 88, No. 1259, Entry 68635 (1985).

144.074   Primary–cosmic–ray energy spectrum up to 50 TeV derived from sea–level muon measurements.
O. C. Allkofer, W. D. Dau, H. Jokisch, G. Klemke, R. C. Uhr, G. Bella, Y. Oren.
Phys. Rev. D, Vol. 31, No. 7, p. 1557 – 1560 (1985). Abstr. in Phys. Abstr., Vol. 88, No. 1259, Entry 68636 (1985).

144.075   Possible contribution of primary gamma–rays to the observed cosmic–ray anisotropy.
V. V. Alexeenko, G. Navarra.
Lett. Nuovo Cimento, Vol. 42, Ser. 2, No. 7, p. 321 – 324 (1985).
Abstr. in Phys. Abstr., Vol. 88, No. 1260, Entry 74026 (1985).

144.076   Longitudinal development of air–shower electrons studied from the arrival time distributions of atmospheric Cerenkov light measured at 900 m above sea level.
N. Inoue, S. Sugawa, Y. Miyazaki, T. Enoki, F. Kakimoto, K. Suga, K. Nishi.
J. Phys. G, Vol. 11, No. 5, p. 657 – 668 (1985). Abstr. in Phys. Abstr., Vol. 88, No. 1260, Entry 74030 (1985).

144.077   Longitudinal development of air–shower electrons studied from the arrival time distributions of atmospheric Cerenkov light measured at 5200 m above sea level.
N. Inoue, T. Kaneko, H. Yoshii, K. Hagiwara, F. Kakimoto, Y. Miyazaki, T. Enoki, K. Suga, K. Nishi, N. Martinic, P. Miranda, L. Siles.
J. Phys. G, Vol. 11, No. 5, p. 669 – 678 (1985). Abstr. in Phys. Abstr., Vol. 88, No. 1260, Entry 74031 (1985).

144.078   Measurement of the fluxes of medium nuclei with an energy of 10 – 25 MeV/nucleon in the near–earth space.
V. V. Bobrovskaya, E. V. Gorchakov, N. L. Grigorov, R. A. Nymmik, C. A. Tretyakova, A. Marin, M. Haiduc, D. Hasegan.
18th International Cosmic Ray Conference, Vol. 2, p. 2 – 5 (1983). – See Abstr. 012.096.

144.079   Charge composition of primary cosmic ray nuclei from oxygen to iron using cellulose nitrate plastic detector.
P. K. Das, T. D. Goswami.
18th International Cosmic Ray Conference, Vol. 2, p. 6 – 9 (1983). – See Abstr. 012.096.

144.080   Elemental composition of medium energy cosmic ray nuclei with $10 \leqslant Z \leqslant 26$.
S. C. Arora, R. Hasan, M. S. Swami.
18th International Cosmic Ray Conference, Vol. 2, p. 10 – 13 (1983). – See Abstr. 012.096.
The elemental abundances of heavy cosmic ray nuclei from neon to iron in a standard energy range (200 – 650 MeV/n) are presented.

144.081   High energy primary iron observed with emulsion chamber.
Y. Sato, I. Ohta, H. Sugimoto, K. Taira, S. Tasaka, N. Tateyama, K. Hoshino.
18th International Cosmic Ray Conference, Vol. 2, p. 14 (1983). Abstract. – See Abstr. 012.096.

144.082   Measurement of cosmic ray nuclei ($Z \geqslant 6$) using CR–39.
G.–x. Ren, Y.–z. Zhou, R.–q. Huang.
18th International Cosmic Ray Conference, Vol. 2, p. 15 (1983). Abstract. – See Abstr. 012.096.

144.083   Elemental composition of cosmic rays from Be to Ni as measured by the French–Danish instrument on HEAO–3.
J. J. Engelmann, P. Goret, E. Juliusson, L. Koch–Miramond, P. Masse, A. Soutoul, B. Byrnak, N. Lund, B. Peters, I. L. Rasmussen, M. Rotenberg, N. J. Westergaard.
18th International Cosmic Ray Conference, Vol. 2, p. 17 – 20 (1983). – See Abstr. 012.096.
The authors report new values for the elemental composition of the galactic cosmic rays between beryllium and nickel in the energy range 0.8 to 25 GeV/nuc.

144.084   The galactic cosmic ray energy spectra as measured by the French–Danish instrument on HEAO–3.
E. Juliusson, J. J. Engelmann, J. Jorrand, L. Koch–Miramond, P. Masse, N. Petrou, N. Lund, I. L. Rasmussen, M. Rotenberg.
18th International Cosmic Ray Conference, Vol. 2, p. 21 – 24 (1983). – See Abstr. 012.096.
The authors report measurements of the energy spectra of the galactic cosmic rays from beryllium to nickel made between 0.7 and 30 GeV/nuc.

144.085   Intensities, energy spectra and interactions of 10 – 30 GeV/n nuclei.
D. J. Fixsen, R. K. Fickle, P. S. Freier, C. J. Waddington.
18th International Cosmic Ray Conference, Vol. 2, p. 25 (1983). Abstract. – See Abstr. 012.096.

144.086   Measurement of heavy cosmic spectra above 20 GeV per nucleon.
J. H. Derrickson, T. A. Parnell, J. W. Watts, J. C. Gregory.
18th International Cosmic Ray Conference, Vol. 2, p. 26 (1983). Abstract. – See Abstr. 012.096.

144.087   Experimental measurement of high energy iron spectrum.
R. E. Streitmatter, V. K. Balasubrahmanyan, B. G. Mauger, J. F. Ormes, S. A. Stephens.
18th International Cosmic Ray Conference, Vol. 2, p. 27 (1983). Abstract. – See Abstr. 012.096.

144.088   The abundances of the elements with $Z > 26$ in the cosmic radiation.
B. Byrnak, N. Lund, I. L. Rasmussen, M. Rotenberg, J. J. Engelmann, P. Goret, E. Juliusson.
18th International Cosmic Ray Conference, Vol. 2, p. 29 – 32 (1983). – See Abstr. 012.096.
The data from the Danish–French HEAO–3 experiment have been analyzed to determine the abundances of the elements heavier than iron.

144.089   Deuterium, ^{3}He, and anomalous H and He nuclei.
R. A. Mewaldt, E. C. Stone.
18th International Cosmic Ray Conference, Vol. 2, p. 42 (1983). Abstract. – See Abstr. 012.096.

144.090   The isotopic composition of helium at high rigidities.
S. P. Jordan, P. Meyer.
18th International Cosmic Ray Conference, Vol. 2, p. 45 – 48 (1983). – See Abstr. 012.096.
A balloon–borne instrument was flown for 12 hours near the geomagnetic equator to measure the relative abundance of the helium isotopes ^{3}He and ^{4}He around 12 GV rigidity.

144.091   A Cerenkov – $\Delta E/\Delta \gamma$ experiment for measuring cosmic–ray isotopes from neon through iron.
A. Buffington, K. Lau, S. Laursen, I. L. Rasmussen, S. M. Schindler, E. C. Stone.
18th International Cosmic Ray Conference, Vol. 2, p. 49 – 52 (1983). – See Abstr. 012.096.
A balloon–borne cosmic–ray experiment has been constructed to measure cosmic–ray isotope masses. It employs a pair of Cerenkov counters and a NaI scintillator stack to determine changes in energy and in Lorentz factor for a traversing or stopping particle.

144.092   Interpretation of manganese isotopes in galactic cosmic rays.
Y. Huang.
18th International Cosmic Ray Conference, Vol. 2, p. 56 – 59 (1983). – See Abstr. 012.096.

This paper describes a new way to determine cosmic ray age and the Mn–54/(Mn–53 + Mn–55) ratio based on the data from HEAO–3 C–2 by assuming that the isotopes of manganese and chromium in cosmic rays have the same energy spectral exponent at the time of their production by primary nuclei.

144.093   The spectrum of high energy cosmic–ray electrons: results and interpretation.
D. Müller, J. Tang.
18th International Cosmic Ray Conference, Vol. 2, p. 60 – 63 (1983). – See Abstr. 012.096.

The data cover the energy range from 5 to 300 GeV. The interstellar differential spectrum can be described by a power law $\sim E^{-1.4}$ below 2 GeV, but attaining a slope $\sim E^{-2.6}$ between 2 and 10 GeV, and steepening further to almost $\sim E^{-3.6}$ between 30 and 300 GeV. The authors interpret the features of this spectrum in the context of models of galactic propagation.

144.094   Observations of high energy electrons beyond 1 TeV.
J. Nishimura, M. Fujii, M. Kato, T. Taira, A. Aizu, Y. Nomura, T. Kobayashi, A. Nishio, M. Kazuno, J. J. Lord, R. J. Wilkes, T. Koss, R. L. Golden.
18th International Cosmic Ray Conference, Vol. 2, p. 64 (1983). Abstract. – See Abstr. 012.096.

144.095   Electron propagation in the leaky box model with a truncated pathlength distribution.
B. G. Mauger, J. F. Ormes.
18th International Cosmic Ray Conference, Vol. 2, p. 65 – 68 (1983). – See Abstr. 012.096.

The authors suggest that a truncated pathlength distribution due to a lack of nearby sources could be responsible for the additional steepening of the electron spectrum.

144.096   Consequences of positron fluxes from densely shrouded sources.
B. G. Mauger, S. A. Stephens.
18th International Cosmic Ray Conference, Vol. 2, p. 69 (1983). Abstract. – See Abstr. 012.096.

144.097   Constraints on accretion powered positron sources.
S. Phinney, D. Eichler.
18th International Cosmic Ray Conference, Vol. 2, p. 70 (1983). Abstract. – See Abstr. 012.096.

144.098   Further analysis of a recent cosmic–ray antiproton experiment.
A. Buffington, S. M. Schindler.
18th International Cosmic Ray Conference, Vol. 2, p. 71 – 74 (1983). – See Abstr. 012.096.

The authors report a method of partial retrieval of pulse size information for the scintillator S_2. The method suffices to separate protons from more highly charged particles.

144.099   Observation of an anti–triton in cosmic rays.
K. M. V. Apparao, N. Durgaprasad, S. A. Stephens, S. Biswas.
18th International Cosmic Ray Conference, Vol. 2, p. 75 – 78 (1983). – See Abstr. 012.096.

The authors report in this paper the observation of a low energy anti–triton recorded in a nuclear emulsion stack exposed to primary cosmic rays.

144.100   Abundance of antiprotons in low energy cosmic rays.
K. M. V. Apparao, N. Durgaprasad, S. A. Stephens, S. Biswas.
18th International Cosmic Ray Conference, Vol. 2, p. 79 (1983). Abstract. – See Abstr. 012.096.

144.101   Energy dependence of the p̄/p ratio in cosmic rays.
R. L. Golden, S. Nunn, S. Horan.
18th International Cosmic Ray Conference, Vol. 2, p. 80 (1983). Abstract. – See Abstr. 012.096.

144.102   On the origin of low energy antiprotons in cosmic rays.
F. A. Agaronian (*F. A. Agaronyan*), S. R. Kelner, Yu. D. Kotov.
18th International Cosmic Ray Conference, Vol. 2, p. 81 – 84 (1983). – See Abstr. 012.096.

Models for the origin of antiprotons are formulated. The results are discussed and the agreement with experimental data is shown.

144.103   On the possible interpretation of antiproton and heavy nuclei flux in low energy cosmic rays.
S. Biswas, N. Durgaprasad, S. Ramadurai, J. S. Yadav.
18th International Cosmic Ray Conference, Vol. 2, p. 85 (1983). Abstract. – See Abstr. 012.096.

144.104   Nonuniform galactic disk model for estimating the interstellar antiproton flux.
L. C. Tan, L. K. Ng.
18th International Cosmic Ray Conference, Vol. 2, p. 86 – 89 (1983). – See Abstr. 012.096.

A cosmic–ray propagation model with a nonuniform galactic disk is found to be able to explain the existing observations of interstellar antiprotons (p̄'s) on the basis of the p̄ secondary production. The model takes into account the generation of p̄'s in the dense molecular hydrogen cloud region and the adiabatic deceleration of p̄'s in the gradient region of cosmic–ray intensity in the galactic disk.

144.105   Equilibrium antiproton spectra of some existing cosmic–ray propagation models.
L. C. Tan, L. K. Ng.
18th International Cosmic Ray Conference, Vol. 2, p. 90 – 93 (1983). – See Abstr. 012.096.

The equilibrium antiproton (p̄) spectra predicted for the usual leaky box model, the closed galaxy model of Rasmussen and Peters (1975) and the revised closed galaxy model of Peters and Westergaard (1976) have been recalculated. It is found that the effect of the p̄ non–annihilation inelastic interaction is significant even for the leaky box model. The existence of the low–energy kinematic cut–off in the p̄ production spectrum does not appear in the p̄ equilibrium spectra.

144.106   Energy dependence of annihilation and non–annihilation cross sections of p̄'s and implications for cosmic ray p̄ origin models.
T. K. Gaisser, R. J. Protheroe.
18th International Cosmic Ray Conference, Vol. 2, p. 94 (1983). – See Abstr. 012.096.

144.107   Observations of large abundances of low–energy antiprotons.
T. Bowen, E. W. Jenkins, J. J. Jones, A. E. Pifer, G. H. Sembroski.
18th International Cosmic Ray Conference, Vol. 2, p. 96 – 98 (1983). – See Abstr. 012.096.

The authors have observed a few low–energy p̄'s in a magnetic spectrometer at mountain altitude which indicate surprisingly high p̄/p ratios.

144.108   Analysis of spectral indices of γ– and μ–components in energy region > 1 TeV and a model of nucleon spectrum of primary cosmic rays.
N. V. Sokolskaya, A. Ya. Varkovitskaya, V. I. Zatsepin.
18th International Cosmic Ray Conference, Vol. 2, p. 101 – 104 (1983). – See Abstr. 012.096.

Available experimental data on spectral indices of muon and γ–components of cosmic rays at various atmospheric depths were analysed. The analysis indicates a change of the spectrum of primary nucleon flux. Since primary nucleons are mainly protons, this conclusion concerns proton component. A model of primary proton spectrum with a bend at $E_p = 5\,\text{TeV}$ is suggested.

144.109 Compositions of cosmic ray particles at energies greater than 10^{13}eV.
T. H. Burnett, S. Dake, M. Fuki, J. C. Gregory, T. Hayashi,
R. Holynski, J. Iwai, W. V. Jones, A. Jurak, J. J. Lord,
O. Miyamura, H. Oda, T. Ogata, T. A. Parnell, T. Tabuki,
Y. Takahashi, T. Tominaga, J. Watts, R. J. Wilkes, W. Wolter,
B. Wosiek.
18th International Cosmic Ray Conference, Vol. 2, p. 105 (1983).
Abstract. – See Abstr. 012.096.

144.110 10^{14}eV iron nuclei in cosmic rays.
R. K. Sood.
18th International Cosmic Ray Conference, Vol. 2, p. 109 (1983).
Abstract. – See Abstr. 012.096.

144.111 Composition of primary cosmic rays at energies $\sim 10^{15}$eV from the study of delayed hadrons in air showers at sea level.
A. I. Mincer, H. T. Freudenreich, J. A. Goodman,
S. C. Tonwar, G. B. Yodh, R. W. Ellsworth, D. Berley.
18th International Cosmic Ray Conference, Vol. 2, p. 110 (1983).
Abstract. – See Abstr. 012.096.

144.112 Mass composition of the primary cosmic radiation in the energy interval $10^{15} - 10^{16}$eV, derived from the Tien Shan EAS data.
J. N. Stamenov, S. Z. Ushev, V. D. Janminchev
(*V. D. Yanminchev*), N. M. Nikolskaja (*N. M. Nikol'skaya*),
V. P. Pavljuchenko (*V. P. Pavlyuchenko*).
18th International Cosmic Ray Conference, Vol. 2, p. 111 – 114
(1983). – See Abstr. 012.096.

The relative contribution values of the five main primary nuclei groups were obtained and their systematical distortions estimated.

144.113 Mass composition of the primary cosmic radiation in the energy interval $10^{16} - 10^{17}$eV, derived from the Akeno EAS data.
S. I. Nikolsky (*S. I. Nikol'skij*), J. N. Stamenov.
18th International Cosmic Ray Conference, Vol. 2, p. 115 – 118
(1983). – See Abstr. 012.096.

It is shown on the base of the analysed Tien Shan and Akeno EAS fluctuations data, that in the interval $10^{15} - 10^{17}$eV the relative proton contribution is not smaller than 40%, what is in contradiction with the predictions of the galactic diffusion models.

144.114 About the possibility of the primary mass composition investigation in the energy interval $10^{15} - 10^{18}$eV with the help of the method of the EAS fluctuation distribution shape analysis.
S. I. Nikolsky (*S. I. Nikol'skij*), J. N. Stamenov.
18th International Cosmic Ray Conference, Vol. 2, p. 119 – 122
(1983). – See Abstr. 012.096.

For each energy interval $(E_i – E_k)$ $[(E_k/E_i) < 10^2]$ exists an optimal observation level, where the accuracy of the primary mass composition estimations is principal the best.

144.115 Density spectrum of the EAS Cerenkov radiation and primary energy spectrum.
N. N. Efimov, V. F. Sokurov.
18th International Cosmic Ray Conference, Vol. 2, p. 123 – 126
(1983). – See Abstr. 012.096.

Data on density spectrum of the EAS Cerenkov radiation with exposition of 301 hours are given.

144.116 Cosmic ray energy spectrum above 10^{17}eV.
L. Horton, C. B. A. McCusker, L. S. Peak, J. Ulrichs,
M. M. Winn.
18th International Cosmic Ray Conference, Vol. 2, p. 128 – 130
(1983). – See Abstr. 012.096.

The authors present a revised energy spectrum of showers recorded by the S.U.G.A.R. array. They also examine the statisti-

cal significance of the apparent flattening of the spectrum for cosmic ray energies above 10^{19}eV.

144.117 The existence of a shower excess from the direction of the Crab Nebula.
T. Dzikowski, J. Gawin, B. Grochalska, J. Korejwo,
J. Wdowczyk.
18th International Cosmic Ray Conference, Vol. 2, p. 132 – 135
(1983). – See Abstr. 012.096.

Analysis of an enlarged set of EAS data has confirmed the existence of an excess of showers from the direction of the Crab Pulsar. The properties of the showers are summarised, the most important results being that the showers seem to be slightly deficient in muons and possibly absorbed slightly faster than normal EAS.

144.118 Measurement of the arrival directions of small air showers.
C. Morello, G. Navarra, P. Vallania.
18th International Cosmic Ray Conference, Vol. 2, p. 137 – 139
(1983). – See Abstr. 012.096.

Cosmic ray anisotropies are studied by measuring the arrival directions of small air showers ($E_0 \sim 3 \times 10^{13}$eV) at mountain altitude.

144.119 Measurements of cosmic ray variations in Chacaltaya.
C. Morello, G. Navarra, P. Miranda, N. Martinic,
A. Turtelli.
18th International Cosmic Ray Conference, Vol. 2, p. 140 (1983).
Abstract. – See Abstr. 012.096.

144.120 Small scale anisotropies in cosmic rays near 10^{16}eV.
P. R. Gerhardy, R. W. Clay.
18th International Cosmic Ray Conference, Vol. 2, p. 142 – 143
(1983). – See Abstr. 012.096.

Data from three years of recording by the Buckland Park EAS array have been used to produce intensity maps of the sky for energies near 10^{16}eV. These maps are analysed for any small scale effects such as may be produced by local particle sources or gamma ray sources. No significant anisotropy is found and an upper limit to the flux from Cen A is set at 10^{-14} photons cm^{-2}s^{-1}.

144.121 Cosmic ray arrival direction measurements: $10^{15} - 10^{17}$eV.
S. M. Astley, R. N. Coy, J. Lloyd–Evans, M. Patel,
R. J. O. Reid, A. A. Watson.
18th International Cosmic Ray Conference, Vol. 2, p. 144 (1983).
Abstract. – See Abstr. 012.096.

144.122 Anisotropy of cosmic rays of superhigh energies.
N. N. Efimov, A. A. Mikhailov (*A. A. Mikhajlov*),
M. I. Pravdin.
18th International Cosmic Ray Conference, Vol. 2, p. 149 – 152
(1983). – See Abstr. 012.096.

Anisotropy analysis results of 14000 showers with energy $E_0 > 10^{17}$eV registered at the Yakutsk EAS array for 1974 – 1982 are given. Together with traditional method the data are analyzed by a new method. The analysis results show that the particles up to extremely high energies are rather galactic ones. It's supposed that higher than 2×10^{19}eV a flux of particles from the Virgo supercluster side is surplus.

144.123 Arrival directions of large EAS in the southern hemisphere.
L. Horton, C. B. A. McCusker, L. S. Peak, J. Ulrichs,
M. M. Winn.
18th International Cosmic Ray Conference, Vol. 2, p. 153 – 156
(1983). – See Abstr. 012.096.

The arrival directions of the axes of air–showers, as observed by the S. U. G. A. R. array, are analysed in an attempt to find the sources of the radiation. The authors do not find any statistically significant departure from isotropy within the declination range $+30°$ to $-90°$.

144.124   Arrival direction and origin of multi–Joule cosmic rays.
C. Cunningham, J. Lloyd–Evans, R. J. O. Reid,
A. A. Watson.
18th International Cosmic Ray Conference, Vol. 2, p. 157 – 160
(1983). – See Abstr. 012.096.

Arrival direction data on cosmic ray primaries with energy above 4×10^{19}eV, observed at Haverah Park, have been updated to May 1982. Above this energy, which is approximately the Greisen–Zatsepin cut–off, the authors have observed 35 events. The amplitude of the first harmonic in right ascension is $(57 \pm 20)\%$ with a maximum at $190 \pm 25°$, chance probability $= 0.06$. Nine of the events have $E > 10^{20}$eV.

144.125   Spectra of cosmic rays and Galaxy's background emission in a diffusion model of cosmic rays distribution.
V. A. Dogiel (*V. A. Dogel'*), A. V. Uryson.
18th International Cosmic Ray Conference, Vol. 2, p. 170 – 173
(1983). – See Abstr. 012.096.

The problem of the influence of cosmic ray sources' nonuniformity on cosmic ray spectra is investigated.

144.126   On a possibility of galactic origin of cosmic rays with energies up to 10^{19}eV.
V. S. Berezinsky (*V. S. Berezinskij*), A. A. Mikhailov
(*A. A. Mikhajlov*).
18th International Cosmic Ray Conference, Vol. 2, p. 174 – 177
(1983). – See Abstr. 012.096.

The authors assume a realistic magnetic field in the halo and study the way of diminishing of anisotropy.

144.127   Models for cosmic ray propagation at energies $10^8 - 10^{14}$eV.
R. Silberberg, C. H. Tsao, M. M. Shapiro, J. R. Letaw.
18th International Cosmic Ray Conference, Vol. 2, p. 179 – 182
(1983). – See Abstr. 012.096.

Difficulties with the standard leaky box model are outlined. The multiple cloud model is discussed and shown to be consistent with observations.

144.128   Local superbubble model of cosmic ray propagation.
R. E. Streitmatter, V. K. Balasubrahmanyan,
J. F. Ormes, R. J. Protheroe.
18th International Cosmic Ray Conference, Vol. 2, p. 183 – 186
(1983). – See Abstr. 012.096.

The authors explore the consequences for cosmic ray phenomena of the solar system being inside a superbubble.

144.129   On the localized nature of the galactic cosmic rays.
J. F. Ormes.
18th International Cosmic Ray Conference, Vol. 2, p. 187 – 190
(1983). – See Abstr. 012.096.

The author explores the consequences of a low value of the diffusion coefficient and of various values of δ. He then addresses consequences for the propagation of cosmic rays, in particular their localization near their point of origin sufficient to explain the role over the escape of low energies.

144.130   Mean lifetime of cosmic rays in the dynamical halo model.
I. Lerche, R. Schlickeiser.
18th International Cosmic Ray Conference, Vol. 2, p. 191 (1983).
Abstract. – See Abstr. 012.096.

144.131   Evidence for an energy dependence in the short pathlengths of the cosmic ray pathlength distribution.
M. Garcia–Munoz, T. G. Guzik, J. A. Simpson, J. P. Wefel.
18th International Cosmic Ray Conference, Vol. 2, p. 210 – 213
(1983). – See Abstr. 012.096.

The energy dependence of the short pathlengths in the cosmic ray pathlength distribution is investigated in a galactic propagation plus solar modulation calculation.

144.132   The diffusion coefficient of cosmic rays derived from the shape of individual cosmic ray spectra.
M. Simon, K. D. Mathis.
18th International Cosmic Ray Conference, Vol. 2, p. 215 – 218
(1983). – See Abstr. 012.096.

The energy dependence of the mean escape length $\lambda_{esc}(E)$ of cosmic ray particles in interstellar space has been investigated by analyzing the shape of the cosmic ray spectra. The dependence seems to steepen with energy, and $\lambda_{esc} \sim R^{-(0.65-0.75)}$ is more likely than $R^{-0.5}$ above roughly 15 GeV/nuc.

144.133   Acceleration and propagation of galactic cosmic rays: implications of HEAO–3 data.
J. F. Ormes, R. J. Protheroe.
18th International Cosmic Ray Conference, Vol. 2, p. 221 – 224
(1983). – See Abstr. 012.096.

The authors re–examine the energy dependence of the mean escape length of cosmic rays from the Galaxy in the light of recent measurements of cosmic ray abundances. Implications for cosmic ray acceleration and galactic confinement are discussed.

144.134   Implications of the ratio $(61 \leqslant Z \leqslant 75)/(76 \leqslant Z \leqslant 83)$ on the origin and propagation of theultra–heavy cosmic rays.
C. H. Tsao, R. Silberberg, J. H. Adams Jr., J. R. Letaw.
18th International Cosmic Ray Conference, Vol. 2, p. 225 – 228
(1983). – See Abstr. 012.096.

144.135   Photon damping in cosmic ray acceleration in active galactic nuclei.
S. A. Colgate.
18th International Cosmic Ray Conference, Vol. 2, p. 230 – 233
(1983). – See Abstr. 012.096.

The usual assumption of the acceleration of ultra high energy cosmic rays, $\geqslant 10^{18}$eV in quasars, Seyfert galaxies, and other active galactic nuclei is challenged on the basis of the photon interactions with the accelerated nucleons.

144.136   The isotropy of ultra high energy cosmic rays and multiple supernova I galactic source.
S. A. Colgate.
18th International Cosmic Ray Conference, Vol. 2, p. 238 – 240
(1983). – See Abstr. 012.096.

The possibility of supernova events as a source for ultra high energy cosmic rays are considered. The time spread of arrival of 10^{20}eV protons is 100 to 400 years at 10 to 20 kpc and the angular spread is ± 15 to $\pm 30°$ depending upon the galactic field configuration. The time spread is sufficient to include several to a dozen type I SN. This is enough events and angular spread to include the observed data.

144.137   Coupled hydromagnetic wave excitation and cosmic ray acceleration at interstellar shocks.
M. A. Lee.
18th International Cosmic Ray Conference, Vol. 2, p. 241 (1983).
Abstract. – See Abstr. 012.096.

144.138   Cosmic ray source spectra produced in supernova blast waves.
H. Moraal, W. I. Axford.
18th International Cosmic Ray Conference, Vol. 2, p. 242 – 245
(1983). – See Abstr. 012.096.

An attempt is made to calculate spectra of both primary and secondary cosmic rays, produced by a typical supernova blast wave whose strength evolves from an initial high Mach number M_0 to a final Mach number $M_f = 1$. It is shown that the primary spectra are nearly power laws over a wide range of momentum, and that adiabatic cooling effects may account for the high antiproton flux in the kinetic energy range $0.1 \lesssim T \lesssim 10$ GeV.

144.139   Shock–wave acceleration out of the hot interstellar medium: effect of charge states on composition.
C. Cesarsky, R. Rothenflug, M. Cassé.
18th International Cosmic Ray Conference, Vol. 2, p. 246 (1983).
Abstract. – See Abstr. 012.096.

144.140 On the theory of cosmic ray mediated shocks with variable compression ratio.
D. Eichler.
18th International Cosmic Ray Conference, Vol. 2, p. 247 – 250 (1983). – See Abstr. 012.096.

Solutions are obtained for cosmic–ray–mediated shock profiles that include a) the thermal pressure of the preshock gas and b) the variation of the compression ratio caused by escape of highest–energy cosmic rays. These must be taken into account in order to have a self–consistent picture of cosmic ray mediation.

144.141 Oblique MHD shock waves modified by cosmic rays.
G. M. Webb.
18th International Cosmic Ray Conference, Vol. 2, p. 251 – 254 (1983). – See Abstr. 012.096.

The time asymptotic structure of an oblique MHD shock wave significantly modified by the back reaction from diffusive acceleration of cosmic rays is investigated. Examples are given.

144.142 Formation of cosmic ray spectrum by shock wave ensemble in the Galaxy.
E. G. Berezhko, G. F. Krymsky (*G. F. Krymskij*).
18th International Cosmic Ray Conference, Vol. 2, p. 255 – 258 (1983). – See Abstr. 012.096.

The authors show that there's a principal possibility of acceleration of particles accelerated by a single shock wave with the help of joint (collective) action of an ensemble of all the shock waves in Galaxy.

144.143 Spectrum of accelerated particles and their influence on shock front structure.
E. G. Berezhko, V. K. Yelshin (*V. K. Elshin*), G. F. Krymsky (*G. F. Krymskij*), Yu. A. Romaschenko (*Yu. A. Romashchenko*).
18th International Cosmic Ray Conference, Vol. 2, p. 259 – 262 (1983). – See Abstr. 012.096.

A process of heating and acceleration of charged particles by collisionless shock waves is considered taking into account their opposite influence on shock front structure. Based on numerical solution of Boltzmann's kinetic equation the distribution function was obtained which allowed to determine the absolute number of accelerated particles and parts of the wave energy passing into these particles.

144.144 Cosmic–ray acceleration and transport in stellar winds with terminal shocks.
G. M. Webb, W. I. Axford, M. A. Forman.
18th International Cosmic Ray Conference, Vol. 2, p. 263 – 266 (1983). – See Abstr. 012.096.

The propagation of cosmic–rays in a stellar wind with a terminal shock is investigated by means of monoenergetic, steady-state, spherically–symmetric solutions of the cosmic–ray equation of transport. The authors show the redistribution of monoenergetic galactic particles in both momentum and position, the streaming pattern, and the energy transfer pattern between the wind and the cosmic–rays. Corresponding results for the case of monoenergetic injection within the cavity (e.g. from the star) are also discussed.

144.145 Time–dependent shock acceleration: approximations and exact solutions.
M. A. Forman, L. O'C. Drury.
18th International Cosmic Ray Conference, Vol. 2, p. 267 – 270 (1983). – See Abstr. 012.096.

The time–dependent spectrum of particles accelerated by a shock is derived exactly for the case where the diffusion coefficient is a power–law in momentum, p. A simple approximate time–dependent spectrum which appears to be quite accurate near the cutoff energy is also presented.

144.146 Monte Carlo simulation of steady state shock structure including cosmic ray mediation and particle escape.
D. C. Ellison, F. C. Jones, D. Eichler.
18th International Cosmic Ray Conference, Vol. 2, p. 271 – 274 (1983). – See Abstr. 012.096.

Results are reported of a Monte Carlo simulation that includes both the back reaction of accelerated particles on the inflowing plasma producing a smoothing of the shock transition, and the free escape of particles allowing arbitrarily large overall compression ratios in high Mach number, steady state shocks. Energy spectra and estimates of the proportion of thermal ions accelerated to high energy are obtained.

144.147 Propagation and origin of energetic cosmic rays, $10^{14} - 10^{19}$ eV.
R. Silberberg, C. H. Tsao, J. R. Letaw, M. M. Shapiro.
18th International Cosmic Ray Conference, Vol. 2, p. 283 – 286 (1983). – See Abstr. 012.096.

The authors explore observational tests of alternative origins for the ultra–high–energy cosmic–ray particles (UHEP) at 10^{14} to 10^{19} eV. They describe several noteworthy features of the UHEP. Then, they outline a pulsar model of origin consistent with those attributes.

144.148 Scaling from Jupiter to pulsars and the acceleration of cosmic ray particles by pulsars. II.
C. Y. Fan, H. R. Hang.
18th International Cosmic Ray Conference, Vol. 2, p. 287 – 290 (1983). – See Abstr. 012.096.

The authors present observational evidences supporting the theory on acceleration of cosmic ray particles by pulsars.

144.149 Can cosmic rays be accelerated in collapsing molecular clouds?
V. A. Dogiel (V. A. Dogel'), P. Kiraly, I. Freedman, A. W. Wolfendale.
18th International Cosmic Ray Conference, Vol. 2, p. 292 – 295 (1983). – See Abstr. 012.096.

The present paper examines the probability of cosmic ray acceleration during cloud collapse, if one assumes that some giant molecular clouds give more γ–rays than would be expected.

144.150 Are young supernova remnants the acceleration sites of cosmic ray electrons?
S. P. Reynolds, R. A. Chevalier.
18th International Cosmic Ray Conference, Vol. 2, p. 297 – 300 (1983). – See Abstr. 012.096.

144.151 Cosmic ray acceleration efficiency by supernova remnants.
G. F. Krymsky (*G. F. Krymskij*), S. I. Petukhov.
18th International Cosmic Ray Conference, Vol. 2, p. 301 – 304 (1983). – See Abstr. 012.096.

Supernova remnants can play a double role in the formation of cosmic ray features: (1) to transform background plasma particles into cosmic rays; (2) to change energy of already existing cosmic rays. The authors describe characteristic consequences of cosmic ray acceleration models by shock waves from supernova remnants and their comparison with radioastronomical observations.

144.152 Onion–shell model of cosmic ray acceleration in supernova remnants.
T. J. Bogdan, H. J. Völk.
18th International Cosmic Ray Conference, Vol. 2, p. 305 (1983). Abstract. – See Abstr. 012.096.

144.153 Acceleration of electrons in supernova remnants.
R. Cowsik, S. Sarkar.
18th International Cosmic Ray Conference, Vol. 2, p. 306 – 309 (1983). – See Abstr. 012.096.

By using the transport coefficients and magnetic field strength variations in supernova remnants derived on the basis of hydro-dynamical calculations the authors are able to derive consistently their radio properties such as the turn–on, their spectra and their evolution. Also when these expand and finally merge into the interstellar medium, they contain enough high energy electrons to contribute significantly to the galactic cosmic rays.

144.154   Blast waves with cosmic ray pressure: the steep energy spectrum case.
R. A. Chevalier.
18th International Cosmic Ray Conference, Vol. 2, p. 314 – 317 (1983). – See Abstr. 012.096.

The effects of cosmic ray pressure on the dynamics of self–similar spherical blast waves are investigated on the assumptions that the ratio of cosmic ray pressure to total pressure at the shock front is constant, that the momentum distribution function of the cosmic rays is a power law, and that the cosmic rays and thermal gas evolve as independent adiabatic fluids in the postshock flow.

144.155   Statistical acceleration of electrons in the galactic halo.
V. A. Dogiel (*V. A. Dogel'*), V. M. Kovalenko.
18th International Cosmic Ray Conference, Vol. 2, p. 318 – 321 (1983). – See Abstr. 012.096.

The problem of shock wave acceleration of relativistic electrons is investigated. It is shown that adiabatic losses behind a shock front transform the electron spectrum significantly. The spectral index of radio emission of the halo is calculated for the case when a shock wave is moving inside the halo.

144.156   Two–stage acceleration and propagation at low energy.
J. P. Meyer.
18th International Cosmic Ray Conference, Vol. 2, p. 322 (1983). Abstract. – See Abstr. 012.096.

144.157   Which types of stars are the dominant cosmic ray injectors?
J. P. Meyer.
18th International Cosmic Ray Conference, Vol. 2, p. 323 (1983). Abstract. – See Abstr. 012.096.

144.158   Ionization effects shaping the elemental cosmic–ray source composition.
R. Rothenflug, M. Cassé, M. Arnaud.
18th International Cosmic Ray Conference, Vol. 2, p. 324 (1983). Abstract. – See Abstr. 012.096.

144.159 ^{60}Fe: an unusual cosmic–ray clock?
M. Cassé, A. Soutoul.
18th International Cosmic Ray Conference, Vol. 2, p. 327 (1983). Abstract. – See Abstr. 012.096.

144.160   The source spectra and mean path–length of primary high energy cosmic rays.
A. Soutoul.
18th International Cosmic Ray Conference, Vol. 2, p. 328 (1983). Abstract. – See Abstr. 012.096.

144.161   Rapid X–ray burster as a source of cosmic rays.
L. M. Singh, H. L. Duorah.
18th International Cosmic Ray Conference, Vol. 2, p. 329 – 332 (1983). – See Abstr. 012.096.

The potential difference developed between the centre of the polar cap and its outer edge in a highly magnetised neutron star like the MXB 1730–335, a long period transient, is found to accelerate charged particles to cosmic ray energies. For a charge $Z = 50$, and $\gamma = 10 - 100$, the maximum energy attainable is found to be $10^{21 - 22}$eV.

144.162   Cosmogenic nuclide ^{10}Be abundance in the ice cores and the time variation of cosmic rays.
G. E. Kocharov, A. N. Konstantinov.
18th International Cosmic Ray Conference, Vol. 2, p. 333 – 336 (1983). – See Abstr. 012.096.

The time–dependence of cosmic ray intensity for the last 30,000 years is obtained based on the experimental data on ^{10}Be concentration in ice cores. The variation of cosmic ray intensity caused by solar modulation has been obtained. The analysis of radiocarbon and Be–10 data allows to indicate the supernova flare about 15,000 years ago.

144.163   Beryllium–10 in deep sea sediments and cosmic ray intensity variations.
P. Sharma, S. K. Bhattacharya, B. L. K. Somayajulu.
18th International Cosmic Ray Conference, Vol. 2, p. 337 – 340 (1983). – See Abstr. 012.096.

Based on Be–10 measurements on well dated Pacific sediment cores, the CR intensity variations are shown to be within $\pm 50\%$ (at the 95% confidence level) during the past 2 m.y.

144.164   Size dependence determination of cosmogenic radionuclides in black, metallic spherules from marine sediments.
K. Yamakoshi, H. Ohashi, M. Imamura.
18th International Cosmic Ray Conference, Vol. 2, p. 341 (1983). – See Abstr. 012.096.

144.165   Time dependence of cosmic ray intensity during 8000 years.
V. A. Dergachev, G. E. Kocharov, V. M. Ostryakov.
18th International Cosmic Ray Conference, Vol. 2, p. 342 – 345 (1983). – See Abstr. 012.096.

During the considered time interval more than 10 periods of decreased solar activity like Maunder minimum are established.

144.166   The time– and space–averaged cosmic ray flux experienced by meteorites.
J. C. Barton.
18th International Cosmic Ray Conference, Vol. 2, p. 346 – 349 (1983). – See Abstr. 012.096.

An attempt is made to assess the uncertainties in the intensity and spectrum of the flux responsible for the production of cosmogenic nuclides in meteorites.

144.167   Are cosmic ray isotopic "anomalies" due to the solar system being anomalous?
D. N. Schramm.
18th International Cosmic Ray Conference, Vol. 2, p. 361 (1983). Abstract. – See Abstr. 012.096.

144.168   The parameters of irradiation by accelerated particles during the different stages of the solar system matter evolution.
A. K. Lavrukhina, R. I. Kuznetsova.
18th International Cosmic Ray Conference, Vol. 2, p. 362 – 365 (1983). – See Abstr. 012.096.

The authors made the calculations for the model based on an idea that an explosion of cosmic objects (supernova and active galactic nuclei) leads simultaneously to the explosive burning of H, He, C, O(Ne), to the collective acceleration of particles up to relativistic energies by hydrodynamic shock waves and to the spallation reactions.

144.169   On the problem of weak modulation of GCR few million years ago.
S. Tokár, P. Povinec.
18th International Cosmic Ray Conference, Vol. 2, p. 381 – 384 (1983). – See Abstr. 012.096.

Monte Carlo method for description of hadronic cascade in matter has been developed for calculation of ^{53}Mn production rate in lunar soil. Only a weak modulation of GCR for the period 2–6 My ago has been obtained.

144.170 ^{22}Na and ^{26}Al in Luna 24 samples.
P. Povinec, A. K. Lavrukhina, G. K. Ustinova.
18th International Cosmic Ray Conference, Vol. 2, p. 385 – 388 (1983). – See Abstr. 012.096.

A comparison of theoretical and experimental results of ^{22}Na and ^{26}Al measurements in lunar samples shows that the average intensity of galactic cosmic rays has been during the last million years constant within $\pm 20\%$.

144.171 The origin of the ultra high energy cosmic rays.
D. N. Schramm, C. T. Hill.
18th International Cosmic Ray Conference, Vol. 2, p. 393 – 396
(1983). – See Abstr. 012.096.
Very distant cosmological sources can generate significant
structure above 10^{19}eV. Local sources are not required and may
be problematic.

**144.172 Testing baryon–symmetric cosmologies on a super-
galactic scale.**
G. Tarlé, P. B. Price, M. H. Salamon, S. P. Ahlen.
18th International Cosmic Ray Conference, Vol. 2, p. 397 (1983).
Abstract. – See Abstr. 012.096.

144.173 Cosmic ray modulation in the coherent regime.
J. A. Earl.
18th International Cosmic Ray Conference, Vol. 3, p. 1 (1983).
Abstract. – See Abstr. 012.096.

144.174 Effect of electric field on the cosmic ray propagation.
K. Nagashima, K. Munakata.
18th International Cosmic Ray Conference, Vol. 3, p. 2 – 5
(1983). – See Abstr. 012.096.
The propagation of galactic cosmic rays through the heliosphere is affected by the electric field induced by the solar
wind. The numerical solution of the steady state transport equation is obtained and the influence of electric field on the cosmic
ray propagation is estimated.

**144.175 How do cosmic rays change their energy in the solar
wind?**
F. C. Jones.
18th International Cosmic Ray Conference, Vol. 3, p. 6 – 8
(1983). – See Abstr. 012.096.
The author has derived the diffusion–convection (modulation)
equation directly from the Boltzmann equation employing a minimum number of assumptions about the scattering process.

**144.176 Galactic cosmic ray modulation by solar wind in the
spherically symmetric nonstationary diffusive model.**
G. N. Guniava, B. D. Naskidashvili, N. Kh. Shatashvili.
18th International Cosmic Ray Conference, Vol. 3, p. 9 – 11
(1983). – See Abstr. 012.096.
Calculations of the temporal deformation of the spatial energetic distribution of galactic cosmic rays in the interplanetary
space has been carried out on the basis of the transfer equation
of galactic cosmic rays. It is shown that in time the enrichment of
galactic cosmic rays spectrum by the high–energy particles takes
place.

**144.177 22–year variation of the solar diurnal anisotropy in terms
of diffusion convection model.**
K. Nagashima, K. Munakata.
18th International Cosmic Ray Conference, Vol. 3, p. 12 – 15
(1983). – See Abstr. 012.096.
It has been reported that following the polarity reversal of the
polar magnetic field of the Sun in 1969, the solar diurnal variation of cosmic rays changed its phase toward earlier hours from
18h LT. This phenomenon, known as the 22–year variation of the
diurnal variation, is theoretically studied.

144.178 The cosmic ray transport equation revisited.
R. A. Burger, J. J. Henning, H. Moraal.
18th International Cosmic Ray Conference, Vol. 3, p. 16 (1983).
Abstract. – See Abstr. 012.096.

**144.179 Nonstationary modulation of cosmic rays for constant
radial gradient.**
A. G. Zusmanovich, O. N. Kryakunova.
18th International Cosmic Ray Conference, Vol. 3, p. 19 – 22
(1983). – See Abstr. 012.096.
The nonstationary equation of cosmic ray transport in the
interplanetary space was solved for the case when the radial
gradient of cosmic rays intensity is independent on the distance
from the Sun. The energy spectra of the modulated cosmic ray
intensity were calculated for the stationary and nonstationary
cases. The lag time of cosmic rays for particles with different
energy is determined.

**144.180 On the energetic "hysteresis" in the galactic cosmic ray
intensity.**
M. B. Krainev (*M. B. Krajnev*), Yu. I. Stozhkov,
T. N. Charakhch'yan.
18th International Cosmic Ray Conference, Vol. 3, p. 23 – 26
(1983). – See Abstr. 012.096.
It is shown that the break of close correlation between the time
behaviour of high and low energy galactic cosmic ray intensity
during the periods of the inversion of the general magnetic field
of the Sun can be explained by the reduction of the regular
interplanetary magnetic field intensity.

**144.181 Proton, helium and electron spectra from 1965 through
1979 as tests of modulation theory.**
P. Evenson, M. Garcia–Munoz, P. Meyer, K. R. Pyle,
J. A. Simpson.
18th International Cosmic Ray Conference, Vol. 3, p. 27 – 30
(1983). – See Abstr. 012.096.
This work studies quantitatively the degree to which the predictions of conventional modulation theory agree with the long-
term experimental data. The authors conclude that (1) to a large
extent the modulation process is independent of the sign of the
particle charge, and (2) the assumption of steady state is a fairly
good approximation throughout the long–term modulation period studied.

144.182 Solar modulation of cosmic ray electrons 1978 – 1983.
P. Evenson, P. Meyer.
18th International Cosmic Ray Conference, Vol. 3, p. 31 – 34
(1983). – See Abstr. 012.096.
The steplike onset of modulation of the positive particles in
1979 is also seen in the electrons indicating that the basic modulation mechanism is not sensitve to the charge of the particle. The
details of the hysteresis loops observed in the previous solar
maximum by Burger and Swanenburg (1973) are not reproduced
at the present solar maximum.

**144.183 Asymmetries in the modulation of protons and helium
nuclei over two solar cycles.**
W. R. Webber, J. C. Kish, D. A. Schrier.
18th International Cosmic Ray Conference, Vol. 3, p. 35 – 38
(1983). – See Abstr. 012.096.
Measurements of proton and helium spectra over two cycles of
solar modulation show large asymmetries in modulation. These
effects are apparent at sunspot minimum conditions in 1965 and
1977, where a much larger intensity is observed in 1977, and also
at sunspot maximum conditions where a much lower intensity is
observed in 1981 relative to 1970. The details of this asymmetry
are discussed and possible causes are considered.

**144.184 Anomalous He in the heliosphere: the 1972 – 77 vs. the
1965 solar minimum.**
M. Garcia–Munoz, K. R. Pyle, J. A. Simpson.
18th International Cosmic Ray Conference, Vol. 3, p. 39 – 42
(1983). – See Abstr. 012.096.
The anomalous ^{4}He component, He(A), has a higher sensitivity to solar modulation than the galactic cosmic ray nuclei of
similar energies. Analysis shows that the absence of He(A) in
1965 could be explained as the consequence of a greater level of
solar modulation at low energies in 1965 than in most of the
1972 – 77 solar minimum.

**144.185 Anomalous component of cosmic rays in the strato-
sphere.**
A. K. Svirzhevskaya, N. S. Svirzhevsky (*N. S. Svirzhevskij*),
Yu. I. Stozhkov, T. N. Charakhch'yan.
18th International Cosmic Ray Conference, Vol. 3, p. 43 – 46
(1983). – See Abstr. 012.096.
The shape change of the cosmic ray absorption curves in the
stratosphere was observed after 1972 both at middle and at high

latitudes. The change is supposed to be due to the penetration of the low energy particles into the atmosphere.

144.186   Solar modulation of medium energy (160 – 759 MeV/n) multiply charged cosmic ray nuclei with $10 \leqslant Z \leqslant 26$ in the interplanetary region.
S. C. Arora, R. Hasan, M. S. Swami.
18th International Cosmic Ray Conference, Vol. 3, p. 47 – 50 (1983). – See Abstr. 012.096.

144.187   The features of galactic cosmic ray modulation by solar activity processes.
N. N. Kontor, G. P. Lyubimov, T. G. Khotilovskaya, E. A. Chuchkov.
18th International Cosmic Ray Conference, Vol. 3, p. 51 – 54 (1983). – See Abstr. 012.096.

The data obtained from neutron monitors and space probes (>30 MeV protons) during solar activity cycles 19, 20, 21 are used to study the modulation of galactic cosmic rays. The character of the modulation is shown to be determined by the occurrence frequency of strong solar flares. Arguments are presented for unsteady–state model of galactic cosmic ray modulation by shock waves from strong solar flares.

144.188   Cosmic ray gradients in the outer heliosphere.
W. Fillius, B. Wake, W.-H. Ip, I. Axford.
18th International Cosmic Ray Conference, Vol. 3, p. 55 – 58 (1983). – See Abstr. 012.096.

Pioneer 10 and 11 spacecraft are now probing the outer heliosphere. The authors have used the UCSD instruments on board to study the gradient, and to look at the time and spatial variations of the cosmic ray intensities.

144.189   Spatial and temporal variability of the B/C ratio.
M. Garcia–Munoz, R. B. McKibben, K. R. Pyle, J. A. Simpson, R. Zamow.
18th International Cosmic Ray Conference, Vol. 3, p. 61 (1983). Abstract. – See Abstr. 012.096.

144.190   Radial gradients of galactic cosmic rays measured by Pioneer 10 from $1 – 29$ AU through the period of solar maximum.
R. B. McKibben, K. R. Pyle, J. A. Simpson.
18th International Cosmic Ray Conference, Vol. 3, p. 63 – 66 (1983). – See Abstr. 012.096.

The authors find that at solar maximum, in contrast to solar minimum, essentially all of the modulation of low energy cosmic rays occurs outside $R = 29$ AU. The combined effects of solar flare shocks coalescing in the far outer heliosphere may account for this dramatic change in the radial dependence of modulation.

144.191   Galactic cosmic ray modulation in the inner heliosphere by solar flare disturbances.
C. J. Hatton, G. A. Bowe.
18th International Cosmic Ray Conference, Vol. 3, p. 68 – 71 (1983). – See Abstr. 012.096.

The authors develop a model to describe the observed intensity at various heliocentric distances as measured by Pioneers 10 and 11.

144.192   A 50 day periodicity in the cosmic ray anisotropy.
W. Fillius, W.-H. Ip, I. Axford.
18th International Cosmic Ray Conference, Vol. 3, p. 72 – 75 (1983). – See Abstr. 012.096.

The authors report on a modulation of cosmic ray anisotropy with a period of more or less than 50 days.

144.193   Observation of the long term cosmic ray modulation beween 1 and 25 AU.
F. B. McDonald, L. Burlaga, T. T. von Rosenvinge, M. Van Hollebeke.
18th International Cosmic Ray Conference, Vol. 3, p. 76 (1983). Abstract. – See Abstr. 012.096.

144.194   Peculiarities of galactic cosmic ray modulation in 1970 – 1974.
I. S. Samsonov, N. P. Chirkov, Z. N. Samsonova.
18th International Cosmic Ray Conference, Vol. 3, p. 77 – 80 (1983). – See Abstr. 012.096.

The authors try to explain the observed anomalous phenomena in galactic cosmic rays by evolution of coronal holes on solar activity cycles.

144.195   11–year cycle solar modulation of cosmic ray intensity inferred from ^{14}C content variation in dated tree rings.
C. Y. Fan, T. M. Chen, S. X. Yun, K. M. Dai.
18th International Cosmic Ray Conference, Vol. 3, p. 82 – 85 (1983). – See Abstr. 012.096.

144.196   Solar activity and radiocarbon abundance in the earth's atmosphere.
N. I. Akatova, G. E. Kocharov.
18th International Cosmic Ray Conference, Vol. 3, p. 86 – 89 (1983). – See Abstr. 012.096.

The radiocarbon production rate and cosmic ray intensity caused by sunspot activity for the period 1874 – 1969 are calculated. Empirical connections between them and solar activity data are determined. Comparison of experimental and calculated data leads to the conclusion about the existence of an additional source of cosmic ray intensity variation.

144.197   Features of long term cosmic ray variations in recent periods and relation with solar activity.
P. K. Pandey, A. K. Jain, S. K. Garde.
18th International Cosmic Ray Conference, Vol. 3, p. 91 – 94 (1983). – See Abstr. 012.096.

The continued decrease of cosmic ray intensity in 1981 – 82 is found to be related with the decreasing trend in sunspot number and the increase in the value of A_p Index, which is contrary to the established ideas. The results are discussed to explain the observational features of the long term modulation of these parameters.

144.198   On the "anomalous" phenomena in the galactic cosmic ray intensity in the periods of the inversion of general magnetic field of the Sun.
M. B. Krainev (*M. B. Krajnev*), Yu. I. Stozhkov, T. N. Charakhch'yan.
18th International Cosmic Ray Conference, Vol. 3, p. 95 – 98 (1983). – See Abstr. 012.096.

The unusual phenomena in the galactic cosmic ray intensity are considered which took place during three periods (including the last one, 1979 – 81) of the inversion of general magnetic field of the Sun. Some restrictions are imposed on possible ways of the interpretation of these effects.

144.199   Cosmic ray modulation at solar activity minima.
A. G. Zusmanovich, V. V. Satsuk, L. F. Churunova.
18th International Cosmic Ray Conference, Vol. 3, p. 102 – 105 (1983). – See Abstr. 012.096.

The values of the cosmic rays modulation at solar activity minima 1954, 1965 and 1976 were determined by the neutron monitor data. The values of the absolute modulation coefficient during these periods and the form of unmodulated cosmic rays energy spectrum were determined using these results and the data of direct measurements of the protons spectra out of the Earth's atmosphere.

144.200   On the possible nature of anomalies in chemical composition and low–energy cosmic ray spectrum.
V. M. Dvornikov.
18th International Cosmic Ray Conference, Vol. 3, p. 110 – 113 (1983). – See Abstr. 012.096.

On the basis of the analysis of trajectory and energy variation of charged particles in electromagnetic fields of a decelerating solar wind, a calculation was performed of modulated energy spectra for nuclei of helium, carbon and oxygen in the energy region $10^{-3} – 1$ GeV/nuc.

144.201 Observed and theoretical characteristics of Forbush decreases in the inner and outer heliosphere.
R. Gall, B. Thomas, H. Durand.
18th International Cosmic Ray Conference, Vol. 3, p. 117 (1983).
– See Abstr. 012.096.

144.202 Modulation effects due to a change of the general solar magnetic field.
A. G. Zusmanovich.
18th International Cosmic Ray Conference, Vol. 3, p. 118 – 121 (1983). – See Abstr. 012.096.

Analytic and numcerical solutions of the equation of the cosmic rays transport in the interplanetary space with a particle drift in the regular field are compared. It is shown that the amplitude of the effect connected with a particle drift in the regular magnetic field is about 20 – 30% from the convective–diffusion modulation at the solar activity minimum when this effect is highest.

144.203 On the cosmic ray kinetics in the anisotropic random magnetic field.
L. I. Dorman, M. E. Katz (*M. E. Kats*), S. F. Nosov, Yu. I. Fedorov, B. A. Shakhov, M. Stehlík.
18th International Cosmic Ray Conference, Vol. 3, p. 127 – 130 (1983). – See Abstr. 012.096.

The motion of charged cosmic ray particles in the strong large–scale magnetic field with random walks of force–lines is investigated. A kinetic equation for the particle distribution in the interplanetary space is derived.

144.204 The acceleration of the anomalous cosmic ray component.
L. A. Fisk, M. A. Lee.
18th International Cosmic Ray Conference, Vol. 3, p. 135 (1983). Abstract. – See Abstr. 012.096.

144.205 Acceleration of nonrelativistic electrons in interplanetary space.
M. F. Bakhareva.
18th International Cosmic Ray Conference, Vol. 3, p. 136 – 139 (1983). – See Abstr. 012.096.

The author studies the possibility of the Alfvén acceleration of the electrons with the mean energy ~ 50 keV by assuming that the interplanetary turbulence at $\sim 10^{-4}$ Hz plays the role of magnetic pumping. The energy spectrum of particles is calculated with taking into account scattering, convection, and adiabatic deceleration. This mechanism is shown to be able of producing the fluxes of nonrelativistic electrons observed near the Earth's orbit.

144.206 Particle acceleration by flare–induced interplanetary shocks.
G. Wibberenz, W. Scholz, H. Kunow.
18th International Cosmic Ray Conference, Vol. 3, p. 140 (1983). Abstract. – See Abstr. 012.096.

144.207 Large–scale solar wind disturbances in events associated with particle acceleration up to relativistic energies in interplanetary medium.
A. T. Filippov, N. P. Chirkov, S. A. Starodubtsev, A. S. Niskovskikh.
18th International Cosmic Ray Conference, Vol. 3, p. 141 – 144 (1983). – See Abstr. 012.096.

Spectral characteristics of galactic cosmic ray intensity during powerful events in July 1959, November 1960, August 1972 and October 1981 are investigated when particle acceleration up to relativistic energies was observed. Cosmic ray intensity variations with 6 – 8, 10 – 12, 18 – 20, 32 – 35, 43 – 58, 64 – 80 hour periods were found. The nature of variations is associated with large–scale solar wind disturbances with similar periods.

144.208 Turbulent dynamo α–effect in the problem of cosmic ray acceleration.
L. L. Kichatinov.
18th International Cosmic Ray Conference, Vol. 3, p. 145 – 148 (1983). – See Abstr. 012.096.

The CR propagation in a helically turbulent cosmic plasma is considered. The CR acceleration mechanism, more effective than second–order Fermi acceleration, is found to act in such a medium in the presence of a large–scale magnetic field. This acceleration is caused by an electric field parallel (or anti–parallel) to large–scale magnetic field. The electric field is generated by small–scale helical motions of the cosmic plasma (α–effect). This suggests that CR acceleration and cosmic magnetic field generation may be caused by the same physical effect.

144.209 Some questions of particle acceleration between approaching fronts of shock waves in interplanetary medium.
A. S. Niskovskikh, N. P. Chirkov, A. T. Filippov.
18th International Cosmic Ray Conference, Vol. 3, p. 149 – 151 (1983). – See Abstr. 012.096.

144.210 The effects of adiabatic deceleration and shock lifetime on energetic storm particle events.
M. A. Forman.
18th International Cosmic Ray Conference, Vol. 3, p. 153 (1983). Abstract. – See Abstr. 012.096.

144.211 Anisotropies of 35 – 56 keV ions associated with interplanetary shocks.
T. R. Sanderson, R. Reinhard, K.–P. Wenzel.
18th International Cosmic Ray Conference, Vol. 3, p. 156 – 159 (1983). – See Abstr. 012.096.

The authors present preliminary results of a survey of ~ 40 interplanetary shocks observed on ISEE–3, using 35 – 56 keV 3–dimensional proton anisotropy parameters. The results are interpreted as evidence for Fermi acceleration associated with quasi–parallel shocks and an electric field drift acceleration with quasi–perpendicular shocks.

144.212 A statistical study of interplanetary shock associated proton intensity increases.
R. Reinhard, P. van Nes, T. R. Sanderson, K.–P. Wenzel, E. J. Smith, B. T. Tsurutani.
18th International Cosmic Ray Conference, Vol. 3, p. 160 – 163 (1983). – See Abstr. 012.096.

The authors find that the events with the highest fluxes of 35 – 56 keV protons are associated with shocks which are quasi–parallel and originate close to the central meridian of the Sun and conclude that they are produced by a first order Fermi acceleration process.

144.213 Energetic particle spectra upstream and downstream of interplanetary shock waves.
M. Scholer, B. Klecker, D. Hovestadt, G. Gloeckler.
18th International Cosmic Ray Conference, Vol. 3, p. 166 (1983). Abstract. – See Abstr. 012.096.

144.214 Anisotropies in ESP events.
T. T. von Rosenvinge, C. Paizis.
18th International Cosmic Ray Conference, Vol. 3, p. 167 (1983). Abstract. – See Abstr. 012.096.

144.215 Features of the low energy proton spectra due to variations of parameters of propagation and statistical acceleration of particles.
M. F. Bakhareva, M. A. Zel'dovich, Yu. I. Logachev.
18th International Cosmic Ray Conference, Vol. 3, p. 168 – 171 (1983). – See Abstr. 012.096.

The proton energy spectra were calculated on the basis of solution of the transport equation with different parameters of propagation and statistical acceleration of particles. The parameters were varied to provide the best fit of the calculated energy spectra with some measurements of 0.5 – 10 MeV proton spectra in the corotating streams at 1 AU. It is shown that the cases of negative radial intensity gradient observed at r ⩽ 1 AU are consistent with statistical acceleration in the interplanetary space.

144.216   Acceleration of particles in the magnetotail.
R. L. Xu, C. Y. Fan, G. Gloeckler, L. S. Ma–Sung,
D. Hovestadt.
18th International Cosmic Ray Conference, Vol. 3, p. 173 – 176
(1983). – See Abstr. 012.096.

The authors present evidences for the association of particle bursts with the neutral sheet. They found that the degree of asymmetry for the ion distribution is also lesser for lower energy ions. They propose two mechanisms of acceleration for the particles.

144.217   Energy spectra of upstream protons in energy interval 23 – 480 keV measured in December 1976.
R. N. Bacilova (*R. N. Basilova*), I. A. Kurillo,
N. A. Mamontova, S. P. Rumin, K. N. Sharvina.
18th International Cosmic Ray Conference, Vol. 3, p. 178 – 181
(1983). – See Abstr. 012.096.

The results of measurements of 300 energy spectra of upstream protons are presented.

144.218   27–day recurrences of enhanced solar diurnal variations in 1973 – 1974 and its relation to high–speed streams from coronal holes.
S. Mori, Y. Sugie, S. Matsuzaki, T. Chiba.
18th International Cosmic Ray Conference, Vol. 3, p. 186 (1983).
Abstract. – See Abstr. 012.096.

144.219   On the degree of correlation between the 27–day variations in cosmic rays and in solar activity.
S. V. Vernov, E. A. Voronina, M. S. Grigoryan, A. I. Sladkova,
T. N. Charakhch'yan, G. A. Bazilevskaya, E. S. Vernova,
M. I. Tyasto.
18th International Cosmic Ray Conference, Vol. 3, p. 187 – 190
(1983). – See Abstr. 012.096.

A correlation analysis is made for the power spectra of the 15 – 50–day variations in galactic cosmic rays and in solar activity indices using the data for 1958 – 1979. The results obtained allow to conclude that the parameters of the cosmic ray modulation associated with the Sun's rotation were changed in 1970 after polarity reversal of the Sun's general magnetic field.

144.220   27–day recurrent effects in cosmic ray anisotropies during solar cycle 20.
K. N. Mishra, S. K. Nigam, R. P. Mishra, J. P. Shukla,
R. P. Mishra.
18th International Cosmic Ray Conference, Vol. 3, p. 191 – 194
(1983). – See Abstr. 012.096.

The harmonic components of galactic cosmic rays and its intensity, ΣK_p index and solar wind velocity have been analysed by Chree's superposed epoch method for maximum and minimum period of solar cycle 20. The amplitude and phase of all the significant harmonics of the galactic cosmic ray intensity show strong recurrent tendency with a periodicity of 27, 13.5, and 4.5 days along with other parameters in both phases of solar activity.

144.221   Diurnal and semi–diurnal anisotropy during the Forbush decrease.
D. S. Rana, R. S. Yadav, Badruddin, N. R. Yadav.
18th International Cosmic Ray Conference, Vol. 3, p. 195 (1983).
Abstract. – See Abstr. 012.096.

144.222   Cosmic ray intensity decrease in July 1982.
A. I. Kuz'min, V. P. Mamrukova, A. N. Prikhod'ko,
G. V. Skripin, I. A. Transky (*I. A. Transkij*).
18th International Cosmic Ray Conference, Vol. 3, p. 202 – 205
(1983). – See Abstr. 012.096.

The authors investigated cosmic ray distribution in various energy ranges during strong Forbush–decreases on July 11 – 12, 1982. Cosmic rays accelerated up to E ~ 250 GeV. The stable third harmonic of distribution of anomalously large amplitude is found.

144.223   Anisotropic intensity waves observed underground at Mawson with a proportional counter telescope.
R. M. Jacklyn, M. A. Pomerantz.
18th International Cosmic Ray Conference, Vol. 3, p. 206 – 209
(1983). – See Abstr. 012.096.

144.224   Anomalous anisotropy of cosmic rays in September 1982.
A. I. Kuz'min, V. M. Migunov, A. N. Prikhod'ko,
G. V. Skripin, I. A. Transky (*I. A. Transkij*).
18th International Cosmic Ray Conference, Vol. 3, p. 210 – 212
(1983). – See Abstr. 012.096.

Anomalies both in value and phase of the first and second harmonics of diurnal variation in the period of minimum intensity value are found.

144.225   The relations of the SSC and FD and their correlation with parameter of solar wind and IMF.
B. Y. Zhu, M. Wada.
18th International Cosmic Ray Conference, Vol. 3, p. 213 – 216
(1983). – See Abstr. 012.096.

Some characteristics of solar wind and interplanetary magnetic field at the time of SSC are investigated according to the range of Forbush decrease accompanied at the same SSC.

144.226   Cosmic ray increases during severe geomagnetic storms.
P. Tanskanen, H. Kananen, S. Kudo, M. Wada,
H. Komori.
18th International Cosmic Ray Conference, Vol. 3, p. 217 – 220
(1983). – See Abstr. 012.096.

144.227   Solar wind stream interfaces and transient decreases in cosmic ray intensity.
D. P. Tiwari, R. P. Mishra, A. P. Mishra, R. L. Singh.
18th International Cosmic Ray Conference, Vol. 3, p. 221 – 224
(1983). – See Abstr. 012.096.

The simultaneous changes in cosmic ray intensity in relation to solar wind stream interfaces have been analysed for the period 1965 – 1979.

144.228   The effect of equatorial coronal holes on galactic cosmic ray intensity.
R. L. Singh, D. P. Tiwari, S. K. Nigam, K. N. Mishra.
18th International Cosmic Ray Conference, Vol. 3, p. 225 – 228
(1983). – See Abstr. 012.096.

It has been observed that after 3 days of the central meridian passage of coronal holes the cosmic ray nucleonic intensity decreases ≈ 1%.

144.229   Effect of solar wind disturbances on galactic cosmic ray intensity depression.
K. N. Mishra, U. Singh, A. P. Mishra, D. P. Tiwari,
R. P. Mishra, R. L. Singh.
18th International Cosmic Ray Conference, Vol. 3, p. 229 – 231
(1983). – See Abstr. 012.096.

The time variation in the ground based nucleonic cosmic ray intensity depressions have been studied in association with solar wind disturbances.

144.230   Differences in time profiles of Forbush decreases observed by neutron monitors and ion chambers at low latitudes.
L. S. Chuang, M. Kusunose, M. Wada.
18th International Cosmic Ray Conference, Vol. 3, p. 232 (1983).
Abstract. – See Abstr. 012.096.

144.231   Forbush decrease of July 1982 observed on board a jet plane.
M. Okano, M. Wada.
18th International Cosmic Ray Conference, Vol. 3, p. 233 – 235
(1983). – See Abstr. 012.096.

144.232 Energy dependence of Forbush decreases observed by the Turku hadron spectrometer.
M. Lumme, M. Nieminen, J. Peltonen, J. J. Torsti,
E. Vainikka, E. Valtonen.
18th International Cosmic Ray Conference, Vol. 3, p. 237 – 240 (1983). – See Abstr. 012.096.

144.233 Forbush–effect and the variations of interplanetary magnetic field and solar wind velocity.
G. P. Lyubimov, N. V. Pereslegina, V. I. Tulupov,
E. A. Chuchkov.
18th International Cosmic Ray Conference, Vol. 3, p. 241 – 244 (1983). – See Abstr. 012.096.
The relationships between Forbush–decreases in galactic cosmic ray intensity and the variations in solar wind velocity and in the intensity and orientation of the interplanetary magnetic field are studied. The parameters are shown to be determined by the nature, form, and structure inherent to flare ejection or plasma stream.

144.234 Forbush–decrease and cosmic ray fluctuations in July 1982.
N. P. Chirkov, A. T. Filippov, V. L. Yanchukovsky
(*V. L. Yanchukovskij*).
18th International Cosmic Ray Conference, Vol. 3, p. 245 – 248 (1983). – See Abstr. 012.096.

144.235 Analysis of cosmic ray pitch–angle anisotropy during the Forbush–effect in June 1972 by the method of spectrographic global survey.
V. M. Dvornikov, V. E. Sdobnov, A. V. Sergeev.
18th International Cosmic Ray Conference, Vol. 3, p. 249 – 252 (1983). – See Abstr. 012.096.
The authors demonstrate the presence of a strong CR pitch–angle anisotropy in interplanetary space. The phase of pitch–angle anisotropy was used to derive information about the IMF orientation.

144.236 27–day recurrence in the cosmic ray north–south anisotropy at high rigidities.
V. H. Regener, R. H. S. John, D. B. Swinson.
18th International Cosmic Ray Conference, Vol. 3, p. 253 – 256 (1983). – See Abstr. 012.096.
There is a strong correlation between the N–S anisotropy and the IMF components in the ecliptic plane. This correlation shows a strong recurrence at 27 day intervals, with an absence of significant recurrence at other time intervals. The authors conclude that the N–S anisotropy arises from the B × ∇N drift, where B is the IMF vector and ∇N is the radial heliocentric cosmic ray density gradient.

144.237 Unusual periodicities in the cosmic ray intensity during the Forbush decrease of July 13/14, 1982.
H. Debrunner, E. O. Flückiger, A. Golliez,
H. Neuenschwander, M. Schubnell, G. Sebor.
18th International Cosmic Ray Conference, Vol. 3, p. 257 (1983). – See Abstr. 012.096.

144.238 Pulsations of electron flux intensity with a period 2 – 3 hours at geosynchronous orbit.
I. P. Bezrodnykh, Yu. G. Shafer.
18th International Cosmic Ray Conference, Vol. 3, p. 258 – 261 (1983). – See Abstr. 012.096.
The experimental data on electron flux pulsations with a period 2 – 3 hours with energy 1 – 1.5 MeV at geosynchronous orbit are presented. It's shown that plunging of geomagnetosphere into high–speed solar plasma stream precedes electron flux pulsations.

144.239 Cosmic ray intensity variations and solar wind velocity.
K. Fujimoto, H. Kojima, K. Murakami.
18th International Cosmic Ray Conference, Vol. 3, p. 267 – 269 (1983). – See Abstr. 012.096.
The relation of cosmic ray intensity to solar wind velocity is analyzed. The correlation between cosmic ray intensity and solar wind velocity is statistically significant. The regression coefficients obtained on yearly basis depend on sunspot number and are ~–0.8 and ~–0.2% per 100 km/s at the solar maximum and minimum, respectively.

144.240 Study of solar flare associated cosmic ray decreases during 1975 – 79.
A. K. Jain, P. K. Pandey, S. P. Pathak, S. P. Agrawal,
R. S. Yadav.
18th International Cosmic Ray Conference, Vol. 3, p. 270 – 273 (1983). – See Abstr. 012.096.
It is found that out of 174 major solar flares, which are associated with SSC during 1975 – 79, 119 of them produce significant world–wide decrease in cosmic ray intensity. These flares occur in a helio–longitude between 60° east to 30° west.

144.241 The low frequency fluctuations of the cosmic ray intensity.
L. I. Dorman, M. E. Katz (*M. E. Kats*), S. F. Nosov,
Yu. I. Fedorov, B. A. Shakhov, M. Stehlík.
18th International Cosmic Ray Conference, Vol. 3, p. 282 – 285 (1983). – See Abstr. 012.096.
The cosmic ray intensity fluctuations are investigated in the limit of very low frequencies, which are caused by the large–scale changes in the structure of the interplanetary medium. The expressions in both the approximation of the anisotropic diffusion and drift approximation are derived.

144.242 On the theory of cosmic ray fluctuations in the strong magnetic field.
L. I. Dorman, M. E. Katz (*M. E. Kats*), S. F. Nosov,
Yu. I. Fedorov, B. A. Shakhov, M. Stehlík.
18th International Cosmic Ray Conference, Vol. 3, p. 286 – 288 (1983). – See Abstr. 012.096.
The fluctuations of the cosmic ray distribution function in the interplanetary magnetic field are calculated on the base of the drift kinetic equation for the exact distribution funtion. The case of the isotropic magnetic field turbulence is investigated. The relations between the spectral characteristics of the cosmic ray distribution function and the random magnetic field are obtained.

144.243 Harmonics of the cosmic ray diurnal variation: a comparison of theory and observation.
J. W. Bieber, M. A. Pomerantz, C. H. Tsao.
18th International Cosmic Ray Conference, Vol. 3, p. 289 – 292 (1983). – See Abstr. 012.096.
The various observed harmonics of the cosmic ray diurnal variation may be understood on a unified basis if the free space cosmic ray anisotropy is non–sinusoidal in form. The major objective of this report is to emphasize that such a non–sinusoidal anisotropy is a natural result of the normal interplanetary transport processes of pitch angle scattering and adiabatic focusing in a spatially varying mean field.

144.244 Rigidity spectrum of semi–diurnal anisotropy of cosmic rays.
K. Fujimoto, S. Mori, H. Ueno, K. Nagashima.
18th International Cosmic Ray Conference, Vol. 3, p. 295 – 298 (1983). – See Abstr. 012.096.
The rigidity spectrum of the cosmic ray anisotropy responsible for solar semi–diurnal variation is examined in wide rigidity range of 10 ~ 600 GV. The anisotropy is in the direction of 3 hr solar time, and has the rigidity spectrum with its maximum around 100 GV.

144.245 Diurnal effect in galactic cosmic ray intensity according to the stratospheric measurements.
G. A. Asatryan, Yu. I. Stozhkov.
18th International Cosmic Ray Conference, Vol. 3, p. 299 – 302 (1983). – See Abstr. 012.096.
The stratospheric measurements of cosmic ray intensity at the latitude with the geomagnetic cutoff rigidity R_c = 7.6 GV show the existence of the diurnal wave with the amplitude ~ 3%. It

may be connected with the strong pitch–angle anisotropy of galactic cosmic rays in the interplanetary space.

144.246   Recent trends in the average behaviour of the daily variation of cosmic rays.

S. P. Pathak, P. K. Shrivastava, D. S. Awasthi, S. P. Agrawal.
18th International Cosmic Ray Conference, Vol. 3, p. 304 – 307 (1983). – See Abstr. 012.096.

Time profiles of the first three harmonics, during recent period of 1980 – 82, are studied. The deduced monthly average variations do not indicate any substantial changes in the amplitude and phase of the cosmic ray daily variations. The invariance of the second and third harmonics is in agreement with that expected from the earlier observations. However, the invariance of first harmonic during 1980 – 82 is not in accordance with that expected from earlier results.

144.247   Statistical study of correlated changes between the first three harmonics of daily variations of cosmic rays on a day–to–day basis.

V. P. Khare, D. K. Jadhav, A. K. Tiwari.
18th International Cosmic Ray Conference, Vol. 3, p. 308 – 311 (1983). – See Abstr. 012.096.

It is found that with the increase in diurnal amplitude neither the diurnal phase nor the semi–diurnal/tri–diurnal amplitude changes very significantly. However, an increasing trend in the semi–diurnal amplitude is observed.

144.248   Study of anomalous change in cosmic ray diurnal variation during Feb. – Mar. 1977.

P. K. Shrivastava, V. P. Khare, D. K. Jadhav.
18th International Cosmic Ray Conference, Vol. 3, p. 312 – 315 (1983). – See Abstr. 012.096.

It is shown that the anomalous behaviour of low diurnal amplitude observed during February – March 1977 is universal, and is not due to any Forbush decrease or any universal time–associated cosmic ray effects.

144.249   Study of 11 and 22 year periodic variation of cosmic ray diurnal anisotropy.

S. P. Agrawal, M. Bercovitch.
18th International Cosmic Ray Conference, Vol. 3, p. 316 – 319 (1983). – See Abstr. 012.096.

The authors present here a survey of the diurnal variation from 1955 to 1979, extending the period studied up to the present solar maximum. The authors find that the direction of the 22 year component which makes an angle of 89° with the 22 year mean anisotropy direction, lies 162° east of the sun–earth line, deviating significantly from the interplanetary magnetic field direction.

144.250   Harmonics of daily variation of galactic cosmic rays during 1968 – 1979.

S. P. Agrawal, S. P. Pathak, B. L. Mishra.
18th International Cosmic Ray Conference, Vol. 3, p. 320 – 323 (1983). – See Abstr. 012.096.

The time of maximum for II and III harmonics is found to be constant during the whole interval of 1968 – 79, whereas it changes significantly in the case of the first harmonic of the daily variation. The observations indicate considerable influence of the interplanetary regions perpendicular to the ecliptic plane.

144.251   Cosmic ray power spectral variations 2. Ambient power.

D. Venkatesan, L. J. Lanzerotti, S. P. Agrawal.
18th International Cosmic Ray Conference, Vol. 3, p. 324 – 327 (1983). – See Abstr. 012.096.

The power spectral technique has been used to study the ambient power for each solar rotation during the interval 1964 – 1977. The long term variability of the ambient power and its maximum and minimum excursion with the sunspot activity is observed and interpreted. It is suggested that the ambient power so derived need to be subtracted from the observed power under the daily variation peaks to obtain the proper amplitudes of the daily variation of cosmic rays.

144.252   22–year recurrence tendency of diurnal anisotropy during maximum solar activity periods observed on quiet days.

S. Kumar.
18th International Cosmic Ray Conference, Vol. 3, p. 328 (1983). Abstract. – See Abstr. 012.096.

144.253   Solar semi–diurnal variations of cosmic ray nucleonic components.

N. Yahagi, H. Takahashi, S. Mori, K. Nagashima.
18th International Cosmic Ray Conference, Vol. 3, p. 329 – 332 (1983). – See Abstr. 012.096.

Average characteristics of the solar semi–diurnal variation, particularly its spectral dependence are examined. The anisotropy has the amplitude of 0.28% at 100 GV with a positive exponent ($\gamma = 1.0$) up to $\cong 200$ GV in the 3.2 hr direction.

144.254   Energy spectrum of the semidiurnal variation of cosmic rays.

A. G. Zusmanovich, L. A. Mirkin.
18th International Cosmic Ray Conference, Vol. 3, p. 333 – 336 (1983). – See Abstr. 012.096.

The average amplitudes of semidiurnal variation of cosmic ray intensity increases up to 50 – 70 GV and after that it decreases with increasing of the particle energy. The cosmic rays gradient perpendicular to the ecliptic plane is calculated.

144.255   Study of semidiurnal variation of cosmic rays during days of low and high diurnal amplitude wave trains.

D. K. Jadhav, M. Shrivastava, A. K. Tiwari, P. K. Shrivastava.
18th International Cosmic Ray Conference, Vol. 3, p. 337 – 340 (1983). – See Abstr. 012.096.

144.256   Variation of the diurnal and semi–diurnal anisotropy on quiet days during the recent period up to 1980.

S. Kumar, R. S. Yadav, S. P. Agrawal.
18th International Cosmic Ray Conference, Vol. 3, p. 341 (1983). Abstract. – See Abstr. 012.096.

144.257   Interplanetary magnetic field sector structure and second harmonic of cosmic ray intensity.

S. K. Nigam, K. N. Mishra, D. P. Tiwari, R. P. Mishra, R. L. Singh.
18th International Cosmic Ray Conference, Vol. 3, p. 342 – 345 (1983). – See Abstr. 012.096.

144.258   Multiplicity determined upper cutoff rigidity of the second harmonic daily variation of cosmic ray intensity during 1968 – 1973.

B. L. Mishra, S. P. Pathak, S. P. Agrawal.
18th International Cosmic Ray Conference, Vol. 3, p. 346 – 349 (1983). – See Abstr. 012.096.

144.259   Lunar–diurnal variations of high–energy cosmic rays.

B. D. Naskidashvili, L. Kh. Shatashvili.
18th International Cosmic Ray Conference, Vol. 3, p. 350 – 353 (1983). – See Abstr. 012.096.

The lunar–diurnal variations of cosmic ray neutron and rigid components in a wide energy range are analysed. The results show the existence of increase of part of lunar–diurnal variations in intensity changes of high–energy μ–mesons.

144.260   The behaviour of the cosmic ray equatorial anisotropy inside fast solar wind streams ejected by coronal holes.

N. Iucci, M. Parisi, M. Storini, G. Villoresi.
18th International Cosmic Ray Conference, Vol. 3, p. 354 – 357 (1983). – See Abstr. 012.096.

The authors analyse the behaviour of the equatorial anisotropy during the days in which the earth is inside broad RHSSs in the years 1973 – 1974.

144.261 Cosmic ray north–south asymmetry related with the latitudinal angular distance of the earth from the heliospheric current sheet.
Y. Munakata, K. Hakamada, S. Mori, S. Yasue, M. Ichinose.
18th International Cosmic Ray Conference, Vol. 3, p. 358 – 361 (1983). – See Abstr. 012.096.

The relation between the north–south asymmetry of cosmic ray intensity for the wide range of rigidity ($\sim 15 - \sim 150$ GV in median primary rigidity) and the latitudinal angular distance (λ) of the earth from the heliospheric current sheet is examined during the period from Apr. 1976 to Aug. 1977. It is found that the amplitude of the north–south asymmetry of cosmic ray intensity is fairly correlated with sin λ.

144.262 Heliolongitudinal asymmetry of solar activity and cosmic ray distribution.
N. P. Chirkov, V. G. Grigoryev (*V. G. Grigor'ev*).
18th International Cosmic Ray Conference, Vol. 3, p. 362 – 365 (1983). – See Abstr. 012.096.

Galactic cosmic ray currents behaviour during 44 Forbush–decreases in 1968 – 1974 are considered. The obtained results show the presence of definite current structures in interplanetary medium related with solar flare heliocoordinates.

144.263 Polarity of the solar polar magnetic field and cosmic ray anisotropy.
R. S. Yadav, Badruddin.
18th International Cosmic Ray Conference, Vol. 3, p. 366 – 369 (1983). – See Abstr. 012.096.

It is found that the amplitude of the diurnal anisotropy is large during the period 1972 – 76, with a positive orientation of the northern solar magnetic pole, as compared to the period 1964 – 68. The phase of the diurnal anisotropy is found to shift to earlier hours during the period 1972 – 76, where the magnetic pole becomes positive, as compared to the period 1964 – 68.

144.264 Cosmic ray anisotropy variation on solar activity cycle.
N. G. Kravtsov, P. A. Krivoshapkin, A. I. Kuz'min, G. V. Skripin.
18th International Cosmic Ray Conference, Vol. 3, p. 370 – 372 (1983). – See Abstr. 012.096.

On data of muon intensity registration in Yakutsk the cosmic ray anisotropy for 1958 – 1981 in dependence upon the solar activity cycle is investigated. The presence of the 22–d and 11–th year anisotropy variation is found. The 22–d variation is explained by the large–scale drift of cosmic rays in the interplanetary medium when there is a neutral sheet and 11–th one is caused by the cosmic ray diffusion.

144.265 Cosmic ray sidereal diurnal variation of galactic origin observed by neutron monitors.
K. Nagashima, Y. Ishida, S. Mori, I. Morishita.
18th International Cosmic Ray Conference, Vol. 3, p. 375 – 378 (1983). – See Abstr. 012.096.

The sidereal diurnal variations of cosmic rays were analyzed using 620 station–years of neutron monitor data during the period of 1958 – 1979. It was found that the observed sidereal variations averaged over the period for all the stations in the northern hemisphere is different from the corresponding variation in the southern hemisphere.

144.266 Seasonal variation of sidereal daily variation produced by modulation of galactic cosmic ray anisotropy in heliomagnetosphere.
K. Nagashima, I. Morishita, S.–i. Yasue.
18th International Cosmic Ray Conference, Vol. 3, p. 379 – 382 (1983). – See Abstr. 012.096.

The authors present a general formulation of the seasonal variations, based on the frequency modulation theory, and discuss general characteristics of the variations caused by the cosmic ray deflection in the model heliomagnetosphere.

144.267 The sidereal diurnal variation measured underground in London.
T. Thambyahpillai.
18th International Cosmic Ray Conference, Vol. 3, p. 383 – 386 (1983). – See Abstr. 012.096.

Data from vertical as well as inclined telescopes operating at a depth of 60 m.w.e. in London are presented and it is shown that in the case of the vertical telescopes the existence of a 22–year variation is confirmed within the limits of statistical uncertainty.

144.268 Solar and sidereal time variations of cosmic rays at 220 hg cm^{-2} underground at Matsushiro.
S. Yasue, S. Mori, S. Sagisaka, M. Ichinose.
18th International Cosmic Ray Conference, Vol. 3, p. 387 – 390 (1983). – See Abstr. 012.096.

144.269 Further evidence for a bi–directional anisotropy of galactic origin observed at moderate depths underground.
R. M. Jacklyn, M. L. Duldig.
18th International Cosmic Ray Conference, Vol. 3, p. 391 – 394 (1983). – See Abstr. 012.096.

144.270 The sidereal variation deep underground in Tasmania – an updating report.
J. E. Humble, A. G. Fenton, K. B. Fenton.
18th International Cosmic Ray Conference, Vol. 3, p. 396 (1983). Abstract. – See Abstr. 012.096.

144.271 Influence of statistical cosmic ray scattering on sidereal daily variation.
S. Yasue, I. Morishita, K. Nagashima.
18th International Cosmic Ray Conference, Vol. 3, p. 397 – 400 (1983). – See Abstr. 012.096.

144.272 The sector–dependent sidereal diurnal variation produced from solar anisotropy of cosmic rays.
K. Nagashima, R. Tatsuoka, S. Matsuzaki.
18th International Cosmic Ray Conference, Vol. 3, p. 401 – 402 (1983). – See Abstr. 012.096.

The sidereal variation is obtained with a phase of 6.57 hours or 18.57 hours.

144.273 Sidereal variations at 34 hg cm^{-2} underground and interplanetary magnetic field directions.
S. Mori, S. Yasue, M. Ichinose, S. Sagisaka, K. Chino, S. Sugie.
18th International Cosmic Ray Conference, Vol. 3, p. 403 – 406 (1983). – See Abstr. 012.096.

The authors present the results of observations made during the period 1975 – 81.

144.274 Study of sidereal time daily variation of cosmic rays during 1973 – 78 by neutron and meson monitors.
D. S. Awasthi, S. P. Agrawal.
18th International Cosmic Ray Conference, Vol. 3, p. 407 – 410 (1983). – See Abstr. 012.096.

144.275 Vertical cutoff rigidities for selected cosmic ray stations for epoch 1980.0.
M. A. Shea, D. F. Smart, L. C. Gentile.
18th International Cosmic Ray Conference, Vol. 3, p. 411 – 414 (1983). – See Abstr. 012.096.

The authors found that in general the vertical cutoff rigidities are decreasing in Latin and South America, in Southern Africa and in India; increases are noted in Europe and Japan.

144.276 A world grid of calculated cosmic ray vertical cutoff rigidities for 1980.0.
M. A. Shea, D. F. Smart.
18th International Cosmic Ray Conference, Vol. 3, p. 415 – 418 (1983). – See Abstr. 012.096.

Significant increases ($\geqslant 0.30$ GV) were noted in the North Atlantic Ocean area with similar decreases found in the South Atlantic.

144.277 Geomagnetic transmission functions for a 400 km altitude satellite.
D. F. Smart, M. A. Shea.
18th International Cosmic Ray Conference, Vol. 3, p. 419 – 422 (1983). – See Abstr. 012.096.

144.278 The cosmic ray equator determined using the international geomagnetic reference field for 1980.0.
M. A. Shea, D. F. Smart, L. C. Gentile.
18th International Cosmic Ray Conference, Vol. 3, p. 423 – 426 (1983). – See Abstr. 012.096.

In comparing new results with the cosmic ray equators calculated for 1955 and 1965 the authors find a steady and consistent northerly shift in latitude of the cosmic ray equator between longitudes 265° E and 10° E with a maximum change of ~4° in latitude for geographic longitudes between 310° E and 325° E.

144.279 Changes in asymptotic directions for various geomagnetic storm conditions.
E. O. Flückiger, D. F. Smart, M. A. Shea.
18th International Cosmic Ray Conference, Vol. 3, p. 427 – 430 (1983). – See Abstr. 012.096.

The authors demonstrate the effect of a geomagnetic storm on the asymptotic directions of cosmic ray particles.

144.280 Procedure for estimating the change in asymptotic directions at a mid–latitude cosmic ray station resulting from a local change in cosmic ray cutoff rigidity.
E. O. Flückiger, D. F. Smart, M. A. Shea.
18th International Cosmic Ray Conference, Vol. 3, p. 431 – 434 (1983). – See Abstr. 012.096.

Recent studies have shown that at low and mid–latitudes the asymptotic directions of cosmic ray particles during perturbed geomagnetic conditions can be deduced with considerable accuracy from the asymptotic directions in the quiescent field, if the associated change in the cutoff rigidity is known.

144.281 Cosmic ray access to satellites from large zenith angles.
J. E. Humble, D. F. Smart, M. A. Shea.
18th International Cosmic Ray Conference, Vol. 3, p. 442 – 445 (1983). – See Abstr. 012.096.

Results are presented from a comprehensive survey of the largest accessible zenith angles for a range of altitudes and geographic locations. The search has disclosed that primary particles are able to reach a satellite at an altitude of 1250 km from zenith angles as large as 178°.

144.282 Space distribution of cosmic rays in the lower atmosphere.
A. M. Altukhov, E. V. Gorchakov, P. P. Ignatiev
(*P. P. Ignat'ev*), T. E. Shvidkovskaya.
18th International Cosmic Ray Conference, Vol. 3, p. 452 – 455 (1983). – See Abstr. 012.096.

Some results of airborne measurements of charged particles and gamma–rays in broad intervals of latitudes and longitudes at altitudes of 0.5 – 11 km over the USSR territory are presented.

144.283 Time variations in the longitudinal and latitudinal cosmic ray asymmetry in the lower atmosphere.
A. M. Altukhov, A. S. Isaev, V. A. Rogozhin,
V. M. Senchyenok, T. N. Charakhch'yan.
18th International Cosmic Ray Conference, Vol. 3, p. 456 – 459 (1983). – See Abstr. 012.096.

The anomalies in the longitudinal and latitudinal distribution of cosmic rays in the lower atmosphere ($X > 300$ g/cm^2) and the time variations are discussed on the basis of the data of aircraft and balloon measurements.

144.284 Geomagnetic field effect on cosmic rays in the earth's atmosphere.
L. I. Dorman, N. I. Pakhomov.
18th International Cosmic Ray Conference, Vol. 3, p. 516 – 519 (1983). – See Abstr. 012.096.

The propagation of cosmic rays in the earth's atmosphere is calculated, with the geomagnetic field effect taken into account.

144.285 Selectivity to primary particle energy and the north–south asymmetry of zonal cosmic ray modulation.
T. N. Charakhch'yan.
18th International Cosmic Ray Conference, Vol. 3, p. 523 – 526 (1983). – See Abstr. 012.096.

The energy spectrum and the north–south asymmetry effects of the new type of variations, the zonal cosmic ray modulation, are discussed.

144.286 Energy distribution of parent nucleons for muons observed in a large–zenith–angle muon telescope array at Ottawa.
A. K. Das, M. Bercovitch.
18th International Cosmic Ray Conference, Vol. 3, p. 534 – 537 (1983). – See Abstr. 012.096.

144.287 Multiplicity response function of the double neutron monitor at Turku.
M. Lumme, M. Nieminen, J. Peltonen, J. J. Torsti,
E. Vainikka, E. Valtonen.
18th International Cosmic Ray Conference, Vol. 3, p. 538 – 541 (1983). – See Abstr. 012.096.

A Monte Carlo method has been applied to the simulation of atmospheric hadron cascades in the primary rigidity range from 1 GV up to 10^{4}GV.

144.288 Annual cosmic ray variations at various atmospheric levels.
T. N. Charakhch'yan, V. P. Okhlopkov, L. S. Okhlopkova.
18th International Cosmic Ray Conference, Vol. 3, p. 547 – 550 (1983). – See Abstr. 012.096.

In the stratosphere, the annual variation amplitude is 2 – 4% and the variations are probably due to solar activity. In the lower atmosphere (at pressures above 300 g/cm^2) the annual variation shows two clear periods: (1) variation in 1961 – 1970 with an amplitude of up to 6 – 7% due to the effects of zonal cosmic ray modulation and (2) variation in 1971 – 1980 with an amplitude of about 2% due mainly to temperature effects.

144.289 The features of the dynamics of the 240–day cosmic ray variation.
V. P. Okhlopkov.
18th International Cosmic Ray Conference, Vol. 3, p. 551 – 554 (1983). – See Abstr. 012.096.

The 240–day variation has been found to appear during individual periods only. The variation amplitude is 2 – 3% at altitudes of about 30 km and 5 – 7% in the lower atmosphere. These features may be displayed of the zonal modulation of cosmic rays.

144.290 Passage of the current sheet and cosmic ray modulation.
Badruddin, R. S. Yadav, N. R. Yadav.
18th International Cosmic Ray Conference, Vol. 3, p. 559 – 562 (1983). – See Abstr. 012.096.

It is found that there is a decrease in cosmic ray intensity after a sector boundary crossing. The periods 1966 – 68 and 1971 – 73 are observed.

144.291 Spectral–time analysis of cosmic ray intensity and solar activity during decreasing branch of the solar cycle.
E. S. Vernova, N. G. Ptitsyna, M. I. Tyasto.
18th International Cosmic Ray Conference, Vol. 3, p. 563 – 566 (1983). – See Abstr. 012.096.

Spectral composition of cosmic ray intensity and of solar activity at different frequency range is observed from 1958 to 1963. The spectral–time diagram is the result of inverse Fourier transform according to subsequently extracted parts of the data spectrum analyzed.

144.292 Cosmic ray decreases and cold plasma in the heliosphere.
A. Geranios, H. Rosenbauer.
18th International Cosmic Ray Conference, Vol. 4, p. 206 – 209 (1983). – See Abstr. 012.096.

The observed magnetically closed structures in the solar wind, containing low temperature interplanetary plasma, are identified and examined in relation to the cosmic ray flux decreases, as measured by high–latitude neutron monitors in the earth. A new model is proposed. Small–amplitude cosmic ray decreases seem to be related to the observed magnetically closed regions in the solar wind.

144.293 An estimate of the flux of free quarks in high energy cosmic radiation.
C. B. A. McCusker.
18th International Cosmic Ray Conference, Vol. 5, p. 78 – 80 (1983). – See Abstr. 012.096.

144.294 Search for 4/3·e charged leptons in cosmic rays.
T. Wada, Y. Yamashita, H. Kawanishi, K. Ieyoshi, I. Yamamoto, K. Imaeda.
18th International Cosmic Ray Conference, Vol. 5, p. 85 – 88 (1983). – See Abstr. 012.096.

144.295 A cosmic ray experiment to detect neutral particles.
A. Subramanian, S. D. Verma.
18th International Cosmic Ray Conference, Vol. 5, p. 93 – 96 (1983). – See Abstr. 012.096.
It is proposed to reinvestigate the results reported by Cowan et al. (1965 – 1970) on the possible existence of a hitherto unknown neutral primary radiation in cosmic rays.

144.296 On coordination of primary proton flux with the flux of hadrons without shower accompany deep in the atmosphere.
R. A. Nam, S. I. Nikolsky (*S. I. Nikol'skij*), A. P. Chubenko, V. I. Yakovlev.
18th International Cosmic Ray Conference, Vol. 5, p. 336 – 339 (1983). – See Abstr. 012.096.

144.297 Evaluation of the hadron energy spectra in the atmosphere.
J. Kempa, J. Wdowczyk.
18th International Cosmic Ray Conference, Vol. 5, p. 340 – 343 (1983). – See Abstr. 012.096.
It is shown that the agreement between experimental data and predictions can be obtained only if the average mass of the primaries does not increase significantly up to few times 10^{15}eV.

144.298 Calculation of neutron proton flux ratio at the different depths of atmosphere.
D. P. Bhattacharyya.
18th International Cosmic Ray Conference, Vol. 5, p. 360 – 363 (1983). – See Abstr. 012.096.
The neutron–proton flux ratio at different atmospheric depths has been calculated from the recently estimated primary cosmic ray spectrum.

144.299 Altitude variation of the derived electron photon intensity.
D. P. Bhattacharyya, A. Mukhopadhyay.
18th International Cosmic Ray Conference, Vol. 5, p. 364 – 367 (1983). – See Abstr. 012.096.
The photon spectra at the atmospheric depths 260, 400 and 650 g/cm^2 estimated from the recently measured primary spectrum are in accord with experimental data for energies up to 10 TeV.

144.300 On cascade studies and on the propagation of cosmic radiation in the atmosphere.
Z. Strugalski.
18th International Cosmic Ray Conference, Vol. 5, p. 376 – 379 (1983). – See Abstr. 012.096.

144.301 Propagation of cosmic rays in the atmosphere and energy deposition in detectors.
C. H. Tsao, R. Silberberg, J. H. Adams Jr., J. R. Letaw.
18th International Cosmic Ray Conference, Vol. 5, p. 380 – 383 (1983). – See Abstr. 012.096.
The cross section equations and data on the composition and energy spectra of cosmic rays have been combined to calculate the effects of nuclear transformations and energy losses of cosmic rays in the atmosphere. The energy deposition spectra of the cosmic ray nuclei in silicon are presented for times of solar minimum and solar maximum.

144.302 Derivation of the energy spectra of cosmic ray hadrons at different depths of the atmosphere.
D. P. Bhattacharyya.
18th International Cosmic Ray Conference, Vol. 5, p. 384 – 387 (1983). – See Abstr. 012.096.

144.303 Primary radiation and lateral muon development in E.A.S.
J. Procureur.
18th International Cosmic Ray Conference, Vol. 6, p. 58 – 61 (1983). – See Abstr. 012.096.
The fluctuations of the lateral muon development are analysed in order to obtain information on the chemical nature of the primary component. Comparison with experimental results from Tien–Shan favours the mixed composition: proton 50%, light nuclei 20%, heavy nuclei 30%.

144.304 On the method of primary cosmic ray nuclear composition determination by investigation of EAS hadron energy spectra.
A. G. Dybovy, N. M. Nesterova.
18th International Cosmic Ray Conference, Vol. 6, p. 82 – 85 (1983). – See Abstr. 012.096.
A method of primary cosmic ray nuclear composition determination is described by examination of the EAS hadron spectra at energy close to the initial energy per nucleon E_0/A.

144.305 Fluctuations in depth of shower maximum and the mass composition of cosmic rays near 10^{19}eV.
R. Walker, A. A. Watson.
18th International Cosmic Ray Conference, Vol. 6, p. 114 – 117 (1983). – See Abstr. 012.096.
Measurements of the fluctuation of the depth of shower maximum are briefly reviewed. It is shown that the hypothesis of a pure iron beam at 10^{19}eV can be strongly rejected in favour of one in which the proton fraction is at least 40% and probably higher.

144.306 Properties of hadron interactions at extremely high energies in the light of cosmic ray data and recent accelerator observations.
J. Wdowczyk, A. W. Wolfendale.
18th International Cosmic Ray Conference, Vol. 6, p. 189 – 192 (1983). – See Abstr. 012.096.
The problem of the nature of the primary cosmic rays, and the manner of their interaction, in the region of 10^{15}eV, is examined. The authors examine recent mass measurements, indirect studies and new interaction data CERN (p$\bar{\text{p}}$ interactions corresponding to $\cong 2 \times 10^{14}$eV) and conclude that if the trend is extrapolated much of the cosmic ray data at 10^{15}eV can be explained without invoking a large increase in the iron component in the primary beam.

144.307 The method for the study of primary cosmic ray mass composition at EAS arrays with large calorimeter area.
T. V. Danilova, A. D. Erlykin.
18th International Cosmic Ray Conference, Vol. 6, p. 262 – 265 (1983). – See Abstr. 012.096.
The new method is proposed for the analysis of the mass composition of primary cosmic rays.

144.308 On the interpretation of DEIS and MUTRON data of near–horizontal muon spectrum and charge ratio.
A. K. Das, A. K. De.
18th International Cosmic Ray Conference, Vol. 7, p. 5 – 8 (1983). – See Abstr. 012.096.
DEIS and MUTRON data of near–horizontal muon spectrum and charge ratio have been interpreted here on a phenomenological model for the propagation of cosmic rays through the atmosphere.

144.309   Experimental determination of the charge ratio of low energy cosmic ray muons.
K. P. Singhal.
18th International Cosmic Ray Conference, Vol. 7, p. 27 – 30 (1983). – See Abstr. 012.096.
Experimental determination of the charge ratio of cosmic ray muons at different energies is of considerable interest, because such measurements reflect the nature and multiplicities of the parent particles produced in the primary interactions.

144.310   Multiple muon events in the Kolar Gold Field experiment.
M. R. Krishnaswamy, M. G. K. Menon, N. K. Mondal, V. S. Narasimham, Y. Hayashi, N. Ito, S. Kawakami, S. Miyake.
18th International Cosmic Ray Conference, Vol. 7, p. 52 (1983). Abstract. – See Abstr. 012.096.

144.311   Analysis of multiple muons in deep underground detectors.
T. K. Gaisser, T. Stanev.
18th International Cosmic Ray Conference, Vol. 7, p. 53 (1983). Abstract. – See Abstr. 012.096.

144.312   A study of the multiple muon phenomena by Monte Carlo method.
Y. Muraki, A. Okada, P. Bianco.
18th International Cosmic Ray Conference, Vol. 7, p. 54 – 57 (1983). – See Abstr. 012.096.
Monte Carlo simulations are made in order to determine the cosmic ray composition between 10^{14}eV and 10^{15}eV. It is shown that an underground muon detector with the total area 10,000 m^2 will be able to measure the cosmic ray composition between 10^{16}eV and 10^{17}eV. High energy astronomical gamma rays will be possibly detected by this system, selecting mu–less air showers.

144.313   Calculation of cosmic ray neutrino flux.
T. K. Gaisser, T. Stanev, S. A. Bludman, H. Lee.
18th International Cosmic Ray Conference, Vol. 7, p. 91 – 94 (1983). – See Abstr. 012.096.
The authors report first results of new calculations of the flux of neutrinos produced by cosmic rays in the atmosphere. They have taken account of effects of the geomagnetic cutoff and of solar modulation separately for upward and downward going neutrinos of both electron and muon flavor with energies from 200 MeV to 10 GeV. The closely related flux of muons is also calculated as a function of depth in the atmosphere.

144.314   Results on cosmic ray neutrino interactions from the KGF experiment.
M. R. Krishnaswamy, M. G. K. Menon, N. K. Mondal, V. S. Narasimham, B. V. Sreekantan, Y. Hayashi, N. Ito, S. Kawakami, S. Miyake.
18th International Cosmic Ray Conference, Vol. 7, p. 95 (1983). Abstract. – See Abstr. 012.096.

144.315   Study of arrival time correlations of horizontal cosmic ray particles.
V. D. Ashitkov, T. M. Kirina, A. P. Klimakov, R. P. Kokoulin, A. A. Petrukhin.
18th International Cosmic Ray Conference, Vol. 7, p. 129 – 132 (1983). – See Abstr. 012.096.
Limits on the flux of cosmic ray particles correlated in time and arrival direction with high energy muons have been estimated for time intervals within ± 26 ms.

144.316   Observations of the Crab pulsar near $10^{15} - 10^{16}$eV.
J. Boone, R. Cady, G. L. Cassiday, J. W. Elbert, E. C. Loh, P. V. Sokolsky, D. Steck, S. Wasserbaech.
18th International Cosmic Ray Conference, Vol. 9, p. 57 – 60 (1983). – See Abstr. 012.096.
During 1980 December, a 3.1 σ excess of Čerenkov light flashes was observed within an angle of about 3.5° from the Crab nebula. The estimated threshold energy of the detected showers

was 10^{15}eV. No excess was detected in observations made during 1981 February.

144.317   Studies of low energy cosmic rays – the anomalous component.
W. R. Webber, A. C. Cummings, E. C. Stone.
18th International Cosmic Ray Conference, Vol. 9, p. 95 – 98 (1983). – See Abstr. 012.096.
The authors have determined the energy spectra of the anomalous cosmic ray species He, N, O, and Ne. They find that the anomalous particles are singly ionized.

144.318   Energy spectra and charge ratios of primary cosmic ray nuclei in the intermediate energy region.
P. K. Das, T. D. Goswami.
18th International Cosmic Ray Conference, Vol. 9, p. 99 – 102 (1983). – See Abstr. 012.096.
The energy spectra and the charge ratios of the primary cosmic ray nuclei C, O, Ne, Mg and Si in the energy interval 150 to 350 MeV/n have been estimated by using a balloon exposed cellulose nitrate plastic detector over Fort Churchill, Canada.

144.319   Observation of cosmic ray heavy nuclei above 5 GeV/n.
Y. Fukada, I. Kondo, Y. Hatano, T. Saito, Y. Kobayashi, M. Miyajima, M. Noma, H. Sakurai, H. Oda.
18th International Cosmic Ray Conference, Vol. 9, p. 103 – 105 (1983). – See Abstr. 012.096.
Observation of cosmic ray heavy nuclei with energies above 5 GeV/nuc. were performed by balloon–borne detector systems in 1980 and 1981. Results from these flights were analysed in terms of the chemical abundance of cosmic ray heavy nuclei above 10 GV rigidity. The ratio of the secondary nuclei to primary nuclei at the top of the atmosphere is derived.

144.320   Cosmic ray elemental abundances for $26 \leqslant Z \leqslant 42$ measured on HEAO–3.
W. R. Binns, D. P. Grossman, M. H. Israel, M. D. Jones, J. Klarmann, T. L. Garrard, E. C. Stone, R. K. Fickle, C. J. Waddington.
18th International Cosmic Ray Conference, Vol. 9, p. 106 – 109 (1983). – See Abstr. 012.096.
The earlier study of the abundances of even–Z nuclei in the interval of atomic number (Z) $26 \leqslant Z \leqslant 42$ has been extended using an expanded data set, with careful attention to potential sources of Z–dependent bias in the data. The results will be compared with individual element abundances expected from various source abundances, including solar system abundances and those from specific nucleosynthesis models, both with and without first–ionization–potential biases.

144.321   Ultra–heavy cosmic ray fluxes from Ariel 6.
P. H. Fowler, M. R. W. Masheder, R. T. Moses, R. N. F. Walker, A. Worley.
18th International Cosmic Ray Conference, Vol. 9, p. 110 – 114 (1983). – See Abstr. 012.096.

144.322   Cosmic–ray abundances of the even charge elements from $_{50}$Sn to $_{58}$Ce measured on HEAO–3.
E. C. Stone, T. L. Garrard, K. E. Krombel, W. R. Binns, M. H. Israel, J. Klarmann, N. R. Brewster, R. K. Fickle, C. J. Waddington.
18th International Cosmic Ray Conference, Vol. 9, p. 115 – 118 (1983). – See Abstr. 012.096.
Elements with even atomic number (Z) in the interval $50 \leqslant Z \leqslant 58$ have been resolved in the cosmic radiation using the Heavy Nuclei Experiment on the HEAO–3 satellite.

144.323   The cosmic ray abundances of the platinum–lead elements as measured on HEAO–3.
D. J. Fixsen, C. J. Waddington, W. R. Binns, M. H. Israel, J. Klarmann, T. L. Garrard, B. J. Newport, E. C. Stone.
18th International Cosmic Ray Conference, Vol. 9, p. 119 – 122 (1983). – See Abstr. 012.096.
Data from the HEAO 3 Heavy Nuclei Experiment are used to establish abundances, relative to iron, of elements in the charge

ranges of $75 \leqslant Z \leqslant 79$ (platinum) and $80 \leqslant Z \leqslant 83$ (lead), as well as the ratio of "secondary" elements, in the $62 \leqslant Z \leqslant 74$ range, to the primary lead–platinum elements.

144.324 Energy spectra of ultraheavy cosmic rays: results from HEAO–3.
M. H. Israel, M. D. Jones, J. Klarmann, W. R. Binns, T. L. Garrard, E. C. Stone, R. K. Fickle, C. J. Waddington.
18th International Cosmic Ray Conference, Vol. 9, p. 123 – 126 (1983). – See Abstr. 012.096.
The HEAO–3 Heavy Nuclei Experiment measures cosmic–ray energy directly in the interval 400 to ~ 1200 MeV/amu. Geomagnetic cutoffs can also be derived up to ~ 15 GV. The authors present preliminary rigidity spectra of various ultraheavy cosmic–ray elements relative to iron.

144.325 Can we get the cosmic ray actinide abundance from the study of tracks in meteorites?
C. Perron, P. Pellas.
18th International Cosmic Ray Conference, Vol. 9, p. 127 – 130 (1983). – See Abstr. 012.096.
The feasibility of a cosmic ray actinide abundance measurement by means of tracks in meteorites is discussed in the light of recent studies of high energy U ion tracks in meteoritic minerals.

144.326 The elemental and isotopic composition of quiet time low energy cosmic rays.
B. Smith, F. B. McDonald.
18th International Cosmic Ray Conference, Vol. 9, p. 131 – 134 (1983). – See Abstr. 012.096.
Isotopic abundances are derived from multidimensional analysis of double dE/dX modes made with two 150 μm ΔE detectors and a 3000 μm stopping E detector.

144.327 The isotopic composition of the cosmic rays at energies above 2 GeV/n.
B. Byrnak, N.–Y. Herrström, N. Lund, B. Peters, I. L. Rasmussen, M. Rotenberg, N. J. Westergaard, P. Ferrando, P. Goret, L. Koch–Miramond, A. Soutoul.
18th International Cosmic Ray Conference, Vol. 9, p. 135 – 138 (1983). – See Abstr. 012.096.
The authors have analyzed a high quality data set from the Danish–French cosmic ray experiment on the HEAO–3 satellite. They have determined accurate cut–off patterns and performed the isotopic analysis for a number of elements. The most important result is the persistence of the Ne–22 excess at particle energies between 2 and 8 GeV/nucleon.

144.328 Mean masses of cosmic ray heavy elements: results from HEAO 3 C–2 at 2.5 GeV/n.
P. Goret, J. J. Engelmann, P. Ferrando, L. Koch–Miramond, N. Petrou, A. Soutoul, B. Byrnak, N.–Y. Herrstroem, N. Lund, B. Peters, I. L. Rasmussen, M. Rotenberg, N. J. Westergaard.
18th International Cosmic Ray Conference, Vol. 9, p. 139 – 142 (1983). – See Abstr. 012.096.
The authors outline a new method of isotope analysis and give mean mass results for $Z \geqslant 6$ at 2.5 GeV/n.

144.329 Nitrogen and boron isotopes around 2 GeV/n from the French–Danish instrument on HEAO–3.
A. Soutoul, J. J. Engelmann, P. Goret, J. Jorrand, L. Koch–Miramond, P. Masse, N. Petrou, N.–Y. Herrstroem, N. Lund.
18th International Cosmic Ray Conference, Vol. 9, p. 143 – 146 (1983). – See Abstr. 012.096.
The authors describe a new method of isotope analysis and for computing mean masses of particles collected at cut off rigidities. They determine the mean masses of nitrogen and boron at momentum per nucleon ~ 3 GeV/c.n.

144.330 The abundance of the radioactive isotope ^{26}Al in galactic cosmic rays.
M. E. Wiedenbeck.
18th International Cosmic Ray Conference, Vol. 9, p. 147 – 150 (1983). – See Abstr. 012.096.
Satellite observations of the isotopic composition of aluminum in low energy cosmic rays ($E/M \cong 200$ MeV/amu) have been used to determine the abundance of the unstable isotope ^{26}Al ($T_{1/2} = 0.87$ Myr).

144.331 The nitrogen abundance in the cosmic ray source.
W. R. Webber.
18th International Cosmic Ray Conference, Vol. 9, p. 151 – 154 (1983). – See Abstr. 012.096.
The author utilizes new cosmic ray data on the isotopic and charge abundance of N nuclei, along with new cross section data for the fragmentation of O into N, to re–examine the question of the nitrogen abundance in the cosmic ray source.

144.332 Cosmic rays from Wolf–Rayet stars.
M. Cassé, A. Maeder.
18th International Cosmic Ray Conference, Vol. 9, p. 159 – 162 (1983). – See Abstr. 012.096.
Wolf–Rayet stars possess stellar winds having high density and high velocity. They seem to be well suited to accelerate their own material to cosmic ray energies. The authors discuss a few delicate points attached to the WR scenario.

144.333 Detection of cosmic ray electrons above 10^{14}eV using gamma ray observatories.
S. A. Stephens.
18th International Cosmic Ray Conference, Vol. 9, p. 163 – 166 (1983). – See Abstr. 012.096.
A quantitative evaluation of high energy gamma ray observatories for the study of cosmic ray electrons is made. It is shown that one may be able to set useful upper limits to the flux of electrons by making use of the high energy gamma ray detector in the GRO.

144.334 Upper limit to antiproton flux in cosmic radiation above 100 GeV using muon charge ratio.
S. A. Stephens.
18th International Cosmic Ray Conference, Vol. 9, p. 167 – 170 (1983). – See Abstr. 012.096.
The author has estimated upper limits to the fraction of antiprotons in cosmic radiation from the observed charge ratio of muons at sea–level.

144.335 Adiabatic expansion of cosmic ray sources and the consequences for secondary antiprotons.
B. G. Mauger, S. A. Stephens.
18th International Cosmic Ray Conference, Vol. 9, p. 171 – 174 (1983). – See Abstr. 012.096.
Adiabatic energy losses due to the supernova expansion will increase the flux of low energy antiprotons. The authors examine the antiproton flux from such sources.

144.336 Primary cosmic ray spectrum above 10^{12}eV from Cerenkov light images.
P. K. MacKeown, M. F. Cawley, J. Clear, D. J. Fegan, R. C. Lamb, K. E. Turver, T. C. Weekes.
18th International Cosmic Ray Conference, Vol. 9, p. 175 – 178 (1983). – See Abstr. 012.096.

144.337 Energy spectra of cascades produced by primary cosmic rays in emulsion chambers exposed in the stratosphere.
V. G. Abulova, L. A. Hein, K. V. Mandritskaya, G. P. Sazhina, N. V. Sokolskaya, E. S. Troshina, A. Ya. Varkovitskaya, E. A. Zamchalova, V. I. Zatsepin.
18th International Cosmic Ray Conference, Vol. 9, p. 179 – 182 (1983). – See Abstr. 012.096.
About 1000 cascades were registered. For 135 events with energy higher than 3 TeV the type of primary particle is determined (proton, α–particle, heavy nucleus).

144.338 Composition of primary cosmic rays at energies $10^{15} - 10^{16}$eV. Analysis of low energy muons in air showers.
G. B. Yodh, J. A. Goodman, S. C. Tonwar, R. W. Ellsworth.
18th International Cosmic Ray Conference, Vol. 9, p. 183 – 186 (1983). – See Abstr. 012.096.

The authors present a comparison of observations with the expectations from simulation of showers and experimental conditions and discuss the sensitivity of the results to composition of primary cosmic rays.

144.339 Composition of primary cosmic rays at energies $\sim 10^{15} - 10^{16}$eV. Analysis of high energy muons in air showers.
G. B. Yodh, J. A. Goodman, S. C. Tonwar, R. W. Ellsworth.
18th International Cosmic Ray Conference, Vol. 9, p. 187 – 190 (1983). – See Abstr. 012.096.

The authors compare experimental data on high energy muons with expectations from their simulations.

144.340 Composition of primary cosmic rays at $10^{14} - 10^{16}$eV.
B. S. Acharya, M. V. S. Rao, K. Sivaprasad, B. V. Sreekantan.
18th International Cosmic Ray Conference, Vol. 9, p. 191 – 194 (1983). – See Abstr. 012.096.

The average mass number of primary cosmic rays reduces from ~ 10 to ~ 1 at around 2×10^{15}eV. The results of other EAS experiments which suggest a different composition are also discussed.

144.341 Energy spectrum of primary cosmic rays in the $10^{15} - 10^{16}$eV energy range.
G. B. Khristiansen, Yu. A. Fomin, N. A. Aliev, T. A. Alimov, N. Kh. Khakimov, M. K. Kakhkharov, B. M. Machmudov, N. R. Rakhimova, R. T. Tashpulatov.
18th International Cosmic Ray Conference, Vol. 9, p. 195 – 197 (1983). – See Abstr. 012.096.

The main feature of the experiment was that the showers were selected using the readings of Cerenkov detectors which are directly relevant to the primary energies of showers, rather than the readings of charged–particle detectors. The authors constructed the integral energy spectrum of cosmic rays in the $10^{15} - 10^{16}$eV energy range.

144.342 Primary energy spectrum between $10^{14.5}$ and 10^{18}eV estimated from electron and muon size spectra at Akeno.
T. Hara, N. Hayashida, M. Honda, F. Ishikawa, K. Kamata, T. Kifune, Y. Mizumoto, M. Nagano, G. Tanahashi, M. Teshima.
18th International Cosmic Ray Conference, Vol. 9, p. 198 – 201 (1983). – See Abstr. 012.096.

Size spectra of electrons and muons are obtained from Akeno AS experiment. Primary spectrum estimated from both spectra agrees with each other.

144.343 The cosmic ray spectrum at $E > 10^{17}$eV.
B. Cady, G. L. Cassiday, J. W. Elbert, P. R. Gerhardy, E. C. Loh, Y. Mizumoto, M. H. Salamon, P. Sokolsky, D. Steck.
18th International Cosmic Ray Conference, Vol. 9, p. 202 – 205 (1983). – See Abstr. 012.096.

Extensive air showers have been recorded for approximately 1000 hours during 1981 and 1982. At $E > 10^{17}$eV the integral spectrum slope and intensity obtained are $\gamma = 2.04 \pm .05$ and $I(>E) = 2.02 \pm .17 \times 10^{-10}\text{m}^{-2}\text{sr}^{-1}\text{s}^{-1}$ respectively.

144.344 Estimation of the EAS primary energy and cosmic ray energy spectrum above 3×10^{17}eV.
D. D. Krasil'nikov, M. N. D'yakonov, A. A. Ivanov, V. A. Kolosov, I. Ye. Sleptsov (*I. E. Sleptsov*).
18th International Cosmic Ray Conference, Vol. 9, p. 206 (1983). Abstract. – See Abstr. 012.096.

144.345 On estimating the energy of giant air–shower primaries.
A. J. Bower, G. Cunningham, J. Linsley, J. C. Perrett, R. J. O. Reid, A. A. Watson.
18th International Cosmic Ray Conference, Vol. 9, p. 207 – 210 (1983). – See Abstr. 012.096.

The principal methods currently used to evaluate the energy of giant air–shower primaries detected by particle detector arrays are discussed in the light of the conflict between the recently reported spectrum from Yakutsk and the results of experiments at Haverah Park and Volcano Ranch.

144.346 EAS arrival direction and muon to electron ratio.
T. Hara, N. Hayashida, M. Honda, K. Kamata, T. Kifune, M. Nagano, G. Tanahashi, M. Teshima.
18th International Cosmic Ray Conference, Vol. 9, p. 211 – 214 (1983). – See Abstr. 012.096.

The anisotropy of the ratio of muon to electron size is looked for. The showers with rich muon content seem to be preferentially from the direction of about 250° in right ascension.

144.347 Celestial arrival directions of primary cosmic rays of median energy 2×10^{15}eV.
M. M. Abdullah, F. Ashton, J. Fatemi.
18th International Cosmic Ray Conference, Vol. 9, p. 215 – 218 (1983). – See Abstr. 012.096.

An analysis of the arrival directions of 3,575 extensive air showers of median energy 2×10^{15}eV shows that the arrival directions are consistent with an isotropic distribution in galactic coordinates to the precision of the present measurements.

144.348 Anisotropy of intensity of cosmic rays with $E_0 > 10^{17}$eV.
D. D. Krasilnikov, A. A. Ivanov, V. A. Kolosov, A. D. Krasilnikov, K. N. Makarov, V. N. Pavlov, I. Ye. Sleptsov (*I. E. Sleptsov*), F. K. Shamsutdinova, G. G. Struchkov, T. A. Yegorov (*T. A. Egorov*), V. P. Yegorova (*V. P. Egorova*).
18th International Cosmic Ray Conference, Vol. 9, p. 223 – 226 (1983). – See Abstr. 012.096.

The sharp changes of anisotropy parameters with energy E_0 and the intensity specific correlations with galactic frame structure are observed, which indicate probable galactic origin of cosmic rays of $E_0 \lesssim 10^{20}$eV.

144.349 Search for a primary cosmic ray anisotropy near 10^{18}eV.
B. Cady, G. L. Cassiday, J. W. Elbert, P. R. Gerhardy, E. C. Loh, Y. Mizumoto, P. Sokolsky, D. Steck.
18th International Cosmic Ray Conference, Vol. 9, p. 227 – 230 (1983). – See Abstr. 012.096.

The results include weak support for increased cosmic ray intensities at lower galactic latitudes for showers below 10^{19}eV.

144.350 The energy spectrum of cosmic ray electrons in radio loop III.
V. A. Dogiel (*V. A. Dogel'*), C. J. Mayer, J. L. Osborne.
18th International Cosmic Ray Conference, Vol. 9, p. 231 – 234 (1983). – See Abstr. 012.096.

Assuming that loop III is the shell of an old supernova remnant the authors investigate models that account both for the radio spectral index and brightness temperature while remaining consistent with soft X–ray surveys of this region of the sky.

144.351 Stability of hydrostatic equilibrium in the galactic halo: cosmic ray effect.
V. D. Kuznetsov, V. S. Ptuskin.
18th International Cosmic Ray Conference, Vol. 9, p. 235 – 238 (1983). – See Abstr. 012.096.

144.352 On the production of ^{50}V by cosmic ray spallation reactions.
N. Prantzos, H. Reeves, M. Cassé.
18th International Cosmic Ray Conference, Vol. 9, p. 239 – 242 (1983). – See Abstr. 012.096.
For the galactic cosmic ray fluence the authors obtain only 7% of the observed solar system abundance of ^{50}V.

144.353 Diffusion of cosmic rays from local galactic sources.
A. Ghosh, L. I. Dorman, V. S. Ptuskin.
18th International Cosmic Ray Conference, Vol. 9, p. 243 – 246 (1983). – See Abstr. 012.096.
Cosmic ray propagation in the vicinities of 1 kpc from the Sun is considered. The data on the $10^{12} - 10^{15}$eV particle anisotropy, on the 10^{12}eV electron spectrum, and on temporal cosmic ray variations are analysed.

144.354 Magnetohydrodynamic turbulence in the cosmic ray confinement regions of the Galaxy.
A. M. Bykov, I. N. Toptygin.
18th International Cosmic Ray Conference, Vol. 9, p. 247 – 250 (1983). – See Abstr. 012.096.
Observational data and theoretical models of magneto-hydrodynamic turbulence generation in the interstellar medium and supernova remnants are considered. The role of shock waves and cosmic rays for turbulence generation is discussed.

144.355 The propagation of cosmic rays in the Galaxy: further evidence for a ”nested leaky box”.
J. Tang, D. Müller.
18th International Cosmic Ray Conference, Vol. 9, p. 251 – 254 (1983). – See Abstr. 012.096.
A propagation model in form of a double leaky box is presented. The model requires few adjustable parameters but assumes that all primary cosmic ray species are accelerated in form of a common power law in momentum above 100 MeV/c. The authors show that this model describes correctly the observed nuclear composition and the energy spectra of electrons and nuclei.

144.356 Quantitative estimates of the effects of cross section uncertainties on the derivation of GCR source composition.
G. F. Hinshaw, M. E. Wiedenbeck.
18th International Cosmic Ray Conference, Vol. 9, p. 263 – 266 (1983). – See Abstr. 012.096.
The authors have found that under several sets of reasonable assumptions about the magnitudes and correlations of cross section uncertainties it is possible to easily obtain estimates of the resulting source composition uncertainties.

144.357 The cosmic ray source composition.
S. H. Margolis.
18th International Cosmic Ray Conference, Vol. 9, p. 267 – 270 (1983). – See Abstr. 012.096.
The cosmic ray source composition is shown to be related to the observed cosmic ray abundances by a set of linear equations which are valid for either ad hoc pathlength distributions or quantitiative solutions to galactic propagation models.

144.358 The energy dependence of the secondary to primary ratio.
S. H. Margolis, R. W. Bussard.
18th International Cosmic Ray Conference, Vol. 9, p. 271 – 274 (1983). – See Abstr. 012.096.
The energy dependence of the secondary to primary ratios in the cosmic rays can be explained by a model accounting for the scattering of the particles by self–generated Alfvén waves. This model predicts an energy–dependent truncation of the pathlength distribution.

144.359 Energy spectra of cosmic rays at source as derived from the French–Danish experiment on HEAO–3.
L. Koch–Miramond, J. J. Engelmann, P. Goret, E. Juliusson, P. Masse, A. Soutoul, C. Perron, N. Lund, I. L. Rasmussen.
18th International Cosmic Ray Conference, Vol. 9, p. 275 – 278 (1983). – See Abstr. 012.096.
Cosmic ray spectra at source have been derived from the spectra observed near Earth. Calculations have been performed in the framework of the leaky box model. Carbon and oxygen do not seem to have the same spectral shape at source.

144.360 Abundances of 'secondary' elements among the ultra-heavy cosmic rays – results from HEAO-3.
J. Klarmann, W. R. Binns, M. H. Israel, S. H. Margolis, T. L. Garrard, E. C. Stone, N. R. Brewster, D. J. Fixsen, C. J. Waddington.
18th International Cosmic Ray Conference, Vol. 9, p. 279 – 282 (1983). – See Abstr. 012.096.
The elements with atomic number (Z) in the intervals $44 \leqslant Z \leqslant 48$ and $62 \leqslant Z \leqslant 74$ arriving at earth are expected to have significant secondary components. The results are consistent with solar system abundances modified for first ionization potential with possibly some enhancement of the r to s ratio.

144.361 The ultraheavy cosmic rays: propagation and selective acceleration.
S. H. Margolis, J. B. Blake.
18th International Cosmic Ray Conference, Vol. 9, p. 283 – 286 (1983). – See Abstr. 012.096.
The purpose of this work is not to derive source abundances, but to distinguish the effects of propagation from nucleosynthesis and preferential acceleration where possible.

144.362 Cosmic ray acceleration efficiency in shock waves with wave dissipation.
H. J. Völk, L. O'C. Drury, J. F. McKenzie.
18th International Cosmic Ray Conference, Vol. 9, p. 287 – 288 (1983). – See Abstr. 012.096.
This paper outlines an attempt to provide more realistic estimates of the efficiency of diffusive acceleration of cosmic rays in shock waves.

144.363 Estimates of galactic cosmic ray spectra at low energies.
W.–H. Ip, W. I. Axford.
18th International Cosmic Ray Conference, Vol. 9, p. 289 – 292 (1983). – See Abstr. 012.096.
A simple model including shock acceleration, energy loss, and propagation loss is employed to calculate the energy spectra of cosmic ray protons, alpha particles, and electrons between 1 MeV/nuc and 100 GeV/nuc. On the basis of these theoretical results, the energy density and pressure density of the cosmic ray particles are also computed.

144.364 Cosmic ray sources and propagation.
W. Kundt.
18th International Cosmic Ray Conference, Vol. 9, p. 297 – 300 (1983). – See Abstr. 012.096.
It is argued that cosmic rays are essentially of galactic origin, up to the highest observed energies. Their ionic component is boosted by young binary neutron stars (grindstone effect); their electron–positron component comes mainly from young pulsars. Injection into the magnetic flux tubes of the galactic disk occurs through the heads of relativistic jets, via field–line crossing, and may be rigidity–dependent.

144.365 Cosmic rays from the galactic centre.
M. Giler, T. Wibig.
18th International Cosmic Ray Conference, Vol. 9, p. 301 – 304 (1983). – See Abstr. 012.096.
The authors consider the possibility that cosmic rays may originate in the galactic centre. They restrain their attention mainly to the problem of mean grammage and grammage distribution of cosmic ray particles.

144.366 Correlation of source abundances of ultraheavy cosmic rays with first ionization potential – results from HEAO–3.
M. H. Israel, W. R. Binns, D. P. Grossman, J. Klarmann, S. H. Margolis, E. C. Stone, T. L. Garrard, K. E. Krombel, N. R. Brewster, R. K. Fickle, C. J. Waddington.
18th International Cosmic Ray Conference, Vol. 9, p. 305 – 308 (1983). – See Abstr. 012.096.
The HEAO–3 heavy nuclei experiment yields the abundances of many individual elements of even atomic number (Z) in the interval $26 \leqslant Z \leqslant 58$. The cosmic–ray–source (CRS) abundances, which have been derived from these observations using standard propagation models is compared with solar–system (SS) abundances. The correlation of the CRS/SS ratio with first ionization potential is presented.

144.367 Particle charge evolution during acceleration and implications on mass and charge spectra.
J. Pérez–Peraza, J. Martinell, A. Villareal.
18th International Cosmic Ray Conference, Vol. 9, p. 309 – 312 (1983). – See Abstr. 012.096.

144.368 Cosmic rays acceleration by large scale compressible and incompressible motions of plasma.
A. M. Bykov, I. N. Toptygin.
18th International Cosmic Ray Conference, Vol. 9, p. 313 – 316 (1983). – See Abstr. 012.096.
Cosmic rays (CR) acceleration by large scale motion with account of effectively particles scattering by small scale turbulence, magnetic fields and shock waves is considered. Some features of particle acceleration by compressible and incompressible motions are discussed. It is shown that this acceleration mechanism plays an important role for CR acceleration problems in young and old supernova remnants.

144.369 Temporal ^{10}Be variations.
J. Beer, U. Siegenthaler, H. Oeschger, M. Andrée, G. Bonani, M. Suter, W. Wölfli, R. C. Finkel, C. C. Langway.
18th International Cosmic Ray Conference, Vol. 9, p. 317 – 320 (1983). – See Abstr. 012.096.
The authors have studied ^{10}Be during the period 1200 – 1800 A.D. including the so–called Maunder minimum (1640 – 1715 A.D.). First measurements on an Antarctic ice core by Raisbeck et al. (1981) showed evidence for enhanced ^{10}Be production rate during the Maunder minimum.

144.370 Tree ring radiocarbon and sea sediment thermoluminescence variations.
M. R. Attolini, G. Bonino, G. Cini–Castagnoli, M. Galli.
18th International Cosmic Ray Conference, Vol. 9, p. 321 – 324 (1983). – See Abstr. 012.096.

144.371 Cyclic variation analysis of time series.
M. R. Attolini, S. Cecchini, M. Galli.
18th International Cosmic Ray Conference, Vol. 9, p. 441 – 446 (1983). – See Abstr. 012.096.
The authors present some criteria that can be applied either to test a time series for the existence of non random oscillations of a given period or to find out which non random oscillations are present in a time series, according to a fixed confidence limit. As an example they show a scanning in time and period length for non–random oscillations, performed on the time series of Climax cosmic ray neutron 5–day averages, from 1953 to 1979.

144.372 Study of the albedo of cosmic ray neutron component at energies below 1 MeV.
Ya. L. Blokh, S. I. Rogovaya.
18th International Cosmic Ray Conference, Vol. 9, p. 448 – 451 (1983). – See Abstr. 012.096.
The report deals with the properties of cosmic ray neutron reflectances at energies below 1 MeV from semi–infinite and finite scatterers.

144.373 22–year variation of cosmic ray intensity in the stratosphere.
E. A. Chebakova, V. A. Likhoded, E. V. Kolomeets, V. M. Erkhov, N. V. Stekolnikov.
18th International Cosmic Ray Conference, Vol. 10, p. 1 – 4 (1983). – See Abstr. 012.096.
Stratosphere data were obtained during the period 1958 – 1982. Analysis of experimental data revealed 11–year wave of cosmic ray intensity.

144.374 Solutions of the two–dimensional steady–state cosmic ray transport equation in interplanetary space.
M. S. Potgieter, H. Moraal.
18th International Cosmic Ray Conference, Vol. 10, p. 5 – 8 (1983). – See Abstr. 012.096.
The authors present further numerical solutions of the two–dimensional steady–state cosmic ray transport equation in a three–dimensional axisymmetric heliosphere. They illustrate the effect of the inclusion of particle drift and compare the results with some experimental data.

144.375 Application of time–reversed method in C. R. modulation.
J. Kóta.
18th International Cosmic Ray Conference, Vol. 10, p. 9 (1983). Abstract. – See Abstr. 012.096.

144.376 Transition to subsonic streaming in nonlinear model for galactic comic ray modulation by solar wind.
V. Kh. Babayan, L. I. Dorman.
18th International Cosmic Ray Conference, Vol. 10, p. 10 – 13 (1983). – See Abstr. 012.096.
The studies of solar wind interaction with interstellar medium are made considering the following five possible factors affecting the propagation of supersonic solar wind: (1) static pressure of interstellar gas, (2) pressure of galactic magnetic field, (3) dynamic pressure of interstellar gas due to peculiar motion of solar system, (4) charge exchange of solar wind protons with interstellar neutral hydrogen, (5) pressure of galactic cosmic rays.

144.377 The effects of inversion of total solar magnetic and N–S asymmetry of solar activity in cosmic ray intensity.
E. V. Kolomeets, A. Kh. Lyakhona, N. V. Stekolnikov.
18th International Cosmic Ray Conference, Vol. 10, p. 14 – 17 (1983). – See Abstr. 012.096.
N–S asymmetry was examined on the basis of experimental data obtained at the various depths of the Earth atmosphere over south and north poles. The amplitude of asymmetry was shown to increase with the depth. The paper presents the data on N–S asymmetry of solar activity for the period 1954 – 1980.

144.378 Temporal and energetic characteristics of additional flux of particles in the stratosphere.
F. B. Aitbaev, B. N. Dyusembaev, V. I. Erkhov, E. V. Kolomeets, V. A. Likhoded, N. V. Stekolnikov.
18th International Cosmic Ray Conference, Vol. 10, p. 18 – 21 (1983). – See Abstr. 012.096.
Additional flux of particles was shown to be present in the atmosphere during 1972 – 1980 and absent in 1962 – 1971.

144.379 Integral radial cosmic–ray gradients in the solar system from 1972 – 1982.
W. R. Webber, J. A. Lockwood.
18th International Cosmic Ray Conference, Vol. 10, p. 22 – 25 (1983). – See Abstr. 012.096.
The cosmic–ray telescope data on IMP 8, Voyager 1 and 2 and Pioneer 10 have been used to determine the integral radial cosmic–ray intensity gradients for energies greater than 60 MeV/nucleon.

144.380 Suprathermal H$^+$ and He^{++} in the distant geomagnetic tail: ISEE–3 observations.
G. Gloeckler, F. M. Ipavich, D. Hovestadt, B. Klecker, M. Scholer, C. Y. Fan.
18th International Cosmic Ray Conference, Vol. 10, p. 26 – 29 (1983). – See Abstr. 012.096.

144.381 Short and long term variations of the anomalous component.
A. C. Cummings, R. A. Mewaldt, W. R. Webber.
18th International Cosmic Ray Conference, Vol. 10, p. 30 – 33 (1983). – See Abstr. 012.096.

The authors have used data from the cosmic ray experiment (CRS) on Voyagers 1 and 2 to examine anomalous O and He in the time period from launch in 1977 to mid–1982.

144.382 Energetic He$^+$ and He^{++} in the earth's outer radiation belt.
L. Ma Sung, G. Gloeckler, D. Hovestadt, B. Klecker.
18th International Cosmic Ray Conference, Vol. 10, p. 34 – 37 (1983). – See Abstr. 012.096.

The authors have examined the characteristics of He^{++} and He$^+$ ions during 10 equatorial passes of ISEE–1. Large abundances of He$^+$ ions (He$^+$:He^{++} $\sim$ 0.5 – 1.0) are discovered during a number of these passes.

144.383 Comparative analysis of various solar activity indices describing the long–term cosmic ray variations.
R. T. Gushchina, N. S. Kaminer, A. V. Yakovlev.
18th International Cosmic Ray Conference, Vol. 10, p. 38 – 41 (1983). – See Abstr. 012.096.

144.384 Galactic cosmic ray variations in the heliosphere according to radioactivity of meteorites.
M. Alania (*M. V. Alaniya*), L. I. Dorman, A. K. Lavrukhina, G. Ustinova.
18th International Cosmic Ray Conference, Vol. 10, p. 42 – 45 (1983). – See Abstr. 012.096.

The results of the investigation of cosmogenic radionuclides in meteorites testify to the fact that the galactic cosmic ray modulation range in the heliosphere is not spherical, but it is compressed in the meridional plane, mainly, within the heliolatitudinal region of $\pm 40°$.

144.385 Further studies of eleven year variation of cosmic rays.
H. S. Ahluwalia, T. Majeed, J. R. Asbridge, S. J. Bame.
18th International Cosmic Ray Conference, Vol. 10, p. 46 (1983). Abstract. – See Abstr. 012.096.

144.386 Quasi–quinquennial variation in cosmic ray intensity.
J. A. Otaola, R. P. Enríquez.
18th International Cosmic Ray Conference, Vol. 10, p. 47 – 50 (1983). – See Abstr. 012.096.

Clear evidence for the existence of a quasi–quinquennial variation in cosmic ray intensity is shown by an analysis of monthly averages of the neutron monitor intensity observed at various observatories over the period 1958 – 1980. The variation is well correlated with a corresponding variation in solar activity. It is suggested that a small change in the heliospheric boundary location of approximately 0.7 AU could produce this variation.

144.387 Twenty–two year cycle in the cosmic ray quasi–quinquennnial variation.
R. P. Enríquez, J. A. Otaola.
18th International Cosmic Ray Conference, Vol. 10, p. 51 – 54 (1983). – See Abstr. 012.096.

It is shown, from the data obtained during two complete sunspot cycles, that a 22 yr variation in the phase of the cosmic ray quasi–quinquennial wave exists. The changes of this variation in cosmic ray intensity as a function of a corresponding solar activity variation show that the phase lag between these two waves is changing over the solar activity cycle. The results are interpreted in terms of the solar magnetic field inversion.

144.388 Long term changes in the solar diurnal variation of cosmic rays and cosmic ray density gradients at high rigidities.
D. B. Swinson.
18th International Cosmic Ray Conference, Vol. 10, p. 55 – 58 (1983). – See Abstr. 012.096.

Analysis of the solar diurnal variation up to 1982 shows that the phase has returned to the later phase observed during the 1960's following the reversal of the sun's polar magnetic field again in 1980 – 81. The data are also examined to search for the presence of a cosmic ray density gradient perpendicular to the ecliptic plane during this period.

144.389 Comparison of the latitude distributions of cosmic rays for the periods of solar minimum during 1954, 1965 and 1976.
A. J. Van der Walt, P. H. Stoker, B. C. Raubenheimer.
18th International Cosmic Ray Conference, Vol. 10, p. 59 – 62 (1983). – See Abstr. 012.096.

Using a new curve fitting technique, the 22 year effect on cosmic ray differential response functions during the periods of cosmic ray maximum as found by Potgieter et al. (1980), is confirmed for sea–level latitude survey data.

144.390 Diffusion approximation in cosmic ray kinetics.
V. Kh. Shogenov.
18th International Cosmic Ray Conference, Vol. 10, p. 67 – 70 (1983). – See Abstr. 012.096.

The equation of cosmic ray diffusion is derived from the Boltzmann kinetic equation. The magnetic inhomogeneities which scatter cosmic rays are treated to be certain moving scatterers.

144.391 Quasistationary asymmetry of the galactic cosmic ray density distribution in the heliosphere.
G. K. Ustinova.
18th International Cosmic Ray Conference, Vol. 10, p. 71 – 74 (1983). – See Abstr. 012.096.

The results of the investigation of cosmogenic radionuclides in meteorites evidence that the 11–year periods of the maximum north–south asymmetry in the galactic cosmic ray density distribution in the heliosphere are replaced by the 11–year periods of the minimum one, and, on the average over a million years some residual asymmetry, apparently, exists.

144.392 Long–period changes of large–scale cosmic ray fluctuations.
Ya. L. Blokh, A. M. Lemberger, I. Ya. Libin, K. F. Yudakhin, I. M. Belkin.
18th International Cosmic Ray Conference, Vol. 10, p. 75 – 77 (1983). – See Abstr. 012.096.

Results of one– and many–dimensional spectral analyses of large–scale cosmic ray fluctuations with periods exceeding 2 hours in 1966 – 1969 and 1977 – 1981 are presented.

144.393 Development of the concepts concerning the global cosmic ray modulation in interplanetary space.
I. V. Dorman.
18th International Cosmic Ray Conference, Vol. 10, p. 82 – 86 (1983). – See Abstr. 012.096.

Different models for global modulation of cosmic rays in interplanetary space are reviewed critically.

144.394 Cosmic ray gradient and anisotropy according to the "Prognoz–3" an "Mars–7" data.
Yu. V. Mineev, E. S. Spirkova.
18th International Cosmic Ray Conference, Vol. 10, p. 87 – 90 (1983). – See Abstr. 012.096.

Four recurrent cosmic ray increases are analysed. It is established that the increases are connected with the interplanetary medium parameters. In the four increases the electron flux gradient at a distance as high as 1.5 a.u. is negative and equal to $\sim 300\%$ per a.u. These protons are moving to the solar system from the interstellar medium.

144.395 The effect of the particle drift in cosmic ray anisotropy.
M. V. Alania (*M. V. Alaniya*), R. G. Aslamazashvili,
T. V. Djapiashvili (*T. V. Dzhapiashvili*), V. S. Tkemaladze.
18th International Cosmic Ray Conference, Vol. 10, p. 91 – 94
(1983). – See Abstr. 012.096.

It is shown that the effect of the particle drift in the sectors of
the regular magnetic field of the Sun is manifested single-
valuedly for the minimum of solar activity. The 22–year change
of the anisotropy phase connected with the global sign change of
the magnetic field of the Sun is revealed.

**144.396 Energetic particle events associated with Forbush de-
creases.**
C. Bagnoli, C. Paizis, N. Iucci, M. Parisi, M. Storini,
G. Villoresi, T. T. Von Rosenvinge.
18th International Cosmic Ray Conference, Vol. 10, p. 95 – 98
(1983). – See Abstr. 012.096.

The authors investigate the influence of the magnetic perturbations producing Forbush decreases on the low–energy particle
fluxes.

**144.397 Diffusive first–order Fermi acceleration at quasi-
parallel interplanetary shocks: injection of thermal ions.**
D. C. Ellison.
18th International Cosmic Ray Conference, Vol. 10, p. 108 – 111
(1983). – See Abstr. 012.096.

The acceleration process is modeled with a Monte Carlo simulation where the particle acceleration time is comparable to the
shock lifetime and the scattering mfp is proportional to rigidity.
Good agreement between the simulation and observations suggests that diffusive quasi–parallel shocks can accelerate thermal
ions directly from the solar wind.

**144.398 Characteristics of particle events associated with
interplanetary shocks.**
K.-P. Wenzel, R. Reinhard, T. R. Sanderson, E. T. Sarris.
18th International Cosmic Ray Conference, Vol. 10, p. 112 – 115
(1983). – See Abstr. 012.096.

The authors present observations of the intensity–time profiles
and the three–dimensional anisotropy distributions of
35 – 56 keV protons observed during two interplanetary shock
events on ISEE–3. They discuss the observations in the light of
current particle acceleration models.

**144.399 ISEE–3 observations of 35 – 1600 keV protons and low
frequency waves upstream of an interplanetary shock.**
T. R. Sanderson, K.-P. Wenzel, E. J. Smith, B. T. Tsurutani.
18th International Cosmic Ray Conference, Vol. 10, p. 116 – 119
(1983). – See Abstr. 012.096.

For several hours prior to the arrival at the spacecraft of the
quasi–parallel shock of April 5, 1979, the authors observed enhanced fluxes of 35 – 1600 keV protons and large amplitude, low
frequency waves. During most of this period the spectrum was
dominated by particles with energies of ~ 200 keV. The r.m.s.
amplitude of the waves is well correlated with the proton intensity
at these energies, suggesting a close relation between the waves
and the particles.

**144.400 Steplike and shock related changes in the long term cos-
mic ray modulation.**
P. J. Ankiewicz, P. H. Stoker, H. Moraal.
18th International Cosmic Ray Conference, Vol. 10, p. 120 – 123
(1983). – See Abstr. 012.096.

Steplike changes in the modulation of cosmic rays can be identified when the counting rates of neutron monitors at high and
low cutoff rigidities are compared. It is suggested that these steplike changes are caused when cosmic rays are accelerated by a
series of outward moving shock waves during transport of the
cosmic rays from the heliospheric boundary to Earth.

**144.401 A technique to study the cosmic ray anisotropy during
the days of Forbush decreases.**
R. S. Yadav, D. S.'Rana.
18th International Cosmic Ray Conference, Vol. 10, p. 128 – 131
(1983). – See Abstr. 012.096.

The authors report a new technique for separating the world–
wide component of a Forbush decrease (Fd) from the pressure
corrected hourly cosmic ray intensity data which provides an
exact shape of Fd profiles. This effort will certainly elucidate the
basic modulation mechanism of Fds.

**144.402 The best–fit rigidity spectra for the first and second zonal
harmonic components of the spherical harmonic analysis
of the neutron monitor data.**
H. Takahashi, N. Yahagi, T. Chiba.
18th International Cosmic Ray Conference, Vol. 10, p. 132 – 135
(1983). – See Abstr. 012.096.

**144.403 Investigation of the cosmic ray north–south anisotropy
by means of the three–dimensional analysis method.**
H. Takahashi, N. Yahagi, T. Chiba.
18th International Cosmic Ray Conference, Vol. 10, p. 136 – 139
(1983). – See Abstr. 012.096.

**144.404 Semi–annual variation of the cosmic ray equator–pole
anisotropy.**
H. Takahashi, N. Yahagi, T. Chiba.
18th International Cosmic Ray Conference, Vol. 10, p. 148 – 151
(1983). – See Abstr. 012.096.

**144.405 The recurrent variations of cosmic ray intensity and
anisotropy.**
N. A. Nachkebia (*N. A. Nachkebiya*), L. Kh. Shatashvili.
18th International Cosmic Ray Conference, Vol. 10, p. 152 – 155
(1983). – See Abstr. 012.096.

The interpretation of the studied cosmic ray variations is that
the sporadic Forbush decreases originate as a result of the interactions of cosmic rays with the separate fluxes of the localized
formations in solar wind while the recurrent Forbush decreases
are formed by the long–lived quasistationary enhanced fluxes of
solar wind, whose dimensions were more extended in the interplanetary space.

**144.406 Cosmic ray intensity gradients in the radial distance
1 – 13 AU as determined from a comparative study of
observations by spacecraft Voyagers 1 and 2, and earth–orbiting
satellite IMP–8.**
D. Venkatesan, R. B. Decker, S. M. Krimigis.
18th International Cosmic Ray Conference, Vol. 10, p. 156 – 159
(1983). – See Abstr. 012.096.

144.407 Forbush decreases at high rigidities.
G. Erdös, J. Kóta.
18th International Cosmic Ray Conference, Vol. 10, p. 160 – 163
(1983). – See Abstr. 012.096.

The method of calculating energy losses along regular trajectories in a model Interplanetary Magnetic Field is extended to
describe Forbush events at high rigidities. Computed intensity
profiles of 70 GV protons are presented for both halves of the
22–year solar magnetic cycle.

**144.408 The rigidity dependence of the July 13, 1982, Forbush
decrease.**
A. G. Fenton, K. B. Fenton, J. E. Humble.
18th International Cosmic Ray Conference, Vol. 10, p. 164 – 167
(1983). – See Abstr. 012.096.

Cosmic ray data are given and the solar and interplanetary
conditions during this event are examined.

144.409 On the nature of cosmic ray Forbush effect.
M. V. Alania (*M. V. Alaniya*), D. P. Bochikashvili.
18th International Cosmic Ray Conference, Vol. 10, p. 168 – 171
(1983). – See Abstr. 012.096.

The authors show that the shock waves from the chromospheric flares in the first approximation can be divided mainly

into two types by the data of cosmic ray Forbush effect: 1) the shock waves that terminated their development before the Earth's orbit and whose cavities have no sharp boundaries and 2) the shock waves that have sharp boundaries and extend intensively not only in the region of the Earth's orbit, but at large distances from the Sun too.

144.410 Time evolution of very rapid cosmic ray variations during quiet and perturbed period.
M. R. Attolini, S. Cecchini, C. Cavani, M. Galli.
18th International Cosmic Ray Conference, Vol. 10, p. 172 (1983). Abstract. – See Abstr. 012.096.

144.411 Dynamic spectral analysis of cosmic ray fluctuations on time scales 1c/d to 1c/30d.
M. R. Attolini, S. Cecchini, M. Galli.
18th International Cosmic Ray Conference, Vol. 10, p. 173 (1983). Abstract. – See Abstr. 012.096.

144.412 A search for heliosphere pulsation in the range 1c/yr to 1c/10 yr.
M. R. Attolini, S. Cecchini, M. Galli.
18th International Cosmic Ray Conference, Vol. 10, p. 174 – 177 (1983). – See Abstr. 012.096.
The authors investigate the possibility of detecting significant signature of long term variations and of other periodicities at frequency greater than 1c/yr, by analysing the time evolution of the recurrence index of CR fluctuations.

144.413 Cosmic ray anisotropy relevant to high–velocity solar wind fluxes.
N. S. Kaminer, A. E. Kuzmicheva, N. V. Mymrina.
18th International Cosmic Ray Conference, Vol. 10, p. 178 – 181 (1983). – See Abstr. 012.096.
The behaviour of galactic cosmic rays in the periods when the Earth traverses high–velocity solar wind fluxes (HVF) is examined. The magnetic field in HVF is shown to be of more regular character than in steady solar wind.

144.414 Solar diurnal variation of cosmic rays related to the interplanetary magnetic field.
M. Ichinose, K. Nagashima, I. Morishita, Y. Ishida.
18th International Cosmic Ray Conference, Vol. 10, p. 182 – 185 (1983). – See Abstr. 012.096.

144.415 Long–term changes in the solar diurnal variation.
A. G. Fenton, J. E. Humble, T. Thambyahpillai.
18th International Cosmic Ray Conference, Vol. 10, p. 186 – 189 (1983). – See Abstr. 012.096.

144.416 Persistent cosmic ray solar diurnal variations observed during 1973 – 79.
J. F. Riker, H. S. Ahluwalia.
18th International Cosmic Ray Conference, Vol. 10, p. 190 – 193 (1983). – See Abstr. 012.096.
Hourly cosmic ray data obtained with University of New Mexico underground vertical muon telescopes at Embudo are Fourier analyzed for the period 1973 – 79. The diurnal variation is shown to be composed of persistent and random components.

144.417 Diurnal anisotropy of cosmic rays inferred from the data of the Elbrus spectrograph.
L. M. Baisultanova, S. P. Daova, V. Kh. Shogenov.
18th International Cosmic Ray Conference, Vol. 10, p. 194 – 196 (1983). – See Abstr. 012.096.
The data obtained with the Elbrus spectrograph are used to study the diurnal cosmic ray anisotropy. The amplitude and phases of the diurnal anisotropy are presented. The results obtained are compared with theoretical conclusions.

144.418 Cosmic ray anisotropy and its temporal variations.
G. A. Gonchar, E. V. Kolomeets, N. V. Slyunyaeva.
18th International Cosmic Ray Conference, Vol. 10, p. 197 – 200 (1983). – See Abstr. 012.096.
The direction of anisotropy is shown to be a parameter most sensitive to the variations of solar magnetic field.

144.419 Discrimination of cosmic ray anisotropy components using crossed telescopes.
A. V. Belov, L. I. Dorman, E. A. Eroshenko, V. A. Oleneva.
18th International Cosmic Ray Conference, Vol. 10, p. 201 – 204 (1983). – See Abstr. 012.096.
The variation of the north–south cosmic ray asymmetry in 1973 and the solar–diurnal variation parameters for the anisotropy enhancement in August – September 1974 are found.

144.420 A search for the directional distribution of the cosmic ray intensity by means of super neutron monitors.
T. Chiba, H. Takahashi.
18th International Cosmic Ray Conference, Vol. 10, p. 205 – 208 (1983). – See Abstr. 012.096.
The authors found that there exists a cosmic ray flow of different direction from the familiar cosmic ray flow of 18 hr direction. A preliminary interpretation and discussion of the result are presented.

144.421 Harmonic and spectral analyses of the cosmic ray data.
J. F. Riker, H. S. Ahluwalia.
18th International Cosmic Ray Conference, Vol. 10, p. 209 – 212 (1983). – See Abstr. 012.096.
Hourly cosmic ray data, obtained with underground vertical muon telescopes at Embudo during the period 1973 – 79, are analyzed. Harmonic and spectral analyses are employed to obtain the amplitudes and phases for the solar diurnal variation.

144.422 Solar and sidereal daily waves around maximum solar activity.
G. Benkó, K. Kecskeméty, J. Kóta, E. Merényi, A. Somogyi, A. Varga.
18th International Cosmic Ray Conference, Vol. 10, p. 213 (1983). Abstract. – See Abstr. 012.096.

144.423 The effect of the interplanetary magnetic field on sidereal variations at 46 hg/cm² – a 22 year cycle?
J. E. Humble, A. G. Fenton.
18th International Cosmic Ray Conference, Vol. 10, p. 214 – 217 (1983). – See Abstr. 012.096.
Data from the Hobart underground detectors have been analysed as a function of the inferred direction of the interplanetary magnetic field. There are significant differences between the sidereal variations observed in both field directions in the period 1965 to 1977 and the variations observed in years outside that range.

144.424 Altitude and latitude variations of various cosmic ray components observed by airborne detectors.
K. Takahashi, A. Inoue, S. Kudo, M. Okano.
18th International Cosmic Ray Conference, Vol. 10, p. 218 – 220 (1983). – See Abstr. 012.096.

144.425 Geomagnetic effect observed by the French–Danish cosmic ray instrument on HEAO–3 satellite.
N. Petrou, J. Jorrand.
18th International Cosmic Ray Conference, Vol. 10, p. 221 – 224 (1983). – See Abstr. 012.096.
The authors confirm with the HEAO–3–C2 data their predictions of the shadow cone effect ($70° \leqslant$ zenith $\leqslant 110°$).

144.426 About the relative importance of the Earth's magnetic field to formation of the observed long–period cosmic ray variations.
A. A. Shadov.
18th International Cosmic Ray Conference, Vol. 10, p. 230 – 232 (1983). – See Abstr. 012.096.
The long–period cosmic ray variations produced in the part by the variations of the Earth's magnetic field are studied. The results obtained are compared with the studies of the relative contents of stable and unstable cosmogenic isotopes in the Earth atmosphere.

144.427   Secondary protons ($E_p \geqslant 500$ MeV) under the radiation belts in the equatorial region.
A. A. Gusev, G. I. Pugacheva.
18th International Cosmic Ray Conference, Vol. 10, p. 241 – 244 (1983). – See Abstr. 012.096.

The proton fluxes in excess of the primaries in the equatorial region have been registered in a series of experiments. It is shown that the effect can be associated with the secondary albedo protons.

144.428   Results of latitude observations with a without–lead neutron monitor.
I. V. Dorman, I. Ya. Libin, V. K. Korotkov, Yu. Kh. Khamukov.
18th International Cosmic Ray Conference, Vol. 10, p. 255 – 258 (1983). – See Abstr. 012.096.

The results are presented of the July, 1982 expedition measurements of the latitude dependence of cosmic ray intensity with a without–lead monitor in the 0.5 – 8.1 GV geomagnetic cutoff rigidity range.

144.429   On the temperature effect of the cosmic ray neutron component.
Yu. Ya. Krest'yannikov, V. E. Sdobnov.
18th International Cosmic Ray Conference, Vol. 10, p. 259 – 262 (1983). – See Abstr. 012.096.

144.430   The calculation of cosmic–ray integral and differential response functions and their standard errors.
A. J. Van der Walt.
18th International Cosmic Ray Conference, Vol. 10, p. 287 – 290 (1983). – See Abstr. 012.096.

144.431   Methodical principles for studying cosmic ray variations.
I. V. Dorman.
18th International Cosmic Ray Conference, Vol. 10, p. 291 – 294 (1983). – See Abstr. 012.096.

Consideration is given to the main development stages, the present state, and the development prospects of the methodical principles for studying cosmic ray variations on the basis of the data of ground–based and underground measurements.

144.432   The quasi–two–year variations of galactic cosmic rays by the data of neutron and hard components of the world network of observation stations.
T. V. Djapiashvili (*T. V. Dzhapiashvili*), O. G. Rogava, L. Kh. Shatashvili.
18th International Cosmic Ray Conference, Vol. 10, p. 295 – 298 (1983). – See Abstr. 012.096.

The results of the complex investigation of the quasi–two–year variations of galactic cosmic ray (CR) intensity and anisotropy, of the barometric coefficient of CR neutron component, solar activity, solar wind velocity, geomagnetic field, pressure and temperature of the Earth atmosphere are presented.

144.433   Search for particles with the charge 4e/3 at mountain altitudes.
G. L. Bashindzhagyan, L. I. Sarycheva, N. B. Sinev, V. V. Avakyan, V. S. Eganov.
18th International Cosmic Ray Conference, Vol. 11, p. 16 – 19 (1983). – See Abstr. 012.096.

The present experiment is a continuous search for the singly–charged and fractionally–charged massive particles in the flux of cosmic ray single hadrons. The results support further search for new particles in cosmic rays.

144.434   Search for long–lived massive particles in cosmic ray air showers of energies $10^5 - 10^7$GeV.
A. I. Mincer, H. T. Freudenreich, J. A. Goodman, G. B. Yodh, R. W. Ellsworth, D. Berley.
18th International Cosmic Ray Conference, Vol. 11, p. 20 – 23 (1983). – See Abstr. 012.096.

A comparison is made between data on large–signal, single counter, delayed events obtained at sea level and a Monte Carlo simulation.

144.435   Search for the decays of hypothetic penetrating heavy hadrons in comic rays.
Yu. N. Bazhutov.
18th International Cosmic Ray Conference, Vol. 11, p. 24 – 26 (1983). – See Abstr. 012.096.

The author made an attempt to seek for the decay of long–lived ($\tau > 10^{-6}$s) penetrating heavy (10 GeV < M < 40 GeV) hadrons whose absorption path is $L = (80 \pm 25)$ m.w.e.

144.436   Muon densities of extensive air showers induced by $10^{15} - 2 \times 10^{16}$eV gamma–rays.
M. Samorski, W. Stamm.
18th International Cosmic Ray Conference, Vol. 11, p. 244 – 247 (1983). – See Abstr. 012.096.

Extensive air showers induced by $10^{15} - 2 \times 10^{16}$eV gamma–rays from Cygnus X–3 have been detected with the EAS experiment at Kiel. The muon densities in such showers are not essentially smaller than in normal showers induced by charged primary particles.

144.437   Primary composition determination between 10^5 and 10^8GeV.
P. K. F. Grieder.
18th International Cosmic Ray Conference, Vol. 11, p. 323 – 325 (1983). – See Abstr. 012.096.

The author has been carrying out a large number of Monte Carlo simulations of extensive air showers in an attempt to solve the problem of the primary composition between 10^5 and 10^8GeV. First results of the calculations are presented and compared with experimental data. From these it is concluded that iron is not the dominating component in the primary energy range between 10^6 and 10^7GeV.

144.438   The possibility of searching for cosmic rays in the energy range above 10^{20}eV.
L. G. Dedenko, G. A. Gusev, M. A. Markov, I. M. Zheleznykh.
18th International Cosmic Ray Conference, Vol. 11, p. 420 – 423 (1983). – See Abstr. 012.096.

If the superhigh mass particles predicted by the theories of the "grand unification" really exist then their interactions can create the hadrons and leptons with the energies above 10^{20}eV. The authors suggest to use the radio emission reflected and scattered by the Earth surface and then by the Earth's ionosphere to detect EAS in the energy range above 10^{20}eV.

144.439   New EAS project to study primary composition at $\geqslant 10^{15}$eV.
S. Kino, A. Nakanishi, S. Miono, T. Kitajima, T. Yanagita, M. Kawamoto, K. Asakimori, M. Hazama, K. Mizushima, T. Nakatsuka, N. Ohmori, M. Kusunose.
18th International Cosmic Ray Conference, Vol. 11, p. 424 – 427 (1983). – See Abstr. 012.096.

Observation of the high energy component in EAS has been expected to confirm the primary cosmic ray composition in the EAS range. To remove the various defects of the past experiments, the authors attempt to install the thermoluminescence counter system in the AS array at Mt. Norikura.

144.440   Function of spatial separation distribution and energy spectrum of muons in groups at the depth 850 hg cm^{-2} underground.
V. N. Bakatanov, E. V. Budko, A. E. Chudakov, V. A. Dogujaev, A. R. Mikhelev, Yu. V. Novosel'tsev, M. V. Novosel'tseva, V. I. Petkov, Yu. V. Sten'kin, O. V. Suvorova, A. V. Voevodsky.
18th International Cosmic Ray Conference, Vol. 11, p. 453 – 456 (1983). – See Abstr. 012.096.

144.441   High energy muon groups deep underground.
Yu. N. Vavilov, L. G. Dedenko.
18th International Cosmic Ray Conference, Vol. 11, p. 457 (1983). Abstract. – See Abstr. 012.096.

144.442 On the high–energy multiple muon phenomena.
K. Mizutani, T. Sato.
18th International Cosmic Ray Conference, Vol. 11, p. 458 (1983). Abstract. – See Abstr. 012.096.

144.443 The energy spectrum of cosmic ray neutrinos at sea level.
K. Mitsui, Y. Minorikawa.
18th International Cosmic Ray Conference, Vol. 11, p. 475 – 478 (1983). – See Abstr. 012.096.

144.444 The neutrino spectra in the atmosphere.
A. V. Butkevich, I. M. Zheleznykh, L. G. Dedenko.
18th International Cosmic Ray Conference, Vol. 11, p. 479 (1983). Abstract. – See Abstr. 012.096.

144.445 Composition of cosmic rays.
E. Juliusson.
18th International Cosmic Ray Conference, Vol. 12, p. 117 – 134 (1983). – See Abstr. 012.096.

This report covers the sessions OG1, 2 and 3, of the conference: "The composition of cosmic rays." That includes a wide range of topics, namely elemental composition and energy spectra including the ultra heavy nuclei, isotopic composition and measurements of electrons and antiparticles.

144.446 Spectra, anisotropies and composition of cosmic rays above 1000 GeV.
J. Linsley.
18th International Cosmic Ray Conference, Vol. 12, p. 135 – 191 (1983). – See Abstr. 012.096.

Updated summaries are given of results on (1) the equal-energy primary composition from charge–resolved energy spectra, (2) the all–particle energy spectrum, (3) N_{max} and N_μ spectra, and (4) harmonic components of the sidereal time variation. A new calorimetric all–particle spectrum is obtained. A model in which a rigidity cutoff of low energy cosmic rays is masked by proton enhancement beginning before the all–particle knee seems plausible.

144.447 Galactic propagation of cosmic rays.
R. Schlickeiser.
18th International Cosmic Ray Conference, Vol. 12, p. 193 – 206 (1983). – See Abstr. 012.096.

This paper summarizes the highlights of 32 papers presented in sessions OG3, OG5 and OG8 of the conference which were concerned with the propagation of cosmic ray electrons, nucleons and antiprotons in galaxies.

144.448 Cosmic ray sources and acceleration.
J. L. Osborne.
18th International Cosmic Ray Conference, Vol. 12, p. 207 – 217 (1983). – See Abstr. 012.096.

This report is concerned with the 35 papers in the OG6 sessions of the conference. Cosmic ray composition, shock wave acceleration, acceleration in supernova remnants (again mainly concerned with shock waves), stellar winds and pulsar acceleration are considered.

144.449 Secular and sidereal variations of cosmic rays.
J. N. Goswami.
18th International Cosmic Ray Conference, Vol. 12, p. 219 – 233 (1983). – See Abstr. 012.096.

The author discusses some of the salient points that emerged from the presentations in the following sessions of this conference: Time and spatial variations in cosmic rays. Solar cycle modulation. Sidereal variations.

144.450 New particles and baryon non–conservation.
P. Bareyre.
18th International Cosmic Ray Conference, Vol. 12, p. 299 – 315 (1983). – See Abstr. 012.096.

This paper covers the sessions HE2 (new particles) and HE7 (baryon non–conservation) of the conference for both theory and observations.

144.451 Rapporteur paper of the section HE 4, HE 6 (*of the conference*).
S. A. Slavatinsky (*S. A. Slavatinskij*).
18th International Cosmic Ray Conference, Vol. 12, p. 317 – 339 (1983). – See Abstr. 012.096.

Energy dependence of inelastic cross–section, scaling violation, primary composition, electromagnetic cascade study, air and dense matter, and theory of high energy interaction are presented.

144.452 Family phenomena and particle interactions. Rapporteur talk on HE 5 (*of the conference*).
T. Yuda.
18th International Cosmic Ray Conference, Vol. 12, p. 341 – 356 (1983). – See Abstr. 012.096.

The author introduces emulsion chamber experiments, and then reviews the problem of possible scaling violation in the fragmentation region, rising cross sections, high P_t jets, exotic events, primary composition and so on.

144.453 Rapporteur paper for sessions MG1, MG3, and MG4 (*of the conference*): modulation theory, interplanetary propagation, and interplanetary acceleration.
F. C. Jones.
18th International Cosmic Ray Conference, Vol. 12, p. 379   388 (1983). – See Abstr. 012.096.

144.454 Cosmic rays as probes of heliosphere and magnetospheres: rapporteur talk on MG–2 and –5 (*of the conference*).
M. Wada, J. A. Simpson.
18th International Cosmic Ray Conference, Vol. 12, p. 389 – 402 (1983). – See Abstr. 012.096.

A general survey of the solar cycle variation is given. Then the author deals with (1) gradient, specially radial, (2) anomalous components, helium and oxygen, (3) spectra, protons, alpha particles, electrons, response function, (4) hysteresis, time, energetic, component, (5) solar magnetic field and its reversal, (6) periodicity, other than well known periods, and (7) magnetospheric phenomena, specially earth magnetosphere.

144.455 Short–term isotropic recurrent and anisotropic variations of cosmic ray intensity.
S. P. Agrawal.
18th International Cosmic Ray Conference, Vol. 12, p. 403 – 424 (1983). – See Abstr. 012.096.

This paper presents a review of the results presented in various sessions of the conference relating to the phenomena of short–term variations of galactic cosmic ray intensity, in the energy range of 1 GeV to ≈ 1000 GeV.

144.456 Geomagnetic and atmospheric effects.
P. H. Stoker.
18th International Cosmic Ray Conference, Vol. 12, p. 425 – 433 (1983). – See Abstr. 012.096.

Rapporteur talk of the sessions MG10 and 11 of the conference. The author discusses the question: why are we interested in geomagnetic and atmospheric effects of cosmic rays?

144.457 Behaviour of high energy interactions from structure of small and medium air showers.
M. V. S. Rao.
18th International Cosmic Ray Conference, Vol. 12, p. 449 – 474 (1983). – See Abstr. 012.096.

This rapporteur paper reviews papers presented in sessions EA1.1 and EA3 of the conference dealing with the structure of air showers in the size range $10^4 - 10^7$ particles and the behaviour of high energy interactions derived from measurements of this structure. The results on the electron, muon and hadron components are reviewed and deliberations on the behaviour of high energy interactions are discussed.

144.458   Study of large air showers through electron, muon and hadron components and optical emissions.
M. Nagano.
18th International Cosmic Ray Conference, Vol. 12, p. 475 – 493 (1983). – See Abstr. 012.096.

Rapporteur paper dealing with the structure of large air showers, cascade studies, optical and radio emission, and other topics presented in the sessions EA1.2, EA2, EA4 and EA5 of the conference.

144.459   Muons and neutrinos.
J. W. Elbert.
18th International Cosmic Ray Conference, Vol. 12, p. 495 – 514 (1983). – See Abstr. 012.096.

Contents: The muon spectrum and charge ratio. Prompt muons. Multiple muon events. Muon interactions. Atmospheric neutrino and neutrino oscillation calculations. Calculations of expected extraterrestrial neutrino fluxes. Neutrino observations and analyses of neutrino data.

144.460   Accreting binaries, diffuse radio emission, and galactic cosmic rays.
D. J. Helfand, R. H. Becker.
Bull. Am. Astron. Soc., Vol. 17, No. 2, p. 566 (1985). Abstract. – See Abstr. 010.065.

144.461   Evidence for a high–energy cosmic–ray spectrum cutoff.
R. M. Baltrusaitis, R. Cady, G. L. Cassiday, R. Cooper, J. W. Elbert, P. R. Gerhardy, S. Ko, E. C. Loh, Y. Mizumoto, M. Salamon, P. Sokolsky, D. Steck.
Phys. Rev. Lett., Vol. 54, No. 16, p. 1875 – 1877 (1985).

The authors report a measurement of the ultrahigh–energy cosmic–ray spectrum using an atmospheric fluorescence technique for extensive–air–shower detection. The differential spectrum between 0.1 and 10 EeV (1 EeV = 10^{18}eV) is well fitted by a power law with slope 2.94 ± 0.02. Above 10 EeV evidence is presented for the development of a spectral "bump" followed by a cutoff at 70 EeV.

144.462   The propagation of ultraheavy cosmic rays.
N. R. Brewster.
Diss. Abstr. Int., Sect. B, Vol. 45, No. 11, p. 3533 (1985). Thesis, University of Minnesota, 150 pp. (1984). Order No. DA8501855.

Astrophysics of cosmic rays.
See Abstr. 003.061.

Precision observations of cosmic rays in Yakutsk.
See Abstr. 003.168.

Perspectives on space and astrophysical plasma physics.
See Abstr. 013.001.

The role of stimulated processes in cosmic annihilation–line sources. II.
See Abstr. 022.113.

Production of long–lived cosmogenic nuclei and their applications.
See Abstr. 022.145.

Accelerator mass spectrometry at the Rehovot Pelletron tandem: measurements of abundances of cosmogenic radioisotopes and future prospects.
See Abstr. 022.146.

Analysis of tracks in the stacked film track detector.
See Abstr. 022.150.

Proton–proton total cross–section at $E_{c.m.}$ = 15 to 150 TeV from cosmic–ray data.
See Abstr. 022.157.

Total and partial inelastic cross sections of proton–nucleus reactions.
See Abstr. 022.159.

Fragmentation of 710 and 1050 MeV/nuc Fe nuclei in CH_2 and C targets – isotopic cross sections for H targets.
See Abstr. 022.160.

Fragmentation of ~ 500 MeV/nuc Ne and O nuclei and CH_2, C and H targets – charge and isotopic cross sections.
See Abstr. 022.161.

How sensitive can underground cosmic ray experiments be in measuring neutrino oscillation parameters?
See Abstr. 022.163.

An accelerator test of semi–empirical cross–sections.
See Abstr. 022.165.

Interactions of 200 GeV gold nuclei in light elements.
See Abstr. 022.166.

Determination of coupling coefficients of neutrons incident at arbitrary angles.
See Abstr. 022.168.

Coupling functions and meteorological coefficients for the detectors of the Elbrus spectrograph.
See Abstr. 022.169.

Multiple muons in DUMAND and their interpretation.
See Abstr. 022.170.

Preliminary results on cosmic muons with NUSEX experiment at the Mt. Blanc Laboratory.
See Abstr. 022.171.

The picture of nucleus–nucleus interactions evolving from QCD models and data on hadron–nucleus collisions.
See Abstr. 022.173.

Observations of small air showers at Mt. Norikura.
See Abstr. 034.007.

Semi–automated, three–dimensional measurement of etched tracks in solid–state nuclear track detectors.
See Abstr. 034.124.

Nagoya cosmic–ray muon spectrometer.
See Abstr. 034.129.

A precision measurement of the cosmic ray muon angular distribution underground.
See Abstr. 034.130.

The investigation of muon groups at large zenith angles with the Aragats spark calorimeter.
See Abstr. 034.133.

Effect of the Mott cross section on charge identification in the HEAO–3 Heavy Cosmic Ray Experiment.
See Abstr. 035.012.

Results of a Transmediterranean balloon flight experiment designed to evaluate the high resolution characteristics of CR–39 in cosmic ray studies.
See Abstr. 035.105.

The non–Z^2 response of the heavy nuclei cosmic ray detector on HEAO–3.
See Abstr. 035.116.

Use of relativistic rise in ionization chambers for measurement of high energy heavy nuclei.
See Abstr. 035.117.

An instrument employing electronic counters and an emulsion chamber for studying heavy cosmic ray interactions (JACEE–3).
See Abstr. 035.118.

Measurement of heavy cosmic rays above 20 GeV/n in passive detectors: preliminary results from JACEE–3.
See Abstr. 035.119.

The ultra heavy cosmic ray experiment on the NASA Long Duration Exposure Facility (LDEF).
See Abstr. 035.121.

Gamma–ray astronomy and the study of electrons $> 10^{13}$eV and bursts using an array of γ–ray detectors in space platform.
See Abstr. 036.139.

Optical lightnings as a background in searches for cosmic X– and gamma–ray bursts through atmospheric fluorescence.
See Abstr. 036.192.

Precision of the spectrographic method.
See Abstr. 036.196.

The simplest versions of the global spectrographic method.
See Abstr. 036.197.

The radiation situation in space and its modification by geomagnetic field and shielding.
See Abstr. 051.025.

LET–distributions and doses of HZE radiation components at near–earth orbits.
See Abstr. 051.026.

Summary of current radiation dosimetry results on manned spacecraft.
See Abstr. 051.027.

Cosmic ray exposure factors for shuttle altitudes derived from calculated cut–off rigidities.
See Abstr. 051.028.

LDEF–1 and beyond.
See Abstr. 051.030.

Zeitabhängige Lösungen für die Stoßbeschleunigung von energetischen Teilchen.
See Abstr. 061.038.

Electronic cosmic ray monopole searches.
See Abstr. 061.091.

Acoustic detection of monopoles.
See Abstr. 061.093.

First results from the Chicago–Fermilab–Michigan cosmic ray magnetic monopole detector.
See Abstr. 061.097.

A limit on the flux of heavily–ionizing penetrating particles in cosmic rays at large zenith angles.
See Abstr. 061.120.

Search for baryon decay catalysed by the Rubakov interaction of slow magnetic monopoles.
See Abstr. 061.123.

Self–similar modified shock structures.
See Abstr. 062.091.

The back action of accelerated particles on shock–front structure.
See Abstr. 062.131.

Acceleration to ultra high energies in magnetospheres of young pulsars.
See Abstr. 067.179.

Frictional acceleration of high–energy particles in pulsar vicinity.
See Abstr. 067.180.

Consequences of neutrino production in the Cygnus X–3 system.
See Abstr. 067.193.

Long–time changes of solar and geomagnetic activity and cosmic ray density.
See Abstr. 072.084.

Solar activity in the periods of anomalous cosmic ray variations (solar maxima of cycles 19 – 21).
See Abstr. 072.088.

Using high energy cosmic particles to probe the coronal magnetic field.
See Abstr. 075.024.

Characteristics of Forbush–type cosmic ray intensity decreases observed during solar cycle 21.
See Abstr. 078.004.

Understanding the heliosphere and its energetic particles.
See Abstr. 078.081.

^{81}Kr production rate in the atmosphere.
See Abstr. 082.079.

The reaction of F–region of the ionosphere on cosmic rays.
See Abstr. 083.022.

Determination of the ionization by cosmic rays during daytime.
See Abstr. 083.023.

Effects of sudden commencements on particle intensities at synchronous orbit.
See Abstr. 084.128.

Probable contribution to cosmic rays by Jupiter.
See Abstr. 099.081.

On identification of ultraheavy cosmic nuclei in extraterrestrial olivines.
See Abstr. 105.010.

On the record of galactic cosmic ray flux and exposure histories of iron meteorites.
See Abstr. 105.151.

Cosmic ray induced thermoluminescence in meteorites.
See Abstr. 105.244.

Isotope production in meteorites and cosmic ray variation.
See Abstr. 105.245.

Cosmic–ray picture of the heliosphere.
See Abstr. 106.050.

The configuration of a heliospheric current sheet and some features of its evolution over the cycle of solar activity.
See Abstr. 106.071.

Coupled hydromagnetic wave excitation and ion acceleration at interplanetary shocks.
See Abstr. 106.074.

The stellar upper mass limit in an OB association and implications for cosmic rays and cosmic abundances.
See Abstr. 107.032.

Ultra–high–energy gamma–rays and cosmic rays from accreting degenerate stars.
See Abstr. 117.233.

Something strange from Cygnus X–3.
See Abstr. 117.363.

Adiabatic deceleration of secondary antiprotons in the envelopes of supernova exploding in dense clouds.
See Abstr. 125.049.

Desorption from interstellar grains.
See Abstr. 131.070.

The physical structure of the dark cloud B5.
See Abstr. 131.071.

Cosmic ray ionization in cold clouds.
See Abstr. 131.351.

Correlation between X–ray luminosity and the deviation from the main sequence colour diagram for stellar X–ray sources.
See Abstr. 142.064.

X rays from cosmic ray interactions.
See Abstr. 142.066.

Discussion on zero–point field acceleration and applicability to cosmic rays, X–ray background and energetics of the intergalactic medium.
See Abstr. 142.068.

COS–B gamma–ray sources and interstellar gas in the first galactic quadrant.
See Abstr. 143.040.

Is Cygnus X–3 a source of gamma rays or of new particles?
See Abstr. 143.054.

The radial distribution of gamma rays and cosmic rays in the outer Galaxy.
See Abstr. 143.094.

The production of galactic gamma–rays above 35 MeV.
See Abstr. 143.095.

High–energy gamma rays and the large–scale distribution of gas and cosmic rays.
See Abstr. 155.035.

High–energy galactic phenomena and the interstellar medium.
See Abstr. 155.037.

The results of observations of the galactic disk for very high energy at the Tien–Shan High Altitude Station.
See Abstr. 155.080.

Cosmic γ rays and the mass of gas in the Galaxy.
See Abstr. 155.091.

The galactic gamma–ray distribution: implications for galactic structure and the radial cosmic–ray gradient.
See Abstr. 155.101.

On the structure and mass of the galactic halo.
See Abstr. 155.104.

Spectral index variations in the galactic radio continuum emission.
See Abstr. 155.158.

The Large Magellanic Cloud: a test for gamma ray and cosmic ray origin theories.
See Abstr. 156.027.

Ultra–high energy cosmic ray production by current disruption in active galactic nuclei.
See Abstr. 158.014.

The development of the electromagnetic cascade showers initiated by high–energy electrons and gammay–rays in the hot photon gas.
See Abstr. 158.283.

The excess charge hypothesis, a possible key for the highest energy cosmic rays springing out of early contracting protoclouds.
See Abstr. 161.341.

Stellar Systems, Galaxy, Extragalactic Objects, Cosmology

151 Stellar Systems (Kinematics, Dynamics)

151.001 Gravitational N–body problem and the validity of concepts of the gas model.
E. Bettwieser, D. Sugimoto.
Mon. Not. R. Astron. Soc., Vol. 212, No. 1, p. 189 – 195 (1985).

This paper concerns the relation between N–body systems and continuum models of stellar dynamics. A 1000–body calculation is examined for the validity of concepts used in gaseous models simulating N–body systems. Moments of velocity distribution are defined with respect to subsystems (10 shells). When moments are averaged in time and over a shell large enough, heat flows down the gradient in velocity dispersion. Redistribution of heat is a mechanism which contributes to core–collapse and core–halo formation. The coefficient of heat conductivity is in fair agreement with simple theoretical considerations.

151.002 Dynamical rules for barred spiral galaxies.
P. J. Teuben, R. H. Sanders.
Mon. Not. R. Astron. Soc., Vol. 212, No. 2, p. 257 – 273 (1985).
With a correction in Vol. 214, No. 2, p. 319, plate 1 (1985).

Numerical integrations of 2D–orbits demonstrate that the principal family of periodic orbits in strong barred potentials is the prograde parallel family. All other families have orbit density distributions which are inconsistent with the imposed potential. Some important implications are discussed: (1) bars tumble in the sense of galactic rotation; (2) corotation must lie outside the bar distortion; (3) strong rapidly tumbling bars are cold, and, conversely, strong slowly tumbling bars are hot and (4) the magnitude of the bar distortion is limited to axial ratios of about 1:5, due to the onset of stochasticity well inside corotation. A small population of stars on stochastic orbits may form the more nearly axisymmetric lens structure seen in some barred galaxies.

151.003 A re–examination of some bar–driven models of spiral structure.
M. P. Schwarz.
Mon. Not. R. Astron. Soc., Vol. 212, No. 3, p. 677 – 686 (1985).

Some numerical experiments by Schempp of gas response to a rotating bar potential are repeated using a particle scheme to simulate the gas dynamics. Schempp's results are shown to be dependent on the disruptive way he turns on the bar. Quasi–static models unaffected by transients are computed. The shocks which develop are in significantly different positions from those of Schempp, and the bar rotation rate is shown to have no effect on whether the morphology is ringed or spiral. Furthermore the spiral structure is not confined to inside corotation, but extends outward past the outer Lindblad resonance. It is shown that inner rings and straight dust lanes require different sorts of bar field for their formation.

151.004 On the formation of spiral structure in gas discs through tidal interaction – I. Direct encounters.
S.-A. Sørensen.
Mon. Not. R. Astron. Soc., Vol. 212, No. 3, p. 723 – 736 (1985).

This paper investigates the waves which are formed when a thin gas disc in a smooth axisymmetric potential is perturbed. The perturbation is introduced through tidal interaction with an external body moving in the plane of the disc. The model is investigated using numerical techniques which follow the formation of large–scale hyperbolic spirals. These are identified as the propagating fronts of epicyclic waves. Over an area comparable to the visual image of a galaxy the spirals change from the hyperbolic form towards an equiangular appearance. Predictions by analytical models were found to be in good agreement with the results.

151.005 Testing for triaxiality with kinematic data.
J. Binney.
Mon. Not. R. Astron. Soc., Vol. 212, No. 4, p. 767 – 781 (1985).

Simple kinematical models of elliptical galaxies are used to show that if elliptical galaxies are triaxial, rotational velocities along their apparent major axes must be associated with similar motions along their apparent minor axes. The relative magnitudes of these velocities indicate the degree to which elliptical galaxies deviate from the classical picture of them as oblate figures of rotation. An analysis of the observational data for 10 galaxies suggests that a typical elliptical galaxy is probably far from being axisymmetric, but rather has a middle–axis length about equal to the average of the lengths of the longest and shortest axes.

151.006 Periodic orbits in a triaxial galaxy.
H. Robe.
Astron. Astrophys., Vol. 142, No. 2, p. 351 – 354 (1985).

The general properties of perturbed differential autonomous systems are used to investigate and prove the existence of periodic orbits in a non rotating triaxial model elliptical galaxy. Explicitly, the author gives an easy, complete and rigorous inventory of the regular periodic orbits of a three dimensional harmonic oscillator with 1:1:1 resonances perturbed by small linear and quartic terms.

151.007 Changes of entropy and entropy flux induced by spiral structure in galaxies.
B. Fuchs.
Astron. Astrophys., Vol. 142, No. 2, p. 487 – 490 (1985).

Changes of entropy and entropy flux induced by the occurrence of spiral density waves in the disks of normal spiral galaxies are discussed. It is shown that the entropy of the galactic disk is lowered by the appearance of spiral structure in the regions between the Lindblad and corotation resonances. The spiral waves are assumed to be tightly wound so that the WKBJ approximation may be applied. The excess entropy associated with the spiral waves is identified as the energy density of the waves divided by the kinetic temperature of the galactic disk. The swing amplifier and WASER mechanisms to amplify spiral density waves are discussed regarding the entropy flux which carries continuously entropy out of the regions of spiral structure towards the outer Lindblad resonance region.

151.008 Application of a new gasdynamics code to gas flow problems in disk galaxies.
G. D. van Albada.
Astron. Astrophys., Vol. 142, No. 2, p. 491 – 496 (1985).

Recent studies of the gas flow in disk galaxies by different authors have produced a wide variety of often discrepant results. The gasdynamics code used in these calculations, which is generally different from one author to the next, appears to be instrumental in producing these discrepancies. In this paper a relatively new code is used to recalculate some of the models in the literature. The code has been tested very extensively and appears to be a suitable choice for modelling the flow of an ideal, isothermal gas. Though the real interstellar medium is far more complex than that, the computed gas flows appear to represent the observations quite well.

151.009   On the existence of periodic orbits in a nearly axisymmetric galaxy.
N. Caranicolas.
Astron. Nachr., Vol. 306, No. 1, p. 17 – 20 (1985).

Assuming that the potential field on the plane of symmetry of a nearly axisymmetric galaxy is a polynomial of the fourth degree, the author studies the conditions of existence and stability of the main types of periodic orbits. He verifies the theoretical results by numerical calculations.

151.010   A collapsing plane model of a stellar system with finite radius and its stability.
S. N. Nuritdinov.
Pis'ma Astron. Zh., Tom 11, No. 2, p. 89 – 93 (1985). In Russian. English translation in Sov. Astron. Lett., Vol. 11.

An exact self–consistent non–linear model of a collapsing system without pressure in the form of a non–rotating circular disk is constructed in the Newtonian approximation. The study of stability against non–symmetric perturbations shows that for untrivial types of harmonics the Jeans power instability takes place that violates the process of collapse and leads to fragmentation of the disk into separate condensations.

151.011   Tidal interactions between the Galaxy and the Magellanic Clouds.
N. Ramamani, T. M. Singh, S. M. Alladin.
The Milky Way galaxy, p. 477 – 478 (1985). – See Abstr. 012.007 (IAU Symp. No. 106).

The merging time and the disruption time in a binary galaxy system are analytically obtained under the adiabatic approximation. Applications are made to the Galaxy–LMC (Large Magellanic Cloud) pair.

151.012   Dynamical evolution of the galactic disk.
R. Wielen, B. Fuchs.
The Milky Way galaxy, p. 481 – 490 (1985). – See Abstr. 012.007 (IAU Symp. No. 106).

After some general remarks on the dynamical evolution of the galactic disk, the authors review mechanisms which may affect the velocities of disk stars: stochastic heating, deflections, adiabatic cooling or heating. They compare the observed velocities of nearby disk stars with theoretical predictions based on the diffusion of stellar orbits.

151.013   Evolution of the vertical structure of galactic disks.
J. V. Villumsen.
The Milky Way galaxy, p. 491 – 492 (1985). – See Abstr. 012.007 (IAU Symp. No. 106).

151.014   Heating of stellar disks by massive gas clouds.
C. G. Lacey.
The Milky Way galaxy, p. 493 – 495 (1985). – See Abstr. 012.007 (IAU Symp. No. 106).

The author has analytically calculated the evolution of the three components of velocity dispersion of the stars in a galactic disk when these are scattered by massive gas clouds, in a generalization of Spitzer & Schwarzschild's (1953) calculation.

151.015   Stellar velocity distribution in the presence of scattering massive clouds.
W. Renz.
The Milky Way galaxy, p. 497 – 498 (1985). – See Abstr. 012.007 (IAU Symp. No. 106).

151.016   Warps and heavy halos.
L. S. Sparke.
The Milky Way galaxy, p. 499 – 502 (1985). – See Abstr. 012.007 (IAU Symp. No. 106).

The paper discusses how an axisymmetric (but not spherical) heavy halo affects the warping modes of a galactic disk.

151.017   3–dimensional particle simulation of vertical oscillations in gas discs.
T. C. Johns, A. H. Nelson.
The Milky Way galaxy, p. 503 – 504 (1985). – See Abstr. 012.007 (IAU Symp. No. 106).

151.018   Two–fluid gravitational instabilities in a galactic disk.
C. J. Jog.
The Milky Way galaxy, p. 505 – 507 (1985). – See Abstr. 012.007 (IAU Symp. No. 106).

The author formulates and solves the hydrodynamic equations describing an azimuthally symmetric galactic disk as a two–fluid system.

151.019   Collective phenomena in a multi–component gravitating system.
A. M. Fridman, A. G. Morozov, J. Palous, A. Piskunov, V. L. Polyachenko.
The Milky Way galaxy, p. 509 – 510 (1985). – See Abstr. 012.007 (IAU Symp. No. 106).

The paper aims at a demonstration of the principal differences between the oscillation spectra of multi–component systems and a one–component medium. The character of the mutual motions of components appears then to be of importance. Three cases are considered.

151.020   Formation and maintenance of spiral structure in galaxies.
C. C. Lin, G. Bertin.
The Milky Way galaxy, p. 513 – 532 (1985). – See Abstr. 012.007 (IAU Symp. No. 106).

The formation and maintenance of spiral structure in galaxies is discussed in terms of spiral modes with a general perception consistent with the hypothesis of quasi–stationary spiral structure. The latter is explained in some detail to give it the original proper perspective and to contrast it with recent studies of non–stationary spiral structures. The authors emphasize the possible coexistence of fast–evolving spiral features and slow–evolving spiral grand designs. Spiral modes are described in terms of three categories of propagating waves. The reason is given why isolated non–barred galaxies with spiral grand design are found to be mostly two–armed. Some suggestions are made for future research.

151.021   Two–dimensional calculations of tightly wound spiral shocks.
A. H. Nelson, T. Johns, M. Tosa.
The Milky Way galaxy, p. 533 – 534 (1985). – See Abstr. 012.007 (IAU Symp. No. 106).

Recently it has been possible to perform computer calculations with sufficiently high resolution to obtain the response of the gas in galaxies to a tightly wound spiral potential due to a stellar density wave. This paper reports the early results of a 2–dimensional calculation, and compares them with 1–dimensional results.

151.022   Oscillations of rotating gas disks: p–modes, g–modes, and r–modes.
M. Iye.
The Milky Way galaxy, p. 535 – 536 (1985). – See Abstr. 012.007 (IAU Symp. No. 106).

Short–wavelength adiabatic oscillations of an infinitesimally thin, rotating and self–gravitating gas disk in dynamical equilibrium are studied. The problem is formulated in terms of the theory of oscillations of gas spheres (i.e. stars) in order to help galactic and stellar seismologists towards their mutual understanding of their rather similar subjects.

151.023   The stabilizing effects of haloes and spiral structure.
H. Robe.
The Milky Way galaxy, p. 537 – 538 (1985). – See Abstr. 012.007 (IAU Symp. No. 106).

The equilibrium and the stability of homogeneous gaseous and uniformly rotating MacLaurin spheroids imbedded in a rigid, homogeneous and spherical stellar halo are considered.

151.024   New N–body experiments on the spiral structure of galaxies.
K. O. Thielheim, H. Wolff.
The Milky Way galaxy, p. 539 – 540 (1985). – See Abstr. 012.007 (IAU Symp. No. 106).

151.025   Velocity dispersion and the stability of galactic disks.
J. Kormendy.
The Milky Way galaxy, p. 541 (1985). Abstract. – See Abstr. 012.007 (IAU Symp. No. 106).

151.026   Periodic orbits relevant for the formation of inner rings in barred galaxies.
M. Michalodimitrakis, C. Terzides.
The Milky Way galaxy, p. 543 – 544 (1985). – See Abstr. 012.007 (IAU Symp. No. 106).

151.027   On the 3–kpc arm: waves excited at the resonance by an oval distortion in the central region.
C. Yuan.
The Milky Way galaxy, p. 545 – 546 (1985). – See Abstr. 012.007 (IAU Symp. No. 106).

151.028   A barred galaxy: the inside viewpoint.
G. D. van Albada.
The Milky Way galaxy, p. 547 – 549 (1985). – See Abstr. 012.007 (IAU Symp. No. 106).
The observed gas distribution and kinematics in the central 4 kpc of our Galaxy show many aspects reminiscent of barred spirals. The author describes a collection of $(1,v)$ maps obtained from a gas flow model of a barred spiral.

151.029   Stochastic star formation and spiral structure.
P. E. Seiden.
The Milky Way galaxy, p. 551 – 558 (1985). – See Abstr. 012.007 (IAU Symp. No. 106).

151.030   Spiral structure in galaxies: large–scale stochastic self–organization of interstellar matter and young stars.
J. V. Feitzinger.
The Milky Way galaxy, p. 559 – 560 (1985). – See Abstr. 012.007 (IAU Symp. No. 106).

151.031   Star formation in a density–wave–dominated, cloudy interstellar medium.
W. W. Roberts Jr., M. A. Hausman.
The Milky Way galaxy, p. 567 – 569 (1985). – See Abstr. 012.007 (IAU Symp. No. 106).
The authors present a model of the interstellar medium in spiral galaxies. The ISM is simulated by a system of particles, representing gas clouds. It is a probabilistic N–body system in which "cloud" particles orbit ballistically, undergo dissipative collisions with other clouds, and experience velocity–boosting interactions with expanding supernova remnants.

151.032   Spiral tracers and prestellar incubation periods in a cloudy interstellar medium.
M. A. Hausman, W. W. Roberts Jr.
The Milky Way galaxy, p. 571 – 572 (1985). – See Abstr. 012.007 (IAU Symp. No. 106).

151.033   Secular evolution in galaxies.
A. May, C. A. Norman, T. S. van Albada.
The Milky Way galaxy, p. 613 – 614 (1985). – See Abstr. 012.007 (IAU Symp. No. 106).

151.034   Gas dynamics and disk–galaxy evolution.
R. G. Carlberg.
The Milky Way galaxy, p. 615 – 619 (1985). – See Abstr. 012.007 (IAU Symp. No. 106).
Gas dynamics plays a vital role in the formation and ongoing evolution of a disk galaxy through its ability to have dissipative collisions. In this paper the simplest possible gas physics which provides a description of a gaseous component is incorporated

into N–body experiments to investigate the formation of a disk galaxy.

151.035   Anisotropic models of elliptical galaxies. II. The case of flattened isophotes.
R. Bacon.
Astron. Astrophys., Vol. 143, No. 1, p. 84 – 93 (1985).
The anisotropic models built in Paper I are applied to elliptical galaxies, with good photometric and kinematic data. Special emphasis is given to detailed testing of the isotropy – including assessment of the (small) effect of the inclination of the galaxy, and the sometimes (dramatic) effect of the seeing limitations. In good agreement with previous global tests, it is found that anisotropy appears only in bright ellipticals, and it is suggested that it comes from the merger building of cD class galaxies. Accurate M/L ratios are derived, and found to remain within a small range independent of the absolute magnitude.

151.036   On the dynamical evolution of a spheroidal cluster.
G. Som Sunder, R. K. Kochhar.
Mon. Not. R. Astron. Soc., Vol. 213, No. 2, p. 381 – 388 (1985).
Tensor virial equations are used to investigate the dynamical evolution of a cluster of mass points characterized by a spheroidal (homogeneous or heterogeneous) distribution of matter and an isotropic distribution of velocities. It is known that if the total energy of the system is positive, it will disperse to infinity. The authors show that in the case of negative total energy, the system executes finite amplitude oscillations between oblate and prolate shapes.

151.037   Evolution of barred galaxies by dynamical friction.
M. D. Weinberg.
Mon. Not. R. Astron. Soc., Vol. 213, No. 2, p. 451 – 471 (1985).
The transfer of angular momentum between the bar and halo components of a galaxy is computed using both analytic dynamical friction theory and 'semi–restricted' n–body simulations. The two methods yield results which are in qualitative agreement and demonstrate that dynamical friction can exert strong torques on galactic bars. A rotating rigid bar with typical parameters initially ending at corotation slows down in several bar rotation times in the presence of an isothermal halo. This rapid angular momentum transfer implies that the corotation radius must be far beyond the end of the bar unless: (1) angular momentum is added to the bar (i.e. by the disc) allowing an equilibrium bar pattern speed to be reached; (2) the galactic halo is small or non–existent; or (3) the bar is weak.

151.038   A gas model for multicomponent star clusters: method and comparison with Fokker–Planck theory.
E. Bettwieser, S. Inagaki.
Mon. Not. R. Astron. Soc., Vol. 213, No. 2, p. 473 – 483 (1985).
A multi–mass stellar system is simulated by integrating numerically a set of four gas–model equations for each component. The components interact by mutual gravitational interaction, by transfer of kinetic energy and by dynamical friction effects. Pre–collapse calculations of two–component systems are presented and compared with the results obtained by direct integration of the fundamental Fokker–Planck equation. The gas model shows a higher degree of equipartition of the kinetic energies. Later–phase evolution is somewhat slower as compared with results of the Fokker–Planck theory. The qualitative behaviour of evolution is, however, in fair agreement.

151.039   The role of initial binaries in cluster evolution: direct N–body calculations.
G. Giannone, D. Molteni.
Astron. Astrophys., Vol. 143, No. 2, p. 321 – 326 (1985).
The presence of initial binary systems in stellar clusters seems to have a dominant effect on the dynamical evolution of the whole cluster. The authors discuss simulations, carried out with a direct N–body method for several models with $N = 300$ equal mass objects, with 20% of initial binaries and different binary binding energy. These calculations show that binaries with binding energy, h, in the range 5 – 10 times the mean kinetic energy interact strongly with the field stars and between themselves. The

main result of these interactions is an enhanced expansion of the core and of the whole cluster. Binary–binary interactions occur in central regions, while single star–binary ones mainly occur in outer regions. The former produce core heating and are mostly responsible of the escaping stars, while the latter contribute to a strong expansion of the halo.

151.040   Distant encounters between disk galaxies and the origin of S0 spirals.
V. Icke.
Astron. Astrophys., Vol. 144, No. 1, p. 115 – 123 (1985).
  Simulations of the interactions between galaxies have hitherto emphasized the galactic stellar component. The author presents numerical hydrodynamic calculations of the fate of a disk of gas in a galaxy perturbed gravitationally by an intruder flying by in a hyperbolic orbit. He considers the possibility that distant galaxy encounters are responsible for making S0 galaxies.

151.041   Dynamical consequences of star collisions for core–envelope structure in red giants.
Y. Tuchman.
Astrophys. J., Vol. 288, No. 1, p. 248 – 251 (1985).
  The possibility of a spatial core–envelope separation in a red giant caused by its collision with a main–sequence star is discussed. This separation, which is shown to occur, at least, in collisions with an impact parameter less than $\sim 50\,R_\odot$ and with an impact velocity less than $360\,\mathrm{km\,s^{-1}}$, leads to the dispersion of the entire envelope with typical velocities of several hundreds of kilometers per second. It is suggested that this mechanism for the dispersion of stellar material might play an important role in the central part of galaxies and in other high star density environments.

151.042   Direct physical collisions and the evolution of dense stellar systems.
H. E. Kandrup.
Astrophys. Space Sci., Vol. 109, No. 2, p. 215 – 230 (1985).
  This paper formulates kinetic equations to describe the destruction of stars via direct physical collisions, assumed characterized by a constant geometric cross section, which are valid for spherically–symmetric systems both in Newtonian gravity and general relativity.

151.043   Gas dynamics of flow past galaxies.
T. J. Gaetz.
News Lett. Astron. Soc. N.Y., Vol. 2, No. 4, p. 34 (1983). Abstract. – See Abstr. 010.242.

151.044   Gas dynamics of flow past galaxies – radiative cooling.
T. J. Gaetz.
News Lett. Astron. Soc. N.Y., Vol. 2, No. 5, p. 28 (1984). Abstract. – See Abstr. 010.243.

151.045   The effect of dynamical friction on orbits: the case of a particle orbiting a central point mass embedded in a massless stellar system.
J. B. Hoffer.
Astrophys. J., Vol. 289, No. 1, p. 193 – 197 (1985).
  The effects of dynamical friction on the orbit of a particle about a central, massive object are both analytically and numerically investigated. It is assumed that the constant–density stellar background providing the friction has no local velocity dispersion and does not contribute to the central force keeping the particle in orbit. Approximate expressions relating the semimajor axis of the decaying circular orbit and time are found for both nonrotating and rotating stellar systems. The general validity of these relations is supported by a computer solution of the equations of motion for several circular and eccentric cases.

151.046   Comparison of resonant orbits. 3. Resonance 1:1 – some theoretical results.
P. Andrle.
Bull. Astron. Inst. Czech., Vol. 36, No. 3, p. 150 – 152 (1985).
  The properties of a spherical gravitational field of a stellar system with the disturbing term $2\beta\xi^2 z^2$ are studied. Since the

general expression of a formal third integral was found earlier (Andrle, 1979) the resonance 1/1 is studied only and the theoretical invariant curve and boundary of orbits were derived.

151.047   Gasdynamical calculations of preferred planes in prolate and triaxial galaxies. I. Case of no figure rotation.
A. Habe, S. Ikeuchi.
Astrophys. J., Vol. 289, No. 2, p. 540 – 555 (1985).
  The gas motions in prolate and triaxial galaxies are studied numerically with a "smoothed particle" method, in order to confirm the presence of the preferred planes predicted by single–particle orbital integrations. Initially, a gaseous ring is set with its angular momentum vector inclined to the principal axes, and its dynamical behavior is pursued for $\sim 2.0 \times 10^9\,\mathrm{yr}$. In a prolate potential the tilted gas ring precesses around the symmetry axis, and loses the normal components of its angular momentum within about one precession time ($\sim 10^9\,\mathrm{yr}$). As a result, the gas settles into the equatorial plane of the potential. In the triaxial potential, when the initial angle between the angular momentum and the longest or shortest axis of the potential is small, the gas ring precesses around the corresponding axis with decreasing angular momentum, and finally settles into the preferred plane which is normal to this axis.

151.048   A self–consistent box–shaped galaxy.
A. May, T. S. Van Albada, C. A. Norman.
Mon. Not. R. Astron. Soc., Vol. 214, No. 1, p. 131 – 136 (1985).
  The authors describe an N–body model of a box–shaped galaxy which is self–consistent and in equilibrium. The model has a cylindrical pattern of rotation.

151.049   Evolution of the velocity distribution in galactic disks.
J. V. Villumsen.
Astrophys. J., Vol. 290, No. 1, p. 75 – 85 (1985).
  This paper presents a numerical investigation of the evolution of the three–dimensional velocity distribution and of the vertical density distribution in galactic disks. The physical mechanism is the scattering of stars off giant molecular clouds (GMCs) which heat up the stellar populations. A population of GMCs is embedded in the fixed potential of a constant scale height exponential disk and a nearly isothermal halo. The equations of motion are integrated directly. The functional form and growth rate for the velocity dispersions are in fairly good agreement with published theoretical predictions based on the Fokker–Planck equation.

151.050   On the problem of relaxation of stellar systems.
V. G. Gurzadyan, G. K. Savvidin.
Dokl. Akad. Nauk SSSR. Ser. Mat. Fiz., Tom 277, No. 1, p. 69 – 73 (1984). In Russian. Abstr. in Ref. Zh., 51. Astron., 2.51.742 (1985).

151.051   Erratum: "Some evolutionary consequences of shock–induced star formation" [Astron. Astrophys., Vol. 138, No. 1, p. 201 – 204 (1984)].
M. Nepveu, M. Robnik.
Astron. Astrophys., Vol. 145, No. 2, p. 482 (1985). See Abstr. 38.151.023.

151.052   The effect of high–order nonlinearity on the modulation instability of spiral density waves.
V. I. Korchagin, P. I. Korchagin.
Astron. Zh., Tom 62, Vyp. 2, p. 202 – 208 (1985). In Russian. English translation in Sov. Astron., Vol. 29, No. 2.
  The evolution of the amplitude of a tightly wound spiral density wave is considered with taking into account the fifth–power nonlinearity of the amplitude. The limits of the amplitude tolerance of the weak nonlinear description are determined. It is shown that the fifth order of nonlinearity leads to amplitude dependence of the group velocity. The effect of the fifth–order nonlinearity is significant for a spiral wave with wave number near the minimum of a dispersion curve or for a marginally stable disk.

151.053 Local stability criterion for gaseous subsystems of flat galaxies.
A. G. Morozov.
Astron. Zh., Tom 62, Vyp. 2, p. 209 – 217 (1985). In Russian. English translation in Sov. Astron., Vol. 29, No. 2.

The local stability criterion is obtained according to which the velocity dispersion of gas clouds c_g, necessary to stabilize a heterogeneous, differentially rotating gaseous subsystem of a flat galaxy against all nonaxisymmetric disturbances in its plane, should be $c_g = [1 + f(d\Omega/dr, d\sigma_0/dr)]c_0$ (the estimates for the solar galactic vicinity yield $f = 1.6$), $c_0 = \pi G\sigma_0/\varkappa$ being the velocity dispersion sufficient to stabilize axisymmetric disturbances, Ω, σ_0 – the angular rotational velocity and the surface density of the gaseous subsystem, $\varkappa$ – the epicyclic frequency.

151.054 Influence of nonlinearity and dispersion on the evolution of packets of spiral density waves.
V. I. Korchagin, P. I. Korchagin.
Sov. Astron., Vol. 28, No. 4, p. 476 – 477 (1984). English translation of 38.151.009.

151.055 Model integrated B–V colors for metal–poor stellar systems.
O. K. Sil'chenko.
Sov. Astron. Lett., Vol. 10, No. 4, p. 209 – 211 (1984). English translation of 38.151.001.

151.056 The origin of inner rings in barred spiral galaxies.
M. P. Schwarz.
Proc. Astron. Soc. Aust., Vol. 5, No. 4, p. 464 – 467 (1984).

In numerical simulations of gas flow in a bar field the author has shown that a ring can form just inside corotation (CR) if the pattern speed is low enough (Schwarz 1979, 1984). The ring will appear to be at the ends of the bar if, as generally believed, the bar terminates near CR. The rings thus formed have sharp corners on the bar major axis as does that in NGC 1433. Since no other convincing mechanism for the formation of inner rings has been proposed, the author attempts to understand what causes the rings in the gas simulations, and what affects their shape.

151.057 The three–body problem in stellar dynamics.
P. Hut.
The Big Bang and Georges Lemaitre, p. 239 – 255 (1984). – See Abstr. 012.043.

The role of the three–body scattering problem in stellar dynamics is investigated. Three–body interactions exchange energy and angular momentum between internal degrees of freedom of binary stars, and external degrees of freedom of both the single stars and the center of mass motion of binaries. In this way three–body scattering can significantly influence the evolution of a star cluster. The scattering of single stars off tight binary stars on average increases the binding energy of these binaries. The energy released heats up the star cluster. Quantitative results are presented to illustrate the role of three–body scattering as a heat source in stellar dynamics.

151.058 Dynamics of elliptical galaxies and the $r^{1/4}$ luminosity law.
G. Bertin, M. Stiavelli.
ESA Spec. Publ., ESA SP–207, p. 111 – 113 (1984). – See Abstr. 012.044.

Observations of elliptical galaxies identify two main features that still lack a proper interpretation: (1) Ellipticals are mostly pressure supported systems. (2) Their luminosity profiles follow an empirical law with striking regularity. Theoretical considerations and important N–body simulations suggest that these galaxies should be described by a distribution function dependent on three integrals of the motion. The authors illustrate the properties of a family of self–consistent equilibrium models that is found to provide a natural dynamical basis for the luminosity law of elliptical galaxies.

151.059 On the morphology of spiral modes.
G. Bertin, C. C. Lin, S. A. Lowe.
ESA Spec. Publ., ESA SP–207, p. 115 – 120 (1984). – See Abstr. 012.044.

A dynamical basis for Hubble's morphology classification of galaxies is given by combining the existing theory for normal spirals composed of two trailing waves with a similar theory for open spirals composed of one leading wave and one trailing wave. Thus the authors identify the two classes of spiral modes which form the prototypes for the two prongs of the tuning fork in the Hubble classification. In addition, they recognize the existence of transition spirals and bar–driven spirals as explained in a previous paper. The relevant physical regimes of the dynamical parameters are those for which modes are moderately growing, consistent with the hypothesis of Quasi–Stationary Spiral Structure.

151.060 On a spectral theory of the excitation mechanism of galactic spirals.
C. C. Lin, R. P. Thurstans.
ESA Spec. Publ., ESA SP–207, p. 121 – 130 (1984). – See Abstr. 012.044.

The spectral theory is developed in full to bring out the relationship between swinging wave packets and exponentially growing wave trains in the conventional infinite sheet model with uniform differential rotation. The theoretical results obtained are compared with those presented in the preceding paper, which is based on the local dispersion relationship for wave trains. General agreement is found. It is pointed out that there may be coexistence of quasi–stationary grand designs with transient spiral features that evolve at a rate comparable to that of shear. The quasi–stationary state may be expected only for certain regimes of dynamical parameters.

151.061 Drift–accretion theory of the spiral structure of galaxies.
Y. N. Eliseev (*Yu. N. Eliseev*), K. N. Stepanov.
ESA Spec. Publ., ESA SP–207, p. 215 – 222 (1984). – See Abstr. 012.044.

The process of the spiral structure formation in galaxies with a bar–like nucleus due to matter accretion onto the disk is considered. The matter flow patterns in the galactic plane in the corotation resonance region are constructed. The separatrices in these patterns are similar to the spiral arms of some types of galaxies. It is shown that high density regions – spiral arms – arise near the separatrices as a result of the matter accretion onto the disk. The explanation of main peculiarities of spiral design is given. The analogue with wave–particle nonlinear interaction in plasma is pointed out.

151.062 N–body simulations of the peculiar ring–tail galaxy NGC 5614–15.
M. Capaccioli, G. Malvasi.
Mem. Soc. Astron. Ital., Vol. 55, No. 3, p. 443 – 450 (1984). – See Abstr. 012.049.

The authors discuss two possible mechanisms to account for the formation of rings in galaxies. The first one is an improved version of the "central collision" mechanism discussed by Lynds (1976) and others. To the classical scheme the authors add tow aspects of great importance: the dynamical friction and a realistic model of mass distribution inside disk galaxies. They also propose a new model of interaction leading to the formation of ring galaxies, which is based on "tidal interaction" with a low mass satellite. This is a very efficient mechanism and seems to have a great probability to happen. Both schemes are applied to investigate the case of the system NGC 5614–15.

151.063 Struttura a spirale: dinamica interna e interazioni con l'ambiente.
G. Bertin.
Mem. Soc. Astron. Ital., Vol. 55, No. 3, p. 451 – 455 (1984). – See Abstr. 012.049.

Is spiral structure essentially a natural product of internal dynamics or is it mostly driven from the outside? Comments are made on these issues by stressing the role of internal dynamics and the advantage of a semi–empirical approach based on quasi–stationary spiral structure as a working hypothesis.

151.064   Hydrostatic models of gas in clusters in unsteady state with respect to an irregular force–field.
K. A. Sidorov.
Astrofizika, Tom 22, Vyp. 2, p. 303 – 313 (1985). In Russian. English translation in Astrophysics, Vol. 22, No. 2.

The hydrostatic distribution of gas in a spherical system in steady state with respect to the regular force–field but unsteady with respect to the irregular one is considered. This system has a velocity distribution with the mean–square radial velocity component larger than the mean–square transverse component. It is likely that clusters of galaxies have such a structure. The relationship between gas and galaxy densities achieved by Cavaliere and Fusco–Femiano (1976) is valid for isothermal gas in the considered system. The hydrostatic equilibrium of gas is destroyed for a cluster with a very strong asymmetry of velocity distribution. The surface brightness of X–ray radiation of the gas is calculated.

151.065   Structure of box–shaped bulges and other spheroidal components.
J. Binney, M. Petrou.
Mon. Not. R. Astron. Soc., Vol. 214, No. 3, p. 449 – 462 (1985).

The authors use a systematic approach to the choice of distribution functions to construct models of box–shaped spheroidal components. They argue that box–shaped bulges differ from elliptical spheroidal components in that in box–shaped systems, only a particular two–dimensional subset of the full three–dimensional continuum of possible orbits is heavily populated. This conjecture is tested by means of two families of distribution functions tailored to yield models with the same radial density profile as Henon's isochrone model. The rotation properties of models of the two families differ principally in that box–shaped models 'rotate on cylinders', in contrast to the more nearly spheroidal rotation characteristic of the elliptical models. This distinction is expected on general grounds and is in agreement with observation. It is argued that distribution functions of the type that generate box–shaped systems can arise either through the cannibalism of small galaxies by large, or through mergers of favourably inclined pairs of galaxies.

151.066   Distribution functions for spherical galaxies.
D. Merritt.
Mon. Not. R. Astron. Soc., Vol. 214, No. 3, p. 25P – 28P (1985).

A family of anisotropic distribution functions is derived that correspond exactly to the spherical density law $\varrho(r) \propto r^{-2}(1+r)^{-2}$, suggested by Jaffe for describing giant elliptical galaxies. The family is characterized by a single free parameter that specifies the degree of velocity anisotropy, from isotropic to completely radial. Both the distribution functions, and the equations describing the radial variation of the velocity dispersion, can be expressed in terms of simple functions.

151.067   Approximate rotation curve solutions for the evolution of a viscous protogalactic disk.
W. K. Brown.
Astrophys. Space Sci., Vol. 111, No. 1, p. 139 – 155 (1985).

The problem of evolution of viscous protogalactic disks is examined. A modification of traditional turbulent viscosity theory is proposed which is based on the premise that the eddies are so small and numerous that each eddy can be considered to be a large molecule. The main result is a family of rotation curves that, as the system evolves, serially reproduces most types of observed rotation curves. With the use of Seiden's star formation theory, the present model produces an exponential–like luminosity profile whenever stars form and the viscous action ceases.

151.068   Collisionless counterparts of Riemann ellipsoids: the self–consistent ellipsoidal model with oblique rotation.
B. P. Kondrat'ev.
Astrofizika, Tom 21, Vyp. 3, p. 499 – 521 (1984). In Russian. English translation in Astrophysics, Vol. 21, No. 3.

A new self–consistent stellar dynamics model of a rotating barred system is constructed. The case where the angular velocity vector $\vec{\Omega}$ does not coincide with a principal symmetry axis but lies in one of the principal planes of the ellipsoid is considered. The properties of individual stellar orbits are investigated and the condition for preservation of the outer elliptical boundary is derived.

151.069   Dynamical evolution in galactic disks.
R. G. Carlberg, J. A. Sellwood.
Astrophys. J., Vol. 292, No. 1, p. 79 – 89 (1985).

The authors derive the basic equation governing the response of a flat stellar disk to *any* transient–perturbing potential. They derive an expression for the induced changes in the phase–space distribution function, finding them to be second order in the perturbing potential. The result is evaluated for a model transient spiral and a continuous sequence of spirals. The predictions are in good agreement with the age/velocity–dispersion relation observed in N–body experiments, and it is concluded that recurrent transient spirals naturally account for that observed for the solar neighborhood stars.

151.070   Linear oscillations of isotropic stellar systems. II. Radial modes of energy–truncated models.
Y. Sobouti.
Astron. Astrophys., Vol. 147, No. 1, p. 61 – 66 (1985). Contrib. No. 12 Biruni Obs.

The eigenvalue problem for the first order radial perturbation of four energy–truncated distributions is solved. Wooley's and Michie–King's models are among the examples analyzed. All eigenvalues of all models with finite masses and radii are real. There is no indication of instability in dynamical time scales.

151.071   The structure and dynamics of ringed galaxies.
R. J. Buta.
Univ. Tex., Publ. Astron., No. 23, 11 + 699 pp. (1984).

The major conclusion of this dissertation is that most of the observations favor the "resonance hypothesis" for the origin of these rings in normal galaxies. This hypothesis interprets the rings as concentrations of stars and gas that have developed via secular effects near the well–known dynamical Lindblad resonances expected to occur in systems which include a small non–axisymmetric component in the gravitational potential. Only in very few cases could alternative hypotheses, such as nuclear explosions or tidal interactions with a companion, be viable explanations for the structure and dynamics of galactic rings.

151.072   Black holes and the shapes of galaxies.
C. A. Norman, A. May, T. S. van Albada.
Space Telesc. Sci. Inst., Prepr. Ser., No. 44, 47 pp. (1985). To appear in Astrophys. J.

151.073   Bifurcations and stability in three–dimensional systems.
G. Contopoulos.
ESO Sci. Prepr., No. 357, 17 pp. (1984).

The author reports some results from a study of simple perturbed three dimensional galactic systems. He finds the stability types and bifurcations of the main families of periodic orbits in such systems. Complex instability is a general phenomenon that introduces stochasticity without bifurcations. It seems that in general 3–D systems there are no infinite sequences of period doubling bifurcations. Collisions of bifurcations lead to a change of the interconnections between the various families of periodic orbits.

151.074   Triaxial galaxies containing massive black holes or central density cusps.
O. E. Gerhard, J. Binney.
MPA Rep., No. 175, 61 pp. (1985).

The box orbits that form the backbone of a triaxial elliptical galaxy carry stars arbitrarily close to the centre. In this paper the authors investigate how these orbits are affected if either (1) a massive black hole lurks at the centre, or (2) the stellar density becomes arbitrarily large near the centre.

151.075 Jacobi integral for the system of two extended configurations (binary galaxies) and an infinitesimal particle (a star).
P. V. Subrahmanyam.
Indian J. Pure Appl. Math., Vol. 14, No. 11, p. 1357 – 1361 (1983). = Nizamiah Rangapur Obs. Dep. Astron., Osmania Univ., Repr. No. 116.

Using the restricted three body problem theory the equations of motion of an infinitesimal particle (a star) moving in the gravitational field of two extended configurations (binary galaxies) are derived. The expression for the surfaces of zero relative velocity is also derived.

151.076 Dynamical evolution of two–component stellar systems: gravitational focusing effect.
S. Inagaki, P. Wiyanto.
Contrib. Bosscha Obs., No. 77, 7 pp. (1982). Paper presented at the IAU Second Asia–Pacific Regional Meeting, Bandung 24 – 29 August 1981.

To examine the effect of gravitational focusing, the dynamical evolution of two–component stellar systems is followed. This is done by solving the Fokker–Planck equation numerically. The results show that the gravitational focusing is not so effective as to inhibit the mass segregation instability.

151.077 Equipartition in multicomponent gravitational systems.
S. Inagaki, W. C. Saslaw.
Astrophys. J., Vol. 292, No. 2, p. 339 – 347 (1985).

The authors examine the role of energy equipartition and mass segregation in spherical dynamical systems having a large number of mass species. The mass spectra are power laws with a fairly wide range of steepness and cutoffs. The authors discuss the necessity and sufficiency of conditions for different types of local and global equipartition and compute Fokker–Planck simulations of various models. The results suggest that globular clusters, galactic nuclei, and clusters of galaxies will all evolve faster and produce denser cores than has been estimated from analyses with just one or two mass species. In some cases the evolution can speed up by an order of magnitude.

151.078 Die Verteilung von Blasen in Galaxien.
J. V. Feitzinger, P. E. Seiden.
Mitt. Astron. Ges., Nr. 63, p. 166 (1985). Abstract. Submitted to Astron. Astrophys. – See Abstr. 012.063.

151.079 Stern–Gas–Reibung in Galaxienkernen. Eine numerische Methode zur Berechnung gekoppelter Gas– und Stellardynamik.
R. Spurzem, T. Langbein.
Mitt. Astron. Ges., Nr. 63, p. 171 (1985). – See Abstr. 012.063.

151.080 Gas dynamics in a polar ring galaxy.
A. Habe, S. Ikeuchi.
Astron. Her., Vol. 78, No. 6, p. 156 – 162 (1985). In Japanese.

151.081 Evolution of stellar systems supplied by external stars: a model for galactic nuclei.
V. I. Dokuchaev, L. M. Ozernoj.
Pis'ma Astron. Zh., Tom 11, No. 5, p. 335 – 342 (1985). In Russian. English translation in Sov. Astron. Lett., Vol. 11.

Extended rarefied environments around the core of a non–isothermic galactic nucleus can supply the core by both energies and masses of external stars due to relaxation mechanisms. These factors can influence considerably the secular evolution of the core when competing with usual star evaporation from it. Conditions are found under which external environments influence the core evolution much more than star evaporation. This results in expansion of the core instead of its collapse.

151.082 Equilibrium non–rotating magnetized disks as a plane analog to the n = 3 polytrope.
G. S. Bisnovatyj–Kogan, Z. F. Seidov.
Pis'ma Astron. Zh., Tom 11, No. 5, p. 395 – 399 (1985). In Russian. English translation in Sov. Astron. Lett., Vol. 11.

Numerical models are constructed for a magnetized non–rotating disk which is in equilibrium due to balance between magnetic and gravitational forces. The solutions obtained turned out to be unstable against short–wave perturbations.

151.083 What happens to a galaxy in a cluster?
R. H. Miller.
Bull. Am. Astron. Soc., Vol. 16, No. 4, p. 876 (1984). Abstract. – See Abstr. 010.062.

151.084 Self–consistent models of ellipticals with $R^{1/4}$ luminosity law.
G. Bertin, M. Stiavelli.
Bull. Am. Astron. Soc., Vol. 16, No. 4, p. 889 (1984). Abstract. – See Abstr. 010.062.

151.085 New stochastic star formation models of disk galaxy evolution.
N. F. Comins, D. Balser.
Bull. Am. Astron. Soc., Vol. 16, No. 4, p. 911 (1984). Abstract. – See Abstr. 010.062.

151.086 Numerical experiments on the oscillations of axisymmetric galaxies.
P. O. Vandervoort, R. H. Miller, D. E. Welty, B. F. Smith.
Bull. Am. Astron. Soc., Vol. 16, No. 4, p. 911 – 912 (1984). Abstract. – See Abstr. 010.062.

151.087 Models of axisymmetric stellar systems.
J. L. Bishop.
Bull. Am. Astron. Soc., Vol. 16, No. 4, p. 912 (1984). Abstract. – See Abstr. 010.062.

151.088 Gravitational instabilities and the formation of molecular cloud complexes in spiral arms.
S. A. Balbus, L. L. Cowie.
Bull. Am. Astron. Soc., Vol. 16, No. 4, p. 919 – 920 (1984). Abstract. – See Abstr. 010.062.

151.089 Collisional coalescence of binaries in globular clusters.
J. G. Hills.
Bull. Am. Astron. Soc., Vol. 16, No. 4, p. 946 (1984). Abstract. – See Abstr. 010.062.

151.090 Can a moderately massive black hole reverse core collapse in globular clusters?
M. J. Duncan.
Bull. Am. Astron. Soc., Vol. 16, No. 4, p. 946 (1984). Abstract. – See Abstr. 010.062.

151.091 Evolution of stellar systems including tidally captured binaries.
J. P. Ostriker, T. S. Statler, H. M. Lee.
Bull. Am. Astron. Soc., Vol. 16, No. 4, p. 946 (1984). Abstract. – See Abstr. 010.062.

151.092 The formation and early dynamical evolution of bound stellar systems.
D. S. Dearborn, M. Margulis, C. J. Lada.
Bull. Am. Astron. Soc., Vol. 16, No. 4, p. 958 (1984). Abstract. – See Abstr. 010.062.

151.093 Viscous evolution of protogalactic disk using a modified turbulent viscosity equation.
W. K. Brown.
Bull. Am. Astron. Soc., Vol. 16, No. 4, p. 990 (1984). Abstract. – See Abstr. 010.062.

151.094 A new method for constructing spherical galaxy models with anisotropic velocity dispersions.
D. Merritt.
Bull. Am. Astron. Soc., Vol. 16, No. 4, p. 991 (1984). Abstract. – See Abstr. 010.062.

151.095   Radial and non–radial oscillations of isotropic stellar systems – solutions of Antonov's equation.
Y. Sobouti.
Bull. Am. Astron. Soc., Vol. 16, No. 4, p. 997 (1984). Abstract. – See Abstr. 010.062.

151.096   Spherical stellar systems with spheroidal velocity distributions.
D. Merritt.
Astron. J., Vol. 90, No. 6, p. 1027 – 1037 (1985).

A new method is described for deriving families of anisotropic distribution functions consistent with any spherically symmetric density profile. The algorithm is straightforward and sufficiently simple that analytic solutions can often be obtained. The models are isotropic in the center, and become either radially or tangentially anisotropic at large radii. Superposition of two or more solutions with different degrees of anisotropy allows one to construct models with almost any desired velocity structure. The algorithm is applied to a model galaxy with a de Vaucouleurs density law, and families of line–of–sight velocity dispersion profiles are obtained.

151.097   Tidal interactions in binary stellar systems.
S. M. Alladin, N. Ramamani, T. M. Singh.
J. Astrophys. Astron., Vol. 6, No. 1, p. 5 – 19 (1985).

The tidal interactions in binary stellar systems are studied under the assumption that the orbital motion of the binary is negligible in comparison with the stellar motion. Energy changes are derived. These are used to obtain simple analytical expressions for the rates of disruption and merging. This method gives an appropriate value for the Roche density. Comparison is made with earlier results obtained under the simplifying assumption that stellar motion is negligible in comparison with the orbital motion of the binary and its implications are discussed.

151.098   Should a self–gravitating system relax towards an isothermal distribution?
H. E. Kandrup.
Astrophys. Space Sci., Vol. 112, No. 2, p. 215 – 223 (1985).

Physical effects ordinarily neglected suggest that, even ignoring three– and higher–body "collisions", a self–gravitating system of stars, such as a globular cluster, does not necessarily "want" to relax completely towards an isothermal distribution. Even if one neglects evaporation and the gravothermal instability, one might anticipate deviations from a Maxwellian distribution of velocities manifest on a time scale $t_s \sim (\log N)t_R$, where t_R is the ordinary binary relaxation time and N is the number of stars.

151.099   Resonant periodic orbits in a bisymmetrical potential.
N. Caranicolas.
Astrophys. Space Sci., Vol. 112, No. 2, p. 367 – 380 (1985).

The author discusses properties of several families of resonant periodic orbits in a general, time–independent, two–dimensional potential field, symmetric with respect to both axes x and y. He classifies the different cases by varying the parameters of the problem. In order to verify the theoretical results he also presents some numerical examples.

151.100   Precollapse evolution of globular clusters.
L. Spitzer Jr.
Dynamics of star clusters, p. 109 – 137 (1985). – See Abstr. 012.070 (IAU Symp. No. 113).

The general tendency of large–N clusters towards a combination of collapse and disruption has been known for some years. The present paper discusses the progress made in this field during the past decade; the topics considered are essentially those related to the precollapse phase until n, the number of stars per unit volume, approaches its peak value. The extensive research in these fields prior to 1974 is mostly omitted from the present review.

151.101   Dynamical evolution of globular clusters after core collapse.
D. C. Heggie.
Dynamics of star clusters, p. 139 – 160 (1985). – See Abstr. 012.070 (IAU Symp. No. 113).

This review describes work on the evolution of a stellar system during the phase which starts at the end of core collapse. It begins with an account of the models of Hénon, Goodman, and Inagaki and Lynden–Bell, as well as evaporative models, and modifications to these models which are needed in the core. Next, these models are related to more detailed numerical calculations of gaseous models, Fokker–Planck models, N–body calculations, etc., and some problems for further work in these directions are outlined. The review concludes with a discussion of the relation between theoretical models and observations of the surface density profiles and statistics of actual globular clusters.

151.102   Direct Fokker–Planck calculations.
H. Cohn.
Dynamics of star clusters, p. 161 – 178 (1985). – See Abstr. 012.070 (IAU Symp. No. 113).

The past decade has seen the development of powerful numerical methods for studying star cluster evolution by direct integration of the Fokker–Planck equation. Cohn's basic algorithm for spherical systems of identical point masses and its application to the study of core collapse is reviewed. Merritt's extension of this method to treat systems containing a mass spectrum, and Goodman's extensions to include strong scattering and rotation are discussed. Results from direct Fokker–Planck computations of core collapse are presented to illustrate the method. Calculations of pre– and post–collapse evolution with a central black hole and with heating by hard binaries are reported.

151.103   Close encounters.
J. Goodman.
Dynamics of star clusters, p. 179 – 187 (1985). – See Abstr. 012.070 (IAU Symp. No. 113).

The mathematical treatment of encounters is discussed briefly, and it is pointed out that in a spherical system with an isotropic distribution function and equal–mass stars, the relevant equations can be analytically orbit–averaged. Even in small systems, close encounters have little effect on core collapse. Near a black hole, however, close encounters may cause an accumulation of stars on large, very eccentric orbits.

151.104   Dynamical evolution of multi–component clusters.
S. Inagaki.
Dynamics of star clusters, p. 189 – 205 (1985). – See Abstr. 012.070 (IAU Symp. No. 113).

Presence of stars with disparate masses causes great differences in the dynamical evolution of star clusters from the evolution of single component clusters. One remarkable effect is acceleration of the evolution. Another effect is destabilization or stabilization. In two–component clusters equipartition at the cluster centre is nearly achieved if Spitzer's (1969) condition is satisfied. In multi–component clusters equipartition at the cluster centre is achieved if either the range of stellar mass is very narrow or the mass spectrum is very steep. Global equipartition is never achieved. Post–collapse evolution of multi–component clusters is discussed briefly and some remaining problems are presented.

151.105   Gaseous models.
D. Sugimoto.
Dynamics of star clusters, p. 207 – 218 (1985). – See Abstr. 012.070 (IAU Symp. No. 113).

It is possible to understand the physics of self–gravitating systems in terms of gaseous models in so far that their global natures and effects of self–gravity are concerned. Here is summarized what is known in idealized gaseous models. They include gravothermal collapse/expansion in linear and non–linear regimes, and post–collapse evolution with gravothermal oscillation. Also discussed are their relations with discrete systems and with treatment in statistical mechanics.

151.106 Gravothermal oscillations.
E. Bettwieser.
Dynamics of star clusters, p. 219 – 230 (1985). – See Abstr. 012.070 (IAU Symp. No. 113).

Stellardynamical systems are modelled by gaseous spheres. The properties of an isothermal equilibrium configuration with a central singularity are discussed in the context of post–collapse evolution. For the numerical calculations heating effects are taken into account. Since the equilibrium configuration is gravothermally unstable a minute emission of energy is sufficient to turn gravothermal contraction into gravothermal expansion. In such a case equilibrium is approached by oscillations of the core.

151.107 Binary formation and interactions with field stars.
P. Hut.
Dynamics of star clusters, p. 231 – 249 (1985). – See Abstr. 012.070 (IAU Symp. No. 113).

Binaries provide an energy source in dense stellar systems. Exothermic gravitational interactions in star clusters can play a role similar to that of nuclear reactions in single stars. These gravitational interactions can be modeled in a laboratory setting, in the form of numerical binary–single star and binary–binary scattering experiments. Gravitational cross sections obtained this way can be applied to model star cluster evolution, just as nuclear cross sections are used as input data in stellar evolution calculations. References are given to detailed descriptions of gravitational cross sections, and a useful new example of an application is given.

151.108 Direct N–body calculations.
S. J. Aarseth.
Dynamics of star clusters, p. 251 – 259 (1985). – See Abstr. 012.070 (IAU Symp. No. 113).

The main principles for direct integration of large point–mass systems are outlined. Most particles are advanced by the Ahmad–Cohen neighbour scheme, using fourth–order force polynomials and individual time–steps. Close encounters and persistent binaries are handled by two–body regularization, whereas extreme triple and quadruple configurations are treated as unperturbed systems by special regularization techniques. As an illustration, some recent results of an isolated system containing 1000 particles of unequal mass are discussed, with special emphasis on the post–collapse phase.

151.109 A unified N–body and statistical treatment of stellar dynamics.
A. P. Lightman, S. L. W. McMillan.
Dynamics of star clusters, p. 261 – 274 (1985). – See Abstr. 012.070 (IAU Symp. No. 113).

The authors summarize the methods of a new ''hybrid'' computer code for stellar dynamics. All particles in the inner spatial region are followed exactly via a direct N–body code and all particles in the outer spatial region are treated statistically via a distribution function and Fokker–Planck type methods. An intermediate region, with features of both, allows exchange of particles and energy between the outer and inner regions. The authors apply their code to the period just before core collapse and just after and summarize the results.

151.110 Direct N–body simulations with a recursive center of mass reduction and regularization.
J. G. Jernigan.
Dynamics of star clusters, p. 275 – 284 (1985). – See Abstr. 012.070 (IAU Symp. No. 113).

The author reports on an investigation of the core collapse of an initially bound system of 1024 point masses based on a new numerical method which incorporates a recursive center of mass reduction of the N–body system and includes regularization of the two body encounters. A tree which organizes the point masses into a logarithmic hierarchy is the primary feature of the numerical algorithm. Simulations are calculated for several core collapse times in an effort to follow the end point evolution of the system. The point–like nature of the interactions is maintained throughout the simulations.

151.111 Dynamical instabilities in spherical stellar systems.
J. Barnes.
Dynamics of star clusters, p. 297 – 299 (1985). – See Abstr. 012.070 (IAU Symp. No. 113).

Equilibrium spherical stellar systems exhibiting instabilities on a dynamical timescale were first studied by Hénon (1973), using a spherically symmetric N–body code. The author has re–examined Hénon's models using an improved code which includes non–radial forces to quadrupole order. In addition to the radial instability reported by Hénon, two new non–radial instabilities are also observed. These instabilities, which are driven by fluctuations of the mean field, offer some analogies to the well–known dynamical instabilities of a cold disk of stars.

151.112 Why are stellar systems anisotropic?
E. Bettwieser, K. J. Fricke, R. Spurzem.
Dynamics of star clusters, p. 301 – 303 (1985). – See Abstr. 012.070 (IAU Symp. No. 113).

Spherical stellar systems show during their secular evolution the development of velocity anisotropy in their halo (cf. e.g. Hénon, 1971). The present study examines the general reasons for generation of anisotropy in stellar systems by means of a gaseous star cluster model including anisotropy.

151.113 A statistical treatment of low–N systems.
S. Casertano.
Dynamics of star clusters, p. 305 – 308 (1985). – See Abstr. 012.070 (IAU Symp. No. 113).

The long–term evolution of N–body systems with high N $(>10^4)$ cannot yet be studied by direct integration of the equations of motion. Instead, the study of systems with smaller N can give insight into the higher N limit in two ways. First, it is possible to measure the dependence on N of all quantities of interest for a range of N–values, and extrapolate the results to higher N. Second, even a very high–N system will eventually undergo core collapse, at which point the central dynamics will be dominated by a subset of stars with small N (10 – 100).

151.114 Functional integrals in stellar dynamics.
E. Dagan.
Dynamics of star clusters, p. 313 – 315 (1985). – See Abstr. 012.070 (IAU Symp. No. 113).

In the late 70s, the functional integral formulation equivalent to the Fokker–Planck equation was worked out (Graham). The author applies this functional integral formulation to gravitationally interacting systems, whose dynamics may be analyzed by separating the forces operating on a particle into a mean field force and fluctuations due to random collisions at intermediate range (scattering at small angles).

151.115 Multi–component models for the structure and evolution of spherical stellar systems.
E. Davoust.
Dynamics of star clusters, p. 317 – 319 (1985). – See Abstr. 012.070 (IAU Symp. No. 113).

The author presents stationary analytical models for the structure of spherical stellar systems with several mass components. Evolution is simulated by increasing the central velocity of escape. The author compares his models to the results of numerical simulations. An observable kinematical effect of mass segregation is pointed out.

151.116 The role of hard binaries in cluster evolution.
G. Giannone, D. Molteni.
Dynamics of star clusters, p. 321 (1985). Abstract. – See Abstr. 012.070 (IAU Symp. No. 113).

151.117 Collisional stellar dynamics in general relativity: an overview.
H. E. Kandrup.
Dynamics of star clusters, p. 323 – 325 (1985). – See Abstr. 012.070 (IAU Symp. No. 113).

Recently, Israel and Kandrup (1984) have formulated a new, manifestly covariant approach to non–equilibrium statistical mechanics in classical general relativity. The object here is to indicate how that formalism may be used to construct a theory of 'collisional' stellar dynamics, valid for a collection of point mass stars in the limit that incoherent radiative effects may be neglected.

151.118 3–integral models for globular clusters.
R. Lupton, J. Gunn.
Dynamics of star clusters, p. 327 – 330 (1985). – See Abstr. 012.070 (IAU Symp. No. 113).

In order to represent rotating clusters with an anisotropic velocity dispersion the authors took $f = C(e^{-\beta E}-1)e^{-\beta\gamma I3}e^{-\beta\bar{\Omega}J3}$ and approximated the third integral I_3 by the total angular momentum, J^2. The term in J_3 leads to solid body rotation in the core. This rotation leads to flattened isopotentials which means that J^2 is not an integral of the motion.

151.119 The early stages of post–collapse cluster evolution.
S. L. W. McMillan.
Dynamics of star clusters, p. 331 – 333 (1985). – See Abstr. 012.070 (IAU Symp. No. 113).

The author presents here some recent (and very preliminary) findings from a study of the early stages of the post–core–collapse evolution of an isolated cluster of identical point "stars". The method used to follow the behavior of the system is the unified N–body/statistical treatment described in detail by McMillan and Lightman (1984) and by Lightman and McMillan (1985).

151.120 Numerical simulations of encounters of hard binaries.
S. Mikkola.
Dynamics of star clusters, p. 335 – 338 (1985). – See Abstr. 012.070 (IAU Symp. No. 113).

Results from numerical integrations of random binary–binary encounters have been used to obtain various cross–sections and outcome distributions for the four–body scattering. The initial orbital elements were chosen randomly except the Kepler–energies for which various selected values were used.

151.121 Collisional relaxation: a new approach.
G. Severne, M. Luwel.
Dynamics of star clusters, p. 339 – 341 (1985). – See Abstr. 012.070 (IAU Symp. No. 113).

Recent numerical simulations for 1–dimensional systems have shown that the relaxation time due to encounters is far shorter than the generally accepted estimate. The analysis of encounters presented here is characterized by the retention of periodic trajectories in the mean field. The kinetic equation obtained yields a relaxation time scale in qualitative agreement with the simulations. The analysis can be extended to the 3–dimensional case, and preliminary results predict here also a reduction of the relaxation time.

151.122 Tidal effects on globular clusters.
P. Seitzer.
Dynamics of star clusters, p. 343 – 346 (1985). – See Abstr. 012.070 (IAU Symp. No. 113).

The author followed 1000 test particles orbiting in the outer half of a cluster for 10 cluster orbits about the Galaxy. Escape was arbitrarily defined as when a particle went beyond 3 times the expected tidal radius as computed at apogalacticon. Models were computed for different cluster orbital eccentricities e = 0.0, 0.3, 0.5, and 0.7 to check for the dependence of observed tidal radius on eccentricity.

151.123 Physical interactions between stars.
J. P. Ostriker.
Dynamics of star clusters, p. 347 – 360 (1985). – See Abstr. 012.070 (IAU Symp. No. 113).

Physical arguments are presented to show that two–body, tidal–capture binaries should form in abundance ($N_b \sim 10^3$) during the evolution of globular clusters by the time that core collapse begins. Interactions amongst these binaries and with core single stars will cause ejections from the cluster which pump energy into the system producing a bounce and re–expansion. Detailed numerical Fokker–Planck evolutionary calculations

presented here confirm this scenario and indicate that this process is likely to be the dominant energy input for most clusters.

151.124 Monte–Carlo calculations.
J. S. Stodółkiewicz.
Dynamics of star clusters, p. 361 – 372 (1985). – See Abstr. 012.070 (IAU Symp. No. 113).

The evolution of a nonisolated globular cluster is presented. The binaries (both, tidally captured and formed in three–body interactions), outflow of mass from stellar envelopes and shocks are considered as sources of energy in the cluster.

151.125 Monte Carlo simulations of the 2+1 dimensional Fokker–Planck equation: spherical star clusters containing massive, central black holes.
S. L. Shapiro.
Dynamics of star clusters, p. 373 – 413 (1985). – See Abstr. 012.070 (IAU Symp. No. 113).

The historical development is sketched and the salient features of the physical solution and its observational consequences are summarized. A new 2+1 dimensional Monte Carlo simulation code has been developed in appreciable generality to solve the time–dependent Fokker–Planck equation for this problem. Numerical results obtained with this Monte Carlo scheme for the dynamical structure and evolution of globular star clusters and dense galactic nuclei containing massive black holes are reviewed. Recent dynamical integrations of the Einstein field equations for spherical, collisionless (Vlasov) systems in General Relativity suggest a possible origin for the supermassive black holes believed to power quasars and active galactic nuclei. This scenario is discussed briefly.

151.126 Can a moderately massive black hole reverse core collapse?
M. J. Duncan.
Dynamics of star clusters, p. 415 – 417 (1985). – See Abstr. 012.070 (IAU Symp. No. 113).

N–body simulations show that the heat flux from the cusp of stars surrounding an object of mass $M \sim 10 - 100\ M_\odot$ has a profound effect on the evolution of a cluster of N stars of mass $m \sim 1\ M_\odot$. A cluster containing an object which is $\gtrsim 5\%$ of its total mass expands, transferring most of the cluster binding energy to a few stars deep in the cusp within several central relaxation times. The results suggest that core collapse in a globular cluster will be reversed when the core mass is reduced to $\sim 10\ M_\odot$. This result does not depend on whether or not stars are disrupted by the black hole.

151.127 Two body capture in large N body systems.
M. Giersz.
Dynamics of star clusters, p. 419 – 420 (1985). – See Abstr. 012.070 (IAU Symp. No. 113).

The author has considered the rates of formation of binaries in the following processes: two–body tidal encounters, three–body encounters, self–hardening of soft binaries formed in three–body encounters and two–body encounters connected with gravitational wave radiation. He has considered also the cooling and the heating connected with formation of binaries and the subsequent interactions of binaries with single stars.

151.128 Tidal stripping and disruption of globular clusters.
T. S. van Albada, T. R. Bontekoe.
Dynamics of star clusters, p. 423 – 425 (1985). – See Abstr. 012.070 (IAU Symp. No. 113).

Tidal stripping and disruption of globular clusters may be responsible for the absence of low density clusters in the inner region of the Galaxy. The authors have studied these processes by integrating orbits of stars while the cluster is moving through a spherically symmetric galactic potential with constant circular velocity ($V_{cir} = 220$ km/s). The response of the cluster to the tidal field of the Galaxy is calculated in a selfconsistent manner with a collisionless N–body code with N = 5000.

151.129 Dynamics of open star clusters.
  R. Wielen.
Dynamics of star clusters, p. 449 – 462 (1985). – See Abstr. 012.070 (IAU Symp. No. 113).
  The dynamical evolution of open star clusters has been studied successfully by numerical N–body simulations. The author compares in detail the theoretically predicted lifetimes with those derived from the observed age distribution of open clusters. Rare but efficient encounters between clusters and giant molecular clouds are probably responsible for the sudden disruption of many open clusters. Massive black holes from the galactic corona would, on the average, affect only old open clusters.

151.130 The dynamical evolution of young open clusters.
  M. Margulis, C. J. Lada, D. Dearborn.
Dynamics of star clusters, p. 463 – 465 (1985). – See Abstr. 012.070 (IAU Symp. No. 113).
  Using numerical N–body calculations the authors have simulated the dynamical evolution of young clusters as they emerge from molecular clouds.

151.131 N–body simulations of realistic open clusters.
  E. Terlevich.
Dynamics of star clusters, p. 471 – 475 (1985). – See Abstr. 012.070 (IAU Symp. No. 113).
  N–body simulations of dynamical evolution of open clusters have been computed with the purpose of comparing them with observations. Most of the models contain 1000 bodies with masses following a power–law mass function of slope $\alpha = -2.75$ and mean mass $0.5\,M_\odot$. Neutron stars or white dwarfs (depending on the initial stellar mass) are generated by instantaneous changes in individual masses, when stars reach the end of their main sequence life. Close approaches between particles are treated by a two–body regularization technique that allows to follow binary evolution in detail.

151.132 On the dynamics of open clusters.
  V. A. Ambartsumian (*V. A. Ambartsumyan*).
Dynamics of star clusters, p. 521 – 524 (1985). – See Abstr. 012.070 (IAU Symp. No. 113).
  This paper is an English translation of a classic paper by Ambartsumyan [Orig.: Uch. Zap. L.G.U. No. 22, p. 19 (1938)], in which he showed for the first time that the evolution of an isolated self–gravitating star cluster under the influence of two-body relaxation does not cease with the establishment of a maxwellian velocity distribution.

151.133 Most probable phase distribution in spherical star systems and conditions for its existence.
V. A. Antonov.
Dynamics of star clusters, p. 525 – 540 (1985). – See Abstr. 012.070 (IAU Symp. No. 113).
  This paper is an English translation of that seminal paper by Antonov which led eventually to our present understanding of the gravothermal instability. [Orig.: Vest. Leningr. Univ., Tom 7, p. 135 (1962)].

151.134 Sheared-pancake instability and protocluster fragmentation.
P. I. Kolykhalov, S. F. Shandarin.
Sov. Astron. Lett., Vol. 10, No. 6, p. 342 – 345 (1984). English translation of 38.151.040.

151.135 Some kinetic effects in inhomogeneous gravitating systems.
B. S. Sagintaev, O. V. Chumak.
Tr. Astrofiz. Inst. Alma–Ata, Tom 40, p. 12 – 20 (1983). In Russian.

151.136 Phase model of a galaxy with triaxial velocity distribution of stars.
I. L. Genkin, L. M. Genkina.
Tr. Astrofiz. Inst. Alma–Ata, Tom 40, p. 21 – 31 (1983). In Russian.

151.137 Some models of gravitating systems with variable mass.
  E. A. Malkov.
Tr. Astrofiz. Inst. Alma–Ata, Tom 40, p. 66 – 71 (1983). In Russian.

151.138 On the dynamics of gravitating systems of variable composition and mass.
M. D. Minglibaev.
Tr. Astrofiz. Inst. Alma–Ata, Tom 40, p. 72 – 79 (1983). In Russian.

151.139 On the stability of the ring structure of peculiar galaxies.
  A. A. Bekov.
Tr. Astrofiz. Inst. Alma–Ata, Tom 40, p. 80 – 86 (1983). In Russian.

151.140 Around which axis do elliptical galaxies rotate?
  B. P. Kondrat'ev.
Tr. Astrofiz. Inst. Alma–Ata, Tom 43, p. 23 – 37 (1984). In Russian.

151.141 Stochastic diffusion of an ensemble of circular orbits.
  B. S. Sagintaev, O. V. Chumak.
Tr. Astrofiz. Inst. Alma–Ata, Tom 43, p. 51 – 58 (1984). In Russian.

151.142 The law of stellar velocity distribution as discontinuity function of phase coordinates.
I. L. Genkin, L. M. Genkina.
Tr. Astrofiz. Inst. Alma–Ata, Tom 43, p. 82 – 91 (1984). In Russian.

151.143 The relation between the velocity and mass distributions. The role of collisionless relaxation processes.
L. I. Vinokurov, A. V. Kats, V. M. Kontorovich.
J. Stat. Phys., Vol. 38, No. 1 – 2, p. 217 – 229 (1985). Abstr. in Phys. Abstr., Vol. 88, No. 1257, Entry 57602 (1985).

151.144 Models of disks of galaxies: comparison with observational data.
A. V. Zasov, A. G. Morozov.
Astron. Zh., Tom 62, Vyp. 3, p. 475 – 481 (1985). In Russian. English translation in Sov. Astron., Vol. 29, No. 3.
  The results of numerical experiments of modelling 3–dimensional galactic disks are described and compared with observations.

151.145 Evolution of rotating star clusters at the stage of inelastic collisions. Gas–stellar disc dynamics.
M. M. Romanova.
Astron. Zh., Tom 62, Vyp. 3, p. 491 – 499 (1985). In Russian. English translation in Sov. Astron., Vol. 29, No. 3.
  The dynamics of a gas–stellar disc in a dense star cluster of small ellipticity ($\varepsilon \lesssim 0.3$), undergoing inelastic contact collisions, is investigated. The disc–radius time variation in an evolving cluster is calculated. During cluster contraction the disc radius is shown to decrease; moreover, the disc contraction rate exceeds that of the cluster during most of the cluster's evolution time. The disc parameters at the time of Ostriker and Peebles instability are calculated.

151.146 Instability of a nonlinearly pulsating model of a stellar system. Volume perturbations against the background of Einstein's nonequilibrium sphere.
S. N. Nuritdinov.
Astron. Zh., Tom 62, Vyp. 3, p. 506 – 517 (1985). In Russian. English translation in Sov. Astron., Vol. 29, No. 3.
  The stability with respect to volume perturbations is studied for a nonlinearly pulsating model with elliptic orbits of stars; the basis of the model is Einstein's equilibrium sphere. The corresponding complete exact dispersion relation is derived. The large–scale harmonic $N = 3$, $m = 1$ ("egg–shaped" perturbations) is analysed in detail. It is shown that in the nonequilibrium model anisotropic instability appears beginning from the value of

the ratio of radial and total transversal components by energy equal to ≈ 0.826.

151.147   Are cloud–cloud collisions necessary for global spiral structure?
W. W. Roberts Jr., G. R. Stewart.
Bull. Am. Astron. Soc., Vol. 17, No. 2, p. 547 (1985). Abstract. – See Abstr. 010.065.

151.148   Hydrodynamic theory of galactic shocks in a clumpy ISM.
G. R. Stewart, W. W. Roberts Jr.
Bull. Am. Astron. Soc., Vol. 17, No. 2, p. 547 – 548 (1985). Abstract. – See Abstr. 010.065.

151.149   Self–consistent equilibrium models of perfect ellipsoids.
T. S. Statler.
Bull. Am. Astron. Soc., Vol. 17, No. 2, p. 562 (1985). Abstract. – See Abstr. 010.065.

151.150   Compact configurations within small evolving groups of galaxies.
G. A. Mamon.
Bull. Am. Astron. Soc., Vol. 17, No. 2, p. 601 (1985). Abstract. – See Abstr. 010.065.

151.151   Clustering of galaxies in explosion scenarios.
S. Saarinen, A. Dekel, B. J. Carr.
Bull. Am. Astron. Soc., Vol. 17, No. 2, p. 602 (1985). Abstract. – See Abstr. 010.065.

151.152   Triple encounters involving finite–sized stars.
S. L. W. McMillan.
Bull. Am. Astron. Soc., Vol. 17, No. 2, p. 603 (1985). Abstract. – See Abstr. 010.065.

151.153   Stellar collisions: a numerical study.
W. Benz, J. G. Hills.
Bull. Am. Astron. Soc., Vol. 17, No. 2, p. 603 (1985). Abstract. – See Abstr. 010.065.

151.154   A model of protogalactic cloud collapse.
W. K. Brown.
Bull. Am. Astron. Soc., Vol. 17, No. 2, p. 611 (1985). Abstract. – See Abstr. 010.065.

151.155   Dynamical friction on a satellite of a disk galaxy.
G. G. Byrd, S. Saarinen, M. J. Valtonen.
Bull. Am. Astron. Soc., Vol. 17, No. 2, p. 623 (1985). Abstract. – See Abstr. 010.066.

151.156   Threshold conditions for unstable and/or chaotic orbits in Hénon–Heiles and related potentials.
K. A. Innanen.
Bull. Am. Astron. Soc., Vol. 17, No. 2, p. 623 (1985). Abstract. – See Abstr. 010.066.

151.157   Stability of spherical galaxies.
D. Merritt, L. A. Aguilar.
Bull. Am. Astron. Soc., Vol. 17, No. 2, p. 623 (1985). Abstract. – See Abstr. 010.066.

151.158   On the number of effective integrals in galactic models.
P. Magnenat.
Celest. Mech., Vol. 35, No. 4, p. 329 – 342 (1985).
Three different numerical techniques are tested to determine the number of integrals of motion in dynamical systems with three degrees of freedom: (1) The computation of the whole set of Lyapunov Characteristic Exponents (LCE), (2) The triple sections in the configurations space, (3) The Stine–Noid box–counting technique. As an application, the LCE procedure is applied to a triaxial elliptical galaxy model. Contrary to similar 2–dimensional systems, this 3–dimensional one presents noticeable zones in the phase space without any non–classical integral.

151.159   On N–body simulations of disk–galaxy evolution.
Yu. N. Mishurov.
Sov. Astron., Vol. 28, No. 6, p. 628 – 630 (1984). English translation of 38.151.093.

151.160   The structure and dynamics of ringed galaxies.
R. J. Buta.
Diss. Abstr. Int., Sect. B, Vol. 45, No. 8, p. 2580 (1985). Thesis, University of Texas, 1022 pp. (1984). Order No. DA8421673.

151.161   Dynamics of gas disks in triaxial galaxies.
T. Y. Steiman–Cameron.
Diss. Abstr. Int., Sect. B, Vol. 45, No. 9, p. 2958 (1985). Thesis, Indiana University, 285 pp. (1984). Order No. DA8426685.

151.162   Waves and gravitational instability in a rotating polytropic cylinder.
S. Ishibashi, H. Ando.
Theoretical aspects on structure, activity, and evolution of galaxies: III, p. 3 – 10 (1985). – See Abstr. 012.110.
The authors briefly review the various kinds of waves appearing in a rotating system, and then they investigate the nature of these waves and also the gravitational instability for a series of rotating polytropic cylinders. In particular, they focus on the unstable mode. Their numerical results are tried to be interpreted using the local theory developed by Ando (1984).

151.163   Global instability of stellar disks with Kuzmin's density distribution.
T. Fujiwara, S. Hozumi.
Theoretical aspects on structure, activity, and evolution of galaxies: III, p. 11 – 16 (1985). – See Abstr. 012.110.
The authors determined unstable global modes of stellar disks by solving the linearized collisionless Boltzmann equation as an initial value problem. The models examined are the Kuzmin disks with two types of the distribution function: one has local stability parameter Q rising with radius and the other has nearly constant Q. It is found that the growth rate correlates well with the value of Q at the center of the disk.

151.164   Global instability of stellar disks with bulge–like components.
S. Hozumi, T. Fujiwara, M. T. Nishida.
Theoretical aspects on structure, activity, and evolution of galaxies: III, p. 17 – 22 (1985). – See Abstr. 012.110.
The authors investigate the effects of a bulge–like component on the stability of stellar disks against linear two–armed modes. A series of mode calculations have been made for the disk with a bulge of different length scales. The results show that a compact bulge is effective in reducing the growth rate of instabilities, even if it is not so massive as an often assumed halo.

151.165   Interchange model of disk galaxies with gas infall.
Y. D. Tanaka, S. Ikeuchi, A. Habe.
Theoretical aspects on structure, activity, and evolution of galaxies: III, p. 33 – 40 (1985). – See Abstr. 012.110.
Assuming that a disk galaxy is composed of a general pervasive (ambient) medium, small clouds, molecular clouds and stars, its evolution is studied through examining interchange processes including gas infall.

151.166   Dissipative structures in galaxies – effects of advection.
T. Nozakura, S. Ikeuchi.
Theoretical aspects on structure, activity, and evolution of galaxies: III, p. 46 – 48 (1985). – See Abstr. 012.110.
Effects of advection on the dissipative structures of interstellar matter are investigated using 1D spiral coordinates. The results of simulation suggest that there could be dissipative structures even if the center–of–mass–velocity has a strong spatial structure.

151.167 N–body simulation of giant molecular clouds in a galaxy.
M. Fukunaga, M. Tosa.
Theoretical aspects on structure, activity, and evolution of galaxies: III, p. 49 – 56 (1985). – See Abstr. 012.110.

The authors performed N–body simulations of particles of thin disk layer in general background potential in order to see how the giant molecular clouds acquire random motion from pure circular rotation in centrifugal balance with the galactic gravitational field through mutual gravitational encounters. The simulation shows that the increase of velocity dispersion of random motion agrees with that of an infinitesimally thin gas layer caused by viscous dissipation of mechanical energy of rotation.

151.168 Evolution of globular clusters after a gravothermal catastrophe.
S. Inagaki.
Theoretical aspects on structure, activity, and evolution of galaxies: III, p. 57 – 64 (1985). – See Abstr. 012.110.

The evolution of globular cluster cores is studied by modifying the simple homology model. Binaries formed by two–body tidal dissipational encounters as well as by three–body encounters are taken into account. It is shown that escape of stars due to encounters between a tidally captured binary and a single star causes the reexpansion of the cluster cores.

151.169 Mass exchange of galaxies by a tidal encounter.
M. T. Nishida, K.–i. Wakamatsu.
Theoretical aspects on structure, activity, and evolution of galaxies: III, p. 89 – 91 (1985). – See Abstr. 012.110.

The authors simulate tidal encounters between two galaxies whose gravitational potentials are determined by massive extended spheres and investigate the mass exchange process due to those tidal encounters.

151.170 A further study of solution to three–dimensional Poisson equation for spirals.
H.–g. Bi, Q.–h. Peng, Z.–y. Li.
Acta Astron. Sin., Vol. 25, No. 4, p. 356 – 364 (1984). In Chinese.

151.171 The penetration of a dwarf galaxy into an elliptical galaxy.
G.–x. Song.
Acta Astron. Sin., Vol. 26, No. 1, p. 87 – 91 (1985). In Chinese.

In this paper a numerical simulation of the penetration of a dwarf galaxy into an elliptical galaxy has been done. The ratio of the mass of the dwarf to that of the elliptical galaxy is 1/5 and the direction of motion of the dwarf is similar to that of rotation in the elliptical galaxy. The result has shown that different from the case in spirals without halo, the distortion induced in the elliptical galaxy by the tidal forces of the dwarf is smaller.

151.172 Spiral solitons in flat gaseous disks of galaxies.
M. G. Abramyan.
Astrofizika, Tom 22, Vyp. 3, p. 487 – 503 (1985). In Russian. English translation in Astrophysics, Vol. 22, No. 3.

The theory of non–linear perturbations of a self–gravitating thin rotating gaseous disk is developed taking into consideration terms of five–degree nonlinearity. The critical value of the surface polytropic index, below which develops a break–up instability of the disk, decreases by a few percent. The correction to the amplitude and small changes of a cubic soliton depend on its group velocity. The question of stability of the solitons is investigated.

151.173 Evolution of a massive binary in a star field.
A. S. Baranov.
Sov. Astron., Vol. 28, No. 6, p. 642 – 648 (1984). English translation of 38.118.047.

151.174 Contemporary dynamical problems – a possible contribution to their solution by galaxy photometry.
L. Martinet.
New aspects of galaxy photometry, p. 121 – 130 (1985). – See Abstr. 012.114.

This paper discusses some of the most recent developments in galactic dynamics, in particular those for which the contribution of galaxy photometry may be useful in the future. The paper is divided into three parts: (1) Triaxiality in elliptical galaxies (2) Stability of disks and structure of bars (3) Interactions between components of galaxies.

151.175 Can shells help to distinguish prolate from oblate elliptical galaxies?
C. Dupraz, F. Combes.
New aspects of galaxy photometry, p. 151 – 154 (1985). – See Abstr. 012.114.

The authors present 3–body simulations of a galaxy merger involving a true elliptical potential, either oblate or prolate (ellipticity in density of E3.5). It is shown that the shell systems formed in the two cases are quite different: shells are indeed a mean of revealing the unknown 3D shape of elliptical galaxies.

151.176 Constraints on the potential of barred galaxies from surface photometry and direct photography.
E. Athanassoula.
New aspects of galaxy photometry, p. 155 – 160 (1985). – See Abstr. 012.114.

This paper outlines a method whereby the gravitational potential of a barred galaxy can be derived from surface photometry and from an analysis of the shape of dust lanes in these galaxies.

151.177 N–body simulations and galaxy photometry.
O. E. Gerhard.
New aspects of galaxy photometry, p. 175 – 182 (1985). – See Abstr. 012.114.

N–body models are now a standard tool to improve our understanding of galaxies. Since photometric and kinematic observations of a galaxy do not in general very strongly constrain a dynamical model, the main use of N–body simulations in this respect is to answer qualitative questions about the nature of possible equilibria. This is illustrated by presenting a model with strong radial ellipticity gradients and one with intrinsic principal axes twists. Another important application of N–body methods is in the study of transient phenomena.

151.178 Constructing distribution functions for galaxies with application to M87.
A. J. Newton.
New aspects of galaxy photometry, p. 279 – 282 (1985). – See Abstr. 012.114.

The distribution of orbits in a galaxy is described by the phase–space distribution function $f(x,v,t)$. The task is to construct a distribution function which yields surface brightness and kinematics compatible with the observations. The author discusses a technique that has been developed for producing distribution functions for spherical systems, and its application to the nearly-spherical elliptical M87. The extension of this technique to a wider class of stellar systems is also briefly considered.

151.179 Stochastic stellar orbits in galaxies.
O. E. Gerhard.
New aspects of galaxy photometry, p. 329 – 330 (1985). – See Abstr. 012.114.

The author has studied the appearance of stochastic orbits caused by perturbations of two–dimensional integrable galaxy potentials, using a canonically invariant form of a method due to Melnikov.

151.180 The velocity fields of barred spirals.
D. Pfenniger.
New aspects of galaxy photometry, p. 341 – 342 (1985). – See Abstr. 012.114.

Schwarzschild's method (1979) is used for constructing self–consistent 2–D models of a well pronounced and fast rotating SB galaxy.

151.181   **Flat rotation curves and exponential luminosity profiles from a model of galaxy formation.**
W. K. Brown.
Proceedings of the Southwest Regional Conference for Astronomy and Astrophysics, Vol. 10, p. 49 (1985). Abstract. – See Abstr. 012.125.

The dynamics of stellar and planetary systems. Summary of the RAS specialist discussion, held 1984 October 12 at the Scientific Societies' Lecture Theatre, Savile Row.
See Abstr. 010.722.

Particle simulation of plasmas and stellar systems.
See Abstr. 014.043.

Photoelectric meridian circles for planetary theory and stellar dynamics.
See Abstr. 041.020.

The dynamics of bodies with variable masses.
See Abstr. 042.026.

Periodic orbits.
See Abstr. 042.028.

Energy conserving, arbitrary order numerical solutions of the N–body problem.
See Abstr. 042.088.

The manifold structure for collision and for hyperbolic–parabolic orbits in the n–body problem.
See Abstr. 042.093.

Birkhoff normalization and resonances in dynamics.
See Abstr. 042.099.

Nonlinear spiral density waves: an inviscid theory.
See Abstr. 062.112.

Equilibrium models of differentially–rotating polytropic cylinders.
See Abstr. 065.017.

On the role of rotation and magnetic field in the structure and energetics of AGN.
See Abstr. 067.083.

The collapse of dense star clusters to supermassive black holes: the origin of quasars and AGNs.
See Abstr. 067.101.

Effect of dynamical friction on the escape of a supermassive black hole from a galaxy.
See Abstr. 067.118.

Periodic star formation activity and color change of galaxies.
See Abstr. 131.365.

Behavior of interstellar gas clouds in galaxy encounters.
See Abstr. 131.366.

Molekülwolken und die Lebensdauer offener Sternhaufen.
See Abstr. 153.032.

The structure and internal kinematics of open clusters.
See Abstr. 153.045.

Report of IAU Commission 37: Star clusters and associations (*Amas stellaires et associations*).
See Abstr. 153.060.

Stellar systems, star clusters and globular clusters.
See Abstr. 154.019.

Rotation and flattening of globular clusters.
See Abstr. 154.065.

Minimum of the eccentricity of the galactic globular cluster orbits.
See Abstr. 154.066.

Black hole remnants in globular clusters.
See Abstr. 154.067.

What next? Priorities in theory and observations. Summary of the panel discussion.
See Abstr. 154.068.

The velocity–age relation for B and A stars.
See Abstr. 155.060.

Report of IAU Commission 33: Structure and dynamics of the galactic system (*Structure et dynamique du système galactique*).
See Abstr. 155.193.

A collision between the Large and Small Magellanic Clouds 2×10^8 years ago.
See Abstr. 156.006.

Tidal models of the Magellanic system.
See Abstr. 156.013.

Kinematics and dynamics of the haloes of supergiant galaxies.
See Abstr. 157.004.

Global modal analysis of disk galaxies: application to the S0 galaxy NGC 3115.
See Abstr. 157.048.

Properties of barred spiral galaxies.
See Abstr. 157.063.

An analysis of observations of the streaming velocities in the bulge of M31.
See Abstr. 157.070.

The plume phenomenon in barred spirals.
See Abstr. 157.108.

Blast wave formation of the extended stellar shells surrounding elliptical galaxies.
See Abstr. 157.118.

Dynamics of the bulge of the lenticular galaxy NGC 7814.
See Abstr. 157.142.

The spiral structure of galaxies.
See Abstr. 157.210.

The dynamics of tri–axis elliptical Milky Way systems.
See Abstr. 157.212.

Can ram pressure distort spiral arms?
See Abstr. 157.249.

The three–dimensional shape of galaxies.
See Abstr. 157.260.

Dynamics of ellipticals: the case for 2–dimensional photometry.
See Abstr. 157.261.

Dark halos around late–type galaxies.
See Abstr. 157.264.

Rotation and velocity dispersion in the stellar component of NGC 1316 (Fornax A).
See Abstr. 158.009.

Slippery evidence on masses in the Local Group.
See Abstr. 160.008.

Simulations of galaxy cluster relaxation.
See Abstr. 160.011.

The intergalactic H I cloud in Leo: a simple modeling of the Spitzer–Baade collision event.
See Abstr. 160.030.

The nature of orbits of multiple nuclei near brightest cluster galaxies.
See Abstr. 160.058.

On the infall velocity towards Virgo.
See Abstr. 160.137.

Brightest member luminosities and dynamical times of galaxy clusters.
See Abstr. 160.158.

On massive neutrino halos and galactic structure.
See Abstr. 161.229.

152 Stellar Associations

152.001   The initial mass function in the R–associations CMa R1, Mon R1 and Mon R2 from radio data.
T. B. Pyatunina.
Pis'ma Astron. Zh., Tom 11, No. 1, p. 27 – 36 (1985). In Russian. English translation in Sov. Astron. Lett., Vol.11.

Results of a search for compact radio sources in the R–associations CMa R1 and Mon R1 are given. The number of sources found in the association Mon R1 is approximately equal to the expected number of background extragalactic radio sources. In the association CMa R1 seven radio sources of small angular diameter with flux greater than 30 mJy are found, two of which probably are background sources. A comparison of optical and radio data on the association CMa R1 and previously published data on the association Mon R2 make it possible to estimate the initial mass function for associations under study: $\xi(M) \propto M^{-2.7\pm0.7}$ for stars with $M \sim 10M_\odot$.

152.002   Multicolour UBVRi photographic and photoelectric photometry of members and possible members of T–associations T1 and T3 Tauri.
U. A. Nurmanova.
Perem. Zvezdy, Tom 21, No. 6, p. 777 – 805 (1983). In Russian.

The results of UBVRi photometry of members and possible members of T–associations T1 and T3 Tauri beginning from 1974 are given. The method by which the observations were made is described.

152.003   Time–dependent star formation in OB associations.
C. Doom, J. P. De Greve, C. de Loore.
Astrophys. J., Vol. 290, No. 1, p. 185 – 190 (1985).

The authors show that in the OB associations Per OB1 and Cen OB1, the lower mass stars were formed first, the most massive stars being $10 - 20 \times 10^6$yr younger than the lower mass stars. It is shown that this result cannot be due to selection effects. The consequences of this time–dependent star formation are briefly discussed.

152.004   A search for OB associations near long–period Cepheids. III. U Carinae,   XZ Carinae,   QY Centauri, VX Crucis, and AA Normae.
S. van den Bergh, P. F. Younger, D. G. Turner.
Astrophys. J., Suppl. Ser., Vol. 57, No. 4, p. 743 – 750 (1985). With plates 13 – 17.

UBV photometry is presented for fields surrounding five long–period Cepheids. None of these objects appear to be located in an OB association.

152.005   X–ray spectra of stars in the Cygnus OB–2 association.
F. R. Harnden Jr.
Bull. Am. Astron. Soc., Vol. 16, No. 4, p. 948 (1984). Abstract. – See Abstr. 010.062.

152.006   On the dynamical evolution of Trapezium–type stellar systems.
T. S. Kozhanov.
Tr. Astrofiz. Inst. Alma–Ata, Tom 40, p. 3 – 11 (1983). In Russian.

152.007   Stellar kinematics in star–forming regions.
R. D. Mathieu, D. W. Latham, L. Hartmann.
Stellar radial velocities, p. 263 – 264 (1985). – See Abstr. 012.095 (IAU Colloq. No. 88).

High–precision radial–velocity studies of three young, star–forming regions – λ Orionis, NGC 2264 and Taurus–Auriga – are briefly described.

152.008   Radial velocities of F and G dwarfs in the Orion cluster region.
J.–C. Mermilliod, M. Mayor.
Stellar radial velocities, p. 367 – 370 (1985). – See Abstr. 012.095 (IAU Colloq. No. 88).

Rotational velocities have been measured with CORAVEL for 38 stars in the Orion Ic association. The distribution of $v \sin i$ for the F– and G–type dwarfs in the Orion Ic association, in a post T Tauri stage (age around 4×10^6y), does not present the bimodal characteristics observed in the rotation of the stars in older open clusters like the Pleiades.

152.009   Time–dependent energetics of OB associations and gamma–ray sources.
C. Doom, T. Montmerle.
18th International Cosmic Ray Conference, Vol. 1, p. 152 (1983). Abstract. – See Abstr. 012.096.

152.010   The chain of three Hα stars in Vulpecula.
V. M. Petrosyan.
Astrofizika, Tom 22, Vyp. 3, p. 635 – 636 (1985). In Russian. English translation in Astrophysics, Vol. 22, No. 3.

Three new Hα stars were discovered inside of a dark nebula in Vulpecula. These stars form a Trapezium type system.

Early–type stars in OB associations in the infrared. II. A discussion of density and temperature distributions in stellar winds.
See Abstr. 112.044.

***UBV* photometry of new Hα stars in the association Orion OB1d.**
See Abstr. 113.030.

Remote OB stars in Puppis.
See Abstr. 114.119.

An empirical Hγ luminosity calibration for class V – III stars.
See Abstr. 115.003.

The evolutionary status of OB stars with peculiar nitrogen spectra.
See Abstr. 115.011.

Trapezium–type multiple systems.
See Abstr. 118.027.

A search for long–period Cepheids in associations.
See Abstr. 122.073.

New flare stars in the Mon I association.
See Abstr. 122.120.

Verification of Cepheids in open clusters and associations as distance indicators.
See Abstr. 122.158.

New variable stars in the region of the association T4 Cygni.
See Abstr. 123.004.

Structure and origin of the Cygnus superbubble.
See Abstr. 131.056.

Structure and evolution of superbubbles in the stratified gas distribution.
See Abstr. 131.364.

X–ray observations of the runaway stars HD 206327 and 26 Cephei and of the λ^1Orionis region.
See Abstr. 142.029.

Report of IAU Commission 37: Star clusters and associations (*Amas stellaires et associations*).
See Abstr. 153.060.

Age and processes of star formation in very young LMC associations.
See Abstr. 156.017.

The structure of the spiral arm S4 in Andromeda galaxy. (A contribution to the theory of density waves).
See Abstr. 157.126.

153 Open Clusters

153.001   Ultraviolet studies of the young populous cluster NGC 2100 in the LMC.
E. Böhm–Vitense, P. Hodge, C. Proffitt.
NASA Conf. Publ., NASA CP–2349, p. 191 – 195 (1984). – See Abstr. 012.001.

Stars in the populous young cluster NGC 2100 were observed with IUE in the low resolution mode. Color excesses, effective temperatures and luminosities were determined for eight stars in the cluster. A comparison of observed and model atmosphere energy distributions shows that the Nandy et al. (1981) average LMC extinction curve does not give good matches for the cluster stars. The authors find masses between 15 and 30 solar masses for their stars and an age of about 6×10^6years. Two stars, B27 and C1, may have an age of up to 12×10^6years.

153.002   *uvby* β photometry of southern clusters – VI. NGC 3766.
     R. R. Shobbrook.
Mon. Not. R. Astron. Soc., Vol. 212, No. 3, p. 591 – 599 (1985).

Strömgren four colour and Hβ photometry has been obtained for 39 early to mid–B stars in the young southern cluster NGC 3766. The cluster contains an unusually large number of Be stars, the proportion between magnitudes 6.5 and 11.5 being about 30 per cent. Estimates from both *uvby*β and *UBV* data indicate a reddening $E(b-y)$ of 0.15 mag, corresponding to $E(B-V)$ 0.20. The distance modulus obtained from both photometric systems is 11.4, giving a distance of 1.9 kpc.

153.003   Mapping of the Beta Cepheid instability strip: the cluster NGC 6231.
L. A. Balona, C. A. Engelbrecht.
Mon. Not. R. Astron. Soc., Vol. 212, No. 4, p. 889 – 897 (1985).

Intensive photometry of 17 early–B stars in the young open cluster NGC 6231 was conducted in order to map the β Cephei instability strip near the zero–age main sequence. The authors report the discovery of two certain β Cephei stars in addition to the four already known. One probable and two possible variables of this class were also found.

153.004   *UBV* three–colour photometric parameters of four galactic clusters near Cassiopeia.
R. P. Fenkart, A. Schröder.
Astron. Astrophys., Suppl. Ser., Vol. 59, No. 1, p. 83 – 86 (1985).

Interstellar reddening and absorption, distance, earliest spectral type and probable physical membership have been determined for four galactic clusters near Cassiopeia: NGC 7510, Ba 2, Ba 3 and NGC 7654, according to Becker's (1963) method A, using both colour–magnitude diagrams.

153.005   Membership and photometric abundances of red evolved stars in open clusters.
J. J. Clariá.
Astron. Astrophys., Suppl. Ser., Vol. 59, No. 2, p. 195 – 204 (1985).

Results from *UBV* and DDO intermediate–band photometry are presented for 31 red evolved stars in eleven open clusters. The likelihood of membership for each observed star is evaluated by using two independent photometric criteria. Three of the clusters studied (NGC 2232, NGC 2264 and NGC 2451) probably do not contain red evolved stars. The photometric abundances range between [Fe/H] = –0.2 in NGC 2343 and Ruprecht 46 to about [Fe/H] = +0.3 in NGC 2546.

153.006   *uvby* and Hβ photometry of the open cluster IC 4996.
     E. J. Alfaro, A. J. Delgado, J. M. García–Pelayo,
R. Garrido, M. Sáez.
Astron. Astrophys., Suppl. Ser., Vol. 59, No. 3, p. 441– 447 (1985).

Photoelectric *uvby* and Hβ observations have been obtained for 15 stars in the field of the open cluster IC 4996. An average colour excess of $E(b-y) = 0.473$ is estimated for six observed members of the cluster, in good agreement with previous results. A true distance modulus of 11.43 and an age of 7.5×10^6years is calculated for IC 4996.

153.007   NGC 2112 – a forgotten old open cluster.
     T. Richtler.
Astron. Astrophys., Suppl. Ser., Vol. 59, No. 3, p. 491 – 495 (1985).

A first color–magnitude diagram of the open cluster NGC 2112 is presented. Although the photometry does not reach the main sequence and thus no firm conclusion concerning the age is possible, it turns out that the cluster has to be placed among the oldest open clusters known. Preliminary values for distance and reddening are 800 pc and $0^m\!.5$, respectively. Strömgren data of a few stars indicate a metallicity about –1 dex.

153.008 Systematic search for members of the Hyades Supercluster. VI. The lower main sequence.
O. J. Eggen.
Astron. J., Vol. 90, No. 1, p. 74 – 79 (1985).

From a complete sample of stars with visual magnitude brighter than 15 and annual proper motion larger than 0.5 arcsec, 50 members of the Hyades Supercluster have been identified by their proper motion. The resulting main sequence is represented, for $M_{bol} > +5.5$ mag, by $M_{bol} = 4.51 + 4.31(R–I)$ with a dispersion of 0.15 mag. The supercluster stars may represent 15% of the star density near the Sun. Two incomplete samples of red dwarfs, formed by Vyssotsky and by Weis, are also examined for supercluster stars and add 34 members.

153.009 A systematic search for members of the Hyades Supercluster. V. The red giants.
O. J. Eggen.
Astron. J., Vol. 90, No. 2, p. 333 – 340 (1985).

The direction of the proper motion identifies 44 bright, red giants as members of the Hyades Supercluster and the available radial–velocity observations do not eliminate any of these from membership. The supercluster contains the known variables R Hya, RR Sco, π^1Gru, R Lyr, VZ Cam, AD Cet, and the two carbon variables TW Hor and NP Pup. Available intermediate– band photometry gives luminosities for the G– and K–type stars that are in reasonable agreement with the astrometric values and, with DDO photometry, indicates a large degree of consistency in the metal abundance of G– and K–type members.

153.010 Investigation of the initial mass spectrum of open star clusters.
B. Stecklum.
Astron. Nachr., Vol. 306, No. 2, p. 45 – 61 (1985). = Mitt. Univ.–Sternw. Jena, Nr. 165.

The mass spectra of 228 open star clusters were derived by comparison of colour–magnitude diagrams with evolutionary tracks. The application to binary stars showed the reliability of the mass determination. The derived mass spectra were fitted by power laws as well as exponential laws. The present investigation revealed a correlation of the slope of the mass spectra with the cluster age, whereas a detected correlation of the slope with galactocentric distance is slight. The results suggest that the slope of the mass spectrum increases with increasing cluster age and galactocentric distance. These findings are discussed with respect to their reasons and previous results concerning open clusters and field stars.

153.011 A photometric study of the open cluster M39.
V. Mohan, R. Sagar.
Mon. Not. R. Astron. Soc., Vol. 213, No. 2, p. 337 – 344 (1985).

Photoelectric UBV magnitudes of 50 stars in the field of M39 have been obtained for most of which previous photoelectric data were not available. These include 36 new members given by Platais (1984). The reddening for the cluster has been found to be 0.03 mag. A distance of 300 pc has been estimated for the cluster. The age of the cluster lies between 2.0×10^8 and 4.0×10^8 yr.

153.012 The Pleiades cluster. II. Does it contain pre–main– sequence stars?
M. Breger.
Astron. Astrophys., Vol. 143, No. 2, p. 455 – 457 (1985).

The suspected existence of pre–main–sequence stars in the Pleiades cluster is based on measurements of abnormal polarization as well as the position of many faint stars above the zero–age main sequence. New polarimetric measurements do not confirm reported abnormal wavelength dependences of polarization.

153.013 Cousins VRI photometry of the Hyades, Coma, and M67.
B. J. Taylor, M. D. Joner.
Astron. J., Vol. 90, No. 3, 479 – 487 (1985).

For little–evolved stars in the Hyades, Coma, and M67, the authors present Cousins VRI photometry which includes and expands on results published in other systems by Taylor. The observing and reduction procedures and error analysis are discussed in some detail, and evidence is presented that the transformations to the Cousins system are satisfactory and the results are internally consistent. Comparisons with data published by Mendoza are made.

153.014 On the distance to the open cluster Lyngå 6.
E. R. Anderson, B. F. Madore, M. H. Pedreros.
Cepheids: theory and observations, p. 203 – 204 (1985). – See Abstr. 012.027. (IAU Colloq. No. 82).

153.015 Evolution of low–mass stars in the Alpha Persei cluster.
J. R. Stauffer, L. W. Hartmann, J. N. Burnham, B. F. Jones.
Astrophys. J., Vol. 289, No. 1, p. 247 – 261 (1985).

The authors present a photometric and spectroscopic study of low–mass members of the α Persei cluster. New relative proper motions have been obtained for 4000 stars in a $1°2 \times 1°2$ region of the α Persei open cluster. The survey extends to $V \approx 16.5$ mag. Optical photometry and high–dispersion spectroscopy of the possible cluster members have also been obtained. The new photometry shows an apparent pre–main sequence (PMS), but it is not yet possible to accurately determine the PMS turn–on point. Projected rotational velocities derived from the echelle spectra indicate that nearly 50% of the stars observed that are later than G2 have 25 km s^{-1} $< v \sin i < 150$ km s^{-1}. The large rotational velocities among low–mass stars in young clusters are ascribed to spin–up during contraction to the main sequence. The evolution of stellar rotation in open clusters is outlined. Considerable differences in the evolution of G dwarfs and K dwarfs are noted.

153.016 The Einstein soft X–ray survey of the Pleiades.
J.–P. Caillault, D. J. Helfand.
Astrophys. J., Vol. 289, No. 1, p. 279 – 299 (1985).

The authors have conducted an extensive soft X–ray survey of the Pleiades open cluster using the imaging instruments on board the Einstein Observatory. A total of 61 X–ray sources were detected in the 4 square degree survey region, of which 44 can be identified with cluster members of spectral types B – M. The results of this survey are used together with data for the Orion Nebula and the Hyades to examine in detail the following topics: (1) X–ray emission from cluster stars compared and contrasted with that of field stars, (2) the coronal activity–rotation relation for late–type stars, and (3) stellar X–ray flares – their frequency, strength, and time scales.

153.017 The occurrence of peculiar stars in clusters and visual systems.
H. A. Abt.
The MK process and stellar classification, p. 369 – 371 (1984). – See Abstr. 012.033.

153.018 CCD photometry of Galactic clusters containing Cepheid variables – I. Lyngå 6.
A. R. Walker.
Mon. Not. R. Astron. Soc., Vol. 213, No. 4, p. 889 – 897 (1985).

A colour–magnitude diagram extending to $V = 20$ ($M_V \cong 4.5$) is given for the cluster Lyngå 6. The true distance modulus is found to be 11.15 ± 0.3, assuming $E(B–V) = 1.34$. The 10.8–day Cepheid TW Nor, situated near the cluster centre, has $\langle M_V \rangle = -3.8 \pm 0.3$ and appears to be a cluster member.

153.019 CCD photometry of galactic clusters containing Cepheid variables. II. NGC 6067.
A. R. Walker.
Mon. Not. R. Astron. Soc., Vol. 214, No. 1, p. 45 – 53 (1985).

A colour–magnitude diagram extending to $V = 19$ is given for 1019 stars in the open cluster NGC 6067. The true distance modulus is found to be 11.05 ± 0.10 mag assuming $E(B–V) = 0.35$. A colour–magnitude diagram for an adjacent field shows that the Norma star cloud is at a similar distance to NGC 6067. The Cepheids CPD $-53°7400$A ($p = 11.3$ day) and HD 144972 ($p = 3.8$ day) have $\langle M_V \rangle = -3.77 \pm 0.10$ and $\langle M_V \rangle = -3.27 \pm 0.10$, respectively, if they are cluster members.

153.020   UBV photographic photometry of Ruprecht 36.
B. C. Reed.
J. R. Astron. Soc. Can., Vol. 79, No. 1, p. 1 – 8 (1985).
Photographic UBV photometry is presented for 78 stars in the region of the small open star cluster Ruprecht 36. Consistent with Vogt and Moffat (1972) the author finds a colour excess E_{B-V} of $0^{m}.13$ and a distance of 2.1 kpc.

153.021   New photometric data for the red giants in the open cluster NGC 5822: membership, chemical composition and physical properties.
J. J. Clariá, E. Lapasset.
Mon. Not. R. Astron. Soc., Vol. 214, No. 2, p. 229 – 240 (1985).
New photoelectric UBV, DDO, and CMT_1T_2 data of 14 G and K stars in the field of the open cluster NGC 5822 are presented. By applying two independent photometric criteria 11 stars are found to have a high probability of being cluster giants, while three remaining stars are almost certainly red field objects. A reddening $E(B-V) = 0.15 \pm 0.01$, a distance modulus of $V_0 - M_V = 8.77 \pm 0.18$, and an age of about $6 - 8 \times 10^8$yr have been derived. NGC 5822 has a mean ultraviolet excess of $\delta(U-B) = 0.00 \pm 0.01$, with respect to the field K giants, and a mean cyanogen anomaly of $\delta CN = +0.031 \pm 0.008$, both implying a nearly solar iron abundance. The effective temperatures, surface gravities and masses of the red giants have also been determined.

153.022   Mg II emission of late main sequence stars in open clusters.
S. Catalano, E. Marilli.
Mem. Soc. Astron. Ital., Vol. 55, No. 3, p. 417 – 421 (1984). – See Abstr. 012.049.
IUE spectra of the Mg II h and k lines of main sequence stars in the Hyades and in the ζ Her moving group are presented. The emission luminosity in these lines shows a power law dependence as function of the mass, with slope not dependent on the age.

153.023   Faint stellar photometry in clusters. II. NGC 6791 and NGC 6535.
B. J. Anthony–Twarog, B. A. Twarog.
Astrophys. J., Vol. 291, No. 2, p. 595 – 610 (1985).
Video camera photometry of the clusters NGC 6791 and NGC 6535 obtained with the KPNO 2.1 m telescope is presented. For the open cluster NGC 6791, the new data provide improved delineation of the cluster color–magnitude diagram in the region of the turnoff and fainter. NGC 6791 is found to be intermediate in age between NGC 188 and Melotte 66. Its apparent distance modulus is $(m-M) = 13.5$ using the $Y = 0.3$ Yale isochrones and $(m-M) = 13.2$ using the $Y = 0.2$ isochrones of VandenBerg. For the sparse globular NGC 6535, the photometry reveals a color–magnitude diagram essentially identical to that of M13, with an age comparable to that found for other globulars. No evidence is found for a significant population of blue stragglers in the cluster core.

153.024   Photometry and age of star clusters in M33.
A. S. Sharov, V. M. Lyutyj.
Sov. Astron. Lett., Vol. 10, No. 5, p. 273 – 276 (1984). English translation of 38.153.016.

153.025   Steps towards the abundance scale. I. The abundance of heavy elements in the Hyades cluster.
R. Cayrel, G. Cayrel de Strobel, B. Campbell.
Astron. Astrophys., Vol. 146, No. 2, p. 249 – 259 (1985).
High signal–to–noise ratio coudé spectra of twelve Hyades dwarfs, plus comparison solar spectra, have been analyzed to obtain accurate abundances for metals (principally iron) in the Hyades. Temperatures relative to the Sun were derived from Hα wing profiles or broadband colors, and cluster stars were selected to bracket the solar temperature. Equivalent widths of metal lines have been obtained with a profile fitting technique that accounts for blends. The resulting iron abundances are anomalously low for two Hyades dwarfs, apparently due to high levels of chromospheric activity in these stars. For the remaining 10 stars the mean abundance is $[Fe/H]_{Hyades} = +0.12 \pm 0.03$. Abundances for ten other elements are also determined.

153.026   Ultraviolet studies of stars in the populous cluster NGC 2100 in the Large Magellanic Cloud.
E. Böhm–Vitense, P. Hodge, C. Proffitt.
Astrophys. J., Vol. 292, No. 1, p. 130 – 142 (1985). With plates 1 – 2.
The authors have studied the ultraviolet energy distributions of the B stars in the LMC cluster NGC 2100. The ultraviolet extinction law was studied for the LMC cluster. The T_{eff}–luminosity diagrams were determined. The brightest stars in the clusters are evolved stars on the "horizontal" supergiant branch. No P Cygni profiles indicating mass loss were recognized for these supergiants.

153.027   The young open cluster Stock 16: an example of star formation in an elephant trunk?
D. G. Turner.
Astrophys. J., Vol. 292, No. 1, p. 148 – 154 (1985). With plates 3 – 4.
UBV photometry is presented for 33 stars in the young open cluster Stock 16 lying in the H II region RCW 75, and these observations are used to obtain new values for the reddening $(E_{B-V} = 0.49 \pm 0.01)$, distance $(1.90 \pm 0.08$ kpc), and age $(3 - 5 \times 10^6$yr) of the cluster. The relationship of Stock 16 to nearby members of Cen OB1 and Cen R1 is discussed, and arguments are presented for the origin of this sparsely populated cluster from a now–truncated elephant–trunk structure visible in RCW 75.

153.028   Einstein X–ray survey of the Pleiades: the dependence of X–ray emission on stellar age.
G. Micela, S. Sciortino, S. Serio, G. S. Vaiana, J. Bookbinder, L. Golub, F. R. Harnden Jr., R. Rosner.
Astrophys. J., Vol. 292, No. 1, p. 172 – 180 (1985).
Two $1° \times 1°$ fields of the Pleiades region, containing 78 cluster members within a limiting magnitude of 14 mag and centered on two of the most luminous stars of the cluster (20 Tau and 17 Tau) have been observed with the Einstein Observatory at a mean detection threshold of $10^{29.5}$ergs s^{-1}. The authors derive the maximum–likelihood X–ray luminosity functions for the Pleiades G and K stars in the cluster, and show that for the G stars, the Pleiades X–ray luminosity function is significantly brighter than the corresponding function for Hyades G dwarf stars. This finding indicates a dependence of X–ray luminosity on stellar age, which is confirmed by comparison of the same data with median X–ray luminosities of pre–main–sequence and local–disk population dwarf G stars.

153.029   Active star formation in NGC 2264.
P. R. Schwartz, H. A. Thronson Jr., S. F. Odenwald, W. Glaccum, R. F. Loewenstein, G. Wolf.
Astrophys. J., Vol. 292, No. 1, p. 231 – 237 (1985).
The region of NGC 2264 near the cone nebula is the site of active star formation in a rotating ring seen nearly edge on as a two lobed source. Allen's infrared source (IRS 1) surrounds a B3 V star still embedded in the southern lobe of the cloud. The northern lobe, IRS 2, also probably contains young stars.

153.030   Metallicities and distances of galactic clusters as determined from UBV–data. II. The metallicities and distances of 38 open clusters.
L. M. Cameron.
Astron. Astrophys., Vol. 147, No. 1, p. 39 – 46 (1985).
A method is presented for calculating simultaneously both the ultraviolet excess and the reddening of open clusters. The ultraviolet excess is then transformed to an [Fe/H]–value. The reddening is used to determine the distances of the clusters incorporating a ratio of total–to–selective absorption that is a function of interstellar extinction and dependent on the spectral type. Furthermore, the changes in the V–magnitude due to abundance effects are also considered. Thus, metallicities and distances of 38 open clusters are given. Finally, composite diagrams of the clusters in

both color–magnitude diagrams are presented to give an idea of the dependence of the cluster main–sequences on the metallicity.

153.031 Metallicities and distances of galactic clusters as determined from UBV–data. III. Ages and abundance gradients of open clusters.
L. M. Cameron.
Astron. Astrophys., Vol. 147, No. 1, p. 47 – 53 (1985).

The metallicities and distances of 38 open clusters are used to estimate the variation of metal abundance along the galactic disc. The resulting gradient amounts to $d[Fe/H]/dR = -0.11 \pm 0.02$ kpc^{-1}. No abundance gradient perpendicular to the galactic plane can be detected using open clusters with $z < 800$ pc. The ages of the clusters are determined using the turn–off point method in the H–R diagram. They cover a span of 5.1 billion years. It is shown that young clusters are placed in the galactic plane whereas the older clusters preferentially lie at higher z–values. No firm relationship between the cluster ages and their metallicity can be deduced. Finally, it is argued that the overall cluster–metallicity ought to be deduced from main–sequence stars rather than from giants.

153.032 Molekülwolken und die Lebensdauer offener Sternhaufen.
R. Wielen, H.–J. Kuhrau.
Mitt. Astron. Ges., Nr. 63, p. 178 (1985). – See Abstr. 012.063.

153.033 Auswahl von Mitgliedern offener Sternhaufen zur Messung der Eigenbewegung der Haufen mit dem Astrometriesatelliten HIPPARCOS.
C. Dettbarn, R. Wielen.
Mitt. Astron. Ges., Nr. 63, p. 178 (1985). Abstract. – See Abstr. 012.063.

153.034 A radial velocity study of open clusters containing blue stragglers.
M. J. Bolte, M. Mateo.
Bull. Am. Astron. Soc., Vol. 16, No. 4, p. 947 (1984). Abstract. – See Abstr. 010.062.

153.035 Synthetic cluster color–magnitude diagrams.
J. H. Jones, I. S. Kaufman, P. J. Flower.
Bull. Am. Astron. Soc., Vol. 16, No. 4, p. 967 (1984). Abstract. – See Abstr. 010.062.

153.036 The metallicities of M5, M67, and M71 as determined from K giant spectra.
D. Burstein, S. M. Faber, J. J. Gonzalez, A. Spaenhauer.
Bull. Am. Astron. Soc., Vol. 16, No. 4, p. 968 (1984). Abstract. – See Abstr. 010.062.

153.037 Ultraviolet observations of the extinction in the galactic cluster, Trumpler 37.
G. C. Clayton, E. L. Fitzpatrick, B. D. Savage.
Bull. Am. Astron. Soc., Vol. 16, No. 4, p. 997 (1984). Abstract. – See Abstr. 010.062.

153.038 The abundance of lithium in the old galactic cluster N7789.
C. A. Pilachowski.
Bull. Am. Astron. Soc., Vol. 16, No. 4, p. 997 (1984). Abstract. – See Abstr. 010.062.

153.039 A systematic search for members of the Hyades Supercluster. IV. The metallic–line stars and ultrashort–period Cepheids.
O. J. Eggen.
Astron. J., Vol. 90, No. 6, p. 1046 – 1059 (1985).

The stars in the Bright Star Catalogue with $\beta = 2.70 - 2.88$ ($T_e \sim 6700 - 8100$K, spectral type A4 to F1) contain 127 members of the Hyades Supercluster, chosen on the basis of the direction of their proper motion. The available radial velocities confirm supercluster membership, within their uncertainty, for most of these stars. The older stars with $\beta > 2.5$ mag are Am and/or ultrashort–period Cepheids (USPC) with only a half–dozen exceptions, whereas in the younger population the Am stars and USPC represent only some 25% of the sample. Outside the center of the Hyades cluster, the largest concentration of supercluster stars is in a region of 30 pc radius between 60 and 80 pc above the Sun, where 30% of the expected stars in the temperature range discussed here are supercluster members.

153.040 A study of the open cluster NGC 2374.
G. S. D. Babu.
J. Astrophys. Astron., Vol. 6, No. 1, p. 61 – 70 (1985).

The results of modified objective grating observations and photoelectric as well as photographic photometry of the open cluster NGC 2374 are presented. The cluster contains at least twenty stars as definite members down to $m_v \approx 15$ mag. There is a uniform extinction of $E(B-V) = 0.175$ mag and the distance is 1.2 ± 0.1 kpc. The most likely age of this cluster is 7.5×10^7 years.

153.041 Integrated magnitudes and colors of open clusters.
N. M. Spassova (*N. M. Spasova*), P. V. Baev.
Astrophys. Space Sci., Vol. 112, No. 1, p. 111 – 123 (1985).

The integrated UBV characteristics of 50 galactic clusters are discussed. On the basis of a comparison between the empirically obtained dependences and their theoretical counterparts (Searle et al., 1973) a suggestion has been made that the number of the massive stars (with masses $> 10 \; M_\odot$) in the initial mass function must be considerably greater. Evolutionary effects on integrated parameters and their dependences have been discussed. An attempt at an evolutionary interpretation of the integrated luminosity function for galactic clusters, obtained by Starikova (1962) in the light of the self–obtained dependences has been made.

153.042 Far–infrared observations of young clusters embedded in the R Coronae Australis and Rho Ophiuchi dark clouds.
B. A. Wilking, P. M. Harvey, M. Joy, A. R. Hyland, T. J. Jones.
Astrophys. J., Vol. 293, No. 1, p. 165 – 177 (1985).

The authors have made multicolor far–infrared maps in two nearby dark clouds, R Coronae Australis and ϱ Ophiuchi, in order to investigate the individual contribution of low–mass stars to the energetics and dynamics of the surrounding gas and dust. They have detected emission from cool dust associated with five low–mass stars in CrA and four in ϱ Oph; their far–infrared luminosities range from $2 \; L_\odot$ up to $40 \; L_\odot$. When an estimate of the bolometric luminosity was possible, it was found that typically more than 50% of the star's energy was radiated longward of 20 µm.

153.043 The star–formation history of very young clusters.
S. W. Stahler.
Astrophys. J., Vol. 293, No. 1, p. 207 – 215 (1985).

The popular idea that star formation has proceeded sequentially from lowest to highest mass members in open clusters is examined critically. Focusing on the prototypical young clusters NGC 2264 and NGC 6530, it is shown that, although the basic concept of continuous star formation is amply supported by observations, the quantitative extension of this idea to sequential formation is not. It is argued that the practice of assigning pre–main–sequence ages to all cluster members leads to a pattern of stellar births that is physically implausible. Any cluster which contains an appreciable main–sequence population will show apparent sequential formation, regardless of its true history, if contraction ages are assigned to all of its stars. It is concluded that the observational evidence is consistent with the simple picture that all stellar masses form over the same broad interval of time in a given cluster.

153.044 Stellar complexes. Review.
Yu. N. Efremov.
Vestn. Akad. Nauk SSSR, No. 12, p. 56 – 66 (1984). In Russian.
From Ref. Zh., 51. Astron., 6.51.667 (1985).

153.045   The structure and internal kinematics of open clusters.
R. D. Mathieu.
Dynamics of star clusters, p. 427 – 448 (1985). – See Abstr. 012.070 (IAU Symp. No. 113).

Open clusters provide an opportunity for observational study of several important facets of stellar dynamics which cannot easily be addressed in globular clusters for either fundamental or practical reasons. In particular, open clusters are case studies for: (1) The dynamics of small–N stellar systems. (2) The effects of a large stellar–mass spectrum. (3) A spectrum of cluster ages. (4) The effects of the external environment. Finally, open clusters have the advantages that many lie within 1 kpc of the sun.

153.046   The dense stellar cores of giant H II regions.
A. F. J. Moffat, W. Seggewiss, M. M. Shara.
Dynamics of star clusters, p. 467 – 469 (1985). – See Abstr. 012.070 (IAU Symp. No. 113).

The authors explore the bright central diffuse objects of the closest visible giant H II regions 30 Dor (LMC) and NGC 3603 (Galaxy), with the aid of CCD imagery in 1″ seeing. Both cenral objects are interpreted as dense cluster cores composed of normal stars with masses ranging up to $\approx 100\,M_\odot$. Each object appears to contain at least one close massive binary. Being significantly denser, the core of NGC 3603 seems to be dynamically more evolved than the core of 30 Dor.

153.047   A proper motion study of the Pleiades cluster.
F. van Leeuwen.
Dynamics of star clusters, p. 477 – 480 (1985). – See Abstr. 012.070 (IAU Symp. No. 113).

The first results of an extensive proper motion study of the Pleiades cluster are presented. A total of 166 exposures covering a 3 by 3 degrees area are now incorporated. The accuracies of the centennial proper motions range from 0″015 in the central region to 0″2 in the outermost region.

153.048   Proper motion studies of stars in and around open clusters.
F. van Leeuwen.
Dynamics of star clusters, p. 579 – 606 (1985). – See Abstr. 012.070 (IAU Symp. No. 113).

A compilation of proper motion studies of stars in and around open clusters is presented. It can serve as a reference to cluster member selections, studies of cluster dynamics or as a guide to further improvement of the data presently available.

153.049   Alpha Tauri CD: another Hyades binary.
K. M. Cudworth.
Publ. Astron. Soc. Pac., Vol. 97, No. 590, p. 348 (1985).

The proper motion and a very approximate parallax of α Tau C are consistent with Hyades membership. Thus α Tau CD is most likely a previously unrecognized binary in the Hyades cluster. The period, however, is probably very long.

153.050   Ultraviolet and X–ray observations of NGC 2264.
T. Simon, W. Cash, T. P. Snow Jr.
Astrophys. J., Vol. 293, No. 2, p. 542 – 550 (1985).

Pre–main–sequence stars in the galactic cluster NGC 2264 were observed with the IUE to search for ultraviolet chromospheric and transition region emission lines and with the Einstein Observatory to search for coronal X–ray emission. Fifteen cluster members (and four other stars) of spectral type B – K, ranging in age from 10^6 to 10^7yr, were observed in the ultraviolet. Chromospheric emission was detected in the faint Hα emission–line star W79 (NX Mon). The observations are consistent with the generally accepted idea that chromospheric activity of late–type stars subsides with age, but they are subject to uncertainties in circumstellar and interstellar extinction. Seven luminous X–ray sources (log $L_x \gtrsim 31.4$) were found. Except for W189 and W107, the likely optical counterparts of the X–ray sources are inconspicuous stars fainter than 14th visual magnitude.

153.051   The Hyades: membership and convergent point from radial velocities.
R. P. Stefanik, D. W. Latham.
Stellar radial velocities, p. 213 – 222 (1985). – See Abstr. 012.095 (IAU Colloq. No. 88).

Multiple radial velocities of 212 proper motion and/or photometric Hyades candidates have been obtained in the magnitude range $5 < V < 15$ with an accuracy of ± 0.7 km s^{-1} for an individual observation. These observations have been used to establish cluster membership, to identify binary star candidates, to derive a convergent point solution and to determine the cluster distance.

153.052   Radial velocity studies of the internal kinematics of open clusters.
R. D. Mathieu.
Stellar radial velocities, p. 249 – 256 (1985). – See Abstr. 012.095 (IAU Colloq. No. 88).

The study of the internal kinematics of open clusters based on radial velocity measurements is reviewed. In particular, the effect of binaries on the determination of open cluster velocity dispersions is discussed. Studies of the internal kinematics of the clusters M11 and M67 are reviewed in some detail.

153.053   The binary population of the Hyades.
R. D. Mathieu, R. P. Stefanik, D. W. Latham.
Stellar radial velocities, p. 385 – 388 (1985). – See Abstr. 012.095 (IAU Colloq. No. 88).

The authors have made a detailed analysis of the binary population of the Hyades, using radial–velocity, visual and proper–motion data in the literature and new multiple high–precision ($\sigma = 0.85$ km/sec) radial–velocity observations of over 100 Hyads with $6 < V < 14$. This report presents the analysis on the 95 Hyads found by van Bueren (1952) with $6 < V < 9$ (0.9 to 1.7 $M_\odot$).

153.054   The extinction law in the open cluster NGC 457 and the intrinsic energy distribution of ϕ Cas (F0 Ia).
P. Rosenzweig, N. D. Morrison.
Bull. Am. Astron. Soc., Vol. 17, No. 2, p. 557 (1985). Abstract. – See Abstr. 010.065.

153.055   An anomalous intracluster reddening law in NGC 281.
H. H. Guetter, D. G. Turner, J. N. Scrimger.
Bull. Am. Astron. Soc., Vol. 17, No. 2, p. 557 – 558 (1985). Abstract. – See Abstr. 010.065.

153.056   How old is NGC 188?
D. S. Adler, R. T. Rood.
Bull. Am. Astron. Soc., Vol. 17, No. 2, p. 593 (1985). Abstract. – See Abstr. 010.065.

153.057   uvby photometry of the main sequence of M67.
B. A. Twarog.
Bull. Am. Astron. Soc., Vol. 17, No. 2, p. 593 (1985). Abstract. – See Abstr. 010.065.

153.058   The distant galactic anticenter cluster, Berkeley 21.
K. A. Janes.
Bull. Am. Astron. Soc., Vol. 17, No. 2, p. 593 (1985). Abstract. – See Abstr. 010.065.

153.059   The spatial distribution of spectroscopic binaries and blue stragglers in the open cluster M67.
R. D. Mathieu, D. W. Latham.
Bull. Am. Astron. Soc., Vol. 17, No. 2, p. 602 (1985). Abstract. – See Abstr. 010.065.

153.060   Report of IAU Commission 37: Star clusters and associations (*Amas stellaires et associations*).
K. C. Freeman.
Trans. IAU, Vol. XIXA, p. 521 – 545 (1985). – See Abstr. 003.046.

153.061   The determination of frequencies and mass characteristics of binary stars in open clusters.
J. P. Dabrowski.
Diss. Abstr. Int., Sect. B, Vol. 45, No. 3, p. 900 (1984). Thesis, University of Pittsburgh, 80 pp. (1983). Order No. DA8411621.

153.062   The structure, internal kinematics and dynamics of open star clusters.
R. D. Mathieu.
Diss. Abstr. Int., Sect. B, Vol. 45, No. 3, p. 902 (1984). Thesis, University of California, 304 pp. (1983). Order No. DA8413499.

153.063   On determining the membership to open clusters from relative proper motions.
J.-l. Zhao, K.-p. Tian.
Acta Astron. Sin., Vol. 26, No. 2, p. 152 – 161 (1985). In Chinese.

Results obtained from the determination of membership to 42 open clusters are summarized and discussed in this paper. It is shown from analysis of the results that the mathematical model, which was developed by Vasilevskis and has been used to determine membership to open clusters, is quite reasonable. The major axis of the distribution of proper motions of field stars has a preferential orientation, which is parallel to the galactic plane. The possibility of application of the statistical method is dependent on the proper motion distributions of members and field stars in a stellar field.

153.064   A kinematic and photometric study of the old cluster M67.
V. N. Frolov.
Izv. Glav. Astron. Obs. Pulkovo, Astrometr. Astrofiz., No. 202, p. 105 – 115 (1984). In Russian.

M67 proper motions of stars within a $130' \times 130'$ area centered on NGC 2682 (M67) were derived using 5 plate pairs. The plate material was obtained with the Pulkovo Normal Astrograph, the mean epoch difference being 62 years. Proper motions of 1104 stars to $B = 15\overset{m}{.}8$ were determined. UBV magnitudes of the stars were determined. It was found that 332 of the measured stars belong to the cluster and 57 are probable members. The membership of blue stragglers and red giants is discussed.

153.065   The 2 Mon star formation region. The cluster region NGC 2244.
N. G. Guseva.
Astrofizika, Tom 22, Vyp. 3, p. 505 – 514 (1985). In Russian. English translation in Astrophysics, Vol. 22, No. 3.

The dust matter has been concluded to be absent within the cast bound of maximal emission of Rosette. It has been shown that if the dust shell surrounding the cluster nucleus exists in the emission region the upper limit of its optical thickness will not exceed $A_v = 0\overset{m}{.}3$.

Computer–based catalogue of open–cluster data.
See Abstr. 002.002.

Sky catalogue 2000.0. Volume 2: Double stars, variable stars and nonstellar objects.
See Abstr. 002.019.

Photometric detection of open cluster binaries.
See Abstr. 036.132.

Radial velocities for 28 southern young open clusters.
See Abstr. 111.022.

Stellar winds in the young galactic cluster NGC 6530.
See Abstr. 112.005.

Main sequence B stars with strong winds.
See Abstr. 112.006.

The surface morphology of solar–type Hyades stars.
See Abstr. 112.109.

The intrinsic UV colors of O stars.
See Abstr. 113.046.

Peculiar stars in the Pleiades group.
See Abstr. 114.025.

An empirical Hγ luminosity calibration for class V – III stars.
See Abstr. 115.003.

A new Hγ–absolute magnitude calibration.
See Abstr. 115.007.

Metallicities and distances of galactic clusters as determined from *UBV*–data. I. The effects of metallicity and reddening on the colors of main–sequence stars.
See Abstr. 115.010.

Spectral types and luminosities of supergiants in open clusters.
See Abstr. 115.022.

Short time scale periodicity in Hα emission from the main–sequence star H II 1883.
See Abstr. 116.015.

An FK Comae–like star in NGC 188.
See Abstr. 116.038.

The massive near–contact binary system V348 Carinae (HD 90707) in IC 2581.
See Abstr. 117.061.

Spectroscopic orbits for 16 more binaries in the Hyades field.
See Abstr. 120.008.

A search for young stellar chromospheres in NGC 2264.
See Abstr. 121.005.

Fundamental parameters of Cepheids.
See Abstr. 122.036.

The zero–point of the Cepheid luminosity scale from a calibration of the luminosities of early–type stars.
See Abstr. 122.070.

Are the Cepheids in cluster nuclei a rare breed?
See Abstr. 122.072.

Verification of Cepheids in open clusters and associations as distance indicators.
See Abstr. 122.158.

Flare stars in the Pleiades.
See Abstr. 122.177.

The energy spectrum of flares of UV Cet–type stars and physical meaning of several statistical characteristics of these stars.
See Abstr. 122.215.

New variable stars found in southern open clusters.
See Abstr. 123.031.

The UV extinction laws in two very young clusters, NGC 6530 in the Galaxy and NGC 2100 in the LMC.
See Abstr. 131.002.

Interstellar extinction in the nucleus of h Per.
See Abstr. 131.008.

A survey of molecular clouds associated with young open star clusters.
See Abstr. 131.039.

Formation of OB clusters: CO, NH_3, and H_2O observations of the distant H II region complex in S128.
See Abstr. 131.182.

Die interstellare Extinktionskurve bei O– und B–Sternen im Rosettennebel.
See Abstr. 131.209.

Interstellar reddening law towards the nucleus of h Per.
See Abstr. 131.267.

Possible consequences of gas accretion for the initial mass function of star clusters.
See Abstr. 131.278.

The formation and early dynamical evolution of bound stellar systems.
See Abstr. 151.092.

Dynamics of open star clusters.
See Abstr. 151.129.

The dynamical evolution of young open clusters.
See Abstr. 151.130.

N–body simulations of realistic open clusters.
See Abstr. 151.131.

On the dynamics of open clusters.
See Abstr. 151.132.

Stellar kinematics in star–forming regions.
See Abstr. 152.007.

Radial velocities of F and G dwarfs in the Orion cluster region.
See Abstr. 152.008.

Stellar systems, star clusters and globular clusters.
See Abstr. 154.019.

What next? Priorities in theory and observations. Summary of the panel discussion.
See Abstr. 154.068.

The stellar disk component: distribution, motions, age and stellar composition.
See Abstr. 155.016.

Carbon stars and S stars near open clusters. A statistical approach.
See Abstr. 155.168.

Ellipticities of star clusters in the Small Magellanic Cloud.
See Abstr. 156.022.

Color–magnitude diagrams of eleven Magellanic Cloud clusters and the Magellanic Cloud distance.
See Abstr. 156.023.

The LMC star cluster NGC 1777 and its neighboring field.
See Abstr. 156.024.

Observed dynamical parameters of star clusters in the SMC.
See Abstr. 156.025.

154 Globular Clusters

154.001   IUE observations of blue horizontal branch globular clusters and UV–bright stars.
B. M. Altner, B. Holton, T. A. Matilsky.
NASA Conf. Publ., NASA CP–2349, p. 179 – 182 (1984). – See Abstr. 012.001.

Recent theoretical attempts to understand the later evolution of Population II stars have met with some success in accounting for the positions of UV–bright stars in globular cluster color magnitude diagrams. The existence of two distinct sub–classes of UV–bright stars has been suggested, based on the luminosity above the horizontal branch. IUE observations of individual UV–bright stars in several globular clusters do indeed show a separation into two luminosity groups. Whereas spectra of resolved hot sources found in the cores of seven clusters lead to believe that they are also UV–bright stars, unlike the stars outside the core they fall mostly in the low luminosity group.

154.002   X–raying the dynamics of globular clusters.
J. E. Grindlay.
News Lett. Astron. Soc. N.Y., Vol. 2, No. 7, p. 13 – 14 (1985). Abstract. – See Abstr. 010.241.

154.003   Photometry, proper motions, and membership in the globular cluster M71.
K. M. Cudworth.
Astron. J., Vol. 90, No. 1, p. 65 – 73 (1985). With plate 8.

New photographic photometry and proper motions have been obtained for over 350 stars with $V \lesssim 16$ in the region of the globular cluster M71 (NGC 6838). The photometry has been calibrated via new photoelectric observations which resolve the zero–point discrepancy of previous sequences. The proper motions have been used to derive cluster membership probabilities. The horizontal branch and lower giant branch very closely resemble those of 47 Tuc, but the upper part of the giant branch is slightly brighter and bluer in M71. The cluster space velocity is well determined and is characteristic of a disk–population object: ~ 50 km s^{-1}.

154.004   Reddening, distance modulus and age of the globular cluster NGC 6121 (M4) from the properties of RR Lyrae variables.
F. Caputo, V. Castellani, M. L. Quarta.
Astron. Astrophys., Vol. 143, No. 1, p. 8 – 12 (1985).

It is shown that pulsational properties of RR Lyrae variables in globular clusters can be used to put theoretical constraints on the values of cluster reddening and distance modulus. By requiring that the HR diagram location of pulsators agrees with the period distribution observed and with the theoretical boundaries of the instability strip, reddening and distance modulus of the globular cluster M4 are derived as a (slow) function of the pulsator masses. Thus, a best guess is presented for the cluster age ($t = 12.2$ billion years), some evidence for a non–canonical evolutionary having been taken into account.

154.005   The globular cluster Palomar 13.
S. Ortolani, L. Rosino, A. Sandage.
Astron. J., Vol. 90, No. 3, 473 – 478, 567 – 568 (1985).

B and V magnitudes of 122 stars in the globular cluster Palomar 13 have been determined by examining, with a PDS microphotometer, plates obtained at the prime focus of the Hale telescope. The new data improve and extend to the upper main sequence the preliminary results published by Ciatti et al. (1965) concerning the c–m diagram and the luminosity function of the cluster.

154.006   A search for millisecond pulsar candidates in globular clusters.
T. Hamilton, D. Helfand, R. Becker.
News Lett. Astron. Soc. N.Y., Vol. 2, No. 4, p. 10 (1983). Abstract. – See Abstr. 010.242.

154.007 Observational constraints on the ages and abundances of old stellar populations.
D. Burstein.
Publ. Astron. Soc. Pac., Vol. 97, No. 588, p. 89 – 103 (1985). Invited paper presented at the Symposium on Stellar Populations at the 95th Annual Scientific Meeting of the Astronomical Society of the Pacific, University of California, Santa Cruz, July 1984.

Intrinsic differences in the integrated spectra of old stellar populations have been chiefly found in the strengths of CNO-based absorption–line features. Differences have also been found in the strengths of the Balmer lines, relative to both the CNO-based features and to the broad–band indices. These differences are summarized here, and at least four different kinds of "old" stellar populations can be identified from the present data: galactic globular clusters, the brighter M31 globular clusters, the nucleus of M32, and the nucleus of M31. The ages of stars in giant elliptical galaxies and in the brighter M31 globular clusters lie between 5 and 15 billion years.

154.008 Homogenized integrated _UBVRI_ colors for galactic globular clusters.
B. C. Reed.
Publ. Astron. Soc. Pac., Vol. 97, No. 588, p. 120 – 125 (1985).

Using the data homogenization technique of Reed and FitzGerald (1982), some 1400 photoelectric observations of integrated _UBVRI_ colors for 114 galactic globular clusters have been reduced to the _UBV_ and _VRI_ systems of van den Bergh (1967) and Hamuy (1984), respectively. Use is made of all mutually common obervations.

154.009 The giant, asymptotic, and horizontal branches of globular clusters. II. Photographic photometry of the metal-poor clusters M15, M92 and NGC 5466.
R. Buonanno, C. E. Corsi, F. Fusi Pecci.
Astron. Astrophys., Vol. 145, No. 1, p. 97 – 117 (1985).

Photographic photometry has been presented in a series of papers on the very metal–poor clusters M15, M92, and NGC 5466. From the analysis and comparison of the C–M diagrams the following conclusions are drawn: (1) The HB's of the three clusters are different. (2) The giant and subgiant branches of the three clusters are narrow. (3) The ratios of the number of HB, AGB, and RGB stars are in good agreement with the evolutionary models taking into account semiconvection and mass loss. (4) The detailed morphology of the HB's in the three clusters is discussed under the assumption that they have the same [Fe/H]–values, the same helium abundance and age. (5) No satisfactory explanation for the presence of the gap in M15 has been found yet. (6) The interpretation deduced from the study of these three clusters has been applied to the comparison with other clusters, and the problem of "the second parameter" in the classification of globular clusters is dealt with. (7) Structure and separation of the main branches and luminosity functions are analysed. (8) Variables and UV–bright stars are discussed.

154.010 Two mid–giant branch variable stars im M15.
Y.–h. Chu, C. M. Clement, H. Sawyer Hogg, T. R. Wells.
Cepheids: theory and observations, p. 264 – 267 (1985). – See Abstr. 012.027. (IAU Colloq. No. 82).

High precision photographic photometry indicates that two stars lying on the giant branch in the C–M diagram of M15 are small amplitude (~ 0.2 mag) variables. The two stars are Kustner 64 and 152. The existing data is not sufficient for period determination.

154.011 Structural parameters and masses for three old LMC clusters.
R. A. W. Elson, K. C. Freeman.
Astrophys. J., Vol. 288, No. 2, p. 521 – 530 (1985).

This paper presents luminosity profiles for three old LMC clusters, NGC 1835, NGC 2210, and NGC 2257. King models are fitted to the profiles, the core and tidal radii are determined, and the total mass of each cluster is deduced. Total luminosities and mass–to–light ratios are also derived. Lower limits on tidal mass–to–light ratios are 0.18 ± 0.03, 011 ± 0.02, and 0.56 ± 0.09

for NGC 1835, NGC 2210, and NGC 2257, respectively. These values are up to 3 times larger than previous determinations. The values for NGC 1835 and NGC 2210 are still somewhat smaller than mass–to–light ratios of Milky Way globulars and it is suggested that this reflects a difference in age.

154.012 The extended giant branches of intermediate age globular clusters in the Magellanic Clouds. IV.
M. Aaronson, J. Mould.
Astrophys. J., Vol. 288, No. 2, p. 551 – 557 (1985). With plates 15 – 18.

A complete survey is now available for asymptotic giant branch stars in the rich star clusters of the Magellanic Clouds. Clusters younger than approximately 8 billion years have carbon stars at the tip of the giant branch, produced by the third dredge-up mechanism. Clusters younger than approximately 0.8 billion years have giant branches populated by M stars. It is suggested that in stars of this mass range thermal pulses have not commenced before mass loss completely erodes the stellar envelope. Cluster stars of 5 $M_\odot$ turnoff suffer of order 80% mass loss in the course of their evolution.

154.013 Variable stars in the globular cluster M92.
Z. I. Kadla, N. V. Yablokova, A. N. Gerashchenko, N. Spasova.
Perem. Zvezdy, Tom 21, No. 6, p. 827 – 830 (1983). In Russian.

A search for variable stars in the central region of the globular cluster M92 revealed the presence of a number of previously undetected variables. Data for 11 of these stars, which showed light variations exceeding $0^m\!.3$ on available plates are given in a table. In the catalogue Sawyer–Hogg (1973) the sign of the X-coordinate of variable 12 is incorrect. Evidently the star corresponds to variable 10 as listed by Nassau (1938). The variability of the Nassau variable 16 is confirmed.

154.014 Variable stars in the globular cluster M13. III. Short–period variables.
T. S. Ruseva, R. M. Rusev.
Perem. Zvezdy, Tom 22, No. 1, p. 49 – 63 (1983). In Russian.

New results of photographic observations are given for seven variable stars in the globular cluster M13. More accurate data for the elements of the light curves in the photometric B system are obtained for W Virginis–type stars V1, V2, V6 and for RR Lyrae type stars V5, V7, V8, V9. The period changing of the stars is studied on the basis of the total published observational material. The period–luminosity, period–amplitude and period–colour relations are constructed. The position of the variables in the instability strip is shown and the pulsational masses and radii of V1 and V6 are determined.

154.015 Period variations of RR Lyrae–type variable stars in the globular cluster M22.
Eh. K. Marinchev.
Perem. Zvezdy, Tom 22, No. 1, p. 65 – 84 (1983). In Russian.

Period changes of 12 RR Lyrae variables were studied on the basis of 238 plates of the Moscow collection taken during 1960, 1961, 1970, 1971 – 1981. The maxima of variables are summarized for the JD interval (2441100–44900). 8 red long–period variables were also studied.

154.016 Observations of variable stars in the globular cluster M15.
A. F. Gordenko, A. V. Klabukova, N. N. Fashchevskij.
Odes. univ. Odessa, 51 pp. (1984). In Russian. Abstr. in Ref. Zh., 51. Astron., 2.51.565 (1985).

154.017 M62: an RR Lyrae–rich, UV–bright galactic globular cluster.
V. Caloi, V. Castellani, M. Tarenghi.
Astron. Astrophys., Vol. 145, No. 2, p. 286 – 289 (1985).

The central region of the globular cluster M62 was observed with IUE. The authors confirm that, in the UV, M62 is the brightest cluster observed up to now in the galaxy. This RR Lyrae–rich system is another example of an UV–bright cluster with a horizontal branch not confined to the blue. The interpretation of UV excesses from spiral nuclei and elliptical galaxies has to take into account the occurrence of these UV emitters in old, metal–poor populations.

154.018   Photoelectric photometry of star clusters in M31. X.
A. S. Sharov, V. M. Lyutyj, V. F. Esipov.
Sov. Astron. Lett., Vol. 10, No. 4, p. 243 (1984). English translation of 38.154.015.

154.019   Stellar systems, star clusters and globular clusters.
V. Castellani.
Mem. Soc. Astron. Ital., Vol. 55, No. 3, p. 399 – 415 (1984). – See Abstr. 012.049.
Contents: Stellar systems and the universe. Galactic clusters. Globular clusters (evolution, helium content, ages, the problem of the second parameter, gaps in the HR diagram, extragalactic clusters).

154.020   Globular clusters and the galactic halo.
L. Angeletti, R. Capuzzo–Dolcetta, P. Giannone.
Mem. Soc. Astron. Ital., Vol. 55, No. 3, p. 423 (1984). Abstract. – See Abstr. 012.049.

154.021   Search for (globular) clusters in nearby galaxies: a progress report.
F. Bònoli, L. Federici.
Mem. Soc. Astron. Ital., Vol. 55, No. 3, p. 425 – 427 (1984). – See Abstr. 012.049.
A long term project on the search for globular clusters in external galaxies has been undertaken at the Dipartimento di Astronomia dell'Università degli Studi di Bologna. Here the status and the future development of this search are briefly presented.

154.022   C0422–213 (Eridanus): a second intermediate metal abundance globular cluster in the outer galactic halo.
G. S. Da Costa.
Astrophys. J., Vol. 291, No. 1, p. 230 – 236 (1985). With plates 6 – 7.
The CTIO 4 m prime focus CCD system has been used to determine a color–magnitude diagram for the distant stellar system in Eridanus (C0422–213). From the magnitude of the horizontal–branch stars, a distance modulus of 19.55 is derived placing the cluster in the outer halo, 84 kpc from the galactic center. The C–M diagram reveals a red horizontal branch while the giant branch parameters suggest $[Fe/H] = -1.35 \pm 0.2$ for the cluster. Thus C0422–213 joins Pal 14 as an intermediate metal abundance cluster in the outer galactic halo.

154.023   The globular cluster NGC 6712.
J. A. Frogel.
Astrophys. J., Vol. 291, No. 2, p. 581 – 585 (1985).
New infrared photometry is presented for 15 red giants, including five variables, in the metal–rich globular cluster NGC 6712. From infrared color–magnitude diagrams a value of $[Fe/H]_{IR} = -0.92$ is derived, identical to that for NGC 6171 and in close agreement with optical determinations of the abundance. Nonetheless, the CO indices of the NGC 6712 giants are quite strong – about 0.04 mag greater than for stars in 47 Tucanae.

154.024   Ellipticities of "disk" and "halo" globular clusters in the SMC.
E. Kontizas, D. Dialetis, T. Prokakis, M. Kontizas.
Astron. Astrophys., Vol. 146, No. 2, p. 293 – 296 (1985).
The projected ellipticities of twenty–four "disk" and "halo" globular clusters of various ages and positions in the SMC have been determined using isodensity contours. B and V plates, taken with the 1.2 m U.K. Schmidt telescope in Australia, were scanned at the National Observatory of Athens with an isodensitometer. The derived ellipticities $e = (1-b/a)$ have shown that the globular clusters of the SMC are more elliptical than their counterparts of the Large Magellanic Cloud and our Galaxy. The comparison of the observed distribution of the ellipticities for both cluster types "disk" and "halo" do not show a statistically significant difference; therefore the observations do not support age dependence.

154.025   A *UBVRI* photoelectric sequence in 47 Tuc.
G. Alcaino, W. Liller.
Astron. Astrophys., Vol. 146, No. 2, p. 389 – 391 (1985).
The authors present a *UBVRI* photoelectric sequence of 24 stars in the magnitude range $8.3 < V < 15.1$ in the globular cluster 47 Tuc. Fourteen of these stars have been observed in *UBV* by Hesser and Hartwick (1977) and six by Lee (1977). The results presented are in excellent agreement with both of those investigations.

154.026   Erratum: "Eight new variable stars in the globular cluster M15" [Astron. Tsirk., No. 1314, p. 1 – 2 (1984)].
Z. I. Kadla, A. N. Gerashchenko, N. V. Yablokova, N. Spasova.
Astron. Tsirk., No. 1342, p. 8 (1984). See Abstr. 38.154.070.

154.027   Deep photometry of globular clusters. V. Age derivations and their implications for galactic evolution.
R. G. Gratton.
Astron. Astrophys., Vol. 147, No. 1, p. 169 – 177 (1985).
The photometric data presently available on main sequence stars in globular clusters are examined. The total sample is composed of 26 clusters. The derivation of a consistent set of cluster ages, with two distinct distance scales, suggests a picture of the Galaxy formation. The collapse in the inner region of the Galaxy was probably quite fast, originating the nearly coeval system of the inner halo globular clusters. A mean age $t = 14.6 \pm 0.4$ Gyr is obtained from a comparison with VandenBerg isochrones. In the outer halo, collapse and cluster formation took probably a longer time, and clusters continued to form for at least 3 – 5 Gyr.

154.028   Results of the galactocentric orbit calculation for the globular cluster NGC 5466.
S. Ninković.
Publ. Astron. Opservatorije Beogr., No. 33, p. 32 – 35 (1985). – See Abstr. 012.061.
The obtained results clearly indicate an orbit, highly inclined to the galactic plane ($80° \pm 10°$) and highly elongated (eccentricity >0.7).

154.029   Images in the rocket ultraviolet: the initial helium abundance and distance modulus of the globular cluster M5 from photometry of horizontal–branch stars.
R. C. Bohlin, R. H. Cornett, J. K. Hill, A. M. Smith, T. P. Stecher.
Astrophys. J., Vol. 292, No. 2, p. 687 – 695 (1985). With plate 6.
The globular cluster M5 (NGC 5904) was observed in the ultraviolet with a rocket–borne telescope to obtain images at effective wavelengths for hot stars near 1540 Å and 2360 Å with bandpasses of 340 Å and 990 Å. The authors have used the 2360 Å magnitude along with the V magnitude to determine effective temperatures and bolometric corrections for 50 stars using model atmospheres. Applying the horizontal–branch models of Sweigart and Gross for $Z = 0.001$, they determine best fit model parameters $Y = 0.21$, distance modulus 14.39, and helium core mass $0.492\ M_\odot$.

154.030   Radial velocities of stars in globular clusters: a look into ω Cen and 47 Tuc.
M. Mayor, G. Meylan.
Messenger, No. 40, p. 1 – 4 (1985).

154.031   Spectroscopy of horizontal branch stars in NGC 6752.
V. Caloi.
Messenger, No. 40, p. 14 – 16 (1985).

154.032   On the helium abundance in globular cluster stars.
Z. I. Kadla, A. N. Gerashchenko, N. V. Yablokova.
Pis'ma Astron. Zh., Tom 11, No. 5, p. 343 – 350 (1985). In Russian. English translation in Sov. Astron. Lett., Vol. 11.
Several stars from globular clusters were observed with a scanner at the 6–m telescope. The helium abundance of two horizontal–branch stars from M13 is found to be nearly normal.

154.033 Galactic shocks and the lifetime of globular clusters.
L. A. Aguilar, P. Hut, J. P. Ostriker.
Bull. Am. Astron. Soc., Vol. 16, No. 4, p. 911 (1984). Abstract. –
See Abstr. 010.062.

154.034 Spectroscopy of the brightest giants in Palomar 5.
G. H. Smith.
Bull. Am. Astron. Soc., Vol. 16, No. 4, p. 947 (1984). Abstract. –
See Abstr. 010.062.

154.035 Color–magnitude diagrams for NGC 7006 and NGC 6229.
J. G. Cohen.
Bull. Am. Astron. Soc., Vol. 16, No. 4, p. 948 (1984). Abstract. –
See Abstr. 010.062.

154.036 IRAS observations of M22.
F. C. Gillett, J. R. Houck, W. L. Rice.
Bull. Am. Astron. Soc., Vol. 16, No. 4, p. 948 (1984). Abstract. –
See Abstr. 010.062.

154.037 The dynamics of the globular cluster M2.
C. Pryor, R. D. McClure, M. Fletcher,
F. D. A. Hartwick, J. Kormendy.
Bull. Am. Astron. Soc., Vol. 16, No. 4, p. 966 (1984). Abstract. –
See Abstr. 010.062.

154.038 Spectroscopy of the NGC 5128 (Cen A) globular clusters.
J. E. Hesser, H. C. Harris, G. L. H. Harris.
Bull. Am. Astron. Soc., Vol. 16, No. 4, p. 967 (1984). Abstract. –
See Abstr. 010.062.

154.039 Rotation and velocity dispersion in the cores of M15 and M92.
J. F. Kielkopf, J. M. Lattis.
Bull. Am. Astron. Soc., Vol. 16, No. 4, p. 967 (1984). Abstract. –
See Abstr. 010.062.

154.040 The variable stars in the globular cluster NGC 6569.
M. L. Hazen–Liller.
Bull. Am. Astron. Soc., Vol. 16, No. 4, p. 967 (1984). Abstract. –
See Abstr. 010.062.

154.041 B–V concentric aperture photometry of globular clusters.
C. J. Peterson.
Bull. Am. Astron. Soc., Vol. 16, No. 4, p. 967 (1984). Abstract. –
See Abstr. 010.062.

154.042 Outlying BHB stars in globular clusters.
N. Sanduleak.
Bull. Am. Astron. Soc., Vol. 16, No. 4, p. 968 (1984). Abstract. –
See Abstr. 010.062.

154.043 The age of the LMC red globular cluster NGC 2213.
G. S. Da Costa, J. R. Mould, M. D. Crawford.
Bull. Am. Astron. Soc., Vol. 16, No. 4, p. 968 (1984). Abstract. –
See Abstr. 010.062.

154.044 The globular cluster M22.
K. M. Cudworth.
Bull. Am. Astron. Soc., Vol. 16, No. 4, p. 968 (1984). Abstract. –
See Abstr. 010.062.

154.045 Radial velocities of remote globular clusters and the mass of the Galaxy.
R. C. Peterson.
Bull. Am. Astron. Soc., Vol. 16, No. 4, p. 969 – 970 (1984). Abstract. – See Abstr. 010.062.

154.046 Globular clusters.
I. R. King.
Sci. Am., Vol. 252, No. 6, p. 66 – 73 (1985).

Globular clusters are dense crowds of ancient stars bound together by their own gravitation. For decades the study of clusters has yielded insights into the evolution of stars, of galaxies and of the universe as a whole.

154.047 CO band strengths for giants in galactic globular clusters. Are they related to the second parameter problem?
F. Caputo.
Astron. Astrophys., Vol. 147, No. 2, p. 317 – 320 (1985).

Data of CO band strengths for giant stars in galactic globular clusters have been investigated. The results appear compatible with CO abundance following the "second parameter" necessary to account for the observed HB morphology in very–blue metal-intermediate globular clusters. Nevertheless, the contribution of other mechanisms (e.g. stellar rotation, non canonical helium-core mass) cannot be excluded, mainly for the two anomalous blue clusters NGC 288 and M10.

154.048 Integrated spectral types from image–tube spectra for 90 galactic globular clusters.
S. J. Shawl, J. E. Hesser.
Bull. Am. Astron. Soc., Vol. 17, No. 1, p. 515 (1985). Abstract. –
See Abstr. 010.064.

154.049 The early stages of post–collapse globular cluster evolution.
S. L. W. McMillan.
Bull. Am. Astron. Soc., Vol. 17, No. 1, p. 515 (1985). Abstract. –
See Abstr. 010.064.

154.050 Observed surface densities in globular clusters.
I. R. King.
Dynamics of star clusters, p. 1 – 17 (1985). – See Abstr. 012.070
(IAU Symp. No. 113).

This review of observations is oriented toward the problem of core collapse in globular clusters. After a brief discussion of methods of determining surface densities, recent results are cited which show that the central brightness peak seen in M15 also appears in several other clusters. The central peaks of brightness are interpreted as the result of core collapse, which theories have repeatedly predicted. If binaries stabilize the collapsed core, the cluster should follow a Hénon model. In two of the three cases tested this model is a reasonable fit. Remaining theoretical questions are listed, as are observational needs.

154.051 Radial velocities and proper motions of globular cluster stars.
R. Lupton, J. E. Gunn, R. F. Griffin.
Dynamics of star clusters, p. 19 – 31 (1985). – See Abstr. 012.070
(IAU Symp. No. 113).

The authors describe techniques and results for obtaining radial velocities and proper motions for individual stars in globular clusters. One set of these results for the nearby northern clusters M92 and M13 are discussed in terms of new dynamical models and distance estimators.

154.052 Chemical gradients in globular clusters.
K. C. Freeman.
Dynamics of star clusters, p. 33 – 42 (1985). – See Abstr. 012.070
(IAU Symp. No. 113).

First the author describes briefly some older work on chemical gradients in globular clusters. Then he discusses some more recent data on the chemical and kinematical properties of 47 Tuc and Omega Cen, that leave us in little doubt about the reality of such chemical gradients.

154.053 X–raying the dynamics of globular clusters.
J. E. Grindlay.
Dynamics of star clusters, p. 43 – 61 (1985). – See Abstr. 012.070
(IAU Symp. No. 113).

The X–ray surveys conducted with the Einstein Observatory are reviewed and the results derived for the luminosity function, masses and nature of the compact X–ray sources are discussed. The evidence for the compact binary nature of the sources is now overwhelming, but long–term X–ray variability studies previously reported may suggest that some of the systems are in fact triple systems with distant companions. Possible relationships between the initial mass function, stellar density and cluster evolution are discussed, and the arguments that the ostensibly similar compact X–ray sources in the galactic bulge are remnants of a population of globular clusters disrupted by giant molecular clouds are updated.

154.054   A high spatial resolution investigation of the core of some dynamically evolved globular clusters.
M. Aurière, J. P. Cordoni, O. Le Fèvre, A. Terzan.
Dynamics of star clusters, p. 63 – 64 (1985). – See Abstr. 012.070 (IAU Symp. No. 113).
A high spatial resolution photographic study of the core of some concentrated globular clusters yields new results concerning M15 and M30.

154.055   Dynamical modeling of M13 proper motions.
K. Cudworth, D. N. C. Lin, K.–S. Oh.
Dynamics of star clusters, p. 65 – 67 (1985). – See Abstr. 012.070 (IAU Symp. No. 113).
Proper motion data of M13 cluster members are very useful for the investigation of the cluster's internal dynamics. The authors introduce a maximum likelihood method to fit these data in terms of conventional King–Michie models.

154.056   The dynamics of 47 Tucanae.
G. S. Da Costa, K. C. Freeman.
Dynamics of star clusters, p. 69 – 72 (1985). – See Abstr. 012.070 (IAU Symp. No. 113).
Observations made at Las Campanas Observatory and at the Anglo–Australian Observatory have been used to determine line–of–sight velocities for member stars in the globular cluster 47 Tucanae. The inner parts of the cluster show appreciable differential rotation. In contrast to M3, "thermal equilibrium" multimass models can only reproduce the observed velocity dispersion values by including a substantial amount of "dark matter"; i.e. unlike M3, there is "missing mass" in 47 Tuc.

154.057   A search for post–collapse cores.
S. Djorgovski, H. Penner.
Dynamics of star clusters, p. 73 – 76 (1985). – See Abstr. 012.070 (IAU Symp. No. 113).
The authors report the preliminary results of a surface photometry survey of globular cores. Two new cores with post-collapse morphology have been found, and two possible candidates. The estimated fraction of clusters with this core morphology in the Galaxy is only a few percent. Most clusters do not show a morphology similar to the predictions of the central black hole models.

154.058   A deep luminosity function for 47 Tucanae.
W. E. Harris, J. E. Hesser.
Dynamics of star clusters, p. 81 – 83 (1985). – See Abstr. 012.070 (IAU Symp. No. 113).
CCD photometry in B and V reaching B(lim) $\cong$ 25 has been employed to obtain the luminosity function and color–magnitude diagram for the main sequence of 47 Tuc. For $5 < M_v < 10$ the authors find that its LF is essentially flat ($\Delta \log n / \Delta m \sim 0$). The CMD is successfully matched by isochrones with [Fe/H] = –0.5 and t $\cong 15 \times 10^9$y.

154.059   Where's the cusp? (Or CCD photometry of globular cluster cores).
P. M. Lugger, H. Cohn, J. E. Grindlay.
Dynamics of star clusters, p. 89 – 92 (1985). – See Abstr. 012.070 (IAU Symp. No. 113).
UBVR CCD surface photometry for a sample of 11 globular clusters has been obtained at the Kitt Peak National Observatory #1 0.9 m telescope as part of a major study of globular cluster core structure. These data are being used to test for the presence of central cusps in the surface brightness profiles, which are expected to be present in clusters which have already undergone core collapse (Cohn and Hut 1984).

154.060   Rotational field and velocity dispersion in globular clusters: ω Cen and 47 Tuc.
G. Meylan, M. Mayor.
Dynamics of star clusters, p. 93 – 96 (1985). – See Abstr. 012.070 (IAU Symp. No. 113).
Thanks to numerous and accurate radial velocity measurements obtained with CORAVEL, the rotational velocity fields, the spatial velocity dispersions from the centre to the edge and the ordered to random motion ratios are deduced for two globular clusters: ω Cen and 47 Tuc. Masses through virial theorem and M/L ratios are also given.

154.061   BV concentric aperture photometry of globular clusters.
C. J. Peterson.
Dynamics of star clusters, p. 97 – 98 (1985). – See Abstr. 012.070 (IAU Symp. No. 113).
Multi–aperture photometric observations are being used for a reassessment of cluster central surface brightnesses, cluster total magnitudes, and cluster structural parameters as defined in the self–consistent dynamical models of King (1966; Peterson and King 1975 and Peterson 1976).

154.062   An extension of the search for spectroscopic binaries in M3.
C. P. Pryor, D. W. Latham, M. L. Hazen–Liller.
Dynamics of star clusters, p. 99 – 101 (1985). – See Abstr. 012.070 (IAU Symp. No. 113).
The authors have obtained 295 new radial velocities for the 112 giants in the globular cluster M3. The velocities have been combined with the Gunn and Griffin data in order to search for radial velocity variations over a time span of ten years. The authors find no convincing evidence that any of the giants observed are spectroscopic binaries with one notable exception, von Zeipel 164, which is believed to be the first spectroscopic binary to be found in a globular cluster. Modelling of the velocity variations that would be expected in the data for a variety of binary populations confirms Gunn and Griffin's conclusion that binaries with separations of less than 10 AU must occur much less frequently among the giants of M3 than among the population I field stars.

154.063   A search for cataclysmic binaries in the globular cluster M3.
M. M. Shara, A. F. J. Moffat, D. A. Hanes.
Dynamics of star clusters, p. 103 – 104 (1985). – See Abstr. 012.070 (IAU Symp. No. 113).
Deep Hβ narrowband and broadband images of M3 have been electronically blinked to search for cataclysmic binaries. Tests of the method on a known, faint cataclysmic enable to set limits on the sensitivity of the technique. No bright ($M_B < 6$) emission-line (equivalent width > 12 Å) cataclysmic binaries exist in M3 between 4 and 30 core radii from the center. Low luminosity globular X–ray sources could still be weak–lined (E.W. < 12 Å) and bright ($M_B \cong +5$ like some old novae) or strong–lined (E.W. $\cong 60$ Å) and faint ($M_B > 7$ like dwarf novae).

154.064   Possible characteristics of supernova induced enrichment of globular clusters.
G. H. Smith.
Dynamics of star clusters, p. 105 – 108 (1985). – See Abstr. 012.070 (IAU Symp. No. 113).
The aim of the paper is to investigate whether properties of NGC 6752 might be consistent with supernova enrichment. Both the free and adiabatic expansion phases of supernova shells (Spitzer 1968) are discussed in this context.

154.065   Rotation and flattening of globular clusters.
S. M. Fall, C. S. Frenk.
Dynamics of star clusters, p. 285 – 296 (1985). – See Abstr. 012.070 (IAU Symp. No. 113).
The authors review the methods for measuring ellipticities and the results that have emerged from such studies. Their main purpose, however, is to discuss the processes that determine the shapes of globular clusters and the ways in which they change with time.

154.066 Minimum of the eccentricity of the galactic globular cluster orbits.
J. Colin.
Dynamics of star clusters, p. 309 – 312 (1985). – See Abstr. 012.070 (IAU Symp. No. 113).

The author determines if the orbits of the galactic globular clusters can be circular, using a spherical logarithmic potential. If they are not, he calculates their perigalactic distances and deduces a pseudo–eccentricity. Using a numerical and geometrical method, he obtains the minimum of all the possible pseudo–eccentricities for each cluster. The corresponding perigalactic distances are compared with the values given by Innanen et al. (1983) with a different method.

154.067 Black hole remnants in globular clusters.
R. B. Larson.
Dynamics of star clusters, p. 421 – 422 (1985). – See Abstr. 012.070 (IAU Symp. No. 113).

154.068 What next? Priorities in theory and observations. Summary of the panel discussion.
I. R. King, L. Spitzer Jr., A. Toomre, S. D. Tremaine, T. S. van Albada, S. D. M. White.
Dynamics of star clusters, p. 499 – 509 (1985). – See Abstr. 012.070 (IAU Symp. No. 113).

154.069 Structure parameters of galactic globular clusters.
R. F. Webbink.
Dynamics of star clusters, p. 541 – 577 (1985). – See Abstr. 012.070 (IAU Symp. No. 113).

Observed and derived structure parameters are tabulated for 154 galactic globular clusters, 7 dwarf spheroidal satellites of the Galaxy, and 6 globular clusters in the Fornax dwarf spheroidal. Observational parameters listed include equatorial coordinates, apparent level of the horizontal branch, reddening, subgiant branch color at the horizontal branch level, limiting and core angular radii, integrated magnitudes, and central surface brightnesses. Derived parameters include galactic coordinates, heliocentric and galactocentric distance, metallicity, limiting and core radii, central relaxation time scale, central mass density, central velocity dispersion, and central escape velocity.

154.070 _UBVRI_ multi–aperture photometry for 71 globular clusters in our own Galaxy.
D. A. Hanes, J. P. Brodie.
Mon. Not. R. Astron. Soc., Vol. 214, No. 4, p. 491 – 517 (1985). With microfiche MN 214/1.

The authors present the results of a programme of multi–aperture photoelectric photometry in the Cousins _UBVRI_ system for the integrated light of 71 globular clusters in our Galaxy. The apertures number nine and range in diameter from 9.6 to 270 arcsec. The authors present an analysis of the uncertainties introduced by all sources of error in studies of this nature. Amongst other things, the authors conclude that there is no evidence for significant colour gradients within globular clusters, with the sole exception of the cluster NGC 104 (47 Tuc). They discuss a simplified technique for determining the colour of globular clusters independently of a direct measurement of the sky brightness near the cluster; the method requires simply that there be no strong radial gradient of colour in the cluster and explicitly compensates for the line–of–sight field and sky contamination. The authors present the complete photometric data in tabular form.

154.071 Statistical simulation of the Oosterhoff effect.
A. M. Ehjgenson, O. S. Yatsyk.
Astrofizika, Tom 22, Vyp. 2, p. 411 – 419 (1985). In Russian. English translation in Astrophysics, Vol. 22, No. 2.

Statistical simulation methods are used to show that the Oosterhoff effect can be explained by probability considerations without any physical mechanism.

154.072 The globular cluster system of the Galaxy. IV. The halo and disk subsystems.
R. Zinn.
Astrophys. J., Vol. 293, No. 2, p. 424 – 444 (1985).

The spatial distributions, kinematics, and metallicities of 121 globular clusters are examined for evidence of distinct subpopulations in the cluster system. The clusters more metal–poor than [Fe/H] = –0.8 constitute the familiar halo population. They have an essentially spherical distribution about the galactic center, a small rotational velocity ($v_{rot} = 50 \pm 23$ km s^{-1}), and a large velocity dispersion ($\sigma = 114$ km s^{-1}). The clusters more metal–rich than [Fe/H] = –0.8 have the properties of a disk system: a highly flattened spatial distribution, $v_{rot} = 152 \pm 29$ km s^{-1}, and $\sigma = 71$ km s^{-1}. Possible metallicity gradients and the luminosity functions are studied for both cluster populations. The scale height of the disk clusters is about 500 pc. Cluster ages and the implications of the present results for the early evolution of the Galaxy are also discussed.

154.073 Deep CCD photometry in globular clusters. III. M15.
G. G. Fahlman, H. B. Richer, D. A. VandenBerg.
Astrophys. J., Suppl. Ser., Vol. 58, No. 2, p. 225 – 254 (1985). With plate 18.

CCD photometry in U, B, and V is presented for a $5' \times 3'$ field in M15. A new color–magnitude diagram is presented, and a detailed comparison with the most recent VandenBerg and Bell isochrones, including an oxygen–enhanced sequence, leads to a cluster age of 16 ± 3 Gyr. New determinations of the cluster reddening, ultraviolet excess, and distance modulus are given, and the cluster luminosity function and field star population are studied in some detail.

154.074 Internal dynamics of globular clusters.
K. C. Freeman.
Stellar radial velocities, p. 241 – 248 (1985). – See Abstr. 012.095 (IAU Colloq. No. 88).

This paper discusses the importance of stellar radial velocities for globular cluster dynamics by reviewing recent studies of M3, 47 Tuc, and ω Cen.

154.075 Dynamics of NGC 188.
H. C. Harris, R. D. McClure.
Stellar radial velocities, p. 257 – 262 (1985). – See Abstr. 012.095 (IAU Colloq. No. 88).

Radial velocities have been obtained of 48 stars (33 member giants) in NGC 188 to investigate the internal dynamics. The velocity dispersion drops from the center outward, but the observed central dispersion is undoubtedly inflated by undetected binaries. The binaries concentrate toward the center. The dispersion in the outer ring indicates a cluster mass in agreement with the mass of visible stars.

154.076 Spectroscopy of globular clusters in the remote halo.
R. C. Peterson.
Stellar radial velocities, p. 363 – 364 (1985). Abstract. – See Abstr. 012.095 (IAU Colloq. No. 88).

154.077 The velocity dispersion of M4 horizontal–branch stars.
R. C. Peterson.
Stellar radial velocities, p. 365 – 366 (1985). Abstract. – See Abstr. 012.095 (IAU Colloq. No. 88).

154.078 Heavy remnants in globular cluster cores.
M. Mayor, G. Meylan.
Stellar radial velocities, p. 393 – 396 (1985). – See Abstr. 012.095 (IAU Colloq. No. 88).

Radial velocity measurements of 169 giants in 47 Tuc and of 298 giants in ω Cen are used to derive the rotation $V(r, z)$ and the velocity dispersion $\sigma(r)$ in these globular clusters.

154.079   Chemical abundances of field horizontal–branch stars of globular clusters.
V. G. Klochkova, V. E. Panchuk.
Astron. Zh., Tom 62, Vyp. 3, p. 552–557 (1985). In Russian. English translation in Sov. Astron., Vol. 29, No. 3.

The chemical composition is obtained for four stars, which are members of the horizontal branch of globular clusters, HD 74721, 86986, 109995, 161817. Within the accuracy of the model atmosphere method the chemical abundances in these stars are equal.

154.080   Globular clusters: homogeneous entities?
P. L. Cottrell.
South. Stars, Vol. 31, No. 2, p. 142–152 (1985).

Globular clusters have come under much more scrutiny during the last 5–10 years. These clusters can no longer be regarded as dead remnants of the early evolution of galaxies, but vital elements in the understanding not only of galactic evolution, but stellar evolution in a crowded environment. In this paper the author reviews some spectroscopic observations and analyses of globular cluster stars.

154.081   A single mass dynamical model for NGC 5139 (ω Cen).
B. J. Jarvis.
Bull. Am. Astron. Soc., Vol. 17, No. 2, p. 552–553 (1985). Abstract. – See Abstr. 010.065.

154.082   Putting the horizontal back in the horizontal branch.
D. A. Crocker, R. T. Rood.
Bull. Am. Astron. Soc., Vol. 17, No. 2, p. 555–556 (1985). Abstract. – See Abstr. 010.065.

154.083   CCD photometry of the E3 star cluster: evidence for main–sequence binaries in a globular cluster?
J. E. Hesser, R. D. McClure, P. B. Stetson, L. L. Stryker.
Bull. Am. Astron. Soc., Vol. 17, No. 2, p. 593 (1985). Abstract. – See Abstr. 010.065.

154.084   Refined photoelectric sequences for NGC 6712 and 6723.
D. H. Martins, D. A. Fraquelli.
Bull. Am. Astron. Soc., Vol. 17, No. 2, p. 594 (1985). Abstract. – See Abstr. 010.065.

154.085   Blue straggler stars in NGC 5466.
J. M. Nemec, H. C. Harris.
Bull. Am. Astron. Soc., Vol. 17, No. 2, p. 602 (1985). Abstract. – See Abstr. 010.065.

154.086   CCD photometry of the core structure of globular clusters.
P. M. Lugger, H. Cohn, J. E. Grindlay.
Bull. Am. Astron. Soc., Vol. 17, No. 2, p. 602–603 (1985). Abstract. – See Abstr. 010.065.

154.087   Core oscillations in globular clusters.
H. Cohn, P. Hut.
Bull. Am. Astron. Soc., Vol. 17, No. 2, p. 603 (1985). Abstract. – See Abstr. 010.065.

154.088   Measurements of the metallicities and radial velocities of globular clusters from the infrared Ca II lines.
R. J. Zinn, T. E. Armandroff.
Bull. Am. Astron. Soc., Vol. 17, No. 2, p. 603 (1985). Abstract. – See Abstr. 010.065.

154.089   Bimodal cyanogen distributions in moderately metal–poor globular clusters. I. Carbon and nitrogen abundances for six M5 giants.
G. E. Langer, R. P. Kraft, E. D. Friel.
Publ. Astron. Soc. Pac., Vol. 97, No. 591, p. 373–381 (1985). = Lick Obs. Bull., No. 1004.

The authors have fit synthetic spectra to low–resolution spectrophotometric scans in order to determine carbon and nitrogen abundances for three "CN–strong" and three "CN–weak" red giants in the globular cluster M5. Although the $\lambda3883$ CN band strengths are "bimodal", all six stars have very similar luminosities and effective temperatures and are presumably in the same evolutionary state. It is found that $\langle[C/Fe]\rangle_{CN-strong} - \langle[C/Fe]\rangle_{CN-weak} = -0.36$ and $\langle[N/Fe]\rangle_{CN-strong} - \langle[N/Fe]\rangle_{CN-weak} = +0.40$. However, there is no significant difference in the total $(C+N)$ abundance.

154.090   Bimodal cyanogen distributions in moderately metal–poor globular clusters. II. Evidence for "mixing" and "saturation".
G. E. Langer.
Publ. Astron. Soc. Pac., Vol. 97, No. 591, p. 382–392 (1985). = Lick Obs. Bull., No. 1003.

This paper describes how the bimodal cyanogen distributions that are observed in the moderately metal–poor ($[Fe/H] \approx -1.5$) globular clusters M5, M3, M13, and NGC 6752 might be interpreted within the framework of a simple mixing scenario in which $C \rightarrow N$ nucleosynthesis proceeds at different rates in otherwise similar stars.

154.091   Erratum: "The evolution of the galactic globular clusters. I. Metal abundance calibrations" [J. Korean Astron. Soc., Vol. 17, No. 2, p. 69–103 (1984)].
S.–W. Lee, N.–K. Park.
J. Korean Astron. Soc., Vol. 18, No. 1, p. 33–40 (1985). See Abstr. 38.154.060.

154.092   Globular clusters and RR Lyrae variables.
Z. I. Kadla, A. N. Gerashchenko.
Izv. Glav. Astron. Obs. Pulkovo, Astrometr. Astrofiz., No. 202, p. 83–104 (1984). In Russian.

Investigations of several globular clusters revealed the presence of a considerable number of previously undiscovered RR Lyrae variables located mainly in the central region of these objects. A more detailed study of the "missing" variables was made.

154.093   A remark on the paper of V. S. Popov "A determination of the distance to the object NGC 5272 using variable stars" [Izv. Glav. Astron. Obs. Pulkovo, Astrometr. Astrofiz., No. 199, p. 102 (1982). In Russian].
Z. I. Kadla.
Izv. Glav. Astron. Obs. Pulkovo, Astrometr. Astrofiz., No. 202, p. 116–117 (1984). In Russian. See Abstr. 32.154.045.

Sky catalogue 2000.0. Volume 2: Double stars, variable stars and nonstellar objects.
See Abstr. 002.019.

Space Telescope observations of globular clusters.
See Abstr. 051.085.

Theoretical constraints on far UV radiation from old Population II stellar systems.
See Abstr. 065.006.

Horizontal–branch sequences for globular cluster RR Lyrae stars.
See Abstr. 065.069.

The distribution of the projected density of stars near a Reissner–Nordström black hole in a globular cluster.
See Abstr. 067.103.

A high dispersion analysis of a giant star in 47 Tucanae.
See Abstr. 114.021.

UV–bright stars in galactic globular clusters, their far–UV spectra and their contribution to the globular cluster luminosity.
See Abstr. 114.023.

Observations of an sdO star in the globular cluster M22.
See Abstr. 114.040.

The rotation of horizontal–branch stars. III. Members of the globular cluster M4.
See Abstr. 116.021.

Are galactic bulge X–ray burst sources leftovers of disrupted globular clusters?
See Abstr. 117.046.

Galactic bulge X–ray burst sources from disrupted globular clusters?
See Abstr. 117.308.

X–ray luminous white dwarf binaries in globular clusters.
See Abstr. 117.309.

Three–body interactions and cataclysmic binaries in globular clusters.
See Abstr. 117.310.

The evolution of highly compact binaries in globular clusters.
See Abstr. 117.311.

CCD based B and V lightcurves for the eclipsing binary NJL 5 in Omega Centauri.
See Abstr. 119.030.

The discovery of a spectroscopic binary in M3.
See Abstr. 120.033.

CCD observations of the RR Lyrae stars in NGC 2210 and the distance to the Large Magellanic Cloud.
See Abstr. 122.006.

A period–luminosity relation for Mira variables in globular clusters and its impact on the distance scale.
See Abstr. 122.014.

Variations of the period of 35 variables in the globular cluster M15.
See Abstr. 122.018.

Double–mode RR Lyrae stars in M15: reanalysis, and experiments with simulated photometry.
See Abstr. 122.023.

The RR Lyrae stars in and around the LMC globular cluster NGC 2257.
See Abstr. 122.032.

The secular period behavior of 38 RR Lyrae stars in the LMC globular cluster NGC 2257.
See Abstr. 122.033.

Population II Cepheids.
See Abstr. 122.076.

The two Pop II Cepheids in the globular cluster Messier 10.
See Abstr. 122.081.

Eight Pop II Cepheids recently identified in globular clusters.
See Abstr. 122.082.

Comparison of the pulsation properties of the RR Lyrae stars in ω Centauri with those of classical Cepheids.
See Abstr. 122.084.

The long term behaviour of two variables in the globular cluster M56.
See Abstr. 122.086.

Photographic observations of the semiregular variable star Z Sagittae (V1, NGC 6838).
See Abstr. 122.094.

Photoelectric photometry of the red variable HD 157010.
See Abstr. 122.111.

Pulsation masses of globular cluster RR Lyrae variables.
See Abstr. 122.168.

Mehrfarben–Beobachtungen von V18, V19, V20, V21, V22, V23 im Kugelhaufen M3.
See Abstr. 123.012.

On the nature of M28 V7.
See Abstr. 123.023.

A search for millisecond pulsars in globular clusters.
See Abstr. 126.039.

On the expected number of white dwarfs in globular clusters.
See Abstr. 126.101.

Globular cluster origin of X–ray bursters.
See Abstr. 142.020.

The nature of the low–luminosity globular cluster X–ray sources.
See Abstr. 142.028.

X–ray sources in globular clusters and in the plane of the Galaxy.
See Abstr. 142.061.

Model integrated $B-V$ colors for metal–poor stellar systems.
See Abstr. 151.055.

Equipartition in multicomponent gravitational systems.
See Abstr. 151.077.

Collisional coalescence of binaries in globular clusters.
See Abstr. 151.089.

Can a moderately massive black hole reverse core collapse in globular clusters?
See Abstr. 151.090.

Evolution of stellar systems including tidally captured binaries.
See Abstr. 151.091.

Precollapse evolution of globular clusters.
See Abstr. 151.100.

Dynamical evolution of globular clusters after core collapse.
See Abstr. 151.101.

Direct Fokker–Planck calculations.
See Abstr. 151.102.

Dynamical evolution of multi–component clusters.
See Abstr. 151.104.

3–integral models for globular clusters.
See Abstr. 151.118.

Tidal effects on globular clusters.
See Abstr. 151.122.

Physical interactions between stars.
See Abstr. 151.123.

Monte–Carlo calculations.
See Abstr. 151.124.

Monte Carlo simulations of the 2+1 dimensional Fokker–Planck equation: spherical star clusters containing massive, central black holes.
See Abstr. 151.125.

Can a moderately massive black hole reverse core collapse?
See Abstr. 151.126.

Tidal stripping and disruption of globular clusters.
See Abstr. 151.128.

Evolution of globular clusters after a gravothermal catastrophe.
See Abstr. 151.168.

Faint stellar photometry in clusters. II. NGC 6791 and NGC 6535.
See Abstr. 153.023.

Synthetic cluster color–magnitude diagrams.
See Abstr. 153.035.

The metallicities of M5, M67, and M71 as determined from K giant spectra.
See Abstr. 153.036.

Report of IAU Commission 37: Star clusters and associations (*Amas stellaires et associations*).
See Abstr. 153.060.

On the structure and mass of the galactic halo.
See Abstr. 155.104.

Contribution to the study of the structure and dynamics of the galactic halo.
See Abstr. 155.113.

The kinematics of Population II stars.
See Abstr. 155.114.

Observed dynamical parameters of star clusters in the SMC.
See Abstr. 156.025.

The evolutionary connection between S and C stars: evidence from star clusters and the Magellanic Clouds.
See Abstr. 156.028.

Reddening of globular clusters in M31.
See Abstr. 157.057.

NGC 3557 and its globular clusters.
See Abstr. 157.059.

The M31 globular cluster system.
See Abstr. 157.067.

The missing bulge globular clusters in M31: new optical candidates.
See Abstr. 157.088.

Globular clusters in galaxies beyond the Local Group. IV. The elliptical galaxies NGC 524 and 1052.
See Abstr. 157.119.

Abundances of giant stars in the Draco and Ursa Minor dwarf galaxies.
See Abstr. 157.124.

Globular clusters in spiral galaxies.
See Abstr. 157.165.

Globular clusters and the distance to M87.
See Abstr. 158.082.

Globular clusters as extragalactic distance indicators.
See Abstr. 161.098.

155 Galaxy

155.001 Galactic absorption lines measured from the IUE low dispersion spectra of active galaxies.
C.–C. Wu, J. C. Blades, A. Boggess, D. G. York.
NASA Conf. Publ., NASA CP–2349, p. 123 – 126 (1984). – See Abstr. 012.001.

Galactic absorption lines were detected and measured from the low dispersion spectra of 10 active galaxies. In the galactic absorption line systems, while C II λ1335 is commonly seen, C IV λ1550 is detected with confidence only in high latitude objects and Fairall 9 which has a latitude close to 60°. The C IV to C II ratio is about one for the high latitude objects; and lower for the low latitude objects. This is consistent with the picture that C IV and C II exist mostly in the halo and disk, respectively. In the quasar absorption line systems, C II is rarely seen while C IV is generally strong. This suggests that the halo of intervening galaxies are large compared to their disk.

155.002 Carbon deflagrating supernovae and the chemical history of the solar neighbourhood.
F. Matteucci, A. Tornambé.
Astron. Astrophys., Vol. 142, No. 1, p. 13 – 20 (1985).

Yields of some heavy elements per stellar generation have been computed, by taking into account the chemical enrichment coming from carbon–deflagration in single intermediate mass stars (Type I 1/2SN) and supernova explosion in massive stars (Type IISN). It has also been assumed that the mass range of stars becoming Type I 1/2SNe is a function of the initial stellar metal content, favouring the occurrence of these SNe in the first stellar generations. These yields, which have been found to decrease from the first stellar generations to the late ones, have been incorporated in a chemical evolution model for the solar neighbourhood under different assumptions concerning the initial mass function. By means of this model, the evolution of the abundances of C^{12}, O^{16}, and Fe^{56} has been followed.

155.003 Giant magnetic features near the core of our Galaxy.
M. Morris.
Phys. Today, Vol. 38, No. 1, p. S7 – S8 (1985). Short review paper. – See Abstr. 013.003.

155.004 On the distribution of stars in the Orion spiral arm.
T. S. Basharina, E. D. Pavlovskaya, A. A. Filippova.
Astron. Zh., Tom 62, Vyp. 1, p. 29 – 33 (1985). In Russian. English translation in Sov. Astron., Vol. 29, No. 1.

The structure of the Orion arm has been studied by numerical experiments supposing that the distribution of the stars along the line of sight is a mixture of two normal laws. The parameters values have been obtained. It has been proved that the density functions of stars along the line of sight in the Orion spiral arm are asymmetric curves which are steeper at the external edge of the arm.

155.005 Models of the mass distribution of the Galaxy.
M. Schmidt.
The Milky Way galaxy, p. 75 – 84 (1985). – See Abstr. 012.007 (IAU Symp. No. 106).

The author reviews briefly published mass models that have approximate exponential disks and flat rotation curves. Then, he reports on some test models to investigate what range of model parameters is permissible for a given rotation curve. Finally, the author comments on some of the properties of the dark corona which is postulated to explain the flat rotation curve.

155.006 A model of our Galaxy.
U. Haud, M. Jôeveer, J. Einasto.
The Milky Way galaxy, p. 85 – 94 (1985). – See Abstr. 012.007 (IAU Symp. No. 106).
The authors present a new model of the Galaxy. It has been constructed using the most recent data available.

155.007 The galactic constants – an interim report.
F. J. Kerr.
The Milky Way galaxy, p. 97 – 98 (1985). – See Abstr. 012.007 (IAU Symp. No. 106).

155.008 Galactic rotation outside the solar circle.
R. Chini.
The Milky Way galaxy, p. 101 – 104 (1985). – See Abstr. 012.007 (IAU Symp. No. 106).
New radial velocities and optical distances for partly very distant H II regions give evidence for a rising rotation curve in the outer parts of the Galaxy.

155.009 Young galactic clusters and the rotation curve of our galaxy.
J. Hron, H. M. Maitzen.
The Milky Way galaxy, p. 105 – 106 (1985). – See Abstr. 012.007 (IAU Symp. No. 106).

155.010 How well do we know the rotation curve of our galaxy?
P. Pismis.
The Milky Way galaxy, p. 107 – 108 (1985). – See Abstr. 012.007 (IAU Symp. No. 106).

155.011 Galactic rotation and velocity fields.
W. L. H. Shuter, A. Gill.
The Milky Way galaxy, p. 109 – 110 (1985). – See Abstr. 012.007 (IAU Symp. No. 106).

155.012 The old population.
K. C. Freeman.
The Milky Way galaxy, p. 113 – 122 (1985). – See Abstr. 012.007 (IAU Symp. No. 106).
The author reviews the kinematical and chemical properties of the old population of our Galaxy. Comparison is made with the properties of the old population in the other disk galaxies.

155.013 Near–infrared studies of the Milky Way.
H. Okuda.
The Milky Way galaxy, p. 123 – 126 (1985). – See Abstr. 012.007 (IAU Symp. No. 106).
The inner region of the Galaxy has been explored by means of near–infrared observations; the distribution and population of the stars are studied from the near–infrared brightness mapping and star counts in the Milky Way, while the magnetic–field configuration is probed by the near–infrared polarimetry.

155.014 Infrared scanning of the galactic bulge.
R. M. Catchpole, P. A. Whitelock, I. S. Glass.
The Milky Way galaxy, p. 127 – 128 (1985). – See Abstr. 012.007 (IAU Symp. No. 106).
The authors find values of the interstellar absorption from J, H and K scans of a 6.8×200 arcmin strip of sky extending from the Sgr I field to the galactic centre. The visual absorption, excluding dark clouds, increases from 3 to 30 magnitudes in this region.

155.015 Infrared studies of the stellar population in Baade's Window.
R. A. Ruelas–Mayorga, A. R. Hyland, T. J. Jones.
The Milky Way galaxy, p. 129 – 130 (1985). – See Abstr. 012.007 (IAU Symp. No. 106).
An IR–scan of the Baade's Window ($l \cong 0°0$, $b \cong -4°0$) area has been obtained. The Cumulative Counts Function at 2.2 µm down to K $\cong +13.5$ was formed by combining 1.9–m telescope scans with Anglo Australian Telescope scans.

155.016 The stellar disk component: distribution, motions, age and stellar composition.
G. Lyngå.
The Milky Way galaxy, p. 133 – 142 (1985). – See Abstr. 012.007 (IAU Symp. No. 106).
The distribution of the stellar disk component is studied with particular emphasis on the properties of open clusters. A scale height of 80 pc is found for clusters younger than 10^9 years while the scale height is 200 pc for older clusters. A radial metallicity gradient in the disk is confirmed but there seem to be significant abundance variations apart from this. There are radial gradients in age as well as in linear diameter. Kinematically, there are a number of regions in the Milky Way with systematic radial velocity residuals.

155.017 The kinematics of nearby stars and large–scale radial motion in the Galaxy.
S. V. M. Clube.
The Milky Way galaxy, p. 145 – 147 (1985). – See Abstr. 012.007 (IAU Symp. No. 106).
The concept of star streams originally due to Kapteyn is revived. It is suggested that the existence of Stream II may have been overlooked, yet it is of crucial importance in specifying the true local standard of rest.

155.018 The galactic radial gradient of velocity dispersion.
M. Mayor, E. Oblak.
The Milky Way galaxy, p. 149 – 150 (1985). – See Abstr. 012.007 (IAU Symp. No. 106).

155.019 Stellar chemical–abundance gradient in the direction of the South Galactic Pole – preliminary results.
C. F. Trefzger, J. W. Pel, A. Blaauw.
The Milky Way galaxy, p. 151 – 152 (1985). – See Abstr. 012.007 (IAU Symp. No. 106).
Using the Walraven VBLUW photometric system, the authors have studied the metal content of 89 F and G stars in the Galactic South Pole field SA141.

155.020 Star counts, local density and K_z force.
B. Strömgren.
The Milky Way galaxy, p. 153 – 160 (1985). – See Abstr. 012.007 (IAU Symp. No. 106).
The approach by Bahcall and Soneira to the determination of galactic parameters through the use of star counts is referred to, and tests of the Bahcall–Soneira Galaxy model are discussed. The determination of the local mass density by Hill, Hilditch and Barnes through studies of A and F stars in the region of the North Galactic Pole is briefly outlined. The author discusses the determination of the galactic force K_z and the local mass density. Three uvbyβ surveys and their role in connection with the K_z problem are presented.

155.021 The stellar distribution in the galactic spheroid.
G. Gilmore.
The Milky Way galaxy, p. 161 – 162 (1985). – See Abstr. 012.007 (IAU Symp. No. 106).

155.022 Studies of O – F5 stars at the galactic poles.
R. W. Hilditch, A. D. McFadzean, G. Hill, J. V. Barnes.
The Milky Way galaxy, p. 163 – 164 (1985). – See Abstr. 012.007 (IAU Symp. No. 106).
The authors report progress on a spectroscopic and photometric programme devoted to the study of the dynamics of O – F5 stars within 15° of the North and South Galactic Poles. The aims of the programme are to test dynamical and chemical evolution models of the Galaxy by establishing velocity dispersions as a function of z–distance for stars of different population groups.

155.023   Galaxy population structure from proper motions.
N. Reid, C. A. Murray.
The Milky Way galaxy, p. 165 – 166 (1985). – See Abstr. 012.007
(IAU Symp. No. 106).
The authors are currently engaged in astrometry of Kapteyn
Selected Area photographic plates covering the period 1908 to
the present day using the GALAXY measuring engine at RGO.
The intention is to use the apparent motions to investigate the
stellar velocity structure within a kiloparsec of the Sun.

155.024   Survey of galactic H I emission at $|b| \leqslant 20°$.
W. B. Burton, P. te Lintel Hekkert.
The Milky Way galaxy, p. 171 – 172 (1985). – See Abstr. 012.007
(IAU Symp. No. 106).
The observations cover the part of the sky north of $\delta = -40°$
in the latitude range b = −20° to at least +20°; in some regions
the survey extends to b = +33°. The longitude coverage in the
equator ranges from l = 340° to 270°. The data represent an
improvement over existing large–scale H I surveys.

**155.025   The vertical distribution of galactic H I: the Arecibo–
Green Bank survey.**
T. M. Bania, F. J. Lockman.
The Milky Way galaxy, p. 173 (1985). – See Abstr. 012.007 (IAU
Symp. No. 106).

155.026   New light on the corrugation phenomenon in our galaxy.
J. V. Feitzinger, J. Spicker.
The Milky Way galaxy, p. 175 – 178 (1985). – See Abstr. 012.007
(IAU Symp. No. 106).
This investigation presents a total picture of the well–known
corrugation–phenomenon for the (heliocentric) longitude range
$10° \leqslant l \leqslant 240°$ as derived from H I–studies.

**155.027   H I at the outer edge of the Galaxy and its implications
for galactic rotation.**
P. D. Jackson.
The Milky Way galaxy, p. 179 – 180 (1985). – See Abstr. 012.007
(IAU Symp. No. 106).

155.028   The electron density in the plane of the Galaxy.
J. M. Weisberg, J. M. Rankin, V. Boriakoff.
The Milky Way galaxy, p. 181 (1985). – See Abstr. 012.007 (IAU
Symp. No. 106).

155.029   CO survey of the southern Milky Way.
R. S. Cohen, P. Thaddeus, L. Bronfman.
The Milky Way galaxy, p. 2, 199 – 202 (1985). – See Abstr.
012.007 (IAU Symp. No. 106).

**155.030   Distribution of CO in the southern Milky Way and
large–scale structure in the Galaxy.**
W. H. McCutcheon, B. J. Robinson, R. N. Manchester,
J. B. Whiteoak.
The Milky Way galaxy, p. 203 – 204 (1985). – See Abstr. 012.007
(IAU Symp. No. 106).
The southern galactic–plane region, in the ranges
$294° \leqslant l \leqslant 358°$, $-0°.075 \leqslant b \leqslant 0°.075$, has been surveyed in the
in the J = 1–0 line of ^{12}CO with a sampling interval of 3′.

155.031   A CO (2–1) survey of the southern Milky Way.
H. van de Stadt, F. P. Israel, T. de Graauw,
C. P. de Vries, J. Brand, H. J. Habing, J. Wouterloot.
The Milky Way galaxy, p. 205 – 206 (1985). – See Abstr. 012.007
(IAU Symp. No. 106).
The survey consists of three parts: 1. a survey of the galactic
plane (b = 0°) in the range l = 270 – 355°; 2. a survey of 88 dark
clouds with and without associated nebulosity; 3. a survey of 47
bright H II–region complexes.

155.032   The carbon monoxide distribution in the inner Galaxy.
T. M. Bania.
The Milky Way galaxy, p. 207 – 208 (1985). – See Abstr. 012.007
(IAU Symp. No. 106).
The latitude distribution of the emission from the ^{12}CO
J = 1→0 rotational transition has been surveyed for the region

$350° \leqslant l \leqslant 25°$ at b = 0′, ±10′ and ±20′. Most of the ^{12}CO
emission in the inner Galaxy, the region extending from the
galactic center to 4 kpc radius, is produced by three large and
massive objects: the nuclear disk/bar, the 3–kpc arm and the
"+135 km s^{-1} feature".

155.033   Outer–Galaxy molecular clouds.
M. L. Kutner, K. N. Mead.
The Milky Way galaxy, p. 209 – 210 (1985). – See Abstr. 012.007
(IAU Symp. No. 106).

155.034   CH in the Galaxy.
L. E. B. Johansson.
The Milky Way galaxy, p. 211 – 212 (1985). – See Abstr. 012.007
(IAU Symp. No. 106).
The distribution of CH in the Galaxy has been investigated via
its main–line transition in the $^2\pi_{1/2}$, J = 1/2 ground–state
Λ–doublet at 3335 MHz. The galactic plane was observed with a
spacing between adjacent points of 2°.5 in the longitude ranges
10° – 60° and 310° – 350°.

**155.035   High–energy gamma rays and the large–scale distribu-
tion of gas and cosmic rays.**
W. Hermsen, J. B. G. M. Bloemen.
The Milky Way galaxy, p. 213 – 218 (1985). – See Abstr. 012.007
(IAU Symp. No. 106).
The COS–B gamma–ray survey is compared with ^{12}CO and
H I surveys in a region containing the Orion complex and in the
outer Galaxy. The observed gamma–ray intensities in the Orion
region (100 MeV < E < 5 GeV) can be ascribed to the interaction of uniformly distributed cosmic rays with the interstellar gas.
In the outer Galaxy H I column–density maps in three galactocentric distance ranges are used in combination with COS–B
gamma–ray data to determine the radial distribution of the
gamma–ray emissivity. A strong decrease in the cosmic–ray electron density and a near constancy of the nuclear component is
found.

155.036   Large–scale mapping of the Galaxy by IRAS.
T. N. Gautier III, M. G. Hauser.
The Milky Way galaxy, p. 219 – 222 (1985). – See Abstr. 012.007
(IAU Symp. No. 106).
The extremely high sensitivity of the IRAS instrument for the
detection of interstellar material in the survey mode is illustrated
in terms of visual extinction and dust and gas column densities.

**155.037   High–energy galactic phenomena and the interstellar
medium.**
C. J. Cesarsky.
The Milky Way galaxy, p. 225 – 233 (1985). – See Abstr. 012.007
(IAU Symp. No. 106).
Contents: Gamma rays and the galactic distribution of cosmic
rays. Gamma–ray sources: a new galactic population? Cosmic-
ray acceleration by interstellar shocks. Supernovae and the interstellar medium.

**155.038   The high–latitude distribution of galactic gamma rays
and possible evidence for a gamma–ray halo.**
A. W. Strong.
The Milky Way galaxy, p. 235 – 236 (1985). – See Abstr. 012.007
(IAU Symp. No. 106).

155.039   On the ultraviolet background radiation of the Galaxy.
A. Zvereva, A. Severny (*A. Severnyj*), L. Granitsky
(*L. Granitskij*), G. Courtès, P. Cruvellier, C. T. Hua.
The Milky Way galaxy, p. 237 – 238 (1985). – See Abstr. 012.007
(IAU Symp. No. 106).
Observations of the far–UV spectrum (1300 – 1800 Å) of the
sky background at different galactic latitudes are presented. The
measurements were made with the photoelectric spectrometer
("Galaktika") on board the "Prognoz–6" satellite. The highly
elongated orbit of the satellite permitted corrections for scattered
L_α–light. The contribution of stars was eliminated using fluxes
measured in the TD–I and OAO–2 experiments.

155.040   Radio continuum emission of the Milky Way and nearby galaxies.
R. Beck, W. Reich.
The Milky Way galaxy, p. 239 – 244 (1985). – See Abstr. 012.007 (IAU Symp. No. 106).
The radio continuum emission of the Milky Way and nearby galaxies can be decomposed into a central region, a clumpy "thin disk", concentrated in the spiral arms, and a smooth "thick disk" (or flattened "halo"). The emissivity ratio of the two disks seems to be related to the magnetic field properties: Galaxies with strong radio spiral arms reveal a highly ordered field following the arm direction, while galaxies with diffuse disks contain a less ordered, smoothly distributed field. The degree of uniformity of the field seems to correlate with the total optical luminosity. The average magnetic field in the Milky Way is weak and turbulent compared to most of the nearby galaxies observed so far.

155.041   Loop I (the North Polar Spur) region – a quasi radio halo.
J. Milogradov–Turin.
The Milky Way galaxy, p. 245 – 246 (1985). – See Abstr. 012.007 (IAU Symp. No. 106).

155.042   Modelling the galactic contribution to the Faraday rotation of radiation from extra–galactic sources.
B. J. Brett.
The Milky Way galaxy, p. 249 – 250 (1985). – See Abstr. 012.007 (IAU Symp. No. 106).

155.043   A bisymmetric spiral magnetic field in the Milky Way.
Y. Sofue, M. Fujimoto.
The Milky Way galaxy, p. 251 – 252 (1985). – See Abstr. 012.007 (IAU Symp. No. 106).
The distribution of Faraday rotation measure of extragalactic radio sources shows that a large–scale magnetic field in the Galaxy is oriented along the spiral arms. The field lines change direction from one arm to the next in the inter–arm region.

155.044   Spiral structure of the Milky Way and external galaxies.
D. M. Elmegreen.
The Milky Way galaxy, p. 255 – 272 (1985). – See Abstr. 012.007 (IAU Symp. No. 106).
The spiral structure of the Milky Way galaxy has always been somewhat elusive because of our internal vantage point. The review presents methods and data for determining the overall pattern, and summarizes various models that have been proposed. Observations of spirals in external galaxies are also discussed, because they can provide insight into the spiral structure of the Milky Way.

155.045   Which kind of spiral structure can fit the observed gradient of vertex deviation?
E. Oblak, M. Crézé.
The Milky Way galaxy, p. 273 – 274 (1985). – See Abstr. 012.007 (IAU Symp. No. 106).
The authors analyse the distribution of peculiar velocities of stars within 200 pc of the sun in terms of space variations of the velocity ellipsoid.

155.046   Determination of galactic spiral structure at radio-frequencies.
H. S. Liszt.
The Milky Way galaxy, p. 283 – 300 (1985). – See Abstr. 012.007 (IAU Symp. No. 106).
The author considers some results of attempts to trace the pattern of galactic spiral structure in H II regions, H I, and CO. There is really no adequate method available for solving this problem, a fact reflected in the lack of consensus regarding the "correct" spiral pattern. The newly–begun process of deriving galactic structure in CO seems to be recapitulating the history laid down by H I observers.

155.047   Giant clouds and star–forming regions as spiral–arm tracers.
B. G. Elmegreen.
The Milky Way galaxy, p. 301 – 302 (1985). – See Abstr. 012.007 (IAU Symp. No. 106).
A variety of observations suggest that clouds of $10^6 - 10^7 M_\odot$ and extended regions of star formation are the best tracers of spiral structure in the Milky Way.

155.048   The largest molecular complexes in the first galactic quadrant.
T. M. Dame, B. G. Elmegreen, R. S. Cohen, P. Thaddeus.
The Milky Way galaxy, p. 303 – 304 (1985). – See Abstr. 012.007 (IAU Symp. No. 106).
The authors have used the Columbia University CO survey of the first galactic quadrant to determine the locations and physical properties of the largest molecular complexes in the inner Galaxy.

155.049   Interpretation of the apparent anomalies of the galactic structure.
T. Jaakkola, N. Holsti, P. Teerikorpi.
The Milky Way galaxy, p. 305 – 307 (1985). – See Abstr. 012.007 (IAU Symp. No. 106).

155.050   A 10–GHz radio–continuum survey of the galactic–plane region at the Nobeyama Radio Observatory – a complex region at $l = 22° – 25°$.
Y. Sofue, H. Hirabayashi, K. Akabane, M. Inoue, T. Handa, N. Nakai.
The Milky Way galaxy, p. 323 – 324 (1985). – See Abstr. 012.007 (IAU Symp. No. 106).
Preliminary results of a 10–GHz radio–continuum survey of the galactic–plane region using the 45–m telescope at NRO are presented. An extensive study of a complex region at $22° \leqslant l \leqslant 25°, |b| \lesssim 1°$ has been made.

155.051   The galactic nucleus.
J. H. Oort.
The Milky Way galaxy, p. 349 – 365 (1985). – See Abstr. 012.007 (IAU Symp. No. 106).
Contents: The mini–spiral. A central black hole? The compact radio source. The source IRS 16. GCX. The positron–electron annihilation line at 511 keV. Energetics and source of ionization in the central few parsecs. Supernova remnant close to the centre. Summary concerning the compact nucleus. The wider surroundings of the centre. H II regions within $0°.25$ (40 pc) from the centre. Molecular clouds. The $+40$ km s^{-1} cloud. Other phenomena of the gas in the central region. Expanding features? Asymmetry and low rotation of the bulk of the molecular gas. The tilt of the gas layer.

155.052   Recombination–line observations of the galactic centre.
J. H. van Gorkom, U. J. Schwarz, J. D. Bregman.
The Milky Way galaxy, p. 371 – 376 (1985). – See Abstr. 012.007 (IAU Symp. No. 106).
Aperture–synthesis observations of the H76α and H110α recombination lines are presented for the inner (3pc) region of the galactic nucleus. The large line width measured with single dishes (Pauls et al., 1974) is caused by well–ordered large–scale motions of the ionized gas. The velocities of the Ne II clumps (Lacy et al., 1980) fit well into the authors' smooth velocity field and smooth intensity distribution. The authors suggest therefore that the cloud picture of the Ne II gas is (at least partly) invalid.

155.053   Phenomena at the galactic centre – a massive black hole?
M. J. Rees.
The Milky Way galaxy, p. 379 – 384 (1985). – See Abstr. 012.007 (IAU Symp. No. 106).
The manifestations of activity at the galactic centre that are unambiguously non–thermal in character are: the γ–ray annihilation line ($\sim 10^{38}$erg s^{-1}) and the compact radio source ($\sim 10^{34}$erg s^{-1}). The author comments briefly on these two phenomena, and also suggests an interpretation of the remarkable pseudo–spiral structures revealed by the Ne II infrared and the radio–continuum maps. These phenomena relate to the old ques-

tion of whether our Galaxy has ever experienced a more violent phase, leaving a massive collapsed remnant.

155.054   The Milky Way: a halo, a corona, or both?
K. S. de Boer.
The Milky Way galaxy, p. 415 – 420 (1985). – See Abstr. 012.007 (IAU Symp. No. 106).

155.055   The chemical evolution of the Galaxy.
B. A. Twarog.
The Milky Way galaxy, p. 587 – 596 (1985). – See Abstr. 012.007 (IAU Symp. No. 106).

The review is restricted to those areas, both observational and theoretical, where a consensus of opinion appears to be emerging on some of the major components of chemical evolution of the Galaxy, particularly those which deviate from the simple, closed models.

155.056   Kinematical and chemical evolution of the galactic disk.
C. G. Lacey, S. M. Fall.
The Milky Way galaxy, p. 597 – 599 (1985). – See Abstr. 012.007 (IAU Symp. No. 106).

The authors have calculated models for the kinematical and chemical evolution of the galactic disk that include the infall of metal–free gas and the stochastic acceleration of disk stars.

155.057   The formation and early evolution of the Milky Way.
S. M. Fall.
The Milky Way galaxy, p. 603 – 610 (1985). – See Abstr. 012.007 (IAU Symp. No. 106).

The purpose of the article is to review several topics concerning the collapse phase in the evolution of the Milky Way.

155.058   The global properties of the Galaxy. III. Maps of the ^{12}CO emission in the first quadrant of the Galaxy.
G. R. Knapp, A. A. Stark, R. W. Wilson.
Astron. J., Vol. 90, No. 2, p. 254 – 300 (1985).

A survey of the ^{12}CO emission between longitudes 4° and 90° of the Galaxy is described. The survey consists of strip maps in latitude at 38 galactic longitudes, spaced at equal intervals of sin l. The maps are sampled at intervals of $\Delta b = 2$ arcmin with velocity resolutions of 0.65 and 2.6 km/s. The rms noise level of the observations is mostly better than 0.3K in the 2.6 km/s channels. The data are presented as latitude–velocity maps, as latitude–averaged profiles, and as a latitude–averaged velocity–sin l map.

155.059   A UV survey of the galactic plane.
M. Golay, N. Cramer, D. Huguenin, B. Nicolet, A. Blecha.
Astrophys. Space Sci., Vol. 109, No. 1, p. 191 – 212 (1985).

The present paper is the introduction to a systematic analysis of 123 6–degree fields near the galactic plane, recorded in the medium ultraviolet ($\lambda \approx 2000$ Å) by the balloon–borne experiment SCAP 2000 of the Laboratoire d'Astronomie Spatiale, Marseille, and Geneva Observatory. The available data are presented and the general properties of the images are briefly discussed. It is shown that the high selectivity of the UV passband regarding spectral type, together with the strong interstellar extinction at that wavelength, provide the necessary conditions for an efficient application of Wolf's method to study the distribution of interstellar matter in the solar neighbourhood. The results of a fast analysis of the available data are presented here.

155.060   The velocity–age relation for B and A stars.
J. Palouš, A. E. Piskunov (*A. Eh. Piskunov*).
Astron. Astrophys., Vol. 143, No. 1, p. 102 – 107 (1985).

The mean velocity and velocity dispersion versus age relations were investigated for B and A stars. The authors found significant differences between the motion of the B stars younger than 5×10^7yr and the motion of moderately old A stars with ages between 4×10^8 and 6×10^8yr. This may be evidence of the systematic motion of Gould's belt relative to the Local Standard of Rest. The authors have observed the initial relaxation phase of the velocity ellipsoid to be about 2×10^8yr; the initial shape of the velocity ellipsoid is given. The subsequent increase in the velocity dispersion in the age interval of 2×10^8 to 10^9yr was not confirmed by the data.

155.061   Recent observations of the galactic center.
J. Inatani.
Grand unified theories and cosmology, p. 122 – 130 (1984). Abstr. in Phys. Abstr., Vol. 88, No. 1251, Entry 24537 (1985). – See Abstr. 012.020.

155.062   New light on faint stars – VII. Luminosity and mass distributions in two high galactic latitude fields.
G. Gilmore, N. Reid, P. Hewett.
Mon. Not. R. Astron. Soc., Vol. 213, No. 2, p. 257 – 278 (1985).

The stellar distribution in apparent V magnitude and $B–V$ colour has been determined for complete samples of ~ 10000 stars in 11.5 square degrees towards $(l, b) = (0°, -90°)$, and of ~ 28000 stars in 17 square degrees towards $(l, b) = (37°, -51°)$. $V–I$ colour data for the second sample are discussed here. From these data the authors derive the structural parameters of the Galactic spheroid, and the form of the stellar luminosity function in the solar neighbourhood.

155.063   Comparisons of the Standard Galaxy Model with observations in two fields.
J. N. Bahcall, K. U. Ratnatunga.
Mon. Not. R. Astron. Soc., Vol. 213, No. 2, p. 39P – 41P (1985).

The Bahcall–Soneira Galaxy model is in good agreement with the observations of Gilmore, Reid and Hewett in the two fields in which they have reported their results.

155.064   Parametrization of the flow of halo high–velocity clouds.
A. Kaelble, K. S. de Boer, M. Grewing.
Astron. Astrophys., Vol. 143, No. 2, p. 408 – 412 (1985).

The authors have analysed the velocities of the sample of high–velocity clouds detected by Giovanelli (1980). The sample can be made consistent with a flow of neutral gas clouds from the Milky Way halo to the disk, which is characterized by a z–velocity between 0 and -100 km s^{-1}, and a galactocentric velocity of between 0 and -100 km s^{-1}, while the cool material in the halo, being between 2 and 5 kpc from the disk, participates only to a limited extent in the differential rotation of the Milky Way. The overall behaviour suggests a galactic origin, supporting the hypothesis that the HVCs are the cooling part of the Galactic Fountain circulation. The inflow rate is about $3 \times 10^{-3} M_\odot \text{kpc}^{-2} \text{yr}^{-1}$ as required by galactic evolution models but the gas may be of near solar composition.

155.065   A model of the radio continuum filaments in the galactic center.
P. J. Quinn, G. J. Sussman.
Astrophys. J., Vol. 288, No. 1, p. 377 – 384 (1985).

High–resolution radio continuum studies of the galactic center have revealed several spiral–like structures within the central 2 pc. The authors show that one of these structures can be modeled as the infall and tidal breakup of a small molecular cloud under the influence of the gravitational field of the center of the galaxy and a dissipative medium. The derived model parameters and model geometry imply the ejection of material near the escape velocity and the existence of a diffuse interstellar medium with a temperature of order 10^6K.

155.066   A region of low absorption in the galactic plane near $l = 102°$.
F. Chromey.
News Lett. Astron. Soc. N.Y., Vol. 2, No. 4, p. 19 (1983). Abstract. – See Abstr. 010.242.

155.067   VLA observations of the galactic center.
F. Yusef–Zadeh, M. Morris, D. Chance.
News Lett. Astron. Soc. N.Y., Vol. 2, No. 5, p. 11 – 12 (1984). Abstract. – See Abstr. 010.243.

155.068 The 157μ [C II] emission of the Galaxy.
G. Stacey, P. Viscuso, C. Fuller, N. Kurtz,
M. Harwit.
News Lett. Astron. Soc. N.Y., Vol. 2, No. 5, p. 14 (1984). Abstract. – See Abstr. 010.243.

155.069 Small scale magnetic field fluctuations in the Galaxy.
J. H. Simonetti, J. M. Cordes, S. R. Spangler.
News Lett. Astron. Soc. N.Y., Vol. 2, No. 5, p. 15 (1984). Abstract. – See Abstr. 010.243.

155.070 The Basel high–latitude survey of the Galaxy. I. Star count analysis from galaxy models.
R. Buser, U. Kaeser.
Astron. Astrophys., Vol. 145, No. 1, p. 1 – 18 (1985).
The authors have analyzed star count data to $m_v \lesssim 18^m$ in fifteen fields of the Basel Palomar–Schmidt–based high–latitude survey, using a large number of parameterized two–component models of the Galaxy consisting of an exponential disk and an (oblate) spheroid halo. Least–squares calculations were formed to investigate best–fitting models and place constraints on parameter values. The present approach interprets the data in terms of luminosity functions for kinematically distinct disk and halo populations. The authors' best models reproduce the observations to within an accuracy of better than $\sim 15\%$ for most fields.

155.071 Distribution of stellar radiation in the Galaxy.
N. Oda.
Astron. Astrophys., Vol. 145, No. 1, p. 45 – 49 (1985).
A model distribution of stellar radiation is constructed for the southern Milky Way and is improved for the northern Milky Way, by reference to recent balloon observations of the extended galactic emission at 2.4 μm. The model accounts for a local radiation density and well approximates the distributions of extended galactic light in the B and V bands, taking into account the contribution of scattered light. With a reasonable assumption that the 2.4 μm volume emissivity of the so–called 5 kpc ring is dominated mainly by M I and M III stars, it is concluded that $\sim 80\%$ of the 2.4 μm volume emissivity of the ring component is ascribed to M III stars. With this assignment, the near–infrared source count observations (Kawara et al., 1982) are explained, and it it demonstrated that the near–infrared observations of the extended galactic emission is consistent with the bimodal star formation model (Güsten and Mezger, 1982).

155.072 Wide–field kinematic Hα observations of the Milky Way with a scanning and imaging Fabry–Perot: preliminary results.
J. Caplan, J.–M. Perrin, J.–P. Sivan.
Astron. Astrophys., Vol. 145, No. 1, p. 221 – 226 (1985).
The authors have developed a new method for the large–scale kinematic study of galactic Hα. A scanning Fabry–Perot interferometer is used with a wide–field optical system, and the interference rings are recorded with a photon–counting television camera. The Fabry–Perot scan is synchronized with the photon–counting acquisition. Data reduction produces a sequence of n wide–field monochromatic images, with a radial velocity spacing given by the fraction $1/n$ of the interferometer's free spectral range. Reduction is effected by "photon sorting", a method which differs from that employed for Taurus by Atherton et al. (1982). This procedure has been successfully tested on the sky. Twenty five monochromatic images of a $15° \times 17°5$ field in Orion have been reconstituted over a range of ± 174 km s^{-1} with respect to the rest wavelength of Hα, with a step of 14.5 km s^{-1}.

155.073 *RGU*–photometry of a starfield in the direction to Scutum (S1').
S. Karaali, W. Becker, R. P. Fenkart.
Astron. Astrophys., Suppl. Ser., Vol. 60, No. 1, p. 75 – 85 (1985).
2647 stars down to a limiting magnitude in G of 17^m7 have been measured in a starfield of 0.167 square degrees in the direction to Scutum on 48″ Palomar Schmidt plates taken in the *RGU*–system. The data have been combined with those obtained earlier in larger fields treated by Becker (1962) and Karaali (1979) in

order to derive far–reaching reddening and space density functions in this direction.

155.074 The local system of early type stars. Spatial extent and kinematics.
T. N. G. Westin.
Astron. Astrophys., Suppl. Ser., Vol. 60, No. 1, p. 99 – 134 (1985).
Bright early type O – A0 stars in different age groups have been used to determine the spatial extent and the kinematic properties of the Gould's belt system. An inclination of 19° to the galactic plane was found for the Gould's belt. The line of nodes of the Gould's belt was found to be oriented almost exactly in the direction of galactic rotation. The extinction in the galactic plane is very irregular. It is shown that both the young and the old stars outside the limits of Gould's belt show the same kinematic characteristics as the nearby stars older than 6×10^7 years; they all seem to follow pure circular differential rotation.

155.075 The application of three–colour photometry on galactic starfields with strong interstellar extinction (Aquila II).
W. Becker, H. Steppe.
Astron. Astrophys., Suppl. Ser., Vol. 60, No. 1, p. 169 – 181 (1985). In German.
The authors report on first experiences from applying the method of three–colour photometry on Milky Way fields with strong interstellar extinction. Since the apparently sparsely populated field Aquila II is at a distance of only $2°4$ from Aquila I, a field with large apparent star density, a comparison of the extremes is tempting and possible. 1125 stars on 24 48″ Palomar Schmidt plates (8 in each colour R, G and U) within an area of 1.76 square degrees have been measured down to a limiting magnitude of 17^m8 in G. To make the distribution of the other stars somewhat more transparent it is advisable to determine the reddening function, at least provisionally according to the classical method by means of spectral types which are known for 74 stars. It shows monotonously increasing reddening with increasing apparent distance modulus. The density function of the late–type giants is almost constant, but the densities are smaller by a factor of about 10 than the ones in Aquila I.

155.076 Galactic chemical evolution and nucleocosmochronology: analytic quadratic models.
D. D. Clayton.
Astrophys. J., Vol. 288, No. 2, p. 569 – 574 (1985).
This paper studies models of the chemical evolution of the Galaxy for a star formation rate proportional to the square of the mass of gas. The first objective is to find analytic solutions to the gas mass and star mass for time–dependent rates of gaseous infall onto the disk. The second objective is to compare these quadratic models to models having linear star formation rates.

155.077 Interactions between the continuum sources in the galactic center and their immediate molecular environment.
P. T. P. Ho, J. M. Jackson, A. H. Barrett, J. T. Armstrong.
Astrophys. J., Vol. 288, No. 2, p. 575 – 579 (1985).
New VLA maps of the galactic center at 6 and 2 cm reveal thermal H II regions at the boundaries of nonthermal structures. In addition to star formation immediately east of SgrA–East, there is a cluster of OB stars $\sim 5'$ south of SgrA–West, and a nonthermal "wisp" $\sim 4'$ south of SgrA–West. This "wisp" may be part of a heretofore undetected supernova remnant in the galactic center. The juxtaposition of nonthermal and thermal continuum emission, as well as the apparent embedding of the H II regions in the molecular condensations, suggest strong interactions between galactic center continuum sources with their molecular environment.

155.078 RR Lyrae stars and the distant galactic halo: distribution, chemical composition, kinematics, and dynamics.
A. Saha.
Astrophys. J., Vol. 289, No. 1, p. 310 – 319 (1985).
The author here uses his earlier work on RR Lyrae stars in the galactic halo to probe the stellar population of the outer halo and to deduce the mass distribution in the far reaches of the Galaxy.

The space densities of RR Lyrae stars are traced out to 33 kpc from the galactic center. When combined with data from other surveys, the run of RR Lyrae star densities covering a very wide range of galactocentric distances is available. The chemical abundances of these probe stars show a very wide spread of metallicity, [Fe/H] from -1.0 to -2.2. However, no net abundance gradient is seen in the far halo. From the radial velocities, the kinematics of these stars has been studied. The mass of the Galaxy within 25 kpc is derived via dynamical arguments to be about $3 \times 10^{11} M_\odot$.

155.079   Giant molecular clouds in the Galaxy. II. Characteristics of discrete features.
D. B. Sanders, N. Z. Scoville, P. M. Solomon.
Astrophys. J., Vol. 289, No. 1, p. 373 – 387 (1985).

A sample of 315 discrete CO emission features with diameters larger than 10 pc have been identified from a uniform CO survey in l and b of the first galactic quadrant. The longitude and velocity of strong emission features are correlated with giant H II regions and may be associated with armlike spiral structure. Cooler clouds are distributed much more widely throughout the disk. The radial distribution of cloud number density is similar to the distribution of CO emissivity – that is, a ringlike concentration at $R = 4 - 8$ kpc with a relatively sharp inner edge and a gradual falloff toward large R. In the inner Galaxy ($R > 2$ kpc) 85% of the H_2 mass is in ~ 6000 clouds larger than 22 pc diameter with $M_{cloud} > 10^5 M_\odot$. Their total mass of $\sim 3 \times 10^9 M_\odot$ represents the largest reservoir of interstellar matter in the inner Galaxy.

155.080   The results of observations of the galactic disk for very high energy at the Tien–Shan High Altitude Station.
D. B. Mukanov, V. P. Fomin.
Izv. Krymskoj Astrofiz. Obs., Tom 69, p. 64 – 66 (1984). In Russian. English translation in Bull. Crimean Astrophys. Obs., Vol. 69.

Results of observation of the galactic disk for energy $E > 3 \times 10^{12}$eV by using the atmospheric Čerenkov technique are presented. It is shown that the count rate of showers is lower in the direction to the galactic equator $|b^{II}| < 1.5°$. The amplitude of decrease is 0.8% of the cosmic–ray background for the confidence level being about 95%.

155.081   Observed properties of the "central parsec" of the Galaxy as manifestation of a permanent star formation process.
I. S. Shklovskij.
Pis'ma Astron. Zh., Tom 11, No. 3, p. 163 – 168 (1985). In Russian. English translation in Sov. Astron. Lett., Vol. 11.

Specific features of the galactic center, observed in different spectral regions (the cold molecular gas–dust ring, sources of ionizing UV radiation, the Sgr A* radio source, peculiar morphology of plasma which emits thermal radio and IR radiation in this region) can be explained by a permanent star formation process.

155.082   Solar motion and galactic rotation obtained from the proper motions of faint stars in areas with galaxies.
S. P. Rybka.
Kinematika Fiz. Nebesn. Tel, Tom 1, No. 2, p. 37 – 42 (1985). In Russian.

The maximum likelihood method is used to divide 2463 faint stars in 30 areas with galaxies into two groups at different distances. The Ogorodnikov–Milne model of a three–dimensional differential centroid velocity field is applied to the proper motions of these stars. The kinematics of distant stars (~ 1700 pc) is shown to have departures from the Oort–Lindblad model.

155.083   Annihilation radiation from the galactic center: positrons in dust?
W. H. Zurek.
Astrophys. J., Vol. 289, No. 2, p. 603 – 608 (1985).

HEAO 3 autumn 1979 γ–ray observations of the galactic center indicate that the 2γ 0.511 MeV annihilation line contains more than 2 times as many photons as can be accounted for by standard models, in which positron annihilation takes place in an ionized interstellar gas. The author discusses positron annihilation in the interstellar medium and shows that a significant increase of the line intensity, sufficient to explain the observed spectrum, may arise from positron annihilation on grains of interstellar dust. Approximate analytic expressions for the rates of various annihilation processes occurring in the interstellar medium are given.

155.084   The fraction of high velocity dispersion H I in the Galaxy.
S. R. Kulkarni, M. Fich.
Astrophys. J., Vol. 289, No. 2, p. 792 – 802 (1985).

It has been recently suggested that a significant fraction of the neutral hydrogen clouds in our Galaxy have a cloud–cloud velocity dispersion of 35 km s^{-1} (Radhakrishnan and Srinivasan). Using H I emission data, the authors find that, indeed, a detectable amount of higher than normal velocity–dispersion H I exists. However, the fraction of H I in these clouds is about an order of magnitude less than that suggested by Radhakrishnan and Srinivasan. This "fast" H I constitutes less than 5% of the volume density of the H I in the galactic plane. It probably constitutes $\leqslant 20\%$ of the total mass of galactic H I but contains most of the kinetic energy of the ISM.

155.085   The 157 micron [C II] luminosity of the Galaxy. II. The presence of knotlike features in the [C II] emission.
G. J. Stacey, P. J. Viscuso, C. E. Fuller, N. T. Kurtz.
Astrophys. J., Vol. 289, No. 2, p. 803 – 806 (1985).

The authors have augmented their previous measurement of the diffuse 157 µm emission of singly ionized carbon from the galactic plane by measurements at two additional galactic longitudes. The present results indicate that the total [C II] flux from the Galaxy is $\sim 6 \times 10^7 L_\odot$ – a factor of 8 lower than the previous estimate. It is likely that the earlier measurement at $l^{II} = 8°0$ was due to a knot in the [C II] emission. The [C II] emission probably arises at the edges of molecular clouds.

155.086   Cosmic gamma rays and the mass of molecular hydrogen in the Galaxy.
C. L. Bhat, M. R. Issa, B. P. Houston, C. J. Mayer, A. W. Wolfendale.
Gas in the interstellar medium, p. 39 – 54 (1984). – See Abstr. 012.032.

It is shown that the γ–ray inferred total gas densities in the Galaxy are considerably less than those implied by conventional analyses of the ^{12}CO line. Specifically, the γ–ray analysis implies a total mass of H_2 in the inner Galaxy of only $\cong 6 \times 10^8 M_\odot$; molecular hydrogen therefore ceases to be the predominant component in this region. Explanations for the previous overestimate are put forward as are pieces of evidence from a variety of sources favouring the new estimate.

155.087   Maximum mass of objects that constitute unseen disk material.
J. N. Bahcall, P. Hut, S. Tremaine.
Astrophys. J., Vol. 290, No. 1, p. 15 – 20 (1985).

The existence of wide binaries in the Galaxy implies that the mass of the unseen disk objects, which provide about half of the local mass density (the Oort limit), must be less than 2 $M_\odot$.

155.088   Chemical evolution of the galactic disk with radial gas flows.
C. G. Lacey, S. M. Fall.
Astrophys. J., Vol. 290, No. 1, p. 154 – 170 (1985).

The authors present a series of models for the chemical evolution of the galactic disk with radial inflows of the gas at a velocity that is constant in time but may vary with galactocentric radius. The models also include the infall of metal–free gas from outside the disk at a rate that decays exponentially with time, and a star formation rate that varies as the mth power of the surface density of the gas layer, with $m = 1$ or 2. Chemical evolution is treated in the approximation of instantaneous recycling, with a constant initial mass function. The models are compared with the observed age–metallicity relation and metallicity distribution for stars in the solar neighborhood and with the observed radial variations of the metallicity, the star formation rate, and the surface densities of gas and stars.

155.089 Molecular clouds in the Carina arm.
R. S. Cohen, D. A. Grabelsky, J. May, L. Bronfman, H. Alvarez, P. Thaddeus.
Astrophys. J., Lett. Ed., Vol. 290, No. 1, p. L15 – L20 (1985).

From a new survey of the 2.6 mm line of CO in the southern Milky Way, the authors have been able to identify 37 molecular clouds along the Carina arm from $l = 282°$ to $336°$ with masses generally greater than $10^5 M_\odot$. The clouds lie approximately every 700 pc along a spiral segment that is nearly 25 kpc long and has a pitch of about $10°$. The total mass of these clouds is $40 \times 10^6 M_\odot$. The abrupt tangent point in molecular clouds at $l = 280°$ and the characteristic loop structure in the l–v diagram are unmistakable evidence of a CO spiral arm in Carina.

155.090 How much H_2 in the Milky Way?
L. Blitz.
Nature, Vol. 314, No. 6011, p. 495 – 496 (1985).

155.091 Cosmic γ rays and the mass of gas in the Galaxy.
C. L. Bhat, M. R. Issa, B. P. Houston, C. J. Mayer, A. W. Wolfendale.
Nature, Vol. 314, No. 6011, p. 511 – 515 (1985).

The authors make the case that the cosmic–ray particles that produce cosmic γ–rays in the 100–MeV region can penetrate most of the dense clouds of gas that seem to contain much of the mass of gas in the Galaxy. The authors determine a probable form for cosmic–ray intensity as a function of galactic position and use the measured γ–ray intensities to derive the surface density of gas as a function of position. Their estimate of $\sim 6 \times 10^8 M_\odot$ for the total mass of molecular hydrogen compared with $1 \times 10^9 M_\odot$ for atomic hydrogen in the inner Galaxy is a significant change in the contemporary perspective.

155.092 Black hole at the galactic centre.
J. Maddox.
Nature, Vol. 315, No. 6015, p. 93 (1985).

The latest accurate measurements of the size of the galactic core at centimetre radio–wavelengths leave no reasonable escape from the conclusion that there is a black hole at the centre (see Abstr. 155.093).

155.093 On the size of the galactic centre compact radio source: diameter < 20 AU.
K. Y. Lo, D. C. Backer, R. D. Ekers, K. I. Kellermann, M. Reid, J. M. Moran.
Nature, Vol. 315, No. 6015, p. 124 – 126 (1985).

The authors have re–observed Sgr A* at $\lambda 3.6$ cm and $\lambda 1.35$ cm with many VLBI baselines. These observations set an upper limit of 20 AU to the diameter of Sgr A* at $\lambda 1.35$ cm and reveal for the first time an elongated structure at $\lambda 3.6$ cm, with the position angle of the long axis almost parallel to the rotation axis of the Galaxy. Observations of the central 0.2 pc at radio and other wavelengths are best explained by a single massive collapsed object at the galactic centre.

155.094 Galactic spiral structure: statistical analysis of the space distribution of bright stars and open clusters.
E. D. Pavlovskaya, A. A. Suchkov.
Sov. Astron., Vol. 28, No. 4, p. 389 – 393 (1984). English translation of 38.155.002.

155.095 Chemical composition and evolution of the Galaxy.
V. I. Slysh.
Zemlya Vselennaya, No. 1, p. 9 – 15 (1985). In Russian.

155.096 New limits on the surface density of M dwarfs. I. Photographic survey and preliminary CCD data.
P. C. Boeshaar, J. A. Tyson.
Astron. J., Vol. 90, No. 5, p. 817 – 822 (1985). With plates 64 – 66.

Several observations have suggested a massive, dark component to galaxies (the "missing mass" controversy). Very–low–mass stars near the hydrogen–burning limit ($0.07\ M_\odot$) have long been a favorite theoretical candidate for the missing mass. The authors show that the FOCAS faint red star counts do not support this hypothesis for the disk and that the stellar luminosity function at the faint end of the main sequence (for the lowest–mass stars, $M < 0.1\ M_\odot$) most likely decreases with decreasing absolute luminosity. This faint red star search was done out to 40 – 100 pc from the Sun; it implies small or negligible dynamical consequences arising from intrinsically faint stars in the disk. This extreme M–dwarf search is being continued to beyond 2 kpc (halo) with the new CCD survey on the CTIO 4–m telescope. Multicolor CCD data on the Lowell 1.8–m telescope for a complete sequence of M dwarfs to dM9 is presented. The 0.7 – 1 µm color of the most extreme M dwarfs reaches a limiting value.

155.097 Some constraints on magnetic fields in galactic spiral arms.
J. Spicker, J. V. Feitzinger.
ESA Spec. Publ., ESA SP–207, p. 225 – 228 (1984). – See Abstr. 012.044.

After a short summary of existing contradicting theories and observations of the longitudinal and vertical components of the magnetic fields in galactic spiral arms, the authors present some constraints on the nature of this field in the light of recent investigations of the rolling motion phenomenon. These vertical structures of H I spiral arms are explained in the context of the galactic fountain model and of vertical magnetic field components. The corrugation structure of the spiral arms also sheds light on the magnetic field structure. Simple energetic requirements for the field are derived. Various components of galactic spiral arms show equal scale properties, which can be understood in terms of the energy equipartition.

155.098 High–energy X–ray observations of the galactic center region.
F. K. Knight, W. N. Johnson III, J. D. Kurfess, M. S. Strickman.
Astrophys. J., Vol. 290, No. 2, p. 557 – 567 (1985).

The authors present results from high–energy X–ray observations (18 – 250 keV) of the region within $5°$ of the galactic center on 1977 November 24 with a balloon–borne scintillation detector. The detector had a point source 3 σ detection threshold of 40 µJy (2 – 10 keV) for a Crab–like spectrum. The authors consider the known 2 – 10 keV sources to determine which are strongest above 200 keV and find an acceptable fit to their data using the bulge sources GX 3 + 1 and GX 5–1, the binary GX 1 + 4, the transient A1743–322, and A1742–294. More than one of these sources must vary to produce the variation in the γ–ray continuum slope and integrated flux (60 – 250 keV) observed 2 years later by HEAO 3.

155.099 The nonthermal radio emission of the galactic center.
Ch. N. Seitnepesov, A. Kh. Khanberdiev.
Astrofizika, Tom 22, Vyp. 2, p. 293 – 301 (1985). In Russian. English translation in Astrophysics, Vol. 22, No. 2.

Different models of the activity of the nonthermal radio source in the galactic center are analyzed. The profiles of the galactic nonthermal radio emission at 85 and 150 MHz on the basis of the three–component model of observed galactic radio emission "Stationary nucleus – radio disc – loop objects" are calculated. The observed profiles are compared with the calculated profiles. The observed angle distribution of the nonthermal galactic radio emission is in agreement with the model stationary source of cosmic electrons in the galactic center. A model of the origin of the galactic cosmic electrons is considered in which the major part of cosmic electrons is generated during the periodic activity of the galactic nucleus.

155.100 Kinematics of K giants in the outer galactic halo.
K. U. Ratnatunga, K. C. Freeman.
Astrophys. J., Vol. 291, No. 1, p. 260 – 269 (1985).

The authors describe a technique for finding field K giants in the outer galactic halo, 10 – 40 kpc from the galactic center. A sample of over 150 distant halo giants was located from an automated objective–prism survey of three high galactic latitude fields, each of 20 square degrees. Radial velocities of these giants show that (1) the outer halo is at most slowly rotating and (2) the line–of–sight velocity dispersion in our two best studied fields is approximately constant with distance from the Sun. The data on the run of velocity dispersion with distance in the halo fits well to a simple kinematical model in which the components of the velocity dispersion in cylindrical polar coordinates (with origin at the galactic center) remain constant everywhere, and equal to the anisotropic values seen in the solar neighborhood.

155.101 The galactic gamma–ray distribution: implications for galactic structure and the radial cosmic–ray gradient.
A. K. Harding, F. W. Stecker.
Astrophys. J., Vol. 291, No. 2, p. 471 – 478 (1985).

The authors have derived the radial distribution of γ–ray emissivity in the Galaxy from flux–longitude profiles, using both the final *SAS 2* results, and the recently corrected *COS B* results and analyzing the northern and southern galactic regions separately. The recent CO surveys of the southern hemisphere are used in conjunction with the northern hemisphere CO data to derive the radial distribution of cosmic rays on both sides of the Galaxy. It is found that, in addition to the "5 kpc ring" of enhanced emission, there is evidence from the asymmetry in the radial distributions for spiral features which are consistent with those derived from the distribution of bright H II regions. There is evidence for a strong increase in the cosmic–ray flux in the inner Galaxy.

155.102 OH/IR stars within 50 parsecs of the galactic center.
A. Winnberg, B. Baud, H. E. Matthews, H. J. Habing, F. M. Olnon.
Astrophys. J., Lett. Ed., Vol. 291, No. 2, p. L45 – L50 (1985).

A search for OH/IR stars close to the galactic center has been initiated using the VLA at 18 cm. Thirty–three stars were found in a $34' \times 34'$ field centered on Sgr A West. The strong concentration toward the galactic plane and the bulk rotation around the nucleus reported by Habing et al. in 1983 in a 2° field around the galactic center are confirmed. The circumstellar shell expansion velocities are higher than found at larger distances from the galactic center. The OH/IR stars are strongly concentrated toward the continuum source Sgr A West, suggesting a pronounced increase in the density of stars toward the galactic center.

155.103 The stellar population of the galactic nuclear bulge.
A. E. Whitford.
Publ. Astron. Soc. Pac., Vol. 97, No. 589, p. 205 – 213 (1985). = Lick Obs. Bull., No. 1001.

Line strengths in the spectra of bulge giants show a considerable spread of metallicity. The dominant component is super metal rich. A minority fraction of the giants is mildly metal poor, as are the RR Lyrae variables. Spectral surveys have identified all the late M giants in several bulge areas. These giants show the IR features noted in studies of the integrated light of galaxies, but are neither as cool nor as luminous as local giants of the same spectral type. Broad–band colors of bulge giants indicate other differences relative to solar–neighborhood prototypes that have traditionally been used in modeling galaxy populations.

155.104 On the structure and mass of the galactic halo.
S. Ninković.
Astrophys. Space Sci., Vol. 110, No. 2, p. 379 – 391 (1985).

On the basis of a globular cluster study a crude estimate of the total mass of the galactic halo within 20 kpc from the centre is done. It gives a minimal halo mass of the order of $10^{10} M_\odot$, yielding possiblities for a mass as large as $\sim 10^{11} M_\odot$. The content of the interstellar matter in the halo is estimated, too. It is found that the gas content is a few percents the minimal mass, the gas temperature is very high – about 1×10^6K, the magnetic field weak – about 0.25 nT. A weak nonthermal radio emission might be expected from such a halo.

155.105 Phenomenological model of the galactic core.
N. S. Kardashev.
Astrophys. Space Phys. Rev., Vol. 4, p. 287 – 315 (1985). – See Abstr. 003.020. Revised and extended English translation of 34.155.061.

A model of the galactic center is constructed, based on the observational data in the radio, infrared, X–ray, and γ–ray regions. The hypothesis that this center contains an extremely massive binary system – a compact cluster of stars and a black hole or a binary hole – is analyzed. The masses of the components of the system are $\sim 10^6 M_\odot$, the rotational period is ~ 6000 yr, and the luminosity is $\sim 3 \times 10^7 L_\odot$. The radiation energy is drawn either from the rotational energy of the black hole or from the accretion of gas clouds (debris from obliterated stars) onto the black hole. The main parameters of the compact radio source are determined and models of the regions in which relativistic $e^\pm$ particles are accelerated and of the region in which the 511–keV annihilation line is emitted are analyzed.

155.106 Far–infrared and submillimeter survey of the galactic center and nearby galactic plane.
M. F. Campbell, D. W. Niles, W. F. Hoffmann, H. A. Thronson, K. Shivanandan, M. Greenhouse, P. W. Woida, M. Kanskar.
Adv. Space Res., Vol. 5, No. 3, p. 3 – 6 (1985). – See Abstr. 012.059.

Maps are presented with 12' resolution of the galactic center and adjacent galactic plane, from $l^{II} = 359°$ to $l^{II} = 5°$. The data were obtained with the Steward Observatory cryogenically–cooled, balloon–borne telescope. The data are from channels filtered for a bandpass of 70 µm $< \lambda <$ 110 µm and for a long-pass of $\lambda > 80$ µm. For the typical effective temperature of 25K of a galactic H II region at this spatial resolution, the effective wavelengths of the channels are 93 µm and 145 µm. Continuous emission is mapped along the galactic plane in both wavelengths. The maps are discussed in detail.

155.107 Balloon–borne far–infrared spectrophotometry of the galactic center region.
S. Drapatz, L. Haser, R. Hofmann, N. Oda, R. Wilczek.
Adv. Space Res., Vol. 5, No. 3, p. 7 – 10 (1985). – See Abstr. 012.059.

Far–infrared observations of the galactic center have been carried through with the MPE 1m balloon–borne telescope "Golden Dragon". The measurements are composed of photometric scanning (33 – 95 µm) of the inner $4' \times 4'$ and low resolution spectroscopy ($\Delta\nu = 10$ cm^{-1}) of the center and of a position approximately 1.5' to the north. A Mars spectrum has been obtained for calibration. The spatial resolution of the photometry map is increased using the Maximum Entropy Method and the resulting map is compared to other observations in the same and other spectral regions. A clear asymmetry in the ring–like structure around the center indicates the presence of noncircular motions. The shape of the spectra is fairly smooth with at least no prominent dust features. A simple modelling shows a drastic increase of column density within 2 pc from the center and a modest drop over the next 3 pc to the north.

155.108 Wind from the centre of the Galaxy and its implications.
L. M. Ozernoj.
Astron. Tsirk., No. 1342, p. 1 – 5 (1984). In Russian.

Recent observations of broad He I and H I lines from IRS 16 located at the galactic centre indicate strongly an outflow of matter from it. The source of the wind cannot be a supermassive black hole.

155.109 Some results of a stellar–statistical investigation on the basis of proper motions of bright stars.
E. V. Khrutskaya.
Astron. Tsirk., No. 1347, p. 6 – 7 (1984). In Russian.

155.110 The vertical potential of the Galaxy.
W. B. Wilson.
South. Stars, Vol. 31, No. 1, p. 4 – 10 (1984).

The standard method of deriving the form of the gravitational potential of the Galaxy from kinematic data of stars in the solar neighbourhood is described. A new technique based on phase–angles is proposed. It makes no assumptions regarding the distribution of energies of the stars and so may not be very powerful. However, it is able to cope with the selection effects usually encountered in samples and this allows wide application.

155.111 A symbiotic model for the source in the center of the Galaxy.
L. M. Ozernoj.
Astron. Tsirk., No. 1349, p. 1 – 5 (1984). In Russian.

As it has been shown earlier, both the H I + He I wind and a significant part of the total radiation of the IRS 16 centre can be attributed to a superstar (SS) of the mass $\sim 300\ M_\odot$, while both gamma continuum radiation and positrons for the 0.511 MeV annihilation line from the galactic centre can be most naturally explained by a moderate–mass black hole (MMBH) of the mass $10 - 300\ M_\odot$. It is argued that, if the IRS 16 centre is spatially separated from the compact radio source Sgr A, the symbiotic object consisting of an SS and an MMBH seems to be able to explain the most principal features of the galactic centre emission.

155.112 The Massachusetts–Stony Brook galactic plane CO survey: disk and spiral arm molecular cloud populations.
P. M. Solomon, D. B. Sanders, A. R. Rivolo.
Astrophys. J., Lett. Ed., Vol. 292, No. 1, p. L19 – L24 (1985).

The authors present a preliminary analysis of a new high–resolution CO survey of the galactic disk which can detect and measure essentially all molecular clouds and cloud components in the inner Galaxy with size greater than 10 pc. In the region $20° < l < 50°$, approximately 2000 emission centers are identified. The authors find two populations which separate according to temperature. The disk population of cold molecular cores contains about three–quarters of the total number of cores, and must be widespread in the Galaxy both in and out of spiral arms. The spiral arm population of warm molecular cores contains about one–quarter of the population with one–half of the emission and is very closely associated with radio H II regions.

155.113 Contribution to the study of the structure and dynamics of the galactic halo.
S. Ninković.
Publ. Astron. Opservatorije Beogr., No. 33, p. 31 (1985). Abstract of a thesis. – See Abstr. 012.061.

No evidence that globular cluster masses exceed $10^5 - 10^6 M_\odot$ was found. A possible exception is ω Cen, whose mass could attain an order of $10^7 M_\odot$. Eccentricities of globular cluster galactocentric orbits are greater than 0.3; it seems that the total mass of the galactic halo within 25 kpc from the galactic centre does not exceed $1 \times 10^{11} M_\odot$.

155.114 The kinematics of Population II stars.
S. D. M. White.
Prepr. Steward Obs., No. 566, 18 pp. (1985).

Simple dynamical models are made to study the implications of recent kinematic data for Population II stars in the galactic halo. The models assume a scale–free distribution of stars orbiting within a spherical potential with a flat rotation curve. This potential is consistent with the observed properties of the globular cluster system. A highly flattened star distribution can reproduce the velocity dispersion behavior found by Ratnatunga and Freeman (1985) in a variety of high latitude fields. In particular it can reproduce the small dispersion they observed out to 25 kpc towards the galactic pole. These data appear to require the spatial distribution of metal–poor Population II stars to be quite different from that of the globular clusters.

155.115 Detection of galactic ^{26}Al gamma radiation by the SMM spectrometer.
G. H. Share, R. L. Kinzer, J. D. Kurfess, D. J. Forrest, E. L. Chupp, E. Rieger.
Astrophys. J., Lett. Ed., Vol. 292, No. 2, p. L61 – L65 (1985).

The gamma–ray spectrometer on the SMM satellite has detected a line near 1.81 MeV with a significance of $> 5\ \sigma$ in each of 3 yr when the galactic center traversed the broad aperture of the instrument in 1980, 1981, and 1982. There was no significant variation in intensity from year to year. The galactic center/anticenter intensity ratio is > 2.5, and the center of the emission is consistent with the location of the galactic center. The measured energy of the line is 1.804 ± 0.004 MeV. These measurements are consistent with the detection of a narrow γ–ray line from interstellar ^{26}Al by HEAO 3 in 1979/1980.

155.116 A possible galactic positron annihilation medium: neutral hydrogen.
B. L. Brown.
Astrophys. J., Lett. Ed., Vol. 292, No. 2, p. L67 – L70 (1985).

Previous investigators have ruled out neutral hydrogen as a possible medium for prompt annihilation of galactic positrons principally because the expected two–photon positronium spectrum is wider than the observed line width. The small fraction of incident positrons that directly annihilate and produce a narrow line was previously thought to be insignificant. Though the fraction is small, it is important in determining the line shape. Detailed fits to the galactic center annihilation spectra are consistent with prompt positron annihilation in a neutral hydrogen medium.

155.117 The gaseous galactic halo.
K. S. de Boer.
Mitt. Astron. Ges., Nr. 63, p. 21 – 33 (1985). – See Abstr. 012.063.

The observational evidence for the existence of a gaseous halo with coronal gas around the Milky Way will be critically reviewed. Highly ionized gas can be produced in a variety of ways in the halo. The models calculated have sufficient freedom to reach agreement with the observations of C IV and Si IV ions. The velocities observed for neutral halo clouds provide boundary conditions for the dynamics and origin of gas in the galactic halo. Possible further reasearch is discussed.

155.118 The chemical evolution of the Milky Way.
R. Güsten.
Mitt. Astron. Ges., Nr. 63, p. 49 – 64 (1985). – See Abstr. 012.063.

After a brief introduction into the evolution scenario of spiral galaxies, the observational evidence for temporal and spatial abundance variations across the galactic disk is discussed. The theoretical framework of chemical evolution models is summarised, followed by a critical discussion of available constraints. Special solutions applying to the solar vicinity and explaining the galactic abundance gradients are presented.

155.119 Morphology of galactic dust and gas based on IRAS and other data.
E. R. Deul, H. Walker, W. B. Burton.
Mitt. Astron. Ges., Nr. 63, p. 163 – 164 (1985). – See Abstr. 012.063.

155.120 Die Wellblechstruktur der Milchstraße.
J. V. Feitzinger, J. Spicker.
Mitt. Astron. Ges., Nr. 63, p. 165 (1985). Abstract. Submitted to Mon. Not. R. Astron. Soc. – See Abstr. 012.063.

155.121 Die projizierte Dunkelwolkenverteilung der Milchstraße.
J. V. Feitzinger, J. A. Stüwe.
Mitt. Astron. Ges., Nr. 63, p. 165 (1985). Abstract. – See Abstr. 012.063.

155.122 Novae in the galactic halo?
H. W. Duerbeck.
Mitt. Astron. Ges., Nr. 63, p. 190 (1985). – See Abstr. 012.063.

155.123 Steps towards a rotation curve of the southern outer Galaxy.
J. Brand, L. Blitz, J. Wouterloot.
Mitt. Astron. Ges., Nr. 63, p. 207 – 210 (1985). – See Abstr. 012.063.

An extensive study aimed at deriving the rotation curve of the outer Galaxy is reported. Preliminary results show that the rotation curve in that section of the Galaxy stays flat out to 18 kpc.

155.124   Ultraviolet absorption by highly ionized halo gas near the galactic center.
B. D. Savage, D. Massa.
Bull. Am. Astron. Soc., Vol. 16, No. 4, p. 910 (1984). Abstract. – See Abstr. 010.062.

155.125   A low frequency high resolution galactic plane survey.
N. Kassim.
Bull. Am. Astron. Soc., Vol. 16, No. 4, p. 910 (1984). Abstract. – See Abstr. 010.062.

155.126   Massachusetts–Stony Brook galactic plane CO survey: the galactic disk rotation curve.
D. P. Clemens.
Bull. Am. Astron. Soc., Vol. 16, No. 4, p. 911 (1984). Abstract. – See Abstr. 010.062.

155.127   Clark Lake observations of the galactic center region.
T. N. LaRosa, N. Kassim.
Bull. Am. Astron. Soc., Vol. 16, No. 4, p. 911 (1984). Abstract. – See Abstr. 010.062.

155.128   The Massachusetts–Stony Brook CO survey of the galactic plane: disk and spiral arm molecular cloud populations.
P. M. Solomon, D. B. Sanders, A. R. Rivolo.
Bull. Am. Astron. Soc., Vol. 16, No. 4, p. 919 (1984). Abstract. – See Abstr. 010.062.

155.129   Gamma ray imaging of the Vela region and galactic plane at 2 – 20 MeV.
T. O'Neill, O. T. Tümer, A. Zych, R. S. White.
Bull. Am. Astron. Soc., Vol. 16, No. 4, p. 934 (1984). Abstract. – See Abstr. 010.062.

155.130   Annihilation in the galactic center: positrons on dust?
W. H. Zurek.
Bull. Am. Astron. Soc., Vol. 16, No. 4, p. 935 (1984). Abstract. – See Abstr. 010.062.

155.131   Observations of the CO J = 7→6 line at 370 μm toward the galactic center.
A. Harris, D. T. Jaffe, M. Silber, R. Genzel.
Bull. Am. Astron. Soc., Vol. 16, No. 4, p. 936 (1984). Abstract. – See Abstr. 010.062.

155.132   A model for the annihilation radiation from the galactic center.
A. Mastichiadis, K. Brecher, A. P. Marscher.
Bull. Am. Astron. Soc., Vol. 16, No. 4, p. 963 (1984). Abstract. – See Abstr. 010.062.

155.133   A galactic component of the far infrared background.
F. J. Low, G. Neugebauer, T. N. Gautier III, F. C. Gillett.
Bull. Am. Astron. Soc., Vol. 16, No. 4, p. 968 (1984). Abstract. – See Abstr. 010.062.

155.134   A spiral structure model for the diffuse far–infrared emission from the Galaxy.
T. Hewagama, E. Dwek.
Bull. Am. Astron. Soc., Vol. 16, No. 4, p. 969 (1984). Abstract. – See Abstr. 010.062.

155.135   Kinematics of the galactic spheroid.
R. F. G. Wyse, G. Gilmore.
Bull. Am. Astron. Soc., Vol. 16, No. 4, p. 969 (1984). Abstract. – See Abstr. 010.062.

155.136   Kinematics and chemical properties of the old disk.
J. R. Lewis, K. C. Freeman.
Bull. Am. Astron. Soc., Vol. 16, No. 4, p. 969 (1984). Abstract. – See Abstr. 010.062.

155.137   Ne II observations of the galactic center.
E. Serabyn, J. H. Lacy.
Bull. Am. Astron. Soc., Vol. 16, No. 4, p. 980 (1984). Abstract. – See Abstr. 010.062.

155.138   The galactic ultraviolet radiation field.
R. C. Henry, W. B. Landsman.
Bull. Am. Astron. Soc., Vol. 16, No. 4, p. 980 (1984). Abstract. – See Abstr. 010.062.

155.139   Further observations of H and He in the very center of the Galaxy.
K. Krisciunas, T. R. Geballe, I. Gatley, R. Wade.
Bull. Am. Astron. Soc., Vol. 16, No. 4, p. 980 (1984). Abstract. – See Abstr. 010.062.

155.140   Mapping molecular hydrogen emission from the galactic center.
I. Gatley, R. Wade, T. J. Jones, A. R. Hyland.
Bull. Am. Astron. Soc., Vol. 16, No. 4, p. 980 (1984). Abstract. – See Abstr. 010.062.

155.141   Molecular clouds in the Carina arm.
D. A. Grabelsky, R. S. Cohen, L. Bronfman, H. Alvarez, J. May, P. Thaddeus.
Bull. Am. Astron. Soc., Vol. 16, No. 4, p. 980 – 981 (1984). Abstract. – See Abstr. 010.062.

155.142   The thickness of the galactic H I layer.
F. J. Lockman, L. M. Hobbs, J. M. Shull.
Bull. Am. Astron. Soc., Vol. 16, No. 4, p. 981 (1984). Abstract. – See Abstr. 010.062.

155.143   Mass distribution in the galactic centre.
M. K. Crawford, R. Genzel, A. I. Harris, D. T. Jaffe, J. H. Lacy, J. B. Lugten, E. Serabyn, C. H. Townes.
Nature, Vol. 315, No. 6019, p. 467 – 470 (1985).
New infrared and submillimetre spectroscopic measurements of the gas dynamics in the central 10 pc of the Galaxy make a convincing case that the mass distribution at the centre of the Galaxy is more concentrated than a spherical isothermal stellar cluster. The measurements fit a point mass of $\sim 4 \times 10^6 M_\odot$ but are also consistent with a cluster where stellar density decreases with radius (R) at least as fast as $R^{-2.7}$, or a combination of a point mass and a stellar cluster. The dynamical information combined with previous 2–μm observations favour a large point mass, which is presumably a massive black hole.

155.144   A constraint on the past star formation rate from the kinematics of nearby stars.
H. Meusinger.
Astron. Nachr., Vol. 306, No. 3, p. 123 – 127 (1985).
The age dependence of the velocity dispersions $\sigma_v(\tau)$ of the nearby stars is considered under distinct assumptions on the history of the star formation rate. The velocity distributions $N_{|U|}$, $N_{|V|}$, $N_{|W|}$ are derived from this by means of the diffusion theory using the standard case by Wielen and Fuchs (1983) and are compared with the observed distributions of a kinematically unbiased sample of nearby stars. The best agreement is found for a nearly constant star formation rate.

155.145   On the structure of the large–scale magnetic field of the Galaxy.
Ch. Seitnepesov, A. Kh. Khanberdiev.
Izv. Akad. Nauk TSSR. Ser. fiz.-tekh., khim. geol. nauk, No. 3, p. 14 – 19 (1984). In Russian. Abstr. in Ref. Zh., 51. Astron., 5.51.620 (1985).

155.146 The galactic magnetic field and a monopole halo.
D. F. Chernoff, S. L. Shapiro, I. Wasserman.
Bull. Am. Astron. Soc., Vol. 17, No. 1, p. 518 (1985). Abstract. – See Abstr. 010.064.

155.147 The interstellar medium in the outer Galaxy.
M. Fich.
Bull. Am. Astron. Soc., Vol. 17, No. 1, p. 520 (1985). Abstract. – See Abstr. 010.064.

155.148 The vertical structure of galactic spiral arms: the velocity asymmetries in H I observations.
J. V. Feitzinger, J. Spicker.
Mon. Not. R. Astron. Soc., Vol. 214, No. 4, p. 539 – 558 (1985).
On the basis of recently published H I data, and new radial velocity fields from improved rotation curves, expansion terms and spiral density–wave perturbations, the authors show that the so–called rolling motion phenomenon is only partly explained by geometric effects. They use an improved method and a new warp model to correct the observed velocity gradients for these apparent rolling motions. In all spiral arms, real velocity gradients perpendicular to the galactic plane are detected, with typical values of $\pm 20\,\mathrm{km\,s^{-1}kpc^{-1}}$ and values reaching $-75\,\mathrm{km\,s^{-1}kpc^{-1}}$ in some localized regions. The phenomenon is also observed in CO measurements. The results agree well with previous H I work.

155.149 [Ne II] observations of the galactic center: evidence for a massive black hole.
E. Serabyn, J. H. Lacy.
Astrophys. J., Vol. 293, No. 2, p. 445 – 458 (1985).
The velocity field within 2 pc of the galactic center has been investigated by observing the [Ne II] 12.8 µm fine–structure line emission from Sgr A West. The observations show regular variations in velocity along several of the most prominent radio continuum features, indicating that at least two of the features are large–scale flows of ionized gas. The central mass distribution is inferred from an analysis of the observed velocity field. It is concluded that the central mass cannot be due to an isothermal stellar cluster. A substantial fraction of the mass must be centrally more concentrated, most likely in a massive black hole of $3 - 4 \times 10^6 M_\odot$. The source of excitation for the ionized gas is also discussed.

155.150 Structure of Sagittarius C observed at radio frequencies.
H. S. Liszt.
Astrophys. J., Lett. Ed., Vol. 293, No. 2, p. L65 – L67 (1985). With plate L4.
Two large radio continuum arcs cross the galactic equator in the vicinity of the galactic center. One arc at longitude $+0°2$ brightens considerably over a 50 pc region nearest Sgr A, to which it is apparently connected by a remarkable set of curved filaments. The internal arc structure near Sgr A exhibits a unique series of parallel streamers. Here it is shown that Sgr C, located near the base of the second arc at $l \approx 359°45$, strongly resembles Sgr A and is connected to a filament which forms the brightest part of that arc.

155.151 Lynden–Bell statistics and velocity distribution of stars in the solar neighbourhood.
I. L. Genkin, A. D. Miroshnik.
Tr. Astrofiz. Inst. Alma–Ata, Tom 40, p. 87 – 90 (1983). In Russian.

155.152 A new halo survey.
B. W. Carney, D. W. Latham.
Stellar radial velocities, p. 139 – 150 (1985). – See Abstr. 012.095 (IAU Colloq. No. 88).
The authors describe a new survey of proper motion stars now nearing completion. They briefly discuss the survey's completeness and the data analysis methods. Preliminary results are presented for the motion of the Local Standard of Rest (LSR) with respect to the most extreme velocity stars and the possible presence of a metallicity gradient amongst the halo stars.

155.153 Current programs of local galactic structure and evolution.
J. Andersen, B. Nordström.
Stellar radial velocities, p. 171 – 176 (1985). – See Abstr. 012.095 (IAU Colloq. No. 88).
The authors present a progress report on some current radial–velocity observing programs aiming to provide complete data for selected samples of stars covering the whole sky. The velocities are based on ESO coudé spectra as well as CORAVEL observations obtained in both hemispheres. Currently, a survey of some 4000 dwarf F stars from Olsen's (1983) $uvby\beta$ photometric survey is near completion. The data will be used to study the velocity dispersion of these stars as a function of age and metal abundance from a kinematically unbiased sample.

155.154 Kinematics of halo red giants.
B. W. Carney, D. W. Latham, V. Straižys.
Stellar radial velocities, p. 177 – 181 (1985). – See Abstr. 012.095 (IAU Colloq. No. 88).
The authors discuss results from radial velocity data for a sample of metal–poor kinematically–unbiased red giants. They determine the motions of the stars with respect to the Local Standard of Rest, and estimate their velocity dispersions.

155.155 The local galactic escape velocity.
B. W. Carney, D. W. Latham, R. C. Peterson.
Stellar radial velocities, p. 187 – 188 (1985). Abstract. – See Abstr. 012.095 (IAU Colloq. No. 88).

155.156 Largescale problems of galactic structure and dynamics.
K. C. Freeman.
Stellar radial velocities, p. 223 – 229 (1985). – See Abstr. 012.095 (IAU Colloq. No. 88).
This review discusses some galactic problems for which stellar radial velocities are important. They include the total matter density near the Sun, the kinematics of moving groups, the kinematics of the old disk stars and the kinematics of the galactic halo.

155.157 Diffusion characteristics of electrons in the Galaxy.
A. Goned.
18th International Cosmic Ray Conference, Vol. 1, p. 178 – 181 (1983). – See Abstr. 012.096.
The radial distribution of electrons in the Galaxy in the range 50 – 5000 MeV has been derived from gamma ray data.

155.158 Spectral index variations in the galactic radio continuum emission.
C. J. Mayer, J. L. Osborne, C. G. T. Haslam, W. Reich.
18th International Cosmic Ray Conference, Vol. 2, p. 166 – 169 (1983). – See Abstr. 012.096.
Radio continuum surveys of the background sky temperature are an important aspect of cosmic ray research since they provide direct evidence of cosmic ray electrons in the range 0.1 – 10 GeV distributed throughout the Galaxy.

155.159 Gamma rays of 1 – 30 MeV from the galactic center region.
T. O'Neill, B. Dayton, J. L. Long, E. Zanrosso, A. D. Zych, R. S. White.
18th International Cosmic Ray Conference, Vol. 9, p. 45 – 48 (1983). – See Abstr. 012.096.
Preliminary results are reported for gamma rays from the galactic center region at 1 – 30 MeV. The observations were made with the UCR double Compton scatter gamma ray telescope. A sky map was obtained by projecting each gamma ray event circle onto the plane of the sky. Weights for the various directions were determined from Monte Carlo calculations.

155.160 Galactic hydrogen distribution and the production of galactic gamma–rays.
K. H. Chik, E. C. M. Young.
18th International Cosmic Ray Conference, Vol. 9, p. 85 – 88 (1983). – See Abstr. 012.096.
By using a simple expanding–ring model, the authors simulate the effect of non–circular motions on the gas distributions

derived from emission line profiles. They work out the radial density distributions of H I and H_2 by an unfolding procedure that does not require the use of a rotation curve. The galactic gamma–ray production is discussed in the light of the derived distributions.

155.161   Spiral structure of the Galaxy as outlined by diffuse nebulae in the solar neighbourhood.
V. S. Avedisova.
Pis'ma Astron. Zh., Tom 11, No. 6, p. 448 – 454 (1985). In Russian. English translation in Sov. Astron. Lett., Vol. 11.

Spiral arms in the Galaxy were traced by 255 emission nebulae with known photometric distances. Some parameters of the spiral structure were found. Pitch angles of three arms identified with the Local, Perseus and Sagittarius arms range between 18° and 21°. The radio source W51, a giant star formation region turned out to be located in the extension of the Local arm.

155.162   The distribution and motions of peculiar red giants.
R. M. Catchpole, M. W. Feast.
Cool stars with excesses of heavy elements, p. 113 – 133 (1985). – See Abstr. 012.101.

The principal interest in studying the distributions and motions of peculiar red giants (in our own and other galaxies) is in the information that this gives us on the population types (ages, masses) and evolutionary phases of these stars. This paper reviews recent population studies of carbon stars, S stars, Mira variables, P giants, barium stars, and CH stars in the Galaxy and in the Magellanic Clouds.

155.163   The Catalogue of Stellar Groups. II. Cool peculiar stars.
D. Egret, M. Jaschek.
Cool stars with excesses of heavy elements, p. 135 – 140 (1985). – See Abstr. 012.101.

The "Catalogue of Stellar Groups" is a set of files listing all the kinds of stars with known peculiarities. A first volume covering the earlier types has already been published. Preliminary lists for about thirty groups of cool peculiar stars are now available. Histograms in visual magnitudes and plots of the galactic distribution are presented here for carbon stars, S stars, CH stars, CN stars, and barium stars.

155.164   Kinematics of late–type giants.
F. Ochsenbein, M. Holtzer.
Cool stars with excesses of heavy elements, p. 141 – 146 (1985). – See Abstr. 012.101.

The well defined data found in the Bright Star Catalog ($m_v \leqslant 6^{m}\!.50$) were used to derive the main kinematical parameters of stars grouped according to spectral classification. The monotonic variation of the dispersion of the residual velocities σ_{res} among main sequence stars is stressed and the incidence of high–velocity stars in late–type giants is discussed.

155.165   Kinematics and spatial density of main sequence SMR (*super–metal–rich*) stars.
M. Grenon.
Cool stars with excesses of heavy elements, p. 147 – 152 (1985). – See Abstr. 012.101.

A population of super–metal–rich stars is searched among nearby and proper–motion stars. The metallicities, T_{eff}, M_v are obtained photometrically and lead to estimates of the age and space density of the SMR subpopulation. Its origin is deduced from the evolutionary and kinematical properties of the SMR candidates.

155.166   Some statistics on barium stars.
M. Holtzer.
Cool stars with excesses of heavy elements, p. 153 – 157 (1985). – See Abstr. 012.101.

The 170 certain Ba stars and the 140 marginal Ba stars contained in the "Catalog of Stellar Groups" were compared to a sample of 200 giants of the same spectral type, G5 to K5, taken from the "Bright Star Catalog". It is shown that the statistics of the Ba star group suffers from observational and classification biases. The marginal Ba stars and the certain Ba stars seems to be homogeneous in both age and population type.

155.167   Carbon stars and the center of our Galaxy.
M. F. McCarthy, V. M. Blanco.
Cool stars with excesses of heavy elements, p. 159 – 161 (1985). – See Abstr. 012.101.

Red giant stars in the galactic center region are studied on infrared sensitive plates taken with a grism and the 4 m CTIO reflector. The authors find M5 to M9 stars abundant and carbon and S stars almost non existent in the direction of the galactic center. In two regions (Baade's window at NGC 6522 and in Sgr I) the authors have counted more than 1000 very cool M giants and no C or S stars.

155.168   Carbon stars and S stars near open clusters. A statistical approach.
U. G. Jørgensen.
Cool stars with excesses of heavy elements, p. 181 – 184 (1985). – See Abstr. 012.101.

The positions of 3218 C stars, 741 S stars and 1140 open clusters have been compared. 59 C stars and 15 S stars have been found to lie in the line of sight to open clusters. This is no more than could be expected from a random distribution of the stars.

155.169   The effect of incompleteness among nearby stars.
A. R. Upgren.
Bull. Am. Astron. Soc., Vol. 17, No. 2, p. 553 (1985). Abstract. – See Abstr. 010.065.

155.170   A photometric survey of two high galactic latitude fields.
E. D. Friel, R. P. Kraft, K. M. Cudworth.
Bull. Am. Astron. Soc., Vol. 17, No. 2, p. 560 (1985). Abstract. – See Abstr. 010.065.

155.171   On the density of the spheroid in the solar neighborhood.
S. Casertano, J. N. Bahcall.
Bull. Am. Astron. Soc., Vol. 17, No. 2, p. 560 – 561 (1985). Abstract. – See Abstr. 010.065.

155.172   Haystacks, needles and carbon stars.
M. F. McCarthy, V. M. Blanco.
Bull. Am. Astron. Soc., Vol. 17, No. 2, p. 561 (1985). Abstract. – See Abstr. 010.065.

155.173   Far–IR and sub–mm diffuse emission and extended sources in the central regions of the Galaxy.
M. F. Campbell, D. W. Niles, M. Kanskar, W. F. Hoffmann, M. Greenhouse, P. W. Woida, H. A. Thronson, K. Shivanandan.
Bull. Am. Astron. Soc., Vol. 17, No. 2, p. 561 (1985). Abstract. – See Abstr. 010.065.

155.174   Low frequency aperture synthesis observations of the galactic center region: possible evidence for Seyfert–like activity.
T. N. LaRosa, N. E. Kassim.
Bull. Am. Astron. Soc., Vol. 17, No. 2, p. 561 (1985). Abstract. – See Abstr. 010.065.

155.175   Photometric imaging of the galactic center source complex with the Goddard 10–micron array camera.
D. Y. Gezari, R. Tresch–Fienberg, G. G. Fazio, W. F. Hoffmann, I. Gatley, G. Lamb, P. Shu, C. McCreight.
Bull. Am. Astron. Soc., Vol. 17, No. 2, p. 561 (1985). Abstract. – See Abstr. 010.065.

155.176   Modeling the galactic bulge at infrared wavelengths.
R. W. Garwood, T. J. Jones.
Bull. Am. Astron. Soc., Vol. 17, No. 2, p. 561 (1985). Abstract. – See Abstr. 010.065.

155.177 The local galactic escape velocity.
B. W. Carney, D. W. Latham.
Bull. Am. Astron. Soc., Vol. 17, No. 2, p. 562 (1985). Abstract. –
See Abstr. 010.065.

155.178 The distance to the galactic center.
V. M. Blanco.
Bull. Am. Astron. Soc., Vol. 17, No. 2, p. 562 (1985). Abstract. –
See Abstr. 010.065.

155.179 The total amount of matter in the solar vicinity.
J. N. Bahcall.
Bull. Am. Astron. Soc., Vol. 17, No. 2, p. 581 (1985). Abstract. –
See Abstr. 010.065.

155.180 The mass distribution in the galactic spheroid.
G. Gilmore.
Bull. Am. Astron. Soc., Vol. 17, No. 2, p. 584 – 585 (1985). Abstract. – See Abstr. 010.065.

155.181 Giant molecular clouds in the Galaxy: the Massachusetts – Stony Brook CO galactic plane survey.
P. M. Solomon.
Bull. Am. Astron. Soc., Vol. 17, No. 2, p. 606 (1985). Abstract. –
See Abstr. 010.065.

155.182 The large–scale distribution of molecules in the northern galactic disk.
D. P. Clemens, N. Z. Scoville, D. B. Sanders.
Bull. Am. Astron. Soc., Vol. 17, No. 2, p. 606 (1985). Abstract. –
See Abstr. 010.065.

155.183 A carbon monoxide survey of the galactic center.
M. Bitran, H. Alvarez, R. S. Cohen, P. Thaddeus.
Bull. Am. Astron. Soc., Vol. 17, No. 2, p. 607 (1985). Abstract. –
See Abstr. 010.065.

155.184 The scale height of molecular clouds in the Carina arm.
D. A. Grabelsky, R. S. Cohen, L. Bronfman, J. May, P. Thaddeus.
Bull. Am. Astron. Soc., Vol. 17, No. 2, p. 607 (1985). Abstract. –
See Abstr. 010.065.

155.185 A deep CO survey of molecular clouds in the fourth galactic quadrant.
L. Bronfman, R. S. Cohen, P. Thaddeus, H. Alvarez.
Bull. Am. Astron. Soc., Vol. 17, No. 2, p. 607 (1985). Abstract. –
See Abstr. 010.065.

155.186 The Bell Laboratories CO survey.
A. A. Stark, J. Bally, G. R. Knapp, A. C. Krahnert, A. A. Penzias, R. W. Wilson.
Bull. Am. Astron. Soc., Vol. 17, No. 2, p. 613 (1985). Abstract. –
See Abstr. 010.065.

155.187 Spiral arm and interarm regions in the Bell Laboratories CO survey.
A. C. Krahnert, A. A. Stark, J. Bally, G. R. Knapp, A. A. Penzias, R. W. Wilson.
Bull. Am. Astron. Soc., Vol. 17, No. 2, p. 613 (1985). Abstract. –
See Abstr. 010.065.

155.188 The Massachusetts–Stony Brook CO survey of the inner galaxy.
W. M. Irvine, D. P. Clemens, D. B. Sanders, N. Z. Scoville.
Bull. Am. Astron. Soc., Vol. 17, No. 2, p. 615 (1985). Abstract. –
See Abstr. 010.065.

155.189 "Windows" in the galactic plane.
W. H. Waller, D. P. Clemens, D. B. Sanders, N. Z. Scoville.
Bull. Am. Astron. Soc., Vol. 17, No. 2, p. 615 (1985). Abstract. –
See Abstr. 010.065.

155.190 Massachusetts–Stony Brook CO galactic plane survey: large scale distribution of molecular clouds in the Galaxy.
J. W. Barrett, P. M. Solomon.
Bull. Am. Astron. Soc., Vol. 17, No. 2, p. 615 (1985). Abstract. –
See Abstr. 010.065.

155.191 CO(1–0) survey at high galactic latitudes.
E. A. Lada, L. Magnani, L. Blitz.
Bull. Am. Astron. Soc., Vol. 17, No. 2, p. 616 (1985). Abstract. –
See Abstr. 010.065.

155.192 Correlations between the distributions of stars and of the intensity of continuous radio emission in the region of the north polar spur.
G. B. Anisimova.
Sov. Astron., Vol. 28, No. 6, p. 718 – 719 (1984). English translation of 38.155.073.

155.193 Report of IAU Commission 33: Structure and dynamics of the galactic system (*Structure et dynamique du système galactique*).
R. Wielen.
Trans. IAU, Vol. XIXA, p. 397 – 435 (1985). – See Abstr. 003.046.

155.194 A survey for galactic plane ultraviolet–excess objects: space densities for white dwarfs, subdwarfs, and cataclysmic variables.
R. A. Downes.
Diss. Abstr. Int., Sect. B, Vol. 45, No. 9, p. 2955 – 2956 (1985).
Thesis, University of California, 163 pp. (1984). Order No. DA8428506.

155.195 The galactic rotation curve from an OH survey.
I. G. Kolesnik, L. V. Yurevich.
Astrofizika, Tom 22, Vyp. 3, p. 461 – 472 (1985). In Russian. English translation in Astrophysics, Vol. 22, No. 3.

The galactic rotation curve up to 16 kpc has been determined using the relation between parameters of OH molecular absorption features of clouds and distance to the clouds. The rotation curve obtained follows the kinematic behaviour of the Galaxy which has been determined from other interstellar molecules. From the position of the central peak in the rotation curve the sun's galactocentric distance was estimated to be ~ 8.5 kpc. The mass of the Galaxy interior to 100 kpc from the galactic center is $\sim 3 \times 10^{12} M_{\odot}$, as was derived from the rotational velocity distribution.

155.196 Scans across Sagittarius A at 1.38 cm.
A. B. Berlin, Yu. N. Parijskij, N. S. Soboleva.
Sov. Astron., Vol. 28, No. 6, p. 609 – 612 (1984). English translation of 38.141.025.

155.197 The Galaxy scene and Quasat.
C. A. Norman.
ESA Spec. Publ., ESA SP–213, p. 11 – 12 (1984). – See Abstr. 012.124.

Current Galaxy research is reviewed with emphasis on aspects relevant to the Quasat project.

A galactic center molecular cloud atlas.
See Abstr. 002.106.

The Galaxy.
See Abstr. 003.132.

The center of the Galaxy.
See Abstr. 003.158.

The Milky Way from antiquity to modern times.
See Abstr. 004.005.

Kapteyn and statistical astronomy.
See Abstr. 004.006.

Studies of the Milky Way 1850 – 1930: some highlights.
See Abstr. 004.007.

The discovery of the spiral arms of the Milky Way.
See Abstr. 004.008.

Serendipity in the Galaxy: the galactic warp and the galactic nucleus.
See Abstr. 004.199.

Cratering theories bombarded.
See Abstr. 011.008.

Positrons in a simulated low–density galactic environment: recent experimental results.
See Abstr. 022.178.

Astrophysical and geophysical observations with PIRAMIG/Salyut 7 experiment.
See Abstr. 035.049.

Galactic magnetic fields and magnetic monopoles.
See Abstr. 061.080.

Monopoles and the galactic magnetic field.
See Abstr. 061.086.

The plasma physics of magnetic monopoles in the Galaxy.
See Abstr. 061.087.

Solar abundances and the role of nucleogenesis in low–to–medium mass stars in the Galaxy.
See Abstr. 061.148.

Report of IAU Commission 21: Light of the night sky (*Lumière du ciel nocturne*).
See Abstr. 082.096.

Cometary impacts, molecular clouds, and the motion of the Sun perpendicular to the galactic plane.
See Abstr. 107.013.

Radial velocities of blue horizontal branch field stars.
See Abstr. 111.001.

The velocity dispersion of carbon stars at the north galactic pole.
See Abstr. 111.012.

Radial velocities of late–type Population II stars.
See Abstr. 111.032.

A new radial velocity survey at the NGP.
See Abstr. 111.045.

Radial velocity and four–color measures of halo A–type stars.
See Abstr. 111.046.

A photometric search for nearby stars in the NLTT catalogue.
See Abstr. 111.049.

Defining a bias–free sample of nearby stars.
See Abstr. 111.052.

RGU three–colour photometry of a starfield in anticentre–direction (A6).
See Abstr. 113.004.

Galactic structure photoelectric standards.
See Abstr. 113.038.

CCD stellar sequences in galactic structure fields.
See Abstr. 113.039.

Analysis of three field halo stars and the chemical evolution of the Galaxy.
See Abstr. 114.036.

An extremely metal–poor star with *r*–process overabundances.
See Abstr. 114.037.

Abundances in field dwarf stars. II. Carbon and nitrogen abundances.
See Abstr. 114.045.

Discovery of carbon stars in the galactic bulge.
See Abstr. 114.073.

Red horizontal–branch stars in the Galactic disk.
See Abstr. 114.082.

Strong–lined G dwarfs in the Galactic disk.
See Abstr. 114.083.

Extremely metal–deficient red giants. III. Chemical abundance patterns in field halo giants.
See Abstr. 114.092.

Discovery of carbon stars in the direction of the galactic center.
See Abstr. 114.107.

A southern survey for extremely metal–poor stars.
See Abstr. 114.111.

The discovery and frequency of barium stars.
See Abstr. 114.146.

M supergiants in the Milky Way and the Magellanic Clouds: colors, spectral types, and luminosities.
See Abstr. 115.004.

The luminosity function of the halo main sequence in the solar neighbourhood.
See Abstr. 115.028.

Metal abundances of RR Lyrae variables in selected galactic star fields. IV. The Lick Astrograph field RR I (MWF 361) in Serpens and Ophiuchus.
See Abstr. 122.021.

The infrared distance scale: the Galaxy and the Magellanic Clouds.
See Abstr. 122.075.

The color excesses of classical Cepheids and the structure of our Galaxy.
See Abstr. 122.212.

Novae and galactic chemical evolution.
See Abstr. 124.013.

The galactic reddening law: the evidence from $uvby\beta$ photometry of B stars.
See Abstr. 131.019.

Associations between neutral and ionized gas in Sgr A.
See Abstr. 131.020.

A high resolution H I absorption spectrum of Sgr A*.
See Abstr. 131.021.

Temperature determinations in molecular clouds of the galactic center.
See Abstr. 131.025.

On the abundance of deuterated molecules in the Galaxy.
See Abstr. 131.030.

Small–scale structure and motions in the interstellar gas.
See Abstr. 131.032.

Atomic hydrogen towards 3C 10.
See Abstr. 131.033.

Interferometric observations of the small–scale structure of galactic neutral hydrogen.
See Abstr. 131.034.

Interstellar sodium and galactic structure. A high–resolution survey.
See Abstr. 131.035.

Molecular–cloud clusters and chains.
See Abstr. 131.036.

Three large molecular complexes in Norma.
See Abstr. 131.037.

Star formation in the Orion Arm.
See Abstr. 131.038.

A new study of the neutral hydrogen in Gould's Belt.
See Abstr. 131.040.

Fine structure of molecular clouds within 1 minute of arc of the galactic center.
See Abstr. 131.041.

Highlights of high–velocity clouds.
See Abstr. 131.042.

A deep survey for high–velocity clouds.
See Abstr. 131.043.

A survey for high–velocity clouds in the inner galaxy.
See Abstr. 131.044.

Observations of high–velocity clouds colliding in the anticenter.
See Abstr. 131.045.

Chemical composition of interstellar material.
See Abstr. 131.047.

Optical spiral structure at $l = 30° - 70°$. II. The distribution of interstellar extinction.
See Abstr. 131.051.

$[^{12}C]/[^{13}C]$ ratios from formaldehyde in the inner galactic disk.
See Abstr. 131.058.

High–resolution surveys of the Sagittarius A molecular cloud complex in ammonia, carbon monoxide, and isocyanic acid.
See Abstr. 131.064.

Gamma–ray emission from atomic and molecular gas at intermediate galactic latitudes.
See Abstr. 131.089.

Anomalous motions of H I clouds.
See Abstr. 131.091.

Constraints on the sites of nitrogen nucleosynthesis from $^{15}NH_3$–observations.
See Abstr. 131.093.

Correlation of infrared dust emission, galaxy counts, and hydrogen column density.
See Abstr. 131.104.

Distances of high–velocity H I clouds in the galactic corona.
See Abstr. 131.116.

High–latitude H I structure and the soft X–ray background.
See Abstr. 131.119.

The interstellar radiation field and the production of inverse–Compton gamma rays in the Galaxy.
See Abstr. 131.128.

The space distribution and internal structure of the Local dust arm in Cygnus.
See Abstr. 131.150.

Observation of neutral hydrogen 21–cm absorption toward double sources.
See Abstr. 131.170.

Star formation in the turbulent protogalaxy.
See Abstr. 131.178.

An investigation of the interstellar extinction. II. Towards the mid–infrared sources in the galactic centre.
See Abstr. 131.188.

Far–infrared and submillimeter astronomical spectroscopy.
See Abstr. 131.214.

A comparison of [S II] and Hα radial velocities, intensities, and line widths in the galactic background.
See Abstr. 131.236.

Radio recombination lines from the galactic plane in Cygnus.
See Abstr. 131.268.

Optical absorption from the high–velocity neutral hydrogen complex C in the spectrum of the RR Lyrae star BT Draconis.
See Abstr. 131.279.

The molecular cloud size spectrum in the outer Galaxy.
See Abstr. 131.282.

A new general survey of high–velocity neutral hydrogen in the southern hemisphere.
See Abstr. 131.296.

Kinematic distances to cool H I clouds.
See Abstr. 131.313.

Galactic H I in directions of low total column density.
See Abstr. 131.327.

Interstellar clouds near the sun.
See Abstr. 131.328.

The hydrocarbon ring C_3H_2 is ubiquitous in the Galaxy.
See Abstr. 131.329.

The largest molecular clouds in the galactic plane.
See Abstr. 131.340.

A study of CO emission from molecular clouds in the northern Milky Way.
See Abstr. 131.341.

A CO survey of molecular clouds in Taurus, Auriga, and Perseus.
See Abstr. 131.342.

The stellar distribution at $l = 245° \; b = 0°$ in Puppis.
See Abstr. 131.359.

Studies of galactic H I in 21–centimeter absorption.
See Abstr. 131.360.

On the problem of the shape of expanding supershells of neutral hydrogen.
See Abstr. 131.368.

H II regions in the Galaxy and Magellanic Clouds.
See Abstr. 132.029.

A long filament in the faint galactic Hα background.
See Abstr. 132.031.

Far–infrared survey of the galactic disc. II. The sources.
See Abstr. 133.003.

The Stockert 2.7 GHz radio continuum survey of the northern galactic plane.
See Abstr. 141.030.

EINSTEIN observations of the soft X–ray background close to the galactic plane.
See Abstr. 142.021.

Ultrasoft X–ray background observations towards the north galactic pole.
See Abstr. 142.040.

Long term X–ray observations of galactic bulge sources.
See Abstr. 142.041.

Einstein observations of the soft X–ray background close to the galactic plane.
See Abstr. 142.045.

Spectra and positions of galactic center high–energy X–ray sources.
See Abstr. 142.046.

X–ray sources in globular clusters and in the plane of the Galaxy.
See Abstr. 142.061.

X rays from cosmic ray interactions.
See Abstr. 142.066.

The galactic plane: a source of 1000 GeV gamma rays.
See Abstr. 143.002.

The gamma–ray burst spatial distribution $\log N(>S)$ vs $\log S$ and $N(>S, l^{II}, b^{II})$ vs S.
See Abstr. 143.016.

The gamma–ray deficit toward the galactic center.
See Abstr. 143.022.

The radial distribution of gamma rays and cosmic rays in the outer Galaxy.
See Abstr. 143.094.

Upper limits to ultra–high energy ($>10^{15}$eV) γ–ray fluxes from regions within the Galaxy.
See Abstr. 143.099.

Diffusion of the galactic cosmic rays in the vicinity of the solar system.
See Abstr. 144.017.

Cosmic ray sources and confinement in the Galaxy.
See Abstr. 144.037.

The origin of cosmic rays.
See Abstr. 144.053.

Cosmic–ray propagation in the solar neighborhood.
See Abstr. 144.059.

High–energy electrons in the nonuniform galactic disk model.
See Abstr. 144.063.

Spectra of cosmic rays and Galaxy's background emission in a diffusion model of cosmic rays distribution.
See Abstr. 144.125.

On a possibility of galactic origin of cosmic rays with energies up to 10^{19}eV.
See Abstr. 144.126.

Local superbubble model of cosmic ray propagation.
See Abstr. 144.128.

Statistical acceleration of electrons in the galactic halo.
See Abstr. 144.155.

Anisotropy of intensity of cosmic rays with $E_0 > 10^{17}$eV.
See Abstr. 144.348.

Stability of hydrostatic equilibrium in the galactic halo: cosmic ray effect.
See Abstr. 144.351.

Cosmic rays from the galactic centre.
See Abstr. 144.365.

Tidal interactions between the Galaxy and the Magellanic Clouds.
See Abstr. 151.011.

Dynamical evolution of the galactic disk.
See Abstr. 151.012.

Warps and heavy halos.
See Abstr. 151.016.

Two–fluid gravitational instabilities in a galactic disk.
See Abstr. 151.018.

On the 3–kpc arm: waves excited at the resonance by an oval distortion in the central region.
See Abstr. 151.027.

A barred galaxy: the inside viewpoint.
See Abstr. 151.028.

Stochastic star formation and spiral structure.
See Abstr. 151.029.

Tidal stripping and disruption of globular clusters.
See Abstr. 151.128.

UBV three–colour photometric parameters of four galactic clusters near Cassiopeia.
See Abstr. 153.004.

Metallicities and distances of galactic clusters as determined from UBV–data. III. Ages and abundance gradients of open clusters.
See Abstr. 153.031.

Globular clusters and the galactic halo.
See Abstr. 154.020.

C0422–213 (Eridanus): a second intermediate metal abundance globular cluster in the outer galactic halo.
See Abstr. 154.022.

Deep photometry of globular clusters. V. Age derivations and their implications for galactic evolution.
See Abstr. 154.027.

Results of the galactocentric orbit calculation for the globular cluster NGC 5466.
See Abstr. 154.028.

Radial velocities of remote globular clusters and the mass of the Galaxy.
See Abstr. 154.045.

The globular cluster system of the Galaxy. IV. The halo and disk subsystems.
See Abstr. 154.072.

Comparison of CO in the Galaxy and the Magellanic Clouds.
See Abstr. 156.005.

A comparison of the Andromeda and Milky Way galaxies.
See Abstr. 157.021.

Radial distributions of constituents in M33, the Galaxy, and M51.
See Abstr. 157.022.

Self–regulating galaxy formation. I. H II disk and Lyman–alpha pressure.
See Abstr. 157.065.

An analysis of the distribution of faint galaxies at the south galactic pole.
See Abstr. 157.109.

The structure of the spiral arm S4 in Andromeda galaxy. (A contribution to the theory of density waves).
See Abstr. 157.126.

Problems of galactic and extragalactic CO photometry.
See Abstr. 157.258.

Predictions of distance within the Local Group.
See Abstr. 160.001.

Magnetohydrodynamic effects on the growth of condensations in an expanding Universe and the formation of galaxies.
See Abstr. 161.051.

156 Magellanic Clouds

156.001 Magellanic Cloud objects.
J. B. Hutchings.
NASA Conf. Publ., NASA CP–2349, p. 32 – 41 (1984). – See Abstr. 012.001.

In this review, the considerable contributions of IUE to understanding the composition, stellar evolution, interstellar medium, and extinction of the Clouds are discussed and compared with those for the Galaxy. Future directions for UV work are suggested.

156.002 Highly ionized gas associated with the Small Magellanic Cloud.
E. L. Fitzpatrick, B. D. Savage.
NASA Conf. Publ., NASA CP–2349, p. 119 – 122 (1984). – See Abstr. 012.001.

High dispersion IUE spectra of SMC stars are examined to search for evidence of a widespread distribution of highly ionized gas. For all of the stars, absorption by C IV and/or Si IV is found near 160 km s^{-1}, and is attributed to nebular material. In addition, the stars show absorption by the high ion lines in the range 100 – 130 km s^{-1}. This absorption might indicate the presence of global distributions of these ions, such as a hot phase of the interstellar medium or a halo–like region.

156.003 *BVI* reddenings of Magellanic Cloud Cepheids.
J. A. R. Caldwell, I. M. Coulson.
Mon. Not. R. Astron. Soc., Vol. 212, No. 4, p. 879 – 888 (1985). With a correction in Vol. 214, No. 4, p. 639 (1985).

Photoelectric *BVI*$_c$ reddenings of 29 LMC and 48 SMC Cepheids are discussed. A mean *E(B–V)* of 0.074 mag is found in the LMC and 0.054 mag in the SMC, with a small dispersion. This would imply an intrinsic reddening in each Cloud of about 0.04 mag above the Galactic foreground. The dependence of these results on the deficiency of heavy elements in the Clouds is discussed.

156.004 Bright red variables of large amplitude in the Magellanic Clouds.
T. Lloyd Evans.
Mon. Not. R. Astron. Soc., Vol. 212, No. 4, p. 955 – 973 (1985).

Periods and epochs of maximum light are given for 17 red variables in the Magellanic Clouds. Spectroscopic observations of six SMC variables of large amplitude confirm membership and show that a wide range of carbon and heavy element enrichment occurs: in particular two stars are of type CS. A new discussion of the Harvard data for bright red variables in both Magellanic Clouds confirm the presence of two groups, of high and intermediate mass, and indicates that stars in the latter group have

$M \leqslant 6 M_\odot$. The large amplitude variables of high luminosity follow a linear relation in the $\log P - M_{bol}$ diagram, with a slope steeper than that found for ordinary Mira variables in the LMC.

156.005 Comparison of CO in the Galaxy and the Magellanic Clouds.
F. P. Israel, T. de Graauw, H. van de Stadt, C. P. de Vries.
The Milky Way galaxy, p. 333 – 334 (1985). – See Abstr. 012.007 (IAU Symp. No. 106).

156.006 A collision between the Large and Small Magellanic Clouds 2×10^8 years ago.
M. Fujimoto, T. Murai.
The Milky Way galaxy, p. 471 – 475 (1985). – See Abstr. 012.007 (IAU Symp. No. 106).

A number of orbits are obtained for the Large and Small Magellanic Clouds (LMC and SMC) revolving around a model Galaxy with a massive halo. It is suggested that the SMC approached the LMC as close as 3 to 7 kpc about 200 million years ago, if these clouds have been in a binary state for the past 10^{10} years, and the Magellanic Stream is due to the gravitational interaction among the triple system of the Galaxy, LMC, and SMC.

156.007 Nonequilibrium models for LMC supernova remnants.
J. Hughes.
News Lett. Astron. Soc. N.Y., Vol. 2, No. 5, p. 13 (1984). Abstract. – See Abstr. 010.243.

156.008 The X–ray source populations of the Magellanic Clouds.
D. J. Helfand.
News Lett. Astron. Soc. N.Y., Vol. 2, No. 5, p. 24 – 25 (1984). Abstract. – See Abstr. 010.243.

156.009 New distances to the Magellanic Clouds.
N. Visvanathan.
Cepheids: theory and observations, p. 223 – 224 (1985). – See Abstr. 012.027. (IAU Colloq. No. 82).

The mean phase magnitudes at the IV waveband (1.05 micron) of thirteen Cepheids in the SMC and the LMC and nine Cepheids in groups and clusters in the Galaxy, are used in conjunction with periods, to construct P–L(IV) relations in these galaxies. The slopes and the dispersions of the relations are nearly the same. The authors derives a distance modulus of 19.11 ± 0.07 for the SMC and 18.82 ± 0.07 for the LMC.

156.010 Interstellar dust in the Large Magellanic Cloud.
G. C. Clayton, P. G. Martin.
Astrophys. J., Vol. 288, No. 2, p. 558 – 568 (1985).

This paper presents extensive observations of the properties of interstellar dust in the Large Magellanic Cloud. Extinction and polarization analysis of 12 reddened stars spread across the LMC shows that in the optical and infrared the dust characteristics are remarkably similar to the Galaxy. The wavelength dependence of the polarization and the polarization efficiency, $p/E(B-V)$, are also comparable to galactic values. On average, the ultraviolet extinction in the LMC is also different from that in the Galaxy, having proportionately higher far–ultraviolet extinction and a weaker 2200 Å bump. The dust–to–gas ratio, $E(B-V)/N_H$ in the LMC is several times lower than the galactic value. A relationship between ultraviolet extinction properties and heavy element abundances in the Galaxy, LMC, and SMC is noted.

156.011 The clouds of Magellan.
D. Mathewson.
Sci. Am., Vol. 252, No. 4, p. 84 – 92 (1985).

The nearest neighbors of our galaxy have had a turbulent history. Close encounters with each other and with the Milky Way have split the smaller cloud and drawn out an enormous stream of hydrogen gas.

156.012 The diffuse interstellar medium in the Magellanic Clouds.
K. S. de Boer.
Gas in the interstellar medium, p. 1 – 6 (1984). – See Abstr. 012.032.

156.013 Tidal models of the Magellanic system.
R. A. Gingold.
Proc. Astron. Soc. Aust., Vol. 5, No. 4, p. 469 – 471 (1984).

Of the attempts to model the Magellanic system as the result of tidal interactions, only the works of Lin and Lynden–Bell (1982) and Murai and Fujimoto (1981) can be considered to be successful. In this paper it is argued that even these calculations do not unequivocally establish tidal interactions as being responsible for the origin of the Magellanic Stream, though they may well be important in giving rise to some features of the system.

156.014 A sample of long–period variables in the bar of the Large Magellanic Cloud and evidence for a recent burst of star formation.
P. R. Wood, M. S. Bessell, G. Paltoglou.
Astrophys. J., Vol. 290, No. 2, p. 477 – 486 (1985).

Infrared JHK photometry and spectral types have been obtained for a sample of long–period variables in, and to the north of, the bar of the LMC. It is found that the carbon stars and the oxygen–rich stars obey different period–luminosity relations. The period–luminosity distribution of the long–period variables in the central bar of the LMC suggests that (1) the bulk of the stars in this region have main–sequence masses $M_{ms} \lesssim 1.6\,M_\odot$ and ages $\gtrsim 10^9$yr, and (2) there is a distinct group of stars in this region with $M_{ms} \sim 5 – 6\,M_\odot$ and age $\sim 6 \times 10^7$yr. The period and luminosity distributions of the Cepheid variables in the central LMC bar also suggest that a burst of star formation occurred in the bar of the LMC $\sim 6 \times 10^7$yr ago.

156.015 Evidence of a hot population in the SMC/LMC bridge detected by the very wide field camera of Spacelab–1.
M. Viton, M. J. P. Sivan, M. G. Courtes, M. A. Gary, R. Decher.
Adv. Space Res., Vol. 5, No. 3, p. 207 – 210 (1985). – See Abstr. 012.059.

Deep 66° field photographs of the sky have been taken by the SL–1 very wide field camera at 1650, 1930 and 2530 Å, with a limiting magnitude of 9.3 at 1930 Å. A 1.2×2.4 kpc ultraviolet extension of the Shapley's wing of the Small Magellanic Cloud is revealed.

156.016 Ultraviolet interstellar absorption toward stars in the Small Magellanic Cloud. IV. Highly ionized gas associated with the Small Magellanic Cloud.
E. L. Fitzpatrick, B. D. Savage.
Astrophys. J., Vol. 292, No. 1, p. 122 – 129 (1985).

High–dispersion IUE satellite spectra of seven stars in the Small Magellanic Cloud (SMC) are examined to study the properties of interstellar C IV and Si IV absorption in the SMC. Absorption by C IV or Si IV or both is found near 160 km s^{-1} for all the stars. The velocity and the relative C IV and Si IV strengths suggest UV–photoionized nebular gas as the origin of this absorption. In addition, the stars show absorption by C IV and, sometimes, Si IV in the velocity range 100 – 130 km s^{-1}. The relative C IV and Si IV strengths indicate anomalous ionization. Possible global origins are considered for this absorption, including a hot phase of the SMC interstellar medium and a circum–SMC distribution of highly ionized gas.

156.017 Age and processes of star formation in very young LMC associations.
I. Tarrab.
Inst. Astrophys. Paris, Pré–Publ., No. 95, 32 pp. (1985). To appear in Astron. Astrophys.

156.018 Extreme Be stars in the Small Magellanic Cloud.
C. D. Garmany, P. S. Conti, P. Massey.
Bull. Am. Astron. Soc., Vol. 16, No. 4, p. 896 (1984). Abstract. – See Abstr. 010.062.

156.019 Regional ultraviolet extinction variations in the Large Magellanic Cloud.
E. L. Fitzpatrick.
Bull. Am. Astron. Soc., Vol. 16, No. 4, p. 915 (1984). Abstract. – See Abstr. 010.062.

156.020 A statistical study of the supernova remnants in the Large Magellanic Cloud.
J. P. Hughes, D. J. Helfand.
Bull. Am. Astron. Soc., Vol. 16, No. 4, p. 927 (1984). Abstract. – See Abstr. 010.062.

156.021 O–type stars in the Magellanic Clouds.
P. S. Conti, C. D. Garmany, P. Massey.
Bull. Am. Astron. Soc., Vol. 16, No. 4, p. 948 (1984). Abstract. – See Abstr. 010.062.

156.022 Ellipticities of star clusters in the Small Magellanic Cloud.
F. R. Chromey, H. Albers.
Bull. Am. Astron. Soc., Vol. 16, No. 4, p. 948 (1984). Abstract. – See Abstr. 010.062.

156.023 Color–magnitude diagrams of eleven Magellanic Cloud clusters and the Magellanic Cloud distance.
R. A. Schommer, E. W. Olszewski, M. Aaronson.
Bull. Am. Astron. Soc., Vol. 16, No. 4, p. 967 (1984). Abstract. – See Abstr. 010.062.

156.024 The LMC star cluster NGC 1777 and its neighboring field.
J. M. Nemec, J. R. Mould, J. Kristian.
Bull. Am. Astron. Soc., Vol. 17, No. 1, p. 518 (1985). Abstract. – See Abstr. 010.064.

156.025 Observed dynamical parameters of star clusters in the SMC.
E. Kontizas, M. Kontizas.
Dynamics of star clusters, p. 85 – 87 (1985). – See Abstr. 012.070 (IAU Symp. No. 113).

The ellipticities of the disk SMC clusters are plotted versus their masses. It can be seen that there is a linear correlation of mass–ellipticity. The orientation of the major axes and the mass–ellipticity relation permit to suggest that rotation should be considered as an important mechanism of the clusters' flatness.

156.026 CORAVEL radial velocities of supergiants in the Small Magellanic Cloud.
E. Maurice, N. Martin, L. Prévot, E. Rebeirot.
Stellar radial velocities, p. 265 – 268 (1985). – See Abstr. 012.095 (IAU Colloq. No. 88).

A total of 253 late–type supergiants in the SMC were observed with CORAVEL between January 1981 and October 1984. The limiting magnitude was $V \approx 14.5$. Accurate radial velocities were obtained for 195 SMC member stars of which 4 are Cepheids, 16 F to K type stars and 175 K to M type stars.

156.027 The Large Magellanic Cloud: a test for gamma ray and cosmic ray origin theories.
B. P. Houston, P. A. Riley, A. W. Wolfendale.
18th International Cosmic Ray Conference, Vol. 1, p. 89 – 92 (1983). – See Abstr. 012.096.

The authors discuss the flux of high energy gamma rays and the radio synchrotron radiation which would be expected from the Large Magellanic Cloud on the basis of a galactic model of cosmic ray origin.

156.028 The evolutionary connection between S and C stars: evidence from star clusters and the Magellanic Clouds.
T. Lloyd Evans.
Cool stars with excesses of heavy elements, p. 163 – 166 (1985). – See Abstr. 012.101.

Observations of intermediate–age clusters in the Magellanic Clouds indicate an evolutionary sequence, $M \rightarrow S \rightarrow C$, for stars on the asymptotic giant branch. Trends with age of the clusters are generally in agreement with theoretical prediction but the lifetime of the S star stage appears to be longer than expected. CS stars have been found among high luminosity field variables and can be accounted for by the envelope–burning process. The presence of faint S stars in ω Cen, where they may be interpreted as the result of a high primordial heavy–element abundance, shows that S stars are not necessarily all self–enriched in s–process material.

156.029 Kinematics of carbon stars in the Large Magellanic Cloud.
H. B. Richer, M. W. Ovenden, G. D. Joslin, B. E. Westerlund, N. Olander.
Cool stars with excesses of heavy elements, p. 167 – 170 (1985). – See Abstr. 012.101.

Published radial velocities of H II regions, luminous early– and late–type stars, and planetary nebulae, are consistent with a rotation law for the Large Magellanic Cloud $V \propto R$ for $R < 5.5°$. New radial velocities for some of the carbon stars in the WORC Catalogue extend the determination of rotation velocities to $R \cong 7°$. These observations are consistent with $V \propto R$ for $R < 5.5°$, with a rapid decrease of V for greater distances. If these results are interpreted as indicating solid–body rotation of a finite disc, and if a true distance modulus of 18.6 is assumed for the LMC, the mass of the LMC is found to be $\sim 2.6 \times 10^{10} M_\odot$.

156.030 A complete CO survey of the Small and Large Magellanic Clouds.
R. S. Cohen, J. Montani, M. Rubio.
Bull. Am. Astron. Soc., Vol. 17, No. 2, p. 548 (1985). Abstract. – See Abstr. 010.065.

156.031 A 1/4 keV X–ray survey of the Large Magellanic Cloud with HEAO–1.
J. A. Nousek, K. P. Singh, D. N. Burrows, G. P. Garmire.
Bull. Am. Astron. Soc., Vol. 17, No. 2, p. 586 (1985). Abstract. – See Abstr. 010.065.

156.032 Interstellar dust in the Large Magellanic Cloud.
J. C. Clayton.
Diss. Abstr. Int., Sect. B, Vol. 45, No. 9, p. 2955 (1985). Thesis, University of Toronto (1984).

156.033 IRAS observations of the Magellanic Clouds.
P. B. W. Schwering.
New aspects of galaxy photometry, p. 115 – 117 (1985). – See Abstr. 012.114.

Infrared maps obtained with IRAS at 60 μm for both Magellanic Clouds are presented.

156.034 On the structure and kinematics of the Small Magellanic Cloud.
E. Maurice, N. Martin, L. Prévot, E. Rebeirot.
New aspects of galaxy photometry, p. 241 – 244 (1985). – See Abstr. 012.114.

Observations with CORAVEL provided radial velocities of 195 SMC member–stars of late spectral type. This paper reports preliminary results from a kinematical study of the SMC based on these stellar velocities.

A catalog of small Magellanic Cloud objects in progress.
See Abstr. 002.067.

A catalog and atlas for the Small Magellanic Cloud.
See Abstr. 002.081.

Nomenclature for objects in the Magellanic Clouds.
See Abstr. 002.115.

Astrophysical and geophysical observations with PIRAMIG/ Salyut 7 experiment.
See Abstr. 035.049.

An astronomical application of deconvolution by the maximum entropy method.
See Abstr. 036.179.

Current problems in horizontal–branch theory: some implications for RR Lyrae variables.
See Abstr. 065.027.

The spectroscopic antics of S 18/SMC.
See Abstr. 112.009.

Spektroskopische und photometrische Untersuchungen der B–Überriesen mit zirkumstellarem Staub in den Magellanschen Wolken.
See Abstr. 112.023.

The hybrid spectrum of the LMC hypergiant R126.
See Abstr. 112.036.

Circumstellar shells in the Large Magellanic Cloud.
See Abstr. 112.056.

On the problem of the luminous emission line stars.
See Abstr. 112.058.

Stellar winds from hot stars in the Magellanic Clouds.
See Abstr. 112.137.

An unstable Ofpe star in the LMC.
See Abstr. 114.058.

M supergiants in the Milky Way and the Magellanic Clouds: colors, spectral types, and luminosities.
See Abstr. 115.004.

R 136a and the central object in the giant H II region NGC 3603 resolved by holographic speckle interferometry.
See Abstr. 115.016.

IUE and optical observations of LMC X–ray binaries.
See Abstr. 117.015.

On fast X–ray rotators with long–term periodicities.
See Abstr. 117.144.

The first IUE observations of LMC X–1 (star 32).
See Abstr. 117.164.

Wolf–Rayet stars in the Magellanic Clouds. III. The WO4 + O4 V binary Sk 188 in the SMC.
See Abstr. 117.187.

On the number of W–R stars in the Large Magellanic Cloud.
See Abstr. 117.188.

Wolf–Rayet stars in the Magellanic Clouds. III. The WO4 + O4 V binary Sk 188 in the SMC.
See Abstr. 117.195.

On fast X–ray rotators with long–term periodicities.
See Abstr. 117.228.

The optical orbit of the X–ray pulsar binary 0535–668 (= A0538–66).
See Abstr. 117.384.

CCD observations of the RR Lyrae stars in NGC 2210 and the distance to the Large Magellanic Cloud.
See Abstr. 122.006.

Period change in Magellanic Cloud Cepheids.
See Abstr. 122.007.

A period–luminosity relation for Mira variables in globular clusters and its impact on the distance scale.
See Abstr. 122.014.

Distances to Magellanic Clouds from observations of Cepheids at 1.05 microns.
See Abstr. 122.030.

The RR Lyrae stars in and around the LMC globular cluster NGC 2257.
See Abstr. 122.032.

The secular period behavior of 38 RR Lyrae stars in the LMC globular cluster NGC 2257.
See Abstr. 122.033.

Infrared observations of galactic Cepheids.
See Abstr. 122.047.

Magellanic Cloud and other extragalactic Cepheids: some current topics.
See Abstr. 122.067.

Cepheid variables as extra–galactic distance indicators.
See Abstr. 122.068.

Leavitt variables: the brightest Cepheid variables and their implications for the distance scale.
See Abstr. 122.074.

The infrared distance scale: the Galaxy and the Magellanic Clouds.
See Abstr. 122.075.

A survey for red variables in the LMC. I.
See Abstr. 122.124.

On the type of variability of the star H 42 in the Andromeda galaxy.
See Abstr. 122.156.

Rectification of nine epochs of photometry for the LMC Cepheid HV 5541.
See Abstr. 122.161.

CORAVEL radial–velocity curves for Cepheids in the Large Magellanic Cloud.
See Abstr. 122.187.

UBVRI observations of Magellanic supergiants and Cepheids.
See Abstr. 122.209.

A model for SNR evolution in a cloudy medium and its application to N49.
See Abstr. 125.007.

On the nature of the supernova remnant 0540–69.3 in the Large Magellanic Cloud.
See Abstr. 125.012.

Supernova remnant searches at Molonglo.
See Abstr. 125.037.

An evolutionary history for the Crablike, pulsar–powered supernova remnant 0540–69.3.
See Abstr. 125.041.

Optical spectrophotometric study of SNRs in the LMC.
See Abstr. 125.060.

Permitted oxygen lines in supernova remnant spectra.
See Abstr. 125.064.

Supernova remnants in the Magellanic Clouds. III.
See Abstr. 125.087.

N63A – an unusual supernova remnant in the LMC.
See Abstr. 125.096.

X–ray studies of the supernova remnant N132D.
See Abstr. 125.097.

An optical synchrotron nebula around the X–ray pulsar 0540–693 in the Large Magellanic Cloud.
See Abstr. 125.122.

Optical pulsations in the Large Magellanic Cloud remnant 0540–69.3.
See Abstr. 126.013.

Optical pulsar in LMC remnant 0540–69.3.
See Abstr. 126.076.

The UV extinction laws in two very young clusters, NGC 6530 in the Galaxy and NGC 2100 in the LMC.
See Abstr. 131.002.

IDS and CCD study of the LMC excited blobs.
See Abstr. 131.069.

High resolution observations of interstellar Na I absorption towards Large Magellanic Cloud stars: I. R 116, R 127 and R 128 with the ESO coudé échelle spectrometer.
See Abstr. 131.190.

A new general survey of high–velocity neutral hydrogen in the southern hemisphere.
See Abstr. 131.296.

Ultraviolet interstellar absorption toward stars in the Small Magellanic Cloud.
See Abstr. 131.357.

Comments on the origins and variations of carbon and nitrogen in the ISM of galaxies as evident from IUE spectroscopy of H II regions.
See Abstr. 132.001.

IUE spectroscopy of extragalactic H II regions.
See Abstr. 132.002.

Discovery of a very bright compact object N88A in the Small Magellanic Cloud.
See Abstr. 132.025.

H II regions in the Galaxy and Magellanic Clouds.
See Abstr. 132.029.

The halo of the 30 Doradus nebula.
See Abstr. 132.030.

Detection of molecular hydrogen in the Small Magellanic Cloud H II region N81.
See Abstr. 132.039.

Spectrophotometry and chemical composition of the 30 Doradus nebula.
See Abstr. 132.044.

IUE observations of planetary nebulae in the Magellanic Clouds.
See Abstr. 134.001.

New planetary nebulae in the Small Magellanic Cloud.
See Abstr. 134.013.

Chemical abundances in planetary nebulae in the Galaxy and in the Magellanic Clouds.
See Abstr. 134.027.

Masses for the central stars of three planetary nebulae in the Magellanic Clouds.
See Abstr. 134.071.

Further spectroscopy of LMC X–ray candidates.
See Abstr. 142.022.

First observation of ultra–high–energy γ rays from LMC X–4.
See Abstr. 143.032.

Search for light flashes from a gamma ray burst source.
See Abstr. 143.034.

Corequake and shock heating model of the 5 March 1979 γ–ray burst.
See Abstr. 143.067.

Tidal interactions between the Galaxy and the Magellanic Clouds.
See Abstr. 151.011.

Ultraviolet studies of the young populous cluster NGC 2100 in the LMC.
See Abstr. 153.001.

Ultraviolet studies of stars in the populous cluster NGC 2100 in the Large Magellanic Cloud.
See Abstr. 153.026.

Structural parameters and masses for three old LMC clusters.
See Abstr. 154.011.

The extended giant branches of intermediate age globular clusters in the Magellanic Clouds. IV.
See Abstr. 154.012.

Ellipticities of "disk" and "halo" globular clusters in the SMC.
See Abstr. 154.024.

The age of the LMC red globular cluster NGC 2213.
See Abstr. 154.043.

The distribution and motions of peculiar red giants.
See Abstr. 155.162.

Globular clusters as extragalactic distance indicators.
See Abstr. 161.098.

157 Normal Galaxies (Structure, Evolution, Pairs, etc.)

157.001 Supernova remnants and the carbon abundance in M33.
W. P. Blair, J. C. Raymond.
NASA Conf. Publ., NASA CP–2349, p. 103 – 106 (1984). – See Abstr. 012.001.

The authors have combined IUE and optical spectral data on a supernova remnant in M33 in order to determine the abundance of carbon in the interstellar gas of M33. They have used new shock model calculations to guide the interpretation and find that a substantially lower value of the ratio C/O than has been assumed previously for M33 may be necessary to explain the data. However, some inconsistencies in the observed and predicted line intensities indicate that the spectra may be partly contaminated by H II region emission or that the preshock gas has been ionized by nearby stars.

157.002 The usual hot stars in unusual galaxies.
S. A. Lamb, J. S. Gallagher, M. J. Hjellming, D. A. Hunter.
NASA Conf. Publ., NASA CP–2349, p. 115 – 118 (1984). – See Abstr. 012.001.

Two hot amorphous galaxies, NGC 1705 and NGC 1800 were successfully observed with the IUE when the authors obtained low dispersion, short wavelength data. These spectra, together with new optical data, indicate that the OB stellar content of these systems is quite like that of the normal OB complexes found in spiral and regular Irr galaxies.

157.003 Radio continuum observation of NGC 4631 at 327 MHz.
S. Sukumar, T. Velusamy.
Mon. Not. R. Astron. Soc., Vol. 212, No. 2, p. 367 – 373 (1985).

The edge–on spiral galaxy NGC 4631 has been mapped at 327 MHz with a resolution of 60×60 arcsec^2 using the Ooty Synthesis Radio Telescope. Faint emission extending up to 7.5 kpc above the plane of the galaxy has been detected, and the spectral index distribution in the disc and halo has been derived by comparing the 327–MHz map with high–frequency maps at 610, 1412 and 10700 MHz. The variation of the low–frequency spectral index (327 – 610 MHz) is consistent with that for a dynamical radio halo, indicating the presence of a galactic wind in NGC 4631.

157.004 Kinematics and dynamics of the haloes of supergiant galaxies.
D. Carter, I. Inglis, R. S. Ellis, G. Efstathiou, J. G. Godwin.
Mon. Not. R. Astron. Soc., Vol. 212, No. 2, p. 471 – 488 (1985).

The authors present measurements of the internal kinematics of the haloes of three supergiant (cD or giant dumb–bell) galaxies in moderately rich clusters. They find that these galaxies have velocity dispersion profiles which are flat or which rise with increasing radius. The observations can equally well be explained by two simple models. The first class has isotropic velocity dispersions and a two–component mass distribution: a bright component with a density distribution which follows a King model, and a dark, isothermal component with a density distribution which is slightly more concentrated than that of the galaxies in the cluster as a whole. The second class has predominantly tangential motions in the halo of the galaxy, and no dark component. In common with most luminous elliptical galaxies these supergiant galaxies are not flattened by rotation. In the two dumb–bells studied the relative velocities between the nuclei are comparable to the stellar velocity dispersions in the cD envelopes.

157.005   A method for determining the spatial correlation function from a sample with partial distance information.
S. Phillipps.
Mon. Not. R. Astron. Soc., Vol. 212, No. 3, p. 657 – 661 (1985).

The projected surface density, Σ, of galaxies seen around a galaxy of known distance is a particularly simple angular statistic. Here, the relationship between Σ and the spatial correlation function ξ is derived and it is shown that this enables ξ to be calculated in a simple way from a sample of galaxies with only partial distance information.

157.006   The redshifts of double galaxies.
G. Picchio, G. Tanzella–Nitti.
Astron. Astrophys., Vol. 142, No. 1, p. 21 – 30 (1985).

The authors have discussed the reliability of redshift measurements and the classification according to the dynamical environment and morphology for a sample of 587 double galaxies in order to produce and discuss a large set of projected orbital velocity differences ΔV. A critical comparison of the redshifts published in the literature has been made on the basis of large Redshift Catalogues. The environment classification of the full sample – cluster, group, nearby companions, isolated pairs – has been performed using literature data and POSS red print inspections. The velocity distribution $N(\Delta V)$ of isolated pairs is significantly different from the one derived for binaries located in groups and clusters.

157.007   The direction of rotation of the spirals in 109 galaxies. NGC 3786 and NGC 5426 – the galaxies with leading spiral arms in interacting systems.
I. I. Pasha.
Pis'ma Astron. Zh., Tom 11, No. 1, p. 3 – 9 (1985). In Russian. English translation in Sov. Astron. Lett., Vol. 11.

The results on determinations of the sense of rotation of the spirals in 132 galaxies are presented. 107 spirals have revealed trailing arms while NGC 3786 and NGC 5426, both in pairs with their companion spirals NGC 3788 and NGC 5427, have shown leading arms. A total of 190 objects constitute the current sample of galaxies with determined sense of spiral rotation. Only four of them are inferred to have leading arms, evidently due to dynamical interaction with companion galaxies. A mechanism of excitation of the leading pattern is suggested and discussed.

157.008   H I line studies of galaxies. IV. Distance moduli of 468 disk galaxies.
L. Bottinelli, L. Gouguenheim, G. Paturel, G. de Vaucouleurs.
Astron. Astrophys., Suppl. Ser., Vol. 59, No. 1, p. 43 – 57 (1985).

The radial velocities, V, line widths, W, and integrated fluxes, S, in the catalogues of Fisher and Tully (1981) and of the authors (Paper I) are compared for ~ 300 galaxies in common. After correction of the FT line widths for resolution effects, the V and W parameters are in excellent systematic agreement. A catalogue of distance moduli of 468 disk galaxies ($-2 \leqslant T \leqslant 8$) derived from the T–F data via the revised B–band Tully–Fisher relation of Paper II and homogeneous with the catalogue of 822 objects in Paper III is given.

157.009   Detailed surface photometry of 36 E–S0 galaxies.
R. Michard.
Astron. Astrophys., Suppl. Ser., Vol. 59, No. 2, p. 205 – 228 (1985).

Complete surface photometry has been obtained for 36 E–S0 galaxies, 33 being members of the Virgo E cluster, on the basis of plates mostly taken with the 1.20 m, f/6, telescope of the Observatoire de Haute–Provence. The data are compared with the results of previous surveys. The results are given as tabulations of the parameters describing ellipses fitted to the isophotes. The axis ratio c/a and major axis position angle P, are tabulated down to a level of B brightness $\mu_0 = 25$ to 26. The "radius" of isophotes $r = (ac)^{1/2}$ is given down to $\mu_0 = 27$ to 28. A preliminary discussion of the data is presented. The $\mu_0(r)$ relations for E galaxies are compared to the $r^{1/4}$ interpolation formula, and it is shown that systematic deviations are observed and are similar for galaxies of similar luminosities.

157.010   H I synthesis observations of the peculiar galaxy NGC 3718 and its companion NGC 3729.
U. J. Schwarz.
Astron. Astrophys., Vol. 142, No. 2, p. 273 – 288 (1985).

NGC 3718 is an early type galaxy with a very peculiar optical appearance. The H I–distribution is very complex in the angular coordinates–velocity space. It can, however, be explained relatively simply by a strong warp, with parts of the orbits seen almost edge–on. The companion galaxy NGC 3729 is also detected in hydrogen. It has a systemic velocity 78 km s^{-1} higher than NGC 3718. Mass estimates of both galaxies are given and limits for the sum of masses of the galaxy pair are derived. There is weak evidence of an extended spiral arm or tidal bridge. The most likely explanation is that the warp is caused by captured gas that moves in a family of stable orbits, which are possible in a tumbling tri–axial potential (v. Albada et al., 1982, Merrit and de Zeeuw, 1983).

157.011   Wolf–Rayet stars in "lazy" galaxies: a statistical approach.
D. Kunth, M. Joubert.
Astron. Astrophys., Vol. 142, No. 2, p. 411 – 420 (1985).

Lazy galaxies are blue emission–line galaxies forming stars by intermittent short bursts. Spectra of 45 such galaxies have been reanalyzed to search for Wolf–Rayet emission at 4686 Å. WR emission has been positively detected in 1 galaxy and suspected in 14 others. The data suggest a mass to light ratio of about 1 for the lazy galaxies. It is suggested that in lazy galaxies and giant extragalactic H II regions, WR stars evolve from main sequence stars less massive than 60 $M_\odot$.

157.012   Orientation of double galaxies on the celestial sphere.
I. G. Plaksina.
Probl. Kosm. Fiz., Vyp. 19, p. 118 – 121 (1984). In Russian. – See Abstr. 003.002.

157.013   Mass models of three southern late–type dwarf spirals.
C. Carignan.
The Milky Way galaxy, p. 95 – 96 (1985). – See Abstr. 012.007 (IAU Symp. No. 106).

157.014   Molecular clouds in external galaxies.
J. S. Young.
The Milky Way galaxy, p. 183 – 191 (1985). – See Abstr. 012.007 (IAU Symp. No. 106).

The author has been conducting a large observational program investigating the molecular contents of galaxies. The aims of this program are to determine (1) the radial distributions of molecular gas in particular galaxies, (2) the relative CO content in galaxies as a function of Hubble type and luminosity, (3) the relative confinement of molecular clouds to spiral arms, and (4) the CO contents of active galaxy nuclei.

157.015   Comments on the distribution of molecules in spiral galaxies.
L. J. Rickard, P. Palmer.
The Milky Way galaxy, p. 193 – 194 (1985). – See Abstr. 012.007 (IAU Symp. No. 106).

157.016   Distribution and motions of CO in M51.
G. Rydbeck, Å. Hjalmarson, O. Rydbeck.
The Milky Way galaxy, p. 195 – 196 (1985). – See Abstr. 012.007 (IAU Symp. No. 106).

The authors present results (partly preliminary) of an extensive map (73 positions) of CO ($J = 1$–0) emission in M51.

157.017 CO (2–1) observations of Maffei 2.
A. I. Sargent, E. C. Sutton, C. R. Masson,
T. G. Phillips, K.–Y. Lo.
The Milky Way galaxy, p. 197 – 198 (1985). – See Abstr. 012.007
(IAU Symp. No. 106).

157.018 The magnetic–field structure and dynamics of NGC 253.
M. Urbanik, U. Klein, R. Beck, R. Wielebinski.
The Milky Way galaxy, p. 247 – 248 (1985). – See Abstr. 012.007
(IAU Symp. No. 106).
Total–power and polarization observations of NGC 253 at
10.7 GHz have been performed with the 100–m MPIfR radio
telescope. The observed arm/interarm polarization contrasts are
discussed in the context of possible field configurations in spiral
arms.

157.019 Spiral structure and kinematics of H I and H II in external galaxies.
R. J. Allen, P. D. Atherton, R. P. J. Tilanus.
The Milky Way galaxy, p. 275 – 279 (1985). – See Abstr. 012.007
(IAU Symp. No. 106).
The authors present new observations of the distribution and
kinematics of the Hβ emission in the spiral arms of M83, obtained with the TAURUS imaging Fabry–Pérot system. These
results, when combined with observations of the neutral and
molecular components with sufficiently high resolution, should
contribute to the understanding of the time sequence by which
stars form out of the interstellar gas.

157.020 A systematic study of M81.
F. Bash.
The Milky Way galaxy, p. 281 – 282 (1985). – See Abstr. 012.007
(IAU Symp. No. 106).
The author describes a series of observational studies of M81
which are presently underway. He hopes that these studies will
allow improved understanding of this one, simple, density–wave
spiral galaxy.

157.021 A comparison of the Andromeda and Milky Way galaxies.
P. Hodge.
The Milky Way galaxy, p. 423 – 430 (1985). – See Abstr. 012.007
(IAU Symp. No. 106).
A comparison of some of the basic properties of M31 and the
Milky Way indicates that in almost every respect M31 is larger
than the Galaxy. It is more luminous, redder, more massive, and
of earlier Hubble type. A detailed comparison of the spiral structure, based on optical tracers, for comparable areas in the outer
parts of each galaxy shows differences in the arm spacings, in
density enhancement, and in pitch angle.

157.022 Radial distributions of constituents in M33, the Galaxy, and M51.
E. M. Berkhuijsen, U. Klein.
The Milky Way galaxy, p. 431 – 434 (1985). – See Abstr. 012.007
(IAU Symp. No. 106).
The radial distributions of the surface brightness or column
density of thermal and nonthermal radio emission, far–infrared
emission, blue light, H I and CO in the Sc galaxies M33 and M51
are compared with the corresponding distributions in the Galaxy.
Information on the variation of the absorption at Hα and on the
variation of the abundance ratio O/H is also shown.

157.023 A coordinated radio and optical survey of M31.
R. A. M. Walterbos, R. C. Kennicutt.
The Milky Way galaxy, p. 435 – 436 (1985). – See Abstr. 012.007
(IAU Symp. No. 106).

157.024 Distribution and motions of H I in M31.
E. Brinks, W. B. Burton.
The Milky Way galaxy, p. 437 – 442 (1985). – See Abstr. 012.007
(IAU Symp. No. 106).

157.025 Distribution and motion of CO in M31.
A. A. Stark.
The Milky Way galaxy, p. 445 – 450 (1985). – See Abstr. 012.007
(IAU Symp. No. 106).

157.026 Large–scale maps of M31 at middle and far infrared wavelengths.
H. J. Habing.
The Milky Way galaxy, p. 8, 451 – 456 (1985). – See Abstr.
012.007 (IAU Symp. No. 106).
The satellite IRAS was in operation from January 27 until
November 22, 1983. Its main goals were to carry out a survey of
the sky at 12, 25, 60, and 100 micron and to make observations
around selected positions achieving higher sensitivity or better
angular resolution than provided in the survey. Among the very
first targets of these pointed observations was the Andromeda
galaxy, M31, because it was considered of prime importance and
was about to leave the observing window. Here the author reports briefly on crude maps made from the first measurements.

157.027 The infrared excess of galaxies.
T. de Jong.
Observatory, Vol. 105, No. 1064, p. 2 – 3 (1985). Abstract. – See
Abstr. 010.721.

157.028 IRAS observations of interacting galaxies.
R. D. Joseph.
Observatory, Vol. 105, No. 1064, p. 3 (1985). Abstract. – See
Abstr. 010.721.

157.029 CCD surface photometry of galaxies with dynamical data. I. NGC 3379, M87, and NGC 1052.
L. E. Davis, M. Cawson, R. L. Davies, G. Illingworth.
Astron. J., Vol. 90, No. 2, p. 169 – 182 (1985).
U, B, and R CCD surface photometry for M87 and NGC 1052,
and B and R CCD surface photometry for NGC 3379 is presented. Ellipses were fit to the U, B, and R surface–brightness
data at each radius to produce a major–axis luminosity profile,
an ellipticity and position–angle profile, and major axis U–R and
B–R color gradients for each galaxy. The surface photometry
spans radii greater than or equal to those reached by the available
kinematic data. Comparison of the luminosity and ellipticity profiles suggests that changes in the slope of the luminosity profile
are related to changes in the ellipticity profiles for NGC 3379 and
NGC 1052. In the case of NGC 3379, the form of the rotation
curve may also be related to changes in the ellipticity and luminosity gradients.

157.030 A catalog of dusty elliptical galaxies.
K. Ebneter, B. Balick.
Astron. J., Vol. 90, No. 2, p. 183 – 191 (1985). With plate 14.
A catalogue of early–type galaxies containing anomalous
amounts and/or distributions of dust is compiled from the literature. The biases inherent in the catalogue are discussed. New
CCD observations of selected dusty elliptical galaxies are presented that help to clarify the configuration of the dust. The
morphological types, radio properties, radio–source orientations,
neutral–hydrogen contents, and kinematics of the catalogue galaxies are reviewed, and the possible correlations between the
presence and distribution of dust are compared to other attributes of these galaxies.

157.031 The nucleus of the giant spiral M101.
K. Davidson, R. M. Humphreys, C. Blaha.
Astron. J., Vol. 90, No. 2, p. 192 – 196 (1985). With plate 15.
High–resolution spectroscopic observations of Hα in the nucleus and other emission knots in the nuclear region of M101
reveal a velocity gradient of about 17 km s^{-1}arcsec^{-1} across the
nuclear region. The Hα contour map of the nucleus itself shows
an obvious east–west velocity gradient. The velocity profile of the
central condensation has a FWHM of 47 km s^{-1} and an rms
dispersion of 22 km s^{-1}. The authors find a mass of only
$10^{6.7} M_\odot$ for the central condensation. The emission–line spectrum and chemical composition of the nucleus are discussed.

157.032   The stellar content of the A496 cD galaxy.
E. A. Valentijn, A. F. M. Moorwood.
Astron. Astrophys., Vol. 143, No. 1, p. 46 – 53 (1985).

In addition to already published radial $B-V$ colour profiles of the A496 cD galaxy the authors have determined the infrared (J, H, K) radial colour distribution of this galaxy. While in $B-V$ the halo becomes bluer at increasing radii, it becomes redder in the visual–infrared colours. The combined visual–infrared data lead to a unique model for both the main sequence and giant stars.

157.033   The flattening distribution of lenticular galaxies.
R. E. De Souza, G. Vettolani, G. Chincarini.
Astron. Astrophys., Vol. 143, No. 1, p. 143 – 147 (1985).

Using the data lists in the RC2 and UGC catalogues the authors show that the distribution of flattening of lenticular galaxies is related to the galaxian density of the environment. If the effect is real it can be due to a higher probability of finding lenticulars with large bulge–to–disk diameter ratios in regions of low density.

157.034   H I observations of galaxies in nearby groups.
W. K. Huchtmeier, J. H. Seiradakis.
Astron. Astrophys., Vol. 143, No. 1, p. 216 – 225 (1985).

Neutral hydrogen observations of 44 nearby galaxies are presented. This project was undertaken in order to search for extended H I haloes. Also it is possible from these data to investigate the effects of incomplete mapping on derived parameters. The derived H I mass is a relatively strong function of the completeness of the mapping, whereas the line width seems to be a rather independent parameter, as long as the dynamically important part of the galaxy is sufficiently covered. This also applies to spot observations of the central position of a galaxy (central profile), as long as the antenna beam covers the galaxy adequately, which is the case for the present sample (optical diameter to antenna beam <2.3). The dwarf galaxy, DDO 154, was found to have a particularly extended H I envelope.

157.035   Photometric investigations of compact galaxies.
F. Börngen, A. T. Kalloghlian (*A. T. Kalloglyan*).
Astron. Nachr., Vol. 306, No. 2, p. 81 – 89 (1985).

No exact definition of compact galaxies with regard to their degree of concentration exists up to now. 169 objects visually selected by Zwicky respectively Börngen and Kalloghlian in the field ZwCl 1710.4 + 6401 are analysed photometrically and morphologically. The brightness profiles are determined by means of equidensities calibrated into intensities. The different parts of these profiles are investigated with regard to their utility of an objective description of compact galaxies. Whereas the halo is not suitable, a classification is practicable by means of the gradient G of the middle part of the profile. Especially, it is shown that the ($\lg I_{max}$, m)-diagram is well suited to represent the compact galaxies. In an interval of some magnitudes in which the density of star images is not yet saturated the positions of stars and compact galaxies differ significantly. Extreme compact galaxies are difficult or not at all discernible from stars.

157.036   Extragalactic dust. I. NGC 7070A.
N. Brosch, J. M. Greenberg, P. J. Grosbøl.
Astron. Astrophys., Vol. 143, No. 2, p. 399 – 407 (1985).

The authors have obtained deep U, B, V and R plates of the lenticular galaxy NGC 7070A at the prime focus of the 3.6 m ESO reflector. Some properties are derived of the dust particles within the dark lanes which are seen projected against the main body of the galaxy. The typical size of a dust grain there is probably somewhat smaller than that usually taken as canonical for galactic dust. The dust lanes are attributed to the accretion of a small disk galaxy, containing at least $\sim 4 \times 10^7 M_\odot$ of gas and dust, by the lenticular galaxy. This accretion event, $\sim 10^9$yr in the past, produced also luminous tails from NGC 7070A towards a small group of nearby galaxies and incomplete shells around the S0 galaxy. This is probably the first case of a disk galaxy showing shells, dust lanes and tails at the same time.

157.037   The statistical distribution of the neutral–hydrogen content of elliptical galaxies.
G. R. Knapp, E. L. Turner, P. E. Cunniffe.
Astron. J., Vol. 90, No. 3, 454 – 468 (1985).

The authors have examined the form of the distribution function for the relative H I content, M_{HI}/L_B, of elliptical galaxies, using a data set derived from all recent H I observations of ellipticals in the literature. The data set contains 152 galaxies; 23 of these have been detected in the H I line. The detected galaxies are shown to be more H I rich on average than the galaxies in the whole sample. A method for recovering the intrinsic distribution of M_{HI}/L_B, using both detection and upper–limit data, is described, and the data are shown to be consistent with a shallow power–law differential distribution $N \sim (M_{HI}/L_B)^{-1.5}$. This distribution is quite different from that for spirals. The result suggests that the gas has an external origin.

157.038   The interacting spiral galaxy NGC 5395.
N. A. Sharp, W. C. Keel.
Astron. J., Vol. 90, No. 3, 469 – 472, 563 – 566 (1985).

The authors present spectroscopic and imaging observations of NGC 5395, and images of its companion NGC 5394. These results show quite clearly that the spiral arms of NGC 5395 are trailing, as is usual, despite claims by Pasha and Smirnov (1982) that NGC 5395 is a leading–arm galaxy. However, NGC 5395 shows some rather unusual structure, and its smaller companion NGC 5394 shows a very bright stellar nucleus with the possibility of asymmetric extended Hα emission.

157.039   The chemistry of galaxies. I. The nature of giant extragalactic H II regions.
M. L. McCall, P. M. Rybski, G. A. Shields.
Astrophys. J., Suppl. Ser., Vol. 57, No. 1, p. 1 – 62 (1985).

Spectrophotometric observations of 99 H II regions in 20 spiral and irregular galaxies are presented and discussed. Morphological and theoretical analyses of the spectra are made in order to study the dimensionality of the spectral sequence, sequencing parameters, chemical abundances, and the overall physical structure of the H II regions and embedded OB associations.

157.040   Elliptische Galaxien mit Schalen: NGC 3923.
F. Bertola.
Sterne Weltraum, 24. Jahrg., Nr. 1, p. 28 – 29 (1985).

157.041   NGC 4258 (M106).
F. Bertola.
Sterne Weltraum, 24. Jahrg., Nr. 2, p. 74 – 75 (1985).

157.042   NGC 7252.
F. Bertola.
Sterne Weltraum, 24. Jahrg., Nr. 3, p. 148 – 149 (1985).

157.043   The seven dwarfs.
D. H. Smith.
Sky Telesc., Vol. 69, No. 3, p. 216 – 217 (1985).

157.044   Distribution and motions of atomic hydrogen in lenticular galaxies. IV. A ring of H I around NGC 4262.
N. Krumm, W. van Driel, H. van Woerden.
Astron. Astrophys., Vol. 144, No. 1, p. 202 – 210 (1985).

The distribution of H I in the normal, barred lenticular galaxy NGC 4262 in the Virgo Cluster has been mapped at a resolution of $13'' \times 51''$. The gas is confined to a narrow ring, apparently in circular rotation about the galaxy center. The ring has a diameter about twice the optical diameter of the galaxy, and is inclined by at least 50° (and possibly as much as 120°) to the optical disk. The rotation velocity of the ring implies that atomic hydrogen accounts for far less than 1% of the total mass, and that the total mass–to–light ratio is near 20. The local mass–to–light ratio in the ring is at least 50 to 80, giving evidence for a massive, "invisible" envelope or halo surrounding the galaxy. The authors consider two reasonable explanations for the presence of gas in NGC 4262, namely that it is the remnant of the primordial proto–galaxy cloud, or of a captured, gas–rich dwarf or inter-galactic H I cloud.

157.045 Separations of H II regions in galaxies.
E. Braunsfurth, J. V. Feitzinger.
Astron. Astrophys., Vol. 144, No. 1, p. 215 – 219 (1985).

Separations of H II regions in a sample of 44 late type galaxies are determined. The distribution function of the linear separa-tions follows an exponential law. This distribution function may be used as secondary standard for distance determinations of galaxies. Typical values of the separations are of the order of 300 pc; they reflect an important scale length in the process of star formation.

157.046 Double galaxy investigations. III. The differential red-shift distribution and emission–line correlations.
W. G. Tifft.
Astrophys. J., Vol. 288, No. 1, p. 65 – 72 (1985).

The distribution of redshift differences for a sample of 250 close double galaxies is shown to depend strongly upon the emis-sion properties of the component galaxies and to some extent upon interaction morphology. The distribution is not signifi-cantly dependent either upon physical or angular separation or upon the magnitude differential. The ΔV distribution can be decomposed into a sharply peaked component with dispersion ~ 100 km s^{-1} characterized by emission objects, and a broad component with dispersion ~ 300 km s^{-1} characterized by ab-sorption spectra. Within these two components there appears to be a dependence of emission strength, and to a lesser extent interaction morphology, upon physical separation in the range 5 – 20 kpc.

157.047 Variations in the strength of spectral features in spheroidal galactic systems. I. The method and CN band variation.
A. Wirth.
Astrophys. J., Vol. 288, No. 1, p. 132 – 137 (1985).

Moderate–resolution, high signal–to–noise spectra of elliptical galaxies are used to study the relation of the strength of the CN bands to other properties of the galaxy. A new technique for the measurement of the strength of spectral features is described. Large intergalaxy CN variations are seen. Further, the CN band strength is shown to be not well correlated with either the strength of other spectral features or with the total luminosity. Two tentative explanations for this effect are proposed.

157.048 Global modal analysis of disk galaxies: application to the S0 galaxy NGC 3115.
T. Ueda, M. Noguchi, M. Iye, S. Aoki.
Astrophys. J., Vol. 288, No. 1, p. 196 – 200 (1985).

The global modal analysis of oscillations is applied to a series of models for NGC 3115. All the models in this series are con-sistent with the observed rotation curve and the observed velocity dispersion curve of NGC 3115 but have different halo mass frac-tions. The apparent lack of any spiral structure in this galaxy implies that the disk is stable, and this requirement gives a loose lower limit for the halo mass fraction of 30% of the total mass.

157.049 On the spatial distribution of Population II stars in Sb and later type galaxies.
J. N. Bahcall, N. D. Kylafis.
Astrophys. J., Vol. 288, No. 1, p. 252 – 258 (1985).

The authors show that surface brightness measurements must be accurate to about 0.1 mag at 1% of the visual sky brightness in order to determine the spatial distribution of Population II stars in Sb galaxies with bulge–to–disk ratios like those in NGC 891 and in galaxies of later Hubble type. They fit the obser-vations of van der Kruit and Searle of the edge–on galaxy NGC 891 with two different models: (1) a thin disk and a de Vaucouleurs spheroid and (2) a thin disk plus a thicker disk. The observations are well described by both models.

157.050 Erratum: "An objective classification system for spiral galaxies. I. The two dominant dimensions" [Astrophys. J., Vol. 278, No. 1, p. 61 – 80 (1984)].
B. C. Whitmore.
Astrophys. J., Vol. 288, No. 1, p. 410 (1985). See Abstr. 37.157.137.

157.051 Wolf–Rayet stars in nearby galaxies: tracers of the most massive stars.
P. Massey.
Publ. Astron. Soc. Pac., Vol. 97, No. 587, p. 5 – 24 (1985).

The properties and evolution of massive stars ($\gtrsim 20\,M_\odot$) are reviewed, along with methods of determining the massive star content of nearby galaxies. Owing to their short lifetimes, the frequency of various types of O stars provide a direct measure of the instantaneous initial mass function (IMF); the relative total number reflects the present–day star formation rate (SFR). As most Wolf–Rayet (W–R) stars are descended from only the most massive O–type stars, and are relatively easy to find by emission-line surveys, they can be used as probes of the massive star content in other galaxies. Recent studies of the W–R content of several Local Group galaxies are reviewed; these data indicate that factors of ~ 3 differences exist in the present–day massive star content of galaxies of similar metallicity. There is some evi-dence that these differences are due in large part to differences in the IMF rather than in the SFR.

157.052 The anomalous arms of the spiral galaxy NGC 4258.
J. R. Roy, R. Arsenault, L. Noreau.
Publ. Astron. Soc. Pac., Vol. 97, No. 587, p. 32 – 36 (1985).

The nature of the optical emission of the southern "anomalous arms" of the spiral galaxy NGC 4258 is investigated. High-resolution spectral scans at Hα have been obtained with a Fabry–Pérot spectrometer. No spectral line was detected, indicating that the emission of these arms is probably mostly continuum. A search for high optical polarization with polarizing filters and a photon–counting camera resulted in an upper polarization limit of 5% (1 σ), eliminating synchrotron radiation as an important contribution to the optical emission of the arms. It remains to be established whether the optical anomalous arms are made of late–type stars or perhaps emitters of nonthermal continuum.

157.053 The luminosity function of galaxies in ^{12}CO emission.
F. Verter.
News Lett. Astron. Soc. N.Y., Vol. 2, No. 5, p. 7 – 8 (1984). Ab-stract. – See Abstr. 010.243.

157.054 Large–scale configuration of magnetic fields in spiral galaxies.
Y. Sofue, U. Klein, R. Beck, R. Wielebinski.
Astron. Astrophys., Vol. 144, No. 2, p. 257 – 260 (1985).

The authors propose a simple method to derive the large–scale configuration of magnetic field in a disk galaxy from the charac-teristic variation of linear polarization angle along major and minor axes. The method tests two assumed models of field distri-bution, a bisymmetric–open spiral or a ring field configuration, by inferring a Faraday rotation distribution along major and minor axes of the galaxy at a single frequency. Deep measure-ments of linear polarization at 5 GHz were made for ten spiral galaxies using the 100 m telescope. Distributions of the polarized intensity and polarization angle along the major and minor axes were obtained for three galaxies NGC 253, NGC 2903, and IC 342. Adding to the data in the literature, it is concluded that spiral galaxies seem to have either a ring or a bisymmetric spiral magnetic field.

157.055 Distribution and kinematics of CO in spiral galaxy M51.
G. Rydbeck, Å. Hjalmarson, O. E. H. Rydbeck.
Astron. Astrophys., Vol. 144, No. 2, p. 282 – 294 (1985).

The authors report results from a continuing in–depth study, with high spatial resolution (33″), of the CO $J = 1$–0 emission from the Sc galaxy M51. In this paper the first 75 observed posi-tions are studied.

157.056   New carbon stars in spheroidal galaxies. I. Sculptor, Carina, Leo I and Leo II systems.
M. Azzopardi, J. Lequeux, B. E. Westerlund.
Astron. Astrophys., Vol. 144, No. 2, p. 388 – 394 (1985).

An objective grating survey of the dwarf spheroidal galaxies Sculptor, Carina, Leo I, and Leo II was made with the ESO and CFH telescopes. It yielded a number of new carbon stars in all these galaxies, more than doubling the number of identified objects.

157.057   Reddening of globular clusters in M31.
M. Iye, O.–G. Richter.
Astron. Astrophys., Vol. 144, No. 2, p. 471 – 478 (1985).

Three homogeneous sets of photometric data for confirmed and suspected globular clusters in M31 are used to examine the differential reddening of these objects due to absorption in the disk of M31. The interstellar medium in the disk of M31 is found to have an average reddening ratio $E_{U-B}/E_{B-V} = 1.01 \pm 0.11$, which is different from the standard value 0.72 of our Galaxy. A more detailed analysis of the colours of the globular clusters in M31 shows that either there exists a (radial) colour gradient for the whole cluster system, or the reddening ratio in the inner part of the disk differs from the outer part where the galactic value is well reproduced. The ratio of total to selective absorption in the inner part of the disk of M31 may also be different from that observed in the solar neighbourhood. In agreement with previous studies, the NW side of the disk of M31 is confirmed to be the near side.

157.058   The two superposed galaxies of NGC 3314.
F. Schweizer, N. Thonnard.
Publ. Astron. Soc. Pac., Vol. 97, No. 588, p. 104 – 109 (1985).

Spectroscopic observations of NGC 3314 are presented. Long-slit spectrograms clearly show two sets of emission lines stemming from the two superposed galaxies. The foreground galaxy, NGC 3314a, is a type SBbc or SBc spiral of absolute magnitude $M_B \approx -20$ and has a radial velocity $cz_\odot = 2835 \pm 15$ km s^{-1}. The background galaxy, NGC 3314b, is probably an Sb of $M_B \approx -20.5$ with $cz_\odot = 4641 \pm 6$ km s^{-1}. The availabe evidence suggests that at least NGC 3314a, and probably also NGC 3314b, are members of the Hydra I cluster of galaxies ($cz_\odot = 3699$ km s^{-1}, $\sigma_v = 681$ km s^{-1}), although the alternative hypothesis that one of them or both are field galaxies cannot presently be ruled out.

157.059   NGC 3557 and its globular clusters.
C. L. Morbey, R. D. McClure.
Publ. Astron. Soc. Pac., Vol. 97, No. 588, p. 110 – 117 (1985).

A 50–minute CTIO CCD frame shows that the giant elliptical galaxy NGC 3557 has a luminosity structure very much like the first-ranked galaxies of poor clusters. An extended luminous envelope surrounds the galaxy but it is less luminous than the typical envelope of a cD galaxy in a rich cluster. Only a marginal excess of faint star–like images brighter than the limiting magnitude, $m_B = 25.5$, can be detected. This suggests the existence of a sparse globular cluster population with a specific frequency $S = 0.8 \pm 0.6$, which is unusually low for an elliptical $M_B = -22.55$.

157.060   Mass–luminosity ratio in edge–on spirals.
A. Meisels.
Astron. Astrophys., Vol. 145, No. 1, p. 135 – 138 (1985).

The author presents a study of the five galaxies with the lowest mass to light ratios in the sample of galaxies of Rubin et al. with optically measured rotation curves. UBV multiaperture photometry was performed on these Sb–Sc galaxies, in order to find their local M/L behavior. M/L_B was found to be less than unity in the center of only one galaxy, NGC 4605. Another galaxy, NGC 1421, shows a decrease of local M/L between 0.2 to 0.5 of de Vaucouleurs' radius. This is, probably, not the result of non-constant internal absorption in edge–on spirals. It is suggested that the low mass-to–light ratios might be a result of wrong distances to the two galaxies mentioned above.

157.061   The distance to M33 based on BVRI CCD observations of Cepheids.
W. L. Freedman.
Cepheids: theory and observations, p. 225 – 227 (1985). – See Abstr. 012.027. (IAU Colloq. No. 82).

157.062   The distances to nearby galaxies from near–infrared photometry of Cepheids.
C. W. McAlary, D. L. Welch.
Cepheids: theory and observations, p. 228 – 231 (1985). – See Abstr. 012.027. (IAU Colloq. No. 82).

Near–infrared photometry of Cepheid variables in the Local Group galaxies, NGC 6822, IC 1613, and M33, and in the M81 Group galaxy, NGC 2403 has been used to determine new, independent distances to these objects, which are almost unaffected by dust extinction and by differences in metallicities among the galaxies.

157.063   Properties of barred spiral galaxies.
B. G. Elmegreen, D. M. Elmegreen.
Astrophys. J., Vol. 288, No. 2, p. 438 – 455 (1985).

Blue and near–infrared surface photometry of 15 barred spiral galaxies reveals that early Hubble types tend to have bars with uniform intensities along their lengths and stellar spiral arms with radially decreasing amplitudes, and that late Hubble types tend to have bars with exponential–like intensity profiles and constant or radially increasing stellar spiral arm amplitudes. Fourier decompositions of the bar azimuthal profiles are used to derive relative bar luminosities. A compilation of previous photometric and kinematic results also reveals that bars in early–type galaxies tend to be larger than bars in late–type galaxies, relative to both the galaxy size and the length of the rising part of the velocity curve. The implications of the observations for theories of bar and spiral arm formation in galaxies are discussed.

157.064   The evolution of spiral galaxies and uncertainties in interpreting galaxy counts.
C. R. King, R. S. Ellis.
Astrophys. J., Vol. 288, No. 2, p. 456 – 464 (1985).

Evolutionary models are discussed with respect to the interpretation of faint galaxy counts. The authors review the K-corrections with available UV data, and the necessary assumptions and model input parameters. In particular they explore the possibility that the bulges of spiral galaxies have evolved in a form similar to that expected for elliptical galaxies. The inclusion of bulge evolution predicts significantly more galaxies and may be consistent with the steepening count–magnitude slope at faint magnitudes.

157.065   Self–regulating galaxy formation. I. H II disk and Lyman–alpha pressure.
D. P. Cox.
Astrophys. J., Vol. 288, No. 2, p. 465 – 480 (1985).

Assuming a simple model for the behavior of interstellar material during formation of a disk galaxy, coupled with the lowest order description of infall, a scenario is developed for self-regulated disk galaxy formation. Radiation pressure, particularly that of Lyman–α is an essential component, maintaining an inflated disk and stopping infall when only a small fraction of the overall perturbation has joined the disk. The model leads naturally to galaxies with a rich circumgalactic environment and flat rotation curves. The model seems to apply most naturally to large spirals with radii from 4 to 20 kpc and average column (or surface) densities from 100 to 500 $M_\odot$pc^{-2}. A detailed description of the formation of a galaxy similar to the Milky Way galaxy is included.

157.066   The dependence of CO content on morphological type and luminosity for spiral galaxies in the Virgo cluster.
J. S. Young, N. Z. Scoville, E. Brady.
Astrophys. J., Vol. 288, No. 2, p. 487 – 493 (1985).

From CO observations of the centers of 25 spiral galaxies in the Virgo cluster, the authors confirm the relation found previously for Sc field galaxies, that the CO luminosity in the central 5 kpc diameter is proportional to the B luminosity in the same region. They also find for both the early– and late–type galaxies that the high–luminosity objects of a given type have more CO in their centers than do the low–luminosity galaxies of the same type. The CO luminosities in the centers of the 18 Virgo galaxies in which CO emission was detected show a strong correlation with the total optical luminosities, and therefore central galaxy density, such that more molecular emission is observed in galaxies with higher mass densities. The maximum CO content observed is similar for early *and* late spiral types.

157.067 The M31 globular cluster system.
D. Crampton, A. P. Cowley, D. Schade, P. Chayer.
Astrophys. J., Vol. 288, No. 2, p. 494 – 513 (1985).

One hundred and nine new globular cluster candidates around M31 have been identified on CFHT prime–focus slitless spectra plates. All previously recognized clusters were examined on the plates, and a compilation made of 509 objects which have the appearance and spectra of globular clusters. Of these objects, 408 have colors typical of globular clusters in the Galaxy. Methods were devised to estimate the core radii and intrinsic colors of most of the 509 clusters. The characteristics of the globular cluster systems of M31 and the Galaxy are compared and are shown to be very similar.

157.068 The colours and Byurakan classification of galaxies.
V. G. Malumyan.
Astrofizika, Tom 22, Vyp. 1, p. 25 – 30 (1985). In Russian. English translation in Astrophysics, Vol. 22, No. 1.

It is shown that galaxies of Byurakan classes 2 and 5 which belong to morphological subtypes Sa, Sab, Sb and Sbc (including SB galaxies), on average, are somewhat bluer than galaxies of classes 3 and 4 of the same subtypes. Among the galaxies of classes 2.5 and 2s the objects with ultraviolet excess radiation occur many times more often than among galaxies of classes 3 and 4. Sa, Sab, Sb and Sbc galaxies of classes 3 and 4 have nearly the same mean B–V and U–B. Sc and Scd galaxies of class 4 as well as of classes 2 and 5, on average, are bluer than Sc and Scd galaxies of class 3. In comparison with the latter, among Sc and Scd galaxies of classes 2, 4 and 5 there is a large excess of objects with negative U–B. These facts as well as the properties of activity of galaxies of classes 2, 4, 5, and 2s, which were known earlier, indicate active processes occurring in galaxies with diffuse, split, starlike or semistellar nuclei.

157.069 Investigation of the galaxies from the Byurakan classification at a frequency of 102 MHz. II.
V. S. Artyukh, M. A. Ogannisyan.
Astrofizika, Tom 22, Vyp. 1, p. 211 – 213 (1985). In Russian. English translation in Astrophysics, Vol. 22, No. 1.

The results of the observations of interplanetary scintillation at 102 MHz of 197 galaxies from the Byurakan classification are presented.

157.070 An analysis of observations of the streaming velocities in the bulge of M31.
P. Teuben, E. L. Turner, M. Schwarzschild.
Astrophys. J., Vol. 289, No. 1, p. 58 – 66 (1985).

McElroy has recently provided an extensive set of radial velocity measurements for the stellar mean motions in the bulge of M31. These velocities show a surprisingly bumpy pattern. It is frequently postulated that the absorption clouds in the bulge are the cause of these observed velocity bumps. The analysis presented here shows that the velocity residuals derived for the inner bulge of M31 are not likely to be caused by the effects of absorption clouds. The statistics of the velocity residuals suggest that they are caused mainly by systematic observational errors that are not accessible to the objective internal error analysis. Requirements on the accuracy of radial velocity data for an analysis of the dynamics of M31 are outlined.

157.071 Rotation velocities of 16 Sa galaxies and a comparison of Sa, Sb, and Sc rotation properties.
V. C. Rubin, D. Burstein, W. K. Ford Jr., N. Thonnard.
Astrophys. J., Vol. 289, No. 1, p. 81 – 104 (1985). With plates 2 – 3.

Rotational velocities are presented for 16 Sa galaxies; for 11 of these, velocities extend to beyond 66% of R_{25}. The Sa rotation curves show the same progression as do the Sb and Sc rotation curves, from low central gradient, low rotational velocity for Sa's of low luminosity, to high central gradient, high rotational velocity for Sa's of high luminosity. There is a marked similarity of form (but not of amplitude) of rotation curves for galaxies with morphologies as different as large–bulged Sa's and small–bulged Sc's. Parameters for the Sa galaxies are compared with those for Sb and Sc galaxies. The maximum rotation velocities V_{max} for Sa galaxies are higher than those of Sb and Sc galaxies of equivalent blue or infrared magnitude. The Tully–Fisher correlation of luminosity with V_{max} has slopes near 10 for Sa, Sb, and Sc galaxies separately, but with zero–point displacements.

157.072 The inner disk of NGC 253.
N. Z. Scoville, B. T. Soifer, G. Neugebauer, J. S. Young, K. Matthews, J. Yerka.
Astrophys. J., Vol. 289, No. 1, p. 129 – 140 (1985).

The stellar and interstellar matter distributions in the luminous Sc galaxy NGC 253 are analyzed using maps of the near–infrared continuum ($\lambda = 1 - 2$ µm) and the millimeter CO emission line. The stellar disk traced out in the near–infrared exhibits two components: a nuclear peak of diameter $\sim 20''$ (~ 300 pc) and an extended, inner disk of diameter $\sim 360''$ (~ 6 kpc). The 2 µm brightness distribution mapped over the full extent of the inner disk is dominated by a barlike feature at position angle $17°$ east of the major axis extending to $\pm 120''$ on opposite sides of the nucleus. The derived abundance of molecular gas in the inner disk exceeds that of H I by a factor of 10 and exhibits a much steeper falloff with radius. The total mass of H_2 inside $R = 4$ kpc is $\sim 2 \times 10^9 M_\odot$, which is $\sim 7\%$ of the dynamical mass at the same radius.

157.073 M supergiants in Local Group irregular galaxies: metallicities and distances.
J. H. Elias, J. A. Frogel.
Astrophys. J., Vol. 289, No. 1, p. 141 – 149 (1985).

JHK photometry of confirmed and suspected red supergiants in NGC 3109, NGC 6822, IC 1613, Sextans A, and DDO 210 together with CO and H_2O indices of some of the brighter stars has been done. The colors and spectral type distributions are discussed and metallicities are derived. The absolute K magnitudes of the brightest red supergiants become brighter as the parent galaxy luminosity increases. The scatter about the mean relation is ~ 0.1 mag, allowing an estimate of 25.34 for the true distance modulus of NGC 3109.

157.074 CO(2 – 1) observations of the nucleus of Maffei 2.
A. I. Sargent, E. C. Sutton, C. R. Masson, K. Y. Lo, T. G. Phillips.
Astrophys. J., Vol. 289, No. 1, p. 150 – 154 (1985). With plate 4.

Observations of the core of the spiral galaxy Maffei 2 have been made in the $J = 2 - 1$ line of ^{12}CO, with a spatial resolution of $30''$. Emission in this transition is confined to an area not much greater than 740 pc in extent. It is shown that enhancement of the line strength at the nucleus of the galaxy is unlikely to be due solely to a temperature gradient between the core and the disk. Star formation may have taken place quite recently in Maffei 2, perhaps as the result of interaction between Maffei 1 and Maffei 2.

157.075 The cross–section of the spiral arm S4 in the Andromeda galaxy.
Yu. N. Efremov.
Pis'ma Astron. Zh., Tom 11, No. 3, p. 169 – 177 (1985). In Russian. English translation in Sov. Astron. Lett., Vol. 11.

For the S4 arm of the M31 galaxy it is found that the regions of H_2 maximum density coincide with those of H I in between OB75 and OB82, and the brightest H II regions are located along the inner border of the stellar arm. The structure of the S4 arm between OB75 and OB82 is in accord with the density wave theory.

157.076 **Distribution of surface brightness across the spiral arm S4 in the Andromeda galaxy.**
Ts. B. Georgiev, G. R. Ivanov.
Pis'ma Astron. Zh., Tom 11, No. 3, p. 178 – 183 (1985). In Russian. English translation in Sov. Astron. Lett., Vol. 11.

Surface brightness in the UBV system across the spiral arm S4 of the galaxy M31 is investigated. The maximum brightness is observed near the inner edge of the arm.

157.077 **Star formation in grand design and flocculent spiral galaxies.**
W. Romanishin.
Astrophys. J., Vol. 289, No. 2, p. 570 – 573 (1985).

The author examines the colors and neutral hydrogen contents of spiral galaxies which have been classified in the Elmegreen and Elmegreen arm morphology system. At the same revised Hubble type, spiral galaxies with regular global arm patterns (the grand design spirals) are bluer than spiral galaxies lacking such patterns (the flocculent spirals) by a small, but statistically significant amount (~ 0.05 in B–V and ~ 0.15 in U–V). The neutral hydrogen contents of the two groups are roughly similar. The color differences are interpreted in terms of different star formation histories in spiral galaxies with and without grand design spiral arms.

157.078 **A grism search of Messier 33 for emission–line objects.**
B. Bohannan, P. S. Conti, P. Massey.
Astron. J., Vol. 90, No. 4, p. 600 – 605, 690 (1985).

Using grism plates, the authors have surveyed Messier 33 for star–like objects with strong emission lines. Of the 82 objects found, most are H II regions, probably excited by a single star. Four are Wolf–Rayet stars and five are tentatively identified as supernovae remnants or planetary nebulae.

157.079 **8 – 13 μm spectrophotometry of galaxies. V. The nuclei of five spiral galaxies.**
P. F. Roche, D. K. Aitken.
Mon. Not. R. Astron. Soc., Vol. 213, No. 4, p. 789 – 797 (1985).

Spectra at 8 – 13 μm of the nuclei of five nearby bright spiral galaxies are presented. Four of the galaxies, IC 342, NGC 5195, NGC 5236 and NGC 253 display prominent spectral structure arising from gas and dust emission. They resemble the spectra of other galaxy nuclei undergoing vigorous nuclear star formation, and it is argued that the strong emission structure arises from the interaction of many H II regions in the central regions of the galaxies. NGC 4736 shows no evidence for emission in the dust and gas features seen in almost all starburst galaxies observed to date. It has been classed as a LINER, and the 10–μm spectrum resembles those of the Seyfert galaxies.

157.080 **Penetrating the forest: the distance to Maffei 1.**
R. J. Buta, M. L. McCall.
Mercury, Vol. 13, No. 5, p. 147 (1985).

157.081 **Surface brightness of isolated and double galaxies.**
I. D. Karachentsev, V. E. Karachentseva,
A. L. Shcherbanovksij.
Astrofiz. Issled. Izv. Spets. Astrofiz. Obs., Tom 19, p. 3 – 13 (1985). In Russian. English translation in Bull. Spec. Astrophys. Obs. – North Caucasus.

For isolated and for double galaxies reductions were made of apparent magnitudes to the photometric system of Holmberg and of angular diameters to the standard isophote $25^m/\square''(a_{25})$. The surface brightness in this system was determined for 902 single galaxies and for 980 pair components. Dependences of the mean surface brightness on morphologic type, apparent axial ratio and on other parameters of galaxies are considered.

157.082 **Morphology of spiral galaxies. I. General properties.**
P. J. Grosbøl.
Astron. Astrophys., Suppl. Ser., Vol. 60, No. 2, p. 261 – 276 (1985).

Until recently either large samples of galaxies have been examined visually or smaller sets have been analyzed quantitatively. It is, however, now feasible to analyze large samples of galaxies in an objective way using modern image processing facilities. In the present paper these techniques have been used to derive quantitative information about the internal structure of spiral galaxies. The selection of the galaxies, the material used, and the reduction procedure are described. A catalogue of the derived quantities and their general properties are presented. Finally, the distributions of disk parameters and isophotal magnitudes are discussed.

157.083 **Systematic errors in Zwicky's magnitudes.**
G. Fasano.
Astron. Astrophys., Suppl. Ser., Vol. 60, No. 2, p. 285 – 292 (1985).

Zwicky's magnitudes m_{Zw} are compared with the B_T magnitudes given by Sandage and Tammann in the Revised Shapley–Ames Catalogue of Bright Galaxies. The slope of the well known linear correlation between the correction $\Delta B = (B_T - m_{Zw})$ and the Zwicky surface brightness, is found to depend on declination and strongly on morphological type. On the contrary, the dependence of ΔB on the volume number of the Zwicky catalogue is explained as due to the above mentioned dependence on declination. Errors of the Zwicky magnitudes corrected for these effects are expected to be of about $0^m.24$, independently of the morphological type.

157.084 **A radio continuum survey of Sbc spiral galaxies at 1.465 GHz.**
E. Hummel, A. Pedlar, J. M. van der Hulst, R. D. Davies.
Astron. Astrophys., Suppl. Ser., Vol. 60, No. 2, p. 293 – 313 (1985).

The radio continuum emission at 1.465 GHz of 88 Sbc spiral galaxies was observed with the Very Large Array. The angular resolution of the observations is $\sim 15''$ and the detection limits are ~ 1 mJy for point–like sources and 5 to 10 mJy for extended sources. A detection rate of 82% was obtained. The survey provides data on the total intensity and on the two–dimensional brightness distribution of the radio emission of the galaxies observed.

157.085 **A useful list of redshift calibrators.**
G. Tanzella–Nitti, P. Fontanelli.
Astron. Astrophys., Suppl. Ser., Vol. 60, No. 2, p. 343 – 347 (1985).

The authors have extracted from the Catalogue of Radial Velocities of Galaxies and its updated extension described in Tanzella–Nitti (1984) a list of galaxies with highly consistent independent 21–cm radial velocity determinations. The selection is also based on a critical comparison of different redshifts, environmental and morphological properties of the objects. A final sample of 34 galaxies is presented as a suitable list of galaxies that can be used as redshift calibrators.

157.086 **Detailed X–ray observations of M83.**
G. Trinchieri, G. Fabbiano, G. G. C. Palumbo.
Astrophys. J., Vol. 290, No. 1, p. 96 – 107 (1985). With plate 1.

M83 is detected in X–rays with the Einstein Observatory with an X–ray luminosity of $\sim 6 \times 10^{39}$erg s^{-1} (0.5 – 3.0 keV, for a distance of 3.75 Mpc). The extent and shape of the X–ray emission are comparable to those in the optical (blue band), although the inner 2' region shows a relative X–ray excess. High–resolution X–ray observations show no strong correlation between the X–ray emission and the spiral arm pattern. Six bright sources, each with $L_x > 1 \times 10^{38}$erg s^{-1}, are detected in the plane and near the nuclear region of M83. Their high X–ray luminosities suggest that they are close accreting binaries. The starburst nucleus of M83 is detected with $L_x \approx 2 \times 10^{39}$erg s^{-1}.

157.087 **Ionized gas in the center of M31.**
G. H. Jacoby, H. Ford, R. Ciardullo.
Astrophys. J., Vol. 290, No. 1, p. 136 – 139 (1985). With plate 2.

The authors present Hα + [N II], [S II], and [O III] CCD images of the nuclear bulge of M31 which show a striking spiral pattern of ionized gas extending to within a few parsecs of the nucleus. The mass of ionized gas is approximately 1500 $M_\odot$ and can easily be provided by mass lost from evolving stars. The ionization source for this gas is discussed. The geometry of the apparent spiral arms implies that much of the gas is in a plane which is tipped with respect to the disk of M31.

157.088 The missing bulge globular clusters in M31: new optical candidates.
A. Wirth, L. L. Smarr, T. L. Bruno.
Astrophys. J., Vol. 290, No. 1, p. 140 – 153 (1985).
The authors use a new method to attack the question of the "missing" globular clusters in the bulge of M31. Image processing techniques were used on 13 videocamera fields to obtain an accurate photometric census of stellar objects in M31's bulge down to a limiting B magnitude of 21. This luminosity distribution is compared with the Bahcall–Soneira model of galactic foreground stars. The authors find a statistically significant excess of bright images in the luminosity range of globular clusters of M31's distance. If these candidates prove to be globular clusters, they would double the number of known globular clusters in the surveyed region.

157.089 Infrared photometry and the comparative stellar content of dwarf spheroidals in the galactic halo.
M. Aaronson, J. Mould.
Astrophys. J., Vol. 290, No. 1, p. 191 – 210 (1985).
The authors report infrared JHK photometry for carbon stars in the Ursa Minor, Draco, Leo I and Leo II dwarf spheroidal galaxies, and for a number of oxygen–rich giants in these systems as well. In combination with previously published data, the results enable a detailed comparison to be made of the red stellar content in all seven nearby spheroidals. The origin of the carbon stars, which are the most luminous stars in these systems, is investigated. Furthermore, the authors demonstrate the existence of a well–defined abundance–absolute magnitude relation among the seven dwarfs, which joins smoothly onto the abundance–magnitude relation for more luminous early–type galaxies.

157.090 Unidentified *IRAS* sources: ultrahigh–luminosity galaxies.
J. R. Houck, D. P. Schneider, G. E. Danielson,
C. A. Beichman, C. J. Lonsdale, G. Neugebauer, B. T. Soifer.
Astrophys. J., Lett. Ed., Vol. 290, No. 1, p. L5 – L8 (1985). With plate L1.
Optical imaging and spectroscopy has been obtained for six of the high galactic latitude infrared sources recently reported by Houck and colleagues from the *IRAS* survey to have no obvious optical counterparts on the POSS prints. All are identified with visually faint galaxies that have total luminosities in the range $5 \times 10^{11} L_\odot$ to $5 \times 10^{12} L_\odot$. This luminosity emerges virtually entirely in the infrared. The origin of the luminosity, which is one to two orders of magnitude greater than that of "normal" galaxies, is not known at this time.

157.091 Certain cosmological considerations connected with binary galaxies.
A. N. Tomov.
Dokl. Bolg. Akad. Nauk, Vol. 37, No. 6, p. 703 – 705 (1984).
Abstr. in Ref. Zh., 51. Astron., 2.51.966 (1985).

157.092 Gas in elliptical galaxies.
A. C. Fabian.
Nature, Vol. 314, No. 6007, p. 130 (1985).

157.093 A very bright water vapour maser source in the galaxy NGC 3079.
A. D. Haschick, W. A. Baan.
Nature, Vol. 314, No. 6007, p. 144 – 146 (1985).
This letter reports on a survey for 22.24 GHz water maser emission from spiral galaxies displaying continuum radio emission, which is an indication of star formation activity. The authors have detected the most luminous H_2O maser source yet reported, $\sim 500 L_\odot$, in the galaxy NGC 3079. OH absorption was also detected in this galaxy and it is suggested that maser amplification of the nuclear radio source causes the observable H_2O emission.

157.094 The ultraviolet spectra of normal elliptical galaxies: a population synthesis approach.
R. Nesci, G. C. Perola.
Astron. Astrophys., Vol. 145, No. 2, p. 296 – 304 (1985).
To explain the upturn shortward of $\lambda 2000$ Å observed in the spectra of elliptical galaxies, the authors have considered three kinds of hot stellar objects, HB stars, accreting white dwarfs in binary systems and young stars. By using a population synthesis technique the authors have been able to reproduce in all three cases the shape of the ultraviolet spectra and to define rather narrow constraints which anyone of these populations of objects must fulfil.

157.095 Star formation in early–type galaxies.
P. Véron, M.–P. Véron–Cetty.
Astron. Astrophys., Vol. 145, No. 2, p. 433 – 438 (1985).
A photometric and spectroscopic study of a complete sample of giant elliptical and S0 galaxies shows a well defined correlation between the $U–B$ and $B–V$ colours of these galaxies. The three bluest galaxies in the sample have stronger Balmer absorption lines in their spectra than the mean elliptical revealing the presence of a component of hot stars; the smaller UV excess seen in many other galaxies may have the same cause. Independent evidence suggests that young stars are sometimes present in early–type galaxies and that the hot stars the authors have observed could be young rather than old. This suggests that recent star formation is more common in early–type galaxies than previously thought. Amorphous galaxies could be early–type galaxies in which very recent star formation has occurred.

157.096 S0 galaxies and the influence of the gas content on the radio continuum emission from a disk component.
E. Hummel, C. G. Kotanyi.
Astron. Astrophys., Vol. 145, No. 2, p. 475 – 478 (1985).
Four S0 galaxies, NGC 3115, 3607, 4111, and 4762, were observed with the Very Large Array at 1.46 GHz in an attempt to detect radio continuum emission from their disk component. No disk emission was detected and the emissivity is at least a factor 10 lower than in spiral galaxies. This is probably due to the very low gas content of these galaxies, showing the influence of the gas on the radio emission (magnetic field and sources of relativistic electrons).

157.097 How common are "dust–lanes" in early–type galaxies?
E. M. Sadler, O. E. Gerhard.
Mon. Not. R. Astron. Soc., Vol. 214, No. 2, p. 177 – 187 (1985).
Despite much recent interest in the properties of elliptical galaxies with dust–lanes, no attempt has yet been made to determine the incidence of such galaxies among the elliptical population as a whole. Using a complete sample of early–type galaxies, and accounting as far as possible for selection effects, the true fraction of galaxies with dust is estimated to be ~ 40 per cent for nearby ellipticals and somewhat higher than this for S0s.

157.098 Luminosity function of E and S0 galaxies.
J. Chołoniewski.
Mon. Not. R. Astron. Soc., Vol. 214, No. 2, p. 189 – 195 (1985).
The author obtains a luminosity function (LF) for E and S0 galaxies from a new sample of 248 southern galaxies (Sadler 1984). He finds that the parameters describing this LF in the Schechter form are $M_* = -21.7$ and $\alpha = -1.35$ ($H_0 = 50$). In his computations the author takes into account the inhomogeneity of galaxy distribution which is large and changes considerably the shape of LF. The incompleteness of the sample and the infall on to the Virgo supercluster are also taken into consideration.

157.099 Bivariate luminosity function of E and S0 galaxies.
J. Chołoniewski.
Mon. Not. R. Astron. Soc., Vol. 214, No. 2, p. 197 – 202 (1985).
A function which describes the joint distribution of luminosity and radius of galaxies – the bivariate luminosity function (BLF) is defined. A simple analytical formula for the shape of BLF is proposed and fitted to the data for E and S0 galaxies from the sample of Sadler (1984).

157.100   A giant black hole at the centre of a dwarf galaxy.
C. M. Gaskell.
Nature, Vol. 314, No. 6013, p. 672 (1985).

This note comments on the recent hypothesis, based on an analysis of velocity dispersions, that the centre of the dwarf elliptical galaxy M32 may contain a massive ($\sim 5 \times 10^5 M_\odot$) black hole.

157.101   Powerful extragalactic masers.
W. A. Baan.
Nature, Vol. 315, No. 6014, p. 26 – 31 (1985).

Two strong extragalactic OH masers have been detected in the galaxies NGC 3690 and Mrk 231 which are in the same class as the megamaser source IC 4553 (Arp 220). The amplification model for the background continuum proposed for IC 4553 can account for most of the powerful OH maser lines and for part of the H_2O emission. Most masing galaxies have sufficient infrared flux to pump the masing regions, except that the H_2O pumping occur at a much higher conversion efficiency than the OH pumping.

157.102   Color correlations between paired galaxies: the Holmberg effect.
V. V. Dëmin (*V. V. Demin*), A. V. Zasov, É. A. Dibaj
(*Eh. A. Dibaj*), A. N. Tomov.
Sov. Astron., Vol. 28, No. 4, p. 367 – 371 (1984). English translation of 38.157.018.

157.103   Star–formation bursts in the central regions of Sc galaxies.
O. K. Sil'chenko.
Sov. Astron., Vol. 28, No. 4, p. 372 – 374 (1984). English translation of 38.157.019.

157.104   103–MHz observations of the Andromeda nebula.
V. S. Artyukh, M. A. Ogannisyan.
Sov. Astron., Vol. 28, No. 4, p. 375 – 378 (1984). English translation of 38.157.020.

157.105   Dwarf satellite galaxies and radiative decay of elementary particles composing massive galactic halos.
A. L. Melott.
Sov. Astron., Vol. 28, No. 4, p. 478 – 479 (1984). English translation of 38.157.021.

157.106   On the cigar–shaped ring galaxies.
B. A. Vorontsov–Vel'yaminov, V. A. Dostal',
V. G. Metlov.
Sov. Astron. Lett., Vol. 10, No. 4, p. 205 – 208 (1984). English translation of 38.157.001.

157.107   Rotation and mass of 37 paired galaxies.
I. D. Karachentsev, V. A. Mineva.
Sov. Astron. Lett., Vol. 10, No. 4, p. 235 – 239 (1984). English translation of 38.157.043.

157.108   The plume phenomenon in barred spirals.
R. Buta.
Proc. Astron. Soc. Aust., Vol. 5, No. 4, p. 472 – 478 (1984).

Barred spirals with ring types in their light distribution shows frequently "plumes", features located symmetrically just off their leading ends of the bar. The author examines the frequency and nature of these phenomena.

157.109   An analysis of the distribution of faint galaxies at the south galactic pole.
R. J. Dodd, H. T. MacGillivray.
Proc. Astron. Soc. Aust., Vol. 5, No. 4, p. 505 – 507 (1984).

The distribution of faint galaxies in a deep sample, down to a limiting magnitude of B $\sim$ 21.5 in a region of some 30 square degrees of sky at the south galactic pole, is investigated. On application of an algorithm for the detection of filamentary structure in the distribution of the galaxies, a negative result is obtained.

157.110   Multiaperture infrared photometry of the nuclei of spiral galaxies.
P. J. Cizdziel, C. G. Wynn–Williams, E. E. Becklin.
Astron. J., Vol. 90, No. 5, p. 731 – 735 (1985).

Results of a multiaperture, 2.2– and 10–μm photometric survey of the nuclei of 25 nearby, low–inclination spiral galaxies are presented. The 10–μm nuclear luminosities, attributed to thermal reradiation by dust in regions of active star formation, and 10–μm/2.2–μm nuclear flux ratios are examined as a function of the slope of the 2.2–μm growth curve. No correlation is seen, suggesting the contribution of young, red giants and supergiants to the near–infrared light of spiral nuclei is small. These results indicate the 2.2–μm light of galaxies in the sample is not radically altered by starbursts.

157.111   A search for hydroxyl masers in M33.
J. D. Fix, R. L. Mutel.
Astron. J., Vol. 90, No. 5, p. 736 – 737 (1985).

The authors have used the NRAO Very Large Array to search for emission from the $^2\pi_{3/2}$ ground–state transitions of hydroxyl in the direction of the nearby Sc galaxy M33. They detected no hydroxyl masers and conclude that there are no type I masers in M33 that are as luminous as the brightest type I masers in our galaxy.

157.112   CO abundances and star formation in the three irregular galaxies NGC 4449, NGC 4214, and NGC 3738.
L. J. Tacconi, J. S. Young.
Astrophys. J., Vol. 290, No. 2, p. 602 – 608 (1985).

CO emission has been detected in the central 1.3 kpc of the three irregular galaxies NGC 4449, NGC 4214, and NGC 3738, and also in a giant H II complex in NGC 4449 using the 14 m telescope of the Five College Radio Astronomy Observatory. The CO distribution in NGC 4449 appears clumped, with stronger emission found in one H II complex than in the central region of the galaxy. Comparing the irregulars with the luminous Scd galaxy IC 342, all of which are roughly at the same distance, the central 50″ of the irregulars have 2 – 5 times less blue luminosity than the central 50″ in IC 342 but have factors of 40 – 70 less CO.

157.113   Photographic photometry of the main galactic system VV 242.
V. P. Reshetnikov, L. A. Vinnikova.
Vestn. Leningr. Univ., Mat. Mekh. Astron., No. 19, p. 90 – 93 (1984). In Russian. Abstr. in Ref. Zh., 51. Astron., 3.51.877 (1985).

157.114   The UV energy distribution of the bright spiral NGC 4258, and the presence of an intermediately hot stellar component.
R. Barbon, M. Capaccioli, G. Longo.
Mem. Soc. Astron. Ital., Vol. 55, No. 3, p. 429 – 442 (1984). – See Abstr. 012.049.

The 1250 – 3055 Å (UV) energy distribution in the nuclear region of the Sbc galaxy NGC 4258 has been obtained from two International Ultraviolet Explorer spectra. It shows a minimum around 2600 Å followed by a turn–up toward the ultraviolet. Using qualitative arguments the authors have temptatively identified two UV components corresponding to black–body temperatures of about 15,000 and 30,000K respectively. They are interpreted as due to hot young stars describing successive phases of stellar formation bursts. The same two hot components are also found in the energy distributions of ellipticals, especially those which are very bright in the UV. This fact is discussed in the context of the hypothesis that ellipticals still retain the capability to form stars.

157.115   High–resolution surface photometry of elliptical galaxies.
T. R. Lauer.
Astrophys. J., Suppl. Ser., Vol. 57, No. 3, p. 473 – 502 (1985). = Lick Obs. Bull., No. 987.

High–resolution CCD surface photometry profiles have been obtained for 42 nearby elliptical and S0 galaxies as a first step to investigate their central structure and core properties. Accuracy of the central surface brightnesses is to better than 0.02 mag rms, and central resolution is limited by atmospheric seeing. A hybrid Fourier deconvolution procedure has been developed to correct the surface photometry for seeing in a model–independent way. This procedure yields surface photometry profiles of slightly sub–arc second resolution.

157.116 Rossby solitons in a galactic disk.
V. I. Korchagin, V. I. Petviashvili.
Pis'ma Astron. Zh., Tom 11, No. 4, p. 298 – 301 (1985). In Russian. English translation in Sov. Astron. Lett., Vol. 11.

A nonlinear equation describing the low–frequency waves in a galactic disk which do not perturb the gravitational field is obtained. This equation has a solution in the form of undamped Rossby solitons – cyclones or anticyclones.

157.117 *IUE* observations of amorphous hot galaxies.
S. A. Lamb, J. S. Gallagher III, M. S. Hjellming, D. A. Hunter.
Astrophys. J., Vol. 291, No. 1, p. 63 – 71 (1985).

Blue amorphous galaxies are star–forming, irregularlike systems which lack the spatially distinct OB stellar groups that are characteristic of most late–type galaxies. The authors have obtained short–wavelength *IUE* spectra of the amorphous galaxies NGC 1705 and NGC 1800. They find that NGC 1705 contains a normal mix of OB stars, which is consistent with the nearly constant recent star–formation rate inferred from new optical data. NGC 1800 is likely to have similar properties, and blue galaxies with amorphous structures thus do not show evidence for anomalies in stellar populations.

157.118 Blast wave formation of the extended stellar shells surrounding elliptical galaxies.
R. E. Williams, W. A. Christiansen.
Astrophys. J., Vol. 291, No. 1, p. 80 – 87 (1985). With plate 1.

The existence of stellar shells at large distances from isolated elliptical galaxies is explained in terms of a blast wave associated with an active nucleus phase early in the history of the galaxy. The blast wave sweeps the initial interstellar medium out of the galaxy into an expanding shell which radiatively cools behind its leading shock front. Cooling of the shell following turnoff of the nucleus activity, which keeps the shell photoionized, leads to a brief epoch of star formation which is terminated by heating of the shell from supernovae and UV radiation from massive stars. The stars so formed would spend much of their time near apogalacteum, thus taking on the appearance of a shell.

157.119 Globular clusters in galaxies beyond the Local Group. IV. The elliptical galaxies NGC 524 and 1052.
W. E. Harris, D. A. Hanes.
Astrophys. J., Vol. 291, No. 1, p. 147 – 151 (1985). With plates 2 – 4.

The authors describe the global properties (total population and spatial structures, derived by starcounts) of the globular cluster systems in two large E/S0 galaxies: NGC 524, the central giant in the compact cluster CfA 13; and NGC 1052, the only major elliptical in the loose southern group HG 44 = Cetus I. In NGC 1052, the halo cluster system is small and more compact than normal. The globular cluster system in NGC 524 is somewhat more populous and structurally more extended.

157.120 Distribution of the magnetic field in spiral galaxies.
A. A. Ruzmajkin, D. D. Sokolov, A. M. Shukurov.
Inst. prikl. mat. Akad. Nauk SSSR. Prepr., No. 117, 29 pp. (1984). In Russian. Abstr. in Ref. Zh., 51. Astron., 4.51.644 (1985).

157.121 Wolf–Rayet stars in NGC 6822 and IC 1613.
T. E. Armandroff, P. Massey.
Astrophys. J., Vol. 291, No. 2, p. 685 – 692 (1985). With plates 14 – 19.

The authors have surveyed the Local–Group irregular galaxies IC 1613 and NGC 6822 for Wolf–Rayet stars using CCD exposures through interference filters. By using crowded–field point–spread–function algorithms, they were able to identify with various degrees of significance 12 W–R candidates in NGC 6822, of which 7 are extremely probable, and 8 candidates in IC 1613, 5 of which are extremely probable. All of the W–R stars in NGC 6822 are likely WN, while two W–R stars in IC 1613 are of WC type. Tests on two previously surveyed fields of M33 revealed all the known W–R stars. NGC 6822 most resembles the SMC in its surface density and type of W–R stars, while IC 1613 resembles the LMC.

157.122 Far–infrared spectroscopy of galaxies: the 158 micron C^+ line and the energy balance of molecular clouds.
M. K. Crawford, R. Genzel, C. H. Townes, D. M. Watson.
Astrophys. J., Vol. 291, No. 2, p. 755 – 771 (1985).

The authors have detected the 158 µm ground–state fine–structure line of singly ionized carbon in six gas–rich galaxies and mapped its spatial distribution in M82, NGC 1068, and M83. The [C II] line is bright toward the centers of all observed galaxies, containing about 0.5% of the bolometric luminosity. Spatial distributions and velocity profiles of the [C II] line are very similar to the 2.6 mm $J = 1 \rightarrow 0$ emission of CO. The ratio of integrated line intensities of [C II] to CO is constant, for [C II] intensities which vary by a factor of 100 in the extragalactic and galactic sources studied. The [C II] line is a tracer of photo–dissociation regions in molecular clouds. The excitation mechanism, line intensities, the overall energy balance and the physical state of these regions in the various galaxy environments are discussed.

157.123 Discovery of the first S star in NGC 6822.
M. Aaronson, J. Mould, K. H. Cook.
Astrophys. J., Lett. Ed., Vol. 291, No. 2, p. L41 – L44 (1985). With plate L2.

A spectroscopic survey of the red giants located in a $1\rlap.'5 \times 2\rlap.'5$ patch of NGC 6822 has led to the discovery of an S star, the first to be identified in a system lying beyond the Milky Way's outer halo. Seven M stars and eight carbon stars have also been spectroscopically confirmed. The S star spectrum shows prominent ZrO and LaO features, while TiO, C_2, and CN bands are weak or absent. Infrared photometry yields $M_{bol} \approx -5.15$ mag, a value similar to that of the only known pure S star in the Magellanic Clouds.

157.124 Abundances of giant stars in the Draco and Ursa Minor dwarf galaxies.
R. A. Bell.
Publ. Astron. Soc. Pac., Vol. 97, No. 589, p. 219 – 228 (1985).

A number of synthetic spectra have been compared with Stetson's (1984) low–dispersion spectra of stars in the Draco and Ursa Minor dwarf galaxies. Spectral indices have been used to deduce the overall metal abundances of the stars, as indicated by the strengths of the H and K lines, as well as the carbon abundances, from the G bands. Similar analyses have been carried out for stars in the globular clusters M3 and M92. Stars in both dwarf spheroidals have the same mean overall abundances as the M92 stars. Carbon is depleted relative to the other metals in both the Draco and Ursa Minor stars ([C/M] = –0.5), but the depletion is not as great as in M92 where [C/M] = –0.8. The abundances deduced from the Draco color–magnitude diagram are higher than those deduced from the M92 diagram, a factor which may be due to higher CNO abundances.

157.125 The deconvolved nuclear structure of M31.
G. Parmeggiani, F. Zavatti, O. Bendinelli.
Astrophys. Space Sci., Vol. 110, No. 2, p. 351 – 355 (1985).

The authors present the nuclear structure of M31 derived from the ground–based CCD observations of Kent (1983). The digital deconvolution from PSF has been performed taking into account the ellipticity. The comparison with old photographic (Johnson, 1961) and extra atmospheric (Light et al., 1974) measurements shows a satisfactory agreement in view of the uncertainties of the various types of observations.

157.126   The structure of the spiral arm S4 in Andromeda galaxy. (A contribution to the theory of density waves).
G. R. Ivanov.
Astrophys. Space Sci., Vol. 110, No. 2, p. 357 – 369 (1985).

The UBV photometry of 690 stars in the spiral arm S4 and the U magnitudes of 120 stars in the spiral arm S6 with the help of the 2 m RCC telescope of the Rozhen Observatory at the Bulgarian Academy of Sciences, has been used to obtain the colour–magnitude and colour–colour diagrams across the arms. The age gradient across the S4 arm has been found. The author has evaluated the velocity of star formation propagation across the arm S4 $\omega_\perp \approx 60\,\mathrm{km\,s^{-1}}$, pattern frequency $\Omega_p \approx 14\,\mathrm{km\,s^{-1}kpc^{-1}}$ and corotation radius $R_c \approx 20$ kpc. A value of pattern frequency $\Omega_p = 12\,\mathrm{km\,s^{-1}kpc^{-1}}$ and $R_c = 12$ kpc of our Galaxy has been obtained from the age distribution of the open clusters and cepheids across the Carina–Sagittarius arm. The spiral structure of M31 has been compared with that of our Galaxy.

157.127   Radio observations of H I in UGC 2885, the largest identified Sc galaxy.
P. R. Roelfsema, R. J. Allen.
Astron. Astrophys., Vol. 146, No. 2, p. 213 – 222 (1985).

Observations of the distribution and motions of H I in the giant Sc galaxy UGC 2885 confirm that it is an extremely regular and symmetric system. Spiral structure is clearly present in H I, corresponding closely to that seen in the optical. Analysis of the velocity data results in a rotation curve which agrees with the optical results. The authors present a simple interpretation for the observed spiral structure and rotation curve in terms of a single dominant two–armed spiral mode. The extent of the spiral arms indicates a corotation radius of approximately 45 kpc, with the spiral wave extending from the inner Lindblad resonance through corotation to the outer Lindblad resonance.

157.128   Star formation and chemical abundances in the blue compact galaxy ESO 338–IG04.
N. Bergvall.
Astron. Astrophys., Vol. 146, No. 2, p. 269 – 281 (1985).

New broadband $UBVRI$ photometry, spectroscopy and direct photographic (B and V) and CCD (Hα and Gunn I) observations are presented for the blue compact galaxy ESO 338–IG04. A low resolution IUE spectrum is reanalyzed. The galaxy, which is undergoing a burst of star formation, is rather luminous for its type, $M_B = -19.4$, and has extremely blue colours. Hot stars are found to be the most important ionization source. A dynamical mass estimate yields $M_{tot} \cong 10^9 M_\odot$. The internal extinction and the chemical abundances are found to be low.

157.129   Near–infrared mapping of spiral galaxies. I. The Sb galaxy NGC 2841: annular structure in the young disc component.
M. Prieto, E. Battaner, C. Sanchez, J. Beckman.
Astron. Astrophys., Vol. 146, No. 2, p. 297 – 302 (1985).

The authors have mapped NGC 2841 in the J and K near–infrared bands out to some 10 kpc from its centre. Maps and radial profiles in μ_J, μ_K, and $J–K$ are presented, and profiles in $B–J$ and $B–K$ derived. The bulge and disc are very well distinguished, but there is also an indication of a bluer ring in the disc between 2 and 7 kpc galactocentric distance. This annular region of enhanced star formation rate coincides with the presence of an enhanced H_2 distribution, derived from CO measurements by Young and Scoville (1982).

157.130   Spectral observations of low surface–brightness galaxies in the M81 group region.
I. D. Karachentsev, V. E. Karachentseva.
Astrofizika, Tom 21, Vyp. 3, p. 641 – 644 (1984). In Russian. English translation in Astrophysics, Vol. 21, No. 3.

Radial velocities are presented for six galaxies nearby M81. Four of them are dwarf members of the M81 group.

157.131   On the dust absorption in the Andromeda galaxy.
G. R. Ivanov, V. K. Golev.
Astron. Tsirk., No. 1352, p. 1 – 3 (1984). In Russian.

157.132   The stellar magnitude (age) gradient across the spiral arm S4 in the region of the star complex NGC 206 in the Andromeda galaxy.
I. A. Rakitin.
Astron. Tsirk., No. 1352, p. 3 – 4 (1984). In Russian.

157.133   An interpretation of the photometric profile of M31 along its major axis.
Yu. N. Efremov, G. R. Ivanov.
Astron. Tsirk., No. 1352, p. 4 – 6 (1984). In Russian.

157.134   An interpretation of the diagram "magnitude–distance from the arm edge" from the point of view of the density wave theory.
Yu. N. Efremov.
Astron. Tsirk., No. 1352, p. 6 – 8 (1984). In Russian.

157.135   Nitrogen abundance in the nuclei of SAbc and SAc galaxies.
A. R. Petrosyan.
Astron. Tsirk., No. 1355, p. 6 – 8 (1984). In Russian.

157.136   An empirical model for the M87 jet.
I. S. Shklovskij.
Sov. Astron., Vol. 28, No. 5, p. 489 – 491 (1984). English translation of 38.157.095.

157.137   A catalog of the rotation curves of normal galaxies.
G. A. Kyazumov.
Sov. Astron., Vol. 28, No. 5, p. 496 – 499 (1984). English translation of 38.157.097.

157.138   Digital surface photometry of galaxies toward a quantitative classification. IV. Principle component analysis of surface–photometric parameters.
M. Watanabe, K. Kodaira, S. Okamura.
Astrophys. J., Vol. 292, No. 1, p. 72 – 78 (1985).

Principal component analysis is applied to five samples composed of galaxies in the Virgo Cluster and the Ursa Major clouds: a standard sample consisting of 201 galaxies, a Virgo sample, an Ursa Major sample, an elliptical sample, and a disk sample. The standard sample includes galaxies of all morphological types. Included in the analysis are four surface–photometric parameters – diameter, magnitude, mean surface brightness, and mean concentration index – derived in the previous papers of this series. The results for these three samples indicate that both elliptical galaxies and disk galaxies lie in the same plane in the space of the four surface–photometric parameters.

157.139   The cores of elliptical galaxies.
T. R. Lauer.
Astrophys. J., Vol. 292, No. 1, p. 104 – 121 (1985). = Lick Obs. Bull., No. 1000.

Core radii and central surface brightnesses are measured from seeing–deconvolved high–resolution CCD surface photometry profiles for 42 nearby elliptical galaxies presented in an earlier paper. Analysis of the apparent core radii and seeing profiles indicates that 14 galaxies have resolvable cores between 1″5 and 5″0. None of the resolved cores can be described by isothermal King model cores. Analysis of the core parameters finds that the central luminosity density, ϱ_c, acts as a second parameter, along with luminosity L, in determining core structure. Central mass–to–light ratios are calculated from the core parameters. A brief discussion is presented on implications of the results for current theories on the formation of elliptical galaxies.

157.140   Brightness profiles of the cores of bulges and elliptical galaxies.
J. Kormendy.
Astrophys. J., Lett. Ed., Vol. 292, No. 1, p. L9 – L13 (1985).

High–resolution surface photometry of the cores of bulges and elliptical galaxies has been obtained with a CCD camera at the CFH Telescope. With the resulting excellent scale and seeing (0″45 – 1″0), many cores are well resolved. Only a few elliptical galaxies are found to have core profiles consistent with a projected isothermal model. Generally these are first–ranked galaxies in rich clusters, e.g., the cD galaxy NGC 6166. Most ellipticals have cores which are nonisothermal by a remarkably consistent small amount. These profiles are more gently curved than an isothermal. Bulges of disk galaxies have profiles which rise still more steeply toward the center. Core profile shape correlates with luminosity.

157.141 Radio continuum and far–infrared emission from spiral galaxies: a close correlation.
T. de Jong, U. Klein, R. Wielebinski, E. Wunderlich.
Astron. Astrophys., Vol. 147, No. 1, p. L6 – L9 (1985).

A total number of 91 galaxies, mostly spirals, have been observed at $\lambda6.3$ cm with the 100m telescope of the MPIfR. For most of these galaxies IRAS 60 and 100 μm flux densities have been published. Plotting far–infrared against radio flux densities the authors find that both quantities are closely correlated. A further analysis reveals that the predominantly non–thermal radio emission of spiral galaxies is intimately connected with the young stellar population; supernova explosions of massive stars are the dominant source of relativistic electrons in the disks of spiral galaxies.

157.142 Dynamics of the bulge of the lenticular galaxy NGC 7814.
R. Bacon.
Astron. Astrophys., Vol. 147, No. 1, p. L16 – L18 (1985).

A detailed isotropic model is applied to the bulge of the lenticular galaxy NGC 7814, using photometric data from Van der Kruit and Searle (1982) and kinematic data from Kormendy and Illingworth (1982). This model gives a good fit of all the 103 observed velocities and velocity dispersions. The only free parameter of this model is the blue mass–to–light ratio, found to be 7 in solar units.

157.143 The structure of the elliptical galaxy IC 4296.
S. Younis, J. Meaburn, P. Stewart.
Astron. Astrophys., Vol. 147, No. 1, p. 178 – 183 (1985).

Best fit ellipses have been fitted with a high degree of accuracy to the isophotes of the elliptical galaxy IC 4296 obtained from J and R photographs with the SERC 1.2 m Schmidt camera and the Anglo Australian Telescope. Changes of ellipticity, of position angle of the major axes and of the centroids of these ellipses as a function of distance from the nucleus have been found. J and R aperture photometry was also performed on the same data arrays. The nuclear regions are shown to be very red compared with the halo. The fast rotation of IC 4296 is discussed in relation to IC 4299.

157.144 How common are "dust–lanes" in early–type galaxies?
E. M. Sadler, O. E. Gerhard.
ESO Sci. Prepr., No. 361, 27 pp. (1985). To be published in Mon. Not. R. Astron. Soc.

157.145 Nitrogen and oxygen evolution in dwarf irregular galaxies.
F. Matteucci, M. Tosi.
ESO Sci. Prepr., No. 372, 33 pp. (1985). Submitted to Mon. Not. R. Astron. Soc.

157.146 Kiso survey for ultraviolet–excess galaxies. II.
B. Takase, N. Miyauchi–Isobe.
Ann. Tokyo Astron. Obs., Second Ser., Vol. 20, No. 3, p. 237 – 281 (1985).

Presented are the second lists and identification charts of the ultraviolet–excess galaxies which have been detected on the multi–color plates taken with the Kiso Schmidt telescope for 10 survey fields. In a sky area of some 300 square degrees 556 objects were catalogued down to a photographic magnitude of about 17.

157.147 H I observations of supermassive spiral galaxies.
B. M. Lewis.
Astrophys. J., Vol. 292, No. 2, p. 451 – 459 (1985).

H I observations are presented for 13 supermassive spirals. The 20% profile width of UGC 9255 (NGC 5619) at 750 km s^{-1} is among the largest reliably recorded for any galaxy and is the largest for an Sb. Comparison with the other well–observed objects with wide profiles, NGC 4594 and NGC 5635, shows that it has a diameter 50% larger than the first and a more than 1.5 mag greater brightness than the second. In relation to a comparison group of supermassive galaxies, however, UGC 9255 is normal.

157.148 Warped dust lanes in elliptical galaxies: transient or stationary phenomena?
F. Bertola, G. Galletta, W. W. Zeilinger.
Astrophys. J., Lett. Ed., Vol. 292, No. 2, p. L51 – L55 (1985).

Spectroscopic observations of NGC 5128 show that the stellar component possesses a velocity gradient along its major axis with a total velocity difference of 160 km s^{-1}. This velocity, together with a velocity dispersion of 140 km s^{-1}, places NGC 5128 among the fast rotating elliptical galaxies. The motion of the warped dust lane with respect to that of the stars is direct. Prograde motions are also found in the warp of the E4 Anon 0151–498, while the E3.5 galaxy NGC 5363 shows a warped dust lane with retrograde motion.

157.149 Carbon monoxide isotope ratios in galactic centers and disks.
L. J. Rickard, L. Blitz.
Astrophys. J., Lett. Ed., Vol. 292, No. 2, p. L57 – L60 (1985).

The authors report observations of millimeter wavelength emission of CO and ^{13}CO in six spiral galaxies, two previously unreported at ^{13}CO. For each galaxy, ^{13}CO emission was observed both from the center and from the disk. The ratio of integrated intensities, $P(CO)/P(^{13}CO)$, varies markedly from galactic center to galactic center, from galactic disk to galactic disk, and from point to point within a given galaxy. These variations indicate that the use of a constant $P(CO)\rightarrow\sigma(H_2)$ conversion to derive H_2 surface densities from extragalactic CO observations can introduce major errors.

157.150 A new high resolution hydrogen line survey of M33.
E. R. Deul, J. M. van der Hulst.
Mitt. Astron. Ges., Nr. 63, p. 168 – 169 (1985). – See Abstr. 012.063.

157.151 Cepheids in the dwarf galaxies Sextans A, Sextans B, and WLM.
A. Sandage, G. A. Carlson.
Bull. Am. Astron. Soc., Vol. 16, No. 4, p. 880 – 881 (1984). Abstract. – See Abstr. 010.062.

157.152 NGC 4782 + 4783: supermassive binary or unbound colliders?
K. D. Borne, J. G. Hoessel.
Bull. Am. Astron. Soc., Vol. 16, No. 4, p. 881 (1984). Abstract. – See Abstr. 010.062.

157.153 A remarkable correlation between IRAS and radio continuum fluxes from disk galaxies.
G. Helou, B. T. Soifer, M. Rowan–Robinson.
Bull. Am. Astron. Soc., Vol. 16, No. 4, p. 888 (1984). Abstract. – See Abstr. 010.062.

157.154 Distances to Local Group galaxies based on BVRI CCD photometry of Cepheids.
W. L. Freedman.
Bull. Am. Astron. Soc., Vol. 16, No. 4, p. 888 (1984). Abstract. – See Abstr. 010.062.

157.155 Isophotal diameters of cluster spirals.
M. E. Cornell, M. Aaronson, G. D. Bothun, J. R. Mould.
Bull. Am. Astron. Soc., Vol. 16, No. 4, p. 888 (1984). Abstract. – See Abstr. 010.062.

157.156 CNO abundance variations in galaxies.
R. J. Dufour.
Bull. Am. Astron. Soc., Vol. 16, No. 4, p. 888 – 889 (1984). Abstract. – See Abstr. 010.062.

157.157   **Fine structure in elliptical and merger galaxies.**
F. Schweizer, W. K. Ford Jr.
Bull. Am. Astron. Soc., Vol. 16, No. 4, p. 889 (1984). Abstract. –
See Abstr. 010.062.

157.158   **Stellar population in the outskirts of elliptical galaxies.**
W. A. Baum, B. Thomsen, B. L. Morgan.
Bull. Am. Astron. Soc., Vol. 16, No. 4, p. 889 (1984). Abstract. –
See Abstr. 010.062.

157.159   **Radial velocities of stars in dwarf elliptical galaxies.**
P. O. Seitzer, J. A. Frogel.
Bull. Am. Astron. Soc., Vol. 16, No. 4, p. 889 (1984). Abstract. –
See Abstr. 010.062.

157.160   **The red stellar content of nearby galaxies.**
K. H. Cook, M. Aaronson, J. Norris.
Bull. Am. Astron. Soc., Vol. 16, No. 4, p. 889 – 890 (1984). Abstract. – See Abstr. 010.062.

157.161   **Main sequence CCD photometry in the Ursa Minor dwarf.**
E. W. Olszewski, M. Aaronson.
Bull. Am. Astron. Soc., Vol. 16, No. 4, p. 890 (1984). Abstract. –
See Abstr. 010.062.

157.162   **CO emission from IRAS galaxies.**
J. S. Young, J. Kenney, S. D. Lord, F. P. Schloerb.
Bull. Am. Astron. Soc., Vol. 16, No. 4, p. 936 (1984). Abstract. –
See Abstr. 010.062.

157.163   **The main sequence turn–off of the Draco dwarf spheroidal galaxy.**
B. W. Carney, P. O. Seitzer.
Bull. Am. Astron. Soc., Vol. 16, No. 4, p. 947 (1984). Abstract. –
See Abstr. 010.062.

157.164   **Massive stars in nearby galaxies: a search for Wolf–Rayet stars in M31 and NGC 300.**
P. Massey, T. E. Armandroff, P. S. Conti.
Bull. Am. Astron. Soc., Vol. 16, No. 4, p. 948 (1984). Abstract. –
See Abstr. 010.062.

157.165   **Globular clusters in spiral galaxies.**
H. C. Harris, J. E. Hesser, G. D. Bothun.
Bull. Am. Astron. Soc., Vol. 16, No. 4, p. 949 (1984). Abstract. –
See Abstr. 010.062.

157.166   **Observations of gas flow in barred galaxies.**
M. P. Ondrechen.
Bull. Am. Astron. Soc., Vol. 16, No. 4, p. 949 (1984). Abstract. –
See Abstr. 010.062.

157.167   **Ultraviolet observations of galaxies with recent star formation.**
G. E. Miller, C.–C. Wu.
Bull. Am. Astron. Soc., Vol. 16, No. 4, p. 949 (1984). Abstract. –
See Abstr. 010.062.

157.168   **Star–formation rates of interacting galaxies.**
H. A. Bushouse.
Bull. Am. Astron. Soc., Vol. 16, No. 4, p. 949 (1984). Abstract. –
See Abstr. 010.062.

157.169   **Star formation in grand design and flocculent spiral galaxies.**
W. Romanishin.
Bull. Am. Astron. Soc., Vol. 16, No. 4, p. 949 (1984). Abstract. –
See Abstr. 010.062.

157.170   **Neutral hydrogen observations of NGC 3079.**
J. A. Irwin, A. R. Taylor, E. R. Seaquist.
Bull. Am. Astron. Soc., Vol. 16, No. 4, p. 950 (1984). Abstract. –
See Abstr. 010.062.

157.171   **CCD photometry of galaxies.**
M. J. Pierce, R. B. Tully.
Bull. Am. Astron. Soc., Vol. 16, No. 4, p. 950 (1984). Abstract. –
See Abstr. 010.062.

157.172   **Observations of stellar spiral arms in old galactic disks.**
R. C. Kennicutt, B. K. Edgar.
Bull. Am. Astron. Soc., Vol. 16, No. 4, p. 950 (1984). Abstract. –
See Abstr. 010.062.

157.173   **IRAS maps of NGC 6822.**
W. L. Rice, F. C. Gillett.
Bull. Am. Astron. Soc., Vol. 16, No. 4, p. 950 (1984). Abstract. –
See Abstr. 010.062.

157.174   **Hα velocity field of NGC 4449.**
E. M. Malumuth, T. B. Williams, R. A. Schommer.
Bull. Am. Astron. Soc., Vol. 16, No. 4, p. 950 (1984). Abstract. –
See Abstr. 010.062.

157.175   **The distance to M31 from H(1.6 µm) photometry of its Cepheids.**
D. L. Welch, C. W. McAlary, R. A. McLaren, B. F. Madore.
Bull. Am. Astron. Soc., Vol. 16, No. 4, p. 951 (1984). Abstract. –
See Abstr. 010.062.

157.176   **Differential spectral synthesis of S0 disks.**
M. D. Gregg.
Bull. Am. Astron. Soc., Vol. 16, No. 4, p. 951 (1984). Abstract. –
See Abstr. 010.062.

157.177   **Enhancement of galaxy spectra using median filters.**
J. W. Sulentic.
Bull. Am. Astron. Soc., Vol. 16, No. 4, p. 953 (1984). Abstract. –
See Abstr. 010.062.

157.178   **Spatially–resolved high–velocity outflow from the nucleus of M51.**
G. Cecil.
Bull. Am. Astron. Soc., Vol. 16, No. 4, p. 956 (1984). Abstract. –
See Abstr. 010.062.

157.179   **Radio structure of the elliptical galaxy MCG 5–4–18.**
J. M. Wrobel, R. S. Simon.
Bull. Am. Astron. Soc., Vol. 16, No. 4, p. 956 (1984). Abstract. –
See Abstr. 010.062.

157.180   **Companions of radio elliptical galaxies.**
K. A. Kingham.
Bull. Am. Astron. Soc., Vol. 16, No. 4, p. 956 – 957 (1984). Abstract. – See Abstr. 010.062.

157.181   **Multicolor near–infrared mapping of star–forming regions in interacting galaxies.**
M. Joy, P. M. Harvey.
Bull. Am. Astron. Soc., Vol. 16, No. 4, p. 960 (1984). Abstract. –
See Abstr. 010.062.

157.182   **H I in NGC 7241: evidence for a hidden companion.**
R. Giovanelli, M. P. Haynes.
Bull. Am. Astron. Soc., Vol. 16, No. 4, p. 961 (1984). Abstract. –
See Abstr. 010.062.

157.183   **The discovery of an optical tail in NGC 7236/37.**
J. J. E. Hayes, N. F. Comins.
Bull. Am. Astron. Soc., Vol. 16, No. 4, p. 962 (1984). Abstract. –
See Abstr. 010.062.

157.184   **The evolution of spiral galaxies and uncertainties in interpreting galaxy counts.**
C. R. King, R. S. Ellis.
Bull. Am. Astron. Soc., Vol. 16, No. 4, p. 963 – 964 (1984). Abstract. – See Abstr. 010.062.

157.185   A VLA study of M81.
F. Bash, M. Kaufman.
Bull. Am. Astron. Soc., Vol. 16, No. 4, p. 976 (1984). Abstract. – See Abstr. 010.062.

157.186   The chemical composition gradient in M81.
D. R. Garnett, G. A. Shields.
Bull. Am. Astron. Soc., Vol. 16, No. 4, p. 976 (1984). Abstract. – See Abstr. 010.062.

157.187 ^{12}CO and ^{13}CO emission from the galaxy NGC 5055.
D. Chance, J. Bally, B. G. Elmegreen.
Bull. Am. Astron. Soc., Vol. 16, No. 4, p. 977 (1984). Abstract. – See Abstr. 010.062.

157.188   Aperture synthesis CO observations of M51.
K.-Y. Lo, G. L. Berge, M. J. Claussen,
G. M. Heiligman, J. Keene, R. B. Leighton, C. R. Masson,
A. T. Moffet, T. G. Phillips, A. I. Sargent, S. L. Scott,
N. Z. Scoville, D. M. Watson, D. P. Woody.
Bull. Am. Astron. Soc., Vol. 16, No. 4, p. 977 (1984). Abstract. – See Abstr. 010.062.

157.189   Hα observations of novae in M31.
R. Ciardullo, H. C. Ford, G. H. Jacoby, J. D. Neill,
A. Shafter.
Bull. Am. Astron. Soc., Vol. 16, No. 4, p. 977 (1984). Abstract. – See Abstr. 010.062.

157.190   Images of M31 in the rocket ultraviolet: the central bulge.
T. P. Stecher, R. C. Bohlin, R. H. Cornett, J. K. Hill, R. S. Hill,
R. W. O'Connell.
Bull. Am. Astron. Soc., Vol. 16, No. 4, p. 977 (1984). Abstract. – See Abstr. 010.062.

157.191   Two color rocket ultraviolet images of M33.
W. B. Landsman, R. C. Bohlin, R. H. Cornett,
J. K. Hill, T. P. Stecher.
Bull. Am. Astron. Soc., Vol. 16, No. 4, p. 977 (1984). Abstract. – See Abstr. 010.062.

157.192   Observations of two novae in M87.
S. van den Bergh, C. Pritchet.
Bull. Am. Astron. Soc., Vol. 16, No. 4, p. 989 (1984). Abstract. – See Abstr. 010.062.

157.193   Optical properties of IRAS galaxies in the Coma and Hercules clusters.
R. L. Seaman, S. E. Strom, E. T. Young.
Bull. Am. Astron. Soc., Vol. 16, No. 4, p. 989 – 990 (1984). Abstract. – See Abstr. 010.062.

157.194   A late–forming galaxy.
W. C. Keel.
Bull. Am. Astron. Soc., Vol. 16, No. 4, p. 990 (1984). Abstract. – See Abstr. 010.062.

157.195   Ring galaxies: a radio and optical investigation of closely interacting systems.
N. Jeske, M. A. Stevens, M. Davis, C. Heiles.
Bull. Am. Astron. Soc., Vol. 16, No. 4, p. 990 (1984). Abstract. – See Abstr. 010.062.

157.196   The X–ray emission from normal early–type galaxies: a statistical analysis.
G. Trinchieri, G. Fabbiano.
Bull. Am. Astron. Soc., Vol. 16, No. 4, p. 990 (1984). Abstract. – See Abstr. 010.062.

157.197   New insights into the physical properties and evolution of galaxies.
A. F. Davidsen.
Bull. Am. Astron. Soc., Vol. 16, No. 4, p. 1011 (1984). Abstract. – See Abstr. 010.062.

157.198   The brightest stars in nearby galaxies. V. Cepheids and the brightest stars in the dwarf galaxy Sextans B compared with those in Sextans A.
A. Sandage, G. Carlson.
Astron. J., Vol. 90, No. 6, p. 1019 – 1026 (1985). With plates 81 – 83.

Photometry of the brightest resolved red and blue stars and of four of seven Cepheids found in the dwarf galaxy Sextans B is compared with similar data for Sextans A. The Cepheids in Sextans B have periods between 8 and 28 days, compared with 10 to 25 days for the five Cepheids known in Sextans A. Combining the data for the nine Cepheids of known period shows both galaxies to be at the same distance of $(m–M)_{AB} = 26.2 \pm \sim 0.2$. The derived absolute magnitudes of the three brightest stars are $\langle M_V(3) \rangle = -7.26$ and -7.92 for the *red supergiants* (at random phase in their variability) for Sextans B and A, respectively, and $\langle M_B(3) \rangle = -6.75$ and -7.53 for the *blue supergiants*, again in B and A, respectively.

157.199   Analysis of high–resolution velocity fields – NGC 6503.
G. A. van Moorsel, D. C. Wells.
Astron. J., Vol. 90, No. 6, p. 1038 – 1045 (1985). With plate 84.

A nonlinear least–squares fitting routine is introduced to analyze velocity fields of galaxies, and to determine the orbital parameters of these galaxies. The new method is shown to require less unnecessary user interaction than previous methods do, and lead to more accurate and unambiguous results. The new method is applied to the velocity field of NGC 6503, which is shown to be globally very regular. On the other hand, interpretation of the residual velocity field indicates that there are regions in this galaxy that show substantial noncircular velocities.

157.200   Peculiar motions and star formation in the interacting galaxy complex Mk 171 = NGC 3690 + IC 694.
R. Augarde, J. Lequeux.
Astron. Astrophys., Vol. 147, No. 2, p. 273 – 280 (1985).

Spectroscopic observations in the visible and in the far ultraviolet are combined with other existing information to interpret the properties of the exceptionally active extended star formation in Mk 171. The authors show that this system corresponds to the interaction, and probably to the direct collision, of two galaxies with a total mass of $\lesssim 4 \times 10^{10} M_\odot$. All the properties of the system are best accounted for by the formation of O stars only in the approximate mass range 20 – 35 $M_\odot$, which has started about 1.5×10^7 yr ago. The total mass of stars formed is of the order of $5 \times 10^7 M_\odot$, to be compared with a total mass of atomic hydrogen of $1.4 \times 10^9 M_\odot$ and a mass of molecular gas of perhaps $10^{10} M_\odot$.

157.201   Discovery of the first S star in NGC 6822.
M. Aaronson, J. Mould, K. H. Cook.
Prepr. Steward Obs., No. 563, 17 pp. (1985). To appear in Astrophys. J., Lett. Ed.

157.202   Hot coronae around early–type galaxies.
W. Forman, C. Jones, W. Tucker.
Astrophys. J., Vol. 293, No. 1, p. 102 – 119 (1985).

The analysis of the X–ray emission from a sample of 55 bright early–type galaxies shows that hot gaseous coronae are a common and perhaps ubiquitous feature of such systems. The X–ray emission can be explained most naturally as thermal bremsstrahlung from hot gas ($kT \approx 0.5 – 1.5$ keV) which may be accumulated from mass loss during normal stellar evolution. The authors describe the X–ray properties of these coronae (gas mass, temperature, and extent) and discuss their implications for the presence of massive dark halos around individual early–type galaxies. They find total masses up to $5 \times 10^{12} M_\odot$ and mass-to-light ratios up to ~ 100 (in solar units) for these galaxies.

157.203   Some possible regularities in the missing mass problem.
J. N. Bahcall, S. Casertano.
Astrophys. J., Lett. Ed., Vol. 293, No. 1, p. L7 – L10 (1985).

The unseen matter in a sample of spiral galaxies exhibits simple regularities and characteristic numerical values.

157.204   Infrared observations of dwarf elliptical galaxies in the Virgo cluster.
H. Zinnecker, R. D. Cannon, T. G. Hawarden, H. T. MacGillivray.
Bull. Am. Astron. Soc., Vol. 17, No. 1, p. 522 (1985). Abstract. – See Abstr. 010.064.

157.205   Diameters of galaxies inside and outside clusters.
A. Bosma.
The Virgo cluster of galaxies, p. 425 – 426 (1985). – See Abstr. 012.069.

The author examines the following hypothesis: spiral galaxies in clusters are smaller than those in the field due to truncation. Evidence for this comes from comparing spirals in Virgo with field spirals.

157.206   Core and tidal radii of the Carina dwarf spheroidal galaxy from UK Schmidt Telescope plates.
P. J. Godwin.
Dynamics of star clusters, p. 77 – 79 (1985). – See Abstr. 012.070 (IAU Symp. No. 113).

Core and tidal radii of the Carina dwarf galaxy are determined by fitting King dynamical models to number count radial profiles, derived from COSMOS data. These values are compared with those of the other six known Local Group dwarf spheroidals.

157.207   Hydroxyl absorption in NGC 520, NGC 2623, and NGC 6240.
W. A. Baan, A. D. Haschick, D. Buckley, J. T. Schmelz.
Astrophys. J., Vol. 293, No. 2, p. 394 – 399 (1985).

Extragalactic absorption of hydroxyl is reported for the galaxies NGC 520, NGC 2623, and NGC 6240. H I absorption has also been detected in these galaxies. The properties of the H I and OH absorption features are compared for these galaxies and for other galaxies with absorption in both OH and H I. The characteristics of extragalactic OH absorption lines are consistent with the existence of dense molecular disks in the inner parts of the galaxies. An asymmetry in a number of extragalactic OH absorption lines suggests a peculiar velocity structure for the absorbing molecular disk.

157.208   Radio detection of historical supernovae and H II regions in M83.
J. J. Cowan, D. Branch.
Astrophys. J., Vol. 293, No. 2, p. 400 – 406 (1985). With plate 23.

VLA maps of the galaxy M83 (NGC 5236) at 6 and 20 cm reveal the presence of both nonthermal and thermal sources, lying predominantly along the inner edges of the optical spiral arms. A radio source coincident with the optical position of supernova 1957d is found to have a nonthermal spectrum; thus SN 1957d is confirmed as the first supernova of intermediate age ($\sim 10 - 300$ yr) to be detected at any wavelength. Three other nonthermal sources are probably supernova remnants. A composite "radio light curve" for supernovae and young supernova remnants of known age is pesented and discussed. Eight thermal sources also are listed, and five of these are identified with optical H II regions in M83.

157.209   Surface photometry of the Sculptor Group galaxies: NGC 7793, NGC 247, and NGC 300.
C. Carignan.
Astrophys. J., Suppl. Ser., Vol. 58, No. 1, p. 107 – 124 (1985). With plates 1 – 3.

Surface photometry from Schmidt plates of the three nearly pure disk galaxies NGC 7793, NGC 247, and NGC 300 is presented. In each case, the luminosity profile shows an almost purely exponential decrease out to the detection limit. The three galaxies cover an exceptionally wide range of disk surface brightness. The color maps reveal no color gradient. The luminosity profiles decomposed into red and blue pixel profiles show that the surface brightness of young blue stars falls off more gradually than for the old disk.

157.210   The spiral structure of galaxies.
E. Athanassoula.
Phys. Rep., Vol. 114, No. 5 – 6, p. 319 – 403 (1984). Abstr. in Phys. Abstr., Vol. 88, No. 1255, Entry 46209 (1985).

157.211   Velocity field and orientation of isolated galaxies in the Perseus and Pisces constellations.
N. N. Pavlova.
Tr. Astrofiz. Inst. Alma–Ata, Tom 40, p. 34 – 40 (1983). In Russian.

157.212   The dynamics of tri–axis elliptical Milky Way systems.
T. de Zeeuw.
Ned. Tijdschr. Natuurkd. A, Vol. A51, No. 1, p. 41 – 44 (1985). In Dutch. Abstr. in Phys. Abstr., Vol. 88, No. 1258, Entry 63499 (1985).

157.213   A determination of the radial velocities of galaxies.
V. S. Popov, G. D. Polyakova.
Stellar radial velocities, p. 397 – 398 (1985). – See Abstr. 012.095 (IAU Colloq. No. 88).

Fourteen galaxies were studied using 24 spetra, obtained with the 6 meter telescope of the Special Astrophysical Observatory of the USSR Academy of Sciences. Results of radial velocity determinations are presented here.

157.214   Individual masses of galaxies in pairs.
I. D. Karachentsev.
Astron. Zh., Tom 62, Vyp. 3, p. 417 – 431 (1985). In Russian. English translation in Sov. Astron., Vol. 29, No. 3.

A list of individual mass estimations is presented for 227 components of isolated pairs of galaxies from the catalogue compiled by the author. For the mass determination, the following data are used: rotation curves of galaxies, the 21–cm line profile width, the radial–velocity dispersion in the central parts of galaxies.

157.215   Gas equilibrium conditions in the gravitational field of massive central galaxies in clusters.
E. V. Volkov.
Astron. Zh., Tom 62, Vyp. 3, p. 450 – 458 (1985). In Russian. English translation in Sov. Astron., Vol. 29, No. 3.

The restrictions imposed by thermal conductivity on the gas distribution in a cD galaxy's gravitational field in the framework of the hydrostatic model are considered. A comparison of the possibility of existence of the hydrostatic regime in cD galaxies and giant ellipticals is carried out. It is shown that for the gas parameters, suggested earlier when interpreting the X–ray data on cD galaxies, steep temperature and density gradients must exist in the galaxy's center. The analysis of the gas density gradient far away from the center of M87 indicates the existence of an extended halo.

157.216   Carbon and M stars in NGC 205.
H. B. Richer, D. R. Crabtree, C. J. Pritchet.
Cool stars with excesses of heavy elements, p. 171 – 174 (1985). – See Abstr. 012.101.

Broadband *VRI* and narrowband CCD frames were obtained at CFHT of the central region of NGC 205, the dwarf elliptical companion of the Andromeda galaxy. The narrowband images isolate TiO and CN and colors formed from the photometry of these images allow carbon and M stars to be unambiguously separated. Among the stars with reliable photometry, 7 carbon stars were found together with 21 late M stars with spectral types as cool as $\sim$ M5. The presence of carbon stars in NGC 205 implies that star formation has been an ongoing process in this dwarf elliptical galaxy.

157.217   Carbon stars in the Carina dwarf galaxy.
R. D. Cannon.
Cool stars with excesses of heavy elements, p. 175 – 179 (1985). – See Abstr. 012.101.

This note summarises recently obtained data on carbon stars in the Carina dwarf spheroidal galaxy, and the implications for the origin and evolution both of carbon stars and of dwarf galaxies.

157.218 The link between the X–ray and the radio emission of spiral and irregular galaxies.
G. Fabbiano, G. Trinchieri.
Bull. Am. Astron. Soc., Vol. 17, No. 2, p. 548 (1985). Abstract. – See Abstr. 010.065.

157.219 Optical spectrophotometry of IRAS galaxies.
M. Simon.
Bull. Am. Astron. Soc., Vol. 17, No. 2, p. 548 (1985). Abstract. – See Abstr. 010.065.

157.220 CO maps of IR bright galaxies.
J. S. Young, F. P. Schloerb, J. Kenney, S. Lord.
Bull. Am. Astron. Soc., Vol. 17, No. 2, p. 548 (1985). Abstract. – See Abstr. 010.065.

157.221 CO map of the Sc galaxy IC 342.
L. Sage, P. M. Solomon.
Bull. Am. Astron. Soc., Vol. 17, No. 2, p. 548 (1985). Abstract. – See Abstr. 010.065.

157.222 An Arecibo survey for extragalactic hydroxyl.
J. T. Schmelz, W. A. Baan, A. D. Haschick.
Bull. Am. Astron. Soc., Vol. 17, No. 2, p. 549 (1985). Abstract. – See Abstr. 010.065.

157.223 Arecibo observations of IRAS galaxies at 18 and 21 cm wavelengths.
J. M. Dickey, R. Garwood, G. Helou, J. Houck.
Bull. Am. Astron. Soc., Vol. 17, No. 2, p. 579 (1985). Abstract. – See Abstr. 010.065.

157.224 Faint object CCD survey.
P. Seitzer, J. A. Tyson.
Bull. Am. Astron. Soc., Vol. 17, No. 2, p. 580 (1985). Abstract. – See Abstr. 010.065.

157.225 Re–assessing the primordial helium abundance: observations of NGC 4861 and CG1116 + 51.
H. L. Dinerstein, G. A. Shields.
Bull. Am. Astron. Soc., Vol. 17, No. 2, p. 581 (1985). Abstract. – See Abstr. 010.065.

157.226 Hot gas and massive halos around early type galaxies.
W. Forman, C. Jones.
Bull. Am. Astron. Soc., Vol. 17, No. 2, p. 586 (1985). Abstract. – See Abstr. 010.065.

157.227 Detailed X–ray study of NGC 4636.
G. Trinchieri, G. Fabbiano, J. Schwarz, V. Stanger.
Bull. Am. Astron. Soc., Vol. 17, No. 2, p. 586 (1985). Abstract. – See Abstr. 010.065.

157.228 Spectroscopy of shells around elliptical galaxies.
W. D. Pence.
Bull. Am. Astron. Soc., Vol. 17, No. 2, p. 586 (1985). Abstract. – See Abstr. 010.065.

157.229 Evidence for scalloping in other galaxies.
H. Karimabadi, L. Blitz.
Bull. Am. Astron. Soc., Vol. 17, No. 2, p. 586 (1985). Abstract. – See Abstr. 010.065.

157.230 Evidence for supernova regulation of metal enrichment in disk galaxies.
R. F. Wyse, J. Silk.
Bull. Am. Astron. Soc., Vol. 17, No. 2, p. 586 – 587 (1985). Abstract. – See Abstr. 010.065.

157.231 Mass accretion rates for giant elliptical galaxies.
W. Tucker, R. Rosner.
Bull. Am. Astron. Soc., Vol. 17, No. 2, p. 600 (1985). Abstract. – See Abstr. 010.065.

157.232 Observational signatures of tidal friction in interacting binary galaxies.
K. D. Borne, J. G. Hoessel.
Bull. Am. Astron. Soc., Vol. 17, No. 2, p. 601 (1985). Abstract. – See Abstr. 010.065.

157.233 Observations of galaxies at high redshift.
R. G. Kron.
Bull. Am. Astron. Soc., Vol. 17, No. 2, p. 605 (1985). Abstract. – See Abstr. 010.065.

157.234 Giant molecular clouds in M31.
L. Blitz, R. Mathieu.
Bull. Am. Astron. Soc., Vol. 17, No. 2, p. 607 – 608 (1985). Abstract. – See Abstr. 010.065.

157.235 Extended X–ray emission from early type galaxies.
J. Schwarz, G. Trinchieri, V. Stanger.
Bull. Am. Astron. Soc., Vol. 17, No. 2, p. 611 (1985). Abstract. – See Abstr. 010.065.

157.236 The CM relation for spiral bulges and their stellar populations.
C. R. King.
Bull. Am. Astron. Soc., Vol. 17, No. 2, p. 612 (1985). Abstract. – See Abstr. 010.065.

157.237 A barred spiral undergoing a collision – neutral hydrogen aperture synthesis observations of NGC 1313.
W. L. Peters.
Bull. Am. Astron. Soc., Vol. 17, No. 2, p. 612 (1985). Abstract. – See Abstr. 010.065.

157.238 Sub–millimeter observations of spiral galaxies.
T. L. Roellig, M. W. Werner, C. G. Seab, E. E. Becklin, C. Impey, J. Davidson, M. Dragovan.
Bull. Am. Astron. Soc., Vol. 17, No. 2, p. 612 (1985). Abstract. – See Abstr. 010.065.

157.239 Star formation rates in disk galaxies: comparison of far–infrared and H–alpha emission.
C. J. Lonsdale, G. Helou.
Bull. Am. Astron. Soc., Vol. 17, No. 2, p. 612 (1985). Abstract. – See Abstr. 010.065.

157.240 The detailed gas distribution of NGC 6946.
L. J. Tacconi, J. S. Young.
Bull. Am. Astron. Soc., Vol. 17, No. 2, p. 613 (1985). Abstract. – See Abstr. 010.065.

157.241 The correlation of CO and Hα in M51.
S. Lord, J. S. Young, S. E. Strom.
Bull. Am. Astron. Soc., Vol. 17, No. 2, p. 613 (1985). Abstract. – See Abstr. 010.065.

157.242 A detailed study of dust in NGC 205 and NGC 185.
J. S. Price, S. F. Mason, E. D. Loh, E. Spillar, G. L. Grasdalen.
Bull. Am. Astron. Soc., Vol. 17, No. 2, p. 613 (1985). Abstract. – See Abstr. 010.065.

157.243 Photometric characteristics of paired E and S0 galaxies.
V. V. Demin.
Sov. Astron., Vol. 28, No. 6, p. 622 – 627 (1984). English translation of 38.157.199.

157.244 Report of IAU Commission 28: Galaxies (*Galaxies*).
V. C. Rubin.
Trans. IAU, Vol. XIXA, p. 313 – 352 (1985). – See Abstr. 003.046.

157.245   A comparative study of dust and young stars in three small galaxies.
J. S. Price.
Diss. Abstr. Int., Sect. B, Vol. 45, No. 8, p. 2581 (1985). Thesis, University of Wyoming, 168 pp. (1984). Order No. DA8425936.

157.246   Gas dynamics of the barred spiral galaxy NGC 3359.
J. R. Ball.
Diss. Abstr. Int., Sect. B, Vol. 45, No. 9, p. 2955 (1985). Thesis, University of Florida, 299 pp. (1984). Order No. DA8429182.

157.247   Current problems of determining distances to galaxies.
A. Sandage, G. A. Tammann.
Supernovae as distance indicators, p. 1 – 13 (1985). – See Abstr. 012.107.

Traditional methods of determining distances to remote galaxies are reviewed to show their relative strengths and weaknesses. Methods that use the luminosity function of globular clusters are rejected as being precise at only the $\pm 1^m$ level. Similar rejection is made of H II region angular diameters. Methods of more promise, in inverse order of precision, are (1) the luminosity function of Sc I galaxies, (2) the linewidth–absolute magnitude (Tully–Fisher) method, (3) use of brightest red and blue supergiants, and (4) use of Type I supernovae. Application of each favors $H_0 \cong 50$ km s^{-1}Mpc^{-1} for the Hubble constant.

157.248   Investigations of interacting galaxies.
B. A. Vorontsov–Vel'yaminov.
Main directions of astronomical investigations at the Moscow University, p. 90 – 95 (1985). In Russian. – See Abstr. 003.049.

157.249   Can ram pressure distort spiral arms?
M. Iye.
Theoretical aspects on structure, activity, and evolution of galaxies: III, p. 74 – 78 (1985). – See Abstr. 012.110.

Dynamical effects of ram pressure exerted on a spiral galaxy running through the intracluster plasma are studied. The ram pressure component parallel to the disk plane of the galaxy can distort the spiral structure of interstellar gas asymmetrically even if this component is one order of magnitude smaller than that required to cause the stripping process. An evaluation for the case of NGC 4254 in the Virgo cluster, however, indicates that this ram pressure distortion is hardly sufficient to explain the observed asymmetry of spiral arms of this galaxy.

157.250   On cepheids in the galaxy M31.
O. A. Mel'nikov.
Izv. Glav. Astron. Obs. Pulkovo, Astrometr. Astrofiz., No. 201, p. 104 – 109 (1985). In Russian.

Using known observations by W. Baade, who gave photometric parameters of variable stars, in particular the cepheids in M31, the following dependencies have been calculated on the basis of 17 long–period cepheids: $<V^0> = $ f(lg P) (inclination –2.40) and $<(B–V)^0> = $ f(lg P) (inclination +0.30). Wesselink's dependence and the surface brightness–color and radius–period relations for cepheids in the Galaxy were used for determining the period–luminosity relation $P \sim L$: $M_V = -1.84$–2.37 lg P. This inclination coincides with that obtained using visual magnitudes. Known M_V and $<V^0>$ lead to a distance modulus of M31 equal to 24.34.

157.251   A spectrophotometric investigation of a superassociation in the galaxy NGC 2820 A.
N. K. Andreasyan, Eh. E. Khachikyan.
Astrofizika, Tom 22, Vyp. 3, p. 441 – 448 (1985). In Russian. English translation in Astrophysics, Vol. 22, No. 3.

The results of spectral and densitometric investigations of the galaxy NGC 2820 A with UV excess are presented. On U, B, V photos it appears to be an irregular galaxy with two condensations. A spectrophotometric investigation of the brightest one – the object Mark 108 – has been carried out. The physical parameters in it are nearly the same as in normal H II regions and there is some underabundance of heavy elements in Mark 108 compared with them.

157.252   On the separation of components of radiation of variable extragalactic sources.
V. A. Hagen–Thorn.
Astrofizika, Tom 22, Vyp. 3, p. 449 – 460 (1985). In Russian. English translation in Astrophysics, Vol. 22, No. 3.

Within the framework of a two–component model of variable extragalactic sources (normal galaxy + variable point source with constant spectral energy distribution) a method is proposed of separation of the components of its radiation. The method is illustrated by several examples of its application.

157.253   Results from surface photometry of extensive samples of galaxies.
S. Okamura.
New aspects of galaxy photometry, p. 39 – 42 (1985). – See Abstr. 012.114.

A summary of the results from photographic photometry of 261 galaxies, obtained at the Kiso Schmidt telescope, is presented.

157.254   Accuracy and improvements in galaxy photometry: why and how.
M. Capaccioli.
New aspects of galaxy photometry, p. 53 – 66 (1985). – See Abstr. 012.114.

This paper discusses the present state of surface photometry of elliptical galaxies and considers the future improvements in photometric accuracy, which are necessary for a more detailed understanding of the physical parameters of these systems.

157.255   Photometry and radio continuum observations of galaxies.
R.–J. Dettmar, R. Beck, R. Wielebinski.
New aspects of galaxy photometry, p. 79 – 84 (1985). – See Abstr. 012.114.

This paper summarizes briefly how a comparison of optical and radio continuum observations opens new insight into the physics of active star formation regions and the interstellar medium of spiral galaxies.

157.256   Galaxy photometry in the infrared.
A. F. M. Moorwood.
New aspects of galaxy photometry, p. 85 – 94 (1985). – See Abstr. 012.114.

This paper briefly reviews applications and results of infrared photometry in galaxy research. Emphasis is placed primarily on recent work in the areas of star formation and cosmology.

157.257   Galaxy photometry and X–ray astronomy.
A. C. Fabian.
New aspects of galaxy photometry, p. 95 – 100 (1985). – See Abstr. 012.114.

Many nearby normal galaxies have been detected as X–ray sources by the Einstein Observatory. The X–ray emission from the normal stellar component is presumably negligible but populations of X–ray binaries and supernova remnants as well as any hot interstellar medium were observable. The emission from late–type galaxies, which is mostly due to X–ray binaries, is discussed here, then a more detailed account is given of the discovery and implications of substantial amounts of ~ 10 million degree interstellar gas in early–type galaxies. The density of much of this hot gas is high enough that a cooling flow may occur with consequent high–pressure star formation.

157.258   Problems of galactic and extragalactic CO photometry.
F. P. Israel.
New aspects of galaxy photometry, p. 101 – 110 (1985). – See Abstr. 012.114.

Extragalactic CO studies yield information on the distribution and kinematics of dense and usually large molecular cloud complexes as well as information on the overall molecular content of galaxies. This paper reviews our present knowledge of the molecular cloud content of the Galaxy and of nearby galaxies.

157.259 A study of star formation in Shapley–Ames spiral galaxies based on IRAS far–infrared observations.
K. Brink, T. de Jong.
New aspects of galaxy photometry, p. 111 – 114 (1985). – See Abstr. 012.114.

IRAS far–infrared observations of an optically complete sample of galaxies are used to compare star formation rates in spiral galaxies in the field, in small groups and in the Virgo cluster.

157.260 The three–dimensional shape of galaxies.
F. Bertola.
New aspects of galaxy photometry, p. 131 – 136 (1985). – See Abstr. 012.114.

This paper discusses observational tests of the intrinsic three–dimensional shape of elliptical galaxies and reviews briefly indications for departure from oblateness in disk galaxies.

157.261 Dynamics of ellipticals: the case for 2–dimensional photometry.
R. Bacon, G. Monnet.
New aspects of galaxy photometry, p. 137 – 144 (1985). – See Abstr. 012.114.

Observed mean rotation curves $V(r)$ and velocity dispersion curves $\sigma(r)$ for elliptical galaxies are compared with predictions from isotropic oblate models.

157.262 Fine structure in elliptical galaxies.
F. Schweizer, W. K. Ford Jr.
New aspects of galaxy photometry, p. 145 – 150 (1985). – See Abstr. 012.114.

The authors have been conducting a survey of 36 field giant elliptical galaxies with the aim of finding fine structure in them. Emphasis is on structures that may be signatures of merger activity, and on correlations between them and the presence of isophotal twists, color anomalies, and abnormal M/L ratios. The authors discuss here the analysis of the direct images obtained in the survey. First they describe a digital masking technique that is closely related to the photographic unsharp masking. Then they briefly present some statistics from the field–elliptical survey and interpret the results.

157.263 Blue and near–infrared surface photometry of 50 barred and nonbarred spiral galaxies.
B. G. Elmegreen, D. M. Elmegreen.
New aspects of galaxy photometry, p. 161 – 166 (1985). – See Abstr. 012.114.

Fifty spiral galaxies of various Hubble types and spiral arm classes were observed with the 1.2 m Palomar Schmidt telescope in blue (4360 Å) and near–infrared (8250 Å) passbands. Surface brightnesses were derived from the plates using a microdensitometer. This paper presents a summary of the major results from an analysis of the spiral structure of these galaxies.

157.264 Dark halos around late–type galaxies.
G. Comte.
New aspects of galaxy photometry, p. 169 – 174 (1985). – See Abstr. 012.114.

In order to investigate the mass–to–light ratio in irregular Magellanic galaxies, a sample of 21 Sdm, Sm, Im objects, for which high–resolution velocity fields observed in 21 cm line are available in the literature, has been selected. The rotation curves of these galaxies were analysed using two–component models consisting of disks and surrounding spherical dark halos. Implications for the particle mass of hypothetical massive neutrinos are discussed.

157.265 Galaxy photometry and cosmology.
G. Bruzual A.
New aspects of galaxy photometry, p. 187 – 192 (1985). – See Abstr. 012.114.

A brief review is presented of the current understanding of galaxy photometry as related to cosmology. Attention is paid to the conclusions about spectral evolution of galaxies that can be derived from the deepest faint galaxy samples currently available. Some typical results are discussed.

157.266 Far–UV to infrared photometric star formation tracers.
B. Rocca–Volmerange.
New aspects of galaxy photometry, p. 199 – 210 (1985). – See Abstr. 012.114.

A short review of quantitative photometric tracers of star formation and their limits is given. The author separates distance–dependent indicators of current star formation rates (SFR) from those (colors and equivalent widths) which, distance independent, yield access to the present/past SFR ratio. Extinction and initial mass function are considered. A comparison of SFR values from various sources shows agreement at a local (per unit mass) but not global scale while no discrepancy appears for present/past SFR ratio.

157.267 Evolution of spiral galaxies in a cluster environment.
B. Guiderdoni.
New aspects of galaxy photometry, p. 211 – 214 (1985). – See Abstr. 012.114.

The H I deficiency and the integrated colors of spiral galaxies in the Virgo Cluster are studied by means of a model of photometric evolution.

157.268 Galaxy photometry at faint light levels – interaction with the environment.
D. Carter.
New aspects of galaxy photometry, p. 215 – 222 (1985). – See Abstr. 012.114.

Galaxies interact with their environment in many ways, and many of these interactions produce features and effects at low surface brightnesses which can be diagnostics for those interactions. This paper discusses the morphology of some of these features (tidal tails, loops, shells, ripples, dust lanes) and their implications for possible interactions of galaxies with neighbouring galaxies or intergalactic matter.

157.269 Malmquist bias in "Tully–Fisher" relation.
L. Bottinelli, L. Gouguenheim, G. Paturel, P. Teerikorpi.
New aspects of galaxy photometry, p. 223 – 226 (1985). – See Abstr. 012.114.

An analysis of the Tully–Fisher relation for 41 spiral galaxies leads to an unbiased estimate of the Hubble constant $H_0 = 72 \pm 3$ km s^{-1}Mpc^{-1}.

157.270 Possible evidence for evolution in the photometric properties of nearby galaxies.
R. B. Tully.
New aspects of galaxy photometry, p. 237 (1985). Abstract. – See Abstr. 012.114.

157.271 Photographic surface photometry of the Andromeda galaxy.
R. A. M. Walterbos, R. C. Kennicutt Jr.
New aspects of galaxy photometry, p. 245 – 248 (1985). – See Abstr. 012.114.

The authors obtained calibrated photographic plates in U, B, V, R, and Hα of M31, using the Burrell Schmidt telescope at Kitt Peak and the Palomar Schmidt. The plates have been scanned at Leiden with the ASTROSCAN densitometer at a resolution of 4″. Here the authors discuss some of the specific reduction techniques used to process the large amount of optical data, and a few preliminary results regarding the warp and the interstellar dust distribution are presented.

157.272 H I holes in the interstellar medium of Messier 31.
E. Brinks.
New aspects of galaxy photometry, p. 249 – 252 (1985). – See Abstr. 012.114.

Recently a high resolution (24″ × 36″) survey of the Andromeda galaxy in the 21–cm line of neutral hydrogen has led to the detection of structures in the interstellar medium which can best be referred to as H I holes. The H I holes in M31 show a clear resemblance to the H I shells and supershells which have been found in our Galaxy and to the superbubbles in the LMC. This leads to the conclusion that H I holes seem to be quite a common phenomenon in spiral galaxies.

157.273   High resolution CO observations in M31.
F. Boulanger, J. Bystedt, F. Casoli, F. Combes.
New aspects of galaxy photometry, p. 253 – 256 (1985). – See Abstr. 012.114.

CO observations towards M31 made with the Onsala 20 m telescope, reveal a high degree of inhomogeneity of the emission on the 100 pc scale. When comparing the CO observations made towards the same spiral arm but with lower resolution, two kinds of emission appear: with the small beam (33″) the authors detected a narrow and bright component, corresponding to giant molecular complexes; a fainter and broader emission appears only at low resolution, or when adjacent spectra are averaged. This wider emission is interpreted as coming from a lot of small clouds spread all over the beam and thus sharing all the velocities expected from the H I emission.

157.274   Morphological investigation of elliptical galaxies.
S. Djorgovski, M. Davis, S. Kent.
New aspects of galaxy photometry, p. 257 – 262 (1985). – See Abstr. 012.114.

The authors describe the two large surface photometry surveys of elliptical galaxies, completed at Mt. Hopkins and Lick observatories. These comprise the largest such surveys of early–type galaxies to this date, and will provide a large set of uniform quality data for a variety of statistical and dynamical studies. The authors discuss briefly some results from the Mt. Hopkins survey.

157.275   A comparison of the ellipticity profiles of E and S0 galaxies.
R. Michard.
New aspects of galaxy photometry, p. 263 – 264 (1985). Abstract. – See Abstr. 012.114.

157.276   The lenticular NGC 3115: a standard for galaxy photometry.
M. Capaccioli, E. V. Held, J.-L. Nieto.
New aspects of galaxy photometry, p. 265 – 268 (1985). – See Abstr. 012.114.

Preliminary results of extensive high–resolution photographic surface photometry in the B band are reported for the lenticular galaxy NGC 3115.

157.277   Dust in elliptical galaxies – how often, how much?
E. M. Sadler, O. E. Gerhard.
New aspects of galaxy photometry, p. 269 – 274 (1985). – See Abstr. 012.114.

The authors report results showing that dust–lanes are a common feature of nearby elliptical galaxies. About 40% of ellipticals have dust–lanes with diameters of a few kpc and dust masses of $10^4 - 10^6 M_\odot$. This result has some possible consequences for galaxy surface photometry, since dust–lanes too small to be seen on photographic plates may produce both ″twisting″ of inner isophotes and off–centre nuclei.

157.278   Liller's photometry revisited.
M. Capaccioli, R. Rampazzo.
New aspects of galaxy photometry, p. 275 – 278 (1985). – See Abstr. 012.114.

This paper reports on the determination of corrections to Liller's (1960 and 1966) B band surface photometry of 22 elliptical galaxies in the Virgo cluster.

157.279   Radio studies of peculiar galaxies at 10.7 GHz.
M. Urbanik, R. Gräve, U. Klein.
New aspects of galaxy photometry, p. 283 – 286 (1985). – See Abstr. 012.114.

Four objects from Arp's (1966) ″Atlas of Peculiar Galaxies″ have been mapped at 10.7 GHz using the Effelsberg 100–m radio telescope. The galaxies observed represent two distinct classes of interaction with their companions. NGC 520 and NGC 3448 represent highly disturbed objects with long optical tails. The two

other objects, NGC 2276 and NGC 3627, are probably interacting galaxies within looser groups. The integrated 10.7 GHz flux densities of all four galaxies were used to compute the average disk brightness temperature and the radio index R defined as the ratio of the radio to the optical luminosities.

157.280   The velocity field of NGC 1365.
S. Jörsäter.
New aspects of galaxy photometry, p. 287 – 288 (1985). – See Abstr. 012.114.

The velocity field of the barred spiral NGC 1365 has been mapped using a large number of optical spectra.

157.281   Morphology and radial velocity of galaxies using color–color diagrams.
G. I. Thompson, K. Nandy.
New aspects of galaxy photometry, p. 289 – 292 (1985). – See Abstr. 012.114.

It is known that the position of a galaxy in a color–color diagram contains information on its morphology and radial velocity. The authors report on a program to measure colors of faint galaxies on Schmidt and AAT plates, preferably to the plate limit, to add another dimension to work that has already been done on the distribution of faint galaxies.

157.282   The z–distribution of radio continuum emission in NGC 891.
R. J. Allen, F. X. Hu.
New aspects of galaxy photometry, p. 293 – 296 (1985). – See Abstr. 012.114.

The authors have combined the most recent observations at 6–cm made by Strong and Allen (1982) with the highest resolution 21–cm data obtained from new H I synthesis observations by Sancisi for constructing a simple two–component model for the z–distribution of radio emission from the galaxy NGC 891.

157.283   An H I and optical study of the gas poor Virgo cluster spiral NGC 4571.
J. M. van der Hulst, R. C. Kennicutt.
New aspects of galaxy photometry, p. 303 – 308 (1985). – See Abstr. 012.114.

Results of VLA observations and photographic surface photometry are presented for the spiral galaxy NGC 4571 in the Virgo cluster. Radial distributions of the blue surface brightness, H I column density and the surface density of H II regions are shown. The very low total gas content and the low star formation rate are confirmed. Possible gas depletion mechanisms are briefly outlined.

157.284   Mass–to–light ratio of elliptical galaxies.
R. Bacon.
New aspects of galaxy photometry, p. 317 – 318 (1985). – See Abstr. 012.114.

Mass–to–light ratios (M/L_B) have been calculated for 197 elliptical galaxies. No correlation could be found between M/L_B and the total luminosity.

157.285   Deconvolution of the elliptical nucleus of M31.
O. Bendinelli, G. Parmeggiani, F. Zavatti.
New aspects of galaxy photometry, p. 319 – 320 (1985). – See Abstr. 012.114.

The authors present a deconvolved brightness distribution of the nucleus of M31 obtained from Stratoscope II data.

157.286   Distance of galaxies from the method of ″sosies″.
L. Bottinelli, L. Gouguenheim, G. Paturel, P. Teerikorpi, I. Vauglin.
New aspects of galaxy photometry, p. 321 – 322 (1985). – See Abstr. 012.114.

From a large body of data the authors have selected ″sosie″ galaxies having the same apparent axis ratio, the same morphological type and the same velocity dispersion. Applying the Faber–Jackson relation and using 31 calibrating galaxies in the Virgo cluster, distance moduli for 97 galaxies have been estimated.

157.287   Mass to luminosity ratio of bars in spiral galaxies.
M. F. Duval, G. Monnet.
New aspects of galaxy photometry, p. 323 – 324 (1985). – See Abstr. 012.114.

Mass–to–light ratios of the bars have been derived for 5 galaxies ranging in type from SBb to SBm.

157.288   Systematic errors in Zwicky's magnitudes.
G. Fasano.
New aspects of galaxy photometry, p. 325 – 328 (1985). – See Abstr. 012.114.

Zwicky's magnitudes m_{zw} in his "Catalogue of Galaxies and Clusters of Galaxies" (1961 – 1968) are compared with B_T magnitudes given by Sandage and Tammann in the "Revised Shapley–Ames Catalog of Bright Galaxies" (1981).

157.289   Surface photometry and detailed structure of the elliptical galaxy NGC 1199 and its compact companions.
J.-L. Nieto, F. Deschamps, A. Maury, T. Roudier.
New aspects of galaxy photometry, p. 339 – 340 (1985). – See Abstr. 012.114.

A photometric analysis of deep plates in blue light of the elliptical galaxy NGC 1199 and its high–redshift SW companion galaxy shows that the companion is strongly obscured by interstellar dust located in the dust lane of NGC 1199. It is concluded that the observations are consistent with a cosmological distance–redshift relation for NGC 1199 and its SW companion galaxy.

157.290   Absorbing matter in E and S0 galaxies.
P. Poulain.
New aspects of galaxy photometry, p. 343 – 344 (1985). – See Abstr. 012.114.

157.291   A peculiarity in the photometric profiles of some S0 galaxies.
F. Simien, R. Michard.
New aspects of galaxy photometry, p. 345 – 346 (1985). – See Abstr. 012.114.

The authors present B band surface brightness profiles of four highly–inclined S0 galaxies.

157.292   Star formation in irregular galaxies.
D. Hunter, S. Wolff.
Mercury, Vol. 14, No. 3, p. 76 – 79 (1985).

Catalog of CO observations of galaxies.
See Abstr. 002.012.

Sky catalogue 2000.0. Volume 2: Double stars, variable stars and nonstellar objects.
See Abstr. 002.019.

Atlas of dwarf galaxies in the region of M81 group.
See Abstr. 002.024.

A new version of the Abastumani compiled catalogue of galaxies on magnetic tape.
See Abstr. 002.042.

Supplement to the detailed bibliography on the surface photometry of galaxies.
See Abstr. 002.044.

An atlas of selected galaxies, with illustrations of photometric analyses.
See Abstr. 002.047.

A catalog of stellar velocity dispersions. I. Compilation and standard galaxies.
See Abstr. 002.048.

A compilation of UBVRI photometry for galaxies in the ESO/Uppsala catalogue.
See Abstr. 002.054.

A catalog of stellar velocity dispersions.
See Abstr. 002.076.

Nomenclature for objects in the galaxy M33.
See Abstr. 002.114.

A compilation of *UBVRI* photometry for galaxies in the ESO/Uppsala catalogue.
See Abstr. 002.121.

The universe of galaxies. Readings from Scientific American.
See Abstr. 003.007.

Galassie e quasars.
See Abstr. 003.114.

Surface photometry from the ESO/SRC Atlas.
See Abstr. 036.005.

The production of a 16–mm film of M31.
See Abstr. 036.009.

Correlation analyses of deep galaxy samples – III.
See Abstr. 036.067.

Automatic classification of galaxies into morphological types.
See Abstr. 036.182.

Galaxies seen through gravitational lenses.
See Abstr. 036.206.

Tests of the Fourier quotient method.
See Abstr. 036.210.

On using a space telescope to detect faint galaxies.
See Abstr. 036.214.

New trends in CCD photometry of galaxies.
See Abstr. 036.218.

A new filter system for CCD surface photometry.
See Abstr. 036.219.

Electronography: performances, results, expectations.
See Abstr. 036.220.

Food for the photometrists – faint galaxies revealed.
See Abstr. 036.221.

Computer–linked microdensitometers: performance, results and expectations.
See Abstr. 036.222.

High resolution in galaxy photometry and imaging.
See Abstr. 036.223.

Comparison of photographic and CCD surface photometry of galaxies.
See Abstr. 036.225.

Photographic photometry of 16000 galaxies on ESO blue and red survey plates.
See Abstr. 036.226.

Some methods for photometry of galaxies in clusters.
See Abstr. 036.227.

Photometric calibration and first results obtained with the UFT experiment on board the ASTRON satellite.
See Abstr. 051.066.

Neutral fermions as the missing mass matter in galactic halos.
See Abstr. 061.045.

The formation and evolution of carbon stars.
See Abstr. 065.062.

Spectroscopy of the winds from Hubble–Sandage stars in M31 and M33.
See Abstr. 112.073.

Spectroscopy of the winds from Hubble–Sandage stars in M31 and M33.
See Abstr. 112.115.

Spectral classification of individual stars in nearby galaxies.
See Abstr. 114.059.

Double–mode RR Lyrae stars in the Draco dwarf galaxy.
See Abstr. 122.022.

Magellanic Cloud and other extragalactic Cepheids: some current topics.
See Abstr. 122.067.

The search, discovery, and use of Cepheid variables as distance indicators: past procedures and future strategies.
See Abstr. 122.171.

Spectroscopy of anomalous Cepheids in the Sculptor dwarf galaxy.
See Abstr. 122.198.

Observations of two novae in M87.
See Abstr. 124.008.

Nova shells. II. Calibration of the distance scale using novae.
See Abstr. 124.009.

Observations of 14 supernovae.
See Abstr. 125.021.

Improved optical spectrophotometry of supernova remnants in M33.
See Abstr. 125.024.

Why are there no type II supernovae in irregular galaxies?
See Abstr. 125.047.

Optical and UV spectra of supernova remnants in M33.
See Abstr. 125.065.

Radio observations of historical, extragalactic supernovae.
See Abstr. 125.109.

Detecting supernova remnants in external galaxies.
See Abstr. 125.111.

Angular diameter determinations of radio supernovae and the distance scale.
See Abstr. 125.112.

The supernova of 1885 in M31.
See Abstr. 125.222.

Supernova 1983g and the distance to NGC 4753.
See Abstr. 125.281.

Far–infrared and submillimeter astronomical spectroscopy.
See Abstr. 131.214.

Periodic star formation activity and color change of galaxies.
See Abstr. 131.365.

Comments on the origins and variations of carbon and nitrogen in the ISM of galaxies as evident from IUE spectroscopy of H II regions.
See Abstr. 132.001.

IUE spectroscopy of extragalactic H II regions.
See Abstr. 132.002.

Structural similarities among nearby giant H II regions.
See Abstr. 132.006.

Gas and star content and spatial distribution in the giant extragalactic H II region Tol 89.
See Abstr. 132.014.

The form of the initial mass function in an H II complex in NGC 6946.
See Abstr. 132.019.

Spatial variations in the physical conditions in the giant extragalactic H II region NGC 5471.
See Abstr. 132.036.

Kinematic properties of three giant H II regions.
See Abstr. 132.054.

JHKL observations of *IRAS* sources – II.
See Abstr. 133.002.

Spectroscopic verification of suspected planetary nebulae. II.
See Abstr. 134.022.

Dynamical rules for barred spiral galaxies.
See Abstr. 151.002.

Testing for triaxiality with kinematic data.
See Abstr. 151.005.

Periodic orbits in a triaxial galaxy.
See Abstr. 151.006.

Changes of entropy and entropy flux induced by spiral structure in galaxies.
See Abstr. 151.007.

Application of a new gasdynamics code to gas flow problems in disk galaxies.
See Abstr. 151.008.

A barred galaxy: the inside viewpoint.
See Abstr. 151.028.

Anisotropic models of elliptical galaxies. II. The case of flattened isophotes.
See Abstr. 151.035.

Dynamical consequences of star collisions for core–envelope structure in red giants.
See Abstr. 151.041.

Gasdynamical calculations of preferred planes in prolate and triaxial galaxies. I. Case of no figure rotation.
See Abstr. 151.047.

On the morphology of spiral modes.
See Abstr. 151.059.

N–body simulations of the peculiar ring–tail galaxy NGC 5614–15.
See Abstr. 151.062.

Structure of box–shaped bulges and other spheroidal components.
See Abstr. 151.065.

Distribution functions for spherical galaxies.
See Abstr. 151.066.

Dynamical evolution in galactic disks.
See Abstr. 151.069.

The structure and dynamics of ringed galaxies.
See Abstr. 151.071.

Black holes and the shapes of galaxies.
See Abstr. 151.072.

Equipartition in multicomponent gravitational systems.
See Abstr. 151.077.

Gas dynamics in a polar ring galaxy.
See Abstr. 151.080.

Evolution of stellar systems supplied by external stars: a model for galactic nuclei.
See Abstr. 151.081.

The structure and dynamics of ringed galaxies.
See Abstr. 151.160.

Dissipative structures in galaxies – effects of advection.
See Abstr. 151.166.

N–body simulation of giant molecular clouds in a galaxy.
See Abstr. 151.167.

Contemporary dynamical problems – a possible contribution to their solution by galaxy photometry.
See Abstr. 151.174.

Can shells help to distinguish prolate from oblate elliptical galaxies?
See Abstr. 151.175.

Constraints on the potential of barred galaxies from surface photometry and direct photography.
See Abstr. 151.176.

N–body simulations and galaxy photometry.
See Abstr. 151.177.

Constructing distribution functions for galaxies with application to M87.
See Abstr. 151.178.

The velocity fields of barred spirals.
See Abstr. 151.180.

Observational constraints on the ages and abundances of old stellar populations.
See Abstr. 154.007.

Search for (globular) clusters in nearby galaxies: a progress report.
See Abstr. 154.021.

Spectroscopy of the NGC 5128 (Cen A) globular clusters.
See Abstr. 154.038.

The old population.
See Abstr. 155.012.

Radio continuum emission of the Milky Way and nearby galaxies.
See Abstr. 155.040.

Spiral structure of the Milky Way and external galaxies.
See Abstr. 155.044.

Chemical evolution of the galactic disk with radial gas flows.
See Abstr. 155.088.

The stellar population of the galactic nuclear bulge.
See Abstr. 155.103.

VLBI observations of the nucleus of M87 at two epochs.
See Abstr. 158.070.

NGC 2110 – a Seyfert and X–ray elliptical(?) galaxy with displaced kinematic and light centers.
See Abstr. 158.071.

Parsec–scale radio emission from the E/S0 galaxy NGC 3894.
See Abstr. 158.080.

M87: round and slow.
See Abstr. 158.094.

The effects of interactions on spiral galaxies. I. Nuclear activity and star formation.
See Abstr. 158.116.

The orientations of the rotation axes of radio galaxies. I. Radio morphologies of bright elliptical galaxies.
See Abstr. 158.132.

The nuclear activity of interacting galaxies.
See Abstr. 158.181.

Optical and infrared colors of high redshift galaxies.
See Abstr. 158.235.

Galactic mergers, starburst galaxies, quasar activity and massive binary black holes.
See Abstr. 158.244.

Bubbles and jets in the center of M51.
See Abstr. 158.250.

The nature of the luminous feature between Markarian 205 and NGC 4319.
See Abstr. 159.024.

On the extendedness of faint ultraviolet excess quasar candidates.
See Abstr. 159.034.

Evidence for underlying galaxies in a complete sample of variable quasars.
See Abstr. 159.049.

Quasars near NGC 1097.
See Abstr. 159.115.

Predictions of distance within the Local Group.
See Abstr. 160.001.

Redshifts of galaxies in the Fornax cluster.
See Abstr. 160.004.

The intergalactic H I cloud in Leo: a simple modeling of the Spitzer–Baade collision event.
See Abstr. 160.030.

Neutral hydrogen in the M96 group: evidence for a giant intergalactic ring.
See Abstr. 160.031.

H I properties of dwarf irregular galaxies in the Virgo Cluster.
See Abstr. 160.036.

An investigation of dwarf galaxies in the Virgo cluster.
See Abstr. 160.048.

Superclusters, cellular structures and pairs of galaxies.
See Abstr. 160.052.

A catalog of radio, optical, and infrared observations of spiral galaxies in clusters.
See Abstr. 160.054.

Dynamics of luminous galaxies. II. Surface photometry and velocity dispersions of brightest cluster members.
See Abstr. 160.057.

Analysis of groups of galaxies with accurate redshifts.
See Abstr. 160.059.

The infrared color–magnitude and color–gas content relations for cluster spirals.
See Abstr. 160.061.

Gas deficiency in cluster galaxies: a comparison of nine clusters.
See Abstr. 160.077.

H I mapping of galaxies in the Hercules cluster.
See Abstr. 160.078.

The evolution of galaxies in clusters. IV. Photometry of 10 low–redshift clusters.
See Abstr. 160.081.

A new determination of the peculiar Virgocentric velocity of the Local Group of galaxies.
See Abstr. 160.098.

H I observations of Virgo cluster galaxies.
See Abstr. 160.107.

The H I deficiency of the S0 galaxies in the Virgo cluster.
See Abstr. 160.108.

H I deficiency in the Virgo cluster.
See Abstr. 160.109.

The H I distribution in spiral galaxies in the Virgo cluster.
See Abstr. 160.110.

Neutral hydrogen observations at the VLA.
See Abstr. 160.111.

A comparison of the gas deficiency of galaxies in Virgo and other nearby clusters.
See Abstr. 160.112.

Rotation curves of hydrogen deficient galaxies in Virgo.
See Abstr. 160.113.

H I deficiency and photometric evolution of spiral galaxies in the Virgo cluster.
See Abstr. 160.114.

An H I and optical study of the gas poor Virgo cluster spiral NGC 4571.
See Abstr. 160.115.

Radio continuum and H I emission from the spiral galaxies in the Virgo cluster area.
See Abstr. 160.116.

H I observations of late–type Virgo dwarfs.
See Abstr. 160.117.

IRAS observations of Virgo cluster galaxies.
See Abstr. 160.118.

An IRAS survey of bright spirals in Virgo.
See Abstr. 160.119.

A near–infrared survey of bright Virgo spiral galaxies.
See Abstr. 160.120.

Infrared observations of dwarf elliptical galaxies in the Virgo cluster.
See Abstr. 160.121.

Molecular clouds in spiral galaxies in the Virgo cluster.
See Abstr. 160.122.

Molecular clouds in H I–deficient Virgo spirals.
See Abstr. 160.123.

The Virgo cluster redshift survey.
See Abstr. 160.124.

Global structure of Virgo cluster galaxies.
See Abstr. 160.125.

An analysis of the geometrical properties for galaxies in the Virgo cluster core.
See Abstr. 160.126.

Optical comparisons of Virgo cluster galaxies and field galaxies.
See Abstr. 160.127.

Search for young stars in Virgo dwarf ellipticals.
See Abstr. 160.129.

Applications of a nuclear spectroscopic survey of Virgo spiral galaxies.
See Abstr. 160.130.

X–ray survey of the Virgo cluster and comparison to field galaxies.
See Abstr. 160.131.

H I observations of the Virgo cluster galaxies.
See Abstr. 160.159.

Influence of the K–effect on the colours and the $\lambda_0 4000$ break amplitudes of galaxies in distant clusters.
See Abstr. 160.160.

The nuclear radio sources in the elliptical galaxies NGC 3309 and NGC 3311 in the cluster Abell 1060.
See Abstr. 160.162.

Cosmological observations of galaxies: number counts.
See Abstr. 161.001.

Optical multicolors: a poor person's z machine for galaxies.
See Abstr. 161.076.

Theory and interpretation of quantized extragalactic redshifts.
See Abstr. 161.080.

A higher limit to the mean galaxy mass allowed by gravitational lens image distortion.
See Abstr. 161.120.

Estimation of galactic masses using the zero energy–momentum cosmological principle.
See Abstr. 161.136.

Galaxy formation revisited.
See Abstr. 161.137.

Statistical comparison of galaxy formation models: the bispectrum.
See Abstr. 161.179.

On neutrino halos of galaxies or clusters of galaxies and their restriction on neutrino mass.
See Abstr. 161.228.

On massive neutrino halos and galactic structure.
See Abstr. 161.229.

The extragalactic distance scale. I. Corrections to fundamental observables.
See Abstr. 161.237.

Where do galaxies form?
See Abstr. 161.258.

Dwarf galaxies as a clue to cosmological dark matter.
See Abstr. 161.348.

Infrared background light from galaxies.
See Abstr. 161.393.

158 Active Galaxies (Seyfert Galaxies, BL Lacertae Objects, Radio Galaxies)

158.001 Ultraviolet and X–ray observations of active galactic nuclei: constraints on models of the broad emission line region.
R. Mushotzky.
NASA Conf. Publ., NASA CP–2349, p. 42 – 50 (1984). – See Abstr. 012.001.

The author is considering the extensive ultraviolet spectra of broad line active nuclei and the large body of X–ray photometry and spectroscopy of these objects and the constraints that these data place on the photoionization models of the broad emission line region of these objects. Many active galactic nuclei, in particular quasars and Seyfert I's, are characterized by strong, broad emission lines from permitted transitions in abundant elements. In the simplest models these lines originate in clouds located at some characteristic distance from the central ionizing source in a quasi–spherical distribution. The author concentrates on how to derive the observed line strengths and line ratios.

158.002 Structure of the Seyfert broad line region and absorption region of NGC 4151.
G. E. Bromage, A. Boksenberg, J. Clavel,
M. H. Demoulin–Ulrich, A. Elvius, M. V. Penston,
G. C. Perola, M. Pettini, M. A. J. Snijders, E. G. Tanzi,
M. Tarenghi.
NASA Conf. Publ., NASA CP–2349, p. 127 – 128 (1984). – See Abstr. 012.001.

The authors have monitored the Seyfert nucleus of NGC 4151 for six years, including periods of intensive study of the complex variations in the ultraviolet spectrum with IUE, and some co–ordinated optical and X–ray observations. Here they briefly review important IUE results relating to the structure of the broad line region and its physical conditions. Their detailed analyses and results are published in a series of papers.

158.003 Broad emission features in active galactic nuclei.
H. Netzer, B. J. Wills, D. Wills, W. Wamsteker.
NASA Conf. Publ., NASA CP–2349, p. 129 – 132 (1984). – See Abstr. 012.001.

The authors have carried out simultaneous ultraviolet and optical observations of several Seyfert 1 galaxies and QSOs in an attempt to understand some broad emission features between 2000 and 5000 Å. Most of the emission is due to Fe II and Balmer continuum. Improved model calculations are used to estimate the flux in these components and to compare the strengths of the ultraviolet and optical Fe II lines. The spectral fitting shows that the ultraviolet Fe II lines are much stronger than previously believed.

158.004 Multifrequency observations of BL Lac objects and violently variable quasars.
J. N. Bregman, A. E. Glassgold, P. J. Huggins, A. L. Kinney.
NASA Conf. Publ., NASA CP–2349, p. 135 – 138 (1984). – See Abstr. 012.001.

The authors report on the status of their program to obtain simultaneous multifrequency spectra of BL Lac objects and violently variable quasars from the X–ray through radio wavebands. They illustrate the importance of the ultraviolet data in defining the overall shape of optically thin synchrotron emission, its relation to the X–ray emission, and the determination of basic physical parameters of the continuum emitting region.

158.005 Very recent IUE observations of 2 BL Lacertae objects.
C. M. Urry, Y. Kondo, K. R. H. Hackney,
R. L. Hackney, S. L. Mufson, W. Wisniewski.
NASA Conf. Publ., NASA CP–2349, p. 139 – 142 (1984). – See Abstr. 012.001.

IUE observations of two BL Lacertae objects, consisting of 2 spectra of 1218 + 304 and 12 spectra of Mrk 421, are presented. The observations indicate an intensity decrease of $\sim 25\%$ on a timescale of one month, with little if any associated spectral change. With one exception, the spectra of these two objects show no discrete features, and the continua are well fit by power law models. The one exception is a long SWP exposure of Mrk 421, from 1984 March 9, which shows a broad emission feature near ~ 1580 Å. The validity of this feature has not yet been established.

158.006 Coordinated multifrequency observations of the BL Lac objects Mrk 180, Mrk 421, and Mrk 501.
S. L. Mufson, D. J. Hutter, K. R. Hackney, R. L. Hackney,
Y. Kondo, C. M. Urry, W. Z. Wisniewski, M. F. Aller,
H. D. Aller.
NASA Conf. Publ., NASA CP–2349, p. 143 – 146 (1984). – See Abstr. 012.001.

Over the past several years the authors have been investigating the properties of the three BL Lac objects Mrk 180, Mrk 421, and Mrk 501 using data gathered at ultraviolet, X–ray, optical, and radio wavelengths. These three objects are relatively nearby ($z = .046, .031,$ and $.034$, respectively) composite systems in which a moderately luminous ($L = 10^{44}$ergs/s) BL Lac object is found at the core of an elliptical galaxy (Ulrich 1978; Ulrich et al. 1975).

158.007 Fluorescent excitation of Fe II, Mn II, Ti II, N I lines by C IV, N V, O VI,... emission lines in the spectra of symbiotic stars & Seyfert galaxies.
D. P. Gilra.
NASA Conf. Publ., NASA CP–2349, p. 147 (1984). Abstract. – See Abstr. 012.001.

158.008 A model for the variable continuum of the Seyfert I galaxy NGC 5548.
W. Wamsteker, P. Benvenuti, A. W. Harris, A. Talavera,
C. Gry, A. Cassatella, B. Hassall, R. Gilmozzi, P. Barr,
J. C. Clavel, H. Netzer, B. J. Wills, D. Wills.
NASA Conf. Publ., NASA CP–2349, p. 148 – 151 (1984). – See Abstr. 012.001.

The variations of NGC 5548 over the full period of IUE observing until February 1984 are analyzed. The results indicate that the amplitude of the variation is largest at the shorter UV wavelengths. The implications of this wavelength dependence are of importance for the continuum modeling. A fit to a simultaneously obtained UV and optical spectrum in July 1981, based on the Fe II fluorescence model of Netzer and Wills (1983) is shown to give an extremely good match with the data.

158.009 Rotation and velocity dispersion in the stellar component of NGC 1316 (Fornax A).
A. Bosma, R. M. Smith, K. J. Wellington.
Mon. Not. R. Astron. Soc., Vol. 212, No. 2, p. 301 – 307 (1985).

Observations have been made of the stellar kinematics in NGC 1316, the galaxy associated with the strong radio source Fornax A. The galaxy is rotating rapidly about its photometric minor axis. A jump in the radial velocity of approximately 30 km s^{-1} is found along the major axis at the position of the first ripple in the luminosity profile. The velocity dispersion decreases from a nuclear value of 230 km s^{-1} to about 180 km s^{-1} at the first ripple, remaining approximately constant at larger radii. The data are consistent with the galaxy being an oblate spheroid flattened by rotation, with isotropic velocity dispersion. The presence of the radial velocity step agrees with the hypothesis that NGC 1316 is the result of at least one merger between a disc and an elliptical galaxy.

158.010 The nuclear colours of weakly active elliptical galaxies.
C. R. Jenkins, C. P. Blackman.
Mon. Not. R. Astron. Soc., Vol. 212, No. 2, p. 385 – 394 (1985).

The authors have obtained long–slit spectroscopy of 15 elliptical radio galaxies. Most nuclear regions are redder than the surrounding regions by a larger amount than expected from the colour–aperture effect, although ~20 per cent of the nuclei are bluer. Evidence from the magnesium index suggests that the reddening is due to abundance effects, and the colours then imply excess metallicities by factors $\cong 1.5$. It seems that the apparent excess redness over the standard colour aperture effect only shows that the colour gradient is generally steeper in the nucleus than previously thought, rather than a distinct stellar population being responsible. The blue nuclei may contain either a power-law continuum source, or young stars.

158.011   Soft gamma–ray production in active galactic nuclei.
M. Sikora, M. Zbyszewska.
Mon. Not. R. Astron. Soc., Vol. 212, No. 3, p. 553 – 564 (1985).

The authors extend recent studies by Lightman, Svensson and others on the pair–equilibrium states of mildly relativistic thermal plasma, including magnetic fields and optical sources. Resulting constraints on luminosities and proton densities together with observational data allow to select an accretion scenario in which a high–energy spectrum, similar to that of NGC 4151, is produced. It is shown that soft γ–ray production via thermal bremsstrahlung can occur in the central region of the two–temperature, geometrically thick part of the disc. On the other hand, the power–law X–ray spectrum is expected to be generated in the intermediate region due to Comptonization of optical photons coming from an outer, geometrically thin part of the disc. Implications for the relation between quasars and Seyfert galaxies are discussed.

158.012   The intensities of the sulphur III lines and the ionization mechanisms in Liners.
A. I. Díaz, B. E. J. Pagel, I. R. G. Wilson.
Mon. Not. R. Astron. Soc., Vol. 212, No. 3, p. 737 – 749 (1985).

New near–infrared spectrophotometric observations show that the relative strengths of the sulphur emission lines [S III] $\lambda\lambda$ 9069, 9532 Å with respect to Hα constitute a powerful diagnostic to distinguish between shock and photoionization mechanisms in Liners. These observations strongly support the idea, already favoured on the basis of abundance arguments, that Liners and Seyferts form a continuous sequence, both types of objects being photoionized, but with the ionization parameter being lower in Liners. This photoionization could be due either to the presence of a central nonthermal source or to recent star formation.

158.013   Is the jet in M87 magnetically confined?
P. E. Hardee.
Unstable current systems and plasma instabilities in astrophysics, p. 439 – 443 (1985). – See Abstr. 012.002 (IAU Symp. No. 107).

Calculations of the minimum pressure in the M87 jet show that the jet pressure is an order or magnitude in excess of the thermal pressure in the interstellar medium calculated from the X–ray emission. The continuous energy input required to produce the jet emission is incompatible with a freely expanding jet. On the other hand, the uniform jet expansion, the jet emission and the jet polarization can be explained if the jet is self–confined by $\mathbf{j} \times \mathbf{B}$ forces.

158.014   Ultra–high energy cosmic ray production by current disruption in active galactic nuclei.
G. Lake, R. E. Pudritz.
Unstable current systems and plasma instabilities in astrophysics, p. 471 – 473 (1985). – See Abstr. 012.002 (IAU Symp. No. 107).

Electrodynamic models for the activity of galactic nuclei are shown to be current systems which can be examined in terms of equivalent circuits. The resulting inductive circuit which describes the coupling of the generator (black–hole and accretion disk) to the distant load (jet plasma) is prone to various instabilities. The authors consider the disruption of this current system and propose that ultra–energetic cosmic rays ($E \sim 10^{19} - 10^{21}$ eV) could be produced during the discharges, which occur at distances of $\sim 10^{16} - 10^{18}$ cm from the central massive hole ($M \sim 10^8 M_\odot$). Such discharges will also produce variable γ–ray and X–ray activity and the authors discuss observations of Cen A in this regard.

158.015   Distribution and motions of atomic hydrogen in lenticular galaxies. III. NGC 3998.
G. R. Knapp, W. van Driel, H. van Woerden.
Astron. Astrophys., Vol. 142, No. 1, p. 1 – 8 (1985).

Synthesis observations of H I in the active S0 galaxy NGC 3998 show that the gas is in a polar ring, inclined about 80° to the optical major axis. The position angle of the ionized gas in the inner galaxy is intermediate between those of H I and the light, suggesting differential precession of the gas structure. Weak extensions on the radio continuum emission from the nucleus are roughly perpendicular to the inner gas distribution. The circular velocity derived from the H I is normal for a galaxy of this luminosity; however, the ratio of central bulge velocity dispersion to circular velocity is anomalously high. This, and the extreme central light concentration, suggest that the galaxy contains a central massive object.

158.016   Multiple hotspots in extragalactic radio sources.
A. G. Williams, S. F. Gull.
Nature, Vol. 313, No. 5997, p. 34 – 36 (1985).

Recent observations with sub–kpc resolution show that the hotspots in powerful extragalactic radio sources often exhibit internal double structure. One subcomponent is usually extremely compact, of size <1 kpc, whereas the other, which is more diffuse, often appears to point away from the smaller component and is usually further from the nucleus. The authors present a model of a jet with variable direction that is able to account for this structure, interpreting the compact subcomponent as the initial point of impact of the jet with the intergalactic medium and the more diffuse subcomponent as a "splatter–spot" formed by jet material deflected asymmetrically by the compact subcomponent.

158.017   A Seyfert galaxy joins the jet set.
B. J. Wills.
Nature, Vol. 313, No. 6005, p. 741 (1985).

This note comments on recent observations of a jet–like feature in the core of the Seyfert galaxy NGC 4151.

158.018   Narrow and variable lines in the ultraviolet spectrum of the Seyfert galaxy NGC 4151.
M. H. Ulrich, A. Altamore, A. Boksenberg, G. E. Bromage, J. Clavel, A. Elvius, M. V. Penston, G. C. Perola, M. A. J. Snijders.
Nature, Vol. 313, No. 6005, p. 747 – 751 (1985).

The Seyfert galaxy NGC 4151 observed when the nucleus is at a minimum has two emission lines of full width at half maximum <7 and 16 Å respectively and varying in intensity by a factor of three in 10 days. These lines are too narrow to be emitted by the whole broad–line region and must arise instead from two localized regions which have a special excitation mechanism, possibly a two–sided jet.

158.019   A radio and optical study of the radio galaxy PKS 0511–48.
R. M. Smith, J. G. Robertson.
Mon. Not. R. Astron. Soc., Vol. 212, No. 4, p. 809 – 816 (1985).

The radio galaxy PKS 0511–48 has been mapped with the Fleurs Synthesis Telescope at 1415 MHz and the Molonglo Observatory Synthesis Telescope at 843 MHz. The source is resolved into at least three unrelated components. A central double source is apparently associated with a tidally disrupted galaxy with a long, bright, curved filament. Spectroscopic observations show this galaxy to have an emission–line spectrum typical of high–ionization narrow–line radio galaxies. The morphology and velocity structure of the filament are consistent with a tidal origin for such a feature.

158.020   The optical luminosity function of Seyfert 1 nuclei.
F.–z. Cheng, L. Danese, G. De Zotti, A. Franceschini.
Mon. Not. R. Astron. Soc., Vol. 212, No. 4, p. 857 – 871 (1985).

The authors have collected from the literature all confirmed Seyfert 1 and 1.5 galaxies in the area covered by the first nine Markarian lists and have estimated their nuclear magnitudes in two independent ways. They have then defined a 'homogeneous' subsample suitable for deriving the luminosity function with the method described by Neyman & Scott. The resulting luminosity function matches remarkably well that of optically selected QSOs at $M_B \cong -23$ ($H_0 = 50$), flattens out somewhat at $M_B \cong -21.5$, but does not show clear evidence of a levelling off down to $M_B \cong -18$; it also compares favourably with the local luminosity function of X–ray–selected active galactic nuclei.

158.021   Erratum: "The optical variability of two Seyfert galaxies: Arakelian 120 and Markarian 231" [Astron. Astrophys., Suppl. Ser., Vol. 35, p. 387 – 389 (1979)].
H. R. Miller.
Astron. Astrophys., Suppl. Ser., Vol. 59, No. 2, p. 367 (1985). See Abstr. 25.158.044.

158.022   High resolution radio observations of low luminosity radio galaxies.
P. Parma, R. D. Ekers, R. Fanti.
Astron. Astrophys., Suppl. Ser., Vol. 59, No. 3, p. 511 – 521 (1985).

New maps for seven low luminosity radio galaxies, obtained with the Westerbork Synthesis Radio Telescope at the frequencies 0.6, 1.4 and 5.0 GHz are presented. For two of them a VLA map at 1.4 GHz, with a resolution of about 3″, is also given. Two, possibly three, of the sources show radio structure of the Head–Tail or Wide–Angle–Tail type. Two others show pronounced S (or rotational type) distortions.

158.023   Very luminous, dust–shrouded galaxies.
H. L. Shipman.
Phys. Today, Vol. 38, No. 1, p. S9 – S10 (1985). Short review paper. – See Abstr. 013.003.

158.024   Arp 220 revealed.
R. E. Schild.
Sky Telesc., Vol. 69, No. 1, p. 24 – 25 (1985).

158.025   Radio source counts below 100 μJy predicted from IRAS 60 μ counts.
P. Biermann, A. Eckart, A. Witzel.
Astron. Astrophys., Vol. 142, No. 2, p. L23 – L24 (1985).

Given the good correlation between 60 μ flux density and 6 cm flux density for galaxies the authors predict the location in the log N – log S curve where the radio source counts for weak sources become dominated by galaxies and again approach the $S^{-1.5}$ law. This happens near 30 to 100 μJy for 6 cm in agreement with a P(D) analysis of Fomalont et al.

158.026   The variable extragalactic object 3C 446.
C. Barbieri, S. Cristiani, S. Omizzolo, G. Romano.
Astron. Astrophys., Vol. 142, No. 2, p. 316 – 320 (1985).

The authors report recent photometric and spectroscopic data on 3C 446, including a very active phase in 1983. A discussion of all the available data confirms the existence of temporal scales in the optical variations of 1540 and 2130 days (640 and 890 days, respectively in the object's rest frame), with the first one corresponding to a greater number of maxima. During the brightening, the emission lines vanish and 3C 446 becomes a typical BL Lac object.

158.027   The structure of the broad–line region of the galaxy NGC 4151.
I. I. Shevchenko.
Pis'ma Astron. Zh., Tom 11, No. 2, p. 83 – 88 (1985). In Russian. English translation in Sov. Astron. Lett., Vol. 11.

It is shown that as applied to the broad–line region of the nucleus of the Seyfert galaxy NGC 4151 the model of homogeneous distribution of emission–line clouds around the source of ionizing continuum predicts the existence of a time delay of the maximum of a flare of the nucleus in the Hα line in relation to the maximum of its flare in the continuum and the absence of any time delay of the maxima of flares in other Balmer lines.

158.028   Are spiral nuclei "active" in the radio continuum?
P. T. P. Ho, J. L. Turner.
The Milky Way galaxy, p. 377 – 378 (1985). – See Abstr. 012.007 (IAU Symp. No. 106).

Using the VLA, the authors obtained matched–array continuum observations at 6 and 2 cm. An angular resolution of ~1″ and an rms sensitivity of ~0.05 mJy were achieved for a sample of 17 nearby spiral galaxies. Spectral–index maps derived for the nuclear regions reveal a mixture of thermal and nonthermal activity.

158.029   The remarkable infrared galaxy Arp 220.
M. Rowan–Robinson.
Observatory, Vol. 105, No. 1064, p. 3 (1985). Abstract. – See Abstr. 010.721.

158.030   X–ray luminosity and the hydrogen–line ratios of quasars and Seyfert galaxies.
G. A. Kriss.
Astron. J., Vol. 90, No. 1, p. 1 – 4 (1985).

New observations of the Lyα/Hα ratio in a set of X–ray–selected active galactic nuclei and an archival study of IUE observations of Lyα in low–redshift quasars and Seyfert galaxies have been used to form a large sample (49 objects) for studying the influence of soft X–rays on the enhancement of Balmer emission in the broad–line region. In common models of broad–line clouds the ratio Lyα/Hα is expected to depend weakly on the ratio of ionizing ultraviolet luminosity to X–ray luminosity (L_{UV}/L_X). If the Lyα luminosity is used as a measure of L_{UV}, a weak dependence of Lyα/Hα on the X–ray luminosity is found (valid at the 94% confidence level).

158.031   A complete sample of intermediate–strength radio sources selected from the GB/GB2 1400–MHz surveys. V. VLA observations of very extended and confused sources.
J. Machalski, J. J. Condon.
Astron. J., Vol. 90, No. 1, p. 5 – 29 (1985). With plates 1 – 2.

Very extended or confused GB/GB2 sources in complete samples with $0.20 \leqslant S < 2.0$ Jy at 1400 MHz were observed with the VLA in its C configuration at 1465 MHz. Low–resolution maps, source component parameters, and new optical identifications are presented. These low–resolution data complement the high–resolution data already published.

158.032   Alignment of radio and optical polarization with VLBI structure.
R. Rusk, E. R. Seaquist.
Astron. J., Vol. 90, No. 1, p. 30 – 38 (1985).

The authors have examined the available data on VLBI structural position angles, and radio and optical polarization position angles to determine whether these angles are correlated. There is a strong trend for the radio E vector to lie normal and for the optical E vector to lie parallel to the VLBI structural axis.

158.033   Footprints of an active galactic nucleus: the nuclear zone of NGC 1068.
B. Balick, T. Heckman.
Astron. J., Vol. 90, No. 2, p. 197 – 203 (1985).

NGC 1068 is a spiral galaxy that hosts an extremely luminous active nucleus imbedded in an unusually bright gaseous medium. The galaxy is nearby and, unlike Cen A, its near–nuclear region is plainly visible. Therefore the galaxy presents an exceptionally propitious opportunity to separate clues to the origin of nuclear activity from the disruptions resulting from this activity. The many observational data, including new, deep emission–line images and long–slit spectra, are reviewed with this aim. NGC 1068 appears to be a hybrid Seyfert/starburst galaxy.

158.034   The narrow line region of active galaxies – I. Nuclear [O III] profiles.
M. Whittle.
Mon. Not. R. Astron. Soc., Vol. 213, No. 1, p. 1 – 31 (1985).

High–dispersion [O III] profiles of 42 southern hemisphere active and H II galaxies are presented as the first part of a study of the narrow line region (NLR) of active galaxies. A parameter scheme based on area measurements is defined which characterizes the width, asymmetry and kurtosis of different parts of the profile. Techniques for estimating errors and resolution corrections are described. A comparison with parameters defined by a cut at constant height shows the area parameters to be less sensitive to noise, more stable to the effects of poor resolution and more sensitive to information contained in the base and wings of the profile.

158.035   The narrow line region of active galaxies – II. Relations between [O III] profile shape and other properties.
M. Whittle.
Mon. Not. R. Astron. Soc., Vol. 213, No. 1, p. 33 – 58 (1985).

Data from paper I have been combined with most previously published [O III] profile information to form a large sample of line width, asymmetry and kurtosis measurements for Seyfert galaxies. This sample has been used to investigate systematically the dependence of the [O III] profile shape on other properties.

158.036   Redshifts of radio sources in the 1–Jy sample.
J. R. Allington–Smith, S. J. Lilly, M. S. Longair.
Mon. Not. R. Astron. Soc., Vol. 213, No. 1, p. 243 – 255 (1985).

Low–resolution optical spectra have yielded firm redshifts for 12 galaxies and 1 QSO identified with radio sources in the 1–Jy sample. In the case of the radio galaxy $1043+37A$ ($z = 0.789$), an asymmetric extended region of [O II] $\lambda 3727$ emission was discovered. This feature extends to about 70 kpc from the galaxy nucleus and displays differential velocities of about 600 km s^{-1}. It is found that the 1–Jy radio galaxies have line strengths that are about a factor 2 weaker than those of the 3C galaxies in the same redshift interval.

158.037   Dust in central regions of five Seyfert galaxies.
M. Šolc.
Acta Univ. Carol. Math. Phys., Vol. 25, No. 2, p. 43 – 56 (1984).

On the basis of multicolour $UBVRIJHKL$ photometry done by Lebofsky and Rieke (1980) for five Seyfert galaxies III Zw2, 3C 120, NGC 1275, 4151 and 5548, the temperature and total mass of the interstellar dust, surrounding in a shell the galaxy nucleus, is studied. The estimated masses of the dust in the shell range at maximum to $10^3 M_\odot$, the shell radii range at most some tens of parsec and dust temperatures are spread from 400K to 800K.

158.038   The form of $\log N - \log S$ relation in the domain of strong flux densities.
V. R. Amirkhanyan.
Astrophys. Space Sci., Vol. 108, No. 2, p. 319 – 322 (1985).

The slope formed in the curves of counts in a strong flux density region can be explained by a constraint on the flux density of radio sources and the spread of their spectral indices.

158.039   Theoretical properties of radio–galaxies. I. Velocities and transit times.
L. Zaninetti.
Astrophys. Space Sci., Vol. 108, No. 2, p. 401 – 407 (1985).

A complete sample of symmetric double radio–galaxies has been investigated by inferring the propagation of a jet along the radio–axis and computing velocities with the aid of different models. A best linear fit between radio–luminosity and velocity is proposed in the range $10^{40} \text{ergs s}^{-1} <$ radio–luminosity $< 10^{45} \text{ergs s}^{-1}$.

158.040   Optical spectroscopy of the oval distortion NGC 4151.
H. Schulz.
Astron. Astrophys., Vol. 143, No. 1, p. 29 – 34 (1985).

Image tube spectra have been obtained from emission line regions within the bar of NGC 4151 at about 1′ distance from the nucleus. The line widths [$(50 \pm 30) \text{ km s}^{-1}$], the electron densities ($N_e < 500 \text{ cm}^{-3}$) and the appearance on photographs of different color suggest that these regions are normal H II regions. Significant radial streaming motions ($\sim 20 \text{ km s}^{-1}$) of the H II regions are detected after constraining the circular velocities ($\sim 150 \text{ km s}^{-1}$) by 21–cm–data. The velocity vectors are interpreted by oval streamlines of the gas which are caused by the gravitational potential of the bar. Finally, some morphological peculiarities in the bar of NGC 4151 are emphasized.

158.041   Beams in head–tail galaxies.
W. A. Baan, M. R. McKee.
Astron. Astrophys., Vol. 143, No. 1, p. 136 – 142 (1985).

The authors reconsider the ram pressure interaction of a continuous beam and the intracluster medium and discuss the characteristic beam configuration in its simplest form. Subsequently they apply the results to three head–tail galaxies and consider the general behavior of the beams in the context of observed fluid dynamical phenomena.

158.042   The radio structure of 3C 130 interpreted with a dynamical model.
W. J. Jägers, M. H. K. de Grijp.
Astron. Astrophys., Vol. 143, No. 1, p. 176 – 181 (1985).

The radio structure of the giant radio source 3C 130 is interpreted in terms of the dynamical model of Blandford and Icke. The overall edge–darkened double structure of this 1.2 Mpc source is disfigured by several symmetric wiggles. Assuming that the wiggles are produced by orbital interaction between the parent galaxy and a nearby companion the radio structure of 3C 130 is well reproduced. The effects of buoyancy on the structure are also taken into account. Together with the observed wiggling, the model also predicts the widening of the source as a function of distance from the parent galaxy. The best fit orbital parameters require a jet velocity of $\sim 120 \text{ km s}^{-1}$, an orbital period of $\sim 1.3 \times 10^9 \text{yr}$ and an age of $\sim 2 \times 10^9 \text{yr}$ for the radio source. They provide an additional indication for the existence of slow jets.

158.043   Optically thin synchrotron emission and Type C_L radio source spectra.
J. A. Roberts.
Astron. Astrophys., Vol. 143, No. 1, p. 231 – 232 (1985).

Contrary to the suggestion of Junhan et al. (1984) Type C_L radio spectra cannot result from optically thin synchrotron emission even if the source is non–uniform.

158.044   Orientation correlation of radio double sources.
L.–z. Fang, Y.–q. Chu, X.–f. Zhu.
Astron. Astrophys., Vol. 143, No. 1, p. 241 – 243 (1985).

A possible difference between the orientation correlations in adiabatic and isothermal scenarios of clustering in the universe has been discussed. Statistical results confirm the existence of the orientation correlation for radio double sources. The scale of such correlation does not appear different from that of clusters of galaxies, namely, it is not inconsistent with the adiabatic scenarios.

158.045   Faint *IRAS* galaxies: a new species in the extragalactic zoo.
D. A. Allen, P. F. Roche, R. P. Norris.
Mon. Not. R. Astron. Soc., Vol. 213, No. 2, p. 67P – 74P (1985).

The authors report optical spectroscopy and near–infrared photometry of 19 optically faint galaxies which have been discovered as intense 100 μm sources by *IRAS*. All have low–excitation emission lines with redshifts up to 0.2, and some exceed $10^{12} L_\odot$ in their infrared luminosities. There is no evidence of Seyfert–like spectra, but instead starburst activity to power the infrared emission. Many of the galaxies are clearly disturbed or show evidence of interaction with neighbours; the highest redshift source consists of a trio of emission galaxies.

158.046   Compact steep spectrum 3CR radio sources. VLBI observations at 18 cm.
C. Fanti, R. Fanti, P. Parma, R. T. Schilizzi,
W. J. M. van Breugel.
Astron. Astrophys., Vol. 143, No. 2, p. 292 – 306 (1985).

The first results are presented of a program to investigate the kiloparsec sized radio structure of a representative sample of compact steep spectrum (CSS) sources from the 3CR catalogue. The overall sizes of these CSS sources range from about 100 mas to 1 or 2″, corresponding to linear sizes of the order of 1 to

10 kpc. The morphological classification ranges from double, to "core–jet", to complex; CSS quasars are generally "core–jets" or complex, while CSS radio galaxies are doubles, although not necessarily simple doubles.

158.047 Infrared emission from dust in the Seyfert galaxy NGC 4151.
A. J. Mayes, G. Pearce, A. Evans.
Astron. Astrophys., Vol. 143, No. 2, p. 327 – 333 (1985).

The non–variable infrared radiation from the nucleus of NGC 4151 is discussed in terms of radiation from circumnuclear dust heated by nuclear radiation. The dust is modelled by a spherical shell and by a torus, both consisting of silicate and graphite dust grains similar to those found in the Galaxy. The model predictions are compared with the observations in an attempt to determine some parameters of the circumnuclear dust. The comparison indicates a spherical shell rather than a torus with a silicate to graphite dust mass ratio of 90:10, an inner radius of about 4 pc and an outer radius of $\geqslant 20$ pc.

158.048 Photoionization models for Liners: gas distribution and abundances.
L. Binette.
Astron. Astrophys., Vol. 143, No. 2, p. 334 – 346 (1985).

The sum of recently published spectral data on Liners is re–analysed to reveal improved correlations. In the most successful diagram which involves [O III]/(Hα/3) and [O I]/[O III], three quarters of the objects taken from magnitude limited complete samples and extending over 1.7 dex in [O I]/[O III] form a linear band which is only 0.3 dex wide. An important property of this diagram is that the correlation is unaffected by reddening or geometrical parameters. Assuming photoionization by a non–stellar object, new models have been computed to represent a continuum in the ionization parameter space. The brightest objects, as measured by the Hα equivalent widths, have a much lower [O I]/Hβ ratio which is consistent with the ionized clouds being optically thin or with the presence of nuclear H II regions.

158.049 The NGC 4593 group of active galaxies.
W. Kollatschny, K. J. Fricke.
Astron. Astrophys., Vol. 143, No. 2, p. 393 – 399 (1985).

The optical properties of the Seyfert 1 galaxy NGC 4593 are described in detail on the basis of spectra from different epochs. NGC 4593 is found to be a member of a group of galaxies. Almost all of the members of this group detected show strong emission line activity. The implications of this finding for the origin and evolution of galactic nuclear activity are briefly discussed.

158.050 An extended extragalactic radio source behind the Milky Way.
J. H. Seiradakis, W. Reich, W. Sieber, H. Kühr, R. Schlickeiser.
Astron. Astrophys., Vol. 143, No. 2, p. 478 – 480 (1985).

Observations at 2.7 and 4.8 GHz are presented of an unusual and so far undetected, bar–like structure in the galactic plane at $l = 65°$. The source extends over $0°\!.7$ by $0°\!.15$, is polarized and has a steep spectrum. It consists of two outer components and a flatter spectrum central bulge. At 10.7 GHz the authors discovered an 8 mJy source in the centre, lying symmetrically between the two outer components. VLA observations at 4.885 GHz showed that this is compact with flux density 3.4 ± 0.5 mJy, indicating a flat or inverted spectrum. These characteristics suggest that it is an extragalactic radio source seen in projection through the plane of the Galaxy.

158.051 CCD photometry of Seyfert galaxy NGC 1068.
R. Schild, R. Tresch–Fienberg, J. Huchra.
Astron. J., Vol. 90, No. 3, 441 – 449, 561 – 562 (1985).

CCD observations of NGC 1068 at the F. L. Whipple Observatory 0.61–m and MMT telescopes at B, V, and R wavelengths are analyzed for color and brightness features of the Seyfert galaxy nucleus and spiral arms. The differential brightness and color profiles of the galaxy show the presence of very young star–forming regions dominating the spiral structure out to

$r = 18$ arcsec, surrounded by a region of abnormally strong dust absorption. These two features cause the previously noted discontinuity between "inner" and "outer" spiral structure. 9% of the luminosity within a 10–arcsec radius arises from the $R = 12.74$ mag semistellar nucleus. An arc of bright optical knots may be related to the proposed radio jet and may link Seyfert activity in the nucleus to star–formation activity in the disk.

158.052 Mysteries of cosmic jets.
D. H. Smith.
Sky Telesc., Vol. 69, No. 3, p. 213 (1985).

158.053 The effect of local galaxy density on the production of powerful radio sources by early–type galaxies.
T. M. Heckman, T. J. Carty, G. D. Bothun.
Astrophys. J., Vol. 288, No. 1, p. 122 – 131 (1985).

The authors have quantitatively analyzed the POSS prints and a set of CCD images obtained at KPNO in order to investigate the local galaxy density around samples of 47 radio–loud and 46 radio–quiet elliptical and lenticular galaxies. The radio sources studied are dominated by steep–spectrum components, not by compact, flat–spectrum ones. The local galaxy density has been measured by weighting the companion galaxies according to their relative size (or luminosity) and/or projected proximity. The primary conclusion is that all measures of average local galaxy density (applied to both the large POSS data set and smaller CCD data set) are larger (by at least a factor of 2 – 3) for the radio–loud galaxies. The statistical significance levels of these results are very high (typically > 99.9%). It is argued that the evidence that galaxy interactions foster nuclear activity is now strong and may apply to the whole "zoo" of active extragalactic objects (nuclear starburst galaxies, Seyfert galaxies, Liners, radio galaxies, quasars).

158.054 Recent spectral variations in the active nucleus of NGC 1566.
D. Alloin, D. Pelat, M. Phillips, M. Whittle.
Astrophys. J., Vol. 288, No. 1, p. 205 – 220 (1985).

The authors present spectrophotometric observations monitoring both the nuclear Hβ and Hα emission–line profiles and the 3500 – 7000 Å continuum of the Seyfert galaxy NGC 1566 over a 14 month period from 1980 December to 1982 January. During this time the spectrum changed from one closely resembling a Seyfert 2 to one closely resembling a Seyfert 1, with most of the change taking place in 4 months. The nuclear spectrum is decomposed into an underlying stellar component and a varying nonstellar component. The implications of the observed variations are considered for both the broad–line region and the narrow–line clouds.

158.055 A radio study of 3C 390.3 – a precessing jet?
P. Alexander.
Mon. Not. R. Astron. Soc., Vol. 213, No. 3, p. 743 – 752 (1985).

Observations of the classical double radio galaxy 3C 390.3 have been made to determine the spectral variations over the source at high radio frequencies. The corresponding spectral ages and the derived magnetic field directions are used to deduce the flow pattern of the radio–emitting plasma within the Np head. The results imply that (1) there is acceleration of material outside the hotspots in the Np head; (2) the beam has changed direction by $\sim 5°$ in the past 10^6yr or so; (3) there is little spectral variation over the southern lobe.

158.056 Markarian 1388 – a high–ionization narrow–line Seyfert galaxy.
D. E. Osterbrock.
Publ. Astron. Soc. Pac., Vol. 97, No. 587, p. 25 – 29 (1985). = Lick Obs. Bull., No. 997.

The spectrum of the Seyfert galaxy Mrk 1388 is described. It is unusual in having strong high–ionization lines (in particular, of [Fe VII] and [Fe X]) and a strong featureless continuum, but narrow H I lines with essentially the same widths as the forbidden lines. It thus combines characteristics usually found separately in Seyfert 1 and Seyfert 2 galaxies.

158.057   Extinction in the central regions of M82.
A. Moneti, W. J. Forrest, J. L. Pipher,
C. E. Woodward.
News Lett. Astron. Soc. N.Y., Vol. 2, No. 5, p. 9 – 10 (1984).
Abstract. – See Abstr. 010.243.

158.058   Een microscopische kijk op het grootste object.
P. D. Barthel, W. J. Jägers,
F. H. P. M. van Roermund.
Zenit, 12. Jaarg., No. 3, p. 92 – 94 (1985).

158.059   The nuclear radio sources in the elliptical galaxies NGC 3309 and NGC 3311 in the cluster Abell 1060.
P. O. Lindblad, S. Jörsäter, Å. Sandqvist.
Astron. Astrophys., Vol. 144, No. 2, p. 496 – 501 (1985).
VLA observations in the 6–cm continuum of the Abell 1060 cluster of galaxies have revealed a radio jet in one of the dominant central elliptical galaxies, NGC 3309. The total flux density is 4.8 mJy and the total minimum equipartition energy is 5×10^{33}erg. A weak radio source, with a total 6–cm flux density of 0.3 mJy, has barely been detected in the other dominant central cD galaxy, NGC 3311. Optical observations of NGC 3309 and NGC 3311 have also been obtained using the ESO 3.6 m and the Danish 1.5 m telescopes on La Silla.

158.060   A dynamical halo around the edge–on galaxy NGC 4631.
W. Werner.
Astron. Astrophys., Vol. 144, No. 2, p. 502 – 505 (1985).
Investigations of radio continuum observations at four frequencies of the edge–on galaxy NGC 4631 indicate that frequency dependence of the radio emission above the disk supports a dynamical halo model.

158.061   The radio structure and host galaxy of 3C 459.
J. S. Ulvestad.
Astrophys. J., Vol. 288, No. 2, p. 514 – 520 (1985).
The radio galaxy 3C 459 has been mapped at 2, 6, and 20 cm using the VLA. It shows a triple structure of total extent 29 kpc, with the western lobe being 5.5 times as far from the core as the eastern lobe. The compact radio core has a steep spectrum and may be similar to the optically unidentified steep–spectrum core sources. Analysis of published absorption–line data yields a spectral type of F for the host galaxy, much earlier than expected for a typical elliptical galaxy and the earliest type known for a powerful radio galaxy. Despite the early spectral type, the colors are not consistent with a normal spiral galaxy.

158.062   Echelle spectroscopy of the Seyfert 1 galaxy Markarian 231.
R. J. Rudy, C. B. Foltz, J. T. Stocke.
Astrophys. J., Vol. 288, No. 2, p. 531 – 534 (1985).
This paper presents high-resolution (~ 10 km s^{-1}) spectroscopy of the Na D absorptions associated with two of the three absorption–line systems present in the spectrum of the Seyfert 1 galaxy Markarian 231. These features are blueshifted by several thousand km s^{-1} relative to the emission lines and show the shape, displacement, and extent in velocity space characteristic of broad absorption–line QSOs. This indicates that the mechanism which accelerates the gas in those high–luminosity objects may be operating in Mrk 231 as well. Arguments are presented which suggest the gas producing the system I absorptions is located well outside the broad emission–line region.

158.063   The near–ultraviolet spectrum of the high–redshift BL Lacertae object 0215 + 015.
J. C. Blades, R. W. Hunstead, H. S. Murdoch, M. Pettini.
Astrophys. J., Vol. 288, No. 2, p. 580 – 594 (1985).
Spectra of the luminous high–redshift BL Lac object 0215 + 015 have been obtained with the IUE from 2300 to 3150 Å and with the AAT from 3040 to 3870 Å. A total of 66 absorption lines were detected, 50 of which can be identified in the six definite and one probable redshift systems ranging from $z_{abs} = 1.254$ to $z_{abs} = 1.719$. Strong unidentified lines found shortward of 3310 Å are assumed to be members of a sparse Lyα forest. This is the first reported detection of a Lyα forest in a BL Lac object. The intrinsic redshift of 0215 + 015 is likely to be near 1.7. From a detailed column density comparison it is shown that the $z_{abs} = 1.345$ system is most probably an intervening galaxy, with the sight line intercepting both disk– and halo–type material.

158.064   The bursting behavior of the BL Lacertae object B2 1308 + 326.
S. L. Mufson, W. A. Stein, W. Z. Wiśniewski, J. T. Pollock, H. D. Aller, M. F. Aller.
Astrophys. J., Vol. 288, No. 2, p. 718 – 724 (1985).
Visual and radio observations of the BL Lac object B2 1308 + 326 have been accumulated for a period of several years. In this source, repetitive optical–near–infrared outbursts are observed which can have luminosities exceeding 10^{48}ergs s^{-1}. In the radio, bursting behavior is also observed. However, the radio bursts are less energetic and come from larger regions. It is argued that the photometric properties of B2 1308 + 326 are the result of nonthermal emission by an object or objects with $M > 10^8 M_\odot$ confined to a region of dimension $R \lesssim 10^{17}$cm.

158.065   IUE observations of the Seyfert 1.9 galaxy Markarian 423.
R. J. Rudy, R. D. Cohen, R. C. Puetter.
Astrophys. J., Lett. Ed., Vol. 288, No. 2, p. L29 – L32 (1985).
The authors present the first ultraviolet observations of a Seyfert 1.9 galaxy. The galaxy, Markarian 423, displays a moderately strong ultraviolet continuum in the 1200 – 2000 Å region, a strong narrow component of Lyα, and no detectable broad lines. Optical observations show little evidence for variation from the 1979 – 1980 measurements of Osterbrock. The absence of any broad ultraviolet lines indicates that the gas which gives rise to the broad emission lines is significantly reddened. The nonstellar continuum and the narrow–line emissions, however, are only slightly reddened.

158.066   Polarimetric and photometric studies of BL Lac. Analysis of observational data. I.
V. A. Hagen–Thorn, S. G. Marchenko, V. A. Yakovleva.
Astrofizika, Tom 22, Vyp. 1, p. 5 – 14 (1985). In Russian. English translation in Astrophysics, Vol. 22, No. 1.
An analysis is given of the results of polarimetric and photometric observations (1969 – 1982) of BL Lac published (1984). Photometric data suggest a two–component model of the object: a giant elliptical galaxy plus a point source. The point source varies in intensity while its energy distribution is invariable. Polarimetric data indicate that BL Lac has a preferred angle; the highest polarization always has directions close to this preferred direction. A strong variability of polarization on time scales of 1 hour was observed but only twice in 316 cases.

158.067   Photometric and polarimetric investigations of two BL Lac–type objects.
S. G. Marchenko.
Astrofizika, Tom 22, Vyp. 1, p. 15 – 23 (1985). In Russian. English translation in Astrophysics, Vol. 22, No. 1.
The results of photometric and polarimetric observations of BL Lac–type objects OI 090.4 in 1979 – 1982 and B2 1418 + 54 in 1980 – 1982 are given. The brightness and polarization variability for both objects on time scales from several years to several days are observed. The comparison of polarization parameters suggests the existence of preferred direction of polarization for both B2 1418 + 54 ($\theta_0 = 120°$) and OI 090.4 ($\theta_0 = 50°$). Some wavelength dependence of polarization parameters for OI 090.4 is found.

158.068   Observations of the Hβ region in some broad–line objects.
M. De Robertis.
Astrophys. J., Vol. 289, No. 1, p. 67 – 80 (1985).
Data are presented of the Hβ region in 27 broad–line objects (Seyfert 1 nuclei and QSOs) at a resolution of ≤ 5 Å. A technique is discussed for removing the Fe II emission from the Hβ region and is applied to the data. While most of the "cleaned" broad–

line profiles are symmetric, both blue and redward asymmetries persist in some of the objects. Despite the apparent diversity in the broad Hβ profile shapes, excellent correlations are found between the line width ratios at different intensities for the entire sample. This suggests that there may be a similar acceleration mechanism or geometry, or both operating in both the broad–and narrow–line regions. Further, there is an excellent correlation between the nonthermal nuclear magnitude and the Hβ line luminosity.

158.069 The hydrogen line spectra of narrow–line radio galaxies.
G. J. Ferland, D. E. Osterbrock.
Astrophys. J., Vol. 289, No. 1, p. 105 – 108 (1985). = Lick Obs. Bull., No. 994.

The authors report the results of the first detection of Lyα in a narrow–line galaxy. Nearly simultaneous optical and UV observations of 3C 192 and 3C 223 allow the measurement of both Balmer and Lyman decrements. These line ratios are approximate functions of the interstellar reddening and of a parameter which is proportional to the amount of H I collisional excitation present. Both galaxies have intrinsic Balmer and Lyman decrements which are significantly steeper than case B, suggesting that the gas is photoionized by a fairly hard X–ray continuum.

158.070 VLBI observations of the nucleus of M87 at two epochs.
J. H. M. M. Schmitt, M. J. Reid.
Astrophys. J., Vol. 289, No. 1, p. 120 – 123 (1985).

The authors present VLBI hybrid maps at 18 cm wavelength of the nucleus of M87 at epochs 1980.12 and 1982.27. The differences between these two maps are very slight. The authors could not detect internal proper motions at a $2\,\sigma$ confidence level of 0.3c. For a relativistic beaming model in which one sees the same material radiating at both epochs, the outflow velocity must exceed about 0.6c and the jet must be aligned to better than $12°$ with respect to the line of sight.

158.071 NGC 2110 – a Seyfert and X–ray elliptical(?) galaxy with displaced kinematic and light centers.
A. S. Wilson, J. A. Baldwin.
Astrophys. J., Vol. 289, No. 1, p. 124 – 128 (1985).

The authors report kinematic and imaging observations of NGC 2110 and find that the interstellar gas is orbiting with projected rotation axis parallel to the minor axis of the stellar light. The rotation curve is steep close to the center and flat farther out. However, the nucleus is displaced by ~ 220 pc from the apparent mass center of the galaxy. This last result is discussed in terms of models invoking differential obscuration of stars and gas, non–Keplerian motion of the gas, or a large mass of unseen material which dominates the rotation curve.

158.072 A model of the polarization position–angle swings in BL Lacertae objects.
A. Königl, A. R. Choudhuri.
Astrophys. J., Vol. 289, No. 1, p. 188 – 192 (1985).

The polarization position–angle swings that have been measured in a number of BL Lacertae objects and highly variable quasars are interpreted in terms of shocks waves which illuminate (by enhanced synchrotron radiation) successive transverse cross sections of a magnetized, relativistic jet. The jet is assumed to have a nonaxisymmetric magnetic field configuration. For a jet that is viewed at a small angle to the axis, the passage of a shock will give rise to an apparent rotation of the polarization position angle whose amplitude can be substantially larger than $180°$. The effects of freely propagating shocks are compared with those of bow shocks which form in front of dense obstacles in the jet, and specific applications to 0727–115 and BL Lacertae are considered.

158.073 Observations of extragalactic radio sources at 8.2 and 13.5 mm wavelengths.
V. A. Efanov, I. G. Moiseev, N. S. Nesterov.
Izv. Krymskoj Astrofiz. Obs., Tom 69, p. 78 – 90 (1984). In Russian. English translation in Bull. Crimean Astrophys. Obs., Vol. 69.

The results of observations of 95 extragalactic radio sources at 8.2 and 13.5 mm wavelength obtained in 1979 – 1982 are given. The observed emission variations of sources at millimeter waves are discussed briefly.

158.074 The multi–faceted active galaxy PKS 0521–36.
I. J. Danziger, P. A. Shaver, A. F. M. Moorwood, R. A. E. Fosbury, W. M. Goss, R. D. Ekers.
Messenger, No. 39, p. 20 – 22 (1985).

158.075 Explosionsartige Sternentstehung in der irregulären Galaxie M82.
P. Biermann.
Sterne Weltraum, 24. Jahrg., Nr. 4, p. 202 – 204 (1985).

158.076 Interferometric observations of the radio source 3C 134 in the decameter wave band.
A. V. Men', S. Ya. Braude, S. L. Rashkovskij, I. S. Fal'kovich, N. K. Sharykin, V. A. Shepelev, A. D. Khristenko.
Astron. Zh., Tom 62, Vyp. 1, p. 38 – 42 (1985). In Russian. English translation in Sov. Astron., Vol. 29, No. 1.

Results are presented of the observation of the radio source 3C 134 made by means of the wide–band decameter radio interferometer URAN–1. It is shown that the angular structure of this object does not appreciably differ from its radio brightness distribution at meter and decimeter waves. The analysis of radio interferometric measurements performered in the band of 20 to 1425 MHz has permitted to estimate the angular size of the source, which at the 50 per cent intensity level has been found equal to $37'' \pm 7''$.

158.077 New evidence for photoionization as the dominant excitation mechanism in Liners.
A. V. Filippenko.
Astrophys. J., Vol. 289, No. 2, p. 475 – 489 (1985).

Optical spectrophotometry of the nearby QSO MR 2251–178 and the powerful radio sources PKS 1718–649 (a classical Liner) and Pictor A (a type 1 Seyfert galaxy with features of Liners) is presented. There is strong evidence that the emitting gas in these objects is photoionized by nonstellar radiation. In all three nuclei, on the other hand, the great strength of [O III] λ4363 relative to [O III] λ5007 implies $T_e > 50{,}000$K, which is incompatible with photoionized low–density clouds. The dilemma vanishes if relatively dense clouds ($n_e \gtrsim 10^6 - 10^7 \mathrm{cm}^{-3}$) exist in the narrow–line regions. This paper provides unambiguous evidence for such high densities: it is shown that in each object the width of forbidden lines increases with their critical density for collisional deexcitation over the range $10^3 \lesssim n_e(\mathrm{crit}) \lesssim 10^7 \mathrm{cm}^{-3}$.

158.078 Sub–millijansky 1.4 GHz source counts and multicolor studies of weak radio galaxy populations.
R. A. Windhorst, G. K. Miley, F. N. Owen, R. G. Kron, D. C. Koo.
Astrophys. J., Vol. 289, No. 2, p. 494 – 513 (1985). With plates 5 – 7.

A deep survey has been carried out with the VLA at 21 cm to a noise level of 45 μJy. The region chosen was the Leiden–Berkeley Deep Survey (LBDS) area Lynx.2, known to be devoid of radio sources stronger than 10 mJy. A catalog of 124 sources is presented, of which 93 form a sample complete to the 5 σ level. The normalized differential source counts show a flattening below 5 mJy. Optical data are used to probe the nature of the sub–millijansky radio source population. Deep identifications and calibrated photometry are made on prime–focus plates obtained in four bands with the Kitt Peak 4 m Mayall telescope. In the complete sample 41 radio sources are reliably identified. It is shown that for radio sources with $1 < S_{1.4} < 10$ mJy a blue radio galaxy population becomes increasingly important; these often have a peculiar optical morphology indicative of interacting or merging galaxies. Below 1 mJy the VLA sources consist of some red galaxies, but the majority are faint blue galaxies mostly fainter than $V \approx 20$.

158.079   Extragalactic high–velocity clouds: VLA observations of the broad neutral hydrogen absorption in the radio galaxy 3C 293.
A. D. Haschick, W. A. Baan.
Astrophys. J., Vol. 289, No. 2, p. 574 – 581 (1985).

The radio galaxy 3C 293 has been mapped using the VLA at a resolution of $1\rlap{.}''5 \times 1\rlap{.}''3$ in order to study the broad neutral hydrogen absorption centered in its spectrum at a velocity of 13,500 km s^{-1}. The spectral–line maps showed that one component of the absorbing material has a steep velocity gradient of ~ 83 km s^{-1}arcsec^{-1} across the nuclear radio source. It is proposed that the absorbing gas associated with the strongest features is part of a rapidly rotating disk surrounding the nucleus of the galaxy. In addition, several high–velocity features indicate that gas is falling onto or is being expelled from the nucleus of 3C 293 with velocities up to 300 km s^{-1}.

158.080   Parsec–scale radio emission from the E/S0 galaxy NGC 3894.
J. M. Wrobel, D. L. Jones, D. B. Shaffer.
Astrophys. J., Vol. 289, No. 2, p. 598 – 602 (1985).

The radio continuum emission from the E/S0 galaxy NGC 3894 (1146 + 596) was mapped with VLBI techniques at 5 GHz. The hybrid map shows elongated structure on parsec scales which is closely aligned with the galaxy's isophotal minor axis. Although collimated on a parsec scale, the radio emission is amorphous in appearance on a 30–kpc scale. This large–scale emission may be inherently uncollimated, perhaps because of disruption of an intrinsically weak beam by its interaction with NGC 3894's interstellar medium, or by the interaction of NGC 3894 with its binary companion galaxy NGC 3895.

158.081   A radio–quiet emission–line jet in the type 2 Seyfert galaxy NGC 7682.
W. C. Keel.
Astron. J., Vol. 90, No. 4, p. 577 – 581, 689 (1985).

Narrowband imaging in Hα + [N II] shows a jet–like feature emerging from the nucleus of NGC 7682. The feature is 10″ in apparent length and less than $1\rlap{.}''5$ in width. For $H_0 = 50$, the length projects to 3 kpc, and the Hα luminosity of the jet is about 10^{39}erg s^{-1}. Spectroscopy shows weak emission most likely due to [N II] at $+ 275$ km s^{-1} relative to the nucleus. VLA–B maps at 6 cm show no emission from the feature above ~ 1 mJy. A consistent picture may be made of this as a "fossil" jet now seen as a string of H II regions formed of entrained interstellar matter after an earlier radio jet turned off $\sim 10^7$years ago, representing either conventional (OB–star powered) regions or cooling jet matter.

158.082   Globular clusters and the distance to M87.
S. van den Bergh, C. Pritchet, C. Grillmair.
Astron. J., Vol. 90, No. 4, p. 595 – 599 (1985).

The luminosity function of the globular clusters surrounding M87 (NGC 4486) has been determined down to $B = 25.4$. The luminosity function possesses a turnover at $B = 25.0 \pm 0.3$ (estimated uncertainty), i.e., 0.4 mag above the plate limit. Assuming that this peak occurs at the same absolute magnitude as that of the luminosity of Local Group globulars yields a distance modulus $(m–M)_0 = 31.43 \pm 0.3$, which corresponds to a distance 19 ± 3 Mpc. After correction for Virgocentric infall, a value of 68 ± 10 km s^{-1}Mpc^{-1} is obtained for the global Hubble parameter. The luminosity function of globular clusters surrounding M87 is not well fitted by a Gaussian with σ = 1.2 (which fits the data for Local Group globular clusters quite well). The authors find that the radial distribution of globular clusters near the core of M87 is much shallower than that of the light in M87. The total number of globular clusters associated with M87 is estimated to be $\sim 2 \times 10^4$.

158.083   8 – 13 μm spectrophotometry of galaxies. IV. Six more Seyferts and 3C345.
D. K. Aitken, P. F. Roche.
Mon. Not. R. Astron. Soc., Vol. 213, No. 4, p. 777 – 788 (1985).

8 – 13 μm spectra of a further six Seyfert nuclei and the quasar 3C345 are presented, together with those of 4 more starburst nuclei in a companion paper. With other published work such spectra now exist for 21 active and 13 starburst nuclei, and this division is clearly distinguished by the form of their 10–μm spectra. The active galaxies are typified by power–law spectra with a tendency for the spectral index to steepen with the progression quasar – Seyfert 1 – Seyfert 2 and with silicate absorption sometimes occurring in Seyfert 2's. In contrast the starburst galaxy spectra are dominated by emission from the unidentified dust–associated features and frequently show the [Ne II] fine structure line.

158.084   MERLIN observations of OH masers around the active galactic nucleus IC 4553.
R. P. Norris, W. A. Baan, A. D. Haschick, P. J. Diamond, R. S. Booth.
Mon. Not. R. Astron. Soc., Vol. 213, No. 4, p. 821 – 831 (1985).

IC 4553, a peculiar galaxy at a redshift 0.02, contains the most luminous and the most distant OH masers yet discovered. The authors present 18–cm MERLIN observations of both the OH masers and the nuclear continuum source, and find that the masers amplify the image of the nucleus. The nucleus is a core–dominated triple source similar to those seen in other active galactic nuclei, whilst the masers consist of molecular clouds in the plane of the galaxy. The OH populations of these clouds are probably inverted by far–infrared radiation from a recent or current starburst. The masers represent a new tool with which to study active galactic nuclei, and the observations indicate that the triple continuum source represents the interaction of a pair of relativistic jets with a rotating interstellar medium.

158.085   Warmers: the missing link between Starburst and Seyfert galaxies.
R. Terlevich, J. Melnick.
Mon. Not. R. Astron. Soc., Vol. 213, No. 4, p. 841 – 856 (1985).

Recent observational and theoretical work has shown that, as a consequence of mass loss in the form of stellar winds in the last stages of their evolution, massive stars can reach effective temperatures of more than 100000K, being observed as extreme WC or WO Wolf–Rayet stars. This paper examines the effect of these hot stars (which the authors call Warmers) in the evolution of metal–rich giant H II regions. The authors show that as a function of age, the emission–line spectrum of the H II region evolves first into a type 2 Seyfert spectrum and then into a Liner–type spectrum. The authors suggest that many of the active nuclei classified as type 2 Seyferts and Liners are not associated with a non–thermal power source but rather with violent star formation activity at high metal abundance.

158.086   CCD magnitudes of 3C radio sources.
S. A. Eales.
Mon. Not. R. Astron. Soc., Vol. 213, No. 4, p. 899 – 904 (1985).

As a by–product of an investigation of the environments of powerful radio sources, R magnitudes have been estimated from CCD images for 23 3C radio galaxies and quasars. Although not as accurate as those measured during rigorous photometric investigations, they are considerably more so than estimates from the Palomar Sky Survey or from deep photographic plates. An accurate position has been obtained for the radio galaxy 3C 268.3. The Hubble diagram of 3C radio galaxies is briefly discussed.

158.087   The variability and radio structure of 3C 159.
I. W. A. Browne, F. Mantovani, T. W. B. Muxlow, L. Padrielli, J. D. Romney.
Mon. Not. R. Astron. Soc., Vol. 213, No. 4, p. 945 – 951 (1985).

The 408–MHz flux density of the high–luminosity radio galaxy 3C 159 shows strong evidence for variability. Intrinsic variability or variability due to propagation effects in the interstellar medium would imply the existence of a strong, very compact, component in the radio structure, whereas MERLIN and VLBI observations show no evidence for such a component. The authors conclude that the most likely explanation of these apparently contradictory observations is that the flux density changes are caused by a combination of exceptionally high linear polarization and variable ionospheric Faraday rotation.

158.088 Radio and optical observations of the radio galaxy 0831 + 557.
N. D. Whyborn, I. W. A. Browne, P. N. Wilkinson,
R. W. Porcas, H. Spinrad.
Mon. Not. R. Astron. Soc., Vol. 214, No. 1, p. 55 – 65 (1985).

Radio and optical observations of the high–luminosity radio galaxy 0831 + 557 are presented. MERLIN and VLBI maps, ranging in resolution from 1.7 arcsec to 15 milliarcsec, reveal low–brightness, steep–spectrum emission of overall extent 11 arcsec, and a compact core, with two components separated by 160 milliarcsec each exhibiting a spectral peak. These components combine with the extended emission to give an overall radio spectrum which is that between 178 MHz and 2.7 GHz. The optical appearance and spectrum of 0831 + 557, however, show little trace of a non–thermal continuum nor the broad emission lines normally associated with such flat–spectrum radio sources. Some properties of the source can be accounted for within the framework of relativistic beaming models but others point to it being an intrinsically unusual object. It may be a member of the class of compact VLBI doubles.

158.089 Hotspots in radio galaxies: a comparison with hydrodynamic simulations.
M. D. Smith, M. L. Norman, K.-H. A. Winkler, L. Smarr.
Mon. Not. R. Astron. Soc., Vol. 214, No. 1, p. 67 – 85 (1985).

Surface brightness distributions are derived for the beam cap of an axisymmetric supersonic jet ramming through its confining medium. The detailed structure and time–evolution of the working surface is computed numerically with a high–resolution gas dynamics code employing a special technique for accurately tracking and preserving the jet boundary. The synchrotron emissivity is assumed to be proportional to the square of the gas pressure inside the jet, and zero outside. The geometry of the beam cap is found to change on a dynamical time–scale, producing an enormous variety of hotspot shapes in a single evolution. A classification of well–resolved hotspots in radio galaxies is made with which the derived contour maps are compared. Several quite common features of observed hotspots are reproduced.

158.090 Recent star formation in interacting galaxies. II. Super starbursts in merging galaxies.
R. D. Joseph, G. S. Wright.
Mon. Not. R. Astron. Soc., Vol. 214, No. 1, p. 87 – 95 (1985).

The subset of galaxy-galaxy interactions which have resulted in a merger are, as a class, ultraluminous IR galaxies. Their IR luminosities span a narrow range which overlaps with the most luminous Seyfert galaxies. However, in contrast with Seyfert galaxies, the available optical, IR, and radio properties of mergers show no evidence for a compact non–thermal central source, and are easily understood in terms of a burst of star formation of extraordinary intensity and spatial extent: they are "super starbursts". The authors argue that super starbursts occur in the evolution of most mergers, and discuss the implications of super starbursts for the suggestion that mergers evolve into elliptical galaxies.

158.091 Non–stellar radiation in radio galaxies at 3.5 μm.
S. J. Lilly, M. S. Longair, M. Miller.
Mon. Not. R. Astron. Soc., Vol. 214, No. 1, p. 109 – 118 (1985).

$JHKLL'$ observations of 24 3CR radio galaxies forming a statistically complete sample at redshifts between 0.01 and 0.20 are presented and discussed. Over a third of the sample show significant non–stellar radiation components at 3.5 μm, but only those galaxies with strong optical emission lines are found to have infrared excesses. The remaining galaxies in which the emission lines are weak or absent have continua from 1 to 3.5 μm that are purely stellar in origin. In the galaxies with excesses, the strength of the excess is loosely correlated with the strength of the Hβ line and with the nuclear core radio luminosity and X–ray luminosity in the sense that the more luminous objects have stronger infrared excesses. It is inferred that the narrow–line objects probably have steeper non–stellar continua than the broad–line radio galaxies, and it is suggested that the non–stellar emission in the narrow–lined objects is produced by thermally reradiating dust associated with gas in the radio galaxies.

158.092 Shock formation of the broad emission–line regions in QSOs and active galactic nuclei.
J. J. Perry, J. E. Dyson.
Mon. Not. R. Astron. Soc., Vol. 213, No. 3, p. 665 – 710 (1985).

The authors present a model for the formation of the BLR (*broad emission–line regions*) of QSOs and AGNs which accounts simultaneously for the formation, structure and velocities of the clouds. These emerge as a natural consequence of the interaction of optically thin supersonic flows with the dense stellar cluster which is assumed to exist in the immediate circumquasar environment. Low density gas near a very luminous source must be hot with $T_e \approx 10^7$–10^8K. So long as such gas flows supersonically, stand–off shocks inevitably form around all obstacles to the flow. The high–pressure, superheated gas in the shocked layer cools rapidly by inverse–Compton cooling off the radiation field. If the obstacles are large enough so that the flow time around them is longer than the cooling time, cold condensations – which are identified with the BLR clouds – form in the shocked gas. Both supernovae shells and groups of stars with strong stellar winds are shown to provide such cooling obstacles. The clouds are continuously generated and need no external confinement mechanism.

158.093 The structure of three compact double radio sources at 5 GHz.
R. L. Mutel, M. W. Hodges, R. B. Phillips.
Astrophys. J., Vol. 290, No. 1, p. 86 – 93 (1985).

The compact double sources 1518 + 047, CTD 93, and 2050 + 364 have been mapped at 5 GHz using a five–element VLBI array. Comparison with previous maps at 1.6 GHz shows that individual components of each double have nearly identical steep spectral indices, consistent with the models in which both components originate on opposite sides of a central energy source. Individual components often show complex structure, including the "head–tail" morphology characteristic of luminous, extended double sources.

158.094 M87: round and slow.
L. Dones, S. D. M. White.
Astrophys. J., Vol. 290, No. 1, p. 94 – 95 (1985).

The authors show that the X–ray emitting gas in M87 is flowing through the galaxy at less than 10 km s^{-1}. This rules out models in which motion of the gas explains the asymmetric radio continuum and Hβ emission in the central regions.

158.095 The infrared and radio morphology of the "hot–spot" galaxy NGC 2903.
C. G. Wynn–Williams, E. E. Becklin.
Astrophys. J., Vol. 290, No. 1, p. 108 – 115 (1985).

New maps of the "hot–spot" galaxy NGC 2903 made at the IRTF and VLA show that while the "starburst" model for the activity in the central 0.6 kpc is valid, the star–forming activity is definitely not confined to the hot–spots themselves. Comparison of 10–μm and free–free flux densities of this and other galaxies suggests there is an extra component of 10–μm emission from galaxy nuclei above that expected from Lyα heating. It is suggested that this emission may arise from small dust grains that are temporarily heated to high temperatures by visible or UV photons.

158.096 $10^{12}L_\odot$ starbursts and shocked molecular hydrogen in the colliding galaxies Arp 220 (= IC 4553) and NGC 6240.
G. H. Rieke, R. M. Cutri, J. H. Black, W. F. Kailey,
C. W. McAlary, M. J. Lebofsky, R. Elston.
Astrophys. J., Vol. 290, No. 1, p. 116 – 124 (1985).

The authors present infrared spectroscopy and photometry and optical spectroscopy of the exceedingly luminous interacting galaxies Arp 220 (= IC 4553) and NGC 6240. These galaxies are the sites of exceedingly powerful bursts of star formation, as shown by (1) the luminosity and extent of the stellar component seen near 2 μm; (2) the large population of red supergiants required to account for the depth of the stellar CO absorption; (3) the absorption and emission features in the spectrum longward of 3 μm; and (4) the extended luminosity source indicated by the spatial extent of the galaxies at 10 μm. Both galaxies also show strong lines from shocked molecular hydrogen.

158.097   Ca II emission in I Zwicky 1.
S. E. Persson, P. J. McGregor.
Astrophys. J., Vol. 290, No. 1, p. 125 – 129 (1985).

Spectrophotometric data between 6000 and 10,000 Å are presented of the nucleus of the Seyfert 1 galaxy I Zwicky 1. The brightest emission lines in this spectral region are the λ8446 line of O I, the λλ8498, 8542, and 8662 lines of Ca II (the infrared triplet), and a strong blend of low–excitation lines near H I Pa9. A few parameters of the broad–line region (BLR) that control the Ca II emission are discussed. Similarities between I Zw 1 and Mrk 231, another Ca II emitter, are noted and together contrasted with other active galaxies. The deduced physical conditions in the Ca II emission zones within the BLR of these two active galaxies are shown to be similar to low–ionization zones in certain galactic emission–line objects.

158.098   X–ray spectrum and variability of 3C 120.
J. P. Halpern.
Astrophys. J., Vol. 290, No. 1, p. 130 – 135 (1985).

The X–ray luminosity of 3C 120 is variable by a factor of 2.5 on time scales of days to months. The spectral slope of the $2 - 10$ keV X–rays changes systematically in the sense that higher intensity states are steeper. Simple power laws fitted to the data are consistent with an energy index α of 0.51 ± 0.13 at minimum intensity and 0.83 ± 0.03 at maximum, with a continuous variation in between. The results are interpreted in terms of beamed synchrotron emission from a relativistic jet, a model which has been invoked previously for the spectra of BL Lac objects.

158.099   Theoretical properties of radio–galaxies. II. Luminosity and energetics.
L. Zaninetti.
Astrophys. Space Sci., Vol. 109, No. 2, p. 287 – 292 (1985).

The author deduces, following a dimensional approach, the theoretical luminosity of a radio–galaxy and makes a confrontation with the observed one as deduced from a large sample of symmetric double radio–galaxies. Equalizing the two luminosities, he obtains the density and the energetics function of the intrinsic radio–luminosities.

158.100   Hot intercloud gas in the nuclei of Seyfert galaxies.
A. S. Zentsova.
Astrophys. Space Sci., Vol. 109, No. 2, p. 327 – 332 (1985).

Physical conditions are found for a hot intercloud gas in the nuclei of Seyfert galaxies. The gas temperature is determined by photoionization and Compton–scattering of the shortwave radiation of the nucleus. Using observational data for the coronal emission line [Fe X] λ6374 Å, the gas density $n = 10^4 \mathrm{cm}^{-3}$ and temperature $T = 10^6$K, typical for the distance 2 pc from the central source, are obtained. It is shown that the intercloud gas is in the state of accretion by the nucleus with a rate $\dot{M} \cong 10^{-2} M_\odot \mathrm{yr}^{-1}$.

158.101   Periodicity in the BL Lac object OJ 287.
L. Carrasco, D. Dultzin–Hacyan, I. Cruz–Gonzalez.
Nature, Vol. 314, No. 6007, p. 146 – 148 (1985).

Photometric observations of OJ 287 reveal periodic variability in timescales of a few minutes. It is concluded that the results are compatible with the presence of "hotspots" in a thick accretion disk spiraling into a Kerr's supermassive black hole and that the central source is seen nearly pole–on.

158.102   A 15.7–min periodicity in OJ 287.
E. Valtaoja, H. Lehto, P. Teerikorpi, T. Korhonen, M. Valtonen, H. Teräsranta, E. Salonen, S. Urpo, M. Tiuri, V. Piirola, W. C. Saslaw.
Nature, Vol. 314, No. 6007, p. 148 – 149 (1985).

The authors report here the detection of a 15.7–min periodic flux variation in OJ 287 in April 1981 at 37 GHz. The source has been subsequently monitored for 265 h at 22 and 37 GHz, as well as in optical photometry, and the same period has been seen in simultaneous observations. The peak–peak amplitude of the variation is 1 – 10% of the total intensity. Other periods may also exist.

158.103   Infrared Seyferts: a new population of active galaxies?
M. H. K. de Grijp, G. K. Miley, J. Lub, T. de Jong.
Nature, Vol. 314, No. 6008, p. 240 – 242 (1985).

The authors present a new method for finding hitherto unknown Seyferts based on their mid–infrared spectra and they show, using data from IRAS, that this method has a success rate of ~70%. One Seyfert, so discovered, contains the most luminous infrared nucleus known so far.

158.104   NGC 1808: a nearby galaxy with a faint Seyfert nucleus.
M.–P. Véron–Cetty, P. Véron.
Astron. Astrophys., Vol. 145, No. 2, p. 425 – 429 (1985).

Spectroscopic observations show that the nucleus of NGC 1808 has some Seyfert characteristics, namely broad (FWHM 550 km s^{-1}) emission lines with a relatively strong [N II] λ6583 line. This nucleus is embedded in an extended and very bright emission line region, ionized by hot stars and rotating with the galaxy, which is not obviously genetically related to the Seyfert nucleus itself.

158.105   UBV photometry of some galaxies with UV excess.
V. S. Tamazyan.
Soobshch. Byurakan. Obs., Vyp. 54, p. 80 – 88 (1983). In Russian.

Detailed photographic UBV photometry of the galaxies Nos. 18, 19, 23, 31 and 300 from lists by Kazaryan are carried out. It is shown that the UV excess in the galaxies Nos. 18, 23 and 300 is concentrated in nuclear regions but in the galaxy No. 19 is widespread. The galaxy No. 31 by photometric data is close to normal elliptical galaxies.

158.106   Cosmological evolution of linear sizes of radio galaxies.
V. K. Kapahi.
Mon. Not. R. Astron. Soc., Vol. 214, No. 2, p. 19P – 23P (1985).

The angular size–redshift relation is investigated for complete samples of radio galaxies of a constant radio luminosity by combining data on relatively nearby galaxies in strong and intermediate source surveys with distant galaxies in the recent Leiden–Berkeley deep radio and optical survey. The data are shown to provide strong evidence for linear–size evolution independent of any luminosity–size relationship. It also appears unlikely that the contribution of compact steep–spectrum sources at large redshifts can obviate the need for size evolution.

158.107   Low frequency variability and interstellar focusing.
T. V. Cawthorne, B. J. Rickett.
Nature, Vol. 315, No. 6014, p. 40 – 42 (1985).

The flux variations of extragalactic radio sources at low radio frequencies have proved very difficult to explain as changes in the luminosity of a simple synchrotron source. It has been suggested recently that the observed flux changes may be the result of a focusing effect of irregularities in the local interstellar medium. The authors present evidence from a sample of radio sources, monitored over five years at Bologna, which suggests that, in accordance with this explanation, low frequency variability is a more common feature of sources at low galactic latitudes.

158.108   The nucleus of the Seyfert galaxy NGC 4151: a search for 1–day variability in the broad emission–line components.
N. G. Bochkarëv.
Sov. Astron. Lett., Vol. 10, No. 4, p. 239 – 242 (1984). English translation of 38.158.048.

158.109   Variability of the radiation of active galactic nuclei.
V. M. Lyutyj.
Zemlya Vselennaya, No. 2, p. 46 – 51 (1985). In Russian.

158.110 The fading of the Seyfert galaxy F–9.
W. Kollatschny, K. J. Fricke.
Astron. Astrophys., Vol. 146, No. 1, p. L11 – L14 (1985).
Optical variability of the once extremely bright Seyfert 1 galaxy F–9 is discussed. The optical Fe II–lines and Balmer–lines changed in a different way. The spectrum of F–9 is approaching that of a Seyfert 2 galaxy.

158.111 Observations of emission line galaxies. III. The broad Balmer line profiles of 12 Seyfert–1 galaxies.
P. Rafanelli.
Astron. Astrophys., Vol. 146, No. 1, p. 17 – 24 (1985).
Spectrophotometric observations of a sample of 12 compact Seyfert–1 galaxies have been obtained with resolutions of 6 Å and 9 Å in the spectral range 3800 – 7250 Å. The high signal to noise ratio of these spectra has allowed to measure the fluxes in the Balmer lines up to H_δ with typical errors of the order of 10% for H_α, H_β, H_γ and 20% for H_δ. A linear correlation between the $\log[F(H_\gamma)/F(H_\beta)]_{BC}$ and $\log[F(H_\delta)/F(H_\beta)]_{BC}$ ratios, which nearly corresponds to the galactic reddening line, has been found. The absolute value of the asymmetry index at 20% and 50% intensity level of the H_β broad component was found to be correlated with the $[F(H_\gamma)/F(H_\beta)]_{BC}$ ratio providing a new constraint on future models of the broad emission–line region.

158.112 A deep Westerbork survey of areas with multicolor Mayall 4 m plates. III. Photometry and spectroscopy of faint source identifications.
R. G. Kron, D. C. Koo, R. Windhorst.
Astron. Astrophys., Vol. 146, No. 1, p. 38 – 58 (1985).
The paper aims to contribute to the empirical understanding of faint radio galaxies at cosmological distances. The authors intend to achieve this by photometric and spectroscopic measurements of the optical identifications of very weak radio sources found in a recent deep survey with the Westerbork Synthesis Radio Telescope. Among 302 sources in the complete sample with $S_{1.4GHz} \gtrsim 0.6$ mJy, 171 reliable optical identifications were found on deep, multicolor Mayall 4 m plates. The authors discuss the characteristics of the radio galaxy sample by combining the optical and radio data in ways that enable the radio galaxies to be classified. For the giant elliptical class, it is investigated whether the data show evidence for evolution of the optical spectra. The authors discuss the quasars and the radio stars. The bivariate radio–optical flux density distribution is used to put constraints on the epoch–dependent radio luminosity function of weak radio galaxies.

158.113 Alignment of southern radio sources.
G. L. White, W. B. McAdam, I. G. Jones.
Proc. Astron. Soc. Aust., Vol. 5, No. 4, p. 507 – 510 (1984).

158.114 IC 4553: masers around an active galactic nucleus.
R. P. Norris.
Proc. Astron. Soc. Aust., Vol. 5, No. 4, p. 514 – 516 (1984).
The author describes radio, optical, and infrared observations of IC 4553. The observations indicate that at the centre of the galaxy is a Seyfert–like active galactic nucleus, which appears double on optical plates only because of the dense dust lane which bisects the nuclear region. This indicates that IC 4553 is an edge–on Seyfert–like galaxy. The OH maser radiation is a result of stimulation of the OH along the line of sight within the galactic plane by the strong far–infrared radiation from the nucleus.

158.115 Steep–spectrum radio sources in clusters of galaxies – the southern sample.
O. B. Slee, J. E. Reynolds.
Proc. Astron. Soc. Aust., Vol. 5, No. 4, p. 516 – 529 (1984).
In order to clarify the distribution of spectral index in sources with unusually steep spectral indices $\alpha < -1.5$ and to compare their radio, optical and X–ray morphologies the authors selected 11 very–steep–spectrum sources from the cluster sample for detailed observation with the VLA at 1.465 and 4.885 GHz. Six of these sources south of declination $-10°$ were also satisfactory observable with the Molonglo Observatory synthesis telescope

(MOST) at 0.843 GHz. This paper is concerned with the properties of the southern sample.

158.116 The effects of interactions on spiral galaxies. I. Nuclear activity and star formation.
W. C. Keel, R. C. Kennicutt Jr., E. Hummel, J. M. van der Hulst.
Astron. J., Vol. 90, No. 5, p. 708 – 730 (1985). With plates 60 – 63.
The authors have obtained spectra of the nuclei of 161 interacting spirals, and compared their properties to those of a complete sample of field spirals. They find that the interacting systems show: (1) Significantly higher levels of emission, in equivalent width and luminosity. (2) More active and extensive nuclear star formation, over all Hubble types. (3) More frequent Seyfert nuclei among close pairs, but fewer among very disturbed (post-encounter) systems. (4) Fewer low–ionization nuclei, which have presumably been turned into Seyferts or masked by nuclear H II regions. The higher incidence of Seyferts in the complete sample supports the notion of central compact objects' being common but normally quiescent in luminous spirals, while their deficiency among distorted pairs suggests that these objects may become gas starved.

158.117 Compact radio sources in the 3C catalog.
T. J. Pearson, R. A. Perley, A. C. S. Readhead.
Astron. J., Vol. 90, No. 5, p. 738 – 755 (1985).
Maps of 36 compact sources from the 3C catalog have been made with the VLA at 6 cm wavelength, with a resolution of $0\rlap{.}''4$. Most of the sources are "steep–spectrum compact sources". They show a variety of structures; several appear to be small "classical double" sources, while others are more complex.

158.118 Determination of physical parameters of jets in radiogalaxies.
G. Grosso, A. Ferrari.
ESA Spec. Publ., ESA SP–207, p. 239 – 242 (1984). – See Abstr. 012.044.
High–resolution observations of radiojets by the Very Large Array have been statistically collected by Bridle and Perley in a recent review. Their list allows in principle the determination of the physical parameters for a large sample of jets in the attempt to define a "reference standard". In this note the authors present preliminary results of such a study and comment on the intrinsic limitations of presently available informations.

158.119 MHD turbulence in radiogalaxies.
L. Zaninetti, G. Pelletier.
ESA Spec. Publ., ESA SP–207, p. 243 – 246 (1984). – See Abstr. 012.044.
The authors investigate the efficiency of turbulence produced by large hydrodynamic instabilities such as Kelvin–Helmholtz and internal shear to make radio–jets shine. Assuming that most of the power of a turbulent cascade is absorbed by radiating particles, the authors deduce the contribution of hydrodynamic instabilities to the luminosity. The efficiency of the turbulent cascade is very sensitive to the variation of the Mach number and this could explain the morphological differences between the low luminosity sources with supersonic flow and the high luminosity sources with hypersonic flow. The theoretical luminosity derived is applied to a sample of extragalactic radiosources.

158.120 Jets in galaxies.
A. Ferrari.
ESA Spec. Publ., ESA SP–207, p. 259 – 273 (1984). – See Abstr. 012.044.
Powerful extragalactic radio sources consist of two extended regions separated by kilo or megaparsecs containing magnetic fields and relativistic particles, and linked by jets to a central compact radio source located in the nucleus of the associated (optical) galaxy. These jets are collimated plasma streams transferring mass, momentum, energy and magnetic flux from the nuclei to the extended components. In this review the study of the dynamics of origin and propagation and of the radiation proper-

ties of jets is presented as a challenging subject for plasma physics.

158.121 Small scale structure and the spinar model for active galactic nuclei.
E. Asseo.
ESA Spec. Publ., ESA SP–207, p. 303 – 306 (1984). – See Abstr. 012.044.

The author considers the properties of strong, low frequency electromagnetic waves associated to the magnetic oblique rotator model for AGN. In the presence of a large scale magnetic field in the surroundings of AGN, the otherwise catastrophic Weibel instability is inefficient. Then a slower damping mechanism, associated to the radiation reaction of particles moving in and interacting with the strong waves, may lead to the formation of small scale (VLBI) structures. This may also explain some VLBI features observed in the jets from extended extragalactic radio sources, in case a precollimation process on larger scales is available.

158.122 Optical emission–line gas associated with the radio source 3C 277.3.
W. van Breugel, G. Miley, T. Heckman, H. Butcher, A. Bridle.
Astrophys. J., Vol. 290, No. 2, p. 496 – 516 (1985). With plates 10 – 12.

The authors present the results of a detailed optical and radio investigation of the radio galaxy 3C 277.3. First, they have used the VLA to obtain radio maps at several wavelengths. Second, using a CCD on the 4 m telescope at KPNO they have obtained narrow–band images in sub–arcsecond seeing which show the detailed morphology of the line–emitting gas. Third, using long slit spectroscopes on the 4 m telescope, they have obtained kinematic information about the gas at various positions and in several position angles. Fourth, using an aperture spectroscope on the 2.3 m telescope of Steward Observatory, they have observed the southern jet in order to study the density and temperature of the line–emitting gas. These data demonstrate conclusively the existence of morphological and energetic relationships between the radio plasma and the emission–line gas. The authors discuss the observations in terms of twin radio jets which propagate through an inhomogeneous (cloudy) medium. The results support a picture in which the emission–line gas is *local* material which is excited and accelerated by the radio source.

158.123 Spectropolarimetry of Seyfert nuclei.
G. D. Schmidt, J. S. Miller.
Astrophys. J., Vol. 290, No. 2, p. 517 – 530 (1985).

The authors present the results of new spectropolarimetric observations of the Seyfert galaxies Mrk 3, Mrk 231, Mrk 486, NGC 3227, and NGC 6814. The extensive measurements, obtained with a spectral resolution (~ 10 Å) far better than in previous studies, are analyzed and compared with the accompanying high–quality flux spectra to discern both the origin of polarization and, where possible, the structure of the nuclear environment.

158.124 Magnetic field structures in active compact radio sources.
T. W. Jones, L. Rudnick, H. D. Aller, M. F. Aller,
P. E. Hodge, R. L. Fiedler.
Astrophys. J., Vol. 290, No. 2, p. 627 – 636 (1985).

The authors present the analysis of simultaneous multifrequency linear polarimetry data between 1.4 GHz and 90 GHz for about 20 active, compact radio sources at six epochs from 1977 December to 1980 July. The general polarization characteristics of these sources can be well described in terms of magnetic fields which are largely turbulent and slightly anisotropic. The magnetic field symmetry axes are generally aligned with the source structural axes on the milli–arcsecond scale. Observed polarization variations and in particular "rotator" polarization events can be produced in this model as a consequence of "random walks" generated through evolution of the turbulent magnetic field.

158.125 The local luminosity function of Seyfert 1 nuclei.
F. Z. Cheng, L. Danese, G. De Zotti,
A. Franceschini.
Mem. Soc. Astron. Ital., Vol. 55, No. 3, p. 597 – 608 (1984). – See Abstr. 012.050.

The authors present a new derivation of the local luminosity function of Seyfert 1 to 1.9 nuclei. Nuclear magnitudes have been estimated using two independent methods. The luminosity function matches remarkably well that of optically selected QSOs at $M_B \cong -23$ ($H_0 = 50$) and compares very favourably with the local luminosity function of X–ray selected active galactic nuclei.

158.126 A search for "dwarf" Seyfert 1 nuclei. I. The initial data and results.
A. V. Filippenko, W. L. W. Sargent.
Astrophys. J., Suppl. Ser., Vol. 57, No. 3, p. 503 – 522 (1985).

A sensitive search for low–luminosity Seyfert 1 nuclei is described and illustrated. Compared with the forbidden emission, Hα exhibits broad wings in 19 – 28 of 75 objects, including at least five E or E/S0 galaxies. Of the 26 low–ionization nuclear emission–line regions (Liners) also discussed by Heckman, 8 – 12 show broad Hα. [O I] $\lambda6300$ is broader than the [S II] $\lambda\lambda6716$, 6731 lines in about one–third of the Liners, implying the existence of clouds having very different densities. The faint end of the luminosity function for active galactic nuclei must be much more populated than was previously believed.

158.127 Spectra of galaxies with UV continuum. V.
B. E. Markarian, L. K. Erastova, V. A. Lipovetskij,
D. A. Stepanyan, A. I. Shapovalova.
Astrofizika, Tom 22, Vyp. 2, p. 215 – 227 (1985). In Russian. English translation in Astrophysics, Vol. 22, No. 2.

The results of spectroscopic observations of 88 objects from the lists of galaxies with UV continuum are presented. For most of the investigated objects the presence of emission lines in red and green parts of the spectra is established. The redshifts and luminosities for all of these galaxies are also determined. Four galaxies have Seyfert features in their spectra. Two of them – Mark 670 and Mark 822 – are Seyferts 2, while two other – Mark 864 and Mark 1400 – are Seyfert 1 galaxies.

158.128 On some new ultraviolet galaxies with jets.
A. R. Petrosyan, K. A. Saakyan, Eh. E. Khachikyan.
Astrofizika, Tom 22, Vyp. 2, p. 229 – 238 (1985). In Russian. English translation in Astrophysics, Vol. 22, No. 2.

A morphological investigation of the UV galaxies Mark 423, 739, 773 and 984 with jets has been carried out. Some of their parameters are estimated. For one of them, Mark 984, a spectral investigation of the nucleus and two condensations in the jet are carried out. Some peculiarities of galaxies with jets are noted. The great majority of galaxies with jets show Seyfert characteristics.

158.129 Spectral observations of the galaxy M82. II.
A. S. Amirkhanyan, V. A. Hagen–Thorn,
V. P. Reshetnikov.
Astrofizika, Tom 22, Vyp. 2, p. 239 – 246 (1985). In Russian. English translation in Astrophysics, Vol. 22, No. 2.

The results of spectral observations of the peculiar galaxy M82 are given. The emission line splitting in spectra of the south filaments is studied. Electron densities in the filamentary structure and radial velocity curves for six positions are found. New data allow to return to the explanation of observed polarization in terms of scattering of radiation of central galaxy regions by dust and free electrons.

158.130 Similarity between the extended components of radio galaxies and plerion–type supernova remnants.
B. V. Komberg, M. A. Smirnov.
Astrofizika, Tom 22, Vyp. 2, p. 257 – 266 (1985). In Russian. English translation in Astrophysics, Vol. 22, No. 2.

In order to have a more representative sample of extended components of double radio galaxies, surface radio brightness – size functions at different frequencies and for different redshift ranges are obtained. The mean slope of these functions is 2.5 ± 0.28, which is close to the value for plerion–type SNR

(2.35±0.20). The coincidence is obviously a manifestation of a continued supply of extended components by high–energy particles from the nuclei. The 1400 MHz subsample for radio galaxies with $z \leqslant 0.1$ including 55 extended components is used to plot a similar dependence and in fact, it may help to estimate the distances to unidentified near radio sources.

158.131 On the nature of variability of radiation from active galactic nuclei.
V. G. Gorbatskij.
Astrofizika, Tom 22, Vyp. 2, p. 267 – 271 (1985). In Russian. English translation in Astrophysics, Vol. 22, No. 2.

Observed fast variability of X–ray radiation from active nuclei is considered. Assuming the hypothesis of ejection of relativistic electron emitters from the active nucleus the strength of the magnetic field and relativistic electron density in the flare region are estimated.

158.132 The orientations of the rotation axes of radio galaxies. I. Radio morphologies of bright elliptical galaxies.
M. Birkinshaw, R. L. Davies.
Astrophys. J., Vol. 291, No. 1, p. 32 – 44 (1985).

The results of VLA observations of 47 bright elliptical galaxies from a sample of 54 galaxies for dynamical study are presented, and the structures of the extended sources are described. No evidence is found for a unique alignment of the radio axes and minor axes of the host galaxies. The distribution of misalignment angles is consistent with being uniform between 0° and 50°, with larger misalignments being significantly less likely. No significant correlations are found between the radio power or misalignment angle and any of the intrinsic kinematic parameters.

158.133 Synchrotron aging in the lobes of luminous radio galaxies.
S. T. Myers, S. R. Spangler.
Astrophys. J., Vol. 291, No. 1, p. 52 – 62 (1985).

The authors have made a study of the gradient of the 1.4 – 5.0 GHz spectral index observed in five luminous 3C radio galaxies. The observations have been compared with a model in which the radiation is due to an isotropic ensemble of electrons subject to synchrotron radiation losses (synchrotron aging). The speeds of separation of the hot spots and lobe material are estimated. The results are consistent with the beam model, in which the lobe material is left behind, by a hot spot advancing through the intergalactic medium at speeds $\sim 10^4 \mathrm{km\ s^{-1}}$.

158.134 *IUE* observations of a starburst disk and the detectability of high redshift galaxies.
D. W. Weedman, D. P. Huenemoerder.
Astrophys. J., Vol. 291, No. 1, p. 72 – 79 (1985).

The high–surface–brightness inner disk of NGC 1068 has been observed with the *IUE* to determine the ultraviolet spectrum and surface brightness profile. This disk is explained by an intense starburst, requiring formation of $1 - 2$ massive stars $\mathrm{yr^{-1}}$, so it is a good prototype for seeking at early epochs in the universe. Calculations of isophotal diameter as a function of redshift are presented. It is found that this disk could be resolved to $z \approx 0.3$ from the ground or to $z \approx 2.0$ with the Hubble Space Telescope.

158.135 Observational evidence for the radiative acceleration of broad–line clouds in Seyfert 1 galaxies and quasars.
C. M. Gaskell.
Astrophys. J., Vol. 291, No. 1, p. 112 – 116 (1985).

It is shown that the relative strength of the broad Hβ emission line in Seyfert 1 galaxies depends on the widths of the lines, being very weak for objects with narrow lines. The equivalent width of Fe II λ4570, on the other hand, is independent of velocity width. Simple calculations show that Hβ is collisionally suppressed at much lower densities or optical depths than the optical Fe II lines. The correlation of Hβ emissivity with cloud velocity provides strong support for the radiative acceleration models developed by Mathews and Blumenthal. It is suggested that the cause of object–to–object variations is the pressure of the hot confining intercloud medium.

158.136 Observations from 1 to 20 microns of low–luminosity active galaxies.
A. Lawrence, M. Ward, M. Elvis, G. Fabbiano, S. P. Willner, N. P. Carleton, A. Longmore.
Astrophys. J., Vol. 291, No. 1, p. 117 – 127 (1985).

The authors present and discuss $1 - 20$ μm photometry of Liners (low–ionization nuclear emission–line regions), star burst nuclei, and high–excitation (mostly Seyfert 2) nuclei. Liners have strong, flattish $10 - 20$ μm excesses but are dominated by a stellar population through $1 - 5$ μm. By contrast, most type 2 Seyfert galaxies and all starburst nuclei have flux distributions that are flat through $1 - 5$ μm, and steeply rising through $10 - 20$ μm. The $1 - 5$ μm flux distributions of type 2 Seyfert galaxies and starburst nuclei can be explained by mixtures of stellar emission, recombination radiation, and hot dust, although a nonthermal component is likely in a few objects.

158.137 The physics of extragalactic radio sources.
D. S. De Young.
Phys. Rep., Vol. 111, No. 6, p. 373 – 408 (1984). Abstr. in Phys. Abstr., Vol. 88, No. 1252, Entry 30078 (1985).

158.138 Multifrequency observations of blazars. I. The shape of the 1 micron to 2 millimeter continuum.
W. K. Gear, E. I. Robson, P. A. R. Ade, M. J. Griffin, L. M. J. Brown, M. G. Smith, I. G. Nolt, J. V. Radostitz, G. Veeder, L. Lebofsky.
Astrophys. J., Vol. 291, No. 2, p. 511 – 517 (1985).

The authors present near–simultaneous measurements in 11 wavebands between 1 μm and 2 mm of a sample of 13 "blazars" (BL Lac objects and optically violent variable quasars). Most of the sources have very flat millimeter/ submillimeter spectra up to the highest observed frequency. However, 3C 279 and 3C 446 show evidence of turnovers in their submillimeter spectra. The $1 - 4$ μm spectra can be characterized by simple power laws, all steeper than –0.9; several sources show evidence of spectral breaks in the $10 - 20$ μm region. It is shown that the spectral properties are consistent with synchrotron emission from relativistic jets aligned close to the line of sight.

158.139 Kinematics and ionization of extended ionized gas in active galaxies. I. The X–ray luminous galaxies NGC 2110, NGC 5506, and MCG –5–23–16.
A. S. Wilson, J. A. Baldwin, J. S. Ulvestad.
Astrophys. J., Vol. 291, No. 2, p. 627 – 654 (1985). With plates 9 – 13.

Direct imaging with CCDs and long–slit spectroscopy with an IPCS and SIT Vidicons have been used to map the emission–line ratios, profiles, and velocity fields over the extended narrow–line regions in three nearby Seyfert galaxies. NGC 2110 (type E3) and NGC 5506 (type IO?) show high–excitation gaseous nebulosities extending over ~ 2 kpc, while the gas in MCG –5–23–16 is spatially unresolved. Each nebulosity is photoionized by an ultraviolet source with a power–law spectrum. An extensive discussion of the global kinematics of the emitting gas in the three galaxies is presented and models of the gas motions are constructed. In all three galaxies, the line profiles near the centers exhibit blue wings, suggesting outflow, and are more centrally peaked than Gaussians.

158.140 X–ray emission from E and S0 galaxies with compact nuclear radio sources.
L. L. Dressel, A. S. Wilson.
Astrophys. J., Vol. 291, No. 2, p. 668 – 676 (1985).

Using the Einstein Observatory IPC, the authors have searched for X–ray emission from 13 elliptical and S0 galaxies with compact nuclear radio sources. Five galaxies were detected with $L_x \approx 3 \times 10^{40} - 3 \times 10^{41} \mathrm{erg\ s^{-1}}$. Combining the results with other published X–ray measurements, it is found that radio–emitting E and S0 galaxies have a higher X–ray luminosity per unit optical luminosity than their radio–quiet counterparts. The authors discuss possible emission mechanisms and models for the resolved source (NGC 1407), for the most powerful source (NGC 3998), and for the class of E/S0 galaxies with active nuclei as a whole.

158.141   The effects of stellar–absorption features on the broad–line profiles of Seyfert 1 galaxies.
D. M. Crenshaw, B. M. Peterson.
Astrophys. J., Vol. 291, No. 2, p. 677 – 684 (1985).

The authors demonstrate that in Seyfert 1 galaxy spectra in which the contribution of the host galaxy is significant, the stellar absorption features can introduce structure and asymmetries into the broad–line profiles. Spectroscopic observations through large apertures of a small sample of Seyfert 1 galaxies reveal that light from stars can contribute up to 70% of the total continuum flux at Hβ. It is shown that, to first order, the stellar contribution to the spectra can be removed by subtracting a suitably scaled spectrum of the dwarf elliptical galaxy M32.

158.142   The nucleus of M82 at radio and X–ray bands: discovery of a new radio population of supernova candidates.
P. P. Kronberg, P. Biermann, F. R. Schwab.
Astrophys. J., Vol. 291, No. 2, p. 693 – 707 (1985).

Detailed VLA radio maps are presented of the continuum radio emission in M82 and the X–ray emission from the HRI of the Einstein X–ray satellite. The authors have discovered more than 40 new radio sources within 300 pc of M82's nucleus. It is argued that nearly all of these sources are radio supernovae or supernova remnants and their radio luminosity distribution is derived. X–ray maps, both from reprocessed Einstein data and from EXOSAT reveal that most of the X–ray emission does not come from the radio/IR complex, but rather from the inner filamentary region seen in Hα and [O II] λ3727 photographs. Its distribution is perpendicular to the stellar distribution in M82. A discussion of the apparent correlation between radio continuum and far–infrared luminosities in "starburst" galaxies is included.

158.143   A gravitational origin for the broad emission line profiles in quasars and Seyfert galaxies: time variation.
T. J. Carroll.
Mon. Not. R. Astron. Soc., Vol. 214, No. 3, p. 321 – 326 (1985).

The response of the broad emission line profiles observed in Seyfert galaxies to a change in the ionizing continuum luminosity is examined within the framework of a gravitational kinematical model for the motions of the emitting material. Particular attention is paid to the evidence from NGC 4151.

158.144   *UBVRI* observations of BL Lacertae objects.
M. Moles, J. M. Garcia–Pelayo, J. Masegosa,
A. Aparicio.
Astrophys. J., Suppl. Ser., Vol. 58, No. 2, p. 255 – 263 (1985).

A total of 85 *UBVRI* observations of BL Lac objects is presented. For most of them, an average continuum can be traced by means of well–defined color indices. These continua are always well reproduced by power laws, strongly supporting the hypothesis of a synchrotron origin of the radiation output, as well as the conclusion that there is no dust in such objects. The analysis of the variability leads to a rough estimate of the value of the magnetic field, which is of the same order for all the objects, around 3×10^{-5}G.

158.145   The interaction between beams and their cocoons.
P. A. Hughes, A. J. Allen.
Mon. Not. R. Astron. Soc., Vol. 214, No. 3, p. 399 – 404 (1985).

It is demonstrated that large–scale density structure may exist within the cocoons of FRII radio sources, due to instability of the working surface of these sources. The authors suggest that this structure is responsible for multiple deflections of the beam, and that it is convected into the beam path by a large–scale circulation in the cocoon, so that the pattern of deflections changes on a time–scale of 10^6yr. This effect can explain the slight lack of collinearity of nulceus and hotspots, beam deflection such as is inferred for Cygnus A, and multiple hotspot structure.

158.146   *JHKL* properties of emission–line galaxies.
I. S. Glass, A. F. M. Moorwood.
Mon. Not. R. Astron. Soc., Vol. 214, No. 3, p. 429 – 447 (1985).

Multi–aperture *JHKL* colours on the new SAAO photometric system are presented for samples of Seyfert 1, Seyfert 2 and H II–region galaxies. Comparisons are made with the colours of "ordinary" galaxies obtained using the same system.

158.147   Radio Seyferts and the forbidden–line region.
A. Pedlar, J. E. Dyson, S. W. Unger.
Mon. Not. R. Astron. Soc., Vol. 214, No. 3, p. 463 – 473 (1985).

The authors describe a model in which the expansion of the radio–emitting components in Seyfert 2 nuclei compresses and accelerates the ambient medium by means of shockwaves. When the expansion velocity of the component has decreased to ~ 400 km s^{-1}, the compressed gas cools and has the properties attributed to the forbidden–line region. This readily accounts for the intimate relationship between the radio continuum emission and the forbidden–line region.

158.148   Sulphur III lines and the excitation mechanism in NGC 1052.
A. I. Díaz, B. E. J. Pagel, E. Terlevich.
Mon. Not. R. Astron. Soc., Vol. 214, No. 3, p. 41P – 45P (1985).

The authors have measured the intensities of [S III] λλ9069, 9532 in the nucleus of the classic Liner elliptical galaxy NGC 1052. Their strength favours photoionization as opposed to shock–heating models, which is in agreement with recent [O III] electron temperature data for NGC 1052 and with [S III] data for some other Liners.

158.149   Inhomogeneous synchrotron–self–Compton models and the problem of relativistic beaming of BL Lac objects.
G. Ghisellini, L. Maraschi, A. Treves.
Astron. Astrophys., Vol. 146, No. 2, p. 204 – 212 (1985).

A formalism for the computation of the synchrotron–self–Compton (SSC) spectrum produced by a wide class of elongated inhomogeneous sources (jets) is described. The relativistic electron density, the maximum electron energy and the magnetic field are assumed to be power law functions of a radial coordinate. The general characteristics of the resulting spectra for various choices of the radial dependences of the physical quantities are discussed. The formalism is applied to the BL Lac objects PKS 2155–304 and PKS 0537–441 and the results are compared with those of the homogeneous SSC model.

158.150   Linear polarization at λ49 cm of 27 double radio sources.
R. G. Conway, R. G. Strom.
Astron. Astrophys., Vol. 146, No. 2, p. 392 – 394 (1985).

Using new observations of the λ49 cm polarization of 27 double radio sources, the authors have reevaluated the depolarization characteristics of a flux–limited sample of 46 classical double sources drawn from the 3CR catalogue. The electron density has been determined for 86 source components assuming that depolarization is due to internal Faraday rotation. The authors obtain values between 200 and 3000 m^{-3} with a frequency distribution which varies approximately as the inverse of electron density. In an investigation of "polarization–asymmetry" in component pairs, they find a slight positive correlation between the electron densities in the components of sources from their sample, but note that there are some sources in which the densities differ by up to an order of magnitude.

158.151   Spectra of galaxies with UV continuum. IV.
B. E. Markarian, V. A. Lipovetskij, D. A. Stepanyan.
Astrofizika, Tom 21, Vyp. 3, p. 419 – 431 (1984). In Russian. English translation in Astrophysics, Vol. 21, No. 3.

The results of spectral observations of 75 UV continuum galaxies are presented. The presence of emission lines in the spectra of 72 of them is established. Two galaxies, Mark 1388 and 1447, possibly relate to Sy2 and Sy1.5 respectively. Eight objects, Mark 522, 539, 540, 596, 940, 948, 971 and 998, have spectral features specific for active galactic nuclei and need further detail investigation.

158.152 A spectral investigation of the galaxy Markarian 111.
A. N. Burenkov, A. R. Petrosyan, K. A. Saakyan, Eh. E. Khachikyan.
Astrofizika, Tom 21, Vyp. 3, p. 433 – 443 (1984). In Russian. English translation in Astrophysics, Vol. 21, No. 3.

On the basis of spectra a spectrometric and spectrophotometric investigation of the galaxy Markarian 111 is carried out. It is shown that the southern condensation observed in the galaxy is a companion situated before Markarian 111. By means of relative intensities of observed emission lines are estimated: values of T_e and n_e, content of some elements (He, O, N, S) in the nucleus, in a condensation situated in the spiral arm of Markarian 111 and in its companion. It is shown that the ionization sources of the gas in these objects are young stars of spectral types O9 – B0.

158.153 Luminosity function of faint galaxies with UV continuum.
D. A. Stepanyan.
Astrofizika, Tom 21, Vyp. 3, p. 445 – 459 (1984). In Russian. English translation in Astrophysics, Vol. 21, No. 3.

The space density of faint galaxies with UV continuum of the second Byurakan survey is determined. The average space density of galaxies with UV continuum for the luminosity interval $-16^m\!.5$ to $-21^m\!.5$ is equal to 0.18 of the total space density of field galaxies for the same interval of absolute magnitudes. The space density of low luminosity galaxies with UV continuum is very high. In the interval $-12^m\!.5$ to $15^m\!.5$ it makes up 0.23 Mpc^{-3}.

158.154 A possible mechanism of the kinetic energy dissipation of plasma streams in radio galaxies.
V. N. Morozov.
Astrofizika, Tom 21, Vyp. 3, p. 475 – 486 (1984). In Russian. English translation in Astrophysics, Vol. 21, No. 3.

Based on the Blandford–Rees model the kinetic energy dissipation of plasma streams in radio galaxies is considered. The energy of the resulting magnetohydrodynamic turbulence is estimated using the quasi–linear theory. The estimates show that the resulting turbulence may be a source of energy which compensates energy losses due to radiation and adiabatic cooling by means of relativistic particles.

158.155 Brightness distribution of synchrotron radiation in the dipole magnetic field and the nature of double radio sources.
Yu. L. Zyskin, A. A. Stepanyan.
Astrofizika, Tom 21, Vyp. 3, p. 487 – 497 (1984). In Russian. English translation in Astrophysics, Vol. 21, No. 3.

It is suggested that the lobes of double radio sources are only a small part of the dipolar magnetic field. The calculations of synchrotron radiation of high–energy particles show that an observer can see only two radio emitting spots in spite of axisymmetrical distribution of particles. The following features of double radio sources may be accounted for in the suggested model: a) the duality of the sources, b) the tendency of the parent galaxy to be near the axis of the source, c) the size dependence on the frequency observed, etc.

158.156 Exosat observations of active galactic nuclei.
G. Branduardi–Raymont, S. J. Bell–Burnell, B. Kellett, H. Fink, D. Molteni, I. McHardy.
Adv. Space Res., Vol. 5, No. 3, p. 129 – 131 (1985). – See Abstr. 012.059.

Exosat observations of the type 1 Seyfert galaxies NGC 3783 and NGC 5548 are reported. Contrary to previous findings, the X–ray spectrum of NGC 3783 is not heavily cutoff at low energies. This suggests that the conditions of the broad–line region in NGC 3783 are different from those in which the earlier data were obtained. The authors speculate on the presence of a two–component spectrum in NGC 5548, under the assumption that the galactic obscuration in its direction is what is measured on a large scale by H I surveys.

158.157 Optical variability of the object 1055 + 567.
V. P. Goranskij.
Astron. Tsirk., No. 1345, p. 8 (1984). In Russian.

158.158 IC 4553 = Arp 220.
F. Bertola.
Sterne Weltraum, 24. Jahrg., Nr. 5, p. 265 (1985).

158.159 Linearly polarized emission of variable radio sources. VI. ”Rotators”: imitation of rotation.
V. N. Kuril’chik, N. T. Ashimbaeva.
Astron. Tsirk., No. 1343, p. 1 – 3 (1984). In Russian.

158.160 Polarization and photometric study of BL Lac. Results of observations.
V. A. Hagen–Thorn, S. G. Marchenko, V. A. Yakovleva.
Sov. Astron., Vol. 28, No. 5, p. 538 – 544 (1984). English translation of 38.158.051.

158.161 Consequences of hot gas in the broad–line region of active galactic nuclei.
T. Kallman, R. Mushotzky.
Astrophys. J., Vol. 292, No. 1, p. 49 – 57 (1985).

The authors discuss models for hot gas in the broad–line region of active galactic nuclei. The results of the two–phase equilibrium models for confinement of broad–line clouds by Compton–heated gas are used to show that high–luminosity quasars are expected to show that Fe XXVI Lyα line absorption which will be observable with spectrometers such as those planned for the future X–ray spectroscopy experiment. Two–phase equilibrium models also predict that the gas in the broad–line clouds and the confining medium may be Compton thick. The implications for polarization and variability are discussed.

158.162 The ultraviolet spectra of active galaxies with weak optical Fe II lines.
H. Netzer, W. Wamsteker, B. J. Wills, D. Wills.
Astrophys. J., Vol. 292, No. 1, p. 143 – 147 (1985).

The authors have obtained simultaneous optical and *IUE* spectra of two Seyfert galaxies that show very weak optical Fe II lines: Mrk 290 and Mrk 915. *IUE* archival spectra of the radio galaxy 3C 390.3 are also compared with optical data obtained during the same year. Strong, distinct ultraviolet features are observed in the spectrum of Mrk 290. A fit to all spectral features between 1900 and 5500 Å includes several thousand Fe II lines, many Balmer lines, Balmer continuum, and metallic lines. Evidence for strong ultraviolet Fe II lines is found also in 3C 390.3, and the case of Mrk 915 is inconclusive. This paper demonstrates several new aspects of the spectra of active galactic nuclei, especially the large amount of energy emitted by Fe II lines and the possible importance of dust or very high density clouds or both in such objects.

158.163 The variability of the spectrum of Arakelian 120. II. Evidence for a small broad line emitting region.
B. M. Peterson, K. A. Meyers, E. R. Capriotti, C. B. Foltz, B. J. Wilkes, H. R. Miller.
Astrophys. J., Vol. 292, No. 1, p. 164 – 171 (1985).

Four years of spectroscopic and photometric monitoring of the Seyfert 1 galaxy Akn 120 reveal that the broad Hβ line flux changes on a time scale indistinguishable from that of the optical continuum, which sets an upper limit of ~ 30 lt–day on the size of the broad–line region. The high luminosity of the Akn 120 continuum source thus implies a very high radiation parameter, significantly larger than the values used in successful models of the emission–line region, even if the electron density is of order 10^9cm^{-3}. A possible solution to this problem is discussed.

158.164 The short wide angle tail radio source in NGC 4874.
L. Feretti, G. Giovannini.
Astron. Astrophys., Vol. 147, No. 1, p. L13 – L15 (1985).

Radio maps at 1.5 and 4.9 GHz are presented of the radio source associated with NGC 4874, one of the two dominant galaxies of the Coma cluster. This radio source of the WAT type, has a very small linear size and is all contained within the optical galaxy. The existence of a very short WAT source implies that either the role of interstellar medium in bending the radio structure is not negligible, or the transition between interstellar and intergalactic medium occurs very close to the center of the galaxy, or other mechanisms are responsible for the wide angle tail phenomenon.

158.165   Orientation effects and the reddening of the lines and continua of Seyfert galaxies.
G. De Zotti, C. M. Gaskell.
Astron. Astrophys., Vol. 147, No. 1, p. 1 – 12 (1985).

The authors present a compilation of axial ratios and published broad– and narrow–line Balmer decrements for 109 Seyfert galaxies of all classes. The reddening law they deduce for the broad–line region is identical to that found by Cheng et al. (1983) from broad–band colours of Seyferts and the distribution of dust is consistent with that producing the reddening in normal spiral galaxies. It is found that broad and narrow–line decrements are not well correlated. There is some evidence that the reddening of Seyfert 1 and Seyfert 1.5 galaxies increases with decreasing luminosity with $\Delta E(B–V) \cong +0.06\Delta M_B$ for $b/a = 0.6$. The authors' prediction of reddenings from orientations are well correlated with published estimates of the reddening based on the $\lambda 2200$ absorption feature.

158.166   Non–thermal models of the X–ray emission in Seyfert galaxies: the case of NGC 4151.
I. Piro, G. C. Perola, E. Massaro.
Astron. Astrophys., Vol. 147, No. 1, p. 22 – 34 (1985).

The radiative properties of non–thermal models (synchrotron and inverse Compton with a spherically symmetric distribution of electrons) of the X–ray emission of NGC 4151 are discussed taking into account all the available observational constraints, in particular the lack of a sizeable spectral variability (at least in the range 2 – 50 keV) during both the brightening and the dimming phase of flare–like events (Mushotzky et al., 1978)

158.167   Similar structures in the outbursts of OJ 287 in 1972 and 1983.
A. Sillanpää, P. Teerikorpi, S. Haarala, T. Korhonen,
Yu. S. Efimov, N. M. Shakhovskoy (*N. M. Shakhovskoj*).
Astron. Astrophys., Vol. 147, No. 1, p. 67 – 70 (1985).

Optical observations during the outburst of OJ 287 in 1983 are presented. On the basis of the authors' observations the object reached $m_v = 12.64$ in January 10, 1983. The declining tails of the light curves during the outbursts in 1972 and 1983 seem to be quite similar.

158.168   Bubbles and jets in the center of M51.
H. C. Ford, P. C. Crane, G. H. Jacoby, D. G. Lawrie,
J. M. van der Hulst.
Space Telesc. Sci. Inst., Prepr. Ser., No. 38, 73 pp. (1985). To appear in Astrophys. J.

158.169   IRAS observations of Seyfert galaxies.
G. K. Miley, G. Neugebauer, B. T. Soifer.
Space Telesc. Sci. Inst., Prepr. Ser., No. 45, 16 pp. (1985). To appear in Astrophys. J., Lett. Ed.

158.170   NGC 1808: a nearby galaxy with a faint Seyfert nucleus.
M.–P. Véron–Cetty, P. Véron.
ESO Sci. Prepr., No. 355, 20 pp. (1984). Submitted to Astron. Astrophys.

158.171   WSRT and VLA observations of very steep spectrum radio galaxies in clusters.
J. Roland, R. J. Hanisch, P. Véron, E. Fomalont.
ESO Sci. Prepr., No. 363, 39 pp. (1985). To be published in Astron. Astrophys.

158.172   *JHKL* properties of emission–line galaxies.
I. S. Glass, A. F. M. Moorwood.
ESO Sci. Prepr., No. 364, 36 pp. (1985). To appear in Mon. Not. R. Astron. Soc.

158.173   Spectra of the two brightest objects in the amorphous galaxy NGC 1569: superluminous young star clusters – or stars in a nearby peculiar galaxy?
H. Arp, A. Sandage.
MPA Rep., No. 173, 29 pp. (1985). Submitted to Astron. J.

158.174   Surveys of local AGN's.
P. Véron.
Obs. Haute Provence, Pré–Publ. No. 1, 12 pp. (1985). Paper presented at the conference on "Structure and evolution of active galactic nuclei", Trieste, 1985.

The author reviews the optical, X–ray and infrared surveys of local AGN's (*Active Galactic Nuclei*), discussing the limitations and biases of each of them. He concludes that no single survey may lead to a complete sample but that all of them must be used simultaneously to build a meaningful luminosity function.

158.175   A statistical study of the relationship between galaxy interactions and nuclear activity.
R. M. Cutri, C. W. McAlary.
Prepr. Steward Obs., No. 569, 63 pp. (1985). To appear in Astrophys. J.

158.176   CCD spectrocopy of the active galaxies NGC 5506 and NGC 7314.
S. L. Morris, M. J. Ward.
Prepr. Steward Obs., No. 572, 12 pp. (1985). Submitted to Mon. Not. R. Astron. Soc.

158.177   The optical and radio properties of X–ray selected BL Lac objects.
J. T. Stocke, J. Liebert, G. D. Schmidt, I. M. Gioia,
T. Maccacaro, R. E. Schild, D. Maccagni, H. C. Arp.
Prepr. Steward Obs., No. 577, 51 pp. (1985).

The eight BL Lac objects from the HEAO–1 A–2 all–sky survey and from the Einstein Medium Sensitivity Survey (MSS) form a flux–limited complete X–ray selected sample. The optical and radio properties of the MSS BL Lacs are presented and compared with those of the HEAO–1 A–2 sample and with those of radio–selected BL Lacs. The X–ray selected BL Lacs possess smaller polarized fractions and less violent optical variability than radio–selected BL Lacs. These properties are consistent with the substantial starlight fraction seen in the optical spectra of a majority of these objects. This starlight allows a determination of definite redshifts for two of four MSS BL Lacs and a probable redshift for a third. Despite the differences in characteristics between the X–ray selected and radio–selected samples, the authors conclude that these eight objects possess most of the basic qualities of BL Lacs and should be considered members of that class.

158.178   The association of optical emission with the inner Centaurus A jet.
J. Brodie, S. Bowyer.
Astrophys. J., Vol. 292, No. 2, p. 447 – 450 (1985).

Optical observations of the inner region of the Centaurus A jet show that the size of the optical feature located at the site of X–ray/radio knot B corresponds in length and breadth to the X–ray and radio dimensions. The physical association of optical emission with the jet is thus confirmed. The data reveal a core/halo structure in the optical knot which is similar to both the radio and the X–ray morphology. This is interpreted in terms of a cocoon of emission surrounding the beam of the jet which is due to backflow from the working surface.

158.179   Optical and radio observations for the BL Lacertae objects 1219 + 28, 0851 + 202, and 1400 + 162.
D. Weistrop, D. B. Shaffer, P. Hintzen, W. Romanishin.
Astrophys. J., Vol. 292, No. 2, p. 614 – 619 (1985).

Radio and visible wavelength observations have been obtained for the BL Lacertae objects 0851 + 202 (OJ 287), 1219 + 28 (ON 231, W Com), and 1400 + 162 (4C 16.39, OQ 100). Spectra of 1219 + 28 show two emission features which give a redshift $z = 0.102$. Deep imaging of 1219 + 28 indicates the presence of an elliptical nebulosity underlying the point source. The observations are consistent with the nebulosity being an associated elliptical galaxy. VLBI maps of 1219+28 at 2.3 GHz, 5.0 GHz, and 8.3 GHz show a central source plus a jet with at least three components. Radio observations of 1400+162 were also obtained using the VLBI network. Imaging and photometry of 0851+202 (OJ 287) are also reported. The observations show no evidence of an underlying nebulosity.

158.180 Outflow in the nucleus of the Seyfert I galaxy NGC 3783.
W. Wamsteker, P. Barr.
Astrophys. J., Lett. Ed., Vol. 292, No. 2, p. L45 – L49 (1985).

Averaged low–resolution IUE spectra of the Seyfert I galaxy NGC 3783 at two brightness levels are presented. The wavelength interval 1250 – 1400 Å, which is essentially free of emission lines in active galaxies, contains two absorption lines at 1303 Å and 1335 Å. The line at 1303 Å shows considerable variations with continuum brightness. This is interpreted as variations in the Si III (4) metastable absorption at 1294 – 1305 Å, with an outflow velocity of ~ 3000 km s^{-1}. This indicates the presence of outflow of material, which is optically thin to the variable ionizing UV continuum.

158.181 The nuclear activity of interacting galaxies.
O. Dahari.
Astrophys. J., Suppl. Ser., Vol. 57, No. 4, p. 643 – 664 (1985) = Lick Obs. Bull., No. 996.

A search for active galactic nuclei among interacting galaxies is reported. A sample of 167 systems of interacting and asymmetric galaxies was observed spectrophotometrically in the spectral range 4700 – 7100 Å. The results are compared with a sample of isolated galaxies. It is found that (1) there are no Seyfert nuclei in elliptical or dwarf irregular galaxies of the sample; (2) there is an excess of Seyfert nuclei among interacting spirals, but it is only at the 90% confidence level; (3) this excess becomes statistically significant (98%) when only *strongly* interacting spirals are included (four new Seyfert nuclei are presented); (4) in the subgroup of galaxies with extreme tidal distortions, no Seyfert nuclei were found.

158.182 Broad–band polarization observations of active compact radio sources.
L. Rudnick, T. W. Jones, H. D. Aller, M. F. Aller, P. E. Hodge, F. N. Owen, R. L. Fiedler, J. J. Puschell, R. C. Bignell.
Astrophys. J., Suppl. Ser., Vol. 57, No. 4, p. 693 – 709 (1985).

The authors have obtained simultaneous multifrequency linear polarimetry data between 1.4 GHz and 90 GHz for about 20 active, compact radio sources at six epochs from 1977 December to 1980 July. Analysis of these data, making use of complementary rotation measure, VLBI, and short–time scale information, is presented in a separate paper.

158.183 Jet Strukturen in Seyfert und Star Burst Galaxien.
G. F. O. Schnur, T. Schmidt–Kaler.
Mitt. Astron. Ges., Nr. 63, p. 167 (1985). – See Abstr. 012.063.

158.184 Star formation in the cooling flows of M87 and NGC 1275.
R. E. White III, C. L. Sarazin.
Bull. Am. Astron. Soc., Vol. 16, No. 4, p. 881 – 882 (1984). Abstract. – See Abstr. 010.062.

158.185 Infrared Seyferts: a new population of active galaxies.
G. K. Miley, R. de Grijp, J. Lub, T. de Jong.
Bull. Am. Astron. Soc., Vol. 16, No. 4, p. 915 (1984). Abstract. – See Abstr. 010.062.

158.186 Infrared photometry of Seyfert 1.8 and 1.9 galaxies.
R. J. Rudy, J. Rodriguez–Espinosa.
Bull. Am. Astron. Soc., Vol. 16, No. 4, p. 915 – 916 (1984). Abstract. – See Abstr. 010.062.

158.187 The IRAS galaxy 0421+040P06: an active spiral galaxy with extended radio lobes.
C. G. Wynn–Williams, C. A. Beichman, C. J. Lonsdale, S. E. Persson, J. N. Heasley, G. K. Miley, B. T. Soifer, G. Neugebauer, E. E. Becklin, J. R. Houck.
Bull. Am. Astron. Soc., Vol. 16, No. 4, p. 916 (1984). Abstract. – See Abstr. 010.062.

158.188 Optical and infrared observations of hot IRAS galaxies.
W. F. Kailey, M. J. Lebofsky.
Bull. Am. Astron. Soc., Vol. 16, No. 4, p. 916 (1984). Abstract. – See Abstr. 010.062.

158.189 The velocity field in the active galaxy NGC 7582.
S. L. Morris, M. J. Ward, M. Whittle, A. S. Wilson, K. Taylor.
Bull. Am. Astron. Soc., Vol. 16, No. 4, p. 916 (1984). Abstract. – See Abstr. 010.062.

158.190 The influence of galaxy interactions on nuclear activity.
R. M. Cutri.
Bull. Am. Astron. Soc., Vol. 16, No. 4, p. 916 – 917 (1984). Abstract. – See Abstr. 010.062.

158.191 H I absorption in active spiral nuclei: predominance of infalling clouds.
J. M. Dickey.
Bull. Am. Astron. Soc., Vol. 16, No. 4, p. 917 (1984). Abstract. – See Abstr. 010.062.

158.192 Multi–frequency VLBI and VLA maps of compact highly polarized radio sources.
R. E. Rusk, E. R. Seaquist, J. L. Yen, R. A. Perley.
Bull. Am. Astron. Soc., Vol. 16, No. 4, p. 917 (1984). Abstract. – See Abstr. 010.062.

158.193 The orientations of the radio and rotation axes of bright elliptical galaxies.
M. Birkinshaw, R. L. Davies.
Bull. Am. Astron. Soc., Vol. 16, No. 4, p. 917 – 918 (1984). Abstract. – See Abstr. 010.062.

158.194 Circular polarization in simple compact extragalactic radio sources.
A. Y. S. Cheng, A. G. Pacholczyk, K. H. Cook.
Bull. Am. Astron. Soc., Vol. 16, No. 4, p. 918 (1984). Abstract. – See Abstr. 010.062.

158.195 X–ray spectroscopy of BL Lacertae objects.
C. M. Urry, R. F. Mushotzky, S. S. Holt.
Bull. Am. Astron. Soc., Vol. 16, No. 4, p. 931 (1984). Abstract. – See Abstr. 010.062.

158.196 Soft X–ray spectra of optically selected Seyfert galaxies.
J. Kruper, C. R. Canizares.
Bull. Am. Astron. Soc., Vol. 16, No. 4, p. 931 (1984). Abstract. – See Abstr. 010.062.

158.197 Einstein X–ray spectra of BL Lac objects.
G. M. Madejski, D. A. Schwartz.
Bull. Am. Astron. Soc., Vol. 16, No. 4, p. 931 – 932 (1984). Abstract. See Abstr. 010.062.

158.198 X–ray line emission from 3C 120.
R. Petre, R. F. Mushotzky, T. Kallman.
Bull. Am. Astron. Soc., Vol. 16, No. 4, p. 932 (1984). Abstract. – See Abstr. 010.062.

158.199 EXOSAT observations of the radio galaxy 3C 390.3.
R. A. Shafer, M. J. Ward, P. Barr.
Bull. Am. Astron. Soc., Vol. 16, No. 4, p. 932 (1984). Abstract. – See Abstr. 010.062.

158.200 SSC models in spherical geometries.
D. L. Band, J. E. Grindlay.
Bull. Am. Astron. Soc., Vol. 16, No. 4, p. 933 (1984). Abstract. –
See Abstr. 010.062.

158.201 Cen A X–ray activity, 1969 – 1979.
J. Terrell.
Bull. Am. Astron. Soc., Vol. 16, No. 4, p. 933 (1984). Abstract. –
See Abstr. 010.062.

158.202 Long–term X–ray monitoring of 4 active galactic nuclei.
W. A. Snyder, K. S. Wood.
Bull. Am. Astron. Soc., Vol. 16, No. 4, p. 933 (1984). Abstract. –
See Abstr. 010.062.

158.203 A model for starbursts in galaxies.
C. Struck–Marcell, J. M. Scalo.
Bull. Am. Astron. Soc., Vol. 16, No. 4, p. 949 (1984). Abstract. –
See Abstr. 010.062.

158.204 High–speed polarimetry of BL Lac.
R. L. Moore, G. D. Schmidt, S. C. West.
Bull. Am. Astron. Soc., Vol. 16, No. 4, p. 951 (1984). Abstract. –
See Abstr. 010.062.

158.205 Interpretation of recent outbursts in BL Lac.
P. A Hughes, H. D. Aller, M. F. Aller.
Bull. Am. Astron. Soc., Vol. 16, No. 4, p. 951 (1984). Abstract. –
See Abstr. 010.062.

158.206 Milliarcsecond polarization structure of several BL Lac objects and quasars.
D. C. Gabuzda, D. H. Roberts, J. F. C. Wardle.
Bull. Am. Astron. Soc., Vol. 16, No. 4, p. 951 – 952 (1984). Abstract. – See Abstr. 010.062.

158.207 Optical polarimetry of BL Lacertae objects and violent variable quasars.
M. L. Sitko, W. A. Stein, G. D. Schmidt.
Bull. Am. Astron. Soc., Vol. 16, No. 4, p. 952 (1984). Abstract. –
See Abstr. 010.062.

158.208 Multifrequency radio observations of four BL Lacertae objects.
T. J. Balonek, C. P. O'Dea, W. A. Dent, W. M. Kinzel.
Bull. Am. Astron. Soc., Vol. 16, No. 4, p. 952 (1984). Abstract. –
See Abstr. 010.062.

158.209 The fate of He II Lyα emission in Seyfert galaxies and quasars.
R. G. Eastman, G. M. MacAlpine.
Bull. Am. Astron. Soc., Vol. 16, No. 4, p. 953 (1984). Abstract. –
See Abstr. 010.062.

158.210 Location of dust in active galactic nuclei.
N. P. Carleton.
Bull. Am. Astron. Soc., Vol. 16, No. 4, p. 954 (1984). Abstract. –
See Abstr. 010.062.

158.211 A search for optical emission from hot spots in radio galaxies.
A. Chokshi, E. L. Wright.
Bull. Am. Astron. Soc., Vol. 16, No. 4, p. 954 (1984). Abstract. –
See Abstr. 010.062.

158.212 The interaction of interstellar clouds with a "quasar wind" and its relation to the NELRs (*narrow emission line regions*) of Seyfert galaxies.
A. V. R. Schiano.
Bull. Am. Astron. Soc., Vol. 16, No. 4, p. 955 (1984). Abstract. –
See Abstr. 010.062.

158.213 The jet in M87: an unstable jet in a cooling inflow atmosphere?
D. M. Sumi, L. L. Smarr.
Bull. Am. Astron. Soc., Vol. 16, No. 4, p. 955 (1984). Abstract. –
See Abstr. 010.062.

158.214 Water masers as starburst tracers in external galaxies.
M. J. Claussen, K.–Y. Lo, G. M. Heiligman.
Bull. Am. Astron. Soc., Vol. 16, No. 4, p. 955 (1984). Abstract. –
See Abstr. 010.062.

158.215 Do extragalactic compact radio sources rotate?
L. F. Brown, J. F. C. Wardle.
Bull. Am. Astron. Soc., Vol. 16, No. 4, p. 955 (1984). Abstract. –
See Abstr. 010.062.

158.216 Superluminal resupply of a radio lobe in 3C 395.
J. H. Spencer, J. A. Waak, K. J. Johnston,
R. S. Simon.
Bull. Am. Astron. Soc., Vol. 16, No. 4, p. 956 (1984). Abstract. –
See Abstr. 010.062.

158.217 Constraints on bent beams in narrow angle tail radio sources.
C. P. O'Dea.
Bull. Am. Astron. Soc., Vol. 16, No. 4, p. 956 (1984). Abstract. –
See Abstr. 010.062.

158.218 A digital surface photometric study of NGC 5128.
R. L. Pennington.
Bull. Am. Astron. Soc., Vol. 16, No. 4, p. 956 (1984). Abstract. –
See Abstr. 010.062.

158.219 cm–wavelength linear polarizations and spectra of superluminal radio sources.
M. F. Aller, H. D. Aller.
Bull. Am. Astron. Soc., Vol. 16, No. 4, p. 957 (1984). Abstract. –
See Abstr. 010.062.

158.220 An unusual ionized gas structure in the Seyfert galaxy Markarian 315.
J. W. MacKenty.
Bull. Am. Astron. Soc., Vol. 16, No. 4, p. 957 (1984). Abstract. –
See Abstr. 010.062.

158.221 The Seyfert 1 inside NGC 1068.
R. R. J. Antonucci, J. S. Miller.
Bull. Am. Astron. Soc., Vol. 16, No. 4, p. 957 (1984). Abstract. –
See Abstr. 010.062.

158.222 Observations of upper–level hydrogen lines in Seyfert 1 galaxies.
C. W. McAlary, G. H. Rieke.
Bull. Am. Astron. Soc., Vol. 16, No. 4, p. 957 (1984). Abstract. –
See Abstr. 010.062.

158.223 Far–infrared spectroscopy of M82.
J. B. Lugten, D. M. Watson, M. K. Crawford,
R. Genzel.
Bull. Am. Astron. Soc., Vol. 16, No. 4, p. 976 (1984). Abstract. –
See Abstr. 010.062.

158.224 Mrk 1388 and other high–ionization narrow–line Seyfert galaxies.
D. E. Osterbrock, R. W. Pogge.
Bull. Am. Astron. Soc., Vol. 16, No. 4, p. 987 (1984). Abstract. –
See Abstr. 010.062.

158.225 An extensive search for low–luminosity Seyfert 1 nuclei.
A. V. Filippenko, W. L. W. Sargent.
Bull. Am. Astron. Soc., Vol. 16, No. 4, p. 987 (1984). Abstract. –
See Abstr. 010.062.

158.226 Speckle observations of the nucleus of NGC 1068.
M. Malkan, E. K. Hege, P. A. Strittmatter.
Bull. Am. Astron. Soc., Vol. 16, No. 4, p. 987 (1984). Abstract. –
See Abstr. 010.062.

158.227 Comparison of the widths of braod emission lines in Seyfert galaxies and quasars.
C.–C. Wu, C. A. Grady, A. Boggess.
Bull. Am. Astron. Soc., Vol. 16, No. 4, p. 987 – 988 (1984). Abstract. – See Abstr. 010.062.

158.228 The broad emission line profiles of Seyfert 1 galaxies.
D. M. Crenshaw.
Bull. Am. Astron. Soc., Vol. 16, No. 4, p. 988 (1984). Abstract. –
See Abstr. 010.062.

158.229 CCD observations of AGN: the use of O I λ8446, [S III] $\lambda\lambda$9532, 9069 and the Paschen series as diagnostics.
M. J. Ward, S. L. Morris.
Bull. Am. Astron. Soc., Vol. 16, No. 4, p. 988 (1984). Abstract. –
See Abstr. 010.062.

158.230 High resolution, long slit spectroscopy of NGC 7469.
M. M. De Robertis, R. W. Pogge.
Bull. Am. Astron. Soc., Vol. 16, No. 4, p. 988 (1984). Abstract. –
See Abstr. 010.062.

158.231 Rapid variations in the UV emission lines of the Seyfert 1 galaxy MCG –2–58–22.
R. L. Ptak, R. E. Stoner.
Bull. Am. Astron. Soc., Vol. 16, No. 4, p. 988 (1984). Abstract. –
See Abstr. 010.062.

158.232 Evidence for supermassive stars in three Seyfert–1 nuclei from variability in IUE spectra.
R. E. Stoner, R. L. Ptak.
Bull. Am. Astron. Soc., Vol. 16, No. 4, p. 988 – 989 (1984). Abstract. – See Abstr. 010.062.

158.233 A statistical study of the galaxian environment of Seyfert galaxies.
T. Fuentes–Williams, J. T. Stocke.
Bull. Am. Astron. Soc., Vol. 16, No. 4, p. 988 – 989 (1984). Abstract. – See Abstr. 010.062.

158.234 Reddening estimates to the Seyfert 1 galaxy II Zw 136.
M. A. DiSanti.
Bull. Am. Astron. Soc., Vol. 16, No. 4, p. 989 (1984). Abstract. –
See Abstr. 010.062.

158.235 Optical and infrared colors of high redshift galaxies.
P. R. Eisenhardt, M. J. Lebofsky.
Bull. Am. Astron. Soc., Vol. 16, No. 4, p. 990 (1984). Abstract. –
See Abstr. 010.062.

158.236 Constraints on the broad–line region of active galaxies imposed by emission–line variability.
B. M. Peterson.
Bull. Am. Astron. Soc., Vol. 16, No. 4, p. 1007 (1984). Abstract. – See Abstr. 010.062.

158.237 High dynamic range VLBI observations of NGC 6251.
D. L. Jones.
Bull. Am. Astron. Soc., Vol. 16, No. 4, p. 1008 (1984). Abstract. – See Abstr. 010.062.

158.238 VLBI observations of the jet in 3C 111.
R. P. Linfield, S. C. Unwin.
Bull. Am. Astron. Soc., Vol. 16, No. 4, p. 1008 (1984). Abstract. – See Abstr. 010.062.

158.239 VLBI monitoring of superluminal motion in BL Lac.
R. L. Mutel, R. B. Phillips.
Bull. Am. Astron. Soc., Vol. 16, No. 4, p. 1009 (1984). Abstract. – See Abstr. 010.062.

158.240 Sideways shocks in the radio jet of Centaurus A.
J. O. Burns, D. A. Clarke.
Bull. Am. Astron. Soc., Vol. 16, No. 4, p. 1009 (1984). Abstract. – See Abstr. 010.062.

158.241 The association of optical emission with the inner Cen A jet.
J. P. Brodie, S. Bowyer.
Bull. Am. Astron. Soc., Vol. 16, No. 4, p. 1009 – 1010 (1984). Abstract. – See Abstr. 010.062.

158.242 An optical investigation of a jet–cloud interaction in the galaxy cluster A 194.
P. J. McCarthy, J. P. Brodie, S. Bowyer.
Bull. Am. Astron. Soc., Vol. 16, No. 4, p. 1010 (1984). Abstract. – See Abstr. 010.062.

158.243 The milliarcsecond structure of NRAO 150.
M. W. Hodges, R. B. Phillips.
Bull. Am. Astron. Soc., Vol. 16, No. 4, p. 1010 (1984). Abstract. – See Abstr. 010.062.

158.244 Galactic mergers, starburst galaxies, quasar activity and massive binary black holes.
C. M. Gaskell.
Nature, Vol. 315, No. 6018, p. 386 (1985).

Many quasar–like objects show evidence for massive binary black holes. The recent discovery of a massive ($5 \times 10^6 M_\odot$) object in the centre of the local group dwarf elliptical M32 greatly raises the probability of forming such binaries through galactic mergers. The author argues that the enhancement of all kinds of activity (quasar–like activity and star formation) in galaxies with companions is not so much a consequence of tidal interaction between the massive galaxies as the result of collisions with their dwarf satellites.

158.245 EXOSAT observations of a 2,000–s intensity dip in Seyfert galaxy NGC 4151.
D. R. Whitehouse, A. M. Cruise.
Nature, Vol. 315, No. 6020, p. 554 – 555 (1985).

During a 5–h X–ray observation of NGC 4151 the authors observed a significant drop in intensity lasting $\sim 2,000$ s. Simultaneous intensity monitoring of a quasar with coordinates close to those of NGC 4151 revealed no variability in its flux, allowing instrumental causes of the event to be discounted. The short timescale of the intensity dip imposes limits of $2 \times 10^7 M_\odot \leqslant M \leqslant 3 \times 10^8 M_\odot$ on the mass of the central black hole.

158.246 The GB/GB2 sample of 122 intermediate–strength radio galaxies.
J. Machalski, J. J. Condon.
Astron. J., Vol. 90, No. 6, p. 973 – 986 (1985).

A sample of 122 galaxies has been selected from the GB/GB2 intermediate–strength radio sources recently mapped with the VLA by Machalski and Condon. Seven observational parameters were obtained for each galaxy: 1400–MHz flux density, low– and high–frequency spectral indices, radio morphological type, radio angular size, optical R magnitude, and optical color index (B–R). An eighth parameter, optical morphological type, was determined for the brighter galaxies only. Linear sizes, total luminosities, and core luminosities of the radio sources were derived from the data. Results of a statistical analysis of this sample are presented, too.

158.247 The wide angle tailed radio source NGC 2329 in the cluster A569.
L. Feretti, G. Giovannini, L. Gregorini, L. Padrielli, J. Roland, E. A. Valentijn.
Astron. Astrophys., Vol. 147, No. 2, p. 321 – 327 (1985).

The authors present WSRT observations at 0.6, 1.4 and 5 GHz of the wide angle tail source associated with NGC 2329, the dominant galaxy of the cluster A 569. Since NGC 2329 is observed to be associated with an X–ray source, it is likely to possess an extended gaseous X–ray halo. The authors suggest that it is possible to describe the observed structure of the wide angle tailed source as due to large scale pressure asymmetries resulting from the combination of radial gas accretion on a subsonically moving galaxy.

158.248   Extragalactic variable radio sources.
P. A. Sturrock.
Astrophys. J., Vol. 293, No. 1, p. 52 – 55 (1985).

The purpose of this article is to consider radio emission from a plasmoid that expands in a non–spherically symmetric manner and to compare the properties of this model with the observational results from highly variable quasars and radio galaxies.

158.249   Minkowski's object: a starburst triggered by a radio jet.
W. van Breugel, A. V. Filippenko, T. Heckman, G. Miley.
Astrophys. J., Vol. 293, No. 1, p. 83 – 93 (1985). With plates 8 – 9.

Radio and optical imaging, as well as optical spectrophotometry, are reported for "Minkowski's object", a peculiar blue object near the elliptical galaxy NGC 541 in the cluster Abell 194. The observations show that it is a region with bright optical line and continuum emission associated with a radio jet emanating from NGC 541. The radio and optical properties may be ascribed to the collision of the jet with relatively dense extranuclear gas. The origin of this gas could be related to its apparent location in a galactic bridge connecting NGC 541 with the nearby dumbbell system NGC 545/NGC 547. A careful study of the excitation mechanism for the line emission leads to the conclusion that Minkowski's object is an irregular galaxy currently undergoing a vigorous burst of star formation that was triggered by the radio jet.

158.250   Bubbles and jets in the center of M51.
H. C. Ford, P. C. Crane, G. H. Jacoby, D. G. Lawrie, J. M. van der Hulst.
Astrophys. J., Vol. 293, No. 1, p. 132 – 147 (1985). With plates 15 – 21

The authors present monochromatic $H\alpha + [N\ II]$ pictures and high–resolution VLA 20 cm and 6 cm radio maps which show a bright, filled cloud and a large ring paired across the nucleus of M51. Spectrophotometry shows that the nuclear emission–line characteristics change abruptly at the edges of the ring and the cloud, and that the cloud has internal velocities larger than the nucleus and a spectrum similar to some galactic supernova remnants. The authors derive the physical characteristics of the nuclear emission regions and then use these to discuss the ionization and excitation of the ring and cloud. They conclude that the nuclear phenomena are the result of an active nucleus which has created a pair of bubbles in the disk. The bubbles may have their origin in a bidirectional jet emanating from the nucleus and interacting with the disk of M51.

158.251   The *IRAS* galaxy 0421 + 040P06: an active spiral (?) galaxy with extended radio lobes.
C. Beichman, C. G. Wynn–Williams, C. J. Lonsdale, S. E. Persson, J. N. Heasley, G. K. Miley, B. T. Soifer, G. Neugebauer, E. E. Becklin, J. R. Houck.
Astrophys. J., Vol. 293, No. 1, p. 148 – 153 (1985). With plate 22.

The infrared bright galaxy 0421 + 040P06 detected by *IRAS* at 25 and 60 µm has been studied at optical, infrared, and radio wavelengths. It is a luminous galaxy with apparent spiral structure emitting $1 \times 10^{11} L_\odot$ from far–infrared to optical wavelengths, assuming $H_0 = 75$ km s^{-1}Mpc^{-1}. Optical spectroscopy reveals a Seyfert 2 emission–line spectrum. The radio observations reveal the presence of a nonthermal source that, at 6 cm, shows a prominent double lobed structure 20 – 30 kpc in size extending beyond the optical confines of the galaxy. The radio source is 3 – 10 times larger than structures previously seen in spiral galaxies.

158.252   *IRAS* observations of Seyfert galaxies.
G. K. Miley, G. Neugebauer, B. T. Soifer.
Astrophys. J., Lett. Ed., Vol. 293, No. 1, p. L11 – L14 (1985).

IRAS measurements at 25, 60, and 100 µm have been used to analyze the infrared properties of Seyfert galaxies from the Markarian and NGC catalogs. One hundred sixteen of 186 Seyfert galaxies were detected. About 50% of all Seyfert galaxies in the sample have 60 µm luminosities in excess of $\sim 10^{10} L_\odot$, and the mean 60 µm luminosity increases with the optical B absolute magnitude. The luminosity functions of the Seyfert 1 and Seyfert 2 galaxies appear quite similar. The infrared measurements provide a measure of the bolometric luminosity of the Seyfert galaxies but do not discriminate between the physical processes involved.

158.253   Far–infrared intense–field phenomena in active galactic nuclei.
H. R. Reiss.
Infrared Phys., Vol. 25, No. 1/2, p. 521 – 524 (1985). – See Abstr. 012.065.

A major feature of many active galactic nuclei is that the bulk of their luminosity occurs in the FIR. This fact, combined with the small size of the objects, means that nonperturbative intense–field behavior is likely to exist at long wavelengths, even though visible and UV behavior is perturbative. The FIR can then cause unusual spectral phenomena. One example is significant broadening of spectral lines in quasars, and another is the absence of line spectra in BL Lac objects.

158.254   1E 1402.3 + 0416.
IAU Circ., No. 4033 (1985).

158.255   NGC 4151.
IAU Circ., No. 4036 (1985).

158.256   B2 1506 + 34.
IAU Circ., No. 4037 (1985).

158.257   NGC 3031.
IAU Circ., Nos. 4044, 4050 (1985).

158.258   NGC 4051.
IAU Circ., No. 4054 (1985).

158.259   H 0323 + 022.
IAU Circ., No. 4059 (1985).

158.260   NGC 6212.
IAU Circ., Nos. 4060, 4068, 4073 (1985).

158.261   OJ 287.
IAU Circ., No. 4070 (1985).

158.262   OH maser in Markarian 273.
IAU Circ., No. 4074 (1985).

158.263   Discovery of new variable radio sources in the nucleus of the nearby galaxy Messier 82.
P. P. Kronberg, R. A. Sramek.
Science, Vol. 227, No. 4682, p. 28 – 31 (1985).

158.264   Luminosity and color evolution of high–redshift radio galaxies.
S. Djorgovski, H. Spinrad, J. Marr.
Bull. Am. Astron. Soc., Vol. 17, No. 1, p. 515 – 516 (1985). Abstract. – See Abstr. 010.064.

158.265   Galaxy activity and environment – a statistical study.
N. Jeske, S. Djorgovski, M. Davis.
Bull. Am. Astron. Soc., Vol. 17, No. 1, p. 516 (1985). Abstract. – See Abstr. 010.064.

158.266 Photometric decomposition of multiple–nuclei cD galaxies.
T. R. Lauer, J. E. Gunn.
Bull. Am. Astron. Soc., Vol. 17, No. 1, p. 516 (1985). Abstract. – See Abstr. 010.064.

158.267 Hot spots in radio galaxies: a comparison with hydrodynamic simulations.
M. D. Smith, M. L. Norman, K.-H. A. Winkler, L. Smarr.
Bull. Am. Astron. Soc., Vol. 17, No. 1, p. 516 (1985). Abstract. – See Abstr. 010.064.

158.268 Nature of the irregular galaxy NGC 5253.
N. Caldwell, M. M. Phillips.
Bull. Am. Astron. Soc., Vol. 17, No. 1, p. 517 (1985). Abstract. – See Abstr. 010.064.

158.269 Minkowski's object: a starburst triggered by a radio jet.
W. van Breugel, A. Filippenko, T. Heckman,
G. Miley.
Bull. Am. Astron. Soc., Vol. 17, No. 1, p. 517 (1985). Abstract. – See Abstr. 010.064.

158.270 The nuclear mass of type I Seyfert galaxies.
R. Liu.
High energy astrophysics and cosmology, p. 389 – 393 (1983). – See Abstr. 012.068.

This paper investigates the statistical relation between X–ray luminosities and total masses in the nuclei of Seyfert I galaxies.

158.271 Optical counterparts of the radio source Virgo A (= M87).
J.-L. Nieto, G. Lelièvre.
The Virgo cluster of galaxies, p. 311 – 320 (1985). – See Abstr. 012.069.

The authors present some physical properties of the optical counterparts of the radio source Virgo A. This paper is a compilation of results from a series of optical and UV observations of M87 (1981–1984) essentially at high resolution.

158.272 NGC 4151: a Seyfert 2 in a deep photometric minimum.
V. M. Lyutyj, V. L. Oknyanskij, K. K. Chuvaev.
Sov. Astron. Lett., Vol. 10, No. 6, p. 335 – 336 (1984). English translation of 38.158.110.

158.273 Distant, medium–luminosity radio galaxies near 3C 273.
V. V. Vitkovskij, W. Reich, N. S. Soboleva,
A. V. Temirova.
Sov. Astron. Lett., Vol. 10, No. 6, p. 337 – 339 (1984). English translation of 38.158.111.

158.274 U,B,V photometry of an interesting Seyfert galaxy.
M. A. Kazaryan, V. S. Tamazyan.
Sov. Astron. Lett., Vol. 10, No. 6, p. 340 – 342 (1984). English translation of 38.158.112.

158.275 The lag between continuum and line flares, and the spatial structure of Seyfert broad–line regions.
I. I. Shevchenko.
Sov. Astron. Lett., Vol. 10, No. 6, p. 377 – 379 (1984). English translation of 38.158.244.

158.276 Coordinated multisite observations of the variability of BL Lac.
C. Brindle, J. H. Hough, J. A. Bailey, D. J. Axon, H. Schulz,
S. Kikuchi, J. T. McGraw, W. J. Wisniewski, G. Fontaine,
D. Nadesu, G. Clayton, E. Anderson, R. F. Jameson, R. Smith,
R. E. Wallis.
Mon. Not. R. Astron. Soc., Vol. 214, No. 4, p. 619 – 638 (1985). With microfiche MN 214/1.

The authors present the results of multisite observations of the optical polarization of BL Lac, together with some flux measurements, at a time when BL Lac was relatively faint. The level of polarization varies in a manner similar to that reported by Moore et al., although the position angle behaviour is quite different,
varying only a small amount from a mean of 28°. Models are presented to account for the observed polarization behaviour of BL Lac. The best model is one consisting of two polarized components, one having a variable polarization, intensity and random position angle, and the other a fixed polarization, intensity and position angle. The authors tentatively identify this latter component as being associated with the inner region of the optical jet of BL Lac.

158.277 A radio and optical study of a jet/cloud interaction in the galaxy cluster A194.
J. P. Brodie, S. Bowyer, P. McCarthy.
Astrophys. J., Lett. Ed., Vol. 293, No. 2, p. L59 – L63 (1985). With plates L2 – L3.

The cluster of galaxies Abell 194 has within it the strong radio source PKS 0123–016. Spectroscopic and photometric data obtained in the field of this source have been combined with VLA maps to reveal a remarkable example of a jet/cloud interaction at the site of the peculiar galaxy known as Minkowski's object, which lies in the path of one of the radio jets discovered in this region. The optical results can be used to rule out shock heating and power–law photoionization for this source and indicate a stellar origin for the ionizing flux. The authors suggest that Minkowski's object was produced by a burst of star formation triggered by the impact of the jet of an obstructing cloud in its path.

158.278 Active galactic nuclei.
N. Dallaporta.
Sir Arthur Eddington Centenary Symposium. Vol. 1: Relativistic astrophysics and cosmology, p. 1 – 27 (1984). – See Abstr. 012.080.

Observational aspects of activity in Seyfert galaxies, radio galaxies, BL Lac objects and quasars are reviewed with emphasis on an emerging unified theoretical understanding of the various activity phenomena.

158.279 The physics of active stellar systems.
J. Kuijpers.
Ned. Tijdschr. Natuurkd. A, Vol. A51, No. 1, p. 21 – 26 (1985). In Dutch. Abstr. in Phys. Abstr., Vol. 88, No. 1258, Entry 63497 (1985).

158.280 Energy phenomena in the cores of Milky Way systems.
A. G. de Bruyn, G. K. Miley.
Ned. Tijdschr. Natuurkd. A, Vol. A51, No. 1, p. 26 – 30 (1985). In Dutch. Abstr. in Phys. Abstr., Vol. 88, No. 1258, Entry 63498 (1985).

158.281 X–ray variability of BL Lac objects: H2155–304 and PKS 0548–322.
P. C. Agrawal, K. P. Singh, G. R. Riegler.
18th International Cosmic Ray Conference, Vol. 1, p. 3 – 6 (1983). – See Abstr. 012.096.

High spatial resolution observations confirm the identification of these X–ray sources with the BL Lac objects and give a source size $\lesssim 2''$. Variations in the X–ray intensity and spectrum of these sources were detected over time scales of a few hours.

158.282 Gamma ray observations of active galactic nuclei as a test of black hole accretion models.
L. Brassani, A. J. Dean, S. Sembay.
18th International Cosmic Ray Conference, Vol. 1, p. 85 – 88 (1983). – See Abstr. 012.096.

Recent γ–ray observations of active galaxies suggest that the major fraction of the total luminosity is emitted in the high energy region of the electromagnetic spectrum. Analysis of the variability timescale – bolometric luminosity relation for about sixty objects, including Seyferts, quasars and BL Lacs, shows that a number of active galaxies violate the classical Eddington limit. A more realistic formulation of this limit is discussed in the context of black hole accretion models for active galaxies.

158.283   The development of the electromagnetic cascade showers initiated by high–energy electrons and gammay–rays in the hot photon gas.
F. A. Aharonian (*F. A. Agaronyan*), V. V. Vardanian.
18th International Cosmic Ray Conference, Vol. 1, p. 153 – 156 (1983). – See Abstr. 012.096.

The development of the electromagnetic cascade shower, occurring at high energy electrons and gamma–rays passage through the hot plasma in the nuclei of active galaxies and quasars is considered.

158.284   The global radio continuum of nearby galaxies.
R. Wielebinski.
18th International Cosmic Ray Conference, Vol. 2, p. 161 – 164 (1983). – See Abstr. 012.096.

The study of the radio continuum from our Milky Way has now been in progress for over 50 years. The radio continuum of normal nearby galaxies is expected to be some variant of what is seen in our own Galaxy. A summary of all results available so far is given. Emphasis is placed on recent data which have close connection with the studies of propagation processes of cosmic rays in galaxies.

158.285   Spectral and temporal constraints on active galactic nuclei.
A. C. Fabian.
18th International Cosmic Ray Conference, Vol. 12, p. 109 – 116 (1983). – See Abstr. 012.096.

Many active galaxies are found to vary in luminosity. Such variations are valuable indicators of source parameters and radiation mechanisms. The author attempts to show how one might read these indications, and, in particular, show how they suggest that electron–positron plasmas are formed.

158.286   Investigations of radio objects with continuous optical spectra. The results of four–colour electrophotometric observations.
G. M. Beskin, V. M. Lyutyj, S. I. Neizvestnyj, S. A. Pustil'nik, V. F. Shvartsman.
Astron. Zh., Tom 62, Vyp. 3, p. 432 – 449 (1985). In Russian. English translation in Sov. Astron., Vol. 29, No. 3.

The results of *UBVR* photometry of 30 radio objects with continuous optical spectra (ROCOSes) are reported. The observations were performed using five telescopes during the years 1979 – 1982; 54 values have been obtained of *U, B, V* magnitudes and 26 of *R* magnitude. Colours for 16 ROCOSes have been obtained for the first time. An appendix contains the results of *UBV* photometry of the BL Lac object OJ 287 during the years 1976 – 1982 (24 measurements).

158.287   On the dependence between the luminosities of Seyfert galaxies and those of their active nuclei.
Eh. A. Dibaj, A. V. Zasov.
Astron. Zh., Tom 62, Vyp. 3, p. 468 – 474 (1985). In Russian. English translation in Sov. Astron., Vol. 29, No. 3.

Absolute magnitudes of active nuclei of some Seyfert galaxies are determined from photoelectric photometry. There exists a well–defined dependence between the luminosity (optical or bolometric) of the nucleus and the total luminosity of the parent galaxy. This is apparently not a result of observational selection only; it rather reflects a real interdependence between the nuclear activity and the properties of the galaxy as a whole. There possibly exists a sharp jump of nonstellar nuclear luminosity between normal and Seyfert galaxies.

158.288   Investigation of radio galaxies from the Bologna survey with the radio telescope RATAN–600.
V. G. Malumyan.
Pis'ma Astron. Zh., Tom 11, No. 6, p. 415 – 419 (1985). In Russian. English translation in Sov. Astron. Lett., Vol. 11.

15 radio galaxies from the Bologna survey have been observed with the radio telescope RATAN–600 at 3.95 GHz. Among these galaxies not only radio galaxies having small angular sizes but some objects of large angular sizes ($\geqslant 40''$) as well display a flat radio spectrum.

158.289   Line–emitting clouds in broad–line regions of active galactic nuclei: their geometry and phase functions.
I. I. Shevchenko.
Pis'ma Astron. Zh., Tom 11, No. 6, p. 432 – 437 (1985). In Russian. English translation in Sov. Astron. Lett., Vol. 11.

For a spherical cloud photoionized by a distant point source a phase function which describes the anisotropy of the cloud radiation in a given line is derived. The derived formulae may be useful in investigations of the structure of broad–line regions of active galactic nuclei. The effect of cloud geometry on the phase function is discussed.

158.290   Energetics and radiation spectrum of the plasma jet from the M87 galaxy nucleus.
A. A. Galeev.
Pis'ma Astron. Zh., Tom 11, No. 6, p. 438 – 443 (1985). In Russian. English translation in Sov. Astron. Lett., Vol. 11.

The theory of resonant acceleration of electrons by magnetosonic waves which are generated by the beam of ions reflected from the shock front is used to calculate parameters of the plasma jet from the M87 galaxy nucleus and its radiation spectrum. It is shown that the drastic break in the optical radiation spectrum results from insufficient pitch–angle scattering of electrons accelerated preferentially along the magnetic field lines. The X–ray spectrum flattens due to pitch–angle scattering of electrons by MHD turbulence of the plasma jet.

158.291   A polarization survey of compact radio sources.
R. E. Rusk, E. R. Seaquist.
Bull. Am. Astron. Soc., Vol. 17, No. 2, p. 576 (1985). Abstract. – See Abstr. 010.065.

158.292   Reinterpretation of the "jet" in M87.
M. L. White.
Bull. Am. Astron. Soc., Vol. 17, No. 2, p. 576 (1985). Abstract. – See Abstr. 010.065.

158.293   Evolution of the outburst characteristics of BL Lac.
E. B. Waltman, J. H. Spencer, K. J. Johnston, P. E. Angerhofer, D. R. Florkowski, D. N. Matsakis, D. D. McCarthy.
Bull. Am. Astron. Soc., Vol. 17, No. 2, p. 577 (1985). Abstract. – See Abstr. 010.065.

158.294   Star formation in galaxies: the active dwarfs.
H. A. Thronson, C. M. Telesco.
Bull. Am. Astron. Soc., Vol. 17, No. 2, p. 577 (1985). Abstract. – See Abstr. 010.065.

158.295   First results from an imaging survey of Markarian Seyfert galaxies.
J. W. MacKenty.
Bull. Am. Astron. Soc., Vol. 17, No. 2, p. 577 (1985). Abstract. – See Abstr. 010.065.

158.296   Ultraviolet emission line and continuum variability in the broad line radio galaxies 3C 120, 3C 382, and 3C 390.3.
G. A. Reichert, C.–C. Wu, A. Boggess, J. B. Oke.
Bull. Am. Astron. Soc., Vol. 17, No. 2, p. 578 (1985). Abstract. – See Abstr. 010.065.

158.297   Deep 21 cm radio imaging of 1 degree fields around 12 core–dominated radio sources.
R. S. Simon, J. H. Spencer, K. J. Johnston.
Bull. Am. Astron. Soc., Vol. 17, No. 2, p. 578 (1985). Abstract. – See Abstr. 010.065.

158.298   The wind from a starburst galaxy nucleus.
A. W. Clegg, R. A. Chevalier.
Bull. Am. Astron. Soc., Vol. 17, No. 2, p. 587 (1985). Abstract. – See Abstr. 010.065.

158.299 **Dust in the emission line gas in Markarian 3.**
S. B. Kraemer.
Bull. Am. Astron. Soc., Vol. 17, No. 2, p. 587 (1985). Abstract. –
See Abstr. 010.065.

158.300 **An analysis of the narrow–line profiles in Seyfert 2 galaxies.**
M. M. De Robertis, D. E. Osterbrock.
Bull. Am. Astron. Soc., Vol. 17, No. 2, p. 587 (1985). Abstract. –
See Abstr. 010.065.

158.301 **Emission–line kinematics in AGN: evidence for a universal M–L relation?**
A. Wandel.
Bull. Am. Astron. Soc., Vol. 17, No. 2, p. 588 (1985). Abstract. –
See Abstr. 010.065.

158.302 **Rapid X–ray variability in NGC 3031 (M81).**
P. Barr, P. Giommi, W. Wamsteker, R. Gilmozzi,
R. F. Mushotzky.
Bull. Am. Astron. Soc., Vol. 17, No. 2, p. 608 (1985). Abstract. –
See Abstr. 010.065.

158.303 **X–ray and optical observations of four X–ray selected BL Lac objects.**
P. Giommi, P. Barr, D. Maccagni, B. Garilli, I. M. Gioia,
T. Maccacaro, R. Schild, J. Stocke.
Bull. Am. Astron. Soc., Vol. 17, No. 2, p. 608 (1985). Abstract. –
See Abstr. 010.065.

158.304 **H0323 + 022: a remarkably variable BL Lac object.**
E. D. Feigelson, H. Bradt, J. McClintock,
R. Remillard, S. Tapia, B. Geldzahler, K. Johnston,
W. Romanishin, P. A. Wehinger, S. Wyckoff, G. Madejski,
D. A. Schwartz, J. Thorstensen, B. E. Schaefer.
Bull. Am. Astron. Soc., Vol. 17, No. 2, p. 608 (1985). Abstract. –
See Abstr. 010.065.

158.305 **Newly discovered BL Lacertae objects identified as bright X–ray source counterparts by the HEAO–1 scanning modulation collimator.**
D. A. Schwartz, W. Roberts, S. Murray, J. Huchra,
R. Remillard, H. Bradt, J. McClintock, I. Tuohy, D. Buckley,
S. Tapia, E. Feigelson, J. Schmelz.
Bull. Am. Astron. Soc., Vol. 17, No. 2, p. 608 (1985). Abstract. –
See Abstr. 010.065.

158.306 **Models for low–radio–frequency variability in the blazar AO 0235 + 164.**
S. L. O'Dell, J. J. Broderick, B. Dennison, K. J. Mitchell,
D. R. Altschuler, J. J. Condon, H. E. Payne.
Bull. Am. Astron. Soc., Vol. 17, No. 2, p. 609 (1985). Abstract. –
See Abstr. 010.065.

158.307 **On the relationship between cold gas and nuclear activity in nearby early–type galaxies.**
M. J. Wardle, G. R. Knapp.
Bull. Am. Astron. Soc., Vol. 17, No. 2, p. 611 (1985). Abstract. –
See Abstr. 010.065.

158.308 **High resolution CCD images of M87.**
T. R. Lauer, J. Kormendy.
Bull. Am. Astron. Soc., Vol. 17, No. 2, p. 611 (1985). Abstract. –
See Abstr. 010.065.

158.309 **Low frequency radio observations of Seyfert galaxies.**
S. G. Neff, R. J. Hanisch, N. E. Kassim,
A. S. Wilson.
Bull. Am. Astron. Soc., Vol. 17, No. 2, p. 612 (1985). Abstract. –
See Abstr. 010.065.

158.310 **Decameter fine structure in 4C 21.53.**
V. P. Bovkun, I. N. Zhuk.
Sov. Astron., Vol. 28, No. 6, p. 648 – 651 (1984). English translation of 38.158.280.

158.311 **Linear polarization and total flux density of silicon monoxide masers and active extragalactic objects at millimeter wavelengths.**
R. E. Barvainis.
Diss. Abstr. Int., Sect. B, Vol. 45, No. 1, p. 228 – 229 (1984).
Thesis, University of Massachusetts, 154 pp. (1984). Order No. DA8410258.

158.312 **The radio continuum properties of bright E/S0 galaxies which contain compact radio core sources.**
J. M. Wrobel.
Diss. Abstr. Int., Sect. B, Vol. 45, No. 2, p. 585 (1984). Thesis, University of Toronto (1983).

158.313 **Dynamical models for radio jets.**
J. Siah.
Diss. Abstr. Int., Sect. B, Vol. 45, No. 5, p. 1502 (1984). Thesis, University of Pennsylvania, 146 pp. (1984). Order No. DA8417362.

158.314 **Continuum distributions of active galactic nuclei.**
I. Cruz–Gonzalez.
Diss. Abstr. Int., Sect. B, Vol. 45, No. 6, p. 1810 (1984). Thesis, Harvard University, 367 pp. (1984). Order No. DA8419443.

158.315 **Physical conditions in low–luminosity active galactic nuclei.**
A. V. Filippenko.
Diss. Abstr. Int., Sect. B, Vol. 45, No. 6, p. 1810 (1984). Thesis, California Institute of Technology, 313 pp. (1984). Order No. DA8420807.

158.316 **Rapid X–ray variability of active galaxies.**
A. F. Tennant Jr.
Diss. Abstr. Int., Sect. B, Vol. 45, No. 6, p. 1812 (1984). Thesis, University of Maryland (1983).

158.317 **X–ray and ultraviolet observations of BL Lacertae objects.**
C. M. Urry.
Diss. Abstr. Int., Sect. B, Vol. 45, No. 6, p. 1812 (1984). Thesis, Johns Hopkins University, 365 pp. (1984). Order No. DA8421235.

158.318 **VLBI observations of compact double radio sources.**
M. W. Hodges.
Diss. Abstr. Int., Sect. B, Vol. 45, No. 7, p. 2200 (1985). Thesis, University of Iowa, 97 pp. (1984). Order No. DA8423567.

158.319 **Frequent observations of extragalactic compact sources at 24 GHz.**
F. T. Haddock.
Diss. Abstr. Int., Sect. B, Vol. 45, No. 12, p. 3847 (1985). Thesis, University of Michigan, 143 pp. (1984). Order No. DA8502830.

158.320 **Electron–positron pairs in active galactic nuclei.**
M. Kusunose, F. Takahara.
Theoretical aspects on structure, activity, and evolution of galaxies: III, p. 95 – 101 (1985). – See Abstr. 012.110.

Three topics on electron–positron pairs in a relativistic plasma are presented in connection with active galactic nuclei. Firstly, the properties of optically thin pair equilibrium states are summarized. Secondly, pair production in a hot accretion plasma is discussed. Finally, the authors present preliminary numerical results of time development of the pair concentration.

158.321 **Spectrophotometry and morphology of galaxies with UV excess. VI.**
M. A. Kazaryan, Eh. S. Kazaryan.
Astrofizika, Tom 22, Vyp. 3, p. 431 – 440 (1985). In Russian. English translation in Astrophysics, Vol. 22, No. 3.

The results of spectrophotometry and morphology of galaxies No. 73, 125 and 229 from the lists (1979) are presented. The masses of the gaseous component of these galaxies are obtained. It has been established that galaxy No. 73 is of Sy 2 type and by its physical properties is similar to the Sy 2 type galaxies Mark 744 and 1066. Though the galaxy No. 125 by its several properties is similar to No. 73, it is, however, evidently in a later stage of evolution than galaxy No. 73.

158.322   Observing the galaxy evolution at high redshifts.
S. Djorgovski, H. Spinrad, J. Marr.
New aspects of galaxy photometry, p. 193 – 198 (1985). – See Abstr. 012.114.

Magnitudes and colors for a sample of distant 3CR radio–galaxies, with redshifts up to $z = 1.82$, show very dramatic evidence for luminosity and color evolution of these objects. Cosmological differences are small compared to the detected evolutionary effects. Elongated shapes and very extended (both in space and velocity) [O II] emission lines suggest violent merging activity at the early epochs, and possibly mark the beginnings of activity of powerful radio–galaxies.

158.323   Phenomenology of extragalactic radio jets.
A. H. Bridle.
Physics of energy transport in extragalactic radio sources, p. 1 – 19 (1984). – See Abstr. 012.116.

The use of the term "jet" is critically reviewed. "Jets" occur often, in a wide variety of extragalactic radio sources, and with properties well correlated both with the total luminosities of the sources and with the relative prominence of their compact radio cores. The one sided jets in powerful sources with symmetrical double lobes pose some acute problems. The evidence bearing on particle acceleration and on the 3–D magnetic field structures in radio jets is briefly summarized.

158.324   3C 120: a continuous link between moving features and a large scale radio jet.
R. C. Walker.
Physics of energy transport in extragalactic radio sources, p. 20 – 24 (1984). – See Abstr. 012.116.

VLBI monitoring observations at 6 cm show features in the parsec–scale radio core of the galaxy 3C 120 that are moving at apparent velocities of a few times the speed of light. Lower resolution VLBI observations at 18 cm and VLA observations with resolutions between 0″.1 and 12″ show a continuous connection between these moving features and 100–kpc scale jet and lobe structures. These data provide the best evidence available, although still indirect, that the large scale, linear structures usually referred to as jets actually contain moving material.

158.325   Radio jets in strong core classical doubles.
J. O. Burns.
Physics of energy transport in extragalactic radio sources, p. 25 – 29 (1984). – See Abstr. 012.116.

The properties of radio jets in classical double sources are described. Particular attention is paid to jet statistics, clumpiness, efficiencies, and confinement.

158.326   Pieces of jets and other thoughts on flip–flops.
L. Rudnick.
Physics of energy transport in extragalactic radio sources, p. 35 – 38 (1984). – See Abstr. 012.116.

The author briefly discusses the status of the flip–flop model and introduces the use of isolated pieces of jets as one possible signature. Study of these jet pieces is also useful independent of the flip–flop, and may help to distinguish between discontinuous energy supplies and intermittent illumination as explanations of gaps in high surface brightness structures. Several areas are suggested for further observational and theoretical work.

158.327   The radio jets of 3C 449.
T. Cornwell, R. Perley.
Physics of energy transport in extragalactic radio sources, p. 39 – 46 (1984). – See Abstr. 012.116.

Multifrequency VLA observations of 3C 449 have revealed the presence of steep gradients in rotation measure in the jets. Consequently, previous estimates of the jet thermal density and velocity from the depolarization are inaccurate. The rotation measure can be attributed to a foreground screen, probably in the galaxy itself

since it is an extended X–ray source. The same X–ray emitting gas can easily confine the jets. Some estimates of the jet velocity can be made but all are subject to large uncertainties.

158.328   Polarization structure of the jets in NGC 1265.
C. P. O'Dea, F. N. Owen.
Physics of energy transport in extragalactic radio sources, p. 47 – 56 (1984). – See Abstr. 012.116.

Arcsecond resolution VLA observations of the polarization structure of the jets in NGC 1265 at 21, 6, and 2 cm are presented and discussed.

158.329   Multiconfiguration VLA maps of Cygnus A.
J. Dreher, R. Perley, J. Cowan.
Physics of energy transport in extragalactic radio sources, p. 57 – 62 (1984). – See Abstr. 012.116.

Maps of Cygnus A at 1.4 and 5 GHz made with multiple configurations of the VLA show a number of interesting features: a jet, a possible counterjet, complex fine structures in the lobes, and very hard edges to the lobes. The ratio of jet to core powers is similar to that of other powerful sources. An upper limit to the jet/counterjet brightness ratio is 4. The hard edges to the lobes suggest that ram confinement may be occurring on the sides as well as the fronts of the lobes. Overall, the lobes and the hotspots both exhibit a strong "S" symmetry.

158.330   Jet turn–on and confinement – indications from IC 4296.
G. V. Bicknell.
Physics of energy transport in extragalactic radio sources, p. 63 (1984). Abstract. – See Abstr. 012.116.

158.331   Wide angle tail radio galaxies.
F. N. Owen.
Physics of energy transport in extragalactic radio sources, p. 89 (1984). Abstract. – See Abstr. 012.116.

158.332   Interpretation of radio polarization data.
R. A. Laing.
Physics of energy transport in extragalactic radio sources, p. 90 – 98 (1984). – See Abstr. 012.116.

The author reviews the interpretation of radio polarization data, partly because misinterpretations are evident in the current literature, and partly because it is now known more about the possible Faraday–rotating and depolarizing phases around radio sources.

158.333   VLBI polarimetry – first results.
J. F. C. Wardle, D. H. Roberts.
Physics of energy transport in extragalactic radio sources, p. 99 (1984). Abstract. – See Abstr. 012.116.

158.334   Energy transport in radio sources: efficient or not?
D. S. De Young.
Physics of energy transport in extragalactic radio sources, p. 100 – 108 (1984). – See Abstr. 012.116.

The question of the flow velocity of the material which powers the extended and compact radio sources is considered, as this issue is crucial to energy transport in radio sources. The various velocity indicators are reviewed and their implications examined. The principal conclusions are that there is strong evidence for non–relativistic flow on the large scales, and that there are serious problems with relativistic flow in compact sources.

158.335   An argument for jet velocities > 0.1 c in powerful doubles.
J. W. Dreher.
Physics of energy transport in extragalactic radio sources, p. 109 – 113 (1984). – See Abstr. 012.116.

The L/cT method is used to estimate the jet velocities for 12 edge–brightened doubles. Using plausible model parameters, the estimates for L/cT are all much larger than unity, which is impossible. The model parameters can be adjusted to give a conservative lower limit to L/cT. These limits are of order 0.1, leading to lower limits on the jet velocities of $\sim 0.1\ c$.

158.336 Jet asymmetries in radio galaxies with dust lanes.
R. A. Laing.
Physics of energy transport in extragalactic radio sources, p. 119 (1984). Abstract. – See Abstr. 012.116.

158.337 Semi–dynamical models of superluminal radio sources.
K. R. Lind, R. D. Blandford.
Physics of energy transport in extragalactic radio sources, p. 121 (1984). Abstract. – See Abstr. 012.116.

158.338 Symmetries of jets and hot spots.
R. A. Laing.
Physics of energy transport in extragalactic radio sources, p. 128 (1984). Abstract. – See Abstr. 012.116.

158.339 Beam inefficiencies and extended source morphologies.
W. A. Christiansen.
Physics of energy transport in extragalactic radio sources, p. 129 – 132 (1984). – See Abstr. 012.116.
One of the most striking aspects of the structural morphology of extended radio sources possessing one–sided jets is the apparent symmetry of the two radio lobes in conjunction with the strong asymmetry of the jet itself. A plausible interpretation of this peculiar morphology within the context of symmetric, continuous beam models for radio source evolution is that the observed radio jet delineates regions where beam transport of energy is "inefficient". An analysis of hot spot and lobe luminosity ratios indicates that, at least for sources in which the jet luminosity is within an order of magnitude of the lobe luminosity, the implied energetics of two–sided continuous beams with asymmetric transport inefficiencies are not consistent with observations.

158.340 Menu for an all–purpose source model.
A. H. Bridle.
Physics of energy transport in extragalactic radio sources, p. 135 – 143 (1984). – See Abstr. 012.116.
Evening discussion concerning energy transport models in extragalactic sources.

158.341 Turbulence, entrainment and magnetic fields.
D. S. De Young.
Physics of energy transport in extragalactic radio sources, p. 202 – 210 (1984). – See Abstr. 012.116.
Evidence for the presence of turbulence in extended radio sources is reviewed, and its effects on the physics of energy transport in these objects is discussed. Turbulent entrainment of the interstellar medium in radio galaxies is suggested as the origin of the optical emission lines, and results of some numerical experiments on entrainment are given which suggest that most entrainment occurs largely near the leading edge of a radio source and that its extent decreases with increasing Mach number. Turbulent amplification of magnetic fields is also reviewed.

158.342 Sidewise shocks in the Centaurus A radio jet.
J. O. Burns, D. Clarke, E. D. Feigelson,
E. J. Schreier.
Physics of energy transport in extragalactic radio sources, p. 255 – 259 (1984). – See Abstr. 012.116.
New high resolution ($1'' \times 0\rlap{.}''3$), high dynamic range (13,000:1) observations at 2 and 6 cm of Cen A have revealed the presence of side–to–side limb brightening in the inner 700 pc of the radio jet. This edge brightening combined with the magnetic field orientation, the spectral index, and the internal pressure of the knots suggest the presence of transverse shocks in the jet.

158.343 An observer's perspective – I.
P. N. Wilkinson.
Physics of energy transport in extragalactic radio sources, p. 272 – 275 (1984). – See Abstr. 012.116.
The author reminds the progress since, for example, the IAU Symposium on Extragalactic Radio Sources in Albuquerque in 1981. This is largely because the full VLA has come on stream. The author stresses that the observations for finer–scale structure

are improving steadily and that VLA–quality results on some small–scale jets should be with us quite soon.

158.344 An observer's perspective – II.
R. A. Laing.
Physics of energy transport in extragalactic radio sources, p. 276 – 279 (1984). – See Abstr. 012.116.
The author summarises what has been done so far as concerns observations and understanding the physics of extragalactic radio sources.

158.345 A theorist's perspective – I.
P. J. Wiita.
Physics of energy transport in extragalactic radio sources, p. 285 – 291 (1984). – See Abstr. 012.116.
The author starts with a very brief history of extragalactic radio astronomy, discusses some exciting observations and reviews theoretical trends in this field.

158.346 A theorist's perspective – II.
A. Königl.
Physics of energy transport in extragalactic radio sources, p. 292 – 296 (1984). – See Abstr. 012.116.
The author reviews and discusses three qualitatively new approaches to the modeling of energy transport in extragalactic radio jets.

158.347 Some aspects of active galactic nuclei.
A. C. Fabian.
ESA Spec. Publ., ESA SP–213, p. 7 – 9 (1984). – See Abstr. 012.124.
A brief summary of current models for energy sources in active galactic nuclei is presented.

158.348 The environment of active galactic nuclei.
G. Miley.
ESA Spec. Publ., ESA SP–213, p. 131 – 137 (1984). – See Abstr. 012.124.
Quasat will provide observations of radio jets propagating through the broad line regions of active nuclei. Here evidence is presented that the radio emitting relativistic plasma and thermal gas in the broad line region interact with each other. The detailed effects produced in the radio and optical by such interactions on a larger scale are described and it is concluded that parsec–scale interactions will produce similar effects observable with Quasat.

158.349 The compact radio emission regions in active galactic nuclei.
A. C. S. Readhead.
ESA Spec. Publ., ESA SP–213, p. 139 – 145 (1984). – See Abstr. 012.124.
The most powerful extragalactic objects can be divided into three main classes according to morphology and spectral index, viz. extended steep spectrum objects, in which the central galaxy is straddled by two large lobes of emission; compact steep spectrum objects; and compact flat spectrum objects. The structure of the compact radio emission regions in the nuclei of these objects can only be studied by very long baseline interferometry. The present state of VLBI observations of active nuclei is reviewed, and the expected impact of Quasat is discussed.

158.350 Strong source VLBI survey at 1.35 cm.
C. R. Lawrence, A. C. S. Readhead, R. P. Linfield,
R. A. Preston, R. T. Schilizzi, R. W. Porcas, R. Booth,
B. F. Burke.
ESA Spec. Publ., ESA SP–213, p. 147 – 148 (1984). – See Abstr. 012.124.
The authors present the first results of a VLBI survey at 1.35 cm. Fringes were detected on the Bonn–OVRO baseline for 25 out of 26 sources, and 14 had mean visibilities greater than 50%.

158.351 Testing theories of superluminal motion.
P. A. G. Scheuer.
ESA Spec. Publ., ESA SP–213, p. 149 – 151 (1984). – See Abstr. 012.124.

The basic theories of superluminal motion have changed little, but a few comparatively recent changes are reviewed. The impact of Quasat in advancing possible test of relativistic beaming theories is then discussed.

158.352 VLBI in space and X–ray observations of compact sources in active galactic nuclei.
P. Biermann.
ESA Spec. Publ., ESA SP–213, p. 157 – 159 (1984). – See Abstr. 012.124.

The possible impact of space VLBI (Quasat) and sensitive new X–ray satellites (EXOSAT, ROSAT) on our understanding of the relativistic emission from compact extragalactic sources is outlined.

Sky catalogue 2000.0. Volume 2: Double stars, variable stars and nonstellar objects.
See Abstr. 002.019.

Decametric survey of discrete sources in the northern sky. IX. Source catalogue in the declination range 52° to 60°.
See Abstr. 002.033.

Supplement to the detailed bibliography on the surface photometry of galaxies.
See Abstr. 002.044.

An atlas of selected galaxies, with illustrations of photometric analyses.
See Abstr. 002.047.

A catalogue of quasars and active nuclei (2nd edition).
See Abstr. 002.055.

Catalogue of radial velocities of galaxies with emission lines.
See Abstr. 002.095.

Accretion power in astrophysics.
See Abstr. 003.012.

Galassie e quasars.
See Abstr. 003.114.

The Universe, bit by bit.
See Abstr. 014.042.

The study of astrophysical interactions with the ESA Photon Counting Detector.
See Abstr. 035.052.

Improved continuum definition in high–background IUE images.
See Abstr. 036.002.

Optimal enhancement of features in digital spectra.
See Abstr. 036.036.

Arcsecond positions for milliarcsecond VLBI nuclei of extragalactic radio sources. III. 74 sources.
See Abstr. 041.012.

Nonlinear effects and the limitation of electron streaming instabilities in astrophysics.
See Abstr. 062.020.

MHD equilibrium and instabilities in extragalactic jets.
See Abstr. 062.025.

Bursting particle acceleration in radio jets.
See Abstr. 062.026.

Instabilities in astrophysical jets: disease and cure.
See Abstr. 062.027.

Current systems in radio jets.
See Abstr. 062.028.

Quasi–two–dimensional cosmic jets.
See Abstr. 062.032.

Relativistic plasmas and active galactic nuclei.
See Abstr. 062.039.

Magnetic vortex tubes, jets and nonthermal sources.
See Abstr. 062.059.

Force–free equilibria of magnetized jets.
See Abstr. 062.066.

Power–law X–ray and gamma–ray emission from relativistic thermal plasmas.
See Abstr. 062.069.

Turbulence and angular momentum in extragalactic radio jets. I. Linear stability analysis.
See Abstr. 062.081.

Electron–positron pairs in a mildly relativistic plasma in active galactic nuclei.
See Abstr. 062.096.

Hydrodynamics of astrophysical jets.
See Abstr. 062.097.

Viscous jets.
See Abstr. 062.098.

An analytic model for hydromagnetospheres.
See Abstr. 062.099.

The stability of confined radio jets: the role of reflection modes.
See Abstr. 062.113.

Cosmic jets.
See Abstr. 062.134.

Electron temperatures of astrophysical plasmas from the [O III] line ratio.
See Abstr. 062.144.

Jets with entrained clouds – I. Hydrodynamic simulations and magnetic field structure.
See Abstr. 062.166.

MHD beam solitons: incompressible flow.
See Abstr. 062.182.

Solutions for magnetic fields and surface electric fields produced by charged particle beams in which electrons (or electrons and positrons) have drift velocities with respect to protons.
See Abstr. 062.185.

Constraints on the properties of bent beams.
See Abstr. 062.195.

Linear analysis of jet stability.
See Abstr. 062.196.

Why dominant cluster jets are different.
See Abstr. 062.198.

Formation and propagation of magnetized radio jets.
See Abstr. 062.201.

Force–free equilibria of magnetized jets.
See Abstr. 062.202.

Collimation of intermediate scale motion entrained by nuclear jets.
See Abstr. 062.203.

The physics of particle acceleration in radio galaxies.
See Abstr. 062.204.

The relationship between laboratory and astrophysical jets.
See Abstr. 062.205.

Magnetic energy dissipation as the source of synchrotron emission in jets.
See Abstr. 062.207.

Laboratory electron beam simulation of cosmic radio jets.
See Abstr. 062.208.

Nonthermal, optically thin e^+e^- pair production and the universal X-ray spectrum of active galactic nuclei.
See Abstr. 063.065.

Theoretical models of polarization structure in compact radio jets.
See Abstr. 063.066.

Radiation from relativistic extragalactic jets interacting with interstellar matter.
See Abstr. 063.067.

Special relativistic effects in rapidly moving plasmoids.
See Abstr. 066.159.

Analytic structure of cosmic radio jets: a preliminary investigation.
See Abstr. 067.054.

Wind interactions above accretion discs: a model for broad-line regions and collimated outflow.
See Abstr. 067.061.

On the role of rotation and magnetic field in the structure and energetics of AGN.
See Abstr. 067.083.

Two-phase accretion model for emission-line regions in quasars and active galactic nuclei.
See Abstr. 067.097.

The theory of an object with a compact core and its applications.
See Abstr. 067.139.

Tidal effects: explosive disruption of stars by a giant black hole.
See Abstr. 067.169.

The evolution of the stellar and interstellar medium components of active galactic nuclei.
See Abstr. 067.189.

Shock acceleration in active galactic nuclei and quasars.
See Abstr. 067.190.

Two temperature accretion disks in active galactic nuclei.
See Abstr. 067.208.

Circumstellar dust distributions around variable stars.
See Abstr. 112.032.

Observations of the emission from some stars at millimeter wavelengths.
See Abstr. 116.040.

Observations of two novae in M87.
See Abstr. 124.008.

Discovery of an entire population of radio supernova candidates in the nucleus of Messier 82.
See Abstr. 125.110.

Refractive scintillation: a cause of compact source variability?
See Abstr. 126.084.

Low-frequency variability of pulsars.
See Abstr. 131.082.

On recombination lines of excited hydrogen: 1. High emission measure H II regions. 2. Radio galaxies and quasars.
See Abstr. 132.011.

Zum Verhältnis der Infrarot- zu Radiokontinuumstrahlung in H II-Gebieten und Galaxien.
See Abstr. 132.052.

Nuclear H II regions in galaxies with emission lines.
See Abstr. 132.068.

Results from the IRAS survey.
See Abstr. 133.001.

Spectroscopic verification of suspected planetary nebulae. I.
See Abstr. 134.021.

Determination of flux densities of radio sources using broad-band radiometers of the radio telescope RATAN-600.
See Abstr. 141.005.

Frequency-dependence of the statistics of radio sources.
See Abstr. 141.008.

Absolute measurements of the radio radiation of Cassiopeia A, Cygnus A and the Crab nebula in the 60 – 100 cm range.
See Abstr. 141.012.

Observations of the Ooty radio sources at 102 MHz.
See Abstr. 141.014.

Observations of Ooty radio sources at RATAN-600.
See Abstr. 141.015.

On the variability of some Parkes radio sources.
See Abstr. 141.021.

A catalog of radio sources to be occulted by comets P/Halley and P/Giacobini-Zinner.
See Abstr. 141.022.

5 GHz source counts from the MG survey.
See Abstr. 141.033.

Report of IAU Commission 40: Radio astronomy (*Radio astronomie*).
See Abstr. 141.039.

Recent optical identifications of HEAO-1 X-ray sources.
See Abstr. 142.034.

NGC 6212: an elliptical galaxy with a highly active nucleus.
See Abstr. 142.049.

Search for gamma ray lines from the Seyfert galaxy NGC 1275 and the COS B source 2CG 195+04.
See Abstr. 143.090.

New developments in high energy astrophysics.
See Abstr. 143.100.

Diffusion and convection of relativistic electrons in non-thermal sources: steady-state solutions.
See Abstr. 144.016.

Photon damping in cosmic ray acceleration in active galactic nuclei.
See Abstr. 144.135.

The energy spectrum of cosmic ray electrons in radio loop III.
See Abstr. 144.350.

Black holes and the shapes of galaxies.
See Abstr. 151.072.

Stern–Gas–Reibung in Galaxienkernen. Eine numerische Methode
zur Berechnung gekoppelter Gas– und Stellardynamik.
See Abstr. 151.079.

Equilibrium non–rotating magnetized disks as a plane analog to the
n = 3 polytrope.
See Abstr. 151.082.

Monte Carlo simulations of the 2 + 1 dimensional Fokker–Planck
equation: spherical star clusters containing massive, central black
holes.
See Abstr. 151.125.

Galactic absorption lines measured from the IUE low dispersion
spectra of active galaxies.
See Abstr. 155.001.

IRAS observations of interacting galaxies.
See Abstr. 157.028.

The inner disk of NGC 253.
See Abstr. 157.072.

8 – 13 μm spectrophotometry of galaxies. V. The nuclei of five
spiral galaxies.
See Abstr. 157.079.

A very bright water vapour maser source in the galaxy NGC 3079.
See Abstr. 157.093.

Blast wave formation of the extended stellar shells surrounding
elliptical galaxies.
See Abstr. 157.118.

Far–infrared spectroscopy of galaxies: the 158 micron C^+ line and
the energy balance of molecular clouds.
See Abstr. 157.122.

Radio continuum and far–infrared emission from spiral galaxies: a
close correlation.
See Abstr. 157.141.

Spatially–resolved high–velocity outflow from the nucleus of M51.
See Abstr. 157.178.

Radio structure of the elliptical galaxy MCG 5–4–18.
See Abstr. 157.179.

Companions of radio elliptical galaxies.
See Abstr. 157.180.

Report of IAU Commission 28: Galaxies (*Galaxies*).
See Abstr. 157.244.

A spectrophotometric investigation of a superassociation in the
galaxy NGC 2820 A.
See Abstr. 157.251.

On the separation of components of radiation of variable extragalactic sources.
See Abstr. 157.252.

Radio studies of peculiar galaxies at 10.7 GHz.
See Abstr. 157.279.

Interplanetary scintillation observations of 57 flat–spectrum
sources at 327 MHz.
See Abstr. 159.004.

Infrared study and classification of optically faint steep–spectrum
radio sources.
See Abstr. 159.005.

Diffuse shock acceleration and quasar photospheres.
See Abstr. 159.020.

Kinematics of the broad–line emission gas in quasars and Seyfert
nuclei: line profiles in a gravitational model.
See Abstr. 159.021.

Broad emission features in QSOs and active galactic nuclei.
II. New observations and theory of Fe II and H I emission.
See Abstr. 159.023.

Models for high–frequency radio to infrared outbursts in extragalactic sources.
See Abstr. 159.076.

The ultraviolet excess in quasars and AGNs: where have all the
stars gone?
See Abstr. 159.098.

Free–free emission from high density braod–line clouds.
See Abstr. 159.100.

Spectroscopy of possible QSO–galaxy interactions.
See Abstr. 159.114.

Quasar and related problems – some recent developments.
See Abstr. 159.118.

The multiple images of the quasar 0957 + 561.
See Abstr. 159.122.

The gravitationally lensed quasar 0957 + 561: VLA observations
and mass models.
See Abstr. 159.123.

The evolution of X–ray radiation of quasars.
See Abstr. 159.128.

Structure of the nuclei of the Seyfert galaxy NGC 1275 and the
quasar 3C 345.
See Abstr. 159.129.

The luminosity function of quasars and low luminosity active galactic nuclei.
See Abstr. 159.134.

The kinematics of the emission line gas in quasars and active galactic nuclei.
See Abstr. 159.141.

Observations of quasars and active nuclei of galaxies.
See Abstr. 159.143.

The nature of the IGM.
See Abstr. 160.003.

Statistics of emission–line galaxies in rich clusters.
See Abstr. 160.029.

X–ray emission possibly coincident with the radio tail of
PKS 0301–123.
See Abstr. 160.062.

Abell 400 and the 3C 75 radio jets.
See Abstr. 160.088.

The nature of radio sources in compact groups of galaxies.
See Abstr. 160.090.

VLA observations of 57 sources in clusters of galaxies.
See Abstr. 160.096.

The global properties of a representative sample of 51 narrow–angle–tail radio sources in the directions of Abell clusters.
See Abstr. 160.097.

Radio continuum emission in the Virgo cluster.
See Abstr. 160.106.

H I observations of Virgo cluster galaxies.
See Abstr. 160.107.

The H I distribution in spiral galaxies in the Virgo cluster.
See Abstr. 160.110.

Neutral hydrogen observations at the VLA.
See Abstr. 160.111.

Radio continuum and H I emission from the spiral galaxies in the Virgo cluster area.
See Abstr. 160.116.

IRAS observations of Virgo cluster galaxies.
See Abstr. 160.118.

An IRAS survey of bright spirals in Virgo.
See Abstr. 160.119.

A near–infrared survey of bright Virgo spiral galaxies.
See Abstr. 160.120.

The Virgo cluster redshift survey.
See Abstr. 160.124.

Global structure of Virgo cluster galaxies.
See Abstr. 160.125.

Applications of a nuclear spectroscopic survey of Virgo spiral galaxies.
See Abstr. 160.130.

X–ray survey of the Virgo cluster and comparison to field galaxies.
See Abstr. 160.131.

EXOSAT observations of the Virgo cluster.
See Abstr. 160.132.

Morphology and energetics of narrow angle tail radio sources.
See Abstr. 160.156.

The nuclear radio sources in the elliptical galaxies NGC 3309 and NGC 3311 in the cluster Abell 1060.
See Abstr. 160.162.

Cloud–beam interactions: bends, lobes and superluminals.
See Abstr. 160.163.

The alignment of distant radio sources.
See Abstr. 161.003.

A comparative study of cosmological evolution in the radio, optical and X–ray bands.
See Abstr. 161.074.

On log N $\sim$ log S.
See Abstr. 161.394.

Cosmology.
See Abstr. 161.400.

159 Quasi-stellar Objects

159.001   The UV spectra of intermediate redshift quasars.
A. L. Kinney, P. J. Huggins, J. N. Bregman, A. E. Glassgold.
NASA Conf. Publ., NASA CP–2349, p. 133 – 134 (1984). – See Abstr. 012.001.

The authors have studied the line and continuum properties of 21 intermediate redshift quasars in the ultraviolet and X–ray wavebands. The main results are: (1) The mean UV slope for this sample is –0.99 ± 0.11. (2) The mean slope from 1200 Å to 2 keV is –1.45 ± 0.03. (3) Two thirds of the objects have UV slopes that extrapolate to more than two sigma above the X–ray flux. (4) For 11 objects, the mean Lα/Hβ ratio is 8.3 ± 1.4. (5) The covering factor is estimated to be about 0.17.

159.002   The close QSO pair Q1548 + 114A, B.
P. A. Shaver, J. G. Robertson.
Mon. Not. R. Astron. Soc., Vol. 212, No. 1, p. 15P – 20P (1985).

IPCS spectroscopic and CCD imaging observations have been made of the QSOs Q1548 + 114A and B, which are separated by 4.8 arcsec and have redshifts of 0.44 and 1.90. There is no evidence either for line–locking in the absorption spectrum of B or for associated absorption in B at the redshift of A, both of which had previously been reported. Galactic H and K absorption is present in both spectra. This remains the closest pair of distinct QSOs, although the number of QSOs has increased tenfold since its discovery, and its existence is consistent with a random distribution of QSOs on the sky.

159.003   Molecular hydrogen in the $z = 2.811$ absorbing material toward the quasar PKS 0528–250.
S. A. Levshakov, D. A. Varshalovich.
Mon. Not. R. Astron. Soc., Vol. 212, No. 3, p. 517 – 521 (1985).

Among the previously unidentified absorption features in the spectrum of the quasar PKS 0528–250 obtained by Morton et al. 1980, the authors have found tentative evidence for H_2 lines at the same redshift ($z = 2.811$) as that of the well–known absorption–line system. They have estimated the column density of molecules from the curve of growth derived for this system by Morton et al., $N(H_2) = (2.9 \pm 0.3) \times 10^{16} \mathrm{cm}^{-2}$; the fraction of H_2 appears to be about 2×10^{-5}. The ortho– to para–H_2 ratio corresponds to the excitation temperature of $T_{ex}(H_2) = 300 \pm 150 K$, though the upper rotational levels $J \geqslant 3$ turn out to be overpopulated. The results obtained seem to provide substantial evidence for the existence of molecular hydrogen at the early cosmological epoch with $z \sim 3$.

159.004   Interplanetary scintillation observations of 57 flat–spectrum sources at 327 MHz.
U. Vijayanarasimha, S. Ananthakrishnan, G. Swarup.
Mon. Not. R. Astron. Soc., Vol. 212, No. 3, p. 601 – 608 (1985).

The authors describe interplanetary scintillation observations of an unbiased sample of 57 flat–spectrum radio sources at 327 MHz using the Ooty Radio Telescope (Swarup et al. 1971). The observations were carried out during 1981 – 82 and allowed detection of compact scintillating components of flux density $S_{obs} \gtrsim 50$ mJy and angular size between < 50 to 500 milliarcsec.

The results obtained are compared with the predictions of the Orr & Browne (1982) model.

159.005   Infrared study and classification of optically faint steep–spectrum radio sources.
D. Walsh, M. J. Lebofsky, G. H. Rieke, D. Shone, R. Elston.
Mon. Not. R. Astron. Soc., Vol. 212, No. 3, p. 631 – 643 (1985).

The acquisition of CCD frames and extensive near–infrared photometry for radio sources selected from the Jodrell Bank 966–MHz survey has shown that objects not visible on the POSS can be classified. As predicted, many of these sources are high–redshift galaxies, but a substantial number appear to be compact and variable, suggesting that they are related to QSOs. The infra–red colours of the objects classified as QSOs are extremely red, lending support to the QSO classification. The large number of QSOs found in this steep–spectrum radio sample brings into question the traditional assumption that such objects will be exclusively high–redshift galaxies.

159.006   New evidence for keV X–ray variability of QSO 0241 + 622.
S. Mereghetti, G. F. Bignami, P. A. Caraveo.
Astron. Astrophys., Vol. 142, No. 1, p. 37 – 40 (1985).

The X–ray quasar QSO $0241 + 622$ was observed twice by the Einstein Observatory IPC instrument in 1979. Comparing the two observations corrected for the instrumental response, a variability of a factor 2 on a time span of ~ 6 months is found for this $\sim 10^{-11}$ erg cm^{-2}s^{-1} source.

159.007   The quasars 1038 + 528 A and B.
J. M. Marcaide, I. I. Shapiro, B. E. Corey,
W. D. Cotton, M. V. Gorenstein, A. E. E. Rogers,
J. D. Romney, R. E. Schild, L. Bååth, N. Bartel, N. L. Cohen,
T. A. Clark, R. A. Preston, M. I. Ratner, A. R. Whitney.
Astron. Astrophys., Vol. 142, No. 1, p. 71 – 84 (1985).

The authors observed the quasars $1038 + 528$ A and B with VLBI at $\lambda 2.8$, $\lambda 3.6$, $\lambda 13$ and $\lambda 18$ cm at various times between November 1979 and March 1981. The compact radio structure of the A quasar ($z = 0.678$) is of the "core–jet" morphological type with a jet whose brightness temperature is only about 1% of that of the core at $\lambda 13$ cm and extends beyond 50 mas from the core with the same orientation as a radio feature of the core. The jet has a steep spectral index and is not visible at shorter wavelengths. The morphology of the B quasar ($z = 2.296$) is also of the "core–jet" type with a very short jet (~ 6 mas or ~ 70 pc in extent) and with a spectral index that becomes very steep at the tail. The B quasar has remained static during the span of time studied, whereas a new component may have emerged from the core of the A quasar.

159.008   Pavo XD–10, an X–ray QSO with extended optical structure.
R. Gilmozzi, J. V. Wall, P. G. Murdin, P. R. Jorden,
D. J. Thorne, I. G. van Breda, J. A. Peacock.
Nature, Vol. 313, No. 6003, p. 557 – 559 (1985).

Deep CCD images of the quasi–stellar object (QSO) identified with the Einstein X–ray source Pavo XD–10 show an optical structure that is dramatically extended. The authors discuss the morphology, the photometry from the CCD data, and the limited spectral information which indicates a redshift (z) of 0.72 for the system. If the extended structure represents a galaxy hosting the $z = 0.72$ QSO, the galaxy luminosity is 2 mag brighter than the host galaxies detected for QSOs at lower redshifts.

159.009   Quasar birthpangs observed?
C. M. Gaskell.
Nature, Vol. 313, No. 6004, p. 628 (1985).

159.010   Autocorrelation functions of the absorption spectra of quasars.
I. E. Val'tts.
Astron. Zh., Tom 62, Vyp. 1, p. 19 – 28 (1985). In Russian. English translation in Sov. Astron., Vol. 29, No. 1.

The autocorrelation function method is suggested to select absorption line systems in quasar spectra without using a refer-ence spectrum. The efficiency of the method is illustrated with 4 objects as an example ($1226 + 105$, 4C 24.61, $0824 + 110$ and Q 0453–423).

159.011   The redshift distribution law of quasars revisited.
S. Depaquit, J.–C. Pecker, J.–P. Vigier.
Astron. Nachr., Vol. 306, No. 1, p. 7 – 15 (1985).

Can observational selection effects, tied with the existence of strong lines, explain the observed $\log(1 + z)$ periodicity of the quasar's histogram, as sometimes claimed? A first approach shows that one must distinguish radio from optical quasars. In the former case spectroscopic selection due to strong lines is investigated from the study of homogeneous samples. Then, the role played by pairs of strong lines in the redshift's determination is evaluated and it shows that the spectroscopic selection cannot explain the observed periodicity of the radio quasar's histogram. The study of the complete sample confirms this conclusion.

159.012   On the optical variability of the QSO PKS 2128–123.
M. Moles, J. García–Pelayo, J. Masegosa,
A. Aparicio.
Astron. J., Vol. 90, No. 1, p. 39 – 42 (1985).

$UBVRI$ observations of the QSO PKS 2128–123 show it as a moderate–amplitude variable object in all bands except the I one, where it appears as highly variable with ΔI up to $1\overset{m}{.}69$. From the detected changes in brightness, the estimated variability parame-ter is found to be about 1 day.

159.013   Relativistic beaming and the optical magnitudes of QSOs.
I. W. A. Browne, A. E. Wright.
Mon. Not. R. Astron. Soc., Vol. 213, No. 1, p. 97 – 102 (1985).

The authors have compared the optical magnitudes of weak, lobe–dominated QSOs and strong, core–dominated QSOs. They find differences in the average magnitudes of the two classes of objects which imply at least one magnitude of enhancement of the optical continuum radiation in core–dominated sources if the Orr & Browne unified scheme of QSOs is correct. Some further observational tests of the scheme are suggested.

159.014   Common lines in the rest–frame absorption–line spectra of QSOs?
Y. P. Varshni, D. Singh.
Astrophys. Space Sci., Vol. 109, No. 1, p. 149 – 153 (1985).

Libby et al. (1984) have studied the absorption–line data for 13 QSOs in the rest–frames of the QSOs. It is shown that the number of groups in which 5 lines or more lie within a wavelength interval of 1.0 Å found by these authors is insignificantly different from that that would be expected from chance coincidences. Conse-quently, there is no evidence that the rest–frame wavelengths at which these groups occur have any special significance.

159.015   Millimetre observations of optically selected quasars.
E. I. Robson, W. K. Gear, M. G. Smith,
P. A. R. Ade, I. G. Nolt.
Mon. Not. R. Astron. Soc., Vol. 213, No. 2, p. 355 – 364 (1985).

The authors report observations of a sample of optically se-lected quasars at wavelengths close to one millimetre. The qua-sars were selected on the basis of (1) optical and near infrared colours, (2) optical luminosity. No objects were detected from either sample despite previous claimed detections of apparently similar objects by other authors. The constrained continuum spectral shape of the sources is inconsistent with a homogeneous synchrotron model, but consistent with an inhomogeneous model and the two–component model of Ennis et al.

159.016   Maximum cosmic redshift of quasars at $Z > 3$.
W. H. Sorrell.
Mon. Not. R. Astron. Soc., Vol. 213, No. 2, p. 389 – 398 (1985).

Timing arguments and current theories of galaxy evolution are used to calculate the maximum cosmological redshift Z_{max} below which quasars first formed in the nuclei of galaxies. On the as-sumption that the mass density of the universe is dominated by massive neutrinos ~ 10 eV of density $\Omega_\nu \sim 1$ (Einstein–de Sitter cosmology), it is found that galaxies first formed at redshift

$Z_f \sim 10$, whereas luminous ($\sim 10^{48}$erg s^{-1}) quasars formed first in cD galaxies at $Z \lesssim Z_{max} \sim 5$. For a dust–free universe, the space density of high–redshift quasars would show a real decrease at redshifts $Z > 4.7$.

159.017 Spectroscopy of quasar candidates from searches of UK Schmidt Telescope objective prism plates.
A. Savage, R. G. Clowes, R. D. Cannon, K. Cheung, M. G. Smith, A. Boksenberg, J. V. Wall.
Mon. Not. R. Astron. Soc., Vol. 213, No. 2, p. 485 – 490 (1985). With microfiche MN 213/1.

In the course of a number of observing campaigns over several years, slit spectra have been obtained for an inhomogeneous set of 126 quasar candidates from searches of five UKST low-dispersion prism plates. One–hundred and eight of the candidates are confirmed as quasars, and 45 previously unpublished red-shifts are presented here. Nine objects were found to be galactic stars and a further nine objects have continuous spectra or incon-clusive data. Ninety–nine of the objects with slit spectroscopy, which were described from the prism searches as 'probable' qua-sars, are confirmed as quasars, while 72 per cent of the more dubious candidates have been subsequently confirmed as qua-sars.

159.018 A study of quasar pairs.
P. A. Shaver.
Astron. Astrophys., Vol. 143, No. 2, p. 451 – 454 (1985).

A study of QSO pairs from large samples shows that the sky distribution of QSOs of different redshifts is random over scales from arcminutes to tens of degrees. Limits have been placed on brightness and colour changes in the background members of close QSO pairs which could be caused by matter associated with the foreground QSOs.

159.019 Evidence for beamed radiation from QSOs.
J. B. Hutchings, A. C. Gower.
Astron. J., Vol. 90, No. 3, 405 – 409 (1985).

The authors have performed deep radio–continuum detection observations with the VLA on a representative sample of opti-cally and X–ray selected QSOs of low redshift. Eight of 27 objects are detected as radio–continuum sources with core luminosities down to 10^{21}W/Hz. A further 15 have higher detection limits. The detected radio–source structures are all small, probably one-sided, and complex. The optical properties of the objects indicate that the ratio of nuclear to host galaxy luminosity is about 5 times higher in the radio–emitting objects, and that their nuclear lumi-nosities are about 10 times higher. The authors use these and related statistics to argue that the optical and radio–continuum emission from these quasars is beamed towards us.

159.020 Diffuse shock acceleration and quasar photospheres.
J. N. Bregman.
Astrophys. J., Vol. 288, No. 1, p. 32 – 42 (1985).

Diffusive shock acceleration, successful in explaining power-law particle distributions for cosmic rays and the solar wind, is adapted to the quasar environment. The author uses the Laplace transform technique to find exact solutions to the kinetic trans-port equations when the effects of synchrotron losses are in-cluded. Below a critical energy, the distribution function of parti-cles has the same power–law energy dependence as the solution without synchrotron losses. Above this energy, the distribution function is cut off exponentially. The emergent synchrotron spec-trum is calculated for cases in which the shock region is resolved and unresolved by an observer. These calculations are compared with existing observations of quasars and BL Lac objects and are found to be consistent with them.

159.021 Kinematics of the broad–line emission gas in quasars and Seyfert nuclei: line profiles in a gravitational model.
T. J. Carroll, J. Kwan.
Astrophys. J., Vol. 288, No. 1, p. 73 – 81 (1985).

The parabolic–orbital model reported by Kwan and Carroll in 1982 for the broad emission–line regions of quasars and Seyfert galaxies is extended and improved to include the effects of a finite infalling cloud number and size. Disruption of the cloud by the differential angular momentum across its face results in a large velocity dispersion among the resulting fragment cloudlets. The drag of a confining intercloud wind is also considered. It is dem-onstrated that the gravitational model can reproduce the ob-served broadness of He I λ5876 relative to Hβ. Profiles for Lyα, Mg II λ2798, C IV λ1549, and C III] λ1909 are also generated and compared to existing observations.

159.022 Emission–line QSOs in the region of the Hercules cluster of galaxies.
E. M. Burbidge, H. E. Smith, V. T. Junkkarinen, A. A. Hoag.
Astrophys. J., Vol. 288, No. 1, p. 82 – 93 (1985). With plates 1 – 2.

Emission–line QSO candidates found by slitless spectroscopy on and off the field of the Hercules cluster of galaxies have been studied by slit spectrophotometry carried out with the Lick and KPNO image dissector scanners. In two 1 deg^2 fields centered on the cluster Abell 2151 and on a nearby off–cluster region, 16 and 9 candidate QSOs, respectively, were found, to a blue continuum magnitude limit of about 20.5. Of these, 14 were confirmed as QSOs in the cluster field, with $0.89 < z < 3.24$, and 6 in the off–cluster field, with $1.52 < z < 3.00$. One candidate QSO was found to be an unresolved low–redshift intergalactic H II region or emission–line galaxy. Accurate positions, finding charts, red-shifts, and descriptions of the spectra are presented.

159.023 Broad emission features in QSOs and active galactic nuclei. II. New observations and theory of Fe II and H I emission.
B. J. Wills, H. Netzer, D. Wills.
Astrophys. J., Vol. 288, No. 1, p. 94 – 116 (1985).

The authors have observed seven low and intermediate redshift QSOs and calculated new models to fit all spectral features be-tween rest wavelengths of about 1800 and 5500 Å. A grid of photoionization models has been used to calculate the strengths of the Balmer continuum and of more than 45 Balmer lines. The authors have also extended previous calculations of the Fe$^+$ atom and have included about 3000 Fe II lines in the models. The models are compared with observed spectra to derive strengths of Fe II emission and other features, showing that Fe II emission is much stronger than previously imagined. Consequences for the physical conditions in the broad–emission–line clouds of QSOs and active galactic nuclei are discussed.

159.024 The nature of the luminous feature between Markarian 205 and NGC 4319.
G. Cecil, A. Stockton.
Astrophys. J., Vol. 288, No. 1, p. 201 – 204 (1985). With plates 5 – 6.

Deep CCD images confirm that a real feature is present at the position of the luminous connection originally reported by Arp. Considerable structure exists in the envelope surrounding Mrk 205, including apparent tidal tails from both the QSO and its compact companion 3".3 to the NNE. The luminous ''connection'' has a complex morphology, reflecting its com-posite character: the part near Mrk 205 seems to be largely a portion of the tail from the companion, while the remaining part is a continuation of a tangential feature on the east side of Mrk 205. The ''connection'' appears to consist of material associ-ated solely with Mrk 205 and its companion.

159.025 PG 0946 + 301: a low–redshift, broad absorption line QSO.
B. J. Wilkes.
Astrophys. J., Lett. Ed., Vol. 288, No. 1, p. L1 – L3 (1985).

The discovery of a new broad absorption line (BAL) QSO with the lowest emisson–line redshift to date ($z = 1.216$) is reported. The data provide evidence for complex broad absorption up to velocities $\sim 14,500$ km s^{-1} relative to the emission–line center.

159.026 Quasar evolution.
D. Weedman.
News Lett. Astron. Soc. N.Y., Vol. 2, No. 4, p. 5 – 6 (1983). Ab-stract. – See Abstr. 010.242.

159.027   Are Ly–α clouds detectable in emission at 21 cm?
M. Vietri.
Astron. Astrophys., Vol. 144, No. 2, p. L17 – L19 (1985).

The author considers two hypotheses concerning the nature of the objects which produce the Ly–α absorption features in the optical spectra of QSOs (Sargent et al. 1980): first, he argues that Oort's (1981) suggestion that such features arise in superclusters is not ruled out by the lack of correlations of the Ly–α lines in the spectra of two QSOs close on the sky (Sargent et al., 1982), due to the filamentary nature of superclusters. He suggests that such superclusters may be detectable as 21 cm emission lines at very high redshifts. Second, he proposes that, if the lines are due to a new species of objects (Sargent et al., 1980, Ostriker and Ikeuchi, 1983), maser action may take place, which might make these objects detectable as high–redshift 21 cm line emitters.

159.028   Uniformity of quasars in the chronometric cosmology.
I. E. Segal, J. F. Nicoll.
Astron. Astrophys., Vol. 144, No. 2, p. L23 – L26 (1985).

Bright quasars form a standard candle within the framework of the chronometric cosmology, without evolution. The standard deviation of the portion of the luminosity function observed in the Palomar Bright Quasar Survey at redshifts above 0.1 is estimated to be less than 1/2 mag., with due allowance for the observational magnitude cutoff bias by an optimal nonparametric statistical method (known as ROBUST). The inclusion of deeper complete samples as reported by Schmidt and Green does not affect this value. Apparent luminosity evolution within the framework of a Friedmann cosmology is shown to be as predicted by the chronometric cosmology.

159.029   Recent photographic monitoring of 3C 273 in blue light.
G. J. Corso, J. Schultz, B. Purcell, G. Garino, A. Dey.
Publ. Astron. Soc. Pac., Vol. 97, No. 588, p. 118 – 119 (1985).

Photographic photometry of 3C 273 on blue plates taken with the 1–meter reflector of Northwestern's Lindheimer Astronomical Research Center during the 1983 and 1984 observing seasons are reported.

159.030   High redshift Mg II and Fe II absorption in QSO spectra from known C IV systems.
P. Boissé, J. Bergeron.
Astron. Astrophys., Vol. 145, No. 1, p. 59 – 69 (1985).

The authors present observations in the red of Mg II and Fe II absorptions in 18 QSOs known to contain C IV absorption systems. They have found an anti–correlation between the strength of the C IV or Mg II absorption lines and the degree of excitation of the absorbers. This partly accounts for the rareness of Mg II systems as compared to C IV. Physical properties of the absorbers are derived from a curve of growth analysis. The authors have not found any evidence for a strong cosmological evolution in the number density of Mg II absorption lines when comparing their results to data obtained at lower redshift. It is shown that a two–phase model for the absorbers is compatible with a different cosmological evolution for the low and high excitation systems.

159.031   VLBI monitoring of the superluminal quasar 3C 273, 1977 – 1982.
S. C. Unwin, M. H. Cohen, J. A. Biretta, T. J. Pearson, G. A. Seielstad, R. C. Walker, R. S. Simon, R. P. Linfield.
Astrophys. J., Vol. 289, No. 1, p. 109 – 119 (1985).

The compact radio structure of the quasar 3C 273 has been monitored with a VLBI array at 5.0 and 10.7 GHz, at 6 month intervals during 1977 – 1982. On each hybrid map there is a bright "core" at the east end of the source, and several compact "knots" extending to the west. Superluminal motion of at least two knots is visible at both frequencies, with apparent speeds $v \approx 6c$ (for $H_0 = 100$ km s^{-1}Mpc^{-1}). These knots decay with half–lives of $1 - 2$ years as they move out from the core; their positions are frequency–independent. The authors deduce lower limits to the Doppler factors δ of the knots, by comparing the predicted inverse Compton X–ray emission from a simple homogeneous sphere model with the observed flux; at least one knot must be moving relativistically with $\delta_{min} \gtrsim 2.5$.

159.032   Serendipitous discovery of a high redshift quasar.
M. Azzopardi.
Messenger, No. 39, p. 12 – 13 (1985).

159.033   Observations of high redshift Mg II and Fe II absorption lines in QSO spectra.
P. Boissé, J. Bergeron.
Messenger, No. 39, p. 22 – 25 (1985).

159.034   On the extendedness of faint ultraviolet excess quasar candidates.
G. Edwards, E. F. Borra, E. Hardy.
Astrophys. J., Vol. 289, No. 2, p. 446 – 450 (1985).

The authors have conducted a search for UV excess quasars based on deep U, J, and F plates taken with the CFH telescope. The plates were fully digitized and analyzed with automated software. After applying a technique for separating stars from galaxies, the authors came upon the surprising result that a significant fraction of the quasar candidates are not starlike. They discuss briefly the possible nature of these objects, concluding that a simple explanation is that they are compact blue galaxies.

159.035   The far–ultraviolet continuum of quasars and the universe at $z > 4$.
H. Netzer.
Astrophys. J., Vol. 289, No. 2, p. 451 – 456 (1985). = Lick Obs. Bull., No. 995.

New observations and line measurements in quasars present a servere energy budget problem and a challenge to present models of such objects. The observed ultraviolet continuum is too faint and too steep to produce the observed emission–line flux, if no or very little reddening is present. The assumption of no reddening in the lines leads to a complete change of view of these objects and suggests extremely high densities or very large column densities. Emission–line reddening helps to ease this problem but suggests from equivalent width considerations that the intrinsic 1200 Å continuum is brighter than observed. Continuum reddening and/ or special geometry could explain this. These new ideas have important implications for the physical conditions in quasars and the visibility of all objects with $z \gtrsim 5$.

159.036   On the statistics of quasar alignments.
M. G. Edmunds, G. H. George.
Mon. Not. R. Astron. Soc., Vol. 213, No. 4, p. 905 – 915 (1985).

The authors investigate the statistics of chance three–point alignments in random clustered and unclustered fields of points, and apply such analysis to two well–searched published quasar fields from the UK Schmidt telescope. They discuss the statistical detection limits for a sub–population of physically aligned triplets, and find no evidence (within the statistical limits) of such a population being present. In appendices the authors give formulae for boundary corrections in triplet searches, and an approximate expectation value for higher order alignments in random fields.

159.037   Cerenkov line emission. A new interpretation of the emission lines of quasars.
T. Kiang.
Ir. Astron. J., Vol. 17, No. 1, p. 11 – 15 (1985). – See Abstr. 012.034.

159.038   Search for "antipode" images of quasars.
V. S. Lebedev, I. A. Lebedeva.
Astrofiz. Issled. Izv. Spets. Astrofiz. Obs., Tom 19, p. 14 – 15 (1985). In Russian. English translation in Bull. Spec. Astrophys. Obs. – North Caucasus.

An analysis is done of quasar distributions on the celestial sphere for the purpose of detecting the surplus number of objects in diametrically reverse directions which can be interpreted as the existence of "antipode" images of the objects in the topologically complex Universe. A negative result is obtained.

159.039   Search for periodicity in the space distribution of quasars.
V. S. Lebedev, I. A. Lebedeva.
Astrofiz. Issled. Izv. Spets. Astrofiz. Obs., Tom 19, p. 16 – 21 (1985). In Russian. English translation in Bull. Spec. Astrophys. Obs. – North Caucasus.

No significant periodicity is detected in quasar distributions due to various arguments (z, $\ln(1+z)$, $1-(1+z)^{-1/2}$ etc.). The spectral analysis method and the quasar catalogue of Hewitt and Burbidge (1980) were used. The presence of some details in the power spectra is connected with a narrow peak ($\Delta z \approx 0.1$) near $z \approx 2$ in the quasar distribution.

159.040   A $0\rlap{.}''25$ jet in the quasar 3C 446.
R. S. Simon, K. J. Johnston, J. H. Spencer.
Astrophys. J., Vol. 290, No. 1, p. 66 – 74 (1985).

The authors have mapped the quasar 3C 446 with VLBI techniques at 18 cm with a resolution of 10×25 milli–arcsec (mas). The radio structure of this quasar exhibits a core–jet morphology, with the length of the jet exceeding $0\rlap{.}''25$. VLBI observations at 2.8 cm reveal that the 1 mas core of 3C 446 has a rising radio spectrum from 18 to 2.8 cm. A simple synchrotron–emission model of the radio emission from 3C 446, taking into account both the observed X–ray flux and the rapid variability of the radio flux density, implies that bulk relativistic motion may be occurring in the core.

159.041   Revised V/V_{max} test with incompleteness–compensated factors and its application to X–ray quasar sample.
H.-q. Zhang, Z.-w. Liu.
Astrophys. Space Sci., Vol. 109, No. 2, p. 333 – 344 (1985).

Both density– and luminosity–evolution of quasars proceed simultaneously. If only one of them is considered, the value of $\langle V/V_{max} \rangle$ would be too high. Here the authors present an idea that the quasar luminosity–redshift curve can be obtained by statistics, from which the incompleteness–compensated factor $K(Z|l_{min})$ for each source is given, and then the V/V_{max} test for density law can be carried out. The authors use this revised V/V_{max} test in investigating 3CR and 4C samples including the 3CR sources detected in X–ray. The result is that the space density distribution of quasars is relative to q_0.

159.042   New quasars with $z = 3.4$ and 3.7 and the surface density of very high redshift quasars.
C. Hazard, R. McMahon.
Nature, Vol. 314, No. 6008, p. 238 – 240 (1985).

The authors describe here the discovery on a UKST low–dispersion IIIaF objective prism plate of a QSO (quasistellar object) with $z = 3.7$, which is the second highest redshift QSO known and, moreover, lies in the same region in which a QSO with $z = 3.61$ has already been reported. The authors also note a QSO with $z = 3.4$. When compared with the results of a search for QSOs with $2.7 < z < 3.3$ on the same plate, these observations do not support suggestions of a $z \sim 3.5$ cut–off in the QSO distribution, but rather support the suggestion of a steady decline from $z \sim 2$.

159.043   Direct estimate of Hubble's constant from the double quasar.
C. M. Gaskell.
Nature, Vol. 314, No. 6010, p. 402 (1985).

159.044   Quasar 4C 39.25 is not contracting.
J. M. Marcaide, N. Bartel, M. V. Gorenstein, I. I. Shapiro, B. E. Corey, A. E. E. Rogers, J. C. Webber, T. A. Clark, J. D. Romney, R. A. Preston.
Nature, Vol. 314, No. 6010, p. 424 (1985).

Recently, Shaffer suggested that the quasar 4C 39.25 may have been contracting superluminally in the period 1979 – 1982. Here, based on a map of this source made from VLBI observations, in 1983 at $\lambda 3.6$ cm the authors conclude that this conjecture is not correct. They find three distinct components in the structure, two of which are separated by 2.0 mas, whereas the third, presumably new and not previously reported, is situated between the other two.

159.045   The radio jet of the quasar 3C 273.
C. Flatters, R. G. Conway.
Nature, Vol. 314, No. 6010, p. 425 – 426 (1985).

The authors present new MERLIN observations of the brightness and polarization of the radio jet of 3C 273 at a resolution of 0.35 arc s. One of the most marked features of the new map, the high polarization found within the head of the source, is hard to explain. If the motion is indeed fast, then relativistic aberration should be taken into account; it is suggested that this leads to a natural explanation of the high observed polarization.

159.046   Combined MERLIN/VLA observations of the super-luminal quasar 3C 179.
D. L. Shone, R. W. Porcas, J. A. Zensus.
Nature, Vol. 314, No. 6012, p. 603 – 604 (1985).

MERLIN at Jodrell Bank and the VLA in New Mexico are complementary instruments, representing the latest generation of aperture–synthesis radio telescopes. MERLIN provides a factor of four greater resolution than VLA, whereas the shorter spacings of the latter provide essential information on large–scale structure. The authors present here the first map to be produced by combining data from MERLIN and the VLA. The new map, of the quasar 3C 179 (redshift, $z = 0.846$), is a considerable improvement on maps made from the individual arrays and reveals greater detail, particularly in the quasars's jet, which seems to be splitting or disrupting.

159.047   Photoionization models for QSO absorption–line regions.
R. B. Gruenwald, S. M. V. Aldrovandi.
Astron. Astrophys., Vol. 145, No. 2, p. 324 – 330 (1985).

Photoionization models are applied to the regions which produce the absorption spectra of QSOs. The absorbing material is assumed to be located according to the statistical studies of Weymann et al. (1979). Observational data for intervening systems can be reproduced by clouds ionized by the integrated radiation field of all QSOs; in this case, the obtained density and mass of the clouds are small ($< 10\ M_\odot$). The observational data can also be reproduced for the regions ejected and ionized by the object as well as for absorbing systems and associated QSOs belonging to the same cluster.

159.048   A search for H I and OH absorption in high redshift quasars.
G. J. de Waard, R. G. Strom, G. K. Miley.
Astron. Astrophys., Vol. 145, No. 2, p. 479 – 481 (1985).

A search has been made for redshifted H I and OH absorption in high redshift quasars using the Westerbork Synthesis Radio Telescope at 0.6 GHz. The investigation was carried out near $z = 1.33$ (H I) and $z = 1.74$ (OH), close to the emission line redshifts of most of the objects observed. Seven quasars, 0202–172, 1331+170, 1356+022, 1615+028, 1756+237, 2005+403, and 2158+101 were investigated, but in no case could absorption be detected. Upper limits to absorption features of 1 to 2 percent were obtained, and the results are briefly discussed.

159.049   Evidence for underlying galaxies in a complete sample of variable quasars.
M. R. S. Hawkins, L. Woltjer.
Mon. Not. R. Astron. Soc., Vol. 214, No. 2, p. 241 – 249 (1985).

A complete sample of quasars has been selected on the basis of optical variability. The objects are observed to become redder at faint magnitudes, both in the colour–magnitude diagram of the sample as a whole, and in the colour of individual objects as they vary. It is concluded that the faint end of the sample is dominated by low–luminosity objects, and the reddening is interpreted as being caused by the presence of an underlying red galaxy which progressively contaminates the integrated colour as the blue nuclear component becomes weaker. Some constraints on the nature of the underlying galaxy are discussed.

159.050 The sensitivity of the V/V_m test to uncertainties in observed parameters.
G. L. White.
Proc. Astron. Soc. Aust., Vol. 5, No. 4, p. 529 – 531 (1984).

Schmidt and Rowan–Robinson (1968) developed a simple and concise test for the uniformity of the distribution of quasars (QSOs) based on QSO luminosity and occupied volume. The luminosity–volume test (V/V_m test) is undertaken by computing the ratio of co–moving volume, V, enclosed by the distance of the QSO (assuming redshift to be cosmological) to the volume, V_m, enclosed by the QSO if it were to be on the very limit of visibility. This paper draws attention to the sensitivity of the V/V_m statistic to various parameters and illustrates these using a complete sample of QSOs from the first Molonglo deep survey (Robertson 1977).

159.051 2237+0305: a new and unusual gravitational lens.
J. Huchra, M. Gorenstein, S. Kent, I. Shapiro, G. Smith, E. Horine, R. Perley.
Astron. J., Vol. 90, No. 5, p. 691 – 696 (1985). With plate 58.

The authors have discovered a new gravitational lens system which consists of a quasar at $z = 1.695$ nearly centered on a 15 mag spiral galaxy, $2237+0305$, at $z = 0.0394$. At $2''$ resolution, only a single optical image of the quasar is visible; its centroid is located within approximately $0\overset{''}{.}3$ of the center of the galaxy. "Snapshot" observations at the VLA yielded no detectable radio radiation, placing an upper limit of about 0.5 mJy on the combined flux density of the galaxy and quasar at $\lambda = 6$ cm. The authors discuss a simple gravitational lens model that accounts for these observations.

159.052 Theoretical quasar emission–line ratios. VII. Energy–balance models for finite hydrogen slabs.
E. N. Hubbard, R. C. Puetter.
Astrophys. J., Vol. 290, No. 2, p. 394 – 410 (1985).

The authors present energy–balance calculations for finite, isobaric, hydrogen–slab quasar emission–line clouds which incorporate probabilistic radiative transfer (RT) in all lines and bound–free continua of a five–level plus continuum model hydrogen atom. They discuss the resulting line ratios, regions of line formation, level populations, and range of applicability of the models. The calculations neglect the effects of heavy elements and secondary ionizations from primary photoelectrons. An appendix presents a detailed comparison of differences between four different methods of treating the radiative transfer: (1) the probabilistic method, (2) an exact RT solution, (3) an effective two–level $\varrho = p_e$ energy–balance calulation, and (4) a $\varrho = p_e$ calculation explicitly incorporating indirect photon creation and destruction paths.

159.053 Multifrequency radio VLBI observations of the superluminal, low–frequency variable quasar NRAO 140.
A. P. Marscher, J. J. Broderick.
Astrophys. J., Vol. 290, No. 2, p. 735 – 741 (1985).

The authors present VLBI maps of the quasar NRAO 140 at three wavelengths: 18, 6, and 2.8 cm. The source consists of a jetlike structure delineated by a nearly colinear series of components which are progressively more compact toward the northwestern end of the source. The multifrequency observations reaffirm the previously reported Compton problem and superluminal motion. One of the components is separating from the "core" at a rate of 0.15 milliarcsec yr^{-1}, which translates to an apparent velocity between $4c$ and $13c$, depending on the values of H_0 and q_0.

159.054 Faint emission–line quasars and the number density for $2 \leqslant z < 2.5$.
D. W. Weedman.
Astrophys. J., Suppl. Ser., Vol. 57, No. 3, p. 523 – 533 (1985).

A list is presented of 189 faint emission–line quasars with magnitudes between 19 and 22. The quasars are in 20 fields spread around the sky, observed with the CFH Telescope. Limits to the number as a function of magnitude for $2.0 \leqslant z < 2.5$ are determined using objects with redshifts determined from two lines and using single–line objects in which the line could be Lyα.

The results show that quasar numbers in this redshift interval increase more slowly with magnitude after 20 mag than they do for quasars brighter than 20 mag. A list is also given of close quasar–galaxy and quasar–quasar pairs.

159.055 The connection between the activity manifestations of the quasi–stellar object 3C 345 in optical and radio ranges.
M. K. Babadzhanyants, E. T. Belokon', V. L. Gorokhov.
Astrofizika, Tom 22, Vyp. 2, p. 247 – 255 (1985). In Russian. English translation in Astrophysics, Vol. 22, No. 2.

Cross–correlation analysis is made of data on optical and radio variability of 3C 345 on large time scales (hundred days). A high degree of correlation is found between radio outbursts and the "slow" (characteristic time ~ 1 yr) component of optical brightness changes; radio outbursts were delayed for 700 – 1000 days. Comparison of the time delay with apparent "superluminal" velocity of the radio components of 3C 345 allows to estimate the distance from the nucleus at which the radio emission arises (~ 0.7 milliarcsec).

159.056 The ultraviolet spectra of intermediate–redshift quasars.
A. L. Kinney, P. J. Huggins, J. N. Bregman, A. E. Glassgold.
Astrophys. J., Vol. 291, No. 1, p. 128 – 135 (1985).

IUE observations of the Lyα line flux and the UV continuum are presented for a sample of 21 quasars of intermediate redshift. Power–law fits to the continuum data yield a mean spectral index of –1.0. The strength of Lyα is found to be well correlated with the luminosity at 1450 Å. Lyα/Hβ line ratios range from 2.7 to 18, with a mean of 8.3. The ratio of the number of Lyα to ionizing photons yields an estimate of 0.17 for the mean covering factor of the sample.

159.057 X–ray emission from red quasars.
J. N. Bregman, A. E. Glassgold, P. J. Huggins, A. L. Kinney.
Astrophys. J., Vol. 291, No. 2, p. 505 – 510 (1985).

A dozen red quasars were observed with the Einstein Observatory in order to determine their X–ray properties. The observations show that for all these sources, the infrared–optical continuum is so steep that when extrapolated to higher frequencies, it passes orders of magnitude below the measured X–ray flux. The X–ray emission is better correlated with the radio than with the infrared flux. The physical conditions in the emitting regions are studied and the consistency of the observed lines and continua with photoionization models is discussed.

159.058 Detection of a supernova in the host galaxy of the QSO 1059+730.
B. Campbell, C. Christian, C. Pritchet, P. Hickson.
Astrophys. J., Lett. Ed., Vol. 291, No. 2, p. L37 – L39 (1985). With plate L1.

A blue stellar object has been discovered within the fuzz surrounding the quasar $1059+730$. This object was observed in multiple CCD frames taken on one night, but it is not visible in other images of the QSO. The authors propose that the object is probably a supernova event in the host galaxy of the quasar. The detection of this supernova provides a consistency argument that QSOs are at their cosmological distances and demonstrates that the envelopes surrounding QSOs contain stars and possibly star formation.

159.059 Evidence for axisymmetric emission regions in quasars with broad absorption lines.
V. P. Grinin.
Sov. Astron. Lett., Vol. 10, No. 5, p. 269 – 273 (1984). English translation of 38.159.040.

159.060 The infrared emission of quasars.
J. C. Carvalho.
Astrophys. Space Sci., Vol. 111, No. 1, p. 97 – 102 (1985).

A model of two compact components (one contributes at radio and submillimeter wavelength and the other is very small and emits at the infrared region) is proposed to interpret the spectrum of quasi–stellar objects at radio and infrared wavelengths. The physical parameters are estimated from the observed data of 3C 273 for the author's model and for the model which is derived by assuming the infrared emission as an extrapolation of the synchrotron radio source. It is proved that the author's model can explain the observational spectra, while the other model has some difficulties to interpret them completely.

159.061   New quasi–stellar objects. V.
        B. E. Markarian, D. A. Stepanyan, V. A. Lipovetskij.
Astron. Tsirk., No. 1346, p. 7 – 8 (1984). In Russian.

159.062   Der leuchtkräftigste Quasar: S5 0014 + 81.
        H. Kühr.
Sterne Weltraum, 24. Jahrg., Nr. 5, p. 260 – 264 (1985).

159.063   Redshift evolution of the Lyman–line–absorbing clouds in quasar spectra.
B. Atwood, J. A. Baldwin, R. F. Carswell.
Astrophys. J., Vol. 292, No. 1, p. 58 – 71 (1985).
        The results of an analysis of the absorption lines measured on an echelle spectrum of the $z = 3.12$ quasar Q0420–388 are presented. The spectral resolution is sufficiently high (33 km s^{-1}) that most of the Lyman absorption lines are resolved, and so velocity dispersions and column densities have been obtained by profile fitting. By comparing the properties of the Lyman absorption systems at redshifts $z \approx 2$ to those at $z \approx 3$ the authors find that, apart from the numbers of detectable clouds, there is no evidence for redshift evolution in the distribution function for any measured quantity. Simple models in which a hot intergalactic medium pressure confined a comoving population of clouds are consistent with the available data.

159.064   Observations of 3C 273 with high north–south resolution.
J. A. Biretta, M. H. Cohen, H. E. Hardebeck, P. Kaufmann, Z. Abraham, A. A. Perfetto, E. Scalise Jr., R. E. Schaal, P. M. Silva.
Astrophys. J., Lett. Ed., Vol. 292, No. 1, p. L5 – L8 (1985).
        The authors present the first VLBI maps of 3C 273 with high north–south resolution. A strong, nonmonotonic curvature is found in the jet at projected radii $\leqslant$ 5 pc. Measurements of the core size show that bulk relativistic motion in the core is not required for consistency with the observed X–ray flux.

159.065   Observations of the low redshift BAL QSO PG 1700 + 518: limits on the fraction of QSOs with BALs at low redshift and the physical conditions in the BAL region.
D. A. Turnshek, C. B. Foltz, R. J. Weymann, O. L. Lupie, R. G. McMahon, B. M. Peterson.
Space Telesc. Sci. Inst., Prepr. Ser., No. 41, 18 pp. (1985).
        The authors present IUE and optical observations demonstrating that the $z_{em} = 0.29$ QSO PG 1700 + 518 is a broad absorption line (BAL) QSO. This object is unique because of its low redshift and because it is one of only two BAL QSOs which exhibit moderate strength Mg II BALs. The discovery of a low redshift BAL QSO is noteworthy because, barring systematics between redshift and the presence of appreciable Mg II BALs in BAL QSOs, it suggests that the fraction of QSOs with BALs at low redshifts is comparable to the observed fraction at high redshifts. This has implications for the evolutionary history of the QSO and its BAL region.

159.066   Evidence for underlying galaxies in a complete sample of variable quasars.
M. R. S. Hawkins, L. Woltjer.
ESO Sci. Prepr., No. 359, 20 pp. (1985). To appear in Mon. Not. R. Astron. Soc.

159.067   The gravitational lens effect and the surface density of quasars near foreground galaxies.
E. J. Zuiderwijk.
ESO Sci. Prepr., No. 371, 40 pp. (1985). To appear in Mon. Not. R. Astron. Soc.

159.068   Variability and the nature of QSO optical–infrared continua.
R. M. Cutri, W. Z. Wisniewski, G. H. Rieke, M. J. Lebofsky.
Prepr. Steward Obs., No. 570, 21 pp. (1985). To appear in Astrophys. J.

159.069   Broad absorption line quasars ("BALQSOs").
        R. J. Weymann, D. A. Turnshek, W. A. Christiansen.
Prepr. Steward Obs., No. 574, 17 pp. (1985). To appear in "Proceedings of the Santa Cruz Summer Workshop in Astrophysics: Quasars and active galactic nuclei", July 1984.

159.070   High resolution radio observations of the most luminous quasar.
H. Kühr, J. T. Stocke, P. A. Strittmatter, N. Bartel, A. Eckart, C. Schalinski, A. Witzel, P. Biermann.
Prepr. Steward Obs., No. 575, 17 pp. (1985).
        The extremely high energy output of S5 0014 + 81 of 5×10^{48}erg/s and the presence of a high column density metal absorption system at $z = 1.11$ in the optical spectrum suggest that a gravitational lens might be present. Indeed, a weak second component was found with the VLA at 6 cm, 0.6 arcsec south of the unresolved main component. Further VLA and VLBI observations clearly detect only the one bright source but with an elongation in the direction of the second component. Therefore, the radio structure is probably a jet like phenomenon and does not provide evidence for or against a gravitational lens which might have boosted the observed flux density of the quasar.

159.071   The diversity of soft X–ray spectra in quasars.
        M. Elvis, B. J. Wilkes, H. Tananbaum.
Astrophys. J., Vol. 292, No. 2, p. 357 – 361 (1985).
        The authors present Einstein IPC spectra for three quasars (NAB 0205 + 024, B2 1028 + 313, PG 1211 + 143). These spectra cover the soft X–ray band 0.1 – 4.0 keV. Power law fits to these spectra have best–fit energy indices of 1.2, 0.6, and 2.2 respectively. These diverse power–law slopes contrast strongly with the suggestion of a uniform 0.7 value for Seyfert galaxies at higher energies (2 – 100 keV). The implications of this diversity for the X–ray background, and X–ray continuum emission mechanisms are discussed.

159.072   High–resolution spectroscopy of selected absorption lines toward quasi–stellar objects. II. The metal–to–hydrogen ratio in a "metal–free" cloud toward S5 0014 + 81.
F. H. Chaffee Jr., C. B. Foltz, H.–J. Röser, R. J. Weymann, D. W. Latham.
Astrophys. J., Vol. 292, No. 2, p. 362 – 370 (1985).
        The authors have obtained 1 Å resolution spectra with the MMT spectrograph of the QSO S5 0014 + 81 ($z_{em} = 3.42$) and have selected a strong Lyman absorption–line system ($z_{abs} = 3.32$), in which absorption lines down to Ly14 can be detected, for detailed study at higher resolution. Observations of H I, C II, C IV, O VI and Si III in this system are reported. A calculation of the ionization equilibrium of material photoionized by integrated QSO radiation is used to infer that the metal–to–hydrogen ratio for the observed cloud is approximately $10^{-2.7}$ of the solar value if the Si III detection is real. Implications of this result for the so–called "metal–free" clouds in quasar absorption systems are discussed.

159.073   Quasars on plates taken with a Schmidt telescope.
        U. Haug.
Bull. Inf. Cent. Données Stellaires, No. 28, p. 75 – 76 (1985). – See Abstr. 012.062.
        First results are reported to study visually selected quasar candidates on Schmidt plates with the help of PDS scans.

159.074   VLA observations of quasars with "dogleg" radio structure.
J. T. Stocke, J. O. Burns, W. A. Christiansen.
Prepr. Steward Obs., No. 581, 56 pp. (1985).
        The authors present high resolution and high dynamic range maps of four quasars chosen because previous work revealed

them to have non–collinear radio structure. Contrary to previous suggestions, the detailed radio morphology of these quasars does not resemble Fanaroff and Riley class 1 radio galaxies (head–tails or wide–angle–tails). Rather these sources are best described as bent classical doubles with edge brightened extended lobes and no evidence for a smoothly changing angle of flight but rather indications that all the bending in these sources took place abruptly. This is what the authors refer to herein as "dogleg" morphology. Two models are presented to account for the "dogleg" morphology; one of which, an inelastic collision between one of the radio lobes and a nearby galaxy halo or intergalactic cloud, seems to account reasonably well for the observed morphological details.

159.075   **Quasi–stellar objects – a personal view.**
E. M. Burbidge.
Bull. Am. Astron. Soc., Vol. 16, No. 4, p. 875 (1984). Abstract. – See Abstr. 010.062.

159.076   **Models for high–frequency radio to infrared outbursts in extragalactic sources.**
A. P. Marscher, W. K. Gear.
Bull. Am. Astron. Soc., Vol. 16, No. 4, p. 917 (1984). Abstract. – See Abstr. 010.062.

159.077   **Active extragalactic sources: nearly simultaneous observations from 20 cm to 1400 Å.**
R. Landau.
Bull. Am. Astron. Soc., Vol. 16, No. 4, p. 917 (1984). Abstract. – See Abstr. 010.062.

159.078   **VLA observations of faint optically selected quasars.**
H. L. Marshall.
Bull. Am. Astron. Soc., Vol. 16, No. 4, p. 918 (1984). Abstract. – See Abstr. 010.062.

159.079   **First Einstein IPC results on X–ray spectra of quasars.**
B. J. Wilkes, M. Elvis, H. Tananbaum.
Bull. Am. Astron. Soc., Vol. 16, No. 4, p. 931 (1984). Abstract. – See Abstr. 010.062.

159.080   **Multifrequency observations of the QSO PG 1211 + 143.**
J. Bechtold, M. Elvis, R. F. Green, G. Fabbiano.
Bull. Am. Astron. Soc., Vol. 16, No. 4, p. 931 (1984). Abstract. – See Abstr. 010.062.

159.081   **The X–ray to optical luminosity ratio of quasars.**
G. A. Kriss, C. R. Canizares.
Bull. Am. Astron. Soc., Vol. 16, No. 4, p. 932 (1984). Abstract. – See Abstr. 010.062.

159.082   **New observations of the X–ray discovered QSO–galaxy pair 1E 0104.2 + 3153.**
I. M. Gioia, T. Maccacaro, R. Schild, P. Giommi, J. T. Stocke.
Bull. Am. Astron. Soc., Vol. 16, No. 4, p. 932 (1984). Abstract. – See Abstr. 010.062.

159.083   **Balloon observations of hard X–rays from 3C 273.**
D. Venkatesan, S. V. Damle, P. K. Kunte, S. Naranan, B. V. Sreekantan.
Bull. Am. Astron. Soc., Vol. 16, No. 4, p. 933 (1984). Abstract. – See Abstr. 010.062.

159.084   **A flare in the optical spectrum of 3C 273.**
A. C. Sadun.
Bull. Am. Astron. Soc., Vol. 16, No. 4, p. 952 (1984). Abstract. – See Abstr. 010.062.

159.085   **The optical polarization properties of the quasar 3C 345.**
P. S. Smith, T. J. Balonek, P. A. Heckert, R. Elston.
Bull. Am. Astron. Soc., Vol. 16, No. 4, p. 952 – 953 (1984). Abstract. – See Abstr. 010.062.

159.086   **VLA observations of the physically largest QSO radio sources.**
P. Hintzen, F. N. Owen.
Bull. Am. Astron. Soc., Vol. 16, No. 4, p. 953 (1984). Abstract. – See Abstr. 010.062.

159.087   **Radio structure and kinematics of the core of 3C 279.**
S. C. Unwin, J. A. Biretta.
Bull. Am. Astron. Soc., Vol. 16, No. 4, p. 953 (1984). Abstract. – See Abstr. 010.062.

159.088   **2237 + 0305: a new and unusual gravitational lens.**
J. Huchra, M. Gorenstein, E. Horine, S. Kent, R. A. Perley, G. H. Smith.
Bull. Am. Astron. Soc., Vol. 16, No. 4, p. 1005 (1984). Abstract. – See Abstr. 010.062.

159.089   **Are there two rich clusters in the line of sight to the QSO G1556?**
R. J. Weymann, C. B. Foltz, S. A. Shectman, C. Price, T. Boroson.
Bull. Am. Astron. Soc., Vol. 16, No. 4, p. 1005 (1984). Abstract. – See Abstr. 010.062.

159.090   **Detection of a galaxy at z = 0.373 responsible for Mg II absorption in the z = 1.24 QSO 4C 55.27.**
J. S. Miller, R. Goodrich, S. A. Stephens.
Bull. Am. Astron. Soc., Vol. 16, No. 4, p. 1005 – 1006 (1984). Abstract. – See Abstr. 010.062.

159.091   **Observations of an interesting new broad–absorption–line QSO: H0302 + 170.**
C. B. Foltz, R. J. Weymann, C. Hazard, D. A. Turnshek.
Bull. Am. Astron. Soc., Vol. 16, No. 4, p. 1006 (1984). Abstract. – See Abstr. 010.062.

159.092   **On the detectability of metal–line transitions in Lyman alpha forest absorption–line systems.**
F. H. Chaffee Jr., C. B. Foltz, J. Bechtold, R. J. Weymann.
Bull. Am. Astron. Soc., Vol. 16, No. 4, p. 1006 (1984). Abstract. – See Abstr. 010.062.

159.093   **Observations of the QSO PKS 0119–046 and the relative velocities of absorption systems with $z_a > z_e$.**
V. T. Junkkarinen.
Bull. Am. Astron. Soc., Vol. 16, No. 4, p. 1006 (1984). Abstract. – See Abstr. 010.062.

159.094   **An imaging study of the nebulosities surrounding low redshift QSOs.**
E. P. Smith, T. M. Heckman, W. Romanishin, G. D. Bothun, B. Balick.
Bull. Am. Astron. Soc., Vol. 16, No. 4, p. 1006 – 1007 (1984). Abstract. – See Abstr. 010.062.

159.095   **Narrow–line regions and the different classes of quasars.**
C. M. Gaskell.
Bull. Am. Astron. Soc., Vol. 16, No. 4, p. 1007 (1984). Abstract. – See Abstr. 010.062.

159.096   **Emission line variations in 3C 446.**
S. A. Stephens, J. S. Miller.
Bull. Am. Astron. Soc., Vol. 16, No. 4, p. 1007 (1984). Abstract. – See Abstr. 010.062.

159.097   **Lyα variability in the high luminosity quasar 3C 446.**
J. N. Bregman, A. E. Glassgold, P. J. Huggins, A. L. Kinney.
Bull. Am. Astron. Soc., Vol. 16, No. 4, p. 1007 (1984). Abstract. – See Abstr. 010.062.

159.098 The ultraviolet excess in quasars and AGNs: where have all the stars gone?
D. Kazanas.
Bull. Am. Astron. Soc., Vol. 16, No. 4, p. 1007 – 1008 (1984).
Abstract. – See Abstr. 010.062.

159.099 QSO Lyman continuum radiation and the ultraviolet extragalactic background.
D. Tytler.
Bull. Am. Astron. Soc., Vol. 16, No. 4, p. 1008 (1984). Abstract. – See Abstr. 010.062.

159.100 Free–free emission from high density braod–line clouds.
R. C. Puetter, E. N. Hubbard.
Bull. Am. Astron. Soc., Vol. 16, No. 4, p. 1008 (1984). Abstract. – See Abstr. 010.062.

159.101 VLBI observations of 3C 273 with high north–south resolution.
J. A. Biretta, M. H. Cohen, P. Kaufmann.
Bull. Am. Astron. Soc., Vol. 16, No. 4, p. 1008 (1984). Abstract. – See Abstr. 010.062.

159.102 High redshift quasars have more bent radio emission.
P. D. Barthel, G. K. Miley.
Bull. Am. Astron. Soc., Vol. 16, No. 4, p. 1009 (1984). Abstract. – See Abstr. 010.062.

159.103 Radio structure in the central components of double–lobed quasars.
D. H. Hough, A. C. S. Readhead.
Bull. Am. Astron. Soc., Vol. 16, No. 4, p. 1010 (1984). Abstract. – See Abstr. 010.062.

159.104 Detection of 21 cm absorption at z = 2.04.
A. M. Wolfe, M. M. Davis, D. A. Turnshek,
H. E. Smith, R. Cohen, F. H. Briggs.
Bull. Am. Astron. Soc., Vol. 16, No. 4, p. 1014 (1984). Abstract. – See Abstr. 010.062.

159.105 Quasar candidates from grens plates.
D. Crampton, D. Schade, A. P. Cowley.
Astron. J., Vol. 90, No. 6, p. 987 – 997 (1985).
Positions and magnitudes are given for 439 quasar candidates discoverd on CFHT blue grens plates, as well as for some white dwarfs and intergalactic H II regions. Spectroscopic observations with the MMT indicate that 70% of the candidates are quasars, and most of the remaining objects are white dwarfs or hot subluminous stars. The MMT observations also reveal that about 68% of the quasars have redshifts less than $z = 1.7$. The redshifts of quasars estimated from their grens spectra peak sharply at $z = 2.2$.

159.106 QSO evolution in the interaction model.
M. De Robertis.
Astron. J., Vol. 90, No. 6, p. 998 – 1003 (1985).
QSO evolution is investigated according to the interaction hypothesis in which activity results from an interaction between two galaxies resulting in the transfer of gas onto a supermassive black hole at the center of at least one participant. Explicit models presented here for interactions in cluster environments show that a peak QSO population can be formed in this way at $z \approx 2 - 3$. Substantial density evolution is expected in such models. At smaller redshifts, QSOs should be found primarily in poor clusters or groups. Probability estimates provided by this model are consistent with local estimates for the observed number of QSOs per interaction.

159.107 Submilliarcsecond VLBI observations of the close pair GC 1342 + 662 and GC 1342 + 663.
D. D. Morabito.
Astron. J., Vol. 90, No. 6, p. 1004 – 1006 (1985).
Differential VLBI has been simultaneously performed on the source pair GC 1342 + 662 and GC 1342 + 663 (4.4 arcmin separation) at 2.29 GHz on the Goldstone/Madrid baseline. These measurements were acquired on two separate observing sessions: 30 December 1982 and 14 May 1983. The difference in arclength separation of GC 1342 + 662 relative to GC 1342 + 663 between the two epochs was 29 ± 31 μarcsec. The arclength differences of the relative position measurements between measurement epochs of GC 1342 + 662 relative to GC 1342 + 663 were -283 ± 36 μarcsec in right ascension and 141 ± 24 μarcsec in declination.

159.108 An X–ray Hubble diagram for quasi–stellar objects.
A. K. Sapre, V. D. Mishra.
J. Astrophys. Astron., Vol. 6, No. 1, p. 49 – 56 (1985).
To form the Hubble diagram for quasi–stellar objects (QSOs), the authors have made use of the recently published data on X–ray fluxes of 159 QSOs observed from the Einstein Observatory. When the optical, radio and X–ray selection effects are removed, keeping only the intrinsically brighter sources, the authors obtain a sample of 16 QSOs having a small dispersion in X–ray luminosities, a statistically significant linear correlation between $(\log f_x, \log cz)$ pairs and a slope $A = -1.906 \pm 0.061$ of the linear regression of $\log f_x$ on $\log cz$. This slope is consistent, at a confidence level of 95 per cent or greater, with the slope of -2.0 expected theoretically based on the assumption that the redshifts of QSOs are cosmological in nature.

159.109 Emission line redshift distribution of QSOs.
Y.–y. Zhou, Z.–g. Deng, H.–j. Dai.
Astrophys. Space Sci., Vol. 112, No. 1, p. 93 – 110 (1985).
The authors study the influence of the selection effect in the identification of emission lines on the redshift distribution of QSOs more thoroughly than in the previous paper (Zhou et al., 1983). They assume that the QSO's redshift is cosmological, adopt the standard model, and consider the selection effect due to the redshift identification, the limiting apparent magnitude in the observation and the evolutionary effect of QSOs, and compute the emission line redshift distribution for the so–called optically selected QSOs discovered by objective prism, grating prism technique alone, the QSOs discovered by positional methods or by color technique and for all QSOs, respectively. The results of computation agree with the observations very well.

159.110 More spectroscopy of the fuzz around QSOs: additional evidence for two types of QSO.
T. A. Boroson, S. E. Persson, J. B. Oke.
Astrophys. J., Vol. 293, No. 1, p. 120 – 131 (1985).
The nebulosities around five optically selected (by UV excess) luminous QSOs have been observed spectroscopically. 3C 273, which has been observed off–nucleus before, has nebulosity dominated by a stellar continuum. TON 202, a steep–spectrum radio source, has nebulosity dominated by strong Balmer and [O III] emission lines. In addition, the spectra of the nuclear regions of 12 low–luminosity QSOs are presented. Nebulosity spectra of these objects were obtained earlier. A sample of 24 objects with surrounding nebulosities is statistically analyzed. It is shown that these objects can be separated into two groups, defined by the equivalent width of the [O III] λ5007 emission line in the nebulosity. The differences in the fundamental properties of these two groups are worked out and the implications for theoretical models of active galactic nuclei are discussed.

159.111 PKS 0537–441.
IAU Circ., Nos. 4027, 4030, 4060 (1985).

159.112 QSO 1156 + 295.
IAU Circ., Nos. 4053, 4057, 4076 (1985).

159.113 PG 1115 + 080A.
IAU Circ., No. 4055 (1985).

159.114 Spectroscopy of possible QSO–galaxy interactions.
S. Djorgovski, H. Spinrad.
Bull. Am. Astron. Soc., Vol. 17, No. 1, p. 516 (1985). Abstract. – See Abstr. 010.064.

159.115 Quasars near NGC 1097.
M. F. Barnothy, J. M. Barnothy.
Bull. Am. Astron. Soc., Vol. 17, No. 1, p. 518 (1985). Abstract. –
See Abstr. 010.064.

159.116 A test for a gravitational lens hypothesis.
H. Kühr.
Bull. Am. Astron. Soc., Vol. 17, No. 1, p. 521 (1985). Abstract. –
See Abstr. 010.064.

159.117 The identification of high redshift QSO's.
B. A. Peterson, M. R. S. Hawkins.
Bull. Am. Astron. Soc., Vol. 17, No. 1, p. 521 – 522 (1985). Abstract. – See Abstr. 010.064.

159.118 Quasar and related problems – some recent developments.
S. Tsuruta.
High energy astrophysics and cosmology, p. 351 – 371 (1983). –
See Abstr. 012.068.

The pertinent observational parameters of quasars and other active galactic nuclei are summarized and models for these objects based on supermassive black holes are reviewed.

159.119 Luminosity evolution of QSOs.
S. Gong, C. Xia.
High energy astrophysics and cosmology, p. 372 – 379 (1983). –
See Abstr. 012.068.

Using a sample of QSOs as standard candles the authors derive an improved deceleration parameter q_0 by means of the exact formula of luminosity distance of a Friedmann universe model. Then from the QSOs catalogue of Hewitt and Burbidge, they derive the statistical magnitude–redshift relation for QSOs. Finally the luminosity of QSOs as a function of redshift z or time t and their respective derivatives are calculated.

159.120 The statistical analysis of Lα absorption lines from a uniform sample of nine high redshift QSOs.
Z. Zou, J. Chen, Y. Bian, X. Tang, Z. Cui.
High energy astrophysics and cosmology, p. 380 – 388 (1983). –
See Abstr. 012.068.

Observations of Lα absorption systems in the spectra of 9 quasars are used for a statistical analysis of the absorbing clouds. This analysis attempts at deciding whether the clouds have been ejected by the quasars or are distributed cosmologically and are not related with the quasars.

159.121 Quasargalaxien.
J. Fried.
Sterne Weltraum, 24. Jahrg., Nr. 6, p. 315 – 317 (1985).

159.122 The multiple images of the quasar 0957 + 561.
D. H. Roberts, P. E. Greenfield, J. N. Hewitt,
B. F. Burke, A. K. Dupree.
Astrophys. J., Vol. 293, No. 2, p. 356 – 369 (1985).

The gravitationally image double quasar 0957 + 561 has been observed at an angular resolution of 0″.3 using the full VLA at 6 cm wavelength. These observations demonstrate structure in the B radio quasar component and in the radio source G which lies close to the nucleus of primary lens galaxy G1. When combined with earlier observations, the data show that most of the flux in the G radio source is probably due to a nuclear source intrinsic to the lens galaxy G1. The radio properties of this galaxy are similar to those of M87. The structure of the B quasar region strongly suggests that an additional image of the arc second radio jet has been found. No unambiguous evidence for the third quasar image B2 was found near B or near the galaxy G1, but limits are obtained on the intensity of B2 as a function of its separation from the nucleus of the lens galaxy G1.

159.123 The gravitationally lensed quasar 0957 + 561: VLA observations and mass models.
P. E. Greenfield, D. H. Roberts, B. F. Burke.
Astrophys. J., Vol. 293, No. 2, p. 370 – 386 (1985).

The paper summarizes the results of observations of 0957 + 561 taken over several years with the VLA at wavelengths of 2 cm, 6 cm, 18 cm, and 20 cm. The radio observations narrow the range of acceptable models for the mass distribution of the lens, and improved models are presented. These models require a very high mass–to–light ratio for the central part of the foreground cluster of galaxies. It is also shown that the lensing material cannot be distributed in the same way as the visible matter in the cluster members.

159.124 Magnesium, iron, and calcium in the $z = 0.39498$ 21 centimeter absorber of PKS 1229–021.
F. H. Briggs, D. A. Turnshek, J. Schaeffer, A. M. Wolfe.
Astrophys. J., Vol. 293, No. 2, p. 387 – 393 (1985).

Sensitive observations of the Fe II, Mg I, Mg II, and Ca II lines in the spectrum of PKS 1229–021 at $z = 0.39498$ with the Multiple Mirror Telescope confirm previous deductions about the velocity structure of the absorber that were based on curve–of–growth analysis of optical data and the width of the 21 cm absorption profile. The velocity and opacity structure conforms to a multicloud picture where our line of sight to the QSO ($z_{em} \approx 1.042$) encounters both the disk and the halo of an intervening galaxy.

159.125 On a method of search for periodicities in redshift distribution of quasars.
O. A. Pushkarev.
Tr. Astrofiz. Inst. Alma–Ata, Tom 43, p. 67 – 81 (1984). In Russian.

159.126 A statistical X–ray QSOs classification.
A. Coradini, F. Giovannelli, M. L. Polimene.
18th International Cosmic Ray Conference, Vol. 1, p. 35 – 38 (1983). – See Abstr. 012.096.

An attempt to classify a QSOs sample measured by Einstein satellite has been done. The used method (G–Mode) allows to perform such a classification without any "a priori" knowledge of the taxonomic units.

159.127 The high energy X–ray spectrum of 3C 273.
R. Staubert, M. Bezler, E. Kendziorra, W. Pietsch,
C. Reppin, J. Trümper, W. Voges.
18th International Cosmic Ray Conference, Vol. 9, p. 13 (1983). Abstract. – See Abstr. 012.096.

159.128 The evolution of X–ray radiation of quasars.
H.–q. Zhang, R.–l. Liu, H.–s. Yang.
18th International Cosmic Ray Conference, Vol. 9, p. 14 – 17 (1983). – See Abstr. 012.096.

Statistical studies of the evolution of X–ray radiation of 273 objects (235 quasars and 38 type 1 Seyfert galaxies) in the 0.5 – 4.5 keV band observed from the Einstein Satellite are carried out. The relationships of luminosities in various wave bands with redshift z and among themselves are studied. The implications of quasar properties are analysed.

159.129 Structure of the nuclei of the Seyfert galaxy NGC 1275 and the quasar 3C 345.
L. I. Matveenko, I. I. K. Pauliny-Toth, V. I. Kostenko,
J. D. Romney, L. B. Bååth.
Pis'ma Astron. Zh., Tom 11, No. 6, p. 420 – 431 (1985). In Russian. English translation in Sov. Astron. Lett., Vol. 11.

The structure of the quasar 3C 345 and the Seyfert galaxy NGC 1275 has been studied with a global VLBI network at 18 cm wavelength.

159.130 Observations of Mg II emission and absorption in broad absorption line QSOs.
V. T. Junkkarinen.
Bull. Am. Astron. Soc., Vol. 17, No. 2, p. 575 (1985). Abstract. –
See Abstr. 010.065.

159.131 Spectroscopy of a rotating gas cloud 100 kpc in diameter surrounding the QSO 3C275.1.
P. Hintzen, J. Stocke.
Bull. Am. Astron. Soc., Vol. 17, No. 2, p. 576 (1985). Abstract. –
See Abstr. 010.065.

159.132 Further results on QSO soft X–ray spectra.
B. J. Wilkes, M. Elvis.
Bull. Am. Astron. Soc., Vol. 17, No. 2, p. 576 (1985). Abstract. –
See Abstr. 010.065.

159.133 X–ray to IR continua of PG quasars.
M. Elvis, J. Bechtold, R. Green.
Bull. Am. Astron. Soc., Vol. 17, No. 2, p. 576 (1985). Abstract. –
See Abstr. 010.065.

159.134 The luminosity function of quasars and low luminosity active galactic nuclei.
H. L. Marshall.
Bull. Am. Astron. Soc., Vol. 17, No. 2, p. 577 (1985). Abstract. –
See Abstr. 010.065.

159.135 The effects of electron scattering opacity in the broad emission line regions of quasars.
T. Kallman, J. Krolik.
Bull. Am. Astron. Soc., Vol. 17, No. 2, p. 587 (1985). Abstract. –
See Abstr. 010.065.

159.136 The far and extreme ultraviolet spectra of QSOs.
U. J. Sofia, F. C. Bruhweiler, M. Kafatos.
Bull. Am. Astron. Soc., Vol. 17, No. 2, p. 588 (1985). Abstract. –
See Abstr. 010.065.

159.137 18 cm VLBI observations of the quasar NRAO 140 during and after a low–frequency outburst.
A. P. Marscher, J. J. Broderick, N. Bartel, L. Padrielli,
J. D. Romney.
Bull. Am. Astron. Soc., Vol. 17, No. 2, p. 608 (1985). Abstract. –
See Abstr. 010.065.

159.138 Multi–epoch VLBI observations of the quasar 4C39.25: superluminal motion sandwiched by stationary structure.
D. B. Shaffer, A. P. Marscher.
Bull. Am. Astron. Soc., Vol. 17, No. 2, p. 609 (1985). Abstract. –
See Abstr. 010.065.

159.139 Does quasar nebulosity represent normal host galaxies?
B. V. Komberg.
Sov. Astron., Vol. 28, No. 6, p. 613 – 616 (1984). English translation of 38.159.123.

159.140 Where are the dead quasars?
M. M. Waldrop.
Science, Vol. 228, No. 4704, p. 1185 – 1186 (1985).
Two recent observations suggest that many "ordinary" galaxies – including our own – contain ultramassive black holes.

159.141 The kinematics of the emission line gas in quasars and active galactic nuclei.
T. J. Carroll.
Diss. Abstr. Int., Sect. B, Vol. 45, No. 1, p. 229 (1984). Thesis, University of Massachusetts, 143 pp. (1984). Order No. DA8410270.

159.142 Photoionization models of the Mg II and C IV selected clouds observed in absorption in QSO spectra.
J. R. Schaeffer.
Diss. Abstr. Int., Sect. B, Vol. 45, No. 2, p. 585 (1984). Thesis, University of Pittsburgh, 124 pp. (1983). Order No. DA8411773.

159.143 Observations of quasars and active nuclei of galaxies.
Eh. A. Dibaj.
Main directions of astronomical investigations at the Moscow University, p. 95 – 97 (1985). In Russian. – See Abstr. 003.049.

159.144 Optical monitoring of the quasar 3C 351 in blue light.
G. J. Corso, J. Schultz, T. Pfaff, A. Dey.
Publ. Astron. Soc. Pac., Vol. 97, No. 591, p. 393 – 394 (1985).
Iris photometry is reported for blue photographic images of the quasar 3C 351 made in 1983 and 1984 with the 1–meter reflector of Northwestern's Lindheimer Astronomical Research Center.

159.145 A flare in the optical spectrum of 3C 273.
A. C. Sadun.
Publ. Astron. Soc. Pac., Vol. 97, No. 591, p. 395 – 396 (1985).
Flaring has been observed in 3C 273 in the $BVRI$ spectrum during January 1983. The increase in luminosity is especially pronounced at longer wavelengths and amounts to as much as half a magnitude increase in brightness over quiescent values. These observations complement and are consistent with similar observations in the millimeter to IR made at about the same time by Robson et al.

159.146 Quasar redshift distribution and selection effects.
Y.–y. Zhou, Z.–g. Deng, Z.–l. Zou.
Chin. Astron. Astrophys., Vol. 9, No. 1, p. 20 – 26 (1985). English translation of 38.159.125.

159.147 Variable blue objects in Kapteyn Selected Area 94.
K.–l. Huang, P. Usher.
Chin. Astron. Astrophys., Vol. 9, No. 2, p. 95 – 97 (1985). English translation of Acta Astron. Sin., Vol. 25, No. 4, p. 321 – 324 (1984).
The authors present a list of ultraviolet excess objects in the Palomar Schmidt field centred on Kapteyn Selected Area 94. All these objects were selected on the basis of optical variability. They can serve as quasar candidates.

159.148 An analysis on the three dimensional distribution of QSOs.
Z.–g. Deng, Y.–z. Liu, S.–l. Cao.
Acta Astron. Sin., Vol. 25, No. 4, p. 348 – 355 (1984). In Chinese.
The authors have analysed the three dimensional distribution of QSOs by a Monte Carlo method, in which they compared the nearest distance distribution expected by randomly distributing QSOs with the distribution given by a Monte Carlo calculation. The results of the analysis showed that the three dimensional distribution of QSOs is highly clumped. This picture is consistent with the existence of cluster of QSOs and the giant voids.

159.149 Optical identifications of QSOs with flat radio spectra.
Y.–q. Chu, X.–f. Zhu, H. Butcher.
Acta Astron. Sin., Vol. 26, No. 1, p. 76 – 86 (1985). In Chinese.
Optical identifications of 35 "empty fields" in a radio sample with flat spectra have been made using the TV and CCD camera at the Kitt Peak National Observatory 4–m telescope. Based on close radio–optical position coincidence, 29 sources are identified with stellar objects. The new identified objects have more steep radio–optical spectra indices. The statistical results suggest a correlation of the radio and optical luminosities.

159.150 Photometry of quasar host galaxies and cosmological implications.
T. Gehren.
New aspects of galaxy photometry, p. 227 – 236 (1985). – See Abstr. 012.114.
Surface photometry of sky–limited red photographic and CCD observations, corrected for galactic extinction, the K term and seeing image degradation, reveals the decomposition of low–redshift quasar images into a central point source and an extended underlying nebulosity. The investigation of the statistical properties of a well–resolved subsample of these nebulosities shows that (1) the underlying nebulosities are in fact the host galaxies of quasar nuclei, (2) quasar host galaxies show strong evidence for luminosity evolution, (3) quasar host galaxies in many cases appear to be heavily distorted by interaction with faint companion galaxies.

159.151 Observations of large scale jets in quasars and the sidedness problem.
J. F. C. Wardle, R. I. Potash.
Physics of energy transport in extragalactic radio sources, p. 30 – 34 (1984). – See Abstr. 012.116.
The authors have found large scale jets in each of the eight largest radio sources from a complete sample of 4C quasars. In

each case there is no visible counter jet. It is argued that the observed limits on jet–counter jet ratios are incompatible with differential Doppler boosting, and that these jets are intrinsically one–sided.

159.152 Bending in the first few hundred parsecs.
P. N. Wilkinson, T. J. Cornwell, A. J. Kus, A. C. S. Readhead, T. J. Pearson.
Physics of energy transport in extragalactic radio sources, p. 76 – 82 (1984). – See Abstr. 012.116.

VLBI maps of jets in the nuclei of the quasars 3C 309.1 and 3C 380 show sharp apparent bends on the 100 pc scale. At present the authors cannot say whether these bends are greatly enhanced by projection or what their underlying cause is. However, progress can be expected on understanding this phenomenon within a few years.

159.153 Distorted quasars: can sharp bends in radio structures be attributed to localized collisions?
W. A. Christiansen, J. O. Burns, J. T. Stocke.
Physics of energy transport in extragalactic radio sources, p. 83 – 88 (1984). – See Abstr. 012.116.

The radio structures of many distorted quasars seem to follow a distinctive pattern consisting of a "classical double" configuration with a single sharp bend. Thus, they resemble the shape of a dog's leg. The authors suggest that these dogleg structures are the result of collisions between the radio ejecta of the quasars and clumps of high density gas in the vicinity of the quasar. Specifically, the possibility that the "clumps" might be the ISM's or gaseous halos of companion galaxies is examined and found to be plausible.

159.154 Jet speed.
L. Rudnick.
Physics of energy transport in extragalactic radio sources, p. 114 – 118 (1984). – See Abstr. 012.116.

Line emission has been detected from the optical counterparts of the extended radio QSO 0812+02 (Wehinger et al., 1984). At the radio/optical hot spot, the velocity is measured to be 1200 km s^{-1} relative to the QSO nucleus. Spatially asymmetric line emissions argues against a relativistic explanation for the one–sided radio jet. Observations such as these hold great promise for measurements of velocities and momentum fluxes in radio sources, although a better understanding of the interactions between thermal and relativistic material are needed.

159.155 Evolution of superluminal radio components in 3C 345.
J. A. Biretta.
Physics of energy transport in extragalactic radio sources, p. 120 (1984). Abstract. – See Abstr. 012.116.

Sky catalogue 2000.0. Volume 2: Double stars, variable stars and nonstellar objects.
See Abstr. 002.019.

A catalogue of quasars and active nuclei (2nd edition).
See Abstr. 002.055.

Accretion power in astrophysics.
See Abstr. 003.012.

Galassie e quasars.
See Abstr. 003.114.

Discovery of quasars.
See Abstr. 004.189.

Berichte aus den Workshops.
See Abstr. 011.024.

The CCD/transit instrument – scientific programs.
See Abstr. 032.030.

The study of astrophysical interactions with the ESA Photon Counting Detector.
See Abstr. 035.052.

Automated analysis of objective–prism spectra. I. Quasar detection.
See Abstr. 036.068.

Effects of source structure on position determinations made with VLBI observations.
See Abstr. 041.048.

Quasar energy from frozen fusion via massive neutrinos?
See Abstr. 061.028.

Quasi–two–dimensional cosmic jets.
See Abstr. 062.032.

Force–free equilibria of magnetized jets.
See Abstr. 062.066.

The stability of confined radio jets: the role of reflection modes.
See Abstr. 062.113.

Solutions for magnetic fields and surface electric fields produced by charged particle beams in which electrons (or electrons and positrons) have drift velocities with respect to protons.
See Abstr. 062.185.

Line fluorescence in astrophysics.
See Abstr. 063.045.

Minilensing of multiply imaged quasars: flux variations and vanishing of images.
See Abstr. 066.047.

Special relativistic effects in rapidly moving plasmoids.
See Abstr. 066.159.

Close–up on gravitational lensing: the gravitational mirages.
See Abstr. 066.187.

Two–phase accretion model for emission–line regions in quasars and active galactic nuclei.
See Abstr. 067.097.

X–ray induced stellar winds: a mass supply for QSO's?
See Abstr. 067.108.

Magnetic buoyancy in QSO accretion disks: the underlying physics and analytic approximations.
See Abstr. 067.124.

Accretion disk model for quasar PG1211+143.
See Abstr. 067.186.

Limb darkening and polarization of accretion disks.
See Abstr. 067.187.

Shock acceleration in active galactic nuclei and quasars.
See Abstr. 067.190.

A deep optical survey of a small region in Aquarius – II. Stellar spectroscopy.
See Abstr. 114.015.

Interstellar cloud phase transitions: effects of metal abundances, grains and X–rays.
See Abstr. 131.072.

Thermal phases of interstellar and quasar gas.
See Abstr. 131.073.

On the abundance of metals and the ionization state in absorbing clouds toward QSOs.
See Abstr. 131.193.

On recombination lines of excited hydrogen: 1. High emission measure H II regions. 2. Radio galaxies and quasars.
See Abstr. 132.011.

Results from the IRAS survey.
See Abstr. 133.001.

Determination of flux densities of radio sources using broad–band radiometers of the radio telescope RATAN–600.
See Abstr. 141.005.

Observations of the Ooty radio sources at 102 MHz.
See Abstr. 141.014.

Observations of Ooty radio sources at RATAN–600.
See Abstr. 141.015.

On the variability of some Parkes radio sources.
See Abstr. 141.021.

Report of IAU Commission 40: Radio astronomy (*Radio astronomie*).
See Abstr. 141.039.

Are there gamma–ray loud quasars?
See Abstr. 143.075.

New developments in high energy astrophysics.
See Abstr. 143.100.

Equilibrium non–rotating magnetized disks as a plane analog to the $n = 3$ polytrope.
See Abstr. 151.082.

Monte Carlo simulations of the $2 + 1$ dimensional Fokker–Planck equation: spherical star clusters containing massive, central black holes.
See Abstr. 151.125.

Galactic absorption lines measured from the IUE low dispersion spectra of active galaxies.
See Abstr. 155.001.

Report of IAU Commission 28: Galaxies (*Galaxies*).
See Abstr. 157.244.

Ultraviolet and X–ray observations of active galactic nuclei: constraints on models of the broad emission line region.
See Abstr. 158.001.

Broad emission features in active galactic nuclei.
See Abstr. 158.003.

Multifrequency observations of BL Lac objects and violently variable quasars.
See Abstr. 158.004.

Fluorescent excitation of Fe II, Mn II, Ti II, N I lines by C IV, N V, O VI,... emission lines in the spectra of symbiotic stars & Seyfert galaxies.
See Abstr. 158.007.

Soft gamma–ray production in active galactic nuclei.
See Abstr. 158.011.

Multiple hotspots in extragalactic radio sources.
See Abstr. 158.016.

The variable extragalactic object 3C 446.
See Abstr. 158.026.

X–ray luminosity and the hydrogen–line ratios of quasars and Seyfert galaxies.
See Abstr. 158.030.

Alignment of radio and optical polarization with VLBI structure.
See Abstr. 158.032.

Redshifts of radio sources in the 1–Jy sample.
See Abstr. 158.036.

Orientation correlation of radio double sources.
See Abstr. 158.044.

Photoionization models for Liners: gas distribution and abundances.
See Abstr. 158.048.

Mysteries of cosmic jets.
See Abstr. 158.052.

Observations of the Hβ region in some broad–line objects.
See Abstr. 158.068.

New evidence for photoionization as the dominant excitation mechanism in Liners.
See Abstr. 158.077.

$8 - 13\,\mu m$ spectrophotometry of galaxies. IV. Six more Seyferts and 3C345.
See Abstr. 158.083.

CCD magnitudes of 3C radio sources.
See Abstr. 158.086.

Shock formation of the broad emission–line regions in QSOs and active galactic nuclei.
See Abstr. 158.092.

A deep Westerbork survey of areas with multicolor Mayall 4 m plates. III. Photometry and spectroscopy of faint source identifications.
See Abstr. 158.112.

Magnetic field structures in active compact radio sources.
See Abstr. 158.124.

Observational evidence for the radiative acceleration of broad–line clouds in Seyfert 1 galaxies and quasars.
See Abstr. 158.135.

Multifrequency observations of blazars. I. The shape of the 1 micron to 2 millimeter continuum.
See Abstr. 158.138.

A gravitational origin for the broad emission line profiles in quasars and Seyfert galaxies: time variation.
See Abstr. 158.143.

Linear polarization at λ49 cm of 27 double radio sources.
See Abstr. 158.150.

Optical variability of the object $1055 + 567$.
See Abstr. 158.157.

Consequences of hot gas in the broad–line region of active galactic nuclei.
See Abstr. 158.161.

Broad–band polarization observations of active compact radio sources.
See Abstr. 158.182.

Long–term X–ray monitoring of 4 active galactic nuclei.
See Abstr. 158.202.

Milliarcsecond polarization structure of several BL Lac objects and quasars.
See Abstr. 158.206.

Optical polarimetry of BL Lacertae objects and violent variable quasars.
See Abstr. 158.207.

The fate of He II Lyα emission in Seyfert galaxies and quasars.
See Abstr. 158.209.

cm–wavelength linear polarizations and spectra of superluminal radio sources.
See Abstr. 158.219.

Comparison of the widths of braod emission lines in Seyfert galaxies and quasars.
See Abstr. 158.227.

Galactic mergers, starburst galaxies, quasar activity and massive binary black holes.
See Abstr. 158.244.

Extragalactic variable radio sources.
See Abstr. 158.248.

Active galactic nuclei.
See Abstr. 158.278.

The physics of active stellar systems.
See Abstr. 158.279.

The development of the electromagnetic cascade showers initiated by high–energy electrons and gammay–rays in the hot photon gas.
See Abstr. 158.283.

Frequent observations of extragalactic compact sources at 24 GHz.
See Abstr. 158.319.

Radio jets in strong core classical doubles.
See Abstr. 158.325.

Interpretation of radio polarization data.
See Abstr. 158.332.

VLBI polarimetry – first results.
See Abstr. 158.333.

Energy transport in radio sources: efficient or not?
See Abstr. 158.334.

Symmetries of jets and hot spots.
See Abstr. 158.338.

Menu for an all–purpose source model.
See Abstr. 158.340.

An observer's perspective – II.
See Abstr. 158.344.

A theorist's perspective – I.
See Abstr. 158.345.

A theorist's perspective – II.
See Abstr. 158.346.

Some aspects of active galactic nuclei.
See Abstr. 158.347.

The compact radio emission regions in active galactic nuclei.
See Abstr. 158.349.

Strong source VLBI survey at 1.35 cm.
See Abstr. 158.350.

Testing theories of superluminal motion.
See Abstr. 158.351.

VLBI in space and X–ray observations of compact sources in active galactic nuclei.
See Abstr. 158.352.

The nature of the IGM.
See Abstr. 160.003.

Quasat, gravitational lenses, and studies of the interstellar, intra-cluster and intergalactic medium.
See Abstr. 160.164.

The alignment of distant radio sources.
See Abstr. 161.003.

A comparative study of cosmological evolution in the radio, optical and X–ray bands.
See Abstr. 161.074.

Counterimages in closed elliptical Friedmann universes.
See Abstr. 161.082.

Gravitational lensing effects of vacuum strings: exact solutions.
See Abstr. 161.092.

On model–dependent bounds on H_0 from gravitational images: application to Q0957 + 561A,B.
See Abstr. 161.097.

Quasar–galaxy associations with discordant redshifts as a topological effect. I. Two–dimensional study.
See Abstr. 161.160.

On the use of measured time delays in gravitational lenses to determine the Hubble constant.
See Abstr. 161.162.

Dark matter and formation of large–scale structure in the universe – the test by distribution of quasars.
See Abstr. 161.191.

The "forest" of absorption lines in QSO spectra and the structure of the Universe.
See Abstr. 161.194.

The statistics of gravitational lenses. II. Apparent evolution in the quasars' luminosity function.
See Abstr. 161.232.

Report of IAU Commission 47: Cosmology (*Cosmologie*).
See Abstr. 161.358.

On log N ~ log S.
See Abstr. 161.394.

160 Galaxy Groups, Clusters of Galaxies, Superclusters, Intergalactic Matter

160.001 Predictions of distance within the Local Group.
R. Mishra.
Mon. Not. R. Astron. Soc., Vol. 212, No. 1, p. 163 – 180 (1985).

The author assumes that the Local Group of galaxies is an isolated collection of objects, of which only the Milky Way and M31 have dynamically significant masses. He further demands that the galaxies in the Local Group emerged from near its barycentre at the Big Bang. This allows to find the paths these objects have taken to reach their present positions. These positions are uncertain because the distances are not known, however, paths exist for only a limited range of present distances to the objects. The author finds what these ranges are as a function of the age of the Universe and the ratio of the Milky Way to M31 masses. Results for 21 objects are shown. These results can be compared to measurements to obtain limits on the age of the Universe and the mass ratio. The results for NGC 6822 show that the mass of the Milky Way is at least half that of M31.

160.002 Multi–object spectroscopy of the distant cluster AC 103.
R. M. Sharples, R. S. Ellis, W. J. Couch, P. M. Gray.
Mon. Not. R. Astron. Soc., Vol. 212, No. 3, p. 687 – 707 (1985).

The authors have obtained intermediate dispersion spectra for a sample of faint galaxies in the field of the $z = 0.31$ cluster AC 103, using a fibre–optic coupler at the Anglo–Australian Telescope. The distribution of blue and red cluster members in the colour–absolute magnitude diagram is in good agreement with that predicted from a knowledge of galaxy populations in low–redshift clusters. Two of the blue galaxies exhibit unusually strong Balmer absorption lines. A global mass–to–light ratio of 90 ± 24 solar units is derived. The redshift distribution of the 15 non–members is consistent with field models incorporating a mild amount of luminosity evolution.

160.003 The nature of the IGM.
W. Kundt, M. Krause.
Astron. Astrophys., Vol. 142, No. 1, p. 150 – 156 (1985).

The authors interpret the concentric systems of luminous arcs seen around many large galaxies as (partial) filamentary shells ejected at supersonic speeds by their nucleus during active stages. The filaments develop a core–halo structure, with hydrogen-deficient cores and extended metal–poor halos. Such giant filamentary shells can at the same time explain the sharp absorption line systems in quasar spectra, both $Ly\alpha$ and metal lines, and the X–ray findings of a metal–enriched intracluster medium. The proposed filaments coexist in pressure equilibrium with a galactic halo and the ambient hot IGM.

160.004 Redshifts of galaxies in the Fornax cluster.
O.–G. Richter, E. M. Sadler.
Astron. Astrophys., Suppl. Ser., Vol. 59, No. 3, p. 433 – 440 (1985).

The authors present new redshifts for 51 galaxies in the region of the Fornax cluster of galaxies. A cross–comparison of results for galaxies with previously measured velocities is also included.

160.005 Dispersion of radial velocities in the Local Supercluster.
E. Giraud, J.–P. Vigier.
C. R. Acad. Sci., Sér. II, Tome 300, No. 1, p. 9 – 12 (1985). In French.

After determination of the distances of 228 spiral galaxies through the Tully–Fisher method the authors have tested various empirical models in order to minimize the differences between radial and redshift velocities in the Local Supercluster.

160.006 Westerbork bekijkt de Virgo–cluster.
C. Kotanyi, R. D. Ekers.
Zenit, 12. Jaarg., No. 1, p. 22 – 24 (1985).

160.007 Radio emission of Abell clusters of galaxies with redshifts from 0.02 to 0.075 at 102.5 MHz. II. Observations of clusters northward from the Galaxy's plane.
A. G. Gubanov.
Astron. Zh., Tom 62, Vyp. 1, p. 1 – 7 (1985). In Russian. English translation in Sov. Astron., Vol. 29, No. 1.

Results of observations at 102.5 MHz of 99 Abell clusters of galaxies with redshifts from 0.02 to 0.075 which supplement observations of the sample of clusters from Paper I (1983) are presented.

160.008 Slippery evidence on masses in the Local Group.
D. Lynden–Bell.
The Milky Way galaxy, p. 461 – 469 (1985). – See Abstr. 012.007 (IAU Symp. No. 106).

The light of the Local Group is dominated by that of M31 and the Milky Way, so the author assumes that they and their haloes dominate and these masses will in turn determine the dynamics of the Group. The author's aim is to determine their masses.

160.009 A survey of galaxies in the field of A194.
R. J. Nemiroff, C. Ftaclas, M. F. Struble.
Astron. J., Vol. 90, No. 2, p. 163 – 168 (1985).

The authors have used glass plate copies of the Palomar Observatory Sky Survey to obtain a diameter–limited survey of galaxies in the field of the cluster Abell 194. The distribution on the sky of the survey galaxies is flattened with an axial ratio of 0.72 and a position angle of 128° east of north, orthogonal to the line of bright galaxies that characterizes the cluster center. The authors find a statistically significant deviation between the cluster galaxy axial–ratio distribution and a random one.

160.010 An optical search for the intergalactic H I cloud in Leo.
E. J. Kibblewhite, M. G. M. Cawson, M. J. Disney, S. Phillipps.
Mon. Not. R. Astron. Soc., Vol. 213, No. 1, p. 111 – 115 (1985).

An optical search has been made for the large intergalactic H I cloud discovered from Arecibo by Schneider et al. A very deep red UKSTU plate of the area has been scanned by the APM machine and deep CCD frames of a small area near a peak in the H I emission have been acquired. No extended emission is found at the limiting surface brightness of the photographic material and no excess of stars above that expected from our Galaxy is found in the CCD data. However, due to the extreme size of the H I cloud, the upper limit on the total luminosity is that of a dwarf galaxy, $M_B \gtrsim -18$. Thus a highly extended very low surface brightness galaxy can not be ruled out, at present.

160.011 Simulations of galaxy cluster relaxation.
S. Yabushita, A. J. Allen.
Mon. Not. R. Astron. Soc., Vol. 213, No. 1, p. 117 – 127 (1985).

A parameter study of N–body simulations of cluster and supercluster relaxation is presented. Ellipsoidal protoclusters lead to ellipsoidal clusters, whilst flat 'pancakes' can only be associated with supercluster scales. Poor clusters may have less obvious cores than rich, with irregular outer structure and the possible formation of bound, orbiting subgroups. A new scheme for fitting the spatial distribution is used to show its variation with relaxation route.

160.012 CCD photometry of a cluster of galaxies at a redshift of 0.57.
W. J. Couch, T. Shanks, W. D. Pence.
Mon. Not. R. Astron. Soc., Vol. 213, No. 1, p. 215 – 241 (1985).

In a search of deep, prime–focus 4m–telescope plates, a very faint cluster of galaxies has been found at 0055–279 in the region of the South Galactic Pole. The authors present CCD $BVRI$ photometry of this cluster together with spectroscopic measurements of its first–ranked member which show the cluster to be at a redshift of $z = 0.57$. The cluster is of Bautz–Morgan type I, is moderately concentrated, and is half as rich as the nearby Coma cluster with an Abell richness class of 1. An analysis of the colours of the cluster's brightest galaxy and of the cluster population as a whole is presented.

160.013 X–ray scattering by intergalactic dust.
A. Evans, G. A. Norwell, M. F. Bode.
Mon. Not. R. Astron. Soc., Vol. 213, No. 1, p. 1P – 6P (1985).

The authors suggest that, if the local density of intergalactic dust is as high as has been suggested by some authors, suitable extragalactic X–ray sources should have scattered haloes comparable with those now being detected around galactic X–ray sources. The results of a search for such haloes could provide useful model–independent constraints on the amount and origin of intergalactic dust.

160.014 Cell structure and selective extinction in the Jagellonian field.
T. Grabinska, M. Zabierowski.
Nuovo Cimento B, Vol. 82B, Ser. 2, No. 2, p. 235 – 245 (1984).
Abstr. in Phys. Abstr., Vol. 88, No. 1248, Entry 9908 (1985).

160.015 Thermal and dynamical evolution of intergalactic clouds.
M. Umemura, S. Ikeuchi.
Prog. Theor. Phys., Vol. 72, No. 1, p. 47 – 62 (1984). Abstr. in Phys. Abstr., Vol. 88, No. 1250, Entry 18922 (1985).

160.016 Evidence for cannibalism in the 0004.8–3450 cluster of galaxies.
H. V. Capelato, G. des Forêts, A. Mazure, E. Salvador–Sole.
Astrophys. Space Sci., Vol. 108, No. 2, p. 363 – 368 (1985).

An analysis of the luminosity segregation in the 0004.8–3450 cluster of galaxies is made which reinforces the assumption of merging mechanisns.

160.017 No missing mass in clusters of galaxies?
M. J. Valtonen, K. A. Innanen, T.–y. Huang, S. Saarinen.
Astron. Astrophys., Vol. 143, No. 1, p. 182 – 187 (1985).

Models of clusters of galaxies are constructed where the galaxies initially form in a flat disk or a 'pancake' and subsequently fall towards the center of the cluster. At the center a massive galaxy (a few times $10^{14} M_\odot$) is located with a companion galaxy. The evolution of the clusters is followed through a large fraction of the Hubble time. At the end of the evolution the clusters appear to possess large amounts of 'missing mass', up to a factor of 10 more than the true mass of the cluster. It is suggested that the observed clusters of galaxies may have gone through similar evolution and therefore may not possess as much cluster 'missing mass' as the direct application of the virial equation would predict.

160.018 Galaxy groups: virial quantities and galaxy population.
M. Mezzetti, A. Pisani, G. Giuricin, F. Mardirossian.
Astron. Astrophys., Vol. 143, No. 1, p. 188 – 193 (1985).

Some statistical properties of the galaxy groups identified by Geller and Huchra (1983) have been examined, taking into account contamination and incompleteness in luminosity. Different weighting procedures do not significantly affect the mass–to–light ratio whose median turns out to be about $400\ M_\odot/L_\odot$ ($H_0 = 100\ \mathrm{km\ s^{-1} Mpc^{-1}}$). On the other hand, virial radii are affected by weighting; this fact, together with the behaviour of the mean pairwise separation among spirals and ellipticals and/or lenticulars, gives evidence of morphological segregation.

160.019 Taxonomical analysis of superclusters – I. The Hercules and Perseus superclusters.
M. Moles, A. del Olmo, J. Perea.
Mon. Not. R. Astron. Soc., Vol. 213, No. 2, p. 365 – 380 (1985).

The structure of the Hercules and Perseus superclusters has been analysed by the non–hierarchical–descendent taxonomical method. In the case of Hercules, well–defined groups are present. The membership of the galaxies with respect to the groups is given, as well as the physical parameters for the groups. A mor-

phological type–redshift effect is present. For Perseus, the internal structure is much less well defined, as it appears as a filamentary aggregate without a clear subgrouping. Tentatively some apparent groups in it are noted and its physical parameters calculated.

160.020 Studies of the Virgo cluster. IV. An atlas of Virgo cluster spiral galaxies: the luminosity range within a given spiral type.
A. Sandage, B. Binggeli, G. A. Tammann.
Astron. J., Vol. 90, No. 3, p. 395 – 404, 536 – 554 (1985).

Photographs enlarged to a common scale are given for 131 spirals of types Sab to Sdm, together with a few Sm III's that form the link of this paper to Paper III. The illustrated galaxies occur within a $\sim 6°$ radius of the Virgo cluster center and have been assigned probable membership in the Virgo cluster (Paper II). Variations of angular size and B_T magnitude among galaxies of a given Hubble type and van den Bergh luminosity class span the range of a factor of ~ 5 in diameter and ~ 3.5 mag in B_T. The luminosity functions of individual types and classes are broad, but cut off at both the bright and faint limits, confirming earlier suggestions that no dwarf spirals exist. The present, nearly complete, distance–limited sample of cluster spirals is used to show that the Λ index of de Vaucouleurs for these Virgo cluster galaxies has an intrinsic dispersion of $\sigma(M) \cong 1.0$ mag at a given Λ value.

160.021 Optical studies of X–ray clusters of galaxies. IV. Velocity dispersions for the cD clusters A496 and A2052.
H. Quintana, J. Melnick, L. Infante, B. Thomas.
Astron. J., Vol. 90, No. 3, 410 – 413, 555 – 556 (1985).

The authors present spectroscopic observations carried out at CTIO and La Silla for these two regular, elliptically dominated, rich Abell clusters with central cD galaxies. They derive velocity dispersions of $657 (+104, -72)\ \mathrm{kms^{-1}}$ for A496 and $576 (+83, -60)\ \mathrm{kms^{-1}}$ for A2052. They briefly discuss some of the consequences of these values for the proposed interpretations of the X–ray characteristics of both clusters.

160.022 Clusters of galaxies from the Shane–Wirtanen counts.
S. A. Shectman.
Astrophys. J., Suppl. Ser., Vol. 57, No. 1, p. 77 – 90 (1985).

A sample of 646 clusters of galaxies has been selected from the Shane–Wirtanen counts in 10' bins by finding local density maxima above a threshold value, after lightly smoothing the data to reduce the effect of the sampling grid. The procedure finds 70% of Abell clusters at distance class 4, and 10% at distance class 5, but only 40% of the Shane–Wirtanen clusters are members of the Abell catalog. A sample of 97 redshifts of Shane–Wirtanen clusters exhibits the same distribution as Abell distance class $D \leqslant 4$ clusters. The Shane–Wirtanen clusters have a higher space density than Abell clusters and may provide a clearer picture of the spatial distribution of clusters.

160.023 The probability generating function for galaxy clustering.
R. Schaeffer.
Astron. Astrophys., Vol. 144, No. 1, p. L1 – L4 (1985).

The generating function for the galaxy distribution probabilities is obtained in a closed form with the help of simple and rather natural assumptions. It allows to calculate a large number of statistical features relative to galaxies: N–body clustering properties, probability of voids, correlations among the large clusters. A few first applications are presented.

160.024 Mass segregation and the rank correlation of luminosity with projected distance in clusters of galaxies.
H. Smith Jr.
Astrophys. J., Vol. 288, No. 1, p. 117 – 121 (1985). = Dep. Astron., Univ. Florida, Contrib. No. 70.

Simulated data from 64–body numerical experiments have been used to test the suitability of the rank correlation coefficient of luminosity with projected distance as a measure of mass segregation. In situations in which the individual masses or mass–to–

light ratios are known, this coefficient is a relatively sensitive indicator of mass segregation. However, in the more realistic situation in which there is an appreciable scatter of the individual values around the mean mass–luminosity relation, it loses much of its sensitivity. For this reason the absence of significant luminosity segregation is not necessarily a strong argument against a cluster's mass being contained entirely in its member galaxies.

160.025   Discovery of a large intergalactic H I cloud in the M96 group.
S. E. Schneider, G. Helou, E. E. Salpeter, Y. Terzian.
News Lett. Astron. Soc. N.Y., Vol. 2, No. 4, p. 16 (1983). Abstract. – See Abstr. 010.242.

160.026   Verre clusters van sterrenstelsels: het zicht is 0.7 miljard lichtjaar.
R. Tiller.
Zenit, 12. Jaarg., No. 3, p. 101 – 103 (1985).

160.027   Iron abundance in galaxy clusters.
R. Rothenflug, M. Arnaud.
Astron. Astrophys., Vol. 144, No. 2, p. 431 – 442 (1985).
Up to now, there are 22 clusters of galaxies for which reliable measures of both gas temperature and iron line equivalent width are available. The authors derive the variation of the iron line equivalent width with the temperature. The cluster iron abundance can be deduced from the comparison between the measurements and the computations. Its value appears to be universal, 0.53 ± 0.03 (1σ) solar value. Gas masses in clusters, and thus iron masses are shown to vary as $N_0^{2.4 \pm 0.6}$, where N_0 is the central number of galaxies. It is confirmed that gas masses are proportional to virial masses, with $M_{\text{virial}} = 10\, M_{\text{gas}} = 1.5 \times 10^4 M_{\text{iron}}$. Implications of such relations in terms of observable gas quantities, galaxy formation and iron origin are briefly discussed.

160.028   The anisotropy of the spatial orientations of galaxies in the Local Supercluster.
H. T. MacGillivray, R. J. Dodd.
Astron. Astrophys., Vol. 145, No. 1, p. 269 – 274 (1985).
The authors find a tendency for galaxies of type Sab–Sc to be less concentrated towards the Local Supercluster plane than spirals of other types, to preferentially show the non–random alignment effect and to show a higher preponderance of winding direction "s" over "reversed s". Furthermore, among these intermediate–type spirals there is evidence for a dependence of spiral winding direction on Supergalactic hemisphere. The results are discussed in the context of theories for the formation of galaxies.

160.029   Statistics of emission–line galaxies in rich clusters.
A. Dressler, I. B. Thompson, S. A. Shectman.
Astrophys. J., Vol. 288, No. 2, p. 481 – 486 (1985).
The authors tabulate and analyze the emission–line characteristics of a sample of 1095 galaxies in rich clusters and 173 field galaxies. It is found that the emission–line frequency is much higher in field galaxies than in cluster galaxies. Relatively strong emission is found in the central regions of 31% of the field galaxies, but in only 7% of the cluster galaxies. Active galactic nuclei occur at a frequency of $\sim 5\%$ in the field sample but only $\sim 1\%$ in the cluster sample, the same ratio as found for the emission–line galaxies. It is concluded that much of this difference must be attributed to environmental influences.

160.030   The intergalactic H I cloud in Leo: a simple modeling of the Spitzer Baade collision event.
H. J. Rood, B. A. Williams.
Astrophys. J., Vol. 288, No. 2, p. 535 – 550 (1985).
Results of a controlled experiment suggest that the recently discovered intergalactic cloud of neutral hydrogen (H I) in Leo is the product of a Spitzer-Baade (gas–sweeping) collision most probably between the large spiral galaxy NGC 3368 and the spiral predecessor of the SB0 galaxy NGC 3384. NGC 3368 and NGC 3384 are members of the group Materne GWa and lie at equal angular distances on either side of the cloud center. Traces of two optical rings in NGC 3368 and an arc or spiral arm in

NGC 3384 have been observed. These are precisely the features found in computer–simulated collisions by Theys and Spiegel and by A. Toomre. Observational data have guided the construction of a simple physical model for the gas–sweeping collision. The physical processes governing the subsequent evolution of the intergalactic cloud are also examined.

160.031   Neutral hydrogen in the M96 group: evidence for a giant intergalactic ring.
S. E. Schneider.
Astrophys. J., Lett. Ed., Vol. 288, No. 2, p. L33 – L35 (1985). With plate L1.
Further Arecibo observations of the region around the intergalactic cloud in Leo have uncovered several smaller isolated clouds. Altogether, the H I lies primarily along a 200 kpc diameter ring around the galaxies M105 (E0) and NGC 3384 (SB0). The radial velocities around the ring follow a Keplerian orbit pattern with a period of 4×10^9yr.

160.032   Relaxation and tidal stripping in rich clusters of galaxies. III. Growth of a massive central galaxy.
D. Merritt.
Astrophys. J., Vol. 289, No. 1, p. 18 – 32 (1985).
The rate at which a massive galaxy ("cannibal") grows by capturing other galaxies at the center of a rich, relaxed cluster is calculated. It is shown that the orbital decay preceding capture tends to leave the distribution of orbital velocities isotropic. As a result, most captures occur from nearly radial orbits, and relatively few from circular orbits. The capture rate is initially very low, due to the paucity of low–velocity galaxies, and to the fact that orbital decay times are comparable to a Hubble time. Encounters between galaxies further inhibit their orbital decay. It is suggested that most cD galaxies formed relatively rapidly, during the collapse and virialization of compact groups or poor clusters, and not during the quieter postcollapse stages as previous authors have advocated.

160.033   Dispersion of electromagnetic waves by the hot intergalactic plasma.
X. Barcons, R. Lapiedra.
Astrophys. J., Vol. 289, No. 1, p. 33 – 36 (1985).
The authors consider some features of the (hypothetical) hot intergalactic plasma, in the framework of the Friedmann–Robertson–Walker universe. First, they obtain a relation between redshift and temperature of this gas which differs strongly from the nonrelativistic model. Then they reformulate the problem of the time delay of the electromagnetic waves traveling in the hot intergalactic plasma. The differences between relativistic and nonrelativistic calculations are emphasized.

160.034   Der Lokale Superhaufen und seine Umgebung. Zur Struktur des nahen Universums.
J. Materne.
Sterne Weltraum, 24. Jahrg., Nr. 4, p. 194 – 201 (1985).

160.035   Origin of redshift differentials in galaxy groups.
G. G. Byrd, M. J. Valtonen.
Astrophys. J., Vol. 289, No. 2, p. 535 – 539 (1985).
Sulentic has found a statistically significant excess number of higher redshift companions relative to the brightest galaxy in spiral dominated groups in Huchra and Geller's catalog of galaxy groups. Of the explanations discussed, he favors non–Doppler redshifts in the spiral groups. Instead, the authors propose that the excess results from most of the group population being composed of unbound expanding members and from a mistaken tendency to pick a galaxy of brighter apparent magnitude from the nearer portion of the expanding group population rather than the true primary, which would generally be in the middle of the population. Using the known galaxy number versus luminosity distribution, and the assumption that most spiral group members are unbound, the expected redshift excess for the spiral dominated groups is calculated.

160.036 H I properties of dwarf irregular galaxies in the Virgo Cluster.
G. L. Hoffman, G. Helou, E. E. Salpeter, A. Sandage.
Astrophys. J., Lett. Ed., Vol. 289, No. 2, p. L15 – L18 (1985).

A neutral hydrogen survey was carried out at Arecibo on 91 dwarf irregular galaxies, with morphological types Sdm through Im, in and around the direction of the Virgo Cluster. Only nine of these were found to be background galaxies, and 19 remain undetected, i.e., most of the candidate galaxies are indeed members of the Virgo Cluster or Supercluster. The distribution of positions and systemic velocities (compared with large spirals) shows no evidence for mass segregation. Statistics on H I masses agree partially with a simple stochastic star formation model.

160.037 The velocity field of the rich Abell cluster A1146.
J. Melnick, H. Quintana.
Astron. J., Vol. 90, No. 4, p. 575 – 576, 688 (1985).

The authors report radial–velocity measurements for 33 galaxies in the very rich cluster A1146 that complement and improve previously reported values. They derive a projected velocity dispersion of 1040 km s^{-1}, which is considerably lower than a previous estimate. This new value is consistent with the X–ray luminosity of A1146.

160.038 A new application of Binggeli's test for large–scale alignment of clusters of galaxies.
M. F. Struble, P. J. E. Peebles.
Astron. J., Vol. 90, No. 4, p. 582 – 589 (1985).

The authors have repeated and extended Binggeli's test for possible large–scale alignment of clusters of galaxies. They find marginally statistically significant evidence that clusters and their first–ranked galaxies tend to point toward neighbors, which is in the direction of Binggeli's result, but the effect is small and at a level that might be influenced by systematic errors. The authors find no pronounced correlation of anisotropy with cluster separation. The authors also looked for a tendency of clusters or their dominant galaxies to be parallel, independent for their relative positions. The results are consistent with isotropy. As an upper bound on a real tendency for clusters or their dominant members to point toward nearby clusters, the authors conclude that the mean value of the relative position angle can not differ from isotropy (45°) by more than 5°.

160.039 First ranked galaxies in groups and clusters.
S. P. Bhavsar, J. D. Barrow.
Mon. Not. R. Astron. Soc., Vol. 213, No. 4, p. 857 – 869 (1985).

The small scatter in the luminosities of the brightest galaxies in clusters has been a topic of much debate. It has been argued that these galaxies are either special objects or the tail–end of a statistical distribution. In 1928, Fisher and Tippett derived the general form a distribution of extreme sample values should take, independent of the parent distribution from which they are drawn. The authors compare this asymptotic form with the distribution of first ranked cluster members and conclude that these galaxies are not the extreme members of a statistical population. On the other hand, comparison of first ranked members of 'loose' groups with the extreme value distributions shows that these galaxies are consistent with their being the tail–end of a statistical distribution.

160.040 CCD observations of a rich distant cluster of galaxies (1046 + 35) containing a classical double radio source.
S. A. Eales.
Mon. Not. R. Astron. Soc., Vol. 214, No. 1, p. 27 – 32 (1985).

CCD imaging observations are presented of a newly discovered cluster of galaxies at RA 10^{h}46^{m}03^s, Dec 35°47'28" (1950.0). The cluster is of Richness Class I, has an estimated redshift of 0.35, and probably contains a classical double radio source.

160.041 Reheating of the intergalactic medium at z > 10.
H. M. P. Couchman.
Mon. Not. R. Astron. Soc., Vol. 214, No. 1, p. 137 – 159 (1985).

The reheating and reionization of the intergalactic medium by photoionizing radiation is considered. It is shown that in a low gas density universe a modest input, even at $z \cong 100$, can reionize the gas. The temperature of the intergalactic medium can be significantly affected even if the gas subsequently recombines. Over the range of redshifts considered, $0 \leqslant z \leqslant 100$, the difference is small for gas recombining at $z \sim 100$, becoming much larger for small redshifts. The molecular hydrogen abundance after recombination is also calculated. Results are presented for an $\Omega = 1$ universe with $\Omega_g = 0.05$, $n_{He}/n_H = 1/12$ and $h_0 = 1/2$.

160.042 The extragalactic H I cloud in Leo.
M. J. Pierce, R. B. Tully.
Astron. J., Vol. 90, No. 3, 450 – 453 (1985).

Wide–field CCD images are used to provide a surface–brightness limit to the B passband flux of $\mu_B = 30.2$ mag arcsec^{-2} for the extragalactic H I cloud in Leo. Corresponding upper limits of $M(H\,I)/L_B > 10\,M_\odot/L_\odot$ and $M(\text{virial})/L_B > 700\,M_\odot/L_\odot$ are obtained. These limits are discussed in terms of the two general hypotheses: is the cloud a stable or transient configuration? It seems unlikely that such an extended cloud could endure for a Hubble time in the tidal field associated with the group.

160.043 Are galaxies more strongly correlated than clusters?
A. S. Szalay, D. N. Schramm.
Nature, Vol. 314, No. 6013, p. 718 – 719 (1985).

One of the most powerful tools used in attempts to understand the structure of the universe is the correlation function $\xi(r)$, the excess probability over random that there are two objects separated by a distance r. In particular, distributions of galaxies and of clusters of galaxies have been investigated using this parameter. The authors show that if the amplitudes of the cluster–cluster correlation function is made dimensionless, systematic changes with cluster richness vanish, implying scale invariance in the clustering process. The dimensionless galaxy–galaxy correlation seems stronger, implying gravitational enhancement on smaller scales.

160.044 Energy content of non–isothermal X–ray plasma in clusters of galaxies.
D. Gerbal, G. Mathez, J.–L. Monin, E. Salvador–Sole.
Astron. Astrophys., Vol. 146, No. 1, p. 119 – 122 (1985).

The analysis of the X–ray emission of the Intra Cluster Medium (ICM) by hydrostatic–isothermal models leads to the so–called scale height problem (Jones and Forman, 1984): the mean energy per unit mass is higher in the ICM than in the galaxies. In this paper the authors investigate what happens when the isothermal hypothesis is abandoned.

160.045 Dynamics of clusters of galaxies.
D. Carter, P. F. Teague.
Proc. Astron. Soc. Aust., Vol. 5, No. 4, p. 502 – 505 (1984).

Clusters of galaxies appear to contain much more mass than just that of the galaxies which provide most of their optical luminosity. There are two important ways to investigate the structure of their gravitational potential wells, firstly by an analysis of the radial velocities of the cluster galaxies, and secondly by a study of the X–ray emission from the hot gas which forms part of the intra–cluster medium. The authors present new samples of radial velocity data for three clusters, and discuss some simple types of models of the X–ray sources.

160.046 Clusters and superclusters.
J. H. Oort.
The Big Bang and Georges Lemaitre, p. 299 – 314 (1984). – See Abstr. 012.043.

This paper summarizes presently available evidence for the existence of galaxy superclusters in the observable universe.

160.047 Strong evidence for metagalactic shock waves at redshifts $z \sim 2-3$.
L. M. Ozernoy (*L. M. Ozernoj*), V. V. Chernomordik.
The Big Bang and Georges Lemaitre, p. 327 (1984). Abstract. – See Abstr. 012.043.

160.048   An investigation of dwarf galaxies in the Virgo cluster.
G. D. Bothun, J. R. Mould, A. Wirth, N. Caldwell.
Astron. J., Vol. 90, No. 5, p. 697 – 707 (1985). With plate 59.

The authors have obtained 21–cm H I observations of a sample of 32 dwarf irregular (dI) and 12 dwarf elliptical (dE) galaxies that are located in the Virgo cluster. Altogether, 18 of 32 dIs were detected in H I, but none of the dEs were detected at a sensitivity level of $M_{HI} = 2 - 3 \times 10^6 M_\odot$. The detected dIs have $M_{HI} > 3 \times 10^7 M_\odot$. This disparity in H I content between dIs and dEs effectively dispels the possibility that the dEs are presently in a stage of quiescence (hibernation), between bursts of star formation. In order to supplement the 21–cm data, the authors have acquired optical spectroscopy, CCD images, and infrared photometry for a limited subsample of these dwarfs. They discuss the general H I, infrared, and dynamical properties of the dwarfs and show how these properties relate to the hypothesis previously outlined, namely that dEs may be stripped dIs.

160.049   Plasma in clusters of galaxies.
A. Michalec.
ESA Spec. Publ., ESA SP–207, p. 223 – 224 (1984). – See Abstr. 012.044.

Einstein observations used by Quintana and Melnick seem to support the power–law relation between the X–ray luminosity and the velocity dispersion proposed by Solinger and Tucker for rich clusters of galaxies. Quintana and Melnick investigated a sample of 31 clusters without any corrections for the redshifts. The author inspected the redshifts of individual clusters in their sample and found that the power law relation is strongly altered when divided into two subgroups with different redshifts. The author concludes that, due to the small samples statistics, and the selection effects it is not possible to draw definite conclusions as to the nature of physical processes in the intergalactic plasma from the "observed" index of the Solinger and Tucker relation.

160.050   MKW 10: a group of galaxies with a compact core.
B. A. Williams.
Astrophys. J., Vol. 290, No. 2, p. 462 – 476 (1985). With plates 5 – 9.

MKW 10 is a poor cluster of galaxies containing nine galaxies with $m_{pg} \leqslant 15.5$, five of which appear to form a compact subsystem. In this paper, H I line observations of the galaxies in MKW 10 are presented along with observations of spiral galaxies that lie within 1°5 of the group. The physical compactness of the subsystem is analyzed by examining the observable effects of tidal interactions and collisions on the optical and H I properties of the spiral galaxies in the group. The dynamical stability of MKW 10 and of a larger ($\sim 3°$ in diameter) feature, which includes this group, is investigated. The time scale of dynamical friction and the expected collision rate are estimated and used to predict the degree of dynamical evolution that may have occurred in the group.

160.051   X–ray observations of possible binary clusters of galaxies.
M. P. Ulmer, R. G. Cruddace, M. P. Kowalski.
Astrophys. J., Vol. 290, No. 2, p. 551 – 556 (1985).

The authors have made X–ray measurements of six pairs of clusters of galaxies and have found two new systems in which the two clusters may have a physical association. One system in particular, A2204/A2210, deserves further study, in addition to the more well known system A399/A401.

160.052   Superclusters, cellular structures and pairs of galaxies.
B. I. Fesenko.
Gor'k. gos. ped. inst. Gor'kij, 17 pp. (1984). In Russian. Abstr. in Ref. Zh., 51. Astron., 3.51.888 (1985).

160.053   Statistical and dynamical properties of galaxy groups.
M. Mezzetti.
Mem. Soc. Astron. Ital., Vol. 55, No. 3, p. 615 – 618 (1984). – See Abstr. 012.050.

Loose galaxy groups contain a substantial fraction of all galaxies. Their study may give information on the density of the universe, on the origin and development of its structures and on the interactions between galaxies and their environment. From the analysis of the dynamical, morphological and photometric properties of groups it turns out that most of them are not yet relaxed, their properties show continuity with those of clusters, cannibalism is not common and $\Omega \sim 0.2$.

160.054   A catalog of radio, optical, and infrared observations of spiral galaxies in clusters.
G. D. Bothun, M. Aaronson, B. Schommer, J. Mould, J. Huchra, W. T. Sullivan III.
Astrophys. J., Suppl. Ser., Vol. 57, No. 3, p. 423 – 472 (1985).

The authors present the results of a major observational program on the luminosities, colors, and gas contents of spiral galaxies in clusters of galaxies. These data have been used as part of a detailed investigation into the nature of cluster spirals and for revisions of the distance scale using the infrared Tully–Fisher relation. The observational strategies, reduction procedures, and sources of error are briefly discussed. The data include 21 cm H I observations, $UBVR$ multiaperture photometry, and H–band photometry of serveral hundred spiral galaxies in 10 clusters.

160.055   Correlations of optical characteristics of Abell clusters of galaxies.
M. Kalinkov, K. Stavrev, I. Kyneva.
Pis'ma Astron. Zh., Tom 11, No. 4, p. 243 – 250 (1985). In Russian. English translation in Sov. Astron. Lett., Vol. 11.

A sample of 420 Abell clusters with measured redshifts has been compiled. The correlations and linear regressions between lg z values, magnitudes of first– and tenth–rank galaxies, cluster richness, BM type, and apparent areas occupied by cluster and first–rank galaxies have been determined.

160.056   About a self–consistent model of rich clusters of galaxies. I. Galaxy component of the cluster.
M. V. Konyukov.
Astrofizika, Tom 22, Vyp. 2, p. 273 – 285 (1985). In Russian. English translation in Astrophysics, Vol. 22, No. 2.

It has been shown that the distribution function of the galaxy component of a cluster is determined by the solution of a boundary problem for the gravitational potential of a self–consistent field. The distribution function is determined by two main parameters. An algorithm of the solution is built up and used to obtain the region of parameters for which a solution exists. This way is extended to a cluster of galaxies with a central body (for example cD–galaxies). A process is suggested to estimate the model parameters using optical observations of clusters of galaxies.

160.057   Dynamics of luminous galaxies. II. Surface photometry and velocity dispersions of brightest cluster members.
E. M. Malumuth, R. P. Kirshner.
Astrophys. J., Vol. 291, No. 1, p. 8 – 31 (1985).

Velocity dispersions for 46 galaxies and CCD surface photometry for 27 galaxies have been obtain using the McGraw–Hill Observatory 1.3 m and the KPNO No. 1 0.9 m telescopes. The results provide a greatly improved set of data for investigating the brightest galaxies in galaxy clusters. It is found that the brightest cluster members (BCMs) are substantially brighter than predicted from their velocity dispersions and the $L \propto \sigma^4$ relation for elliptical galaxies. The data for the BCMs are discussed in terms of the merger hypothesis for cD formation.

160.058   The nature of orbits of multiple nuclei near brightest cluster galaxies.
J. L. Tonry.
Astrophys. J., Vol. 291, No. 1, p. 45 – 51 (1985).

Multiple nuclei in brightest cluster galaxies have two traits that are difficult to understand: (1) They are often moving very rapidly with respect to the central galaxy, and (2) their frequency suggests an immense rate of cannibalism if they are bound in circular orbits. The properties and evolution of orbits in a model rich cluster of galaxies are examined in order to determine the implications and consequences of these traits. Both of these observations can be explained if more than 50% of multiple nuclei are on eccentric orbits within the core of the cluster, rising a few hundred kpc from the center. A consequence of this is that the density distribution of galaxies in a cluster will have a cusp in the center.

160.059   Analysis of groups of galaxies with accurate redshifts.
H. Arp, J. W. Sulentic.
Astrophys. J., Vol. 291, No. 1, p. 88 – 111 (1985).

Redshift measures at 21 cm wavelength have been made on over 100 galaxies in more than 40 different groups with the Arecibo radio telescope. These groups generally consist of a large spiral galaxy with one or more companions. This list of galaxies is supplemented with over 160 galaxies in more than 40 groups with a dominant galaxy that is brighter than 11.8 mag. It is shown that the companion galaxies in these groups have significantly higher redshifts than the brightest galaxy in the group. The most accurate redshifts available in the Local Group (M31) and in the M81 group are analyzed. It is shown that all 21 of the best–known physical companions have significantly higher redshifts than the central galaxies. It is suggested that these companions contain a component of nonvelocity redshift. A detailed analysis of the tendency toward quantization in the differential redshift between the dominant galaxy and its companions is also presented.

160.060   Il quartetto di Stephan.
R. Monella.
Orione, Vol. 5, N. 2, p. 17 – 20 (1985).

160.061   The infrared color–magnitude and color–gas content relations for cluster spirals.
G. D. Bothun, J. Mould, R. A. Schommer, M. Aaronson.
Astrophys. J., Vol. 291, No. 2, p. 586 – 594 (1985).

Using multiaperture optical and infrared magnitudes of 164 spirals in nearby galaxy clusters, the authors explore the utility of the spiral color–magnitude relation as an accurate distance indicator. In combination with H I line widths, they find there to be a well–defined relationship between the observed $(B–H)_{-0.5}$ color of spirals and their observed 20% line width. However, it is shown that the spiral color–magnitude relation may be impossible to calibrate due to the lack of a well–defined zero point. Therefore the $B–H, H$ relation cannot be considered as an accurate distance indicator. Finally, the authors compare the H I content of these cluster spirals to the Local Supercluster sample. They can find no evidence that any of the cluster spirals that have been used in establishing the IR Tully–Fisher distance scale are H I deficient with respect to field spirals.

160.062   X–ray emission possibly coincident with the radio tail of PKS 0301–123.
J. O. Burns, E. R. Nelson, R. A. White, S. A. Gregory.
Astrophys. J., Vol. 291, No. 2, p. 611 – 620 (1985). With plate 8.

X–ray, radio, and optical observations are reported for a very poor cluster of galaxies containing the radio source PKS 0301–123. A long X–ray observation using the Einstein Observatory resulted in the serendipitous discovery of a peculiar extended X–ray morphology associated with the poor cluster. The X–ray emission is extended in the same direction and is approximately the same length as the radio–tailed source, PKS 0301–123, which has been mapped at 6 and 20 cm with the VLA. Optical redshift observations confirm the identification of the X–ray and radio emission with the poor cluster at $z = 0.1$. Possible radiation mechanisms responsible for both the X–ray and radio emission are considered.

160.063   An X–ray study of the Centaurus cluster of galaxies using Einstein.
T. Matilsky, C. Jones, W. Forman.
Astrophys. J., Vol. 291, No. 2, p. 621 – 626 (1985).

Einstein IPC observations of the core of the Centaurus cluster of galaxies have been analyzed to map the 0.5 – 3.5 keV surface brightness and temperature of the intracluster gas. The emission is centered on NGC 4696, the elliptical galaxy believed to be at or near the dynamical center of the cluster. The observations are used to measure the total mass distribution around the central region. It is found that surrounding NGC 4696, like M87 at the

center of the Virgo Cluster, is a dark, massive halo, with a gravitating mass of $\sim 2 \times 10^{13} M_\odot$ out to a radius of about 20'. The elliptical galaxy NGC 4709, at the core of a more distant cluster, is also detected with a luminosity of 2×10^{40}ergs s^{-1}.

160.064   The structure of the Virgo cluster of galaxies.
K. I. Tanaka.
Publ. Astron. Soc. Jpn., Vol. 37, No. 1, p. 133 – 154 (1985).

A new detailed analysis of galaxies in the Virgo area shows that the conventional Virgo cluster comprises two different clouds of galaxies: the Virgo cluster I and the Southern cloud II by our definition. The Virgo cluster I centered at 12^h27^m6, $+13°07'$ (1950), has a mean radial velocity of $\langle V_0 \rangle \cong +980 \pm 60$ km s^{-1} and there is no significant velocity difference between elliptical–lenticular and spiral–irregular galaxies in the Virgo cluster I. The Southern cloud II is centered at about 12^h25^m, $+7°30'$ (1950) and all types of galaxies in the cloud are diffusely distributed, while elliptical galaxies in the Virgo cluster I are concentrated to the center. The mean radial velocity within a radius of $3°5$ of the Southern cloud II is $\langle V_0 \rangle \cong +1240 \pm 80$ km s^{-1}, which is significantly higher than the mean velocity of the Virgo cluster I.

160.065   The H I content of spirals in galaxy groups.
G. Giuricin, F. Mardirossian, M. Mezzetti.
Astron. Astrophys., Vol. 146, No. 2, p. 317 – 324 (1985).

The authors have analyzed the H I content of ~ 230 spiral and irregular members of Geller und Huchra's (1983) galaxy groups. In contrast with some earlier findings it has been found that the H I content – as expressed by two distance–independent parameters M_H/L (H I mass–to–luminosity ratio) and σ_H (mean H I surface density) – depends neither on the distance from the respective group center nor on the ratio between this distance and the mean pairwise separation of all group members. Furthermore, no variation of the H I properties as a function of the group compactness has been evidenced. A discussion of general conditions for ram pressure and turbulent viscous stripping allows to fix an upper limit of $\sim 10^{-4}$ particles per cm^3 for the average number density of the intragroup medium.

160.066   Gas and dust emission in galaxy clusters in the radio wave region.
V. K. Khersonskij, N. V. Voshchinnikov.
Astrofizika, Tom 21, Vyp. 3, p. 461 – 474 (1984). In Russian. English translation in Astrophysics, Vol. 21, No. 3.

The consequences of the hypothesis on the presence of dust grains in the intergalactic medium of galaxy clusters are considered. Gas and dust emission in the range $\lambda = 0.04 - 50$ cm in the Coma and Perseus galaxy clusters are calculated. It is shown that the dust emission cannot compensate the Sunyaev–Zel'dovich effect in the centimeter wavelength region. In the submillimeter wavelength region the galaxy clusters may be bright sources of emission.

160.067   Exosat observations of the Perseus cluster.
G. Branduardi–Raymont, B. Kellett, A. C. Fabian, T. McGlynn, G. Manzo, A. Peacock.
Adv. Space Res., Vol. 5, No. 3, p. 133 – 136 (1985). – See Abstr. 012.059.

Exosat CMA and GSPC observations of the Perseus cluster of galaxies are reported. The soft X–ray image data confirm the earlier Einstein Observatory results in terms of the density distribution radially outward from NGC 1275. The authors have used the GSPC spectral data at higher energies to study the strength of the X–ray continuum and of the Fe line emission over the cluster. An asymmetry is found in the distribution of the hot gas around NGC 1275, which is consistent with the E–W direction of the ellipticity discovered at soft X–rays by the Einstein Observatory.

160.068   Nearby superclusters of rich clusters of galaxies.
M. Kalinkov, K. Stavrev, I. Kaneva.
Astron. Tsirk., No. 1340, p. 4 – 7 (1984). In Russian.

160.069 Programme "Northern Cone". I. The results of redshift measurements for 36 distant rich clusters of galaxies with the 6–m telescope.
A. I. Kopylov, T. S. Fetisova, V. F. Shvartsman.
Astron. Tsirk., No. 1344, p. 1 – 4 (1984). In Russian.

This paper contains the results of redshift measurements of the very rich compact clusters of galaxies which are inside the "Northern Cone" (galactic latitude $b^{II} \geqslant 60°$) and which have indirect estimates of z according to Leire and van den Bergh as $0.10 \leqslant z_{LB} \leqslant 0.25$.

160.070 Programme "Northern Cone". II. Spatial correlation function of rich compact clusters of galaxies indside the Cone.
A. I. Kopylov, D. Yu. Kuznetsov, T. S. Fetisova, V. F. Shvartsman.
Astron. Tsirk., No. 1347, p. 1 – 5 (1984). In Russian.

160.071 Rich galaxy clusters may result from large–scale motions inside superclusters.
S. F. Shandarin, A. A. Klypin.
Sov. Astron., Vol. 28, No. 5, p. 491 – 495 (1984). English translation of 38.160.030.

160.072 Far–ultraviolet background observations at high galactic latitude. I. The Coma Cluster.
J. B. Holberg, H. B. Barber.
Astrophys. J., Vol. 292, No. 1, p. 16 – 21 (1985).

A series of high–galactic latitude observations made with the Voyager 2 ultraviolet spectrometer are used to place stringent upper limits on the far–ultraviolet emissions from the central regions of the Coma Cluster of galaxies. In particular, consideration of the $912 – 1150$ Å region yields lower limits upon the radiative decay lifetimes of any neutrino species with masses in the range $22.1 – 27.8$ eV, assuming massless decay products. Limits on the radiative decay lifetimes of such neutrinos are found to vary from 2.4×10^{25} to 7.1×10^{24}s, respectively, as a function of neutrino rest mass.

160.073 Cluster redshifts in five suspected superclusters.
R. Ciardullo, H. Ford, R. Harms.
Space Telesc. Sci. Inst., Prepr. Ser., No. 35, 44 pp. (1985). To appear in Astrophys. J.

160.074 The dynamics of four multiple–nucleus brightest cluster galaxies.
J. G. Hoessel, K. D. Borne, D. P. Schneider.
Space Telesc. Sci. Inst., Prepr. Ser., No. 37, 36 pp. (1985). To be published in Astrophys. J.

160.075 Multi–object spectroscopy of the distant cluster AC 103.
R. M. Sharples, R. S. Ellis, W. J. Couch, P. M. Gray.
Anglo–Aust. Obs., Prepr., No. 200, 36 pp. (1985). Submitted to Mon. Not. R. Astron. Soc.

160.076 Optical studies of X–ray clusters of galaxies. IV. Velocity dispersions for the cD clusters A496 and A2052.
H. Quintana, J. Melnick, L. Infante, B. Thomas.
Astrophys. Prepr. Ser., No. 6, 16 pp. (1984).

The authors present spectroscopic observations carried out at CTIO and La Silla for two regular, elliptically dominated, rich Abell clusters with central cD galaxies. They derive velocity dispersions of $657\,(+104, -72)$ kms^{-1} for A496 and $576\,(+83, -60)$ kms^{-1} for A2052. The authors briefly discuss some of the consequences of these values for the proposed interpretations of the X–ray characteristics of both clusters.

160.077 Gas deficiency in cluster galaxies: a comparison of nine clusters.
R. Giovanelli, M. P. Haynes.
Astrophys. J., Vol. 292, No. 2, p. 404 – 425 (1985).

Using a combined sample of previously published results and new high–sensitivity observations, the neutral hydrogen contents of galaxies in nine clusters have been compared with those of isolated galaxies. Of the nine clusters, six show substantial H I deficiency, to varying degrees. The three clusters whose galaxies contain normal H I masses contain a higher proportion of spirals, and are not X–ray sources. In the clusters which show H I deficiency, the depletion of H I gas, exceeding factors of 10, strongly correlates with the radial distance of a galaxy from the cluster center. The data lead to the conclusion that the occurrence of H I deficiency is correlated with the presence of a hot X–ray intracluster medium, and that while perhaps not responsible for the complete mechanism of morphological segregation in clusters, an ongoing interaction process is active through the cores of X–ray clusters.

160.078 H I mapping of galaxies in the Hercules cluster.
E. E. Salpeter, J. M. Dickey.
Astrophys. J., Vol. 292, No. 2, p. 426 – 440 (1985).

The authors have made an H I survey, with the VLA, of three fields centered on the Hercules cluster of galaxies (A2151). They have detections (or possible detections) for 31 optically known galaxies, of which 13 are optically bright Zwicky galaxies. For 16 of the 31 detected galaxies the center position is different for the emission in different frequency channels, i.e., there is a rotation signature. The core of A2151, although elongated and clumpy, is probably a single dynamic unit, and the dispersion of systemic velocities decreases with increasing distance from the cluster core.

160.079 The physical implications of an isothermal model for the hot intracluster medium.
M. J. Henriksen, R. F. Mushotzky.
Astrophys. J., Vol. 292, No. 2, p. 441 – 446 (1985).

The authors have used X–ray fluxes from *HEAO 1* and *Einstein Observatory* observations of clusters of galaxies to constrain the parameter β in the isothermal surface brightness profile. This parameter is found to have values primarily between 0.50 and 0.75 for 15 clusters. These values are compared with those derived from isothermal model fits to the X–ray images. The paper then addresses the physical constraints imposed on this model by the existing optical data, the implied gas mass, and the gas contribution to the binding cluster mass. It is concluded that the isothermal model is a nonphysical model, although it has proved to be a useful empirical characterization for clusters.

160.080 Galaxy clustering and the method of voids.
A. J. S. Hamilton.
Astrophys. J., Lett. Ed., Vol. 292, No. 2, p. L35 – L39 (1985).

The probability of finding a void in a random volume in a distribution of galaxies is related to the generating function of the sequence of correlation amplitudes. This letter describes methods for measuring this function from volume–limited or apparent–magnitude–limited data and shows how to deconvolve the observed function to infer an underlying positive definite continuum density distribution $f(\varrho)$, and hence a consistent sequence of correlation amplitudes, which are proportional to the irreducible moments of the continuum distribution.

160.081 The evolution of galaxies in clusters. IV. Photometry of 10 low–redshift clusters.
H. R. Butcher, A. Oemler Jr.
Astrophys. J., Suppl. Ser., Vol. 57, No. 4, p. 665 – 691 (1985).

Photographic colors and magnitudes and morphological types are presented for galaxies in the cores of 10 nearby clusters of galaxies. In the typical cluster, the sample includes all galaxies within a radius of 1.5 Mpc (assuming $H_0 = 50$ km s^{-1}Mpc^{-1}) of the cluster center and brighter than $J = 17.5$. The accuracy of the photometry varies with cluster, but most magnitudes are accurate to 0.10 mag, and typical errors in the colors are of the same order. A determination of the density of field galaxies has also been made using existing data, extended with new measurements in two fields toward the galactic polar caps.

160.082 Studies of distant clusters of galaxies.
A. Dressler, J. E. Gunn.
Bull. Am. Astron. Soc., Vol. 16, No. 4, p. 881 (1984). Abstract. - See Abstr. 010.062.

160.083   Statistics of emission–line galaxies in rich clusters.
A. Dressler, I. B. Thompson, S. A. Shectman.
Bull. Am. Astron. Soc., Vol. 16, No. 4, p. 881 (1984). Abstract. – See Abstr. 010.062.

160.084   Survey of nearby Abell clusters for narrow emission–line galaxies: A 347 and A 1367.
C. Moss, M. Whittle.
Bull. Am. Astron. Soc., Vol. 16, No. 4, p. 881 (1984). Abstract. – See Abstr. 010.062.

160.085   The luminosity function of Abell 168.
W. R. Oegerle, J. G. Hoessel, R. M. Ernst.
Bull. Am. Astron. Soc., Vol. 16, No. 4, p. 881 (1984). Abstract. – See Abstr. 010.062.

160.086   X–ray spectra of low luminosity clusters of galaxies.
M. J. Henriksen, R. F. Mushotzky.
Bull. Am. Astron. Soc., Vol. 16, No. 4, p. 882 (1984). Abstract. – See Abstr. 010.062.

160.087   H I deficient galaxies in X–ray clusters.
M. P. Haynes, C. A. Magri, R. Giovanelli.
Bull. Am. Astron. Soc., Vol. 16, No. 4, p. 882 (1984). Abstract. – See Abstr. 010.062.

160.088   Abell 400 and the 3C 75 radio jets.
F. N. Owen, J. A. Eilek, C. P. O'Dea, M. Inoue, R. A. White.
Bull. Am. Astron. Soc., Vol. 16, No. 4, p. 882 (1984). Abstract. – See Abstr. 010.062.

160.089   An extensive ICM in rich clusters.
R. J. Hanisch, M. P. Ulmer.
Bull. Am. Astron. Soc., Vol. 16, No. 4, p. 882 (1984). Abstract. – See Abstr. 010.062.

160.090   The nature of radio sources in compact groups of galaxies.
T. K. Menon, P. Hickson.
Bull. Am. Astron. Soc., Vol. 16, No. 4, p. 961 (1984). Abstract. – See Abstr. 010.062.

160.091   X–ray observations of the mass distribution in the cluster of galaxies A754.
D. Fabricant, P. Gorenstein, T. C. Beers, M. Kurtz, M. Geller.
Bull. Am. Astron. Soc., Vol. 16, No. 4, p. 961 (1984). Abstract. – See Abstr. 010.062.

160.092   Origin of redshift asymmetry in galaxy groups.
G. G. Byrd, M. J. Valtonen.
Bull. Am. Astron. Soc., Vol. 16, No. 4, p. 962 (1984). Abstract. – See Abstr. 010.062.

160.093   Microcomputer simultations of galaxy clusters redshift observations.
R. L. Negroń, J. O. Burns.
Bull. Am. Astron. Soc., Vol. 16, No. 4, p. 962 (1984). Abstract. – See Abstr. 010.062.

160.094   Intergalactic molecule formation at high redshift.
M.-M. Mac Low, J. M. Shull.
Bull. Am. Astron. Soc., Vol. 16, No. 4, p. 962 (1984). Abstract. – See Abstr. 010.062.

160.095   VLA observations of intergalactic H I in the M96 group.
S. E. Schneider, E. E. Salpeter, Y. Terzian.
Bull. Am. Astron. Soc., Vol. 16, No. 4, p. 991 (1984). Abstract. – See Abstr. 010.062.

160.096   VLA observations of 57 sources in clusters of galaxies.
C. P. O'Dea, F. N. Owen.
Astron. J., Vol. 90, No. 6, p. 927 – 953 (1985). With plates 74 – 80.

VLA observations of 57 radio sources in the directions of clusters of galaxies are presented. The data include observations of 41 narrow–angle–tail (NAT) sources, nine wide–angle–tail (WAT) sources and seven sources with complex morphology. Twin–jet structure is found in $\sim 75\%$ of the NATs. Contour plots of the sources are presented. Optical finding charts for 25 previously unidentified sources are given. Brief comments on the properties of the individual radio sources are made.

160.097   The global properties of a representative sample of 51 narrow–angle–tail radio sources in the directions of Abell clusters.
C. P. O'Dea, F. N. Owen.
Astron. J., Vol. 90, No. 6, p. 954 – 972 (1985).

The global and statistical properties of a representative heterogeneous sample of 51 narrow–angle–tail (NAT) sources in Abell clusters, as well as those of a homogeneous subsample of 39 NATs, are investigated. NATs are equally likely to be found in clusters of any Bautz–Morgan type. The observed distribution of projected distances from the cluster centers is consistent with that expected for an isothermal King model with a core radius of 170 kpc ($H_0 = 75$ km s^{-1}Mpc^{-1}), and does not depend on relative optical dominance. The present results are consistent with the hypothesis that the morphology of NATs is due to the galaxy's motion through the intracluster medium but the results do not distinguish between the detailed models.

160.098   A new determination of the peculiar Virgocentric velocity of the Local Group of galaxies.
J. A. de Freitas Pacheco.
Astron. J., Vol. 90, No. 6, p. 1007 – 1011 (1985).

The author presents a new determination of the peculiar velocity of the Local Group of galaxies towards the Virgo cluster center. The study is based on a sample of 67 E galaxies that are members of groups and clusters, with radial velocities in the range $1000 \leqslant V \leqslant 7000$ km s^{-1}. The relation between luminosity and effective radius (half–light) was used as a distance indicator. The derived Virgocentric velocity is $V_p = 200 \pm 50$ km s^{-1}.

160.099   H I clouds in the Sculptor and Local Groups.
H. Arp.
Astron. J., Vol. 90, No. 6, p. 1012 – 1018 (1985).

Previous measures of low–redshift (-400 km s$^{-1} \lesssim z \lesssim +300$ km s^{-1}) hydrogen clouds over the sky are discussed. It is shown that there are H I clouds in the directions of the Sculptor Group galaxies, NGC 55 and NGC 300, which are distinguished from the Magellanic Stream in scale, redshift, and distribution on the sky. These H I clouds have redshift differences of -220 km s$^{-1} \lesssim z \lesssim +100$ km s^{-1} from the Sculptor Group galaxies with which they are associated. The hydrogen clouds in the Local and Sculptor Groups are examined for possible connection with the high–redshift quasars, which were recently suggested to be members of the two groups. Finally, the physical situation of the Local and Sculptor Groups with respect to other nearby galaxies and groups is considered.

160.100   Photometry of groups of galaxies. I.
G. M. Richter, J. Vennik.
Astron. Nachr., Vol. 306, No. 3, p. 107 – 116 (1985).

Photometric data (magnitudes, radii, profiles) for galaxies in the field of 7 nearby groups of galaxies are measured by photographic surface photometry. Most of them are dwarf galaxies.

160.101   Simple redshift calibration of Abell clusters of galaxies.
M. Kalinkov, K. Y. Stavrev, I. F. Kuneva.
Dokl. Bolg. Akad. Nauk, Vol. 37, No. 8, p. 987 – 990 (1984).

160.102 Cluster redshift in five suspected superclusters.
R. Ciardullo, H. Ford, R. Harms.
Astrophys. J., Vol. 293, No. 1, p. 69 – 82 (1985). With plates
1 – 7.

Redshift surveys for rich superclusters were carried out in five
regions of the sky containing surface–density enhancements of
Abell clusters. While several superclusters are identified,
projection effects dominate each field, and no system contains
more than five rich clusters. The system 2206–22, though a region
of exceedingly high Abell cluster surface density, appears to be a
remarkable superposition of 23 rich clusters almost uniformly
distributed in redshift space between $0.08 < z < 0.24$. The new
redshifts significantly increase the three–dimensional information available for the distance class 5 and 6 Abell clusters and
allow an estimate of the spatial correlation function around rich
superclusters.

**160.103 The dynamics of four multiple–nuclei brightest cluster
galaxies.**
J. G. Hoessel, K. D. Borne, D. P. Schneider.
Astrophys. J., Vol. 293, No. 1, p. 94 – 101 (1985). With plates
10 – 14.

Roughly 50% of all first–ranked cluster galaxies are seen with
two or more nuclei within 20 kpc of the luminosity center of the
the system ($H_0 = 50$ km s^{-1}Mpc^{-1}). Spectroscopic measurements and CCD imaging are presented for the nuclear components of four multiple–nucleus systems: those found in Abell
clusters 400, 671, 1185, and 2052. The spectroscopic and imaging
data for the systems in the two clusters A400 and A671 suggest
that their multiplicity is no more than a projection effect. The
shapes of the contours around the nuclear components in A1185
and A2052 are consistent with physical association among those
nuclei, a conjecture supported by the spectroscopic data. This
physical association has probably resulted from the action of
dynamical friction and the subsequent orbital decay of cluster
galaxies whose trajectories brought them close to the first–ranked
galaxy.

**160.104 Percolation as descriptive statistics: methods and
perspectives.**
A. A. Klypin.
Inst. prikl. mat. Akad. Nauk SSSR. Prepr., No. 167, 28 pp.
(1984). In Russian. Abstr. in Ref. Zh., 51. Astron., 6.51.741
(1985).

160.105 Stephans Quintett im 8–Zöller.
M. Costantino.
Sterne Weltraum, 24. Jahrg., Nr. 6, p. 336 (1985).

160.106 Radio continuum emission in the Virgo cluster.
C. Kotanyi.
The Virgo cluster of galaxies, p. 13 – 22 (1985). – See Abstr.
012.069.

No diffuse intergalactic radio emission has been found at a
level tenfold below that of Coma C (Kotanyi, 1981). Many discrete sources not associated with optical galaxies are detected in
synthesis observations. No narrow–angle head–tail sources were
found in the Virgo cluster. M87 may show features found in
wide–angle tail sources. Undersized radio continuum and H I
disks are only seen within a fraction of the core radius of the
Virgo cluster.

160.107 H I observations of Virgo cluster galaxies.
W. K. Huchtmeier.
The Virgo cluster of galaxies, p. 23 – 35 (1985). – See Abstr.
012.069.

H I–observations of 456 galaxies in the Virgo cluster area
within 10° of M87 are discussed and compared with a sample of
nearby galaxies ($v_0 < 500$ km/s). The total H I–mass ($M_{H\,I}$) and
the relative H I–content ($M_{H\,I}/L_B$ and σ_H) for a given morphological type are the same in both data samples with the exception
of the inner 3° of the Virgo cluster where a pronounced

H I–deficiency is observed. Application of the Tully–Fisher relation for deriving differential distances in that field is discussed.

**160.108 The H I deficiency of the S0 galaxies in the Virgo
cluster.**
C. Balkowski, P. Chamaraux, P. Fontanelli.
The Virgo cluster of galaxies, p. 37 – 43 (1985). – See Abstr.
012.069.

The authors analyse the H I contents of a large sample of S0
galaxies measured at Arecibo with a high sensitivity, mainly by
Giovanardi et al. (1983) and located within and outside the Virgo
cluster. As a result, the Virgo S0 galaxies have a deficiency factor
DF > 2.3 and most probably a much larger value. The estimation of the stripping of cluster galaxies by the ram pressure of the
intracluster medium shows that the deficiency of the Virgo cluster
S0 galaxies can be caused by this process.

160.109 H I deficiency in the Virgo cluster.
M. P. Haynes.
The Virgo cluster of galaxies, p. 45 – 50 (1985). – See Abstr.
012.069.

The author summarizes the comparative studies of both the
H I content and H I extents of Virgo and isolated galaxies undertaken by himself, R. Giovanelli and J. N. Hewitt which can now
be combined with other results in order to investigate the pattern
of H I deficiency seen in the Virgo cluster.

**160.110 The H I distribution in spiral galaxies in the Virgo
cluster.**
R. H. Warmels.
The Virgo cluster of galaxies, p. 51 – 60 (1985). – See Abstr.
012.069.

Neutral hydrogen observations with the Westerbork Synthesis
Radio Telescope have revealed the H I distributions in 36 spiral
galaxies in the Virgo cluster area. For these galaxies, with morphological types ranging from Sab to Sdm, the author presents
the face–on corrected H I diameter at the surface density level of
1 M$_\odot$pc^{-2} and the asymmetry of the H I disk at this level. The
results show that, for a given optical diameter, the isophotal H I
sizes of Virgo spirals become smaller with decreasing distance to
the cluster core. In addition, galaxies which are located in the
core have more asymmetric H I disks than galaxies at the periphery.

160.111 Neutral hydrogen observations at the VLA.
J. van Gorkom, C. Kotanyi.
The Virgo cluster of galaxies, p. 61 – 66 (1985). – See Abstr.
012.069.

A survey of the neutral hydrogen line emission from bright
spiral galaxies in the Virgo cluster has been done using the Very
Large Array (VLA). The first results on 10 of the brightest spirals
close to the cluster centre have been presented previously
(van Gorkom et al., 1984). They show that the H I disks of spirals
within a radius of about 3° from M87 are anomalously small,
those outside being normal. The change occurs rather sharply at
that radius, suggesting that the dominant mechanism is ram–
pressure stripping of the interstellar gas by the diffuse halo of
M87. Here the authors present preliminary data for the full sample, which is nearly complete to a limiting optical magnitude.

**160.112 A comparison of the gas deficiency of galaxies in Virgo
and other nearby clusters.**
R. Giovanelli.
The Virgo cluster of galaxies, p. 67 – 70 (1985). – See Abstr.
012.069.

The author reports the summary of a comparative study of the
H I content of galaxies in nine nearby clusters, including Virgo.
The data base consists of 458 galaxies of type Sa or later for
which high quality 21 cm line data are available. The galaxies
belong to the clusters A262, Cancer, A1367, Virgo, A1656,
Z 1400+0949, A2147, A2151 and Pegasus. They were compared
with a sample of 324 isolated galaxies observed by Haynes and
Giovanelli (1984).

160.113   Rotation curves of hydrogen deficient galaxies in Virgo.
  G. Chincarini, R. E. de Souza.
The Virgo cluster of galaxies, p. 71 – 79 (1985). – See Abstr. 012.069.

Galaxies inside clusters are more likely to suffer perturbations due to environmental effects. An important question is whether these environmental effects are sufficiently strong to disturb the mass distribution of cluster galaxies. In that case it is possible, at least in principle, to detect these disturbances by comparing the rotation curves of galaxies in clusters with isolated galaxies. An early answer to this matter was given by Rubin (1982) based on a sample of eight member galaxies of the Cancer and Pegasus I clusters. The authors' data in Virgo agree with her statement, indicating that the rotation curves of cluster galaxies (Virgo and A1060) are similar to those of non–cluster galaxies.

160.114   H I deficiency and photometric evolution of spiral galaxies in the Virgo cluster.
  B. Guiderdoni, B. Rocca–Volmerange.
The Virgo cluster of galaxies, p. 81 – 89 (1985). – See Abstr. 012.069.

From a compilation of H I data of 107 spirals in the Virgo cluster, with their available UBV, far–UV and Hα photometry, the authors examine the spatial distribution of H I deficient galaxies and the correlation of the deficiency with morphological type and colors. The history of star formation is studied by means of a model of photometric evolution, together with a consistent dynamical scenario.

160.115   An H I and optical study of the gas poor Virgo cluster spiral NGC 4571.
  R. C. Kennicutt, J. M. van der Hulst.
The Virgo cluster of galaxies, p. 91 – 94 (1985). – See Abstr. 012.069.

Several studies have indicated that the galaxies in the Virgo cluster core region have low H I contents and low star formation rates compared to galaxies of similar morphological type in the field. Kennicutt (1983) identified a few galaxies with extreme properties: galaxies with morphological type Sc, but colors, H I contents and star formation rates which are more typical for field Sb galaxies. The authors chose one such anemic galaxy, NGC 4571, for a detailed study using the VLA to study the H I and the Mt. Lemmon, Mt. Palomar and Kitt Peak facilities to obtain optical continuum and Hα data.

160.116   Radio continuum and H I emission from the spiral galaxies in the Virgo cluster area.
  G. Gavazzi.
The Virgo cluster of galaxies, p. 95 – 99 (1985). – See Abstr. 012.069.

The author shows that the statistical method used to derive luminosity functions (from radio data to X–ray) can be profitably applied also to H I data. He applies the method to show quantitatively the amount of H I deficiency among the spiral galaxies in the Virgo area without invoking the deficiency parameter. Secondly he presents some evidence that spiral galaxies in Virgo might show a correlation between H I content and radio continuum luminosity.

160.117   H I observations of late–type Virgo dwarfs.
  G. L. Hoffman.
The Virgo cluster of galaxies, p. 101 – 108 (1985). – See Abstr. 012.069.

The Las Campanas optical survey discussed by Sandage and Binggeli includes 242 galaxies of types Sdm and later which are probable members of the Virgo cluster. In collaboration with G. Helou, E. E. Salpeter and A. Sandage, the author is in the midst of an H I detection survey of all these galaxies, and reports here results for the 91 galaxies observed so far at Arecibo.

160.118   IRAS observations of Virgo cluster galaxies.
  T. de Jong.
The Virgo cluster of galaxies, p. 111 – 123 (1985). – See Abstr. 012.069.

The author analyses and discusses infrared data of all 88 galaxies detected by IRAS in the core of the Virgo cluster. Two elliptical galaxies, M84 and M87, are detected by IRAS. Masses, removal times and destruction mechanisms of dust in these ellipticals are briefly discussed. Apart from ellipticals, which are not detected in the field sample, the infrared detection probability of galaxies in the Virgo cluster and in the field is similar. Star formation rates in early–type spirals (Sa – Sbc) are more than twice smaller in the Virgo cluster than in the field.

160.119   An IRAS survey of bright spirals in Virgo.
  G. Helou.
The Virgo cluster of galaxies, p. 125 (1985). Abstract. – See Abstr. 012.069.

160.120   A near–infrared survey of bright Virgo spiral galaxies.
  R. D. Joseph, T. G. Hawarden, I. Gatley.
The Virgo cluster of galaxies, p. 127 – 133 (1985). – See Abstr. 012.069.

The nuclei of many spiral galaxies exhibit large far–IR luminosities. This "IR activity" is commonly interpreted in terms of a recent burst of star formation. One of the recent attempts to address this question was an IR (10 μm) survey by Becklin et al. (1983) of the 60 spiral galaxies in the Virgo cluster. The authors chose to repeat the survey in the JHKL wavebands. Preliminary analysis of their data indicates that IR excesses are common, but not universal and the IR excesses tend to correlate with the stellar density in the nucleus.

160.121   Infrared observations of dwarf elliptical galaxies in the Virgo cluster.
  H. Zinnecker, R. D. Cannon, T. G. Hawarden,
  H. T. MacGillivray.
The Virgo cluster of galaxies, p. 135 – 150 (1985). – See Abstr. 012.069.

The authors have obtained JHK photometry of a sample of 10 "nucleated" dwarf elliptical galaxies (dE, N) in the Virgo cluster. The galaxies have been selected from the list of Bingelli et al. (1984). It was found that the J–H vs. H–K colour–colour diagram of the dE, N sample is distinctly different from the corresponding diagrams of both the central regions of giant elliptical galaxies and metal–poor galactic globular clusters. The data suggest that the stellar content of dE, N galaxies may be similar to that of fairly metal–rich globular clusters and to that of the outermost parts of giant ellipticals. An optical–infrared (B–H) colour for 8 of the 10 dE, N galaxies was also deduced.

160.122   Molecular clouds in spiral galaxies in the Virgo cluster.
  J. S. Young.
The Virgo cluster of galaxies, p. 151 – 164 (1985). – See Abstr. 012.069.

CO observations of the centers of 25 spirals in the Virgo cluster have been made using the 14 meter telescopes of the FCRAO (Young et al., 1985). It is found that the optical luminosity in the central 5 kpc diameter is proportional to the first power of the CO luminosity in the same region. The major axis distributions of CO emission in 8 spiral galaxies in the Virgo cluster have also been mapped (Kenney and Young, 1984). The shapes of the CO distributions with radius in the Virgo cluster and the total CO luminosities agree well with those found in field galaxies, suggesting that the molecular contents of the galaxies in Virgo are not significantly different from field galaxies of the same type and luminosity.

160.123   Molecular clouds in H I–deficient Virgo spirals.
  J. D. Kenney, J. S. Young.
The Virgo cluster of galaxies, p. 165 – 177 (1985). – See Abstr. 012.069.

The authors have mapped ^{12}CO along the major axis of 8 bright spiral galaxies in the Virgo cluster. Comparisons of the CO diameters with the H I and optical diameters, and of the CO line fluxes with the H I line fluxes for a sample of Virgo and non–Virgo galaxies show that the molecular gas contents are roughly normal in Virgo spirals, including the severely H I–deficient galaxies NGC 4569 and NGC 4579. Apparently, some process has removed the low density atomic gas, while leaving the high density molecular gas unscathed. The authors show that this is consistent with ram–pressure stripping of H I, and discuss the fate of molecular clouds in a galaxy which has lost most of its atomic gas.

160.124   The Virgo cluster redshift survey.
    J. P. Huchra.
The Virgo cluster of galaxies, p. 181 – 200 (1985). – See Abstr. 012.069.

The author has collected velocities for 99% of the 471 galaxies brighter than $m_{pg} = 15.5$ and within 6° of the center of the Virgo cluster. There is significant correlated velocity and spatial structure in the core of the cluster. There exist individual clumps of galaxies associated with both M87 (NGC 4486) and M49 (NGC 4472). The cluster luminosity function is derived with background galaxies removed assuming that all objects with $v < 3000$ km s^{-1} are members.

160.125   Global structure of Virgo cluster galaxies.
    S. Okamura.
The Virgo cluster of galaxies, p. 201 – 215 (1985). – See Abstr. 012.069.

The author summarizes the results of photometric analyses applied to the two samples of Virgo cluster galaxies, sample of 213 bright galaxies and sample of 64 faint galaxies. Emphasis is put on a proposed quantitative classification system based upon global luminosity distribution and characteristics of faint (dwarf) galaxies compared with those of bright galaxies.

160.126   An analysis of the geometrical properties for galaxies in the Virgo cluster core.
H. T. MacGillivray, R. J. Dodd.
The Virgo cluster of galaxies, p. 217 – 225 (1985). – See Abstr. 012.069.

The authors have examined the geometrical properties for galaxies in a magnitude–limited sample in a field of 25 square degrees of sky centred on the Virgo cluster. The aim has been to carry out a search for systematic, non–random trends of the type detected in the Local Supercluster. No global effect of alignment of Virgo galaxies along the Local Supercluster plane has been detected, in agreement with previous observations. However, galaxies do show a strong effect of alignment with respect to the elliptical galaxy NGC 4406 (M86) in the sense that they are preferentially oriented with their major planes perpendicular to the direction vector to NGC 4406.

160.127   Optical comparisons of Virgo cluster galaxies and field galaxies.
R. C. Kennicutt Jr.
The Virgo cluster of galaxies, p. 227 – 238 (1985). – See Abstr. 012.069.

The author summarizes what little is known about the optical properties of Virgo cluster versus field spirals. The available statistics indicate that at least some of the Virgo members are different, but not enough is known to unambiguously assign a cause. The author also tries to point out some areas in which future observations would be most helpful.

160.128   Morphological and physical characteristics of the Virgo cluster: first results from the Las Campanas photographic survey.
A. Sandage, B. Binggeli, G. A. Tammann.
The Virgo cluster of galaxies, p. 239 – 295 (1985). – See Abstr. 012.069.

The Las Campanas du Pont 2.5 m telescope has been used for an extensive photographic survey of the Virgo cluster, the basic results which appear in a "Catalog of 2096 galaxies in the Virgo cluster area". The Catalog data are currently been used to investigate the structural properties of Virgo cluster galaxies, and the Virgo cluster as a whole. This paper reviews the first results of these investigations. Three topics are discussed after a short description of the Catalog. (1) Morphological and physical properties of the dwarfs; (2) luminosity functions for the various morphological types; and (3) the projected spatial and velocity distributions.

160.129   Search for young stars in Virgo dwarf ellipticals.
    L. Vigroux, M. Lachieze Rey, J. Souviron, J. P. Vader.
The Virgo cluster of galaxies, p. 297 – 300 (1985). – See Abstr. 012.069.

The authors are analyzing about 10 more dwarf ellipticals and it may be premature to draw any firm conclusions. However they can already say that: Young stars can be found in several dwarf ellipticals but it is not a general feature of these galaxies. Interaction with a larger spiral may trigger a burst of star formation. This could be the origin of young stars in NGC 205 and NGC 4627. However young stars can also be found in galaxies isolated or loosely bound such as NGC 185 and M100 DW5.

160.130   Applications of a nuclear spectroscopic survey of Virgo spiral galaxies.
J. R. Stauffer.
The Virgo cluster of galaxies, p. 301 – 310 (1985). – See Abstr. 012.069.

The author comments on two recent Virgo papers published by other authors. His database for these comments is the nuclear spectrophotometric survey of Virgo and field spirals that he obtained for his thesis (Stauffer 1982, 1983). He ends his paper with a brief remark concerning the distance of NGC 4569.

160.131   X–ray survey of the Virgo cluster and comparison to field galaxies.
W. Forman, C. Jones, M. DeFaccio.
The Virgo cluster of galaxies, p. 323 – 343 (1985). – See Abstr. 012.069.

The observations presented in this review include about 60 galaxies in the Virgo cluster listed in the Shapley–Ames Catalog and about 80 field galaxies. M87 and its dominance of the cluster in X–rays is described – it is about one hundred times more luminous in X–rays than any other cluster member and it possesses a massive, dark halo. Further, the X–ray emission from the early type galaxies (E, S0, and Sa) in Virgo and the field is discussed. The last section of this review compares the early and late type galaxies in the field to those in the Virgo cluster and discusses the possible effects of the Virgo cluster environment on their X–ray properties.

160.132   EXOSAT observations of the Virgo cluster.
    A. Smith, G. Stewart.
The Virgo cluster of galaxies, p. 345 – 355 (1985). – See Abstr. 012.069.

Reported here are the observations of the Virgo cluster made with the Medium Energy experiment on board EXOSAT. The aims of the observations were to seek further evidence for the presence of the hard very extended object as well as to study M87 and its environment.

160.133   Dynamics and evolution of the Virgo Supercluster.
    A. Yahil.
The Virgo cluster of galaxies, p. 359 – 373 (1985). – See Abstr. 012.069.

Existing observations are all consistent with a standard baryonic universe with a cosmological density factor $\Omega_0 = 0.1 - 0.2$. This includes a comparison of the abundances of the light elements with those predicted from standard big–bang nucleosynthesis, as well as virial estimates of the mass content of galaxies with their massive halos. To that is added the determination of Ω_0 from the Virgocentric infall; it provides an estimate of the mean density on scales of tens of Mpc, both in and out of galaxies, and whether baryonic or not. N–body simulations support the standard interpretation of a baryonic universe.

160.134   Dynamics of the Local Supercluster and the Virgo cluster.
E. J. Shaya, R. B. Tully.
The Virgo cluster of galaxies, p. 375 – 384 (1985). – See Abstr. 012.069.

The paper gives a discussion of the velocity fields and the galaxy distributions in the Local Supercluster, with special emphasis given to determinations of the mass and the dynamical history of the Virgo cluster. The authors find: (1) There is presently a large flux of predominantly spiral galaxies falling into the Virgo cluster. (2) The infall velocities of galaxies near the cluster are consistent with the virial theorem mass determinations if the universe is between $10-15$ Gyrs. (3) The virial theorem mass determined from Virgo spirals is only 2.2 times the mass determined from the Virgo ellipticals. (4) The expansion of galaxies out to recessional velocities of 3000 km s^{-1} is relatively undisturbed aside from the Virgo cluster, and spherically symmetric about the cluster. (5) Motions deviating from the Virgo influenced flow are usually less than 200 km s^{-1}.

160.135   The termination of galaxy formation by cluster formation.
R. B. Tully.
The Virgo cluster of galaxies, p. 385 – 390 (1985). – See Abstr. 012.069.

This review contains a Hawaiian perspective of aspects of the evolution of the Virgo Cluster and the effect of such an environment on the formation and evolution of galaxies. There is an attempt to draw connections between five fragments of information and ideas: (1) the distribution of galaxies in the supercluster about Virgo is very clumpy, (2) at least the outer parts of spiral galaxies probably formed relatively recently, (3) both galaxies and larger–scale structure within the supercluster might be in, or near to, a tidally limited regime, (4) large–scale tidal fields can dictate the morphologies of galaxies, and (5) a color–magnitude diagram might contain information on the recent evolution of the cluster.

160.136   The Local Group velocity and the Virgo cluster.
R. D. Davies, L. Staveley–Smith.
The Virgo cluster of galaxies, p. 391 – 396 (1985). – See Abstr. 012.069.

The authors discuss their measurements of the Local Group velocity determined relative to the backdrop of galaxies extending out to 5000 km s^{-1}. This is compared with the discrepant result of Rubin, Thonnard, Ford & Roberts (1976) and with the measurement of the dipole component of the cosmic microwave background which gives the Local Group velocity relative to the microwave background reference frame at $z \cong 1000$. The authors' results agree with this latter derivation.

160.137   On the infall velocity towards Virgo.
R. C. Kraan–Korteweg.
The Virgo cluster of galaxies, p. 397 – 406 (1985). – See Abstr. 012.069.

An investigation of the parameters which determine a self-consistent Virgocentric infall model reveals two stable special regions, viz. (1) a region where the observed radial velocities v_0 nearly reflect the unperturbed Hubble velocities v_H and which lends itself to a determination of the Hubble constant H_0; and (2) a region where the peculiar radial velocities Δv, caused by the infall model, are about equal to the infall velocity v_{vc}; this region is particularly suited for a determination of v_{vc}. Spiral galaxies within these regions with known 21 cm–line widths are used to determine values of $H_0 = 58$ km s^{-1}Mpc^{-1} and $v_{vc} = 210$ km s^{-1}.

160.138   The historical evolution of the Virgo cluster distance modulus.
O.–G. Richter.
The Virgo cluster of galaxies, p. 409 – 411 (1985). – See Abstr. 012.069.

The author has attempted to collect all the many distance modulus determinations for the Virgo cluster that are widely scattered in the literature. However, the resulting list is probably incomplete. Nevertheless, it should serve as a guide to the relevant literature for the many different distance indicators that are potentially applicable to the Virgo cluster.

160.139   Distance and kinematics of the Virgo cluster.
G. de Vaucouleurs.
The Virgo cluster of galaxies, p. 413 – 416 (1985). – See Abstr. 012.069.

The distances, structure and kinematics of various components of the Virgo cluster are reviewed. The mean distance moduli of the Virgo E cluster and the Virgo S cloud, from seven methods each, are $\mu_0 = 30.40 \pm 0.10$ and 30.70 ± 0.10.

160.140   The Virgo cluster distance corrections to global parameters.
R. B. Tully, P. Fouqué.
The Virgo cluster of galaxies, p. 417 – 424 (1985). – See Abstr. 012.069.

Outside of the local and nearest groups, distances of individual galaxies rest on their global properties, calibrated in a statistical way.

160.141   A comparison of the Virgo cluster with (one) other nearby cluster(s).
O.–G. Richter.
The Virgo cluster of galaxies, p. 427 – 432 (1985). – See Abstr. 012.069.

The purpose of this note is to draw attention to some other clusters of galaxies that are very similar to the Virgo cluster and may provide a better playground for some observational studies. Furthermore, the detailed knowledge about properties in common could perhaps lead to a better understanding of the Virgo cluster.

160.142   The far background of the Virgo cluster.
G. Reaves.
The Virgo cluster of galaxies, p. 433 – 436 (1985). – See Abstr. 012.069.

The author studies, whether it is possible to put a limit on the intergalactic extinction within the Virgo cluster by comparing the projected areal density of very faint – and thus very distant – galaxies and clusters of galaxies within the area of the cluster, with the density in regions surrounding the cluster. A large area, from $\alpha = 11^h00^m$ to 14^h00^m and from $\delta = +5°$ to $+20°$ (1950.0), was chosen for study. The Virgo cluster occupies the central 15% of this approximately 675 sq. deg. region.

160.143   Future needs for observations of the Virgo cluster. Panel discussion.
H. Arp, W. Forman, R. Giovanelli, R. Kennicutt, A. Sandage, B. Tully, M. Roberts.
The Virgo cluster of galaxies, p. 438 – 468 (1985). – See Abstr. 012.069.

160.144   Radial velocity dispersion in nonstationary clusters of galaxies.
E. A. Malkov.
Tr. Astrofiz. Inst. Alma–Ata, Tom 40, p. 32 – 33 (1983). In Russian.

160.145   Between the galaxies.
S. E. Schneider, Y. Terzian.
Am. Sci., Vol. 72, No. 6, p. 574 – 581 (1984). Abstr. in Phys. Abstr., Vol. 88, No. 1257, Entry 57869 (1985).

160.146   Gravitational lens effects of superclusters.
T. A. Oosterloo.
Ned. Tijdschr. Natuurkd. A, Vol. A51, No. 1, p. 36 – 37 (1985). In Dutch. Abstr. in Phys. Abstr., Vol. 88, No. 1258, Entry 63512 (1985).

160.147   A study of the large–scale structure of the distribution of galaxies in and around the Cancer cluster.
M. D. Bicay, R. Giovanelli.
Bull. Am. Astron. Soc., Vol. 17, No. 2, p. 579 (1985). Abstract. – See Abstr. 010.065.

160.148   The distribution of mass and light in the poor cluster MKW4.
G. A. Kriss, E. M. Malumuth.
Bull. Am. Astron. Soc., Vol. 17, No. 2, p. 579 (1985). Abstract. – See Abstr. 010.065.

160.149   Morphological segregation in the Pisces–Perseus supercluster.
R. Giovanelli, M. P. Haynes, G. L. Chincarini.
Bull. Am. Astron. Soc., Vol. 17, No. 2, p. 581 – 582 (1985). Abstract. – See Abstr. 010.065.

160.150   Cooling flows in clusters of galaxies.
L. Cowie.
Bull. Am. Astron. Soc., Vol. 17, No. 2, p. 600 (1985). Abstract. – See Abstr. 010.065.

160.151   The X–ray structure of the 3C 295 cluster: a cooling flow at a redshift of 0.5.
M. J. Henriksen, J. P. Henry.
Bull. Am. Astron. Soc., Vol. 17, No. 2, p. 601 (1985). Abstract. – See Abstr. 010.065.

160.152   A dynamical study of the NGC 4005 group.
B. A. Williams.
Bull. Am. Astron. Soc., Vol. 17, No. 2, p. 601 (1985). Abstract. – See Abstr. 010.065.

160.153   CCD photometry of the cluster of galaxies Abell 1689.
D. H. Gudehus, D. J. Hegyi.
Bull. Am. Astron. Soc., Vol. 17, No. 2, p. 601 (1985). Abstract. – See Abstr. 010.065.

160.154   A search for additional intergalactic neutral hydrogen in the M96 group.
S. E. Schneider, E. E. Salpeter, Y. Terzian.
Bull. Am. Astron. Soc., Vol. 17, No. 2, p. 602 (1985). Abstract. – See Abstr. 010.065.

160.155   Some results from two–dimensional simulation of neutrino–dominated universes.
A. L. Melott.
Sov. Astron., Vol. 28, No. 6, p. 631 – 635 (1984). English translation of 38.160.075.

160.156   Morphology and energetics of narrow angle tail radio sources.
C. P. O'Dea.
Diss. Abstr. Int., Sect. B, Vol. 45, No. 10, p. 3256 – 3257 (1985). Thesis, University of Massachusetts, 469 pp. (1984). Order No. DA8500112.

160.157   On a self–consistent model of rich clusters of galaxies. II. The plasma component of a cluster.
M. V. Konyukov.
Astrofizika, Tom 22, Vyp. 3, p. 473 – 485 (1985). In Russian. English translation in Astrophysics, Vol. 22, No. 3.
It has been shown that the distribution of hydrodynamic parameters of cluster plasma is determined by the solution of a boundary problem for the gravitational potential of a self–consistent field. A model of the plasma component is determined by three main parameters. The equation for distribution of ions in a cluster under diffusion equilibrium of ions is obtained. The main types of solutions for the plasma component are determined. An algorithm of the solution of the problem for a self–consistent model cluster with ions is built up.

160.158   Brightest member luminosities and dynamical times of galaxy clusters.
G. Giuricin, F. Mardirossian, M. Mezzetti.
New aspects of galaxy photometry, p. 167 – 168 (1985). – See Abstr. 012.114.

160.159   H I observations of the Virgo cluster galaxies.
W. K. Huchtmeier.
New aspects of galaxy photometry, p. 299 – 302 (1985). – See Abstr. 012.114.
Global properties of a complete sample of Virgo cluster galaxies are derived containing all spiral and irregular galaxies brighter than $m_B = 14.2$ within $10°$ of M87 and a large number of galaxies brighter than $m_B = 14.95$ in that area. Total H I masses, radial velocities and the Tully–Fisher relation are determined.

160.160   Influence of the K–effect on the colours and the $\lambda_0 4000$ break amplitudes of galaxies in distant clusters.
E. Laurikainen, T. Jaakkola.
New aspects of galaxy photometry, p. 309 – 312 (1985). – See Abstr. 012.114.
The colour distributions of the Coma cluster and DC 0329–52 shifted to $z = 0.39$ and $z = 1.0$ are simulated, in regard to the influence of the K–term on the morphological and colour distributions of 50 brightest galaxies. A good fit with the colours of actual distant clusters is obtained, except that the observed colours are slightly redder than the predicted ones. Also Spinrad's data on the z–dependence of the $\lambda_0 4000$ break amplitude can be understood via the K–effect. The results do not support the hypothesis of cosmological evolution of galaxies.

160.161   Relations among optical, radio and X–ray properties of Abell clusters of galaxies.
M. Kalinkov, K. Stavrev, I. Kuneva.
New aspects of galaxy photometry, p. 331 – 334 (1985). – See Abstr. 012.114.
A data bank, containing optical, radio and X–ray data for all Abell clusters of galaxies has been compiled. The relations between various parameters, presenting global or specific characteristics of the clusters have been investigated. Some examples of these relations are presented here.

160.162   The nuclear radio sources in the elliptical galaxies NGC 3309 and NGC 3311 in the cluster Abell 1060.
P. O. Lindblad, S. Jörsäter, Å. Sandqvist.
New aspects of galaxy photometry, p. 337 – 338 (1985). – See Abstr. 012.114.
The authors have obtained VLA 6 cm continuum observations, as well as CCD photometry in various Gunn and Johnson filters for the giant elliptical galaxies NGC 3309 and NGC 3311 at the center of the Abell 1060 cluster. A jet–like radio source is coincident with the optical nucleus of NGC 3309. A brief discussion of the implications of these data for a possible accretion flow in the central part of Abell 1060 is given.

160.163   Cloud–beam interactions: bends, lobes and superluminals.
R. N. Henriksen.
Physics of energy transport in extragalactic radio sources, p. 122 – 127 (1984). – See Abstr. 012.116.
There is growing evidence for the interaction of the energy–momentum beams believed to power the radio "jets" and both intra and inter galactic "clouds". The author clarifies qualitatively the various possible types of beam–cloud interaction. Finally, he sketches how these basic interactions might produce apparent superluminal motion.

160.164   Quasat, gravitational lenses, and studies of the interstellar, intracluster and intergalactic medium.
B. F. Burke.
ESA Spec. Publ., ESA SP–213, p. 153 – 155 (1984). – See Abstr. 012.124.
The refractive effects of the intervening medium between a radio source and the observer can be studied by orbiting VLBI, and Quasat will lead to significant advances in our understanding of these effects.

Atlas of dwarf galaxies in the region of M81 group.
See Abstr. 002.024.

Correlation analyses of deep galaxy samples – III.
See Abstr. 036.067.

Some methods for photometry of galaxies in clusters.
See Abstr. 036.227.

X–ray astronomy and plasma astrophysics.
See Abstr. 062.090.

Gravitational lensing and galaxy shape.
See Abstr. 066.060.

Cepheid variables as extra–galactic distance indicators.
See Abstr. 122.068.

Radio lines of heavy elements from hot gas in the remnants of supernova explosions and clusters of galaxies.
See Abstr. 125.031.

Millimeter–wavelength lines of heavy elements predicted from the hot gas in supernova remnants and galaxy clusters.
See Abstr. 125.035.

Endpoints of stellar evolution: X–ray surveys of the Local Group.
See Abstr. 142.037.

Discussion on zero–point field acceleration and applicability to cosmic rays, X–ray background and energetics of the intergalactic medium.
See Abstr. 142.068.

New developments in high energy astrophysics.
See Abstr. 143.100.

Anisotropy of cosmic rays of superhigh energies.
See Abstr. 144.122.

Equipartition in multicomponent gravitational systems.
See Abstr. 151.077.

The relation between the velocity and mass distributions. The role of collisionless relaxation processes.
See Abstr. 151.143.

Compact configurations within small evolving groups of galaxies.
See Abstr. 151.150.

Kinematics and dynamics of the haloes of supergiant galaxies.
See Abstr. 157.004.

A method for determining the spatial correlation function from a sample with partial distance information.
See Abstr. 157.005.

The redshifts of double galaxies.
See Abstr. 157.006.

The stellar content of the A496 cD galaxy.
See Abstr. 157.032.

Distribution and motions of atomic hydrogen in lenticular galaxies. IV. A ring of H I around NGC 4262.
See Abstr. 157.044.

The dependence of CO content on morphological type and luminosity for spiral galaxies in the Virgo cluster.
See Abstr. 157.066.

Globular clusters in galaxies beyond the Local Group. IV. The elliptical galaxies NGC 524 and 1052.
See Abstr. 157.119.

The structure of the elliptical galaxy IC 4296.
See Abstr. 157.143.

Distances to Local Group galaxies based on BVRI CCD photometry of Cepheids.
See Abstr. 157.154.

Isophotal diameters of cluster spirals.
See Abstr. 157.155.

Optical properties of IRAS galaxies in the Coma and Hercules clusters.
See Abstr. 157.193.

Infrared observations of dwarf elliptical galaxies in the Virgo cluster.
See Abstr. 157.204.

Diameters of galaxies inside and outside clusters.
See Abstr. 157.205.

Surface photometry of the Sculptor Group galaxies: NGC 7793, NGC 247, and NGC 300.
See Abstr. 157.209.

Gas equilibrium conditions in the gravitational field of massive central galaxies in clusters.
See Abstr. 157.215.

Mass accretion rates for giant elliptical galaxies.
See Abstr. 157.231.

Report of IAU Commission 28: Galaxies (*Galaxies*).
See Abstr. 157.244.

Can ram pressure distort spiral arms?
See Abstr. 157.249.

Evolution of spiral galaxies in a cluster environment.
See Abstr. 157.267.

Galaxy photometry at faint light levels – interaction with the environment.
See Abstr. 157.268.

An H I and optical study of the gas poor Virgo cluster spiral NGC 4571.
See Abstr. 157.283.

Beams in head–tail galaxies.
See Abstr. 158.041.

The NGC 4593 group of active galaxies.
See Abstr. 158.049.

The effect of local galaxy density on the production of powerful radio sources by early–type galaxies.
See Abstr. 158.053.

The nuclear radio sources in the elliptical galaxies NGC 3309 and NGC 3311 in the cluster Abell 1060.
See Abstr. 158.059.

Globular clusters and the distance to M87.
See Abstr. 158.082.

M87: round and slow.
See Abstr. 158.094.

Steep–spectrum radio sources in clusters of galaxies – the southern sample.
See Abstr. 158.115.

The short wide angle tail radio source in NGC 4874.
See Abstr. 158.164.

WSRT and VLA observations of very steep spectrum radio galaxies in clusters.
See Abstr. 158.171.

The nuclear activity of interacting galaxies.
See Abstr. 158.181.

Star formation in the cooling flows of M87 and NGC 1275.
See Abstr. 158.184.

An optical investigation of a jet–cloud interaction in the galaxy cluster A 194.
See Abstr. 158.242.

The wide angle tailed radio source NGC 2329 in the cluster A569.
See Abstr. 158.247.

Minkowski's object: a starburst triggered by a radio jet.
See Abstr. 158.249.

Minkowski's object: a starburst triggered by a radio jet.
See Abstr. 158.269.

Optical counterparts of the radio source Virgo A (= M87).
See Abstr. 158.271.

A radio and optical study of a jet/cloud interaction in the galaxy cluster A194.
See Abstr. 158.277.

Wide angle tail radio galaxies.
See Abstr. 158.331.

Emission–line QSOs in the region of the Hercules cluster of galaxies.
See Abstr. 159.022.

Are Ly–α clouds detectable in emission at 21 cm?
See Abstr. 159.027.

The statistical analysis of Lα absorption lines from a uniform sample of nine high redshift QSOs.
See Abstr. 159.120.

Photometry of quasar host galaxies and cosmological implications.
See Abstr. 159.150.

Large–scale structure in the Universe.
See Abstr. 161.014.

Constraints on decaying particles from Virgo infall.
See Abstr. 161.015.

On the existence of cosmological evolutionary effects. II. Influence of the K–effect on the colours and other parameters of galaxies in distant clusters.
See Abstr. 161.053.

Current status of the missing mass problem.
See Abstr. 161.061.

Extragalactic dust and near–infrared cosmic background.
See Abstr. 161.081.

The distribution of voids.
See Abstr. 161.089.

On percolation as a cosmological test.
See Abstr. 161.090.

Cosmological density fluctuations and large–scale structure: from N–point correlation functions to the probability distribution.
See Abstr. 161.096.

Galaxy and cluster formation in a neutrino universe.
See Abstr. 161.123.

Quasar–galaxy associations with discordant redshifts as a topological effect. I. Two–dimensional study.
See Abstr. 161.160.

The evolution of large–scale structure in a universe dominated by cold dark matter.
See Abstr. 161.178.

Statistical comparison of galaxy formation models: the bispectrum.
See Abstr. 161.179.

The distance scale: present status and future prospects.
See Abstr. 161.195.

Status of quantized extragalactic redshifts.
See Abstr. 161.199.

Redshifts of emission line galaxies near the Bootes void.
See Abstr. 161.200.

Some very large scale structure in the distribution of Abell galaxy clusters.
See Abstr. 161.201.

On neutrino halos of galaxies or clusters of galaxies and their restriction on neutrino mass.
See Abstr. 161.228.

On the false fluctuations of the cosmic background caused by the dust and gas emission in galaxy clusters.
See Abstr. 161.253.

Profiles of galaxy clusters in cosmological scenarios.
See Abstr. 161.344.

Report of IAU Commission 47: Cosmology (*Cosmologie*).
See Abstr. 161.358.

The large scale structure of the universe.
See Abstr. 161.359.

161 Universe, Cosmology, Background Radiation

161.001   Cosmological observations of galaxies: number counts.
A. W. Sievers, J. J. Perry, G. F. R. Ellis.
Mon. Not. R. Astron. Soc., Vol. 212, No. 1, p. 197 – 209 (1985).
A simple model of selection effects and the diameter–luminosity distribution of elliptical galaxies is used to investigate the effect of various parameters on optical number counts of galaxies. It is shown that in addition to knowledge of the detection limit, point–spread effects, and angular resolution limits, the number counts are significantly affected by the nature of the diameter–luminosity distribution of the galaxies.

161.002   Millisecond pulsars, gravitational waves and the inflationary universe.
L. M. Krauss.
Nature, Vol. 313, No. 5997, p. 32 – 33 (1985).
Signals from millisecond pulsars, being exceptionally accurate clocks, are sensitive to a cosmic background of gravitational waves. Present measurements imply that the energy density in such waves in an octave bandwidth near 10^{-7} Hz must be $\lesssim 10^{-3}$ of the closure density. The author investigates the sensitivity of a millisecond pulsar timing signal to the background of gravitational waves expected from an inflationary phase of expansion during the early universe.

161.003   The alignment of distant radio sources.
V. K. Kapahi, R. Subrahmanyan, A. K. Singal.
Nature, Vol. 313, No. 6002, p. 463 – 465 (1985).
The relative orientation of the axes of extragalactic extended radio sources was first investigated by Willson, who noted that in the 3CR sample there was a tendency for the axes to lie parallel to each other for sources separated by $\lesssim 10°$ on the sky. Using two larger samples, Sanders has recently concluded that this tendency to align is present at a high confidence level ($> 3\sigma$) only for distant ($z > 0.4$) sources. The authors have attempted to verify the effect using two more independent and larger source samples and find no significant tendency for axis alignment. They also investigated a sample formed by adding well–observed radio quasars to 3CR sources and found that, even in this sample, the effect shows up only at a marginally significant level ($< 2\sigma$).

161.004   Green's functions in Robertson–Walker–space.
G. Süßmann, H.-D. Radecke.
Z. Naturforsch., A, Band 40a, Heft 1, p. 96 – 97 (1985).
The wave equations for the metric of the cosmological standard model and their solutions in form of Green's functions are given.

161.005   The new inflationary universe.
A. H. Guth.
Phys. Today, Vol. 38, No. 1, p. S6 – S7 (1985). Short review paper. – See Abstr. 013.003.

161.006   Relics of creation.
P. Davies.
Sky Telesc., Vol. 69, No. 2, p. 112 – 115 (1985).

161.007   On population III stars.
S. Hayakawa.
Astron. Her., Vol. 78, No. 2, p. 32 – 37 (1985). In Japanese.

161.008   Cosmology of magnetic monopoles.
M. Izawa, M. Morikawa.
Astron. Her., Vol. 78, No. 2, p. 38 – 44 (1985). In Japanese.

161.009   Statistical fluctuations in an inflationary universe.
Y. Rephaeli.
Astron. Astrophys., Vol. 142, No. 2, p. L29 – L31 (1985).
The viability of statistical fluctuations as a source for the structure in the Universe is considered in the context of inflationary universe models.

161.010   The dynamics of spatially homogeneous axisymmetrical cosmological models with a primordial magnetic field.
V. A. Ruban.
Astron. Zh., Tom 62, Vyp. 1, p. 8 – 18 (1985). In Russian. English translation in Sov. Astron., Vol. 29, No. 1.
It is found that for all axisymmetrical cosmological models with a "primordial" magnetic field degenerate dynamics without an initial vacuum stage as well as specific matter singularities caused by the gravitation of matter are typical. New kinds of matter singularities are pointed out.

161.011   Observer dependence of past light cone initial data in cosmology.
G. Dautcourt.
Astron. Nachr., Vol. 306, No. 1, p. 1 – 6 (1985).
The author discusses some properties of the initial data introduced in previous articles to study the past light cone initial value problem in cosmology. A parametrization of the conformal cone metric (the geometrical part of the initial data) is proposed which is closely related to astronomical observations of image distortions of distant extragalactic objects. The quantities used here as initial data have simple transformation properties under a Lorentz rotation of the observer.

161.012   Erratum: "Bianchi–I universes of Brans–Dicke theory for the vacuum" [Astron. Nachr., Vol. 304, No. 4, p. 163 – 166 (1983). In German.]
U. Dyllong.
Astron. Nachr., Vol. 306, No. 1, p. 44 (1985). See Abstr. 34.161.034.

161.013   Supposed history of our Galaxy.
W. Iwanowska.
The Milky Way galaxy, p. 611 – 612 (1985). – See Abstr. 012.007 (IAU Symp. No. 106).

161.014   Large–scale structure in the Universe.
A. Heavens.
Mon. Not. R. Astron. Soc., Vol. 213, No. 1, p. 143 – 155 (1985).
Models containing the essential features of the adiabatic theory of galaxy formation are investigated. A cellular structure of uniform two–dimensional sheets of galaxies, with voids in between, is found to provide a natural explanation for the discrepancy between the observed clustering length (where the two–point correlation function is unity) and the expected scale of large–scale structure in a universe dominated by light neutrinos. An analysis of the growth of surface density perturbations within expanding two–dimensional 'pancakes' is shown to lead to structure on scales up to the scale of galaxy clusters, provided that galaxies are formed by local fragmentation of pancakes, and the pancaking epoch is rather early ($z_p \gtrsim 3$). It is also shown that rejection of the neutrino pancake theory on the basis of comparison of the correlation function with existing numerical simulations is premature.

161.015   Constraints on decaying particles from Virgo infall.
G. Efstathiou.
Mon. Not. R. Astron. Soc., Vol. 213, No. 1, p. 29P – 34P (1985).
Decaying particles have been proposed to reconcile inflationary models of the early Universe with observations which suggest that the Universe is open. The bulk of the present mass–density is assumed to be in relativistic particles resulting from the non–radiative decay of massive weakly interacting particles. The decays must have occurred recently $(1 + z_d) < 5$, otherwise the cores of rich galaxy clusters would not be in virial equilibrium. However, this model is not compatible with recent estimates of our peculiar velocity relative to the Virgo cluster unless $(1 + z_d) \gtrsim 20$.

161.016 Does cosmology imply a Dirac neutrino mass?
D. Fargion, M. G. Shepkin.
Phys. Lett. B, Vol. 146B, No. 1 – 2, p. 46 – 50 (1984). Abstr. in
Phys. Abstr., Vol. 88, No. 1248, Entry 9963 (1985).

161.017 Analytic model for phase transitions in the early Universe.
L. Diosi, B. Keszthelyi, B. Lukacs, G. Paal.
Report KFKI–1984–84, Hungarian Acad. Sci., Budapest, Hungary, 36 pp. (1984). Abstr. in Phys. Abstr., Vol. 88, No. 1248, Entry 9965 (1985).

161.018 Towards a supersymmetric cosmology.
J. Ellis, K. Enqvist, G. Gelmini, C. Kounnas, A. Masiero, D. V. Nanopoulos, A. Y. Smirnov.
Phys. Lett. B, Vol. 147B, No. 1 – 3, p. 27 – 33 (1984). Abstr. in
Phys. Abstr., Vol. 88, No. 1249, Entry 10210 (1985).

161.019 Quantum fluctuations as the source of classical gravitational perturbations in inflationary universe models.
R. H. Brandenberger.
Nucl. Phys. B, Part. Phys., Vol. B245, No. 2, p. 328 – 342 (1984).
Abstr. in Phys. Abstr., Vol. 88, No. 1249, Entry 10217 (1985).

161.020 A model for the early universe and the connection between gravitation and the quantum nature of matter.
E. J. Sternglass.
Lett. Nuovo Cimento, Vol. 41, Ser. 2, No. 6, p. 203 – 208 (1984).
Abstr. in Phys. Abstr., Vol. 88, No. 1249, Entry 10225 (1985).

161.021 Restriction on amount of antimatter in the early Universe from $\bar{p}$–^{4}He reaction data.
Yu. A. Batusov, I. V. Falomkin, G. B. Pontecorvo, M. G. Sapozhnikov, C. Guaraldo, A. Maggiora, M. Vascon, G. Zanella, G. Bendiscioli, V. Filippini, E. Lodi Rizzini, A. Rotondi, A. Zenoni, F. Balestra, M. P. Bussa, L. Busso, L. Ferrero, G. Gervino, D. Panzieri, G. Piragino, F. Tosello.
Lett. Nuovo Cimento, Vol. 41, Ser. 2, No. 7, p. 223 – 226 (1984).
Abstr. in Phys. Abstr., Vol. 88, No. 1249, Entry 10857 (1985).

161.022 Analysis of Coleman–Weinberg potentials in orbit space: general method and application to the inflationary universe.
J. S. Kim, C. W. Kim.
Nucl. Phys. B, Part. Phys., Vol. B244, No. 2, p. 523 – 540 (1984).
Abstr. in Phys. Abstr., Vol. 88, No. 1249, Entry 14802 (1985).

161.023 Constraints on generalized inflationary cosmologies.
L. F. Abbott, M. B. Wise.
Nucl. Phys. B, Part. Phys., Vol. B244, No. 2, p. 541 – 548 (1984).
Abstr. in Phys. Abstr., Vol. 88, No. 1249, Entry 14803 (1985).

161.024 The Kasner type Universe in new general relativity.
M. Fukui, J. Masukawa, O. Miyamura.
Prog. Theor. Phys., Vol. 71, No. 6, p. 1213 – 1220 (1984). Abstr.
in Phys. Abstr., Vol. 88, No. 1249, Entry 14808 (1985).

161.025 Quantum Kaluza–Klein cosmologies. I.
Z. C. Wu.
Phys. Lett. B, Vol. 146B, No. 5, p. 307 – 310 (1984). Abstr. in
Phys. Abstr., Vol. 88, No. 1249, Entry 14809 (1985).

161.026 Majorons: a simultaneous solution to the large and small scale dark matter problems.
G. Gelmini, D. N. Schramm, J. W. F. Valle.
Phys. Lett. B, Vol. 146B, No. 5, p. 311 – 317 (1984). Abstr. in
Phys. Abstr., Vol. 88, No. 1249, Entry 14810 (1985).

161.027 The future of the Universe.
N. Prantzos, M. Casse.
Recherche, Vol. 15, No. 156, p. 838 – 848 (1984). In French.
Abstr. in Phys. Abstr., Vol. 88, No. 1249, Entry 14811 (1985).

161.028 The Einstein static universe, associated potentials and the equation of motion.
H. P. Berg, J. Tarski.
Rev. Roum. Phys., Vol. 29, No. 8, p. 711 – 719 (1984). Abstr. in
Phys. Abstr., Vol. 88, No. 1250, Entry 15011 (1985).

161.029 Cosmological solutions of Bi–metric theories of gravitation.
D.-E. Liebscher, J. Mücket.
Ann. Phys. (Leipzig), Vol. 41, No. 3, p. 172 – 178 (1984). Abstr.
in Phys. Abstr., Vol. 88, No. 1250, Entry 15013 (1985).

161.030 Cosmological problems for spontaneously broken supergravity.
A. S. Goncharov, A. D. Linde.
Phys. Lett. B, Vol. 147B, No. 4 – 5, p. 279 – 283 (1984). Abstr. in
Phys. Abstr., Vol. 88, No. 1250, Entry 15035 (1985).

161.031 The scale of extragalactic distances.
L. Bottinelli, L. Gouguenheim, G. Paturel.
Ann. Phys. (Paris), Vol. 9, No. 3, p. 503 – 533 (1984). Abstr. in
Phys. Abstr., Vol. 88, No. 1250, Entry 18655 (1985).

161.032 Cosmological coincidences and the Dirac–cosmology.
H.–J. Treder.
Ann. Phys. (Leipzig), Vol. 41, No. 3, p. 208 – 217 (1984). Abstr.
in Phys. Abstr., Vol. 88, No. 1250, Entry 18950 (1985).

161.033 A magnetohydrodynamic Universe.
S. Prakash.
Curr. Sci., Vol. 53, No. 5, p. 252 – 253 (1984). Abstr. in Phys.
Abstr., Vol. 88, No. 1250, Entry 18954 (1985).

161.034 Quantum effects, inflationary model, and the observable Universe.
M. Demianski.
Int. J. Theor. Phys., Vol. 23, No. 8, p. 713 – 723 (1984). Abstr. in
Phys. Abstr., Vol. 88, No. 1250, Entry 18955 (1985). – See Abstr.
012.015.

161.035 Can inflation explain the second law of thermodynamics?
D. N. Page.
Int. J. Theor. Phys., Vol. 23, No. 8, p. 725 – 733 (1984). Abstr. in
Phys. Abstr., Vol. 88, No. 1250, Entry 18956 (1985). – See Abstr.
012.015.

161.036 Kantowski–Sachs cosmological models approaching isotropy.
E. Weber.
J. Math. Phys., Vol. 25, No. 11, p. 3279 – 3285 (1984). Abstr. in
Phys. Abstr., Vol. 88, No. 1250, Entry 18957 (1985).

161.037 A cosmological solution of Einstein's equations.
P. S. Wesson.
J. Math. Phys., Vol. 25, No. 11, p. 3297 – 3298 (1984). Abstr. in
Phys. Abstr., Vol. 88, No. 1250, Entry 18958 (1985).

161.038 Nonequilibrium processes and primordial nucleosynthesis.
T. Rothman, R. A. Matzner.
Lett. Nuovo Cimento, Vol. 41, Ser. 2, No. 3, p. 65 – 68 (1984).
Abstr. in Phys. Abstr., Vol. 88, No. 1250, Entry 18959 (1985).

161.039 The role of spin in cosmological models.
M. L. Bedran, E. P. Vasconcellos–Vaidya.
Lett. Nuovo Cimento, Vol. 41, Ser. 2, No. 3, p. 73 – 74 (1984).
Abstr. in Phys. Abstr., Vol. 88, No. 1250, Entry 18960 (1985).

161.040 The edge of spacetime.
F. Hawking.
New Sci., Vol. 103, No. 1417, p. 10 – 14 (1984). Abstr. in Phys.
Abstr., Vol. 88, No. 1250, Entry 18961 (1985).

161.041 Statistical formulation of electromagnetism in the expanding Universe and arrow of time.
T. Futamase, T. Matsuda.
Prog. Theor. Phys., Vol. 72, No. 1, p. 38 – 46 (1984). Abstr. in Phys. Abstr., Vol. 88, No. 1250, Entry 18962 (1985).

161.042 Dynamics of higher dimensional Universe models.
H. Sato.
Prog. Theor. Phys., Vol. 72, No. 1, p. 98 – 105 (1984). Abstr. in Phys. Abstr., Vol. 88, No. 1250, Entry 18963 (1985).

161.043 An effective action functional for the inflationary cosmology.
A. Aurilia, G. Denardo, F. Legovini, E. Spallucci.
Phys. Lett. B, Vol. 147B, No. 4 – 5, p. 258 – 262 (1984). Abstr. in Phys. Abstr., Vol. 88, No. 1250, Entry 18964 (1985).

161.044 Supersymmetric inflation, baryon asymmetry and the gravitino problem.
B. A. Ovrut, P. J. Steinhardt.
Phys. Lett. B, Vol. 147B, No. 4 – 5, p. 263 – 268 (1984). Abstr. in Phys. Abstr., Vol. 88, No. 1250, Entry 18965 (1985).

161.045 On the absence of particle creation in the hot Friedmann Universe.
Yu. P. Goncharov.
Phys. Lett. B, Vol. 147B, No. 4 – 5, p. 269 – 272 (1984). Abstr. in Phys. Abstr., Vol. 88, No. 1250, Entry 18966 (1985).

161.046 Supercooling, entropy production, and bubble kinetics in the quark–hadron phase transition in the early universe.
T. DeGrand, K. Kajantie.
Phys. Lett. B, Vol. 147B, Nos. 4 – 5, p. 273 – 278 (1984). Abstr. in Phys. Abstr., Vol. 88, No. 1250, Entry 18967 (1985).

161.047 Cascading inflationary universe and the origin of density fluctuations.
Pham Quang Hung.
Phys. Rev. D, Vol. 30, No. 8, p. 1637 – 1648 (1984). Abstr. in Phys. Abstr., Vol. 88, No. 1250, Entry 18969 (1985).

161.048 Nucleosynthesis in anisotropic cosmologies revisited.
T. Rothman, R. Matzner.
Phys. Rev. D, Vol. 30, No. 8, p. 1649 – 1668 (1984). Abstr. in Phys. Abstr., Vol. 88, No. 1250, Entry 18970 (1985).

161.049 Uniqueness of the Friedmann–Lemaitre–Robertson–Walker universes.
B. Mashhoon, M. H. Partovi.
Phys. Rev. D, Vol. 30, No. 8, p. 1839 – 1842 (1984). Abstr. in Phys. Abstr., Vol. 88, No. 1250, Entry 18971 (1985).

161.050 Inhomogeneous generalisation of Bianchi type–VI_0 cosmological model of perfect fluid distribution in general relativity.
S. R. Roy, S. Narain.
Astrophys. Space Sci., Vol. 108, No. 1, p. 195 – 201 (1985).
Einstein's equations have been solved for perfect fluid distribution. The solution obtained may be considered as inhomogeneous generalisation of Bianchi type–VI_0 cosmological model.

161.051 Magnetohydrodynamic effects on the growth of condensations in an expanding Universe and the formation of galaxies.
C. R. Evans, A. J. Fennelly.
Astrophys. Space Sci., Vol. 109, No. 1, p. 15 – 44 (1985).
The authors review evidence for the existence of the Galaxy's (Milky Way's) magnetic field and the evidence for an inter-galactic magnetic field permeating an ionized inter–galactic medium. The magnetohydrodynamics of such an inter–galactic medium during the post–recombination expansion and the effects of MHD on the formation of galaxies are then considered: isotropic cosmologies, the interaction of magnetic fields with isotropic backgrounds, detailed MHD perturbation computations, and an analysis of related MHD galaxy–formation studies.

161.052 Oscillations in the primeval ambiplasma in the Universe.
B. Singh, G. L. Kalra.
Astrophys. Space Sci., Vol. 109, No. 1, p. 63 – 75 (1985).
This paper discusses the propagation of waves in the early Universe simulated by rarefied, homogeneous ambiplasma (a mixed matter–antimatter ionized gas) placed in a uniform magnetic field. A rich variety of oscillations appear due to enhanced degrees of freedom in the ambiplasma. These oscillations have been analyzed for the propagation vector along and across the direction of magnetic field both in a symmetric and asymmetric ambiplasma. It is found that the oscillations do not grow in different situations which have been discussed. Dispersion curves have been drawn and various cut–offs and resonances have been found out. The study is concluded by highlighting some characteristic features of the waves in the ambiplasma.

161.053 On the existence of cosmological evolutionary effects. II. Influence of the K–effect on the colours and other parameters of galaxies in distant clusters.
E. Laurikainen, T. Jaakkola.
Astrophys. Space Sci., Vol. 109, No. 1, p. 111 – 122 (1985).
Parameters of the distant galaxy clusters of 3C295 ($z = 0.46$) and Cl 0024 + 1654 ($z = 0.39$) are compared with the predictions made using galaxies of the local clusters Coma ($z = 0.023$) and DC 0329–52 ($z = 0.057$) taking the K–effect into account. The distributions of colour and morphological type, and the amplitudes F^+/F^- of the $\lambda_0 4000$ discontinuity are examined and no evidence for evolution of the galaxies and the clusters can be seen.

161.054 Electromagnetic field in the structure of space–time.
S. Selak.
Astrophys. Space Sci., Vol. 109, No. 1, p. 123 – 130 (1985).
The author makes an attempt to show that electromagnetic field appears to be a phenomenon caused by the presence of a dilatational degree of freedom which is attributed to the whole matter of the Universe. For this attempt he used Weyl's generalization of the Riemannian geometry and tried to show that Einstein's basic objection to the Weyl theory can in this case be removed.

161.055 Evolution of adiabatic perturbations in a photino–dominated universe.
R. Valdarnini.
Astron. Astrophys., Vol. 143, No. 1, p. 94 – 98 (1985).
The author presents the results of numerical computations for the linear evolution of adiabatic perturbations in a photino (X) dominated universe. The integration runs from early redshift ($z_{in} = 10^9$) down to recombination. The X perturbations are treated according to a Boltzmann–Liouville approach, while matter and radiation perturbations are treated according to an ideal fluid approximation. A comparison of the observed clustering with the computed mass variances shows that a photino–dominated universe with $m_x \cong 2\,keV$ and $n = 0$ or $n = 1$ as spectral index for the initial X perturbation, might be favoured over a neutrino–dominated universe, where only spectra with $n > 2$ would be allowed.

161.056 Relations for a Kaluza–Klein cosmology with variable rest mass.
P. S. Wesson.
Astron. Astrophys., Vol. 143, No. 1, p. 233 – 234 (1985).
The components of the Einstein tensor and other relations are given for a cosmological model that is homogeneous and isotropic in a 5D theory of gravity of the Kaluza–Klein type. These relations are particularly relevant to the embedding for general relativity with variable rest mass proposed by Wesson (1983), and may also be of use in other Kaluza–Klein theories.

161.057 A simple realisation of the inflationary Universe scenario in SU(1,1) supergravity.
A. S. Goncharov, A. D. Linde.
Classical Quantum Gravity, Vol. 1, No. 6, p. L75 – L79 (1984). Abstr. in Phys. Abstr., Vol. 88, No. 1251, Entry 19176 (1985).

161.058   **Nonlinear growth of density contrast and formation of galaxies in a neutrino–dominant Universe.**
S. Ikeuchi, M. Umemura.
Prog. Theor. Phys., Vol. 72, No. 2, p. 216 – 232 (1984). Abstr. in Phys. Abstr., Vol. 88, No. 1251, Entry 24530 (1985).

161.059   **Scenarios of galaxy formation.**
H. Sato.
Grand unified theories and cosmology, p. 275 – 286 (1984). Abstr. in Phys. Abstr., Vol. 88, No. 1251, Entry 24538 (1985). – See Abstr. 012.020.

161.060   **Galaxy formation theory and large scale structure in the Universe.**
S. Ikeuchi.
Grand unified theories and cosmology, p. 287 – 302 (1984). Abstr. in Phys. Abstr., Vol. 88, No. 1251, Entry 24539 (1985). – See Abstr. 012.020.

161.061   **Current status of the missing mass problem.**
F. Takahara.
Grand unified theories and cosmology, p. 148 – 168 (1984). Abstr. in Phys. Abstr., Vol. 88, No. 1251, Entry 24542 (1985). – See Abstr. 012.020.

161.062   **Infrared observation of the early Universe.**
T. Matsumoto.
Grand unified theories and cosmology, p. 131 – 147 (1984). Abstr. in Phys. Abstr., Vol. 88, No. 1251, Entry 24563 (1985). – See Abstr. 012.020.

161.063   **Coleman–Weinberg model in Einstein space–time.**
G. Denardo, E. Spallucci.
Nuovo Cimento A, Vol. 83A, Ser. 2, No. 1, p. 35 – 49 (1984). Abstr. in Phys. Abstr., Vol. 88, No. 1251, Entry 24570 (1985).

161.064   **The invisible universe.**
J. D. Barrow, J. Silk.
New Sci., Vol. 103, No. 1419, p. 28 – 30 (1984). Abstr. in Phys. Abstr., Vol. 88, No. 1251, Entry 24571 (1985).

161.065   **Singularities in a stochastically predictable universe.**
J. Gruszczak, M. Heller.
Phys. Lett. A, Vol. 106A, No. 1 – 2, p. 13 – 16 (1984). Abstr. in Phys. Abstr., Vol. 88, No. 1251, Entry 24572 (1985).

161.066   **Singularities and the Hausdorff condition.**
W. Rudnicki.
Phys. Lett. A, Vol. 106A, No. 1 – 2, p. 17 – 18 (1984). Abstr. in Phys. Abstr., Vol. 88, No. 1251, Entry 24573 (1985).

161.067   **A bound on inflationary energy density from the isotropy of the microwave background.**
D. H. Lyth.
Phys. Lett. B, Vol. 147B, No. 6, p. 403 – 404 (1984). Abstr. in Phys. Abstr., Vol. 88, No. 1251, Entry 24574 (1985).

161.068   **Extended structures at intermediate scales in an inflationary cosmology.**
G. Lazarides, Q. Shafi.
Phys. Lett. B, Vol. 148B, No. 1 – 3, p. 35 – 38 (1984). Abstr. in Phys. Abstr., Vol. 88, No. 1251, Entry 24575 (1985).

161.069   **Higher–dimensional cosmologies.**
D. Lorenz–Petzold.
Phys. Lett. B, Vol. 148B, No. 1 – 3, p. 43 – 47 (1984). Abstr. in Phys. Abstr., Vol. 88, No. 1251, Entry 24576 (1985).

161.070   **Cosmological models based on the Nordström theory.**
N. K. Mohammed.
ZANCO, Ser. A, Vol. 8, No. 4, p. 23 – 31 (1983). Abstr. in Phys. Abstr., Vol. 88, No. 1251, Entry 24578 (1985).

161.071   **Inflationary universe model.**
K. Sato.
Grand unified theories and cosmology, p. 303 – 348 (1984). Abstr. in Phys. Abstr., Vol. 88, No. 1251, Entry 24581 (1985). – See Abstr. 012.020.

161.072   **Problems in the perturbation analysis of a universe dominated by a coherent scalar field.**
M. Sasaki.
Grand unified theories and cosmology, p. 349 – 358 (1984). Abstr. in Phys. Abstr., Vol. 88, No. 1251, Entry 24582 (1985). – See Abstr. 012.020.

161.073   **Comments on the chaotic inflation.**
H. Kodama.
Grand unified theories and cosmology, p. 359 – 375 (1984). Abstr. in Phys. Abstr., Vol. 88, No. 1251, Entry 24583 (1985). – See Abstr. 012.020.

161.074   **A comparative study of cosmological evolution in the radio, optical and X–ray bands.**
L. Danese, G. De Zotti, A. Franceschini.
Astron. Astrophys., Vol. 143, No. 2, p. 277 – 291 (1985).
The authors have carried out, using all the available data, a comparative study of the cosmological evolution of steep– and "flat"–spectrum radio sources, optically selected QSOs and X–ray selected active galactic nuclei, in the framework of both luminosity dependent density evolution models and luminosity evolution models. Models of the former type provide a good fit to all the observations with evolution functions strongly increasing with luminosity; the steepest dependence on luminosity is found for optical QSOs, the shallowest for "flat"–spectrum radio sources. Good fits can also be obtained in the framework of pure luminosity evolution models.

161.075   **H II regions and the value of H_0.**
P. Teerikorpi.
Astron. Astrophys., Vol. 143, No. 2, p. 469 – 474 (1985).
An approximately linear D_3 vs. LC relation from H II region diameter data of Kennicutt for late–type galaxies is discussed and found to provide useful complementary information on the Hubble constant H_0 within a local velocity field model.

161.076   **Optical multicolors: a poor person's z machine for galaxies.**
D. C. Koo.
Astron. J., Vol. 90, No. 3, 418 – 440, 560 (1985).
The author explores the feasibility, limits, accuracy, and uncertainties of using optical broadband colors in the form of two–color plots to measure the redshifts of faint galaxies. Comparison of redshifts from slit spectra and 4–m photographic $UBVI$ photometry of 100 faint galaxies confirms that colors do yield redshifts. For redshifts less than 0.35, two–thirds of the color redshifts agree with the spectroscopic values to within ± 0.04 in z; for higher redshifts to z of 0.6, the errors increase to ± 0.06. These accuracies were achieved for blue as well as red galaxies, and ensure that multicolors can serve as a poor person's redshift machine to study distant galaxies in clusters and the field.

161.077   **The hidden dimensions of spacetime.**
D. Z. Freedman, P. van Nieuwenhuizen.
Sci. Am., Vol. 252, No. 3, p. 62 – 69 (1985).
Spacetime, usually thought of as four–dimensional, may have as many as seven extra dimensions. Eleven–dimensional structures now under study might give a unified account of the four basic forces of nature.

161.078   **Betazerfall und das Alter des Universums.**
H. V. Klapdor.
Sterne Weltraum, 24. Jahrg., Nr. 3, p. 132 – 139 (1985).

161.079 Inhomogeneous cosmology. II. Linearly polarized gravitational waves.
P. J. Adams, R. W. Hellings, R. L. Zimmerman.
Astrophys. J., Vol. 288, No. 1, p. 14 – 21 (1985).

This is the second in a series of papers whose purpose is to investigate the possible strength of a cosmic background of gravitational radiation. The authors present a metric which may be used to describe exact plane gravitational waves propagating over a homogeneous background spacetime, and they find solutions representing waves of linear polarization over backgrounds of Bianchi types I, III, V, and VI, as well as over a background with a smooth global inhomogeneity. In all these cases an inhomogeneity in the structure of the initial cosmic singularity evolves into gravitational waves propagating over a homogeneous background spacetime.

161.080 Theory and interpretation of quantized extragalactic redshifts.
W. J. Cocke.
Astrophys. J., Vol. 288, No. 1, p. 22 – 28 (1985).

A consistent 3–dimensional form of a preliminary theory of extragalactic redshift quantization is developed. The results are applied to the dynamics of an isolated galaxy component, by means of a Schrödinger equation. The interpretation of the wave function is discussed, and it is possible to associate the coefficients of the expansion of a wave function in terms of redshift eigenfunctions with observed redshift components.

161.081 Extragalactic dust and near–infrared cosmic background.
P. de Bernardis, S. Masi, A. Malagoli, F. Melchiorri.
Astrophys. J., Vol. 288, No. 1, p. 29 – 31 (1985).

The possible existence of a near–infrared background ($1 < \lambda < 20$ µm) has been recently suggested by rocket observations. The authors discuss the effects of the interaction between such a background and extragalactic dust. Primordial dust ($40 < z < 100$) cannot exceed $n_0 < 10^{-21}\mathrm{cm}^{-3}$, in order to avoid a complete absorption of the background. A far–infrared diffuse radiation would be generated by dust heated by the background. Such radiation could be consistent with the recent observations of Gush in the $500 – 700$ µm region.

161.082 Counterimages in closed elliptical Friedmann universes.
J. V. Narlikar, T. R. Seshadri.
Astrophys. J., Vol. 288, No. 1, p. 43 – 49 (1985).

It is shown that the different connectivity implied by the elliptical version of a closed Friedmann model allows two images of a distant astronomical object to be seen, provided the deceleration parameter q_0 of the Friedmann model exceeds unity. Of the two images the "direct image" is along the shortest route light track. If the redshift of the direct image exceeds $(1.5q_0-1)(q_0-1)^{-2}$ then a second "counterimage" should be visible at the diametrically antipodal position. The direct image has a maximum possible redshift $z_{max}(q_0)$, and it is suggested that the apparent cutoff in the redshifts of QSOs may be due to this effect.

161.083 Pancakes and the formation of galaxies in a universe dominated by collisionless particles.
P. R. Shapiro, C. Struck-Marcell.
Astrophys. J., Suppl. Ser., Vol. 57, No. 2, p. 205 – 239 (1985).

Detailed, numerical hydrodynamical calculations are presented of the coupled growth of (baryon)–(collisionless particle) pancakes in a post–recombination Friedmann universe dominated by finite mass, stable, weakly interacting elementary particles, such as neutrinos, gravitinos, photinos, or axions. Pancakes such as these arise during the cosmological growth of primordial, adiabatic density fluctuations and may be the sites of galaxy formation. The evolution of individual pancakes is calculated by numerically solving the hydrodynamical conservation equations for the baryonic fluid, together with the Poisson equation for the combined (baryon)–(collisionless particle) gravitational potential and the Vlasov equation for the collisionless particles, all in a homogeneous, isotropically expanding universe described by the Robertson–Walker metric.

161.084 Numerical techniques for large cosmological N–body simulations.
G. Efstathiou, M. Davis, C. S. Frenk, S. D. M. White.
Astrophys. J., Suppl. Ser., Vol. 57, No. 2, p. 241 – 260 (1985).

The authors discuss techniques for carrying out large N–body simulations of the growth of structure in an expanding universe; both particle mesh schemes and schemes with more accurate interparticle forces are considered. The authors investigate force smoothing and anisotropy, integration efficiency, energy and momentum conservation, and the effects of boundaries and finite particle number. They present an algorithm for setting up a representation of any desired initial spectrum of growing mode density perturbations, for example those predicted in neutrino– or axion–dominated universes. Test simulations illustrate the performance and range of validity of the methods.

161.085 A hierarchical explosion scheme for the origin of cosmological structure.
B. J. Carr, S. Ikeuchi.
Mon. Not. R. Astron. Soc., Vol. 213, No. 3, p. 497 – 518 (1985).

The explosions of pregalactic stars in the period $10 < z < 10^3$, when Compton cooling dominates, could initiate a bootstrap process in which ever–larger shells of swept–up gas successively fragment into more exploding stars. The authors derive a simple relationship between the amplifications occurring at each stage and infer that the shell mass, in the absence of shell mergers, would tend to a characteristic value which depends only on the redshift and explosive efficiency. The formation of a few pregalactic stars could lead to the present cosmological structure.

161.086 Mechanisms for biased galaxy formation.
M. J. Rees.
Mon. Not. R. Astron. Soc., Vol. 213, No. 3, p. 75P – 81P (1985).

A 'flat' universe with $\Omega = 1$, attractive on theoretical grounds, can only be reconciled with virial–type analyses if galaxies are more 'clumped' than the overall mass distribution even on scales up to $10 – 20$ Mpc. Formation of conspicuous galaxies must be favoured in high–density regions, and inhibited in voids. Mechanisms are discussed for bringing this about, particularly in the context of the 'cold dark matter' cosmogony. There are two forms in which energy from early galaxies could provide a pressure that prevents condensation of further galaxies $\gtrsim 20$ Mpc away: (1) fast particles, or (2) UV radiation pressure, coupled to the intergalactic gas via Lyman α scattering. If the universe has $\Omega = 1$, with baryons contributing $\Omega_b = 0.1$, then up to 90 per cent of these baryons could remain uncondensed in intergalactic space or be in systems with $M/L > 100$ solar units.

161.087 The cosmological constant in the McCrea–Milne cosmological scheme.
V. G. Gurzadyan.
Observatory, Vol. 105, No. 1065, p. 42 – 43 (1985).

A natural way of introducing the cosmological constant into the equations of motion in the McCrea–Milne approach is demonstrated.

161.088 On the way of understanding Particle–In–Cell simulations of gravitational clustering.
F. R. Bouchet, J.–C. Adam, R. Pellat.
Astron. Astrophys., Vol. 144, No. 2, p. 413 – 426 (1985).

This paper is the first in a projected series on an attempt to study, define, and improve the range and the conditions of validity of the numerical simulations of gravitational collapse. The authors have built a 2–D Particle–In–Cell code to address the problem of collisionless gravitational clustering. Owing to the unmatched resolution obtained by using a 512^2 grid, they have clearly assessed the role of the two scales present in their simulations, the Jeans length and the grid–spacing. They discovered that numerical effects perturb an unexpectedly large range, and report on the very interesting behaviour of very small scale perturbations. Furthermore, they have computed velocity moments of the two–point correlation function up to the third order.

161.089 The distribution of voids.
G. Vettolani, R. E. de Souza, B. Marano,
G. Chincarini.
Astron. Astrophys., Vol. 144, No. 2, p. 506 – 513 (1985).

After determining the location and estimating the volume of regions which are void of galaxies in a sample complete to $m_z \leqslant 14.5$, the authors compare the observed volume-distribution with the distribution estimated in simulations generated according to the model of Soneira–Peebles. The numerical analysis and a rather approximated analytical approach show that (1) the voids observed in the sample used do not contradict a hierarchical distribution of matter in the Universe and (2) even the presence of large voids as the one claimed in Bootes could be explained in a hierarchical realization.

161.090 On percolation as a cosmological test.
A. Dekel, M. J. West.
Astrophys. J., Vol. 288, No. 2, p. 411 – 417 (1985).

The authors point out difficulties in the use of percolation as a complementary statistic for the galaxy clustering pattern by studying simple toy models and dynamical N–body models that represent the competing clustering scenarios. The percolation properties are found not to be very sensitive to the presence of "pancakes" and "strings" once they are clumpy, and hence they do not distinguish properly between models that are very different. It is also shown that the dependence of the percolation properties on the sampling parameters is sensitive to the same topological properties that the test ought to measure.

161.091 Radiation–like imperfect fluid cosmologies.
A. A. Coley, B. O. J. Tupper.
Astrophys. J., Vol. 288, No. 2, p. 418 – 421 (1985).

Imperfect fluid solutions of the Einstein field equations are presented in which the geometrical part is formally identical to the $k = 0$, $R(t) = t^{1/2}$ radiation–like FRW line element. These exact solutions include some which are self–similar.

161.092 Gravitational lensing effects of vacuum strings: exact solutions.
J. R. Gott III.
Astrophys. J., Vol. 288, No. 2, p. 422 – 427 (1985).

Exact interior and exterior solutions to Einstein's field equations are obtained for vacuum strings. A maximum mass per unit length for a string is found: $\mu_{max} = 6.73 \times 10^{27} \mathrm{g\,cm}^{-1}$. Grand unified vacuum strings with $\mu \sim 10^{24} \mathrm{g\,cm}^{-1}$, consistent with the observed isotropy of the microwave background and large enough to promote galaxy formation, would produce equal brightness double images of QSOs with separations of up to 6′. Formulae for lensing probabilities, image splittings, time delays, etc., are derived for strings in a reasonable cosmological setting.

161.093 A comment on Dirac large number hypothesis cosmology.
N. T. Bishop.
Astrophys. J., Vol. 289, No. 1, p. 1 (1985).

Results on Dirac cosmology presented in a recent paper by Julg (1983) are criticized.

161.094 Can a neutrino–dominated universe be rejected?
A. L. Melott.
Astrophys. J., Vol. 289, No. 1, p. 2 – 4 (1985).

The clustering properties and large–scale motion in a standard neutrino–pancake model are shown to be consistent with observation for a steep primordial density perturbation spectrum. The autocorrelation properties of QSO Lyα absorbers are shown to be consistent with the behavior of matter in numerical simulations of neutrino–dominated universes. The importance of beginning simulations at a highly linear stage is emphasized.

161.095 Tycho's supernova and the Hubble constant.
G. de Vaucouleurs.
Astrophys. J., Vol. 289, No. 1, p. 5 – 9 (1985).

Recent determinations of the distance of 3C 10, the remnant of SN 1572, are reviewed; all indicate distances smaller than earlier estimates suggested. With the average distance modulus $\mu_0 = 12.5 \pm 0.2$, the apparent magnitude at maximum of SN 1572, corrected for absorption, leads to an absolute magnitude $M_{pg} = -18.85 \pm 0.3$. The mean corrected magnitudes of four well–observed Type I supernovae in the Virgo Cluster and five in the Coma Cluster, lead to distances of 12.2 – 13.7 and 68 – 80 Mpc, respectively, according to whether or not M_{pg} depends on rate of decay. With these distances and the mean radial velocities corrected for the peculiar motion of the Local Group, the corresponding values of the Hubble ratio are in the range 89 – 100 and 92 – 108 km s^{-1}Mpc^{-1}, or 97 ± 20 on the average.

161.096 Cosmological density fluctuations and large–scale structure: from N–point correlation functions to the probability distribution.
J. N. Fry.
Astrophys. J., Vol. 289, No. 1, p. 10 – 17 (1985).

Knowledge of N–point correlation functions for all N allows one to invert and obtain the probability distribution of mass fluctuations in a fixed volume. This method is applied to the hierarchical sequence of higher order correlations with dimensionless amplitudes suggested by the BBGKY equations. The resulting distribution is significantly non–Gaussian, even for quite small mean square fluctuations. The qualitative and to some degree quantitative results are to a large degree independent of the exact sequence of amplitudes. An ensemble of such models compared with N–body simulations fails to account for the low–density frequency distribution.

161.097 On model–dependent bounds on H_0 from gravitational images: application to Q0957 + 561A,B.
E. E. Falco, M. V. Gorenstein, I. I. Shapiro.
Astrophys. J., Lett. Ed., Vol. 289, No. 1, p. L1 – L4 (1985).

The authors consider the images of a background source formed by a foreground mass distribution in a matter–dominated Friedmann cosmology. They present a transformation that leaves observable properties of these gravitational images invariant. Using a compound mass distribution to represent a galaxy–cluster lens, the authors can account for the positions and relative magnifications of the A and B images of Q0957 + 561. Since mass cannot be negative, the transformation applied to this lens model yields bounds on the Hubble constant, given the difference $\Delta\tau_{BA}$ (in years) between the propagation times associated with the two images: $0 < H_0 \leqslant (98 \pm 3)\Delta\tau_{BA}^{-1}$ km s^{-1}Mpc^{-1}. Temporal variations in the quasar's luminosity always appear first in the A image. Measurements of any such variations, reliably correlated between images, would determine $\Delta\tau_{BA}$.

161.098 Globular clusters as extragalactic distance indicators.
G. Paturel.
C. R. Acad. Sci., Sér. II, Tome 300, No. 12, p. 569 – 572 (1985).
In French.

New relationships are found between observable parameters and the absolute magnitude for globular clusters in the Magellanic Clouds. Such relationships could be useful for extragalactic distance determination.

161.099 Aufbau und Entwicklung des Weltalls. VII. Zeugen der kosmischen Frühgeschichte.
D.-E. Liebscher.
Sterne, 61. Band, Heft 1, p. 11 – 20 (1985).

161.100 On the dimensional cosmological principles.
L. K. Chi.
Astrophys. J., Vol. 289, No. 2, p. 443 – 445 (1985).

The dimensional cosmological principles proposed by Wesson require that the density, pressure, and mass of cosmological models be functions of the dimensionless variables which are themselves combinations of the gravitational constant, the speed of light, and the spacetime coordinates. The space coordinate is not the comoving coordinate. In this paper, the dimensional cosmological principle and the dimensional perfect cosmological principle are reformulated by using the comoving coordinate.

161.101   Dark matter and inflation.
L. M. Krauss.
Gen. Relativ. Gravitation, Vol. 17, No. 1, p. 89 – 94 (1985).

The author demonstrates that dark matter consisting of any type or types of stable weakly interacting elementary particle is incompatible with the minimal predictions of inflation, based on present observations of galaxy clustering, and assuming galaxies are good tracers of mass in the universe.

161.102   Universal rotation: how large can it be?
J. D. Barrow, R. Juszkiewicz, D. H. Sonoda.
Mon. Not. R. Astron. Soc., Vol. 213, No. 4, p. 917 – 943 (1985).

The authors present a detailed analysis of the expected temperature distribution of the cosmic microwave background radiation in flat, open and closed universes possessing small anisotropies. They predict the most general temperature patterns on the sky and their associated angular correlation functions. The authors use these results to evaluate the largest level of cosmic vorticity compatible with existing observations of the dipole and quadrupole fluctuations in the microwave background. This analysis extends previous work by employing the quadrupole observations and calculating all observable quantities.

161.103   The polarization of the microwave background in open universes.
B. W. Tolman.
Astrophys. J., Vol. 290, No. 1, p. 1 – 11 (1985).

The polarization and anisotropy of the microwave background radiation in open universes with an expansion anisotropy, arising from either a cosmological anisotropy or density inhomogeneities on large scales, are computed by a direct, or Monte Carlo, simulation of the problem. The simulation includes accurate numerical data for the ionization of matter during both the decoupling and the reheated eras. Besides the well–known result that the radiation anisotropy is distorted and focused by the spatial curvature into a single very small, intense spot, it is found that the scattering of this spot during the reheated epoch generates features in the polarization of the radiation which cover fully one–half of the sky.

161.104   Carrier of "hidden mass" in models of an inflationary universe.
P. D. Nasel'skij, A. G. Polnarev.
Inst. kosm. issled. Akad. Nauk SSSR. Prepr., No. 913, 20 pp. (1984). In Russian. Abstr. in Ref. Zh., 51. Astron., 1.51.1218 (1985).

161.105   The dependence of baryon asymmetry of the universe on the equation of state.
Yu. A. Beletskij, A. I. Bugrij, A. A. Trushevskij.
Inst. teor. fiz. Akad. Nauk USSR. Prepr., No. 77 E, 17 pp. (1984). Abstr. in Ref. Zh., 51. Astron., 1.51.1220 (1985).

161.106   Conformal anomalies and generation of massless particles in a Friedmann universe.
I. L. Bukhbinder, S. D. Odintsov.
Izv. vuzov. Fiz., Tom 27, No. 8, p. 54 – 58 (1984). In Russian. Abstr. in Ref. Zh., 51. Astron., 1.51.1221 (1985).

161.107   Large–scale structure of the universe: three–dimensional model.
L. V. Rozhanskij, S. F. Shandarin.
Inst. prikl. mat. Akad. Nauk SSSR. Prepr., No. 79, 26 pp. (1984). In Russian. Abstr. in Ref. Zh., 51. Astron., 1.51.1227 (1985).

161.108   Quantum dynamics of an isotropic cosmological model.
D. E. Burlankov, V. N. Dutyshev, A. A. Kochnev.
Zh. Ehksp. Teor. Fiz., Tom 87, No. 9, p. 705 – 716 (1984). In Russian. Abstr. in Ref. Zh., 51. Astron., 1.51.1234 (1985).

161.109   On two variants in the scenario of evolution of the universe.
M. E. Gertsenshtejn.
Izv. vuzov. Fiz., Tom 27, No. 6, p. 103 – 106 (1984). In Russian. Abstr. in Ref. Zh., 51. Astron., 1.51.1237 (1985).

161.110   De Sitter initial state of the universe as a result of asymptotic disappearance of gravitational interactions of matter.
M. A. Markov, V. F. Mukhanov.
Pis'ma ZhTEF, Tom 40, No. 6, p. 265 – 268 (1984). In Russian. Abstr. in Ref. Zh., 51. Astron., 1.51.1238 (1985).

161.111   Spontaneous violation with C–invariance in a Friedmann model of the universe with matter.
F. Sh. Zaripov.
Izv. vuzov. Fiz., Tom 27, No. 7, p. 22 – 26 (1984). In Russian. Abstr. in Ref. Zh., 51. Astron., 1.51.1239 (1985).

161.112   Classical models of a topological transition in cosmology.
M. Yu. Konstantinov, V. N. Mel'nikov.
Izv. vuzov. Fiz., Tom 27, No. 8, p. 32 – 37 (1984). In Russian. Abstr. in Ref. Zh., 51. Astron., 1.51.1240 (1985).

161.113   Magnetic monopole solutions of a non–Abelian gauge theory in an expanding universe.
V. K. Shigolev.
Izv. vuzov. Fiz., Tom 27, No. 8, p. 67 – 71 (1984). In Russian. Abstr. in Ref. Zh., 51. Astron., 1.51.1241 (1985).

161.114   Gas–dynamical model of the universe.
V. T. Sarychev.
Izv. vuzov. Fiz., Tomsk, 44 pp. (1984). In Russian. Abstr. in Ref. Zh., 51. Astron., 1.51.1254 (1985).

161.115   Comment on the BDT–Bianchi type–I stiff matter solution.
D. Lorenz–Petzold.
Astrophys. Space Sci., Vol. 109, No. 2, p. 387 – 391 (1985).

The author wishes to point out that the Brans–Dicke–stiff matter solution recently given by Ram and Singh (1984) is incorrect, in so far as the corresponding field equations have not been solved completely. The 'new' GRT–Bianchi type–I stiff matter solution does not exist.

161.116   Comment on the Brans–Dicke–Bianchi type–III vacuum solution.
D. Lorenz–Petzold.
Astrophys. Space Sci., Vol. 109, No. 2, p. 411 (1985).

The author wishes to point out that the Brans–Dicke–Bianchi type–III vacuum solution recently given by Tiwari and Singh (1984) is not new. Moreover, the solution given has no correct Einstein limit, contrary to what is claimed by these authors. The Ellis–MacCallum vacuum solution in the Einstein case can be obtained from the Brans–Dicke solution first given by Lorenz–Petzold (1984).

161.117   Experimental test of cosmological theories.
F. Melchiorri, B. Olivo–Melchiorri.
Nauk. i chelovechestvo, 1984: Dostupno i tochno o gl. v mirov. nauk. Mezhdunar. ezhegod. Moskva, p. 229 – 233 (1984). In Russian. Abstr. in Ref. Zh., 51. Astron., 2.51.968 (1985).

161.118   Anisotropic cosmological model in a new scalar–tensor theory.
A. Banerjee, A. K. G. de Oliveira, N. O. Santos.
Gen. Relativ. Gravitation, Vol. 17, No. 4, p. 371 – 380 (1985).

The authors study a stiff fluid Bianchi type I cosmological model in a scalar–tensor theory where the gravitational constant depends on the strength of the scalar field. Near the singularity the matter density goes to infinity at a faster rate than the amount of shear, which itself becomes infinitely large at this limit. Exact solutions in some special cases are obtained.

161.119   The nature of the initial singularity.
G. F. R. Ellis, W. L. Roque.
Gen. Relativ. Gravitation, Vol. 17, No. 4, p. 397 – 406 (1985).

The authors consider the initial singularity in the universe as isotropic and with a strength the same as that of the singularity in the density function. They study the effect of this initial singularity on a particle falling into it in the future at a high speed. This gives a precise understanding of the fate of particles falling into the final singularity in the future of a $K = +1$ Friedmann–Lemaitre–Robertson–Walker universe.

161.120 A higher limit to the mean galaxy mass allowed by gravitational lens image distortion.
S. Phillipps.
Nature, Vol. 314, No. 6013, p. 721 (1985).

Tyson et al., in a recent discussion of the gravitational lens distortion, by foreground galaxies, of the images of distant ("background") galaxies, claim that their non–detection of this effect implies a low mean mass for the foreground galaxies. It would require that the cosmological mass density in the form of galaxies is $\Omega_{gal} < 0.03$. This letter shows that by including the effects of galaxy clustering the limit is raised to $\Omega_{gal} \lesssim 0.2$.

161.121 Generation of entropy perturbations by primary adiabatic ones in the early Universe.
B. V. Vajner, A. A. Gusev.
Astron. Zh., Tom 62, Vyp. 2, p. 193 – 201 (1985). In Russian. English translation in Sov. Astron., Vol. 29, No. 2.

The generation of entropy perturbations by primary adiabatic ones in the early Universe is considered, provided that the baryon asymmetry is explained by a decay of heavy X–bosons with violation of the CP–invariance and of the conservation law of baryon charge. It is shown that in the standard Universe it is possible to obtain the entropy perturbations similar to the adiabatic spectrum. A model of the Universe with second–kind phase transitions caused by spontaneous symmetry breaking is constructed. The spectra of primary adiabatic and following entropy perturbations and their role in creation of large–scale structure of the Universe are discussed.

161.122 The "hidden mass" in the Universe: neutrinos or supermassive stable particles?
N. A. Zabotin, P. D. Nasel'skij.
Astron. Zh., Tom 62, Vyp. 2, p. 410 – 412 (1985). In Russian. English translation in Sov. Astron., Vol. 29, No. 2.

The analysis of the data on the small–scale anisotropy of the relic electromagnetic radiation leads to a conclusion on the advantages of the models of the Universe with supermassive carriers of its "hidden mass".

161.123 Galaxy and cluster formation in a neutrino universe.
A. G. Doroshkevich.
Sov. Astron., Vol. 28, No. 4, p. 378 – 383 (1984). English translation of 38.161.024.

161.124 Heavy neutrinos and the evolution of primordial gravitational perturbations.
I. K. Rozgacheva.
Sov. Astron., Vol. 28, No. 4, p. 384 – 389 (1984). English translation of 38.161.025.

161.125 60 years of the theory of the expanding universe.
P. Ya. Polubarinova–Kochina, V. I. Khlebnikov.
Zemlya Vselennaya, No. 1, p. 74 – 85 (1985). In Russian.

161.126 Physics and cosmology: some interactions.
W. H. McCrea.
The Big Bang and Georges Lemaitre, p. 3 – 22 (1984). – See Abstr. 012.043.

The purpose of this paper is to sketch in some of the background to Lemaitre's cosmology, to recall its main features, briefly to review the development of observational cosmology since the time when Lemaitre proposed his model, and then to note some sequels to his ideas in some of the most recent models.

161.127 Impact of Lemaitre's ideas on modern cosmology.
P. J. E. Peebles.
The Big Bang and Georges Lemaitre, p. 23 – 30 (1984). – See Abstr. 012.043.

This paper recalls some aspects of the history of the discovery of the expansion of the universe and assesses the present

status of some of Lemaitre's main ideas in physical cosmology.

161.128 Massive neutrinos and photinos in cosmology and galactic astronomy.
D. W. Sciama.
The Big Bang and Georges Lemaitre, p. 31 – 41 (1984). – See Abstr. 012.043.

The hot Primeval Atom would have produced pairs of neutrinos and of photinos (if they exist), many of which would have survived to the present day. If these particles have non–zero rest–mass they might dominate the universe, providing it with the critical density, and also individual galaxies, providing them with their missing mass. This hypothesis might be tested by searching for the photons which these particles would be expected to emit.

161.129 The primordial nucleosynthesis.
J. Audouze.
The Big Bang and Georges Lemaitre, p. 43 – 61 (1984). – See Abstr. 012.043.

This paper first reviews the attempts to determine the primordial abundances of the lightest elements which can be formed by the Big Bang nucleosynthesis. It then presents a summary of the Standard Big Bang nucleosynthesis. In particular, it is shown that from the abundances of D, ^{3}He and ^{7}Li a baryonic cosmological parameter of $0.10(+0.11, -0.09)$ can be derived. Furthermore, the He primordial abundance is consistent with the three different types of neutrinos which are presently observed.

161.130 A generalization of the Lemaitre models.
A. Krasiński.
The Big Bang and Georges Lemaitre, p. 63 – 73 (1984). – See Abstr. 012.043.

A generalization of the Friedmann–Lemaitre–Robertson–Walker (FLRW) models is obtained by weakening the assumptions under which they are derived from Einstein's relativity. It is assumed that each section t = const. is homogeneous and isotropic while the spacetime itself does not necessarily have any symmetry. In this universe, some spatial sections may be open while others will be closed. Its geometrical picture is presented and its physical properties are discussed.

161.131 Is the universe homogeneous on a large scale?
B. Mashhoon.
The Big Bang and Georges Lemaitre, p. 75 – 81 (1984). – See Abstr. 012.043.

It is proposed to replace the cosmological principle of spatial homogeneity by the assumption that the (matter and radiation) content of the universe is on the average uniform, i.e., the equation of state of the system is everywhere the same. Einstein's equations allow then solutions which differ from the FLRW models by the existence of radial inhomogeneities due to shear. As a first step toward a general study of inhomogeneities, local models with radial inhomogeneities have been developed and the observational quantities for such models have been determined.

161.132 Vacuum inhomogeneous cosmological models.
J.–L. Hanquin.
The Big Bang and Georges Lemaitre, p. 83 – 92 (1984). – See Abstr. 012.043.

The author presents some results concerning the vacuum cosmological models which admit a 2–dimensional Abelian group of isometries; the classification of these space–times based on the topological nature of their space–like hypersurfaces and on their time evolution is presented.

161.133 The significance of Newtonian cosmology.
D. Galletto, B. Barberis.
The Big Bang and Georges Lemaitre, p. 93 – 102 (1984). – See Abstr. 012.043.

Starting from the hypotheses that the physical space is Euclidean, that the universe is infinite and homogeneous and that with regard to our Galaxy its behaviour is isotropic, without resorting to Newton's law of gravitation the authors deduce Hubble's law, the law of motion of a typical galaxy, the equation of evolution of the universe, and that the force at a distance exerted between any two galaxies is expressed by Newton's law of gravitation.

161.134 Numerical simulation of evolution of a multi-dimensional Higgs field in the new inflationary scenario.
K. Sato, H. Kodama.
The Big Bang and Georges Lemaitre, p. 103 – 112 (1984). – See Abstr. 012.043.
Numerical simulation of the evolution of the adjoint Higgs field in the new inflationary universe scenario based on the SU(5) GUT model with the Coleman–Weinberg potential is carried out by introducing a suitable viscosity term in the equation of motion.

161.135 Primordial nucleosynthesis and nuclear reaction rates uncertainties.
P. Delbourgo–Salvador, E. Vangioni–Flam, G. Malinie, J. Audouze.
The Big Bang and Georges Lemaitre, p. 113 – 122 (1984). – See Abstr. 012.043.
This paper investigates the implications of uncertainties in the nuclear reaction rates used for calculating primordial abundances in the Standard Big–Bang models. It is concluded that the experimental uncertainties cannot account for discrepancies between theoretical and observed abundances.

161.136 Estimation of galactic masses using the zero energy-momentum cosmological principle.
D. K. Callebaut, N. Van den Bergh, P. Wils.
The Big Bang and Georges Lemaitre, p. 139 – 148 (1984). – See Abstr. 012.043.
The cosmological zero energy–momentum principle is briefly explained. It is then applied in a quasi–Newtonian model to derive Jeans' instability criterion, with and without electrical charges. The calculation of galaxy masses from it yields reasonable agreement if the estimated values for the temperature in galaxies are used.

161.137 Galaxy formation revisited.
J. Silk.
The Big Bang and Georges Lemaitre, p. 279 – 297 (1984). – See Abstr. 012.043.
The statistical properties of galaxies are used to infer the distribution of specific binding energy with surface density for a range of Hubble types. It is argued that the inferred locus characterizes physical conditions at the end of the dissipative phase of protogalactic evolution. Cosmological fluctuations in density provide the initial conditions at the onset of galaxy formation, and the critical surface density for radiative cooling to have occurred in protogalactic gas clouds provides a necessary lower bound for star formation. If gas is supported in, and eventually driven from, galactic potential wells by energy input from forming and dying stars, many of the statistical correlations found for galaxies can be understood.

161.138 Cosmology, galactic astronomy and elementary particle physics.
D. W. Sciama.
ESA Spec. Publ., ESA SP–207, p. 171 – 178 (1984). – See Abstr. 012.044.
Cosmology, galactic astronomy and elementary particle physics are now closely interwoven. Despite the many existing uncertainties, each of these subjects appears to provide basic information for the others, and cannot be understood without taking the others into account. This recent development is illustrated by brief discussions of the inflationary universe, the primordial synthesis of baryons and the light elements, and the possible roles of massive neutrinos, photinos and gravitinos in providing the "missing" matter in the universe as a whole, in galaxy clusters and in individual galaxies.

161.139 Conserved quantities and radiative aspects of asymptotically anti–de Sitter space–times.
A. Magnon.
ESA Spec. Publ., ESA SP–207, p. 237 – 238 (1984). – See Abstr. 012.044.
A new framework to describe the structure of asymptotically anti–de Sitter space–times is constructed and the resulting definition of conserved quantities is compared and contrasted with that proposed by Abbot and Deser.

161.140 Galaxy formation.
J. Silk.
ESA Spec. Publ., ESA SP–207, p. 327 – 340 (1984). – See Abstr. 012.044.
Implications of the isotropy of the cosmic microwave background on large and small angular scales for galaxy formation are reviewed. In the context of primeval adiabatic fluctuations, one is compelled to consider a universe dominated by cold, weakly interacting non–baryonic matter. One of the more interesting candidates for this matter is the massive photino, and a possible signature of photino annihilation in our galactic halo involves production of cosmic ray antiprotons. If the density is near its closure value, it is necessary to invoke a biasing mechanism for suppressing galaxy formation throughout most of the universe in order to reconcile the dark matter density with the lower astronomical determinations of the mean cosmological density. A mechanism is described utilizing the onset of primordial massive star formation to strip gaseous protogalaxies, only the densest, early collapsing systems forming luminous galaxies.

161.141 Shell crossings and the Tolman model.
C. Hellaby, K. Lake.
Astrophys. J., Vol. 290, No. 2, p. 381 – 387 (1985).
The authors consider the problem of shell crossings and regular maxima in the Tolman model. The necessary and sufficient conditions which guarantee no shell crossings will arise in Tolman models are derived, and the authors show explicitly that a Tolman model (in general, with a surface layer) may contain both elliptic and hyperbolic regions without developing any shell crossings and without the hyperbolic regions recollapsing. This finding is contrary to the recent hypothesis of Zel'dovich and Grishchuk.

161.142 The angular distance between stars in a Brans–Dicke scalar–tensor theory.
R. K. Kadyev, M. Ablaeva.
Investigations of nuclear physics and physics of elementary particles, p. 4 – 9 (1984). In Russian. Abstr. in Ref. Zh., 51. Astron., 3.51.186 (1985). – See Abstr. 003.016.

161.143 On a possible solution of the problem of initial perturbations in cosmology.
B. L. Spokojnyj.
Pis'ma ZhEhTF, Tom 40, No. 8, p. 354 – 356 (1984). In Russian. Abstr. in Ref. Zh., 51. Astron., 3.51.913 (1985).

161.144 Generalized Lie super–algebra and supergravitation with positive cosmological constant.
M. A. Vasil'ev.
Pis'ma ZhEhTF, Tom 40, No. 10, p. 437 – 440 (1984). In Russian. Abstr. in Ref. Zh., 51. Astron., 3.51.926 (1985).

161.145 On rotation of the universe and redshift.
V. D. Lyakhovets.
Univ. druzhby narodov. Moskva, 5 pp. (1984). In Russian. Abstr. in Ref. Zh., 51. Astron., 3.51.942 (1985).

161.146 The distribution of distant galaxies.
D. Nanni, D. Trevese, A. Vignato.
Mem. Soc. Astron. Ital., Vol. 55, No. 3, p. 609 – 613 (1984). – See Abstr. 012.050.

161.147 Imprints on the cosmic background radiation due to primordial fluctuations in universe models dominated by non–baryonic dark matter.
F. Lucchin.
Mem. Soc. Astron. Ital., Vol. 55, No. 3, p. 625 – 628 (1984). – See Abstr. 012.050.

Several detailed calculations of the CBR temperature fluctuations have been worked out for universe models dominated by non–baryonic weakly interacting particles. The comparison with the observed fluctuations strongly constrains all possible cosmological scenarios and favours a universe model dominated by "cold" dark matter (axions, photinos,...).

161.148 Evolution of a very early universe with polarized vacuum.
V. G. Gurzadyan, A. A. Kocharyan, S. G. Matinyan.
Astrofizika, Tom 22, Vyp. 2, p. 287 – 292 (1985). In Russian. English translation in Astrophysics, Vol. 22, No. 2.

The Einstein equations with one–loop quantum corrections describing the polarization of vacuum of physical fields are investigated. It is shown that within a time interval, when only one–loop corrections for a strong gravitational field are sufficient, the equations have the only stable solution of the Friedmann type.

161.149 Classical Kaluza–Klein cosmology for a torus space with a cosmological constant and matter.
Y. Tosa.
Phys. Rev. D, Vol. 30, No. 10, p. 2054 – 2060 (1984). Abstr. in Phys. Abstr., Vol. 88, No. 1252, Entry 24810 (1985).

161.150 How can heavy neutrinos dominate the Universe?
P. B. Pal.
Phys. Rev. D, Vol. 30, No. 10, p. 2100 – 2104 (1984). Abstr. in Phys. Abstr., Vol. 88, No. 1252, Entry 24811 (1985).

161.151 Quantum mechanics in the tunneling universe.
V. A. Rubakov.
Phys. Lett. B, Vol. 148B, No. 4 – 5, p. 280 – 286 (1984). Abstr. in Phys. Abstr., Vol. 88, No. 1252, Entry 24815 (1985).

161.152 Viscosity and the monopole density of the Universe.
L. Diosi, B. Keszthelyi, B. Lukacs, G. Paal.
Acta Phys. Pol., Ser. B, Vol. B15, No. 10, p. 909 – 918 (1984). Abstr. in Phys. Abstr., Vol. 88, No. 1252, Entry 30098 (1985).

161.153 Higher derivatives in quantum cosmology. I. The isotropic case.
S. W. Hawking, J. C. Luttrell.
Nucl. Phys. B, Part. Phys., Vol. B247, No. 1, p. 250 – 260 (1984). Abstr. in Phys. Abstr., Vol. 88, No. 1252, Entry 30101 (1985).

161.154 Gauge–invariant density fluctuations in a cold dark matter universe.
T. Moody.
Phys. Lett. B, Vol. 149B, No. 4 – 5, p. 328 – 332 (1984). Abstr. in Phys. Abstr., Vol. 88, No. 1252, Entry 30104 (1985).

161.155 Generation of isothermal density perturbations in an inflationary universe.
A. D. Linde.
Pis'ma ZhEhTF, Tom 40, No. 11, p. 496 – 498 (1984). In Russian. Abstr. in Ref. Zh., 51. Astron., 4.51.778 (1985).

161.156 Motion of a free particle in a Friedmann world.
Z. F. Seidov.
Dokl. Akad. Nauk AzSSR, Tom 40, No. 7, p. 51 – 54 (1984). In Russian. Abstr. in Ref. Zh., 51. Astron., 4.51.793 (1985).

161.157 Cosmology of magnetic monopoles.
M. Izawa.
Astron. Her., Vol. 78, No. 5, p. 138 – 144 (1985). In Japanese.

161.158 On the cosmological parameters.
S. J. M. Stoelinga.
Astrophys. J., Vol. 291, No. 2, p. 396 – 398 (1985).

From the Einstein mass–energy equivalence, a simple Newtonian relationship concerning the main cosmological parameters is suggested which corresponds well to recent observationally obtained values.

161.159 Explosions in the early universe.
E. T. Vishniac, J. P. Ostriker, E. Bertschinger.
Astrophys. J., Vol. 291, No. 2, p. 399 – 416 (1985).

The authors have constructed a series of simulations of spherically symmetric explosions in an otherwise unperturbed cosmology. The numerical hydrodynamics code includes effects due to electron conductivity, radiative cooling, gravity, ion viscosity, and the difference between the electron and ion temperatures. The initial redshift of the explosions has been varied between a redshift of 50 and a redshift of 2. The initial energies of the explosions range from 10^{59}ergs to 10^{61}ergs, appropriate to the energy input from quasars or primeval galaxies. The authors discuss the evolution of these explosions in terms of the quantities most clearly related to observable effects on the intergalactic medium, including their size, total thermal and kinetic energies, shell temperatures and densities, and mean interior pressure as a function of epoch.

161.160 Quasar–galaxy associations with discordant redshifts as a topological effect. I. Two–dimensional study.
H. V. Fagundes.
Astrophys. J., Vol. 291, No. 2, p. 450 – 459 (1985).

This paper discusses the hypothesis that a quasar–galaxy association with discordant redshifts may be the multiple image of a single source, produced by rays emitted along paths of different lengths. This is allowed by the multiply–connected topologies of Friedmann's *closed* models of negative spatial curvature. In this first part of a two–paper sequence the author deals with the problem in the relatively simple context of a hyperbolic two–dimensional space. This physically unrealistic case is used for deriving some qualitative observational consequences and for elucidating the mathematical method, which applies the tesselations of hyperbolic spaces. A fully three–dimensional treatment will be given in a following paper.

161.161 Low–frequency measurements of the cosmic background radiation spectrum.
G. F. Smoot, G. De Amici, S. D. Friedman, C. Witebsky, G. Sironi, G. Bonelli, N. Mandolesi, S. Cortiglioni, G. Morigi, R. B. Partridge, L. Danese, G. De Zotti.
Astrophys. J., Lett. Ed., Vol. 291, No. 2, p. L23 – L27 (1985).

The long-wavelength spectrum of the cosmic background radiation has been measured at five wavelengths (0.33, 0.9, 3.0, 6.3, 12.0 cm). These measurements represent a continuation of the work reported by Smoot and colleagues in 1983. The combined results have a weighted average of 2.73 ± 0.05K. They limit the possible Compton distortion of the cosmic background radiation spectrum to less than 6%.

161.162 On the use of measured time delays in gravitational lenses to determine the Hubble constant.
C. Alcock, N. Anderson.
Astrophys. J., Lett. Ed., Vol. 291, No. 2, p. L29 – L32 (1985).

Gravitational lenses are rare in the known samples of quasars, indicating that the conditions involved in their formation are unusual. In particular, the distribution of matter along the light rays from the observer through the deflector to the quasar may be very different from mean conditions. The authors show that reasonable deviations in the density of matter along the beams can significantly alter the relationship between time delays and the Hubble constant (H_0) and conclude that gravitational lenses are not promising estimators of H_0.

161.163 The specific physical character of the expansion of the Universe.
P. Voráček.
Astrophys. Space Sci., Vol. 111, No. 1, p. 103 – 121 (1985).

In some few steps a model describing the physical character of the cosmological expansion of the Friedmann–Robertson–Walker closed Universe is developed. Due to the fact that the rate of the cosmic time is changing, the conclusion is drawn that a hidden expansion is superposed on the generally known expansion of the Universe. The resulting picture – the bi–expansive model – is brought into connection with the voids in the Universe. The expansion is described as a 'motion' in a spatial dimension orthogonal to our space – the 4–dimensional sphere, and some characteristic properties of such a 'motion', compatible with the special and general relativity, are outlined.

161.164 Density perturbations in a two–fluid component universe.
S.–p. Xiang.
Astrophys. Space Sci., Vol. 111, No. 1, p. 171 – 178 (1985).

In a two–fluid component universe consisting of visible matter and neutrinos, the developed features of perturbations in the two components are quite different. If the densities ϱ_1 and ϱ_2, and the Jeans lengths λ_{1J} and λ_{2J} of the two components satisfy the relations $\varrho_1 \ll \varrho_2$, $\lambda_{1J} \ll \lambda_{2J}$, the developed inhomogeneities in the non–dominant component 1 are larger than those in the dominant component 2. Moreover, the increase of perturbations is in some situations not monotonous but oscillatory, and such oscillations in the two components are contrary.

161.165 Fluctuation evolution in a two–component dark–matter model.
R. Valdarnini, S. A. Bonometto.
Astron. Astrophys., Vol. 146, No. 2, p. 235 – 241 (1985).

The authors investigate the evolution of adiabatic perturbations in a scenario dominated by dark matter made of two kinds of collisionless quanta of different masses. These masses are taken around 10 eV (massive ν) and in the keV range (X–particles, tentatively photinos). ν–dominated and X–dominated cases are considered. While the ν component moderately alters the dynamics of X–dominated models, it is found that the presence of X–particles can lead to a drastic modification of the density fluctuation spectra, at the beginning of non–linear stages. It is found that ν and X spectra become equal before non–linear collapses start. The possibility of double pancaking (dissipative and non–dissipative) over two different scales seems to have to be discarded, in the frame of the present models.

161.166 Structure of the universe.
Ya. B. Zel'dovich.
Astrophys. Space Phys. Rev., Vol. 3, p. 1 – 34 (1984). – See Abstr. 003.019. Revised and extended English translation of 34.161.276.

(1). Introduction. General state of cosmology. (2). The initial stages of perturbation and the relativistic theory. (3). Structure of the universe in the adiabatic theory. (4). Neutrino universe. Neutrino pancakes. (5). A brief discussion of supernovae. (6). Entropy (isothermal) perturbations. (7). Other perturbations, other particles. (8). Some remarks on the observational data, on the comparison of these data with the theory, and on the unsolved problems. (9). Conclusion. Appendix: Mathematical structure of the theory.

161.167 Dark matter and formation of large–scale structure in the universe – composition of dark matter and clustering scenario.
L. Fang, S. Xiang, S. Li.
Sci. Sin., Ser. A, Vol. 28, No. 3, p. 301 – 310 (1985).

One can obtain knowledge on the composition and the abundance of dark matter from the distributions and the redshifts of visible objects. On the other hand, the dark matter plays an important role in the formation of various scale structure. The dark matter originated from the problem of "missing mass". The dark matter was taken as a passive factor in cosmology. It has become, recently, a dominant factor in the problem of the formation of large–scale structures.

161.168 Unser Teil des Universums.
F. Mossig.
Orion, 43. Jahrg., Nr. 208, p. 83 – 84 (1985).

161.169 A stroll in the realm of cosmology.
W. Davidson.
South. Stars, Vol. 31, No. 1, p. 77 – 84 (1984).

A review is made of the cosmological problem with emphasis on the need to resolve certain difficulties in the comparison between theory and observation before the current big–bang model of the expanding universe can be generally accepted.

161.170 Recent heavy–particle decay in a matter–dominated universe.
K.–A. Olive, D. Seckel, E. Vishniac.
Astrophys. J., Vol. 292, No. 1, p. 1 – 11 (1985).

The cold–matter scenario for galaxy formation solves the dark–matter problem very nicely on small scales corresponding to galaxies and clusters of galaxies. It is, however, difficult to reconcile with a universe with an Einstein–deSitter value of $\Omega = 1$. It is shown here that cold matter and $\Omega = 1$ can be made compatible while retaining the feature that the universe is matter–dominated today. This is done by means of heavy (cold) particles whose decay subsequently leads to the unbinding of a large fraction of lighter clustered matter.

161.171 Constraint on the amplitude of isothermal perturbations imposed by an isotropy of the cosmic microwave background radiation.
Y. Suto, K. Sato, H. Kodama.
Astrophys. J., Lett. Ed., Vol. 292, No. 1, p. L1 – L4 (1985).

The evolution of initially isothermal perturbations produced in a radiation–dominated universe is investigated until the time of recombination. The ratio of the amplitude of baryon irregularities to that of radiation generated in the course of the evolution of the universe is determined. Combining it with the observed upper bound on the anisotropy of the cosmic background radiation, the authors obtain a severe constraint on the amplitude of initially isothermal fluctuations at the time of recombination. This constraint poses a serious difficulty for the theory of galaxy formation based on isothermal perturbations.

161.172 The boundary conditions of the universe.
S. W. Hawking.
Priroda, No. 4, p. 21 – 27 (1985). In Russian.

161.173 The Universe as laboratory of elementary particles.
M. Yu. Khlopov.
Priroda, No. 5, p. 20 – 29 (1985). In Russian.

161.174 A note on the formation of matter in the early universe. Hypothesis of basic monopole balance. With brief introduction to magnetic quark theory.
T. Gimse.
Theor. Pap., Vol. 3, Nr. 2, p. 27 – 42 (1985).

In this paper, by means of the magnetic monopole quark theory, a new proposal dealing with the problem of the formation of matter/antimatter in the early universe is presented.

161.175 Particle–mesh simulations of clustering in cosmology.
F. R. Bouchet, H. E. Kandrup.
Inst. Astrophys. Paris, Pré–Publ., No. 97, 13 pp. (1985). To appear in Astrophys. J.

161.176 Deuterium and ^{3}He formation by supernovae of the first generation.
A. K. Lavrukhina, R. I. Kuznetsova.
Izv. Akad. Nauk SSSR. Ser. fiz., Tom 48, No. 11, p. 2080 – 2082 (1984). In Russian. Abstr. in Ref. Zh., 51. Astron., 3.51.905 (1985).

161.177 Biased galaxy formation in a universe dominated by cold dark matter.
R. Schaeffer, J. Silk.
Astrophys. J., Vol. 292, No. 2, p. 319 – 329 (1985).

The hypothesis that galaxies form from rare fluctuations in a universe dominated by cold, weakly interacting dark matter is shown to be in reasonable accord with most observed properties of the distribution of galaxies. The authors calculate the mass distribution of galaxies and the cluster multiplicity function for a model in which galaxy formation is biased by the assumption that galaxies form only at a redshift in excess of 3. They determine the fraction of gaseous matter that, at the epoch of galaxy formation, has not already collapsed on smaller scales. On scales which have turned around recently, gas infall to dark–matter potential wells is shown to result in observable diffuse X–ray emission. The galaxy distribution on these scales is found to be highly aniso-tropic, building up into large–scale filaments or pancakes.

161.178 The evolution of large–scale structure in a universe dominated by cold dark matter.
M. Davis, G. Efstathiou, C. S. Frenk, S. D. M. White.
Astrophys. J., Vol. 292, No. 2, p. 371 – 394 (1985).

The authors present the results of numerical simulations of nonlinear gravitational clustering in universes dominated by weakly interacting, "cold" dark matter (e.g., axions or photinos). These studies employ a high resolution N–body code with periodic boundary conditions and 32,768 particles. The calculations followed the evolution of ensembles of models with $\Omega = 1$ and $\Omega < 1$ from the initial conditions predicted for a "constant curvature" primordial fluctuation spectrum. It is concluded that an $\Omega = 1$ model is quite unacceptable if galaxies trace the mass distribution and that models with $\Omega \approx 0.2$, while better, still do not provide a fully acceptable match to observation. An alternative hypothesis is that galaxies formed only at high peaks of the initial density field. It is shown that this leads to models which can agree with observations quite well even for $\Omega = 1$.

161.179 Statistical comparison of galaxy formation models: the bispectrum.
J. N. Fry, A. L. Melott.
Astrophys. J., Vol. 292, No. 2, p. 395 – 403 (1985).

To discriminate between galaxy–formation models that give similar two–point correlation functions, the authors look to higher moments, specifically the three–point function in trans-form space or bispectrum, for an ensemble of numerical simula-tions. They find different behaviors of this statistics that can be used to distinguish between classes of models. Up to the uncer-tainty in identifying the galaxy distribution with the mass distri-bution in the models, the results seem to prefer the isothermal or the cold–particle models. No model, however, agrees in all detail with the observational bispectrum of the Shane–Wirtanen galaxy catalog.

161.180 Stability of certain spatially homogeneous cosmological models.
J. D. Barrow, D. H. Sonoda.
Gen. Relativ. Gravitation, Vol. 17, No. 5, p. 409 – 415 (1985).

The authors investigate the general behavior of spatially ho-mogeneous cosmological models at large times. Using existing techniques from stability theory they discover that some known exact solutions are asymptotically stable despite possessing spe-cial properties. Some consequences of these results are discussed.

161.181 Comment on the Bianchi type–V vacuum model.
D. Lorenz–Petzold, V. Joseph.
Gen. Relativ. Gravitation, Vol. 17, No. 5, p. 509 – 513 (1985).

It is shown that the nonlocally rotationally symmetric Bianchi type–V vacuum solution recently discussed by Nayak is identical, after trivial transformations, with the solution found some years ago by Joseph. In addition some calculations and remarks con-cerning the integration of the null geodesic equations are given.

161.182 Early universe and GUT.
M. Yoshimura.
Cosmology of the early universe, p. 1 – 21 (1984). – See Abstr. 003.021.

Recent developments in applications of grand unified theories (GUT) to the early universe are reviewed. The classical puzzles of the standard cosmology (baryon asymmetry, horizon-homogeneity, flatness, monopole overproduction) may be solved by GUT cosmology.

161.183 Vacuum viscosity and entropy generation in quantum gravitational processes in the early universe.
B. L. Hu.
Cosmology of the early universe, p. 23 – 44 (1984). – See Abstr. 003.021.

Entropy generation in quantum gravitational processes due to vacuum polarization and particle production in the early universe is discussed. The quantum processes of spontaneous and induced particle production from the vacuum and n–particle state as well as the classical process of non–adiabatic frequency shifts of nor-mal modes of the system are described in the context of an adiabatic formulation of finite temperature quantum field theory and the thermodynamics of relativistic imperfect fluids.

161.184 Phase transitions in supersymmetric grand unified mod-els.
E. W. Kolb, S.–Y. Pi, S. Raby.
Cosmology of the early universe, p. 45 – 70 (1984). – See Abstr. 003.021.

Cosmological consequences of supersymmetric grand unified models are reviewed. The development of the phase transition in the early universe associated with the spontaneous breaking of supersymmetry is studied. In models based on the Witten-O'Raifeartaigh potential, the universe experiences a long de–Sitter phase which may solve several cosmological conundrums. Problems associated with the phase transition are discussed.

161.185 Gravitational effects of bubbles' collision in the very early universe.
Z. C. Wu.
Cosmology of the early universe, p. 71 – 90 (1984). – See Abstr. 003.021.

First order phase transition occurred in the very early universe. This process is described by the formation, expansion and colli-sion of Coleman's bubbles. The author investigates the gravita-tional effects in collisions of bubbles.

161.186 Massive neutrino and cosmology.
T. Lu.
Cosmology of the early universe, p. 91 – 128 (1984). – See Abstr. 003.021.

The progress of experimental efforts and theoretical predic-tions on neutrino rest–mass is reviewed, and cosmological impli-cations of massive neutrinos, especially clustering phenomena in the universe, are discussed in some detail.

161.187 The large–scale structure in the universe.
L. Z. Fang.
Cosmology of the early universe, p. 129 – 164 (1984). – See Abstr. 003.021.

After reviewing observational evidence for large–scale inho-mogeneity in the distribution of visible matter in the universe, this paper discusses evidence for dark matter and studies clustering phenomena in a universe dominated by dark matter.

161.188 Phase transition and inflationary universe model.
K. Sato.
Cosmology of the early universe, p. 165 – 232 (1984). – See Abstr. 003.021.

Contents: (1) First order phase transition and inflationary uni-verse model. (2) Details of the original inflationary model and its difficulties. (3) New inflationary model.

161.189 Spatially homogeneous dynamics: a unified picture.
R. T. Jantzen.
Cosmology of the early universe, p. 233 – 305 (1984). – See Abstr. 003.021.

The Einstein equations for a perfect fluid spatially homogeneous spacetime are studied in a unified manner by retaining the generality of certain parameters whose discrete values correspond to the various Bianchi types of spatial homogeneity. A parameter dependent decomposition of the metric variables adapted to the symmetry breaking effects of the nonabelian Bianchi types on the "free dynamics" leads to a reduction of the equations of motion for those variables to a 2–dimensional time dependent Hamiltonian system containing various time dependent potentials which are explicitly described and diagrammed.

161.190 Das Modell des inflationären Universums.
G. Börner, N. Straumann.
Phys. Bl., 41. Jahrg., Heft 6, p. 146 – 151 (1985).

161.191 Dark matter and formation of large–scale structure in the universe – the test by distribution of quasars.
L. Fang, Y. Chu, X. Zhu.
Sci. Sin., Ser. A, Vol. 28, No. 5, p. 517 – 523 (1985).

Based on their study of the formation of large scale structure in the universe, the authors expect that: (1) the distribution of quasars should differ from that of galaxies because it has no strong inhomogeneity on the scale of $10 - 100$ Mpc; (2) the distributions of quasars with $z > 2$ and $z < 2$ should differ from each other. Various analyses on quasar distribution are consistent with these predictions. Particularly, the nearest neighbor test for the complete quasar sample given by Savage and Bolton clearly shows that the distribution of $z > 2$ quasars is rather homogeneous while the $z < 2$ quasars have a tendency to clustering.

161.192 Asymptotically anti–de Sitter spaces.
M. Henneaux, C. Teitelboim.
Commun. Math. Phys., Vol. 98, No. 3, p. 391 – 424 (1985).

The authors investigate in detail in this paper the dynamics of the gravitational field when the cosmological constant is negative.

161.193 Anisotropy of the relic electromagnetic radiation generated by an isotropic gravitational wave background with flat spectrum.
A. A. Starobinskij.
Pis'ma Astron. Zh., Tom 11, No. 5, p. 323 – 330 (1985). In Russian. English translation in Sov. Astron. Lett., Vol. 11.

The temperature anisotropy of background electromagnetic radiation generated by an isotropic stochastic relic gravitational wave background with flat initial spectrum is calculated. For the angles $\theta > 2°$ the angular dependence of the anisotropy correlation function practically coincides with the known dependence for the case of adiabatic perturbations with flat spectrum. The anisotropy is small for $\theta < 1°$. Upper limits on the amplitude of gravitational waves and the expected value of the quadrupole anisotropy are obtained.

161.194 The "forest" of absorption lines in QSO spectra and the structure of the Universe.
A. G. Doroshkevich, J. P. Mücket.
Pis'ma Astron. Zh., Tom 11, No. 5, p. 331 – 334 (1985). In Russian. English translation in Sov. Astron. Lett., Vol. 11.

The problem of the "forest" of absorption lines in the spectra of quasars is discussed in the frame of the Λ–theory of formation and evolution of the structure of the Universe. It is assumed that the hidden mass is connected with neutrino–like particles (possibly unstable) with a rest mass of the order of $60 - 100$ eV. An observational test for the hypothesis proposed is discussed.

161.195 The distance scale: present status and future prospects.
M. Aaronson, J. R. Mould.
Bull. Am. Astron. Soc., Vol. 16, No. 4, p. 875 (1984). Abstract. – See Abstr. 010.062.

161.196 Cosmological lessons from the very early universe.
M. S. Turner.
Bull. Am. Astron. Soc., Vol. 16, No. 4, p. 875 – 876 (1984). Abstract. – See Abstr. 010.062.

161.197 Dark matter in the universe: astrophysical constraints and effects on galaxy formation.
G. R. Blumenthal.
Bull. Am. Astron. Soc., Vol. 16, No. 4, p. 918 (1984). Abstract. – See Abstr. 010.062.

161.198 The cosmic antineutrino background from gravitational collapse.
S. E. Woosley, J. R. Wilson, R. Mayle.
Bull. Am. Astron. Soc., Vol. 16, No. 4, p. 935 (1984). Abstract. – See Abstr. 010.062.

161.199 Status of quantized extragalactic redshifts.
W. G. Tifft, W. J. Cocke.
Bull. Am. Astron. Soc., Vol. 16, No. 4, p. 950 (1984). Abstract. – See Abstr. 010.062.

161.200 Redshifts of emission line galaxies near the Bootes void.
S. A. Gregory, W. G. Tifft, R. P. Kirshner, J. W. Moody.
Bull. Am. Astron. Soc., Vol. 16, No. 4, p. 961 (1984). Abstract. – See Abstr. 010.062.

161.201 Some very large scale structure in the distribution of Abell galaxy clusters.
D. J. Batuski, J. O. Burns.
Bull. Am. Astron. Soc., Vol. 16, No. 4, p. 962 (1984). Abstract. – See Abstr. 010.062.

161.202 A precise measurement of the microwave background temperature from observations of interstellar CN.
D. M. Meyer, M. Jura.
Bull. Am. Astron. Soc., Vol. 16, No. 4, p. 1014 – 1015 (1984). Abstract. – See Abstr. 010.062.

161.203 Molecules in the early universe.
S. Lepp, A. Dalgarno, J. M. Shull.
Bull. Am. Astron. Soc., Vol. 16, No. 4, p. 1015 (1984). Abstract. – See Abstr. 010.062.

161.204 Determining H_0 and q_0 from supernova spectra.
R. V. Wagoner.
Bull. Am. Astron. Soc., Vol. 16, No. 4, p. 1015 (1984). Abstract. – See Abstr. 010.062.

161.205 Graphical display of R(t) for Friedmann universes with arbitrary cosmological constant.
J. E. Felten, R. Isaacman.
Bull. Am. Astron. Soc., Vol. 16, No. 4, p. 1015 (1984). Abstract. – See Abstr. 010.062.

161.206 Microuniverses in a FIB cosmology.
J. M. Barnothy.
Bull. Am. Astron. Soc., Vol. 16, No. 4, p. 1016 (1984). Abstract. – See Abstr. 010.062.

161.207 Death of a star: birth of a universe.
M. R. Bernstein.
Bull. Am. Astron. Soc., Vol. 16, No. 4, p. 1016 (1984). Abstract. – See Abstr. 010.062.

161.208 Espacio–tiempo, gravitación, cosmología.
P. Lustgarten.
Bol. Acad. Cienc. Fis. Mat. Nat., Tomo 43, Nos. 131 – 132, p. 32 – 228 (1983).

161.209   Erratum: "Nonconservation of baryons in cosmology – revisited" [J. Astrophys. Astron., Vol. 5, No. 1, p. 67 – 78 (1984)].
J. V. Narlikar.
J. Astrophys. Astron., Vol. 6, No. 1, p. 75 (1985). See Abstr. 37.161.288.

161.210   Are galaxies due to accretion?
H.-E. Fröhlich.
Astron. Nachr., Vol. 306, No. 3, p. 101 – 105 (1985).

Burnt–out Population III remnants are thought to have acted as protogalactic seeds. Once tepid pancakes have formed such seeds ($\approx 10^6 M_\odot$) could be subject to considerable growing by quasi–stationary accretion.

161.211   Some magnetofluid cosmological models of plane symmetry.
S. Prakash.
Astrophys. Space Sci., Vol. 111, No. 2, p. 383 – 388 (1985).

Considering the cylindrically–symmetric metric of Marder, some plane–symmetric magnetofluid cosmological models have been derived. Various physical and geometrical properties of the models have been discussed.

161.212   Bianchi type–VI_0 perfect fluid cosmological models with magnetic field.
S. R. Roy, J. P. Singh, S. Narain.
Astrophys. Space Sci., Vol. 111, No. 2, p. 389 – 397 (1985).

The authors discuss Bianchi type–VI_0 cosmological models with perfect fluid distribution and magnetic field directed along the axial direction. Einstein's equations have been solved for the condition that the expansion scalar θ bears a constant ratio to the anisotropy in the direction of a space–like unit vector λi.

161.213   Density perturbations in a universe consisting of fluid plus collisionless neutrino gas.
S.–p. Xiang.
Astrophys. Space Sci., Vol. 112, No. 1, p. 83 – 91 (1985).

In a two–component universe which consists of fluid (visible matter) plus collisionless massive neutrino gas (dark matter) the remarkable difference between the developed inhomogeneities in two components could be formed after the decoupling time. The non–dominant visible component in the universe is strongly clustered especially on smaller scales, while the distribution of the dominant dark matter (neutrinos) is fairly uniform.

161.214   Exact cosmological solutions in the scalar–tensor theory with cosmological constant.
L. O. Pimentel.
Astrophys. Space Sci., Vol. 112, No. 1, p. 175 – 183 (1985).

Under the assumption of a power–law between the expansion factor of the Universe, and the scalar field ($\phi a^n = c = \text{const.}$) tensor theory with cosmological constant are reduced to quadrature. Several exact solutions are obtained, among them inflationary universes that have barotropic equation of state.

161.215   The microwave background and (m, Z) relations in a tilted cosmological model.
D. R. Matravers, M. S. Madsen, D. L. Vogel.
Astrophys. Space Sci., Vol. 112, No. 1, p. 193 – 202 (1985).

The structure of the cosmic microwave background temperature is studied in the context of a Bianchi type–V tilted cosmological model. First integrals of the equations for the null geodesics are found by use of the symmetries of the model, enabling celestial temperature distribution to be found. The quadrupole and dipole moments are calculated for some models, suggesting that the observed anisotropy in the cosmic microwave background can be understood in the context of a Bianchi type–V model of the Universe. The apparent magnitude–redshift relations are also calculated for these models.

161.216   Large–scale structure of the Universe in the framework of a model equation of nonlinear diffusion.
S. N. Gurbatov, A. I. Saichev, S. F. Shandarin.
Inst. prikl. mat. Akad. Nauk SSSR. Prepr., No. 152, 27 pp. (1984). In Russian. Abstr. in Ref. Zh., 51. Astron., 5.51.777 (1985).

161.217   Propagation of a spherical shock wave in an expanding Universe taking gravitation into account.
Ya. M. Kazhdan.
Inst. prikl. mat. Akad. Nauk SSSR. Prepr., No. 141, 24 pp. (1984). In Russian. Abstr. in Ref. Zh., 51. Astron., 5.51.780 (1985).

161.218   On possible spontaneous compactification leading to zero cosmological constant.
G. V. Lavrelashvili, P. G. Tinyakov.
Yad. Fiz., Tom 41, No. 1, p. 271 – 277 (1985). In Russian. Abstr. in Ref. Zh., 51. Astron., 5.51.782 (1985).

161.219   Some solutions of equations of supergravity at the background of a cosmological model.
A. T. Il'ichev, N. R. Sibgatullin.
Vestn. MGU. Mat. Mekh., No. 1, p. 83 – 86 (1985). In Russian. Abstr. in Ref. Zh., 51. Astron., 5.51.784 (1985).

161.220   Theory of evolution of the Universe as a consequence of cosmological postulates.
I. L. Rozental'.
Inst. kosm. issled. Akad. Nauk SSSR. Prepr., No. 909, 12 pp. (1984). In Russian. Abstr. in Ref. Zh., 51. Astron., 5.51.787 (1985).

161.221   Small–scale angular fluctuations in the microwave background radiation and the existence of isolated large–scale structures in the universe.
L. J. Goicoechea, J. L. Sanz.
Astrophys. J., Vol. 293, No. 1, p. 17 – 24 (1985).

The relative fluctuation of the present temperature associated with the microwave background radiation (MBR) on a small angular scale can be related for a general inhomogeneous cosmological model to the kinematical quantities, their gradients, and the Weyl tensor through the geodesic deviation equation. The authors apply this result to calculate the induction of temperature fluctuations in the MBR by a spherically symmetric cluster (or void) of matter or radiation or both, considered as a perturbation in a flat Friedmann universe, with negligible pressure.

161.222   Scale–invariant density perturbations, anisotropy of the cosmic microwave background, and large–scale peculiar velocity field.
N. Vittorio, J. Silk.
Astrophys. J., Lett. Ed., Vol. 293, No. 1, p. L1 – L5 (1985).

The large–scale peculiar velocity field and the large– and intermediate–angular scale anisotropy of the cosmic microwave background are studied in inflationary cosmological models of critical density and containing primordial scale–invariant adiabatic density perturbations. Comparison with recent observations by de Vaucouleurs and Peters provides tentative support for a cold dark matter scenario in which the dark matter is not appreciably less clustered than the luminous galaxy distribution.

161.223   The dark night–sky riddle: a "paradox" that resisted solution.
E. R. Harrison.
Science, Vol. 226, No. 4677, p. 941 – 945 (1984).

The riddle of a dark night sky, now known as "Olbers's paradox", can be traced back to Thomas Digges in 1576. Since the time of Edmund Halley (1721) the riddle of a dark night sky in an infinite universe uniformly populated with stars has been regarded as a paradox. Constant emphasis on the paradoxical aspect of the problem of darkness at night, however, leads to a one–sided interpretation of the riddle. Calling the phenomenon a "paradox" distorts the historical perspective, and consequently one incorrectly attributes the origin of the riddle to Edmund

Halley. Also it distorts the cosmological perspective and quite probably has greatly delayed the solution of the riddle.

161.224 The clustering of matter in a neutrino–dominated universe.
T. Lu, L. Fang.
High energy astrophysics and cosmology, p. 405 – 424 (1983). – See Abstr. 012.068.

The effects of a non–zero mass of neutrinos on the formation of large–scale inhomogeneities leading to galaxy superclusters and voids in the universe are investigated.

161.225 Deceleration of the expanding universe.
L. Fang, F. Hu.
High energy astrophysics and cosmology, p. 425 – 445 (1983). – See Abstr. 012.068.

The various methods of determining the deceleration parameter q_0 are reviewed. The Hubble diagram method for galaxies and various subsets of quasars consistently gives $q_0 > 1/2$, the mean mass density method always gives $q_0 < 1/2$. A possible resolution of the discrepancy invoking massive neutrinos is outlined.

161.226 The large–scale inhomogeneities in the universe.
L. Fang.
High energy astrophysics and cosmology, p. 446 – 460 (1983). – See Abstr. 012.068.

The large–scale distribution of both luminous and dark matter in the universe is discussed. It is shown that the fundamental problem in cosmic clustering is no longer to search the mechanism of forming a large density perturbation, but to explain the difference between the small total density perturbation and the large apparent amplitude in the inhomogeneity of visible objects.

161.227 The rest–mass of the neutrino and the cosmic clustering phenomena.
T. Lu, P. Zhu.
High energy astrophysics and cosmology, p. 461 – 466 (1983). – See Abstr. 012.068.

161.228 On neutrino halos of galaxies or clusters of galaxies and their restriction on neutrino mass.
W. Huang, C. Xu, C. Qing, Z. He.
High energy astrophysics and cosmology, p. 467 – 472 (1983). – See Abstr. 012.068.

161.229 On massive neutrino halos and galactic structure.
A. Crollalanza, J. G. Gao, R. Ruffini.
High energy astrophysics and cosmology, p. 473 – 480 (1983). – See Abstr. 012.068.

The aim of this paper is to clarify, by using some simplified models, the interplay of "visible matter" (plasma, atoms, molecules, etc.) and massive neutrino halos in a galactic structure. In particular, the effect of the central concentration of baryon matter on the neutrino distribution is calculated.

161.230 A de Sitter–Friedmann phase transition model.
H. Guo, Z. Zou.
High energy astrophysics and cosmology, p. 489 – 494 (1983). – See Abstr. 012.068.

It is shown that a new type of singularity–free cosmological model can be established by means of a simple finite temperature $\lambda\phi^4$ model. This model takes a series of important quantum and statistical effects into account, such as spontaneous symmetry breaking, trace anomaly and particle creation, and symmetry restoration at high temperature through phase transition.

161.231 The microwave background anisotropy and the homogeneity of the universe.
V. F. Mukhanov, G. V. Chibisov.
Sov. Astron. Lett., Vol. 10, No. 6, p. 374 – 376 (1984). English translation of 38.161.326.

161.232 The statistics of gravitational lenses. II. Apparent evolution in the quasars' luminosity function.
M. Vietri.
Astrophys. J., Vol. 293, No. 2, p. 343 – 355 (1985).

The author studies the statistical effects on the luminosity function of quasars due to gravitational lensing in an inhomogeneous universe. He uses a modified version of the approach developed in Vietri and Ostriker, which naturally includes cosmological effects, energy conservation, a realistic mass distribution within galaxies, a realistic mass spectrum for galaxies, amplification bias, lensing by galaxies ($\equiv$ maxilensing) and by stars ($\equiv$ minilensing); the influence of galaxy clusters and superclusters is neglected. The main conclusions are: (1) If minilensing is neglected, the effect of gravitational lensing is negligible (~ 0.005 of all quasars are amplified). (2) If minilensing is included, the intrinsic evolution of quasars is less than demanded by the V/V_{max} test.

161.233 Erratum: "The dependence on distance and redshift of the velocity vectors of the Sun, the Galaxy, and the Local Group with respect to different extragalactic frames of reference" [Astrophys. J., Vol. 287, No. 1, p. 1 – 16 (1984)].
G. de Vaucouleurs, W. L. Peters.
Astrophys. J., Vol. 293, No. 2, p. 616 (1985). See Abstr. 38.161.363.

161.234 Big bang photosynthesis and pregalactic nucleosynthesis of light elements.
J. Audouze, D. Lindley, J. Silk.
Astrophys. J., Lett. Ed., Vol. 293, No. 2, p. L53 – L57 (1985).

Two nonstandard scenarios for pregalactic synthesis of the light elements (^2H, ^3He, ^4He, and ^7Li) are developed. Big bang photosynthesis occurs if energetic photons, produced by the decay of massive neutrinos or gravitinos, partially photodisintegrate ^4He (formed in the standard hot big bang) to produce ^2H and ^3He. In this case, primordial nucleosynthesis no longer constrains the baryon density of the universe, or the number of neutrino species. Alternatively, one may dispense partially or completely with the hot big bang and produce the light elements by bombardment of primordial gas, provided that ^4He is synthesized by a later generation of massive stars.

161.235 The self–similar evolution of holes in an Einstein–de Sitter universe.
E. Bertschinger.
Astrophys. J., Suppl. Ser., Vol. 58, No. 1, p. 1 – 37 (1985).

Similarity solutions are presented for the nonlinear evolution of isolated, spherical, negative density perturbations in an Einstein–de Sitter universe. The solutions are characterized by a void surrounded by an overdense shell, both expanding in comoving coordinates. The solutions may apply to the giant voids observed in the distribution of galaxies.

161.236 Self–similar secondary infall and accretion in an Einstein–de Sitter universe.
E. Bertschinger.
Astrophys. J., Suppl. Ser., Vol. 58, No. 1, p. 39 – 66 (1985).

Similarity solutions are presented for the nonlinear evolution of a "hatbox" positive density perturbation in an Einstein–de Sitter universe. After the initial collapse and violent relaxation, secondary infall of bound shells creates a $\varrho \propto r^{-2.25}$ halo. Collisionless particles turning around in the halo create caustics similar to those seen around elliptical galaxies. The solutions may apply to the formation of massive galactic halos.

161.237 The extragalactic distance scale. I. Corrections to fundamental observables.
R. B. Tully, P. Fouqué.
Astrophys. J., Suppl. Ser., Vol. 58, No. 1, p. 67 – 80 (1985).

Five pairwise permutations of four observable parameters that describe the global properties of galaxies – blue magnitudes, infrared magnitudes, diameters, and H I line widths – can be used to estimate the distances of individual spiral systems. Before observables are useful, however, they must be corrected for aperture, projection, and absorption effects. Tests for these effects are described that involve minimization of the scatter in four *distance–independent* combinations of the observable parameters.

161.238 Cosmological production of Kaluza–Klein monopoles.
J. A. Harvey, E. W. Kolb, M. J. Perry.
Phys. Lett. B, Vol. 149B, No. 6, p. 465 – 469 (1984). Abstr. in Phys. Abstr., Vol. 88, No. 1253, Entry 30299 (1985).

161.239 The cosmic distance scale.
P. Hodge.
Am. Sci., Vol. 72, No. 5, p. 474 – 482 (1984). Abstr. in Phys. Abstr., Vol. 88, No. 1253, Entry 35886 (1985).

161.240 Non–Abelian isometries of multidimensional universes.
F. Mansouri, L. Witten.
Found. Phys., Vol. 14, No. 11, p. 1095 – 1106 (1984). Abstr. in Phys. Abstr., Vol. 88, No. 1253, Entry 35892 (1985).

161.241 A mass particle in a rotating homogeneous universe with an electromagnetic field.
L. K. Patel, B. M. Pandya.
J. Math. Phys. Sci., Vol. 18, No. 2, p. 197 – 203 (1984). Abstr. in Phys. Abstr., Vol. 88, No. 1253, Entry 35893 (1985).

161.242 Abundance of light elements and lepton asymmetry.
N. C. Rana.
Nuovo Cimento B, Vol. 84B, Ser. 11, No. 1, p. 53 – 64 (1984). Abstr. in Phys. Abstr., Vol. 88, No. 1253, Entry 35896 (1985).

161.243 Some rotating, time–dependent Bianchi type-IX cosmologies with heat flow.
E. Sviestins.
Gen. Relativ. Gravitation, Vol. 17, No. 6, p. 521 – 523 (1985).

This note reports on a study of cosmologies which constitute explicit, rotating, and expanding solutions of Einstein's field equations, with spacelike, timelike, or null–like homogeneous hypersurfaces of Bianchi type IX, and the source of which is a non–thermalized perfect fluid.

161.244 The time–symmetric initial value problem for a homogeneous, anisotropic, empty, closed universe, using Regge calculus.
R. M. Williams.
Gen. Relativ. Gravitation, Vol. 17, No. 6, p. 559 – 571 (1985).

Models for a homogeneous, anisotropic, empty, closed universe at a particular moment of time are constructed from sets of empty tetrahedra. It is shown using Regge calculus that only two such models have a moment of time symmetry.

161.245 The current state of the inflationary universe.
P. J. Steinhardt.
Comments Nucl. Part. Phys., Vol. 12, No. 5 – 6, p. 273 – 286 (1984). Abstr. in Phys. Abstr., Vol. 88, No. 1254, Entry 40773 (1985).

161.246 Spontaneous breaking of the baryon–antibaryon symmetry in a non–equilibrium process.
K. Tennakone.
J. Phys. G, Vol. 11, No. 1, p. 15 – 20 (1985). Abstr. in Phys. Abstr., Vol. 88, No. 1254, Entry 40774 (1985).

161.247 Kaluza–Klein–Bianchi–Kantowski–Sachs cosmologies.
D. Lorenz–Petzold.
Phys. Lett. B, Vol. 149B, No. 1 – 3, p. 79 – 81 (1984). Abstr. in Phys. Abstr., Vol. 88, No. 1254, Entry 40776 (1985).

161.248 Particle creation effect on $M_4 \times S^7$ Kaluza–Klein cosmologies.
T. Koikawa, K. Maeda.
Phys. Lett. B, Vol. 149B, No. 1 – 3, p. 82 – 86 (1984). Abstr. in Phys. Abstr., Vol. 88, No. 1254, Entry 40777 (1985).

161.249 Quantum Kaluza–Klein cosmologies. II.
X. M. Hu, Z. C. Wu.
Phys. Lett. B, Vol. 149B, No. 1 – 3, p. 87 – 89 (1984). Abstr. in Phys. Abstr., Vol. 88, No. 1254, Entry 40778 (1985).

161.250 Effect of particle creation on Kaluza–Klein cosmologies.
K. Maeda.
Phys. Rev. D, Vol. 30, No. 12, p. 2482 – 2494 (1984). Abstr. in Phys. Abstr., Vol. 88, No. 1254, Entry 40779 (1985).

161.251 Perfect–fluid higher–dimensional cosmologies.
D. Sahdev.
Phys. Rev. D, Vol. 30, No. 12, p. 2495 – 2507 (1984). Abstr. in Phys. Abstr., Vol. 88, No. 1254, Entry 40780 (1985).

161.252 A neutrino universe with viscosity.
B. Modak.
Pramāna, Vol. 23, No. 6, p. 809 – 814 (1984). Abstr. in Phys. Abstr., Vol. 88, No. 1254, Entry 40781 (1985).

161.253 On the false fluctuations of the cosmic background caused by the dust and gas emission in galaxy clusters.
V. K. Khersonskij, N. V. Voshchinnikov.
Astrophys. Lett., Vol. 24, No. 4, p. 217 – 224 (1985).

The expected value and spectral dependence of the false fluctuations of the cosmic background radiation connected with the gas and dust emission in distant galaxy clusters are considered. For typical clusters in the epoch with redshift $z = 1$ to $z = 5$ the gas and dust emission in the range 0.2 – 50.0 cm is computed. It is found that the expected values of the false fluctuations can reach $|\Delta T/T| \gtrsim 10^{-5}$. In the millimeter wavelength range $\Delta T/T > 0$, caused by the dust emission, while in the centimeter wavelength range $\Delta T/T < 0$, due to the Sunyaev–Zel'dovich effect. Angular scales of the false fluctuations of order $5' - 10'$ are close to the scales of the expected cosmological fluctuations. The search for the latter currently have sensitivities of $|\Delta T/T| \approx 10^{-4} - 10^{-5}$. It is shown that the false fluctuations can be separated from the true ones because of their essentially different spectral dependence.

161.254 Surface brightness, galaxy evolution and cosmology.
S. Phillipps.
Astrophys. Lett., Vol. 24, No. 4, p. 225 – 230 (1985).

A cosmological test recently suggested by Thomsen and Frandsen which employs the observed correlation between surface brightness and scale size for elliptical galaxies to avoid problems due to dynamical evolution, has been extended to make use of arbitrary galaxy samples. The explicit dependence on stellar evolution is also demonstrated and, assuming this evolution to be calculable, an equation for the deceleration parameter q_0 entirely in terms of observables is derived. The test is then applied to two available samples.

161.255 Generalized cosmological models.
A. Krasiński.
Sir Arthur Eddington Centenary Symposium. Vol. 1: Relativistic astrophysics and cosmology, p. 45 – 63 (1984). – See Abstr. 012.080.

Two classes of generalizations of the Friedman–Lemaître models within Einstein's theory of relativity are presented. One is obtained by assuming that the spacetime is a congruence of homogeneous and isotropic hypersurfaces whose orthogonal trajectories coincide with matter world–lines – without assuming that the spacetime inherits these symmetries. A second class is obtained by assuming that the spatial sections are conformal to flat spaces. The field equations are in this case reduced to three linear partial differential equations of first order for two functions.

161.256 The de Sitter universe and "projective relativity".
G. Arcidiacono.
Sir Arthur Eddington Centenary Symposium. Vol. 1: Relativistic astrophysics and cosmology, p. 64 – 88 (1984). – See Abstr. 012.080.

Contents: (1) The special projective relativity. (2) The law of the cosmic expansion. (3) Study of the projective red–shift law.

(4) The de Sitter–Castelnuovo universe and mechanics. (5) Study of the projective Laplacian and D'Alembertian. (6) Magnetohydrodynamics and cosmology. (7) Study of the magnetohydrodynamics energy tensor. (8) The general projective relativity. (9) The scalar–tensor gravitational field. (10) The vector–tensor gravitational field.

161.257   Perturbative approach to self consistent cosmologies.
P. Nardone.
Sir Arthur Eddington Centenary Symposium. Vol. 1: Relativistic astrophysics and cosmology, p. 244 – 252 (1984). – See Abstr. 012.080.
The author presents a perturbative approach to the equations describing the behaviour of a scalar field in a self–consistent Robertson–Walker universe. This approach throws a new light on the threshold mass which governs the instability of Minkowsky space and the existence of possible self–supporting cosmologies.

161.258   Where do galaxies form?
A. S. Szalay.
Proceedings of the XIth International Conference on Neutrino Physics and Astrophysics, p. 204 – 213 (1984). – See Abstr. 012.081.
The present theories of galaxy formation are reviewed. The relation between peculiar velocities and the correlation function of galaxies points to the possibility that galaxies do not form uniformly everywhere. None of the present theories of galaxy formation can account for this fact in a natural way. Theoretical implications of several observations, such as Lyman–α clouds, and correlations of faint galaxies, are discussed.

161.259   Radiation and five–dimensional cosmology.
R. B. Mann, D. E. Vincent.
Phys. Lett. A, Vol. 107A, No. 2, p. 75 – 78 (1985). Abstr. in Phys. Abstr., Vol. 88, No. 1255, Entry 40989 (1985).

161.260   A mechanism for reducing the value of the cosmological constant.
L. F. Abbott.
Phys. Lett. B, Vol. 150B, No. 6, p. 427 – 430 (1985). Abstr. in Phys. Abstr., Vol. 88, No. 1255, Entry 41016 (1985).

161.261   Anisotropic Kaluza–Klein cosmologies.
J. Demaret, J.–L. Hanquin.
Phys. Rev. D, Vol. 31, No. 2, p. 258 – 261 (1985). Abstr. in Phys. Abstr., Vol. 88, No. 1255, Entry 41017 (1985).

161.262   Does a phase transition in the early universe produce the conditions needed for inflation?
G. F. Mazenko, W. G. Unruh, R. M. Wald.
Phys. Rev. D, Vol. 31, No. 2, p. 273 – 282 (1985). Abstr. in Phys. Abstr., Vol. 88, No. 1255, Entry 41020 (1985).

161.263   Superstring cosmology.
E. Alvarez.
Phys. Rev. D, Vol. 31, No. 2, p. 418 – 421 (1985). Abstr. in Phys. Abstr., Vol. 88, No. 1255, Entry 41021 (1985).

161.264   Quantum effects in "mixmaster universe".
S. A. Pritomanov.
Phys. Lett. A, Vol. 107A, No. 1, p. 33 – 35 (1985). Abstr. in Phys. Abstr., Vol. 88, No. 1255, Entry 46247 (1985).

161.265   Confinement transition in cosmology.
S. A. Bonometto, L. Sokolowski.
Phys. Lett. A, Vol. 107A, No. 5, p. 210 – 214 (1985). Abstr. in Phys. Abstr., Vol. 88, No. 1255, Entry 46248 (1985).

161.266   Can primordial black holes solve the overproduction problem of monopoles?
M. Izawa, K. Sato.
Prog. Theor. Phys., Vol. 72, No. 4, p. 768 – 781 (1984). Abstr. in Phys. Abstr., Vol. 88, No. 1255, Entry 46249 (1985).

161.267   Entropy production in the inflationary universe.
M. Morikawa, M. Sasaki.
Prog. Theor. Phys., Vol. 72, No. 4, p. 782 – 798 (1984). Abstr. in Phys. Abstr., Vol. 88, No. 1255, Entry 46250 (1985).

161.268   Inflation in Kaluza–Klein cosmology.
Y. Okada.
Phys. Lett. B, Vol. 150B, No. 1 – 3, p. 103 – 106 (1985). Abstr. in Phys. Abstr., Vol. 88, No. 1255, Entry 46252 (1985).

161.269   Observable consequences of dynamical compactification in Kaluza–Klein cosmology.
T. Koikawa, M. Yoshimura.
Phys. Lett. B, Vol. 150B, No. 1 – 3, p. 107 – 112 (1985). Abstr. in Phys. Abstr., Vol. 88, No. 1255, Entry 46253 (1985).

161.270   Spontaneous compactification and singularities in inner space.
B. T. McInnes.
Phys. Lett. B, Vol. 150B, No. 1 – 3, p. 113 – 114 (1985). Abstr. in Phys. Abstr., Vol. 88, No. 1255, Entry 46254 (1985).

161.271   Limits on inflationary models of the universe.
S. W. Hawking.
Phys. Lett. B, Vol. 150B, No. 5, p. 339 – 341 (1985). Abstr. in Phys. Abstr., Vol. 88, No. 1255, Entry 46255 (1985).

161.272   Fate of order parameter in the inflationary universe.
M. Sakagami, A. Hosoya.
Phys. Lett. B, Vol. 150B, No. 5, p. 342 – 346 (1985). Abstr. in Phys. Abstr., Vol. 88, No. 1255, Entry 46256 (1985).

161.273   Anisotropic Universe with non–linear photons.
E. von Ruckert.
Rev. Bras. Fis., Vol. 13, No. 4, p. 721 – 734 (1983). Abstr. in Phys. Abstr., Vol. 88, No. 1255, Entry 46257 (1985).

161.274   Isotropic singularities in cosmological models.
S. W. Goode, J. Wainwright.
Classical Quantum Gravity, Vol. 2, No. 1, p. 99 – 115 (1985). Abstr. in Phys. Abstr., Vol. 88, No. 1256, Entry 51810 (1985).

161.275   Void in the neutrino dominated expanding universe.
J.–g. Gao, T.–z. Shen, H.–s. Zhang.
Kexue Tongbao (Beijing), Vol. 29, No. 11, p. 1466 – 1468 (1984). Abstr. in Phys. Abstr., Vol. 88, No. 1256, Entry 51811 (1985).

161.276   Numerical calculations of minisuperspace cosmological models.
S. W. Hawking, Z. C. Wu.
Phys. Lett. B, Vol. 151B, No. 1, p. 15 – 20 (1985). Abstr. in Phys. Abstr., Vol. 88, No. 1256, Entry 51812 (1985).

161.277   Anisotropic supergravity cosmologies.
D. Lorenz–Petzold.
Phys. Lett. B, Vol. 151B, No. 2, p. 105 – 110 (1985). Abstr. in Phys. Abstr., Vol. 88, No. 1256, Entry 51813 (1985).

161.278   A maximally symmetric no–scale inflationary universe.
C. Kounnas, M. Quiros.
Phys. Lett. B, Vol. 151B, No. 3 – 4, p. 189 – 194 (1985). Abstr. in Phys. Abstr., Vol. 88, No. 1256, Entry 51814 (1985).

161.279   The Universe's parameters.
T. X. Thuan.
Recherche, Vol. 16, No. 162, p. 52 – 63 (1985). In French. Abstr. in Phys. Abstr., Vol. 88, No. 1256, Entry 51815 (1985).

161.280   Conformal models of a universe filled with radiation and dust.
T. S. Kozhanov, Eh. G. Mychelkin.
Tr. Astrofiz. Inst. Alma–Ata, Tom 43, p. 38 – 50 (1984). In Russian.

161.281   **Mach's principle. I. Initial state of the Universe.**
B. L. Altshuler.
Int. J. Theor. Phys., Vol. 24, No. 1, p. 99 – 107 (1985). Abstr. in
Phys. Abstr., Vol. 88, No. 1257, Entry 52027 (1985).

161.282   **Mach's principle. II. Realization of Dirac's hypothesis in
Brans–Dicke theory with cosmological term.**
B. L. Altshuler.
Int. J. Theor. Phys., Vol. 24, No. 1, p. 109 – 118 (1985). Abstr. in
Phys. Abstr., Vol. 88, No. 1257, Entry 52068 (1985).

161.283   **On the stability of de Sitter space in the presence of
particle production.**
P. R. Anderson, R. Holman.
Phys. Lett. B, Vol. 151B, No. 5 – 6, p. 339 – 342 (1985). Abstr. in
Phys. Abstr., Vol. 88, No. 1257, Entry 52029 (1985).

161.284   **Scalar perturbations at the reheating phase in a new
inflationary universe.**
M. Den, K. Tomita.
Prog. Theor. Phys., Vol. 72, No. 5, p. 989 – 999 (1984). Abstr. in
Phys. Abstr., Vol. 88, No. 1257, Entry 52069 (1985).

161.285   **Kaluza–Klein cosmology.**
C. Wetterich.
Monopole '83, p. 117 – 123 (1984). Abstr. in Phys. Abstr.,
Vol. 88, No. 1257, Entry 52088 (1985). – See Abstr. 012.088.

161.286   **Primordial inflation with flat supergravity potentials.**
G. Gelmini, C. Kounnas, D. V. Nanopoulos.
Nucl. Phys. B, Part. Phys., Vol. B250, No. 1, p. 177 – 198 (1985).
Abstr. in Phys. Abstr., Vol. 88, No. 1257, Entry 52109 (1985).

161.287   **TCP, quantum gravity, the cosmological constant and all
that. . .**
T. Banks.
Nucl. Phys. B, Part. Phys., Vol. B249, No. 2, p. 332 – 360 (1985).
Abstr. in Phys. Abstr., Vol. 88, No. 1257, Entry 57926 (1985).

161.288   **A new mechanism for baryogenesis.**
I. Affleck, M. Dine.
Nucl. Phys. B, Part. Phys., Vol. B249, No. 2, p. 361 – 380 (1985).
Abstr. in Phys. Abstr., Vol. 88, No. 1257, Entry 57927 (1985).

161.289   **On the stochasticity in relativistic cosmology.**
I. M. Khalatnikov, E. M. Lifshitz (*E. M. Lifshits*),
K. M. Khanin, L. N. Shchur, Ya. G. Sinai.
J. Stat. Phys., Vol. 38, No. 1 – 2, p. 97 – 114 (1985). Abstr. in
Phys. Abstr., Vol. 88, No. 1257, Entry 57929 (1985).

161.290   **Bianchi type–I cosmological models in Schwinger's sca-
lar tensor theory.**
T. Singh, T. Singh.
Nuovo Cimento B, Vol. 84B, Ser. 11, No. 2, p. 134 – 140 (1984).
Abstr. in Phys. Abstr., Vol. 88, No. 1257, Entry 57932 (1985).

161.291   **The projective geometry of simple cosmological models.**
T. R. Hurd.
Proc. R. Soc. London, Ser. A, Vol. 397, No. 1813, p. 233 – 243
(1985). Abstr. in Phys. Abstr., Vol. 88, No. 1257, Entry 57933
(1985).

161.292   **Kaluza–Klein cosmologies and inflation. II.**
R. B. Abbott, S. D. Ellis, S. M. Barr.
Phys. Rev. D, Vol. 31, No. 4, p. 673 – 680 (1985). Abstr. in Phys.
Abstr., Vol. 88, No. 1257, Entry 57934 (1985).

161.293   **Decaying particles do not "heat up" the Universe.**
R. J. Scherrer, M. S. Turner.
Phys. Rev. D, Vol. 31, No. 4, p. 681 – 688 (1985). Abstr. in Phys.
Abstr., Vol. 88, No. 1257, Entry 57935 (1985).

161.294   **Feynman rules for finite–temperature Green's functions
in an expanding universe.**
G. Semenoff, N. Weiss.
Phys. Rev. D, Vol. 31, No. 4, p. 689 – 698 (1985). Abstr. in Phys.
Abstr., Vol. 88, No. 1257, Entry 57936 (1985).

161.295   **Evolution equation for the Higgs field in an expanding
universe.**
G. Semenoff, N. Weiss.
Phys. Rev. D, Vol. 31, No. 4, p. 699 – 703 (1985). Abstr. in Phys.
Abstr., Vol. 88, No. 1257, Entry 57937 (1985).

161.296   **Scalar electrodynamics in Robertson–Walker universes.**
L. H. Ford.
Phys. Rev. D, Vol. 31, No. 4, p. 704 – 709 (1985). Abstr. in Phys.
Abstr., Vol. 88, No. 1257, Entry 57938 (1985).

161.297   **Quantum instability of de Sitter spacetime.**
L. H. Ford.
Phys. Rev. D, Vol. 31, No. 4, p. 710 – 717 (1985). Abstr. in Phys.
Abstr., Vol. 88, No. 1257, Entry 57939 (1985).

161.298   **Inflation and reheating in Bianchi type–IX cosmology.**
A. Zardecki.
Phys. Rev. D, Vol. 31, No. 4, p. 718 – 724 (1985). Abstr. in Phys.
Abstr., Vol. 88, No. 1257, Entry 57940 (1985).

161.299   **Weak and strong quantum vacua in Bianchi type I uni-
verses.**
M. Castagnino, F. D. Mazzitelli.
Phys. Rev. D, Vol. 31, No. 4, p. 742 – 753 (1985). Abstr. in Phys.
Abstr., Vol. 88, No. 1257, Entry 57941 (1985).

161.300   **Particle creation in de Sitter space.**
E. Mottola.
Phys. Rev. D, Vol. 31, No. 4, p. 754 – 766 (1985). Abstr. in Phys.
Abstr., Vol. 88, No. 1257, Entry 57942 (1985).

161.301   **Higher–dimensional extensions of Bianchi type I cos-
mologies.**
D. Lorenz–Petzold.
Phys. Rev. D, Vol. 31, No. 4, p. 929 – 931 (1985). Abstr. in Phys.
Abstr., Vol. 88, No. 1257, Entry 57943 (1985).

161.302   **Thermodynamic fluctuations and the monopole density
of the early Universe.**
L. Diosi, B. Lukacs.
Report KFKI–1984–102, Hungarian Acad. Sci., Budapest, Hun-
gary, 15 pp. (1984). Abstr. in Phys. Abstr., Vol. 88, No. 1257,
Entry 57948 (1985).

161.303   **Progress and prospects for the inflationary universe.**
P. J. Steinhardt.
Monopole '83, p. 85 – 106 (1984). Abstr. in Phys. Abstr., Vol. 88,
No. 1257, Entry 57960 (1985). – See Abstr. 012.088.

161.304   **Primordial inflation and the monopole problem.**
K. A. Olive, D. Seckel.
Monopole '83, p. 107 – 116 (1984). Abstr. in Phys. Abstr.,
Vol. 88, No. 1257, Entry 57961 (1985). – See Abstr. 012.088.

161.305   **Magnetic monopoles in a magnetic universe.**
D. Fargion.
Monopole '83, p. 161 – 168 (1984). Abstr. in Phys. Abstr.,
Vol. 88, No. 1257, Entry 57962 (1985). – See Abstr. 012.088.

161.306   **New inhomogeneous viscous–fluid cosmologies.**
D. Papadopoulos, J. L. Sanz.
Lett. Nuovo Cimento, Vol. 42, Ser. 2, No. 5, p. 215 – 220
(1985). Abstr. in Phys. Abstr., Vol. 88, No. 1258, Entry 63543
(1985).

161.307   The development of questions about the Universe.
F. A. Bais.
Ned. Tijdschr. Natuurkd. A, Vol. A51, No. 1, p. 12 – 14 (1985).
In Dutch. Abstr. in Phys. Abstr., Vol. 88, No. 1258, Entry 63544
(1985).

161.308   Astronomical observations relating to the Universe question.
P. Katgert.
Ned. Tijdschr. Natuurkd. A, Vol. A51, No. 1, p. 15 – 17 (1985).
In Dutch. Abstr. in Phys. Abstr., Vol. 88, No. 1258, Entry 63545
(1985).

161.309   The formation of medium scale structures in the Universe.
V. Icke.
Ned. Tijdschr. Natuurkd. A, Vol. A51, No. 1, p. 18 – 19 (1985).
In Dutch. Abstr. in Phys. Abstr., Vol. 88, No. 1258, Entry 63546
(1985).

161.310   Nonlinear evolution of negative density perturbations in a radiation–dominated Universe.
Y. Suto, K. Sato, H. Sato.
Prog. Theor. Phys., Vol. 72, No. 6, p. 1137 – 1145 (1984). Abstr.
in Phys. Abstr., Vol. 88, No. 1258, Entry 63547 (1985).

161.311   Constraints on baryon and lepton number asymmetries of the early Universe from primordial nucleosynthesis.
N. Terasawa, K. Sato.
Prog. Theor. Phys., Vol. 72, No. 6, p. 1262 – 1265 (1984). Abstr.
in Phys. Abstr., Vol. 88, No. 1258, Entry 63548 (1985).

161.312   Questions about the gravitational instability of axion density fluctuations in a axion dominated Universe.
M. Sasaki.
Prog. Theor. Phys., Vol. 72, No. 6, p. 1266 – 1269 (1984). Abstr.
in Phys. Abstr., Vol. 88, No. 1258, Entry 63549 (1985).

161.313   Antimatter in the Universe.
N. C. Rana, A. W. Wolfendale.
Antiproton 1984, p. 235 – 242 (1985). Abstr. in Phys. Abstr.,
Vol. 88, No. 1258, Entry 63550 (1985). – See Abstr. 012.092.

161.314   Particle physics and cosmology.
P. J. E. Peebles.
Antiproton 1984, p. 243 – 250 (1985). Abstr. in Phys. Abstr.,
Vol. 88, No. 1258, Entry 63551 (1985). – See Abstr. 012.092.

161.315   Isotropic homogeneous universe with viscous fluid.
N. O. Santos, R. S. Dias, A. Banerjee.
J. Math. Phys., Vol. 26, No. 4, p. 878 – 881 (1985). Abstr. in
Phys. Abstr., Vol. 88, No. 1259, Entry 63773 (1985).

161.316   Cosmologies   with   quasiregular   singularities. I. Spacetimes and test waves.
D. A. Konkowski, T. M. Helliwell, L. C. Shepley.
Phys. Rev. D, Vol. 31, No. 6, p. 1178 – 1194 (1985). Abstr. in
Phys. Abstr., Vol. 88, No. 1259, Entry 63775 (1985).

161.317   Cosmologies with quasiregular singularities. II. Stability considerations.
D. A. Konkowski, T. M. Helliwell.
Phys. Rev. D, Vol. 31, No. 6, p. 1195 – 1211 (1985). Abstr. in
Phys. Abstr., Vol. 88, No. 1259, Entry 63776 (1985).

161.318   Cosmology with decaying particles.
M. S. Turner.
Phys. Rev. D, Vol. 31, No. 6, p. 1212 – 1224 (1985). Abstr. in
Phys. Abstr., Vol. 88, No. 1259, Entry 63791 (1985).

161.319   Realization of new inflation.
A. Albrecht, R. H. Brandenberger.
Phys. Rev. D, Vol. 31, No. 6, p. 1225 – 1231 (1985). Abstr. in
Phys. Abstr., Vol. 88, No. 1259, Entry 63792 (1985).

161.320   Global–symmetry evolution in axion cosmologies.
R. Holman, T. W. Kephart.
Phys. Rev. D, Vol. 31, No. 6, p. 1232 – 1235 (1985). Abstr. in
Phys. Abstr., Vol. 88, No. 1259, Entry 63793 (1985).

161.321   Quantum cosmology with a positive–definite action.
G. T. Horowitz.
Phys. Rev. D, Vol. 31, No. 6, p. 1169 – 1177 (1985). Abstr. in
Phys. Abstr., Vol. 88, No. 1259, Entry 63805 (1985).

161.322   Primordial two–component inflation.
K. Enqvist, D. V. Nanopoulos.
Nucl. Phys. B, Part. Phys., Vol. B252, No. 3, p. 508 – 522 (1985).
Abstr. in Phys. Abstr., Vol. 88, No. 1259, Entry 69004 (1985).

161.323   Vacuum tension effects on the evolution of domain walls in the early Universe.
A. Aurilia, G. Denardo, F. Legovini, E. Spallucci.
Nucl. Phys. B, Part. Phys., Vol. B252, No. 3, p. 523 – 537 (1985).
Abstr. in Phys. Abstr., Vol. 88, No. 1259, Entry 69005 (1985).

161.324   Cosmological models in eleven–dimensional supergravity.
J. Demaret, J.–L. Hanquin, M. Henneaux, P. Spindel.
Nucl. Phys. B, Part. Phys., Vol. B252, No. 3, p. 538 – 560 (1985).
Abstr. in Phys. Abstr., Vol. 88, No. 1259, Entry 69006 (1985).

161.325   The cosmic field tensor in bimetric general relativity.
D. B. Kerrighan.
Found. Phys., Vol. 15, No. 3, p. 379 – 386 (1985). Abstr. in Phys.
Abstr., Vol. 88, No. 1259, Entry 69007 (1985).

161.326   Inflation with higher dimensional gravity.
Q. Shafi, C. Wetterich.
Phys. Lett. B, Vol. 152B, No. 1 – 2, p. 51 – 55 (1985). Abstr. in
Phys. Abstr., Vol. 88, No. 1259, Entry 69012 (1985).

161.327   Quark–hadron phase transition in Bianchi type I cosmology.
E. Giannetto, P. Salucci.
Phys. Lett. B, Vol. 152B, No. 3 – 4, p. 171 – 174 (1985). Abstr. in
Phys. Abstr., Vol. 88, No. 1259, Entry 69013 (1985).

161.328   SU(N,1) inflation.
E. Ellis, K. Enqvist, D. V. Nanopoulos, K. A. Olive,
M. Srednicki.
Phys. Lett. B, Vol. 152B, No. 3 – 4, p. 175 – 180 (1985). Abstr. in
Phys. Abstr., Vol. 88, No. 1259, Entry 69014 (1985).

161.329   Cosmological perturbation theory.
H. Kodama, M. Sasaki.
Prog. Theor. Phys., Suppl., No. 78, p. 1 – 166 (1984). Abstr. in
Phys. Abstr., Vol. 88, No. 1259, Entry 69015 (1985).

161.330   Large scale inhomogeneities and the cosmological principle.
B. Lukacs, A. Meszaros.
Report KFKI–1984–130, Hungarian Acad. Sci., Budapest, Hungary, 24 pp. (1984). Abstr. in Phys. Abstr., Vol. 88, No. 1259,
Entry 69016 (1985).

161.331   The distance scale of the Universe.
B. Takase.
Oyo Buturi, Vol. 53, No. 10, p. 842 – 848 (1984). In Japanese.
Abstr. in Phys. Abstr., Vol. 88, No. 1260, Entry 74089 (1985).

161.332   Magnetic Brans–Dicke–Bianchi type–I cosmological models.
D. Lorenz–Petzold.
Acta Phys. Pol., Ser. B, Vol. B16, No. 3, p. 165 – 170 (1985).
Abstr. in Phys. Abstr., Vol. 88, No. 1260, Entry 74296 (1985).

161.333 Exact Brans–Dicke–Bianchi type–VI$_h$ solutions.
D. Lorenz–Petzold.
Acta Phys. Pol., Ser. B, Vol. B16, No. 3, p. 171 – 178 (1985).
Abstr. in Phys. Abstr., Vol. 88, No. 1260, Entry 74297 (1985).

161.334 Cosmological models in Kaluza–Klein theories.
R. Bergamini, R. Capovilla.
Lett. Nuovo Cimento, Vol. 42, Ser. 2, No. 7, p. 367 – 370 (1985).
Abstr. in Phys. Abstr., Vol. 88, No. 1260, Entry 74309 (1985).

161.335 Higher–dimensional perfect fluid cosmologies.
D. Lorenz–Petzold.
Phys. Lett. B, Vol. 153B, No. 3, p. 134 – 136 (1985). Abstr. in
Phys. Abstr., Vol. 88, No. 1260, Entry 74311 (1985).

161.336 Quark confinement and phase transition in the early Universe.
V. V. Dixit, J. Lodenquai.
Phys. Lett. B, Vol. 153B, No. 4 – 5, p. 240 – 242 (1985). Abstr. in
Phys. Abstr., Vol. 88, No. 1260, Entry 74312 (1985).

161.337 Cosmic superstrings.
E. Witten.
Phys. Lett. B, Vol. 153B, No. 4 – 5, p. 243 – 246 (1985). Abstr. in
Phys. Abstr., Vol. 88, No. 1260, Entry 74313 (1985).

161.338 Extended thermodynamics in the early Universe.
B. Lukacs, K. Martinas, T. Pacher.
Report KFKI–1985–03, Hungarian Acad. Sci., Budapest, Hungary, 28 pp. (1985). Abstr. in Phys. Abstr., Vol. 88, No. 1260, Entry 74314 (1985).

161.339 Homogeneous shear–free fluid distribution with spherical symmetry.
B. Modak.
Indian J. Phys., Part B, Vol. 59B, No. 1, p. 48 – 52 (1985). Abstr. in Phys. Abstr., Vol. 88, No. 1258, Entry 63542 (1985).

161.340 Primordial helium abundance and grand unification schemes.
J. Pasupathy, S. Ramadurai.
18th International Cosmic Ray Conference, Vol. 5, p. 509 (1983). Abstract. – See Abstr. 012.096.

161.341 The excess charge hypothesis, a possible key for the highest energy cosmic rays springing out of early contracting protoclouds.
E. Fischhoff.
18th International Cosmic Ray Conference, Vol. 9, p. 326 – 329 (1983). – See Abstr. 012.096.
The author investigates the trajectory physical characteristics of an ejected particle from the moment of its escape to present cosmic time in an expanding and red–shifting universe as a function of usual cosmological data, ejection cosmic time and given cloud parameters. These results are matched with the physical and evolution parameters of protoclouds consistent with current knowledge and/or theory.

161.342 Radiation backgrounds and their cosmological implications.
J. V. Narlikar.
18th International Cosmic Ray Conference, Vol. 12, p. 1 – 20 (1983). – See Abstr. 012.096.
The information content of radiation backgrounds at wavelengths ranging from radio waves to gamma rays is reviewed, within the context of the standard big bang cosmology. It is shown that the various backgrounds provide useful inputs and constraints on the physical features of the universe such as the existence and growth of large scale inhomogeneities like galaxies, clusters and superclusters, the overall density of the universe, the photon to baryon ratio and the extent of antimatter in the universe, etc.

161.343 Gravitational instability in a multi–component expanding medium.
L. V. Solov'eva, I. S. Nurgaliev.
Astron. Zh., Tom 62, Vyp. 3, p. 459 – 467 (1985). In Russian. English translation in Sov. Astron., Vol. 29, No. 3.
Gravitational instability in a medium consisting of two or more components in the expanding Universe is considered in the framework of the Newtonian theory. The system of equations for density perturbations is solved in short–wave and long–wave approximations.

161.344 Profiles of galaxy clusters in cosmological scenarios.
M. J. West, A. Dekel, A. Oemler Jr.
Bull. Am. Astron. Soc., Vol. 17, No. 2, p. 579 (1985). Abstract. – See Abstr. 010.065.

161.345 Cosmic nucleosynthesis and nonlinear inhomogeneities.
J. Centrella, R. A. Matzner, T. Rothman, J. Wilson.
Bull. Am. Astron. Soc., Vol. 17, No. 2, p. 580 (1985). Abstract. – See Abstr. 010.065.

161.346 Numerical simulation of self–similar clustering of galaxies in an expanding universe.
D. H. Porter, J. G. Jernigan.
Bull. Am. Astron. Soc., Vol. 17, No. 2, p. 580 – 581 (1985). Abstract. – See Abstr. 010.065.

161.347 Limits on small–scale fluctuations in the microwave background radiation.
H. M. Martin, R. B. Partridge, M. I. Ratner.
Bull. Am. Astron. Soc., Vol. 17, No. 2, p. 581 (1985). Abstract. – See Abstr. 010.065.

161.348 Dwarf galaxies as a clue to cosmological dark matter.
A. Dekel, J. Silk.
Bull. Am. Astron. Soc., Vol. 17, No. 2, p. 581 (1985). Abstract. – See Abstr. 010.065.

161.349 Quark nuggets: a candidate for cold dark matter.
C. Alcock, E. Farhi.
Bull. Am. Astron. Soc., Vol. 17, No. 2, p. 582 (1985). Abstract. – See Abstr. 010.065.

161.350 Flux conservation by an isolated Schwarzschild gravitational lens.
Y. Avni, I. Shulami.
Bull. Am. Astron. Soc., Vol. 17, No. 2, p. 582 (1985). Abstract. – See Abstr. 010.065.

161.351 The multipole harmonics of the unresolved–source and cosmic microwave backgrounds.
V. A. Korolev, M. V. Sazhin.
Sov. Astron., Vol. 28, No. 6, p. 616 – 621 (1984). English translation of 38.161.327.

161.352 Quantum field theory methods and inflationary universe models.
R. H. Brandenberger.
Rev. Mod. Phys., Vol. 57, No. 1, p. 1 – 60 (1985).
The paper reviews the theory of inflationary universe models, giving particular emphasis to the question of origin and growth of energy–density fluctuations in these new cosmologies.

161.353 Why do galaxies exist?
M. M. Waldrop.
Science, Vol. 228, No. 4702, p. 978 – 980 (1985).
Theorists can answer that question in remarkable detail if they postulate a universe with "cold" dark matter.

161.354 The 3 K background radiation: observational and theoretical status.
R. Fabbri.
Gravitation, geometry and relativistic physics, p. 249 – 264 (1984). – See Abstr. 012.103.
This review surveys recent observational and theoretical efforts to use the detailed properties of the cosmic microwave back-

ground, i.e. spectral distortions, angular anisotropies and polarization, as sensitive probes of the large scale structure of the universe.

161.355   Newtonian and relativistic Bianchi I models of the universe.
B. Barberis, D. Galletto.
Gravitation, geometry and relativistic physics, p. 290 – 293 (1984). – See Abstr. 012.103.

161.356   The cosmological constant.
A. Blanchard, F. X. Désert.
Gravitation, geometry and relativistic physics, p. 294 – 301 (1984). – See Abstr. 012.103.

In this paper the authors recall briefly the ''zoology'' of dust-filled universes with a positive cosmological constant Λ. They analyse present observational limits and point out the absence of any convincing argument against a non–vanishing Λ along with important consequences for cosmology. Some possible observational tests able to improve existing limits are proposed.

161.357   The inflationary universe: a primer.
R. Hakim.
Gravitation, geometry and relativistic physics, p. 302 – 332 (1984). – See Abstr. 012.103.

A summary of the basic features of inflationary universe models is presented. A bibliographic appendix lists research papers published before May 1984.

161.358   Report of IAU Commission 47: Cosmology (*Cosmologie*).
J. Audouze.
Trans. IAU, Vol. XIXA, p. 655 – 694 (1985). – See Abstr. 003.046.

161.359   The large scale structure of the universe.
P. Crane.
Nucl. Phys. B, Part. Phys., Vol. B252, No. 1 – 2, p. 1 – 10 (1985). – See Abstr. 012.105.

Observational data on galaxies and clusters of galaxies are reviewed in the context of how they relate to the large scale structure of the universe. Various methods of interpreting these data are discussed.

161.360   Primordial nucleosynthesis – a window on the early universe.
G. Steigman.
Nucl. Phys. B, Part. Phys., Vol. B252, No. 1 – 2, p. 11 – 23 (1985). – See Abstr. 012.105.

Big bang nucleosynthesis provides a unique probe of the early evolution of the universe. By confronting the predictions of element synthesis in the standard hot big bang model with the observed abundances of the light elements, one may test the consistency of the standard model and derive valuable constraints on parameters of particle physics and cosmology.

161.361   On the nature of the baryon asymmetry.
F. W. Stecker.
Nucl. Phys. B, Part. Phys., Vol. B252, No. 1 – 2, p. 25 – 36 (1985). – See Abstr. 012.105.

This paper examines the question as to whether the baryon asymmetry in the universe is a locally varying or universally fixed number. In particular, it concetrates on the question as to the existence of a possible matter–antimatter domain structure for the universe deriving from a GUT with spontaneous CP symmetry breaking. The paper reviews theoretical considerations and astrophysical data and tests relating to this fundamental question.

161.362   Galaxies and dark matter.
A. Zee.
Nucl. Phys. B, Part. Phys., Vol. B252, No. 1 – 2, p. 37 – 51 (1985). – See Abstr. 012.105.

This paper reviews some aspects of the theory of galaxy formation, which are related to the possible existence of dark matter

composed of elementary particles (massive neutrinos, axions, gravitinos, etc.).

161.363   Phase transitions and dark matter problems.
D. N. Schramm.
Nucl. Phys. B, Part. Phys., Vol. B252, No. 1 – 2, p. 53 – 71 (1985). – See Abstr. 012.105.

The possible relationships between phase transitions in the early universe and dark matter problems are discussed. It is shown that there are at least 3 distinct cosmological dark matter problems (1) halos; (2) galaxy formation and clustering; and (3) $\Omega = 1$, each emphasizing different attributes for the dark matter. At least some of the dark matter must be baryonic but if problems 2 and 3 are real they seem to also require non–baryonic material. However, if seeds are generated at the quark–hadron-chiral symmetry transition then alternatives to the standard scenarios may occur.

161.364   Massive relic neutrinos – a status report.
G. Steigman.
Nucl. Phys. B, Part. Phys., Vol. B252, No. 1 – 2, p. 73 – 80 (1985). – See Abstr. 012.105.

Massive relic neutrinos are the most natural exotic candidate for the dark mass in the universe. With modest masses (tens of eV), relic neutrinos can easily supply a critical mass density. Such neutrinos, however, interfere with galaxy formation and are most likely excluded by the data. Recent work suggests that unstable neutrinos may rescue the massive relic neutrino scenario.

161.365   The origin of cosmological density fluctuations.
B. J. Carr.
Nucl. Phys. B, Part. Phys., Vol. B252, No. 1 – 2, p. 81 – 111 (1985). – See Abstr. 012.105.

The density fluctuations required to explain the large–scale cosmological structure may have arisen spontaneously as a result of a phase transition in the early universe. There are several ways in which such fluctuations may have been produced, and they could have a variety of spectra, so one should not necessarily expect all features of the large–scale structure to derive from a simple power law spectrum. Some features may even result from astrophysical amplification mechanisms rather than gravitational instability.

161.366   Formation of galaxies.
A. S. Szalay.
Nucl. Phys. B, Part. Phys., Vol. B252, No. 1 – 2, p. 113 – 125 (1985). – See Abstr. 012.105.

The present theories of galaxy formation are reviewed. The relation between peculiar velocities and the correlation function of galaxies points to the possibility that galaxies do not form uniformly everywhere. Scale invariant properties of the cluster-cluster correlations are discussed. Comparing the correlation functions in a dimensionless way, galaxies appear to be stronger clustered, in contrast with the comparison of the dimensional amplitudes of the correlation functions. Theoretical implications of observations of Lyman–α clouds and of correlations of faint galaxies are discussed.

161.367   Inflation–driving scalar field.
S.-y. Pi.
Nucl. Phys. B, Part. Phys., Vol. B252, No. 1 – 2, p. 127 – 140 (1985). – See Abstr. 012.105.

The quantum mechanical behavior of the Higgs field which drives inflation of the early universe is discussed. The standard picture of the new inflationary universe cannot be realized in the minimal SU(5) model. A realistic model in which inflation is driven by a complex gauge singlet Higgs field containing the axion is presented.

161.368   Quantum origin of the universe.
A. Vilenkin.
Nucl. Phys. B, Part. Phys., Vol. B252, No. 1 – 2, p. 141 – 151 (1985). – See Abstr. 012.105.

This paper presents a cosmological model in which the universe is created by quantum tunneling from ''nothing'', where by ''nothing'' a state with no classical space time is implied.

161.369 Recent progress in the inflationary universe scenario.
A. D. Linde.
Nucl. Phys. B, Part. Phys., Vol. B252, No. 1 – 2, p. 153 – 159 (1985). – See Abstr. 012.105.

It is argued that the most natural realization of the idea of inflation can be achieved in the context of the chaotic inflation scenario. Two different versions of this scenario are discussed.

161.370 Monopoles and cosmology.
G. Lazarides.
Nucl. Phys. B, Part. Phys., Vol. B252, No. 1 – 2, p. 207 – 216 (1985). – See Abstr. 012.105.

The production of grand unified monopoles in the very early universe as well as their subsequent evolution is summarized. It is argued that grand unified theories and standard cosmology tend to lead to too many magnetic monopoles in the universe. Finally, the paper reviews some of the mechanisms suggested, so far, for solving this primordial monopole problem and stresses the existence of non–inflationary models which may lead to a measurable monopole flux.

161.371 Evolution of a system of cosmic strings.
T. W. B. Kibble.
Nucl. Phys. B, Part. Phys., Vol. B252, No. 1 – 2, p. 227 – 244 (1985). – See Abstr. 012.105.

The formation and destruction of closed loops in a system of strings is discussed. Coupled equations are set up to describe the evolution of the system and it is shown that it must either evolve to a scaling solution where the persistence length is proportional to time or to a situation where the strings come to dominate the energy density. The first case provides a very attractive cosmological scenario in which the loops generate the density fluctuation needed for galaxy formation.

161.372 The invisible axion, the gravitino and cosmology.
J. E. Kim.
Nucl. Phys. B, Part. Phys., Vol. B252, No. 1 – 2, p. 245 – 259 (1985). – See Abstr. 012.105.

The author reviews the invisible axion, and also discuss the gravitino problem in cosmology. It is pointed out that the upper bound of the Peccei–Quinn symmetry breaking scale depends on the particle physics models and can easily go up to 10^{14}GeV. The gravitino with about a 100 GeV mass is not favoured from the observed abundances of light nuclei.

161.373 Cosmic supergravity.
D. V. Nanopoulos.
Nucl. Phys. B, Part. Phys., Vol. B252, No. 1 – 2, p. 295 – 308 (1985). – See Abstr. 012.105.

Supergravity seems to play a very important role in particle physics. This paper is devoted to recent cosmological applications of supergravity which indicate that supergravity is of prime importance also in cosmology.

161.374 Kaluza–Klein cosmology and the inflationary universe.
C. Wetterich.
Nucl. Phys. B, Part. Phys., Vol. B252, No. 1 – 2, p. 309 – 320 (1985). – See Abstr. 012.105.

This paper discusses the problem whether a realistic inflationary scenario can be incorporated into higher–dimensional cosmological space–time theories.

161.375 The dimensional reduction transition.
E. Kolb.
Nucl. Phys. B, Part. Phys., Vol. B252, No. 1 – 2, p. 321 – 330 (1985). – See Abstr. 012.105.

If there are more than four space–time dimensions, dimensions small today may have been dynamically important in the evolution of the early universe. The transition to four space–time dimensions may have produced physically significant phenomena observable today.

161.376 Les symétries de l'Univers.
J.–M. Souriau.
C. R. Acad. Sci., Sér. Gén., Vie Sci., Tome 2, No. 3, p. 213 – 242 (1985).

161.377 Relativistic cosmologies with closed, locally homogeneous spatial sections.
H. V. Fagundes.
Phys. Rev. Lett., Vol. 54, No. 11, p. 1200 – 1202 (1985).

The homogeneous Bianchi and Kantowski–Sachs metrics of relativistic cosmology are investigated through their correspondence with recent geometrical results of Thurston. These allow a partial classification of the topologies for closed, locally homogeneous spaces according to Thurston's eight geometric types.

161.378 Evolution of cosmic strings.
A. Albrecht, N. Turok.
Phys. Rev. Lett., Vol. 54, No. 16, p. 1868 – 1871 (1985).

Results of numerical simulations of the evolution of a network of interacting cosmic strings in an expanding universe are presented and compared with a simple model. The results lend weight to earlier speculations about the evolution of cosmic strings.

161.379 Simple realization of the inflationary expansion of the universe.
G. F. Mazenko.
Phys. Rev. Lett., Vol. 54, No. 19, p. 2163 – 2165 (1985).

161.380 Microwave background anisotropy and decaying–particle models for a flat universe.
N. Vittorio, J. Silk.
Phys. Rev. Lett., Vol. 54, No. 20, p. 2269 – 2273 (1985).

The fine–scale anisotropy of the cosmic microwave background radiation, induced by primordial scale–invariant adiabatic density fluctuations, has been studied in flat cosmological models dominated by relativistic particles from the recent decay of a massive relic–particle species. The authors find that, if the relic–particle species consists of massive, unstable neutrinos, there is appreciable, and probably excessive, fine–scale anisotropy in the cosmic microwave background.

161.381 Can the universe be charged?
S. Orito, M. Yoshimura.
Phys. Rev. Lett., Vol. 54, No. 22, p. 2457 – 2460 (1985).

The authors point out a possibility of temporal charge nonconservation in cosmological models based on more–dimensional theories a là Kaluza–Klein. A net excess charge may then have been created before the extra dimensional space was compactified. Cosmological implications of the excess charge are discussed.

161.382 Improved astronomical limits on the neutrino mass.
J. Madsen, R. Epstein.
Phys. Rev. Lett., Vol. 54, No. 25, p. 2720 – 2722 (1985).

Lower limits for the neutrino mass are obtained if one flavor of neutrino is predominantly responsible for the dark matter in galaxies. From rotation curves of spiral galaxies, it is found that the neutrino mass must exceed 35 $h^{1/2}$eV (where h is the Hubble constant in units of 100 km s^{-1}Mpc^{-1}) if the velocity distribution of halo neutrinos is isotropic. If radial dispersion dominates, limits are slightly weakened. These results do not rely on a specific form of the phase–space distribution of halo neutrinos.

161.383 The cosmic microwave background temperature at 2.64 and 1.32 millimeters.
D. M. Meyer.
Diss. Abstr. Int., Sect. B, Vol. 45, No. 6, p. 1811 (1984). Thesis, University of California, 157 pp. (1984). Order No. DA8420213.

161.384 Measurement of the large–scale anisotropy of the cosmic background radiation at 3 mm.
G. L. Epstein.
Diss. Abstr. Int., Sect. B, Vol. 45, No. 9, p. 2956 (1985). Thesis, University of California, 150 pp. (1984). Order No. DA8426951.

161.385 A measurement of the intensity of the cosmic background radiation at 3.0 cm.
S. D. Friedman.
Diss. Abstr. Int., Sect. B, Vol. 45, No. 9, p. 2956 (1985). Thesis, University of California, 134 pp. (1984). Order No. DA8426965.

161.386 Integral constraints on stress energy perturbations in general relativity and applications to cosmology.
J. Traschen.
Diss. Abstr. Int., Sect. B, Vol. 45, No. 12, p. 3848 (1985). Thesis, Harvard University, 137 pp. (1984). Order No. DA8503585.

161.387 Bianchi type I cosmological models with perfect fluid and magnetic field.
S. R. Roy, S. Narain, J. P. Singh.
Aust. J. Phys., Vol. 38, No. 2, p. 239 – 248 (1985).

The authors investigate Bianchi type I cosmological models containing a perfect fluid and with an incident magnetic field directed along the x axis. The solutions obtained represent an expansion scalar θ bearing a constant ratio to the anisotropy in the direction of a space–like unit vector.

161.388 Fragmentation of a pre–galactic gas cloud and initial stellar mass function.
Y. Yoshii, H. Saio.
Theoretical aspects on structure, activity, and evolution of galaxies: III, p. 109 – 114 (1985). – See Abstr. 012.110.

The initial mass function (IMF) for the zero–metal stars is derived based on a mechanism such that radiation from protostars formed in a pregalactic gas cloud interacts with the surrounding gas and raises the Jeans mass at the end of successive fragmentation processes. The derived shape of the IMF has a peak. In particular, the slope toward the massive part of the IMF is steeper than the Salpeter–like IMF. The mass at the peak of the IMF is $4 - 10\ M_{\odot}$, depending on the assumed mass–luminosity relation for protostars.

161.389 Thermal–chemical instability in pre–galactic gas clouds.
Y. Sabano, M. Tosa.
Theoretical aspects on structure, activity, and evolution of galaxies: III, p. 115 – 120 (1985). – See Abstr. 012.110.

The growth of thermal–chemical instabilities in the pre–galactic medium is followed by numerical simulations of gas dynamics. Results show that a collapsing primordial gas cloud breaks into self–gravitating subcondensations with the mass of normal stars, suggesting that population III objects can be formed as normal stars.

161.390 Dark matter and formation of subgalactic objects.
M. Umemura, S. Ikeuchi.
Theoretical aspects on structure, activity, and evolution of galaxies: III, p. 123 – 128 (1985). – See Abstr. 012.110.

The authors consider the Einstein–de Sitter universe dominated by both hot and cold dark matter. It is investigated which hybrid model ensures the formation of subgalactic objects such as dwarf galaxies and Lyα clouds. Some observable characteristics of subgalactic objects and galaxies are discussed.

161.391 Dimension of singularity and correlation of density in an expanding universe.
T. Nakamura.
Theoretical aspects on structure, activity, and evolution of galaxies: III, p. 129 (1985). Abstract. – See Abstr. 012.110.

161.392 Isothermal perturbations and anisotropy of the cosmic background radiation.
Y. Suto, K. Sato, H. Kodama.
Theoretical aspects on structure, activity, and evolution of galaxies: III, p. 130 – 137 (1985). – See Abstr. 012.110.

The authors study the quantitative behavior of evolution of the initially isothermal perturbations and show that the isothermal scenario is also severely limited by the cosmic background radiation isotropy.

161.393 Infrared background light from galaxies.
F. Takahara, Y. Yoshii.
Theoretical aspects on structure, activity, and evolution of galaxies: III, p. 138 – 142 (1985). – See Abstr. 012.110.

The spectral energy distribution of the background light from galaxies is synthesized in optical and infrared wavelength regions. The infrared background turns out to be dominated by the contribution from elliptical galaxies. Comparing the calculated spectra with the recent measurements of the infrared background by Matsumoto et al., the authors try to set constraints on the evolution of galaxies.

161.394 On log N $\sim$ log S.
K. Aizu.
Theoretical aspects on structure, activity, and evolution of galaxies: III, p. 143 – 151 (1985). – See Abstr. 012.110.

Recent deep surveys of extragalactic radio sources are reviewed and several attempts of model calculations are discussed.

161.395 Formation of difference in distribution between visible and invisible matters in the expanding universe.
L.-z. Fang, S.-p. Xiang.
Chin. Astron. Astrophys., Vol. 9, No. 1, p. 66 – 70 (1985). English translation of Acta Astrophys. Sin., Vol. 4, No. 4, p. 313 – 321 (1984).

The visible and invisible matters in the universe have rather different density distributions. The former is obvious clustered on the scales of galaxies, clusters and superclusters, while the latter is rather more uniform. The authors discuss the possibility of this difference forming during the stage of Jeans clustering. If the invisible matter is mainly neutrino with a finite rest–mass, then the quasi–uniform distribution of the invisible matter can be explained.

161.396 De Sitter – invariant vacuum state and temperature Green function.
M.-y. Zhou, L.-f. Chen, H.-y. Guo.
Chin. Astron. Astrophys., Vol. 9, No. 1, p. 71 – 75 (1985). English translation of Acta Astrophys. Sin., Vol. 4, No. 4, p. 322 – 328 (1984).

Using the method of canonical quantization in the static spacetime, the authors calculated the temperature Green function of the real scalar field of the de Sitter spacetime.

161.397 Energy level perturbation of hydrogen in the Robertson–Walker metric.
S.-w. Zhang, Y.-g. Shen.
Chin. Astron. Astrophys., Vol. 9, No. 1, p. 76 – 79 (1985). English translation of Acta Astrophys. Sin., Vol. 4, No. 4, p. 329 – 332 (1984).

The authors calculated the perturbations in the energy level of the hydrogen atom in the four states $^1S_{1/2}$, $^2S_{1/2}$, $^2P_{3/2}$ and $^2P_{1/2}$, caused by the background curvature of the Robertson–Walker metric. They found that it is only when the radius of curvature is less than 10^{-7}cm that an energy level split of the size of the Lamb shift can occur between $^2S_{1/2}$ and $^2P_{1/2}$.

161.398 Exact solutions in cosmology.
M. A. H. MacCallum.
Solutions of Einstein's equations: techniques and results, p. 334 – 366 (1984). – See Abstr. 012.112.

The aim of this paper is to explain why and how exact solutions of Einstein's equations have been used to advance our understanding of cosmology, to review the models which have been used, and to illustrate their use by describing some recent work and current problems. In particular, a detailed survey of recent work on inhomogeneous solutions is presented.

161.399 Quantum cosmogonic model.
M. F. Khodyachikh.
Astrofizika, Tom 22, Vyp. 3, p. 619 – 631 (1985). In Russian. English translation in Astrophysics, Vol. 22, No. 3.

An expression for the total mechanical energy of a compressing homogeneous sphere with due regard for pressures and rotations has been obtained. The exact solution of Schrödinger's equation describing stationary states of a body–quanton has been found. The division reactions of quantons are considered. The quantons might turn into macrobodies in the process of disintegration. The obtained formulas allow to calculate models of formation of concrete systems of objects observed in the Universe.

161.400   Cosmology.
H. van der Laan.
ESA Spec. Publ., ESA SP–213, p. 3 – 6 (1984). – See Abstr. 012.124.
A brief outline of current cosmological problems related to the population of strong radio sources and to the microwave background is presented.

General relativity.
See Abstr. 003.009.

Cosmologia: teoria cosmogaláctica e galáxcentrismo.
See Abstr. 003.011.

Superforce. The search for a Grand Unified Theory of nature.
See Abstr. 003.079.

The hidden universe.
See Abstr. 003.082.

The creation of matter: the universe from beginning to end.
See Abstr. 003.092.

Le destin ultime de l'univers.
See Abstr. 003.111.

Universe.
See Abstr. 003.115.

Ursprung und Zukunft des Weltalls. Pflanzen – Tiere – Menschen.
See Abstr. 003.152.

Building the universe.
See Abstr. 003.176.

Im Augenblick der Schöpfung.
See Abstr. 003.178.

The moment of creation: big bang physics from before the first millisecond to the present universe.
See Abstr. 003.179.

Evolution of matter and energy on a cosmic and planetary scale.
See Abstr. 003.194.

Eddington's "Expanding Universe".
See Abstr. 004.108.

Das Werden des Kosmos. Von der Erfahrung der zeitlichen Dimension astronomischer Objekte im 18. Jahrhundert.
See Abstr. 004.111.

Discovery of the 3K radiation.
See Abstr. 004.190.

Discovery of the cosmic microwave background.
See Abstr. 004.191.

Unsmoothing the Universe.
See Abstr. 011.003.

Particles and gravity. Proceedings of the Eighth Johns Hopkins Workshop of Current Problems in Particle Theory, held in Baltimore, 20 – 22 June 1984.
See Abstr. 012.120.

Contemporary cosmology and philosophy. (Some aspects of relationships and possibilities of overcoming the present difficulties).
See Abstr. 015.034.

Finite Euclidean and non–Euclidean geometry with applications to the finite pendulum and the polygonal harmonic motion. A first step to finite cosmology.
See Abstr. 021.011.

Prospects for determining the cosmological helium–3 to helium–4 ratio via absorption–line studies of the local interstellar medium.
See Abstr. 036.099.

Suche nach Neutrino–Oszillationen an Kernreaktoren.
See Abstr. 061.001.

The steady–state free–electron population of free space.
See Abstr. 061.004.

Exact solution to the Einstein–Yang–Mills equation.
See Abstr. 061.005.

Mirror neutrinos in standard cosmology.
See Abstr. 061.006.

Origin of the chemical elements.
See Abstr. 061.008.

Effective Lagrangian for massless pair production by an oscillating field.
See Abstr. 061.009.

Global and gauge symmetries in finite temperature sypersymmetric theories.
See Abstr. 061.010.

Electrodisintegration and electrocapture in primordial nucleosynthesis.
See Abstr. 061.013.

The shadow world of superstring theories.
See Abstr. 061.025.

Astration of cosmological deuterium.
See Abstr. 061.029.

Cosmological and experimental constraints on the tau neutrino.
See Abstr. 061.032.

A difficulty with evasion of a cosmological limit on massive neutrinos.
See Abstr. 061.033.

Effects of nuclear uncertainties and chemical evolution on the Standard Big Bang nucleosynthesis.
See Abstr. 061.037.

The quest for the origin of the elements.
See Abstr. 061.048.

Interaction between antiprotons and helium and astrophysics.
See Abstr. 061.049.

Incoherent radiation in an n–dimensional space.
See Abstr. 061.053.

Non–gauge real vector multiplets coupled to the $N = 1$ supergravity.
See Abstr. 061.054.

The magnetic monopole.
See Abstr. 061.055.

Large number coincidences and unification of the parameters underlying elementary particles astrophysics and cosmology.
See Abstr. 061.058.

Lepton–number violation in cosmology and astrophysics.
See Abstr. 061.060.

A stochastic Lie–isotopic approach to the foundations of classical electrodynamics.
See Abstr. 061.063.

A model for neutrino decays.
See Abstr. 061.066.

A new mechanism of gauge symmetry breaking induced by gravity and a naturally vanishing cosmological constant.
See Abstr. 061.067.

Superconducting strings in axion models.
See Abstr. 061.068.

Neutron capture processes in astrophysics.
See Abstr. 061.074.

Superheavy magnetic monopoles and the standard cosmology.
See Abstr. 061.079.

Monopoles in 1983.
See Abstr. 061.082.

Monopole compatibility with cosmology.
See Abstr. 061.083.

Reducing the monopole abundance.
See Abstr. 061.084.

Monopolonium.
See Abstr. 061.088.

Some possible tests of the inapplicability of Pauli's exclusion principle.
See Abstr. 061.095.

Nontrivial anisotropic supergravity cosmological solution.
See Abstr. 061.098.

Capture processes and element synthesis in the Universe.
See Abstr. 061.099.

Evolution of monopoles in big bang universe.
See Abstr. 061.114.

Monopolonium.
See Abstr. 061.115.

Chaotic behaviour and the Einstein equations.
See Abstr. 066.007.

Cosmic no–hair theorems.
See Abstr. 066.008.

Cosmic censorship, persistent curvature and asymptotic causal pathology.
See Abstr. 066.019.

Scalar gravitation and cosmology.
See Abstr. 066.023.

On the generalizations of Nariais GRT–solution in the Brans–Dicke theory.
See Abstr. 066.024.

de Sitter invariant gravity coupled with matter and its cosmological consequences.
See Abstr. 066.029.

Static space–time in general scalar tensor theory.
See Abstr. 066.030.

Comment on the BDT–Bianchi type–VI$_0$ stiff matter solution.
See Abstr. 066.033.

Comment on the GRT–LRS Bianchi type V solutions.
See Abstr. 066.034.

Einstein–Kahler solutions in some cosmological space–times.
See Abstr. 066.036.

The use of generating techniques for space–times with two non–null commuting Killing vectors in vacuum and stiff perfect fluid cosmological models.
See Abstr. 066.037.

A new formulation of gravitational lens theory, time–delay, and Fermat's principle.
See Abstr. 066.042.

Gravitational amplification of light by a non–static and thick lens.
See Abstr. 066.044.

Numerical analysis of one–soliton solutions on a Bianchi type–II background.
See Abstr. 066.048.

Spontaneously broken symmetries in gravitational fields.
See Abstr. 066.051.

Gravity and grand unified theories.
See Abstr. 066.056.

Gravitational lensing and galaxy shape.
See Abstr. 066.060.

Time and singularity.
See Abstr. 066.074.

Algebraic programming in general relativity and cosmology.
See Abstr. 066.075.

Pioneer 10 search for gravitational waves – limits on a possible isotropic cosmic background of radiation in the microhertz region.
See Abstr. 066.076.

On negative time dilation, the yttrium effect and applications to chemistry, physics and cosmology.
See Abstr. 066.077.

Role of Maxwellian distribution in an expanding universe.
See Abstr. 066.078.

Gravitation and irreversibility.
See Abstr. 066.079.

On the connection between the cosmological constant and the gravitational constant in a theory of induced gravity.
See Abstr. 066.080.

On Kaluza–Klein theories.
See Abstr. 066.092.

On the formulation of the positive energy theorem in Kaluza–Klein theories.
See Abstr. 066.093.

Note on cosmic censorship.
See Abstr. 066.101.

Thermal gravitational radiation from stellar objects and its possible detection.
See Abstr. 066.104.

Einstein's gravity with massive gravitons.
See Abstr. 066.106.

The Einstein–Rosen gravitational waves and cosmology.
See Abstr. 066.111.

Does a cosmic censor exist?
See Abstr. 066.112.

Gravitational waves in an expanding universe.
See Abstr. 066.113.

A constraint on physical quantum states in cosmological space–times.
See Abstr. 066.118.

On the scalar–tetradic theories of gravitation: cosmology and gravitational radiation.
See Abstr. 066.119.

Propagation of gravitational waves in Robertson–Walker backgrounds.
See Abstr. 066.121.

Unorthodox curvature interactions, black holes, and inflation.
See Abstr. 066.125.

Severe limitations upon the ultracentrifuge gravitational redshift experiment.
See Abstr. 066.126.

Naked singularities and causality violation.
See Abstr. 066.128.

Comments on the source of Gödel–type metrics.
See Abstr. 066.136.

Infinitesimal null isotropy and Robertson–Walker metrics.
See Abstr. 066.153.

Some simple type D solutions to the Einstein equations with sources.
See Abstr. 066.154.

Casimir energy in quantum gravity in $R^1 \times T^3$ space–time.
See Abstr. 066.155.

General relativity: progress, problems, and prospects.
See Abstr. 066.158.

Close–up on gravitational lensing: the gravitational mirages.
See Abstr. 066.187.

Exact Brans–Dicke–Bianchi solutions.
See Abstr. 066.213.

Conformal structure of a Schwarzschild black hole immersed in a Friedmann universe.
See Abstr. 067.072.

Report of IAU Commission 21: Light of the night sky (*Lumière du ciel nocturne*).
See Abstr. 082.096.

Les étoiles pauvres en métaux.
See Abstr. 114.127.

The type I supernovae absolute magnitude brightness decline rate relation and the Hubble constant.
See Abstr. 125.030.

Supernovae and the distance scale.
See Abstr. 125.032.

Hubble's constant and exploding carbon–oxygen white dwarf models for Type I supernovae.
See Abstr. 125.033.

Angular diameter determinations of radio supernovae and the distance scale.
See Abstr. 125.112.

Optical supernovae and the Hubble constant.
See Abstr. 125.115.

Type I supernovae as standard candles.
See Abstr. 125.116.

Physical models for Type I supernovae and the distance scale.
See Abstr. 125.120.

Probing the early universe with the millisecond pulsar.
See Abstr. 126.087.

New developments in high energy astrophysics.
See Abstr. 143.100.

The origin of the ultra high energy cosmic rays.
See Abstr. 144.171.

Testing baryon–symmetric cosmologies on a supergalactic scale.
See Abstr. 144.172.

Approximate rotation curve solutions for the evolution of a viscous protogalactic disk.
See Abstr. 151.067.

Deep photometry of globular clusters. V. Age derivations and their implications for galactic evolution.
See Abstr. 154.027.

A method for determining the spatial correlation function from a sample with partial distance information.
See Abstr. 157.005.

The evolution of spiral galaxies and uncertainties in interpreting galaxy counts.
See Abstr. 157.064.

Self–regulating galaxy formation. I. H II disk and Lyman–alpha pressure.
See Abstr. 157.065.

Certain cosmological considerations connected with binary galaxies.
See Abstr. 157.091.

An analysis of the distribution of faint galaxies at the south galactic pole.
See Abstr. 157.109.

Some possible regularities in the missing mass problem.
See Abstr. 157.203.

Re–assessing the primordial helium abundance: observations of NGC 4861 and CG 1116 + 51.
See Abstr. 157.225.

Hot gas and massive halos around early type galaxies.
See Abstr. 157.226.

Current problems of determining distances to galaxies.
See Abstr. 157.247.

Galaxy photometry and cosmology.
See Abstr. 157.265.

The form of $\log N - \log S$ relation in the domain of strong flux densities.
See Abstr. 158.038.

Orientation correlation of radio double sources.
See Abstr. 158.044.

Sub–millijansky 1.4 GHz source counts and multicolor studies of weak radio galaxy populations.
See Abstr. 158.078.

CCD magnitudes of 3C radio sources.
See Abstr. 158.086.

Cosmological evolution of linear sizes of radio galaxies.
See Abstr. 158.106.

Observing the galaxy evolution at high redshifts.
See Abstr. 158.322.

Maximum cosmic redshift of quasars at $Z > 3$.
See Abstr. 159.016.

A study of quasar pairs.
See Abstr. 159.018.

Uniformity of quasars in the chronometric cosmology.
See Abstr. 159.028.

The far–ultraviolet continuum of quasars and the universe at $z > 4$.
See Abstr. 159.035.

Direct estimate of Hubble's constant from the double quasar.
See Abstr. 159.043.

The sensitivity of the V/V_m test to uncertainties in observed parameters.
See Abstr. 159.050.

QSO evolution in the interaction model.
See Abstr. 159.106.

An X–ray Hubble diagram for quasi–stellar objects.
See Abstr. 159.108.

Luminosity evolution of QSOs.
See Abstr. 159.119.

Predictions of distance within the Local Group.
See Abstr. 160.001.

Cell structure and selective extinction in the Jagellonian field.
See Abstr. 160.014.

No missing mass in clusters of galaxies?
See Abstr. 160.017.

Galaxy groups: virial quantities and galaxy population.
See Abstr. 160.018.

Reheating of the intergalactic medium at $z > 10$.
See Abstr. 160.041.

Are galaxies more strongly correlated than clusters?
See Abstr. 160.043.

Statistical and dynamical properties of galaxy groups.
See Abstr. 160.053.

Galaxy clustering and the method of voids.
See Abstr. 160.080.

Dynamics and evolution of the Virgo Supercluster.
See Abstr. 160.133.

Dynamics of the Local Supercluster and the Virgo cluster.
See Abstr. 160.134.

The termination of galaxy formation by cluster formation.
See Abstr. 160.135.

On the infall velocity towards Virgo.
See Abstr. 160.137.

Between the galaxies.
See Abstr. 160.145.

A study of the large–scale structure of the distribution of galaxies in and around the Cancer cluster.
See Abstr. 160.147.

Some results from two–dimensional simulation of neutrino–dominated universes.
See Abstr. 160.155.

Influence of the K–effect on the colours and the $\lambda_0 4000$ break amplitudes of galaxies in distant clusters.
See Abstr. 160.160.

Author Index

The authors are listed in alphabetical order according to the initial letter following the first names. Author names which are not transliterated according to our scheme are indicated by †.

Aab, O. E. †
117.296
Aab, O. Eh.
114.024 .076
117.296
Aaman, J. E.
066.004
Aannestad, P. A.
126.081
132.033
Aanstoos, J. V.
036.180
Aaronson, M.
065.062
111.012
154.012
156.023
157.089 .123 .155 .160 .161
   .201
160.054 .061
161.195
Aarseth, S. J.
151.108
Abadie, D.
022.090
Abalakin, V. K.
082.045
Abbasov, A. R.
077.035
Abbott, D. C.
002.122
064.024 .032 .104
116.012
Abbott, L. F.
161.023 .260
Abbott, R. B.
161.292
Abdellatif, R. A.
062.077
Abdil'din, M. M.
066.146
Abdulla, A. Y.
042.030
Abdullah, M. M.
144.347
Abdulova, R. V.
002.095
Abdusamatov, Kh. I.
072.023
Abe, O.
066.155
Abell, R.
034.161
Abhyankar, K. D.
002.034 .035
063.068 .069
120.024
Ablaeva, M.
161.142
Ables, H. D.
111.005
Abraham, Z.
159.064
Abrahamian, A. V.
034.131 .133
Abramenko, V. I.
077.020
Abramov, V. I.
033.011 .012
Abramyan, L. Eh.
033.017
Abramyan, M. G.
151.172
Abt, H. A.
036.188
114.057
120.034
153.017

Abuladze, O. P.
122.089
Abulova, V. G.
144.337
Acharya, B. S.
144.340
Acharya, Y. B.
035.026
Acker, A.
009.005
114.145
131.155
134.007 .011 .040
Ackerman, M.
082.033
Acton, I. I.
035.124
Acton, L. W.
072.069
073.127 .204
076.013
Acuna, M. H.
100.027
Adam, J.
045.008
134.005
Adam, J. A.
062.122
Adam, J. C.
062.160
Adam, J.-C.
161.088
Adam, M. G.
072.049
Adam, V.
035.073
Adamoli, G.
013.090
100.044
Adams, F.
131.256
Adams, J. H.
035.122
Adams, M.
113.061
Adams, P. J.
161.079
Adams Jr., J. H.
051.026
144.134 .301
Adamson, A. J.
111.045
Ade, P. A. R.
158.138
159.015
Adeishvili, T. G. †
083.014
Adejshvili, T. G.
083.014
Adelman, C. J.
002.025
Adelman, S. J.
002.025
036.127
114.019 .115
122.136
Adler, D. S.
153.056
Adzhyan, G. S.
065.035
Afanas'ev, P. M.
033.043
Afanas'eva, P. M.
002.032
Affleck, I.
161.288
Africano, J. L.
117.362

Africano, J. L.
134.057
Agafonov, M. I.
125.048
Agafonova, I. I.
099.042
100.048
Agaronian, F. A. †
144.102
Agaronyan, F. A.
143.068
144.102
158.283
Aggarwal, K. M.
022.059 .143
Agnese, A. G.
066.157
Agnetta, G.
035.091
Agrawal, P. C.
035.090
113.042
117.111 .261
118.049
143.100
158.281
Agrawal, S. P.
073.191
144.240 .246 .249 .250 .251
   .256 .258 .274 .455
Agrinier, B.
035.091
Aguilar, L. A.
151.157
154.033
Agus, S.
079.141
Aharonian, F. A. †
143.068
158.283
A'Hearn, M.
103.341 .342
A'Hearn, M. F.
102.001
103.107
Ahlen, S. P.
061.090 .134
144.172
Ahluwalia, H. S.
144.385 .416 .421
Ahmad, I. A.
119.001 .077 .097
Ahmad, J.
036.181
Ahmed, S. A. S.
080.124
Ahrens, T. J.
022.126
Aichelburg, P. C.
067.149
Aihara, Y.
091.016
Aikawa, T.
065.076
122.025 .064 .195
Aikman, G. C. L.
111.041
Aitbaev, F. B.
144.378
Aitken, D. K.
098.074
103.122
131.188
157.079
158.083
Aizu, A.
144.094

Aizu, K.
161.394
Ajello, J. M.
106.015
Ajmanov, A. K.
034.115
074.129
Ajmanova, G. K.
036.160
073.182
Ajzenberg-Selove, F.
061.022
Akabane, K.
132.042
155.050
Akasofu, S.-I.
072.002
075.015
084.010
085.006
144.054
Akatova, N. I.
144.196
Ake, T. B.
114.147
117.209
119.002 .079
120.016
Ake III, T. B.
119.004
120.003
121.006
Akim, E. L. †
093.044
Akim, Eh. L.
093.020 .044
Akimov, L. A.
105.207
Akin'yan, S. T.
077.024
Akioka, T.
061.122
Akita, K.
073.079
Akiyama, H.
035.028
079.141
Akopova, A. B.
144.070
Aksenov, A. N.
099.038
Aksenov, E. P.
009.032
Aksnes, K.
099.011 .057
Al-Mufti, S.
015.036 .037
022.101 .103
Al-Naimiy, H. M. K.
119.012
Alam, B.
022.131
Alania, M. V. †
072.088
144.395 .409
Alania, M. †
144.384
Alaniya, I. F.
122.089
Alaniya, M. V.
072.088
144.046 .384 .395 .409
Alaverdyan, G. B.
065.035
Albee, A. L.
003.032
051.023

Falchi

Falchi, A.
073.054
Falciani, R.
013.047
073.054
Falco, E. E.
036.206
161.097
Falgarone, E.
131.018
Falin, J. L.
036.166 .168 .169
041.030 .044
051.088 .091
Fal'kovich, I. S.
158.076
Fall, S. M.
154.065
155.056 .057 .088
Falomkin, I. V.
161.021
Fan, C. Y.
126.094
144.148 .195 .216 .380
Fang, C.
002.010
072.120
073.239
Fang, L.
067.139
161.167 .191 .224 .225 .226
Fang, L. Z.
003.021
161.187
Fang, L.-z.
158.044
161.395
Fang, T.
081.026
Fanning, A. E.
004.050
Fanti, C.
158.046
Fanti, R.
158.022 .046
Faraggiana, R.
114.056
Farago, M.
035.086
Fargion, D.
061.006
161.016 .305
Farhi, E.
161.349
Farhoomand, J.
022.093
Farilla, A.
036.164
Farinella, P.
098.022 .040 .071 .072
Farnik, F.
073.094
074.025
076.006
Fasakhova, M. A.
077.044
Fasano, G.
002.030
157.083 .288
Fashchevskij, N. N.
122.018
154.016
Fassino, B.
002.097
Fast, V. G.
104.027
Fatemi, I
144.347
Fathauer, R. W.
082.093
Fauci, F.
126.029 .127 .146
Faulkner, D. J.
134.031
Faulkner, D. R.
112.144
119.087
Faulkner, J.
041.012
117.019 .327
Favale, A. J.
035.094

Favin, S.
084.007
Fawcett, B. C.
073.127
Fay, R. A.
084.043
Fazio, G. G.
012.059
035.030 .060
051.062
112.060
132.007 .055
155.175
Feast, M. W.
008.165
115.014
122.067
155.162
Fechtig, H.
008.339
106.028 .057 .063
Fedder, J.
106.061
Fedder, J. A.
103.187
Federer, W.
022.140
Federici, L.
154.021
Federman, S. R.
131.074 .146
Fedorov, P. N.
032.054
082.102
Fedorov, Yu. I.
078.061
144.203 .241 .242
Fedorova, A. V.
117.054
Fedorova, R. T.
032.054
Fedoseev, E. N.
044.004
Fedyaev, Yu. A.
021.012
Fedyanin, M. R.
082.036
Fegan, D. J.
034.081 .128
061.119 .123
117.112
126.113
143.047 .048 .091
144.336
Fegley, B.
105.083
107.019
Fegley Jr., B.
105.274
Fehrenbach, C.
007.009
011.022
111.034
Feibelman, W. A.
112.084
134.002 .069 .074
Feierman, B.
115.017
Feigelson, E.
158.305
Feigelson, E. D.
021.017
036.201 .203
121.013 .024
158.304 .342
Feijth, H.
032.003
124.223
Feinstein, A.
002.060
Feitzinger, J. V.
002.068
151.030 .078
155.026 .097 .120 .121 .148
157.045
Feix, M. R.
065.084
Fejes, I.
036.028
Fejgin, F. Z.
084.103
Fekel, F. C.
111.043

Fekel, F. C.
112.072
120.037
123.009
Fekel Jr., F. C.
120.045
Feklistova, T. H. †
114.051
Feklistova, T. Kh.
022.051
114.051
Fel'dman, B. Ya.
093.044
Feldman, P. A.
126.025
Feldman, P. D.
035.129
099.003 .060
102.001
103.015
113.041
Feldman, U.
071.006
073.001 .136 .155 .206
074.050
Fel'dman, V. I.
105.037 .201
Feldman, W. C.
106.072
Felenbok, P.
121.014
Felici, F.
035.040
Fellberg, G.
035.082
093.026 .027
Felli, M.
112.069
Felten, J. E.
161.205
Fenimore, E. E.
051.005
067.027
143.010 .019 .025 .035 .051
.052 .053 .059 .064 .065
Fenkart, R. P.
002.010 .102
113.004
153.004
155.073
Fennell, J. F.
035.108
Fennelly, A. J.
161.051
Fennelly, J. A.
076.044
Fenton, A. G.
078.004 .005 .056 .083
144.025 .270 .408 .415 .423
Fenton, K. B.
078.004 .005 .056
143.034
144.270 .408
Feretti, L.
158.164 .247
Ferguson, D. H.
117.238
Ferguson, H. C.
035.063
Ferland, G. J.
158.069
Ferlet, R.
112.129
122.101 .108 .137
131.190 .191
Ferluga, S.
119.020
Fernandes, M.
119.100
Fernandez, F.
035.105
Fernandez-Figueroa, M. J.
114.028
117.094
Fernie, J. D.
012.027
119.066
Fernley, J. A.
122.047
Ferrandi, G.
035.074
Ferrando, P.
144.327 .328

Field

Ferrara, S.
012.118
Ferrari, A.
062.025 .032
064.048 .054
158.118 .120
Ferrari, J. A.
066.023
Ferrari, V.
066.025
Ferrari d'Occhieppo, K.
015.002
Ferrari-Toniolo, M.
117.005 .043
132.024
133.005 .008
Ferraz-Mello, S.
042.031 .071 .072
097.005
Ferrer, S.
042.043
Ferreri, W.
098.007
103.025
Ferrero, L.
161.021
Ferri, G.
034.150
Ferris, J. P.
099.091
Ferris, T.
003.088
Ferro Fontan, C.
067.147
Ferronsky, S. V.
042.036
Ferronsky, V. I.
042.036
Fertel, J. H.
099.044
Fertig, J.
103.914
Fesen, R. A.
125.056 .072 .074 .089
131.006
134.015
Fesenko, B. I.
160.052
Fesenkova, L. V.
015.041
Festa, A.
051.118
Festou, M.
103.035 .104
Festou, M. C.
035.019
099.003 .060
102.018 .043
103.101 .103 .105 .106
Fetisova, T. S.
160.069 .070
Feuchter, C. A.
014.034
Feyerabend, P. K.
004.098
Fiala, A. D.
080.132
Fialkov, V. A.
034.110
Ficarra, A.
141.003
Fich, M.
112.085
131.282
132.066
155.084 .147
Fichtel, C. E.
035.093 .094
143.085
Fickle, R. K.
022.166
035.116
144.085 .320 .322 .324 .366
Fiedler, R. A.
131.161 .223
Fiedler, R. L.
062.182
158.124 .182
Field, D.
131.063
Field, G.
062.135

Field

Field, G. B.
003.089
Figer, A.
099.021
Figueras, F.
113.006
Fijalkow, E.
065.084
Fikani, M. M.
143.050
Filice, J. P.
074.123 .160
Filimonova, T. K.
104.042
Filipenko, V. I.
033.010
Filip'ev, G. K.
122.090
Filippenko, A.
158.269
Filippenko, A. V.
117.084
158.077 .126 .225 .249 .315
Filippini, V.
161.021
Filippov, A. E.
045.026
Filippov, A. T.
144.207 .209 .234
Filippova, A. A.
155.004
Fillius, W.
144.004 .188 .192
Finch, M. L.
051.107
Findlay, J. W.
004.183
Finger, G.
082.077
Fink, D.
022.146
Fink, H.
158.156
Fink, U.
103.601
131.247
Finkel, R. C.
144.369
Finkenzeller, U.
036.096
Finley, D. S.
126.006
Finman, L. C.
035.137
Finney, B. R.
015.021 .026
Fiorentini, G.
061.070
Fiorini, E.
022.171
Firmanyuk, B. N.
122.173
Firneis, M. G.
009.044
Firsova, L. A.
098.106
Fischbach, E.
080.030
Fischel, D.
002.025
Fischer, G.
008.528
Fischer, H.
032.017
045.021
Fischer, J.
112.069
131.293 .332
Fischhoff, E.
161.341
Fisenko, A. V.
105.028 .032 .200
Fisenko, M. I.
036.071
Fisher, D. E.
105.214
Fisher, G. H.
073.089 .090 .091 .213 .235
Fisher, R.
074.097
116.039
Fisher, R. R.
034.049 .102

Fisher, R. R.
074.023 .117 .134 .145 .146
.147
Fisher, W. A.
111.045
117.060 .200
119.009
Fishkova, L. M.
082.070
Fishman, G. J.
035.003
126.083
143.017 .038
Fisk, L. A.
144.029 .204
Fitton, B.
082.005 .065
FitzGerald, M. P.
114.119
Fitzpatrick, E. L.
131.357
153.037
156.002 .016 .019
Fix, J. D.
112.102
131.249
157.111
Fixsen, D. J.
144.085 .323 .360
Flaherty, F. J.
012.102
Flaih, H. A.
119.012
Flatters, C.
159.045
Fleischmann, H. H.
062.072
Fleming, T. A.
126.079
133.013
Flerov, G. N.
105.243 .246
Fletcher, J. M.
034.125
Fletcher, M.
154.037
Flin, P.
119.071
Floquet, M.
120.013 .017
Florence, W. B.
098.095
Florenskij, K. P.
091.031
Florides, P. S.
066.010 .069
Florinskij, A. A.
073.170
Florkowski, D.
117.231
Florkowski, D. R.
158.293
Florsch, A.
002.067
012.067
036.118
111.033
Florsch, J.
002.067
Flower, D. R.
022.088
131.111 .130
134.027
Flower, P. J.
153.035
Fludra, A.
073.051
Flueckiger, E. O.
078.022 .036 .046 .073
144.237 .279 .280
Flury, W.
051.118
Flynn, E. R.
061.022
Flynn, R. W.
014.040
Foglietti, V.
034.114
Foing, B. H.
076.008
116.023
Fokin, A. B.
065.075

Fokin, A. B.
122.079
Foltz, C. B.
131.193
158.062 .163
159.065 .072 .089 .091 .092
Fomalont, E.
158.171
Fomichev, V. V.
012.040
073.094 .167
077.024
Fomin, S. S.
076.023
Fomin, V. A.
033.043
Fomin, V. P.
143.027 .036
155.080
Fomin, Yu. A.
144.341
Fong, R.
036.067
Fontaine, G.
008.504
126.003 .033 .089 .120
158.276
Fontan, C. F.
067.156
Fontanelli, P.
157.085
160.108
Fontenla, J. M.
073.132
Fonti, S.
031.035 .036
Forbes, D.
119.089
131.051
Forbes, D. W.
122.059
Forbes, E. G.
004.151
Forbes, T. C.
073.032
Forbes, T. G.
073.106
Ford, H.
157.087
160.073 .102
Ford, H. C.
157.189
158.168 .250
Ford, L. H.
161.296 .297
Ford, V. L.
125.087
Ford Jr., W. K.
157.071 .157 .262
Foresta Martin, F.
094.042
Forkert, T.
033.001
Forlani, A.
062.173
Forman, M. A.
144.144 .145 .210
Forman, W.
157.202 .226
160.063 .131 .143
Formisano, V.
074.075
084.040
102.025
Fornasari, L.
077.011
Forrest, D. J.
035.135
073.146 .192
076.033 .039
078.036 .037 .073
117.237
143.012 .013 .026
155.115
Forrest, W. J.
034.023 .027
035.060 .061
158.057
Fort, B.
036.218
Fosbury, R. A. E.
158.074

Frazier

Foukal, P.
072.104
Foukal, P. V.
071.043
Fountain, G. H.
035.045
Fountain, W.
035.119
Fouque, P.
160.140
161.237
Fowler, A. C.
091.065
Fowler, P. H.
035.101 .113
144.321
Fowler, W. A.
061.046 .048 .092
065.065
Fox, K.
022.073
093.052
100.034
Fox, P.
080.060
Frabel, P.
035.091
Fracassini, M.
126.022 .055 .056
Fracastoro, M. G.
005.005
007.030
013.037
118.058
Fraczyk, P.
036.027
Fradkin, M. I.
073.143
078.035
143.073 .083
Fraedrich, D.
022.149
Fraknoi, A.
002.040
003.090
006.002 .004 .008 .010 .021
.023
014.037
Francaviglia, M.
066.048
Franceschini, A.
158.020 .125
161.074
Franchuk, N. G.
077.036
Francis, G. W.
052.065
Franco, M. L.
131.068 .267
Francois, P.
114.128
Frank, J.
003.012
117.064
Frank, L. A.
084.097
Franklin, F.
007.004
099.011 .057
Franklin, K. L.
004.193
Fransson, C.
125.113
Frantzen, H.-P.
041.022
Franz, O. G.
098.014
118.051
Fraquelli, D. A.
118.040
154.084
Fraser, B. J.
011.007
Frater, R. H.
033.014
Frauenholz, R. B.
045.001
Fraundorf, P.
106.065
Frazer, J.
116.036 .051
Frazier, R. M.
105.255

Hunt

Hunt, G. C.
125.004
141.040
Hunt, G. E.
012.026 .071
051.063
091.018 .026 .050
099.075
100.035
Hunt, J. J.
012.044 .045
Hunten, D. M.
099.069
101.008
Hunter, D.
157.292
Hunter, D. A.
132.006
157.002 .117
Hunter, J. H.
067.123
Hunter, W. R.
034.092 .118
Huntress Jr., W. T.
131.100
Hur, R. C.
144.045
Hurd, T. R.
161.291
Hurford, G. J.
072.070
073.039
074.104
Hurley, K.
051.039
125.090
143.006 .024 .025 .035 .043
   .056 .064 .065
Hurukawa, K.
098.056 .057
Hurwitz, M.
036.099
Huston, T. J.
105.117
Hut, P.
012.070
117.310
151.057 .107
154.033 .087
155.087
Hutcheon, I. D.
105.049 .139
Hutcheon, I. E.
105.101 .102
Hutchings, J. B.
035.064
067.176
117.083 .200 .384 .385
119.093
142.022
156.001
159.019
Hutchinson, B.
004.141
Hutchison, R.
013.054
105.047 .103
Huters, A. F.
117.153
Hutsemekers, D.
112.086
123.002
Hutter, D. J.
036.199 .200
118.051
158.006
Hyde, F. W.
013.017
Hyde, T.
022.034
091.021
Hyland, A. R.
112.077
153.042
155.015 .140
Hyman, M.
105.154
Hyun, J. J.
073.245

Iafrate, L.
035.029

Ianna, P. A.
032.044
111.048 .050
118.042 .045
120.021
Iarocci, R.
022.171
Ibadinov, Kh. I.
102.029
Ibadov, S.
102.030 .040
Ibanez Cabanell, J. M.
067.085
Ibanez S., M. H.
062.074
Ibanoglu, C.
117.250
119.013 .064 .085
Iben Jr., I.
065.005 .011 .090
067.016
Ibragimov, K. Yu.
099.038
Ibrahim, F. S.
062.149
Ichikawa, M.
009.012
033.056
Ichinose, M.
144.261 .268 .273 .414
Icke, V.
151.040
161.309
Ierkic, H. M.
099.067
Ieyoshi, K.
144.294
Igarashi, G.
081.039
Igenbergs, E.
035.036
Iglseder, H.
035.036
Ignatenko, K. I.
105.200 .227
Ignat'ev, P. P.
144.282
Ignat'ev, S. B.
034.110
Ignatiev, P. P. †
144.282
Ignatova, R.
105.010
Ignatova, S. P.
093.026
Ihle, W.
021.005
Iida, S.
034.129
Ikeda, Y.
105.104 .169
Ikeuchi, S.
131.365
151.047 .080 .165 .166
160.015
161.058 .060 .085 .390
Ikhsanov, R. N.
072.116 .121
075.043
Ikromov, A.
042.058
Il'ichev, A. T.
161.219
Iliev, L. H. †
113.031
Iliev, L. Kh.
113.031 .071
Il'in, V. D.
084.127
Il'ina, A. N.
084.127
Iliyin, V. D. †
084.127
Iliyina, A. N. †
084.127
Ill, M.
082.014
Illes-Almar, E.
082.015
Illing, R. M. E.
073.176
074.011 .122 .152

Illingworth, G.
157.029
Ilmas, M.
117.115
Ilovaisky, S. A.
117.186 .370
143.024
Ilyasov, Yu. P.
131.288
Imaeda, K.
144.294
Imamura, J. N.
065.103
067.143
Imamura, M.
144.164
Imbert, M.
111.002
119.007
122.187
Imbriale, W. A.
051.053
Imhof, W. L.
143.035
Imhoff, C. L.
008.699
121.001 .006 .045 .053 .062
Impey, C.
157.238
Imshennik, V. S.
003.119
065.037
Inagaki, S.
151.038 .076 .077 .104 .168
Inasaridze, R. Ya.
031.006
Inatani, J.
033.047
131.041
155.061
Incandela, J. R.
061.097
Indorato, L.
004.002
Indrawan, S.
079.141
Infante, L.
160.021 .076
Ingersoll, A. P.
097.020
Inglis, I.
157.004
Ingraham, R. L.
014.041
Ingvarsson, S. I.
119.112
120.028
Innanen, K. A.
042.062 .063
151.156
160.017
Innis, J. L.
114.099
117.139
Inoue, A.
082.081
144.424
Inoue, H.
067.011
117.050
142.011 .016 .031
Inoue, K.
046.020
Inoue, M.
132.042
155.050
160.088
Inoue, N.
144.076 .077
Inoue, T.
042.017
Intriligator, D. S.
013.006
022.112
074.085
093.006
Inzhelevskaya, I. V.
084.106
Ioakimidis, N. I.
042.106
Ioffe, Z. M.
102.032
104.016

Issa

Ionson, J. A.
074.003 .084
Iorio Fili, D.
031.039
Ioshpa, B. A.
075.027 .028 .030
Ip, W.-H.
013.075
100.013
102.024
103.601
144.188 .192 .363
Ipatov, A. P.
117.031
Ipatov, A. V.
141.035
Ipatova, L. P.
072.046
077.034
Ipavich, F. M.
078.033 .071 .072
106.076
144.380
Ipser, J. R.
065.059
Irvine, W. M.
131.105 .143 .215 .315 .319
   .329 .334
132.026 .048
155.188
Irwin, A. W.
064.058
Irwin, J. A.
157.170
Irwin, M. J.
036.068 .100 .153
Isaacman, R.
161.205
Isaev, A. S.
144.283
Ishibashi, S.
131.366
151.162
Ishida, G.
118.028
Ishida, K.-i.
117.160
Ishida, Y.
144.265 .414
Ishikawa, F.
144.342
Ishikawa, T.
002.056
Ishkov, V. N.
073.094
Ishwar, B.
042.070
Islam, J. N.
003.111
012.011
066.014 .015
Isles, J. E.
119.111
Isobe, S.
036.147
079.141
117.160
Isobe, T.
036.201
Isogai, M.
052.017
Israel, F. P.
132.039
155.031
156.005
157.258
Israel, G.
003.018
Israel, M.
013.001
Israel, M. H.
022.166
034.135
035.116 .117
144.320 .322 .323 .324 .360
   .366
Israel, W.
066.112 .158
Issa, M. R.
142.067
144.031
155.086 .091

Subject Index

The Subject Index provides an alphabetical subject key to the papers abstracted in *Astronomy and Astrophysics Abstracts*. In order to explain the structure of this Index some general remarks on the construction principles of our key words will follow.

Whenever possible a key word was formed in such a way as to combine two different descriptors characterizing an abstracted article. Such a combination represents the contents of a paper more precisely than both components regarded as single entries. As an example we take the two terms Minor Planets and Orbits and the combination

Minor Planets
Orbits

Efforts were made to choose combinations which can be inverted in order to increase the usefulness of the Index. That means the two components represent terms of equal status which both serve as primary entries. In the given example we find the two entries

Minor Planets
Orbits

and

Orbits
Minor Planets

But there are also two-term key words which cannot be reversed. The conditions for non-inversibility are: the secondary descriptor is either a very specific term (e.g. LTE Analyses) or a very general one (e.g. Models). Thus we distinguish two kinds of descriptors, the primary ones, serving as direct entries and the secondary ones as auxiliary entries to the Index. Our intention is to choose comprehensive as well as significant formulations.

Some further aids to the use of this Subject Index are given by the following concordances:

Asteroids	see Minor Planets
Earth: Magnetic Field	see Geomagnetic Field
Galactic...	see also Galaxies, Galaxy
Lunar...	see also Moon
Planetary...	see also Planets
Solar...	see also Sun
Stellar...	see also Stars

Efforts were made to standardize the key words. This idea led to the development of our vocabulary of astronomical and astrophysical terms. From time to time we enlarged our vocabulary and at present version 09.85 consists of 2,303 items.

The list of key words was examined in order to standardize the key words and to confine some of the synonymous entries. For example, according to our vocabulary the key words Catalogues, Positions, Star Catalogues, and Star Positions were admissible terms, likewise the combinations

Catalogues
Star Positions

Star Catalogues
Positions

Star Catalogues
Star Positions

The user is – in any case – kindly requested to look for synonymous entries, because further references to the topic looked for might exist elsewhere in the Index under another current astronomical term.

Some key words concerning comets, novae, and solar eclipses are sorted by a special routine. Comets are listed according to their Roman numeral designation or preliminary designation, novae according to their stellar constellation and solar eclipses according to the date of the event.

Due to the introduction of a special object index containing a list of single objects, special object designations (e.g. SS433, BL Lacertae, Algol, etc) have been removed from the subject index and transferred to the object index (see SS433, BL Lac, β Per).

Exceptions have been made where major planets and their moons, comets, novae, and supernovae are concerned, which are given in the subject index only. Some objects (name or catalog designation) are contained in both indexes

Andromeda Galaxy	M31, NGC 224
Magellanic Clouds	LMC, SMC
Pleiades	M45
Hyades	Melotte 25
Crab Nebula	M1

In order to obtain the most comprehensive information possible the use of both indexes is recommended.

Starting with Volume 30 of *Astronomy and Astrophysics Abstracts,* this Index is given in a four-column arrangement. The overall size further has been optimized by suppression of a repetition of the first key word in a two-term combination. Thus, the primary term is printed only once (in semi-boldface) for all the following secondary key words which belong to this header.

The following List of Key Words might serve as a further guidance to the use of this Subject Index. This list, a subset of our vocabulary 09.85, consists of 1,190 entries and comprises all one-term key words and all primary terms of the two-term ones actually used in this Subject Index.

List of Key Words

A Dwarfs
A Stars
A Supergiants
Absolute Magnitudes
Absorption Lines
Absorption Spectra
Accretion
Accretion Disks
Achondrites
Acoustic Waves
Active Galactic Nuclei
Active Galaxies
Ae Stars
Aerosols
Air Showers
Airglow
Albedo
Alfven Waves
Algol Systems
AM Herculis Stars
Am Stars
Andromeda Galaxy
Angular Diameters
Angular Momentum
Annual Parallaxes
Antimatter
Antineutrinos
Ap Stars
Aperture Synthesis
Apollo Objects
Apsidal Motion
Aquarids
Archaeoastronomy
Artificial Satellites
Asteroidal Belt
Astrodynamics
Astrographic Catalogues
Astrographs
Astrolabes
Astrometric Binaries
Astrometric Plates
Astrometric Reflectors
Astrometry
Astronomical Geodesy
Astronomical Instruments
Astronomical Optics
Aten-Type Objects
Atlases
Atmospheres
Atomic Clocks
Atomic Processes
Aurorae
Auroral Arcs
Auxiliary Instrumentation
Azimuth Determination

B Stars
B Subdwarfs
B Supergiants
Background Radiation
Balloons
Balmer Decrement
Balmer Lines
Barium Stars
Barred Spirals
Baryons
Be Stars
Beta Cephei Stars
Beta CMa Stars
Beta Lyrae Stars
Binaries
Binary Pulsars
Bipolar Nebulae
Bipolar Outflows
BL Herculis Stars
BL Lacertae Objects
Black Holes
Black-Body Radiation
Blue Galaxies
Blue Stars

Blue Stragglers
Bolometric Corrections
Bolometric Magnitudes
Bp Stars
Bremsstrahlung
Bright Galaxies
Bright Stars
Brightness Temperature
Broad-Band Photometry
BV Photometry
BY Draconis Stars

C-M Diagrams
Callisto
Cameras
Carbon Stars
Cataclysmic Binaries
Cataclysmic Variables
Catalogues
CCD Cameras
CCD Detectors
CCD Observations
CCD Photometry
Celestial Bodies
Celestial Mechanics
Celestial Objects
Central Stars
Cepheids
Cerenkov Radiation
CH Stars
Chandler Wobble
Chemical Elements
Chondrites
Chondrules
Circumstellar Clouds
Circumstellar Disks
Circumstellar Dust
Circumstellar Envelopes
Circumstellar Grains
Circumstellar Matter
Circumstellar Shells
Clocks
Close Binaries
Clusters of Galaxies
CN Stars
Cocoon Stars
Collapse
Collapsed Objects
Collapsed Stars
Collapsing Clouds
Collapsing Stars
Color Excesses
Color Indices
Color-Luminosity Relation
Colors
Cometary Atmospheres
Cometary Comae
Cometary Dust
Cometary Ionospheres
Cometary Nebulae
Cometary Nuclei
Cometary Plasma
Cometary Tails
Comets
Compact Galaxies
Compact Objects
Compton Effect
Compton Scattering
Contact Binaries
Convection
Convective Zones
Cool Giants
Cool Stars
Cosmic Dust
Cosmic Rays
Cosmochemistry
Cosmochronology
Cosmogony
Cosmological Constant
Cosmology

Crab Nebula
Curves-of-Growth
Cyclotron Radiation
Cygnus Loop

Dark Clouds
Dark Matter
Dark Nebulae
Data Bases
Data Centers
Data Processing
DDO Photometry
Degenerate Dwarfs
Degenerate Stars
Delta Cephei Stars
Delta Scuti Stars
Dense Clouds
Dense Matter
Densities
Density
Density Waves
Detector Arrays
Detectors
Diameter
Diameters
Differential Rotation
Diffuse Nebulae
Disk Galaxies
Distance
Distance Indicators
Distance Moduli
Distance Scale
Distances
Draconids
Dust
Dust Clouds
Dust Grains
Dust Nebulae
Dust Shells
Dwarf Cepheids
Dwarf Galaxies
Dwarf Novae
Dwarfs
Dynamical Parallaxes
Dynamo Theory

Early Universe
Early-Stage Stars
Early-Type Galaxies
Early-Type Stars
Early-Type Supergiants
Earth
Earth Atmosphere
Earth Core
Earth Interior
Earth Ionosphere
Earth Magnetosphere
Earth Mantle
Earth Mesosphere
Earth Radiation Belts
Earth Rotation
Earth Stratosphere
Earth Thermosphere
Earth Troposphere
Earth-Moon System
Echelle Spectra
Echelle Spectrometers
Eclipses
Eclipsing Binaries
Ecliptic
Effective Temperature
Effective Temperatures
Electron Densities
Electron Temperatures
Electronographic Cameras
Electrophotometers
Elementary Particles
Ellipsoidal Binaries
Elliptical Galaxies
Emission Lines

Emission Nebulae
Emission-Line Galaxies
Emission-Line Objects
Emission-Line Stars
Enceladus
Ephemerides
Equilibrium Figures
Eruptive Variables
Europa
Europa Surface
Extinction
Extragalactic Objects
Extraterrestrial Intelligence
Extraterrestrial Life
Extreme UV
Extreme UV Spectra

F Dwarfs
F Stars
F Supergiants
Fabry-Perot Interferometry
Fabry-Perot Spectrometers
Faint Galaxies
Faint Stars
Fiber Optics
Field Theories
Filamentary Nebulae
Filters
Fireballs
FK4
FK4 Stars
FK5
FK5 System
Flare Stars
Fluid Dynamics
Fluid Spheres
Flux Densities
Forbidden Lines
Forbush Effect
Four-Body Problem
Four-Color Photometry
Fourier Spectrometers
Fraunhofer Lines
Frequency Standards
FU Orionis Stars
Fundamental Astrometry
Fundamental Stars

G Dwarfs
G Giants
G Stars
G Supergiants
Galactic Anticenter
Galactic Bars
Galactic Center
Galactic Corona
Galactic Disk
Galactic Disks
Galactic Halo
Galactic Halos
Galactic Magnetic Field
Galactic Nuclei
Galactic Nucleus
Galactic Orbits
Galactic Plane
Galactic Poles
Galactic Structure
Galactic Wind
Galactic Winds
Galaxies
Galaxy
Galaxy Clustering
Galaxy Counts
Galaxy Evolution
Galaxy Formation
Galaxy Groups
Galaxy Mergers
Galaxy Pairs
Galaxy Voids
Galilean Satellites

Object Index

The usual bibliographic information accompanying the papers in the fields of astronomy and astrophysics certainly allows for their exact identification but the selection process of information retrieval is facilitated by supplementary information. Phenomena of all kinds, methods of observation and reduction as well as properties of classes of objects are referred to in the Subject Index.

Beginning with Vol. 39 *Astronomy and Astrophysics Abstracts* will present an additional index of astronomical objects dealt with in papers compiled. Basically all individual objects appearing in the title and the author's abstract are listed, but objects treated in the paper proper have also been included into the index. The number of single objects listed usually has been limited to 20. Papers discussing more than 20 individual objects have not been evaluated in the Object Index.

It is recommended to use the Object Index together with the key words of the Subject Index. Individual comets, novae, supernovae, major planets and their satellites are exclusively referred to in the Subject Index. A number of familiar astronomical objects will continue to appear in the Subject Index with their common designation. These objects are listed in the Object Index with their standard acronyms:

M31, NGC 224	Andromeda Galaxy
LMC, SMC	Magellanic Clouds
M45	Pleiades
Melotte 25	Hyades
M1	Crab Nebula

Novae identified with variables are referred to in the Object Index with their variable name:

GK Per	Nova Persei 1901

Astronomy and Astrophysics Abstracts supports the attempts of the IAU in standardizing the nomenclature of astronomical objects. Acronyms used follow closely the recommendations of "The First Dictionary of the Nomenclature of Celestial Objects" (A. Fernandez, M.-C. Lortet, F. Spite; Astron. Astrophys., Suppl. Ser., Vol. 52, No. 4 (1983)).

In sorting astronomical objects deviations from the lexicographic sequence are unavoidable in many cases. Sorting of the Object Index is based on the IBM standard routine SORT, which sorts blanks followed by special characters, Latin small and capital letters, and finally numbers. The three-letter mnemonic code of the 88 stellar constellations adopted by the IAU defines the highest rank within the sorting hierarchy. This implies that all stellar and non-stellar objects, whose designations explicitly refer to a constellation, are listed under the constellation's abbreviation. Stars within a constellation are sorted according to their Greek letter, followed by the standard variable nomen-

clature, and finally the Flamsteed numbers as well as other catalogue designations:

Cyg A
Cyg OB2
Cyg X–3
α Cyg
χ Cyg
BF Cyg
NML Cyg
Z Cyg
V367 Cyg
V1727 Cyg
6 Cyg

Designations without reference to stellar constellations are treated – as a rule – by sorting letters before numbers. An exception has been made for the minor planets. These objects have been retained in a particular section following after the letter 'Z'. The internal sequence is (1) permanently numbered planets, (2) objects with provisional designations, and (3) Palomar-Leiden objects:

(9) Metis
(29) Amphitrite
1979 VA
1984 HA$_1$
6344 P–L

Efforts were made to sort objects from individual catalogues according to the internal order of these catalogues. In Durchmusterungen the stars are arranged from the equator to the pole, taking the northward direction first. Object designations which contain coordinates are sorted first according to the usual catalogue abbreviation (B2, HMS, PKS, ZwCl, etc.) or object class designator (GX, PSR, SNR, etc.) and then according to the first coordinate. If only a position is given, the object is referred to by the number sequence. Objects with catalogue designations beginning with a number (1E, 3C, 4U, etc.) are listed under the respective number at the end of the Index. Note that Roman numbers are treated as Latin capital letters. For instance, objects from the II Zw and V Zw lists have to be searched under the letter I respectively V. A consequence of the applied sorting routine is, e.g., that Messier objects will be found at the end of the list of objects beginning with the letter 'M'.

Attempts have been made to avoid ambiguous acronyms. For instance, in the astronomical literature the acronym VB is used to designate (1) planetary nebula listed by Van den Bergh, (2) Bologna variables, (3) Van Bueren's list of Hyades stars, (4) Van Biesbroeck's list of faint stars. This ambiguity is partially resolved by using 'Van Bueren ...' for objects in (3), and 'Van Biesbroeck ...' for those in (4). With respect to the problem of multiple designations for a single object, a concordance relation is beyond the scope of the present Object Index. The user is therefore advised to search for possible alternative designations of a particular object. For example, the supergiant α Ori may also appear under BD $+7°1055$ or HD 39801, or the radio source Virgo A may be referred under $1218+1240$, 3C 274, NGC 4486 or M87.

ASTRONOMY AND ASTROPHYSICS ABSTRACTS

A Publication of the Astronomisches Rechen-Institut Heidelberg

Editors: S. Böhme, U. Esser, W. Fricke, H. Hefele, I. Heinrich,
W. Hofmann, D. Krahn, V. R. Matas, L. D. Schmadel, G. Zech

Volume 1	Literature 1969, Part 1	X+ 435 pp. (1969)
Volume 2	Literature 1969, Part 2	X+ 516 pp. (1970)
Volume 3	Literature 1970, Part 1	X+ 490 pp. (1970)
Volume 4	Literature 1970, Part 2	X+ 562 pp. (1971)
Volume 5	Literature 1971, Part 1	X+ 505 pp. (1971)
Volume 6	Literature 1971, Part 2	X+ 560 pp. (1972)
Volume 7	Literature 1972, Part 1	X+ 526 pp. (1972)
Volume 8	Literature 1972, Part 2	X+ 594 pp. (1973)
Volume 9	Literature 1973, Part 1	X+ 610 pp. (1973)
Volume 10	Literature 1973, Part 2	X+ 661 pp. (1974)
Volume 11	Literature 1974, Part 1	X+ 579 pp. (1974)
Volume 12	Literature 1974, Part 2	X+ 699 pp. (1975)
Volume 13	Literature 1975, Part 1	X+ 632 pp. (1975)
Volume 14	Literature 1975, Part 2	X+ 747 pp. (1976)
Volume 15/16	Author and Subject Indexes to Volumes 1–10	
	Literature 1969–1973	VII+ 655 pp. (1976)
Volume 17	Literature 1976, Part 1	XII+ 645 pp. (1976)
Volume 18	Literature 1976, Part 2	X+ 859 pp. (1977)
Volume 19	Literature 1977, Part 1	X+ 732 pp. (1977)
Volume 20	Literature 1977, Part 2	X+ 786 pp. (1978)
Volume 21	Literature 1978, Part 1	X+ 834 pp. (1978)
Volume 22	Literature 1978, Part 2	X+ 849 pp. (1979)
Volume 23/24	Author and Subject Indexes to Volumes 11–14 and 17–22	
	Literature 1974–1978	VIII+1127 pp. (1979)
Volume 25	Literature 1979, Part 1	X+ 872 pp. (1979)
Volume 26	Literature 1979, Part 2	X+ 794 pp. (1980)
Volume 27	Literature 1980, Part 1	X+ 939 pp. (1980)
Volume 28	Literature 1980, Part 2	X+ 841 pp. (1981)
Volume 29	Literature 1981, Part 1	X+ 853 pp. (1981)
Volume 30	Literature 1981, Part 2	X+ 792 pp. (1982)
Volume 31	Literature 1982, Part 1	X+ 776 pp. (1982)
Volume 32	Literature 1982, Part 2	X+ 848 pp. (1983)
Volume 33	Literature 1983, Part 1	X+ 815 pp. (1983)
Volume 34	Literature 1983, Part 2	X+ 921 pp. (1984)
Volume 35/36	Author and Subject Indexes to Volumes 25–34	
	Literature 1979–1983	XIII+ 891 pp. (1984)
Volume 37	Literature 1984, Part 1	X+ 937 pp. (1984)
Volume 38	Literature 1984, Part 2	X+ 919 pp. (1985)
Volume 39	Literature 1985, Part 1	X+1149 pp. (1985)

Published for Astronomisches Rechen-Institut by
Springer-Verlag Berlin Heidelberg New York Tokyo